| | | | | | | | | | 18 VIIIA |
|---|---|---|---|---|---|---|---|---|---|
| | | | | 13 IIIA | 14 IVA | 15 VA | 16 VIA | 17 VIIA | 2 **He** 4.002602 |
| | | | | 5 **B** 10.811 | 6 **C** 12.0107 | 7 **N** 14.00674 | 8 **O** 15.9994 | 9 **F** 18.9984032 | 10 **Ne** 20.1797 |
| 10 | 11 IB | 12 IIB | | 13 **Al** 26.981538 | 14 **Si** 28.0855 | 15 **P** 30.973762 | 16 **S** 32.066 | 17 **Cl** 35.4527 | 18 **Ar** 39.948 |
| 28 **Ni** 58.6934 | 29 **Cu** 63.546 | 30 **Zn** 65.39 | 31 **Ga** 69.723 | 32 **Ge** 72.61 | 33 **As** 74.92160 | 34 **Se** 78.96 | 35 **Br** 79.904 | 36 **Kr** 83.80 |
| 46 **Pd** 106.42 | 47 **Ag** 107.8682 | 48 **Cd** 112.411 | 49 **In** 114.818 | 50 **Sn** 118.710 | 51 **Sb** 121.760 | 52 **Te** 127.60 | 53 **I** 126.90447 | 54 **Xe** 131.29 |
| 78 **Pt** 195.078 | 79 **Au** 196.96655 | 80 **Hg** 200.59 | 81 **Tl** 204.3833 | 82 **Pb** 207.2 | 83 **Bi** 208.98038 | 84 **Po** (209) | 85 **At** (210) | 86 **Rn** (222) |
| 110 **Uun** (269) | 111 **Uuu** (272) | 112 **Uub** (277) | | 114 **Uuq** (289) | | 116 **Uuh** (289) | | 118 **Uuo** (293) |

## Inner-Transition Metals

| 63 **Eu** 151.964 | 64 **Gd** 157.25 | 65 **Tb** 158.92534 | 66 **Dy** 162.50 | 67 **Ho** 164.93032 | 68 **Er** 167.26 | 69 **Tm** 168.93421 | 70 **Yb** 173.04 | 71 **Lu** 174.967 |
|---|---|---|---|---|---|---|---|---|
| 95 **Am** (243) | 96 **Cm** (247) | 97 **Bk** (247) | 98 **Cf** (251) | 99 **Es** (252) | 100 **Fm** (257) | 101 **Md** (258) | 102 **No** (259) | 103 **Lr** (262) |

# General Chemistry

# General Chemistry

**SEVENTH EDITION**

## Darrell D. Ebbing
WAYNE STATE UNIVERSITY

## Steven D. Gammon
UNIVERSITY OF IDAHO

HOUGHTON MIFFLIN COMPANY   BOSTON  NEW YORK

*Editor-in-Chief:* Kathi Prancan
*Senior Sponsoring Editor:* Richard Stratton
*Associate Editor:* Sara Wise
*Senior Project Editor:* Nancy Blodget
*Editorial Assistant:* Kristin Penta
*Senior Production/Design Coordinator:* Jill Haber
*Manufacturing Manager:* Florence Cadran
*Executive Marketing Manager:* Andy Fisher

**Warning:** This book contains descriptions of chemical reactions and photographs of experiments that are potentially dangerous and harmful if undertaken without proper supervision, equipment, and safety precautions. DO NOT attempt to perform these experiments relying solely on the information presented in this text.

**Photo credits:** A list of credits precedes the index.

**Cover image:** Kennan Harvey. © Stone Images

*Printed in the U.S.A.*

Library of Congress Control Number: 2001131490

ISBN: 0-618-118373

Instructor's Annotated Edition ISBN: 0-618-118381

123456789-VH-05 04 03 02 01

# Contents in Brief

## Basics of Chemistry

## Atomic and Molecular Structure

## States of Matter and Solutions

## Chemical Reactions and Equilibrium

## Nuclear Chemistry and Chemistry of the Elements

# Contents

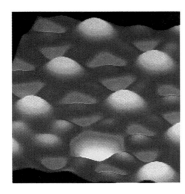

# 5 Gaseous State

# 6 Thermochemistry

# 9   Ionic and Covalent Bonding

# 10   Molecular Geometry and Chemical Bonding Theory

# 11    States of Matter; Liquids and Solids    440

# 12    Solutions    504

# 13   Materials of Technology

# 14   Rates of Reaction

# 17    Acid–Base Equilibria                                    723

# 18    Solubility and Complex-Ion Equilibria                  768

# 21 Nuclear Chemistry

# 22 Chemistry of the Main-Group Elements

# 23 The Transition Elements and Coordination Compounds

# Essays

## A Chemist Looks At

**LIFE SCIENCE**

**MATERIALS**

**ENVIRONMENT**

**FRONTIERS**

**EVERYDAY LIFE**

# Instrumental Methods

# Preface

## To the Instructor

In the preface to the first edition, we wrote, "Scientists delve into the molecular machinery of the biological cell and examine bits of material from the planets of the solar system. The challenge for the instructors of introductory chemistry is to capture the excitement of these discoveries [of chemistry] while giving students a solid understanding of the basic principles and facts. The challenge for the students is to be receptive to a new way of thinking, which will allow them to be caught up in the excitement of discovery." From the very first edition of this text, our aims have always been to help instructors capture the excitement of chemistry and to teach students to "think chemistry." Here are some of the features of the text that we feel are especially important in achieving these goals.

**Clear, Lucid Explanations of Chemical Concepts**  From the first edition, we have placed the highest priority on writing clear, lucid explanations of chemical concepts. We have strived to relate abstract concepts to specific real-world events and have presented topics in a logical, yet flexible, order. With succeeding editions, incorporating suggestions from instructors and students, we have refined the writing.

**Coherent Problem-Solving Approach**  With the first edition, we presented a coherent problem-solving approach that involved worked-out *Examples* coupled with *Exercises* and corresponding end-of-chapter *Problems*. This approach received an enormously positive response. In later editions, while refining the pedagogy of these examples and problems, we added *Problem Strategies* to underscore how one thinks through a problem solution.

Then, in the sixth edition, we added two new study aids, which met with an enthusiastic response: *Concept Checks* and *Conceptual Problems*. We wrote these problems to force students to think about the chemical concepts involved, rather than focus on obtaining a numerical result. In this new edition, we have strengthened this feature of the book. We now cover many more topics than in the previous edition, and we have added artwork to many more of these problems. The artwork is an important element of the pedagogy, helping the student visualize the concepts.

**Media Activities to Help You Integrate Our Electronic Resources into Your Course**  With this edition, we have added *Media Activities* to the problems section of the book to help you integrate our electronic resources into your course work and assignments. A media activity might ask the student to view a particular topic in the *Visualizations* segment of the Student CD or to work through a topic in the *Understanding Concepts* activity of that CD. The *Visualizations* are videos and animations whose purpose is to help the student visualize a topic in chemistry. The *Understanding Concepts* activities are in-depth explorations of a particular concept or problem-solving approach. At the end of each set of questions for a chapter, the *Media Activities* direct the student to take a practice test that is available on the Ebbing/Gammon web site.

**Chapter Essays Showcasing Chemistry as a Modern Science** With this edition, we have refocused and strengthened our *A Chemist Looks At* essays. In this process, we have dropped many essays from the previous edition and have added many new ones. In doing this, we chose essay topics that we thought would engage students' interest while at the same time showcase the chemistry involved. Most of the topics concern up-to-date research. With this edition, we have also adopted *icons* to describe the content area (materials, environment, everyday life, frontiers, and life science) of these essays.

The new essays show students that chemistry is a vibrant, constantly changing science that has relevance for our modern world. The essay "Removing Caffeine from Coffee," for example, describes the removal of caffeine from coffee using supercritical carbon dioxide instead of environmentally problematical organic solvents. Some other essays are "Nitric Oxide and Biological Signaling," which describes the importance of a simple chemical compound in human biology, and "Seeing Molecules React," which emphasizes that chemists are more and more able to work with single molecules.

**A New Illustration Program with an Emphasis on Molecular Concepts** Most of us are strongly visual in our learning. When you see something, you tend to remember it. For this new edition, we went over each piece of art, asking how we might improve it or where we might add art to improve student comprehension. One area that we focused on was the presentation of chemistry at the molecular level. We start building the molecular "story" in Chapter 1, and by Chapter 2 we have developed the molecular view and have integrated it into the problem-solving apparatus as well as into the text discussions. We continue in the following chapters to use the molecular view to strengthen chemical concepts. In Chapter 3, on stoichiometry, for example, a Concept Check asks the student to visualize the concept of limiting reactants in terms of molecular models.

**New Chapters on Materials and Polymers** We added a chapter on materials to this new edition—Chapter 13, *Materials of Technology*. This chapter includes much new material on nonmetals, such as the fullerenes and nanotechnology, ceramics, and composites. We also added a chapter (Chapter 25) on polymers, one of the most important materials of modern technology. This chapter consists of two parts, one on synthetic polymers (including a section on conducting polymers), the other on biological polymers.

**New Combined Chapter on the Main-Group Elements** Chapter 22 now combines the discussion of the main-group elements, which was formerly divided into two chapters. This format allows for the seamless discussion of the chemistry of these elements. It includes a section on the chemistry of hydrogen.

# To the Student

You are about to embark on a fascinating intellectual adventure. You are surrounded by chemical materials and substances, and if you are even a bit curious, you have asked yourself questions about these things. As you look at the print in this book, light particles, or photons, scatter from the ink, some passing into your eye. What chemistry happens when these photons strike the retina at the back of your eye? Perhaps you are drinking a cup of decaffeinated coffee or a cola soda that contains some of the caffeine taken from the coffee. How do they remove the caffeine without removing the flavor (at least not *all* of it)?

Of course, any new subject has its concepts and its ways of looking at things that may, at first, seem strange. Chemistry, dealing as it does with all the materials of the universe, from living things to the objects of the heavens, has its share of complexity. But its central concept is simple: Everything is composed of a selection (from about one hundred different kinds) of atoms chemically combined as molecules or other clumps of matter. As abstract as such objects might seem, chemists now have supermicroscopes that can see these atoms and molecules.

Having studied and taught chemistry for some years, your authors are well aware of the problems students encounter in their study of chemistry. In writing this book, we have constantly striven to relate chemical concepts to specific things in the real world, and we have written an abundance of instructional aids to help you grapple with the ideas presented.

# Features of the Seventh Edition

Each individual learns in a different way. For that reason, we have incorporated a number of different features into the text to help you master the subject. We hope that by becoming familiar with these features, which are listed below, you will be able to tailor a study program that meets your particular needs.

**Chapter Theme**   We begin each chapter with a theme—something specific that reveals the real-world relevance of the chapter topic. For example, we open Chapter 2, *Atoms, Molecules, and Ions,* with a discussion of sodium, chlorine, and sodium chloride. This chapter theme then leads naturally into a series of questions (for instance, "How do we explain the differences in properties of different forms of matter?"), which we answer later in the chapter.

**Vocabulary**   Chemistry uses words in a precise way, and it is important that you develop a vocabulary of terms in order to read and communicate the subject effectively. When a new, important word is introduced in the text, we have flagged it by putting it in boldface type. The definition of that word will generally follow in the same sentence in italic type. All of these words are collected at the end of the chapter in the list of *Important Terms.* They also appear, along with a few other words, in the *Glossary* at the end of the book, and on the CD-ROM. In addition, on the student web site, you can use the *Flashcard* program or print them out to practice learning these terms.

**Concept Checks and Conceptual Problems**   Many of the questions that you encounter daily begin with: "What do you think will happen if...?" or "Why is this choice the better one?" These questions are not asking for a numerical answer; rather, they are asking you to apply your conceptual knowledge to a problem. To answer this type of question, you need to think critically and have a mastery of chemical concepts and ideas. Merely knowing a set of formulas and memorizing a series of steps to a problem will not help you obtain the answer. In many cases, your conceptual understanding of chemistry will be what you remember and apply later in life.

To assist you in mastering the concepts presented in this text, we have placed *Concept Checks* in the body of every chapter and a section of *Conceptual Problems* at the end of each chapter. Answers to the *Concept Checks* and the odd-numbered *Conceptual Problems* are provided at the end of the book. More detailed discussions of the *Concept Checks* are included on the student web site.

**Problem-Solving Program**     Problem solving is an important part of chemistry. By solving problems yourself, you become involved with the subject, and by being involved with the subject, you learn it. But problem solving is like learning to swim or play a musical instrument; it becomes easy only with practice. Chemistry builds one concept on another, and fact on fact. The secret of problem solving in chemistry is to know what you learned earlier so well that when you approach a new problem, you know how to put the pieces together.

Recognizing the importance of problem solving in chemistry, we felt the burden could be much reduced if we followed a consistent problem-solving program. We introduce each problem-solving skill with an *Example,* in which you are led through the reasoning involved in working out a particular type of problem. Most of these Examples include a *Problem Strategy* that underscores the thinking process involved in solving the problem. Each Example is accompanied by an *Exercise,* which is a similar problem that you can try. The answers to these Exercises are at the end of the book. At the end of each Exercise is a list of corresponding end-of-chapter *Practice Problems.* Try some of these to gain mastery of that problem-solving skill. Answers to odd-numbered problems are provided at the end of the book.

We have also provided three other sources of problems for you to try: the *Problem Solving* questions on the CD-ROM, the *ACE* self-quizzing on the student web site, and the on-line homework problems (also on the web site).

**Checklist for Review**     When it comes to reviewing, students generally develop their own techniques. We have tried to accommodate these differences by presenting various review possibilities. For example, you may find that the list of *Important Terms* is useful, not only because it is a list of new words, but also because as you look over the words you see the structure of the chapter. As you mentally note this structure, try to recall the ideas associated with the words. You may also choose to use the *Flashcard* program on the web site to review these terms. Many chapters also introduce one or more mathematical equations to be used in problem solving. In the chapter, these are shaded in color; then, in the Checklist for Review, they are listed as *Key Equations.* The *Summary of Facts and Concepts* presents a verbal summary of the chapter. Study this, and as you go over each statement, try to flesh out points. Finally, we present a list of *Operational Skills.* This is a summary of the chapter's problem-solving skills. Each operational skill tells you what information is needed and what is to be solved for in a given type of problem. Each operational skill also refers back to the Examples that discuss that problem-solving skill.

**End-of-Chapter Questions and Problems**     This section begins with *Review Questions,* which have been designed to test your understanding of the chapter ideas. Generally, they can be answered by straightforward recall or by simple extension of the chapter material. After these questions, we have added a section of *Conceptual Problems.* After these problems, we have listed several sections of problems to help you gain mastery of problem-solving skills. The *Practice Problems* are keyed to a particular topic by heading; the *General Problems* are not. The *Cumulative-Skills Problems* require you to combine several skills, often from previous chapters. These problems all appear in matched pairs for additional practice. The questions under *Media Activities* refer you to the Student CD and the Ebbing/Gammon web site.

# Complete Instructional Package

This text is complemented by a complete package of print and electronic ancillaries.

## For Students and Instructors

**Student Technology Package**   This package is available with all new texts and is also available for sale separately. The package includes:

**Technology guide.**   A handy guide to the technology resources available to accompany the seventh edition.

*General Chemistry Interactive 5.0* **Student CD.**   A highly interactive CD-ROM that supports the goals of the seventh edition by helping students visualize molecular behavior, explore important concepts, and practice working problems. The CD runs on both Macintosh and Windows platforms.

**Passkey.**   A unique 13-digit code allowing students access to web site materials; the passkey is valid for 12 months from sign-up.

**SMARTHINKING™ on-line tutoring.**   Passkey required. Web-based tutoring by qualified *e*-structors at www.Smarthinking.com. Available Fall 2001.

**Student web site (http://college.hmco.com/, select "chemistry").**   Passkey required. Includes Houghton Mifflin's ACE self-quizzing, interactive molecules, movies, flashcards of key terms and concepts, and links to other useful sites.

**Houghton Mifflin's Eduspace™ on-line homework system.**   Passkey required. The on-line homework system provides students with questions that are keyed directly to the text content. Individual student progress is tracked and may be viewed, managed, and downloaded by instructors.

**eBook**   The complete text of *General Chemistry*, Seventh Edition is available in electronic format for selected eBook readers. Contact your Houghton Mifflin representative for details.

**Conceptual Guide to** *General Chemistry*, Darrell D. Ebbing, Wayne State University, and Steven D. Gammon, University of Idaho. This guide is intended to underscore and clarify conceptual aspects of the text. It contains step-by-step solutions to the in-chapter Concept Checks and the end-of-chapter Conceptual Problems and a "Concept Target," which summarizes the concept or concepts that are addressed in the problem.

**Experiments in** *General Chemistry*, R.A.D. Wentworth, Indiana University. Forty experiments parallel the material found in the textbook. Each lab exercise has a pre-lab assignment, background information, clear instructions for performing the experiment, and a convenient section for reporting results and observations. An instructor's resource manual is also available. Several experiments in this edition incorporate the use of computers and the Internet.

**Student Solutions Manual,** David Bookin, Mount San Jacinto College, Darrell D. Ebbing, Wayne State University, and Steven D. Gammon, University of Idaho. This manual contains detailed solutions to the in-chapter exercises and odd-numbered end-of-chapter problems. It also contains answers to the Review Questions. All solutions have been checked by an independent reviewer to ensure accuracy.

**Study Guide for *General Chemistry*,** Larry K. Krannich, University of Alabama at Birmingham. Each chapter of the study guide reinforces the students' understanding of concepts and operational skills presented in the text. It includes the following features for each chapter: a list of key terms and their definitions, a diagnostic test with answers, a summary of major concepts and operational skills, additional practice problems and their solutions, and a chapter post-test with answers.

**Qualitative Analysis and Ionic Equilibrium,** George H. Schenk, Wayne State University. This laboratory manual presents a traditional qualitative analysis scheme that can be used with the text or can stand alone. It stresses the chemistry of metal ions and anions. Each exercise has a preliminary report and a final report for pre-lab and post-lab activities.

## For Instructors

**Instructor's Annotated Edition,** Darrell D. Ebbing, Wayne State University and Steven D. Gammon, University of Idaho. The IAE comprises the student text and a program of annotations to assist the instructor in syllabus and lecture preparation. Many of the annotations refer to material such as transparencies, lab experiments, video, and electronic ancillaries available with this text. Others suggest ways to integrate other instructional media with your use of this text.

**Instructor's Resource Manual,** Darrell D. Ebbing, Wayne State University, and Steven D. Gammon, University of Idaho. This manual offers information about chapter essays, suggestions for alternate sequencing of topics, short chapter descriptions, a master list of operational skills, correlation of cumulative-skills problems with text topics, alternative examples for lectures, suggested lecture demonstrations, and a list of overhead transparencies.

**HMClassPrep™ Package.**   This Instructor CD-ROM includes PowerPoint slides of text figures, photos, concept checks, and lecture outlines, along with videos, animations, and an on-line Instructor's Resource Manual. The assets can be accessed from the CD-ROM by chapter or asset type. They can be customized to help instructors create lectures and in-class activities in a flexible manner.

**HMTesting Computerized Testing System.**   This system offers a flexible test-editing program and a comprehensive gradebook to allow easy administration and tracking. Instructors can administer tests via a network server or the Web. The HMTesting database can produce multiple-choice or free-response tests. Instructors can customize questions based on chapter, question format, and specific topics.

**Printed Test Bank,** Ronald Ragsdale, University of Utah. The printed test bank contains more than 2000 multiple-choice questions organized by chapter. Over 400 of the questions are new to this edition.

**Transparencies.**   More than 250 full-color transparencies of figures, tables, and photographs selected from the text are provided.

**Instructor web site at college.hmco.com.** This web site, created exclusively for instructors, allows access to all student web site resources plus downloadable PowerPoint slides, additional instructor and classroom resources, and relevant links.

**Complete Solutions Manual,** David Bookin, Mount San Jacinto College, Darrell D. Ebbing, Wayne State University, and Steven D. Gammon, University of Idaho. This complete version of the *Student Solutions Manual* contains detailed solutions to the in-chapter exercises and *all* end-of-chapter problems, as well as the answers to the Concept Checks and Review Questions. This supplement is intended for the instructor's convenience and for those who wish their students to have solutions to all problems.

**Instructor's Resource Manual to the Lab Manual,** R.A.D. Wentworth, Indiana University. This manual provides instructors with sample results to all pre- and post-lab activities in *Experiments in General Chemistry.*

**Content for Course Management Software** is available through *WebCT* and *Blackboard.com.* These two distributed learning systems allow you to create a virtual classroom without any knowledge of HTML. Features include assessment tools and a gradebook, on-line file exchange between you and your students, and on-line syllabi and course descriptions. The customized *General Chemistry* cartridges feature quizzes, study materials, and exercises related to the text.

**Laserdisc.** The laserdisc provides 34 videoclips of lecture demonstrations, line art from the text, and animations of difficult concepts in chemistry. Lecture manager software is available for IBM and Macintosh platforms. A print instructor's guide, complete with barcodes, frame references, and cross-references to the text, is available.

**Video Lecture Demonstrations,** John Luoma, Cleveland State University; John J. Fortman and Rubin Battino, Wright State University; Patricia L. Samuel, Boston University; and Paul Kelter, University of North Carolina at Greensboro. This series of four videotapes (A–D) contains 130 lecture demonstrations. Series C demonstrations appear on the laserdisc.

# Acknowledgments

We especially want to thank Richard Stratton, Senior Sponsoring Editor, and Sara Wise, Associate Editor, for helping us to achieve our goals for this revision. Besides contributing ideas, insights, and inspiration, they assembled an outstanding group of people to participate in this project. We also wish to thank Nancy Blodget, Senior Project Editor, and Jill Haber, Senior Production/Design Coordinator, for an excellent job in carrying out the monumental task of melding the final manuscript, line art, and photos into a finished book.

This revision involved a complete reworking of the art program. We were fortunate to have John Murdzek lead in this critical area. John went through the previous edition of the text, critically evaluating every piece of art. His excellent suggestions have substantially improved the pedagogy and impact of the art program. Jessyca Broekman returned again as art editor for the book. She had the task of turning the hundreds of new art ideas into reality. She executed this colossal task with enthusiasm and innovation. We were pleased also to have Naomi Kornhauser again as the photography editor. Naomi has the exceptional talent of finding the perfect picture when presented with even a vague request. She also coordinated the setup photography. Douglas Sawyer and Steve Borick, both of Scottsdale Community College, did the actual photography for the setup photos. Often this required original and innovative work, and we were fortunate to have them on our team.

The technology component of the text package is crucial. We thank Sue Warne, Instructional Technology Sponsor at Houghton Mifflin, for coordinating the development of an outstanding package of technology-based materials to accompany the text. We thank Muhonjia Khaminwa and John Murdzek for their work on the multimedia package, and Jessica Fisher Willson and Jessica Yuan for their work on the web site. We also wish to thank Sharon Hutchison, Professor of Chemistry at the University of Idaho, for her critical work in the development of the content for the on-line testing package.

The revision of the text required a corresponding revision of the ancillary items. We thank Marisa Papile, Editorial Assistant, for her work in overseeing the ancillary revision process. Andy Fisher, Executive Marketing Manager, has been with this project since the beginning of the revision. We greatly appreciate his suggestions for improvements and his infectious enthusiasm for the project. We also thank Henry Rachlin for designing a book that is exceptionally pleasing to look at while soundly emphasizing the pedagogy of the text.

Darrell wishes to thank his wife, Jean, and children, Julie, Linda, and Russell, for their continued support and encouragement over many years of writing. Steve thanks his wife, Jodi, his two children, Katie and Andrew, and his parents, Judy and Dick, for their support and for helping him keep a perspective on the important things in life.

The development of any revision would be impossible without the help of reviewers and accuracy checkers. We are enormously grateful to the following people for giving their time and ideas to this edition of *General Chemistry:*

Carey Bissonnette, *University of Waterloo*

Bob Belford, *West Virginia University*

Conrad Bergo, *East Stroudsburg University*

Aaron Brown, *Ventura College*

Tim Champion, *Johnson C. Smith University*

Paul Cohen, *College of New Jersey*

Lee Coombs, *California Polytechnic State University*

Jack Cummins, *Metro State College*

William M. Davis, *The University of Texas at Brownsville*

Earline F. Dikeman, *Kansas State University*

Evelyn S. Erenrich, *Rutgers University*

Greg Ferrence, *Illinois State University*

Renee Gittler, *Penn State Lehigh Valley*

Brian Glaser, *Black Hawk College*

David Grainger, *Colorado State University*

Christopher Grayce, *University of California at Irvine*

John M. Halpin, *New York University*

Carol Handy, *Portland Community College*

Daniel Haworth, *Marquette University*

Gregory Kent Haynes, *Morgan State University*

Robert Henry, *Tarrant County College*

Grant Holder, *Appalachian State University*

Andrew Jorgensen, *University of Toledo*

Kirk Kawagoe, *Fresno City College*

David Kort, *Mississippi State University*

Charles Kosky, *Borough of Manhattan Community College*

Jeffrey Kovac, *University of Tennessee at Knoxville*

Art Landis, *Emporia State University*

Richard Langley, *Stephen F. Austin State University*

Robert Mentore, *Ramapo College*

Joyce Miller, *San Jacinto College (South)*

Bob Morris, *Ball State University*

John Nash, *Purdue University*

Deborah Nycz, *Broward Community College*

Michael A. Quinlan, *University of Southern California*

Joe Rorke, *College of DuPage*

John Schaumloffel, *University of Massachusetts at Dartmouth*

Vernon Thielmann, *Southwest Missouri State University*

Jennifer Travers, *Oregon State University*

Gershon Vincow, *Syracuse University*

Donald Wirz, *University of California at Riverside*

Pete Witt, *Midlands Technical College*

Kim Woodrum, *University of Kentucky*

DARRELL D. EBBING
STEVEN D. GAMMON

# Features of *General Chemistry*, Seventh Edition

The Seventh Edition of **General Chemistry** retains all of the qualities that have made it a trusted and authoritative first-year text. Its hallmark ability to teach quantitative **problem solving** is complemented by new features that focus on key themes of **visualization** and **conceptual understanding**.

## VISUALIZATION

**Figure 4.14**
**Decomposition reaction.**
The decomposition reaction of mercury(II) oxide into its elements, mercury and oxygen.

New molecular blowups help students connect the macroscopic to the molecular level.

4.5 Oxidation–Reduction Reactions **153**

tion is one in which one or more atoms change oxidation number, implying that there has been a transfer of electrons.

Consider the combustion of calcium metal in oxygen gas (Figure 4.11).

$$2Ca(s) + O_2(g) \longrightarrow 2CaO(s)$$

This is an oxidation–reduction reaction. To see this, you assign oxidation numbers to the atoms in the equation and then note that the atoms change oxidation number during the reaction.

The oxidation number of an atom in an element is always zero. Thus, Ca in the metal and O in $O_2$ have oxidation numbers of zero. Another rule follows from the definition of oxidation number: The oxidation number of an atom that exists in a substance as a monatomic ion equals the charge on that ion. So the oxidation number of Ca in CaO is $+2$ (the charge on $Ca^{2+}$), and the oxidation number of O in CaO is $-2$ (the charge on $O^{2-}$). To emphasize these oxidation numbers in an equation, we will write them above the atomic symbols in the formulas.

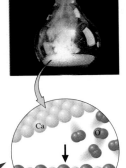

**Figure 4.11**
**The burning of calcium metal in oxygen.** The burning calcium emits a red-orange flame.

**Figure 11.3**
**Sublimation of iodine.**
*Left:* The beaker contains iodine crystals, $I_2$; a dish of ice rests on top of the beaker. *Right:* Iodine has an appreciable vapor pressure even below its melting point (114°C); thus, when heated carefully the solid sublimes without melting. The vapor deposits as crystals on the cool underside of the dish.

Sublimation can be used to purify solids that readily vaporize. Figure 11.3 shows a simple way to purify iodine by sublimation. Impure iodine is heated in a beaker so that it vaporizes, leaving nonvolatile impurities behind. The vapor crystallizes on the bottom surface of a dish containing ice that rests on top of the beaker. Freeze-drying of foods is a commercial application of sublimation. Brewed coffee, for example, is frozen and placed in a vacuum to remove water vapor. The ice continues to sublime until it is all gone, leaving freeze-dried coffee. Most freeze-dried foods are easily reconstituted by adding water.

**Condensation** is *the change of a gas to either the liquid or the solid state* (the change of vapor to solid is sometimes called *deposition*). Dew is liquid water formed

Solid $I_2$

Gaseous $I_2$

### Hydrogen Bonding

CH₃F

CH₃OH

**Figure 11.23**
**Fluoromethane and methanol.** Space-filling molecular models. *(Top)* fluoromethane, CH₃F; *(bottom)* methanol, CH₃OH.

It is interesting to compare fluoromethane, $CH_3F$, and methanol, $CH_3OH$ (Figure 11.23). They have about the same molecular weight (34 for $CH_3F$ and 32 for $CH_3OH$) and about the same dipole moment (1.81 D for $CH_3F$ and 1.70 D for $CH_3OH$). You might expect these substances to have about the same intermolecular attractive forces and therefore about the same boiling points. In fact, the boiling points are quite different. Fluoromethane boils at $-78°C$ and is a gas under normal conditions. Methanol boils at 65°C and is normally a liquid. There is apparently an intermolecular attraction in methanol that is not present in fluoromethane.

We have already seen that the properties of water and glycerol cannot be explained in terms of van der Waals forces alone. What water, glycerol, and methanol have in common is one or more —OH groups.

H—O—H

water

glycerol

methanol

> **Many chemical structures are now depicted in multiple ways to help students make the leap from symbolic to visual representations.**

---

**Figure 7.3**
**Water wave (ripple).** The wavelength (λ) is the distance between any two adjacent identical points of a wave, such as two adjacent peaks or two adjacent troughs.

> **Margin icons key students to visual resources on the student CD-ROM.**

**Visualization:**
**Electormagnetic Wave**

The **frequency** of a wave is *the number of wavelengths of that wave that pass a fixed point in one unit of time* (usually one second). For example, imagine you are anchored in a small boat on a pond when a stone is dropped into the water. Waves travel outward from this point and move past your boat. The number of wavelengths that pass you in one second is the frequency of that wave. Frequency is denoted by

---

As you can see, $\Delta U$ does not differ a great deal from $\Delta H$. This is the case in most reactions, so that the heat of reaction at constant pressure is approximately equal to the change of internal energy.

**Figure 6.9**
**Pressure–volume work.** In this experiment, we replace the pressure of the atmosphere by a piston-and-weight assembly of equal pressure. As sodium metal reacts with water, the hydrogen gas evolved pushes the piston and weight upward (compare *before* and *after*). It requires work to raise the piston and weight upward in a gravitational field.

> **Revised diagrams convey chemical principles clearly and effectively.**

# CONCEPTUAL UNDERSTANDING

Concept Checks throughout the chapters have been revised to further challenge the student to learn the ideas underlying chemistry. A new icon identifies conceptually focused material.

CONCEPT CHECK 4.1

LiI($s$) and $CH_3OH$($l$) are each introduced into a separate beaker containing water. Using the drawings shown here, label each beaker with the appropriate compound and indicate whether you would expect each substance to be a strong electrolyte, weak electrolyte, or nonelectrolyte.

 Conceptual Problems

**5.23** Using the concepts developed in this chapter, explain the following observations.
a. Automobile tires are flatter on cold days.
b. You are not supposed to dispose of aerosol cans in a fire.
c. The lid of a water bottle pops off when the bottle sits in the sun.
d. A balloon pops when you squeeze it.

**5.24** You have three identical flasks, each containing equal amounts of $N_2$, $O_2$, and He. The volume of the $N_2$ flask is doubled, the $O_2$ flask volume is halved, and the He flask volume is reduced to one-third of the original volume. Rank the flasks from highest to lowest pressure both before and after the volume is changed and indicate by what factor the pressure has changed.

**5.25** Consider the following gas container equipped with a movable piston.

a. By what factor (increase by 1, decrease by 1.5, etc.) would you change the pressure if you wanted to change the volume from volume C to volume D?
b. If the piston were moved from volume C to volume A, by what factor would the pressure change?
c. By what factor would you change the temperature in order to change from volume C to volume B?
d. If you increased the number of moles of gas in the container by a factor of 2, by what factors would the pressure and the volume change?

**5.26** A 3.00-L flask containing 2.0 mol of $O_2$ and 1.0 mol of $N_2$ is in a room that is at 22.0°C.
a. How much (what fraction) of the total pressure in the flask is due to the $N_2$?
b. The flask is cooled and the pressure drops. What happens, if anything, to the mole fraction of the $O_2$ at the lower temperature?

c. 1.0 L of liquid water is introduced into the flask containing both gases. The pressure is then measured about 45 minutes later. Would you expect the measured pressure to be higher or lower?
d. Given the information in this problem and the conditions in part c, would it be possible to calculate the pressure in the flask after the introduction of the water? If it is not possible with the given information, what further information would you need to accomplish this task?

**5.27** Consider the following setup, which shows identical containers connected by a tube with a valve that is presently closed. The container on the left has 1.0 mol of $H_2$ gas; the container on the right has 1.0 mol of $O_2$.

Conceptual Problems at the end of each chapter reinforce principles by asking nonquantitative questions about the material.

*Note:* Acceptable answers to some of these questions might be "both" or "neither one."
a. Which container has the greatest density of gas?
b. Which container has molecules that are moving at a faster average molecular speed?
c. Which container has more molecules?
d. If the valve is opened, will the pressure in each of the containers change? If it does, how will it change (increase, decrease, or no change)?
e. 2.0 mol of Ar is added to the system with the valve open. What fraction of the total pressure will be due to the $H_2$?

**5.28** Two identical He-filled balloons, each with a volume of 20 L, are allowed to rise into the atmosphere. One rises to an altitude of 3000 m while the other rises to 6000 m.
a. Assuming that the balloons are at the same temperature, which balloon has the greater volume?
b. What information would you need in order to calculate the volume of each of the balloons at their respective heights?

## Solid State

A solid is a nearly incompressible state of matter with a well-defined shape, because the units (atoms, molecules, or ions) making up the solid are in close contact and in fixed positions or sites. In the next section, we will look at the kinds of forces holding the units together in different types of solids. In later sections, we will look at crystalline solids and their structure.

### 11.6 Classification of Solids by Type of Attraction of Units

> Margin icons key students to conceptual activities on the student CD-ROM.

**Understanding Concepts: Classification of Solids**

A solid consists of structural units—atoms, molecules, or ions—that are attracted to one another strongly enough to give a rigid substance. One way to classify solids is by the type of force holding the structural units together. In some cases, these forces are intermolecular and hold molecules together. In other cases, these forces are chemical bonds (metallic, ionic, or covalent) and hold atoms together. From this point of view, then, there are four different types of solids: molecular, metallic, ionic, and covalent. It should be noted, however, that a given ionic bond may have considerable covalent character or vice versa, so the distinction between ionic and covalent solids is not a precise one.

---

### A Checklist for Review

#### Important Terms

| | | | |
|---|---|---|---|
| ionic bond (9.1) | Lewis electron-dot | octet rule (9.4) | resonance description (9.7) |
| Lewis electron-dot |   formula (9.4) | single bond (9.4) | formal charge (9.9) |
|   symbol (9.1) | bonding pair (9.4) | double bond (9.4) | bond length (bond |
| lattice energy (9.1) | lone (nonbonding) | triple bond (9.4) |   distance) (9.10) |
| ionic radius (9.3) |   pair (9.4) | polar covalent bond (9.5) | covalent radii (9.10) |
| isoelectronic (9.3) | coordinate covalent | electronegativity (9.5) | bond order (9.10) |
| covalent bond (p. 362) |   bond (9.4) | delocalized bonding (9.7) | bond energy (9.11) |

> A narrative **Summary of Facts and Concepts** at the end of each chapter recaps the important concepts in each chapter.

#### Summary of Facts and Concepts

An *ionic bond* is a strong attractive force holding ions together. An ionic bond can form between two atoms by the transfer of electrons from the valence shell of one atom to the valence shell of the other. Many similar ions attract one another to form a crystalline solid, in which positive ions are surrounded by negative ions and negative ions are surrounded by positive ions. As a result, ionic solids are typically high-melting solids. Monatomic cations of the main-group elements have charges equal to the group number (or in some cases, the group number minus two). Monatomic anions of the main-group elements have charges equal to the group number minus eight.

A *covalent bond* is a strong attractive force that holds two atoms together by their sharing of electrons. These bonding electrons are attracted simultaneously to both atomic nuclei, and they spend part of the time near one atom and part of the time near the other. If an electron pair is not equally shared, the bond is *polar*. This polarity results from the difference in *electronegativities* of the atoms—that is, in the unequal abilities of the atoms to draw bonding electrons to themselves.

*Lewis electron-dot formulas* are simple representations of the valence-shell electrons of atoms in molecules and ions. You can apply simple rules to draw these formulas. In molecules with *delocalized bonding,* it is not possible to describe accurately the electron distribution with a single Lewis formula. For these molecules, you must use *resonance*. Although the atoms in Lewis formulas often satisfy the octet rule, *exceptions to the octet rule* are not uncommon. You can obtain the Lewis formulas for these exceptions by following the rules for writing Lewis formulas. The concept of *formal charge* will often help you decide which of several Lewis formulas gives the best description of a molecule or ion.

*Bond lengths* can be estimated from the *covalent radii* of atoms. Bond length depends on *bond order;* as the bond order increases, the bond length decreases. The A—B *bond energy* is the average enthalpy change when an A—B bond is broken. You can use bond energies to estimate $\Delta H$ for gas-phase reactions.

#### Operational Skills

*Note:* A periodic table is needed for skills 1, 2, 3, 5, and 6.

1. **Using Lewis symbols to represent ionic bond formation**   Given a metallic and a nonmetallic main-group element, use Lewis symbols to represent the transfer of electrons to form ions of noble-gas configurations. **(EXAMPLE 9.1)**
2. **Writing electron configurations of ions**   Given an ion, write the electron configuration. For an ion of a main-group element, give the Lewis symbol. **(EXAMPLES 9.2, 9.3)**
3. **Using periodic trends to obtain relative ionic radii** Given a series of ions, arrange them in order of increasing ionic radius. **(EXAMPLE 9.4)**

5. **Writing Lewis formulas**   Given the molecular formula of a simple compound or ion, write the Lewis electron-dot formula. **(EXAMPLES 9.6, 9.7, 9.8, 9.10)**
6. **Writing resonance formulas**   Given a simple molecule with delocalized bonding, write the resonance description. **(EXAMPLE 9.9)**
7. **Using formal charges to determine the best Lewis formula**   Given two or more Lewis formulas, use formal charges to determine which formula best describes the electron distribution or gives the most plausible molecular structure. **(EXAMPLE 9.11)**
8. **Relating bond order and bond length**   Know the relationship between bond order and bond length. **(EXAMPLE 9.12)**

# PROBLEM SOLVING

In-text **Examples** guide students through the logic of solving a type of problem.

**Problem Strategies** outline the thinking that underlies the numerical solution of the problem. The **Solution** then applies that thinking to a particular problem.

Each example is followed by an Exercise to give students a chance to practice what they have just seen worked out. A **Reference to End-of-Chapter Problems** directs students to other problems of this type.

New margin icons key students to similar problems on the student CD-ROM for further practice.

### EXAMPLE 3.3

**Calculating the Mass of an Atom or Molecule**

a. What is the mass in grams of a chlorine atom, Cl?

b. What is the mass in grams of a hydrogen chloride molecule, HCl?

#### PROBLEM STRATEGY

Because each mole of substance contains Avogadro's number of units of that substance, dividing the mass of a substance in one mole (molar mass) by Avogadro's number gives the mass of that unit of substance.

#### SOLUTION

a. The atomic weight of Cl is 35.5 amu, so the molar mass of Cl is 35.5 g/mol. Dividing 35.5 g (per mole) by $6.02 \times 10^{23}$ (Avogadro's number) gives the mass of one atom.

$$\text{Mass of a Cl atom} = \frac{35.5 \text{ g}}{6.02 \times 10^{23}} = 5.90 \times 10^{-23} \text{ g}$$

b. The molecular weight of HCl equals the AW of H plus the AW of Cl, or 1.01 amu + 35.5 amu = 36.5 amu. Therefore, 1 mol HCl contains 36.5 g HCl and

$$\text{Mass of an HCl molecule} = \frac{36.5 \text{ g}}{6.02 \times 10^{23}} = 6.06 \times 10^{-23} \text{ g}$$

Problem Solving:
Determining the Mass of a
Sample of Atoms

### EXERCISE 3.3

a. What is the mass in grams of a calcium atom, Ca? b. What is the mass in grams of an ethanol molecule, $C_2H_5OH$?

See Problems 3.25, 3.26, 3.27, and 3.28.

---

End-of-chapter **Practice Problems** are keyed to particular topics by a heading. In addition, there are **General Problems** and **Cumulative-Skills Problems**, which require students to combine several skills.

## Practice Problems

### Phase Transitions

**11.29** Identify the phase transition occurring in each of the following.

a. The water level in an aquarium tank falls continuously (the tank has no leak).

b. A mixture of scrambled eggs placed in a cold vacuum chamber slowly turns to a powdery solid.

c. Chlorine gas is passed into a very cold test tube where it turns to a yellow liquid.

d. When carbon dioxide gas under pressure exits from a small orifice, it turns to a white "snow."

the time the iodine began to melt until the iodine was completely melted. What is the heat of fusion per mole of iodine?

**11.34** A 35.8-g sample of cadmium metal was melted by an electric heater providing 4.66 J/s of heat. If it took 6.92 min from the time the metal began to melt until it was completely melted, what is the heat of fusion per mole of cadmium?

**11.35** Isopropyl alcohol, $CH_3CHOHCH_3$, is used in rubbing alcohol mixtures. Alcohol on the skin cools by evaporation. How much heat is absorbed by the alcohol if 5.00 g evaporates? The heat of vaporization of isopropyl alcohol is 42.1 kJ/mol.

---

New **Media Activities** direct students to electronic resources and ask questions related to activities on the student CD-ROM and web site.

## Media Activities

**11.123** Open the Changes of State *Understanding Concepts* activity on the student CD to review phase changes and heating curves.

a. When a solid converts to a liquid and then to a gas, describe what happens during each step in terms of the intermolecular forces.

b. Rationalize why $\Delta H_{vap}$ for water is over six times larger than $\Delta H_{fus}$.

c. What values of $\Delta H$ would you need to calculate in order to

forces present in $SO_2$, $N_2H_4$, $O_3$, and $SP_6$. Try as many additional structures provided by the program as necessary in order to master the concepts.

**11.126** Run the Unit Cells and Crystal Packing *Visualization* on the student CD.

a. Which unit cell—simple cubic, face-centered cubic, or body-centered cubic—is a closest-packed arrangement?

b. Which unit cell—simple cubic, face-centered cubic, or body-centered cubic—contains the most atoms?

An STM image of the atomic structure of gallium arsenide, GaAs (Ga, blue; As, red).

# Materials of Technology

In the summer of 2000, scientists at Bell Laboratories and Oxford University reported that they had constructed a motor out of DNA, the chemical material of the genetic code. This molecular-scale motor used DNA as both a structural material and as a fuel to run the motor. Earlier, Harvard University chemists had constructed molecular-scale tweezers out of carbon-atom tubes, showing that they could manipulate clusters of polystyrene molecules, and NASA scientists have studied the possibility of constructing a molecular-scale gear assembly (see Figure 13.1). These scientists' efforts represent some of the many ongoing investigations into "nanotechnology," in

A new chapter on **Materials** includes cutting-edge developments and applications of particular value to students in science-related majors.

Each chapter begins with a **chapter theme**, a real-world topic that raises a question that leads into the chapter topic.

---

## A CHEMIST LOOKS AT

### Removing Caffeine from Coffee

EVERYDAY LIFE

Legend has it that a goat herder named Kaldi, in the ancient country of Abyssinia (now Ethiopia, in northeast Africa), discovered the pleasant effect of coffee beans when he noted that his goats danced wildly about after eating the shiny leaves and red berries of a small tree. Later he tried eating the whole berries himself, and he soon found that he was dancing with his goats. Whether the legend is true or not, it does effectively describe the stimulant effect of this berry from the coffee tree. As we now know, the active principle of coffee, which today is obtained by roasting the berries' green pits, or beans as they are called, is caffeine, a bitter-tasting white substance with the formula $C_8H_{10}N_4O_2$. (Figure 11.14 shows the structural formula of caffeine.)

For those who like the taste of roasted coffee but don't want the caffeine, decaffeinated coffee is available. A German chemist, Ludwig Roselius, first made "decaf" coffee about 1900 by extracting the caffeine from green coffee beans with the solvent chloroform, $CHCl_3$. Later,

"A Chemist Looks At" essays relate chemical concepts to real-world applications. New icons key students to particular areas of interest: medicine, health, frontiers of science, the environment, and everyday life.

**Figure 11.14**
**The caffeine molecule.** Ball-and-stick model of caffeine, $C_8H_{10}N_4O_2$.

day, though, most of the commercial decaffeinated coffee produced (Figure 11.15) uses *supercritical* carbon dioxide as the extracting fluid.

# A new suite of robust technology tools accompanies the text to augment teaching and learning.

The **Ebbing/Gammon web sites** for students and instructors enrich, enhance, and facilitate teaching and learning by taking the text's three main themes—problem solving, understanding concepts, and visualization—to an interactive level. A unique passkey that enables students to access the site is available with all new student texts and is also sold separately. Instructor passkeys are provided upon adoption.

The **online homework system tools** allow instructors to create homework assignments with a wealth of exercises keyed to the text. Students can submit homework assignments online. The online homework system takes the drudgery out of homework grading and tracking, freeing instructors' time to help students in other ways. Instructors can also use its powerful course management features to monitor, analyze, and download student assignments and progress.

**ACE self-quizzes** boost learning and confidence. Students receive immediate feedback and can e-mail quiz results to instructors.

**Electronic Flash Cards** allow students to study chemistry terms, equations, and key reactions.

**Media Activities** tie it all together, helping students use the CD-ROM and web site to enhance learning.

## HMClassPrep™ CD-ROM

HMClassPrep is a cross-platform CD-ROM that provides a multitude of text-specific resources to facilitate class preparation for instructors and enhance the classroom experience for students. These customizable resources can be accessed from the CD-ROM by chapter or asset type to help professors create lectures and in-class activities easily and quickly.

**General Chemistry Interactive 5.0** student CD-ROM offers topic summaries, conceptual enhancement, dynamic visualization, problem-solving aids, and more in a highly interactive environment.

# About the Instructor's Annotated Edition

This *Instructor's Annotated Edition* identifies various instructional materials that will support student learning of various topics in *General Chemistry*. The IAE comprises the student text and a program of annotations, printed in red, to assist the instructor in syllabus and lecture preparation. These notes do not appear in the student edition of the text. Other marginal material, printed in blue, appears in both the student's and the instructor's versions.

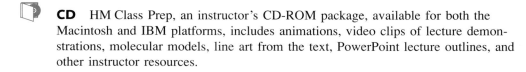 **CD** HM Class Prep, an instructor's CD-ROM package, available for both the Macintosh and IBM platforms, includes animations, video clips of lecture demonstrations, molecular models, line art from the text, PowerPoint lecture outlines, and other instructor resources.

 **Video Lecture Demonstrations** Videos of lecture demonstrations are available on four sets of videocassettes (Series A, Series B, Series C, and Series D). The demonstrations on Series C are also found on the CD and the laserdisc. Each video reference is denoted by a special icon depending on whether the video occurs only on the videotape or on the CD and laserdisc as well.

 **Laserdisc** A laserdisc is available containing the 34 lecture demonstrations on videocassette C, the animations found on the CD, and the line art from the text.

**References to *Experiments in General Chemistry*** Marginal references to the experiments in this laboratory manual appear near related text material.

**References to Lecture Demonstrations** References are given to literature sources and to descriptions in the *Instructor's Resource Manual* for demonstrations that you can use in your lectures to dramatize or emphasize a point.

**Transparencies** The instructional package includes over 250 overhead transparencies selected from the text. These figures can also be accessed on the instructor's CD.

**Alternate Examples** The *Instructor's Resource Manual* contains one or more alternate examples for each of the examples in the text.

**Comments** The annotation at the beginning of each chapter comments briefly on the chapter, giving some ideas for alternate placement in the syllabus or ideas for abbreviation of the material. Other marginal notes present information that might be useful in lectures or as asides.

Optical fibers that use light for data transmission.

# Chemistry and Measurement

I n 1964 Barnett Rosenberg and his coworkers at Michigan State University were studying the effects of electricity on bacterial growth. They inserted platinum electrodes, or electrical connections, into a live bacterial culture and allowed an electric current to pass. After 1 to 2 hours, they noted that cell division in the bacteria stopped. The researchers were very surprised by this result, but even more surprised by the explanation. They were able to show that cell division was inhibited by a substance containing platinum, produced from the platinum electrodes by the electric current. A substance such as this one, the researchers thought, might be useful as an anticancer drug, because cancer involves runaway cell division. Later research confirmed this view, and today the platinum-containing substance *cisplatin* is a leading anticancer drug (Figure 1.1).

This story illustrates three significant reasons to study chemistry. First, chemistry has important practical applications. The development of lifesaving drugs is one, and a complete list would touch upon most areas of modern technology.

1

**Figure 1.1**
**Barnett Rosenberg.**
Discoverer of the anticancer
activity of cisplatin.

▶ Because the material of this chapter is likely to be more or less familiar to students, you might wish to confine your lecture to a few main points, while stressing to students that they should read the chapter carefully.

Second, chemistry is an intellectual enterprise, a way of explaining our material world. When Rosenberg and his coworkers saw that cell division in the bacteria had ceased, they systematically looked for the chemical substance that caused it to cease. They sought a chemical explanation for the occurrence.

Finally, chemistry figures prominently in other fields. Rosenberg's experiment began as a problem in biology; through the application of chemistry it led to an advance in medicine. Whatever your career plans, you will find your knowledge of chemistry is a useful intellectual tool for making important decisions.

# An Introduction to Chemistry

All of the objects around you—this book, your pen or pencil, and the things of nature such as rocks, water, and plant and animal substances—constitute the *matter* of the universe. Each of the particular kinds of matter, such as a certain kind of paper or plastic or metal, is referred to as a *material*. We can define chemistry as the science of the composition and structure of materials and of the changes that materials undergo.

▶ Tyrian purple is one of the few known natural compounds of bromine; it is a bromine derivative of indigo, another natural dye. The bromine atoms are the large red-brown atoms in Figure 1.2.

One chemist may hope that by understanding certain materials he or she will be able to find a cure for a disease or a solution for an environmental ill. Another chemist may simply want to understand a phenomenon. Because chemistry deals with all materials, it is a subject of enormous breadth. It would be difficult to exaggerate the influence of chemistry on modern science and technology or on our ideas about our planet and the universe. In the section that follows, we will take a brief glimpse at modern chemistry and see some of the ways it has influenced technology, science, and modern thought.

## 1.1  Modern Chemistry: A Brief Glimpse

For thousands of years, human beings have fashioned natural materials into useful products. Modern chemistry certainly has its roots in this endeavor. After the discovery of fire, people began to notice changes in certain rocks and minerals exposed to high temperatures. From these observations came the development of ceramics, glass, and metals, which today are among our most useful materials. Dyes and medicines were other early products obtained from natural substances. For example, the ancient

**Figure 1.2**
**Molecular models of
Tyrian purple and aniline.**
Tyrian purple *(top)* is a dye
that was obtained by the early
Phoenicians from a species of
sea snail. The dye was
eventually synthesized from
aniline *(bottom)*.

▶ Liquid crystals and liquid-
crystal displays are described
in the essay at the end of
Section 11.7.

**Figure 1.3** *(left)*
**A watch that uses a
liquid-crystal display.**
These liquid-crystal displays
are common in watches and
laptop computers.

**Figure 1.4** *(right)*
**Model of a molecule that
forms a liquid crystal.**
Note that the molecule has a
rodlike shape.

Phoenicians extracted a bright purple dye, known as Tyrian purple, from a species of sea snail. One ounce of Tyrian purple required over 200,000 snails. Because of its brilliant hue and scarcity, the dye became the choice of royalty.

Although chemistry has its roots in early technology, chemistry as a field of study based on scientific principles came into being only in the latter part of the eighteenth century. Chemists began to look at the precise quantities of substances they used in their experiments. From this work came the central principle of modern chemistry: the materials around us are composed of exceedingly small particles called *atoms,* and the precise arrangement of these atoms into *molecules* or more complicated structures accounts for the many different characteristics of materials. Once chemists understood this central principle, they could begin to fashion molecules to order. They could *synthesize* molecules; that is, they could build large molecules from small ones. Tyrian purple, for example, was eventually synthesized from the simpler molecule aniline; see Figure 1.2. Chemists could also correlate molecular structure with the characteristics of materials and so begin to fashion materials with special characteristics.

The liquid-crystal displays (LCDs) that you see in watches, calculators, and similar devices (Figure 1.3) are an example of an application that depends on the special characteristics of materials. The liquid crystals used in these displays are a form of matter intermediate in characteristics between those of liquids and those of solid crystals; hence the name. These liquid crystals are composed of rodlike molecules that tend to align themselves something like the wood matches in a matchbox. The liquid crystals are held between thin plates that align the molecules in a particular direction, giving the normal light-gray background of the display. These plates are covered with small electrodes, and when any one of them is electrified, the nearby molecules of the liquid crystal are realigned so that they point in a new direction, changing the gray in that area to black. Figure 1.4 shows a model of one of the molecules that forms a liquid crystal; note the rodlike shape of the molecule. Chemists have designed many similar molecules for liquid-crystal applications. ◀

Chemists continue to develop new materials and to discover new properties of old ones. Electronics and communications, for example, have been completely transformed by technological advances in materials. Optical-fiber cables have replaced long-distance telephone cables made of copper wire. Optical fibers are fine threads of extremely pure glass. Because of their purity, these fibers can transmit laser light pulses for miles compared with only a few inches in ordinary glass. Not only is

**Figure 1.5** *(left)*
**Comparison of optical-fiber and copper-wire cables.** The fine optical fiber (in the upper center part of the photograph) can carry much more information than copper-wire cables.

**Figure 1.6** *(right)*
**A gallium-arsenide chip.** Using this chip, IBM scientists demonstrated that they could transform light pulses into electronic computer signals at the rate of 3 billion bits of information a second (equivalent to reading 40 encyclopedia volumes a second).

**Figure 1.7**
**A computer-generated model of a fragment of a DNA molecule.** DNA contains the hereditary information of an organism that is passed on from one generation to the next.

optical-fiber cable cheaper and less bulky than copper cable carrying the same information (Figure 1.5), but by using different colors of light, optical-fiber cable can carry voice, data, and video information at the same time. At the ends of an optical-fiber cable, devices using other new materials convert the light pulses to electrical signals and back, while computer chips constructed from still other materials process the signals (Figure 1.6).

Chemistry has also affected the way we think of the world around us. For example, biochemists and molecular biologists—scientists who study the molecular basis of living organisms—have made a remarkable finding: all forms of life appear to share many of the same molecules and molecular processes. Consider the information of inheritance, the genetic information that is passed on from one generation of organism to the next. Individual organisms, whether bacteria or human beings, store this information in a particular kind of molecule called deoxyribonucleic acid, or DNA for short (Figure 1.7).

DNA consists of two intertwined molecular chains; each chain consists of links of four different types of molecular pieces, or bases. Just as you record information on a page by stringing together characters (letters, numbers, spaces, and so on), an organism stores the information for reproducing itself in the order of these bases in its DNA. In a multicellular organism, such as a human being, every cell contains the same DNA.

The atomic theory of matter, which forms the basis of modern chemistry, was the work of the British chemist John Dalton (1766–1844). Throughout his life, Dalton maintained an interest in the science of weather and climate. This interest led Dalton to study the atmosphere and to speculate on its fundamental structure, which eventually led him to his atomic theory. If Dalton were alive today, he might well be conducting experiments designed to answer some of the current questions in atmospheric chemistry: How dramatically is the earth's climate being changed by human activity and the burning of fossil fuels? What are the chemical processes involved in the depletion of ozone (a form of oxygen) over Antarctica and how will this depletion affect life on earth?

One of our first projects will be to look at this central concept of chemistry, the atomic theory of matter. We will do that in the next chapter, but first we must lay the groundwork for this discussion. We will need some basic vocabulary to talk about science and to describe materials; then we will need to discuss measurement and units, because measurement is critical for quantitative work.

## 1.2 Experiment and Explanation

Experiment and explanation are the heart of chemical research. A chemist makes observations under circumstances in which variables, such as temperature and amounts of substances, can be controlled. An **experiment** is *an observation of natural phenomena carried out in a controlled manner so that the results can be duplicated and rational conclusions obtained.* In the chapter opening it was mentioned that Rosenberg studied the effects of electricity on bacterial growth. Temperature and amounts of nutrients in a given volume of bacterial medium are important variables in such experiments. Unless these variables are controlled, the work cannot be duplicated, nor can any reasonable conclusion be drawn.

After a series of experiments, perhaps a researcher sees some relationship or regularity in the results. For instance, Rosenberg noted that in each experiment in which an electric current was passed through a bacterial culture by means of platinum electrodes, the bacteria ceased dividing. If the regularity or relationship is fundamental and we can state it simply, we call it a law. A **law** is *a concise statement or mathematical equation about a fundamental relationship or regularity of nature.* An example is the law of conservation of mass, which says that the mass, or quantity of matter, remains constant during any chemical change.

At some point in a research project, a scientist tries to make sense of the results by devising an explanation. Explanations help us organize knowledge and predict future events. A **hypothesis** is *a tentative explanation of some regularity of nature.* Having seen that bacteria ceased to divide when an electric current from platinum electrodes passed through the culture, Rosenberg was eventually able to propose the hypothesis that certain platinum compounds were responsible. If a hypothesis is to be useful, it should suggest new experiments that become tests of the hypothesis. Rosenberg could test his hypothesis by looking for the platinum compound and testing for its ability to inhibit cell division.

If a hypothesis successfully passes many tests, it becomes known as a theory. A **theory** is *a tested explanation of basic natural phenomena.* An example is the molecular theory of gases—the theory that all gases are composed of very small particles called molecules. This theory has withstood many tests and has been fruitful in suggesting many experiments. Note that we cannot prove a theory absolutely. It is always possible that further experiments will show the theory to be limited or that someone will develop a better theory. For example, the physics of the motion of objects devised by Isaac Newton withstood experimental tests for more than two centuries, until physicists discovered that the equations do not hold for objects moving near the speed of light. Later physicists showed that very small objects also do not follow Newton's equations. Both discoveries resulted in revolutionary developments in physics. The first led to the theory of relativity; the second, to quantum mechanics, which has had an immense impact on chemistry.

▶ Video no. 1 (Series D), "Color Spectrum."

The two aspects of science, experiment and explanation, are closely related. A scientist performs experiments and observes some regularity; someone explains this regularity and proposes more experiments; and so on. From his experiments, Rosenberg explained that certain platinum compounds inhibit cell division. This explanation led him to do new experiments on the anticancer activity of these compounds.

The *general* process of advancing scientific knowledge through observation; the framing of laws, hypotheses, or theories; and the conducting of more experiments is called the *scientific method* (Figure 1.8). It is not a method for carrying out a *specific* research program, because the design of experiments and the explanation of results draw on the creativity and individuality of a researcher.

**Figure 1.8**
**A representation of the scientific method.** The flow diagram shows the general steps in the scientific method. At the right, Rosenberg's work on the development of an anticancer drug illustrates the steps.

**General Steps**

**Experiments**

**Results**

**Hypothesis**

**Further experiments** devised based on hypothesis

Negative results lead to modification or rejection of hypothesis and formulation of new hypothesis

Positive results support hypothesis

A **theory** follows after results consistently support a hypothesis

**Further experiments**

**Rosenberg's Work**

Platinum electrodes are inserted into a live bacterial culture. Variables controlled:
- amount of nutrients in a given volume of bacterial medium
- temperature
- time

Bacteria ceased dividing.

Certain platinum compounds inhibit cell division.

Look for platinum compounds in bacterial culture.
Further test platinum compounds' ability to inhibit cell division.

Cisplatin, recovered from bacterial culture. When cisplatin is added to a new culture, the bacteria cease dividing.

Certain platinum compounds inhibit cell division.

Experiments to determine the anticancer activity of platinum compounds.

# The Birth of the Post-it Note®

**EVERYDAY LIFE**

Have you ever used a Post-it and wondered where the idea for those little sticky notes came from? You have a chemist to thank for its invention. The story of the Post-it Note is one that illustrates how the creativity and insights of a scientist can result in a product that is as common in the office as the stapler or pen.

In the early 1970s, Art Fry, a 3M scientist, was standing in the choir at his church trying to keep track of all the little bits of paper that marked the music selections for the service. During the service, a number of the markers fell out of the music, making him lose his place. While standing in front of the congregation, he realized that he needed a bookmark that would stick to the book, wouldn't hurt the book, and could be easily detached. To make his plan work, he required an adhesive that would not *permanently* stick things together. Finding the appropriate adhesive was not as simple as it may seem, because most adhesives at that time were created to stick things together permanently.

Still thinking about his problem the next day, Fry consulted a colleague, Spencer Silver, who was studying adhesives at the 3M research labs. The study consisted of conducting a series of tests on a range of adhesives to determine the strength of the bond they formed. One of the adhesives that Silver created for the study was an adhesive that always remained sticky. Fry recognized that this adhesive was just what he needed for his bookmark. His first bookmark, invented the day after the initial idea, consisted of a strip of Silver's tacky adhesive applied to the edge of a piece of paper.

Part of Fry's job description at 3M was to spend time working on creative ideas such as his bookmark. As a result he continued to experiment with the bookmark to improve its properties of sticking, detaching, and not hurting the surface to which it was attached. One day, while working on some paperwork, he wrote a question on one of his experimental strips of paper and sent it to his boss stuck to the top of a file folder. His boss then answered the question on the note and returned it attached to some other documents. During a later discussion over coffee, Fry and his boss realized that they had invented a new way for people to communicate: the Post-it Note was born. The Post-it Note is one of the top five office products sold in the United States.

## 1.3 Law of Conservation of Mass

Modern chemistry emerged in the eighteenth century, when chemists began to use the balance systematically as a tool in research. Balances measure **mass,** which is *the quantity of matter in a material* (Figure 1.9). **Matter** is the general term for the material things around us; we can define it as *whatever occupies space and can be perceived by our senses.*

Antoine Lavoisier (1743–1794), a French chemist, was one of the first to insist on the use of the balance in chemical research. By weighing substances before and after chemical change, he demonstrated the **law of conservation of mass,** which states that *the total mass remains constant during a chemical change (chemical reaction).* ◄

In a series of experiments, Lavoisier applied the law of conservation of mass to clarify the phenomenon of burning, or combustion. He showed that when a material burns, a component of air (which he called oxygen) combines chemically with the material. For example, when the liquid metal mercury is heated in air, it burns or combines with oxygen to give a red-orange substance, whose modern name is mercury(II) oxide. We can represent the chemical change as follows:

► Chemical reactions may involve a gain or loss of heat and other forms of energy. According to Einstein, mass and energy are equivalent. Thus, when energy is lost as heat, mass is also lost, but changes of mass in chemical reactions (billionths of a gram) are too small to detect.

$$\text{Mercury} + \text{oxygen} \longrightarrow \text{mercury(II) oxide}$$

The arrow means "is changed to." See Figure 1.10.

**Figure 1.9**
**Laboratory balances.** *Left:* A simple two-pan balance. Material to be weighed is placed on the left pan and is balanced with weights of known mass placed on the right pan. *Right:* A modern single-pan balance. The mass of the material on the pan appears on the digital readout.

**Figure 1.10**
**Heating mercury metal in air.** *Left:* Mercury metal reacts with oxygen to yield mercury(II) oxide. The color of the oxide varies from red to yellow depending on the particle size. *Right:* When you heat mercury(II) oxide, it decomposes to mercury and oxygen gas.

By strongly heating the red-orange substance, Lavoisier was able to decompose it to yield the original mercury and oxygen gas. The following example illustrates how the law of conservation of mass can be used to study this reaction.

► An alternate example is provided in the Instructor's Resource Manual.

### EXAMPLE 1.1

## Using the Law of Conservation of Mass

You heat 2.53 grams of metallic mercury in air, which produces 2.73 grams of a red-orange residue. Assume that the chemical change is the reaction of the metal with oxygen in air.

$$\text{Mercury} + \text{oxygen} \longrightarrow \text{red-orange residue}$$

What is the mass of oxygen that reacts? When you strongly heat the red-orange residue, it decomposes to give back the mercury and release the oxygen, which you collect. What is the mass of oxygen you collect?

 **PROBLEM STRATEGY**

You apply the law of conservation of mass to the reaction. According to this law, the total mass remains constant during a chemical reaction; that is,

Mass of substances before reaction = mass of substances after reaction

**SOLUTION**

From the law of conservation of mass,

Mass of mercury + mass of oxygen = mass of red-orange residue

Substituting, you obtain

2.53 grams + mass of oxygen = 2.73 grams

or

Mass of oxygen = (2.73 − 2.53) grams = **0.20 grams**

The mass of oxygen collected when the red-orange residue decomposes equals the mass of oxygen that originally reacted **(0.20 grams).**

---

**EXERCISE 1.1**

You place 1.85 grams of wood in a vessel with 9.45 grams of air and seal the vessel. Then you heat the vessel strongly so that the wood burns. In burning, the wood yields ash and gases. After the experiment, you weigh the ash and find that its mass is 0.28 grams. What is the mass of the gases in the vessel at the end of the experiment?

---

See Problems 1.29, 1.30, 1.31, and 1.32.

Lavoisier set out his views on chemistry in his book, *Traité Élémentaire de Chimie* (Basic Treatise on Chemistry), in 1789. The book was very influential, especially among younger chemists, and set the stage for modern chemistry.

Before leaving this section, you should note the distinction between the terms *mass* and *weight* in precise usage. The weight of an object is the force of gravity exerted on it. The weight is proportional to the mass of the object divided by the square of the distance between the center of mass of the object and that of the earth. ◄ Because the earth is slightly flattened at the poles, an object weighs more at the North Pole, where it is closer to the center of the earth, than at the equator. The mass of an object is the same wherever it is measured.

► The force of gravity $F$ between objects whose masses are $m_1$ and $m_2$ is $Gm_1m_2/r^2$, where $G$ is the gravitational constant and $r$ is the distance between the centers of mass of the two objects.

## 1.4   Matter: Physical State and Chemical Constitution

► Experiments in General Chemistry, 1A, "Identification of an Unknown Compound."

We describe iron as a silvery-colored metal that melts at 1535°C (2795°F). Once we have collected enough descriptive information about many different kinds of matter, patterns emerge that suggest ways of classifying it. There are two principal ways of

Solid

Liquid

Gas

**Figure 1.11**
**Molecular representations of solid, liquid, and gas.**
The top beaker contains a solid with a molecular view of the solid; the molecular view depicts the closely packed, immobile atoms that make up the solid structure. The middle beaker contains a liquid with a molecular view of the liquid; the molecular view depicts atoms that are close together but moving freely. The bottom beaker contains a gas with a molecular view of the gas; the molecular view depicts atoms that are far apart and moving freely.

**Understanding Concepts:
States and Classification
of Matter**

classifying matter: by its physical state as a solid, liquid, or gas and by its chemical constitution as an element, compound, or mixture.

## Solids, Liquids, and Gases

Commonly, a given kind of matter exists in different physical forms under different conditions. Water, for example, exists as ice (solid water), as liquid water, and as steam (gaseous water) (Figure 1.11). The main identifying characteristic of solids is their rigidity: they tend to maintain their shapes when subjected to outside forces. Liquids and gases, however, are *fluids;* that is, they flow easily and change their shapes in response to slight outside forces.

What distinguishes a gas from a liquid is the characteristic of *compressibility* (and its opposite, *expansibility*). A gas is easily compressible, whereas a liquid is not. You can put more and more air into a tire, which increases only slightly in volume. In fact, a given quantity of gas can fill a container of almost any size. A small quantity would expand to fill the container; a larger quantity could be compressed to fill the same space. By contrast, if you were to try to force more liquid water into a closed glass bottle that was already full of water, it would burst.

These two characteristics, rigidity (or fluidity) and compressibility (or expansibility), can be used to frame definitions of the three states of matter:

**solid**  *the form of matter characterized by rigidity;* a solid is relatively incompressible and has fixed shape and volume.

**liquid**  *the form of matter that is a relatively incompressible fluid;* a liquid has a fixed volume but no fixed shape.

**gas**  *the form of matter that is an easily compressible fluid;* a given quantity of gas will fit into a container of almost any size and shape.

The term *vapor* is often used to refer to the gaseous state of any kind of matter that normally exists as a liquid or a solid.

*The three forms of matter—solid, liquid, and gas—*are referred to as the **states of matter.** Their characteristics are summarized in Table 1.1.

## Elements, Compounds, and Mixtures

To understand how matter is classified by its chemical constitution, we must first distinguish between physical and chemical changes and between physical and chemical properties. A **physical change** is *a change in the form of matter but not in its chemical identity.* Changes of physical state are examples of physical changes. The process

**Table 1.1
Characteristics of the States of Matter**

| State of Matter | Fluidity or Rigidity | Compressibility |
|---|---|---|
| Solid | Rigid | Very low |
| Liquid | Fluid | Very low |
| Gas | Fluid | High |

**Figure 1.12**
**Separation by distillation.**
You can separate an easily
vaporized liquid from another
substance by distillation.

Thermometer

Coolant
water out

Distillation
flask

Condenser

Coolant
water in

Receiver

of dissolving one material in another is a further example of a physical change. For instance, you can dissolve sodium chloride (table salt) in water. The result is a clear liquid, like pure water, though many of its other characteristics are different from those of pure water. The water and sodium chloride in this liquid retain their chemical identities and can be separated by some method that depends on physical changes.

*Distillation* is one way to separate the sodium chloride and water components of this liquid. You place the liquid in a flask to which a device called a *condenser* is attached (see Figure 1.12). The liquid in the flask is heated to bring it to a boil. (Boiling entails the formation of bubbles of the vapor in the body of the liquid.) Water vapor forms and passes from the flask into the cooled condenser, where the vapor changes back to liquid water. The liquid water is collected in another flask, called a *receiver.* The original flask now contains the solid sodium chloride. Thus, by means of physical changes (the change of liquid water to vapor and back to liquid), you have separated the sodium chloride and water that you had earlier mixed together.

A **chemical change,** or **chemical reaction,** is *a change in which one or more kinds of matter are transformed into a new kind of matter or several new kinds of matter.* The rusting of iron, during which iron combines with oxygen in air to form a new material called rust, is a chemical change. The original materials (iron and oxygen) combine chemically and cannot be separated by any physical means. To recover the iron and oxygen from rust requires a chemical change or a series of chemical changes.

We characterize or identify a material by its various properties, which may be either physical or chemical. A **physical property** is *a characteristic that can be observed for a material without changing its chemical identity.* Examples are physical state (solid, liquid, or gas), melting point, and color. A **chemical property** is *a*

**Figure 1.13**
**Reaction of sodium with water.** Sodium metal flits around the water surface as it reacts briskly, giving off hydrogen gas. The other product is sodium hydroxide, which changes a substance added to the water (phenolphthalein) from colorless to pink.

▶ Demonstration of reaction of hydrogen with oxygen: Summerlin and Ealy, *Chemical Demonstrations,* Vol. 1 (Washington, D.C.: Am. Chem. Soc., 1985), no. 108, p. 173.

▶ Video Series C, "Oxygen, Hydrogen, Soap Bubbles, and Balloons."

▶ In Chapter 2, we will redefine an element in terms of atoms.

▶ It is now known that some compounds do not follow the law of definite proportions. These nonstoichiometric compounds, as they are called, are described briefly in Chapter 11.

▶ Chromatography, another example of a physical method used to separate mixtures, is described in the essay at the end of this section.

*characteristic of a material involving its chemical change.* A chemical property of iron is its ability to react with oxygen to produce rust.

**Substances**  The various materials we see around us are either substances or mixtures of substances. A **substance** is *a kind of matter that cannot be separated into other kinds of matter by any physical process.* Earlier you saw that, when sodium chloride is dissolved in water, it is possible to separate the sodium chloride from the water by the physical process of distillation. However, sodium chloride is itself a substance and cannot be separated by physical processes into new materials. Similarly, pure water is a substance.

No matter what its source, a substance always has the same characteristic properties. Sodium is a solid metal having a melting point of 98°C. The metal also reacts vigorously with water (Figure 1.13). No matter how sodium is prepared, it always has these properties. Similarly, whether sodium chloride is obtained by burning sodium in chlorine or from seawater, it is a white solid melting at 801°C.

---

**EXERCISE 1.2**

Potassium is a soft, silvery-colored metal that melts at 64°C. It reacts vigorously with water, with oxygen, and with chlorine. Identify all of the physical properties given in this description. Identify all of the chemical properties given.

**See Problems 1.39, 1.40, 1.41, and 1.42.**

---

**Elements**  Millions of substances have been characterized by chemists. Of these, a very small number are known as elements, from which all other substances are made. Lavoisier was the first to establish an experimentally useful definition of an element. He defined an **element** as *a substance that cannot be decomposed by any chemical reaction into simpler substances.* In 1789 Lavoisier listed 33 substances as elements, of which more than 20 are still so regarded. Today 111 elements are known. Some elements are shown in Figure 1.14. ◀

**Compounds**  Most substances are compounds. A **compound** is *a substance composed of two or more elements chemically combined.* By the end of the eighteenth century, Lavoisier and others had examined many compounds and showed that all of them were composed of the elements in definite proportions by mass. Joseph Louis Proust (1754–1826), by his painstaking work, convinced the majority of chemists of the general validity of the **law of definite proportions** (also known as the **law of constant composition**): *a pure compound, whatever its source, always contains definite or constant proportions of the elements by mass.* For example, 1.0000 gram of sodium chloride always contains 0.3934 gram of sodium and 0.6066 gram of chlorine, chemically combined. Sodium chloride has definite proportions of sodium and chlorine; that is, it has constant or definite composition. ◀

**Mixtures**  Most of the materials around us are mixtures. A **mixture** is *a material that can be separated by physical means into two or more substances.* Unlike a pure compound, a mixture has variable composition. When you dissolve sodium chloride in water, you obtain a mixture; its composition depends on the relative amount of sodium chloride dissolved. You can separate the mixture by the physical process of distillation. ◀

**Figure 1.14**
**Some elements.** *Center:*
Sulfur. *From upper right,*
*clockwise:* Arsenic, iodine,
magnesium, bismuth, mercury.

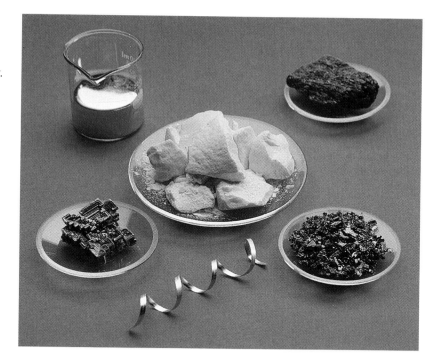

▶ Demonstration of separating
iron from cereal: Summerlin
et al., *Chemical Demonstrations,*
Vol. 2 (Washington, D.C.: Am.
Chem. Soc., 1987), no. 34, p. 62.
(Alternatively, magnetically
separate a mixture of sulfur and
iron filings.)

**Figure 1.15**
**A heterogeneous mixture.**
*Left:* The mixture on the
watch glass consists of
potassium dichromate (orange
crystals) and iron filings. *Right:*
A magnet separates the iron
filings from the mixture.

Mixtures are classified into two types. A **heterogeneous mixture** is *a mixture that consists of physically distinct parts, each with different properties.* Figure 1.15 shows a heterogeneous mixture of potassium dichromate and iron filings. Another example is salt and sugar that have been stirred together. If you were to look closely, you would see the separate crystals of sugar and salt. A **homogeneous mixture** (also known as a **solution**) is *a mixture that is uniform in its properties throughout given samples.* When sodium chloride is dissolved in water, you obtain a homogeneous mixture, or solution. Air is a gaseous solution, principally of two elementary substances, nitrogen and oxygen, which are physically mixed but not chemically combined.

A **phase** is *one of several different homogeneous materials present in the portion of matter under study.* A heterogeneous mixture of salt and sugar is said to be composed

**Figure 1.16**
**Relationships among elements, compounds, and mixtures.** Mixtures can be separated by physical processes into substances, and substances can be combined physically into mixtures. Compounds can be separated by chemical reactions into their elements, and elements can be combined chemically to form compounds.

▶ **Experiments in General Chemistry, 1B, "Separation of a Mixture by Paper Chromatography."**

of two different phases; one of the phases is salt, the other is sugar. Similarly, ice cubes in water is said to be composed of two phases; one phase is ice, the other is liquid water. Ice floating in a solution of sodium chloride in water also consists of two phases, ice and the liquid solution. Note that a phase may be either a pure substance in a particular state or a solution in a particular state (solid, liquid, or gaseous). Also, the portion of matter under consideration may consist of several phases of the same substance or several phases of different substances.

Figure 1.16 summarizes the relationships among elements, compounds, and mixtures. Materials are either substances or mixtures. Substances can be mixed by physical processes, and other physical processes can be used to separate the mixtures into substances. Substances are either elements or compounds. Elements may react chemically to yield compounds, and compounds may be decomposed by chemical reactions into elements.

▶ **Video no. 20 (Series D), "Paper Chromatography."**

### CONCEPT CHECK 1.1

Matter can be represented as being composed of individual units. For example, the smallest individual unit of matter can be represented as a single circle, •, and chemical combinations of these units of matter as connected circles, ••, with each element represented by a different color. Using this model, place the appropriate label—element, compound, or mixture—on each container.

      A                B                C

# Separation of Mixtures by Chromatography

*Chromatography* is a group of similar separation techniques. Each depends on how fast a substance moves, in a stream of gas or liquid, past a stationary phase to which the substance may be slightly attracted. An example is provided by a simple experiment in *paper chromatography* (see Figure 1.17). In this experiment, a line of ink is

*(continued)*

**Figure 1.17**
**An illustration of paper chromatography.** A line of ink has been drawn along the lower edge of a sheet of paper. The dyes in the ink separate as a solution of methanol and water creeps up the paper.

**Figure 1.18**
**Column chromatography.**
*(A)* A solution containing substances to be separated is poured into the top of a column, which contains powdered chalk. *(B)* Pure liquid is added to the column, and the substances begin to separate into bands. *(C)* The substances separate further on the column. Each substance is collected in a separate flask as it comes off the column.

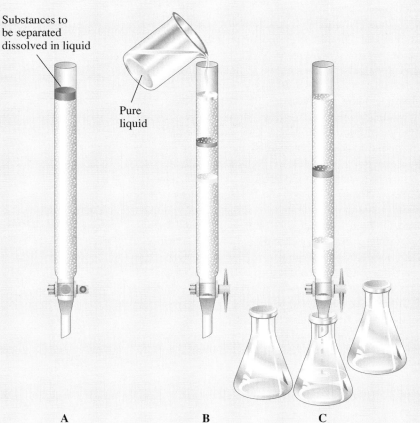

Substances to be separated dissolved in liquid

Pure liquid

A          B          C

drawn near one edge of a sheet of paper, and the paper is placed upright with this edge in a solution of methanol and water. As the solution creeps up the paper, the ink moves upward, separating into a series of different-colored bands that correspond to the different dyes in the ink. Each dye is attracted to the wet paper fibers but with different strengths of attraction. As the solution moves upward, the dyes less strongly attracted to the paper fibers move more rapidly.

The Russian botanist Mikhail Tswett was the first to understand the basis of chromatography and to apply it systematically as a method of separation. In 1906 Tswett separated pigments in plant leaves by *column chromatography.* He first dissolved the pigments from the leaves in petroleum ether, a liquid similar to gasoline. After packing a glass tube or column with powdered chalk, he poured the solution of pigments into the top of the column (see Figure 1.18). When he washed the column by pouring in more petroleum ether, it began to show distinct yellow and green bands. These bands, each containing a pure pigment, became well separated as they moved down the column, so that the pure pigments could be obtained. The name chromatography originates from this early separation of colored substances (the stem *chromato-* means "color"), although the technique is not limited to colored substances.

*Gas chromatography* (GC) is a more recent separation method. Here the moving stream is a gaseous mixture of vaporized substances plus a gas such as helium, which is called the *carrier.* The stationary material is either a solid or a liquid adhering to a solid, packed in a column. As the gas passes through the column, substances in the mixture are attracted differently to the stationary column packing and thus are separated. Gas chromatography is a rapid, small-scale method of separating mixtures. It is also important in the analysis of mixtures because the time it takes for a substance at a given temperature to travel through the column to a detector (called the *retention time*) is fixed. You can therefore use retention times to help identify substances. Figure 1.19 shows a gas chromatograph and a portion of a computer plot *(chromatogram).* Each peak on the chromatogram corresponds to a specific substance. The peaks were automatically recorded by the instrument as the different substances in the mixture passed the detector. Chemists have analyzed complicated mixtures by gas chromatography. Analysis of chocolate, for example, shows that it contains over 800 flavor compounds.

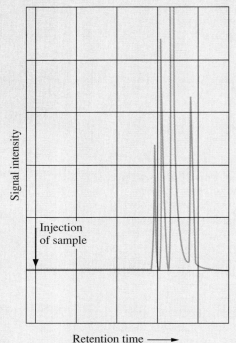

**Figure 1.19**
**Gas chromatography.** *Top:* A modern gas chromatograph. *Bottom:* This is a chromatogram of a hexane mixture, showing its separation into four different substances. Such hexane mixtures occur in gasoline; hexane is also used as a solvent to extract the oil from certain vegetable seeds.

# Physical Measurements

Chemists characterize and identify substances by their particular properties. To determine many of these properties requires physical measurements. Cisplatin, the substance featured in the chapter opening, is a yellow substance whose solubility in water (the amount that dissolves in a given quantity of water) is 0.252 gram in 100 grams of water. You can use the solubility, as well as other physical properties, to identify cisplatin. In a modern chemical laboratory, measurements often are complex, but many experiments begin with simple measurements of mass, volume, time, and so forth. In the next few sections, we will look at the measurement process. We first consider the concept of *precision* of a measurement.

## 1.5 Measurement and Significant Figures

▶ **2.54 centimeters = 1 inch**

Measurement is the comparison of a physical quantity to be measured with a **unit** of measurement—that is, with *a fixed standard of measurement*. On a centimeter scale, the centimeter unit is the standard of comparison. ◀ A steel rod that measures 9.12 times the centimeter unit has a length of 9.12 centimeters. To record the measurement, you must be careful to give both the *measured number* (9.12) and the *unit* (centimeters).

If you repeat a particular measurement, you usually do not obtain precisely the same result, because each measurement is subject to experimental error. The measured values vary slightly from one another. Suppose you perform a series of identical measurements of a quantity. The term **precision** refers to *the closeness of the set of values obtained from identical measurements of a quantity*. **Accuracy** is a related term; it refers to *the closeness of a single measurement to its true value*. To illustrate the idea of precision, consider a simple measuring device, the centimeter ruler. In Figure 1.20, a steel rod has been placed near a ruler subdivided into tenths of a centimeter. You can see that the rod measures just over 9.1 cm (cm = centimeter). With care, it is possible to estimate by eye to hundredths of a centimeter. Here you might give the measurement as 9.12 cm. Suppose you measure the length of this rod twice more. You find the values to be 9.11 cm and 9.13 cm. Thus, you record the length of the rod as being somewhere between 9.11 cm and 9.13 cm. The spread of values indicates the precision with which a measurement can be made by this centimeter ruler. ◀

▶ **Measurements that are of high precision are usually accurate. It is possible, however, to have a systematic error in a measurement. Suppose that, in calibrating a ruler, the first centimeter is made too small by 0.1 cm. Then, although the measurements of length on this ruler are still precise to 0.01 cm, they are accurate to only 0.1 cm.**

To indicate the precision of a measured number (or result of calculations on measured numbers), we often use the concept of significant figures. **Significant figures** are *those digits in a measured number (or result of a calculation with measured numbers) that include all certain digits plus a final one having some uncertainty*. When you

**Figure 1.20**
**Precision of measurement with a centimeter ruler.**
The length of the rod is just over 9.1 cm. On successive measurements, we estimate the length by eye at 9.12, 9.11, and 9.13 cm. We record the length as between 9.11 cm and 9.13 cm.

measured the rod, you obtained the values 9.12 cm, 9.11 cm, and 9.13 cm. You could report the result as the average, 9.12 cm. The first two digits (9.1) are certain; the next digit (2) is estimated, so it has some uncertainty. It would be incorrect to write 9.120 cm for the length of the rod. This would say that the last digit (0) has some uncertainty but that the other digits (9.12) are certain, which is not true.

## Number of Significant Figures

**Number of significant figures** refers to *the number of digits reported for the value of a measured or calculated quantity, indicating the precision of the value.* Thus, there are three significant figures in 9.12 cm, whereas 9.123 cm has four. To count the number of significant figures in a given measured quantity, you observe the following rules:

1. All digits are significant except zeros at the beginning of the number and possibly terminal zeros (one or more zeros at the end of a number). Thus, 9.12 cm, 0.912 cm, and 0.00912 cm all contain three significant figures.

2. Terminal zeros ending at the right of the decimal point are significant. Each of the following has three significant figures: 9.00 cm, 9.10 cm, 90.0 cm.

3. Terminal zeros in a number without an explicit decimal point may or may not be significant. If someone gives a measurement as 900 cm, you do not know whether one, two, or three significant figures are intended. If the person writes 900. cm (note the decimal point), the zeros are significant. More generally, you can remove any uncertainty in such cases by expressing the measurement in scientific notation.

**Scientific notation** is *the representation of a number in the form A × 10ⁿ, where A is a number with a single nonzero digit to the left of the decimal point and n is an integer, or whole number.* In scientific notation, the measurement 900 cm precise to two significant figures is written $9.0 \times 10^2$ cm. If precise to three significant figures, it is written $9.00 \times 10^2$ cm. Scientific notation is also convenient for expressing very large or very small quantities. It is much easier (and simplifies calculations) to write the speed of light as $3.00 \times 10^8$ (rather than as 300,000,000) meters per second. ◄

▶ **See Appendix A for a review of scientific notation.**

## Significant Figures in Calculations

Measurements are often used in calculations. How do you determine the correct number of significant figures to report for the answer to a calculation? The following are two rules that we use:

1. **Multiplication and division.** When multiplying or dividing measured quantities, give as many significant figures in the answer as there are in the measurement with *the least number of significant figures.*

2. **Addition and subtraction.** When adding or subtracting measured quantities, give the same number of decimal places in the answer as there are in the measurement with *the least number of decimal places.*

Let us see how you apply these rules. Suppose you have a substance believed to be cisplatin, and, in an effort to establish its identity, you measure its solubility (the

**Figure 1.21**
**Significant figures and calculators.**   Not all of the figures that appear on a calculator display are significant. In performing the calculation 100.0 × 0.0634 ÷ 25.31, the calculator display shows 0.2504938. We would report the answer as 0.250, however, because the factor 0.0634 has the least number of significant figures (three).

amount that dissolves in a given quantity of water). You find that 0.0634 gram of the substance dissolves in 25.31 grams of water. The amount dissolving in 100.0 grams is

$$100.0 \text{ grams of water} \times \frac{0.0634 \text{ gram cisplatin}}{25.31 \text{ grams of water}}$$

Performing the arithmetic on a pocket calculator (Figure 1.21), you get 0.2504938 for the numerical part of the answer (100.0 × 0.0634 ÷ 25.31). It would be incorrect to give this number as the final answer. The measurement 0.0634 gram has the least number of significant figures (three). Therefore, you report the answer to three significant figures—that is, 0.250 gram.

Now consider the addition of 184.2 grams and 2.324 grams. On a calculator, you find that 184.2 + 2.324 = 186.524. But because the quantity 184.2 grams has the least number of decimal places—one, whereas 2.324 grams has three—the answer is 186.5 grams.

## Exact Numbers

So far we have discussed only numbers that involve uncertainties. However, you will also encounter exact numbers. An **exact number** is *a number that arises when you count items or sometimes when you define a unit*. For example, when you say there are 9 coins in a bottle, you mean exactly 9, not 8.9 or 9.1. Also, when you say there are 12 inches to a foot, you mean exactly 12. Similarly, the inch is defined to be exactly 2.54 centimeters. The conventions of significant figures do not apply to exact numbers. Thus, the 2.54 in the expression "1 inch equals 2.54 centimeters" should not be interpreted as a measured number with three significant figures. In effect, the 2.54 has an infinite number of significant figures, but of course it would be impossible to write out an infinite number of digits. You should make a mental note of any numbers in a calculation that are exact, because they have no effect on the number of significant figures in a calculation. The number of significant figures in a calculation result depends only on the numbers of significant figures in quantities having uncertainties. For example, suppose you want the total mass of 9 coins when each coin has a mass of 3.0 grams. The calculation is

$$3.0 \text{ grams} \times 9 = 27 \text{ grams}$$

You report the answer to two significant figures because 3.0 grams has two significant figures. The number 9 is exact and does not determine the number of significant figures.

## Rounding

In reporting the solubility of your substance in 100.0 grams of water as 0.250 gram, you *rounded* the number you read off the calculator (0.2504938). **Rounding** is *the procedure of dropping nonsignificant digits in a calculation result and adjusting the last digit reported*. The general procedure is as follows. Look at the leftmost digit to be dropped.

1. If this digit is 5 or greater, add 1 to the last digit to be retained and drop all digits farther to the right. Thus, rounding 1.2151 to three significant figures gives 1.22.

2. If this digit is less than 5, simply drop it and all digits farther to the right. Rounding 1.2143 to three significant figures gives 1.21.

In doing a calculation of two or more steps, it is desirable to retain nonsignificant digits for intermediate answers. This ensures that accumulated small errors from rounding do not appear in the final result. If you use a calculator, you can simply enter numbers one after the other, performing each arithmetic operation and rounding only the final answer. ◀ To keep track of the correct number of significant figures, you may want to record intermediate answers with a line under the last significant figure, as shown in the solution to part d of the following example.

▶ **The answers for Exercises and Problems will follow this practice. For most Example Problems where intermediate answers are reported, all answers, intermediate and final, will be rounded.**

▶ **An alternate example is provided in the Instructor's Resource Manual.**

## EXAMPLE 1.2

## Using Significant Figures in Calculations

Perform the following calculations and round the answers to the correct number of significant figures (units of measurement have been omitted).

a. $\dfrac{2.568 \times 5.8}{4.186}$

b. $5.41 - 0.398$

c. $3.38 - 3.01$

d. $4.18 - 58.16 \times (3.38 - 3.01)$

 **PROBLEM STRATEGY**

Parts a, b, and c are relatively straightforward; use the rule for significant figures that applies in each case. In a multistep calculation as in d, proceed step by step, applying the relevant rule to each step. First, do the operations within parentheses. By convention, you perform multiplications and divisions before additions and subtractions. Do not round intermediate answers, but do keep track of the least significant digit, say by underlining it.

**SOLUTION**

a. The factor 5.8 has the fewest significant figures; therefore, the answer should be reported to two significant figures. Round the answer to **3.6.** b. The number with the least number of decimal places is 5.41. Therefore, round the answer to two decimal places, to **5.01.** c. The answer is **0.37.** Note how you have lost one significant figure in the subtraction. d. You first do the subtraction within parentheses, underlining the least significant digits.

$$4.1\underline{8} - 58.1\underline{6} \times (3.3\underline{8} - 3.0\underline{1}) = 4.1\underline{8} - 58.1\underline{6} \times 0.3\underline{7}$$

Following convention, you do the multiplication before the subtraction.

$$4.1\underline{8} - 58.1\underline{6} \times 0.3\underline{7} = 4.1\underline{8} - 2\underline{1}.5192 = -1\underline{7}.3392$$

The final answer is **−17.**

**Problem Solving: Significant Figures in Mathematical Operations**

## EXERCISE 1.3

Give answers to the following arithmetic setups. Round to the correct number of significant figures.

a. $\dfrac{5.61 \times 7.891}{9.1}$

b. $8.91 - 6.435$

c. $6.81 - 6.730$

d. $38.91 \times (6.81 - 6.730)$

See Problems 1.53 and 1.54.

## CONCEPT CHECK 1.2

a. When you report your weight to someone, how many significant figures do you typically use?

b. What is your weight with two significant figures?

c. Indicate your weight and the number of significant figures you would obtain if you weighed yourself on a truck scale that can measure in 50-kg or 100-lb increments.

# 1.6 SI Units

The first measurements were probably based on the human body (the length of the foot, for example). In time, fixed standards developed, but these varied from place to place. Each country or government (and often each trade) adopted its own units. As science became more quantitative in the seventeenth and eighteenth centuries, scientists found that the lack of standard units was a problem. ◄ They began to seek a simple, international system of measurement. In 1791 a study committee of the French Academy of Sciences devised such a system. Called the *metric system,* it became the official system of measurement for France and was soon used by scientists throughout the world. Most nations have since adopted the metric system or, at least, have set a schedule for changing to it.

▶ In the system of units that Lavoisier used in the eighteenth century, there were 9216 grains to the pound (the *livre*). English chemists of the same period used a system in which there were 7000 grains to the pound—unless they were trained as apothecaries, in which case there were 5760 grains to the pound!

## SI Base Units and SI Prefixes

In 1960 the General Conference of Weights and Measures adopted the **International System** of units (or **SI,** after the French *le Système International d'Unités*), which is *a particular choice of metric units.* This system has seven **SI base units,** *the SI units from which all others can be derived.* Table 1.2 lists these base units and the symbols used to represent them. In this chapter, we will discuss four base quantities: length, mass, time, and temperature. ◄

One advantage of any metric system is that it is a decimal system. In SI, a larger or smaller unit for a physical quantity is indicated by an **SI prefix,** which is *a prefix used in the International System to indicate a power of ten.* For example, the base

▶ The amount of substance is discussed in Chapter 3, and the electric current (ampere) is introduced in Chapter 20. Luminous intensity will not be used in this book.

**Table 1.2
SI Base Units**

| Quantity | Unit | Symbol |
| --- | --- | --- |
| Length | meter | m |
| Mass | kilogram | kg |
| Time | second | s |
| Temperature | kelvin | K |
| Amount of substance | mole | mol |
| Electric current | ampere | A |
| Luminous intensity | candela | cd |

**Table 1.3**
**SI Prefixes***

| Multiple | Prefix | Symbol |
|----------|--------|--------|
| $10^{18}$ | exa | E |
| $10^{15}$ | peta | P |
| $10^{12}$ | tera | T |
| $10^{9}$ | giga | G |
| $10^{6}$ | mega | M |
| $10^{3}$ | kilo | k |
| $10^{2}$ | hecto | h |
| 10 | deka | da |
| $10^{-1}$ | deci | d |
| $10^{-2}$ | centi | c |
| $10^{-3}$ | milli | m |
| $10^{-6}$ | micro | $\mu$† |
| $10^{-9}$ | nano | n |
| $10^{-12}$ | pico | p |
| $10^{-15}$ | femto | f |
| $10^{-18}$ | atto | a |

*The units highlighted in color are the ones most commonly used.
†Greek letter mu, pronounced "mew."

unit of length in SI is the meter (somewhat longer than a yard), and $10^{-2}$ meter is called a *centi*meter. Thus 2.54 centimeters equals $2.54 \times 10^{-2}$ meters. The SI prefixes are listed in Table 1.3. Only those shown in color will be used in this book: *mega-* ($10^{6}$), *kilo-* ($10^{3}$), *deci-* ($10^{-1}$), *centi-* ($10^{-2}$), *milli-* ($10^{-3}$), *micro-* ($10^{-6}$), *nano-* ($10^{-9}$), and *pico-* ($10^{-12}$).

## Length, Mass, and Time

▶ The meter was originally defined in terms of a standard platinum–iridium bar kept at Sèvres, France. In 1983, the meter was defined as the distance traveled by light in a vacuum in 1/299,792,458 seconds.

The **meter (m)** is *the SI base unit of length.* ◀ By combining it with one of the SI prefixes, you can get a unit of appropriate size for any length measurement. For the very small lengths used in chemistry, the nanometer (nm; 1 nanometer = $10^{-9}$ m) or the picometer (pm; 1 picometer = $10^{-12}$ m) is an acceptable SI unit. *A non-SI unit of length* traditionally used by chemists is the **angstrom (Å),** which equals $10^{-10}$ m. (An oxygen atom, one of the minute particles of which the substance oxygen is composed, has a diameter of about 1.3 Å. If you could place oxygen atoms adjacent to one another, you could line up over 75 million of them in 1 cm.)

▶ The present standard of mass is the platinum–iridium kilogram mass kept at the International Bureau of Weights and Measures in Sèvres, France.

The **kilogram (kg)** is *the SI base unit of mass,* equal to about 2.2 pounds. ◀ This is an unusual base unit in that it contains a prefix. In forming other SI mass units, prefixes are added to the word *gram* (g) to give units such as the *milli*gram (mg; 1 mg = $10^{-3}$ g).

The **second (s)** is *the SI base unit of time* (Figure 1.22). Combining this unit with prefixes such as *milli-, micro-, nano-,* and *pico-,* you create units appropriate for measuring very rapid events. For example, computer workstations can add two 10-digit

**Figure 1.22**
**NIST F-1 cesium fountain clock.**   First put into service in 1999 at the National Institute of Standards and Technology (NIST) in Boulder, Colorado, this clock represents a new generation of super accurate atomic clocks. The clock is so precise that it only loses one second every 20 million years.

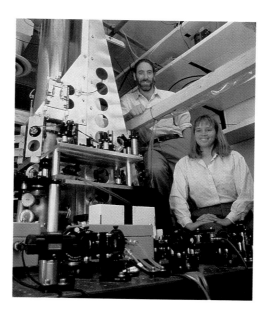

numbers in roughly a nanosecond; a supercomputer can perform the same calculation in about a picosecond. The time required for the fastest chemical processes is about a picosecond. When you measure times much longer than a few hundred seconds, you revert to *minutes* and *hours,* an obvious exception to the prefix–base format of the International System.

---

**EXERCISE 1.4**

Express the following quantities using an SI prefix and a base unit. For instance, $1.6 \times 10^{-6}$ m $= 1.6$ $\mu$m. A quantity such as 0.000168 g could be written 0.168 mg or 168 $\mu$g.

a. $1.84 \times 10^{-9}$ m        b. $5.67 \times 10^{-12}$ s        c. $7.85 \times 10^{-3}$ g

d. $9.7 \times 10^{3}$ m        e. 0.000732 s        f. 0.000000000154 m

---

See Problems 1.57 and 1.58.

## Temperature

Temperature is difficult to define precisely, but we all have an intuitive idea of what we mean by it. It is a measure of "hotness." A hot object placed next to a cold one becomes cooler, while the cold object becomes hotter. Heat energy passes from a hot object to a cold one, and the quantity of heat passed between the objects depends on the difference in temperature between the two. Therefore, temperature and heat are different, but related, concepts.

A thermometer is a device for measuring temperature. The common type consists of a glass capillary containing a column of liquid whose length varies with temperature. A scale alongside the capillary gives a measure of the temperature. The **Celsius scale** (formerly the centigrade scale) is *the temperature scale in general scientific use*. On this scale, the freezing point of water is 0°C and the boiling point of water at normal barometric pressure is 100°C. However, *the SI base unit of temperature* is the **kelvin (K),** a

**Figure 1.23**
**Comparison of temperature scales.** Room temperature is about 293 K, 20°C, and 68°F. Water freezes at 273.15 K, 0°C, and 32°F. Water boils under normal pressure at 373.15 K, 100°C, and 212°F.

▶ Note that the degree sign (°) is not used with the Kelvin scale, and the unit of temperature is simply the kelvin (not capitalized).

unit on an *absolute temperature* scale. ◄ On any absolute scale, the lowest temperature that can be attained theoretically is zero. The Celsius and the Kelvin scales have equal-size units, but 0°C is equivalent to 273.15 K. Thus, it is easy to convert from one scale to the other, using the formula

$$T_K = \left(t_C \times \frac{1\ K}{1°C}\right) + 273.15\ K$$

where $T_K$ is the temperature in kelvins and $t_C$ is the temperature in degrees Celsius. A temperature of 20°C (about room temperature) equals 293 K.

The Fahrenheit scale is at present the common temperature scale in the United States. Figure 1.23 compares Kelvin, Celsius, and Fahrenheit scales. As the figure shows, 0°C is the same as 32°F (both exact), and 100°C corresponds to 212°F (both exact). Therefore, there are $212 - 32 = 180$ Fahrenheit degrees in the range of 100 Celsius degrees. That is, there are exactly 9 Fahrenheit degrees for every 5 Celsius degrees. Knowing this, and knowing that 0°C equals 32°F, we can derive a formula to convert degrees Celsius to degrees Fahrenheit. ◄

▶ These two temperature conversion formulas are often written K = °C + 273.15 and °F = (1.8 × °C) + 32. Although these yield the correct number, they do not take into account the units.

$$t_F = \left(t_C \times \frac{9°F}{5°C}\right) + 32°F$$

where $t_F$ is the temperature in Fahrenheit and $t_C$ is the temperature in Celsius. By rearranging this, we can obtain a formula for converting degrees Fahrenheit to degrees Celsius:

$$t_C = \frac{5°C}{9°F} \times (t_F - 32°F)$$

▶ An alternate example is provided in the Instructor's Resource Manual.

### EXAMPLE 1.3

## Converting from One Temperature Scale to Another

The hottest place on record in North America is Death Valley in California. It reached a temperature of 134°F in 1913. What is this temperature reading in degrees Celsius? in kelvins?

### SOLUTION

Substituting, we find that

$$t_C = \frac{5°C}{9°F} \times (t_F - 32°F) = \frac{5°C}{9°F} \times (134°F - 32°F) = \mathbf{56.7°C}$$

In kelvins,

$$T_K = \left(t_C \times \frac{1\,K}{1°C}\right) + 273.15\,K = \left(56.7°C \times \frac{1\,K}{1°C}\right) + 273.15\,K = \mathbf{329.9\,K}$$

### EXERCISE 1.5

a. A person with a fever has a temperature of 102.5°F. What is this temperature in degrees Celsius? b. A cooling mixture of dry ice and isopropyl alcohol has a temperature of −78°C. What is this temperature in kelvins?

**See Problems 1.61, 1.62, 1.63, and 1.64.**

 **CONCEPT CHECK 1.3**

a. Estimate and express the length of your leg in an appropriate metric unit.

b. What would be a reasonable height, in meters, of a three-story building?

c. How would you be feeling if your body temperature was 39°C?

d. Would you be comfortable sitting in a room at 23°C in a short sleeved shirt?

## 1.7 Derived Units

▶ Experiments in General Chemistry, 1C, "Some Measurements of Mass and Volume."

Once base units have been defined for a system of measurement, you can derive other units from them. You do this by using the base units in equations that define other physical quantities. For example, *area* is defined as length times length. Therefore,

$$\text{SI unit of area} = (\text{SI unit of length}) \times (\text{SI unit of length})$$

From this, you see that the SI unit of area is meter × meter, or $m^2$. Similarly, *speed* is defined as the rate of change of distance with time; that is, speed = distance/time. So

$$\text{SI unit of speed} = \frac{\text{SI unit of distance}}{\text{SI unit of time}}$$

The SI unit of speed is meters per second (that is, meters divided by seconds). The unit is symbolized m/s or $m \cdot s^{-1}$. The unit of speed is an example of an **SI derived unit,** which is *a unit derived by combining SI base units.* Table 1.4 defines a number

**Table 1.4
Derived Units**

| Quantity | Definition of Quantity | SI Unit |
|---|---|---|
| Area | Length squared | $m^2$ |
| Volume | Length cubed | $m^3$ |
| Density | Mass per unit volume | $kg/m^3$ |
| Speed | Distance traveled per unit time | $m/s$ |
| Acceleration | Speed changed per unit time | $m/s^2$ |
| Force | Mass times acceleration of object | $kg \cdot m/s^2$ (= newton, N) |
| Pressure | Force per unit area | $kg/(m \cdot s^2)$ (= pascal, Pa) |
| Energy | Force times distance traveled | $kg \cdot m^2/s^2$ (= joule, J) |

of derived units. Volume and density are discussed in this section; pressure and energy are discussed later (in Sections 5.1 and 6.1, respectively).

## Volume

*Volume* is defined as length cubed and has the SI unit of cubic meter ($m^3$). This is too large a unit for normal laboratory work, so we use either cubic decimeters ($dm^3$) or cubic centimeters ($cm^3$, also written cc). Traditionally, chemists have used the **liter (L),** which is *a unit of volume equal to a cubic decimeter* (approximately one quart). In fact, most laboratory glassware (Figure 1.24) is calibrated in liters or milliliters (1000 mL = 1 L). Because 1 dm equals 10 cm, a cubic decimeter, or one liter, equals $(10 \text{ cm})^3 = 1000 \text{ cm}^3$. Therefore, a milliliter equals a cubic centimeter. In summary

$$1 \text{ L} = 1 \text{ dm}^3 \quad \text{and} \quad 1 \text{ mL} = 1 \text{ cm}^3$$

## Density

The **density** of an object is its *mass per unit volume*. You can express this as

$$d = \frac{m}{V}$$

where $d$ is the density, $m$ is the mass, and $V$ is the volume. Suppose an object has a mass of 15.0 g and a volume of 10.0 $cm^3$. Substituting, you find that

$$d = \frac{15.0 \text{ g}}{10.0 \text{ cm}^3} = 1.50 \text{ g/cm}^3$$

The density of the object is 1.50 $g/cm^3$ (or 1.50 $g \cdot cm^{-3}$).

Density is an important characteristic property of a material. Water, for example, has a density of 1.000 $g/cm^3$ at 4°C and a density of 0.998 $g/cm^3$ at 20°C. Lead has a density of 11.3 $g/cm^3$ at 20°C. (Figures 1.25 and 1.26 dramatically show some relative densities.) ◄ Oxygen gas has a density of $1.33 \times 10^{-3}$ $g/cm^3$ at normal

▶ Demonstration of relative liquid densities, Bassam Z. Shakhashiri, *Chemical Demonstrations,* Vol. 3 (Madison: University of Wisconsin Press, 1989), no. 9.2, pp. 229–233.

▶ Demonstration of relative gas densities, Summerlin and Ealy, *Chemical Demonstrations,* Vol. 1 (Washington, D.C.: Am. Chem. Soc., 1985), no. 2, p. 7.

▶ The density of solid materials on earth ranges from about 1 $g/cm^3$ to 22.5 $g/cm^3$ (osmium metal). In the interior of certain stars, the density of matter is truly staggering. Black neutron stars—stars composed of neutrons, or atomic cores compressed by gravity to a superdense state—have densities of about $10^{15}$ $g/cm^3$.

**Figure 1.24**
**Some laboratory glassware.** *Left to right:* A 600-mL beaker; a 100-mL graduated cylinder; a 250-mL volumetric flask; a 250-mL Erlenmeyer flask. *Front:* A 5-mL pipet. A graduated cylinder is used to measure approximate volumes of liquids. A pipet is calibrated to deliver a specified volume of liquid when filled to an etched line with a suction bulb. The pipet shown is accurate to about 0.01 mL. A volumetric flask is filled to an etched line and is used in making up specific volumes of solutions. The flask shown is accurate to about 0.1 mL.

pressure and 20°C. (Like other gases under normal conditions, oxygen has a density that is about 1000 times smaller than those of liquids and solids.) Because the density is characteristic of a substance, it can be helpful in identifying it. Example 1.4 illustrates this point. Density can also be useful in determining whether a substance is pure. Consider a gold bar whose purity is questioned. The metals likely to be mixed with gold, such as silver or copper, have lower densities than gold. Therefore, an adulterated (impure) gold bar can be expected to be far less dense than pure gold.

**Figure 1.25** *(left)*
**The relative densities of copper and mercury.** Copper floats on mercury because the density of copper is less than that of mercury. (This is a copper penny; recent pennies are copper-clad zinc.)

**Figure 1.26** *(right)*
**Relative densities of some liquids.** Shown are three liquids (dyed so that they will show up clearly): *ortho*-xylene, water, and 1,1,1-trichloroethane. Water (blue layer) is less dense than 1,1,1-trichloroethane (colorless) and floats on it. *Ortho*-xylene (top, golden layer) is less dense than water and floats on it.

▶ An alternate example is provided in the Instructor's Resource Manual.

---

**EXAMPLE 1.4**

## Calculating the Density of a Substance

A colorless liquid, used as a solvent (a liquid that dissolves other substances), is believed to be one of the following:

| Substance | Density (in g/mL) |
|---|---|
| *n*-butyl alcohol | 0.810 |
| ethylene glycol | 1.114 |
| isopropyl alcohol | 0.785 |
| toluene | 0.866 |

To identify the substance, a chemist determined its density. By pouring a sample of the liquid into a graduated cylinder, she found that the volume was 35.1 mL. She also found that the sample weighed 30.5 g. What was the density of the liquid? What was the substance?

**SOLUTION**

Recall that density equals mass per volume (mass divided by volume). You substitute 30.5 g for the mass and 35.1 mL for the volume.

$$d = \frac{m}{V} = \frac{30.5 \text{ g}}{35.1 \text{ mL}} = \textbf{0.869 g/mL}$$

The density of the liquid equals that of **toluene** (within experimental error).

---

**EXERCISE 1.6**

A piece of metal wire has a volume of 20.2 cm$^3$ and a mass of 159 g. What is the density of the metal? The metal is either manganese, iron, or nickel, which have densities of 7.21 g/cm$^3$, 7.87 g/cm$^3$, and 8.90 g/cm$^3$, respectively. From which metal is the wire made?

---

**See Problems 1.65, 1.66, 1.67, and 1.68.**

▶ Video no. 2 (Series D), "The Six Layers" and no. 3 (Series D), "Diet vs. Regular Soda."

In addition to characterizing a substance, the density provides a useful relationship between mass and volume. For example, suppose an experiment calls for a certain mass of liquid. Rather than weigh the liquid on a balance, you might instead measure out the corresponding volume. Example 1.5 illustrates this idea.

▶ An alternate example is provided in the Instructor's Resource Manual.

---

**EXAMPLE 1.5**

## Using the Density to Relate Mass and Volume

An experiment requires 43.7 g of isopropyl alcohol. Instead of measuring out the sample on a balance, a chemist dispenses the liquid into a graduated cylinder. The density of isopropyl alcohol is 0.785 g/mL. What volume of isopropyl alcohol should he use?

**SOLUTION**

You rearrange the formula defining the density to obtain the volume.

$$V = \frac{m}{d}$$

Then you substitute into this formula:

$$V = \frac{43.7 \text{ g}}{0.785 \text{ g/mL}} = \textbf{55.7 mL}$$

---

**EXERCISE 1.7**

Ethanol (grain alcohol) has a density of 0.789 $g/cm^3$. What volume of ethanol must be poured into a graduated cylinder to equal 30.3 g?

---

See Problems 1.69, 1.70, 1.71, and 1.72.

 **CONCEPT CHECK 1.4**

You are working in the office of a precious metal buyer. A miner brings you a nugget of metal that he claims is gold. You suspect that the metal is a form of "fool's gold," called marcasite, that is composed of iron and sulfur. In the back of your office, you have a chunk of pure gold. What simple experiments could you perform to decide whether or not the miner's nugget is gold?

# 1.8   Units and Dimensional Analysis (Factor-Label Method)

In performing numerical calculations with physical quantities, it is good practice to enter each quantity as a number with its associated unit. Both the numbers and the units are then carried through the indicated algebraic operations. The advantages of this are twofold:

1. The units for the answer will come out of the calculations.

2. If you make an error in arranging factors in the calculation (for example, if you use the wrong formula), this will become apparent because the final units will be nonsense.

**Dimensional analysis** (or the **factor-label method**) is *the method of calculation in which one carries along the units for quantities.* As an illustration, suppose you want to find the volume $V$ of a cube, given $s$, the length of a side of the cube. Because $V = s^3$, if $s = 5.00$ cm, you find that $V = (5.00 \text{ cm})^3 = 5.00^3 \text{ cm}^3$. There is no guesswork about the unit of volume here; it is cubic centimeters ($cm^3$).

Suppose, however, that you wish to express the volume in liters (L), a metric unit that equals $10^3$ cubic centimeters (approximately one quart). You can write this equality as

$$1 \text{ L} = 10^3 \text{ cm}^3$$

If you divide both sides of the equality by the right-hand quantity, you get

$$\frac{1 \text{ L}}{10^3 \text{ cm}^3} = \frac{10^3 \cancel{\text{cm}^3}}{10^3 \cancel{\text{cm}^3}} = 1$$

Observe that you treat units in the same way as algebraic quantities. Note too that the right-hand side now equals 1 and no units are associated with it. Because it is always possible to multiply any quantity by 1 without changing that quantity, you can multiply the previous expression for the volume by the factor 1 L/$10^3$ cm$^3$ without changing the actual volume. You are changing only the way you express this volume:

$$V = 5.00^3 \text{ cm}^3 \times \underbrace{\frac{1 \text{ L}}{10^3 \text{ cm}^3}}_{\substack{\text{converts} \\ \text{cm}^3 \text{ to L}}} = 125 \times 10^{-3} \text{ L} = 0.125 \text{ L}$$

The ratio 1 L/$10^3$ cm$^3$ is called a **conversion factor** because it is *a factor equal to 1 that converts a quantity expressed in one unit to a quantity expressed in another unit.* ◀ Note that the numbers in this conversion factor are *exact*, because 1 L equals exactly 1000 cm$^3$. Such exact conversion factors do not affect the number of significant figures in an arithmetic result. In the previous calculation, the quantity 5.00 cm (the measured length of the side) does determine or limit the number of significant figures.

The next two examples illustrate how the conversion-factor method may be used to convert one metric unit to another.

▶ It takes more room to explain conversion factors than it does to use them. With practice, you will be able to write the final conversion step without the intermediate algebraic manipulations outlined here.

▶ An alternate example is provided in the Instructor's Resource Manual.

---

### EXAMPLE 1.6

## Converting Units: Metric Unit to Metric Unit

Nitrogen gas is the major component of air. A sample of nitrogen gas in a glass bulb weighs 243 mg. What is this mass in SI base units of mass (kilograms)?

 **PROBLEM STRATEGY**

Do the conversion in two steps. First convert milligrams to grams; then convert grams to kilograms. To convert from milligrams to grams, note that the prefix *milli-* means $10^{-3}$. To convert from grams to kilograms, note that the prefix *kilo-* means $10^3$.

**SOLUTION**

Since 1 mg = $10^{-3}$ g, you can write

$$243 \text{ mg} \times \frac{10^{-3} \text{ g}}{1 \text{ mg}} = 2.43 \times 10^{-1} \text{ g}$$

Then, because the prefix *kilo-* means $10^3$, you write

$$1 \text{ kg} = 10^3 \text{ g}$$

and

$$2.43 \times 10^{-1} \text{ g} \times \frac{1 \text{ kg}}{10^3 \text{ g}} = \mathbf{2.43 \times 10^{-4} \text{ kg}}$$

Note, however, that you can do the two conversions in one step, as follows:

$$243 \text{ mg} \times \underbrace{\frac{10^{-3} \text{ g}}{1 \text{ mg}}}_{\substack{\text{converts} \\ \text{mg to g}}} \times \underbrace{\frac{1 \text{ kg}}{10^3 \text{ g}}}_{\substack{\text{converts} \\ \text{g to kg}}} = 2.43 \times 10^{-4} \text{ kg}$$

121 pm

► An alternate example is provided in the Instructor's Resource Manual.

## EXERCISE 1.8

The oxygen molecule (the smallest particle of oxygen gas) consists of two oxygen atoms a distance of 121 pm apart. How many millimeters is this distance?

See Problems 1.73, 1.74, 1.75, and 1.76.

## EXAMPLE 1.7

## Converting Units: Metric Volume to Metric Volume

The world's oceans contain approximately $1.35 \times 10^9$ km$^3$ of water. What is this volume in liters?

 **PROBLEM STRATEGY**

Convert cubic kilometers to cubic decimeters, which are equivalent to liters. Do this in two steps, first cubic kilometers to cubic meters, then cubic meters to cubic decimeters. (The steps may be combined.) For a conversion factor such as that for cubic kilometers to cubic meters, note that $10^3$ m $= 1$ km, so the factor is $(10^3$ m/1 km$)^3$.

**SOLUTION**

$$1.35 \times 10^9 \text{ km}^3 \times \left(\frac{10^3 \text{ m}}{1 \text{ km}}\right)^3 \times \left(\frac{1 \text{ dm}}{10^{-1} \text{ m}}\right)^3 = 1.35 \times 10^{21} \text{ dm}^3$$

$\underbrace{\qquad}_{\text{converts km}^3 \text{ to m}^3}$ $\underbrace{\qquad}_{\text{converts m}^3 \text{ to dm}^3}$

Because a cubic decimeter is equal to a liter, the volume of the oceans is

$$\textbf{1.35} \times \textbf{10}^{\textbf{21}} \textbf{ L}$$

## EXERCISE 1.9

A large crystal is constructed by stacking small, identical pieces of crystal, much as you construct a brick wall by stacking bricks. A unit cell is the smallest such piece from which a crystal can be made. The unit cell of a crystal of gold metal has a volume of 67.6 Å$^3$. What is this volume in cubic decimeters?

See Problems 1.77 and 1.78.

The conversion-factor method can be used to convert any unit to another unit, provided a conversion equation exists between the two units. Relationships between certain U.S. units and metric units are given in Table 1.5. You can use these to convert between U.S. and metric units. Suppose you wish to convert 0.547 lb to grams. From Table 1.5, note that 1 lb = 0.4536 kg, or 1 lb = 453.6 g, so the conversion factor from pounds to grams is 453.6 g/1 lb. Therefore,

$$0.547 \text{ lb} \times \frac{453.6 \text{ g}}{1 \text{ lb}} = 248 \text{ g}$$

The next example illustrates a conversion requiring several steps.

**Table 1.5**
**Relationships of Some U.S. and Metric Units**

| Length | Mass | Volume |
|---|---|---|
| 1 in = 2.54 cm (exact) | 1 lb = 0.4536 kg | 1 qt = 0.9464 L |
| 1 yd = 0.9144 m (exact) | 1 lb = 16 oz (exact) | 4 qt = 1 gal (exact) |
| 1 mi = 1.609 km | 1 oz = 28.35 g | |
| 1 mi = 5280 ft (exact) | | |

▶ Two alternate examples are provided in the Instructor's Resource Manual.

---

**EXAMPLE 1.8**

## Converting Units: Any Unit to Another Unit

How many centimeters are there in 6.51 miles? Use the exact definitions 1 mi = 5280 ft, 1 ft = 12 in, and 1 in = 2.54 cm.

**SOLUTION**

This problem involves several conversions, which can be done all at once. From the definitions, you obtain the following conversion factors:

$$1 = \frac{5280 \text{ ft}}{1 \text{ mi}} \qquad 1 = \frac{12 \text{ in}}{1 \text{ ft}} \qquad 1 = \frac{2.54 \text{ cm}}{1 \text{ in}}$$

Then,

$$6.51 \text{ mi} \times \underbrace{\frac{5280 \text{ ft}}{1 \text{ mi}}}_{\substack{\text{converts} \\ \text{mi to ft}}} \times \underbrace{\frac{12 \text{ in}}{1 \text{ ft}}}_{\substack{\text{converts} \\ \text{ft to in}}} \times \underbrace{\frac{2.54 \text{ cm}}{1 \text{ in}}}_{\substack{\text{converts} \\ \text{in to cm}}} = \mathbf{1.05 \times 10^6 \text{ cm}}$$

All of the conversion factors are exact, so the number of significant figures in the result is determined by the number of significant figures in 6.51 mi.

**EXERCISE 1.10**

Using the definitions 1 in = 2.54 cm and 1 yd = 36 in (both exact), obtain the conversion factor for yards to meters. How many meters are there in 3.54 yd?

See Problems 1.79, 1.80, 1.81, and 1.82.

# A Checklist for Review

## Important Terms

*Note:* The number in parentheses denotes the section in which the term is defined.

experiment (1.2)
law (1.2)
hypothesis (1.2)
theory (1.2)
mass (1.3)
matter (1.3)
law of conservation
of mass (1.3)
solid (1.4)
liquid (1.4)
gas (1.4)
states of matter (1.4)
physical change (1.4)

chemical change (chemical
reaction) (1.4)
physical property (1.4)
chemical property (1.4)
substance (1.4)
element (1.4)
compound (1.4)
law of definite proportions
(law of constant
composition) (1.4)
mixture (1.4)
heterogeneous mixture (1.4)
homogeneous mixture
(solution) (1.4)

phase (1.4)
unit (1.5)
precision (1.5)
accuracy (1.5)
significant figures (1.5)
number of significant
figures (1.5)
scientific notation (1.5)
exact number (1.5)
rounding (1.5)
International System
(SI) (1.6)
SI base units (1.6)

SI prefix (1.6)
meter (m) (1.6)
angstrom (Å) (1.6)
kilogram (kg) (1.6)
second (s) (1.6)
Celsius scale (1.6)
kelvin (K) (1.6)
SI derived unit (1.7)
liter (L) (1.7)
density (1.7)
dimensional analysis (factor-
label method) (1.8)
conversion factor (1.8)

## Key Equations

$$T_K = \left(t_C \times \frac{1\ K}{1°C}\right) + 273.15\ K$$

$$t_C = \frac{5°C}{9°F} \times (t_F - 32°F)$$

$$d = m/V$$

## Summary of Facts and Concepts

Chemistry is an experimental science in that the facts of chemistry are obtained by *experiment*. These facts are systematized and explained by *theory*, and theory suggests more experiments. The *scientific method* involves this interplay, in which the body of accepted knowledge grows as it is tested by experiment.

Chemistry emerged as a *quantitative science* with the work of the eighteenth-century French chemist Antoine Lavoisier. He made use of the idea that the mass, or quantity of matter, remains constant during a chemical reaction *(law of conservation of mass)*.

Matter may be classified by its physical state as a *gas, liquid,* or *solid.* Matter may also be classified by its chemical constitution as an *element, compound,* or *mixture.* Materials are either substances or mixtures of substances. Substances are either elements or compounds, which are composed of two or more elements. Mixtures can be separated into substances by physical processes, but compounds can be separated into elements only by chemical reactions.

A quantitative science requires the making of measurements. Any measurement has limited *precision,* which you convey by writing the measured number to a certain number of *significant figures.* There are many different systems of measurement, but scientists generally use the metric system. The International System (SI) uses a particular selection of metric units. It employs seven *base units* combined with *prefixes* to obtain units of various size. Units for other quantities are derived from these. To obtain a *derived unit* in SI for a quantity such as the volume or density, you merely substitute base units into a defining equation for the quantity.

*Dimensional analysis* is a technique of calculating with physical quantities in which units are included and treated in the same way as numbers. You can thus obtain the *conversion factor* needed to express a quantity in new units.

## Operational Skills

1. **Using the law of conservation of mass** Given the masses of all substances in a chemical reaction except one, calculate the mass of this one substance. **(EXAMPLE 1.1)**

2. **Using significant figures in calculations** Given an arithmetic setup, report the answer to the correct number of significant figures and round it properly. **(EXAMPLE 1.2)**

3. **Converting from one temperature scale to another** Given a temperature reading on one scale, convert it to another scale—Celsius, Kelvin, or Fahrenheit. **(EXAMPLE 1.3)**

4. **Calculating the density of a substance** Given the mass and volume of a substance, calculate the density. **(EXAMPLE 1.4)**

5. **Using the density to relate mass and volume** Given the mass and density of a substance, calculate the volume; or given the volume and density, calculate the mass. **(EXAMPLE 1.5)**

6. **Converting units** Given an equation relating one unit to another (or a series of such equations), convert a measurement expressed in one unit to a new unit. **(EXAMPLES 1.6, 1.7, AND 1.8)**

# Review Questions

**1.1** Discuss some ways in which chemistry has changed technology. Give one or more examples of how chemistry has affected another science.

**1.2** Define the terms *experiment* and *theory.* How are theory and experiment related? What is a hypothesis?

**1.3** Illustrate the steps in the scientific method using Rosenberg's discovery of the anticancer activity of cisplatin.

**1.4** Define the terms *matter* and *mass.* What is the difference between mass and weight?

**1.5** State the law of conservation of mass. Describe how you might demonstrate this law.

**1.6** A chemical reaction is often accompanied by definite changes in appearance. For example, heat and light may be emitted, and there may be a change of color of the substances. Figure 1.10 shows the reactions of the metal mercury with oxygen in air. Describe the changes that occur.

**1.7** Characterize gases, liquids, and solids in terms of compressibility and fluidity.

**1.8** Choose a substance and give several of its physical properties and several of its chemical properties.

**1.9** Give examples of an element, a compound, a heterogeneous mixture, and a homogeneous mixture.

**1.10** What phases or states of matter are present in a glass of bubbling carbonated beverage that contains ice cubes?

**1.11** What distinguishes an element from a compound? Can a compound also be an element?

**1.12** What is meant by the precision of a measurement? How is it indicated?

**1.13** Two rules are used to decide how to round the result of a calculation to the correct number of significant figures. Use a calculation to illustrate each rule. Explain how you obtained the number of significant figures in the answers.

**1.14** Distinguish between a measured number and an exact number. Give examples of each.

**1.15** How does the International System (SI) obtain units of different size from a given unit? How does the International System obtain units for all possible physical quantities from only seven base units?

**1.16** What is an absolute temperature scale? How are degrees Celsius related to kelvin units?

**1.17** Define *density.* Describe some uses of density.

**1.18** Why should units be carried along with numbers in a calculation?

#  Conceptual Problems

*Key:* These problems are designed to check your understanding of the concepts associated with some of the main topics presented in the chapter. A strong conceptual understanding of chemistry is the foundation for both applying chemical knowledge and solving chemical problems. These problems vary in level of difficulty and often can be used as a basis for group discussion.

**1.19**

a. Sodium metal is partially melted. What are the two phases present?

b. A sample of sand is composed of granules of quartz (silicon dioxide) and seashells (calcium carbonate). The sand is mixed with water. What phases are present?

**1.20** A material is believed to be a compound. Suppose you have several samples of this material obtained from various places around the world. Comment on what you would expect to find upon observing the melting point and color for each sample. What would you expect to find upon determining the elemental composition for each sample?

**1.21** You need a thermometer that is accurate to ±5°C to conduct some experiments in the temperature range of 0°C to 100°C. You find one in your lab drawer that has lost its markings.
a. What experiments could you do to make sure your thermometer is suitable for your experiments?
b. Assuming that the thermometer works, what procedure could you follow to put a scale on your thermometer that has the desired accuracy?

**1.22** You are teaching a class of second graders some chemistry, and they are confused about an object's mass versus its density. Keeping in mind that most second graders don't understand fractions, how would you explain these two ideas and illustrate how they differ? What examples would you use as a part of your explanation?

**1.23** Say you live in a climate where the temperature ranges from −100°F to 20°F and you want to define a new temperature scale, YS (YS is the "Your Scale" temperature scale), which defines this range as 0.0°YS to 100.0°YS.
a. Come up with an equation that would allow you to convert between °F and °YS.
b. Using your equation, what would be the temperature in °F if it were 66°YS?

**1.24** You are presented with a piece of metal in a jar. It is your job to determine what the metal is. What are some physical properties that you could measure in order to determine the type of metal? You suspect that the metal might be sodium; what are some chemical properties that you could investigate? (See Section 1.4 for some ideas.)

**1.25** You have two identical boxes with interior dimensions of 8.0 cm × 8.0 cm × 8.0 cm. You completely fill one of the boxes with wooden spheres that are 1.6 cm in diameter. The other box gets filled with wooden cubes that are 1.6 cm on each edge. After putting the lid on the filled boxes, you then measure the density of each. Which one is more dense?

**1.26** Consider the following compounds and their densities.

| Substance | Density (g/mL) | Substance | Density (g/mL) |
|---|---|---|---|
| Isopropyl alcohol | 0.785 | Toluene | 0.866 |
| n-Butyl alcohol | 0.810 | Ethylene glycol | 1.114 |

You create a column of the liquids in a glass cylinder with the most dense material on the bottom layer and the least dense on the top. You do not allow the liquids to mix.
a. First you drop a plastic bead that has a density of 0.24 g/cm$^3$ into the column. What do you expect to observe?
b. Next you drop a different plastic bead that has a volume of 0.043 mL and a mass of $3.92 \times 10^{-2}$ g into the column. What would you expect to observe in this case?
c. You drop another bead into the column and observe that it makes it all the way to the bottom of the column. What can you conclude about the density of this bead?

**1.27**
a. Which of the following items have a mass of about 1 g?
a grain of sand
a paper clip
a nickel
a 5.0-gallon bucket of water
a brick
a car
b. What is the approximate mass (using SI mass units) of each of the items in part a?

**1.28** What is the length of the nail reported to the correct number of significant figures?

## Practice Problems

*Key:* These problems are for practice in applying problem-solving skills. They are divided by topic, and some are keyed to exercises (see the ends of the exercises). The problems are arranged in matched pairs; the odd-numbered problem of each pair is listed first and its answer is given in the back of the book.

### Conservation of Mass

**1.29** A 15.9-g sample of sodium carbonate is added to a solution of acetic acid weighing 20.0 g. The two substances react, releasing carbon dioxide gas to the atmosphere. After reaction, the contents of the reaction vessel weighs 29.3 g. What is the mass of carbon dioxide given off during the reaction?

**1.30** Some iron wire weighing 5.6 g is placed in a beaker and covered with 15.0 g of dilute hydrochloric acid. The acid reacts with the metal and gives off hydrogen gas, which escapes into the surrounding air. After reaction, the contents of the beaker weighs 20.4 g. What is the mass of hydrogen gas produced by the reaction?

**1.31**  Zinc metal reacts with yellow crystals of sulfur in a fiery reaction to produce a white powder of zinc sulfide. A chemist determines that 65.4 g of zinc reacts with 32.1 g of sulfur. How many grams of zinc sulfide could be produced from 20.0 g of zinc metal?

**1.32**  Aluminum metal reacts with bromine, a red-brown liquid with a noxious odor. The reaction is vigorous and produces aluminum bromide, a white crystalline substance. A sample of 27.0 g of aluminum yields 266.7 g of aluminum bromide. How many grams of bromine react with 15.0 g of aluminum?

## Solids, Liquids, and Gases

**1.33**  Give the normal state (solid, liquid, or gas) of each of the following.
a. sodium hydrogen carbonate (baking soda)
b. isopropyl alcohol (rubbing alcohol)
c. nitrogen
d. copper

**1.34**  Give the normal state (solid, liquid, or gas) of each of the following.
a. potassium hydrogen tartrate (cream of tartar)
b. oxygen
c. carbon (diamond)
d. bromine

## Chemical and Physical Changes; Properties of Substances

**1.35**  Which of the following are physical changes and which are chemical changes?
a. melting of sodium chloride
b. pulverizing of rock salt
c. burning of sulfur
d. dissolving of salt in water

**1.36**  For each of the following, decide whether a physical or a chemical change is involved.
a. dissolving of sugar in water
b. rusting of iron
c. burning of wood
d. evaporation of alcohol

**1.37**  A sample of mercury(II) oxide was heated to produce mercury metal and oxygen gas. Then the liquid mercury was cooled to $-40°C$, where it solidified. A glowing wood splint was thrust into the oxygen, and the splint burst into flame. Identify each physical change and each chemical change.

**1.38**  Solid iodine, contaminated with salt, was heated until the iodine vaporized. The violet vapor of iodine was then cooled to yield the pure solid. Solid iodine and zinc metal powder were mixed and ignited to give a white powder. Identify each physical change and each chemical change.

**1.39**  The following are properties of substances. Decide whether each is a physical property or a chemical property.
a. Chlorine gas liquefies at $-35°C$ under normal pressure.
b. Hydrogen burns in chlorine gas.
c. Bromine boils at $-7.2°C$.
d. Lithium is a soft, silvery-colored metal.
e. Iron rusts in an atmosphere of moist air.

**1.40**  Decide whether each of the following is a physical property or a chemical property of the substance.
a. Salt substitute, potassium chloride, dissolves in water.
b. Seashells, calcium carbonate, fizz when immersed in vinegar.
c. The gas hydrogen sulfide smells like rotten eggs.
d. Fine steel wool (Fe) can be burned in air.
e. Pure water freezes at $0°C$.

**1.41**  Iodine is a solid having somewhat lustrous, blue-black crystals. The crystals vaporize readily to a violet-colored gas. Iodine combines with many metals. For example, aluminum combines with iodine to give aluminum iodide. Identify the physical and the chemical properties of iodine that are cited.

**1.42**  Mercury(II) oxide is an orange-red solid with a density of $11.1 \ g/cm^3$. It decomposes when heated to give mercury and oxygen. The compound is insoluble in water (does not dissolve in water). Identify the physical and the chemical properties of mercury(II) oxide that are cited.

## Elements, Compounds, and Mixtures

**1.43**  Consider the following separations of materials. State whether a physical process or a chemical reaction is involved in each separation.
a. Sodium chloride is obtained from seawater by evaporation of the water.
b. Mercury is obtained by heating the substance mercury(II) oxide; oxygen is also obtained.
c. Pure water is obtained from ocean water by evaporating the water, then condensing it.
d. Iron is produced from an iron ore that contains the substance iron(III) oxide.
e. Gold is obtained from river sand by panning (allowing the heavy metal to settle in flowing water).

**1.44**  All of the following processes involve a separation of either a mixture into substances or a compound into elements. For each, decide whether a physical process or a chemical reaction is required.
a. Sodium metal is obtained from the substance sodium chloride.
b. Iron filings are separated from sand by using a magnet.
c. Sugar crystals are separated from a sugar syrup by evaporation of water.
d. Fine crystals of silver chloride are separated from a suspension of the crystals in water.

e. Copper is produced when zinc metal is placed in a solution of copper(II) sulfate, a compound.

**1.45**   Label each of the following as a substance, a heterogeneous mixture, or a solution.
a. seawater          b. sulfur
c. fluorine          d. beach sand

**1.46**   Indicate whether each of the following materials is a substance, a heterogeneous mixture, or a solution.
a. milk              b. bromine
c. gasoline          d. aluminum

**1.47**   Which of the following are pure substances and which are mixtures? For each, list all of the different phases present.
a. bromine liquid and its vapor
b. paint, containing a liquid solution and a dispersed solid pigment
c. partially molten iron
d. baking powder containing sodium hydrogen carbonate and potassium hydrogen tartrate

**1.48**   Which of the following are pure substances and which are mixtures? For each, list all of the different phases present.
a. a sugar solution with sugar crystals at the bottom
b. ink containing a liquid solution with fine particles of carbon
c. a sand containing quartz (silicon dioxide) and calcite (calcium carbonate)
d. liquid water and steam at 100°C

## Significant Figures

**1.49**   How many significant figures are there in each of the following measurements?
a. 73.0000 g          b. 0.0503 kg
c. 6.300 cm           d. 0.80090 m
e. $5.10 \times 10^{-7}$ m      f. 2.001 s

**1.50**   How many significant figures are there in each of the following measurements?
a. 130.0 kg           b. 0.05930 g
c. 0.224800 m         d. 1008 s
e. $4.380 \times 10^{-8}$ m     f. $9.100 \times 10^4$ cm

**1.51**   The circumference of the earth at the equator is 40,000 km. This value is precise to two significant figures. Write this in scientific notation to express correctly the number of significant figures.

**1.52**   The astronomical unit equals the mean distance between the earth and the sun. This distance is 150,000,000 km, which is precise to three significant figures. Express this in scientific notation to the correct number of significant figures.

**1.53**   Do the indicated arithmetic and give the answer to the correct number of significant figures.
a. $\dfrac{8.71 \times 0.0301}{0.056}$

b. $0.71 + 81.8$
c. $934 \times 0.00435 + 107$
d. $(847.89 - 847.73) \times 14673$

**1.54**   Do the indicated arithmetic and give the answer to the correct number of significant figures.
a. $\dfrac{0.871 \times 0.23}{5.871}$

b. $8.937 - 8.930$
c. $8.937 + 8.930$
d. $0.00015 \times 54.6 + 1.002$

**1.55**   One sphere has a radius of 5.10 cm; another has a radius of 5.00 cm. What is the difference in volume (in cubic centimeters) between the two spheres? Give the answer to the correct number of significant figures. The volume of a sphere is $(4/3)\pi r^3$, where $\pi = 3.1416$ and $r$ is the radius.

**1.56**   A solid cylinder of iron of circular cross section with a radius of 1.500 cm has a ruler etched along its length. What is the volume of iron contained between the marks labeled 3.10 cm and 3.50 cm? The volume of a cylinder is $\pi r^2 l$, where $\pi = 3.1416$, $r$ is the radius, and $l$ is the length.

## SI Units

**1.57**   Write the following measurements, without scientific notation, using the appropriate SI prefix.
a. $5.89 \times 10^{-12}$ s      b. 0.2010 m
c. $2.560 \times 10^{-9}$        d. $6.05 \times 10^3$ m

**1.58**   Write the following measurements, without scientific notation, using the appropriate SI prefix.
a. $4.851 \times 10^{-6}$ g      b. $3.16 \times 10^{-2}$ m
c. $2.591 \times 10^{-9}$ s      d. $8.93 \times 10^{-12}$ g

**1.59**   Using scientific notation, convert:
a. 6.15 ps to s              b. 3.781 $\mu$m to m
c. 1.546 Å to m             d. 9.7 mg to g

**1.60**   Using scientific notation, convert:
a. 6.20 km to m             b. 1.98 ns to s
c. 2.54 cm to m             d. 5.23 $\mu$g to g

## Temperature Conversion

**1.61**   Convert:
a. 68°F to degrees Celsius
b. −11°F to degrees Celsius
c. 26°C to degrees Fahrenheit
d. −70°C to degrees Fahrenheit

**1.62**   Convert:
a. 121°F to degrees Celsius
b. −7°F to degrees Celsius
c. −41°C to degrees Fahrenheit
d. 58°C to degrees Fahrenheit

**1.63**   Salt and ice are stirred together to give a mixture to freeze ice cream. The temperature of the mixture is $-21.1°C$. What is this temperature in degrees Fahrenheit?

**1.64**   Liquid nitrogen can be used for the quick freezing of foods. The liquid boils at $-196°C$. What is this temperature in degrees Fahrenheit?

## Density

**1.65**   A certain sample of the mineral galena (lead sulfide) weighs 12.4 g and has a volume of 1.64 cm$^3$. What is the density of galena?

**1.66**   A flask contains 25.0 mL of diethyl ether weighing 17.84 g. What is the density of the ether?

**1.67**   A liquid with a volume of 8.5 mL has a mass of 6.74 g. The liquid is either octane, ethanol, or benzene, the densities of which are 0.702 g/cm$^3$, 0.789 g/cm$^3$, and 0.879 g/cm$^3$, respectively. What is the identity of the liquid?

**1.68**   A mineral sample has a mass of 31.5 g and a volume of 3.9 cm$^3$. The mineral is either sphalerite (density = 4.0 g/cm$^3$), cassiterite (density = 6.99 g/cm$^3$), or cinnabar (density = 8.10 g/cm$^3$). Which is it?

**1.69**   Platinum has a density of 21.4 g/cm$^3$. What is the mass of 5.9 cm$^3$ of this metal?

**1.70**   What is the mass of a 22.7-mL sample of gasoline, which has a density of 0.70 g/cm$^3$?

**1.71**   Ethanol has a density of 0.789 g/cm$^3$. What volume must be poured into a graduated cylinder to give 19.8 g of alcohol?

**1.72**   Bromine is a red-brown liquid with a density of 3.10 g/mL. A sample of bromine weighing 88.5 g occupies what volume?

## Unit Conversions

**1.73**   Sodium hydrogen carbonate, known commercially as baking soda, reacts with acidic materials such as vinegar to release carbon dioxide gas. An experiment calls for 0.480 kg of sodium hydrogen carbonate. Express this mass in milligrams.

**1.74**   The acidic constituent in vinegar is acetic acid. A 10.0-mL sample of a certain vinegar contains 483 mg of acetic acid. What is this mass of acetic acid expressed in micrograms?

**1.75**   The different colors of light have different wavelengths. The human eye is most sensitive to light whose wavelength is 555 nm (greenish-yellow). What is this wavelength in millimeters?

**1.76**   Water consists of molecules (groups of atoms). A water molecule has two hydrogen atoms, each connected to an oxygen atom. The distance between any one hydrogen atom and the oxygen atom is 0.96 Å. What is this distance in millimeters?

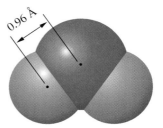

0.96 Å

**1.77**   The total amount of fresh water on earth is estimated to be $3.73 \times 10^8$ km$^3$. What is this volume in cubic meters? in liters?

**1.78**   A submicroscopic particle suspended in a solution has a volume of 1.4 $\mu$m$^3$. What is this volume in liters?

**1.79**   How many grams are there in 3.58 short tons? Note that 1 g = 0.03527 oz (ounces avoirdupois), 1 lb (pound) = 16 oz, and 1 short ton = 2000 lb. (These relations are exact.)

**1.80**   The calorie, the Btu (British thermal unit), and the joule are units of energy; 1 calorie = 4.184 joules (exact), and 1 Btu = 252.0 calories. Convert 2.17 Btu to joules.

**1.81**   The first measurement of sea depth was made in 1840 in the central South Atlantic, where a plummet was lowered 2425 fathoms. What is this depth in meters? Note that 1 fathom = 6 ft, 1 ft = 12 in, and 1 in = $2.54 \times 10^{-2}$ m. (These relations are exact.)

**1.82**   The estimated amount of recoverable oil from the field at Prudhoe Bay in Alaska is $9.6 \times 10^9$ barrels. What is this amount of oil in cubic meters? One barrel = 42 gal (exact), 1 gal = 4 qt (exact), and 1 qt = $9.46 \times 10^{-4}$ m$^3$.

## General Problems

*Key*:  These problems provide more practice but are not divided by topic or keyed to exercises. Each odd-numbered problem and the even-numbered problem that follows it are similar; answers to odd-numbered problems are given in the back of the book.

**1.83**   Sodium metal reacts vigorously with water. A piece of sodium weighing 19.70 g was added to a beaker containing 126.22 g of water. During reaction, hydrogen gas was produced and bubbled from the solution. The solution, containing sodium hydroxide, weighed 145.06 g. How many grams of hydrogen gas were produced?

**1.84** An antacid tablet weighing 0.853 g contained calcium carbonate as the active ingredient, in addition to an inert binder. When an acid solution weighing 56.519 g was added to the tablet, carbon dioxide gas was released, producing a fizz. The resulting solution weighed 57.152 g. How many grams of carbon dioxide were produced?

**1.85** When a mixture of aluminum powder and iron(III) oxide is ignited, it produces molten iron and aluminum oxide. In an experiment, 5.40 g of aluminum was mixed with 18.50 g of iron(III) oxide. At the end of the reaction, the mixture contained 11.17 g of iron, 10.20 g of aluminum oxide, and an undetermined amount of unreacted iron(III) oxide. No aluminum was left. What is the mass of the iron(III) oxide?

**1.86** When chlorine gas is bubbled into a solution of sodium bromide, the sodium bromide reacts to give bromine, a red-brown liquid, and sodium chloride (ordinary table salt). A solution was made by dissolving 20.6 g of sodium bromide in 100.0 g of water. After passing chlorine through the solution, investigators analyzed the mixture. It contained 16.0 g of bromine and 11.7 g of sodium chloride. How many grams of chlorine reacted?

**1.87** A beaker weighed 53.10 g. To the beaker was added 5.348 g of iron pellets and 56.1 g of hydrochloric acid. What was the total mass of the beaker and the mixture (before reaction)? Express the answer to the correct number of significant figures.

**1.88** A graduated cylinder weighed 68.1 g. To the cylinder was added 58.2 g of water and 5.318 g of sodium chloride. What was the total mass of the cylinder and the solution? Express the answer to the correct number of significant figures.

**1.89** Identify each of the following elements from its properties under normal conditions. (See Table 2.1.)
a. red-brown liquid      c. soft, yellow metal
b. white, waxy solid      d. soft, black solid

**1.90** Identify each of the following elements from its properties under normal conditions. (See Table 2.1.)
a. reddish metal
b. brittle, yellow solid
c. greenish-yellow gas
d. silver-colored liquid metal

**1.91** Analyses of several samples of a material containing only iron and oxygen gave the following results. Could this material be a compound?

| | Mass of Sample | Mass of Iron | Mass of Oxygen |
|---|---|---|---|
| Sample A | 1.518 g | 1.094 g | 0.424 g |
| Sample B | 2.056 g | 1.449 g | 0.607 g |
| Sample C | 1.873 g | 1.335 g | 0.538 g |

**1.92** A red-orange solid contains only mercury and oxygen. Analyses of three different samples gave the following results. Are the data consistent with the hypothesis that the material is a compound?

| | Mass of Sample | Mass of Mercury | Mass of Oxygen |
|---|---|---|---|
| Sample A | 1.0410 g | 0.9641 g | 0.0769 g |
| Sample B | 1.5434 g | 1.4293 g | 0.1141 g |
| Sample C | 1.2183 g | 1.1283 g | 0.0900 g |

**1.93** A cubic box measures 39.3 cm on an edge. What is the volume of the box in cubic centimeters? Express the answer to the correct number of significant figures.

**1.94** A cylinder with circular cross section has a radius of 2.45 cm and a height of 56.32 cm. What is the volume of the cylinder? Express the answer to the correct number of significant figures.

**1.95** An aquarium has a rectangular cross section that is 47.8 in by 12.5 in; it is 19.5 in high. How many U.S. gallons does the aquarium contain? One U.S. gallon equals exactly 231 $in^3$.

**1.96** A spherical tank has a radius of 150.0 in. Calculate the volume of the tank in cubic inches; then convert this to Imperial gallons. The volume of a sphere is $(4/3)\,\pi r^3$, where $r$ is the radius. One Imperial gallon equals 277.4 $in^3$.

**1.97** Obtain the difference in volume between two spheres, one of radius 5.61 cm, the other of radius 5.85 cm. The volume $V$ of a sphere is $(4/3)\,\pi r^3$, where $r$ is the radius. Give the result to the correct number of significant figures.

**1.98** What is the difference in surface area between two circles, one of radius 7.98 cm, the other of radius 8.50 cm? The surface area of a circle of radius $r$ is $\pi r^2$. Obtain the result to the correct number of significant figures.

**1.99** Perform the following arithmetic setups and express the answers to the correct number of significant figures.
a. $\dfrac{56.1 - 51.1}{6.58}$      b. $\dfrac{56.1 + 51.1}{6.58}$
c. $(9.1 + 8.6) \times 26.91$      d. $0.0065 \times 3.21 + 0.0911$

**1.100** Perform the following arithmetic setups and report the answers to the correct number of significant figures.
a. $\dfrac{9.345 - 9.005}{9.811}$      b. $\dfrac{9.345 + 9.005}{9.811}$
c. $(8.12 + 7.53) \times 3.71$      d. $0.71 \times 0.36 + 17.36$

**1.101** For each of the following, write the measurement in terms of an appropriate prefix and base unit.
a. The mass of calcium per milliliter in a sample of blood serum is 0.0912 g.
b. The radius of an oxygen atom is about 0.000000000066 m.

c. A particular red blood cell measures 0.0000071 m.

d. The wavelength of a certain ultraviolet radiation is 0.000000056 m.

**1.102** For each of the following, write the measurement in terms of an appropriate prefix and base unit.

a. The mass of magnesium per milliliter in a sample of blood serum is 0.0186 g.

b. The radius of a carbon atom is about 0.000000000077 m.

c. The hemoglobin molecule, a component of red blood cells, is 0.0000000065 m in diameter.

d. The wavelength of a certain infrared radiation is 0.00000085 m.

**1.103** Write each of the following in terms of the SI base unit (that is, express the prefix as the power of ten).

a. 1.07 ps      b. 5.8 $\mu$m

c. 319 nm      d. 15.3 ms

**1.104** Write each of the following in terms of the SI base unit (that is, express the prefix as the power of ten).

a. 7.3 mK      b. 275 pm

c. 19.6 ms      d. 45 $\mu$m

**1.105** Tungsten metal, which is used in lightbulb filaments, has the highest melting point of any metal (3410°C). What is this melting point in degrees Fahrenheit?

**1.106** Titanium metal is used in aerospace alloys to add strength and corrosion resistance. Titanium melts at 1677°C. What is this temperature in degrees Fahrenheit?

**1.107** Calcium carbonate, a white powder used in toothpastes, antacids, and other preparations, decomposes when heated to about 825°C. What is this temperature in degrees Fahrenheit?

**1.108** Sodium hydrogen carbonate (baking soda) starts to decompose to sodium carbonate (soda ash) at about 50°C. What is this temperature in degrees Fahrenheit?

**1.109** Gallium metal can be melted by the heat of one's hand. Its melting point is 29.8°C. What is this temperature in kelvin units? in degrees Fahrenheit?

**1.110** Mercury metal is liquid at normal temperatures but freezes at −38.9°C. What is this temperature in kelvin units? in degrees Fahrenheit?

**1.111** Zinc metal can be purified by distillation (transforming the liquid metal to vapor, then condensing the vapor back to liquid). The metal boils at normal atmospheric pressure at 1666°F. What is this temperature in degrees Celsius? in kelvin units?

**1.112** Iodine is a bluish-black solid. It forms a violet-colored vapor when heated. The solid melts at 236°F. What is this temperature in degrees Celsius? in kelvin units?

**1.113** The density of magnesium metal (used in fireworks) is 1.74 g/cm³. Express this density in SI units (kg/m³).

**1.114** Vanadium metal is added to steel to impart strength. The density of vanadium is 5.96 g/cm³. Express this in SI units (kg/m³).

**1.115** The density of quartz mineral was determined by adding a weighed piece to a graduated cylinder containing 51.2 mL water. After the quartz was submerged, the water level was 65.7 mL. The quartz piece weighed 38.4 g. What was the density of the quartz?

**1.116** Hematite (iron ore) weighing 70.7 g was placed in a flask whose volume was 53.2 mL. The flask with hematite was then carefully filled with water and weighed. The hematite and water weighed 109.3 g. The density of the water was 0.997 g/cm³. What was the density of the hematite?

**1.117** Some bottles of colorless liquids were being labeled when the technicians accidentally mixed them up and lost track of their contents. A 15.0-mL sample withdrawn from one bottle weighed 22.3 g. The technicians knew that the liquid was either acetone, benzene, chloroform, or carbon tetrachloride (which have densities of 0.792 g/cm³, 0.899 g/cm³, 1.489 g/cm³, and 1.595 g/cm³, respectively). What was the identity of the liquid?

**1.118** A solid will float on any liquid that is more dense than it is. The volume of a piece of calcite weighing 35.6 g is 12.9 cm³. On which of the following liquids will the calcite float: carbon tetrachloride (density = 1.60 g/cm³), methylene bromide (density = 2.50 g/cm³), tetrabromoethane (density = 2.96 g/cm³), methylene iodide (density = 3.33 g/cm³)?

**1.119** Platinum metal is used in jewelry; it is also used in automobile catalytic converters. What is the mass of a cube of platinum 4.40 cm on an edge? The density of platinum is 21.4 g/cm³.

**1.120** Ultrapure silicon is used to make solid-state devices, such as computer chips. What is the mass of a circular cylinder of silicon that is 12.40 cm long and has a radius of 4.00 cm? The density of silicon is 2.33 g/cm³.

**1.121** Vinegar contains acetic acid (about 5% by mass). Pure acetic acid has a strong vinegar smell but is corrosive to the skin. What volume of pure acetic acid has a mass of 35.00 g? The density of acetic acid is 1.053 g/mL.

**1.122** Ethyl acetate has a characteristic fruity odor and is used as a solvent in paint lacquers and perfumes. An experiment requires 0.021 kg of ethyl acetate. What volume is this (in liters)? The density of ethyl acetate is 0.902 g/mL.

**1.123** Convert:

a. 8.45 kg to micrograms

b. 318 $\mu$s to milliseconds

c. 93 km to nanometers

d. 37.1 mm to centimeters

**1.124** Convert:
a. 239 Å to micrometers
b. 19.6 kg to milligrams
c. 24.8 cm to millimeters
d. 4.3 ns to microseconds

**1.125** Convert:
a. 5.91 kg of chrome yellow to milligrams
b. 753 mg of vitamin A to micrograms
c. 90.1 MHz (megahertz), the wavelength of an FM signal, to kilohertz
d. 498 mJ (the joule, J, is a unit of energy) to kilojoules

**1.126** Convert:
a. 7.19 $\mu$g of cyanocobalamin (vitamin $B_{12}$) to milligrams
b. 104 pm, the radius of a sulfur atom, to angstroms
c. 0.010 mm, the diameter of a typical blood capillary, to centimeters
d. 0.0605 kPa (the pascal, Pa, is a unit of pressure) to centipascals

**1.127** The largest of the Great Lakes is Lake Superior, which has a volume of 12,230 $km^3$. What is this volume in liters?

**1.128** The average flow of the Niagara River is 0.477 $km^3$ of water per day. What is this volume in liters?

**1.129** A room measures 10.0 ft × 11.0 ft and is 9.0 ft high. What is its volume in liters?

**1.130** A cylindrical settling tank is 6.0 ft deep and has a radius of 15.0 ft. What is the volume of the tank in liters?

**1.131** The masses of diamonds and gems are measured in carats. A carat is defined as 200 mg. If a jeweler has 275 carats of diamonds, how many grams does she have?

**1.132** Recent world production of gold was 49.6 × $10^6$ troy ounces. One troy ounce equals 31.10 g. What is the world production of gold in metric tons ($10^6$ g)?

## Cumulative-Skills Problems

*Note:* These problems require two or more operational skills you learned in this chapter. In later chapters, the problems under this heading will combine skills introduced in previous chapters with those given in the current one.

**1.133** When 10.0 g of marble chips (calcium carbonate) are treated with 50.0 mL of hydrochloric acid (density 1.096 g/mL), the marble dissolves, giving a solution and releasing carbon dioxide gas. The solution weighs 60.4 g. How many liters of carbon dioxide gas are released? The density of the gas is 1.798 g/L.

**1.134** Zinc ore (zinc sulfide) is treated with sulfuric acid, leaving a solution with some undissolved bits of material and releasing hydrogen sulfide gas. If 10.8 g of zinc ore is treated with 50.0 mL of sulfuric acid (density 1.153 g/mL), 65.1 g of solution and undissolved material remains. In addition, hydrogen sulfide (density 1.393 g/L) is evolved. What is the volume (in liters) of this gas?

**1.135** A steel sphere has a radius of 1.58 in. If this steel has a density of 7.88 $g/cm^3$, what is the mass of the sphere in grams?

**1.136** A weather balloon filled with helium has a diameter of 3.00 ft. What is the mass in grams of the helium in the balloon at 21°C and normal pressure? The density of helium under these conditions is 0.166 g/L.

**1.137** The land area of Greenland is 840,000 $mi^2$, with only 132,000 $mi^2$ free of perpetual ice. The average thickness of this ice is 5000 ft. Estimate the mass of the ice (assume two significant figures). The density of ice is 0.917 $g/cm^3$.

**1.138** Antarctica, almost completely covered in ice, has an area of 5,500,000 $mi^2$ with an average height of 7500 ft. Without the ice, the height would be only 1500 ft. Estimate the mass of this ice (two significant figures). The density of ice is 0.917 $g/cm^3$.

**1.139** A sample of an ethanol–water solution has a volume of 54.2 $cm^3$ and a mass of 49.6 g. What is the percentage of ethanol (by mass) in the solution? (Assume that there is no change in volume when the pure compounds are mixed.) The density of ethanol is 0.789 $g/cm^3$ and that of water is 0.998 $g/cm^3$. Alcoholic beverages are rated in *proof,* which is a measure of the relative amount of ethanol in the beverage. Pure ethanol is exactly 200 proof; a solution that is 50% ethanol by volume is exactly 100 proof. What is the proof of the given ethanol–water solution?

**1.140** You have a piece of gold jewelry weighing 9.35 g. Its volume is 0.654 $cm^3$. Assume that the metal is an alloy (mixture) of gold and silver, which have densities of 19.3 $g/cm^3$ and 10.5 $g/cm^3$, respectively. Also assume that there is no change in volume when the pure metals are mixed. Calculate the percentage of gold (by mass) in the alloy. The relative amount of gold in an alloy is measured in *karats.* Pure gold is 24 karats; an alloy of 50% gold is 12 karats. State the proportion of gold in the jewelry in karats.

**1.141** A sample of vermilion-colored mineral was weighed in air, then weighed again while suspended in water. An object is buoyed up by the mass of the fluid displaced by the object. In air, the mineral weighed 18.49 g; in water, it weighed 16.21 g. The densities of air and water are 1.205 g/L and 0.9982 $g/cm^3$, respectively. What is the density of the mineral?

**1.142**   A sample of a bright blue mineral was weighed in air, then weighed again while suspended in water. An object is buoyed up by the mass of the fluid displaced by the object. In air, the mineral weighed 7.35 g; in water, it weighed 5.40 g. The densities of air and water are 1.205 g/L and 0.9982 g/cm$^3$, respectively. What is the density of the mineral?

**1.143**   A student gently drops an object weighing 15.8 g into an open vessel that is full of ethanol, so that a volume of ethanol spills out equal to the volume of the object. The exper-imenter now finds that the vessel and its contents weigh 10.5 g more than the vessel full of ethanol only. The density of ethanol is 0.789 g/cm$^3$. What is the density of the object?

**1.144**   An experimenter places a piece of a solid metal weighing 255 g into a graduated cylinder, which she then fills with liquid mercury. After weighing the cylinder and its contents, she removes the solid metal and fills the cylinder with mercury. She now finds that the cylinder and its contents weigh 101 g less than before. The density of mercury is 13.6 g/cm$^3$. What is the density of the solid metal?

# Media Activities

*Key:* The media activities are designed to direct you to electronic resources that will help you master the material or serve as a reference. The activities contain questions to help you develop a more thorough understanding of important principles and concepts.

**1.145**   Open Problem Solving for Chapter 1 on the student CD. In this case, Significant Figures in Mathematical Operations has been selected, so enter 2 or 3 in the box for the number of questions, and then click Begin to work a problem. Perform the following calculations to the correct number of significant figures.
a. 12 + 12.34
b. (0.12 × 345) − 15.2

**1.146**   Go to the Molecule Library on the student web site (select Ebbing/Gammon General Chemistry from www.college.hmco.com/chemistry). Select a water molecule by clicking on H$_2$O. From the Display option in the menu, choose Ball & Stick.
a. Examine the molecular model and determine which of the balls represent H atoms.
b. Experiment with the other Display options and determine which one gives you the best sense of the relative size of the H and O atoms in a water molecule (they are not the same size). Which element is larger, O or H?

c. On paper, draw a ball-and-stick model of an H$_2$ molecule. Use the Molecule Library to check your answer.

**1.147**   Watch the Density Experiment *Visualization.*
a. Given the reason stated for the observed behavior, what assumption is being made about the volume of liquid in each of the cans of soda?
b. What assumptions are being made about the total volume of the cans?
c. Without changing the volume of diet cola, how could you make the diet cola can sink?

**1.148**   Open the Periodic Table on the student CD and look up the density of sodium, Na, and potassium, K (the units are g/cm$^3$).
a. If you placed 10 grams of each of these metals in a liquid with a density of 1.00 g/cm$^3$, what would you observe?
b. Plastics that are being recycled are often separated on the basis of density: less dense plastics float, while the denser ones sink. If you wanted to separate Na and K by the same technique, what density liquid would be required?
c. What is the volume of a 10 g piece of Na?

**1.149**   For review, take the ACE practice test for Chapter 1 on the student web site.

Iron atoms arranged in circle on copper metal surface by scanning tunneling microscope probe.

# Atoms, Molecules, and Ions

**S**odium is a soft, silvery metal. This metal cannot be handled with bare fingers, because it reacts with any moisture on the skin, causing a burn. Chlorine is a poisonous, greenish-yellow gas with a choking odor. When molten sodium is placed into an atmosphere of chlorine, a dramatic reaction occurs: the sodium bursts into flame and a white, crystalline powder forms (see Figure 2.1). What is particularly interesting about this reaction is that this chemical combination of two toxic substances produces sodium chloride, the substance that we commonly call table salt.

Sodium metal and chlorine gas are particular forms of matter. In burning, they undergo a chemical change—a chemical reaction—in which these forms of matter change to a form of matter with different chemical and physical properties. How do we explain the differences in properties of different forms of matter? And how do we explain chemical reactions such as the burning of sodium metal in chlorine gas? This chapter and the next take an introductory look at these basic questions in chemistry. In later chapters we will develop the concepts introduced here.

## CONTENTS

**Figure 2.1**
**Reaction of sodium and chlorine.** *Left:* Sodium metal and chlorine gas. *Right:* A small piece of molten sodium burning in a flask of chlorine. The product is sodium chloride, common table salt.

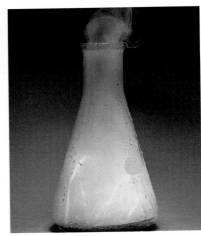

# Atomic Theory and Atomic Structure

▶ Video no. 1 (Series A), "Reactions of Alkali Metals with Halogen Gases"; video no. 2 (Series B), "Metals and Chlorine."

As noted in Chapter 1, all matter is composed of atoms. In this first part of Chapter 2, we explore the atomic theory of matter and the structure of atoms. We also examine the *periodic table*, which is an organizational chart designed to highlight similarities among the elements.

## 2.1   Atomic Theory of Matter

▶ This chapter introduces basic concepts needed in the course. Required lecture time will depend on the background of the students and the weight the instructor wants to put on certain topics, such as naming of compounds.

As we noted in Chapter 1, Lavoisier laid the experimental foundation of modern chemistry. But the British chemist John Dalton (1766–1844) provided the basic theory: all matter—whether element, compound, or mixture—is composed of small particles called atoms. The postulates, or basic statements, of Dalton's theory are presented in this section. Note that the terms *element, compound,* and *chemical reaction,* which were defined in Chapter 1 in terms of matter as we normally see it, are redefined here by the postulates of Dalton's theory in terms of atoms.

▶ Demonstration of the reaction of sodium and chlorine: Summerlin et al., *Chemical Demonstrations,* Vol. 2 (Washington, D.C.: Am. Chem. Soc., 1987), no. 30, p. 56.

### Postulates of Dalton's Atomic Theory

The main points of Dalton's **atomic theory,** *an explanation of the structure of matter in terms of different combinations of very small particles,* are given by the following postulates:

1. All matter is composed of indivisible atoms. An **atom** is *an extremely small particle of matter that retains its identity during chemical reactions.* (See Figure 2.2.)

2. An **element** is *a type of matter composed of only one kind of atom,* each atom of a given kind having the same properties. Mass is one such property. Thus, the atoms of a given element have a characteristic mass. (We will need to revise this definition of element in Section 2.3 so that it is stated in modern terms.) ◀

▶ As you will see in Section 2.4, it is the *average* mass of an atom that is characteristic of each element on earth.

3. A **compound** is *a type of matter composed of atoms of two or more elements chemically combined in fixed proportions.* The relative numbers of any two kinds of atoms in a compound occur in simple ratios. Water, for example, a compound

**Figure 2.2**
**Image of iodine atoms on a platinum metal surface.** This computer-screen image was obtained by a scanning tunneling microscope (discussed in Chapter 7); the color was added to the image by computer. Iodine atoms are the large peaks with pink tops. Note the "vacancy" in the array of iodine atoms. A scale shows the size of the atoms in nanometers.

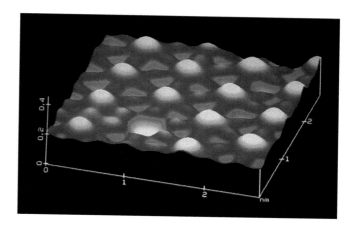

of the elements hydrogen and oxygen, consists of hydrogen and oxygen atoms in the ratio of 2 to 1.

4. A **chemical reaction** consists of *the rearrangement of the atoms present in the reacting substances to give new chemical combinations present in the substances formed by the reaction.* Atoms are not created, destroyed, or broken into smaller particles by any chemical reaction.

Today we know that atoms are not truly indivisible; they are themselves made up of particles. Nevertheless, Dalton's postulates are essentially correct.

## Atomic Symbols and Models

It is convenient to use symbols for the atoms of the different elements. An **atomic symbol** is *a one- or two-letter notation used to represent an atom corresponding to a particular element.* ◄ Typically, the atomic symbol consists of the first letter, capitalized, from the name of the element, sometimes with an additional letter from the name in lowercase. For example, chlorine has the symbol Cl. Other symbols are derived from a name in another language (usually Latin). Sodium is given the symbol Na from its Latin name, *natrium.* Symbols of selected elements are listed in Table 2.1.

► The atomic names (temporary) given to newly discovered elements are derived from the atomic number (discussed in Section 2.3) using the numerical roots: *nil* (0), *un* (1), *bi* (2), *tri* (3), *quad* (4), *pent* (5), *hex* (6), *sept* (7), *oct* (8), and *enn* (9) with the ending *-ium*. For example, the element with atomic number 118 is ununoctium; its atomic symbol is Uuo.

Early in the development of his atomic theory, Dalton built spheres to represent atoms and used combinations of these spheres to represent compounds. Chemists continue to use this idea of representing atoms by three-dimensional models, but because we now know so much more about atoms and molecules than Dalton did, these models have become more refined. Section 2.6 describes these models in more detail.

## Deductions from Dalton's Atomic Theory

Note how atomic theory explains the difference between an element and a compound. Atomic theory also explains two laws we considered earlier. One of these is the law of conservation of mass, which states that the total mass remains constant during a chemical reaction. By postulate 2, every atom has a definite mass. Because a chemical reaction only rearranges the chemical combinations of atoms (postulate 4), the mass must remain constant. The other law explained by atomic theory is the law of definite proportions (constant composition). Postulate 3 defines a compound as a type of matter containing the atoms of two or more elements in definite proportions.

**Table 2.1**
**Some Common Elements**

| Name of Element | Atomic Symbol | Physical Appearance of Element* |
|---|---|---|
| Aluminum | Al | Silvery-white metal |
| Barium | Ba | Silvery-white metal |
| Bromine | Br | Reddish-brown liquid |
| Calcium | Ca | Silvery-white metal |
| Carbon | C | |
| Graphite | | Soft, black solid |
| Diamond | | Hard, colorless crystal |
| Chlorine | Cl | Greenish-yellow gas |
| Chromium | Cr | Silvery-white metal |
| Cobalt | Co | Silvery-white metal |
| Copper | Cu (from *cuprum*) | Reddish metal |
| Fluorine | F | Pale yellow gas |
| Gold | Au (from *aurum*) | Pale yellow metal |
| Helium | He | Colorless gas |
| Hydrogen | H | Colorless gas |
| Iodine | I | Bluish-black solid |
| Iron | Fe (from *ferrum*) | Silvery-white metal |
| Lead | Pb (from *plumbum*) | Bluish-white metal |
| Magnesium | Mg | Silvery-white metal |
| Manganese | Mn | Gray-white metal |
| Mercury | Hg (from *hydrargyrum*) | Silvery-white liquid metal |
| Neon | Ne | Colorless gas |
| Nickel | Ni | Silvery-white metal |
| Nitrogen | N | Colorless gas |
| Oxygen | O | Colorless gas |
| Phosphorus (white) | P | Yellowish-white, waxy solid |
| Potassium | K (from *kalium*) | Soft, silvery-white metal |
| Silicon | Si | Gray, lustrous solid |
| Silver | Ag (from *argentum*) | Silvery-white metal |
| Sodium | Na (from *natrium*) | Soft, silvery-white metal |
| Sulfur | S | Yellow solid |
| Tin | Sn (from *stannum*) | Silvery-white metal |
| Zinc | Zn | Bluish-white metal |

*Common form of the element under normal conditions.

Because the atoms have definite mass, compounds must have the elements in definite proportions by mass.

A good theory should not only explain known facts and laws but also predict new ones. The **law of multiple proportions,** deduced by Dalton from his atomic theory, states that *when two elements form more than one compound, the masses of one element in these compounds for a fixed mass of the other element are in ratios of small whole numbers.* To illustrate this law, consider the following. If we take a fixed mass of carbon, 1.000 gram, and react it with oxygen, we end up with two compounds: one that contains 1.3321 grams of oxygen for each 1.000 gram of carbon, and one that contains 2.6642 grams of oxygen per 1.000 gram of carbon. Note that the ratio of the amounts of oxygen in the two compounds is 2 to 1 (2.6642 ÷ 1.3321), which indicates that there must be twice as much oxygen in the second compound. Applying atomic theory, if we assume that the compound that has 1.3321 grams of oxygen to 1.000 gram of carbon is CO (the combination of one carbon atom and one oxygen atom), then the compound that contains twice as much oxygen per 1.000 gram of carbon must be $CO_2$ (the combination of one carbon atom and two oxygen atoms). The deduction of the law of multiple proportions from atomic theory was important in convincing chemists of the validity of the theory.

 **CONCEPT CHECK 2.1**

Like Dalton, chemists continue to model atoms using spheres. Modern models are usually drawn with a computer and use different colors to represent atoms of different elements. Which of the models below represents $CO_2$?

## 2.2   The Structure of the Atom

Although Dalton had postulated that atoms were indivisible particles, experiments conducted around the beginning of the present century showed that atoms themselves consist of particles. These experiments showed that an atom consists of two kinds of particles: a **nucleus,** *the atom's central core, which is positively charged and contains most of the atom's mass,* and one or more electrons. An **electron** is *a very light, negatively charged particle that exists in the region around the atom's positively charged nucleus.*

### Discovery of the Electron

 ▶ Video no. 2 (Series A), "Electron Beam."

In 1897 the British physicist J. J. Thomson (Figure 2.3) conducted a series of experiments that showed that atoms were not indivisible particles. Figure 2.4 shows an experimental apparatus similar to the one used by Thomson. In this apparatus, two electrodes from a high-voltage source are sealed into a glass tube from which the air has been evacuated. The negative electrode is called the *cathode;* the positive one, the *anode.* When the high-voltage current is turned on, the glass tube emits a greenish light. Experiments showed that this greenish light is caused by the interaction of the glass with *cathode rays,* which are rays that originate from the cathode.

▶ Video Series C, "Cathode ray tube."

After the cathode rays leave the negative electrode, they move toward the anode, where some rays pass through a hole to form a beam (Figure 2.4). This beam bends

**Figure 2.3**
**Joseph John Thomson (1856–1940).** J. J. Thomson's scientific ability was recognized early with his appointment as professor of physics in the Cavendish Laboratory at Cambridge University when he was not quite 28 years old. Soon after this appointment, Thomson began research on the discharge of electricity through gases. This work culminated in 1897 with the discovery of the electron. Thomson was awarded the Nobel Prize in physics in 1906.

away from the negatively charged plate and toward the positively charged plate. (Cathode rays are not directly visible, but do cause certain materials such as zinc sulfide to glow so you can observe them.) Figure 2.5 shows a similar experiment, in which cathode rays are seen to bend when a magnet is brought toward them. Thomson showed that the characteristics of cathode rays are independent of the material making up the cathode. From such evidence, he concluded that a cathode ray consists of a beam of negatively charged particles (or electrons) and that electrons are constituents of all matter. ◄

From his experiments, Thomson could also calculate the ratio of the electron's mass, $m_e$, to its electric charge, $e$. He could not obtain either the mass or the charge separately, however. In 1909 the U.S. physicist Robert Millikan performed a series of ingenious experiments in which he obtained the charge on the electron by observing how a charged drop of oil falls in the presence and in the absence of an electric field (Figure 2.6). From this type of experiment, the charge on the electron is found to be $1.602 \times 10^{-19}$ coulombs (the *coulomb,* abbreviated C, is a unit of electric charge). If you use this charge with the most recent value of the mass-to-charge ratio of the

► A television tube uses the deflection of cathode rays by magnetic fields. A beam of cathode rays is directed toward a coated screen on the front of the tube, where by varying the magnetism generated by electromagnetic coils, the beam traces a luminescent image.

**Figure 2.4**
**Formation of cathode rays.** Cathode rays leave the cathode, or negative electrode, and are accelerated toward the anode, or positive electrode. Some of the rays pass through the hole in the anode to form a beam, which then is bent by the electric plates in the tube.

**Figure 2.5**
**Bending of cathode rays by a magnet.** *Left:* The cathode-ray beam travels from right to left. It is visible where it falls on a zinc-sulfide screen. *Center:* The beam of negative particles bends downward as the south pole of the magnet is brought toward it. *Right:* When the magnet is turned around, the beam bends in the opposite direction.

**Figure 2.6**
**Millikan's oil-drop experiment.** An atomizer, or spray bottle, introduces a fine mist of oil drops into the top chamber. Several drops happen to fall through a small hole into the lower chamber, where the experimenter follows the motion of one drop with a microscope. Some of these drops have picked up one or more electrons as a result of friction in the atomizer and have become negatively charged. A negatively charged drop will be attracted upward when the experimenter turns on a current to the electric plates. The drop's upward speed (obtained by timing its rise) is related to its mass-to-charge ratio, from which you can calculate the charge on the electron.

electron, you obtain an electron mass of $9.109 \times 10^{-31}$ kg, which is more than 1800 times smaller than the mass of the lightest atom (hydrogen). This shows quite clearly that the electron is indeed a subatomic particle.

## The Nuclear Model of the Atom

Ernest Rutherford (1871–1937), a British physicist, put forth the idea of the *nuclear model* of the atom in 1911, based on experiments done in his laboratory by Hans Geiger and Ernest Marsden. These scientists observed the effect of bombarding thin gold foil (and other metal foils) with alpha radiation from radioactive substances such as uranium (Figure 2.7). Rutherford had already shown that alpha rays consist of positively charged particles. Geiger and Marsden found that most of the alpha particles passed through a metal foil as though nothing were there, but a few (about 1 in 8000) were scattered at large angles and sometimes almost backward.

According to Rutherford's model, most of the mass of the atom (99.95% or more) is concentrated in a positively charged center, or nucleus, around which the negatively charged electrons move. Although most of the mass of an atom is in its nucleus, the nucleus occupies only a very small portion of the space of the atom. Nuclei have diameters of about $10^{-15}$ m ($10^{-3}$ pm), whereas atomic diameters are about $10^{-10}$ m (100 pm), a hundred thousand times larger. If you were to use a golf ball to represent the nucleus, the atom would be about 3 miles in diameter!

The nuclear model easily explains the results of bombarding gold and other metal foils with alpha particles. Alpha particles are much lighter than gold atoms. (Alpha

Charged plate (+)

Charged oil drop

Charged plate (—)

Spray bottle

Microscope

Window

Viewing chamber

**Figure 2.7**
**Alpha-particle scattering from metal foils.** Alpha radiation is produced by a radioactive source and formed into a beam by a lead plate with a hole in it. (Lead absorbs the radiation.) Scattered alpha particles are made visible by a zinc sulfide screen, which emits tiny flashes where particles strike it. A movable microscope is used for viewing the flashes.

**Figure 2.8**
**Representation of the scattering of alpha particles by a gold foil.** Most of the alpha particles pass through the foil barely deflected. A few, however, collide with gold nuclei and are deflected at large angles. (The relative sizes of nuclei are smaller than can be drawn here.)

**Visualization: Gold Foil Experiment**

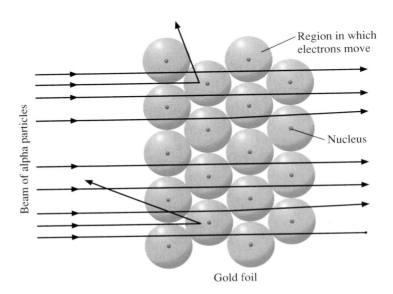

particles are helium nuclei.) Most of the alpha particles pass through the metal atoms of the foil, undeflected by the lightweight electrons. When an alpha particle does happen to hit a metal-atom nucleus, however, it is scattered at a wide angle because it is deflected by the massive, positively charged nucleus (Figure 2.8).

**CONCEPT CHECK 2.2**

What would be a feasible model for the atom if Geiger and Marsden had found that 7999 out of 8000 alpha particles were deflected back at the alpha-particle source?

## 2.3 Nuclear Structure; Isotopes

The nucleus of an atom also has a structure; the nucleus is composed of two different kinds of particles, protons and neutrons. The type of alpha-particle scattering experiment that led to the nuclear model of the atom was also instrumental in clarifying this structure of the nucleus.

An important property of the nucleus is its electric charge. It is possible to deduce the value of the positive charge of a nucleus by analyzing the distribution of alpha particles scattered from a metal foil; other experiments also provide the nuclear charge. ◄ From such experiments, researchers discovered that each element has a unique nuclear charge that is an integer multiple of the magnitude of the electron charge. This integer, which is characteristic of an element, is called the *atomic number* (Z). A hydrogen atom nucleus, whose magnitude of charge equals that of the electron, has the smallest atomic number (1).

► The simplest way to obtain the nuclear charge is by analyzing the x rays emitted by the element when irradiated with cathode rays. This is discussed in the essay at the end of Section 8.2.

In 1911, Rutherford discovered that hydrogen nuclei, or what we now call protons, form when alpha particles strike some of the lighter elements, such as nitrogen. A **proton** is *a nuclear particle having a positive charge equal to that of the electron and a mass more than 1800 times that of the electron.* When an alpha particle collides with a nitrogen atom, a proton may be knocked out of the nitrogen nucleus. The protons in a nucleus give the nucleus its positive charge.

The **atomic number** (Z) is therefore *the number of protons in the nucleus of an atom.* Because you can determine the atomic number experimentally, you can determine unambiguously whether or not a sample is a pure element or if perhaps you have discovered a new element. We can now state the definition of an element with more precision than we could in Section 2.1. An **element** is *a substance whose atoms all have the same atomic number.* The inside back cover of the book lists the elements and their atomic numbers (see the Table of Atomic Numbers and Atomic Weights).

The neutron was also discovered by alpha-particle scattering experiments. When beryllium metal is irradiated with alpha rays, a strongly penetrating radiation is obtained from the metal. In 1932 the British physicist James Chadwick (1891–1974) showed that this penetrating radiation consists of neutral particles, or neutrons. The **neutron** is *a nuclear particle having a mass almost identical to that of the proton but no electric charge.* When beryllium nuclei are struck by alpha particles, neutrons are knocked out. Table 2.2 compares the masses and charges of the electron and the two nuclear particles, the proton and the neutron.

Now consider the nucleus of some atom, say of the naturally occurring sodium atom. The nucleus contains 11 protons and 12 neutrons. Thus, the charge on the sodium nucleus is $+11e$, which we usually write as simply $+11$, meaning $+11$ units of electron charge $e$. Similarly, the nucleus of a naturally occurring aluminum atom contains 13 protons and 14 neutrons; the charge on the nucleus is $+13$.

We characterize a nucleus by its atomic number (Z) and its mass number (A). The **mass number** (A) is *the total number of protons and neutrons in a nucleus.* The nucleus of the naturally occurring sodium atom has an atomic number of 11 and a mass number of 23 (11 + 12).

**Table 2.2**
**Properties of the Electron, Proton, and Neutron**

| Particle | Mass (kg) | Charge (C) | Mass (amu)* | Charge (e) |
|---|---|---|---|---|
| Electron | $9.10939 \times 10^{-31}$ | $-1.60218 \times 10^{-19}$ | 0.00055 | $-1$ |
| Proton | $1.67262 \times 10^{-27}$ | $+1.60218 \times 10^{-19}$ | 1.00728 | $+1$ |
| Neutron | $1.67493 \times 10^{-27}$ | 0 | 1.00866 | 0 |

*The atomic mass unit (amu) equals $1.66054 \times 10^{-27}$ kg; it is defined in Section 2.4.

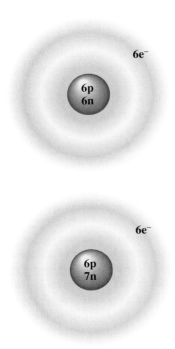

**Figure 2.9**
**A representation of two isotopes of carbon.** The drawing shows the basic particles making up the carbon-12 and carbon-13 isotopes. (The relative sizes of the nuclei are much exaggerated here.)

▶ Isotopes were first suspected in about 1912 when chemically identical elements with different atomic masses were found in radioactive materials. The most convincing evidence, however, came from the mass spectrometer, discussed in Section 2.4.

Problem Solving: Writing Symbols for Atoms

A **nuclide** is *an atom characterized by a definite atomic number and mass number.* The shorthand notation for any nuclide consists of the symbol of the element with the atomic number written as a subscript on the left and the mass number as a superscript on the left. You write the *nuclide symbol* for the naturally occurring sodium nuclide as follows:

$$\text{Mass number} \longrightarrow {}^{23}_{11}\text{Na} \longleftarrow \text{Atomic number}$$

An atom is normally electrically neutral, so it has as many electrons about its nucleus as the nucleus has protons; that is, the number of electrons in a neutral atom equals its atomic number. A sodium atom has a nucleus of charge $+11$, and around this nucleus are 11 electrons (with a charge of $-11$, giving the atom a charge of 0).

All nuclei of the atoms of a particular element have the same atomic number, but the nuclei may have different mass numbers. **Isotopes** are *atoms whose nuclei have the same atomic number but different mass numbers; that is, the nuclei have the same number of protons but different numbers of neutrons.* Sodium has only one naturally occurring isotope, which we denote by the same symbol we used for the nuclide (${}^{23}_{11}\text{Na}$); we also call the isotope sodium-23. Naturally occurring oxygen is a mixture of isotopes; it contains 99.759% oxygen-16, 0.037% oxygen-17, and 0.204% oxygen-18. ◀

Figure 2.9 may help you to visualize the relationship among the different subatomic particles in the isotopes of an element. The size of the nucleus as represented in the figure is very much exaggerated in order to show the numbers of protons and neutrons. In fact, the space taken up by the nucleus is minuscule compared with the region occupied by electrons. Electrons, although extremely light particles, move throughout relatively large diffuse regions, or "shells," about the nucleus of an atom.

---

**EXAMPLE 2.1**

**Writing Nuclide Symbols**

▶ An alternate example is provided in the Instructor's Resource Manual.

What is the nuclide symbol for a nucleus that contains 38 protons and 50 neutrons?

**SOLUTION**

From the Table of Atomic Numbers and Atomic Weights on the inside back cover of this book, you will note that the element with atomic number 38 (the number of protons in the nucleus) is strontium, symbol Sr. The mass number is $38 + 50 = 88$. The symbol is ${}^{88}_{38}\text{Sr}$.

---

**EXERCISE 2.1**

A nucleus consists of 17 protons and 18 neutrons. What is its nuclide symbol?

See Problems 2.39 and 2.40.

---

## 2.4 Atomic Weights

A central feature of Dalton's atomic theory was the idea that an atom of an element has a characteristic mass. Now we know that a naturally occurring element may be a mixture of isotopes, each isotope having its own characteristic mass. However, the percentages of the different isotopes in most naturally occurring elements have remained essentially constant over time and in most cases are independent of the origin of the element. Thus, what Dalton actually calculated were *average* atomic masses (actually average *relative* masses, as we will discuss). Since you normally work with naturally occurring mixtures of elements (whether in pure form or as compounds), it is indeed these average atomic masses that you need in chemical work. ◀

▶ Some variation of isotopic composition occurs in a number of elements, and this limits the significant figures in their average atomic masses.

## Relative Atomic Masses

Dalton could not weigh individual atoms. What he could do was find the average mass of one atom relative to the average mass of another. We will refer to these relative atomic masses as *atomic weights*. To see how Dalton obtained his atomic weights, imagine that you burn hydrogen gas in oxygen. The product of this reaction is water, a compound of hydrogen and oxygen. By experiment, you find that 1.0000 gram of hydrogen reacts with 7.9367 g of oxygen to form water. To obtain the atomic weight of oxygen (relative to hydrogen), you need to know the relative numbers of hydrogen atoms and oxygen atoms in water. As you might imagine, Dalton had difficulty with this. Today we know that water contains two atoms of hydrogen for every atom of oxygen. So the atomic weight of oxygen is $2 \times 7.9367 = 15.873$ times that of the mass of the average hydrogen atom. ◀

▶ Dalton had assumed the formula for water to be HO. Using this formula and accurate data, he would have obtained 7.9367 for the relative atomic weight of oxygen.

## Atomic Mass Units

Dalton's hydrogen-based atomic weight scale was eventually replaced by a scale based on oxygen and then, in 1961, by the present carbon-12 mass scale. This scale depends on measurements of atomic mass by an instrument called a *mass spectrometer* (Figure 2.10), which we will describe briefly later in this section. You make accurate measurements of mass on this instrument by comparing the mass of an atom to the mass of a particular atom chosen as a standard. On the present atomic weight scale, the carbon-12 isotope is chosen as the standard and is arbitrarily assigned a mass of exactly 12 atomic mass units. One **atomic mass unit (amu)** is, therefore, *a mass unit equal to exactly one-twelfth the mass of a carbon-12 atom.*

**Figure 2.10**
**A mass spectrometer.**
This instrument measures the masses of atoms and molecules.

**Figure 2.11**

**Diagram of a simple mass spectrometer, showing the separation of neon isotopes.** Neon gas enters an evacuated chamber, where neon atoms form positive ions when they collide with electrons. Positively charged neon atoms, $Ne^+$, are accelerated from this region by the negative grid and pass between the poles of a magnet. The beam of positively charged atoms is split into three beams by the magnetic field according to the mass-to-charge ratios. The three beams then travel to a detector at the end of the tube. (The detector is shown here as a photographic plate; in modern spectrometers, the detector is electronic, and the mass positions are recorded on a computer.)

▶ Neon was the first element to be separated into isotopes. In 1913 J. J. Thomson found the mass of neon to be 20 amu, but he also found a less-abundant mass at 22 amu, which he thought was a contaminant. Later, with improved apparatus, F. W. Aston showed that most elements are mixtures of isotopes.

On this modern scale, the **atomic weight** of an element is *the average atomic mass for the naturally occurring element, expressed in atomic mass units.* A complete table of atomic weights appears on the inside back cover of this book. (Although the use of the term *weight* in *atomic weight* is not strictly correct, it is sanctioned by convention to mean "average atomic mass.")

## Mass Spectrometry and Atomic Weights

Earlier we noted that you use a mass spectrometer to obtain accurate masses of atoms. The earliest mass spectrometers measured the mass-to-charge ratios of positively charged atoms using the same ideas that Thomson used to study electrons. Modern instruments may use other techniques, but all measure the mass-to-charge ratios of positively charged atoms (and molecules as well). Figure 2.11 shows a simplified sketch of a mass spectrometer that employs Thomson's ideas.

In the mass spectrometer shown in Figure 2.11, neon gas passes through an inlet tube into a chamber where the neon atoms collide with electrons from an electron beam (cathode rays). The force of a collision can knock an electron from a neon atom. The positive neon atoms produced this way are drawn toward a negative grid, and some of them pass through slits to form a beam of positive electricity. This beam then travels through a magnetic field (from a magnet, whose poles are above and below the positive beam). The magnetic field deflects the positively charged atoms in the positive beam according to their mass-to-charge ratios. Just as a heavy car is buffeted less by a strong wind than a light car is, the most massive positively charged atoms are deflected the least by the magnetic field. For this reason, each positively charged atom arrives at a definite position at the end of the tube depending on its mass-to-charge ratio. The positively charged neon atoms split into three beams corresponding to the three different isotopes of neon. Each atom has a charge of $+1$ but a mass number of either 20, 21, or 22. ◀

You obtain the mass of each neon isotope from the magnitude of deflection of its positively charged atoms. The neon isotopes have the following masses: neon-20, 19.992 amu; neon-21, 20.994 amu; and neon-22, 21.991 amu. You can also obtain the percentage of the total number of neon atoms that are neon-20, for example, from the relative number of positively charged neon-20 atoms reaching the detector at the end

Fractional abundance

18   19   20   21   22   23

Atomic mass (amu)

**Figure 2.12**
**The mass spectrum of neon.** Neon is separated into its isotopes Ne-20, Ne-21, and Ne-22. The height at each mass peak is proportional to the fraction of that isotope in the element.

▶ An alternate example is provided in the Instructor's Resource Manual.

▶ Experiments in General Chemistry, 2, "Isotopes and Mass Spectrometry."

of the tube. Figure 2.12 shows the *mass spectrum* of neon, which is a chart recording that shows the relative numbers of atoms for various masses.

The mass spectrum gives us all of the information needed to calculate the atomic weight: the masses of isotopes and their relative numbers, or fractional abundances. The **fractional abundance** of an isotope is *the fraction of the total number of atoms that is composed of a particular isotope.* The fractional abundances of the neon isotopes in naturally occurring neon are neon-20, 0.9051; neon-21, 0.0027; and neon-22, 0.0922.

You calculate the atomic weight of an element by multiplying each isotopic mass by its fractional abundance and summing the values. If you do that for neon using the data given here, you will obtain 20.179 amu. The next example illustrates the calculation in full for chromium.

---

**EXAMPLE 2.2**

## Determining Atomic Weight from Isotopic Masses and Fractional Abundances

Chromium, Cr, has the following isotopic masses and fractional abundances:

| Mass Number | Isotopic Mass (amu) | Fractional Abundance |
|---|---|---|
| 50 | 49.9461 | 0.0435 |
| 52 | 51.9405 | 0.8379 |
| 53 | 52.9407 | 0.0950 |
| 54 | 53.9389 | 0.0236 |

What is the atomic weight of chromium?

 **PROBLEM STRATEGY**

The type of average used to calculate the atomic weight of an element is similar to the "average" an instructor might use to obtain a student's final grade in a class. Suppose the student has a total exam grade of 76 and a total laboratory grade of 84. The instructor decides to give a weight of 70% to the exams and 30% to the laboratory. How would the instructor calculate the final grade? He or she would multiply each type of grade by its weight factor and add the results:

$$(76 \times 0.70) + (84 \times 0.30) = 78$$

The final grade is closer to the exam grade, because the instructor chose to give the exam grade greater weight.

**SOLUTION**

Multiply each isotopic mass by its fractional abundance, then sum:

$$49.9461 \text{ amu} \times 0.0435 = \phantom{0}2.17 \text{ amu}$$
$$51.9405 \text{ amu} \times 0.8379 = 43.52 \text{ amu}$$
$$52.9407 \text{ amu} \times 0.0950 = \phantom{0}5.03 \text{ amu}$$
$$53.9389 \text{ amu} \times 0.0236 = \underline{\phantom{0}1.27 \text{ amu}}$$
$$51.99 \text{ amu}$$

The atomic weight of chromium is **51.99 amu.**

**EXERCISE 2.2**

Chlorine consists of the following isotopes:

| Isotope | Isotopic Mass (amu) | Fractional Abundance |
|---------|---------------------|----------------------|
| Chlorine-35 | 34.96885 | 0.75771 |
| Chlorine-37 | 36.96590 | 0.24229 |

What is the atomic weight of chlorine?

See Problems 2.43, 2.44, 2.45, and 2.46.

## 2.5 Periodic Table of the Elements

▶ The relationship between the periodic table and electron configurations and the periodicity of atomic properties are discussed in Chapter 8.

In 1869 the Russian chemist Dmitri Mendeleev (1834–1907; Figure 2.13) and the German chemist J. Lothar Meyer (1830–1895), working independently, made similar discoveries. They found that when they arranged the elements in order of atomic weight, they could place them in horizontal rows, one row under the other, so that the elements in each vertical column have similar properties. *A tabular arrangement of elements in rows and columns, highlighting the regular repetition of properties of the elements,* is called a **periodic table.**

Eventually, more accurate determinations of atomic weights revealed discrepancies in this ordering of the elements. However, in the early part of this century, it was shown that the elements are characterized by their atomic numbers, rather than atomic weights. When the elements in the periodic table are ordered by atomic number, such discrepancies vanish.

A modern version of the periodic table, with the elements arranged by atomic number, is shown in Figure 2.14 (see also inside front cover). Each entry lists the atomic number, atomic symbol, and atomic weight of an element. This is a convenient way of tabulating such information, and you should become familiar with using the periodic table

**Figure 2.13**
**Dmitri Ivanovich Mendeleev (1834–1907).** Mendeleev constructed a periodic table as part of his effort to systematize chemistry. He received many international honors for his work, but his reception in czarist Russia was mixed. He had pushed for political reforms and made many enemies as a result.

**Figure 2.14**
**A modern form of the periodic table.** This table is also given on the inside front cover of the book. The placement in the main body of the table of the two partial rows at the bottom is indicated by an asterisk (*) and a double asterisk (**).

for that purpose. As we develop the subject matter of chemistry throughout the text, you will see how the periodic table can become even more useful.

## Periods and Groups

The basic structure of the periodic table is its division into rows and columns, or periods and groups. A **period** consists of *the elements in any one horizontal row of the periodic table.* A **group** consists of *the elements in any one column of the periodic table.*

The first period of elements is short, consisting of only hydrogen (H) and helium (He). The second period has 8 elements, beginning with lithium (Li) and ending with neon (Ne). There is then another period of 8 elements, and this is followed by a period having 18 elements, beginning with potassium (K) and ending with krypton (Kr). The fifth period also has 18 elements. The sixth period actually consists of 32 elements, but in order for the row to fit on a page, part of it appears at the bottom of the table. Otherwise the table would have to be expanded, with the additional elements placed after lanthanum (La, atomic number 57). The seventh period, though not complete, also has some of its elements placed as a row at the bottom of the table.

The groups are usually numbered. The numbering frequently seen in North America labels the groups with Roman numerals and A's and B's. In Europe a similar convention has been used, but some columns have the A's and B's interchanged. To eliminate this confusion, the International Union of Pure and Applied Chemistry (IUPAC) suggested a convention in which the columns are numbered 1 to 18. Figure 2.14 shows the traditional North American and the IUPAC conventions. When we refer to an element by its periodic group, we will use the traditional North American convention. The A groups are called *main-group* (or *representative*) *elements;* the B groups are called *transition elements.* The two rows of elements at the bottom of the table are called *inner-transition elements* (the first row is referred to as the *lanthanides;* the second row, as the *actinides*).

As noted earlier, the elements in any one group have similar properties. For example, the elements in Group IA, often known as the *alkali metals,* are soft metals that react easily with water. (Hydrogen, a gas, is an exception and might better be put in a group by itself.) Sodium is an alkali metal. So is potassium. The Group VIIA elements, known as *halogens,* are also reactive elements. Chlorine is a halogen. We have already noted its vigorous reaction with sodium. Bromine, which is a red-brown liquid, is another halogen. It too reacts vigorously with sodium.

## Metals, Nonmetals, and Metalloids

The elements of the periodic table in Figure 2.14 are divided by a heavy "staircase" line into metals on the left and nonmetals on the right. A **metal** is *a substance or mixture that has a characteristic luster, or shine, and is generally a good conductor of heat and electricity.* Except for mercury, the metallic elements are solids at room temperature (about 20°C). They are more or less *malleable* (can be hammered into sheets) and *ductile* (can be drawn into wire).

A **nonmetal** is *an element that does not exhibit the characteristics of a metal.* Most of the nonmetals are gases (for example, chlorine and oxygen) or solids (for example, phosphorus and sulfur). The solid nonmetals are usually hard, brittle substances. Bromine is the only liquid nonmetal.

Most of the elements bordering the staircase line in the periodic table (Figure 2.14) are metalloids, or semimetals. A **metalloid,** or **semimetal,** is *an element having both metallic and nonmetallic properties.* These elements, such as silicon (Si) and germanium (Ge), are usually good *semiconductors*—elements that, when pure, are poor conductors of electricity at room temperature but become moderately good conductors at higher temperatures. ◄

▶ When these pure semiconductor elements have small amounts of certain other elements added to them (a process called *doping*), they become very good conductors of electricity. Semiconductors are the critical materials in solid-state electronic devices. (Chapter 13)

---

### EXERCISE 2.3

By referring to the periodic table (Figure 2.14 or inside front cover), identify the group and period to which each of the following elements belongs. Then decide whether the element is a metal, nonmetal, or metalloid.

a. Se    b. Cs    c. Fe    d. Cu    e. Br

---

See Problems 2.49 and 2.50.

 **CONCEPT CHECK 2.3**

Consider the elements He, Ne, and Ar. Can you come up with a reason why they are in the same group in the periodic table?

# Thirty Seconds on the Island of Stability

**FRONTIERS**

It appears that scientists have now been on the "Island of Stability"—for at least 30 seconds. In recent years, the island of stability has attracted nuclear scientists the way Mount Everest has attracted mountain climbers. Like mountain climbers, scientists have wanted to surmount their own tallest peak—discovery of the relatively stable, superheavy nuclide at the peak of stability that is surrounded by foothills comprised of less stable nuclides. Discovering the superheavy nuclide at the peak would also mean discovering a new element—a task that has had its allure since Lavoisier gave currency to the concept of elements in the late eighteenth century.

Theoretical physicists predicted the existence of the island of stability, centered at element 114 with mass number 298, in the 1960s. The term *stability* refers here to that of the atomic nucleus. An unstable nucleus tends to fall apart by radioactive decay—a piece spontaneously flies off the nucleus, leaving a different one behind. Approximately 275 nuclides are completely stable, or nonradioactive. All of these nuclides have atomic numbers (or numbers of protons) no greater than 83 (the atomic number for the element bismuth). Beyond bismuth, all elements are radioactive and become increasingly unstable. In fact, none of the original, primeval elements past uranium (element 92)—the *transuranium* elements—exists any longer; they have long since vanished by radioactive decay. Scientists have made transuranium elements in the laboratory, however. In 1996, for example, German scientists succeeded in making a few atoms of element 112, with mass number 277; each of these atoms lasted only about 240 microseconds.

In December 1998, Russian scientists at the laboratory in Dubna, north of Moscow, and their American collaborators from Lawrence Livermore Laboratory in California produced one atom of element 114 (provisionally named ununquadium, Uuq), of mass number 298, that lasted about 30 seconds. (While a 30-second lifetime may seem brief, it is many times longer than that seen in the previously discovered element 112.) The Russian scientists bombarded a target of plutonium-244 with calcium-48, a rare isotope, for 34 days. We can represent the result by a nuclear equation, which is similar to a chemical equation (described at the end of the chapter):

$$^{48}_{20}\text{Ca} + {}^{244}_{94}\text{Pu} \longrightarrow {}^{289}_{114}\text{Uuq} + {}^{1}_{0}\text{n}$$

The nuclide symbols used here were introduced in Section 2.3. A high-speed calcium-48 atom smashes into a plutonium-244 target where, as the arrow denotes, it yields new products, an atom of element 114 (mass number 289) and a neutron, denoted $^{1}_{0}\text{n}$.

The discovery of new elements has not abated. Late in May 1999, scientists at Lawrence Berkeley National Laboratory in Berkeley, California, reported the discovery of two new elements, 116 (ununhexium, Uuh) and 118 (ununoctium, Uuo), which they produced by bombarding a lead-208 target with krypton-86 nuclei:

$$^{86}_{36}\text{Kr} + {}^{208}_{82}\text{Pb} \longrightarrow {}^{293}_{118}\text{Uuo} + {}^{1}_{0}\text{n}$$

Figure 2.15 shows scientists working on the *Berkeley gas-filled separator,* which they used to isolate the new element from the target. Element 118 decayed first to a nuclide of 116, then to a nuclide of 114. Element 118 would presumably be a noble-gas element and occupy the position under radon in the periodic table. Elements 113, 115, and 117 have yet to be discovered, but when found will complete the seventh row of the periodic table.

**Figure 2.15**
**Berkeley gas-filled separator.** Nuclear physicist Victor Ninov *(left)* and nuclear chemist Kenneth Gregorich work on the apparatus used to separate the new element 118.

# Chemical Substances: Formulas and Names

Atomic theory has developed steadily since Dalton's time and has become the cornerstone of chemistry. It results in an enormous simplification: all of the millions of compounds we know today are composed of the atoms of just a few elements. Now we look more closely at how we describe the composition and structure of chemical substances in terms of atoms.

## 2.6 Chemical Formulas; Molecular and Ionic Substances

The **chemical formula** of a substance is *a notation that uses atomic symbols with numerical subscripts to convey the relative proportions of atoms of the different elements in the substance.* Consider the formula of aluminum oxide, $Al_2O_3$. This means that the compound is composed of aluminum atoms and oxygen atoms in the ratio $2:3$. Consider the formula for sodium chloride, NaCl. When no subscript is written for a symbol, it is assumed to be 1. Therefore, the formula NaCl means that the compound is composed of sodium atoms and chlorine atoms in the ratio $1:1$.

Additional information may be conveyed by different kinds of chemical formulas. To understand this, we need to look briefly at two main types of substances: molecular and ionic.

### Molecular Substances

A **molecule** is *a definite group of atoms that are chemically bonded together—that is, tightly connected by attractive forces.* The nature of these strong forces is discussed in Chapters 9 and 10. A *molecular substance* is a substance that is composed of molecules all of which are alike. The molecules in such a substance are so small that even extremely minute samples contain tremendous numbers of them. One billionth ($10^{-9}$) of a drop of water, for example, contains about 2 trillion ($2 \times 10^{12}$) water molecules. ◄

A **molecular formula** *gives the exact number of different atoms of an element in a molecule.* The hydrogen peroxide molecule contains two hydrogen atoms and two oxygen atoms chemically bonded. Therefore, its molecular formula is $H_2O_2$. Other simple molecular substances are water, $H_2O$; ammonia, $NH_3$; carbon dioxide, $CO_2$; and ethanol (ethyl alcohol), $C_2H_6O$. In Chapter 3, you will see how we determine such formulas.

The atoms in a molecule are not simply piled together randomly. Rather, the atoms are chemically bonded in a definite way. A *structural formula* is a chemical formula that shows how the atoms are bonded to one another in a molecule. For example, it is known that each of the hydrogen atoms in the water molecule is bonded to the oxygen atom. Thus, the structural formula of water is H—O—H. A line joining two atomic symbols in such a formula represents the chemical bond connecting the atoms. Figure 2.16 shows some structural formulas. Structural formulas are sometimes condensed in writing. For example, the structural formula of ethanol may be written $CH_3CH_2OH$ or $C_2H_5OH$, depending on the detail you want to convey.

The atoms in a molecule are not only connected in definite ways but exhibit definite spatial arrangements as well. Chemists often construct molecular models as an aid in visualizing the shapes and sizes of molecules. Figure 2.16 shows molecular models for several compounds. While the ball-and-stick type of model shows the bonds and bond angles clearly, the space-filling type gives a more realistic feeling of the space occupied by the atoms. Chemists also use computer-generated models of molecules, which can be produced in a variety of forms.

► Another way to understand the large numbers of molecules involved in relatively small quantities of matter is to consider 1 g of water (about one-fifth teaspoon). It contains $3.3 \times 10^{22}$ water molecules. If you had a penny for every molecule in this quantity of water, the height of your stack of pennies would be about 300 million times the distance from the earth to the sun.

**Figure 2.16**
**Examples of molecular and structural formulas and molecular models.**
Three common molecules—water, ammonia, and ethanol—are shown.

|  | Water | Ammonia | Ethanol |
|---|---|---|---|
| Molecular formula | $H_2O$ | $NH_3$ | $C_2H_6O$ |
| Structural formula | H—O—H | H—N—H \ H | $\begin{matrix} & H & H \\ & | & | \\ H- & C- & C- & O-H \\ & | & | \\ & H & H \end{matrix}$ |
| Molecular model (ball-and-stick type) |  |  |  |
| Molecular model (space-filling type) |  |  |  |

Some elements are molecular substances and are represented by molecular formulas. Chlorine, for example, is a molecular substance and has the formula $Cl_2$, each molecule being composed of two chlorine atoms bonded together. Sulfur consists of molecules composed of eight atoms; its molecular formula is $S_8$. Helium and neon are composed of isolated atoms; their formulas are He and Ne, respectively. Other elements, such as carbon (in the form of graphite or diamond), do not have a simple molecular structure but consist of a very large, indefinite number of atoms bonded together. These elements are represented simply by their atomic symbols. (A recently discovered form of carbon called buckminsterfullerene has the molecular formula $C_{60}$.) Models of some elementary substances are shown in Figure 2.17.

An important class of molecular substances is the polymers. **Polymers** are *very large molecules that are made up of a number of smaller molecules repeatedly linked together.* **Monomers** are *the small molecules that are linked together to form the polymer.* A good analogy for the formation of polymers from monomers is that of making a chain of paper clips. Say you have boxes of red, blue, and yellow paper clips. Each paper clip represents a monomer. You can make a large chain of paper clips, representing the polymer, with a variety of patterns and repeating units. One chain might involve a repeating pattern of two red, three yellow, and one blue. Another chain might consist of only blue paper clips. Just like the paper clips, the creation of a particular polymer in a laboratory involves controlling both the monomers that are being linked together and the linkage pattern.

**Figure 2.17**
**Molecular models of some elementary substances.** *Left to right:* Chlorine, $Cl_2$, white phosphorus, $P_4$, and sulfur, $S_8$.

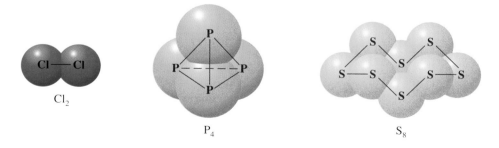

$Cl_2$          $P_4$          $S_8$

**Figure 2.18**
**$F_2C{=}CF_2$ monomer and Teflon®.** *Left:* Model of the monomer used to make Teflon®, $CF_2CF_2$. *Center:* Model showing linkage of $CF_2CF_2$ monomers that make Teflon®. *Right:* Pan with Teflon® coating.

Polymers are both natural and synthetic. Hard plastic soda bottles are made from chemically linking two different monomers in an alternating pattern. Wool and silk are natural polymers of amino acids linked by peptide bonds. Nylon® for fabrics, Kevlar® for bulletproof vests, and Nomex® for flame-retardant clothing all contain a CONH linkage. Plastics and rubbers are also polymers that are made from carbon- and hydrogen-containing monomers. Even the Teflon® coating on cookware is a polymer that is the result of linking $CF_2CF_2$ monomers (Figure 2.18). From these examples, it is obvious that polymers are very important molecular materials that we use every day in a wide variety of applications. For more on the chemistry of polymers, refer to Chapter 25.

## Ionic Substances

Although many substances are molecular, others are composed of ions (pronounced "eye′-ons"). An **ion** is *an electrically charged particle obtained from an atom or chemically bonded group of atoms by adding or removing electrons.* Sodium chloride is a substance made up of ions.

Although isolated atoms are normally electrically neutral and therefore contain equal numbers of positive and negative charges, during the formation of certain compounds atoms can become ions. Metal atoms tend to lose electrons, whereas nonmetals tend to gain electrons. When a metal atom such as sodium and a nonmetal atom such as chlorine approach one another, an electron can transfer from the metal atom to the nonmetal atom to produce ions.

An atom that picks up an extra electron becomes *a negatively charged ion,* called an **anion** (pronounced "an′-ion"). An atom that loses an electron becomes *a positively charged ion,* called a **cation** ("cat′-ion"). A sodium atom, for example, can lose an electron to form a sodium cation (denoted $Na^+$). A chlorine atom can gain an electron to form a chloride anion (denoted $Cl^-$). A calcium atom can lose two electrons to form a calcium cation, denoted $Ca^{2+}$. Note that the positive-two charge on the ion is indicated by a superscript $2+$.

Some ions consist of two or more atoms chemically bonded but having an excess or deficiency of electrons so that the unit has an electric charge. An example is the sulfate ion, $SO_4{}^{2-}$. The superscript $2-$ indicates an excess of two electrons on the group of atoms.

An **ionic compound** is *a compound composed of cations and anions.* Sodium chloride consists of equal numbers of sodium ions, $Na^+$, and chloride ions, $Cl^-$. The strong attraction between positive and negative charges holds the ions together in a

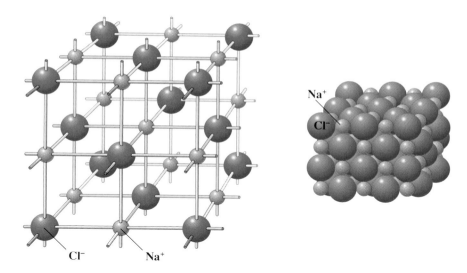

**Figure 2.19**
**The sodium chloride crystal.** *Left:* A photograph showing crystals of sodium chloride. *Center:* A model of a portion of a crystal detailing the regular arrangement of sodium ions and chloride ions. Each sodium ion is surrounded by six chloride ions, and each chloride ion is surrounded by six sodium ions. *Right:* A model highlighting the packing arrangement of the sodium and chloride ions in a solid sodium chloride crystal.

regular arrangement in space. For example, in sodium chloride, each $Na^+$ ion is surrounded by six $Cl^-$ ions, and each $Cl^-$ ion is surrounded by six $Na^+$ ions. The result is a *crystal,* which is a kind of solid having a regular three-dimensional arrangement of atoms, molecules, or (as in the case of sodium chloride) ions. Figure 2.19 shows sodium chloride crystals and two types of models used to depict the arrangement of the ions in the crystal. The number of ions in an individual sodium chloride crystal determines the size of the crystal.

The formula of an ionic compound is written by giving the smallest possible integer number of different ions in the substance, except that the charges on the ions are omitted so that the formulas merely indicate the atoms involved. For example, sodium chloride contains equal numbers of $Na^+$ and $Cl^-$ ions. The formula is written NaCl (not $Na^+Cl^-$). Iron(III) sulfate is a compound consisting of iron(III) ions, $Fe^{3+}$, and sulfate ions, $SO_4^{2-}$, in the ratio 2 : 3. The formula is written $Fe_2(SO_4)_3$, in which parentheses enclose the formula of an ion composed of more than one atom (again, omitting ion charges); parentheses are used only when there are two or more such ions.

Although ionic substances do not contain molecules, we can speak of the smallest unit of such a substance. The **formula unit** of a substance is *the group of atoms or ions explicitly symbolized in the formula.* For example, the formula unit of water, $H_2O$, is the $H_2O$ molecule. The formula unit of iron(III) sulfate, $Fe_2(SO_4)_3$, consists of two $Fe^{3+}$ ions and three $SO_4^{2-}$ ions. The formula unit is the smallest unit of such substances.

All substances, including ionic compounds, are electrically neutral. You can use this fact to obtain the formula of an ionic compound, given the formulas of the ions. This is illustrated in the following example.

▶ An alternate example is provided in the Instructor's Resource Manual.

**EXAMPLE 2.3**

## Writing an Ionic Formula, Given the Ions

a. Chromium(III) oxide is used as a green paint pigment (Figure 2.20). It is a compound composed of $Cr^{3+}$ and $O^{2-}$ ions. What is the formula of chromium(III) oxide?

**Figure 2.20**
**Chromium(III) oxide.** The compound is used as a green paint pigment.

b. Strontium oxide is a compound composed of $Sr^{2+}$ and $O^{2-}$ ions. Write the formula of this compound.

 **PROBLEM STRATEGY**

Because a compound is neutral, the sum of positive and negative charges equals zero. Consider the ionic compound $CaCl_2$, which consists of one $Ca^{2+}$ ion and two $Cl^-$ ions. The sum of the charges is:

$$\underbrace{1 \times (+2)}_{\text{Positive charge}} + \underbrace{2 \times (-1)}_{\text{Negative charge}} = 0$$

Note that the number of calcium ions in $CaCl_2$ equals the magnitude of charge on the chloride ion (1), whereas the number of chloride ions in $CaCl_2$ equals the magnitude of charge on the calcium ion (2). In general, you use the magnitude of charge on one ion to obtain the subscript for the other ion. You may need to simplify the formula you obtain this way so that it expresses the simplest ratio of ions.

**SOLUTION**

a. You can achieve electrical neutrality by taking as many cations as there are units of charge on the anion and as many anions as there are units of charge on the cation. Two $Cr^{3+}$ ions have a total charge of 6+, and three $O^{2-}$ ions have a total charge of 6−, giving the combination a net charge of zero. The simplest ratio of $Cr^{3+}$ to $O^{2-}$ is 2 : 3, and the formula is **$Cr_2O_3$.** Note that the charge (without its sign) on one ion becomes the subscript of the other ion.

$$Cr^{③+} \diagdown O^{②-} \qquad \text{or} \qquad Cr_2O_3$$

b. You see that equal numbers of $Sr^{2+}$ and $O^{2-}$ ions will give a neutral compound. Thus, the formula is SrO. If you use the units of charge to find the subscripts, you get

$$Sr^{②+} \diagdown O^{②-} \qquad \text{or} \qquad Sr_2O_2$$

The final formula is **SrO,** because this gives the simplest ratio of ions.

**Figure 2.21**
**Potassium chromate.** Many compounds of chromium have bright colors, which is the origin of the name of the element. It comes from the Greek word *chroma,* meaning "color."

**EXERCISE 2.4**

Potassium chromate is an important compound of chromium (Figure 2.21). It is composed of $K^+$ and $CrO_4^{2-}$ ions. Write the formula of the compound.

See Problems 2.67 and 2.68.

## 2.7 Organic Compounds

An important class of *molecular substances that contain carbon combined with other elements, such as hydrogen, oxygen, and nitrogen* is **organic compounds.** Organic chemistry is the area of chemistry that is concerned with these compounds (Chapter 24 is devoted to this topic). Historically, organic compounds were restricted to those that could be produced only from living entities and were thought to contain a "vital force" based on their natural origin. The concept of the vital force

**Figure 2.22**
**Molecular model of urea (CH₄N₂O).**   Urea was the first organic molecule deliberately synthesized by a chemist.

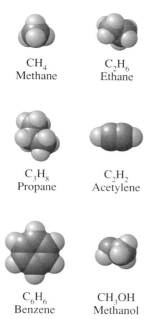

$CH_4$
Methane

$C_2H_6$
Ethane

$C_3H_8$
Propane

$C_2H_2$
Acetylene

$C_6H_6$
Benzene

$CH_3OH$
Methanol

$CH_3CH_2OCH_2CH_3$
Diethyl ether

was disproved in 1828 when a German chemist, Friedrick Wöhler, synthesized urea (a molecular compound in human urine, $CH_4N_2O$, Figure 2.22) from the molecular compounds ammonia ($NH_3$) and cyanic acid ($HNCO$). His work clearly demonstrated that a given compound is exactly the same whether it comes from a living entity or is synthesized.

Organic compounds make up the majority of all known compounds. Since 1957, more than 13 million (60%) of the recorded substances in an international materials registry have been listed as organic. You encounter organic compounds in both living and nonliving materials every day. The proteins, amino acids, enzymes, and DNA that make up your body are all either individual organic molecules or contain organic molecules. Table sugar, peanut oil, antibiotic medicines, and methanol (windshield washer) are all examples of organic molecules as well. Organic chemistry and the compounds produced by the reactions of organic molecules are probably responsible for the majority of the materials that currently surround you as you read this page of text.

The simplest organic compounds are hydrocarbons. **Hydrocarbons** are *those compounds containing only hydrogen and carbon.* Common examples include methane ($CH_4$), ethane ($C_2H_6$), propane ($C_3H_8$), acetylene ($C_2H_2$), and benzene ($C_6H_6$). Hydrocarbons are used extensively as sources of energy for heating our homes, for powering internal combustion engines, and for generating electricity. They also are the starting materials for most plastics. Much of the mobility and comfort of our current civilization is built on the low cost and availability of hydrocarbons.

The chemistry of organic molecules is often determined by groups of atoms in the molecule that have characteristic chemical properties. A **functional group** is a *reactive portion of a molecule that undergoes predictable reactions.* When you use the term *alcohol* when referring to a molecular compound, you actually are indicating a molecule that contains an —OH functional group. Methyl alcohol has the chemical formula $CH_3OH$. The term *ether* indicates that an organic molecule contains an oxygen atom between two carbon atoms as in diethyl ether ($CH_3CH_2OCH_2CH_3$). Table 2.3 contains a few examples of organic functional groups along with example compounds.

Chapter 24 contains a much more extensive treatment of organic chemistry. Due to its importance and its relation to living organisms, a substantial part of your future chemistry experience will likely be in the field of organic chemistry.

**CONCEPT CHECK 2.4**

Identify the following compounds as being a hydrocarbon, an alcohol, an ether, or a carboxylic acid.

(a)          (b)          (c)          (d)

**Table 2.3**
**Examples of Organic Functional Groups**

| Functional Group | Name of Functional Group | Example Molecule | Common Use |
|---|---|---|---|
| —OH | Alcohol | Methyl alcohol ($CH_3OH$) | Windshield washer |
| —O— | Ether | Dimethyl ether ($CH_3OCH_3$) | Solvent |
| —COOH | Carboxylic acid | Acetic acid ($CH_3COOH$) | Acid in vinegar |

## 2.8 Naming Simple Compounds

Before the structural basis of chemical substances became established, compounds were named after people, places, or particular characteristics. Examples are Glauber's salt (sodium sulfate, discovered by J. R. Glauber), sal ammoniac (ammonium chloride, named after the ancient Egyptian deity Ammon from the temple near which the substance was made), and washing soda (sodium carbonate, used for softening wash water). Today several million compounds are known and thousands of new ones are discovered every year. Without a system for naming compounds, coping with this multitude of substances would be a hopeless task. **Chemical nomenclature** is *the systematic naming of chemical compounds.*

If a compound is not classified as organic, as discussed in Section 2.7, it must be inorganic. **Inorganic compounds** are *composed of elements other than carbon.* A few notable exceptions to this classification scheme include carbon monoxide, carbon dioxide, carbonates, and the cyanides; all contain carbon and yet are generally considered to be inorganic.

In this section, we discuss the nomenclature of some simple inorganic compounds. We first look at the naming of ionic compounds. Then, we look at the naming of some simple molecular compounds, including binary molecular compounds (molecular compounds of two elements) and acids. Finally, we look at hydrates of ionic compounds. These substances contain water molecules in loose association with ionic compounds.

### Ionic Compounds

Ionic compounds, as we saw in the previous section, are substances composed of ions. Most ionic compounds contain metal and nonmetal atoms; for example, NaCl. (The ammonium salts, such as $NH_4Cl$, are a prominent exception.) You name an ionic compound by giving the name of the cation followed by the name of the anion. For example,

potassium sulfate

cation name    anion name

Before you can name ionic compounds, you need to be able to write and name ions.

**Table 2.4**
**Common Monatomic Ions of the Main-Group Elements***

|  | IA | IIA | IIIA | IVA | VA | VIA | VIIA |
|---|---|---|---|---|---|---|---|
| Period 1 |  |  |  |  |  |  | $H^-$ |
| Period 2 | $Li^+$ | $Be^{2+}$ | B | C | $N^{3-}$ | $O^{2-}$ | $F^-$ |
| Period 3 | $Na^+$ | $Mg^{2+}$ | $Al^{3+}$ | Si | P | $S^{2-}$ | $Cl^-$ |
| Period 4 | $K^+$ | $Ca^{2+}$ | $Ga^{3+}$ | Ge | As | $Se^{2-}$ | $Br^-$ |
| Period 5 | $Rb^+$ | $Sr^{2+}$ | $In^{3+}$ | $Sn^{2+}$ | Sb | $Te^{2-}$ | $I^-$ |
| Period 6 | $Cs^+$ | $Ba^{2+}$ | $Tl^+, Tl^{3+}$ | $Pb^{2+}$ | $Bi^{3+}$ |  |  |

*Elements shown in color do not normally form compounds having monatomic ions.

The simplest ions are monatomic. A **monatomic ion** is *an ion formed from a single atom.* Table 2.4 lists common monatomic ions of the main-group elements. You may want to look at the table while you read first the rules for predicting the charges on such ions and then the rules for naming the monatomic ions.

### Rules for Predicting the Charges on Monatomic Ions

1. Most of the main-group metallic elements have one monatomic cation with a charge equal to the group number in the periodic table (the Roman numeral). Example: aluminum, in Group IIIA, has a monatomic ion $Al^{3+}$.

2. Some metallic elements of high atomic number are exceptions to the previous rule. These elements have common cations with a charge equal to the group number minus 2, in addition to having a cation with a charge equal to the group number. Example: The common ion of lead is $Pb^{2+}$. (The group number is 4; the charge is 4 − 2.) In addition to compounds containing $Pb^{2+}$, some lead compounds contain $Pb^{4+}$.

3. Most transition elements form more than one monatomic cation, each with a different charge. Most of these elements have one ion with a charge of +2. Example: Iron has common cations $Fe^{2+}$ and $Fe^{3+}$. Copper has common cations $Cu^+$ and $Cu^{2+}$.

4. The charge on a monatomic anion for a nonmetallic main-group element equals the group number minus 8. Example: Oxygen has the monatomic anion $O^{2-}$. (The group number is 6; the charge is 6 − 8.)

▶ The Roman numeral actually denotes the *oxidation state,* or *oxidation number,* of the atom in the compound. For a monatomic ion, the oxidation state equals the charge. Otherwise, the oxidation state is a hypothetical charge assigned in accordance with certain rules; see Section 4.5.

### Rules for Naming Monatomic Ions

1. Monatomic cations are named after the element if there is only one such ion. Example: $Al^{3+}$ is called aluminum ion; $Na^+$ is called sodium ion.

2. If there is more than one monatomic cation of an element, Rule 1 is not sufficient. The *Stock system* of nomenclature names the cations after the element, as in Rule 1, but follows this by a Roman numeral in parentheses denoting the charge on the ion. Example: $Fe^{2+}$ is called iron(II) ion and $Fe^{3+}$ is called iron(III) ion. ◀

**Table 2.5**
**Common Cations of the Transition Elements**

| Ion | Ion Name |
|-----|----------|
| $Cr^{3+}$ | Chromium(III) or chromic |
| $Mn^{2+}$ | Manganese(II) or manganous |
| $Fe^{2+}$ | Iron(II) or ferrous |
| $Fe^{3+}$ | Iron(III) or ferric |
| $Co^{2+}$ | Cobalt(II) or cobaltous |
| $Ni^{2+}$ | Nickel(II) or nickel |
| $Cu^{2+}$ | Copper(II) or cupric |
| $Zn^{2+}$ | Zinc |
| $Ag^{+}$ | Silver |
| $Cd^{2+}$ | Cadmium |
| $Hg^{2+}$ | Mercury(II) or mercuric |

In an older system of nomenclature, such ions are named by adding the suffixes -*ous* and -*ic* to a stem name of the element (which may be from the Latin) to indicate the ions of lower and higher charge, respectively. Example: $Fe^{2+}$ is called ferrous ion; $Fe^{3+}$, ferric ion. $Cu^{+}$ is called cuprous ion; $Cu^{2+}$, cupric ion.

Table 2.5 lists some common cations of the transition elements. Most of these elements have more than one ion, so require the Stock nomenclature system or the older suffix system. A few, such as zinc, have only a single ion that is normally encountered, and you usually name them by just the metal name. You would not be wrong, however, if, for example, you named $Zn^{2+}$ as zinc(II) ion.

3. The names of the monatomic anions are obtained from a stem name of the element followed by the suffix -*ide*. Example: $Br^{-}$ is called bromide ion, from the stem name *brom-* for bromine and the suffix -*ide*.

A **polyatomic ion** is *an ion consisting of two or more atoms chemically bonded together and carrying a net electric charge.* Table 2.6 lists some common polyatomic ions. The first two are cations ($Hg_2^{2+}$ and $NH_4^{+}$); the rest are anions. There are no simple rules for writing the formulas of such ions, although you will find it helpful to note a few points.

Most of the ions in Table 2.6 are *oxoanions* (also called *oxyanions*), which consist of oxygen with another element (called the characteristic or central element). Sulfur, for example, forms the oxoanions sulfate ion, $SO_4^{2-}$, and sulfite ion, $SO_3^{2-}$. The oxoanions in the table are grouped by the characteristic element. For instance, the sulfur ions occur as a group at the end of the table.

Note that the names of the oxoanions have a stem name from the characteristic element, plus a suffix -*ate* or -*ite*. These suffixes denote the relative number of oxygen atoms in the oxoanions of a given characteristic element. The name of the oxoanion with the greater number of oxygen atoms has the suffix -*ate*; the name of the oxoanion with the lesser number of oxygen atoms has the suffix -*ite*. The sulfite ion, $SO_3^{2-}$, and sulfate ion, $SO_4^{2-}$, are examples. Another example are the two oxoanions of nitrogen listed in Table 2.6:

**Table 2.6**
**Some Common Polyatomic Ions**

| Name | Formula | Name | Formula |
|---|---|---|---|
| Mercury(I) or mercurous | $Hg_2^{2+}$ | Nitrite | $NO_2^-$ |
| Ammonium | $NH_4^+$ | Nitrate | $NO_3^-$ |
| Cyanide | $CN^-$ | Hydroxide | $OH^-$ |
| Carbonate | $CO_3^{2-}$ | Peroxide | $O_2^{2-}$ |
| Hydrogen carbonate (or bicarbonate) | $HCO_3^-$ | Phosphate | $PO_4^{3-}$ |
| | | Monohydrogen phosphate | $HPO_4^{2-}$ |
| Acetate | $C_2H_3O_2^-$ | Dihydrogen phosphate | $H_2PO_4^-$ |
| Oxalate | $C_2O_4^{2-}$ | Sulfite | $SO_3^{2-}$ |
| Hypochlorite | $ClO^-$ | Sulfate | $SO_4^{2-}$ |
| Chlorite | $ClO_2^-$ | Hydrogen sulfite (or bisulfite) | $HSO_3^-$ |
| Chlorate | $ClO_3^-$ | | |
| Perchlorate | $ClO_4^-$ | Hydrogen sulfate (or bisulfate) | $HSO_4^-$ |
| Chromate | $CrO_4^{2-}$ | | |
| Dichromate | $Cr_2O_7^{2-}$ | Thiosulfate | $S_2O_3^{2-}$ |
| Permanganate | $MnO_4^-$ | | |

$$NO_2^- \quad \text{nitr\underline{ite} ion}$$
$$NO_3^- \quad \text{nitr\underline{ate} ion}$$

Unfortunately, the suffixes do not tell you the actual number of oxygen atoms in the oxoanion, only the relative number. However, if you were given the formulas of the two anions, you could name them.

In some cases, the two suffixes -*ite* and -*ate* are not enough to distinguish among the possible oxoanions of a given characteristic element. The oxoanions of chlorine are an example. Table 2.6 lists four oxoanions of chlorine: $ClO^-$, $ClO_2^-$, $ClO_3^-$, and $ClO_4^-$. In such cases, two prefixes *hypo-* and *per-* are used in addition to the two suffixes. The two oxoanions with the least number of oxygen atoms ($ClO^-$ and $ClO_2^-$) are named using the suffix -*ite*, and the prefix *hypo-* is added to the one of these two ions with the fewer oxygen atoms:

$$ClO^- \quad \text{\underline{hypo}chlor\underline{ite} ion}$$
$$ClO_2^- \quad \text{chlor\underline{ite} ion}$$

The two oxoanions with the greatest number of oxygen atoms ($ClO_3^-$ and $ClO_4^-$) are named using the suffix -*ate*, and the prefix *per-* is added to the one of these two ions with the greater number of oxygen atoms:

$$ClO_3^- \quad \text{chlor\underline{ate} ion}$$
$$ClO_4^- \quad \text{\underline{per}chlor\underline{ate} ion}$$

Some of the polyatomic ions in Table 2.6 are oxoanions bonded to one or more hydrogen ions ($H^+$). They are sometimes referred to as *acid anions,* because acids are substances that provide $H^+$ ions. As an example, monohydrogen phosphate ion, $HPO_4^{2-}$, is essentially a phosphate ion ($PO_4^{3-}$) to which a hydrogen ion ($H^+$) has bonded. The prefix *mono-*, from the Greek, means "one." Similarly, *di-* is a prefix from the Greek meaning two, so dihydrogen phosphate ion is a phosphate ion to which two hydrogen ions have bonded. In an older terminology, ions such as hydrogen carbonate and hydrogen sulfate were called bicarbonate and bisulfate, respectively.

The last anion in the table is thiosulfate ion, $S_2O_3^{2-}$. The prefix *thio-* means that an oxygen atom in the root ion name (sulfate, $SO_4^{2-}$) has been replaced by a sulfur atom.

There are only a few common polyatomic cations. Those listed in Table 2.6 are the mercury(I) ion (also called mercurous ion) and the ammonium ion. Mercury(I) ion is one of the few common metal ions that is not monatomic; its formula is $Hg_2^{2+}$. The ion charge indicated in parentheses is the charge per metal atom. The ammonium ion, $NH_4^+$, is one of the few common cations composed of only nonmetal atoms.

Now that we have discussed the naming of ions, let us look at the naming of ionic compounds. Example 2.4 illustrates how you name an ionic compound given its formula.

▶ An alternate example is provided in the Instructor's Resource Manual.

## EXAMPLE 2.4

## Naming an Ionic Compound from Its Formula

Name the following: a. $Mg_3N_2$, b. $CrSO_4$.

### ⬧ PROBLEM STRATEGY

Note that the compounds in a and b contain both metal and nonmetal atoms, so we expect them to be ionic compounds. The first thing to do is write the formulas of the cations and anions in the compound, then name the ions. Look first for any monatomic cations and anions whose charges are predictable. Use the rules given in the text to name the ions. For those metals having more than one cation, you will need to deduce the charge on the metal ion (say the ion of Cr in $CrSO_4$) from the charge on the anion. You will need to memorize or have available the formulas and names of the polyatomic anions to be able to write them from the formula of the compound.

### SOLUTION

a. Magnesium, a Group IIA metal, is expected to form only a 2+ ion ($Mg^{2+}$, the magnesium ion). Nitrogen (Group VA) is expected to form an anion of charge equal to the group number minus 8 ($N^{3-}$, the nitride ion). You can check that these ions would give the formula $Mg_3N_2$. The name of the compound is **magnesium nitride** (the name of the cation followed by the name of the anion).

b. Chromium is a transition element and, like most such elements, has more than one monatomic ion. You can find the charge on the Cr ion if you know the formula of the anion. From Table 2.6, you see that the $SO_4$ in $CrSO_4$ refers to the anion $SO_4^{2-}$ (the sulfate ion). Therefore, the Cr cation must be $Cr^{2+}$ to give electrical neutrality. The name of $Cr^{2+}$ is chromium(II) ion, so the name of the compound is **chromium(II) sulfate.**

### EXERCISE 2.5

Write the names of the following compounds: a. CaO, b. $PbCrO_4$.

Problem Solving:
Naming Compounds Containing
Polyatomic Ions

See Problems 2.69 and 2.70.

---

### EXAMPLE 2.5

## Writing the Formula from the Name of an Ionic Compound

Write formulas for the following compounds: a. iron(II) phosphate b. titanium(IV) oxide.

 **PROBLEM STRATEGY**

From the name of the compound, obtain the name of the ions, then write the formulas of the ions. Finally, use the method of Example 2.3 to obtain the formula of the compound from the formulas of the ions.

**SOLUTION**

a. Iron(II) phosphate contains the iron(II) ion, $Fe^{2+}$, and the phosphate ion, $PO_4^{3-}$. (You will need to have memorized the formula of the phosphate ion, or have access to Table 2.6.) Once you know the formulas of the ions, you use the method of Example 2.3 to obtain the formula. The formula is **$Fe_3(PO_4)_2$.**

b. Titanium(IV) oxide is composed of titanium(IV) ions, $Ti^{4+}$, and oxide ions, $O^{2-}$. The formula of titanium(IV) oxide is **$TiO_2$.**

---

### EXERCISE 2.6

A compound has the name thallium(III) nitrate. What is its formula? The symbol of thallium is Tl.

---

See Problems 2.71 and 2.72.

### Binary Molecular Compounds

A **binary compound** is *a compound composed of only two elements.* Binary compounds composed of a metal and a nonmetal are usually ionic and are named as ionic compounds, as we have just discussed. Binary compounds composed of two nonmetals or metalloids are usually molecular and are named using a prefix system. In this system, you name the two elements using the order given by the formula of the compound.

**Order of Elements in the Formula**   The order of elements in the formula of a binary molecular compound is established by convention. By this convention, the nonmetal or metalloid occurring first in the following sequence is written first in the formula of the compound.

| Element | B | Si C | Sb As P N | H | Te Se S | I Br Cl | O F |
|---------|---|------|-----------|---|---------|---------|-----|
| Group | IIIA | IVA | VA | | VIA | VIIA | |

You can reproduce this order easily if you have a periodic table available. You arrange the nonmetals and metalloids in the order of the groups of the periodic table, listing the elements from the bottom of the group upward. Then you place H between Groups VA and VIA, and move O so that it is just before F.

This order places the nonmetals and metalloids approximately in order of increasing nonmetallic character. Thus, in the formula, the first element is the more metallic and the second element is the more nonmetallic, similar to the situation with binary ionic compounds. For example, the compound whose molecule contains three fluorine atoms and one nitrogen atom is written $NF_3$, not $F_3N$.

**Table 2.7**
**Greek Prefixes for Naming Compounds**

| Number | Prefix |
|--------|--------|
| 1 | mono- |
| 2 | di- |
| 3 | tri- |
| 4 | tetra- |
| 5 | penta- |
| 6 | hexa- |
| 7 | hepta- |
| 8 | octa- |
| 9 | nona- |
| 10 | deca- |

**Rules for Naming Binary Molecular Compounds**   Now let us look at the rules for naming binary molecular compounds by the *prefix system.*

1. The name of the compound has the elements in the order given in the formula.

2. You name the first element using the exact element name.

3. You name the second element by writing the stem name of the element with the suffix *-ide* (as if the element occurred as the anion).

4. You add a prefix, derived from the Greek, to each element name to denote the subscript of the element in the formula. (The Greek prefixes are listed in Table 2.7.) Generally, the prefix *mono-* is not used, unless it is needed to distinguish two compounds of the same two elements.

Consider $N_2O_3$. This is a binary molecular compound, as you would predict because N and O are nonmetals. According to Rule 1, you name it after the elements, N before O, following the order of elements in the formula. By Rule 2, the N is named exactly as the element (nitrogen), and by Rule 3, the O is named as the anion (oxide). The compound is a nitrogen oxide. Finally, you add prefixes to denote the subscripts in the formula (Rule 4). As Table 2.7 shows, the prefix for two is *di-*, and the prefix for three is *tri-*. The name of the compound is <u>di</u>nitrogen <u>tri</u>oxide.

Here are some examples to illustrate how the prefix *mono-* is used. There is only one compound of hydrogen and chlorine: HCl. It is called hydrogen chloride, not monohydrogen monochloride, since the prefix *mono-* is not generally written. On the other hand, there are two common compounds of carbon and oxygen: CO and $CO_2$. They are both carbon oxides. To distinguish one from the other, we name them carbon monoxide and carbon dioxide, respectively.

Here are some other examples of prefix names for binary molecular compounds.

| | | | |
|---|---|---|---|
| $SF_4$ | sulfur tetrafluoride | $SF_6$ | sulfur hexafluoride |
| $ClO_2$ | chlorine dioxide | $Cl_2O_7$ | dichlorine hept(a)oxide |

The final vowel in a prefix is often dropped before a vowel in a stem name, for ease in pronunciation. Tetraoxide, for instance, becomes tetroxide.

A few compounds have older, well-established names. The compound $H_2S$ would be named dihydrogen sulfide by the prefix system, but is commonly called hydrogen sulfide. NO is still called nitric oxide, although its prefix name is nitrogen monoxide. The names water and ammonia for $H_2O$ and $NH_3$, respectively, are very old names, and the prefix system is rarely used for them.

▶ An alternate example is provided in the Instructor's Resource Manual.

**EXAMPLE 2.6**

### Naming a Binary Compound from Its Formula

Name the following compounds: a. $N_2O_4$, b. $P_4O_6$.

**SOLUTION**

These compounds are molecular, so we name them using Greek prefixes, naming the elements in the same order as in the formulas: a. **dinitrogen tetroxide,** b. **tetraphosphorus hexoxide.**

**EXERCISE 2.7**

Name the following compounds: a. $Cl_2O_6$, b. $PCl_3$, c. $PCl_5$.

See Problems 2.75 and 2.76.

▶ An alternate example is provided in the Instructor's Resource Manual.

**EXAMPLE 2.7**

## Writing the Formula from the Name of a Binary Compound

Give the formulas of the following compounds: a. disulfur dichloride, b. tetraphosphorus trisulfide.

**SOLUTION**

You change the names of the elements to symbols and translate the prefixes to subscripts. The formulas are a. $S_2Cl_2$ and b. $P_4S_3$.

**EXERCISE 2.8**

Give formulas for the following compounds: a. carbon disulfide, b. sulfur trioxide.

See Problems 2.77 and 2.78.

**EXAMPLE 2.8**

## Naming a Binary Chemical Compound from Its Molecular Model

Name the following chemical compounds:

(a)                    (b)

 **PROBLEM STRATEGY**

After recognizing that these are molecular compounds, start by writing down the elements using subscripts to denote how many atoms of each element occur in the compound. Next, determine which element should be written first in the chemical formula. (The more metallic element is written first.)

**SOLUTION**

a.  The elements in the compound are O and N. There are two O atoms for each N atom, so the chemical formula could be $O_2N$. Applying the rule for writing the more metallic element first, the formula is rearranged to yield the correct formula, $NO_2$. Using Greek prefixes and naming the elements in the order in which they appear, we get the correct chemical formula: **nitrogen dioxide.**

b.  Following the same procedure as part a, we get **chlorine monofluoride.**

**EXERCISE 2.9**

Using the molecular models, name the following chemical compounds:

(a)                    (b)

See Problems 2.79 and 2.80.

**Figure 2.23**
**Molecular model of nitric acid.** The photograph shows a computer-screen image of the model.

## Acids and Corresponding Anions

Acids are an important class of compounds, and we will look closely at them in the next chapter. Here we want merely to see how we name these compounds and how they are related to the anions that we encounter in some ionic compounds. For our present purposes, an *acid* is a molecular compound that yields hydrogen ions, $H^+$, and an anion for each acid molecule when the acid dissolves in water. An example is nitric acid, $HNO_3$. The $HNO_3$ molecule yields one $H^+$ ion and one nitrate ion, $NO_3^-$, in aqueous (water) solution.

Nitric acid, $HNO_3$, is an oxoacid, or oxyacid (Figure 2.23). An **oxoacid** is *an acid containing hydrogen, oxygen, and another element (often called the central atom).* In water the oxoacid molecule yields one or more hydrogen ions, $H^+$, and an oxoanion. The names of the oxoacids are related to the names of the corresponding oxoanions. If you know the name of the oxoanion, you can obtain the name of the corresponding acid by replacing the suffix as follows:

| Anion Suffix | Acid Suffix |
|---|---|
| *-ite* | *-ous* |
| *-ate* | *-ic* |

For example, the acid name corresponding to sulfite ion is sulfurous acid; the acid name corresponding to sulfate ion is sulfuric acid. Table 2.8 lists some oxoanions and their corresponding oxoacids.

Some binary compounds of hydrogen and nonmetals yield acidic solutions when dissolved in water. These *solutions* are named like compounds by using the prefix *hydro-* and the suffix *-ic* with the stem name of the nonmetal, followed by the word *acid.* We denote the solution by the formula of the binary compound followed by (*aq*) for aqueous (water) solution. The corresponding binary compound can be distinguished from the solution by appending the state of the compound to the formula.

**Table 2.8**
**Some Oxoanions and Their Corresponding Oxoacids**

| Oxoanion | | Oxoacid | |
|---|---|---|---|
| $CO_3^{2-}$ | Carbon*ate ion* | $H_2CO_3$ | Carbon*ic acid* |
| $NO_2^-$ | Nitr*ite ion* | $HNO_2$ | Nitr*ous acid* |
| $NO_3^-$ | Nitr*ate ion* | $HNO_3$ | Nitr*ic acid* |
| $PO_4^{3-}$ | Phosph*ate ion* | $H_3PO_4$ | Phosphor*ic acid* |
| $SO_3^{2-}$ | Sulf*ite ion* | $H_2SO_3$ | Sulfur*ous acid* |
| $SO_4^{2-}$ | Sulf*ate ion* | $H_2SO_4$ | Sulfur*ic acid* |
| $ClO^-$ | *Hypo*chlor*ite ion* | $HClO$ | *Hypo*chlor*ous acid* |
| $ClO_2^-$ | Chlor*ite ion* | $HClO_2$ | Chlor*ous acid* |
| $ClO_3^-$ | Chlor*ate ion* | $HClO_3$ | Chlor*ic acid* |
| $ClO_4^-$ | *Per*chlor*ate ion* | $HClO_4$ | *Per*chlor*ic acid* |

Thus, when hydrogen chloride gas, HCl($g$), is dissolved in water, it forms hydrochloric acid, HCl($aq$). Here are some other examples:

| Binary Compound | Acid Solution |
|---|---|
| HBr($g$), hydrogen bromide | *hydro*bromic *acid*, HBr($aq$) |
| HF($l$), hydrogen fluoride | *hydro*fluoric *acid*, HF($aq$) |

► An alternate example is provided in the Instructor's Resource Manual.

### EXAMPLE 2.9

## Writing the Name and Formula of an Anion from the Acid

Selenium has an oxoacid, $H_2SeO_4$, called selenic acid. What is the formula and name of the corresponding anion?

### SOLUTION

When you remove two $H^+$ ions from $H_2SeO_4$, you obtain the $SeO_4{}^{2-}$ ion. You name the ion from the acid by replacing *-ic* with *-ate*. The anion is called the **selenate ion.**

### EXERCISE 2.10

What is the name and formula of the anion corresponding to perbromic acid, $HBrO_4$?

See Problems 2.81 and 2.82.

### Hydrates

A **hydrate** is *a compound that contains water molecules weakly bound in its crystals*. These substances are often obtained by evaporating an aqueous solution of the compound. Consider copper(II) sulfate. When an aqueous solution of this substance is evaporated, blue crystals form in which each formula unit of copper(II) sulfate, $CuSO_4$, is associated with five molecules of water. The formula of the hydrate is written $CuSO_4 \cdot 5H_2O$, where a centered dot separates $CuSO_4$ and $5H_2O$. When the blue crystals of the hydrate are heated, the water is driven off, leaving behind white crystals of copper(II) sulfate without associated water, a substance called *anhydrous* copper(II) sulfate (see Figure 2.24).

Hydrates are named from the anhydrous compound, followed by the word *hydrate* with a prefix to indicate the number of water molecules per formula unit of the compound. For example, $CuSO_4 \cdot 5H_2O$ is known as copper(II) sulfate pentahydrate.

► Demonstration of the dehydration of copper(II) sulfate pentahydrate: Summerlin et al., *Chemical Demonstrations,* Vol. 2 (Washington, D.C.: Am. Chem. Soc., 1987), no. 38, p. 69.

► An alternate example is provided in the Instructor's Resource Manual.

### EXAMPLE 2.10

## Naming a Hydrate from Its Formula

Epsom salts has the formula $MgSO_4 \cdot 7H_2O$. What is the chemical name of the substance?

### SOLUTION

$MgSO_4$ is magnesium sulfate. $MgSO_4 \cdot 7H_2O$ is **magnesium sulfate heptahydrate.**

**Figure 2.24**
**Copper(II) sulfate.** The hydrate $CuSO_4 \cdot 5H_2O$ is blue; the anhydrous compound, $CuSO_4$, is white.

### EXERCISE 2.11

Washing soda has the formula $Na_2CO_3 \cdot 10H_2O$. What is the chemical name of this substance?

See Problems 2.83 and 2.84.

► An alternate example is provided in the Instructor's Resource Manual.

### EXAMPLE 2.11

## Writing the Formula from the Name of a Hydrate

The mineral gypsum has the chemical name calcium sulfate dihydrate. What is the chemical formula of this substance?

**SOLUTION**

Calcium sulfate is composed of calcium ions ($Ca^{2+}$) and sulfate ions ($SO_4^{2-}$), so the formula of the anhydrous compound is $CaSO_4$. Since the mineral is a *di*hydrate, the formula of the compound is **$CaSO_4 \cdot 2H_2O$.**

### EXERCISE 2.12

Photographers' hypo, used to fix negatives during the development process, is sodium thiosulfate pentahydrate. What is the chemical formula of this compound?

See Problems 2.85 and 2.86.

 **CONCEPT CHECK 2.5**

You take a job with the U.S. Environmental Protection Agency (EPA) inspecting college chemistry laboratories. On the first day of the job while inspecting Generic University you encounter a bottle with the formula $Al_2Q_3$ that was used as an unknown compound in an experiment. Before you send the compound off to the EPA lab for analysis you want to narrow down the possibilities for element Q. What are the likely (real element) candidates for element Q?

# Chemical Reactions: Equations

An important feature of atomic theory is its explanation of a chemical reaction as a rearrangement of the atoms of substances. The reaction of sodium metal with chlorine gas described in the chapter opening involves the rearrangement of the atoms of sodium and chlorine to give the new combination of atoms in sodium chloride. Such a rearrangement of atoms is conveniently represented by a chemical equation, which uses chemical formulas.

## 2.9 Writing Chemical Equations

A **chemical equation** is *the symbolic representation of a chemical reaction in terms of chemical formulas.* For example, the burning of sodium in chlorine to produce sodium chloride is written

$$2Na + Cl_2 \longrightarrow 2NaCl$$

The formulas on the left side of an equation (before the arrow) represent the reactants; a **reactant** is *a starting substance in a chemical reaction.* The arrow means "react to form" or "yield." The formulas on the right side represent the products; a **product** is *a substance that results from a reaction.* Note the *coefficient* of 2 in front of the formula NaCl. Coefficients such as this give the relative number of molecules or formula units involved in the reaction. Coefficients of 1 are usually understood, not written.

In many cases, it is useful to indicate the states or phases of the substances in an equation. You do this by placing appropriate labels indicating the phases within parentheses following the formulas of the substances. You use the following phase labels:

$(g)$ = gas, $(l)$ = liquid, $(s)$ = solid, $(aq)$ = aqueous (water) solution

When you use these labels, the previous equation becomes

$$2Na(s) + Cl_2(g) \longrightarrow 2NaCl(s)$$

You can also indicate in an equation the conditions under which a reaction takes place. If the reactants are heated to make the reaction go, you can indicate this by putting the symbol $\Delta$ (capital Greek delta) over the arrow. For example, the equation

$$2NaNO_3(s) \xrightarrow{\Delta} 2NaNO_2(s) + O_2(g)$$

indicates that solid sodium nitrate, $NaNO_3$, decomposes when heated to give solid sodium nitrite, $NaNO_2$, and oxygen gas, $O_2$.

When an aqueous solution of hydrogen peroxide, $H_2O_2$, comes in contact with platinum metal, Pt, the hydrogen peroxide decomposes into water and oxygen gas. The platinum acts as a *catalyst,* a substance that speeds up a reaction without undergoing any net change itself. You write the equation for this reaction as follows. The catalyst, Pt, is written over the arrow. ◄

▶ **Platinum is a silvery-white metal used for jewelry. It is also valuable as a catalyst for many reactions, including those that occur in the catalytic converters of automobiles.**

$$2H_2O_2(aq) \xrightarrow{Pt} 2H_2O(l) + O_2(g)$$

You also can combine the symbol $\Delta$ with the formula of a catalyst, placing one above the arrow and the other below the arrow.

**Figure 2.25**
**Representation of the reaction of methane with oxygen.** Molecular models represent the reaction of $CH_4$ with $O_2$ to give $CO_2$ and $H_2O$.

## 2.10  Balancing Chemical Equations

**Understanding Concepts:**
**Conservation of Mass and**
**Balancing Equations**

When the coefficients in a chemical equation are correctly given, the numbers of atoms of each element are equal on both sides of the arrow. The equation is then said to be *balanced*. That a chemical equation should be balanced follows from atomic theory. A chemical reaction involves simply a recombination of the atoms; none are destroyed and none are created. Consider the burning of natural gas, which is composed mostly of methane, $CH_4$. Using atomic theory, you describe this as the chemical reaction of one molecule of methane, $CH_4$, with two molecules of oxygen, $O_2$, to form one molecule of carbon dioxide, $CO_2$, and two molecules of water, $H_2O$.

$$CH_4 \quad + \quad 2O_2 \quad \longrightarrow \quad CO_2 \quad + \quad 2H_2O$$
one molecule  +  two molecules  react to  one molecule  +  two molecules
of methane        of oxygen        form    of carbon dioxide        of water

Figure 2.25 represents the reaction in terms of molecular models.

Before you can write a balanced chemical equation for a reaction, you must determine *by experiment* those substances that are reactants and those that are products. You must also determine the formulas of each substance. Once you know these things, you can write the balanced chemical equation.

As an example, consider the burning of propane gas (Figure 2.26). By experiment, you determine that propane reacts with oxygen in air to give carbon dioxide and water. You also determine from experiment that the formulas of propane, oxygen, carbon dioxide, and water are $C_3H_8$, $O_2$, $CO_2$, and $H_2O$, respectively. Then you can write

$$C_3H_8 + O_2 \longrightarrow CO_2 + H_2O$$

This equation is not balanced because the coefficients that give the relative number of molecules involved have not yet been determined. *To balance the equation, you select coefficients that will make the numbers of atoms of each element equal on both sides of the equation.*

Because there are three carbon atoms on the left side of the equation ($C_3H_8$), you must have three carbon atoms on the right. This is achieved by writing 3 for the coefficient of $CO_2$.

$$\underline{1}C_3H_8 + O_2 \longrightarrow \underline{3}CO_2 + H_2O$$

(We have written the coefficient for $C_3H_8$ to show that it is now determined; normally when the coefficient is 1, it is omitted.) Similarly, there are eight hydrogen atoms on the left (in $C_3H_8$), so you write 4 for the coefficient of $H_2O$.

$$\underline{1}C_3H_8 + O_2 \longrightarrow \underline{3}CO_2 + \underline{4}H_2O$$

The coefficients on the right are now determined, and there are ten oxygen atoms on this side of the equation: 6 from the three $CO_2$ molecules and 4 from the four $H_2O$

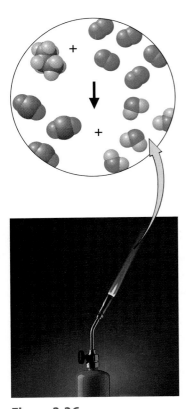

**Figure 2.26**
**The burning of propane gas.** Propane gas, $C_3H_8$, from the tank burns by reacting with oxygen, $O_2$, in air to give carbon dioxide, $CO_2$, and water, $H_2O$.

molecules. You now write 5 for the coefficient of $O_2$. We drop the coefficient 1 from $C_3H_8$ and have the balanced equation.

$$C_3H_8 + 5O_2 \longrightarrow 3CO_2 + 4H_2O$$

It is a good idea to check your work by counting the number of atoms of each element on the left side and then on the right side of the equation.

The combustion of propane can be equally well represented by

$$2C_3H_8 + 10O_2 \longrightarrow 6CO_2 + 8H_2O$$

in which the coefficients of the previous equation have been doubled. Usually, however, it is preferable to *write the coefficients so that they are the smallest whole numbers possible.* When we refer to a balanced equation, we will assume that this is the case unless otherwise specified.

▶ **Balancing equations this way is relatively quick for all but the most complicated equations. A special method for such equations is described in Chapters 4 and 20.**

The method we have outlined—called *balancing by inspection*—is essentially a trial-and-error method. ◀ It can be made easier, however, by observing the following rule:

> Balance first the atoms for elements that occur in only one substance on each side of the equation.

The usefulness of this rule is illustrated in the next example. Remember that in any formula, such as $Fe_2(SO_4)_3$, a subscript to the right of the parentheses multiplies *each* subscript within the parentheses. Thus, $Fe_2(SO_4)_3$ represents $4 \times 3$ oxygen atoms. Remember also that you cannot change a subscript in any formula; only the coefficients can be altered to balance an equation.

---

### EXAMPLE 2.12

## Balancing Simple Equations

▶ **An alternate example is provided in the Instructor's Resource Manual.**

Balance the following equations.

a. $H_3PO_3 \longrightarrow H_3PO_4 + PH_3$      b. $Ca + H_2O \longrightarrow Ca(OH)_2 + H_2$

c. $Fe_2(SO_4)_3 + NH_3 + H_2O \longrightarrow Fe(OH)_3 + (NH_4)_2SO_4$

 **PROBLEM STRATEGY**

Look at the chemical equation for atoms of elements that occur in only one substance on each side of the equation. Begin by balancing the equation in one of these atoms. Use the number of atoms on the left side of the arrow as the coefficient of the substance containing that element on the right side, and vice versa. After balancing one element in an equation, the rest become easier.

**SOLUTION**

a. Oxygen occurs in just one of the products ($H_3PO_4$). It is therefore easiest to balance O atoms first. To do this, note that $H_3PO_3$ has three O atoms; use 3 as the coefficient of $H_3PO_4$ on the right side. There are four O atoms in $H_3PO_4$ on the right side; use 4 as the coefficient of $H_3PO_3$ on the left side.

$$\underline{4}H_3PO_3 \longrightarrow \underline{3}H_3PO_4 + PH_3$$

This equation is now also balanced in P and H atoms. Thus, the balanced equation is

$$4H_3PO_3 \longrightarrow 3H_3PO_4 + PH_3$$

b. The equation is balanced in Ca atoms as it stands.

$$\underline{1}Ca + H_2O \longrightarrow \underline{1}Ca(OH)_2 + H_2$$

O atoms occur in only one reactant and in only one product, so they are balanced next.

$$\underline{1}Ca + \underline{2}H_2O \longrightarrow \underline{1}Ca(OH)_2 + H_2$$

The equation is now also balanced in H atoms. Thus, the answer is

$$Ca + 2H_2O \longrightarrow Ca(OH)_2 + H_2$$

c. To balance the equation in Fe atoms, you write

$$\underline{1}Fe_2(SO_4)_3 + NH_3 + H_2O \longrightarrow \underline{2}Fe(OH)_3 + (NH_4)_2SO_4$$

You balance the S atoms by placing the coefficient 3 for $(NH_4)_2SO_4$.

$$\underline{1}Fe_2(SO_4)_3 + NH_3 + H_2O \longrightarrow \underline{2}Fe(OH)_3 + \underline{3}(NH_4)_2SO_4$$

Now you balance the N atoms.

$$\underline{1}Fe_2(SO_4)_3 + \underline{6}NH_3 + H_2O \longrightarrow \underline{2}Fe(OH)_3 + \underline{3}(NH_4)_2SO_4$$

To balance the O atoms, you first count the number of O atoms on the right (18). Then you count the number of O atoms in substances on the left with known coefficients. There are 12 O's in $Fe_2(SO_4)_3$; hence, the number of remaining O's (in $H_2O$) must be $18 - 12 = 6$.

$$\underline{1}Fe_2(SO_4)_3 + \underline{6}NH_3 + \underline{6}H_2O \longrightarrow \underline{2}Fe(OH)_3 + \underline{3}(NH_4)_2SO_4$$

Finally, note that the equation is now balanced in H atoms. The answer is

$$Fe_2(SO_4)_3 + 6NH_3 + 6H_2O \longrightarrow 2Fe(OH)_3 + 3(NH_4)_2SO_4$$

Check each final equation by counting the number of atoms of each element on both sides of the equations.

### EXERCISE 2.13

**Problem Solving: Balancing a Chemical Equation**

Find the coefficients that balance the following equations.

a. $O_2 + PCl_3 \longrightarrow POCl_3$

b. $P_4 + N_2O \longrightarrow P_4O_6 + N_2$

c. $As_2S_3 + O_2 \longrightarrow As_2O_3 + SO_2$

d. $Ca_3(PO_4)_2 + H_3PO_4 \longrightarrow Ca(H_2PO_4)_2$

See Problems 2.89 and 2.90.

# A Checklist for Review

## Important Terms

atomic theory (2.1)
atom (2.1)
element (2.1, 2.3)
compound (2.1)
chemical reaction (2.1)
atomic symbol (2.1)
law of multiple proportions (2.1)
nucleus (2.2)
electron (2.2)
proton (2.3)
atomic number (Z) (2.3)
neutron (2.3)

mass number (A) (2.3)
nuclide (2.3)
isotope (2.3)
atomic mass unit (amu) (2.4)
atomic weight (2.4)
fractional abundance (2.4)
periodic table (2.5)
period (of periodic table) (2.5)
group (of periodic table) (2.5)
metal (2.5)

nonmetal (2.5)
metalloid (semimetal) (2.5)
chemical formula (2.6)
molecule (2.6)
molecular formula (2.6)
polymer (2.6)
monomer (2.6)
ion (2.6)
anion (2.6)
cation (2.6)
ionic compound (2.6)
formula unit (2.6)
organic compound (2.7)

hydrocarbon (2.7)
functional group (2.7)
chemical nomenclature (2.7)
inorganic compound (2.8)
monatomic ion (2.8)
polyatomic ion (2.8)
binary compound (2.8)
oxoacid (2.8)
hydrate (2.8)
chemical equation (2.9)
reactant (2.9)
product (2.9)

## Summary of Facts and Concepts

*Atomic theory* is central to chemistry. According to this theory, all matter is composed of small particles, or *atoms.* Each *element* is composed of the same kind of atom, and a *compound* is composed of two or more elements chemically combined in fixed proportions. A *chemical reaction* consists of the rearrangement of the atoms present in the reacting substances to give new chemical combinations present in the substances formed by the reaction. Although Dalton considered atoms to be the ultimate particles of matter, we now know that atoms themselves have structure. An atom has a *nucleus* and *electrons*. These atomic particles were discovered by J. J. Thomson and Ernest Rutherford.

Thomson established that cathode rays consist of negatively charged particles, or electrons, that are constituents of all atoms. Thomson measured the mass-to-charge ratio of the electron, and later Millikan measured its charge. From these measurements, the electron was found to be more than 1800 times lighter than the lightest atom.

Rutherford proposed the nuclear model of the atom to account for the results of experiments in which alpha particles were scattered from metal foils. According to this model, the atom consists of a central core, or nucleus, around which the electrons exist. The nucleus has most of the mass of the atom and consists of *protons* (with a positive charge) and *neutrons* (with no charge). Each chemically distinct atom has a nucleus with a specific number of protons (*atomic number*), and around the nucleus in the neutral atom are an equal number of electrons. The number of protons plus neutrons in a nucleus equals the *mass number*. Atoms whose nuclei have the same number of protons but different number of neutrons are called *isotopes.*

The *atomic weight* can be calculated from the isotopic masses and *fractional abundances* of the isotopes in a naturally occurring element. These data can be determined by the use of a mass spectrometer, which separates ions according to their mass-to-charge ratios.

The elements can be arranged in rows and columns by atomic number to form the *periodic table.* Elements in a given *group* (column) have similar properties. (A *period* is a row in the periodic table.) Elements on the left and at the center of the table are *metals;* those on the right are *nonmetals.*

A *chemical formula* is a notation used to convey the relative proportions of the atoms of the different elements in a substance. If the substance is molecular, the formula gives the precise number of each kind of atom in the molecule. If the substance is ionic, the formula gives the relative number of different ions in the compound.

Chemical nomenclature is the systematic naming of compounds based on their formulas or structures. Rules are given for naming *ionic compounds, binary molecular compounds, acids,* and *hydrates.*

A chemical reaction occurs when the atoms in substances rearrange and combine into new substances. We represent a reaction by a *chemical equation,* writing a chemical formula for each reactant and product. The coefficients in the equation indicate the relative numbers of reactant and product molecules or formula units. Once the reactants and products and their formulas have been determined by experiment, we determine the coefficients by *balancing* the numbers of each kind of atom on both sides of the equation.

## Operational Skills

1. **Writing nuclide symbols** Given the number of protons and neutrons in a nucleus, write its nuclide symbol. **(EXAMPLE 2.1)**

2. **Determining atomic weight from isotopic masses and fractional abundances** Given the isotopic masses (in atomic mass units) and fractional isotopic abundances for a naturally occurring element, calculate its atomic weight. **(EXAMPLE 2.2)**

3. **Writing an ionic formula, given the ions** Given the formulas of a cation and an anion, write the formula of the ionic compound of these ions. **(EXAMPLE 2.3)**

4. **Writing the name of a compound from its formula, or vice versa** Given the formula of a simple compound (ionic, binary molecular, acid, or hydrate), write the name

**(EXAMPLES 2.4, 2.6, 2.10)**, or vice versa **(EXAMPLES 2.5, 2.7, 2.11)**.

5. **Writing the name of a binary molecular compound from its molecular model** Given the molecular model of a binary compound, write the name. **(EXAMPLE 2.8)**

6. **Writing the name and formula of an anion from the acid** Given the name and formula of an oxoacid, write the name and formula of the oxoanion; or from the name and formula of the oxoanion, write the formula and name of the oxoacid. **(EXAMPLE 2.9)**

7. **Balancing simple equations** Given the formulas of the reactants and products in a chemical reaction, obtain the coefficients of the balanced equation. **(EXAMPLE 2.12)**

## Review Questions

**2.1** Describe atomic theory and discuss how it explains the great variety of different substances. How does it explain chemical reactions?

**2.2** Two compounds of iron and chlorine, A and B, contain 1.270 g and 1.904 g of chlorine, respectively, for each gram of iron. Show that these amounts are in the ratio 2 : 3. Is this consistent with the law of multiple proportions? Explain.

**2.3** Explain the operation of a cathode-ray tube. Describe the deflection of cathode rays by electrically charged plates placed within the cathode-ray tube. What does this imply about cathode rays?

**2.4** Explain Millikan's oil-drop experiment.

**2.5** Describe the nuclear model of the atom. How does this model explain the results of alpha-particle scattering from metal foils?

**2.6** What are the different kinds of particles in the atom's nucleus? Compare their properties with each other and with those of an electron.

**2.7** Describe how protons and neutrons were discovered to be constituents of nuclei.

**2.8** Oxygen consists of three different _____, each having eight protons but different numbers of neutrons.

**2.9** Describe how Dalton obtained relative atomic masses.

**2.10** Briefly explain how a mass spectrometer works. What kinds of information does one obtain from the instrument?

**2.11** Define the term *atomic weight*. Why might the values of atomic weights on a planet elsewhere in the universe be different from those on earth?

**2.12** What is the name of the element in Group IVA and Period 5?

**2.13** Cite some properties that are characteristic of a metal.

**2.14** Ethane consists of molecules with two atoms of carbon and six atoms of hydrogen. Write the molecular formula for ethane.

**2.15** What is the difference between a molecular formula and a structural formula?

**2.16** What is the fundamental difference between an organic substance and an inorganic substance? Write chemical formulas of three inorganic molecules that contain carbon.

**2.17** Give an example of a binary compound that is ionic. Give an example of a binary compound that is molecular.

**2.18** Which of the following models represent a(n):

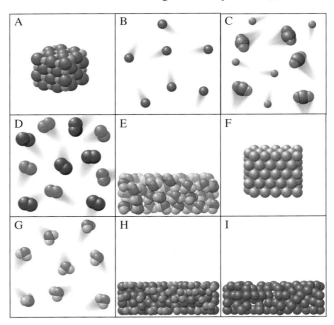

a. element
b. compound
c. mixture
d. ionic solid
e. gas made up of an element and a compound
f. mixture of elements
g. solid element
h. solid
i. liquid

**2.19** The compounds CuCl and $CuCl_2$ were formerly called cuprous chloride and cupric chloride, respectively. What are their names using the Stock system of nomenclature? What are the advantages of the Stock system of nomenclature over the former one?

**2.20** Explain what is meant by the term *balanced chemical equation*.

## Conceptual Problems

**2.21** One of the early models of the atom proposed that atoms were wispy balls of positive charge with the electrons evenly distributed throughout. What would you expect to observe if you conducted Rutherford's experiment and the atom had this structure?

**2.22** A friend is trying to balance the following equation:

$$N_2 + H_2 \longrightarrow NH_3$$

He presents you with his version of the "balanced" equation:

$$N + H_3 \longrightarrow NH_3$$

You immediately recognize that he has committed a serious error; however, he argues that there is nothing wrong, since the equation is balanced. What reason can you give to convince him that his "method" of balancing the equation is flawed?

**2.23** Given that the periodic table is an organizational scheme for the elements, what might be some other logical ways in which to group the elements that would provide meaningful chemical information in a periodic table of your own devising?

**2.24** You discover a new set of polyatomic anions that has the newly discovered element "X" combined with oxygen. Since you made the discovery, you get to choose the names for these new polyatomic ions. In developing the names, you want to pay attention to convention. A colleague in your lab has come up with a name that she really likes for one of the ions. How would you name the following in a way consistent with her name?

| Formula | Name |
|---|---|
| $XO_4{}^{2-}$ | |
| $XO_3{}^{2-}$ | |
| $XO_2{}^{2-}$ | excite |
| $XO^{2-}$ | |

**2.25** You have the mythical metal element "X" that can exist as $X^+$, $X^{2+}$, and $X^{5+}$ ions.
a. What would be the chemical formula for compounds formed from the combination of each of the X ions and $SO_4{}^{2-}$?
b. If the name of the element X is exy, what would be the names of each of the compounds from part a of this problem?

**2.26** Match the molecular model with the correct chemical formula: $CH_3OH$, $NH_3$, $KCl$, $H_2O$.

(a)   (b)   (c)   (d)

**2.27** Consider a hypothetical case in which the charge on a proton is twice that of an electron. Using this hypothetical case, and the fact that atoms maintain a charge of 0, how many protons, neutrons, and electrons would a potassium-39 atom contain?

**2.28** Currently, the atomic mass unit (amu) is based on being exactly one-twelfth the mass of a carbon-12 atom and is equal to $1.66 \times 10^{-27}$ kg.
a. If the amu were based on sodium-23 with a mass equal to exactly 1/23 of the mass of a sodium-23 atom, would the mass of the amu be different?
b. If the new mass of the amu based on sodium-23 is $1.67 \times 10^{-27}$ kg, how would the mass of a hydrogen atom, in amu, compare with the current mass of a hydrogen atom in amu?

# Practice Problems

**2.29** What is the name of the element represented by each of the following atomic symbols?
a. Ne    b. Zn    c. Ag    d. Mg

**2.30** For each atomic symbol, give the name of the element.
a. Ca    b. Pb    c. Hg    d. Sn

**2.31** Give the atomic symbol for each of the following elements.
a. potassium    b. sulfur    c. iron    d. manganese

**2.32** Give the atomic symbol for each of the following elements.
a. cobalt    b. aluminum    c. neon    d. titanium

## Electrons, Protons, and Neutrons

**2.33** A student has determined the mass-to-charge ratio for an electron to be $5.64 \times 10^{-12}$ kg/C. In another experiment, using Millikan's oil-drop apparatus, he found the charge on the electron to be $1.605 \times 10^{-19}$ C. What would be the mass of the electron, according to these data?

**2.34** The mass-to-charge ratio for the positive ion $F^+$ is $1.97 \times 10^{-7}$ kg/C. Using the value of $1.602 \times 10^{-19}$ C for the charge on the ion, calculate the mass of the fluorine atom. (The mass of the electron is negligible compared with that of the ion, so the ion mass is essentially the atomic mass.)

**2.35** The following table gives the number of protons and neutrons in the nuclei of various atoms. Which atom is the isotope of atom A? Which atom has the same mass number as atom A?

|        | Protons | Neutrons |
|--------|---------|----------|
| Atom A | 17      | 19       |
| Atom B | 16      | 19       |
| Atom C | 18      | 18       |
| Atom D | 17      | 20       |

**2.36** The following table gives the number of protons and neutrons in the nuclei of various atoms. Which atom is the isotope of atom A? Which atom has the same mass number as atom A?

|        | Protons | Neutrons |
|--------|---------|----------|
| Atom A | 31      | 39       |
| Atom B | 32      | 38       |
| Atom C | 38      | 50       |
| Atom D | 31      | 38       |

**2.37** Naturally occurring chlorine is a mixture of the isotopes Cl-35 and Cl-37. How many protons and how many neutrons are there in each isotope? How many electrons are there in the neutral atoms?

**2.38** Naturally occurring lithium is a mixture of $^{6}_{3}Li$ and $^{7}_{3}Li$. Give the number of protons, neutrons, and electrons in the neutral atom of each isotope.

**2.39** What is the nuclide symbol for the nucleus that contains 14 protons and 14 neutrons?

**2.40** An atom contains 17 protons and 19 neutrons. What is the nuclide symbol for the nucleus?

## Atomic Weights

**2.41** Ammonia is a gas with a characteristic pungent odor. It is sold as a water solution for use in household cleaning. The gas is a compound of nitrogen and hydrogen in the atomic ratio 1 : 3. A sample of ammonia contains 7.933 g N and 1.712 g H. What is the atomic mass of N relative to H?

**2.42** Hydrogen sulfide is a gas with the odor of rotten eggs. The gas can sometimes be detected in automobile exhaust. It is a compound of hydrogen and sulfur in the atomic ratio 2 : 1. A sample of hydrogen sulfide contains 0.587 g H and 9.330 g S. What is the atomic mass of S relative to H?

**2.43** Calculate the atomic weight of boron, B, from the following data:

| Isotope | Isotopic Mass (amu) | Fractional Abundance |
|---------|---------------------|----------------------|
| B-10    | 10.013              | 0.1978               |
| B-11    | 11.009              | 0.8022               |

**2.44** An element has two naturally occurring isotopes with the following masses and abundances:

| Isotopic Mass (amu) | Fractional Abundance |
|---------------------|----------------------|
| 84.9118             | 0.7215               |
| 86.9092             | 0.2785               |

What is the atomic weight of this element? What is the identity of the element?

**2.45** An element has three naturally occurring isotopes with the following masses and abundances:

| Isotopic Mass (amu) | Fractional Abundance  |
|---------------------|-----------------------|
| 38.964              | 0.9326                |
| 39.964              | $1.000 \times 10^{-4}$ |
| 40.962              | 0.0673                |

Calculate the atomic weight of this element. What is the identity of the element?

**2.46** An element has three naturally occurring isotopes with the following masses and abundances:

| Isotopic Mass (amu) | Fractional Abundance |
|---|---|
| 27.977 | 0.9221 |
| 28.976 | 0.0470 |
| 29.974 | 0.0309 |

Calculate the atomic weight of this element. What is the identity of the element?

**2.47**  While traveling to a distant universe, you discover the hypothetical element, "X." You obtain a representative sample of the element and discover that it is made up of two isotopes, X-23 and X-25. To help your science team calculate the atomic mass of the substance, you send the following drawing of your sample with your report.

In the report, you also inform the science team that the gold atoms are X-23, which have an isotopic mass of 23.02 amu, and the green atoms are X-25, which have an isotopic mass of 25.147 amu. What is the atomic mass of element X?

**2.48**  While roaming a parallel universe, you discover the hypothetical element, "Z." You obtain a representative sample of the element and discover that it is made up of two isotopes, Z-47 and Z-51. To help your science team calculate the atomic mass of the substance, you send the following drawing of your sample with your report.

In the report, you also inform the science team that the blue atoms are Z-47, which have an isotopic mass of 47.510 amu, and the orange atoms are Z-51, which have an isotopic mass of 51.126 amu. What is the atomic mass of element Z?

## Periodic Table

**2.49**  Identify the group and period for each of the following. Refer to the periodic table (Figure 2.14 or inside front cover). Label each as a metal, nonmetal, or metalloid.
a. C      b. Po      c. Cr      d. Mg      e. B

**2.50**  Refer to the periodic table (Figure 2.14 or inside front cover) and obtain the group and period for each of the following elements. Also determine whether the element is a metal, nonmetal, or metalloid.
a. Si      b. F      c. Ca      d. Co      e. Xe

**2.51**  Refer to the periodic table (Figure 2.14 or inside front cover) and answer the following questions.
a. What Group VIA element is a metalloid?
b. What is the Group IIIA element in Period 3?

**2.52**  Refer to the periodic table (Figure 2.14 or inside front cover) and answer the following questions.
a. What Group VA element is a metal?
b. What is the Group IIA element in Period 3?

**2.53**  Give one example (atomic symbol and name) for each of the following.
a. a main-group (representative) element in the second period
b. an alkali metal
c. a transition element in the fourth period
d. a lanthanide element

**2.54**  Give one example (atomic symbol and name) for each of the following.
a. a transition element in the sixth period
b. a halogen
c. a main-group (representative) element in the third period
d. an actinide element

## Molecular and Ionic Substances

**2.55**  The normal form of the element sulfur is a brittle, yellow solid. This is a molecular substance, $S_8$. If this solid is vaporized, it first forms $S_8$ molecules; but at high temperature, $S_2$ molecules are formed. How do the molecules of the solid sulfur and of the hot vapor differ? How are the molecules alike?

**2.56**  White phosphorus is available in sticks, which have a waxy appearance. This is a molecular substance, $P_4$. When this solid is vaporized, it first forms $P_4$ molecules; but at high temperature, $P_2$ molecules are formed. How do the molecules of white phosphorus and those of the hot vapor differ? How are the molecules alike?

**2.57**  A 1.50-g sample of nitrous oxide (an anesthetic, sometimes called laughing gas) contains $2.05 \times 10^{22}$ $N_2O$ molecules. How many nitrogen atoms are in this sample? How many nitrogen atoms are in 1.00 g of nitrous oxide?

**2.58**  Nitric acid is composed of $HNO_3$ molecules. A sample weighing 3.50 g contains $3.34 \times 10^{22}$ $HNO_3$ molecules. How many nitrogen atoms are in this sample? How many oxygen atoms are in 4.50 g of nitric acid?

**2.59**  A sample of ammonia, $NH_3$, contains $3.3 \times 10^{21}$ hydrogen atoms. How many $NH_3$ molecules are in this sample?

**2.60**  A sample of ethanol (ethyl alcohol), $C_2H_5OH$, contains $4.2 \times 10^{23}$ hydrogen atoms. How many $C_2H_5OH$ molecules are in this sample?

**2.61**  Give the molecular formula for each of the following structural formulas.
a.

c.

isopropyl alcohol

d.

$$Cl—P—Cl$$

phosphorus trichloride

**2.62**  What molecular formula corresponds to each of the following structural formulas?

a.

glycerol

b.

$$H—Si—Si—H$$

disilane

c.

$$H—N—O—H$$

hydroxylamine

d.

$$F—S—F$$

sulfur tetrafluoride

**2.63**  Write the molecular formula for each of the following compounds represented by molecular models.

(a)            (b)            (c)

**2.64**  Write the molecular formula for each of the following compounds represented by molecular models.

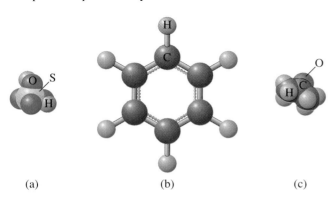

(a)            (b)            (c)

**2.65**  Iron(III) nitrate has the formula $Fe(NO_3)_3$. What is the ratio of iron atoms to oxygen atoms in this compound?

**2.66**  Ammonium phosphate, $(NH_4)_3PO_4$, has how many hydrogen atoms for each oxygen atom?

**2.67**  Write the formula for the compound of each of the following pairs of ions.
a.  $Fe^{3+}$ and $CN^-$        b.  $K^+$ and $SO_4^{2-}$
c.  $Li^+$ and $N^{3-}$          d.  $Ca^{2+}$ and $P^{3-}$

**2.68**  For each of the following pairs of ions, write the formula of the corresponding compound.
a.  $Co^{2+}$ and $Br^-$        b.  $NH_4^+$ and $SO_4^{2-}$
c.  $Na^+$ and $PO_4^{3-}$      d.  $Fe^{3+}$ and $NO_3^-$

## Chemical Nomenclature

**2.69**  Name the following compounds.
a.  $Na_2SO_4$         b.  $CaO$
c.  $CuCl$             d.  $Cr_2O_3$

**2.70**  Name the following compounds.
a.  $Na_2O$            b.  $Mn_2O_3$
c.  $NH_4HCO_3$        d.  $Cu(NO_3)_2$

**2.71**  Write the formulas of:
a.  lead(II) dichromate
b.  barium hydrogen carbonate
c.  cesium sulfide
d.  iron(II) acetate

**2.72**  Write the formulas of:
a.  sodium thiosulfate
b.  copper(I) hydroxide
c.  calcium hydrogen carbonate
d.  nickel(II) nitride

**2.73**  For each of the following binary compounds, decide whether the compound is expected to be ionic or molecular.
a.  $SeF_4$            b.  $LiBr$
c.  $SiF_4$            d.  $Cs_2O$

**2.74**  For each of the following binary compounds, decide whether the compound is expected to be ionic or molecular.
a.  $BaO$              b.  $As_4O_6$
c.  $ClF_3$            d.  $CuCl$

**2.75**  Give systematic names to the following binary compounds.
a.  $N_2O$             b.  $P_4O_{10}$
c.  $AsCl_3$           d.  $Cl_2O_7$

**2.76**  Give systematic names to the following binary compounds.
a.  $N_2F_2$           b.  $NCl_3$
c.  $N_2O_5$           d.  $As_4O_6$

**2.77**  Write the formulas of the following compounds.
a.  nitrogen tribromide
b.  xenon tetroxide

c. oxygen difluoride
d. dichlorine pentoxide

**2.78** Write the formulas of the following compounds.
a. chlorine trifluoride
b. dichlorine monoxide
c. dinitrogen tetrafluoride
d. phosphorus pentafluoride

**2.79** Write the systematic name for each of the following compounds represented by a molecular model.

(a)            (b)            (c)

**2.80** Write the systematic name for each of the following molecules represented by a molecular model.

(a)            (b)            (c)

**2.81** Give the name and formula of the acid corresponding to each of the following oxoanions.
a. bromate ion, $BrO_3^-$
b. hyponitrite ion, $N_2O_2^{2-}$
c. disulfite ion, $S_2O_5^{2-}$
d. arsenate ion, $AsO_4^{3-}$

**2.82** Give the name and formula of the acid corresponding to each of the following oxoanions.
a. selenite ion, $SeO_3^{2-}$
b. sulfite ion, $SO_3^{2-}$
c. hypoiodite ion, $IO^-$
d. diphosphate ion, $P_2O_7^{4-}$

**2.83** Glauber's salt has the formula $Na_2SO_4 \cdot 10H_2O$. What is the chemical name of this substance?

**2.84** Emerald-green crystals of the substance $NiSO_4 \cdot 6H_2O$ are used in nickel plating. What is the chemical name of this compound?

**2.85** Iron(II) sulfate heptahydrate is a blue-green, crystalline compound used to prepare other iron compounds. What is the formula of iron(II) sulfate heptahydrate?

**2.86** Cobalt(II) chloride hexahydrate has a pink color. It loses water on heating and changes to a blue-colored compound. What is the formula of cobalt(II) chloride hexahydrate?

## Chemical Equations

**2.87** For the balanced chemical equation $Pb(NO_3)_2 + K_2CO_3 \longrightarrow PbCO_3 + 2KNO_3$, how many oxygen atoms are on the left side?

**2.88** In the equation $2PbS + O_2 \longrightarrow 2PbO + 2SO_2$, how many oxygen atoms are there on the right side? Is the equation balanced as written?

**2.89** Balance the following equations.
a. $Sn + NaOH \longrightarrow Na_2SnO_2 + H_2$
b. $Al + Fe_3O_4 \longrightarrow Al_2O_3 + Fe$
c. $CH_3OH + O_2 \longrightarrow CO_2 + H_2O$
d. $P_4O_{10} + H_2O \longrightarrow H_3PO_4$
e. $PCl_5 + H_2O \longrightarrow H_3PO_4 + HCl$

**2.90** Balance the following equations.
a. $Cl_2O_7 + H_2O \longrightarrow HClO_4$
b. $MnO_2 + HCl \longrightarrow MnCl_2 + Cl_2 + H_2O$
c. $Na_2S_2O_3 + I_2 \longrightarrow NaI + Na_2S_4O_6$
d. $Al_4C_3 + H_2O \longrightarrow Al(OH)_3 + CH_4$
e. $NO_2 + H_2O \longrightarrow HNO_3 + NO$

**2.91** Solid calcium phosphate and aqueous sulfuric acid solution react to give calcium sulfate, which comes out of the solution as a solid. The other product is phosphoric acid, which remains in solution. Write a balanced equation for the reaction using complete formulas for the compounds with phase labels.

**2.92** Solid potassium metal reacts with water, giving a solution of potassium hydroxide and releasing hydrogen gas. Write a balanced equation for the reaction using complete formulas for the compounds with phase labels.

**2.93** An aqueous solution of ammonium chloride and barium hydroxide is heated, and the compounds react to give off ammonia gas. Barium chloride solution and water are also products. Write a balanced equation for the reaction using complete formulas for the compounds with phase labels; indicate that the reactants are heated.

**2.94** Lead metal is produced by heating solid lead(II) sulfide with solid lead(II) sulfate, resulting in liquid lead and sulfur dioxide gas. Write a balanced equation for the reaction using complete formulas for the compounds with phase labels; indicate that the reactants are heated.

## General Problems

**2.95** Two samples of different compounds of nitrogen and oxygen have the following compositions. Show that the compounds follow the law of multiple proportions. What is the ratio of oxygen in the two compounds for a fixed amount of nitrogen?

|            | Amount N  | Amount O  |
|------------|-----------|-----------|
| Compound A | 1.206 g   | 2.755 g   |
| Compound B | 1.651 g   | 4.714 g   |

**2.96** Two samples of different compounds of sulfur and oxygen have the following compositions. Show that the compounds follow the law of multiple proportions. What is the ratio of oxygen in the two compounds for a fixed amount of sulfur?

|            | Amount S  | Amount O  |
|------------|-----------|-----------|
| Compound A | 1.210 g   | 1.811 g   |
| Compound B | 1.783 g   | 1.779 g   |

**2.97** In a series of oil-drop experiments, the charges measured on the oil drops were $-3.20 \times 10^{-19}$ C, $-6.40 \times 10^{-19}$ C, $-9.60 \times 10^{-19}$ C, and $-1.12 \times 10^{-18}$ C. What is the smallest difference in charge between any two drops? If this is assumed to be the charge on the electron, how many excess electrons are there on each drop?

**2.98** In a hypothetical universe, an oil-drop experiment gave the following measurements of charges on oil drops: $-5.55 \times 10^{-19}$ C, $-9.25 \times 10^{-19}$ C, $-1.11 \times 10^{-18}$ C, and $-1.48 \times 10^{-18}$ C. Assume that the smallest difference in charge equals the unit of negative charge in this universe. What is the value of this unit of charge? How many units of excess negative charge are there on each oil drop?

**2.99** Compounds of europium, Eu, are used to make color television screens. The europium nucleus has a charge of +63. How many electrons are there in the neutral atom? in the $Eu^{3+}$ ion?

**2.100** Cesium, Cs, is used in photoelectric cells ("electric eyes"). The cesium nucleus has a charge of +55. What is the number of electrons in the neutral atom? in the $Cs^+$ ion?

**2.101** A nucleus of mass number 80 contains 45 neutrons. An atomic ion of this element has 36 electrons in it. Write the symbol for this atomic ion (give the symbol for the nucleus and give the ionic charge as a right superscript).

**2.102** One isotope of a metallic element has mass number 119 and has 69 neutrons in the nucleus. An atomic ion has 48 electrons. Write the symbol for this ion (give the symbol for the nucleus and give the ionic charge as a right superscript).

**2.103** Silver has two naturally occurring isotopes, one of mass 106.91 amu and the other of mass 108.90 amu. Find the fractional abundances for these two isotopes. The atomic weight is 107.87 amu.

**2.104** Obtain the fractional abundances for the two naturally occurring isotopes of copper. The masses of the isotopes are $_{29}^{63}Cu$, 62.9298 amu; $_{29}^{65}Cu$, 64.9278 amu. The atomic weight is 63.546 amu.

**2.105** Identify the following elements, giving their name and atomic symbol.
a. a nonmetal that is normally a liquid
b. a normally gaseous element in Group IA
c. a transition element in Group VB, Period 5
d. the halogen in Period 2

**2.106** Identify the following elements, giving their name and atomic symbol.
a. a normally gaseous element in Group VIA
b. a metal that is normally a liquid
c. a main-group element in Group IIIA, Period 2
d. the alkali metal in Period 4

**2.107** Give the names of the following ions.
a. $Pb^{2+}$     b. $Pb^{4+}$     c. $Cu^+$     d. $Cu^{2+}$

**2.108** Give the names of the following ions.
a. $Mn^{2+}$     b. $Ni^{2+}$     c. $Co^{2+}$     d. $Co^{3+}$

**2.109** Write formulas for all the ionic compounds that can be formed by these ions: $Na^+$, $Ni^{2+}$, $SO_4^{2-}$, and $Cl^-$.

**2.110** Write formulas for all the ionic compounds that can be formed by these ions: $Ca^{2+}$, $Cr^{3+}$, $O^{2-}$, and $NO_3^-$.

**2.111** Name the following compounds.
a. $Sn_3(PO_4)_2$     b. $NH_4NO_2$
c. $Mg(OH)_2$     d. $CrSO_4$

**2.112** Name the following compounds.
a. $Cu(NO_2)_2$     b. $(NH_4)_3P$
c. $Na_2SO_4$     d. $HgCl_2$

**2.113** Give the formulas for the following compounds.
a. mercury(I) sulfide          b. cobalt(III) sulfite
c. ammonium dichromate     d. aluminum sulfide

**2.114** Give the formulas for the following compounds.
a. hydrogen peroxide          b. aluminum phosphate
c. lead(IV) nitride          d. boron trifluoride

**2.115** Name the following molecular compounds.
a. $AsBr_3$     b. $H_2Se$     c. $P_2O_5$     d. $SiO_2$

**2.116** Name the following molecular compounds.
a. $ClF_4$     b. $CS_2$     c. $NF_3$     d. $SF_6$

**2.117** Balance the following equations.
a. $C_2H_6 + O_2 \longrightarrow CO_2 + H_2O$
b. $P_4O_6 + H_2O \longrightarrow H_3PO_3$
c. $KClO_3 \longrightarrow KCl + KClO_4$

d. $(NH_4)_2SO_4 + NaOH \longrightarrow NH_3 + H_2O + Na_2SO_4$

e. $NBr_3 + NaOH \longrightarrow N_2 + NaBr + HOBr$

**2.118** Balance the following equations.

a. $NaOH + H_3PO_4 \longrightarrow Na_3PO_4 + H_2O$

b. $SiCl_4 + H_2O \longrightarrow SiO_2 + HCl$

c. $Ca_3(PO_4)_2 + C \longrightarrow Ca_3P_2 + CO$

d. $H_2S + O_2 \longrightarrow SO_2 + H_2O$

e. $N_2O_5 \longrightarrow NO_2 + O_2$

**2.119** A monatomic ion has a charge of $+2$. The nucleus of the ion has a mass number of 62. The number of neutrons in the nucleus is 1.21 times that of the number of protons. How many electrons are in the ion? What is the name of the element?

**2.120** A monatomic ion has a charge of $+3$. The nucleus of the ion has a mass number of 27. The number of neutrons in the nucleus is 2.08 times that of the number of protons. How many electrons are in the ion? What is the name of the element?

**2.121** Natural carbon, which has an atomic weight of 12.011 amu, consists of carbon-12 and carbon-13 isotopes. Given that the mass of carbon-13 is 13.00335 amu, what would be the average atomic mass (in amu) of a carbon sample prepared by mixing equal numbers of carbon atoms from a sample of natural carbon and a sample of pure carbon-13?

**2.122** Natural chlorine, which has an atomic weight of 35.4527 amu, consists of chlorine-35 and chlorine-37 isotopes. Given that the mass of chlorine-35 is 34.96885 amu, what is the average atomic mass (in amu) of a chlorine sample prepared by mixing equal numbers of chlorine atoms from a sample of natural chlorine and a sample of pure chlorine-35?

## Cumulative-Skills Problems

**2.123** There are $2.619 \times 10^{22}$ atoms in 1.000 g of sodium. Assume that sodium atoms are spheres of radius 1.86 Å and that they are lined up side by side. How many miles in length is the line of sodium atoms?

**2.124** There are $1.699 \times 10^{22}$ atoms in 1.000 g of chlorine. Assume that chlorine atoms are spheres of radius 0.99 Å and that they are lined up side by side. How many miles in length is the line of chlorine atoms?

**2.125** A sample of green crystals of nickel(II) sulfate heptahydrate was heated carefully to produce the bluish-green nickel(II) sulfate hexahydrate. What are the formulas of the hydrates? If 8.753 g of the heptahydrate produces 8.192 g of the hexahydrate, how many grams of anhydrous nickel(II) sulfate could be obtained?

**2.126** Cobalt(II) sulfate heptahydrate has pink-colored crystals. When heated carefully, it produces cobalt(II) sulfate monohydrate, which has red crystals. What are the formulas of these hydrates? If 3.548 g of the heptahydrate yields 2.184 g of the monohydrate, how many grams of the anhydrous cobalt(II) sulfate could be obtained?

**2.127** A sample of metallic element X, weighing 3.177 g, combines with 0.6015 L of $O_2$ gas (at normal pressure and 20.0°C) to form the metal oxide with the formula XO. If the density of $O_2$ gas under these conditions is 1.330 g/L, what is the mass of this oxygen? The atomic weight of oxygen is 15.9994 amu. What is the atomic weight of X? What is the identity of X?

**2.128** A sample of metallic element X, weighing 4.315 g, combines with 0.4810 L of $Cl_2$ gas (at normal pressure and 20.0°C) to form the metal chloride with the formula XCl. If the density of $Cl_2$ gas under these conditions is 2.948 g/L, what is the mass of the chlorine? The atomic weight of chlorine is 35.453 amu. What is the atomic weight of X? What is the identity of X?

## Media Activities

**2.129** Open the Law of Definite Proportions *Understanding Concepts* simulation on the student web site. Pick ammonia, $NH_3$, from the list of molecules and analyze two different masses. (a) Looking at the analysis section of the activity, what numerical values did not change as you changed the mass of $NH_3$? (b) If all the numbers changed, from the first analysis to the next, what impact would this have on Dalton's Atomic Theory?

**2.130** Open the Gold Foil Experiment *Visualization* on the student CD. (a) What would you observe if electrons were equally massive as protons? (b) What insight on atomic structure did this experiment provide?

**2.131** Write the correct formula for each of the following ionic compounds: ammonium phosphate, calcium nitrate, lead (II) perchlorate, and sodium sulfide. Check your answers by running the Ionic Compound Formulas *Understanding Concepts* simulation on the student web site.

**2.132** Write the correct name and formula from combinations of the following ions: $Ba^{2+}$ and $S^{2-}$, $Fe^{3+}$ and $NO_3^-$, and $Al^{3+}$ and $CO_3^{2-}$. Check your answers by running the Ionic Compound Formulas *Understanding Concepts* simulation on the student web site.

**2.133** Open the Balancing Equations *Understanding Concepts* activity on the student CD. Balance each of the reactions.

**2.134** For review, take the ACE practice test for Chapter 2 on the student web site.

# 3

Reaction of zinc and iodine to produce zinc iodide.

# Calculations with Chemical Formulas and Equations

Acetic acid (ah-see′-tik acid) is a colorless liquid with a sharp, vinegary odor. In fact, vinegar contains acetic acid, which accounts for vinegar's odor and sour taste. The name vinegar derives from the French word *vinaigre*, meaning "sour wine." Vinegar results from the fermentation of wine or cider by certain bacteria (Figure 3.1). These bacteria require oxygen, and the overall chemical change is the reaction of ethanol (alcohol) in wine with oxygen to give acetic acid.

Laboratory preparation of acetic acid may also start from ethanol, which reacts with oxygen in two steps. First, ethanol reacts with oxygen to yield a compound called acetaldehyde, in addition to water. In the second step, the acetaldehyde reacts with more oxygen to produce acetic acid. (The human body also produces acetaldehyde and then acetic acid from alcohol, as it attempts to eliminate alcohol from the system.)

**Figure 3.1**
**Preparing gourmet vinegar.** In the presence of air, acetic bacteria ferment wine or cider to vinegar.

▶ This chapter develops the mole concept and related calculations.

This chapter focuses on two basic questions, which we can illustrate using these compounds: How do you determine the chemical formula of a substance such as acetic acid (or acetaldehyde)? How much acetic acid can you prepare from a given quantity of ethanol (or a given quantity of acetaldehyde)? These types of questions are very important in chemistry. You must know the formulas of all the substances involved in a reaction before you can write the chemical equation, and you need the balanced chemical equation to determine the quantitative relationships among the different substances in the reaction. We begin by discussing how you relate number of atoms or molecules to grams of substance, because this is the key to answering both questions.

# Mass and Moles of Substance

You buy a quantity of groceries in several ways. You often purchase items such as oranges and lemons by counting out a particular number. Some things, such as eggs and soda, can be purchased in a "package" that represents a known quantity; for example, a dozen or a case. Bulk foods, such as peanuts or hard candy, are usually purchased by mass, because it is too tedious to count them out, and because we know that a given mass yields a certain quantity of the item. All three of these methods are used by chemists to determine the quantity of matter: counting, using a package that represents a quantity, and measuring the mass. For a chemist, it is a relatively easy matter to weigh a substance to obtain the mass. However, the number of atoms or molecules in even a seemingly minute amount is much too large to count. (You may recall from Section 2.6 that a billionth of a drop of water contains $2 \times 10^{12}$ $H_2O$ molecules.) Nevertheless, chemists are interested in knowing such numbers. How many atoms of carbon are there in one molecule of acetic acid? How many molecules of acetic acid can be obtained from one molecule of ethanol? Before we look at how chemists solve this problem of measuring numbers of atoms, molecules, and ions, we must look at the concept of molecular weight and formula weight and introduce the package chemists call the mole.

# 3.1 Molecular Weight and Formula Weight

In Chapter 2 (Section 2.4), we discussed the concept of atomic weight. We can easily extend this idea to include molecular weight. The **molecular weight** (MW) of a substance is *the sum of the atomic weights of all the atoms in a molecule of the substance.*

It is, therefore, the average mass of a molecule of that substance, expressed in atomic mass units. For example, the molecular weight of water, $H_2O$, is 18.0 amu ($2 \times 1.0$ amu from two H atoms plus 16.0 amu from one O atom). If the molecular formula for the substance is not known, you can determine the molecular weight experimentally by means of a mass spectrometer. In later chapters we will discuss simple, inexpensive methods of molecular-weight determination.

The **formula weight** (FW) of a substance is *the sum of the atomic weights of all atoms in a formula unit of the compound,* whether molecular or not. Sodium chloride, with the formula unit NaCl, has a formula weight of 58.44 amu (22.99 amu from Na plus 35.45 amu from Cl). NaCl is ionic, so strictly speaking the expression "molecular weight of NaCl" has no meaning. On the other hand, the molecular weight and the formula weight calculated from the molecular formula of a substance are identical. ◀

▶ Some chemists use the term *molecular weight* in a less strict sense for ionic as well as molecular compounds.

▶ An alternate example is provided in the Instructor's Resource Manual.

### EXAMPLE 3.1

## Calculating the Formula Weight from a Formula

Calculate the formula weight of each of the following to three significant figures, using a table of atomic weights (AW): a. chloroform, $CHCl_3$; b. iron(III) sulfate, $Fe_2(SO_4)_3$.

 **PROBLEM STRATEGY**

Obtain the sum of the atomic weights for the atoms in each formula. Remember in part b that the number of atoms of each element within parentheses in the formula should be multiplied by the subscript to the parentheses.

**SOLUTION**

a. The calculation is

$$
\begin{array}{lll}
1 \times \text{AW of C} = & & 12.0 \text{ amu} \\
1 \times \text{AW of H} = & & 1.0 \text{ amu} \\
3 \times \text{AW of Cl} = 3 \times 35.45 \text{ amu} = & \underline{106.4 \text{ amu}} \\
\text{FW of } CHCl_3 = & & 119.4 \text{ amu}
\end{array}
$$

The answer rounded to three significant figures is **119 amu.**

b. The calculation is

$$
\begin{array}{lll}
2 \times \text{AW of Fe} = & 2 \times 55.8 \text{ amu} = 111.6 \text{ amu} \\
3 \times \text{AW of S} = & 3 \times 32.1 \text{ amu} = \phantom{0}96.3 \text{ amu} \\
3 \times 4 \times \text{AW of O} = & 12 \times 16.00 \text{ amu} = \underline{192.0 \text{ amu}} \\
\text{FW of } Fe_2(SO_4)_3 = & 399.9 \text{ amu}
\end{array}
$$

The answer rounded to three significant figures is **$4.00 \times 10^2$ amu.**

### EXERCISE 3.1

Calculate the formula weights of the following compounds, using a table of atomic weights. Give the answers to three significant figures. a. nitrogen dioxide, $NO_2$; b. glucose, $C_6H_{12}O_6$; c. sodium hydroxide, NaOH; d. magnesium hydroxide, $Mg(OH)_2$.

See Problems 3.19 and 3.20.

---

**EXAMPLE 3.2**

## Calculating the Formula Weight from Molecular Models

For the following two compounds, write the molecular formula and calculate the formula weight to four significant figures:

(a)     (b)

**SOLUTION**

a. This molecular model is of a molecule that is composed of two O and two H atoms. For inorganic compounds, the elements in a chemical formula are written in order such that the most metallic element is listed first. (Even though H is not metallic, it is positioned in the periodic table in such a way that it is considered to be a metal when writing formulas.) Hence, the chemical formula is $H_2O_2$. Using the same approach as Example 3.1, calculating the formula weight yields **34.02 amu.**

b. This molecular model represents a molecule made up of one N atom, three O atoms, and one H atom. The chemical formula is then $HNO_3$. The formula weight is **63.01 amu.**

---

**EXERCISE 3.2**

For the following compounds, write the molecular formula and calculate the formula weight to three significant figures.

(a)     (b)

---

See Problems 3.21 and 3.22.

# 3.2 The Mole Concept

When we prepare a compound industrially or even study a reaction in the laboratory, we deal with tremendous numbers of molecules or ions. Suppose you wish to prepare acetic acid, starting from 10.0 g of ethanol. This small sample (less than 3 teaspoonsful) contains $1.31 \times 10^{23}$ molecules, a truly staggering number. Imagine a device that counts molecules at the rate of one million per second. It would take more than four billion years—nearly the age of the earth—for this device to count that many molecules! Chemists have adopted the *mole concept* as a convenient way to deal with the enormous numbers of molecules or ions in the samples they work with.

## Definition of Mole and Molar Mass

▶ **Demonstration of mole concept:** show a mole of various substances.

A **mole** (symbol **mol**) is defined as *the quantity of a given substance that contains as many molecules or formula units as the number of atoms in exactly 12 g of carbon-12.* One mole of ethanol, for example, contains the same number of ethanol molecules as there are carbon atoms in 12 g of carbon-12.

*The number of atoms in a 12-g sample of carbon-12* is called **Avogadro's number** (to which we give the symbol $N_A$). Recent measurements of this number give the value $6.0221367 \times 10^{23}$, which to three significant figures is $6.02 \times 10^{23}$.

A mole of a substance contains Avogadro's number ($6.02 \times 10^{23}$) of molecules (or formula units). The term *mole,* like a dozen or a gross, thus refers to a particular number of things. A dozen eggs equals 12 eggs, a gross of pencils equals 144 pencils, and a mole of ethanol equals $6.02 \times 10^{23}$ ethanol molecules.

▶ **Sodium carbonate, $Na_2CO_3$, is a white, crystalline solid known commercially as soda ash. Large amounts of soda ash are used in the manufacture of glass. The hydrated compound, $Na_2CO_3 \cdot 10H_2O$, is known as washing soda.**

In using the term *mole* for ionic substances, we mean the number of formula units of the substance. For example, a mole of sodium carbonate, $Na_2CO_3$, is a quantity containing $6.02 \times 10^{23}$ $Na_2CO_3$ units. But each formula unit of $Na_2CO_3$ contains two $Na^+$ ions and one $CO_3^{2-}$ ion. Therefore, a mole of $Na_2CO_3$ also contains $2 \times 6.02 \times 10^{23}$ $Na^+$ ions and $1 \times 6.02 \times 10^{23}$ $CO_3^{2-}$ ions. ◀

When using the term *mole,* it is important to specify the formula of the unit to avoid any misunderstanding. For example, a mole of oxygen atoms (with the formula O) contains $6.02 \times 10^{23}$ O atoms. A mole of oxygen molecules (formula $O_2$) contains $6.02 \times 10^{23}$ $O_2$ molecules—that is, $2 \times 6.02 \times 10^{23}$ O atoms.

The **molar mass** of a substance is *the mass of one mole of the substance.* Carbon-12 has a molar mass of exactly 12 g/mol, by definition.

For all substances, the molar mass in grams per mole is numerically equal to the formula weight in atomic mass units.

Ethanol, whose molecular formula is $C_2H_6O$ (frequently written as the condensed structural formula $C_2H_5OH$), has a molecular weight of 46.1 amu and a molar mass of 46.1 g/mol. Figure 3.2 shows molar amounts of different substances.

Ethanol
$C_2H_5OH$

**Figure 3.2**
**One mole each of various substances.** *Clockwise from top left:* 1-octanol ($C_8H_{17}OH$); mercury(II) iodide ($HgI_2$); methanol ($CH_3OH$); sulfur ($S_8$).

► An alternate example is provided in the Instructor's Resource Manual.

---

## EXAMPLE 3.3

### Calculating the Mass of an Atom or Molecule

a. What is the mass in grams of a chlorine atom, Cl?

b. What is the mass in grams of a hydrogen chloride molecule, HCl?

 **PROBLEM STRATEGY**

Because each mole of substance contains Avogadro's number of units of that substance, dividing the mass of a substance in one mole (molar mass) by Avogadro's number gives the mass of that unit of substance.

**SOLUTION**

a. The atomic weight of Cl is 35.5 amu, so the molar mass of Cl is 35.5 g/mol. Dividing 35.5 g (per mole) by $6.02 \times 10^{23}$ (Avogadro's number) gives the mass of one atom.

$$\text{Mass of a Cl atom} = \frac{35.5 \text{ g}}{6.02 \times 10^{23}} = \mathbf{5.90 \times 10^{-23} \text{ g}}$$

b. The molecular weight of HCl equals the AW of H plus the AW of Cl, or 1.01 amu + 35.5 amu = 36.5 amu. Therefore, 1 mol HCl contains 36.5 g HCl and

$$\text{Mass of an HCl molecule} = \frac{36.5 \text{ g}}{6.02 \times 10^{23}} = \mathbf{6.06 \times 10^{-23} \text{ g}}$$

---

## EXERCISE 3.3

a. What is the mass in grams of a calcium atom, Ca? b. What is the mass in grams of an ethanol molecule, $C_2H_5OH$?

---

See Problems 3.25, 3.26, 3.27, and 3.28.

**Problem Solving: Determining the Mass of a Sample of Atoms**

## Mole Calculations

Now that you know how to find the mass of one mole of substance, there are two important questions to ask. First, how much does a given number of moles of a substance weigh? Second, how many moles of a given formula unit does a given mass of substance contain? Both questions are easily answered using dimensional analysis, or the conversion-factor method. ◄

To illustrate, consider the conversion of grams of ethanol, $C_2H_5OH$, to moles of ethanol. The molar mass of ethanol is 46.1 g/mol, so we write

$$1 \text{ mol } C_2H_5OH = 46.1 \text{ g } C_2H_5OH$$

Thus, the factor converting grams of ethanol to moles of ethanol is 1 mol $C_2H_5OH$/46.1 g $C_2H_5OH$. To convert moles of ethanol to grams of ethanol, we simply invert the conversion factor (46.1 g $C_2H_5OH$/1 mol $C_2H_5OH$). Note that the unit you are converting *from* is on the bottom of the conversion factor; the unit you are converting *to* is on the top.

► Alternatively, because the molar mass is the mass per mole, you can relate mass and moles by means of the formula

**Molar mass = mass/moles**

Again, suppose you are going to prepare acetic acid from 10.0 g of ethanol, $C_2H_5OH$. How many moles of $C_2H_5OH$ is this? You convert 10.0 g $C_2H_5OH$ to moles $C_2H_5OH$ by multiplying by the appropriate conversion factor.

$$10.0 \text{ g } \cancel{C_2H_5OH} \times \frac{1 \text{ mol } C_2H_5OH}{46.1 \text{ g } \cancel{C_2H_5OH}} = 0.217 \text{ mol } C_2H_5OH$$

The following examples further illustrate this conversion-factor technique.

▶ An alternate example is provided in the Instructor's Resource Manual.

▶ Video no. 4 (Series B), "The Purple Cloud."

**Figure 3.3**
**Reaction of zinc and iodine.** Heat from the reaction of the elements causes some iodine to vaporize (violet vapor).

**EXAMPLE 3.4**

## Converting Moles of Substance to Grams

Zinc iodide, $ZnI_2$, can be prepared by the direct combination of elements (Figure 3.3). A chemist determines from the amounts of elements that 0.0654 mol $ZnI_2$ can form. How many grams of zinc iodide is this?

 **PROBLEM STRATEGY**

Use the formula weight to write the factor that converts from mol $ZnI_2$ to g $ZnI_2$. Note that the unit you are converting from (mol $ZnI_2$) is on the bottom of the conversion factor, and the unit you are converting to (g $ZnI_2$) is on the top.

**SOLUTION**

The molar mass of $ZnI_2$ is 319 g/mol. (The formula weight is 319 amu, which is obtained by summing the atomic weights in the formula.) Therefore,

$$0.0654 \text{ } \cancel{\text{mol } ZnI_2} \times \frac{319 \text{ g } ZnI_2}{1 \text{ } \cancel{\text{mol } ZnI_2}} = \textbf{20.9 g } \textbf{ZnI}_2$$

**EXERCISE 3.4**

Hydrogen peroxide, $H_2O_2$, is a colorless liquid. A concentrated solution of it is used as a source of oxygen for rocket propellant fuels. Dilute aqueous solutions are used as a bleach. Analysis of a solution shows that it contains 0.909 mol $H_2O_2$ in 1.00 L of solution. What is the mass of hydrogen peroxide in this volume of solution?

See Problems 3.29, 3.30, 3.31, and 3.32.

**Figure 3.4**
**Preparation of lead(II) chromate.** When lead(II) nitrate solution (colorless) is added to potassium chromate solution (clear yellow), bright yellow solid lead(II) chromate forms (giving a cloudlike formation of fine crystals).

**EXAMPLE 3.5**

## Converting Grams of Substance to Moles

Lead(II) chromate, $PbCrO_4$, is a yellow paint pigment (called chrome yellow) prepared by a precipitation reaction (Figure 3.4). In a preparation, 45.6 g of lead(II) chromate is obtained as a precipitate. How many moles of $PbCrO_4$ is this?

**SOLUTION**

The molar mass of $PbCrO_4$ is 323 g/mol. That is,

$$1 \text{ mol } PbCrO_4 = 323 \text{ g } PbCrO_4$$

Therefore,

$$45.6 \text{ g PbCrO}_4 \times \frac{1 \text{ mol PbCrO}_4}{323 \text{ g PbCrO}_4} = \textbf{0.141 mol PbCrO}_4$$

## EXERCISE 3.5

Nitric acid, $HNO_3$, is a colorless, corrosive liquid used in the manufacture of nitrogen fertilizers and explosives. In an experiment to develop new explosives for mining operations, a 28.5-g sample of nitric acid was poured into a beaker. How many moles of $HNO_3$ are there in this sample of nitric acid?

See Problems 3.33 and 3.34.

## EXAMPLE 3.6

### Calculating the Number of Molecules in a Given Mass

How many molecules are there in a 3.46-g sample of hydrogen chloride, HCl?

 **PROBLEM STRATEGY**

The number of molecules in a sample is related to moles of compound (1 mol HCl = $6.02 \times 10^{23}$ HCl molecules). Therefore, if you first convert grams HCl to moles, then you can convert moles to number of molecules.

**SOLUTION**

Here is the calculation:

$$3.46 \text{ g HCl} \times \frac{1 \text{ mol HCl}}{36.5 \text{ g HCl}} \times \frac{6.02 \times 10^{23} \text{ HCl molecules}}{1 \text{ mol HCl}}$$
$$= \textbf{5.71} \times \textbf{10}^{22} \textbf{ HCl molecules}$$

Note how the units in the numerator of a factor are canceled by the units in the denominator of the following factor.

## EXERCISE 3.6

Hydrogen cyanide, HCN, is a volatile, colorless liquid with the odor of certain fruit pits (such as peach and cherry pits). The compound is highly poisonous. How many molecules are there in 56 mg HCN, the average toxic dose?

See Problems 3.37, 3.38, 3.39, and 3.40.

Problem Solving: Molar Mass and Number of Molecules

 **CONCEPT CHECK 3.1**

You have 1.5 moles of tricycles.

a. How many moles of seats do you have?

b. How many moles of tires do you have?

c. How could you use parts a and b as an analogy to teach a friend about the number of moles of $OH^-$ ions in 1.5 moles of $Mg(OH)_2$?

# Determining Chemical Formulas

When a chemist has discovered a new compound, the first question to answer is, What is the formula? To answer, you begin by analyzing the compound to determine amounts of the elements for a given amount of compound. This is conveniently expressed as **percentage composition**—that is, as *the mass percentages of each element in the compound.* You then determine the formula from this percentage composition. If the compound is a molecular substance, you must also find the molecular weight of the compound in order to determine the molecular formula.

The next section describes the calculation of mass percentages. Then, in two following sections, we describe how to determine a chemical formula.

## 3.3   Mass Percentages from the Formula

Suppose that $A$ is a part of something—that is, part of a whole. It could be an element in a compound or one substance in a mixture. We define the **mass percentage** of $A$ as *the parts of* $A$ *per hundred parts of the total, by mass.* That is,

$$\text{Mass } \% \ A = \frac{\text{mass of } A \text{ in the whole}}{\text{mass of the whole}} \times 100\%$$

You can look at the mass percentage of $A$ as the number of grams of $A$ in 100 g of the whole.

The next example will provide practice with the concept of mass percentage. In this example we will start with a compound (formaldehyde, $CH_2O$) whose formula is given and obtain the percentage composition.

▶ An alternate example is provided in the Instructor's Resource Manual.

---

**EXAMPLE 3.7**

### Calculating the Percentage Composition from the Formula

Formaldehyde, $CH_2O$, is a toxic gas with a pungent odor. Large quantities are consumed in the manufacture of plastics (Figure 3.5), and a water solution of the compound is used to preserve biological specimens. Calculate the mass percentages of the elements in formaldehyde (give answers to three significant figures).

**Figure 3.5**
**Preparing resorcinol–formaldehyde plastic.**
*Left:* The clear solution contains formaldehyde, $CH_2O$, and resorcinol, $C_6H_4(OH)_2$. The formation of the plastic is started by adding several drops of potassium hydroxide. *Right:* The red plastic has formed in the beaker.

**PROBLEM STRATEGY**

To calculate mass percentage, you need the mass of an element in a given mass of compound. You can get this information by interpreting the formula in molar terms and then converting moles to masses, using a table of atomic weights. Thus, 1 mol $CH_2O$ has a mass of 30.0 g and contains 1 mol C (12.0 g), 2 mol H (2 × 1.01 g), and 1 mol O (16.0 g). You divide each mass of element by the molar mass, then multiply by 100, to obtain the mass percentage.

**SOLUTION**

Here are the calculations:

$$\% \text{ C} = \frac{12.0 \text{ g}}{30.0 \text{ g}} \times 100\% = \textbf{40.0\%}$$

$$\% \text{ H} = \frac{2 \times 1.01 \text{ g}}{30.0 \text{ g}} \times 100\% = \textbf{6.73\%}$$

You can calculate the percentage of O in the same way, but it can also be found by subtracting the percentages of C and H from 100%:

$$\% \text{ O} = 100\% - (40.0\% + 6.73\%) = \textbf{53.3\%}$$

**Problem Solving:
Calculating Mass Percent**

**EXERCISE 3.7**

Ammonium nitrate, $NH_4NO_3$, which is prepared from nitric acid, is used as a nitrogen fertilizer. Calculate the mass percentages of the elements in ammonium nitrate (to three significant figures).

See Problems 3.49, 3.50, 3.51, and 3.52.

▶ An alternate example is provided in the Instructor's Resource Manual.

**EXAMPLE 3.8**

## Calculating the Mass of an Element in a Given Mass of Compound

How many grams of carbon are there in 83.5 g of formaldehyde, $CH_2O$? Use the percentage composition obtained in the previous example (40.0% C, 6.73% H, 53.3% O).

**SOLUTION**

$CH_2O$ is 40.0% C, so the mass of carbon in 83.5 g $CH_2O$ is

$$83.5 \text{ g} \times 0.400 = \textbf{33.4 g}$$

Formaldehyde
$CH_2O$

**EXERCISE 3.8**

How many grams of nitrogen, N, are there in a fertilizer containing 48.5 g of ammonium nitrate and no other nitrogen-containing compound? See Exercise 3.7 for the percentage composition of $NH_4NO_3$.

See Problems 3.53 and 3.54.

# 3.4 Elemental Analysis: Percentages of Carbon, Hydrogen, and Oxygen

▶ **You might want to begin Section 3.4 with this simpler example. An iron oxide was heated in a stream of hydrogen gas. If 1.534 g of the oxide yields 1.110 g of iron, what is the percentage composition? (The formula is $Fe_3O_4$.)**

Suppose you have a newly discovered compound whose formula you wish to determine. The first step is to obtain its percentage composition. As an example, consider the determination of the percentages of carbon, hydrogen, and oxygen in compounds containing only these three elements. The basic idea is this: You burn a sample of the compound of known mass and get $CO_2$ and $H_2O$. Next you relate the masses of $CO_2$ and $H_2O$ to the masses of carbon and hydrogen. Then you calculate the mass percentages of C and H. You find the mass percentage of O by subtracting the mass percentages of C and H from 100.

Figure 3.6 shows an apparatus used to find the amount of carbon and hydrogen in a compound. The compound is burned in a stream of oxygen gas. The vapor of the compound and its combustion products pass over copper pellets coated with copper(II) oxide, CuO, which supplies additional oxygen and ensures that the compound is completely burned. As a result of the combustion, every mole of carbon (C) in the compound ends up as a mole of carbon dioxide ($CO_2$), and every mole of hydrogen (H) ends up as one-half mole of water ($H_2O$). The water is collected by a drying agent, a substance that has a strong affinity for water. The carbon dioxide is collected by chemical reaction with sodium hydroxide, NaOH. ◀ By weighing the U-tubes containing the drying agent and the sodium hydroxide before and after combustion, it is possible to determine the masses of water and carbon dioxide produced. From these data, you can calculate the percentage composition of the compound.

▶ **Sodium hydroxide reacts with carbon dioxide according to the following equations:**

$$NaOH + CO_2 \longrightarrow NaHCO_3$$
$$2NaOH + CO_2 \longrightarrow Na_2CO_3 + H_2O$$

The chapter opened with a discussion of acetic acid. The next example shows how to determine the percentage composition of this substance from combustion data. We will use this percentage composition later in Example 3.12 to determine the formula of acetic acid.

Oxygen in →

Copper(II) oxide (CuO) pellets

Excess oxygen out →

Furnace —

Sample dish

Drying agent (traps $H_2O$)

Sodium hydroxide (traps $CO_2$)

**Figure 3.6**
**Combustion method for determining the percentages of carbon and hydrogen in a compound.** The compound is placed in the sample dish and is heated by the furnace. Vapor of the compound burns in $O_2$ in the presence of CuO pellets, giving $CO_2$ and $H_2O$. The water vapor is collected by a drying agent, and $CO_2$ combines with the sodium hydroxide. Amounts of $CO_2$ and $H_2O$ are obtained by weighing the U-tubes before and after combustion.

▶ An alternate example is provided in the Instructor's Resource Manual.

---

### EXAMPLE 3.9

## Calculating the Percentages of C and H by Combustion

Acetic acid contains only C, H, and O. A 4.24-mg sample of acetic acid is completely burned. It gives 6.21 mg of carbon dioxide and 2.54 mg of water. What is the mass percentage of each element in acetic acid?

 **PROBLEM STRATEGY**

You first convert the mass of $CO_2$ to moles of $CO_2$. Then you convert this to moles of C, noting that 1 mol C produces 1 mol $CO_2$. Finally, you convert to mass of C. Similarly, you convert the mass of $H_2O$ to mol $H_2O$, then to mol H, and finally to mass of H. (Remember that 1 mol $H_2O$ produces 2 mol H.) Once you have the masses of C and H, you can calculate the mass percentages. Subtract from 100% to get %O.

**SOLUTION**

Following is the calculation of grams C:

$$6.21 \times 10^{-3} \text{ g CO}_2 \times \frac{1 \text{ mol CO}_2}{44.0 \text{ g CO}_2} \times \frac{1 \text{ mol C}}{1 \text{ mol CO}_2} \times \frac{12.0 \text{ g C}}{1 \text{ mol C}}$$
$$= 1.69 \times 10^{-3} \text{ g C (or 1.69 mg C)}$$

For hydrogen, you note that 1 mol $H_2O$ yields 2 mol H, so you write

$$2.54 \times 10^{-3} \text{ g H}_2\text{O} \times \frac{1 \text{ mol H}_2\text{O}}{18.0 \text{ g H}_2\text{O}} \times \frac{2 \text{ mol H}}{1 \text{ mol H}_2\text{O}} \times \frac{1.01 \text{ g H}}{1 \text{ mol H}}$$
$$= 2.85 \times 10^{-4} \text{ g H (or 0.285 mg H)}$$

You can now calculate the mass percentages of C and H in acetic acid.

$$\text{Mass \% C} = \frac{1.69 \text{ mg}}{4.24 \text{ mg}} \times 100\% = 39.9\%$$

$$\text{Mass \% H} = \frac{0.285 \text{ mg}}{4.24 \text{ mg}} \times 100\% = 6.72\%$$

You find the mass percentage of oxygen by subtracting the sum of these percentages from 100%:

$$\text{Mass \% O} = 100\% - (39.9\% + 6.72\%) = 53.4\%$$

Thus, the percentage composition of acetic acid is **39.9% C, 6.7% H, and 53.4% O.**

---

### EXERCISE 3.9

A 3.87-mg sample of ascorbic acid (vitamin C) gives 5.80 mg $CO_2$ and 1.58 mg $H_2O$ on combustion. What is the percentage composition of this compound (the mass percentage of each element)? Ascorbic acid contains only C, H, and O.

---

See Problems 3.55 and 3.56.

**CONCEPT CHECK 3.2**

You perform combustion analysis on a compound that contains only C and H.

a. Considering the fact that the combustion products $CO_2$ and $H_2O$ are colorless, how can you tell if some of the product got trapped in the CuO pellets (see Figure 3.6)?

b. Would your calculated results of mass percentage of C and H be affected if some of the combustion products got trapped in the CuO pellets? If your answer is yes, how might your results differ from the expected values for the compound?

## 3.5 Determining Formulas

The percentage composition of a compound leads directly to its empirical formula. An **empirical formula** (or **simplest formula**) for a compound is *the formula of a substance written with the smallest integer (whole number) subscripts.* For most ionic substances, the empirical formula is the formula of the compound. ◀ This is often not the case for molecular substances. For example, hydrogen peroxide has the molecular formula $H_2O_2$. The molecular formula, you may recall, tells you the precise number of atoms of different elements in a molecule of the substance. The empirical formula, however, merely tells you the ratio of numbers of atoms in the compound. The empirical formula of hydrogen peroxide ($H_2O_2$) is HO.

Compounds with different molecular formulas can have the same empirical formula, and such substances will have the same percentage composition. An example is acetylene, $C_2H_2$, and benzene, $C_6H_6$. Acetylene is a gas used as a fuel and also in welding. Benzene, in contrast, is a liquid that is used in the manufacture of plastics and is a component of gasoline (Figure 3.7). Table 3.1 illustrates how these two com-

▶ The formula of sodium peroxide, an ionic compound of $Na^+$ and $O_2^{2-}$, is $Na_2O_2$. Its empirical formula is NaO.

**Figure 3.7**
**Acetylene and benzene.**
*Left:* Acetylene is a colorless gas that when burned in oxygen produces a hot flame that can be used for welding. *Right:* Benzene is a colorless liquid that is an essential component in the manufacture of many petrochemicals.

**Table 3.1**
**Molecular Models of Two Compounds That Have the Empirical Formula CH.**
Although benzene and acetylene have the same empirical formula, they do not have the same molecular formula or structure.

| Compound | Empirical Formula | Molecular Formula | Molecular Model |
|---|---|---|---|
| Acetylene | CH | $C_2H_2$ | |
| Benzene | CH | $C_6H_6$ | |

pounds, with the same empirical formula, but different molecular formulas, also have different chemical structures. Because the empirical formulas of acetylene and benzene are the same, they have the same percentage composition: 92.3% C and 7.7% H, by mass.

To obtain the molecular formula of a substance, you need two pieces of information: (1) as in the previous section, the percentage composition, from which the empirical formula can be determined; and (2) the molecular weight. The molecular weight allows you to choose the correct multiple of the empirical formula for the molecular formula. We will illustrate these steps in the next three examples.

## Empirical Formula from the Composition

The empirical formula of a compound shows the ratios of numbers of atoms in the compound. You can find this formula from the composition of the compound by converting masses of the elements to moles. The next two examples show these calculations in detail.

▶ An alternate example is provided in the Instructor's Resource Manual.

**EXAMPLE 3.10**

### Determining the Empirical Formula from Masses of Elements (Binary Compound)

A compound of nitrogen and oxygen is analyzed, and a sample weighing 1.587 g is found to contain 0.483 g N and 1.104 g O. What is the empirical formula of the compound?

# Mass Spectrometry and Molecular Formula

Sophisticated instruments have become indispensable in modern chemical analysis and research. One such instrument is the mass spectrometer. The mass spectrum of a compound is used in combination with other information to identify a substance or to obtain the molecular formula and structure of a newly prepared compound.

Positive ions of molecules, like those of atoms, can be generated by bombarding the gas, or vapor of the substance, with electrons. The mass spectra of molecules, however, are usually much more complicated than those of atoms. One reason is that the molecular ions produced often break into fragments, giving several different kinds of positive ions. Consider the $CH_2Cl_2$ molecule (methylene chloride). When this molecule is struck by a high-energy electron, a positive ion $CH_2Cl_2^+$ may form.

$$CH_2Cl_2 + e^- \longrightarrow CH_2Cl_2^+ + 2e^-$$

The $CH_2Cl_2^+$ ion gains a great deal of energy from the collision of $CH_2Cl_2$ with the electron, and the ion frequently loses this energy by breaking into smaller pieces. One way is

$$CH_2Cl_2^+ \longrightarrow CH_2Cl^+ + Cl$$

Thus, the original molecule, even one as simple as $CH_2Cl_2$, can give rise to a number of ions.

The second reason for the complexity of the mass spectrum of a molecular substance is that many of the atoms in any ion can occur with different isotopic mass, so each ion often has many peaks. The mass spectrum of methylene chloride, $CH_2Cl_2$, is shown in Figure 3.10. Fourteen peaks are clearly visible. A larger molecule can give an even more complicated spectrum.

Because of the complexity of the mass spectrum, it can be used as a "fingerprint" in identifying a compound. Only methylene chloride has exactly the spectrum shown in Figure 3.10. Thus, by comparing the mass spectrum of an unknown substance with those in a catalog of mass spectra of known compounds, you can determine its identity. The more information you have about the compound, the shorter the search through the catalog of spectra.

The mass spectrum itself contains a wealth of information about molecular structure. Some experience is needed to analyze the spectrum of a compound, but you can get an idea of how it is done by looking at Figure 3.10. Suppose you do not know the identity of the compound. The most intense peaks at the greatest mass often correspond to the ion from the original molecule and give you the molecular weight. Thus, you would expect the peaks at mass 84 and mass 86 to be from the original molecular ion, so the molecular weight is approximately 84 to 86.

An elemental analysis is also possible. The two most intense peaks in Figure 3.10 are at 84 amu and 49 amu. They differ by 35 amu. Perhaps the original mol-

 **PROBLEM STRATEGY**

You convert masses of each element to moles, because these quantities will be proportional to the subscripts in the empirical formula. To obtain the smallest integers from the moles, you divide each by the smallest one. If the results are all whole numbers, they will be the subscripts in the formula. (Otherwise, you will need to multiply by some factor, as illustrated in Example 3.11.)

**SOLUTION**

You convert the masses to moles:

$$0.483 \text{ g N} \times \frac{1 \text{ mol N}}{14.0 \text{ g N}} = 0.0345 \text{ mol N}$$

$$1.104 \text{ g O} \times \frac{1 \text{ mol O}}{16.00 \text{ g O}} = 0.06900 \text{ mol O}$$

Now divide each mole number by the smaller one (0.0345 mol). For N, you get 1.00; for O, you get 2.00. Thus, the ratio of number of N atoms to the number of O atoms is 1 to 2. Hence, the empirical formula is **NO$_2$**.

**Figure 3.10**
**Mass spectrum of methylene chloride, $CH_2Cl_2$.** The lines at higher mass correspond to ions of the original molecule. Several lines occur because of the presence of different isotopes in the ion.

ecule that gives the peak at 84 amu contains a chlorine-35 atom, which is lost to give the peak at 49 amu. If this is true, you should expect a weaker peak at mass 86, corresponding to the original molecular ion with chlorine-37 in place of a chlorine-35 atom. This is indeed what you see.

The relative heights of the peaks, which depend on the natural abundances of atoms, are also important, because they give you additional information about the elements present. The relative heights can also tell you how many atoms of a given element are in the original molecule. Naturally occurring chlorine is 75.8%

chlorine-35 and 24.2% chlorine-37. If the original molecule contained only one Cl atom, the peaks at 84 amu and 86 amu would be in the ratio 0.758 : 0.242. That is, the peak at 86 amu would be about one-third the height of the one at 84 amu. In fact, the relative height is twice this value. This means that the molecular ion contains two chlorine atoms, because the chance that any such ion contains one chlorine-37 atom is then twice as great. Thus, simply by comparing relative peak heights, you can both confirm the presence of particular elements and obtain the molecular formula.

---

**EXERCISE 3.10**

A sample of compound weighing 83.5 g contains 33.4 g of sulfur. The rest is oxygen. What is the empirical formula?

See Problems 3.57 and 3.58.

---

► An alternate example is provided in the Instructor's Resource Manual.

**EXAMPLE 3.11**

## Determining the Empirical Formula from Percentage Composition (General)

Chromium forms compounds of various colors. (The word *chromium* comes from the Greek *khroma,* meaning "color"; see Figure 3.8.) Sodium dichromate is the most important commercial chromium compound, from which many other chromium compounds are manufactured. It is a bright orange, crystalline substance. An analysis of sodium dichromate gives the following mass percentages: 17.5% Na, 39.7% Cr, and 42.8% O. What is the

**Figure 3.8**
**Chromium compounds of different colors.**
*Clockwise from top:*
Potassium chromate, $K_2CrO_4$; chromium(VI) oxide, $CrO_3$; chromium(III) sulfate, $Cr_2(SO_4)_3$; chromium metal; potassium dichromate, $K_2Cr_2O_7$; chromium(III) oxide, $Cr_2O_3$.

empirical formula of this compound? (Sodium dichromate is ionic, so it has no molecular formula.)

 **PROBLEM STRATEGY**

Assume for the purposes of this calculation that you have 100.0 g of substance. Then the mass of each element in the sample equals the numerical value of the percentage. For example, the quantity of sodium in sodium dichromate is 17.5 g, since the substance is 17.5% Na. Now convert the masses to moles and divide each mole number by the smallest. In this example, you do not obtain a series of integers, or whole numbers, from this division. You will need to find a whole-number factor to multiply these results by to obtain integers. Normally this factor will be 2 or 3, though it might be larger.

**SOLUTION**

Of the 100.0 g of sodium dichromate, 17.5 g is Na, 39.7 g is Cr, and 42.8 g is O. You convert these amounts to moles.

$$17.5 \text{ g Na} \times \frac{1 \text{ mol Na}}{23.0 \text{ g Na}} = 0.761 \text{ mol Na}$$

$$39.7 \text{ g Cr} \times \frac{1 \text{ mol Cr}}{52.0 \text{ g Cr}} = 0.763 \text{ mol Cr}$$

$$42.8 \text{ g O} \times \frac{1 \text{ mol O}}{16.0 \text{ g O}} = 2.68 \text{ mol O}$$

Now you divide all the mole numbers by the smallest one.

$$\text{For Na: } \frac{0.761 \text{ mol}}{0.761 \text{ mol}} = 1.00$$

$$\text{For Cr: } \frac{0.763 \text{ mol}}{0.761 \text{ mol}} = 1.00$$

$$\text{For O: } \frac{2.68 \text{ mol}}{0.761 \text{ mol}} = 3.52$$

You must decide whether these numbers are integers (whole numbers), within experimental error. If you round off the last digit, which is subject to experimental error, you get $Na_{1.0}Cr_{1.0}O_{3.5}$. In this case, the subscripts are not all integers. However, they can be made into integers by multiplying each one by 2; you get $Na_{2.0}Cr_{2.0}O_{7.0}$. Thus, the empirical formula is $\mathbf{Na_2Cr_2O_7}$.

**EXERCISE 3.11**

Benzoic acid is a white, crystalline powder used as a food preservative. The compound contains 68.8% C, 5.0% H, and 26.2% O, by mass. What is its empirical formula?

See Problems 3.59, 3.60, 3.61, and 3.62.

## Molecular Formula from Empirical Formula

The molecular formula of a compound is a multiple of its empirical formula. For example, the molecular formula of acetylene, $C_2H_2$, is equivalent to $(CH)_2$, and the molecular formula of benzene, $C_6H_6$, is equivalent to $(CH)_6$. Therefore, the molecular weight is some multiple of the empirical formula weight, which is obtained by summing the atomic weights of the atoms in the empirical formula. For any molecular compound, you can write

$$\text{Molecular weight} = n \times \text{empirical formula weight}$$

where $n$ is the number of empirical formula units in the molecule. You get the molecular formula by multiplying the subscripts of the empirical formula by $n$, which you calculate from the equation

$$n = \frac{\text{molecular weight}}{\text{empirical formula weight}}$$

Once you determine the empirical formula for a compound, you can calculate its empirical formula weight. If you have an experimental determination of its molecular weight, you can calculate $n$ and then the molecular formula. The next example illustrates how you use percentage composition and molecular weight to determine the molecular formula of acetic acid.

▶ Two alternate examples are provided in the Instructor's Resource Manual.

### EXAMPLE 3.12

## Determining the Molecular Formula from Percentage Composition and Molecular Weight

In Example 3.9, we found the percentage composition of acetic acid to be 39.9% C, 6.7% H, and 53.4% O. Determine the empirical formula. The molecular weight of acetic acid was determined by experiment to be 60.0 amu. What is its molecular formula?

### SOLUTION

A sample of 100.0 g of acetic acid contains 39.9 g C, 6.7 g H, and 53.4 g O. Converting these masses to moles gives 3.33 mol C, 6.6 mol H, and 3.34 mol O. Dividing the mole numbers by the smallest one gives 1.00 for C, 2.0 for H, and 1.00 for O. **The empirical formula of acetic acid is $CH_2O$.** (You may have noted that the percentage composition of acetic acid is, within experimental error, the same as that of formaldehyde—see Example 3.7—so they must have the same empirical formula.) The empirical formula weight is 30.0 amu. Dividing the empirical formula weight into the molecular weight gives the number by which the subscripts in $CH_2O$ must be multiplied.

$$n = \frac{\text{molecular weight}}{\text{empirical formula weight}} = \frac{60.0 \text{ amu}}{30.0 \text{ amu}} = 2.00$$

**The molecular formula of acetic acid is $(CH_2O)_2$, or $C_2H_4O_2$.**

### EXERCISE 3.12

The percentage composition of acetaldehyde is 54.5% C, 9.2% H, and 36.3% O, and its molecular weight is 44 amu. Obtain the molecular formula of acetaldehyde.

See Problems 3.65, 3.66, 3.67, and 3.68.

▶ The condensed structural formula for acetaldehyde is **CH₃CHO**. Its structural formula is

$$
\begin{array}{ccc}
& H & O \\
& | & \| \\
H - & C - & C - H \\
& | & \\
& H &
\end{array}
$$

Determining this structural formula requires additional information.

The formula of acetic acid is often written $HC_2H_3O_2$ to indicate that one of the hydrogen atoms is acidic (lost easily) while the other three are not (Figure 3.9). Now that you know the formulas of acetic acid and acetaldehyde (determined from the data in Exercise 3.12 to be $C_2H_4O$), you can write the equations for the preparation of acetic acid described in the chapter opener. ◀ The first step consists of reacting ethanol with oxygen to obtain acetaldehyde and water. If you write the chemical equation and balance it, you obtain

$$2C_2H_5OH + O_2 \longrightarrow 2C_2H_4O + 2H_2O$$

$$\text{ethanol} \qquad\qquad\qquad \text{acetaldehyde}$$

In practice, the reaction is carried out with the reactants as gases (gas phase) at about 400°C using silver as a catalyst.

The second step consists of reacting acetaldehyde with oxygen to obtain acetic acid. Acetaldehyde liquid is mixed with a catalyst—manganese(II) acetate—and air is bubbled through it. The balanced equation is

$$2C_2H_4O + O_2 \longrightarrow 2HC_2H_3O_2$$

$$\text{acetaldehyde} \qquad\qquad \text{acetic acid}$$

Once you have this balanced equation, you are in a position to answer quantitative questions such as, How much acetic acid can you obtain from a 10.0-g sample of acetaldehyde? You will see how to answer such questions in the next sections.

**Figure 3.9**
**Molecular model of acetic acid.** The hydrogen atom (blue atom at the top right) attached to the oxygen atom (red atom) is easily lost, so it is acidic. The three hydrogen atoms attached to the carbon atom (black atom at the left) are not acidic.

## CONCEPT CHECK 3.3

A friend has some questions about empirical formulas and molecular formulas. You can assume that he is good at performing the calculations.

a. For a problem that asked him to determine the empirical formula, he came up with the answer $C_2H_8O_2$. Is this a possible answer to the problem? If not, what guidance would you offer your friend?

b. For another problem he came up with the answer $C_{1.5}H_4$ as the empirical formula. Is this answer correct? Once again, if it isn't correct, what could you do to help your friend?

c. Since you have been a big help, your friend asks one more question. He completed a problem of the same type as Example 3.12. His answers indicate that the compound had an empirical formula of $C_3H_8O$ and the molecular formula $C_3H_8O$. Is this result possible?

# Stoichiometry: Quantitative Relations in Chemical Reactions

In Chapter 2, we described a chemical equation as a representation of what occurs when molecules react. We will now study chemical equations more closely to answer questions about the stoichiometry of reactions. **Stoichiometry** (pronounced "stoy-key-om′-e-tree") is *the calculation of the quantities of reactants and products involved in a chemical reaction*. It is based on the chemical equation and on the relationship between mass and moles. Such calculations are fundamental to most quantitative work in chemistry. In the next sections, we will use the industrial Haber process for the production of ammonia to illustrate stoichiometric calculations.

## 3.6 Molar Interpretation of a Chemical Equation

In the Haber process for producing ammonia, $NH_3$, nitrogen (from the atmosphere) reacts with hydrogen at high temperature and pressure.

$$N_2(g) + 3H_2(g) \longrightarrow 2NH_3(g)$$

Hydrogen is usually obtained from natural gas or petroleum and so is relatively expensive. For this reason, the price of hydrogen partly determines the price of ammonia. Thus, an important question to answer is, How much hydrogen is required to give a particular quantity of ammonia? For example, how much hydrogen would be needed to produce one ton (907 kg) of ammonia? Similar kinds of questions arise throughout chemical research and industry.

To answer such quantitative questions, you must first look at the balanced chemical equation: one $N_2$ molecule and three $H_2$ molecules react to produce two $NH_3$ molecules (see models above). A similar statement involving multiples of these numbers of molecules is also correct. For example, $6.02 \times 10^{23}$ $N_2$ molecules react with $3 \times 6.02 \times 10^{23}$ $H_2$ molecules, giving $2 \times 6.02 \times 10^{23}$ $NH_3$ molecules. This last statement can be put in molar terminology: one mole of $N_2$ reacts with three moles of $H_2$ to give two moles of $NH_3$.

▶ Video no. 4 (Series D), "Hydrogen/Oxygen Balloons," no. 5 (Series D), "The Exploding Pringles Can," no. 6 (Series D), "Ammonium Dichromate Volcano," and no. 7 (Series D), "Talesnick's Production of HCl."

> You may interpret a chemical equation either in terms of numbers of molecules (or ions or formula units) or in terms of numbers of moles, depending on your needs.

Because moles can be converted to mass, you can also give a mass interpretation of a chemical equation. The molar masses of $N_2$, $H_2$, and $NH_3$ are 28.0, 2.02, and 17.0 g/mol, respectively. Therefore 28.0 g of $N_2$ reacts with $3 \times 2.02$ g of $H_2$ to yield $2 \times 17.0$ g of $NH_3$.

We summarize these three interpretations as follows:

| $N_2$ | + | $3H_2$ | $\longrightarrow$ | $2NH_3$ | |
|---|---|---|---|---|---|
| 1 molecule $N_2$ | + | 3 molecules $H_2$ | $\longrightarrow$ | 2 molecules $NH_3$ | (molecular interpretation) |
| 1 mol $N_2$ | + | 3 mol $H_2$ | $\longrightarrow$ | 2 mol $NH_3$ | (molar interpretation) |
| 28.0 g $N_2$ | + | $3 \times 2.02$ g $H_2$ | $\longrightarrow$ | $2 \times 17.0$ g $NH_3$ | (mass interpretation) |

Suppose you ask how many grams of atmospheric nitrogen will react with 6.06 g ($3 \times 2.02$ g) of hydrogen. You see from the last equation that the answer is 28.0 g $N_2$. We formulated this question for one mole of atmospheric nitrogen. Recalling the question posed earlier, you may ask how much hydrogen (in kg) is needed to yield 907 kg of ammonia in the Haber process. The solution to this problem depends on the fact that *the number of moles involved in a reaction is proportional to the coefficients in the balanced chemical equation.* In the next section, we will describe a procedure for solving such problems.

**EXERCISE 3.13**

In an industrial process, hydrogen chloride, HCl, is prepared by burning hydrogen gas, $H_2$, in an atmosphere of chlorine, $Cl_2$. Write the chemical equation for the reaction. Below the equation, give the molecular, molar, and mass interpretations.

See Problems 3.69 and 3.70.

## 3.7 Amounts of Substances in a Chemical Reaction

You see from the preceding discussion that a balanced chemical equation relates the amounts of substances in a reaction. The coefficients in the equation can be given a molar interpretation, and using this interpretation you can, for example, calculate the moles of product obtained from any given moles of reactant. Also, you can extend this type of calculation to answer questions about masses of reactants and products.

Again consider the Haber process for producing ammonia gas. Suppose you have a mixture of $H_2$ and $N_2$, and 4.8 mol $H_2$ in this mixture reacts with $N_2$ to produce $NH_3$. How many moles of $NH_3$ can you produce from this quantity of $H_2$? In asking questions like this, you assume that the mixture contains a sufficient quantity of the other reactant ($N_2$, in this case).

▶ **This conversion factor simply expresses the fact that the mole ratio of $NH_3$ to $H_2$ in the reaction is 2 to 3.**

The balanced chemical equation for the Haber process tells you that 3 mol $H_2$ produce 2 mol $NH_3$. You can express this as a conversion factor: ◀

$$\underbrace{\frac{2 \text{ mol NH}_3}{3 \text{ mol H}_2}}_{\substack{\text{Converts from} \\ \text{mol H}_2 \text{ to mol NH}_3}}$$

Multiplying any quantity of $H_2$ by this conversion factor mathematically converts that quantity of $H_2$ to the quantity of $NH_3$ as specified by the balanced chemical equation. Note that you write the conversion factor with the quantity you are converting *from* on the bottom (3 mol $H_2$), and the quantity you are converting *to* on the top (2 mol $NH_3$).

To calculate the quantity of $NH_3$ produced from 4.8 mol $H_2$, you write 4.8 mol $H_2$ and multiply this by the preceding conversion factor:

$$4.8 \text{ mol H}_2 \times \frac{2 \text{ mol NH}_3}{3 \text{ mol H}_2} = 3.2 \text{ mol NH}_3$$

Note how the unit mol $H_2$ cancels to give the answer in mol $NH_3$.

In this calculation, you converted moles of reactant to moles of product.

$$\text{Moles reactant} \longrightarrow \text{moles product}$$

It is just as easy to calculate the moles of reactant needed to obtain the specified moles of product. You mathematically convert moles of product to moles of reactant.

$$\text{Moles product} \longrightarrow \text{moles reactant}$$

To set up the conversion factor, you refer to the balanced chemical equation and place the quantity you are converting from on the bottom and the quantity you are converting to on the top.

$$\underbrace{\frac{3 \text{ mol H}_2}{2 \text{ mol NH}_3}}_{\substack{\text{Converts from} \\ \text{mol NH}_3 \text{ to mol H}_2}}$$

Now consider the problem asked in the previous section: How much hydrogen (in kg) is needed to yield 907 kg of ammonia by the Haber process? The balanced chemical equation directly relates moles of substances, not masses. Therefore, you must first convert the mass of ammonia to moles of ammonia, then convert moles of ammonia to moles of hydrogen. Finally, you convert moles of hydrogen to mass of hydrogen.

$$\text{Mass NH}_3 \longrightarrow \text{mol NH}_3 \longrightarrow \text{mol H}_2 \longrightarrow \text{mass H}_2$$

You learned how to convert mass to moles, and vice versa, in Section 3.2.

The calculation to convert 907 kg $NH_3$, or $9.07 \times 10^5$ g $NH_3$, to mol $NH_3$ is as follows:

$$9.07 \times 10^5 \text{ g NH}_3 \times \underbrace{\frac{1 \text{ mol NH}_3}{17.0 \text{ g NH}_3}}_{\substack{\text{Converts from} \\ \text{g NH}_3 \text{ to mol NH}_3}} = 5.34 \times 10^4 \text{ mol NH}_3$$

Now you convert from moles $NH_3$ to moles $H_2$.

$$5.34 \times 10^4 \text{ mol NH}_3 \times \underbrace{\frac{3 \text{ mol H}_2}{2 \text{ mol NH}_3}}_{\substack{\text{Converts from} \\ \text{mol NH}_3 \text{ to mol H}_2}} = 8.01 \times 10^4 \text{ mol H}_2$$

Finally, you convert moles $H_2$ to grams $H_2$.

$$8.01 \times 10^4 \text{ mol H}_2 \times \underbrace{\frac{2.02 \text{ g H}_2}{1 \text{ mol H}_2}}_{\substack{\text{Converts from} \\ \text{mol H}_2 \text{ to g H}_2}} = 1.62 \times 10^5 \text{ g H}_2$$

The result says that to produce 907 kg $NH_3$, you need $1.62 \times 10^5$ g $H_2$, or 162 kg $H_2$.

Once you feel comfortable with the individual conversions, you can do this type of calculation in a single step by multiplying successively by conversion factors, as follows:

$$9.07 \times 10^5 \text{ g NH}_3 \times \frac{1 \text{ mol NH}_3}{17.0 \text{ g NH}_3} \times \frac{3 \text{ mol H}_2}{2 \text{ mol NH}_3} \times \frac{2.02 \text{ g H}_2}{1 \text{ mol H}_2} =$$
$$1.62 \times 10^5 \text{ g H}_2 \text{ (or 162 kg H}_2)$$

Note that the unit in the denominator of each conversion factor cancels the unit in the numerator of the preceding factor. Figure 3.11 illustrates this calculation diagrammatically.

The following two examples illustrate additional variations of this type of calculation.

**Figure 3.11**
**Steps in a stoichiometric calculation.** You convert the mass of substance A in a reaction to moles of substance A, then to moles of another substance B, and finally to mass of substance B.

► An alternate example is provided in the Instructor's Resource Manual.

**Figure 3.12**
**Hematite.** The name of this iron mineral stems from the Greek word for blood, which alludes to the color of certain forms of the mineral.

**Figure 3.13**
**Preparation of chlorine.**
Concentrated hydrochloric acid was added to manganese dioxide in the beaker. Note the formation of yellowish-green gas (chlorine).

## EXAMPLE 3.13

### Relating the Quantity of Reactant to Quantity of Product

Hematite, $Fe_2O_3$, is an important ore of iron; see Figure 3.12. (An ore is a natural substance from which the metal can be profitably obtained.) The free metal is obtained by reacting hematite with carbon monoxide, CO, in a blast furnace. Carbon monoxide is formed in the furnace by partial combustion of carbon. The reaction is

$$Fe_2O_3(s) + 3CO(g) \longrightarrow 2Fe(s) + 3CO_2(g)$$

How many grams of iron can be produced from 1.00 kg $Fe_2O_3$?

 **PROBLEM STRATEGY**

This calculation involves the conversion of a quantity of $Fe_2O_3$ to a quantity of Fe. In performing this calculation, we assume that there is enough CO for the complete reaction of the $Fe_2O_3$. An essential feature of this type of calculation is that you must use the information from the balanced chemical equation to convert from the moles of a given substance to the moles of another substance. Therefore, you first convert the mass of $Fe_2O_3$ (1.00 kg $Fe_2O_3$ = $1.0 \times 10^3$ g $Fe_2O_3$) to moles of $Fe_2O_3$. Then, using the relationship from the balanced chemical equation, you convert moles of $Fe_2O_3$ to moles of Fe. Finally, you convert the moles of Fe to grams of Fe.

**SOLUTION**

The calculation is as follows:

$$1.00 \times 10^3 \text{ g } Fe_2O_3 \times \frac{1 \text{ mol } Fe_2O_3}{160 \text{ g } Fe_2O_3} \times \frac{2 \text{ mol } Fe}{1 \text{ mol } Fe_2O_3} \times \frac{55.8 \text{ g } Fe}{1 \text{ mol } Fe} = \textbf{698 g Fe}$$

## EXERCISE 3.14

Sodium is a soft, reactive metal that instantly reacts with water to give hydrogen gas and a solution of sodium hydroxide, NaOH. How many grams of sodium metal are needed to give 7.81 g of hydrogen by this reaction? (Remember to write the balanced equation first.)

*See Problems 3.75, 3.76, 3.77, and 3.78.*

## EXAMPLE 3.14

### Relating the Quantities of Two Reactants (or Two Products)

Today chlorine is prepared from sodium chloride by electrochemical decomposition. Formerly chlorine was produced by heating hydrochloric acid with pyrolusite (manganese dioxide, $MnO_2$), a common manganese ore. Small amounts of chlorine may be prepared in the laboratory by the same reaction (see Figure 3.13):

$$4HCl(aq) + MnO_2(s) \longrightarrow 2H_2O(l) + MnCl_2(aq) + Cl_2(g)$$

How many grams of HCl react with 5.00 g of manganese dioxide, according to this equation?

▶ An alternate example is
provided in the Instructor's
Resource Manual.

**Problem Solving:
Chemical Stoichiometry**

**SOLUTION**

You write what is given (5.00 g $MnO_2$) and convert this to moles, then to moles of what is desired (mol HCl). Finally you convert this to mass (g HCl). The calculation is:

$$5.00 \text{ g } MnO_2 \times \frac{1 \text{ mol } MnO_2}{86.9 \text{ g } MnO_2} \times \frac{4 \text{ mol HCl}}{1 \text{ mol } MnO_2} \times \frac{36.5 \text{ g HCl}}{1 \text{ mol HCl}} = \textbf{8.40 g HCl}$$

**EXERCISE 3.15**

Sphalerite is a zinc sulfide (ZnS) mineral and an important commercial source of zinc metal. The first step in the processing of the ore consists of heating the sulfide with oxygen to give zinc oxide, ZnO, and sulfur dioxide, $SO_2$. How many kilograms of oxygen gas combine with $5.00 \times 10^3$ g of zinc sulfide in this reaction? (You must first write the balanced chemical equation.)

*See Problems 3.79 and 3.80.*

**EXERCISE 3.16**

The British chemist Joseph Priestley prepared oxygen in 1774 by heating mercury(II) oxide, HgO. Mercury metal is the other product. If 6.47 g of oxygen is collected, how many grams of mercury metal are also produced?

*See Problems 3.81 and 3.82.*

 **CONCEPT CHECK 3.4**

The main reaction of a charcoal grill is $C(s) + O_2(g) \longrightarrow CO_2(g)$. Which of the statements below are incorrect? Why?

a. 1 atom of carbon reacts with 1 molecule of oxygen to produce 1 molecule of $CO_2$.

b. 1 g of C reacts with 1 g of $O_2$ to produce 2 grams of $CO_2$.

c. 1 g of C reacts with 0.5 g of $O_2$ to produce 1 g of $CO_2$.

d. 12 g of C reacts with 32 g of $O_2$ to produce 44 g of $CO_2$.

e. 1 mol of C reacts with 1 mol of $O_2$ to produce 1 mol of $CO_2$.

f. 1 mol of C reacts with 0.5 mol of $O_2$ to produce 1 mol of $CO_2$.

## 3.8 Limiting Reactant; Theoretical and Percentage Yields

▶ **Experiments in General Chemistry, 3A, "The Empirical Formula of an Oxide."**

Often reactants are added to a reaction vessel in amounts different from the molar proportions given by the chemical equation. In such cases, only one of the reactants may be completely consumed at the end of the reaction, whereas some amounts of other reactants will remain unreacted. The **limiting reactant** (or **limiting reagent**) is *the reactant that is entirely consumed when a reaction goes to completion.* A reactant that is not completely consumed is often referred to as an *excess reactant.* Once one of the reactants is used up, the reaction stops. This means that:

> The moles of product are always determined by the starting moles of limiting reactant.

A couple of analogies may help you understand the limiting reactant problem. Suppose you want to make some cheese sandwiches. Each is made from two slices of bread and a slice of cheese. Let's write that in the form of a chemical equation:

$$2 \text{ slices bread } + 1 \text{ slice cheese } \longrightarrow 1 \text{ sandwich}$$

You look in the kitchen and see that you have six slices of bread and two slices of cheese. The six slices of bread would be enough to make three sandwiches if you had enough cheese. The two slices of cheese would be enough to make two sandwiches if you had enough bread. How many sandwiches can you make? What you will find is that once you have made two sandwiches, you will be out of cheese (Figure 3.14). Cheese is the "limiting reactant" in the language of chemistry. If you look at each "reactant," and ask how much "product" you can make from it, the reactant that limits your amount of product is called the limiting reactant.

Here is another analogy. Suppose you are supervising the assembly of automobiles. Your plant has in stock 300 steering wheels and 900 tires, plus an excess of

**Figure 3.14**
**Limiting reactant analogy using cheese sandwiches.** Start with six slices of bread, two slices of cheese, and the sandwich-making equation. Even though you have extra bread you are limited to making two sandwiches by the amount of cheese you have on hand. Cheese is the limiting reactant.

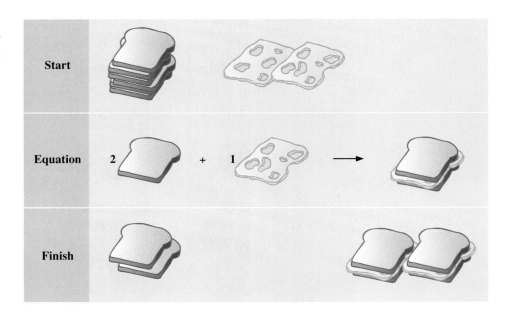

every other needed component. How many autos can you assemble from this stock? Here is the "balanced equation" for the auto assembly:

$$1 \text{ steering wheel } + 4 \text{ tires } + \text{ other components } \longrightarrow 1 \text{ auto}$$

One way to solve this problem is to calculate the number of autos that you could assemble from each component. In this case, you only need to concentrate on the effects of the steering wheels and tires since production will not be limited by the other, excess components. Looking at the equation, you can determine that from 300 steering wheels, you could assemble 300 autos; from 900 tires you could assemble $900 \div 4 = 225$ autos.

How many autos can you produce, 225 or 300? Note that by the time you have assembled 225 autos, you will have exhausted your stock of tires, so no more autos can be assembled. Tires are the "limiting reactant," since they limit the number of automobiles that you can actually assemble (225).

You can set up these calculations in the same way for a chemical reaction. First, you calculate the numbers of autos you could assemble from the number of steering wheels and from the number of tires available, and then you compare the results. From the balanced equation, you can see that one steering wheel is equivalent to one auto, and four tires are equivalent to one auto. Therefore:

$$300 \text{ steering wheels} \times \frac{1 \text{ auto}}{1 \text{ steering wheel}} = 300 \text{ autos}$$

$$900 \text{ tires} \times \frac{1 \text{ auto}}{4 \text{ tires}} = 225 \text{ autos}$$

Comparing the numbers of autos produced (225 autos versus 300 autos), you conclude that tires are the limiting component. (The component producing the fewer autos is the limiting component.) Note that apart from the units (or factor labels), the calculation is identical with what was done previously.

Now consider a chemical reaction, the burning of hydrogen in oxygen.

$$2H_2(g) + O_2(g) \longrightarrow 2H_2O(g)$$

Suppose you put 1 mol $H_2$ and 1 mol $O_2$ into a reaction vessel. How many moles of $H_2O$ will be produced? First, you note that 2 mol $H_2$ produces 2 mol $H_2O$ and that 1 mol $O_2$ produces 2 mol $H_2O$. Now you calculate the moles of $H_2O$ that you could produce from each quantity of reactant, assuming that there is sufficient other reactant.

$$1 \text{ mol } H_2 \times \frac{2 \text{ mol } H_2O}{2 \text{ mol } H_2} = 1 \text{ mol } H_2O$$

$$1 \text{ mol } O_2 \times \frac{2 \text{ mol } H_2O}{1 \text{ mol } O_2} = 2 \text{ mol } H_2O$$

Comparing these results, you see that hydrogen, $H_2$, yields the least amount of product, so it must be the limiting reactant. By the time 1 mol $H_2O$ is produced, all of the hydrogen is used up; the reaction stops. Oxygen, $O_2$, is the excess reactant.

We can summarize the limiting-reactant problem as follows. Suppose you are given the amounts of reactants added to a vessel, and you wish to calculate the amount of product obtained when the reaction is complete. Unless you know that the reactants have been added in the molar proportions given by the chemical equation, the problem is twofold: (1) you must first identify the limiting reactant; (2) you then calculate the amount of product from the amount of limiting reactant. The next examples illustrate the steps.

► An alternate example is provided in the Instructor's Resource Manual.

### EXAMPLE 3.15

## Calculating with a Limiting Reactant (Involving Moles)

Zinc metal reacts with hydrochloric acid by the following reaction:

$$Zn(s) + 2HCl(aq) \longrightarrow ZnCl_2(aq) + H_2(g)$$

If 0.30 mol Zn is added to hydrochloric acid containing 0.52 mol HCl, how many moles of $H_2$ are produced?

 **PROBLEM STRATEGY**

STEP 1 Which is the limiting reactant? To answer this, using the relationship from the balanced chemical equation, you take each reactant in turn and ask how much product ($H_2$) would be obtained if each were totally consumed. The reactant that gives the smaller amount of product is the limiting reactant. (Remember how you obtained the limiting component in the auto-assembly analogy.)

STEP 2 You obtain the amount of product actually obtained from the amount of limiting reactant.

**SOLUTION**

STEP 1

$$0.30 \text{ mol Zn} \times \frac{1 \text{ mol } H_2}{1 \text{ mol Zn}} = 0.30 \text{ mol } H_2$$

$$0.52 \text{ mol HCl} \times \frac{1 \text{ mol } H_2}{2 \text{ mol HCl}} = 0.26 \text{ mol } H_2$$

You see that hydrochloric acid must be the limiting reactant and that some zinc must be left unconsumed. (Zinc is the excess reactant.)

STEP 2 Since HCl is the limiting reactant, the amount of $H_2$ produced must be **0.26 mol.**

### EXERCISE 3.17

Aluminum chloride, $AlCl_3$, is used as a catalyst in various industrial reactions. It is prepared from hydrogen chloride gas and aluminum metal shavings.

$$2Al(s) + 6HCl(g) \longrightarrow 2AlCl_3(s) + 3H_2(g)$$

Suppose a reaction vessel contains 0.15 mol Al and 0.35 mol HCl. How many moles of $AlCl_3$ can be prepared from this mixture?

*See Problems 3.83 and 3.84.*

► An alternate example is provided in the Instructor's Resource Manual.

### EXAMPLE 3.16

## Calculating with a Limiting Reactant (Involving Masses)

In a process for producing acetic acid, oxygen gas is bubbled into acetaldehyde, $CH_3CHO$, containing manganese(II) acetate (catalyst) under pressure at 60°C.

$$2CH_3CHO(l) + O_2(g) \longrightarrow 2HC_2H_3O_2(l)$$

In a laboratory test of this reaction, 20.0 g $CH_3CHO$ and 10.0 g $O_2$ were put into a reaction vessel. a. How many grams of acetic acid can be produced by this reaction from these amounts of reactants? b. How many grams of the excess reactant remain after the reaction is complete?

### ▶ PROBLEM STRATEGY

a. This part is similar to the preceding example, but now you must convert grams of each reactant (acetaldehyde and oxygen) to moles of product (acetic acid). From these results, you decide which is the limiting reactant and the moles of product obtained, which you convert to grams of product.

b. From the moles of product, you calculate the grams of the excess reactant needed in the reaction. Then you subtract this quantity from the grams of the excess reactant available to find the quantity of excess reactant remaining.

### SOLUTION

a. How much acetic acid is produced?

STEP 1   To determine which reactant is limiting, you convert grams of each reactant (20.0 g $CH_3CHO$ and 10.0 g $O_2$) to moles of product, $HC_2H_3O_2$. Acetaldehyde has a molar mass of 44.1 g/mol, and oxygen has a molar mass of 32.0 g/mol.

$$20.0 \text{ g } CH_3CHO \times \frac{1 \text{ mol } CH_3CHO}{44.1 \text{ g } CH_3CHO} \times \frac{2 \text{ mol } HC_2H_3O_2}{2 \text{ mol } CH_3CHO} = 0.454 \text{ mol } HC_2H_3O_2$$

$$10.0 \text{ g } O_2 \times \frac{1 \text{ mol } O_2}{32.0 \text{ g } O_2} \times \frac{2 \text{ mol } HC_2H_3O_2}{1 \text{ mol } O_2} = 0.625 \text{ mol } HC_2H_3O_2$$

Thus, acetaldehyde, $CH_3CHO$, is the limiting reactant, so 0.454 mol $HC_2H_3O_2$ was produced.

STEP 2   You convert 0.454 mol $HC_2H_3O_2$ to grams of $HC_2H_3O_2$.

$$0.454 \text{ mol } HC_2H_3O_2 \times \frac{60.1 \text{ g } HC_2H_3O_2}{1 \text{ mol } HC_2H_3O_2} = \textbf{27.3 g } HC_2H_3O_2$$

b. How much of the excess reactant (oxygen) was left over? You convert the moles of acetic acid to grams of oxygen (the quantity of oxygen needed to produce this amount of acetic acid).

$$0.454 \text{ mol } HC_2H_3O_2 \times \frac{1 \text{ mol } O_2}{2 \text{ mol } HC_2H_3O_2} \times \frac{32.0 \text{ g } O_2}{1 \text{ mol } O_2} = 7.26 \text{ g } O_2$$

You started with 10.0 g $O_2$, so the quantity remaining is

$$(10.0 - 7.26) \text{ g } O_2 = \textbf{2.7 g } O_2 \qquad \text{(mass remaining)}$$

**Figure 3.15**
**Reaction of zinc and sulfur.**   When a hot nail is stuck into a pile of zinc and sulfur, a fiery reaction occurs and zinc sulfide forms.

**Problem Solving: Stoichiometry: Limiting Reactant**

### EXERCISE 3.18

In an experiment, 7.36 g of zinc was heated with 6.45 g of sulfur (Figure 3.15). Assume that these substances react according to the equation

$$8Zn + S_8 \longrightarrow 8ZnS$$

What amount of zinc sulfide was produced?

See Problems 3.85 and 3.86.

The **theoretical yield** of product is *the maximum amount of product that can be obtained by a reaction from given amounts of reactants*. It is the amount that you calculate from the stoichiometry based on the limiting reactant. In Example 3.16, the theoretical yield of acetic acid is 27.3 g. In practice, the *actual yield* of a product may be much less for several possible reasons. First, some product may be lost during the process of separating it from the final reaction mixture. Second, there may be other, competing reactions that occur simultaneously with the reactant on which the theoretical yield is based. Finally, many reactions appear to stop before they reach completion; they give mixtures of reactants and products. ◄

▶ Such reactions reach chemical equilibrium. We will discuss equilibrium quantitatively in Chapter 15.

It is important to know the actual yield from a reaction in order to make economic decisions about a preparation method. The reactants for a given method may not be too costly per kilogram, but if the actual yield is very low, the final cost can be very high. The **percentage yield** of product is *the actual yield (experimentally determined) expressed as a percentage of the theoretical yield (calculated)*.

$$\text{Percentage yield} = \frac{\text{actual yield}}{\text{theoretical yield}} \times 100\%$$

To illustrate the calculation of percentage yield, recall that the theoretical yield of acetic acid calculated in Example 3.16 was 27.3 g. If the actual yield of acetic acid obtained in an experiment, using the amounts of reactants given in Example 3.16, is 23.8 g, then

$$\text{Percentage yield of } HC_2H_3O_2 = \frac{23.8 \ g}{27.3 \ g} \times 100\% = 87.2\%$$

---

**EXERCISE 3.19**

▶ Two alternate exercises are provided in the Instructor's Resource Manual.

New industrial plants for acetic acid react liquid methanol with carbon monoxide in the presence of a catalyst.

$$CH_3OH(l) + CO(g) \longrightarrow HC_2H_3O_2(l)$$

In an experiment, 15.0 g of methanol and 10.0 g of carbon monoxide were placed in a reaction vessel. What is the theoretical yield of acetic acid? If the actual yield is 19.1 g, what is the percentage yield?

---

**See Problems 3.89 and 3.90.**

 **CONCEPT CHECK 3.5**

You perform the hypothetical reaction of an element, $X_2(g)$, with another element, $Y(g)$, to produce $XY(g)$.

a. Write the balanced chemical equation for the reaction.

b. If $X_2$ and $Y$ were mixed in the quantities shown in the container on the next page and allowed to react, which of the following is the correct representation of the contents of the container after the reaction has occurred?

c. Using the information presented in part b, identify the limiting reactant.

**Before reaction:**

**After reaction:**

● = Atom X

◉ = Atom Y

Option #1    or    Option #2    or    Option #3

## A Checklist for Review

### Important Terms

molecular weight (3.1)
formula weight (3.1)
mole (mol) (3.2)
Avogadro's number
   ($N_A$) (3.2)

molar mass (3.2)
percentage composition
   (p. 98)
mass percentage (3.3)

empirical (simplest)
   formula (3.5)
stoichiometry (p. 107)

limiting reactant
   (reagent) (3.8)
theoretical yield (3.8)
percentage yield (3.8)

### Key Equations

$$\text{Mass \% } A = \frac{\text{mass of } A \text{ in the whole}}{\text{mass of the whole}} \times 100\%$$

$$\text{Percentage yield} = \frac{\text{actual yield}}{\text{theoretical yield}} \times 100\%$$

$$n = \frac{\text{molecular weight}}{\text{empirical formula weight}}$$

### Summary of Facts and Concepts

A *formula weight* equals the sum of the atomic weights of the atoms in the formula of a compound. If the formula corresponds to that of a molecule, this sum of atomic weights equals the *molecular weight* of the compound. The mass of *Avogadro's number* ($6.02 \times 10^{23}$) of formula units—that is, the mass of one *mole* of substance—equals the mass in grams that corresponds to the numerical value of the formula weight in amu. This mass is called the *molar mass*.

The *empirical formula (simplest formula)* of a compound is obtained from the *percentage composition* of the substance, which is expressed as *mass percentages* of the elements. To calculate the empirical formula, you convert mass percentages

to ratios of moles, which, when expressed in smallest whole numbers, give the subscripts in the formula. A molecular formula is a multiple of the empirical formula; this multiple is determined from the experimental value of the molecular weight.

A chemical equation may be interpreted in terms of moles of reactants and products, as well as in terms of molecules. Using this *molar interpretation,* you can convert from the mass of one substance in a chemical equation to the mass of another. The maximum amount of product from a reaction is determined by the *limiting reactant,* the reactant that is completely used up; the other reactants are in excess.

## Operational Skills

*Note:* A table of atomic weights is necessary for most of these skills.

1. **Calculating the formula weight from a formula or molecular model** Given the formula of a compound and a table of atomic weights, calculate the formula weight. **(EXAMPLES 3.1, 3.2)**
2. **Calculating the mass of an atom or molecule** Using the molar mass and Avogadro's number, calculate the mass of an atom or molecule in grams. **(EXAMPLE 3.3)**
3. **Converting moles of substance to grams, and vice versa** Given the moles of a compound with a known formula, calculate the mass **(EXAMPLE 3.4)**. Or, given the mass of a compound with a known formula, calculate the moles. **(EXAMPLE 3.5)**
4. **Calculating the number of molecules in a given mass** Given the mass of a sample of a molecular substance and its formula, calculate the number of molecules in the sample. **(EXAMPLE 3.6)**
5. **Calculating the percentage composition from the formula** Given the formula of a compound, calculate the mass percentages of the elements in it. **(EXAMPLE 3.7)**
6. **Calculating the mass of an element in a given mass of compound** Given the mass percentages of elements in a given mass of a compound, calculate the mass of any element. **(EXAMPLE 3.8)**
7. **Calculating the percentages of C and H by combustion** Given the masses of $CO_2$ and $H_2O$ obtained from the combustion of a known mass of a compound of C, H, and O, compute the mass percentage of each element. **(EXAMPLE 3.9)**
8. **Determining the empirical formula from percentage composition** Given the masses of elements in a known mass of compound, or given its percentage composition, obtain the empirical formula. **(EXAMPLES 3.10, 3.11)**
9. **Determining the molecular formula from percentage composition and molecular weight** Given the empirical formula and molecular weight of a substance, obtain its molecular formula. **(EXAMPLE 3.12)**
10. **Relating quantities in a chemical equation** Given a chemical equation and the amount of one substance, calculate the amount of another substance involved in the reaction. **(EXAMPLES 3.13, 3.14)**
11. **Calculating with a limiting reactant** Given the amounts of reactants and the chemical equation, find the limiting reactant; then calculate the amount of a product. **(EXAMPLES 3.15, 3.16)**

## Review Questions

**3.1** What is the difference between a formula weight and a molecular weight? Could a given substance have both a formula weight and a molecular weight?

**3.2** Describe in words how to obtain the formula weight of a compound from the formula.

**3.3** One mole of $N_2$ contains how many $N_2$ molecules? How many N atoms are there in one mole of $N_2$? One mole of iron(III) sulfate, $Fe_2(SO_4)_3$, contains how many moles of $SO_4^{2-}$ ions? how many moles of O atoms?

**3.4** Explain what is involved in determining the composition of a compound of C, H, and O by combustion.

**3.5** Explain what is involved in obtaining the empirical formula from the percentage composition.

**3.6** A substance has the molecular formula $C_6H_{12}O_2$. What is its empirical formula?

**3.7** Hydrogen peroxide has the empirical formula HO and an empirical formula weight of 17.0 amu. If the molecular weight is 34.0 amu, what is the molecular formula?

**3.8** Describe in words the meaning of the equation

$$CH_4 + 2O_2 \longrightarrow CO_2 + 2H_2O$$

using a molecular, a molar, and then a mass interpretation.

**3.9** Explain how a chemical equation can be used to relate the masses of different substances involved in a reaction.

**3.10** What is a limiting reactant in a reaction mixture? Explain how it determines the amount of product.

**3.11** Come up with some examples of limiting reactants that use the concept but don't involve chemical reactions.

**3.12** Explain why it is impossible to have a theoretical yield of more than 100%.

## Conceptual Problems

**3.13** You react nitrogen and hydrogen in a container to produce ammonia, $NH_3(g)$. The following figure depicts the contents of the container after the reaction is complete.

= $NH_3$

= $N_2$

a. Write a balanced chemical equation for the reaction.
b. What is the limiting reactant?
c. How many molecules of the limiting reactant would you need to add to the container in order to have a complete reaction (convert all reactants to products)?

**3.14** Propane, $C_3H_8$, is the fuel of choice in a gas barbecue. When burning, the balanced equation is

$$C_3H_8 + 5O_2 \longrightarrow 3CO_2 + 4H_2O$$

a. What is the limiting reactant when cooking with a gas grill?
b. If the grill will not light and you know that you have an ample flow of propane to the burner, what is the limiting reactant?
c. When using a gas grill you can sometimes turn the gas up to the point at which the flame becomes yellow and smokey. In terms of the chemical reaction, what is happening?

**3.15** A critical point to master in becoming proficient at solving problems is evaluating whether or not your answer is reasonable. A friend asks you to look over her homework to see if she has done the calculations correctly. Shown below are descriptions of some of her answers. Without using your calculator or doing calculations on paper, see if you can judge the answers below as being reasonable or ones that will require her to go back and work the problems again.

a. 0.33 mole of an element has a mass of $1.0 \times 10^{-3}$ g.
b. The mass of one molecule of water is $1.80 \times 10^{-10}$ g.
c. There are $3.01 \times 10^{23}$ atoms of Na in 0.500 mol of Na.
d. The molar mass of $CO_2$ is 44.0 kg/mol.

**3.16** An exciting, and often loud, chemical demonstration involves the simple reaction of hydrogen gas and oxygen gas to produce water vapor:

$$2H_2(g) + O_2(g) \longrightarrow 2H_2O(g)$$

The reaction is carried out in soap bubbles or balloons that are filled with the reactant gases. We get the reaction to proceed by igniting the bubbles or balloons. The more $H_2O$ that is formed during the reaction, the bigger the bang. Explain the following observations.

a. A bubble containing just $H_2$ makes a quiet "fffft" sound when ignited.
b. When a bubble containing equal amounts of $H_2$ and $O_2$ is ignited, a sizable bang results.
c. When a bubble containing a ratio of 2 to 1 in the amounts of $H_2$ and $O_2$ is ignited, the loudest bang results.
d. When a bubble containing just $O_2$ is ignited, virtually no sound is made.

**3.17** High cost and limited availability of a reactant often dictate which reactant is limiting in a particular process. Identify the limiting reactant when running the reactions below and come up with a reason to support your decision.

a. Burning charcoal on a grill:

$$C(s) + O_2(g) \longrightarrow CO_2(g)$$

b. Burning a chunk of Mg in water:

$$Mg(s) + 2H_2O(l) \longrightarrow Mg(OH)_2(aq) + H_2(g)$$

c. The Haber process of ammonia production:

$$3H_2(g) + N_2(g) \longrightarrow 2NH_3(g)$$

**3.18** A few hydrogen and oxygen molecules are introduced into a container in the quantities depicted in the following drawing. The gases are then ignited by a spark causing them to react and form $H_2O$.

= $H_2$

= $O_2$

a. What is the maximum number of water molecules that can be formed in the chemical reaction?
b. Draw a molecular level representation of the container's contents after the chemical reaction.

# Practice Problems

## Formula Weights and Mole Calculations

**3.19**    Find the formula weights of the following substances to three significant figures.
a.  methanol, $CH_3OH$
b.  phosphorus trichloride, $PCl_3$
c.  potassium carbonate, $K_2CO_3$
d.  nickel phosphate, $Ni_3(PO_4)_2$

**3.20**    Find the formula weights of the following substances to three significant figures.
a.  acetic acid, $HC_2H_3O_2$
b.  phosphorus pentachloride, $PCl_5$
c.  potassium sulfate, $K_2SO_4$
d.  calcium hydroxide, $Ca(OH)_2$

**3.21**    Calculate the formula weight of the following molecules to three significant figures.

(a)                    (b)

**3.22**    Calculate the formula weight of the following molecules to three significant figures.

(a)                    (b)

**3.23**    Ammonium nitrate, $NH_4NO_3$, is used as a nitrogen fertilizer and in explosives. What is the molar mass of $NH_4NO_3$?

**3.24**    Phosphoric acid, $H_3PO_4$, is used to make phosphate fertilizers and detergents and is also used in carbonated beverages. What is the molar mass of $H_3PO_4$?

**3.25**    Calculate the mass (in grams) of each of the following species.
a.  Na atom                    b.  S atom
c.  $CH_3Cl$ molecule          d.  $Hg(NO_3)_2$ formula unit

**3.26**    Calculate the mass (in grams) of each of the following species.
a.  Fe atom                    b.  F atom
c.  $N_2O$ molecule            d.  $Al(OH)_3$ formula unit

**3.27**    Diethyl ether, $(C_2H_5)_2O$, commonly known as ether, is used as an anesthetic. What is the mass in grams of a molecule of diethyl ether?

**3.28**    Glycerol, $C_3H_8O_3$, is used as a moistening agent for candy and is the starting material for nitroglycerin. Calculate the mass of a glycerol molecule in grams.

**3.29**    Calculate the mass in grams of the following.
a.  0.15 mol Na                b.  0.594 mol S
c.  2.78 mol $CH_2Cl_2$        d.  38 mol $Na_2SO_3$

**3.30**    Calculate the mass in grams of the following.
a.  0.205 mol Fe               b.  0.79 mol F
c.  5.8 mol $CO_2$             d.  48.1 mol $K_2CrO_4$

**3.31**    Boric acid, $H_3BO_3$, is a mild antiseptic and is often used as an eyewash. A sample contains 0.543 mol $H_3BO_3$. What is the mass of boric acid in the sample?

**3.32**    Carbon disulfide, $CS_2$, is a colorless, highly flammable liquid used in the manufacture of rayon and cellophane. A sample contains 0.0205 mol $CS_2$. Calculate the mass of carbon disulfide in the sample.

**3.33**    Obtain the moles of substance in the following.
a.  2.86 g C                   b.  7.05 g $Cl_2$
c.  76 g $C_4H_{10}$           d.  26.2 g $Li_2CO_3$

**3.34**    Obtain the moles of substance in the following.
a.  2.57 g As                  b.  7.83 g $S_8$
c.  41.4 g $N_2H_4$            d.  227 g $Al_2(SO_4)_3$

**3.35**    Calcium sulfate, $CaSO_4$, is a white, crystalline powder. Gypsum is a mineral, or natural substance, that is a hydrate of calcium sulfate. A 1.000-g sample of gypsum contains 0.791 g $CaSO_4$. How many moles of $CaSO_4$ are there in this sample? Assuming that the rest of the sample is water, how many moles of $H_2O$ are there in the sample? Show that the result is consistent with the formula $CaSO_4 \cdot 2H_2O$.

**3.36**    A 1.547-g sample of blue copper(II) sulfate pentahydrate, $CuSO_4 \cdot 5H_2O$, is heated carefully to drive off the water. The white crystals of $CuSO_4$ that are left behind have a mass of 0.989 g. How many moles of $H_2O$ were in the original sample? Show that the relative molar amounts of $CuSO_4$ and $H_2O$ agree with the formula of the hydrate.

**3.37**    Calculate the following.
a.  number of atoms in 8.21 g Li
b.  number of atoms in 32.0 g $Br_2$
c.  number of molecules in 62 g $NH_3$
d.  number of formula units in 201 g $PbCrO_4$
e.  number of $SO_4^{2-}$ ions in 14.3 g $Cr_2(SO_4)_3$

**3.38** Calculate the following.
a. number of atoms in 25.7 g Al
b. number of atoms in 8.71 g $I_2$
c. number of molecules in 14.9 g $N_2O_5$
d. number of formula units in 3.31 g $NaClO_4$
e. number of $Ca^{2+}$ ions in 3.91 g $Ca_3(PO_4)_2$

**3.39** Carbon tetrachloride is a colorless liquid used in the manufacture of fluorocarbons and as an industrial solvent. How many molecules are there in 7.58 mg of carbon tetrachloride?

**3.40** Chlorine trifluoride is a colorless, reactive gas used in nuclear fuel reprocessing. How many molecules are there in a 5.88-mg sample of chlorine trifluoride?

## Mass Percentage

**3.41** A 1.836-g sample of coal contains 1.584 g C. Calculate the mass percentage of C in the coal.

**3.42** A 5.69-g aqueous solution of isopropyl alcohol contains 3.98 g of isopropyl alcohol. What is the mass percentage of isopropyl alcohol in the solution?

**3.43** Phosphorus oxychloride is the starting compound for preparing substances used as flame retardants for plastics. An 8.53-mg sample of phosphorus oxychloride contains 1.72 mg of phosphorus. What is the mass percentage of phosphorus in the compound?

**3.44** Ethyl mercaptan is an odorous substance added to natural gas to make leaks easily detectable. A sample of ethyl mercaptan weighing 3.17 mg contains 1.64 mg of sulfur. What is the mass percentage of sulfur in the substance?

**3.45** A fertilizer is advertised as containing 15.8% nitrogen (by mass). How much nitrogen is there in 4.15 kg of fertilizer?

**3.46** Seawater contains 0.0065% (by mass) of bromine. How many grams of bromine are there in 1.00 L of seawater? The density of seawater is 1.025 g/cm$^3$.

**3.47** A sample of an alloy of aluminum contains 0.0898 mol Al and 0.0381 mol Mg. What are the mass percentages of Al and Mg in the alloy?

**3.48** A sample of gas mixture from a neon sign contains 0.0856 mol Ne and 0.0254 mol Kr. What are the mass percentages of Ne and Kr in the gas mixture?

## Chemical Formulas

**3.49** Calculate the percentage composition for each of the following compounds (three significant figures).
a. CO          b. $CO_2$
c. $NaH_2PO_4$   d. $Co(NO_3)_2$

**3.50** Calculate the percentage composition for each of the following compounds (three significant figures).
a. NO          b. $H_2O$
c. $KClO_4$      d. $Mn(NO_2)_2$

**3.51** Calculate the mass percentage of each element in toluene, represented by the following molecular model.

**3.52** Calculate the mass percentage of each element in 2-propanol, represented by the following molecular model.

**3.53** Which contains more carbon, 6.01 g of glucose, $C_6H_{12}O_6$, or 5.85 g of ethanol, $C_2H_6O$?

**3.54**  Which contains more sulfur, 40.8 g of calcium sulfate, $CaSO_4$, or 35.2 g of sodium sulfite, $Na_2SO_3$?

**3.55**  Ethylene glycol is used as an automobile antifreeze and in the manufacture of polyester fibers. The name glycol stems from the sweet taste of this poisonous compound. Combustion of 6.38 mg of ethylene glycol gives 9.06 mg $CO_2$ and 5.58 mg $H_2O$. The compound contains only C, H, and O. What are the mass percentages of the elements in ethylene glycol?

**3.56**  Phenol, commonly known as carbolic acid, was used by Joseph Lister as an antiseptic for surgery in 1865. Its principal use today is in the manufacture of phenolic resins and plastics. Combustion of 5.23 mg of phenol yields 14.67 mg $CO_2$ and 3.01 mg $H_2O$. Phenol contains only C, H, and O. What is the percentage of each element in this substance?

**3.57**  An oxide of osmium (symbol Os) is a pale yellow solid. If 2.89 g of the compound contains 2.16 g of osmium, what is its empirical formula?

**3.58**  An oxide of tungsten (symbol W) is a bright yellow solid. If 5.34 g of the compound contains 4.23 g of tungsten, what is its empirical formula?

**3.59**  Potassium manganate is a dark green, crystalline substance whose composition is 39.6% K, 27.9% Mn, and 32.5% O, by mass. What is its empirical formula?

**3.60**  Hydroquinone, used as a photographic developer, is 65.4% C, 5.5% H, and 29.1% O, by mass. What is the empirical formula of hydroquinone?

**3.61**  Acrylic acid, used in the manufacture of acrylic plastics, has the composition 50.0% C, 5.6% H, and 44.4% O. What is its empirical formula?

**3.62**  Malonic acid is used in the manufacture of barbiturates (sleeping pills). The composition of the acid is 34.6% C, 3.9% H, and 61.5% O. What is malonic acid's empirical formula?

**3.63**  Two compounds have the same composition: 92.25% C and 7.75% H.
a.  Obtain the empirical formula corresponding to this composition.
b.  One of the compounds has a molecular weight of 52.03 amu; the other, of 78.05 amu. Obtain the molecular formulas of both compounds.

**3.64**  Two compounds have the same composition: 85.62% C and 14.38% H.
a.  Obtain the empirical formula corresponding to this composition.
b.  One of the compounds has a molecular weight of 28.03 amu; the other, of 56.06 amu. Obtain the molecular formulas of both compounds.

**3.65**  Putrescine, a substance produced by decaying animals, has the empirical formula $C_2H_6N$. Several determinations of molecular weight give values in the range of 87 to 90 amu. Find the molecular formula of putrescine.

**3.66**  Compounds of boron with hydrogen are called boranes. One of these boranes has the empirical formula $BH_3$ and a molecular weight of 28 amu. What is its molecular formula?

**3.67**  Oxalic acid is a toxic substance used by laundries to remove rust stains. Its composition is 26.7% C, 2.2% H, and 71.1% O (by mass), and its molecular weight is 90 amu. What is its molecular formula?

**3.68**  Adipic acid is used in the manufacture of nylon. The composition of the acid is 49.3% C, 6.9% H, and 43.8% O (by mass), and the molecular weight is 146 amu. What is the molecular formula?

## Stoichiometry: Quantitative Relations in Reactions

**3.69**  Ethylene, $C_2H_4$, burns in oxygen to give carbon dioxide, $CO_2$, and water. Write the equation for the reaction, giving molecular, molar, and mass interpretations below the equation.

**3.70**  Hydrogen sulfide gas, $H_2S$, burns in oxygen to give sulfur dioxide, $SO_2$, and water. Write the equation for the reaction, giving molecular, molar, and mass interpretations below the equation.

**3.71**  Butane, $C_4H_{10}$, burns with the oxygen in air to give carbon dioxide and water.

$$2C_4H_{10}(g) + 13O_2(g) \longrightarrow 8CO_2(g) + 10H_2O(g)$$

What is the amount (in moles) of carbon dioxide produced from 0.30 mol $C_4H_{10}$?

**3.72**  Ethanol, $C_2H_5OH$, burns with the oxygen in air to give carbon dioxide and water.

$$C_2H_5OH(l) + 3O_2(g) \longrightarrow 2CO_2(g) + 3H_2O(l)$$

What is the amount (in moles) of water produced from 0.51 mol $C_2H_5OH$?

**3.73**  Iron in the form of fine wire burns in oxygen to form iron(III) oxide.

$$4Fe(s) + 3O_2(g) \longrightarrow 2Fe_2O_3(s)$$

How many moles of $O_2$ are needed to produce 5.21 mol $Fe_2O_3$?

**3.74** Nickel(II) chloride reacts with sodium phosphate to precipitate nickel(II) phosphate.

$$3NiCl_2(aq) + 2Na_3PO_4(aq) \longrightarrow Ni_3(PO_4)_2(s) + 6NaCl(aq)$$

How many moles of nickel(II) chloride are needed to produce 0.479 mol nickel(II) phosphate?

**3.75** Nitric acid, $HNO_3$, is manufactured by the Ostwald process, in which nitrogen dioxide, $NO_2$, reacts with water.

$$3NO_2(g) + H_2O(l) \longrightarrow HNO_3(aq) + NO(g)$$

How many grams of nitrogen dioxide are required in this reaction to produce 7.50 g $HNO_3$?

**3.76** White phosphorus, $P_4$, is prepared by fusing calcium phosphate, $Ca_3(PO_4)_2$, with carbon, C, and sand, $SiO_2$, in an electric furnace.

$$2Ca_3(PO_4)_2(s) + 6SiO_2(s) + 10C(s) \longrightarrow$$
$$P_4(g) + 6CaSiO_3(l) + 10CO(g)$$

How many grams of calcium phosphate are required to give 10.0 g of phosphorus?

**3.77** Tungsten metal, W, is used to make incandescent bulb filaments. The metal is produced from the yellow tungsten(VI) oxide, $WO_3$, by reaction with hydrogen.

$$WO_3(s) + 3H_2(g) \longrightarrow W(s) + 3H_2O(g)$$

How many grams of tungsten can be obtained from 4.81 kg of hydrogen with excess tungsten(VI) oxide?

**3.78** Acrylonitrile, $C_3H_3N$, is the starting material for the production of a kind of synthetic fiber (acrylics). It can be made from propylene, $C_3H_6$, by reaction with nitric oxide, NO.

$$4C_3H_6(g) + 6NO(g) \longrightarrow 4C_3H_3N(g) + 6H_2O(g) + N_2(g)$$

How many grams of acrylonitrile are obtained from 651 kg of propylene and excess NO?

**3.79** The following reaction, depicted using molecular models, is used to make carbon tetrachloride, $CCl_4$, a solvent and starting material for the manufacture of fluorocarbon refrigerants and aerosol propellants.

Calculate the number of grams of carbon disulfide, $CS_2$, needed for a laboratory-scale reaction with 62.7 g of chlorine, $Cl_2$.

**3.80** Using the following reaction (depicted using molecular models), large quantities of ammonia are burned in the presence of a platinum catalyst to give nitric oxide, as the first step in the preparation of nitric acid.

Suppose a vessel contains 12.2 g of $NH_3$, how many grams of $O_2$ are needed for a complete reaction?

**3.81** When dinitrogen pentoxide, $N_2O_5$, a white solid, is heated, it decomposes to nitrogen dioxide and oxygen.

$$2N_2O_5(s) \xrightarrow{\Delta} 4NO_2(g) + O_2(g)$$

If a sample of $N_2O_5$ produces 1.315 g $O_2$, how many grams of $NO_2$ are formed?

**3.82** Copper metal reacts with nitric acid. Assume that the reaction is

$$3Cu(s) + 8HNO_3(aq) \longrightarrow$$
$$3Cu(NO_3)_2(aq) + 2NO(g) + 4H_2O(l)$$

If 5.92 g $Cu(NO_3)_2$ is eventually obtained, how many grams of nitric oxide, NO, would have formed also, according to the preceding equation?

## Limiting Reactant; Theoretical and Percentage Yields

**3.83** Potassium superoxide, $KO_2$, is used in rebreathing gas masks to generate oxygen.

$$4KO_2(s) + 2H_2O(l) \longrightarrow 4KOH(s) + 3O_2(g)$$

If a reaction vessel contains 0.25 mol $KO_2$ and 0.15 mol $H_2O$, what is the limiting reactant? How many moles of oxygen can be produced?

**3.84** Solutions of sodium hypochlorite, NaClO, are sold as a bleach (such as Clorox). They are prepared by the reaction of chlorine with sodium hydroxide.

$$2NaOH(aq) + Cl_2(g) \longrightarrow$$
$$NaCl(aq) + NaClO(aq) + H_2O(l)$$

If you have 1.23 mol of NaOH in solution and 1.47 mol of $Cl_2$ gas available to react, which is the limiting reactant? How many moles of $NO(g)$ could be obtained?

**3.85** Methanol, $CH_3OH$, is prepared industrially from the gas-phase catalytic balanced reaction that has been depicted here using molecular models.

In a laboratory test, a reaction vessel was filled with 35.4 g CO and 10.2 g $H_2$. How many grams of methanol would be produced in a complete reaction? Which reactant remains unconsumed at the end of the reaction? How many grams of it remain?

**3.86** Carbon disulfide, $CS_2$, burns in oxygen. Complete combustion gives the balanced reaction that has been depicted here using molecular models.

Calculate the grams of sulfur dioxide, $SO_2$, produced when a mixture of 30.0 g of carbon disulfide and 35.0 g of oxygen reacts. Which reactant remains unconsumed at the end of the combustion? How many grams remain?

**3.87** Titanium, which is used to make airplane engines and frames, can be obtained from titanium tetrachloride, which in turn is obtained from titanium dioxide by the following process:

$$3TiO_2(s) + 4C(s) + 6Cl_2(g) \longrightarrow$$
$$3TiCl_4(g) + 2CO_2(g) + 2CO(g)$$

A vessel contains 4.15 g $TiO_2$, 5.67 g C, and 6.78 g $Cl_2$. Suppose the reaction goes to completion as written. How many grams of titanium tetrachloride can be produced?

## General Problems

**3.91** Caffeine, the stimulant in coffee and tea, has the molecular formula $C_8H_{10}N_4O_2$. Calculate the mass percentage of each element in the substance. Give the answers to three significant figures.

**3.92** Morphine, a narcotic substance obtained from opium, has the molecular formula $C_{17}H_{19}NO_3$. What is the mass percentage of each element in morphine (to three significant figures)?

**3.93** A moth repellent, *para*-dichlorobenzene, has the composition 49.1% C, 2.7% H, and 48.2% Cl. Its molecular weight is 147 amu. What is its molecular formula?

**3.88** Hydrogen cyanide, HCN, is prepared from ammonia, air, and natural gas ($CH_4$) by the following process:

$$2NH_3(g) + 3O_2(g) + 2CH_4(g) \xrightarrow{Pt} 2HCN(g) + 6H_2O(g)$$

Hydrogen cyanide is used to prepare sodium cyanide, which is used in part to obtain gold from gold-containing rock. If a reaction vessel contains 11.5 g $NH_3$, 12.0 g $O_2$, and 10.5 g $CH_4$, what is the maximum mass in grams of hydrogen cyanide that could be made, assuming the reaction goes to completion as written?

**3.89** Aspirin (acetylsalicylic acid) is prepared by heating salicylic acid, $C_7H_6O_3$, with acetic anhydride, $C_4H_6O_3$. The other product is acetic acid, $C_2H_4O_2$.

$$C_7H_6O_3 + C_4H_6O_3 \longrightarrow C_9H_8O_4 + C_2H_4O_2$$

What is the theoretical yield (in grams) of aspirin, $C_9H_8O_4$, when 2.00 g of salicylic acid is heated with 4.00 g of acetic anhydride? If the actual yield of aspirin is 1.98 g, what is the percentage yield?

**3.90** Methyl salicylate (oil of wintergreen) is prepared by heating salicylic acid, $C_7H_6O_3$, with methanol, $CH_3OH$.

$$C_7H_6O_3 + CH_3OH \longrightarrow C_8H_8O_3 + H_2O$$

In an experiment, 1.50 g of salicylic acid is reacted with 11.20 g of methanol. The yield of methyl salicylate, $C_8H_8O_3$, is 1.31 g. What is the percentage yield?

**3.94** Sorbic acid is added to food as a mold inhibitor. Its composition is 64.3% C, 7.2% H, and 28.5% O, and its molecular weight is 112 amu. What is its molecular formula?

**3.95** Thiophene is a liquid compound of the elements C, H, and S. A sample of thiophene weighing 7.96 mg was burned in oxygen, giving 16.65 mg $CO_2$. Another sample was subjected to a series of reactions that transformed all of the sulfur in the compound to barium sulfate. If 4.31 mg of thiophene gave 11.96 mg of barium sulfate, what is the empirical formula of thiophene? Its molecular weight is 84 amu. What is its molecular formula?

**3.96** Aniline, a starting compound for urethane plastic foams, consists of C, H, and N. Combustion of such compounds yields $CO_2$, $H_2O$, and $N_2$ as products. If the combustion of 9.71 mg of aniline yields 6.63 mg $H_2O$ and 1.46 mg $N_2$, what is its empirical formula? The molecular weight of aniline is 93 amu. What is its molecular formula?

**3.97** A sample of limestone (containing calcium carbonate, $CaCO_3$) weighing 438 mg is treated with oxalic acid, $H_2C_2O_4$, to give calcium oxalate, $CaC_2O_4$.

$$CaCO_3(s) + H_2C_2O_4(aq) \longrightarrow$$
$$CaC_2O_4(s) + H_2O(l) + CO_2(g)$$

The mass of the calcium oxalate is 472 mg. What is the mass percentage of calcium carbonate in this limestone?

**3.98** A titanium ore contains rutile ($TiO_2$) plus some iron oxide and silica. When it is heated with carbon in the presence of chlorine, titanium tetrachloride, $TiCl_4$, is formed.

$$TiO_2(s) + C(s) + 2Cl_2(g) \longrightarrow TiCl_4(g) + CO_2(g)$$

Titanium tetrachloride, a liquid, can be distilled from the mixture. If 35.4 g of titanium tetrachloride is recovered from 17.4 g of crude ore, what is the mass percentage of $TiO_2$ in the ore (assuming all $TiO_2$ reacts)?

**3.99** Ethylene oxide, $C_2H_4O$, is made by the oxidation of ethylene, $C_2H_4$.

$$2C_2H_4(g) + O_2(g) \longrightarrow 2C_2H_4O(g)$$

Ethylene oxide is used to make ethylene glycol for automobile antifreeze. In a pilot study, 10.6 g of ethylene gave 9.91 g of ethylene oxide. What is the percentage yield of ethylene oxide?

**3.100** Nitrobenzene, $C_6H_5NO_2$, an important raw material for the dye industry, is prepared from benzene, $C_6H_6$, and nitric acid, $HNO_3$.

$$C_6H_6(l) + HNO_3(l) \longrightarrow C_6H_5NO_2(l) + H_2O(l)$$

When 22.4 g of benzene and an excess of $HNO_3$ are used, what is the theoretical yield of nitrobenzene? If 28.7 g of nitrobenzene is recovered, what is the percentage yield?

**3.101** Zinc metal can be obtained from zinc oxide, ZnO, by reaction at high temperature with carbon monoxide, CO.

$$ZnO(s) + CO(g) \longrightarrow Zn(s) + CO_2(g)$$

The carbon monoxide is obtained from carbon.

$$2C(s) + O_2(g) \longrightarrow 2CO(g)$$

What is the maximum amount of zinc that can be obtained from 75.0 g of zinc oxide and 50.0 g of carbon?

**3.102** Hydrogen cyanide, HCN, can be made by a two-step process. First, ammonia is reacted with $O_2$ to give nitric oxide, NO.

$$4NH_3(g) + 5O_2(g) \longrightarrow 4NO(g) + 6H_2O(g)$$

Then nitric oxide is reacted with methane, $CH_4$.

$$2NO(g) + 2CH_4(g) \longrightarrow 2HCN(g) + 2H_2O(g) + H_2(g)$$

When 24.2 g of ammonia and 25.1 g of methane are used, how many grams of hydrogen cyanide can be produced?

**3.103** Calcium carbide, $CaC_2$, used to produce acetylene, $C_2H_2$, is prepared by heating calcium oxide, CaO, and carbon, C, to high temperature.

$$CaO(s) + 3C(s) \longrightarrow CaC_2(s) + CO(g)$$

If a mixture contains 1.15 kg of each reactant, how many grams of calcium carbide can be prepared?

**3.104** A mixture consisting of 12.8 g of calcium fluoride, $CaF_2$, and 13.2 g of sulfuric acid, $H_2SO_4$, is heated to drive off hydrogen fluoride, HF.

$$CaF_2(s) + H_2SO_4(l) \longrightarrow 2HF(g) + CaSO_4(s)$$

What is the maximum number of grams of hydrogen fluoride that can be obtained?

**3.105** Alloys, or metallic mixtures, of mercury with another metal are called amalgams. Sodium in sodium amalgam reacts with water. (Mercury does not.)

$$2Na(s) + 2H_2O(l) \longrightarrow 2NaOH(aq) + H_2(g)$$

If a 15.23-g sample of sodium amalgam evolves 0.108 g of hydrogen, what is the percentage of sodium in the amalgam?

**3.106** A sample of sandstone consists of silica, $SiO_2$, and calcite, $CaCO_3$. When the sandstone is heated, calcium carbonate, $CaCO_3$, decomposes into calcium oxide, CaO, and carbon dioxide.

$$CaCO_3(s) \longrightarrow CaO(s) + CO_2(g)$$

What is the percentage of silica in the sandstone if 18.7 mg of the rock yields 3.95 mg of carbon dioxide?

# Cumulative-Skills Problems

**3.107**    A 0.500-g mixture of $Cu_2O$ and $CuO$ contains 0.425 g Cu. What is the mass of $CuO$ in the mixture?

**3.108**    A mixture of $Fe_2O_3$ and $FeO$ was found to contain 72.00% Fe by mass. What is the mass of $Fe_2O_3$ in 0.500 g of this mixture?

**3.109**    Hemoglobin is the oxygen-carrying molecule of red blood cells, consisting of a protein and a nonprotein substance. The nonprotein substance is called heme. A sample of heme weighing 35.2 mg contains 3.19 mg of iron. If a heme molecule contains one atom of iron, what is the molecular weight of heme?

**3.110**    Penicillin V was treated chemically to convert sulfur to barium sulfate, $BaSO_4$. An 8.19-mg sample of penicillin V gave 5.46 mg $BaSO_4$. What is the percentage of sulfur in penicillin V? If there is one sulfur atom in the molecule, what is the molecular weight?

**3.111**    A 3.41-g sample of a metallic element, M, reacts completely with 0.0158 mol of a gas, $X_2$, to form 4.52 g MX. What are the identities of M and X?

**3.112**    1.92 g $M^+$ ion reacts with 0.158 mol $X^-$ ion to produce a compound, $MX_2$, which is 86.8% X by mass. What are the identities of $M^+$ and $X^-$?

**3.113**    An alloy of iron (54.7%), nickel (45.0%), and manganese (0.3%) has a density of 8.17 $g/cm^3$. How many iron atoms are there in a block of alloy measuring 10.0 cm × 20.0 cm × 15.0 cm?

**3.114**    An alloy of iron (71.0%), cobalt (12.0%), and molybdenum (17.0%) has a density of 8.20 $g/cm^3$. How many cobalt atoms are there in a cylinder with a radius of 2.50 cm and a length of 10.0 cm?

 # Media Activities

**3.115**    Watch the Ammonium Dichromate Volcano *Visualization* on the student CD.
a. Determine the balanced chemical equation, assuming that ammonium dichromate is the only reactant and that chromium (III) oxide, nitrogen gas, and water vapor are the products.
b. If you started with 25 g of ammonium dichromate, what mass of chromium(III) oxide would you expect to find on the watch glass after the reaction?

**3.116**    Open the Limiting Reactants *Understanding Concepts* activity on the student CD and select the combustion of acetylene example.
a. If two molecules of acetylene are reacted with four molecules of oxygen, what is the limiting reactant?
b. How many $H_2O$ molecules can form?
c. How many molecules of $O_2$ are required to react completely with two molecules of acetylene?

**3.117**    Open the Limiting Reactants *Understanding Concepts* activity on the student CD and select the combustion of hydrazine example. After you have balanced the equation, calculate the number of grams of nitrogen dioxide produced when 15 g of hydrazine is reacted with 15 g of oxygen.

**3.118**    Open the Conservation of Mass and Balancing Equations *Understanding Concepts* activity on the student CD.

a. Using the combustion of ammonia example, balance the equation with two different sets of coefficients.
b. Even though the numbers have changed in your two examples of balancing the equation, what has remained the same? (*Hint:* Compare the mole ratios of substances in the reaction.) Why is this important?

**3.119**    Go to the Molecule Library on the student web site and view the structures and chemical formulas of 1, 1-dichloroethane and 1,2-dichloroethane.
a. What is the principal difference between these two molecules?
b. How does the percent C, H, and Cl, compare for each of these molecules?

**3.120**    Open the Oxygen, Hydrogen, Soap Bubbles, and Balloons *Visualization* on the student CD.
a. If you haven't done so, complete Conceptual Problem 3.14.
b. Calculate the mass of water produced during the chemical reaction depicted in the video if 0.05 g of hydrogen was ignited in 0.2 g of oxygen.
c. What state (solid, liquid, or gas) is the water in when the reaction is completed?

**3.121**    For review, take the ACE practice test for Chapter 3 on the student web site.

Beakers with precipitates of lead(II) iodide and mercury(II) iodide.

# Chemical Reactions

Chemical reactions are the heart of chemistry. Some reactions, such as those accompanying a forest fire or the explosion of dynamite, are quite dramatic. Others are much less obvious, although all chemical reactions must involve detectable change. A chemical reaction involves a change from reactant substances to product substances, and the product substances will have physical and chemical properties different from those of the reactants.

Figure 4.1 shows an experimenter adding a colorless solution of potassium iodide, KI, to a colorless solution of lead(II) nitrate, $Pb(NO_3)_2$. What you see is the formation of a cloud of bright yellow crystals where the two solutions have come into contact, clear evidence of a chemical reaction. The bright yellow crystals are lead(II) iodide, $PbI_2$, one of the reaction products. We call a solid that forms during a chemical reaction in solution a *precipitate;* the reaction is a precipitation reaction.

In this chapter, we will discuss the major types of chemical reactions, including precipitation reactions. Some of the most important reactions we will describe involve ions in aqueous (water) solution. Therefore, we will first look at these ions and see how we represent by chemical equations the reactions involving ions in aqueous solution.

Some questions we will answer are: What is the evidence for ions in solution? How do we write chemical equations for reactions involving ions? How can we classify and describe the many reactions we observe so that we can begin to understand them? What is the quantitative description of solutions and reactions in solution?

**CONTENTS**

**Figure 4.1**
**Reaction of potassium iodide solution and lead(II) nitrate solution.** The reactant solutions are colorless, but one of the products, lead(II) iodide, forms as a yellow precipitate.

# Ions in Aqueous Solution

You probably have heard that you should not operate electrical equipment while standing in water. And you may have read a murder mystery in which the victim was electrocuted when an electrical appliance "accidentally" fell into his or her bath water. Actually, if the water were truly pure, the person would be safe from electrocution, because pure water is a nonconductor of electricity. Bath water or water as it flows from the faucet, however, is a *solution* of water with small amounts of dissolved substances in it, and these dissolved substances make the solution an electrical conductor. This allows an electric current to flow from an electrical appliance to the human body. Let us look at the nature of such solutions.

## 4.1 Ionic Theory of Solutions and Solubility Rules

Chemists began studying the electrical behavior of substances in the early nineteenth century, and they knew that you could make pure water electrically conducting by dissolving certain substances in it. In 1884, the young Swedish chemist Svante Arrhenius proposed the *ionic theory of solutions* to account for this conductivity. He said that certain substances produce freely moving ions when they dissolve in water, and these ions conduct an electric current in an aqueous solution. ◄

▶ Arrhenius submitted his ionic theory as part of his doctoral dissertation to the faculty at Uppsala, Sweden, in 1884. It was not well received and he barely passed. In 1903, however, he was awarded the Nobel Prize in chemistry for this theory.

Suppose you dissolve sodium chloride, NaCl, in water. From our discussion in Section 2.6, you may remember that sodium chloride is an ionic solid consisting of sodium ions, $Na^+$, and chloride ions, $Cl^-$, held in a regular, fixed array. When you dissolve solid sodium chloride in water, the $Na^+$ and $Cl^-$ ions go into solution as freely moving ions. Now suppose you dip electric wires that are connected to the poles of a battery into a solution of sodium chloride. The wire that connects to the positive pole of the battery attracts the negatively charged chloride ions in solution, because of their opposite charges. Similarly, the wire connected to the negative pole of the battery attracts the positively charged sodium ions in solution (Figure 4.2). Thus, the ions in the solution begin to move, and these moving charges form the electric current in the solution. (In a wire, it is moving electrons that constitute the electric current.)

**Figure 4.2**
**Motion of ions in solution.** Ions are in fixed positions in a crystal. During the solution process, however, ions leave the crystal and become freely moving. Note that $Na^+$ ions (small gray spheres) are attracted to the negative wire, whereas $Cl^-$ ions (large green spheres) are attracted to the positive wire.

▶ Demonstration of conductivity of liquid versus frozen solution: Summerlin et al., *Chemical Demonstrations*, Vol. 2 (Washington, D.C.: Am. Chem. Soc., 1987), no. 71, p. 132.

▶ Video no. 3 (Series A), "Electrical Conductivity of Aqueous Solutions."

Now consider pure water. Water consists of molecules, each of which is electrically neutral. Since each molecule carries no net electric charge, it carries no overall electric charge when it moves. Thus, pure water is a nonconductor of electricity.

In summary, although water is itself nonconducting, it has the ability to dissolve various substances, some of which go into solution as freely moving ions. An aqueous solution of ions is electrically conducting.

## Electrolytes and Nonelectrolytes

We can divide the substances that dissolve in water into two broad classes, electrolytes and nonelectrolytes. An **electrolyte** is *a substance that dissolves in water to give an electrically conducting solution.* Sodium chloride, table salt, is an example of an electrolyte. When most ionic substances dissolve in water, ions that were in fixed sites in the crystalline solid go into the surrounding aqueous solution, where they are free to move about. The resulting solution is conducting because the moving ions form an electric current. Thus, in general, *ionic solids that dissolve in water are electrolytes.*

Not all electrolytes are ionic substances. Certain molecular substances dissolve in water to form ions. The resulting solution is electrically conducting, and so we say that the molecular substance is an electrolyte. An example is hydrogen chloride gas, $HCl(g)$, which is a molecular substance. Hydrogen chloride gas dissolves in water, giving $HCl(aq)$, which in turn produces hydrogen ions, $H^+$, and chloride ions, $Cl^-$, in aqueous solution. (The solution of $H^+$ and $Cl^-$ ions is called hydrochloric acid.)

$$HCl(aq) \xrightarrow{H_2O} H^+(aq) + Cl^-(aq)$$

We will look more closely at molecular electrolytes, such as HCl, at the end of this section.

A **nonelectrolyte** is *a substance that dissolves in water to give a nonconducting or very poorly conducting solution.* A common example is sucrose, $C_{12}H_{22}O_{11}$, which is ordinary table sugar. Another example is methanol, $CH_3OH$, a compound used in car window washer solution. Both of these are molecular substances. The solution process occurs because molecules of the substance mix with molecules of water. Molecules are electrically neutral and cannot carry an electric current, so the solution is electrically nonconducting.

## Observing the Electrical Conductivity of a Solution

Figure 4.3 shows a simple apparatus that allows you to observe the ability of a solution to conduct an electric current. The apparatus has two electrodes, here they are flat metal plates, dipping into the solution in a beaker. One electrode connects directly to a battery through a wire. The other electrode connects by a wire to a light bulb that connects with another wire to the other side of the battery. For an electrical current to flow from the battery, there must be a complete circuit, which allows the current to flow from the positive pole of the battery through the circuit to the negative pole of the battery. To have a complete circuit, the solution in the beaker must conduct electricity, as the wires do. If the solution is conducting, the circuit is complete and the bulb lights. If the solution is nonconducting, the circuit is incomplete and the bulb does not light.

The beaker shown on the left side of Figure 4.3 contains pure water. Because the bulb is not lit, we conclude that pure water is a nonconductor (or very poor conductor) of electricity, which is what we expect from our earlier discussion. The beaker

**Figure 4.3**
**Testing the electrical conductivity of a solution.**
*Left:* Pure water does not conduct; therefore the bulb does not light. *Right:* A solution of sodium chloride allows the current to pass through it, and the bulb lights.

▶ Video Series C, "Conductivities of Aqueous Solutions."

shown on the right side of Figure 4.3 contains a solution of sodium chloride in water. In this case, the bulb burns brightly, showing that the solution is a very good conductor of electricity, due to the movement of ions in the solution.

How brightly the bulb lights tells you whether the solution is a very good conductor (contains a "strong" electrolyte) or only a moderately good conductor (contains a "weak" electrolyte). The solution of sodium chloride burns brightly, so we conclude that sodium chloride is a strong electrolyte. Let us look more closely at such strong and weak electrolytes.

## Strong and Weak Electrolytes

When electrolytes dissolve in water they produce ions, but to varying extents. A **strong electrolyte** is *an electrolyte that exists in solution almost entirely as ions.* Most ionic solids that dissolve in water do so by going into the solution almost completely as ions, so they are strong electrolytes. An example is sodium chloride. We can represent the dissolution of sodium chloride in water by the following equation:

$$NaCl(s) \xrightarrow{H_2O} Na^+(aq) + Cl^-(aq)$$

▶ Video no. 23 (Series D), "Conductivity Apparatus."

A **weak electrolyte** is *an electrolyte that dissolves in water to give a relatively small percentage of ions.* These are generally molecular substances. Ammonia, $NH_3$, is an example. Pure ammonia is a gas that readily dissolves in water and goes into solution as ammonia molecules, $NH_3(aq)$. When you buy "ammonia" in the grocery store, you are buying an aqueous solution of ammonia. Ammonia molecules react with water to form ammonium ions, $NH_4^+$, and hydroxide ions, $OH^-$.

$$NH_3(aq) + H_2O(l) \longrightarrow NH_4^+(aq) + OH^-(aq)$$

However, these ions, $NH_4^+ + OH^-$, react with each other to give back ammonia molecules and water molecules.

$$NH_4^+(aq) + OH^-(aq) \longrightarrow NH_3(aq) + H_2O(l)$$

Both reactions, the original one and its reverse, occur constantly and simultaneously. We denote this situation by writing a single equation with a double arrow:

$$NH_3(aq) + H_2O(l) \rightleftharpoons NH_4^+(aq) + OH^-(aq)$$

**Figure 4.4**
**Comparing strong and weak electrolytes.** The apparatus is similar to that in Figure 4.3, but this time strong and weak electrolytes are compared. The solution on the left is of HCl (a strong electrolyte) and that on the right is of $NH_3$ (a weak electrolyte). Note how much more brightly the bulb on the left burns compared with that on the right.

As a result of this forward and reverse reaction, just a small percentage of the $NH_3$ molecules (about 3%) have reacted at any given moment to form ions. Thus, ammonia is a weak electrolyte. The electrical conductivities of a strong and a weak electrolyte are contrasted in Figure 4.4.

Most soluble molecular substances are either nonelectrolytes or weak electrolytes. An exception is hydrogen chloride gas, $HCl(g)$, which, as noted earlier, dissolves in water to produce hydrogen ions and chloride ions. The ions that form in this case hardly react at all with each other, so we represent the reaction by an equation with a single arrow:

$$HCl(aq) \longrightarrow H^+(aq) + Cl^-(aq)$$

Since hydrogen chloride dissolves to give almost entirely ions, hydrogen chloride (or hydrochloric acid) is a strong electrolyte, unlike ammonia.

### Solubility Rules

It is clear from the preceding discussion that substances vary widely in their *solubility,* or ability to dissolve, in water. Some compounds, such as sodium chloride and ethyl alcohol ($CH_3CH_2OH$), dissolve readily and are said to be *soluble*. Others, such as calcium carbonate (which occurs naturally as limestone and marble) and benzene ($C_6H_6$), have quite limited solubilities and are thus said to be *insoluble*.

Soluble ionic compounds form solutions that contain many ions and therefore are strong electrolytes. To predict the solubility of ionic compounds, chemists have developed solubility rules. Table 4.1 lists eight solubility rules for ionic compounds. These rules apply to most of the common ionic compounds that we will discuss in this course. Example 4.1 illustrates how to use the rules.

**EXAMPLE 4.1**

## Using the Solubility Rules

Determine whether the following compounds are soluble or insoluble in water.

a. $Hg_2Cl_2$     b. KI     c. lead(II) nitrate

**PROBLEM STRATEGY**

You should refer to Table 4.1 for the solubility rules of ionic compounds to see which compounds are soluble in water. The soluble ionic compounds are those that readily dissolve in water. The insoluble compounds hardly dissolve at all.

**SOLUTION**

a. According to Rule 3 in Table 4.1, most compounds that contain chloride, $Cl^-$, are soluble. However, $Hg_2Cl_2$ is listed as one of the exceptions to this rule so it does not dissolve in water. Therefore, **$Hg_2Cl_2$ is not soluble in water.**

b. According to both Rule 1, Group IA compounds are soluble, and Rule 3, most iodides are soluble, KI is expected to be soluble. Therefore, **KI is soluble in water.**

c. According to Rule 2, compounds containing nitrates, $NO_3^-$, are soluble. Since there are no exceptions to this rule, **lead(II) nitrate, $Pb(NO_3)_2$, is soluble in water.**

---

**EXERCISE 4.1**

Determine whether the following compounds are soluble or insoluble in water.

a. NaBr    b. $Ba(OH)_2$    c. calcium carbonate

---

See Problems 4.21 and 4.22.

Let us summarize the main points in this section. Compounds that dissolve in water are soluble; those that dissolve little, or not at all, are insoluble. Soluble substances are either electrolytes or nonelectrolytes. Nonelectrolytes form noncon-

**Table 4.1**
**Solubility Rules for Ionic Compounds**

| Rule | Applies to | Statement | Exceptions |
|------|-----------|-----------|-----------|
| 1 | $Li^+$, $Na^+$, $K^+$, $NH_4^+$ | Group IA and ammonium compounds are soluble. | — |
| 2 | $C_2H_3O_2^-$, $NO_3^-$ | Acetates and nitrates are soluble. | — |
| 3 | $Cl^-$, $Br^-$, $I^-$ | Most chlorides, bromides, and iodides are soluble. | $AgCl$, $Hg_2Cl_2$, $PbCl_2$, $AgBr$, $HgBr_2$, $Hg_2Br_2$, $PbBr_2$, $AgI$, $HgI_2$, $Hg_2I_2$, $PbI_2$ |
| 4 | $SO_4^{2-}$ | Most sulfates are soluble. | $CaSO_4$, $SrSO_4$, $BaSO_4$, $Ag_2SO_4$, $Hg_2SO_4$, $PbSO_4$ |
| 5 | $CO_3^{2-}$ | Most carbonates are insoluble. | Group IA carbonates, $(NH_4)_2CO_3$ |
| 6 | $PO_4^{3-}$ | Most phosphates are insoluble. | Group IA phosphates, $(NH_4)_3PO_4$ |
| 7 | $S^{2-}$ | Most sulfides are insoluble. | Group IA sulfides, $(NH_4)_2S$ |
| 8 | $OH^-$ | Most hydroxides are insoluble. | Group IA hydroxides, $Ca(OH)_2$, $Sr(OH)_2$, $Ba(OH)_2$ |

ducting aqueous solutions because they dissolve completely as molecules. Electrolytes form electrically conducting solutions in water because they dissolve to give ions in solution. Electrolytes can be strong or weak. Almost all soluble ionic substances are strong electrolytes. Soluble molecular substances usually are nonelectrolytes or weak electrolytes; the latter solution consists primarily of molecules, but has a small percentage of ions. Ammonia, $NH_3$, is an example of a molecular substance that is a weak electrolyte. A few molecular substances (such as HCl) dissolve almost entirely as ions in the solution and are therefore strong electrolytes. The solubility rules can be used to predict the solubility of ionic compounds in water.

 **CONCEPT CHECK 4.1**

LiI(*s*) and $CH_3OH$(*l*) are each introduced into a separate beaker containing water. Using the drawings shown here, label each beaker with the appropriate compound and indicate whether you would expect each substance to be a strong electrolyte, weak electrolyte, or nonelectrolyte.

# 4.2 Molecular and Ionic Equations

We use chemical equations to help us describe chemical reactions. For a reaction involving ions, we have a choice of chemical equations, depending on the kind of information we want to convey. We can represent such a reaction by a *molecular equation*, a *complete ionic equation*, or a *net ionic equation*.

To illustrate these different kinds of equations, consider the preparation of pre-cipitated calcium carbonate, $CaCO_3$. This white, fine powdery compound is used as a paper filler to brighten and retain ink, as an antacid as in the trade-named Tums, and as a mild abrasive in toothpastes. One way to prepare this compound is to react calcium hydroxide, $Ca(OH)_2$, with sodium carbonate, $Na_2CO_3$. Let us look at the different ways to write the equation for this reaction.

## Molecular Equations

You could write the equation for this reaction as follows:

$$Ca(OH)_2(aq) + Na_2CO_3(aq) \longrightarrow CaCO_3(s) + 2NaOH(aq)$$

We call this a **molecular equation,** which is *a chemical equation in which the reactants and products are written as if they were molecular substances, even though they may actually exist in solution as ions.* The molecular equation is useful because it is explicit about what the reactant solutions are and what products you obtain. The equation says that you add aqueous solutions of calcium hydroxide and sodium carbonate to the reaction vessel. As a result of the reaction, the insoluble, white calcium carbonate solid forms in the solution; that is, calcium carbonate precipitates. After you remove the precipitate, you are left with a solution of sodium hydroxide. The molecular equation closely describes what you actually do in the laboratory or in an industrial process.

## Complete Ionic Equations

Although a molecular equation is useful in describing the actual reactant and product substances, it does not tell you what is happening at the level of ions. That is, it does not give you an ionic-theory interpretation of the reaction. Because this kind of information is useful, you often need to rewrite the molecular equation as an ionic equation.

Again, consider the reaction of calcium hydroxide, $Ca(OH)_2$, and sodium carbonate, $Na_2CO_3$. Both are soluble ionic substances and therefore strong electrolytes; when they dissolve in water, they go into solution as ions. Each formula unit of $Ca(OH)_2$ forms one $Ca^{2+}$ ion and two $OH^-$ ions in solution. If you want to emphasize that the solution contains freely moving ions, it would be better to write $Ca^{2+}(aq)$ + $2OH^-(aq)$ in place of $Ca(OH)_2(aq)$. Similarly, each formula unit of $Na_2CO_3$ forms two $Na^+$ ions and one $CO_3^{2-}$ ion in solution, and you would emphasize this by writing $2Na^+(aq)$ + $CO_3^{2-}(aq)$ in place of $Na_2CO_3(aq)$. The reactant side of the equation becomes

$$Ca^{2+}(aq) + 2OH^-(aq) + 2Na^+(aq) + CO_3^{2-}(aq) \longrightarrow$$

Thus, the reaction mixture begins as a solution of four different kinds of ions.

Now let us look at the product side of the equation. One product is the precipitate $CaCO_3(s)$. According to the solubility rules, this is an insoluble ionic compound so it will exist in water as a solid. We leave the formula as $CaCO_3(s)$ to convey this information in the equation. On the other hand, NaOH is a soluble ionic substance and therefore a strong electrolyte; it dissolves in aqueous solution to give the freely moving ions, which we denote by writing $Na^+(aq)$ + $OH^-(aq)$. The complete equation is

$$Ca^{2+}(aq) + 2OH^-(aq) + 2Na^+(aq) + CO_3^{2-}(aq) \longrightarrow$$
$$CaCO_3(s) + 2Na^+(aq) + 2OH^-(aq)$$

The purpose of such a *complete ionic equation* is to represent each substance by its predominant form in the reaction mixture. For example, if the substance is a soluble ionic compound, it probably dissolves as individual ions (so it is a strong electrolyte). In a complete ionic equation, you represent the compound as separate ions. If the substance is a weak electrolyte, it is present in solution primarily as molecules, so you represent it by its molecular formula. If the substance is an insoluble ionic compound, you represent it by the formula of the compound, not by the formulas of the separate ions in solution.

Thus, a **complete ionic equation** is *a chemical equation in which strong electrolytes (such as soluble ionic compounds) are written as separate ions in the solution.* You represent other reactants and products by the formulas of the compounds, indicating any soluble substance by *(aq)* after its formula and any insoluble solid substance by *(s)* after its formula.

## Net Ionic Equations

In the complete ionic equation representing the reaction of calcium hydroxide and sodium carbonate, some ions ($OH^-$ and $Na^+$) appear on both sides of the equation. This means that nothing happens to these ions as the reaction occurs. They are called spectator ions. A **spectator ion** is *an ion in an ionic equation that does not take part in the reaction.* You can cancel such ions from both sides to express the essential reaction that occurs.

$$Ca^{2+}(aq) + 2\cancel{OH^-(aq)} + 2\cancel{Na^+(aq)} + CO_3^{2-}(aq) \longrightarrow$$
$$CaCO_3(s) + 2\cancel{Na^+(aq)} + 2\cancel{OH^-(aq)}$$

The resulting equation is

$$Ca^{2+}(aq) + CO_3^{2-}(aq) \longrightarrow CaCO_3(s)$$

This is the **net ionic equation,** *an ionic equation from which spectator ions have been canceled.* It shows that the reaction that actually occurs at the ionic level is between calcium ions and carbonate ions to form solid calcium carbonate.

From the net ionic equation, you can see that mixing any solution of calcium ion with any solution of carbonate ion will give you this same reaction. For example, the strong electrolyte calcium nitrate, $Ca(NO_3)$, dissolves readily in water to provide a source of calcium ions. (According to the solubility rules, calcium nitrate is soluble, so it goes into solution as $Ca^{2+}$ and $NO_3^-$ ions.) Similarly, the strong electrolyte potassium carbonate, $K_2CO_3$, dissolves readily in water to provide a source of carbonate ions. (According to the solubility rules, potassium carbonate is soluble, so it goes into solution as $K^+$ and $CO_3^{2-}$ ions.) When you mix solutions of these two compounds, you obtain a solution of calcium ions and carbonate ions, which react to form the insoluble calcium carbonate. The other product is potassium nitrate, a soluble ionic compound, and therefore a strong electrolyte. The molecular equation representing the reaction is

$$Ca(NO_3)_2(aq) + K_2CO_3(aq) \longrightarrow CaCO_3(s) + 2KNO_3(aq)$$

You obtain the complete ionic equation from this molecular equation by rewriting each of the soluble ionic compounds as ions, but retaining the formula for the precipitate $CaCO_3(s)$:

$$Ca^{2+}(aq) + 2\cancel{NO_3^-(aq)} + 2\cancel{K^+(aq)} + CO_3^{2-}(aq) \longrightarrow$$
$$CaCO_3(s) + 2\cancel{K^+(aq)} + 2\cancel{NO_3^-(aq)}$$

**Figure 4.5**
**Limestone formations.** It is believed that most limestone formed as a precipitate of calcium carbonate (and other carbonates) from seawater. The photograph shows limestone formations at Bryce Point, Bryce Canyon National Park, Utah. More than sixty million years ago, this area was covered by seawater.

The net ionic equation is

$$Ca^{2+}(aq) + CO_3{}^{2-}(aq) \longrightarrow CaCO_3(s)$$

Note that the net ionic equation is identical to the one obtained from the reaction of $Ca(OH)_2$ and $Na_2CO_3$. The essential reaction is the same whether you mix solutions of calcium hydroxide and sodium carbonate or solutions of calcium nitrate and potassium carbonate.

The value of the net ionic equation is its generality. For example, seawater contains $Ca^{2+}$ and $CO_3{}^{2-}$ ions from various sources. Whatever the sources of these ions, you expect them to react to form a precipitate of calcium carbonate. In seawater, this precipitate results in sediments of calcium carbonate, which eventually form limestone (Figure 4.5).

► An alternate example is provided in the Instructor's Resource Manual.

## EXAMPLE 4.2

## Writing Net Ionic Equations

Write a net ionic equation for each of the following molecular equations.

a. $2HClO_4(aq) + Ca(OH)_2(aq) \longrightarrow Ca(ClO_4)_2(aq) + 2H_2O(l)$
   Perchloric acid, $HClO_4$, is a strong electrolyte, forming $H^+$ and $ClO_4{}^-$ ions in solution. $Ca(ClO_4)_2$ is a soluble ionic compound.

b. $HC_2H_3O_2(aq) + NaOH(aq) \longrightarrow NaC_2H_3O_2(aq) + H_2O(l)$
   Acetic acid, $HC_2H_3O_2$, is a molecular substance and a weak electrolyte.

 **PROBLEM STRATEGY**

You will need to convert the molecular equation to the complete ionic equation, then cancel spectator ions to obtain the net ionic equation. For each ionic compound in the reaction, use the solubility rules to determine if the compound will be soluble (in the solution as ions) or insoluble (present as an undissolved solid). For the complete ionic equation, represent all of the strong electrolytes by their separate ions in solution, adding (*aq*) after the formula of each. Retain the formulas of the other compounds. An ionic compound should have (*aq*) after its formula if it is soluble or (*s*) if it is insoluble.

**SOLUTION**

a. According to the solubility rules presented in Table 4.1 and the problem statement, $Ca(OH)_2$ and $Ca(ClO_4)_2$ are soluble ionic compounds so they are strong electrolytes. The problem statement notes that $HClO_4$ is also a strong electrolyte. You write each strong electrolyte in the form of separate ions. Water, $H_2O$, is a nonelectrolyte (or very weak electrolyte), so you retain its molecular formula. The complete ionic equation is

$$2H^+(aq) + 2\cancel{ClO_4^-(aq)} + \cancel{Ca^{2+}(aq)} + 2OH^-(aq) \longrightarrow$$
$$\cancel{Ca^{2+}(aq)} + 2\cancel{ClO_4^-(aq)} + 2H_2O(l)$$

After canceling spectator ions and dividing by 2, you get the following net ionic equation:

$$H^+(aq) + OH^-(aq) \longrightarrow H_2O(l)$$

b. According to the solubility rules, $NaOH$ and $NaC_2H_3O_2$ are soluble ionic compounds, so they are strong electrolytes. The problem statement notes that $HC_2H_3O_2$ is a weak electrolyte, which you write by its molecular formula. Water, $H_2O$, is a nonelectrolyte, so you retain its molecular formula also. The complete ionic equation is

$$HC_2H_3O_2(aq) + \cancel{Na^+(aq)} + OH^-(aq) \longrightarrow \cancel{Na^+(aq)} + C_2H_3O_2^-(aq) + H_2O(l)$$

and the net ionic equation is

$$HC_2H_3O_2(aq) + OH^-(aq) \longrightarrow C_2H_3O_2^-(aq) + H_2O(l)$$

---

**EXERCISE 4.2**

Write complete ionic and net ionic equations for each of the following molecular equations.

a. $2HNO_3(aq) + Mg(OH)_2(s) \longrightarrow 2H_2O(l) + Mg(NO_3)_2(aq)$
Nitric acid, $HNO_3$, is a strong electrolyte.

b. $Pb(NO_3)_2(aq) + Na_2SO_4(aq) \longrightarrow PbSO_4(s) + 2NaNO_3(aq)$

---

See Problems 4.25 and 4.26.

# Types of Chemical Reactions

▶ Demonstration of precipitation: Summerlin and Ealy, *Chemical Demonstrations,* Vol. 1 (Washington, D.C.: Am. Chem. Soc., 1985), no. 12–14, pp. 23–25; Summerlin et al., *Chemical Demonstrations,* Vol. 2 (Washington, D.C.: Am. Chem. Soc., 1987), no. 65–66, pp. 121–125.

Among the several million known substances, many millions of chemical reactions are possible. Beginning students are often bewildered by the possibilities. How can I know when two substances will react when they are mixed? How can I predict the products? Although it is not possible to give completely general answers to these questions, it is possible to make sense of chemical reactions. Most of the reactions we will study belong to one of three types:

1. Precipitation reactions. In these reactions, you mix solutions of two ionic substances and a solid ionic substance (a precipitate) forms.

2. Acid–base reactions. An acid substance reacts with a substance called a base. Such reactions involve the transfer of a proton between reactants.

3. Oxidation–reduction reactions. These involve the transfer of electrons between reactants.

We will look at each of these types of reactions. By the time you finish this chapter, you should feel much more comfortable with the descriptions of chemical reactions that you will encounter in this course.

## 4.3 Precipitation Reactions

▶ Video no. 4 (Series A), "Precipitation Reactions."

▶ Animation: Precipitation

▶ Video Series D, "Chemical Casserole," "The Ammonia Fountain," and "Fun with Limewater."

In the previous section, we used a precipitation reaction to illustrate how to convert a molecular equation to an ionic equation. A precipitation reaction occurs in aqueous solution because one product is insoluble. A **precipitate** is *an insoluble solid compound formed during a chemical reaction in solution.* To predict whether a precipitate will form when you mix two solutions of ionic compounds, you need to know whether any of the potential products that might form are insoluble or not. This is another application of the solubility rules (Section 4.1).

### Predicting Precipitation Reactions

Now let us see how you would go about predicting whether a precipitation reaction will occur. Suppose you mix together solutions of magnesium chloride, $MgCl_2$, and silver nitrate, $AgNO_3$. You can write the potential reactants as follows:

$$MgCl_2 + AgNO_3 \longrightarrow$$

How can you tell if a reaction will occur, and if it does, what products to expect?

When you write a precipitation reaction as a molecular equation, the reaction has the form of an exchange reaction. An **exchange** (or **metathesis**) **reaction** is *a reaction between compounds that, when written as a molecular equation, appears to involve the exchange of parts between the two reactants.* In a precipitation reaction, the anions exchange between the two cations (or vice versa).

▶ Demonstration of metathesis reaction between solid $Pb(NO_3)_2$ and solid KI: Summerlin and Ealy, *Chemical Demonstrations,* Vol. 1 (Washington, D.C.: Am. Chem. Soc., 1985), no. 75, p. 121.

Let us momentarily assume that a reaction does occur between magnesium chloride and silver nitrate. If you exchange the anions, you get silver chloride and magnesium nitrate. Once you figure out the formulas of these potential products, you can write the molecular equation. The formulas are $AgCl$ and $Mg(NO_3)_2$. (If you do not recall how to write the formula of an ionic compound, given the ions, refer back to Example 2.3.) The balanced equation, assuming there is a reaction, is

$$MgCl_2 + 2AgNO_3 \longrightarrow 2AgCl + Mg(NO_3)_2$$

Let us verify that $MgCl_2$ and $AgNO_3$ are soluble and then check the solubilities of the products. Rule 3 in Table 4.1 says that chlorides are soluble, with certain exceptions, which do not include magnesium chloride. Thus, we predict that magnesium chloride is soluble. Rule 2 indicates that nitrates are soluble, so $AgNO_3$ is soluble as well. The potential products are silver chloride and magnesium nitrate. According to Rule 3, silver chloride is one of the exceptions to the general solubility of chlorides. Therefore, we predict that the silver chloride is insoluble. Magnesium nitrate is soluble according to Rule 2.

Now we can append the appropriate phase labels to the compounds in the preceding equation.

$$MgCl_2(aq) + 2AgNO_3(aq) \longrightarrow 2AgCl(s) + Mg(NO_3)_2(aq)$$

We predict that reaction occurs because silver chloride is insoluble and precipitates from the reaction mixture. Figure 4.6 shows the formation of the white silver chloride from this reaction. If you separate the precipitate from the solution by pouring it through filter paper, the solution that passes through (the filtrate) contains magnesium nitrate which you could obtain by evaporating the water. The molecular equation is a summary of the actual reactants and products in the reaction. ◄

To see the reaction that occurs on an ionic level, you need to rewrite the molecular equation as a net ionic equation. You first write the strong electrolytes (here soluble ionic compounds) in the form of ions, leaving the formula of the precipitate unchanged.

$$\mathrm{Mg^{2+}(aq)} + \mathrm{2Cl^-}(aq) + \mathrm{2Ag^+}(aq) + \mathrm{2NO_3^-(aq)} \longrightarrow$$
$$\mathrm{2AgCl}(s) + \mathrm{Mg^{2+}(aq)} + \mathrm{2NO_3^-(aq)}$$

After canceling spectator ions and reducing the coefficients to the smallest whole numbers, you obtain the net ionic equation:

$$\mathrm{Ag^+}(aq) + \mathrm{Cl^-}(aq) \longrightarrow \mathrm{AgCl}(s)$$

This equation represents the essential reaction that occurs: $\mathrm{Ag^+}$ ions and $\mathrm{Cl^-}$ ions in aqueous solution react to form solid silver chloride.

If silver chloride were soluble, a reaction would not have occurred. When you first mix solutions of $\mathrm{MgCl_2}(aq)$ and $\mathrm{AgNO_3}(aq)$, you obtain a solution of four ions: $\mathrm{Mg^{2+}}(aq)$, $\mathrm{Cl^-}(aq)$, $\mathrm{Ag^+}(aq)$, and $\mathrm{NO_3^-}(aq)$. If no precipitate formed, you would end up simply with a solution of these four ions. However, because $\mathrm{Ag^+}$ and $\mathrm{Cl^-}$ react to give the precipitate AgCl, the effect is to remove these ions from the reaction mixture as an insoluble compound and leave behind a solution of $\mathrm{Mg(NO_3)_2}$.

► The reactants $MgCl_2$ and $AgNO_3$ must be added in correct amounts; otherwise, the excess reactant will remain along with the product $Mg(NO_3)_2$.

► Demonstration of variation of conductivity during a metathesis reaction (decrease of ions): Summerlin et al., *Chemical Demonstrations,* Vol. 2 (Washington, D.C.: Am. Chem. Soc., 1987), no. 70, pp. 130–131.

**Figure 4.6**
**Reaction of magnesium chloride and silver nitrate.**
Magnesium chloride solution is added to a beaker of silver nitrate solution. A white precipitate of silver chloride forms.

---

### EXAMPLE 4.3

## Deciding Whether Precipitation Occurs

For each of the following, decide whether a precipitation reaction occurs. If it does, write the balanced molecular equation, and then the net ionic equation. If no reaction occurs, write the compounds followed by an arrow and then *NR* (no reaction).

a. Aqueous solutions of sodium chloride and iron(II) nitrate are mixed.

b. Aqueous solutions of aluminum sulfate and sodium hydroxide are mixed.

### ▨ PROBLEM STRATEGY

Start by writing the formulas of the compounds that are mixed. (If you have trouble with this, see Examples 2.3 and 2.4.) Then, assuming momentarily that the compounds do react, write the exchange reaction. Make sure that you write the correct formulas of the products. Using the solubility rules, append phase labels to each formula in the equation. If one of the products is a precipitate, reaction occurs; otherwise, no reaction occurs.

### SOLUTION

a. The formulas of the compounds are NaCl and $\mathrm{Fe(NO_3)_2}$. Exchanging anions, you get sodium nitrate, $\mathrm{NaNO_3}$, and iron(II) chloride, $\mathrm{FeCl_2}$. The equation for the exchange reaction is

$$\mathrm{NaCl} + \mathrm{Fe(NO_3)_2} \longrightarrow \mathrm{NaNO_3} + \mathrm{FeCl_2} \quad \text{(not balanced)}$$

▶ Two alternate examples are provided in the Instructor's Resource Manual.

Referring to Table 4.1, note that NaCl and $NaNO_3$ are soluble (Rule 1). Also, iron(II) nitrate is soluble (Rule 2), and iron(II) chloride is soluble. (Rule 3 says that chlorides are soluble with some exceptions, none of which include $FeCl_2$.) Since there is no precipitate, no reaction occurs. You obtain simply an aqueous solution of the four different ions ($Na^+$, $Cl^-$, $Fe^{2+}$, and $NO_3^-$). For the answer, we write

$$NaCl(aq) + Fe(NO_3)_2(aq) \longrightarrow NR$$

b. The formulas of the compounds are $Al_2(SO_4)_3$ and NaOH. Exchanging anions, you get aluminum hydroxide, $Al(OH)_3$, and sodium sulfate, $Na_2SO_4$. The equation for the exchange reaction is

$$Al_2(SO_4)_3 + NaOH \longrightarrow Al(OH)_3 + Na_2SO_4 \quad \text{(not balanced)}$$

From Table 4.1, you see that $Al_2(SO_4)_3$ is soluble (Rule 4), NaOH and $Na_2SO_4$ are soluble (Rule 1), and $Al(OH)_3$ is insoluble (Rule 8). Thus, aluminum hydroxide precipitates. The balanced molecular equation with phase labels is

$$Al_2(SO_4)_3(aq) + 6NaOH(aq) \longrightarrow 2Al(OH)_3(s) + 3Na_2SO_4(aq)$$

To get the net ionic equation, you write the strong electrolytes (here, soluble ionic compounds) as ions in aqueous solution and cancel spectator ions.

$$2Al^{3+}(aq) + 3\cancel{SO_4^{2-}(aq)} + 6\cancel{Na^+(aq)} + 6OH^-(aq) \longrightarrow$$
$$2Al(OH)_3(s) + 6\cancel{Na^+(aq)} + 3\cancel{SO_4^{2-}(aq)}$$

The net ionic equation is

$$Al^{3+}(aq) + 3OH^-(aq) \longrightarrow Al(OH)_3(s)$$

Thus, aluminum ion reacts with hydroxide ion to precipitate aluminum hydroxide.

### EXERCISE 4.3

You mix aqueous solutions of sodium iodide and lead(II) acetate. If a reaction occurs, write the balanced molecular equation and the net ionic equation. If no reaction occurs, indicate this by writing the formulas of the compounds and an arrow followed by *NR*.

See Problems 4.29, 4.30, 4.31, and 4.32.

**Problem Solving: Predicting Reaction Products**

 ### CONCEPT CHECK 4.2

Your lab partner tells you that she mixed two solutions that contain ions. You analyze the solution and find that it contains the ions and precipitate shown in the beaker.

a. Write the molecular equation for the reaction.

b. Write the complete ionic equation for the reaction.

c. Write the net ionic equation for the reaction.

$Na^+(aq)$

$C_2H_3O_2^-(aq)$

$SrSO_4(s)$

# 4.4 Acid–Base Reactions

Acids and bases are some of the most important electrolytes. You can recognize acids and bases by some simple properties. Acids have a sour taste. Solutions of bases, on the other hand, have a bitter taste and a soapy feel. (Of course, you should never taste laboratory chemicals.) Some examples of acids are acetic acid, present in vinegar; citric acid, a constituent of lemon juice; and hydrochloric acid, found in the digestive fluid of the stomach. An example of a base is aqueous ammonia, often used as a household cleaner. Table 4.2 lists further examples; see also Figure 4.7.

Another simple property of acids and bases is their ability to cause color changes in certain dyes. An **acid–base indicator** is *a dye used to distinguish between acidic and basic solutions by means of the color changes it undergoes in these solutions.* Such dyes are common in natural materials. The amber color of tea, for example, is

▶ Demonstration of acid–base indicators: Bassam Z. Shakhashiri, *Chemical Demonstrations*, Vol. 3 (Madison: University of Wisconsin Press, 1989), no. 8.1–8.3, pp. 33–49; Summerlin and Eary, *Chemical Demonstrations*, Vol. 1 (Washington, D.C.: Am. Chem. Soc., 1985), no. 22–25, pp. 37–40.

**Figure 4.7**
**Household acids and bases.** Shown are a variety of household products that are either acids or bases.

**Table 4.2**
**Common Acids and Bases**

| Name | Formula | Remarks |
|---|---|---|
| *Acids:* | | |
| Acetic acid | $HC_2H_3O_2$ | Found in vinegar |
| Acetylsalicylic acid | $HC_9H_7O_4$ | Aspirin |
| Ascorbic acid | $H_2C_6H_6O_6$ | Vitamin C |
| Citric acid | $H_3C_6H_5O_7$ | Found in lemon juice |
| Hydrochloric acid | HCl | Found in gastric juice (digestive fluid in stomach) |
| Sulfuric acid | $H_2SO_4$ | Battery acid |
| *Bases:* | | |
| Ammonia | $NH_3$ | Aqueous solution used as a household cleaner |
| Calcium hydroxide | $Ca(OH)_2$ | Slaked lime (used in mortar for building construction) |
| Magnesium hydroxide | $Mg(OH)_2$ | Milk of magnesia (antacid and laxative) |
| Sodium hydroxide | NaOH | Drain cleaners, oven cleaners |

**Figure 4.8**
**Red cabbage juice as an acid–base indicator.**
*Left:* Preparation of red cabbage juice. The beaker (green solution) contains red cabbage juice and sodium bicarbonate (baking soda). *Right:* Red cabbage juice has been added from a pipet to solutions in the beakers. These solutions vary in acidity from highly acidic on the left to highly basic on the right.

▶ Demonstration of indicators from plants: Bassam Z. Shakhashiri, *Chemical Demonstrations,* Vol. 3 (Madison: University of Wisconsin Press, 1989), no. 8.4, pp. 50–57.

lightened by the addition of lemon juice (citric acid). Red cabbage juice changes to green then yellow when a base is added (Figure 4.8). The green and yellow colors change back to red when an acid is added. Litmus is a common laboratory acid–base indicator. This dye, produced from certain species of lichens, turns red in acidic solution and blue in basic solution. Phenolphthalein (fee′ nol thay′ leen), another laboratory acid–base indicator, is colorless in acidic solution and pink in basic solution.

An important chemical characteristic of acids and bases is the way they react with one another. To understand these acid–base reactions, we need to have precise definitions of the terms *acid* and *base*.

## Definitions of Acid and Base

▶ Demonstration of properties of acids and bases: Bassam Z. Shakhashiri, *Chemical Demonstrations,* Vol. 3 (Madison: University of Wisconsin Press, 1989), no. 8.5, p. 58.

When Arrhenius developed his ionic theory of solutions, he also gave the classic definitions of acids and bases. According to Arrhenius, an **acid** is *a substance that produces hydrogen ions, $H^+$, when it dissolves in water.* An example is nitric acid, $HNO_3$, a molecular substance that dissolves in water to give $H^+$ and $NO_3^-$.

$$HNO_3(aq) \xrightarrow{H_2O} H^+(aq) + NO_3^-(aq)$$

▶ Video no. 37 (Series D), "Red Cabbage Juice."

According to Arrhenius, a **base** is *a substance that produces hydroxide ions, $OH^-$, when it dissolves in water.* Sodium hydroxide, $NaOH$, is an ionic solid that dissolves in water to give hydroxide ions. Thus, sodium hydroxide is a base.

$$NaOH(s) \xrightarrow{H_2O} Na^+(aq) + OH^-(aq)$$

The molecular substance ammonia, $NH_3$, is a base in the Arrhenius view, because it yields hydroxide ions when it reacts with water. ◀

▶ Solutions of ammonia are sometimes called ammonium hydroxide and given the formula $NH_4OH(aq)$, based on analogy with NaOH. No $NH_4OH$ molecule or compound has ever been found, however.

 +  ⇌  +

$$NH_3(aq) + H_2O(l) \rightleftharpoons NH_4^+(aq) + OH^-(aq)$$

Although the Arrhenius concept of acids and bases is useful, it is somewhat limited. For example, it tends to single out the $OH^-$ ion as the source of base character, when other ions or molecules can play a similar role. In 1923, Johannes N. Brønsted and Thomas M. Lowry independently noted that many reactions involve nothing more than the transfer of a proton ($H^+$) between reactants, and they realized that they could use this idea to expand the definitions of acids and bases to describe a large class of chemical reactions. In this view, acid–base reactions are *proton-transfer reactions.*

Consider the preceding reaction. It involves the transfer of a proton from the water molecule, $H_2O$, to the ammonia molecule, $NH_3$.

$$NH_3(aq) + H_2O(l) \rightleftharpoons NH_4^+(aq) + OH^-(aq)$$

Once the proton, $H^+$, has left the $H_2O$ molecule, it leaves behind an $OH^-$ ion. When the $H^+$ adds to $NH_3$, an $NH_4^+$ ion results. $H_2O$ is said to *donate* a proton to $NH_3$, and $NH_3$ is said to *accept* a proton from $H_2O$.

Brønsted and Lowry defined an **acid** as *the species (molecule or ion) that donates a proton to another species in a proton-transfer reaction.* They defined a **base** as *the species (molecule or ion) that accepts a proton in a proton-transfer reaction.* In the reaction of ammonia with water, the $H_2O$ molecule is the acid, because it donates a proton. The $NH_3$ molecule is a base, because it accepts a proton.

$$\underset{\text{base}}{NH_3}(aq) + \underset{\text{acid}}{H_2O}(l) \rightleftharpoons NH_4^+(aq) + OH^-(aq)$$

The dissolution of nitric acid, $HNO_3$, in water is actually a proton-transfer reaction, although the following equation, which we used as an illustration of an Arrhenius acid, does not spell that out.

$$HNO_3(aq) \xrightarrow{H_2O} H^+(aq) + NO_3^-(aq)$$

To see that this is a proton-transfer reaction, we need to clarify the structure of the hydrogen ion, $H^+(aq)$. This ion consists of a proton ($H^+$) in association with water molecules, which is what ($aq$) means. This is not a weak association, however, because the proton (or hydrogen nucleus) would be expected to attract electrons strongly to itself. In fact, the $H^+(aq)$ ion might be better thought of as a proton chemically bonded to a water molecule to give the $H_3O^+$ ion, with other water molecules less strongly associated with this ion, which we represent by the phase-labeled formula $H_3O^+(aq)$. Written in this form, we usually call this the *hydronium ion.* It is important to understand that the hydrogen ion, $H^+(aq)$, and the hydronium ion, $H_3O^+(aq)$, represent precisely the same physical ion. For simplicity, we often write the formula for this ion as $H^+(aq)$, but when we want to be explicit about the proton-transfer aspect of a reaction, we write the formula as $H_3O^+(aq)$. ◄

Now let us rewrite the preceding equation by replacing $H^+(aq)$ by $H_3O^+(aq)$. To maintain a balanced equation, we will also need to add $H_2O(l)$ to the left side.

$$HNO_3(aq) + H_2O(l) \longrightarrow H_3O^+(aq) + NO_3^-(aq)$$

► The Brønsted–Lowry concept is treated in detail in Chapter 16.

► Animation: Proton transfer.

► The $H^+(aq)$ ion has variable structure in solution. In solids, there is evidence for $[H(H_2O)_n]^+$, where $n$ is 1, 2, 3, 4, or 6.

Note that the reaction involves simply a transfer of a proton ($H^+$) from $HNO_3$ to $H_2O$:

$$\overset{\overset{\displaystyle H^+}{\frown}}{HNO_3(aq)} + \underset{base}{H_2O(l)} \longrightarrow NO_3^-(aq) + H_3O^+(aq)$$

The $HNO_3$ molecule is an acid (proton donor) and $H_2O$ is a base (proton acceptor). Note that $H_2O$ may function as an acid or a base, depending on the other reactant.

The Arrhenius definitions and those of Brønsted and Lowry are essentially equivalent for aqueous solutions, although their points of view are different. For instance, sodium hydroxide and ammonia are bases in the Arrhenius view because they increase the percentage of $OH^-$ ion in the aqueous solution. They are bases in the Brønsted–Lowry view because they provide species ($OH^-$ from the strong electrolyte sodium hydroxide and $NH_3$ from ammonia) that can accept protons. ◀

## Strong and Weak Acids and Bases

Acids and bases are classified as strong or weak, depending on whether they are strong or weak electrolytes. A **strong acid** is *an acid that ionizes completely in water; it is a strong electrolyte.* Hydrochloric acid, $HCl(aq)$, and nitric acid, $HNO_3(aq)$, are examples of strong acids. Using the hydronium ion notation, we write the respective equations as follows:

$$HCl(aq) + H_2O(l) \longrightarrow H_3O^+(aq) + Cl^-(aq)$$

$$HNO_3(aq) + H_2O(l) \longrightarrow H_3O^+(aq) + NO_3^-(aq)$$

Table 4.3 lists six common strong acids. Most of the other acids we will discuss are weak acids.

A **weak acid** is *an acid that only partly ionizes in water; it is a weak electrolyte.* An example of a weak acid is hydrocyanic acid, $HCN(aq)$. The hydrogen cyanide molecule, $HCN$, reacts with water to produce a small percentage of ions in solution.

$$HCN(aq) + H_2O(l) \rightleftharpoons H_3O^+(aq) + CN^-(aq)$$

A **strong base** is *a base that is present in aqueous solution entirely as ions, one of which is $OH^-$; it is a strong electrolyte.* The ionic compound sodium hydroxide, $NaOH$, is an example of a strong base. It dissolves in water as $Na^+$ and $OH^-$.

$$NaOH(s) \xrightarrow{H_2O} Na^+(aq) + OH^-(aq)$$

▶ **We will discuss the Brønsted–Lowry concept of acids and bases more thoroughly in Chapter 16.**

▶ Demonstration of the conductivity of strong and weak acids: Bassam Z. Shakhashiri, *Chemical Demonstrations,* Vol. 3 (Madison: University of Wisconsin Press, 1989), no. 8.21, pp. 140–145.

**Table 4.3**
**Common Strong Acids and Bases**

| Strong Acids | Strong Bases |
|---|---|
| $HClO_4$ | $LiOH$ |
| $H_2SO_4$ | $NaOH$ |
| $HI$ | $KOH$ |
| $HBr$ | $Ca(OH)_2$ |
| $HCl$ | $Sr(OH)_2$ |
| $HNO_3$ | $Ba(OH)_2$ |

The hydroxides of Groups IA and IIA elements, except for beryllium hydroxide, are strong bases (see Table 4.3).

A **weak base** is *a base that is only partly ionized in water; it is a weak electrolyte.* Ammonia, $NH_3$, is an example.

$$NH_3(aq) + H_2O(l) \rightleftharpoons NH_4^+(aq) + OH^-(aq)$$

You will find it important to be able to identify an acid or base as strong or weak. When you write an ionic equation, you represent strong acids and bases by the ions they form and weak acids and bases by the formulas of the compounds. The next example will give you some practice identifying acids and bases as strong or weak.

▶ An alternate example is provided in the Instructor's Resource Manual.

---
**EXAMPLE 4.4**

## Classifying Acids and Bases as Strong or Weak

Identify each of the following compounds as a strong or weak acid or base:

a. LiOH  b. $HC_2H_3O_2$  c. HBr  d. $HNO_2$

**SOLUTION**

Refer to Table 4.3 for the common strong acids and bases. You can assume that other acids and bases are weak.

a. As noted in Table 4.3, **LiOH is a strong base.**

b. Acetic acid, $HC_2H_3O_2$, is not one of the strong acids listed in Table 4.3; therefore, we assume **$HC_2H_3O_2$ is a weak acid.**

c. As noted in Table 4.3, **HBr is a strong acid.**

d. Nitrous acid, $HNO_2$, is not one of the strong acids listed in Table 4.3; therefore, we assume **$HNO_2$ is a weak acid.**

---
**EXERCISE 4.4**

Label each of the following as a strong or weak acid or base:

a. $H_3PO_4$  b. HClO  c. $HClO_4$  d. $Sr(OH)_2$

---
See Problems 4.33 and 4.34.

### Neutralization Reactions

▶ Demonstration of neutralization of acid by milk of magnesia: Summerlin et al., *Chemical Demonstrations,* Vol. 2 (Washington, D.C.: Am. Chem. Soc., 1987), no. 90, p. 171.

One of the chemical properties of acids and bases is that they neutralize one another. A **neutralization reaction** is *a reaction of an acid and a base that results in an ionic compound and possibly water.* When a base is added to an acid solution, the acid is said to be neutralized. *The ionic compound that is a product of a neutralization reaction* is called a **salt**. Most ionic compounds other than hydroxides and oxides are salts; that is, they can be obtained from neutralization reactions such as the following:

$$\underset{\text{acid}}{2HCl(aq)} + \underset{\text{base}}{Ca(OH)_2(aq)} \longrightarrow \underset{\text{salt}}{CaCl_2(aq)} + 2H_2O(l)$$

$$\underset{\text{acid}}{HCN(aq)} + \underset{\text{base}}{KOH(aq)} \longrightarrow \underset{\text{salt}}{KCN(aq)} + H_2O(l)$$

The salt formed in a neutralization reaction consists of cations obtained from the base and anions obtained from the acid. In the first example, the base is $Ca(OH)_2$, which supplies $Ca^{2+}$ cations; the acid is HCl, which supplies $Cl^-$ anions. The salt contains $Ca^{2+}$ and $Cl^-$ ions ($CaCl_2$).

We wrote these reactions as molecular equations. Written in this form, the equations make explicit the reactant compounds and the salts produced. However, to discuss the essential reactions that occur, you need to write the net ionic equations. You will see that the net ionic equation for each acid–base reaction involves the transfer of a proton.

The first reaction involves a strong acid, HCl($aq$), and a strong base, $Ca(OH)_2$($aq$). Both $Ca(OH)_2$ and the product $CaCl_2$, being soluble ionic compounds, are strong electrolytes. Writing these strong electrolytes in the form of ions gives the following complete ionic equation:

$$2H^+(aq) + 2\cancel{Cl^-(aq)} + \cancel{Ca^{2+}(aq)} + 2OH^-(aq) \longrightarrow \cancel{Ca^{2+}(aq)} + 2\cancel{Cl^-(aq)} + 2H_2O(l)$$

Canceling spectator ions and dividing by 2 gives the net ionic equation:

$$H^+(aq) + OH^-(aq) \xrightarrow{\ H^+\ } H_2O(l)$$

Note the transfer of a proton from the hydrogen ion (or hydronium ion, $H_3O^+$) to the hydroxide ion.

The second reaction involves HCN($aq$), a weak acid, and KOH($aq$), a strong base; the product is KCN, a strong electrolyte. The net ionic equation is:

$$HCN(aq) + OH^-(aq) \xrightarrow{\ H^+\ } CN^-(aq) + H_2O(l)$$

Note the proton transfer, characteristic of an acid–base reaction.

In each of these examples, hydroxide ions latch strongly onto protons to form water. Because water is a very stable substance, it effectively provides the driving force of the reaction.

Although water is one of the products in most neutralization reactions, the reaction of an acid with the base ammonia provides a prominent exception. Consider the reaction of sulfuric acid with ammonia:

$$\underset{\text{acid}}{H_2SO_4(aq)} + \underset{\text{base}}{2NH_3(aq)} \longrightarrow \underset{\text{salt}}{(NH_4)_2SO_4(aq)}$$

The net ionic equation is

$$H^+(aq) + NH_3(aq) \xrightarrow{\ H^+\ } NH_4^+(aq)$$

Note the proton transfer, a hallmark of an acid–base reaction. In this case, $NH_3$ molecules latch onto protons and form relatively stable $NH_4^+$ ions.

---

### EXAMPLE 4.5

## Writing an Equation for a Neutralization

Write the molecular equation and then the net ionic equation for the neutralization of nitrous acid, $HNO_2$, by sodium hydroxide, NaOH, both in aqueous solution. Use an arrow with $H^+$ over it to show the proton transfer.

▶ An alternate example is provided in the Instructor's Resource Manual.

**PROBLEM STRATEGY**

Start by writing the acid and base reactants and the salt product. Recall that the salt consists of cations from the base and anions from the acid. Note also that water is frequently a product; if it is, you will not be able to balance the equation without it. Once you have the balanced molecular equation, write the complete ionic equation. Do that by representing any strong electrolytes by the formulas of their ions. Finally, write the net ionic equation by canceling any spectator ions, and from it note the proton transfer.

**SOLUTION**

The salt consists of the cation from the base ($Na^+$) and the anion from the acid ($NO_2^-$); its formula is $NaNO_2$. You will need to add $H_2O$ as a product to complete and balance the molecular equation:

$$HNO_2(aq) + NaOH(aq) \longrightarrow NaNO_2(aq) + H_2O(l) \quad \textbf{(molecular equation)}$$

Note that NaOH (a strong base) and $NaNO_2$ (a soluble ionic substance) are strong electrolytes; $HNO_2$ is a weak electrolyte (it is not one of the strong acids in Table 4.3). You write both NaOH and $NaNO_2$ in the form of ions. The complete ionic equation is

$$HNO_2(aq) + \cancel{Na^+(aq)} + OH^-(aq) \longrightarrow \cancel{Na^+(aq)} + NO_2^-(aq) + H_2O(l)$$

The net ionic equation is

$$HNO_2(aq) + OH^-(aq) \longrightarrow NO_2^-(aq) + H_2O(l)$$

Note that a proton is transferred from $HNO_2$ to the $OH^-$ ion to yield the products:

$$\overset{\overset{\displaystyle H^+}{\curvearrowright}}{HNO_2(aq)} + OH^-(aq) \longrightarrow NO_2^-(aq) + H_2O(l) \quad \textbf{(net ionic equation)}$$

**EXERCISE 4.5**

Write the molecular equation and the net ionic equation for the neutralization of hydrocyanic acid, HCN, by lithium hydroxide, LiOH, both in aqueous solution.

See Problems 4.35, 4.36, 4.37, and 4.38.

Acids such as HCl and $HNO_3$ that have only one acidic hydrogen atom per acid molecule are called *monoprotic* acids. A **polyprotic acid** is *an acid that yields two or more acidic hydrogens per molecule*. Phosphoric acid is an example of a triprotic acid. By reacting this acid with different amounts of a base, you can obtain a series of salts:

$$H_3PO_4(aq) + NaOH(aq) \longrightarrow NaH_2PO_4(aq) + H_2O(l)$$
$$H_3PO_4(aq) + 2NaOH(aq) \longrightarrow Na_2HPO_4(aq) + 2H_2O(l)$$
$$H_3PO_4(aq) + 3NaOH(aq) \longrightarrow Na_3PO_4(aq) + 3H_2O(l)$$

Salts such as $NaH_2PO_4$ and $Na_2HPO_4$ that have acidic hydrogen atoms and can undergo neutralization with bases are called *acid salts*.

▶ Video no. 36 (Series D), "Bicarb./Hydrochloric Acid Volcano."

▶ Demonstration of reaction of acids with carbonates: Bassam Z. Shakhashiri, *Chemical Demonstrations,* Vol. 3 (Madison: University of Wisconsin Press, 1989), no. 8.11, pp. 96–99.

---

### EXERCISE 4.6

Write molecular and net ionic equations for the successive neutralizations of each of the acidic hydrogens of sulfuric acid with potassium hydroxide. (That is, write equations for the reaction of sulfuric acid with KOH to give the acid salt and for the reaction of the acid salt with more KOH to give potassium sulfate.)

See Problems 4.39, 4.40, 4.41, and 4.42.

---

### Acid–Base Reactions with Gas Formation

Certain salts, notably carbonates, sulfites, and sulfides, react with acids to form a gaseous product. Consider the reaction of sodium carbonate with hydrochloric acid. The molecular equation for the reaction is

$$\underset{\text{carbonate}}{Na_2CO_3(aq)} + \underset{\text{acid}}{2HCl(aq)} \longrightarrow \underset{\text{salt}}{2NaCl(aq)} + H_2O(l) + CO_2(g)$$

Here, a carbonate (sodium carbonate) reacts with an acid (hydrochloric acid) to produce a salt (sodium chloride), water, and carbon dioxide gas. A similar reaction is shown in Figure 4.9, in which baking soda (sodium hydrogen carbonate) reacts with the acetic acid in vinegar. Note the bubbles of carbon dioxide gas that evolve during the reaction. This reaction of a carbonate with an acid is the basis of a simple test for carbonate minerals. When you treat a carbonate mineral or rock, such as limestone, with hydrochloric acid, the material fizzes, as bubbles of odorless carbon dioxide form.

It is useful to consider the preceding reaction as an exchange, or metathesis, reaction. Recall that in an exchange reaction, you obtain the products from the reactants by exchanging the anions (or the cations) between the two reactants. In this case, we interchange the carbonate ion with the chloride ion and we obtain the following:

$$Na_2CO_3(aq) + 2HCl(aq) \longrightarrow 2NaCl(aq) + H_2CO_3(aq)$$

The last product shown in the equation is carbonic acid, $H_2CO_3$. (You obtain the formula of carbonic acid by noting that you need to associate two $H^+$ ions with the carbonate ion, $CO_3^{2-}$, to obtain a neutral compound.) Carbonic acid is unstable and decomposes to water and carbon dioxide gas. The overall result is the equation we wrote earlier:

$$Na_2CO_3(aq) + 2HCl(aq) \longrightarrow 2NaCl(aq) + \underbrace{H_2O(l) + CO_2(g)}_{H_2CO_3(aq)}$$

The net ionic equation for this reaction is

$$CO_3^{2-}(aq) + 2H^+(aq) \longrightarrow H_2O(l) + CO_2(g)$$

The carbonate ion reacts with hydrogen ion from the acid. If you write the hydrogen ion as $H_3O^+$ (hydronium ion), you can see also that the reaction involves a proton transfer:

$$\overset{2H^+}{\overgroup{CO_3^{2-}(aq) + 2H_3O^+(aq)}} \longrightarrow H_2CO_3(aq) + 2H_2O(l) \longrightarrow 3H_2O(l) + CO_2(g)$$

From the broader perspective of the Brønsted–Lowry view, this is an acid–base reaction.

Sulfites behave similarly to carbonates. When a sulfite, such as sodium sulfite ($Na_2SO_3$), reacts with an acid, sulfur dioxide ($SO_2$) is a product. Sulfides, such as sodium sulfide ($Na_2S$), react with acids to produce hydrogen sulfide gas. The reaction can be viewed as a simple exchange reaction. The reactions are summarized in Table 4.4. ◀

**Figure 4.9**
**Reaction of a carbonate with an acid.** Baking soda (sodium hydrogen carbonate) reacts with acetic acid in vinegar to evolve bubbles of carbon dioxide.

▶ $CO_3^{2-}$, $S^{2-}$, and $SO_3^{2-}$ are bases. Thus the net equations involve a basic anion with $H^+$, an acid.

**Table 4.4**
**Some Ionic Compounds That Evolve Gases When Treated with Acids**

| Ionic Compound | Gas | Example |
|---|---|---|
| Carbonate ($CO_3^{2-}$) $CO_2$ | $CO_2$ | $Na_2CO_3 + 2HCl \longrightarrow 2NaCl + H_2O +$ |
| Sulfite ($SO_3^{2-}$) | $SO_2$ | $Na_2SO_3 + 2HCl \longrightarrow 2NaCl + H_2O + SO_2$ |

▶ An alternate example is provided in the Instructor's Resource Manual.

▶ Experiments in General Chemistry, 4B, "Ionic Reactions in Aqueous Solutions."

**EXAMPLE 4.6**

## Writing an Equation for a Reaction with Gas Formation

Write the molecular equation and the net ionic equation for the reaction of zinc sulfide with hydrochloric acid.

**SOLUTION**

First write the reactants, noting from Table 4.1 that most sulfides are insoluble (except those of the IA metals and $(NH_4)_2S$).

$$ZnS(s) + HCl(aq) \longrightarrow$$

If you look at this as an exchange reaction, the products are $ZnCl_2$ and $H_2S$. Since chlorides are soluble (with some exceptions, not including $ZnCl_2$), we write the following balanced equation.

$$\textbf{ZnS}(s) + \textbf{2HCl}(aq) \longrightarrow \textbf{ZnCl}_2(aq) + \textbf{H}_2\textbf{S}(g)$$

The complete ionic equation is

$$ZnS(s) + 2H^+(aq) + 2Cl^-(aq) \longrightarrow Zn^{2+}(aq) + 2Cl^-(aq) + H_2S(g)$$

and the net ionic equation is

$$\textbf{ZnS}(s) + \textbf{2H}^+(aq) \longrightarrow \textbf{Zn}^{2+}(aq) + \textbf{H}_2\textbf{S}(g)$$

**EXERCISE 4.7**

Write the molecular equation and the net ionic equation for the reaction of calcium carbonate with nitric acid.

See Problems 4.43, 4.44, 4.45, and 4.46.

 **CONCEPT CHECK 4.3**

At times, we want to generalize the formula of certain important chemical substances; acids and bases fall into this category. Given the following reactions, try to identify the acids, bases, and some examples of what the general symbols (M and $A^-$) represent.

a. $MOH(s) \longrightarrow M^+(aq) + OH^-(aq)$

b. $HA(aq) + H_2O(l) \rightleftharpoons H_3O^+(aq) + A^-(aq)$

c. $H_2A(aq) + 2H_2O(l) \rightleftharpoons H_3O^+(aq) + HA^-(aq)$

d. For parts a through c, come up with real examples for M and A.

## 4.5 Oxidation–Reduction Reactions

▶ Video no. 24 (Series D), "The Gold Penny."

▶ Animation: Oxidation–reduction reaction.

▶ Video no. 27 (Series D), "Complexes of Vanadium."

In the two preceding sections, we described precipitation reactions (reactions producing a precipitate) and acid–base reactions (reactions involving proton transfer). Here we discuss the third major class of reactions, oxidation–reduction reactions, which are reactions involving a transfer of electrons from one species to another.

As a simple example of an oxidation–reduction reaction, let us look at what happens when you dip an iron nail into a blue solution of copper(II) sulfate (Figure 4.10). What you see is that the iron nail becomes coated with a reddish-brown tinge of metallic copper. The molecular equation for this reaction is

$$Fe(s) + CuSO_4(aq) \longrightarrow FeSO_4(aq) + Cu(s)$$

From this equation, you can obtain the net ionic equation, which shows that iron metal reacts with copper(II) ion to produce iron(II) ion and copper metal.

$$Fe(s) + Cu^{2+}(aq) \longrightarrow Fe^{2+}(aq) + Cu(s)$$

The electron-transfer aspect of the reaction is apparent from this equation. Note that each iron atom in the metal loses two electrons to form an iron(II) ion, and each copper(II) ion gains two electrons to form a copper atom in the metal. The net effect is that two electrons are transferred from each iron atom in the metal to each copper(II) ion.

The concept of *oxidation numbers* was developed as a simple way of keeping track of electrons in a reaction. Using oxidation numbers, you can determine whether or not electrons have been transferred from one atom to another. If electrons have been transferred, an oxidation–reduction reaction has occurred.

### Oxidation Numbers

We define the **oxidation number** (or **oxidation state**) of an atom in a substance as *the actual charge of the atom if it exists as a monatomic ion, or a hypothetical charge assigned to the atom in the substance by simple rules.* An oxidation–reduction reac-

**Figure 4.10**
**Reaction of iron with $Cu^{2+}$(aq).** *Left:* Iron nail and copper(II) sulfate solution, which has a blue color. *Center:* Fe reacts with $Cu^{2+}$(aq) to yield $Fe^{2+}$(aq) and Cu(s). *Right:* The copper metal plates out on the nail.

tion is one in which one or more atoms change oxidation number, implying that there has been a transfer of electrons.

Consider the combustion of calcium metal in oxygen gas (Figure 4.11).

$$2Ca(s) + O_2(g) \longrightarrow 2CaO(s)$$

This is an oxidation–reduction reaction. To see this, you assign oxidation numbers to the atoms in the equation and then note that the atoms change oxidation number during the reaction.

The oxidation number of an atom in an element is always zero. Thus, Ca in the metal and O in $O_2$ have oxidation numbers of zero. Another rule follows from the definition of oxidation number: The oxidation number of an atom that exists in a substance as a monatomic ion equals the charge on that ion. So the oxidation number of Ca in CaO is $+2$ (the charge on $Ca^{2+}$), and the oxidation number of O in CaO is $-2$ (the charge on $O^{2-}$). To emphasize these oxidation numbers in an equation, we will write them above the atomic symbols in the formulas.

$$\overset{0}{2Ca}(s) + \overset{0}{O_2}(g) \longrightarrow \overset{+2\ -2}{2CaO}(s)$$

From this, you see that the Ca and O atoms change in oxidation number during the reaction. In effect, each calcium atom in the metal loses two electrons to form $Ca^{2+}$ ions, and each oxygen atom in $O_2$ gains two electrons to form $O^{2-}$ ions. The net result is a transfer of electrons from calcium to oxygen, so this reaction is an oxidation–reduction reaction. In other words, an **oxidation–reduction reaction** (or **redox reaction**) is *a reaction in which electrons are transferred between species or in which atoms change oxidation number.*

Note that calcium has gained in oxidation number from 0 to $+2$. (Each calcium atom loses two electrons.) We say that calcium has been *oxidized*. Oxygen, on the other hand, has decreased in oxidation number from 0 to $-2$. (Each oxygen atom gains two electrons.) We say that oxygen has been *reduced*. An oxidation–reduction reaction always involves both oxidation (the loss of electrons) and reduction (the gain of electrons).

Formerly, the term *oxidation* meant "reaction with oxygen." The current definition greatly enlarges the meaning of this term. Consider the reaction of calcium metal with chlorine gas (Figure 4.12); the reaction looks similar to the burning of calcium in oxygen. The chemical equation is

$$\overset{0}{Ca}(s) + \overset{0}{Cl_2}(g) \longrightarrow \overset{+2\ -1}{CaCl_2}(s)$$

In this reaction, the calcium atom is oxidized, because it increases in oxidation number (from 0 to $+2$, as in the previous equation). Chlorine is reduced; it decreases in oxidation number from 0 to $-1$. This is clearly an oxidation–reduction reaction that does not involve oxygen.

## Oxidation-Number Rules

So far, we have used two rules for obtaining oxidation numbers: (1) the oxidation number of an atom in an element is zero, and (2) the oxidation number of an atom in a monatomic ion equals the charge on the ion. These and several other rules for assigning oxidation numbers are given in Table 4.5.

In molecular substances, we use these rules to give the *approximate* charges on the atoms. Consider the molecule $SO_2$. Oxygen atoms tend to attract electrons, pulling them from other atoms (sulfur in the case of $SO_2$). As a result, an oxygen atom in

**Figure 4.11**
**The burning of calcium metal in oxygen.** The burning calcium emits a red-orange flame.

**Figure 4.12**
**The burning of calcium metal in chlorine.** The reaction appears similar to the burning of calcium in oxygen.

$SO_2$ takes on a negative charge relative to the sulfur atom. The magnitude of the charge on an oxygen atom in a molecule is not a full $-2$ charge as in the $O^{2-}$ ion. However, it is convenient to assign an oxidation number of $-2$ to oxygen in $SO_2$ (and in most other compounds of oxygen) to help us express the approximate charge distribution in the molecule. Rule 3 in Table 4.5 says that an oxygen atom has an oxidation number of $-2$ in most of its compounds.

Rules 4 and 5 are similar in that they tell you what to expect for the oxidation number of certain elements in their compounds. Rule 4, for instance, says that hydrogen has an oxidation number of $+1$ in most of its compounds.

Rule 6 states that the sum of the oxidation numbers of the atoms in a compound is zero. This rule follows from the interpretation of oxidation numbers as (hypothetical) charges on the atoms. Because any compound is electrically neutral, the sum of the charges on its atoms must be zero. This rule is easily extended to ions: the sum of the oxidation numbers (hypothetical charges) of the atoms in a polyatomic ion equals the charge on the ion.

You can use Rule 6 to obtain the oxidation number of one atom in a compound or ion, if you know the oxidation numbers of the other atoms in the compound or ion. Consider the $SO_2$ molecule. According to Rule 6,

$$\text{(Oxidation number of S)} + 2 \times \text{(oxidation number of O)} = 0$$

or

$$\text{(Oxidation number of S)} + 2 \times (-2) = 0.$$

Therefore,

$$\text{Oxidation number of S (in } SO_2) = -2 \times (-2) = +4$$

**Table 4.5**
**Rules for Assigning Oxidation Numbers**

| Rule | Applies to | Statement |
|---|---|---|
| 1 | Elements | The oxidation number of an atom in an element is zero. |
| 2 | Monatomic ions | The oxidation number of an atom in a monatomic ion equals the charge on the ion. |
| 3 | Oxygen | The oxidation number of oxygen is $-2$ in most of its compounds. (An exception is O in $H_2O_2$ and other peroxides, where the oxidation number is $-1$.) |
| 4 | Hydrogen | The oxidation number of hydrogen is $+1$ in most of its compounds. (The oxidation number of hydrogen is $-1$ in binary compounds with a metal, such as $CaH_2$.) |
| 5 | Halogens | The oxidation number of fluorine is $-1$ in all of its compounds. Each of the other halogens (Cl, Br, I) has an oxidation number of $-1$ in binary compounds, except when the other element is another halogen above it in the periodic table or the other element is oxygen. |
| 6 | Compounds and ions | The sum of the oxidation numbers of the atoms in a compound is zero. The sum of the oxidation numbers of the atoms in a polyatomic ion equals the charge on the ion. |

The next example illustrates the use of the rules in Table 4.5 to assign the oxidation numbers to all atoms in a compound or ion.

► Two alternate examples are provided in the Instructor's Resource Manual.

### EXAMPLE 4.7

## Assigning Oxidation Numbers

Use the rules from Table 4.5 to obtain the oxidation number of the chlorine atom in each of the following: (a) $HClO_4$ (perchloric acid), (b) $ClO_3^-$ (chlorate ion).

 **PROBLEM STRATEGY**

In each case, write the expression for the sum of the oxidation numbers, equating this to zero for a compound or to the charge for an ion (Rule 6). Now, use Rules 2 to 5 to substitute oxidation numbers for particular atoms, such as $-2$ for oxygen and $+1$ for hydrogen, and solve for the unknown oxidation number (Cl in this example).

**SOLUTION**

a. For perchloric acid, Rule 6 gives the equation

$$\text{(Oxidation number of H)} + \text{(oxidation number of Cl)} + 4 \times \text{(oxidation number of O)} = 0$$

Using Rules 3 and 4, you obtain

$$(+1) + \text{(oxidation number of Cl)} + 4 \times (-2) = 0$$

Therefore,

$$\text{Oxidation number of Cl (in } HClO_4) = -(+1) - 4 \times (-2) = \mathbf{+7}$$

b. For the chlorate ion, Rule 6 gives the equation

$$\text{(Oxidation number of Cl)} + 3 \times \text{(oxidation number of O)} = -1$$

Using Rule 3, you obtain

$$\text{(Oxidation number of Cl)} + 3 \times (-2) = -1$$

Therefore,

$$\text{Oxidation number of Cl (in } ClO_3^-) = -1 - 3 \times (-2) = \mathbf{+5}$$

### EXERCISE 4.8

Obtain the oxidation numbers of the atoms in each of the following: (a) potassium dichromate, $K_2Cr_2O_7$, (b) permanganate ion, $MnO_4^-$.

**Problem Solving: Assigning Oxidation States**

See Problems 4.47, 4.48, 4.49, and 4.50.

## Describing Oxidation–Reduction Reactions

We use special terminology to describe oxidation–reduction reactions. To illustrate this, we will look again at the reaction of iron with copper(II) sulfate. The net ionic equation is

$$\overset{0}{Fe}(s) + \overset{+2}{Cu^{2+}}(aq) \longrightarrow \overset{+2}{Fe^{2+}}(aq) + \overset{0}{Cu}(s)$$

We can write this reaction in terms of two half-reactions. A **half-reaction** is *one of two parts of an oxidation–reduction reaction, one part of which involves a loss of electrons (or increase of oxidation number) and the other a gain of electrons (or decrease of oxidation number).* The half-reactions for the preceding equation are

$$\overset{0}{Fe}(s) \longrightarrow \overset{+2}{Fe^{2+}}(aq) + 2e^- \qquad \text{(electrons lost by Fe)}$$

$$\overset{+2}{Cu^{2+}}(aq) + 2e^- \longrightarrow \overset{0}{Cu}(s) \qquad \text{(electrons gained by } Cu^{2+})$$

**Oxidation** is *the half-reaction in which there is a loss of electrons by a species (or an increase of oxidation number of an atom).* **Reduction** is *the half-reaction in which there is a gain of electrons by a species (or a decrease in the oxidation number of an atom).* Thus, the equation $Fe(s) \longrightarrow Fe^{2+}(aq) + 2e^-$ represents the oxidation half-reaction, and the equation $Cu^{2+}(aq) + 2e^- \longrightarrow Cu(s)$ represents the reduction half-reaction.

Recall that a species that is *oxidized* loses electrons (or contains an atom that increases in oxidation number) and a species that is *reduced* gains electrons (or contains an atom that decreases in oxidation number). An **oxidizing agent** is *a species that oxidizes another species; it is itself reduced.* Similarly, a **reducing agent** is *a species that reduces another species; it is itself oxidized.* In our example reaction, the copper(II) ion is the oxidizing agent, whereas iron metal is the reducing agent.

The relationships among these terms are shown in the following diagram for the reaction of iron with copper(II) ion.

## Some Common Oxidation–Reduction Reactions

Many of the oxidation–reduction reactions we will discuss in this book fall into the following simple categories:

1. Combination reactions

2. Decomposition reactions

3. Displacement reactions

4. Combustion reactions

We will describe examples of each of these in this section.

▶ Video Series C, "Zinc and iodine."

**Combination Reactions** A **combination reaction** is *a reaction in which two substances combine to form a third substance.* Not all combination reactions are oxidation–reduction reactions. However, the simplest cases are those in which two elements react to form a compound; these are clearly oxidation–reduction reactions. In Chapter 2, we discussed the reaction of sodium metal and chlorine gas (Figure 4.13).

$$2Na(s) + Cl_2(g) \longrightarrow 2NaCl(s)$$

**Figure 4.13**
**Combination reaction.**   *Left:* Sodium metal and chlorine gas. *Right:* The spectacular combination reaction of sodium and chlorine.

Antimony and chlorine also combine in a fiery reaction.

$$2Sb + 3Cl_2 \longrightarrow 2SbCl_3$$

Other combination reactions involve compounds as reactants. For example,

$$CaO(s) + SO_2(g) \longrightarrow CaSO_3(s)$$

If you check the oxidation numbers, you will see that this is not an oxidation–reduction reaction.

▶ Video no. 19 (Series A),
"Ammonium Dichromate
Volcano."

▶ Video Series C,
"Ammonium Dichromate
Volcano."

**Decomposition Reactions**   A **decomposition reaction** is *a reaction in which a single compound reacts to give two or more substances.* Often these reactions occur when the temperature is raised. In Chapter 1, we described the decomposition of mercury(II) oxide into its elements when the compound is heated (Figure 4.14). This is an oxidation–reduction reaction.

$$2HgO(s) \xrightarrow{\Delta} 2Hg(l) + O_2(g)$$

A common method of preparing oxygen in the laboratory consists of heating potassium chlorate with manganese(IV) oxide (manganese dioxide) as a catalyst.

$$2KClO_3(s) \xrightarrow[MnO_2]{\Delta} 2KCl(s) + 3O_2(g)$$

In this reaction, a compound decomposes into another compound and an element; it is an oxidation–reduction reaction.

Not all decomposition reactions are of the oxidation–reduction type. For example, when you heat calcium carbonate (limestone) to a high enough temperature, it decomposes into calcium oxide and carbon dioxide.

$$CaCO_3(s) \xrightarrow{\Delta} CaO(s) + CO_2(g)$$

Check the oxidation numbers to verify that this is not an oxidation–reduction reaction.

**Figure 4.14**
**Decomposition reaction.**
The decomposition reaction of mercury(II) oxide into its elements, mercury and oxygen.

 **Demonstration of metal/metal salt displacement reactions:** Summerlin and Ealy, *Chemical Demonstrations,* Vol. 1 (Washington, D.C.: Am. Chem. Soc., 1985), no. 96, p. 151.

 **Video Series C,** "Reactions of metals with acids."

 **Demonstration of a displacement reaction:** Summerlin et al., *Chemical Demonstrations,* Vol. 2 (Washington, D.C.: Am. Chem. Soc., 1987), no. 105, p. 107.

 **Video no. 8 (Series D),** "Sodium and Potassium in Water."

**Displacement Reactions** A **displacement reaction** (also called a **single-replacement reaction**) is *a reaction in which an element reacts with a compound, displacing an element from it*. Since these reactions involve an element and one of its compounds, these must be oxidation–reduction reactions. An example is the reaction that occurs when you dip a copper metal strip into a solution of silver nitrate.

$$Cu(s) + 2AgNO_3(aq) \longrightarrow Cu(NO_3)_2(aq) + 2Ag(s)$$

From the molecular equation, it appears that copper displaces silver in silver nitrate, producing crystals of silver metal and a greenish-blue solution of copper(II) nitrate. The net ionic equation, however, shows that the reaction involves the transfer of electrons from copper metal to silver ion:

$$Cu(s) + 2Ag^+(aq) \longrightarrow Cu^{2+}(aq) + 2Ag(s)$$

When you dip a zinc metal strip into an acid solution, bubbles of hydrogen form on the metal and escape from the solution (Figure 4.15).

$$Zn(s) + 2HCl(aq) \longrightarrow ZnCl_2(aq) + H_2(g)$$

Zinc displaces hydrogen in the acid, producing zinc chloride solution and hydrogen gas. The net ionic equation is

$$Zn(s) + 2H^+(aq) \longrightarrow Zn^{2+}(aq) + H_2(g)$$

Whether a reaction occurs between a given element and a monatomic ion depends on the relative ease with which the two species gain or lose electrons. Table 4.6 shows the activity series of the elements, a listing of the elements in decreasing order of their ease of losing electrons during reactions in aqueous solution. The metals listed at the top are the strongest reducing agents (they lose electrons easily); those at the bottom, the weakest. A free element reacts with the monatomic ion of another element if the free element is above the other element in the activity series.

**Figure 4.15**
**Displacement reaction.** Displacement reaction of zinc metal and hydrochloric acid. Hydrogen gas formed in the reaction bubbles from the metal surface that dips into the acid.

**Table 4.6**
**Activity Series of the Elements**

React vigorously with acidic solutions to give $H_2$
$\left\{\begin{array}{l} \text{Li} \\ \text{K} \\ \text{Ba} \\ \text{Ca} \\ \text{Na} \end{array}\right\}$
React vigorously with liquid water to give $H_2$

React with acids to give $H_2$
$\left\{\begin{array}{l} \text{Mg} \\ \text{Al} \\ \text{Zn} \\ \text{Cr} \\ \text{Fe} \\ \text{Cd} \end{array}\right\}$
React slowly with liquid water but readily with steam to give $H_2$

Co
Ni
Sn
Pb

$H_2$

Do not react with acids to give $H_2$*
$\left\{\begin{array}{l} \text{Cu} \\ \text{Hg} \\ \text{Ag} \\ \text{Au} \end{array}\right.$

*Cu, Hg, and Ag react with $HNO_3$ but do not produce $H_2$. In these reactions, the metal is oxidized to the metal ion, and $NO_3^-$ ion is reduced to $NO_2$ or other nitrogen species.

**Figure 4.16**
**Combustion reaction.** The combustion of iron wool. Iron reacts with oxygen in the air to produce iron(III) oxide, $Fe_2O_3$. The reaction is similar to the rusting of iron but is much faster.

Consider this reaction:

$$2K(s) + 2H^+(aq) \longrightarrow 2K^+(aq) + H_2(g)$$

You would expect this reaction to proceed as written, because potassium metal (K) is well above hydrogen in the activity series. In fact, potassium metal reacts violently with water, which contains only a very small percentage of $H^+$ ions. Imagine the reaction of potassium metal with a strong acid like HCl!

**Combustion Reactions** A **combustion reaction** is *a reaction in which a substance reacts with oxygen, usually with the rapid release of heat to produce a flame.* The products include one or more oxides. Oxygen changes oxidation number from 0 to −2, so combustions are oxidation–reduction reactions.

Organic compounds usually burn in oxygen or air to yield carbon dioxide. If the compound contains hydrogen (as most do), water is also a product. For instance, butane ($C_4H_{10}$) burns in air as follows:

$$2C_4H_{10}(g) + 13O_2(g) \longrightarrow 8CO_2(g) + 10H_2O(g)$$

▶ **Video no. 3 (Series B), "Does Steel Wool Burn?"**

Many metals burn in air as well. Although chunks of iron do not burn readily in air, iron wool, which consists of fine strands of iron, does (Figure 4.16). The increased surface area of the metal in iron wool allows oxygen from air to react quickly with it.

$$4Fe(s) + 3O_2(g) \longrightarrow 2Fe_2O_3(s)$$

The rusting of iron wool is essentially a similar reaction, although slower (Figure 4.17).

## 4.6 Balancing Simple Oxidation–Reduction Equations

▶ **Balancing equations was discussed in Section 2.10.**

Oxidation–reduction reactions can often be quite difficult to balance. ◀ Some are so complex in fact that chemists have written computer programs to accomplish the task. In this section, we will develop a method for balancing simple oxidation–reduction reactions that can later be generalized for far more complex reactions. One of the advantages of using this technique for even simple reactions is that you focus on what makes oxidation–reduction reactions different from other reaction types. See Chapter 20 for a comprehensive treatment of this topic.

At first glance, the equation representing the reaction of zinc metal with silver(I) ions in solution might appear to be balanced.

$$Zn(s) + Ag^+(aq) \longrightarrow Zn^{2+}(aq) + Ag(s)$$

However, because a balanced chemical equation must have a charge balance as well as a mass balance, this equation is not balanced: it has a total charge of $+1$ for the reactants and $+2$ for the products. In a simple case like this, you might balance the equation by inspection, or you could apply a different approach, called the half-reaction method.

## Half-Reaction Method Applied to Simple Oxidation–Reduction Equations

The *half-reaction method* consists of first separating the equation into two half-reactions, one for oxidation, the other for reduction. You balance each half-reaction, then combine them to obtain a balanced oxidation–reduction reaction. Here is an illustration of the process. First we identify the species being oxidized and reduced and assign the appropriate oxidation states.

$$\overset{0}{Zn}(s) + \overset{+1}{Ag^+}(aq) \longrightarrow \overset{+2}{Zn^{2+}}(aq) + \overset{0}{Ag}(s)$$

Next, write the half-reactions in an unbalanced form.

$$Zn \longrightarrow Zn^{2+} \qquad \text{(oxidation)}$$
$$Ag^+ \longrightarrow Ag \qquad \text{(reduction)}$$

Next, balance the charge in each equation by adding electrons to the more positive side to create balanced half-reactions. Following this procedure, the balanced half-reactions are:

$$Zn \longrightarrow Zn^{2+} + 2e^- \qquad \text{(oxidation half-reaction)}$$
$$Ag^+ + e^- \longrightarrow Ag \qquad \text{(reduction half-reaction)}$$

Note that the number of electrons that Zn loses during the oxidation process (two) exceeds the number of electrons gained by $Ag^+$ during the reduction (one). Since, according to the reduction half-reaction, each $Ag^+$ is capable of gaining only one electron, we need to double the amount of $Ag^+$ in order for it to accept all of the electrons produced by Zn during oxidation. To meet this goal and obtain the balanced oxidation–reduction reaction, we multiply each half-reaction by a factor (integer) so that when we add them together, the electrons cancel. We multiply the first equation by 1 (the number of electrons in the second half-reaction) and multiply the second equation by 2 (the number of electrons in the first half-reaction).

$$1 \times (Zn \longrightarrow Zn^{2+} + 2e^-)$$
$$\underline{2 \times (Ag^+ + e^- \longrightarrow Ag)}$$
$$Zn + 2Ag^+ + \cancel{2e^-} \longrightarrow Zn^{2+} + 2Ag + \cancel{2e^-}$$

The electrons cancel, which finally yields the balanced oxidation–reduction equation:

$$Zn(s) + 2Ag^+(aq) \longrightarrow Zn^{2+}(aq) + 2Ag(s)$$

**Figure 4.17**
**Rusting of iron wool.** The temperature rises perceptibly during the rusting, showing that heat is released by the reaction just as if the iron were being burned (see Figure 4.16). Note the temperature increase of the thermometer wrapped with moist iron wool.

Example 4.8 further illustrates this technique.

---

**EXAMPLE 4.8**

## Balancing Simple Oxidation–Reduction Reactions by the Half-Reaction Method

Consider a more difficult problem, the combination (oxidation–reduction) reaction of magnesium metal and nitrogen gas:

$$Mg(s) + N_2(g) \longrightarrow Mg_3N_2(s)$$

Apply the half-reaction method to balance this equation.

 **PROBLEM STRATEGY**

Start by identifying the species undergoing oxidation and reduction and assigning oxidation numbers. Then write the two balanced half-reactions, keeping in mind that you add the electrons to the more positive side of the reaction. Next, multiply each of the half-reactions by a whole number that will cancel the electrons on each side of the equation. Finally, add the half-reactions together to yield the balanced equation.

**SOLUTION**

Identify the oxidation states of the elements:

$$\overset{0}{Mg}(s) + \overset{0}{N_2}(g) \longrightarrow \overset{+2\ -3}{Mg_3N_2}(s)$$

In this problem, a molecular compound, nitrogen ($N_2$), is undergoing reduction. When a species undergoing reduction or oxidation is a molecule, write the formula of the molecule (do not split it up) in the half-reaction. Also, make sure that both the mass and the charge are balanced. (Note the $6e^-$ required to balance the charge due to the $2N^{3-}$.)

$$Mg \longrightarrow Mg^{2+} + 2e^- \qquad \text{(balanced oxidation half-reaction)}$$
$$N_2 + 6e^- \longrightarrow 2N^{3-} \qquad \text{(balanced reduction half-reaction)}$$

We now need to multiply each half-reaction by a factor that will cancel the electrons.

$$3 \times (Mg \longrightarrow Mg^{2+} + 2e^-)$$
$$1 \times (N_2 + 6e^- \longrightarrow 2N^{3-})$$
$$\overline{3Mg + N_2 + \cancel{6e^-} \longrightarrow 3Mg^{2+} + 2N^{3-} + \cancel{6e^-}}$$

Therefore, the balanced combination (oxidation–reduction) reaction is

$$\mathbf{3Mg + N_2 \longrightarrow 3Mg^{2+} + 2N^{3-}}$$

From inspecting the coefficients in this reaction, looking at the original equation, and knowing that $Mg^{2+}$ and $N^{3-}$ will combine to form an ionic compound ($Mg_3N_2$), we can rewrite the equation in the following form:

$$3Mg(s) + N_2(g) \longrightarrow Mg_3N_2(s)$$

---

**EXERCISE 4.9**

Use the half-reaction method to balance the equation $Ca(s) + Cl_2(g) \longrightarrow CaCl_2(s)$.

---

See Problems 4.57 and 4.58.

# Working with Solutions

The majority of the chemical reactions discussed in this chapter take place in solution. This is because the reaction between two solid reactants often proceeds very slowly or not at all. In a solid, the molecules or ions in a crystal tend to occupy approximately fixed positions, so that the chance of two molecules or ions coming together to react is small. In liquid solutions, reactant molecules are free to move throughout the liquid; therefore, reaction is much faster. When you run reactions in liquid solutions, it is convenient to dispense the amounts of reactants by measuring out volumes of reactant solutions. In the next two sections, we will discuss calculations involved in making up solutions, and in Section 4.10 we will describe stoichiometric calculations involving such solutions.

## 4.7   Molar Concentration

When we dissolve a substance in a liquid, we call the substance the *solute* and the liquid the *solvent*. Consider ammonia solutions. Ammonia gas dissolves readily in water, and aqueous ammonia solutions are often used in the laboratory. In such solutions, ammonia gas is the solute and water is the solvent.

The general term *concentration* refers to the quantity of solute in a standard quantity of solution. Qualitatively, we say that a solution is *dilute* when the solute concentration is low and *concentrated* when the solute concentration is high. Usually these terms are used in a comparative sense and do not refer to a specific concentration. We say that one solution is more dilute, or less concentrated, than another. However, for commercially available solutions, the term *concentrated* refers to the maximum, or near maximum, concentration available. For example, concentrated aqueous ammonia contains about 28% $NH_3$ by mass.

In this example, we expressed the concentration quantitatively by giving the mass percentage of solute—that is, the mass of solute in 100 g of solution. However, we need a unit of concentration that is convenient for dispensing reactants in solution, such as one that specifies moles of solute per solution volume.

**Molar concentration,** or **molarity (*M*),** is defined as *the moles of solute dissolved in one liter (cubic decimeter) of solution.*

$$\text{Molarity } (M) = \frac{\text{moles of solute}}{\text{liters of solution}}$$

An aqueous solution that is 0.15 *M* $NH_3$ (read this as "0.15 molar $NH_3$") contains 0.15 mol $NH_3$ per liter of solution. If you want to prepare a solution that is, for

**Figure 4.18**
**Preparing a 0.200 *M* CuSO₄ solution.** *Left:* 0.0500 mol CuSO₄ · 5H₂O
(12.48 g) is weighed on a platform balance. *Center:* The copper(II) sulfate
pentahydrate is transferred carefully to the volumetric flask. *Right:* Water is added to
bring the solution level to the mark on the neck of the 250-mL volumetric flask. The
molarity is 0.0500 mol/0.250 L = 0.200 *M*.

example, 0.200 *M* CuSO₄, you place 0.200 mol CuSO₄ in a 1.000-L volumetric
flask, or a proportional amount in a flask of a different size (Figure 4.18). You add
a small quantity of water to dissolve the CuSO₄. Then you fill the flask with additional water to the mark on the neck and mix the solution. The following example
shows how to calculate the molarity of a solution given the mass of solute and the
volume of solution.

### EXAMPLE 4.9

## Calculating Molarity from Mass and Volume

A sample of NaNO₃ weighing 0.38 g is placed in a 50.0-mL volumetric flask.
The flask is then filled with water to the mark on the neck, dissolving the solid.
What is the molarity of the resulting solution?

 **PROBLEM STRATEGY**

▶ An alternate example is
provided in the Instructor's
Resource Manual.

To calculate the molarity, you need the moles of solute. Therefore, you first convert grams NaNO₃ to moles. The molarity equals the moles of solute divided
by the liters of solution.

**SOLUTION**

You find that 0.38 g NaNO₃ is $4.\underline{4}7 \times 10^{-3}$ mol NaNO₃; the last significant figure is underlined. The volume of solution is 50.0 mL, or $50.0 \times 10^{-3}$ L, so the
molarity is

$$\text{Molarity} = \frac{4.\underline{4}7 \times 10^{-3} \text{ mol NaNO}_3}{50.0 \times 10^{-3} \text{ L soln}} = \mathbf{0.089 \ \textit{M} \ NaNO_3}$$

**Problem Solving: Calculations of Molarity**

### EXERCISE 4.10

A sample of sodium chloride, NaCl, weighing 0.0678 g is placed in a 25.0-mL volumetric flask. Enough water is added to dissolve the NaCl, and then the flask is filled to the mark with water and carefully shaken to mix the contents. What is the molarity of the resulting solution?

*See Problems 4.59, 4.60, 4.61, and 4.62.*

The advantage of molarity as a concentration unit is that, for a solution of known molarity, the amount of solute is related to the volume of solution. Rather than having to weigh out a specified mass of substance, you can instead measure out a definite volume of solution of the substance, which is usually easier. As the following example illustrates, molarity can be used as a factor for converting from moles of solute to liters of solution, and vice versa.

► An alternate example is provided in the Instructor's Resource Manual.

### EXAMPLE 4.10

## Using Molarity as a Conversion Factor

An experiment calls for the addition to a reaction vessel of 0.184 g of sodium hydroxide, NaOH, in aqueous solution. How many milliliters of 0.150 *M* NaOH should be added?

 **PROBLEM STRATEGY**

You first need to convert grams NaOH to moles NaOH, because molarity relates moles of solute to volume of solution. Then, you convert moles NaOH to liters of solution, using the molarity as a conversion factor. Here, 0.150 *M* means that 1 L of solution contains 0.150 moles of solute, so the conversion factor is:

$$\underbrace{\frac{1 \text{ L soln}}{0.150 \text{ mol NaOH}}}_{\substack{\text{Converts} \\ \text{mol NaOH to L soln}}}$$

**SOLUTION**

Here is the calculation. (The molar mass of NaOH is 40.0 g/mol.)

$$0.184 \text{ g NaOH} \times \frac{1 \text{ mol NaOH}}{40.0 \text{ g NaOH}} \times \frac{1 \text{ L soln}}{0.150 \text{ mol NaOH}}$$
$$= 3.07 \times 10^{-2} \text{ L soln (or 30.7 mL)}$$

You need to add **30.7 mL** of 0.150 *M* NaOH solution to the reaction vessel.

### EXERCISE 4.11

How many milliliters of 0.163 *M* NaCl are required to give 0.0958 g of sodium chloride?

*See Problems 4.63, 4.64, 4.65, and 4.66.*

**Problem Solving: Concentration and Volume**

▶ Experiments in General Chemistry, 4A, "Conductivity of Aqueous Solutions."

**EXERCISE 4.12**

How many moles of sodium chloride should be put in a 50.0-mL volumetric flask to give a 0.15 $M$ NaCl solution when the flask is filled to the mark with water? How many grams of NaCl is this?

See Problems 4.67, 4.68, 4.69, and 4.70.

## 4.8 Diluting Solutions

Add water (solvent)

**Figure 4.19**
**Molecular view of the dilution process.** *Left:* A molecular view of a solution of $Cl_2$ dissolved in water. *Right:* The solution after performing a dilution by adding water. Note how the number of moles of $Cl_2$ in the container does not change when perfoming the dilution, only the concentration changes. In this particular case, the concentration of $Cl_2$ drops to half of the starting concentration because the volume was doubled.

Commercially available aqueous ammonia (28.0% $NH_3$) is 14.8 $M$ $NH_3$. Suppose, however, that you want a solution that is 1.00 $M$ $NH_3$. You need to dilute the concentrated solution with a definite quantity of water. For this purpose, you must know the relationship between the molarity of the solution before dilution (the *initial molarity*) and that after dilution (the *final molarity*).

To obtain this relationship, first recall the equation defining molarity:

$$\text{Molarity} = \frac{\text{moles of solute}}{\text{liters of solution}}$$

You can rearrange this to give

$$\text{Moles of solute} = \text{molarity} \times \text{liters of solution}$$

The product of molarity and the volume (in liters) gives the moles of solute in the solution. Writing $M_i$ for the initial molar concentration and $V_i$ for the initial volume of solution, you get

$$\text{Moles of solute} = M_i \times V_i$$

When the solution is diluted by adding more water, the concentration and volume change to $M_f$ (the final molar concentration) and $V_f$ (the final volume), and the moles of solute equals

$$\text{Moles of solute} = M_f \times V_f$$

Because the moles of solute has not changed during the dilution (Figure 4.19),

$$M_i \times V_i = M_f \times V_f$$

(Note: You can use any volume units, but both $V_i$ and $V_f$ must be in the same units.)

The next example illustrates how you can use this formula to find the volume of a concentrated solution needed to prepare a given volume of dilute solution.

**EXAMPLE 4.11**

### Diluting a Solution

You are given a solution of 14.8 $M$ $NH_3$. How many milliliters of this solution do you require to give 100.0 mL of 1.00 $M$ $NH_3$ when diluted (Figure 4.20)?

**SOLUTION**

You know the final volume (100.0 mL), final concentration (1.00 $M$), and initial concentration (14.8 $M$). You write the dilution formula and rearrange it to give the initial volume.

▶ An alternate example is provided in the Instructor's Resource Manual.

$$M_iV_i = M_fV_f$$

$$V_i = \frac{M_fV_f}{M_i}$$

Now you substitute the known values into the right side of the equation.

$$V_i = \frac{1.00\,M \times 100.0\ \text{mL}}{14.8\,M} = \textbf{6.76 mL}$$

**Problem Solving: Concentration and Volume II**

### EXERCISE 4.13

You have a solution that is 1.5 $M$ $H_2SO_4$ (sulfuric acid). How many milliliters of this acid do you need to prepare 100.0 mL of 0.18 $M$ $H_2SO_4$?

See Problems 4.71 and 4.72.

**Figure 4.20**
**Preparing a solution by diluting a concentrated one.** A volume of concentrated ammonia, similar to the one in the beaker, has been added to the volumetric flask. Here water is being added from the plastic squeeze bottle to bring the volume up to the mark on the flask.

### CONCEPT CHECK 4.4

Consider the following beakers. Each contains a solution of the hypothetical atom X.

a. Arrange the beakers in order of increasing concentration of X.

b. Without adding or removing X, what specific things could you do to make the concentrations of X equal in each beaker? (*Hint:* Think about dilutions.)

# Quantitative Analysis

Analytical chemistry deals with the determination of composition of materials—that is, the analysis of materials. The materials that one might analyze include air, water, food, hair, body fluids, pharmaceutical preparations, and so forth. The analysis of materials is divided into qualitative and quantitative analysis. *Qualitative analysis* involves the identification of substances or species present in a material. For instance, you might determine that a sample of water contains lead(II) ion. **Quantitative analysis,** which we will discuss in the last sections of this chapter, involves *the determination of the amount of a substance or species present in a material.* In a quantitative analysis, you might determine that the amount of lead(II) ion in a sample of water is 0.067 mg/L.

## 4.9   Gravimetric Analysis

**Gravimetric analysis** is *a type of quantitative analysis in which the amount of a species in a material is determined by converting the species to a product that can be isolated completely and weighed.* Precipitation reactions are frequently used in gravimetric analyses. In these reactions, you determine the amount of an ionic species by precipitating it from solution. The precipitate, or solid formed in the reaction, is then filtered from the solution, dried, and weighed. The advantages of a gravimetric analysis are its simplicity (at least in theory) and its accuracy. The chief disadvantage is that it requires meticulous, time-consuming work. Because of this, whenever possible, chemists use modern instrumental methods.

As an example of a gravimetric analysis, consider the problem of determining the amount of lead in a sample of drinking water. Lead, if it occurs in the water, probably exists as the lead(II) ion, $Pb^{2+}$. Lead(II) sulfate is a very insoluble compound of lead(II) ion. When sodium sulfate, $Na_2SO_4$, is added to a solution containing $Pb^{2+}$, lead(II) sulfate precipitates (that is, $PbSO_4$ comes out of the solution as a fine, crystalline solid). If you assume that the lead is present in solution as lead(II) nitrate, you can write the following equation for the reaction:

$$Na_2SO_4(aq) + Pb(NO_3)_2(aq) \longrightarrow 2NaNO_3(aq) + PbSO_4(s)$$

You can separate the white precipitate of lead(II) sulfate from the solution by filtration. Then you dry and weigh the precipitate. Figure 4.21 shows the laboratory setup used in a similar analysis.

▶ An alternate example is provided in the Instructor's Resource Manual.

### EXAMPLE 4.12

### Determining the Amount of a Species by Gravimetric Analysis

A 1.000-L sample of polluted water was analyzed for lead(II) ion, $Pb^{2+}$, by adding an excess of sodium sulfate to it. The mass of lead(II) sulfate that precipitated was 229.8 mg. What is the mass of lead in a liter of the water? Give the answer as milligrams of lead per liter of solution.

**Figure 4.21**
**Gravimetric analysis for barium ion.**   *Left:* A solution of potassium chromate (yellow) is poured down a stirring rod into a solution containing an unknown amount of barium ion, $Ba^{2+}$. The yellow precipitate that forms is barium chromate, $BaCrO_4$. *Right:* The solution is filtered by pouring it into a crucible containing a porous glass partition. Afterward, the crucible is heated to dry the barium chromate. By weighing the crucible before and afterward, you can determine the mass of precipitate.

**PROBLEM STRATEGY**

All of the lead in the water solution is precipitated as lead(II) sulfate, $PbSO_4$. If you determine the percentage of lead in $PbSO_4$, you can calculate the quantity of lead in the water sample.

**SOLUTION**

Following Example 3.7, you obtain the mass percentage of Pb in $PbSO_4$ by dividing the molar mass of Pb by the molar mass of $PbSO_4$, then multiplying by 100%:

$$\%Pb = \frac{207.2 \text{ g/mol}}{303.3 \text{ g/mol}} \times 100\% = 68.32\%$$

Therefore, the 1.000-L sample of water contains

$$\text{Amount Pb in sample} = 229.8 \text{ mg } PbSO_4 \times 0.6832 = 157.0 \text{ mg Pb}$$

The water sample contains **157.0 mg Pb per liter.**

---

**EXERCISE 4.14**

You are given a sample of limestone, which is mostly $CaCO_3$, to determine the mass percentage of Ca in the rock. You dissolve the limestone in hydrochloric acid, which gives a solution of calcium chloride. Then you precipitate the calcium ion in solution by adding sodium oxalate, $Na_2C_2O_4$. The precipitate is calcium oxalate, $CaC_2O_4$. You find that a sample of limestone weighing 128.3 mg gives 140.2 mg of $CaC_2O_4$. What is the mass percentage of calcium in the limestone?

---

See Problems 4.73 and 4.74.

## 4.10  Volumetric Analysis

As you saw earlier, you can use molarity as a conversion factor, and in this way you can calculate the volume of solution that is equivalent to a given mass of solute (see Example 4.10). This means that you can replace mass measurements in solution reactions by volume measurements. In the next example, we look at the volumes of solutions involved in a given reaction.

▶ An alternate example is provided in the Instructor's Resource Manual.

▶ Video no. 6 (Series A), "Acid–Base Titration."

**EXAMPLE 4.13**

### Calculating the Volume of Reactant Solution Needed

Consider the reaction of sulfuric acid, $H_2SO_4$, with sodium hydroxide, NaOH.

$$H_2SO_4(aq) + 2NaOH(aq) \longrightarrow 2H_2O(l) + Na_2SO_4(aq)$$

Suppose a beaker contains 35.0 mL of 0.175 $M$ $H_2SO_4$. How many milliliters of 0.250 $M$ NaOH must be added to react completely with the sulfuric acid?

**PROBLEM STRATEGY**

You convert from 35.0 mL (or $35.0 \times 10^{-3}$ L) $H_2SO_4$ solution to moles $H_2SO_4$ (using the molarity of $H_2SO_4$), then to moles NaOH (from the chemical equation). Finally, you convert this to volume of NaOH solution (using the molarity of NaOH).

**SOLUTION**

The calculation is as follows:

$$35.0 \times 10^{-3} \text{ L } \cancel{H_2SO_4 \text{ soln}} \times \frac{0.175 \text{ mol } \cancel{H_2SO_4}}{1 \text{ L } \cancel{H_2SO_4 \text{ soln}}} \times \frac{2 \text{ mol } \cancel{NaOH}}{1 \text{ mol } \cancel{H_2SO_4}} \times$$

$$\frac{1 \text{ L NaOH soln}}{0.250 \text{ mol } \cancel{NaOH}} = 4.90 \times 10^{-2} \text{ L NaOH soln (or 49.0 mL NaOH soln)}$$

Thus, 35.0 mL of 0.175 $M$ sulfuric acid solution reacts with **49.0 mL** of 0.250 $M$ sodium hydroxide solution.

---

**EXERCISE 4.15**

Nickel sulfate, $NiSO_4$, reacts with sodium phosphate, $Na_3PO_4$, to give a pale yellow-green precipitate of nickel phosphate, $Ni_3(PO_4)_2$, and a solution of sodium sulfate, $Na_2SO_4$.

$$3NiSO_4(aq) + 2Na_3PO_4(aq) \longrightarrow Ni_3(PO_4)_2(s) + 3Na_2SO_4(aq)$$

How many milliliters of 0.375 $M$ $NiSO_4$ will react with 45.7 mL of 0.265 $M$ $Na_3PO_4$?

---

See Problems 4.79 and 4.80.

An important method for determining the amount of a particular substance is based on measuring the volume of reactant solution. Suppose substance $A$ reacts in solution with substance $B$. If you know the volume and concentration of a solution of $B$ that just reacts with substance $A$ in a sample, you can determine the amount of $A$. **Titration** is *a procedure for determining the amount of substance* A *by adding a carefully measured volume of a solution with known concentration of* B *until the reaction of* A *and* B *is just complete.* **Volumetric analysis** is *a method of analysis based on titration.*

Figure 4.22 shows a flask containing hydrochloric acid with an unknown amount of HCl being titrated with sodium hydroxide solution, NaOH, of known molarity. The reaction is

$$NaOH(aq) + HCl(aq) \longrightarrow NaCl(aq) + H_2O(l)$$

▶ **An indicator is a substance that undergoes a color change when a reaction approaches completion. See Section 4.4.**

To the HCl solution are added a few drops of phenolphthalein indicator. ◀ Phenolphthalein is colorless in the hydrochloric acid but turns pink at the completion of the reaction of NaOH with HCl. Sodium hydroxide with a concentration of 0.207 $M$ is contained in a *buret,* a glass tube graduated to measure the volume of liquid delivered from the stopcock. The solution in the buret is added to the HCl in the flask until the phenolphthalein just changes from colorless to pink. At this point, the reaction is complete and the volume of NaOH that reacts with the HCl is read from the buret. This volume is then used to obtain the mass of HCl in the original solution.

**Figure 4.22**
**Titration of an unknown amount of HCl with NaOH.** *Left:* The flask contains HCl and a few drops of phenolphthalein indicator; the buret contains 0.207 *M* NaOH (the buret reading is 44.97 mL). *Center:* NaOH was added to the solution in the flask until a persistent faint pink color was reached, marking the endpoint of the titration (the buret reading is 49.44 mL). The amount of HCl can be determined from the volume of NaOH used (4.47 mL); see Example 4.14. *Right:* The addition of several drops of NaOH solution beyond the endpoint gives a deep pink color.

▶ An alternate example is provided in the Instructor's Resource Manual.

▶ Experiments in General Chemistry, 4C, "How Much Acetic Acid Is in Vinegar?"

**EXAMPLE 4.14**

## Calculating the Quantity of Substance in a Titrated Solution

A flask contains a solution with an unknown amount of HCl. This solution is titrated with 0.207 *M* NaOH. It takes 4.47 mL NaOH to complete the reaction. What is the mass of the HCl?

### PROBLEM STRATEGY

You convert the volume of NaOH ($4.47 \times 10^{-3}$ L NaOH solution) to moles NaOH (from the molarity of NaOH). Then you convert moles NaOH to moles HCl (from the chemical equation). Finally, you convert moles HCl to grams HCl.

### SOLUTION

The calculation is as follows:

$$4.47 \times 10^{-3} \text{ L NaOH soln} \times \frac{0.207 \text{ mol NaOH}}{1 \text{ L NaOH soln}} \times \frac{1 \text{ mol HCl}}{1 \text{ mol NaOH}} \times \frac{36.5 \text{ g HCl}}{1 \text{ mol HCl}}$$

$$= \textbf{0.0338 g HCl}$$

**EXERCISE 4.16**

A 5.00-g sample of vinegar is titrated with 0.108 *M* NaOH. If the vinegar requires 39.1 mL of the NaOH solution for complete reaction, what is the mass percentage of acetic acid, $HC_2H_3O_2$, in the vinegar? The reaction is

$$HC_2H_3O_2(aq) + NaOH(aq) \longrightarrow NaC_2H_3O_2(aq) + H_2O(l)$$

See Problems 4.81 and 4.82.

**CONCEPT CHECK 4.5**

Consider three flasks, each containing 0.10 mol of acid. You need to learn something about the acids in each of the flasks, so you perform titration using an NaOH solution. Here are the results of the experiment:

Flask A    10 mL of NaOH required for neutralization

Flask B    20 mL of NaOH required for neutralization

Flask C    30 mL of NaOH required for neutralization

a. What have you learned about each of these acids from performing the experiment?

b. Could you use the results of this experiment to determine the concentration of the NaOH? If not, what assumption about the molecular formulas of the acids would allow you to make the concentration determination?

# A Checklist for Review

## Important Terms

electrolyte (4.1)
nonelectrolyte (4.1)
strong electrolyte (4.1)
weak electrolyte (4.1)
molecular equation (4.2)
complete ionic equation (4.2)
spectator ion (4.2)
net ionic equation (4.2)
precipitate (4.3)
exchange (metathesis) reaction (4.3)
acid–base indicator (4.4)

acid (Arrhenius) (4.4)
base (Arrhenius) (4.4)
acid (Brønsted–Lowry) (4.4)
base (Brønsted–Lowry) (4.4)
strong acid (4.4)
weak acid (4.4)
strong base (4.4)
weak base (4.4)
neutralization reaction (4.4)
salt (4.4)
polyprotic acid (4.4)

oxidation number (oxidation state) (4.5)
oxidation–reduction reaction (redox reaction) (4.5)
half-reaction (4.5)
oxidation (4.5)
reduction (4.5)
oxidizing agent (4.5)
reducing agent (4.5)
combination reaction (4.5)

decomposition reaction (4.5)
displacement reaction (single-replacement reaction) (4.5)
combustion reaction (4.5)
molar concentration (molarity) ($M$) (4.7)
quantitative analysis (p. 39)
gravimetric analysis (4.9)
titration (4.10)
volumetric analysis (4.10)

## Key Equations

$$\text{Molarity } (M) = \frac{\text{moles of solute}}{\text{liters of solution}}$$

$$M_i \times V_i = M_f \times V_f$$

## Summary of Facts and Concepts

Reactions often involve ions in aqueous solution. Many of the compounds in such reactions are *electrolytes,* which are substances that dissolve in water to give ions. Electrolytes that exist in solution almost entirely as ions are called *strong electrolytes.* Electrolytes that dissolve in water to give a relatively small percentage of ions are called *weak electrolytes.* The *solubility rules* can be used to predict the extent to which an ionic compound will dissolve in water. Most soluble ionic compounds are strong electrolytes.

We can represent a reaction involving ions in one of three different ways, depending on what information we want to convey. A *molecular equation* is one in which substances are written as if they were molecular, even though they are ionic. This equation closely describes what you actually do in the laboratory. However, this equation does not describe what is happening at the level of ions and molecules. For that purpose, we rewrite the molecular equation as a *complete ionic equation* by replacing the formulas for strong electrolytes by their ion formulas. If you cancel *spectator ions* from the complete ionic equation, you obtain the *net ionic equation.*

Most of the important reactions we will consider in this course can be divided into three major classes: (1) precipitation reactions, (2) acid–base reactions, and (3) oxidation–reduction reactions. A *precipitation reaction* occurs in aqueous solution because one product is insoluble. You can decide whether two ionic compounds will result in a precipitation reaction, if you know from solubility rules that one of the potential products is insoluble.

*Acids* are substances that yield hydrogen ions in aqueous solution or donate protons. *Bases* are substances that yield hydroxide ions in aqueous solution or accept protons. These *acid–base reactions* are proton-transfer reactions. In this chapter, we covered neutralization reactions (reactions of acids and bases to yield salts) and reactions of certain salts with acids to yield a gas.

*Oxidation–reduction reactions* are reactions involving a transfer of electrons from one species to another or a change in the oxidation number of atoms. The concept of *oxidation numbers* helps us describe this type of reaction. The atom that increases in oxidation number is said to undergo *oxidation;* the atom that decreases in oxidation number is said to undergo *reduction.* Oxidation and reduction must occur together in a reaction. Many oxidation–reduction reactions fall into the following categories: combination reactions, decomposition reactions, displacement reactions, and combustion reactions. Oxidation–reduction reactions can be balanced by the *half-reaction method.*

*Molar concentration,* or *molarity,* is the moles of solute in a liter of solution. Knowing the molarity allows you to calculate the amount of solute in any volume of solution. Because the moles of solute are constant during the *dilution of a solution,* you can determine to what volume to dilute a concentrated solution to give one of desired molarity.

*Quantitative analysis* involves the determination of the amount of a species in a material. In *gravimetric analysis,* you determine the amount of a species by converting it to a product you can weigh. In *volumetric analysis,* you determine the amount of a species by titration. *Titration* is a method of chemical analysis in which you measure the volume of solution of known molarity that reacts with a compound of unknown amount. You determine the amount of the compound from this volume of solution.

## Operational Skills

1. **Using the solubility rules** Given the formula of an ionic compound, predict its solubility in water. **(EXAMPLE 4.1)**
2. **Writing net ionic equations** Given a molecular equation, write the corresponding net ionic equation. **(EXAMPLE 4.2)**
3. **Deciding whether precipitation occurs** Using solubility rules, decide whether two soluble ionic compounds react to form a precipitate. If they do, write the net ionic equation. **(EXAMPLE 4.3)**
4. **Classifying acids and bases as strong or weak** Given the formula of an acid or a base, classify it as strong or weak. **(EXAMPLE 4.4)**
5. **Writing an equation for a neutralization** Given an acid and a base, write the molecular equation and then the net ionic equation for the neutralization reaction. **(EXAMPLE 4.5)**
6. **Writing an equation for a reaction with gas formation** Given the reaction between a carbonate, sulfite, or sulfide and an acid, write the molecular and the net ionic equations. **(EXAMPLE 4.6)**
7. **Assigning oxidation numbers** Given the formula of a simple compound or ion, obtain the oxidation numbers of the atoms, using the rules for assigning oxidation numbers. **(EXAMPLE 4.7)**
8. **Balancing simple oxidation–reduction reactions by the half-reaction method** Given an oxidation–reduction reaction, balance it. **(EXAMPLE 4.8)**
9. **Calculating molarity from mass and volume** Given the mass of the solute and the volume of the solution, calculate the molarity. **(EXAMPLE 4.9)**
10. **Using molarity as a conversion factor** Given the volume and molarity of a solution, calculate the amount of solute. Or, given the amount of solute and the molarity of a solution, calculate the volume. **(EXAMPLE 4.10)**
11. **Diluting a solution** Calculate the volume of solution of known molarity required to make a specified volume of solution with different molarity. **(EXAMPLE 4.11)**
12. **Determining the amount of a species by gravimetric analysis** Given the amount of a precipitate in a gravimetric analysis, calculate the amount of a related species. **(EXAMPLE 4.12)**
13. **Calculating the volume of reactant solution needed** Given the chemical equation, calculate the volume of solution of known molarity of one substance that just reacts with a given volume of solution of another substance. **(EXAMPLE 4.13)**
14. **Calculating the quantity of substance in a titrated solution** Calculate the mass of one substance that reacts with a given volume of solution of known molarity of another substance. **(EXAMPLE 4.14)**

# Review Questions

**4.1**    Explain why some electrolyte solutions are strongly conducting, whereas others are weakly conducting.

**4.2**    Define the terms *strong electrolyte* and *weak electrolyte.* Give an example of each.

**4.3**    Explain the terms *soluble* and *insoluble.* Use the solubility rules to write the formula of an insoluble ionic compound.

**4.4**    What are the advantages and disadvantages of using a molecular equation to represent an ionic reaction?

**4.5**    What is a *spectator ion?* Illustrate with a complete ionic reaction.

**4.6**    What is a net ionic equation? What is the value in using a net ionic equation? Give an example.

**4.7**    What are the major types of chemical reactions? Give a brief description and an example of each.

**4.8**    Describe in words how you would prepare pure crystalline AgCl and NaNO₃ from solid AgNO₃ and solid NaCl.

**4.9**    Give an example of a neutralization reaction. Label the acid, base, and salt.

**4.10**    Give an example of a polyprotic acid and write equations for the successive neutralizations of the acidic hydrogen atoms of the acid molecule to produce a series of salts.

**4.11**    Why must oxidation and reduction occur together in a reaction?

**4.12**    Give an example of a displacement reaction. What is the oxidizing agent? What is the reducing agent?

**4.13**    Why is the product of molar concentration and volume constant for a dilution problem?

**4.14**    Describe how the amount of sodium hydroxide in a mixture can be determined by titration with hydrochloric acid of known molarity.

# Conceptual Problems

**4.15**    You need to perform gravimetric analysis of a water sample in order to determine the amount of $Ag^+$ present.
a. List three aqueous solutions that would be suitable for mixing with the sample to perform the analysis.
b. Would adding $KNO_3(aq)$ allow you to perform the analysis?
c. Assume you have performed the analysis and the silver solid that formed is moderately soluble. How might this affect your analysis results?

**4.16**    In this problem you need to draw two pictures of solutions in beakers at different points in time. Time zero ($t = 0$) will be the *hypothetical* instant at which the reactants dissolve in the solution (*if* they dissolve) *before* they react. Time after mixing ($t > 0$) will be the time required to allow sufficient interaction of the materials. For now, we assume that insoluble

solids have no ions in solution and do not worry about representing the stoichiometric amounts of the dissolved ions. Here is an example: Solid NaCl and solid AgNO₃ are added to a beaker containing 250 mL of water.

Note that we are not showing the $H_2O$ and we are representing only the ions and solids in solution. Using the same conditions as the example (adding the solids to $H_2O$), draw pictures of the following:
a. solid lead(II) nitrate and solid ammonium chloride at $t = 0$ and $t > 0$
b. FeS($s$) and NaNO₃($s$) at $t = 0$ and $t > 0$
c. solid lithium iodide and solid sodium carbonate at $t = 0$ and $t > 0$

**4.17**    You come across a beaker that contains water, aqueous ammonium acetate, and a precipitate of calcium phosphate.
a. Write the balanced molecular equation for a reaction between two solutions containing ions that could produce this solution.
b. Write the complete ionic equation for the reaction in part a.
c. Write the net ionic equation for the reaction in part a.

**4.18** .    Three acid samples are prepared for titration by 0.01 *M* NaOH:
1.  Sample 1 is prepared by dissolving 0.01 mol of HCl in 50 mL of water.
2.  Sample 2 is prepared by dissolving 0.01 mol of HCl in 60 mL of water.
3.  Sample 3 is prepared by dissolving 0.01 mol of HCl in 70 mL of water.

$$t = 0 \qquad\qquad t > 0$$

a. Without performing a formal calculation, compare the concentrations of the three acid samples (rank them from highest to lowest).

b. When performing the titration, which sample, if any, will require the largest volume of the 0.01 *M* NaOH for neutralization?

**4.19** Would you expect a precipitation reaction between an ionic compound that is an electrolyte and an ionic compound that is a nonelectrolyte? Justify your answer.

**4.20** Equal quantities of the hypothetical strong acid HX, weak acid HA, and weak base BZ, are each added to a separate beaker of water, producing the solutions depicted in the drawings. In the drawings, the relative amounts of each substance present in the solution (neglecting the water) are shown. Identify the acid or base that was used to produce each of the solutions (HX, HA, or BZ).

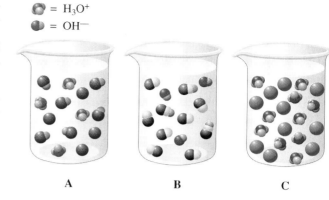

## Practice Problems

### Solubility Rules

**4.21** Using solubility rules, predict the solubility in water of the following ionic compounds.
a. PbS
b. AgNO₃
c. Na₂CO₃
d. CaI₂

**4.22** Using solubility rules, predict the solubility in water of the following ionic compounds.
a. Al(OH)₃
b. Li₃P
c. NH₄Br
d. NaOH

**4.23** Using solubility rules, decide whether the following ionic solids are soluble or insoluble in water. If they are soluble, indicate what ions you would expect to be present in solution.
a. AgBr
b. Li₂SO₄
c. Ca₃(PO₄)₂
d. Na₂CO₃

**4.24** Using solubility rules, decide whether the following ionic solids are soluble or insoluble in water. If they are soluble, indicate what ions you would expect to be present in solution.
a. (NH₄)₂SO₄
b. BaCO₃
c. PbSO₄
d. Ca(NO₃)₂

### Ionic Equations

**4.25** Write net ionic equations for the following molecular equations. HBr is a strong electrolyte.
a. $HBr(aq) + KOH(aq) \longrightarrow KBr(aq) + H_2O(l)$
b. $AgNO_3(aq) + NaBr(aq) \longrightarrow AgBr(s) + NaNO_3(aq)$
c. $CaS(aq) + 2HBr(aq) \longrightarrow CaBr_2(aq) + H_2S(g)$
d. $NaOH(aq) + NH_4Br(aq) \longrightarrow$
$$NaBr(aq) + NH_3(g) + H_2O(l)$$

**4.26** Write net ionic equations for the following molecular equations. HBr is a strong electrolyte.
a. $HBr(aq) + NH_3(aq) \longrightarrow NH_4Br(aq)$
b. $2HBr(aq) + Ba(OH)_2(aq) \longrightarrow 2H_2O(l) + BaBr_2(aq)$
c. $Pb(NO_3)_2(aq) + 2NaBr(aq) \longrightarrow PbBr_2(s) + 2NaNO_3(aq)$
d. $MgCO_3(s) + H_2SO_4(aq) \longrightarrow$
$$MgSO_4(aq) + H_2O(l) + CO_2(g)$$

**4.27** Lead(II) nitrate solution and sodium sulfate solution are mixed. Crystals of lead sulfate come out of solution, leaving a solution of sodium nitrate. Write the molecular equation and the net ionic equation for the reaction.

**4.28** Potassium carbonate solution reacts with aqueous hydrobromic acid to give a solution of potassium bromide, carbon dioxide gas, and water. Write the molecular equation and the net ionic equation for the reaction.

### Precipitation

**4.29** Write the molecular equation and the net ionic equation for each of the following aqueous reactions. If no reaction occurs, write *NR* after the arrow.
a. $FeSO_4 + NaCl \longrightarrow$
b. $Na_2CO_3 + MgBr_2 \longrightarrow$
c. $MgSO_4 + NaOH \longrightarrow$
d. $NiCl_2 + NaBr \longrightarrow$

**4.30** Write the molecular equation and the net ionic equation for each of the following aqueous reactions. If no reaction occurs, write *NR* after the arrow.
a. $AgNO_3 + NaI \longrightarrow$
b. $Ba(NO_3)_2 + K_2SO_4 \longrightarrow$
c. $Mg(NO_3)_2 + K_2SO_4 \longrightarrow$
d. $CaCl_2 + Al(NO_3)_3 \longrightarrow$

**4.31** For each of the following, write molecular and net ionic equations for any precipitation reaction that occurs. If no reaction occurs, indicate this.
a. Solutions of barium nitrate and lithium sulfate are mixed.
b. Solutions of sodium bromide and calcium nitrate are mixed.
c. Solutions of aluminum sulfate and sodium hydroxide are mixed.
d. Solutions of calcium bromide and sodium phosphate are mixed.

**4.32** For each of the following, write molecular and net ionic equations for any precipitation reaction that occurs. If no reaction occurs, indicate this.
a. Zinc chloride and sodium sulfide are dissolved in water.
b. Sodium sulfide and calcium chloride are dissolved in water.
c. Magnesium sulfate and potassium iodide are dissolved in water.
d. Magnesium sulfate and potassium carbonate are dissolved in water.

## Strong and Weak Acids and Bases

**4.33** Classify each of the following as a strong or weak acid or base.
a. HF     b. KOH     c. $HClO_4$     d. HIO

**4.34** Classify each of the following as a strong or weak acid or base.
a. HBrO     b. HCNO     c. $Sr(OH)_2$     d. HI

## Neutralization Reactions

**4.35** Complete and balance each of the following molecular equations (in aqueous solution); include phase labels. Then, for each, write the net ionic equation.
a. $NaOH + HNO_3 \longrightarrow$
b. $HCl + Ba(OH)_2 \longrightarrow$
c. $HC_2H_3O_2 + Ca(OH)_2 \longrightarrow$
d. $NH_3 + HNO_3 \longrightarrow$

**4.36** Complete and balance each of the following molecular equations (in aqueous solution); include phase labels. Then, for each, write the net ionic equation.
a. $Al(OH)_3 + HCl \longrightarrow$
b. $HBr + Sr(OH)_2 \longrightarrow$
c. $Ba(OH)_2 + HC_2H_3O_2 \longrightarrow$
d. $HNO_3 + KOH \longrightarrow$

**4.37** For each of the following, write the molecular equation, including phase labels. Then write the net ionic equation. Note that the salts formed in these reactions are soluble.
a. the neutralization of hydrobromic acid with calcium hydroxide solution
b. the reaction of solid aluminum hydroxide with nitric acid
c. the reaction of aqueous hydrogen cyanide with calcium hydroxide solution

d. the neutralization of lithium hydroxide solution by aqueous hydrogen cyanide

**4.38** For each of the following, write the molecular equation, including phase labels. Then write the net ionic equation. Note that the salts formed in these reactions are soluble.
a. the neutralization of lithium hydroxide solution by aqueous perchloric acid
b. the reaction of barium hydroxide solution and aqueous nitrous acid
c. the reaction of sodium hydroxide solution and aqueous nitrous acid
d. the neutralization of aqueous hydrogen cyanide by aqueous strontium hydroxide

**4.39** Complete the right side of each of the following molecular equations. Then write the net ionic equations. Assume all salts formed are soluble. Acid salts are possible.
a. $2KOH(aq) + H_3PO_4(aq) \longrightarrow$
b. $3H_2SO_4(aq) + 2Al(OH)_3(s) \longrightarrow$
c. $2HC_2H_3O_2(aq) + Ca(OH)_2(aq) \longrightarrow$
d. $H_2SO_3(aq) + NaOH(aq) \longrightarrow$

**4.40** Complete the right side of each of the following molecular equations. Then write the net ionic equations. Assume all salts formed are soluble. Acid salts are possible.
a. $Ca(OH)_2(aq) + 2H_2SO_4(aq) \longrightarrow$
b. $2H_3PO_4(aq) + Ca(OH)_2(aq) \longrightarrow$
c. $NaOH(aq) + H_2SO_4(aq) \longrightarrow$
d. $Sr(OH)_2(aq) + 2H_2CO_3(aq) \longrightarrow$

**4.41** Write molecular and net ionic equations for the successive neutralizations of each acidic hydrogen of sulfurous acid by aqueous calcium hydroxide. $CaSO_3$ is insoluble; the acid salt is soluble.

**4.42** Write molecular and net ionic equations for the successive neutralizations of each acidic hydrogen of phosphoric acid by calcium hydroxide solution. $Ca_3(PO_4)_2$ is insoluble; assume that the acid salts are soluble.

## Reactions Evolving a Gas

**4.43** The following reactions occur in aqueous solution. Complete and balance the molecular equations using phase labels. Then write the net ionic equations.
a. $CaS + HBr \longrightarrow$
b. $MgCO_3 + HNO_3 \longrightarrow$
c. $K_2SO_3 + H_2SO_4 \longrightarrow$

**4.44** The following reactions occur in aqueous solution. Complete and balance the molecular equations using phase labels. Then write the net ionic equations.
a. $BaCO_3 + HNO_3 \longrightarrow$
b. $K_2S + HCl \longrightarrow$
c. $CaSO_3(s) + HI \longrightarrow$

**4.45** Write the molecular equation and the net ionic equation for the reaction of solid iron(II) sulfide and hydrochloric acid. Add phase labels.

**4.46** Write the molecular equation and the net ionic equation for the reaction of solid barium carbonate and hydrogen bromide in aqueous solution. Add phase labels.

## Oxidation Numbers

**4.47** Obtain the oxidation number for the element noted in each of the following.
a. Ga in $Ga_2O_3$  b. Nb in $NbO_2$
c. Br in $KBrO_4$  d. Mn in $K_2MnO_4$

**4.48** Obtain the oxidation number for the element noted in each of the following.
a. Cr in $CrO_3$  b. Hg in $Hg_2Cl_2$
c. Ga in $Ga(OH)_3$  d. P in $Na_3PO_4$

**4.49** Obtain the oxidation number for the element noted in each of the following.
a. N in $NH_2^-$  b. I in $IO_3^-$
c. Al in $Al(OH)_4^-$  d. P in $H_2PO_4^-$

**4.50** Obtain the oxidation number for the element noted in each of the following.
a. N in $NO_2^-$  b. Cr in $CrO_4^{2-}$
c. Zn in $Zn(OH)_4^{2-}$  d. As in $H_2AsO_3^-$

**4.51** Determine the oxidation numbers of all the elements in each of the following compounds. (*Hint:* Look at the ions present.)
a. $Mn(ClO_3)_2$  b. $Fe_2(CrO_4)_3$
c. $HgCr_2O_7$  d. $Co_3(PO_4)_2$

**4.52** Determine the oxidation numbers of all the elements in each of the following compounds. (*Hint:* Look at the ions present.)
a. $Hg_2(BrO_3)_2$  b. $Cr_2(SO_4)_3$
c. $CoSeO_4$  d. $Cu_2SO_3$

## Describing Oxidation–Reduction Reactions

**4.53** In the following reactions, label the oxidizing agent and the reducing agent.
a. $P_4(s) + 5O_2(g) \longrightarrow P_4O_{10}(s)$
b. $Co(s) + Cl_2(g) \longrightarrow CoCl_2(s)$

**4.54** In the following reactions, label the oxidizing agent and the reducing agent.
a. $ZnO(s) + C(s) \longrightarrow Zn(g) + CO(g)$
b. $8Fe(s) + S_8(s) \longrightarrow 8FeS(s)$

**4.55** In the following reactions, label the oxidizing agent and the reducing agent.
a. $2Al(s) + 3F_2(g) \longrightarrow 2AlF_3(s)$

b. $Hg^{2+}(aq) + NO_2^-(aq) + H_2O(l) \longrightarrow$
$$Hg(s) + 2H^+(aq) + NO_3^-(aq)$$

**4.56** In the following reactions, label the oxidizing agent and the reducing agent.
a. $Fe_2O_3(s) + 3CO(g) \longrightarrow 2Fe(s) + 3CO_2(g)$
b. $PbS(s) + 4H_2O_2(aq) \longrightarrow PbSO_4(s) + 4H_2O(l)$

## Balancing Oxidation–Reduction Reactions

**4.57** Balance the following oxidation–reduction reactions by the half-reaction method.
a. $CuCl_2(aq) + Al(s) \longrightarrow AlCl_3(aq) + Cu(s)$
b. $Cr^{3+}(aq) + Zn(s) \longrightarrow Cr(s) + Zn^{2+}(aq)$

**4.58** Balance the following oxidation–reduction reactions by the half-reaction method.
a. $FeI_3(aq) + Mg(s) \longrightarrow Fe(s) + MgI_2(aq)$
b. $H_2(g) + Ag^+(aq) \longrightarrow Ag(s) + H^+(aq)$

## Molarity

**4.59** A sample of 0.0256 mol of iron(III) chloride, $FeCl_3$, was dissolved in water to give 25.0 mL of solution. What is the molarity of the solution?

**4.60** A 50.0-mL volume of $AgNO_3$ solution contains 0.0285 mol $AgNO_3$ (silver nitrate). What is the molarity of the solution?

**4.61** An aqueous solution is made from 0.798 g of potassium permanganate, $KMnO_4$. If the volume of solution is 50.0 mL, what is the molarity of $KMnO_4$ in the solution?

**4.62** A sample of oxalic acid, $H_2C_2O_4$, weighing 1.192 g is placed in a 100.0-mL volumetric flask, which is then filled to the mark with water. What is the molarity of the solution?

**4.63** What volume of 0.120 $M$ $CuSO_4$ is required to give 0.150 mol of copper(II) sulfate, $CuSO_4$?

**4.64** How many milliliters of 0.126 $M$ $HClO_4$ (perchloric acid) are required to give 0.00752 mol $HClO_4$?

**4.65** An experiment calls for 0.0353 g of potassium hydroxide, KOH. How many milliliters of 0.0176 $M$ KOH are required?

**4.66** What is the volume (in milliliters) of 0.215 $M$ $H_2SO_4$ (sulfuric acid) containing 0.949 g $H_2SO_4$?

**4.67** Heme, obtained from red blood cells, binds oxygen, $O_2$. How many moles of heme are there in 75 mL of 0.0019 $M$ heme solution?

**4.68** Insulin is a hormone that controls the use of glucose in the body. How many moles of insulin are required to make up 28 mL of 0.0048 $M$ insulin solution?

**4.69** How many grams of sodium dichromate, $Na_2Cr_2O_7$, should be added to a 50.0-mL volumetric flask to prepare 0.025 $M$ $Na_2Cr_2O_7$ when the flask is filled to the mark with water?

**4.70** Describe how you would prepare $2.50 \times 10^2$ mL of 0.20 $M$ $Na_2SO_4$. What mass (in grams) of sodium sulfate, $Na_2SO_4$, is needed?

**4.71** You wish to prepare 0.12 $M$ $HNO_3$ from a stock solution of nitric acid that is 15.8 $M$. How many milliliters of the stock solution do you require to make up 1.00 L of 0.12 $M$ $HNO_3$?

**4.72** A chemist wants to prepare 0.25 $M$ HCl. Commercial hydrochloric acid is 12.4 $M$. How many milliliters of the commercial acid does the chemist require to make up 1.50 L of the dilute acid?

## Gravimetric Analysis

**4.73** A chemist added an excess of sodium sulfate to a solution of a soluble barium compound to precipitate all of the barium ion as barium sulfate, $BaSO_4$. How many grams of barium ion are in a 458-mg sample of the barium compound if a solution of the sample gave 513 mg $BaSO_4$ precipitate? What is the mass percentage of barium in the compound?

**4.74** A soluble iodide was dissolved in water. Then an excess of silver nitrate, $AgNO_3$, was added to precipitate all of the iodide ion as silver iodide, AgI. If 1.545 g of the soluble iodide gave 2.185 g of silver iodide, how many grams of iodine are in the sample of soluble iodide? What is the mass percentage of iodine, I, in the compound?

**4.75** Copper has compounds with copper(I) ion or copper(II) ion. A compound of copper and chlorine was treated with a solution of silver nitrate, $AgNO_3$, to convert the chloride ion in the compound to a precipitate of AgCl. A 59.40-mg sample of the copper compound gave 86.00 mg AgCl.
a. Calculate the percentage of chlorine in the copper compound.
b. Decide whether the formula of the compound is CuCl or $CuCl_2$.

**4.76** Gold has compounds containing gold(I) ion or gold(III) ion. A compound of gold and chlorine was treated with a solution of silver nitrate, $AgNO_3$, to convert the chloride ion in the compound to a precipitate of AgCl. A 162.7-mg sample of the gold compound gave 100.3 mg AgCl.
a. Calculate the percentage of the chlorine in the gold compound.
b. Decide whether the formula of the compound is AuCl or $AuCl_3$.

**4.77** A compound of iron and chlorine is soluble in water. An excess of silver nitrate was added to precipitate the chloride ion as silver chloride. If a 134.8-mg sample of the compound gave 304.8 mg AgCl, what is the formula of the compound?

**4.78** A 1.345-g sample of a compound of barium and oxygen was dissolved in hydrochloric acid to give a solution of barium ion, which was then precipitated with an excess of potassium chromate to give 2.012 g of barium chromate, $BaCrO_4$. What is the formula of the compound?

## Volumetric Analysis

**4.79** What volume of 0.250 $M$ $HNO_3$ (nitric acid) reacts with 42.4 mL of 0.150 $M$ $Na_2CO_3$ (sodium carbonate) in the following reaction?

$$2HNO_3(aq) + Na_2CO_3(aq) \longrightarrow$$
$$2NaNO_3(aq) + H_2O(l) + CO_2(g)$$

**4.80** A flask contains 49.8 mL of 0.150 $M$ $Ca(OH)_2$ (calcium hydroxide). How many milliliters of 0.350 $M$ $Na_2CO_3$ (sodium carbonate) are required to react completely with the calcium hydroxide in the following reaction?

$$Na_2CO_3(aq) + Ca(OH)_2(aq) \longrightarrow CaCO_3(s) + 2NaOH(aq)$$

**4.81** How many milliliters of 0.150 $M$ $H_2SO_4$ (sulfuric acid) are required to react with 4.10 g of sodium hydrogen carbonate, $NaHCO_3$, according to the following equation?

$$H_2SO_4(aq) + 2NaHCO_3(aq) \longrightarrow$$
$$Na_2SO_4(aq) + 2H_2O(l) + 2CO_2(g)$$

**4.82** How many milliliters of 0.250 $M$ $KMnO_4$ are needed to react with 3.36 g of iron(II) sulfate, $FeSO_4$? The reaction is as follows:

$$10FeSO_4(aq) + 2KMnO_4(aq) + 8H_2SO_4(aq) \longrightarrow$$
$$5Fe_2(SO_4)_3(aq) + 2MnSO_4(aq) + K_2SO_4(aq) + 8H_2O(l)$$

**4.83** A solution of hydrogen peroxide, $H_2O_2$, is titrated with a solution of potassium permanganate, $KMnO_4$. The reaction is

$$5H_2O_2(aq) + 2KMnO_4(aq) + 3H_2SO_4(aq) \longrightarrow$$
$$5O_2(g) + 2MnSO_4(aq) + K_2SO_4(aq) + 8H_2O(l)$$

It requires 51.7 mL of 0.145 $M$ $KMnO_4$ to titrate 20.0 g of the solution of hydrogen peroxide. What is the mass percentage of $H_2O_2$ in the solution?

**4.84** A 3.33-g sample of iron ore is transformed to a solution of iron(II) sulfate, $FeSO_4$, and this solution is titrated with 0.150 $M$ $K_2Cr_2O_7$ (potassium dichromate). If it requires 41.4 mL of potassium dichromate solution to titrate the iron(II) sulfate solution, what is the percentage of iron in the ore? The reaction is

$$6FeSO_4(aq) + K_2Cr_2O_7(aq) + 7H_2SO_4(aq) \longrightarrow$$
$$3Fe_2(SO_4)_3(aq) + Cr_2(SO_4)_3(aq) + 7H_2O(l) + K_2SO_4(aq)$$

# General Problems

**4.85** Magnesium metal reacts with hydrobromic acid to produce hydrogen gas and a solution of magnesium bromide. Write the molecular equation for this reaction. Then write the corresponding net ionic equation.

**4.86** Aluminum metal reacts with perchloric acid to produce hydrogen gas and a solution of aluminum perchlorate. Write the molecular equation for this reaction. Then write the corresponding net ionic equation.

**4.87** Nickel(II) sulfate solution reacts with lithium hydroxide solution to produce a precipitate of nickel(II) hydroxide and a solution of lithium sulfate. Write the molecular equation for this reaction. Then write the corresponding net ionic equation.

**4.88** Potassium sulfate solution reacts with barium bromide solution to produce a precipitate of barium sulfate and a solution of potassium bromide. Write the molecular equation for this reaction. Then write the corresponding net ionic equation.

**4.89** Decide whether a reaction occurs for each of the following. If it does not, write *NR* after the arrow. If it does, write the balanced molecular equation; then write the net ionic equation.
a. $LiOH + HCN \longrightarrow$
b. $Li_2CO_3 + HNO_3 \longrightarrow$
c. $LiCl + AgNO_3 \longrightarrow$
d. $LiCl + MgSO_4 \longrightarrow$

**4.90** Decide whether a reaction occurs for each of the following. If it does not, write *NR* after the arrow. If it does, write the balanced molecular equation; then write the net ionic equation.
a. $Al(OH)_3 + HNO_3 \longrightarrow$
b. $FeS + HClO_4 \longrightarrow$
c. $CaCl_2 + NaNO_3 \longrightarrow$
d. $MgSO_4 + Ba(NO_3)_2 \longrightarrow$

**4.91** Complete and balance each of the following molecular equations, including phase labels, if a reaction occurs. Then write the net ionic equation. If no reaction occurs, write *NR* after the arrow.
a. $Sr(OH)_2 + HC_2H_3O_2 \longrightarrow$
b. $NH_4I + CsCl \longrightarrow$
c. $NaNO_3 + CsCl \longrightarrow$
d. $NH_4I + AgNO_3 \longrightarrow$

**4.92** Complete and balance each of the following molecular equations, including phase labels, if a reaction occurs. Then write the net ionic equation. If no reaction occurs, write *NR* after the arrow.
a. $HClO_4 + BaCO_3 \longrightarrow$
b. $H_2CO_3 + Sr(OH)_2 \longrightarrow$

c. $K_3PO_4 + MgCl_2 \longrightarrow$
d. $FeSO_4 + MgCl_2 \longrightarrow$

**4.93** Describe in words how you would do each of the following preparations. Then give the molecular equation for each preparation.
a. $CuCl_2(s)$ from $CuSO_4(s)$
b. $Ca(C_2H_3O_2)_2(s)$ from $CaCO_3(s)$
c. $NaNO_3(s)$ from $Na_2SO_3(s)$
d. $MgCl_2(s)$ from $Mg(OH)_2(s)$

**4.94** Describe in words how you would do each of the following preparations. Then give the molecular equation for each preparation.
a. $MgCl_2(s)$ from $MgCO_3(s)$
b. $NaNO_3(s)$ from $NaCl(s)$
c. $Al(OH)_3(s)$ from $Al(NO_3)_3(s)$
d. $HCl(aq)$ from $H_2SO_4(aq)$

**4.95** Classify each of the following reactions as a combination reaction, decomposition reaction, displacement reaction, or combustion reaction.
a. When they are heated, ammonium dichromate crystals, $(NH_4)_2Cr_2O_7$, decompose to give nitrogen, water vapor, and solid chromium(III) oxide, $Cr_2O_3$.
b. When aqueous ammonium nitrite, $NH_4NO_2$, is heated, it gives nitrogen and water vapor.
c. When gaseous ammonia, $NH_3$, reacts with hydrogen chloride gas, HCl, fine crystals of ammonium chloride, $NH_4Cl$, are formed.
d. Aluminum added to an aqueous solution of sulfuric acid, $H_2SO_4$, forms a solution of aluminum sulfate, $Al_2(SO_4)_3$. Hydrogen gas is released.

**4.96** Classify each of the following reactions as a combination reaction, decomposition reaction, displacement reaction, or combustion reaction.
a. When solid calcium oxide, CaO, is exposed to gaseous sulfur trioxide, $SO_3$, solid calcium sulfate, $CaSO_4$, is formed.
b. Calcium metal (solid) reacts with water to produce a solution of calcium hydroxide, $Ca(OH)_2$, and hydrogen gas.
c. When solid sodium hydrogen sulfite, $NaHSO_3$, is heated, solid sodium sulfite, $Na_2SO_3$, sulfur dioxide gas, $SO_2$, and water vapor are formed.
d. Magnesium reacts with bromine to give magnesium bromide, $MgBr_2$.

**4.97** Consider the reaction of all pairs of the following compounds in water solution: $Ba(OH)_2$, $Pb(NO_3)_2$, $H_2SO_4$, $NaNO_3$, $MgSO_4$.
a. Which pair (or pairs) forms one insoluble compound and one soluble compound (not water)?
b. Which pair (or pairs) forms two insoluble compounds?

c. Which pair (or pairs) forms one insoluble compound and water?

**4.98** Consider the reaction of all pairs of the following compounds in water solution: $Sr(OH)_2$, $AgNO_3$, $H_3PO_4$, $KNO_3$, $CuSO_4$.
a. Which pair (or pairs) forms one insoluble compound and one soluble compound (not water)?
b. Which pair (or pairs) forms two insoluble compounds?
c. Which pair (or pairs) forms one insoluble compound and water?

**4.99** An aqueous solution contains 4.50 g of calcium chloride, $CaCl_2$, per liter. What is the molarity of $CaCl_2$? When calcium chloride dissolves in water, the calcium ions, $Ca^{2+}$, and chloride ions, $Cl^-$, in the crystal go into the solution. What is the molarity of each ion in the solution?

**4.100** An aqueous solution contains 3.45 g of iron(III) sulfate, $Fe_2(SO_4)_3$, per liter. What is the molarity of $Fe_2(SO_4)_3$? When the compound dissolves in water, the $Fe^{3+}$ ions and $SO_4^{2-}$ ions in the crystal go into the solution. What is the molar concentration of each ion in the solution?

**4.101** A stock solution of potassium dichromate, $K_2Cr_2O_7$, is made by dissolving 89.3 g of the compound in 1.00 L of solution. How many milliliters of this solution are required to prepare 1.00 L of 0.100 $M$ $K_2Cr_2O_7$?

**4.102** A 71.2-g sample of oxalic acid, $H_2C_2O_4$, was dissolved in 1.00 L of solution. How would you prepare 1.00 L of 0.150 $M$ $H_2C_2O_4$ from this solution?

**4.103** A solution contains 6.00% (by mass) NaBr (sodium bromide). The density of the solution is 1.046 g/cm$^3$. What is the molarity of NaBr?

**4.104** An aqueous solution contains 4.00% $NH_3$ (ammonia) by mass. The density of the aqueous ammonia is 0.979 g/mL. What is the molarity of $NH_3$ in the solution?

**4.105** A barium mineral was dissolved in hydrochloric acid to give a solution of barium ion. An excess of potassium sulfate was added to 50.0 mL of the solution, and 1.128 g of barium sulfate precipitate formed. Assume that the original solution was barium chloride. What was the molarity of $BaCl_2$ in this solution?

**4.106** Bone was dissolved in hydrochloric acid, giving 50.0 mL of solution containing calcium chloride, $CaCl_2$. To precipitate the calcium ion from the resulting solution, an excess of potassium oxalate was added. The precipitate of calcium oxalate, $CaC_2O_4$, weighed 1.437 g. What was the molarity of $CaCl_2$ in the solution?

**4.107** You have a sample of a rat poison whose active ingredient is thallium(I) sulfate. You analyze this sample for the mass percentage of active ingredient by adding potassium iodide to precipitate yellow thallium(I) iodide. If the sample of

rat poison weighed 759.0 mg and you obtained 212.2 mg of the dry precipitate, what is the mass percentage of the thallium(I) sulfate in the rat poison?

**4.108** An antacid tablet has calcium carbonate as the active ingredient; other ingredients include a starch binder. You dissolve the tablet in hydrochloric acid and filter off insoluble material. You add potassium oxalate to the filtrate (containing calcium ion) to precipitate calcium oxalate. If a tablet weighing 0.680 g gave 0.6332 g of calcium oxalate, what is the mass percentage of active ingredient in the tablet?

**4.109** A sample of $CuSO_4 \cdot 5H_2O$ was heated to 110°C, where it lost water and gave another hydrate of copper(II) ion that contains 32.50% Cu. A 98.77-mg sample of this new hydrate gave 116.66 mg of barium sulfate precipitate when treated with a barium nitrate solution. What is the formula of the new hydrate?

**4.110** A sample of $CuSO_4 \cdot 5H_2O$ was heated to 100°C, where it lost water and gave another hydrate of copper(II) ion that contained 29.76% Cu. An 85.42-mg sample of this new hydrate gave 93.33 mg of barium sulfate precipitate when treated with a barium nitrate solution. What is the formula of the new hydrate?

**4.111** A water-soluble compound of gold and chlorine is treated with silver nitrate to convert the chlorine completely to silver chloride, AgCl. In an experiment, 328 mg of the compound gave 464 mg of silver chloride. Calculate the percentage of Cl in the compound. What is its empirical formula?

**4.112** A solution of scandium chloride was treated with silver nitrate. The chlorine in the scandium compound was converted to silver chloride, AgCl. A 58.9-mg sample of scandium chloride gave 167.4 mg of silver chloride. What are the mass percentages of Sc and Cl in scandium chloride? What is its empirical formula?

**4.113** A 0.608-g sample of fertilizer contained nitrogen as ammonium sulfate, $(NH_4)_2SO_4$. It was analyzed for nitrogen by heating with sodium hydroxide.

$$(NH_4)_2SO_4(s) + 2NaOH(aq) \longrightarrow$$
$$Na_2SO_4(aq) + 2H_2O(l) + 2NH_3(g)$$

The ammonia was collected in 46.3 mL of 0.213 $M$ HCl (hydrochloric acid), with which it reacted.

$$NH_3(g) + HCl(aq) \longrightarrow NH_4Cl(aq)$$

This solution was titrated for excess hydrochloric acid with 44.3 mL of 0.128 $M$ NaOH.

$$NaOH(aq) + HCl(aq) \longrightarrow NaCl(aq) + H_2O(l)$$

What is the percentage of nitrogen in the fertilizer?

**4.114** An antacid tablet contains sodium hydrogen carbonate, $NaHCO_3$, and inert ingredients. A 0.500-g sample of powdered tablet was mixed with 50.0 mL of 0.190 $M$ HCl (hydrochloric acid). The mixture was allowed to stand until it reacted.

$$NaHCO_3(s) + HCl(aq) \longrightarrow NaCl(aq) + H_2O(l) + CO_2(g)$$

The excess hydrochloric acid was titrated with 47.1 mL of 0.128 $M$ NaOH (sodium hydroxide).

$$HCl(aq) + NaOH(aq) \longrightarrow NaCl(aq) + H_2O(l)$$

What is the percentage of sodium hydrogen carbonate in the antacid?

## Cumulative-Skills Problems

**4.115**   Lead(II) nitrate reacts with cesium sulfate in an aqueous precipitation reaction. What are the formulas of lead(II) nitrate and cesium sulfate? Write the molecular equation and net ionic equation for the reaction. What are the names of the products? Give the molecular equation for another reaction that produces the same precipitate.

**4.116**   Silver nitrate reacts with strontium chloride in an aqueous precipitation reaction. What are the formulas of silver nitrate and strontium chloride? Write the molecular equation and net ionic equation for the reaction. What are the names of the products? Give the molecular equation for another reaction that produces the same precipitate.

**4.117**   Elemental bromine is the source of bromine compounds. The element is produced from certain brine solutions that occur naturally. These brines are essentially solutions of calcium bromide that, when treated with chlorine gas, yield bromine in a displacement reaction. What are the molecular equation and net ionic equation for the reaction? A solution containing 40.0 g of calcium bromide requires 14.2 g of chlorine to react completely with it, and 22.2 g of calcium chloride is produced in addition to whatever bromine is obtained. How many grams of calcium bromide are required to produce 10.0 pounds of bromine?

**4.118**   Barium carbonate is the source of barium compounds. It is produced in an aqueous precipitation reaction from barium sulfide and sodium carbonate. (Barium sulfide is a soluble compound obtained by heating the mineral barite, which is barium sulfate, with carbon.) What are the molecular equation and net ionic equation for the precipitation reaction? A solution containing 33.9 g of barium sulfide requires 21.2 g of sodium carbonate to react completely with it, and 15.6 g of sodium sulfide is produced in addition to whatever barium carbonate is obtained. How many grams of barium sulfide are required to produce 10.0 tons of barium carbonate? (One ton equals 2000 pounds.)

**4.119**   Mercury(II) nitrate is treated with hydrogen sulfide, $H_2S$, forming a precipitate and a solution. Write the molecular equation and the net ionic equation for the reaction. An acid is formed; is it strong or weak? Name each of the products. If 81.15 g of mercury(II) nitrate and 8.52 g of hydrogen sulfide are mixed in 550.0 g of water to form 58.16 g of precipitate, what is the mass of the solution after the reaction?

**4.120**   Mercury(II) nitrate is treated with hydrogen sulfide, $H_2S$, forming a precipitate and a solution. Write the molecular equation and the net ionic equation for the reaction. An acid is formed; is it strong or weak? Name each of the products. If 65.65 g of mercury(II) nitrate and 4.26 g of hydrogen sulfide are mixed in 490.0 g of water to form 54.16 g of precipitate, what is the mass of the solution after the reaction?

**4.121**   Iron forms a sulfide with the approximate formula $Fe_7S_8$. Assume that the oxidation state of sulfur is $-2$, and that iron atoms exist in both $+2$ and $+3$ oxidation states. What is the ratio of Fe(II) atoms to Fe(III) atoms in this compound?

**4.122**   A transition metal X forms an oxide of formula $X_2O_3$. It is found that only 50% of X atoms in this compound are in the $+3$ oxidation state. The only other stable oxidation states of X are $+2$ and $+5$. What percentage of X atoms are in the $+2$ oxidation state in this compound?

**4.123**   What volume of a solution of ethanol, $C_2H_6O$, that is 94.0% ethanol by mass contains 0.200 mol $C_2H_6O$? The density of the solution is 0.807 g/mL.

**4.124**   What volume of a solution of ethylene glycol, $C_2H_6O_2$, that is 56.0% ethylene glycol by mass contains 0.350 mol $C_2H_6O_2$? The density of the solution is 1.072 g/mL.

**4.125**   A 10.0-mL sample of potassium iodide solution was analyzed by adding an excess of silver nitrate solution to produce silver iodide crystals, which were filtered from the solution.

$$KI(aq) + AgNO_3(aq) \longrightarrow KNO_3(aq) + AgI(s)$$

If 2.183 g of silver iodide was obtained, what was the molarity of the original KI solution?

**4.126**   A 25.0-mL sample of sodium sulfate solution was analyzed by adding an excess of barium chloride solution to produce barium sulfate crystals, which were filtered from the solution.

$$Na_2SO_4(aq) + BaCl_2(aq) \longrightarrow 2NaCl(aq) + BaSO_4(s)$$

If 5.719 g of barium sulfate was obtained, what was the molarity of the original $Na_2SO_4$ solution?

**4.127**   A metal, M, was converted to the sulfate, $M_2(SO_4)_3$. Then a solution of the sulfate was treated with barium chloride to give barium sulfate crystals, which were filtered off.

$$M_2(SO_4)_3(aq) + 3BaCl_2(aq) \longrightarrow 2MCl_3(aq) + 3BaSO_4(s)$$

If 1.200 g of the metal gave 6.026 g of barium sulfate, what is the atomic weight of the metal? What is the metal?

**4.128**   A metal, M, was converted to the chloride, $MCl_2$. Then a solution of the chloride was treated with silver nitrate to give silver chloride crystals, which were filtered from the solution.

$$MCl_2(aq) + 2AgNO_3(aq) \longrightarrow M(NO_3)_2(aq) + 2AgCl(s)$$

If 2.434 g of the metal gave 7.964 g of silver chloride, what is the atomic weight of the metal? What is the metal?

**4.129** Phosphoric acid is prepared by dissolving phosphorus(V) oxide, $P_4O_{10}$, in water. What is the balanced equation for this reaction? How many grams of $P_4O_{10}$ are required to make 1.50 L of aqueous solution containing 5.00% phosphoric acid by mass? The density of the solution is 1.025 g/mL.

**4.130** Iron(III) chloride can be prepared by reacting iron metal with chlorine. What is the balanced equation for this reaction? How many grams of iron are required to make 2.50 L of aqueous solution containing 9.00% iron(III) chloride by mass? The density of the solution is 1.067 g/mL.

**4.131** An alloy of aluminum and magnesium was treated with sodium hydroxide solution, in which only aluminum reacts.

$$2Al(s) + 2NaOH(aq) + 6H_2O(l) \longrightarrow$$
$$2NaAl(OH)_4(aq) + 3H_2(g)$$

If a sample of alloy weighing 1.118 g gave 0.1068 g of hydrogen, what is the percentage of aluminum in the alloy?

**4.132** An alloy of iron and carbon was treated with sulfuric acid, in which only iron reacts.

$$2Fe(s) + 3H_2SO_4(aq) \longrightarrow Fe_2(SO_4)_3(aq) + 3H_2(g)$$

If a sample of alloy weighing 2.358 g gave 0.1352 g of hydrogen, what is the percentage of iron in the alloy?

**4.133** Determine the volume of sulfuric acid solution needed to prepare 37.4 g of aluminum sulfate, $Al_2(SO_4)_3$, by the reaction

$$2Al(s) + 3H_2SO_4(aq) \longrightarrow Al_2(SO_4)_3(aq) + 3H_2(g)$$

The sulfuric acid solution, whose density is 1.104 g/mL, contains 15.0% $H_2SO_4$ by mass.

**4.134** Determine the volume of sodium hydroxide solution needed to prepare 26.2 g sodium phosphate, $Na_3PO_4$, by the reaction

$$3NaOH(aq) + H_3PO_4(aq) \longrightarrow Na_3PO_4(aq) + 3H_2O(l)$$

The sodium hydroxide solution, whose density is 1.133 g/mL, contains 12.0% NaOH by mass.

**4.135** The active ingredients of an antacid tablet contained only magnesium hydroxide and aluminum hydroxide. Complete neutralization of a sample of the active ingredients required 48.5 mL of 0.187 $M$ hydrochloric acid. The chloride salts from this neutralization were obtained by evaporation of the filtrate from the titration; they weighed 0.4200 g. What was the percentage by mass of magnesium hydroxide in the active ingredients of the antacid tablet?

**4.136** The active ingredients in an antacid tablet contained only calcium carbonate and magnesium carbonate. Complete reaction of a sample of the active ingredients required 41.33 mL of 0.08750 $M$ hydrochloric acid. The chloride salts from the reaction were obtained by evaporation of the filtrate from this titration; they weighed 0.1900 g. What was the percentage by mass of the calcium carbonate in the active ingredients of the antacid tablet?

## Media Activities

**4.137** Open the Precipitation Reactions *Understanding Concepts* activity and go through the introduction and example.
a. What does the blue ball in the animation represent?
b. What ion is missing from the animation? c. Consider the reaction between four formula units of $Pb(NO_3)_2(aq)$ and three formula units of $CaCl_2(aq)$. After the reaction is complete, how many and which type of ions remain in solution?

**4.138** Open the Exercises in the Precipitation Reactions *Understanding Concepts* activity and work at least two problems.
a. For each of the problems that you work, write the net ionic equation.
b. Write the net ionic equation for the reaction of $PbNO_3(aq)$ and $CaCl_2(aq)$.

**4.139** Open the Dissolution of a Solid in a Liquid *Visualization*.
a. If the process depicted in the animation were allowed to continue, describe what you would eventually see.
b. If the crystal were $MgCl_2$ instead of NaCl, how would the animation compare?
c. If the crystal were $PbCl_2$, how would the animation compare?

**4.140** Open the Conductivities of Aqueous Solutions *Visualization* and play the demonstration illustrating the differences in conductivity between strong, weak, and nonelectrolytes in water.
a. Determine which solution was the strongest electrolyte and explain why this is the case.
b. Based on the conductivity results, would you classify the amino acid alanine as a strong or weak acid?

**4.141** Open the following Problem Solving activities in Chapter 4 that refer to molarity calculations: *Calculation and Molarity, Concentration and Volume,* and *Concentration and Volume II.* Do three to four practice problems from each problem set.
a. Calculate the $Cl^-$ concentration in a 0.10 $M$ NaCl solution.
b. Calculate the $Cl^-$ concentration in a 0.10 $M$ $CaCl_2$ solution.

**4.142** Run the Zinc and Iodine *Visualization* to observe a vigorous oxidation–reduction reaction. The product of the reaction is zinc iodide.
a. Write and balance the equation using the half-reaction method outlined in Section 4.6.
b. Identify the oxidizing agent and reducing agent in the reaction.

**4.143** For review, take the ACE practice test for Chapter 4 on the student web site.

A chemical reaction producing gas bubbles in solution.

# The Gaseous State

Gases have several characteristics that distinguish them from liquids and solids. You can compress gases into smaller and smaller volumes. Anyone who has transported gases such as oxygen compressed in steel cylinders can appreciate this (Figure 5.1). Also, for gases, unlike for liquids and solids, you can relate pressure, volume, temperature, and molar amount of substance with fair accuracy by one simple equation, the ideal gas law. Suppose you wish to determine the amount of oxygen in a tank of compressed gas. You measure the pressure (with a gauge), the temperature, and the volume of gas (volume of the tank). Then, using the ideal gas law, you can calculate the amount of gas in the tank.

Kinetic-molecular theory describes a gas as composed of molecules in constant motion. This theory helps explain the simple relationship that exists among the pressure, volume, temperature, and amount of a gas. Kinetic theory has also enhanced our understanding of the flow of fluids, the transmission of sound, and the conduction of heat.

This chapter introduces the empirical gas laws, such as the ideal gas law, and the kinetic-molecular theory that explains these laws. After finishing the chapter, you will be able to answer questions such as, How many grams of oxygen are there in a 50.0-L tank at 21°C when the oxygen pressure is 15.7 atmospheres? (An atmosphere is a unit of pressure equal to that of the normal atmosphere.) You will also be able to calculate the average speed of an oxygen molecule in this tank (479 m/s, or 1071 mi/hr).

## CONTENTS

**183**

# Gas Laws

▶ This chapter may be covered just before the chapter on liquids and solids (Chapter 11), if you prefer.

Most substances composed of small molecules are gases under normal conditions or else are easily vaporized liquids. Table 5.1 lists selected gaseous substances and some of their properties.

In the first part of this chapter, we will examine the quantitative relationships, or empirical laws, governing gases. First, however, you need to understand the concept of pressure.

## 5.1 Gas Pressure and Its Measurement

▶ **Animations:** Visualizing molecular motion (single molecule); Visualizing molecular motion (many molecules).

▶ **Video no. 16 (Series D), "Boiling Water with Ice."**

▶ **Demonstration with mercury barometers:** Bassam Z. Shakhashiri, *Chemical Demonstrations*, Vol. 2 (Madison: University of Wisconsin Press, 1985), no. 5.2, p. 9.

**Pressure** is defined as *the force exerted per unit area of surface.* A coin resting on a table exerts a force, and therefore a pressure, downward on the table due to gravity. The air above the table exerts an additional pressure on the table, because the air is also being pulled downward by gravity.

To obtain the SI unit of pressure and a feeling for its size, let us calculate the pressure on a table from a perfectly flat coin with a radius and mass equal to that of a new penny (9.3 mm in radius and 2.5 g). The force exerted by the coin from gravity equals the mass of the coin times the constant acceleration of gravity. *Acceleration* is the change of speed per unit time, so the SI unit of acceleration is meters per second per second, abbreviated $m/s^2$. The constant acceleration of gravity is $9.81 \ m/s^2$, and the force on the coin due to gravity is

$$\text{Force} = \text{mass} \times \text{constant acceleration of gravity}$$
$$= (2.5 \times 10^{-3} \ kg) \times (9.81 \ m/s^2) = 2.5 \times 10^{-2} \ kg \cdot m/s^2$$

**Figure 5.1**
**Cylinders of gas.** Gases such as oxygen and nitrogen can be transported as compressed gases in steel cylinders. Large volumes of gas at normal pressures can be compressed into a small volume. Note the pressure gauges.

**Table 5.1**
**Properties of Selected Gases**

| Name | Formula | Color | Odor | Toxicity |
|---|---|---|---|---|
| Ammonia | $NH_3$ | Colorless | Penetrating | Toxic |
| Carbon dioxide | $CO_2$ | Colorless | Odorless | Nontoxic |
| Carbon monoxide | $CO$ | Colorless | Odorless | Very toxic |
| Chlorine | $Cl_2$ | Pale green | Irritating | Very toxic |
| Helium | $He$ | Colorless | Odorless | Nontoxic |
| Hydrogen | $H_2$ | Colorless | Odorless | Nontoxic |
| Hydrogen chloride | $HCl$ | Colorless | Irritating | Corrosive |
| Hydrogen sulfide | $H_2S$ | Colorless | Foul | Very toxic |
| Methane | $CH_4$ | Colorless | Odorless | Nontoxic |
| Neon | $Ne$ | Colorless | Odorless | Nontoxic |
| Nitrogen | $N_2$ | Colorless | Odorless | Nontoxic |
| Nitrogen dioxide | $NO_2$ | Red-brown | Irritating | Very toxic |
| Oxygen | $O_2$ | Colorless | Odorless | Nontoxic |
| Sulfur dioxide | $SO_2$ | Colorless | Choking | Toxic |

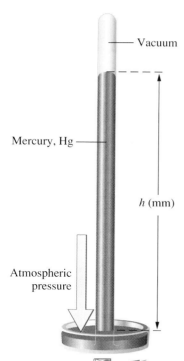

**Figure 5.2**
**A mercury barometer.**
The height $h$ is proportional to the barometric pressure. For that reason, the pressure is often given as the height of the mercury column, in units of millimeters of mercury, mmHg.

The cross-sectional area of the coin is $\pi \times (\text{radius})^2 = 3.14 \times (9.3 \times 10^{-3} \text{ m})^2 = 2.7 \times 10^{-4} \text{ m}^2$. Therefore,

$$\text{Pressure} = \frac{\text{force}}{\text{area}} = \frac{2.5 \times 10^{-2} \text{ kg} \cdot \text{m/s}^2}{2.7 \times 10^{-4} \text{ m}^2} = 93 \text{ kg/(m} \cdot \text{s}^2)$$

*The SI unit of pressure, $kg/(m \cdot s^2)$,* is given the name **pascal (Pa),** after the French physicist Blaise Pascal (1623–1662), who studied fluid pressure. Note that the pressure exerted by a coin the size and mass of a penny is approximately 100 Pa. The pressure exerted by the atmosphere is about 1000 times larger, or about 100,000 Pa. Thus, the pascal is an extremely small unit.

Chemists have traditionally used two other units of pressure, based on the mercury barometer. A **barometer** is *a device for measuring the pressure of the atmosphere.* The mercury barometer consists of a glass tube about one meter long, filled with mercury and inverted in a dish of the same liquid metal (Figure 5.2). At sea level the mercury in the tube falls to a height of about 760 mm above the level in the dish. This height is a direct measure of the atmospheric pressure (Figure 5.3). ◄ Air pressure downward on the surface of the mercury in the dish is transmitted through the liquid and exerts a pressure upward at the base of the mercury column, supporting it. A mercury column placed in a sealed flask, as in Figure 5.4, measures the gas pressure in the flask. It acts as a **manometer,** *a device that measures the pressure of a gas or liquid in a vessel.*

The unit **millimeters of mercury (mmHg),** also called the **torr** (after Evangelista Torricelli, who invented the mercury barometer in 1643), is *a unit of pressure equal to that exerted by a column of mercury 1 mm high at 0.00°C.* The **atmosphere (atm)** is a related *unit of pressure equal to exactly 760 mmHg.*

The general relationship between the pressure $P$ and the height $h$ of a liquid column in a barometer or manometer is

$$P = gdh$$

**Figure 5.3** *(left)*
**Atmospheric pressure from air mass.** The force of gravity on the mass of a column of air exerts a pressure at the earth's surface. This pressure varies slightly with weather, but is approximately 101 kPa (760 mmHg, or 1 atm).

**Figure 5.4** *(right)*
**A flask equipped with a closed-tube manometer.** The gas pressure in the flask is proportional to the difference in heights between the liquid levels in the manometer, $\Delta h$.

▶ Atmospheric, or barometric, pressure depends not only on altitude but also on weather conditions. The "highs" and "lows" of weather reports refer to high- and low-pressure air masses. A high is associated with fair weather; a low brings unsettled weather, or storms.

▶ Video Series C, "Collapsing Can."

**Table 5.2**
**Important Units of Pressure**

| Unit | Relationship or Definition |
|---|---|
| Pascal (Pa) | $kg/(m \cdot s^2)$ |
| Atmosphere (atm) | $1\ atm = 1.01325 \times 10^5\ Pa \approx 100\ kPa$ |
| mmHg, or torr | $760\ mmHg = 1\ atm$ |

Here $g$ is the constant acceleration of gravity (9.81 m/s$^2$) and $d$ is the density of the liquid in the manometer. If $g$, $d$, and $h$ are in SI units, the pressure is given in pascals. Table 5.2 summarizes the relationships among the various units of pressure.

**EXAMPLE 5.1**

## Converting Units of Pressure

The pressure of a gas in a flask is measured to be 797.7 mmHg using a mercury-filled manometer as in Figure 5.4. What is this pressure in pascals and atmospheres?

 **PROBLEM STRATEGY**

This is a unit conversion problem. To convert to the appropriate units, use the conversion factors presented in Table 5.2 and the techniques presented in Chapter 1. Keep in mind that the conversion 760 mmHg = 1 atm is exact.

**SOLUTION**

Conversion to pascals:

$$797.7\ \text{mmHg} \times \frac{1.01325 \times 10^5\ \text{Pa}}{760\ \text{mmHg}} = \textbf{1.064} \times \textbf{10}^5\ \textbf{Pa}$$

Conversion to atmospheres:

$$797.7\ \text{mmHg} \times \frac{1\ \text{atm}}{760\ \text{mmHg}} = \textbf{1.050 atm}$$

**EXERCISE 5.1**

A gas in a container had a measured pressure of 57 kPa. Calculate the pressure in units of atm and mmHg.

See Problems 5.31 and 5.32.

 **CONCEPT CHECK 5.1**

Suppose that you set up two barometers like the one shown in Figure 5.2. In one of the barometers you use mercury, and in the other you use water. Which of the barometers would have a higher column of liquid, the one with Hg or H$_2$O? Explain your answer.

# 5.2  Empirical Gas Laws

All gases under moderate conditions behave quite simply with respect to pressure, temperature, volume, and molar amount. By holding any two of these physical properties constant, it is possible to show a simple relationship between the other two. The discovery of these quantitative relationships, the empirical gas laws, occurred from the mid-seventeenth to the mid-nineteenth century.

### Boyle's Law: Relating Volume and Pressure

▶ Video no. 7 (Series A), "Boyle's Law."

▶ Demonstration of Boyle's law: Bassam Z. Shakhashiri, *Chemical Demonstrations,* Vol. 2 (Madison: University of Wisconsin Press, 1985), no. 5.4, pp. 14–19; Summerlin et al., *Chemical Demonstrations,* Vol. 2 (Washington, D.C.: Am. Chem. Soc., 1987), no. 11, p. 22.

One characteristic property of a gas is its *compressibility*—its ability to be squeezed into a smaller volume by the application of pressure. By comparison, liquids and solids are relatively incompressible. The compressibility of gases was first studied quantitatively by Robert Boyle in 1661. When he poured mercury into the open end of a J-shaped tube, the volume of the enclosed gas decreased (Figure 5.5). Each addition of mercury increased the pressure on the gas, decreasing its volume. From such experiments, he formulated the law now known by his name. According to **Boyle's law,** *the volume of a sample of gas at a given temperature varies inversely with the applied pressure.* That is, $V \propto 1/P$, where $V$ is the volume, $P$ is the pressure, and $\propto$ means "is proportional to." Thus, if the pressure is doubled, the volume is halved.

Boyle's law can also be expressed in the form of an equation. Putting pressure and volume on the same side of the equation, you can write

> Boyle's law:
> $$PV = \text{constant} \qquad \text{(for a given amount of gas at fixed temperature)}$$

▶ Animations: Boyle's law; Microscopic illustration of Boyle's law.

That is, for a given amount of gas at a fixed temperature, the pressure times the volume equals a constant. Table 5.3 gives some pressure and volume data for 1.000 g $O_2$ at 0°C. Note that the product of the pressure and volume is nearly constant. By

**Figure 5.5**
**Boyle's experiment.**
The volume of the gas at normal atmospheric pressure (760 mmHg) is 100 mL. When the pressure is doubled by adding 760 mm of mercury, the volume is halved (to 50 mL). Tripling the pressure decreases the volume to one-third of the original (to 33 mL).

**Table 5.3**
**Pressure–Volume Data for 1.000 g $O_2$ at 0°C**

| P (atm) | V (L) | PV |
|---------|-------|-----|
| 0.2500 | 2.801 | 0.7002 |
| 0.5000 | 1.400 | 0.7000 |
| 0.7500 | 0.9333 | 0.7000 |
| 1.000 | 0.6998 | 0.6998 |
| 2.000 | 0.3495 | 0.6990 |
| 3.000 | 0.2328 | 0.6984 |
| 4.000 | 0.1744 | 0.6976 |
| 5.000 | 0.1394 | 0.6970 |

*Increasing pressure* ↓    *Increasing volume* ↑

plotting the volume of the oxygen at different pressures (as shown in Figure 5.6), you obtain a graph showing the inverse relationship of $P$ and $V$.

You can use Boyle's law to calculate the volume occupied by a gas when the pressure changes. Consider the 50.0-L tank of oxygen mentioned in the chapter opening. The pressure of gas in the tank is 15.7 atm at 21°C. What volume of oxygen can you get from the tank at 21°C if the atmospheric pressure is 1.00 atm? You can write $P_i$ and $V_i$ for the initial pressure (15.7 atm) and initial volume (50.0 L), and $P_f$ and $V_f$ for the final pressure (1.00 atm) and final volume (to be determined). Because the temperature does not change, the product of the pressure and volume remains constant. Thus, you can write

$$P_f V_f = P_i V_i$$

Dividing both sides of the equation by $P_f$ gives

$$V_f = V_i \times \frac{P_i}{P_f}$$

Substituting into this equation,

$$V_f = 50.0 \text{ L} \times \frac{15.7 \text{ atm}}{1.00 \text{ atm}} = 785 \text{ L}$$

**Figure 5.6**
**Gas pressure–volume relationship.** *(A)* Plot of volume vs. pressure for a sample of oxygen. The volume (of 1.000 g $O_2$ at 0°C) decreases with increasing pressure. When the pressure is doubled (from 0.50 atm to 1.00 atm), the volume is halved (from 1.40 L to 0.70 L). *(B)* Plot of 1/$V$ vs. pressure (at constant temperature) for the same sample. The straight line indicates that volume varies inversely with pressure.

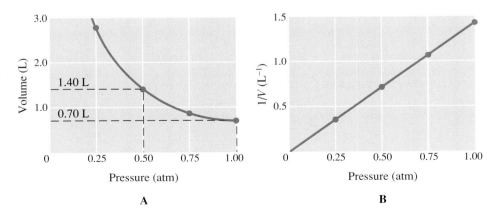

Note that the initial volume is multiplied by a ratio of pressures. You know the oxygen gas is changing to a lower pressure and will therefore expand, so this ratio will be greater than 1. The final volume, 785 L, is that occupied by all of the gas at 1.00 atm pressure. However, because the tank holds 50.0 L, only this volume of gas remains in the tank. The volume that escapes is $(785 - 50.0)$ L $= 735$ L.

◀ An alternate example is provided in the Instructor's Resource Manual.

◀ Experiments in General Chemistry, 5A, "Boyle's Law and the Empty Space in Air."

◀ Demonstration of thermal expansion of gases: Bassam Z. Shakhashiri, *Chemical Demonstrations*, Vol. 2 (Madison: University of Wisconsin Press, 1985), no. 5.6, pp. 24–27; Summerlin et al., *Chemical Demonstrations*, Vol. 2 (Washington, D.C.: Am. Chem. Soc., 1987), no. 12, p. 23.

**Problem Solving: Boyle's Law**

### EXAMPLE 5.2

## Using Boyle's Law

A volume of air occupying 12.0 dm$^3$ at 98.9 kPa is compressed to a pressure of 119.0 kPa. The temperature remains constant. What is the new volume?

#### PROBLEM STRATEGY

Putting the data for the problem in tabular form, you see what data you have and what you must find. This will suggest the method of solution.

$$V_i = 12.0 \text{ dm}^3 \qquad P_i = 98.9 \text{ kPa} \left. \right\} \begin{array}{l} \text{Temperature and} \\ \text{moles remain} \\ \text{constant.} \end{array}$$
$$V_f = \quad ? \qquad P_f = 119.0 \text{ kPa}$$

Because $P$ and $V$ vary but temperature and moles are constant, you use Boyle's law.

#### SOLUTION

Application of Boyle's law gives

$$V_f = V_i \times \frac{P_i}{P_f} = 12.0 \text{ dm}^3 \times \frac{98.9 \text{ kPa}}{119.0 \text{ kPa}} = \mathbf{9.97 \text{ dm}^3}$$

Note that the pressure on the gas increases, so the gas is compressed (gives a smaller volume), and the ratio of pressures is less than 1. Thus, you could get the above result by simply multiplying the volume by the ratio of pressures, choosing the ratio so it is less than 1. (If you had expected the gas volume to increase, you would have chosen the ratio to be greater than 1.)

#### EXERCISE 5.2

A volume of carbon dioxide gas, $CO_2$, equal to 20.0 L was collected at 23°C and 1.00 atm pressure. What would be the volume of carbon dioxide collected at 23°C and 0.830 atm?

**See Problems 5.33, 5.34, 5.35, and 5.36.**

Before leaving the subject of Boyle's law, we should note that the pressure–volume product for a gas is not precisely constant. You can see this from the *PV* data given in Table 5.3 for oxygen. In fact, all gases follow Boyle's law at low to moderate pressures but deviate from this law at high pressures. The extent of deviation depends on the gas. We will return to this point at the end of the chapter.

◀ Demonstration of Charles's law: Bassam Z. Shakhashiri, *Chemical Demonstrations*, Vol. 2 (Madison: University of Wisconsin Press, 1985), no. 5.7, pp. 28–32.

### Charles's Law: Relating Volume and Temperature

Temperature also affects gas volume. When you immerse an air-filled balloon in liquid nitrogen ($-196$°C), the balloon shrinks (Figure 5.7). After the balloon is removed

**Figure 5.7**
**Effect of temperature on volume of a gas.**
A balloon immersed in liquid nitrogen shrinks because the air inside contracts in volume. When the balloon is removed from the liquid nitrogen, the air inside warms up and the balloon expands to its original size.

from the liquid nitrogen, it returns to its original size. A gas contracts when cooled and expands when heated.

One of the first quantitative observations of gases at different temperatures was made by Jacques Alexandre Charles in 1787. Charles was a French physicist and a pioneer in hot-air and hydrogen-filled balloons. ◄ Later, John Dalton (in 1801) and Joseph Louis Gay-Lussac (in 1802) continued these kinds of experiments, which showed that a sample of gas at a fixed pressure increases in volume *linearly* with temperature. By "linearly," we mean that if we plot the volume occupied by a given sample of gas at various temperatures, we get a straight line (Figure 5.8).

When you extend the straight lines in Figure 5.8 from the last experimental point toward lower temperatures—that is, when you *extrapolate* the straight lines backward—you find that they all intersect at a common point. This point occurs at a temperature of −273.15°C, where the graph indicates a volume of zero. This

► The first ascent of a hot-air balloon carrying people was made on November 21, 1783. A few days later, Jacques Alexandre Charles made an ascent in a hydrogen-filled balloon. On landing, the balloon was attacked and torn to shreds by terrified peasants armed with pitchforks.

**Figure 5.8**
**Linear relationship of gas volume and temperature at constant pressure.**
The graph shows gas volume versus temperature for a given mass of gas at 1.00 atm pressure. This linear relationship is independent of amount or kind of gas. Note that all lines extrapolate to −273°C at zero volume.

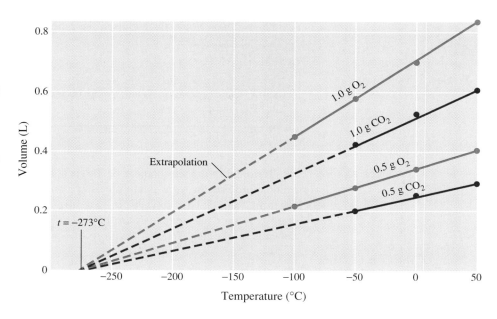

▶ Video Series C, "Liquid Nitrogen and Balloons."

▶ Animation:
Balloon submersion in liquid N₂.

▶ Video no. 18 (Series D),
"Birthday Candle Experiment."

▶ The mathematical equation
of a straight line, $y = mx + b$,
is discussed in Appendix A.

▶ Animation:
Microscopic illustration of
Charles's law.

▶ The Kelvin temperature scale
and a more formal equation for
temperature conversion were
discussed in Section 1.6.

▶ Video no. 15 (Series D),
"Egg in a Bottle."

▶ Animation:
Charles's law.

▶ Video no. 17 (Series D),
"Aluminum Can Implosion."

seems to say that if the substances remain gaseous, the volumes occupied will be zero at −273.15°C. This could not happen, however; all gases liquefy before they reach this temperature, and Charles's law does not apply to liquids. These extrapolations do show that we can express the volume variation of a gas with temperature more simply by choosing a different thermometer scale. Let us see how to do this.

The fact that the volume occupied by a gas varies linearly with degrees Celsius can be expressed mathematically by the following equation:

$$V = bt + a$$

where $t$ is the temperature in degrees Celsius and $a$ and $b$ are constants that determine the straight line. ◀ You can eliminate the constant $a$ by observing that $V = 0$ at $t = -273.15$ for any gas. Substituting into the preceding equation, you get

$$0 = b(-273.15) + a \quad \text{or} \quad a = 273.15b$$

The equation for the volume can now be rewritten:

$$V = bt + 273.15b = b(t + 273.15)$$

Suppose you use a temperature scale equal to degrees Celsius plus 273.15, which you may recognize as the *Kelvin scale* (an absolute scale). ◀

$$K = {}^{\circ}C + 273.15$$

If you write $T$ for the temperature on the Kelvin scale, you obtain

$$V = bT$$

This is **Charles's law,** which we can state as follows: *the volume occupied by any sample of gas at a constant pressure is directly proportional to the absolute temperature*. Thus, doubling the absolute temperature of a gas doubles its volume.

Charles's law can be rearranged into a form that is very useful for computation.

Charles's law:

$$\frac{V}{T} = \text{constant} \qquad \text{(for a given amount of gas at a fixed pressure)}$$

This equation says that the volume divided by the absolute temperature, for a sample of gas at a fixed pressure, remains constant.

Consider a sample of gas at a fixed pressure, and suppose the temperature changes from its initial value $T_i$ to a final value $T_f$. How does the volume change? Because the volume divided by absolute temperature is constant, you can write

$$\frac{V_f}{T_f} = \frac{V_i}{T_i}$$

Or, rearranging slightly,

$$V_f = V_i \times \frac{T_f}{T_i}$$

Note that to obtain the final volume, the initial volume is multiplied by a ratio of absolute temperatures. The next example illustrates a calculation involving a change in the temperature of a gas.

▶ An alternate example is provided in the Instructor's Resource Manual.

---

**EXAMPLE 5.3**

## Using Charles's Law

Earlier we found that the total volume of oxygen that can be obtained from a particular tank at 1.00 atm and 21°C is 785 L (including the volume remaining in the tank). What would be this volume of oxygen if the temperature had been 28°C?

 **PROBLEM STRATEGY**

You must first convert temperatures to the Kelvin scale. Then, list the data in tabular form, so you can easily see what varies and what remains constant. You apply Charles's law when $V$ and $T$ vary and $P$ is constant.

**SOLUTION**

$$T_i = (21 + 273)K = 294\ K$$
$$T_f = (28 + 273)K = 301\ K$$

Following is the data table:

$$V_i = 785\ L \qquad P_i = 1.00\ \text{atm} \qquad T_i = 294\ K$$
$$V_f = ? \qquad P_f = 1.00\ \text{atm} \qquad T_f = 301\ K$$

Note that $T$ varies and $P$ remains constant, so $V$ must change. These are the conditions needed to apply Charles's law.

$$V_f = V_i \times \frac{T_f}{T_i} = 785\ L \times \frac{301\ K}{294\ K} = \textbf{804 L}$$

Note that the temperature increases, so you expect the volume to increase. This means that the ratio of absolute temperatures is greater than 1. You could get the foregoing result by simply multiplying the volume by the ratio of absolute temperatures, choosing the ratio so that it is greater than 1. (If you had expected the gas volume to decrease, you would have chosen the ratio to be less than 1.)

---

**EXERCISE 5.3**

If you expect a chemical reaction to produce 4.38 dm³ of oxygen, $O_2$, at 19°C and 101 kPa, what would be the volume at 25°C and 101 kPa?

---

See Problems 5.39, 5.40, 5.41, and 5.42.

**Problem Solving: Charles's Law**

Although most gases follow Charles's law fairly well, they deviate from it at high pressures and low temperatures.

### Combined Gas Law: Relating Volume, Temperature, and Pressure

Boyle's law ($V \propto 1/P$) and Charles's law ($V \propto T$) can be combined and expressed in a single statement: the volume occupied by a given amount of gas is proportional to the absolute temperature divided by the pressure ($V \propto T/P$). You can write this as an equation.

$$V = \text{constant} \times \frac{T}{P} \quad \text{or} \quad \frac{PV}{T} = \text{constant} \qquad \text{(for a given amount of gas)}$$

Consider a problem in which you wish to calculate the final volume of a gas when the pressure and temperature are changed. $PV/T$ is constant for a given amount of gas, so you can write

$$\frac{P_f V_f}{T_f} = \frac{P_i V_i}{T_i}$$

which rearranges to

$$V_f = V_i \times \frac{P_i}{P_f} \times \frac{T_f}{T_i}$$

Thus, the final volume is obtained by multiplying the initial volume by ratios of pressures and absolute temperatures.

▶ An alternate example is provided in the Instructor's Resource Manual.

---

### EXAMPLE 5.4

## Using the Combined Gas Law

Modern determination of %N in an organic compound is an automated version of one developed by the French chemist Jean-Baptiste Dumas in 1830. The Dumas method uses hot copper(II) oxide to oxidize C and H in the compound to $CO_2$ and $H_2O$ (both are trapped chemically), and to convert N in the compound to $N_2$ gas (Figure 5.9). Using the Dumas method, 39.8 mg of caffeine gives 10.1 cm$^3$ of nitrogen gas at 23°C and 746 mmHg. What is the volume of nitrogen at 0°C and 760 mmHg?

**Figure 5.9**
**Oxidation of an organic compound with copper(II) oxide.** *Left:* Molecular model of caffeine, an organic compound. *Center:* A chemist heats caffeine with copper(II) oxide, CuO. CuO oxidizes C and H in the compound to $CO_2$ and $H_2O$, and converts N to $N_2$. *Right:* Copper metal produced by the reaction.

### SOLUTION

You first express the temperatures in kelvins:

$$T_i = (23 + 273) \text{ K} = 296 \text{ K}$$
$$T_f = (0 + 273) \text{ K} = 273 \text{ K}$$

Now you put the data for the problem in tabular form:

| | | |
|---|---|---|
| $V_i = 10.1 \text{ cm}^3$ | $P_i = 746 \text{ mmHg}$ | $T_i = 296 \text{ K}$ |
| $V_f = \quad ?$ | $P_f = 760 \text{ mmHg}$ | $T_f = 273 \text{ K}$ |

From the data, you see that you need to use Boyle's and Charles's laws combined to find how $V$ varies as $P$ and $T$ change. Note that the increase in pressure decreases the volume, making the pressure ratio less than 1. The decrease in temperature also decreases the volume, and the ratio of absolute temperatures is less than 1.

$$V_f = V_i \times \frac{P_i}{P_f} \times \frac{T_f}{T_i} = 10.1 \text{ cm}^3 \times \frac{746 \text{ mmHg}}{760 \text{ mmHg}} \times \frac{273 \text{ K}}{296 \text{ K}} = \textbf{9.14 cm}^3$$

### EXERCISE 5.4

A balloon contains 5.41 dm$^3$ of helium, He, at 24°C and 101.5 kPa. Suppose the gas in the balloon is heated to 35°C. If the helium pressure is now 102.8 kPa, what is the volume of the gas?

**See Problems 5.45 and 5.46.**

 **CONCEPT CHECK 5.2**

To conduct some experiments, a 10.0-L flask equipped with a movable plunger, as illustrated here, is filled with enough H$_2$ gas to come to a pressure of 20 atm.

a. In the first experiment, we decrease the temperature in the flask by 10°C and then increase the volume. Predict how the pressure in the flask changes during each of these events and, if possible, how the final pressure compares to your starting pressure.

b. Once again we start with the pressure in the flask at 20 atm. The flask is then heated 10°C followed by a volume decrease. Predict how the pressure in the flask changes during each of these events and, if possible, how the final pressure compares to your starting pressure.

▶ Video no. 9 (Series A), "Stoichiometry: Gay-Lussac's Law of Combining Volumes."

▶ Demonstration of Avogadro's law: Bassam Z. Shakhashiri, *Chemical Demonstrations,* Vol. 2 (Madison: University of Wisconsin Press, 1985), no. 5.11, pp. 44–47.

## Avogadro's Law: Relating Volume and Amount

In 1808 the French chemist Joseph Louis Gay-Lussac (1778–1850) concluded from experiments on gas reactions that the volumes of reactant gases at the same pressure and temperature are in ratios of small whole numbers *(the law of combining volumes).* For example, two volumes of hydrogen react with one volume of oxygen gas to produce water.

$$2H_2(g) + O_2(g) \longrightarrow 2H_2O(g)$$
$$\text{2 volumes} \quad \text{1 volume}$$

Three years later, the Italian chemist Amedeo Avogadro (1776–1856) interpreted the law of combining volumes in terms of what we now call **Avogadro's law:** *equal volumes of any two gases at the same temperature and pressure contain the same number of molecules.* Thus, two volumes of hydrogen contain twice the number of molecules as in one volume of oxygen, in agreement with the chemical equation for the reaction.

One mole of any gas contains the same number of molecules (Avogadro's number = $6.02 \times 10^{23}$) and by Avogadro's law must occupy the same volume at a given temperature and pressure. This *volume of one mole of gas* is called the **molar gas volume, $V_m$.** Volumes of gases are often compared at **standard temperature and pressure (STP),** *the reference conditions for gases chosen by convention to be 0°C and 1 atm pressure.* At STP, the molar gas volume is found to be 22.4 L/mol (Figure 5.10).

**Figure 5.10**
**The molar volume of a gas.**   The box has a volume of 22.4 L, equal to the molar volume of a gas at STP. The basketball is shown for comparison.

We can re-express Avogadro's law as follows: the molar gas volume at a given temperature and pressure is a specific constant independent of the nature of the gas.

> Avogadro's law:
> $V_m$ = specific constant (= 22.4 L/mol at STP)
> (depending on $T$ and $P$ but independent of the gas)

Table 5.4 lists the molar volumes of several gases. The values agree within about 2% of the value expected from the empirical gas laws (the "ideal gas" value). We will discuss the reason for the deviations from this value at the end of this chapter, in the section on real gases.

**Table 5.4**
**Molar Volumes of Several Gases**

| Gas | Molar Volume (L) at 0.0°C, 1.00 atm |
|---|---|
| He | 22.40 |
| $H_2$ | 22.43 |
| $O_2$ | 22.39 |
| $CO_2$ | 22.29 |
| $NH_3$ | 22.09 |
| Ideal gas* | 22.41 |

*An ideal gas follows the empirical gas laws.

# 5.3   The Ideal Gas Law

▶ Animation:
   The ideal gas law.

In the previous section, we discussed the empirical gas laws. Here we will show that these laws can be combined into one equation, called the *ideal gas equation*. Earlier we combined Boyle's law and Charles's law into the equation

$$V = \text{constant} \times \frac{T}{P}$$   (for a given amount of gas)

# Nitric Oxide Gas and Biological Signaling

In 1998, the Nobel committee awarded its prize in physiology or medicine to three scientists for the astounding discovery that nitric oxide gas, NO, functions as the signaling agent between biological cells in a wide variety of chemical processes. Until this discovery, biochemists had thought that the major chemical reactions in a cell always involved very large molecules. Now they discovered that a simple gas, NO, could have a central role in cell chemistry.

Prizewinners Robert Furchgott and Louis Ignarro, independently, unraveled the role of nitric oxide in blood-pressure regulation. Cells in the lining of arteries detect increased blood pressure and respond by producing nitric oxide. NO rapidly diffuses through the artery wall to cells in the surrounding muscle tissue. In response, the muscle tissue relaxes, the blood vessel expands, and the blood pressure drops.

In a related discovery, another prizewinner, Ferid Murad, explained how nitroglycerin works to alleviate the intense chest pain of an angina attack, which results from reduced blood flow to the heart muscle as a result of partial blockage of arteries by plaque. Physicians have prescribed nitroglycerin for angina for more than a century, knowing only that it works. Murad found that nitroglycerin breaks down in the body to form nitric oxide,

which relaxes the arteries, allowing greater blood flow to the heart. Alfred Nobel, were he alive today, would no doubt be stunned by this news. Nobel, who established the prizes bearing his name, made his fortune in the nineteenth century from his invention of dynamite, a mixture of nitroglycerin with clay that tamed the otherwise haz-

**Figure 5.11**
**Nitroglycerin patch.** Nitroglycerin has been used for more than a hundred years to treat and prevent angina attacks. Here a patient wears a patch that dispenses nitroglycerin steadily over a period of time.

This "constant" is independent of the temperature and pressure but does depend on the amount of gas. For one mole, the constant will have a specific value, which we will denote as $R$. The molar volume, $V_m$, is

$$V_m = R \times \frac{T}{P}$$

According to Avogadro's law, the molar volume at a specific value of $T$ and $P$ is a constant independent of the nature of the gas, and this implies that $R$ is a constant independent of the gas. The **molar gas constant, $R$,** is *the constant of proportionality relating the molar volume of a gas to $T/P$.* Values of $R$ in various units are given in Table 5.5.

The preceding equation can be written for $n$ moles of gas by multiplying both sides by $n$.

$$\underbrace{nV_m}_{V} = \frac{nRT}{P} \quad \text{or} \quad PV = nRT$$

Because $V_m$ is the volume per mole, $nV_m$ is the total volume $V$. *The equation $PV = nRT$, which combines all of the gas laws,* is called the **ideal gas law.**

**Table 5.5**
**Molar Gas Constant in Various Units**

| Value of $R$ |
| --- |
| 0.082058 L • atm/(K • mol) |
| 8.3145 J/(K • mol)* |
| 8.3145 kg • m$^2$/(s$^2$ • K • mol) |
| 8.3145 kPa • dm$^3$/(K • mol) |
| 1.9872 cal/(K • mol)* |

*The units of pressure times volume are the units of energy— for example, joules (J) or calories (cal).

ardous explosive. When Nobel had heart trouble, his physician recommended that he eat a small quantity of nitroglycerin; he refused. In a letter, he wrote "It is ironical that I am now ordered by my physician to eat nitroglycerin." Today, a patient may take either nitroglycerin pills (containing tenths of a milligram of compound in a stabilized mixture) for occasional use or can use a chest patch to dispense nitroglycerin continuously to the skin, where it is absorbed (Figure 5.11).

Research papers on the biological role of nitric oxide now number in the tens of thousands. For example,

scientists have found that white blood cells use nitric oxide in a kind of chemical warfare. These cells emit concentrated clouds of NO that surround bacterial or tumor cells, killing them by interfering with certain cell processes. Researchers have also discovered that nitric oxide plays a role in penile erection. Pharmacologists found that the drug Viagra (Figure 5.12) assists in the action of NO in dilating arteries, leading to erection.

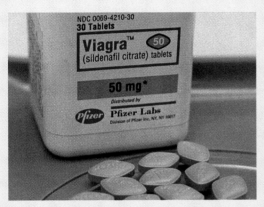

**Figure 5.12**
**Viagra.** Viagra is marketed as an aid to erectile dysfunction. Sexual stimulation produces nitric oxide, which in turn aids an enzyme in the production of a substance called cGMP that relaxes arterial muscles, allowing blood to flow to the penis. Viagra enhances the action of NO by inhibiting an enzyme that degrades cGMP.

Ideal gas law:

$$PV = nRT$$

The ideal gas law includes all the information contained in Boyle's, Charles's, and Avogadro's laws. In fact, starting with the ideal gas law, you can derive any of the other gas laws.

## EXAMPLE 5.5

### Deriving Empirical Gas Laws from the Ideal Gas Law

Prove the following statement: the pressure of a given amount of gas at a fixed volume is proportional to the absolute temperature. This is sometimes called *Amontons's law*. In 1702 Guillaume Amontons constructed a thermometer based on measurement of the pressure of a fixed volume of air. The principle he employed is now used in special gas thermometers to establish the Kelvin scale.

► An alternate example is provided in the Instructor's Resource Manual.

► Video no. 8 (Series A), "Amontons's Law."

**SOLUTION**

From the ideal gas law,

$$PV = nRT$$

Solving for $P$, you get

$$P = \left(\frac{nR}{V}\right)T$$

Note that everything in parentheses in this equation is constant. Therefore, you can write

$$P = \text{constant} \times T$$

Or, expressing this as a proportion, you get

$$P \propto T$$

Boyle's law and Charles's law follow from the ideal gas law by a similar derivation.

### EXERCISE 5.5

Show that the moles of gas are proportional to the pressure for constant volume and temperature.

See Problems 5.49 and 5.50.

The limitations that apply to Boyle's, Charles's, and Avogadro's laws also apply to the ideal gas law. That is, the ideal gas law is most accurate for low to moderate pressures and for temperatures that are not too low. ◀

## Calculations Using the Ideal Gas Law

The type of problem to which Boyle's and Charles's laws are applied involves a change in conditions ($P$, $V$, or $T$) of a gas. The ideal gas law allows us to solve another type of problem: given any three of the quantities $P$, $V$, $n$, and $T$, calculate the unknown quantity. Example 5.6 illustrates such a problem.

### EXAMPLE 5.6

## Using the Ideal Gas Law

Answer the question asked in the chapter opening: how many grams of oxygen, $O_2$, are there in a 50.0-L tank at 21°C when the oxygen pressure is 15.7 atm?

 **PROBLEM STRATEGY**

In asking for the mass of oxygen, we are in effect asking for moles of gas, $n$, because mass and moles are directly related. The problem gives $P$, $V$, and $T$, so you can use the ideal gas law to solve for $n$. The proper value to use for $R$ depends on the units of $P$ and $V$. Here they are in atmospheres and liters, respectively, so you use $R = 0.0821 \text{ L} \cdot \text{atm}/(\text{K} \cdot \text{mol})$ from Table 5.5. Once you have solved for $n$, convert to mass by using the molar mass of oxygen.

**SOLUTION**

The data given in the problem are

| Variable | Value |
|----------|-------|
| $P$ | 15.7 atm |
| $V$ | 50.0 L |
| $T$ | (21 + 273) K = 294 K |
| $n$ | ? |

Solving the ideal gas law for $n$ gives

$$n = \frac{PV}{RT}$$

Substituting the data gives

$$n = \frac{15.7 \text{ atm} \times 50.0 \text{ L}}{0.0821 \text{ L} \cdot \text{atm}/(\text{K} \cdot \text{mol}) \times 294 \text{ K}} = 32.5 \text{ mol}$$

and converting moles to mass of oxygen yields

$$32.5 \text{ mol } O_2 \times \frac{32.0 \text{ g } O_2}{1 \text{ mol } O_2} = \mathbf{1.04 \times 10^3 \text{ g } O_2}$$

**Problem Solving: Ideal Gas Law**

---

**EXERCISE 5.6**

What is the pressure in a 50.0-L tank that contains 3.03 kg of oxygen, $O_2$, at 23°C?

---

See Problems 5.51, 5.52, 5.53, 5.54, 5.55, and 5.56.

**Figure 5.13**
**Hot-air ballooning.**
A propane gas burner on board a balloon heats the air. The heated air expands, occupying a larger volume; it therefore has a lower density than the surrounding air. The hot air and balloon rise.

## Gas Density; Molecular-Weight Determination

The density of a substance is its mass divided by volume, and because the volume of a gas varies with temperature and pressure, the density of a gas also varies with temperature and pressure (Figure 5.13). In the next example, we use the ideal gas law to calculate the density of a gas at any temperature and pressure. We do this by calculating the moles, and from this the mass, in a liter of gas at the given temperature and pressure; the mass per liter of gas is its density. At the end of this section, we will give an alternative method of solving this problem, using a formula relating density to temperature, pressure, and molecular weight.

---

**EXAMPLE 5.7**

## Calculating Gas Density

What is the density of oxygen, $O_2$, in grams per liter at 25°C and 0.850 atm?

 **PROBLEM STRATEGY**

The density of a gas is often expressed in g/L. Using the ideal gas law you can calculate the moles of $O_2$ in 1 L of $O_2$. Next, you can convert the moles of $O_2$ to a mass of $O_2$, keeping in mind that this mass is the amount of $O_2$ per liter of $O_2$ (g/1 L), which is the density.

▶ An alternate example is provided in the Instructor's Resource Manual.

**SOLUTION**

The data given are

| Variable | Value |
|----------|-------|
| $P$ | 0.850 atm |
| $V$ | 1 L (exact value) |
| $T$ | (25 + 273) K = 298 K |
| $n$ | ? |

Therefore,

$$n = \frac{PV}{RT} = \frac{0.850 \text{ atm} \times 1 \text{ L}}{0.0821 \text{ L} \cdot \text{atm}/(\text{K} \cdot \text{mol}) \times 298 \text{ K}} = 0.0347 \text{ mol}$$

Now convert mol $O_2$ to grams

$$0.0347 \text{ mol } O_2 \times \frac{32.0 \text{ g } O_2}{1 \text{ mol } O_2} = 1.11 \text{ g } O_2$$

Therefore, the density of $O_2$ at 25°C and 0.850 atm is **1.11 g/L.**

Suppose you wanted the density of $Cl_2$, instead of $O_2$, at this $T$ and $P$. Only the conversion of moles to mass is affected. The previous calculation becomes

$$0.0347 \text{ mol } Cl_2 \times \frac{70.9 \text{ g } Cl_2}{1 \text{ mol } Cl_2} = 2.46 \text{ g } Cl_2$$

The density of $Cl_2$ at 25°C and 0.850 atm is 2.46 g/L. Note that the density of a gas is directly proportional to its molecular weight.

---

**EXERCISE 5.7**

Calculate the density of helium, He, in grams per liter at 21°C and 752 mmHg. The density of air under these conditions is 1.188 g/L. What is the difference in mass between 1 liter of air and 1 liter of helium? (This mass difference is equivalent to the buoyant, or lifting, force of helium per liter.)

See Problems 5.57, 5.58, 5.59, and 5.60.

---

From the previous example, we see that the density of a gas is directly proportional to its molecular weight. Chlorine, whose molecular weight is more than twice that of oxygen or nitrogen, the major components of air, is thus more than twice as dense as air. Figure 5.14 shows bromine gas being poured into a beaker. Note that the reddish-brown gas hugs the bottom of the beaker, because it is denser than air.

The relation between density and molecular weight of a gas suggests that you could use a measurement of gas density to determine its molecular weight. In fact, gas-density (vapor-density) measurements provided one of the first methods of determining molecular weight. The method was worked out by the French chemist Jean-Baptiste André Dumas in 1826. It can be applied to any substance that can be vaporized without decomposing.

As an illustration, consider the determination of the molecular weight of halothane, an inhalation anesthetic. The density of halothane vapor at 71°C (344 K) and 768 mmHg (1.01 atm) is 7.05 g/L. To obtain the molecular weight, you calculate

**Figure 5.14**
**A gas whose density is greater than that of air.**
The reddish-brown gas being poured from the flask is bromine. Note how the gas, which is denser than air, hugs the bottom of the beaker.

the moles of vapor in a given volume. The molar mass equals mass divided by the moles in the same volume. From the density, you see that one liter of vapor has a mass of 7.05 g. The moles in this volume are obtained from the ideal gas law.

$$n = \frac{PV}{RT} = \frac{1.01 \ \text{atm} \times 1 \ L}{0.0821 \ L \cdot \text{atm}/(K \cdot \text{mol}) \times 344 \ K} = 0.0358 \ \text{mol}$$

Therefore, the molar mass, $M_m$, is

$$M_m = \frac{m}{n} = \frac{7.05 \ \text{g}}{0.0358 \ \text{mol}} = 197 \ \text{g/mol}$$

▶ Dumas tried to obtain the atomic weight of sulfur from vapor-density measurements. His results were difficult to interpret, however. Sulfur, as we now know, exists as $S_8$ molecules in the solid and in the liquid. In the vapor, however, the molecules are not only $S_8$ but $S_6$ and smaller species as well. Above 800°C, the vapor is mostly $S_2$, until about 2000°C, at which point sulfur exists mostly as a monatomic gas. These changes in molecular form can be correlated with the density measurements.

Thus, the molecular weight is 197 amu. ◀

If you look at what we have done, you will see that all you need to have is the mass of vapor in *any* given volume (for a given $T$ and $P$). An explicit determination of density is not necessary. You determine the moles of vapor from the volume using the ideal gas law, then determine the molar mass by dividing the mass by moles for this volume. The next example illustrates this. At the end of this section, we will derive an explicit formula for the molecular weight in terms of the density of a gas, which will provide an alternative way to solve this problem.

**EXAMPLE 5.8**

**Determining the Molecular Weight of a Vapor**

▶ An alternate example is provided in the Instructor's Resource Manual.

You are given a quantity of a volatile (easily vaporized) liquid substance and asked to determine its molecular weight. To do that, you place a sample of the liquid in a 200.0-mL flask and submerge the flask in boiling water to vaporize all of the liquid (Figure 5.15). The vapor fills the flask and the excess escapes. Now you cool the flask so that all the vapor in it condenses, and you weigh the flask and liquid. After you subtract the mass of the flask, you find that the mass of the substance in the flask is 0.970 g. The temperature of the boiling water was 99°C; the barometric pressure was 733 mmHg. What is the molecular weight of the substance?

**Figure 5.15**
**Finding the vapor density of a substance.** After the flask fills with the vapor of the substance at the temperature of the boiling water, you allow the flask to cool so that the vapor condenses. You obtain the mass of the substance (which was vapor) by weighing the flask and substance, then subtracting the mass of the empty flask. The pressure of the vapor equals the barometric pressure, and the temperature equals that of the boiling water.

Thermometer

Boiling water

Vapor of sample

Sample
(should evaporate
completely)

Electric hot plate

▶ Demonstration of the determination of molecular weight of a vapor: Bassam Z. Shakhashiri, *Chemical Demonstrations*, Vol. 2 (Madison: University of Wisconsin Press, 1985), no. 5.13, pp. 51–54.

### PROBLEM STRATEGY

You can calculate the moles of vapor from the ideal gas law and then calculate the molar mass by dividing mass by moles. Note that the temperature of the vapor is the same as the temperature of the boiling water and that the pressure of the vapor equals the barometric pressure.

### SOLUTION

You have the following data for the vapor.

| Variable | Value |
|---|---|
| $P$ | $733 \text{ mmHg} \times \dfrac{1 \text{ atm}}{760 \text{ mmHg}} = 0.964 \text{ atm}$ |
| $V$ | $200.0 \text{ mL} = 0.2000 \text{ L}$ |
| $T$ | $(99 + 273) \text{ K} = 372 \text{ K}$ |
| $n$ | ? |

From the ideal gas law, $PV = nRT$, you obtain

$$n = \frac{PV}{RT} = \frac{0.964 \text{ atm} \times 0.2000 \text{ L}}{0.0821 \text{ L} \cdot \text{atm}/(K \cdot \text{mol}) \times 372 \text{ K}}$$

$$n = 0.00631 \text{ mol}$$

Dividing the mass of the vapor by moles gives you the mass per mole (the molar mass).

$$\text{Molar mass} = \frac{\text{grams vapor}}{\text{moles vapor}} = \frac{0.970 \text{ g}}{0.00631 \text{ mol}} = 154 \text{ g/mol}$$

Thus, the molecular weight is **154 amu.**

## EXERCISE 5.8

A sample of a gaseous substance at 25°C and 0.862 atm has a density of 2.26 g/L. What is the molecular weight of the substance?

See Problems 5.61, 5.62, 5.63, and 5.64.

From the ideal gas law, you can obtain an explicit relationship between the molecular weight and density of a gas. Recall that the molar mass ($M_m = m/n$) when expressed in grams per mole is numerically equal to the molecular weight. If you substitute $n = m/M_m$ into the ideal gas law, $PV = nRT$, you obtain

$$PV = \frac{m}{M_m} RT \qquad \text{or} \qquad PM_m = \frac{m}{V} RT$$

But $m/V$ equals the density, $d$. Substituting this gives

$$PM_m = dRT$$

We can illustrate the use of this equation by solving Examples 5.7 and 5.8 again. In Example 5.7, we asked, "What is the density of oxygen, $O_2$, in grams per liter at 25°C (298 K) and 0.850 atm?" You rearrange the previous equation to give an explicit formula for the density and then substitute into this formula.

$$d = \frac{PM_m}{RT} = \frac{0.850 \text{ atm} \times 32.0 \text{ g/mol}}{0.0821 \text{ L} \cdot \text{atm}/(K \cdot \text{mol}) \times 298 \, K} = 1.11 \text{ g/L}$$

Note that by giving $M_m$ in g/mol and $R$ in L · atm/(K · mol), you obtain the density in g/L.

The question in Example 5.8 is equivalent to the following. What is the molecular weight of a substance weighing 0.970 g whose vapor occupies 200.0 mL (0.2000 L) at 99°C (372 K) and 733 mmHg (0.964 atm)? The density of the vapor is 0.970 g/0.2000 L = 4.85 g/L. To obtain the molecular weight, you rearrange the equation $PM_m = dRT$ to give an explicit formula for $M_m$, then substitute into it.

$$M_m = \frac{dRT}{P} = \frac{4.85 \text{ g/L} \times 0.0821 \, L \cdot \text{atm}/(K \cdot \text{mol}) \times 372 \, K}{0.964 \text{ atm}} = 154 \text{ g/mol}$$

 **CONCEPT CHECK 5.3**

Three 3.0-L flasks, each at a pressure of 878 mmHg, are in a room. The flasks contain He, Ar, and Xe, respectively.

a. Which of the flasks contains the most atoms of gas?

b. Which of the flasks has the greatest density of gas?

c. If the He flask was heated and the Ar flask cooled, which of the three flasks would be at the highest pressure?

d. If the temperature of the He was lowered while the Xe was raised, which of the three flasks would have the greatest number of moles of gas?

## 5.4 Stoichiometry Problems Involving Gas Volumes

▶ Experiments in General Chemistry, 5B, "The Decomposition of Potassium Chlorate."

In Chapter 3 you learned how to find the mass of one substance in a chemical reaction from the mass of another substance in the reaction. Now that you know how to use the ideal gas law, we can extend these types of problems to include gas volumes.

Consider the following reaction, which is often used to generate small quantities of oxygen gas:

$$2KClO_3(s) \xrightarrow[\text{MnO}_2]{\Delta} 2KCl(s) + 3O_2(g)$$

Suppose you heat 0.0100 mol of potassium chlorate, $KClO_3$, in a test tube. How many liters of oxygen can you produce at 298 K and 1.02 atm?

You solve such a problem by breaking it into two problems, one involving stoichiometry and the other involving the ideal gas law. You note that 2 mol $KClO_3$ yields 3 mol $O_2$. Therefore,

$$0.0100 \text{ mol } KClO_3 \times \frac{3 \text{ mol } O_2}{2 \text{ mol } KClO_3} = 0.0150 \text{ mol } O_2$$

Now that you have the moles of oxygen produced, you can use the ideal gas law to calculate the volume of oxygen under the conditions given. You rearrange the ideal gas law, $PV = nRT$, and solve for the volume.

$$V = \frac{nRT}{P}$$

Then you substitute for $n$, $T$, and $P$. Because the pressure is given in units of atmospheres, you choose the value of $R$ in units of $L \cdot atm/(K \cdot mol)$. The answer comes out in liters.

$$V = \frac{0.0150 \text{ mol} \times 0.0821 \text{ L} \cdot atm/(K \cdot mol) \times 298 \text{ K}}{1.02 \text{ atm}} = 0.360 \text{ L}$$

The next example further illustrates this method.

▶ An alternate example is provided in the Instructor's Resource Manual.

### EXAMPLE 5.9

### Solving Stoichiometry Problems Involving Gas Volumes

Automobiles are being equipped with air bags that inflate on collision to protect the occupants from injury (Figure 5.16). Many such air bags are inflated

**Figure 5.16**
**Automobile air bag.**
Automobiles are being equipped with air bags that inflate during a front-end collision. Most air bags are inflated with nitrogen obtained from the rapid reaction of sodium azide, $NaN_3$ (the azide ion is $N_3^-$).

with nitrogen, $N_2$, using the rapid reaction of sodium azide, $NaN_3$, and iron(III) oxide, $Fe_2O_3$, which is initiated by a spark. The overall reaction is

$$6NaN_3(s) + Fe_2O_3(s) \longrightarrow 3Na_2O(s) + 2Fe(s) + 9N_2(g)$$

How many grams of sodium azide would be required to provide 75.0 L of nitrogen gas at 25°C and 748 mmHg?

 **PROBLEM STRATEGY**

The chemical equation relates the moles of $NaN_3$ to moles of $N_2$. You can use the ideal gas law to relate the volume of $N_2$ to moles of $N_2$ and hence moles of $NaN_3$. Then you can obtain the grams of $NaN_3$ from the moles of $NaN_3$.

**SOLUTION**

Here are the available data:

| Variable | Value |
|---|---|
| $P$ | $748 \text{ mmHg} \times \dfrac{1 \text{ atm}}{760 \text{ mmHg}} = 0.984 \text{ atm}$ |
| $V$ | 75.0 L |
| $T$ | $(25 + 273) \text{ K} = 298 \text{ K}$ |
| $n$ | ? |

Note that the pressure was converted from mmHg to atmospheres and the temperature from degrees Celsius to kelvins. Rearrange the ideal gas law to obtain $n$.

$$n = \frac{PV}{RT}$$

Substituting from the data gives you the moles of $N_2$.

$$n = \frac{0.984 \text{ atm} \times 75.0 \text{ L}}{0.0821 \text{ L} \cdot \text{atm}/(K \cdot \text{mol}) \times 298 \text{ K}} = 3.02 \text{ mol}$$

From the moles of $N_2$, you use the chemical equation to obtain moles of $NaN_3$.

$$3.02 \text{ mol } N_2 \times \frac{6 \text{ mol } NaN_3}{9 \text{ mol } N_2} = 2.01 \text{ mol } NaN_3$$

Now you can calculate the grams of $NaN_3$. The molecular weight of $NaN_3$ is 65.01 amu.

$$2.01 \text{ mol } NaN_3 \times \frac{65.01 \text{ g } NaN_3}{1 \text{ mol } NaN_3} = \textbf{131 g } NaN_3$$

---

**EXERCISE 5.9**

How many liters of chlorine gas, $Cl_2$, can be obtained at 40°C and 787 mmHg from 9.41 g of hydrogen chloride, HCl, according to the following equation?

$$2KMnO_4(s) + 16HCl(aq) \longrightarrow 8H_2O(l) + 2KCl(aq) + 2MnCl_2(aq) + 5Cl_2(g)$$

See Problems 5.67, 5.68, 5.69, 5.70, 5.71, and 5.72.

# 5.5 Gas Mixtures; Law of Partial Pressures

▶ Demonstration of Dalton's law of partial pressure: Bassam Z. Shakhashiri, *Chemical Demonstrations*, Vol. 2 (Madison: University of Wisconsin Press, 1985), no. 5.10, pp. 41–43.

**Figure 5.17**
**An illustration of Dalton's law of partial pressures.**
The valve connecting the flasks and the valve at the funnel are opened so that flask A fills with mineral oil. The helium flows into flask B (having the same volume as flask A), where it mixes with hydrogen. Each gas exerts the pressure it would exert if the other were not there.

While studying the composition of air, John Dalton concluded in 1801 that each gas in a mixture of unreactive gases acts, as far as its pressure is concerned, as though it were the only gas in the mixture. To illustrate, consider two 1-L flasks, one filled with helium to a pressure of 152 mmHg at a given temperature and the other filled with hydrogen to a pressure of 608 mmHg at the same temperature. Suppose all of the helium in the one flask is put in with the hydrogen in the other flask (see Figure 5.17). After the gases are mixed in one flask, each gas occupies a volume of one liter, just as before, and has the same temperature.

According to Dalton, each gas exerts the same pressure it would exert if it were the only gas in the flask. Thus, the pressure exerted by helium in the mixture is 152 mmHg, and the pressure exerted by hydrogen in the mixture is 608 mmHg. The total pressure exerted by the gases in the mixture is 152 mmHg + 608 mmHg = 760 mmHg.

## Partial Pressures and Mole Fractions

*The pressure exerted by a particular gas in a mixture* is the **partial pressure** of that gas. The partial pressure of helium in the preceding mixture is 152 mmHg; the partial pressure of hydrogen in the mixture is 608 mmHg. According to **Dalton's law of partial pressures,** *the sum of the partial pressures of all the different gases in a mixture is equal to the total pressure of the mixture.*

If you let $P$ be the total pressure and $P_A$, $P_B$, $P_C$, ... be the partial pressures of the component gases in a mixture, the law of partial pressures can be written as

Add oil to funnel until Flask A is filled

Pressure = 152 mmHg He

Pressure = 608 mmHg $H_2$

Oil

Pressure = 152 mmHg He + 608 mmHg $H_2$ = 760 mmHg

Flask A

Flask B

Flask A

Flask B

**BEFORE MIXING**

**AFTER MIXING**

Dalton's law of partial pressures:

$$P = P_A + P_B + P_C + \cdots$$

The individual partial pressures follow the ideal gas law. For component $A$,

$$P_A V = n_A RT$$

where $n_A$ is the number of moles of component $A$.

The composition of a gas mixture is often described in terms of the mole fractions of component gases. The **mole fraction** of a component gas is *the fraction of moles of that component in the total moles of gas mixture* (or the fraction of molecules that are component molecules). Because the pressure of a gas is proportional to moles, for fixed volume and temperature ($P = nRT/V \propto n$), the mole fraction also equals the partial pressure divided by total pressure.

$$\text{Mole fraction of } A = \frac{n_A}{n} = \frac{P_A}{P}$$

Mole percent equals mole fraction $\times$ 100. Mole percent is equivalent to the percentage of the molecules that are component molecules.

► Two alternate examples are provided in the Instructor's Resource Manual.

## EXAMPLE 5.10

## Calculating Partial Pressures and Mole Fractions of a Gas in a Mixture

A 1.00-L sample of dry air at 25°C and 786 mmHg contains 0.925 g $N_2$, plus other gases including oxygen, argon, and carbon dioxide. a. What is the partial pressure (in mmHg) of $N_2$ in the air sample? b. What is the mole fraction and mole percent of $N_2$ in the mixture?

**SOLUTION**

a. Each gas in a mixture follows the ideal gas law. To calculate the partial pressure of $N_2$, you convert 0.923 g $N_2$ to moles $N_2$.

$$0.923 \text{ g } N_2 \times \frac{1 \text{ mol } N_2}{28.0 \text{ g } N_2} = 0.0330 \text{ mol } N_2$$

You substitute into the ideal gas law (noting that 25°C is 298 K).

$$P_{N_2} = \frac{n_{N_2} RT}{V}$$

$$= \frac{0.0330 \text{ mol} \times 0.0821 \text{ L} \cdot \text{atm}/(\text{K} \cdot \text{mol}) \times 298 \text{ K}}{1.00 \text{ L}}$$

$$= 0.807 \text{ atm } (\mathbf{= 613 \text{ mmHg}})$$

b. The mole fraction of $N_2$ in air is

$$\text{Mole fraction of } N_2 = \frac{P_{N_2}}{P} = \frac{613 \text{ mmHg}}{786 \text{ mmHg}} = \mathbf{0.780}$$

Air contains **78.0 mole percent** of $N_2$.

**Problem Solving: Dalton's Law**

### EXERCISE 5.10

A 10.0-L flask contains 1.031 g $O_2$ and 0.572 g $CO_2$ at 18°C. What are the partial pressures of oxygen and carbon dioxide? What is the total pressure? What is the mole fraction of oxygen in the mixture?

**See Problems 5.75, 5.76, 5.77, and 5.78.**

 **CONCEPT CHECK 5.4**

A flask equipped with a valve contains 3.0 mol of $H_2$ gas. You introduce 3.0 mol of Ar gas into the flask via the valve and then seal the flask.

a. What happens to the pressure of just the $H_2$ gas in the flask after the introduction of the Ar? If it changes, by what factor does it do so?

b. How do the pressures of the Ar and the $H_2$ in the flask compare?

c. How does the total pressure in the flask relate to the pressures of the two gases?

### Collecting Gases over Water

A useful application of the law of partial pressures arises when you collect gases over water (a method used for gases that do not dissolve appreciably in water). Figure 5.18 shows how a gas, produced by chemical reaction in the flask, is collected by leading it to an inverted tube, where it displaces water. As gas bubbles through the water, the gas picks up molecules of water vapor that mix with it. The partial pressure of water vapor in the gas mixture in the collection tube depends only on the temperature. This

**Figure 5.18**
**Collection of gas over water.** Hydrogen, prepared by the reaction of zinc with hydrochloric acid, is led to an inverted tube initially filled with water. When the gas-collection tube is adjusted so that the water level in the tube is at the same height as the level in the beaker, the gas pressure in the tube equals the barometric pressure (769 mmHg). The total gas pressure equals the sum of the partial pressure of the hydrogen (752 mmHg) and the vapor pressure of water (17 mmHg). For clarity, hydrochloric acid is shown in light brown and hydrogen gas in light blue.

Hydrochloric acid

Hydrogen (partial pressure = 752 mmHg) with water vapor (partial pressure = 17 mmHg)

Water at 19°C

Zinc

**Table 5.6**
**Vapor Pressure of Water at Various Temperatures**

| Temperature (°C) | Pressure (mmHg) | Temperature (°C) | Pressure (mmHg) |
|---|---|---|---|
| 0 | 4.6 | 27 | 26.7 |
| 5 | 6.5 | 28 | 28.3 |
| 10 | 9.2 | 29 | 30.0 |
| 11 | 9.8 | 30 | 31.8 |
| 12 | 10.5 | 35 | 42.2 |
| 13 | 11.2 | 40 | 55.3 |
| 14 | 12.0 | 45 | 71.9 |
| 15 | 12.8 | 50 | 92.5 |
| 16 | 13.6 | 55 | 118.0 |
| 17 | 14.5 | 60 | 149.4 |
| 18 | 15.5 | 65 | 187.5 |
| 19 | 16.5 | 70 | 233.7 |
| 20 | 17.5 | 75 | 289.1 |
| 21 | 18.7 | 80 | 355.1 |
| 22 | 19.8 | 85 | 433.6 |
| 23 | 21.1 | 90 | 525.8 |
| 24 | 22.4 | 95 | 633.9 |
| 25 | 23.8 | 100 | 760.0 |
| 26 | 25.2 | 105 | 906.1 |

▶ **Vapor pressure is the maximum partial pressure of the vapor in the presence of the liquid. It is defined more precisely in Chapter 11.**

partial pressure of water vapor is called the *vapor pressure* of water. ◀ Values of the vapor pressure of water at various temperatures are listed in Table 5.6. The following example shows how to find the partial pressure and then the mass of the collected gas.

▶ **An alternate example is provided in the Instructor's Resource Manual.**

**EXAMPLE 5.11**

## Calculating the Amount of Gas Collected over Water

Hydrogen gas is produced by the reaction of hydrochloric acid, HCl, on zinc metal.

$$2HCl(aq) + Zn(s) \longrightarrow ZnCl_2(aq) + H_2(g)$$

The gas is collected over water. If 156 mL of gas is collected at 19°C (two significant figures) and 769 mmHg total pressure, what is the mass of hydrogen collected?

▶ **Video no. 10 (Series A), "Stoichiometry: The Generation of Oxygen Gas."**

 **PROBLEM STRATEGY**

The gas collected is hydrogen mixed with water vapor. To obtain the amount of hydrogen, you must first find its partial pressure in the mixture, using Dalton's law (Step 1). Then you can calculate the moles of hydrogen from the ideal gas law (Step 2). Finally, you can obtain the mass of hydrogen from the moles of hydrogen (Step 3).

**SOLUTION**

**STEP 1** The vapor pressure of water at 19°C is 16.5 mmHg. From Dalton's law of partial pressures, you know that the total gas pressure equals the partial pressure of hydrogen, $P_{H_2}$, plus the partial pressure of water, $P_{H_2O}$.

$$P = P_{H_2} + P_{H_2O}$$

Substituting and solving for the partial pressure of hydrogen, you get

$$P_{H_2} = P - P_{H_2O} = (769 - 16.5) \text{ mmHg} = 752 \text{ mmHg}$$

**STEP 2** Now you can use the ideal gas law to find the moles of hydrogen collected. The data are

| Variable | Value |
|---|---|
| $P$ | $752 \text{ mmHg} \times \dfrac{1 \text{ atm}}{760 \text{ mmHg}} = 0.989 \text{ atm}$ |
| $V$ | $156 \text{ mL} = 0.156 \text{ L}$ |
| $T$ | $(20 + 273) \text{ K} = 293 \text{ K}$ |
| $n$ | ? |

From the ideal gas law, $PV = nRT$, you have

$$n = \frac{PV}{RT} = \frac{0.989 \text{ atm} \times 0.156 \text{ L}}{0.0821 \text{ L} \cdot \text{atm}/(\text{K} \cdot \text{mol}) \times 293 \text{ K}} = 0.00641 \text{ mol}$$

**STEP 3** You convert moles of $H_2$ to grams of $H_2$.

$$0.00641 \text{ mol H}_2 \times \frac{2.02 \text{ g H}_2}{1 \text{ mol H}_2} = \mathbf{0.0129 \text{ g H}_2}$$

---

**EXERCISE 5.11**

Oxygen can be prepared by heating potassium chlorate, $KClO_3$, with manganese dioxide as a catalyst. The reaction is

$$2KClO_3(s) \xrightarrow[\text{MnO}_2]{\Delta} 2KCl(s) + 3O_2(g)$$

How many moles of $O_2$ would be obtained from 1.300 g $KClO_3$? If this amount of $O_2$ were collected over water at 23°C and at a total pressure of 745 mmHg, what volume would it occupy?

---

See Problems 5.81 and 5.82.

# Kinetic-Molecular Theory

In the following sections, you will see how the interpretation of a gas in terms of the **kinetic-molecular theory** (or simply **kinetic theory**) leads to the ideal gas law. *According to this theory, a gas consists of molecules in constant random motion.* The word *kinetic* describes something in motion. Kinetic energy, $E_k$, is the energy associated with the motion of an object of mass $m$. From physics,

$$E_k = \tfrac{1}{2}m \times (\text{speed})^2$$

We will use the concept of kinetic energy in describing the kinetic theory.

# 5.6   Kinetic Theory of an Ideal Gas

**Figure 5.19**
**Kinetic-theory model of gas pressure.** According to kinetic theory, gas pressure is the result of the bombardment of the container walls by constantly moving molecules.

▶ Waterston's paper, with the reviewers' comments, was discovered in the Royal Society's files by Lord Rayleigh in 1892. (Rayleigh codiscovered argon and received the 1904 Nobel Prize in physics.)

▶ Animation:
Kinetic molecular theory/ heat transfer.

▶ The statements that there are no intermolecular forces and that the volume of molecules is negligible are simplifications that lead to the ideal gas law. But intermolecular forces are needed to explain how we get the liquid state from the gaseous state; it is intermolecular forces that hold molecules together in the liquid state.

Our present explanation of gas pressure is that it results from the continual bombardment of the container walls by constantly moving molecules (Figure 5.19). This kinetic interpretation of gas pressure was first put forth in 1676 by Robert Hooke, who had been an assistant of Boyle. Hooke did not pursue this idea, however, so the generally accepted interpretation of gas pressure remained the one given by Isaac Newton, a contemporary of Hooke.

According to Newton, the pressure of a gas was due to the mutual repulsions of the gas particles (molecules). These repulsions pushed the molecules against the walls of the gas container, much as coiled springs packed tightly in a box would push against the walls of the box. This interpretation continued to be the dominant view of gas pressure until the mid-nineteenth century.

Despite the dominance of Newton's view, some people followed the kinetic interpretation. In 1738, Daniel Bernoulli, a Swiss mathematician and physicist, gave a quantitative explanation of Boyle's law using the kinetic interpretation. He even suggested that molecules move faster at higher temperatures, in order to explain Amontons's experiments on the temperature dependence of gas volume and pressure. However, Bernoulli's paper attracted little notice. A similar kinetic interpretation of gases was submitted for publication to the Royal Society of London in 1848 by John James Waterston. His paper was rejected as "nothing but nonsense." ◀

Soon afterward, the scientific climate for the kinetic view improved. The kinetic theory of gases was developed by a number of influential physicists, including James Clerk Maxwell (1859) and Ludwig Boltzmann (in the 1870s). Throughout the last half of the nineteenth century, research continued on the kinetic theory, making it a cornerstone of our present view of molecular substances.

## Postulates of Kinetic Theory

Physical theories are often given in terms of *postulates:* the basic statements from which all conclusions or predictions of a theory are deduced. The postulates are accepted as long as the predictions from the theory agree with experiment. If a particular prediction did not agree with experiment, we would limit the area to which the theory applies, modify the postulates, or start over with a new theory.

The kinetic theory of an *ideal gas* (a gas that follows the ideal gas law) is based on five postulates.

POSTULATE 1   *Gases are composed of molecules whose size is negligible compared with the average distance between them.* Most of the volume occupied by a gas is empty space. This means that you can usually ignore the volume occupied by the molecules.

POSTULATE 2   *Molecules move randomly in straight lines in all directions and at various speeds.* This means that properties of a gas that depend on the motion of molecules, such as pressure, will be the same in all directions.

POSTULATE 3   *The forces of attraction or repulsion between two molecules (intermolecular forces) in a gas are very weak or negligible, except when they collide.* ◀ This means that a molecule will continue moving in a straight line with undiminished speed until it collides with another gas molecule or with the walls of the container.

**Figure 5.20**
**Elastic collision of steel balls.** *Left:* The steel ball at the left is lifted to give it energy. *Right:* When the steel ball is released, the energy is transmitted through elastic collisions to the ball on the right. Note that the ball at the right rises as high as the original ball was lifted.

**POSTULATE 4** *When molecules collide with one another, the collisions are elastic.* In an elastic collision, the total kinetic energy remains constant; no kinetic energy is lost. To understand the difference between an elastic and an inelastic collision, compare the collision of two hard steel spheres with the collision of two masses of putty. The collision of steel spheres is nearly elastic (that is, the spheres bounce off each other and continue moving), but that of putty is not (Figure 5.20). Postulate 4 says that unless the kinetic energy of molecules is removed from a gas—for example, as heat—the molecules will forever move with the same average kinetic energy per molecule.

**POSTULATE 5** *The average kinetic energy of a molecule is proportional to the absolute temperature.* This postulate establishes what we mean by temperature from a molecular point of view: the higher the temperature, the greater the molecular kinetic energy.

## Qualitative Interpretation of the Gas Laws

According to the kinetic theory, the pressure of a gas results from the bombardment of container walls by molecules. Both the concentration of molecules (number per unit volume) and the average speed of the molecules are factors in determining this pressure. Molecular concentration and average speed determine the frequency of collisions with the wall. Average molecular speed determines the average force of a collision.

Let's consider how these terms apply to Avogadro's law, which relates the volume of gas to the moles of gas at constant (fixed) temperature and pressure. Constant temperature means that the average kinetic energy of a molecule is constant (Postulate 5). Therefore, the average molecular speed and thus the average molecular force from collision remain constant. Gas pressure depends on the force and frequency of molecular collisions with container walls. Consider the container depicted in Figure 5.21*A* where the pressure of the gas molecules in the container equals atmospheric pressure. What happens when you introduce more gas molecules into the container? More moles of gas increase the number of molecules per unit volume (concentration) and therefore increase the frequency of collisions with the container walls. If the volume of the container were fixed, the pressure would increase (Figure 5.21*B*). However, Avogadro's law requires that the pressure remain constant, so the container volume expands until the concentration and, therefore, frequency of collisions with the container walls decreases (Figure 5.20*C*). The result is that when you increase the number of moles of gas in a container at a constant pressure and temperature, the volume increases.

The same concepts can be used to discuss Charles's law. Suppose you increase the volume of a gas while holding both the temperature and number of moles of gas

A          B          C

**Figure 5.21**
**Molecular description of Avogadro's law.** *(A)* A container where the gas molecules are at atmospheric pressure ($P_{inside} = P_{atm}$). The pressure in the container is due to the force and frequency of molecular collisions with the container walls. *(B)* The container after increasing the number of gas molecules while not allowing the piston to move. Due to the greater concentration of gas (more moles of gas in the same volume), the frequency of collisions of the gas molecules with the walls of the container has increased, causing the pressure to increase ($P_{inside} > P_{atm}$). *(C)* Container after the molecules are allowed to move the piston and increase the volume. The concentration of gas molecules and frequency of collisions with the container walls have decreased, and the pressure of the gas molecules inside the container is again equal to atmospheric pressure ($P_{inside} = P_{atm}$).

A          B

**Figure 5.22**
**Molecular description of Charles's law.** *(A)* A fixed number of moles of gas in a container at room temperature. The pressure in the container is due to the force and frequency of molecular collisions with the container walls. *(B)* Container after the volume has been doubled, resulting in a decrease in the concentration (number of molecules per unit of volume), which leads to a reduction in the frequency of molecular collisions with the container walls, thereby reducing the pressure.

constant (Figure 5.22). This decreases the concentration of molecules and the frequency of collisions per unit wall area; the pressure must decrease. Thus, when you increase the volume of a gas while keeping the temperature constant, the pressure decreases.

Now consider Charles's law. If you raise the temperature, you increase the average molecular speed. The average force of a collision increases. If all other factors remained fixed, the pressure would increase. For the pressure to remain constant as it does in Charles's law, it is necessary for the volume to increase so that the number of molecules per unit volume decreases and the frequency of collisions decreases. Thus, when you raise the temperature of a gas while keeping the pressure constant, the volume increases.

## The Ideal Gas Law from Kinetic Theory

One of the most important features of kinetic theory is its explanation of the ideal gas law. To show how you can get the ideal gas law from kinetic theory, we will first find an expression for the pressure of a gas. ◄

> ► This derivation is essentially the one given by Daniel Bernoulli.

According to kinetic theory, the pressure of a gas, $P$, will be proportional to the frequency of molecular collisions with a surface and to the average force exerted by a molecule in collision.

$$P \propto \text{frequency of collisions} \times \text{average force}$$

The average force exerted by a molecule during a collision depends on its mass $m$ and its average speed $u$—that is, on its average momentum $mu$. In other words, the greater the mass of the molecule and the faster it is moving, the greater the force exerted during collision. The frequency of collisions is also proportional to the average speed $u$, because the faster a molecule is moving, the more often it strikes the container walls. Frequency of collisions is inversely proportional to the gas volume $V$, because the larger the volume, the less often a given molecule strikes the container walls. Finally, the frequency of collisions is proportional to the number of molecules $N$ in the gas volume. Putting these factors together gives

$$P \propto \left( u \times \frac{1}{V} \times N \right) \times mu$$

Bringing the volume to the left side, you get

$$PV \propto Nmu^2$$

▶ Recall that kinetic energy is defined as $\frac{1}{2}m$ multiplied by (speed)$^2$.

Because the average kinetic energy of a molecule of mass $m$ and average speed $u$ is $\frac{1}{2}mu^2$, $PV$ is proportional to the average kinetic energy of a molecule. ◀ Moreover, the average kinetic energy is proportional to the absolute temperature (Postulate 5). Noting that the number of molecules, $N$, is proportional to the moles of molecules, $n$, you have

$$PV \propto nT$$

You can write this as an equation by inserting a constant of proportionality, $R$, which you can identify as the molar gas constant.

$$PV = nRT$$

The next two sections give additional deductions from kinetic theory.

## 5.7 Molecular Speeds; Diffusion and Effusion

The principal tenet of kinetic theory is that molecules are in constant random motion. Now we will look at the speeds of molecules and at some conclusions of kinetic theory regarding molecular speeds.

### Molecular Speeds

According to kinetic theory, the speeds of molecules in a gas vary over a range of values. The British physicist James Clerk Maxwell (1831–1879) showed theoretically—and it has since been demonstrated experimentally—that molecular speeds are distributed as shown in Figure 5.23. This distribution of speeds depends on the temperature. At any temperature, the molecular speeds vary widely, but most are close to the average speed, which is close to the speed corresponding to the maximum in the distribution curve. As the temperature increases, the average speed increases.

▶ According to kinetic theory, the total kinetic energy of a mole of any gas equals $\frac{3}{2}RT$. The average kinetic energy of a molecule, which by definition is $\frac{1}{2}mu^2$, is obtained by dividing $\frac{3}{2}RT$ by Avogadro's number, $N_A$. Therefore, the kinetic energy is $\frac{1}{2}mu^2 = 3RT/(2N_A)$. Hence $u^2 = 3RT/(N_A m)$. Or, noting that $N_A m$ equals the molar mass $M_m$, you get $u^2 = 3RT/M_m$, from which you get the text equation.

The **root-mean-square (rms) molecular speed**, $u$, is *a type of average molecular speed, equal to the speed of a molecule having the average molecular kinetic energy*. It is given by the following formula:

$$u = \sqrt{\frac{3RT}{M_m}}$$

where $R$ is the molar gas constant, $T$ is the absolute temperature, and $M_m$ is the molar mass for the gas. This result follows from Postulate 5 of kinetic theory. ◀ Note that of two gases, the one with the higher molar mass will have the lower rms speed.

**Figure 5.23**
**Maxwell's distribution of molecular speeds.** The distributions of speeds of $H_2$ molecules are shown for 0°C and 500°C. Note that the speed corresponding to the maximum in the curve (the most probable speed) increases with temperature.

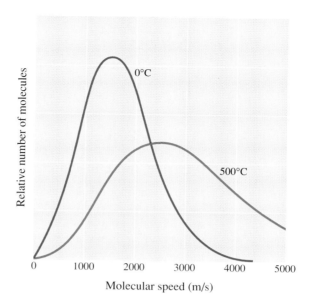

In applying this equation, *care must be taken to use consistent units.* If SI units are used for $R$ (= 8.31 kg · m²/(s² · K · mol)), $T$ (K), and $M_m$ (kg/mol), as in the following example, the rms speed will be in meters per second. Note that in these units, $H_2$, whose molecular weight is 2.02 amu, has a molar mass of 2.02 × $10^{-3}$ kg/mol.

Values of the rms speed calculated from this formula indicate that molecular speeds are astonishingly high. For example, the rms speed of $H_2$ molecules at 20°C is $1.90 \times 10^3$ m/s (over 4000 mi/hr).

▶ An alternate example is provided in the Instructor's Resource Manual.

## EXAMPLE 5.12

### Calculating the rms Speed of Gas Molecules

Calculate the rms speed of $O_2$ molecules in a tank at 21°C and 15.7 atm. See Table 5.5 for the appropriate value of $R$.

**SOLUTION**

The rms molecular speed is independent of pressure but does depend on the absolute temperature, which is (21 + 273) K = 294 K. To calculate $u$, it is best to use SI units throughout. In these units, the molar mass of $O_2$ is 32.0 × $10^{-3}$ kg/mol, and $R$ = 8.31 kg · m²/(s² · K · mol). Hence,

$$u = \left(\frac{3 \times 8.31 \text{ kg} \cdot \text{m}^2/(\text{s}^2 \cdot \text{K} \cdot \text{mol}) \times 294 \text{ K}}{32.0 \times 10^{-3} \text{ kg/mol}}\right)^{1/2} = \textbf{479 m/s}$$

**Problem Solving: Root Mean Square Velocity**

## EXERCISE 5.12

What is the rms speed (in m/s) of a carbon tetrachloride molecule at 22°C?

See Problems 5.83, 5.84, 5.85, and 5.86.

**Figure 5.24**
**Kinetic-theory model of gas diffusion.**   A vessel contains two different gases. For clarity, the figure shows the motion of only one molecule. The path of the molecule is chaotic, because of the constant collisions with other molecules. After some time, the "green" molecule has by its random motion mixed with the "red" gas. Similar motions of other molecules eventually result in the complete mixing of the two gases.

---

**EXERCISE 5.13**

At what temperature do hydrogen molecules, $H_2$, have the same rms speed as nitrogen molecules, $N_2$, at 455°C? At what temperature do hydrogen molecules have the same average kinetic energy?

See Problems 5.87 and 5.88.

---

### Diffusion and Effusion

The pleasant odor of apple pie baking in the oven quickly draws people to the kitchen. The spread of an odor is easily explained by kinetic theory. All molecules are in constant, chaotic motion, and eventually a cluster of molecules of a particular substance will spread out to occupy a larger and larger space. Gaseous **diffusion** is *the process whereby a gas spreads out through another gas to occupy the space uniformly.* Figure 5.24 is a kinetic-theory explanation of the diffusion of two gases. An individual molecule moves chaotically, as it is buffeted by other molecules. After sufficient time, this constant chaotic motion of molecules results in a complete mixing of the gases. Figure 5.25 demonstrates the diffusion of ammonia gas through air.

When you think about the kinetic-theory calculations of molecular speed, you might ask why diffusion is not even much faster than it is. Why does it take minutes for the gas to diffuse throughout a room when the molecules are moving at perhaps a thousand miles per hour? This was, in fact, one of the first criticisms of kinetic theory. The answer is simply that a molecule never travels very far in one direction (at ordinary pressures) before it collides with another molecule and moves off in another direction. If you could trace out the path of an individual molecule, it would be a zigzagging trail. For a molecule to cross a room, it has to travel many times the straight-line distance.

Although the rate of diffusion certainly depends in part on the average molecular speed, the effect of molecular collisions makes the theoretical picture a bit complicated. *Effusion,* like diffusion, is a process involving the flow of a gas but is theoretically much simpler.

If you place a container of gas in a vacuum and then make a very small hole in the container, the gas molecules escape through the hole at the same speed they had in the container (Figure 5.26). *The process in which a gas flows through a small hole in a container* is called **effusion.** It was first studied by Thomas Graham, who discovered in 1846 that the rate of effusion of a gas is inversely proportional to the square root of its density. Today we usually state **Graham's law of effusion** in terms of molecular weight: *the rate of effusion of gas molecules from a particular hole is inversely proportional to the square root of the molecular weight of the gas at constant temperature and pressure.*

**Figure 5.25**
**Gaseous diffusion.**
*Left:* Concentrated aqueous ammonia in the beaker evolves ammonia gas into the glass tube, which contains a strip of wet indicator paper. *Right:* The indicator changes color as the ammonia gas diffuses upward through the air in the tube.

Let us consider a kinetic theory analysis of an effusion experiment. Suppose the hole in the container is made small enough so that the gas molecules continue to move randomly (rather than moving together as they would in a wind). When a molecule happens to encounter the hole, it leaves the container. The collection of molecules leaving the container by chance encounters with the hole constitutes the effusing gas. All you have to consider for effusion is the rate at which molecules encounter the hole in the container.

The rate of effusion of molecules from a container depends on three factors: the cross-sectional area of the hole (the larger it is, the more likely molecules are to escape); the number of molecules per unit volume (the more crowded the molecules are, the more likely they are to encounter the hole); and the average molecular speed (the faster the molecules are moving, the sooner they will find the hole and escape).

If you compare the effusion of different gases from the same container, at the same temperature and pressure, the first two factors will be the same. The average molecular speeds will be different, however.

**Figure 5.26**
**Model of gaseous effusion.**  According to kinetic theory, gas molecules are in constant random motion. When a molecule on the left side of the box happens to hit the pinhole in the partition between the two parts of the box, it passes (or effuses) to the right side. The rate of effusion depends on the speed of the molecules— the faster the molecules move, the more likely they are to encounter the pinhole and pass from the left side of the box to the right side.

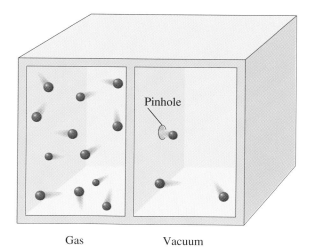

Gas                    Vacuum

**Figure 5.27**
**The hydrogen fountain.**
The demonstrator places a beaker containing hydrogen gas over the porous clay container. Hydrogen effuses into the porous container faster than air effuses out. As a result, the pressure inside the porous container and the flask connected to it increases, forcing colored water out the side tube as a stream.

▶ Demonstration of rates of effusion: Bassam Z. Shakhashiri, *Chemical Demonstrations*, Vol. 2 (Madison: University of Wisconsin Press, 1985), no. 5.14, pp. 55–58, and no. 5.18, pp. 72–74.

Because the average molecular speed essentially equals $\sqrt{3RT/M_m}$, where $M_m$ is the molar mass, the rate of effusion is proportional to $M_m^{1/2}$. That is, the rate of effusion is inversely proportional to the square root of the molar mass (or molecular weight), as Graham's law states. The derivation of Graham's law from kinetic theory was considered a triumph of the theory and greatly strengthened confidence in its validity.

Graham's law of effusion:

$$\text{Rate of effusion of molecules} \propto \frac{1}{\sqrt{M_m}} \quad \text{(for the same container at constant } T \text{ and } P\text{)}$$

The hydrogen fountain, shown in Figure 5.27, is dependent on the differences in rates of effusion of gases. (Can you explain why a helium-filled balloon loses pressure after a few hours?)

▶ An alternate example is provided in the Instructor's Resource Manual.

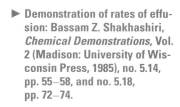

$CO_2$
$M_m = 44.0$ g/mol

$SO_2$
$M_m = 64.1$ g/mol

**EXAMPLE 5.13**

## Calculating the Ratio of Effusion Rates of Gases

Calculate the ratio of effusion rates of molecules of carbon dioxide, $CO_2$, and sulfur dioxide, $SO_2$, from the same container and at the same temperature and pressure.

**SOLUTION**

The two rates of effusion are inversely proportional to the square roots of their molar masses, so you can write

$$\frac{\text{Rate of effusion of CO}_2}{\text{Rate of effusion of SO}_2} = \sqrt{\frac{M_m(SO_2)}{M_m(CO_2)}}$$

where $M_m(SO_2)$ is the molar mass of $SO_2$ (64.1 g/mol) and $M_m(CO_2)$ is the molar mass of $CO_2$ (44.0 g/mol). Substituting these molar masses into the formula gives

$$\frac{\text{Rate of effusion of CO}_2}{\text{Rate of effusion of SO}_2} = \sqrt{\frac{64.1 \text{ g/mol}}{44.0 \text{ g/mol}}} = \mathbf{1.21}$$

In other words, carbon dioxide effuses 1.21 times faster than sulfur dioxide (because $CO_2$ molecules move 1.21 times faster on average than $SO_2$ molecules).

### EXERCISE 5.14

If it takes 3.52 s for 10.0 mL of helium to effuse through a hole in a container at a particular temperature and pressure, how long would it take for 10.0 mL of oxygen, $O_2$, to effuse from the same container at the same temperature and pressure? (Note that the rate of effusion can be given in terms of volume of gas effused per second.)

See Problems 5.89, 5.90, 5.91, and 5.92.

### EXERCISE 5.15

If it takes 4.67 times as long for a particular gas to effuse as it takes hydrogen under the same conditions, what is the molecular weight of the gas? (Note that the rate of effusion is inversely proportional to the time it takes for a gas to effuse.)

See Problems 5.93 and 5.94.

Graham's law has practical application in the preparation of fuel rods for nuclear fission reactors. Such reactors depend on the fact that the uranium-235 nucleus undergoes fission (splits) when bombarded with neutrons. When the nucleus splits, several neutrons are emitted and a large amount of energy is liberated. These neutrons bombard more uranium-235 nuclei, and the process continues with the evolution of more energy. However, natural uranium consists of 99.27% uranium-238 (which does not undergo fission) and only 0.72% uranium-235 (which does undergo fission). A uranium fuel rod must contain about 3% uranium-235 to sustain the nuclear reaction.

▶ Uranium hexafluoride vapor has a density of 13.0 g/L at 57°C (the temperature at which the solid sublimes). This is more than 12 times as dense as air at the same temperature.

To increase the percentage of uranium-235 in a sample of uranium (a process called *enrichment*), one first prepares uranium hexafluoride, $UF_6$, a white, crystalline solid that is easily vaporized. Uranium hexafluoride vapor is allowed to pass through a series of porous membranes. Each membrane has many small holes through which the vapor can effuse. Because the $UF_6$ molecules with the lighter isotope of uranium travel about 0.4% faster than the $UF_6$ molecules with the heavier isotope, the gas that passes through first is somewhat richer in uranium-235. When this vapor passes through another membrane, the uranium-235 vapor becomes further concentrated. It takes many effusion stages to reach the necessary enrichment; for complete separation of the isotopes, as required for bomb-grade uranium, many thousands of effusion stages are needed (Figure 5.28 on the following page).

 **CONCEPT CHECK 5.5**

Consider the experimental apparatus shown. In this setup, each round flask contains a gas and the long tube contains no gas (that is, it is a vacuum).

a. We use 1.0 mol of He for experiment X and 1.0 mol of Ar for experiment Y. If both valves are opened at the same time, which gas would you expect to reach the end of the long tube first?

b. If you wanted the Ar to reach the end of the long tube at the same time as the He, what experimental condition (that is, you cannot change the equipment) could you change to make this happen?

**Figure 5.28**
**A plant for separating uranium isotopes by effusion.** These large metal vessels at the Oak Ridge Gaseous Diffusion Plant contain porous membranes through which uranium hexafluoride vapor effuses. After many effusion stages, the vapor becomes more concentrated in the uranium-235 isotope.

## 5.8 Real Gases

In Section 5.2, we found that, contrary to Boyle's law, the pressure–volume product for $O_2$ at 0°C was not quite constant, particularly at high pressures (see $PV$ column in Table 5.3). Experiments show that the ideal gas law describes the behavior of real gases quite well at low pressures and moderate temperatures, but not at high pressures and low temperatures.

Figure 5.29 shows the behavior of the pressure–volume product at various pressures for several gases. The gases deviate noticeably from Boyle's law (ideal gas) behavior at high pressures. Also, the deviations differ for each kind of gas. We can explain why this is so by examining the postulates of kinetic theory, from which the ideal gas law can be derived.

Postulate 1 of kinetic theory says that the volume of space occupied by molecules is negligible compared with the total gas volume. To derive the ideal gas law from kinetic theory, we must assume that each molecule is free to move throughout the entire gas volume, $V$. At low pressures, where the volume of individual molecules is negligible compared with the total volume available (Figure 5.30A), the ideal gas law is a good approximation. At higher pressures, where the volume of individual molecules becomes important (Figure 5.30B), the space through which a molecule can move is significantly different from $V$.

Postulate 3 says that the forces of attraction between molecules (intermolecular forces) in a gas are very weak or negligible. This is a good approximation at low pressures, where molecules tend to be far apart, because these forces diminish rap-

**Figure 5.29**
**Pressure–volume product of gases at different pressures.** The bottom graph shows the pressure–volume product of one mole of various gases at 0°C and at different pressures. The top graph shows values at low pressure. For an ideal gas, the pressure–volume product is constant.

**A**          **B**

**Figure 5.30**
**Effect of molecular volume at high pressure.** *(A)* At low pressures, the volume of molecules is a small fraction of the total volume and can be neglected, as in the ideal gas law. *(B)* At high pressures, the volume of molecules is a significant fraction of the total volume and cannot be neglected. The ideal gas law is no longer a good approximation.

idly as the distance between molecules increases. Intermolecular forces become significant at higher pressures, though, because the molecules tend to be close together. Because of these intermolecular forces, the actual pressure of a gas is less than that predicted by ideal gas behavior. As a molecule begins to collide with a wall surface, neighboring molecules pull this colliding molecule slightly away from the wall, giving a reduced pressure (Figure 5.31).

The Dutch physicist J. D. van der Waals (1837–1923) was the first to account for these deviations of a real gas from ideal gas behavior. The **van der Waals equation** is *an equation similar to the ideal gas law, but includes two constants, a and b, to account for deviations from ideal behavior.*

$$\left(P + \frac{n^2a}{V^2}\right)(V - nb) = nRT$$

The constants $a$ and $b$ are chosen to fit experiment as closely as possible. Table 5.7 gives values of van der Waals constants for several gases.

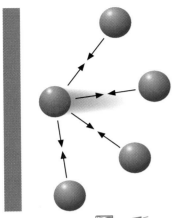

**Figure 5.31**
**Effect of intermolecular attractions on gas pressure.** Gas pressure on a wall is due to molecules colliding with it. Here a molecule (shown in red) about to collide with the wall is attracted away from the wall by the weak attractive forces of neighboring molecules (intermolecular forces). As a result, the pressure exerted by the gas is less than that expected in the absence of those forces.

**Table 5.7**
**van der Waals Constants for Some Gases**

| Gas | $a$ L$^2$ • atm/mol$^2$ | $b$ L/mol |
|---|---|---|
| Carbon dioxide, $CO_2$ | 3.658 | 0.04286 |
| Ethane, $C_2H_6$ | 5.570 | 0.06499 |
| Ethanol, $C_2H_5OH$ | 12.56 | 0.08710 |
| Helium, He | 0.0346 | 0.0238 |
| Hydrogen, $H_2$ | 0.2453 | 0.02651 |
| Oxygen, $O_2$ | 1.382 | 0.03186 |
| Sulfur dioxide, $SO_2$ | 6.865 | 0.05679 |
| Water, $H_2O$ | 5.537 | 0.03049 |

(From Lide, David R., Ed., in *CRC Handbook of Chemistry and Physics*, 74th ed., pp. 6–48. Copyright CRC Press Inc., Boca Raton, Florida, 1993. With the permission of CRC Press, Inc.)

You can obtain the van der Waals equation from the ideal gas law, $PV = nRT$, by making the following replacements:

**Ideal Gas Equation**          **van der Waals Equation**

$V$          becomes          $V - nb$

$P$          becomes          $P + n^2a/V^2$

To obtain these substitutions, van der Waals reasoned as follows. The volume available for a molecule to move in equals the gas volume, $V$, minus the volume occupied by molecules. So he replaced $V$ in the ideal gas law with $V - nb$, where $nb$ represents the volume occupied by $n$ moles of molecules. Then he noted that the total force of attraction on any molecule about to hit a wall is proportional to the concentration of neighboring molecules, $n/V$. However, the number of molecules about to hit the wall per unit wall area is also proportional to the concentration, $n/V$. Therefore, the force per unit wall area, or pressure, is reduced from that assumed in the ideal gas law by a factor proportional to $n^2/V^2$. Letting $a$ be the proportionality constant, we can write

$$P(\text{actual}) = P(\text{ideal}) - n^2a/V^2$$

or

$$P(\text{ideal}) = P(\text{actual}) + n^2a/V^2$$

In other words, you replace $P$ in the ideal gas law by $P + n^2a/V^2$.

The next example illustrates the use of the van der Waals equation to calculate the pressure exerted by a gas under given conditions.

▶ An alternate example is provided in the Instructor's Resource Manual.

**EXAMPLE 5.14**

**Using the van der Waals Equation**

If sulfur dioxide were an ideal gas, the pressure at 0.0°C exerted by 1.000 mole occupying 22.41 L would be 1.000 atm (22.41 L is the molar volume of an ideal gas at STP). Use the van der Waals equation to estimate the pressure of this volume of 1.000 mol $SO_2$ at 0.0°C. See Table 5.7 for values of $a$ and $b$.

### PROBLEM STRATEGY

You first rearrange the van der Waals equation to give $P$ in terms of the other variables.

$$P = \frac{nRT}{V - nb} - \frac{n^2 a}{V^2}$$

Now you substitute $R = 0.08206$ L · atm/(K · mol), $T = 273.2$ K, $V = 22.41$ L, $a = 6.865$ L$^2$ · atm/mol$^2$, and $b = 0.05679$ L/mol.

### SOLUTION

$$P = \frac{1.000 \text{ mol} \times 0.08206 \text{ L} \cdot \text{atm/(K} \cdot \text{mol}) \times 273.2 \text{ K}}{22.41 \text{ L} - (1.000 \text{ mol} \times 0.05679 \text{ L/mol})} -$$

$$\frac{(1.000 \text{ mol})^2 \times 6.865 \text{ L}^2 \cdot \text{atm/mol}^2}{(22.41 \text{ L})^2}$$

$$= 1.003 \text{ atm} - 0.014 \text{ atm} = \textbf{0.989 atm}$$

Note that the first term, $nRT/(V - nb)$, gives the pressure corrected for molecular volume. This pressure, 1.003 atm, is slightly higher than the ideal value, 1.000 atm, because the volume through which a molecule is free to move, $V - nb$, is smaller than what is assumed for an ideal gas ($V$). Therefore, the concentration of molecules in this volume is greater, and the number of collisions per unit surface area is expected to be greater. However, the last term, $n^2 a/V^2$, corrects for intermolecular forces, which reduce the force of the collisions, and this reduces the pressure to 0.989 atm.

### EXERCISE 5.16

Use the van der Waals equation to calculate the pressure of 1.000 mol ethane, $C_2H_6$, that has a volume of 22.41 L at 0.0°C. Compare the result with the value predicted by the ideal gas law.

See Problems 5.95 and 5.96.

### CONCEPT CHECK 5.6

A 1.00-L container is filled with an ideal gas and the recorded pressure is 350 atm. We then put the same amount of a real gas into the container and measure the pressure.

a. If the real gas molecules occupy a relatively small volume and have large intermolecular attractions, how would you expect the pressures of the two gases to compare?

b. If the real gas molecules occupy a relatively large volume and there are negligible intermolecular attractions, how would you expect the pressures of the two gases to compare?

c. If the real gas molecules occupy a relatively large volume and have large intermolecular attractions, how would you expect the pressures of the two gases to compare?

# Carbon Dioxide Gas and the Greenhouse Effect

**ENVIRONMENT**

Yes, I know, I shouldn't have left that chocolate bar on the seat of my car, which I parked on a sunny street near the university. You and I know that the interior of such a car can become quite hot. Have you ever wondered why? If the sun's heat energy can get into the car through its window glass, why can't it leave equally well?

Parked cars are similar to greenhouses, which people have used for centuries to protect plants from cold weather. The Swedish chemist Svante Arrhenius, who you may remember from the previous chapter (for his theory of ionic solutions), realized that carbon dioxide in the atmosphere acts like the glass in a greenhouse. His calculations, published in 1898, were the first to show how sensitive the temperature of the earth might be to the percentage of carbon dioxide in the atmosphere. Arrhenius's reference to the earth's atmospheric envelope, with its carbon dioxide, as a "hothouse" has evolved into our present-day term *greenhouse effect*.

Here is the explanation of the greenhouse effect. The principal gases in the atmosphere are oxygen, $O_2$, and nitrogen, $N_2$. These gases are transparent to visible light from the sun, and when this light reaches the surface of the earth, it is absorbed and converted to heat. This heat causes atoms in the earth's surface to vibrate, which then radiate the heat energy as infrared radiation, or heat rays (Figure 5.32). Neither oxygen nor nitrogen absorb infrared radiation, so if the earth's atmosphere contained only these gases, the infrared radiation, or heat rays, would simply escape into outer space; there would be no greenhouse effect. However, other gases in the atmosphere, especially carbon dioxide, do absorb infrared radiation, and it is this absorption that warms the atmosphere and eventually also the earth's surface, giving us a greenhouse effect. The greater the percentage of carbon dioxide in the atmosphere, the warmer the earth should be. (Glass, acting like carbon dioxide, allows visible light to pass into a greenhouse, but absorbs infrared radiation, effectively trapping the heat.)

The greenhouse effect has become associated with the concept of *global warming*. Records kept since the nineteenth century clearly show that the amount of carbon dioxide in the atmosphere has increased dramatically, from about 284 ppm (parts per million) in 1830 to about 370 ppm in 2000 (Figure 5.33). (Parts per million is equivalent to mole percent $\times 10^{-4}$.) Scientists

**Figure 5.32**
**Greenhouse effect of certain gases in the atmosphere.**
The atmosphere is transparent to sunlight, except for radiation in the far ultraviolet. When sunlight reaches the earth's surface, it is absorbed. The heated surface then radiates the energy back as infrared radiation. The major components of the atmosphere, $O_2$ and $N_2$, are transparent to this radiation, but gases such as $CO_2$, $H_2O$, $CH_4$, and chlorofluorocarbons, do absorb infrared rays. These substances in the atmosphere then reradiate the infrared rays, with the result that a significant fraction of the radiation returns to earth. In effect, these gases trap the radiation, acting like the glass on a greenhouse.

believe that our burning of fossil fuels (coal, oil, and natural gas) is responsible for this change. Moreover, temperature measurements made at the surface of the earth during the past 100 years indicate an average temperature increase of between 0.4°C and 0.8°C. Thus, many climate scientists believe that human activities are at least partially responsible for global warming and, hence, they predict drastic changes in climate. But to complicate matters, satellite measurements of upper atmosphere temperatures made during the last 20 years do not show an increase in temperature. So, a number of climate scientists question the idea of human-induced global warming, claiming that what we are seeing is a natural climate variation.

In January 2000, the U.S. National Research Council (NRC) released a report of a panel that was trying to reconcile these views. They concluded that the increase in surface temperature is real. But John M. Wallace, professor of atmospheric sciences at the University of Washington and chair of the NRC panel, said, "The rapid increase in the Earth's surface temperature over the past 20 years is not necessarily representative of how the atmosphere is responding to long-term, human-induced changes. . . ." He cautioned against politicizing a scientific investigation and noted that "We still don't have all the answers." Climate is enormously complicated, depending not only on the concentration of carbon dioxide in the atmosphere (greenhouse effect), but also on dust particles (which reflect sunlight), natural variation in radiation from the sun, and many other variables. Despite the complications of climate research, the majority of scientists do believe that human activities are affecting our climate.

**Figure 5.33**
**Concentrations of carbon dioxide in the atmosphere (average annual values).**
Concentrations of carbon dioxide in the atmosphere have been increasing steadily since about 1830. The blue curve is derived from ice core samples (containing air bubbles) obtained from Antarctica. The purple curve is derived from direct measurements of atmospheric carbon dioxide collected since 1958 at Mauna Loa, Hawaii. (*Source of blue curve:* D.M. Etheridge, L.P. Steele, R.L. Langenfelds, R.J. Francey, J.M. Barnola and V.I. Morgan. 1998. Historical $CO_2$ records from the Law Dome DE08, DE08-2, and DSS ice cores. In Trends: A Compendium of Data on Global Change. Carbon Dioxide Information Analysis Center, Oak Ridge National Laboratory, U.S. Department of Energy, Oak Ridge, Tenn., U.S.A. *Source of purple curve:* C.D. Keeling and T.P. Whorf. 1999. Atmospheric $CO_2$ records from sites in the SIO air sampling network. In Trends: A Compendium of Data on Global Change. Carbon Dioxide Information Analysis Center, Oak Ridge National Laboratory, U.S. Department of Energy, Oak Ridge, Tenn., U.S.A.)

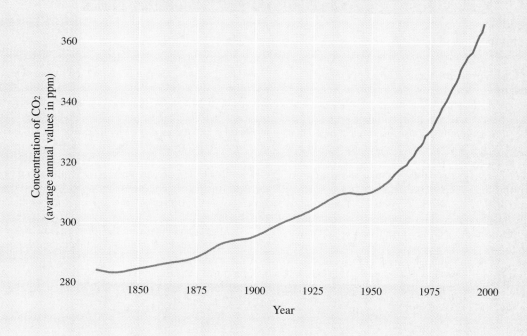

# A Checklist for Review

## Important Terms

pressure (5.1)
pascal (Pa) (5.1)
barometer (5.1)
manometer (5.1)
millimeters of mercury
   (mmHg or torr) (5.1)
atmosphere (atm) (5.1)
Boyle's law (5.2)

Charles's law (5.2)
Avogadro's law (5.2)
molar gas volume ($V_m$) (5.2)
standard temperature and
   pressure (STP) (5.2)
molar gas constant ($R$) (5.3)
ideal gas law (5.3)
partial pressure (5.5)

Dalton's law of partial
   pressures (5.5)
mole fraction (5.5)
kinetic-molecular theory
   of gases (kinetic theory)
   (p. 82)
root-mean-square (rms)
   molecular speed (5.7)

diffusion (5.7)
effusion (5.7)
Graham's law of effusion
   (5.7)
van der Waals equation
   (5.8)

## Key Equations

$PV = \text{constant}$     (constant $n, T$)

$\dfrac{V}{T} = \text{constant}$     (constant $n, P$)

$V_m = \text{specific constant}$
(depending on $T, P$; independent of gas)

$PV = nRT$

$PM_m = dRT$

$P = P_A + P_B + P_C + \ldots$

Mole fraction of A $= \dfrac{n_A}{n} = \dfrac{P_A}{P}$

$u = \sqrt{\dfrac{3RT}{M_m}}$

Rate of effusion $\propto \dfrac{1}{\sqrt{M_m}}$

(same container at constant $T, P$)

$\left(P + \dfrac{n^2a}{V^2}\right)(V - nb) = nRT$

## Summary of Facts and Concepts

The *pressure* of a gas equals the force exerted per unit area of surface. It is measured by a *manometer* in units of *pascals, millimeters of mercury,* or *atmospheres.*

Gases at low to moderate pressures and moderate temperatures follow the same simple relationships, or gas laws. Thus, for a given amount of gas at constant temperature, the volume varies inversely with pressure *(Boyle's law)*. Also, for a given amount of gas at constant pressure, the volume is directly proportional to the absolute temperature *(Charles's law)*. These two laws, together with *Avogadro's law* (equal volumes of any two gases at the same temperature and pressure contain the same number of molecules), can be formulated as one equation, *PV = nRT (ideal gas law)*. This equation gives the relationship among *P, V, n,* and *T* for a gas. It also relates these quantities for each component in a gas mixture. The total gas pressure, *P,* equals the sum of the partial pressures of each component *(law of partial pressures)*.

The ideal gas law can be explained by the *kinetic-molecular theory.* We define an ideal gas as consisting of molecules with negligible volume that are in constant random motion. In the ideal gas model, there are no intermolecular forces between molecules, and the average kinetic energy of molecules is proportional to the absolute temperature. From kinetic theory, one can show that the *rms molecular speed* equals $\sqrt{3RT/M_m}$. Given this result, one can derive *Graham's law of effusion:* the rate of effusion of gas molecules under identical conditions is inversely proportional to the square root of the molecular weight.

Real gases deviate from the ideal gas law at high pressure and low temperature. From kinetic theory, we expect these deviations, because real molecules do have volume and intermolecular forces do exist. These two factors are partially accounted for in the *van der Waals equation.*

## Operational Skills

1. **Converting units of pressure**   Given the pressure of a gas in one unit of pressure (Pa, mmHg, or atm), convert to another unit of pressure.   **(EXAMPLE 5.1)**
2. **Using the empirical gas laws**   Given an initial volume occupied by a gas, calculate the final volume when the pressure changes at fixed temperature **(EXAMPLE 5.2)**; when the temperature changes at fixed pressure **(EXAMPLE 5.3)**; and when both pressure and temperature change **(EXAMPLE 5.4)**.
3. **Deriving empirical gas laws from the ideal gas law**   Starting from the ideal gas law, derive the relationship between any two variables.   **(EXAMPLE 5.5)**
4. **Using the ideal gas law**   Given any three of the variables $P$, $V$, $T$, and $n$ for a gas, calculate the fourth from the ideal gas law.   **(EXAMPLE 5.6)**
5. **Relating gas density and molecular weight**   Given the molecular weight, calculate the density of a gas for a particular temperature and pressure **(EXAMPLE 5.7)**; or, given the gas density, calculate the molecular weight **(EXAMPLE 5.8)**.
6. **Solving stoichiometry problems involving gas volumes**   Given the volume (or mass) of one substance in a reaction, calculate the mass (or volume) of another produced or used up.   **(EXAMPLE 5.9)**
7. **Calculating partial pressures and mole fractions of a gas in a mixture**   Given the masses of gases in a mixture, calculate the partial pressures and mole fractions.   **(EXAMPLE 5.10)**
8. **Calculating the amount of gas collected over water**   Given the volume, total pressure, and temperature of gas collected over water, calculate the mass of the dry gas.   **(EXAMPLE 5.11)**
9. **Calculating the rms speed of gas molecules**   Given the molecular weight and temperature of a gas, calculate the rms molecular speed.   **(EXAMPLE 5.12)**
10. **Calculating the ratio of effusion rates of gases**   Given the molecular weights of two gases, calculate the ratio of rates of effusion **(EXAMPLE 5.13)**; or, given the relative effusion rates of a known and an unknown gas, obtain the molecular weight of the unknown gas **(EXERCISE 5.15)**.
11. **Using the van der Waals equation**   Given $n$, $T$, $V$, and the van der Waals constants $a$ and $b$ for a gas, calculate the pressure from the van der Waals equation.   **(EXAMPLE 5.14)**

## Review Questions

**5.1**   Define *pressure*. From the definition, obtain the SI unit of pressure in terms of SI base units.

**5.2**   For what purpose is a manometer used? How does it work?

**5.3**   What variables determine the height of the liquid in a manometer?

**5.4**   Starting with Boyle's law (stated as an equation), obtain an equation for the final volume occupied by a gas from the initial volume when the pressure is changed at constant temperature.

**5.5**   The volume occupied by a gas depends linearly on degrees Celsius at constant pressure, but it is not directly proportional to degrees Celsius. However, it is directly proportional to kelvins. What is the difference between a linear relationship and a direct proportion?

**5.6**   Explain how you would set up an absolute temperature scale based on the Fahrenheit scale, knowing that absolute zero is −273.15°C.

**5.7**   Starting with Charles's law (stated as an equation), obtain an equation for the final volume of a gas from its initial volume when the temperature is changed at constant pressure.

**5.8**   State Avogadro's law in words. How does this law explain the law of combining volumes? Use the gas reaction

$$N_2 + 3H_2 \longrightarrow 2NH_3$$

as an example in your explanation.

**5.9**   What are the standard conditions for comparing gas volumes?

**5.10**   What does the term *molar gas volume* mean? What is the molar gas volume (in liters) at STP for an ideal gas?

**5.11**   Starting from Boyle's, Charles's, and Avogadro's laws, obtain the ideal gas law, $PV = nRT$.

**5.12**   What variables are needed to describe a gas that obeys the ideal gas law? What are the SI units for each variable?

**5.13**   What is the value of $R$ in units of L · mmHg/ (K · mol)?

**5.14**   The ideal gas law relates four variables. An empirical gas law relates two variables, assuming the other two are constant. How many empirical gas laws can be obtained? Give statements of each.

**5.15**   Give the postulates of kinetic theory and state any evidence that supports them.

**5.16** Explain Boyle's law in terms of the kinetic theory.

**5.17** What is the origin of gas pressure, according to kinetic theory?

**5.18** How does the rms molecular speed depend on absolute temperature? on molar volume?

**5.19** Explain why a gas appears to diffuse more slowly than average molecular speeds might suggest.

**5.20** What is effusion? Why does a gas whose molecules have smaller mass effuse faster than one whose molecules have larger mass?

**5.21** Under what conditions does the behavior of a real gas begin to differ significantly from the ideal gas law?

**5.22** What is the physical meaning of the $a$ and $b$ constants in the van der Waals equation?

 ## Conceptual Problems

**5.23** Using the concepts developed in this chapter, explain the following observations.
a. Automobile tires are flatter on cold days.
b. You are not supposed to dispose of aerosol cans in a fire.
c. The lid of a water bottle pops off when the bottle sits in the sun.
d. A balloon pops when you squeeze it.

**5.24** You have three identical flasks, each containing equal amounts of $N_2$, $O_2$, and He. The volume of the $N_2$ flask is doubled, the $O_2$ flask volume is halved, and the He flask volume is reduced to one-third of the original volume. Rank the flasks from highest to lowest pressure both before and after the volume is changed and indicate by what factor the pressure has changed.

**5.25** Consider the following gas container equipped with a movable piston.

a. By what factor (increase by 1, decrease by 1.5, etc.) would you change the pressure if you wanted the volume to change from volume C to volume D?
b. If the piston were moved from volume C to volume A, by what factor would the pressure change?
c. By what factor would you change the temperature in order to change from volume C to volume B?
d. If you increased the number of moles of gas in the container by a factor of 2, by what factors would the pressure and the volume change?

**5.26** A 3.00-L flask containing 2.0 mol of $O_2$ and 1.0 mol of $N_2$ is in a room that is at 22.0°C.
a. How much (what fraction) of the total pressure in the flask is due to the $N_2$?
b. The flask is cooled and the pressure drops. What happens, if anything, to the mole fraction of the $O_2$ at the lower temperature?

c. 1.0 L of liquid water is introduced into the flask containing both gases. The pressure is then measured about 45 minutes later. Would you expect the measured pressure to be higher or lower?
d. Given the information in this problem and the conditions in part c, would it be possible to calculate the pressure in the flask after the introduction of the water? If it is not possible with the given information, what further information would you need to accomplish this task?

**5.27** Consider the following setup, which shows identical containers connected by a tube with a valve that is presently closed. The container on the left has 1.0 mol of $H_2$ gas; the container on the right has 1.0 mol of $O_2$.

*Note:* Acceptable answers to some of these questions might be "both" or "neither one."
a. Which container has the greatest density of gas?
b. Which container has molecules that are moving at a faster average molecular speed?
c. Which container has more molecules?
d. If the valve is opened, will the pressure in each of the containers change? If it does, how will it change (increase, decrease, or no change)?
e. 2.0 mol of Ar is added to the system with the valve open. What fraction of the total pressure will be due to the $H_2$?

**5.28** Two identical He-filled balloons, each with a volume of 20 L, are allowed to rise into the atmosphere. One rises to an altitude of 3000 m while the other rises to 6000 m.
a. Assuming that the balloons are at the same temperature, which balloon has the greater volume?
b. What information would you need in order to calculate the volume of each of the balloons at their respective heights?

**5.29**  You have a balloon that contains $O_2$. What could you do to the balloon in order to double the volume? Be specific in your answers; for example, you could increase the number of moles of $O_2$ by a factor of 2.

**5.30**  Three 25.0-L flasks are placed next to each other on a shelf in a chemistry stockroom. The first flask contains He at a pressure of 1.0 atm, the second contains Xe at 1.50 atm, and the

third contains $F_2$ and has a label that says 2.0 mol $F_2$. Consider the following questions about these flasks.
a. Which flask has the greatest number of moles of gas?
b. If you wanted each of the flasks to be at the same pressure as the He flask, what general things could you do to the other two containers to make this happen?

# Practice Problems

*Note:*  In these problems, the final zeros given in temperatures and pressures (for example, 20°C, 760 mmHg) are significant figures.

## Units of Pressure

**5.31**  A gas in a closed-tube manometer (Figure 5.4) has a measured pressure of 0.047 atm. Calculate the pressure in mmHg.

**5.32**  The barometric pressure measured outside an airplane at 9 km (30,000 ft) was 259 mmHg. Calculate the pressure in kPa.

## Empirical Gas Laws

**5.33**  Suppose you had a 3.15-L sample of neon gas at 21°C and a pressure of 0.951 atm. What would be the volume of this gas if the pressure were increased to 1.292 atm while the temperature remained constant?

**5.34**  You fill a balloon with helium gas to a volume of 2.68 L at 23°C and 789 mmHg. Now you release the balloon. What would be the volume of helium if its pressure changes to 556 mmHg but the temperature is unchanged?

**5.35**  You have a tank of argon gas at 19.8 atm pressure at 19°C. The volume of argon in the tank is 50.0 L. What would be the volume of this gas if you allowed it to expand to the pressure of the surrounding air (0.974 atm)? Assume the temperature remains constant.

**5.36**  A diving bell is a container open at the bottom. As the bell descends, the water level inside changes so that the pressure inside equals the pressure outside. Initially, the volume of air is $8.58 \text{ m}^3$ at 1.020 atm and 20°C. What is the volume at 1.584 atm and 20°C?

**5.37**  A McLeod gauge measures low gas pressures by compressing a known volume of the gas at constant temperature. If $315 \text{ cm}^3$ of gas is compressed to a volume of $0.0457 \text{ cm}^3$ under a pressure of 2.51 kPa, what was the original gas pressure?

**5.38**  If $456 \text{ dm}^3$ of krypton at 101 kPa and 21°C is compressed into a $27.0\text{-dm}^3$ tank at the same temperature, what is the pressure of krypton in the tank?

**5.39**  A sample of nitrogen gas at 18°C and 760 mmHg has a volume of 2.67 mL. What is the volume at 0°C and 1 atm of pressure?

**5.40**  A mole of gas at 0°C and 760 mmHg occupies 22.41 L. What is the volume at 20°C and 760 mmHg?

**5.41**  Helium gas, He, at 22°C and 1.00 atm occupied a vessel whose volume was 2.54 L. What volume would this gas occupy if it were cooled to liquid-nitrogen temperature (−197°C)?

**5.42**  An experiment called for 4.83 L of sulfur dioxide, $SO_2$, at 0°C and 1.00 atm. What would be the volume of this gas at 25°C and 1.00 atm?

**5.43**  A vessel containing $39.5 \text{ cm}^3$ of helium gas at 25°C and 106 kPa was inverted and placed in cold ethanol. As the gas contracted, ethanol was forced into the vessel to maintain the same pressure of helium. If this required $18.8 \text{ cm}^3$ of ethanol, what was the final temperature of the helium?

**5.44**  A sample of $62.3 \text{ cm}^3$ of argon gas at 18°C was contained at a pressure of 155 kPa in a J-shaped tube with mercury, as in Figure 5.5. Later the temperature changed. When the mercury level was adjusted to give the same pressure of argon, the gas volume changed to $46.7 \text{ cm}^3$. What was the final temperature of the argon?

**5.45**  A bacterial culture isolated from sewage produced 35.5 mL of methane, $CH_4$, at 31°C and 753 mmHg. What is the volume of this methane at standard temperature and pressure (0°C, 760 mmHg)?

**5.46**  Pantothenic acid is a B vitamin. Using the Dumas method, you find that a sample weighing 71.6 mg gives 3.84 mL of nitrogen gas at 23°C and 785 mmHg. What is the volume of nitrogen at STP?

**5.47**  In the presence of a platinum catalyst, ammonia, $NH_3$, burns in oxygen, $O_2$, to give nitric oxide, NO, and water vapor. How many volumes of nitric oxide are obtained from one volume of ammonia, assuming each gas is at the same temperature and pressure?

**5.48** Methanol, $CH_3OH$, can be produced in industrial plants by reacting carbon dioxide with hydrogen in the presence of a catalyst. Water is the other product. How many volumes of hydrogen are required for each volume of carbon dioxide when each gas is at the same temperature and pressure?

## Ideal Gas Law

**5.49** Starting from the ideal gas law, prove that the volume of a mole of gas is inversely proportional to the pressure at constant temperature (Boyle's law).

**5.50** Starting from the ideal gas law, prove that the volume of a mole of gas is directly proportional to the absolute temperature at constant pressure (Charles's law).

**5.51** A cylinder of oxygen gas contains 91.3 g $O_2$. If the volume of the cylinder is 8.58 L, what is the pressure of the $O_2$ if the gas temperature is 21°C?

**5.52** In an experiment, you fill a heavy-walled 5.00-L flask with methane gas, $CH_4$. If the flask contains 6.68 g of methane at 19°C, what is the gas pressure?

**5.53** An experiment calls for 3.50 mol of chlorine, $Cl_2$. What volume would this be if the gas volume is measured at 34°C and 3.00 atm?

**5.54** According to your calculations, a reaction should yield 5.67 g of oxygen, $O_2$. What do you expect the volume to be at 23°C and 0.985 atm?

**5.55** The maximum safe pressure that a certain 4.00-L vessel can hold is 3.50 atm. If the vessel contains 0.410 mol of gas, what is the maximum temperature (in degrees Celsius) to which this vessel can be subjected?

**5.56** A 2.50-L flask was used to collect a 5.65-g sample of propane gas, $C_3H_8$. After the sample was collected, the gas pressure was found to be 741 mmHg. What was the temperature of the propane in the flask?

**5.57** What is the density of ammonia gas, $NH_3$, at 31°C and 751 mmHg? Obtain the density in grams per liter.

**5.58** Calculate the density of hydrogen sulfide gas, $H_2S$, at 56°C and 967 mmHg. Obtain the density in grams per liter.

**5.59** Butane, $C_4H_{10}$, is an easily liquefied gaseous fuel. Calculate the density of butane gas at 0.897 atm and 23°C. Give the answer in grams per liter.

**5.60** Chloroform, $CHCl_3$, is a volatile (easily vaporized) liquid solvent. Calculate the density of chloroform vapor at 98°C and 797 mmHg. Give the answer in grams per liter.

**5.61** A chemist vaporized a liquid compound and determined its density. If the density of the vapor at 90°C and 753 mmHg is 1.585 g/L, what is the molecular weight of the compound?

**5.62** You vaporize a liquid substance at 100°C and 755 mmHg. The volume of 0.548 g of vapor is 237 mL. What is the molecular weight of the substance?

**5.63** A 1.28-g sample of a colorless liquid was vaporized in a 250-mL flask at 121°C and 786 mmHg. What is the molecular weight of this substance?

**5.64** A 2.30-g sample of white solid was vaporized in a 345-mL vessel. If the vapor has a pressure of 985 mmHg at 148°C, what is the molecular weight of the solid?

**5.65** Ammonium chloride, $NH_4Cl$, is a white solid. When heated to 325°C, it gives a vapor that is a mixture of ammonia and hydrogen chloride.

$$NH_4Cl(s) \longrightarrow NH_3(g) + HCl(g)$$

Suppose someone contends that the vapor consists of $NH_4Cl$ molecules rather than a mixture of $NH_3$ and HCl. Could you decide between these alternative views on the basis of gas-density measurements? Explain.

**5.66** Phosphorus pentachloride, $PCl_5$, is a white solid that sublimes (vaporizes without melting) at about 100°C. At higher temperatures, the $PCl_5$ vapor decomposes to give phosphorus trichloride, $PCl_3$, and chlorine, $Cl_2$.

$$PCl_5(g) \longrightarrow PCl_3(g) + Cl_2(g)$$

How could gas-density measurements help to establish that $PCl_5$ vapor is decomposing?

## Stoichiometry with Gas Volumes

**5.67** Calcium carbide reacts with water to produce acetylene gas, $C_2H_2$.

$$CaC_2(s) + 2H_2O(l) \longrightarrow Ca(OH)_2(aq) + C_2H_2(g)$$

Calculate the volume (in liters) of acetylene produced at 26°C and 684 mmHg from 0.050 mol $CaC_2$ and excess $H_2O$.

**5.68** Magnesium metal reacts with hydrochloric acid to produce hydrogen gas, $H_2$.

$$Mg(s) + 2HCl(aq) \longrightarrow MgCl_2(aq) + H_2(g)$$

Calculate the volume (in liters) of hydrogen produced at 28°C and 665 mmHg from 0.0840 mol Mg and excess HCl.

**5.69** Lithium hydroxide, LiOH, is used in spacecraft to recondition the air by absorbing the carbon dioxide exhaled by astronauts. The reaction is

$$2LiOH(s) + CO_2(g) \longrightarrow Li_2CO_3(s) + H_2O(l)$$

What volume of carbon dioxide gas at 21°C and 781 mmHg could be absorbed by 327 g of lithium hydroxide?

**5.70** Magnesium burns in air to produce magnesium oxide, MgO, and magnesium nitride, $Mg_3N_2$. Magnesium nitride reacts with water to give ammonia.

$$Mg_3N_2(s) + 6H_2O(l) \longrightarrow 3Mg(OH)_2(s) + 2NH_3(g)$$

What volume of ammonia gas at 24°C and 753 mmHg will be produced from 4.56 g of magnesium nitride?

**5.71**   Urea, $NH_2CONH_2$, is a nitrogen fertilizer that is manufactured from ammonia and carbon dioxide.

$$2NH_3(g) + CO_2(g) \longrightarrow NH_2CONH_2(aq) + H_2O(l)$$

What volume of ammonia at 25°C and 1.50 atm is needed to produce 908 g (2 lb) of urea?

**5.72**   Nitric acid is produced from nitric oxide, NO, which in turn is prepared from ammonia by the Ostwald process:

$$4NH_3(g) + 5O_2(g) \longrightarrow 4NO(g) + 6H_2O(g)$$

What volume of oxygen at 35°C and 2.15 atm is needed to produce 50.0 g of nitric oxide?

**5.73**   Ammonium sulfate is used as a nitrogen and sulfur fertilizer. It is produced by reacting ammonia with sulfuric acid. Write the balanced equation for the reaction of gaseous ammonia with sulfuric acid solution. What volume (in liters) of ammonia at 15°C and 1.15 atm is required to produce 150.0 g of ammonium sulfate?

**5.74**   Sodium hydrogen carbonate is also known as baking soda. When this compound is heated, it decomposes to sodium carbonate, carbon dioxide, and water vapor. Write the balanced equation for this reaction. What volume (in liters) of carbon dioxide gas at 75°C and 756 mmHg will be produced from 26.8 g of sodium hydrogen carbonate?

## Gas Mixtures

**5.75**   Calculate the total pressure (in atm) of a mixture of 0.0200 mol of helium, He, and 0.0100 mol of hydrogen, $H_2$, in a 2.50-L flask at 10°C. Assume ideal gas behavior.

**5.76**   Calculate the total pressure (in atm) of a mixture of 0.0300 mol of helium, He, and 0.0200 mol of oxygen, $O_2$, in a 4.00-L flask at 20°C. Assume ideal gas behavior.

**5.77**   A 200.0-mL flask contains 1.03 mg $O_2$ and 0.56 mg He at 15°C. Calculate the partial pressures of oxygen and of helium in the flask. What is the total pressure?

**5.78**   The atmosphere in a sealed diving bell contained oxygen and helium. If the gas mixture has 0.200 atm of oxygen and a total pressure of 3.00 atm, calculate the mass of helium in 10.0 L of the gas mixture at 20°C.

**5.79**   The gas from a certain volcano had the following composition in mole percent (that is, mole fraction × 100): 65.0% $CO_2$, 25.0% $H_2$, 5.4% HCl, 2.8% HF, 1.7% $SO_2$, and 0.1% $H_2S$. What would be the partial pressure of each of these gases if the total pressure of volcanic gas were 760 mmHg?

**5.80**   In a series of experiments, the U.S. Navy developed an undersea habitat. In one experiment, the mole percent composition of the atmosphere in the undersea habitat was 79.0% He, 17.0% $N_2$, and 4.0% $O_2$. What will the partial pressure of each

gas be when the habitat is 58.8 m below sea level, where the pressure is 6.91 atm?

**5.81**   Formic acid, $HCHO_2$, is a convenient source of small quantities of carbon monoxide. When warmed with sulfuric acid, formic acid decomposes to give CO gas.

$$HCHO_2(l) \longrightarrow H_2O(l) + CO(g)$$

If 3.85 L of carbon monoxide was collected over water at 25°C and 689 mmHg, how many grams of formic acid were consumed?

**5.82**   An aqueous solution of ammonium nitrite, $NH_4NO_2$, decomposes when heated to give off nitrogen, $N_2$.

$$NH_4NO_2(s) \longrightarrow 2H_2O(g) + N_2(g)$$

This reaction may be used to prepare pure nitrogen. How many grams of ammonium nitrite must have reacted if 3.75 $dm^3$ of nitrogen gas was collected over water at 19°C and 97.8 kPa?

## Molecular Speeds; Effusion

**5.83**   Calculate the rms speeds of $N_2$ molecules at 25°C and at 125°C. Sketch approximate curves of the molecular speed distributions of $N_2$ at 25°C and at 125°C.

**5.84**   Calculate the rms speed of $Br_2$ molecules at 23°C and 1.00 atm. What is the rms speed of $Br_2$ at 23°C and 1.50 atm?

**5.85**   Uranium hexafluoride, $UF_6$, is a white solid that sublimes (vaporizes without melting) at 57°C under normal atmospheric pressure. The compound is used to separate uranium isotopes by effusion. What is the rms speed (in m/s) of a uranium hexafluoride molecule at 57°C?

**5.86**   For a spacecraft or a molecule to leave the moon, it must reach the escape velocity (speed) of the moon, which is 2.37 km/s. The average daytime temperature of the moon's surface is 365 K. What is the rms speed (in m/s) of a hydrogen molecule at this temperature? How does this compare with the escape velocity?

**5.87**   At what temperature would $CO_2$ molecules have an rms speed equal to that of $H_2$ molecules at 25°C?

**5.88**   At what temperature does the rms speed of $O_2$ molecules equal 400. m/s?

**5.89**   What is the ratio of rates of effusion of $N_2$ and $O_2$ under the same conditions?

**5.90**   Obtain the ratio of rates of effusion of $H_2$ and $H_2Te$ under the same conditions.

**5.91**   If 0.10 mol of $I_2$ vapor can effuse from an opening in a heated vessel in 52 s, how long will it take 0.10 mol $H_2$ to effuse under the same conditions?

**5.92**   If it takes 10.6 hours for 1.00 L of nitrogen, $N_2$, to effuse through the pores in a balloon, how long would it take for 1.00 L of helium, He, to effuse under the same conditions?

**5.93** If 4.83 mL of an unknown gas effuses through a hole in a plate in the same time it takes 9.23 mL of argon, Ar, to effuse through the same hole under the same conditions, what is the molecular weight of the unknown gas?

**5.94** A given volume of nitrogen, $N_2$, required 68.3 s to effuse from a hole in a chamber. Under the same conditions, another gas required 85.6 s for the same volume to effuse. What is the molecular weight of this gas?

## van der Waals Equation

**5.95** Calculate the pressure of ethanol vapor, $C_2H_5OH(g)$, at 82.0°C if 1.000 mol $C_2H_5OH(g)$ occupies 30.00 L. Use the van der Waals equation (see Table 5.7 for data). Compare with the result from the ideal gas law.

**5.96** Calculate the pressure of water vapor at 120.0°C if 1.000 mol of water vapor occupies 32.50 L. Use the van der Waals equation (see Table 5.7 for data). Compare with the result from the ideal gas law.

**5.97** Calculate the molar volume of ethane at 1.00 atm and 0°C and at 10.0 atm and 0°C, using the van der Waals equation. The van der Waals constants are given in Table 5.7. To simplify, note that the term $n^2a/V^2$ is small compared with $P$. Hence, it may be approximated with negligible error by substituting $nRT/P$ from the ideal gas law for $V$ in this term. Then the van der Waals equation can be solved for the volume. Compare the results with the values predicted by the ideal gas law.

**5.98** Calculate the molar volume of oxygen at 1.00 atm and 0°C and at 10.0 atm and 0°C, using the van der Waals equation. The van der Waals constants are given in Table 5.7. (See the note on solving the equation given in Problem 5.97.) Compare the results with the values predicted by the ideal gas law. Also compare with the values obtained from Table 5.3.

## General Problems

**5.99** A glass tumbler containing 243 cm$^3$ of air at $1.00 \times 10^2$ kPa (the barometric pressure) and 20°C is turned upside down and immersed in a body of water to a depth of 20.5 m. The air in the glass is compressed by the weight of water above it. Calculate the volume of air in the glass, assuming the temperature and barometric pressure have not changed.

**5.100** The density of air at 20°C and 1.00 atm is 1.205 g/L. If this air were compressed at the same temperature to equal the pressure at 30.0 m below sea level, what would be its density? Assume the barometric pressure is constant at 1.00 atm. The density of seawater is 1.025 g/cm$^3$.

**5.101** A flask contains 157 mL of argon at 21°C and 738 mmHg. What is the volume of gas, corrected to STP?

**5.102** A steel bottle contains 12.0 L of a gas at 11.0 atm and 20°C. What is the volume of gas at STP?

**5.103** A balloon containing 5.0 dm$^3$ of gas at 14°C and 100.0 kPa rises to an altitude of 2000. m, where the temperature is 20°C. The pressure of gas in the balloon is now 79.0 kPa. What is the volume of gas in the balloon?

**5.104** A volume of air is taken from the earth's surface, at 15°C and 1.00 atm, to the stratosphere, where the temperature is −20°C and the pressure is $1.00 \times 10^{-3}$ atm. By what factor is the volume increased?

**5.105** A radioactive metal atom decays (goes to another kind of atom) by emitting an alpha particle (He$^{2+}$ ion). The alpha particles are collected as helium gas. A sample of helium with a volume of 12.05 mL was obtained at 765 mmHg and 23°C. How many atoms decayed during the period of the experiment?

**5.106** The combustion method used to analyze for carbon and hydrogen can be adapted to give percentage N by collecting the nitrogen from combustion of the compound as $N_2$. A sample of a compound weighing 8.75 mg gave 1.77 mL $N_2$ at 25°C and 749 mmHg. What is the percentage N in the compound?

**5.107** Dry air at STP has a density of 1.2929 g/L. Calculate the average molecular weight of air from the density.

**5.108** A hydrocarbon gas has a density of 1.22 g/L at 20°C and 1.00 atm. An analysis gives 80.0% C and 20.0% H. What is the molecular formula?

**5.109** A person exhales about $5.8 \times 10^2$ L of carbon dioxide per day (at STP). The carbon dioxide exhaled by an astronaut is absorbed from the air of a space capsule by reaction with lithium hydroxide, LiOH.

$$2\text{LiOH}(s) + \text{CO}_2(g) \longrightarrow \text{Li}_2\text{CO}_3(s) + \text{H}_2\text{O}(l)$$

How many grams of lithium hydroxide are required per astronaut per day?

**5.110** Pyruvic acid, $HC_3H_3O_3$, is involved in cell metabolism. It can be assayed for (that is, the amount of it determined) by using a yeast enzyme. The enzyme makes the following reaction go to completion:

$$\text{HC}_3\text{H}_3\text{O}_3(aq) \longrightarrow \text{C}_2\text{H}_4\text{O}(aq) + \text{CO}_2(g)$$

If a sample containing pyruvic acid gives 20.3 mL of carbon dioxide gas, $CO_2$, at 349 mmHg and 30°C, how many grams of pyruvic acid are there in the sample?

**5.111** Liquid oxygen was first prepared by heating potassium chlorate, $KClO_3$, in a closed vessel to obtain oxygen at high pressure. The oxygen was cooled until it liquefied.

$$2KClO_3(s) \longrightarrow 2KCl(s) + 3O_2(g)$$

If 170. g of potassium chlorate reacts in a 2.50-L vessel, which was initially evacuated, what pressure of oxygen will be attained when the temperature is finally cooled to 25°C? Use the preceding chemical equation and ignore the volume of solid product.

**5.112** Raoul Pictet, the Swiss physicist who first liquefied oxygen, attempted to liquefy hydrogen. He heated potassium formate, $KCHO_2$, with KOH in a closed 2.50-L vessel.

$$KCHO_2(s) + KOH(s) \longrightarrow K_2CO_3(s) + H_2(g)$$

If 75.0 g of potassium formate reacts in a 2.50-L vessel, which was initially evacuated, what pressure of hydrogen will be attained when the temperature is finally cooled to 21°C? Use the preceding chemical equation and ignore the volume of solid product.

**5.113** A 24.9-mL volume of hydrochloric acid reacts completely with 55.0 mL of aqueous $Na_2CO_3$. The reaction is

$$2HCl(aq) + Na_2CO_3(aq) \longrightarrow$$
$$CO_2(g) + H_2O(l) + 2NaCl(aq)$$

The volume of $CO_2$ formed is 141 mL at 27°C and 742 mmHg. What is the molarity of the HCl solution?

**5.114** An 18.6-mL volume of hydrochloric acid reacts completely with a solid sample of $MgCO_3$. The reaction is

$$2HCl(aq) + MgCO_3(s) \longrightarrow$$
$$CO_2(g) + H_2O(l) + MgCl_2(aq)$$

The volume of $CO_2$ formed is 159 mL at 23°C and 731 mmHg. What is the molarity of the HCl solution?

**5.115** A 41.41-mL sample of a 0.1250 $M$ acid reacts with an excess of $Na_2CO_3$ to form 150.0 mL $CO_2$ at 646 mmHg and 27°C. If the acid is either HCl or $H_2SO_4$, which is it?

**5.116** A 48.90-mL sample of a 0.2040 $M$ acid reacts with an excess of $Na_2CO_3$ to form 125.0 mL $CO_2$ at 722 mmHg and 17°C. If the acid is either HCl or $H_2SO_4$, which is it?

**5.117** If the rms speed of $NH_3$ molecules is found to be 0.510 km/s, what is the temperature (in degrees Celsius)?

**5.118** If the rms speed of He atoms in the exosphere (highest region of the atmosphere) is $3.53 \times 10^3$ m/s, what is the temperature (in kelvins)?

**5.119** Calculate the ratio of rates of effusion of $^{235}UF_6$ and $^{238}UF_6$, where $^{235}U$ and $^{238}U$ are isotopes of uranium. The atomic masses are $^{235}U$, 235.04 amu; $^{238}U$, 238.05 amu; $^{19}F$ (the only naturally occurring isotope), 18.998 amu. Carry five significant figures in the calculation.

**5.120** Hydrogen has two stable isotopes, $^1H$ and $^2H$, with atomic masses of 1.0078 amu and 2.0141 amu, respectively. Ordinary hydrogen gas, $H_2$, is a mixture consisting mostly of $^1H_2$ and $^1H^2H$. Calculate the ratio of rates of effusion of $^1H_2$ and $^1H^2H$ under the same conditions.

**5.121** A 1.000-g sample of an unknown gas at 0°C gives the following data:

| $P$(atm) | $V$(L) |
|---|---|
| 0.2500 | 3.1908 |
| 0.5000 | 1.5928 |
| 0.7500 | 1.0601 |
| 1.0000 | 0.7930 |

Use these data to calculate the value of the molar mass at each of the given pressures from the ideal gas law (we will call this the "apparent molar mass" at this pressure). Plot the apparent molar masses against pressure and extrapolate to find the molar mass at zero pressure. Because the ideal gas law is most accurate at low pressures, this extrapolation will give an accurate value for the molar mass. What is the accurate molar mass?

**5.122** Plot the data given in Table 5.3 for oxygen at 0°C to obtain an accurate molar mass for $O_2$. To do this, calculate a value of the molar mass at each of the given pressures from the ideal gas law (we will call this the "apparent molar mass" at this pressure). On a graph show the apparent molar mass versus the pressure and extrapolate to find the molar mass at zero pressure. Because the ideal gas law is most accurate at low pressures, this extrapolation will give an accurate value for the molar mass. What is the accurate molar mass?

**5.123** Carbon monoxide, CO, and oxygen, $O_2$, react according to

$$2CO(g) + O_2(g) \longrightarrow 2CO_2(g)$$

Assuming that the reaction takes place and goes to completion, determine what substances remain and what their partial pressures are after the valve is opened in the apparatus represented in the accompanying figure. Also assume that the temperature is fixed at 300 K.

**5.124** Suppose the apparatus shown in the figure accompanying Problem 5.123 contains $H_2$ at 0.500 atm in the left vessel separated from $O_2$ at 1.00 atm in the other vessel. The valve is then opened. If $H_2$ and $O_2$ react to give $H_2O$ when the temperature is fixed at 533 K, what substances remain and what are their partial pressures after reaction?

## Cumulative-Skills Problems

**5.125**    A sample of natural gas is 85.2% methane, $CH_4$, and 14.8% ethane, $C_2H_6$, by mass. What is the density of this mixture at 18°C and 771 mmHg?

**5.126**    A sample of a breathing mixture for divers contained 34.3% helium, He; 51.7% nitrogen, $N_2$; and 14.0% oxygen, $O_2$ (by mass). What is the density of this mixture at 22°C and 755 mmHg?

**5.127**    A sample of sodium peroxide, $Na_2O_2$, was reacted with an excess of water.

$$2Na_2O_2(s) + 2H_2O(l) \longrightarrow 4NaOH(aq) + O_2(g)$$

All of the sodium peroxide reacted, and the oxygen was collected over water at 21°C. The barometric pressure was 771 mmHg. The apparatus was similar to that shown in Figure 5.18. However, the level of water inside the tube was 25.0 cm above the level of water outside the tube. If the volume of gas in the tube is 31.0 mL, how many grams of sodium peroxide were in the sample?

**5.128**    A sample of zinc metal was reacted with an excess of hydrochloric acid.

$$Zn(s) + 2HCl(aq) \longrightarrow ZnCl_2(aq) + H_2(g)$$

All of the zinc reacted, and the hydrogen gas was collected over water at 18°C; the barometric pressure was 751 mmHg. The apparatus was similar to that shown in Figure 5.18, but the level of water inside the tube was 31.0 cm above the level outside the tube. If the volume of gas in the tube is 22.1 mL, how many grams of zinc were there in the sample?

**5.129**    A mixture contained calcium carbonate, $CaCO_3$, and magnesium carbonate, $MgCO_3$. A sample of this mixture weighing 7.85 g was reacted with excess hydrochloric acid. The reactions are

$$CaCO_3(g) + 2HCl(aq) \longrightarrow CaCl_2(aq) + H_2O(l) + CO_2(g)$$
$$MgCO_3(s) + 2HCl(aq) \longrightarrow MgCl_2(aq) + H_2O(l) + CO_2(g)$$

If the sample reacted completely and produced 1.94 L of carbon dioxide, $CO_2$, at 25°C and 785 mmHg, what were the percentages of $CaCO_3$ and $MgCO_3$ in the mixture?

**5.130**    A mixture contained zinc sulfide, ZnS, and lead sulfide, PbS. A sample of the mixture weighing 6.12 g was reacted with an excess of hydrochloric acid. The reactions are

$$ZnS(s) + 2HCl(aq) \longrightarrow ZnCl_2(aq) + H_2S(g)$$
$$PbS(s) + 2HCl(aq) \longrightarrow PbCl_2(aq) + H_2S(g)$$

If the sample reacted completely and produced 1.049 L of hydrogen sulfide, $H_2S$, at 23°C and 745 mmHg, what were the percentages of ZnS and PbS in the mixture?

**5.131**    A mixture of $N_2$ and Ne contains equal moles of each gas and has a total mass of 10.0 g. What is the density of this gas mixture at 500 K and 15.00 atm? Assume ideal gas behavior.

**5.132**    A mixture of Ne and Ar gases at 350 K contains twice as many moles of Ne as of Ar and has a total mass of 50.0 g. If the density of the mixture is 4.00 g/L, what is the partial pressure of Ne?

**5.133**    An ideal gas with a density of 3.00 g/L has a pressure of 675 mmHg at 25°C. What is the root-mean-square speed of the molecules of this gas?

**5.134**    The root-mean-square speed of the molecules of an ideal gas at 25°C and a pressure of 2.50 atm is $5.00 \times 10^2$ km/s. What is the density of this gas?

## Media Activities

**5.135**    Open the Liquid Nitrogen and Balloons *Visualization*.
a. Explain why the balloon collapsed when Liquid $N_2$ was poured on the balloon in terms of the force of collisions per unit area.
b. Why did the balloon return to its original shape at the end of the video?
c. View the Balloon Submersion in Liquid $N_2$ *Visualization* and see if your answer in part a is consistent with what is presented in this animation.

**5.136**    Open the Diffusion of Gases *Visualization*.
a. Explain why the $NH_3(g)$ and $HCl(g)$ molecules did not meet in the middle of the tube.
b. If the animation were allowed to continue for significantly longer, how would the molecules be distributed in the tube?

**5.137**    Open the Collapsing Can *Visualization*.

a. Assuming only $H_2O(g)$ is in the can, what is the pressure of water vapor inside the can before the can is sealed?
b. What happens to the gas molecules inside the can as the can is immersed in cold water?
c. Upon immersion in the water, what happens to the pressure inside the can relative to the pressure outside of the can?
d. Why did the can collapse?

**5.138**    Open the Molecular Motion (many molecules) *Visualization*.
a. List three things you could do to this system that would decrease the pressure in the box.
b. For each of your answers in part a, what common feature were you trying to control in order to reduce the pressure?

**5.139**    For review, take the ACE practice test for Chapter 5 on the student web site.

Ammonium dichromate decomposes with the release of heat in a fiery reaction.

# Thermochemistry

Nearly all chemical reactions involve either the release or the absorption of heat, a form of energy. The burning of coal and gasoline are dramatic examples of chemical reactions in which a great deal of heat is released. Such reactions are important sources of warmth and energy. Chemical reactions that absorb heat are usually less dramatic. The reaction of barium hydroxide with an ammonium salt is an exception. If crystals of barium hydroxide octahydrate, $Ba(OH)_2 \cdot 8H_2O$, are mixed with crystals of ammonium nitrate, $NH_4NO_3$, in a flask, the solids form first a slush, then a liquid. Because the reaction mixture absorbs heat from the surroundings, the flask feels cool. It soon becomes so cold that if it is set in a puddle of water on a board, the water freezes; the board can then be inverted with the flask frozen to it (Figure 6.1).

In this chapter, we will be concerned with the quantity of heat released or absorbed in a chemical reaction. We will address several questions: How do you measure the quantity of heat released or absorbed by a chemical reaction? To what extent can you relate the quantity of heat involved in a given reaction to the quantities of heat in other reactions? And how can you use this information?

## CONTENTS

**Figure 6.1**
**A reaction that absorbs heat.** Two crystalline substances, barium hydroxide octahydrate and an ammonium salt, are mixed thoroughly in a flask. Then the flask, which feels quite cold to the touch, is set in a puddle of water on a board. In a couple of minutes, the flask and board are frozen solidly together. The board can then be inverted with the flask frozen to it.

# Understanding Heats of Reaction

**Visualization: Ammonium Dichromate Volcano**

▶ Demonstration of an endothermic reaction: Bassam Z. Shakhashiri, *Chemical Demonstrations,* Vol. 1 (Madison: University of Wisconsin Press, 1983), no. 1.3, pp. 10–12; Summerlin and Ealy, *Chemical Demonstrations,* Vol. 1 (Washington, D.C.: Am. Chem. Soc., 1985), no. 27, p. 44.

▶ These questions concern chemical equilibrium and will be discussed in Chapter 19.

*Thermodynamics* is the science of the relationships between heat and other forms of energy. *Thermochemistry* is one area of thermodynamics. It concerns the study of the quantity of heat absorbed or evolved (given off) by chemical reactions. An example of a heat-evolving reaction is the burning of fuel. There may be practical reasons why you want to know the quantity of heat evolved during the burning of a fuel: you could calculate the cost of the fuel per unit of heat energy produced; you could calculate the quantity of heat obtained per unit mass of rocket fuel; and so forth. But there are also theoretical reasons for wanting to know the quantity of heat involved in a reaction. For example, knowing such values, you are able to calculate the amount of energy needed to break a particular kind of chemical bond and so learn something about the strength of that bond. Heat measurements also provide data needed to determine whether a particular chemical reaction occurs and, if so, to what extent. ◀

We have just used the terms *energy* and *heat,* assuming you have some idea of what they mean. But to proceed, you will need precise definitions of these and other terms. In the following section, we will define energy and its different forms and introduce units of energy.

## 6.1 Energy and Its Units

▶ This chapter gives the student an introduction to energy, which can be useful in discussing bonding. However, you can delay discussion of this chapter and treat it with Chapter 19 as a unit on thermodynamics.

▶ Video no. 11 (Series A), "An Endothermic Reaction."

We can define **energy** briefly as *the potential or capacity to move matter.* According to this definition, energy is not a material thing but rather a property of matter. Energy exists in different forms that can be interconverted. You can see the relationship of a given form of energy to the motion of matter by following its interconversions into different forms.

Consider the interconversions of energy in a steam-driven electrical generator. A fuel is burned to heat water and generate steam. The steam expands against a piston (or turbine), which is connected to a drive shaft that turns an electrical coil in a magnetic field. Electricity is generated in the coil. The fuel contains chemical energy, which is converted to heat. Part of the heat is then converted to motion of the drive

**Figure 6.2**
**Conversion of light energy to kinetic energy.** *Left:* The winning solar-powered vehicle from the 1999 Sunrayce, a race featuring vehicles designed and built by North American universities. The panel of photovoltaic cells is at the rear of the vehicle. *Right:* Solar-powered lawn mower. The panel of photovoltaic cells is on the handlebar.

▶ Video Series C, "Ammonium Chloride and Barium Hydroxide."

shaft, and this motion is converted to electrical energy. The electrical energy could be used to run a motor, transforming the electrical energy back to the energy of motion. Or you could send the electricity into a lightbulb, converting electrical energy to heat energy and light energy. Photovoltaic cells can convert light back to electricity, which could be used to run a motor that can move matter (Figure 6.2). These examples show that energy can exist in different forms, including heat, light, and electrical energy, and these different forms can be interconverted. You also see their relationship to the energy of motion.

In this chapter, we will be especially concerned with the energy of substances, or chemical energy, and its transformation during chemical reaction into heat energy. To prepare for this, we will first explore the quantitative meaning of the energy of motion *(kinetic energy)*. Then we will look at the concepts of *potential energy* and of the *internal energy* of substances, which is defined in terms of the kinetic and potential energies of the particles making up the substance.

## Kinetic Energy; Units of Energy

**Kinetic energy** is *the energy associated with an object by virtue of its motion.* An object of mass $m$ and speed or velocity $v$ has kinetic energy $E_k$ equal to

$$E_k = \tfrac{1}{2}mv^2$$

This formula shows that the kinetic energy of an object depends on both its mass and its speed. A heavy object can move more slowly than a light object and still have the same kinetic energy. ◀

▶ In the previous chapter, we used the symbol *u* for *average* molecular speed. Here *v* is the speed of an individual object or particle.

Consider the kinetic energy of a person whose mass is 59.0 kg and whose speed is 26.8 m/s. (This is equivalent to a person with a mass of 130 lb traveling in an automobile going 60 miles per hour.) You substitute the mass and speed into the formula.

$$E_k = \tfrac{1}{2} \times (59.0 \text{ kg}) \times (26.8 \text{ m/s})^2 = 2.12 \times 10^4 \text{ kg} \cdot \text{m}^2/\text{s}^2$$

Note that the unit of energy comes out of the calculation. Because you substituted SI units of mass and speed, you obtain the SI unit of energy. *The SI unit of energy, $kg \cdot m^2/s^2$, is given the name* **joule (J)** (pronounced "jewl") after the English physicist James Prescott Joule (1818–1889), who studied the energy concept. You see that a person weighing 130 lb and traveling 60 miles per hour has a kinetic energy equal to $2.12 \times 10^4$ J, or 21.2 kJ (21.2 kilojoules).

The joule is an extremely small unit. To appreciate its size, note that the *watt* is a measure of the quantity of energy used per unit time and equals 1 joule per second. A 100-watt bulb, for example, uses 100 joules of energy every second. A kilowatt-hour, the unit by which electric energy is sold, equals 3600 kilowatt-seconds (there are 3600 seconds in 1 hour), or 3.6 million joules. A household might use something like 1000 kilowatt-hours (3.6 billion joules) of electricity in a month.

The **calorie (cal)** is *a non-SI unit of energy commonly used by chemists, originally defined as the amount of energy required to raise the temperature of one gram of water by one degree Celsius.* This is only an approximate definition, however, because we now know that the energy needed to heat water depends slightly on the temperature of the water. In 1925 the calorie was defined in terms of the joule:

$$1 \text{ cal} = 4.184 \text{ J} \qquad \text{(exact definition)}$$

A person weighing 130 lb and traveling 60 miles per hour has a kinetic energy of

$$2.12 \times 10^4 \text{ J} \times \frac{1 \text{ cal}}{4.184 \text{ J}} = 5.07 \times 10^3 \text{ cal (5.07 kcal)}$$

▶ An alternate example is provided in the Instructor's Resource Manual.

## EXAMPLE 6.1

## Calculating Kinetic Energy

A good pitcher can throw a baseball so that it travels between 60 and 80 miles per hour. A regulation baseball weighing 143 g (0.143 kg) travels 75 miles per hour (33.5 m/s). What is the kinetic energy of this baseball in joules? in calories?

**SOLUTION**

You substitute into the formula $E_k = \frac{1}{2}mv^2$ using SI units.

$$E_k = \frac{1}{2} \times 0.143 \text{ kg} \times (33.5 \text{ m/s})^2 = \textbf{80.2 J}$$

To obtain the kinetic energy in calories, you multiply by the conversion factor 1 cal/4.184 J.

$$80.2 \, \cancel{J} \times \frac{1 \text{ cal}}{4.184 \, \cancel{J}} = \textbf{19.2 cal}$$

## EXERCISE 6.1

An electron, whose mass is $9.11 \times 10^{-31}$ kg, is accelerated by a positive charge to a speed of $5.0 \times 10^6$ m/s. What is the kinetic energy of the electron in joules? in calories?

See Problems 6.35, 6.36, 6.37, and 6.38.

**Figure 6.3**
**Potential energy and kinetic energy.** Water at the top of the dam has potential energy. As the water falls over the dam, this potential energy is converted to kinetic energy.

## Potential Energy

**Potential energy** is *the energy an object has by virtue of its position in a field of force.* For example, water at the top of a dam has potential energy (in addition to whatever kinetic energy it may possess), because the water is at a relatively high position in the gravitational force field of the earth. You can calculate this potential energy of the water from the formula $E_p = mgh$. Here $E_p$ is the potential energy of a quantity of water at the top of the dam, $m$ is the mass of the water, $g$ is the constant acceleration of gravity, and $h$ is the height of the water measured from some standard level. The choice of this standard level is arbitrary, because only *differences* of potential energy are important in any physical situation. It is convenient to choose the standard level to be the surface of the earth. As a quantity of water falls over the dam, its potential energy decreases from $mgh$ at the top of the dam to zero at the earth's surface.

The potential energy of the water at the top of the dam is converted to kinetic energy when the water falls to a lower level. As the water falls, it moves more quickly. The potential energy decreases and the kinetic energy increases. Figure 6.3 shows the potential energy of water being converted to kinetic energy as the water falls over a dam.

## Internal Energy

Consider the total energy of a quantity of water as it moves over the dam. This water as a whole has kinetic energy and potential energy. However, we know that water is made up of molecules, which are made up of smaller particles, electrons and nuclei. Each of these particles also has kinetic energy and potential energy. *The sum of the kinetic and potential energies of the particles making up a substance* is referred to as the **internal energy,** $U$, of the substance. Therefore, the total energy, $E_{tot}$, of a quantity of water equals the sum of its kinetic and potential energies as a whole $(E_k + E_p)$ plus its internal energy.

$$E_{tot} = E_k + E_p + U$$

Normally when you study a substance in the laboratory, the substance is at rest in a vessel. Its kinetic energy as a whole is zero. Moreover, its potential energy as a whole is constant, and you can take it to be zero. In this case, the total energy of the substance equals its internal energy, $U$.

### Law of Conservation of Energy

We have discussed situations in which one form of energy can be converted into another form of energy. For example, when water falls over a dam, potential energy is converted into kinetic energy. Some of the kinetic energy of the water may also be converted into random molecular motion—that is, into internal energy of the water. The total energy, $E_{tot}$, of the water, however, remains constant, equal to the sum of the kinetic energy, $E_k$, the potential energy, $E_p$, and the internal energy, $U$, of the water.

This result can be stated more generally as the **law of conservation of energy:** *energy may be converted from one form to another, but the total quantity of energy remains constant.*

### CONCEPT CHECK 6.1

A solar-powered water pump has photovoltaic cells that protrude from top panels. These cells collect energy from sunlight, storing it momentarily in a battery, which later runs an electric motor that pumps water up to a storage tank on a hill. What energy conversions are involved in using sunlight to pump water into the storage tank?

► Animation: Kinetic-molecular theory/heat transfer.

## 6.2 Heat of Reaction

In the chapter opening, we mentioned chemical reactions (such as the burning of coal) that evolve, or release, heat. We also described a reaction that absorbs heat. Both types of reaction involve a *heat of reaction.* To understand this concept, you need to know what is meant by a thermodynamic *system* and its *surroundings* and to have a precise definition of the term *heat.*

Suppose that you are interested in studying the change of a thermodynamic property (such as internal energy) during a physical or chemical change. *The substance or mixture of substances under study in which a change occurs* is called the **thermodynamic system** (or simply **system**). The **surroundings** are *everything in the vicinity of the thermodynamic system* (Figure 6.4).

### Definition of Heat

**Heat** is defined as *the energy that flows into or out of a system because of a difference in temperature between the thermodynamic system and its surroundings.* As long as a system and its surroundings are in thermal contact (that is, they are not thermally insulated from one another), energy (heat) flows between them to establish temperature equality, or *thermal equilibrium.* Heat flows from a region of higher temperature to one of lower temperature; once the temperatures become equal, heat flow stops. Note that once heat flows into a system, it appears in the system as an increase in its internal energy. You would not say that the system has heat, because heat is only an energy flow.

We can explain this flow of energy between two regions of different temperatures in terms of kinetic-molecular theory. Imagine two vessels in contact, each containing oxygen gas and the one on the left being hotter (Figure 6.5). According to kinetic theory, the average speed of molecules in the hotter gas is greater than that of molecules in the colder gas. But as the molecules in their random motions collide with the ves-

**Figure 6.4**
**Illustration of a thermodynamic system.** The *system* consists of the portion of the universe that we choose to study; in this case, it is a solution of $Ba(OH)_2$ and $NH_4NO_3$. Everything else, including the reaction vessel, constitutes the *surroundings.*

**Figure 6.5
A kinetic-theory explanation of heat.** The vessel on the left contains oxygen molecules at a higher temperature than the oxygen molecules on the right. Molecules collide with the vessel walls, thereby losing or gaining energy. The faster molecules tend to slow down, and the slower molecules tend to speed up. The net result is that energy is transferred through the vessel walls from the hot gas to the cold gas. We call this energy transfer *heat*.

▶ Demonstration of an exothermic reaction (hand-warmer reaction): Summerlin et al., *Chemical Demonstrations,* Vol. 2 (Washington, D.C.: Am. Chem. Soc., 1987), no. 54, pp. 99–100.

Surroundings

Heat → System ← Heat

**Endothermic**

Surroundings

Heat ← System → Heat

**Exothermic**

sel walls, they lose energy to or gain energy from the walls. The faster molecules tend to slow down, while the slower molecules tend to speed up. Eventually, the average speeds of the molecules in the two vessels (and therefore the temperatures of the two gases) become equal. The net result is that energy has transferred through the vessel walls from the hot gas to the cold gas; that is, heat has flowed from the hotter vessel to the cooler one.

Heat and temperature are sometimes confused. The distinction is clear in the kinetic picture of a gas. According to kinetic theory, the absolute temperature of a gas is directly proportional to the average kinetic energy of the molecules. When you add heat to a gas, you increase its internal energy, and therefore its total kinetic energy. This increase in kinetic energy will be distributed over the molecules in the sample. Therefore, the increase in average kinetic energy per molecule (and therefore the increase in temperature) depends on the size of the gas sample. A given quantity of heat will raise the temperature of a sample more if the sample is small.

Heat is denoted by the symbol $q$. The algebraic sign of $q$ is chosen to be positive if heat is absorbed by the system and negative if heat is evolved. The sign of $q$ can be remembered this way: when heat is absorbed by a system, energy is *added* to it; $q$ is assigned a positive quantity. On the other hand, when heat is evolved by a system, energy is *subtracted* from it; $q$ is assigned a negative number.

## Heat of Reaction

Consider a system in which a chemical reaction occurs. Before the reaction, the system and surroundings are at the same temperature, say 25°C. When the reaction starts, however, the temperature changes.

Suppose the temperature of the system falls. In that case, heat flows from the surroundings into the system. When the reaction stops, heat continues to flow until the system returns to the temperature of its surroundings (at 25°C). Heat has flowed into the system; that is, the system has absorbed heat and $q$ is positive.

Suppose, on the other hand, that the temperature of the system rises. In this case, heat flows from the system to the surroundings. When the reaction stops, heat continues to flow until the system returns to the temperature of its surroundings (at 25°C). In this case, heat has flowed out of the system; that is, the system has evolved heat and $q$ is negative.

The **heat of reaction** (at a given temperature) is *the value of q required to return a system to the given temperature at the completion of the reaction.*

Chemical reactions or physical changes are classified as exothermic or endothermic. An **exothermic process** is *a chemical reaction or a physical change in which*

**Figure 6.6**
**An exothermic process.**
*Left:* The test tube contains anhydrous copper(II) sulfate (white crystals) and a thermometer that registers 26.1°C. The graduated cylinder contains water. *Right:* Water from the graduated cylinder has been added to the test tube, where the copper(II) sulfate forms hydrated copper(II) ion (blue color). The thermometer now registers 90.2°C, because the hydration process is exothermic.

heat is evolved ($q$ is negative); see Figure 6.6. An **endothermic process** is *a chemical reaction or a physical change in which heat is absorbed* ($q$ is positive). Experimentally, you note that in the exothermic reaction, the reaction flask initially warms; in the endothermic reaction, the reaction flask initially cools. We can summarize as follows:

| Type of Reaction | Experimental Effect Noted | Result on System | Sign of $q$ |
|---|---|---|---|
| Endothermic | Reaction vessel cools (heat is absorbed) | Energy added | + |
| Exothermic | Reaction vessel warms (heat is evolved) | Energy subtracted | − |

Suppose that in an experiment 1 mol of methane burns in oxygen and evolves 890 kJ of heat: $CH_4(g) + 2O_2(g) \longrightarrow CO_2(g) + 2H_2O(l)$. The reaction is exothermic. Therefore the heat of reaction, $q$, is −890 kJ.

Or consider the reaction described in the chapter opening, in which crystals of barium hydroxide octahydrate, $Ba(OH)_2 \cdot 8H_2O$, react with crystals of ammonium nitrate, $NH_4NO_3$.

$$Ba(OH)_2 \cdot 8H_2O(s) + 2NH_4NO_3(s) \longrightarrow 2NH_3(g) + 10H_2O(l) + Ba(NO_3)_2(aq)$$

When 1 mol $Ba(OH)_2 \cdot 8H_2O$ reacts with 2 mol $NH_4NO_3$, the reaction mixture absorbs 170.4 kJ of heat. The reaction is endothermic. Therefore the heat of reaction, $q$, is +170.4 kJ.

▶ Video no. 12 (Series B), "Potassium Permanganate and Glycerine; The Thermite Reaction (exothermic reactions)."

▶ Video Series C, "Thermite Reaction."

▶ Video no. 9 (Series D), "Sterno and Non-Burning Money"; no. 10 (Series D), "Tesla Coil Explosion"; no. 11 (Series D), "Chromic Acid"; no. 12 (Series D), "Magnesium and Dry Ice"; no. 13 (Series D), "Chemical Cold Pack"; and no. 14 (Series D), "Thermite."

**EXERCISE 6.2**

Ammonia burns in the presence of a platinum catalyst to give nitric oxide, NO.

$$4NH_3(g) + 5O_2(g) \xrightarrow{\text{Pt}} 4NO(g) + 6H_2O(l)$$

In an experiment, 4 mol $NH_3$ is burned and evolves 1170 kJ of heat. Is the reaction endothermic or exothermic? What is the value of $q$?

See Problems 6.39, 6.40, 6.41, and 6.42.

# 6.3   Enthalpy and Enthalpy Change

The heat absorbed or evolved by a reaction depends on the conditions under which the reaction occurs. Usually, a reaction takes place in a vessel open to the atmosphere, and therefore at the constant pressure of the atmosphere. We will assume that this is the case and write the heat of reaction as $q_p$, the subscript $p$ indicating that the process occurs at constant pressure.

## Enthalpy

There is a property of substances called enthalpy (en′-thal-py) that is related to the heat of reaction $q_p$. **Enthalpy** (denoted $H$) is *an extensive property of a substance that can be used to obtain the heat absorbed or evolved in a chemical reaction.* (An extensive property is a property that depends on the amount of substance. Other examples of extensive properties are mass and volume.)

Enthalpy is a state function. A **state function** is *a property of a system that depends only on its present state, which is determined by variables such as temperature and pressure, and is independent of any previous history of the system.* This means that a change in enthalpy does not depend on how the change was made, but only on the initial state and final state of the system. An analogy may clarify the point. Suppose you are hiking in mountainous terrain. You start walking from campsite A, which is at an altitude of 1200 ft above sea level according to your pocket altimeter, and you take the curving, graded path upward to campsite B (Figure 6.7). Alternatively, you could have

**Figure 6.7**
**An analogy to illustrate a state function.**   The two campsites differ by 3600 ft in altitude. This difference in altitude is independent of the path taken between the campsites. The distance traveled and the things encountered along the way, however, do depend on the path taken. Altitude here is analogous to a thermodynamic state function.

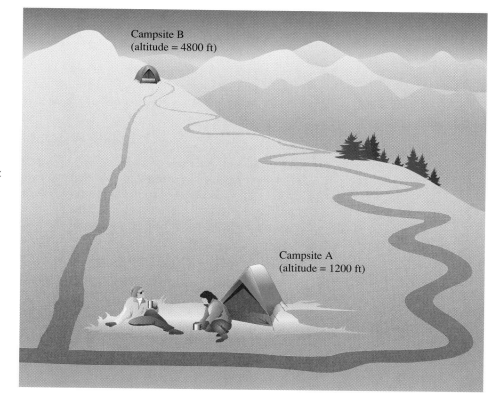

Campsite B
(altitude = 4800 ft)

Campsite A
(altitude = 1200 ft)

**Figure 6.8**
**An enthalpy diagram.**
When 2 mol Na(s) and 2 mol H$_2$O(l) react to give 2 mol NaOH(aq) and 1 mol H$_2$(g), 367.5 kJ of heat is released, and the enthalpy of the system decreases by 367.5 kJ.

taken a more direct, but more difficult, route to campsite B. The distances traveled are different; the distance by the graded path is longer than by the more direct route. However, the altitude (height above sea level) of campsite B is independent of how you got there (independent of any history). If you find the altitude of campsite B to be 4800 ft after arriving by the graded path, the altitude will be 4800 ft by any other path (assuming your altimeter is accurate, of course). Moreover, the difference in altitude of the two campsites is independent of the route you take from one to the other. Altitude is analogous to a state function, whereas distance traveled is not.

## Enthalpy of Reaction

Consider a chemical reaction system. At first, the enthalpy of the system is that of the reactants. But as the reaction proceeds, the enthalpy changes and finally becomes equal to that of the products. *The change in enthalpy for a reaction at a given temperature and pressure* (called the **enthalpy of reaction**) is obtained by subtracting the enthalpy of the reactants from the enthalpy of the products. We will use the symbol $\Delta$ (meaning "change in") and write the change in enthalpy as $\Delta H$. You apply the $\Delta$ notation by taking the final value and subtracting the initial value. Thus $\Delta H = H_{fiina} - H_{initial}$. Since you start from reactants and end with products, the enthalpy of reaction is

$$\Delta H = H(\text{products}) - H(\text{reactants})$$

Because $H$ is a state function, the value of $\Delta H$ is independent of the details of the reaction. It depends only on the initial state (the reactants) and the final state (the products).

The key relation in this chapter is that between enthalpy change and heat of reaction:

$$\Delta H = q_p$$

*The enthalpy of reaction equals the heat of reaction at constant pressure.*

To illustrate the concepts we have introduced, consider the reaction at 25°C of sodium metal and water, carried out in a beaker open to the atmosphere at 1.00 atm pressure.

$$2Na(s) + 2H_2O(l) \longrightarrow 2NaOH(aq) + H_2(g)$$

The metal and water react vigorously and heat evolves. Experiment shows that 2 mol of sodium metal reacts with 2 mol of water to evolve 367.5 kJ of heat. Because heat evolves, the reaction is exothermic, and you write $q_p = -367.5$ kJ. Therefore, the enthalpy of reaction, or change of enthalpy for the reaction, is $\Delta H = -367.5$ kJ. Figure 6.8 shows an *enthalpy diagram* for this reaction.

## Enthalpy and Internal Energy

In the preceding discussion, we noted that the enthalpy change equals the heat of reaction at constant pressure. This will be sufficient for the purpose of this chapter, which is to introduce the concepts of heat of reaction and enthalpy change. Later we will look at thermodynamics in more detail. Still, it is useful at this point to note briefly the relationship of enthalpy to internal energy.

The enthalpy, $H$, is defined precisely as the internal energy, $U$, plus pressure, $P$, times volume, $V$.

$$H = U + PV$$

To obtain some understanding of this equation, consider a reaction system at constant pressure $P$. We will label the initial quantities (those for reactants) with a subscript $i$ and the final quantities (those for products) with a subscript $f$. Then

$$\Delta H = H_f - H_i = (U_f + PV_f) - (U_i + PV_i)$$

Collecting the internal-energy terms and the pressure-volume terms, you can rewrite this as

$$\Delta H = (U_f - U_i) + P(V_f - V_i) = \Delta U + P\Delta V$$

You write $\Delta U$ for $U_f - U_i$ and $\Delta V$ for $V_f - V_i$ and rearrange this as follows:

$$\Delta U = \Delta H - P\Delta V$$

The last term in this equation $(-P\Delta V)$ is the energy required by the system to change volume against the constant pressure of the atmosphere. (The minus sign means that energy is required to increase the volume of the system.) This required energy is called the *pressure–volume work*. Thus, the equation says that the internal energy of the system changes in two ways. It changes because energy leaves or enters the system as heat $(\Delta H)$, and it changes because the system increases or decreases in volume against the constant pressure of the atmosphere (which requires energy $-P\Delta V$).

Consider a specific reaction. When 2 mol of sodium metal and 2 mol of water react in a beaker, 1 mol of hydrogen gas forms and heat evolves. If you measured this heat, you would find it to be $-367.5$ kJ, indicating that 367.5 kJ of energy has left the system in the form of heat.

Because hydrogen gas forms during the reaction, the volume of the system increases. To expand, the system must push back the atmosphere, and this requires energy equal to the pressure–volume work. It may be easier to see this pressure–volume work if you replace the constant pressure of the atmosphere by an equivalent pressure from a piston-and-weight assembly, as in Figure 6.9. When hydrogen gas is released during the reaction, it pushes upward on the piston and raises the weight. It requires energy to lift a weight upward in a gravitational field. If you calculate this pressure–volume work at 25°C and 1.00 atm pressure, you find that it is $-P\Delta V = -2.5$ kJ.

In the sodium–water reaction, the internal energy changes by $-367.5$ kJ because heat evolves and changes by $-2.5$ kJ because pressure–volume work is done. The total change of internal energy is

$$\Delta U = \Delta H - P\Delta V = -367.5 \text{ kJ} - 2.5 \text{ kJ} = -370.0 \text{ kJ}$$

As you can see, $\Delta U$ does not differ a great deal from $\Delta H$. This is the case in most reactions, so that the heat of reaction at constant pressure is approximately equal to the change of internal energy.

**Figure 6.9**
**Pressure–volume work.**
In this experiment, we replace the pressure of the atmosphere by a piston-and-weight assembly of equal pressure. As sodium metal reacts with water, the hydrogen gas evolved pushes the piston and weight upward (compare *before* and *after*). It requires work to raise the piston and weight upward in a gravitational field.

## 6.4 Thermochemical Equations

We will often find it convenient to write the enthalpy of reaction, $\Delta H$, with the chemical equation. A **thermochemical equation** is *the chemical equation for a reaction (including phase labels) in which the equation is given a molar interpretation, and the enthalpy of reaction for these molar amounts is written directly after the equation.* For the reaction of sodium and water, you would write

$$2Na(s) + 2H_2O(l) \longrightarrow 2NaOH(aq) + H_2(g); \Delta H = -367.5 \text{ kJ}$$

This equation says that 2 mol of sodium reacts with 2 mol of water to produce 2 mol of sodium hydroxide and 1 mol of hydrogen gas, and 367.5 kJ of heat evolves.

Note that we specify that the thermochemical equation includes phase labels. This is important because the enthalpy change, $\Delta H$, depends on the phase of the substances. Consider the reaction of hydrogen and oxygen to produce water. If the product is water vapor, 2 mol of $H_2$ burn to release 483.7 kJ of heat.

$$2H_2(g) + O_2(g) \longrightarrow 2H_2O(g); \Delta H = -483.7 \text{ kJ}$$

On the other hand, if the product is liquid water, the heat released is 571.7 kJ.

$$2H_2(g) + O_2(g) \longrightarrow 2H_2O(l); \Delta H = -571.7 \text{ kJ}$$

▶ It takes 44.0 kJ of heat to vaporize 1 mol of liquid water at 25°C.

In this case, additional heat is released when water vapor condenses to liquid. ◀

▶ An alternate example is provided in the Instructor's Resource Manual.

### EXAMPLE 6.2

### Writing Thermochemical Equations

Aqueous sodium hydrogen carbonate solution (baking soda solution) reacts with hydrochloric acid to produce aqueous sodium chloride, water, and carbon dioxide gas. The reaction absorbs 11.8 kJ of heat at constant pressure for each mole of sodium hydrogen carbonate. Write the thermochemical equation for the reaction.

### SOLUTION

You first write the balanced chemical equation.

$$NaHCO_3(aq) + HCl(aq) \longrightarrow NaCl(aq) + H_2O(l) + CO_2(g)$$

Because the reaction absorbs heat, the enthalpy of reaction for molar amounts of this equation is $+11.8$ kJ. The thermochemical equation is

$$\textbf{NaHCO}_3\textbf{(aq) + HCl(aq)} \longrightarrow \textbf{NaCl(aq) + H}_2\textbf{O(l) + CO}_2\textbf{(g); } \boldsymbol{\Delta H = +11.8 \text{ kJ}}$$

### EXERCISE 6.3

A propellant for rockets is obtained by mixing the liquids hydrazine, $N_2H_4$, and dinitrogen tetroxide, $N_2O_4$. These compounds react to give gaseous nitrogen, $N_2$, and water vapor, evolving 1049 kJ of heat at constant pressure when 1 mol $N_2O_4$ reacts. Write the thermochemical equation for this reaction.

**See Problems 6.43 and 6.44.**

The following are two important rules for manipulating thermochemical equations:

1. When a thermochemical equation is multiplied by any factor, the value of $\Delta H$ for the new equation is obtained by multiplying the value of $\Delta H$ in the original equation by that same factor.

2. When a chemical equation is reversed, the value of $\Delta H$ is reversed in sign.

Consider the thermochemical equation for the synthesis of ammonia.

$$N_2(g) + 3H_2(g) \longrightarrow 2NH_3(g); \Delta H = -91.8 \text{ kJ}$$

Suppose you want the thermochemical equation to show what happens when twice as many moles of nitrogen and hydrogen react to produce ammonia. Because double the amounts of substances are present, the enthalpy of reaction is doubled (enthalpy is an extensive quantity). Doubling the previous equation, you obtain

$$2N_2(g) + 6H_2(g) \longrightarrow 4NH_3(g); \Delta H = -184 \text{ kJ}$$

Suppose you reverse the first equation we wrote for the synthesis of ammonia. Then the reaction is the dissociation of 2 mol of ammonia into its elements. The thermochemical equation is

$$2NH_3(g) \longrightarrow N_2(g) + 3H_2(g); \Delta H = +91.8 \text{ kJ}$$

► **If you use the molar interpretation of a chemical equation, there is nothing unreasonable about using such coefficients as $\frac{1}{2}$ and $\frac{3}{2}$.**

If you want to express this in terms of 1 mol of ammonia, you simply multiply this equation by a factor of $\frac{1}{2}$. ◄

The next example further illustrates the use of these rules of thermochemical equations.

► **An alternate example is provided in the Instructor's Resource Manual.**

### EXAMPLE 6.3

## Manipulating Thermochemical Equations

When 2 mol $H_2(g)$ and 1 mol $O_2(g)$ react to give liquid water, 572 kJ of heat evolves.

$$2H_2(g) + O_2(g) \longrightarrow 2H_2O(l); \Delta H = -572 \text{ kJ}$$

Write this equation for 1 mol of liquid water. Give the reverse equation, in which 1 mol of liquid water dissociates into hydrogen and oxygen.

### SOLUTION

You multiply the coefficients and $\Delta H$ by $\frac{1}{2}$:

$$H_2(g) + \tfrac{1}{2}O_2(g) \longrightarrow H_2O(l); \Delta H = -286 \text{ kJ}$$

Reversing the equation, you get

$$H_2O(l) \longrightarrow H_2(g) + \tfrac{1}{2}O_2(g); \Delta H = +286 \text{ kJ}$$

### EXERCISE 6.4

a. Write the thermochemical equation for the reaction described in Exercise 6.3 for the case involving 1 mol $N_2H_4$. b. Write the thermochemical equation for the reverse of the reaction described in Exercise 6.3.

**See Problems 6.45, 6.46, 6.47, and 6.48.**

# Lucifers and Other Matches

**EVERYDAY LIFE**

Samuel Jones, an Englishman, patented one of the first kinds of matches in 1828. It consisted of a glass bead containing sulfuric acid surrounded by a coating of sugar with an oxidizing agent. You ignited the match by breaking the bead using a pair of pliers or, if you were more daring, your teeth. This action released the acid, which ignited an exothermic reaction in the surrounding combustible materials.

Later Jones began to market a friction match discovered, but not patented, by John Walker. Walker, who had been experimenting with explosives, discovered this match one day when he tried to remove a small glob of a dried mixture from a stick. He rubbed the stick on the floor and was surprised when it burst into flame. Jones called his matches "Lucifers." They were well named; when lighted, they gave off a shower of sparks and smoky fumes with the acrid odor of sulfur dioxide. Jones had every box inscribed with the warning "Persons whose lungs are delicate should by no means use Lucifers."

A few years later, a Frenchman, Charles Sauria, invented the white phosphorus match, which became an immediate success. When rubbed on a rough surface, the match lighted easily, without hazardous sparks, and smelled better than Lucifers. The match head contained white phosphorus, an oxidizing agent, and glue. White phosphorus is a yellowish-white, waxy substance, often sold in the form of sticks looking something like fat crayons. Unlike crayons, though, white phosphorus ignites spontaneously in air (so it is generally stored under water). The glue in the match mixture had two purposes: it protected the white phosphorus from air, and it held the match mixture firmly together. The white phosphorus match had one serious drawback. White phosphorus is quite poisonous. Workers in match factories often began to show the agonizing symptoms of "phossy jaw," from white phosphorus poisoning, in which the jawbone disintegrates. The manufacture of white phosphorus matches was outlawed in the early 1900s.

The head of the "strike-anywhere" match, which you can buy today at any grocery store, contains the relatively nontoxic tetraphosphorus trisulfide, $P_4S_3$, and potassium chlorate, $KClO_3$ (Figure 6.10). By rubbing the match head against a surface, you create enough heat by friction to ignite the match material. Tetraphosphorus trisulfide then burns in air in a very exothermic reaction.

$$P_4S_3(s) + 8O_2(g) \longrightarrow P_4O_{10}(s) + 3SO_2(g);$$
$$\Delta H = -3677 \text{ kJ}$$

Safety matches have a head containing mostly an oxidizing agent and require a striking surface containing nonpoisonous red phosphorus.

**Figure 6.10**
**Strike-anywhere matches.** *Left:* The head of the match contains tetraphosphorus trisulfide and potassium chlorate. *Right:* The substances in the match head react when frictional heat ignites the mixture.

**CONCEPT CHECK 6.2**

Natural gas consists primarily of methane, $CH_4$. It is used in a process called *steam reforming* to prepare a gaseous mixture of carbon monoxide and hydrogen for industrial use.

$$CH_4(g) + H_2O(g) \longrightarrow CO(g) + 3H_2(g); \Delta H = 206 \text{ kJ}$$

The reverse reaction, the reaction of carbon monoxide and hydrogen, has been explored as a way to prepare methane (synthetic natural gas). Which of the following are exothermic? Of these, which one is the most exothermic?

a. $CH_4(g) + H_2O(g) \longrightarrow CO(g) + 3H_2(g)$

b. $2CH_4(g) + 2H_2O(g) \longrightarrow 2CO(g) + 6H_2(g)$

c. $CO(g) + 3H_2(g) \longrightarrow CH_4(g) + H_2O(g)$

d. $2CO(g) + 6H_2(g) \longrightarrow 2CH_4(g) + 2H_2O(g)$

# 6.5  Applying Stoichiometry to Heats of Reaction

As you might expect, the quantity of heat obtained from a reaction will depend on the amount of reactants. We can extend the method used to solve stoichiometry problems, described in Chapter 3, to problems involving the quantity of heat.

Consider the reaction of methane, $CH_4$ (the principal constituent of natural gas), burning in oxygen at constant pressure. How much heat could you obtain from 10.0 g of methane, assuming you have an excess of oxygen? You can answer this question if you know the enthalpy change for the reaction of 1 mol of methane. The thermochemical equation is

$$CH_4(g) + 2O_2(g) \longrightarrow CO_2(g) + 2H_2O(l); \Delta H = -890.3 \text{ kJ}$$

The calculation involves the following conversions:

> ▶ If you wanted the result in kilocalories, you could convert this answer as follows:
>
> $556 \text{ kJ} \times \dfrac{1 \text{ kcal}}{4.184 \text{ kJ}} = 133 \text{ kcal}$

Grams of $CH_4$ $\longrightarrow$ moles of $CH_4$ $\longrightarrow$ kilojoules of heat

$$10.0 \text{ g } CH_4 \times \frac{1 \text{ mol } CH_4}{16.0 \text{ g } CH_4} \times \frac{-890.3 \text{ kJ}}{1 \text{ mol } CH_4} = -556 \text{ kJ}$$

That is, 10.0 g of methane burns in excess oxygen to evolve 556 kJ of heat. ◀

### EXAMPLE 6.4

## Calculating the Heat of Reaction from the Stoichiometry

> ▶ Two alternate examples are provided in the Instructor's Resource Manual.

How much heat is evolved when 907 kg of ammonia is produced according to the following equation? (Assume that the reaction occurs at constant pressure.)

$$N_2(g) + 3H_2(g) \longrightarrow 2NH_3(g); \Delta H = -91.8 \text{ kJ}$$

 **PROBLEM STRATEGY**

The calculation involves converting grams of $NH_3$ to moles of $NH_3$ and then to kilojoules of heat.

Grams of $NH_3$ $\longrightarrow$ moles of $NH_3$ $\longrightarrow$ kilojoules of heat

You obtain the conversion factor for the second step from the thermochemical equation, which says that the production of 2 mol $NH_3$ is accompanied by $q_p = -91.8$ kJ.

**SOLUTION**

Note that 907 kg equals $9.07 \times 10^5$ g. Hence,

$$9.07 \times 10^5 \text{ g } NH_3 \times \frac{1 \text{ mol } NH_3}{17.0 \text{ g } NH_3} \times \frac{-91.8 \text{ kJ}}{2 \text{ mol } NH_3} = -2.45 \times 10^6 \text{ kJ}$$

Thus, **$2.45 \times 10^6$ kJ of heat evolves.**

---

**EXERCISE 6.5**

How much heat evolves when 10.0 g of hydrazine reacts according to the reaction described in Exercise 6.3?

---

See Problems 6.49, 6.50, 6.51, and 6.52.

## 6.6 Measuring Heats of Reaction

So far we have introduced the concept of heat of reaction and have related it to the enthalpy change. We also showed how to display the enthalpy change in a thermochemical equation, and from this equation how to calculate the heat of reaction for any given amount of substance. Now that you have a firm idea of what heats of reaction are, how would you measure them?

First, we need to look at the heat required to raise the temperature of a substance, because a thermochemical measurement is based on the relationship between heat and temperature change. The heat required to raise the temperature of a substance is called its *heat capacity.*

### Heat Capacity and Specific Heat

Heat is required to raise the temperature of a given amount of substance, and the quantity of heat depends on the temperature change. The **heat capacity** (*C*) of a sample of substance is *the quantity of heat needed to raise the temperature of the sample of substance one degree Celsius (or one kelvin).* Changing the temperature of the sample from an initial temperature $t_i$ to a final temperature $t_f$ requires heat equal to

$$q = C\Delta t$$

▶ The heat capacity will depend on whether the process is constant-pressure or constant-volume. We will assume a constant-pressure process unless otherwise stated.

where $\Delta t$ is the change of temperature and equals $t_f - t_i$. ◀

Suppose a piece of iron requires 6.70 J of heat to raise the temperature by one degree Celsius. Its heat capacity is therefore 6.70 J/°C. The quantity of heat required to raise the temperature of the piece of iron from 25.0°C to 35.0°C is

$$q = C\Delta t = (6.70 \text{ J/°C}) \times (35.0°C - 25.0°C) = 67.0 \text{ J}$$

**Table 6.1**
**Specific Heats and Molar Heat Capacities of Some Substances***

| Substance | Specific Heat J/(g · °C) | Molar Heat Capacity J/(mol · °C) |
|---|---|---|
| Aluminum, Al | 0.901 | 24.3 |
| Copper, Cu | 0.384 | 24.4 |
| Ethanol, $C_2H_5OH$ | 2.43 | 112.2 |
| Iron, Fe | 0.449 | 25.1 |
| Water, $H_2O$ | 4.18 | 75.3 |

*Values are for 25°C.

Heat capacity is directly proportional to the amount of substance. Often heat capacities are listed for molar amounts of substances. The *molar heat capacity* of a substance is its heat capacity for one mole of substance.

Heat capacities are also compared for one-gram amounts of substances. The **specific heat capacity** (or simply **specific heat**) is *the quantity of heat required to raise the temperature of one gram of a substance by one degree Celsius (or one kelvin) at constant pressure.* To find the heat $q$ required to raise the temperature of a sample, you multiply the specific heat of the substance, $s$, by the mass in grams, $m$, and the temperature change, $\Delta t$.

$$q = s \times m \times \Delta t$$

The specific heats and molar heat capacities of a few substances are listed in Table 6.1. Values depend somewhat on temperature, and those listed are for 25°C. Water has a specific heat of 4.18 J/(g · °C)—that is, 4.18 joules per gram per degree Celsius. In terms of calories, the specific heat of water is 1.00 cal/(g · °C). Example 6.5 illustrates the use of the preceding equation.

▶ An alternate example is provided in the Instructor's Resource Manual.

**EXAMPLE 6.5**

## Relating Heat and Specific Heat

Calculate the heat absorbed by 15.0 g of water to raise its temperature from 20.0°C to 50.0°C (at constant pressure). The specific heat of water is 4.18 J/(g · °C).

**SOLUTION**

You substitute into the equation

$$q = s \times m \times \Delta t$$

The temperature change is

$$\Delta t = t_f - t_i = 50.0°C - 20.0°C = +30.0°C$$

Therefore,

$$q = 4.18 \text{ J/(g · °C)} \times 15.0 \text{ g} \times (+30.0°C) = \mathbf{1.88 \times 10^3 \text{ J}}$$

---

**EXERCISE 6.6**

Iron metal has a specific heat of 0.449 J/(g · °C). How much heat is transferred to a 5.00-g piece of iron, initially at 20.0°C, when it is placed in a pot of boiling water? Assume that the temperature of the water is 100.0°C and that the water remains at this temperature, which is the final temperature of the iron.

**See Problems 6.55 and 6.56.**

---

## Measurement of Heat of Reaction

You measure the heat of reaction in a **calorimeter,** *a device used to measure the heat absorbed or evolved during a physical or chemical change.* The device can be as simple as the apparatus sketched in Figure 6.11, which consists of an insulated container (for example, a pair of polystyrene coffee cups) with a thermometer. More elaborate calorimeters are employed when precise measurements are needed for research, although the basic idea remains the same—to measure temperature changes under controlled circumstances and relate these temperature changes to heat.

The coffee-cup calorimeter shown in Figure 6.11 is a constant-pressure calorimeter. The heat of the reaction is calculated from the temperature change caused by the

**Figure 6.11**
**A simple coffee-cup calorimeter.** This calorimeter is made of two nested polystyrene coffee cups. The outer cup helps to insulate the reaction mixture from its surroundings. After the reactants are added to the inner cup, the calorimeter is covered to reduce heat loss by evaporation and convection. The heat of reaction is determined by noting the temperature rise or fall.

Thermometer

HCl (*aq*)

2 polystyrene coffee cups

NaOH (*aq*)

**Figure 6.12**
**A bomb calorimeter.** The heat of a reaction involving a gas (here the heat of combustion of graphite) is conveniently determined in a sealed vessel called a bomb. The reaction is started by an ignition coil running through the graphite sample.

reaction, and since this is a constant-pressure process, the heat can be directly related to the enthalpy change, $\Delta H$. Research versions of a constant-pressure calorimeter are available, and these are used when gases are not involved.

For reactions involving gases, a *bomb calorimeter* is generally used (Figure 6.12). Consider the heat of combustion of graphite, the form of carbon used in the "lead" of a pencil. To measure the heat released when graphite burns in oxygen, a sample of graphite is placed in a small cup in the calorimeter. The graphite is surrounded by oxygen, and the graphite and oxygen are sealed in a steel vessel, or bomb. An electrical circuit is activated to start the burning of the graphite. The bomb is surrounded by water in an insulated container, and the heat of reaction is calculated from the temperature change of the calorimeter caused by the reaction.

Because the reaction in a bomb calorimeter occurs in a closed vessel, the pressure does not generally remain constant. Rather the volume remains constant, and under these conditions the heat of reaction does not in general equal $\Delta H$; a small correction is usually needed. However, this correction is negligible when the reaction does not involve gases or when the number of moles of reactant gas equals the number of moles of product gas, as in the combustion of graphite to carbon dioxide. ◄

The next example describes the calculations needed to obtain $\Delta H$ from calorimetric measurements.

▶ At the end of Section 6.3, we noted that $\Delta H = \Delta U + P\Delta V$. The heat at constant volume equals $\Delta U$. To obtain $\Delta H$, you must add a correction, $P\Delta V$. The change in volume is significant only if there are changes in gas volumes. From the ideal gas law, the molar volume of a gas is $RT/P$. If the moles of gas that have reacted is $n_i$ and the moles of gas produced is $n_f$, then $P\Delta V$ equals essentially $RT \times (n_f - n_i)$, or $RT\Delta n$, where $\Delta n$ refers to the change of moles of gas. To obtain the correction in joules, we use $R = 8.31$ J/(K · mol). Thus, $RT$ at 25°C (298 K) is 8.31 J/(K · mol) $\times$ 298 K = $2.48 \times 10^3$ J/mol. The final result is $\Delta H = \Delta U + (2.48$ kJ$) \times \Delta n$.

▶ An alternate example is provided in the Instructor's Resource Manual.

---

**EXAMPLE 6.6**

### Calculating $\Delta H$ from Calorimetric Data

Suppose 0.562 g of graphite is placed in a calorimeter with an excess of oxygen at 25.00°C and 1 atm pressure (Figure 6.12). Excess $O_2$ ensures that all carbon burns to form $CO_2$. The graphite is ignited, and it burns according to the equation

$$C(graphite) + O_2(g) \longrightarrow CO_2(g)$$

On reaction, the calorimeter temperature rises from 25.00°C to 25.89°C. The heat capacity of the calorimeter and its contents was determined in a separate experiment to be 20.7 kJ/°C. What is the heat of reaction at 25.00°C and 1 atm pressure? Express the answer as a thermochemical equation.

### PROBLEM STRATEGY

**STEP 1**    You obtain the heat released by the graphite sample when it burns in the calorimeter. You reason as follows. The heat released by the reaction is absorbed by the calorimeter and its contents. Let $q_{rxn}$ be the quantity of heat from the reaction mixture, and let $C_{cal}$ be the heat capacity of the calorimeter and contents. The quantity of heat absorbed by the calorimeter is $C_{cal}\Delta t$. This will have the same magnitude as $q_{rxn}$, but the opposite sign: $q_{rxn} = -C_{cal}\Delta t$. Substituting values of $C_{cal}$ and $\Delta t$, you obtain the heat $q_{rxn}$ from the sample of C (0.562 g). The factor $(q_{rxn}/0.562 \text{ g C})$ converts

$$\text{Grams C} \longrightarrow \text{kJ heat}$$

**STEP 2**    The $\Delta H$ of reaction is the heat obtained from 1 mol C. Imagine you start with this amount of C and convert (mathematically) to grams C, then to kJ heat ($\Delta H$ for the reaction). The conversion steps are:

$$\text{Moles C} \longrightarrow \text{grams C} \longrightarrow \text{kJ heat } (\Delta H)$$

### SOLUTION

**STEP 1**    The heat from the graphite sample is

$$q_{rxn} = -C_{cal}\Delta t = -20.7 \text{ kJ/°C} \times (25.89°C - 25.00°C)$$
$$= -20.7 \text{ kJ/°C} \times 0.89°C = -18.4 \text{ kJ}$$

An extra figure is retained in $q_{rxn}$ for further computation. The negative sign indicates the reaction is exothermic, as expected for a combustion. The factor to convert grams C to kJ heat is $-18.4 \text{ kJ}/0.562 \text{ g C}$.

**STEP 2**    The conversion of 1 mol C to kJ heat for 1 mol ($\Delta H$) is

$$1 \text{ mol C} \times \frac{12 \text{ g C}}{1 \text{ mol C}} \times \frac{-18.4 \text{ kJ}}{0.562 \text{ g C}} = -3.9 \times 10^2 \text{ kJ}$$

(The final answer has been rounded to two significant figures.) When 1 mol of carbon burns, $3.9 \times 10^2$ kJ of heat is released. You can summarize the results by the thermochemical equation:

$$\textbf{C(graphite)} + \textbf{O}_2\textbf{(g)} \longrightarrow \textbf{CO}_2\textbf{(g); } \Delta H = -3.9 \times 10^2 \textbf{ kJ}$$

### EXERCISE 6.7

Suppose 33 mL of 1.20 $M$ HCl is added to 42 mL of a solution containing excess sodium hydroxide, NaOH, in a coffee-cup calorimeter. The solution temperature, originally 25.0°C, rises to 31.8°C. Give the enthalpy change, $\Delta H$, for the reaction

$$\text{HCl}(aq) + \text{NaOH}(aq) \longrightarrow \text{NaCl}(aq) + \text{H}_2\text{O}(l)$$

Express the answer as a thermochemical equation. For simplicity, assume that the heat capacity and the density of the final solution in the cup are those of water. (In more accurate work, these values must be determined.) Also assume that the total volume of the solution equals the sum of the volumes of HCl(*aq*) and NaOH(*aq*).

See Problems 6.59, 6.60, 6.61, and 6.62.

# Using Heats of Reaction

In the first part of this chapter, we looked at the basic properties of the heat of reaction and how to measure it. Now we want to find how heats of reaction can be used. We will see that the $\Delta H$ for one reaction can be obtained from the $\Delta H$'s of other reactions. This means that we can tabulate a small number of values and use them to calculate others.

## 6.7 Hess's Law

▶ Experiments in General Chemistry, 6, "Thermochemistry and Hess's law." 

Enthalpy, you may recall, is a state function (Section 6.3). This means that the enthalpy change for a chemical reaction is independent of the path by which the products are obtained. In 1840, the Russian chemist Germain Henri Hess, a professor at the University of St. Petersburg, discovered this result by experiment. **Hess's law of heat summation** states that *for a chemical equation that can be written as the sum of two or more steps, the enthalpy change for the overall equation equals the sum of the enthalpy changes for the individual steps.* In other words, no matter how you go from given reactants to products (whether in one step or several), the enthalpy change for the overall chemical change is the same.

To understand Hess's law fully and to see how you can use it, consider a simple example. Suppose you would like to find the enthalpy change for the combustion of graphite (carbon) to carbon monoxide.

$$2C(\text{graphite}) + O_2(g) \longrightarrow 2CO(g)$$

The direct determination of this enthalpy change is very difficult, because once carbon monoxide forms it reacts further with oxygen to yield carbon dioxide. If you do the experiment in an excess of oxygen, you obtain only carbon dioxide, and the enthalpy change is the heat of complete combustion of graphite. On the other hand, if you do the experiment in a limited quantity of oxygen, you obtain a mixture of carbon monoxide and carbon dioxide, and the heat of reaction is a value appropriate for a mixture of these products. How can you obtain the enthalpy change for the preparation of pure carbon monoxide from graphite and oxygen?

The answer is to apply Hess's law. To do this, imagine that the combustion of graphite to carbon monoxide takes place in two separate steps:

$$2C(\text{graphite}) + 2O_2(g) \longrightarrow 2CO_2(g) \qquad \text{(first step)}$$
$$2CO_2(g) \longrightarrow 2CO(g) + O_2(g) \qquad \text{(second step)}$$

In the first step, you burn 2 mol of graphite in 2 mol of oxygen to produce 2 mol of carbon dioxide. In the second step, you decompose this carbon dioxide to give 2 mol of carbon monoxide and 1 mol of oxygen. The net result is the combustion of 2 mol of graphite in 1 mol of oxygen to give 2 mol of carbon monoxide. You can obtain this result by adding the two steps, canceling out 2 mol $CO_2$ and 1 mol $O_2$ on both sides of the equation.

$$2C(graphite) + \cancel{2}O_2(g) \longrightarrow \cancel{2CO_2(g)}$$
$$\underline{\cancel{2CO_2(g)}} \qquad\qquad \longrightarrow 2CO(g) + \cancel{O_2(g)}$$
$$2C(graphite) + O_2(g) \longrightarrow 2CO(g)$$

According to Hess's law, the enthalpy change for the overall equation (which is the equation you want) equals the sum of the enthalpy changes for the two steps. Now you need to determine the enthalpy changes for the separate steps.

You can determine the enthalpy change for the first step by simply burning graphite in an excess of oxygen, as described in Example 6.6. The result is $\Delta H = -393.5$ kJ per mole of $CO_2$ formed. For 2 mol $CO_2$, you multiply by 2.

$$2C(graphite) + 2O_2(g) \longrightarrow 2CO_2(g); \Delta H = (-393.5 \text{ kJ}) \times (2)$$

The second step, the decomposition of carbon dioxide, is not an easy experiment. However, the reverse of this decomposition is simply the combustion of carbon monoxide. You could determine the $\Delta H$ for that combustion by burning carbon monoxide in an excess of oxygen. The experiment is similar to the one for the combustion of graphite to carbon dioxide.

$$2CO(g) + O_2(g) \longrightarrow 2CO_2(g); \Delta H = -566.0 \text{ kJ}$$

From the properties of thermochemical equations (Section 6.4), you know that the enthalpy change for the reverse reaction is simply $(-1)$ times the original reaction.

$$2CO_2(g) \longrightarrow 2CO(g) + O_2(g); \Delta H = (-566.0 \text{ kJ}) \times (-1)$$

If you now add these two steps and add their enthalpy changes, you obtain the chemical equation and the enthalpy change for the combustion of carbon monoxide, which is what you wanted.

$$2C(graphite) + 2O_2(g) \longrightarrow 2CO_2(g) \qquad \Delta H_1 = (-393.5 \text{ kJ}) \times (2)$$
$$\underline{2CO_2(g) \qquad\qquad \longrightarrow 2CO(g) + O_2(g)} \qquad \underline{\Delta H_2 = (-566.0 \text{ kJ}) \times (-1)}$$
$$2C(graphite) + O_2(g) \longrightarrow 2CO(g) \qquad \Delta H_3 = -221.0 \text{ kJ}$$

You see that the combustion of 2 mol of graphite to give 2 mol of carbon monoxide has an enthalpy change of $-221.0$ kJ. Figure 6.13 gives an enthalpy diagram showing the relationship among the enthalpy changes for this calculation.

This calculation illustrates how you can use Hess's law to obtain the enthalpy change for a reaction that is difficult to determine by direct experiment. Hess's law is more generally useful, however, in that it allows you to calculate the enthalpy change for one reaction from the values for others, whatever their source. Suppose you are given the following data: ◄

▶ **We have followed common convention in writing S for the formula of sulfur in Equations 1 and 3, although the molecular formula is $S_8$. Doing this gives much simpler equations. You might try to do the calculations again, rewriting the equations using $S_8$ for sulfur. Remember to multiply the value of $\Delta H$ in each case by 8.**

$$S(s) + O_2(g) \longrightarrow SO_2(g); \Delta H = -297 \text{ kJ} \qquad\qquad \textbf{(1)}$$
$$2SO_3(g) \longrightarrow 2SO_2(g) + O_2(g); \Delta H = 198 \text{ kJ} \qquad\qquad \textbf{(2)}$$

How could you use these data to obtain the enthalpy change for the following equation?

$$2S(s) + 3O_2(g) \longrightarrow 2SO_3(g) \qquad\qquad \textbf{(3)}$$

You need to multiply Equations 1 and 2 by factors (perhaps reversing one or both equations) so that when you add them together you obtain Equation 3. You can usually guess what you need to do to the first two equations to obtain the third one. Note that

**Figure 6.13**
**Enthalpy diagram illustrating Hess's law.**
The diagram shows two different ways to go from graphite and oxygen (reactants) to carbon monoxide (products). Going by way of reactions 1 and 2 is equivalent to the direct reaction 3.

Reactants:
2 mol C (graphite) +
2 mol $O_2(g)$

③ $\Delta H_3 = -221.0$ kJ

Products:
2 mol CO($g$) + 1 mol $O_2(g)$

① $\Delta H_1 = -787.0$ kJ

② $\Delta H_2 = +566.0$ kJ

Enthalpy, $H$ (kJ)

2 mol $CO_2(g)$

Equation 3 has a coefficient of 2 for S($s$). This suggests that you should multiply Equation 1 by 2 (and multiply the $\Delta H$ by 2). Note also that $SO_3(g)$ in Equation 3 is on the right-hand side. This suggests that you should reverse Equation 2 (and multiply the $\Delta H$ by $-1$). Here is the calculation:

$$2S(s) + 2O_2(g) \longrightarrow 2\cancel{SO_2(g)} \qquad \Delta H = (-297 \text{ kJ}) \times (2)$$
$$\underline{2\cancel{SO_2(g)} + O_2(g) \longrightarrow 2SO_3(g) \qquad \Delta H = (198 \text{ kJ}) \times (-1)}$$
$$2S(s) + 3O_2(g) \longrightarrow 2SO_3(g) \qquad \Delta H = -792 \text{ kJ}$$

The next example gives another illustration of how Hess's law can be used to calculate the enthalpy change for a reaction from the enthalpy values for other reactions. In this case, the problem involves three equations from which we obtain a fourth. Although this problem is somewhat more complicated than the one we just did, the basic procedure is the same: Compare the coefficients in the equations. See what factors you need to multiply the equations whose $\Delta H$'s you know to obtain the equation you want. Check your results to be sure that the equations (and the $\Delta H$'s) add up properly.

▶ An alternate example is provided in the Instructor's Resource Manual.

---
**EXAMPLE 6.7**

## Applying Hess's Law

What is the enthalpy of reaction, $\Delta H$, for the formation of tungsten carbide, WC, from the elements? (Tungsten carbide is very hard and is used to make cutting tools and rock drills.)

$$W(s) + C(\text{graphite}) \longrightarrow WC(s)$$

The enthalpy change for this reaction is difficult to measure directly, because the reaction occurs at 1400°C. However, the heats of combustion of the elements and of tungsten carbide can be measured easily:

$$2W(s) + 3O_2(g) \longrightarrow 2WO_3(s); \Delta H = -1680.6 \text{ kJ} \qquad \textbf{(1)}$$
$$C(\text{graphite}) + O_2(g) \longrightarrow CO_2(g); \Delta H = -393.5 \text{ kJ} \qquad \textbf{(2)}$$
$$2WC(s) + 5O_2(g) \longrightarrow 2WO_3(s) + 2CO_2(g); \Delta H = -2391.6 \text{ kJ} \qquad \textbf{(3)}$$

**PROBLEM STRATEGY**

You need to multiply Equations 1, 2, and 3 by factors so that when you add the three equations you obtain the desired equation for the formation of WC(s). To obtain these factors, compare Equations 1, 2, and 3 in turn with the desired equation. For instance, note that Equation 1 has 2W(s) on the left side, whereas the desired equation has W(s). Therefore, you multiply Equation 1 (and its $\Delta H$) by $\frac{1}{2}$.

**SOLUTION**

Multiplying Equation 1 by $\frac{1}{2}$, you obtain

$$W(s) + \tfrac{3}{2}O_2(g) \longrightarrow WO_3(s); \Delta H = \tfrac{1}{2} \times (-1680.6 \text{ kJ}) = -840.3 \text{ kJ}$$

Compare Equation 2 with the desired equation. Both have C(graphite) on the left side; therefore, you leave Equation 2 as it is. Now, compare Equation 3 with the desired equation. Equation 3 has 2WC(s) on the left side, whereas the desired equation has WC(s) on the right side. Hence, you reverse Equation 3 and multiply it (and its $\Delta H$) by $\frac{1}{2}$.

$$WO_3(s) + CO_2(g) \longrightarrow WC(s) + \tfrac{5}{2}O_2(g); \Delta H = -\tfrac{1}{2} \times (-2391.6 \text{ kJ}) = 1195.8 \text{ kJ}$$

Note that the $\Delta H$ is obtained by multiplying the value for Equation 3 by $-\frac{1}{2}$. Now these three equations and the corresponding $\Delta H$'s are added together.

$$
\begin{array}{ll}
W(s) + \tfrac{3}{2}\cancel{O_2(g)} \longrightarrow \cancel{WO_3(s)} & \Delta H = -840.3 \text{ kJ} \\
C(\text{graphite}) + \cancel{O_2(g)} \longrightarrow \cancel{CO_2(g)} & \Delta H = -393.5 \text{ kJ} \\
\cancel{WO_3(s)} + \cancel{CO_2(g)} \longrightarrow WC(s) + \tfrac{5}{2}\cancel{O_2(g)} & \Delta H = 1195.8 \text{ kJ} \\
\hline
W(s) + C(\text{graphite}) \longrightarrow WC(s) & \mathbf{\Delta H = -38.0 \text{ kJ}}
\end{array}
$$

**Problem Solving: Hess's Law**

### EXERCISE 6.8

Manganese metal can be obtained by reaction of manganese dioxide with aluminum.

$$4Al(s) + 3MnO_2(s) \longrightarrow 2Al_2O_3(s) + 3Mn(s)$$

What is $\Delta H$ for this reaction? Use the following data:

$$2Al(s) + \tfrac{3}{2}O_2(g) \longrightarrow Al_2O_3(s); \Delta H = -1676 \text{ kJ}$$

$$Mn(s) + O_2(g) \longrightarrow MnO_2(s); \Delta H = -521 \text{ kJ}$$

See Problems 6.63, 6.64, 6.65, and 6.66.

 **CONCEPT CHECK 6.3**

The heat of fusion (also called heat of melting), $\Delta H_{fus}$, of ice is the enthalpy change for

$$H_2O(s) \longrightarrow H_2O(l); \Delta H_{fus}$$

Similarly, the heat of vaporization, $\Delta H_{vap}$, of liquid water is the enthalpy change for

$$H_2O(l) \longrightarrow H_2O(g); \Delta H_{vap}$$

How is the heat of sublimation, $\Delta H_{sub}$, the enthalpy change for the reaction

$$H_2O(s) \longrightarrow H_2O(g);\ \Delta H_{sub}$$

related to $\Delta H_{fus}$ and $\Delta H_{vap}$?

# 6.8   Standard Enthalpies of Formation

Because Hess's law relates the enthalpy changes of some reactions to the enthalpy changes of others, we only need to tabulate the enthalpy changes of certain types of reactions. We also generally list enthalpy changes only for certain standard thermodynamic conditions (which are not identical to the standard conditions for gases, STP).

The term **standard state** refers to *the standard thermodynamic conditions chosen for substances when listing or comparing thermodynamic data: 1 atm pressure and the specified temperature (usually 25°C).* ◄ These standard conditions are indicated by a superscript degree sign (°). The enthalpy change for a reaction in which reactants in their standard states yield products in their standard states is denoted $\Delta H°$ ("delta H degree," but often read as "delta H zero"). The quantity $\Delta H°$ is called the *standard enthalpy of reaction.*

► The International Union of Pure and Applied Chemistry (IUPAC) recommends that the standard pressure be 1 bar ($1 \times 10^5$ Pa). Thermodynamic tables are becoming available for 1 bar pressure, and in the future such tables will probably replace those for 1 atm.

As we will show, it is sufficient to tabulate just the enthalpy changes for formation reactions—that is, for reactions in which compounds are formed from their elements. To specify the formation reaction precisely, however, we must specify the exact form of each element.

Some elements exist in the same physical state (gas, liquid, or solid) in two or more distinct forms. For example, oxygen in any of the physical states occurs both as dioxygen (commonly called simply oxygen), with $O_2$ molecules, and as ozone, with $O_3$ molecules. Dioxygen gas is odorless; ozone gas has a characteristic pungent odor. Solid carbon has two principal crystalline forms: graphite and diamond. Graphite is a soft, black, crystalline substance; diamond is a hard, usually colorless crystal. The elements oxygen and carbon are said to exist in different allotropic forms. An **allotrope** is *one of two or more distinct forms of an element in the same physical state* (Figure 6.14).

► Although the reference form is usually the stablest allotrope of an element, the choice is essentially arbitrary as long as one is consistent.

The **reference form** of an element for the purpose of specifying the formation reaction is usually *the stablest form (physical state and allotrope) of the element under standard thermodynamic conditions.* The reference form of oxygen at 25°C is $O_2(g)$; the reference form of carbon at 25°C is graphite. ◄

**Figure 6.14**
**Allotropes of sulfur.**
*Left:* An evaporating dish contains rhombic sulfur, the stable form of the element at room temperature. *Right:* When this sulfur is melted, then cooled, it forms long needles of monoclinic sulfur, another allotrope. At room temperature, monoclinic sulfur will slowly change back to rhombic sulfur. Both forms contain the molecule $S_8$, depicted by the model.

**Table 6.2**
**Standard Enthalpies of Formation (at 25°C)***

| Formula | $\Delta H_f^\circ$ (kJ/mol) | Formula | $\Delta H_f^\circ$ (kJ/mol) |
|---|---|---|---|
| $e^-(g)$ | 0 | $C_2H_4(g)$ | 52.5 |
| **Hydrogen** | | $C_2H_6(g)$ | −84.7 |
| $H^+(aq)$ | 0 | $C_6H_6(l)$ | 49.0 |
| $H(g)$ | 218.0 | $HCHO(g)$ | −116 |
| $H_2(g)$ | 0 | $CH_3OH(l)$ | −238.6 |
| **Sodium** | | $CS_2(g)$ | 117 |
| $Na^+(g)$ | 609.8 | $CS_2(l)$ | 87.9 |
| $Na^+(aq)$ | −239.7 | $HCN(g)$ | 135 |
| $Na(g)$ | 107.8 | $HCN(l)$ | 105 |
| $Na(s)$ | 0 | $CCl_4(g)$ | −96.0 |
| $NaCl(s)$ | −411.1 | $CCl_4(l)$ | −139 |
| $NaHCO_3(s)$ | −947.7 | $CH_3CHO(g)$ | −166 |
| $Na_2CO_3(s)$ | −1130.8 | $C_2H_5OH(l)$ | −277.6 |
| **Calcium** | | **Silicon** | |
| $Ca^{2+}(aq)$ | −543.0 | $Si(s)$ | 0 |
| $Ca(s)$ | 0 | $SiO_2(s)$ | −910.9 |
| $CaO(s)$ | −635.1 | $SiF_4(g)$ | −1548 |
| $CaCO_3(s)$ (calcite) | −1206.9 | **Lead** | |
| **Carbon** | | $Pb(s)$ | 0 |
| $C(g)$ | 715.0 | $PbO(s)$ | −219 |
| $C$(graphite) | 0 | $PbS(s)$ | −98.3 |
| $C$(diamond) | 1.9 | **Nitrogen** | |
| $CO(g)$ | −110.5 | $N(g)$ | 473 |
| $CO_2(g)$ | −393.5 | $N_2(g)$ | 0 |
| $HCO_3^-(aq)$ | −691.1 | $NH_3(g)$ | −45.9 |
| $CH_4(g)$ | −74.9 | | *(continued)* |

Table 6.2 lists standard enthalpies of formation of substances (a longer table is given in Appendix C). The **standard enthalpy of formation** (also called the **standard heat of formation**) of a substance, denoted $\Delta H_f^\circ$, is *the enthalpy change for the formation of one mole of the substance in its standard state from its elements in their reference form and in their standard states.*

To understand this definition, consider the standard enthalpy of formation of liquid water. Note that the stablest forms of hydrogen and oxygen at 1 atm and 25°C are $H_2(g)$ and $O_2(g)$, respectively. These are therefore the reference forms of the elements. You write the formation reaction for 1 mol of liquid water as follows:

$$H_2(g) + \tfrac{1}{2}O_2(g) \longrightarrow H_2O(l)$$

The standard enthalpy change for this reaction is −285.8 kJ per mole of $H_2O$. Therefore, the thermochemical equation is

$$H_2(g) + \tfrac{1}{2}O_2(g) \longrightarrow H_2O(l); \ \Delta H_f^\circ = -285.8 \text{ kJ}$$

**Table 6.2   (Continued)**

| Formula | $\Delta H_f^\circ$ (kJ/mol) | Formula | $\Delta H_f^\circ$ (kJ/mol) |
|---|---|---|---|
| $NH_4^+(aq)$ | $-132.8$ | $HF(g)$ | $-273$ |
| $NO(g)$ | $90.3$ | *Chlorine* | |
| $NO_2(g)$ | $33.2$ | $Cl^-(aq)$ | $-167.5$ |
| $HNO_3(aq)$ | $-206.6$ | $Cl(g)$ | $121.0$ |
| *Oxygen* | | $Cl_2(g)$ | $0$ |
| $O(g)$ | $249.2$ | $HCl(g)$ | $-92.3$ |
| $O_2(g)$ | $0$ | *Bromine* | |
| $O_3(g)$ | $143$ | $Br^-(g)$ | $-218.9$ |
| $OH^-(aq)$ | $-229.9$ | $Br^-(aq)$ | $-120.9$ |
| $H_2O(g)$ | $-241.8$ | $Br_2(l)$ | $0$ |
| $H_2O(l)$ | $-285.8$ | *Iodine* | |
| *Sulfur* | | $I^-(g)$ | $-194.7$ |
| $S(g)$ | $279$ | $I^-(aq)$ | $-55.9$ |
| $S_2(g)$ | $129$ | $I_2(s)$ | $0$ |
| $S_8$(rhombic) | $0$ | *Silver* | |
| $S_8$(monoclinic) | $2$ | $Ag^+(g)$ | $1026.4$ |
| $SO_2(g)$ | $-296.8$ | $Ag^+(aq)$ | $105.9$ |
| $H_2S(g)$ | $-20$ | $Ag(s)$ | $0$ |
| *Fluorine* | | $AgF(s)$ | $-203$ |
| $F^-(g)$ | $-255.6$ | $AgCl(s)$ | $-127.0$ |
| $F^-(aq)$ | $-329.1$ | $AgBr(s)$ | $-99.5$ |
| $F_2(g)$ | $0$ | $AgI(s)$ | $-62.4$ |

*See Appendix C for additional values.

The values of standard enthalpies of formation listed in Table 6.2 and in other tables are determined by direct measurement in some cases and by applying Hess's law in others. Oxides, such as water, can often be determined by direct calorimetric measurement of the combustion reaction. If you look back at Example 6.7, you will see an illustration of how Hess's law can be used to obtain the enthalpy of formation of tungsten carbide, WC.

Note that the standard enthalpy of formation of an element will depend on the form of the element. For example, the $\Delta H_f^\circ$ for diamond equals the enthalpy change from the stablest form of carbon (graphite) to diamond. The thermochemical equation is

$$C(\text{graphite}) \longrightarrow C(\text{diamond}); \Delta H_f^\circ = 1.9 \text{ kJ}$$

On the other hand, the $\Delta H_f^\circ$ for graphite equals zero. Note the values of $\Delta H_f^\circ$ for the elements listed in Table 6.2; the reference forms have zero values.

Now let us see how to use standard enthalpies of formation (listed in Table 6.2) to find the standard enthalpy change for a reaction. We will first look at this problem from the point of view of Hess's law. But when we are done, we will note a pattern in the result, which will allow us to state a simple formula for solving this type of problem.

Consider the equation

$$CH_4(g) + 4Cl_2(g) \longrightarrow CCl_4(l) + 4HCl(g); \Delta H° = ?$$

From Table 6.2 you find the enthalpies of formation for $CH_4(g)$, $CCl_4(l)$, and $HCl(g)$. You can then write the following thermochemical equations:

$$C(\text{graphite}) + 2H_2(g) \longrightarrow CH_4(g); \Delta H_f° = -74.9 \text{ kJ} \qquad \textbf{(1)}$$

$$C(\text{graphite}) + 2Cl_2(g) \longrightarrow CCl_4(l); \Delta H_f° = -139 \text{ kJ} \qquad \textbf{(2)}$$

$$\tfrac{1}{2}H_2(g) + \tfrac{1}{2}Cl_2(g) \longrightarrow HCl(g); \Delta H_f° = -92.3 \text{ kJ} \qquad \textbf{(3)}$$

You now apply Hess's law. Since you want $CH_4$ to appear on the left, and $CCl_4$ and 4HCl on the right, you reverse Equation 1 and add Equation 2 and 4 × Equation 3.

$$
\begin{array}{ll}
CH_4(g) \longrightarrow \cancel{C(\text{graphite})} + \cancel{2H_2(g)} & (-74.9 \text{ kJ}) \times (-1) \\
\cancel{C(\text{graphite})} + 2Cl_2(g) \longrightarrow CCl_4(l) & (-139 \text{ kJ}) \times (1) \\
\cancel{2H_2(g)} + 2Cl_2(g) \longrightarrow 4HCl(g) & (-92.3 \text{ kJ}) \times (4) \\
\hline
CH_4(g) + 4Cl_2(g) \longrightarrow CCl_4(l) + 4HCl(g) & \Delta H° = -433 \text{ kJ}
\end{array}
$$

The setup of this calculation can be greatly simplified once you closely examine what you are doing. Note that the $\Delta H_f°$ for each compound is multiplied by its coefficient in the chemical equation whose $\Delta H°$ you are calculating. Moreover, the $\Delta H_f°$ for each reactant is multiplied by a negative sign. You can symbolize the enthalpy of formation of a substance by writing the formula in parentheses following $\Delta H_f°$. Then your calculation can be written as follows:

$$\Delta H° = [\Delta H_f°(CCl_4) + 4 \, \Delta H_f°(HCl)] - [\Delta H_f°(CH_4) + 4 \, \Delta H_f°(Cl_2)]$$

$$= [(-139) + 4(-92.3)] \text{ kJ} - [(-74.9) + 4(0)] \text{ kJ}$$

$$= -433 \text{ kJ}$$

In general, you can calculate the $\Delta H°$ for a reaction by the equation

$$\Delta H° = \Sigma \, n \, \Delta H_f°(\text{products}) - \Sigma \, m \, \Delta H_f°(\text{reactants})$$

Here $\Sigma$ is the mathematical symbol meaning "the sum of," and $m$ and $n$ are the coefficients of the substances in the chemical equation.

The next two examples illustrate the calculation of enthalpies of reaction from standard enthalpies of formation.

► An alternate example is provided in the Instructor's Resource Manual.

---

### EXAMPLE 6.8

## Calculating the Heat of Phase Transition from Standard Enthalpies of Formation

Use values of $\Delta H_f°$ to calculate the heat of vaporization, $\Delta H_{vap}°$, of carbon disulfide at 25°C. The vaporization process is

$$CS_2(l) \longrightarrow CS_2(g)$$

**PROBLEM STRATEGY**

The vaporization process can be treated like a chemical reaction, with $CS_2(l)$ the "reactant" and $CS_2(g)$ the "product." It is convenient to record the values of $\Delta H_f°$ under the formulas in the equation, multiplying them by the coefficients

$CS_2$

in the equation. You calculate $\Delta H^\circ_{vap}$ by subtracting values for the "reactant" from values for the "product."

**SOLUTION**

Here is the equation for the vaporization, with values of $\Delta H^\circ_f$ multiplied by coefficients (here, all 1's).

$$CS_2(l) \longrightarrow CS_2(g)$$
$$1(88) \qquad\qquad 1(117) \quad (kJ)$$

The calculation is

$$\Delta H^\circ_{vap} = \Sigma\, n\, \Delta H^\circ_f(\text{products}) - \Sigma\, m\, \Delta H^\circ_f(\text{reactants})$$
$$= \Delta H^\circ_f[CS_2(g)] - \Delta H^\circ_f[CS_2(l)]$$
$$= (117 - 88)\ kJ = \mathbf{29\ kJ}$$

---

**EXERCISE 6.9**

Calculate the heat of vaporization, $\Delta H^\circ_{vap}$, of water, using standard enthalpies of formation (Table 6.2).

---

See Problems 6.69 and 6.70.

**EXAMPLE 6.9**

## Calculating the Enthalpy of Reaction from Standard Enthalpies of Formation

Large quantities of ammonia are used to prepare nitric acid. The first step consists of the catalytic oxidation of ammonia to nitric oxide, NO.

$$4NH_3(g) + 5O_2(g) \xrightarrow{\text{Pt}} 4NO(g) + 6H_2O(g)$$

What is the standard enthalpy change for this reaction? Use Table 6.2 for data.

 **PROBLEM STRATEGY**

You record the values of $\Delta H^\circ_f$ under the formulas in the equation, multiplying them by the coefficients in the equation. You calculate $\Delta H^\circ$ by subtracting values for the reactants from values for the products.

**SOLUTION**

Here is the equation with the $\Delta H^\circ_f$'s recorded beneath it:

$$4NH_3(g) + 5O_2(g) \longrightarrow 4NO(g) + 6H_2O(g)$$
$$4(-45.9) \qquad 5(0) \qquad\qquad 4(90.3) \qquad 6(-241.8) \quad (kJ)$$

You can ignore the catalyst because its only effect is to speed up the reaction. The calculation is

$$\Delta H^\circ = \Sigma\, n\, \Delta H^\circ_f(\text{products}) - \Sigma\, m\, \Delta H^\circ_f(\text{reactants})$$
$$= [4\, \Delta H^\circ_f(NO) + 6\, \Delta H^\circ_f(H_2O)] - [4\, \Delta H^\circ_f(NH_3) + 5\, \Delta H^\circ_f(O_2)]$$
$$= [4(90.3) + 6(-241.8)]\ kJ - [4(-45.9) + 5(0)]\ kJ$$
$$= \mathbf{-906\ kJ}$$

▶ An alternate example is provided in the Instructor's Resource Manual.

**Problem Solving: Enthalpy from Standard Enthalpies of Formation**

Be very careful of arithmetical signs—they are a likely source of mistakes. Also pay particular attention to the state of each substance. Here, for example, you must use the $\Delta H_f^\circ$ for $H_2O(g)$, not for $H_2O(l)$.

---

**EXERCISE 6.10**

Calculate the enthalpy change for the following reaction:

$$3NO_2(g) + H_2O(l) \longrightarrow 2HNO_3(aq) + NO(g)$$

Use standard enthalpies of formation.

---

See Problems 6.71, 6.72, 6.73, and 6.74.

Enthalpies of formation can also be defined for ions. In this case, because it is not possible to make thermal measurements on individual ions, we must arbitrarily define the standard enthalpy of formation of one ion as zero. Then values for all other ions can be deduced from calorimetric data. By convention, the standard enthalpy of formation of $H^+(aq)$ is taken as zero. Values of $\Delta H_f^\circ$ for some ions are given in Table 6.2.

---

**EXERCISE 6.11**

Calculate the standard enthalpy change for the reaction of an aqueous solution of barium hydroxide, $Ba(OH)_2$, with an aqueous solution of ammonium nitrate, $NH_4NO_3$, at 25°C. (Figure 6.1 illustrated this reaction using solids instead of solutions.) The complete ionic equation is

$$Ba^{2+}(aq) + 2OH^-(aq) + 2NH_4^+(aq) + NO_3^-(aq) \longrightarrow$$
$$2NH_3(g) + 2H_2O(l) + Ba^{2+}(aq) + 2NO_3^-(aq)$$

---

See Problems 6.75 and 6.76.

## 6.9 Fuels—Foods, Commercial Fuels, and Rocket Fuels

A *fuel* is any substance that is burned or similarly reacted to provide heat and other forms of energy. The earliest use of fuels for heat came with the control of fire, which was achieved about 750,000 years ago. This major advance allowed the human species to migrate from tropical savannas and eventually to inhabit most of the earth. Through cooking, fire also increased the variety of edible food supplies and provided some protection of the food from bacterial decay. During the mid-eighteenth century, the discovery of the steam engine, which converts the chemical energy latent in fuels to mechanical energy, ushered in the Industrial Revolution. Today fuels not only heat our homes and move our cars but are absolutely necessary for every facet of modern technology. For example, fuels generate the electricity required for our modern computing and communications technologies, and they propel the rocket engines that make possible our explorations of outer space. In this section we will look at foods as fuels; at fossil fuels (which include gas, oil, and coal); at coal gasification and liquefaction; and at rocket fuels.

## Foods as Fuels

Foods fill three needs of the body: they supply substances for the growth and repair of tissue, they supply substances for the synthesis of compounds used in the regulation of body processes, and they supply energy. About 80% of the energy we need is for heat. The rest is used for muscular action, chemical processes, and other body processes. ◄

▶ The human body requires about as much energy in a day as does a 100-watt lightbulb.

The body generates energy from food by the same overall process as combustion, so the overall enthalpy change is the same as the heat of combustion, which can be determined in a calorimeter. You can get some idea of the energy available from carbohydrate foods by looking at a typical one, glucose ($C_6H_{12}O_6$). The thermochemical equation for the combustion of glucose is

$$C_6H_{12}O_6(s) + 6O_2(g) \longrightarrow 6CO_2(g) + 6H_2O(l); \ \Delta H^\circ = -2803 \text{ kJ}$$

One gram of glucose yields 15.6 kJ (3.73 kcal) of heat when burned.

A representative fat is glyceryl trimyristate, $C_{45}H_{86}O_6$. The equation for its combustion is

$$C_{45}H_{86}O_6(s) + \tfrac{127}{2}O_2(g) \longrightarrow 45CO_2(g) + 43H_2O(l); \ \Delta H^\circ = -27,820 \text{ kJ}$$

▶ In the popular literature of nutrition, the kilocalorie is referred to as the Calorie (capital C). Thus, these values are given as 4.0 Calories and 9.0 Calories.

One gram of this fat yields 38.5 kJ (9.20 kcal) of heat when burned. The average values quoted for carbohydrates and fats are 4.0 kcal/g and 9.0 kcal/g, respectively. ◄ Note that fats contain more than twice the fuel value per gram as do carbohydrates. Thus, by storing its fuel as fat, the body can store more fuel for a given mass of body tissue.

## Fossil Fuels

All of the fossil fuels in existence today were created millions of years ago when aquatic plants and animals were buried and compressed by layers of sediment at the bottoms of swamps and seas. Over time this organic matter was converted by bacterial decay and pressure to petroleum (oil), gas, and coal, which today are our major fuel sources. Figure 6.15 gives the percentages of the total energy consumed in the United States from various sources. Together, the fossil fuels account for nearly 90% of the total.

**Figure 6.15
Sources of energy consumed in the United States (1996).** Data are from *Monthly Energy Review On-line*, http://www.eia.doe.gov.

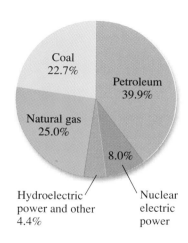

Coal 22.7%

Petroleum 39.9%

Natural gas 25.0%

8.0%

Hydroelectric power and other 4.4%

Nuclear electric power

Anthracite, or hard coal, the oldest variety of coal, was laid down as long as 250 million years ago and may contain over 80% carbon. Bituminous coal, a younger variety of coal, has between 45% and 65% carbon. Fuel values of coals are rated in Btu's (British thermal units) per pound, which are essentially heats of combustion per pound of coal. A typical value is 13,200 Btu/lb. A Btu equals 1054 J, so 1 Btu/lb equals 2.32 J/g. Therefore, the combustion of coal in oxygen yields about 30.6 kJ/g. You can compare this value with the heat of combustion of pure carbon (graphite).

$$C(graphite) + O_2(g) \longrightarrow CO_2(g); \Delta H° = -393.5 \text{ kJ}$$

The value given in the equation is for 1 mol (12.0 g) of carbon. Per gram, you get 32.8 kJ/g, which is comparable with the values obtained for coal.

Natural gas and petroleum together account for nearly three-quarters of the fossil fuels consumed per year. They are very convenient fluid fuels, being easily transportable and having no ash. Purified natural gas is primarily methane, $CH_4$, but it also contains small amounts of ethane, $C_2H_6$; propane, $C_3H_8$; and butane, $C_4H_{10}$. We would expect the fuel values of natural gas to be close to that for the heat of combustion of methane:

$$CH_4(g) + 2O_2(g) \longrightarrow CO_2(g) + 2H_2O(g); \Delta H° = -802 \text{ kJ}$$

This value of $\Delta H°$ is equivalent to 50.1 kJ per gram of fuel.

Petroleum is a very complicated mixture of compounds. Gasoline, which is obtained from petroleum by chemical and physical processes, contains many different hydrocarbons (compounds of carbon and hydrogen). One such hydrocarbon is octane, $C_8H_{18}$. The combustion of octane evolves 5074 kJ of heat per mole.

$$C_8H_{18}(l) + \tfrac{25}{2}O_2(g) \longrightarrow 8CO_2(g) + 9H_2O(g); \Delta H° = -5074 \text{ kJ}$$

This value of $\Delta H°$ is equivalent to 44.4 kJ/g. These combustion values indicate another reason why the fluid fossil fuels are popular: they release more heat per gram than coal does.

## Coal Gasification and Liquefaction

The major problem with petroleum and natural gas as fuels is their relative short supply. It has been estimated that petroleum supplies will be 80% depleted by about the year 2030. Natural-gas supplies may be depleted even sooner.

Coal supplies, on the other hand, are sufficient to last several more centuries. This abundance has spurred much research into developing commercial methods for converting coal to the more easily handled liquid and gaseous fuels. Most of these methods begin by converting coal to carbon monoxide, CO. One way involves the water–gas reaction.

$$C(s) + H_2O(g) \longrightarrow CO(g) + H_2(g)$$

▶ The methanation reaction uses a nickel catalyst. The Fisher–Tropsch process for making gasoline from CO and $H_2$ employs a cobalt catalyst.

In this reaction, steam is passed over hot coal. Once a mixture of carbon monoxide and hydrogen is obtained, it can be transformed by various reactions into useful products. For example, in the methanation reaction (discussed at length in Chapter 15), this mixture is reacted over a catalyst to give methane.

$$CO(g) + 3H_2(g) \longrightarrow CH_4(g) + H_2O(g)$$

Different catalysts and different reaction conditions result in liquid fuels. An added advantage of coal gasification and coal liquefaction is that sulfur, normally present in fossil fuels, can be removed during the process. The burning of sulfur-containing coal is a major source of air pollution and acid rain.

## Rocket Fuels

Rockets are self-contained missiles propelled by the ejection of gases from an orifice. Usually these are hot gases propelled from the rocket by the reaction of a fuel with an oxidizer. Rockets are believed to have originated with the Chinese—perhaps before the thirteenth century, which is when they began to appear in Europe. However, it was not until the twentieth century that rocket propulsion began to be studied seriously, and since World War II rockets have become major weapons. Space exploration with satellites propelled by rocket engines began in 1957 with the Russian satellite *Sputnik I.* Today weather and communications satellites are regularly put into orbit about the earth using rocket engines.

One of the factors determining which fuel and oxidizer to use is the mass of the fuel and oxidizer required. We have already seen that natural gas and gasoline have higher fuel values per gram than does coal. The difference is caused by the higher hydrogen content of natural gas and gasoline. Hydrogen is the element of lowest density, and at the same time it reacts exothermically with oxygen to give water. You might expect hydrogen and oxygen to be an ideal fuel–oxidizer combination. The thermochemical equation for the combustion of hydrogen is

$$H_2(g) + \tfrac{1}{2}O_2(g) \longrightarrow H_2O(g); \Delta H^\circ = -242 \text{ kJ}$$

This value of $\Delta H^\circ$ is equivalent to 120 kJ/g of fuel ($H_2$) compared with 50 kJ/g of methane. The second and third stages of the *Saturn V* launch vehicle that sent a three-man Apollo crew to the moon used a hydrogen/oxygen system. The launch vehicle contained liquid hydrogen (boiling at $-253°C$) and liquid oxygen, or LOX (boiling at $-183°C$).

The first stage of liftoff used kerosene and oxygen, and an unbelievable 550 metric tons ($550 \times 10^3$ kg) of kerosene were burned in 2.5 minutes! It is interesting to calculate the average rate of energy production in this 2.5-minute interval. Kerosene is approximately $C_{12}H_{26}$. The thermochemical equation is

$$C_{12}H_{26}(l) + \tfrac{37}{2}O_2(g) \longrightarrow 12CO_2(g) + 13H_2O(g); \Delta H^\circ = -7513 \text{ kJ}$$

This value of $\Delta H^\circ$ is equivalent to 44.1 kJ/g. Thus, $550 \times 10^6$ g of fuel generated $2.43 \times 10^{10}$ kJ in 150 s (2.5 min). Each second, the average energy produced was $1.62 \times 10^{11}$ J. This is equivalent to $1.62 \times 10^{11}$ watts, or 217 million horsepower (1 horsepower equals 745.7 watts, or J/s).

The landing module for the Apollo mission used a fuel made of hydrazine, $N_2H_4$, and a derivative of hydrazine. The oxidizer was dinitrogen tetroxide, $N_2O_4$. These substances are normally liquids and therefore easier to store than liquid hydrogen and oxygen. The reaction of the oxidizer with hydrazine is

$$2N_2H_4(l) + N_2O_4(l) \longrightarrow 3N_2(g) + 4H_2O(g); \Delta H^\circ = -1049 \text{ kJ}$$

Solid propellants are also used as rocket fuels. The mixture used in the booster rockets of the *Columbia* space shuttle (Figure 6.16) was a fuel containing aluminum metal powder. An oxidizer of ammonium perchlorate, $NH_4ClO_4$, was mixed with the fuel.

**Figure 6.16**
**The launching of the**
*Columbia* **space shuttle.**
The solid fuel for the booster
rockets is a mixture of
aluminum metal powder
and other materials with
ammonium perchlorate
as the oxidizer. A cloud of
aluminum oxide forms as
the rockets burn.

<h1>A Checklist for Review</h1>

## Important Terms

energy (6.1)
kinetic energy (6.1)
joule (J) (6.1)
calorie (cal) (6.1)
potential energy (6.1)
internal energy (6.1)
law of conservation of
  energy (6.1)

thermodynamic system (or
  system) (6.2)
surroundings (6.2)
heat (6.2)
heat of reaction (6.2)
exothermic process (6.2)
endothermic process (6.2)
enthalpy (6.3)

state function (6.3)
enthalpy of reaction (6.3)
thermochemical equation
  (6.4)
heat capacity (6.6)
specific heat capacity
  (specific heat) (6.6)
calorimeter (6.6)

Hess's law of heat standard
  summation (6.7)
standard state (6.8)
allotrope (6.8)
reference form (6.8)
standard enthalpy of
  formation (standard heat
  of formation) (6.8)

## Key Equations

$E_k = \frac{1}{2}mv^2$

1 cal = 4.184 J

$\Delta H = q_p$

$q = C \, \Delta t$

$q = s \times m \times \Delta t$

$\Delta H^\circ = \Sigma \, n \, \Delta H_f^\circ(\text{products}) - \Sigma \, m \, \Delta H_f^\circ(\text{reactants})$

## Summary of Facts and Concepts

*Energy* exists in various forms, including *kinetic energy* and *potential energy.* The SI unit of energy is the *joule* (1 calorie = 4.184 joules). The *internal energy* of a substance is the sum of the kinetic energies and potential energies of the particles making up the substance. According to the *law of conservation of energy,* the total quantity of energy remains constant.

Reactions absorb or evolve definite quantities of heat under given conditions. At constant pressure, this *heat of reaction* is the *enthalpy of reaction,* $\Delta H$. The chemical equation plus $\Delta H$ for molar amounts of reactants is referred to as the *thermochemical equation.* With it, you can calculate the heat for any

amount of substance by *applying stoichiometry to heats of reaction.* One measures the heat of reaction in a *calorimeter.* Direct calorimetric determination of the heat of reaction requires a reaction that goes to completion without other reactions occurring at the same time. Otherwise, the heat or enthalpy of reaction is determined indirectly from other enthalpies of reaction by using *Hess's law of heat summation.* Thermochemical data are conveniently tabulated as *enthalpies of formation.* If you know the values for each substance in an equation, you can easily compute the enthalpy of reaction. As an application of thermochemistry, the last section of the chapter discusses fuels.

## Operational Skills

1. **Calculating kinetic energy**   Given the mass and speed of an object, calculate the kinetic energy.   **(EXAMPLE 6.1)**
2. **Writing thermochemical equations**   Given a chemical equation, states of substances, and the quantity of heat absorbed or evolved for molar amounts, write the thermochemical equation.   **(EXAMPLE 6.2)**
3. **Manipulating thermochemical equations**   Given a thermochemical equation, write the thermochemical equation for different multiples of the coefficients or for the reverse reaction.   **(EXAMPLE 6.3)**
4. **Calculating the heat of reaction from the stoichiometry**   Given the value of $\Delta H$ for a chemical equation, calculate the heat of reaction for a given mass of reactant or product.   **(EXAMPLE 6.4)**
5. **Relating heat and specific heat**   Given any three of the quantities $q$, $s$, $m$, and $\Delta t$, calculate the fourth one. **(EXAMPLE 6.5)**
6. **Calculating $\Delta H$ from calorimetric data**   Given the amounts of reactants and the temperature change of a calorimeter of specified heat capacity, calculate the heat of reaction.   **(EXAMPLE 6.6)**
7. **Applying Hess's law**   Given a set of reactions with enthalpy changes, calculate $\Delta H$ for a reaction obtained from these other reactions by using Hess's law.   **(EXAMPLE 6.7)**
8. **Calculating the heat of phase transition from standard enthalpies of formation**   Given a table of standard enthalpies of formation, calculate the heat of phase transition.   **(EXAMPLE 6.8)**
9. **Calculating the enthalpy of reaction from standard enthalpies of formation**   Given a table of standard enthalpies of formation, calculate the enthalpy of reaction. **(EXAMPLE 6.9)**

## Review Questions

**6.1**   Define *energy, kinetic energy, potential energy,* and *internal energy.*

**6.2**   Define the joule in terms of SI base units.

**6.3**   What is the original definition of the calorie? What is the present definition?

**6.4**   Describe the interconversions of potential and kinetic energy in a moving pendulum. A moving pendulum eventually comes to rest. Has the energy been lost? If not, what has happened to it?

**6.5**   Suppose heat flows into a vessel containing a gas. As the heat flows into the gas, what happens to the gas molecules? What happens to the internal energy of the gas?

**6.6**   Define an *exothermic* reaction and an *endothermic* reaction. Give an example of each.

**6.7**   The internal energy of a substance is a state function. What does this mean?

**6.8**   Under what condition is the enthalpy change equal to the heat of reaction?

**6.9**   How does the enthalpy change for an endothermic reaction occurring at constant pressure?

**6.10**   Why is it important to give the states of the reactants and products when giving an equation for $\Delta H$?

**6.11**   If an equation for a reaction is doubled and then reversed, how is the value of $\Delta H$ changed?

**6.12**   Consider the reaction of methane, $CH_4$, with oxygen, $O_2$, discussed in Section 6.5. How would you set up the calculation if the problem had been to compute the heat if 10.0 g $H_2O$ were produced (instead of 10.0 g $CH_4$ reacted)?

**6.13**   Define the heat capacity of a substance. Define the specific heat of a substance.

**6.14**   Describe a simple calorimeter. What measurements are needed to determine the heat of reaction?

**6.15**   What property of enthalpy provides the basis of Hess's law? Explain.

**6.16**   You discover that you cannot carry out a particular reaction for which you would like the enthalpy change. Does this mean that you will be unable to obtain this enthalpy change? Explain.

**6.17**   What is meant by the thermodynamic standard state?

**6.18**   What is meant by the reference form of an element? What is the standard enthalpy of formation of an element in its reference form?

**6.19**   What is meant by the standard enthalpy of formation of a substance?

**6.20**   Write the chemical equation for the formation reaction of $H_2S(g)$.

**6.21**   Is the following reaction the appropriate one to use in determining the enthalpy of formation of methane, $CH_4(g)$? Why or why not?

$$C(g) + 4H(g) \longrightarrow CH_4(g)$$

**6.22**   What is a fuel? What are the fossil fuels?

**6.23**   Give chemical equations for the conversion of carbon in coal to methane, $CH_4$.

**6.24**   List some rocket fuels and corresponding oxidizers. Give thermochemical equations for the exothermic reactions of these fuels with the oxidizers.

# Conceptual Problems

**6.25** A small car is traveling at twice the speed of a larger car, which has twice the mass of the smaller car. Which car has the greater kinetic energy? (Or do they both have the same kinetic energy?)

**6.26** The equation for the combustion of butane, $C_4H_{10}$, is

$$C_4H_{10}(g) + \tfrac{13}{2}O_2(g) \longrightarrow 4CO_2(g) + 5H_2O(g)$$

Which one of the following generates the least heat? Why?
a. burning one mole of butane
b. reacting one mole of oxygen with excess butane
c. producing one mole of carbon dioxide by burning butane
d. producing one mole of water by burning butane

**6.27** What is the enthalpy change for the preparation of one mole of liquid water from the elements, given the following equations?

$$H_2(g) + \tfrac{1}{2}O_2(g) \longrightarrow H_2O(g); \Delta H_f$$
$$H_2O(l) \longrightarrow H_2O(g); \Delta H_{vap}$$

**6.28** A block of aluminum and a block of iron, both having the same mass, are removed from a freezer and placed outside on a warm day. When the same quantity of heat has flowed into each block, which block will be warmer? Assume that neither block has yet reached the outside temperature. (See Table 6.1 for the specific heats of the metals.)

**6.29** You have two samples of different metals, metal A and metal B, each having the same mass. You heat both metals to 95°C and then place each one into separate beakers containing the same quantity of water at 25°C.

a. You measure the temperatures of the water in the two beakers when each metal has cooled by 10°C and find that the temperature of the water with metal A is higher than the temperature of the water with metal B. Which metal has the greater specific heat? Explain.
b. After waiting a period of time, the temperature of the water in each beaker rises to a maximum value. In which beaker does the water rise to the higher value, the one with metal A or the one with metal B? Explain.

**6.30** Consider the reactions of silver metal, $Ag(s)$, with each of the halogens: fluorine, $F_2(g)$, chlorine, $Cl_2(g)$, and bromine, $Br_2(l)$. What chapter data could you use to decide which reaction is most exothermic? Which reaction is that?

**6.31** Tetraphosphorus trisulfide, $P_4S_3$, burns in excess oxygen to give tetraphosphorus decoxide, $P_4O_{10}$, and sulfur dioxide, $SO_2$. Suppose you have measured the enthalpy change for this reaction. How could you use it to obtain the enthalpy of formation of $P_4S_3$? What other data do you need?

**6.32** A soluble salt, $MX_2$, is added to water in a beaker. The equation for the dissolving of the salt is:

$$MX_2(s) \longrightarrow M^{2+}(aq) + 2X^-(aq); \quad \Delta H > 0$$

a. Immediately after the salt dissolves, is the solution warmer or colder?
b. Indicate the direction of heat flow, in or out of the beaker, while the salt dissolves.
c. After the salt dissolves and the water returns to room temperature, what is the value of $q$ for the system?

# Practice Problems

## Energy and Its Units

**6.33** Methane, $CH_4$, is a major component of marsh gas. When 0.5000 mol methane burns to produce carbon dioxide and liquid water, $-445.2$ kJ of heat is released. What is this heat in kilocalories?

**6.34** Hydrogen sulfide, $H_2S$, is produced during decomposition of organic matter. When 0.5000 mol $H_2S$ burns to produce $SO_2(g)$ and $H_2O(l)$, $-270.7$ kJ of heat is released. What is this heat in kilocalories?

**6.35** A car whose mass is $4.85 \times 10^3$ lb is traveling at a speed of 57 miles per hour. What is the kinetic energy of the car in joules? in calories? See Table 1.5 for conversion factors.

**6.36** A bullet weighing 235 grains is moving at a speed of $2.52 \times 10^3$ ft/s. Calculate the kinetic energy of the bullet in joules and in calories. One grain equals 0.0648 g.

**6.37** Chlorine dioxide, $ClO_2$, is a reddish-yellow gas used in bleaching paper pulp. The average speed of a $ClO_2$ molecule at 25°C is 306 m/s. What is the kinetic energy (in joules) of a $ClO_2$ molecule moving at this speed?

**6.38** Nitrous oxide, $N_2O$, has been used as a dental anesthetic. The average speed of an $N_2O$ molecule at 25°C is 379 m/s. Calculate the kinetic energy (in joules) of an $N_2O$ molecule traveling at this speed.

## Heat of Reaction

**6.39** The process of dissolving ammonium nitrate, $NH_4NO_3$, in water is an endothermic process. What is the sign of $q$? If you were to add some ammonium nitrate to water in a flask, would you expect the flask to feel warm or cool?

**6.40** The decomposition of ozone, $O_3$, to oxygen, $O_2$, is an exothermic reaction. What is the sign of $q$? If you were to touch a flask in which ozone is decomposing to oxygen, would you expect the flask to feel warm or cool?

**6.41** Nitric acid, a source of many nitrogen compounds, is produced from nitrogen dioxide. An old process for making nitrogen dioxide employed nitrogen and oxygen.

$$N_2(g) + 2O_2(g) \longrightarrow 2NO_2(g)$$

The reaction absorbs 66.4 kJ of heat per 2 mol $NO_2$ produced. Is the reaction endothermic or exothermic? What is the value of $q$?

**6.42** Hydrogen cyanide is used in the manufacture of clear plastics such as Lucite and Plexiglas. It is prepared from ammonia and natural gas ($CH_4$).

$$2NH_3(g) + 3O_2(g) + 2CH_4(g) \longrightarrow 2HCN(g) + 6H_2O(g)$$

The reaction evolves 939 kJ of heat per 2 mol HCN formed. Is the reaction endothermic or exothermic? What is the value of $q$?

## Thermochemical Equations

**6.43** When 1 mol of iron metal reacts with hydrochloric acid at constant temperature and pressure to produce hydrogen gas and aqueous iron(II) chloride, 87.9 kJ of heat evolves. Write a thermochemical equation for this reaction.

**6.44** When 2 mol of potassium chlorate crystals decompose to potassium chloride crystals and oxygen gas at constant temperature and pressure, 44.7 kJ of heat is given off. Write a thermochemical equation for this reaction.

**6.45** When white phosphorus burns in air, it produces phosphorus(V) oxide.

$$P_4(s) + 5O_2(g) \longrightarrow P_4O_{10}(s); \Delta H = -2940 \text{ kJ}$$

What is $\Delta H$ for the following equation?

$$P_4O_{10}(s) \longrightarrow P_4(s) + 5O_2(g)$$

$P_4$

**6.46** Carbon disulfide burns in air, producing carbon dioxide and sulfur dioxide.

$$CS_2(l) + 3O_2(g) \longrightarrow CO_2(g) + 2SO_2(g); \Delta H = -1075 \text{ kJ}$$

What is $\Delta H$ for the following equation?

$$\tfrac{1}{2}CS_2(l) + \tfrac{3}{2}O_2(g) \longrightarrow \tfrac{1}{2}CO_2(g) + SO_2(g)$$

$CS_2$

**6.47** Phosphoric acid, $H_3PO_4$, can be prepared by the reaction of phosphorus(V) oxide, $P_4O_{10}$, with water.

$$\tfrac{1}{4}P_4O_{10}(s) + \tfrac{3}{2}H_2O(l) \longrightarrow H_3PO_4(aq); \Delta H = -113.2 \text{ kJ}$$

What is $\Delta H$ for the reaction involving 1 mol of $P_4O_{10}$?

$$P_4O_{10}(s) + 6H_2O(l) \longrightarrow 4H_3PO_4(aq)$$

**6.48** With a platinum catalyst, ammonia will burn in oxygen to give nitric oxide, NO.

$$4NH_3(g) + 5O_2(g) \longrightarrow 4NO(g) + 6H_2O(g);$$
$$\Delta H = -906 \text{ kJ}$$

What is the enthalpy change for the following reaction?

$$NO(g) + \tfrac{3}{2}H_2O(g) \longrightarrow NH_3(g) + \tfrac{5}{4}O_2(g)$$

## Stoichiometry of Reaction Heats

**6.49** Colorless nitric oxide, NO, combines with oxygen to form nitrogen dioxide, $NO_2$, a brown gas.

$$2NO(g) + O_2(g) \longrightarrow 2NO_2(g); \Delta H = -114 \text{ kJ}$$

What is the enthalpy change per gram of nitric oxide?

**6.50** Hydrogen, $H_2$, is used as a rocket fuel. The hydrogen is burned in oxygen to produce water vapor.

$$2H_2(g) + O_2(g) \longrightarrow 2H_2O(g); \Delta H = -484 \text{ kJ}$$

What is the enthalpy change per gram of hydrogen?

**6.51** Ammonia burns in the presence of a copper catalyst to form nitrogen gas.

$$4NH_3(g) + 3O_2(g) \longrightarrow 2N_2(g) + 6H_2O(g);$$
$$\Delta H = -1267 \text{ kJ}$$

What is the enthalpy change to burn 35.8 g of ammonia?

**6.52** Hydrogen sulfide, $H_2S$, is a foul-smelling gas. It burns to form sulfur dioxide.

$$2H_2S(g) + 3O_2(g) \longrightarrow 2SO_2(g) + 2H_2O(g);$$
$$\Delta H = -1037 \text{ kJ}$$

Calculate the enthalpy change to burn 26.7 g of hydrogen sulfide.

**6.53** Propane, $C_3H_8$, is a common fuel gas. Use the following to calculate the grams of propane you would need to provide 369 kJ of heat.

$$C_3H_8(g) + 5O_2(g) \longrightarrow 3CO_2(g) + 4H_2O(g);$$
$$\Delta H = -2044 \text{ kJ}$$

**6.54** Ethanol, $C_2H_5OH$, is mixed with gasoline and sold as gasohol. Use the following to calculate the grams of ethanol needed to provide 293 kJ of heat.

$$C_2H_5OH(l) + 3O_2(g) \longrightarrow 2CO_2(g) + 3H_2O(g);$$
$$\Delta H = -1235 \text{ kJ}$$

## Heat Capacity and Calorimetry

**6.55** You wish to heat water to make coffee. How much heat (in joules) must be used to raise the temperature of 0.180 kg of tap water (enough for one cup of coffee) from 19°C to 96°C (near the ideal brewing temperature)? Assume the specific heat is that of pure water, 4.18 J/(g · °C).

**6.56** An iron skillet weighing 1.63 kg is heated on a stove to 178°C. Suppose the skillet is cooled to room temperature, 21°C. How much heat energy (in joules) must be removed to effect this cooling? The specific heat of iron is 0.449 J/(g · °C).

**6.57** When steam condenses to liquid water, 2.26 kJ of heat is released per gram. The heat from 168 g of steam is used to heat a room containing $6.44 \times 10^4$ g of air (20 ft × 12 ft × 8 ft). The specific heat of air at normal pressure is 1.015 J/(g · °C). What is the change in air temperature, assuming the heat from the steam is all absorbed by air?

**6.58** When ice at 0°C melts to liquid water at 0°C, it absorbs 0.334 kJ of heat per gram. Suppose the heat needed to melt 38.0 g of ice is absorbed from the water contained in a glass. If this water has a mass of 0.210 kg and a temperature of 21.0°C, what is the final temperature of the water? (Note that you will also have 38.0 g of water at 0°C from the ice.)

**6.59** When 15.3 g of sodium nitrate, $NaNO_3$, was dissolved in water in a calorimeter, the temperature fell from 25.00°C to 21.56°C. If the heat capacity of the solution and the calorimeter is 1071 J/°C, what is the enthalpy change when 1 mol of sodium nitrate dissolves in water? The solution process is

$$NaNO_3(s) \longrightarrow Na^+(aq) + NO_3^-(aq); \Delta H = ?$$

**6.60** When 23.6 g of calcium chloride, $CaCl_2$, was dissolved in water in a calorimeter, the temperature rose from 25.0°C to 38.7°C. If the heat capacity of the solution and the calorimeter is 1258 J/°C, what is the enthalpy change when 1 mol of calcium chloride dissolves in water? The solution process is

$$CaCl_2(s) \longrightarrow Ca^{2+}(aq) + 2Cl^-(aq); \Delta H = ?$$

**6.61** A sample of ethanol, $C_2H_5OH$, weighing 2.84 g was burned in an excess of oxygen in a bomb calorimeter. The temperature of the calorimeter rose from 25.00°C to 33.73°C. If the heat capacity of the calorimeter and contents was 9.63 kJ/°C, what is the value of $q$ for burning 1 mol of ethanol at constant volume and 25.00°C? The reaction is

$$C_2H_5OH(l) + 3O_2(g) \longrightarrow 2CO_2(g) + 3H_2O(l)$$

**6.62** A sample of benzene, $C_6H_6$, weighing 3.51 g was burned in an excess of oxygen in a bomb calorimeter. The temperature of the calorimeter rose from 25.00°C to 37.18°C. If the heat capacity of the calorimeter and contents was 12.05 kJ/°C, what is the value of $q$ for burning 1 mol of benzene at constant volume and 25.00°C? The reaction is

$$C_6H_6(l) + \tfrac{15}{2}O_2(g) \longrightarrow 6CO_2(g) + 3H_2O(l)$$

## Hess's Law

**6.63** Hydrazine, $N_2H_4$, is a colorless liquid used as a rocket fuel. What is the enthalpy change for the process in which hydrazine is formed from its elements?

$$N_2(g) + 2H_2(g) \longrightarrow N_2H_4(l)$$

Use the following reactions and enthalpy changes:

$$N_2H_4(l) + O_2(g) \longrightarrow N_2(g) + 2H_2O(l); \Delta H = -622.2 \text{ kJ}$$
$$H_2(g) + \tfrac{1}{2}O_2(g) \longrightarrow H_2O(l); \Delta H = -285.8 \text{ kJ}$$

**6.64** Hydrogen peroxide, $H_2O_2$, is a colorless liquid whose solutions are used as a bleach and an antiseptic. $H_2O_2$ can be prepared in a process whose overall change is

$$H_2(g) + O_2(g) \longrightarrow H_2O_2(l)$$

Calculate the enthalpy change using the following data:

$$H_2O_2(l) \longrightarrow H_2O(l) + \tfrac{1}{2}O_2(g); \Delta H = -98.0 \text{ kJ}$$
$$2H_2(g) + O_2(g) \longrightarrow 2H_2O(l); \Delta H = -571.6 \text{ kJ}$$

**6.65** Ammonia will burn in the presence of a platinum catalyst to produce nitric oxide, NO.

$$4NH_3(g) + 5O_2(g) \longrightarrow 4NO(g) + 6H_2O(g)$$

What is the heat of reaction at constant pressure? Use the following thermochemical equations:

$$N_2(g) + O_2(g) \longrightarrow 2NO(g); \Delta H = 180.6 \text{ kJ}$$
$$N_2(g) + 3H_2(g) \longrightarrow 2NH_3(g); \Delta H = -91.8 \text{ kJ}$$
$$2H_2(g) + O_2(g) \longrightarrow 2H_2O(g); \Delta H = -483.7 \text{ kJ}$$

**6.66** Hydrogen cyanide is a highly poisonous, volatile liquid. It can be prepared by the reaction

$$CH_4(g) + NH_3(g) \longrightarrow HCN(g) + 3H_2(g)$$

What is the heat of reaction at constant pressure? Use the following thermochemical equations:

$$N_2(g) + 3H_2(g) \longrightarrow 2NH_3(g); \Delta H = -91.8 \text{ kJ}$$
$$C(\text{graphite}) + 2H_2(g) \longrightarrow CH_4(g); \Delta H = -74.9 \text{ kJ}$$
$$H_2(g) + 2C(\text{graphite}) + N_2(g) \longrightarrow 2HCN(g);$$
$$\Delta H = 270.3 \text{ kJ}$$

**6.67** Compounds with carbon–carbon double bonds, such as ethylene, $C_2H_4$, add hydrogen in a reaction called hydrogenation.

$$C_2H_4(g) + H_2(g) \longrightarrow C_2H_6(g)$$

Calculate the enthalpy change for this reaction, using the following combustion data:

$$C_2H_4(g) + 3O_2(g) \longrightarrow 2CO_2(g) + 2H_2O(l);$$
$$\Delta H = -1401 \text{ kJ}$$
$$C_2H_6(g) + \tfrac{7}{2}O_2(g) \longrightarrow 2CO_2(g) + 3H_2O(l);$$
$$\Delta H = -1550 \text{ kJ}$$
$$H_2(g) + \tfrac{1}{2}O_2(g) \longrightarrow H_2O(l); \Delta H = -286 \text{ kJ}$$

**6.68** Acetic acid, $CH_3COOH$, is contained in vinegar. Suppose acetic acid was formed from its elements, according to the following equation:

$$2C(\text{graphite}) + 2H_2(g) + O_2(g) \longrightarrow CH_3COOH(l)$$

Find the enthalpy change, $\Delta H$, for this reaction, using the following data:

$$CH_3COOH(l) + 2O_2(g) \longrightarrow 2CO_2(g) + 2H_2O(l);$$
$$\Delta H = -871 \text{ kJ}$$
$$C(\text{graphite}) + O_2(g) \longrightarrow CO_2(g); \Delta H = -394 \text{ kJ}$$
$$H_2(g) + \tfrac{1}{2}O_2(g) \longrightarrow H_2O(l); \Delta H = -286 \text{ kJ}$$

## Standard Enthalpies of Formation

**6.69** The cooling effect of alcohol on the skin is due to its evaporation. Calculate the heat of vaporization of ethanol (ethyl alcohol), $C_2H_5OH$.

$$C_2H_5OH(l) \longrightarrow C_2H_5OH(g); \Delta H° = ?$$

The standard enthalpy of formation of $C_2H_5OH(l)$ is $-277.6$ kJ/mol and that of $C_2H_5OH(g)$ is $-235.4$ kJ/mol.

**6.70** Carbon tetrachloride, $CCl_4$, is a liquid used as an industrial solvent and in the preparation of fluorocarbons. What is the heat of vaporization of carbon tetrachloride?

$$CCl_4(l) \longrightarrow CCl_4(g); \Delta H° = ?$$

Use standard enthalpies of formation (Table 6.2).

**6.71** Hydrogen sulfide gas is a poisonous gas with the odor of rotten eggs. It occurs in natural gas and is produced during the decay of organic matter, which contains sulfur. The gas burns in oxygen as follows:

$$2H_2S(g) + 3O_2(g) \longrightarrow 2H_2O(l) + 2SO_2(g)$$

Calculate the standard enthalpy change for this reaction using standard enthalpies of formation.

**6.72** Carbon disulfide is a colorless liquid. When pure, it is nearly odorless, but the commercial product smells vile. Carbon disulfide is used in the manufacture of rayon and cellophane. The liquid burns as follows:

$$CS_2(l) + 3O_2(g) \longrightarrow CO_2(g) + 2SO_2(g)$$

Calculate the standard enthalpy change for this reaction using standard enthalpies of formation.

**6.73** Iron is obtained from iron ore by reduction with carbon monoxide. The overall reaction is

$$Fe_2O_3(s) + 3CO(g) \longrightarrow 2Fe(s) + 3CO_2(g)$$

Calculate the standard enthalpy change for this equation. See Appendix C for data.

**6.74** The first step in the preparation of lead from its ore (galena, PbS) consists of roasting the ore.

$$2PbS(s) + 3O_2(g) \longrightarrow 2SO_2(g) + 2PbO(s)$$

Calculate the standard enthalpy change for this reaction, using enthalpies of formation (see Appendix C).

**6.75** Hydrogen chloride gas dissolves in water to form hydrochloric acid (an ionic solution).

$$HCl(g) \xrightarrow{H_2O} H^+(aq) + Cl^-(aq)$$

Find $\Delta H°$ for the above reaction. The data are given in Table 6.2.

**6.76** Carbon dioxide from the atmosphere "weathers," or dissolves, limestone ($CaCO_3$) by the reaction

$$CaCO_3(s) + CO_2(g) + H_2O(l) \longrightarrow Ca^{2+}(aq) + 2HCO_3^-(aq)$$

Obtain $\Delta H°$ for this reaction. See Table 6.2 for the data.

**6.77** The Group IIA carbonates decompose when heated. For example,

$$MgCO_3(s) \longrightarrow MgO(s) + CO_2(g)$$

Use enthalpies of formation (Appendix C) and calculate the heat required to decompose 10.0 g of magnesium carbonate.

**6.78** The Group IIA carbonates decompose when heated. For example,

$$BaCO_3(s) \longrightarrow BaO(s) + CO_2(g)$$

Use enthalpies of formation (Appendix C) and calculate the heat required to decompose 10.0 g of barium carbonate.

# General Problems

**6.79** The energy, $E$, needed to move an object a distance $d$ by applying a force $F$ is $E = F \times d$. What must be the SI unit of force if this equation is to be consistent with the SI unit of energy for $E$?

**6.80** The potential energy of an object in the gravitational field of the earth is $E_p = mgh$. What must be the SI unit of $g$ if this equation is to be consistent with the SI unit of energy for $E_p$?

**6.81** Liquid hydrogen peroxide has been used as a propellant for rockets. Hydrogen peroxide decomposes into oxygen and water, giving off heat energy equal to 686 Btu per pound of propellant. What is this energy in joules per gram of hydrogen peroxide? (1 Btu = 252 cal; see also Table 1.5.)

**6.82** Hydrogen is an ideal fuel in many respects; for example, the product of its combustion, water, is nonpolluting. The heat given off in burning hydrogen to gaseous water is $5.16 \times 10^4$ Btu per pound. What is this heat energy in joules per gram? (1 Btu = 252 cal; see also Table 1.5.)

**6.83** Niagara Falls has a height of 167 ft (American Falls). What is the potential energy in joules of 1.00 lb of water at the top of the falls if we take water at the bottom to have a potential energy of zero? What would be the speed of this water at the bottom of the falls if we neglect friction during the descent of the water?

**6.84** Any object, be it a space satellite or a molecule, must attain an initial upward velocity of at least 11.2 km/s in order to escape the gravitational attraction of the earth. What would be the kinetic energy in joules of a satellite weighing 2354 lb that has the speed equal to this escape velocity of 11.2 km/s?

**6.85** When calcium carbonate, $CaCO_3$ (the major constituent of limestone and seashells), is heated, it decomposes to calcium oxide (quicklime).

$$CaCO_3(s) \longrightarrow CaO(s) + CO_2(g); \Delta H = 178.3 \text{ kJ}$$

How much heat is required to decompose 21.3 g of calcium carbonate?

**6.86** Calcium oxide (quicklime) reacts with water to produce calcium hydroxide (slaked lime).

$$CaO(s) + H_2O(l) \longrightarrow Ca(OH)_2(s); \Delta H = -65.2 \text{ kJ}$$

The heat released by this reaction is sufficient to ignite paper. How much heat is released when 24.5 g of calcium oxide reacts?

**6.87** Formic acid, $HCHO_2$, was first discovered in ants (*formica* is Latin for "ant"). In an experiment, 5.48 g of formic acid was burned at constant pressure.

$$2HCHO_2(l) + O_2(g) \longrightarrow 2CO_2(g) + 2H_2O(l)$$

If 30.3 kJ of heat evolved, what is $\Delta H$ per mole of formic acid?

**6.88** Acetic acid, $HC_2H_3O_2$, is the sour constituent of vinegar (*acetum* is Latin for "vinegar"). In an experiment, 3.58 g of acetic acid was burned.

$$HC_2H_3O_2(l) + 2O_2(g) \longrightarrow 2CO_2(g) + 2H_2O(l)$$

If 52.0 kJ of heat evolved, what is $\Delta H$ per mole of acetic acid?

**6.89** A piece of lead of mass 121.6 g was heated by an electrical coil. From the resistance of the coil, the current, and the time the current flowed, it was calculated that 235 J of heat was added to the lead. The temperature of the lead rose from 20.4°C to 35.5°C. What is the specific heat of the lead?

**6.90** The specific heat of copper metal was determined by putting a piece of the metal weighing 35.4 g in hot water. The quantity of heat absorbed by the metal was calculated to be 47.0 J from the temperature drop of the water. What was the specific heat of the metal if the temperature of the metal rose 3.45°C?

**6.91** A 50.0-g sample of water at 100.00°C was placed in an insulated cup. Then 25.3 g of zinc metal at 25.00°C was added to the water. The temperature of the water dropped to 96.68°C. What is the specific heat of zinc?

**6.92** A 19.6-g sample of a metal was heated to 61.67°C. When the metal was placed into 26.7 g of water in a calorimeter, the temperature of the water increased from 25.00°C to 30.00°C. What is the specific heat of the metal?

**6.93** In a calorimetric experiment, 6.48 g of lithium hydroxide, LiOH, was dissolved in water. The temperature of the calorimeter rose from 25.00°C to 36.66°C. What is $\Delta H$ for the solution process?

$$LiOH(s) \longrightarrow Li^+(aq) + OH^-(aq)$$

The heat capacity of the calorimeter and its contents is 547 J/°C.

**6.94** When 21.45 g of potassium nitrate, $KNO_3$, was dissolved in water in a calorimeter, the temperature fell from 25.00°C to 14.14°C. What is the $\Delta H$ for the solution process?

$$KNO_3(s) \longrightarrow K^+(aq) + NO_3^-(aq)$$

The heat capacity of the calorimeter and its contents is 682 J/°C.

**6.95** A 10.00-g sample of acetic acid, $HC_2H_3O_2$, was burned in a bomb calorimeter in an excess of oxygen.

$$HC_2H_3O_2(l) + 2O_2(g) \longrightarrow 2CO_2(g) + 2H_2O(l)$$

The temperature of the calorimeter rose from 25.00°C to 35.81°C. If the heat capacity of the calorimeter and its contents is 13.43 kJ/°C, what is the enthalpy change for the reaction?

**6.96** The sugar arabinose, $C_5H_{10}O_5$, is burned completely in oxygen in a calorimeter.

$$C_5H_{10}O_5(s) + 5O_2(g) \longrightarrow 5CO_2(g) + 5H_2O(l)$$

Burning a 0.548-g sample caused the temperature to rise from 20.00°C to 20.54°C. The heat capacity of the calorimeter and its contents is 15.8 kJ/°C. Calculate $\Delta H$ for the combustion reaction per mole of arabinose.

**6.97** Hydrogen sulfide, $H_2S$, is a poisonous gas with the odor of rotten eggs. The reaction for the formation of $H_2S$ from the elements is

$$H_2(g) + \tfrac{1}{8}S_8(rhombic) \longrightarrow H_2S(g)$$

Use Hess's law to obtain the enthalpy change for this reaction from the following enthalpy changes:

$$H_2S(g) + \tfrac{3}{2}O_2(g) \longrightarrow H_2O(g) + SO_2(g); \Delta H = -519 \text{ kJ}$$
$$H_2(g) + \tfrac{1}{2}O_2(g) \longrightarrow H_2O(g); \Delta H = -242 \text{ kJ}$$
$$\tfrac{1}{8}S_8(rhombic) + O_2(g) \longrightarrow SO_2(g); \Delta H = -297 \text{ kJ}$$

**6.98** Ethylene glycol, $HOCH_2CH_2OH$, is used as antifreeze. It is produced from ethylene oxide, $C_2H_4O$, by the reaction

$$C_2H_4O(g) + H_2O(l) \longrightarrow HOCH_2CH_2OH(l)$$

Use Hess's law to obtain the enthalpy change for this reaction from the following enthalpy changes:

$$2C_2H_4O(g) + 5O_2(g) \longrightarrow 4CO_2(g) + 4H_2O(l);$$
$$\Delta H = -2612.2 \text{ kJ}$$
$$HOCH_2CH_2OH(l) + \tfrac{5}{2}O_2(g) \longrightarrow 2CO_2(g) + 3H_2O(l);$$
$$\Delta H = -1189.8 \text{ kJ}$$

**6.99** Hydrogen, $H_2$, is prepared by *steam reforming,* in which hydrocarbons are reacted with steam. For $CH_4$,

$$CH_4(g) + H_2O(g) \longrightarrow CO(g) + 3H_2(g)$$

Calculate the enthalpy change $\Delta H°$ for this reaction, using standard enthalpies of formation.

**6.100** Hydrogen is prepared from natural gas (mainly methane, $CH_4$) by partial oxidation.

$$2CH_4(g) + O_2(g) \longrightarrow 2CO(g) + 4H_2(g)$$

Calculate the enthalpy change $\Delta H°$ for this reaction, using standard enthalpies of formation.

**6.101** Calcium oxide, CaO, is prepared by heating calcium carbonate (from limestone and seashells).

$$CaCO_3(s) \longrightarrow CaO(s) + CO_2(g)$$

Calculate the standard enthalpy of reaction, using enthalpies of formation. The $\Delta H_f°$ of CaO(s) is $-635$ kJ/mol. Other values are given in Table 6.2.

**6.102** Sodium carbonate, $Na_2CO_3$, is used to manufacture glass. It is obtained from sodium hydrogen carbonate, $NaHCO_3$, by heating.

$$2NaHCO_3(s) \longrightarrow Na_2CO_3(s) + H_2O(g) + CO_2(g)$$

Calculate the standard enthalpy of reaction, using enthalpies of formation (Table 6.2).

**6.103** Calculate the heat released when 2.000 L $O_2$ with a density of 1.11 g/L at 25°C reacts with an excess of hydrogen to form liquid water at 25°C.

**6.104**    Calculate the heat released when 4.000 L $Cl_2$ with a density of 2.46 g/L at 25°C reacts with an excess of sodium metal to form solid sodium chloride at 25°C.

**6.105**    Sucrose, $C_{12}H_{22}O_{11}$, is common table sugar. The enthalpy change at 25°C and 1 atm for the complete burning of 1 mol of sucrose in oxygen to give $CO_2(g)$ and $H_2O(l)$ is −5641 kJ. From this and from data given in Table 6.2, calculate the standard enthalpy of formation of sucrose.

**6.106**    Acetone, $CH_3COCH_3$, is a liquid solvent. The enthalpy change at 25°C and 1 atm for the complete burning of 1 mol of acetone in oxygen to give $CO_2(g)$ and $H_2O(l)$ is −1791 kJ. From this and from data given in Table 6.2, calculate the standard enthalpy of formation of acetone.

**6.107**    Ammonium nitrate is an oxidizing agent and can give rise to explosive mixtures. A mixture of 2.00 mol of powdered aluminum and 3.00 mol of ammonium nitrate crystals reacts exothermically, yielding nitrogen gas, water vapor, and aluminum oxide. How many grams of the mixture are required to provide 245 kJ of heat? See Appendix C for data.

**6.108**    The thermite reaction is a very exothermic reaction; it has been used to produce liquid iron for welding. A mixture of 2 mol of powdered aluminum metal and 1 mol of iron(III) oxide yields liquid iron and solid aluminum oxide. How many grams of the mixture are needed to produce 348 kJ of heat? See Appendix C for data.

## Cumulative-Skills Problems

**6.109**    What will be the final temperature of a mixture made from 25.0 g of water at 15.0°C, from 45.0 g of water at 50.0°C, and from 15.0 g of water at 37.0°C?

**6.110**    What will be the final temperature of a mixture made from equal masses of the following: water at 25.0°C, ethanol at 35.5°C, and iron at 95°C?

**6.111**    Graphite is burned in oxygen to give carbon monoxide and carbon dioxide. If the product mixture is 33% CO and 67% $CO_2$ by mass, what is the heat from the combustion of 1.00 g of graphite?

**6.112**    A sample of natural gas is 80.0% $CH_4$ and 20.0% $C_2H_6$ by mass. What is the heat from the combustion of 1.00 g of this mixture? Assume the products are $CO_2(g)$ and $H_2O(l)$.

**6.113**    A sample containing 2.00 mol of graphite reacts completely with a limited quantity of oxygen at 25°C and 1.0 atm pressure, producing 481 kJ of heat and a mixture of CO and $CO_2$. Calculate the masses of CO and $CO_2$ produced.

**6.114**    A 10.0-g sample of a mixture of $CH_4$ and $C_2H_4$ reacts with oxygen at 25°C and 1 atm to produce $CO_2(g)$ and $H_2O(l)$. If the reaction produces 520 kJ of heat, what is the mass percentage of $CH_4$ in the mixture?

**6.115**    How much heat is released when a mixture containing 10.0 g $NH_3$ and 20.0 g $O_2$ reacts by the equation

$$4NH_3(g) + 5O_2(g) \longrightarrow 4NO(g) + 6H_2O(g);$$
$$\Delta H° = -906 \text{ kJ}$$

**6.116**    How much heat is released when a mixture containing 10.0 g $CS_2$ and 10.0 g $Cl_2$ reacts by the equation

$$CS_2(g) + 3Cl_2(g) \longrightarrow S_2Cl_2(g) + CCl_4(g);$$
$$\Delta H° = -232 \text{ kJ}$$

**6.117**    Consider the Haber process:

$$N_2(g) + 3H_2(g) \longrightarrow 2NH_3(g); \Delta H° = -91.8 \text{ kJ}$$

The density of ammonia at 25°C and 1.00 atm is 0.696 g/L. The density of nitrogen, $N_2$, is 1.145 g/L, and the molar heat capacity is 29.12 J/(mol · °C). (a) How much heat is evolved in the production of 1.00 L of ammonia at 25°C and 1.00 atm? (b) What percentage of this heat is required to heat the nitrogen required for this reaction (0.500 L) from 25°C to 400°C, the temperature at which the Haber process is run?

**6.118**    An industrial process for manufacturing sulfuric acid, $H_2SO_4$, uses hydrogen sulfide, $H_2S$, from the purification of natural gas. In the first step of this process, the hydrogen sulfide is burned to obtain sulfur dioxide, $SO_2$.

$$2H_2S(g) + 3O_2(g) \longrightarrow 2H_2O(l) + 2SO_2(g);$$
$$\Delta H° = -1125 \text{ kJ}$$

The density of sulfur dioxide at 25°C and 1.00 atm is 2.62 g/L, and the molar heat capacity is 30.2 J/(mol · °C). (a) How much heat would be evolved in producing 1.00 L of $SO_2$ at 25°C and 1.00 atm? (b) Suppose heat from this reaction is used to heat 1.00 L of $SO_2$ from 25°C and 1.00 atm to 500°C for its use in the next step of the process. What percentage of the heat evolved is required for this?

**6.119**    The carbon dioxide exhaled in the breath of astronauts is often removed from the spacecraft by reaction with lithium hydroxide.

$$2LiOH(s) + CO_2(g) \longrightarrow Li_2CO_3(s) + H_2O(l)$$

Estimate the grams of lithium hydroxide required per astronaut per day. Assume that each astronaut requires $2.50 \times 10^3$ kcal of energy per day. Further assume that this energy can be equated to the heat of combustion of a quantity of glucose,

$C_6H_{12}O_6$, to $CO_2(g)$ and $H_2O(l)$. From the amount of glucose required to give $2.50 \times 10^3$ kcal of heat, calculate the amount of $CO_2$ produced and hence the amount of LiOH required. The $\Delta H_f^\circ$ for glucose$(s)$ is $-1273$ kJ/mol.

**6.120** A rebreathing gas mask contains potassium superoxide, $KO_2$, which reacts with moisture in the breath to give oxygen.

$$4KO_2(s) + 2H_2O(l) \longrightarrow 4KOH(s) + 3O_2(g)$$

Estimate the grams of potassium superoxide required to supply a person's oxygen needs for one hour. Assume a person requires $1.00 \times 10^2$ kcal of energy for this time period. Further assume that this energy can be equated to the heat of combustion of a quantity of glucose, $C_6H_{12}O_6$, to $CO_2(g)$ and $H_2O(l)$. From the amount of glucose required to give $1.00 \times 10^2$ kcal of heat, calculate the amount of oxygen consumed and hence the amount of $KO_2$ required. The $\Delta H_f^\circ$ for glucose$(s)$ is $-1273$ kJ/mol.

 ## Media Activities

**6.121** Watch the Thermite Reaction *Visualization* on the student CD.
a. Using the Standard Enthalpy of Formation Values in Appendix C, calculate $\Delta H$ for the thermite reaction $Fe_2O_3(s) + 2Al(s) \rightarrow Al_2O_3(s) + 2Fe(s)$.
b. Calculate the amount of heat given off when 5.00 grams of aluminum is completely reacted with an excess of iron(III) oxide.
c. If all of the heat from the reaction in part a was absorbed by 2.00 L of water a 20.0°C, what would be the final temperature of the water?

**6.122** Play the Heat of Reaction *Visualization* on the student CD and make note of the reactants and both the initial and final temperatures of the experiment.
a. Write a balanced chemical equation for the reaction.
b. If there was 100.00 g of solution in the calorimeter during the reaction, calculate the amount of heat evolved.

c. If $\Delta H$ for the reaction is $-54$ kJ/mol, how many moles of acid were consumed during the experiment?

**6.123** Open the Calorimetry Laboratory *Understanding Concepts* simulation on the student web site.
a. Determine the heat capacity of the calorimeter by performing an experiment where you burn (ignite) a known quantity of benzoic acid ($\Delta H = 26.43$ kJ/g), and note the temperature change ($\Delta T$) of the calorimeter.
b. Conduct an experiment using 10 grams of pepperoni pizza and 10 grams of peanut butter. If you were trying to add weight, which of the two foods would you choose on a per gram basis?
c. Using your answer from part a and performing an additional experiment, determine $\Delta H$ in kJ/g for potato chips.

**6.124** For review, take the ACE practice test for Chapter 6 on the student web site.

# 7

The colored flames from several metal compounds.

# Quantum Theory of the Atom

According to Rutherford's model (Section 2.2), an atom consists of a nucleus many times smaller than the atom itself, with electrons occupying the remaining space. How are the electrons distributed in this space? Or, we might ask, what are the electrons doing in the atom?

The answer was to come from an unexpected area: the study of colored flames. When metal compounds burn in a flame, they emit bright colors (Figure 7.1). The spectacular colors of fireworks are due to the burning of metal compounds. Lithium and strontium compounds give a deep red color; barium compounds, a green color; and copper compounds, a bluish-green color.

Although the red flames of lithium and strontium appear similar, the light from each can be resolved (separated) by means of a prism into distinctly different colors. This resolution easily distinguishes the two elements. A prism disperses the colors of

**Figure 7.1**
**Flame tests of Groups IA and IIA elements.** A wire loop containing a sample of a metal compound is placed in a flame. *Left to right:* flames of lithium (red), sodium (yellow), strontium (red), and calcium (orange).

▶ This chapter can immediately follow Chapter 2, if you wish.

**Visualization: Flame Tests**
**Visualization: H₂ Line Spectrum**

▶ Video no. 7 (Series B), "Pyrotechnics: The Chemistry of Fireworks."

white light just as small raindrops spread the colors of sunlight into a rainbow or spectrum. But the light from a flame, when passed through a prism, reveals something other than a rainbow. Instead of the continuous range of color from red to yellow to violet, the spectrum from a strontium flame, for example, shows a cluster of red lines and blue lines against a black background. The spectrum of lithium is different, showing a red line, a yellow line, and two blue lines against a black background. (See Figure 7.2.)

Each element, in fact, has a characteristic line spectrum because of the emission of light from atoms in the hot gas. The spectra can be used to identify elements. How is it that each atom emits particular colors of light? What does a line spectrum tell us about the structure of an atom? If you know something about the structures of atoms, can you explain the formation of ions and molecules? We will answer these questions in this and the next few chapters.

# Light Waves, Photons, and the Bohr Theory

▶ Video Series C, "Flame Tests."

▶ Video no. 13 (Series A), "Flame Tests."

In Chapter 2 we looked at the *basic structure* of atoms and we introduced the concept of a chemical bond. To understand the formation of a chemical bond between atoms, however, you need to know something about the *electronic structure* of atoms. The present theory of the electronic structure of atoms started with an explanation of the colored light produced in hot gases and flames. Before we can discuss this, we need to describe the nature of light.

## 7.1 The Wave Nature of Light

▶ Demonstration of line spectra: Have students observe atomic line spectra from flames, gas discharge tubes, and incandescent bulbs using diffraction gratings made of acetate film in slide mounts (available from Edmund Scientific Company).

If you drop a stone into one end of a quiet pond, the impact of the stone with the water starts an up-and-down motion of the water surface. This up-and-down motion travels outward from where the stone hit; it is a familiar example of a wave. A *wave* is a continuously repeating change or oscillation in matter or in a physical field. Light is also a wave. It consists of oscillations in electric and magnetic fields that can travel through space. Visible light, x rays, and radio waves are all forms of *electromagnetic radiation*.

## EMISSION (LINE) SPECTRA

**Figure 7.2**
**Emission (line) spectra of some elements.** The lines correspond to visible light emitted by atoms. (Wavelengths of lines are given in nanometers.)

▶ **Animation: Electromagnetic wave.**

You characterize a wave by its wavelength and frequency. The **wavelength,** denoted by the Greek letter $\lambda$ (lambda), is *the distance between any two adjacent identical points of a wave*. Thus, the wavelength is the distance between two adjacent peaks or troughs of a wave. Figure 7.3 shows a cross section of a water wave at a given moment, with the wavelength ($\lambda$) identified. Radio waves have wavelengths from approximately 100 mm to several hundred meters. Visible light has much shorter wavelengths, about $10^{-6}$ m. Wavelengths of visible light are often given in nanometers (1 nm = $10^{-9}$ m). For example, light of wavelength $5.55 \times 10^{-7}$ m, the greenish-yellow light to which the human eye is most sensitive, equals 555 nm.

**Figure 7.3**
**Water wave (ripple).** The wavelength (λ) is the distance between any two adjacent identical points of a wave, such as two adjacent peaks or two adjacent troughs.

Visualization:
Electromagnetic Wave

The **frequency** of a wave is *the number of wavelengths of that wave that pass a fixed point in one unit of time* (usually one second). For example, imagine you are anchored in a small boat on a pond when a stone is dropped into the water. Waves travel outward from this point and move past your boat. The number of wavelengths that pass you in one second is the frequency of that wave. Frequency is denoted by the Greek letter $\nu$ (nu, pronounced "new"). The unit of frequency is /s, or s$^{-1}$, also called the *hertz* (Hz).

The wavelength and frequency of a wave are related to each other. Figure 7.4 shows two waves, each traveling from left to right at the same speed; that is, each wave moves the same total length in 1s. The top wave, however, has a wavelength twice that of the bottom wave. In 1s, four complete wavelengths of the top wave move left to right from the origin. It has a frequency of 4/s, or 4 Hz. In the same time, eight complete wavelengths of the bottom wave move left to right from the origin. It has a frequency of 8/s, or 8 Hz. Note that for two waves traveling with a given speed, wavelength and frequency are inversely related: the greater the wavelength, the lower the frequency, and vice versa. In general, with a wave of frequency $\nu$ and wavelength $\lambda$, there are $\nu$ wavelengths, each of length $\lambda$, that pass a fixed point every second. The product $\nu\lambda$ is the total length of the wave that has passed the point in 1s. This length of wave per second is the speed of the wave. For light of speed $c$

$$c = \nu\lambda$$

**Figure 7.4**
**Relation between wavelength and frequency.** Both waves are traveling at the same speed. The top wave has a wavelength twice that of the bottom wave. The bottom wave, however, has twice the frequency of the top wave.

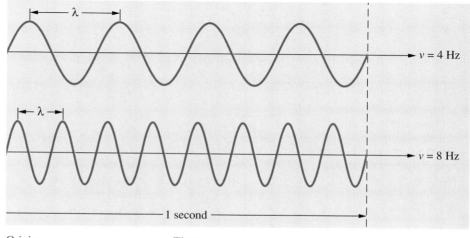

The speed of light waves in a vacuum is a constant and is independent of wavelength or frequency. This speed is $3.00 \times 10^8$ m/s, which is the value for $c$ that we use in the following examples. ◄

► To understand how fast the
speed of light is, it might help
to realize that it takes only 2.5 s
for radar waves (which travel
at the speed of light) to leave
earth, bounce off the moon, and
return—a total distance of
478,000 miles.

► An alternate example is
provided in the Instructor's
Resource Manual.

## EXAMPLE 7.1

## Obtaining the Wavelength of Light from Its Frequency

What is the wavelength of the yellow sodium emission, which has a frequency of $5.09 \times 10^{14}$/s?

### SOLUTION

The frequency and wavelength are related by the formula $c = \nu\lambda$. You rearrange this formula to give

$$\lambda = \frac{c}{\nu}$$

in which $c$ is the speed of light ($3.00 \times 10^8$ m/s). Substituting,

$$\lambda = \frac{3.00 \times 10^8 \text{ m/s}}{5.09 \times 10^{14}/\text{s}} = 5.89 \times 10^{-7} \text{ m, or } \mathbf{589 \text{ nm}}$$

### EXERCISE 7.1

The frequency of the strong red line in the spectrum of potassium is $3.91 \times 10^{14}$/s. What is the wavelength of this light in nanometers?

See Problems 7.27 and 7.28.

► An alternate example is
provided in the Instructor's
Resource Manual.

## EXAMPLE 7.2

## Obtaining the Frequency of Light from Its Wavelength

What is the frequency of violet light with a wavelength of 408 nm?

### SOLUTION

You rearrange the equation relating frequency and wavelength to give

$$\nu = \frac{c}{\lambda}$$

Substituting for $\lambda$ (408 nm = $408 \times 10^{-9}$ m) gives

$$\nu = \frac{3.00 \times 10^8 \text{ m/s}}{408 \times 10^{-9} \text{ m}} = \mathbf{7.35 \times 10^{14}/\text{s}}$$

► In 1854, Kirchhoff found that
each element has a unique
spectrum. Later, Bunsen and
Kirchhoff developed the prism
spectroscope and used it to
confirm their discovery of two
new elements, cesium (in 1860)
and rubidium (in 1861).

### EXERCISE 7.2

The element cesium was discovered in 1860 by Robert Bunsen and Gustav Kirchhoff, who found two bright blue lines in the spectrum of a substance isolated from a mineral water. One of the spectral lines of cesium has a wavelength of 456 nm. What is its frequency? ◄

See Problems 7.29 and 7.30.

**Figure 7.5**
**The electromagnetic spectrum.** Divisions between regions are not defined precisely.

*The range of frequencies or wavelengths of electromagnetic radiation* is called the **electromagnetic spectrum,** shown in Figure 7.5. Visible light extends from the violet end of the spectrum, which has a wavelength of about 400 nm, to the red end, with a wavelength of less than 800 nm. Beyond these extremes, electromagnetic radiation is not visible to the human eye. Infrared radiation has wavelengths greater than 800 nm (greater than the wavelength of red light), and ultraviolet radiation has wavelengths less than 400 nm (less than the wavelength of violet light).

### CONCEPT CHECK 7.1

Laser light of a specific frequency falls on a crystal that converts this light into one with double the original frequency. How is the wavelength of this frequency-doubled light related to the wavelength of the original laser light? Suppose the original laser light was red. In which region of the spectrum would the frequency-doubled light be? (If this is the visible region, what color is the light?)

## 7.2 Quantum Effects and Photons

Isaac Newton, who studied the properties of light in the seventeenth century, believed that light consisted of a beam of particles. In 1801, however, British physicist Thomas Young showed that light, like waves, could be diffracted. *Diffraction* is a property of waves in which the waves spread out when they encounter an obstruction or small hole about the size of the wavelength. You can observe diffraction by viewing a light source through a hole—for example, a streetlight through a mesh curtain. The image of the streetlight is blurred by diffraction.

By the early part of the twentieth century, the wave theory of light appeared to be well entrenched. But in 1905 the German physicist Albert Einstein (1879–1955; emigrated to the U.S. in 1933) discovered that he could explain a phenomenon known as the *photoelectric effect* by postulating that light had both wave and particle properties. Einstein based this idea on the work of the German physicist Max Planck (1858–1947).

## Planck's Quantization of Energy

▶ Max Planck was professor of physics at the University of Berlin when he did this research. He received the Nobel Prize in physics for it in 1918.

In 1900 Max Planck found a theoretical formula that exactly describes the intensity of light of various frequencies emitted by a hot solid at different temperatures. ◀ Earlier, others had shown experimentally that the light of maximum intensity from a hot solid varies in a definite way with temperature. A solid glows red at 750°C, then white as the temperature increases to 1200°C. At the lower temperature, chiefly red light is emitted. As the temperature increases, more yellow and blue light become mixed with the red, giving white light. ◀

▶ Radiation emitted from the human body and warm objects is mostly infrared, which is detected by burglar alarms, military night-vision scopes, and similar equipment.

According to Planck, the atoms of the solid oscillate, or vibrate, with a definite frequency $v$, depending on the solid. But in order to reproduce the results of experiments on glowing solids, he found it necessary to accept a strange idea. An atom could have only certain energies of vibration, $E$, those allowed by the formula

$$E = nhv, \qquad n = 1, 2, 3, \ldots$$

where $h$ is a constant, now called **Planck's constant,** *a physical constant with the value 6.63 × 10$^{-34}$ J · s.* The value of $n$ must be 1 or 2 or some other whole number. Thus, the only energies a vibrating atom can have are $hv$, $2hv$, $3hv$, and so forth.

The numbers symbolized by $n$ are called *quantum numbers.* The vibrational energies of the atoms are said to be *quantized;* that is, the possible energies are limited to certain values. Quantization of energy seems contradicted by everyday experience. Imagine the same concept being applied to the energy of a moving automobile. Quantization of a car's energy would mean that only certain speeds were possible. A car could travel at, say, 10, 20, or 30 miles per hour (mi/hr) but not at 12 or 25 mi/hr or any other intermediate speeds! For car energies, this is unreasonable. But for atoms, quantization is the rule.

Visualization:
Photoelectric Effect (yellow)
Photoelectric Effect (blue)

## Photoelectric Effect

Planck himself was uneasy with the quantization assumption and tried unsuccessfully to eliminate it from his theory. Albert Einstein, on the other hand, boldly extended Planck's work to include the structure of light itself. Einstein reasoned that if a vibrating atom changed energy, say from $3hv$ to $2hv$, it would decrease in energy by $hv$, and this energy would be emitted as a bit (or quantum) of light energy. He therefore postulated that light consists of quanta (now called **photons**), or *particles of electromagnetic energy, with energy E proportional to the observed frequency of the light:* ◀

▶ Albert Einstein obtained his Ph.D. in 1905. In the same year he published four ground-breaking research papers: one on the photoelectric effect (for which he received the Nobel Prize in physics in 1921); two on special relativity; and one on Brownian motion (which led to experiments that tested kinetic-molecular theory and ended remaining doubts of the existence of atoms and molecules).

$$E = hv$$

In 1905 Einstein used this photon concept to explain the photoelectric effect.

The **photoelectric effect** is *the ejection of electrons from the surface of a metal or from another material when light shines on it* (see Figure 7.6). Electrons are ejected, however, only when the frequency of light exceeds a certain *threshold value* characteristic of the particular metal. For example, although violet light will cause potassium metal to eject electrons, no amount of red light (which has a lower frequency) has any effect.

To explain this dependence of the photoelectric effect on the frequency, Einstein assumed that an electron is ejected from a metal when it is struck by a single photon. Therefore, this photon must have at least enough energy to remove the electron from the attractive forces of the metal. No matter how many photons strike the metal, if no single one has sufficient energy, an electron cannot be ejected. A photon of red light has insufficient energy to remove an electron from potassium. But a photon corre-

▶ Animations: Photoelectric effect.

**Figure 7.6**
**The photoelectric effect.**
Light shines on a metal surface, knocking out electrons. The metal surface is contained in an evacuated tube, which allows the ejected electrons to be accelerated to a positively charged plate. As long as light of sufficient frequency shines on the metal, free electrons are produced and a current flows through the tube. (The current is measured by an ammeter.) When the light is turned off, the current stops flowing.

Light rays

Released electron is attracted to positive wire

Positive wire or plate

Light-sensitive metal surface

Battery    Ammeter

sponding to the threshold frequency has just enough energy, and at higher frequencies it has more than enough energy. When the photon hits the metal, its energy $h\nu$ is taken up by the electron. The photon ceases to exist as a particle; it is said to be *absorbed*.

The wave and particle pictures of light should be regarded as complementary views of the same physical entity. This is called the *wave–particle duality* of light. The equation $E = h\nu$ displays this duality; $E$ is the energy of a light particle or photon, and $\nu$ is the frequency of the associated wave. Neither the wave nor the particle view alone is a complete description of light.

▶ An alternate example is provided in the Instructor's Resource Manual.

**EXAMPLE 7.3**

## Calculating the Energy of a Photon

The red spectral line of lithium occurs at 671 nm ($6.71 \times 10^{-7}$ m). Calculate the energy of one photon of this light.

**SOLUTION**

The frequency of this light is

$$\nu = \frac{c}{\lambda} = \frac{3.00 \times 10^8 \text{ m/s}}{6.71 \times 10^{-7} \text{ m}} = 4.47 \times 10^{14}/\text{s}$$

Hence, the energy of one photon is

$$E = h\nu = 6.63 \times 10^{-34} \text{ J} \cdot \text{s} \times 4.47 \times 10^{14}/\text{s} = \mathbf{2.96 \times 10^{-19} \text{ J}}$$

**Problem Solving:**
**Frequency and Energy of**
**Electromagnetic Radiation**

**EXERCISE 7.3**

The following are representative wavelengths in the infrared, ultraviolet, and x-ray regions of the electromagnetic spectrum, respectively: $1.0 \times 10^{-6}$ m, $1.0 \times 10^{-8}$ m, and $1.0 \times 10^{-10}$ m. What is the energy of a photon of each radiation? Which has the greatest amount of energy per photon? Which has the least?

See Problems 7.35, 7.36, 7.37, and 7.38.

# Zapping Hamburger with Gamma Rays

**EVERYDAY LIFE**

In 1993 in Seattle, Washington, four children died and hundreds of people became sick from food poisoning when they ate undercooked hamburgers containing a dangerous strain of a normally harmless bacteria, *Escherichia coli*. A similar bout of food poisoning from hamburgers containing the dangerous strain of *E. coli* occurred in the summer of 1997; 17 people became ill. This time the tainted hamburger was traced to a meat processor in Nebraska, which immediately recalled 25 million pounds of ground beef. Within months, the U.S. Food and Drug Administration approved the irradiation of red meat by gamma rays to kill harmful bacteria. Gamma rays are a form of electromagnetic radiation similar to x rays, but the photons of gamma rays have higher energy.

The idea of using high-energy radiation to disinfect foods is not new (Figure 7.7). It has been known for about 50 years that high-energy radiation kills bacteria and molds in foods, prolonging their shelf life. Gamma rays kill organisms by breaking up the DNA molecules within their cells. This DNA holds the information for producing vital cell proteins.

Currently, the gamma rays used in food irradiation come from the radioactive decay of cobalt-60. It is perhaps this association with radioactivity that has slowed the acceptance of irradiation as a food disinfectant. However, food is not made radioactive by irradiating it

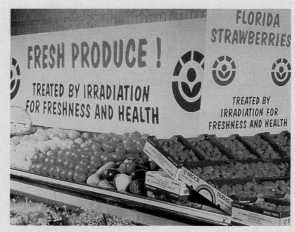

**Figure 7.7**
**Gamma irradiation of foods.** Here, fresh produce has been irradiated to reduce spoilage and to ensure that the produce does not contain harmful bacteria. Now, meat can also be irradiated.

with gamma rays. The two major concerns about food irradiation—whether a food's nutritional value might be significantly diminished and whether decomposition products might cause cancer—appear to be without foundation. Irradiation of foods with gamma rays may soon join pasteurization as a way to protect our food supply from harmful bacteria.

## 7.3 The Bohr Theory of the Hydrogen Atom

According to Rutherford's nuclear model, the atom consists of a nucleus with most of the mass of the atom and a positive charge, around which move enough electrons to make the atom electrically neutral. But this model, offered in 1911, posed a dilemma. Using the then-current theory, one could show that an electrically charged particle (such as an electron) that revolves around a center would continuously lose energy as electromagnetic radiation. As an electron in an atom lost energy, it would spiral into the nucleus (in about $10^{-10}$ s, according to available theory). The stability of the atom could not be explained.

A solution to this theoretical dilemma was found in 1913 by Niels Bohr (1885–1962), a Danish physicist, who at the time was working with Rutherford (see Figure 7.8). Using the work of Planck and Einstein, Bohr applied a new theory to the simplest atom, hydrogen. Before we look at Bohr's theory, we need to consider the line spectra of atoms, which we looked at briefly in the chapter opening.

**Figure 7.8**
**Niels Bohr (1885–1962).**
After Bohr developed his
quantum theory of the
hydrogen atom, he used his
ideas to explain the periodic
behavior of the elements.
Later, when the new quantum
mechanics was discovered by
Schrödinger and Heisenberg,
Bohr spent much of his time
developing its philosophical
basis. He received the Nobel
Prize in physics in 1922.

▶ Animation: Refraction of white
light.

▶ Video no. 21 (Series D), Chemi-
luminescence in a Bottle;" no.
22 (Series D), "UV Lamp and
Credit Cards."

## Atomic Line Spectra

As described in the previous section, a heated solid emits light. A heated tungsten
filament in an ordinary lightbulb is a typical example. With a prism we can spread
out the light from a bulb to give a **continuous spectrum**—that is, *a spectrum containing light of all wavelengths,* like that of a rainbow (see Figure 7.9). The light emitted by a heated gas, however, yields different results. Rather than a continuous spectrum, with all colors of the rainbow, we obtain a **line spectrum**—*a spectrum showing
only certain colors or specific wavelengths of light.* When the light from a hydrogen
gas discharge tube (in which an electrical discharge heats hydrogen gas) is separated
into its components by a prism, it gives a spectrum of lines, each line corresponding
to light of a given wavelength. The light produced in the discharge tube is emitted
by hydrogen atoms. Figure 7.2 shows the line spectrum of the hydrogen atom, as well
as line spectra of other atoms.

The line spectrum of the hydrogen atom is especially simple. In the visible region,
it consists of only four lines (a red, a blue-green, a blue, and a violet), although others
appear in the infrared and ultraviolet regions. In 1885 J. J. Balmer showed that the wavelengths $\lambda$ in the visible spectrum of hydrogen could be reproduced by a simple formula:

$$\frac{1}{\lambda} = 1.097 \times 10^7/\text{m}\left(\frac{1}{2^2} - \frac{1}{n^2}\right)$$

Here $n$ is some whole number (integer) greater than 2. By substituting $n = 3$, for
example, and calculating $1/\lambda$ and then $\lambda$, one finds $\lambda = 6.56 \times 10^{-7}$ m, or 656 nm,
a wavelength corresponding to red light. The wavelengths of the other lines in the
hydrogen atom visible spectrum are obtained by successively substituting $n = 4$,
$n = 5$, and $n = 6$.

## Bohr's Postulates

Bohr set down the following postulates to account for (1) the stability of the hydrogen atom (that the atom exists and its electron does not continuously radiate energy
and spiral into the nucleus) and (2) the line spectrum of the atom.

**1. Energy-level Postulate** An electron can have only *specific energy values in an
atom,* which are called its **energy levels.** Therefore, the atom itself can have only specific total energy values.

**Figure 7.9**
**Dispersion of white light
by a prism.** White light,
entering at the left, strikes a
prism, which disperses the
light into a continuous
spectrum of wavelengths.

Bohr borrowed the idea of quantization of energy from Planck. Bohr, however, devised a rule for this quantization that could be applied to the motion of an electron in an atom. From this he derived the following formula for the energy levels of the electron in the hydrogen atom:

$$E = -\frac{R_H}{n^2} \qquad n = 1, 2, 3, \ldots \infty \qquad \text{(for H atom)}$$

where $R_H$ is a constant (expressed in energy units) with the value $2.179 \times 10^{-18}$ J. Different values of the possible energies of the electron are obtained by putting in different values of $n$, which can have only the integral values 1, 2, 3, and so forth (up to infinity, $\infty$). Here $n$ is called the *principal quantum number*. The diagram in Figure 7.10 shows the energy levels of the electron in the H atom. ◀

▶ **The energies have negative values because the energy of the separated nucleus and electron is taken to be zero. As the nucleus and electron come together to form a stable state of the atom, energy is released and the energy becomes less than zero, or negative.**

**2. Transitions Between Energy Levels**   An electron in an atom can change energy only by going from one energy level to another energy level. By so doing, the electron undergoes a *transition*.

We explain the emission of light by atoms to give a line spectrum as follows: An electron in a higher energy level (initial energy level $E_i$) undergoes a transition to a lower energy level (final energy level, $E_f$). (See Figure 7.10.) In this process, the electron loses energy, which is emitted as a photon. In other words, the final energy of the electron plus the energy of the photon equals the initial energy of the electron (from the law of conservation of energy):

$$E_f + h\nu = E_i$$

**Figure 7.10**
**Energy-level diagram for the electron in the hydrogen atom.**   Energy is plotted on the vertical axis (in fractional multiples of $R_H$). The arrow represents an electron transition (discussed in Postulate 2) from level $n = 4$ to level $n = 2$. Light of wavelength 486 nm (blue-green) is emitted. (See Example 7.4 for the calculation of this wavelength.)

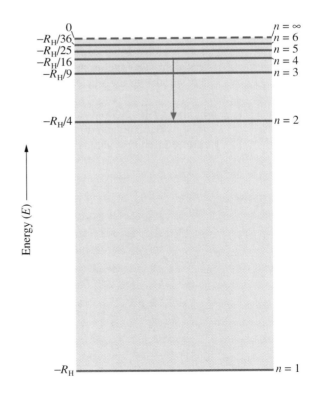

When you rearrange this, you obtain

Energy of emitted photon $= h\nu = E_i - E_f$

In this postulate, Bohr used Einstein's photon concept to explain the line spectra of atoms. By substituting values of the energy levels of the electron in the H atom, which he had derived, into the preceding equation, he was able to reproduce Balmer's formula exactly. Moreover, he was able to predict all of the lines in the spectrum of the hydrogen atom, including those in the infrared and ultraviolet regions.

To show how Bohr obtained Balmer's formula, write $n_i$ for the principal quantum number of the initial energy level, and $n_f$ for the principal quantum number of the final energy level. Then, from Postulate 1,

$$E_i = -\frac{R_H}{n_i^2} \quad \text{and} \quad E_f = -\frac{R_H}{n_f^2}$$

You substitute these into the previous equation (the one shaded green), which states that the energy of the emitted photon, $h\nu$, equals $E_i - E_f$.

$$h\nu = E_i - E_f = \left(-\frac{R_H}{n_i^2}\right) - \left(-\frac{R_H}{n_f^2}\right)$$

That is,

$$h\nu = R_H\left(\frac{1}{n_f^2} - \frac{1}{n_i^2}\right)$$

Now recalling that $\nu = c/\lambda$, you can rewrite this as

$$\frac{1}{\lambda} = \frac{R_H}{hc}\left(\frac{1}{n_f^2} - \frac{1}{n_i^2}\right)$$

By substituting $R_H = 2.179 \times 10^{-18}$ J, $h = 6.626 \times 10^{-34}$ J·s, and $c = 2.998 \times 10^8$ m/s, you find that $R_H/hc = 1.097 \times 10^7$/m, which is the constant given in the Balmer formula. In Balmer's formula, the quantum number $n_f$ is 2. This means that Balmer's formula gives wavelengths that occur when electrons in H atoms undergo transitions from energy levels $n_i > 2$ to level $n_f = 2$. If you change $n_f$ to other integers, you obtain different series of lines (or wavelengths) for the spectrum of the H atom (see Figure 7.11).

▶ An alternate example is provided in the Instructor's Resource Manual.

## EXAMPLE 7.4

## Determining the Wavelength or Frequency of a Hydrogen Atom Transition

What is the wavelength of light emitted when the electron in a hydrogen atom undergoes a transition from energy level $n = 4$ to level $n = 2$?

 **PROBLEM STRATEGY**

You start with the formula for the energy levels of the H atom, $E = -R_H/n^2$, and obtain the energy change for the transition. This energy difference will equal the energy of the photon, from which you can calculate frequency, then wavelength,

**Figure 7.11**
**Transitions of the electron in the hydrogen atom.**
The diagram shows the Lyman, Balmer, and Paschen series of transitions that occur for $n_f = 1, 2,$ and 3, respectively.

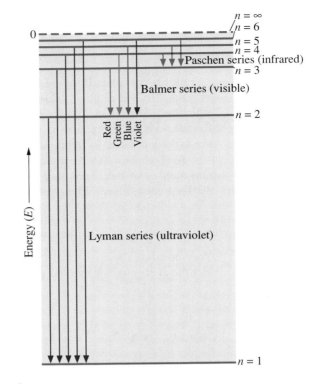

of the emitted light. (Although you could do this problem by "plugging into" the previously derived equation for $1/\lambda$, the method followed here requires you to remember only key formulas, shown in color in the text.)

**SOLUTION**

From the formula for the energy levels, you know that

$$E_i = \frac{-R_H}{4^2} = \frac{-R_H}{16} \quad \text{and} \quad E_f = \frac{-R_H}{2^2} = \frac{-R_H}{4}$$

You subtract the lower value from the higher value, to get a positive result. (The energy of the photon is positive. If your result is negative, it means you calculated $E_f - E_i$, rather than $E_i - E_f$. Simply reverse the subtraction to obtain a positive result.) Because this result equals the energy of the photon, you equate it to $h\nu$:

$$\left(\frac{-R_H}{16}\right) - \left(\frac{-R_H}{4}\right) = \frac{-4R_H + 16R_H}{64} = \frac{-R_H + 4R_H}{16} = \frac{3R_H}{16} = h\nu$$

The frequency of the light emitted is

$$\nu = \frac{3R_H}{16\,h} = \frac{3}{16} \times \frac{2.179 \times 10^{-18}\,\text{J}}{6.63 \times 10^{-34}\,\text{J} \cdot \text{s}} = 6.17 \times 10^{14}/\text{s}$$

Since $\lambda = c/\nu$,

$$\lambda = \frac{3.00 \times 10^8\,\text{m/s}}{6.17 \times 10^{14}/\text{s}} = 4.86 \times 10^{-7}\,\text{m, or \textbf{486 nm}}$$

The color is blue-green (see Figure 7.10).

---

**EXERCISE 7.4**

Calculate the wavelength of light emitted from the hydrogen atom when the electron undergoes a transition from level $n = 3$ to level $n = 1$.

---

See Problems 7.41, 7.42, 7.43, and 7.44.

▶ Experiments in General Chemistry, 7, "The Absorption Spectrum of Cobalt(II) Chloride."

   According to Bohr's theory, the *emission* of light from an atom occurs when an electron undergoes a transition from an upper energy level to a lower one. To complete the explanation, we need to describe how the electron gets into the upper level prior to emission. Normally, the electron in a hydrogen atom exists in its lowest, or $n = 1$, level. To get into a higher energy level, the electron must gain energy, or be *excited*. One way this can happen is through the collision of two hydrogen atoms. During this collision, some of the kinetic energy of one atom can be gained by the electron of another atom, thereby boosting, or exciting, the electron from the $n = 1$ level to a higher energy level. The excitation of atoms and the subsequent emission of light are most likely to occur in a hot gas, where atoms have large kinetic energies.

   Bohr's theory explains not only the emission but also the *absorption* of light. When an electron in the hydrogen atom undergoes a transition from $n = 3$ to $n = 2$, a photon of red light (wavelength 656 nm) is emitted. When red light of wavelength 656 nm shines on a hydrogen atom in the $n = 2$ level, a photon can be absorbed. If the photon is absorbed, the energy is gained by the electron, which undergoes a transition to the $n = 3$ level. (This is the reverse of the emission process we just discussed.) Materials that have a color, such as dyed textiles and painted walls, appear colored because of the absorption of light. For example, when white light falls on a substance that absorbs red light, the color components that are not absorbed, the yellow and blue light, are reflected. The substance appears blue-green.

   Postulates 1 and 2 hold for atoms other than hydrogen, except that the energy levels cannot be obtained by a simple formula. However, if you know the wavelength of the emitted light, you can relate it to $\nu$ and then to the difference in energy levels of the atom. The energy levels of atoms have been experimentally determined in this way.

---

**EXERCISE 7.5**

What is the difference in energy levels of the sodium atom if emitted light has a wavelength of 589 nm?

---

See Problems 7.47 and 7.48.

**CONCEPT CHECK 7.2**

An atom has a line spectrum consisting of a red line and a blue line. Assume that each line corresponds to a transition between two adjacent energy levels. Sketch an energy-level diagram with three energy levels that might explain this line spectrum, indicating the transitions on this diagram. Consider the transition from the highest energy level on this diagram to the lowest energy level. How would you describe the color or region of the spectrum corresponding to this transition?

# Lasers and Compact Disc Players

**MATERIALS**

Lasers are sources of intense, highly directed beams of *monochromatic* light—light of very narrow wavelength range. Figure 7.12 shows a laser being used in an industrial research project. The word *laser* is an acronym meaning *l*ight *a*mplification by *s*timulated *e*mission of *r*adiation. Many different kinds of lasers now exist, but the general principle of a laser can be understood by looking at the ruby laser, the first type constructed (in 1960).

Ruby is aluminum oxide containing a small concentration of chromium(III) ions, $Cr^{3+}$, in place of some aluminum ions. The electron transitions in a ruby laser are those of $Cr^{3+}$ ions in solid $Al_2O_3$. An energy-level diagram of $Cr^{3+}$ in ruby is shown in Figure 7.13. Most of the $Cr^{3+}$ ions are initially in the lowest energy level (level 1). If you shine light of wavelength 545 nm on a ruby crystal, the light is absorbed and $Cr^{3+}$ ions undergo transitions from level 1 to level 3. A few of these ions in level 3 emit photons and return to level 1, but most of them undergo *radiationless transitions* to level 2. In these transitions, the ions lose energy as heat to the ruby crystal, rather than emit photons.

The $Cr^{3+}$ ion in level 2, like any species in a level other than the lowest, is unstable and in time will undergo a transition to level 1. Thus, you expect the $Cr^{3+}$ ions in level 2 to undergo transitions to level 1 with the

**Figure 7.13**
**Energy levels of the chromium(III) ion in ruby.**
The energy levels have been numbered by increasing energy from 1 to 3. (Level 3 is broadened in the solid state.) When chromium(III) ions absorb the 545-nm portion of the light from a flash lamp, electrons in the ions first undergo transitions to level 3 and then radiationless transitions to level 2. (The energy is lost as heat.) Electrons accumulate in level 2 until more exist in level 2 than in level 1. Then a photon from a spontaneous emission from level 2 to level 1 can stimulate the emission from other atoms in level 2. These photons move back and forth between reflective surfaces, stimulating additional emissions and forming a laser pulse at 694 nm.

**Figure 7.12**
**Lasers in industrial research.** A laser beam is being used in a research project to design a metal-cutting tool.

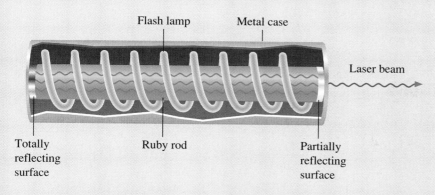

**Figure 7.14**
**A ruby laser.** A flash lamp encircles a ruby rod. Light of 545 nm (green) from the flash "pumps" electrons from level 1 to level 3, then level 2, from which stimulated emission forms a laser pulse at 694 nm (red). The stimulated emission bounces back and forth between reflective surfaces at the ends of the ruby rod, building up a coherent laser beam. One end has a partially reflective surface to allow the laser beam to exit from the ruby.

*spontaneous emission* of photons. (The wavelength of this emission is 694 nm.) However, this spontaneous emission of $Cr^{3+}$ is relatively slow. If you flash a ruby rod with a bright light at 545 nm, most of the $Cr^{3+}$ ions end up in level 2 for perhaps a fraction of a millisecond (which is quite a long time, compared with that of most excited states of atoms—they exist only about $10^{-8}$ s). This buildup of many excited species is crucial to the operation of a laser. If these excited ions can be triggered to emit simultaneously, or nearly simultaneously, an intense emission will be obtained.

The process of *stimulated emission* is ideal for this triggering. When a photon corresponding to 694 nm encounters a $Cr^{3+}$ ion in level 2, it stimulates the ion to undergo the transition from level 2 to level 1. The ion emits a photon corresponding to exactly the same wavelength as the original photon. In place of just one photon, there are now two photons, the original one and the one obtained by stimulated emission. The net effect is to increase the intensity of the light at this wavelength. Thus, a weak light at 694 nm can be amplified by stimulated emission of the excited ruby.

We can now explain the operation of a ruby laser. A sketch of the laser is shown in Figure 7.14. It consists of a ruby rod a centimeter or less in diameter and 2 cm to 10 cm long. The ruby rod is silvered at one end to provide a completely reflective surface; the other end has a partially reflective surface. A coiled flash-lamp tube is placed around the ruby rod. When the flash lamp is discharged, a bright light is emitted, and light at 545 nm (green) is absorbed by the ruby. Most of the $Cr^{3+}$ ions end up in level 2. A few of these ions spontaneously emit photons corre-

sponding to 694 nm (red light), and these photons stimulate other ions to emit. As the photons are reflected back and forth between the reflective surfaces, more and more ions are stimulated to emit, and the light quickly builds up in intensity until it passes through the partially reflective surface as a pulse of laser light at 694 nm.

Laser light is *coherent*. This means that the waves forming the beam are all in phase; that is, the waves have their maxima and minima at the same points in space and time. The property of coherence of a laser beam is used in compact disc (CD) audio players (Figure 7.15). Music is encoded on the disc in the form of pits, or indentations, on a spiral track. When the disc is played, a small laser beam scans the track and is reflected back to a detector. Light reflected from an indentation is out of phase with light from the laser and interferes with it. Because of this interference, the reflected beam is diminished in intensity and gives a diminished detector signal. Fluctuations in the signal are then converted to sound.

Other properties of laser light are used in different applications. The ability to focus intense light on a spot is used in the surgical correction of a detached retina in the eye. In effect, the laser beam is used to do "spot welding" of the retina. Laser printers also employ the intensity of a laser beam. These printers follow the principle of photocopiers but use a computer to direct the laser light in a pattern of dots to form an image. In chemical research, laser beams provide intense monochromatic light to locate energy levels in molecules, to study the products of very fast chemical reactions, and to analyze samples for small amounts of substances.

*(continued)*

**Figure 7.15**
**How a compact disc player works.** Music or other information is encoded in the form of pits on a compact disc. Lenses direct laser light to the disc, which reflects it back. Light reflected from a pit is out of phase with light from the laser, and these two waves interfere, reducing the intensity of the wave. As the disc moves around, light is reflected from either the main surface at full intensity (on) or from a pit at reduced intensity (off). A light-sensitive detector converts this reflected light to a digital signal (a series of on and off pulses), and a digital-to-analog converter changes this signal to an analog signal that the speakers can accept.

# Quantum Mechanics and Quantum Numbers

Bohr's theory firmly established the concept of atomic energy levels. It was unsuccessful, however, in accounting for the details of atomic structure and in predicting energy levels for atoms other than hydrogen. Further understanding of atomic structure required other theoretical developments.

## 7.4 Quantum Mechanics

Current ideas about atomic structure depend on the principles of *quantum mechanics*, a theory that applies to submicroscopic (that is, extremely small) particles of matter, such as electrons. The first clue to how to proceed in developing this theory came with the discovery of the de Broglie relation.

## de Broglie Relation

According to Einstein, light has not only wave properties, which we characterize by frequency and wavelength, but also particle properties. For example, a particle of light, the photon, has a definite energy $E = h\nu$. One can also show that the photon has momentum. (The momentum of a particle is the product of its mass and speed.) This momentum, $mc$, is related to the wavelength of the light: $mc = h/\lambda$ or $\lambda = h/mc$. ◄

▶ A photon has a rest mass of zero but a relativistic mass $m$ as a result of its motion. Einstein's equation $E = mc^2$ relates this relativistic mass to the energy of the photon, which also equals $h\nu$. Therefore, $mc^2 = h\nu$, or $mc = h\nu/c = h/\lambda$.

In 1923 the French physicist Louis de Broglie reasoned that if light (considered as a wave) exhibits particle aspects, then perhaps particles of matter show characteristics of waves under the proper circumstances. He therefore postulated that a particle of matter of mass $m$ and speed $v$ has an associated wavelength, by analogy with light:

$$\lambda = \frac{h}{mv}$$

The equation $\lambda = h/mv$ is called the **de Broglie relation.**

If matter has wave properties, why are they not commonly observed? Calculation using the de Broglie relation shows that a baseball (0.145 kg) moving at about 60 mi/hr (27 m/s) has a wavelength of about $10^{-34}$ m, a value so incredibly small that such waves cannot be detected. On the other hand, electrons moving at experimentally accessible speeds have wavelengths the size of a few hundred picometers (1 pm = $10^{-12}$ m). Under the proper circumstances, the wave character of electrons should be observable.

The wave property of electrons was first demonstrated in 1927 by C. Davisson and L. H. Germer in the United States and by George Paget Thomson (son of J. J. Thomson) in Britain. They showed that a beam of electrons, just like x rays, can be diffracted by a crystal. The German physicist Ernst Ruska used this wave property to construct the first *electron microscope* in 1933; he shared the 1986 Nobel Prize in physics for this work. A modern instrument is shown in Figure 7.16. The *resolving power,* or ability to distinguish detail, of a microscope that uses waves depends on their wavelength. To resolve detail the size of several hundred picometers, we need a wavelength on that order. X rays have wavelengths in this range, but so far no

**Figure 7.16**
**Scanning electron microscope.** This microscope can resolve details down to 1 nm in a sample. The operator places the sample inside the chamber at the left and views the image on the video screens.

**Figure 7.17**
**Scanning electron microscope image.** This image is of a wasp's head. Color has been added by computer for contrast in discerning different parts of the image.

practical means have been found for focusing them. Electrons, on the other hand, are readily focused with electric and magnetic fields. Figure 7.17 shows a photograph taken with an electron microscope, displaying the detail that is possible with this instrument. ◄

▶ In 1986, Thomas H. Newman collected a $1000 prize offered 26 years earlier by Richard Feynman, physics Nobel Prize winner. Newman, a student of R. Fabian Pease at Stanford University, satisfied Feynman's challenge to reduce a normal printed page to 1/25,000th of its linear dimensions. Newman used a specially constructed electron-beam tool to etch the first page of *A Tale of Two Cities* onto an area 5.6 $\mu$m by 5.6 $\mu$m. The image could be read with an electron microscope, as Feynman required.

▶ An alternate example is provided in the Instructor's Resource Manual.

**EXAMPLE 7.5**

## Applying the de Broglie Relation

a. Calculate the wavelength (in meters) of the wave associated with a 1.00-kg mass moving at 1.00 km/hr. b. What is the wavelength (in picometers) associated with an electron, whose mass is $9.11 \times 10^{-31}$ kg, traveling at a speed of $4.19 \times 10^6$ m/s? (This speed can be attained by an electron accelerated between two charged plates differing by 50.0 volts; voltages in the kilovolt range are used in electron microscopes.)

**SOLUTION**

a. A speed $v$ of 1.00 km/hr equals

$$1.00 \, \frac{km}{hr} \times \frac{1 \, hr}{3600 \, s} \times \frac{10^3 \, m}{1 \, km} = 0.278 \, m/s$$

Substituting quantities (all expressed in SI units for consistency), you get

$$\lambda = \frac{h}{mv} = \frac{6.63 \times 10^{-34} \, J \cdot s}{1.00 \, kg \times 0.278 \, m/s} = \mathbf{2.38 \times 10^{-33} \, m}$$

b. $\lambda = \dfrac{h}{mv} = \dfrac{6.63 \times 10^{-34} \, J \cdot s}{9.11 \times 10^{-31} \, kg \times 4.19 \times 10^6 \, m/s} = 1.74 \times 10^{-10} \, m = \mathbf{174 \, pm}$

**EXERCISE 7.6**

Calculate the wavelength (in picometers) associated with an electron traveling at a speed of $2.19 \times 10^6$ m/s.

See Problems 7.49 and 7.50.

### CONCEPT CHECK 7.3

A proton is approximately 2000 times heavier than an electron. How would the speeds of these particles compare if their corresponding wavelengths were about equal?

## Wave Functions

De Broglie's relation applies quantitatively only to particles in a force-free environment. It cannot be applied directly to an electron in an atom, where the electron is subject to the attractive force of the nucleus. But in 1926 Erwin Schrödinger, guided by de Broglie's work, devised a theory that could be used to find the wave properties of electrons in atoms and molecules. *The branch of physics that mathematically describes the wave properties of submicroscopic particles* is called **quantum mechanics** or **wave mechanics.** ◄

▶ Schrödinger received the Nobel Prize in physics in 1933 for his wave formulation of quantum mechanics. Werner Heisenberg won the Nobel Prize the previous year for his matrix-algebra formulation of quantum mechanics. The two formulations yield identical results.

Without going into the mathematics of quantum mechanics here, we will discuss some of the most important conclusions of the theory. In particular, quantum mechanics alters the way we think about the motion of particles. Our usual concept of motion comes from what we see in the everyday world. We might, for instance, visually follow a ball that has been thrown. The path of the ball is given by its position and velocity (or momentum) at various times. We are therefore conditioned to think in terms of a continuous path for moving objects. In Bohr's theory, the electron was thought of as moving about, or orbiting, the nucleus in the way the earth orbits the sun. Quantum mechanics vastly changes this view of motion. We can no longer think of an electron as having a precise orbit in an atom. To describe such an orbit, we would have to know the exact position of the electron at various times and exactly how long it would take it to travel to a nearby position in the orbit. That is, at any moment we would have to know not only the precise position but also the precise momentum (mass times speed) of the electron.

In 1927 Werner Heisenberg showed from quantum mechanics that it is impossible to know simultaneously, with absolute precision, both the position and the momentum of a particle such as an electron. Heisenberg's **uncertainty principle** is *a relation that states that the product of the uncertainty in position and the uncertainty in momentum of a particle can be no smaller than Planck's constant divided by* $4\pi$. Thus, letting $\Delta x$ be the uncertainty in the $x$ coordinate of the particle and letting $\Delta p_x$ be the uncertainty in the momentum in the $x$ direction, we have

$$(\Delta x)(\Delta p_x) \geq \frac{h}{4\pi}$$

There are similar relations in the $y$ and $z$ directions. The uncertainty principle says that the more precisely you know the position (the smaller $\Delta x$, $\Delta y$, and $\Delta z$), the less well you know the momentum of the particle (the larger $\Delta p_x$, $\Delta p_y$, and $\Delta p_z$). In other words, if you know very well where a particle is, you cannot know where it is going!

The uncertainty principle is only significant for particles of very small mass such as electrons. You can see this by noting that the momentum equals mass times velocity, so $p_x = mv_x$. The preceding relation becomes

$$(\Delta x)(\Delta v_x) \geq \frac{h}{4\pi m}$$

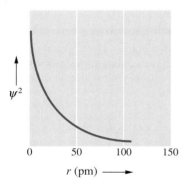

**Figure 7.18**
**Plot of $\psi^2$ for the lowest energy level of the hydrogen atom.** The square of the wave function is plotted versus the distance, $r$, from the nucleus.

For dust particles and baseballs, where $m$ is relatively large, the term on the right becomes nearly zero, and the uncertainties of position and velocity are quite small. The path of a baseball has meaning. For electrons, however, the uncertainties in position and momentum are normally quite large. We cannot describe the electron in an atom as moving in a definite orbit.

Although quantum mechanics does not allow us to describe the electron in the hydrogen atom as moving in an orbit, it does allow us to make *statistical* statements about where we would find the electron if we were to look for it. For example, we can obtain the *probability* of finding an electron at a certain point in a hydrogen atom. Although we cannot say that an electron will definitely be at a particular position at a given time, we can say that the electron is likely (or not likely) to be at this position.

Information about a particle in a given energy level (such as an electron in an atom) is contained in a mathematical expression called a *wave function,* denoted by the Greek letter psi, $\psi$. The wave function is obtained by solving an equation of quantum mechanics (Schrödinger's equation). Its square, $\psi^2$, gives the probability of finding the particle within a region of space.

The wave function and its square, $\psi^2$, have values for all locations about a nucleus. Figure 7.18 shows values of $\psi^2$ for the electron in the lowest energy level of the hydrogen atom along a line starting from the nucleus. Note that $\psi^2$ is large near the nucleus ($r = 0$), indicating that the electron is most likely to be found in this region. The value of $\psi^2$ decreases rapidly as the distance from the nucleus increases, but $\psi^2$ never goes to exactly zero, although the probability does become extremely small at large distances from the nucleus. This means that an atom does not have a definite boundary, unlike in the Bohr model of the atom.

Figure 7.19 shows another view of this electron probability. The graph plots the probability of finding the electron in different spherical shells at particular distances from the nucleus, rather than the probability at a point. Even though the probability of finding the electron at a point near the nucleus is high, the volume of any shell there is small. Therefore, the probability of finding the electron *within a shell* is greatest at some distance from the nucleus. This distance just happens to equal the radius that Bohr calculated for an electron orbit in his model.

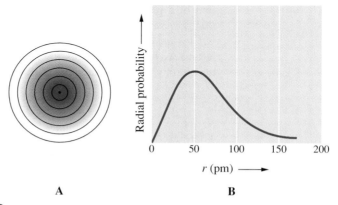

A                    B

**Figure 7.19**
**Probability of finding an electron in a spherical shell about the nucleus.**
*(A)* The diagram shows the probability density for an electron in a hydrogen atom. The region is marked off in shells about the nucleus. *(B)* The graph shows the probability of finding the electron within shells at various distances from the nucleus (radial probability). The curve exhibits a maximum, which means that the radial probability is greatest for a given distance from the nucleus.

# Scanning Tunneling Microscopy

The 1986 Nobel Prize in physics went to three physicists for their work in developing microscopes for viewing extremely small objects. Half of the prize went to Ernst Ruska for the development of the electron microscope, described earlier. The other half was awarded to Gerd Binnig and Heinrich Rohrer, at IBM's research laboratory in Zurich, Switzerland, for their invention of the *scanning tunneling microscope* in 1981. This instrument makes possible the viewing of atoms and molecules on a solid surface (Figure 7.20).

The scanning tunneling microscope is based on the concept of quantum mechanical *tunneling,* which in turn depends on the probability interpretation of quantum mechanics. Consider a hydrogen atom, which consists of an electron about a proton (call it A), and imagine another proton (called B) some distance from the first. In classical terms, if we were to move the electron from the region of proton A to the region of proton B, energy would have to be supplied to remove the electron from the attractive field of proton A. Quantum mechanics, however, gives us a different picture. The probability of the electron in the hydrogen atom being at a location far from proton A, say near proton B, is very small but not zero. In effect, this means that the electron, which nominally belongs to proton A, may find itself near proton B without extra energy having been supplied. The electron is said to have *tunneled* from one atom to another.

The scanning tunneling microscope consists of a tungsten metal needle with an extremely fine point (the probe) placed close to the sample to be viewed (Figure

7.21). If the probe is close enough to the sample, electrons can tunnel from the probe to the sample. The probability for this can be increased by having a small voltage applied between the probe and sample. Electrons tunneling from the probe to the sample give rise to a measurable electric current. The magnitude of the current depends on the distance between the probe and the sample (as well as on the wave function of the atom in the sample). By adjusting this distance, the current can be maintained at a fixed value. As the probe scans the sample, it moves toward or away from the sample to maintain a fixed current; in effect, the probe follows the contours of the sample.

The probe must move incredibly small distances, and this is done by an ingenious mechanism. Certain solids (*piezoelectric* crystals) generate a voltage when their length is changed by small pressure differences (as when a phonograph needle rides over the grooves of a record). The reverse also occurs. Small voltage variations

*(continued)*

**Figure 7.21**
**The scanning tunneling microscope.** A tunneling current flows between the probe and the sample when there is a small voltage between them. A feedback circuit, which provides this voltage, senses the current and varies the voltage on a piezoelectric rod (z-drive) in order to keep the distance constant between the probe and sample. A computer has been programmed to provide voltages to the x-drive and the y-drive to move the probe over the surface of the sample.

**Figure 7.20**
**Scanning tunneling microscope image of benzene molecules on a metal surface.**
Benzene molecules, $C_6H_6$, are arranged in a regular array on a rhodium metal surface.

applied to a piezoelectric rod can generate corresponding small changes in its length. In the tunneling microscope, a variable voltage is applied to a piezoelectric rod to shorten or lengthen it. Because the probe is attached to the rod, the probe is moved toward or away from the sample. In this way the distance of the probe to the sample is adjusted by a voltage in order to maintain a constant tunneling current as the probe scans the sample.

Researchers routinely use the scanning tunneling microscope to study the arrangement of atoms and molecules on surfaces. Lately, scientists have also used the microscope probe to move atoms about a surface in an effort to construct miniature devices. Figure 7.22 shows 48 iron atoms that were arranged in a circle on a copper surface. Each of the iron atoms looks like a sharp mountain peak rising from a plain. Inside this "quantum corral" (as it has come to be called), you can see the wavelike distribution of electrons trapped within. Here is graphic proof of the wavelike nature of electrons!

**Figure 7.22**
**Quantum corral.** IBM scientists used a scanning tunneling microscope probe to arrange 48 iron atoms in a circle on a copper metal surface. The iron atoms appear as the tall peaks in the diagram. Note the wavelike ripple of electrons trapped within the circle of atoms (called a *quantum corral*). [IBM Research Division, Almaden Research Center; research done by Donald M. Eigler and coworkers.]

## 7.5 Quantum Numbers and Atomic Orbitals

▶ Three different quantum numbers are needed because there are three dimensions to space.

According to quantum mechanics, each electron in an atom is described by four different quantum numbers, three of which ($n$, $l$, and $m_l$) specify the wave function that gives the probability of finding the electron at various points in space. ◀ *A wave function for an electron in an atom* is called an **atomic orbital.** An atomic orbital is pictured qualitatively by describing the region of space where there is high probability of finding the electrons. The atomic orbital so pictured has a definite shape. A fourth quantum number ($m_s$) refers to a magnetic property of electrons called *spin*. We first look at quantum numbers, then at atomic orbitals.

### Quantum Numbers

The allowed values and general meaning of each of the four quantum numbers of an electron in an atom are as follows:

**1. Principal Quantum Number ($n$)** *This quantum number is the one on which the energy of an electron in an atom principally depends; it can have any positive value: 1, 2, 3, and so on.* The energy of an electron in an atom depends *principally* on $n$. The smaller $n$ is, the lower the energy. In the case of the hydrogen atom or single-electron atomic ions, such as $Li^{2+}$ and $He^+$, $n$ is the only quantum number determining the energy (which is given by Bohr's formula, discussed in Section 7.3). For other atoms, the energy also depends to a slight extent on the $l$ quantum number.

The *size* of an orbital also depends on *n*. The larger the value of *n* is, the larger the orbital. Orbitals of the same quantum state *n* are said to belong to the same *shell*. Shells are sometimes designated by the following letters:

| **Letter** | *K* | *L* | *M* | *N* ... |
|---|---|---|---|---|
| *n* | 1 | 2 | 3 | 4 ... |

### 2. Angular Momentum Quantum Number (*l*)   (Also Called *Azimuthal Quantum Number*)

*This quantum number distinguishes orbitals of given n having different shapes; it can have any integer value from 0 to n − 1.* Within each shell of quantum number *n*, there are *n* different kinds of orbitals, each with a distinctive shape denoted by an *l* quantum number. For example, if an electron has a principal quantum number of 3, the possible values for *l* are 0, 1, and 2. Thus, within the *M* shell (*n* = 3), there are three kinds of orbitals, each having a different shape for the region where the electron is most likely to be found. These orbital shapes will be discussed later in this section.

Although the energy of an orbital is principally determined by the *n* quantum number, the energy also depends somewhat on the *l* quantum number (except for the H atom). For a given *n*, the energy of an orbital increases with *l*.

Orbitals of the same *n* but different *l* are said to belong to different *subshells* of a given shell. The different subshells are usually denoted by letters as follows:

| **Letter** | *s* | *p* | *d* | *f* | *g* ... |
|---|---|---|---|---|---|
| *l* | 0 | 1 | 2 | 3 | 4 ... |

To denote a subshell within a particular shell, we write the value of the *n* quantum number for the shell, followed by the letter designation for the subshell. For example, 2*p* denotes a subshell with quantum numbers *n* = 2 and *l* = 1. ◄

► The rather odd choice of letter symbols for *l* quantum numbers survives from old spectroscopic terminology (describing the lines in a spectrum as *s*harp, *p*rincipal, *d*iffuse, and *f*undamental).

### 3. Magnetic Quantum Number (*m*$_l$)

*This quantum number distinguishes orbitals of given n and l—that is, of given energy and shape but having a different orientation in space; the allowed values are the integers from −l to +l.* For *l* = 0 (*s* subshell), the allowed *m*$_l$ quantum number is 0 only; there is only one orbital in the *s* subshell. For *l* = 1 (*p* subshell), *m*$_l$ = −1, 0, and +1; there are three different orbitals in the *p* subshell. The orbitals have the same shape but different orientations in space. In addition, each orbital of a given subshell has the same energy. Note that there are 2*l* + 1 orbitals in each subshell of quantum number *l*.

### 4. Spin Quantum Number (*m*$_s$)

*This quantum number refers to the two possible orientations of the spin axis of an electron; possible values are* $+\frac{1}{2}$ *and* $-\frac{1}{2}$. An electron acts as though it were spinning on its axis like the earth. Such an electron spin would give rise to a circulating electric charge that would generate a magnetic field. In this way, an electron behaves like a small bar magnet, with a north and a south pole. ◄

► Electron spin will be discussed further in Section 8.1.

Table 7.1 lists the permissible quantum numbers for all orbitals through the *n* = 4 shell. These values follow from the rules just given. Energies for the orbitals are shown in Figure 7.23 for the hydrogen atom. Note that all orbitals with the same principal quantum number *n* have the same energy. For atoms with more than one electron, however, only orbitals in the same subshell (denoted by a given *n* and *l*) have the same energy. We will have more to say about orbital energies in Chapter 8.

**Table 7.1**
**Permissible Values of Quantum Numbers for Atomic Orbitals**

| $n$ | $l$ | $m_l$* | Subshell Notation | Number of Orbitals in the Subshell |
|---|---|---|---|---|
| 1 | 0 | 0 | $1s$ | 1 |
| 2 | 0 | 0 | $2s$ | 1 |
| 2 | 1 | $-1, 0, +1$ | $2p$ | 3 |
| 3 | 0 | 0 | $3s$ | 1 |
| 3 | 1 | $-1, 0, +1$ | $3p$ | 3 |
| 3 | 2 | $-2, -1, 0, +1, +2$ | $3d$ | 5 |
| 4 | 0 | 0 | $4s$ | 1 |
| 4 | 1 | $-1, 0, +1$ | $4p$ | 3 |
| 4 | 2 | $-2, -1, 0, +1, +2$ | $4d$ | 5 |
| 4 | 3 | $-3, -2, -1, 0, +1, +2, +3$ | $4f$ | 7 |

*Any one of the $m_l$ quantum numbers may be associated with the $n$ and $l$ quantum numbers on the same line.

**Figure 7.23**
**Orbital energies of the hydrogen atom.** The lines for each subshell indicate the number of different orbitals of that subshell. (Note break in the energy scale.)

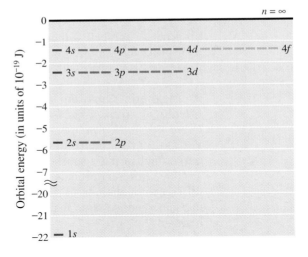

▶ An alternate example is provided in the Instructor's Resource Manual.

**EXAMPLE 7.6**

## Using the Rules for Quantum Numbers

State whether each of the following sets of quantum numbers is permissible for an electron in an atom. If a set is not permissible, explain why.

a. $n = 1,\ l = 1,\ m_l = 0,\ m_s = +\frac{1}{2}$

b. $n = 3,\ l = 1,\ m_l = -2,\ m_s = -\frac{1}{2}$

c. $n = 2,\ l = 1,\ m_l = 0,\ m_s = +\frac{1}{2}$

d. $n = 2,\ l = 0,\ m_l = 0,\ m_s = 1$

**SOLUTION**

a. **Not permissible.** The $l$ quantum number is equal to $n$; it must be less than $n$.

b. **Not permissible.** The magnitude of the $m_l$ quantum number (that is, the $m_l$ value, ignoring its sign) must not be greater than $l$.

c. **Permissible.**

d. **Not permissible.** The $m_s$ quantum number can be only $+\frac{1}{2}$ or $-\frac{1}{2}$.

---

**EXERCISE 7.7**

Explain why each of the following sets of quantum numbers is not permissible for an orbital.

a. $n = 0$, $l = 1$, $m_l = 0$, $m_s = +\frac{1}{2}$

b. $n = 2$, $l = 3$, $m_l = 0$, $m_s = -\frac{1}{2}$

c. $n = 3$, $l = 2$, $m_l = +3$, $m_s = +\frac{1}{2}$

d. $n = 3$, $l = 2$, $m_l = +2$, $m_s = 0$

---

See Problems 7.61 and 7.62.

## Atomic Orbital Shapes

An $s$ orbital has a spherical shape, though specific details of the probability distribution depend on the value of $n$. Figure 7.24 shows cross-sectional representations of the probability distributions of a $1s$ and a $2s$ orbital. The color shading is darker where the electron is more likely to be found. In the case of a $1s$ orbital, the electron is most likely to be found near the nucleus. The shading becomes lighter as the distance from the nucleus increases, indicating that the electron is less likely to be found there.

The orbital does not abruptly end at some particular distance from the nucleus. An atom, therefore, has an indefinite extension, or "size." We can gauge the "size" of the orbital by means of the *99% contour*. The electron has a 99% probability of being found within the space of the 99% contour (the sphere indicated by the dashed line in the diagram).

A $2s$ orbital differs in detail from a $1s$ orbital. The electron in a $2s$ orbital is likely to be found in two regions, one near the nucleus and the other in a spherical shell

**Problem Solving: Electron Subshells**

▶ Animations: *s* orbitals, *p* orbitals, *d* orbitals.

**Figure 7.24 Cross-sectional representations of the probability distributions of *s* orbitals.** In a $1s$ orbital, the probability distribution is largest near the nucleus. In a $2s$ orbital, it is greatest in a spherical shell about the nucleus. Note the relative "size" of the orbitals, indicated by the 99% contours.

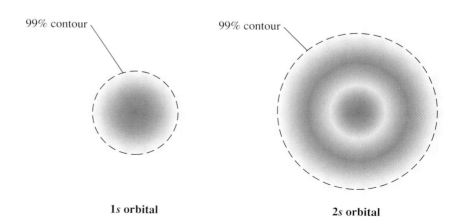

1s orbital          2s orbital

**Figure 7.25**
**Cutaway diagrams showing the spherical shape of s orbitals.** In both diagrams, a segment of each orbital is cut away to reveal the electron distribution of the orbital.

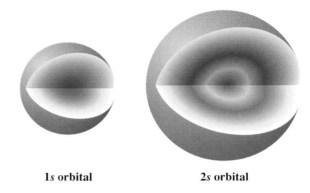

1s orbital          2s orbital

about the nucleus. (The electron is most likely to be here.) The 99% contour shows that the 2s orbital is larger than the 1s orbital.

A cross-sectional diagram cannot portray the three-dimensional aspect of the 1s and 2s atomic orbitals. Figure 7.25 shows cutaway diagrams, which better illustrate this three dimensionality.

There are three p orbitals in each p subshell. All p orbitals have the same basic shape (two lobes arranged along a straight line with the nucleus between the lobes) but differ in their orientations in space. Because the three orbitals are set at right angles to each other, we can show each one as oriented along a different coordinate axis (Figure 7.26). We denote these orbitals as $2p_x$, $2p_y$, and $2p_z$. A $2p_x$ orbital has its greatest electron probability along the x-axis, a $2p_y$ orbital along the y-axis, and a $2p_z$

**Figure 7.26**
**The 2p orbitals.** (A) Electron distribution in the $2p_x$ orbital. Note that it consists of two lobes oriented along the x-axis. (B) Orientations of the three 2p orbitals. The drawings depict the general shape and orientation of the orbitals, but not the detailed electron distribution as in (A).

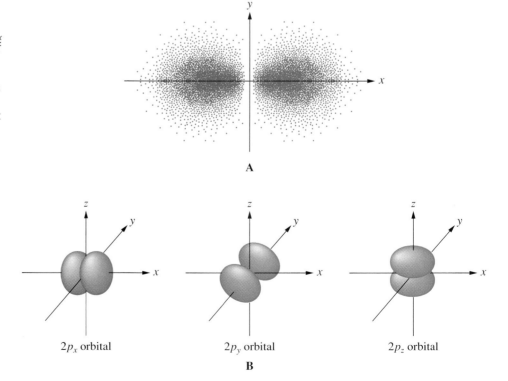

A

$2p_x$ orbital          $2p_y$ orbital          $2p_z$ orbital

B

▶ We will discuss *d* orbitals in Chapter 23, when we look at the bonding in compounds of the transition elements.

orbital along the *z*-axis. Other *p* orbitals, such as 3*p*, have this same general shape, with differences in detail depending on *n*. We will discuss *s* and *p* orbital shapes again in Chapter 10 in reference to chemical bonding.

There are five *d* orbitals, which have more complicated shapes than do *s* and *p* orbitals. These are represented in Figure 7.27.

**Figure 7.27**
**The five 3*d* orbitals.** These are labeled by subscripts, as in $d_{xy}$, that describe their mathematical characteristics.

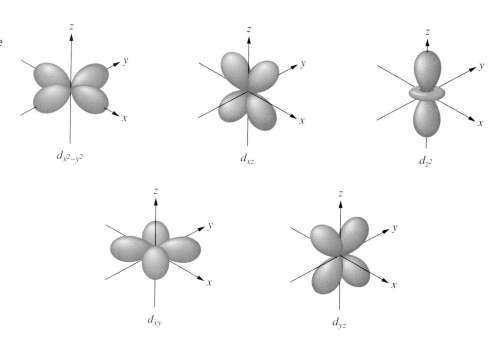

$d_{x^2-y^2}$     $d_{xz}$     $d_{z^2}$

$d_{xy}$     $d_{yz}$

## A Checklist for Review

### Important Terms

wavelength ($\lambda$) (7.1)
frequency ($\nu$) (7.1)
electromagnetic spectrum (7.1)
Planck's constant (7.2)
photons (7.2)

photoelectric effect (7.2)
continuous spectrum (7.3)
line spectrum (7.3)
energy levels (7.3)
de Broglie relation (7.4)

quantum (wave) mechanics (7.4)
uncertainty principle (7.4)
atomic orbital (7.5)
principal quantum number (*n*) (7.5)

angular momentum quantum number (*l*) (7.5)
magnetic quantum number ($m_l$) (7.5)
spin quantum number ($m_s$) (7.5)

### Key Equations

$c = \nu\lambda$

$E = h\nu$

$E = -\dfrac{R_H}{n^2}$     $n = 1, 2, 3, \ldots \infty$     (for H atom)

Energy of emitted photon $= h\nu = E_i - E_f$

$\lambda = \dfrac{h}{mv}$

### Summary of Facts and Concepts

One way to study the electronic structure of the atom is to analyze the electromagnetic radiation that is emitted from an atom. Electromagnetic radiation is characterized by its *wavelength* $\lambda$ and frequency $\nu$, and these quantities are related to the speed of light $c$ ($c = \nu\lambda$).

Einstein showed that light consists of particles *(photons)*, each of energy $E = h\nu$, where *h* is Planck's constant. According to Bohr, electrons in an atom have *energy levels*, and when an electron in a higher energy level drops (or undergoes a transition) to a lower energy level, a photon is emitted. The energy of the photon equals the difference in energy between the two levels.

Electrons and other particles of matter have both particle and wave properties. For a particle of mass $m$ and speed $v$, the wavelength is related to momentum $mv$ by the *de Broglie relation:* $\lambda = h/mv$. The wave properties of a particle are described by a *wave function,* from which we can get the probability of finding the particle in different regions of space.

Each electron in an atom is characterized by four different quantum numbers. The distribution of an electron in space—its *atomic orbital*—is characterized by three of these quantum numbers: the principal quantum number, the angular momentum quantum number, and the magnetic quantum number. The fourth quantum number (spin quantum number) describes the magnetism of the electron.

## Operational Skills

1. **Relating wavelength and frequency of light** Given the frequency of light, calculate the wavelength, or vice versa. **(EXAMPLES 7.1, 7.2)**
2. **Calculating the energy of a photon** Given the frequency or wavelength of light, calculate the energy associated with one photon. **(EXAMPLE 7.3)**
3. **Determining the wavelength or frequency of a hydrogen atom transition** Given the initial and final principal quantum numbers for an electron transition in the hydro-gen atom, calculate the frequency or wavelength of light emitted **(EXAMPLE 7.4)**. You need the value of $R_H$.
4. **Applying the de Broglie relation** Given the mass and speed of a particle, calculate the wavelength of the associated wave. **(EXAMPLE 7.5)**
5. **Using the rules for quantum numbers** Given a set of quantum numbers $n$, $l$, $m_l$, and $m_s$, state whether that set is permissible for an electron. **(EXAMPLE 7.6)**

## Review Questions

**7.1** Give a brief wave description of light. What are two characteristics of light waves?

**7.2** What is the mathematical relationship among the different characteristics of light waves? State the meaning of each of the terms in the equation.

**7.3** Briefly describe the portions of the electromagnetic spectrum, starting with shortest wavelengths and going to longer wavelengths.

**7.4** Planck originated the idea that energies can be quantized. What does the term *quantized* mean? What was Planck trying to explain when he was led to the concept of quantization of energy? Give the formula he arrived at and explain each of the terms in the formula.

**7.5** In your own words, explain the photoelectric effect. How does the photon concept explain this effect?

**7.6** Describe the wave–particle picture of light.

**7.7** Give the equation that relates particle properties of light. Explain the meaning of each symbol in the equation.

**7.8** Physical theory at the time Rutherford proposed his nuclear model of the atom was not able to explain how this model could give a stable atom. Explain the nature of this difficulty.

**7.9** Explain the main features of Bohr's theory. Do these features solve the difficulty alluded to in Question 7.8?

**7.10** Explain the process of emission of light by an atom.

**7.11** Explain the process of absorption of light by an atom.

**7.12** What is the evidence for electron waves? Give a practical application.

**7.13** What kind of information does a wave function give about an electron in an atom?

**7.14** The atom is sometimes said to be similar to a miniature planetary system, with electrons orbiting the nucleus. What does the uncertainty principle have to say about this view of the atom?

**7.15** Bohr described the hydrogen atom as an electron orbiting a hydrogen nucleus. Although certain aspects of his theory are still valid, his theory agreed quantitatively with experiment only in the case of the hydrogen atom. In what way does quantum mechanics change Bohr's original picture of the hydrogen atom?

**7.16** Give the possible values of (a) the principal quantum number, (b) the angular momentum quantum number, (c) the magnetic quantum number, and (d) the spin quantum number.

**7.17** What is the notation for the subshell in which $n = 4$ and $l = 3$? How many orbitals are in this subshell?

**7.18** What is the general shape of an $s$ orbital? of a $p$ orbital?

**Conceptual Problems**

**7.19** Consider two beams of the same yellow light. Imagine that one beam has its wavelength doubled; the other has its frequency doubled. Which of these two beams is then in the ultraviolet region?

**7.20** Some infrared radiation has a wavelength that is 1000 times larger than that of a certain visible light. This visible light has a frequency that is 1000 times smaller than that of some X radiation. How many times more energy is there in a photon of this X radiation than there is in a photon of the infrared radiation?

**7.21** One photon of green light has less than twice the energy of two photons of red light. Consider two hypothetical experiments. In one experiment, potassium metal is exposed to one photon of green light; in another experiment, potassium metal is exposed to two photons of red light. In one of these experiments, no electrons are ejected by the photoelectric effect (no matter how many times this experiment is repeated). In the other experiment, at least one electron was observed to be ejected. What is the maximum number of electrons that could be ejected during this other experiment, one or two?

**7.22** An atom in its ground state absorbs a photon (photon 1), then quickly emits another photon (photon 2). One of these photons corresponds to ultraviolet radiation, whereas the other corresponds to red light. Explain what is happening. Which electromagnetic radiation, ultraviolet or red light, is associated with the emitted photon (photon 2)?

**7.23** Three emission lines involving three energy levels in an atom occur at wavelengths $x$, $1.5x$, and $3.0x$ nanometers. Which wavelength corresponds to the transition from the highest to the lowest of the three energy levels?

**7.24** An atom emits yellow light when an electron makes the transition from the $n = 5$ to the $n = 1$ level. In separate experiments, suppose you bombarded the $n = 1$ level of this atom with red light, yellow light (obtained from the previous emission), and blue light. In which experiment or experiments would the electron be promoted to the $n = 5$ level?

**7.25** Which of the following particles has the longest wavelength?
a. an electron traveling at $x$ meters per second
b. a proton traveling at $x$ meters per second
c. a proton traveling at $2x$ meters per second

**7.26** Imagine a world in which the rule for the $l$ quantum number is that values start with 1 and go up to $n$. The rules for the $n$ and $m_l$ quantum numbers are unchanged from those of our world. Write the quantum numbers for the first two shells (i.e., $n = 1$ and $n = 2$).

## Practice Problems

### Electromagnetic Waves

**7.27** Radio waves in the AM region have frequencies in the range 550 to 1600 kilocycles per second (550 to 1600 kHz). Calculate the wavelength corresponding to a radio wave of frequency $1.365 \times 10^6$/s (that is, 1365 kHz).

**7.28** Microwaves have frequencies in the range $10^9$ to $10^{12}$/s (cycles per second), equivalent to between 1 gigahertz and 1 terahertz. What is the wavelength of microwave radiation whose frequency is $1.258 \times 10^{10}$/s?

**7.29** Light with a wavelength of 478 nm lies in the blue region of the visible spectrum. Calculate the frequency of this light.

**7.30** Calculate the frequency associated with light of wavelength 656 nm. (This corresponds to one of the wavelengths of light emitted by the hydrogen atom.)

**7.31** At its closest approach, Mars is 56 million km from earth. How long would it take to send a radio message from a space probe of Mars to earth when the planets are at this closest distance?

**7.32** The space probe *Pioneer 11* was launched April 5, 1973, and reached Jupiter in December 1974, traveling a distance of 998 million km. How long did it take an electromagnetic signal to travel to earth from *Pioneer 11* when it was near Jupiter?

**7.33** The meter was defined in 1963 as the length equal to 1,650,763.73 wavelengths of the orange-red radiation emitted by the krypton-86 atom (the meter has since been redefined). What is the wavelength of this transition? What is the frequency?

**7.34** The second is defined as the time it takes for 9,192,631,770 wavelengths of a certain transition of the cesium-133 atom to pass a fixed point. What is the frequency of this electromagnetic radiation? What is the wavelength?

### Photons

**7.35** What is the energy of a photon corresponding to radio waves of frequency $1.365 \times 10^6$/s?

**7.36** What is the energy of a photon corresponding to microwave radiation of frequency $1.258 \times 10^{10}$/s?

**7.37** The green line in the atomic spectrum of thallium has a wavelength of 535 nm. Calculate the energy of a photon of this light.

**7.38** Indium compounds give a blue-violet flame test. The atomic emission responsible for this blue-violet color has a wavelength of 451 nm. Obtain the energy of a single photon of this wavelength.

**7.39** A particular transition of the rubidium atom emits light whose frequency is $3.84 \times 10^{14}$ Hz. (Hz is the abbreviation for *hertz,* which is equivalent to the unit /s, or $s^{-1}$.) Is this light in the visible spectrum? If so, what is the color of the light? See Figure 7.5.

**7.40** Barium atoms have a particular transition that emits light of frequency $5.41 \times 10^{14}$ Hz. (Hz is the abbreviation for *hertz,* which is equivalent to the unit /s, or $s^{-1}$.) Is this light in the visible spectrum? If so, what is the color of the light? See Figure 7.5.

## Bohr Theory

**7.41** An electron in a hydrogen atom in the level $n = 5$ undergoes a transition to level $n = 3$. What is the frequency of the emitted radiation?

**7.42** Calculate the frequency of electromagnetic radiation emitted by the hydrogen atom in the electron transition from $n = 4$ to $n = 3$.

**7.43** The first line of the Lyman series of the hydrogen atom emission results from a transition from the $n = 2$ level to the $n = 1$ level. What is the wavelength of the emitted photon? Using Figure 7.5, describe the region of the electromagnetic spectrum in which this emission lies.

**7.44** What is the wavelength of the electromagnetic radiation emitted from a hydrogen atom when the electron undergoes the transition $n = 5$ to $n = 4$? In what region of the spectrum does this line occur? See Figure 7.5.

**7.45** Calculate the shortest wavelength of the electromagnetic radiation emitted by the hydrogen atom in undergoing a transition from the $n = 6$ level.

**7.46** Calculate the longest wavelength of the electromagnetic radiation emitted by the hydrogen atom in undergoing a transition from the $n = 6$ level.

**7.47** What is the difference in energy between the two levels responsible for the violet emission line of the calcium atom at 422.7 nm?

**7.48** What is the difference in energy between the two levels responsible for the ultraviolet emission line of the magnesium atom at 285.2 nm?

## de Broglie Waves

*Note:* Masses of the electron, proton, and neutron are listed on the inside back cover of this book.

**7.49** What is the wavelength of a neutron traveling at a speed of 4.15 km/s? (Neutrons of these speeds are obtained from a nuclear pile.)

**7.50** What is the wavelength of a proton traveling at a speed of 6.21 km/s? What would be the region of the spectrum for electromagnetic radiation of this wavelength?

**7.51** At what speed must an electron travel to have a wavelength of 10.0 pm?

**7.52** At what speed must a neutron travel to have a wavelength of 10.0 pm?

**7.53** What is the de Broglie wavelength of a 145-g baseball traveling at 30.0 m/s (67.1 mph)? Is the wavelength much smaller or much larger than the diameter of an atom (on the order of 100 pm)?

**7.54** What is the de Broglie wavelength of an oxygen molecule, $O_2$, traveling at 521 m/s? Is the wavelength much smaller or much larger than the diameter of an atom (on the order of 100 pm)?

## Atomic Orbitals

**7.55** If the $n$ quantum number of an atomic orbital is 4, what are the possible values of $l$? If the $l$ quantum number is 3, what are the possible values of $m_l$?

**7.56** The $n$ quantum number of an atomic orbital is 5. What are the possible values of $l$? What are the possible values of $m_l$ if the $l$ quantum number is 4?

**7.57** How many subshells are there in the $M$ shell? How many orbitals are there in the $f$ subshell?

**7.58** How many subshells are there in the $N$ shell? How many orbitals are there in the $g$ subshell?

**7.59** Give the notation (using letter designations for $l$) for the subshells denoted by the following quantum numbers.
a. $n = 6, l = 2$
b. $n = 5, l = 4$
c. $n = 4, l = 3$
d. $n = 6, l = 1$

**7.60** Give the notation (using letter designations for $l$) for the subshells denoted by the following quantum numbers.
a. $n = 3, l = 1$
b. $n = 4, l = 2$
c. $n = 4, l = 0$
d. $n = 5, l = 3$

**7.61** Explain why each of the following sets of quantum numbers would not be permissible for an electron, according to the rules for quantum numbers.
a. $n = 1, l = 0, m_l = 0, m_s = +1$
b. $n = 1, l = 3, m_l = +3, m_s = +\frac{1}{2}$
c. $n = 3, l = 2, m_l = +3, m_s = -\frac{1}{2}$
d. $n = 0, l = 1, m_l = 0, m_s = +\frac{1}{2}$
e. $n = 2, l = 1, m_l = -1, m_s = +\frac{3}{2}$

**7.62** State which of the following sets of quantum numbers would be possible and which impossible for an electron in an atom.
a. $n = 0, l = 0, m_l = 0, m_s = +\frac{1}{2}$
b. $n = 1, l = 1, m_l = 0, m_s = +\frac{1}{2}$
c. $n = 1, l = 0, m_l = 0, m_s = -\frac{1}{2}$
d. $n = 2, l = 1, m_l = -2, m_s = +\frac{1}{2}$
e. $n = 2, l = 1, m_l = -1, m_s = +\frac{1}{2}$

## General Problems

**7.63** The blue line of the strontium atom emission has a wavelength of 461 nm. What is the frequency of this light? What is the energy of a photon of this light?

**7.64** The barium atom has an emission with wavelength 554 nm (green). Calculate the frequency of this light and the energy of a photon of this light.

**7.65** The energy of a photon is $4.10 \times 10^{-19}$ J. What is the wavelength of the corresponding light? What is the color of this light?

**7.66** The energy of a photon is $3.34 \times 10^{-19}$ J. What is the wavelength of the corresponding light? What is the color of this light?

**7.67** The photoelectric work function of a metal is the minimum energy needed to eject an electron by irradiating the metal with light. For calcium, this work function equals $4.34 \times 10^{-19}$ J. What is the minimum frequency of light for the photoelectric effect in calcium?

**7.68** The photoelectric work function for magnesium is $5.90 \times 10^{-19}$ J. (The work function is the minimum energy needed to eject an electron from the metal by irradiating it with light.) Calculate the minimum frequency of light required to eject electrons from magnesium.

**7.69** Light of wavelength 345 nm shines on a piece of calcium metal. What is the speed of the ejected electron? (Light energy greater than that of the work function of calcium ends up as kinetic energy of the ejected electron. See Problem 7.67 for the definition of work function and its value for calcium.)

**7.70** Light of wavelength 285 nm shines on a piece of magnesium metal. What is the speed of the ejected electron? (Light energy greater than that of the work function of magnesium ends up as kinetic energy of the ejected electron. See Problem 7.68 for the definition of work function and its value for magnesium.)

**7.71** Calculate the wavelength of the Balmer line of the hydrogen spectrum in which the initial $n$ quantum number is 5 and the final $n$ quantum number is 2.

**7.72** Calculate the wavelength of the Balmer line of the hydrogen spectrum in which the initial $n$ quantum number is 3 and the final $n$ quantum number is 2.

**7.73** One of the lines in the Balmer series of the hydrogen atom emission spectrum is at 397 nm. It results from a transition from an upper energy level to $n = 2$. What is the principal quantum number of the upper level?

**7.74** A line of the Lyman series of the hydrogen atom spectrum has the wavelength $9.50 \times 10^{-8}$ m. It results from a transition from an upper energy level to $n = 1$. What is the principal quantum number of the upper level?

**7.75** A hydrogen-like ion has a nucleus of charge $+Ze$ and a single electron outside this nucleus. The energy levels of these ions are $-Z^2R_H/n^2$ (where $Z$ = atomic number). Calculate the wavelength of the transition from $n = 3$ to $n = 2$ for $He^+$, a hydrogen-like ion. In what region of the spectrum does this emission occur?

**7.76** What is the wavelength of the transition from $n = 4$ to $n = 3$ for $Li^{2+}$? In what region of the spectrum does this emission occur? $Li^{2+}$ is a hydrogen-like ion. Such an ion has a nucleus of charge $+Ze$ and a single electron outside this nucleus. The energy levels of the ion are $-Z^2R_H/n^2$, where $Z$ is the atomic number.

**7.77** An electron microscope employs a beam of electrons to obtain an image of an object. What energy must be imparted to each electron of the beam to obtain a wavelength of 10.0 pm? Obtain the energy in electron volts (eV) (1 eV = $1.602 \times 10^{-19}$ J).

**7.78** Neutrons are used to obtain images of the hydrogen atoms in molecules. What energy must be imparted to each neutron in a neutron beam to obtain a wavelength of 10.0 pm? Obtain the energy in electron volts (eV) (1 eV = $1.602 \times 10^{-19}$ J).

**7.79** What is the number of different orbitals in each of the following subshells?
a. 3d    b. 4f    c. 4p    d. 5s

**7.80** What is the number of different orbitals in each of the following subshells?
a. 5f    b. 5g    c. 6s    d. 5p

**7.81** List the possible subshells for the $n = 6$ shell.

**7.82** List the possible subshells for the $n = 7$ shell.

## Cumulative-Skills Problems

**7.83** The energy required to dissociate the $Cl_2$ molecule to Cl atoms is 239 kJ/mol $Cl_2$. If the dissociation of a $Cl_2$ molecule were accomplished by the absorption of a single photon whose energy was exactly the quantity required, what would be its wavelength (in meters)?

**7.84** The energy required to dissociate the $H_2$ molecule to H atoms is 432 kJ/mol $H_2$. If the dissociation of an $H_2$ molecule were accomplished by the absorption of a single photon whose energy was exactly the quantity required, what would be its wavelength (in meters)?

**7.85** A microwave oven heats by radiating food with microwave radiation, which is absorbed by the food and converted to heat. Suppose an oven's radiation wavelength is 12.5 cm. A container with 0.250 L of water was placed in the oven, and the temperature of the water rose from 20.0°C to 100.0°C. How many photons of this microwave radiation were required? Assume that all the energy from the radiation was used to raise the temperature of the water.

**7.86** Warm objects emit electromagnetic radiation in the infrared region. Heat lamps employ this principle to generate infrared radiation. Water absorbs infrared radiation with wavelengths near 2.80 $\mu$m. Suppose this radiation is absorbed by the water and converted to heat. A 1.00-L sample of water absorbs infrared radiation, and its temperature increases from 20.0°C to 30.0°C. How many photons of this radiation are used to heat the water?

**7.87** Light with a wavelength of 425 nm fell on a potassium surface, and electrons were ejected at a speed of $4.88 \times 10^5$ m/s. What energy was expended in removing an electron from the metal? Express the answer in joules (per electron) and in kilojoules per mole (of electrons).

**7.88** Light with a wavelength of 405 nm fell on a strontium surface, and electrons were ejected. If the speed of an ejected electron is $3.36 \times 10^5$ m/s, what energy was expended in removing the electron from the metal? Express the answer in joules (per electron) and in kilojoules per mole (of electrons).

**7.89** When an electron is accelerated by a voltage difference, the kinetic energy acquired by the electron equals the voltage times the charge on the electron. Thus, one volt imparts a kinetic energy of $1.602 \times 10^{-19}$ volt-coulombs, which equals $1.602 \times 10^{-19}$ J. What is the wavelength associated with electrons accelerated by $4.00 \times 10^3$ volts?

**7.90** What is the wavelength for electrons accelerated by $1.00 \times 10^4$ volts? When an electron is accelerated by a voltage difference, the kinetic energy acquired by the electron equals the voltage times the charge on the electron. Thus, one volt imparts a kinetic energy of $1.602 \times 10^{-19}$ volt-coulombs, or $1.602 \times 10^{-19}$ J.

## Media Activities

**7.91** Watch the Photoelectric Effect *Visualization* on the student CD.
a. Which of the colors of light shown in this *Visualization* has higher energy photons?
b. Which of the colors of light shown in this *Visualization* has the longest wavelength?
c. Calculate the energy of a photon of blue light and a photon of red light.

**7.92** View the Flame Tests *Visualization* of sodium acetate and copper(II) chloride on the student CD.
a. What is the purpose of the flame?
b. Explain why we see the flame turn a different color when each salt is introduced into the flame.
c. Why does each of the salts give off a different color of light when heated?
d. Which of the salts emits higher energy photons of visible light?

**7.93** Play the Refraction of White Light and the $H_2$ Line Spectrum *Visualizations* on the student CD.
a. Using the Bohr model, explain why all the colors of visible light are not present in the $H_2$ emission spectrum.
b. Which line in the hydrogen emission spectrum is of highest energy?

c. The four lines in the hydrogen emission spectrum correspond to the $n = 6$ to $n = 2$, $n = 5$ to $n = 2$, $n = 4$ to $n = 2$, and $n = 3$ to $n = 2$ electronic transitions in the Bohr model. Match the electronic transitions to the wavelengths listed in the $H_2$ line spectrum animation, and verify that these four electronic transitions produce emissions at the four wavelengths listed on the *Visualization*.

**7.94** Open the Hydrogen Emission Spectrum *Understanding Concepts* simulation on the student web site. Once you have worked with the simulation, test your understanding of the concepts by taking the Self Assessment. Using the Rydberg equation, calculate the energy of an electron in a hydrogen atom making a transition from the $n = 4$ to the $n = 1$ level.

**7.95** View the 1s orbital *Visualization* on the student CD.
a. How would the shape of a 1s orbital on H and $He^+$ compare?
b. Compare the size of a 1s orbital for ground state H and ground state $He^+$.
c. Which electron would be of higher energy, a ground state 1s electron in H or a ground state 1s electron in $He^+$?

**7.96** For review, take the ACE practice test for Chapter 7 on the student web site.

Sodium metal reacts rapidly with water, producing hydrogen gas, which catches fire.

# Electron Configurations and Periodicity

**M**arie Curie, a Polish-born French chemist, and her husband, Pierre, announced the discovery of radium in 1898 (Figure 8.1). They had separated a very radioactive mixture from pitchblende, an ore of uranium. This mixture was primarily a compound of barium. When the mixture was heated in a flame, however, it gave a new atomic line spectrum, in addition to the spectrum for barium. The Curies based their discovery of a new element on this finding. It took them four more years to obtain a pure compound of radium. Radium, like uranium, is a radioactive element. But in most of its chemical and physical properties, radium is similar to the nonradioactive element barium. It was this similarity that made the final separation of the new element so difficult.

Chemists had long known that groups of elements have similar properties. In 1869 Dmitri Mendeleev found that when the elements were arranged in a particular way, they fell into columns, with elements in the same column displaying similar properties. Thus, Mendeleev placed beryllium, calcium, strontium, and barium in one column. Now, with the Curies' discovery, radium was added to this column.

**Figure 8.1**
**Marie Sklodowska Curie
(1867–1934).** Marie Sklodowska Curie, born in Warsaw, Poland, began her doctoral work with Henri Becquerel soon after he discovered the spontaneous radiation emitted by uranium salts. She found this radiation to be an atomic property and coined the word *radioactivity* for it. In 1903 the Curies and Becquerel were awarded the Nobel Prize in physics for their discovery of radioactivity. Three years later, Pierre Curie was killed in a carriage accident. Marie Curie continued their work on radium and in 1911 was awarded the Nobel Prize in chemistry for the discovery of polonium and radium and the isolation of pure radium metal. This was the first time a scientist had received two Nobel awards. (Since then two others have been so honored.)

▶ In the preceding chapter, we introduced atomic orbitals and quantum numbers. Here we look at the electron configurations of atoms and the periodic behavior of the elements.

Mendeleev's arrangement of the elements, the *periodic table*, was originally based on the observed chemical and physical properties of the elements and their compounds. We now explain this arrangement in terms of the electronic structure of atoms. In this chapter we will look at this electronic structure and its relationship to the periodic table of elements.

# Electronic Structure of Atoms

In Chapter 7 we found that an electron in an atom has four quantum numbers—$n$, $l$, $m_l$, and $m_s$—associated with it. The first three quantum numbers characterize the orbital that describes the region of space where an electron is most likely to be found; we say that the electron "occupies" this orbital. The spin quantum number, $m_s$, describes the spin orientation of an electron. In the first section, we will look further at electron spin; then we will discuss how electrons are distributed among the possible orbitals.

## 8.1 Electron Spin and the Pauli Exclusion Principle

Otto Stern and Walther Gerlach first observed electron spin magnetism in 1921. They directed a beam of silver atoms into the field of a specially designed magnet. The same experiment can be done with hydrogen atoms. The beam of hydrogen atoms is split into two by the magnetic field; half of the atoms are bent in one direction and half in the other (see Figure 8.2). The fact that the atoms are affected by the laboratory magnet shows that they themselves act as magnets.

The beam of hydrogen atoms is split into two because the electron in each atom behaves as a tiny magnet with only two possible orientations. In effect, the electron acts as though it were a ball of spinning charge (Figure 8.3), and like a circulating

**Figure 8.2**
**The Stern–Gerlach experiment.** The diagram shows the experiment using hydrogen atoms (simpler to interpret theoretically), although the original experiment employed silver atoms. A beam of hydrogen atoms (shown in blue) is split into two by a nonuniform magnetic field. One beam consists of atoms each with an electron having $m_s = +\frac{1}{2}$; the other beam consists of atoms each having an electron with $m_s = -\frac{1}{2}$.

▶ Protons and many nuclei also have spin. See the Instrumental Methods essay at the end of this section.

electric charge the electron would create a magnetic field. Electron spin, however, is subject to a quantum restriction on the possible directions of the spin axis. The resulting directions of spin magnetism correspond to spin quantum numbers $m_s = +\frac{1}{2}$ and $m_s = -\frac{1}{2}$. ◀

## Electron Configurations and Orbital Diagrams

An **electron configuration** of an atom is *a particular distribution of electrons among available subshells.* The notation for a configuration lists the subshell symbols, one after the other, with a superscript giving the number of electrons in that subshell. For example, a configuration of the lithium atom (atomic number 3) with two electrons in the $1s$ subshell and one electron in the $2s$ subshell is written $1s^2 2s^1$.

The notation for a configuration gives the number of electrons in each subshell, but we use *a diagram to show how the orbitals of a subshell are occupied by electrons.* It is called an **orbital diagram.** An orbital is represented by a circle. Each group of orbitals in a subshell is labeled by its subshell notation. An electron in an orbital is shown by an arrow; the arrow points up when $m_s = +\frac{1}{2}$ and down when $m_s = -\frac{1}{2}$. The orbital diagram

$$1s \quad 2s \qquad 2p$$

shows the electronic structure of an atom in which there are two electrons in the $1s$ subshell, or orbital (one electron with $m_s = +\frac{1}{2}$, the other with $m_s = -\frac{1}{2}$); two electrons in the $2s$ subshell ($m_s = +\frac{1}{2}$, $m_s = -\frac{1}{2}$); and one electron in the $2p$ subshell ($m_s = +\frac{1}{2}$). The electron configuration is $1s^2 2s^2 2p^1$.

**Figure 8.3**
**A representation of electron spin.** The two possible spin orientations are indicated by the models. By convention, the spin direction is given as shown by the large arrow on the spin axis. Electrons behave as tiny bar magnets, as shown in the figure.

### Pauli Exclusion Principle

Not all of the conceivable arrangements of electrons among the orbitals of an atom are physically possible. The **Pauli exclusion principle,** which summarizes experimental observations, states that *no two electrons in an atom can have the same four quantum numbers.* If one electron in an atom has the quantum numbers $n = 1$, $l = 0$, $m_l = 0$, and $m_s = +\frac{1}{2}$, no other electron can have these same quantum numbers. In other words, you cannot place two electrons with the same value of $m_s$ in a $1s$ orbital. The orbital diagram

$1s$

is not a possible arrangement of electrons.

Because there are only two possible values of $m_s$, an orbital can hold no more than two electrons—and then only if the two electrons have different spin quantum numbers. In an orbital diagram, an orbital with two electrons must be written with arrows pointing in opposite directions. The two electrons are said to have opposite spins. We can restate the Pauli exclusion principle:

> Pauli exclusion principle: An orbital can hold at most two electrons, and then only if the electrons have opposite spins.

You can see that each subshell holds a maximum of twice as many electrons as the number of orbitals in the subshell. Thus, a $2p$ subshell, which has three orbitals (with $m_l = -1$, 0, and $+1$), can hold a maximum of six electrons. The maximum number of electrons in various subshells is given in the following table.

| Subshell | Number of Orbitals | Maximum Number of Electrons |
|---|---|---|
| $s \ (l = 0)$ | 1 | 2 |
| $p \ (l = 1)$ | 3 | 6 |
| $d \ (l = 2)$ | 5 | 10 |
| $f \ (l = 3)$ | 7 | 14 |

▶ An alternate example is provided in the Instructor's Resource Manual.

### EXAMPLE 8.1

### Applying the Pauli Exclusion Principle

Which of the following orbital diagrams or electron configurations are possible and which are impossible, according to the Pauli exclusion principle? Explain.

a. ⬆⬇  ⬆⬇  ⬆◯◯
$\phantom{}$ $1s$  $2s$  $\phantom{xx}2p$

b. ⬆⬇  ⬆⬆⬇  ◯◯◯
$\phantom{}$ $1s$  $2s$  $\phantom{xx}2p$

c. ⬆⬇  ⬆  ⬆⬆◯◯
$\phantom{}$ $1s$  $2s$  $\phantom{xx}2p$

d. $1s^3 2s^1$

e. $1s^2 2s^1 2p^7$

f. $1s^2 2s^2 2p^6 3s^2 3p^6 3d^8 4s^2$

**SOLUTION**

a. **Possible** orbital diagram. b. **Impossible** orbital diagram; there are three electrons in the $2s$ orbital. c. **Impossible** orbital diagram; there are two electrons in a $2p$ orbital with the same spin. d. **Impossible** electron configuration; there are three electrons in the $1s$ subshell (one orbital). e. **Impossible** electron configuration; there are seven electrons in the $2p$ subshell (which can hold only six electrons). f. **Possible.** Note that the $3d$ subshell can hold as many as ten electrons.

---

**EXERCISE 8.1**

Look at the following orbital diagrams and electron configurations. Which are possible and which are not, according to the Pauli exclusion principle? Explain.

a.
   $1s$  $2s$  $2p$

b.
   $1s$  $2s$  $2p$

c.
   $1s$  $2s$  $2p$

d. $1s^2 2s^2 2p^4$

e. $1s^2 2s^4 2p^2$

f. $1s^2 2s^2 2p^6 3s^2 3p^{10} 3d^{10}$

---

See Problems 8.33, 8.34, 8.35, and 8.36.

**INSTRUMENTAL METHODS**

# Nuclear Magnetic Resonance (NMR)

You have just seen that electrons have a spin and as a result behave like tiny magnets. Protons and neutrons similarly have spins. Therefore, depending on the arrangement of protons and neutrons, a nucleus could have spin. A nucleus with spin will act like a bar magnet, similar to the electron although many times smaller in magnitude. Examples of nuclei with spin are hydrogen-1 (proton), carbon-13 (but carbon-12, the most abundant nuclide of carbon, has no spin), and fluorine-19.

Although nuclear magnetism is much smaller than that of electrons, with the correct equipment it is easily seen and in fact forms the basis of *nuclear magnetic resonance* (NMR) spectroscopy, one of the most important methods for determining molecular structure.

*(continued)*

**Figure 8.4**
**Magnetic resonance imaging.** A patient's head is placed in a large magnet and subjected to a radio pulse. Proton spin transitions give rise to a radio wave emission that can be analyzed electronically and converted by computer to a two-dimensional image of a plane portion of the brain.

Nuclear magnetism also forms the basis of the medical diagnostic tool, *magnetic resonance imaging,* or MRI (Figure 8.4).

The essential features of NMR can be seen if you consider the proton. Like the electron, the proton has two spin states. In the absence of a magnetic field, these spin states have the same energy, but in the field of a strong magnet (external field), they have different energies. The state in which the proton magnetism is aligned with the external field, so the south pole of the proton magnet faces the north pole of the external magnet, will have lower energy. The state in which the proton magnet is turned 180°, with its south pole facing the south pole of the external magnet, will have higher energy. Now if a proton in the lower spin state is irradiated with electromagnetic waves of the proper frequency (in which the photon has energy equal to the difference in energy of the spin states), the proton will change to the higher spin state. The frequency absorbed by the proton depends on the magnitude of the magnetic field. For the magnets used in these instruments, the radiation lies in the radio-frequency range. Frequencies commonly used are 60 MHz and 100 MHz, which are in the FM region; frequencies up to 600 MHz are available.

Figure 8.5 shows a simplified NMR spectrometer. It consists of a sample in the field of a variable electromagnet and near two coils, one a radio wave transmitter and the other a receiver coil perpendicular to the transmitter coil (so that the receiver will not pick up the signal from the transmitter). Suppose the transmitter radiates waves of 100 MHz. If the sample absorbs these radio waves, protons will undergo transitions from the lower to the higher

spin state. Once protons are in the higher energy state, they tend to lose energy, going back to the lower spin state and radiating 100-MHz radio waves. Thus, the sample acts like a transmitter, but one with coils in various directions, so the signal can be detected by the receiver coil.

In general, the sample will not absorb at the chosen frequency. But you can change the energy difference between spin states, and therefore the frequency that is absorbed by the sample, by increasing or decreasing the magnitude of the external magnetic field using small coils on the magnet pole faces. In so doing you can, in effect, "tune" the sample, or bring it into "resonance" with the transmitter frequency. (Alternatively, many modern instruments vary the transmitter frequency to obtain resonance.)

Because each proton in a substance is surrounded by electrons that have their own magnetic fields, the magnetic environment of a proton depends to some extent on the bonding, or chemical, environment in which the proton is involved. So, the external magnetic field needed to bring a proton into resonance with the 100 MHz radiation, say, varies with the chemical (bonding) environment. A spectrum is produced by recording the magnetic field (or frequency) at which the protons in a molecule produce a resonance signal. Figure 8.6 shows a high-resolution NMR spectrum of ethanol.

Ethanol, whose molecular structure is

$$H-O-\underset{\underset{H}{|}}{\overset{\overset{H}{|}}{C}}-\underset{\underset{H}{|}}{\overset{\overset{H}{|}}{C}}-H$$

has protons in three different chemical environments: a proton bonded to an oxygen atom (H—O—), the protons in a —CH$_2$— group, and the protons in a —CH$_3$ group. Each of these three types of protons shows up as a peak or, as we will see, a group of peaks in the NMR spectrum. The position of the center of a group of peaks for a given proton relative to some standard is referred to as the *chemical shift* for that proton, because it depends on the chemical environment of the proton. By measuring the area under the peaks for a given type of proton (usually done electronically by the spectrometer), a researcher can discover the number of protons of that type, effectively giving the chemist structural information about the molecule.

As we noted earlier, a given proton may give rise to several peaks in a high-resolution spectrum. (At low-resolution, these separate peaks may appear as only one for each proton.) For example, the protons for the —CH$_2$— group of ethanol give four closely spaced peaks. These peaks arise because of the interaction of the proton spins on this group with those protons on the adjacent —CH$_3$ group. The number of peaks (four) for the —CH$_2$— group is one more than the number of protons on the adjacent group (three).

NMR spectroscopy is one of the most important tools a chemist has in determining the identity of a substance and its molecular structure. The method is quick and can yield such information as the chemical bonding environment of the protons and the number of protons having a given environment. It can also give information about other nuclei, such as carbon-13.

**Figure 8.5**
**Nuclear magnetic resonance experiment.** A sample is placed in a magnetic field (represented by a blue arrow in the direction of the field). The magnetic field can be varied by changing the electric current in the magnetic coil. A radio transmitter radiates waves, represented by the double-headed red arrow. The waves are absorbed by the sample when the radio frequency corresponds to the difference in energy of the nuclear spin states in the sample. When the nuclear spins of the sample go back to the ground state, radio waves are emitted from the sample in various directions. The receiver coil picks up those waves from the sample represented by the double-headed green arrow. (Source: John D. Roberts, *Nuclear Magnetic Resonance*, McGraw-Hill, 1959, p. 14, Figure 1-5. Reproduced with permission.)

**Figure 8.6**
**High-resolution NMR spectrum of ethanol, CH$_3$CH$_2$OH.** The protons of a given type occur as a peak or closely spaced group of peaks. The position of the center of such a group of peaks relative to some standard is referred to as the *chemical shift*. The peaks within a group arise from the interaction of the spins between protons of one type with those of another type.

## 8.2 Building-Up Principle and the Periodic Table

Every atom has an infinite number of possible electron configurations. The configuration associated with the lowest energy level of the atom corresponds to a quantum-mechanical state called the *ground state*. Other configurations correspond to *excited states,* associated with energy levels other than the lowest. For example, the ground state of the sodium atom is known from experiment to have the electron configuration $1s^2 2s^2 2p^6 3s^1$. The electron configuration $1s^2 2s^2 2p^6 3p^1$ represents an excited state of the sodium atom. ◄

▶ The transition of the sodium atom from the excited state $1s^2 2s^2 2p^6 3p^1$ to the ground state $1s^2 2s^2 2p^6 3s^1$ is accompanied by the emission of yellow light at 589 nm. Excited states of an atom are needed to describe its spectrum.

The chemical properties of an atom are related primarily to the electron configuration of its ground state. Table 8.1 lists the experimentally determined ground-state electron configurations of atoms $Z = 1$ to $Z = 36$. (A complete table appears in Appendix D.)

### Building-Up Principle (Aufbau Principle)

Most of the configurations in Table 8.1 can be explained in terms of the **building-up principle** (or **Aufbau principle**), *a scheme used to reproduce the electron configurations of the ground states of atoms by successively filling subshells with electrons in a specific order (the building-up order).* Following this principle, you obtain the

**Table 8.1**
**Ground-State Electron Configurations of Atoms Z = 1 to 36***

| Z | Element | Configuration | Z | Element | Configuration |
|---|---------|---------------|---|---------|---------------|
| 1 | H | $1s^1$ | 19 | K | $1s^2 2s^2 2p^6 3s^2 3p^6 4s^1$ |
| 2 | He | $1s^2$ | 20 | Ca | $1s^2 2s^2 2p^6 3s^2 3p^6 4s^2$ |
| 3 | Li | $1s^2 2s^1$ | 21 | Sc | $1s^2 2s^2 2p^6 3s^2 3p^6 3d^1 4s^2$ |
| 4 | Be | $1s^2 2s^2$ | 22 | Ti | $1s^2 2s^2 2p^6 3s^2 3p^6 3d^2 4s^2$ |
| 5 | B | $1s^2 2s^2 2p^1$ | 23 | V | $1s^2 2s^2 2p^6 3s^2 3p^6 3d^3 4s^2$ |
| 6 | C | $1s^2 2s^2 2p^2$ | 24 | Cr | $1s^2 2s^2 2p^6 3s^2 3p^6 3d^5 4s^1$ |
| 7 | N | $1s^2 2s^2 2p^3$ | 25 | Mn | $1s^2 2s^2 2p^6 3s^2 3p^6 3d^5 4s^2$ |
| 8 | O | $1s^2 2s^2 2p^4$ | 26 | Fe | $1s^2 2s^2 2p^6 3s^2 3p^6 3d^6 4s^2$ |
| 9 | F | $1s^2 2s^2 2p^5$ | 27 | Co | $1s^2 2s^2 2p^6 3s^2 3p^6 3d^7 4s^2$ |
| 10 | Ne | $1s^2 2s^2 2p^6$ | 28 | Ni | $1s^2 2s^2 2p^6 3s^2 3p^6 3d^8 4s^2$ |
| 11 | Na | $1s^2 2s^2 2p^6 3s^1$ | 29 | Cu | $1s^2 2s^2 2p^6 3s^2 3p^6 3d^{10} 4s^1$ |
| 12 | Mg | $1s^2 2s^2 2p^6 3s^2$ | 30 | Zn | $1s^2 2s^2 2p^6 3s^2 3p^6 3d^{10} 4s^2$ |
| 13 | Al | $1s^2 2s^2 2p^6 3s^2 3p^1$ | 31 | Ga | $1s^2 2s^2 2p^6 3s^2 3p^6 3d^{10} 4s^2 4p^1$ |
| 14 | Si | $1s^2 2s^2 2p^6 3s^2 3p^2$ | 32 | Ge | $1s^2 2s^2 2p^6 3s^2 3p^6 3d^{10} 4s^2 4p^2$ |
| 15 | P | $1s^2 2s^2 2p^6 3s^2 3p^3$ | 33 | As | $1s^2 2s^2 2p^6 3s^2 3p^6 3d^{10} 4s^2 4p^3$ |
| 16 | S | $1s^2 2s^2 2p^6 3s^2 3p^4$ | 34 | Se | $1s^2 2s^2 2p^6 3s^2 3p^6 3d^{10} 4s^2 4p^4$ |
| 17 | Cl | $1s^2 2s^2 2p^6 3s^2 3p^5$ | 35 | Br | $1s^2 2s^2 2p^6 3s^2 3p^6 3d^{10} 4s^2 4p^5$ |
| 18 | Ar | $1s^2 2s^2 2p^6 3s^2 3p^6$ | 36 | Kr | $1s^2 2s^2 2p^6 3s^2 3p^6 3d^{10} 4s^2 4p^6$ |

*A complete table appears in Appendix D.

▶ The quantum numbers and characteristics of orbitals were discussed in Section 7.5.

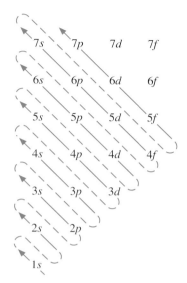

**Figure 8.7**
**A mnemonic diagram for the building-up order (diagonal rule).** You obtain this diagram by writing the subshell(s) in rows, each row having subshell(s) of given $n$. Within each row, you arrange the subshell(s) by increasing $l$. (You can stop after writing the $nf$ subshell(s), since no known elements contain $g$ or higher subshell(s).) Now, starting with the $1s$ subshell(s), draw a series of diagonals, as shown. The building-up order is the order in which these diagonals strike the subshell(s).

electron configuration of an atom by successively filling subshells in the following order: $1s$, $2s$, $2p$, $3s$, $3p$, $4s$, $3d$, $4p$, $5s$, $4d$, $5p$, $6s$, $4f$, $5d$, $6p$, $7s$, $5f$. This order reproduces the experimentally determined electron configurations (with some exceptions, which we will discuss later). You need not memorize this order. As you will see, you can very easily obtain it from the periodic table. (You can also reproduce this order from the mnemonic diagram shown in Figure 8.7.)

The building-up order corresponds for the most part to increasing energy of the subshells. You might expect this. By filling orbitals of lowest energy first, you usually get the lowest total energy (ground state) of the atom. Recall that the energy of an orbital depends only on the quantum numbers $n$ and $l$. ◀ (The energy of the H atom, however, depends only on $n$.) Orbitals with the same $n$ and $l$ but different $m_l$— that is, different orbitals of the same subshell—have the same energy. The energy depends primarily on $n$, increasing with its value. For example, a $3s$ orbital has greater energy than a $2s$ orbital. Except for the H atom, the energies of orbitals with the same $n$ increase with the $l$ quantum number. A $3p$ orbital has slightly greater energy than a $3s$ orbital. The orbital of lowest energy is $1s$; next higher are $2s$ and $2p$, then $3s$ and $3p$. The order of these subshells by energy, from $1s$ to $3p$, follows the building-up order as listed earlier.

When subshells have nearly the same energy, however, the building-up order is not strictly determined by the order of their energies. The ground-state configurations, which we are trying to predict by the building-up order, are determined by the *total* energies of the atoms. The total energy of an atom depends not only on the energies of the subshells but also on the energies of interaction among the different subshells. It so happens that for all elements with $Z = 21$ or greater, the energy of the $3d$ subshell is lower than the energy of the $4s$ subshell (Figure 8.8), which is opposite to the building-up order. You need the building-up order to predict the electron configurations of the ground states of atoms.

Now you can see how to reproduce the electron configurations of Table 8.1 using the building-up principle. Remember that the number of electrons in a neutral atom equals the atomic number $Z$. (The nuclear charge is $+Z$.) In the case of the simplest

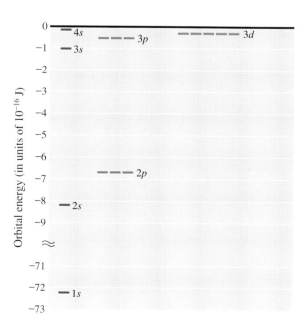

**Figure 8.8**
**Orbital energies for the scandium atom ($Z = 21$).** Note that in the scandium atom, unlike the hydrogen atom, the subshells for each $n$ are spread apart in energy. Thus, the $2p$ energy is above the $2s$. Similarly, the $n = 3$ subshells are spread to give the order $3s < 3p < 3d$. The $3d$ subshell energy is now just below the $4s$. (Values for this figure were calculated from theory by Charlotte F. Fischer, Vanderbilt University.)

atom, hydrogen ($Z = 1$), you obtain the ground state by placing the single electron into the $1s$ orbital, giving the configuration $1s^1$ (this is read as "one-ess-one"). Now you go to helium ($Z = 2$). The first electron goes into the $1s$ orbital, as in hydrogen, followed by the second electron, because any orbital can hold two electrons. The configuration is $1s^2$. Filling the $n = 1$ shell creates a very stable configuration and, as a result, helium is chemically unreactive.

You continue this way through the elements, each time increasing $Z$ by 1 and adding another electron. You obtain the configuration of an atom from that of the preceding element by adding an electron into the next available orbital, following the building-up order. In lithium ($Z = 3$), the first two electrons give the configuration $1s^2$, like helium, but the third electron goes into the next higher orbital in the building-up order, because the $1s$ orbital is now filled. This gives the configuration $1s^2 2s^1$. In beryllium ($Z = 4$), the fourth electron fills the $2s$ orbital, giving the configuration $1s^2 2s^2$.

Using the abbreviation [He] for $1s^2$, the configurations are

$$Z = 3 \quad \text{lithium} \quad 1s^2 2s^1 \quad \text{or} \quad [\text{He}]2s^1$$
$$Z = 4 \quad \text{beryllium} \quad 1s^2 2s^2 \quad \text{or} \quad [\text{He}]2s^2$$

With boron ($Z = 5$), the electrons begin filling the $2p$ subshell. You get

$$Z = 5 \quad \text{boron} \quad 1s^2 2s^2 2p^1 \quad \text{or} \quad [\text{He}]2s^2 2p^1$$
$$Z = 6 \quad \text{carbon} \quad 1s^2 2s^2 2p^2 \quad \text{or} \quad [\text{He}]2s^2 2p^2$$
$$\vdots$$
$$Z = 10 \quad \text{neon} \quad 1s^2 2s^2 2p^6 \quad \text{or} \quad [\text{He}]2s^2 2p^6$$

Having filled the $2p$ subshell, you again find a particularly stable configuration. Neon is chemically unreactive as a result.

With sodium ($Z = 11$), the $3s$ orbital begins to fill. Using the abbreviation [Ne] for $1s^2 2s^2 2p^6$, you have

$$Z = 11 \quad \text{sodium} \quad 1s^2 2s^2 2p^6 3s^1 \quad \text{or} \quad [\text{Ne}]3s^1$$
$$Z = 12 \quad \text{magnesium} \quad 1s^2 2s^2 2p^6 3s^2 \quad \text{or} \quad [\text{Ne}]3s^2$$

Then the $3p$ subshell begins to fill.

$$Z = 13 \quad \text{aluminum} \quad 1s^2 2s^2 2p^6 3s^2 3p^1 \quad \text{or} \quad [\text{Ne}]3s^2 3p^1$$
$$\vdots$$
$$Z = 18 \quad \text{argon} \quad 1s^2 2s^2 2p^6 3s^2 3p^6 \quad \text{or} \quad [\text{Ne}]3s^2 3p^6$$

With the $3p$ subshell filled, a stable configuration has been attained; argon is an unreactive element.

▶ Although the building-up order reproduces the ground-state electron configurations, it has no other significance. The order by $n$ (and then by $l$ within a given $n$), however, generally places the most easily ionized orbitals at the far right. For example, the electron configuration of Fe (in order by shells) is $1s^2 2s^2 2p^6 3s^2 3p^6 3d^6 4s^2$. The $4s$ electrons ionize first.

Now the $4s$ orbital begins to fill. You get $[\text{Ar}]4s^1$ for potassium ($Z = 19$) and $[\text{Ar}]4s^2$ for calcium ($Z = 20$) ($[\text{Ar}] = 1s^2 2s^2 2p^6 3s^2 3p^6$). At this point the $3d$ subshell begins to fill. You get $[\text{Ar}]3d^1 4s^2$ for scandium ($Z = 21$), $[\text{Ar}]3d^2 4s^2$ for titanium ($Z = 22$), and $[\text{Ar}]3d^3 4s^2$ for vanadium ($Z = 23$). Note that we have written the configurations with subshells arranged in order by shells. This generally places the subshells in order by energy and puts the subshells involved in chemical reactions at the far right.

Let us skip to zinc ($Z = 30$). The $3d$ subshell has filled; the configuration is $[\text{Ar}]3d^{10} 4s^2$. Now the $4p$ subshell begins to fill, starting with gallium ($Z = 31$), configuration $[\text{Ar}]3d^{10} 4s^2 4p^1$, and ending with krypton ($Z = 36$), configuration $[\text{Ar}]3d^{10} 4s^2 4p^6$.

## Electron Configurations and the Periodic Table

By this time you can see a pattern develop among the ground-state electron configurations of the atoms. This pattern explains the periodic table, which was briefly described in Section 2.5. Consider helium, neon, argon, and krypton, elements in Group VIIIA of the periodic table. Neon, argon, and krypton have configurations in which a $p$ subshell has just filled. (Helium has a filled $1s$ subshell; no $1p$ subshell is possible.)

| | |
|---|---|
| helium | $1s^2$ |
| neon | $1s^2 2s^2 2p^6$ |
| argon | $1s^2 2s^2 2p^6 3s^2 3p^6$ |
| krypton | $1s^2 2s^2 2p^6 3s^2 3p^6 3d^{10} 4s^2 4p^6$ |

These elements are the first members of the group called *noble gases* because of their relative unreactivity.

Look now at the configurations of beryllium, magnesium, and calcium, members of the group of *alkaline earth metals* (Group IIA), which are similar, moderately reactive elements.

| | | | |
|---|---|---|---|
| beryllium | $1s^2 2s^2$ | or | $[\text{He}]2s^2$ |
| magnesium | $1s^2 2s^2 2p^6 3s^2$ | or | $[\text{Ne}]3s^2$ |
| calcium | $1s^2 2s^2 2p^6 3s^2 3p^6 4s^2$ | or | $[\text{Ar}]4s^2$ |

Each of these configurations consists of a **noble-gas core,** that is, *an inner-shell configuration corresponding to one of the noble gases,* plus two outer electrons with an $ns^2$ configuration.

The elements boron, aluminum, and gallium (Group IIIA) also have similarities. Their configurations are

| | | | |
|---|---|---|---|
| boron | $1s^2 2s^2 2p^1$ | or | $[\text{He}]2s^2 2p^1$ |
| aluminum | $1s^2 2s^2 2p^6 3s^2 3p^1$ | or | $[\text{Ne}]3s^2 3p^1$ |
| gallium | $1s^2 2s^2 2p^6 3s^2 3p^6 3d^{10} 4s^2 4p^1$ | or | $[\text{Ar}]3d^{10} 4s^2 4p^1$ |

Boron and aluminum have noble-gas cores plus three electrons with the configuration $ns^2 np^1$. Gallium has an additional filled $3d$ subshell. *The noble-gas core together with $(n-1)d^{10}$ electrons* is often referred to as a **pseudo-noble-gas core,** because these electrons usually are not involved in chemical reactions.

*An electron in an atom outside the noble-gas or pseudo-noble-gas core* is called a **valence electron.** Such electrons are primarily involved in chemical reactions, and similarities among the configurations of valence electrons (the *valence-shell configurations*) account for similarities of the chemical properties among groups of elements.

Figure 8.9 shows a periodic table with the valence-shell configurations included. Note the similarity in electron configuration within any group (column) of elements. This similarity explains what chemists since Mendeleev have known—the properties of elements in any group are similar.

The *main-group* (or *representative*) *elements* all have valence-shell configurations $ns^a np^b$, with some choice of $a$ and $b$. ($b$ could be equal to 0.) In other words, the outer $s$ or $p$ subshell is being filled. Similarly, in the *d-block transition elements* (often called simply *transition elements* or *transition metals*), a $d$ subshell is being filled. In the *f-block transition elements* (or *inner-transition elements*), an $f$ subshell is being filled. (See Figure 8.9 or Appendix D for the configurations of these elements.)

**Figure 8.9**
**A periodic table.** This table shows the valence-shell configurations of the elements.

▶ There does not appear to be a simple explanation that is generally agreed on for why these configurations do not follow the building-up principle.

▶ More exceptions occur among the heavier transition elements, where the outer subshells are very close together. We must concede that simplicity was not the uppermost concern in the construction of the universe!

## Exceptions to the Building-Up Principle

As we have said, the building-up principle reproduces most of the ground-state configurations correctly. There are some exceptions, however, and chromium ($Z = 24$) is the first we encounter. The building-up principle predicts the configuration $[Ar]3d^4 4s^2$, though the correct one is found experimentally to be $[Ar]3d^5 4s^1$. These two configurations are actually very close in total energy because of the closeness in energies of the $3d$ and $4s$ orbitals (note Figure 8.8). For that reason, small effects can influence which of the two configurations is actually lower in energy. Copper ($Z = 29$) is another exception to the building-up principle, which predicts the configuration $[Ar]3d^9 4s^2$, although experiment shows the ground-state configuration to be $[Ar]3d^{10} 4s^1$.

We need not dwell on these exceptions beyond noting that they occur. The point to remember is that the configuration predicted by the building-up principle is very close in energy to the ground-state configuration (if it is not the ground state). Most of the qualitative conclusions regarding the chemistry of an element are not materially affected by arguing from the configuration given by the building-up principle. ◀

# X Rays, Atomic Numbers, and Orbital Structure (Photoelectron Spectroscopy)

In 1913 Henry G. J. Moseley, a student of Rutherford, used the technique of *x-ray spectroscopy* (just discovered by Max von Laue) to determine the atomic numbers of the elements. X rays are produced in a cathode-ray tube when the electron beam (cathode ray) falls on a metal target. The explanation for the production of x rays is as follows: When an electron in the cathode ray hits a metal atom in the target, it can (if it has sufficient energy) knock an electron from an inner shell of the atom. This produces a metal ion with an electron missing from an inner orbital. The electron configuration is unstable, and an electron from an orbital of higher energy drops into the half-filled orbital and a photon is emitted. The photon corresponds to electromagnetic radiation in the x-ray region.

The energies of the inner orbitals of an atom and the energy changes between them depend on the nuclear charge, $+Z$. Therefore, the photon energies $h\nu$ and the frequencies $\nu$ of emitted x rays depend on the atomic number $Z$ of the metal atom in the target. Figure 8.10 shows the x-ray spectra Moseley obtained with various metal targets. The direct dependence of the x-ray spectrum on atomic number provides an unequivocal way of deciding whether a substance is a pure element or not.

A related technique, *x-ray photoelectron spectroscopy*, experimentally confirms our theoretical view of the orbital structure of the atom. Instead of irradiating a sample with an electron beam and analyzing the frequencies of emitted x rays, you irradiate a sample with x rays and analyze the kinetic energies of ejected electrons. In other words, you observe the *photoelectric effect* on the sample (see Section 7.2).

As an example of photoelectron spectroscopy, consider a sample of neon gas (Ne atoms). Suppose the sample is irradiated with x rays of a specific frequency great enough to remove a $1s$ electron from the neon atom. Part of the energy of the x-ray photon, $h\nu$, is used to remove the electron from the atom. (This is the *ionization energy, I.E.,* for that electron.) The remaining energy appears as kinetic energy, $E_k$, of the ejected electron. From the law of conservation of energy, you can write

$$E_k = h\nu - I.E.$$

Because $h\nu$ is fixed, $E_k$ will depend linearly on *I.E.*, the ionization energy.

If you look at the electrons ejected from neon, you find that they have kinetic energies related to the ionization energies from all possible orbitals ($1s$, $2s$, and $2p$) in the atom. When you scan the various kinetic energies of ejected electrons, you see a spectrum with peaks corresponding to the different occupied orbitals (see Figure 8.11A). These ionization energies are approximately equal to the positive values of the orbital energies (Figure 8.11B), so this spectrum provides direct experimental verification of the discrete energy levels associated with the electrons of the atom.

*(continued)*

**Figure 8.10**
**X-ray spectra of the elements calcium to zinc, obtained by Moseley.** Each line results from an emission of given wavelength. Because of the volatility of zinc, Moseley used brass (a copper–zinc alloy) to observe the spectrum of zinc. Note the copper lines in brass. Also note how the lines progress to the right (indicating increasing wavelength or decreasing energy difference) with decreasing atomic number. [From J. J. Lagowski, *The Structure of Atoms* (Boston: Houghton Mifflin, 1964), Figure 26, p. 80. Used by permission.]

**Figure 8.11**
**X-ray photoelectron spectrum of neon.** *(A)* Each peak shows the energy of the ejected electrons. The energy of each x-ray photon is $200.9 \times 10^{-18}$ J. *(B)* The energy-level diagram for neon. The transitions that occur in the photoelectron spectrum are to $n = \infty$, where the energy = 0.

 **CONCEPT CHECK 8.1**

Imagine a world in which the Pauli principle is "No more than one electron can occupy an atomic orbital, irrespective of its spin." How many elements would there be in the second row of the periodic table, assuming that nothing else is different about this world?

# 8.3 Writing Electron Configurations Using the Periodic Table

▶ See Judith Strong, "The Periodic Table and Electron Configurations," *J. Chem. Educ.* 1986, *63*, 834.

To discuss bonding and the chemistry of the elements coherently, you must be able to reproduce the atomic configurations with ease, following the building-up principle. All you need is some facility in recall of the building-up order of subshells.

One approach is to recall the structure of the periodic table. Because that structure is basic, it offers a sound way to remember the building-up order. There is a definite pattern to the order of filling of the subshells as you go through the elements in the periodic table, and from this you can write the building-up order. Figure 8.12 shows a periodic table stressing this pattern. For example, in the violet-colored area, an $ns$ subshell is being filled. In the blue-colored area, an $np$ subshell is being filled. The value of $n$ is obtained from the period (row) number. In the red area, an $(n - 1)d$ subshell is being filled.

You read the building-up order by starting with the first period, in which the $1s$ subshell is being filled. In the second period, you have $2s$ (violet area); then, staying in the same period but jumping across, you have $2p$ (blue area). In the third period, you have $3s$ and $3p$; in the fourth period, $4s$ (violet area), $3d$ (red area), and $4p$ (blue area). This pattern should become clear enough to visualize with a periodic table that is not labeled with the subshells, such as the one on the inside front cover of this

s-block elements

p-block elements

d-block elements

f-block elements

**Figure 8.12**
**A periodic table illustrating the building-up order.**   The colored areas of elements show the different subshells that are filling with those elements.

▶ An alternate example is provided in the Instructor's Resource Manual.

book. The detailed method is illustrated in the next example. (An alternative way to reconstruct the building-up order of subshells is to write the diagram in Figure 8.7.)

---

### EXAMPLE 8.2

## Determining the Configuration of an Atom Using the Building-Up Principle

Use the building-up principle to obtain the configuration for the ground state of the gallium atom ($Z = 31$). Give the configuration in complete form (do not abbreviate for the core). What is the valence-shell configuration?

### PROBLEM STRATEGY

Write the subshells in their building-up order. You can use any periodic table to do this. All you have to remember is the pattern in which subshells are filled as you progress through the table (Figure 8.12). Go through the periods, starting with hydrogen, writing the subshells that are being filled, and stopping with the element whose configuration you want (gallium in this example). Then you distribute the electrons (equal to the atomic number of the element) to the subshells. The valence-shell configuration includes only the subshells outside the noble-gas or pseudo-noble-gas core.

### SOLUTION

From a periodic table, you get the following building-up order:

|        | $1s$  | $2s\ 2p$ | $3s\ 3p$ | $4s\ 3d\ 4p$ |
|--------|-------|----------|----------|--------------|
| Period: | first | second   | third    | fourth       |

Now you fill the subshells with electrons, remembering that you have a total of 31 electrons to distribute. You get

$$1s^2 2s^2 2p^6 3s^2 3p^6 4s^2 3d^{10} 4p^1$$

Or, if you rearrange the subshells by shells, you write

$$1s^2 2s^2 2p^6 3s^2 3p^6 3d^{10} 4s^2 4p^1$$

The valence-shell configuration is $4s^2 4p^1$.

**Problem Solving:
Electron Configurations**

---

**EXERCISE 8.2**

Use the building-up principle to obtain the electron configuration for the ground state of the manganese atom ($Z = 25$).

See Problems 8.39, 8.40, 8.41, and 8.42.

---

In many cases, you need only the configuration of the outer electrons. You can determine this from the position of the element in the periodic table. Recall that the valence-shell configuration of a main-group element is $ns^a np^b$, where $n$, the principal quantum number of the outer shell, also equals the period number for the element. The total number of valence electrons, which equals $a + b$, can be obtained from the group number. For example, gallium is in Period 4, so $n = 4$. It is in Group IIIA, so the number of valence electrons is 3. This gives the valence-shell configuration $4s^2 4p^1$. The configuration of outer shells of a transition element is obtained in a similar fashion. The next example gives the details.

---

▶ An alternate example is provided in the Instructor's Resource Manual.

**EXAMPLE 8.3**

## Determining the Configuration of an Atom Using the Period and Group Numbers

What are the configurations for the outer electrons of a. tellurium, $Z = 52$, and b. nickel, $Z = 28$?

**SOLUTION**

a. You locate tellurium in a periodic table and find it to be in Period 5, Group VIA. Thus, it is a main-group element, and the outer subshells are $5s$ and $5p$. These subshells contain six electrons, because the group is VIA. The valence-shell configuration is $\mathbf{5s^2 5p^4}$. b. Nickel is a Period 4 transition element, in which the general form of the outer-shell configuration is $3d^{a-2} 4s^2$. To determine a, you note that it equals the Roman numeral group number up to iron (8). After that you count Co as 9 and Ni as 10. Hence, the outer-shell configuration is $\mathbf{3d^8 4s^2}$.

---

**EXERCISE 8.3**

Using the periodic table on the inside front cover, write the valence-shell configuration of arsenic (As).

See Problems 8.43, 8.44, 8.45, and 8.46.

---

**EXERCISE 8.4**

The lead atom has the ground-state configuration $[Xe]4f^{14}5d^{10}6s^2 6p^2$. Find the period and group for this element. From its position in the periodic table, would you classify lead as a main-group, a transition, or an inner-transition element?

See Problems 8.47 and 8.48.

**CONCEPT CHECK 8.2**

Two elements in Period 3 are adjacent to one another in the periodic table. The ground-state atom of one element has only *s* electrons in its valence shell; the other one has at least one *p* electron in its valence shell. Identify the elements.

# 8.4 Orbital Diagrams of Atoms; Hund's Rule

In discussing the ground states of atoms, we have not yet described how the electrons are arranged within each subshell. There may be several different ways of arranging electrons in a particular configuration. Consider the carbon atom ($Z = 6$) with the ground-state configuration $1s^2 2s^2 2p^2$. Three possible arrangements are given in the following orbital diagrams.

These orbital diagrams show different states of the carbon atom. Each state has a different energy and, as you will see, different magnetic characteristics.

## Hund's Rule

In about 1927, Friedrich Hund discovered an empirical rule determining the lowest-energy arrangement of electrons in a subshell. **Hund's rule** states that *the lowest-energy arrangement of electrons in a subshell is obtained by putting electrons into separate orbitals of the subshell with the same spin before pairing electrons.* Let us see how this would apply to the carbon atom, whose ground-state configuration is $1s^2 2s^2 2p^2$. The first four electrons go into the $1s$ and $2s$ orbitals.

The next two electrons go into separate $2p$ orbitals, with both electrons having the same spin, following Hund's rule.

We see that the orbital diagram corresponding to the lowest energy is the one we previously labeled Diagram 1.

To apply Hund's rule to the oxygen atom, whose ground-state configuration is $1s^2 2s^2 2p^4$, we place the first seven electrons as follows:

**Table 8.2**
**Orbital Diagrams for the Ground States of Atoms from $Z = 1$ to $Z = 10$**

| Atom | Z | Configuration | 1s | 2s | 2p |
|------|---|---------------|----|----|----|
| Hydrogen | 1 | $1s^1$ | ⊙ | ◯ | ◯◯◯ |
| Helium | 2 | $1s^2$ | ⊙ | ◯ | ◯◯◯ |
| Lithium | 3 | $1s^2 2s^1$ | ⊙ | ⊙ | ◯◯◯ |
| Beryllium | 4 | $1s^2 2s^2$ | ⊙ | ⊙ | ◯◯◯ |
| Boron | 5 | $1s^2 2s^2 2p^1$ | ⊙ | ⊙ | ⊙◯◯ |
| Carbon | 6 | $1s^2 2s^2 2p^2$ | ⊙ | ⊙ | ⊙⊙◯ |
| Nitrogen | 7 | $1s^2 2s^2 2p^3$ | ⊙ | ⊙ | ⊙⊙⊙ |
| Oxygen | 8 | $1s^2 2s^2 2p^4$ | ⊙ | ⊙ | ⊙⊙⊙ |
| Fluorine | 9 | $1s^2 2s^2 2p^5$ | ⊙ | ⊙ | ⊙⊙⊙ |
| Neon | 10 | $1s^2 2s^2 2p^6$ | ⊙ | ⊙ | ⊙⊙⊙ |

The last electron is paired with one of the $2p$ electrons to give a doubly occupied orbital. The orbital diagram for the ground state of the oxygen atom is

$$1s \quad 2s \quad 2p$$

In the following example, Hund's rule is used to determine the orbital diagram for the ground state of a more complicated atom. Table 8.2 gives orbital diagrams for the ground states of the first ten elements.

► An alternate example is provided in the Instructor's Resource Manual.

**EXAMPLE 8.4**

## Applying Hund's Rule

Write an orbital diagram for the ground state of the iron atom.

 **PROBLEM STRATEGY**

First obtain the electron configuration, as described in Example 8.2. Then draw circles for the orbitals of each subshell. A filled subshell should have doubly occupied orbitals (two electrons with opposite spins). For a partially filled subshell, apply Hund's rule, putting electrons into separate orbitals with the same spin (either all up or all down) before pairing electrons.

**SOLUTION**

The electron configuration of the iron atom is $1s^2 2s^2 2p^6 3s^2 3p^6 3d^6 4s^2$. All the subshells except the $3d$ are filled. In placing the six electrons in the $3d$ subshell, you note that the first five go into separate $3d$ orbitals with their spin arrows in the same direction. The sixth electron must doubly occupy a $3d$ orbital. The orbital diagram is

$$1s \quad 2s \quad 2p \quad 3s \quad 3p \quad 3d \quad 4s$$

You can write this diagram in abbreviated form using [Ar] for the argon-like core of the iron atom:

[Ar] ⬆⬇ ⬆ ⬆ ⬆ ⬆ ⬆⬇
　　　　3d　　　　4s

---

## EXERCISE 8.5

Write an orbital diagram for the ground state of the phosphorus atom ($Z = 15$). Write all orbitals.

---

See Problems 8.49 and 8.50.

### Magnetic Properties of Atoms

The magnetic properties of a substance can reveal certain information about the arrangement of electrons in an atom (or molecule). Although an electron in an atom behaves like a small magnet, the magnetic attractions from two electrons that are opposite in spin cancel each other. As a result, an atom that has only doubly occupied orbitals has no net spin magnetism. However, an atom with *unpaired* electrons— that is, with an excess of one kind of spin—does exhibit a net magnetism.

　　The magnetic properties of an atom can be observed. The most direct way is to determine whether the atomic substance is attracted to the field of a strong magnet. A **paramagnetic substance** is *a substance that is weakly attracted by a magnetic field, and this attraction is generally the result of unpaired electrons.* ◄ For example, sodium vapor has been found experimentally to be paramagnetic. The explanation is that the vapor consists primarily of sodium atoms, each containing an unpaired electron. (The configuration is [Ne]$3s^1$.) A **diamagnetic substance** is *a substance that is not attracted by a magnetic field or is very slightly repelled by such a field. This property generally means that the substance has only paired electrons.* Mercury vapor is found experimentally to be diamagnetic. The explanation is that mercury vapor consists of mercury atoms, with the electron configuration [Xe]$4f^{14}5d^{10}6s^2$, which has only paired electrons.

　　We expect the different orbital diagrams presented at the beginning of this section for the $1s^2 2s^2 2p^2$ configuration of the carbon atom to have different magnetic properties. Diagram 1, predicted by Hund's rule to be the ground state, would give a magnetic atom, whereas the other diagrams would not. If we could prepare a vapor of carbon atoms, it should be attracted to a magnet. (It should be paramagnetic.) It is difficult to prepare a vapor of free carbon atoms in sufficient concentration to observe a result. However, the visible–ultraviolet spectrum of carbon atoms can be obtained easily from dilute vapor. From an analysis of this spectrum, it is possible to show that the ground-state atom is magnetic, which is consistent with the prediction of Hund's rule.

► The strong, permanent magnetism seen in iron objects is called *ferromagnetism* and is due to the cooperative alignment of electron spins in many iron atoms. Paramagnetism is a much weaker effect. Nevertheless, paramagnetic substances can be attracted to a strong magnet. Liquid oxygen is composed of paramagnetic $O_2$ molecules. When poured over a magnet, the liquid clings to the poles. (See Figure 10.30.)

# Periodicity of the Elements

You have seen that the periodic table that Mendeleev discovered in 1869 can be explained by the periodicity of the ground-state electron configurations of the atoms. Now we will look at various aspects of the periodicity of the elements.

# Levitating Frogs and People

**FRONTIERS**

Have you ever seen a magic act in which a person rises above the stage in apparent defiance of gravity? Magicians have often included such acts of "levitation" as part of their performances. Of course, magical levitation uses tricks to deceive you. But levitation can be done, without tricks, using the known laws of electromagnetism.

Recently, researchers wanted to demonstrate that you can levitate almost anything if you have the proper magnetic field. They placed a frog within the magnetic field of a powerful electromagnet (which consists of an electric current flowing through a coil of wire). The frog floated in midair above the coil, at the point where the upward repulsion of the frog from the magnetic field just balanced the downward force of gravity (Figure 8.13).

We don't usually think of things such as frogs as being magnetic. In fact, most materials, including most substances in a biological organism, are diamagnetic. They contain pairs of electrons, in which the magnetism of each electron in a pair offsets the equal and opposite magnetism of the other electron. You might expect diamagnetic materials to be nonmagnetic because of this balance of opposite electron spins. But when you place a diamagnetic material in an external magnetic field, its electrons move so as to induce, or generate, a smaller magnetic field that is opposite in direction to the external field. This results in a repulsive force between the diamagnetic material and the external field. The repulsive force, though generally small, can be easily observed if the external magnet is large enough. Although, with a sufficiently large magnet, it would be possible to levitate a person as well as a frog, the researchers don't plan to do that.

**Figure 8.13**
**Levitation of a frog.** A frog placed in a powerful magnetic field, generated by a current flowing through a water-cooled coil of wire, appears to float in midair in defiance of gravity. The frog contains materials that are diamagnetic (composed of electron pairs), which are repelled by the magnetic field.

# 8.5 Mendeleev's Predictions from the Periodic Table

One of Mendeleev's periodic tables is reproduced in Figure 8.14. Though somewhat different from modern tables, it shows essentially the same arrangement. In this early form of the periodic table, within each column some elements were placed toward the left side and some toward the right. With some exceptions, the elements on a given side have similar properties. Mendeleev left spaces in his periodic table for what he felt were undiscovered elements. There are blank spaces in his row 5, for example, one directly under aluminum and another under silicon (looking at just the elements on the right side of the column). By writing the known elements in this row with their atomic weights, he could determine approximate values (between the known ones) for the missing elements (values in parentheses).

| Cu | Zn | — | — | As | Se | Br |
|---|---|---|---|---|---|---|
| 63 amu | 65 amu | (68 amu) | (72 amu) | 75 amu | 78 amu | 80 amu |

| Reihen | Gruppe I.<br>—<br>$R^2O$ | Gruppe II.<br>—<br>$RO$ | Gruppe III.<br>—<br>$R^2O^3$ | Gruppe IV.<br>$RH^4$<br>$RO^2$ | Gruppe V.<br>$RH^3$<br>$R^2O^5$ | Gruppe VI.<br>$RH^2$<br>$RO^3$ | Gruppe VII.<br>$RH$<br>$R^2O^7$ | Gruppe VIII.<br>—<br>$RO^4$ |
|---|---|---|---|---|---|---|---|---|
| 1 | H = 1 | | | | | | | |
| 2 | Li = 7 | Be = 9,4 | B = 11 | C = 12 | N = 14 | O = 16 | F = 19 | |
| 3 | Na = 23 | Mg = 24 | Al = 27,3 | Si = 28 | P = 31 | S = 32 | Cl = 35,5 | |
| 4 | K = 39 | Ca = 40 | — = 44 | Ti = 48 | V = 51 | Cr = 52 | Mn = 55 | Fe = 56, Co = 59, |
| | | | | | | | | Ni = 59, Cu = 63. |
| 5 | (Cu = 63) | Zn = 65 | — = 68 | — = 72 | As = 75 | Se = 78 | Br = 80 | |
| 6 | Rb = 85 | Sr = 87 | ?Yt = 88 | Zr = 90 | Nb = 94 | Mo = 96 | — = 100 | Ru = 104, Rh =104, |
| | | | | | | | | Pd = 106, Ag = 108. |
| 7 | (Ag = 108) | Cd = 112 | In = 113 | Su = 118 | Sb = 122 | Te = 125 | J = 127 | |
| 8 | Cs = 133 | Ba = 137 | ?Di = 138 | ?Ce = 140 | — | — | — | — — — — |
| 9 | (—) | — | — | | — | — | — | |
| 10 | — | — | ?Er = 178 | ?La = 180 | Ta = 182 | W = 184 | — | Os = 195, Ir = 197, |
| | | | | | | | | Pt = 198, Au = 199. |
| 11 | (Au = 199) | Hg = 200 | Tl =204 | Pb = 207 | Bi = 208 | — | — | — — — — |
| 12 | — | — | — | Th = 231 | — | U = 240 | — | |

**Figure 8.14**
**Mendeleev's periodic
table.**  This one was
published in 1872.

The Group III element directly under aluminum Mendeleev called eka-aluminum, with the symbol Ea. (*Eka* is the Sanskrit word meaning "first"; thus eka-aluminum is the first element under aluminum.) The known Group III elements have oxides of the form $R_2O_3$, so Mendeleev predicted that eka-aluminum would have an oxide with the formula $Ea_2O_3$.

The physical properties of this undiscovered element could be predicted by comparing values for the neighboring known elements. For eka-aluminum Mendeleev predicted a density of 5.9 $g/cm^3$, a low *melting point* (the temperature at which a substance melts), and a high *boiling point* (the temperature at which a substance boils).

In 1874 the French chemist Paul-Émile Lecoq de Boisbaudran found two previously unidentified lines in the atomic spectrum of a sample of sphalerite (a zinc sulfide, ZnS, mineral). Realizing he was on the verge of a discovery, Lecoq de Boisbaudran quickly prepared a large batch of the zinc mineral, from which he isolated a gram of a new element. He called this new element gallium. The properties of gallium were remarkably close to those Mendeleev predicted for eka-aluminum.

| Property | Predicted for<br>Eka-Aluminum | Found for<br>Gallium |
|---|---|---|
| Atomic weight | 68 amu | 69.7 amu |
| Formula of oxide | $Ea_2O_3$ | $Ga_2O_3$ |
| Density of the element | 5.9 $g/cm^3$ | 5.91 $g/cm^3$ |
| Melting point of the element | Low | 30.1°C |
| Boiling point of the element | High | 1983°C |

The predictive power of Mendeleev's periodic table was demonstrated again when scandium (eka-boron) was discovered in 1879 and germanium (eka-silicon) in 1886. Both elements had properties remarkably like those predicted by Mendeleev. These early successes won acceptance for the organizational and predictive power of the periodic table.

## 8.6  Some Periodic Properties

**Understanding Concepts:
Periodic Trends**

The electron configurations of the atoms display a periodic variation with increasing atomic number (nuclear charge). As a result, the elements show periodic variations of physical and chemical behavior. The **periodic law** states that *when the elements are arranged by atomic number, their physical and chemical properties vary periodically.* In this section, we will look at three physical properties of an atom: atomic radius, ionization energy, and electron affinity. These three quantities, especially ionization energy and electron affinity, are important in discussions of chemical bonding (the subject of Chapter 9).

### Atomic Radius

An atom does not have a definite size, because the statistical distribution of electrons does not abruptly end but merely decreases to very small values as the distance from the nucleus increases. This can be seen in the plot of the electron distribution for the argon atom, shown in Figure 8.15. Consequently, atomic size must be defined in a somewhat arbitrary manner, so various measures of atomic size exist. The atomic radii plotted in Figure 8.16 and also represented in Figure 8.17 are *covalent radii*, which are obtained from measurements of distances between the nuclei of atoms in the chemical bonds of molecular substances. (The determination and use of covalent radii are discussed in Chapter 9.)

Figures 8.16 and 8.17 show the following general trends in size of atomic radii:

1. Within each period (horizontal row), the atomic radius tends to decrease with increasing atomic number (nuclear charge). The largest atom in a period is a Group IA atom and the smallest is a noble-gas atom.

2. Within each group (vertical column), the atomic radius tends to increase with the period number.

**Figure 8.15
Electron distribution for the argon atom.**  This is a radial distribution, showing the probability of finding an electron at a given distance from the nucleus. The distribution shows three maxima, for the *K*, *L*, and *M* shells. The outermost maximum occurs at 66 pm; then the distribution falls steadily, becoming negligibly small after several hundred picometers.

Distance from the nucleus (pm)

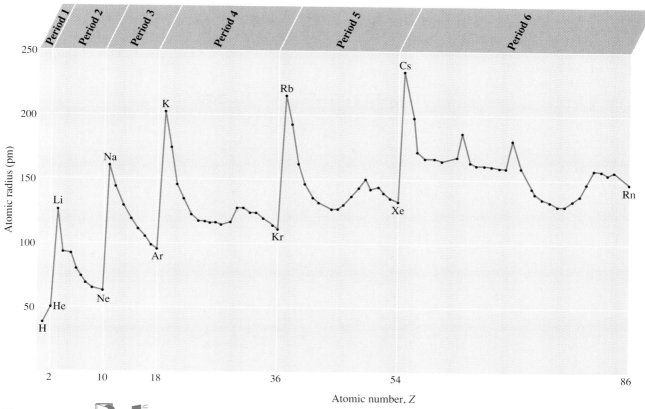

**Figure 8.16**
**Atomic radius (covalent radius) versus atomic number.** Note that the curve is periodic (tends to repeat). Each period of elements begins with the Group IA atom, and the atomic radius tends to decrease until the Group VIIIA atom. (Values for He, Ne, and Ar are estimated because there are no known compounds.)

The atomic radius increases greatly going from any noble-gas atom to the following Group IA atom, giving the curve in Figure 8.16 a saw-tooth appearance. A similar diagram is obtained for other measures of atomic size.

These general trends in atomic radius can be explained if you look at the two factors that primarily determine the size of the outermost orbital. One factor is the principal quantum number $n$ of the orbital; the larger $n$ is, the larger the size of the orbital. The other factor is the effective nuclear charge acting on an electron in the orbital; increasing the effective nuclear charge reduces the size of the orbital by pulling the electrons inward. The **effective nuclear charge** is *the positive charge that an electron experiences from the nucleus, equal to the nuclear charge but reduced by any shielding or screening from any intervening electron distribution.* Consider the effective nuclear charge on the 2s electron in the lithium atom (configuration $1s^2 2s^1$). The nuclear charge is $3e$, but the effect of this charge on the 2s electron is reduced by the distribution of the two 1s electrons lying between the nucleus and the 2s electron (roughly, each core electron reduces the nuclear charge by $1e$).

Consider a given period of elements. The principal quantum number of the outer orbitals remains constant. However, the effective nuclear charge increases, because the nuclear charge increases and the number of core electrons remains constant. Consequently, the size of the outermost orbital and, therefore, the radius of the atom decrease with increasing $Z$ in any period.

Now consider a given column of elements. The effective nuclear charge remains nearly constant (approximately equal to $e$ times the number of valence electrons), but $n$ gets larger. You observe that the atomic radius increases.

**Figure 8.17**
**Representation of atomic radii (covalent radii) of the main-group elements.**
Note the trends within each period and each group.

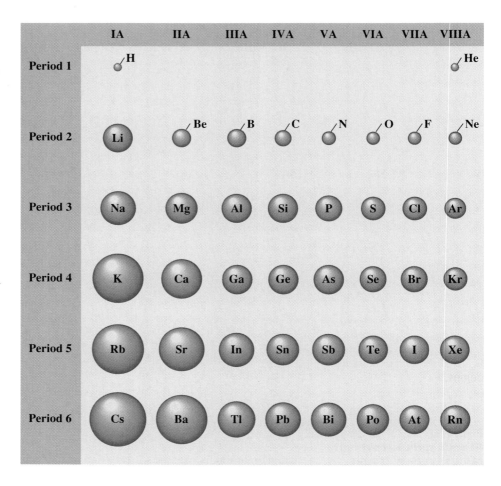

► An alternate example is provided in the Instructor's Resource Manual.

### EXAMPLE 8.5

## Determining Relative Atomic Sizes from Periodic Trends

Refer to a periodic table and use the trends noted for size of atomic radii to arrange the following in order of increasing atomic radius: Al, C, Si.

**SOLUTION**

Note that C is above Si in Group IVA. Therefore, the radius of C is smaller than that of Si (the atomic radius increases going down a group of elements). Note that Al and Si are in the same period. Therefore the radius of Si is smaller than that of Al (radius decreases with Z in a period). Hence, the order of elements by increasing radius is **C, Si, Al.**

### EXERCISE 8.6

Using a periodic table, arrange the following in order of increasing atomic radius: Na, Be, Mg.

See Problems 8.53 and 8.54.

## Ionization Energy

The **first ionization energy** (or **first ionization potential**) of an atom is *the minimum energy needed to remove the highest-energy (that is, the outermost) electron from the neutral atom in the gaseous state.* (When the unqualified term *ionization energy* is used, it generally means first ionization energy.) For the lithium atom, the first ionization energy is the energy needed for the following process (electron configurations are in parentheses):

$$Li(1s^2 2s^1) \longrightarrow Li^+(1s^2) + e^-$$

Values of this energy are usually quoted for one mole of atoms ($6.02 \times 10^{23}$ atoms). The ionization energy of the lithium atom is 520 kJ/mol. ◄

Ionization energies display a periodic variation when plotted against atomic number, as Figure 8.18 shows. Within any period, values tend to increase with atomic number. Thus, the lowest values in a period are found for the Group IA elements *(alkali metals).* It is characteristic of reactive metals such as these to lose electrons easily. The largest ionization energies in any period occur for the noble-gas elements. In other words, a noble-gas atom loses electrons with difficulty, which is partly responsible for the stability of the noble-gas configurations and the unreactivity of the noble gases.

This general trend—increasing ionization energy with atomic number in a given period—can be explained as follows: The energy needed to remove an electron from the outer shell is proportional to the effective nuclear charge divided by the average

► Ionization energies are often measured in electron volts (eV). This is the amount of energy imparted to an electron when it is accelerated through an electrical potential of one volt. One electron volt is approximately 100 kJ/mol (1 eV = 96.5 kJ/mol).

**Figure 8.18**
**Ionization energy versus atomic number.** Note that the values tend to increase within each period, except for small drops in ionization energy at the Group IIIA and VIA elements. Large drops occur when a new period begins.

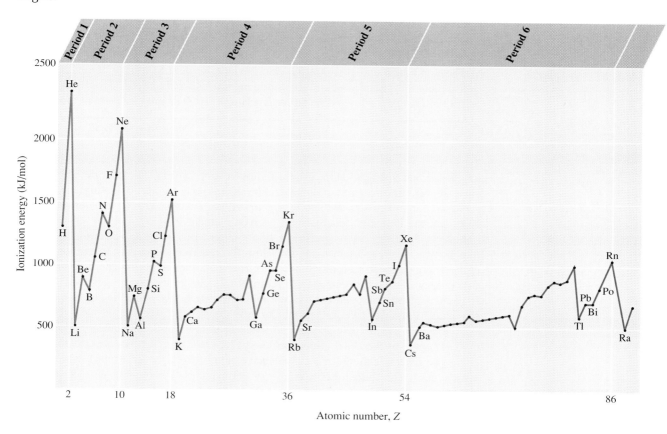

distance between electron and nucleus. (This distance is inversely proportional to the effective nuclear charge.) Hence, the ionization energy is proportional to the square of the effective nuclear charge and increases going across a period.

Small deviations from this general trend occur. A IIIA element ($ns^2np^1$) has smaller ionization energy than the preceding IIA element ($ns^2$). Apparently the $np$ electron of the IIIA element is more easily removed than one of the $ns$ electrons of the preceding IIA element. Also note that a VIA element ($ns^2np^4$) has smaller ionization energy than the preceding VA element. As a result of electron repulsion, it is easier to remove an electron from the doubly occupied $np$ orbital of the VIA element than from a singly occupied orbital of the preceding VA element.

Ionization energies tend to decrease going down any column of main-group elements. This is because atomic size increases going down the column.

▶ An alternate example is provided in the Instructor's Resource Manual.

## EXAMPLE 8.6

### Determining Relative Ionization Energies from Periodic Trends

Using a periodic table only, arrange the following elements in order of increasing ionization energy: Ar, Se, S.

### SOLUTION

Note that Se is below S in Group VIA. Therefore, the ionization energy of Se should be less than that of S. Also, S and Ar are in the same period, with $Z$ increasing from S to Ar. Therefore, the ionization energy of S should be less than that of Ar. Hence the order is **Se, S, Ar.**

### EXERCISE 8.7

The first ionization energy of the chlorine atom is 1251 kJ/mol. Without looking at Figure 8.18, state which of the following values would be the more likely ionization energy for the iodine atom. Explain. a. 1000 kJ/mol. b. 1400 kJ/mol.

See Problems 8.55 and 8.56.

The electrons of an atom can be removed successively. The energies required at each step are known as the *first* ionization energy, the *second* ionization energy, and so forth. Table 8.3 lists the successive ionization energies of the first ten elements. Note that the ionization energies for a given element increase as more electrons are removed. The first and second ionization energies of beryllium (electron configuration $1s^22s^2$) are 899 kJ/mol and 1757 kJ/mol, respectively. The first ionization energy is the energy needed to remove a $2s$ electron from the Be atom. The second ionization energy is the energy needed to remove a $2s$ electron from the positive ion $Be^+$. Its value is greater than that of the first ionization energy because the electron is being removed from a positive ion, which strongly attracts the electron.

Note that there is a large jump in value from the second ionization energy of Be (1757 kJ/mol) to the third ionization energy (14,848 kJ/mol). The second ionization energy corresponds to removing a valence electron (from the $2s$ orbital), which is relatively easy. The third ionization energy corresponds to removing an electron from the core of the atom (from the $1s$ orbital)—that is, from a noble-gas configuration ($1s^2$). A vertical line in Table 8.3 separates the energies needed to remove valence

**Table 8.3**
**Successive Ionization Energies of the First Ten Elements (kJ/mol)***

| Element | First | Second | Third | Fourth | Fifth | Sixth | Seventh |
|---|---|---|---|---|---|---|---|
| H | 1312 | | | | | | |
| He | 2372 | 5250 | | | | | |
| Li | 520 | 7298 | 11,815 | | | | |
| Be | 899 | 1757 | 14,848 | 21,006 | | | |
| B | 801 | 2427 | 3660 | 25,025 | 32,826 | | |
| C | 1086 | 2353 | 4620 | 6222 | 37,829 | 47,276 | |
| N | 1402 | 2857 | 4578 | 7475 | 9445 | 53,265 | 64,358 |
| O | 1314 | 3388 | 5300 | 7469 | 10,989 | 13,326 | 71,333 |
| F | 1681 | 3374 | 6020 | 8407 | 11,022 | 15,164 | 17,867 |
| Ne | 2081 | 3952 | 6122 | 9370 | 12,177 | 15,238 | 19,998 |

*Ionization energies to the right of a vertical line correspond to removal of electrons from the core of the atom.

electrons from those needed to remove core electrons. For each element, a large increase in ionization energy occurs when this line is crossed. The large increase results from the fact that once the valence electrons are removed, a stable noble-gas configuration is obtained. Further ionizations become much more difficult. We will see in Chapter 9 that metal atoms often form compounds by losing valence electrons; core electrons are not significantly involved in the formation of compounds.

## Electron Affinity

When a neutral atom in the gaseous state picks up an electron to form a stable negative ion, energy is released. For example, a chlorine atom can pick up an electron to give a chloride ion, $Cl^-$, and 349 kJ/mol of energy is released. You write the process as follows, with electron configurations noted in parentheses:

$$Cl([Ne]3s^2 3p^5) + e^- \longrightarrow Cl^-([Ne]3s^2 3p^6)$$

The **electron affinity** is *the energy change for the process of adding an electron to a neutral atom in the gaseous state to form a negative ion.* If the negative ion is stable (does not spontaneously disintegrate into the neutral atom and free electron), the energy change for its formation is a negative number. Thus, the electron affinity of Cl is $-349$ kJ/mol. Large negative numbers such as this indicate that a very stable negative ion is formed. Small negative numbers indicate that a less stable ion is formed. Table 8.4 gives the electron affinities of the main-group elements. ◄

▶ Electron affinities are also defined with positive values for the formation of stable ions.

Electron affinities, *E.A.,* have a periodic variation, just as atomic radii and ionization energies do, though somewhat more complicated. To see this variation, consider each of the main groups of elements beginning with Group IA. All of the Group IA elements have small negative electron affinities. When you add an electron to a lithium atom, for example, it goes into the $2s$ orbital to form a moderately stable negative ion, releasing energy.

$$Li(1s^2 2s^1) + e^- \longrightarrow Li^-(1s^2 2s^2); \ E.A. = -60 \text{ kJ/mol}$$

**Table 8.4**
**Electron Affinities of the Main-Group Elements (kJ/mol)\***

| Period | IA | IIIA | IVA | VA | VIA | VIIA |
|---|---|---|---|---|---|---|
| 1 | **H** −73 | | | | | |
| 2 | **Li** −60 | **B** −27 | **C** −122 | **N** 0 | **O** −141 | **F** −328 |
| 3 | **Na** −53 | **Al** −44 | **Si** −134 | **P** −72 | **S** −200 | **Cl** −349 |
| 4 | **K** −48 | **Ga** −30 | **Ge** −120 | **As** −77 | **Se** −195 | **Br** −325 |
| 5 | **Rb** −47 | **In** −30 | **Sn** −121 | **Sb** −101 | **Te** −190 | **I** −295 |
| 6 | **Cs** −45 | **Tl** −30 | **Pb** −110 | **Bi** −110 | **Po** −180 | **At** −270 |

*Atoms of the alkaline earth metals (Group IIA) and the noble gases (Group VIIIA) do not form stable negative ions.

None of the Group IIA elements form stable negative ions; that is, the electron affinities of these elements are positive. Each of the atoms has a filled $ns$ subshell, so that if you were to add an electron, it would have to go into the next higher energy subshell ($np$). Instead of releasing energy, the atom would absorb energy. For example,

$$\text{Be}(1s^2 2s^2) + \text{e}^- \longrightarrow \text{Be}^-(1s^2 2s^2 2p^1); \ E.A. > 0$$

(For a similar reason, the Group VIIIA elements also do not form stable negative ions.)

From Group IIIA to Group VIIA, the added electron goes into the $np$ subshell of the valence shell. With the exception of the Group VA elements, the electron affinities tend toward more negative values as you progress to the right through these elements in any period (more energy is released and stabler negative ions are formed). The electron affinity of a Group VA element is generally less negative than the preceding Group IVA element. Note the trend in electron affinities in the following:

$$\text{B}(1s^2 2s^2 2p^1) + \text{e}^- \longrightarrow \text{B}^-(1s^2 2s^2 2p^2); \ E.A. = -27 \ \text{kJ/mol}$$
$$\text{C}(1s^2 2s^2 2p^2) + \text{e}^- \longrightarrow \text{C}^-(1s^2 2s^2 2p^3); \ E.A. = -122 \ \text{kJ/mol}$$
$$\text{N}(1s^2 2s^2 2p^3) + \text{e}^- \longrightarrow \text{N}^-(1s^2 2s^2 2p^4); \ E.A. = 0 \ \text{kJ/mol}$$
$$\text{O}(1s^2 2s^2 2p^4) + \text{e}^- \longrightarrow \text{O}^-(1s^2 2s^2 2p^5); \ E.A. = -141 \ \text{kJ/mol}$$
$$\text{F}(1s^2 2s^2 2p^5) + \text{e}^- \longrightarrow \text{F}^-(1s^2 2s^2 2p^6); \ E.A. = -328 \ \text{kJ/mol}$$

Broadly speaking, the general trend is toward more negative electron affinities from left to right in any period. Note especially that the Group VIA and Group VIIA elements have the largest negative electron affinities of any of the main-group elements. Therefore, these elements have compounds containing monatomic anions (such as $F^-$ and $O^{2-}$).

**EXERCISE 8.8**

Without looking at Table 8.4 but using the general comments in this section, decide which has the larger negative electron affinity, C or F.

See Problems 8.57 and 8.58.

 **CONCEPT CHECK 8.3**

Given the following information for element E, identify the element's group in the periodic table: The electron affinity of E is positive (that is, it does not form a stable negative ion). The first ionization energy of E is less than the second ionization energy, which in turn is very much less than its third ionization energy.

## 8.7  Periodicity in the Main-Group Elements

▶ **Experiments in General Chemistry, 8, "Solubilities within a Family."**

The chemical and physical properties of the main-group elements clearly display periodic character. For instance, as pointed out in Section 2.5, the metallic elements lie to the left of the "staircase" line in the periodic table (see inside front cover), nonmetals lie to the right, and the metalloids (with intermediate characteristics) lie along this line. So, as you move left to right in any row of the periodic table, the metallic character of the elements decreases. As you progress down a column, however, the elements tend to increase in metallic character.

These variations of metallic–nonmetallic character can be attributed in part to variations in the ionization energies of the corresponding atoms. Elements with low ionization energy tend to be metallic, whereas those with high ionization energy tend to be nonmetallic. As you saw in the previous section, ionization energy is a periodic property, so it is not surprising that the metallic–nonmetallic character of an element is similarly periodic.

The basic–acidic behavior of the oxides of the elements is a good indicator of the metallic–nonmetallic character of the elements. Oxides are classified as basic or acidic depending on their reactions with acids and bases. A **basic oxide** is *an oxide that reacts with acids.* Most metal oxides are basic. An **acidic oxide** is *an oxide that reacts with bases.* Most nonmetal oxides are acidic oxides. An **amphoteric oxide** is *an oxide that has both basic and acidic properties.*

In the following brief descriptions of the main-group elements, we will note the metallic–nonmetallic behavior of the elements, as well as the basic–acidic character of the oxides. Although elements in a given group are expected to be similar, the degree of similarity does vary among the groups. The alkali metals (Group IA) show marked similarities, as do the halogens (Group VIIA). On the other hand, the Group IVA elements range from a nonmetal (carbon) at the top of the column to a metal (lead) at the bottom. In either case, however, the changes from one element in a column to the next lower one are systematic, and the periodic table helps us to correlate these systematic changes.

### Hydrogen ($1s^1$)

Although the electron configuration of hydrogen would seem to place the element in Group IA, its properties are quite different, and it seems best to consider the element

▶ At very high pressures, however, hydrogen is believed to have metallic properties.

as belonging in a group by itself. The element is a colorless gas composed of $H_2$ molecules. ◀

## Group IA Elements, the Alkali Metals ($ns^1$)

The *alkali metals* are soft and reactive, with the reactivities increasing as you move down the column of elements. All of the metals react with water to produce hydrogen.

$$2Li(s) + 2H_2O(l) \longrightarrow 2LiOH(aq) + H_2(g)$$

The vigor of the reaction increases from lithium (moderate) to rubidium (violent). All of the alkali metals form basic oxides with the general formula $R_2O$.

## Group IIA Elements, the Alkaline Earth Metals ($ns^2$)

▶ Demonstration of reactivity of metals with water: see the Instructor's Resource Manual, under lecture demonstrations for Chapter 8.

The alkaline earth metals are also chemically reactive but much less so than the alkali metals. Reactivities increase going down the group. The alkaline earth metals form basic oxides with the general formula RO.

## Group IIIA Elements ($ns^2np^1$)

Groups IA and IIA exhibit only slight increases in metallic character down a column, but with Group IIIA we see a significant increase. The first Group IIIA element, boron, is a metalloid. Other elements in this group—aluminum, gallium, indium, and thallium—are metals. (Gallium is a curious metal; it melts easily in the palm of the hand—see Figure 8.19.)

The oxides in this group have the general formula $R_2O_3$. Boron oxide, $B_2O_3$, is an acidic oxide; aluminum oxide, $Al_2O_3$, and gallium oxide, $Ga_2O_3$, are amphoteric oxides. The change in the oxides from acidic to amphoteric to basic is indicative of an increase in metallic character of the elements.

**Figure 8.19**
**Gallium.** This metal melts from the heat of a hand.

**Figure 8.20**
**Oxides of some Group IVA elements.** Powdered lead monoxide (yellow), lead dioxide (dark brown), tin dioxide (white), and crystalline silicon dioxide (clear quartz).

### Group IVA Elements ($ns^2np^2$)

This group shows the most distinct change in metallic character. It begins with the non-metal carbon, C, followed by the metalloids silicon, Si, and germanium, Ge, and then the metals tin, Sn, and lead, Pb. Both tin and lead were known to the ancients. ◄

All the elements in this group form oxides with the general formula $RO_2$, which progress from acidic to amphoteric. Carbon dioxide, $CO_2$, an acidic oxide, is a gas. (Carbon also forms the monoxide, CO.) Silicon dioxide, $SiO_2$, an acidic oxide, exists as quartz and white sand (particles of quartz). Germanium dioxide, $GeO_2$, is acidic, though less so than silicon dioxide. Tin dioxide, $SnO_2$, an amphoteric oxide, is found as the mineral cassiterite, the principal ore of tin. Lead dioxide, $PbO_2$, is amphoteric. Lead has a more stable monoxide, PbO. Figure 8.20 shows oxides of some Group IVA elements.

► Bronze, one of the first *alloys* (metallic mixtures) used by humans, contains about 90% copper and 10% tin. Bronze melts at a lower temperature than copper but is much harder.

### Group VA Elements ($ns^2np^3$)

The Group VA elements also show the distinct transition from nonmetal (nitrogen, N, and phosphorus, P) to metalloid (arsenic, As, and antimony, Sb) to metal (bismuth, Bi). Nitrogen occurs as a colorless, odorless, relatively unreactive gas with $N_2$ molecules; white phosphorus is a white, waxy solid with $P_4$ molecules. Gray arsenic is a brittle solid with metallic luster; antimony is a brittle solid with a silvery, metallic luster. Bismuth is a hard, lustrous metal with a pinkish tinge.

The Group VA elements form oxides with empirical formulas $R_2O_3$ and $R_2O_5$. In some cases, the molecular formulas are twice these formulas—that is, $R_4O_6$ and $R_4O_{10}$. Nitrogen has the acidic oxides $N_2O_3$ and $N_2O_5$, although it also has other, better known oxides, such as NO. Phosphorus has the acidic oxides $P_4O_6$ and $P_4O_{10}$. Arsenic has the acidic oxides $As_2O_3$ and $As_2O_5$; antimony has the amphoteric oxides $Sb_2O_3$ and $Sb_2O_5$; and bismuth has the basic oxide $Bi_2O_3$.

► Video Series C, "Barking Dogs (White Phosphorus)."

### Group VIA Elements, the Chalcogens ($ns^2np^4$)

These elements, the *chalcogens* (pronounced kal′-ke-jens), show the transition from nonmetal (oxygen, O, sulfur, S, and selenium, Se) to metalloid (tellurium, Te) to metal (polonium, Po). Oxygen occurs as a colorless, odorless gas with $O_2$ molecules. It also has an allotrope, ozone, with molecular formula $O_3$. Sulfur is a brittle, yellow solid with molecular formula $S_8$. Tellurium is a shiny gray, brittle solid; polonium is a silvery metal.

**Figure 8.21**
**Burning sulfur.** The combustion of sulfur in air gives a blue flame. The product is primarily sulfur dioxide, detectable by its acrid odor.

Sulfur, selenium, and tellurium form oxides with the formulas $RO_2$ and $RO_3$. (Sulfur burns in air to form sulfur dioxide; see Figure 8.21.) These oxides, except for $TeO_2$ are acidic; $TeO_2$ is amphoteric. Polonium has an oxide $PoO_2$, which is amphoteric, though more basic in character than $TeO_2$.

### Group VIIA Elements, the Halogens ($ns^2np^5$)

The *halogens* are reactive nonmetals with the general molecular formula $X_2$, where X symbolizes a halogen. Fluorine, $F_2$, is a pale yellow gas; chlorine, $Cl_2$, a pale greenish-yellow gas; bromine, $Br_2$, a reddish-brown liquid; and iodine, $I_2$, a bluish-black solid that has a violet vapor (see Figure 8.22). Little is known about the chemistry of astatine, At, because all isotopes are radioactive with very short half-lives. (The half-life of a radioactive isotope is the time it takes for half of the isotope to decay, or break down, to another element.) It might be expected to be a metalloid.

Each halogen forms several compounds with oxygen; these are generally unstable, acidic oxides.

### Group VIIIA Elements, the Noble Gases ($ns^2np^6$)

The Group VIIIA elements exist as gases consisting of uncombined atoms. For a long time these elements were thought to be chemically inert, because no compounds were known. Then, in the early 1960s several compounds of xenon were prepared. Now compounds are also known for krypton and radon. These elements are known as the *noble gases* because of their relative unreactivity.

### CONCEPT CHECK 8.4

An element is a metalloid that forms an acidic oxide with the formula $R_2O_5$. Identify the element.

**Figure 8.22**
**The halogens.** From left to right, the flasks contain chlorine, bromine, and iodine. (Bromine and iodine were warmed to produce the vapors; bromine is normally a reddish-brown liquid and iodine a bluish-black solid.)

# A Checklist for Review

## Important Terms

electron configuration (8.1)
orbital diagram (8.1)
Pauli exclusion principle (8.1)
building-up (Aufbau) principle (8.2)

noble-gas core (8.2)
pseudo-noble-gas core (8.2)
valence electron (8.2)
Hund's rule (8.4)
paramagnetic substance (8.4)

diamagnetic substance (8.4)
periodic law (8.6)
effective nuclear charge (8.6)
first ionization energy (first ionization potential) (8.6)

electron affinity (8.6)
basic oxide (8.7)
acidic oxide (8.7)
amphoteric oxide (8.7)

## Summary of Facts and Concepts

To understand the similarities that exist among the members of a group of elements, it is necessary to know the *electron configurations* for the ground states of atoms. Only those arrangements of electrons allowed by the *Pauli exclusion principle* are possible. The ground-state configuration of an atom represents the electron arrangement that has the lowest total energy. This arrangement can be reproduced by the *building-up principle* (Aufbau principle), where electrons fill the subshells in particular order (the building-up order) consistent with the Pauli exclusion principle. The arrangement of electrons in partially filled subshells is governed by *Hund's rule*.

Elements in the same group of the periodic table have similar valence-shell configurations. As a result, chemical and physical properties of the elements show periodic behavior. *Atomic radii,* for example, tend to decrease across any period (left to right) and increase down any group (top to bottom). *First ionization energies* tend to increase across a period and decrease down a group. *Electron affinities* of the Group VIA and Group VIIA elements have large negative values.

## Operational Skills

1. **Applying the Pauli exclusion principle**   Given an orbital diagram or electron configuration, decide whether it is possible or not, according to the Pauli exclusion principle.   **(EXAMPLE 8.1)**
2. **Determining the configuration of an atom using the building-up principle**   Given the atomic number of an atom, write the complete electron configuration for the ground state, according to the building-up principle. **(EXAMPLE 8.2)**
3. **Determining the configuration of an atom using the period and group numbers**   Given the period and group for an element, write the configuration of the outer electrons.   **(EXAMPLE 8.3)**
4. **Applying Hund's rule**   Given the electron configuration for the ground state of an atom, write the orbital diagram.   **(EXAMPLE 8.4)**
5. **Applying periodic trends**   Using the known trends and referring to a periodic table, arrange a series of elements in order by atomic radius **(EXAMPLE 8.5)** or ionization energy **(EXAMPLE 8.6)**.

# Review Questions

**8.1**   Describe the experiment of Stern and Gerlach. How are the results for the hydrogen atom explained?

**8.2**   Describe the model of electron spin given in the text. What are the restrictions on electron spin?

**8.3**   How does the Pauli exclusion principle limit the possible electron configurations of an atom?

**8.4**   What is the maximum number of electrons that can occupy a $g$ subshell ($l = 4$)?

**8.5**   List the orbitals in order of increasing orbital energy up to and including $3p$ orbitals.

**8.6**   Define each of the following: noble-gas core, pseudo-noble-gas core, valence electron.

**8.7**   Give two different possible orbital diagrams for the $1s^2 2s^2 2p^4$ configuration of the oxygen atom, one of which should correspond to the ground state. Label the diagram for the ground state.

**8.8** Define the terms *diamagnetic substance* and *paramagnetic substance*. Does the ground-state oxygen atom give a diamagnetic or a paramagnetic substance? Explain.

**8.9** What kind of subshell is being filled in Groups IA and IIA? in Groups IIIA to VIIIA? in the transition elements? in the lanthanides and actinides?

**8.10** How was Mendeleev able to predict the properties of gallium before it was discovered?

**8.11** Describe the major trends that emerge when atomic radii are plotted against atomic number. Describe the trends observed when first ionization energies are plotted against atomic number.

**8.12** What atom has the smallest radius among the alkaline earth elements?

**8.13** What main group in the periodic table has elements with the most negative electron affinities for each period? What electron configurations of neutral atoms have only unstable negative ions?

**8.14** The ions $Na^+$ and $Mg^{2+}$ occur in chemical compounds, but the ions $Na^{2+}$ and $Mg^{3+}$ do not. Explain.

**8.15** Describe the major trends in metallic character observed in the periodic table of the elements.

**8.16** Distinguish between an acidic and a basic oxide. Give examples of each.

**8.17** What is the name of the alkali metal atom with valence-shell configuration $5s^1$?

**8.18** What would you predict for the atomic number of the halogen under astatine in the periodic table?

**8.19** List the elements in Groups IIIA to VIA in the same order as in the periodic table. Label each element as a metal, a metalloid, or a nonmetal. Does each column of elements display the expected trend of increasing metallic characteristics?

**8.20** For the list of elements you made for Question 8.19, note whether the oxides of each element are acidic, basic, or amphoteric.

**8.21** Write an equation for the reaction of potassium metal with water.

**8.22** From what is said in Section 8.7 about Group IIA elements, list some properties of barium.

**8.23** Give the names and formulas of two oxides of carbon.

**8.24** Match each description in the left column with the appropriate element in the right column.

a. A waxy, white solid, normally stored under water — Sulfur
Sodium
b. A yellow solid that burns in air — White phosphorus
c. A reddish-brown liquid — Bromine
d. A soft, light metal that reacts vigorously with water

# Conceptual Problems

**8.25** Suppose that the Pauli exclusion principle were "No more than two electrons can have the same four quantum numbers." What would be the electron configurations of the ground states for the first six elements of the periodic table, assuming that, except for the Pauli principle, the usual building-up principle holds?

**8.26** Imagine a world in which all quantum numbers, except the $l$ quantum number, are as they are in the real world. In this imaginary world, $l$ begins with 1 and goes up to $n$ (the value of the principal quantum number). Assume that the orbitals fill in the order by $n$, then $l$; that is, the first orbital to fill is for $n = 1$, $l = 1$; the next orbital to fill is for $n = 2$, $l = 1$, and so forth. How many elements would there be in the first period of the periodic table?

**8.27** Two elements in Period 5 are adjacent to one another in the periodic table. The ground-state atom of one element has only $s$ electrons in its valence shell; the other has at least one $d$ electron in an unfilled shell. Identify the elements.

**8.28** Two elements are in the same column of the periodic table, one above the other. The ground-state atom of one element has two $s$ electrons in its outer shell, and no $d$ electrons

anywhere in its configuration. The other element has $d$ electrons in its configuration. Identify the elements.

**8.29** You travel to an alternate universe where the atomic orbitals are different from those on earth, but all other aspects of the atoms are the same. In this universe, you find that the first (lowest energy) orbital is filled with three electrons and the second orbital can hold a maximum of nine electrons. You discover an element Z that has five electrons in its atom. Would you expect Z to be more likely to form a cation or an anion? Indicate a possible charge on this ion.

**8.30** Would you expect to find an element having both a very large (positive) first ionization energy and an electron affinity that is much less than zero (large but negative)? Explain.

**8.31** Two elements are in the same group, one following the other. One is a metalloid, the other is a metal. Both form oxides of the formula $RO_2$; the first is acidic, the next is amphoteric. Identify the two elements.

**8.32** A metalloid has an acidic oxide of the formula $R_2O_3$. The element has no oxide of the formula $R_2O_5$. What is the name of the element?

# Practice Problems

## Pauli Exclusion Principle

**8.33**  Which of the following orbital diagrams are allowed by the Pauli exclusion principle? Explain how you arrived at this decision. Give the electron configuration for the allowed ones.

a. (↑↓) (↑↑) (↓↓)(↓)
   $1s$  $2s$  $2p$

b. (↑↓) (↑↓) (↓)(↓)(↓)
   $1s$  $2s$  $2p$

c. (↑↓) (↑↓) (↑)(↑↑)(↑)
   $1s$  $2s$  $2p$

d. (↑↓) (↑↓) (↑↓)(↓)(↓)
   $1s$  $2s$  $2p$

**8.34**  Which of the following orbital diagrams are allowed and which are not allowed by the Pauli exclusion principle? Explain. For those that are allowed, write the electron configuration.

a. (↑↓) (↑) (↑)(↑)(↑)
   $1s$  $2s$  $2p$

b. (↑↓) (↑↓) (↑↑)(↑)(↑)
   $1s$  $2s$  $2p$

c. (↑↓) (↑↓) (↑↓)(↓)(↑)
   $1s$  $2s$  $2p$

d. (↑↓) (↑↓) (↑)(↑)(↑)
   $1s$  $2s$  $2p$

**8.35**  Which of the following electron configurations are possible? Explain why the others are not.
a.  $1s^22s^12p^6$
b.  $1s^22s^22p^63s^13d^6$
c.  $1s^22s^22p^8$
d.  $1s^22s^22p^63s^23d^{12}$

**8.36**  Choose the electron configurations that are possible from among the following. Explain why the others are impossible.
a.  $1s^22s^32p^6$
b.  $1s^22s^22p^4$
c.  $1s^22s^22p^83s^23p^63d^7$
d.  $1s^22s^22p^63s^13d^9$

**8.37**  Write all of the possible orbital diagrams for the electron configuration $1s^22p^1$. (There are six different diagrams.)

**8.38**  Give the different orbital diagrams for the configuration $1s^12p^1$. (There are twelve different diagrams.)

## Building-up Principle and Hund's Rule

**8.39**  Give the electron configuration of the ground state of chlorine, using the building-up principle.

**8.40**  Use the building-up principle to obtain the ground-state configuration of phosphorus.

**8.41**  Use the building-up principle to obtain the electron configuration of the ground state of vanadium.

**8.42**  Give the electron configuration of the ground state of cobalt, using the building-up principle.

**8.43**  Bromine is a Group VIIA element in Period 4. Deduce the valence-shell configuration of bromine.

**8.44**  Bismuth is a Group VA element in Period 6. Write the valence-shell configuration of bismuth.

**8.45**  Cadmium is a Group IIB element in Period 5. What would you expect for the valence-shell configuration of cadmium?

**8.46**  Titanium is a Group IVB element in Period 4. What would you expect for the configuration of outer electrons of titanium?

**8.47**  Thallium has the ground-state configuration $[Xe]4f^{14}5d^{10}6s^26p^1$. Give the group and period for this element. Classify it as a main-group, a $d$-transition, or an $f$-transition element.

**8.48**  The configuration for the ground state of iridium is $[Xe]4f^{14}5d^76s^2$. What are the group and period for this element? Is it a main-group, a $d$-transition, or an $f$-transition element?

**8.49**  Write the orbital diagram for the ground state of nickel. The electron configuration is $[Ar]3d^84s^2$.

**8.50**  Write the orbital diagram for the ground state of terbium. The electron configuration is $[Xe]4f^96s^2$.

**8.51**  Write an orbital diagram for the ground state of the potassium atom. Is the atomic substance diamagnetic or paramagnetic?

**8.52**  Write an orbital diagram for the ground state of the zinc atom. Is the atomic substance diamagnetic or paramagnetic?

## Periodic Trends

**8.53**  Order the following elements by increasing atomic radius according to what you expect from periodic trends: Se, S, As.

**8.54**  Using periodic trends, arrange the following elements in order of increasing atomic radius: F, S, Cl.

**8.55**  Using periodic trends, arrange the following elements by increasing ionization energy: Ar, Na, Cl, Al.

**8.56**  Arrange the following elements in order of increasing ionization energy: Mg, Ca, S. Do not look at Figure 8.18.

**8.57**  From what you know in a general way about electron affinities, state which member of each of the following pairs has the greater negative value: (a) As, Br (b) F, Li.

**8.58**  From what you know in a general way about electron affinities, state which member of each of the following pairs has the greater negative value: (a) Cl, S (b) Se, K.

**8.59**  If potassium perchlorate has the formula $KClO_4$, what formula would you expect for lithium perbromate?

**8.60**  Write the simplest formulas expected for two oxides of tellurium.

## General Problems

**8.61** Write the complete ground-state electron configuration of the strontium atom, Sr, using the building-up principle.

**8.62** Write the complete ground-state electron configuration of the tin atom, Sn, using the building-up principle.

**8.63** Obtain the valence-shell configuration of the polonium atom, Po, using the position of this atom in the periodic table.

**8.64** Obtain the valence-shell configuration of the thallium atom, Tl, using the position of this atom in the periodic table.

**8.65** Write the orbital diagram for the ground state of the arsenic atom. Give all orbitals.

**8.66** Write the orbital diagram for the ground state of the germanium atom. Give all orbitals.

**8.67** For eka-lead, predict the electron configuration, whether the element is a metal or nonmetal, and the formula of an oxide.

**8.68** For eka-bismuth, predict the electron configuration, whether the element is a metal or nonmetal, and the formula of an oxide.

**8.69** From Figure 8.18, predict the first ionization energy of francium (Z = 87).

**8.70** From Figure 8.18, predict the first ionization energy of astatine (Z = 85).

**8.71** Write the orbital diagram corresponding to the ground state of Nb, whose configuration is $[Kr]4d^4 5s^1$.

**8.72** Write the orbital diagram for the ground state of ruthenium. The configuration is $[Kr]4d^7 5s^1$.

**8.73** Match each set of characteristics on the left with an element in the column at the right.

a. A reactive nonmetal; the atom has a large negative electron affinity

b. A soft metal; the atom has low ionization energy

c. A metalloid that forms an oxide of formula $R_2O_3$

d. A chemically unreactive gas

Sodium (Na)
Antimony (Sb)
Argon (Ar)
Chlorine ($Cl_2$)

**8.74** Match each element on the right with a set of characteristics on the left.

a. A reactive, pale yellow gas; the atom has a large negative electron affinity

b. A soft metal that reacts with water to produce hydrogen

c. A metal that forms an oxide of formula $R_2O_3$

d. A colorless gas; the atom has a moderately large negative electron affinity

Oxygen ($O_2$)
Gallium (Ga)
Barium (Ba)
Fluorine ($F_2$)

**8.75** Find the electron configuration of the element with Z = 23. From this, give its group and period in the periodic table. Classify the element as a main-group, a $d$-block transition, or an $f$-block transition element.

**8.76** Find the electron configuration of the element with Z = 33. From this, give its group and period in the periodic table. Is this a main-group, a $d$-block transition, or an $f$-block transition element?

## Cumulative-Skills Problems

**8.77** A 2.50-g sample of barium reacted completely with water. What is the equation for the reaction? How many milliliters of dry $H_2$ evolved at 21°C and 748 mmHg?

**8.78** A sample of cesium metal reacted completely with water, evolving 48.1 mL of dry $H_2$ at 19°C and 768 mmHg. What is the equation for the reaction? What was the mass of cesium in the sample?

**8.79** What is the formula of radium oxide? What is the percentage of radium in this oxide?

**8.80** What is the formula of hydrogen telluride? What is the percentage of tellurium in this compound?

**8.81** How much energy would be required to ionize 5.00 mg of Na(g) atoms to $Na^+(g)$ ions? The first ionization energy of Na atoms is 496 kJ/mol.

**8.82** How much energy is evolved when 2.65 mg of Cl(g) atoms adds electrons to give $Cl^-(g)$ ions?

**8.83** Using the Bohr formula for the energy levels, calculate the energy required to raise the electron in a hydrogen atom from n = 1 to n = ∞. Express the result for 1 mol H atoms. Because the n = ∞ level corresponds to removal of the electron from the atom, this energy equals the ionization energy of the H atom.

**8.84** Calculate the ionization energy of the $He^+$ ion in kJ/mol (this would be the second ionization energy of He). See Problem 8.83. The Bohr formula for the energy levels of an ion consisting of a nucleus of charge $Z$ and a single electron is $-R_H Z^2/n^2$.

**8.85** The lattice energy of an ionic solid such as NaCl is the enthalpy change $\Delta H$ for the process in which the solid changes to ions. For example,

$$NaCl(s) \longrightarrow Na^+(g) + Cl^-(g) \qquad \Delta H = 786 \text{ kJ/mol}$$

Assume that the ionization energy and electron affinity are $\Delta H$ values for the processes defined by those terms. The ionization energy of Na is 496 kJ/mol. Use this, the electron affinity from Table 8.4, and the lattice energy of NaCl to calculate $\Delta H$ for the following process:

$$Na(g) + Cl(g) \longrightarrow NaCl(s)$$

**8.86** Calculate $\Delta H$ for the following process:

$$K(g) + Br(g) \longrightarrow KBr(s)$$

The lattice energy of KBr is 689 kJ/mol, and the ionization energy of K is 419 kJ/mol. The electron affinity of K is given in Table 8.4. See Problem 8.85.

## Media Activities

**8.87** Open the Electron Configuration *Problem Solving* on the student CD and do several practice problems predicting ground-state electron configurations.
a. For each problem, predict the number of unpaired electrons present in the ground state.
b. What is an *excited-state* electron configuration?
c. Explain why there are an infinite number of excited-state electron configurations, but only one ground-state electron configuration.

**8.88** Write orbital diagrams for each of the following atoms in their ground state: Li, O, Cr, Ti. Then, open the Building-Up Principle (Aufbau Principle) *Understanding Concepts* simulation on the student web site and move the slider to verify your answers. Pick several more elements and use the application to test your understanding of the Building-Up Principle.

**8.89** Open the Periodic Trends *Understanding Concepts* activity on the student CD and select Atomic Radius on the first page to investigate periodic trends in atomic radii. Use the periodic table to rank the following atoms from smallest to largest atomic radius: Ne, Sr, He, Br, C, Rb, Ca. Then, using the electronic periodic table on the student web site, verify your answers by clicking on each element to display the atomic radius.

**8.90** Open the Periodic Trends *Understanding Concepts* activity on the student CD and select Ionization Energy on the first page to investigate periodic trends in ionization energy. Use the periodic table to arrange the following elements in order of increasing first ionization energy: Li, He, F, Rb, K. Then, using the electronic periodic table on the student web site, verify your answers by clicking on each element to display the first ionization energy.

**8.91** For review, take the ACE practice test for Chapter 8 on the student web site.

# 9

The shape of snowflakes results from bonding (and intermolecular) forces.

# Ionic and Covalent Bonding

The properties of a substance, such as sodium chloride (Figure 9.1), are determined in part by the chemical bonds that hold the atoms together. A *chemical bond* is a strong attractive force that exists between certain atoms in a substance. In Chapter 2 we described how sodium (a silvery metal) reacts with chlorine (a pale greenish-yellow gas) to produce sodium chloride (table salt, a white solid). The substances in this reaction are quite different, as are their chemical bonds. Sodium chloride, NaCl, consists of $Na^+$ and $Cl^-$ ions held in a regular arrangement, or crystal, by *ionic bonds*. Ionic bonding results from the attractive force of oppositely charged ions.

A second kind of chemical bond is a *covalent bond*. In a covalent bond, two atoms share valence electrons (outer-shell electrons), which are attracted to the positively charged cores of both atoms, thus linking them. For example, chlorine gas consists of $Cl_2$ molecules. You would not expect the two Cl atoms in each $Cl_2$ molecule to acquire the opposite charges required for ionic bonding. Rather, a covalent bond holds the two atoms together. This is consistent with the equal sharing of electrons that you would expect between identical atoms. In most molecules, the atoms are linked by covalent bonds.

*Metallic bonding,* seen in sodium and other metals, represents another important type of bonding. A crystal of sodium metal consists of a regular arrangement of sodium atoms. The valence electrons of these atoms move throughout the crystal, attracted to the positive cores of all $Na^+$ ions. This attraction holds the crystal together.

What determines the type of bonding in each substance? How do you describe the bonding in various substances? In this chapter we will look at some simple, useful concepts of bonding that can help us answer these questions. We will be concerned with ionic and covalent bonds in particular.

**Figure 9.1**
**Sodium chloride crystals.**
Natural crystals of sodium chloride mineral (halite).

# Ionic Bonds

▶ This chapter builds on Chapters 7 and 8; thus, the discussions of ionization energy and electron affinity at the end of Chapter 8 flow smoothly into the discussion of ionic bonding at the beginning of this chapter.

The first explanation of chemical bonding was suggested by the properties of *salts,* substances now known to be ionic. Salts are generally crystalline solids that melt at high temperatures. Sodium chloride, for example, melts at 801°C. A molten salt (the liquid after melting) conducts an electric current. A salt dissolved in water gives a solution that also conducts an electric current. The electrical conductivity of the molten salt and the salt solution results from the motion of ions in the liquids. This suggests the possibility that ions exist in solids, held together by the attraction of opposite charges.

## 9.1  Describing Ionic Bonds

▶ Experiments in General Chemistry, 9A, "The Identity of an Insoluble Precipitate."

An **ionic bond** is *a chemical bond formed by the electrostatic attraction between positive and negative ions.* The bond forms between two atoms when one or more electrons are transferred from the valence shell of one atom to the valence shell of the other. The atom that loses electrons becomes a *cation* (positive ion), and the atom that gains electrons becomes an *anion* (negative ion). Any given ion tends to attract as many neighboring ions of opposite charge as possible. When large numbers of ions gather together, they form an ionic solid. The solid normally has a regular, crystalline structure that allows for the maximum attraction of ions, given their particular sizes.

To understand why ionic bonding occurs, consider the transfer of a valence electron from a sodium atom (electron configuration $[Ne]3s^1$) to the valence shell of a chlorine atom ($[Ne]3s^2 3p^5$). You can represent the electron transfer by the following equation:

$$Na([Ne]3s^1) + Cl([Ne]3s^2 3p^5) \longrightarrow Na^+([Ne]) + Cl^-([Ne]3s^2 3p^6)$$

As a result of the electron transfer, ions are formed, each of which has a noble-gas configuration. The sodium atom has lost its $3s$ electron and has taken on the neon configuration, [Ne]. The chlorine atom has accepted the electron into its $3p$ subshell

▶ Note that the Lewis symbols for Be, B, and C show the maximum number of unpaired electrons to reflect the usual bonding situation and differ from what you might write given the ground-state electron configurations.

▶ An alternate example is provided in the Instructor's Resource Manual.

and has taken on the argon configuration, $[Ne]3s^2 3p^6$. Such noble-gas configurations and the corresponding ions are particularly stable. This stability of the ions accounts in part for the formation of the ionic solid NaCl. Once a cation or anion forms, it attracts ions of opposite charge. Within the sodium chloride crystal, NaCl, every $Na^+$ ion is surrounded by six $Cl^-$ ions, and every $Cl^-$ ion by six $Na^+$ ions. (Figure 2.19 shows the arrangement of ions in the NaCl crystal.)

### Lewis Electron-Dot Symbols

You can simplify the preceding equation for the electron transfer between Na and Cl by writing Lewis electron-dot symbols for the atoms and monatomic ions. A **Lewis electron-dot symbol** is *a symbol in which the electrons in the valence shell of an atom or ion are represented by dots placed around the letter symbol of the element.* Table 9.1 lists Lewis symbols and corresponding valence-shell electron configurations for the atoms of the second and third periods. Note that dots are placed one to each side of a letter symbol until all four sides are occupied. Then the dots are written two to a side until all valence electrons are accounted for. The exact placement of the single dots is immaterial. For example, the single dot in the Lewis symbol for chlorine can be written on any one of the four sides. [This pairing of dots does not always correspond to the pairing of electrons in the ground state. Thus, you write $\cdot \dot{B} \cdot$ for boron, rather than $:B\cdot$, which more closely corresponds to the ground-state configuration $[He]2s^2 2p^1$. The first symbol better reflects boron's chemistry, in which each single electron (single dot) tends to be involved in bond formation.]

The equation representing the transfer of an electron from the sodium atom to the chlorine atom is

$$Na \cdot \overset{\frown}{+} \cdot \overset{\cdot\cdot}{\underset{\cdot\cdot}{Cl}} : \longrightarrow Na^+ + [:\overset{\cdot\cdot}{\underset{\cdot\cdot}{Cl}}:]^-$$

The noble-gas configurations of the ions are apparent from the symbols. No dots are shown for the cation. (All valence electrons have been removed, leaving the noble-gas core.) Eight dots are shown in brackets for the anion (noble-gas configuration $ns^2 np^6$).

### EXAMPLE 9.1

## Using Lewis Symbols to Represent Ionic Bond Formation

Use Lewis electron-dot symbols to represent the transfer of electrons from magnesium to fluorine atoms to form ions with noble-gas configurations.

**Table 9.1**
**Lewis Electron-Dot Symbols for Atoms of the Second and Third Periods**

| Period | IA $ns^1$ | IIA $ns^2$ | IIIA $ns^2 np^1$ | IVA $ns^2 np^2$ | VA $ns^2 np^3$ | VIA $ns^2 np^4$ | VIIA $ns^2 np^5$ | VIIIA $ns^2 np^6$ |
|--------|-----------|-----------|------------------|-----------------|----------------|-----------------|------------------|-------------------|
| Second | Li$\cdot$ | $\cdot$Be$\cdot$ | $\cdot\dot{B}\cdot$ | $\cdot\dot{C}\cdot$ | $:\dot{N}\cdot$ | $:\dot{O}\cdot$ | $:\ddot{F}\cdot$ | $:\ddot{Ne}:$ |
| Third | Na$\cdot$ | $\cdot$Mg$\cdot$ | $\cdot\dot{Al}\cdot$ | $\cdot\dot{Si}\cdot$ | $:\dot{P}\cdot$ | $:\dot{S}\cdot$ | $:\ddot{Cl}\cdot$ | $:\ddot{Ar}:$ |

**SOLUTION**

The Lewis symbols for the atoms are $:\ddot{\text{F}}\cdot$ and $\cdot\text{Mg}\cdot$ (see Table 9.1). The magnesium atom loses two electrons to assume a noble-gas configuration. But because a fluorine atom can accept only one electron to fill its valence shell, two fluorine atoms must take part in the electron transfer. We can represent this electron transfer as follows:

$$:\ddot{\text{F}}\cdot + \cdot\text{Mg}\cdot + \cdot\ddot{\text{F}}: \longrightarrow [:\ddot{\text{F}}:]^- + \text{Mg}^{2+} + [:\ddot{\text{F}}:]^-$$

---

**EXERCISE 9.1**

Represent the transfer of electrons from magnesium to oxygen atoms to assume noble-gas configurations. Use Lewis electron-dot symbols.

---

See Problems 9.29 and 9.30.

## Energy Involved in Ionic Bonding

You have seen in a qualitative way why a sodium atom and a chlorine atom might be expected to form an ionic bond. It is instructive, however, to look at the energy changes involved in ionic bond formation. From this analysis, you can gain further understanding of why certain atoms bond ionically and others do not.

If atoms come together and bond, there should be a net decrease in energy, because the bonded state should be more stable and therefore at a lower energy level. Consider again the formation of an ionic bond between a sodium atom and a chlorine atom. You can think of this as occurring in two steps: (1) An electron is transferred between the two separate atoms to give ions. (2) The ions then attract one another to form an ionic bond. In reality, the transfer of the electron and the formation of an ionic bond occur simultaneously, rather than in discrete steps, as the atoms approach one another. But the *net* quantity of energy involved is the same whether the steps occur one after the other or at the same time.

The first step requires removal of the $3s$ electron from the sodium atom and the addition of this electron to the valence shell of the chlorine atom. Removing the electron from the sodium atom requires energy (the first ionization energy of the sodium atom, which equals 496 kJ/mol). Adding the electron to the chlorine atom releases energy (the electron affinity of the chlorine atom, which equals $-349$ kJ/mol). ◀ It requires more energy to remove an electron from the sodium atom than is gained when the electron is added to the chlorine atom. That is, the formation of ions from the atoms is not in itself energetically favorable. It requires additional energy equal to at least $(496 - 349)$ kJ/mol, or 147 kJ/mol, to form ions (Figure 9.2, Step 1).

When positive and negative ions bond, however, more than enough energy is released to supply this additional requirement. What principally determines the energy released when ions bond is the attraction of oppositely charged ions. You can estimate this energy from *Coulomb's law* if you make the simplifying assumption that the ions are spheres, just touching, with a distance between nuclei equal to that in the NaCl crystal. From experiment, this distance is 282 pm, or $2.82 \times 10^{-10}$ m. According to Coulomb's law, the energy $E$ obtained in bringing two ions with electric charges $Q_1$ and $Q_2$ from infinite separation to a distance $r$ apart is

$$E = \frac{kQ_1Q_2}{r}$$

▶ Ionization energies and electron affinities of atoms were discussed in Section 8.6.

**Figure 9.2**
**Energetics of ionic bonding.** The transfer of an electron from an Na atom to a Cl atom is not in itself energetically favorable; it requires 147 kJ/mol of energy (Step 1). However, 493 kJ of energy is released when these oppositely charged ions come together to form ion pairs (Step 2a). And additional energy (293 kJ) is released when these ion pairs form the solid NaCl crystal (Step 2b). The lattice energy released when one mole each of Na$^+$ and Cl$^-$ ions react to produce NaCl(s) is 786 kJ/mol, and the overall process of NaCl formation is energetically favorable, releasing 639 kJ/mol if one starts with gaseous Na and Cl atoms.

▶ **The energy value −493 kJ/mol is approximate, because of the simplifying assumption we made.**

Here $k$ is a physical constant, equal to $8.99 \times 10^9 \, \text{J} \cdot \text{m/C}^2$ (C is the symbol for coulomb); the charge on Na$^+$ is $+e$ and that on Cl$^-$ is $-e$, where $e$ equals $1.602 \times 10^{-19}$ C. Therefore,

$$E = \frac{-(8.99 \times 10^9 \, \text{J} \cdot \text{m/C}^2) \times (1.602 \times 10^{-19} \, \text{C})^2}{2.82 \times 10^{-10} \, \text{m}} = -8.18 \times 10^{-19} \, \text{J}$$

The minus sign means energy is released. This energy is for the formation of one ion pair. To express this for one mole of NaCl, we multiply by Avogadro's number, $6.02 \times 10^{23}$. We obtain $-493$ kJ/mol for the energy obtained when one mole of Na$^+$ and one mole of Cl$^-$ come together to form NaCl ion pairs (Figure 9.2, Step 2a). ◀

The attraction of oppositely charged ions does not stop with the bonding of pairs of ions. The maximum attraction of ions of opposite charge with the minimum repulsion of ions of the same charge is obtained with the formation of the crystalline solid. Then additional energy is released (Figure 9.2, Step 2b).

The energy released when a crystalline solid forms from ions is related to the lattice energy of the solid. The **lattice energy** is *the change in energy that occurs when an ionic solid is separated into isolated ions in the gas phase.* For sodium chloride, the process is

$$\text{NaCl}(s) \longrightarrow \text{Na}^+(g) + \text{Cl}^-(g)$$

The distances between ions in the crystal are continuously enlarged until the ions are very far apart. The energy required for this process is the lattice energy of NaCl. You can calculate this energy from Coulomb's law, or obtain an experimental value from thermodynamic data. The value, $+786$ kJ/mol, is for just the opposite process of the one we were considering earlier, in which ions come together to bond, forming the solid. When ions come together to bond, this energy is $-786$ kJ/mol (Figure 9.2, Step 2). Consequently, the net energy obtained when gaseous Na and Cl atoms form solid NaCl is $(-786 + 147)$ kJ/mol $= -639$ kJ/mol. The negative sign shows that there has been a net decrease in energy, which you expect when stable bonding has occurred.

From this energy analysis, you can see that two elements bond ionically if the ionization energy of one is sufficiently small and the electron affinity of the other is sufficiently large and negative. This situation exists between a reactive metal (which has low ionization energy) and a reactive nonmetal (which has large negative electron affinity). In general, bonding between a metal and a nonmetal is ionic. This energy analysis also explains why ionic bonding normally results in a solid rather than in ion-pair molecules.

▶ **If you have not covered Hess's law, you may want to omit this subsection.**

### Lattice Energies from the Born–Haber Cycle

The preceding energy analysis requires us to know the lattice energy of solid sodium chloride. Direct experimental determination of the lattice energy of an ionic solid is difficult. However, this quantity can be indirectly determined from experiment by means of a thermochemical "cycle" originated by Max Born and Fritz Haber in 1919 and now called the *Born–Haber cycle*. The reasoning is based on Hess's law.

To obtain the lattice energy of NaCl, you think of solid sodium chloride being formed from the elements by two different routes, as shown in Figure 9.3. In one route, $NaCl(s)$ is formed directly from the elements, $Na(s)$ and $\frac{1}{2}Cl_2(g)$. The enthalpy change for this is $\Delta H_f^{\circ}$, which is given in Table 6.2 as $-411$ kJ per mole of NaCl. The second route consists of five steps:

1. *Sublimation of sodium.* Metallic sodium is vaporized to a gas of sodium atoms. (*Sublimation* is the transformation of a solid to a gas.) The enthalpy change for this process, measured experimentally, is 108 kJ per mole of sodium.

2. *Dissociation of chlorine.* Chlorine molecules are dissociated to atoms. The enthalpy change for this equals the Cl—Cl bond dissociation energy, which is 240 kJ per mole of bonds, or 120 kJ per mole of Cl atoms.

3. *Ionization of sodium.* Sodium atoms are ionized to $Na^+$ ions. The enthalpy change is essentially the ionization energy of atomic sodium, which equals 496 kJ per mole of Na.

4. *Formation of chloride ion.* The electrons from the ionization of sodium atoms are transferred to chlorine atoms. The enthalpy change for this is the electron affinity of atomic chlorine, which equals $-349$ kJ per mole of Cl atoms.

5. *Formation of NaCl(s) from ions.* The ions $Na^+$ and $Cl^-$ formed in Steps 3 and 4 combine to give solid sodium chloride. Because this process is just the reverse of the one corresponding to the lattice energy (breaking the solid into ions), the enthalpy change is the negative of the lattice energy. If we let $U$ be the lattice energy, the enthalpy change for Step 5 is $-U$.

**Figure 9.3**
**Born–Haber cycle for NaCl.** The formation of NaCl(s) from the elements is accomplished by two different routes. The direct route is the formation reaction (shown in boldface), and the enthalpy change is $\Delta H_f^{\circ}$. The indirect route occurs in five steps.

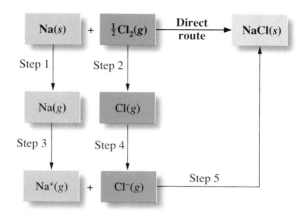

Let us write these five steps and add them. We also add the corresponding enthalpy changes, following Hess's law.

$$
\begin{array}{lll}
\text{Na}(s) & \longrightarrow \text{Na}(g) & \Delta H_1 = \quad 108 \text{ kJ} \\
\tfrac{1}{2}\text{Cl}_2(g) & \longrightarrow \text{Cl}(g) & \Delta H_2 = \quad 120 \text{ kJ} \\
\text{Na}(g) & \longrightarrow \text{Na}^+(g) + e^-(g) & \Delta H_3 = \quad 496 \text{ kJ} \\
\text{Cl}(g) + e^-(g) & \longrightarrow \text{Cl}^-(g) & \Delta H_4 = -349 \text{ kJ} \\
\text{Na}^+(g) + \text{Cl}^-(g) & \longrightarrow \text{NaCl}(s) & \Delta H_5 = -U \\
\hline
\text{Na}(s) + \tfrac{1}{2}\text{Cl}_2(g) & \longrightarrow \text{NaCl}(s) & \Delta H_f^\circ = \quad 375 \text{ kJ} - U
\end{array}
$$

In summing the equations, we have canceled terms that appear on both the left and right sides of the arrows. The final equation is simply the formation reaction for NaCl($s$). Adding the enthalpy changes, we find that the enthalpy change for this formation reaction is $375 \text{ kJ} - U$. But the enthalpy of formation has been determined calorimetrically and equals $-411$ kJ. Equating these two values, we get

$$375 \text{ kJ} - U = -411 \text{ kJ}$$

Solving for $U$ yields the lattice energy of NaCl:

$$U = (375 + 411) \text{ kJ} = 786 \text{ kJ}$$

## Properties of Ionic Substances

Typically, ionic substances are high-melting solids. Sodium chloride, NaCl, ordinary salt, melts at 801°C, and magnesium oxide, MgO, a ceramic, melts at 2800°C. The explanation for the high melting points of these substances is simple.

Small, spherical cations and anions interact by strong bonds that essentially depend on the electrical force of attraction described by Coulomb's law. Large groups of such cations and anions attract one another with strong ionic bonds linking all of the ions together, forming a crystalline solid. When you heat any solid, the atoms or ions of that solid begin to vibrate; and, as you raise the temperature, these atoms or ions vibrate through larger and larger distances. At a high enough temperature, the atoms or ions of the solid may move sufficiently far apart that the solid melts to a liquid (or the solid may simply decompose to different substances). The temperature at which melting occurs depends on the strength of the interaction between the atoms or ions. Typical ionic solids require high temperatures for this process to occur because of the strong interactions between the ions.

Coulomb's law also explains why magnesium oxide, which is composed of ions having double charges ($Mg^{2+}$ and $O^{2-}$), has such a high melting point compared with sodium chloride ($Na^+$ and $Cl^-$). Coulomb interactions depend on the product of the ion charges. For NaCl, this is $1 \times 1$, whereas for MgO, this is $2 \times 2$, or 4 times greater. (Coulomb's law also depends on the distance between charges, or ions, which is about the same in NaCl and MgO.) The much larger Coulomb interaction in MgO requires a much greater temperature to initiate melting.

The liquid melt from an ionic solid consists of ions, and so the liquid conducts an electrical current. If the ionic solid dissolves in a molecular liquid, such as water, the resulting solution consists of ions dispersed among molecules; the solution also conducts an electrical current.

Recently, as the accompanying essay on ionic liquids describes, chemists have prepared ionic substances that behave atypically. Their unusual properties are the result of large, nonspherical cations that lead to especially weak ionic bonding. The melting points of these ionic substances are unusually low; often they are liquids at room temperature (that is, their melting points are below room temperature).

# Ionic Liquids and Green Chemistry

**FRONTIERS**

New substances with peculiar or strange behavior have always intrigued chemists. The German chemist Hennig Brand discovered a white, waxy solid in 1669 that he called "cold fire," because it glowed in the dark—a strange property then and still captivating today. He tried to keep the recipe (lengthy boiling of putrid urine) secret, but people's fascination with the waxy solid, which was white phosphorus, proved irresistible.

Modern chemists continue to expand the boundaries of known materials. They have discovered plastics that conduct electricity like metals and materials that look solid, but are so porous they are almost as light as air. Now chemists have produced room-temperature ionic liquids (Figure 9.4). Most of these are clear, well-behaved substances that look and pour much like water, and like water their strength is as solvents (liquids that dissolve other substances). In fact, they promise to be "super solvents." Given a material—an organic substance, a plastic, or even a rock—researchers believe you will be able to find an ionic liquid that is capable of dissolving it!

**Figure 9.5**
**The ions composing an ionic liquid.** Shown here are space-filling models of the ions composing N-butylpyridinium nitrate. Note the bulky cation and small nitrate anion.

Copyright 2000 American Chemical Society

**Figure 9.4**
**Room-temperature ionic liquid.** Hélène Olivier of the French Petroleum Institute displays a tube containing an ionic liquid (including a blue nickel compound) with a hydrocarbon liquid above it. Typical liquids, such as a hydrocarbon, are molecular substances, whereas ionic liquids are composed of ions.

Compare room-temperature ionic liquids with their more prosaic cousins. Sodium chloride, a typical example, is solid at room temperature and doesn't melt until exposed to temperatures over 800°C. The molten liquid, though clear, is very corrosive (chemically reactive). The high melting point of sodium chloride is easily explained. It consists of small, spherical ions that pack closely together. Thus, the ions interact strongly, giving a solid with a high melting point. Room-temperature ionic liquids, by contrast, consist of large, nonspherical cations with various anions (Figure 9.5). It is the large, bulky cations that keep the ions from packing closely; the large distances between ions result in weak interactions, yielding a substance whose melting point is often well below room temperature.

The demand for *green chemistry,* the commercial production of chemicals using environmentally sound methods, has spurred much of the research into ionic liquids. Many chemical processes use volatile organic solvents. These solvents are liquids that evaporate easily into the surrounding air, where they can contribute to air pollution. Organic solvents are often flammable, too. Ionic liquids are neither volatile nor flammable. In addition to these environmental rewards, however, ionic liquids appear to offer another bonus: The proper choice of ionic liquid may improve the yield and lower the costs of a chemical process.

# 9.2 Electron Configurations of Ions

In the previous section, we described the formation of ions from atoms. Often you can understand what monatomic ions form by looking at the electron configurations of the atoms and deciding what configurations you would expect for the ions.

## Ions of the Main-Group Elements

In Chapter 2, in discussing the naming of ionic compounds, we listed the common monatomic ions of the main-group elements (Table 2.4). Most of the cations are obtained by removing all the valence electrons from the atoms of metallic elements. Once these atoms have lost their valence electrons, they have stable noble-gas or pseudo-noble-gas configurations. The stability of these configurations can be seen by looking at the successive ionization energies of some atoms. Table 9.2 lists the first through the fourth ionization energies of Na, Mg, and Al. The energy needed to remove the first electron from the Na atom is only 496 kJ/mol (first ionization energy). But the energy required to remove another electron (the second ionization energy) is nearly ten times greater (4562 kJ/mol). The electron in this case must be taken from $Na^+$, which has a neon configuration. Magnesium and aluminum atoms are similar. Their valence electrons are easily removed, but the energy needed to take an electron from either of the ions that result ($Mg^{2+}$ and $Al^{3+}$) is extremely high. That is why no compounds are found with ions having charges greater than the group number.

The loss of successive electrons from an atom requires increasingly more energy. Consequently, Group IIIA elements show less tendency to form ionic compounds than do Group IA and IIA elements, which primarily form ionic compounds. Boron, in fact, forms no compounds with $B^{3+}$ ions. The bonding is normally covalent, a topic discussed later in this chapter. However, the tendency to form ions becomes greater going down any column of the periodic table because of decreasing ionization energy. The remaining elements of Group IIIA do form compounds containing 3+ ions.

There is also a tendency for Group IIIA to VA elements of higher periods, particularly Period 6, to form compounds with ions having a positive charge of two less than the group number. Thallium in IIIA, Period 6, has compounds with 1+ ions and compounds with 3+ ions. Ions with charge equal to the group number minus two are

**Table 9.2**
**Ionization Energies of Na, Mg, and Al (in kJ/mol)***

| | Successive Ionization Energies | | | |
|---|---|---|---|---|
| Element | First | Second | Third | Fourth |
| Na | 496 | 4,562 | 6,912 | 9,543 |
| Mg | 738 | 1,451 | 7,733 | 10,540 |
| Al | 578 | 1,817 | 2,745 | 11,577 |

*Energies for the ionization of valence electrons lie to the left of the colored line.

obtained when the $np$ electrons of an atom are lost but the $ns^2$ electrons are retained. For example,

$$Tl([Xe]4f^{14}5d^{10}6s^26p^1) \longrightarrow Tl^+([Xe]4f^{14}5d^{10}6s^2) + e^-$$

Few compounds of $4+$ ions are known because the energy required to form ions is so great. The first three elements of Group IVA—C, Si, and Ge—are nonmetals (or metalloids) and usually form covalent rather than ionic bonds. Tin and lead, however, the fourth and fifth elements of Group IVA, commonly form compounds with $2+$ ions (ionic charge equal to the group number minus two). For example, tin forms tin(II) chloride, $SnCl_2$, which is an ionic compound. It also forms tin(IV) chloride, $SnCl_4$, but this is a covalent, not an ionic, compound. Bismuth, in Group VA, is a metallic element that forms compounds of $Bi^{3+}$ (ionic charge equal to the group number minus two), where only the $6p$ electrons have been lost.

Group VIA and Group VIIA elements, whose atoms have the largest negative electron affinities, would be expected to form monatomic ions by gaining electrons to give noble-gas or pseudo-noble-gas configurations. An atom of a Group VIIA element (valence-shell configuration $ns^2np^5$) picks up one electron to give a $1-$ anion ($ns^2np^6$); examples are $F^-$ and $Cl^-$. (Hydrogen also forms compounds of the $1-$ ion, $H^-$. The hydride ion, $H^-$, has a $1s^2$ configuration, like the noble-gas atom helium.) An atom of a Group VIA element (valence-shell configuration $ns^2np^4$) picks up two electrons to give a $2-$ anion ($ns^2np^6$); examples are $O^{2-}$ and $S^{2-}$. Although the electron affinity of nitrogen ($2s^22p^3$) is zero, the $N^{3-}$ ion ($2s^22p^6$) is stable in the presence of certain positive ions, including $Li^+$ and those of the alkaline earth elements. ◄

To summarize, the common monatomic ions found in compounds of the main-group elements fall into three categories (see Table 2.4):

1. Cations of Groups IA to IIIA having noble-gas or pseudo-noble-gas configurations. The ion charges equal the group numbers.

2. Cations of Groups IIIA to VA having $ns^2$ electron configurations. The ion charges equal the group numbers minus two. Examples are $Tl^+$, $Sn^{2+}$, $Pb^{2+}$, and $Bi^{3+}$.

3. Anions of Groups VA to VIIA having noble-gas or pseudo-noble-gas configurations. The ion charges equal the group numbers minus eight.

► **Interestingly, lithium metal reacts with nitrogen at room temperature to form a layer of lithium nitride, Li₃N, on the metal surface.**

► An alternate example is provided in the Instructor's Resource Manual.

---

### EXAMPLE 9.2

## Writing the Electron Configuration and Lewis Symbol for a Main-Group Ion

Write the electron configuration and the Lewis symbol for $N^{3-}$.

### SOLUTION

The electron configuration of the N atom is $[He]2s^22p^3$. By gaining three electrons, the atom assumes a $3-$ charge and the neon configuration **$[He]2s^22p^6$**. The Lewis symbol is

$$[:\ddot{N}:]^{3-}$$

---

### EXERCISE 9.2

Write the electron configuration and the Lewis symbol for $Ca^{2+}$ and for $S^{2-}$.

---

**See Problems 9.31 and 9.32.**

### EXERCISE 9.3

Write the electron configurations of Pb and $Pb^{2+}$.

See Problems 9.33 and 9.34.

### Polyatomic Ions

Many ions, particularly anions, are polyatomic. Some common polyatomic ions are listed in Table 2.6. The atoms in these ions are held together by covalent bonds, which we will discuss in Section 9.4.

### Transition-Metal Ions

▶ Demonstration of the color of transition-metal ions. Place dishes of aqueous solutions on overhead and project.

Most transition elements form several cations of different charges. For example, iron has the cations $Fe^{2+}$ and $Fe^{3+}$. Neither has a noble-gas configuration; that would require the energetically impossible loss of eight electrons from the neutral atom.

In forming ions in compounds, the atoms of transition elements generally lose the $ns$ electrons first; then they may lose one or more $(n - 1)d$ electrons. The 2+ ions are common for the transition elements and are obtained by the loss of the highest-energy $s$ electrons from the atom. Many transition elements also form 3+ ions by losing one $(n - 1)d$ electron in addition to the two $ns$ electrons. Table 2.5 lists some common transition-metal ions. Many compounds of transition-metal ions are colored because of transitions involving $d$ electrons, whereas the compounds of the main-group elements are usually colorless (Figure 9.6).

### EXAMPLE 9.3

## Writing Electron Configurations of Transition-Metal Ions

Write the electron configurations of $Fe^{2+}$ and $Fe^{3+}$.

▶ An alternate example is provided in the Instructor's Resource Manual.

 **PROBLEM STRATEGY**

After obtaining the electron configuration of the atom (using the building-up principle as in Example 8.2 or Example 8.3), remove $ns$ electrons, then $(n - 1)d$ electrons, until you have the correct positive charge on the ion.

**Figure 9.6**
**Common transition-metal cations in aqueous solution.** *Left to right:* $Cr^{3+}$ (red-violet), $Mn^{2+}$ (pale pink), $Fe^{2+}$ (pale green), $Fe^{3+}$ (pale yellow), $Co^{2+}$ (pink), $Ni^{2+}$ (green), $Cu^{2+}$ (blue), $Zn^{2+}$ (colorless).

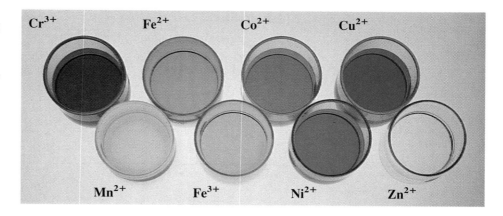

**SOLUTION**

The electron configuration of the Fe atom ($Z = 26$) is $[Ar]3d^64s^2$. To obtain the configuration of $Fe^{2+}$, remove the $4s^2$ electrons. To obtain the configuration of $Fe^{3+}$, also remove a $3d$ electron. **The configuration of $Fe^{2+}$ is $[Ar]3d^6$, and that of $Fe^{3+}$ is $[Ar]3d^5$.**

---

**EXERCISE 9.4**

Write the electron configuration of $Mn^{2+}$.

---

See Problems 9.35 and 9.36.

 **CONCEPT CHECK 9.1**

The following are electron configurations for some ions. Which ones would you expect to see in chemical compounds? State the concept or rule you used to decide for or against any ion.

a. $Fe^{2+}$ $[Ar]3d^44s^2$     b. $N^{2-}$ $[He]2s^22p^5$     c. $Zn^{2+}$ $[Ar]3d^{10}$

d. $Na^{2+}$ $[He]2s^22p^5$     e. $Ca^{2+}$ $[Ne]3s^23p^6$

# 9.3  Ionic Radii

A monatomic ion, like an atom, is a nucleus surrounded by a distribution of electrons. The **ionic radius** is *a measure of the size of the spherical region around the nucleus of an ion within which the electrons are most likely to be found.* As for an atomic radius, defining an ionic radius is somewhat arbitrary, because an electron distribution never abruptly ends. However, if we imagine ions to be spheres of definite size, we can obtain their radii from known distances between nuclei in crystals. (These distances can be determined accurately by observing how crystals diffract x rays.) ◄

To understand how you compute ionic radii, consider the determination of the radius of an $I^-$ ion in a lithium iodide (LiI) crystal. Figure 9.7 depicts a layer of ions in LiI. The distance between adjacent iodine nuclei equals twice the $I^-$ radius. From x-ray diffraction experiments, the iodine–iodine distance is found to be 426 pm (1 pm = $1 \times 10^{-12}$ m). Therefore, the $I^-$ radius in LiI is $\frac{1}{2} \times 426$ pm = 213 pm. Other

► The study of crystal structure by x-ray diffraction is discussed in Chapter 11.

**Figure 9.7**
**Determining the iodide ion radius in the lithium iodide (LiI) crystal.** *(A)* A three-dimensional view of the crystal. *(B)* Cross section through a layer of ions. Iodide ions are assumed to be spheres in contact with one another. The distance between iodine nuclei (426 pm) is determined experimentally. One-half of this distance (213 pm) equals the iodide ion radius.

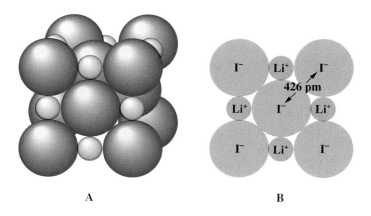

A                                                    B

**Table 9.3**
**Ionic Radii (in pm) of Some Main-Group Elements**

| Period | IA | IIA | IIIA | VIA | VIIA |
|---|---|---|---|---|---|
| 2 | $Li^+$ | $Be^{2+}$ | | $O^{2-}$ | $F^-$ |
| | 60 | 31 | | 140 | 136 |
| 3 | $Na^+$ | $Mg^{2+}$ | $Al^{3+}$ | $S^{2-}$ | $Cl^-$ |
| | 95 | 65 | 50 | 184 | 181 |
| 4 | $K^+$ | $Ca^{2+}$ | $Ga^{3+}$ | $Se^{2-}$ | $Br^-$ |
| | 133 | 99 | 62 | 198 | 195 |
| 5 | $Rb^+$ | $Sr^{2+}$ | $In^{3+}$ | $Te^{2-}$ | $I^-$ |
| | 148 | 113 | 81 | 221 | 216 |
| 6 | $Cs^+$ | $Ba^{2+}$ | $Tl^{3+}$ | | |
| | 169 | 135 | 95 | | |

crystals give approximately the same radius for the $I^-$ ion. Table 9.3, which lists average values of ionic radii obtained from many compounds of the main-group elements, gives 216 pm for the $I^-$ radius.

That you can find values of ionic radii that agree with the known structures of many crystals is strong evidence for the existence of ions in the solid state. Moreover, these values of ionic radii compare with atomic radii in ways that you might expect. For example, you expect a cation to be smaller and an anion to be larger than the corresponding atom (see Figure 9.8).

A cation formed when an atom loses all its valence electrons is smaller than the atom because it has one less shell of electrons. But even when only some of the valence electrons are lost from an atom, the size of the ion is smaller. With fewer electrons in the valence orbitals, the electron–electron repulsion is initially less, so these orbitals can shrink to increase the attraction of the electrons for the nucleus. Similarly, because an anion has more electrons than the atom, the electron–electron repulsion is greater, so the valence orbitals expand. Thus, the anion radius is larger than the atomic radius.

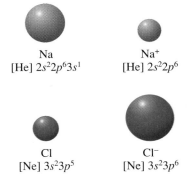

Na
[He] $2s^22p^63s^1$

$Na^+$
[He] $2s^22p^6$

Cl
[Ne] $3s^23p^5$

$Cl^-$
[Ne] $3s^23p^6$

**Figure 9.8**
**Comparison of atomic and ionic radii.** Note that the sodium atom loses its outer shell in forming the $Na^+$ ion. Thus, the cation is smaller than the atom. The $Cl^-$ ion is larger than the Cl atom, because the same nuclear charge holds a greater number of electrons less strongly.

**EXERCISE 9.5**
Which has the larger radius, S or $S^{2-}$? Explain.

See Problems 9.37 and 9.38.

The ionic radii of the main-group elements shown in Table 9.3 follow a regular pattern, just as atomic radii do. *Ionic radii increase down any column because of the addition of electron shells.*

**EXERCISE 9.6**
Without looking at Table 9.3, arrange the following ions in order of increasing ionic radius: $Sr^{2+}$, $Mg^{2+}$, $Ca^{2+}$. (You may use a periodic table.)

See Problems 9.39 and 9.40.

The pattern across a period becomes clear if you look first at the cations, then at the anions. For example, in the third period we have

| Cation | $Na^+$ | $Mg^{2+}$ | $Al^{3+}$ |
|---|---|---|---|
| Radius (pm) | 95 | 65 | 50 |

All of these ions have the neon configuration $1s^2 2s^2 2p^6$, but different nuclear charges; that is, they are isoelectronic. **Isoelectronic** *refers to different species having the same number and configuration of electrons.* To understand the decrease in radius from $Na^+$ to $Al^{3+}$, imagine the nuclear charge (atomic number) of $Na^+$ to increase. With each increase of charge, the orbitals contract due to the greater attractive force of the nucleus. Thus, in any isoelectronic sequence of atomic ions, the ionic radius decreases with increasing atomic number (just as it does for the atoms).

If you look at the anions in the third-period elements, you notice that they are much larger than the cations in the same period. This abrupt increase in ionic radius is due to the fact that the anions $S^{2-}$ and $Cl^-$ have configurations with one more shell of electrons than the cations. And because these anions also constitute an isoelectronic sequence (with argon configuration), the ionic radius decreases with atomic number (as with the atoms):

| Anion | $S^{2-}$ | $Cl^-$ |
|---|---|---|
| Radius (pm) | 184 | 181 |

Thus, *in general, across a period the cations decrease in radius. When you reach the anions, there is an abrupt increase in radius, and then the radius again decreases.*

---

**EXAMPLE 9.4**

▶ An alternate example is provided in the Instructor's Resource Manual.

## Using Periodic Trends to Obtain Relative Ionic Radii

Without looking at Table 9.3, arrange the following ions in order of decreasing ionic radius: $F^-$, $Mg^{2+}$, $O^{2-}$. (You may use a periodic table.)

**SOLUTION**

Note that $F^-$, $Mg^{2+}$, and $O^{2-}$ are isoelectronic. If you arrange them by increasing nuclear charge, they will be in order of decreasing ionic radius. **The order is $O^{2-}$, $F^-$, $Mg^{2+}$.**

**EXERCISE 9.7**

Without looking at Table 9.3, arrange the following ions in order of increasing ionic radius: $Cl^-$, $Ca^{2+}$, $P^{3-}$. (You may use a periodic table.)

See Problems 9.41 and 9.42.

---

# Covalent Bonds

▶ Experiments in General Chemistry, 9B, "From Oil of Wintergreen to Salicylic Acid."

In the preceding sections we looked at ionic substances, which are typically high-melting solids. Many substances, however, are molecular—gases, liquids, or low-melting solids consisting of molecules (Figure 9.9). A molecule is a group of atoms, frequently nonmetal atoms, strongly linked by chemical bonds. Often the forces that hold atoms together in a molecular substance cannot be understood on the basis of the attraction of oppositely charged ions. An obvious example is the molecule $H_2$, in which the two H atoms are held together tightly and no ions are present. In 1916

**Figure 9.9**
**Two molecular substances.**
*Left:* Iodoform, $CHI_3$, is a low-melting, yellow solid (m.p. 120°C). *Right:* Carbon tetrachloride, $CCl_4$, is a colorless liquid.

▶ **G. N. Lewis (1875–1946) was professor of chemistry at the University of California at Berkeley. Well known for work on chemical bonding, he was also noted for his research in molecular spectroscopy and thermodynamics (the study of heat involved in chemical and physical processes).**

Gilbert Newton Lewis proposed that the strong attractive force between two atoms in a molecule results from a **covalent bond,** *a chemical bond formed by the sharing of a pair of electrons between atoms.* ◀ In 1926 Walter Heitler and Fritz London showed that the covalent bond in $H_2$ could be quantitatively explained by the newly discovered theory of quantum mechanics. We will discuss the descriptive aspects of covalent bonding in the following sections

## 9.4 Describing Covalent Bonds

Consider the formation of a covalent bond between two H atoms to give the $H_2$ molecule. As the atoms approach one another, their $1s$ orbitals begin to overlap. Each electron can then occupy the space around both atoms. In other words, the two electrons can be shared by the atoms (Figure 9.10). The electrons are attracted simultaneously by the positive charges of the two hydrogen nuclei. This attraction that bonds the electrons to both nuclei is the force holding the atoms together. Although ions do not exist in $H_2$, the force that holds the atoms together can still be regarded as arising from the attraction of oppositely charged particles: nuclei and electrons.

It is interesting to see how the potential energy of the atoms changes as they approach and then bond. Figure 9.11 shows the potential energy of the atoms for various distances between nuclei. The potential energy of the atoms when they are some distance apart is indicated by a position on the potential-energy curve at the far right. As the atoms approach (moving from right to left on the potential-energy curve), the potential energy gets lower and lower. The decrease in energy is a reflection of the bonding of the atoms. Eventually, as the atoms get close enough, the repulsion of the positive charges on the nuclei becomes larger than the attraction of electrons for nuclei. In other words, the potential energy reaches a minimum value and then increases. The distance between nuclei at this minimum energy is called the *bond length* of $H_2$. It is the normal distance between nuclei in the molecule.

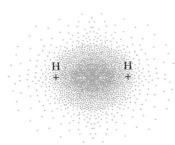

**Figure 9.10**
**The electron probability distribution for the $H_2$ molecule.** The electrons occupy the space around both atoms.

Now imagine the reverse process. You start with the $H_2$ molecule, the atoms at their normal bond length apart (at the minimum of the potential-energy curve). To separate the atoms in the molecule, energy must be added (you move along the curve to the flat portion at the right). The energy that must be added is called the *bond dissociation energy.* The larger the bond dissociation energy, the stronger is the bond.

**Figure 9.11**
**Potential-energy curve for H₂.** The stable molecule occurs at the bond distance corresponding to the minimum in the potential-energy curve.

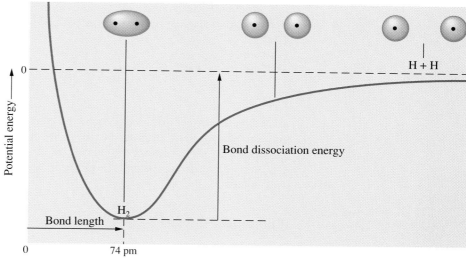

### Lewis Formulas

▶ A systematic scheme for writing Lewis formulas is described in Section 9.6; here we simply introduce Lewis formulas.

You can represent the formation of the covalent bond in $H_2$ from atoms as follows:

$$H\cdot\ +\ \cdot H\ \longrightarrow\ H\!:\!H$$

This uses the Lewis electron-dot symbol for the hydrogen atoms and represents the covalent bond by a pair of dots. Recall that the two electrons from the covalent bond spend part of the time in the region of each atom. In this sense, each atom in $H_2$ has a helium configuration. We can draw a circle about each atom to emphasize this.

The formation of a bond between H and Cl to give an HCl molecule can be represented in a similar way.

$$H\cdot\ +\ \cdot\overset{..}{\underset{..}{Cl}}:\ \longrightarrow\ \left(H\!:\!\overset{..}{\underset{..}{Cl}}:\right)$$

As the two atoms approach each other, unpaired electrons on each atom pair up to form a covalent bond. The pair of electrons is shared by the two atoms. Each atom then acquires a noble-gas configuration of electrons, the H atom having two electrons about it (as in He), and the Cl atom having eight valence electrons about it (as in Ar).

A *formula using dots to represent valence electrons* is called a **Lewis electron-dot formula.** An electron pair represented by a pair of dots in such a formula is either a **bonding pair** (*an electron pair shared between two atoms*) or a **lone,** or **nonbonding, pair** (*an electron pair that remains on one atom and is not shared*). For example,

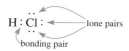

Bonding pairs are often represented by dashes rather than by pairs of dots.

Frequently, the number of covalent bonds formed by an atom equals the number of unpaired electrons shown in its Lewis symbol. Consider the formation of $NH_3$.

$$3H \cdot + \cdot \overset{\displaystyle .}{\underset{\displaystyle .}{N}} : \longrightarrow \begin{matrix} H \\ \overset{..}{H : N :} \\ H \end{matrix}$$

Each bond is formed between an unpaired electron on one atom and an unpaired electron on another atom. In many instances, the number of bonds formed by an atom in Groups IVA to VIIA equals the number of unpaired electrons, which is eight minus the group number. For example, a nitrogen atom (Group VA) forms $8 - 5 = 3$ covalent bonds. ◄

▶ The numbers of unpaired electrons in the Lewis symbols for the atoms of elements in Groups IA and IIIA equal the group numbers. But, except for the first elements of these groups, the atoms usually form ionic bonds.

## Coordinate Covalent Bonds

When bonds form between atoms that both donate an electron, you have

$$A \cdot + \cdot B \longrightarrow A : B$$

However, it is possible for both electrons to come from the same atom. A **coordinate covalent bond** is *a bond formed when both electrons of the bond are donated by one atom:*

$$A + : B \longrightarrow A : B$$

A coordinate covalent bond is not essentially different from other covalent bonds; it involves the sharing of a pair of electrons between two atoms. An example is the formation of the ammonium ion, in which an electron pair on the N atom of $NH_3$ forms a bond with $H^+$.

$$H^+ + : NH_3 \longrightarrow \begin{bmatrix} H \\ \overset{..}{H : N : H} \\ H \end{bmatrix}^+$$

The new N—H bond is clearly identical to the other N—H bonds.

## Octet Rule

In each of the molecules we have described so far, the atoms have obtained noble-gas configurations through the sharing of electrons. Atoms other than hydrogen have obtained eight electrons in their valence shells; hydrogen atoms have obtained two. *The tendency of atoms in molecules to have eight electrons in their valence shells (two for hydrogen atoms)* is known as the **octet rule.** Many of the molecules we will discuss follow the octet rule. Some do not.

## Multiple Bonds

In the molecules we have described so far, each of the bonds has been a **single bond**—that is, *a covalent bond in which a single pair of electrons is shared by two atoms.* But it is possible for atoms to share two or more electron pairs. A **double bond** is *a covalent bond in which two pairs of electrons are shared by two atoms.* A **triple bond** is *a covalent bond in which three pairs of electrons are shared by*

# Chemical Bonds in Nitroglycerin

Nitroglycerin gained a nasty reputation soon after its discovery in 1846. Unless kept cold, the pale yellow, oily liquid detonates from even the mildest vibration. In the French film *The Wages of Fear,* four men agree to drive two trucks loaded with nitroglycerin over mountainous roads from a remote village in South America. They are to drive the nitroglycerin to an out-of-control oil-well fire, where the explosive will be used to close off the well. On the way, they find the road obstructed by a huge boulder that has rolled across it, and they decide to blast the boulder out of their way using some of their explosive cargo. You see one of the men gingerly pouring nitroglycerin from a thermos bottle down a stick and into a hole in the rock. Beads of sweat form on his contorted face, while he tries desperately to suspend his breathing and any extra movement that might set off the nitroglycerin.

Here is the structure of nitroglycerin (also see the molecular model in Figure 9.12):

With just a little jostling, nitroglycerin, $C_3H_5(ONO_2)_3$, can rearrange its atoms to give stable products:

$$4C_3H_5(ONO_2)_3(l) \longrightarrow$$
$$6N_2(g) + 12CO_2(g) + 10H_2O(g) + O_2(g)$$

The stability of the products results from their strong bonds, which are much stronger than those in nitroglycerin. Nitrogen, for example, has a strong nitrogen–nitrogen triple bond, and carbon dioxide has two strong carbon–oxygen double bonds. The explosive force of the reaction results from both the rapid reaction and from the large volume increase on forming gaseous products.

In 1867, the Swedish chemist Alfred Nobel discovered that nitroglycerin behaved better when absorbed on diatomaceous earth, a crumbly rock, giving an explosive mixture that Nobel called *dynamite.* Part of the wealth he made from his explosive factories was left in trust to establish the Nobel Prizes. Although explosives are often associated in people's minds with war, they do have many peacetime uses, including road building, mining, and demolition (Figure 9.13).

*Source:* Adapted from an essay in *Introductory Chemistry,* second edition, by Ebbing and Wentworth (Houghton Mifflin, 1998), p. 308.

**Figure 9.12**
**Molecular model of nitroglycerin.** In this ball-and-stick model, carbon atoms are black, hydrogen atoms are light blue, oxygen atoms are red, and nitrogen atoms are dark blue.

**Figure 9.13**
**Demolition of a building.** Explosives are placed at predetermined positions so that when they are detonated, the building collapses on itself.

Ethylene

Acetylene

*two atoms.* As examples, consider ethylene, $C_2H_4$, and acetylene, $C_2H_2$. Their Lewis formulas are

ethylene

H:C:::C:H    or    H—C≡C—H

acetylene

Note the octet of electrons on each C atom. Double bonds form primarily with C, N, O, and S atoms. Triple bonds form mostly to C and N atoms.

## 9.5 Polar Covalent Bonds; Electronegativity

**Figure 9.14**
**The HCl molecule.**
A molecular model.

A covalent bond involves the sharing of at least one pair of electrons between two atoms. When the atoms are alike, as in the case of the H—H bond of $H_2$, the bonding electrons are shared equally. That is, the electrons spend the same amount of time in the vicinity of each atom. But when the two atoms are of different elements, the bonding electrons need not be shared equally. A **polar covalent bond** (or simply *polar bond*) is *a covalent bond in which the bonding electrons spend more time near one atom than the other.* For example, in the case of the HCl molecule (Figure 9.14), the bonding electrons spend more time near the chlorine atom than the hydrogen atom.

You can consider the polar covalent bond as intermediate between a *nonpolar* covalent bond, as in $H_2$, and an ionic bond, as in NaCl. From this point of view, an ionic bond is simply an extreme example of a polar covalent bond. To illustrate, we can represent the bonding in $H_2$, HCl, and NaCl with electron-dot formulas as follows:

H : H      H : Cl:      Na⁺  : Cl:⁻

Nonpolar covalent     Polar covalent     Ionic

The bonding pairs of electrons are equally shared in $H_2$, unequally shared in HCl, and essentially not shared in NaCl. Thus, it is possible to arrange different bonds to form a gradual transition from nonpolar covalent to ionic.

**Electronegativity** is *a measure of the ability of an atom in a molecule to draw bonding electrons to itself.* Several electronegativity scales have been proposed. In 1934 Robert S. Mulliken suggested on theoretical grounds that the electronegativity (*X*) of an atom be given as half its ionization energy (*I.E.*) minus electron affinity (*E.A.*). ◄

▶ Robert S. Mulliken received the Nobel Prize in chemistry in 1966 for his work on molecular orbital theory (discussed in Chapter 10).

$$X = \frac{I.E. - E.A.}{2}$$

An atom such as fluorine that tends to pick up electrons easily (large negative *E.A.*) and hold on to them strongly (large *I.E.*) has a large electronegativity. On the other hand, an atom such as lithium or cesium that loses electrons readily (small *I.E.*) and has little tendency to gain electrons (small negative or positive *E.A.*) has a small electronegativity. Until recently, only a few electron affinities had been measured. For this reason,

**Figure 9.15**
**Electronegativities of the elements.** The values given are those of Pauling.

▶ Linus Pauling received the Nobel Prize in chemistry in 1954 for his work on the nature of the chemical bond. In 1962 he received the Nobel Peace Prize.

▶ The electronegativity difference in metal–metal bonding would also be small. This bonding frequently involves delocalized *metallic bonding*, briefly described in Section 9.7.

▶ An alternate example is provided in the Instructor's Resource Manual.

Mulliken's scale has had limited utility. A more widely used scale was derived earlier by Linus Pauling from bond energies, which are discussed later in this chapter. Pauling's electronegativity values are given in Figure 9.15. Because electronegativities depend somewhat on the bonds formed, these values are actually average ones. ◀

Fluorine, the most electronegative element, was assigned a value of 4.0 on Pauling's scale. Lithium, at the left end of the same period, has a value of 1.0. Cesium, in the same column but below lithium, has a value of 0.7. *In general, electronegativity increases from left to right and decreases from top to bottom in the periodic table.* Metals are the least electronegative elements (they are *electropositive*) and nonmetals the most electronegative.

The absolute value of the difference in electronegativity of two bonded atoms gives a rough measure of the polarity to be expected in a bond. When this difference is small, the bond is nonpolar. When it is large, the bond is polar (or, if the difference is very large, perhaps ionic). The electronegativity differences for the bonds H—H, H—Cl, and Na—Cl are 0.0, 0.9, and 2.1, respectively, following the expected order. Differences in electronegativity explain why ionic bonds usually form between a metal atom and a nonmetal atom; the electronegativity difference would be largest between these elements. On the other hand, covalent bonds form primarily between two nonmetals because the electronegativity differences are small. ◀

### EXAMPLE 9.5

## Using Electronegativities to Obtain Relative Bond Polarities

Use electronegativity values (Figure 9.15) to arrange the following bonds in order of increasing polarity: P—H, H—O, C—Cl.

### SOLUTION

The absolute values of the electronegativity differences are P—H, 0.0; H—O, 1.4; C—Cl, 0.5. Hence, the order is **P—H, C—Cl, H—O.**

**Problem Solving:**
**Relative Bond Polarities**

---

**EXERCISE 9.8**

Using electronegativities, decide which of the following bonds is most polar:
C—O, C—S, H—Br.

---

See Problems 9.49 and 9.50.

You can use an electronegativity scale to predict the direction in which the electrons shift during bond formation; the electrons are pulled toward the more electronegative atom. For example, consider the H—Cl bond. Because the Cl atom ($X = 3.0$) is more electronegative than the H atom ($X = 2.1$), the bonding electrons in H—Cl are pulled toward Cl. Because the bonding electrons spend most of their time around the Cl atom, that end of the bond acquires a partial negative charge (indicated $\delta-$). The H-atom end of the bond has a partial positive charge ($\delta+$). You can show this as follows:

$$\overset{\delta+ \quad \delta-}{\text{H}-\text{Cl}}$$

The HCl molecule is said to be a *polar molecule*. We will say more about polar molecules when we look at molecular structures in Chapter 10.

## 9.6 Writing Lewis Electron-Dot Formulas

▶ This section describes the rules for writing Lewis formulas; these are then elaborated on in Sections 9.7 to 9.9.

The Lewis electron-dot formula of a molecule is similar to the structural formula in that it shows how atoms are bonded. Bonding electron pairs are indicated either by two dots or by a dash. In addition to the bonding electrons, however, an electron-dot formula shows the positions of lone pairs of electrons, whereas the structural formula does not. Thus, the electron-dot formula is a simple two-dimensional representation of the positions of electrons in a molecule. In the next chapter we will see how to predict the three-dimensional shape of a molecule from the two-dimensional electron-dot formula. In this section we will discuss the steps for writing the electron-dot formula for a molecule made from atoms of the main-group elements. ◀

▶ Lewis formulas do not directly convey information about molecular shape. For example, the Lewis formula of methane, CH₄, is written as the flat (two-dimensional) formula

$$\text{H} \\ \text{H} : \overset{..}{\underset{..}{\text{C}}} : \text{H} \\ \text{H}$$

The actual methane molecule, however, is not flat; it has a three-dimensional structure, as explained in Chapter 10.

Before you can write the Lewis formula of a molecule (or a polyatomic ion), you must know the *skeleton structure* of the molecule. The skeleton structure tells you which atoms are bonded to one another (without regard to whether the bonds are single or not).

Normally the skeleton structure must be found by experiment. For simple molecules, you can often predict the skeleton structure. For instance, many small molecules or polyatomic ions consist of a central atom around which are bonded atoms of greater electronegativity, such as F, Cl, and O. Note the structural formula (and molecular model) of phosphorus oxychloride, POCl₃:

$$\text{Cl} \\ | \\ \text{Cl}-\text{P}-\text{O} \\ | \\ \text{Cl}$$

phosphorus oxychloride
POCl₃

The phosphorus atom is surrounded by more electronegative atoms. (In some cases, H atoms surround a more electronegative atom as in $H_2O$ and $NH_3$, but H cannot be a central atom because it normally forms only one bond.) Similarly, *oxoacids* are substances in which O atoms (and possibly other electronegative atoms) are bonded to a central atom, with one or more H atoms usually bonded to O atoms. An example is chlorosulfonic acid, $HSO_3Cl$:

chlorosulfonic acid
$HSO_3Cl$

Another useful idea for predicting skeleton structures is that molecules or polyatomic ions with symmetrical formulas often have symmetrical structures. For example, disulfur dichloride, $S_2Cl_2$, is symmetrical, with the more electronegative Cl atoms around the S atoms: Cl—S—S—Cl.

Once you know the skeleton structure of a molecule, you can find the Lewis formula by applying the following four steps. These steps allow you to write electron-dot formulas even when the central atom does not follow the octet rule.

**STEP 1:** *Calculate the total number of valence electrons* for the molecule by summing the number of valence electrons (= group number) for each atom. If you are writing the Lewis formula of a polyatomic anion, you *add* the number of negative charges to this total. (For $CO_3^{2-}$ you add 2 because the 2− charge indicates that there are two more electrons than are provided by the neutral atoms.) For a polyatomic cation, you *subtract* the number of positive charges from the total. (For $NH_4^+$ you subtract 1.)

**STEP 2:** *Write the skeleton structure of the molecule or ion,* connecting every bonded pair of atoms by a pair of dots (or a dash).

**STEP 3:** *Distribute electrons to the atoms surrounding the central atom (or atoms)* to satisfy the octet rule for these surrounding atoms.

**STEP 4:** *Distribute the remaining electrons as pairs to the central atom (or atoms),* after subtracting the number of electrons already distributed from the total found in Step 1. If there are fewer than eight electrons on the central atom, this suggests that a multiple bond is present. (Two electrons fewer than an octet suggests a double bond; four fewer suggests a triple bond or two double bonds.) To obtain a multiple bond, move one or two electron pairs (depending on whether the bond is to be double or triple) from a surrounding atom to the bond connecting the central atom. Atoms that often form multiple bonds are C, N, O, and S.

The next several examples illustrate how to write the Lewis electron-dot formula for a small molecule, given the molecular formula.

▶ An alternate example is provided in the Instructor's Resource Manual.

**EXAMPLE 9.6**

## Writing Lewis Formulas (Single Bonds Only)

Sulfur dichloride, $SCl_2$, is a red, fuming liquid used in the manufacture of insecticides. Write the Lewis formula for the molecule.

**PROBLEM STRATEGY**

You follow the four steps: (1) Calculate the total number of valence electrons. (2) Write the skeleton structure with two electrons to each bond between atoms. (3) Distribute electrons to the outer atoms to satisfy the octet rule. (4) Distribute the remaining electrons to the central atom.

**SOLUTION**

The number of valence electrons from an atom equals the group number: 6 for S, 7 for each Cl, for a total of 20 electrons. You expect the skeleton structure to have S as the central atom, with the more electronegative Cl atoms bonded to it. After connecting atoms by electron pairs and distributing electrons to the outer atoms, you have

This accounts for 8 electron pairs, or 16 electrons. Subtracting this from the total number of electrons (20) gives 4 electrons, or 2 electron pairs. You place these on the central atom (S). The final Lewis formula is

$$: \ddot{C}l : \ddot{S} : \ddot{C}l : \quad \text{or} \quad : \ddot{C}l - \ddot{S} - \ddot{C}l :$$

**EXERCISE 9.9**

Dichlorodifluoromethane, $CCl_2F_2$, is a gas used as a refrigerant and aerosol propellant. Write the Lewis formula for $CCl_2F_2$.

*See Problems 9.53 and 9.54.*

$SCl_2$

▶ An alternate example is provided in the Instructor's Resource Manual.

**EXAMPLE 9.7**

## Writing Lewis Formulas (Including Multiple Bonds)

Carbonyl chloride, or phosgene, $COCl_2$, is a highly toxic gas used as a starting material for the preparation of polyurethane plastics. What is the electron-dot formula of $COCl_2$?

**PROBLEM STRATEGY**

You follow the four steps as in the previous example, but after distributing the remaining electrons to the central atom, you find that it does not have an octet. To rectify this, you transfer an electron pair from an outer atom to form a double bond (part of Step 4). Do you move the electron pair from the Cl or the O atom? It helps to remember that C, N, O, and S atoms often form double bonds. In this case, you move the electron pair from the O atom.

**SOLUTION**

The total number of valence electrons is $4 + 6 + (2 \times 7) = 24$. You expect the more electropositive atom, C, to be central, with the O and Cl atoms bonded to it. After distributing electron pairs to these surrounding atoms, you have

$$: \ddot{C}l : C : \ddot{O} : \quad \text{or} \quad : \ddot{C}l - C - \ddot{O} :$$
$$\quad \ \ : \ddot{C}l : \qquad \qquad \qquad : \ddot{C}l :$$

COCl$_2$

This accounts for all 24 valence electrons, leaving only 6 electrons on C. This is two fewer than an octet, so you move a pair of electrons on the O atom to give a carbon–oxygen double bond. The electron-dot formula of COCl$_2$ is

$$:\overset{..}{\underset{..}{C}l}:C::\overset{..}{\underset{..}{O}} \quad \text{or} \quad :\overset{..}{\underset{..}{C}l}-C=\overset{..}{\underset{..}{O}}$$
$$\quad\quad :\overset{}{\underset{..}{C}l}: \quad\quad\quad\quad\quad :\overset{}{\underset{..}{C}l}:$$

Check to see that each atom has an octet.

### EXERCISE 9.10

Write the electron-dot formula of carbon dioxide, $CO_2$.

See Problems 9.55 and 9.56.

► An alternate example is provided in the Instructor's Resource Manual.

### EXAMPLE 9.8

## Writing Lewis Formulas (Ionic Species)

Obtain the electron-dot formula (Lewis formula) of the $BF_4^-$ ion.

### SOLUTION

The total valence electrons provided by the boron and four fluorine atoms is $3 + (4 \times 7) = 31$. Because the anion has a charge of $1-$, it has one more electron than is provided by the neutral atoms. Thus, the total number of valence electrons is 32 (or 16 electron pairs). You assume that the skeleton structure has boron as the central atom, with the more electronegative F atoms bonded to it. After connecting the B and F atoms by bonds and placing electron pairs around the F atoms to satisfy the octet rule, you obtain

$$
\begin{array}{c}
:\overset{..}{F}: \\
:\overset{..}{F}:\overset{..}{B}:\overset{..}{F}: \\
:\overset{..}{F}:
\end{array}
\quad \text{or} \quad
\begin{array}{c}
:\overset{..}{F}: \\
| \\
:\overset{..}{F}-B-\overset{..}{F}: \\
| \\
:\overset{..}{F}:
\end{array}
$$

This uses up all 32 electrons. The charge on the entire ion is indicated by the minus sign written as a superscript to square brackets enclosing the electron-dot formula.

$$
\left[
\begin{array}{c}
:\overset{..}{F}: \\
:\overset{..}{F}:\overset{..}{B}:\overset{..}{F}: \\
:\overset{..}{F}:
\end{array}
\right]^-
\quad \text{or} \quad
\left[
\begin{array}{c}
:\overset{..}{F}: \\
| \\
:\overset{..}{F}-B-\overset{..}{F}: \\
| \\
:\overset{..}{F}:
\end{array}
\right]^-
$$

$BF_4^-$

### EXERCISE 9.11

Write the electron-dot formula of a. the hydronium ion, $H_3O^+$; b. the chlorite ion, $ClO_2^-$.

See Problems 9.57 and 9.58.

### CONCEPT CHECK 9.2

Each of the following may seem, at first glance, to be plausible electron-dot formulas for the molecule $N_2F_2$. Most, however, are incorrect for some reason. What concepts or rules apply to each, either to cast it aside or to keep it as the correct formula?

a. $:\!\overset{..}{\underset{..}{F}}\!:\!\overset{..}{N}\!:\!\overset{..}{N}\!:\!\overset{..}{\underset{..}{F}}\!:$    b. $:\!\overset{..}{\underset{..}{F}}\!:\!\overset{..}{N}\!:\!:\!\overset{..}{N}\!:\!\overset{..}{\underset{..}{F}}\!:$    c. $:\!\overset{..}{\underset{..}{F}}\!:\!:\!\overset{..}{N}\!:\!\overset{..}{N}\!:\!\overset{..}{\underset{..}{F}}\!:$

d. $:\!\overset{..}{\underset{..}{F}}\!:\!\overset{..}{N}\!:\!\overset{..}{N}\!:\!\overset{..}{\underset{..}{F}}\!:$    e. $:\!\overset{..}{\underset{..}{F}}\!:\!\overset{..}{N}\!:\!:\!\overset{..}{F}\!:\!\overset{..}{N}\!:$    f. $:\!\overset{..}{\underset{..}{F}}\!:\!\overset{..}{N}\ \overset{..}{N}\!:\!\overset{..}{\underset{..}{F}}\!:$

## 9.7 Delocalized Bonding: Resonance

We have assumed up to now that the bonding electrons are localized in the region between two atoms. In some cases, however, this assumption does not fit the experimental data. Suppose, for example, that you try to write an electron-dot formula for ozone, $O_3$. You find that you can write two formulas:

    A         and       B

► **The lengths of the two oxygen–oxygen bonds (that is, the distances between the atomic nuclei) are each 128 pm.**

In formula A, the oxygen–oxygen bond on the left is a double bond and the oxygen–oxygen bond on the right is a single bond. In formula B, the situation is just the opposite. Experiment shows, however, that both bonds in $O_3$ are identical. Therefore, neither formula can be correct. ◄

According to theory, one of the bonding pairs in ozone is spread over the region of all three atoms rather than associated with a particular oxygen–oxygen bond. This is called **delocalized bonding,** *a type of bonding in which a bonding pair of electrons is spread over a number of atoms rather than localized between two.* We might symbolically describe the delocalized bonding in ozone as follows:

(For clarity, only the bonding pairs are given.) The broken line indicates a bonding pair of electrons that spans three nuclei rather than only two. In effect, the oxygen–oxygen bond is neither a single bond nor a double bond but an intermediate type.

A single electron-dot formula cannot properly describe delocalized bonding. Instead, a resonance description is often used. According to the **resonance description,** *you describe the electron structure of a molecule having delocalized bonding by writing all possible electron-dot formulas.* These are called the *resonance formulas* of the molecule. The actual electron distribution of the molecule is a composite of these resonance formulas.

The electron structure of ozone can be described in terms of the two resonance formulas presented at the start of this section. By convention, we usually write all of the resonance formulas and connect them by double-headed arrows. For ozone we would write

    A            B

Sodium ions

Valence electrons move
throughout metal

**Figure 9.16**
**Delocalized bonding in
sodium metal.**  The metal
consists of positive sodium
ions in a "sea" of valence
electrons. Valence (bonding)
electrons are free to move
throughout the metal crystal
(beige area).

$CO_3^{2-}$

▶ An alternate example is
provided in the Instructor's
Resource Manual.

Unfortunately, this notation can be misinterpreted. It does not mean that the ozone molecule flips back and forth between two forms. There is only one ozone molecule. The double-headed arrow means that you should form a mental picture of the molecule by fusing the various resonance formulas. The left oxygen–oxygen bond is double in formula A and the right one is double in formula B, so you must picture an electron pair that actually encompasses both bonds.

Attempting to write electron-dot formulas leads you to recognize that delocalized bonding exists in many molecules. Whenever you can write several plausible electron-dot formulas—which often differ merely in the allocation of single and double bonds to the same kinds of atoms (as in ozone)—you can expect delocalized bonding

## EXAMPLE 9.9

### Writing Resonance Formulas

Describe the electron structure of the carbonate ion, $CO_3^{2-}$, in terms of electron-dot formulas.

**SOLUTION**

One possible electron-dot formula for the carbonate ion is

$$\left[\begin{array}{c} :\overset{\cdot\cdot}{O}: \\ \| \\ C \\ :\overset{\cdot\cdot}{O} \quad \overset{\cdot\cdot}{O}: \end{array}\right]^{2-}$$

Because you expect all carbon–oxygen bonds to be equivalent, you must describe the electron structure in resonance terms.

$$\left[\begin{array}{c} :\overset{\cdot\cdot}{O}: \\ C \\ :\overset{\cdot\cdot}{O} \quad \overset{\cdot\cdot}{O}: \end{array}\right]^{2-} \longleftrightarrow \left[\begin{array}{c} :\overset{\cdot\cdot}{O}: \\ C \\ :\overset{\cdot\cdot}{O} \quad \overset{\cdot\cdot}{O}: \end{array}\right]^{2-} \longleftrightarrow \left[\begin{array}{c} :\overset{\cdot\cdot}{O}: \\ C \\ :\overset{\cdot\cdot}{O} \quad \overset{\cdot\cdot}{O}: \end{array}\right]^{2-}$$

You expect one electron pair to be delocalized over the region of all three carbon–oxygen bonds.

## EXERCISE 9.12

Describe the bonding in $NO_3^-$ using resonance formulas.

See Problems 9.59, 9.60, 9.61, and 9.62.

Metals are extreme examples of delocalized bonding. A sodium metal crystal, for example, can be regarded as an array of $Na^+$ ions surrounded by a "sea" of electrons (Figure 9.16). The valence, or bonding, electrons are delocalized over the entire metal crystal. The freedom of these electrons to move throughout the crystal is responsible for the electrical conductivity of a metal.

## 9.8   Exceptions to the Octet Rule

Many molecules composed of atoms of the main-group elements have electronic structures that satisfy the octet rule, but a number of them do not. A few molecules, such as NO, have an odd number of electrons and so cannot satisfy the octet

**Figure 9.17**
**Phosphorus pentafluoride,**
**PF₅.** A molecular model.

rule. Other exceptions to the octet rule fall into two groups—one group of molecules with an atom having fewer than eight valence electrons around it and the other group of molecules with an atom having more than eight valence electrons around it.

The exceptions in which the central atom has more than eight valence electrons around it are fairly numerous. Phosphorus pentafluoride is a simple example. This is a colorless gas of $PF_5$ molecules. Each molecule consists of a phosphorus atom surrounded by the more electronegative fluorine atoms (Figure 9.17). Following the steps outlined in Section 9.6, you arrive at the following electron-dot formula, in which the phosphorus atom has ten valence electrons around it:

The octet rule stems from the fact that the main-group elements in most cases employ only an $ns$ and three $np$ valence-shell orbitals in bonding, and these orbitals hold eight electrons. Elements of the second period are restricted to these orbitals, but from the third period on, the elements also have unfilled $nd$ orbitals, which may be used in bonding. For example, the valence-shell configuration of phosphorus is $3s^2 3p^3$. Using just these $3s$ and $3p$ orbitals, the phosphorus atom can accept only three additional electrons, forming three covalent bonds (as in $PF_3$). However, more bonds can be formed if the empty $3d$ orbitals of the atom are used. If each of the five electrons of the phosphorus atom is paired with unpaired electrons of fluorine atoms, $PF_5$ can be formed. In this way, phosphorus forms both the trifluoride and the pentafluoride. By contrast, nitrogen (which has no available $d$ orbitals in its valence shell) forms only the trifluoride, $NF_3$.

▶ An alternate example is provided in the Instructor's Resource Manual.

### EXAMPLE 9.10

## Writing Lewis Formulas (Exceptions to the Octet Rule)

Xenon, a noble gas, forms a number of compounds. One of these is xenon tetrafluoride, $XeF_4$, a white, crystalline solid first prepared in 1962. What is the electron-dot formula of the $XeF_4$ molecule?

 **PROBLEM STRATEGY**

You follow the four steps, as in Example 9.6. In Step 4, when you distribute the remaining electrons to the central atom, you find that it has more than an octet.

### SOLUTION

There are 8 valence electrons from the Xe atom and 7 from each F atom, for a total of 36 valence electrons. For the skeleton structure, you draw the Xe atom surrounded by the electronegative F atoms. After placing electron pairs on the F atoms to satisfy the octet rule for them, you have

This accounts for 16 pairs, or 32 electrons. A total of 36 electrons is available, so you put an additional $36 - 32 = 4$ electrons (2 pairs) on the Xe atom. The Lewis formula is

---

### EXERCISE 9.13

Sulfur tetrafluoride, $SF_4$, is a colorless gas. Write the electron-dot formula of the $SF_4$ molecule.

---

See Problems 9.63 and 9.64.

The other group of exceptions to the octet rule consists mostly of molecules containing Group IIA or IIIA atoms. Consider boron trifluoride, $BF_3$. The molecule consists of boron surrounded by the much more electronegative fluorine atoms. The total number of valence electrons is $3 + (3 \times 7) = 24$. If you connect boron and fluorine atoms by electron pairs and fill out the fluorine atoms with octets of electrons, you obtain:

$$
\begin{array}{c}
: \ddot{F} : \\
| \\
: \ddot{F} - B \\
| \\
: \ddot{F} :
\end{array}
$$

If you now follow with Step 4 of the procedure outlined in Section 9.6, you note that the boron atom has only six electrons on it. You write the following resonance formulas, each having a double bond:

$$
\begin{array}{ccc}
: \ddot{F} : & : \ddot{F} : & : \ddot{F} : \\
\| & | & | \\
: \ddot{F} - B & \ddot{F} = B & : \ddot{F} - B \\
| & | & \| \\
: \ddot{F} : & : \ddot{F} : & : \ddot{F} :
\end{array}
$$

$: \ddot{F} - B \longleftrightarrow \ddot{F} = B \longleftrightarrow : \ddot{F} - B$

In fact, there is evidence to suggest that the first formula written—the one with all single bonds and in which boron has only six electrons around it—describes the chemistry of boron trifluoride very well. For example, boron trifluoride reacts with molecules having a lone pair, such as with ammonia, $NH_3$, to give the compound $BF_3NH_3$. The reaction is easy to describe in terms of the formula in which boron has only six electrons around it.

$$
\begin{array}{ccc}
: \ddot{F} : \quad H & & : \ddot{F} : \ H \\
| \quad\quad | & & | \quad\quad | \\
: \ddot{F} - B \ + \ : N - H \longrightarrow & : \ddot{F} - B \ : N - H \\
| \quad\quad | & & | \quad\quad | \\
: \ddot{F} : \quad H & & : \ddot{F} : \ H
\end{array}
$$

coordinate covalent bond

In this reaction, a coordinate covalent bond forms between the boron and nitrogen atoms, and the boron achieves an octet of electrons.

The chemistry of $BF_3$ thus appears to support an electron structure with boron having only six electrons around it. No doubt, resonance involving all four of the Lewis formulas that we have drawn best describes the actual electron structure of

▶ The B–F bond length (130 pm) is shorter than expected for a single bond (152 pm), which indicates partial double-bond character. See the discussion in Section 9.10 on bond length and bond order.

boron trifluoride, but the relative importance of the different resonance formulas is not settled. ◀

Other examples of molecules with Group IIIA atoms (such as Al) or Group IIA atoms (such as Be) display electron structures similar to that of boron trifluoride. For example, the $BeF_2$ molecule (found in the vapor over the heated solid $BeF_2$) has the Lewis formula

$$:\ddot{F}:Be:\ddot{F}:$$

Aluminum chloride, $AlCl_3$, offers an interesting study in bonding. At room temperature, the substance is a white, crystalline solid and an ionic compound, as might be expected for a binary compound of a metal and a nonmetal. However, the substance has a relatively low melting point (192°C) for an ionic compound. Apparently this is due to the fact that instead of melting to a liquid of ions, as happens with most ionic solids, the compound forms $Al_2Cl_6$ molecules (Figure 9.18), with Lewis formula

$$\ddot{C}l \quad \ddot{C}l \quad \ddot{C}l$$
$$\quad Al \quad Al$$
$$\ddot{C}l \quad \ddot{C}l \quad \ddot{C}l$$

**Figure 9.18**
**The $Al_2Cl_6$ molecule.**  A molecular model.

Each atom has an octet of electrons around it. Note that two of the Cl atoms are in *bridge* positions, with each Cl atom having two covalent bonds. When this liquid is heated, it vaporizes as $Al_2Cl_6$ molecules. As the vapor is further heated, these molecules break up into $AlCl_3$ molecules. These molecules have an electron structure similar to that of $BF_3$.

---

**EXERCISE 9.14**

Beryllium chloride, $BeCl_2$, is a solid substance consisting of long (essentially infinite) chains of atoms with Cl atoms in bridge positions.

$$\diagdown \overset{\ddot{C}l}{\diagup} \diagdown Be \overset{\diagup}{\diagdown} \overset{\ddot{C}l}{\underset{\ddot{C}l}{}} \diagup Be \diagdown \overset{\ddot{C}l}{\underset{\ddot{C}l}{}} \diagup Be \diagdown$$

However, if the solid is heated, it forms a vapor of $BeCl_2$ molecules. Write the electron-dot formula of the $BeCl_2$ molecule.

**See Problems 9.65 and 9.66.**

---

## 9.9   Formal Charge and Lewis Formulas

Earlier we looked at writing the Lewis formula for carbonyl chloride, $COCl_2$ (Example 9.7). In the solution to that example, we wrote the answer as follows:

$$:\ddot{C}l-C=\ddot{O}$$
$$\quad\quad |$$
$$\quad\quad :\ddot{C}l:$$

You may recall that we had to decide whether to draw the double bond between C and O or between C and Cl. We decided in favor of a carbon–oxygen double bond by applying the simple idea that multiple bonds most likely involve C, N, O, and S atoms. In this section, we will describe how you can use the concept of formal charge to determine the best Lewis formula. Formal charge is also a help in writing the skeleton structure of a molecule.

The **formal charge** of an atom in a Lewis formula is *the hypothetical charge you obtain by assuming that bonding electrons are equally shared between bonded atoms and that the electrons of each lone pair belong completely to one atom.* When you use the Lewis formula that most clearly describes the bonding, the formal charges give you an approximate distribution of electrons in the molecule. Let us see how you would obtain the formal charges, given several possible Lewis formulas of a molecule, and how you would use these formal charges to determine the best Lewis formula.

You begin by writing the possible Lewis formulas. For $COCl_2$, you can write

$$:\ddot{Cl}—C{=}\ddot{O} \qquad \ddot{Cl}{=}C—\ddot{O}: \qquad :\ddot{Cl}—C—\ddot{O}:$$
$$\quad\;\;|\qquad\qquad\;\;|\qquad\qquad\;\;\|$$
$$\quad:\ddot{Cl}:\qquad\quad\;:\ddot{Cl}:\qquad\quad\;:\ddot{Cl}:$$

For each formula, you apply the following *rules for formal charge* to assign the valence electrons to individual atoms:

1. Half of the electrons of a bond are assigned to each atom in the bond (counting each dash as two electrons).

2. Both electrons of a lone pair are assigned to the atom to which the lone pair belongs.

You calculate the formal charge on an atom by taking the number of valence electrons on the free atom (equal to the group number) and subtracting the number of electrons you assigned the atom by the rules of formal charge; that is,

Formal charge $=$ valence electrons on free atom $- \frac{1}{2}$(number of electrons in a bond)
$-$ (number of lone-pair electrons)

As a check on your work, you should note that the sum of the formal charges equals the charge on the molecular species (zero for a neutral molecule).

Consider a Cl atom in the first Lewis formula we wrote for $COCl_2$. You begin by counting the electrons assigned to the Cl atom by the rules of formal charge. You have 1 electron from the single bond and 6 from the lone-pair electrons, for a total of 7. The formal charge of Cl equals the number of valence electrons (7) less the number of assigned electrons (7). This gives a formal charge of 0 for the Cl atom. Similar calculations of formal charge give 0 for each of the other atoms in the Lewis formula.

Now look at the second Lewis formula we wrote for $COCl_2$ and assign the electrons by the rules for formal charge. For the Cl atom having the double bond, you count 2 electrons from the double bond and 4 from the lone pairs, for a total of 6. The formal charge of the Cl atom is $7 - 6 = +1$. For the O atom, you count 1 electron from the single bond and 6 from the lone-pair electrons, for a total of 7. The number of valence electrons on the free O atom is 6, so the formal charge of the O atom is $6 - 7 = -1$. The formal charge of each of the other atoms is 0.

We indicate the formal charges in a Lewis formula by inserting circled numbers near the atoms (writing $+$ for $+1$ and $-$ for $-1$). For the formulas given earlier for $COCl_2$, the formal charges are shown as follows:

$$:\ddot{Cl}—C{=}\ddot{O} \qquad \oplus\ddot{Cl}{=}C—\ddot{O}:\ominus \qquad :\ddot{Cl}—C—\ddot{O}:\ominus$$
$$\quad\;\;|\qquad\qquad\quad\;\;|\qquad\qquad\quad\;\;\|$$
$$\quad:\ddot{Cl}:\qquad\qquad:\ddot{Cl}:\qquad\qquad:\ddot{Cl}:$$
$$\qquad\qquad\qquad\qquad\qquad\qquad\qquad\qquad\qquad\;\;\oplus$$

▶ The rules for computing oxidation number are similar, except that both bonding electrons are assigned to the more electronegative atom.

Now we can discuss two rules that are useful in writing Lewis formulas.

> **RULE A** Whenever you can write several Lewis formulas for a molecule, choose the one having the lowest magnitudes of formal charges.
>
> **RULE B** When two proposed Lewis formulas for a molecule have the same magnitudes of formal charges, choose the one having the negative formal charge on the more electronegative atom.

The second and third formulas for $COCl_2$ have formal charges on the Cl and O atoms. The first formula, however, has zero formal charges for all atoms. So by Rule A, this formula most closely approximates the actual electron distribution. You would write it as the best description by a single Lewis formula. (To be more precise, you could consider a resonance description involving all three Lewis formulas. However, the Lewis formulas would not participate equally in the resonance description; the first one would predominate.)

You can also use formal charges to help you choose the most likely skeleton structure from several possibilities. Consider thionyl chloride, $SOCl_2$. Which of the following Lewis formulas is most plausible? Note that these Lewis formulas have quite different atomic arrangements, so they should not be considered as simply different resonance formulas of the same molecule.

You can verify these formal charges using the method we have outlined here. The last formula has a $+2$ formal charge on the Cl atom, which is larger in magnitude than the formal charges on any atoms in the other two formulas. Therefore, using Rule A, you would not consider the last formula a plausible structure. To choose between the first and second structures, you apply Rule B. According to this rule, you would choose the first structure over the second, because the first one associates a negative formal charge with the more electronegative atom (O), whereas the second structure associates a positive charge with the more electronegative atom. Note that this application of formal charge gives the same result that you would obtain by assuming that the least electronegative atom is the central atom.

Now that you have one Lewis formula with the appropriate skeleton structure for $SOCl_2$, you should explore the possibility of other Lewis formulas with this skeleton structure. You can write a Lewis formula that has a sulfur–oxygen double bond.

$SOCl_2$

This formula has formal charges of zero for each atom; so according to Rule A, it should be closer to the actual electron distribution than the previous formula. (You could write a resonance description involving both formulas.) Note that the sulfur atom has 10 electrons around it. As long as the atom has available *d* orbitals for bonding (as any element in the third and higher periods will have), the atom need not obey the octet rule. The next example further illustrates the application of formal charges in choosing the appropriate Lewis formula.

---

**EXAMPLE 9.11**

## Using Formal Charges to Determine the Best Lewis Formula

Write the Lewis formula that best describes the charge distribution in the sulfuric acid molecule, $H_2SO_4$, according to the rules of formal charge.

 **PROBLEM STRATEGY**

First, draw the skeleton structure of the molecule. Follow the steps for writing an electron-dot formula. Note that you can move electron pairs on outer atoms into bonding regions to obtain additional formulas (having multiple bonds). Use the rules to obtain the formal charges of the atoms in each formula. Decide which formula is the best representation of the electron distribution in the molecule.

**SOLUTION**

Assume a skeleton structure in which the S atom is surrounded by the more electronegative O atoms; the H atoms are then attached to two of the O atoms. The following is a possible Lewis formula having single bonds:

$$
\begin{array}{c}
:\!\ddot{O}\!: \\
| \\
H\!-\!\ddot{O}\!-\!S\!-\!\ddot{O}\!-\!H \\
| \\
:\!\ddot{O}\!:
\end{array}
$$

You calculate the formal charge of each O atom bonded to an H atom as follows: You assign the O atom 2 electrons from the two single bonds and 4 electrons from the two lone pairs, for a total of 6 electrons. Because the number of valence electrons on the O atom is 6, the formal charge is $6 - 6 = 0$. For each of the other two O atoms, you assign 1 electron from the one single bond and 6 electrons from the three lone pairs, for a total of 7. The formal charge is $6 - 7 = -1$. For the S atom, you assign 4 electrons from the four single bonds, for a total of 4. Because the number of valence electrons is 6, the formal charge is $6 - 4 = +2$. The formal charge of each of the H atoms is $1 - 1 = 0$. The Lewis formula with formal charges is

$$
\begin{array}{c}
:\!\ddot{O}\!:^{\ominus} \\
|^{\scriptsize(+2)} \\
H\!-\!\ddot{O}\!-\!S\!-\!\ddot{O}\!-\!H \\
| \\
:\!\ddot{O}\!:^{\ominus}
\end{array}
$$

Note that the sum of the formal charges equals 0, as it should for a neutral molecule.

You can also write a Lewis formula that has zero formal charges for atoms, if you form sulfur–oxygen double bonds:

$$
\begin{array}{c}
:\!O\!: \\
\| \\
H\!-\!\ddot{O}\!-\!S\!-\!\ddot{O}\!-\!H \\
\| \\
:\!O\!:
\end{array}
$$

To the top O atom, you assign 2 electrons from the double bond and 4 electrons from the two lone pairs, for a total of 6 electrons. The formal charge is

$6 - 6 = 0$. To the S atom, you assign a total of 6 electrons from bonds. The formal charge is $6 - 6 = 0$. The formal charges of the other atoms are also 0. Thus, this Lewis formula should be a better representation of the electron distribution in the $H_2SO_4$ molecule.

### EXERCISE 9.15

Write the Lewis formula that best describes the phosphoric acid molecule, $H_3PO_4$.

**See Problems 9.69 and 9.70.**

 **CONCEPT CHECK 9.3**

Which of the models shown below most accurately represents the hydrogen cyanide molecule, HCN? Write the electron-dot formula that most closely agrees with this model. State any concept or rule you used in arriving at your answer.

(a)             (b)             (c)             (d)

## 9.10  Bond Length and Bond Order

**Bond length** (or **bond distance**) is *the distance between the nuclei in a bond.* Bond lengths are determined experimentally using x-ray diffraction or the analysis of molecular spectra. Knowing the bond length in a particular molecule can sometimes provide a clue to the type of bonding present.

In many cases bond lengths for covalent single bonds in compounds can be predicted from covalent radii. **Covalent radii** are *values assigned to atoms in such a way that the sum of covalent radii of atoms A and B predicts an approximate A—B bond length.* Thus, the radius of the Cl atom might be taken to be half the Cl—Cl bond length (198 pm). The covalent radius of Cl would be $\frac{1}{2} \times 198$ pm = 99 pm. Table 9.4 lists single-bond covalent radii for nonmetallic elements. To predict the bond length of C—Cl, you add the covalent radii of the two atoms, C and Cl. You get $(77 + 99)$ pm = 176 pm, which compares favorably with the experimental value of 178 pm found in most compounds.

### EXERCISE 9.16

Estimate the O—H bond length in $H_2O$ from the covalent radii listed in Table 9.4.

**See Problems 9.71, 9.72, 9.73, and 9.74.**

▶ Although a double bond is stronger than a single bond, it is not necessarily less reactive. Ethylene, $CH_2=CH_2$, for example, is more reactive than ethane, $CH_3$—$CH_3$, where carbon atoms are linked through a single bond. The reactivity of ethylene results from the simultaneous formation of a number of strong, single bonds. See Section 9.11.

The **bond order,** defined in terms of the Lewis formula, is *the number of pairs of electrons in a bond.* For example, in C : C the bond order is 1 (single bond); in C : : C the bond order is 2 (double bond). Bond length depends on bond order. As the bond order increases, the bond strength increases and the nuclei are pulled inward, decreasing the bond length. Look at carbon–carbon bonds. The average C—C bond length is 154 pm, whereas C=C is 134 pm long and C≡C is 120 pm long. ◀

**Table 9.4**
**Single-Bond Covalent Radii for Nonmetallic Elements (in pm)**

| | | | | H |
|---|---|---|---|---|
| | | | | 37 |
| B | C | N | O | F |
| 88 | 77 | 70 | 66 | 64 |
| | Si | P | S | Cl |
| | 117 | 110 | 104 | 99 |
| | | As | Se | Br |
| | | 121 | 117 | 114 |
| | | | Te | I |
| | | | 137 | 133 |

▶ An alternate example is provided in the Instructor's Resource Manual.

**EXAMPLE 9.12**
## Relating Bond Order and Bond Length

Consider the molecules $N_2H_4$, $N_2$, and $N_2F_2$. Which molecule has the shortest nitrogen–nitrogen bond? Which has the longest nitrogen–nitrogen bond?

**SOLUTION**

First write the Lewis formulas:

$$\text{H}-\overset{\displaystyle .. }{\underset{\displaystyle \text{H}}{\text{N}}}-\overset{\displaystyle .. }{\underset{\displaystyle \text{H}}{\text{N}}}-\text{H} \qquad :\text{N}\equiv\text{N}: \qquad :\overset{\displaystyle .. }{\underset{\displaystyle .. }{\text{F}}}-\text{N}=\text{N}-\overset{\displaystyle .. }{\underset{\displaystyle .. }{\text{F}}}:$$

$$N_2H_4 \qquad\qquad N_2 \qquad\qquad N_2F_2$$

**The nitrogen–nitrogen bond should be shortest in $N_2$, where it is a triple bond, and longest in $N_2H_4$, where it is a single bond.** (Experimental values for the nitrogen–nitrogen bond lengths are 109 pm for $N_2$, 122 pm for $N_2F_2$, and 147 pm for $N_2H_4$.)

**EXERCISE 9.17**

Formic acid, isolated in 1670, is the irritant in ant bites. The structure of formic acid is

$$\text{H}-\text{C}=\text{O}$$
$$|$$
$$\text{O}$$
$$|$$
$$\text{H}$$

One of the carbon–oxygen bonds has a length of 136 pm; the other is 123 pm long. What is the length of the C=O bond in formic acid?

See Problems 9.75 and 9.76.

## 9.11 Bond Energy

In Section 9.4, when we described the formation of a covalent bond, we introduced the concept of bond dissociation energy, the energy required to break a particular bond in a molecule (see Figure 9.11). The bond dissociation energy is a measure of the strength of a particular bond and is essentially the *enthalpy change* for a gas-phase reaction in which a bond breaks. The enthalpy change, $\Delta H$, is the heat absorbed in a reaction carried out at constant pressure (heat of reaction).

Consider the experimentally determined enthalpy changes for the breaking, or dissociation, of a C—H bond in methane, $CH_4$, and in ethane, $C_2H_6$, in the gas phase:

$$\text{H}-\overset{\displaystyle \text{H}}{\underset{\displaystyle \text{H}}{\text{C}}}-\text{H}(g) \longrightarrow \text{H}-\overset{\displaystyle \text{H}}{\underset{\displaystyle \text{H}}{\text{C}}}(g) + \text{H}(g) \qquad \Delta H = 435\ \text{kJ}$$

$$\text{H}-\overset{\displaystyle \text{H}}{\underset{\displaystyle \text{H}}{\text{C}}}-\overset{\displaystyle \text{H}}{\underset{\displaystyle \text{H}}{\text{C}}}-\text{H}(g) \longrightarrow \text{H}-\overset{\displaystyle \text{H}}{\underset{\displaystyle \text{H}}{\text{C}}}-\overset{\displaystyle \text{H}}{\underset{\displaystyle \text{H}}{\text{C}}}(g) + \text{H}(g) \qquad \Delta H = 410\ \text{kJ}$$

Note that the $\Delta H$ values are approximately the same in the two cases. This suggests that the enthalpy change for the dissociation of a C—H bond may be about the same in other molecules. Comparisons of this sort lead to the conclusion that we can obtain approximate values of energies of various bonds.

We define the A—B **bond energy** (denoted *BE*) as *the average enthalpy change for the breaking of an A—B bond in a molecule in the gas phase.* For example, to calculate a value for the C—H bond energy, or *BE*(C—H), you might look at the experimentally determined enthalpy change for the breaking of all the C—H bonds in methane:

$$CH_4(g) \longrightarrow C(g) + 4H(g); \Delta H = 1662 \text{ kJ}$$

Because four C—H bonds are broken, you obtain an average value for breaking one C—H bond by dividing the enthalpy change for the reaction by 4:

$$BE(\text{C—H}) = \tfrac{1}{4} \times 1662 \text{ kJ} = 416 \text{ kJ}$$

Similar calculations with other molecules, such as ethane, would yield approximately the same values for the C—H bond energy.

Table 9.5 lists values of some bond energies. Because it takes energy to break a bond, bond energies are always positive numbers. When a bond is formed, the energy is equal to the negative of the bond energy (energy is released).

Bond energy is a measure of the strength of a bond: the larger the bond energy, the stronger the chemical bond. Note from Table 9.5 that the bonds C—C, C=C, and C≡C have energies of 346, 602, and 835 kJ/mol, respectively. These numbers

**Table 9.5**
**Bond Energies (in kJ/mol)\***

| | Single Bonds | | | | | | | | |
|---|---|---|---|---|---|---|---|---|---|
| | **H** | **C** | **N** | **O** | **S** | **F** | **Cl** | **Br** | **I** |
| H | 432 | | | | | | | | |
| C | 411 | 346 | | | | | | | |
| N | 386 | 305 | 167 | | | | | | |
| O | 459 | 358 | 201 | 142 | | | | | |
| S | 363 | 272 | — | — | 226 | | | | |
| F | 565 | 485 | 283 | 190 | 284 | 155 | | | |
| Cl | 428 | 327 | 313 | 218 | 255 | 249 | 240 | | |
| Br | 362 | 285 | — | 201 | 217 | 249 | 216 | 190 | |
| I | 295 | 213 | — | 201 | — | 278 | 208 | 175 | 149 |

| Multiple Bonds | | | | | |
|---|---|---|---|---|---|
| C=C | 602 | C=N | 615 | C=O | 745 (799 in CO$_2$) |
| C≡C | 835 | C≡N | 887 | C≡O | 1072 |
| N=N | 418 | N=O | 607 | S=O (in SO$_2$) | 532 |
| N≡N | 942 | O=O | 494 | S=O (in SO$_3$) | 469 |

\*Data are taken from J. E. Huheey, Keiter, and Keiter, *Inorganic Chemistry,* 4th ed. (New York: HarperCollins, 1993), pp. A21–A34.

**Figure 9.19**
**Reaction of methane with chlorine.** Molecular models illustrate the reaction $CH_4 + Cl_2 \longrightarrow CH_3Cl + HCl$.

indicate that the triple bond is stronger than the double bond, which in turn is stronger than the single bond.

You can use this table of bond energies to estimate heats of reaction, or enthalpy changes, $\Delta H$, for gaseous reactions. To illustrate this, find $\Delta H$ for the following reaction (Figure 9.19):

$$CH_4(g) + Cl_2(g) \longrightarrow CH_3Cl(g) + HCl(g)$$

You can *imagine* that the reaction takes place in steps involving the breaking and forming of bonds. Starting with the reactants, you suppose that one C—H bond and the Cl—Cl bond break.

$$
\begin{array}{c}
\quad\;\; H \\
\quad\;\; | \\
H-C-H + Cl-Cl \longrightarrow H-C + H + Cl + Cl \\
\quad\;\; | \qquad\qquad\qquad\qquad\;\; | \\
\quad\;\; H \qquad\qquad\qquad\qquad\;\; H
\end{array}
$$

The enthalpy change is $BE(\text{C—H}) + BE(\text{Cl—Cl})$. Now you reassemble the fragments to give the products.

$$
\begin{array}{c}
\;\; H \qquad\qquad\qquad\qquad\qquad\quad H \\
\;\; | \qquad\qquad\qquad\qquad\qquad\quad | \\
H-C + H + Cl + Cl \longrightarrow H-C-Cl + H-Cl \\
\;\; | \qquad\qquad\qquad\qquad\qquad\quad | \\
\;\; H \qquad\qquad\qquad\qquad\qquad\quad H
\end{array}
$$

In this case, C—Cl and H—Cl bonds are formed, and the enthalpy change equals the negative of the bond energies $-BE(\text{C—Cl}) - BE(\text{H—Cl})$. Substituting bond-energy values from Table 9.5, you get the enthalpy of reaction.

$$
\begin{aligned}
\Delta H &\simeq BE(\text{C—H}) + BE(\text{Cl—Cl}) - BE(\text{C—Cl}) - BE(\text{H—Cl}) \\
&= (411 + 240 - 327 - 428) \text{ kJ} \\
&= -104 \text{ kJ}
\end{aligned}
$$

The negative sign means that heat is released by the reaction. Because the bond-energy concept is only approximate, this value is only approximate (the experimental value is $-101$ kJ).

> In general, the enthalpy of reaction is (approximately) equal to the sum of the bond energies for bonds broken minus the sum of the bond energies for bonds formed.

Rather than calculating a heat of reaction from bond energies, you usually obtain the heat of reaction from thermochemical data, because these are generally known more accurately. If the thermochemical data are not known, however, bond energies can give you an estimate of the heat of reaction.

Bond energies are perhaps of greatest value when you try to explain heats of reaction or understand the relative stabilities of compounds. In general, a reaction is exothermic (gives off heat) if weak bonds are replaced by strong bonds (see Figure 9.20). In the reaction we just discussed, two bonds were broken and replaced by two new, stronger bonds. In the following example, a strong bond (C=C) is replaced by

**Figure 9.20**
**Explosion of nitrogen triiodide–ammonia complex.** The red-brown complex of nitrogen triiodide and ammonia is so sensitive to explosion that it can be detonated with a feather. Nitrogen–iodine single bonds are replaced by very stable nitrogen–nitrogen triple bonds ($N_2$) and iodine–iodine single bonds ($I_2$).

two weaker ones (two C—C bonds). Although the single bond is weaker than the double bond, as you would expect, two C—C bonds *together* create an energetically more stable situation than one C=C bond.

▶ An alternate example is provided in the Instructor's Resource Manual.

**EXAMPLE 9.13**

## Estimating $\Delta H$ from Bond Energies

Polyethylene is formed by linking many ethylene molecules into long chains. Estimate the enthalpy change per mole of ethylene for this reaction (shown below), using bond energies.

$$+ \quad \underset{H}{\overset{H}{>}}C=C\underset{H}{\overset{H}{<}} + \underset{H}{\overset{H}{>}}C=C\underset{H}{\overset{H}{<}} + \underset{H}{\overset{H}{>}}C=C\underset{H}{\overset{H}{<}} + \underset{H}{\overset{H}{>}}C=C\underset{H}{\overset{H}{<}} + \longrightarrow$$

$$\overset{\displaystyle H\;H\;H\;H\;H\;H\;H\;H}{-C-C-C-C-C-C-C-C-}\underset{\displaystyle H\;H\;H\;H\;H\;H\;H\;H}{}$$

**SOLUTION**

Imagine the reaction to involve the breaking of the carbon–carbon double bonds and the formation of carbon–carbon single bonds. For a very long chain, the net result is that for every C=C bond broken, two C—C bonds are formed.

$$\Delta H \approx [602 - (2 \times 346)]\ kJ = \mathbf{-90\ kJ}$$

**EXERCISE 9.18**

Use bond energies to estimate the enthalpy change for the combustion of ethylene, $C_2H_4$, according to the equation

$$C_2H_4(g) + 3O_2(g) \longrightarrow 2CO_2(g) + 2H_2O(g)$$

See Problems 9.77 and 9.78.

# Infrared Spectroscopy and Vibrations of Chemical Bonds

A chemical bond acts like a stiff spring connecting nuclei. As a result, the nuclei in a molecule vibrate, rather than maintaining fixed positions relative to each other. Nuclear vibration is depicted in Figure 9.21, which shows a spring model of HCl.

This vibration of molecules is revealed in their absorption of infrared radiation. (An instrument for observing the absorption of infrared radiation is shown in Figure 9.22.) The frequency of radiation absorbed equals the frequencies of nuclear vibrations. For example, the H—Cl bond vibrates at a frequency of $8.652 \times 10^{13}$ vibrations per second. If radiation of this frequency falls on the molecule, it absorbs the radiation, which is in the infrared region, and begins vibrating more strongly.

The infrared absorption spectrum of a molecule of even moderate size can have a rather complicated appearance. Figure 9.23 shows the infrared (IR) spectrum of ethyl butyrate, a compound present in pineapple flavor. The complicated appearance of the IR spectrum is actually an advantage. Two different compounds are unlikely to have exactly the same IR spectrum. Therefore, the IR spectrum can act as a compound's "fingerprint."

The IR spectrum of a compound can also yield structural information. Suppose you would like to obtain the structural formula of ethyl butyrate. The molecular formula, determined from combustion analysis, is $C_6H_{12}O_2$. Important information about this structure can be obtained from Figure 9.23.

**Figure 9.21**
**Vibration of the HCl molecule.** The vibrating molecule is represented here by a spring model. The atoms in the molecule vibrate; that is, they move constantly back and forth.

You first need to be able to read such a spectrum. Instead of plotting an IR spectrum in frequency units (since the frequencies are very large), you usually would give the frequencies in *wavenumbers*, which are proportional to frequency. To get the wavenumber, you divide the frequency by the speed of light expressed in centimeters per second. For example, HCl absorbs at $(8.652 \times 10^{13} \text{ s}^{-1})/(2.998 \times 10^{10} \text{ cm/s}) = 2886 \text{ cm}^{-1}$ (wavenumbers). Wavenumber, or sometimes wavelength, is plotted along the horizontal axis.

*Percent transmittance*—that is, the percent of radiation that passes through a sample—is plotted on the vertical axis. When a molecule absorbs radiation of a given frequency or wavenumber, this is seen in the spectrum as an inverted spike (peak) at that wavenumber.

*(continued)*

**Figure 9.22**
**A Fourier transform infrared (FTIR) spectrometer.** *Left:* A Nicolet 560 E.S.P. FT-IR spectrometer. *Right:* A high-resolution FTIR spectrum of carbon monoxide.

**Figure 9.23**
**Infrared spectrum of ethyl butyrate.** Note the peaks corresponding to vibrations of C=O and C—O bonds. The molecular structure is shown at the bottom left. [From NIST Mass Spec Data Center, S.E. Stein, director, "IR and Mass Spectra" *in NIST Chemistry WebBook, NIST Standard Reference Database Number 69,* Eds. W.G. Mallard and P.J. Linstrom, February 2000, National Institute of Standards and Technology, Gaithersburg, MD, 20899 (http//webbook. nist.gov). © 1991, 1994, 1996, 1997, 1998, 1999, 2000 copyright by the U.S. Secretary of Commerce on behalf of the United States of America. All rights reserved.]

Certain structural features of molecules appear as absorption peaks in definite regions of the infrared spectrum. For example, the absorption peak at 1730 cm$^{-1}$ is characteristic of the C=O bond. With some knowledge of where various bonds absorb, one can identify other peaks, including that of C—O at 1180 cm$^{-1}$. (Generally, the IR peak for an A—B bond occurs at lower wavenumber than for an A=B bond.) The IR spectrum does not reveal the complete structure, but it provides important clues. Data from other instruments, such as the mass spectrometer (page 104), give additional clues.

Figure 9.24 shows another IR spectrum—this one of a blood sample from a person exposed to carbon monoxide, CO, a poisonous gas. It shows that the CO molecule is chemically attached to hemoglobin, a molecule present in red blood cells and responsible for carrying oxygen. Once hemoglobin bonds to CO, its oxygen-carrying function is lost.

**Figure 9.24**
**Infrared spectrum of carboxyhemoglobin (hemoglobin with CO bonded to it).** The peak at 1953 cm$^{-1}$ corresponds to the vibration of CO bonded to hemoglobin.

**386**

# A Checklist for Review

## Important Terms

ionic bond (9.1)
Lewis electron-dot symbol (9.1)
lattice energy (9.1)
ionic radius (9.3)
isoelectronic (9.3)
covalent bond (p. 362)

Lewis electron-dot formula (9.4)
bonding pair (9.4)
lone (nonbonding) pair (9.4)
coordinate covalent bond (9.4)

octet rule (9.4)
single bond (9.4)
double bond (9.4)
triple bond (9.4)
polar covalent bond (9.5)
electronegativity (9.5)
delocalized bonding (9.7)

resonance description (9.7)
formal charge (9.9)
bond length (bond distance) (9.10)
covalent radii (9.10)
bond order (9.10)
bond energy (9.11)

## Summary of Facts and Concepts

An *ionic bond* is a strong attractive force holding ions together. An ionic bond can form between two atoms by the transfer of electrons from the valence shell of one atom to the valence shell of the other. Many similar ions attract one another to form a crystalline solid, in which positive ions are surrounded by negative ions and negative ions are surrounded by positive ions. As a result, ionic solids are typically high-melting solids. Monatomic cations of the main-group elements have charges equal to the group number (or in some cases, the group number minus two). Monatomic anions of the main-group elements have charges equal to the group number minus eight.

A *covalent bond* is a strong attractive force that holds two atoms together by their sharing of electrons. These bonding electrons are attracted simultaneously to both atomic nuclei, and they spend part of the time near one atom and part of the time near the other. If an electron pair is not equally shared, the bond is *polar*. This polarity results from the difference in *electronegativities* of the atoms—that is, in the unequal abilities of the atoms to draw bonding electrons to themselves.

*Lewis electron-dot formulas* are simple representations of the valence-shell electrons of atoms in molecules and ions. You can apply simple rules to draw these formulas. In molecules with *delocalized bonding*, it is not possible to describe accurately the electron distribution with a single Lewis formula. For these molecules, you must use *resonance*. Although the atoms in Lewis formulas often satisfy the octet rule, *exceptions to the octet rule* are not uncommon. You can obtain the Lewis formulas for these exceptions by following the rules for writing Lewis formulas. The concept of *formal charge* will often help you decide which of several Lewis formulas gives the best description of a molecule or ion.

*Bond lengths* can be estimated from the *covalent radii* of atoms. Bond length depends on *bond order*; as the bond order increases, the bond length decreases. The A—B *bond energy* is the average enthalpy change when an A—B bond is broken. You can use bond energies to estimate $\Delta H$ for gas-phase reactions.

## Operational Skills

*Note:* A periodic table is needed for skills 1, 2, 3, 5, and 6.

1. **Using Lewis symbols to represent ionic bond formation** Given a metallic and a nonmetallic main-group element, use Lewis symbols to represent the transfer of electrons to form ions of noble-gas configurations. **(EXAMPLE 9.1)**
2. **Writing electron configurations of ions** Given an ion, write the electron configuration. For an ion of a main-group element, give the Lewis symbol. **(EXAMPLES 9.2, 9.3)**
3. **Using periodic trends to obtain relative ionic radii** Given a series of ions, arrange them in order of increasing ionic radius. **(EXAMPLE 9.4)**
4. **Using electronegativities to obtain relative bond polarities** Given the electronegativities of the atoms, arrange a series of bonds in order by polarity. **(EXAMPLE 9.5)**
5. **Writing Lewis formulas** Given the molecular formula of a simple compound or ion, write the Lewis electron-dot formula. **(EXAMPLES 9.6, 9.7, 9.8, 9.10)**
6. **Writing resonance formulas** Given a simple molecule with delocalized bonding, write the resonance description. **(EXAMPLE 9.9)**
7. **Using formal charges to determine the best Lewis formula** Given two or more Lewis formulas, use formal charges to determine which formula best describes the electron distribution or gives the most plausible molecular structure. **(EXAMPLE 9.11)**
8. **Relating bond order and bond length** Know the relationship between bond order and bond length. **(EXAMPLE 9.12)**
9. **Estimating $\Delta H$ from bond energies** Given a table of bond energies, estimate the heat of reaction. **(EXAMPLE 9.13)**

# Review Questions

**9.1**   Describe the formation of a sodium chloride crystal from atoms.

**9.2**   Why does sodium chloride normally exist as a crystal rather than as a molecule composed of one cation and one anion?

**9.3**   Explain what energy terms are involved in the formation of an ionic solid from atoms. In what way should these terms change (become larger or smaller) to give the lowest energy possible for the solid?

**9.4**   Define lattice energy for potassium bromide.

**9.5**   Why do most monatomic cations of the main-group elements have a charge equal to the group number? Why do most monatomic anions of these elements have a charge equal to the group number minus eight?

**9.6**   The 2+ ions of transition elements are common. Explain why this might be expected.

**9.7**   Explain how ionic radii are obtained from known distances between nuclei in crystals.

**9.8**   Describe the trends shown by the radii of the monatomic ions for the main-group elements both across a period and down a column.

**9.9**   Describe the formation of a covalent bond in $H_2$ from atoms. What does it mean to say that the bonding electrons are shared by the two atoms?

**9.10**   Draw a potential-energy diagram for a molecule such as $Cl_2$. Indicate the bond length (194 pm) and the bond dissociation energy (240 kJ/mol).

**9.11**   Give an example of a molecule that has a coordinate covalent bond.

**9.12**   The octet rule correctly predicts the Lewis formula of many molecules involving main-group elements. Explain why this is so.

**9.13**   Describe the general trends in electronegativities of the elements in the periodic table both across a period and down a column.

**9.14**   What is the qualitative relationship between bond polarity and electronegativity difference?

**9.15**   What is a resonance description of a molecule? Why is this concept required if we wish to retain Lewis formulas as a description of the electron structure of molecules?

**9.16**   Describe the kinds of exceptions to the octet rule that we encounter in compounds of the main-group elements. Give examples.

**9.17**   What is the relationship between bond order and bond length? Use an example to illustrate it.

**9.18**   Define bond energy. Explain how one can use bond energies to estimate the heat of reaction.

# Conceptual Problems

**9.19**   You land on a distant planet in another universe and find that the $n = 1$ level can hold a maximum of 4 electrons, the $n = 2$ level can hold a maximum of 5 electrons, and the $n = 3$ level can hold a maximum of 3 electrons. Like our universe, protons have a charge of $+1$, electrons have a charge of $-1$, and opposite charges attract. Also, a filled shell results in greater stability of an atom, so the atom tends to gain or lose electrons to give a filled shell. Predict the formula of a compound that results from the reaction of a neutral metal atom X, which has 7 electrons, and a neutral nonmetal atom Y, which has 3 electrons.

**9.20**   Which of the following represent configurations of thallium ions in compounds? Explain your decision in each case.
a. $Tl^{2+}$ $[Xe]4f^{14}5d^{10}6p^1$     b. $Tl^{3+}$ $[Xe]4f^{14}5d^{10}$
c. $Tl^{4+}$ $[Xe]4f^{14}5d^9$     d. $Tl^+$ $[Xe]4f^{14}5d^{10}6s^2$

**9.21**   Examine each of the following electron-dot formulas and decide whether the formula is correct, or whether you could write a formula that better approximates the electron

structure of the molecule. State which concepts or rules you use in each case to arrive at your conclusion.

a. $: \overset{\cdot\cdot}{N} : \overset{\cdot\cdot}{N} :$     b. $: \overset{\cdot\cdot}{Cl} : C ::: N :$

c. $H : C :: \overset{\cdot\cdot}{F}$     d.     $: \overset{\cdot\cdot}{O} :$
    $\quad : \overset{\cdot\cdot}{O} :$          $H : \overset{\cdot\cdot}{O} :: C :: \overset{\cdot\cdot}{O} : H$

**9.22**   For each of the following molecular models, write an appropriate Lewis formula.

(a)          (b)          (c)          (d)

**9.23**   For each of the following molecular formulas, draw the most reasonable skeleton structure.
a. $CH_2Cl_2$     b. $HNO_2$     c. NOF     d. $N_2O_4$

What rule or concept did you use to obtain each structure?

**9.24** Below are a series of resonance formulas for $N_2O$ (nitrous oxide). Rank these in terms of how closely you think each one represents the true electron structure of the molecule. State the rules and concepts you use to do this ranking.

a. $:\ddot{N}-N\equiv O:$     b. $:N\equiv N-\ddot{O}:$     c. $:\ddot{N}=N=\ddot{O}:$

**9.25** Sodium, Na, reacts with element X to form an ionic compound with the formula $Na_3X$.

a. What is the formula of the compound you expect to form when calcium, Ca, reacts with element X?

b. Would you expect this compound to be ionic or molecular?

**9.26** The enthalpy change for each of the following reactions was calculated using bond energies. The bond energies of X—O, Y—O, and Z—O are all equal.

$$X-X + O{=}O \longrightarrow X-O-O-X; \quad \Delta H = -275 \text{ kJ}$$

$$Y-Y + O{=}O \longrightarrow Y-O-O-Y; \quad \Delta H = +275 \text{ kJ}$$

$$Z-Z + O{=}O \longrightarrow Z-O-O-Z; \quad \Delta H = -100 \text{ kJ}$$

a. Rank the bonds X—X, Y—Y, and Z—Z from strongest to weakest.

b. Compare the energies required to completely dissociate each of the products to atoms.

c. If $O_2$ molecules were O—O instead of O=O, how would this change $\Delta H$ for each reaction?

# Practice Problems

## Ionic Bonding

**9.27** Write Lewis symbols for the following:
a. In     b. $In^{3+}$     c. P     d. $P^{3-}$

**9.28** Write Lewis symbols for the following:
a. Ba     b. $Ba^{2+}$     c. I     d. $I^-$

**9.29** Use Lewis symbols to represent the transfer of electrons between the following atoms to form ions with noble-gas configurations:
a. Na and I     b. Ca and Cl

**9.30** Use Lewis symbols to represent the electron transfer between the following atoms to give ions with noble-gas configurations:
a. Mg and I     b. Ca and S

**9.31** For each of the following, write the electron configuration and Lewis symbol:
a. $Ca^{2+}$     b. Se     c. $Se^{2-}$     d. $Br^-$

**9.32** For each of the following, write the electron configuration and Lewis symbol:
a. $Rb^+$     b. I     c. $I^-$     d. $Te^{2-}$

**9.33** Write the electron configurations of Bi and $Bi^{3+}$.

**9.34** Write the electron configurations of Sn and $Sn^{2+}$.

**9.35** Give the electron configurations of $Ni^{2+}$ and $Ni^{3+}$.

**9.36** Give the electron configurations of $Cu^+$ and $Cu^{2+}$.

## Ionic Radii

**9.37** Arrange the members of each of the following pairs in order of increasing radius and explain the order:
a. $I^-$, I     b. $Ca^{2+}$, Ca

**9.38** Arrange the members of each of the following pairs in order of increasing radius and explain the order:
a. Rb, $Rb^+$     b. Se, $Se^{2-}$

**9.39** Without looking at Table 9.3, arrange the following by increasing ionic radius: $Se^{2-}$, $Te^{2-}$, $S^{2-}$. Explain how you arrived at this order. (You may use a periodic table.)

**9.40** Which has the larger radius, $N^{3-}$ or $P^{3-}$? Explain. (You may use a periodic table.)

**9.41** Arrange the following in order of increasing ionic radius: $F^-$, $Na^+$, and $N^{3-}$. Explain this order. (You may use a periodic table.)

**9.42** Arrange the following in order of increasing ionic radius: $I^-$, $Cs^+$, and $Te^{2-}$. Explain this order. (You may use a periodic table.)

## Covalent Bonding

**9.43** Use Lewis symbols to show the reaction of atoms to form phosphine, $PH_3$. Indicate bonding pairs and lone pairs in the electron-dot formula of this compound.

**9.44** Use Lewis symbols to show the reaction of atoms to form hydrogen sulfide. Indicate which electron pairs in the Lewis formula of $H_2S$ are bonding and which are lone pairs.

**9.45** Assuming that the atoms form the normal number of covalent bonds, give the molecular formula of the simplest compound of arsenic and hydrogen atoms.

**9.46** Assuming that the atoms form the normal number of covalent bonds, give the molecular formula of the simplest compound of silicon and chlorine atoms.

## Polar Covalent Bonds; Electronegativity

**9.47** Using a periodic table (not Figure 9.15), arrange the following in order of increasing electronegativity:
a. P, O, N    b. Na, Al, Mg    c. C, Al, Si

**9.48** With the aid of a periodic table (not Figure 9.15), arrange the following in order of increasing electronegativity:
a. Sr, Cs, Ba    b. Ca, Ge, Ga    c. P, As, S

**9.49** Arrange the following bonds in order of increasing polarity using electronegativities of atoms: Si—O, C—Br, As—Br.

**9.50** Decide which of the following bonds is least polar on the basis of electronegativities of atoms: H—Se, P—Cl, N—Cl.

**9.51** Indicate the partial charges for the bonds given in Problem 9.49, using the symbols $\delta^+$ and $\delta^-$.

**9.52** Indicate the partial charges for the bonds given in Problem 9.50, using the symbols $\delta^+$ and $\delta^-$.

## Writing Lewis Formulas

**9.53** Write Lewis formulas for the following molecules:
a. $Br_2$    b. $H_2Se$    c. $NF_3$

**9.54** Write Lewis formulas for the following molecules:
a. BrF    b. $PBr_3$    c. NOF

**9.55** Write Lewis formulas for the following molecules:
a. $P_2$    b. $COBr_2$    c. $HNO_2$

**9.56** Write Lewis formulas for the following molecules:
a. CO    b. BrCN    c. $N_2F_2$

**9.57** Write Lewis formulas for the following ions:
a. $ClO^-$    b. $SnCl_3^-$    c. $S_2^{2-}$

**9.58** Write Lewis formulas for the following ions:
a. $IBr_2^+$    b. $ClF_2^+$    c. $CN^-$

## Resonance

**9.59** Write resonance descriptions for the following:
a. $NO_2^-$    b. $HNO_3$

**9.60** Write resonance descriptions for the following:
a. $FNO_2$    b. $SO_3$

**9.61** Use resonance to describe the electron structure of nitromethane, $CH_3NO_2$. The skeleton structure is

**9.62** Give the resonance description of the formate ion. The skeleton structure is

$$\left[ \begin{array}{c} O \\ | \\ H-C-O \end{array} \right]^-$$

## Exceptions to the Octet Rule

**9.63** Write Lewis formulas for the following:
a. $XeF_2$    b. $SeF_4$    c. $TeF_6$    d. $XeF_5^+$

**9.64** Write Lewis formulas for the following:
a. $I_3^-$    b. $ClF_3$    c. $IF_4^-$    d. $BrF_5$

**9.65** Write Lewis formulas for the following:
a. $BCl_3$    b. $TlCl_2^+$    c. $BeBr_2$

**9.66** Write Lewis formulas for the following:
a. $BeF_2$    b. $BeF_3^-$    c. $AlBr_3$

## Formal Charge and Lewis Formulas

**9.67** Write a Lewis formula for each of the following, assuming that the octet rule holds for the atoms. Then obtain the formal charges of the atoms.
a. $O_3$    b. CO    c. $HNO_3$

**9.68** Write a Lewis formula for each of the following, assuming that the octet rule holds for the atoms. Then obtain the formal charges of the atoms.
a. ClNO    b. $POCl_3$    c. $N_2O$ (NNO)

**9.69** For each of the following, use formal charges to choose the Lewis formula that gives the best description of the electron distribution:
a. $SOF_2$    b. $H_2SO_3$    c. $HClO_2$

**9.70** For each of the following, use formal charges to choose the Lewis formula that gives the best description of the electron distribution:
a. $ClO_2F$    b. $SO_2$    c. $ClO_3^-$

## Bond Length, Bond Order, and Bond Energy

**9.71** Use covalent radii (Table 9.4) to estimate the length of the P—F bond in phosphorus trifluoride, $PF_3$.

**9.72** What do you expect for the B—Cl bond length in boron trichloride, $BCl_3$, on the basis of covalent radii (Table 9.4)?

**9.73** Calculate the bond length for each of the following single bonds, using covalent radii (Table 9.4):
a. C—H    b. S—Cl    c. Br—Cl    d. Si—O

**9.74** Calculate the C—H and C—Cl bond lengths in chloroform, $CHCl_3$, using values for the covalent radii from Table 9.4. How do these values compare with the experimental values: C—H, 107 pm; C—Cl, 177 pm?

**9.75** One of the following compounds has a carbon–nitrogen bond length of 116 pm; the other has a carbon–nitrogen bond length of 147 pm. Match a bond length with each compound.

methylamine          acetonitrile

**9.76** Which of the following two compounds has the shorter carbon–oxygen bond?

methanol          formaldehyde

**9.77** Use bond energies (Table 9.5) to estimate $\Delta H$ for the following gas-phase reaction.

This is called an "addition" reaction, because a compound (HBr) is added across the double bond.

**9.78** A commercial process for preparing ethanol (ethyl alcohol), $C_2H_5OH$, consists of passing ethylene gas, $C_2H_4$, and steam over an acid catalyst (to speed up the reaction). The gas-phase reaction is

Estimate $\Delta H$ for this reaction, using bond energies (Table 9.5).

## General Problems

**9.79** For each of the following pairs of elements, state whether the binary compound formed is likely to be ionic or covalent. Give the formula and name of the compound.
a. Sr, O      b. C, Br      c. Ga, F      d. N, Br

**9.80** For each of the following pairs of elements, state whether the binary compound formed is likely to be ionic or covalent. Give the formula and name of the compound.
a. Na, S      b. Al, F      c. Ca, Cl      d. Si, Br

**9.81** Give the Lewis formula for the arsenate ion, $AsO_4^{3-}$. Write the formula of lead(II) arsenate.

**9.82** Give the Lewis formula for the selenite ion, $SeO_3^{2-}$. Write the formula of aluminum selenite.

**9.83** Iodic acid, $HIO_3$, is a colorless, crystalline compound. What is the electron-dot formula of iodic acid?

**9.84** Selenic acid, $H_2SeO_4$, is a crystalline substance and a strong acid. Write the electron-dot formula of selenic acid.

**9.85** Sodium amide, known commercially as sodamide, is used in preparing indigo, the dye used to color blue jeans. It is an ionic compound with the formula $NaNH_2$. What is the electron-dot formula of the amide anion, $NH_2^-$?

**9.86** Lithium aluminum hydride, $LiAlH_4$, is an important reducing agent (an element or compound that generally has a strong tendency to give up electrons in its chemical reactions). Write the electron-dot formula of the $AlH_4^-$ ion.

**9.87** Nitronium perchlorate, $NO_2ClO_4$, is a reactive salt of the nitronium ion $NO_2^-$. Write the electron-dot formula of $NO_2^+$.

**9.88** Solid phosphorus pentabromide, $PBr_5$, has been shown to have the ionic structure $[PBr_4^+][Br^-]$. Write the electron-dot formula of the $PBr_4^+$ cation.

**9.89** Write electron-dot formulas for the following:
a. $SeOCl_2$      b. $CSe_2$      c. $GaCl_4^-$      d. $C_2^{2-}$

**9.90** Write electron-dot formulas for the following:
a. $POBr_3$      b. $Si_2H_6$      c. $IF_2^+$      d. $NO^+$

**9.91** Write Lewis formulas for the following:
a. $SbCl_3$      b. ICN      c. $ICl_3$      d. $IF_5$

**9.92** Write Lewis formulas for the following:
a. $AlCl_4^-$      b. $AlF_6^{3-}$      c. $BrF_3$      d. $IF_6^+$

**9.93** Give resonance descriptions for the following:
a. $SeO_2$      b. $N_2O_4$

**9.94**    Give resonance descriptions for the following:
a. $O_3$    b. $C_2O_4^{2-}$

**9.95**    The compound $S_2N_2$ has a cyclic structure with alternating sulfur and nitrogen atoms. Draw all resonance formulas in which the atoms obey the octet rule. Of these, select those in which the formal charges of all atoms are closest to zero.

**9.96**    Acetic acid has the structure $CH_3CO(OH)$, in which the OH group is bonded to a C atom. The two carbon–oxygen bonds have different lengths. When an acetic acid molecule loses the H from the OH group to form the acetate ion, the two carbon–oxygen bonds become equal in length. Explain.

**9.97**    The atoms in $N_2O_5$ are connected as follows:

$$\begin{matrix} O & & & O \\ & \diagdown & & \diagup \\ & N-O-N & \\ & \diagup & & \diagdown \\ O & & & O \end{matrix}$$

No attempt has been made here to indicate whether a bond is single or double or whether there is resonance. Obtain the Lewis formula (or formulas). The N—O bond lengths are 118 pm and 136 pm. Indicate the lengths of the bonds in the compound.

**9.98**    Methyl nitrite has the structure

$$\begin{matrix} & & H & & & & \\ & & | & & & & \\ H & - & C & - & O & - & N & - & O \\ & & | & & & & \\ & & H & & & & \end{matrix}$$

No attempt has been made here to indicate whether a bond is single or double or whether there is resonance. Obtain the Lewis formula (or formulas). The N—O bond lengths are 122 pm and 137 pm. Indicate the lengths of the N—O bonds in the compound.

**9.99**    Use bond energies to estimate $\Delta H$ for the reaction

$$H_2(g) + O_2(g) \longrightarrow H_2O_2(g)$$

**9.100**    Use bond energies to estimate $\Delta H$ for the reaction

$$2H_2(g) + N_2(g) \longrightarrow N_2H_4(g)$$

**9.101**    Use bond energies to estimate $\Delta H$ for the reaction

$$N_2F_2(g) + F_2(g) \longrightarrow N_2F_4(g)$$

**9.102**    Use bond energies to estimate $\Delta H$ for the reaction

$$HCN(g) + 2H_2(g) \longrightarrow CH_3NH_2(g)$$

## Cumulative-Skills Problems

**9.103**    Phosphorous acid, $H_3PO_3$, has the structure $(HO)_2PHO$, in which one H atom is bonded to the P atom, and two H atoms are bonded to O atoms. For each bond to an H atom, decide whether it is polar or nonpolar. Assume that only polar-bonded H atoms are acidic. Write the balanced equation for the complete neutralization of phosphorous acid with sodium hydroxide. A 200.0-mL sample of $H_3PO_3$ requires 22.50 mL of 0.1250 $M$ NaOH for complete neutralization. What is the molarity of the $H_3PO_3$ solution?

**9.104**    Hypophosphorous acid, $H_3PO_2$, has the structure $(HO)PH_2O$, in which two H atoms are bonded to the P atom, and one H atom is bonded to an O atom. For each bond to an H atom, decide whether it is polar or nonpolar. Assume that only polar-bonded H atoms are acidic. Write the balanced equation for the complete neutralization of hypophosphorous acid with

sodium hydroxide. A 200.0-mL sample of $H_3PO_2$ requires 22.50 mL of 0.1250 $M$ NaOH for complete neutralization. What is the molarity of the $H_3PO_2$ solution?

**9.105**    An ionic compound has the following composition (by mass): Mg, 10.9%; Cl, 31.8%; O, 57.3%. What are the formula and name of the compound? Write the Lewis formulas for the ions.

**9.106**    An ionic compound has the following composition (by mass): Ca, 30.3%; N, 21.2%; O, 48.5%. What are the formula and name of the compound? Write the Lewis formulas for the ions.

**9.107**    A gaseous compound has the following composition by mass: C, 25.0%; H, 2.1%; F, 39.6%; O, 33.3%. Its molecular weight is 48.0 amu. Write the Lewis formula for the molecule.

**9.108** A liquid compound used in dry cleaning contains 14.5% C and 85.5% Cl by mass and has a molecular weight of 166 amu. Write the Lewis formula for the molecule.

**9.109** A compound of tin and chlorine is a colorless liquid. The vapor has a density of 7.49 g/L at 151°C and 1.00 atm. What is the molecular weight of the compound? Why do you think the compound is molecular and not ionic? Write the Lewis formula for the molecule.

**9.110** A compound of arsenic and fluorine is a gas. A sample weighing 0.100 g occupies 14.2 mL at 23°C and 765 mmHg. What is the molecular weight of the compound? Write the Lewis formula for the molecule.

**9.111** Calculate the enthalpy of reaction for

$$HCN(g) \longrightarrow H(g) + C(g) + N(g)$$

from enthalpies of formation (see Appendix C). Given that the C—H bond energy is 411 kJ/mol, obtain a value for the C≡N bond energy. Compare your result with the value given in Table 9.5.

**9.112** Assume the values of the C—H and C—C bond energies given in Table 9.5. Then, using data given in Appendix C, calculate the C=O bond energy in acetaldehyde,

$$
\begin{array}{c}
\text{H} \quad \text{O} \\
| \quad \quad \| \\
\text{H}-\text{C}-\text{C}-\text{H} \\
| \\
\text{H}
\end{array}
$$

Compare your result with the value given in Table 9.5.

**9.113** According to Pauling, the A—B bond energy is equal to the average of the A—A and B—B bond energies plus an energy contribution from the polar character of the bond:

$$BE(A\text{—}B) = \tfrac{1}{2}[BE(A\text{—}A) + BE(B\text{—}B) + k(X_A - X_B)^2]$$

Here $X_A$ and $X_B$ are the electronegativities of atoms A and B, and $k$ is a constant equal to 98.6 kJ. Assume that the electronegativity of H is 2.1. Use the formula to calculate the electronegativity of oxygen.

**9.114** Because known compounds with N—I bonds tend to be unstable, there are no thermodynamic data available with which to calculate the N—I bond energy. However, we can estimate a value from Pauling's formula relating electronegativities and bond energies (see Problem 9.113). Using Pauling's electronegativities and the bond energies given in Table 9.5, calculate the N—I bond energy.

**9.115** Using Mulliken's formula, calculate a value for the electronegativity of chlorine. Use values of the ionization energy from Figure 8.18 and values of the electron affinity from Table 8.4. Divide this result (in kJ/mol) by 230 to get a value comparable to Pauling's scale.

**9.116** Using Mulliken's formula, calculate a value for the electronegativity of oxygen. Convert the result to a value on Pauling's scale. See Problem 9.115.

## Media Activities

**9.117** Complete the Formation of Ionic Compounds from Ions *Understanding Concepts* activity on the student CD. Write balanced chemical equations for each of the following reactions:
a. Ba + P ⟶
b. Na + S ⟶
c. Mg + N ⟶

**9.118** View the Born–Haber cycle for NaCl(s) *Visualization* on the student CD.
a. Describe the step that releases the most energy during the formation of NaCl(s).

b. Describe the step that requires the most energy during the formation of NaCl(s).

**9.119** Work through the Lewis Dot Structures *Understanding Concepts* simulation on the student web site.
a. Use formal charges to determine the most prominent Lewis structure for $SO_3^{2-}$.
b. Use formal charges to determine the most prominent Lewis structure for $PO_4^{3-}$.

**9.120** For review, take the ACE practice test for Chapter 9 on the student web site.

# 10

A framework model of the vitamin C molecule showing its three-dimensional shape.

# Molecular Geometry and Chemical Bonding Theory

We know from various experiments that molecules have definite structures; that is, the atoms of a molecule have definite positions relative to one another in three-dimensional space. Consider two molecules: boron trifluoride, $BF_3$, and phosphorus trifluoride, $PF_3$. Both have the same general formula, $AX_3$, but their molecular structures are very different (Figure 10.1). Boron trifluoride is a planar, or flat, molecule. The angle between any two B—F bonds is 120°. The geometry or general shape of the $BF_3$ molecule is said to be *trigonal planar* (trigonal because the figure described by imaginary lines connecting the F atoms has three sides). Phosphorus trifluoride is nonplanar, and the angle between any two P—F bonds is 96°. If you imagine lines connecting the fluorine atoms, these lines and those of the P—F

**Figure 10.1**
**Molecular models of BF₃ and PF₃.** Although both molecules have the general formula AX₃, boron trifluoride is planar (flat), whereas phosphorus trifluoride is pyramidal (pyramid shaped).

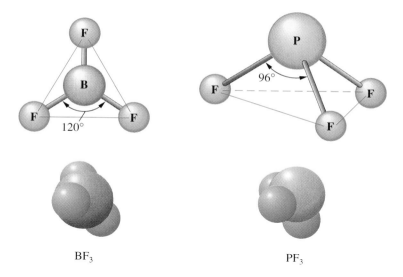

BF₃                              PF₃

▶ This chapter includes more advanced bonding concepts than those presented in the previous chapter. It follows that chapter naturally, but you can postpone or abbreviate the material as you wish.

bonds describe a pyramid with the phosphorus atom at its apex. The geometry of the PF₃ molecule is said to be *trigonal pyramidal*. How do you explain such different molecular geometries?

Subtle differences in structure are also possible. Consider the following molecular structures, which differ only in the arrangement of the atoms about the C=C bond; such molecules are referred to as *isomers*.

*cis*-1,2-dichloroethene          *trans*-1,2-dichloroethene

In the *cis* compound, both H atoms are on the same side of the C=C bond; in the *trans* compound, they are on opposite sides. These structural formulas represent entirely different compounds, as their boiling points clearly demonstrate. *Cis*-1,2-dichloroethene boils at 60°C; *trans*-1,2-dichloroethene boils at 48°C. The differences between such *cis* and *trans* compounds can be quite important. The central molecular event in the detection of light by the human eye involves the transformation of a *cis* compound to its corresponding *trans* compound after the absorption of a photon of light. How do you explain the existence of *cis* and *trans* compounds?

In this chapter we discuss how you explain the geometries of molecules in terms of their electronic structures. We also explore two theories of chemical bonding: valence bond theory and molecular orbital theory.

# Molecular Geometry and Directional Bonding

▶ Experiments in General Chemistry, 10B, "Molecular Architecture of Tetrahedral Molecules."

**Molecular geometry** is *the general shape of a molecule, as determined by the relative positions of the atomic nuclei.* There is a simple model that allows you to predict molecular geometries, or shapes, from Lewis formulas. This valence-shell electron-pair model usually predicts the correct general shape of a molecule. It does not explain chemical bonding, however. For this you must look at a theory, such as valence bond theory, that is based on quantum mechanics. Valence bond theory provides further insight into why bonds form and, at the same time, reveals that bonds have definite directions in space.

## 10.1 The Valence-Shell Electron-Pair Repulsion (VSEPR) Model

▶ "Valence Shell Electron Pair Repulsion Theory," Falcon Software, Wentworth, NH (IBM or compatible).

▶ The acronym VSEPR is pronounced "vesper."

The **valence-shell electron-pair repulsion (VSEPR) model** *predicts the shapes of molecules and ions by assuming that the valence-shell electron pairs are arranged about each atom so that electron pairs are kept as far away from one another as possible, thus minimizing electron-pair repulsions.* ◀ For example, if there are only two electron pairs in the valence shell of an atom, these pairs tend to be at opposite sides of the nucleus, so that repulsion is minimized. This gives a *linear* arrangement of electron pairs; that is, the electron pairs mainly occupy regions of space at an angle of 180° to one another (Figure 10.2).

**Figure 10.2 Arrangement of electron pairs about an atom.** Black lines give the directions of electron pairs about atom A. The blue lines merely help depict the geometric arrangement of electron pairs.

▶ **When you tie different numbers of balloons of the same size together, they assume the different arrangements shown by electron pairs.**

If three electron pairs are in the valence shell of an atom, they tend to be arranged in a plane directed toward the corners of a triangle of equal sides (equilateral triangle). This arrangement is *trigonal planar,* in which the regions of space occupied by electron pairs are directed at 120° angles to one another (Figure 10.2).

Four electron pairs in the valence shell of an atom tend to have a *tetrahedral* arrangement. That is, if you imagine the atom at the center of a regular tetrahedron, each region of space in which an electron pair mainly lies extends toward a corner, or vertex (Figure 10.2). (A regular tetrahedron is a geometrical shape with four faces, each an equilateral triangle. Thus, it has the form of a triangular pyramid.) The regions of space mainly occupied by electron pairs are directed at approximately 109.5° angles to one another.

When you determine the geometry of a molecule experimentally, you locate the positions of the atoms, not the electron pairs. To predict the relative positions of atoms around a given atom using the VSEPR model, you first note the arrangement of valence-shell electron pairs around that central atom. Some of these electron pairs are bonding pairs and some are lone pairs. *The direction in space of the bonding pairs gives you the molecular geometry.* To consider examples of the method, we will first look at molecules $AX_n$ in which the central atom A is surrounded by two, three, or four valence-shell electron pairs. Later we will look at molecules in which A is surrounded by five or six valence-shell electron pairs.

▶ **Video no. 31 (Series D), "Talesnick's Balloon Exercise (VSEPR Theory)."**

### Central Atom with Two, Three, or Four Valence-Shell Electron Pairs

Examples of geometries in which two, three, or four valence-shell electron pairs surround a central atom are shown in Figure 10.3.

▶ **Show large molecular models of different geometries.**

**Two Electron Pairs (Linear Arrangement)**   To find the geometry of the molecule $AX_n$, you first determine the number of valence-shell electron pairs around atom A. You can get this information from the electron-dot formula. For example, following the rules given in Section 9.6, you can find the electron-dot formula for the $BeF_2$ molecule. ◀

▶ **Beryllium fluoride, BeF₂, is normally a solid. The BeF₂ molecule exists in the vapor phase at high temperature.**

$$: \overset{..}{\underset{..}{F}} : Be : \overset{..}{\underset{..}{F}} :$$

There are two electron pairs in the valence shell for beryllium, and the VSEPR model predicts that they will have a linear arrangement (see Figure 10.2). Fluorine atoms are bonded in the same direction as the electron pairs. Hence, the geometry of the $BeF_2$ molecule is *linear*—that is, the atoms are arranged in a straight line (see Figure 10.4).

The VSEPR model can also be applied to molecules with multiple bonds. In this case, each multiple bond is treated as though it were a single electron pair. (All pairs of a multiple bond are required to be in approximately the same region.) To predict the geometry of carbon dioxide, for example, you first write the electron-dot formula.

$$: \overset{..}{O} :: C :: \overset{..}{O} :$$

You have two double bonds, or two electron groups, about the carbon atom, and these are treated as though there were two pairs on carbon. Thus, according to the VSEPR model, the bonds are arranged linearly and the geometry of carbon dioxide is linear.

**Three Electron Pairs (Trigonal Planar Arrangement)**   Let us now predict the geometry of the boron trifluoride molecule, $BF_3$, introduced in the chapter opening. The electron-dot formula is

$$\overset{..}{\underset{}{: F :}}$$
$$: \overset{..}{\underset{..}{F}} : B : \overset{..}{\underset{..}{F}} :$$

$BF_3$

| ELECTRON PAIRS | | | ARRANGEMENT OF PAIRS | MOLECULAR GEOMETRY | | EXAMPLE | |
|---|---|---|---|---|---|---|---|
| Total | Bonding | Lone | | | | | |
| 2 | 2 | 0 | Linear | Linear $AX_2$ | | $BeF_2$ | F—Be—F |
| 3 | 3 | 0 | Trigonal planar | Trigonal planar $AX_3$ | | $BF_3$ | |
| | 2 | 1 | | Bent (or angular) $AX_2$ | —Lone pair | $SO_2$ | |
| 4 | 4 | 0 | Tetrahedral | Tetrahedral $AX_4$ | | $CH_4$ | |
| | 3 | 1 | | Trigonal pyramidal $AX_3$ | | $NH_3$ | |
| | 2 | 2 | | Bent (or angular) $AX_2$ | | $H_2O$ | |

**Figure 10.3**
**Molecular geometries.** Central atom with two, three, or four valence-shell electron pairs. The notation in the column at the right provides a convenient way to represent the shape of a molecule drawn on a flat surface. A straight line represents a bond or electron pair in the plane of the page. A dashed line represents a bond or electron pair extending behind the page. A wedge represents a bond or electron pair extending above (in front of) the page.

**Figure 10.4**
**The beryllium fluoride molecule.** Molecular model.

The three electron pairs on boron have a trigonal planar arrangement, and the molecular geometry, which is determined by the directions of the bonding pairs, is also *trigonal planar* (Figures 10.1 and 10.3).

Sulfur dioxide, $SO_2$, provides an example of a molecule with three electron groups about the S atom; one group is a lone pair. This molecule requires a resonance description, so it gives us a chance to see how the VSEPR model can be applied in such cases. The resonance description can be written

Whichever formula you consider, sulfur has three groups of electrons about it. These groups have a trigonal planar arrangement, so sulfur dioxide is a *bent*, or *angular,* molecule (Figure 10.3). In general, whenever you consider a resonance description, you can use any of the resonance formulas to predict the molecular geometry with the VSEPR model.

Note the difference between the *arrangement of electron pairs* and the *molecular geometry,* or the arrangement of nuclei. What you usually "see" by means of x-ray diffraction and similar methods are the nuclear positions—that is, the molecular geometry. Unseen, but nevertheless important, are the lone pairs, which occupy definite positions about an atom according to the VSEPR model.

### Four Electron Pairs (Tetrahedral Arrangement)

The common and most important case of four electron pairs about the central atom (the octet rule) leads to three different molecular geometries, depending on the number of bonds formed. Note the following examples.

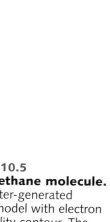

$CH_4$

$NH_3$

$H_2O$

|  | | |
|---|---|---|
| H<br>$\cdot\cdot$<br>H $:$ C $:$ H<br>$\cdot\cdot$<br>H | H<br>$\cdot\cdot$<br>$:$ N $:$ H<br>$\cdot\cdot$<br>H | H<br>$\cdot\cdot$<br>$:$ O $:$ H<br>$\cdot\cdot$ |
| $CH_4$ | $NH_3$ | $H_2O$ |
| **Molecular geometry:** tetrahedral | trigonal pyramidal | bent |

In each of these examples, the electron pairs are arranged tetrahedrally, and two or more atoms are bonded in these tetrahedral directions to give the different geometries (Figure 10.3).

When all electron pairs are bonding, as in methane, $CH_4$, the molecular geometry is *tetrahedral* (Figure 10.5). When three of the pairs are bonding and one pair is nonbonding, as in ammonia, $NH_3$, the molecular geometry is *trigonal pyramidal*. Note that the nitrogen atom is at the apex of the pyramid, and the three hydrogen atoms extend downward to form the triangular base of the pyramid.

Now you can answer the question raised in the chapter opening: Why is the $BF_3$ molecule trigonal planar and the $PF_3$ molecule trigonal pyramidal? The difference occurs because in $BF_3$ there are three pairs of electrons around boron, and in $PF_3$ there are four pairs of electrons around phosphorus. As in $NH_3$, there are three bonding pairs and one lone pair in $PF_3$, which give rise to the trigonal pyramidal geometry.

### Steps in the Prediction of Geometry by the VSEPR Model

Let us summarize the steps to follow in order to predict the geometry of an $AX_n$ molecule or ion by the VSEPR method. (All X atoms of $AX_n$ need not be identical.)

**Figure 10.5**
**The methane molecule.**
Computer-generated frame model with electron probability contour. The frame model shows the tetrahedral arrangement of the bonds.

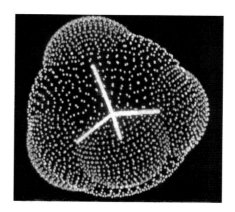

1. Determine from the electron-dot formula how many electron pairs are around the central atom. Count a multiple bond as one pair. (If resonance occurs, use one resonance formula to determine this number.)

2. Arrange the electron pairs as shown in Figure 10.2.

3. Obtain the molecular geometry from the directions of bonding pairs, as shown in Figure 10.3.

 **CONCEPT CHECK 10.1**

An atom in a molecule is surrounded by four pairs of electrons: one lone pair and three bonding pairs. Describe how the four electron pairs are arranged about the atom. How are any three of these pairs arranged in space? What is the geometry about this central atom, taking into account just the bonded atoms?

▶ An alternate example is provided in the Instructor's Resource Manual.

**EXAMPLE 10.1**

### Predicting Molecular Geometries (Two, Three, or Four Electron Pairs)

Predict the geometry of the following molecules or ions, using the VSEPR method: a. $BeCl_2$; b. $NO_2^-$; c. $SiCl_4$.

 **PROBLEM STRATEGY**

After drawing the Lewis formula for a molecule, determine the number of electron groups about the central atom. This number determines the arrangement of the electron groups (Figure 10.2). The geometry is determined by the arrangement of bonds to the central atom (Figure 10.3).

**SOLUTION**

a. Following the steps outlined in Sections 9.6 and 9.8 for writing Lewis formulas, you distribute the valence electrons to the skeleton structure of $BeCl_2$ as follows:

$$: \overset{..}{Cl} : Be : \overset{..}{Cl} :$$

The two pairs on Be have a linear arrangement, indicating a **linear** molecular geometry for $BeCl_2$. (See Figure 10.3; 2 electron pairs, 0 lone pairs.) b. The nitrite ion, $NO_2^-$, has the following resonance description:

$$\left[ : \overset{..}{O} \diagup \overset{\overset{..}{N}}{\phantom{N}} \diagdown\!\!\!\diagdown \overset{..}{O} : \right]^- \longleftrightarrow \left[ : \overset{..}{O} \diagup\!\!\!\diagup \overset{\overset{..}{N}}{\phantom{N}} \diagdown \overset{..}{O} : \right]^-$$

The N atom has three electron groups (one single bond, one double bond, and one lone pair) about it. Therefore the molecular geometry of the $NO_2^-$ ion is **bent**. (See Figure 10.3; 2 bonding pairs, 1 lone pair.) c. The electron-dot formula of $SiCl_4$ is

$$\begin{array}{c} : \overset{..}{Cl} : \\ : \overset{..}{Cl} : \overset{..}{Si} : \overset{..}{Cl} : \\ : \overset{..}{Cl} : \end{array}$$

The molecular geometry is **tetrahedral.** (See Figure 10.3; 4 bonding pairs, 0 lone pairs.)

### EXERCISE 10.1

Use the VSEPR method to predict the geometry of the following ion and molecules: a. $ClO_3^-$; b. $OF_2$; c. $SiF_4$.

See Problems 10.23, 10.24, 10.25, and 10.26.

## Bond Angles and the Effect of Lone Pairs

The VSEPR model allows you to predict the approximate angles between bonds in molecules. For example, it tells you that $CH_4$ should have a tetrahedral geometry and that the H—C—H bond angles should be 109.5° (see Figure 10.2). Because all of the valence-shell electron pairs about the carbon atom are bonding and all of the bonds are alike, you expect the $CH_4$ molecule to have an exact tetrahedral geometry. However, if one or more of the electron pairs is nonbonding or if there are dissimilar bonds, all four valence-shell electron pairs will not be alike. Then you expect the bond angles to deviate from the 109.5° predicted by an exact tetrahedral geometry. This is indeed what you see. Experimentally determined H—A—H bond angles (A is the central atom) in $CH_4$, $CH_3Cl$, $NH_3$, and $H_2O$ are shown in Figure 10.6.

The increase or decrease of bond angles from the ideal values is often predictable. *A lone pair tends to require more space than a corresponding bonding pair.* You can explain this as follows: A lone pair of electrons is attracted to only one atomic core, whereas a bonding pair is attracted to two. As a result, the lone pair is spatially more diffuse, while the bonding pair is drawn more tightly to the nuclei. Consider the trigonal pyramidal molecule $NH_3$. The lone pair on the nitrogen atom requires more space than the bonding pairs. Therefore, the N—H bonds are effectively pushed away from the lone pair, and the H—N—H bond angles become smaller than the tetrahedral value of 109.5°. How much smaller, the VSEPR model cannot tell you. The experimental value of an H—N—H bond angle is a few degrees smaller (107°). But the trigonal pyramidal molecule $PF_3$, mentioned in the chapter opening, has an F—P—F bond angle of 96°, which is significantly smaller than the exact tetrahedral value.

*Multiple bonds require more space than single bonds because of the greater number of electrons.* You therefore expect the C=O bond in the formaldehyde molecule,

**Figure 10.6**
**H—A—H bond angles in some molecules.**
Experimentally determined bond angles are shown for $CH_4$, $CH_3Cl$, $NH_3$, and $H_2O$, represented here by models.

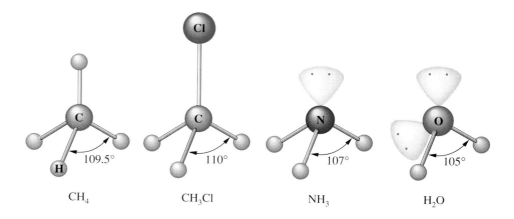

**Figure 10.7**
**H—C—H bond angles in molecules with carbon double bond.** Bond angles are shown for formaldehyde, $CH_2O$, and ethylene, $CH_2CH_2$, represented here by models.

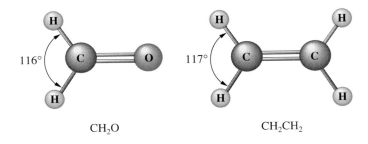

CH₂O CH₂CH₂

$CH_2O$, to require more space than the C—H bonds. You predict that the H—C—H bond angle will be smaller than the 120° seen in an exact trigonal planar geometry, such as the F—B—F angle in $BF_3$. Similarly, you expect the H—C—H bond angles in the ethylene molecule, $CH_2CH_2$, to be smaller than the trigonal planar value. Experimental values are shown in Figure 10.7.

## Central Atom with Five or Six Valence-Shell Electron Pairs

▶ You may decide not to treat five and six electron pairs, particularly if you are short of time.

Five electron pairs tend to have a *trigonal bipyramidal arrangement*. The electron pairs tend to be directed to the corners of a trigonal bipyramid, a figure formed by placing the face of one tetrahedron onto the face of another tetrahedron (see Figure 10.2).

Six electron pairs tend to have an *octahedral arrangement*. The electron pairs tend to be directed to the corners of a regular octahedron, a figure that has eight triangular faces and six vertexes, or corners (see Figure 10.2).

**Five Electron Pairs (Trigonal Bipyramidal Arrangement)** Large atoms like phosphorus can accommodate more than eight valence electrons. The phosphorus atom in phosphorus pentachloride, $PCl_5$, has five electron pairs in its valence shell. With five electron pairs around phosphorus, all bonding, $PCl_5$ should have a *trigonal bipyramidal geometry* (Figure 10.8). Note, however, that the vertexes of the trigonal bipyramid are not all equivalent (that is, the angles between the electron pairs are not all the same). Thus, the directions to which the electron pairs point are not equivalent. Two of the directions, called *axial directions,* form an axis through the central atom (see Figure 10.9). They are 180° apart. The other three directions are called *equatorial directions*. These point toward the corners of the equilateral triangle that lies on a plane through the central atom, perpendicular (at 90°) to the axial directions. The equatorial directions are 120° from each other.

Molecular geometries other than trigonal bipyramidal are possible when one or more of the five electron pairs are lone pairs. Consider the sulfur tetrafluoride molecule, $SF_4$. The Lewis electron-dot formula is

The five electron pairs about sulfur should have a trigonal bipyramidal arrangement. Because the axial and equatorial positions of the electron pairs are not equivalent, you must decide in which of these positions the lone pair appears.

| ELECTRON PAIRS | | | ARRANGEMENT OF PAIRS | MOLECULAR GEOMETRY | EXAMPLE |
|---|---|---|---|---|---|
| **Total** | **Bonding** | **Lone** | | | |

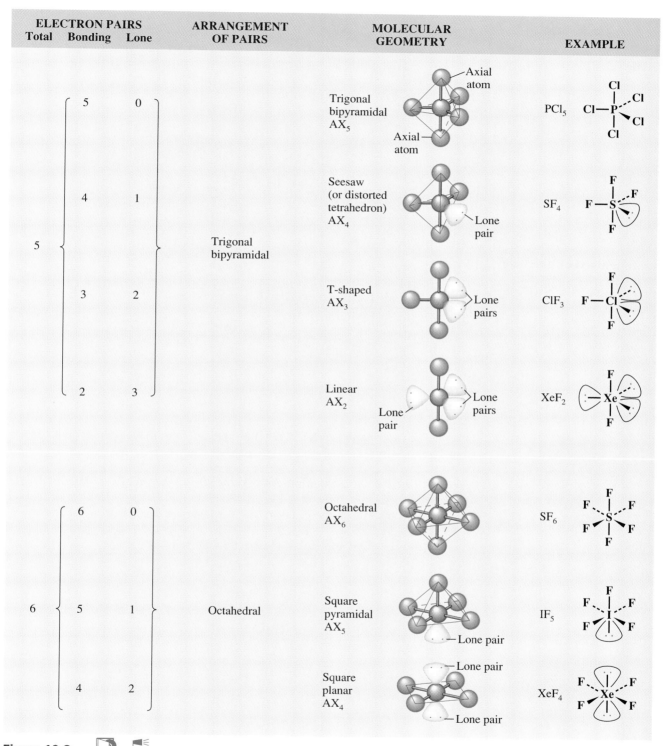

| 5 | | | | | |
| | 5 | 0 | | Trigonal bipyramidal $AX_5$ | $PCl_5$ |
| | 4 | 1 | Trigonal bipyramidal | Seesaw (or distorted tetrahedron) $AX_4$ | $SF_4$ |
| | 3 | 2 | | T-shaped $AX_3$ | $ClF_3$ |
| | 2 | 3 | | Linear $AX_2$ | $XeF_2$ |
| 6 | | | | | |
| | 6 | 0 | | Octahedral $AX_6$ | $SF_6$ |
| | 5 | 1 | Octahedral | Square pyramidal $AX_5$ | $IF_5$ |
| | 4 | 2 | | Square planar $AX_4$ | $XeF_4$ |

**Figure 10.8**
**Molecular geometries.** Central atom with five or six valence-shell electron pairs.

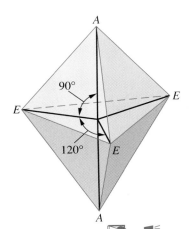

**Figure 10.9**
**Trigonal bipyramidal arrangement of electron pairs.** Electron pairs are directed along the black lines to the vertexes of a trigonal bipyramid. Equatorial directions are labeled *E*, and axial directions are labeled *A*.

The lone pair acts as though it is somewhat "larger" than a bonding pair. Therefore, the total repulsion between all pairs should be lower if the lone pair is in a position that puts it directly adjacent to the smallest number of other pairs.

An electron pair in an equatorial position is directly adjacent to only *two* other pairs—the two axial pairs at 90°. The other two equatorial pairs are farther away at 120°. An electron pair in an axial position is directly adjacent to *three* other pairs—the three equatorial pairs at 90°. The other axial pair is much farther away at 180°.

Thus you expect lone pairs to occupy equatorial positions in the trigonal bipyramidal arrangement. For sulfur tetrafluoride, this gives a *seesaw* (or *distorted tetrahedral*) *geometry* (Figure 10.10).

Chlorine trifluoride, $ClF_3$, has the electron-dot formula

The lone pairs on chlorine occupy two of the equatorial positions of the trigonal bipyramidal arrangement, giving a *T-shaped geometry* (Figure 10.11). The four atoms of the molecule all lie in one plane, with the chlorine nucleus at the intersection of the "T."

Xenon difluoride, $XeF_2$, has the electron-dot formula

The three lone pairs on xenon occupy the equatorial positions of the trigonal bipyramidal arrangement, giving a *linear geometry* (Figure 10.12).

**Six Electron Pairs (Octahedral Arrangement)**   There are six electron pairs (all bonding pairs) about sulfur in sulfur hexafluoride, $SF_6$. Thus, it has an *octahedral geometry*, with sulfur at the center of the octahedron and fluorine atoms at the vertexes (Figure 10.8).

Iodine pentafluoride, $IF_5$, has the electron-dot formula

The lone pair on iodine occupies one of the six equivalent positions in the octahedral arrangement, giving a *square pyramidal geometry* (Figure 10.13). The name derives from the shape formed by drawing lines between atoms.

Xenon tetrafluoride, $XeF_4$, has the electron-dot formula

The two lone pairs on xenon occupy opposing positions in the octahedral arrangement to minimize their repulsion. The result is a *square planar geometry* (Figure 10.14).

**Figure 10.10**
**The sulfur tetrafluoride molecule.** Molecular model.

**Figure 10.11**
**The chlorine trifluoride molecule.** Molecular model.

**Figure 10.12**
**The xenon difluoride molecule.** Molecular model.

**Figure 10.13**
**The iodine pentafluoride molecule.** Molecular model.

**Figure 10.14**
**The xenon tetrafluoride molecule.** Molecular model.

▶ An alternate example is provided in the Instructor's Resource Manual.

▶ Video no. 29 (Series D), "Styrofoam Cup and Acetone," no. 30, "Iodine/Carbon Tetrachloride/ Water."

---

**EXAMPLE 10.2**

**Predicting Molecular Geometries (Five or Six Electron Pairs)**

What do you expect for the geometry of tellurium tetrachloride, $TeCl_4$?

 **PROBLEM STRATEGY**

After drawing the Lewis formula for a molecule, determine the number of electron groups about the central atom. This number determines the arrangement of the electron groups (Figure 10.2). The geometry is determined by the arrangement of bonds to the central atom (Figure 10.8).

**SOLUTION**

First you distribute the valence electrons to the Cl atoms to satisfy the octet rule. Then you allocate the remaining valence electrons to the central atom, Te (following the steps outlined in Section 9.6). The electron-dot formula is

There are five electron pairs in the valence shell of Te in $TeCl_4$. Of these, four are bonding pairs and one is a lone pair. The arrangement of electron pairs is trigonal bipyramidal. You expect the lone pair to occupy an equatorial position, so $TeCl_4$ has a **seesaw** molecular geometry. (See Figure 10.8; 4 bonding pairs, 1 lone pair.)

**EXERCISE 10.2**

According to the VSEPR model, what molecular geometry would you predict for iodine trichloride, $ICl_3$?

See Problems 10.29, 10.30, 10.31, and 10.32.

## 10.2 Dipole Moment and Molecular Geometry

The VSEPR model provides a simple procedure for predicting the geometry of a molecule. However, predictions must be verified by experiment. Information about the geometry of a molecule can sometimes be obtained from an experimental quantity called the dipole moment, which is related to the polarity of the bonds in a molecule. The **dipole moment** is *a quantitative measure of the degree of charge separation in a molecule.* The polarity of a bond, such as that in HCl, is characterized by a separation of electric charge. We can represent this in HCl by indicating partial charges, $\delta^+$ and $\delta^-$, on the atoms.

$$\overset{\delta^+}{H}—\overset{\delta^-}{Cl}$$

Any molecule that has a net separation of charge, as in HCl, has a dipole moment. A molecule in which the distribution of electric charge is equivalent to charges $+\delta$ and $-\delta$ separated by a distance $d$ has a dipole moment equal to $\delta d$. Dipole moments are usually measured in units of *debyes* (D). In SI units, dipole moments are measured in coulomb-meters (C·m), and $1 D = 3.34 \times 10^{-30}$ C·m.

# Left-Handed and Right-Handed Molecules

**EVERYDAY LIFE**

This morning, half asleep, I (D. E.) tried to put my left glove on my right hand. Silly mistake, of course. A person's two hands are similar but not identical. If I look at my hands, say, with both palms toward me, the thumbs point in opposite directions. But, if I now hold the palm of my right hand toward a mirror and compare that image with my actual left hand, I see that the thumbs point in the same direction. A person's left hand looks like the mirror image of his or her right hand, and vice versa. You can mentally superimpose one hand onto the mirror image of the other hand. Yet the two hands are themselves nonsuperimposable. Some molecules have this "handedness" property too, and this possibility gives rise to a subtle kind of isomerism.

The presence of handedness in molecules depends on the fact that the atoms in the molecules occupy specific places in three-dimensional space. In most organic molecules, this property depends on the tetrahedral bonding of the carbon atom. Consider a carbon atom bonded to four different kinds of groups, say, —H, —OH, —$CH_3$, and —COOH. (The resulting molecule is called lactic acid.) Two isomers are possible in which one isomer is the mirror image of the other, but the two isomers are themselves not identical (they are nonsuperimposable) (see Figure 10.15). However, all four groups bonded to this carbon atom must be different to have this isomerism. If you replace the —OH group of lactic acid by, say, an H atom, the molecule and its mirror image become identical. There are no isomers; there is just one kind of molecule.

The two lactic acid isomers shown in Figure 10.15 are labeled D-lactic acid and L-lactic acid (D is for *dextro,* meaning right; L is for *levo,* meaning left). They might be expected to have quite similar properties, and they do. They have identical melting points, boiling points, and

solubilities, for instance. And if we try to prepare lactic acid in the laboratory, we almost always get an equal molar mixture of the D and L isomers. But the biological origins of the molecules are different. Only D-lactic acid occurs in sour milk, whereas only L-lactic acid forms in muscle tissue after exercising. (If you overexercise, lactic acid accumulates in muscle tissue, causing soreness.)

Biological molecules frequently have the handedness character. The chemical substance responsible for the flavor of spearmint is L-carvone. The substance responsible for the flavor of caraway seeds, which you see in some rye bread, is D-carvone (see Figure 10.16). Here are two substances whose molecules differ only in being mirror images of one another. Yet, they have strikingly different flavors: one is minty, the other is pungent aromatic.

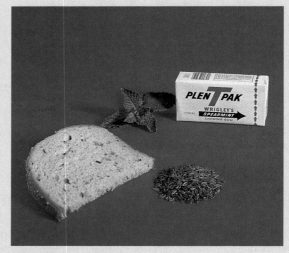

**Figure 10.16**
**Spearmint and caraway.** Spearmint leaves contain L-carvone; caraway seeds contain D-carvone.

**Figure 10.15**
**Isomers of lactic acid.**
Note that the two molecules labeled D-lactic acid and L-lactic acid are mirror images and cannot be superimposed on one another. See Figure 10.3 for an explanation of the notation used to represent the tetrahedral geometry.

HOOC⋯C⋯OH
CH₃
**D-lactic acid**

HO⋯C⋯COOH
CH₃
**L-lactic acid**

—COOH means

$$-C \begin{matrix} O \\ O-H \end{matrix}$$

Metal plates

A

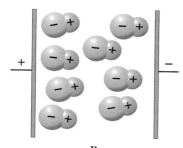

B

**Figure 10.17**
**Alignment of polar molecules by an electric field.** *(A)* A polar substance is placed between metal plates. *(B)* When the plates are connected to an electric voltage (whose direction is shown by the + and − signs), the polar molecules align so that their negative ends point toward the positive plate.

Measurements of dipole moments are based on the fact that polar molecules (molecules having a dipole moment) can be oriented by an electric field. Figure 10.17 shows an electric field generated by charged plates. Note that the polar molecules tend to align themselves so that the negative ends of the molecules point toward the positive plate, and the positive ends point toward the negative plate. This orientation of the molecules affects the *capacitance* of the charged plates (the capacity of the plates to hold a charge). Consequently, measurements of the capacitance of plates separated by different substances can be used to obtain the dipole moments of those substances. Figure 10.18 shows a simple demonstration that distinguishes between a polar liquid (one whose molecules have a dipole moment) and a nonpolar liquid. The polar liquid, but not the nonpolar one, is attracted to a charged rod.

You can sometimes relate the presence or absence of a dipole moment in a molecule to its molecular geometry. For example, consider the carbon dioxide molecule. Each carbon–oxygen bond has a polarity in which the more electronegative oxygen atom has a partial negative charge.

$$\overset{\delta^-}{O}=\overset{2\delta^+}{C}=\overset{\delta^-}{O}$$

We will denote the dipole-moment contribution from each bond (the bond dipole) by an arrow with a positive sign at one end ($\mapsto$). The dipole-moment arrow points from

**Figure 10.18**
**Attraction of a polar liquid to an electrified rod.** *Left:* Water is a polar liquid; it is attracted to the electrically charged rod. The capacitance of the polar liquid (due to the alignment of polar molecules) stabilizes a charge separation induced in the column of liquid by the charged rod. Charges of sign opposite to those on the rod form on the liquid surface near the rod, resulting in a net attraction. *Right:* Carbon tetrachloride, $CCl_4$, is a nonpolar liquid; it is not attracted to the glass rod.

the positive partial charge toward the negative partial charge. Thus, you can rewrite the formula for carbon dioxide as

$$\overset{\longleftrightarrow \ \ \longleftrightarrow}{O=C=O}$$

Each bond dipole, like a force, is a *vector* quantity; that is, it has both magnitude and direction. Like forces, two bond dipoles of equal magnitude but opposite direction cancel each other. (Think of two groups of people in a tug of war. As long as each group pulls on the rope with the same force but in the opposite direction, there is no movement—the net force is zero.) Because the two carbon–oxygen bonds in $CO_2$ are equal but point in opposite directions, they give a net dipole moment of zero for the molecule.

For comparison, consider the water molecule. The bond dipoles point from the hydrogen atoms toward the more electronegative oxygen.

Here, the two bond dipoles do not point directly toward or away from each other. As a result, they add together to give a nonzero dipole moment for the molecule.

The dipole moment of $H_2O$ has been observed to be 1.94 D.

The fact that the water molecule has a dipole moment is excellent experimental evidence for a bent geometry. If the $H_2O$ molecule were linear, the dipole moment would be zero.

The analysis we have just made for two different geometries of $AX_2$ molecules can be extended to other $AX_n$ molecules (in which all X atoms are identical). Table 10.1

**Table 10.1**
**Relationship Between Molecular Geometry and Dipole Moment**

| Formula | Molecular Geometry | Dipole Moment* |
|---------|--------------------|----------------|
| AX | Linear | Can be nonzero |
| $AX_2$ | Linear | Zero |
|  | Bent | Can be nonzero |
| $AX_3$ | Trigonal planar | Zero |
|  | Trigonal pyramidal | Can be nonzero |
|  | T-shaped | Can be nonzero |
| $AX_4$ | Tetrahedral | Zero |
|  | Square planar | Zero |
|  | Seesaw | Can be nonzero |
| $AX_5$ | Trigonal bipyramidal | Zero |
|  | Square pyramidal | Can be nonzero |
| $AX_6$ | Octahedral | Zero |

*All X atoms are assumed to be identical.

summarizes the relationship between molecular geometry and dipole moment. Those geometries in which A—X bonds are directed symmetrically about the central atom (for example, linear, trigonal planar, and tetrahedral) give molecules of zero dipole moment; that is, the molecules are *nonpolar.* Those geometries in which the X atoms tend to be on one side of the molecule (for example, bent and trigonal pyramidal) can have nonzero dipole moments; that is, they can give *polar* molecules.

▶ An alternate example is provided in the Instructor's Resource Manual.

### EXAMPLE 10.3

## Relating Dipole Moment and Molecular Geometry

Each of the following molecules has a nonzero dipole moment. Select the molecular geometry that is consistent with this information. Explain your reasoning.

a. $SO_2$      linear, bent

b. $PH_3$      trigonal planar, trigonal pyramidal

**SOLUTION**

a. In the linear geometry, the S—O bond contributions to the dipole moment would cancel, giving a zero dipole moment. That would not happen in the **bent** geometry; hence, this must be the geometry for the $SO_2$ molecule. b. In the trigonal planar geometry, the bond contributions to the dipole moment would cancel, giving a zero dipole moment. That would not occur in the **trigonal pyramidal** geometry; hence, this is a possible molecular geometry for $PH_3$.

### EXERCISE 10.3

Bromine trifluoride, $BrF_3$, has a nonzero dipole moment. Indicate which of the following geometries are consistent with this information: a. trigonal planar; b. trigonal pyramidal; c. T-shaped.

**See Problems 10.33 and 10.34.**

### EXERCISE 10.4

Which of the following would be expected to have a dipole moment of zero on the basis of symmetry? Explain. a. $SOCl_2$; b. $SiF_4$; c. $OF_2$.

**See Problems 10.35 and 10.36.**

In Example 10.3, the fact that certain molecular geometries necessarily imply a zero dipole moment allowed us to eliminate these geometries when considering a molecule with a nonzero dipole moment. The reverse argument does not follow, however. For example, suppose a molecule of the type $AX_3$ is found to have no measurable dipole moment. It may be that the geometry is trigonal planar. Another possibility, however, is that the geometry is trigonal pyramidal or T-shaped, but the lone pairs on the central atom offset the polarity of the bonds.

You can see this effect of lone pairs on the dipole moment of nitrogen trifluoride, $NF_3$. Judging by the electronegativity difference, you would expect each N—F bond to be quite polar. Yet the dipole moment of nitrogen trifluoride is only 0.2 D. (By contrast, ammonia has a dipole moment of 1.47 D.) The explanation for

**Figure 10.19**
**Explanation of the small dipole moment of NF₃.**
The lone-pair and bond contributions to the dipole moment of NF₃ tend to offset one another.

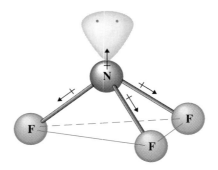

▶ **Can you explain why NH₃ has such a large dipole moment compared with NF₃?**

the small dipole moment in NF₃ appears in Figure 10.19. The lone pair on nitrogen has a dipole-moment contribution that is directed outward from the nucleus, because the electrons are offset from the nuclear center. This dipole-moment contribution thus opposes the N—F bond moments. ◀

 **CONCEPT CHECK 10.2**

> Two molecules, each with the general formula AX₃, have different dipole moments. Molecule Y has a dipole moment of zero, whereas molecule Z has a nonzero dipole moment. From this information, what can you say about the geometries of Y and Z?

## 10.3  Valence Bond Theory

The VSEPR model is usually a satisfactory method for predicting molecular geometries. To understand bonding and electronic structure, however, you must look to quantum mechanics. We will consider two theories stemming from quantum mechanics: valence bond theory and molecular orbital theory. Both use the methods of quantum mechanics but make different simplifying assumptions. In this section, we will look in a qualitative way at the basic ideas involved in **valence bond theory,** *an approximate theory to explain the electron pair or covalent bond by quantum mechanics.*

### Basic Theory

According to valence bond theory, a bond forms between two atoms when the following conditions are met:

1.  An orbital on one atom comes to occupy a portion of the same region of space as an orbital on the other atom. The two orbitals are said to *overlap.*

2.  The total number of electrons in both orbitals is no more than two.

As the orbital of one atom overlaps the orbital of another, the electrons in the orbitals begin to move about both atoms. Because the electrons are attracted to both nuclei at once, they pull the atoms together. Strength of bonding depends on the amount of overlap; the greater the overlap, the greater the bond strength. The two orbitals cannot contain more than two electrons because a given region of space can hold only two electrons (and then only if the electron spins are opposite).

**Figure 10.20**
**Formation of H₂.**
The H—H bond forms when
the 1s orbitals, one from each
atom, overlap.

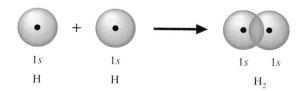

For example, consider the formation of the H₂ molecule from atoms. Each atom has the electron configuration $1s^1$. As the H atoms approach each other, their $1s$ orbitals begin to overlap and a covalent bond forms (Figure 10.20). Valence bond theory also explains why two He atoms (each with electron configuration $1s^2$) do not bond. Suppose two He atoms approach one another and their $1s$ orbitals begin to overlap. Each orbital is doubly occupied, so the sharing of electrons between atoms would place the four valence electrons from the two atoms in the same region. This, of course, could not happen. As the orbitals begin to overlap, each electron pair strongly repels the other. The atoms come together, then fly apart.

Because the strength of bonding depends on orbital overlap, orbitals other than $s$ orbitals bond only in given directions. *Orbitals bond in the directions in which they protrude or point, to obtain maximum overlap.* Consider the bonding between a hydrogen atom and a chlorine atom to give the HCl molecule. A chlorine atom has the electron configuration $[Ne]3s^23p^5$. Of the orbitals in the valence shell of the chlorine atom, three are doubly occupied by electrons and one (a $3p$ orbital) is singly occupied. The bonding of the hydrogen atom has to occur with the singly occupied $3p$ orbital of chlorine. For the strongest bonding to occur, the maximum overlap of orbitals is required. The $1s$ orbital of hydrogen must overlap along the axis of the singly occupied $3p$ orbital of chlorine (see Figure 10.21).

## Hybrid Orbitals

From what has been said, you might expect the number of bonds formed by a given atom to equal the number of unpaired electrons in its valence shell. Chlorine, whose orbital diagram is

has one unpaired electron and forms one bond. Oxygen, whose orbital diagram is

has two unpaired electrons and forms two bonds, as in H₂O.

However, consider the carbon atom, whose orbital diagram is

1s                3p

**Figure 10.21**
**Bonding in HCl.**  The bond
forms by the overlap of a
hydrogen 1s orbital (blue)
along the axis of a chlorine 3p
orbital (green).

You might expect this atom to bond to two hydrogen atoms to form the CH₂ molecule. Although this molecule is known to be present momentarily during some reactions, it is very reactive and cannot be isolated. But methane, CH₄, in which the carbon atom bonds to four hydrogen atoms, is well known. In fact, a carbon atom usually forms four bonds.

► **Animations:** *sp* **hybridization,** ***sp*²** **hybridization, and** ***sp*³** **hybridization.**

You might explain this as follows: Four unpaired electrons are formed when an electron from the $2s$ orbital of the carbon atom is *promoted* (excited) to the vacant $2p$ orbital.

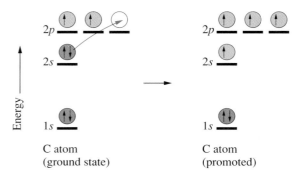

It would require energy to promote the carbon atom this way, but more than enough energy would be obtained from the formation of two additional covalent bonds. One bond would form from the overlap of the carbon $2s$ orbital with a hydrogen $1s$ orbital. Each of the other three bonds would form from a carbon $2p$ orbital and a hydrogen $1s$ orbital.

Experiment shows, however, that the four C—H bonds in methane are identical. ◄ This implies that the carbon orbitals involved in bonding are also equivalent. For this reason, valence bond theory assumes that the four valence orbitals of the carbon atom combine during the bonding process to form four new, but equivalent, hybrid orbitals. **Hybrid orbitals** are *orbitals used to describe bonding that are obtained by taking combinations of atomic orbitals of the isolated atoms.* In this case, a set of hybrid orbitals is constructed from one $s$ orbital and three $p$ orbitals, so they are called $sp^3$ hybrid orbitals. Calculations from theory show that each $sp^3$ hybrid orbital has a large lobe pointing in one direction and a small lobe pointing in the opposite direction. The four $sp^3$ hybrid orbitals point in tetrahedral directions. Figure 10.22*A* shows the shape of a single $sp^3$ orbital; *B* shows a stylized set of four orbitals pointing in tetrahedral directions.

The C—H bonds in methane, $CH_4$, are described by valence bond theory as the overlapping of each $sp^3$ hybrid orbital of the carbon atom with $1s$ orbitals of hydrogen atoms (see Figure 10.22*C*). Thus, the bonds are arranged tetrahedrally, which is

► **Nuclear magnetic resonance (page 315) and infrared spectroscopy (page 385) both show that $CH_4$ has four equivalent C—H bonds.**

**Figure 10.22**
**Spatial arrangement of $sp^3$ hybrid orbitals.**
*(A)* The shape of a single $sp^3$ hybrid orbital. Each orbital consists of two lobes. One lobe is small, but dense, and concentrated near the nucleus. The other lobe is large, but diffuse. Bonding occurs with the large lobe, since it extends farther from the nucleus. *(B)* The four hybrid orbitals are arranged tetrahedrally in space. (Small lobes are omitted here for clarity, and large lobes are stylized and greatly narrowed for ease in depicting the directional bonding.) *(C)* Bonding in $CH_4$. Each C—H bond is formed by the overlap of a $1s$ orbital from a hydrogen atom and an $sp^3$ hybrid orbital of the carbon atom.

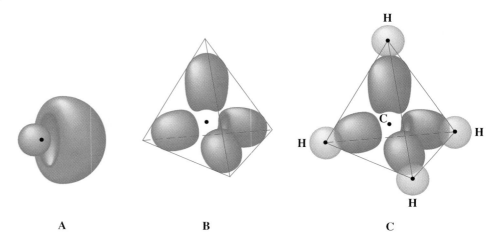

A                    B                    C

predicted by the VSEPR model. You can represent the hybridization of carbon and the bonding of hydrogen to the carbon atom in methane as follows:

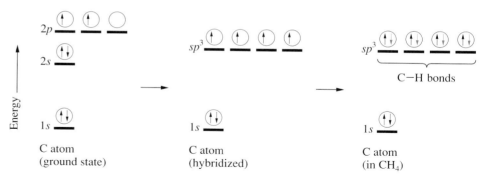

C atom
(ground state)

C atom
(hybridized)

C atom
(in CH$_4$)

**Visualization:** *sp²*
**Hybridization**

Here the blue arrows represent electrons originally belonging to hydrogen atoms.

Hybrid orbitals can be formed from various numbers of atomic orbitals. *The number of hybrid orbitals formed always equals the number of atomic orbitals used.* For example, if you combine an *s* orbital and two *p* orbitals to get a set of equivalent orbitals, you get three hybrid orbitals (called *sp²* hybrid orbitals). A set of hybrid orbitals always has definite directional characteristics. Here, all three *sp²* hybrid orbitals lie in a plane and are directed at 120° angles to one another; that is, they have a trigonal planar arrangement. Some possible hybrid orbitals and their geometric arrangements are listed in Table 10.2 and shown in Figure 10.23. Note that the geometric arrangements of hybrid orbitals are the same as those for electron pairs in the VSEPR model.

Only two of the three *p* orbitals are used to form *sp²* hybrid orbitals. The unhybridized *p* orbital is perpendicular to the plane of the *sp²* hybrid orbitals. Similarly, only one of the three *p* orbitals is used to form *sp* hybrid orbitals. The two unhybridized *p* orbitals are perpendicular to the axis of the *sp* hybrid orbitals and perpendicular to each other. We will use these facts when we discuss multiple bonding in the next section.

Now that you know something about hybrid orbitals, let us develop a general scheme for describing the bonding about any atom (we will call this the central atom). First, notice from Table 10.2 that there is a relationship between type of hybrid orbitals on an atom and the geometric arrangement of those orbitals. If you know one, you can infer the other. Thus, if you know the geometric arrangement, you know what hybrid orbitals to use in the bond description of the central atom. Your first task in describing the bonding, then, is to obtain the geometric arrangement about the central atom.

**Table 10.2**
**Kinds of Hybrid Orbitals**

| Hybrid Orbitals | Geometric Arrangement | Number of Orbitals | Example |
|---|---|---|---|
| *sp* | Linear | 2 | Be in BeF$_2$ |
| *sp²* | Trigonal planar | 3 | B in BF$_3$ |
| *sp³* | Tetrahedral | 4 | C in CH$_4$ |
| *sp³d* | Trigonal bipyramidal | 5 | P in PCl$_5$ |
| *sp³d²* | Octahedral | 6 | S in SF$_6$ |

**Figure 10.23**
**Diagrams of hybrid orbitals showing their spatial arrangements.**
Each lobe shown is one hybrid orbital (small lobes omitted for clarity).

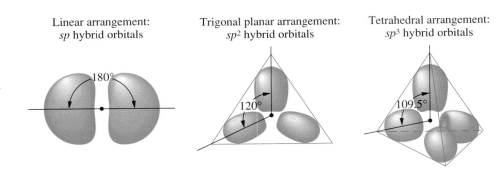

Linear arrangement: *sp* hybrid orbitals

180°

Trigonal planar arrangement: $sp^2$ hybrid orbitals

120°

Tetrahedral arrangement: $sp^3$ hybrid orbitals

109.5°

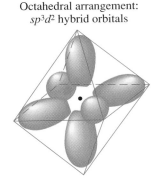

Trigonal bipyramidal arrangement: $sp^3d$ hybrid orbitals

Octahedral arrangement: $sp^3d^2$ hybrid orbitals

In lieu of experimental information about the geometry, you can use the VSEPR model, since it proves to be a fairly good predictor of molecular geometry. To obtain the bonding description about any atom in a molecule, you proceed as follows:

**Step 1:** Write the Lewis electron-dot formula of the molecule.

**Step 2:** From the Lewis formula, use the VSEPR model to obtain the arrangement of electron pairs about this atom.

**Step 3:** From the geometric arrangement of the electron pairs, deduce the type of hybrid orbitals on this atom required for the bonding description (see Table 10.2).

**Step 4:** Assign valence electrons to the hybrid orbitals of this atom one at a time, pairing them only when necessary.

**Step 5:** Form bonds to this atom by overlapping singly occupied orbitals of other atoms with the singly occupied hybrid orbitals of this atom.

As an application of this scheme, let us look at the $BF_3$ molecule and obtain the bond description of the boron atom. Following Step 1, you write the Lewis formula of $BF_3$.

Now you apply the VSEPR model to the boron atom (Step 2). There are three electron pairs about the boron atom, so they are expected to have a planar trigonal arrangement. Looking at Table 10.2 (Step 3), you note that the three $sp^2$ hybrid orbitals have a trigonal planar arrangement. In Step 4, you assign the valence electrons of the boron atom to the hybrid orbitals. Finally, in Step 5, you imagine three fluorine atoms approaching the boron atom. The singly occupied $2p$ orbital on a fluorine atom overlaps one of the $sp^2$ hybrid orbitals on boron, forming a covalent bond. Three such B—F bonds form. Note that one of the $2p$ orbitals of boron remains unhybridized and is unoccupied by electrons. It is oriented perpendicular to the molecular plane. You can summarize these steps using orbital diagrams as follows:

The next example further illustrates how to describe the bonding about an atom in a molecule. To solve problems of this type, you need to know the geometric arrangements of the hybrid orbitals listed in Table 10.2.

## EXAMPLE 10.4

### Applying Valence Bond Theory (Two, Three, or Four Electron Pairs)

Describe the bonding in $H_2O$ according to valence bond theory. Assume that the molecular geometry is the same as given by the VSEPR model.

### PROBLEM STRATEGY

From the Lewis formula for a molecule, determine its geometry about the central atom using the VSEPR model. From this geometry, determine the hybrid orbitals on this atom, assigning its valence electrons to these orbitals one at a time. Finally, form bonds by overlapping these hybrid orbitals with orbitals of the other atoms.

### SOLUTION

The Lewis formula for $H_2O$ is

$$H-\overset{\displaystyle ..}{\underset{\displaystyle |}{O}}:$$
$$H$$

Note that there are four pairs of electrons about the oxygen atom. According to the VSEPR model, these are directed tetrahedrally, and from Table 10.2 you see that you should use $sp^3$ hybrid orbitals. Each O—H bond is formed by the overlap of a $1s$ orbital of a hydrogen atom with one of the singly occupied $sp^3$ hybrid orbitals of the oxygen atom. You can represent the bonding to the oxygen atom in $H_2O$ as follows (see also Figure 10.24):

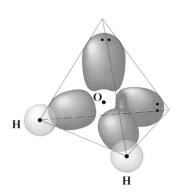

**Figure 10.24**
**Bonding in $H_2O$.** Orbitals on oxygen are $sp^3$ hybridized; bonding is tetrahedral (see Example 10.4).

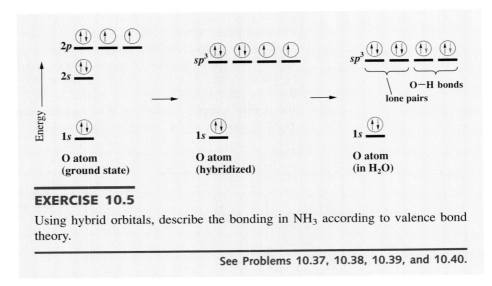

O atom
(ground state)

O atom
(hybridized)

O atom
(in H₂O)

### EXERCISE 10.5

Using hybrid orbitals, describe the bonding in $NH_3$ according to valence bond theory.

See Problems 10.37, 10.38, 10.39, and 10.40.

The actual angle between O—H bonds in $H_2O$ has been experimentally determined to be 104.5°. Because this is close to the tetrahedral angle (109.5°), the description of bonding given in Example 10.4 is essentially correct. To be more precise, you should note that the hybrid orbitals for bonds and lone pairs are not exactly equivalent. The lone pairs are somewhat larger than bonding pairs. Because they take up more space, the lone pairs push the bonding pairs closer together than they are in the exact tetrahedral case.

As the next example illustrates, hybrid orbitals are also useful for describing bonding when the central atom of a molecule is surrounded by more than eight valence electrons.

▶ An alternate example is provided in the Instructor's Resource Manual.

### EXAMPLE 10.5

## Applying Valence Bond Theory (Five or Six Electron Pairs)

Describe the bonding in $XeF_4$ using hybrid orbitals.

 **PROBLEM STRATEGY**

From the Lewis formula for a molecule, determine its geometry about the central atom using the VSEPR model. From this geometry, determine the hybrid orbitals on this atom, assigning its valence electrons to these orbitals one at a time. Finally, form bonds by overlapping these hybrid orbitals with orbitals of the other atoms.

**SOLUTION**

The Lewis formula of $XeF_4$ is

The xenon atom has four single bonds and two lone pairs. It will require six orbitals to describe the bonding. This suggests that you use $sp^3d^2$ hybrid orbitals on xenon (according to Table 10.2). Each fluorine atom (valence-shell configuration $2s^2 2p^5$) has one singly occupied orbital, so you assume that this orbital is used in bond-

# Orbitals—The Glue of Chemical Bonding

**FRONTIERS**

Late in 1999, scientists at Arizona State University for the first time produced clear images of the density of an orbital involved in the chemical bonding between atoms (Figure 10.25). These were not simply computer simulations of orbital densities, but pictures produced from experiment. An orbital is the region of space occupied by an electron in an atom or molecule and, of course, the orbitals between atoms (produced by the overlap of orbitals on separate atoms) are the glue of chemical bonding. Physicists and chemists have had mathematical models of atomic orbitals since the 1920s when quantum mechanics was discovered. The experimental image of the density shown in Figure 10.25 looks like what you would expect for the density of the $3d_{z^2}$ atomic orbital that you can find in textbooks (see Figure 7.27). So finally we can see what, for many years, theory has predicted.

What the ASU scientists did was obtain an extremely accurate electron-density distribution of the copper(I) oxide ($Cu_2O$) crystal, using a special apparatus they devised. As a first approximation, one might expect this compound to be ionic, $(Cu^+)_2O^{2-}$. In fact, the compound is ionic with a covalent component. By subtracting the electron density of the ions from the actual density, the scientists could obtain a picture of the covalent component. This component reveals that some electron density moves from $3d_{z^2}$ orbitals on copper atoms, yielding hybrid orbitals that appear to form copper–copper bonds. The loss of density from the $3d_{z^2}$ orbital shows up in the image as $3d_{z^2}$ "holes" (which are blue in the image).

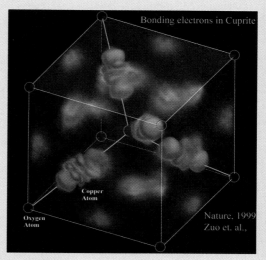

**Figure 10.25**
**Electron-density image of the copper(I) oxide crystal (with densities of ions subtracted).**
Oxygen atoms (marked by circles) are at the corners of a crystal cell, with another oxygen atom at the center. Copper atoms are at the centers of regions of depleted density (shown here in blue). Each of these regions looks very much like what we would expect from a textbook drawing of a $3d_{z^2}$ orbital. Note the dumbbell shape of each region, with a blue collar around the middle. (Compare this with the $3d_{z^2}$ orbital shown in Figure 7.27.) Regions with a build-up of density (shown in pink) are between copper atoms and appear to indicate the formation of copper–copper bonds.

ing. Each Xe—F bond is formed by the overlap of a xenon $sp^3d^2$ hybrid orbital with a singly occupied fluorine $2p$ orbital. You can summarize this as follows:

**Problem Solving:
Localized Electron Model**

**EXERCISE 10.6**

Describe the bonding in $PCl_5$ using hybrid orbitals.

See Problems 10.43, 10.44, 10.45, and 10.46.

## 10.4 Description of Multiple Bonding

In the previous section, we described bonding as the overlap of *one* orbital from each of the bonding atoms. Now we consider the possibility that *more than one* orbital from each bonding atom might overlap, resulting in a multiple bond.

As an example, consider the ethylene molecule.

$$\begin{array}{ccc} H & & H \\ \diagdown & & \diagup \\ & C = C & \\ \diagup & & \diagdown \\ H & & H \end{array}$$

*One hybrid orbital is needed for each bond (whether a single or a multiple bond) and for each lone pair.* Because each carbon atom is bonded to three other atoms and there are no lone pairs, three hybrid orbitals are needed. This suggests the use of $sp^2$ hybrid orbitals on each carbon atom (there are three $sp^2$ hybrid orbitals; see Table 10.2). Thus, during bonding, the $2s$ orbital and two of the $2p$ orbitals of each carbon atom form three hybrid orbitals having trigonal planar orientation. A third $2p$ orbital on each carbon atom remains unhybridized and is perpendicular to the plane of the three $sp^2$ hybrid orbitals.

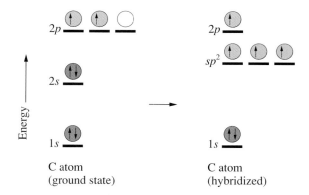

To describe the multiple bonding in ethylene, we must distinguish between two kinds of bonds. A $\sigma$ **(sigma) bond** *has a cylindrical shape about the bond axis.* It is formed either when two $s$ orbitals overlap, as in $H_2$ (Figure 10.26A), or when an orbital with directional character, such as a $p$ orbital or a hybrid orbital, overlaps another orbital *along their axis* (Figure 10.26B). The bonds we discussed in the previous section are $\sigma$ bonds.

A $\pi$ **(pi) bond** *has an electron distribution above and below the bond axis.* It is formed by the *sideways* overlap of two parallel $p$ orbitals (see Figure 10.26C). A sideways overlap will not give so strong a bond as an along-the-axis overlap of two $p$

**Figure 10.26**
**Sigma and pi bonds.**
(A) The formation of a σ bond by the overlap of two s orbitals. (B) A σ bond can also be formed by the overlap of two p orbitals along their axes. (C) When two p orbitals overlap sideways, a π bond is formed.

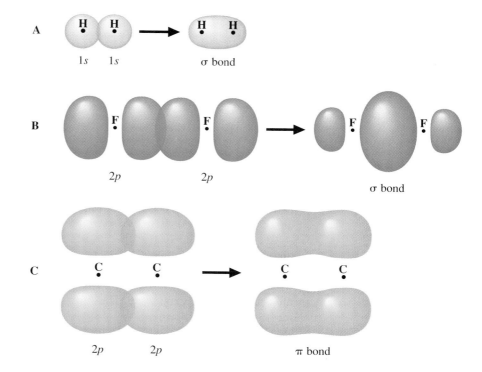

**Figure 10.27**
**Bonding in ethylene.**
(A) The σ-bond framework in ethylene, formed by the overlap of $sp^2$ hybrid orbitals on C atoms and 1s orbitals on H atoms. (B) The formation of the π bonds in ethylene. When the 2p orbitals are perpendicular to one another, there is no overlap and no bond formation. When the two —CH$_2$ groups rotate so that the 2p orbitals are parallel, a π bond forms.

orbitals. A π bond occurs when two parallel orbitals are still available after strong σ bonds have formed.

Now imagine that the separate atoms of ethylene move into their normal molecular positions. Each $sp^2$ hybrid carbon orbital overlaps a 1s orbital of a hydrogen atom or an $sp^2$ hybrid orbital of another carbon atom to form a σ bond (Figure 10.27A). Together, the σ bonds give the molecular framework of ethylene.

As you see from the orbital diagram for the hybridized C atom, a single 2p orbital still remains on each carbon atom. These orbitals are perpendicular to the plane of

A                                    B

**Figure 10.28**
**Bonding in acetylene.**
(A) The σ-bond framework.
(B) Two 2p orbitals on each
carbon atom begin to overlap
(symbolized by lines) to form
two π bonds.

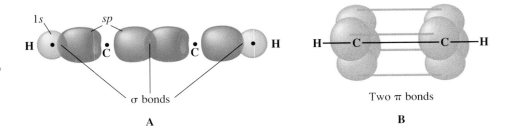

the hybrid orbitals; that is, they are perpendicular to the —CH₂ plane. Note that the two —CH₂ planes can rotate about the carbon–carbon axis without affecting the overlap of the hybrid orbitals. As these planes rotate, the 2p orbitals also rotate. When the —CH₂ planes rotate so that the 2p orbitals become parallel, the orbitals overlap to give a π bond (Figure 10.27B).

You therefore describe the carbon–carbon double bond as one σ bond and one π bond. Note that when the two 2p orbitals are parallel, the two —CH₂ ends of the molecule lie in the same plane. Thus, the formation of a π bond "locks" the two ends into a flat, rigid molecule.

You can describe the triple bonding in acetylene, H—C≡C—H, in similar fashion. Because each carbon atom is bonded to two other atoms and there are no lone pairs, two hybrid orbitals are needed. This suggests sp hybridization (see Table 10.2).

**Visualization:** *sp*
**Hybridization**

These sp hybrid orbitals have a linear arrangement, so the H—C—C—H geometry is linear. Bonds formed by the overlap of these hybrid orbitals are σ bonds. The two 2p orbitals not used to construct hybrid orbitals are perpendicular to the bond axis and to each other. They are used to form two π bonds. Thus, the carbon–carbon triple bond consists of one σ bond and two π bonds (see Figure 10.28).

▶ An alternate example is
provided in the Instructor's
Resource Manual.

**EXAMPLE 10.6**

## Applying Valence Bond Theory (Multiple Bonding)

Describe the bonding on a given N atom in dinitrogen difluoride, N₂F₂, using valence bond theory.

**PROBLEM STRATEGY**

From the Lewis formula, determine the hybrid orbitals on the double-bonded atoms. Do this by counting the number of electron groups about each atom, and from this determine the kind of hybrid orbitals used. Form a $\sigma$ bond between the double-bonded atoms by overlapping a hybrid orbital of one atom with a hybrid orbital of the other atom. Also, form $\sigma$ bonds from these atoms to other atoms. Use any unhybridized $p$ orbitals on the central atoms to form $\pi$ bonds between them.

**SOLUTION**

The electron-dot formula of $N_2F_2$ is

$$\overset{\pi}{\overset{\curvearrowright}{:\ddot{F}-\ddot{N}=\ddot{N}-\ddot{F}:}}$$

Note that a double bond is described as a $\pi$ bond plus a $\sigma$ bond. Hybrid orbitals are needed to describe each $\sigma$ bond and each lone pair (a total of three hybrid orbitals for each N atom). This suggests $sp^2$ hybridization (see Table 10.2). According to this description, one of the $sp^2$ hybrid orbitals is used to form the N—F bond, another to form the $\sigma$ bond of N=N, and the third to hold the lone pair on the N atom. The $2p$ orbitals on each N atom overlap to form the $\pi$ bond of N=N. Hybridization and bonding of the N atoms are shown as follows:

**EXERCISE 10.7**

Describe the bonding on the carbon atom in carbon dioxide, $CO_2$, using valence bond theory.

See Problems 10.47 and 10.48.

▶ **Experiments in General Chemistry, 10A, "Geometric Isomers."**

The $\pi$-bond description of the double bond agrees well with experiment. The *geometric*, or *cis–trans, isomers* of the compound 1,2-dichloroethene (described in the chapter opening) illustrate this. *Isomers* are compounds of the same molecular formula but with different arrangements of the atoms. (The numbers in the name 1,2-dichloroethene refer to the positions of the chlorine atoms; one chlorine atom

**A**

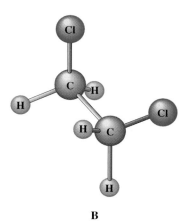

**B**

**Figure 10.29**
**Lack of geometric isomers in 1,2-dichloroethane.**
Because of rotation about the carbon–carbon bond in 1,2-dichloroethane, geometric isomers are not possible. Note that the molecule pictured in (A) can be twisted easily to give the molecule pictured in (B).

is attached to carbon atom 1 and the other to carbon atom 2.) The structures of these isomers of 1,2-dichloroethene are

$$\begin{array}{cc} \underset{H}{\overset{Cl}{\diagdown}}C=C\underset{H}{\overset{Cl}{\diagup}} & \underset{H}{\overset{Cl}{\diagdown}}C=C\underset{Cl}{\overset{H}{\diagup}} \\ \textit{cis}\text{-1,2-dichloroethene} & \textit{trans}\text{-1,2-dichloroethene} \end{array}$$

To transform one isomer into the other, one end of the molecule must be rotated as the other remains fixed. For this to happen, the $\pi$ bond must be broken. Breaking the $\pi$ bond requires considerable energy, so the *cis* and *trans* compounds are not easily interconverted. Contrast this with 1,2-dichloroethane (Figure 10.29), in which the two ends of the molecule can rotate without breaking any bonds. Here isomers corresponding to different spatial orientations of the two chlorine atoms cannot be prepared, because the two ends rotate freely with respect to one another. There is only one compound.

As we noted in the chapter opening, *cis* and *trans* isomers have different properties. The *cis* isomer of 1,2-dichloroethene boils at 60°C; the *trans* compound, at 48°C. These isomers can also be differentiated on the basis of dipole moment. The *trans* compound has no dipole moment because it is symmetrical (the polar C—Cl bonds point in opposite directions and so cancel). However, the *cis* compound has a dipole moment of 1.85 D.

It is possible to convert one isomer to another if sufficient energy is supplied—say by chemical reaction. The role of the conversion of a *cis* isomer to its *trans* isomer in human vision is discussed in the essay at the end of Section 10.7.

---

**EXERCISE 10.8**

Dinitrogen difluoride (see Example 10.6) exists as *cis* and *trans* isomers. Write structural formulas for these isomers and explain (in terms of the valence bond theory of the double bond) why they exist.

---

**See Problems 10.49 and 10.50.**

**CONCEPT CHECK 10.3**

An atom in a molecule has one single bond and one triple bond to other atoms. What hybrid orbitals do you expect for this atom? Describe how you arrive at your answer.

# Molecular Orbital Theory

In the preceding sections, we looked at a simple version of valence bond theory. Although simple valence bond theory satisfactorily describes most of the molecules we encounter, it does not apply to all molecules. For example, according to this theory any molecule with an even number of electrons should be diamagnetic (not attracted to a magnet), because we assume the electrons to be paired and to have opposite spins. In fact, a few molecules with an even number of electrons are paramagnetic (attracted to a magnet), indicating that some of the electrons are not paired. The best-known example of such a paramagnetic molecule is $O_2$. Because of the para-

▶ Video no. 14 (Series A), "Paramagnetism of Liquid Oxygen." For instructions on how to do the demonstration, see Instructor's Resource Manual, Part VIII, "Brief Notes on Suggested Lecture Demonstrations" for Chapter 10.

magnetism of $O_2$, liquid oxygen sticks to a magnet when poured over it (Figure 10.30). Although valence bond theory can be extended to explain the electronic structure of $O_2$, molecular orbital theory (an alternative bonding theory) provides a straightforward explanation of the paramagnetism of $O_2$.

**Molecular orbital theory** is *a theory of the electronic structure of molecules in terms of molecular orbitals, which may spread over several atoms or the entire molecule*. This theory views the electronic structure of molecules to be much like the electronic structure of atoms. Each molecular orbital has a definite energy. To obtain the ground state of a molecule, electrons are put into orbitals of lowest energy, consistent with the Pauli exclusion principle, just as in atoms.

## 10.5 Principles of Molecular Orbital Theory

You can think of a molecular orbital as being formed from a combination of atomic orbitals. As atoms approach each other and their atomic orbitals overlap, molecular orbitals are formed.

### Bonding and Antibonding Orbitals

Consider the $H_2$ molecule. As the atoms approach to form the molecule, their $1s$ orbitals overlap. One molecular orbital is obtained by adding the two $1s$ orbitals (see Figure 10.31). Note that where the atomic orbitals overlap, their values sum to give a larger result. This means that in this molecular orbital, electrons are often found in the region between the two nuclei where the electrons can hold the nuclei together. *Molecular orbitals that are concentrated in regions between nuclei* are called **bonding orbitals.** The bonding orbital in $H_2$, which we have just described, is denoted $\sigma_{1s}$. The $\sigma$ (sigma) means that the molecular orbital has a cylindrical shape about the bond axis. The subscript $1s$ tells us that the molecular orbital was obtained from $1s$ atomic orbitals.

Another molecular orbital is obtained by subtracting the $1s$ orbital on one atom from the $1s$ orbital on the other (Figure 10.31). When the orbitals are subtracted, the resulting values in the region of the overlap are close to zero. This means that in this molecular orbital, the electrons spend little time between the nuclei. *Molecular*

**Figure 10.30**
**Paramagnetism of oxygen, $O_2$.** Liquid oxygen is poured between the poles of a strong magnet. Oxygen adheres to the poles, showing that it is paramagnetic.

**Figure 10.31**
**Formation of bonding and antibonding orbitals from $1s$ orbitals of hydrogen atoms.** When the two $1s$ orbitals overlap, they can either add to give a bonding molecular orbital or subtract to give an antibonding molecular orbital.

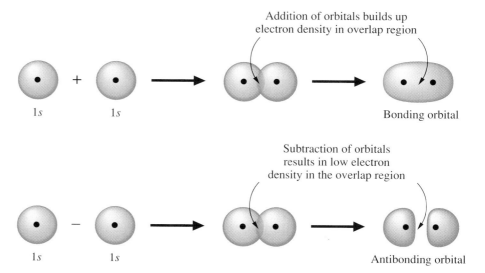

Addition of orbitals builds up electron density in overlap region

$1s$   $1s$

Bonding orbital

Subtraction of orbitals results in low electron density in the overlap region

$1s$   $1s$

Antibonding orbital

**Figure 10.32**
**Relative energies of the 1s orbital of the H atom and the $\sigma_{1s}$ and $\sigma^*_{1s}$ molecular orbitals of H$_2$.**
Arrows denote occupation of the $\sigma_{1s}$ orbital by electrons in the ground state of H$_2$.

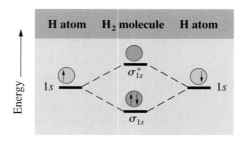

▶ Video Series C, "Magnetic Properties of Liquid Nitrogen and Liquid Oxygen."

*orbitals that are concentrated in regions other than between two nuclei* are called **antibonding orbitals.** The antibonding orbital in H$_2$, which we have just described, is denoted $\sigma^*_{1s}$. The asterisk, which you read as "star," tells us that the molecular orbital is antibonding.

Figure 10.32 shows the energies of the molecular orbitals $\sigma_{1s}$ and $\sigma^*_{1s}$ relative to the atomic orbitals. Energies of the separate atomic orbitals are represented by heavy black lines at the far left and right. The energies of the molecular orbitals are shown by heavy black lines in the center. (These are connected by dashed lines to show which atomic orbitals were used to obtain the molecular orbitals.) Note that the energy of a bonding orbital is less than that of the separate atomic orbitals, whereas the energy of an antibonding orbital is higher.

You obtain the electron configuration for the ground state of H$_2$ by placing the two electrons (one from each atom) into the lower-energy orbital (see Figure 10.32). The orbital diagram is

$$\sigma_{1s} \quad \sigma^*_{1s}$$

and the electron configuration is $(\sigma_{1s})^2$. Because the energy of the two electrons is lower than their energies in the isolated atoms, the H$_2$ molecule is stable.

Configurations involving the $\sigma^*_{1s}$ orbital describe excited states of the molecule. As an example of an excited state, you can write the orbital diagram

$$\sigma_{1s} \quad \sigma^*_{1s}$$

The corresponding electron configuration is $(\sigma_{1s})^1(\sigma^*_{1s})^1$.

You obtain a similar set of orbitals when you consider the approach of two helium atoms. To obtain the ground state of He$_2$, you note that there are four electrons, two from each atom, that would fill the molecular orbitals. Two electrons go into the $\sigma_{1s}$ orbital and two go into the $\sigma^*_{1s}$ orbital. The orbital diagram is

$$\sigma_{1s} \quad \sigma^*_{1s}$$

and the configuration is $(\sigma_{1s})^2(\sigma^*_{1s})^2$. The energy decrease from the bonding electrons is offset by the energy increase from the antibonding electrons. Hence, He$_2$ is not a stable molecule. Molecular orbital theory therefore explains why the element helium exists as a monatomic gas, whereas hydrogen is diatomic.

## Bond Order

The term *bond order* refers to the number of bonds that exist between two atoms. In molecular orbital theory, the bond order of a diatomic molecule is defined as one-half

> ► For a Lewis formula, the bond order equals the number of electron pairs shared between two atoms (see Section 9.10).

the difference between the number of electrons in bonding orbitals, $n_b$, and the number of electrons in antibonding orbitals, $n_a$. ◄

$$\text{Bond order} = \tfrac{1}{2}(n_b - n_a)$$

For $H_2$, which has two bonding electrons,

$$\text{Bond order} = \tfrac{1}{2}(2 - 0) = 1$$

That is, $H_2$ has a single bond. For $He_2$, which has two bonding and two antibonding electrons,

$$\text{Bond order} = \tfrac{1}{2}(2 - 2) = 0$$

Bond orders need not be whole numbers; half-integral bond orders of $\tfrac{1}{2}$, $\tfrac{3}{2}$, and so forth are also possible. For example, the $H_2{}^+$ molecular ion, which is formed in mass spectrometers, has the configuration $(\sigma_{1s})^1$ and a bond order of $\tfrac{1}{2}(1 - 0) = \tfrac{1}{2}$.

## Factors That Determine Orbital Interaction

The $H_2$ and $He_2$ molecules are relatively simple. A $1s$ orbital on one atom interacts with the $1s$ orbital on the other atom. But now let us consider the $Li_2$ molecule. Each Li atom has $1s$ and $2s$ orbitals. Which orbitals interact to form molecular orbitals? To find out, you need to understand the factors that determine orbital interaction.

The strength of the interaction between two atomic orbitals to form molecular orbitals is determined by two factors: (1) the energy difference between the interacting orbitals, and (2) the magnitude of their overlap. *For the interaction to be strong, the energies of the two orbitals must be approximately equal and the overlap must be large.*

From this last statement, you see that when two Li atoms approach one another to form $Li_2$, only like orbitals on the two atoms interact appreciably. The $2s$ orbital of one lithium atom interacts with the $2s$ orbital of the other atom, but the $2s$ orbital from one atom does not interact with the $1s$ orbital of the other atom, because their energies are quite different. Also, because the $2s$ orbitals are outer orbitals, they are able to overlap and interact strongly when the atoms approach. As in $H_2$, these atomic orbitals interact to give a bonding orbital (denoted $\sigma_{2s}$) and an antibonding orbital (denoted $\sigma_{2s}^*$). However, even though the $1s$ orbitals of the two atoms have the same energy, they do not overlap appreciably and so interact weakly. (The difference in energy between $\sigma_{1s}$ and $\sigma_{1s}^*$ is very small.) Figure 10.33 gives the relative energies of the orbitals.

You obtain the ground-state configuration of $Li_2$ by putting six electrons (three from each atom) into the molecular orbitals of lowest energy. The configuration of the diatomic molecule $Li_2$ is

$$Li_2 \qquad (\sigma_{1s})^2(\sigma_{1s}^*)^2(\sigma_{2s})^2$$

> ► However, small shifts in the energy of the core electrons due to bonding can be measured by x-ray photoelectron spectroscopy (page 323).

The $(\sigma_{1s})^2(\sigma_{1s}^*)^2$ part of the configuration is often abbreviated KK (which denotes the K shells, or inner shells, of the two atoms). These electrons do not have a significant effect on bonding. ◄

$$Li_2 \qquad KK(\sigma_{2s})^2$$

In calculating bond order, we can ignore KK (it includes two bonding and two antibonding electrons). Thus, the bond order is 1, as in $H_2$.

In $Be_2$, the energy diagram is similar to that in Figure 10.33. You have eight electrons to distribute (four from each Be atom). The ground-state configuration of $Be_2$ is

$$Be_2 \qquad KK(\sigma_{2s})^2(\sigma_{2s}^*)^2$$

**Figure 10.33**
**The energies of the molecular orbitals of $Li_2$.** Arrows indicate the occupation of these orbitals by electrons in the ground state.

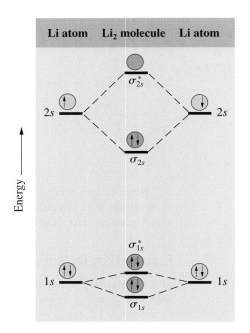

Note that the configuration has two bonding and two antibonding electrons outside the K shells. Thus, the bond order is $\frac{1}{2}(2 - 2) = 0$. No bond is formed, so the $Be_2$ molecule, like the $He_2$ molecule, is unstable.

## 10.6 Electron Configurations of Diatomic Molecules of the Second-Period Elements

The previous section looked at the electron configurations of some simple molecules: $H_2$, $He_2$, $Li_2$, and $Be_2$. These are **homonuclear diatomic molecules**—that is, *molecules composed of two like nuclei.* (**Heteronuclear diatomic molecules** are *molecules composed of two different nuclei*—for example, CO and NO.) To find the electron configurations of other homonuclear diatomic molecules, we need to have additional molecular orbitals.

We have already looked at the formation of molecular orbitals from *s* atomic orbitals. Now we need to consider the formation of molecular orbitals from *p* atomic orbitals. There are two different ways in which $2p$ atomic orbitals can interact. One set of $2p$ orbitals can overlap along their axes to give one bonding and one antibonding $\sigma$ orbital ($\sigma_{2p}$ and $\sigma_{2p}^*$). The other two sets of $2p$ orbitals then overlap sideways to give two bonding and two antibonding $\pi$ orbitals ($\pi_{2p}$ and $\pi_{2p}^*$). See Figure 10.34.

Figure 10.35 shows the relative energies of the molecular orbitals obtained from $2s$ and $2p$ atomic orbitals. This order of molecular orbitals reproduces the known electron configurations of homonuclear diatomic molecules composed of elements in the second row of the periodic table. ◄ The order of filling is

▶ **This order gives the correct number of electrons in subshells, even though in $O_2$ and $F_2$ the energy of $\sigma_{2p}$ is below $\pi_{2p}$.**

$$\sigma_{2s} \quad \sigma_{2s}^* \quad \pi_{2p} \quad \sigma_{2p} \quad \pi_{2p}^* \quad \sigma_{2p}^*$$

Note that there are two orbitals in the $\pi$ subshell and two orbitals in the $\pi^*$ subshell. Because each orbital can hold two electrons, a $\pi$ or $\pi^*$ subshell can hold four electrons. The next example shows how you can use the order of filling to obtain the

**Figure 10.34**
**The different ways in which 2p orbitals can interact.** When the $2p$ orbitals overlap along their axes, they form $\sigma_{2p}$ and $\sigma_{2p}^*$ molecular orbitals. When they overlap sideways, they form $\pi_{2p}$ and $\pi_{2p}^*$ molecular orbitals.

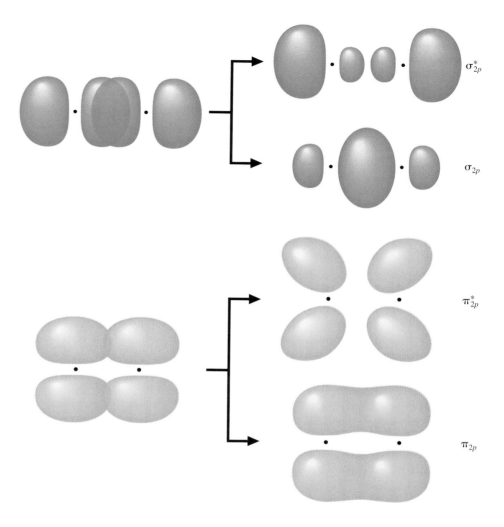

orbital diagram, magnetic character, electron configuration, and bond order of a homonuclear diatomic molecule.

► An alternate example is provided in the Instructor's Resource Manual.

---

**EXAMPLE 10.7**

**Describing Molecular Orbital Configurations (Homonuclear Diatomic Molecules)**

Give the orbital diagram of the $O_2$ molecule. Is the molecular substance diamagnetic or paramagnetic? What is the electron configuration? What is the bond order of $O_2$?

**SOLUTION**

There are 12 valence electrons in $O_2$ (six from each atom), which occupy the molecular orbitals as shown in the following orbital diagram:

$$\mathbf{KK} \; (\uparrow\downarrow) \; (\uparrow\downarrow) \; \underbrace{(\uparrow\downarrow)(\uparrow\downarrow)} \; (\uparrow\downarrow) \; \underbrace{(\uparrow)(\uparrow)} \; (\bigcirc)$$
$$\quad\;\; \sigma_{2s} \quad \sigma_{2s}^* \quad\;\; \pi_{2p} \quad\;\; \sigma_{2p} \quad\; \pi_{2p}^* \quad \sigma_{2p}^*$$

**Figure 10.35**
**Relative energies of molecular orbitals of homonuclear diatomic molecules (excluding K shells).** The arrows show the occupation of molecular orbitals by the valence electrons of $N_2$.

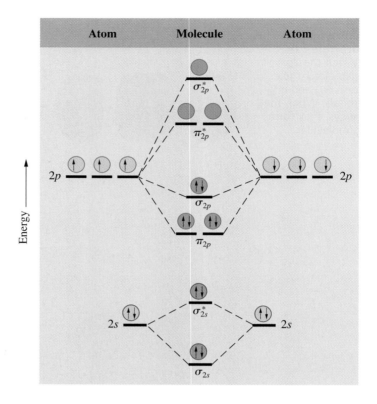

Note that the two electrons in the $\pi_{2p}^*$ subshell must go into different orbitals with their spins in the same direction (Hund's rule). Because there are two unpaired electrons, the molecular substance is **paramagnetic.** The electron configuration is

$$KK(\sigma_{2s})^2(\sigma_{2s}^*)^2(\pi_{2p})^4(\sigma_{2p})^2(\pi_{2p}^*)^2$$

There are eight bonding electrons and four antibonding electrons. Therefore,

$$\text{Bond order} = \tfrac{1}{2}(8 - 4) = \mathbf{2}$$

**EXERCISE 10.9**

The $C_2$ molecule exists in the vapor phase over carbon at high temperature. Describe the molecular orbital structure of this molecule; that is, give the orbital diagram and electron configuration. Would you expect the molecular substance to be diamagnetic or paramagnetic? What is the bond order for $C_2$?

See Problems 10.51 and 10.52.

**Problem Solving:**
**Bond Order**

Table 10.3 compares experimentally determined bond lengths, bond dissociation energies, and magnetic character of the second-period homonuclear diatomic molecules with the bond order calculated from molecular orbital theory, as in the preceding example. Note that as the bond order increases, bond length tends to decrease and bond dissociation energy tends to increase. You should be able to verify that the experimentally determined magnetic character of the molecule, given in the last column of the table, is correctly predicted by molecular orbital theory.

**Table 10.3**
**Theoretical Bond Orders and Experimental Data for the Second-Period Homonuclear Diatomic Molecules**

| Molecule | Bond Order | Bond Length (pm) | Bond Dissociation Energy (kJ/mol) | Magnetic Character |
|---|---|---|---|---|
| $Li_2$ | 1 | 267 | 110 | Diamagnetic |
| $Be_2$ | 0 | * | * | * |
| $B_2$ | 1 | 159 | 290 | Paramagnetic |
| $C_2$ | 2 | 124 | 602 | Diamagnetic |
| $N_2$ | 3 | 110 | 942 | Diamagnetic |
| $O_2$ | 2 | 121 | 494 | Paramagnetic |
| $F_2$ | 1 | 142 | 155 | Diamagnetic |
| $Ne_2$ | 0 | * | * | * |

The symbol * means that no stable molecule has been observed.

When the atoms in a heteronuclear diatomic molecule are close to one another in a row of the periodic table, the molecular orbitals have the same relative order of energies as those for homonuclear diatomic molecules. In this case you can obtain the electron configurations in the same way, as the next example illustrates.

> ► An alternate example is provided in the Instructor's Resource Manual.

**EXAMPLE 10.8**

## Describing Molecular Orbital Configurations (Heteronuclear Diatomic Molecules)

Write the orbital diagram for nitric oxide, NO. What is the bond order of NO?

**SOLUTION**

You assume that the order of filling of orbitals is the same as for homonuclear diatomic molecules. There are 11 valence electrons in NO. Thus, the orbital diagram is

$$KK \; (\uparrow\downarrow) \; (\uparrow\downarrow) \; \underbrace{(\uparrow\downarrow)(\uparrow\downarrow)} \; (\uparrow\downarrow) \; \underbrace{(\uparrow)(\;)} \; (\;)$$
$$\quad\quad \sigma_{2s} \;\; \sigma_{2s}^* \;\;\; \pi_{2p} \quad\; \sigma_{2p} \;\;\; \pi_{2p}^* \;\; \sigma_{2p}^*$$

Because there are eight bonding and three antibonding electrons,

$$\text{Bond order} = \tfrac{1}{2}(8 - 3) = \tfrac{5}{2}$$

**EXERCISE 10.10**

Give the orbital diagram and electron configuration for the carbon monoxide molecule, CO. What is the bond order of CO? Is the molecule diamagnetic or paramagnetic?

See Problems 10.53 and 10.54.

When the two atoms in a heteronuclear diatomic molecule differ appreciably, you can no longer use the scheme appropriate for homonuclear diatomic molecules. The

**Figure 10.36**
**Relative energies of molecular orbitals in the valence shell of HF.**
Arrows show the occupation of orbitals by electrons.

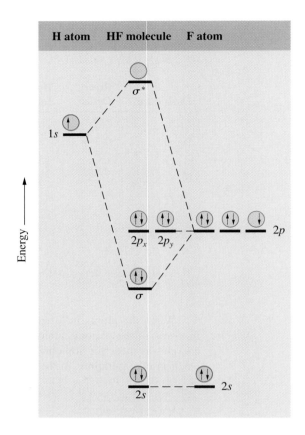

HF molecule is an example. Figure 10.36 shows the relative energies of the molecular orbitals that form. Sigma bonding and antibonding orbitals are formed by combining the $1s$ orbital on the H atom with the $2p$ orbital on the F atom that lies along the bond axis. The $2s$ orbital and the other $2p$ orbitals on fluorine remain as *nonbonding* orbitals (neither bonding nor antibonding). Note that the energy of the $2p$ subshell in fluorine is lower than the energy of the $1s$ orbital in hydrogen, because the fluorine electrons are more tightly held. As a result, the bonding orbital is made up of a greater percentage of the $2p$ fluorine orbital than of the $1s$ hydrogen orbital. This means that electrons in the bonding molecular orbital spend more time in the vicinity of the fluorine atom; that is, the H—F bond is polar, with the F atom having a small negative charge.

## 10.7 Molecular Orbitals and Delocalized Bonding

One of the advantages of using molecular orbital theory is the simple way in which it describes molecules with delocalized bonding. Whereas valence bond theory requires two or more resonance formulas, molecular orbital theory describes the bonding in terms of a single electron configuration.

Consider the ozone molecule as an example. In the previous chapter, we described this molecule in terms of resonance as follows:

$O_3$

**Figure 10.37**
**Hybrid orbitals on oxygen atoms of ozone, $O_3$.**   Each oxygen atom has $sp^2$ hybrid orbitals. One hybrid orbital on the center atom and one hybrid orbital on the left atom overlap to form a single bond at the left. Similar orbitals at the right form another single bond. Each atom has a $\pi$ orbital perpendicular to the plane of these hybrid orbitals.

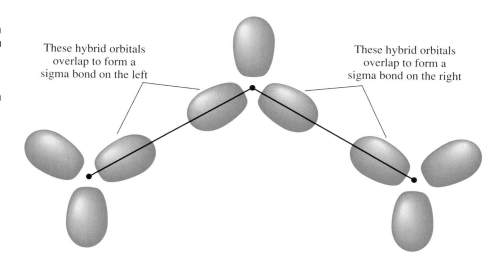

These hybrid orbitals overlap to form a sigma bond on the left

These hybrid orbitals overlap to form a sigma bond on the right

**Visualization: pi Bond**

Some of these electrons are localized either on one atom or between an oxygen–oxygen bond (these are printed in black). The other electrons are delocalized (these are printed in blue). They are the electron pairs that change position in going from one resonance formula to the other.

Each O atom has three localized electron pairs about it, which suggests that each atom uses $sp^2$ hybrid orbitals (see Table 10.2). The overlap of one hybrid orbital on the center O atom with a hybrid orbital on the end O atom at the left forms the left O—O bond (Figure 10.37). The overlap of another hybrid orbital on the center O atom with a hybrid orbital on the end O atom at the right forms the other O—O bond. This leaves one hybrid orbital on the center O atom and two on each of the end O atoms for lone pairs.

After the hybrid orbitals are formed, one unhybridized $p$ orbital remains on each O atom. These three $p$ orbitals are perpendicular to the plane of the molecule and parallel to one another. They can overlap sideways to give three $\pi$ molecular orbitals. (Three interacting atomic orbitals produce three molecular orbitals.) All three will span the entire molecule (Figure 10.38). One of these orbitals turns out to be antibonding; one is nonbonding (that is, it has an energy equal to that of the isolated atoms); and the third is bonding.

The bonding and nonbonding $\pi$ orbitals are both doubly occupied. This agrees with the resonance description, in which one delocalized pair is bonding and the other is a lone pair on the end oxygen atoms.

**Figure 10.38**
**The electron distributions of the $\pi$ molecular orbitals of $O_3$.**   Note that one molecular orbital is nonbonding. It describes a lone pair of electrons that can be on either end-oxygen atom.

Antibonding $\pi$ orbital

Nonbonding $\pi$ orbital

Bonding $\pi$ orbital

# Human Vision

**LIFE SCIENCE**

Human vision, as well as the detection of light by other organisms, is known to involve a compound called 11-*cis*-retinal. The structural formula is shown in Figure 10.39, at the left. This compound, which is derived from vitamin A, is contained in two kinds of cells—rod cells and cone cells—located in the retina of the eye. Rod cells are responsible for vision in dim light. They give the sensation of light or dark but no sensation of color. Cone cells are responsible for color vision, but they require bright light. Color vision is possible because three types of cone cells exist: one type absorbs light in the red region of the spectrum, another absorbs in the green region, and a third absorbs in the blue region. In each of these different cells, 11-*cis*-retinal is attached to a different protein molecule, which affects the region of light that is absorbed by retinal.

When a photon of light is absorbed by 11-*cis*-retinal, it is transformed to 11-*trans*-retinal (see Figure 10.39). This small change at one carbon–carbon double bond results in a large movement of one end of the molecule with respect to the other. In other words, the absorption of a single photon results in a significant change in molecular geometry. This change in shape of retinal also affects the shape of the protein molecule to which it is attached, and *this* change results in a sequence of events, not completely understood, in which an electrical signal is generated.

To understand the *cis–trans* conversion when a photon is absorbed, consider the similar, but simpler, molecule *cis*-1,2-dichloroethene (see structural formula on page 395). The double bond consists of a $\sigma$ bond and a $\pi$ bond. The $\pi$ bond results from two electrons in a $\pi$-type molecular orbital. A corresponding antibonding orbital, $\pi^*$, has higher energy than the $\pi$ orbital and is unoccupied in the ground state. When a photon of light in the ultraviolet region (near 180 nm) is absorbed by *cis*-1,2-dichloroethene, an electron in the $\pi$ orbital undergoes a transition to the $\pi^*$ orbital. Whereas there were previously two electrons in the bonding $\pi$ orbital (contributing one bond), there is now one electron in the bonding orbital and one in the antibonding orbital (with zero net bonding). As a result of the absorption of a photon, the double bond becomes a single bond and rotation about the bond is now possible. The *cis* isomer tends to rotate to the *trans* isomer, which is more stable. Then energy is lost as heat, and the electron in the $\pi^*$ orbital undergoes a transition back to the $\pi$ orbital; the single bond becomes a double bond again. The net result is the conversion of the *cis* isomer to the *trans* isomer.

Retinal differs from dichloroethene in having six double bonds alternating with single bonds instead of only one double bond. The alternating double and single bonds give a rigid structure, and when the middle double bond undergoes the *cis*-to-*trans* change, a large movement of atoms occurs. Another result of a long chain of alternating double and single bonds is a change in the wavelength of light absorbed in the $\pi$-to-$\pi^*$ transition. Dichloroethene absorbs in the ultraviolet region, but retinal attached to its protein absorbs in the visible region. (The attached protein also alters the wavelength of absorbed light.) Many biological compounds that are colored consist of long chains of alternating double and single bonds. $\beta$-Carotene, a yellow pigment in carrots, has this structure. It is broken down in the body to give vitamin A, from which retinal is derived.

**Figure 10.39**

**The *cis–trans* conversion of retinal on absorption of light.** Note the large movement of atoms caused by the rotation about the double bond between carbon atoms 11 and 12.

# Stratospheric Ozone (An Absorber of Ultraviolet Rays)

**ENVIRONMENT**

Ozone (also known as trioxygen, $O_3$) is a reactive form of oxygen having a bent molecular geometry, as you would predict from its resonance formulas:

Ozone forms in the atmosphere whenever $O_2$ is irradiated by ultraviolet light or subjected to electrical discharges. On bright sunny days, you might notice the fresh-air smell of dilute ozone (trioxygen, $O_3$). Around electrical equipment, however, ozone often occurs in higher concentrations and has a disagreeable odor.

Although ozone occurs at ground level and is an ingredient of *photochemical smog,* it is an essential component of the *stratosphere,* which occurs at about 10 to 15 km above ground level. Ozone in the stratosphere absorbs ultraviolet radiation between 200 and 300 nm, which is vitally important to life on earth. Radiation from the sun contains ultraviolet rays that are harmful to the DNA of biological organisms. Oxygen, $O_2$, absorbs the most energetic of these ultraviolet rays in the earth's upper atmosphere, but only ozone in the stratosphere absorbs the remaining ultraviolet radiation that is destructive to life on earth (Figure 10.40).

In 1974, Mario J. Molina and F. Sherwood Rowland expressed concern that chlorofluorocarbons (CFCs), such as $CClF_3$ and $CCl_2F_2$, would be a source of chlorine atoms that could catalyze the decomposition of ozone in the stratosphere, so that the ozone would be destroyed faster than it could be produced. The chlorofluorocarbons are relatively inert compounds used as refrigerants, spray-can propellants, and blowing agents (substances used to produce plastic foams). As a result of their inertness, however, CFCs concentrate in the atmosphere, where they steadily rise into the stratosphere. Once in the stratosphere, ultraviolet light decomposes them to form chlorine atoms, which react with ozone to form ClO and $O_2$. The ClO molecules react with oxygen atoms in the stratosphere to regenerate Cl atoms.

$$Cl(g) + O_3(g) \longrightarrow ClO(g) + O_2(g)$$
$$\underline{ClO(g) + O(g) \longrightarrow Cl(g) + O_2(g)}$$
$$O_3(g) + O(g) \longrightarrow 2O_2(g)$$

The net result is the decomposition of ozone to dioxygen. Chlorine atoms are consumed in the first step but regenerated in the second. Thus, chlorine atoms are not used up, so they function as a catalyst.

In 1985, British researchers reported that the ozone over the Antarctic had been declining precipitously each

*(continued)*

**Figure 10.40**
**The solar spectrum.**
Radiation below 200 nm is absorbed in the upper atmosphere by $O_2$. Ultraviolet radiation between 200 and 300 nm is absorbed by $O_3$ in the stratosphere.

spring for several years, although it returned the following winter (Figure 10.41). Then in August 1987, an expedition was assembled to begin flights through the "ozone hole" with instruments. Researchers found that wherever ozone is depleted in the stratosphere, chlorine monoxide (ClO) appears, as expected if chlorine atoms are catalyzing the depletion (see the previous reactions). The chlorofluorocarbons are the suspected source of these chlorine atoms. Many nations, including the United States, have pledged to phase out the usage of CFCs, and manufacturers are hurrying to substitute safer compounds. Auto manufacturers, for instance, now use hydrofluorocarbons (HFCs), such as $CF_3CH_2F$, which do not contain Cl atoms, as refrigerants in air conditioners.

In October 1995, Sherwood Rowland (University of California, Irvine), Mario Molina (Massachusetts Institute of Technology), and Paul Crutzen (Max Planck Institute, Mainz, Germany) were awarded the Nobel Prize in chemistry for their work on stratospheric ozone depletion. Crutzen studied the effect of nitrogen oxides in catalyzing the decomposition of ozone.

**Figure 10.41**
**Computer map of the South Polar region showing total stratospheric ozone.** Data were obtained by the Total Ozone Mapping Spectrometer (TOMS) aboard the *Nimbus*-7 satellite in September 2000. Total ozone at any point is given in Dobson units, which equal the thickness of total ozone at 1 atm, 0°C in $10^{-2}$ mm. The ozone hole is shown in dark blue, in the region near the South Pole (Antarctica).

## A Checklist for Review

### Important Terms

molecular geometry (p. 396)
valence-shell electron-pair repulsion (VSEPR) model (10.1)
dipole moment (10.2)

valence bond theory (10.3)
hybrid orbitals (10.3)
$\sigma$ (sigma) bond (10.4)
$\pi$ (pi) bond (10.4)

molecular orbital theory (p. 423)
bonding orbitals (10.5)
antibonding orbitals (10.5)

homonuclear diatomic molecules (10.6)
heteronuclear diatomic molecules (10.6)

### Summary of Facts and Concepts

*Molecular geometry* refers to the spatial arrangement of atoms in a molecule. The *valence-shell electron-pair repulsion (VSEPR) model* is a simple model for predicting molecular geometries. It is based on the idea that the valence-shell electron pairs are arranged symmetrically about an atom to minimize electron-pair repulsion. The geometry about an atom is then determined by the directions of the bonding pairs. Information about the geometry of a molecule can sometimes be obtained from the experimentally determined presence or absence of a *dipole moment*.

The bonding and geometry in a molecule can be described in terms of *valence bond theory*. In this theory, a bond is formed by the overlap of orbitals from two atoms. *Hybrid orbitals*, a set of equivalent orbitals formed by combining atomic orbitals, are often needed to describe this bond. Multiple bonds

occur via the overlap of atomic orbitals to give $\sigma$ *bonds* and $\pi$ *bonds*. *Cis–trans* isomers result from the molecular rigidity imposed by a $\pi$ bond.

*Molecular orbital theory* can also be used to explain bonding in molecules. According to this theory, electrons in a molecule occupy orbitals that may spread over the entire molecule. You can think of these molecular orbitals as constructed from atomic orbitals. Thus, when two atoms approach to form a diatomic molecule, the atomic orbitals interact to form *bonding* and *antibonding* molecular orbitals. The electron configuration of a diatomic molecule such as $O_2$ can be predicted from the order of filling of the molecular orbitals. From this configuration, you can predict the bond order and whether the molecular substance is diamagnetic or paramagnetic.

## Operational Skills

1. **Predicting molecular geometries** Given the formula of a simple molecule, predict its geometry, using the VSEPR model. **(EXAMPLES 10.1, 10.2)**
2. **Relating dipole moment and molecular geometry** State what geometries of a molecule $AX_n$ are consistent with the information that the molecule has a nonzero dipole moment. **(EXAMPLE 10.3)**
3. **Applying valence bond theory** Given the formula of a simple molecule, describe its bonding, using valence bond theory. **(EXAMPLE 10.4, 10.5, 10.6)**
4. **Describing molecular orbital configurations** Given the formula of a diatomic molecule obtained from first- or second-period elements, deduce the molecular orbital configuration, the bond order, and whether the molecular substance is diamagnetic or paramagnetic. **(EXAMPLES 10.7, 10.8)**

## Review Questions

**10.1** Describe the main features of the VSEPR model.

**10.2** According to the VSEPR model, what are the arrangements of two, three, four, five, and six valence-shell electron pairs about an atom?

**10.3** Why is a lone pair expected to occupy an equatorial position instead of an axial position in the trigonal bipyramidal arrangement?

**10.4** Why is it possible for a molecule to have polar bonds, yet have a dipole moment of zero?

**10.5** Explain why nitrogen trifluoride has a small dipole moment even though it has polar bonds in a trigonal pyramidal arrangement.

**10.6** Explain in terms of valence bond theory why the orbitals used on an atom give rise to a particular geometry about that atom.

**10.7** What is the angle between two $sp^3$ hybrid orbitals?

**10.8** What is the difference between a sigma bond and a pi bond?

**10.9** Describe the bonding in ethylene, $C_2H_4$, in terms of valence bond theory.

**10.10** How does the valence bond description of a carbon–carbon double bond account for *cis–trans* isomers?

**10.11** What are the differences between a bonding and an antibonding molecular orbital of a diatomic molecule?

**10.12** What factors determine the strength of interaction between two atomic orbitals to form a molecular orbital?

**10.13** Describe the formation of bonding and antibonding molecular orbitals resulting from the interaction of two $2s$ orbitals.

**10.14** Describe the formation of molecular orbitals resulting from the interaction of two $2p$ orbitals.

**10.15** How does molecular orbital theory describe the bonding in the HF molecule?

**10.16** Describe the bonding in $O_3$, using molecular orbital theory. Compare this with the resonance description.

## Conceptual Problems

**10.17** Match the following molecular substances with one of the molecular models (i) to (iv) that correctly depicts the geometry of the corresponding molecule.
a. $SeO_2$    b. $BeCl_2$    c. $PBr_3$    d. $BCl_3$
(i)     (ii)

(iii)     (iv)

**10.18** Suppose that an ethane molecule, $CH_3CH_3$, is broken into two $CH_3$ molecules in such a way that one :$CH_3$ molecule retains the electron pair that was originally the one making up the C—C bond. The other $CH_3$ molecule has two fewer electrons. Imagine that momentarily this $CH_3$ molecule has the geometry it had in the ethane molecule. Describe the electron repulsions present in this molecule and how they would be expected to rearrange its geometry.

**10.19** What are the bond angles predicted by the VSEPR model about the carbon atom in the formate ion, $HCO_2{}^-$? Considering that the bonds to this atom are not identical, would you expect the experimental values to agree precisely with the VSEPR values? How might they differ?

**10.20** An atom in a molecule has two bonds to other atoms and one lone pair. What kind of hybrid orbitals do you expect for this atom? Describe how you arrived at your answer.

**10.21** Two compounds have the same molecular formula, $C_2H_2Br_2$. One has a dipole moment; the other does not. Both compounds react with bromine, $Br_2$, to produce the same compound. This reaction is a generally accepted test for double pound.

bonds, and each bromine atom of $Br_2$ adds to a different atom of the double bond. What is the identity of the original compounds? Describe the argument you use.

**10.22** A neutral molecule is identified as a tetrafluoride, $XF_4$, where X is an unknown atom. If the molecule has a dipole moment of 0.63 D, can you give some possibilities for the identity of X?

## Practice Problems

### The VSEPR Model

**10.23** Predict the shape or geometry of the following molecules, using the VSEPR model.
a. $SCl_2$    b. $CF_4$    c. $COCl_2$    d. $NF_3$

**10.24** Use the electron-pair repulsion model to predict the geometry of the following molecules:
a. $GeCl_2$    b. $AsCl_3$    c. $SO_3$    d. $XeO_4$

**10.25** Predict the geometry of the following ions, using the electron-pair repulsion model.
a. $NO_3^-$    b. $ClO_4^-$    c. $SCN^-$    d. $SO_4^{2-}$

**10.26** Use the VSEPR model to predict the geometry of the following ions:
a. $N_3^-$    b. $BH_4^-$    c. $SO_3^{2-}$    d. $NO_2^-$

**10.27** For each of the following molecules, state the bond angle (or bond angles, as appropriate) that you would expect to see on the central atom based on the simple VSEPR model. Would you expect the actual bond angles to be greater or less than this?
a. $H_2Se$    b. $CF_4$    c. $PH_3$    d. $COF_2$

**10.28** For each of the following molecules, state the bond angle (or bond angles, as appropriate) that you would expect to see on the central atom based on the simple VSEPR model. Would you expect the actual bond angles to be greater or less than this?
a. $NCl_3$    b. $CF_2CF_2$    c. $SiF_4$    d. $OF_2$

**10.29** What geometry is expected for the following molecules, according to the VSEPR model?
a. $PF_5$    b. $BrF_3$    c. $BrF_5$    d. $SCl_4$

**10.30** From the electron-pair repulsion model, predict the geometry of the following molecules:
a. $ClF_5$    b. $SbF_5$    c. $SeF_4$    d. $TeF_6$

**10.31** Predict the geometries of the following ions, using the VSEPR model.
a. $SnCl_5^-$    b. $PF_6^-$    c. $ClF_2^-$    d. $IF_4^-$

**10.32** Name the geometries expected for the following ions, according to the electron-pair repulsion model.
a. $BrF_6^+$    b. $IF_2^-$    c. $ICl_4^-$    d. $IF_4^+$

### Dipole Moment and Molecular Geometry

**10.33** (a) The molecule $AsF_3$ has a dipole moment of 2.59 D. Which of the following geometries are possible: trigonal planar, trigonal pyramidal, or T-shaped? (b) The molecule $H_2S$ has a dipole moment of 0.97 D. Is the geometry linear or bent?

**10.34** (a) The molecule $BrF_3$ has a dipole moment of 1.19 D. Which of the following geometries are possible: trigonal planar, trigonal pyramidal, or T-shaped? (b) The molecule $TeCl_4$ has a dipole moment of 2.54 D. Is the geometry tetrahedral, seesaw, or square planar?

**10.35** Which of the following molecules would be expected to have zero dipole moment on the basis of their geometry?
a. $SeF_2$    b. $NCl_3$    c. $TeF_4$    d. $XeF_4$

**10.36** Which of the following molecules would be expected to have a dipole moment of zero because of symmetry?
a. $BeBr_2$    b. $H_2Se$    c. $AsF_3$    d. $SeF_6$

### Valence Bond Theory

**10.37** What hybrid orbitals would be expected for the central atom in each of the following molecules or ions?
(a)                                (b)

**10.38** What hybrid orbitals would be expected for the central atom in each of the following molecules or ions?
(a)                                (b)

**10.39** What hybrid orbitals would be expected for the central atom in each of the following molecules or ions?
a. $GeF_2$    b. $NO_2^-$

**10.40** What hybrid orbitals would be expected for the central atom in each of the following molecules or ions?
a. $PBr_3$    b. $GeCl_4$

**10.41** (a) Mercury(II) chloride dissolves in water to give poorly conducting solutions, indicating that the compound is largely nonionized in solution—it dissolves as $HgCl_2$ molecules. Describe the bonding of the $HgCl_2$ molecule, using valence bond theory. (b) Phosphorus trichloride, $PCl_3$, is a colorless liquid with a highly irritating vapor. Describe the bonding in the $PCl_3$ molecule, using valence bond theory. Use hybrid orbitals.

**10.42** (a) Nitrogen trifluoride, $NF_3$, is a relatively unreactive, colorless gas. How would you describe the bonding in the $NF_3$ molecule in terms of valence bond theory? Use hybrid orbitals. (b) Silicon tetrafluoride, $SiF_4$, is a colorless gas formed when hydrofluoric acid attacks silica ($SiO_2$) or glass. Describe the bonding in the $SiF_4$ molecule, using valence bond theory.

**10.43** What hybrid orbitals would be expected for the central atom in each of the following?
a. $BrF_5$    b. $BrF_3$    c. $AsCl_5$    d. $ClF_4^+$

**10.44** What hybrid orbitals would be expected for the central atom in each of the following?
a. $PCl_5$    b. $SeF_6$    c $TeF_4$    d. $IF_4^-$

**10.45** Phosphorus pentachloride is normally a white solid. It exists in this state as the ionic compound $[PCl_4^+][PCl_6^-]$. Describe the electron structure of the $PCl_6^-$ ion in terms of valence bond theory.

**10.46** Iodine, $I_2$, dissolves in an aqueous solution of iodide ion, $I^-$, to give the triiodide ion, $I_3^-$. This ion consists of one iodine atom bonded to two others. Describe the bonding of $I_3^-$ in terms of valence bond theory.

**10.47** (a) Formaldehyde, $H_2CO$, is a colorless, pungent gas used to make plastics. Give the valence bond description of the formaldehyde molecule. (Both hydrogen atoms are attached to the carbon atom.) (b) Nitrogen, $N_2$, makes up about 80% of the earth's atmosphere. Give the valence bond description of this molecule.

**10.48** (a) The molecule $HN\!=\!NH$ exists as a transient species in certain reactions. Give the valence bond description of this species. (b) Hydrogen cyanide, HCN, is a very poisonous gas or liquid with the odor of bitter almonds. Give the valence bond description of HCN. (Carbon is the central atom.)

**10.49** The hyponitrite ion, $^-O\!-\!N\!=\!N\!-\!O^-$, exists in solid compounds as the *trans* isomer. Using valence bond theory, explain why *cis–trans* isomers might be expected for this ion. Draw structural formulas of the *cis–trans* isomers.

**10.50** Fumaric acid, $C_4H_4O_4$, occurs in the metabolism of glucose in the cells of plants and animals. It is used commercially in beverages. The structural formula of fumaric acid is

Maleic acid is the *cis* isomer of fumaric acid. Using valence bond theory, explain why these isomers are possible.

## Molecular Orbital Theory

**10.51** Describe the electronic structure of each of the following, using molecular orbital theory. Calculate the bond order of each and decide whether it should be stable. For each, state whether the substance is diamagnetic or paramagnetic.
a. $B_2$    b. $B_2^+$    c. $O_2^-$

**10.52** Use molecular orbital theory to describe the bonding in the following. For each one, find the bond order and decide whether it is stable. Is the substance diamagnetic or paramagnetic?
a. $C_2^+$    b. $Ne_2$    c. $C_2^-$

**10.53** Assume that the cyanide ion, $CN^-$, has molecular orbitals similar to those of a homonuclear diatomic molecule. Write the configuration and bond order of $CN^-$. Is a substance of the ion diamagnetic or paramagnetic?

**10.54** Write the molecular orbital configuration of the diatomic molecule BN. What is the bond order of BN? Is the substance diamagnetic or paramagnetic? Use the order of energies that was given for homonuclear diatomic molecules.

## General Problems

**10.55** Predict the molecular geometry of the following:
a. $SnCl_2$    b. $COF_2$    c. $ICl_2^-$    d. $PCl_6^-$

**10.56** Predict the molecular geometry of the following:
a. HOF    b. $NH_4^+$    c. $PF_5$    d. $ClF_4^+$

**10.57** Which of the following molecules or ions are linear?
a. $SeF_2$    b. $CHCl_3$    c. $SeF_4$    d. $SnF_6^{2-}$

**10.58** Which of the following molecules or ions are trigonal planar?
a. $BBr_3$    b. $AsH_3$    c. $ClF_4^-$    d. $TeF_5^-$

**10.59** Describe the hybrid orbitals used by each carbon atom in the following molecules:

a.
$$\begin{array}{c} H \\ \diagdown \\ C=C-C-OH \\ \diagup \quad | \quad | \\ H \quad H \quad H \end{array}$$
with H's on the first carbon and H, H on the middle carbon

b. $N\equiv C-C\equiv N$

**10.60** Describe the hybrid orbitals used by each nitrogen atom in the following molecules:

a. $H-N-N-H$ (with H, H on top)

b. $N\equiv C-C\equiv N$

**10.61** Explain how the dipole moment could be used to distinguish between the *cis* and *trans* isomers of 1,2-dibromoethene:

Cis  Trans

**10.62** Two compounds have the formula $Pt(NH_3)_2Cl_2$. (Compound B is *cisplatin,* mentioned in the opening to Chapter 1.) They have square planar structures. One is expected to have a dipole moment; the other is not. Which one would have a dipole moment?

(a)  (b)

**10.63** Explain in terms of bonding theory why all four hydrogen atoms of allene, $H_2C=C=CH_2$, cannot lie in the same plane.

**10.64** Explain in terms of bonding theory why all atoms of $H_2C=C=C=CH_2$ must lie in the same plane.

**10.65** What is the molecular orbital configuration of $HeH^+$? Do you expect the ion to be stable?

**10.66** What is the molecular orbital configuration of $He_2^+$? Do you expect the ion to be stable?

**10.67** Calcium carbide, $CaC_2$, consists of $Ca^{2+}$ and $C_2^{2-}$ (acetylide) ions. Write the molecular orbital configuration and bond order of the acetylide ion, $C_2^{2-}$.

**10.68** Sodium peroxide, $Na_2O_2$, consists of $Na^+$ and $O_2^{2-}$ (peroxide) ions. Write the molecular orbital configuration and bond order of the peroxide ion, $O_2^{2-}$.

**10.69** The oxygen–oxygen bond in $O_2^+$ is 112 pm and in $O_2$ is 121 pm. Explain why the bond length in $O_2^+$ is shorter than in $O_2$. Would you expect the bond length in $O_2^-$ to be longer or shorter than that in $O_2$? Why?

**10.70** The nitrogen–nitrogen bond distance in $N_2$ is 109 pm. On the basis of bond orders, would you expect the bond distance in $N_2^+$ to be less than or greater than 109 pm? Answer the same question for $N_2^-$.

**10.71** Using molecular orbital theory, determine the electronic structure of the first excited electronic state of $N_2$. What differences are expected in the properties of the excited state of $N_2$ compared with the same properties of the ground state?

**10.72** The ionization energy of $O_2$ is smaller than the ionization energy of atomic O; the opposite is true for the ionization energies of $N_2$ and atomic N. Explain this behavior in terms of the molecular orbital energy diagrams of $O_2$ and $N_2$.

## Cumulative-Skills Problems

**10.73** A molecular compound is composed of 60.4% Xe, 22.1% O, and 17.5% F, by mass. If the molecular weight is 217.3 amu, what is the molecular formula? What is the Lewis formula? Predict the molecular geometry using the VSEPR model. Describe the bonding, using valence bond theory.

**10.74** A molecular compound is composed of 58.8% Xe, 7.2% O, and 34.0% F, by mass. If the molecular weight is 223 amu, what is the molecular formula? What is the Lewis formula? Predict the molecular geometry using the VSEPR model. Describe the bonding, using valence bond theory.

**10.75** A compound of chlorine and fluorine, $ClF_n$, reacts at about 75°C with uranium metal to produce uranium hexafluoride, $UF_6$, and chlorine monofluoride, $ClF(g)$. A quantity of uranium produced 3.53 g $UF_6$ and 343 mL ClF at 75°C and 2.50 atm. What is the formula (*n*) of the compound? Describe the bonding in the molecule, using valence bond theory.

**10.76** Excess fluorine, $F_2(g)$, reacts at 150°C with bromine, $Br_2(g)$, to give a compound $BrF_n$. If 423 mL $Br_2(g)$ at 150°C and 748 mmHg produced 4.20 g $BrF_n$, what is *n*? Describe the bonding in the molecule, using valence bond theory.

**10.77** Each of the following compounds has a nitrogen–nitrogen bond: $N_2$, $N_2H_4$, $N_2F_2$. Match each compound with one of the following bond lengths: 110 pm, 122 pm, 145 pm. Describe the geometry about one of the N atoms in each compound. What hybrid orbitals are needed to describe the bonding in valence bond theory?

**10.78** The bond length in $C_2$ is 131 pm. Compare this with the bond lengths in $C_2H_2$ (120 pm), $C_2H_4$ (134 pm), and $C_2H_6$ (153 pm). What bond order would you predict for $C_2$ from its bond length? Does this agree with the molecular orbital configuration you would predict for $C_2$?

**10.79** Draw resonance formulas of the nitric acid molecule, $HNO_3$. What is the geometry about the N atom? What is the hybridization on N? Use bond energies and one Lewis formula for $HNO_3$ to estimate $\Delta H_f^\circ$ for $HNO_3(g)$. The actual value of $\Delta H_f^\circ$ for $HNO_3(g)$ is $-135$ kJ/mol, which is lower than the estimated value because of stabilization of $HNO_3$ by resonance.

The *resonance energy* is defined as $\Delta H_f^\circ$(estimated) $- \Delta H_f^\circ$ (actual). What is the resonance energy of $HNO_3$?

**10.80** One resonance formula of benzene, $C_6H_6$, is

What is the other resonance formula? What is the geometry about a carbon atom? What hybridization would be used in valence bond theory to describe the bonding? The $\Delta H_f^\circ$ for $C_6H_6(g)$ is $-83$ kJ/mol; $\Delta H_f^\circ$ for $C(g)$ is 715 kJ/mol. Obtain the resonance energy of benzene. (See Problem 10.79.)

# Media Activities

**10.81** View the Two Electron Pair, Three Electron Pair, and Four Electron Pair *Visualizations* on the student CD.
a. What is the geometry of four electron pairs arranged around an atom?
b. What is the geometry of six electron pairs arranged around an atom?
c. According to the VSEPR model, which geometry results in unequivalent electron pair arrangements? How many electron pairs are in this arrangement?

**10.82** Work through the VSEPR Theory of Molecular Structure *Understanding Concepts* activity on the student CD. Be sure to try several of the exercises presented at the end of this activity. Determine the molecular geometry of $SeO_3$, $IBr_3$, $ICl_2^-$, and $PH_2^+$.

**10.83** Run the $sp^3$, $sp^2$, and $sp$ hybridization *Visualizations* on the student CD. Which hybridization is used to describe the following arrangements of electron pairs around an atom: tetrahedral, trigonal planar, and linear?

**10.84** Open the Diatomic Molecular Orbital Diagrams *Understanding Concepts* simulation on the student web site and answer the following questions:
a. When two atomic orbitals combine, how many molecular orbitals are formed?
b. How many electrons would you place in the molecular orbital diagram for a $B_2$ molecule? Move the slider to check your answer.
c. Move the slider to build a molecular orbital diagram for $Be_2$. Would you expect this to be a stable molecule?
d. Move the slider to build a molecular orbital diagram for $B_2$. What is the bond order of a $B_2$ molecule?

**10.85** Run the Formation of C=C Double Bond in Ethylene *Visualization* on the student CD.
a. Write the Lewis structure of acetylene, $H-C\equiv C-H$.
b. Compare the bonding in acetylene to the bonding in ethylene. What kind of bond is the additional bond in acetylene?
c. How many sigma bonds are in acetylene?

**10.86** For review, take the ACE practice test for Chapter 10 on the student web site.

Microscopic view of nematic liquid crystal film under polarized light.

# States of Matter; Liquids and Solids

Solid carbon dioxide is shown in Figure 11.1. It has an interesting property: at normal pressures it passes directly to the gaseous state without first melting to the liquid. This property, together with the fact that this change occurs at $-78°C$, makes solid carbon dioxide useful for keeping materials very cold. Because solid $CO_2$ cools other objects and does not leave a liquid residue, it is called *dry ice*.

Liquid carbon dioxide is obtained by putting carbon dioxide gas under pressure; it can only exist at pressures greater than 5 atm. In most cases, carbon dioxide is transported as a liquid in high-pressure steel tanks or at low temperature (about $-18°C$) and moderate pressure. When liquid carbon dioxide evaporates (changes to vapor), it absorbs large quantities of heat, cooling as low as $-57°C$. Because of this property, it is used as a refrigerant. Liquid carbon dioxide is often used to freeze foods for grocery stores and fast-food restaurants. If the compressed gas from the evaporating liquid is allowed to expand through a valve, the rapidly cooled vapor forms

► This chapter looks at the states of matter, particularly liquids and solids. You may wish to immediately precede it with Chapter 5, on gases.

solid carbon dioxide "snow." (Most gases cool when they expand in this manner; for example, air cools when it escapes from the valve of an inflated tire.) This carbon dioxide snow is compacted into blocks and is the source of dry ice.

Carbon dioxide, like water and most other pure substances, exists in solid, liquid, and gaseous states and can undergo changes from one state to another. Substances change state under various temperature and pressure conditions. Can we obtain some understanding of these conditions for a change of state? Why, for example, is carbon dioxide normally a gas, whereas water is normally a liquid? What are the conditions under which carbon dioxide gas changes to a liquid or to a solid?

These are some of the questions we will examine in this chapter. They concern certain physical properties that, as you will see, can often be related to the bonding and structure of the liquid and solid states.

## 11.1 Comparison of Gases, Liquids, and Solids

**Figure 11.1**
**Dry ice.** Dry ice is solid carbon dioxide, $CO_2$, whose temperature is about $-78°C$. It passes directly from the solid to the gaseous state. The white plumes are composed of ice crystals ($H_2O$, not $CO_2$) formed by condensation from the cold air.

The different states of matter were defined in Section 1.4, and the gas laws and the kinetic theory of gases were defined in Chapter 5. Here we want to recall the salient features of those discussions and then compare gases, liquids, and solids. We especially want to compare how these states are viewed in terms of kinetic-molecular theory.

Gases are compressible fluids. According to kinetic-molecular theory, gases are composed of molecules or single atoms that are in constant random motion throughout mostly empty space (unless the gas is highly compressed). A gas is easily compressed because the molecules can be pushed into a smaller space. A gas is fluid because individual molecules can move easily relative to one another.

Liquids are relatively incompressible fluids. This state of matter can also be explained by kinetic-molecular theory. According to this theory, the molecules of a liquid are in constant random motion (as in a gas) but are more tightly packed, so there is much less free space. Because the molecules can move relative to one another as in a gas, a liquid can flow (it is fluid). But the lack of empty space explains why a liquid, unlike a gas, is nearly incompressible.

Solids are nearly incompressible and are rigid, not fluid. Kinetic-molecular theory explains this by saying that the particles making up a solid (which may be atoms, molecules, or ions) exist in close contact and (unlike those in a gas or liquid) do not move about but oscillate or vibrate about fixed sites. This explains the rigidity of a solid. And, of course, the compact structure explains its incompressibility. Figure 11.2 compares the gaseous, liquid, and solid states in kinetic-molecular terms.

**Figure 11.2**
**Representation of the states of matter.** In a gas, the molecules are in constant random motion through largely empty space. In a liquid, the molecules are also in constant random motion, but they are more closely packed than in a gas. In a solid, the basic units (atoms, ions, or molecules) are closely packed and vibrate about fixed sites.

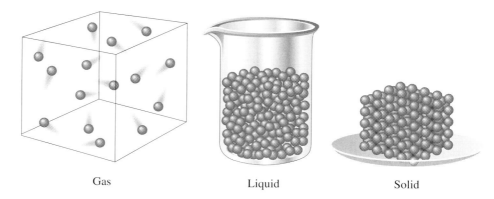

Gas          Liquid          Solid

Recall that gases normally follow closely the ideal gas law, $PV = nRT$. The simplicity of this equation is a result of the nearly negligible forces of interaction between molecules and the nearly negligible molecular size compared with the total volume of gas. Still, to be accurate we should account for both of these quantities, and the van der Waals equation, which we discussed in Section 5.8, is one attempt to do this.

$$(P + n^2a/V^2)(V - nb) = nRT$$

In this equation, the constants $a$ and $b$ depend on the nature of the gas. The constant $a$ depends on the magnitude of the intermolecular forces; the constant $b$ depends on the relative size of the molecules.

No such simple equations exist for the liquid and solid states. The size of the particles making up the liquid and solid states cannot be neglected relative to the total volume or accounted for in a simple way. Similarly, the forces of interaction between particles in the liquid and the solid states cannot be neglected; indeed, these forces of interaction are crucial when you describe the properties of the liquid and solid states.

Some of the most important properties of liquids and solids are concerned with changes of physical state. Therefore, we begin with this topic. In later sections, we will look specifically at the liquid and solid states and see how their properties depend on the forces of attraction between particles.

# Changes of State

In the chapter opener, we discussed the change of solid carbon dioxide (dry ice) directly to a gas. Such *a change of a substance from one state to another* is called a **change of state** or **phase transition.** In the next sections, we will look at the different kinds of phase transitions and the conditions under which they occur. (A *phase* is a homogeneous portion of a system—that is, a given state of either a substance or a solution.)

## 11.2 Phase Transitions

In general, each of the three states of a substance can change into either of the other states. Table 11.1 lists the different kinds of phase transitions.

**Table 11.1**
**Kinds of Phase Transitions**

| Phase Transition | Name | Examples |
|---|---|---|
| Solid ⟶ liquid | Melting, fusion | Melting of snow and ice |
| Solid ⟶ gas | Sublimation | Sublimation of dry ice, freeze-drying of coffee |
| Liquid ⟶ solid | Freezing | Freezing of water or a liquid metal |
| Liquid ⟶ gas | Vaporization | Evaporation of water or refrigerant |
| Gas ⟶ liquid | Condensation, liquefaction | Formation of dew, liquefaction of carbon dioxide |
| Gas ⟶ solid | Condensation, deposition | Formation of frost and snow |

**Melting** is *the change of a solid to the liquid state* (melting is also referred to as *fusion*). For example,

$$H_2O(s) \longrightarrow H_2O(l) \qquad \text{(melting, fusion)}$$
ice, snow    liquid water

**Freezing** is *the change of a liquid to the solid state*. The freezing of liquid water to ice is a common example.

$$H_2O(l) \longrightarrow H_2O(s) \qquad \text{(freezing)}$$
liquid water    ice

The casting of a metal object involves the melting of the metal, then its freezing in a mold.

**Vaporization** is *the change of a solid or a liquid to the vapor.* For example,

$$H_2O(l) \longrightarrow H_2O(g) \qquad \text{(vaporization)}$$
liquid water    water vapor

*The change of a solid directly to the vapor* is specifically referred to as **sublimation.** Although the vapor pressure of many solids is quite low, some (usually molecular solids) have appreciable vapor pressure. Ice, for instance, has a vapor pressure of 4.7 mmHg at 0°C. For this reason, a pile of snow slowly disappears in winter even though the temperature is too low for it to melt. The snow is being changed directly to water vapor.

$$H_2O(s) \longrightarrow H_2O(g) \qquad \text{(sublimation)}$$
ice, snow    water vapor

**Figure 11.3**
**Sublimation of iodine.**
*Left:* The beaker contains iodine crystals, $I_2$; a dish of ice rests on top of the beaker. *Right:* Iodine has an appreciable vapor pressure even below its melting point (114°C); thus, when heated carefully the solid sublimes without melting. The vapor deposits as crystals on the cool underside of the dish.

Sublimation can be used to purify solids that readily vaporize. Figure 11.3 shows a simple way to purify iodine by sublimation. Impure iodine is heated in a beaker so that it vaporizes, leaving nonvolatile impurities behind. The vapor crystallizes on the bottom surface of a dish containing ice that rests on top of the beaker. Freeze-drying of foods is a commercial application of sublimation. Brewed coffee, for example, is frozen and placed in a vacuum to remove water vapor. The ice continues to sublime until it is all gone, leaving freeze-dried coffee. Most freeze-dried foods are easily reconstituted by adding water.

**Condensation** is *the change of a gas to either the liquid or the solid state* (the change of vapor to solid is sometimes called *deposition*). Dew is liquid water formed

Solid $I_2$

Gaseous $I_2$

by condensation of water vapor from the atmosphere. Frost is solid water formed by direct condensation of water vapor from the atmosphere without first forming liquid water. Snow is formed by a similar process in the upper atmosphere.

$$H_2O(g) \longrightarrow H_2O(l) \qquad \text{(condensation)}$$
water vapor    dew

$$H_2O(g) \longrightarrow H_2O(s) \qquad \text{(condensation, deposition)}$$
water vapor    frost, snow

When a substance that is normally a gas, such as carbon dioxide, changes to the liquid state, the phase transition is often referred to as *liquefaction*.

▶ Video no. 15 (Series A), "Vapor Pressure."

▶ Demonstration of the vapor pressure of pure liquids: Bassam Z. Shakhashiri, *Chemical Demonstrations*, Vol. 3 (Madison: University of Wisconsin Press, 1989), no. 9.5A, p. 242.

## Vapor Pressure

Liquids, and even some solids, are continuously vaporizing. If a liquid is in a closed vessel with space above it, a partial pressure of the vapor state builds up in this space. The **vapor pressure** of a liquid is *the partial pressure of the vapor over the liquid, measured at equilibrium at a given temperature.* To understand what we mean by equilibrium, let us look at a simple method for measuring vapor pressure.

You introduce a few drops of water from a medicine dropper into the mercury column of a barometer (Figure 11.4A). Being less dense than mercury, the water rises in the tube to the top of the mercury, where it evaporates, or vaporizes.

To understand the process of vaporization, it is necessary to realize that the molecules in a liquid have a distribution of kinetic energies (Figure 11.5). Those molecules moving away from the surface and toward the vapor phase will escape only if their kinetic energies are greater than a certain minimum value equal to the potential energy from the attraction of molecules in the body of the liquid.

**Figure 11.4**
**Measurement of the vapor pressure of water.**
*(A)* The mercury column drops a distance *h* because of the vapor pressure of water. *(B)* A detail of the upper portion of the tube, showing water molecules evaporating from and condensing on the water surface.

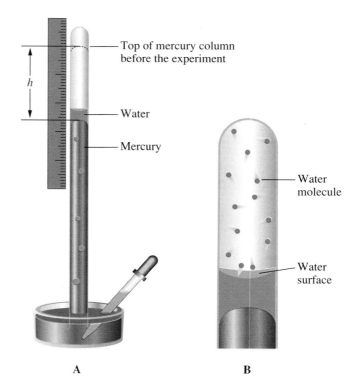

Top of mercury column before the experiment

*h*

Water

Mercury

Water molecule

Water surface

A    B

**Figure 11.5**
**Distribution of kinetic energies of molecules in a liquid.** Distributions of kinetic energies are shown for two different temperatures. The fraction of molecules having kinetic energies greater than the minimum necessary for escape is given by the colored areas (orange for the lower temperature, red area plus orange area at the higher temperature). Note that the fraction of molecules having kinetic energies greater than the minimum value for vaporization increases with temperature.

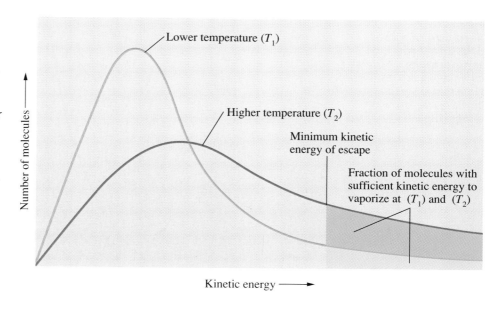

As molecules at the surface gain sufficient kinetic energy (through collisions with neighboring molecules), they escape the liquid and go into the space at the top of the barometer column (see Figure 11.4B). More and more water molecules begin to fill this space. As the number of molecules in the vapor state increases, more and more gaseous molecules collide with the liquid water surface, exerting pressure on it. So, as vaporization of water proceeds, the mercury column moves downward.

Some of the molecules in the vapor collide with the liquid surface and stick; that is, the vapor condenses to liquid. The rate of condensation steadily increases as the number of molecules in the volume of vapor increases, until the rate at which molecules are condensing on the liquid equals the rate at which molecules are vaporizing (Figure 11.6).

**Figure 11.6**
**Rates of vaporization and condensation of a liquid over time.** As vaporization continues at a constant rate, the rate of condensation increases. Eventually, the two rates become equal. The partial pressure of vapor reaches a steady value at a given temperature, which is the vapor pressure of the liquid. The process is said to reach equilibrium.

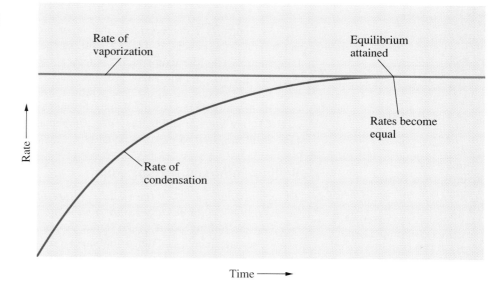

**Figure 11.7**
**Variation of vapor
pressure with
temperature.** Several
common liquids are shown.
Note that the vapor pressure
increases rapidly with
temperature.

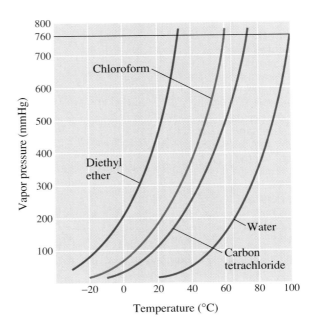

When the rates of vaporization and condensation have become equal, the number of molecules in the vapor at the top of the column in Figure 11.4 stops increasing and remains unchanged. The partial pressure of the vapor (as measured by the change in height of the mercury column) reaches a steady value, which is its vapor pressure. The liquid and vapor are in *equilibrium*. Although the partial pressure of the vapor is unchanging, molecules are still leaving the liquid and coming back to the liquid. For this reason, you speak of this as a *dynamic equilibrium*—one in which the molecular processes (in this case, vaporization and condensation) are continuously occurring. You represent this dynamic equilibrium for the vaporization and condensation of water by the equation with a double arrow.

$$H_2O(l) \rightleftharpoons H_2O(g)$$

As Figure 11.7 demonstrates, the vapor pressure of a substance depends on the temperature. (Appendix B gives a table of the vapor pressures of water at various temperatures.) As the temperature increases, the kinetic energy of molecular motion becomes greater, and the vapor pressure increases. Liquids and solids with relatively high vapor pressure at normal temperatures are said to be *volatile*. Chloroform and carbon tetrachloride are volatile liquids. Naphthalene, $C_{10}H_8$, and *para*-dichlorobenzene, $C_6H_4Cl_2$, are volatile solids: they have appreciable vapor pressures at room temperature. Both are used to make mothballs.

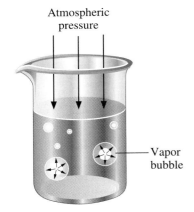

**Figure 11.8
Boiling of a liquid.**
A liquid boils when the vapor
pressure inside a bubble of
vapor equals the external
pressure. The temperature
at which this occurs is called
the boiling point.

## Boiling Point and Melting Point

*The temperature at which the vapor pressure of a liquid equals the pressure exerted on the liquid (atmospheric pressure, unless the vessel containing the liquid is closed)* is called the **boiling point** of the liquid. As the temperature of a liquid is raised, the vapor pressure increases. When the vapor pressure equals the atmospheric pressure, stable bubbles of vapor form within the body of the liquid (Figure 11.8). This process

**Table 11.2**
**Melting Points and Boiling Points (at 1 atm) of Several Substances**

| Name | Type of Solid* | Melting Point, °C | Boiling Point, °C |
|---|---|---|---|
| Neon, Ne | Molecular | −249 | −246 |
| Hydrogen sulfide, $H_2S$ | Molecular | −86 | −61 |
| Chloroform, $CHCl_3$ | Molecular | −64 | 62 |
| Water, $H_2O$ | Molecular | 0 | 100 |
| Acetic acid, $HC_2H_3O_2$ | Molecular | 17 | 118 |
| Mercury, Hg | Metallic | −39 | 357 |
| Sodium, Na | Metallic | 98 | 883 |
| Tungsten, W | Metallic | 3410 | 5660 |
| Cesium chloride, CsCl | Ionic | 645 | 1290 |
| Sodium chloride, NaCl | Ionic | 801 | 1413 |
| Magnesium oxide, MgO | Ionic | 2800 | 3600 |
| Quartz, $SiO_2$ | Covalent network | 1610 | 2230 |
| Diamond, C | Covalent network | 3550 | 4827 |

*Types of solids are discussed in Section 11.6.

is called boiling. Once boiling starts, the temperature of the liquid remains at the boiling point (as long as sufficient heat is supplied).

Because the pressure exerted on a liquid can vary, the boiling point of a liquid can vary. For instance, at 1.00 atm (the average atmospheric pressure at sea level), the boiling point of water is 100°C. But at 0.83 atm (the average atmospheric pressure in Denver, Colorado, at 1609 m above sea level), the boiling point of water is 95°C. The *normal boiling point* of a liquid is the boiling point at 1 atm. ◀

*The temperature at which a pure liquid changes to a crystalline solid, or freezes,* is called the **freezing point.** *The temperature at which a crystalline solid changes to a liquid, or melts,* is called the **melting point;** it is identical to the freezing point. The melting or freezing occurs at the temperature where the liquid and solid are in dynamic equilibrium.

$$\text{Solid} \rightleftharpoons \text{liquid}$$

Unlike boiling points, melting points are affected significantly only by large pressure changes.

Both the melting point and the boiling point are characteristic physical properties of a substance and can be used to help identify it. Table 11.2 gives melting points and boiling points of several substances.

### Heat of Phase Transition

Any change of state involves the addition or removal of energy as heat to or from the substance. A simple experiment shows that this is the case. Suppose you add heat at a constant rate to a beaker containing ice at −20°C. In Figure 11.9, we have

▶ If you want to cook a stew in less time than is possible at atmospheric pressure, you can use a pressure cooker. This is especially valuable at high altitudes, where water boils at a lower temperature. The steam pressure inside the cooker is allowed to build up to a given value before being released by a valve. The stew then boils at a higher temperature and is cooked more quickly.

▶ Video Series C, "Boiling Water with Ice."

**Figure 11.9**
**Heating curve for water.**
Heat is being added at a
constant rate to a system
containing water. Note the
flat regions of the curve.
When heat is added during
a phase transition the
temperature does not
change.

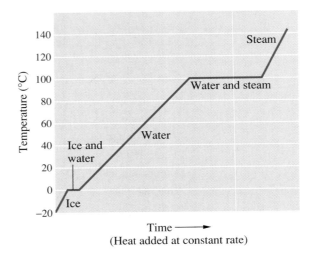

Time ⟶
(Heat added at constant rate)

▶ Demonstration of boiling at
reduced pressure: Summerlin
et al., *Chemical Demonstrations,*
Vol. 2 (Washington, D.C.: Am.
Chem. Soc., 1987), no. 13, p. 24.

▶ This material on heat of phase
transition can be omitted or
abbreviated if you are short
on time.

plotted the temperature of the different phases of water as heat is added. The temperature of the ice begins to rise from −20°C, as you would expect; the addition
of heat normally raises the temperature of a substance. At 0°C, the ice begins to
melt, so that you get a beaker of ice in water. Note the flat region in the curve,
labeled ice and water. Why is this region flat? It means that heat is being added to
the system without a change in temperature; the temperature remains at 0°C. This
temperature, of course, is the melting point of ice. The heat being added is energy
required to melt ice to water at the same temperature. The intermolecular forces
binding water molecules to specific sites in the solid phase must be partially broken to allow water molecules the ability to slide over one another easily, as happens in the liquid state. Note the flat regions for each of the phase transitions.
Because heat is being added at a constant rate, the length of each flat region is proportional to the *heat of phase transition.*

*The heat needed for the melting of a solid* is called the **heat of fusion** (or *enthalpy
of fusion*) and is denoted $\Delta H_{fus}$. For ice, the heat of fusion is 6.01 kJ per mole.

$$H_2O(s) \longrightarrow H_2O(l); \ \Delta H_{fus} = 6.01 \text{ kJ/mol}$$

*The heat needed for the vaporization of a liquid* is called the **heat of vaporization** (or *enthalpy of vaporization*) and is denoted $\Delta H_{vap}$. At 100°C, the heat of vaporization of water is 40.7 kJ per mole. ◀

$$H_2O(l) \longrightarrow H_2O(g); \ \Delta H_{vap} = 40.7 \text{ kJ/mol}$$

▶ The fact that heat is required
for vaporization can be demonstrated by evaporating a dish of
water inside a vessel attached
to a vacuum pump. With the
pressure kept low, water evaporates quickly enough to freeze
the water remaining in the dish.

Note that much more heat is required for vaporization than for melting. Melting
needs only enough energy for the molecules to escape from their sites in the solid.
For vaporization, enough energy must be supplied to break most of the intermolecular attractions.

A refrigerator relies on the cooling effect accompanying vaporization. The mechanism contains an enclosed gas that can be liquefied under pressure, such as ammonia or dichlorodifluoromethane (Freon-12), $CCl_2F_2$. As the liquid is allowed to
evaporate, it absorbs heat and thus cools its surroundings (the interior space of the
refrigerator). Gas from the evaporation is recycled to a compressor and then to a
condenser, where it is liquefied again. Heat leaves the condenser, going into the surrounding air.

**Understanding Concepts:
Changes of State**

▶ Two alternate examples are provided in the Instructor's Resource Manual.

$CCl_2F_2$

## EXAMPLE 11.1

## Calculating the Heat Required for a Phase Change of a Given Mass of Substance

A particular refrigerator cools by evaporating liquefied dichlorodifluoromethane, $CCl_2F_2$. How many kilograms of this liquid must be evaporated to freeze a tray of water at 0°C to ice at 0°C? The mass of the water is 525 g, the heat of fusion of ice is 6.01 kJ/mol, and the heat of vaporization of dichlorodifluoromethane is 17.4 kJ/mol.

### SOLUTION

The heat that must be removed to freeze 525 g of water at 0°C is

$$525 \text{ g } H_2O \times \frac{1 \text{ mol } H_2O}{18.0 \text{ g } H_2O} \times \frac{-6.01 \text{ kJ}}{1 \text{ mol } H_2O} = -175 \text{ kJ}$$

The minus sign indicates that heat energy is taken away from the water. Consequently, the vaporization of dichlorodifluoromethane absorbs 175 kJ of heat. The mass of $CCl_2F_2$ that must be vaporized to absorb this quantity of heat is

$$175 \text{ kJ} \times \frac{1 \text{ mol } CCl_2F_2}{17.4 \text{ kJ}} \times \frac{121 \text{ g } CCl_2F_2}{1 \text{ mol } CCl_2F_2} = 1.22 \times 10^3 \text{ g } CCl_2F_2$$

**1.22 kg of dichlorodifluoromethane must be evaporated.**

### EXERCISE 11.1

The heat of vaporization of ammonia is 23.4 kJ/mol. How much heat is required to vaporize 1.00 kg of ammonia? How many grams of water at 0°C could be frozen to ice at 0°C by the evaporation of this amount of ammonia?

See Problems 11.35, 11.36, 11.37, and 11.38.

 **CONCEPT CHECK 11.1**

$t = 0$

Shown here is a representation of a closed container in which you have just placed 10 L of $H_2O$. In our experiment, we are going to call this starting point in time $t = 0$ and assume that all of the $H_2O$ is in the liquid phase. We have represented a few of the $H_2O$ molecules in the water as dots.

a. Consider a time $t = 1$, at which some time has passed but the system has not reached equilibrium.

　　i. How will the level of the liquid $H_2O$ compare to that at $t = 0$?

　　ii. How will the vapor pressure in the flask compare to that at $t = 0$?

　　iii. How will the number of $H_2O$ molecules in the vapor state compare to that at $t = 0$?

　　iv. How does the rate of evaporation in this system compare to the rate of condensation?

　　v. Draw a picture of the system at $t = 1$.

b. Consider a time $t = 2$, at which enough time has passed for the system to reach equilibrium.

i. How will the level of the liquid $H_2O$ compare to that at $t = 1$?

ii. How will the vapor pressure in the flask compare to that at $t = 1$?

iii. How will the number of $H_2O$ molecules in the vapor state compare to that at $t = 1$?

iv. How does the rate of evaporation in this system compare to the rate of condensation?

v. Draw a picture of the system at $t = 2$.

### Clausius–Clapeyron Equation: Relating Vapor Pressure and Liquid Temperature

We noted earlier that the vapor pressure of a substance depends on temperature. The variation of vapor pressure with temperature of some liquids was given in Figure 11.7. It can be shown that the logarithm of the vapor pressure of a liquid or solid varies with the absolute temperature according to the following approximate relation:

$$\ln P = -\frac{A}{T} + B$$

Here $\ln P$ is the natural logarithm of the vapor pressure, and $A$ and $B$ are positive constants. You can confirm this relation for the liquids shown in Figure 11.7 by replotting the data. If you put $y = \ln P$ and $x = 1/T$, the previous relation becomes

$$y = -Ax + B$$

This means that if you plot $\ln P$ versus $1/T$, you should get a straight line with slope $-A$. The data of Figure 11.7 are replotted this way in Figure 11.10. ◄

▶ **See Appendix A for the graphing of a straight line.**

The previous equation has been derived from thermodynamics, by assuming the vapor behaves like an ideal gas. The result, known as the *Clausius–Clapeyron equation,* shows that the constant $A$ is proportional to the heat of vaporization of the liquid, $\Delta H_{vap}$.

**Figure 11.10**
**Plot of the logarithm of vapor pressure versus 1/T.**
The liquids shown in Figure 11.7 are replotted in this graph. Note the straight-line relationship.

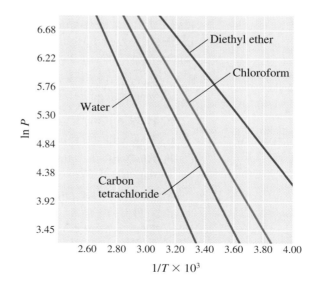

$$\ln P = \frac{-\Delta H_{vap}}{RT} + B$$

A two-point form of the Clausius–Clapeyron equation is very useful for calculations. Let us write the previous equation for two different temperatures:

$$\ln P_2 = \frac{-\Delta H_{vap}}{RT_2} + B$$

$$\ln P_1 = \frac{-\Delta H_{vap}}{RT_1} + B$$

Here $P_1$ is the vapor pressure at absolute temperature $T_1$, and $P_2$ is the vapor pressure at absolute temperature $T_2$. If you subtract the second equation from the first, you get

$$\ln P_2 - \ln P_1 = \frac{-\Delta H_{vap}}{RT_2} - \frac{-\Delta H_{vap}}{RT_1} + B - B$$

which you can write as

$$\ln \frac{P_2}{P_1} = \frac{\Delta H_{vap}}{R} \left( \frac{1}{T_1} - \frac{1}{T_2} \right)$$

The next two examples illustrate the use of this two-point form of the Clausius–Clapeyron equation.

▶ An alternate example is provided in the Instructor's Resource Manual.

## EXAMPLE 11.2

### Calculating the Vapor Pressure at a Given Temperature

Estimate the vapor pressure of water at 85°C. Note that the normal boiling point of water is 100°C and that its heat of vaporization is 40.7 kJ/mol.

**SOLUTION**

The two-point form of the Clausius–Clapeyron equation relates five quantities: $P_1$, $T_1$, $P_2$, $T_2$, and $\Delta H_{vap}$. The normal boiling point is the temperature at which the vapor pressure equals 760 mmHg. Thus, you can let $P_1$ equal 760 mmHg and $T_1$ equal 373 K (100°C). Then $P_2$ would be the vapor pressure at $T_2$, or 358 K (85°C). The heat of vaporization, $\Delta H_{vap}$, is $40.7 \times 10^3$ J/mol. Note that we have expressed this quantity in joules per mole to agree with the units of the molar gas constant, $R$, which equals 8.31 J/(K · mol). You can now substitute into the two-point equation (the equation shaded in color just before this example).

$$\ln \frac{P_2}{760 \text{ mmHg}} = \frac{40.7 \times 10^3 \text{ J/mol}}{8.31 \text{ J/(K · mol)}} \left( \frac{1}{373 \text{ K}} - \frac{1}{358 \text{ K}} \right)$$

$$= (4898 \text{ K}) \times (-1.123 \times 10^{-4}/\text{K}) = -0.550$$

You now want to solve this equation for $P_2$, which is the vapor pressure of water at 85°C. To do so, you take the antilogarithm of both sides of the equation. This gives

$$\frac{P_2}{760 \text{ mmHg}} = \text{antilog}(-0.550)$$

The antilog($-0.550$) is $e^{-0.550} = 0.577$. (Appendix A describes how to obtain the antilogarithm.) Thus,

$$\frac{P_2}{760 \text{ mmHg}} = 0.577$$

Therefore,

$$P_2 = 0.577 \times 760 \text{ mmHg} = \textbf{439 mmHg}$$

This value may be compared with the experimental value of 434 mmHg, given in Appendix B. Differences from the experimental value stem from the fact that the Clausius–Clapeyron equation is an approximate relation.

**Problem Solving: Calculating Vapor Pressure**

---

**EXERCISE 11.2**

Carbon disulfide, $CS_2$, has a normal boiling point of 46°C and a heat of vaporization of 26.8 kJ/mol. What is the vapor pressure of carbon disulfide at 35°C?

---

See Problems 11.41 and 11.42.

---

**EXAMPLE 11.3**

## Calculating the Heat of Vaporization from Vapor Pressures

Calculate the heat of vaporization of diethyl ether (often called simply "ether"), $C_4H_{10}O$, from the following vapor pressures: 400 mmHg at 18°C and 760 mmHg at 35°C. (Each pressure value has three significant figures.)

► An alternate example is provided in the Instructor's Resource Manual.

**SOLUTION**

If you let $P_1$ be 400 mmHg, then $T_1$ is 291 K (18°C), $P_2$ is 760 mmHg, and $T_2$ is 308 K (35°C). You substitute these values into the Clausius–Clapeyron equation.

$$\ln \frac{760 \text{ mmHg}}{400 \text{ mmHg}} = \frac{\Delta H_{vap}}{8.31 \text{ J/(K} \cdot \text{mol)}} \times \left( \frac{1}{291 \text{ K}} - \frac{1}{308 \text{ K}} \right)$$

$$0.642 = (2.28 \times 10^{-5}) \times \Delta H_{vap}/(\text{J/mol})$$

Therefore,

$$\Delta H_{vap} = 2.82 \times 10^4 \text{ J/mol} \quad \textbf{(28.2 kJ/mol)}$$

---

**EXERCISE 11.3**

Selenium tetrafluoride, $SeF_4$, is a colorless liquid. It has a vapor pressure of 757 mmHg at 105°C and 522 mmHg at 95°C. What is the heat of vaporization of selenium tetrafluoride?

---

See Problems 11.43 and 11.44.

# 11.3 Phase Diagrams

As we mentioned in the chapter opening, the solid, liquid, and gaseous states of carbon dioxide exist under different temperature and pressure conditions. A **phase diagram** is *a graphical way to summarize the conditions under which the different states of a substance are stable.*

## Melting-Point Curve

Figure 11.11 is a phase diagram for water. It consists of three curves that divide the diagram into regions labeled "solid," "liquid," and "gas." In each region, the indicated state is stable. Every point on each of the curves indicates experimentally determined temperatures and pressures at which two states are in equilibrium. Thus, the curve *AB*, dividing the solid region from the liquid region, represents the conditions under which the solid and liquid are in equilibrium.

Solid ⇌ liquid

This curve gives the melting points of the solid at various pressures.

Usually, the melting point is only slightly affected by pressure. For this reason, the melting-point curve *AB* is nearly vertical. If a liquid is more dense than the solid, as is the case for water, the melting point decreases with pressure. ◀ The melting-point curve in such cases leans slightly to the left. In the case of ice, the decrease is indeed slight—only 1°C for a pressure increase of 133 atm. For most substances, the liquid state is less dense than the solid. In that case, the melting-point curve leans slightly to the right.

## Vapor-Pressure Curves for the Liquid and the Solid

The curve *AC* that divides the liquid region from the gaseous region gives the vapor pressures of the liquid at various temperatures. It also gives the boiling points of the liquid for various pressures. The boiling point of water at 1 atm is shown on the phase diagram (Figure 11.11).

The curve *AD* that divides the solid region from the gaseous region gives the vapor pressures of the solid at various temperatures. This curve intersects the other curves at point *A*, the **triple point,** which is *the point on a phase diagram representing the temperature and pressure at which three phases of a substance coexist in equilibrium.* For water, the triple point occurs at 0.01°C, 0.00603 atm (4.58 mmHg), and the solid, liquid, and vapor phases coexist. ◀

Suppose a solid is warmed at a pressure below the pressure at the triple point. In a phase diagram, this corresponds to moving along a horizontal line below the triple point. You can see from Figure 11.11 that such a line will intersect curve *AD*, the vapor-pressure curve for the solid. Thus, the solid will pass directly into

▶ The following is a dramatic demonstration of the effect of pressure on the melting of ice: Suspend a block of ice between two chairs. Then loop a length of wire over the block and place weights on the ends of the wire. The pressure of the wire will melt the ice under it. Liquid water will flow over the top of the wire and freeze again, because it is no longer under the pressure of the wire. As a result, the wire will cut through the ice and fall to the floor, but the block of ice will remain as one piece.

▶ Because the triple point for water occurs at a definite temperature, it is used to define the Kelvin temperature scale. The temperature of water at its triple point is defined to be 273.16 K (0.01°C).

**Figure 11.11**
**Phase diagram for water (not to scale).** The curves *AB*, *AC*, and *AD* divide the diagram into regions that give combinations of temperature and pressure for which only one state is stable. Along any curve, the two states from the adjoining regions are in equilibrium.

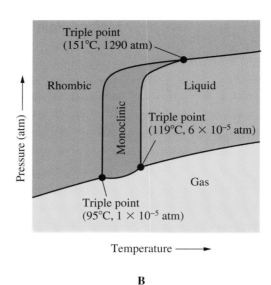

A                                                                    B

**Figure 11.12**
**Phase diagrams for carbon dioxide and sulfur (not to scale).** *(A)* Carbon dioxide. At normal atmospheric pressure (1 atm), the solid sublimes when warmed. *(B)* Sulfur. This substance has a complicated phase diagram with three triple points.

the gas; that is, the solid will sublime. Freeze-drying of a food (or brewed coffee) is accomplished by placing the frozen food in a vacuum (below 0.00603 atm) so that the ice in it sublimes. Because the food can be dried at a lower temperature than if heat-dried, it retains more flavor and can often be reconstituted by simply adding water.

The triple point of carbon dioxide is at $-57°C$ and 5.1 atm (Figure 11.12A). Therefore, the solid sublimes if warmed at any pressure below 5.1 atm. This is why solid carbon dioxide sublimes at normal atmospheric pressure (1 atm). Above 5.1 atm, however, the solid melts if warmed. Sulfur has a more complicated phase diagram (Figure 11.12B). It displays three triple points, one of them involving two different solid forms of sulfur (called rhombic sulfur and monoclinic sulfur), as well as the vapor.

### Critical Temperature and Pressure

Imagine an experiment in which liquid and gaseous carbon dioxide are sealed in a thick-walled glass vessel at 20°C. At this temperature, the liquid is in equilibrium with its vapor at a pressure of 57 atm. You observe that the liquid and vapor are separated by a well-defined boundary, or meniscus. Now suppose the temperature is raised. The vapor pressure increases, and at 30°C it is 71 atm. Then, as the temperature approaches 31°C, a curious thing happens. The meniscus becomes fuzzy and less well defined. At 31°C, the meniscus disappears altogether. Above this temperature and above the critical pressure, there is only one fluid state, called a *supercritical fluid* (see Figure 11.13).

*The temperature above which the liquid state of a substance no longer exists regardless of the pressure* is called the **critical temperature.** For carbon dioxide it is 31°C. *The vapor pressure at the critical temperature* is called the **critical pressure** (73 atm for carbon dioxide). It is the minimum pressure that must be applied to a gas at the critical temperature to liquefy it.

On a phase diagram, the preceding experiment corresponds to following the vapor-pressure curve where the liquid and vapor are in equilibrium. Note that this

**Figure 11.13**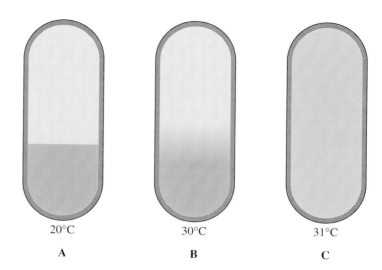
**Observing the critical phenomenon.** *(A)* Carbon dioxide liquid in equilibrium with its vapor at 20°C. *(B)* The same two states in equilibrium at 30°C (just below the critical point); the densities of the liquid and vapor are becoming equal. *(C)* Carbon dioxide at 31°C (the critical temperature); the liquid and vapor now have the same densities— in fact, the distinction between liquid and vapor has disappeared, resulting in what is called a supercritical fluid.

20°C

A

30°C

B

31°C

C

curve in Figure 11.12*A* ends at a point at which the temperature and pressure have their critical values. This is the *critical point*. If you look at the phase diagram for water, you will see that the vapor-pressure curve for the liquid similarly ends, at point *C*, which is the critical point for water. In this case, the critical temperature is 374°C and the critical pressure is 218 atm.

Many important gases cannot be liquefied at room temperature. Nitrogen, for example, has a critical temperature of −147°C. This means the gas cannot be liquefied until the temperature is below −147°C.

▶ An alternate example is provided in the Instructor's Resource Manual.

**EXAMPLE 11.4**

## Relating the Conditions for the Liquefaction of Gases to the Critical Temperature

The critical temperatures of ammonia and nitrogen are 132°C and −147°C, respectively. Explain why ammonia can be liquefied at room temperature by merely compressing the gas to a high enough pressure, whereas the compression of nitrogen requires a low temperature as well.

**SOLUTION**

**The critical temperature of ammonia is well above room temperature.** If ammonia gas at room temperature is compressed, it will liquefy. **Nitrogen, however, has a critical temperature well below room temperature.** It cannot be liquefied by compression unless the temperature is below the critical temperature.

**EXERCISE 11.4**

Describe how you could liquefy the following gases: a. methyl chloride, $CH_3Cl$ (critical point, 144°C, 66 atm); b. oxygen, $O_2$ (critical point, −119°C, 50 atm).

See Problems 11.49 and 11.50.

# Removing Caffeine from Coffee

EVERYDAY LIFE

Legend has it that a goat herder named Kaldi, in the ancient country of Abyssinia (now Ethiopia, in northeast Africa), discovered the pleasant effect of coffee beans when he noted that his goats danced wildly about after eating the shiny leaves and red berries of a small tree. Later he tried eating the whole berries himself, and he soon found that he was dancing with his goats. Whether the legend is true or not, it does effectively describe the stimulant effect of this berry from the coffee tree. As we now know, the active principle of coffee, which today is obtained by roasting the berries' green pits, or beans as they are called, is caffeine, a bitter-tasting white substance with the formula $C_8H_{10}N_4O_2$. (Figure 11.14 shows the structural formula of caffeine.)

For those who like the taste of roasted coffee but don't want the caffeine, decaffeinated coffee is available. A German chemist, Ludwig Roselius, first made "decaf" coffee about 1900 by extracting the caffeine from green coffee beans with the solvent chloroform, $CHCl_3$. Later, commercial processes replaced chloroform with the safer organic solvent methylene chloride, $CH_2Cl_2$. To-

**Figure 11.14**
**The caffeine molecule.** Ball-and-stick model of caffeine, $C_8H_{10}N_4O_2$.

day, though, most of the commercial decaffeinated coffee produced (Figure 11.15) uses *supercritical* carbon dioxide as the extracting fluid.

In a tank of carbon dioxide under pressure (for example, in a $CO_2$ fire extinguisher), the substance nor-

 **CONCEPT CHECK 11.2**

When camping at high altitude, you need to pay particular attention to changes in cooking times for foods that are boiled in water. If you like eggs that are boiled for 10 minutes near sea level, would you have to cook them for a longer or a shorter time at 3200 m to get the egg you like? Be sure to explain your answer.

# Liquid State

Now we will look at some physical properties of liquids. After that, we will explain the experimental values of these properties in terms of intermolecular forces.

## 11.4 Properties of Liquids: Surface Tension and Viscosity

You have seen that molecules tend to escape the liquid state and form a vapor. The vapor pressure is the equilibrium partial pressure of this vapor over the liquid; it increases with temperature. The boiling point is the temperature at which the vapor

**Figure 11.15**
**Decaf coffee.** Coffee is decaffeinated in large plants, such as the one in Houston, Texas.

good solvent for organic substances, but supercritical carbon dioxide readily dissolves many of these substances, including caffeine. It is nontoxic and nonflammable. It also has no effect on the stratospheric ozone layer, whereas methylene chloride does. (See the essay on stratospheric ozone at the end of Chapter 10.) Carbon dioxide does contribute to the greenhouse effect (discussed in the essay at the end of Chaper 5), but the gas once used can be recirculated for solvent use and not vented to the atmosphere.

Supercritical fluids have gained much attention recently because of the possibility of replacing toxic and environmentally less desirable solvents. For example, at the moment, the usual solvent used to dry-clean clothes is perchloroethylene, $CCl_2CCl_2$. Although nonflammable and less toxic than carbon tetrachloride, which was the solvent previously used, perchloroethylene is regulated as an air pollutant under the Clean Air Act. Some scientists have shown that you can dry-clean with supercritical carbon dioxide if you use a special detergent.

Substances other than carbon dioxide have also shown intriguing solvent properties. For example, whereas water under normal conditions dissolves ionic and polar substances, above its critical point (374°C, 217 atm) it becomes an excellent solvent for nonpolar substances. Supercritical water and carbon dioxide promise to replace many toxic or environmentally unfriendly organic solvents.

mally exists as the liquid in equilibrium with its gas phase. But we know from the previous text discussion that above 31°C (88°F), the two phases, gas and liquid, are replaced by a single fluid phase. So, on a hot summer day (above 31°C or 88°F), the carbon dioxide in such a tank is above its critical temperature and pressure and exists as the supercritical fluid.

Supercritical carbon dioxide is a near-ideal solvent. Under normal conditions, carbon dioxide is not a very

pressure equals the pressure applied to the liquid. Both vapor pressure and boiling point are important properties of a liquid. Values of the vapor pressure for some liquids at 20°C are listed in Table 11.3 (column 3). Two additional properties given in Table 11.3 are surface tension and viscosity. All of these properties, as you will see in the next section, depend on intermolecular forces, which are related to molecular structure.

**Table 11.3**
**Properties of Some Liquids at 20°C**

| Substance | Molecular Weight (amu) | Vapor Pressure (mmHg) | Surface Tension (J/m²) | Viscosity (kg/m · s) |
|---|---|---|---|---|
| Water, $H_2O$ | 18 | $1.8 \times 10^1$ | $7.3 \times 10^{-2}$ | $1.0 \times 10^{-3}$ |
| Carbon dioxide, $CO_2$ | 44 | $4.3 \times 10^4$ | $1.2 \times 10^{-3}$ | $7.1 \times 10^{-5}$ |
| Pentane, $C_5H_{12}$ | 72 | $4.4 \times 10^2$ | $1.6 \times 10^{-2}$ | $2.4 \times 10^{-4}$ |
| Glycerol, $C_3H_8O_3$ | 92 | $1.6 \times 10^{-4}$ | $6.3 \times 10^{-2}$ | $1.5 \times 10^0$ |
| Chloroform, $CHCl_3$ | 119 | $1.7 \times 10^2$ | $2.7 \times 10^{-2}$ | $5.8 \times 10^{-4}$ |
| Carbon tetrachloride, $CCl_4$ | 154 | $8.7 \times 10^1$ | $2.7 \times 10^{-2}$ | $9.7 \times 10^{-4}$ |
| Bromoform, $CHBr_3$ | 253 | $3.9 \times 10^0$ | $4.2 \times 10^{-2}$ | $2.0 \times 10^{-3}$ |

**Figure 11.16**
**Explaining surface tension.** Note that a molecule at the surface experiences a net force toward the interior of the liquid, whereas a molecule in the interior experiences no net force.

▶ Demonstration of surface tension: Bassam Z. Shakhashiri, *Chemical Demonstrations,* Vol. 3 (Madison: University of Wisconsin Press, 1989), no. 9.23, p. 301.

**Figure 11.18**
**Demonstration of surface tension of water.** A steel pin will float on the surface of water because of surface tension (*left*). The water surface is depressed and slightly stretched by the pin (*right*), but even more surface would have to be created for the pin to submerge, which takes more energy. The easiest way to float a pin is to place a square of tissue paper, with the pin on top of it, on the prongs of a fork. When the fork is lowered into the water, the tissue and pin float on the surface. The paper soon becomes water-logged and sinks, leaving the pin floating.

**Figure 11.17**
**Water bug on a water surface.** The water bug seems to skate on the surface "skin" of the water, which is actually a result of surface tension.

## Surface Tension

As Figure 11.16 illustrates, a molecule within the body of a liquid tends to be attracted equally in all directions, so that it experiences no net force. On the other hand, a molecule at the surface of a liquid experiences a net attraction by other molecules toward the interior of the liquid. As a result, there is a tendency for the surface area of a liquid to be reduced as much as possible. This explains why falling raindrops are nearly spherical. (The sphere has the smallest surface area for a given volume of any geometrical shape.)

Energy is needed to reverse the tendency toward reduction of surface area in liquids. **Surface tension** is *the energy required to increase the surface area of a liquid by a unit amount.* Column 4 in Table 11.3 lists values of the surface tension of some liquids at 20°C.

The surface tension of a liquid can be affected by dissolved substances. Soaps and detergents, in particular, drastically decrease the surface tension of water. An interesting but simple experiment shows the effect of soap on surface tension. Because of surface tension, a liquid behaves as though it had a skin. Water bugs seem to skitter across this skin as if ice-skating (Figure 11.17). You can actually float a pin on water, if you carefully lay it across the surface (Figure 11.18). If you then put a drop of soap solution onto the water, the soap spreads across the surface and the pin sinks. Water bugs also sink in soapy water.

**Figure 11.19**
**Liquid levels in capillaries.**
*(A)* Capillary rise, due to the attraction of water and glass. The final water level in the capillary is a balance between the force of gravity and the surface tension of water. *(B)* Depression, or lowering, of mercury level in a glass capillary. Unlike water, mercury is not attracted to glass.

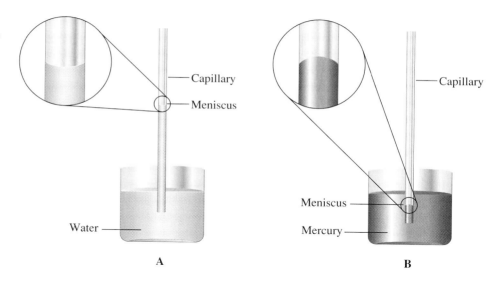

► Demonstration of viscosity: Bassam Z. Shakhashiri, *Chemical Demonstrations,* Vol. 3 (Madison: University of Wisconsin Press, 1989), no. 9.27, p. 313.

**Figure 11.20**
**Comparison of the viscosities of two liquids.** Similar steel balls were dropped simultaneously into two graduated cylinders, the right one containing water and the left one glycerol. A steel ball takes considerably longer to fall through a column of glycerol than through a similar column of water, because glycerol has a greater viscosity than does water.

*Capillary rise* is a phenomenon related to surface tension. When a small-diameter glass tube, or capillary, is placed upright in water, a column of the liquid rises in the tube (Figure 11.19*A*). This capillary rise can be explained in the following way: water molecules happen to be attracted to glass. Because of this attraction, a thin film of water starts to move up the inside of the glass capillary. But in order to reduce the surface area of this film, the water level begins to rise also. The final water level is a balance between the surface tension and the potential energy required to lift the water against gravity. Note that the *meniscus,* or liquid surface within the tube, has its edges curved upward; that is, it has a concave shape.

In the case of mercury, the liquid level in the capillary is lower than for the liquid outside (Figure 11.19*B*). Also, the meniscus has its edges curved downward; that is, it has a convex shape. The difference between Figures 19*B* and 19*A* results from the fact that the attraction between mercury atoms is greater than the attraction of mercury atoms for glass, in contrast to the situation for water and glass.

## Viscosity

**Viscosity** is *the resistance to flow that is exhibited by all liquids and gases.* The viscosity of a liquid can be obtained by measuring the time it takes for a given quantity to flow through a capillary tube or, alternatively, by the time it takes for a steel ball of given radius to fall through a column of liquid (Figure 11.20). Even without such quantitative measurements, you know from experience that the viscosity of syrup is greater than that of water; syrup is a *viscous* liquid. Viscosity is an important characteristic of motor oils. In the United States, the Society of Automotive Engineers (SAE) has established numbers to indicate the viscosity of motor oils at a given temperature (0°F, or −18°C); higher numbers indicate greater viscosities at that temperature. The viscosity of a simple oil increases as the temperature decreases, so an SAE 40 oil that is appropriate for an auto engine in the summer would be too viscous in very cold weather. An oil designated as SAE 10W/40 is an oil mixture that behaves like an SAE 10 oil in the winter (W), but like an SAE 40 oil in the summer. Column 5 in Table 11.3 gives the viscosities (in SI units) of some liquids at 20°C.

# 11.5 Intermolecular Forces; Explaining Liquid Properties

In the preceding section, we described a number of properties of liquids. Now we want to explain these properties. Many of the physical properties of liquids (and certain solids too) can be explained in terms of **intermolecular forces,** *the forces of interaction between molecules.* These forces are normally weakly attractive.

One of the most direct indications of the attraction between molecules is the heat of vaporization of liquids. Consider a substance like neon, which consists of molecules that are single atoms. (There is no tendency for these atoms to bond chemically.) Neon is normally a gas, but it liquefies when the temperature is lowered to $-246°C$ at 1.00 atm pressure. The heat of vaporization of the liquid at this temperature is 1.77 kJ/mol. Some of this energy (0.23 kJ/mol) is needed to push back the atmosphere when the vapor forms. The remaining energy (1.54 kJ/mol) must be supplied to overcome intermolecular attractions. Because a molecule in a liquid is surrounded by several neighbor molecules, this remaining energy is some multiple of a single molecule–molecule interaction. Typically, this multiple is about 5, so we expect the neon–neon interaction energy to be about 0.3 kJ/mol. Other experiments yield a similar value. By comparison, the energy of attraction between two hydrogen atoms in the hydrogen molecule is 432 kJ/mol. Thus, the energy of attraction between neon atoms is about a thousand times smaller than that between atoms in a chemical bond.

Attractive intermolecular forces can be larger than those in neon. For example, chlorine, $Cl_2$, and bromine, $Br_2$, have intermolecular attractive energies of 3.0 kJ/mol and 4.3 kJ/mol, respectively. But even these values are much smaller than bond energies.

Three types of attractive forces are known to exist between neutral molecules: dipole–dipole forces, London (or dispersion) forces, and hydrogen bonding forces. Each of these will be explained in this section. The term **van der Waals forces** is *a general term for those intermolecular forces that include dipole–dipole and London forces.* ◀ Van der Waals forces are the weak attractive forces in a large number of substances, including Ne, $Cl_2$, and $Br_2$, which we just discussed. Hydrogen bonding occurs in substances containing hydrogen atoms bonded to certain very electronegative atoms. Approximate energies of intermolecular attractions are compared with those of chemical bonds in Table 11.4.

▶ **The Dutch physicist Johannes van der Waals (1837–1923) was the first to suggest the importance of intermolecular forces and used the concept to derive his equation for gases. (See van der Waals equation, Section 5.8.)**

**Table 11.4**
**Types of Intermolecular and Chemical Bonding Interactions**

| Type of Interaction | Approximate Energy (kJ/mol) |
|---|---|
| Intermolecular | |
|     Van der Waals (dipole–dipole, London) | 0.1 to 10 |
|     Hydrogen bonding | 10 to 40 |
| Chemical bonding | |
|     Ionic | 100 to 1000 |
|     Covalent | 100 to 1000 |

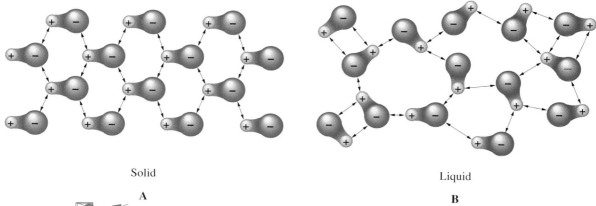

Solid

Liquid

A

B

**Figure 11.21**
**Alignment of polar molecules of HCl.**
(A) Molecules tend to line up in the solid so that positive ends point to negative ends. (B) The normal random motion of molecules in a liquid only partially disrupts this alignment of polar molecules.

▶ Dipole moments were discussed in Section 10.2.

## Dipole–Dipole Forces

Polar molecules can attract one another through dipole–dipole forces. The **dipole–dipole force** is *an attractive intermolecular force resulting from the tendency of polar molecules to align themselves such that the positive end of one molecule is near the negative end of another.* Recall that a polar molecule has a dipole moment as a result of the electronic structure of the molecule. ◀ For example, hydrogen chloride, HCl, is a polar molecule because of the difference in electronegativities of the H and Cl atoms.

$$\overset{\delta+}{H}\text{—}\overset{\delta-}{Cl}$$

Figure 11.21 shows the alignment of polar molecules in the case of HCl. Note that this alignment creates a net attraction between molecules. This attractive force is partly responsible for the fact that hydrogen chloride, a gas at room temperature, becomes a liquid when cooled to $-85°C$. At this temperature, HCl molecules have slowed enough (that is, have a low enough kinetic energy) for the intermolecular forces to hold the molecules in the liquid state.

## London (Dispersion) Forces

In nonpolar molecules, there can be no dipole–dipole force. Yet such substances liquefy, so there must be another type of attractive intermolecular force. Even in hydrogen chloride, calculations show that the dipole–dipole force accounts for less than 20% of the total energy of attraction. What is the nature of the additional attractive force?

In 1930 Fritz London found he could account for a weak attraction between any two molecules. Let us use neon to illustrate his argument. The electrons of this atom move about the nucleus so that over a period of time they have a spherical distribution. The electrons spend as much time on one side of the nucleus as on the other. Experimentally, you find that the atom has no dipole moment. However, at any instant, more electrons may be on one side of the nucleus than on the other. The atom becomes a small, instantaneous dipole, with one side having a partial negative charge, $\delta^-$, and the other side having a partial positive charge, $\delta^+$ (see Figure 11.22A).

Now imagine that another neon atom is close to this first atom—say, next to the partial negative charge that appeared at that instant. This partial negative charge

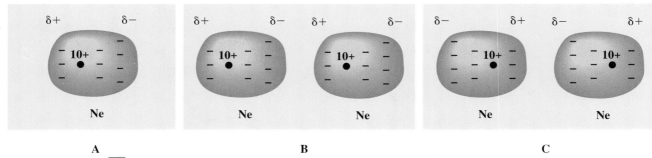

A B C

**Figure 11.22**
**Origin of the London force.** *(A)* At some instant, there are more electrons on one side of a neon atom than on the other. *(B)* If this atom is near another neon atom, the electrons on that atom are repelled. The result is two instantaneous dipoles, which give an attractive force. *(C)* Later the electrons on both atoms have moved, but they tend to move together, which gives an attractive force between the atoms.

repels the electrons in the second atom, so at this instant it too becomes a small dipole (see Figure 11.22B). Note that these two dipoles are oriented with the partial negative charge of one atom next to the partial positive charge of the other atom. Therefore, an attractive force exists between the two atoms.

Electrons are in constant motion, but the motion of electrons on one atom affects the motion of electrons on the other atom. As a result, the *instantaneous dipoles* of the atoms tend to change together, maintaining a net attractive force (compare *B* and *C* of Figure 11.22). Such instantaneous changes of electron distributions occur in all molecules. For this reason, an attractive force always exists between any two molecules.

Thus, you see that **London forces** (also called **dispersion forces**) are *the weak attractive forces between molecules resulting from the small, instantaneous dipoles that occur because of the varying positions of the electrons during their motion about nuclei.*

London forces tend to increase with molecular weight. This is because molecules with larger molecular weight usually have more electrons, and London forces increase in strength with the number of electrons. Also, larger molecular weight often means larger atoms, which are more *polarizable* (more easily distorted to give instantaneous dipoles because the electrons are farther from the nuclei). The relationship of London forces to molecular weight is only approximate, however. For molecules of about the same molecular weight, the more compact one is probably less polarizable, so the London forces are smaller. Consider pentane, isopentane (2-methylbutane), and neopentane (2,2-dimethylpropane).

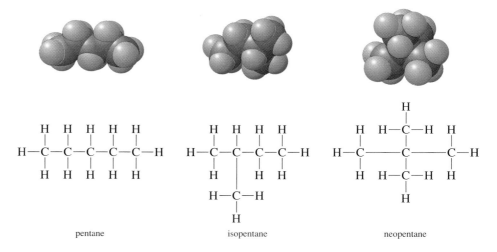

pentane                 isopentane                 neopentane

All have the same molecular formula, $C_5H_{12}$, and thus the same molecular weight, but they differ in the arrangement of the atoms. Pentane is a long chain of carbon atoms to which hydrogen atoms are attached. Isopentane and neopentane, however, have increasingly more compact arrangements of atoms. As a result, you expect London forces to decrease from pentane to isopentane and from isopentane to neopentane. This is what you find in the heats of vaporization. The values are 25.8 kJ/mol, 24.7 kJ/mol, and 22.8 kJ/mol for pentane, isopentane, and neopentane, respectively.

## Van der Waals Forces and the Properties of Liquids

▶ **Video no. 29 (Series D),**
**"Styrofoam Cup and Acetone."**

The vapor pressure of a liquid depends on intermolecular forces, because the ease or difficulty with which a molecule leaves the liquid depends on the strength of its attraction to other molecules. When the intermolecular forces in the liquid are strong, you expect the vapor pressure to be low.

Let us look again at Table 11.3, which lists vapor pressures as well as surface tensions and viscosities of some liquids at 20°C. The intermolecular attractions in these liquids (except for water and glycerol) are due entirely to van der Waals forces. Of the van der Waals forces, the London force is always present and usually dominant. The dipole–dipole force is usually appreciable only in small polar molecules or in large molecules having very large dipole moments. As noted earlier, the dipole–dipole attraction in the small, polar molecule HCl accounts for less than 20% of its interaction energy.

The liquids in Table 11.3 are listed in order of increasing molecular weight. Recall that London forces tend to increase with molecular weight. Therefore, London forces would be expected to increase for these liquids from the top to the bottom of the table. If London forces are the dominant attractive forces in these liquids, the vapor pressures should decrease from top to bottom in the table. And this is what you see, except for water and glycerol. (Their vapor pressures are relatively low, and an additional force, hydrogen bonding, is needed to explain them.)

The normal boiling point of a liquid must depend on intermolecular forces, because it is related to the vapor pressure. Normal boiling points are approximately proportional to the energy of intermolecular attraction. Thus, they are lowest for liquids with the weakest intermolecular forces.

Surface tension also depends on intermolecular forces. Surface tension is the energy needed to increase the surface area of a liquid. To increase the surface area, it is necessary to pull molecules apart against the intermolecular forces of attraction. Thus, surface tension would be expected to increase with the strength of attractive forces. If London forces are the dominant attractive forces in the liquid, surface tension should increase with molecular weight—that is, it should increase from top to bottom in Table 11.3. This is what you see, except for water and glycerol, which have relatively high surface tensions.

The viscosity of a liquid depends in part on intermolecular forces, because increasing the attractive forces between molecules increases the resistance to flow. If London forces are dominant, you should expect the viscosity of the liquids in Table 11.3 to increase from the top of the column to the bottom, which it does (again with the exception of water and glycerol; these liquids have relatively high viscosities). Note, however, that the viscosity also depends on other factors, such as the possibility of molecules tangling together. Liquids with long molecules that tangle together are expected to have high viscosities.

CH₃F

CH₃OH

**Figure 11.23**
**Fluoromethane and methanol.** Space-filling molecular models. *(Top)* fluoromethane, $CH_3F$; *(bottom)* methanol, $CH_3OH$.

## Hydrogen Bonding

It is interesting to compare fluoromethane, $CH_3F$, and methanol, $CH_3OH$ (Figure 11.23). They have about the same molecular weight (34 for $CH_3F$ and 32 for $CH_3OH$) and about the same dipole moment (1.81 D for $CH_3F$ and 1.70 D for $CH_3OH$). You might expect these substances to have about the same intermolecular attractive forces and therefore about the same boiling points. In fact, the boiling points are quite different. Fluoromethane boils at $-78°C$ and is a gas under normal conditions. Methanol boils at 65°C and is normally a liquid. There is apparently an intermolecular attraction in methanol that is not present in fluoromethane.

We have already seen that the properties of water and glycerol cannot be explained in terms of van der Waals forces alone. What water, glycerol, and methanol have in common is one or more —OH groups.

water          glycerol          methanol

Molecules that have the —OH group are subject to an additional attractive force called hydrogen bonding. **Hydrogen bonding** is *a weak to moderate attractive force that exists between a hydrogen atom covalently bonded to a very electronegative atom, X, and a lone pair of electrons on another small, electronegative atom, Y.* Hydrogen bonding is represented in structural drawings by a series of dots.

$$—X—H\cdots Y—$$

Usually, hydrogen bonding is seen in cases where X and Y are the atoms F, O, or N.

Figure 11.24*A* shows a plot of boiling point versus molecular weight for the hydrides (binary compounds with hydrogen) of the Group VIA elements. If London forces were the only intermolecular forces present in this series of compounds, the boiling points should increase regularly with molecular weight. You do see such a regular increase in boiling point for $H_2S$, $H_2Se$, and $H_2Te$. On the other hand, $H_2O$ has a much higher boiling point than you would expect if only London forces were present. This is consistent with the view that hydrogen bonding exists in $H_2O$ but is essentially absent in $H_2S$, $H_2Se$, and $H_2Te$.

Studies of the structure of $H_2O$ in its different physical states show that hydrogen bonding is present. The hydrogen atom of one water molecule is attracted to the electron pair of another water molecule (Figure 11.25). Many $H_2O$ molecules can be linked this way to form clusters of molecules. Similar hydrogen bonding is seen in other molecules containing —O—H groups.

Figure 11.24*B* shows plots of boiling point versus molecular weight for Group VIIA, VA, and IVA hydrides. Hydrogen fluoride, HF, and ammonia, $NH_3$, have particularly high boiling points compared with other hydrides of their periodic group.

**Figure 11.24**
**Boiling point versus molecular weight for hydrides.** (A) Plot for the Group VIA hydrides. Note that the boiling points of $H_2S$, $H_2Se$, and $H_2Te$ increase fairly smoothly, as you would expect if London forces are dominant. $H_2O$ has a much higher boiling point, indicating an additional intermolecular attraction. (B) Plot for the hydrides of Groups VIIA, VA, and IVA.

A

B

**Figure 11.25**
**Hydrogen bonding in water.** (A) The electrons in the O—H bonds of $H_2O$ molecules are attracted to the oxygen atoms, leaving the positively charged protons partially exposed. A proton on one water molecule is attracted to a lone pair on an oxygen atom in another water molecule. (B) Lewis structures of water with hydrogen bonding between molecules.

**Figure 11.26**
**Hydrogen bonding between two biologically important molecules.**
This computer-drawn figure shows the electron densities of two molecules (guanine and cytosine); the electron densities increase as the color changes from blue to green to yellow to red. Note the three H bonds between the two molecules. Each H bond is between an H atom with low electron density (exposing the positive charge on the atom) and the lone electron pair (high density) of either an O or N atom. (The molecules are guanine and cytosine, which are two of the four bases used as codes in DNA; this is explained in Chapter 25.) (Electron density figure taken from G.D. Shusterman, A.J. Shusterman, *J. Chem. Ed.*, 74, 771, 1997.)

**Understanding Concepts: Intermolecular Forces**

▶ Video no. 30 (Series D), "Iodine/Carbon Tetrachloride/Water."

▶ An alternate example is provided in the Instructor's Resource Manual.

Definite structural evidence has established hydrogen bonding in HF. In solid hydrogen fluoride, the structure shows a zigzag arrangement involving hydrogen bonding.

The hydrogen bonding attraction between two water molecules may be explained in part on the basis of the dipole moment of the —O—H bond.

$$\overset{\delta-\quad\ \delta+}{-O-H}$$

The partial positive charge on the hydrogen atom of one water molecule is attracted to the partial negative charge of a lone pair on another water molecule. In addition, however, a hydrogen atom covalently bonded to an electronegative atom appears to be special. The electrons in the —O—H bond are drawn to the O atom, leaving the concentrated positive charge of the proton partially exposed. This concentrated positive charge is strongly attracted to a lone pair of electrons on another O atom. Figure 11.26 shows the hydrogen bonding between two biologically important molecules.

---

**EXAMPLE 11.5**

**Identifying Intermolecular Forces**

What kinds of intermolecular forces (London, dipole–dipole, hydrogen bonding) are expected in the following substances?

a. methane, $CH_4$
b. (see model at right)
c. butanol (butyl alcohol), $CH_3CH_2CH_2CH_2OH$

**SOLUTION**

a. Methane is a nonpolar molecule. Hence, the only intermolecular attractions are **London forces**. b. Trichloromethane is an unsymmetrical molecule with polar bonds. Thus, we expect **dipole–dipole forces,** in addition to **London forces.** c. Butanol has a hydrogen atom attached to an oxygen atom. Therefore, you expect **hydrogen bonding.** Because the molecule is polar (from the O—H bond), you also expect **dipole–dipole forces. London forces** exist too, because such forces exist between all molecules.

## EXERCISE 11.5

List the different intermolecular forces you would expect for each of the following compounds:

a. propanol, $CH_3CH_2CH_2OH$
b. (see model at right)
c. sulfur dioxide, $SO_2$

See Problems 11.55 and 11.56.

## EXAMPLE 11.6

> An alternate example is provided in the Instructor's Resource Manual.

# Determining Relative Vapor Pressure on the Basis of Intermolecular Attraction

For each of the following pairs, choose the substance you expect to have the lower vapor pressure at a given temperature: a. carbon dioxide ($CO_2$) or sulfur dioxide ($SO_2$); b. dimethyl ether ($CH_3OCH_3$) or ethanol ($CH_3CH_2OH$).

## SOLUTION

a. London forces increase roughly with molecular weight. The molecular weights for $SO_2$ and $CO_2$ are 64 amu and 44 amu, respectively. Therefore, the London forces between $SO_2$ molecules should be greater than between $CO_2$ molecules. Moreover, because $SO_2$ is polar but $CO_2$ is not, there are dipole–dipole forces between $SO_2$ molecules but not between $CO_2$ molecules. **You conclude that sulfur dioxide has the lower vapor pressure.** Experimental values of the vapor pressure at 20°C are: $CO_2$, 56.3 atm; $SO_2$, 3.3 atm. b. Dimethyl ether and ethanol have the same molecular formulas but different structural formulas. The structural formulas are

$$
\begin{array}{ccc}
\text{H} & & \text{H} \\
| & & | \\
\text{H}-\text{C}-\text{O}-\text{C}-\text{H} & & \text{H}-\text{C}-\text{C}-\text{O}-\text{H} \\
| & & | \\
\text{H} & & \text{H}
\end{array}
$$

dimethyl ether          ethanol

The molecular weights are equal, and therefore the London forces are approximately the same. However, there will be strong hydrogen bonding in ethanol but not in dimethyl ether. **You expect ethanol to have the lower vapor pressure.** Experimental values of vapor pressure at 20°C are: $CH_3OCH_3$, 4.88 atm; $CH_3CH_2OH$, 0.056 atm.

## EXERCISE 11.6

Arrange the following hydrocarbons in order of increasing vapor pressure: ethane, $C_2H_6$; propane, $C_3H_8$; and butane, $C_4H_{10}$. Explain your answer.

See Problems 11.59 and 11.60.

### EXERCISE 11.7

At the same temperature, methyl chloride, $CH_3Cl$, has a vapor pressure of 1490 mmHg, and ethanol has a vapor pressure of 42 mmHg. Explain why you might expect methyl chloride to have a higher vapor pressure than ethanol, even though methyl chloride has a somewhat larger molecular weight.

**See Problems 11.61 and 11.62.**

### CONCEPT CHECK 11.3

A common misconception is that the following chemical reaction occurs when boiling water:

$$2H_2O(l) \longrightarrow 2H_2(g) + O_2(g)$$

instead of

$$H_2O(l) \longrightarrow H_2O(g)$$

a. What evidence do you have that the second reaction is correct?

b. How would the enthalpy of the wrong reaction compare with that of the correct reaction?

c. How could you calculate the enthalpy change for the wrong reaction (Chapter 6)?

# Solid State

A solid is a nearly incompressible state of matter with a well-defined shape, because the units (atoms, molecules, or ions) making up the solid are in close contact and in fixed positions or sites. In the next section, we will look at the kinds of forces holding the units together in different types of solids. In later sections, we will look at crystalline solids and their structure.

## 11.6 Classification of Solids by Type of Attraction of Units

**Understanding Concepts: Classification of Solids**

A solid consists of structural units—atoms, molecules, or ions—that are attracted to one another strongly enough to give a rigid substance. One way to classify solids is by the type of force holding the structural units together. In some cases, these forces are intermolecular and hold molecules together. In other cases, these forces are chemical bonds (metallic, ionic, or covalent) and hold atoms together. From this point of view, then, there are four different types of solids: molecular, metallic, ionic, and covalent. It should be noted, however, that a given ionic bond may have considerable covalent character or vice versa, so the distinction between ionic and covalent solids is not a precise one.

154 pm

335 pm

142 pm

Diamond

Graphite

**Figure 11.27**
**Structures of diamond and graphite.** Diamond is a three-dimensional network solid; every carbon atom is covalently bonded to four others. Graphite consists of sheets of carbon atoms covalently bonded to form a hexagonal pattern of atoms; the sheets are held together by van der Waals forces.

## Types of Solids

A **molecular solid** is *a solid that consists of atoms or molecules held together by intermolecular forces.* Many solids are of this type. Examples are solid neon, solid water (ice), and solid carbon dioxide (dry ice).

A **metallic solid** is *a solid that consists of positive cores of atoms held together by a surrounding "sea" of electrons (metallic bonding).* In this kind of bonding, positively charged atomic cores are surrounded by delocalized electrons. Examples include iron, copper, and silver.

An **ionic solid** is *a solid that consists of cations and anions held together by the electrical attraction of opposite charges (ionic bonds).* Examples are cesium chloride, sodium chloride, and zinc sulfide (but ZnS has considerable covalent character).

A **covalent network solid** is *a solid that consists of atoms held together in large networks or chains by covalent bonds.* Diamond is an example of a three-dimensional network solid. Every carbon atom in diamond is covalently bonded to four others, so an entire crystal might be considered an immense molecule. It is possible to have similar two-dimensional "sheet" and one-dimensional "chain" molecules, with atoms held together by covalent bonds. Examples are diamond (three-dimensional network), graphite (sheets), and asbestos (chains). Figure 11.27 shows the structures of diamond and graphite.

Table 11.5 summarizes these four types of solids.

▶ An alternate example is provided in the Instructor's Resource Manual.

---

### EXAMPLE 11.7

## Identifying Types of Solids

Which of the four basic types of solids would you expect the following substances to be? a. solid ammonia, $NH_3$; b. cesium, Cs; c. cesium iodide, CsI; d. silicon, Si.

**Table 11.5
Types of Solids**

| Type of Solid | Structural Units | Attractive Forces Between Structural Units | Examples |
|---|---|---|---|
| Molecular | Atoms or molecules | Intermolecular forces | Ne, $H_2O$, $CO_2$ |
| Metallic | Atoms (positive cores surrounded by electron "sea") | Metallic bonding (extreme delocalized bond) | Fe, Cu, Ag |
| Ionic | Ions | Ionic bonding | CsCl, NaCl, ZnS |
| Covalent network | Atoms | Covalent bonding | Diamond, graphite, asbestos |

**SOLUTION**

a. Ammonia has a molecular structure; therefore, it freezes as a **molecular solid.**
b. Cesium is a metal; it is a **metallic solid.** c. Cesium iodide is an ionic substance; it exists as an **ionic solid.** d. Silicon atoms might be expected to form covalent bonds to other silicon atoms, as carbon does in diamond. A **covalent network solid** would result.

**EXERCISE 11.8**

Classify each of the following solids according to the forces of attraction that exist between the structural units: a. zinc; b. sodium iodide, NaI; c. silicon carbide, SiC; d. methane, $CH_4$.

See Problems 11.65, 11.66, 11.67, and 11.68.

### Physical Properties

Many physical properties of a solid can be directly related to its structure. Let us look at several of these properties.

**Melting Point and Structure**   Table 11.2 lists the melting points for various solids and also gives the type of solid. For a solid to melt, the forces holding the structural units in their sites must be overcome, at least partially. In a molecular solid, these forces are weak intermolecular attractions. Thus, molecular solids tend to have low melting points (usually below 300°C). At room temperature, many molecular substances are either liquids (such as water and ethanol) or gases (such as carbon dioxide and ammonia). Note also that the melting points of molecular solids reflect the strengths of intermolecular attractions, just as boiling points do. This can be seen in the melting points of the molecular solids listed in Table 11.2.

For ionic solids and covalent network solids to melt, chemical bonds must be broken. For that reason, the melting points of these types of solids are relatively high. The melting point of the ionic solid sodium chloride is 801°C; that of magnesium oxide is 2800°C. Melting points of covalent network solids are generally quite high. Quartz, for example, melts at 1610°C; diamond, at 3550°C. See Table 11.2. ◄

▶ **Both molecular and covalent network solids have covalent bonds, but none of the covalent bonds are broken during the melting of a molecular solid.**

▶ Lattice energy is the energy needed to separate a crystal into isolated ions in the gaseous state. It represents the strength of attraction of ions in the solid.

The difference between the melting points of sodium chloride (801°C) and magnesium oxide (2800°C) can be explained in terms of the charges on the ions. You expect the melting point to rise with lattice energy. ◀ Lattice energies, however, depend on the product of the ionic charges (and inversely on the distance between charges). For sodium chloride ($Na^+Cl^-$) this ion-charge product is $(+1) \times (-1) = -1$, whereas for magnesium oxide ($Mg^{2+}O^{2-}$) it is $(+2) \times (-2) = -4$. Thus, the lattice energy is greater for magnesium oxide, and this is reflected in its much higher melting point.

Metals often have high melting points, but there is considerable variability. Mercury, a liquid at room temperature, melts at $-39°C$. Tungsten melts at 3410°C, the highest melting point of any metallic element. In general, melting points are low for the Group IA and IIA elements but increase as you move right in the periodic table to the transition metals. The elements in the center of the transition series (such as tungsten) have high melting points. Then, as you move farther to the right, the melting points decrease and are again low for the Group IIB elements.

▶ An alternate example is provided in the Instructor's Resource Manual.

## EXAMPLE 11.8

### Determining Relative Melting Points Based on Types of Solids

Arrange the following elements in order of increasing melting point: silicon, hydrogen, lithium. Explain your reasoning.

### SOLUTION

Hydrogen is a molecular substance ($H_2$) with a very low molecular weight (2 amu). You expect it to have a very low melting point. Lithium is a Group IA element and is expected to be a relatively low-melting metal. Except for mercury, however, all metals are solids below 25°C. So the melting point of lithium is well above that of hydrogen. Silicon might be expected to have a covalent network structure like carbon and to have a high melting point. Therefore, the list of these elements in order of increasing melting point is **hydrogen, lithium, silicon.**

## EXERCISE 11.9

Decide what type of solid is formed for each of the following substances: $C_2H_5OH$, $CH_4$, $CH_3Cl$, $MgSO_4$. On the basis of the type of solid and the expected magnitude of intermolecular forces (for molecular crystals), arrange these substances in order of increasing melting point. Explain your reasoning.

See Problems 11.69 and 11.70.

**Hardness and Structure** Hardness depends on how easily the structural units of a solid can be moved relative to one another and therefore on the strength of attractive forces between the units. Thus, molecular crystals, with weak intermolecular forces, are rather soft compared with ionic crystals, in which the attractive forces are much stronger. A three-dimensional covalent network solid is usually quite hard because of the rigidity given to the structure by strong covalent bonds throughout it. Diamond and silicon carbide (SiC), three-dimensional covalent networks, are among the hardest substances known.

**Figure 11.28**
**Behavior of crystals when struck.** Lead is malleable; rock salt cracks along crystal planes when struck.

We should add that molecular and ionic crystals are generally brittle because they fracture easily along crystal planes. Metallic crystals, by contrast, are *malleable;* that is, they can be shaped by hammering (Figure 11.28).

**Electrical Conductivity and Structure**    One of the characteristic properties of metals is good electrical conductivity. The delocalized valence electrons are easily moved by an electrical field and are responsible for carrying the electric current. By contrast, most covalent and ionic solids are nonconductors, because the electrons are localized to particular atoms or bonds. Ionic substances do become conducting in the liquid state, however, because the ions can move. In an ionic liquid, it is the ions that carry the electric current.

Table 11.6 summarizes the properties we discussed for the different types of solids.

**Table 11.6**
**Properties of the Different Types of Solids**

| Type of Solid | Melting Point | Hardness and Brittleness | Electrical Conductivity |
|---|---|---|---|
| Molecular | Low | Soft and brittle | Nonconducting |
| Metallic | Variable | Variable hardness; malleable | Conducting |
| Ionic | High to very high | Hard and brittle | Nonconducting solid (conducting liquid) |
| Covalent network | Very high | Very hard | Usually nonconducting |

# 11.7  Crystalline Solids; Crystal Lattices and Unit Cells

Solids can be crystalline or amorphous. A **crystalline solid** *is composed of one or more crystals; each crystal has a well-defined ordered structure in three dimensions.* Sodium chloride (table salt) and sucrose (table sugar) are examples of crystalline substances. Metals are usually compact masses of crystals. An **amorphous solid** *has a disordered structure; it lacks the well-defined arrangement of basic units (atoms, molecules, or ions) found in a crystal.* A glass is an amorphous solid obtained by cooling a liquid rapidly enough that its basic units are "frozen" in random positions before they can assume an ordered crystalline arrangement. Common window glass is an example, as is obsidian, a natural glass formed when molten rock from a volcanic eruption cools quickly. ◄

▶ When quartz crystals ($SiO_2$) are melted and then cooled rapidly, they form a glass called silica glass. Though ideal for some purposes, silica glass is difficult to work with because of its high temperature of softening. This temperature can be lowered by adding various oxides, such as $Na_2O$ and $CaO$. In practice, ordinary glass is formed by melting sand (quartz crystals) with sodium carbonate and calcium carbonate. The carbonates decompose to the oxides plus carbon dioxide.

## Crystal Lattices

A crystal is a three-dimensional ordered arrangement of basic units. (The basic unit is either an atom, molecule, or ion, depending on the type of crystal.) The ordered structure of a crystal is conveniently described in terms of a *crystal lattice.* The idea of a crystal lattice is most easily grasped by looking at two-dimensional repeating patterns such as those seen on many fabrics or wallpapers. For simplicity, consider the pattern of *A*'s shown in Figure 11.29*A*. This pattern consists of a repetition of basic units *A*. A lattice of points is obtained by choosing the same point in each basic unit. (The choice of this point is immaterial as long as the same point is chosen in each basic unit.) In the pattern of Figure 11.29, we have chosen the points to be midway on the crossbar of each *A*. The collection of these points constitutes the lattice of the pattern (Figure 11.29*B*). This lattice shows the essential arrangement of the basic units (*A*'s) in the pattern. We could have chosen a different location for the point in each *A*, but we would have obtained the same lattice.

Similarly, we define a **crystal lattice** as *the geometric arrangement of lattice points of a crystal, in which we choose one lattice point at the same location within each of the basic units of the crystal.* Figure 11.30 shows the crystal structure of copper metal. If we place a point at the same location in each copper atom (say at the center of the atom), we obtain the crystal lattice for copper. The crystal lattice and crystal structure are not the same. We can imagine building the crystal structure for copper from its crystal lattice by placing a copper atom at each lattice point. Many different crystals have the same crystal lattice. For instance, the crystal lattice for solid methane, $CH_4$, is the same (except for the distances between points) as that for copper. We obtain the methane crystal structure by placing $CH_4$ molecules at the lattice points. The crystal lattice shows only the arrangement of basic units of the crystal.

**Figure 11.29**
**A two-dimensional pattern.** *(A)* A pattern of *A*'s. A point midway on the crossbar of each *A* has been selected for a lattice point. *(B)* The pattern of lattice points. *(C)* Division of the pattern into unit cells. *(D)* A single unit cell.

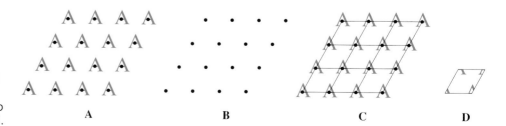

A                    B                    C                    D

**Figure 11.30**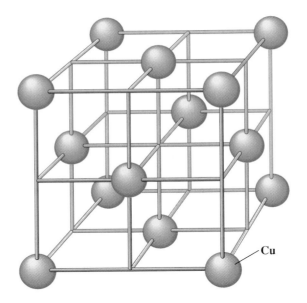
**Crystal structure and crystal lattice of copper.** The copper atoms have been shrunk so that the crystal structure is more visible. The crystal lattice is the geometric arrangement of lattice points, which you can take to be the centers of the atoms. Lines have been drawn to emphasize the geometry of the lattice.

You can imagine dividing a crystal into many equivalent cells, or *unit cells,* so that these cells become the "bricks" from which you mentally construct the crystal. Let us look at the two-dimensional pattern in Figure 11.29*C,* in which we have connected lattice points by straight lines. The parallelograms formed by the straight lines divide the pattern into small cells. These are the smallest possible cells from which you could imagine constructing the pattern by stacking the individual cells—they are the unit cells of the pattern (Figure 11.29*D*).

In the same way, you can divide any crystal lattice into boxlike cells, or unit cells. The **unit cell** of a crystal is *the smallest boxlike unit (each box having faces that are parallelograms) from which you can imagine constructing a crystal by stacking the units in three dimensions.*

There are seven basic shapes possible for unit cells, which give rise to the seven *crystal systems* used to classify crystals. A crystal belonging to a given crystal system has a unit cell with one of the seven shapes shown in Figure 11.31. Each unit-cell shape can be characterized by the angles between the edges of the unit cell and the relative lengths of those edges. A crystal belonging to the cubic system, for example, has 90° angles between the edges of the unit cell, and all edges are of equal length. Table 11.7 summarizes the relationships between the angles and edge lengths for the unit cell of each crystal system.

Most of the crystal systems have more than one possible crystal lattice. A *simple* (or *primitive*) lattice has a unit cell in which there are lattice points only at the corners of the unit cell. Other lattices in the same crystal system have additional lattice points either within the *body* of the unit cell or on *faces* of the unit cell. As an example, we will describe the cubic crystal system in some detail.

### Cubic Unit Cells

The cubic crystal system has three possible cubic unit cells: simple (or primitive) cubic, body-centered cubic, and face-centered cubic. These cells are illustrated in Figure 11.32. A **simple cubic unit cell** is *a cubic unit cell in which lattice points are situated*

**Visualization: Unit Cells and Crystal Packing**

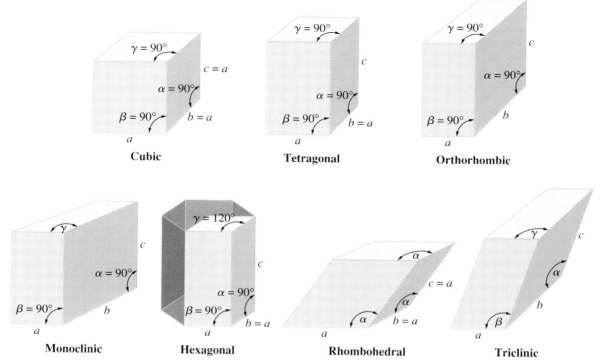

**Figure 11.31**
**Unit-cell shapes of the different crystal systems.**
The unit cells of the seven different crystal systems are displayed, showing the relationships between edge lengths and angles (summarized in Table 11.7).

only at the corners. A **body-centered cubic unit cell** is *a cubic unit cell in which there is a lattice point at the center of the cubic cell in addition to those at the corners.* A **face-centered cubic unit cell** is *a cubic unit cell in which there are lattice points at the centers of each face of the unit cell in addition to those at the corners.*

The simplest crystal structures are those in which there is only a single atom at each lattice point. Most metals are examples. Copper metal has a face-centered cubic unit cell with one copper atom at each lattice point (see Figure 11.30). The unit cells for this and other cubic cells of simple atomic crystals are shown in Figure 11.33.

**Table 11.7**
**The Seven Crystal Systems**

| Crystal System | Edge Length | Angles | Examples |
|---|---|---|---|
| Cubic | $a = b = c$ | $\alpha = \beta = \gamma = 90°$ | $NaCl$, $Cu$ |
| Tetragonal | $a = b \neq c$ | $\alpha = \beta = \gamma = 90°$ | $TiO_2$ (rutile), $Sn$ (white tin) |
| Orthorhombic | $a \neq b \neq c$ | $\alpha = \beta = \gamma = 90°$ | $CaCO_3$ (aragonite), $BaSO_4$ |
| Monoclinic | $a \neq b \neq c$ | $\alpha = \beta = 90°, \gamma \neq 90°$ | $PbCrO_4$ |
| Hexagonal | $a = b \neq c$ | $\alpha = \beta = 90°, \gamma = 120°$ | C (graphite), $ZnO$ |
| Rhombohedral | $a = b = c$ | $\alpha = \beta = \gamma \neq 90°$ | $CaCO_3$ (calcite), $HgS$ (cinnabar) |
| Triclinic | $a \neq b \neq c$ | $\alpha \neq \beta \neq \gamma \neq 90°$ | $K_2Cr_2O_7$, $CuSO_4 \cdot 5H_2O$ |

**Figure 11.32**
**Cubic unit cells.** The simple cubic unit cell has lattice points only at the corners; the body-centered cubic unit cell also has a lattice point at the center of the cell; and the face-centered cubic unit cell has lattice points at the centers of each face in addition to those at the corners.

Simple cubic            Body-centered cubic            Face-centered cubic

**Figure 11.33**
**Space-filling representation of cubic unit cells.** Only that portion of each atom belonging to a unit cell is shown. Note that a corner atom is shared with eight unit cells and a face atom is shared with two.

Simple cubic            Body-centered cubic            Face-centered cubic

Note that only the portions of the atoms actually within the unit cell are shown. For certain applications, you will need to know the number of atoms in a unit cell of such an atomic crystal. The next example shows how you count atoms in a unit cell.

▶ An alternate example is provided in the Instructor's Resource Manual.

**EXAMPLE 11.9**

## Determining the Number of Atoms per Unit Cell

How many atoms are there in the face-centered cubic unit cell of an atomic crystal having one atom at each lattice point? Remember that atoms at the corners and faces of the unit cell are shared with adjoining unit cells (see Figure 11.33).

**SOLUTION**

Each atom at the center of a face is shared with one other unit cell. Hence, only one-half of the atom belongs to a particular unit cell. But there are six faces on a cube. This gives three whole atoms from the faces. Each corner atom is shared among eight unit cells. Hence, only one-eighth of a corner atom belongs to a particular unit cell. But there are eight corners on a cube. Therefore, the corners contribute one whole atom. Thus, there is a total of four atoms in a face-centered cubic unit cell. To summarize,

$$6 \text{ faces} \times \frac{\frac{1}{2} \text{ atom}}{\text{face}} = 3 \text{ atoms}$$

$$8 \text{ corners} \times \frac{\frac{1}{8} \text{ atom}}{\text{corner}} = 1 \text{ atom}$$

Total    **4 atoms**

---

**EXERCISE 11.10**

Figure 11.34 shows solid dots ("atoms") forming a two-dimensional lattice. A unit cell is marked off by the black line. How many "atoms" are there in this unit cell in such a two-dimensional lattice?

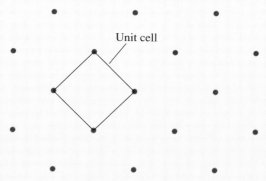

**Figure 11.34**
**A two-dimensional lattice.**

---

See Problems 11.75 and 11.76.

## Crystal Defects

So far we have assumed that crystals have perfect order. In fact, real crystals have various defects or imperfections. These are principally of two kinds: chemical impurities and defects in the formation of the lattice.

Ruby is an example of a crystal with a chemical impurity. The crystal is mainly colorless aluminum oxide, $Al_2O_3$, but occasional aluminum ions, $Al^{3+}$, are replaced by chromium(III) ions, $Cr^{3+}$, which give a red color.

Various lattice defects occur during crystallization. Crystal planes may be misaligned, or sites in the crystal lattice may remain vacant. For example, there might be an equal number of sodium ion vacancies and chloride ion vacancies in a sodium chloride crystal. It is also possible to have an unequal number of cation and anion vacancies in an ionic crystal. For example, iron(II) oxide, FeO, usually crystallizes with some iron(II) sites left unoccupied. Enough of the remaining iron atoms have +3 charges to give an electrically balanced crystal. As a result, there are more oxygen atoms than iron atoms in the crystal. Moreover, the exact composition of the crystal can vary, so the formula FeO is only approximate. Such a compound whose composition varies slightly from its idealized formula is said to be *nonstoichiometric*. Other examples of nonstoichiometric compounds are $Cu_2O$ and $Cu_2S$. Each usually has less copper than expected from the formula. The recently discovered ceramic materials having superconducting properties at relatively high temperatures (page 568) are nonstoichiometric compounds. An example is yttrium barium copper oxide, $YBa_2Cu_3O_{7-x}$, where $x$ is approximately 0.1. It is a compound with oxygen atom vacancies.

# Liquid-Crystal Displays

**EVERYDAY LIFE**

Liquid-crystal displays (LCDs) are common in calculators and wristwatches (see Figure 1.3). Liquid crystals, which are the basis of these displays, generally consist of rodlike molecules in a phase that is intermediate in order between a liquid and a crystalline solid. (Figure 1.4 shows a typical liquid-crystal molecule.) Several different types of liquid-crystal phases exist, differing in the type of molecular order. The *nematic phase* is shown in Figure 11.35. The molecules in this liquid-crystal phase are aligned in one direction, much like the matches in a matchbox. Otherwise, the molecules have random positions, as in a liquid.

Liquid-crystal displays use nematic liquid crystals. The molecular order in a nematic liquid crystal, which results from weak intermolecular forces, is easily disrupted. For this reason, the liquid crystals flow like a liquid. Because of the weakness of the intermolecular forces, the molecules in a nematic phase are easily realigned along new directions.

**Figure 11.35**
**Nematic liquid crystal.** The rodlike molecules are aligned in the same direction, although the positions of the molecules are random.

A liquid-crystal display uses this ease of molecular reorientation to change areas of the display from light to dark, resulting in the patterns you see on the display. The display consists of liquid crystals contained be-

**Figure 11.36**
**A liquid-crystal display (LCD).** *Left:* The display consists of a layer of liquid crystals between glass plates. Sheets of polarizers and transparent electrodes are also included in the "sandwich." *Right:* Portion of one set of electrodes. The electrodes are shaped to form numbers when certain of them have a voltage applied to them. Here, those that combine to form a 3 are shown in gray.

tween glass plates whose interior surfaces are treated to align the molecules in a given direction. The space between the glass plates also contains polarizer sheets (similar to the Polaroid lenses in sunglasses) and transparent electrodes to reorient the molecules (Figure 11.36). When a voltage to a set of electrodes in some area of the display is turned on, the molecules of the liquid crystal in that area reorient along a new direction. When the voltage is turned off, the molecules return to their original orientation.

The purpose of the polarizer sheets is to make the change of molecular orientation visible by using the ability of the liquid-crystal layer to change the plane of polarized light. Light generally consists of waves that are vibrating in various planes. A polarizer filters out all waves except those vibrating in a given plane, resulting in what is called *polarized light*. If you were to hold a second polarizer up to this polarized light, the light would pass or not, depending on the relative orientations of the two polarizers. (You can observe this by taking the lenses from an old pair of Polaroid sunglasses and holding them up so you can look through both lenses at once. When you rotate the second polarizer with respect to the first, you see that more or less light passes through, depending on the relative orientations of the polarizers.)

Figure 11.37 shows how the liquid-crystal display works. Light from a source outside the display passes through the first polarizer, resulting in polarized light, which then passes through the liquid-crystal layer. When the voltage to a set of electrodes in an area of the display is off, the liquid-crystal layer in that area rotates the plane of the polarized light so that it can pass through the second polarizer to the mirror and be reflected back. The display area appears bright. When the voltage is turned on, however, the plane of the polarized light is not rotated. In that case, the light cannot pass through the second polarizer to the mirror, and the display area appears dark. The pattern on the display is created by turning sets of electrodes on and off.

**Figure 11.37**
**How the liquid-crystal display works.** *(A)* Unpolarized light from an external light source (at the right) is filtered by the first polarizer, giving light polarized along the plane indicated by the double-headed arrow. The plane of polarization of the light is rotated by the liquid-crystal layer, so it can pass through the second polarizer to the mirror, where the light is reflected back. This area of the display appears bright. *(B)* The molecules of the liquid-crystal layer are realigned when the electrodes are turned on. This time the liquid crystal layer does not rotate the plane of the polarized light, so it cannot pass through to the mirror. This area of the display appears dark.

# 11.8    Structures of Some Crystalline Solids

We have described the structure of crystals in a general way. Now we want to look in detail at the structure of several crystalline solids that represent the different types: molecular, metallic, ionic, and covalent network.

### Molecular Solids; Closest Packing

The simplest molecular solids are the frozen noble gases—for example, solid neon. In this case, the molecules are single atoms and the intermolecular interactions are London forces. These forces are nondirectional (in contrast to covalent bonding, which is directional), and the maximal attraction is obtained when each atom is surrounded by the largest possible number of other atoms. The problem, then, is simply to find how identical spheres can be packed as tightly as possible into a given space.

To form a layer of close-packed spheres, you place each row of spheres in the crevices of the adjoining rows. This is illustrated in Figure 11.38A. (You may find it helpful to stack identical coins in the manner described here for spheres.) Place the next layer of spheres in the hollows in the first layer. Only half of these hollows can be filled with spheres. Once you have placed a sphere in a given hollow, it partially covers three neighboring hollows (Figure 11.38B) and completely determines the pattern of the layer (Figure 11.38C).

When you come to fill the third layer, you find that you have a choice of sites for spheres, labeled *x* and *y* in Figure 11.38C. Note that each of the *x* sites has a sphere in the first layer directly beneath it, but the *y* sites do not. When the spheres in the third layer are placed in the *x* sites, so that the third layer repeats the first layer, you label this stacking *ABA*. When successive layers are placed so that the spheres of each layer are directly over a layer that is one layer away, you get a stacking that you label *ABABABA* · · · . (This notation indicates that the *A* layers are directly over other *A* layers, whereas the *B* layers are directly over other *B* layers.) The result is a **hexagonal close-packed structure** (hcp), *a crystal structure composed of close-packed atoms (or other units) with the stacking ABABABA · · · ; the structure has a hexagonal unit cell.*

When the spheres in the third layer are placed in *y* sites, so that the third layer is over neither the first nor the second layer, you get a stacking that you label *ABC*. The fourth layer must be over either the *A* or the *B* layer. If subsequent layers are stacked so that they are over the layer two layers below, you get the stacking *ABCAB-*

**Figure 11.38**
**Closest packing of spheres.** *(A)* Each layer is formed by placing spheres in the crevices of the adjoining row. *(B)* When a sphere is placed in a hollow on top of a layer, three other hollows are partially covered. *(C)* There are two types of hollows on top of the second layer, labeled *x* and *y*. The *x* sites are directly above the spheres in the first layer; the *y* sites are not. Spheres of a third layer may occupy either all of the *x* sites or all of the *y* sites.

A

B

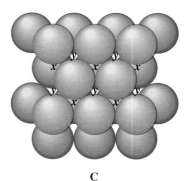

C

**Figure 11.39**
**The cubic close-packed structure has a face-centered cubic lattice.**
*Left:* An "exploded" view of portions of four layers of the cubic close-packed structure. *Right:* These layers form a face-centered cubic unit cell.

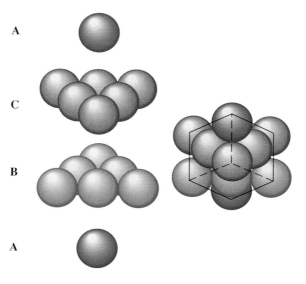

▶ If you pack into a crate as many perfectly round oranges of the same size as possible, they will be close packed. However, they need not be either hexagonal or cubic close packed, because the *A*-, *B*-, and *C*-type layers can be placed at random (except that similar layers cannot be adjacent). A possible arrangement might be *ABCABABABC* ⋯. It is the forces between structural units in a close-packed crystal that determine whether one of the ordered structures, either cubic or hexagonal, will occur.

*CABCA* ⋯ , which results in a **cubic close-packed structure** (ccp), *a crystal structure composed of close-packed atoms (or other units) with the stacking ABCAB-CABCA* ⋯ . ◀

The cubic close-packed lattice is identical to the lattice having a face-centered cubic unit cell. To see this, you take portions of four layers from the cubic close-packed array (Figure 11.39, left). When these are placed together, they form a cube, as shown in Figure 11.39, right.

In any close-packed arrangement, each interior atom is surrounded by 12 nearest-neighbor atoms. *The number of nearest-neighbor atoms of an atom* is called its **coordination number.** Thinking of atoms as hard spheres, one can calculate that the spheres occupy 74% of the space of the crystal. There is no way of packing identical spheres so that an atom has a coordination number greater than 12 or the spheres occupy more than 74% of the space of the crystal. All the noble-gas solids have cubic close-packed crystals except helium, which has a hexagonal close-packed crystal.

A substance composed of polyatomic molecules with approximately spherical shapes might be expected to crystallize in an approximately close-packed structure. Methane, $CH_4$, is an example. Solid methane has a face-centered cubic lattice with one $CH_4$ molecule at each lattice point.

## Metallic Solids

If you were to assume that metallic bonding is completely nondirectional, you would expect metals to crystallize in one of the close-packed structures, as the noble-gas elements do. Indeed, many of the metallic elements do have either cubic or hexagonal close-packed crystals. Copper and silver, for example, are cubic close packed (that is, face-centered cubic with one atom at each lattice point). However, a body-centered cubic arrangement of spheres (having one sphere at each lattice point of a body-centered cubic lattice) occupies almost as great a percentage of space as the close-packed ones. In a body-centered arrangement, 68% of the space is occupied by spheres, compared with the maximal value of 74% in the close-packed arrangement. Each atom in a body-centered lattice has a coordination number of 8, rather than the 12 of each atom in a close-packed lattice.

**Figure 11.40**
**Crystal structures of metals.** Most metals have one of the close-packed structures (hcp = hexagonal close-packed; ccp = cubic close-packed) or a body-centered cubic structure (bcc; manganese has a complicated bcc lattice, with several atoms at each lattice point). Other structures are body-centered tetragonal (bct), orthorhombic (or), and rhombohedral (rh).

| IA | IIA | IIIB | IVB | VB | VIB | VIIB | VIIIB | | | IB | IIB | IIIA | IVA | VA |
|---|---|---|---|---|---|---|---|---|---|---|---|---|---|---|
| Li bcc | Be hcp | | | | | | | | | | | | | |
| Na bcc | Mg hcp | | | | | | | | | | | Al ccp | | |
| K bcc | Ca ccp | Sc hcp | Ti hcp | V bcc | Cr bcc | Mn bcc | Fe bcc | Co hcp | Ni ccp | Cu ccp | Zn hcp | Ga or | | |
| Rb bcc | Sr ccp | Y hcp | Zr hcp | Nb bcc | Mo bcc | Tc hcp | Ru hcp | Rh ccp | Pd ccp | Ag ccp | Cd hcp | In bct | Sn bct | |
| Cs bcc | Ba bcc | La hcp | Hf hcp | Ta bcc | W bcc | Re hcp | Os hcp | Ir ccp | Pt ccp | Au ccp | Hg rh | Tl hcp | Pb ccp | Bi rh |

Figure 11.40 lists the structures of the metallic elements. With few exceptions, they have either a close-packed or a body-centered structure.

## Ionic Solids

The description of ionic crystals is complicated by the fact that you must give the positions in the crystal structure of both the cations and the anions. We will look at the structures of three different cubic crystals with the general formula MX (where M is the metal and X is the nonmetal): cesium chloride (CsCl), sodium chloride (NaCl), and cubic zinc sulfide or zinc blende (ZnS). Many ionic substances of the general formula MX have crystal structures that are one of these types. Once you have learned these structures, you can relate the structures of many other compounds to them.

Cesium chloride consists of positive ions and negative ions of about equal size arranged in a cubic array. Figure 11.41 shows the unit cell of cesium chloride. It shows both a space-filling model (ions are shown as spheres of approximately the correct relative sizes) and a model in which the ions are shrunk in size in order to display more clearly their relative positions. Note that there are chloride ions at each corner of the cube and a cesium ion at the center. Alternatively, the unit cell may be taken to have cesium ions at the corners of a cube and a chloride ion at the center. In either case, the unit cell contains one $Cs^+$ ion and one $Cl^-$ ion. (Figure 11.41 has $\frac{1}{8}$ $Cl^-$ ion at each of eight corners plus one $Cs^+$ ion at the center of the unit cell.)

**Figure 11.41**
**Cesium chloride unit cell.** Cesium chloride consists of interpenetrating simple cubic lattices of $Cs^+$ and $Cl^-$ ions. The figure shows a unit cell with $Cl^-$ ions at the corners of the unit cell and a $Cs^+$ ion at the center. (A space-filling model is on the left; a model with ions shrunk in size to emphasize the structure is on the right.) An alternative unit cell would have $Cs^+$ ions at the corners with a $Cl^-$ ion at the center.

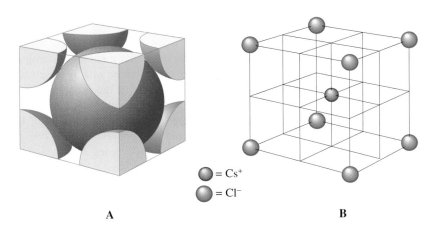

= $Cs^+$
= $Cl^-$

A                    B

**Figure 11.42**
**Sodium chloride unit cell.**
Sodium chloride has a face-centered cubic lattice. *(A)* The unit cell shown here (as a space-filling model) has $Cl^-$ ions at the corners and at the centers of the faces of the cube. The $Cl^-$ ions are much larger than $Na^+$ ions, so the $Cl^-$ ions are nearly touching and have approximately a cubic close-packed structure; the $Na^+$ ions are in cavities in this structure. *(B)* The alternative unit cell shown here has $Na^+$ ions at the corners and the centers of the faces of the unit cell.

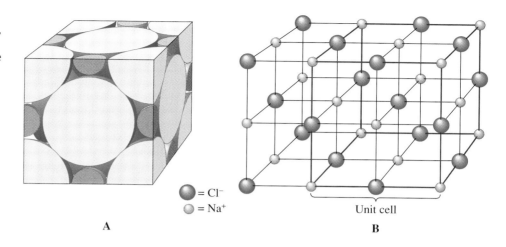

○ = $Cl^-$
○ = $Na^+$

A

Unit cell

B

Ammonium chloride, $NH_4Cl$, and thallium chloride, $TlCl$, are examples of other ionic compounds having this "cesium chloride" structure.

Figure 11.42 shows the unit cell of sodium chloride. It has $Cl^-$ ions at the corners as well as at the center of each cube face. Because the chloride ions are so much larger than the sodium ions, the $Cl^-$ ions are nearly touching and form an approximate cubic close-packed structure of $Cl^-$ ions. The $Na^+$ ions are arranged in cavities of this close-packed structure. Some other compounds that crystallize in this structure are potassium chloride, $KCl$; calcium oxide, $CaO$; and silver chloride, $AgCl$.

Zinc sulfide, $ZnS$, crystallizes in either of two structures. The zinc sulfide mineral having cubic structure is called zinc blende or sphalerite. Zinc sulfide also exists as hexagonal crystals. (The mineral is called wurtzite.) A solid substance that can occur in more than one crystal structure is said to be *polymorphic*. We will discuss only the zinc blende structure (cubic ZnS).

Figure 11.43 shows the unit cell of zinc blende. Sulfide ions are at the corners and at the center of each face of the unit cell. The positions of the zinc ions can be described if you imagine the unit cell divided into eight cubic parts, or subcubes. Zinc

**Figure 11.43**
**Zinc blende (cubic ZnS) unit cell.** Zinc sulfide, ZnS, crystallizes in two different forms, or polymorphs: a hexagonal form (wurtzite) and a face-centered cubic form (zinc blende, or sphalerite), which is shown here. Sulfide ions, $S^{2-}$, are at the corners and at the centers of the faces of the unit cell; $Zn^{2+}$ ions are in alternate subcubes of the unit cell. The unit cell can also be described as having $Zn^{2+}$ ions at each corner and at the center of each face, with $S^{2-}$ ions in alternate subcubes.

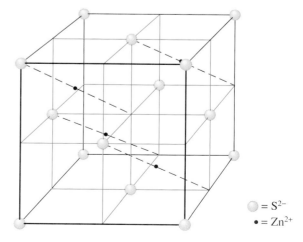

○ = $S^{2-}$
• = $Zn^{2+}$

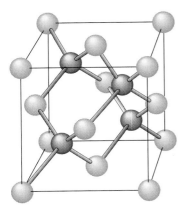

**Figure 11.44**
**Diamond unit cell.** The unit cell of diamond can be obtained from the unit cell of zinc blende by replacing $Zn^{2+}$ ions by carbon atoms (shown here as dark spheres) and $S^{2-}$ ions by carbon atoms (shown here as light spheres). Note that each carbon atom is bonded to four others.

ions are at the centers of alternate subcubes. The number of $S^{2-}$ ions per unit cell is four (8 corner ions $\times \frac{1}{8}$ + 6 face ions $\times \frac{1}{2}$), and the number of $Zn^{2+}$ ions is four (all are completely within the unit cell). Thus, there are equal numbers of $Zn^{2+}$ and $S^{2-}$ ions, as expected by the formula ZnS.

The unit cell of zinc blende can also be taken to have zinc ions at the corners and face centers, with sulfide ions in alternate subcubes. Zinc oxide, ZnO, and beryllium oxide, BeO, also have the "zinc blende" structure.

### Covalent Network Solids

The structures of covalent network solids are determined primarily by the directions of covalent bonds. Diamond is a simple example. It is one allotropic form of the element carbon, in which each carbon atom is covalently bonded to four other carbon atoms in tetrahedral directions to give a three-dimensional covalent network (see Figure 11.27). You can describe the bonding by assuming that every carbon atom is $sp^3$ hybridized. Each C—C bond forms by the overlap of one $sp^3$ hybrid orbital on one carbon atom with an $sp^3$ hybrid orbital on the other atom. The unit cell of diamond is similar to that of zinc blende (Figure 11.43); the $Zn^{2+}$ and $S^{2-}$ ions are replaced by carbon atoms, as in Figure 11.44. The crystal lattice, which is face-centered cubic, can be obtained by placing lattice points at the centers of alternate carbon atoms (either those shown in dark gray or those shown in light gray). Silicon, germanium, and gray tin (a nonmetallic allotrope) have "diamond" structures.

Graphite is another allotrope of carbon. It consists of large, flat sheets of carbon atoms covalently bonded to form hexagonal arrays, and these sheets are stacked one on top of the other (see Figure 11.27). You can describe each sheet in terms of resonance formulas such as the following:

(one resonance formula)

This is just one of many possible resonance formulas, because others can be drawn by moving the double bonds around. This means that the $\pi$ electrons are delocalized over a plane of carbon atoms. You can also describe this delocalization in molecular orbital terms. You imagine that each carbon atom is $sp^2$ hybridized, so each atom has three $sp^2$ hybrid orbitals in the molecular plane and an additional $2p$ orbital perpendicular to this plane. A $\sigma$ bond forms between two carbon atoms when an $sp^2$ hybrid orbital on one atom overlaps an $sp^2$ hybrid orbital on another. The result is a planar $\sigma$-bond framework. The $2p$ orbitals of all the atoms overlap to form $\pi$ orbitals encompassing the entire plane of atoms.

Each carbon–carbon distance within a graphite sheet is 142 pm, which is almost midway between the length of a C—C bond (154 pm) and a C=C bond (134 pm) and which indicates an intermediate bond order (see Figure 11.27). The two-dimensional sheets, or layers, are stacked one on top of the other, and the attraction between sheets results from London forces. The distance between layers is 335 pm,

**Figure 11.45**
**Electron micrograph of graphite.** Note the layer structure.

which is greater than the carbon–carbon bond length, because London forces are weaker than covalent bonding forces.

You can explain several properties of graphite on the basis of its structure. The layer structure, which you see in electron micrographs (Figure 11.45), tends to separate easily, so that sheets of graphite slide over one another. The "lead" in a pencil is graphite. As you write with a pencil, layers of graphite rub off to form the pencil mark. Similarly, the use of graphite to lubricate locks and other closely fitting metal surfaces depends on the ability of graphite particles to move over one another as the layers slide. Another property of graphite is its electrical conductivity. Electrical conductivity is unusual in covalent network solids, because normally the electrons in such bonds are localized. Graphite is a moderately good conductor, however, because delocalization leads to mobile electrons within each carbon atom layer. Figure 11.46 shows a demonstration of the electrical conductivity of graphite.

**CONCEPT CHECK 11.4**

Shown here is a representation of a unit cell for a crystal. The orange balls are atom A, and the black balls are atom B.

**Figure 11.46**
**Demonstration of the electrical conductivity of graphite.** The nonzero reading on the meter shows that the pencil lead (graphite) conducts electricity.

a. What is the chemical formula of the compound that has this unit cell ($A_?B_?$)?

b. Consider the configuration of the A atoms. Is this a cubic unit cell? If so, which type?

# 11.9 Calculations Involving Unit-Cell Dimensions

You can determine the structure and dimensions of a unit cell by diffraction methods, which we will describe briefly in the next section. Once you know the unit-cell dimensions and the structure of a crystal, however, some interesting calculations are possible. For instance, suppose you find that a metallic solid has a face-centered cubic unit cell with all atoms at lattice points, and you determine the edge length of the unit cell. From this unit-cell dimension, you can calculate the volume of the unit cell. Then, knowing the density of the metal, you can calculate the mass of the atoms in the unit cell. Because you know that the unit cell is face-centered cubic with all atoms at lattice points and that such a unit cell has four atoms, you can obtain the mass of an individual atom. This determination of the mass of a single atom gave one of the first accurate determinations of Avogadro's number. The calculations are shown in the next example.

## EXAMPLE 11.10

### Calculating Atomic Mass from Unit-Cell Dimension and Density

X-ray diffraction from crystals provides one of the most accurate ways of determining Avogadro's number. Silver crystallizes in a face-centered cubic lattice with all atoms at the lattice points. The length of an edge of the unit cell was determined by x-ray diffraction to be 408.6 pm (4.086 Å). The density of silver is 10.50 g/cm³. Calculate the mass of a silver atom. Then, using the known value of the atomic weight, calculate Avogadro's number.

 **PROBLEM STRATEGY**

Knowing the edge length of a unit cell, you can calculate the unit-cell volume. Then, from the density, you can find the mass of the unit cell and hence the mass of a silver atom.

**SOLUTION**

You obtain the volume, $V$, of the unit cell by cubing the length of an edge.

$$V = (4.086 \times 10^{-10} \text{ m})^3 = 6.822 \times 10^{-29} \text{ m}^3$$

The density, $d$, of silver in grams per cubic meter is

$$d = 10.50 \frac{\text{g}}{\text{cm}^3} \times \left(\frac{1 \text{ cm}}{10^{-2} \text{ m}}\right)^3 = 1.050 \times 10^7 \text{ g/m}^3$$

Density is mass per volume; hence, the mass of a unit cell equals the density times the volume of the unit cell.

$$\begin{aligned} m &= dV \\ &= 1.050 \times 10^7 \text{ g/m}^3 \times 6.822 \times 10^{-29} \text{ m}^3 \\ &= 7.163 \times 10^{-22} \text{ g} \end{aligned}$$

Because there are four atoms in a face-centered unit cell having one atom at each lattice point (see Example 11.9), the mass of a silver atom is

$$\text{Mass of 1 Ag atom} = \tfrac{1}{4} \times 7.163 \times 10^{-22} \text{ g} = \mathbf{1.791 \times 10^{-22} \text{ g}}$$

The known atomic weight of silver is 107.87 amu. Thus, the molar mass (Avogadro's number of atoms) is 107.87 g/mol, and Avogadro's number, $N_A$, is

$$N_A = \frac{107.87 \text{ g/mol}}{1.791 \times 10^{-22} \text{ g}} = \mathbf{6.023 \times 10^{23}/\text{mol}}$$

## EXERCISE 11.11

Lithium metal has a body-centered cubic structure with all atoms at the lattice points and a unit-cell length of 350.9 pm. Calculate Avogadro's number. The density of lithium is 0.534 g/cm³.

See Problems 11.77 and 11.78.

If you know or assume the structure of an atomic crystal, you can calculate the length of the unit-cell edge from the density of the substance. This is illustrated in the next example. Agreement of this value with that obtained from x-ray diffraction confirms that your view of the structure of the crystal is correct.

► An alternate example is provided in the Instructor's Resource Manual.

Platinum
$d = 21.45$ g/cm$^3$
atomic weight = 195.08 amu

### EXAMPLE 11.11

## Calculating Unit-Cell Dimension from Unit-Cell Type and Density

Platinum crystallizes in a face-centered cubic lattice with all atoms at the lattice points. It has a density of 21.45 g/cm$^3$ and an atomic weight of 195.08 amu. From these data, calculate the length of a unit-cell edge. Compare this with the value of 392.4 pm obtained from x-ray diffraction.

 **PROBLEM STRATEGY**

The mass of an atom, and hence the mass of a unit cell, can be calculated from the atomic weight. Knowing the density and the mass of the unit cell, we can calculate the volume and the edge length of a unit cell.

### SOLUTION

You can use Avogadro's number ($6.022 \times 10^{23}$/mol) to convert the molar mass of platinum (195.08 g/mol) to the mass per atom.

$$\frac{195.08 \text{ g Pt}}{1 \text{ mol Pt}} \times \frac{1 \text{ mol Pt}}{6.022 \times 10^{23} \text{ Pt atoms}} = \frac{3.239 \times 10^{-22} \text{ g Pt}}{1 \text{ Pt atom}}$$

Because there are four atoms per unit cell, the mass per unit cell can be calculated as follows:

$$\frac{3.239 \times 10^{-22} \text{ g Pt}}{1 \text{ Pt atom}} \times \frac{4 \text{ Pt atoms}}{1 \text{ unit cell}} = \frac{1.296 \times 10^{-21} \text{ g}}{1 \text{ unit cell}}$$

The volume of the unit cell is

$$V = \frac{m}{d} = \frac{1.296 \times 10^{-21} \text{ g}}{21.45 \text{ g/cm}^3} = 6.042 \times 10^{-23} \text{ cm}^3$$

If the edge length of the unit cell is denoted as $l$, the volume is $V = l^3$. Hence, the edge length is

$$\begin{aligned} l &= \sqrt[3]{V} \\ &= \sqrt[3]{6.042 \times 10^{-23} \text{ cm}^3} \\ &= 3.924 \times 10^{-8} \text{ cm (or } 3.924 \times 10^{-10} \text{ m)} \end{aligned}$$

**Thus, the edge length is 392.4 pm, which is in excellent agreement with the x-ray diffraction value.**

### EXERCISE 11.12

Potassium metal has a body-centered cubic structure with all atoms at the lattice points. The density of the metal is 0.856 g/cm$^3$. Calculate the edge length of a unit cell.

See Problems 11.79 and 11.80.

Potassium
$d = 0.856$ g/cm$^3$

## 11.10 Determining Crystal Structure by X-Ray Diffraction

The determination of crystal structure by x-ray diffraction is one of the most important ways of determining the structures of molecules. Because of its ordered structure, a crystal consists of repeating planes of the same kind of atom. These planes can act as reflecting surfaces for x rays. When x rays are reflected from these planes, they show a *diffraction pattern,* which can be recorded on a photographic plate as a series of spots (see Figure 11.47). By analyzing the diffraction pattern, you can determine the positions of the atoms in the unit cell of the crystal. Once you have determined the positions of each atom in the unit cell of a molecular solid, you have also found the positions of the atoms in the molecule. ◀

▶ X-ray diffraction has been used to obtain the structure of proteins, which are large molecules essential to life. John Kendrew and Max Perutz received the Nobel Prize in chemistry in 1962 for their x-ray work in determining the structure of myoglobin (MW = 17,600 amu) and hemoglobin (MW = 66,000 amu). Hemoglobin is the oxygen-carrying protein of the red blood cells, and myoglobin is the oxygen-carrying protein in muscle tissue.

To understand how such a diffraction pattern occurs, let us look at the diffraction of two waves. Suppose two waves of the same wavelength come together in phase. By *in phase* we mean that the waves come together so that their peaks (maxima) and troughs (minima) match (see Figure 11.48A). Two waves that come together in phase form a wave with higher peaks and lower troughs than either wave. We say that the waves reinforce each other, or undergo *constructive interference.* The height of an x-ray wave (its *amplitude*) is related to the intensity of the x ray, so constructive interference gives a resultant wave having increased intensity.

Alternatively, suppose two waves of the same wavelength come together out of phase. By *out of phase* we mean that where one wave has its peaks, the other has its troughs (see Figure 11.48B). When two waves come together out of phase, each peak of one wave combines with a trough of the other so that the resultant wave amplitude is smaller than that of either wave (zero, if the two waves have the same amplitudes). We say that the waves undergo *destructive interference.* The resultant wave has decreased intensity.

Now consider the scattering, or reflection, of x rays from a crystal. Figure 11.49 shows two rays: one ray reflected from one plane, the other from another plane. The two rays start out in phase, but because one ray travels farther, they may end up out

**Figure 11.47**
**A crystal diffraction pattern.** This diffraction pattern was obtained by diffracting x rays from a sodium chloride crystal and collecting the image on photographic film. The resulting negative shows dark spots on a lighter background.

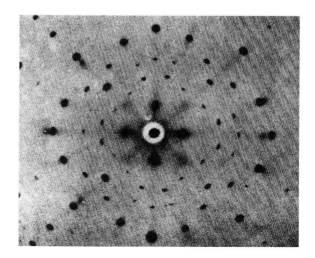

**Figure 11.48**
**Wave interference.**
(A) Two waves constructively interfere when they are in phase (when their peaks and troughs match). The two waves add to give a resultant wave with higher peaks and deeper troughs than either wave. (B) Two waves destructively interfere when they are out of phase (when the peaks of one wave match the troughs of the other wave). The two waves add to give a resultant wave with lower peaks and more shallow troughs than either wave.

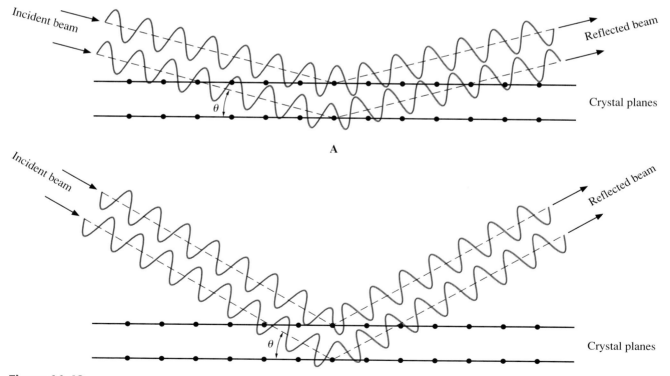

**Figure 11.49**
**Diffraction of x rays from crystal planes.**   (A) At most angles $\theta$, reflected waves are out of phase and destructively interfere. (B) At certain angles $\theta$, however, reflected waves are in phase and constructively interfere.

of phase after reflection. Only at certain angles of reflection do the two rays remain in phase. At some angles the rays constructively interfere (giving dark areas on a negative, corresponding to large x-ray intensity). At other angles the rays destructively interfere (giving light areas on a photographic plate, corresponding to low x-ray intensity). Note the dark and light areas on the photographic negative of the diffraction pattern shown in Figure 11.47.

You can relate the resulting pattern produced by the diffraction to the structure of the crystal. You can determine the type of unit cell and its size; if the solid is composed of molecules, you can determine the position of each atom in the molecule.

## INSTRUMENTAL METHODS

# Automated X-Ray Diffractometry

Max von Laue, a German physicist, was the first to suggest the use of x rays for the determination of crystal structure. Soon afterward, in 1913, the British physicists William Bragg and his son Lawrence developed the method on which modern crystal-structure determination is based. They realized that the atoms in a crystal form reflecting planes for x rays, and from this idea they derived the fundamental equation of crystal-structure determination.

$$n\lambda = 2d \sin \theta, n = 1, 2, 3, \ldots$$

The *Bragg equation* relates the wavelength of x rays, $\lambda$, to the distance between atomic planes, $d$, and the angle of reflection, $\theta$. Note that reflections occur at several angles, corresponding to different integer values of $n$.

A molecular crystal has many different atomic planes, so that it reflects an x-ray beam in many different directions. By analyzing the intensities and angular directions of the reflected beams, you can determine the exact positions of all the atoms in the unit cell of the crystal and therefore obtain the structure of the molecule. The problem of obtaining the x-ray data (intensities and angular directions of the reflections) and then analyzing them, however, is not trivial. Originally, the reflected x rays were recorded on photographic plates. After taking many pictures, the scientist would pore over the negatives, measuring the

densities of the spots and their positions on the plates. Then he or she would work through lengthy and laborious calculations to analyze the data. Even with early computers, the determination of a molecular structure required a year or more.

With the development of electronic x-ray detectors and minicomputers, x-ray diffraction has become automated, so that the time and effort of determining the structure of a molecule have been substantially reduced. Now frequently the most difficult task is preparing a suitable crystal. The crystal should be several tenths of a millimeter in each dimension and without significant defects. Such crystals of protein molecules, for example, can be especially difficult to prepare.

Once a suitable crystal has been obtained, the structure of a molecule of moderate size can often be determined in a day or so. The crystal is mounted on a glass fiber (or in a glass capillary containing an inert gas, if the substance reacts with air) and placed on a pin or spindle within the circular assembly of the x-ray diffractometer (Figure 11.50). The crystal and x-ray detector (placed on the opposite side of the crystal from the x-ray tube) rotate under computer control, while the computer records the intensities and angles of thousands of x-ray reflection spots. After computer analysis of the data, the molecular structure is printed out.

**Figure 11.50**

**Automated x-ray diffractometer.** *(Top)* The single-crystal specimen is mounted on a glass fiber, which is placed on a spindle within the circular assembly of the diffractometer. A new data collection system (left of specimen) reduces data collection time from several days to several hours. *(Bottom)* The schematic diagram shows the diffracted rays being detected by a photograph. In a modern diffractometer, the final data collection is done with a fixed electronic detector, and the crystal is rotated. The data are collected and analyzed by a computer accompanying the diffractometer.

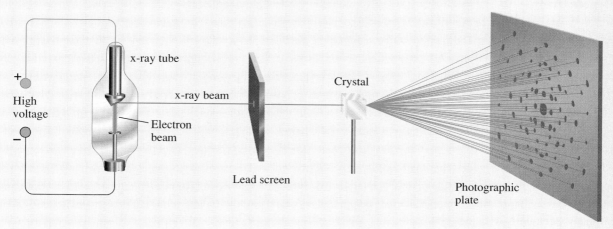

# Water (A Special Substance for Planet Earth)

**ENVIRONMENT**

Water is the only liquid substance (other than petroleum) to be found on earth in significant amounts. This liquid is also readily convertible under conditions on earth to the solid and gaseous forms. Water has several unusual properties that set it apart from other substances. For example, its solid phase, ice, is less dense than liquid water, whereas for most substances the solid phase is more dense than the liquid. In addition, water has an unusually large heat capacity. These are some of the properties that are important in determining conditions favorable to life. In fact, it is difficult to think of life without water.

*(continued)*

**491**

**Figure 11.51**
**The structure of ice.** Oxygen atoms are represented by large red spheres; hydrogen atoms, by small spheres. Each oxygen atom is tetrahedrally surrounded by four hydrogen atoms. Two are close, giving the $H_2O$ molecule. Two are farther away, held by hydrogen bonding (represented by three dots). The distribution of hydrogen atoms in these two types of positions is random.

**Figure 11.52**
**Density of water versus temperature.** Water has a maximum density at 4°C.

The unusual properties of water are largely linked to its ability to form hydrogen bonds. For example, ice is less dense than liquid water because ice has an open, hydrogen-bonded structure (see Figure 11.51). Each oxygen atom in the structure of ice is surrounded tetrahedrally by four hydrogen atoms: two that are close and covalently bonded to give the $H_2O$ molecule, and two others that are farther away and held by hydrogen bonds. The tetrahedral angles give rise to a three-dimensional structure that contains open space. When ice melts, hydrogen bonds break, the ice structure partially disintegrates, and the molecules become more compactly arranged, leading to a more dense liquid.

Even liquid water has significant hydrogen bonding. Only about 15% of the hydrogen bonds are broken when ice melts. You might view the liquid as composed of ice-like clusters in which hydrogen bonds are continually breaking and forming, so clusters disappear and new ones appear. This is sometimes referred to as the "flickering cluster" model of liquid water.

As the temperature rises from 0°C, the clusters tend to break down further, giving an even more compact liquid. Thus, the density rises (see Figure 11.52). However, as is normal in any liquid when the temperature rises, the

molecules begin to move and vibrate faster, and the space occupied by the average molecule increases. For most other liquids, this results in a continuous decrease in density with temperature increase. In water, this normal effect is countered by the density increase due to breaking of hydrogen bonds, but at 4°C the normal effect begins to predominate. Water shows a maximum density at 4°C and becomes less dense at higher temperatures.

The unusually large heat of vaporization of water has an important effect on the earth's weather. Evaporation of the surface waters absorbs over 30% of the solar energy reaching the earth's surface. This energy is released when the water vapor condenses, and thunderstorms and hurricanes may result. In the process, the waters of the earth are circulated and the freshwater sources are replenished. This natural cycle of water from the oceans to freshwater sources and its return to the ocean is called the *hydrologic cycle* (Figure 11.53).

Although evaporation and condensation of water play a dominant role in our weather, other properties are important as well. The exceptionally large heat capacity of bodies of water has an important moderating effect on the surrounding temperature by warming cold air masses in winter and cooling warm air masses in summer. Worldwide, the oceans are most important, but even inland lakes have a pronounced effect. For example, the Great Lakes give Detroit a more moderate winter than cities somewhat farther south but with no lakes nearby.

The fact that ice is less dense than water means that it forms on top of the liquid when freezing occurs. This has far-reaching effects, both for weather and for aquatic

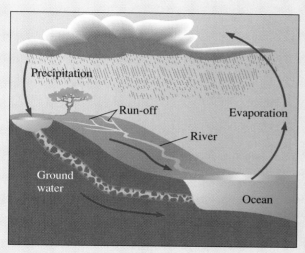

**Figure 11.53**
**The hydrologic cycle.** Ocean water evaporates to form clouds. Then, precipitation (rain and snow) from these clouds replenishes freshwater sources. Water from these sources eventually returns to the oceans via rivers, run-off, or groundwater.

The solvent properties of water are also unusual. Water is both a polar substance and a hydrogen-bonding molecule. As a result, water dissolves many substances, including ionic and polar compounds. These properties make water the premier solvent, biologically and industrially.

As a result of the solvent properties of water, the naturally occurring liquid always contains dissolved materials, particularly ionic substances. *Hard water* contains certain metal ions, such as $Ca^{2+}$ and $Mg^{2+}$. These ions react with soaps, which are sodium salts of stearic acid and similar organic acids, to produce a curdy precipitate of calcium and magnesium salts. This precipitate adheres to clothing and bathtubs (as bathtub ring). Removing $Ca^{2+}$ and $Mg^{2+}$ ions from hard water is referred to as *water softening*.

Water is often softened by *ion exchange*. Ion exchange is a process whereby a water solution is passed through a column of a material that replaces one kind of ion in solution with another kind. Home and commercial water softeners contain cation-exchange resins, which are insoluble macromolecular substances (substances consisting of giant molecules) to which negatively charged groups are chemically bonded. The negative charges are counterbalanced by ions such as $Na^+$. When hard water containing the $Ca^{2+}$ ion passes through a column of this resin, the $Na^+$ ion in the resin is replaced by $Ca^{2+}$ ion.

$$2NaR(s) + Ca^{2+}(aq) \longrightarrow CaR_2(s) + 2Na^+(aq)$$
$$(R^- = \text{anion of exchange resin})$$

The water passing through the column now contains $Na^+$ in place of $Ca^{2+}$ and has been softened. Once the resin has been completely converted to the calcium salt, it can be regenerated by flushing the column with a concentrated solution of NaCl to reverse the previous reaction.

animals. When ice forms on a body of water, it insulates the underlying water from the cold air and limits further freezing. Fish depend on this for winter survival. Consider what would happen to a lake if ice were more dense than water. The ice would freeze from the bottom of the lake upward. Without the insulating effect at the surface, the lake could well freeze solid, killing the fish. Spring thaw would be prolonged, because the insulating effect of the surface water would make it take much longer for the ice at the bottom of a lake to melt.

## A Checklist for Review

### Important Terms

change of state (phase transition) (p. 442)
melting (11.2)
freezing (11.2)
vaporization (11.2)
sublimation (11.2)
condensation (11.2)
vapor pressure (11.2)
boiling point (11.2)
freezing point (11.2)
melting point (11.2)

heat of fusion (11.2)
heat of vaporization (11.2)
phase diagram (11.3)
triple point (11.3)
critical temperature (11.3)
critical pressure (11.3)
surface tension (11.4)
viscosity (11.4)
intermolecular forces (11.5)
van der Waals forces (11.5)
dipole–dipole force (11.5)

London (dispersion) forces (11.5)
hydrogen bonding (11.5)
molecular solid (11.6)
metallic solid (11.6)
ionic solid (11.6)
covalent network solid (11.6)
crystalline solid (11.7)
amorphous solid (11.7)
crystal lattice (11.7)
unit cell (11.7)

simple cubic unit cell (11.7)
body-centered cubic unit cell (11.7)
face-centered cubic unit cell (11.7)
hexagonal close-packed structure (11.8)
cubic close-packed structure (11.8)
coordination number (11.8)

## Key Equation

$$\ln \frac{P_2}{P_1} = \frac{\Delta H_{vap}}{R} \left( \frac{1}{T_1} - \frac{1}{T_2} \right)$$

## Summary of Facts and Concepts

Gases are composed of molecules in constant random motion throughout mostly empty space. This explains why gases are compressible fluids. Liquids are also composed of molecules in constant random motion, but the molecules are more tightly packed. Thus, liquids are incompressible fluids. Solids are composed of atoms, molecules, or ions that are in close contact and oscillate about fixed sites. Thus, solids are incompressible and rigid rather than fluid.

Any state of matter may change to another state. Melting and freezing are examples of such changes of state, or *phase transitions*. Vapor pressure, boiling point, and melting point are properties of substances that involve phase transitions. Vapor pressure increases with temperature; according to the *Clausius–Clapeyron equation,* the logarithm of the vapor pressure varies inversely with the absolute temperature. The conditions under which a given state of a substance exists are graphically summarized in a *phase diagram.*

Like vapor pressure and boiling point, *surface tension* and *viscosity* are important properties of liquids. These properties can be explained in terms of intermolecular forces. The three kinds of attractive intermolecular forces are *dipole–dipole forces, London forces,* and *hydrogen bonding.* London forces are weak attractive forces present in all molecules; London forces tend to increase with increasing molecular weight. Vapor pressure tends to decrease with increasing molecular weight, whereas boiling point, surface tension, and viscosity tend to increase (unless hydrogen bonding is present). Hydrogen bonding occurs in substances containing H atoms bonded to F, O, or N atoms. When hydrogen bonding is present, vapor pressure tends to be lower than otherwise expected, whereas boiling point, surface tension, and viscosity tend to be higher.

Solids can be classified in terms of the type of force between the structural units; there are molecular, metallic, ionic, and covalent network solids. Melting point, hardness, and electrical conductivity are properties that can be related to the structure of the solid.

Solids can be crystalline or amorphous. A crystalline solid has an ordered arrangement of structural units placed at *crystal lattice* points. We may think of a crystal as constructed from *unit cells.* Cubic unit cells are of three kinds: simple cubic, body-centered cubic, and face-centered cubic. One of the most important ways of determining the structure of a crystalline solid is by *x-ray diffraction.*

## Operational Skills

1. **Calculating the heat required for a phase change of a given mass of substance** Given the heat of fusion (or vaporization) of a substance, calculate the amount of heat required to melt (or vaporize) a given quantity of that substance. **(EXAMPLE 11.1)**
2. **Calculating vapor pressures and heats of vaporization** Given the vapor pressure of a liquid at one temperature and its heat of vaporization, calculate the vapor pressure at another temperature **(EXAMPLE 11.2)**. Given the vapor pressures of a liquid at two temperatures, calculate the heat of vaporization **(EXAMPLE 11.3)**.
3. **Relating the conditions for the liquefaction of gases to the critical temperature** Given the critical temperature and pressure of a substance, describe the conditions necessary for liquefying the gaseous substance. **(EXAMPLE 11.4)**
4. **Identifying intermolecular forces** Given the molecular structure, state the kinds of intermolecular forces expected for a substance. **(EXAMPLE 11.5)**
5. **Determining relative vapor pressure on the basis of intermolecular attraction** Given two liquids, decide on the basis of intermolecular forces which has the higher vapor pressure at a given temperature or which has the lower boiling point. **(EXAMPLE 11.6)**
6. **Identifying types of solids** From what you know about the bonding in a solid, classify it as a molecular, metallic, ionic, or covalent network. **(EXAMPLE 11.7)**
7. **Determining the relative melting points based on types of solids** Given a list of substances, arrange them in order of increasing melting point from what you know of their structures. **(EXAMPLE 11.8)**
8. **Determining the number of atoms per unit cell** Given the description of a unit cell, find the number of atoms per cell. **(EXAMPLE 11.9)**
9. **Calculating atomic mass from unit-cell dimension and density** Given the edge length of the unit cell, the crystal structure, and the density of a metal, calculate the mass of a metal atom. **(EXAMPLE 11.10)**
10. **Calculating unit-cell dimension from unit-cell type and density** Given the unit-cell structure, the density, and the atomic weight for an element, calculate the edge length of the unit cell. **(EXAMPLE 11.11)**

# Review Questions

**11.1** List the different phase transitions that are possible and give examples of each.

**11.2** Describe how you could purify iodine by sublimation.

**11.3** Describe vapor pressure in molecular terms. What do we mean by saying it involves a dynamic equilibrium?

**11.4** Explain why 15 g of steam at 100°C melts more ice than 15 g of liquid water at 100°C.

**11.5** Why is the heat of fusion of a substance smaller than its heat of vaporization?

**11.6** Explain why evaporation leads to cooling of the liquid.

**11.7** Describe the behavior of a liquid and its vapor in a closed vessel as the temperature increases.

**11.8** Gases that cannot be liquefied at room temperature merely by compression are called "permanent" gases. How could you liquefy such a gas?

**11.9** The pressure in a cylinder of nitrogen continuously decreases as gas is released from it. On the other hand, a cylinder of propane maintains a constant pressure as propane is released. Explain this difference in behavior.

**11.10** Why does the vapor pressure of a liquid depend on the intermolecular forces?

**11.11** Explain the surface tension of a liquid in molecular terms. How does the surface tension make a liquid act as though it had a "skin"?

**11.12** Explain the origin of the London force that exists between two molecules.

**11.13** Explain what is meant by hydrogen bonding. Describe the hydrogen bonding between two $H_2O$ molecules.

**11.14** Why do molecular substances have relatively low melting points?

**11.15** Describe the distinguishing characteristics of a crystalline solid and an amorphous solid.

**11.16** Describe the face-centered cubic unit cell.

**11.17** Describe the structure of thallium(I) iodide, which has the same structure as cesium chloride.

**11.18** What is the coordination number of $Cs^+$ in CsCl? of $Na^+$ in NaCl? of $Zn^{2+}$ in ZnS?

**11.19** Explain in words how Avogadro's number could be obtained from the unit-cell edge length of a cubic crystal. What other data are required?

**11.20** Explain the production of an x-ray diffraction pattern by a crystal in terms of the interference of waves.

# Conceptual Problems

**11.21** Shown here is a curve of the distribution of kinetic energies of the molecules in a liquid at an arbitrary temperature $T$.

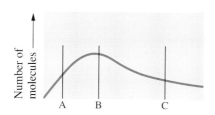

The lines marked A, B, and C represent the point where each of the molecules for three different liquids (liquid A, liquid B, and liquid C) has enough kinetic energy to escape into the gas phase (see Figure 11.5 for more information). Write a brief explanation for each of your answers to the following questions.

a. Which of the molecules—A, B, or C—would have the majority of the molecules in the gas phase at temperature $T$?

b. Which of the molecules—A, B, or C—has the strongest intermolecular attractions?

c. Which of the molecules would have the lowest vapor pressure at temperature $T$?

**11.22** Consider a substance X with a $\Delta H_{vap} = 20.3$ kJ/mol and $\Delta H_{fus} = 9.0$ kJ/mol. The melting point, freezing point, and heat capacities of both the solid and liquid X are identical to those of water.

a. If you place one beaker containing 50 g of X at −10°C and another beaker with 50 g of $H_2O$ at −10°C on a hot plate and start heating them, which material will reach the boiling point first?

b. Which of the materials from part a, X or $H_2O$, would completely boil away first?

c. On a piece of graph paper, draw the heating curve for $H_2O$ and X. How do the heating curves reflect your answers from parts a and b?

**11.23** Using the information presented in this chapter, explain why farmers spray water above and on their fruit trees on still nights when they know the temperature is going to drop below 0°C. (*Hint:* Totally frozen fruit is what the farmers are trying to avoid.)

**11.24** You are presented with three bottles, each containing a different liquid: bottle A, bottle B, and bottle C. Bottle A's label states that it is an ionic compound with a boiling point of 35°C. Bottle B's label states that it is a molecular compound with a boiling point of 29.2°C. Bottle C's label states that it is a molecular compound with a boiling point of 67.1°C.

a. Which of the compounds is most likely to be incorrectly identified?

b. If Bottle A were a molecular compound, which of the compounds has the strongest intermolecular attractions?

c. If Bottle A were a molecular compound, which of the compounds would have the highest vapor pressure?

**11.25**    Shown here is a representation of a unit cell for a crystal. The orange balls are atom A, and the black balls are atom B.

a. What is the chemical formula of the compound that has this unit cell ($A_?B_?$)?

b. Consider the configuration of the A atoms. Is this a cubic unit cell? If so, which type?

## Practice Problems

### Phase Transitions

**11.29**    Identify the phase transition occurring in each of the following.

a. The water level in an aquarium tank falls continuously (the tank has no leak).

b. A mixture of scrambled eggs placed in a cold vacuum chamber slowly turns to a powdery solid.

c. Chlorine gas is passed into a very cold test tube where it turns to a yellow liquid.

d. When carbon dioxide gas under pressure exits from a small orifice, it turns to a white "snow."

e. Molten lava from a volcano cools and turns to solid rock.

**11.30**    Identify the phase transition occurring in each of the following.

a. Mothballs slowly become smaller and eventually disappear.

b. Rubbing alcohol spilled on the palm of the hand feels cool as the volume of liquid decreases.

c. A black deposit of tungsten metal collects on the inside of a lightbulb whose filament is tungsten metal.

d. Raindrops hit a cold metal surface, which becomes covered with ice.

e. Candle wax turns to liquid under the heat of the candle flame.

**11.31**    Use Figure 11.7 to estimate the boiling point of diethyl ether, $(C_2H_5)_2O$, under an external pressure of 350 mmHg.

**11.32**    Use Figure 11.7 to estimate the boiling point of carbon tetrachloride, $CCl_4$, under an external pressure of 350 mmHg.

**11.33**    An electric heater coil provided heat to a 15.5-g sample of iodine, $I_2$, at the rate of 3.48 J/s. It took 4.54 min from the time the iodine began to melt until the iodine was completely melted. What is the heat of fusion per mole of iodine?

**11.26**    Assuming normal winter conditions ($-1.5°C$ and 1.0 atm pressure), consult the phase diagram for water (Figure 11.11) and come up with a reason why ice skates and sleds slide so well on solid water. Keep in mind that sleds and ice skates do not typically slide well on other solid surfaces (concrete and metal, for example).

**11.27**    If you place room temperature water in a well-insulated cup and allow some of the water to evaporate, the temperature of the water in the cup will drop lower than room temperature. Come up with an explanation for this observation.

**11.28**    The heats of vaporization for water and carbon disulfide are 40.7 kJ/mol and 26.8 kJ/mol, respectively. A vapor (steam) burn occurs when the concentrated vapor of a substance condenses on your skin. Which of these substances, water or carbon disulfide, will result in the most severe burn if identical quantities of each vapor at a temperature just above their boiling point came in contact with your skin?

**11.34**    A 35.8-g sample of cadmium metal was melted by an electric heater providing 4.66 J/s of heat. If it took 6.92 min from the time the metal began to melt until it was completely melted, what is the heat of fusion per mole of cadmium?

**11.35**    Isopropyl alcohol, $CH_3CHOHCH_3$, is used in rubbing alcohol mixtures. Alcohol on the skin cools by evaporation. How much heat is absorbed by the alcohol if 5.00 g evaporates? The heat of vaporization of isopropyl alcohol is 42.1 kJ/mol.

**11.36**    Liquid butane, $C_4H_{10}$, is stored in cylinders to be used as a fuel. Suppose 39.3 g of butane gas is removed from a cylinder. How much heat must be provided to vaporize this much gas? The heat of vaporization of butane is 21.3 kJ/mol.

**11.37**    Water at 0°C was placed in a dish inside a vessel maintained at low pressure by a vacuum pump. After a quantity of water had evaporated, the remainder froze. If 9.31 g of ice at 0°C was obtained, how much liquid water must have evaporated? The heat of fusion of water is 6.01 kJ/mol and its heat of vaporization is 44.9 kJ/mol at 0°C.

**11.38**    A quantity of ice at 0.0°C was added to 27.4 g of water at 21.0°C to give water at 0.0°C. How much ice was added? The heat of fusion of water is 6.01 kJ/mol and the specific heat is 4.18 J/(g · °C).

**11.39**    A quantity of ice at 0°C is added to 64.3 g of water in a glass at 55°C. After the ice melted, the temperature of the water in the glass was 15°C. How much ice was added? The heat of fusion of water is 6.01 kJ/mol and the specific heat is 4.18 J/(g · °C).

**11.40**  Steam at 100°C was passed into a flask containing 275 g of water at 21°C, where the steam condensed. How many grams of steam must have condensed if the temperature of the water in the flask was raised to 83°C? The heat of vaporization of water at 100°C is 40.7 kJ/mol and the specific heat is 4.18 J/(g · °C).

**11.41**  Chloroform, $CHCl_3$, a volatile liquid, was once used as an anesthetic but has been replaced by safer compounds. Chloroform boils at 61.7°C and has a heat of vaporization of 31.4 kJ/mol. What is its vapor pressure at 33.0°C?

**11.42**  Methanol, $CH_3OH$, a colorless, volatile liquid, was formerly known as wood alcohol. It boils at 65.0°C and has a heat of vaporization of 37.4 kJ/mol. What is its vapor pressure at 22.0°C?

**11.43**  White phosphorus, $P_4$, is normally a white, waxy solid melting at 44°C to a colorless liquid. The liquid has a vapor pressure of 400.0 mmHg at 251.0°C and 760.0 mmHg at 280.0°C. What is the heat of vaporization of this substance?

**11.44**  Carbon disulfide, $CS_2$, is a volatile, flammable liquid. It has a vapor pressure of 400.0 mmHg at 28.0°C and 760.0 mmHg at 46.5°C. What is the heat of vaporization of this substance?

## Phase Diagrams

**11.45**  Shown here is the phase diagram for compound Z. The triple point of Z is −5.1°C at 3.3 atm and the critical point is 51°C and 99.1 atm.

a. What is the state of Z at position A?
b. If we increase the temperature from the compound at position A to 60°C while holding the pressure constant, what is the state of Z?
c. If we take the compound starting under the conditions of part b and reduce the temperature to 20°C and increase the pressure to 65 atm, what is the state of Z?
d. Would it be possible to make the compound starting under the conditions of part c a solid by increasing just the pressure?

**11.46**  Shown here is the phase diagram for compound X. The triple point of X is −25.1°C at 0.50 atm and the critical point is 22°C and 21.3 atm.

a. What is the state of X at position A?
b. If we decrease the temperature from the compound at position A to −28.2°C while holding the pressure constant, what is the state of X?
c. If we take the compound starting under the conditions of part b and increase the temperature to 15.3°C and decrease the pressure to 0.002 atm, what is the state of X?
d. Would it be possible to make the compound starting under the conditions of part c a solid by increasing just the pressure?

**11.47**  Use graph paper and sketch the phase diagram of oxygen, $O_2$, from the following information: normal melting point, −218°C; normal boiling point, −183°C; triple point, −219°C, 1.10 mmHg; critical point, −118°C, 50.1 atm. Label each phase region on the diagram.

**11.48**  Use graph paper and sketch the phase diagram of argon, Ar, from the following information: normal melting point, −187°C; normal boiling point, −186°C; triple point, −189°C, 0.68 atm; critical point, −122°C, 48 atm. Label each phase region on the diagram.

**11.49**  Which of the following substances can be liquefied by applying pressure at 25°C? For those that cannot, describe the conditions under which they *can* be liquefied.

| Substance | Critical Temperature | Critical Pressure |
|---|---|---|
| Sulfur dioxide, $SO_2$ | 158°C | 78 atm |
| Acetylene, $C_2H_2$ | 36°C | 62 atm |
| Methane, $CH_4$ | −82°C | 46 atm |
| Carbon monoxide, CO | −140°C | 35 atm |

**11.50**  A tank of gas at 21°C has a pressure of 1.0 atm. Using the data in the table, answer the following questions. Explain your answers.
a. If the tank contains carbon tetrafluoride, $CF_4$, is the liquid state also present?
b. If the tank contains butane, $C_4H_{10}$, is the liquid state also present?

| Substance | Boiling Point at 1 atm | Critical Temperature | Critical Pressure |
|---|---|---|---|
| $CF_4$ | −128°C | −46°C | 41 atm |
| $C_4H_{10}$ | −0.5°C | 152°C | 38 atm |

**11.51**     Bromine, $Br_2$, has a triple point at $-7.3°C$ and 44 mmHg and a critical point at $315°C$ and 102 atm. The density of the solid is 3.4 g/cm$^3$, and the density of the liquid is 3.1 g/cm$^3$. Sketch a rough phase diagram of bromine, labeling all important features. Circle the correct word in each of the following sentences (and explain your answers).
a. Bromine vapor at 40 mmHg condenses to the (liquid, solid) when cooled sufficiently.
b. Bromine vapor at 400 mmHg condenses to the (liquid, solid) when cooled sufficiently.

**11.52**     Krypton, Kr, has a triple point at $-169°C$ and 133 mmHg and a critical point at $-63°C$ and 54 atm. The density of the solid is 2.8 g/cm$^3$, and the density of the liquid is 2.4 g/cm$^3$. Sketch a rough phase diagram of krypton. Circle the correct word in each of the following sentences (and explain your answers).
a. Solid krypton at 130 mmHg (melts, sublimes without melting) when the temperature is raised.
b. Solid krypton at 760 mmHg (melts, sublimes without melting) when the temperature is raised.

## Intermolecular Forces and Properties of Liquids

**11.53**     The heats of vaporization of liquid $Cl_2$, liquid $H_2$, and liquid $N_2$ are 20.4 kJ/mol, 0.9 kJ/mol, and 5.6 kJ/mol, respectively. Are the relative values as you would expect? Explain.

**11.54**     The heats of vaporization of liquid $O_2$, liquid Ne, and liquid methanol, $CH_3OH$, are 6.8 kJ/mol, 1.8 kJ/mol, and 34.5 kJ/mol, respectively. Are the relative values as you would expect? Explain.

**11.55**     For each of the following substances, list the kinds of intermolecular forces expected.
a.

b. isopropyl alcohol, $CH_3CHOHCH_3$
c. hydrogen iodide, HI
d. krypton, Kr

**11.56**     Which of the following compounds would you expect to exhibit *only* London forces?
a. carbon tetrachloride, $CCl_4$
b.

c. phosphorus trichloride, $PCl_3$
d. phosphorus pentachloride, $PCl_5$

**11.57**     Arrange the following substances in order of increasing magnitude of the London forces: $SiCl_4$, $CCl_4$, $GeCl_4$.

**11.58**     Arrange the following substances in order of increasing magnitude of the London forces: Ar, He, Kr.

**11.59**     Methane, $CH_4$, reacts with chlorine, $Cl_2$, to produce a series of chlorinated hydrocarbons: methyl chloride ($CH_3Cl$), methylene chloride ($CH_2Cl_2$), chloroform ($CHCl_3$), and carbon tetrachloride ($CCl_4$). Which compound has the lowest vapor pressure at room temperature? Explain.

**11.60**     The halogens form a series of compounds with each other, which are called interhalogens. Examples are bromine chloride (BrCl), iodine bromide (IBr), bromine fluoride (BrF), and chlorine fluoride (ClF). Which compound is expected to have the lowest boiling point at any given pressure? Explain.

**11.61**     Predict the order of increasing vapor pressure at a given temperature for the following compounds:
a. $FCH_2CH_2F$
b. $HOCH_2CH_2OH$
c. $FCH_2CH_2OH$
Explain why you chose this order.

**11.62**     Predict the order of increasing vapor pressure at a given temperature for the following compounds:
a. $CH_3CH_2CH_2CH_2OH$
b. $HOCH_2CH_2CH_2OH$
c. $CH_3CH_2OCH_2CH_3$
Explain why you chose this order.

**11.63**     List the following substances in order of increasing boiling point:

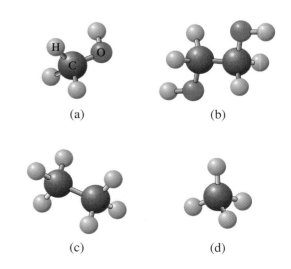

(a)          (b)

(c)          (d)

**11.64** Arrange the following compounds in order of increasing boiling point.

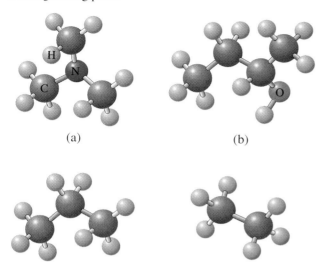

(a)                                        (b)

## Types of Solids

**11.65** Classify each of the following by the type of solid it forms: (a) Na; (b) Fe; (c) B; (d) $H_2O$; (e) KF.

**11.66** Classify each of the following by the type of solid it forms: (a) LiCl; (b) $BaCl_2$; (c) $BCl_3$; (d) $CCl_4$; (e) $NCl_3$.

**11.67** Classify each of the following solid elements as molecular, metallic, ionic, or covalent network.
a. tin, Sn
b. germanium, Ge
c. sulfur, $S_8$
d. iodine, $I_2$

**11.68** Which of the following do you expect to be molecular solids?
a. sodium hydroxide, NaOH
b. solid ethane, $C_2H_6$
c. nickel, Ni
d. solid silane, $SiH_4$

**11.69** Arrange the following compounds in order of increasing melting point.

CaO  (a)

H—C—C—O—C—C—H with H H H H (b)

H—C—C—C—C—O—H with H H H H (c)

KCl (d)

**11.70** Arrange the following substances in order of increasing melting point.

(a)        (b)              (c)            (d)

**11.71** Associate each type of solid in the left-hand column with two of the properties in the right-hand column. Each property may be used more than once.
a. molecular solid
b. ionic solid
c. metallic solid
d. covalent network solid

low-melting
high-melting
brittle
malleable
hard
electrically conducting

**11.72** On the basis of the description given, classify each of the following solids as molecular, metallic, ionic, or covalent network. Explain your answers.
a. a lustrous, yellow solid that conducts electricity
b. a hard, black solid melting at 2350°C to give a nonconducting liquid
c. a nonconducting, pink solid melting at 650°C to give an electrically conducting liquid
d. red crystals having a characteristic odor and melting at 171°C

**11.73** Associate each of the solids Co, LiCl, SiC, and $CHI_3$ with one of the following sets of properties.
a. A white solid melting at 613°C; the liquid is electrically conducting, although the solid is not.
b. A very hard, blackish solid subliming at 2700°C.
c. A yellow solid with a characteristic odor having a melting point of 120°C.
d. A gray, lustrous solid melting at 1495°C; both the solid and liquid are electrical conductors.

**11.74** Associate each of the solids BN, $P_4S_3$, Pb, and $CaCl_2$ with one of the following sets of properties.
a. A bluish-white, lustrous solid melting at 327°C; the solid is soft and malleable.
b. A white solid melting at 772°C; the solid is an electrical nonconductor but dissolves in water to give a conducting solution.
c. A yellowish-green solid melting at 172°C.
d. A very hard, colorless substance melting at about 3000°C.

## Crystal Structure

**11.75** How many atoms are there in a simple cubic unit cell of an atomic crystal in which all atoms are at lattice points?

**11.76** How many atoms are there in a body-centered cubic unit cell of an atomic crystal in which all atoms are at lattice points?

**11.77** Metallic iron has a body-centered cubic lattice with all atoms at lattice points and a unit cell whose edge length is 286.6 pm. The density of iron is 7.87 g/cm³. What is the mass of an iron atom? Compare this value with the value you obtain from the molar mass.

**11.78** Nickel has a face-centered unit cell with all atoms at lattice points and an edge length of 352.4 pm. The density of metallic nickel is 8.91 g/cm³. What is the mass of a nickel atom? From the atomic weight, calculate Avogadro's number.

**11.79** Copper metal has a face-centered cubic structure with all atoms at lattice points and a density of 8.93 g/cm³. Its atomic weight is 63.5 amu. Calculate the edge length of the unit cell.

**11.80** Barium metal has a body-centered cubic lattice with all atoms at lattice points; its density is 3.51 g/cm³. From these data and the atomic weight, calculate the edge length of a unit cell.

**11.81** Gold has cubic crystals whose unit cell has an edge length of 407.9 pm. The density of the metal is 19.3 g/cm³. From these data and the atomic weight, calculate the number of gold atoms in a unit cell, assuming all atoms are at lattice points. What type of cubic lattice does gold have?

**11.82** Chromium forms cubic crystals whose unit cell has an edge length of 288.5 pm. The density of the metal is 7.20 g/cm³. Use these data and the atomic weight to calculate the number of atoms in a unit cell, assuming all atoms are at lattice points. What type of cubic lattice does chromium have?

**11.83** Tungsten has a body-centered cubic lattice with all atoms at the lattice points. The edge length of the unit cell is 316.5 pm. The atomic weight of tungsten is 183.8 amu. Calculate its density.

**11.84** Lead has a face-centered cubic lattice with all atoms at lattice points and a unit-cell edge length of 495.0 pm. Its atomic weight is 207.2 amu. What is the density of lead?

**11.85** Metallic magnesium has a hexagonal close-packed structure and a density of 1.74 g/cm³. Assume magnesium atoms to be spheres of radius $r$. Because magnesium has a close-packed structure, 74.1% of the space is occupied by atoms. Calculate the volume of each atom; then find the atomic radius, $r$. The volume of a sphere is equal to $4\pi r^3/3$.

**11.86** Metallic barium has a body-centered cubic structure (all atoms at the lattice points) and a density of 3.51 g/cm³. Assume barium atoms to be spheres. The spheres in a body-centered array occupy 68.0% of the total space. Find the atomic radius of barium. (See Problem 11.85.)

## General Problems

**11.87** If you leave your car parked outdoors in the winter, you may find frost on the windows in the morning. If you then start the car and let the heater warm the windows, after some minutes the windows will be dry. Describe all of the phase changes that have occurred.

**11.88** Snow forms in the upper atmosphere in a cold air mass that is supersaturated with water vapor. When the snow later falls through a lower, warm air mass, rain forms. When this rain falls on a sunny spot, the drops evaporate. Describe all of the phase changes that have occurred.

**11.89** The percent relative humidity of a sample of air is found as follows: (partial pressure of water vapor/vapor pressure of water) × 100. A sample of air at 21°C was cooled to 15°C, where moisture began to condense as dew. What was the relative humidity of the air at 21°C?

**11.90** A sample of air at 21°C has a relative humidity of 58%. At what temperature will water begin to condense as dew? (See Problem 11.89.)

**11.91** The vapor pressure of benzene is 100.0 mmHg at 26.1°C and 400.0 mmHg at 60.6°C. What is the boiling point of benzene at 760.0 mmHg?

**11.92** The vapor pressure of water is 17.5 mmHg at 20.0°C and 355.1 mmHg at 80.0°C. Calculate the boiling point of water at 760.0 mmHg.

**11.93** Describe the behavior of carbon dioxide gas when compressed at the following temperatures:
a. 20°C        b. −70°C        c. 40°C
The triple point of carbon dioxide is −57°C and 5.1 atm, and the critical point is 31°C and 73 atm.

**11.94** Describe the behavior of iodine vapor when cooled at the following pressures:
a. 120 atm        b. 1 atm        c. 50 mmHg
The triple point of iodine is 114°C and 90.1 mmHg, and the critical point is 512°C and 116 atm.

**11.95** Describe the formation of hydrogen bonds in propanol, $CH_3CH_2CH_2OH$. Represent possible hydrogen bonding structures in propanol by using structural formulas and the conventional notation for a hydrogen bond.

**11.96** Describe the formation of hydrogen bonds in hydrogen peroxide, $H_2O_2$. Represent possible hydrogen bonding structures in hydrogen peroxide by using structural formulas and the conventional notation for a hydrogen bond.

**11.97** Ethylene glycol ($CH_2OHCH_2OH$) is a slightly viscous liquid that boils at 198°C. Pentane ($C_5H_{12}$), which has approximately the same molecular weight as ethylene glycol, is a nonviscous liquid that boils at 36°C. Explain the differences in physical characteristics of these two compounds.

**11.98** Pentylamine, $CH_3CH_2CH_2CH_2CH_2NH_2$, is a liquid that boils at 104°C and has a viscosity of $10 \times 10^{-4}$ kg/(m · s). Triethylamine, $(CH_3CH_2)_3N$, is a liquid that boils at 89°C and has a viscosity of about $4 \times 10^{-4}$ kg/(m · s). Explain the differences in properties of these two compounds.

**11.99** Consider the elements Al, Si, P, and S from the third row of the periodic table. In each case, identify the type of solid the element would form.

**11.100** The elements in Problem 11.99 form the fluorides $AlF_3$, $SiF_4$, $PF_3$, and $SF_4$. In each case, identify the type of solid formed by the fluoride.

**11.101** Decide which substance in each of the following pairs has the lower melting point. Explain how you made each choice.
a. potassium chloride, KCl; or calcium oxide, CaO
b. carbon tetrachloride,

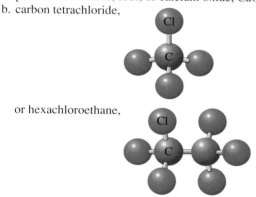

or hexachloroethane,

c. zinc, Zn; or chromium, Cr
d. acetic acid, $CH_3COOH$; or ethyl chloride, $C_2H_5Cl$

**11.102** Decide which substance in each of the following pairs has the lower melting point. Explain how you made each choice.
a. magnesium oxide, MgO; or hexane, $C_6H_{14}$
b. 1-propanol,

or ethylene glycol,

c. silicon, Si; or sodium, Na
d. methane, $CH_4$; or silane, $SiH_4$

**11.103** Iridium metal, Ir, crystallizes in a face-centered cubic (close-packed) structure. The edge length of the unit cell was found by x-ray diffraction to be 383.9 pm. The density of iridium is 22.42 g/cm$^3$. Calculate the mass of an iridium atom. Use Avogadro's number to calculate the atomic weight of iridium.

**11.104** The edge length of the unit cell of tantalum metal, Ta, is 330.6 pm; the unit cell is body-centered cubic (one atom at each lattice point). Tantalum has a density of 16.69 g/cm$^3$. What is the mass of a tantalum atom? Use Avogadro's number to calculate the atomic weight of tantalum.

**11.105** Use your answer to Problem 11.79 to calculate the radius of the copper atom. Assume that copper atoms are spheres. Then note that the spheres on any face of a unit cell touch along the diagonal.

**11.106** Rubidium metal has a body-centered cubic structure (with one atom at each lattice point). The density of the metal is 1.532 g/cm$^3$. From this information and the atomic weight, calculate the edge length of the unit cell. Now assume that rubidium atoms are spheres. Each corner sphere of the unit cell touches the body-centered sphere. Calculate the radius of a rubidium atom.

**11.107** Calculate the percent of volume that is actually occupied by spheres in a body-centered cubic lattice of identical spheres. You can do this by first relating the radius of a sphere, $r$, to the length of an edge of a unit cell, $l$. (Note that the spheres do not touch along an edge, but do touch along a diagonal passing through the body-centered sphere.) Then calculate the volume of a unit cell in terms of $r$. The volume occupied by spheres equals the number of spheres per unit cell times the volume of a sphere ($4\pi r^3/3$).

**11.108** Calculate the percent of volume that is actually occupied by spheres in a face-centered cubic lattice of identical spheres. You can do this by first relating the radius of a sphere, $r$, to the length of an edge of a unit cell, $l$. (Note that the spheres do not touch along an edge, but do touch along the diagonal of a face.) Then calculate the volume of a unit cell in terms of $r$. The volume occupied by spheres equals the number of spheres per unit cell times the volume of a sphere ($4\pi r^3/3$).

**11.109** For the hydrogen halides and the noble gases, we have the following boiling points:

| Halogen Family, °C | Noble Gases, °C |
| --- | --- |
| HF, 19 | Ne, −246 |
| HCl, −115 | Ar, −186 |
| HBr, −67 | Kr, −152 |
| HI, −35 | Xe, −108 |

Account for the following:
a. The general trend in the boiling points of the hydrides and the noble gases.
b. The unusual boiling point of hydrogen fluoride.
c. The observation that the hydrogen halides have boiling points that are significantly higher than the noble gases.

**11.110** For the carbon and nitrogen family hydrides, we have the following boiling points:

| Carbon Family, °C | Nitrogen Family, °C |
|---|---|
| $CH_4$, −164 | $NH_3$, −33 |
| $SiH_4$, −112 | $PH_3$, −88 |
| $GeH_4$, −88 | $AsH_3$, −55 |
| $SnH_4$, −52 | $SbH_3$, −17 |

Account for the following:
a. The general trend in the boiling points of the binary hydrides.
b. The unusual boiling point of ammonia.
c. The observation that the nitrogen family hydrides have boiling points that are notably higher than those of the carbon family.

**11.111** Account for the following observations:
a. Both diamond and silicon carbide are very hard, whereas graphite is both soft and slippery.

b. Carbon dioxide is a gas, whereas silicon dioxide is a high-melting solid.

**11.112** Greater variation exists between the properties of the first and second members of a family in the periodic table than between other members. Discuss this observation for the oxygen family using the following data.

| Element | Boiling Point, °C | Compound | Boiling Point, °C |
|---|---|---|---|
| $O_2$ | −183 | $H_2O$ | 100 |
| $S_8$ | 445 | $H_2S$ | −61 |
| $Se_8$ | 685 | $H_2Se$ | −42 |

**11.113** Use chemical principles to discuss the following observations:
a. $CO_2$ sublimes at −78°C, whereas $SiO_2$ boils at 2200°C.
b. HF boils at 19°C, whereas HCl boils at −85°C.
c. $CF_4$ boils at −128°C, whereas $SiF_4$ boils at −86°C.

**11.114** a. Draw Lewis structures of each of the following compounds: LiH, $NH_3$, $CH_4$, $CO_2$.
b. Which of these has the highest boiling point? Why?
c. Which of these has the lowest boiling point? Why?
d. Which of these has the next-to-highest boiling point? Why?

# Cumulative-Skills Problems

**11.115** The vapor pressure of a volatile liquid can be determined by slowly bubbling a known volume of gas through the liquid at a given temperature and pressure. In an experiment, a 5.40-L sample of nitrogen gas, $N_2$, at 20.0°C and 745 mmHg is bubbled through liquid isopropyl alcohol, $C_3H_8O$, at 20.0°C. Nitrogen containing the vapor of $C_3H_8O$ at its vapor pressure leaves the vessel at 20.0°C and 745 mmHg. It is found that 0.6149 g $C_3H_8O$ has evaporated. How many moles of $N_2$ are in the gas leaving the liquid? How many moles of alcohol are in this gaseous mixture? What is the mole fraction of alcohol vapor in the gaseous mixture? What is the partial pressure of the alcohol in the gaseous mixture? What is the vapor pressure of $C_3H_8O$ at 20.0°C?

**11.116** In an experiment, a sample of 6.35 L of nitrogen at 25.0°C and 768 mmHg is bubbled through liquid acetone, $C_3H_6O$. The gas plus vapor at its equilibrium partial pressure leaves the liquid at the same temperature and pressure. If 6.550 g of acetone has evaporated, what is the vapor pressure of acetone at 25.0°C? See Problem 11.115.

**11.117** How much heat is needed to vaporize 10.0 mL of liquid hydrogen cyanide, HCN, at 25.0°C? The density of the liquid is 0.687 g/mL. Use standard heats of formation, which are given in Appendix C.

**11.118** How much heat is needed to vaporize 20.0 mL of liquid methanol, $CH_3OH$, at 25.0°C? The density of the liquid is 0.787 g/mL. Use standard heats of formation, which are given in Appendix C.

**11.119** How much heat must be added to 12.5 g of solid white phosphorus, $P_4$, at 25.0°C to give the liquid at its melting point, 44.1°C? The heat capacity of solid white phosphorus is 95.4 J/(K · mol); its heat of fusion is 2.63 kJ/mol.

**11.120** How much heat must be added to 25.0 g of solid sodium, Na, at 25.0°C to give the liquid at its melting point, 97.8°C? The heat capacity of solid sodium is 28.2 J/(K · mol), and its heat of fusion is 2.60 kJ/mol.

**11.121** Acetic acid, $CH_3COOH$, forms stable pairs of molecules held together by two hydrogen bonds.

Such molecules—themselves formed by the association of two simpler molecules—are called dimers. The vapor over liquid acetic acid consists of a mixture of monomers (single acetic acid molecules) and dimers. At 100.6°C the total pressure of vapor over liquid acetic acid is 436 mmHg. If the vapor consists of 0.630 mole fraction of the dimer, what are the masses of monomer and dimer in 1.000 L of the vapor? What is the density of the vapor?

**11.122** The total pressure of vapor over liquid acetic acid at 71.3°C is 146 mmHg. If the density of the vapor is 0.702 g/L, what is the mole fraction of dimer in the vapor? See Problem 11.121.

#  Media Activities

**11.123** Open the Changes of State *Understanding Concepts* activity on the student CD to review phase changes and heating curves.

a. When a solid converts to a liquid and then to a gas, describe what happens during each step in terms of the intermolecular forces.

b. Rationalize why $\Delta H_{vap}$ for water is over six times larger than $\Delta H_{fus}$.

c. What values of $\Delta H$ would you need to calculate in order to determine the total amount of heat required to change ice at −20°C to liquid water at 45°C?

**11.124** Work through the Intermolecular Forces *Understanding Concepts* activity on the student CD.

a. What type of intermolecular forces do nonpolar covalent compounds exhibit?

b. How is molecular size related to the strength of London dispersion forces?

c. What three covalent bonds are necessary in order for a covalent compound to exhibit hydrogen bonding?

**11.125** Open the Exercises section in the Intermolecular Forces *Understanding Concepts* activity on the student CD to help you predict the molecular geometry and intermolecular

forces present in $SO_2$, $N_2H_4$, $O_3$, and $SP_6$. Try as many additional structures provided by the program as necessary in order to master the concepts.

**11.126** Run the Unit Cells and Crystal Packing *Visualization* on the student CD.

a. Which unit cell—simple cubic, face-centered cubic, or body-centered cubic—is a closest-packed arrangement?

b. Which unit cell—simple cubic, face-centered cubic, or body-centered cubic—contains the most atoms?

c. Which packing arrangement of the layers in a solid has a face-centered cubic unit cell?

**11.127** Play the Unit Cell Calculations *Understanding Concepts* simulation on the student web site. Calculate the atomic radius of an atom that crystallizes in a structure that has a face-centered cubic unit cell with an edge length of 341 pm.

**11.128** Play the Unit Cells *Understanding Concepts* simulation on the student web site. Calculate the density of a hypothetical element that crystallizes in a body-centered cubic lattice with the atoms at the lattice points. The density of the element is 16.5 g/cm³, and the atomic mass is 41.23 amu.

**11.129** For review, take the ACE practice test for Chapter 11 on the student web site.

Sugar cube dissolving in water.

# Solutions

There are various practical reasons for preparing solutions. For instance, most chemical reactions are run in solution. Also, solutions have particular properties that are useful. When gold is used for jewelry, it is mixed, or alloyed, with a small amount of silver. Gold–silver alloys are not only harder than pure gold, but they also melt at lower temperatures and are therefore easier to cast.

Solubility, the amount of one substance that dissolves in another, varies with temperature and, in some cases, with pressure. The variation of solubility with pressure can be a useful property. Acetylene gas, $C_2H_2$, for example, is used as a fuel in welding torches. It can be transported safely under pressure in cylinders as a solution in acetone, $CH_3COCH_3$. Acetone is a liquid, and at 1 atm pressure 1 L of the liquid dissolves 27 g of acetylene. But at 12 atm, the pressure in a full cylinder, the same quantity of acetone dissolves 320 g of acetylene, so that more can be transported. When the valve on a cylinder is opened, reducing the pressure, acetylene gas comes out of solution.

Another useful property of solutions is their lower melting, or freezing, points compared with those of the major component. We have already mentioned the lowering of the melting point of gold when a small amount of silver is added. The use of ethylene glycol, $CH_2OHCH_2OH$, as an automobile antifreeze depends on the same property. Water containing ethylene glycol freezes at temperatures below the freezing point of pure water.

What determines how much the freezing point of a solution is lowered? How does the solubility of a substance change when conditions such as pressure and temperature change? These are some of the questions we will address in this chapter.

# Solution Formation

▶ This chapter looks at the solution process, concentration units, colligative properties, and colloids. If pressed for time, you might want to devote your attention to concentration units and colligative properties.

When sodium chloride dissolves in water, the resulting uniform dispersion of ions in water is called a solution. In general, a *solution* is a homogeneous mixture of two or more substances, consisting of ions or molecules. A *colloid* is similar in that it appears to be homogeneous like a solution. In fact, however, it consists of comparatively large particles of one substance dispersed throughout another substance or solution.

From the examples of solutions mentioned in the chapter opening, you see that they may be quite varied in their characteristics. To begin our discussion, let us look at the different types of solutions we might encounter. We will discuss colloids at the end of the chapter.

## 12.1  Types of Solutions

Solutions may exist in any of the three states of matter; that is, they may be gases, liquids, or solids. Some examples are listed in Table 12.1. The terms *solute* and *solvent* refer to the components of a solution. The **solute,** *in the case of a solution of a gas or solid dissolved in a liquid, is the gas or solid; in other cases, the solute is the component in smaller amount.* The **solvent,** *in a solution of a gas or solid dissolved in a liquid, is the liquid; in other cases, the solvent is the component in greater amount.* Thus, when sodium chloride is dissolved in water, sodium chloride is the solute and water is the solvent.

### Gaseous Solutions

In general, nonreactive gases or vapors can mix in all proportions to give a gaseous mixture. *Fluids that mix with or dissolve in each other in all proportions* are said to be **miscible fluids.** Gases are thus miscible. (If two fluids do not mix but form two layers, they are said to be *immiscible.* Air, which is a mixture of oxygen, nitrogen, and smaller amounts of other gases, is an example of a gaseous solution.

**Table 12.1**
**Examples of Solutions**

| Solution | State of Matter | Description |
|---|---|---|
| Air | Gas | Homogeneous mixture of gases ($O_2$, $N_2$, others) |
| Soda water | Liquid | Gas ($CO_2$) dissolved in a liquid ($H_2O$) |
| Ethanol in water | Liquid | Liquid solution of two completely miscible liquids |
| Brine | Liquid | Solid (NaCl) dissolved in a liquid ($H_2O$) |
| Potassium–sodium alloy | Liquid | Solution of two solids (K + Na) |
| Dental-filling alloy | Solid | Solution of a liquid (Hg) in a solid (Ag plus other metals) |
| Gold–silver alloy | Solid | Solution of two solids (Au + Ag) |

**Figure 12.1**
**Immiscible and miscible
liquids.** *Left:* Water and
methylene chloride are
immiscible and form two
layers. *Right:* Acetone and
water are miscible liquids; that
is, the two substances dissolve
in each other in all proportions.

## Liquid Solutions

Most liquid solutions are obtained by dissolving a gas, liquid, or solid in some liquid. Soda water, for example, consists of a solution of carbon dioxide gas in water. Acetone, $C_3H_6O$, in water is an example of a liquid–liquid solution. (Immiscible and miscible liquids are shown in Figure 12.1.) Brine is water with sodium chloride (a solid) dissolved in it. Seawater contains both dissolved gases (from air) and solids (mostly sodium chloride).

It is also possible to make a liquid solution by mixing two solids together. Potassium–sodium alloy is an example of this. Both potassium and sodium are solid metals at room temperature, but a liquid solution results when the mixture contains 10% to 50% sodium. ◄

▶ Sodium metal melts at 98°C and potassium metal melts at 63°C. When the two metals are mixed to give a solution that is 20% sodium, for example, the melting point is lowered to −10°C. This is another example of the lowering of the freezing, or melting, point of solutions, which we will discuss later in the chapter. Potassium–sodium alloy is used as a heat-transfer medium in nuclear reactors.

## Solid Solutions

Solid solutions are also possible. In the chapter opening, we mentioned gold–silver alloys. Dental-filling alloy is a solution of mercury (a liquid) in silver (a solid), with small amounts of other metals.

---

**EXERCISE 12.1**

Give an example of a solid solution prepared from a liquid and a solid.

---

See Problems 12.29 and 12.30.

▶ Video no. 16 (Series A),
"Unsaturated, Saturated,
and Supersaturated Solutions."

## CONCEPT CHECK 12.1

Identify the solute(s) and solvent(s) in the following solutions.

a. 80 g of Cr and 5 g of Mo

b. 5 g of $MgCl_2$ dissolved in 1000 g of $H_2O$

c. 39% $N_2$, 41% Ar, and the rest $O_2$

▶ Video no. 17 (Series A),
"Crystallization from Supersaturated Solutions."

# 12.2 Solubility and the Solution Process

**Visualization: Dissolution of a Solid in a Liquid**

**Figure 12.2**
**Solubility equilibrium.**
The solid crystalline phase is in dynamic equilibrium with species (ions or molecules) in a saturated solution. The rate at which species leave the crystals equals the rate at which species return to the crystals.

▶ Video Series C,
"Supersaturated Sodium Acetate."

**Figure 12.3**
**Comparison of unsaturated and saturated solutions.** *Top:* The 30.0-g pile of NaCl will dissolve completely in 100 mL of water at 20°C, giving an unsaturated solution. *Bottom:* When 40.0 g NaCl is stirred into 100 mL $H_2O$, 36.0 g dissolves at 20°C, leaving 4.0 g of the crystalline solid on the bottom of the beaker. This solution is saturated.

The amount of substance that will dissolve in a solvent depends on both the substance and the solvent. We describe the amount that dissolves in terms of solubility.

## Solubility; Saturated Solutions

To understand the concept of solubility, consider the process of dissolving sodium chloride in water. Sodium chloride is an ionic substance and it dissolves in water as $Na^+$ and $Cl^-$ ions. If you could view the dissolving of sodium chloride at the level of ions, you would see a dynamic process. Suppose you stir 40.0 g of sodium chloride crystals into 100 mL of water at 20°C. Sodium ions and chloride ions leave the surface of the crystals and enter the solution. The ions move about at random in the solution and may by chance collide with a crystal and stick, thus returning to the crystalline state. As the sodium chloride continues to dissolve, more ions enter the solution, and the rate at which they return to the crystalline state increases (the more ions in solution, the more likely ions are to collide with the crystals and stick). Eventually, a *dynamic equilibrium* is reached in which the rate at which ions leave the crystals equals the rate at which ions return to the crystals (see Figure 12.2). You write the dynamic equilibrium this way:

$$NaCl(s) \xrightleftharpoons{H_2O} Na^+(aq) + Cl^-(aq)$$

At equilibrium, no more sodium chloride appears to dissolve; 36.0 g has gone into solution, leaving 4.0 g of crystals at the bottom of the vessel. You have a **saturated solution**—that is, *a solution that is in equilibrium with respect to a given dissolved substance.* The solution is saturated with respect to NaCl, and no more NaCl can dissolve. The **solubility** of sodium chloride in water *(the amount that dissolves in a given quantity of water at a given temperature to give a saturated solution)* is 36.0 g/100 mL at 20°C. Note that if you had mixed 30.0 g of sodium chloride with 100 mL of water, all of the crystals would have dissolved. You would have an **unsaturated solution,** *a solution not in equilibrium with respect to a given dissolved substance and in which more of the substance can dissolve.* (Saturated and unsaturated solutions are compared in Figure 12.3.)

+ 30.0 g NaCl 100 mL $H_2O$ = **Unsaturated solution** containing 100 mL $H_2O$ and 30.0 g NaCl

+ 40.0 g NaCl 100 mL $H_2O$ = **Saturated solution** containing 100 mL $H_2O$ and 36.0 g NaCl
The additional 4.0 g NaCl remains undissolved

**Figure 12.4**
**Crystallization from a supersaturated solution of sodium acetate.** *Left:* Crystallization begins to occur when a small crystal of sodium acetate is added. *Center, right:* Within seconds, crystal growth spreads from the original crystal throughout the solution.

Sometimes it is possible to obtain a **supersaturated solution,** *a solution that contains more dissolved substance than a saturated solution does.* For example, the solubility of sodium thiosulfate, $Na_2S_2O_3$, in water at 100°C is 231 g/100 mL. But at room temperature, the solubility is much less—about 50 g/100 mL. Suppose you prepare a solution saturated with sodium thiosulfate at 100°C. You might expect that as the water solution is cooled, sodium thiosulfate would crystallize out. In fact, if the solution is slowly cooled to room temperature, this does not occur. Instead the result is a solution in which 231 g of sodium thiosulfate is dissolved in 100 mL of cold water, compared with the 50 g you would normally expect to find dissolved.

Supersaturated solutions are not in equilibrium with the solid substance. If a small crystal of sodium thiosulfate is added to a supersaturated solution, the excess immediately crystallizes out. Crystallization from a supersaturated solution is usually quite fast and dramatic (see Figure 12.4).

## Factors in Explaining Solubility

The solubilities of substances in one another vary widely. You might find a substance miscible in one solvent but nearly insoluble in another. As a general rule, "like dissolves like." That is, similar substances dissolve one another. Oil is miscible in gasoline. Both are mixtures of hydrocarbon substances (compounds of hydrogen and carbon only). On the other hand, oil does not mix with water. Water is a polar substance, whereas hydrocarbons are not. Why do similar substances dissolve in one another to greater extents than do dissimilar substances? What factors are involved in solubility?

The solubility of one substance in another can be explained in terms of two factors. One is the natural tendency of substances to mix. This is sometimes also referred to as the natural tendency toward disorder. ◄ Figure 12.5A shows a vessel divided into two parts, with oxygen gas on the left and nitrogen gas on the right. If you remove the partition, the molecules of the two gases begin to mix. Ultimately, the molecules

▶ *Entropy* is a measure of disorder. Thus, the tendency toward disorder can be expressed as a tendency toward increasing entropy. The mixing of substances increases entropy. This increase is one of the primary factors leading to the spontaneous formation of a solution.

**Figure 12.5**
**The mixing of gas molecules.** *(A)* A vessel is divided by a removable partition with oxygen gas on the left and nitrogen gas on the right. *(B)* When the partition is removed, molecules of the two gases begin to mix. Eventually, the molecules become thoroughly mixed through their random motions.

become thoroughly mixed through their random motions (Figure 12.5*B*). You might expect a similar mixing of molecules or ions in other types of solutions.

If the process of dissolving one substance in another involved nothing more than simple mixing, you would expect substances to be completely soluble in one another; that is, you would expect substances to be miscible. You know that this is only sometimes the case. Usually, substances have limited solubility in one another. A factor that can limit solubility is the relative forces of attraction between species (molecules or ions). Suppose there are strong attractions between solute species and strong attractions between solvent species, but weak attractions between solute and solvent species. In that case, the strongest attractions are maintained so long as the solute and solvent species do not mix. The lowest energy of the solute–solvent system is obtained then also.

The solubility of a solute in a solvent (that is, the extent of the mixing of the solute and solvent species) depends on a balance between the natural tendency for the solute and solvent species to mix and the tendency for a system to have the lowest energy possible.

## Molecular Solutions

The simplest example of a molecular solution is one gas dissolved in another gas. Air, essentially a solution of oxygen and nitrogen, is an example. The intermolecular forces in gases are weak. The only solubility factor of importance is the natural tendency for molecules to mix (the inclination toward disorder). Gases are therefore miscible.

Substances may be miscible even when the intermolecular forces are not negligible. Consider the solution of the two similar liquid hydrocarbons heptane, $C_7H_{16}$, and octane, $C_8H_{18}$, which are components of gasoline. The intermolecular attractions are due to London forces, and those between heptane and octane molecules are nearly equal to those between octane and octane molecules and heptane and heptane molecules. The different intermolecular attractions are about the same strength, so there are no favored attractions. Octane and heptane molecules tend to move freely through one another. Therefore, the tendency of molecules to mix results in miscibility of the substances.

As a counterexample, consider the mixing of octane with water. There are strong hydrogen bonding forces between water molecules. For octane to mix with water, hydrogen bonds must be broken and replaced by much weaker London forces between water and octane. In this case, the maximum forces of attraction among molecules (and therefore the lower energy) result if the octane and water remain unmixed. Therefore, octane and water are nearly immiscible. (See Figure 12.6.)

The statement "like dissolves like" succinctly expresses these observations. That is, substances with similar intermolecular attractions are usually soluble in one

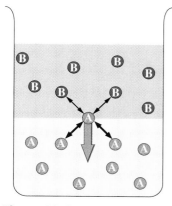

**Figure 12.6**
**The immiscibility of liquids.** Suppose an *A* molecule moves from liquid *A* into liquid *B*. If the intermolecular attraction between two *A* molecules is much stronger than the intermolecular attraction between an *A* molecule and a *B* molecule, the net force of attraction tends to pull the *A* molecule back into liquid *A*. Thus, liquid *A* will be immiscible with liquid *B*.

**Table 12.2**
**Solubilities of Alcohols in Water**

| Name | Formula | Solubility in H₂O (g/100 g H₂O at 20°C) |
|------|---------|------------------------------------------|
| Methanol | $CH_3OH$ | Miscible |
| Ethanol | $CH_3CH_2OH$ | Miscible |
| 1-Propanol | $CH_3CH_2CH_2OH$ | Miscible |
| 1-Butanol | $CH_3CH_2CH_2CH_2OH$ | 7.9 |
| 1-Pentanol | $CH_3CH_2CH_2CH_2CH_2OH$ | 2.7 |
| 1-Hexanol | $CH_3CH_2CH_2CH_2CH_2CH_2OH$ | 0.6 |

another. The two similar hydrocarbons heptane and octane are completely miscible, whereas octane and water (with dissimilar intermolecular attractions) are immiscible.

For the series of alcohols (organic, or carbon-containing, compounds with an —OH group) listed in Table 12.2, the solubility in water decreases from miscible to slightly soluble. Water and alcohols are alike in having —OH groups through which strong hydrogen bonding attractions arise.

(R— = $CH_3$—, $C_2H_5$—, etc.)

The attraction between a methanol molecule, $CH_3OH$, and a water molecule is nearly as strong as that between two methanol molecules or between two water molecules. Methanol and water molecules tend to mix freely. Methanol and water are miscible, as are ethanol and water and 1-propanol and water. However, as the hydrocarbon end, R—, of the alcohol becomes the more prominent portion of the molecule, the alcohol becomes less like water. Now the forces of attraction between alcohol and water molecules are weaker than those between two alcohol molecules or between two water molecules. Therefore, the solubilities of alcohols decrease with increasing length of R.

**Figure 12.7**
**Attraction of water molecules to ions because of the ion–dipole force.**
The O end of H₂O orients toward the cation, whereas an H atom of H₂O orients toward the anion.

---

**EXERCISE 12.2**

Which of the following compounds is likely to be more soluble in water: $C_4H_9OH$ or $C_4H_9SH$? Explain.

See Problems 12.31, 12.32, 12.33, and 12.34.

## Ionic Solutions

Ionic substances differ markedly in their solubilities in water. For example, sodium chloride, NaCl, has a solubility of 36 g per 100 mL of water at room temperature, whereas calcium phosphate, $Ca_3(PO_4)_2$, has a solubility of only 0.002 g per 100 mL of water. In most cases, these differences in solubility can be explained in terms of the different energies of attraction between ions in the crystal and between ions and water.

The energy of attraction between an ion and a water molecule is due to an *ion–dipole force*. Water molecules are polar, so they tend to orient with respect to nearby ions. In the case of a positive ion ($Li^+$, for example), water molecules orient with their oxygen atoms (the negative ends of the molecular dipoles) toward the ion. In the case of a negative ion (for instance, $F^-$), water molecules orient with their hydrogen atoms (the positive ends of the molecular dipoles) toward the ion (see Figure 12.7).

*The attraction of ions for water molecules* is called **hydration.** ◀ Hydration of ions favors the dissolving of an ionic solid in water. Figure 12.8 illustrates the hydration of ions in the dissolving of a lithium fluoride crystal. Ions on the surface become hydrated and then move into the body of the solution as hydrated ions.

If the hydration of ions were the only factor in the solution process, you would expect all ionic solids to be soluble in water. The ions in a crystal, however, are very strongly attracted to one another. Therefore, the solubility of an ionic solid depends not only on the energy of hydration of ions (energy associated with the attraction between ions and water molecules) but also on *lattice energy*, the energy holding ions together in the crystal lattice. Lattice energy works against the solution process, so an ionic solid with relatively large lattice energy is usually insoluble.

Lattice energies depend on the charges on the ions (as well as on the distance between the centers of neighboring positive and negative ions). The greater the magnitude of ion charge, the greater is the lattice energy. ◀ For this reason, you might expect substances with singly charged ions to be comparatively soluble and

▶ Hydration of ions also occurs in crystalline solids. For example, copper(II) sulfate pentahydrate, $CuSO_4 \cdot 5H_2O$, contains five water molecules for each $CuSO_4$ formula unit in the crystal. Substances like this, called hydrates, were discussed in Chapter 2.

▶ According to Coulomb's law, the energy of attraction of two ions is proportional to the product of the ion charges and inversely proportional to the distance between the centers of the ions.

**Figure 12.8**
**The dissolving of lithium fluoride in water.** Ions on the surface of the crystal can hydrate, that is, associate with water molecules; ions at the corners are especially easy to remove because they are held by fewer lattice forces. Ions are completely hydrated in the aqueous phase and move off into the body of the liquid.

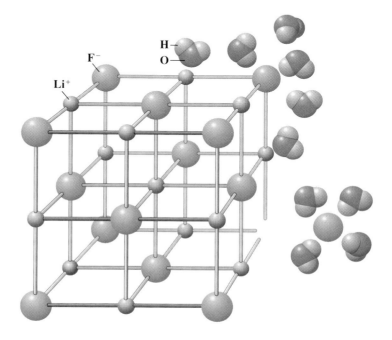

those with multiply charged ions to be less soluble. This is borne out by the fact that compounds of the alkali metal ions (such as $Na^+$ and $K^+$) and ammonium ions ($NH_4^+$) are generally soluble, whereas those of phosphate ions ($PO_4^{3-}$), for example, are generally insoluble.

Lattice energy is also inversely proportional to the distance between neighboring ions, and this distance depends on the sum of the radii of the ions. For example, the lattice energy of magnesium hydroxide, $Mg(OH)_2$, is inversely proportional to the sum of the radii of $Mg^{2+}$ and $OH^-$. In the series of alkaline earth hydroxides—$Mg(OH)_2$, $Ca(OH)_2$, $Sr(OH)_2$, $Ba(OH)_2$—the lattice energy decreases as the radius of the alkaline earth ion increases (from $Mg^{2+}$ to $Ba^{2+}$). If the lattice energy alone determines the trend in solubilities, you should expect the solubility to increase from magnesium hydroxide to barium hydroxide. In fact, this is what you find. Magnesium hydroxide is insoluble in water, and barium hydroxide is soluble. But this is not the whole story. The energy of hydration also depends on ionic radius. A small ion has a concentrated electric charge and a strong electric field that attracts water molecules. Therefore, the energy of hydration is greatest for a small ion such as $Mg^{2+}$ and least for a large ion such as $Ba^{2+}$. If the energy of hydration of ions alone determined the trend in solubilities, you would expect the solubilities to decrease from magnesium hydroxide to barium hydroxide, rather than to increase. (We should also add that the energy of hydration increases with the charge on the ion; energy of hydration is greater for $Mg^{2+}$ than for $Na^+$. Magnesium ion has a larger charge, and also is a smaller ion than $Na^+$.)

The explanation for the observed solubility trend in the alkaline earth hydroxides is that the lattice energy decreases more rapidly in the series $Mg(OH)_2$, $Ca(OH)_2$, $Sr(OH)_2$, and $Ba(OH)_2$ than does the energy of hydration in the series of ions $Mg^{2+}$, $Ca^{2+}$, $Sr^{2+}$, and $Ba^{2+}$. For this reason, the lattice-energy factor dominates this solubility trend.

You see the opposite solubility trend when the energy of hydration decreases more rapidly, so that it dominates the trend. Consider the alkaline earth sulfates. Here the lattice energy depends on the sum of the radius of the cation and the radius of the sulfate ion. Because the sulfate ion, $SO_4^{2-}$, is much larger than the hydroxide ion, $OH^-$, the percent change in lattice energy in going through the series of sulfates from $MgSO_4$ to $BaSO_4$ is smaller than in the hydroxides. The lattice energy changes less, and the energy of hydration of the cation decreases by a greater amount. Now the energy of hydration dominates the solubility trend, and the solubility decreases from magnesium sulfate to barium sulfate. Magnesium sulfate is soluble in water, and barium sulfate is insoluble.

---

### EXERCISE 12.3

Which ion has the larger hydration energy, $Na^+$ or $K^+$?

See Problems 12.35 and 12.36.

---

 ### CONCEPT CHECK 12.2

The hypothetical ionic compound $AB_2$ is very soluble in water. Another hypothetical ionic compound, $CB_2$, is only slightly soluble in water. The lattice energies for these compounds are about the same. Provide an explanation for the solubility difference between these compounds.

# Hemoglobin Solubility and Sickle-Cell Anemia

**LIFE SCIENCE**

Sickle-cell anemia was the first inherited disease shown to have a specific molecular basis. In people with the disease, the red blood cells tend to become elongated (sickle-shaped) when the concentration of oxygen ($O_2$) is low, as it is, for example, in the venous blood supply (see Figure 12.9). Once the red blood cells have sickled, they can no longer function in their normal capacity as oxygen carriers, and they often break apart. Moreover, the sickled cells clog capillaries, interfering with the blood supply to vital organs.

In 1949 Linus Pauling showed that people with sickle-cell anemia have abnormal hemoglobin. Hemoglobin is the substance in red blood cells that carries oxygen. Hemoglobin is normally present in solution within the red blood cells. But in people with sickle-cell anemia, the unoxygenated hemoglobin readily comes out of solution. It produces a fibrous precipitate that deforms the cell, giving it the characteristic sickle shape.

Hemoglobins are large molecules (molecular weight about 64,000 amu) consisting of four protein chains, two of one kind ($\alpha$ chains) and two of another kind

($\beta$ chains). Normal and sickle-cell hemoglobins are almost exactly alike, except that each $\beta$ chain of the hemoglobin responsible for sickle-cell anemia differs from normal hemoglobin in one place. In this place, the normal hemoglobin has the group

$$\begin{array}{c} | \\ CH_2 \\ | \\ CH_2 \\ | \\ HO-C=O \end{array}$$

which helps confer water solubility on the molecule because of the polarity of the group and its ability to form hydrogen bonds. The abnormal hemoglobin has the following hydrocarbon group:

$$\begin{array}{c} | \\ H-C-CH_3 \\ | \\ CH_3 \end{array}$$

Hydrocarbon groups are nonpolar. This small change makes the molecule less water-soluble.

**Figure 12.9**
**Red blood cells.** Scanning electron micrographs of normal (left) and sickled (right) red blood cells. The color is added by computer.

## 12.3 Effects of Temperature and Pressure on Solubility

In general, the solubility of a substance depends on temperature. For example, the solubility of ammonium nitrate in 100 mL of water is 118 g at 0°C and 811 g at 100°C. Pressure may also have an effect on solubility, as you will see.

**Figure 12.10**
**Solubility of some ionic salts at different temperatures.** The solubilities of the salts NaCl, $KNO_3$, and $CuSO_4$ rise with increasing temperature, as is the case with most ionic solids. The solubility of $Ce_2(SeO_4)_3$, however, falls with increasing temperature.

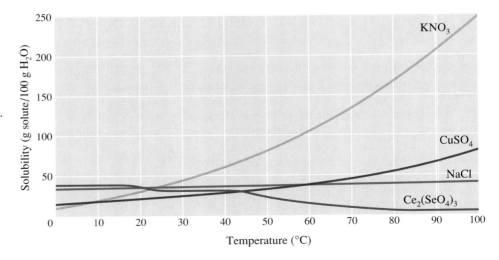

▶ The difficulty of applying Le Chatelier's principle to obtain the temperature dependence of solubility is discussed in G.M. Bodner, *J. Chem. Educ.* 1980, *57*, 117, and R. Treptow, *J. Chem. Educ.* 1984, *61*, 499.

▶ Demonstration of a chemical cold pack: Bassam Z. Shakhashiri, *Chemical Demonstrations*, Vol. 1 (Madison: University of Wisconsin Press, 1983), no. 1.2, pp. 8–9.

## Temperature Change

Most gases become less soluble in water at higher temperatures. The first bubbles that appear when tap water is heated are bubbles of air released as the increasing temperature reduces the solubility of air in water. In contrast, most ionic solids become more soluble in water with rising temperature.

The variations of the solubilities of the salts $KNO_3$, $CuSO_4$, NaCl, and $Ce_2(SeO_4)_3$ are shown in Figure 12.10. Three of the salts show the usual behavior; their solubilities increase with rising temperature. Potassium nitrate, $KNO_3$, changes solubility dramatically from 14 g/100 g $H_2O$ at 0°C to 245 g/100 g $H_2O$ at 100°C. Copper sulfate, $CuSO_4$, shows a moderate increase in solubility over this temperature interval. Sodium chloride, NaCl, increases only slightly in solubility with temperature.

A number of ionic compounds decrease in solubility with increasing temperature. Calcium sulfate, $CaSO_4$, and calcium hydroxide, $Ca(OH)_2$, are common examples. They are slightly soluble compounds that become even less soluble at higher temperatures. Cerium selenate, $Ce_2(SeO_4)_3$, shown in Figure 12.10, is very soluble at 0°C but much less soluble at 100°C.

Heat can be released or absorbed when ionic substances are dissolved in water. In some cases, this *heat of solution* is quite noticeable. When sodium hydroxide is dissolved in water, the solution becomes hot (the solution process is exothermic). On the other hand, when ammonium nitrate is dissolved in water, the solution becomes very cold (the solution process is endothermic). This cooling effect from the dissolving of ammonium nitrate in water is the basis for instant cold packs used in hospitals and elsewhere. An instant cold pack consists of a bag of $NH_4NO_3$ crystals inside a bag of water (Figure 12.11). When the inner bag is broken, $NH_4NO_3$ dissolves in the water. Heat is absorbed, so the bag feels cold. Similarly, hot packs are available, containing either $CaCl_2$ or $MgSO_4$, which dissolve in water with the evolution of heat.

## Pressure Change

In general, pressure change has little effect on the solubility of a liquid or solid in water, but the solubility of a gas is very much affected by pressure. The qualitative effect of a change in pressure on the solubility of a gas can be predicted from Le Chatelier's principle. **Le Chatelier's principle** *states that when a system in equi-*

**Figure 12.11**
**Instant hot and cold compress packs.** The inner bag of the instant hot compress contains calcium chloride, $CaCl_2$. When the inner bag is broken, the $CaCl_2$ dissolves exothermically in water (heat is released). The instant cold compress contains an inner bag of ammonium nitrate, $NH_4NO_3$, which when broken apart allows the $NH_4NO_3$ to dissolve in the water in the outer bag. The solution process is endothermic (heat is absorbed), so the bag feels cold.

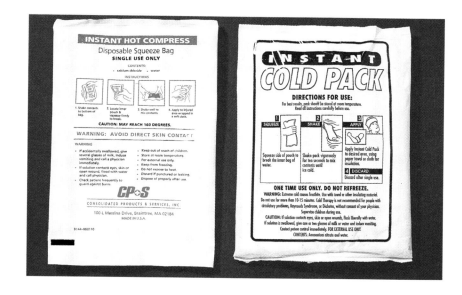

librium is disturbed by a change of temperature, pressure, or concentration variable, the system shifts in equilibrium composition in a way that tends to counteract this change of variable. Let us see how Le Chatelier's principle can predict the effect of a change in pressure on gas solubility.

Imagine a cylindrical vessel that is fitted with a movable piston and contains carbon dioxide gas over its saturated water solution (Figure 12.12). The equilibrium is

$$CO_2(g) \rightleftharpoons CO_2(aq)$$

Suppose you increase the partial pressure of $CO_2$ gas by pushing the piston down. This change of partial pressure, according to Le Chatelier's principle, shifts the equilibrium composition in a way that tends to counteract the pressure increase. From the preceding equation, you see that the partial pressure of $CO_2$ gas decreases if more $CO_2$ dissolves.

$$CO_2(g) \longrightarrow CO_2(aq)$$

The system comes to a new equilibrium, in which more $CO_2$ has dissolved. So you can predict that carbon dioxide is more soluble at higher pressures. Conversely, when the partial pressure of carbon dioxide gas is reduced, its solubility is decreased. A bottle of carbonated beverage fizzes when the cap is removed: as the partial pressure of carbon dioxide is reduced, gas comes out of solution (Figure 12.13). The argument given here for carbon dioxide holds for any gas:

> All gases become more soluble in a liquid at a given temperature when the partial pressure of the gas over the solution is increased.

Piston

Gaseous $CO_2$

Aqueous solution of $CO_2$

**Figure 12.12**
**Effect of pressure on gas solubility.** When the piston is pushed down, increasing the partial pressure of carbon dioxide, more gas dissolves (which tends to reduce the carbon dioxide partial pressure).

## Henry's Law: Relating Pressure to the Solubility of a Gas in a Liquid

The effect of pressure on the solubility of a gas in a liquid can be predicted quantitatively. According to **Henry's law,** *the solubility of a gas is directly proportional to the partial pressure of the gas above the solution.* Expressed mathematically, the law is

$$S = k_H P$$

where $S$ is the solubility of the gas (expressed as mass of solute per unit volume of solvent), $k_H$ is Henry's law constant for the gas for a particular liquid at a given temperature, and $P$ is the partial pressure of the gas. The next example shows how this formula is used.

**Figure 12.13**
**Sudden release of pressure from a carbonated beverage.**
A carbonated beverage is produced by dissolving carbon dioxide in beverage solution under pressure. More carbon dioxide dissolves at the higher pressure than otherwise. When this pressure is suddenly released, carbon dioxide is less soluble, and the excess bubbles from the solution.

▶ An alternate example is provided in the Instructor's Resource Manual.

## EXAMPLE 12.1

## Applying Henry's Law

In the chapter opening, we noted that 27 g of acetylene, $C_2H_2$, dissolves in 1 L of acetone at 1.0 atm pressure. If the partial pressure of acetylene is increased to 12 atm, what is its solubility in acetone?

 **PROBLEM STRATEGY**

Let $S_1$ be the solubility of the gas at partial pressure $P_1$, and let $S_2$ be the solubility at partial pressure $P_2$. Then you can write Henry's law for both pressures.

$$S_1 = k_H P_1$$
$$S_2 = k_H P_2$$

Dividing the second equation by the first, you get

$$\frac{S_2}{S_1} = \frac{k_H P_2}{k_H P_1} \quad \text{or} \quad \frac{S_2}{S_1} = \frac{P_2}{P_1}$$

You can use this relation to find the solubility at one pressure given the solubility at another.

**SOLUTION**

At 1.0 atm partial pressure of acetylene ($P_1$), the solubility $S_1$ is 27 g $C_2H_2$ per liter of acetone. For a partial pressure of 12 atm ($P_2$),

$$\frac{S_2}{27 \text{ g } C_2H_2/L \text{ acetone}} = \frac{12 \text{ atm}}{1.0 \text{ atm}}$$

Then

$$S_2 = \frac{27 \text{ g } C_2H_2}{L \text{ acetone}} \times \frac{12}{1.0} = \frac{3.2 \times 10^2 \text{ g } C_2H_2}{L \text{ acetone}}$$

At 12 atm partial pressure of acetylene, 1 L of acetone dissolves **$3.2 \times 10^2$ g of acetylene.**

## EXERCISE 12.4

A liter of water at 25°C dissolves 0.0404 g $O_2$ when the partial pressure of the oxygen is 1.00 atm. What is the solubility of oxygen from air, in which the partial pressure of $O_2$ is 159 mmHg?

**See Problems 12.39 and 12.40.**

 **CONCEPT CHECK 12.3**

Most fish have a very difficult time surviving at elevations much above 3500 m. How could Henry's law be used to account for this fact?

# Colligative Properties

In the chapter opening, we mentioned that the addition of ethylene glycol, $CH_2OHCH_2OH$, to water lowers the freezing point of water below 0°C. For example, if 0.010 mol of ethylene glycol is added to 1 kg of water, the freezing point is lowered to −0.019°C. The magnitude of freezing-point lowering is directly proportional to the number of ethylene glycol molecules added to a quantity of water. Thus, if you add 0.020 (= 0.010 × 2) mol of ethylene glycol to 1 kg of water, the freezing point is lowered to −0.038°C (= −0.019°C × 2). The same lowering is observed with the addition of other nonelectrolyte substances. For example, an aqueous solution of 0.020 mol of urea, $(NH_2)_2CO$, in 1 kg of water also freezes at −0.038°C.

Freezing-point lowering is a colligative property. **Colligative properties** of solutions are *properties that depend on the concentration of solute molecules or ions in solution but not on the chemical identity of the solute* (whether it is ethylene glycol or urea, for instance). In the next sections, we will discuss several colligative properties, which are expressed quantitatively in terms of various concentration units. We will first look into ways of expressing the concentration of a solution.

## 12.4   Ways of Expressing Concentration

The *concentration* of a solute is the amount of solute dissolved in a given quantity of solvent or solution. The quantity of solvent or solution can be expressed in terms of volume or in terms of mass or molar amount. Thus, there are several ways of expressing the concentration of a solution.

The *molarity* of a solution is the moles of solute in a liter of solution. ◄

▶ **Molarity was discussed in Section 4.7.**

$$\text{Molarity} = \frac{\text{moles of solute}}{\text{liters of solution}}$$

For example, 0.20 mol of ethylene glycol dissolved in enough water to give 2.0 L of solution has a molarity of

$$\frac{0.20 \text{ mol ethylene glycol}}{2.0 \text{ L solution}} = 0.10 \, M \text{ ethylene glycol}$$

The solution is 0.10 molar (denoted $M$). This unit is especially useful when you wish to dispense a given amount of solute, because the unit directly relates the amount of solute to the volume of solution.

Other concentration units are defined in terms of the mass or molar amount of solvent or solution. The most important of these are mass (or weight) percentage of solute, molality, and mole fraction.

### Mass Percentage of Solute

Solution concentration is sometimes expressed in terms of the **mass percentage of solute**—that is, *the percentage by mass of solute contained in a solution.*

$$\text{Mass percentage of solute} = \frac{\text{mass of solute}}{\text{mass of solution}} \times 100\%$$

For example, an aqueous solution that is 3.5% sodium chloride by mass contains 3.5 g of NaCl in 100.0 g of solution. It could be prepared by dissolving 3.5 g of NaCl in 96.5 g of water (100.0 − 3.5 = 96.5).

▶ An alternate example is provided in the Instructor's Resource Manual.

---

**EXAMPLE 12.2**

## Calculating with Mass Percentage of Solute

How would you prepare 425 g of an aqueous solution containing 2.40% by mass of sodium acetate, $NaC_2H_3O_2$?

**SOLUTION**

The mass of sodium acetate in 425 g of solution is

$$\text{Mass of } NaC_2H_3O_2 = 425 \text{ g} \times 0.0240 = 10.2 \text{ g}$$

The quantity of water in the solution is

$$\text{Mass of } H_2O = \text{mass of solution} - \text{mass of } NaC_2H_3O_2 = 425 \text{ g} - 10.2 \text{ g} = 415 \text{ g}$$

**You would prepare the solution by dissolving 10.2 g of sodium acetate in 415 g of water.**

---

**EXERCISE 12.5**

An experiment calls for 35.0 g of hydrochloric acid that is 20.2% HCl by mass. How many grams of HCl is this? How many grams of water?

See Problems 12.41, 12.42, 12.43, and 12.44.

---

## Molality

The **molality** of a solution is *the moles of solute per kilogram of solvent.*

$$\text{Molality} = \frac{\text{moles of solute}}{\text{kilograms of solvent}}$$

For example, 0.20 mol of ethylene glycol dissolved in $2.0 \times 10^3$ g (= 2.0 kg) of water has a molality of

$$\frac{0.20 \text{ mol ethylene glycol}}{2.0 \text{ kg solvent}} = 0.10 \ m \text{ ethylene glycol}$$

That is, the solution is 0.10 molal (denoted $m$). The units of molality and molarity are sometimes confused. Note that molality is defined in terms of *mass of solvent,* and molarity is defined in terms of *volume of solution.*

**Figure 12.14**
**Intravenous feeding of glucose.** Glucose, $C_6H_{12}O_6$, is the principal energy substance of the body and is transported to various cells by the blood. Here glucose is being fed intravenously to a patient.

---

**EXAMPLE 12.3**

## Calculating the Molality of Solute

Glucose, $C_6H_{12}O_6$, is a sugar that occurs in fruits. It is also known as "blood sugar" because it is found in blood and is the body's main source of energy (see Figure 12.14). What is the molality of a solution containing 5.67 g of glucose dissolved in 25.2 g of water?

▶ An alternate example is provided in the Instructor's Resource Manual.

### PROBLEM STRATEGY

You first convert mass of glucose to moles, because the molality equals moles of solute (glucose) divided by mass of solvent (water) in kilograms.

### SOLUTION

The moles of glucose (MW = 180.2 amu) in 5.67 g are found as follows:

$$5.67 \text{ g } C_6H_{12}O_6 \times \frac{1 \text{ mol } C_6H_{12}O_6}{180.2 \text{ g } C_6H_{12}O_6} = 0.0315 \text{ mol } C_6H_{12}O_6$$

The mass of water is 25.2 g, or $25.2 \times 10^{-3}$ kg.

$$\text{Molality} = \frac{0.0315 \text{ mol } C_6H_{12}O_6}{25.2 \times 10^{-3} \text{ kg solvent}} = \mathbf{1.25 \textit{ m } C_6H_{12}O_6}$$

### EXERCISE 12.6

Toluene, $C_6H_5CH_3$, is a liquid compound similar to benzene, $C_6H_6$. It is the starting material for other substances, including trinitrotoluene (TNT). Find the molality of toluene in a solution that contains 35.6 g of toluene and 125 g of benzene.

See Problems 12.45, 12.46, 12.47, and 12.48.

## Mole Fraction

The **mole fraction** of a component substance $A$ ($X_A$) in a solution is defined as *the moles of component substance divided by the total moles of solution* (that is, moles of solute plus solvent).

$$X_A = \frac{\text{moles of substance } A}{\text{total moles of solution}}$$

For example, if a solution is made up of 1 mol of ethylene glycol and 9 mol of water, the total moles of solution are 1 mol + 9 mol = 10 mol. The mole fraction of ethylene glycol is 1/10 = 0.1, and the mole fraction of water is 9/10 = 0.9. Multiplying mole fractions by 100 gives *mole percent*. Hence, this solution is 10 mole percent ethylene glycol and 90 mole percent water. You can also say that 10% of the molecules in the solution are ethylene glycol and 90% are water. The sum of the mole fractions of all the components of a solution equals 1.

▶ An alternate example is provided in the Instructor's Resource Manual.

### EXAMPLE 12.4

## Calculating the Mole Fractions of Components

What are the mole fractions of glucose and water in a solution containing 5.67 g of glucose, $C_6H_{12}O_6$, dissolved in 25.2 g of water?

### SOLUTION

This is the glucose solution described in Example 12.3. There you found that 5.67 g of glucose equals 0.0315 mol of glucose. The moles of water in the solution are

$$25.2 \text{ g } H_2O \times \frac{1 \text{ mol } H_2O}{18.0 \text{ g } H_2O} = 1.40 \text{ mol } H_2O$$

Toluene

Benzene

▶ **An alternate example is provided in the Instructor's Resource Manual.**

Hence, the total moles of solution are

$$1.40 \text{ mol} + 0.0315 \text{ mol} = 1.432 \text{ mol}$$

(You retain an extra figure in the answer for further computation.) Finally, you get

$$\text{Mole fraction glucose} = \frac{0.0315 \text{ mol}}{1.432 \text{ mol}} = \textbf{0.0220}$$

$$\text{Mole fraction water} = \frac{1.40 \text{ mol}}{1.432 \text{ mol}} = \textbf{0.978}$$

The sum of the mole fractions is 1.000.

### EXERCISE 12.7

Calculate the mole fractions of toluene and benzene in the solution described in Exercise 12.6.

See Problems 12.49 and 12.50.

## Conversion of Concentration Units

It is relatively easy to interconvert concentration units when they are expressed in terms of mass or moles of solute and solvent, as the following examples show.

### EXAMPLE 12.5

## Converting Molality to Mole Fractions

An aqueous solution is 0.120 $m$ glucose, $C_6H_{12}O_6$. What are the mole fractions of each component in the solution?

 **PROBLEM STRATEGY**

A 0.120 $m$ glucose solution contains 0.120 mol of glucose in 1.00 kg of water. After converting 1.00 kg $H_2O$ to moles, you can calculate the mole fraction using its definition.

**SOLUTION**

The moles of $H_2O$ in 1.00 kg of water is

$$1.00 \times 10^3 \text{ g } H_2O \times \frac{1 \text{ mol } H_2O}{18.0 \text{ g } H_2O} = 55.6 \text{ mol } H_2O$$

Hence,

$$\text{Mole fraction glucose} = \frac{0.120 \text{ mol}}{(0.120 + 55.6) \text{ mol}} = \textbf{0.00215}$$

$$\text{Mole fraction water} = \frac{55.6 \text{ mol}}{(0.120 + 55.6) \text{ mol}} = \textbf{0.998}$$

### EXERCISE 12.8

A solution is 0.120 $m$ methanol dissolved in ethanol. Calculate the mole fractions of methanol, $CH_3OH$, and ethanol, $C_2H_5OH$, in the solution.

See Problems 12.51 and 12.52.

---

### EXAMPLE 12.6

## Converting Mole Fractions to Molality

A solution is 0.150 mole fraction glucose, $C_6H_{12}O_6$, and 0.850 mole fraction water. What is the molality of glucose in the solution?

 **PROBLEM STRATEGY**

One mole of solution contains 0.150 mol of glucose and 0.850 mol of water. You convert moles of solvent ($H_2O$) to kilograms, then calculate the molality.

**SOLUTION**

The mass of this amount of water is

$$0.850 \ \cancel{mol \ H_2O} \times \frac{18.0 \text{ g } H_2O}{1 \ \cancel{mol \ H_2O}} = 15.3 \text{ g } H_2O \ (= 0.0153 \text{ kg } H_2O)$$

The molality of glucose, $C_6H_{12}O_6$, in the solution is

$$\text{Molality of } C_6H_{12}O_6 = \frac{0.150 \text{ mol } C_6H_{12}O_6}{0.0153 \text{ kg solvent}} = \textbf{9.80 } \textbf{\textit{m} } \textbf{C}_6\textbf{H}_{12}\textbf{O}_6$$

---

### EXERCISE 12.9

A solution is 0.250 mole fraction methanol, $CH_3OH$, and 0.750 mole fraction ethanol, $C_2H_5OH$. What is the molality of methanol in the solution?

---

See Problems 12.53 and 12.54.

To convert molality to molarity and vice versa, you must know the density of the solution. The calculations are described in the next two examples. As the first example shows, molality and molarity are approximately equal in dilute aqueous solutions.

---

### EXAMPLE 12.7

## Converting Molality to Molarity

An aqueous solution is 0.273 $m$ KCl. What is the molar concentration of potassium chloride, KCl? The density of the solution is $1.011 \times 10^3$ g/L.

 **PROBLEM STRATEGY**

The molality tells you that the solution contains 0.273 mol KCl in 1 kg of water. You take the mass of solution that contains 1 kg of water (the mass of 0.273 mol KCl plus $1.000 \times 10^3$ g $H_2O$) and use the density to convert this quantity of solution to volume. Molarity equals moles of solute (KCl) divided by volume of solution in liters.

**SOLUTION**

The mass of potassium chloride in this quantity of solution is

$$0.273 \ \cancel{mol \ KCl} \times \frac{74.6 \text{ g KCl}}{1 \ \cancel{mol \ KCl}} = 20.4 \text{ g KCl}$$

The total mass of the solution equals the mass of water plus the mass of potassium chloride.

$$1.000 \times 10^3 \text{ g} + 20.4 \text{ g} = 1.020 \times 10^3 \text{ g}$$

The volume of the solution can be calculated using the inverse of the density of the solution.

$$\text{Volume of solution} = 1.020 \times 10^3 \text{ g} \times \frac{1 \text{ L}}{1.011 \times 10^3 \text{ g}} = 1.009 \text{ L}$$

Hence, the molarity of the solution is

$$\frac{0.273 \text{ mol KCl}}{1.009 \text{ L solution}} = \mathbf{0.271 \textit{ M} \text{ KCl}}$$

Note that the molarity and the molality of this solution are approximately equal. This happens in cases where the solutions are dilute and the density is about 1 g/mL.

---

### EXERCISE 12.10

Urea, $(NH_2)_2CO$, is used as a fertilizer. What is the molar concentration of an aqueous solution that is 3.42 $m$ urea? The density of the solution is 1.045 g/mL.

**See Problems 12.55 and 12.56.**

---

▶ An alternate example is provided in the Instructor's Resource Manual.

### EXAMPLE 12.8

## Converting Molarity to Molality

An aqueous solution is 0.907 $M$ $Pb(NO_3)_2$. What is the molality of lead nitrate, $Pb(NO_3)_2$, in this solution? The density of the solution is 1.252 g/mL.

 **PROBLEM STRATEGY**

There is 0.907 mol $Pb(NO_3)_2$ per liter of solution. Consider 1 L (= $1.000 \times 10^3$ mL) of solution and calculate its mass, using the density. You can then calculate the mass of lead nitrate and find the mass of water by difference. Now you can calculate the molality.

**SOLUTION**

$$\begin{aligned}
\text{Mass of solution} &= \text{density} \times \text{volume} \\
&= 1.252 \text{ g/mL} \times 1.000 \times 10^3 \text{ mL} \\
&= 1.252 \times 10^3 \text{ g}
\end{aligned}$$

The mass of lead nitrate is

$$0.907 \text{ mol } Pb(NO_3)_2 \times \frac{331.2 \text{ g } Pb(NO_3)_2}{1 \text{ mol } Pb(NO_3)_2} = 3.00 \times 10^2 \text{ g } Pb(NO_3)_2$$

The mass of water is

$$\begin{aligned}
\text{Mass of } H_2O &= \text{mass of solution} - \text{mass of } Pb(NO_3)_2 \\
&= 1.252 \times 10^3 \text{ g} - 3.00 \times 10^2 \text{ g} \\
&= 9.52 \times 10^2 \text{ g} (= 0.952 \text{ kg})
\end{aligned}$$

Hence, the molality of lead nitrate in this solution is

$$\frac{0.907 \text{ mol Pb(NO}_3)_2}{0.952 \text{ kg solvent}} = \textbf{0.953 } \boldsymbol{m} \textbf{ Pb(NO}_3)_2$$

### EXERCISE 12.11

An aqueous solution is 2.00 $M$ urea. The density of the solution is 1.029 g/mL. What is the molal concentration of urea in the solution?

See Problems 12.57 and 12.58.

## 12.5 Vapor Pressure of a Solution

During the nineteenth century, chemists observed that the vapor pressure of a volatile solvent was lowered by addition of a nonvolatile solute. **Vapor-pressure lowering** of a solvent is *a colligative property equal to the vapor pressure of the pure solvent minus the vapor pressure of the solution.* For example, water at 20°C has a vapor pressure of 17.54 mmHg. Ethylene glycol, $CH_2OHCH_2OH$, is a liquid whose vapor pressure at 20°C is relatively low; it can be considered to be nonvolatile compared with water. An aqueous solution containing 0.0100 mole fraction of ethylene glycol has a vapor pressure of 17.36 mmHg. Thus, the vapor-pressure lowering, $\Delta P$, of water is

$$\Delta P = 17.54 \text{ mmHg} - 17.36 \text{ mmHg} = 0.18 \text{ mmHg}$$

Figure 12.15 shows a demonstration of vapor-pressure lowering.

In about 1886, the French chemist François Marie Raoult observed that the partial vapor pressure of solvent over a solution of a nonelectrolyte solute depends on the mole fraction of solvent in the solution. Consider a solution of volatile solvent, $A$, and nonelectrolyte solute, $B$, which may be volatile or nonvolatile. According to **Raoult's law,** *the partial pressure of solvent, $P_A$, over a solution equals the vapor pressure of the pure solvent, $P_A^\circ$, times the mole fraction of solvent, $X_A$, in the solution.*

$$P_A = P_A^\circ X_A$$

If the solute is nonvolatile, $P_A$ is the total vapor pressure of the solution. Because the mole fraction of solvent in a solution is always less than 1, the vapor pressure of the

**Figure 12.15**
**Demonstration of vapor-pressure lowering.** *(A)* Two beakers, each containing water solutions of the same solute, are placed under a bell jar. The solution in the left beaker is less concentrated than that in the right beaker, so its vapor pressure is greater. The partial pressure of vapor in the bell jar is an intermediate value. It is less than the vapor pressure of the solution on the left, but more than that of the solution on the right. As a result, vapor leaves the solution on the left (which becomes more concentrated) and condenses on the solution on the right (which becomes less concentrated). *(B)* After some time, the two solutions become equal in concentration and in vapor pressure.

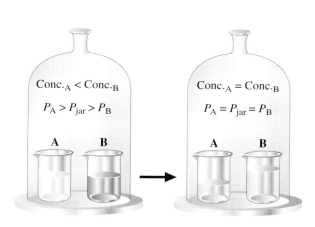

**Figure 12.16**
**Plot of vapor pressures of solutions showing Raoult's law.** Vapor pressures of solvent A for two solutions have been plotted against mole fraction of solvent. In one case (labeled "ideal solution"), the vapor pressure is proportional to the mole fraction of solvent for all mole fractions; it follows Raoult's law for all concentrations of solute. For the "nonideal solution," Raoult's law is followed for low solute concentrations (mole fraction of solvent near 1), but the vapor pressure deviates at other concentrations.

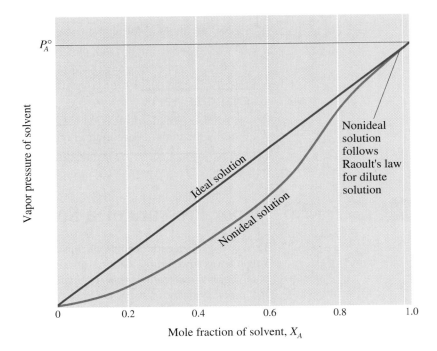

solution of a nonvolatile solute is less than that for the pure solvent; the vapor pressure is lowered. In general, Raoult's law is observed to hold for dilute solutions— that is, solutions in which $X_A$ is close to 1. If the solvent and solute are chemically similar, Raoult's law may hold for all mole fractions. Raoult's law is displayed graphically for two solutions in Figure 12.16.

You can obtain an explicit expression for the vapor-pressure lowering of a solvent in a solution assuming Raoult's law holds and that the solute is a nonvolatile nonelectrolyte. The vapor-pressure lowering, $\Delta P$, is

$$\Delta P = P_A^\circ - P_A$$

Substituting Raoult's law gives

$$\Delta P = P_A^\circ - P_A^\circ X_A = P_A^\circ(1 - X_A)$$

But the sum of the mole fractions of the components of a solution must equal 1; that is, $X_A + X_B = 1$. So $X_B = 1 - X_A$. Therefore,

$$\Delta P = P_A^\circ X_B$$

From this equation, you can see that the vapor-pressure lowering is a colligative property—one that depends on the concentration, but not on the nature, of the solute. Thus, if the mole fraction of ethylene glycol, $X_B$, in an aqueous solution is doubled from 0.010 to 0.020, the vapor-pressure lowering increases from 0.18 mmHg to 0.36 mmHg. Also, because the previous equation does not depend on the characteristics of the solute (other than its being nonvolatile and a nonelectrolyte), a solution that is 0.010 mole fraction urea, $(NH_2)_2CO$, has the same vapor-pressure lowering as one that is 0.010 mole fraction ethylene glycol. The next example illustrates the use of the previous equation.

▶ An alternate example is
provided in the Instructor's
Resource Manual.

## EXAMPLE 12.9

## Calculating Vapor-Pressure Lowering

Calculate the vapor-pressure lowering of water when 5.67 g of glucose, $C_6H_{12}O_6$, is dissolved in 25.2 g of water at 25°C. The vapor pressure of water at 25°C is 23.8 mmHg. What is the vapor pressure of the solution?

**SOLUTION**

This is the glucose solution described in Example 12.4. According to the calculations performed there, the solution is 0.0220 mole fraction glucose. Therefore, the vapor-pressure lowering is

$$\Delta P = P_A^\circ X_B = 23.8 \text{ mmHg} \times 0.0220 = \textbf{0.524 mmHg}$$

The vapor pressure of the solution is

$$P_A = P_A^\circ - \Delta P = (23.8 - 0.524) \text{ mmHg} = \textbf{23.3 mmHg}$$

## EXERCISE 12.12

Naphthalene, $C_{10}H_8$, is used to make mothballs. Suppose a solution is made by dissolving 0.515 g of naphthalene in 60.8 g of chloroform, $CHCl_3$. Calculate the vapor-pressure lowering of chloroform at 20°C from the naphthalene. The vapor pressure of chloroform at 20°C is 156 mmHg. Naphthalene can be assumed to be nonvolatile compared with chloroform. What is the vapor pressure of the solution?

See Problems 12.59 and 12.60.

An *ideal solution* of substances A and B is one in which both substances follow Raoult's law for all values of mole fractions. Such solutions occur when the substances are chemically similar so that the intermolecular forces between A and B molecules are similar to those between two A molecules or between two B molecules. In this case, we are not restricted to the solute being nonvolatile: both the solute and solvent have significant vapor pressure. Therefore, the total vapor pressure over an ideal solution equals the sum of the partial vapor pressures, each of which is given by Raoult's law:

$$P = P_A^\circ X_A + P_B^\circ X_B$$

Solutions of benzene, $C_6H_6$, and toluene, $C_6H_5CH_3$, are ideal. Note the similarity in their structural formulas:

benzene                    toluene

Suppose a solution is 0.70 mole fraction benzene and 0.30 mole fraction toluene. The vapor pressures of pure benzene and pure toluene are 75 mmHg and 22 mmHg, respectively. Hence, the total vapor pressure is

$$P = (75 \text{ mmHg} \times 0.70) + (22 \text{ mmHg} \times 0.30) = 59 \text{ mmHg}$$

You can easily see that the vapor over this solution is richer in the more volatile component (benzene). The partial vapor pressure of benzene over the solution is

$$75 \text{ mmHg} \times 0.70 = 53 \text{ mmHg}$$

Because the total vapor pressure is 59 mmHg, the mole fraction of benzene in the vapor is

$$\frac{53 \text{ mmHg}}{59 \text{ mmHg}} = 0.90$$

The vapor is 0.90 mole fraction benzene, whereas the liquid solution is 0.70 mole fraction benzene. This is a general principle: the vapor over a solution is richer in the more volatile component.

Use is made of this principle to separate a more volatile component from a liquid mixture. If you distill a mixture of benzene and toluene, the vapor and the resulting liquid that distills over (the distillate) will be richer in benzene, the more volatile component. If you then take this distillate and distill it, the vapor and the resulting liquid that comes over will be even richer in benzene. After many such distillations, you can obtain nearly pure benzene. ◄

In practice, instead of actually performing a series of simple distillations to separate the volatile components of a mixture, you perform a single *fractional distillation* using a fractionating column, such as the one shown in Figure 12.17. Vapor from the liquid in the flask condenses on cooler packing material (which might be glass beads) in the lower part of the column. In turn, this liquid (which is slightly richer in benzene) vaporizes and condenses farther up the column, where it is even cooler. The liquid that results is even richer in benzene. The distillation process continues up the column, so the composition of the liquid varies continuously from a mixture of benzene and toluene at the bottom of the column to pure benzene at the top. The temperature also varies continuously up the column, being cooler at the top than at the bottom. The liquid that distills over at first is pure benzene. As the liquid in the distillation flask becomes richer in the less volatile component (toluene), the boiling point of the liquid in the flask increases. Fractional distillation is a common procedure in the laboratory and in industrial processing. For example, it is used to separate the components of petroleum into gasoline, diesel fuel, and so forth. ◄

► Simple distillation was discussed in Section 1.4.

► Demonstration of fractional distillation: Bassam Z. Shakhashiri, *Chemical Demonstrations*, Vol. 3 (Madison: University of Wisconsin Press, 1989), no. 9.10, pp. 258–262.

► Table 24.4 lists the various fractions obtained from petroleum, along with their boiling-point ranges.

 **CONCEPT CHECK 12.4**

You need to boil a water-based solution at a temperature lower than 100°C. What kind of liquid could you add to the water to make this happen?

## 12.6 Boiling-Point Elevation and Freezing-Point Depression

The *normal* boiling point of a liquid is the temperature at which its vapor pressure equals 1 atm. Because the addition of a nonvolatile solute to a liquid reduces its vapor pressure, the temperature must be increased to a value greater than the normal boil-

**Figure 12.17**
**Fractional distillation.**
This is a typical laboratory apparatus for the fractional distillation of a liquid solution to obtain pure liquid components. The fractionating column contains a packing material, such as glass beads, on which vapor condenses and liquid redistills. The stopper on the receiver is loosely fitted to allow ventilation to the atmosphere.

▶ Demonstration of boiling-point elevation: Bassam Z. Shakhashiri, *Chemical Demonstrations*, Vol. 3 (Madison: University of Wisconsin Press, 1989), no. 9.22, pp. 297–300.

▶ Demonstration of freezing-point depression: Bassam Z. Shakhashiri, *Chemical Demonstrations*, Vol. 3 (Madison: University of Wisconsin Press, 1989), no. 9.21, pp. 290–296.

**Understanding Concepts: Colligative Properties**

ing point to achieve a vapor pressure of 1 atm. Figure 12.18 shows the vapor-pressure curve of a solution. The curve is below the vapor-pressure curve of the pure liquid solvent. The **boiling-point elevation,** $\Delta T_b$, is *a colligative property of a solution equal to the boiling point of the solution minus the boiling point of the pure solvent.*

The boiling-point elevation, $\Delta T_b$, is found to be proportional to the molal concentration, $c_m$, of the solution (for dilute solutions).

$$\Delta T_b = K_b c_m$$

The constant of proportionality, $K_b$ (called the *boiling-point-elevation constant*), depends only on the solvent. Table 12.3 lists values of $K_b$, as well as boiling points, for some solvents. Benzene, for example, has a boiling-point-elevation constant of 2.61°C/m. This means that a 0.100 m solution of a nonvolatile, undissociated solute in benzene boils at 0.261°C above the boiling point of pure benzene. Pure benzene boils at 80.2°C, so a 0.100 m solution boils at 80.2°C + 0.261°C = 80.5°C.

Figure 12.18 also shows the effect of a dissolved solute on the freezing point of a solution. Usually it is the pure solvent that freezes out of solution. Sea ice, for example, is almost pure water. For that reason, the vapor-pressure curve for the solid is unchanged. Therefore, the freezing point shifts to a lower temperature. The **freezing-point depression,** $\Delta T_f$, is *a colligative property of a solution equal to the freezing point of the pure solvent minus the freezing point of the solution.* ($\Delta T_f$ is shown in Figure 12.18.)

**Figure 12.18**
**Phase diagram showing the effect of a nonvolatile solute on freezing point and boiling point.** Note that the freezing point ($T_f$) is lowered and the boiling point ($T_b$) is elevated.

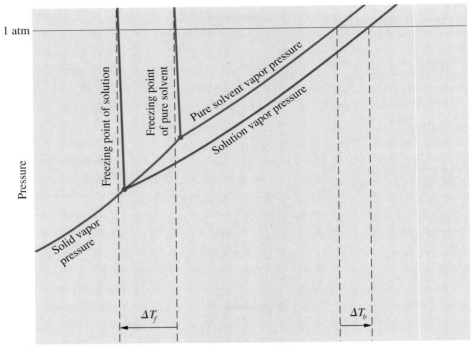

Freezing-point depression, $\Delta T_f$, like boiling-point elevation, is proportional to the molal concentration, $c_m$ (for dilute solutions).

$$\Delta T_f = K_f c_m$$

Here $K_f$ is the *freezing-point-depression constant* and depends only on the solvent. Table 12.3 gives values of $K_f$ for some solvents. The freezing-point-depression constant for benzene is 5.07°C/m. Thus a 0.100 m solution freezes at 0.507°C below the freezing point of pure benzene. Pure benzene freezes at 5.46°C; the freezing point of the solution is 5.46°C − 0.507°C = 4.95°C.

**Table 12.3**
**Boiling-Point-Elevation Constants ($K_b$) and Freezing-Point-Depression Constants ($K_f$)**

| Solvent | Formula | Boiling Point (°C) | Freezing Point (°C) | $K_b$ (°C/m) | $K_f$ (°C/m) |
|---|---|---|---|---|---|
| Acetic acid | $HC_2H_3O_2$ | 118.5 | 16.60 | 3.08 | 3.59 |
| Benzene | $C_6H_6$ | 80.2 | 5.455 | 2.61 | 5.065 |
| Camphor | $C_{10}H_{16}O$ | — | 179.5 | — | 40 |
| Carbon disulfide | $CS_2$ | 46.3 | — | 2.40 | — |
| Cyclohexane | $C_6H_{12}$ | 80.74 | 6.55 | 2.79 | 20.0 |
| Ethanol | $C_2H_5OH$ | 78.3 | — | 1.07 | — |
| Water | $H_2O$ | 100.000 | 0.000 | 0.512 | 1.858 |

[Data are taken from Landolt-Börnstein, 6th ed., *Zahlenwerte und Functionen aus Physik, Chemie, Astronomie, Geophysik, und Technik*, Vol. II, Part IIa (Heidelberg: © Springer-Verlag, 1960), pp. 844–849 and pp. 918–919.]

## EXAMPLE 12.10

### Calculating Boiling-Point Elevation and Freezing-Point Depression

An aqueous solution is 0.0222 *m* glucose. What are the boiling point and the freezing point of this solution?

### SOLUTION

Table 12.3 gives $K_b$ and $K_f$ for water as 0.512°C/*m* and 1.86°C/*m*, respectively. Therefore,

$$\Delta T_b = K_b c_m = 0.512°C/m \times 0.0222\ m = 0.0114°C$$
$$\Delta T_f = K_f c_m = 1.86°C/m \times 0.0222\ m = 0.0413°C$$

The boiling point of the solution is 100.000°C + 0.0114°C = **100.011°C,** and the freezing point is 0.000°C − 0.0413°C = **−0.041°C.** Note that $\Delta T_b$ is added and $\Delta T_f$ is subtracted.

### EXERCISE 12.13

How many grams of ethylene glycol, $CH_2OHCH_2OH$, must be added to 37.8 g of water to give a freezing point of −0.150°C?

See Problems 12.61 and 12.62.

**Figure 12.19**
**Automobile antifreeze mixtures.** The main ingredient of automobile antifreeze mixtures is ethylene glycol, $CH_2OHCH_2OH$. This substance has low vapor pressure, so it does not easily vaporize away. It is relatively cheap, being manufactured from ethylene, $C_2H_4$, which is obtained from petroleum. Ethylene glycol is also used to produce polyester fibers and plastics. The substance has a sweet taste (the name glycol derives from the Greek word *glukus,* meaning "sweet") but is poisonous.

The boiling-point elevation and the freezing-point depression of solutions have a number of practical applications. We mentioned in the chapter opening that ethylene glycol is used in automobile radiators as an antifreeze because it lowers the freezing point of the coolant (Figure 12.19). The same substance also helps prevent the radiator coolant from boiling away by elevating the boiling point. Sodium chloride and calcium chloride are spread on icy roads in the winter to lower the melting point of ice and snow below the temperature of the surrounding air. Salt–ice mixtures are used as freezing mixtures in home ice-cream makers. Melting ice cools the ice-cream mixture to the freezing point of the solution, which is well below the freezing point of pure water.

Colligative properties are also used to obtain molecular weights. Although the mass spectrometer is now often used for routine determinations of the molecular weight of pure substances, colligative properties are still employed to obtain information about the species in solution. Freezing-point depression is often used because it is simple to determine a melting point or a freezing point (Figure 12.20). From the freezing-point lowering, you can calculate the molal concentration, and from the molality, you can obtain the molecular weight. The next two examples illustrate these calculations.

## EXAMPLE 12.11

### Calculating the Molecular Weight of a Solute from Molality

A solution is prepared by dissolving 0.131 g of a substance in 25.4 g of water. The molality of the solution is determined by freezing-point depression to be 0.056 *m*. What is the molecular weight of the substance?

**Figure 12.20**
**Determination of the freezing point of a liquid.**
The liquid is cooled by means of a freezing mixture. To control the rate of temperature decrease, the liquid is separated from the freezing mixture by an air space.

Thermometer

Stirring rod

Air space

Liquid whose freezing point is to be measured

Freezing mixture (ice and salt)

▶ Experiments in General Chemistry, 12A, "Moral Mass from Freezing-Point Depression."

### PROBLEM STRATEGY

Molality equals moles of solute (moles of substance) divided by kilograms of solvent ($H_2O$). By substituting values for molality and kilograms $H_2O$, you can solve for moles of substance. The molar mass of the substance equals mass of substance (0.131 g) divided by moles of substance. The molecular weight (in amu) has the same numerical value as molar mass in g/mol.

### SOLUTION

The molality of an aqueous solution is

$$\text{Molality} = \frac{\text{moles of substance}}{\text{kg } H_2O}$$

Hence, for the given solution,

$$0.056 \, \frac{\text{mol}}{\text{kg } H_2O} = \frac{\text{moles of substance}}{25.4 \times 10^{-3} \text{ kg } H_2O}$$

Rearranging this equation, you get

$$\text{Moles of substance} = 0.056 \, \frac{\text{mol}}{\text{kg } H_2O} \times 25.4 \times 10^{-3} \text{ kg } H_2O = 1.42 \times 10^{-3} \text{ mol}$$

(An extra digit is retained for further computation.) The molar mass of substance equals

$$\text{Molar mass} = \frac{0.131 \text{ g}}{1.42 \times 10^{-3} \text{ mol}} = 92.3 \text{ g/mol}$$

You round the answer to two significant figures. The molecular weight is **92 amu.**

### EXERCISE 12.14

A 0.930-g sample of ascorbic acid (vitamin C) was dissolved in 95.0 g of water. The concentration of ascorbic acid, as determined by freezing-point depression, was 0.0555 $m$. What is the molecular weight of ascorbic acid?

See Problems 12.65 and 12.66.

▶ An alternate example is provided in the Instructor's Resource Manual.

Camphor

### EXAMPLE 12.12

## Calculating the Molecular Weight from Freezing-Point Depression

Camphor is a white solid that melts at 179.5°C. It has been used to determine the molecular weights of organic compounds, because of its unusually large freezing-point-depression constant (40°C/$m$), which allows ordinary thermometers to be used. The organic substance is dissolved in melted camphor, and then the melting point of the solution is determined. a. A 1.07-mg sample of a compound was dissolved in 78.1 mg of camphor. The solution melted at 176.0°C. What is the molecular weight of the compound? b. If the empirical formula of the compound is CH, what is the molecular formula?

 **PROBLEM STRATEGY**

a. From the melting point of the solution, calculate the freezing-point depression, $\Delta T_f$; then solve the equation $\Delta T_f = K_f c_m$ for the molality, $m$. Use the method of the preceding example to obtain the molecular weight, given the mass of the sample and the molality of the solution. b. See Example 3.12 for the relation between number of empirical formula units $n$, empirical formula weight, and molecular weight.

**SOLUTION**

a. The freezing-point lowering is

$$\Delta T_f = (179.5 - 176.0)°C = 3.5°C$$

so the molality of the solution is

$$\frac{\Delta T_f}{K_f} = \frac{3.5°\cancel{C}}{40°\cancel{C}/m} = 0.088 \ m$$

From this you compute the moles of the compound that are dissolved in 78.1 mg of camphor (see Example 12.11).

$$0.088 \ \text{mol/\cancel{kg}} \times 78.1 \times 10^{-6} \ \cancel{kg} = 6.9 \times 10^{-6} \ \text{mol}$$

The molar mass of the compound is

$$M_m = \frac{1.07 \times 10^{-3} \ \text{g}}{6.9 \times 10^{-6} \ \text{mol}} = 1.6 \times 10^2 \ \text{g/mol}$$

The molecular weight is **160 amu** (two significant figures).

b. The empirical formula weight of CH is 13 amu. Therefore, the number of CH units in the molecule is

$$n = \frac{\text{molecular weight}}{\text{empirical formula weight}} = \frac{1.6 \times 10^2 \text{ amu}}{13 \text{ amu}} = 12$$

The molecular formula is $C_{12}H_{12}$.

---

**Problem Solving: Determining the Molar Mass by Freezing-Point Depression**

**EXERCISE 12.15**

A 0.205-g sample of white phosphorus was dissolved in 25.0 g of carbon disulfide, $CS_2$. The boiling-point elevation of the carbon disulfide solution was found to be 0.159°C. What is the molecular weight of the phosphorus in solution? What is the formula of molecular phosphorus?

See Problems 12.67 and 12.68.

---

## 12.7 Osmosis

Certain membranes allow solvent molecules to pass through them but not solute molecules, particularly not those of large molecular weight. Such a membrane is called *semipermeable* and might be an animal bladder, a vegetable tissue, or a piece of cellophane. Figure 12.21 depicts the operation of a semipermeable membrane. **Osmosis** is *the phenomenon of solvent flow through a semipermeable membrane to equalize the solute concentrations on both sides of the membrane.* When two solutions of the same solvent are separated by a semipermeable membrane, solvent molecules migrate through the membrane in both directions. However, the rate of solvent migration is faster from the solution of low concentration to the solution of high concentration.

**Figure 12.21**
**A semipermeable membrane separating water and an aqueous solution of glucose.** Here the membrane is depicted as being similar to a sieve—the pores are too small to allow the glucose molecules to pass. The actual mechanism in particular cases may be more complicated. For example, water may be absorbed on one side of the membrane and diffuse through it to the other side.

Before equilibrium
is reached

After equilibrium
is reached

**Figure 12.22**
**An experiment in osmosis.**
*Left:* Before reaching equilibrium, water molecules pass through the membrane into the glucose solution in the inverted funnel at a greater rate than they return (osmosis). This causes the liquid level in the funnel to rise. *Right:* The column of liquid in the funnel eventually reaches sufficient height to exert a downward pressure, $\pi$ (the osmotic pressure), causing the rates of water molecules entering and leaving the glucose solution to become equal. Once the osmotic pressure, $\pi$, has been reached, osmosis stops, and the liquid level in the funnel has reached its maximum height.

Figure 12.22 shows an experiment that demonstrates osmosis. A concentrated glucose solution is placed in an inverted funnel whose mouth is sealed with a semipermeable membrane. The funnel containing the glucose solution is then placed in a beaker of pure water. As water flows from the beaker through the membrane into the funnel, the liquid level rises in the stem of the funnel.

The glucose solution continues to rise up the funnel stem until the downward pressure exerted by the solution above the membrane eventually stops the upward flow of solvent (water). In general, **osmotic pressure** is *a colligative property of a solution equal to the pressure that, when applied to the solution, just stops osmosis.*

The osmotic pressure $\pi$ of a solution is related to the molar concentration of solute, $M$:

$$\pi = MRT$$

Here $R$ is the gas constant and $T$ is the absolute temperature. There is a formal similarity between this equation for osmotic pressure and the equation for an ideal gas:

$$PV = nRT$$

The molar concentration $M$ of a gas equals $n/V$; therefore, $P = (n/V)RT = MRT$.

To see the order of magnitude of the osmotic pressure of a solution, consider the aqueous solution described in Example 12.10. There you found that a solution that is about 0.02 $m$ has a freezing-point depression of about 0.04°C. The molarity and molality of a dilute aqueous solution are approximately equal (see Example 12.7). Therefore, this solution will also be about 0.02 $M$. Hence, the osmotic pressure at 25°C (298 K) is

$$\pi = MRT$$
$$= 0.02 \; \text{mol/L} \times 0.082 \; \text{L} \cdot \text{atm/(K} \cdot \text{mol)} \times 298 \; \text{K}$$
$$= 0.5 \; \text{atm}$$

If this pressure were to be exerted by a column of a water solution, as in Figure 12.23, the column would have to be more than 4.5 m high.

In most osmotic-pressure experiments, much more dilute solutions are employed. Often osmosis is used to determine the molecular weights of macromolecular or polymeric substances. Polymers are very large molecules generally made up from a simple, repeating unit. A typical molecule of polyethylene, for instance, might have the formula $CH_3(CH_2CH_2)_{2000}CH_3$. Although polymer solutions can be fairly concentrated in terms of grams per liter, on a mole-per-liter basis they may be quite dilute. The freezing-point depression is usually too small to measure, though the osmotic pressure may be appreciable.

▶ An alternate example is provided in the Instructor's Resource Manual.

---

### EXAMPLE 12.13

## Calculating Osmotic Pressure

The formula for low-molecular-weight starch is $(C_6H_{10}O_5)_n$, where $n$ averages $2.00 \times 10^2$. When 0.798 g of starch is dissolved in 100.0 mL of water solution, what is the osmotic pressure at 25°C?

 **PROBLEM STRATEGY**

Obtain the molecular weight of $(C_6H_{10}O_5)_{200}$, and use it to obtain the molarity of the starch solution. Substitute into the formula for the osmotic pressure.

▶ Demonstration of osmotic pressure of a sugar solution: Bassam Z. Shakhashiri, *Chemical Demonstrations*, Vol. 3 (Madison: University of Wisconsin Press, 1989), pp. 286–289.

### SOLUTION

The molecular weight of $(C_6H_{10}O_5)_{200}$ is 32,400 amu. The number of moles in 0.798 g of starch is

$$0.798 \text{ g starch} \times \frac{1 \text{ mol starch}}{32,400 \text{ g starch}} = 2.46 \times 10^{-5} \text{ mol starch}$$

The molarity of the solution is

$$\frac{2.46 \times 10^{-5} \text{ mol}}{0.1000 \text{ L solution}} = 2.46 \times 10^{-4} \text{ mol/L solution}$$

and the osmotic pressure at 25°C is

$$\pi = MRT$$
$$= 2.46 \times 10^{-4} \text{ mol/L} \times 0.0821 \text{ L} \cdot \text{atm}/(K \cdot \text{mol}) \times 298 \text{ K}$$
$$= 6.02 \times 10^{-3} \text{ atm} = 6.02 \times 10^{-3} \text{ atm} \times \frac{760 \text{ mmHg}}{1 \text{ atm}}$$
$$= \textbf{4.58 mmHg}$$

For comparison, you can calculate the freezing-point depression. Assume that the molality is equal to the molarity (this is in effect the case for very dilute aqueous solutions). Then,

$$\Delta T_f = 1.86°\text{C}/m \times 2.46 \times 10^{-4} m = 4.58 \times 10^{-4} °\text{C}$$

which is barely detectable with generally available equipment.

---

### EXERCISE 12.16

Calculate the osmotic pressure at 20°C of an aqueous solution containing 5.0 g of sucrose, $C_{12}H_{22}O_{11}$, in 100.0 mL of solution.

---

See Problems 12.69 and 12.70.

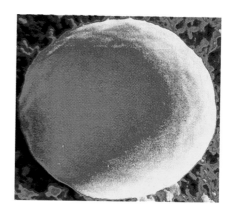

**Figure 12.23**
**The importance of osmotic pressure to cells.**   Shown here are color-enhanced electron micrographs of red blood cells in solutions of various osmotic pressures. *(A)* The osmotic pressure of the solution is greater than that of the cell; the cell has collapsed. *(B)* The osmotic pressure of the solution is equal to that of the cell; the cell has its normal round shape with depressed center. *(C)* The osmotic pressure of the solution is less than that of the cell; the cell has a bloated shape.

▶ **Osmotic pressure appears to be important for the rising of sap in a tree. During the day, water evaporates from the leaves of the tree, so the aqueous solution in the leaves becomes more concentrated and the osmotic pressure increases. Sap flows upward to dilute the water solution in the leaves.**

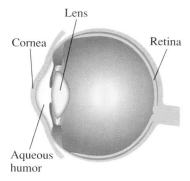

**Figure 12.24**
**Parts of the eye.**   Certain cells in the cornea act as pumps to prevent osmosis of water from the aqueous humor.

Osmosis is important in many biological processes. ◀ A cell might be thought of (simplistically) as an aqueous solution enclosed by a semipermeable membrane. The solution surrounding the cell must have an osmotic pressure equal to that within the cell. Otherwise, water would either leave the cell, dehydrating it, or enter the cell and possibly burst the membrane. For intravenous feeding (adding a nutrient solution to the venous blood supply of a patient), it is necessary that the nutrient solution have exactly the osmotic pressure of blood plasma. If it does not, the blood cells may collapse or burst as a result of osmosis (Figure 12.23).

The body has portions of organs that are at different osmotic pressures, and an active pumping mechanism is required to offset osmosis. For example, cells of the transparent tissue of the exterior eye, the cornea, have a more concentrated optical fluid than does the aqueous humor, a solution just behind the cornea (Figure 12.24). To prevent the cornea from taking up additional water from the aqueous humor, cells that pump water are located in the corneal tissue adjacent to the aqueous humor. Corneas that are to be stored and used for transplants must be removed from the globe of the eye soon after the donor's death. This prevents the clouding that occurs when the pumping mechanism fails (as its energy supply is depleted at death).

The process of *reverse osmosis* has been applied to the problem of purifying water. In particular, the method has been used to *desalinate* ocean water (that is, remove salts from seawater to make drinkable or industrially usable water). In normal osmosis, the solvent flows from a dilute solution through a membrane to a more concentrated solution. By applying a pressure equal to the osmotic pressure to the more concentrated solution, the process of osmosis can be stopped. By applying an even greater pressure, the osmotic process can be reversed. Then, solvent flows from the concentrated solution (which could be ocean water) through a membrane to the more dilute solution (which could be more or less pure water). Figure 12.25 shows a system using reverse osmosis to desalinate ocean water.

**Figure 12.25**
**Desalination system that uses reverse osmosis.**
When ocean water is subjected to a pressure greater than its osmotic pressure, pure water passes through a series of membranes (contained in the blue tubes, lower right), leaving a more concentrated salt solution behind.

 **CONCEPT CHECK 12.5**

Explain why pickles are stored in a brine (salt) solution. What would the pickles look like if they were stored in water?

## 12.8 Colligative Properties of Ionic Solutions

To explain the colligative properties of ionic solutions, you must realize that it is the total concentration of ions, rather than the concentration of ionic substance, that is important. For example, the freezing-point depression of 0.100 $m$ sodium chloride solution is nearly twice that of 0.100 $m$ glucose solution. You can explain this by saying that sodium chloride dissolves in water to form the ions $Na^+$ and $Cl^-$. Each formula unit of NaCl gives two particles. You can write the freezing-point lowering more generally as

$$\Delta T_f = iK_f c_m$$

where $i$ is the number of ions resulting from each formula unit and $c_m$ is the molality computed on the basis of the formula of the ionic compound. The equations for the other colligative properties must be similarly modified by the factor $i$.

Actually, the freezing points of ionic solutions agree with the previous equation only when the solutions are quite dilute. The $i$ values calculated from the freezing-point depression are usually smaller than the number of ions obtained from the formula unit. For example, a 0.029 $m$ aqueous solution of potassium sulfate, $K_2SO_4$, has a freezing point of $-0.14°C$. Hence,

$$i = \frac{\Delta T_f}{K_f c_m} = \frac{0.14°C}{1.86°C/m \times 0.029\ m} = 2.6$$

► The value of $i$ equal to $\Delta T_f / K_f c_m$ is often called the *van't Hoff factor*. Thus, the van't Hoff factor for 0.029 $m$ $K_2SO_4$ is 2.6.

You might have expected a value of 3, on the basis of the fact that $K_2SO_4$ ionizes to give three ions. ◄ At first, this was taken as evidence that salts were not completely ionized in solution. In 1923, however, Peter Debye and Erich Hückel were able to show that the colligative properties of salt solutions could be explained by assuming that the salt is completely ionized in solution but that the *activities,* or effective concentrations, of the ions are less than their actual concentrations as a result of the electrical interactions of the ions in solution. The *Debye–Hückel theory* allows us to calculate these activities. When this is done, excellent agreement is obtained for dilute solutions.

▶ An alternate example is provided in the Instructor's Resource Manual.

---

### EXAMPLE 12.14

### Determining Colligative Properties of Ionic Solutions

Estimate the freezing point of a 0.010 *m* aqueous solution of aluminum sulfate, $Al_2(SO_4)_3$. Assume the value of *i* based on the formula of the compound.

### SOLUTION

When aluminum sulfate, $Al_2(SO_4)_3$, dissolves in water, it dissociates into five ions.

$$Al_2(SO_4)_3(s) \xrightarrow{H_2O} 2Al^{3+}(aq) + 3SO_4{}^{2-}(aq)$$

Therefore, you assume *i* = 5. The freezing-point depression is

$$\Delta T_f = iK_f c_m = 5 \times 1.86°C/m \times 0.010\ m = 0.093°C$$

The estimated freezing point of the solution is 0.000°C − 0.093°C = **−0.093°C.**

---

### EXERCISE 12.17

Estimate the boiling point of a 0.050 *m* aqueous $MgCl_2$ solution. Assume a value of *i* based on the formula.

**See Problems 12.71 and 12.72.**

---

 **CONCEPT CHECK 12.6**

Each of the following substances is dissolved in a separate 10.0-L container of water: 1.5 mol NaCl, 1.3 mol $Na_2SO_4$, 2.0 mol $MgCl_2$, and 2.0 mol KBr. Without doing extensive calculations, rank the boiling points of each of the solutions from highest to lowest.

# Colloid Formation

As we noted at the beginning of the chapter, colloids appear homogeneous like solutions, but they consist of comparatively large particles of one substance dispersed throughout another substance. We will look at colloids in the following section.

# 12.9   Colloids

▶ Demonstration of the Tyndall effect: Bassam Z. Shakhashiri, *Chemical Demonstrations*, Vol. 3 (Madison: University of Wisconsin Press, 1989), pp. 353–357.

A **colloid** is *a dispersion of particles of one substance (the dispersed phase) throughout another substance or solution (the continuous phase).* Fog is an example of a colloid: it consists of very small water droplets (dispersed phase) in air (continuous phase). A colloid differs from a true solution in that the dispersed particles are larger than normal molecules, though they are too small to be seen with a microscope. The particles range from about $1 \times 10^3$ pm to about $2 \times 10^5$ pm in size.

**Figure 12.26**
**A demonstration of the Tyndall effect by a colloid.** A light beam is visible perpendicular to its path only if light is scattered toward the viewer. The vessel on the left contains a colloid, which scatters light. The vessel on the right contains a true solution, which scatters negligible light.

▶ Although all gases and liquids scatter light, the scattering from a pure substance or true solution is quite small and usually not detectable. However, because of the considerable depth of the atmosphere, the scattering of light by air molecules can be seen. The blue color of the sky is due to the fact that blue light is scattered more easily than red light.

▶ Video no. 10 (Series B), "Hot and Cold/Liquid to Solid/Solid to Liquid." (The first demonstration shows the formation of a gel.) 📼

## Tyndall Effect

Although a colloid appears to be homogeneous because the dispersed particles are quite small, it can be distinguished from a true solution by its ability to scatter light. *The scattering of light by colloidal-size particles* is known as the **Tyndall effect.** ◀ For example, the atmosphere appears to be a clear gas, but a ray of sunshine against a dark background shows up many fine dust particles by light scattering. Similarly, when a beam of light is directed through clear gelatin (a colloid, not a true solution), the beam becomes visible by the scattering of light from colloidal gelatin particles. The beam appears as a ray passing through the solution (Figure 12.26). When the same experiment is performed with a true solution, such as an aqueous solution of sodium chloride, the beam of light is not visible.

## Types of Colloids

Colloids are characterized according to the state (solid, liquid, or gas) of the dispersed phase and of the continuous phase. Table 12.4 lists various types of colloids and some examples of each. Fog and smoke are **aerosols,** which are *liquid droplets or solid particles dispersed throughout a gas.* An **emulsion** consists of *liquid droplets dispersed throughout another liquid* (as particles of butterfat are dispersed through homogenized milk). A **sol** consists of *solid particles dispersed in a liquid.*

**Table 12.4**
**Types of Colloids**

| Continuous Phase | Dispersed Phase | Name | Example |
|---|---|---|---|
| Gas | Liquid | Aerosol | Fog, mist |
| Gas | Solid | Aerosol | Smoke |
| Liquid | Gas | Foam | Whipped cream |
| Liquid | Liquid | Emulsion | Mayonnaise (oil dispersed in water) |
| Liquid | Solid | Sol | $AgCl(s)$ dispersed in $H_2O$ |
| Solid | Gas | Foam | Pumice, plastic foams |
| Solid | Liquid | Gel | Jelly, opal (mineral with liquid inclusions) |
| Solid | Solid | Solid sol | Ruby glass (glass with dispersed metal) |

▶ Demonstration of gel formation (canned heat): Bassam Z. Shakhashiri, *Chemical Demonstrations*, Vol. 3 (Madison: University of Wisconsin Press, 1989), pp. 360–361.

## Hydrophilic and Hydrophobic Colloids

Colloids in which the continuous phase is water are divided into two major classes: hydrophilic colloids and hydrophobic colloids. A **hydrophilic colloid** is *a colloid in which there is a strong attraction between the dispersed phase and the continuous phase (water).* Many such colloids consist of macromolecules (very large molecules) dispersed in water. Except for the large size of the dispersed molecules, these colloids are like normal solutions. Protein solutions, such as gelatin in water, are hydrophilic colloids. Gelatin molecules are attracted to water molecules by London forces and hydrogen bonding.

A **hydrophobic colloid** is *a colloid in which there is a lack of attraction between the dispersed phase and the continuous phase (water).* Hydrophobic colloids are basically unstable. Given sufficient time, the dispersed phase aggregates into larger particles. In this behavior, they are quite unlike true solutions and hydrophilic colloids. The time taken to separate may be extremely long, however. A colloid of gold particles in water prepared by Michael Faraday in 1857 is still preserved in the British Museum in London. This colloid is hydrophobic as well as a sol (solid particles dispersed in water).

Hydrophobic sols are often formed when a solid crystallizes rapidly from a chemical reaction or a supersaturated solution. When crystallization occurs rapidly, many centers of crystallization (called *nuclei*) form at the same time. Ions are attracted to these nuclei, and very small crystals form. These small crystals are prevented from settling out by the random thermal motion of the solvent molecules, which continue to buffet them.

You might expect these very small crystals to aggregate into larger crystals because the aggregation would bring ions of opposite charge into contact. However, sol formation appears to happen when, for some reason, each of the small crystals gets a preponderance of one kind of charge on its surface. For example, iron(III) hydroxide forms a colloid because an excess of iron(III) ion ($Fe^{3+}$) is present on the surface, giving each crystal an excess of positive charge. These positively charged crystals repel one another, so aggregation to larger particles is prevented.

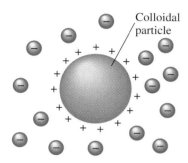

Fe(OH)$_3$ particle surrounded by Cl$^-$ ions

**A**

Fe(OH)$_3$ particle surrounded by PO$_4{}^{3-}$ ions

**B**

**Figure 12.27**
**Layers of ions surrounding charged colloidal particles.** *(A)* A positively charged colloidal particle of iron(III) hydroxide surrounded by chloride ions. *(B)* The same colloidal particle surrounded by phosphate ions. Because these ions gather more closely to the colloidal particles, coagulation is more likely to occur.

## Coagulation

An iron(III) hydroxide sol can be made to aggregate by the addition of an ionic solution, particularly if the solution contains anions with multiple charges (such as phosphate ions, PO$_4{}^{3-}$). **Coagulation** is *the process by which the dispersed phase of a colloid is made to aggregate and thereby separate from the continuous phase.* You can picture what happens in the following way: A positively charged colloidal particle of iron(III) hydroxide gathers a layer of anions around it. The thickness of this layer is determined by the charge on the anions. The greater the magnitude of the negative charge, the more compact is the layer of charge. Phosphate ions, for example, gather more closely to the positively charged colloidal particles than do chloride ions (see Figure 12.27). If the ion layer is gathered close to the colloidal particle, the overall charge is effectively neutralized. In that case, two colloidal particles can approach close enough to aggregate.

The curdling of milk when it sours is another example of coagulation. Milk is a colloidal suspension in which the particles are prevented from aggregating because they have electric charges of the same sign. The ions responsible for the coagulation (curdling) are formed when lactose (milk sugar) ferments to lactic acid. A third example is the coagulation of a colloidal suspension of soil in river water when the water meets the concentrated ionic solution of an ocean. The Mississippi Delta was formed in this way.

---

**EXERCISE 12.18**

Colloidal sulfur particles are negatively charged with thiosulfate ions, $S_2O_3^{2-}$, and other ions on the surface of the sulfur. Indicate which of the following would be most effective in coagulating colloidal sulfur: NaCl, MgCl$_2$, or AlCl$_3$.

---

See Problems 12.77 and 12.78.

## Association Colloids

When molecules or ions that have both a hydrophobic and a hydrophilic end are dispersed in water, they associate, or aggregate, to form colloidal-sized particles, or micelles. A **micelle** is *a colloidal-sized particle formed in water by the association of molecules or ions that each have a hydrophobic end and a hydrophilic end.* The hydrophobic ends point inward toward one another, and the hydrophilic ends are on the outside of the micelle facing the water. *A colloid in which the dispersed phase consists of micelles* is called an **association colloid.**

Ordinary soap in water provides an example of an association colloid. Soap consists of compounds such as sodium stearate, $C_{17}H_{35}COONa$. The stearate ion has a long hydrocarbon end that is hydrophobic (because it is nonpolar) and a carboxyl group ($COO^-$) at the other end that is hydrophilic (because it is ionic).

$$CH_3CH_2CH_2CH_2CH_2CH_2CH_2CH_2CH_2CH_2CH_2CH_2CH_2CH_2CH_2CH_2CH_2C \overset{\displaystyle O}{\underset{\displaystyle O^-}{\big\langle}}$$

hydrophobic end              hydrophilic end

stearate ion

In water solution, the stearate ions associate into micelles in which the hydrocarbon ends point inward toward one another and away from the water, and ionic carboxyl groups are on the outside of the micelle facing the water (Figure 12.28).

The cleansing action of soap occurs because oil and grease can be absorbed into the hydrophobic centers of soap micelles and washed away (Figures 12.29 and 12.30). Synthetic detergents also form association colloids. Sodium lauryl sulfate is a synthetic detergent present in toothpastes and shampoos.

$$CH_3CH_2CH_2CH_2CH_2CH_2CH_2CH_2CH_2CH_2CH_2CH_2OSO_3^-Na^+$$

sodium lauryl sulfate

**Figure 12.28**
**A stearate micelle in a water solution.** Stearate ions associate in groups (micelles), with their hydrocarbon ends pointing inward. The ionic ends, on the outside of the micelle, point into the water solution.

▶ Experiments in General Chemistry, 12B, "Softening Hard Water."

**Figure 12.29**
**The cleansing action of soap.** The hydrocarbon ends of soap ions gather around an oil spot, forming a micelle that can be washed away in the water.

**Figure 12.30**
**Formation of an association colloid with soap.** *Left:* Vegetable oil floating on water (dyed green). *Right:* When the mixture is shaken with soap, an emulsion forms as the oil droplets are absorbed into soap micelles.

It has a hydrophilic sulfate group ($—OSO_3^-$) and a hydrophobic dodecyl group ($C_{12}H_{25}—$, the hydrocarbon end).

The detergent molecules we have discussed so far are classified in the trade as "anionics," because they have a negative charge at the hydrophilic end. Other detergent molecules are "cationics," because they have a positive charge at the hydrophilic end. An example is

Hydrophilic end                                                                 Hydrophobic end

$$H_3C \quad CH_2CH_2CH_2CH_2CH_2CH_2CH_2CH_2CH_2CH_2CH_2CH_2CH_2CH_2CH_3$$
$$\underset{+}{N}$$
$$H_3C \quad CH_2CH_2CH_2CH_2CH_2CH_2CH_2CH_2CH_2CH_2CH_2CH_2CH_2CH_2CH_3$$

hydrophilic end                                                                 hydrophobic end

Many cationic detergents also have germicidal properties and are used in hospital disinfectants and in mouthwashes (Figure 12.31).

 **CONCEPT CHECK 12.7**

If electrodes that are connected to a direct current (DC) source are dipped into a beaker of collodial iron(III) hydroxide, a precipitate collects at the negative electrode. Explain why this happens.

**Figure 12.31**
**Commercial preparations containing cationic detergents.** Many cationic detergents exhibit germicidal properties. For that reason, they are used in certain hospital antiseptics, mouthwashes, and eye-wetting solutions.

# The World's Smallest Test Tubes

Phospholipids are the main constituents of the membranes of living cells. These membranes enclose the chemical substances of a cell and allow these substances to carry on the reactions needed for life processes unimpeded by random events external to the membrane. Recently, chemists have prepared cell-like structures—*lipid vesicles*—from phospholipids and have used them to carry out chemical reactions by combining the contents of two lipid vesicles, with diameters from about 50 nanometers (billionths of a meter) to several micrometers (millionths of a meter). Richard Zare, at Stanford University, who was one of those who described these experiments, has called these lipid vesicles "the world's smallest test tubes."

Like the soap ions, phospholipids have the intriguing property of self-assembly into groups. Soap ions in water naturally group themselves into micelles. Phospholipids, however, have a chemical structure that, while similar to that of soap ions, precludes their assembly into micelles. Rather, they form layer structures. And under the right circumstances, a layer may fold back on itself like the skin of a basketball to enclose a space, forming a vesicle.

Lecithins are typical phospholipids. Lipids are biological substances like fats and oils that are soluble in organic solvents; phospholipids contain phosphate groups. (If you are a label reader, you may have seen soy lecithin listed as an ingredient in food products such as bakery goods.) Here is the general structure of a lecithin:

$$R-\overset{\overset{\textstyle O}{\|}}{C}-O-CH_2$$
$$R'-\overset{\overset{\textstyle O}{\|}}{C}-O-CH$$
$$H_2C-O-\overset{\overset{\textstyle O}{\|}}{\underset{\underset{\textstyle O^-}{\|}}{P}}-O-CH_2-CH_2-\overset{+}{N}(CH_3)_3$$

R and R' denote long hydrocarbon groups, with perhaps 14 to 18 carbon atoms.

Note that, like a soap ion, a phospholipid has a hydrophobic end (the two hydrocarbon groups) and a hydrophilic end (an ionic portion containing a phosphate group, $-PO_4^- -$). Like soap ions, phospholipid molecules in water tend to associate so that their hydrophilic (or ionic) "heads" dip into the water, with their hydrophobic (or hydrocarbon) "tails" pointing away. However, the hydrocarbon tails of a phospholipid molecule are too bulky to associate into micelles and instead form a *bilayer,* a layer two molecules thick. Figure 12.32 shows a model of a cell membrane that displays this bilayer structure, who two phospholipid layers, their hydrocarbon tails pointing inward toward each other. The

*(continued)*

Specific proteins associated with the membrane surface

Specific proteins embedded in phospholipid bilayer

Protein extending fully through the membrane

Phospholipid bilayer composed of polar head groups and hydrophobic tails of phospholipid molecules

**Figure 12.32 A model of a cell membrane.** The membrane consists of a phospholipid bilayer with protein molecules inserted in it.

hydrophilic heads are rendered in blue; the hydrophobic tails are in green. Note that the interior of a membrane (shown in green) is "oily" from the hydrocarbon groups, and therefore repels molecules or ions dissolved in the water solutions outside and inside a cell. (A cell, though, must interact with its environment. Protein molecules embedded in the cell membrane form channels or active pumps to transport specific molecules and ions across the cell membrane.)

In one of their experiments, Zare and his colleagues brought two phospholipid vesicles, each about a micrometer in diameter and containing different reactant substances, together under a special microscope. When the vesicles just touched, the researchers delivered an electrical pulse to them that opened a small pore in each vesicle. The two vesicles coalesced into one, and their contents reacted. The scientists detected the product molecules from the orange fluorescence they gave off when radiated with ultraviolet light. The techniques these researchers are developing should allow scientists to mimic the conditions in biological cells. And lipid vesicles might one day be designed to deliver drugs to specific cells, say cancer cells.

# A Checklist for Review

## Important Terms

solute (12.1)
solvent (12.1)
miscible fluids (12.1)
saturated solution (12.2)
solubility (12.2)
unsaturated solution (12.2)
supersaturated
solution (12.2)
hydration (of ions) (12.2)

Le Chatelier's
principle (12.3)
Henry's law (12.3)
colligative properties
(page 517)
mass percentage of
solute (12.4)
molality (12.4)
mole fraction (12.4)

vapor-pressure
lowering (12.5)
Raoult's law (12.5)
boiling-point elevation (12.6)
freezing-point
depression (12.6)
osmosis (12.7)
osmotic pressure (12.7)
colloid (12.9)

Tyndall effect (12.9)
aerosols (12.9)
emulsion (12.9)
sol (12.9)
hydrophilic colloid (12.9)
hydrophobic colloid (12.9)
coagulation (12.9)
micelle (12.9)
association colloid (12.9)

## Key Equations

$S = k_H P$

Mass percentage of solute $= \dfrac{\text{mass of solute}}{\text{mass of solution}} \times 100\%$

Molality $= \dfrac{\text{moles of solute}}{\text{kilograms of solvent}}$

$X_A = \dfrac{\text{moles of substance } A}{\text{total moles of solution}}$

$P_A = P_A^\circ X_A$

$\Delta P = P_A^\circ X_B$

$\Delta T_b = K_b c_m$

$\Delta T_f = K_f c_m$

$\pi = MRT$

## Summary of Facts and Concepts

*Solutions* are homogeneous mixtures and can be gases, liquids, or solids. Two gases, for example, will mix in all proportions to give a gaseous solution, because gases are *miscible* in one another. Often one substance will dissolve in another only to a limited extent. The maximum amount that dissolves at equilibrium is the *solubility* of that substance. Solubility is explained in terms of the natural tendency toward disorder (say, by the mixing of two substances) and by the tendency to result in the strongest forces of attraction between species. The dissolving of one molecular substance in another is limited when intermolecular forces strongly favor the unmixed substances. When two substances have similar types of intermolecular forces, they tend to be soluble ("like dissolves like"). Ionic substances differ greatly in their solubilities in water, because solubility depends on the relative values of *lattice energy* and *hydration energy.*

Solubilities usually vary with temperature. At higher temperatures, most gases will become less soluble in water, whereas most ionic solids will become more soluble. Pressure has a significant effect on the solubility only of a gas. A gas is more soluble in a liquid if the partial pressure of the gas is increased, which is in agreement with *Le Chatelier's principle.* According to *Henry's law,* the solubility of a gas is directly proportional to the partial pressure of the gas.

*Colligative properties* of solutions depend only on the concentration of solute particles. The concentration may be defined by the *molarity, mass percentage of solute, molality,* or *mole fraction.* One example of a colligative property is the *vapor-pressure lowering* of a volatile solvent by the addition of a nonvolatile solute. According to *Raoult's law,* the vapor pressure of this solution depends on the mole fraction of solvent. Because adding a nonvolatile solute lowers the vapor pressure, the boiling point must be raised to bring the vapor pressure back to 1 atm *(boiling-point elevation).* Such a solution also exhibits a *freezing-point depression.* Boiling-point elevation and freezing-point depression are colligative properties. *Osmosis* is another colligative property. In osmosis, there is a flow of solvent through a semipermeable membrane in order to equalize the concentrations of solutions on the two sides of the membrane. Colligative properties may be used to measure solute concentration. In this way, one can determine the molecular weight of the solute. The colligative properties of an ionic solution depend on the concentration of ions, which you can obtain from the concentration of ionic compound and its formula. Interactions of ions in solution do affect the properties significantly, however.

A *colloid* is a dispersion of particles of one substance (about $1 \times 10^3$ pm to $2 \times 10^5$ pm in size) throughout another.

Colloids can be detected by the *Tyndall effect* (the scattering of light by colloidal-size particles). Colloids are characterized by the state of the dispersed phase and the state of the continuous phase. An aerosol, for example, consists of liquid droplets or solid particles dispersed in a gas. Colloids in water are hydrophilic or hydrophobic. A *hydrophilic colloid* consists of a dispersed phase that is attracted to the water. Many of these colloids have macromolecules dissolved in water. In *hydrophobic colloids,* there is little attraction between the dispersed phase and the water. The colloidal particles may be electrically charged and repel one another, preventing aggregation to larger particles. An ionic solution can neutralize this charge so that the colloid will *coagulate*—that is, aggregate. An *association colloid* consists of molecules with a hydrophobic end and a hydrophilic end dispersed in water. These molecules associate into colloidal-size groups, or *micelles.*

## Operational Skills

1. **Applying Henry's law**  Given the solubility of a gas at one pressure, find its solubility at another pressure. **(EXAMPLE 12.1)**
2. **Calculating solution concentration**  Given the mass percent of solute, state how to prepare a given mass of solution **(EXAMPLE 12.2)**. Given the masses of solute and solvent, find the molality **(EXAMPLE 12.3)** and mole fractions **(EXAMPLE 12.4)**.
3. **Converting concentration units**  Given the molality of a solution, calculate the mole fractions of solute and solvent; and given the mole fractions, calculate the molality **(EXAMPLES 12.5 AND 12.6)**. Given the density, calculate the molarity from the molality, and vice versa **(EXAMPLES 12.7 AND 12.8)**.
4. **Calculating vapor-pressure lowering**  Given the mole fraction of solute in a solution of nonvolatile, undissociated solute and the vapor pressure of pure solvent, calculate the vapor-pressure lowering and vapor pressure of the solution.  **(EXAMPLE 12.9)**
5. **Calculating boiling-point elevation and freezing-point depression**  Given the molality of a solution of nonvolatile, undissociated solute, calculate the boiling-point elevation and freezing-point depression. **(EXAMPLE 12.10)**
6. **Calculating molecular weights**  Given the masses of solvent and solute and the molality of the solution, find the molecular weight of the solute **(EXAMPLE 12.11)**. Given the masses of solvent and solute, the freezing-point depression, and $K_f$, find the molecular weight of the solute **(EXAMPLE 12.12)**.
7. **Calculating osmotic pressure**  Given the molarity and the temperature of a solution, calculate its osmotic pressure.  **(EXAMPLE 12.13)**
8. **Determining colligative properties of ionic solutions**  Given the concentration of ionic compound in a solution, calculate the magnitude of a colligative property; if $i$ is not given, assume the value based on the formula of the ionic compound.  **(EXAMPLE 12.14)**

## Review Questions

**12.1**  Give one example of each: a gaseous solution, a liquid solution, a solid solution.

**12.2**  What are the two factors needed to explain the differences in solubilities of substances?

**12.3**  Explain in terms of intermolecular attractions why octane is immiscible in water.

**12.4**  Explain why ionic substances show a wide range of solubilities in water.

**12.5**  Using the concept of hydration, describe the process of dissolving a sodium chloride crystal in water.

**12.6**  What is the usual solubility behavior of an ionic compound in water when the temperature is raised? Give an example of an exception to this behavior.

**12.7**  Give one example of each: a salt whose heat of solution is exothermic and a salt whose heat of solution is endothermic.

**12.8**  What do you expect to happen to a concentration of dissolved gas in a solution as the solution is heated?

**12.9**  Explain why a carbonated beverage must be stored in a closed container.

**12.10**    Pressure has an effect on the solubility of oxygen in water but a negligible effect on the solubility of sugar in water. Why?

**12.11**    Four ways were discussed to express the concentration of a solute in solution. Identify them and define each concentration unit.

**12.12**    When two beakers containing different concentrations of a solute in water are placed in a closed cabinet for a time, one beaker gains solvent and the other loses it, so that the concentrations of solute in the two beakers become equal. Explain what is happening.

**12.13**    Explain the process of fractional distillation to separate a solution of two liquids into pure components.

**12.14**    Explain why the boiling point of a solution containing a nonvolatile solute is higher than the boiling point of a pure solvent.

**12.15**    List two applications of freezing-point depression.

**12.16**    Explain the process of reverse osmosis to produce drinkable water from ocean water.

**12.17**    One can often see "sunbeams" passing through the less dense portions of clouds. What is the explanation for this?

**12.18**    Give an example of an aerosol, a foam, an emulsion, a sol, and a gel.

**12.19**    Explain on the basis that "like dissolves like" why glycerol, $CH_2OHCHOHCH_2OH$, is miscible in water but benzene, $C_6H_6$, has very limited solubility in water.

**12.20**    Explain how soap removes oil from a fabric.

## Conceptual Problems

**12.21**    Even though the oxygen demands of trout and bass are different, they can exist in the same body of water. However, if the temperature of the water in the summer gets above about 23°C, the trout begin to die, but not the bass. Why is this the case?

**12.22**    You want to purchase a salt to melt snow and ice on your sidewalk. Which one of the following salts would best accomplish your task using the least amount: KCl, $CaCl_2$, $PbS_2$, $MgSO_4$, or AgCl?

**12.23**    Ten grams of the hypothetical ionic compounds XZ and YZ are each placed in a separate 2.0-L beaker of water. XZ completely dissolves, whereas YZ is insoluble. The energy of hydration of the $Y^+$ ion is greater than the $X^+$ ion. Explain this difference in solubility.

**12.24**    Small amounts of a nonvolatile, nonelectrolyte solute and a volatile solute are each dissolved in separate beakers containing 1 kg of water. If the number of moles of each solute is equal:
a. Which solution will have the higher vapor pressure?
b. Which solution will boil at a higher temperature?

**12.25**    A Cottrell precipitator consists of a column containing electrodes that are connected to a high-voltage direct current (DC) source. The Cottrell precipitator is placed in smokestacks to remove smoke particles from the gas discharged from an industrial plant. Explain how you think this works.

**12.26**    Consider the following dilute NaCl(*aq*) solutions.

a. Which one will boil at a higher temperature?
b. Which one will freeze at a lower temperature?
c. If the solutions were separated by a semipermeable membrane that allowed only water to pass, which solution would you expect to show an increase in the concentration of NaCl?

**12.27**    A green leafy salad wilts if left too long in a salad dressing containing vinegar and salt. Explain what happens.

**12.28**    People have proposed towing icebergs to arid parts of the earth as a way to deliver freshwater. Explain why icebergs do not contain salts although they are formed by the freezing of ocean water (i.e., saltwater).

## Practice Problems

### Types of Solutions

**12.29**    Give an example of a liquid solution prepared by dissolving a gas in a liquid.

**12.30**    Give an example of a solid solution prepared from two solids.

### Solubility

**12.31**    Would boric acid, $B(OH)_3$, be more soluble in ethanol, $C_2H_5OH$, or in benzene, $C_6H_6$?

**12.32**    Would naphthalene, $C_{10}H_8$, be more soluble in ethanol, $C_2H_5OH$, or in benzene, $C_6H_6$?

**12.33** Arrange the following substances in order of increasing solubility in hexane, $C_6H_{14}$: $CH_2OHCH_2OH$, $C_{10}H_{22}$, $H_2O$.

**12.34** Indicate which of the following is more soluble in ethanol, $C_2H_5OH$: acetic acid, $CH_3COOH$, or stearic acid, $C_{17}H_{35}COOH$.

**12.35** Which of the following ions would be expected to have the greater energy of hydration, $Mg^{2+}$ or $Al^{3+}$?

**12.36** Which of the following ions would be expected to have the greater energy of hydration, $F^-$ or $Cl^-$?

**12.37** Arrange the following alkaline-earth-metal iodates in order of increasing solubility in water; explain your reasoning: $Ba(IO_3)_2$, $Ca(IO_3)_2$, $Sr(IO_3)_2$, $Mg(IO_3)_2$. Note that $IO_3^-$ is a large anion.

**12.38** Explain the trends in solubility (grams per 100 mL of water) of the alkali-metal fluorides and permanganates.

| | $Li^+$ | $Na^+$ | $K^+$ | $Rb^+$ | $Cs^+$ |
|---|---|---|---|---|---|
| $F^-$ | 0.27 | 4.2 | 92 | 131 | 367 |
| $MnO_4^-$ | 71 | Very soluble | 6.4 | 0.5 | 0.097 |

**12.39** The solubility of carbon dioxide in water is 0.161 g $CO_2$ in 100 mL of water at 20°C and 1.00 atm. A soft drink is carbonated with carbon dioxide gas at 5.50 atm pressure. What is the solubility of carbon dioxide in water at this pressure?

**12.40** Nitrogen, $N_2$, is soluble in blood and can cause intoxication at sufficient concentration. For this reason, the U.S. Navy advises divers using compressed air not to go below 125 feet. The total pressure at this depth is 4.79 atm. If the solubility of nitrogen at 1.00 atm is $1.75 \times 10^{-3}$ g/100 mL of water, and the mole percent of nitrogen in air is 78.1, what is the solubility of nitrogen in water from air at 4.79 atm?

## Solution Concentration

**12.41** How would you prepare 72.5 g of an aqueous solution that is 2.50% potassium iodide, KI, by mass?

**12.42** How would you prepare 455 g of an aqueous solution that is 6.50% sodium sulfate, $Na_2SO_4$, by mass?

**12.43** What mass of solution containing 5.00% potassium iodide, KI, by mass contains 258 mg KI?

**12.44** What mass of solution containing 6.50% sodium sulfate, $Na_2SO_4$, by mass contains 1.50 g $Na_2SO_4$?

**12.45** Vanillin, $C_8H_8O_3$, occurs naturally in vanilla extract and is used as a flavoring agent. A 39.1-mg sample of vanillin was dissolved in 168.5 mg of diphenyl ether, $(C_6H_5)_2O$. What is the molality of vanillin in the solution?

**12.46** Lauryl alcohol, $C_{12}H_{25}OH$, is prepared from coconut oil; it is used to make sodium lauryl sulfate, a synthetic deter-

gent. What is the molality of lauryl alcohol in a solution of 17.1 g lauryl alcohol dissolved in 165 g ethanol, $C_2H_5OH$?

**12.47** Fructose, $C_6H_{12}O_6$, is a sugar occurring in honey and fruits. The sweetest sugar, it is nearly twice as sweet as sucrose (cane or beet sugar). How much water should be added to 3.50 g of fructose to give a 0.125 $m$ solution?

Fructose

**12.48** Caffeine, $C_8H_{10}N_4O_2$, is a stimulant found in tea and coffee. A sample of the substance was dissolved in 45.0 g of chloroform, $CHCl_3$, to give a 0.0946 $m$ solution. How many grams of caffeine were in the sample?

Caffeine

**12.49** A 100.0-g sample of a brand of rubbing alcohol contains 65.0 g of isopropyl alcohol, $C_3H_7OH$, and 35.0 g of water. What is the mole fraction of isopropyl alcohol in the solution? What is the mole fraction of water?

**12.50** An automobile antifreeze solution contains 2.55 kg of ethylene glycol, $CH_2OHCH_2OH$, and 2.00 kg of water. Find the mole fraction of ethylene glycol in this solution. What is the mole fraction of water?

**12.51** A bleaching solution contains sodium hypochlorite, NaClO, dissolved in water. The solution is 0.650 $m$ NaClO. What is the mole fraction of sodium hypochlorite?

**12.52** An antiseptic solution contains hydrogen peroxide, $H_2O_2$, in water. The solution is 0.655 $m$ $H_2O_2$. What is the mole fraction of hydrogen peroxide?

**12.53** Concentrated hydrochloric acid contains 1.00 mol HCl dissolved in 3.31 mol $H_2O$. What is the mole fraction of HCl in concentrated hydrochloric acid? What is the molal concentration of HCl?

**12.54** Concentrated aqueous ammonia contains 1.00 mol $NH_3$ dissolved in 2.44 mol $H_2O$. What is the mole fraction of $NH_3$ in concentrated aqueous ammonia? What is the molal concentration of $NH_3$?

**12.55** Oxalic acid, $H_2C_2O_4$, occurs as the potassium or calcium salt in many plants, including rhubarb and spinach. An aqueous solution of oxalic acid is 0.580 $m$ $H_2C_2O_4$. The density of the solution is 1.022 g/mL. What is the molar concentration?

**12.56** Citric acid, $H_3C_6H_5O_7$, occurs in plants. Lemons contain 5% to 8% citric acid by mass. The acid is added to beverages and candy. An aqueous solution is 0.688 $m$ citric acid. The density is 1.049 g/mL. What is the molar concentration?

**12.57** A solution of vinegar is 0.763 $M$ acetic acid, $HC_2H_3O_2$. The density of the vinegar is 1.004 g/mL. What is the molal concentration of acetic acid?

**12.58** A beverage contains tartaric acid, $H_2C_4H_4O_6$, a substance obtained from grapes during wine making. If the beverage is 0.271 $M$ tartaric acid, what is the molal concentration? The density of the solution is 1.016 g/mL.

## Colligative Properties

**12.59** Calculate the vapor pressure at 35°C of a solution made by dissolving 20.2 g of sucrose, $C_{12}H_{22}O_{11}$, in 70.1 g of water. The vapor pressure of pure water at 35°C is 42.2 mmHg. What is the vapor-pressure lowering of the solution? (Sucrose is nonvolatile.)

**12.60** What is the vapor pressure at 23°C of a solution of 1.20 g of naphthalene, $C_{10}H_8$, in 25.6 g of benzene, $C_6H_6$? The vapor pressure of pure benzene at 23°C is 86.0 mmHg; the vapor pressure of naphthalene can be neglected. Calculate the vapor-pressure lowering of the solution.

**12.61** What is the boiling point of a solution of 0.173 g of glycerol, $C_3H_8O_3$, in 20.0 g of water? What is the freezing point?

**12.62** A solution was prepared by dissolving 0.817 g of sulfur, $S_8$, in 100.0 g of acetic acid, $HC_2H_3O_2$. Calculate the freezing point and boiling point of the solution.

**12.63** An aqueous solution of a molecular compound freezes at −0.086°C. What is the molality of the solution?

**12.64** Urea, $(NH_2)_2CO$, is dissolved in 100.0 g of water. The solution freezes at −0.085°C. How many grams of urea were dissolved to make this solution?

**12.65** A 0.0182-g sample of an unknown substance was dissolved in 2.135 g of benzene. The molality of this solution, determined by freezing-point depression, was 0.0698 $m$. What is the molecular weight of the unknown substance?

**12.66** A solution contains 0.0653 g of a compound in 9.75 g of ethanol. The molality of the solution is 0.0368 $m$. Calculate the molecular weight of the compound.

**12.67** Safrole is contained in oil of sassafras and was once used to flavor root beer. A 2.39-mg sample of safrole was dissolved in 103.0 mg of diphenyl ether. The solution had a melting point of 25.70°C. Calculate the molecular weight of safrole. The freezing point of pure diphenyl ether is 26.84°C, and the freezing-point-depression constant, $K_f$, is 8.00°C/$m$.

**12.68** Butylated hydroxytoluene (BHT) is used as an antioxidant in processed foods. (It prevents fats and oils from becoming rancid.) A solution of 2.500 g of BHT in 100.0 g of benzene had a freezing point of 4.880°C. What is the molecular weight of BHT?

**12.69** Dextran is a polymeric carbohydrate produced by certain bacteria. It is used as a blood plasma substitute. An aqueous solution contains 0.582 g of dextran in 106 mL of solution at 21°C. It has an osmotic pressure of 1.47 mmHg. What is the average molecular weight of the dextran?

**12.70** Arginine vasopressin is a pituitary hormone. It helps regulate the amount of water in the blood by reducing the flow of urine from the kidneys. An aqueous solution containing 21.6 mg of vasopressin in 100.0 mL of solution has an osmotic pressure at 25°C of 3.70 mmHg. What is the molecular weight of the hormone?

**12.71** What is the freezing point of 0.0090 $m$ aqueous calcium chloride, $CaCl_2$? Use the formula of the salt to obtain $i$.

**12.72** What is the freezing point of 0.0091 $m$ aqueous sodium phosphate, $Na_3PO_4$? Use the formula of the salt to obtain $i$.

**12.73** A 0.0140-g sample of an ionic compound with the formula $Cr(NH_3)_5Cl_3$ was dissolved in water to give 25.0 mL of solution at 25°C. The osmotic pressure was determined to be 119 mmHg. How many ions are obtained from each formula unit when the compound is dissolved in water?

**12.74** In a mountainous location, the boiling point of pure water is found to be 95°C. How many grams of sodium chloride must be added to 1 kg of water to bring the boiling point back to 100°C? Assume that $i = 2$.

## Colloids

**12.75** Give the type of colloid (aerosol, foam, emulsion, sol, or gel) that each of the following represents.
a. rain cloud      b. milk of magnesia
c. soapsuds      d. silt in water

**12.76** Give the type of colloid (aerosol, foam, emulsion, sol, or gel) that each of the following represents.
a. ocean spray      b. beaten egg white
c. dust cloud      d. salad dressing

**12.77** Arsenic(III) sulfide forms a sol with a negative charge. Which of the following ionic substances should be most effective in coagulating the sol?
a. KCl      b. $MgCl_2$      c. $Al_2(SO_4)_3$      d. $Na_3PO_4$

**12.78** Aluminum hydroxide forms a positively charged sol. Which of the following ionic substances should be most effective in coagulating the sol?
a. NaCl      b. $CaCl_2$      c. $Fe_2(SO_4)_3$      d. $K_3PO_4$

# General Problems

**12.79** A gaseous mixture consists of 80.0 mole percent $N_2$ and 20.0 mole percent $O_2$ (the approximate composition of air). Suppose water is saturated with the gas mixture at 25°C and 1.00 atm total pressure, and then the gas is expelled from the water by heating. What is the composition in mole fractions of the gas mixture that is expelled? The solubilities of $N_2$ and $O_2$ at 25°C and 1.00 atm are 0.0175 g/L $H_2O$ and 0.0393 g/L $H_2O$, respectively.

**12.80** A natural gas mixture consists of 90.0 mole percent $CH_4$ (methane) and 10.0 mole percent $C_2H_6$ (ethane). Suppose water is saturated with the gas mixture at 20°C and 1.00 atm total pressure, and the gas is then expelled from the water by heating. What is the composition in mole fractions of the gas mixture that is expelled? The solubilities of $CH_4$ and $C_2H_6$ at 20°C and 1.00 atm are 0.023 g/L $H_2O$ and 0.059 g/L $H_2O$, respectively.

**12.81** An aqueous solution is 8.50% ammonium chloride, $NH_4Cl$, by mass. The density of the solution is 1.024 g/mL. What are the molality, mole fraction, and molarity of $NH_4Cl$ in the solution?

**12.82** An aqueous solution is 27.0% lithium chloride, LiCl, by mass. The density of the solution is 1.127 g/mL. What are the molality, mole fraction, and molarity of LiCl in the solution?

**12.83** A 75-g sample of a gaseous fuel mixture contains 0.51 mole fraction propane, $C_3H_8$; the remainder of the mixture is butane, $C_4H_{10}$. What are the masses of propane and butane in the sample?

**12.84** The diving atmosphere used by the U.S. Navy in its undersea Sea-Lab experiments consisted of 0.036 mole fraction $O_2$ and 0.056 mole fraction $N_2$, with helium (He) making up the remainder. What are the masses of nitrogen, oxygen, and helium in a 7.84-g sample of this atmosphere?

**12.85** A liquid solution consists of 0.25 mole fraction ethylene dibromide, $C_2H_4Br_2$, and 0.75 mole fraction propylene dibromide, $C_3H_6Br_2$. Both ethylene dibromide and propylene dibromide are volatile liquids; their vapor pressures at 85°C are 173 mmHg and 127 mmHg, respectively. Assume that each compound follows Raoult's law in the solution. Calculate the total vapor pressure of the solution.

**12.86** What is the total vapor pressure at 20°C of a liquid solution containing 0.25 mole fraction benzene, $C_6H_6$, and 0.75 mole fraction toluene, $C_6H_5CH_3$? Assume that Raoult's law holds for each component of the solution. The vapor pressure of pure benzene at 20°C is 75 mmHg; that of toluene at 20°C is 22 mmHg.

**12.87** A sample of potassium aluminum sulfate 12-hydrate, $KAl(SO_4)_2 \cdot 12H_2O$, containing 118.6 mg is dissolved in 1.000 L of solution. Calculate the following for the solution:
a. The molarity of $KAl(SO_4)_2$.
b. The molarity of $SO_4^{2-}$.
c. The molality of $KAl(SO_4)_2$, assuming that the density of the solution is 1.00 g/mL.

**12.88** A sample of aluminum sulfate 18-hydrate, $Al_2(SO_4)_3 \cdot 18H_2O$, containing 159.3 mg is dissolved in 1.000 L of solution. Calculate the following for the solution:
a. The molarity of $Al_2(SO_4)_3$.
b. The molarity of $SO_4^{2-}$.
c. The molality of $Al_2(SO_4)_3$, assuming that the density of the solution is 1.00 g/mL.

**12.89** Urea, $(NH_2)_2CO$, has been used to melt ice from sidewalks, because the use of salt is harmful to plants. If the saturated aqueous solution contains 44% urea by mass, what is the freezing point? (The answer will be approximate, because the equation in the text applies accurately only to dilute solutions.)

**12.90** Calcium chloride, $CaCl_2$, has been used to melt ice from roadways. Given that the saturated solution is 32% $CaCl_2$ by mass, estimate the freezing point.

**12.91** The osmotic pressure of blood at 37°C is 7.7 atm. A solution that is given intravenously must have the same osmotic pressure as the blood. What should be the molarity of a glucose solution to give an osmotic pressure of 7.7 atm at 37°C?

**12.92** Maltose, $C_{12}H_{22}O_{11}$, is a sugar produced by malting (sprouting) grain. A solution of maltose at 25°C has an osmotic pressure of 5.50 atm. What is the molar concentration of maltose?

**12.93** Which aqueous solution has the lower freezing point, 0.10 $m$ $CaCl_2$ or 0.10 $m$ glucose?

**12.94** Which aqueous solution has the lower boiling point, 0.10 $m$ KCl or 0.10 $m$ $CaCl_2$?

**12.95** Commercially, sulfuric acid is obtained as a 98% solution. If this solution is 18 $M$, what is its density? What is its molality?

**12.96** Phosphoric acid is usually obtained as an 85% phosphoric acid solution. If it is 15 $M$, what is the density of this solution? What is its molality?

**12.97** A compound of manganese, carbon, and oxygen contains 28.17% Mn and 30.80% C. When 0.125 g of this compound is dissolved in 5.38 g of cyclohexane, the solution freezes at 5.28°C. What is the molecular formula of this compound?

**12.98** A compound of cobalt, carbon, and oxygen contains 28.10% C and 34.47% Co. When 0.147 g of this compound is dissolved in 6.72 g of cyclohexane, the solution freezes at 5.23°C. What is the molecular formula of this compound?

**12.99** The carbohydrate digitoxose contains 48.64% carbon and 8.16% hydrogen. The addition of 18.0 g of this compound to 100. g of water gives a solution that has a freezing point of −2.2°C.
a. What is the molecular formula of the compound?
b. What is the molar mass of this compound to the nearest tenth of a gram?

**12.100** Analysis of a compound gave 39.50% C, 2.21% H, and 58.30% Cl. When 0.855 g of this solid was dissolved in 7.50 g of naphthalene, the solution had a freezing point of 78.0°C. The pure solvent freezes at 80.0°C; its molal freezing point constant is 6.8°C/$m$.
a. What is the molecular formula of the compound?
b. What is its molecular mass to the nearest 0.1 g?

**12.101** Fish blood has an osmotic pressure equal to that of seawater. If seawater freezes at −2.3°C, what is the osmotic pressure of the blood at 25°C? To solve this problem, what assumptions must be made?

**12.102** A salt solution has an osmotic pressure of 18 atmospheres at 22°C. What is the freezing point of this solution? What assumptions must be made to solve this problem?

## Cumulative-Skills Problems

**12.103** An experimenter makes up a solution of 0.375 mol $Na_2CO_3$, 0.125 mol $Ca(NO_3)_2$, and 0.200 mol $AgNO_3$ in 2.000 L of water solution. Note any precipitations that occur, writing a balanced equation for each. Then, calculate the molarities of each ion in the solution.

**12.104** An experimenter makes up a solution of 0.625 mol $Na_3PO_4$, 0.100 mol $Ca(NO_3)_2$, and 0.150 mol $AgNO_3$ in 4.000 L of water solution. Note any precipitations that occur, writing a balanced equation for each. Then, calculate the molarities of each ion in the solution.

**12.105** The lattice enthalpy of sodium chloride, $\Delta H°$ for

$$NaCl(s) \longrightarrow Na^+(g) + Cl^-(g)$$

is 787 kJ/mol; the heat of solution in making up 1 $M$ NaCl($aq$) is +4.0 kJ/mol. From these data, obtain the sum of the heats of hydration of $Na^+$ and $Cl^-$. That is, obtain the sum of $\Delta H°$ values for

$$Na^+(g) \longrightarrow Na^+(aq)$$
$$Cl^-(g) \longrightarrow Cl^-(aq)$$

If the heat of hydration of $Cl^-$ is −338 kJ/mol, what is the heat of hydration of $Na^+$?

**12.106** The lattice enthalpy of potassium chloride is 717 kJ/mol; the heat of solution in making up 1 $M$ KCl($aq$) is +18.0 kJ/mol. Using the value for the heat of hydration of $Cl^-$ given in Problem 12.105, obtain the heat of hydration of $K^+$. Compare this with the value you obtained for $Na^+$ in Problem 12.105. Explain the relative values of $Na^+$ and $K^+$.

**12.107** A solution is made up by dissolving 15.0 g $MgSO_4 \cdot 7H_2O$ in 100.0 g of water. What is the molality of $MgSO_4$ in this solution?

**12.108** A solution is made up by dissolving 15.0 g $Na_2CO_3 \cdot 10H_2O$ in 100.0 g of water. What is the molality of $Na_2CO_3$ in this solution?

**12.109** An aqueous solution is 15.0% by mass of copper(II) sulfate pentahydrate, $CuSO_4 \cdot 5H_2O$. What is the molarity of $CuSO_4$ in this solution at 20°C? The density of this solution at 20°C is 1.167 g/mL.

**12.110** An aqueous solution is 20.0% by mass of sodium thiosulfate pentahydrate, $Na_2S_2O_3 \cdot 5H_2O$. What is the molarity of $Na_2S_2O_3$ in this solution at 20°C? The density of this solution at 20°C is 1.174 g/mL.

**12.111**   The freezing point of 0.0830 *m* aqueous acetic acid is −0.159°C. Acetic acid, $HC_2H_3O_2$, is partially dissociated according to the equation

$$HC_2H_3O_2(aq) \rightleftharpoons H^+(aq) + C_2H_3O_2^-(aq)$$

Calculate the percentage of $HC_2H_3O_2$ molecules that are dissociated, assuming the equation for the freezing-point depression holds for the total concentration of molecules and ions in the solution.

**12.112**   The freezing point of 0.109 *m* aqueous formic acid is −0.210°C. Formic acid, $HCHO_2$, is partially dissociated according to the equation

$$HCHO_2(aq) \rightleftharpoons H^+(aq) + CHO_2^-(aq)$$

Calculate the percentage of $HCHO_2$ molecules that are dissociated, assuming the equation for the freezing-point depression

holds for the total concentration of molecules and ions in the solution.

**12.113**   A compound of carbon, hydrogen, and oxygen was burned in oxygen, and 1.000 g of the compound produced 1.434 g $CO_2$ and 0.783 g $H_2O$. In another experiment, 0.1107 g of the compound was dissolved in 25.0 g of water. This solution had a freezing point of −0.0894°C. What is the molecular formula of the compound?

**12.114**   A compound of carbon, hydrogen, and oxygen was burned in oxygen, and 1.000 g of the compound produced 1.418 g $CO_2$ and 0.871 g $H_2O$. In another experiment, 0.1103 g of the compound was dissolved in 45.0 g of water. This solution had a freezing point of −0.0734°C. What is the molecular formula of the compound?

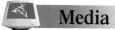 # Media Activities

**12.115**   View the Dissolution of a Solid in a Liquid *Visualization* on the student CD.
a. Describe, on a molecular level, what you would see if the *Visualization* continued to portray an unsaturated solution.
b. Describe, on a macroscopic level, what you would observe in a beaker while the process depicted in the *Visualization* occurs.

**12.116**   Run the Solution Equilibrium *Visualization* on the student CD.
a. At equilibrium, how does the rate of dissolution compare with the rate of solid formation?
b. Imagine that additional water was added to the system depicted in the *Visualization*. What would you observe on both a microscopic and macroscopic level?

c. If you were to evaporate some water from the system depicted in the *Visualization*, what would you observe on both a microscopic and macroscopic level?

**12.117**   Open the Colligative Properties *Understanding Concepts* simulation on the student web site. Move the slider bar to observe how changing the glucose concentrations affects osmotic pressure.
a. Describe, on a molecular level, what is happening when you add glucose.
b. For a real glucose solution, how does the glucose concentration change from the time you initially add the glucose to when the osmotic pressure of the solution becomes constant?

**12.118**   For review, take the ACE practice test for Chapter 12 on the student web site.

# 13

An STM image of the atomic structure of gallium arsenide, GaAs (Ga, blue; As, red).

# Materials of Technology

In the summer of 2000, scientists at Bell Laboratories and Oxford University reported that they had constructed a motor out of DNA, the chemical material of the genetic code. This molecular-scale motor used DNA as both a structural material and as a fuel to run the motor. Earlier, Harvard University chemists had constructed molecular-scale tweezers out of carbon-atom tubes, showing that they could manipulate clusters of polystyrene molecules, and NASA scientists have studied the possibility of constructing a molecular-scale gear assembly (see Figure 13.1). These scientists' efforts represent some of the many ongoing investigations into "nanotechnology," in which one manipulates materials on a molecular scale to create useful devices.

Nanotechnology is only the latest of a series of developments in materials science, in which one studies materials with a view toward developing useful applications. Telecommunications is an example of an area in which the development of

**Figure 13.1**
**Carbon-atom nanotube gear assembly.**   Shown is a computer simulation of two carbon-atom tubes, several nanometers in diameter, with projecting atoms to form two gears.

new materials has had immense impact, leading to rapid change in the technology. Initially, telecommunications was restricted to voice communication by telephone using copper wires to carry a message in the form of an electrical signal. Today, it is just as likely that fibers of pure glass, or fiber-optic cables, carry the message in the form of light pulses. An advantage of fiber-optic cable is that it can carry much more information than a copper-wire cable of similar size. Fiber-optic cables can easily carry television and Internet signals as well as voice messages. At present, fiber-optic communications requires that electrical signals be converted to light pulses then back to electrical signals for processing by, say, a computer. The question scientists are now asking is this: Can the communications processing be done entirely with light, possibly resulting in faster communications and more compact equipment? This, of course, would require new materials with special properties.

In this chapter, we begin by discussing metals and metallurgy (the study of the properties and production of metallic materials). Although metals have been used from earliest times, they are still central to modern technology, and scientists continue to develop new metals with special properties. (See Figure 13.2.) After discussing the bonding in metals, we look at some nonmetallic materials, including the fullerenes, which are new materials made entirely of carbon atoms. We end the chapter by looking at *composites,* materials made of two or more different kinds of materials. Bone is a natural composite of calcium phosphate crystals and a protein called collagen. Calcium phosphate lends structural strength to bone, but by itself is brittle. Collagen binds the calcium phosphate crystals into a stable, somewhat elastic whole. Modern human-made composites try to mimic and extend the properties of such natural composites.

**Figure 13.2**
**Racing bicycle.**   The frame is made of a special high-chromium steel, which is strong but lightweight.

# Metals and Metallurgy

One of the most obvious characteristics of a metal is its luster, or shine. It is certainly this property of metals that initially fascinates us. But metals have other interesting properties. For example, metals are electrical conductors. Over 4 million metric tons ($4 \times 10^9$ kg) of copper are used annually worldwide for electrical purposes alone (Figure 13.3). Metals are also generally good conductors of heat. They feel cold to your touch because heat from your hand is conducted away by the metal. Metals are also more or less malleable and ductile. By *malleable,* we mean you can pound the material into flat sheets; by *ductile,* we mean you can draw it into a wire. Thus we can define a **metal** as *a material that is lustrous (shiny), has high electrical and heat conductivities, and is malleable and ductile.*

A metal need not be a pure element; it can be a compound or mixture. An **alloy** is *a material with metallic properties that is either a compound or a mixture.* If the alloy is a mixture, it may be homogeneous (a solution) or heterogeneous. Most commercial metals are alloys consisting usually of one metal with small quantities of some other metals. Gold jewelry, for example, is made from an alloy that is a solid solution of gold containing some silver. Pure gold is too soft for jewelry; the alloy is much harder.

In the first part of this chapter, we will look at the different sources of the metallic elements and see how we obtain the free metals from their ores. We will also look at the bonding in metals, which can explain some of their properties.

## 13.1 Natural Sources of the Metallic Elements

A major source of metals and their compounds is the earth's crust, the outer solid portion of the planet. Mine shafts go to depths of less than about 4 km (compared with the earth's radius of 6371 km); therefore, we access only the thinnest outer layer of the earth. Table 13.1 lists the ten most abundant elements in the earth's crust and the mass percentage of each element in this part of the earth.

The two most abundant elements in the earth's crust are oxygen and silicon. These elements, along with various metals, are the ingredients of *silicates,* which are the

**Figure 13.3**
**Metals as electrical conductors.** *Left:* A power transmission line; the wire is aluminum. *Right:* Aluminum and copper wire.

**Table 13.1**
**Ten Most Abundant Elements in the Earth's Crust**

| Element | Mass Percentage |
| --- | --- |
| 1.  Oxygen | 46.4 |
| 2.  Silicon | 27.7 |
| 3.  Aluminum | 8.1 |
| 4.  Iron | 5.0 |
| 5.  Calcium | 3.6 |
| 6.  Sodium | 2.8 |
| 7.  Potassium | 2.6 |
| 8.  Magnesium | 2.1 |
| 9.  Titanium | 0.4 |
| 10.  Hydrogen | 0.1 |

▶ **The structure of silicates and aluminosilicates is discussed in Section 13.6.**

chief substances of rocks and clay (Figure 13.4). Clay and many rocks are *alumi-nosilicates,* silicates in which aluminum is an important constituent. These alumi-nosilicates account for the vast quantity of aluminum in the earth's crust. ◀

The commercial sources of most metals are not silicates, however, because generally metals are not easily obtained from them. Despite the abundance and avail-ability of clay, an economical method of producing aluminum from clay has yet to be discovered. The usual compounds from which metals are obtained are oxides (and hydroxides), sulfides, and carbonates. Aluminum, for example, is produced from baux-ite, which is a material containing aluminum oxide hydroxide, $AlO(OH)$, and aluminum hydroxide, $Al(OH)_3$, along with iron oxides and silicates.

Bauxite is technically a *rock;* geologists define a rock as a naturally occurring solid material composed of one or more minerals. A **mineral** is *a naturally occurring inorganic solid substance or solid solution with a definite crystalline structure.* So a mineral might be a definite chemical substance, or it might be a homogeneous solid mixture. One of the constituents of bauxite is gibbsite, which is a mineral form of

**Figure 13.4**
**Pictured Rocks.** The Pictured Rocks along the shore of Lake Superior in Michigan are sandstone cliffs composed principally of silicon and oxygen. The beautiful colors of the cliffs are due to streaks of hematite, or iron(III) oxide.

**Figure 13.5** *(left)*
**Ruby.** Ruby is aluminum oxide, $Al_2O_3$, containing a small percentage of chromium(III) oxide, $Cr_2O_3$.

**Figure 13.6** *(right)*
**Yellow and red ocher.** Yellow and red ocher are natural, earthy forms of iron(III) oxide, $Fe_2O_3$. The colors depend on the crystal form, which can be varied by strong heating.

aluminum hydroxide, $Al(OH)_3$. Corundum is an oxide mineral of aluminum; its formula is $Al_2O_3$. Some of the aluminum atoms in corundum may be replaced by chromium, forming a mineral that is a red solid solution of $Al_2O_3$ and $Cr_2O_3$. When this mineral is of gemstone quality, it is known as ruby (Figure 13.5).

An **ore** is *a rock or mineral from which a metal or nonmetal can be economically produced.* Bauxite is the principal ore of aluminum. Iron, the fourth most abundant element in the earth's crust, has several ores, the most important of which is the mineral hematite. Hematite, which is iron(III) oxide, $Fe_2O_3$, varies in form and color. The earthy, red-colored form is known as red ocher; in addition to its use as an ore, it is used as a paint pigment (Figure 13.6). Metal ores are not evenly distributed over the earth. Table 13.2 lists some of the ores of important metals, along with their country of origin.

**Table 13.2**
**Principal Ores of Some Important Metals**

| Element | Ore | Where Found |
| --- | --- | --- |
| Aluminum, Al | Bauxite, $AlO(OH)$ and $Al(OH)_3$ | Surinam, Jamaica, Guyana |
| Chromium, Cr | Chromite, $FeCr_2O_4$ | Russia, South Africa |
| Copper, Cu | Chalcopyrite, $CuFeS_2$ | USA, Chile, Canada, Kazakhstan |
| Iron, Fe | Hematite, $Fe_2O_3$ | Australia, Ukraine, USA |
| Lead, Pb | Galena, $PbS$ | USA, Australia, Canada |
| Lithium, Li | Spodumene, $LiAlSi_2O_6$ | USA, Canada, Brazil |
| Mercury, Hg | Cinnabar, $HgS$ | Spain, Algeria |
| Nickel, Ni | Pentlandite, $NiS$–$FeS$ mixture | Canada |
| Tin, Sn | Cassiterite, $SnO_2$ | Malaysia, Bolivia |
| Zinc, Zn | Sphalerite, $ZnS$ | Canada, USA, Australia |

# 13.2 Metallurgy

No doubt nuggets of gleaming metal in sand and rock initially fascinated early humans simply because of their appearance. Eventually, however, someone discovered that metals could be fashioned into tools. The first metals used were probably those that occurred naturally in the free state, such as gold. Later, it was found that heating certain rocks in a hot charcoal fire yielded metals. We realize now that these rocks contain metal compounds that can be reduced to the metallic state by reaction with charcoal (carbon) or the carbon monoxide obtained from the partial burning of charcoal. Copper was probably the first metal produced this way. Later, it was discovered that rocks containing copper and tin compounds yielded bronze, the first manufactured alloy. It was from these beginnings that metallurgy arose. **Metallurgy** is *the scientific study of the production of metals from their ores and the making of alloys having various useful properties.* In this section, we will look at the basic steps in the production of a metal from its ore:

1. *Preliminary treatment.* Usually, an ore must first be treated in some way to concentrate its metal-containing portion. Ores are usually mixtures of a mineral containing the metal along with economically worthless rock material that must be discarded. During the preliminary treatment, the metal-containing mineral is separated from these less desirable parts of the ore. It may also be necessary to transform the metal-containing mineral by chemical reaction to a metal compound that is more easily reduced to the free metal.

2. *Reduction.* Unless the metal occurs free, the metal compound obtained from preliminary treatment has to be reduced. Electrolysis or chemical reduction may be used, depending on the metal.

3. *Refining.* Once the free metal is obtained, it may have to be purified before it can be used. This purification process is referred to as metal refining.

## Preliminary Treatment

A metal ore contains varying quantities of economically worthless material along with the mineral containing the metallic element. *The worthless portion of an ore* is called **gangue** (pronounced "gang"). To separate the useful mineral from the gangue, both physical and chemical methods are used, depending on the ore. Panning for gold nuggets is a simple physical separation method based on differences in densities of gold and other minerals. Flushing water over a pan of gold-bearing earth easily washes sand and dirt aside, leaving the more dense grains of gold at the bottom of the pan.

**Flotation** is *a physical method of separating a mineral from the gangue that depends on differences in their wettabilities by a liquid solution* (Figure 13.7). The wetting of solid particles by water depends on the forces of attraction between the particles and water. Certain chemical substances called *flotation agents* are used to coat the metal-bearing mineral particles selectively so they are wet less easily by water than is the gangue. During the flotation process, pulverized ore is mixed with water and a flotation agent. This mixture is then agitated violently with air. The gangue becomes wetted by the water and settles to the bottom of the flotation container. Particles of metal-bearing mineral become coated with the flotation agent. A molecule of this substance has a polar end, which is attracted to the mineral, and a water-repellent (hydrophobic) end. ◀ Mineral particles become coated with molecules of the flotation agent with their hydrophobic ends pointing outward. The coated mineral particles are

▶ The action of a flotation agent has some similarity to detergent action; see Section 12.9.

**Figure 13.7**
**Flotation process for concentrating certain ores.** The ore attaches to bubbles of air and is carried off in the froth. The gangue settles to the bottom of the tank, where it is withdrawn.

Froth containing ore

Compressed air

Stirring blade

Gangue
(rock, sand)

more strongly attracted to air bubbles than to the water solution and so float to the surface, where they are trapped by the froth. The froth is removed at the top of the flotation vessel and the gangue is removed at the bottom. Sulfide ores, such as those of copper, lead, and zinc, are concentrated in this way.

The Bayer process is an example of a chemical method of concentrating an ore in its metal-bearing fraction. The **Bayer process** is *a chemical procedure in which purified aluminum oxide, $Al_2O_3$, is separated from the aluminum ore bauxite.* It depends on the fact that aluminum hydroxide is amphoteric, meaning that it behaves both as a base (reacting with acid to form the metal ion), which is expected of a metal hydroxide, and as an acid (reacting with base to form a hydroxo metal ion). Bauxite, as we noted earlier, contains aluminum hydroxide, $Al(OH)_3$, and aluminum oxide hydroxide, $AlO(OH)$, as well as relatively worthless constituents such as silicate minerals with some iron oxides. When bauxite is mixed with hot, aqueous sodium hydroxide solution, the amphoteric aluminum minerals, along with some silicates, dissolve in the strong base. The aluminum minerals give the aluminate ion, $Al(OH)_4^-$.

$$Al(OH)_3(s) + OH^-(aq) \longrightarrow Al(OH)_4^-(aq)$$
$$AlO(OH)(s) + OH^-(aq) + H_2O(l) \longrightarrow Al(OH)_4^-(aq)$$

Other substances, including the iron oxides, remain undissolved and are filtered off. As the hot solution of sodium aluminate cools, aluminum hydroxide precipitates, leaving the soluble silicates in solution. In practice, the solution is seeded with aluminum hydroxide to start the precipitation.

$$Al(OH)_4^-(aq) \longrightarrow Al(OH)_3(s) + OH^-(aq)$$

Aluminum hydroxide can also be precipitated from the basic solution by acidifying it with carbon dioxide gas (Figure 13.8). The aluminum hydroxide is finally *calcined* (heated strongly in a furnace) to produce aluminum oxide.

$$2Al(OH)_3(s) \xrightarrow{\Delta} Al_2O_3(s) + 3H_2O(g)$$

**Figure 13.8**
**Precipitation of aluminum hydroxide by acidifying an aluminate ion solution.**
*Left:* Carbon dioxide gas is bubbled into a solution of aluminate ion, $Al(OH)_4^-$, containing aluminon dye. *Right:* Aluminum hydroxide, $Al(OH)_3$, precipitates from the acidic solution. The precipitate is normally white but adsorbs the dye to form a pink-colored precipitate, called a *lake*. A standard test for aluminum ion is its precipitation as the hydroxide in the presence of aluminon dye to form this lake.

► Acid rain is discussed in the essay at the end of Section 17.2.

**Figure 13.9**
**Preparation of gallium arsenide, GaAs, semiconductor material.**
The bars in the tube are gallium; the small pieces in the tube near the furnace are arsenic. Gallium arsenide is used to make semiconductor diode lasers for use in fiber optic communications, compact disc players, and laser printers. Also, computer chips are being developed using gallium arsenide, because it conducts an electrical signal five times faster than presently used silicon.

Bauxite is also a source of gallium, the element under aluminum in the periodic table. Gallium is important in preparing gallium arsenide for solid-state devices, including lasers for compact disc players and light-emitting diodes, or LEDs (Figure 13.9). Sodium gallate, $Na[Ga(OH)_4]$, is more soluble than the corresponding aluminum compound, and it remains in the filtrate from the Bayer process after aluminum hydroxide has been filtered off.

Once an ore is concentrated, it may be necessary to convert the mineral to a compound more suitable for reduction. **Roasting** is *the process of heating a mineral in air to obtain the oxide.* Sulfide minerals, such as zinc ore (containing the mineral sphalerite, ZnS), are usually roasted before reducing them to the metal.

$$2ZnS(s) + 3O_2(g) \longrightarrow 2ZnO(s) + 2SO_2(g); \ \Delta H° = -884 \text{ kJ}$$

The roasting process is exothermic, because it is essentially the burning of the sulfide ore. Once the ore has been heated to initiate roasting, additional heating is unnecessary.

One of the products of the roasting of a sulfide ore is sulfur dioxide gas. Until recently, this gas was simply vented to the atmosphere. Sulfur dioxide oxidizes further in polluted air to yield sulfur trioxide, which dissolves in moisture in the air to yield *acid rain.* Considerable damage to the environment has occurred as a result. More and more, ore-roasting facilities use the sulfur dioxide to produce sulfuric acid or otherwise remove it from the gases vented to the atmosphere. ◄

## Reduction

A metal, if it is not already free, must be obtained from one of its compounds by reduction, using either electrolysis or chemical reduction.

**Electrolysis** The reactive main-group metals, such as lithium, are obtained by *electrolysis.* Electrolysis uses an electric current to reduce a metal compound to the metal. Lithium, for example, is obtained commercially by electrolysis of molten lithium chloride, LiCl. (Figure 13.10 shows a simple laboratory setup of the electrolysis process.) Molten lithium chloride is a liquid consisting of lithium ions, $Li^+$, and chloride ions, $Cl^-$. Electrons from the negative pole of a battery flow to an attached electrode and react with lithium ions in the melt, forming lithium metal:

$$Li^+ + e^- \longrightarrow Li(l)$$

**Figure 13.10**
**Electrolysis of molten lithium chloride.**
This sketch shows the basic elements of such an electrolysis. However, in the setup as shown, liquid lithium tends to rise to the surface, where it can catch fire. In the commercial process, the products of the electrolysis are collected in separate containers, and air is not allowed to come in contact with them.

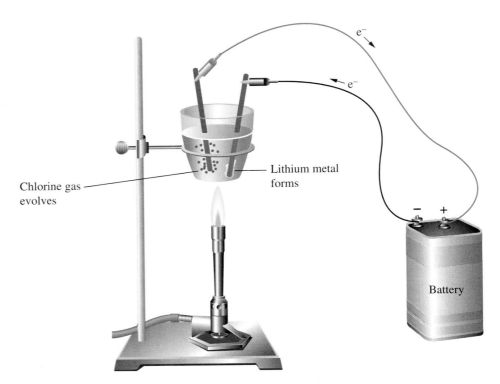

Chlorine gas evolves

Lithium metal forms

Battery

At the other electrode, chloride ions yield chlorine gas and electrons flow back to the battery.

$$Cl^- \longrightarrow \tfrac{1}{2}Cl_2(g) + e^-$$

The net result of these electrode reactions is the production of lithium metal and chlorine from lithium chloride. The electric current supplies the energy for the reaction.

$$LiCl(l) \longrightarrow Li(l) + \tfrac{1}{2}Cl_2(g)$$

Magnesium is another reactive metal obtained by electrolysis of the chloride. Magnesium ion, $Mg^{2+}$, is the third most abundant dissolved ion in the oceans, after $Cl^-$ and $Na^+$. The oceans, therefore, are an essentially inexhaustible supply of magnesium ion, from which the metal can be obtained. The **Dow process,** *a commercial method for isolating magnesium from seawater,* depends on the fact that magnesium ion can be precipitated from aqueous solution by adding a base. Figure 13.11 shows a flowchart for the process. Seashells provide the source of the base (calcium hydroxide). Seashells are principally calcium carbonate, $CaCO_3$, which when heated decomposes to calcium oxide, $CaO$, and carbon dioxide, $CO_2$. Calcium oxide reacts with water to produce calcium hydroxide. The base, calcium hydroxide, provides hydroxide ion that reacts with magnesium ion in seawater to precipitate magnesium hydroxide:

$$Mg^{2+}(aq) + 2OH^-(aq) \longrightarrow Mg(OH)_2(s)$$

The magnesium hydroxide is allowed to settle out in ponds, here it gives a suspension, or slurry, of $Mg(OH)_2$. The slurry is filtered to recover the magnesium hydroxide precipitate, which is then treated with hydrochloric acid to yield magnesium chlo-

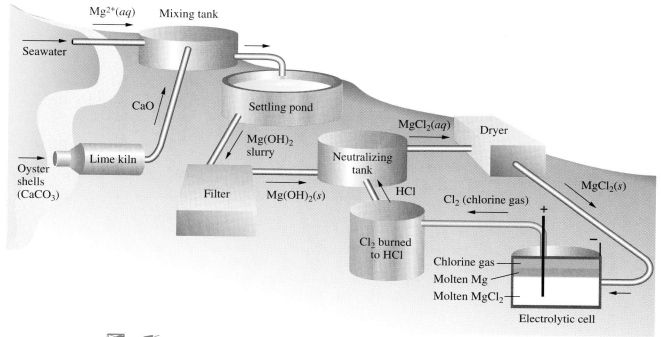

**Figure 13.11**
**Dow process for producing magnesium from seawater.** Oyster shells or other seashells are calcined in a kiln to produce calcium oxide, which when added to seawater precipitates magnesium ion as magnesium hydroxide. This is neutralized with hydrochloric acid to give magnesium chloride. Electrolysis of molten $MgCl_2$ yields magnesium metal and chlorine. Hydrochloric acid can be recovered by burning the chlorine with natural gas.

ride. The dry magnesium chloride salt is melted and electrolyzed at 700°C to yield magnesium metal and chlorine gas.

$$MgCl_2(l) \xrightarrow{\text{electrolysis}} Mg(l) + Cl_2(g)$$

Aluminum is manufactured from the aluminum oxide obtained in the Bayer process using electrolysis. The method for doing this was discovered in 1886 by Charles Martin Hall (then a student at Oberlin College) in the United States and, independently, by Paul Héroult in France. (Both men were 22 years old at the time.) Before this discovery, aluminum was a precious metal, and in the 1850s it was so rare and exotic that the French court used aluminum tableware at royal functions in place of the usual silver and gold utensils.

The **Hall–Héroult process** is *the commercial method for producing aluminum by the electrolysis of a molten mixture of aluminum oxide in cryolite, $Na_3AlF_6$.* Originally, the process used natural cryolite. Today, the Hall–Héroult process uses synthetic cryolite, or sodium hexafluoroaluminate, produced by reacting aluminum hydroxide from the Bayer process with sodium hydroxide and hydrofluoric acid.

$$3NaOH(aq) + Al(OH)_3(s) + 6HF(aq) \longrightarrow Na_3AlF_6(aq) + 6H_2O(l)$$

A Hall–Héroult electrolytic cell consists of a rectangular steel shell lined first with an insulating material, then with carbon (from baked petroleum coke) to form the negative electrode (Figure 13.12). The other (positive) electrode is also made from carbon. The electrolyte consists of molten cryolite, at about 1000°C, into which some aluminum oxide is dissolved. The overall electrolysis reaction is

$$2Al_2O_3(s) + 3C(s) \xrightarrow{\text{electrolysis}} 4Al(l) + 3CO_2(g)$$

**Figure 13.12**
**Hall–Héroult cell for the production of aluminum.**
Aluminum oxide is electrolyzed in molten cryolite (the electrolyte). Molten aluminum forms at the negative electrode (tank lining), where it is periodically withdrawn.

Carbon-lined tank

Carbon electrode

Tank wall (insulator)

Molten aluminum

$Al_2O_3$ in molten cryolite

Carbon dioxide

The positive carbon electrodes are consumed in the electrolysis and must be replaced periodically; aluminum oxide is continually added to the electrolyte bath.

The energy required to produce aluminum by the Hall–Héroult process is considerable—about 14,000 kilowatt-hours per ton of aluminum. (A watt equals one joule per second, so a kilowatt-hour equals 3.6 million joules.) To put this in perspective, note that a typical household might use about 1000 kilowatt-hours per month. Therefore, the energy used to produce one ton of aluminum is sufficient to supply electricity to about 14 homes for one month. Recycling of aluminum would save much of this energy. Only about 7% of the energy required to produce aluminum from aluminum oxide is needed to obtain aluminum metal from used cans and scrap.

**Chemical Reduction**    The cheapest chemical reducing agent is some form of carbon, such as hard coal or coke. Coke is the solid residue left from coal after its volatile constituents have been driven off by heating in the absence of air. Some important metals, including iron and zinc, are obtained by reduction of their compounds with carbon (or carbon monoxide, which is a product of partial oxidation of carbon).

Earlier, we described how zinc is obtained from its sulfide ore by roasting to zinc oxide. (Zinc oxide is a white substance used in many common ointments; see Figure 13.13.) When the oxide is heated strongly with carbon in a *blast furnace,* it is reduced to the metal. Zinc oxide and coke are added at the top of the furnace; a blast

**Figure 13.13**
**Zinc oxide in medicines.**
Zinc oxide is used as an antiseptic and a sunscreen.

**FIGURE 13.14**
**Blast furnace for the reduction of iron ore to iron metal.** Iron ore, coke, and limestone are added at the top of the blast furnace, and heated air enters at the bottom, where it burns the coke. The carbon monoxide from the burning reduces the iron ore to metallic iron.

of air enters at the bottom. Some of the coke burns in the air and heats the material to a temperature above the boiling point of zinc (908°C). At this temperature, zinc oxide reacts with carbon:

$$ZnO(s) + C(s) \longrightarrow Zn(g) + CO(g)$$

Zinc metal vapor leaves at the top of the blast furnace and is condensed to the liquid, which later solidifies.

Iron is also produced by reduction in a blast furnace. A mixture of iron ore (containing an oxide such as $Fe_2O_3$), coke (carbon obtained by heating coal), and limestone (a rock that is mainly calcium carbonate) is added at the top of the furnace (Figure 13.14), and a blast of heated air enters at the bottom. The reactions are somewhat complicated, but essentially coke burns to produce carbon monoxide, which reacts with iron oxides to reduce them to iron and carbon dioxide. For example, the reduction of $Fe_2O_3$ by CO is:

$$Fe_2O_3(s) + 3CO(g) \longrightarrow 2Fe(l) + 3CO_2(g)$$

Molten iron flows to the bottom of the blast furnace, where it is drawn off. Impurities in the iron ore react with calcium oxide obtained from the limestone to produce a glassy material called *slag*. Molten slag collects in a layer floating on the molten iron and is drawn off periodically.

Hydrogen and reactive metals, such as sodium, magnesium, and aluminum, are used as reducing agents when carbon is unsuitable. Tungsten, for example, reacts with carbon to produce tungsten carbide, WC, a relatively unreactive and extremely hard substance. ◄ The metal is prepared from tungsten(VI) oxide, a canary-yellow compound obtained from the processing of tungsten ore. The oxide is reduced by heating it in a stream of hydrogen gas.

▶ Tungsten carbide, WC, is used for cutting tools because of its hardness even at high temperatures.

$$WO_3(s) + 3H_2(g) \longrightarrow W(s) + 3H_2O(g)$$

**Figure 13.15**
**Tungsten filament of an incandescent lightbulb.**
A coiled wire of tungsten becomes white-hot when an electric current flows through it. The wire is enclosed in a glass bulb containing gases that do not react with the tungsten, such as nitrogen and argon. The gases carry the heat away from the wire, which would otherwise overheat and boil away.

One of the most important uses of tungsten metal is in the production of filaments for incandescent bulbs (Figure 13.15). This usage depends on the fact that tungsten has the highest melting point (3410°C) and highest boiling point (5900°C) of any metal. To be useful, the incandescent filament in a lightbulb must not melt and should not vaporize excessively. The tungsten metal filament does slowly vaporize, and you can often see where metal has condensed as a black coating on the inside surface of a burned-out bulb.

## Refining

Often the metal obtained from reduction contains impurities and must be purified, or refined, before it can be used. Zinc, for example, contains lead, cadmium, and iron as impurities. Zinc has a relatively low boiling point (908°C) and is refined by distillation.

Pure nickel metal is used for battery electrodes and as a catalyst. After roasting of the sulfide ore and reduction of the oxide to nickel, the metal may be purified for further use by the Mond process. The **Mond process** is *a chemical procedure for the purification of nickel that depends on the formation and later decomposition of a volatile compound of the metal (nickel tetracarbonyl).* In this process, impure nickel powder is placed in a stream of carbon monoxide gas at 20 atm and 150°C, where it reacts to produce nickel tetracarbonyl vapor, $Ni(CO)_4$. ◄ The vapor is condensed to the liquid (whose normal boiling point is 43°C) and distilled to purify it. The nickel tetracarbonyl vapor is then passed over small pellets of pure nickel heated to 230°C, where it decomposes to deposit more pure nickel and releases carbon monoxide as the other product. The equilibrium reaction involving nickel, carbon monoxide, and nickel tetracarbonyl is

► The transition metals in Groups VB to VIIIB form complexes with CO molecules as the ligands; the complexes are called carbonyls.

$$Ni(s) + 4CO(g) \rightleftharpoons Ni(CO)_4(g)$$

Copper metal used for electrical wire must be quite pure; otherwise its electrical resistance it too large. More than 90% of the copper produced by reduction of copper ores is further refined, or purified, by electrolysis. The electrolysis cell consists of two electrodes, one of impure copper connected to the positive pole of a battery and the other of pure copper connected to the negative pole. The electrodes dip into a solution of copper ions. Copper(II) ions from the solution combine with electrons from the negative pole of the battery, leaving copper atoms that plate out onto the pure copper electrode.

$$Cu^{2+}(aq) + 2e^- \longrightarrow Cu(s)$$

Electrons from the impure copper flow to the positive pole of the battery, leaving copper(II) ions behind that enter the solution. These copper(II) ions replace the ones that plate out onto the pure copper electrode:

$$Cu(s) \longrightarrow Cu^{2+}(aq) + 2e^-$$

The overall result is that copper leaves the impure electrode and plates out onto the pure copper electrode, enlarging it at the expense of the other electrode. ◄

▶ **This method of refining copper is described further at the end of Section 20.10.**

 **CONCEPT CHECK 13.1**

Why must a mineral containing a metal usually be reduced to obtain the free metal?

## 13.3  Bonding in Metals

The special properties of a metal result from its delocalized bonding, in which bonding electrons are spread over a number of atoms. In this section, we will look first at the "electron-sea" model of a metal and then at the molecular orbital theory of bonding in metals.

### Electron-Sea Model of Metals

A very simple picture of a metal depicts an array of positive ions surrounded by a "sea" of valence electrons free to move over the entire metal crystal. When the metal is connected to a source of electric current, the electrons easily move away from the negative side of the electric source and toward the positive side, forming an electric current in the metal. In other words, the metal is a conductor of electric current because of the mobility of the valence electrons. A metal is also a good heat conductor because the mobile electrons can carry additional kinetic energy across the metal. ◄

▶ **The electron-sea model of metallic bonding is described briefly in Section 9.7; see Figure 9.16.**

### Molecular Orbital Theory of Metals

The electron-sea model of metals is a simplified view that accounts in only a qualitative way for properties of a metal such as electrical conductivity. Molecular orbital theory gives a more detailed picture of the bonding in a metal. ◄

▶ **Molecular orbital theory is discussed in Sections 10.5 to 10.7.**

Recall that molecular orbitals form between two atoms when atomic orbitals on the atoms overlap. In some cases, the atomic orbitals on three or more atoms overlap to form molecular orbitals that encompass all of the atoms. These molecular orbitals are said to be *delocalized*. The number of molecular orbitals that form by the overlap of atomic orbitals always equals the number of atomic orbitals. In a metal, the outer orbitals of an enormous number of metal atoms overlap to form an enormous number of molecular orbitals that are delocalized over the metal. As a result, a large number of energy levels are crowded together into "bands." Because of these energy bands, the molecular orbital theory of metals is often referred to as *band theory*.

**Figure 13.16**
**Formation of an energy band in sodium metal.**
Note that the number of energy levels grows until the levels merge into a continuous band of energies. The lower half of the band is occupied by electrons.

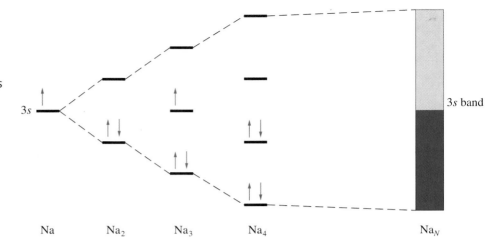

Sodium metal provides a simple illustration of band theory. Imagine that you build a crystal of sodium by bringing sodium atoms together one at a time, and during this process you follow the formation of molecular orbitals and associated energy levels. Each isolated sodium atom has the electron configuration $1s^2 2s^2 2p^6 3s^1$, or [Ne]$3s^1$. When two sodium atoms approach each other, their $3s$ orbitals overlap to form two molecular orbitals (a bonding molecular orbital and an antibonding molecular orbital). The inner-core electrons, represented as [Ne] in the electron configuration, remain essentially nonbonding. Now imagine that a third atom is brought up to this diatomic molecule. The three $3s$ orbitals overlap to form three molecular orbitals, each orbital encompassing the entire $Na_3$ molecule (that is, they are delocalized over the molecule). When a large number $N$ (on the order of Avogadro's number) of sodium atoms have been brought together to form a crystal, the atoms will have formed $N$ molecular orbitals encompassing the entire crystal.

Figure 13.16 shows that at each stage the number of energy levels grows and the energy levels spread until the molecular orbital levels merge into a *band* of essentially continuous energies. We call this the $3s$ band of the sodium metal. A $3s$ band formed from $N$ atoms will have $N$ orbitals that hold a maximum of $2N$ electrons. Because each sodium atom has one valence electron, $N$ atoms will supply $N$ $3s$ electrons and will therefore half-fill the $3s$ band.

Electrons become free to move throughout a crystal when they are excited to unoccupied orbitals of a band. In sodium metal, this requires very little energy; because of the half-filled $3s$ band, unoccupied orbitals lie just above the occupied orbitals of highest energy. When a voltage is applied to a metal crystal, electrons are excited to the unoccupied orbitals and move toward the positive pole of the voltage source.

The explanation of the electrical conductivity of magnesium metal is somewhat more complicated than in the case of sodium. A magnesium atom has the configuration [Ne]$3s^2$. As in sodium, the $3s$ orbitals of magnesium metal overlap to form a $3s$ band. If this were the whole story, you would expect the $2N$ valence electrons of the atoms to completely fill the $3s$ band. Therefore, if a small voltage were applied, the electrons would have no place to go. You would expect magnesium to be an insulator (a nonconducting material), but in fact it is a conductor.

Both sodium and magnesium have unoccupied $3p$ bands formed from unoccupied $3p$ orbitals of the atoms. The presence of this band in sodium has no effect on its electrical conductivity. In magnesium, however, the situation is different, as Figure

**Figure 13.17**
**Formation of 3s and 3p bands in magnesium metal.**  Note that the 3s and 3p bands merge. As a result, the bands are only partially filled.

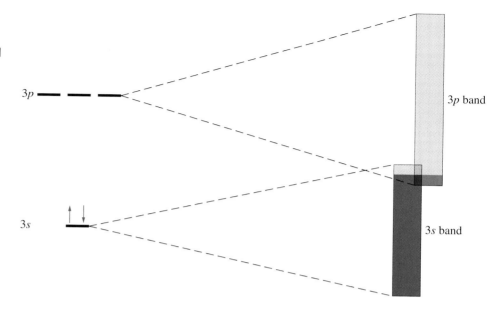

13.17 shows. As the orbitals of the individual atoms interact in forming the metal, the energy levels spread so that the bottom of the 3p band merges with the top of the 3s band. Imagine electrons filling the 3s band. When the electrons reach the energy where the two bands have merged, electrons begin to fill orbitals in both bands. As a result, the 3s and 3p bands of magnesium metal are only partially filled by the time you have accounted for all 2N valence electrons. Therefore, when a voltage is applied to the metal, the highest-energy electrons are easily excited into the unoccupied orbitals, giving an electrical conductor.

 **CONCEPT CHECK 13.2**

A semiconductor is a material with a separation between the filled band and an unfilled band called an energy gap. Would it take more or less energy to make a semiconductor carry current than it would a metal?

# Nonmetallic Materials

Long before humans developed the technology for producing metals, they had begun using nonmetallic materials, such as wood and stone, to make useful tools. Fire was of central importance in extending the range of useful materials. About 20,000 years ago, humans found that they could harden clay by cooking it in a fire; but once they discovered how to increase the temperature of a fire through the use of furnaces, the kinds of materials available to them multiplied quickly. By firing wet clay in furnaces or kilns, they could produce pottery vessels. At even higher temperatures, they could produce glass.

These materials, fired clay products and glass, are still vitally important today, but to these more traditional materials, we can now add a wealth of sophisticated materials with special properties. The new superconductors mentioned in the essay on

# Superconductivity

**MATERIALS**

A *superconductor* is a material that abruptly loses its resistance to an electric current when cooled to a definite characteristic temperature. This means that an electric current will flow in a superconductor without heat loss, unlike the current in a typical conductor. Once a current has been started in a superconducting circuit, it continues to flow indefinitely. Another intriguing property of a superconductor is its perfect diamagnetism. This means that the superconductor completely repels magnetic field lines. Figure 13.18 shows a magnet suspended in midair over one of the newly discovered ceramic superconductors. The magnet seems to levitate—it appears to hover in air in defiance of gravity. In fact, the repulsion of magnetic field lines by the superconductor holds the magnet aloft.

Superconductivity was discovered in 1911 by the Dutch physicist Heike Kamerlingh Onnes soon after he found a way to liquefy helium. By evaporating liquid helium, he could obtain temperatures near absolute zero. Kamerlingh Onnes found that mercury metal suddenly loses all resistance to an electric current when cooled to 4 K. Superconductors first became useful when a niobium metal alloy was found to become superconducting at about 23 K and to remain superconducting even when large currents flow through it. (Many superconducting materials lose their superconductivity when even moderate currents flow through them.) It became possible to construct superconducting magnets with high magnetic fields by starting an electric current in a superconducting circuit. Such magnets are being used in medical magnetic resonance imaging. (See Instrumental Methods: Nuclear Magnetic Resonance in Chapter 8.) Even though expensive liquid helium (more than $10 per gallon) is needed to operate these magnets, they are still cheaper to use than the usual electromagnets because of their much lower power requirements.

**Figure 13.18**
**Levitation of a magnet by a superconductor.**
The magnet (samarium–cobalt alloy) is supported above the ceramic superconductor (approximate formula $YBa_2Cu_3O_7$). The ceramic becomes superconducting when it is cooled by liquid nitrogen.

In 1986, Johannes Georg Bednorz and Karl Alexander Müller of IBM discovered that certain copper oxide ceramic materials became superconducting at 30 K, and within months researchers had found similar materials that become superconducting at 125 K. (They won the 1987 Nobel Prize in physics for their discovery.) This means that superconductors can be operated using a cheap refrigerant, such as liquid nitrogen, at a few cents per gallon. Perhaps materials can be found that are superconducting even at room temperature. What remains is to determine how to fabricate such superconducting materials into wires and similar objects with the ability to carry large currents.

superconductivity are just one example of these new materials. More and more, high-value materials are being designed for specialized applications.

In the following sections, we look at several nonmetallic materials with applications in modern technology. We begin with a discussion of the different allotropes or forms of carbon: diamond, graphite, and the fullerenes (Section 13.4). The fullerenes are recently discovered molecular forms of the element carbon, in which the carbon atoms form hollow balls and tubes that may make them important as catalysts or possibly as drug-delivery materials. Diamond shows promise as a material that might supersede silicon in its role as a material for solid-state electronics. Silicon and diamond can act as semiconductors, which we discuss in Section 13.5. We end the chapter with sections on silicon, silica, and silicates (Section 13.6), ceramics (Section 13.7), and finally composites (Section 13.8).

## 13.4 Diamond, Graphite, the Fullerenes, and Nanotechnology

Until recently, the element carbon was thought to occur in only two allotropic forms: diamond and graphite. Both allotropes are covalent network solids, whose structures we discussed in detail in Section 11.8. In diamond, each carbon atom is tetrahedrally ($sp^3$) bonded to four other carbon atoms (see Figure 11.27 in Section 11.6). The result is a three-dimensional network structure in which every carbon atom is covalently bonded to four other carbon atoms. To move one plane of atoms in the diamond crystal relative to another requires the breaking of many strong carbon–carbon bonds. Because of this, diamond is the hardest material known, and its principal industrial use is as an abrasive (grinding material).

Graphite has an entirely different structure. It has a layer structure, with each layer being attracted to one another by London forces. Within a layer, each carbon atom is covalently bonded to three other atoms, giving a flat layer of carbon-atom hexagons. Here is the resonance formula for a portion of a layer of graphite:

(one resonance formula)

The bonding is $sp^2$ with delocalized $\pi$ bonds. As a result of the delocalized bonding, graphite is an electrical conductor, unlike diamond, which is an insulator.

In 1985, Richard Smalley and Robert Curl of the United States and Harold Kroto of the United Kingdom discovered the first of a series of molecular forms of the element carbon: $C_{60}$, which they called *buckminsterfullerene* or, more casually, the *buckyball*. (See the essay at the end of this section.) Since then, scientists have discovered a series of related carbon-atom molecules, called *fullerenes*. Buckminsterfullerene has a soccer-ball shape, with carbon atoms enclosing a hollow space. Other fullerenes have tube shapes.

### Diamond

Diamond, because of its brilliance and sparkle, is well known as a gemstone. Early chemists, intrigued by its rarity as a natural material, hoped to produce diamond from graphite, the cheaper, black form of carbon. The higher density of diamond compared with that of graphite suggests (by applying Le Chatelier's principle) that high pressure would facilitate the transformation of graphite to diamond. ◄ In 1897, the French chemist Henri Moissan tried to achieve high pressures by quickly solidifying carbon in molten iron. He did find transparent crystals in the solid mixture, but these were

► Le Chatelier's principle was discussed in Section 12.3.

**Figure 13.19**
**Synthetic diamonds.**
The diamonds *(right)* were made by heating graphite *(left)* under high pressure with a catalyst. Pencil "lead" is graphite mixed with clay.

later shown not to be diamond. It was not until 1955 that scientists at General Electric succeeded in synthesizing diamond. Using a transition-metal catalyst, the process required temperatures of about 2000°C and pressure of about 100,000 atm. The synthetic diamonds produced by this process are generally very small and most useful as an abrasive, or grinding, material. The resulting "diamond grit" is used in drill bits and cutting wheels. Gem-quality diamonds have been made, but they cannot compete in price with natural diamonds. Figure 13.19 shows some synthetic diamonds.

More recent research into producing diamond from graphite has concentrated on the process of *chemical vapor deposition*. In general, in chemical vapor deposition, a chemical reaction in a gas or vapor produces a product that can be deposited onto a solid material, or substrate. One procedure for producing diamond films consists of passing a mixture of methane, $CH_4$, and hydrogen, $H_2$, over a hot filament to produce reactive species that deposit diamond onto a solid substrate (Figure 13.20). The methane provides the carbon atoms for the diamond film. Hydrogen gas, it is believed, forms hydrogen atoms on the hot filament, and these hydrogen atoms then play a pivotal role in the formation of diamond, in preference to graphite, which would normally be more stable under the moderate conditions of the process. The hydrogen atoms are thought to abstract hydrogen atoms from methane, forming reactive methyl

**Figure 13.20**
**Diamond-film preparation by chemical vapor deposition (CVD).**
Methane, $CH_4$, and hydrogen, $H_2$, react in the region of a hot filament to produce reactive species that deposit diamond on the substrate.

$CH_4 + H_2$

Hot filament

Substrate with diamond film

Gas out

radicals, $CH_3$, and hydrogen, $H_2$. The methyl radicals diffuse to the substrate surface where they form carbon–carbon bonds, ultimately yielding tiny crystals of diamond. Any graphite that happens to form on the surface of the substrate reacts quickly with hydrogen, and so is etched away, leaving the diamond crystals as a film.

One application of diamond films is as a coating on cutting tools, which depend on the superior hardness of diamond. Another application of diamond films depends on the high thermal conductivity of diamond, which can conduct heat away four times faster than copper metal. Diamond films are now being marketed as a base for microelectronic parts that generate heat. One should be able to pack the electronic components closer on a chip without overheating, resulting in less power loss and higher speeds.

Another application for diamond films depends on the possibility of producing the material for microelectronic components by building up layers of carbon atoms on a diamond film. Although diamond is an electrical insulator, like silicon and other materials, it becomes a conductor when small quantities of other substances, such as boron, are added to it. We say that the diamond has been "doped" and behaves as a *semiconductor*. (We will discuss semiconductors in Section 13.5.) In principle, diamond could supplant silicon as the material for constructing microelectronics devices, and theoretically these devices would be much faster than ones constructed from silicon.

## Graphite

The "lead" in a lead pencil is actually graphite, which occurs naturally as a mineral. It is the layer structure of graphite, with the weak attractive forces between those layers and the delocalized bonding electrons within each layer, that is responsible for the substance's use as the lead in pencils. When you smear graphite over a sheet of paper, you force some of the outer layers of the substance apart, where they adhere to the paper (Figure 13.21). The delocalized electrons, spread over a layer of graphite, give rise to many close energy levels, similar to those in a metal. These energy levels absorb all the wavelengths of visible light, rendering the material black, and giving it its needed contrast on light paper. ◄

▶ See the end of Section 7.3 for a discussion of the relationship between absorption of light and the color observed for a material.

The pencil no doubt has a place in the history of technology, but today most graphite is used for electrodes, for example, to produce aluminum. Although natural graphite is still mined as a material for lead pencils, most of the graphite of modern industry is manufactured by heating coke in an electric furnace to about 3000°C. Coke (which is itself made by heating coal) and other forms of carbon, such as carbon black and charcoal, have a structure that is similar to graphite, but with less order. At high temperatures, the carbon atoms migrate to give crystalline graphite.

**Figure 13.21**
**Writing with a lead pencil.**
When you draw a pencil lead over a sheet of paper, you force outer layers of graphite onto the paper. This depends on the fact that graphite has a layer structure, with weak attractive forces between layers, which can be easily separated.

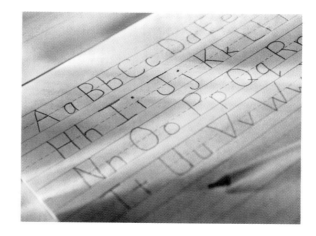

**Figure 13.22**
**Lithium-ion batteries.**
*Left:* Discharge of a lithium-ion battery. Initially, the battery was charged by connecting the two electrodes to an external voltage to draw lithium ions from the $LiCoO_2$ electrode to the graphite electrode, where lithium ions reside in the space between the graphite layers. During discharge, the lithium ions move from the graphite electrode back to the $LiCoO_2$ electrode, which the ions find energetically more favorable. Electrons move in the external circuit in the same direction as the lithium ions to maintain electrical balance in the battery. *Right:* Two commercial lithium-ion batteries.

**Figure 13.23**
**Artificial Heart Valve.**
The valve contains disks coated with pyrolytic carbon. (Pyrolytic carbon is produced by high-temperature decomposition of hydrocarbons; it has a disordered graphite structure.) Pyrolytic carbon gives a smooth surface that impedes clot formation.

The development of lightweight rechargeable batteries as sources of portable energy has become a high priority, and graphite has become one of the most studied materials for use as the negative electrode in these batteries. One of the most successful of these is the lithium-ion battery (Figure 13.22). The battery consists of a positive electrode of lithium cobalt oxide, $LiCoO_2$, or similar compound, with graphite as the other electrode. During charging of the battery, lithium ions ($Li^+$) are pulled by the application of an external voltage from the lithium compound toward the graphite, where the ions penetrate the layer structure. When the battery is discharged, the lithium ions flow from the graphite toward the lithium compound electrode. As an analogy, imagine that you pump water uphill to a reservoir ("charging the battery"). Later, you allow the water to flow downhill, perhaps flowing over a waterwheel, which does some useful work ("discharging the battery"). Electric-charge balance is maintained in the battery during discharge as electrons flow through an external circuit (giving a flow of electricity) from the graphite electrode to the lithium compound electrode. Lithium-ion batteries are lightweight and can go through many charge–discharge cycles. They are being used in cell phones and laptop computers, and they are being studied for use in electric vehicles.

Graphite and carbon materials similar in structure to graphite also have many other uses in modern technology. One of these is in the construction of artificial heart valves. The heart has four valves, which allow the blood to flow in only one direction. When a valve fails, the pumping action of the heart is impaired, and the valve must be surgically repaired or replaced. Artificial heart valves have been made of disks coated with pyrolytic carbon (Figure 13.23). Pyrolytic carbon, produced by high-temperature decomposition of hydrocarbons, has a disordered graphite structure; it gives a sturdy, smooth surface that impedes clot formation.

## Fullerenes and Nanotechnology

Graphite, as we have noted, consists of layers of carbon atoms arranged in hexagons. The *fullerenes* comprise a family of molecules each consisting of a closed cage of carbon atoms (instead of a flat layer of atoms) arranged in pentagons as well as hexagons. The most symmetrical fullerene is buckminsterfullerene, which consists of 12 pentagons and 20 hexagons of carbon atoms. (See the essay at the end of this section, which describes the discovery of buckminsterfullerene.) Other fullerenes consist of graphite-like sheets of carbon atoms rolled into tubes and capped at each end by half of a buckminsterfullerene molecule. These fullerenes can be extremely long tubes of carbon (consisting of a million or more carbon atoms) and are referred to as carbon *nanotubes.*

The unique characteristics of buckminsterfullerene, or buckyball, have sparked interest in possible applications of the substance. A novel feature of the molecule is that the carbon atoms of the cage structure contain just enough empty space to enclose an atom. For example, buckminsterfullerene has been prepared with a helium atom enclosed within the molecule. This must be the smallest helium balloon! Other atoms, including potassium have been encapsulated within the buckminsterfullerene molecule. Possibly, this ability of buckminsterfullerene could be used as a special storage device. Other fullerenes also have spaces within their cage structures, and encapsulation of substances within these materials is being investigated. Buckminsterfullerene also shows the ability to pick up atoms of metals within the spaces between molecules of the crystal. Buckminsterfullerene with potassium atoms in these spaces has shown the properties of a superconductor, in which the electrical resistance of the material drops to zero below about 18 K. (See the previous essay on superconductivity.) The fullerenes also display catalytic behavior. For example, the fullerenes have been shown to act as catalysts in converting methane to ethane and higher hydrocarbons. Thus, they might be useful in converting natural gas (mostly methane) to gasoline (a mixture of various hydrocarbons). Also, it is possible to attach transition-metal atoms (such as osmium) to buckminsterfullerene, and these derivatives might yield unique catalysts, since many transition metals are themselves known catalysts.

Carbon nanotubes, which as we said are giant tubes of carbon atoms, have become an especially active area of research, especially in **nanotechnology** *(an area of technology in which one manipulates materials on a molecular scale to create useful devices).* These nanotubes are especially strong and, depending on the structure, can be as electrically conducting as metals. Chemists at Harvard University have shown that they can make carbon nanotubes bend by applying a voltage to them. They have used this property to construct molecular-scale tweezers, as we mentioned in the chapter opening. Figure 13.24 shows a conceptual drawing of an array of crossed carbon nanotubes that could function as a computer memory device; theoretically it could hold tens of thousands more bits of information in the same space as present memory devices.

**Figure 13.24**
**Conceptual drawing of a carbon-nanotube electronic device.** The nanotubes are molecular-scale tubes of carbon atoms. The device consists of two sets of parallel nanotubes separated from each other by spacers. Each crossing point holds one bit of information (an on or off condition). In the on condition, the top tube is bent toward the bottom tube by applying voltage pulses to the crossing tubes. (Reprinted with permission from Charles M. Lieber *et al., Science,* Vol. 289, No. 94 (2000). Copyright © 2000 American Association for the Advancement of Science.)

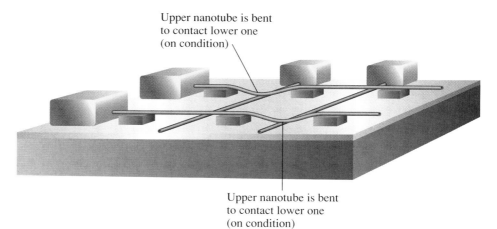

Upper nanotube is bent to contact lower one (on condition)

Upper nanotube is bent to contact lower one (on condition)

# Buckminsterfullerene—A Third Form of Carbon

**MATERIALS**

Until recently, carbon was thought to occur in only two principal forms: diamond and graphite, both network solids. In 1985, Harold W. Kroto (from the University of Sussex in Brighton, England) approached Richard E. Smalley and Robert F. Curl (at Rice University, in Houston, Texas) to do some experiments to simulate the conditions in certain stars to see what sorts of carbon-containing molecules might be produced. The research group at Rice had previously constructed an instrument in which they used an intense laser beam to vaporize solids. The hot vapor produced in this way could then be directed as a molecular beam into a mass spectrometer, where the molecular weights of the species in the vapor could be measured.

The experiments on the vaporization of graphite produced surprising results. Molecular clusters of 2 to 30 carbon atoms were found, as expected; but in addition, the mass spectrum of the vapor consistently showed the presence of a particularly abundant molecule, $C_{60}$. Why was this molecule so stable? Kroto, Smalley, and Curl wrestled with this problem and eventually came to the conclusion that the molecule must be like a piece of graphite sheet that somehow closed back on itself to form a closed-dome structure. Kroto recalled how he once built a cardboard dome with the night sky printed on it for his children; he thought it contained not only hexagons, as in graphite, but also pentagons. Smalley set about trying to construct a model of $C_{60}$ by gluing paper polygons together. He discovered that he could obtain a very stable, closed polygon with 60 vertices by starting with a pentagon and attaching hexagons to each of its five edges. To this bowl-shaped structure he attached more pentagons and hexagons, producing a paper soccer ball stable enough to bounce on the floor. The molecular structure is shown in Figure 13.25. Kroto and Smalley named the molecule buckminsterfullerene, after R. Buckminster Fuller, who studied closed-dome architectural structures constructed from polygons.

In 1990, buckminsterfullerene was prepared in gram quantities. Once the reddish-brown substance was available in sufficient quantity, researchers were able to verify the soccer-ball structure of the $C_{60}$ molecule.

In 1996, the Nobel Prize in chemistry was awarded to professors Curl, Kroto, and Smalley for their discovery.

**Figure 13.25**
**Structure of buckminsterfullerene.**
*Left:* A frame model of $C_{60}$. Carbon atoms are at the corners of each polygon. Each bond is intermediate between a single and a double bond, similar to the bonding in graphite. *Right:* The buckminsterfullerene molecule is often called a "buckyball," because of its soccer-ball appearance.

## 13.5 Semiconductors

Semiconductor materials are the basis of solid-state electronics devices, such as transistors, light-emitting diodes (LEDs), and laser diodes. A CD player, for example, uses transistors for its electronic circuits, as well as a laser diode to "read" a compact disc. (See the essay *Lasers and Compact Disc Players* at the end of Section 7.3.) The basic

**Figure 13.26**
**Effect of doping silicon.**
*(A)* Silicon doped with phosphorus. The extra electrons (denoted e⁻) from phosphorus atoms are free to conduct a current. *(B)* Silicon doped with boron. A bond with a missing electron is equivalent to a positively charged hole, indicated here as a plus sign in a circle.

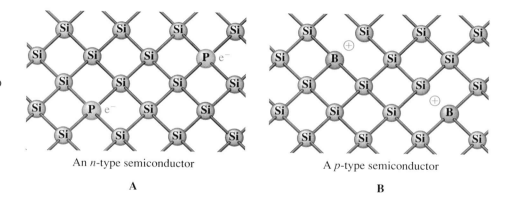

An *n*-type semiconductor

A *p*-type semiconductor

A

B

component of most of these electronic devices is silicon, a metalloid. A metalloid is a **semiconducting element,** which is *an element that has low electrical conductivity at room temperature, but increases in conductivity as the temperature increases.* (This contrasts with the behavior of metals, which are generally good electrical conductors, although their conductivity decreases with increasing temperature.) A striking property of a semiconducting element is that you can very much increase the electrical conductivity of the pure element by adding minute quantities of certain other elements to it, a process that is called *doping.*

To understand what happens when you dope a semiconducting element, let's consider the doping of silicon with phosphorus. You first start with very pure silicon. Phosphorus has five valence electrons, whereas silicon has only four. By doping silicon with phosphorus, you replace a few silicon atoms in the solid with phosphorus atoms. After each phosphorus atom bonds to four silicon atoms, an electron is left over (Figure 13.26*A*). The extra electrons in phosphorus-doped silicon are free to conduct an electric current, so the doped silicon becomes a good conductor. It is called an n-*type semiconductor,* because the current is carried by negative charges (electrons).

Another type of doped semiconductor is a p-*type semiconductor.* You make these by doping the semiconductor element with another element that contains fewer valence electrons. For example, suppose you dope silicon with boron, which has only three valence electrons compared with silicon's four valence electrons. By doping, you replace a few of the silicon atoms in the solid with boron atoms. After a boron atom forms bonds to four silicon atoms, one of the boron–silicon bonds has only one electron in it. In other words, the orbital has a vacancy for an electron. You can think of this vacancy as a positively charged "hole" in the bonding orbital (Figure 13.26*B*). An electron from a neighboring atom can move to occupy this hole. Now, a hole exists on the neighboring atom, and an electron from an atom next to it can move into that hole. This process can continue, and the result is the movement of positively charged holes in the semiconductor. The boron-doped silicon becomes a conductor, with the electrical charge being carried by positively charged holes, hence the name *p*-type semiconductor.

Often a solid-state electronic device depends on the electrical property of a p–n *junction.* A p–n junction consists of a *p*-type semiconductor joined to an *n*-type semiconductor. It can function as a *rectifier,* a device that allows current to flow in one direction but not the other. Imagine a p–n junction with wires connected to it from a source of direct current as shown in Figure 13.27, *top.* Positive holes are continuously generated in the *p*-type semiconductor by the positive wire, and electrons are released into the *n*-type semiconductor from the negative wire. The two charge

**Figure 13.27**
**A _p–n_ junction as a rectifier.** _Top:_ When a source of electricity is connected as shown, an electric current flows. Note the direction of flow of charge carriers (positive holes and electrons). _Bottom:_ When the wire connections are reversed, no current flows. Note the depletion of charge carriers near the junction of the two semiconductors.

_p_-type semiconductor    _n_-type semiconductor

_p_-type semiconductor    _n_-type semiconductor

**KEY**
(+) Positively charged hole     (−) Electron

**Figure 13.28**
**An integrated circuit chip.** The chip is just above the date on the penny. Its base material is silicon; miniature circuits with electronic components are constructed on the chip.

carriers (positive holes and electrons) move toward the junction, where they combine. The result is a continuous flow of current through the _p–n_ junction. Now suppose that the wire connections are reversed, as in Figure 13.27, _bottom._ Positive holes are attracted to the negative wire, and electrons are attracted to the positive wire. The region of the junction quickly becomes depleted in charge carriers, so the _p–n_ junction becomes nonconducting, and no current can flow.

By combining _p_-type and _n_-type semiconductors, you can construct various kinds of _transistors,_ devices for controlling electric signals. Some of the latest computer chips have microscopic electrical circuits integrated with as many as a million transistors per square centimeter of surface area (Figure 13.28).

# 13.6   Silicon, Silica, and Silicates

If we were to list the materials of technology, silicon would certainly be prominent on that list. As we mentioned in the previous section, silicon is the basic material in the semiconductor devices that make up CD players, computers, and other electronics gear. (Silicon Valley, an area in California that is headquarters for many computer companies, takes its name from the element silicon.) But oxygen compounds of silicon also are important to technology. Quartz crystals, a form of silica (also called silicon dioxide, $SiO_2$), for example, are used to control frequencies in radio transmitters and watches. And many practical materials such as cement and bricks are silicate materials. Silicates are compounds of silicon and oxygen with one or more metallic elements. In this section, we will look at silicon, silica, and silicates.

### Silicon

Silicon occurs in the earth's crust, the outermost solid layer of the planet, as compounds with oxygen. About 95% of the earth's crust is silica and silicate rocks and minerals. Elemental silicon is obtained by reducing quartz sand ($SiO_2$) with coke (C) in an electric furnace at 3000°C.

$$SiO_2(l) + 2C(s) \longrightarrow Si(l) + 2CO(g)$$

Silicon has a diamond-like structure (in which silicon atoms bond tetrahedrally to four other silicon atoms). It is a hard, lustrous gray solid and is used to make alloys and solid-state electronic devices.

For the manufacture of solid-state devices, it is necessary to start with extremely pure silicon (no more than $10^{-8}\%$ impurities). You first convert the impure element to silicon tetrachloride, Si, which is a low-boiling liquid (b.p. 58°C) that can be purified by distillation.

$$Si(s) + 2Cl_2(g) \longrightarrow SiCl_4(g)$$

You then reduce the purified silicon tetrachloride by passing the vapor with hydrogen through a hot tube, where pure silicon crystallizes on the surface of a pure silicon rod.

$$SiCl_4(g) + 2H_2(g) \longrightarrow Si(s) + 4HCl(g)$$

The silicon prepared in this process is polycrystalline (that is, it consists of a compact mass of many small crystals). In a final step, this polycrystalline silicon is fashioned into a rod that consists of a single crystal, from which high-purity silicon wafers can be cut (Figure 13.29).

**Figure 13.29**
**High-purity silicon rod and wafers cut from it.**
Silicon wafers form the base material for integrated circuit chips used in solid-state electronic devices.

**Figure 13.30**
**Structure of silica (SiO$_2$).**
*Left:* A silicon atom is bonded tetrahedrally to four oxygen atoms giving an SiO$_4$ tetrahedron. Each of the oxygen atoms on this tetrahedron bonds to silicon atoms on other tetrahedra, giving a three-dimensional structure. *Right:* Shown is a fragment of silica structure with three SiO$_4$ tetrahedra bonded together.

SiO$_4$ tetrahedron                Fragment of silica structure

## Silica (Silicon Dioxide)

Silica (whose formal chemical name is silicon dioxide, SiO$_2$) has several different crystalline forms. The most important of these is quartz, which is a constituent of many rocks. It is the weathering of these rocks that releases quartz particles, a major component of most kinds of sands. Amethyst is a gem form of quartz; it contains a small quantity of Fe$_2$O$_3$, which is believed to be responsible for its purple color. **Silica,** SiO$_2$, is *a covalent network solid in which each silicon atom is covalently bonded in tetrahedral directions to four oxygen atoms; each oxygen atom is in turn bonded to another silicon atom* (Figure 13.30).

Quartz crystals have a very interesting and useful property: they exhibit the *piezoelectric effect*. In a piezoelectric crystal, such as quartz, compression of the crystal in a particular direction causes an electric voltage to develop across it. Such crystals are used in phonograph pickups and microphones to convert sound vibrations to alternating electric currents. The opposite effect is also possible: an alternating electric current applied to a piezoelectric crystal can make it vibrate. When cut to precise dimensions, the crystal responds most strongly to a certain vibrational frequency. Such crystals are used to control the frequency of an alternating electric current. When the alternating current frequency deviates from the natural frequency of the crystal, a feedback mechanism adjusts the alternating current frequency. Quartz crystals are used to control radio and television frequencies, as well as clocks.

Quartz crystals are now produced synthetically to satisfy commercial demand. First, silica is dissolved under pressure in a hot, aqueous solution of sodium hydroxide; then a small "seed" crystal of quartz is added. As the solution cools, quartz deposits from the saturated solution onto the seed crystal to form a large crystal.

## Silicates

A **silicate** is *a compound of silicon and oxygen (with one or more metals) that may be formally regarded as a derivative of silicic acid, H$_4$SiO$_4$ or Si(OH)$_4$.* Silicic acid has never been isolated, although solutions containing anions of the acid are well known. When silica is melted with sodium carbonate, it forms a soluble material referred to commercially as *water glass.* These solutions contain various silicate ions, such as

$$H-O-\underset{\underset{\displaystyle O_-}{|}}{\overset{\overset{\displaystyle O^-}{|}}{Si}}-O-H$$

**Figure 13.31**
**Linking of two SiO$_4$ tetrahedra.**  The SiO$_4$ tetrahedra are linked through a common oxygen atom. Note the negative charges on the O atoms that are not shared by Si atoms.

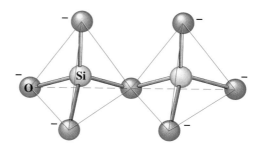

The solution also contains ions with two or more silicon atoms, which form by condensation reactions. A **condensation reaction** is *a reaction in which two molecules or ions are chemically joined by the elimination of a small molecule such as H$_2$O.* For example, the silicic acid anion can react in a condensation reaction with another such anion to form a disilicate anion (anion of two silicon atoms):

$$H-O-\underset{\underset{O_-}{|}}{\overset{\overset{O^-}{|}}{Si}}-O-H \;+\; H-O-\underset{\underset{O_-}{|}}{\overset{\overset{O^-}{|}}{Si}}-O-H \;\longrightarrow\; H-O-\underset{\underset{O_-}{|}}{\overset{\overset{O^-}{|}}{Si}}-O-\underset{\underset{O_-}{|}}{\overset{\overset{O^-}{|}}{Si}}-O-H \;+\; H_2O$$

Silicate anions containing long Si—O—S— . . . chains can result from such condensation reactions. Materials exhibiting chains and networks of this sort of silicon–oxygen bonding are common. The silicate minerals of the earth all have this type of structure.

An enormous variety of silicate minerals exists, but all of their structures have SiO$_4$ tetrahedra as basic units. A few minerals contain SiO$_4^{4-}$ as discrete ions; they are referred to as *orthosilicates*. The bonding in these minerals has considerable covalent character. Zircon, ZrSiO$_4$, is an example of such a silicate mineral; willemite, Zn$_2$SiO$_4$, is another. Zircon is the principal ore of zirconium metal. Because of the brilliance of zircon (a result of its high refractive index, or ability to bend light), colorless crystals of this mineral are used as a diamond-like gem.

Most silicate minerals have more complicated structures in which SiO$_4$ tetrahedra are linked to other SiO$_4$ tetrahedra through common oxygen atoms (Figure 13.31). In such structures, the O atoms attached to only one Si atom have a negative charge, giving silicate anions. These negative charges are balanced by metal atoms in silicate minerals. Beryl, for example, is a beryllium mineral with the formula Be$_3$Al$_2$(SiO$_3$)$_6$. The silicate anion in beryl consists of six SiO$_4$ tetrahedra formed into a cycle, as shown in Figure 13.32 (cyclic silicate anion). Emerald is gem-quality beryl having a deep green color that results from a small quantity of chromium in the crystal. In other silicate minerals, the SiO$_4$ tetrahedra are linked to form anions that are long chains or sheets (Figure 13.32, chain and double-chain silicate anions). Spodumene, LiAl(SiO$_3$)$_2$, an ore of lithium, is a long-chain silicate mineral.

In the aluminosilicate minerals, aluminum atoms substitute for some of the silicon atoms. An **aluminosilicate mineral** is *a mineral consisting of silicate sheets or three-dimensional networks in which some of the SiO$_4$ tetrahedra of a silicate structure have been replaced by AlO$_4$ tetrahedra.* If you imagine a three-dimensional structure such as quartz (SiO$_2$) with some of the SiO$_4$ tetrahedra replaced by AlO$_4$ tetrahedra, you obtain a three-dimensional network aluminosilicate. Because the nucleus of the aluminum atom has one less positive charge than that of the silicon atom, such minerals require another metal atom for charge balance. Orthoclase, which

**Figure 13.32**
**Structures of some silicate minerals.** A number of silicate minerals consist of finite silicate anions, such as $SiO_4^{4-}$ and $Si_2O_7^{6-}$. Beryl contains the cyclic anion $Si_6O_{18}^{12-}$. Other silicate minerals have anions with very long chains or double chains.

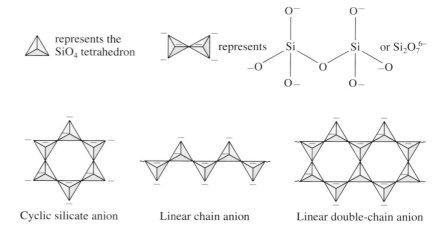

Cyclic silicate anion    Linear chain anion    Linear double-chain anion

occurs in granite, is an example of a three-dimensional network aluminosilicate. The empirical formula of the aluminosilicate anion is $AlSi_3O_8^-$; the formula of orthoclase is $KAlSi_3O_8$.

Figure 13.33 shows several silicate minerals. The one on the left is an example of asbestos, a class of fibrous minerals. In some cases, the fibrous structure results from long-chain silicate anions in the mineral. However, the common asbestos mineral chrysotile is actually a sheet silicate mineral in which the sheets tend to break up into long pieces that roll up into fine tubes. Mica is a sheet silicate mineral that tends to break easily into visible pieces of mineral sheets (Figure 13.33, center). Figure 13.33 (right) shows the three-dimensional aluminosilicate orthoclase, a very hard mineral.

**Figure 13.33**
**Some silicate minerals.** *Left:* Asbestos; note the fibers on its surface. *Center:* Mica; this mineral cleaves easily into sheets. *Right:* Orthoclase, a feldspar mineral; this is a three-dimensional network silicate.

**CONCEPT CHECK 13.3**

For certain chemical reactions, quartz containers are used instead of glass. What would the meniscus of water look like in a quartz test tube? Give an explanation for your answer.

# Silica Aerogels, the Lightest "Solids"

**MATERIALS**

The block with a hazy blue color sitting on the table looks like frozen smoke (Figure 13.34). Though it appears insubstantial, it is fairly robust: you can pick it up and saw it in two. The block is a silica aerogel. Aerogels are the lightest "solid" materials; silica aerogel, for example, has been made with a density that is only about three times that of air. The aerogels are not true solids, however. Rather, they are highly porous networks of a solid generally filled with air. (They are colloids, which we discussed in Section 12.9.) Some aerogels in which the solid structure is silica are 99% air by mass. It is also possible to evacuate the material to create a vacuum within the pores. Ninety percent or more of the air can be removed to give a material that is basically a solid.

The silica aerogels excel as heat insulators, exceeding the insulating ability of polyurethane foams presently used in refrigerators and other appliances (see Figure 13.35). Evacuating the silica aerogel yields a material with the insulating ability of a conventional vacuum bottle. A window made of a sandwich of evacuated silica aerogel between two glass panes could have great potential. Although the silica aerogel would give the window a slightly frosted appearance, this could work well for many applications, say, skylights. Future research might also show us how to produce the aerogel with a more uniform pore structure that would scatter less light and eliminate this frostiness.

How do you make an aerogel? Let's look first at a popular *gel*, Jell-O™. A gel is a liquid-containing network of solid material or macromolecular substance (that is, one composed of very large molecules). Thus, it has a porous structure with the pores filled by liquid. In the case of Jell-O, the network is of the protein gelatin, a macromolecular substance, and the liquid is water, with color and flavorings added. If you leave Jell-O on your counter for an extended period, it shrinks in volume as water evaporates. The shrinkage occurs because of a meniscus (boundary) that forms between water and air. As the water decreases in volume, surface tension (forces) at the meniscus pulls inward on the pore surface through intermolecular forces (mostly hydrogen bonding forces) and the gel structure collapses. Is it possible to remove the liquid in a gel and replace it with air without affecting the pore structure

**Figure 13.34**
**A silica aerogel.** The aerogel is transparent, though it does scatter blue light, giving it a hazy appearance.

of the solid or macromolecular material of the gel? In the 1930s, Steven Kissler at Stanford University discovered how to do this and in the process produced the first aerogel.

Kissler first produced a gel composed of silica in water. He reasoned that if he could eliminate the meniscus that forms as the liquid evaporates, then he could keep the original pore structure of the material and create an aerogel. His solution was to subject the gel to a high enough temperature and pressure to change the liquid to a *supercritical fluid* (the fluid state above the critical point; see the end of Section 11.3.) Above the critical point, there is no distinction between liquid and gas, so there is no meniscus, and the fluid can evaporate without destroying the porous structure of the original gel. The result is a porous structure filled with gas, or aerogel. Kissler's process for making an aerogel, however, was complicated and expensive. More recent research has shown how to simplify the process and make silica aerogel commercially feasible. Research in this area continues and aerogels have now been made from many materials other than silica with properties that may extend the usefulness of aerogels to applications other than as insulators.

*(continued)*

**Figure 13.35**
**Silica aerogel is an insulator.** A gas torch is directed toward a block of silica aerogel. The person doesn't feel any heat on his finger because of the excellent insulating property of the aerogel.

# 13.7 Ceramics

The word *ceramics* comes from a Greek word for pottery, which are objects made by firing clay. Although pottery making is an ancient skill, it is one that continues to flourish today. Fired clay is a major medium for producing objects of art, and fired clay objects including tiles, bricks, and dinnerware are major materials of today's commerce. However, modern research has increased in number the kinds of materials we can make that are chemically similar to these traditional ceramics. Consequently, a broader definition has emerged: **ceramics** are *nonmetallic, inorganic solids that are hard and brittle and usually produced at elevated temperature.* This definition would include glass, although glass is sometimes treated as a separate topic. In this section, we will look first at the traditional ceramics, then describe some recent developments in ceramic materials.

## Traditional Ceramics and Glass

Clay is a natural, earthy mixture of very small crystals of certain silicate sheet minerals. These minerals form by the weathering of granite, the rock that composes the backbone of mountain ranges. Clay minerals easily adsorb water, and wet clay is moldable. The wet platelike crystals adhere to one another to give a *plastic,* or easily deformable, mass. It is this plasticity of wet clay that allows the potter to form useful and artistic shapes.

Once the potter has shaped the wet clay into a vase or other object, he or she allows the clay to dry and then fires it. During the elevated temperatures of the firing, complex irreversible changes occur that result first in a hard, porous material. As the temperature rises, some of the more easily fusible clay minerals begin to melt, forming a glass. The surface tension of the melt draws the liquid and crystals together so that, on cooling, the solid ceramic is more dense than clay fired at lower temperatures and it is no longer porous. It consists of particles suspended in a glass matrix.

Most pottery is *glazed.* After firing, the ceramic object is coated with a water suspension of silica and other oxides; then it is fired again. The oxides melt and react to form a glassy coating, or glaze. In the case of earthenware, the glaze gives the ceramic a nonporous covering. For most pots, glazing simply provides a decorative shell.

Figure 13.36 shows a porcelain vase in which the glaze displays beautiful needle-like crystal formations. The artist carefully formulated a glaze mixture of silica and zinc oxide, in addition to other oxides. He then fired the vase with its glaze, carefully controlling the temperature, until only a few microscopic crystals of zinc oxide remained. These microscopic crystals formed the nuclei on which zinc orthosilicate crystallized.

Pottery is an ancient art, going back thousands of years. Glassmaking is a newer art, dating from about 2500 B.C. Glass is an amorphous, or noncrystalline, solid. When a pure liquid substance freezes, it usually forms a crystalline solid. It is often possible, however, to cool a liquid below its equilibrium freezing point to give a *supercooled* liquid. The term *glass,* in the broadest sense, refers to any supercooled liquid whose viscosity is so high that it has the properties of a solid. Ordinary glass is a silicate produced by fusing silica ($SiO_2$) with other oxides (or substances such as carbonates that yield oxides when heated). Common glass (called soda-lime glass) contains sodium and calcium oxides, in addition to silica.

## Some Recent Ceramic Materials

Research in ceramics has resulted in many new kinds of materials with diverse applications. For example, ceramics and glass have been joined in the production of glass–ceramics. These are made by adding materials to a glass to introduce many crystal nuclei into it. The glass–ceramic stovetop pots shown in Figure 13.37 were made from a lithium aluminosilicate glass to which titanium and zirconium oxides were added. Small crystals of zirconium titanate precipitate in the molten glass and form nuclei on which lithium aluminosilicate crystallizes. The resulting glass–ceramic has a very low thermal expansion, so it does not break easily when subjected to large temperature changes. It remains transparent because the crystals are extremely small and their refractive index is close to that of the glass.

Silicon carbide, SiC, is a ceramic material widely used as an abrasive in grinding wheels and cutting tools. Silicon nitride, $Si_3N_4$, is a newer ceramic. It is available in several crystalline forms, and recently researchers have discovered a high-pressure form that is especially hard. Because of its hardness, silicon nitride has been used in special cutting tools. It also shows some promise as a material for high-temperature engine components. Ceramics have recently been used to make medical

**Figure 13.36**
**A vase with a crystalline glaze.** The crystals are zinc orthosilicate. Artist Stuart Gray has carefully experimented with glaze compositions and firing temperatures to obtain the effect shown.

**Figure 13.37**
**Glass–ceramic cookware.** This transparent glass–ceramic cookware is made by Corning, Inc.

implants and bone replacement materials. High-purity alumina, $Al_2O_3$, is nearly inert in the human body, so its high strength and wear resistance makes it useful for hip replacement devices (Figure 13.38). Although porcelain ceramics have traditionally been used as electrical insulators, some of the newer ceramics are electrical conductors. They are used, for example, in electrodes and heating elements. The ceramic superconductors mentioned in the essay on page 568 are extreme examples of electrical conductors. At low enough temperatures, these ceramic materials lose all electrical resistance.

## 13.8 Composites

A **composite,** as we mentioned in the chapter opening, is a *material constructed of two or more different kinds of materials.* We mentioned bone as a natural composite of calcium phosphate crystals and a protein collagen. Bone, as a composite, is much stronger than either of its components and less subject to fracture than the calcium phosphate structure would be by itself. An example of a commercial composite is epoxy plastic reinforced with carbon fibers; it is used in aircraft parts, golf clubs, fishing rods, and so forth. The carbon fibers are made from textile fibers by pyrolysis (heating the textile until it decomposes). At high temperatures, the carbon fibers become graphitized, or transformed to a more or less crystalline graphite structure. These carbon fibers have great strength (Figure 13.39).

Composites frequently consist of a material with embedded fibers, which in the example just given is epoxy plastic with embedded carbon fibers. The epoxy plastic is strong but subject to fracture. The embedded fibers impede a fracture that might start, thus enhancing the stability of the structure. Another example of a composite is silicon carbide ceramic with embedded silicon carbide fibers. These fibers are made by pyrolysis of a polymer of dimethylsilane, $(CH_3)_2SiH_2$. By itself, silicon carbide ceramic is brittle, like most ceramics. The silicon carbide fibers, however, break up any fracture lines that may form, making the composite much less brittle.

**Figure 13.38**
**Hip Replacement Device.**
Shown here is a hip replacement device with an alumina ceramic joint head.

**Figure 13.39**
**Carbon fibers.** Carbon fibers are made by pyrolyzing textile fibers. They are used to make composites.

# A Checklist for Review

## Important Terms

metal (p. 554)
alloy (p. 554)
mineral (13.1)
ore (13.1)
metallurgy (13.2)
gangue (13.2)

flotation (13.2)
Bayer process (13.2)
roasting (13.2)
Dow process (13.2)
Hall−Héroult process (13.2)
Mond process (13.2)

nanotechnology (13.4)
semiconducting
  element (13.5)
silica (13.6)
silicate (13.6)
condensation reaction (13.6)

aluminosilicate
  mineral (13.6)
ceramics (13.7)
composite (13.8)

## Summary of Facts and Concepts

*Metals* may be pure elements, or they may be *alloys,* which can be either compounds or mixtures. Metallic elements and their compounds are obtained principally from the earth's crust, most of which is composed of metal silicates. The chief sources of metals, however, are not silicates but oxide, carbonate, and sulfide *minerals,* which exist in *ore* deposits widely scattered over the earth.

An important part of *metallurgy* is the production of metals from their ores. This involves three basic steps: *preliminary treatment, reduction,* and *refining.* In preliminary treatment, the ore is concentrated in its metal-containing mineral. The concentrated ore may also require a process such as *roasting,* in which the metal compound is transformed to one that is more easily reduced. In reduction, the metal compound is reduced to the metal, by either electrolysis or chemical reduction. In refining, the metal is purified, or freed of contaminants.

The *electron-sea model* is a simple depiction of a metal as an array of positive ions surrounded by delocalized valence electrons. Molecular orbital theory gives a more detailed picture of the bonding in metals. Because the energy levels in a metal crowd together into bands, this picture of metal bonding is often referred to as *band theory.* According to band theory, the electrons in a crystal become free to move when they are excited to the unoccupied orbitals of a band. In a metal, this requires very little energy, because unoccupied orbitals lie just above the occupied orbitals of highest energy.

Carbon is an important nonmetallic material. It has several allotropes, including diamond, graphite, and the fullerenes. *Diamond* has a three-dimensional network structure in which each carbon atom bonds to four others with $sp^3$ orbitals. Scientists first synthesized diamond from graphite in 1955 using high temperature and pressure. The process produces diamond grit, which is used for drill bits and cutting wheels. More recently, researchers have made diamond films from graphite using *chemical vapor deposition. Graphite* has a layer structure, in which the carbon atoms in each layer bond to three other atoms with $sp^2$ orbitals. Most commercial graphite is produced by heating coke in an electric furnace; it is used to make electrodes. Graphite forms one of the electrodes in the new lithium-ion battery. The *fullerenes* comprise a family of molecules with a closed cage of carbon atoms arranged in pentagons and hexagons. The most symmetrical member is *buckminster-fullerene,* $C_{60}$. Some of the fullerenes consist of tubes of carbon atoms, each end of a tube being capped by half of a buckminsterfullerene molecule; these fullerene tube molecules are referred to as carbon *nanotubes.*

*Semiconducting elements* form the basis of solid-state electronics devices. A striking property of these elements is that they experience large increases in conductivity when *doped* with small quantities of certain other elements. When silicon is doped with phosphorus, it becomes an n-*type semiconductor,* in which an electric current is carried by electrons. However, when silicon is doped with boron, it becomes a p-*type semiconductor,* in which an electric current is carried by positively charged holes. Joining a *p*-type semiconductor to an *n*-type semiconductor gives a *p*−*n* junction, which can function as a *rectifier.*

Silicon is a prominent material of technology. The element is the basic material used in semiconductor devices for solid-state electronics. These devices require silicon of extreme purity. The impure element is converted to silicon tetrachloride, which is purified by distillation, after which the silicon tetrachloride is reduced to pure silicon. *Silica* is chemically known as silicon dioxide; quartz is a common form of silica. It exhibits the *piezoelectric effect,* which is used to control radio, television, and clock frequencies. *Silicates* are formed by *condensation reactions;* an enormous variety of silicate minerals exists. The *aluminosilicate* minerals are silicates in which some $SiO_4$ tetrahedra are replaced by $AlO_4$ tetrahedra.

The word *ceramics* comes from a Greek word for pottery. Traditional ceramics production begins by forming wet clay, then firing it to make a ceramic object such as a pot or sculpture. Glass is an amorphous, or noncrystalline, solid produced by fusing silica with other oxides. Ceramics have been extended to include materials other than fired clay and glass. Examples are the glass−ceramics, used in making stovetop pots, silicon carbide (used as an abrasive), silicon nitride (used in making cutting tools and engine components), and alumina (for replacement of hip joints).

*Composites* are constructed of two or more materials. An example is the composite made of epoxy plastic with embedded carbon fibers. The fibers impede any fracture that might start, thus enhancing the stability of the material.

## Operational Skills

*Note:* The problem-solving skills used in this chapter are discussed in previous chapters.

## Review Questions

**13.1**    What is an alloy? Give an example.

**13.2**    Give four characteristics of a metal.

**13.3**    Define *mineral, rock,* and *ore.* Is bauxite a mineral or a rock? Explain.

**13.4**    Ores are usually metal compounds. What kinds of metal compounds most commonly occur in metal ores?

**13.5**    What are the basic steps in the production of a pure metal from a natural source? Illustrate each step with the preparation of aluminum.

**13.6**    How does flotation separate an ore mineral from impurities?

**13.7**    What is the purpose of roasting zinc ore?

**13.8**    Carbon is often used as a cheap reducing agent, but it is sometimes unsuitable. Give an example in which another reducing agent is preferred.

**13.9**    Briefly describe three different methods for refining a metal.

**13.10**    Write equations for the preparation of magnesium metal from seawater, limestone, and hydrochloric acid.

**13.11**    Draw a flowchart for the preparation of aluminum from its ore. Label each step.

**13.12**    Explain metallic conduction in terms of molecular orbital theory (band theory).

**13.13**    Describe the formation of an energy band in a metal. Assume that the metal is formed by bringing metal atoms together one at a time.

**13.14**    Briefly describe three allotropes of carbon.

**13.15**    Describe two methods of preparing diamond from graphite. What are uses or possible uses of this synthetic diamond?

**13.16**    Describe some commercial uses of graphite.

**13.17**    Describe some possible applications of buckminsterfullerene or of the other fullerenes.

**13.18**    Explain the difference between an *n*-type and a *p*-type semiconductor.

**13.19**    Describe the steps in preparing ultrapure silicon from quartz sand.

**13.20**    Explain how a quartz crystal is used to regulate an alternating current frequency.

**13.21**    Use structural formulas to illustrate the hypothetical condensation reaction of two silicic acid molecules, $Si(OH)_4$.

**13.22**    Describe the structure of the lithium ore spodumene, $LiAl(SiO_3)_2$.

**13.23**    What is an aluminosilicate mineral? Give the name and formula of such a mineral.

**13.24**    Define the term *ceramics*. Give examples of some commercial applications of clay ceramics.

**13.25**    Give some uses or potential uses of the newer (nonclay) ceramics.

**13.26**    Describe the production of glass.

**13.27**    What is a glass–ceramic? Describe an example and give an application.

**13.28**    What is a *composite?* Give an example of a natural composite and an example of a human-made composite.

## Conceptual Problems

**13.29**    Unless zinc is purified, cadmium is normally an impurity in the metal. Why might you expect this to be the case?

**13.30**    Electrolysis is used to obtain some metals from their compounds. List some metals obtained this way.

**13.31**    Aluminum is the third most abundant element (first most abundant metal) in the earth's crust. Does this mean that aluminum ores are widespread and plentiful? Explain.

**13.32**    The text says that the higher density of diamond compared with graphite suggests that the application of high pressure would facilitate the transformation of graphite to diamond. Explain the reasoning behind this statement.

**13.33**    Diamond is an insulator, but when small amounts of boron are added it becomes a conductor. What is the explanation for this change in conduction? What are the electric current carriers?

**13.34**    Is a quartz crystal $(SiO_2)$ a mineral or a rock?

**13.35**    Orthoclase feldspar has the chemical formula $KAlSi_3O_8$. What other cations could replace those in this chemical formula? Write the resulting formula for two such minerals.

**13.36**    Cutting wheels have been made from alumina containing fine fibers of silicon carbide. How do the silicon carbide fibers help alumina in this application?

# Practice Problems

## Metallurgy

**13.37**  Lead metal is purified by electrolysis in a manner similar to that for copper; the electrolyte is lead(II) hexafluorosilicate, $PbSiF_6$. Describe the process in more detail, giving the half-reactions.

**13.38**  Nickel metal can be purified by electrolysis in a manner similar to that for copper; the electrolyte is nickel(II) chloride or sulfate. Describe the process in more detail, giving the half-reactions.

**13.39**  Pure iron is prepared for special purposes by precipitating iron(III) oxide and reducing the dry oxide with hydrogen gas. Write the balanced equation for this reduction of iron(III) oxide.

**13.40**  Manganese is available from its ore pyrolusite, which is manganese(IV) oxide, $MnO_2$. Some manganese metal is prepared from $MnO_2$ by reduction with aluminum metal. Write the balanced equation for this reduction.

**13.41**  How many kilograms of iron can be produced from 2.00 kg of hydrogen, $H_2$, when you reduce iron(III) oxide? (See Problem 13.39.)

**13.42**  How many kilograms of manganese can be produced from 1.00 kg of aluminum, Al, when you reduce manganese(IV) oxide? (See Problem 13.40.)

**13.43**  Using thermodynamic data given in Appendix C, obtain $\Delta H°$ for the roasting of galena, PbS:

$$PbS(s) + \tfrac{3}{2}O_2(g) \longrightarrow PbO(s) + SO_2(g)$$

Is the reaction endothermic or exothermic?

**13.44**  Using thermodynamic data given in Appendix C, obtain $\Delta H°$ for the reduction of pyrolusite, $MnO_2$, by aluminum (see Problem 13.40). Is this reaction endothermic or exothermic?

## Bonding in Metals

**13.45**  Sketch a diagram showing the formation of energy levels from the valence orbitals for K, $K_2$, $K_3$, and $K_n$. On the diagram, place arrows indicating how the electrons fill these energy levels.

**13.46**  Sketch a diagram showing the formation of energy levels from the valence orbitals for Ca, $Ca_2$, $Ca_3$, and $Ca_n$. On the diagram, place arrows indicating how the electrons fill these energy levels.

**13.47**  How many energy levels are there in the valence band of a single crystal of sodium with a mass of 1.00 mg?

**13.48**  How many energy levels are there in the $3p$ band of a single crystal of magnesium with a mass of 1.00 mg?

## Diamond, Graphite, the Fullerenes, and Nanotechnology

**13.49**  Look up the densities of graphite and diamond in a handbook, and compare the values for these allotropes of carbon. According to Le Chatelier's principle, which of these allotropes should be the stabler form at high pressure?

**13.50**  Using data for enthalpies of formation, calculate the standard change of enthalpy when graphite reacts with $O_2$ to form $CO_2$. Do the same calculation but for the reaction of diamond with $O_2$. Which of these values corresponds to the standard enthalpy of formation of $CO_2$?

**13.51**  A mixture of methane, $CH_4$, and hydrogen, $H_2$, is passed over a hot filament to provide a gas containing carbon atoms. If all of the carbon atoms in the methane are converted to diamond film, what volume of methane at STP is needed to produce 1.00 mg of diamond film?

**13.52**  A mixture of ethane, $C_2H_6$, and hydrogen, $H_2$, is passed over a hot filament to provide a gas containing carbon atoms. If all of the carbon atoms in the ethane are converted to diamond film, what volume of methane at STP is needed to produce 1.00 mg of diamond film?

**13.53**  Draw a portion of the structure of the buckminsterfullerene molecule, say, a hexagon and a pentagon of carbon atoms, using one resonance form. Draw another resonance form. Describe the bonding at the carbon atoms in terms of hybrid orbitals.

**13.54**  Draw a portion of the structure of a fullerene nanotube molecule, say, several hexagons, using one resonance form. Draw another resonance form. Describe the bonding at the carbon atoms in terms of hybrid orbitals.

## Semiconductors

**13.55**  Which of the following are *n*-type semiconductors?
a. C(diamond) doped with B
b. Ge doped with As
c. Si doped with Al
d. Ge doped with P

**13.56**  Which of the following are *p*-type semiconductors?
a. C(diamond) doped with B
b. Ge doped with As
c. Si doped with Al
d. Ge doped with P

## Silicon, Silica, and Silicates

**13.57**  Silicon has a diamond-like structure, with each Si atom bonded to four other Si atoms. Describe the bonding in silicon in terms of hybrid orbitals.

**13.58** Silicon carbide, SiC, has a structure in which each Si atom is bonded to four C atoms, and each C atom is bonded to four Si atoms. Describe the bonding in terms of hybrid orbitals.

**13.59** Silicon is prepared by reducing silicon dioxide with carbon at high temperature. How many kilograms of silicon dioxide are required to prepare 5.00 kg of silicon?

**13.60** Silicon is purified by first converting it to silicon tetrachloride, which is then distilled. How many kilograms of silicon are required to prepare 5.00 kg of $SiCl_4$?

**13.61** Draw a portion of the structure of the mineral spodumene, $LiAl(SiO_3)_2$, which is a mineral consisting of a long chain of $SiO_4$ tetrahedra. Use this drawing to verify the empirical formula given.

**13.62** Draw a portion of the structure of the mineral beryl, $Be_3Al_2(SiO_3)_6$, which is a mineral consisting of cyclic silicate anions. Use this drawing to verify the empirical formula given.

## General Problems

**13.63** Zinc is obtained from the sulfide mineral sphalerite, ZnS. First the mineral is roasted to the oxide, which is then reduced with carbon in a blast furnace. Write chemical equations for this process.

**13.64** Lead is obtained from the sulfide mineral galena, PbS. First the mineral is roasted to the oxide, which is then reduced with carbon in a blast furnace. Write chemical equations for this process.

**13.65** A sample of zinc ore contains 87.0% sphalerite, ZnS. How many metric tons of Zn can be obtained from 1.00 metric ton of the zinc ore? (See Problem 13.63 for a discussion of the process.)

**13.66** A sample of lead ore contains 78.0% galena, PbS. How many metric tons of Pb can be obtained from 1.00 metric ton of the lead ore? (See Problem 13.64 for a discussion of the process.)

**13.67** Compare the structure of graphite with that of the buckminsterfullerene. In what ways are they similar? In what ways are they different?

**13.68** Compare the structures of diamond and graphite. How do the structures of these substances explain the hardness of diamond and the electrical conductivity of graphite?

**13.69** Calculate the standard enthalpy change for the reaction of silicon dioxide with carbon to produce silicon:

$$SiO_2(s) + 2C(s) \longrightarrow Si(s) + 2CO(g)$$

**13.70** Calculate the standard enthalpy change for the reaction of zinc sulfide with oxygen to produce zinc oxide:

$$2ZnS(s) + 3O_2 \longrightarrow 2ZnO(s) + 2SO_2(g)$$

## Media Activities

**13.71** Open the Molecule Library on the student web site and look at the models of copper and iron.
a. Do copper and iron have the same packing arrangements of the metal atoms?
b. The unit cell of iron is body-centered cubic. How many atoms are there in this unit cell?

**13.72** Run the Classification of Solids *Understanding Concepts* activity on the student CD. How does the electron-sea model of metals explain the fact that metals generally do not shatter when subjected to blows? What properties of metals can be attributed to the electron-sea model?

**13.73** View the reaction of Cu with Nitric Acid *Visualization* on the student CD. In this *Visualization,* the copper is re-

acting with the nitrate, $NO_3^-$, anion to produce $NO_2$ and $Cu^{2+}$. Which species is undergoing oxidation, and which is being reduced?

**13.74** Open the Molecule Library on the student web site and look at the models of graphite and diamond.
a. Are the diamond and graphite structures the same?
b. Use the structures to explain the fact that diamonds are extremely hard and graphite is a lubricant.
c. Describe, on a molecular level, what happens when a pencil lead (graphite) is used to write on a piece of paper.

**13.75** For review, take the ACE practice test for Chapter 13 on the student web site.

Once started, the reaction of aluminum and iron(III) oxide is fast.

# Rates of Reaction

Chemical reactions require varying lengths of time for completion, depending on the characteristics of the reactants and products and the conditions under which the reaction is run (Figure 14.1). Many reactions are over in a fraction of a second, whereas others can take much longer. If you add barium ion to an aqueous solution of sulfate ion, a precipitate of barium sulfate forms almost immediately. On the other hand, the reactions that occur in a cement mixture as it hardens to concrete require years for completion.

The study of the rate, or speed, of a reaction has important applications. In the manufacture of ammonia from nitrogen and hydrogen, you may wish to know what conditions will help the reaction proceed in a commercially feasible length of time. You may also wish to know whether the nitric oxide, NO, in the exhaust gases of supersonic transports will destroy the ozone in the stratosphere faster than the ozone is produced. Answering these questions requires knowledge of the rates of reactions.

## CONTENTS

**Reaction Rates**
14.1 Definition of Reaction Rate
14.2 Experimental Determination of Rate
14.3 Dependence of Rate on Concentration
14.4 Change of Concentration with Time
14.5 Temperature and Rate; Collision and Transition-State Theories
14.6 Arrhenius Equation
**Reaction Mechanisms**
14.7 Elementary Reactions
14.8 The Rate Law and the Mechanism
14.9 Catalysis

**589**

**Figure 14.1**
**Reaction time.** A chemical reaction takes time. Some reactions are complete in a fraction of a second; others take years. The reaction shown here is complete in less than a minute. The reaction is the formation of a product of formaldehyde with hydrogen sulfite ion: $H_2C{=}O(aq)$ + $HSO_3^-(aq) \longrightarrow$ $H_2C(OH)SO_3^-(aq)$. As the hydrogen sulfite ion is used up, the solution becomes less acidic, then changes to basic. Bromthymol-blue indicator marks the change from acidic to basic by changing from yellow to blue.

▶ A knowledge of kinetics gives the student insight into chemical reactions. However, you can abbreviate the discussion, say, by limiting the treatment to Sections 14.1, 14.3, 14.5, 14.7, and 14.9.

▶ Demonstration of the formaldehyde clock reaction: Summerlin et al., *Chemical Demonstrations*, Vol. 2 (Washington, D.C.: Am. Chem. Soc., 1987), no. 86, pp. 160–161.

# Reaction Rates

▶ Video no. 21 (Series A), "Chemical Kinetics: An Iodine Clock Reaction."

▶ Video Series C, "Oscillating Reaction."

Another reason for studying reaction rates is to understand how chemical reactions occur. By noting how the rate of a reaction is affected by changing conditions, you can sometimes learn the details of what is happening at the molecular level.

A reaction whose rate has been extensively studied under various conditions is the decomposition of dinitrogen pentoxide, $N_2O_5$. When this substance is heated in the gas phase, it decomposes to nitrogen dioxide and oxygen:

$$2N_2O_5(g) \longrightarrow 4NO_2(g) + O_2(g)$$

We will look at this reaction in some detail. The questions we will pose include: How is the rate of a reaction like this measured? What conditions affect the rate of a reaction? How do you express the relationship of rate of a reaction to the variables that affect rate? What happens at the molecular level when $N_2O_5$ decomposes to $NO_2$ and $O_2$?

*Chemical kinetics* is the study of reaction rates, how reaction rates change under varying conditions, and what molecular events occur during the overall reaction. In the first part of this chapter, we will look at reaction rates and the variables that affect them.

What variables affect reaction rates? As we noted in the chapter opening, the rate depends on the characteristics of the reactants in a particular reaction. Some reactions are fast and others are slow, but the rate of any given reaction may be affected by the following factors:

1. *Concentrations of reactants.* Often the rate of reaction increases when the concentration of a reactant is increased. A piece of steel wool burns with some difficulty in air (20% $O_2$) but bursts into a dazzling white flame in pure oxygen. The rate of burning increases with the concentration of $O_2$. In some reactions, however, the rate is unaffected by the concentration of a particular reactant, as long as it is present at some concentration.

▶ Video no. 28 (Series D), "Briggs-Rauscher" (oscillating reactions).

▶ Video no. 13 (Series B), "Dust Explosions."

**Figure 14.2**
**Catalytic decomposition of hydrogen peroxide.** The hydrogen peroxide decomposes rapidly when hydrobromic acid is added to an aqueous solution. One of the products is oxygen gas, which bubbles vigorously from the solution. In addition, some HBr is oxidized to $Br_2$, as can be seen from the red color of the liquid and vapor.

**Figure 14.3**
**Effect of large surface area on the rate of reaction.** Lycopodium powder (from the tiny spores of a club moss) ignites easily to produce a yellow flame. The powder has a large surface area per volume and burns rapidly in air.

2. *Concentration of catalyst.* A **catalyst** is *a substance that increases the rate of reaction without being consumed in the overall reaction.* Because the catalyst is not consumed by the reaction, it does not appear in the balanced chemical equation (although its presence may be indicated by writing its formula over the arrow). A solution of pure hydrogen peroxide, $H_2O_2$, is stable, but when hydrobromic acid, $HBr(aq)$, is added, $H_2O_2$ decomposes rapidly into $H_2O$ and $O_2$ (Figure 14.2).

$$2H_2O_2(aq) \xrightarrow{\text{HBr}(aq)} 2H_2O(l) + O_2(g)$$

Here HBr acts as a catalyst to speed decomposition.

3. *Temperature at which the reaction occurs.* Usually reactions speed up when the temperature increases. It takes less time to boil an egg at sea level than on a mountaintop, where water boils at a lower temperature. Reactions during cooking go faster at higher temperature.

4. *Surface area of a solid reactant or catalyst.* If a reaction involves a solid with a gas or liquid, the surface area of the solid affects the reaction rate. Because the reaction occurs at the surface of the solid, the rate increases with increasing surface area. A wood fire burns faster if the logs are chopped into smaller pieces. Similarly, the surface area of a solid catalyst is important to the rate of reaction. The greater the surface area per unit volume, the faster is the reaction (Figure 14.3).

In the first sections of this chapter, we will look primarily at reactions in the gas phase and in liquid solution where factors 1 to 3 are important. Thus, we will look at the effect of *concentrations* and *temperature* on reaction rates. Before we can explore these factors, however, we need a precise definition of reaction rate.

## 14.1 Definition of Reaction Rate

▶ Demonstration of explosion with lycopodium powder: Bassam Z. Shakhashiri, *Chemical Demonstrations,* Vol. 1 (Madison: University of Wisconsin Press, 1983), no. 1.41, pp. 103–105.

The rate of a reaction is the amount of product formed or the amount of reactant used up per unit of time. So that a rate calculation does not depend on the total quantity of reaction mixture used, you express the rate for a unit volume of the mixture. Therefore, the **reaction rate** is *the increase in molar concentration of product of a reaction per unit time or the decrease in molar concentration of reactant per unit time.* The usual unit of reaction rate is moles per liter per second, $mol/(L \cdot s)$.

Consider the gas-phase reaction discussed in the chapter opening:

$$2N_2O_5(g) \longrightarrow 4NO_2(g) + O_2(g)$$

The rate for this reaction could be found by observing the increase in molar concentration of $O_2$ produced. You denote the molar concentration of a substance by enclosing its formula in square brackets. Thus, $[O_2]$ is the molar concentration of $O_2$. In a given time interval $\Delta t$, the molar concentration of oxygen, $[O_2]$, in the reaction vessel increases by the amount $\Delta[O_2]$. The symbol $\Delta$ (capital Greek delta) means "change in"; you obtain the change by subtracting the initial value from the final value. The rate of the reaction is given by

$$\text{Rate of formation of oxygen} = \frac{\Delta[O_2]}{\Delta t}$$

This equation gives the *average* rate over the time interval $\Delta t$. If the time interval is very short, the equation gives the *instantaneous* rate—that is, the rate at a particular instant of time. The instantaneous rate is also the value of $\Delta[O_2]/\Delta t$ for the tangent at a given instant (the straight line that just touches the curve of concentration versus time at a given point). See Figure 14.4. ◀

▶ In calculus, the rate at a given moment (the instantaneous rate) is given by the derivative $d\,[O_2]/dt$.

To understand the difference between average rate and instantaneous rate, it may help to think of the speed of an automobile. Speed can be defined as the rate of change

**Figure 14.4**
**The instantaneous rate of reaction.** In the reaction $2N_2O_5(g) \longrightarrow 4NO_2(g) + O_2(g)$, the concentration of $O_2$ increases over time. You obtain the instantaneous rate at a given time from the slope of the tangent at the point on the curve corresponding to that time. In this diagram, the slope equals $\Delta[O_2]/\Delta t$ obtained from the tangent.

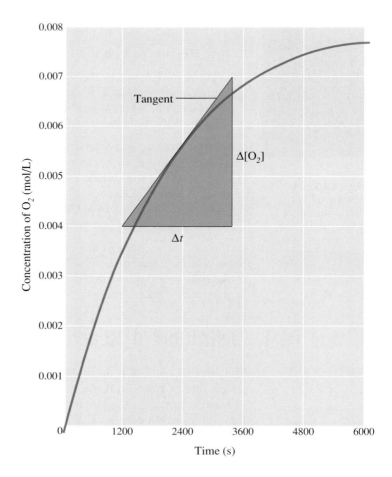

**Figure 14.5**
**Calculation of the average rate.** The average rate of formation of $O_2$ during the decomposition of $N_2O_5$ was calculated during two different time intervals. When the time changes from 600 s to 1200 s, the average rate is $2.5 \times 10^{-6}$ mol/(L · s). Later, when the time changes from 4200 s to 4800 s, the average rate has slowed to $5 \times 10^{-7}$ mol/(L · s). Thus, the rate of the reaction decreases as the reaction proceeds.

| Time | $[O_2]$ |
|------|---------|
| 0 | 0.0000 |
| 600 | 0.0021 |
| 1200 | 0.0036 |
| 1800 | 0.0048 |
| 2400 | 0.0057 |
| 3000 | 0.0063 |
| 3600 | 0.0068 |
| 4200 | 0.0072 |
| 4800 | 0.0075 |
| 5400 | 0.0077 |
| 6000 | 0.0078 |

of position, $x$; that is, speed equals $\Delta x/\Delta t$, where $\Delta x$ is the distance traveled. If an automobile travels 84 miles in 2.0 hours, the average speed over this time interval is 84 mi/2.0 hr = 42 mi/hr. However, at any instant during this interval, the speedometer, which registers instantaneous speed, may read more or less than 42 mi/hr. At some moment on the highway, it may read 55 mi/hr, whereas on a congested city street it may read only 20 mi/hr. The quantity $\Delta x/\Delta t$ becomes more nearly an instantaneous speed as the time interval $\Delta t$ is made smaller.

Figure 14.5 shows the increase in concentration of $O_2$ during the decomposition of $N_2O_5$. It shows the calculation of average rates at two positions on the curve. For example, when the time changes from 600 s to 1200 s ($\Delta t = 600$ s), the $O_2$ concentration increases by 0.0015 mol/L ($= \Delta[O_2]$). Therefore, the average rate $= \Delta[O_2]/\Delta t = $ (0.0015 mol/L)/600 s $= 2.5 \times 10^{-6}$ mol/(L · s). Later, during the time interval from 4200 s to 4800 s, the average rate is $5 \times 10^{-7}$ mol/(L · s). Note that the rate decreases as the reaction proceeds.

Because the amounts of products and reactants are related by stoichiometry, any substance in the reaction can be used to express the rate of reaction. In the case of the decomposition of $N_2O_5$ to $NO_2$ and $O_2$, we gave the rate in terms of the rate of formation of oxygen, $\Delta[O_2]/\Delta t$. However, you can also express it in terms of the rate of decomposition of $N_2O_5$.

$$\text{Rate of decomposition of } N_2O_5 = -\frac{\Delta[N_2O_5]}{\Delta t}$$

Note the negative sign. It always occurs in a rate expression for a reactant in order to indicate a decrease in concentration and to give a positive value for the rate. Thus, because $[N_2O_5]$ decreases, $\Delta[N_2O_5]$ is negative and $-\Delta[N_2O_5]/\Delta t$ is positive.

The rate of decomposition of $N_2O_5$ and the rate of formation of oxygen are easily related. Two moles of $N_2O_5$ decompose for each mole of oxygen formed, so the rate of decomposition of $N_2O_5$ is twice the rate of formation of oxygen. To equate the rates, you must divide the rate of decomposition of $N_2O_5$ by 2 (its coefficient in the balanced chemical equation).

$$\text{Rate of formation of } O_2 = \tfrac{1}{2} \text{ (rate of decomposition of } N_2O_5)$$

or

$$\frac{\Delta[O_2]}{\Delta t} = -\tfrac{1}{2}\frac{\Delta[N_2O_5]}{\Delta t}$$

▶ An alternate example is provided in the Instructor's Resource Manual.

---

### EXAMPLE 14.1

## Relating the Different Ways of Expressing Reaction Rates

Consider the reaction of nitrogen dioxide with fluorine to give nitryl fluoride, $NO_2F$.

$$2NO_2(g) + F_2(g) \longrightarrow 2NO_2F(g)$$

How is the rate of formation of $NO_2F$ related to the rate of reaction of fluorine?

 **PROBLEM STRATEGY**

The rate of disappearance of a reactant X is $-\Delta[X]/\Delta t$; the rate of formation of a product Y is $\Delta[Y]/\Delta t$. You can express the rate of the reaction in terms of either rate expression. These rate expressions can be equated, if you first divide each by the coefficient of the corresponding substance in the chemical equation.

**SOLUTION**

You write

$$\text{Rate of formation of } NO_2F = \frac{\Delta[NO_2F]}{\Delta t}$$

and

$$\text{Rate of reaction of } F_2 = -\frac{\Delta[F_2]}{\Delta t}$$

You divide each rate by the corresponding coefficient and then equate them:

$$\tfrac{1}{2}\frac{\Delta[NO_2F]}{\Delta t} = -\frac{\Delta[F_2]}{\Delta t}$$

---

### EXERCISE 14.1

For the reaction given in Example 14.1, how is the rate of formation of $NO_2F$ related to the rate of reaction of $NO_2$?

**See Problems 14.31 and 14.32.**

▶ An alternate example is
provided in the Instructor's
Resource Manual.

### EXAMPLE 14.2

## Calculating the Average Reaction Rate

Calculate the average rate of decomposition of $N_2O_5$, $-\Delta[N_2O_5]/\Delta t$, by the reaction

$$2N_2O_5(g) \longrightarrow 4NO_2(g) + O_2(g)$$

during the time interval from $t = 600$ s to $t = 1200$ s (regard all time figures as significant). Use the following data:

| Time | $[N_2O_5]$ |
|------|-----------|
| 600 s | $1.24 \times 10^{-2} M$ |
| 1200 s | $0.93 \times 10^{-2} M$ |

 **PROBLEM STRATEGY**

You calculate a $\Delta$ quantity in the rate expression by taking the final value minus the initial value.

**SOLUTION**

Average rate of decomposition of $N_2O_5 = -\dfrac{\Delta[N_2O_5]}{\Delta t} =$

$$-\frac{(0.93 - 1.24) \times 10^{-2} M}{(1200 - 600) \text{ s}} = -\frac{-0.31 \times 10^{-2} M}{600 \text{ s}} = \mathbf{5.2 \times 10^{-6} \, M/s}$$

Note that this rate is twice the rate of formation of $O_2$ in the same time interval (within the experimental error of the value given in the preceding text discussion).

### EXERCISE 14.2

Iodide ion is oxidized by hypochlorite ion in basic solution.

$$I^-(aq) + ClO^-(aq) \longrightarrow Cl^-(aq) + IO^-(aq)$$

In 1.00 $M$ NaOH at 25°C, the iodide-ion concentration (equal to the $ClO^-$ concentration) at different times was as follows:

| Time | $[I^-]$ |
|------|--------|
| 2.00 s | 0.00169 $M$ |
| 8.00 s | 0.00101 $M$ |

Calculate the average rate of reaction of $I^-$ during this time interval.

See Problems 14.35 and 14.36.

 **CONCEPT CHECK 14.1**

Shown here is a plot of the concentration of a reactant D versus time.

a. How do the instantaneous rates at points A and B compare?

b. Is the rate for this reaction constant at all points in time?

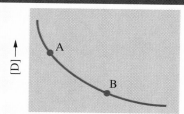

## 14.2 Experimental Determination of Rate

▶ Demonstration of the arsenite–thiosulfate clock reaction: Summerlin and Ealy, *Chemical Demonstrations*, Vol. 1 (Washington, D.C.: Am. Chem. Soc., 1985), no. 51, p. 80.

▶ Ethyl acetate is a liquid with a fruity odor that belongs to a class of organic (carbon-containing) compounds called esters. Unlike a salt, such as sodium acetate, an ester is a covalent compound.

▶ Video no. 34 (Series D), "Lycopodium Combustion."

To obtain the rate of a reaction, you must determine the concentration of a reactant or product during the course of the reaction. One way to do this for a slow reaction is to withdraw samples from the reaction vessel at various times and analyze them. The rate of the reaction of ethyl acetate with water in acidic solution was one of the first to be determined this way. ◀

$$CH_3CH_2O\overset{O}{\overset{\|}{C}}CH_3 + H_2O \xrightarrow{H^+} CH_3CH_2OH + HO\overset{O}{\overset{\|}{C}}CH_3$$

ethyl acetate                ethanol        acetic acid

This reaction is slow, so the amount of acetic acid produced is easily obtained by titration before any significant further reaction occurs.

More convenient are techniques that can continuously follow the progress of a reaction by observing the change in some physical property of the system. These physical methods are often adaptable to fast reactions as well as slow ones. For example, if a gas reaction involves a change in the number of molecules, the pressure of the system changes when the volume and temperature are held constant. By following the pressure change as the reaction proceeds, you can obtain the reaction rate. The decomposition of dinitrogen pentoxide in the gas phase, which we mentioned earlier, has been studied this way. Dinitrogen pentoxide crystals are sealed in a vessel equipped with a manometer (a pressure-measuring device; see Figure 14.6). The vessel is then plunged into a water bath at 45°C, at which temperature the solid vaporizes and the gas decomposes.

$$2N_2O_5(g) \longrightarrow 4NO_2(g) + O_2(g)$$

**Figure 14.6**
**An experiment to follow the concentration of $N_2O_5$ as the decomposition proceeds.** The total pressure is measured during the reaction at 45°C. Pressure values can be related to the concentrations of $N_2O_5$, $NO_2$, and $O_2$ in the flask.

Opening for adding $N_2O_5(s)$

Insulated water bath

45°C

Mercury manometer

Flask containing $N_2O_5(g)$, $NO_2(g)$, and $O_2(g)$

▶ Video Series C, "Reaction Rate and Concentration."

Manometer readings are taken at regular time intervals, and the pressure values are converted to partial pressures or concentrations of $N_2O_5$. The rates of reaction during various time intervals can be calculated as described in the previous section.

Another physical property used to follow the progress of a reaction is color, or the absorption of light by some species. Consider the reaction

$$ClO^-(aq) + I^-(aq) \longrightarrow IO^-(aq) + Cl^-(aq)$$

The hypoiodite ion, $IO^-$, absorbs at the blue end of the spectrum near 400 nm. The intensity of this absorption is proportional to $[IO^-]$, and you can use the absorption to determine the reaction rate. You can also follow the decomposition of $N_2O_5$ from the intensity of the red-brown color of the product $NO_2$.

In these experiments, the intensity of absorption at a particular wavelength is measured by a spectrometer appropriate for the visible region of the spectrum. You are not limited to the use of visible-region spectrometers for the determination of reaction rates, however. Depending on the reaction, other types of instruments are used, including infrared (IR) and nuclear magnetic resonance (NMR) spectrometers, which we described briefly in Instrumental Methods essays in earlier chapters.

# 14.3 Dependence of Rate on Concentration

Experimentally, it has been found that a reaction rate depends on the concentrations of certain reactants as well as the concentration of catalyst, if there is one (Figure 14.7). Consider the reaction of nitrogen dioxide with fluorine to give nitryl fluoride, $NO_2F$.

$$2NO_2(g) + F_2(g) \longrightarrow 2NO_2F(g)$$

The rate of this reaction is observed to be proportional to the concentration of nitrogen dioxide. When the concentration of nitrogen dioxide is doubled, the rate doubles.

Time = 0
A

Time = 20 seconds
B

Time = 40 seconds
C

**Figure 14.7**
**Effect of reactant concentrations on rate of reaction.** *(A)* Both beakers contain the same amounts of reactants: sodium thiosulfate, $Na_2S_2O_3$, in acidic solution and sodium arsenite, $Na_3AsO_3$. However, the beaker on the right contains more water and thus lower concentrations of reactants. *(B)* $Na_2S_2O_3$ decomposes slowly in acidic solution to yield $H_2S$, which reacts quickly with $Na_3AsO_3$ to give a bright yellow precipitate of $As_2S_3$. The time it takes the precipitate to form depends on the decomposition rate of $Na_2S_2O_3$. *(C)* Note that the precipitate forms more slowly in the solution of lower concentrations (the beaker on the right).

▶ Video no. 32 (Series D), "Iodine Clock."

The rate is also proportional to the concentration of fluorine; doubling the concentration of fluorine also doubles the rate.

A **rate law** is *an equation that relates the rate of a reaction to the concentrations of reactants (and catalyst) raised to various powers.* The following equation is the rate law for the foregoing reaction:

$$\text{Rate} = k[NO_2][F_2]$$

Note that in this rate law both reactant concentrations have an exponent of 1. Here $k$, called the **rate constant,** is *a proportionality constant in the relationship between rate and concentrations.* It has a fixed value at any given temperature, but it varies with temperature. Whereas the units of rate are usually given as $mol/(L \cdot s)$, the units of $k$ depend on the form of the rate law. For the previous rate law,

$$k = \frac{\text{rate}}{[NO_2][F_2]}$$

from which you get the following unit for $k$:

$$\frac{mol/(L \cdot s)}{(mol/L)^2} = L/(mol \cdot s)$$

As a more general example, consider the reaction of substances A and B to give D and E, according to the balanced equation

$$a\text{A} + b\text{B} \xrightarrow{\text{C}} d\text{D} + e\text{E} \qquad \text{C} = \text{catalyst}$$

You could write the rate law in the form

$$\text{Rate} = k[A]^m[B]^n[C]^p$$

The exponents $m$, $n$, and $p$ are frequently, but not always, integers. *They must be determined experimentally* and they cannot be obtained simply by looking at the balanced equation. For example, note that the exponents in the equation Rate $= k[NO_2][F_2]$ have no relationship to the coefficients in the balanced equation $2NO_2 + F_2 \longrightarrow 2NO_2F$.

Once you know the rate law for a reaction and have found the value of the rate constant, you can calculate the rate of a reaction for any values of reactant concentrations. As you will see later, knowledge of the rate law is also useful in understanding how the reaction occurs at the molecular level.

## Reaction Order

You can classify a reaction by its orders. The **reaction order** with respect to a given reactant species equals *the exponent of the concentration of that species in the rate law, as determined experimentally.* For the reaction of $NO_2$ with $F_2$ to give $NO_2F$, the reaction is first order with respect to the $NO_2$ because the exponent of $[NO_2]$ in the rate law is 1. Similarly, the reaction is first order with respect to $F_2$.

The *overall order* of a reaction equals the sum of the orders of the reactant species in the rate law. In this example, the overall order is 2; that is, the reaction is second order overall.

Reactions display a variety of reaction orders. Some examples follow:

1. Cyclopropane, $C_3H_6$, has the molecular structure

$$\begin{array}{c} CH_2 \\ \diagup \quad \diagdown \\ H_2C \underline{\qquad} CH_2 \end{array}$$

When heated, the carbon ring (the triangle) opens up to give propylene, $CH_2=CHCH_3$. Because the compounds are isomers (different compounds with the same molecular formula), the reaction is called an *isomerization*.

$$C_3H_6(g) \longrightarrow CH_2=CHCH_3(g)$$

cyclopropane            propylene

The reaction has the experimentally determined rate law

$$\text{Rate} = k[C_3H_6]$$

The reaction is first order in cyclopropane and first order overall.

2. Nitric oxide, NO, reacts with hydrogen according to the equation

$$2NO(g) + 2H_2(g) \longrightarrow N_2(g) + 2H_2O(g)$$

The experimentally determined rate law is

$$\text{Rate} = k[NO]^2[H_2]$$

Thus, the reaction is second order in NO, first order in $H_2$, and third order overall.

3. Acetone, $CH_3COCH_3$, reacts with iodine in acidic solution.

$$CH_3COCH_3(aq) + I_2(aq) \xrightarrow{H^+} CH_3COCH_2I(aq) + HI(aq)$$

The experimentally determined rate law is

$$\text{Rate} = k[CH_3COCH_3][H^+]$$

The reaction is first order in acetone. It is zero order in iodine; that is, the rate law contains the factor $[I_2]^0 = 1$. Therefore, the rate does not depend on the concentration of $I_2$, as long as some concentration of $I_2$ is present. Note that the reaction is first order in the catalyst, $H^+$. Thus, the overall order is 2.

Although reaction orders frequently have whole-number values (particularly 1 or 2), they can be fractional. Zero and negative orders are also possible.

▶ An alternate example is provided in the Instructor's Resource Manual.

## EXAMPLE 14.3

### Determining the Order of Reaction from the Rate Law

Bromide ion is oxidized by bromate ion in acidic solution.

$$5Br^-(aq) + BrO_3^-(aq) + 6H^+(aq) \longrightarrow 3Br_2(aq) + 3H_2O(l)$$

The experimentally determined rate law is

$$\text{Rate} = k[Br^-][BrO_3^-][H^+]^2$$

What is the order of reaction with respect to each reactant species? What is the overall order of reaction?

**SOLUTION**

The reaction is **first order with respect to $Br^-$ and first order with respect to $BrO_3^-$**; it is **second order with respect to $H^+$** (an order with respect to a species equals the exponent of its concentration). **The reaction is fourth order overall** (= 1 + 1 + 2).

---

**EXERCISE 14.3**

What are the reaction orders with respect to each reactant species for the following reaction?

$$NO_2(g) + CO(g) \longrightarrow NO(g) + CO_2(g)$$

Assume the rate law is

$$Rate = k[NO_2]^2$$

What is the overall order?

---

See Problems 14.39, 14.40, 14.41, and 14.42.

 **CONCEPT CHECK 14.2**

Consider the reaction $Q + R \longrightarrow S + T$ and the rate law for the reaction:

$$Rate = k[Q]^0[R]^2$$

a. You run the reaction three times, each time starting with $[R] = 2.0\ M$. For each run you change the starting concentration of $[Q]$: run 1, $[Q] = 0.0\ M$; run 2, $[Q] = 1.0\ M$; run 3, $[Q] = 2.0\ M$. Rank the rate of the three reactions using each of these concentrations.

b. The way the rate law is written in this problem is not typical for expressions containing reactants that are zero order in the rate law. Write the rate law in the more typical fashion.

## Determining the Rate Law

The experimental determination of the rate law for a reaction requires that you find the order of the reaction with respect to each reactant and any catalyst. The *initial-rate method* is a simple way to obtain reaction orders. It consists of doing a series of experiments in which the initial, or starting, concentrations of reactants are varied. Then the initial rates are compared, from which the reaction orders can be deduced.

To see how this method works, again consider the reaction

$$2N_2O_5(g) \longrightarrow 4NO_2(g) + O_2(g)$$

▶ In Figure 14.4, the slope of the tangent to the curve at $t = 0$ equals the initial rate of appearance of $O_2$, which equals one-half the initial rate of disappearance of $N_2O_5$.

You observe this reaction in two experiments. In Experiment 2, the initial concentration of $N_2O_5$ is twice that in Experiment 1. You then note the initial rate of disappearance of $N_2O_5$ in each case. The initial concentrations and corresponding initial rates for the two experiments are given in the following table: ◀

|  | **Initial $N_2O_5$ Concentration** | **Initial Rate of Disappearance of $N_2O_5$** |
|---|---|---|
| Experiment 1 | $1.0 \times 10^{-2}$ mol/L | $4.8 \times 10^{-6}$ mol/(L $\cdot$ s) |
| Experiment 2 | $2.0 \times 10^{-2}$ mol/L | $9.6 \times 10^{-6}$ mol/(L $\cdot$ s) |

The rate law for this reaction will have the concentration of reactant raised to a power $m$.

$$\text{Rate} = k[N_2O_5]^m$$

The value of $m$ (the reaction order) must be determined from the experimental data. Note that when the $N_2O_5$ concentration is doubled, you get a new rate, Rate', given by the following equation:

$$\text{Rate}' = k(2[N_2O_5])^m = 2^m k[N_2O_5]^m$$

This rate is $2^m$ times the original rate.

You can now see how the rate is affected when the concentration is doubled for various choices of $m$. Suppose $m = 2$. You get $2^m = 2^2 = 4$. That is, when the initial concentration is doubled, the rate is multiplied by 4. We summarize the results for various choices of $m$ in Table 14.1.

Suppose you divide the initial rate of reaction of $N_2O_5$ from Experiment 2 by the initial rate from Experiment 1.

$$\frac{9.6 \times 10^{-6} \text{ mol/(L} \cdot \text{s)}}{4.8 \times 10^{-6} \text{ mol/(L} \cdot \text{s)}} = 2.0$$

You see that when the $N_2O_5$ concentration is doubled, the rate is doubled. This corresponds to the case $m = 1$. The rate law must have the form

$$\text{Rate} = k[N_2O_5]$$

You can determine the value of the rate constant $k$ by substituting values of the rate and $N_2O_5$ concentrations from any of the experiments into the rate law. Using values from Experiment 2, you get

$$9.6 \times 10^{-6} \text{ mol/(L} \cdot \text{s)} = k \times 2.0 \times 10^{-2} \text{ mol/L}$$

Hence,

$$k = \frac{9.6 \times 10^{-6}/\text{s}}{2.0 \times 10^{-2}} = 4.8 \times 10^{-4}/\text{s}$$

**Table 14.1**
**Effect on Rate of Doubling the Initial Concentration of Reactant**

| $m$ | **Rate Is Multiplied by:** |
|---|---|
| $-1$ | $\frac{1}{2}$ |
| 0 | 1 |
| 1 | 2 |
| 2 | 4 |

▶ An alternate example is provided in the Instructor's Resource Manual.

---

**EXAMPLE 14.4**

## Determining the Rate Law from Initial Rates

Iodide ion is oxidized in acidic solution to triiodide ion, $I_3^-$, by hydrogen peroxide.

$$H_2O_2(aq) + 3I^-(aq) + 2H^+(aq) \longrightarrow I_3^-(aq) + 2H_2O(l)$$

A series of four experiments was run at different concentrations, and the initial rates of $I_3^-$ formation were determined (see table). a. From these data, obtain the reaction orders with respect to $H_2O_2$, $I^-$, and $H^+$. b. Then find the rate constant.

| | Initial Concentrations (mol/L) | | | Initial Rate |
| | $H_2O_2$ | $I^-$ | $H^+$ | [mol/(L · s)] |
|---|---|---|---|---|
| Exp. 1 | 0.010 | 0.010 | 0.00050 | $1.15 \times 10^{-6}$ |
| Exp. 2 | 0.020 | 0.010 | 0.00050 | $2.30 \times 10^{-6}$ |
| Exp. 3 | 0.010 | 0.020 | 0.00050 | $2.30 \times 10^{-6}$ |
| Exp. 4 | 0.010 | 0.010 | 0.00100 | $1.15 \times 10^{-6}$ |

 **PROBLEM STRATEGY**

You assume that the rate law has the following form

$$\text{Rate} = k[H_2O_2]^m[I^-]^n[H^+]^p$$

and you determine the reaction orders (exponents $m$, $n$, and $p$) by comparing two rate experiments in which all concentrations of reactants but one are held constant.

**SOLUTION**

a. Comparing Experiment 1 and Experiment 2, you see that when the $H_2O_2$ concentration is doubled (with other concentrations constant), the rate is doubled. From Table 14.1, you see that $m = 1$. The reaction is first order in $H_2O_2$.

   To solve a problem such as this in a general way, you approach it algebraically. You write the rate law algebraically for two experiments. (The subscripts denote the experiments.)

$$(\text{Rate})_1 = k[H_2O_2]_1^m[I^-]_1^n[H^+]_1^p$$

$$(\text{Rate})_2 = k[H_2O_2]_2^m[I^-]_2^n[H^+]_2^p$$

Now you divide the second equation by the first.

$$\frac{(\text{Rate})_2}{(\text{Rate})_1} = \frac{k[H_2O_2]_2^m[I^-]_2^n[H^+]_2^p}{k[H_2O_2]_1^m[I^-]_1^n[H^+]_1^p}$$

The rate constant cancels. Grouping the terms, you obtain

$$\frac{(\text{Rate})_2}{(\text{Rate})_1} = \left(\frac{[H_2O_2]_2}{[H_2O_2]_1}\right)^m \left(\frac{[I^-]_2}{[I^-]_1}\right)^n \left(\frac{[H^+]_2}{[H^+]_1}\right)^p$$

Now you substitute values from Experiment 1 and Experiment 2. (In a ratio the units cancel and can be omitted.)

$$\frac{2.30 \times 10^{-6}}{1.15 \times 10^{-6}} = \left[\frac{0.020}{0.010}\right]^m \left[\frac{0.010}{0.010}\right]^n \left[\frac{0.00050}{0.00050}\right]^p$$

This gives $2 = 2^m$, from which you obtain $m = 1$. That is, doubling the $H_2O_2$ concentration doubles the rate.

   The exponents $n$ and $p$ are obtained in the same way. Comparing Experiment 1 and Experiment 3, you see that doubling the $I^-$ concentration (with the other concentrations constant) doubles the rate. Therefore, $n = 1$ (the reaction is first order in $I^-$). Finally, comparing Experiment 1 and Experiment 4, you see that doubling the $H^+$ concentration (holding other concentrations constant) has no effect on the rate. Therefore, $p = 0$ (the reaction is zero order in $H^+$). Because $[H^+]^0 = 1$, the rate law is

$$\text{Rate} = k[H_2O_2][I^-]$$

The reaction orders with respect to $H_2O_2$, $I^-$, and $H^+$, are **1, 1, and 0,** respectively. Note that the orders are *not* related to the coefficients of the overall equation.

b. You calculate the rate constant by substituting values from any of the experiments into the rate law. Using Experiment 1, you obtain

$$1.15 \times 10^{-6}\frac{\cancel{mol}}{\cancel{L}\cdot s} = k \times 0.010\frac{\cancel{mol}}{\cancel{L}} \times 0.010\frac{mol}{L}$$

$$k = \frac{1.15 \times 10^{-6}/s}{0.010 \times 0.010 \times mol/L} = \textbf{1.2} \times \textbf{10}^{-2}\ \textbf{L/(mol} \cdot \textbf{s)}$$

## EXERCISE 14.4

The initial-rate method was applied to the decomposition of nitrogen dioxide.

$$2NO_2(g) \longrightarrow 2NO(g) + O_2(g)$$

It yielded the following results:

| | Initial NO$_2$ Concentration | Initial Rate of Formation of O$_2$ |
|---|---|---|
| Exp. 1 | 0.010 mol/L | $7.1 \times 10^{-5}$ mol/(L · s) |
| Exp. 2 | 0.020 mol/L | $28 \times 10^{-5}$ mol/(L · s) |

Find the rate law and the value of the rate constant with respect to $O_2$ formation.

See Problems 14.43, 14.44, 14.45, 14.46, 14.47, and 14.48.

 **CONCEPT CHECK 14.3**

Rate laws are not restricted to chemical systems; they are used to help describe many "everyday" events. For example, a rate law for tree growth might look something like this:

$$\text{Rate of growth} = (\text{soil type})^w(\text{temperature})^x(\text{light})^y(\text{fertilizer})^z$$

In this equation, like chemical rate equations, the exponents need to be determined by experiment. (Can you think of some other factors?)

a. Say you are a famous physician trying to determine the factors that influence the rate of aging in humans. Develop a rate law, like the one above, that would take into account at least four factors that affect the rate of aging.

b. Explain what you would need to do in order to determine the exponents in your rate law.

c. Consider smoking to be one of the factors in your rate law. You conduct an experiment and find that a person smoking two packs of cigarettes a day quadruples ($4\times$) the rate of aging over that of a one-pack-a-day smoker. Assuming that you could hold all other factors in your rate law constant, what will be the exponent of the smoking term in your rate law?

## 14.4 Change of Concentration with Time

▶ Video no. 33 (Series D), "Old Nassau Reaction." [▭▭]

A rate law tells you how the rate of a reaction depends on reactant concentrations at a particular moment. But often you would like to have a mathematical relationship showing how a reactant concentration changes over a period of time. Such an equation would be directly comparable to the experimental data, which are usually obtained as concentrations at various times. In addition to summarizing the experimental data, this equation would predict concentrations for all times. Using it, you could answer questions such as: How long does it take for this reaction to be 50% complete? to be 90% complete?

Moreover, as you will see, knowing exactly how the concentrations change with time for different rate laws suggests ways of plotting the experimental data on a graph. Graphical plotting provides an alternative to the initial-rate method for determining the rate law.

Using calculus, we can transform a rate law into a mathematical relationship between concentration and time. Because we will work only with the final equations, we need not go into the derivations here. We will look in some depth at first-order reactions and briefly at second-order reactions.

### Concentration–Time Equations

**First-Order Rate Law**  Let us first look at first-order rate laws. The decomposition of nitrogen pentoxide,

$$2N_2O_5(g) \longrightarrow 4NO_2(g) + O_2(g)$$

has the following rate law:

$$\text{Rate} = -\frac{\Delta[N_2O_5]}{\Delta t} = k[N_2O_5]$$

▶ **The general derivation using calculus is as follows. Substituting [A] for [N₂O₅], the rate law becomes**

$$\frac{-d[A]}{dt} = k[A]$$

**You rearrange this to give**

$$\frac{-d[A]}{[A]} = k\, dt$$

**Integrating from time = 0 to time = t,**

$$-\int_{[A]_0}^{[A]_t} \frac{d[A]}{[A]} = k \int_0^t dt$$

**gives**

$$-\{\ln[A]_t - \ln[A]_0\} = k(t - 0)$$

**This can be rearranged to give the equation in the text.**

Using calculus, one can show that such a first-order rate law leads to the following relationship between $N_2O_5$ concentration and time: ◀

$$\ln\frac{[N_2O_5]_t}{[N_2O_5]_0} = -kt$$

Here $[N_2O_5]_t$ is the concentration at time $t$, and $[N_2O_5]_0$ is the initial concentration of $N_2O_5$ (that is, the concentration at $t = 0$). The symbol "ln" denotes the natural logarithm (base $e = 2.718 \ldots$).

This equation enables you to calculate the concentration of $N_2O_5$ at any time, once you are given the initial concentration and the rate constant. Also, you can find the time it takes for the $N_2O_5$ concentration to decrease to a particular value.

More generally, let A be a substance that reacts to give products according to the equation

$$aA \longrightarrow \text{products}$$

where $a$ is the stoichiometric coefficient of reactant A. Suppose that this reaction has a first-order rate law

$$\text{Rate} = -\frac{\Delta[A]}{\Delta t} = k[A]$$

Using calculus, you get the following equation:

$$\ln \frac{[A]_t}{[A]_0} = -kt$$

Here $[A]_t$ is the concentration of reactant A at time $t$, and $[A]_0$ is the initial concentration. The ratio $[A]_t/[A]_0$ is the fraction of reactant remaining at time $t$. The next example illustrates how you work with this equation.

▶ An alternate example is provided in the Instructor's Resource Manual.

### EXAMPLE 14.5

## Using the Concentration–Time Equation for a First-Order Reaction

The decomposition of $N_2O_5$ to $NO_2$ and $O_2$ is first order, with a rate constant of $4.80 \times 10^{-4}$/s at 45°C. a. If the initial concentration is $1.65 \times 10^{-2}$ mol/L, what is the concentration after 825 s? b. How long would it take for the concentration of $N_2O_5$ to decrease to $1.00 \times 10^{-2}$ mol/L from its initial value, given in a?

### SOLUTION

a. You need to use the equation relating concentration to time (the shaded equation given just before this example). You get

$$\ln \frac{[N_2O_5]_t}{1.65 \times 10^{-2} \text{ mol/L}} = -4.80 \times 10^{-4}/\text{s} \times 825 \text{ s} = -0.396$$

To solve for $[N_2O_5]_t$, you take the antilogarithm of both sides. This removes the ln from the left and gives antilog($-0.396$), or $e^{-0.396}$, on the right, which equals 0.673.

$$\frac{[N_2O_5]_t}{1.65 \times 10^{-2} \text{ mol/L}} = 0.673$$

Hence,

$$[N_2O_5]_t = 1.65 \times 10^{-2} \text{ mol/L} \times 0.673 = \textbf{0.0111 mol/L}$$

b. You substitute into the equation relating concentration to time.

$$\ln \frac{1.00 \times 10^{-2} \text{ mol/L}}{1.65 \times 10^{-2} \text{ mol/L}} = -4.80 \times 10^{-4}/\text{s} \times t$$

The left side equals $-5.01$; the right side equals $-4.80 \times 10^{-4}$/s $\times t$. Hence,

$$5.01 = 4.80 \times 10^{-4}/\text{s} \times t$$

Or,

$$t = \frac{5.01}{4.80 \times 10^{-4}/\text{s}} = \textbf{1.04} \times \textbf{10}^3 \textbf{ s} \qquad \textbf{(17.4 min)}$$

**Problem Solving: Integrated First-Order Rate Law**

### EXERCISE 14.5

a. What would be the concentration of dinitrogen pentoxide in the experiment described in Example 14.5 after $6.00 \times 10^2$ s? b. How long would it take for the concentration of $N_2O_5$ to decrease to 10.0% of its initial value?

See Problems 14.49 and 14.50.

**Second-Order Rate Law**  Consider the reaction

$$aA \longrightarrow \text{products}$$

and suppose it has the second-order rate law

$$\text{Rate} = -\frac{\Delta[A]}{\Delta t} = k[A]^2$$

An example is the decomposition of nitrogen dioxide at moderately high temperatures (300°C to 400°C).

$$2NO_2(g) \longrightarrow 2NO(g) + O_2(g)$$

Using calculus, you can obtain the following relationship between the concentration of A and the time.

$$\frac{1}{[A]_t} = kt + \frac{1}{[A]_0}$$

Using this equation, you can calculate the concentration of $NO_2$ at any time during its decomposition if you know the rate constant and the initial concentration. At 330°C, the rate constant for the decomposition of $NO_2$ is 0.775 L/(mol · s). Suppose the initial concentration is 0.0030 mol/L. What is the concentration of $NO_2$ after 645 s? By substituting into the previous equation, you get

$$\frac{1}{[NO_2]_t} = 0.775 \text{ L/(mol · s)} \times 645 \text{ s} + \frac{1}{0.0030 \text{ mol/L}} = 8.3 \times 10^2 \text{ L/mol}$$

If you invert both sides of the equation, you find that $[NO_2]_t = 0.0012$ mol/L. Thus, after 645 s, the concentration of $NO_2$ decreased from 0.0030 mol/L to 0.0012 mol/L.

## Half-Life of a Reaction

As a reaction proceeds, the concentration of a reactant decreases, because it is being consumed. The **half-life,** $t_{1/2}$, of a reaction is *the time it takes for the reactant concentration to decrease to one-half of its initial value.* ◄

▶ The concept of half-life is also used to characterize a radioactive nucleus, whose radioactive decay is a first-order process. This is discussed in Chapter 21.

For a first-order reaction, such as the decomposition of dinitrogen pentoxide, the half-life is independent of the initial concentration. To see this, substitute into the equation

$$\ln \frac{[N_2O_5]_t}{[N_2O_5]_0} = -kt$$

In one half-life, the $N_2O_5$ concentration decreases by one-half, from its initial value, $[N_2O_5]_0$, to the value $[N_2O_5]_t = \frac{1}{2}[N_2O_5]_0$. After substituting, the equation becomes

$$\ln \frac{\frac{1}{2}[N_2O_5]_0}{[N_2O_5]_0} = -kt_{1/2}$$

The expression on the left equals $\ln \frac{1}{2} = -0.693$. Hence,

$$0.693 = kt_{1/2}$$

**Figure 14.8**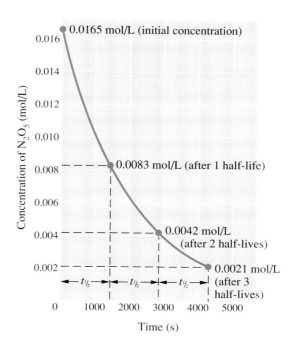

**A graph illustrating that the half-life of a first-order reaction is independent of initial concentration.** In one half-life (1440 s, or 24.0 min), the concentration decreases by one-half, from 0.0165 mol/L to $\frac{1}{2} \times$ 0.0165 mol/L = 0.0083 mol/L. After each succeeding half-life (24.0 min), the concentration decreases by one-half again, from 0.0083 mol/L to $\frac{1}{2} \times$ 0.0083 mol/L = 0.0042 mol/L, then to $\frac{1}{2} \times$ 0.0042 mol/L = 0.0021 mol/L, and so forth.

Solving for the half-life, $t_{1/2}$, you get

$$t_{1/2} = \frac{0.693}{k}$$

Because the rate constant for the decomposition of $N_2O_5$ at 45°C is $4.80 \times 10^{-4}$/s, the half-life is

$$t_{1/2} = \frac{0.693}{4.80 \times 10^{-4}/s} = 1.44 \times 10^3 \text{ s}$$

Thus, the half-life is $1.44 \times 10^3$ s, or 24.0 min.

You see that the half-life of $N_2O_5$ does not depend on the initial concentration of $N_2O_5$. This means that the half-life is the same at any time during the reaction. If the initial concentration is 0.0165 mol/L, after one half-life (24.0 min) the concentration decreases to $\frac{1}{2} \times$ 0.0165 mol/L = 0.0083 mol/L. After another half-life (another 24.0 min), the $N_2O_5$ concentration decreases to $\frac{1}{2} \times$ 0.0083 mol/L = 0.0042 mol/L. Every time one half-life passes, the $N_2O_5$ concentration decreases by one-half again (see Figure 14.8).

The foregoing result for the half-life for the first-order decomposition of $N_2O_5$ is perfectly general. That is, for the general first-order rate law

$$\text{Rate} = -\frac{\Delta[A]}{\Delta t} = k[A]$$

the half-life is related to the rate constant, but is independent of concentration of A.

$$t_{1/2} = \frac{0.693}{k}$$

▶ An alternate example is
provided in the Instructor's
Resource Manual.

## EXAMPLE 14.6

## Relating the Half-Life of a Reaction to the Rate Constant

Sulfuryl chloride, $SO_2Cl_2$, is a colorless, corrosive liquid whose vapor decomposes in a first-order reaction to sulfur dioxide and chlorine.

$$SO_2Cl_2(g) \longrightarrow SO_2(g) + Cl_2(g)$$

At 320°C, the rate constant is $2.20 \times 10^{-5}$/s. a. What is the half-life of $SO_2Cl_2$ vapor at this temperature? b. How long (in hours) would it take for 50.0% of the $SO_2Cl_2$ to decompose? How long would it take for 75.0% of the $SO_2Cl_2$ to decompose?

 **PROBLEM STRATEGY**

You substitute $k = 2.20 \times 10^{-5}$/s into the equation relating $k$ and $t_{1/2}$.

**SOLUTION**

a.
$$t_{1/2} = \frac{0.693}{k} = \frac{0.693}{2.20 \times 10^{-5}/s} = \textbf{3.15} \times \textbf{10}^\textbf{4}\ \textbf{s}$$

b. The half-life is the time required for one-half (50.0%) of the $SO_2Cl_2$ to decompose. This is $3.15 \times 10^4$ s, or **8.75 hr.** After another half-life, one-half of the remaining $SO_2Cl_2$ decomposes. The total decomposed is $\frac{1}{2} + (\frac{1}{2} \times \frac{1}{2}) = \frac{3}{4}$, or 75.0%. The time required is two half-lives, or $2 \times 8.75$ hr = **17.5 hr.**

## EXERCISE 14.6

The isomerization of cyclopropane, $C_3H_6$, to propylene, $CH_2{=}CHCH_3$, is first order in cyclopropane and first order overall. At 1000°C, the rate constant is 9.2/s. What is the half-life of cyclopropane at 1000°C? How long would it take for the concentration of cyclopropane to decrease to 50% of its initial value? to 25% of its initial value?

See Problems 14.53 and 14.54.

It can be shown by reasoning similar to that given previously that the half-life of a second-order rate law, Rate = $k[A]^2$, is $1/(k[A]_0)$. In this case, the half-life depends on initial concentration and becomes larger as time goes on. Consider the decomposition of $NO_2$ at 330°C. It takes 430 s for the concentration to decrease by one-half from 0.0030 mol/L to 0.0015 mol/L. However, it takes 860 s (twice as long) for the concentration to decrease by one-half again. The fact that the half-life changes with time is evidence that the reaction is not first order.

## Graphing of Kinetic Data

Earlier you saw that the order of a reaction can be determined by comparing initial rates for several experiments in which different initial concentrations are used (initial-rate method). It is also possible to determine the order of a reaction by graphical plotting of the data for a particular experiment. The experimental data are plotted in several different ways, first assuming a first-order reaction, then a second-order reaction, and so forth. The order of the reaction is determined by which graph gives the best fit to the experimental data. To illustrate, we will look at how the plotting is done for first-order and second-order reactions and then compare graphs for a specific reaction.

You have seen that the first-order rate law, $-\Delta[A]/\Delta t = k[A]$, gives the following relationship between concentration of A and time:

$$\ln \frac{[A]_t}{[A]_0} = -kt$$

This equation can be rewritten in a slightly different form, which you can identify with the equation of a straight line. Using the property of logarithms that $\log(A/B) = \log A - \log B$, you get

$$\ln[A]_t = -kt + \ln[A]_0$$

A straight line has the mathematical form $y = mx + b$, when $y$ is plotted on the vertical axis against $x$ on the horizontal axis. ◄ Let us now make the following identifications:

$$\underbrace{\ln[A]_t}_{y} = \underbrace{-k}_{m}\ \underbrace{t}_{x} + \underbrace{\ln[A]_0}_{b}$$

► **See Appendix A for a discussion of the mathematics of a straight line.**

This means that if you plot $\log[A]_t$ on the vertical axis against the time $t$ on the horizontal axis, you will get a straight line for a first-order reaction. Figure 14.9 shows a plot of $\ln[N_2O_5]$ at various times during the decomposition reaction. The fact that the points lie on a straight line is confirmation that the rate law is first order.

You can obtain the rate constant for the reaction from the slope, $m$, of the straight line.

$$-k = m \quad \text{or} \quad k = -m$$

**Figure 14.9**
**Plot of $\ln[N_2O_5]$ versus time.** A straight line can be drawn through the experimental points (colored dots). The fact that the straight line fits the experimental data so well confirms that the rate law is first order.

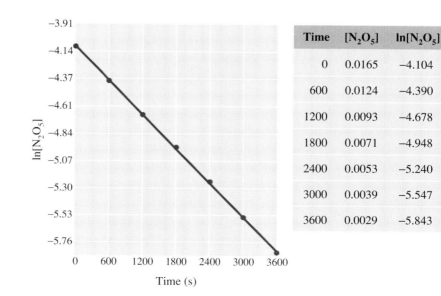

| Time | $[N_2O_5]$ | $\ln[N_2O_5]$ |
|------|-----------|--------------|
| 0 | 0.0165 | −4.104 |
| 600 | 0.0124 | −4.390 |
| 1200 | 0.0093 | −4.678 |
| 1800 | 0.0071 | −4.948 |
| 2400 | 0.0053 | −5.240 |
| 3000 | 0.0039 | −5.547 |
| 3600 | 0.0029 | −5.843 |

You calculate the slope of this curve in the same way you obtained the average rate of reaction from kinetic data (Example 14.2). You select two points far enough apart that when you subtract to obtain $\Delta x$ and $\Delta y$ for the slope, you do not lose significant figures. Using the first and last points in Figure 14.9, you get

$$m = \frac{\Delta y}{\Delta x} = \frac{(-5.843) - (-4.104)}{(3600 - 0) \text{ s}} = \frac{-1.739}{3600 \text{ s}} = -4.83 \times 10^{-4}/\text{s}$$

Therefore, $k = 4.84 \times 10^{-4}/\text{s}$. (Two points were selected directly from the experimental data. In precise work, you would first draw the straight line that best fits the experimental data points and then calculate the slope of this line.)

The second-order rate law, $-\Delta[A]/\Delta t = k[A]^2$, gives the following relationship between concentration of A and time:

$$\underbrace{\frac{1}{[A]_t}}_{y} = \underbrace{kt}_{mx} + \underbrace{\frac{1}{[A]_0}}_{b}$$

In this case, you get a straight line if you plot $1/[A]_t$ on the vertical axis against the time $t$ on the horizontal axis for a second-order reaction.

As an illustration of the determination of reaction order by graphical plotting, consider the following data for the decomposition of $NO_2$ at 330°C.

$$2NO_2(g) \longrightarrow 2NO(g) + O_2(g)$$

The concentrations of $NO_2$ for various times are

| Time (s) | $[NO_2]$ (mol/L) |
|---|---|
| 0 | $1.00 \times 10^{-2}$ |
| 60 | $0.683 \times 10^{-2}$ |
| 120 | $0.518 \times 10^{-2}$ |
| 180 | $0.418 \times 10^{-2}$ |
| 240 | $0.350 \times 10^{-2}$ |
| 300 | $0.301 \times 10^{-2}$ |
| 360 | $0.264 \times 10^{-2}$ |

In Figure 14.10A, we have plotted $\ln [NO_2]$ against $t$, and in Figure 14.10B we have plotted $1/[NO_2]$ against $t$. Only in (B) do the points closely follow a straight line, indicating that the rate law is second order. That is,

$$\text{Rate of disappearance of } NO_2 = -\frac{\Delta[NO_2]}{\Delta t} = k[NO_2]^2$$

You can obtain the rate constant, $k$, for a second-order reaction from the slope of the line, similar to the way you did for a first-order reaction. In this case, however, the slope equals $k$, as you can see from the preceding equation. Choosing the first and last points in Figure 14.10B, you get

$$k = \frac{\Delta y}{\Delta x} = \frac{(379 - 100) \text{ L/mol}}{(360 - 0) \text{ s}} = 0.775 \text{ L/(mol} \cdot \text{s)}$$

Table 14.2 summarizes the relationships discussed in this section for first-order and second-order reactions.

A

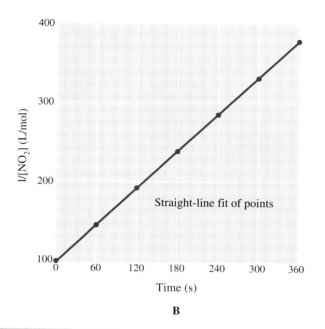

B

| Time (s) | (A) ln [NO$_2$] | (B) 1/[NO$_2$] |
|---|---|---|
| 0 | −4.605 | 100 |
| 60 | −4.984 | 146 |
| 120 | −5.263 | 193 |
| 180 | −5.476 | 239 |
| 240 | −5.656 | 286 |
| 300 | −5.805 | 332 |
| 360 | −5.938 | 379 |

**Figure 14.10**
**Plotting the data for the decomposition of nitrogen dioxide at 330°C.**
(A) Plot of ln[NO$_2$] against time. Note that a straight line does not fit the points well.
(B) Plot of 1/[NO$_2$] against time. Note how closely the points follow the straight line,
indicating that the decomposition is second order.

**Table 14.2**
**Relationships for First-Order and Second-Order Reactions**

| Order | Rate Law | Concentration–Time Equation | Half-Life | Graphical Plot |
|---|---|---|---|---|
| 1 | Rate $= k[A]$ | $\ln \dfrac{[A]_t}{[A]_0} = -kt$ | $0.693/k$ | $\ln [A]$ vs $t$ |
| 2 | Rate $= k[A]^2$ | $\dfrac{1}{[A]_t} = kt + \dfrac{1}{[A]_0}$ | $1/(k[A]_0)$ | $\dfrac{1}{[A]}$ vs $t$ |

**CONCEPT CHECK 14.4**

A reaction believed to be either first or second order has a half-life of 20 s at the beginning of the reaction but a half-life of 40 s some time later. What is the order of the reaction?

## 14.5 Temperature and Rate; Collision and Transition-State Theories

As we noted earlier, the rate of reaction depends on temperature. This shows up in the rate law through the rate constant, which is found to vary with temperature. In most cases, the rate increases with temperature (Figure 14.11). Consider the reaction of nitric oxide, NO, with chlorine, $Cl_2$, to give nitrosyl chloride, NOCl, and chlorine atoms.

$$NO(g) + Cl_2(g) \longrightarrow NOCl(g) + Cl(g)$$

▶ The change in rate constant with temperature varies considerably from one reaction to another. In many cases, the rate of reaction approximately doubles for a 10°C rise, and this is often given as an approximate rule.

The rate constant $k$ for this reaction is $4.9 \times 10^{-6}$ L/(mol · s) at 25°C and $1.5 \times 10^{-5}$ L/(mol · s) at 35°C. In this case the rate constant and therefore the rate are more than tripled for a 10°C rise in temperature. ◀ How do you explain this strong dependence of reaction rate on temperature? To understand it, you need to look at a simple theoretical explanation of reaction rates.

### Collision Theory

Why the rate constant depends on temperature can be explained by collision theory. **Collision theory** of reaction rates is *a theory that assumes that, for reaction to occur, reactant molecules must collide with an energy greater than some minimum value and with the proper orientation. The minimum energy of collision required for two molecules to react* is called the **activation energy,** $E_a$. The value of $E_a$ depends on the particular reaction.

In collision theory, the rate constant for a reaction is given as a product of three factors: (1) $Z$, the collision frequency, (2) $f$, the fraction of collisions having energy

**Figure 14.11**
**Effect of temperature on reaction rate.** *Left:* Each test tube contains potassium permanganate, $KMnO_4$, and oxalic acid, $H_2C_2O_4$, at the same concentrations. Permanganate ion oxidizes oxalic acid to $CO_2$ and $H_2O$. One test tube was placed in a beaker of warm water (40°C); the other was kept at room temperature (20°C). *Right:* After 10 minutes, the test tube at 40°C showed noticeable reaction, whereas the other did not.

greater than the activation energy, and (3) $p$, the fraction of collisions that occur with the reactant molecules properly oriented. Thus,

$$k = Zfp$$

We will discuss each of these factors in turn.

To have a specific reaction to relate the concepts to, consider the gas-phase reaction of NO with $Cl_2$, mentioned previously. This reaction is believed to occur in a single step. An NO molecule collides with a $Cl_2$ molecule. If the collision has sufficient energy and if the molecules are properly oriented, they react to produce NOCl and Cl.

Collision frequency $Z$, the frequency with which the reactant molecules collide, depends on temperature. As you will see, however, this dependence of collision frequency on temperature does not explain why reaction rates usually change greatly with small temperature increases. You can easily explain why the collision frequency depends on temperature. As the temperature rises, the gas molecules move faster and therefore collide more frequently. Thus, collision frequency is proportional to the root-mean-square (rms) molecular speed, which in turn is proportional to the square root of the absolute temperature, according to the kinetic theory of gases. ◀ From kinetic theory, one can show that at 25°C, a 10°C rise in temperature increases the collision frequency by about 2%. If you were to assume that each collision of reactant molecules resulted in reaction, you would conclude that the rate would increase with temperature at the same rate as the collision frequency—that is, by 2% for a 10°C rise in temperature. This clearly does not explain the tripling of the rate (a 300% increase) that you see in the reaction of NO with $Cl_2$ when the temperature is raised from 25°C to 35°C.

▶ According to the kinetic theory of gases, the rms molecular speed equals $\sqrt{3RT/M_m}$ (see Section 5.6).

You see that the collision frequency varies only slowly with temperature. However, $f$, the fraction of molecular collisions having energy greater than the activation energy, changes rapidly in most reactions with even small temperature changes. It can be shown that $f$ is related to the activation energy, $E_a$, this way:

$$f = e^{-E_a/RT}$$

Here $e = 2.718\ldots$ , and $R$ is the gas constant, which equals 8.31 J/(mol · K). For the reaction of NO with $Cl_2$, the activation energy is $8.5 \times 10^4$ J/mol. At 25°C (298 K), the fraction of collisions with sufficient energy for reaction is $1.2 \times 10^{-15}$. Thus, the number of collisions of reactant molecules that actually result in reaction is extremely small. But the frequency of collisions is very large, so the reaction rate, which depends on the product of these quantities, is not small. If the temperature is raised by 10°C to 35°C, the fraction of collisions of NO and $Cl_2$ molecules with sufficient energy for reaction is $3.8 \times 10^{-15}$, over three times larger than the value at 25°C! In other words, the tripling of the reaction rate when the temperature rises 10°C is explained by the temperature dependence of $f$.

From the previous equation relating $f$ to $E_a$, you see that $f$ decreases with increasing values of $E_a$. Because the rate constant depends on $f$, this means that reactions with large activation energies have small rate constants and that reactions with small activation energies have large rate constants.

The reaction rate also depends on $p$, the proper orientation of the reactant molecules when they collide. This factor is independent of temperature changes. You can see why it is important that the reactant molecules be properly oriented by looking in some detail at the reaction of NO with $Cl_2$. Figure 14.12 shows two possible molecular collisions. In Figure 14.12A, the NO and $Cl_2$ molecules collide properly oriented for reaction. The NO molecule approaches with its N atom toward the $Cl_2$ molecule. In addition, the angle of approach is about that expected for the formation of bonds in the product molecule NOCl. In Figure 14.12B, an NO molecule approaches with

**Figure 14.12**
**Importance of molecular orientation in the reaction of NO and $Cl_2$.** *(A)* NO approaches with its N atom toward $Cl_2$, and an N—Cl bond forms. Also, the angle of approach is close to that in the product NOCl. *(B)* NO approaches with its O atom toward $Cl_2$. No N—Cl bond can form, so NO and $Cl_2$ collide and then fly apart.

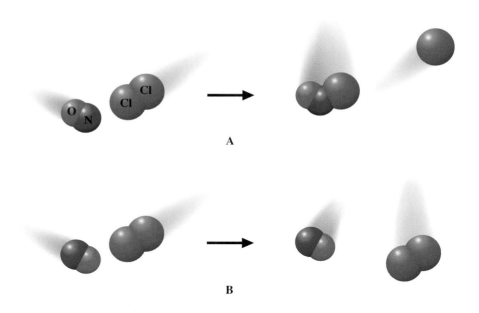

its O atom toward the $Cl_2$ molecule. Because this orientation does not allow the formation of a bond between the N atom and a Cl atom, it is ineffective for reaction. The NO and $Cl_2$ molecules come together and then fly apart. All orientations except those close to that shown in *(A)* are ineffective.

## Transition-State Theory

Collision theory explains some important features of a reaction, but it is limited in that it does not explain the role of activation energy. **Transition-state theory** *explains the reaction resulting from the collision of two molecules in terms of an activated complex.* An **activated complex** (transition state) is *an unstable grouping of atoms that can break up to form products.* We can represent the formation of the activated complex this way:

$$O{=}N + Cl{—}Cl \longrightarrow [O{=}N\text{----}Cl\text{----}Cl]$$

When the molecules come together with proper orientation, an N—Cl bond begins to form. At the same time, the kinetic energy of the collision is absorbed by the activated complex as a vibrational motion of the atoms. This energy becomes concentrated in the bonds denoted by the dashed lines and can flow between them. If, at some moment, sufficient energy becomes concentrated in one of the bonds of the activated complex, that bond breaks or falls apart. Depending on whether the N----Cl or Cl----Cl bond breaks, the activated complex either reverts to the reactants or yields the products.

$$\underset{\text{reactants}}{O{=}N + Cl_2} \longleftarrow \underset{\text{activated complex}}{[O{=}N\text{----}Cl\text{----}Cl]} \longrightarrow \underset{\text{products}}{O{=}N{—}Cl + Cl}$$

## Potential-Energy Diagrams for Reactions

It is instructive to consider a potential-energy diagram for the reaction of NO with $Cl_2$. We can represent this reaction by the equation

$$NO + Cl_2 \longrightarrow NOCl_2^{‡} \longrightarrow NOCl + Cl$$

**Figure 14.13**
**Potential-energy curve (not to scale) for the endothermic reaction**
**NO + Cl$_2$ $\longrightarrow$ NOCl + Cl.**
For NO and Cl$_2$ to react, at least 85 kJ/mol of energy must be supplied by the collision of reactant molecules. Once the activated complex forms, it may break up to products, releasing 2 kJ/mol of energy. The difference, $(85 - 2)$ kJ/mol = 83 kJ/mol, is the heat energy absorbed, $\Delta H$.

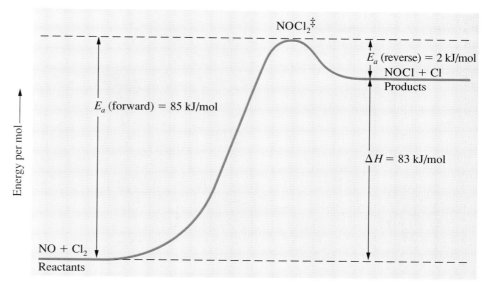

Here NOCl$_2$$^{\ddagger}$ denotes the activated complex. Figure 14.13 shows the change in potential energy (indicated by the solid curve) that occurs during the progress of the reaction. The potential-energy curve starts at the left with the potential energy of the reactants, NO + Cl$_2$. Moving along the curve toward the right, the potential energy increases to a maximum, corresponding to the activated complex. Farther to the right, the potential energy decreases to that of the products, NOCl + Cl.

At the start, the NO and Cl$_2$ molecules have a certain quantity of kinetic energy. The total energy of these molecules equals their potential energy plus their kinetic energy, and it remains constant throughout the reaction (according to the law of conservation of energy). As the reaction progresses (going from left to right in the diagram), the reactants come together. The potential energy increases because the outer electrons of the two molecules repel. The kinetic energy decreases, and the molecules slow down. Only if the reactant molecules have sufficient kinetic energy is it possible for the potential energy to increase to the value for the activated complex. This kinetic energy must be equal to or greater than the difference in energy between the activated complex and the reactant molecules (85 kJ/mol). The energy difference is the activation energy for the forward reaction.

At the maximum in the potential-energy curve, the reactant molecules have come together as the activated complex. When the activated complex breaks up into products, the products go to lower potential energy (by 2 kJ/mol) and gain in kinetic energy. Note that the energy of the products is higher than the energy of the reactants. The difference in energy equals the heat of reaction, $\Delta H$. Because the energy increases, $\Delta H$ is positive and the reaction is endothermic.

Now look at the reverse reaction:

$$NOCl + Cl \longrightarrow NO + Cl_2$$

The activation energy is 2 kJ/mol, the difference in energy of the initial species, NOCl + Cl, and the activated complex. This is a smaller quantity than that for the forward reaction, so the rate constant for the reverse reaction is larger.

Figure 14.14 shows the potential-energy curve for an exothermic reaction. In this case, the energy of the reactants is higher than that of the products, so heat energy is released when the reaction goes in the forward direction.

**Understanding Concepts:**
**Transition States and**
**Activation Energy**

**Figure 14.14**
**Potential-energy curve for an exothermic reaction.**
The energy of the reactants is higher than that of the products, so heat energy is released when the reaction goes in the forward direction.

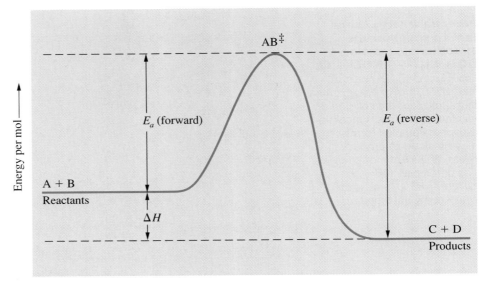

## CONCEPT CHECK 14.5

Consider the following potential energy curves for two different reactions:

a. Which reaction has a higher activation energy for the forward reaction?

b. If both reactions were run at the same temperature and have the same orientation requirements to react, which one would have the larger rate constant?

c. Are these reactions exothermic or endothermic?

## 14.6 Arrhenius Equation

Rate constants for most chemical reactions closely follow an equation of the form

$$k = Ae^{-E_a/RT}$$

*The mathematical equation $k = Ae^{-E_a/RT}$, which expresses the dependence of the rate constant on temperature,* is called the **Arrhenius equation,** after its formulator, the Swedish chemist Svante Arrhenius. ◄ Here $e$ is the base of natural logarithms,

**Figure 14.15**
**Plot of ln $k$ versus $1/T$.**
The logarithm of the rate constant for the decomposition of $N_2O_5$ (data from Table 14.3) is plotted versus $1/T$. A straight line is then fitted to the points; the slope equals $-E_a/(2.303\,R)$.

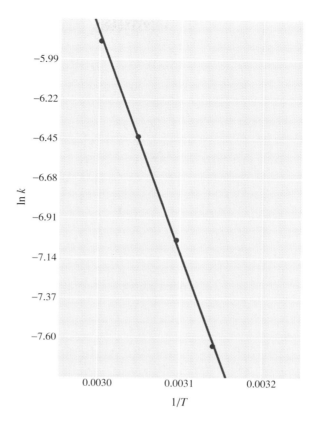

▶ Arrhenius published this equation in 1889 and suggested that molecules must be given enough energy to become "activated" before they could react. Collision and transition-state theories, which enlarged on this concept, were developed later (1920s and 1930s, respectively).

$2.718\ldots$; $E_a$ is the activation energy; $R$ is the gas constant, $8.31$ J/(K · mol); and $T$ is the absolute temperature. *The symbol A in the Arrhenius equation, which is assumed to be a constant,* is called the **frequency factor.** The frequency factor is related to the frequency of collisions with proper orientation ($pZ$). (The frequency factor does have a slight dependence on temperature, as you see from collision theory, but usually it can be ignored.)

It is useful to recast Arrhenius's equation in logarithmic form. Taking the natural logarithm of both sides of the Arrhenius equation gives

$$\ln k = \ln A - \frac{E_a}{RT}$$

Let us make the following identification of symbols:

$$\underbrace{\ln k}_{y} = \underbrace{\ln A}_{b} + \underbrace{\left(\frac{-E_a}{R}\right)}_{m}\underbrace{\left(\frac{1}{T}\right)}_{x}$$

This shows that if you plot $\ln k$ against $1/T$, you should get a straight line. The slope of this line is $-E_a/R$, from which you can obtain the activation energy $E_a$. The intercept is $\ln A$. Figure 14.15 shows a plot of $\ln k$ versus $1/T$ for the data given in Table 14.3. It demonstrates that the points do lie on a straight line.

You can put the previous equation into a form that is useful for computation. First you write the equation for two different absolute temperatures $T_1$ and $T_2$. You

**Table 14.3**
**Rate Constant for the Decomposition of $N_2O_5$ at Various Temperatures**

| Temperature (°C) | $k$ (/s) |
|---|---|
| 45.0 | $4.8 \times 10^{-4}$ |
| 50.0 | $8.8 \times 10^{-4}$ |
| 55.0 | $1.6 \times 10^{-3}$ |
| 60.0 | $2.8 \times 10^{-3}$ |

write $k_1$ for the rate constant at temperature $T_1$ and $k_2$ for the rate constant at temperature $T_2$.

$$\ln k_2 = \ln A - \frac{E_a}{RT_2}$$

$$\ln k_1 = \ln A - \frac{E_a}{RT_1}$$

You eliminate $\ln A$ by subtracting the equations.

$$\ln k_2 - \ln k_1 = -\frac{E_a}{RT_2} + \frac{E_a}{RT_1}$$

or

$$\ln \frac{k_2}{k_1} = \frac{E_a}{R}\left(\frac{1}{T_1} - \frac{1}{T_2}\right)$$

The next example illustrates the use of the Arrhenius equation in this form.

► An alternate example is provided in the Instructor's Resource Manual.

## EXAMPLE 14.7

### Using the Arrhenius Equation

The rate constant for the formation of hydrogen iodide from the elements

$$H_2(g) + I_2(g) \longrightarrow 2HI(g)$$

is $2.7 \times 10^{-4}$ L/(mol · s) at 600 K and $3.5 \times 10^{-3}$ L/(mol · s) at 650 K.
a. Find the activation energy $E_a$. b. Then calculate the rate constant at 700 K.

 **PROBLEM STRATEGY**

a. Substitute the data given in the problem statement into the equation noted just before this example, then solve for $E_a$. b. Use the same equation, but substitute for $k_1$, $T_1$, $T_2$, and $E_a$ obtained in a, and solve for $k_2$.

**SOLUTION**

a.

$$\ln \frac{3.5 \times 10^{-3}}{2.7 \times 10^{-4}} = \frac{E_a}{8.31 \text{ J/(mol · K)}}\left(\frac{1}{600 \text{ K}} - \frac{1}{650 \text{ K}}\right)$$

$$\ln 1.30 \times 10^{1} = 1.11 = \frac{E_a}{8.31 \text{ J/mol}} \times (1.28 \times 10^{-4})$$

Hence,

$$E_a = \frac{2.56 \times 8.31 \text{ J/mol}}{1.28 \times 10^{-4}} = \mathbf{1.66 \times 10^5 \text{ J/mol}}$$

b. Substitute $E_a = 1.66 \times 10^5$ J/mol and

$$k_1 = 2.7 \times 10^{-4} \text{ L/(mol · s)} \qquad (T_1 = 600 \text{ K})$$
$$k_2 = \text{unknown as yet} \qquad (T_2 = 700 \text{ K})$$

You get

$$\ln \frac{k_2}{2.7 \times 10^{-4} \text{ L/(mol} \cdot \text{s)}} = \frac{1.66 \times 10^5 \text{ J/mol}}{8.31 \text{ J/(mol} \cdot \text{K)}} \times \left(\frac{1}{600 \text{ K}} - \frac{1}{700 \text{ K}}\right) = 4.77$$

Taking antilogarithms,

$$\frac{k_2}{2.7 \times 10^{-4} \text{ L/(mol} \cdot \text{s)}} = e^{4.77} = 1.2 \times 10^2$$

Hence,

$$k_2 = (1.2 \times 10^2) \times (2.7 \times 10^{-4}) \text{ L/(mol} \cdot \text{s)} = \mathbf{3.2 \times 10^{-2} \text{ L/(mol} \cdot \text{s)}}$$

**Problem Solving:
Determining
Activation Energy**

### EXERCISE 14.7

Acetaldehyde, $CH_3CHO$, decomposes when heated.

$$CH_3CHO(g) \longrightarrow CH_4(g) + CO(g)$$

The rate constant for the decomposition is $1.05 \times 10^{-3}/(M^{1/2} \cdot \text{s})$ at 759 K and $2.14 \times 10^{-2}/(M^{1/2} \cdot \text{s})$ at 836 K. What is the activation energy for this decomposition? What is the rate constant at 865 K?

**See Problems 14.63, 14.64, 14.65, and 14.66.**

# Reaction Mechanisms

A balanced chemical equation is a description of the overall result of a chemical reaction. However, what actually happens at the molecular level may be more involved than is represented by this single equation. The reaction may take place in several steps. In the next sections, we will examine some reactions and see how the rate law can give us information about these steps, or *elementary reactions.*

## 14.7 Elementary Reactions

Consider the reaction of nitrogen dioxide with carbon monoxide.

$$NO_2(g) + CO(g) \longrightarrow NO(g) + CO_2(g) \quad \text{(net chemical equation)}$$

At temperatures below 500 K, this gas-phase reaction is believed to take place in two steps.

$$NO_2 + NO_2 \longrightarrow NO_3 + NO \quad \text{(elementary reaction)}$$
$$NO_3 + CO \longrightarrow NO_2 + CO_2 \quad \text{(elementary reaction)}$$

Each step, called an **elementary reaction,** is *a single molecular event, such as a collision of molecules, resulting in a reaction. The set of elementary reactions whose overall effect is given by the net chemical equation* is called the **reaction mechanism.**

According to the reaction mechanism just given, two $NO_2$ molecules collide and react to give the product molecule NO and the reaction intermediate $NO_3$. A **reaction intermediate** is *a species produced during a reaction that does not appear in the net equation because it reacts in a subsequent step in the mechanism.* Often the reaction intermediate has a fleeting existence and cannot be isolated from the reaction mixture. The $NO_3$ molecule is known only from its visible light spectrum. It reacts quickly with CO to give the product molecules $NO_2$ and $CO_2$.

The overall chemical equation, which represents the net result of these two elementary reactions in the mechanism, is obtained by adding the steps and canceling species that occur on both sides.

$$NO_2 + NO_2 \longrightarrow NO_3 + NO \qquad \text{(elementary reaction)}$$
$$NO_3 + CO \longrightarrow NO_2 + CO_2 \qquad \text{(elementary reaction)}$$
$$\overline{NO_2 + \cancel{NO_2} + \cancel{NO_3} + CO \longrightarrow \cancel{NO_3} + NO + \cancel{NO_2} + CO_2} \qquad \text{(overall equation)}$$

▶ An alternate example is provided in the Instructor's Resource Manual.

### EXAMPLE 14.8

## Writing the Overall Chemical Equation from a Mechanism

Carbon tetrachloride, $CCl_4$, is obtained by chlorinating methane or an incompletely chlorinated methane such as chloroform, $CHCl_3$. The mechanism for the gas-phase chlorination of $CHCl_3$ is

$$Cl_2 \rightleftharpoons 2Cl \qquad \text{(elementary reaction)}$$
$$Cl + CHCl_3 \longrightarrow HCl + CCl_3 \qquad \text{(elementary reaction)}$$
$$Cl + CCl_3 \longrightarrow CCl_4 \qquad \text{(elementary reaction)}$$

Obtain the net, or overall, chemical equation from this mechanism.

**SOLUTION**

The first step produces two Cl atoms (a reaction intermediate). One Cl atom is used in the second step and another is used in the third step. Thus, all Cl atoms cancel. Similarly, the intermediate $CCl_3$, produced in the second step, is used up in the third step. You can cancel all Cl and $CCl_3$ species.

$$Cl_2 \rightleftharpoons \cancel{2Cl}$$
$$\cancel{Cl} + CHCl_3 \longrightarrow HCl + \cancel{CCl_3}$$
$$\overline{\cancel{Cl} + \cancel{CCl_3} \longrightarrow CCl_4}$$
$$Cl_2 + CHCl_3 \longrightarrow HCl + CCl_4 \qquad \textbf{(overall equation)}$$

### EXERCISE 14.8

The iodide ion catalyzes the decomposition of aqueous hydrogen peroxide, $H_2O_2$. This decomposition is believed to occur in two steps.

$$H_2O_2 + I^- \longrightarrow H_2O + IO^- \qquad \text{(elementary reaction)}$$
$$H_2O_2 + IO^- \longrightarrow H_2O + O_2 + I^- \qquad \text{(elementary reaction)}$$

What is the overall equation representing this decomposition? Note that $IO^-$ is a reaction intermediate. The iodide ion is not an intermediate; it was added to the reaction mixture.

**See Problems 14.69 and 14.70.**

## Molecularity

Elementary reactions are classified according to their molecularity. The **molecularity** is *the number of molecules on the reactant side of an elementary reaction*. A **unimolecular reaction** is *an elementary reaction that involves one reactant molecule;* a **bimolecular reaction** is *an elementary reaction that involves two reactant molecules*. Bimolecular reactions are the most common. Unimolecular reactions are best illustrated by decomposition of some previously excited species. Some gas-phase reactions are thought to occur in a **termolecular reaction,** *an elementary reaction that involves three reactant molecules*. Higher molecularities are not encountered, presumably because the chance of the correct four molecules coming together at once is extremely small.

As an example of a unimolecular reaction, consider the elementary process in which an energetically excited ozone molecule (symbolized by $O_3^*$) spontaneously decomposes.

$$O_3^* \longrightarrow O_2 + O$$

Normally, a molecule in a sample of ozone gas is in a lower energy level. But such a molecule may be excited to a higher level if it collides with another molecule or absorbs a photon. The energy of the excited molecule is distributed among its three nuclei as vibrational energy. After a period of time, this energy becomes redistributed. If by chance most of the energy finds its way to one end oxygen atom, that atom will fly off. In other words, the excited ozone molecule decomposes into an oxygen molecule and an oxygen atom.

All of the steps in the mechanism of the reaction of $NO_2$ with CO, given earlier in this section, are bimolecular. Consider the first step, which involves the reaction of two $NO_2$ molecules.

$$NO_2 + NO_2 \longrightarrow NO_3 + NO$$

When these two $NO_2$ molecules come together, they form an activated complex of the six atoms, $(NO_2)_2$, which immediately breaks into two new molecules, $NO_3$ and NO.

The overall reaction of two atoms—say, two Br atoms—to form a diatomic molecule ($Br_2$) is normally a termolecular process. When two bromine atoms collide, they form an excited bromine molecule, $Br_2^*$. This excited molecule immediately flies apart, re-forming the atoms, unless another atom or molecule is present just at the moment of molecule formation to take away the excess energy. Suppose an argon atom and the two bromine atoms all collide at the same moment.

$$Br + Br + Ar \longrightarrow Br_2 + Ar^*$$

Energy that would have been left with the bromine molecule is now picked up by the argon atom (giving the energized atom $Ar^*$). The bromine molecule is stabilized by being left in a lower energy level.

▶ An alternate example is provided in the Instructor's Resource Manual.

### EXAMPLE 14.9

## Determining the Molecularity of an Elementary Reaction

What is the molecularity of each step in the mechanism described in Example 14.8?

$$Cl_2 \rightleftharpoons 2Cl$$
$$Cl + CHCl_3 \longrightarrow HCl + CCl_3$$
$$Cl + CCl_3 \longrightarrow CCl_4$$

**SOLUTION**

The molecularity of any elementary reaction equals the number of reactant molecules. Thus, the forward part of the first step is **unimolecular;** the reverse of the first step, the second step, and the third step are each **bimolecular.**

---

**EXERCISE 14.9**

The following is an elementary reaction that occurs in the decomposition of ozone in the stratosphere by nitric oxide.

$$O_3 + NO \longrightarrow O_2 + NO_2$$

What is the molecularity of this reaction? That is, is the reaction unimolecular, bimolecular, or termolecular?

---

See Problems 14.71 and 14.72.

## Rate Equation for an Elementary Reaction

There is no necessarily simple relationship between the overall reaction and the rate law that you observe for it. As we stressed before, the rate law must be obtained experimentally. However, when you are dealing with an elementary reaction, the rate does have a simple, predictable form.

> For an elementary reaction, the rate is proportional to the product of the concentration of each reactant molecule.

To understand this, let us look at the different possibilities. Consider the unimolecular elementary reaction

$$A \longrightarrow B + C$$

For each A molecule there is a definite probability, or chance, that it will decompose into B and C molecules. The more A molecules there are in a given volume, the more A molecules that can decompose in that volume per unit time. In other words, the rate of reaction is proportional to the concentration of A.

$$Rate = k[A]$$

Now consider a bimolecular elementary reaction, such as

$$A + B \longrightarrow C + D$$

For the reaction to occur, the reactant molecules A and B must collide. Reaction does not occur with every collision. Nevertheless, the rate of formation of product is proportional to the frequency of molecular collisions, because a definite fraction of the collisions produces reaction. Within a given volume, the frequency of collisions is proportional to the number of A molecules, $n_A$, times the number of B molecules, $n_B$. Furthermore, the concentration of A is proportional to $n_A$, and the concentration of B is proportional to $n_B$. Therefore, the rate of this elementary reaction is proportional to [A][B].

$$Rate = k[A][B]$$

A termolecular elementary reaction has a rate equation that is obtained by similar reasoning. For the elementary reaction

$$A + B + C \longrightarrow D + E$$

the rate is proportional to the concentrations of A, B, and C.

$$\text{Rate} = k[\text{A}][\text{B}][\text{C}]$$

Any reaction you observe is likely to consist of several elementary steps, and the rate law that you find is the combined result of these steps. This is why you cannot predict the rate law by looking at the overall equation.

▶ An alternate example is provided in the Instructor's Resource Manual.

---

**EXAMPLE 14.10**

## Writing the Rate Equation for an Elementary Reaction

Write rate equations for each of the following elementary reactions.

a. Ozone is converted to $O_2$ by NO in a single step.

$$O_3 + NO \longrightarrow O_2 + NO_2$$

b. The recombination of iodine atoms occurs as follows:

$$I + I + M \longrightarrow I_2 + M^*$$

where M is some atom or molecule that absorbs energy from the reaction.

c. An $H_2O$ molecule absorbs energy; some time later enough of this energy flows into one O—H bond to break it.

$$H_2O \longrightarrow H + O—H$$

**SOLUTION**

The rate equation can be written directly from the elementary reaction (but *only* for an elementary reaction).

a. **Rate** = $k[O_3][NO]$

b. **Rate** = $k[I]^2[M]$

c. **Rate** = $k[H_2O]$

---

**EXERCISE 14.10**

Write the rate equation, showing the dependence of rate on concentrations, for the elementary reaction

$$NO_2 + NO_2 \longrightarrow N_2O_4$$

---

See Problems 14.73 and 14.74.

In the next section, we look at the relationship between a reaction mechanism and the observed rate law.

---

## 14.8  The Rate Law and the Mechanism

The mechanism of a reaction cannot be observed directly. A mechanism is devised to explain the experimental observations. It is like the explanation provided by a detective to explain a crime in terms of the clues found. Other explanations may be possible, and further clues may make one of these other explanations seem more

▶ You see the scientific method in operation here, which recalls the discussion in Chapter 1. Experiments have been made from which you determine the rate law. Then a mechanism is devised to explain the rate law. This mechanism in turn suggests more experiments. These may confirm the explanation or may disagree with it. If they disagree, a new mechanism must be devised that explains all of the experimental evidence.

plausible than the currently accepted one. So it is with reaction mechanisms. They are accepted provisionally, with the understanding that further experiments may lead you to accept another mechanism as the more probable explanation. ◀

An important clue in understanding the mechanism of a reaction is the rate law. The reason for its importance is that once you assume a mechanism, you can predict the rate law. If this prediction does not agree with the experimental rate law, the assumed mechanism must be wrong. Take, for example, the overall equation

$$2NO_2(g) + F_2(g) \longrightarrow 2NO_2F(g)$$

If you follow the rate of disappearance of $F_2$, you observe that it is directly proportional to the concentration of $NO_2$ and $F_2$.

$$\text{Rate} = k[NO_2][F_2] \qquad \text{(experimental rate law)}$$

This rate law is a summary of the experimental data. Assume that the reaction occurs in a single elementary reaction.

$$NO_2 + NO_2 + F_2 \longrightarrow NO_2F + NO_2F \quad \text{(elementary reaction)}$$

This, then, is your assumed mechanism. Because this is an elementary reaction, you can immediately write the rate law predicted by it.

$$\text{Rate} = k[NO_2]^2[F_2] \qquad \text{(predicted rate law)}$$

However, this does not agree with experiment, and your assumed mechanism must be discarded. You conclude that the reaction occurs in more than one step.

### Rate-Determining Step

▶ Experiments in General Chemistry, 14, "The Rate of an Iodine Clock Reaction." ⚗

The reaction of $NO_2$ with $F_2$ is believed to occur in the following steps (elementary reactions).

$$NO_2 + F_2 \xrightarrow{k_1} NO_2F + F \qquad \text{(slow step)}$$
$$F + NO_2 \xrightarrow{k_2} NO_2F \qquad \text{(fast step)}$$

The rate constant for each step has been written over the arrow. The net result of the mechanism must be equivalent to the net result of the overall equation. By adding the two steps together, you can see that this is the case.

$$
\begin{array}{l}
NO_2 + F_2 \longrightarrow NO_2F + F \\
\underline{F + NO_2 \longrightarrow NO_2F} \\
2NO_2 + F_2 \longrightarrow 2NO_2F
\end{array}
$$

The F atom is a reaction intermediate.

The mechanism must also be in agreement with the experimental rate law. Let us look at that. Note that the second step is assumed to be much faster than the first, so that as soon as $NO_2$ and $F_2$ react, the F atom that is formed reacts with an $NO_2$ molecule to give another $NO_2F$ molecule. Therefore, the rate of disappearance of $F_2$ (and therefore the rate law) is determined completely by the slow step, or rate-determining step. The **rate-determining step** is *the slowest step in the reaction mechanism.*

To understand better the significance of the rate-determining step, suppose you and a friend want to start a chemistry study group. You and your friend decide to send invitation cards to some students in your class. You write a lengthy note on each card explaining the study group and then sign the card, taking an average of 2.0 minutes per card. Your friend puts the card in an envelope, affixes a computer-printed address label, seals the envelope, stamps it, and places it in a stack for mailing, taking 0.5 minute per card. How long does it take you and your friend to do 100 cards? What is the average time taken per card (*rate* of completing the cards)?

Because you take longer than your friend to do each card, your friend can complete the task for one card while you are working on another (your friend has time left over). It takes you a total of (100 cards × 2.0 min/card) to do your step in the task. When you have finished the last card, your friend still has to place the card in the envelope, and so forth. Therefore, you have to add an additional 0.5 min to the total time.

$$\text{Total time for 100 cards} = (100 \text{ cards} \times 2.\underline{0} \text{ min/card}) + 0.\underline{5} \text{ min}$$
$$= 2\underline{0}0.5 \text{ min}$$

We have underlined the last significant figure.

The rate of completing the invitation cards is

$$\text{Rate} = 2\underline{0}0.5 \text{ min/100 cards} = 2.0 \text{ min/card}$$

The rate is essentially the time it takes you to compose a note and sign a card. The time for your friend to finish the task after you finish is insignificant compared with the total time. The rate for the task equals the rate for the slower step (the rate-determining step).

The rate-determining step in the reaction of $NO_2$ and $F_2$ is the first step in the mechanism, in which $NO_2$ reacts with $F_2$ to produce $NO_2F$ and an F atom. The rate equation for this rate-determining step of the mechanism is

$$\text{Rate} = k_1[NO_2][F_2]$$

This should equal the experimental rate law (otherwise the mechanism cannot be correct), which it does if you equate $k_1$ to $k$ (the experimental rate constant). This agreement is not absolute evidence that the mechanism is correct. However, one can perform experiments to see whether fluorine atoms react very quickly with nitrogen dioxide. Such experiments show that they do.

▶ An alternate example is provided in the Instructor's Resource Manual.

### EXAMPLE 14.11

## Determining the Rate Law from a Mechanism with an Initial Slow Step

Ozone reacts with nitrogen dioxide to produce oxygen and dinitrogen pentoxide.

$$O_3(g) + 2NO_2(g) \longrightarrow O_2(g) + N_2O_5(g)$$

The proposed mechanism is

$$O_3 + NO_2 \longrightarrow NO_3 + O_2 \qquad \text{(slow)}$$
$$NO_3 + NO_2 \longrightarrow N_2O_5 \qquad \text{(fast)}$$

What is the rate law predicted by this mechanism?

### SOLUTION

The first step in the mechanism is rate determining, because its rate is much slower than the second step. We can write the rate law directly from this step.

$$\text{Rate} = k[O_3][NO_2]$$

---

**EXERCISE 14.11**

The iodide ion-catalyzed decomposition of hydrogen peroxide, $H_2O_2$, is believed to follow the mechanism

$$H_2O_2 + I^- \xrightarrow{k_1} H_2O + IO^- \qquad \text{(slow)}$$
$$H_2O_2 + IO^- \xrightarrow{k_2} H_2O + O_2 + I^- \qquad \text{(fast)}$$

What rate law is predicted by this mechanism? Explain.

---

See Problems 14.75 and 14.76.

### Mechanisms with an Initial Fast Step

A somewhat more complicated situation occurs when the rate-determining step follows an initial fast, equilibrium step. The decomposition of dinitrogen pentoxide,

$$2N_2O_5(g) \longrightarrow 4NO_2(g) + O_2(g) \qquad \text{(overall equation)}$$

which we discussed in the chapter opening section, is believed to follow this type of mechanism.

$$N_2O_5 \underset{k_{-1}}{\overset{k_1}{\rightleftharpoons}} NO_2 + NO_3 \qquad \text{(fast, equilibrium)}$$

$$NO_2 + NO_3 \xrightarrow{k_2} NO + NO_2 + O_2 \qquad \text{(slow)}$$
$$NO_3 + NO \xrightarrow{k_3} 2NO_2 \qquad \text{(fast)}$$

To give the overall stoichiometry, you need to multiply the first step by two. Note that there are two reaction intermediates, $NO_3$ and NO. Figure 14.16 represents the mechanism by means of molecular models. Let us show that this mechanism is consistent with the experimentally determined rate law,

$$\text{Rate} = k[N_2O_5]$$

The second step in the mechanism is assumed to be much slower than the other steps and is therefore rate determining. Hence, the rate law predicted from this mechanism is

$$\text{Rate} = k_2[NO_2][NO_3]$$

However, this equation cannot be compared directly with experiment because it is written in terms of the reaction intermediate, $NO_3$. The experimental rate law will be written in terms of substances that occur in the chemical equation, not of reaction intermediates. For purposes of comparison, it is necessary to re-express the rate equation, eliminating $[NO_3]$. To do this, you must look at the first step in the mechanism.

This step is fast and reversible. That is, $N_2O_5$ dissociates rapidly into $NO_2$ and $NO_3$, and these products in turn react to re-form $N_2O_5$. The rate of the forward reaction (dissociation of $N_2O_5$) is

$$\text{Forward rate} = k_1[N_2O_5]$$

and the rate of the reverse reaction (formation of $N_2O_5$ from $NO_2$ and $NO_3$) is

$$\text{Reverse rate} = k_{-1}[NO_2][NO_3]$$

When the reaction first begins, there are no $NO_2$ or $NO_3$ molecules, and the reverse rate is zero. But as $N_2O_5$ dissociates, the concentration of $N_2O_5$ decreases and

**Figure 14.16**
**Representation of the mechanism of decomposition of $N_2O_5$, using molecular models.**
Note that little atomic rearrangement occurs during each step.

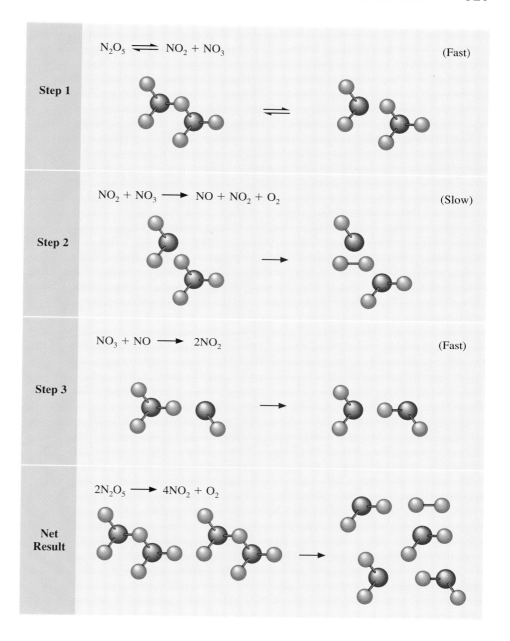

Step 1    $N_2O_5 \rightleftharpoons NO_2 + NO_3$    (Fast)

Step 2    $NO_2 + NO_3 \longrightarrow NO + NO_2 + O_2$    (Slow)

Step 3    $NO_3 + NO \longrightarrow 2NO_2$    (Fast)

Net Result    $2N_2O_5 \longrightarrow 4NO_2 + O_2$

the concentrations of $NO_2$ and $NO_3$ increase. Therefore, the forward rate decreases and the reverse rate increases. Soon the two rates become equal, such that $N_2O_5$ molecules form as often as other $N_2O_5$ molecules dissociate. The first step has reached *dynamic equilibrium*. Because these elementary reactions are much faster than the second step, this equilibrium is reached before any significant reaction by the second step occurs. Moreover, this equilibrium is maintained throughout the reaction. ◄

At equilibrium, the forward and reverse rates are equal, so you can write

▶ Chemical equilibrium is discussed in detail in the next chapter.

$$k_1[N_2O_5] = k_{-1}[NO_2][NO_3]$$

$$[NO_3] = \frac{k_1}{k_{-1}} \frac{[N_2O_5]}{[NO_2]}$$

Substituting into the rate law, you get

$$\text{Rate} = k_2[NO_2][NO_3] = k_2[\cancel{NO_2}] \times \frac{k_1}{k_{-1}} \frac{[N_2O_5]}{[\cancel{NO_2}]}$$

or

$$\text{Rate} = k_2 \frac{k_1}{k_{-1}} [N_2O_5]$$

Thus, if you identify $k_1 k_2 / k_{-1}$ as $k$, you reproduce the experimental rate law.

In addition to correctly predicting the experimental rate law, the mechanism must also be in agreement with the overall equation for the reaction. Although the first step is essentially in equilibrium, the products of this step ($NO_2$ and $NO_3$) are being continuously used in the subsequent steps. Note that each of these subsequent steps uses up a molecule of $NO_3$. For these steps to proceed, the first step must effectively produce two molecules of $NO_3$. Thus, the net result of the mechanism is as follows:

$$
\begin{array}{rcl}
2(N_2O_5 & \longrightarrow & NO_2 + \cancel{NO_3}) \\
\cancel{NO_2} + \cancel{NO_3} & \longrightarrow & \cancel{NO} + NO_2 + O_2 \\
\cancel{NO_3} + \cancel{NO} & \longrightarrow & 2NO_2 \\
\hline
2N_2O_5 & \longrightarrow & 4NO_2 + O_2
\end{array}
$$

The net result of the mechanism is equivalent to the overall equation for the reaction, as it should be.

▶ An alternate example is provided in the Instructor's Resource Manual.

## EXAMPLE 14.12

## Determining the Rate Law from a Mechanism with an Initial Fast, Equilibrium Step

Nitric oxide can be reduced with hydrogen gas to give nitrogen and water vapor.

$$2NO(g) + 2H_2(g) \longrightarrow N_2(g) + 2H_2O(g) \qquad \text{(overall equation)}$$

A proposed mechanism is

$$2NO \underset{k_{-1}}{\overset{k_1}{\rightleftharpoons}} N_2O_2 \qquad \text{(fast, equilibrium)}$$

$$N_2O_2 + H_2 \overset{k_2}{\longrightarrow} N_2O + H_2O \qquad \text{(slow)}$$

$$N_2O + H_2 \overset{k_3}{\longrightarrow} N_2 + H_2O \qquad \text{(fast)}$$

What rate law is predicted by this mechanism?

### PROBLEM STRATEGY

You write the rate equation for the rate-determining (slow) step, just as in the previous example. In this case, however, the equation contains a species, $N_2O_2$, that does not appear in the overall equation for the reaction. You need to eliminate it from the final form of the rate law. Note that the first step is fast and reaches equilibrium. You use this fact to write an expression for $[N_2O_2]$. Then you substitute this expression for $[N_2O_2]$ into the rate equation.

### SOLUTION

According to the rate-determining step,

$$\text{Rate} = k_2[N_2O_2][H_2]$$

You try to eliminate $N_2O_2$ from the rate law by looking at the first step, which is fast and reaches equilibrium. At equilibrium, the forward rate and the reverse rate are equal.

$$k_1[NO]^2 = k_{-1}[N_2O_2]$$

Therefore, $[N_2O_2] = (k_1/k_{-1})[NO]^2$, so

$$\text{Rate} = \frac{k_2 k_1}{k_{-1}}[NO]^2[H_2]$$

Experimentally, you should observe the rate law

$$\textbf{Rate} = k[NO]^2[H_2]$$

where we have replaced the constants $k_2 k_1/k_{-1}$ with $k$, which represents the experimentally observed rate constant.

### EXERCISE 14.12

Nitric oxide, NO, reacts with oxygen to produce nitrogen dioxide.

$$2NO(g) + O_2(g) \longrightarrow 2NO_2(g) \qquad \text{(overall equation)}$$

If the mechanism is

$$NO + O_2 \underset{k_{-1}}{\overset{k_1}{\rightleftharpoons}} NO_3 \qquad \text{(fast, equilibrium)}$$

$$NO_3 + NO \overset{k_2}{\longrightarrow} NO_2 + NO_2 \qquad \text{(slow)}$$

what is the predicted rate law? Remember to express this in terms of substances in the chemical equation.

See Problems 14.77 and 14.78.

 **CONCEPT CHECK 14.6**

You are a chemist in charge of a research laboratory that is trying to increase the reaction rate for the balanced chemical reaction

$$X + 2Y \longrightarrow Z$$

a. One of your researchers comes into your office and states that she has found a material that significantly lowers the activation energy of the reaction. Explain the effect this will have on the rate of the reaction.

b. Another researcher states that after doing some experiments, he has determined that the rate law is

$$\text{Rate} = k[X][Y]$$

Is this possible?

c. Yet another person in the lab reports that the mechanism for the reaction is:

$$2Y \longrightarrow I \qquad \text{(slow)}$$

$$X + I \longrightarrow Z \qquad \text{(fast)}$$

Is the rate law from part b consistent with this mechanism? If not, what should the rate law be?

## 14.9 Catalysis

▶ Video no. 22 (Series A), "Catalysis."

▶ Video no. 35 (Series D), "Effect of Cobalt(II) Chloride."

A catalyst is a substance that has the seemingly miraculous power of speeding up a reaction without being consumed by it. In theory, you could add a catalyst to a reaction mixture and, after the reaction, separate that catalyst and use it over and over again. In practice, there is often some loss of catalyst through other reactions that can occur at the same time.

Catalysts are of enormous importance to the chemical industry, because they allow a reaction to occur with a reasonable rate at a much lower temperature than otherwise; lower temperatures translate into lower energy costs. Moreover, catalysts are often quite specific—they increase the rate of certain reactions, but not others. For instance, an industrial chemist can start with a mixture of carbon monoxide and hydrogen and produce methane gas using one catalyst or produce gasoline using another catalyst. The most remarkable catalysts are enzymes. *Enzymes* are the marvelously selective catalysts employed by biological organisms. A biological cell contains thousands of different enzymes that in effect direct all of the chemical processes that occur in the cell.

How can we explain how a catalyst can influence a reaction without being consumed by it? Briefly, the catalyst must participate in at least one step of a reaction and be regenerated in a later step. Consider the commercial preparation of sulfuric acid, $H_2SO_4$, from sulfur dioxide, $SO_2$. The first step involves the reaction of $SO_2$ with $O_2$ to produce sulfur trioxide, $SO_3$. For this reaction to occur at an economical rate, it requires a catalyst. An early industrial process employed nitric oxide, NO, as the catalyst.

$$2SO_2(g) + O_2(g) \xrightarrow{\text{NO}} 2SO_3(g)$$

Nitric oxide does not appear in the overall equation, but must participate in the reaction mechanism. Here is a proposed mechanism:

$$2NO + O_2 \longrightarrow 2NO_2$$
$$NO_2 + SO_2 \longrightarrow NO + SO_3$$

To obtain the overall reaction, the last step must occur twice each time the first step occurs once. As you can see from the mechanism, two molecules of NO are used up in the first step and are regenerated in the second step.

Note that the catalyst is an active participant in the reaction. But how does this participation explain the increase in speed of the catalyzed reaction over the uncatalyzed reaction? The Arrhenius equation (given at the beginning of Section 14.6) provides an answer. The catalyzed reaction mechanism makes available a reaction path having an increased overall rate of reaction. It increases this rate either by increasing the frequency factor $A$ or, more commonly, by decreasing the activation energy $E_a$. The most dramatic effect comes from decreasing the activation energy, because it occurs as an exponent in the Arrhenius equation.

The depletion of ozone in the stratosphere by Cl atoms provides an example of the lowering of activation energy by a catalyst. Ozone is normally present in the stratosphere and provides protection against biologically destructive, short-wavelength ultraviolet radiation from the sun. Some recent ozone depletion in the stratosphere is believed to result from the Cl-catalyzed decomposition of $O_3$. Cl atoms in the stratosphere originate from the decomposition of chlorofluorocarbons (CFCs), which are compounds manufactured as refrigerants, aerosol propellants, and so forth. These Cl

**Figure 14.17**
**Comparison of activation energies in the uncatalyzed and catalyzed decompositions of ozone.**
The uncatalyzed reaction is $O_3 + O \longrightarrow O_2 + O_2$. Catalysis by Cl atoms provides an alternate pathway with lower activation energy, and therefore a faster reaction.

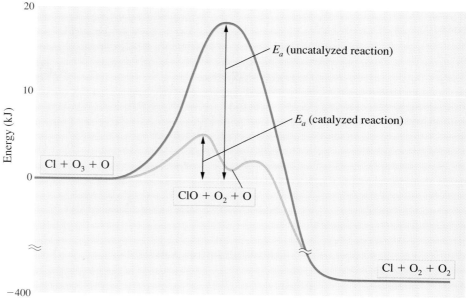

▶ **Ozone depletion in the stratosphere was discussed in an essay at the end of Chapter 10.**

▶ **Video no. 14 (Series B), "Exploding Flasks: Heterogeneous Catalysis."**

▶ **Animation: Homogeneous Catalysis**

atoms react with ozone to form ClO and $O_2$, and the ClO reacts with O atoms (normally in the stratosphere) to produce Cl and $O_2$.

$$\begin{array}{r} Cl(g) + O_3(g) \longrightarrow ClO(g) + O_2(g) \\ ClO(g) + O(g) \longrightarrow Cl(g) + O_2(g) \\ \hline O_3(g) + O(g) \longrightarrow 2O_2(g) \end{array}$$

The net result is the decomposition of ozone with O atoms to produce $O_2$. Figure 14.17 shows the potential-energy curves for the uncatalyzed and the catalyzed reactions. The uncatalyzed reaction has such a large activation energy that its rate is extremely low. The addition of chlorine atoms provides an alternate pathway to the same overall reaction, but has a much lower activation energy and therefore an increased rate. ◀

## Homogeneous Catalysis

The oxidation of sulfur dioxide using nitric oxide as a catalyst is an example of **homogeneous catalysis,** which is *the use of a catalyst in the same phase as the reacting species.* The catalyst NO and the reacting species $SO_2$ and $O_2$ are all in the gas phase. Another example occurs in the oxidation of thallium(I) to thallium(III) by cerium(IV) in aqueous solution.

$$2Ce^{4+}(aq) + Tl^+(aq) \longrightarrow 2Ce^{3+}(aq) + Tl^{3+}(aq)$$

The uncatalyzed reaction is very slow; presumably it involves the collision of three positive ions. The reaction can be catalyzed by manganese(II) ion, however. The mechanism is thought to be

$$\begin{array}{r} Ce^{4+} + Mn^{2+} \longrightarrow Ce^{3+} + Mn^{3+} \\ Ce^{4+} + Mn^{3+} \longrightarrow Ce^{3+} + Mn^{4+} \\ Mn^{4+} + Tl^+ \longrightarrow Tl^{3+} + Mn^{2+} \end{array}$$

Each step is bimolecular.

**Figure 14.18**
**Heterogeneous catalysts.**
*Left:* The chemists shown here are developing platinum, palladium, and ruthenium metal catalysts. *Right:* The catalyst is absorbed on support materials, such as carbon and alumina.

▶ **Animation: Heterogeneous Catalysis**

▶ **Understanding the chemical processes occurring at surfaces is being advanced by new techniques, including x-ray photoelectron spectroscopy and scanning tunneling microscopy.**

## Heterogeneous Catalysis

Some of the most important industrial reactions involve **heterogeneous catalysis**—that is, *the use of a catalyst that exists in a different phase from the reacting species, usually a solid catalyst in contact with a gaseous or liquid solution of reactants* (Figure 14.18). Such surface, or heterogeneous, catalysis is thought to occur by chemical adsorption of the reactants onto the surface of the catalyst. *Adsorption* is the attraction of molecules to a surface. In *physical adsorption,* the attraction is provided by weak intermolecular forces. **Chemisorption,** by contrast, is *the binding of a species to a surface by chemical bonding forces.* It may happen that bonds in the species are broken during chemisorption, and this may provide the basis of catalytic action in certain cases. ◀

An example of heterogeneous catalysis involving chemisorption is provided by *catalytic hydrogenation.* This is the addition of $H_2$ to a compound, such as one with a carbon–carbon double bond, using a catalyst of platinum or nickel metal. Vegetable oils, which contain carbon–carbon double bonds, are changed to solid fats (shortening) when the bonds are catalytically hydrogenated. In the case of ethylene, $C_2H_4$, the equation is

$$
\underset{\text{ethylene}}{\overset{\displaystyle H}{\underset{\displaystyle H}{}}C=C\overset{\displaystyle H}{\underset{\displaystyle H}{}}} + H_2 \xrightarrow{\text{Pt}} \underset{\text{ethane}}{H-\overset{\displaystyle \overset{H}{|}}{\underset{\displaystyle \underset{H}{|}}{C}}-\overset{\displaystyle \overset{H}{|}}{\underset{\displaystyle \underset{H}{|}}{C}}-H}
$$

A mechanism for this reaction is represented by the four steps shown in Figure 14.19. In *(A),* ethylene and hydrogen molecules diffuse to the catalyst surface, where, as shown in *(B),* they undergo chemisorption. The pi electrons of ethylene form bonds to the metal, and hydrogen molecules break into H atoms that bond to the metal. In *(C),* two H atoms migrate to an ethylene molecule bonded to the catalyst, where they react to form ethane. Then, in *(D),* because ethane does not bond to the metal, it diffuses away. Note that the catalyst surface that was used in *(B)* is regenerated in *(D).*

At the beginning of this section, we described the catalytic oxidation of $SO_2$ to $SO_3$, the anhydride of sulfuric acid. In an early process, the homogeneous catalyst NO was used. Today, in the *contact process,* a heterogeneous catalyst, Pt or $V_2O_5$, is used. Surface catalysts are used in the catalytic converters of automobiles to convert substances that would be atmospheric pollutants, such as CO and NO, into harmless substances, such as $CO_2$ and $N_2$ (Figure 14.20).

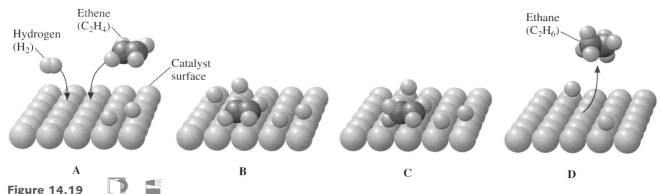

**Figure 14.19**

**Proposed mechanism of catalytic hydrogenation of $C_2H_4$.** *(A)* $C_2H_4$ and $H_2$ molecules diffuse to the catalyst. *(B)* The molecules form bonds to the catalyst surface. (The $H_2$ molecules dissociate to atoms in the process.) *(C)* H atoms migrate to the $C_2H_4$ molecule, where they react to form $C_2H_6$. *(D)* $C_2H_6$ diffuses away from the catalyst.

**Figure 14.20**
**Automobile catalytic converter.** *Top:* Exhaust gases from the automobile engine pass to the exhaust manifold, then to the catalytic converter, where pollutants CO and NO are converted to $CO_2$ and $N_2$. *Bottom:* Cross-sectional views of some automobile catalytic converters, showing the packing material with catalyst on its surface.

## Enzyme Catalysis

▶ Demonstration of the catalytic decomposition of hydrogen peroxide: Summerlin and Ealy, *Chemical Demonstrations*, Vol. 1 (Washington, D.C.: Am. Chem. Soc., 1985), no. 44, p. 71.

Almost all enzymes, the catalysts of biological organisms, are protein molecules with molecular weights ranging to over a million amu. An enzyme has enormous catalytic activity, converting a thousand or so reactant molecules to products in a second. Enzymes are also highly specific, each enzyme acting only on a specific substance, or a specific type of substance, catalyzing it to undergo a particular reaction. For

Substrate
+
Enzyme

Enzyme-substrate complex
leading to new product formation

Products
+
Enzyme

**Figure 14.21**

**Enzyme action (lock-and-key model).** The enzyme has an active site to which the substrate binds to form an enzyme–substrate complex. The active site of the enzyme acts like a lock into which the substrate (key) fits. While bound to the enzyme, the substrate may have bonds weakened or new bonds formed to yield the products, which leave the enzyme.

example, the enzyme sucrase, which is present in the digestive fluid of the small intestine, catalyzes the reaction of sucrose (table sugar) with water to form the simpler sugars glucose and fructose. *The substance whose reaction the enzyme catalyzes* is called the **substrate.** Thus, the enzyme sucrase acts on the substrate sucrose.

Figure 14.21 shows schematically how an enzyme acts. The enzyme molecule is a protein chain that tends to fold into a roughly spherical form with an *active site* at which the substrate molecule binds and the catalysis takes place. The substrate molecule, S, fits into the active site on the enzyme molecule, E, somewhat in the way a key fits into a lock, forming an enzyme–substrate complex, ES. (The lock-and-key model is only a rough approximation, because the active site on an enzyme deforms somewhat to fit the substrate molecule.) In effect, the active site "recognizes" the substrate and gives the enzyme its specificity. On binding to the enzyme, the substrate may have bonds that weaken or new bonds form that help yield the products, P.

$$E + S \longrightarrow ES \longrightarrow E + P$$

The formation of the enzyme–substrate complex provides a new pathway to products with a lower activation energy. Figure 14.22 compares the potential-energy curves of the uncatalyzed and catalyzed reactions. Note the lower activation energy of the catalyzed reaction.

**Figure 14.22**
**Potential-energy curves for the reaction of substrate, S, to products, P.** *(A)* The uncatalyzed reaction of substrate to product. *(B)* The enzyme-catalyzed reaction of substrate to product. Note that the catalyzed reaction provides a pathway with lower activation energy than does the uncatalyzed reaction.

A

B

# Seeing Molecules React

**FRONTIERS**

Exactly what happens when molecules react? Transition-state theory and, more recently, calculations from quantum-mechanical theory have given us insight into how molecules behave in elementary reactions. But can we actually observe *through experiment* how molecules react? Until fairly recently, this seemed an impossible dream. Molecules are so small, the distances that atoms in molecules move to break bonds and form new ones are so miniscule, and the time that it takes a reaction to occur is so fast that to be able to observe molecules while they react seemed well beyond human ability.

Within the last couple of decades, however, chemistry has progressed enormously in its quest to deal with substances at the molecular level. In 1987, Ahmed H. Zewail, professor of chemistry at California Institute of Technology, managed to devise an experimental technique that could follow molecules while they are reacting. In effect, Zewail developed "the world's smallest camera." In the process, he founded an area of research that has come to be called "femtochemistry," because reactions between molecules happen on the scale of femtoseconds. A femtosecond (fs) is $10^{-15}$ s, which is a million times smaller than a billionth of a second!

The experiment involves using a laser, first to send a *pump pulse* to excite a reactant molecule to a higher energy state. (See the essay on lasers at the end of Section 7.3.) Then, a weaker laser pulse or *probe pulse* of properly chosen wavelength is used to detect the molecule as it transforms to products. By varying the time between the pump pulse and the probe pulse, the experimenter can follow a molecule throughout the course of its reaction. The change of wavelength of the probe pulse tells the experimenter something about the change in character of the molecule as it reacts.

As an example of a reaction, Zewail looked at the decomposition of cyclobutane to ethylene:

$$\underset{\text{cyclobutane}}{\begin{array}{c} \text{H} \quad \text{H} \\ | \quad\; | \\ \text{H}-\text{C}-\text{C}-\text{H} \\ \text{H}-\text{C}-\text{C}-\text{H} \\ | \quad\; | \\ \text{H} \quad \text{H} \end{array}} \longrightarrow 2\;\underset{\text{ethylene}}{\begin{array}{c} \text{H} \qquad \text{H} \\ \text{C}=\text{C} \\ \text{H} \qquad \text{H} \end{array}}$$

Figure 14.23 shows two possible potential-energy diagrams for this reaction. The left diagram shows the cyclobutane molecule passing through a transition state

*(continued)*

**Figure 14.23**
**Two possible potential-energy curves for the decomposition of cyclobutane to ethylene** *Left:* The cyclobutane molecule goes to a transition state in which two carbon–carbon bonds are stretched. When the transition state goes over to products, these stretched bonds break. *Right:* The cyclobutane molecule goes to an intermediate molecule in which one carbon–carbon bond has broken. When this intermediate goes over to products, the other carbon–carbon bond breaks. Experiments show that this is what actually happens. (Reprinted with the permission of The Nobel e-Museum from wysiwyg://26/http://www.nobel.se/chemistry/laureates/1999/press.html.)

(at the peak of the potential curve) in which two carbon–carbon bonds have simultaneously stretched. When the transition state goes over to products, these stretched bonds break. The right diagram shows the cyclobutane molecule going first to an intermediate molecule (between the two peaks). In this intermediate molecule, one carbon–carbon bond is broken. When the intermediate molecule goes over to products, another bond breaks, giving two ethylene molecules. Zewail and his coworkers were able to show that cyclobutane decomposed to ethylene by first going to an intermediate molecule (that is, the reaction actually goes by the potential curve on the right in Figure 14.23). The total time of the reaction was 700 fs! In 1999, Zewail (Figure 14.24) won the Nobel Prize in chemistry for his work in femtochemistry.

**Figure 14.24**
**Ahmed H. Zewail** The chemist is shown here with his laser apparatus.

## A Checklist for Review

### Important Terms

catalyst (page 591)
reaction rate (14.1)
rate law (14.3)
rate constant (14.3)
reaction order (14.3)
half-life (14.4)
collision theory (14.5)

activation energy (14.5)
transition-state
   theory (14.5)
activated complex (14.5)
Arrhenius equation (14.6)
frequency factor (14.6)
elementary reaction (14.7)

reaction mechanism (14.7)
reaction intermediate (14.7)
molecularity (14.7)
unimolecular
   reaction (14.7)
bimolecular reaction (14.7)
termolecular reaction (14.7)

rate-determining step (14.8)
homogeneous
   catalysis (14.9)
heterogeneous
   catalysis (14.9)
chemisorption (14.9)
substrate (14.9)

### Key Equations

$$\ln\frac{[A]_t}{[A]_0} = -kt \qquad\qquad t_{1/2} = \frac{0.693}{k} \qquad\qquad \ln\frac{k_2}{k_1} = \frac{E_a}{R}\left(\frac{1}{T_1} - \frac{1}{T_2}\right)$$

### Summary of Facts and Concepts

The *reaction rate* is defined as the increase in moles of product per liter per second (or as the decrease in moles of reactant per liter per second). Rates of reaction are determined by following the change of concentration of a reactant or product, either by chemical analysis or by observing a physical property. It is found that reaction rates are proportional to concentrations of reactants raised to various powers (usually 1 or 2, but they can be fractional or negative). The *rate law* mathematically expresses the relationship between rates and concentrations for a chemical reaction. Although the rate law tells you how the rate of a reaction depends on concentrations at a given moment, it is possible to transform a rate law to show how concentrations

change with time. The *half-life* of a reaction is the time it takes for the reaction concentration to decrease to one-half of its original concentration.

Reaction rates can often double or triple with a 10°C rise in temperature. The effect of temperature on the rate can be explained by *collision theory*. According to this theory, two molecules react after colliding only when the energy of collision is greater than the *activation energy* and when the molecules are properly oriented. It is the rapid increase in the fraction of collisions having energy greater than the activation energy that explains the large temperature dependence of reaction rates. *Transition-state theory* explains reaction rates in terms of the formation of an *activated complex* of the colliding molecules. The *Arrhenius equation* is a mathematical relationship showing the dependence of a rate constant on temperature.

A chemical equation describes the overall result of a chemical reaction that may take place in one or more steps. The steps are called *elementary reactions,* and the set of elementary reactions that describes what is happening at the molecular level in an overall reaction is called the *reaction mechanism.* In some cases, the overall reaction involves a *reaction intermediate*—a species produced in one step but used up in a subsequent one. The rate of the overall reaction depends on the rate of the slowest step (the *rate-determining step*). This rate is proportional to the product of the concentrations of each reactant molecule in that step. If this step involves a reaction intermediate, its con-

centration can be eliminated by using the relationship of the concentrations in the preceding fast, *equilibrium* step.

A *catalyst* is a substance that speeds up a chemical reaction without being consumed by it. The catalyst is used up in one step of the reaction mechanism but regenerated in a later step. Catalytic activity operates by providing a reaction mechanism that has lower activation energy. Catalysis is classified as *homogeneous catalysis* if the substances react within one phase and as *heterogeneous catalysis* if the substances in a gas or liquid phase react at the surface of a solid catalyst. Many industrial reactions involve heterogeneous catalysis. Enzymes are biological catalysts (usually proteins). *Enzyme catalysis* is highly specific, so each enzyme acts on a particular kind of *substrate.*

## Operational Skills

1. **Relating the different ways of expressing reaction rates** Given the balanced equation for a reaction, relate the different possible ways of defining the rate of the reaction. **(EXAMPLE 14.1)**
2. **Calculating the average reaction rate** Given the concentration of reactant or product at two different times, calculate the average rate of reaction over that time interval. **(EXAMPLE 14.2)**
3. **Determining the order of reaction from the rate law** Given an empirical rate law, obtain the orders with respect to each reactant (and catalyst, if any) and the overall order. **(EXAMPLE 14.3)**
4. **Determining the rate law from initial rates** Given initial concentrations and initial-rate data (in which the concentrations of all species are changed one at a time, holding the others constant), find the rate law for the reaction. **(EXAMPLE 14.4)**
5. **Using the concentration-time equation for a first-order reaction** Given the rate constant and initial reactant concentration for a first-order reaction, calculate the reactant concentration after a definite time, or calculate the time it takes for the concentration to decrease to a prescribed value. **(EXAMPLE 14.5)**

6. **Relating the half-life of a reaction to the rate constant** Given the rate constant for a reaction, calculate the half-life. **(EXAMPLE 14.6)**
7. **Using the Arrhenius equation** Given the values of the rate constant for two temperatures, find the activation energy and calculate the rate constant at a third temperature. **(EXAMPLE 14.7)**
8. **Writing the overall chemical equation from a mechanism** Given a mechanism for a reaction, obtain the overall chemical equation. **(EXAMPLE 14.8)**
9. **Determining the molecularity of an elementary reaction** Given an elementary reaction, state the molecularity. **(EXAMPLE 14.9)**
10. **Writing the rate equation for an elementary reaction** Given an elementary reaction, write the rate equation. **(EXAMPLE 14.10)**
11. **Determining the rate law from a mechanism** Given a mechanism with an initial slow step, obtain the rate law **(EXAMPLE 14.11)**. Given a mechanism with an initial fast, equilibrium step, obtain the rate law **(EXAMPLE 14.12)**.

## Review Questions

**14.1** List the four variables or factors that can affect the rate of reaction.

**14.2** Define the rate of reaction of HBr in the following reaction. How is this related to the rate of formation of $Br_2$?

$$4HBr(g) + O_2(g) \longrightarrow 2Br_2(g) + 2H_2O(g)$$

**14.3** Give at least two physical properties that might be used to determine the rate of a reaction.

**14.4** A rate of reaction depends on four variables (Question 14.1). Explain by means of an example how the rate law deals with each of these variables.

**14.5** The exponents in a rate law have no relationship to the coefficients in the overall balanced equation for the reaction. Give an example of a balanced equation and the rate law for a reaction that clearly demonstrates this.

**14.6**    The reaction

$$3I^-(aq) + H_3AsO_4(aq) + 2H^+(aq) \longrightarrow$$
$$I_3^-(aq) + H_3AsO_3(aq) + H_2O(l)$$

is found to be first order with respect to each of the reactants. Write the rate law. What is the overall order?

**14.7**    The rate of a reaction is quadrupled when the concentration of one reactant is doubled. What is the order of the reaction with respect to this reactant?

**14.8**    A rate law is one-half order with respect to a reactant. What is the effect on the rate when the concentration of this reactant is doubled?

**14.9**    The reaction $A(g) \longrightarrow B(g) + C(g)$ is known to be first order in $A(g)$. It takes 25 s for the concentration of $A(g)$ to decrease by one-half of its initial value. How long does it take for the concentration of $A(g)$ to decrease to one-fourth of its initial value? to one-eighth of its initial value?

**14.10**    Compare the half-life equations for a first-order and second-order reaction. For which reaction order is the value of the half-life independent of the reactant concentration?

**14.11**    What two factors determine whether a collision between two reactant molecules will result in reaction?

**14.12**    Sketch a potential-energy diagram for the exothermic, elementary reaction

$$A + B \longrightarrow C + D$$

and on it denote the activation energies for the forward and reverse reactions. Also indicate the reactants, products, and activated complex.

**14.13**    Draw a structural formula for the activated complex in the following reaction:

$$NO_2 + NO_3 \longrightarrow NO + NO_2 + O_2$$

Refer to Figure 14.16, Step 2, for details of this reaction. Use dashed lines for bonds about to form or break. Use a single line for all other bonds.

**14.14**    Rate constants for reactions often follow the Arrhenius equation. Write this equation and then identify each term in it with the corresponding factor or factors from collision theory. Give a physical interpretation of each of those factors.

**14.15**    By means of an example, explain what is meant by the term *reaction intermediate*.

**14.16**    Why is it generally impossible to predict the rate law for a reaction on the basis of the chemical equation only?

**14.17**    The rate law for the reaction

$$2NO_2Cl(g) \longrightarrow 2NO_2(g) + Cl_2(g) \quad \text{(overall equation)}$$

is first order in nitryl chloride, $NO_2Cl$.

$$\text{Rate} = k[NO_2Cl]$$

Explain why the mechanism for this reaction cannot be the single elementary reaction

$$2NO_2Cl \longrightarrow 2NO_2 + Cl_2 \quad \text{(elementary reaction)}$$

**14.18**    There is often one step in a reaction mechanism that is rate determining. What characteristic of such a step makes it rate determining? Explain.

**14.19**    The dissociation of $N_2O_4$ into $NO_2$,

$$N_2O_4(g) \rightleftharpoons 2NO_2(g)$$

is believed to occur in one step. Obtain the concentration of $N_2O_4$ in terms of the concentration of $NO_2$ and the rate constants for the forward and reverse reactions, when the reactions have come to equilibrium.

**14.20**    How does a catalyst speed up a reaction? How can a catalyst be involved in a reaction without being consumed by it?

**14.21**    Compare physical adsorption and chemisorption (chemical adsorption).

**14.22**    Describe the steps in the catalytic hydrogenation of ethylene.

 ## Conceptual Problems

**14.23**    Consider the reaction $3A \longrightarrow 2B + C$.
a. One rate expression for the reaction is

$$\text{Rate of formation of C} = +\frac{\Delta[C]}{\Delta t}$$

Write two other rate expressions for this reaction in this form.
b. Using your two rate expressions, if you calculated the average rate of the reaction over the same time interval, would the rates be equal?
c. If your answer to part b was no, write two rate expressions that would give an equal rate when calculated over the same time interval.

**14.24**    Given the reaction $2A + B \longrightarrow C + 3D$, can you write the rate law for this reaction? If so, write the rate law; if not, why?

**14.25**    You perform some experiments for the reaction $A \longrightarrow B + C$ and determine the rate law has the form

$$\text{Rate} = k[A]^x$$

Calculate the value of exponent $x$ for each of the following.
a. [A] is tripled and you observe no rate change.
b. [A] is doubled and the rate doubles.
c. [A] is tripled and the rate goes up by a factor of 27.

**14.26** A friend of yours runs a reaction and generates the following plot. She explains that in following the reaction, she measured the concentration of a compound that she calls "E."

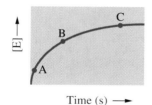

a. Your friend tells you that E is either a reactant or product. Which is it and why?
b. Is the average rate faster between points A and B or B and C? Why?

**14.27** Given the hypothetical plot shown here for the concentration of compound Y versus time, answer the following questions.

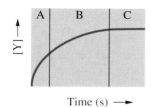

a. In which region of the curve does the rate have a constant value (A, B, or C)?
b. In which region of the curve is the rate the fastest (A, B, or C)?

**14.28** You carry out the following reaction by introducing $N_2O_4$ into an evacuated flask and observing the concentration change of the product over time.

$$N_2O_4(g) \longrightarrow 2NO_2(g)$$

Which one of the curves shown here reflects the data collected for this reaction?

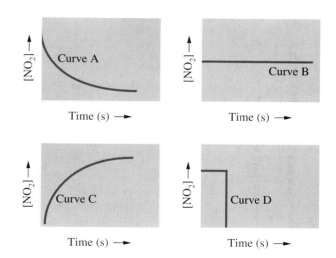

**14.29** You are running the reaction $2A + B \longrightarrow C + 3D$. Your lab partner has conducted the first two experiments to determine the rate law for the reaction. He has recorded the initial rates for these experiments in another data table. Come up with some reactant concentrations for Experiment 3 that will allow you to determine the rate law by measuring the initial rate.

| Experiment Number | Concentration of A ($M$) | Concentration of B ($M$) |
|---|---|---|
| 1 | 1.0 | 1.0 |
| 2 | 2.0 | 1.0 |
| 3 | | |

**14.30** The chemical reaction $A \longrightarrow B + C$ has a rate constant that obeys the Arrhenius equation. Predict what happens to both the rate constant $k$ and the rate of the reaction if the following were to occur.
a. a decrease in temperature
b. an increase in the activation energy of the forward and reverse reactions
c. an increase in both activation energy and temperature

## Practice Problems

### Reaction Rates

**14.31** Relate the rate of decomposition of $NO_2$ to the rate of formation of $O_2$ for the following reaction:

$$2NO_2(g) \longrightarrow 2NO(g) + O_2(g)$$

**14.32** For the reaction of hydrogen with iodine

$$H_2(g) + I_2(g) \longrightarrow 2HI(g)$$

relate the rate of disappearance of hydrogen gas to the rate of formation of hydrogen iodide.

**14.33** To obtain the rate of the reaction

$$5Br^-(aq) + BrO_3^-(aq) + 6H^+(aq) \longrightarrow 3Br_2(aq) + 3H_2O(l)$$

you might follow the $Br^-$ concentration or the $BrO_3^-$ concentration. How are the rates in terms of these species related?

**14.34** To obtain the rate of the reaction

$$3I^-(aq) + H_3AsO_4(aq) + 2H^+(aq) \longrightarrow I_3^-(aq) + H_3AsO_3(aq) + H_2O(l)$$

you might follow the $I^-$ concentration or the $I_3^-$ concentration. How are the rates in terms of these species related?

**14.35**    Ammonium nitrite, $NH_4NO_2$, decomposes in solution, as shown here.

$$NH_4NO_2(aq) \longrightarrow N_2(g) + 2H_2O(l)$$

The concentration of $NH_4^+$ ion at the beginning of an experiment was 0.500 $M$. After 3.00 hours, it was 0.432 $M$. What is the average rate of decomposition of $NH_4NO_2$ in this time interval?

**14.36**    Iron(III) chloride is reduced by tin(II) chloride.

$$2FeCl_3(aq) + SnCl_2(aq) \longrightarrow 2FeCl_2(aq) + SnCl_4(aq)$$

The concentration of $Fe^{3+}$ ion at the beginning of an experiment was 0.03586 $M$. After 4.00 min, it was 0.02638 $M$. What is the average rate of reaction of $FeCl_3$ in this time interval?

**14.37**    Azomethane, $CH_3NNCH_3$, decomposes according to the following equation:

$$CH_3NNCH_3(g) \longrightarrow C_2H_6(g) + N_2(g)$$

The initial concentration of azomethane was $1.50 \times 10^{-2}$ mol/L. After 9.00 min, this concentration decreased to $1.01 \times 10^{-2}$ mol/L. Obtain the average rate of reaction during this time interval. Express the answer in units of mol/(L · s).

**14.38**    Nitrogen dioxide, $NO_2$, decomposes upon heating to form nitric oxide and oxygen according to the following equation:

$$2NO_2(g) \longrightarrow 2NO(g) + O_2(g)$$

At the beginning of an experiment, the concentration of nitrogen dioxide in a reaction vessel was 0.1103 mol/L. After 60.0 s, the concentration decreased to 0.1076 mol/L. What is the average rate of decomposition of $NO_2$ during this time interval, in mol/(L · s)?

## Rate Laws

**14.39**    Hydrogen sulfide is oxidized by chlorine in aqueous solution.

$$H_2S(aq) + Cl_2(aq) \longrightarrow S(s) + 2HCl(aq)$$

The experimental rate law is

$$\text{Rate} = k[H_2S][Cl_2]$$

What is the reaction order with respect to $H_2S$? with respect to $Cl_2$? What is the overall order?

**14.40**    For the reaction of nitric oxide, NO, with chlorine, $Cl_2$,

$$2NO(g) + Cl_2(g) \longrightarrow 2NOCl(g)$$

the observed rate law is

$$\text{Rate} = k[NO]^2[Cl_2]$$

What is the reaction order with respect to nitric oxide? with respect to $Cl_2$? What is the overall order?

**14.41**    Oxalic acid, $H_2C_2O_4$, is oxidized by permanganate ion to $CO_2$ and $H_2O$.

$$2MnO_4^-(aq) + 5H_2C_2O_4(aq) + 6H^+(aq) \longrightarrow$$
$$2Mn^{2+}(aq) + 10CO_2(g) + 8H_2O(l)$$

The rate law is

$$\text{Rate} = k[MnO_4^-][H_2C_2O_4]$$

What is the order with respect to each reactant? What is the overall order?

**14.42**    Iron(II) ion is oxidized by hydrogen peroxide in acidic solution.

$$H_2O_2(aq) + 2Fe^{2+}(aq) + 2H^+(aq) \longrightarrow$$
$$2Fe^{3+}(aq) + 2H_2O(l)$$

The rate law is

$$\text{Rate} = k[H_2O_2][Fe^{2+}]$$

What is the order with respect to each reactant? What is the overall order?

**14.43**    In experiments on the decomposition of azomethane,

$$CH_3NNCH_3(g) \longrightarrow C_2H_6(g) + N_2(g)$$

the following data were obtained:

|  | Initial Concentration of Azomethane | Initial Rate |
|---|---|---|
| Exp. 1 | $1.13 \times 10^{-2}$ $M$ | $2.8 \times 10^{-6}$ $M/s$ |
| Exp. 2 | $2.26 \times 10^{-2}$ $M$ | $5.6 \times 10^{-6}$ $M/s$ |

What is the rate law? What is the value of the rate constant?

**14.44**    Ethylene oxide, $C_2H_4O$, decomposes when heated to give methane and carbon monoxide.

$$C_2H_4O(g) \longrightarrow CH_4(g) + CO(g)$$

The following kinetic data were observed for the reaction at 688 K:

|  | Initial Concentration of Ethylene Oxide | Initial Rate |
|---|---|---|
| Exp. 1 | 0.00272 $M$ | $5.57 \times 10^{-7}$ $M/s$ |
| Exp. 2 | 0.00544 $M$ | $1.11 \times 10^{-6}$ $M/s$ |

Find the rate law and the value of the rate constant for this reaction.

**14.45** Nitric oxide, NO, reacts with hydrogen to give nitrous oxide, $N_2O$, and water.

$$2NO(g) + H_2(g) \longrightarrow N_2O(g) + H_2O(g)$$

In a series of experiments, the following initial rates of disappearance of NO were obtained:

| Initial Concentrations | | Initial Rate of Reaction |
|---|---|---|
| NO | $H_2$ | of NO |
| Exp. 1 | $6.4 \times 10^{-3} M$ | $2.2 \times 10^{-3} M$ | $2.6 \times 10^{-5} M/s$ |
| Exp. 2 | $12.8 \times 10^{-3} M$ | $2.2 \times 10^{-3} M$ | $1.0 \times 10^{-4} M/s$ |
| Exp. 3 | $6.4 \times 10^{-3} M$ | $4.5 \times 10^{-3} M$ | $5.1 \times 10^{-5} M/s$ |

Find the rate law and the value of the rate constant for the reaction of NO.

**14.46** In a kinetic study of the reaction

$$2NO(g) + O_2(g) \longrightarrow 2NO_2(g)$$

the following data were obtained for the initial rates of disappearance of NO:

| Initial Concentrations | | Initial Rate of |
|---|---|---|
| NO | $O_2$ | Reaction of NO |
| Exp. 1 | 0.0125 M | 0.0253 M | 0.0281 M/s |
| Exp. 2 | 0.0250 M | 0.0253 M | 0.112 M/s |
| Exp. 3 | 0.0125 M | 0.0506 M | 0.0561 M/s |

Obtain the rate law. What is the value of the rate constant?

**14.47** Chlorine dioxide, $ClO_2$, is a reddish-yellow gas that is soluble in water. In basic solution it gives $ClO_3^-$ and $ClO_2^-$ ions.

$$2ClO_2(aq) + 2OH^-(aq) \longrightarrow$$
$$ClO_3^-(aq) + ClO_2^-(aq) + H_2O$$

To obtain the rate law for this reaction, the following experiments were run and, for each, the initial rate of reaction of $ClO_2$ was determined. Obtain the rate law and the value of the rate constant.

| Initial Concentrations (mol/L) | | Initial Rate |
|---|---|---|
| $ClO_2$ | $OH^-$ | (mol/(L · s)) |
| Exp. 1 | 0.060 | 0.030 | 0.0248 |
| Exp. 2 | 0.020 | 0.030 | 0.00276 |
| Exp. 3 | 0.020 | 0.090 | 0.00828 |

**14.48** Iodide ion is oxidized to hypoiodite ion, $IO^-$, by hypochlorite ion, $ClO^-$, in basic solution. The equation is

$$I^-(aq) + ClO^-(aq) \xrightarrow{OH^-} IO^-(aq) + Cl^-(aq)$$

The following initial-rate experiments were run and, for each, the initial rate of formation of $IO^-$ was determined. Find the rate law and the value of the rate constant.

| Initial Concentrations (mol/L) | | | Initial Rate |
|---|---|---|---|
| $I^-$ | $ClO^-$ | $OH^-$ | (mol/(L · s)) |
| Exp. 1 | 0.010 | 0.020 | 0.010 | $12.2 \times 10^{-2}$ |
| Exp. 2 | 0.020 | 0.010 | 0.010 | $12.2 \times 10^{-2}$ |
| Exp. 3 | 0.010 | 0.010 | 0.010 | $6.1 \times 10^{-2}$ |
| Exp. 4 | 0.010 | 0.010 | 0.020 | $3.0 \times 10^{-2}$ |

## Change of Concentration with Time; Half-Life

**14.49** Sulfuryl chloride, $SO_2Cl_2$, decomposes when heated.

$$SO_2Cl_2(g) \longrightarrow SO_2(g) + Cl_2(g)$$

In an experiment, the initial concentration of $SO_2Cl_2$ was 0.0248 mol/L. If the rate constant is $2.2 \times 10^{-5}$/s, what is the concentration of $SO_2Cl_2$ after 2.0 hr? The reaction is first order.

**14.50** Cyclopropane, $C_3H_6$, is converted to its isomer propylene, $CH_2=CHCH_3$, when heated. The rate law is first order in cyclopropane, and the rate constant is $6.0 \times 10^{-4}$/s at 500°C. If the initial concentration of cyclopropane is 0.0226 mol/L, what is the concentration after 899 s?

**14.51** Ethyl chloride, $CH_3CH_2Cl$, used to produce tetraethyllead gasoline additive, decomposes, when heated, to give ethylene and hydrogen chloride.

$$CH_3CH_2Cl(g) \longrightarrow C_2H_4(g) + HCl(g)$$

The reaction is first order. In an experiment, the initial concentration of ethyl chloride was 0.00100 M. After heating at 500°C for 155 s, this was reduced to 0.00067 M. What was the concentration of ethyl chloride after a total of 256 s?

**14.52** Cyclobutane, $C_4H_8$, consisting of molecules in which four carbon atoms form a ring, decomposes, when heated, to give ethylene.

$$C_4H_8(g) \longrightarrow 2C_2H_4(g)$$

The reaction is first order. In an experiment, the initial concentration of cyclobutane was 0.00150 M. After heating at 450°C for 455 s, this was reduced to 0.00119 $M$. What was the concentration of cyclobutane after a total of 750 s?

**14.53** Methyl isocyanide, $CH_3NC$, isomerizes, when heated, to give acetonitrile (methyl cyanide), $CH_3CN$.

$$CH_3NC(g) \longrightarrow CH_3CN(g)$$

The reaction is first order. At 230°C, the rate constant for the isomerization is $6.3 \times 10^{-4}$/s. What is the half-life? How long would it take for the concentration of $CH_3NC$ to decrease to 50.0% of its initial value? to 25.0% of its initial value?

**14.54** Dinitrogen pentoxide, $N_2O_5$, decomposes when heated in carbon tetrachloride solvent.

$$N_2O_5 \longrightarrow 2NO_2 + \tfrac{1}{2}O_2(g)$$

If the rate constant for the decomposition of $N_2O_5$ is $6.2 \times 10^{-4}$/min, what is the half-life? (The rate law is first order in $N_2O_5$.) How long would it take for the concentration of $N_2O_5$ to decrease to 25% of its initial value? to 12.5% of its initial value?

**14.55** In the presence of excess thiocyanate ion, $SCN^-$, the following reaction is first order in chromium(III) ion, $Cr^{3+}$; the rate constant is $2.0 \times 10^{-6}$/s.

$$Cr^{3+}(aq) + SCN^-(aq) \longrightarrow Cr(SCN)^{2+}(aq)$$

What is the half-life in hours? How many hours would be required for the initial concentration of $Cr^{3+}$ to decrease to each of the following values: 25.0% left, 12.5% left, 6.25% left, 3.125% left?

**14.56** In the presence of excess thiocyanate ion, $SCN^-$, the following reaction is first order in iron(III) ion, $Fe^{3+}$; the rate constant is 1.27/s.

$$Fe^{3+}(aq) + SCN^-(aq) \longrightarrow Fe(SCN)^{2+}(aq)$$

What is the half-life in seconds? How many seconds would be required for the initial concentration of $Fe^{3+}$ to decrease to each of the following values: 25.0% left, 12.5% left, 6.25% left, 3.125% left? What is the relationship between these times and the half-life?

**14.57** In the presence of excess thiocyanate ion, $SCN^-$, the following reaction is first order in chromium(III) ion, $Cr^{3+}$; the rate constant is $2.0 \times 10^{-6}$/s.

$$Cr^{3+}(aq) + SCN^-(aq) \longrightarrow Cr(SCN)^{2+}(aq)$$

If 85.0% reaction is required to obtain a noticeable color from the formation of the $Cr(SCN)^{2+}$ ion, how many hours are required?

**14.58** In the presence of excess thiocyanate ion, $SCN^-$, the following reaction is first order in iron(III) ion, $Fe^{3+}$; the rate constant is 1.27/s.

$$Fe^{3+}(aq) + SCN^-(aq) \longrightarrow Fe(SCN)^{2+}(aq)$$

If 90.0% reaction is required to obtain a noticeable color from the formation of the $Fe(SCN)^{2+}$ ion, how many seconds are required?

**14.59** Chlorine dioxide oxidizes iodide ion in aqueous solution to iodine; chlorine dioxide is reduced to chlorite ion.

$$2ClO_2(aq) + 2I^-(aq) \longrightarrow 2ClO_2^-(aq) + I_2(aq)$$

The order of the reaction with respect to $ClO_2$ was determined by starting with a large excess of $I^-$, so that its concentration was essentially constant. Then

$$\text{Rate} = k[ClO_2]^m[I^-]^n = k'[ClO_2]^m$$

where $k' = k[I^-]^n$. Determine the order with respect to $ClO_2$ and the rate constant $k'$ by plotting the following data assuming first- and then second-order kinetics. [Data from H. Fukutomi and G. Gordon, *J. Am. Chem. Soc., 89,* 1362 (1967).]

| Time (s) | $[ClO_2]$ (mol/L) |
|---|---|
| 0.00 | $4.77 \times 10^{-4}$ |
| 1.00 | $4.31 \times 10^{-4}$ |
| 2.00 | $3.91 \times 10^{-4}$ |
| 3.00 | $3.53 \times 10^{-4}$ |

**14.60** Methyl acetate, $CH_3COOCH_3$, reacts in basic solution to give acetate ion, $CH_3COO^-$, and methanol, $CH_3OH$.

$$CH_3COOCH_3(aq) + OH^-(aq) \longrightarrow \\ CH_3COO^-(aq) + CH_3OH(aq)$$

The overall order of the reaction was determined by starting with methyl acetate and hydroxide ion at the same concentrations, so $[CH_3COOCH_3] = [OH^-] = x$. Then

$$\text{Rate} = k[CH_3COOCH_3]^m[OH^-]^n = kx^{m+n}$$

Determine the overall order and the value of the rate constant by plotting the following data assuming first- and then second-order kinetics.

| Time (min) | $[CH_3COOCH_3]$ (mol/L) |
|---|---|
| 0.00 | 0.01000 |
| 3.00 | 0.00740 |
| 4.00 | 0.00683 |
| 5.00 | 0.00634 |

## Rate and Temperature

**14.61** Sketch a potential-energy diagram for the reaction of nitric oxide with ozone.

$$NO(g) + O_3(g) \longrightarrow NO_2(g) + O_2(g)$$

The activation energy for the forward reaction is 10 kJ; the $\Delta H°$ is $-200$ kJ. What is the activation energy for the reverse reaction? Label your diagram appropriately.

**14.62** Sketch a potential-energy diagram for the decomposition of nitrous oxide.

$$N_2O(g) \longrightarrow N_2(g) + O(g)$$

The activation energy for the forward reaction is 251 kJ; the $\Delta H°$ is +167 kJ. What is the activation energy for the reverse reaction? Label your diagram appropriately.

**14.63** In a series of experiments on the decomposition of dinitrogen pentoxide, $N_2O_5$, rate constants were determined at two different temperatures. At 35°C, the rate constant was $1.4 \times 10^{-4}$/s; at 45°C, the rate constant was $5.0 \times 10^{-4}$/s. What is the activation energy for this reaction? What is the value of the rate constant at 55°C?

**14.64** The reaction

$$2NOCl(g) \longrightarrow 2NO(g) + Cl_2(g)$$

has rate-constant values for the reaction of NOCl of $9.3 \times 10^{-6}$/s at 350 K and $6.9 \times 10^{-4}$/s at 400 K. Calculate activation energy for the reaction. What is the rate constant at 425 K?

**14.65** The rate of a particular reaction triples when the temperature is increased from 25°C to 35°C. Calculate the activation energy for this reaction.

**14.66** The rate of a particular reaction quadruples when the temperature is increased from 25°C to 35°C. Calculate the activation energy for this reaction.

**14.67** The following values of the rate constant were obtained for the decomposition of nitrogen dioxide at various temperatures. Plot the logarithm of $k$ versus $1/T$ and from the graph obtain the energy of activation.

| Temperature (°C) | $k$ (L/mol · s) |
|---|---|
| 320 | 0.527 |
| 330 | 0.776 |
| 340 | 1.121 |
| 350 | 1.607 |

**14.68** The following values of the rate constant were obtained for the decomposition of hydrogen iodide at various temperatures. Plot the logarithm of $k$ versus $1/T$ and from the graph obtain the energy of activation.

| Temperature (°C) | $k$ (L/mol · s) |
|---|---|
| 440 | $2.69 \times 10^{-3}$ |
| 460 | $6.21 \times 10^{-3}$ |
| 480 | $1.40 \times 10^{-2}$ |
| 500 | $3.93 \times 10^{-2}$ |

## Reaction Mechanisms; Catalysis

**14.69** Nitric oxide, NO, is believed to react with chlorine according to the following mechanism:

$$NO + Cl_2 \rightleftharpoons NOCl_2 \quad \text{(elementary reaction)}$$
$$NOCl_2 + NO \longrightarrow 2NOCl \quad \text{(elementary reaction)}$$

Identify any reaction intermediate. What is the overall equation?

**14.70** The decomposition of ozone is believed to occur in two steps.

$$O_3 \rightleftharpoons O_2 + O \quad \text{(elementary reaction)}$$
$$O_3 + O \longrightarrow 2O_2 \quad \text{(elementary reaction)}$$

Identify any reaction intermediate. What is the overall reaction?

**14.71** Identify the molecularity of each of the following elementary reactions.
a. $NO + O_3 \longrightarrow NO_2 + O_2$
b. $NOCl_2 + NO \longrightarrow 2NOCl$
c. $O_3 \longrightarrow O_2 + O$
d. $H + H + N_2 \longrightarrow H_2 + N_2^*$

**14.72** What is the molecularity of each of the following elementary reactions?
a. $O + O_2 + N_2 \longrightarrow O_3 + N_2^*$
b. $NO_2Cl + Cl \longrightarrow NO_2 + Cl_2$
c. $Cl + H_2 \longrightarrow HCl + H$
d. $CS_2 \longrightarrow CS + S$

**14.73** Write a rate equation, showing the dependence of rate on reactant concentrations, for each of the following elementary reactions.
a. $O_3 \longrightarrow O_2 + O$
b. $NOCl_2 + NO \longrightarrow 2NOCl$

**14.74** Write a rate equation, showing the dependence of rate on reactant concentrations, for each of the following elementary reactions.
a. $CS_2 \longrightarrow CS + S$
b. $CH_3Br + OH^- \longrightarrow CH_3OH + Br^-$

**14.75** The isomerization of cyclopropane, $C_3H_6$, is believed to occur by the mechanism shown in the following equations:

$$C_3H_6 + C_3H_6 \xrightarrow{k_1} C_3H_6 + C_3H_6^* \quad \text{(Step 1)}$$
$$C_3H_6^* \xrightarrow{k_2} CH_2{=}CHCH_3 \quad \text{(Step 2)}$$

Here $C_3H_6^*$ is an excited cyclopropane molecule. At low pressure, Step 1 is much slower than Step 2. Derive the rate law for this mechanism at low pressure. Explain.

**14.76** The thermal decomposition of nitryl chloride, $NO_2Cl$,

$$2NO_2Cl(g) \longrightarrow 2NO_2(g) + Cl_2(g)$$

is thought to occur by the mechanism shown in the following equations:

$$NO_2Cl \xrightarrow{k_1} NO_2 + Cl \quad \text{(slow step)}$$
$$NO_2Cl + Cl \xrightarrow{k_2} NO_2 + Cl_2 \quad \text{(fast step)}$$

What rate law is predicted by this mechanism?

**14.77** The reaction

$$H_2(g) + I_2(g) \longrightarrow 2HI(g)$$

may occur by the following mechanism:

$$I_2 \underset{k_{-1}}{\overset{k_1}{\rightleftharpoons}} 2I \qquad \text{(fast, equilibrium)}$$

$$I + I + H_2 \overset{k_2}{\longrightarrow} 2HI \qquad \text{(slow)}$$

What rate law is predicted by the mechanism?

**14.78** Ozone decomposes to oxygen gas.

$$2O_3(g) \longrightarrow 3O_2(g)$$

A proposed mechanism for this decomposition is

$$O_3 \underset{k_{-1}}{\overset{k_1}{\rightleftharpoons}} O_2 + O \qquad \text{(fast, equilibrium)}$$

$$O_3 + O \overset{k_2}{\longrightarrow} 2O_2 \qquad \text{(slow)}$$

What is the rate law derived from this mechanism?

**14.79** The following is a possible mechanism for a reaction involving hydrogen peroxide in aqueous solution; only a small amount of sodium bromide was added to the reaction mixture:

$$H_2O_2 + Br^- \longrightarrow BrO^- + H_2O$$
$$H_2O_2 + BrO^- \longrightarrow Br^- + H_2O + O_2$$

What is the overall reaction? What species is acting as a catalyst? Are there any reaction intermediates?

**14.80** Consider the following mechanism for a reaction in aqueous solution and indicate the species acting as a catalyst:

$$NH_2NO_2 + OH^- \rightleftharpoons H_2O + NHNO_2^-$$
$$NHNO_2^- \longrightarrow N_2O(g) + OH^-$$

Explain why you believe this species is a catalyst. What is the overall reaction? What substance might be added to the reaction mixture to give the catalytic activity?

## General Problems

**14.81** A study of the decomposition of azomethane,

$$CH_3NNCH_3(g) \longrightarrow C_2H_6(g) + N_2(g)$$

gave the following concentrations of azomethane at various times:

| Time | [CH$_3$NNCH$_3$] |
|---|---|
| 0 min | $1.50 \times 10^{-2}$ M |
| 10 min | $1.29 \times 10^{-2}$ M |
| 20 min | $1.10 \times 10^{-2}$ M |
| 30 min | $0.95 \times 10^{-2}$ M |

Obtain the average rate of decomposition in units of M/s for each time interval.

**14.82** Nitrogen dioxide decomposes when heated.

$$2NO_2(g) \longrightarrow 2NO(g) + O_2(g)$$

During an experiment, the concentration of NO$_2$ varied with time in the following way:

| Time | [NO$_2$] |
|---|---|
| 0.0 min | 0.1103 M |
| 1.0 min | 0.1076 M |
| 2.0 min | 0.1050 M |
| 3.0 min | 0.1026 M |

Obtain the average rate of decomposition of NO$_2$ in units of M/s for each time interval.

**14.83** You can write the rate law for the decomposition of azomethane as

$$\text{Rate} = k[CH_3NNCH_3]^n \qquad \text{or} \qquad k = \frac{\text{rate}}{[CH_3NNCH_3]^n}$$

when $n$ is the order of the reaction. Note that when you divide the rates at various times by the concentrations raised to the correct power $n$, you should get the same number (the rate constant $k$). Verify that the decomposition of azomethane is first order by dividing each average rate in a time interval (obtained in Problem 14.81) by the average concentration in that interval. Note that each calculation gives nearly the same value. Take the average of these values to obtain the rate constant.

**14.84** Use the technique described in Problem 14.83 to verify that the decomposition of nitrogen dioxide is second order. That is, divide each average rate in a time interval (obtained in Problem 14.82) by the square of the average concentration in that interval. Note that each calculation gives nearly the same value. Take the average of these calculated values to obtain the rate constant.

**14.85** Methyl acetate reacts in acidic solution.

$$CH_3COOCH_3 + H_2O \overset{H^+}{\longrightarrow} CH_3OH + CH_3COOH$$

methyl acetate      methanol    acetic acid

The rate law is first order in methyl acetate in acidic solution, and the rate constant at 25°C is $1.26 \times 10^{-4}$/s. How long will it take for 65% of the methyl acetate to react?

**14.86** Benzene diazonium chloride, $C_6H_5NNCl$, decomposes by a first-order rate law.

$$C_6H_5NNCl \longrightarrow C_6H_5Cl + N_2(g)$$

If the rate constant at 20°C is $4.3 \times 10^{-5}$/s, how long will it take for 75% of the compound to decompose?

**14.87** What is the half-life of methyl acetate at 25°C in the acidic solution described in Problem 14.85?

**14.88** What is the half-life of benzene diazonium chloride at 20°C? See Problem 14.86 for data.

**14.89** A compound decomposes by a first-order reaction. If the concentration of the compound is 0.0250 $M$ after 65 s when the initial concentration was 0.0350 $M$, what is the concentration of the compound after 88 s?

**14.90** A compound decomposes by a first-order reaction. The concentration of compound decreases from 0.1180 $M$ to 0.0950 $M$ in 5.2 min. What fraction of the compound remains after 7.1 min?

**14.91** Plot the data given in Problem 14.81 to verify that the decomposition of azomethane is first order. Determine the rate constant from the slope of the straight-line plot of log $[CH_3NNCH_3]$ versus time.

**14.92** The decomposition of aqueous hydrogen peroxide in a given concentration of catalyst yielded the following data:

| Time | 0.0 min | 5.0 min | 10.0 min | 15.0 min |
|---|---|---|---|---|
| $[H_2O_2]$ | 0.1000 $M$ | 0.0804 $M$ | 0.0648 $M$ | 0.0519 $M$ |

Verify that the reaction is first order. Determine the rate constant for the decomposition of $H_2O_2$ (in units of /s) from the slope of the straight-line plot of ln $[H_2O_2]$ versus time.

**14.93** The decomposition of nitrogen dioxide,

$$2NO_2(g) \longrightarrow 2NO(g) + O_2(g)$$

has a rate constant of 0.498 $M$/s at 319°C and a rate constant of 1.81 $M$/s at 354°C. What are the values of the activation energy and the frequency factor for this reaction? What is the rate constant at 420°C?

**14.94** A second-order reaction has a rate constant of $8.7 \times 10^{-4}$/($M \cdot$ s) at 30°C. At 40°C, the rate constant is $1.8 \times 10^{-3}$/($M \cdot$ s). What are the activation energy and frequency factor for this reaction? Predict the value of the rate constant at 45°C.

**14.95** At high temperature, the reaction

$$NO_2(g) + CO(g) \longrightarrow NO(g) + CO_2(g)$$

is thought to occur in a single step. What should be the rate law in that case?

**14.96** Methyl chloride, $CH_3Cl$, reacts in basic solution to give methanol.

$$CH_3Cl + OH^- \longrightarrow CH_3OH + Cl^-$$

This reaction is believed to occur in a single step. If so, what should be the rate law?

**14.97** Nitryl bromide, $NO_2Br$, decomposes into nitrogen dioxide and bromine.

$$2NO_2Br(g) \longrightarrow 2NO_2(g) + Br_2(g)$$

A proposed mechanism is

$$NO_2Br \longrightarrow NO_2 + Br \qquad \text{(slow)}$$
$$NO_2Br + Br \longrightarrow NO_2 + Br_2 \qquad \text{(fast)}$$

Write the rate law predicted by this mechanism.

**14.98** Tertiary butyl chloride reacts in basic solution according to the equation

$$(CH_3)_3CCl + OH^- \longrightarrow (CH_3)_3COH + Cl^-$$

The accepted mechanism for this reaction is

$$(CH_3)_3CCl \longrightarrow (CH_3)_3C^+ + Cl^- \qquad \text{(slow)}$$
$$(CH_3)_3C^+ + OH^- \longrightarrow (CH_3)_3COH \qquad \text{(fast)}$$

What should be the rate law for this reaction?

**14.99** Urea, $(NH_2)_2CO$, can be prepared by heating ammonium cyanate, $NH_4OCN$.

$$NH_4OCN \longrightarrow (NH_2)_2CO$$

This reaction may occur by the following mechanism:

$$NH_4^+ + OCN^- \underset{k_{-1}}{\overset{k_1}{\rightleftharpoons}} NH_3 + HOCN \qquad \text{(fast, equilibrium)}$$
$$NH_3 + HOCN \overset{k_2}{\longrightarrow} (NH_2)_2CO \qquad \text{(slow)}$$

What is the rate law predicted by this mechanism?

**14.100** Acetone reacts with iodine in acidic aqueous solution to give monoiodoacetone.

$$CH_3COCH_3 + I_2 \overset{H^+}{\longrightarrow} CH_3COCH_2I + HI$$
$$\text{acetone} \qquad\qquad\qquad \text{monoiodoacetone}$$

A possible mechanism for this reaction is

Write the rate law that you derive from this mechanism.

**14.101** A study of the gas-phase oxidation of nitrogen monoxide at 25°C and 1.00 atm pressure gave the following results:

$$2NO(g) + O_2(g) \longrightarrow NO_2(g)$$

| | Conc. NO, mol/L | Conc. O$_2$, mol/L | Initial Rate |
|---|---|---|---|
| Exp. 1 | $4.5 \times 10^{-2}$ | $2.2 \times 10^{-2}$ | $0.80 \times 10^{-2}$ mol/(L · s) |
| Exp. 2 | $4.5 \times 10^{-2}$ | $4.5 \times 10^{-2}$ | $1.60 \times 10^{-2}$ mol/(L · s) |
| Exp. 3 | $9.0 \times 10^{-2}$ | $9.0 \times 10^{-2}$ | $1.28 \times 10^{-1}$ mol/(L · s) |
| Exp. 4 | $3.8 \times 10^{-1}$ | $4.6 \times 10^{-3}$ | ? |

a. What is the experimental rate law for the reaction above?
b. What is the initial rate of the reaction in Experiment 4?

**14.102** The reaction of water with CH$_3$Cl in acetone as a solvent is represented by the equation

$$CH_3Cl + H_2O \longrightarrow CH_3OH + HCl$$

The rate of the reaction doubles when the concentration of CH$_3$Cl is doubled and it quadruples when the concentration of H$_2$O is doubled.
a. What is the unit for $k$?
b. Calculate $k$ if CH$_3$OH is formed at a rate of 1.50 $M$/s when [CH$_3$Cl] = [H$_2$O] = 0.40 $M$.

**14.103** The reaction of thioacetamide with water is shown by the equation below:

$$CH_3C(S)NH_2(aq) + H_2O \longrightarrow H_2S(aq) + CH_3C(O)NH_2(aq)$$

The rate of reaction is given by the rate law:

$$Rate = k[H_3O^+][CH_3C(S)NH_2]$$

Consider one liter of solution that is 0.20 $M$ in CH$_3$C(S)NH$_2$ and 0.15 $M$ in HCl at 25°C.
a. For each of the changes listed below, state whether the rate of reaction increases, decreases, or remains the same. Why?
    i. A 4-g sample of NaOH is added to the solution.
    ii. 500 mL of water is added to the solution.
b. For each of the changes listed below, state whether the value of $k$ will increase, decrease, or remain the same. Why?
    i. A catalyst is added to the solution.
    ii. The reaction is carried out at 15°C instead of 25°C.

**14.104** The reaction of methylacetate with water is shown by the equation below:

$$CH_3COOCH_3(aq) + H_2O \longrightarrow CH_3COOH(aq) + CH_3OH(aq)$$

The rate of the reaction is given by the rate law:

$$Rate = k[H_3O^+][CH_3COOCH_3]$$

Consider one liter of solution that is 0.15 $M$ in CH$_3$COOCH$_3$ and 0.015 $M$ in H$_3$O$^+$ at 25°C.
a. For each of the changes listed below, state whether the rate of reaction increases, decreases, or remains the same. Why?
    i. Some concentrated sulfuric acid is added to the solution.
    ii. Water is added to the solution.
b. For each of the changes listed below, state whether the value of $k$ will increase, decrease, or remain the same. Why?
    i. Some concentrated sulfuric acid is added to the solution.
    ii. The reaction is carried out at 35°C instead of 25°C.

**14.105** Draw a potential-energy diagram for an uncatalyzed exothermic reaction. On the same diagram, indicate the change that results on the addition of a catalyst. Discuss the role of a catalyst in changing the rate of reaction.

**14.106** Draw and label the potential-energy curve for the reaction

$$N_2O_4(g) \rightleftharpoons 2NO_2(g); \Delta H = 57 \text{ kJ}$$

The activation energy for the reverse reaction is 23 kJ. Note $\Delta H$ and $E_a$ on the diagram. What is the activation energy for the forward reaction? For which reaction (forward or reverse) will the reaction rate be most sensitive to a temperature increase? Explain.

**14.107** What is meant by the term *rate of a chemical reaction?* Why does the rate of a reaction normally change with time? When does the rate of a chemical reaction equal the rate constant?

**14.108** Briefly discuss the factors that affect the rates of chemical reactions. Which of the factors affect the magnitude of the rate constant? Which factor(s) do not affect the magnitude of the rate constant? Why?

## Cumulative-Skills Problems

**14.109** Dinitrogen pentoxide, N$_2$O$_5$, undergoes first-order decomposition in chloroform solvent to yield NO$_2$ and O$_2$. The rate constant at 45°C is $6.2 \times 10^{-4}$/min. Calculate the volume of O$_2$ obtained from the reaction of 1.00 mol N$_2$O$_5$ at 45°C and 770 mmHg after 20.0 hr.

**14.110** Hydrogen peroxide undergoes a first-order decomposition to water and O$_2$ in aqueous solution. The rate constant at 25°C is $7.40 \times 10^{-4}$/s. Calculate the volume of O$_2$ obtained from the decomposition reaction of 1.00 mol H$_2$O$_2$ at 25°C and 740 mmHg after 20.0 min.

**14.111** Hydrogen peroxide in aqueous solution decomposes by a first-order reaction to water and oxygen. The rate constant for this decomposition is $7.40 \times 10^{-4}$/s. What quantity of heat energy is initially liberated per second from 2.00 L of solution that is $1.50\ M\ H_2O_2$? See Appendix C for data.

**14.112** Nitrogen dioxide reacts with carbon monoxide by the overall equation

$$NO_2(g) + CO(g) \longrightarrow NO(g) + CO_2(g)$$

At a particular temperature, the reaction is second order in $NO_2$ and zero order in CO. The rate constant is 0.515 L/(mol · s). How much heat energy evolves per second initially from 3.50 L of reaction mixture containing $0.0250\ M\ NO_2$? See Appendix C for data. Assume the enthalpy change is constant with temperature.

**14.113** Nitric oxide reacts with oxygen to give nitrogen dioxide.

$$2NO(g) + O_2(g) \longrightarrow 2NO_2(g)$$

The rate law is $-\Delta[NO]/\Delta t = k[NO]^2[O_2]$, where the rate constant is $1.16 \times 10^{-5}\ L^2/(mol^2 \cdot s)$ at 339°C. A vessel contains NO and $O_2$ at 339°C. The initial partial pressures of NO and $O_2$ are 155 mmHg and 345 mmHg, respectively. What is the rate of decrease of partial pressure of NO (in mmHg per second)? (*Hint:* From the ideal gas law, obtain an expression for the molar concentration of a particular gas in terms of its partial pressure.)

**14.114** Nitric oxide reacts with hydrogen as follows:

$$2NO(g) + H_2(g) \longrightarrow N_2O(g) + H_2O(g)$$

The rate law is $-\Delta[H_2]/\Delta t = k[NO]^2[H_2]$, where $k$ is $1.10 \times 10^{-7}\ L^2/(mol^2 \cdot s)$ at 826°C. A vessel contains NO and $H_2$ at 826°C. The partial pressures of NO and $H_2$ are 144 mmHg and 324 mmHg, respectively. What is the rate of decrease of partial pressure of NO? See Problem 14.113.

 ## Media Activities

**14.115** Experiment with the Time Course of a Reaction *Understanding Concepts* simulation on the student web site by answering the Self-Assessment questions. Does the curve depicted in this simulation represent a product of a reactant?

**14.116** Open the Rate Law *Understanding Concepts* activity on the student web site. Conduct several experiments to determine the rate law of the reaction. (You can vary the concentration of the reactants by moving the sliders, and you can display the initial rate of the reaction by clicking on the curve near the origin.) For each individual experiment, be sure to record the [A], [B], [C], and the initial rate. What is the overall order of the reaction?

**14.117** Run the First Order Rates *Problem Solving* activity on the student web site. For the reaction depicted in the *Visual-*ization, what is the initial concentration of $N_2O_3$ if the concentration of $N_2O_3$ is $5.1 \times 10^3\ M$ after 88 seconds have elapsed?

**14.118** Work through the Transition States and Activation Energy *Understanding Concepts* activity on the student CD. At the end of this activity, open up the $H_2$ + Br example and move the slider to step through the reaction.
a. Is the $H_2$ + Br reaction exothermic or endothermic?
b. Calculate the $\Delta H$ for the $H_2$ + Br reaction.
c. What is the formula of the activated complex for the $H_2$ + Br reaction?

**14.119** For review, take the ACE practice test for Chapter 14 on the student web site.

# 15

Oscillating pattern formed by a reaction far from equilibrium.

# Chemical Equilibrium

*C*hemical reactions often seem to stop before they are complete. Such reactions are *reversible*. That is, the original reactants form products, but then the products react with themselves to give back the original reactants. Actually two reactions are occurring, and the eventual result is a mixture of reactants and products, rather than simply a mixture of products.

Consider the gaseous reaction in which carbon monoxide and hydrogen react to produce methane and steam:

$$CO(g) + 3H_2(g) \longrightarrow CH_4(g) + H_2O(g)$$
$$\text{methane}$$

This reaction, which requires a catalyst to occur at a reasonable rate, is called *catalytic methanation*. It is a potentially useful reaction because it can be used to produce methane from coal or organic wastes from which carbon monoxide can be obtained (Figure 15.1). Methane, in turn, can be used as a fuel and as a starting material for the production of organic chemicals. Although, at the moment, natural gas (which is mostly methane) is plentiful, analysts predict that oil and natural gas production will peak sometime in the early part of this century. Then, the demand

**Figure 15.1**
**Coal gasification plant.**
Gaseous fuels can be made from coal in a number of ways. One way (described in the text) is to react coal with steam to produce a mixture of CO and $H_2$, which then reacts by catalytic methanation to yield $CH_4$, the major component of natural gas.

▶ This chapter introduces the concept of equilibrium. The method described for solving equilibrium problems is used uniformly in the following chapters.

for other sources of fuels and organic chemicals, such as coal and organic wastes, will rise.

Catalytic methanation is a reversible reaction, and depending on the reaction conditions, the final reaction mixture will have varying amounts of the products methane and steam, as well as the starting substances carbon monoxide and hydrogen. It is also possible to start with methane and steam and, under the right conditions, form a mixture that is predominantly carbon monoxide and hydrogen. The process is called *steam reforming*.

$$CH_4(g) + H_2O(g) \longrightarrow CO(g) + 3H_2(g)$$

The product mixture of CO and $H_2$ (synthesis gas) is used to prepare a number of industrial chemicals.

The processes of catalytic methanation and steam reforming illustrate the reversibility of chemical reactions. Starting with CO and $H_2$ and using the right conditions, you can form predominantly $CH_4$ and $H_2O$. Starting with $CH_4$ and $H_2O$ and using different conditions, you can obtain a reaction mixture that is predominantly CO and $H_2$. An important question is, What conditions favor the production of $CH_4$ and $H_2O$, and what conditions favor the production of CO and $H_2$?

As noted earlier, certain reactions (such as catalytic methanation) appear to stop before they are complete. The reaction mixture ceases to change in any of its properties and consists of both reactants and products in definite concentrations. Such a reaction mixture is said to have reached *chemical equilibrium*. In earlier chapters, we discussed other types of equilibria, including the equilibrium between a liquid and its vapor and the equilibrium between a solid and its saturated solution. In this chapter, we will see how to determine the composition of a reaction mixture at equilibrium and how to alter this composition by changing the conditions for the reaction.

# Describing Chemical Equilibrium

Many chemical reactions are like the catalytic methanation reaction. Such reactions can be made to go predominantly in one direction or the other depending on the conditions. Let us look more closely at this reversibility and see how to characterize it quantitatively.

## 15.1 Chemical Equilibrium—A Dynamic Equilibrium

▶ The concept of dynamic equilibrium in chemical reactions was briefly mentioned in Section 14.8. Discussions on vapor pressure (Section 11.2) and solubility (Section 12.2) give detailed explanations of the role of dynamic equilibria.

When substances react, they eventually form a mixture of reactants and products in *dynamic equilibrium*. ◀ This dynamic equilibrium consists of a forward reaction, in which substances react to give products, and a reverse reaction, in which products react to give the original reactants. Both forward and reverse reactions occur at the same rate, or speed.

Consider the catalytic methanation reaction discussed in the chapter opening. It consists of forward and reverse reactions, as represented by the chemical equation

$$CO(g) + 3H_2(g) \rightleftharpoons CH_4(g) + H_2O(g)$$

Suppose you put 1.000 mol CO and 3.000 mol $H_2$ into a 10.00-L vessel at 1200 K (927°C). The rate of the reaction of CO and $H_2$ depends on the concentrations of CO and $H_2$. At first these concentrations are large, but as the substances react their concentrations decrease (Figure 15.2). The rate of the forward reaction, which depends on reactant concentrations, is large at first but steadily decreases. On the other hand, the concentrations of $CH_4$ and $H_2O$, which are zero at first, increase with time. The rate of the reverse reaction starts at zero and steadily increases. The forward rate decreases and the reverse rate increases until eventually the rates become equal. When that happens, CO and $H_2$ molecules are formed as fast as they react. The concentrations of reactants and products no longer change, and the reaction mixture has reached *equilibrium*. Figure 15.2 shows how the amounts of substances in the reaction mixture become constant when equilibrium is reached.

**Chemical equilibrium** is *the state reached by a reaction mixture when the rates of forward and reverse reactions have become equal.* If you observe the reaction mixture, you see no net change, although the forward and reverse reactions are continuing. The continuing forward and reverse reactions make the equilibrium a *dynamic* process.

**Figure 15.2**
**Catalytic methanation reaction approaches equilibrium.** (A) The experiment begins with 1.000 mol CO and 3.000 mol $H_2$ in a 10.00-L vessel. Note that the amounts of substances become constant at equilibrium. (B) The forward rate is large at first but steadily decreases, whereas the reverse rate starts at zero and steadily increases. Eventually both rates become equal (at equilibrium).

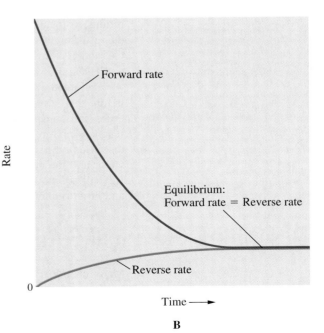

Suppose you place known amounts of reactants in a vessel and let the mixture come to equilibrium. To obtain the composition of the equilibrium mixture, you need to determine the amount of only one of the substances. The amounts of the others can be calculated from the amounts originally placed in the vessel and the equation that represents the reaction. For example, suppose you place 1.000 mol CO and 3.000 mol $H_2$ in a reaction vessel at 1200 K. After equilibrium is reached, you chill the reaction mixture quickly to condense the water vapor to liquid, which you then weigh and find to be 0.387 mol. You calculate the other substances from the stoichiometry of the reaction, as shown in the following example.

▶ An alternate example is provided in the Instructor's Resource Manual.

### EXAMPLE 15.1

## Applying Stoichiometry to an Equilibrium Mixture

Carbon monoxide and hydrogen react according to the following equation:

$$CO(g) + 3H_2(g) \rightleftharpoons CH_4(g) + H_2O(g)$$

When 1.000 mol CO and 3.000 mol $H_2$ are placed in a 10.00-L vessel at 927°C (1200 K) and allowed to come to equilibrium, the mixture is found to contain 0.387 mol $H_2O$. What is the molar composition of the equilibrium mixture? That is, how many moles of each substance are present?

 **PROBLEM STRATEGY**

The problem is essentially one of stoichiometry. It involves initial, or *starting*, amounts of reactants. These amounts *change* as reaction occurs. Later, the reaction comes to *equilibrium*, and you analyze the reaction mixture for the amount of one of the reactants or products. It is convenient to solve this problem by first setting up a table in which you write the starting, change, and equilibrium values of each substance under the balanced equation. This way you can easily see what you have to calculate.

| Amount (mol) | $CO(g) + 3H_2(g) \rightleftharpoons CH_4(g) + H_2O(g)$ |
|---|---|
| Starting | |
| Change | |
| Equilibrium | |

You fill in the starting amounts from the values given in the problem statement. You are not given explicit values for the changes that occur, so you let $x$ be the molar change. That is, each product increases by $x$ moles multiplied by the coefficients of the substances in the balanced equation. Reactants decrease by $x$ moles multiplied by the corresponding coefficients. The decrease is indicated by a negative sign. Equilibrium values are equal to starting values plus the changes.

### SOLUTION

Using the information given in the problem, you set up the following table:

| Amount (mol) | $CO(g)$ | + | $3H_2(g)$ | $\rightleftharpoons$ | $CH_4(g)$ | + | $H_2O(g)$ |
|---|---|---|---|---|---|---|---|
| Starting | 1.000 | | 3.000 | | 0 | | 0 |
| Change | $-x$ | | $-3x$ | | $+x$ | | $+x$ |
| Equilibrium | $1.000 - x$ | | $3.000 - 3x$ | | $x$ | | $x = 0.387$ |

The problem statement gives the equilibrium amount for $H_2O$. This tells you that $x = 0.387$ mol. You calculate equilibrium amounts for other substances from the expressions given in the table, using this value of $x$.

$$\text{Equilibrium amount CO} = (1.000 - x) \text{ mol}$$
$$= (1.000 - 0.387) \text{ mol}$$
$$= 0.613 \text{ mol}$$

$$\text{Equilibrium amount } H_2 = (3.000 - 3x) \text{ mol}$$
$$= (3.000 - 3 \times 0.387) \text{ mol}$$
$$= 1.839 \text{ mol}$$

$$\text{Equilibrium amount } CH_4 = x \text{ mol} = 0.387 \text{ mol}$$

Therefore, the amounts of substances in the equilibrium mixture are **0.613 mol CO, 1.839 mol $H_2$, 0.387 mol $CH_4$, and 0.387 mol $H_2O$.**

---

### EXERCISE 15.1

Synthesis gas (a mixture of CO and $H_2$) is increased in concentration of hydrogen by passing it with steam over a catalyst. This is the so-called water–gas shift reaction. Some of the CO is converted to $CO_2$, which can be removed:

$$CO(g) + H_2O(g) \rightleftharpoons CO_2(g) + H_2(g)$$

Suppose you start with a gaseous mixture containing 1.00 mol CO and 1.00 mol $H_2O$. When equilibrium is reached at 1000°C, the mixture contains 0.43 mol $H_2$. What is the molar composition of the equilibrium mixture?

---

**See Problems 15.21, 15.22, 15.23, 15.24, 15.25, and 15.26.**

 **CONCEPT CHECK 15.1**

---

Two substances A and B react to produce substance C. When reactant A decreases by molar amount $x$, product C increases by molar amount $x$. When reactant B decreases by molar amount $x$, product C increases by molar amount $2x$. Write the chemical equation for the reaction.

## 15.2 The Equilibrium Constant

In the preceding example, we found that when 1.000 mol CO and 3.000 mol $H_2$ react in a 10.00-L vessel by catalytic methanation at 1200 K, they give an equilibrium mixture containing 0.613 mol CO, 1.839 mol $H_2$, 0.387 mol $CH_4$, and 0.387 mol $H_2O$. Let us call this Experiment 1. Now consider a similar experiment, Experiment 2, in which you start with an additional mole of carbon monoxide. That is, you place 2.000 mol CO and 3.000 mol $H_2$ in a 10.00-L vessel at 1200 K. At equilibrium, you find that the vessel contains 1.522 mol CO, 1.566 mol $H_2$, 0.478 mol $CH_4$, and 0.478 mol $H_2O$. What you observe from the results of Experiments 1 and 2 is that the equilibrium composition depends on the amounts of starting substances. Nevertheless, you will see that all of the equilibrium compositions for a reaction at a given temperature are related by a quantity called the *equilibrium constant*.

### Definition of the Equilibrium Constant $K_c$

Consider the reaction

$$aA + bB \rightleftharpoons cC + dD$$

where A, B, C, and D denote reactants and products, and $a$, $b$, $c$, and $d$ are coefficients in the balanced chemical equation. The **equilibrium-constant expression** for a reaction is *an expression obtained by multiplying the concentrations of products, dividing by the concentrations of reactants, and raising each concentration term to a power equal to the coefficient in the chemical equation.* The **equilibrium constant** $K_c$ is *the value obtained for the equilibrium-constant expression when equilibrium concentrations are substituted.* ◄ For the previous reaction, you have

▶ In Section 15.3, you will see that you ignore concentrations of pure liquids and solids when you write the equilibrium-constant expression. The concentrations of these substances are constant and are incorporated into the value of $K_c$.

$$K_c = \frac{[C]^c[D]^d}{[A]^a[B]^b}$$

Here you denote the molar concentration of a substance by writing its formula in square brackets. The subscript $c$ on the equilibrium constant means that it is defined in terms of molar concentrations. The **law of mass action** is *a relation that states that the values of the equilibrium-constant expression $K_c$ are constant for a particular reaction at a given temperature, whatever equilibrium concentrations are substituted.* ◄

▶ "Active mass" was an early term for concentration; hence the term *mass action.*

As the following example illustrates, the equilibrium-constant expression is defined in terms of the balanced chemical equation. If the equation is rewritten with different coefficients, the equilibrium-constant expression will be changed.

### EXAMPLE 15.2

## Writing Equilibrium-Constant Expressions

▶ An alternate example is provided in the Instructor's Resource Manual.

a. Write the equilibrium-constant expression $K_c$ for catalytic methanation.

$$CO(g) + 3H_2(g) \rightleftharpoons CH_4(g) + H_2O(g)$$

b. Write the equilibrium-constant expression $K_c$ for the reverse of the previous reaction.

$$CH_4(g) + H_2O(g) \rightleftharpoons CO(g) + 3H_2(g)$$

c. Write the equilibrium-constant expression $K_c$ for the synthesis of ammonia.

$$N_2(g) + 3H_2(g) \rightleftharpoons 2NH_3(g)$$

d. Write the equilibrium-constant expression $K_c$ when the equation for the previous reaction is written

$$\tfrac{1}{2}N_2(g) + \tfrac{3}{2}H_2(g) \rightleftharpoons NH_3(g)$$

 **PROBLEM STRATEGY**

Write the concentrations of products in the top (numerator) of the equilibrium-constant expression, and write the concentrations of reactants in the bottom (denominator). Raise each concentration term to the power equal to the coefficient of the substance in the chemical equation.

**SOLUTION**

a. The expression for the equilibrium constant is

$$K_c = \frac{[CH_4][H_2O]}{[CO][H_2]^3}$$

Note that concentrations of products are on the top and concentrations of reactants are on the bottom. Also, note that each concentration term is raised to a power equal to the coefficient of the substance in the chemical equation.

b. When the equation is written in reverse order, the expression for $K_c$ is inverted:

$$K_c = \frac{[CO][H_2]^3}{[CH_4][H_2O]}$$

c. The equilibrium constant for $N_2 + 3H_2 \rightleftharpoons 2NH_3$ is

$$K_c = \frac{[NH_3]^2}{[N_2][H_2]^3}$$

d. When the coefficients in the equation in part c are multiplied by $\frac{1}{2}$ to give

$$\tfrac{1}{2}N_2 + \tfrac{3}{2}H_2 \rightleftharpoons NH_3$$

the equilibrium-constant expression becomes

$$K_c = \frac{[NH_3]}{[N_2]^{1/2}[H_2]^{3/2}}$$

which is the square root of the previous expression.

---

**EXERCISE 15.2**

a. Write the equilibrium-constant expression $K_c$ for the equation

$$2NO_2(g) + 7H_2(g) \rightleftharpoons 2NH_3(g) + 4H_2O(g)$$

b. Write the equilibrium-constant expression $K_c$ when this reaction is written

$$NO_2(g) + \tfrac{7}{2}H_2(g) \rightleftharpoons NH_3(g) + 2H_2O(g)$$

---

See Problems 15.27 and 15.28.

**Problem Solving:
Equilibrium Constant**

## Equilibrium: A Kinetics Argument

▶ If you have not covered Chapter 14, you may want to omit this subsection.

▶ Animation: Equilibrium decomposition of $N_2O_4$.

The law of mass action was first stated by the Norwegian chemists Cato Guldberg and Peter Waage in 1867. They were led to this law by a kinetics argument. To understand the argument, consider the decomposition of dinitrogen tetroxide, $N_2O_4$.

Dinitrogen tetroxide is a colorless substance that decomposes, when warmed, to give nitrogen dioxide, $NO_2$, a brown substance. Suppose you start with dinitrogen tetroxide gas and heat it above room temperature. At this higher temperature, $N_2O_4$

decomposes to $NO_2$, which can be noted by the increasing intensity of the brown color of the reaction mixture.

$$N_2O_4(g) \longrightarrow 2NO_2(g)$$

Once some $NO_2$ is produced, it can react to re-form $N_2O_4$. Call the decomposition of $N_2O_4$ the forward reaction and the formation of $N_2O_4$ the reverse reaction. These are elementary reactions, and you can immediately write the rate equations. The rate of the forward reaction is $k_f[N_2O_4]$, where $k_f$ is the rate constant for the forward reaction. The rate of the reverse reaction is $k_r[NO_2]^2$, where $k_r$ is the rate constant for the reverse reaction.

At first, the reaction vessel contains mostly $N_2O_4$, so the concentration of $NO_2$ is very small. Therefore, the rate of the forward reaction is relatively large, and the rate of the reverse reaction is relatively small. However, as the decomposition proceeds, the concentration of $N_2O_4$ decreases, so the forward rate decreases. At the same time, the concentration of $NO_2$ increases, so the reverse rate increases. Eventually the two rates become equal. After that, on average, every time an $N_2O_4$ molecule decomposes, two $NO_2$ molecules recombine. Overall, the concentrations of $N_2O_4$ and $NO_2$ no longer change, although forward and reverse reactions are still occurring. The reaction has reached a dynamic equilibrium.

At equilibrium, you can write: Rate of forward reaction = rate of reverse reaction, or

$$k_f[N_2O_4] = k_r[NO_2]^2$$

Rearranging gives

$$\frac{k_f}{k_r} = \frac{[NO_2]^2}{[N_2O_4]}$$

**Visualization: Equilibrium Decomposition of $N_2O_4$**

The right side is the equilibrium-constant expression for the decomposition of dinitrogen tetroxide. The left side is the ratio of rate constants. Thus, the equilibrium-constant expression is constant for a given temperature, and you can identify the equilibrium constant $K_c$ as the ratio of rate constants for the forward and reverse reactions.

$$K_c = \frac{k_f}{k_r}$$

If the overall reaction occurs by a multistep mechanism, one can show that the equilibrium constant equals a product of ratios of rate constants—one ratio for each step in the mechanism.

## Obtaining Equilibrium Constants for Reactions

▶ Experiments in General Chemistry, 15B, "Determining an Equilibrium Constant."

At the beginning of this section, we gave data from the results of two experiments, Experiments 1 and 2, involving catalytic methanation. By substituting the molar concentrations from these two experiments into the equilibrium-constant expression for the reaction, you can show that you get the same value for both experiments, as you expect from the law of mass action. This value equals $K_c$ for methanation at 1200 K.

▶ It is also possible to determine equilibrium constants from thermochemical data, as described in Section 19.6.

Thus, besides verifying the validity of the law of mass action in this case, you can also see how an equilibrium constant can be obtained from concentration data. ◀

**Experiment 1**  The equilibrium composition is 0.613 mol CO, 1.839 mol $H_2$, 0.387 mol $CH_4$, and 0.387 mol $H_2O$. The volume of the reaction vessel is 10.00 L, so the equilibrium concentration of CO is

$$[CO] = \frac{0.613 \text{ mol}}{10.00 \text{ L}} = 0.0613 \text{ } M$$

Similarly, the other equilibrium concentrations are $[H_2] = 0.1839 \text{ } M$, $[CH_4] = 0.0387 \text{ } M$, and $[H_2O] = 0.0387 \text{ } M$.

Substitute these values into the equilibrium expression for catalytic methanation. (The expression was obtained in Example 15.2.) Although until now we have consistently carried units along with numbers in calculations, it is the usual practice to write equilibrium constants without units. We will follow that practice here. ◀ Substitution of concentrations into the equilibrium-constant expression gives

▶ In thermodynamics, the equilibrium constant is defined in terms of *activities*, rather than concentrations. For an ideal mixture, the activity of a substance is the ratio of its concentration (or partial pressure if a gas) to a standard concentration of 1 *M* (or partial pressure of 1 atm), so that units cancel. Thus, activities have numerical values but no units.

$$K_c = \frac{[CH_4][H_2O]}{[CO][H_2]^3} = \frac{(0.0387)(0.0387)}{(0.0613)(0.1839)^3} = 3.93$$

**Experiment 2**  The equilibrium composition is 1.522 mol CO, 1.566 mol $H_2$, 0.478 mol $CH_4$, and 0.478 mol $H_2O$. Therefore, the concentrations, which are obtained by dividing by 10.00 L, are $[CO] = 0.1522 \text{ } M$, $[H_2] = 0.1566 \text{ } M$, $[CH_4] = 0.0478 \text{ } M$, and $[H_2O] = 0.0478 \text{ } M$. Substituting into the equilibrium-constant expression gives

$$K_c = \frac{[CH_4][H_2O]}{[CO][H_2]^3} = \frac{(0.0478)(0.0478)}{(0.1522)(0.1566)^3} = 3.91$$

Within the precision of the data, these values (3.93 and 3.91) of the equilibrium expression for different starting mixtures of these gases at 1200 K are the same. Moreover, experiment shows that when you start with $CH_4$ and $H_2O$, instead of CO and $H_2$, an equilibrium mixture is again reached that yields the same value of $K_c$ (see Experiment 3, Figure 15.3). You can take the equilibrium constant for catalytic methanation at 1200 K to be 3.92, the average of these values.

**Figure 15.3**
**Some equilibrium compositions for the methanation reaction.**
Different starting concentrations have been used in each experiment (all at 1200 K). Experiments 1 and 2 start with different concentrations of reactants CO and $H_2$. Experiment 3 starts with the products $CH_4$ and $H_2O$. All three experiments yield essentially the same value of $K_c$.

| | Starting concentrations | | Equilibrium concentrations | Calculated value of $K_c$ |
|---|---|---|---|---|
| **Experiment 1** | 0.1000 $M$ CO<br>0.3000 $M$ $H_2$ | → | 0.0613 $M$ CO<br>0.1839 $M$ $H_2$<br>0.0387 $M$ $CH_4$<br>0.0387 $M$ $H_2O$ | $K_c = 3.93$ |
| **Experiment 2** | 0.2000 $M$ CO<br>0.3000 $M$ $H_2$ | → | 0.1522 $M$ CO<br>0.1566 $M$ $H_2$<br>0.0478 $M$ $CH_4$<br>0.0478 $M$ $H_2O$ | $K_c = 3.91$ |
| **Experiment 3** | 0.1000 $M$ $CH_4$<br>0.1000 $M$ $H_2O$ | → | 0.0613 $M$ CO<br>0.1839 $M$ $H_2$<br>0.0387 $M$ $CH_4$<br>0.0387 $M$ $H_2O$ | $K_c = 3.93$ |

The following exercise provides additional practice in evaluating an equilibrium constant from equilibrium compositions that are experimentally determined. Example 15.3 gives the detailed solution of a more complicated problem of this type.

► The reaction described here is used industrially to adjust the ratio of $H_2$ to CO in synthesis gas (mixture of CO and $H_2$). In this process, CO reacts with $H_2O$ to give $H_2$, so the ratio of $H_2$ to CO is increased.

### EXERCISE 15.3

When 1.00 mol each of carbon monoxide and water reach equilibrium at 1000°C in a 10.0-L vessel, the equilibrium mixture contains 0.57 mol CO, 0.57 mol $H_2O$, 0.43 mol $CO_2$, and 0.43 mol $H_2$. Write the chemical equation for the equilibrium. What is the value of $K_c$? ◄

See Problems 15.35 and 15.36.

► An alternate example is provided in the Instructor's Resource Manual.

### EXAMPLE 15.3

## Obtaining an Equilibrium Constant from Reaction Composition

Hydrogen iodide, HI, decomposes at moderate temperatures according to the equation

$$2HI(g) \rightleftharpoons H_2(g) + I_2(g)$$

The amount of $I_2$ in the reaction mixture can be determined from the intensity of the violet color of $I_2$; the more intense the color, the more $I_2$ in the reaction vessel. When 4.00 mol HI was placed in a 5.00-L vessel at 458°C, the equilibrium mixture was found to contain 0.442 mol $I_2$. What is the value of $K_c$ for the decomposition of HI at this temperature?

 **PROBLEM STRATEGY**

You will first need to calculate the molar concentrations of substances, because the equilibrium constant is stated in terms of these quantities. The problem statement gives the starting amount of HI and the equilibrium amount of $I_2$. As in Example 15.1, it is convenient to set up a table of starting, change, and equilibrium concentrations. You use the equilibrium values to calculate the equilibrium constant $K_c$.

**SOLUTION**

To obtain the concentrations of HI and $I_2$, you divide the molar amounts by the volume of the reaction vessel (5.00 L).

$$\text{Starting concentration of HI} = \frac{4.00 \text{ mol}}{5.00 \text{ L}} = 0.800 \text{ } M$$

$$\text{Equilibrium concentration of } I_2 = \frac{0.442 \text{ mol}}{5.00 \text{ L}} = 0.0884 \text{ } M$$

From these values, you set up the following table:

| Concentration ($M$) | 2HI($g$) $\rightleftharpoons$ | H$_2$($g$) + | I$_2$($g$) |
|---|---|---|---|
| Starting | 0.800 | 0 | 0 |
| Change | $-2x$ | $x$ | $x$ |
| Equilibrium | $0.800 - 2x$ | $x$ | $x = 0.0884$ |

The equilibrium concentrations of substances can be evaluated from the expressions given in the last line of this table. You know that the equilibrium concentration of $I_2$ is 0.0884 $M$ and that this equals $x$. Therefore,

$$[HI] = (0.800 - 2x)\,M = (0.800 - 2 \times 0.0884)\,M = 0.623\,M$$
$$[H_2] = x = 0.0884\,M$$

Now you substitute into the equilibrium-constant expression for the reaction. From the chemical equation, you write

$$K_c = \frac{[H_2][I_2]}{[HI]^2}$$

Substituting,

$$K_c = \frac{(0.0884)(0.0884)}{(0.623)^2} = \mathbf{0.0201}$$

---

**EXERCISE 15.4**

Hydrogen sulfide, a colorless gas with a foul odor, dissociates on heating:

$$2H_2S(g) \rightleftharpoons 2H_2(g) + S_2(g)$$

When 0.100 mol $H_2S$ was put into a 10.0-L vessel and heated to 1132°C, it gave an equilibrium mixture containing 0.0285 mol $H_2$. What is the value of $K_c$ at this temperature?

---

**See Problems 15.37, 15.38, 15.39, and 15.40.**

### The Equilibrium Constant $K_p$

In discussing gas-phase equilibria, it is often convenient to write the equilibrium constant in terms of partial pressures of gases rather than concentrations. Note that the concentration of a gas is proportional to its partial pressure at a fixed temperature (Figure 15.4). You can see this by looking at the ideal gas law, $PV = nRT$, and solving for $n/V$, which is the molar concentration of the gas. You get $n/V = P/RT$. In other words, the molar concentration of a gas equals its partial pressure divided by $RT$, which is constant at a given temperature.

**Figure 15.4
The concentration of a gas at a given temperature is proportional to the pressure.**  When the gas is compressed, the pressure increases (as noted from the pressure gauge) and the concentration of the gas (molecules per unit volume) increases.

When you express an equilibrium constant for a gaseous reaction in terms of partial pressures, you call it $K_p$. For catalytic methanation,

$$CO(g) + 3H_2(g) \rightleftharpoons CH_4(g) + H_2O(g)$$

the equilibrium expression in terms of partial pressures becomes

$$K_p = \frac{P_{CH_4}P_{H_2O}}{P_{CO}P_{H_2}{}^3}$$

In general, the value of $K_p$ is different from that of $K_c$. From the relationship for molar concentration $n/V = P/RT$, one can show that

$$K_p = K_c(RT)^{\Delta n}$$

where $\Delta n$ is the sum of the coefficients of *gaseous* products in the chemical equation minus the sum of the coefficients of *gaseous* reactants. For the methanation reaction ($K_c = 3.92$), in which 2 mol of gaseous products ($CH_4 + H_2O$) are obtained from 4 mol of gaseous reactants ($CO + 3H_2$), $\Delta n$ equals $2 - 4 = -2$. The usual unit of partial pressures in $K_p$ is atmospheres; therefore, the value of $R$ is 0.0821 L · atm/(K · mol). Hence,

$$K_p = 3.92 \times (0.0821 \times 1200)^{-2} = 4.04 \times 10^{-4}$$

---

### EXERCISE 15.5

Phosphorus pentachloride dissociates on heating:

$$PCl_5(g) \rightleftharpoons PCl_3(g) + Cl_2(g)$$

If $K_c$ equals $3.26 \times 10^{-2}$ at 191°C, what is $K_p$ at this temperature?

---

**See Problems 15.43, 15.44, 15.45, and 15.46.**

---

## Equilibrium Constant for the Sum of Reactions

It is possible to determine the equilibrium constants for various chemical reactions and then use them to obtain the equilibrium constants of other reactions. A useful rule is as follows: *if a given chemical equation can be obtained by taking the sum of other equations, the equilibrium constant for the given equation equals the product of the equilibrium constants of the other equations.*

As an application of this rule, consider the following reactions at 1200 K:

$$CO(g) + 3H_2(g) \rightleftharpoons CH_4(g) + H_2O(g); K_1 = 3.92 \qquad \text{(Reaction 1)}$$

$$CH_4(g) + 2H_2S(g) \rightleftharpoons CS_2(g) + 4H_2(g); K_2 = 3.3 \times 10^4 \quad \text{(Reaction 2)}$$

When you take the sum of these two equations, you get

$$CO(g) + 2H_2S(g) \rightleftharpoons CS_2(g) + H_2O(g) + H_2(g) \qquad \text{(Reaction 3)}$$

According to the rule, the equilibrium constant for this reaction, $K_3$, is $K_3 = K_1K_2$. This result is easy to verify. Substituting the expressions for $K_1$ and $K_2$ in the product $K_1K_2$ gives

$$K_1K_2 = \frac{[CH_4][H_2O]}{[CO][H_2]^3} \times \frac{[CS_2][H_2]^4}{[CH_4][H_2S]^2} = \frac{[CS_2][H_2O][H_2]}{[CO][H_2S]^2}$$

▶ You may also use equilibrium expressions in terms of pressures in this calculation.

which is the equilibrium-constant expression for Reaction 3. The value of the equilibrium constant is $K_1K_2 = 3.92 \times (3.3 \times 10^4) = 1.3 \times 10^5$. ◀

## 15.3 Heterogeneous Equilibria; Solvents in Homogeneous Equilibria

A **homogeneous equilibrium** is *an equilibrium that involves reactants and products in a single phase.* Catalytic methanation is an example of a homogeneous equilibrium; it involves only gaseous reactants and products. On the other hand, a **heterogeneous equilibrium** is *an equilibrium involving reactants and products in more than one phase.* For example, the reaction of iron metal filings with steam to produce iron oxide, $Fe_3O_4$, and hydrogen involves solid phases, Fe and $Fe_3O_4$, in addition to a gaseous phase.

$$3Fe(s) + 4H_2O(g) \rightleftharpoons Fe_3O_4(s) + 4H_2(g)$$

In writing the equilibrium-constant expression for a heterogeneous equilibrium, you omit concentration terms for pure solids and liquids. For the reaction of iron with steam, you would write

$$K_c = \frac{[H_2]^4}{[H_2O]^4}$$

Concentrations of Fe and $Fe_3O_4$ are omitted, because whereas the concentration of a gas can have various values, the concentration of a pure solid or a pure liquid is a constant at a given temperature.

To understand why such constant concentrations are omitted in writing $K_c$, write the equilibrium-constant expression for the reaction of iron with steam, but include [Fe] and [$Fe_3O_4$]. Call this $K_c'$.

$$K_c' = \frac{[Fe_3O_4][H_2]^4}{[Fe]^3[H_2O]^4}$$

You can rearrange the above equation, putting all of the constant factors on the left side.

$$\underbrace{\frac{[Fe]^3}{[Fe_3O_4]}}_{\text{constant factors}} \times K_c' = \underbrace{\frac{[H_2]^4}{[H_2O]^4}}_{\text{variable factors}}$$

▶ If you use the thermodynamic equilibrium constant, which is defined in terms of activities, you find that the activity of a pure solid or liquid is 1. Therefore, when you write the equilibrium constant, the activities of pure solids and liquids need not be given explicitly.

Note that all terms on the left are constant but that those on the right (the concentrations of $H_2$ and $H_2O$) are variable. The left side equals $K_c$. Thus, in effect, concentrations of pure solids or pure liquids are incorporated in the value of $K_c$. Following similar reasoning, you also omit the concentration of solvent from $K_c$ for a homogeneous reaction, as long as the concentration of solvent remains essentially constant during reaction. ◀

The fact that the concentrations of both Fe and $Fe_3O_4$ do not appear in the equilibrium-constant expression means that the equilibrium is not affected by the amounts of these substances, *as long as some of each is present.*

# Slime Molds and Leopards' Spots

**EVERYDAY LIFE**

A bright yellow splotch highlights a patch of wood chips in a garden. It looks as if a prankster has poured yellow paint over the chips. Next day the yellow splotch has moved a few inches, and now appears to be climbing up the base of a wall by the garden! A biologist would tell you the yellow splotch is a slime mold (Figure 15.5).

Less noticeable are the cellular slime molds. Much of the time they consist of a group of single-celled amoebas searching for bacteria to eat. But when their food supply is gone, the amoebas release a chemical that signals them to clump together, forming a multicelled slug that moves to a dry, sunlit patch where it forms spores that can be blown to better feeding grounds. During the initial stages of aggregation, the amoebas form a distinctive pattern reminiscent of that seen by chemists in an *oscillating reaction,* which occurs as substances diffuse through a gel (Figure 15.6).

The Russian chemist Boris P. Belousov reported one of the first oscillating reactions in 1958, in which the reaction cycled back and forth over time from one color to another. At first, his work was ignored. You might expect a reaction to form products of a given color, with the reaction gradually slowing as it comes to equilibrium, but how could a chemical reaction oscillate, say, from red to blue and back, over a period of hours? Another Russian chemist, Anatol Zhabotinsky, improved the experiment,

leaving no doubt that some chemical reactions do oscillate. And when the chemicals are placed in a gel in a flat dish, the colors appear as a pattern of waves in space rather than as oscillations in time. Chemists have since shown that the Belousov–Zhabotinsky reaction occurs by two different mechanisms, first by one, then by the other. These mechanisms are repeated in space or time, depending on the concentrations of intermediate substances. During the reaction, an indicator changes color depending on which mechanism is active. Although the complete set of elementary steps of the Belousov–Zhabotinsky reaction is complicated, the overall reaction occurs just as you would expect. The initial reactants continue to decrease over time and the final products increase as the substances come to equilibrium.

An oscillating mechanism also appears to be at work in the aggregation phase of cellular slime molds. And these are not the only biological examples of chemical oscillations. One theory of the formation of fur patterns in animals, such as a leopard's spots, assumes that an oscillating reaction occurs in the skin of the embryo, which eventually generates the fur pattern you see in the adult. A similar theory has been used to explain striped patterns on seashells. The rhythmic beat of your heart is an example where the concentrations of substances oscillate in time. Who would have thought that slime molds and leopards' spots had anything in common?

**Figure 15.5**
**A slime mold.** The fruiting body (yellow) of the slime mold *Fuligo septica* appears on a patch of leaves.

**Figure 15.6**
**An oscillating reaction.**
This is a plate showing the wave patterns of the Belousov–Zhabotinsky reaction (in a gel).

► Video no. 28 (Series D), "Briggs-Rauscher" (Oscillating Reactions).

Ni(CO)$_4$

## EXAMPLE 15.4

## Writing $K_c$ for a Reaction with Pure Solids or Liquids

a. Quicklime (calcium oxide, CaO) is prepared by heating a source of calcium carbonate, CaCO$_3$, such as limestone or seashells.

$$CaCO_3(s) \rightleftharpoons CaO(s) + CO_2(g)$$

Write the expression for $K_c$.

b. You can write the equilibrium-constant expression for a physical equilibrium, such as vaporization, as well as for a chemical equilibrium. Write the expression for $K_c$ for the vaporization of water.

$$H_2O(l) \rightleftharpoons H_2O(g)$$

 **PROBLEM STRATEGY**

Write the equilibrium-constant expression excluding terms for pure solids and liquids.

**SOLUTION**

In writing the equilibrium expressions, you ignore pure liquid and solid phases. a. $K_c = [CO_2]$; b. $K_c = [H_2O(g)]$.

## EXERCISE 15.6

The Mond process for purifying nickel involves the formation of nickel tetracarbonyl, Ni(CO)$_4$, a volatile liquid, from nickel metal and carbon monoxide. Carbon monoxide is passed over impure nickel to form nickel carbonyl vapor, which, when heated, decomposes and deposits pure nickel.

$$Ni(s) + 4CO(g) \rightleftharpoons Ni(CO)_4(g)$$

Write the expression for $K_c$ for this reaction.

*See Problems 15.47 and 15.48.*

# Using the Equilibrium Constant

In the preceding sections, we described how a chemical reaction reaches equilibrium and how you can characterize this equilibrium by the equilibrium constant. Now we want to see the ways in which an equilibrium constant can be used to answer important questions. We will look at the following uses:

1. *Qualitatively interpreting the equilibrium constant.* By merely looking at the magnitude of $K_c$, you can tell whether a particular equilibrium favors products or reactants.

2. *Predicting the direction of reaction.* Consider a reaction mixture that is not at equilibrium. By substituting the concentrations of substances that exist in a reaction mixture into an expression similar to the equilibrium constant and comparing with $K_c$, you can predict whether the reaction will proceed toward products or toward reactants (as defined by the way you write the chemical equation).

3. *Calculating equilibrium concentrations.* Once you know the value of $K_c$ for a reaction, you can determine the composition at equilibrium for any set of starting concentrations.

Let us see what meaning we can attach to the value of $K_c$.

## 15.4 Qualitatively Interpreting the Equilibrium Constant

If the value of the equilibrium constant is large, you know immediately that the products are favored at equilibrium. Consider the synthesis of ammonia from its elements.

$$N_2(g) + 3H_2(g) \rightleftharpoons 2NH_3(g)$$

At 25°C the equilibrium constant $K_c$ equals $4.1 \times 10^8$. This means that the numerator (product concentrations) is $4.1 \times 10^8$ times larger than the denominator (reactant concentrations). In other words, at this temperature the reaction favors the formation of ammonia at equilibrium.

You can verify this by calculating one possible equilibrium composition for this reaction. Suppose the equilibrium mixture is 0.010 $M$ in $N_2$ and 0.010 $M$ in $H_2$. From these concentrations you can calculate the concentration of ammonia necessary to give equilibrium. Substitute the concentrations of $N_2$ and $H_2$ and the value of $K_c$ into the equilibrium expression

$$K_c = \frac{[NH_3]^2}{[N_2][H_2]^3}$$

You get

$$\frac{[NH_3]^2}{(0.010)(0.010)^3} = 4.1 \times 10^8$$

Now you can solve for the concentration of ammonia.

$$[NH_3]^2 = 4.1 \times 10^8 \times (0.010)(0.010)^3 = 4.1$$

After you take the square root of both sides of this equation, you find that $[NH_3] = 2.0\, M$. The concentrations of $N_2$ and $H_2$ are both 0.010 $M$, so the amount of ammonia formed at equilibrium is 200 times that of any one reactant. Figure 15.7 shows the burning of methane, another reaction with an enormously large equilibrium constant.

If the value of the equilibrium constant is small, the reactants are favored at equilibrium. As an example, consider the reaction of nitrogen and oxygen to give nitric oxide, NO:

$$N_2(g) + O_2(g) \rightleftharpoons 2NO(g)$$

The equilibrium constant $K_c$ equals $4.6 \times 10^{-31}$ at 25°C. If you assume that the concentrations of $N_2$ and $O_2$ are 1.0 $M$, you find that the concentration of NO is $6.8 \times 10^{-16}\, M$. In this case the equilibrium constant is very small, and the concentration of product is not detectable. Reaction occurs to only a very limited extent. ◄

When the equilibrium constant is neither large nor small (around 1), neither reactants nor products are strongly favored. The equilibrium mixture contains significant amounts of all substances in the reaction. For example, in the case of the methanation reaction, the equilibrium constant $K_c$ equals 3.92 at 1200 K. You found that if

**Figure 15.7**
**Methane (natural gas) reacts with oxygen.** The equilibrium mixture is almost entirely carbon dioxide and water. The equilibrium constant $K_c$ for the reaction $CH_4 + 2O_2 \rightleftharpoons CO_2 + 2H_2O$ is $10^{140}$ at 25°C.

► The equilibrium constant does become large enough at higher temperatures (about 2000°C) to give appreciable amounts of nitric oxide. This is why air ($N_2 + O_2$) in flames and in auto engines becomes a source of the pollutant NO.

you start with 1.000 mol CO and 3.000 mol $H_2$ in a 10.00-L vessel, the equilibrium composition is 0.613 mol CO, 1.839 mol $H_2$, 0.387 mol $CH_4$, and 0.387 mol $H_2O$. Neither the reactants nor the products are predominant.

If $K_c$ for a reaction,

$$\underbrace{aA + bB}_{\text{reactants}} \rightleftharpoons \underbrace{cC + dD}_{\text{products}}$$

is large, the equilibrium mixture is mostly products. If $K_c$ is small, the equilibrium mixture is mostly reactants. When $K_c$ is around 1, the equilibrium mixture contains appreciable amounts of both reactants and products.

---

**EXERCISE 15.7**

The equilibrium constant $K_c$ for the reaction

$$2NO(g) + O_2(g) \rightleftharpoons 2NO_2(g)$$

equals $4.0 \times 10^{13}$ at 25°C. Does the equilibrium mixture contain predominantly reactants or products? If $[NO] = [O_2] = 2.0 \times 10^{-6}\,M$ at equilibrium, what is the equilibrium concentration of $NO_2$?

---

See Problems 15.49, 15.50, 15.51, and 15.52.

## 15.5 Predicting the Direction of Reaction

Suppose a gaseous mixture from an industrial plant has the following composition at 1200 K: 0.0200 $M$ CO, 0.0200 $M$ $H_2$, 0.00100 $M$ $CH_4$, and 0.00100 $M$ $H_2O$. If the mixture is passed over a catalyst at 1200 K, would the reaction

$$CO(g) + 3H_2(g) \rightleftharpoons CH_4(g) + H_2O(g)$$

go toward the right or the left? That is, would the mixture form more $CH_4$ and $H_2O$ in going toward equilibrium, or more CO and $H_2$?

To answer this question, you substitute the concentrations of substances into the *reaction quotient* and compare its value to $K_c$. The **reaction quotient, $Q_c$**, is *an expression that has the same form as the equilibrium-constant expression but whose concentration values are not necessarily those at equilibrium.* For catalytic methanation, the reaction quotient is

$$Q_c = \frac{[CH_4]_i[H_2O]_i}{[CO]_i[H_2]_i^3}$$

where the subscript $i$ indicates concentrations at a particular instant $i$. When you substitute the concentrations of the gaseous mixture described earlier, you get

$$Q_c = \frac{(0.00100)(0.00100)}{(0.0200)(0.0200)^3} = 6.25$$

Recall that the equilibrium constant $K_c$ for catalytic methanation is 3.92 at 1200 K. For the reaction mixture to go to equilibrium, the value of $Q_c$ must decrease from 6.25 to 3.92. This will happen if the reaction goes to the left. In that case, the numerator of $Q_c$ ($[CH_4]_i[H_2O]_i$) will decrease, and the denominator ($[CO]_i[H_2]_i^3$) will increase. Thus, the gaseous mixture will give more CO and $H_2$.

Consider the problem more generally. You are given a reaction mixture that is not yet at equilibrium. You would like to know in what direction the reaction will go as it approaches equilibrium. To answer this, you substitute the concentrations of substances from the mixture into the reaction quotient $Q_c$. Then, you compare $Q_c$ to the equilibrium constant $K_c$.

For the general reaction $a\mathrm{A} + b\mathrm{B} \rightleftharpoons c\mathrm{C} + d\mathrm{D}$,

$$Q_c = \frac{[\mathrm{C}]_i^c[\mathrm{D}]_i^d}{[\mathrm{A}]_i^a[\mathrm{B}]_i^b}$$

Then

If $Q_c > K_c$, the reaction will go to the left.
If $Q_c < K_c$, the reaction will go to the right.
If $Q_c = K_c$, the reaction mixture is at equilibrium.

▶ An alternate example is provided in the Instructor's Resource Manual.

### EXAMPLE 15.5

## Using the Reaction Quotient

A 50.0-L reaction vessel contains 1.00 mol $N_2$, 3.00 mol $H_2$, and 0.500 mol $NH_3$. Will more ammonia, $NH_3$, be formed or will it dissociate when the mixture goes to equilibrium at 400°C? The equation is

$$N_2(g) + 3H_2(g) \rightleftharpoons 2NH_3(g)$$

$K_c$ is 0.500 at 400°C.

 **PROBLEM STRATEGY**

First, write the reaction quotient, $Q_c$, which has the same form as the equilibrium constant expression, but whose concentrations are those for a reaction mixture. Then, compare the value of $Q_c$ with that of $K_c$. If $Q_c$ is greater than $K_c$, the reaction goes in the direction written; otherwise it goes in the opposite direction (unless $Q_c = K_c$, in which case the reaction is at equilibrium).

### SOLUTION

The composition of the gas has been given in terms of moles. You convert these to molar concentrations by dividing by the volume (50.0 L). This gives 0.0200 $M$ $N_2$, 0.0600 $M$ $H_2$, and 0.0100 $M$ $NH_3$. Substituting these concentrations into the reaction quotient gives

$$Q_c = \frac{[\mathrm{NH_3}]_i^2}{[\mathrm{N_2}]_i[\mathrm{H_2}]_i^3} = \frac{(0.0100)^2}{(0.0200)(0.0600)^3} = 23.1$$

Because $Q_c = 23.1$ is greater than $K_c = 0.500$, the reaction will go to the left as it approaches equilibrium. Therefore, **ammonia will dissociate.**

### EXERCISE 15.8

A 10.0-L vessel contains 0.0015 mol $CO_2$ and 0.10 mol CO. If a small amount of carbon is added to this vessel and the temperature raised to 1000°C, will more CO form? The reaction is

$$CO_2(g) + C(s) \rightleftharpoons 2CO(g)$$

**Problem Solving:
Applications of the
Equilibrium Constant**

The value of $K_c$ is 1.17 at 1000°C. Assume that the volume of gas in the vessel is 10.0 L.

---

See Problems 15.53 and 15.54.

## CONCEPT CHECK 15.2

Carbon monoxide and hydrogen react in the presence of a catalyst to form methanol, $CH_3OH$:

$$CO(g) + 2H_2(g) \rightleftharpoons CH_3OH(g)$$

An equilibrium mixture of these three substances is suddenly compressed so that the concentrations of all substances initially double. In what direction does the reaction go as a new equilibrium is attained?

# 15.6 Calculating Equilibrium Concentrations

Once you have determined the equilibrium constant for a reaction, you can use it to calculate the concentrations of substances in an equilibrium mixture. The next example illustrates a simple type of equilibrium problem.

---

### EXAMPLE 15.6

## Obtaining One Equilibrium Concentration Given the Others

A gaseous mixture contains 0.30 mol CO, 0.10 mol $H_2$, and 0.020 mol $H_2O$, plus an unknown amount of $CH_4$, in each liter. This mixture is at equilibrium at 1200 K.

$$CO(g) + 3H_2(g) \rightleftharpoons CH_4(g) + H_2O(g)$$

What is the concentration of $CH_4$ in this mixture? The equilibrium constant $K_c$ equals 3.92.

 **PROBLEM STRATEGY**

Calculate concentrations from moles of substances. Substitute these concentrations into the equilibrium-constant equation and solve for the unknown concentration.

**SOLUTION**

The equilibrium-constant equation is

$$K_c = \frac{[CH_4][H_2O]}{[CO][H_2]^3}$$

Substituting the known concentrations and the value of $K_c$ gives

$$3.92 = \frac{[CH_4](0.020)}{(0.30)(0.10)^3}$$

You can now solve for $[CH_4]$.

$$[CH_4] = \frac{(3.92)(0.30)(0.10)^3}{(0.020)} = 0.059$$

The concentration of $CH_4$ in the mixture is **0.059 mol/L.**

### EXERCISE 15.9

Phosphorus pentachloride gives an equilibrium mixture of $PCl_5$, $PCl_3$, and $Cl_2$ when heated.

$$PCl_5(g) \rightleftharpoons PCl_3(g) + Cl_2(g)$$

A 1.00-L vessel contains an unknown amount of $PCl_5$ and 0.020 mol each of $PCl_3$ and $Cl_2$ at equilibrium at 250°C. How many moles of $PCl_5$ are in the vessel if $K_c$ for this reaction is 0.0415 at 250°C?

*See Problems 15.57 and 15.58.*

Usually you begin a reaction with known starting quantities of substances and want to calculate the quantities at equilibrium. The next example illustrates the steps used to solve this type of problem.

### EXAMPLE 15.7

## Solving an Equilibrium Problem (Involving a Linear Equation in *x*)

▶ An alternate example is provided in the Instructor's Resource Manual.

The reaction

$$CO(g) + H_2O(g) \rightleftharpoons CO_2(g) + H_2(g)$$

is used to increase the ratio of hydrogen in synthesis gas (mixtures of CO and $H_2$). Suppose you start with 1.00 mol each of carbon monoxide and water in a 50.0-L vessel. How many moles of each substance are in the equilibrium mixture at 1000°C? The equilibrium constant $K_c$ at this temperature is 0.58.

 **PROBLEM STRATEGY**

The solution of an equilibrium problem involves three steps. In Step 1, you use the given information to set up a table of starting, change, and equilibrium concentrations. In this problem, you are given starting amounts of reactants. You use these to obtain the starting concentrations. Express the change concentrations in terms of $x$, as we did in Example 15.1. In Step 2, you substitute the equilibrium concentrations in the table into the equilibrium-constant expression and equate this to the value of the equilibrium constant. In Step 3, you solve this equilibrium-constant equation. This example gives an equation that is a perfect square; taking the square root of it gives a linear equation, which is easy to solve. (More generally, an equilibrium equation involves solving an $n$th-order equation for its roots. For $n = 2$, you can use the quadratic formula, as illustrated in Example 15.8. For $n > 2$, it is time to think about using a computer.) ◀

▶ See the "Equilibrium Calculator (Equilib)," by Robert Allendorfer, JCE: Software, Madison, WI (requires MS Windows).

**SOLUTION**

**STEP 1** The *starting concentrations* of CO and $H_2O$ are

$$[CO] = [H_2O] = \frac{1.00 \text{ mol}}{50.0 \text{ L}} = 0.0200 \text{ mol/L}$$

The starting concentrations of the products, $CO_2$ and $H_2$, are 0. The *changes in concentrations* when the mixture goes to equilibrium are not given. However, you can write them all in terms of a single unknown. If you let $x$ be the moles of $CO_2$ formed per liter, then the moles of $H_2$ formed per liter is also $x$. Similarly, $x$ moles each of CO and $H_2O$ are consumed. You write the changes for CO and $H_2O$ as $-x$. You obtain the *equilibrium concentrations* by adding the change in concentrations to the starting concentrations, as shown in the following table:

| Concentrations (*M*) | $CO(g)$ | + | $H_2O(g)$ | $\rightleftharpoons$ | $CO_2(g)$ | + | $H_2(g)$ |
|---|---|---|---|---|---|---|---|
| Starting | 0.0200 | | 0.0200 | | 0 | | 0 |
| Change | $-x$ | | $-x$ | | $+x$ | | $+x$ |
| Equilibrium | $0.0200 - x$ | | $0.0200 - x$ | | $x$ | | $x$ |

**STEP 2** You then substitute the values for the equilibrium concentrations into the equilibrium-constant equation,

$$K_c = \frac{[CO_2][H_2]}{[CO][H_2O]}$$

and you get

$$0.58 = \frac{(x)(x)}{(0.0200 - x)(0.0200 - x)}$$

or

$$0.58 = \frac{x^2}{(0.0200 - x)^2}$$

**STEP 3** You now solve this equilibrium equation for the value of $x$. Note that the right-hand side is a perfect square. If you take the square root of both sides, you get

$$\pm 0.76 = \frac{x}{0.0200 - x}$$

We have written $\pm$ to indicate that you should consider both positive and negative values, because both are mathematically possible. Rearranging the equation gives

$$x = \frac{0.0200 \times 0.76}{1.76} = 0.0086 \quad \text{and} \quad x = \frac{-0.0200 \times 0.76}{0.24} = -0.063$$

You can dismiss the negative value as physically impossible ($x$ can only be positive; it represents the concentration of $CO_2$ formed). If you substitute for $x$ in the last line of the table, the equilibrium concentrations are 0.0114 *M* CO, 0.0114 *M* $H_2O$, 0.0086 *M* $CO_2$, and 0.0086 *M* $H_2$. To find the moles of each substance in the 50.0-L vessel, you multiply the concentrations by the volume of the vessel. For example, the amount of CO is

$$0.0114 \text{ mol/L} \times 50.0 \text{ L} = 0.570 \text{ mol}$$

The equilibrium composition of the reaction mixture is **0.570 mol CO, 0.570 mol H₂O, 0.43 mol CO₂, and 0.43 mol H₂.**

### EXERCISE 15.10

What is the equilibrium composition of a reaction mixture if you start with 0.500 mol each of $H_2$ and $I_2$ in a 1.0-L vessel? The reaction is

$$H_2(g) + I_2(g) \rightleftharpoons 2HI(g) \qquad K_c = 49.7 \text{ at } 458°C$$

See Problems 15.59 and 15.60.

The preceding example illustrates the three steps used to solve for equilibrium concentrations, as summarized in the following list:

1. Set up a table of concentrations (starting, change, and equilibrium expressions in $x$).

2. Substitute the expressions in $x$ for equilibrium concentrations into the equilibrium-constant equation.

3. Solve the equilibrium-constant equation for the values of the equilibrium concentrations.

▶ **A quadratic equation of the form**

$$ax^2 + bx + c = 0$$

**has the solutions**

$$x = \frac{-b \pm \sqrt{b^2 - 4ac}}{2a}$$

**This equation for $x$ is called the *quadratic formula*. (See Appendix A.)**

In the previous example, if you had not started with the same number of moles of reactants, you would not have gotten an equation with a perfect square. In that case you would have had to solve a quadratic equation. The next example illustrates how to solve such an equation. ◀

▶ **An alternate example is provided in the Instructor's Resource Manual.**

### EXAMPLE 15.8

## Solving an Equilibrium Problem (Involving a Quadratic Equation in $x$)

Hydrogen and iodine react according to the equation

$$H_2(g) + I_2(g) \rightleftharpoons 2HI(g)$$

Suppose 1.00 mol $H_2$ and 2.00 mol $I_2$ are placed in a 1.00-L vessel. How many moles of substances are in the gaseous mixture when it comes to equilibrium at 458°C? The equilibrium constant $K_c$ at this temperature is 49.7.

 **PROBLEM STRATEGY**

You follow the three-step approach outlined just before this example. The equilibrium equation will be a quadratic equation, which you can solve by the quadratic formula. (See Appendix A for a discussion of this equation and formula.)

### SOLUTION

**STEP 1**    The concentrations of substances are as follows:

| Concentrations ($M$) | H₂(g) | + | I₂(g) | ⇌ | 2HI(g) |
|---|---|---|---|---|---|
| Starting | 1.00 | | 2.00 | | 0 |
| Change | $-x$ | | $-x$ | | $2x$ |
| Equilibrium | $1.00 - x$ | | $2.00 - x$ | | $2x$ |

Note that the changes in concentrations equal $x$ multiplied by the coefficient of that substance in the balanced chemical equation. The change is negative for a reactant and positive for a product. Equilibrium concentrations equal starting concentrations plus the changes in concentrations.

**STEP 2** Substituting into the equilibrium-constant equation,

$$K_c = \frac{[HI]^2}{[H_2][I_2]}$$

you get

$$49.7 = \frac{(2x)^2}{(1.00 - x)(2.00 - x)}$$

**STEP 3** Because the right-hand side is not a perfect square, you must use the quadratic formula to solve for $x$. The equation rearranges to give

$$(1.00 - x)(2.00 - x) = (2x)^2/49.7 = 0.0805x^2$$

or

$$0.920x^2 - 3.00x + 2.00 = 0$$

Hence,

$$x = \frac{3.00 \pm \sqrt{9.00 - 7.36}}{1.84} = 1.63 \pm 0.70$$

A quadratic equation has two mathematical solutions. You obtain one solution by taking the upper (positive) sign in $\pm$ and the other by taking the lower (negative) sign. You get

$$x = 2.33 \quad \text{and} \quad x = 0.93$$

However, $x = 2.33$ gives a negative value to $1.00 - x$ (the equilibrium concentration of $H_2$), which is physically impossible. Only $x = 0.93$ remains. You substitute this value of $x$ into the last line of the table in Step 1 to get the equilibrium concentrations, and then multiply these by the volume of the vessel (1.00 L) to get the amounts of substances. The last line of the table is rewritten

| Concentrations ($M$) | $H_2(g)$ | $+$ | $I_2(g)$ | $\rightleftharpoons$ | $2HI(g)$ |
|---|---|---|---|---|---|
| Equilibrium | $1.00 - x = 0.07$ | | $2.00 - x = 1.07$ | | $2x = 1.86$ |

The equilibrium composition is **0.07 mol $H_2$, 1.07 mol $I_2$, and 1.86 mol HI.**

---

### EXERCISE 15.11

Phosphorus pentachloride, $PCl_5$, decomposes when heated.

$$PCl_5(g) \rightleftharpoons PCl_3(g) + Cl_2(g)$$

If the initial concentration of $PCl_5$ is 1.00 mol/L, what is the equilibrium composition of the gaseous mixture at 160°C? The equilibrium constant $K_c$ at 160°C is 0.0211.

---

See Problems 15.61 and 15.62.

### ▶ CONCEPT CHECK 15.3

A and B react to produce C according to the following chemical equation:

$$A + B \longrightarrow C$$

Amounts of A and B are added to an equilibrium reaction mixture of A, B, and C such that when equilibrium is again attained, the amounts of A and B are doubled in the same volume. How is the amount of C changed?

# Changing the Reaction Conditions; Le Chatelier's Principle

Obtaining the maximum amount of product from a reaction depends on the proper selection of reaction conditions. By changing these conditions, you can increase or decrease the yield of product. There are three ways to alter the equilibrium composition of a gaseous reaction mixture and possibly increase the yield of product:

1. Changing the concentrations by removing products or adding reactants to the reaction vessel.

2. Changing the partial pressure of gaseous reactants and products by changing the volume.

3. Changing the temperature.

Note that a catalyst cannot alter equilibrium composition, although it can change the rate at which a product is formed and affect the result.

## 15.7  Removing Products or Adding Reactants

▶ Demonstration of the effect of removing a product on chemical equilibrium: Summerlin and Ealy, *Chemical Demonstrations,* Vol. 1 (Washington, D.C.: Am. Chem. Soc., 1985), no. 39, p. 64.

▶ The common statement of Le Chatelier's principle in terms of "stresses" is too vague and may give wrong results. See R.S. Treptow, *J. Chem. Educ.* 1980, *57,* 417–420.

▶ Le Chatelier's principle was introduced in Section 12.3, where it was used to determine the effect of pressure on solubility of gases in liquids.

One way to increase the yield of a desired product is to change concentrations in a reaction mixture by removing a product or adding a reactant. Consider the methanation reaction,

$$CO(g) + 3H_2(g) \rightleftharpoons CH_4(g) + H_2O(g)$$

If you place 1.000 mol CO and 3.000 mol $H_2$ in a 10.00-L reaction vessel, the equilibrium composition at 1200 K is 0.613 mol CO, 1.839 mol $H_2$, 0.387 mol $CH_4$, and 0.387 mol $H_2O$. Can you alter this composition by removing or adding one of the substances to improve the yield of methane?

To answer this question, you can apply **Le Chatelier's principle,** which states that *when a system in chemical equilibrium is disturbed by a change of temperature, pressure, or a concentration, the system shifts in equilibrium composition in a way that tends to counteract this change of variable.* ◀ Suppose you remove a substance from or add a substance to the equilibrium mixture in order to alter the concentration of the substance. Chemical reaction then occurs to partially restore the initial concentration of the removed or added substance. (However, if the concentration of substance cannot be changed, as in the case of a pure solid or liquid reactant or product, changes in amount will have no effect on the equilibrium composition.)

**Table 15.1**
**Effect of Removing Water Vapor from a Methanation Mixture (in a 10.00-L Vessel)**

| Stage of Process | Mol CO | Mol $H_2$ | Mol $CH_4$ | Mol $H_2O$ |
|---|---|---|---|---|
| Original reaction mixure | 0.613 | 1.839 | 0.387 | 0.387 |
| After removing water (before equilibrium) | 0.613 | 1.839 | 0.387 | 0 |
| When equilibrium is reestablished | 0.491 | 1.473 | 0.509 | 0.122 |

▶ Experiments in General
Chemistry, 15A, "Le
Chatelier's Principle."

For example, suppose that water vapor is removed from the reaction vessel containing the equilibrium mixture for methanation. Le Chatelier's principle predicts that net chemical change will occur to partially reinstate the original concentration of water vapor. This means that the methanation reaction momentarily goes in the forward direction,

$$CO(g) + 3H_2(g) \longrightarrow CH_4(g) + H_2O(g)$$

until equilibrium is re-established. Going in the forward direction, the concentrations of both water vapor and methane increase.

A practical way to remove water vapor in this reaction would be to cool the reaction mixture quickly to condense the water. Liquid water could be removed and the gases reheated until equilibrium is again established. The concentration of water vapor would build up as the concentration of methane increases. Table 15.1 lists the amounts of each substance at each stage of this process (carried out in a 10.00-L reaction vessel). Note how the yield of methane has been improved.

It is often useful to add an excess of a cheap reactant in order to force the reaction toward more products. In this way, the more expensive reactant is made to react to a greater extent than it would otherwise.

Consider the ammonia synthesis

$$N_2(g) + 3H_2(g) \rightleftharpoons 2NH_3(g)$$

If you wished to convert as much hydrogen to ammonia as possible, you might increase the concentration of nitrogen. To understand the effect of this, first suppose that a mixture of nitrogen, hydrogen, and ammonia is at equilibrium. If nitrogen is now added to this mixture, the equilibrium is disturbed. According to Le Chatelier's principle, the reaction will now go in the direction that will use up some of the added nitrogen.

$$N_2(g) + 3H_2(g) \longrightarrow 2NH_3(g)$$

Consequently, adding more nitrogen has the effect of converting a greater quantity of hydrogen to ammonia. Figure 15.8 illustrates the effect of adding or removing a reactant on another chemical equilibrium.

You can look at these situations in terms of the reaction quotient, $Q_c$. Consider the methanation reaction, in which

$$Q_c = \frac{[CH_4]_i[H_2O]_i}{[CO]_i[H_2]_i^3}$$

If the reaction mixture is at equilibrium, $Q_c = K_c$. Suppose you remove some $H_2O$ from the equilibrium mixture. Now $Q_c < K_c$, and from what we said in Section 15.5, you know the reaction will proceed in the forward direction to restore equilibrium.

**Figure 15.8**
**Effect on a chemical equilibrium of changing the concentration of a substance.**  The concentration is changed by adding or removing a substance. *(A)* The center beaker contains an orange precipitate of $HgI_2$ in equilibrium with $I^-$ (colorless) and $HgI_4^{2-}$ (pale yellow): $HgI_2(s) + 2I^-(aq) \rightleftharpoons HgI_4^{2-}(aq)$. *(B)* The solution of NaI (right beaker) is added to the equilibrium mixture in the center beaker. By adding $I^-$, the reaction is forced to the right: $HgI_2(s) + 2I^-(aq) \longrightarrow HgI_4^{2-}(aq)$. The orange precipitate of $HgI_2$ disappears. *(C)* A solution of NaClO (left beaker) is now added to the reaction mixture; the NaClO removes $I^-$ by oxidizing it to $IO^-$. By removing $I^-$, the reaction shifts to the left: $HgI_2(s) + 2I^-(aq) \longleftarrow HgI_4^{2-}(aq)$. An orange precipitate of $HgI_2$ appears.

**Understanding Concepts:**
**Equilibrium**

We can now summarize the conclusions from this section:

> When more reactant is added to, or some product is removed from, an equilibrium mixture, thereby changing the concentration of reactant or product, net reaction occurs left to right (that is, in the forward direction) to give a new equilibrium, and more products are produced.

> When more product is added to, or some reactant is removed from, an equilibrium mixture, thereby changing the concentration of reactant or product, net reaction occurs right to left (that is, in the reverse direction) to give a new equilibrium, and more reactants are produced.

**EXAMPLE 15.9**

## Applying Le Chatelier's Principle When a Concentration Is Altered

Predict the direction of reaction when $H_2$ is removed from a mixture (lowering its concentration) in which the following equilibrium has been established:

$$H_2(g) + I_2(g) \rightleftharpoons 2HI(g)$$

**SOLUTION**

When $H_2$ is removed from the reaction mixture, lowering its concentration, the reaction goes in the **reverse direction** (more HI dissociates to $H_2$ and $I_2$) to partially restore the $H_2$ that was removed.

$$H_2(g) + I_2(g) \longleftarrow 2HI(g)$$

---

**EXERCISE 15.12**

Consider each of the following equilibria, which are disturbed as indicated. Predict the direction of reaction.

a. The equilibrium

$$CaCO_3(s) \rightleftharpoons CaO(s) + CO_2(g)$$

is disturbed by increasing the pressure (that is, concentration) of carbon dioxide.

b. The equilibrium

$$2Fe(s) + 3H_2O(g) \rightleftharpoons Fe_2O_3(s) + 3H_2(g)$$

is disturbed by increasing the concentration of hydrogen.

**See Problems 15.65 and 15.66.**

---

A reaction whose equilibrium constant is extremely small remains almost completely as reactants and cannot be shifted to products by adding an excess of one reactant. For example, the reaction

$$CO_2(g) + 2H_2O(g) \rightleftharpoons CH_4(g) + 2O_2(g)$$

has an equilibrium constant $K_c$ equal to $10^{-140}$ at 25°C. This value is so small that the equilibrium mixture practically consists only of carbon dioxide and water. Adding more carbon dioxide has no appreciable effect.

## 15.8 Changing the Pressure and Temperature

▶ Video no. 23 (Series A),
"Gaseous Equilibrium."

The optimum conditions for catalytic methanation involve moderately elevated temperatures and normal to moderately high pressures.

$$CO(g) + 3H_2(g) \xrightarrow[\text{1 atm–100 atm}]{\text{230°C–450°C}} CH_4(g) + H_2O(g)$$

Let us see whether we can gain insight into why these might be the optimum conditions for the reaction.

### Effect of Pressure Change

A pressure change *obtained by changing the volume* can affect the yield of product in a gaseous reaction if the reaction involves a change in total moles of gas. The methanation reaction, $CO + 3H_2 \rightleftharpoons CH_4 + H_2O$, is an example of a change in moles of gas. When the reaction goes in the forward direction, four moles of reactant gas ($CO + 3H_2$) become two moles of product gas ($CH_4 + H_2O$).

To see the effect of such a pressure change, consider what happens when an equilibrium mixture from the methanation reaction is compressed to one-half of its original volume at a fixed temperature (see Figure 15.9). The total pressure is doubled ($PV$ = constant at a fixed temperature, according to Boyle's law, so halving $V$ requires that $P$ double). Because the partial pressures and therefore the concentrations of reactants and products have changed, the mixture is no longer at equilibrium. The direction in which the reaction goes to reestablish equilibrium can be predicted by applying Le Chatelier's principle. Reaction should go in the forward direction,

**Figure 15.9**
**Effect on chemical equilibrium of changing the pressure.** The effect on the methanation reaction is shown; the approximate composition is represented by the proportion of different "molecules" (see key at right). The actual arrangement of molecules in a gas is random. *(A)* The original equilibrium mixture of CO, $H_2$, $CH_4$, and $H_2O$ molecules. *(B)* The gases are compressed to one-half their initial volume, increasing their partial pressures, so that the mixture is no longer at equilibrium. *(C)* Equilibrium is reestablished when the reaction goes in the forward direction: $CO + 3H_2 \longrightarrow CH_4 + H_2O$. In this way, the total number of molecules is reduced, which reduces the initial pressure increase.

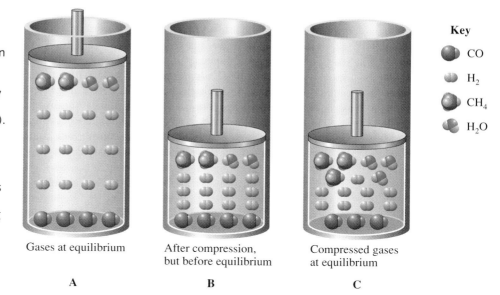

Gases at equilibrium

After compression, but before equilibrium

Compressed gases at equilibrium

Key

CO

$H_2$

$CH_4$

$H_2O$

A

B

C

because then the moles of gas decrease, and the pressure (which is proportional to moles of gas) decreases. In this way, the initial pressure increase is partially reduced.

You find the same result by looking at the reaction quotient $Q_c$. Let [CO], [$H_2$], [$CH_4$], and [$H_2O$] be the molar concentrations at equilibrium for the methanation reaction. When the volume of an equilibrium mixture is halved, the partial pressures and therefore the concentrations are doubled. You obtain the reaction quotient at that moment by replacing each equilibrium concentration by double its value.

$$Q_c = \frac{(2[CH_4])(2[H_2O])}{(2[CO])(2[H_2])^3} = \frac{K_c}{4}$$

Because $Q_c < K_c$, the reaction proceeds in the forward direction.

You can see the quantitative effect of the pressure change by solving the equilibrium problem. Recall that if you put 1.000 mol CO and 3.000 mol $H_2$ into a 10.00-L vessel, the equilibrium composition at 1200 K, where $K_c$ equals 3.92, is 0.613 mol CO, 1.839 mol $H_2$, 0.387 mol $CH_4$, and 0.387 mol $H_2O$. Suppose the volume of the reaction gases is halved so that the initial concentrations are doubled. The temperature remains at 1200 K, so $K_c$ is still 3.92. But if you solve for the new equilibrium composition (following the method of Example 15.8), you obtain 0.495 mol CO, 1.485 mol $H_2$, 0.505 mol $CH_4$, and 0.505 mol $H_2O$. Note that the amount of $CH_4$ has increased from 0.387 mol to 0.505 mol. You conclude that high pressure of reaction gases favors high yields of methane.

To decide the direction of reaction when the pressure of the reaction mixture is increased (say, by decreasing the volume), you ignore liquids and solids. They are not much affected by pressure changes, because they are nearly incompressible. Consider the reaction

$$C(s) + CO_2(g) \rightleftharpoons 2CO(g)$$

The moles *of gas* increase when the reaction goes in the forward direction (1 mol $CO_2$ goes to 2 mol CO). Therefore, when you increase the pressure of the reaction mixture by decreasing its volume, the reaction goes in the reverse direction. The moles of gas decrease, and the initial increase of pressure is partially reduced, as Le Chatelier's principle leads you to expect.

It is important to note that an increase or a decrease in pressure of a gaseous reaction must result in changes of partial pressures of substances in the chemical equation if it is to have an effect on the equilibrium composition. Only changes in partial pressures, or concentrations, of these substances can affect the reaction quotient. Consider the effect of increasing the pressure in the methanation reaction by adding helium gas. Although the total pressure increases, the partial pressures of CO, $H_2$, $CH_4$, and $H_2O$ do not change. Thus, the equilibrium composition is not affected. However, changing the pressure by changing the volume of this system changes the partial pressures of all gases, so the equilibrium composition is affected.

We can summarize these conclusions:

If the pressure is increased by decreasing the volume of a reaction mixture, the reaction shifts in the direction of fewer moles of gas.

▶ An alternate example is provided in the Instructor's Resource Manual.

### EXAMPLE 15.10

### Applying Le Chatelier's Principle When the Pressure Is Altered

Look at each of the following equations and decide whether an increase of pressure obtained by decreasing the volume will increase, decrease, or have no effect on the amounts of products.

a. $CO(g) + Cl_2(g) \rightleftharpoons COCl_2(g)$

b. $2H_2S(g) \rightleftharpoons 2H_2(g) + S_2(g)$

c. $C(graphite) + S_2(g) \rightleftharpoons CS_2(g)$

**SOLUTION**

a. Reaction decreases the number of molecules of gas (from two to one). According to Le Chatelier's principle, an increase of pressure **increases** the amount of product.

b. Reaction increases the number of molecules of gas (from two to three); hence, an increase of pressure **decreases** the amounts of products.

c. Reaction does not change the number of molecules of gas. (You ignore the change in volume due to consumption of solid carbon, because the change in volume of a solid is insignificant. Look only at gas volumes when deciding the effect of pressure change on equilibrium composition.) Pressure change has **no effect.**

### EXERCISE 15.13

Can you increase the amount of product in each of the following reactions by increasing the pressure? Explain.

a. $CO_2(g) + H_2(g) \rightleftharpoons CO(g) + H_2O(g)$

b. $4CuO(s) \rightleftharpoons 2Cu_2O(s) + O_2(g)$

c. $2SO_2(g) + O_2(g) \rightleftharpoons 2SO_3(g)$

See Problems 15.67 and 15.68.

| Table 15.2 Equilibrium Constant for Methanation at Different Temperatures | |
| --- | --- |
| **Temperature (K)** | **$K_c$** |
| 298 | $4.9 \times 10^{27}$ |
| 800 | $1.38 \times 10^5$ |
| 1000 | $2.54 \times 10^2$ |
| 1200 | 3.92 |

## Effect of Temperature Change

Temperature has a profound effect on most reactions. In the first place, reaction rates usually increase with an increase in temperature, meaning that equilibrium is reached sooner. Many gaseous reactions are sluggish or have imperceptible rates at room temperature but speed up enough at higher temperature to become commercially feasible processes.

Second, equilibrium constants vary with temperature. Table 15.2 gives values of $K_c$ for methanation at various temperatures. Note that $K_c$ equals $4.9 \times 10^{27}$ at 298 K. Therefore, an equilibrium mixture at room temperature is mostly methane and water.

Whether you should raise or lower the temperature of a reaction mixture to increase the equilibrium amount of product can be shown by Le Chatelier's principle. Again, consider the methanation reaction,

$$CO(g) + 3H_2(g) \rightleftharpoons CH_4(g) + H_2O(g); \Delta H° = -206.2 \text{ kJ}$$

The value of $\Delta H°$ shows this reaction to be quite exothermic. As products are formed, considerable heat is released. According to Le Chatelier's principle, as the temperature is raised, the reaction shifts to form more reactants, thereby absorbing heat and attempting to counter the increase in temperature.

$$CO(g) + 3H_2(g) \longleftarrow CH_4(g) + H_2O(g) + \text{heat}$$

You predict that the equilibrium constant will be smaller for higher temperatures, in agreement with the values of $K_c$ given in Table 15.2. Figure 15.10 illustrates the effect of temperature on another chemical equilibrium. ◀

The conclusions from Le Chatelier's principle regarding temperature effects on an equilibrium can be summarized this way:

▶ The quantitative effect of temperature on the equilibrium constant is discussed in Chapter 19.

> For an endothermic reaction ($\Delta H°$ positive), the amounts of products are increased at equilibrium by an increase in temperature ($K_c$ is larger at higher $T$). For an exothermic reaction ($\Delta H°$ negative), the amounts of products are increased at equilibrium by a decrease in temperature ($K_c$ is larger at lower $T$).

**Figure 15.10**
**Effect on chemical equilibrium of changing the temperature.**
The reaction shown is endothermic:
$Co(H_2O)_6^{2+}(aq) + 4Cl^-(aq) \rightleftharpoons CoCl_4^{2-}(aq) + 6H_2O(l)$.
At room temperature, the equilibrium mixture is blue, from $CoCl_4^{2-}$. When cooled by the ice bath, the equilibrium mixture turns pink, from $Co(H_2O)_6^{2+}$.

▶ An alternate example is provided in the Instructor's Resource Manual.

## EXAMPLE 15.11

## Applying Le Chatelier's Principle When the Temperature Is Altered

Carbon monoxide is formed when carbon dioxide reacts with solid carbon (graphite).

$$CO_2(g) + C(\text{graphite}) \rightleftharpoons 2CO(g); \Delta H° = 172.5 \text{ kJ}$$

Is a high or low temperature more favorable to the formation of carbon monoxide?

### SOLUTION

The reaction absorbs heat in the forward direction.

$$\text{Heat} + CO_2(g) + C(\text{graphite}) \longrightarrow 2CO(g)$$

As the temperature is raised, reaction occurs in the forward direction, using heat and thereby tending to lower the temperature. Thus, **high temperature** is more favorable to the formation of carbon monoxide. This is one reason why combustions of carbon and organic materials can produce significant amounts of carbon monoxide.

### EXERCISE 15.14

Consider the possibility of converting carbon dioxide to carbon monoxide by the endothermic reaction

$$CO_2(g) + H_2(g) \rightleftharpoons CO(g) + H_2O(g)$$

Is a high or a low temperature more favorable to the production of carbon monoxide? Explain.

See Problems 15.69 and 15.70.

## Choosing the Optimum Conditions for Reaction

You are now in a position to understand the optimum conditions for the methanation reaction. Because the reaction is exothermic, low temperatures should favor high yields of methane; that is, the equilibrium constant is large for low temperature. But gaseous reactions are often very slow at room temperature. In practice, the methanation reaction is run at moderately elevated temperatures (230°C–450°C) in the presence of a nickel catalyst. Under these conditions, the rate of reaction is sufficiently fast but the equilibrium constant is not too small. Because the methanation reaction involves a decrease in moles of gas, the yield of methane should increase as the pressure increases. However, the equilibrium constant is large at the usual operating temperatures, so very high pressures are not needed to obtain economical yields of methane. Pressures of 1 atm to 100 atm are usual for this reaction.

As another example, consider the Haber process for the synthesis of ammonia:

$$N_2(g) + 3H_2(g) \xrightleftharpoons{\text{Fe catalyst}} 2NH_3(g); \Delta H° = -91.8 \text{ kJ}$$

Because the reaction is exothermic, the equilibrium constant is larger for lower temperatures. But the reaction proceeds too slowly at room temperature to be practical,

▶ This technological problem has stimulated a great deal of basic research into understanding how certain bacteria "fix" nitrogen at atmospheric pressure to make $NH_3$. An enzyme called nitrogenase is responsible for $N_2$ fixation in these bacteria. This enzyme contains Fe and Mo, which may play a catalytic role.

even in the presence of the best available catalysts. ◀ The optimum choice of temperature, found experimentally to be about 450°C, is a compromise between an increased rate of reaction at higher temperature and an increased yield of ammonia at lower temperature. Because the formation of ammonia decreases the moles of gases, the yield of product is improved by high pressures. The equilibrium constant $K_c$ is only 0.159 at 450°C, so higher pressures (up to 600 atm) are required for an economical yield of ammonia. Ammonia from the Haber reactor is removed from the reaction mixture by cooling the compressed gases until $NH_3$ liquefies. Unreacted $N_2$ and $H_2$ circulate back to the reactor.

**EXERCISE 15.15**

Consider the reaction

$$2CO_2(g) \rightleftharpoons 2CO(g) + O_2(g); \Delta H° = 566 \text{ kJ}$$

Discuss the temperature and pressure conditions that would give the best yield of carbon monoxide.

*See Problems 15.73 and 15.74.*

## 15.9   Effect of a Catalyst

A *catalyst* is a substance that increases the rate of a reaction but is not consumed by it. The significance of a catalyst can be seen in the reaction of sulfur dioxide with oxygen to give sulfur trioxide.

$$2SO_2(g) + O_2(g) \rightleftharpoons 2SO_3(g)$$

The equilibrium constant $K_c$ for this reaction is $1.7 \times 10^{26}$, which indicates that for all practical purposes the reaction should go almost completely to products. Yet when sulfur is burned in air or oxygen, it forms predominantly $SO_2$ and very little $SO_3$. Oxidation of $SO_2$ to $SO_3$ is simply too slow to give a significant amount of product. However, the rate of the reaction is appreciable in the presence of a platinum or divanadium pentoxide catalyst. The oxidation of $SO_2$ in the presence of a catalyst is the main step in the *contact process* for the industrial production of sulfuric acid, $H_2SO_4$. Sulfur trioxide reacts with water to form sulfuric acid. (In the industrial process, $SO_3$ is dissolved in concentrated $H_2SO_4$, which is then diluted.) ◀

▶ Sulfur dioxide from the combustion of coal and from other sources appears to be a major cause of the marked increase in acidity of rain in the eastern United States in the past few decades. This *acid rain* has been shown to contain sulfuric and nitric acids. The $SO_2$ is oxidized in moist, polluted air to $H_2SO_4$. Acid rain is discussed in the essay on acid rain at the end of Section 17.2.

It is important to understand that *a catalyst has no effect on the equilibrium composition of a reaction mixture. A catalyst merely speeds up the attainment of equilibrium.* For example, suppose you mix 2.00 mol $SO_2$ and 1.00 mol $O_2$ in a 100.0-L vessel. In the absence of a catalyst, these substances appear unreactive. Much later, if you analyze the mixture, you find essentially the same amounts of $SO_2$ and $O_2$. But when a catalyst is added, the rates of both forward and reverse reactions are very much increased. As a result, the reaction mixture comes to equilibrium in a short time. The amounts of $SO_2$, $O_2$, and $SO_3$ can be calculated from the equilibrium constant. You find that the mixture is mostly $SO_3$ (2.00 mol), with only $1.7 \times 10^{-8}$ mol $SO_2$ and $8.4 \times 10^{-9}$ mol $O_2$.

A catalyst is useful for a reaction, such as $2SO_2 + O_2 \rightleftharpoons 2SO_3$, that is normally slow but has a large equilibrium constant. However, if the reaction has an exceedingly small equilibrium constant, a catalyst is of little help. The reaction

$$N_2(g) + O_2(g) \rightleftharpoons 2NO(g)$$

**Figure 15.11    An example of how rates of reaction can affect the kind of product.** *Left:* At time = 0, a bright orange precipitate of tetragonal crystals of $HgI_2$ forms from a concentrated solution of $HgCl_2$ and KI. In the dilute solution in the right beaker, however, the $HgI_2$ precipitates as rhombic crystals with a yellow color. Although less stable, the yellow crystals form faster in the dilute solution. *Right:* In 45 minutes the yellow crystals have changed to the more stable orange crystals.

has been considered for the industrial production of nitric acid. (NO reacts with $O_2$ and $H_2O$ to give nitric acid.) At 25°C, however, the equilibrium constant $K_c$ equals $4.6 \times 10^{-31}$. An equilibrium mixture would contain an extremely small concentration of NO. A catalyst would not provide a significant yield at this temperature; a catalyst merely speeds up the attainment of equilibrium. The equilibrium constant increases as the temperature is raised, so that at 2000°C, air (a mixture of $N_2$ and $O_2$) forms about 0.4% NO at equilibrium. An industrial plant was set up in Norway in 1905 to prepare nitrate fertilizers using this reaction. The plant was eventually made obsolete by the *Ostwald process* for making nitric acid, in which NO is prepared by the oxidation of ammonia. This latter reaction is more economical than the direct reaction of $N_2$ and $O_2$, in part because the equilibrium constant is larger at moderate temperatures.

Although a catalyst cannot affect the composition at true equilibrium, in some cases it can affect the product in a reaction because it affects the rate of one reaction out of several possible reactions. The importance of rate for determining the product in a reaction is illustrated by a simple example in Figure 15.11. Mercury(II) ion reacts with iodide ion to form a precipitate of mercury(II) iodide.

$$Hg^{2+}(aq) + 2I^-(aq) \longrightarrow HgI_2(s)$$

In concentrated solutions, orange tetragonal crystals of $HgI_2$ form. In dilute solution, however, yellow rhombic crystals form faster and are the initial product. Later, when true equilibrium is attained, the orange crystals appear.

The Ostwald process presents an interesting example of the effect of a catalyst in determining a product when several possibilities exist. Two reactions of ammonia with oxygen are possible. The reaction used in the Ostwald process is

$$4NH_3(g) + 5O_2(g) \rightleftharpoons 4NO(g) + 6H_2O(g)$$

▶ **Video Series C, "Ammonia and Oxygen."**

**Figure 15.12**
**Oxidation of ammonia using a copper catalyst.**
The products (formed on the glowing copper coil) are $N_2$ and $H_2O$, whereas a platinum catalyst results in NO and $H_2O$ as products.

However, nitric oxide dissociates to its elements,

$$2NO(g) \rightleftharpoons N_2(g) + O_2(g)$$

and the equilibrium constant for this is quite large ($2.2 \times 10^{30}$ at 25°C), so at true equilibrium the products of the reaction of ammonia with oxygen might be expected to be nitrogen and water.

$$4NH_3(g) + 3O_2(g) \rightleftharpoons 2N_2(g) + 6H_2O(g)$$

Ammonia can be made to burn in oxygen, and $N_2$ and $H_2O$ are the products. (The reaction occurs most readily in the presence of a copper catalyst; see Figure 15.12.) However, what Ostwald discovered was that the first reaction, to form NO from $NH_3$ and $O_2$, is catalyzed by platinum. Therefore, by using this catalyst at moderate temperatures, nitric oxide can be selectively formed. The dissociation of nitric oxide to its elements is normally too slow at moderate temperatures to be significant. Many other examples could be cited in which the products of given reactants depend on the catalyst.

## A Checklist for Review

### Important Terms

chemical equilibrium (15.1)
equilibrium-constant
   expression (15.2)
equilibrium constant (15.2)

law of mass action (15.2)
homogeneous
   equilibrium (15.3)

heterogeneous
   equilibrium (15.3)
reaction quotient (15.5)

Le Chatelier's
   principle (15.7)

### Key Equations

$$K_c = \frac{[C]^c[D]^d}{[A]^a[B]^b} \qquad Q_c = \frac{[C]_i^c[D]_i^d}{[A]_i^a[B]_i^b}$$

## Summary of Facts and Concepts

Chemical equilibrium can be characterized by the *equilibrium constant* $K_c$. In the expression for $K_c$, the concentration of products is in the numerator and the concentration of reactants is in the denominator. Pure liquids and solids are ignored in writing the equilibrium-constant expression. When $K_c$ is very large, the equilibrium mixture is mostly products, and when $K_c$ is very small, the equilibrium mixture is mostly reactants. The reaction quotient $Q_c$ takes the form of the equilibrium-constant expression. If you substitute the concentrations of substances in a reaction mixture into $Q_c$, you can predict the direction the reaction must go to attain equilibrium. You can use $K_c$ to calculate the composition of the reaction mixture at equilibrium, starting from various initial compositions.

The *choice of conditions*, including *catalysts*, can be very important to the success of a reaction. Removing a product from the reaction mixture, for example, shifts the equilibrium composition to give more product. Changing the pressure and temperature can also affect the product yield. *Le Chatelier's principle* is useful in predicting the effect of such changes.

## Operational Skills

1. **Applying stoichiometry to an equilibrium mixture** Given the starting amounts of reactants and the amount of one substance at equilibrium, find the equilibrium composition. **(EXAMPLE 15.1)**
2. **Writing equilibrium-constant expressions** Given the chemical equation, write the equilibrium-constant expression. **(EXAMPLES 15.2, 15.4)**
3. **Obtaining an equilibrium constant from reaction composition** Given the equilibrium composition, find $K_c$. **(EXAMPLE 15.3)**
4. **Using the reaction quotient** Given the concentrations of substances in a reaction mixture, predict the direction of reaction. **(EXAMPLE 15.5)**
5. **Obtaining one equilibrium concentration given the others** Given $K_c$ and all concentrations of substances but one in an equilibrium mixture, calculate the concentration of this one substance. **(EXAMPLE 15.6)**
6. **Solving equilibrium problems** Given the starting composition and $K_c$ of a reaction mixture, calculate the equilibrium composition. **(EXAMPLES 15.7, 15.8)**
7. **Applying Le Chatelier's principle** Given a reaction, use Le Chatelier's principle to decide the effect of adding or removing a substance **(EXAMPLE 15.9)**, changing the pressure **(EXAMPLE 15.10)**, or changing the temperature **(EXAMPLE 15.11)**.

## Review Questions

**15.1** Consider the reaction $N_2O_4(g) \rightleftharpoons 2NO_2(g)$. Draw a graph illustrating the changes of concentrations of $N_2O_4$ and $NO_2$ as equilibrium is approached. Describe how the rates of the forward and reverse reactions change as the mixture approaches dynamic equilibrium. Why is this called a *dynamic* equilibrium?

**15.2** When 1.0 mol each of $H_2(g)$ and $I_2(g)$ are mixed at a certain high temperature, they react to give a final mixture consisting of 0.5 mol each of $H_2(g)$ and $I_2(g)$ and 1.0 mol HI(g). Why do you obtain the same final mixture when you bring 2.0 mol HI(g) to the same temperature?

**15.3** Explain why the equilibrium constant for a gaseous reaction can be written in terms of partial pressures instead of concentrations.

**15.4** Obtain the equilibrium constant for the reaction

$$HCN(aq) \rightleftharpoons H^+(aq) + CN^-(aq)$$

from the following

$$HCN(aq) + OH^-(aq) \rightleftharpoons CN^-(aq) + H_2O(l);$$
$$K_1 = 4.9 \times 10^4$$

$$H_2O(l) \rightleftharpoons H^+(aq) + OH^-(aq); K_2 = 1.0 \times 10^{-14}$$

**15.5** Which of the following reactions involve homogeneous equilibria and which involve heterogeneous equilibria? Explain the difference.
a. $2NO(g) + O_2(g) \rightleftharpoons 2NO_2(g)$
b. $2Cu(NO_3)_2(s) \rightleftharpoons 2CuO(s) + 4NO_2(g) + O_2(g)$
c. $2N_2O(g) \rightleftharpoons 2N_2(g) + O_2(g)$
d. $2NH_3(g) + 3CuO(s) \rightleftharpoons 3H_2O(g) + N_2(g) + 3Cu(s)$

**15.6** Explain why pure liquids and solids can be ignored when writing the equilibrium-constant expression.

**15.7** What qualitative information can you get from the magnitude of the equilibrium constant?

**15.8** What is the reaction quotient? How is it useful?

**15.9** List the possible ways in which you can alter the equilibrium composition of a reaction mixture.

**15.10** Two moles of $H_2$ are mixed with 1 mol $O_2$ at 25°C. No observable reaction takes place, although $K_c$ for the reaction to form water is very large at this temperature. When a piece of platinum is added, however, the gases react rapidly.

Explain the role of platinum in this reaction. How does it affect the equilibrium composition of the reaction mixture?

**15.11**   How is it possible for a catalyst to give products from a reaction mixture that are different from those obtained when no catalyst or a different catalyst is used? Give an example.

**15.12**   List four ways in which the yield of ammonia in the reaction

$$N_2(g) + 3H_2(g) \rightleftharpoons 2NH_3(g); \Delta H° < 0$$

can be improved for a given amount of $H_2$. Explain the principle behind each way.

 ## Conceptual Problems

**15.13**   During an experiment with the Haber process, a researcher put 1 mol $N_2$ and 1 mol $H_2$ into a reaction vessel to observe the equilibrium formation of ammonia, $NH_3$.

$$N_2(g) + 3H_2(g) \rightleftharpoons 2NH_3(g)$$

When these reactants come to equilibrium, assume that $x$ mol $H_2$ react. How many moles of ammonia form?

**15.14**   Suppose liquid water and water vapor exist in equilibrium in a closed container. If you add a small amount of liquid water to the container, how does this affect the amount of water vapor in the container? If, instead, you add a small amount of water vapor to the container, how does this affect the amount of liquid water in the container?

**15.15**   A mixture initially consisting of 2 mol CO and 2 mol $H_2$ comes to equilibrium with methanol, $CH_3OH$, as the product:

$$CO(g) + 2H_2(g) \rightleftharpoons CH_3OH(g)$$

At equilibrium, the mixture will contain which of the following?
a. less than 1 mol $CH_3OH$
b. 1 mol $CH_3OH$
c. more than 1 mol $CH_3OH$ but less than 2 mol
d. 2 mol $CH_3OH$
e. more than 2 mol $CH_3OH$

**15.16**   When a continuous stream of hydrogen gas, $H_2$, passes over hot magnetic iron oxide, $Fe_3O_4$, metallic iron and water vapor form. When a continuous stream of water vapor passes over hot metallic iron, the oxide $Fe_3O_4$ and $H_2$ form. Explain why the reaction goes in one direction in one case but in the reverse direction in the other.

**15.17**   During the commercial preparation of sulfuric acid, sulfur dioxide reacts with oxygen in an exothermic reaction to produce sulfur trioxide. In this step, sulfur dioxide mixed with oxygen-enriched air passes into a reaction tower at about 420°C, where reaction occurs on a vanadium(V) oxide catalyst. Discuss the conditions used in this reaction in terms of its effect on the yield of sulfur trioxide. Are there any other conditions that you might explore in order to increase the yield of sulfur trioxide?

**15.18**   The following reaction is carried out at 500 K in a container equipped with a movable piston.

$$A(g) + B(g) \rightleftharpoons C(g); \quad K_c = 10 \text{ (at 500 K)}$$

After the reaction has reached equilibrium, the container has the composition depicted here.

Suppose the container volume is doubled.
a. How does the equilibrium composition shift?
b. How does the concentration of each of the reactants and the product change? (That is, does the concentration increase, decrease, or stay the same?)

**15.19**   An experimenter introduces 4.0 mol of gas A into a 1.0-L container at 200 K to form product B according to the reaction

$$2A(g) \rightleftharpoons B(g)$$

Using the experimenter's data (one curve is for A, the other is for B), calculate the equilibrium constant at 200 K.

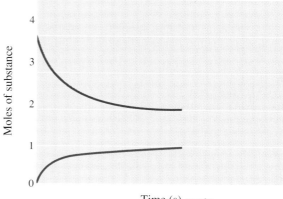

**15.20** At some temperature, a 100-L reaction vessel contains a mixture that is initially 1.00 mol CO and 2.00 mol $H_2$. The vessel also contains a catalyst so that the following equilibrium is attained:

$$CO(g) + 2H_2(g) \rightleftharpoons CH_3OH(g)$$

At equilibrium, the mixture contains 0.100 mol $CH_3OH$. In a later experiment in the same vessel, you start with 1.00 mol $CH_3OH$. How much methanol is there at equilibrium? Explain.

## Practice Problems

### Reaction Stoichiometry

**15.21** You place 1.50 mol of dinitrogen trioxide, $N_2O_3$, into a flask, where it decomposes at 25.0°C and 1.00 atm:

$$N_2O_3(g) \rightleftharpoons NO_2(g) + NO(g)$$

What is the composition of the reaction mixture at equilibrium if it contains 0.45 mol of nitrogen dioxide, $NO_2$?

**15.22** A 1.250-mol sample of phosphorus pentachloride, $PCl_5$, dissociates at 160°C and 1.00 atm to give 0.169 mol of phosphorus trichloride, $PCl_3$, at equilibrium.

$$PCl_5(g) \rightleftharpoons PCl_3(g) + Cl_2(g)$$

What is the composition of the final reaction mixture?

**15.23** Nitric oxide, NO, reacts with bromine, $Br_2$, to give nitrosyl bromide, NOBr.

$$2NO(g) + Br_2(g) \rightleftharpoons 2NOBr(g)$$

A sample of 0.0655 mol NO with 0.0328 mol $Br_2$ gives an equilibrium mixture containing 0.0389 mol NOBr. What is the composition of the equilibrium mixture?

**15.24** You place 0.750 mol of nitrogen, $N_2$, and 2.250 mol of hydrogen, $H_2$, into a reaction vessel at 450°C and 10.0 atm. The reaction is

$$N_2(g) + 3H_2(g) \rightleftharpoons 2NH_3(g)$$

What is the composition of the equilibrium mixture if you obtain 0.060 mol of ammonia, $NH_3$, from it?

**15.25** Methanol, $CH_3OH$, formerly known as wood alcohol, is manufactured commercially by the following reaction:

$$CO(g) + 2H_2(g) \rightleftharpoons CH_3OH(g)$$

A 1.500-L vessel was filled with 0.1500 mol CO and 0.3000 mol $H_2$. When this mixture came to equilibrium at 500 K, the vessel contained 0.1187 mol CO. How many moles of each substance were in the vessel?

**15.26** In the contact process, sulfuric acid is manufactured by first oxidizing $SO_2$ to $SO_3$, which is then reacted with water. The reaction of $SO_2$ with $O_2$ is

$$2SO_2(g) + O_2(g) \rightleftharpoons 2SO_3(g)$$

A 2.000-L flask was filled with 0.0400 mol $SO_2$ and 0.0200 mol $O_2$. At equilibrium at 900 K, the flask contained 0.0296 mol $SO_3$. How many moles of each substance were in the flask at equilibrium?

### The Equilibrium Constant and its Evaluation

**15.27** Write equilibrium-constant expressions $K_c$ for each of the following reactions.
a. $N_2O_3(g) \rightleftharpoons NO_2(g) + NO(g)$
b. $2H_2S(g) \rightleftharpoons 2H_2(g) + S_2(g)$
c. $2NO(g) + O_2(g) \rightleftharpoons 2NO_2(g)$
d. $PCl_3(g) + 3NH_3(g) \rightleftharpoons P(NH_2)_3(g) + 3HCl(g)$

**15.28** Write equilibrium-constant expressions $K_c$ for each of the following reactions.
a. $N_2H_4(g) \rightleftharpoons N_2(g) + 2H_2(g)$
b. $2NOCl(g) \rightleftharpoons 2NO(g) + Cl_2(g)$
c. $2NO(g) + 2H_2(g) \rightleftharpoons N_2(g) + 2H_2O(g)$
d. $4HCl(g) + O_2(g) \rightleftharpoons 2Cl_2(g) + 2H_2O(g)$

**15.29** The equilibrium-constant expression for a gas reaction is

$$K_c = \frac{[H_2O]^2[SO_2]^2}{[H_2S]^2[O_2]^3}$$

Write the balanced chemical equation corresponding to this expression.

**15.30** The equilibrium-constant expression for a gas reaction is

$$K_c = \frac{[CO_2]^3[H_2O]^4}{[C_3H_8][O_2]^5}$$

Write the balanced chemical equation corresponding to this expression.

**15.31** The equilibrium-constant expression for a reaction is

$$K_c = \frac{[NO_2]^4[O_2]}{[N_2O_5]^2}$$

What is the equilibrium-constant expression when the equation for this reaction is halved and then reversed?

**15.32** The equilibrium-constant expression for a reaction is

$$K_c = \frac{[NH_3]^4[O_2]^5}{[NO]^4[H_2O]^6}$$

What is the equilibrium-constant expression when the equation for this reaction is halved and then reversed?

**15.33** The equilibrium constant $K_c$ for the equation

$$2HI(g) \rightleftharpoons H_2(g) + I_2(g)$$

at 425°C is 1.84. What is the value of $K_c$ for the following equation?

$$H_2(g) + I_2(g) \rightleftharpoons 2HI(g)$$

**15.34** The equilibrium constant $K_c$ for the equation

$$CS_2(g) + 4H_2(g) \rightleftharpoons CH_4(g) + 2H_2S(g)$$

at 900°C is 27.8. What is the value of $K_c$ for the following equation?

$$\tfrac{1}{2}CS_2(g) + 2H_2(g) \rightleftharpoons \tfrac{1}{2}CH_4(g) + H_2S(g)$$

**15.35** A 5.00-L vessel contained 0.0185 mol of phosphorus trichloride, 0.0158 mol of phosphorus pentachloride, and 0.0870 mol of chlorine at 230°C in an equilibrium mixture. Calculate the value of $K_c$ for the reaction

$$PCl_3(g) + Cl_2(g) \rightleftharpoons PCl_5(g)$$

**15.36** An 8.00-L reaction vessel at 491°C contained 0.650 mol $H_2$, 0.275 mol $I_2$, and 3.000 mol HI. Assuming that the substances are at equilibrium, find the value of $K_c$ at 491°C for the reaction of hydrogen and iodine to give hydrogen iodide. The equation is

$$H_2(g) + I_2(g) \rightleftharpoons 2HI(g)$$

**15.37** Obtain the value of $K_c$ for the following reaction at 500 K:

$$CO(g) + 2H_2(g) \rightleftharpoons CH_3OH(g)$$

Use the data given in Problem 15.25.

**15.38** Obtain the value of $K_c$ for the following reaction at 900 K:

$$2SO_2(g) + O_2(g) \rightleftharpoons 2SO_3(g)$$

Use the data given in Problem 15.26.

**15.39** At 77°C, 2.00 mol of nitrosyl bromide, NOBr, placed in a 1.00-L flask dissociates to the extent of 9.4%; that is, for each mole of NOBr before reaction, $(1.000 - 0.094)$ mol NOBr remains after dissociation. Calculate the value of $K_c$ for the dissociation reaction

$$2NOBr(g) \rightleftharpoons 2NO(g) + Br_2(g)$$

**15.40** A 2.00-mol sample of nitrogen dioxide was placed in an 80.0-L vessel. At 200°C, the nitrogen dioxide was 6.0% decomposed according to the equation

$$2NO_2(g) \rightleftharpoons 2NO(g) + O_2(g)$$

Calculate the value of $K_c$ for this reaction at 200°C. (See Problem 15.39.)

**15.41** Write equilibrium-constant expressions $K_p$ for each of the following reactions:
a. $H_2(g) + Br_2(g) \rightleftharpoons 2HBr(g)$
b. $CS_2(g) + 4H_2(g) \rightleftharpoons CH_4(g) + 2H_2S(g)$
c. $4HCl(g) + O_2(g) \rightleftharpoons 2H_2O(g) + 2Cl_2(g)$
d. $CO(g) + 2H_2(g) \rightleftharpoons CH_3OH(g)$

**15.42** Write equilibrium-constant expressions $K_p$ for each of the following reactions:
a. $N_2O_4(g) \rightleftharpoons 2NO_2(g)$
b. $2NO(g) + Br_2(g) \rightleftharpoons 2NOBr(g)$
c. $2SO_2(g) + O_2(g) \rightleftharpoons 2SO_3(g)$
d. $4NH_3(g) + 5O_2(g) \rightleftharpoons 4NO(g) + 6H_2O(g)$

**15.43** The value of $K_c$ for the following reaction at 900°C is 0.28.

$$CS_2(g) + 4H_2(g) \rightleftharpoons CH_4(g) + 2H_2S(g)$$

What is $K_p$ at this temperature?

**15.44** The equilibrium constant $K_c$ equals 10.5 for the following reaction at 227°C.

$$CO(g) + 2H_2(g) \rightleftharpoons CH_3OH(g)$$

What is the value of $K_p$ at this temperature?

**15.45** The reaction

$$SO_2(g) + \tfrac{1}{2}O_2(g) \rightleftharpoons SO_3(g)$$

has $K_p$ equal to 6.55 at 627°C. What is the value of $K_c$ at this temperature?

**15.46** Fluorine, $F_2$, dissociates into atoms on heating.

$$\tfrac{1}{2}F_2(g) \rightleftharpoons F(g)$$

The value of $K_p$ at 842°C is $7.55 \times 10^{-2}$. What is the value of $K_c$ at this temperature?

**15.47** Write the expression for the equilibrium constant $K_c$ for each of the following equations:
a. $C(s) + CO_2(g) \rightleftharpoons 2CO(g)$
b. $FeO(s) + CO(g) \rightleftharpoons Fe(s) + CO_2(g)$
c. $Na_2CO_3(s) + SO_2(g) + \tfrac{1}{2}O_2(g) \rightleftharpoons Na_2SO_4(s) + CO_2(g)$
d. $PbI_2(s) \rightleftharpoons Pb^{2+}(aq) + 2I^-(aq)$

**15.48** For each of the following equations, give the expression for the equilibrium constant $K_c$:
a. $NH_4Cl(s) \rightleftharpoons NH_3(g) + HCl(g)$
b. $C(s) + 2N_2O(g) \rightleftharpoons CO_2(g) + 2N_2(g)$

c. $2NaHCO_3(s) \rightleftharpoons Na_2CO_3(s) + H_2O(g) + CO_2(g)$
d. $Fe^{3+}(aq) + 3OH^-(aq) \rightleftharpoons Fe(OH)_3(s)$

## Using the Equilibrium Constant

**15.49** On the basis of the value of $K_c$, decide whether or not you expect nearly complete reaction at equilibrium for each of the following:
a. $N_2(g) + O_2(g) \rightleftharpoons 2NO(g); K_c = 4.6 \times 10^{-31}$
b. $C_2H_4(g) + H_2(g) \rightleftharpoons C_2H_6(g); K_c = 1.3 \times 10^{21}$

**15.50** Would either of the following reactions go almost completely to product at equilibrium?
a. $2NO(g) + 2H_2(g) \rightleftharpoons N_2(g) + 2H_2O(g); K_c = 6.5 \times 10^{113}$
b. $COCl_2(g) \rightleftharpoons CO(g) + Cl_2(g); K_c = 3.6 \times 10^{-16}$

**15.51** Hydrogen fluoride decomposes according to the following equation:

$$2HF(g) \rightleftharpoons H_2(g) + F_2(g)$$

The value of $K_c$ at room temperature is $1.0 \times 10^{-95}$. From the magnitude of $K_c$, do you think the decomposition occurs to any great extent at room temperature? If an equilibrium mixture in a 1.0-L vessel contains 1.0 mol HF, what is the amount of $H_2$ formed? Does this result agree with what you expect from the magnitude of $K_c$?

**15.52** Suppose sulfur dioxide reacts with oxygen at 25°C.

$$2SO_2(g) + O_2(g) \rightleftharpoons 2SO_3(g)$$

The equilibrium constant $K_c$ equals $8.0 \times 10^{35}$ at this temperature. From the magnitude of $K_c$, do you think this reaction occurs to any great extent when equilibrium is reached at room temperature? If an equilibrium mixture is $1.0\ M$ $SO_3$ and has equal concentrations of $SO_2$ and $O_2$, what is the concentration of $SO_2$ in the mixture? Does this result agree with what you expect from the magnitude of $K_c$?

**15.53** The following reaction has an equilibrium constant $K_c$ equal to $3.07 \times 10^{-4}$ at 24°C.

$$2NOBr(g) \rightleftharpoons 2NO(g) + Br_2(g)$$

For each of the following compositions, decide whether the reaction mixture is at equilibrium. If it is not, decide which direction the reaction should go.
a. $[NOBr] = 0.0720\ M$, $[NO] = 0.0162\ M$,
   $[Br_2] = 0.0123\ M$
b. $[NOBr] = 0.121\ M$, $[NO] = 0.0159\ M$,
   $[Br_2] = 0.0139\ M$
c. $[NOBr] = 0.103\ M$, $[NO] = 0.0134\ M$,
   $[Br_2] = 0.0181\ M$
d. $[NOBr] = 0.0472\ M$, $[NO] = 0.0121\ M$,
   $[Br_2] = 0.0105\ M$

**15.54** The following reaction has an equilibrium constant $K_c$ equal to 3.59 at 900°C.

$$CH_4(g) + 2H_2S(g) \rightleftharpoons CS_2(g) + 4H_2(g)$$

For each of the following compositions, decide whether the reaction mixture is at equilibrium. If it is not, decide which direction the reaction should go.
a. $[CH_4] = 1.26\ M$, $[H_2S] = 1.32\ M$,
   $[CS_2] = 1.43\ M$, $[H_2] = 1.12\ M$
b. $[CH_4] = 1.25\ M$, $[H_2S] = 1.52\ M$,
   $[CS_2] = 1.15\ M$, $[H_2] = 1.73\ M$
c. $[CH_4] = 1.20\ M$, $[H_2S] = 1.31\ M$,
   $[CS_2] = 1.15\ M$, $[H_2] = 1.85\ M$
d. $[CH_4] = 1.56\ M$, $[H_2S] = 1.43\ M$,
   $[CS_2] = 1.23\ M$, $[H_2] = 1.91\ M$

**15.55** Methanol, $CH_3OH$, is manufactured industrially by the reaction

$$CO(g) + 2H_2(g) \rightleftharpoons CH_3OH(g)$$

A gaseous mixture at 500 K is $0.020\ M$ $CH_3OH$, $0.10\ M$ CO, and $0.10\ M$ $H_2$. What will be the direction of reaction if this mixture goes to equilibrium? The equilibrium constant $K_c$ equals 10.5 at 500 K.

**15.56** Sulfur trioxide, used to manufacture sulfuric acid, is obtained commercially from sulfur dioxide.

$$2SO_2(g) + O_2(g) \rightleftharpoons 2SO_3(g)$$

The equilibrium constant $K_c$ for this reaction is $4.17 \times 10^{-2}$ at 727°C. What is the direction of reaction when a mixture that is $0.20\ M$ $SO_2$, $0.10\ M$ $O_2$, and $0.40\ M$ $SO_3$ approaches equilibrium?

**15.57** Phosgene, $COCl_2$, used in the manufacture of polyurethane plastics, is prepared from CO and $Cl_2$.

$$CO(g) + Cl_2(g) \rightleftharpoons COCl_2(g)$$

An equilibrium mixture at 395°C contains 0.012 mol CO and 0.025 mol $Cl_2$ per liter, as well as $COCl_2$. If $K_c$ at 395°C is $1.23 \times 10^3$, what is the concentration of $COCl_2$?

**15.58** Nitric oxide, NO, is formed in automobile exhaust by the reaction of $N_2$ and $O_2$ (from air).

$$N_2(g) + O_2(g) \rightleftharpoons 2NO(g)$$

The equilibrium constant $K_c$ is 0.0025 at 2127°C. If an equilibrium mixture at this temperature contains 0.023 mol $N_2$ and 0.031 mol $O_2$ per liter, what is the concentration of NO?

**15.59** Iodine and bromine react to give iodine monobromide, IBr.

$$I_2(g) + Br_2(g) \rightleftharpoons 2IBr(g)$$

What is the equilibrium composition of a mixture at 150°C that initially contained 0.0015 mol each of iodine and bromine in a 5.0-L vessel? The equilibrium constant $K_c$ for this reaction at 150°C is $1.2 \times 10^2$.

**15.60** Initially a mixture contains 0.850 mol each of $N_2$ and $O_2$ in an 8.00-L vessel. Find the composition of the mixture when equilibrium is reached at 3900°C. The reaction is

$$N_2(g) + O_2(g) \rightleftharpoons 2NO(g)$$

and $K_c = 0.0123$ at 3900°C.

**15.61** The equilibrium constant $K_c$ for the reaction

$$PCl_3(g) + Cl_2(g) \rightleftharpoons PCl_5(g)$$

equals 49 at 230°C. If 0.500 mol each of phosphorus trichloride and chlorine are added to a 5.0-L reaction vessel, what is the equilibrium composition of the mixture at 230°C?

**15.62** Calculate the composition of the gaseous mixture obtained when 1.00 mol of carbon dioxide is exposed to hot carbon at 800°C in a 1.00-L vessel. The equilibrium constant $K_c$ at 800°C is 14.0 for the reaction

$$CO_2(g) + C(s) \rightleftharpoons 2CO(g)$$

**15.63** Suppose 1.000 mol CO and 3.000 mol $H_2$ are put in a 10.00-L vessel at 1200 K. The equilibrium constant $K_c$ for

$$CO(g) + 3H_2(g) \rightleftharpoons CH_4(g) + H_2O(g)$$

equals 3.92. Find the equilibrium composition of the reaction mixture.

**15.64** The equilibrium constant $K_c$ for the reaction

$$N_2(g) + 3H_2(g) \rightleftharpoons 2NH_3(g)$$

at 450°C is 0.159. Calculate the equilibrium composition when 1.00 mol $N_2$ is mixed with 3.00 mol $H_2$ in a 2.00-L vessel.

## Le Chatelier's Principle

**15.65** Consider the equilibrium

$$FeO(s) + CO(g) \rightleftharpoons Fe(s) + CO_2(g)$$

When carbon dioxide is removed from the equilibrium mixture (say, by passing the gases through water to absorb $CO_2$), what is the direction of net reaction as the new equilibrium is achieved?

**15.66** (a) Predict the direction of reaction when chlorine gas is added to an equilibrium mixture of $PCl_3$, $PCl_5$, and $Cl_2$. The reaction is

$$PCl_3(g) + Cl_2(g) \rightleftharpoons PCl_5(g)$$

(b) What is the direction of reaction when chlorine gas is removed from an equilibrium mixture of these gases?

**15.67** What would you expect to be the effect of an increase of pressure on each of the following reactions? Would the pressure change cause reaction to go to the right or left?
a. $CH_4(g) + 2S_2(g) \rightleftharpoons CS_2(g) + 2H_2S(g)$
b. $H_2(g) + Br_2(g) \rightleftharpoons 2HBr(g)$
c. $CO_2(g) + C(s) \rightleftharpoons 2CO(g)$

**15.68** Indicate whether either an increase or a decrease of pressure obtained by changing the volume would increase the amount of product in each of the following reactions.
a. $CO(g) + 2H_2(g) \rightleftharpoons CH_3OH(g)$
b. $2SO_2(g) + O_2(g) \rightleftharpoons 2SO_3(g)$
c. $N_2O_4(g) \rightleftharpoons 2NO_2(g)$

**15.69** Methanol is prepared industrially from synthesis gas (CO and $H_2$).

$$CO(g) + 2H_2(g) \rightleftharpoons CH_3OH(g); \Delta H° = -21.7\ kcal$$

Would the fraction of methanol obtained at equilibrium be increased by raising the temperature? Explain.

**15.70** One way of preparing hydrogen is by the decomposition of water.

$$2H_2O(g) \rightleftharpoons 2H_2(g) + O_2(g); \Delta H° = 484\ kJ$$

Would you expect the decomposition to be favorable at high or low temperature? Explain.

**15.71** Use thermochemical data (Appendix C) to decide whether the equilibrium constant for the following reaction will increase or decrease with temperature.

$$2NO_2(g) + 7H_2(g) \rightleftharpoons 2NH_3(g) + 4H_2O(g)$$

**15.72** Use thermochemical data (Appendix C) to decide whether the equilibrium constant for the following reaction will increase or decrease with temperature.

$$CH_4(g) + 2H_2S(g) \rightleftharpoons CS_2(g) + 4H_2(g)$$

**15.73** What would you expect to be the general temperature and pressure conditions for an optimum yield of nitric oxide, NO, by the oxidation of ammonia?

$$4NH_3(g) + 5O_2(g) \rightleftharpoons 4NO(g) + 6H_2O(g); \Delta H° < 0$$

**15.74** Predict the general temperature and pressure conditions for the optimum conversion of ethylene ($C_2H_4$) to ethane ($C_2H_6$).

$$C_2H_4(g) + H_2(g) \rightleftharpoons C_2H_6(g); \Delta H° < 0$$

# General Problems

**15.75** A mixture of carbon monoxide, hydrogen, and methanol, $CH_3OH$, is at equilibrium according to the equation

$$CO(g) + 2H_2(g) \rightleftharpoons CH_3OH(g)$$

At 250°C, the mixture is 0.096 $M$ CO, 0.191 $M$ $H_2$, and 0.015 $M$ $CH_3OH$. What is $K_c$ for this reaction at 250°C?

**15.76** An equilibrium mixture of $SO_3$, $SO_2$, and $O_2$ at 727°C is 0.0160 $M$ $SO_3$, 0.0056 $M$ $SO_2$, and 0.0021 $M$ $O_2$. What is the value of $K_c$ for the following reaction?

$$SO_2(g) + \tfrac{1}{2}O_2(g) \rightleftharpoons SO_3(g)$$

**15.77** At 850°C and 1.000 atm pressure, a gaseous mixture of carbon monoxide and carbon dioxide in equilibrium with solid carbon is 90.55% CO by mass.

$$C(s) + CO_2(g) \rightleftharpoons 2CO(g)$$

Calculate $K_c$ for this reaction at 850°C.

**15.78** An equilibrium mixture of dinitrogen tetroxide, $N_2O_4$, and nitrogen dioxide, $NO_2$, is 65.8% $NO_2$ by mass at 1.00 atm pressure and 25°C. Calculate $K_c$ at 25°C for the reaction

$$N_2O_4(g) \rightleftharpoons 2NO_2(g)$$

**15.79** A 2.00-L vessel contains 1.00 mol $N_2$, 1.00 mol $H_2$, and 2.00 mol $NH_3$. What is the direction of reaction (forward or reverse) needed to attain equilibrium at 400°C? The equilibrium constant $K_c$ for the reaction

$$N_2(g) + 3H_2(g) \rightleftharpoons 2NH_3(g)$$

is 0.51 at 400°C.

**15.80** A vessel originally contained 0.200 mol iodine monobromide (IBr), 0.0010 mol $I_2$, and 0.0010 mol $Br_2$. The equilibrium constant $K_c$ for the reaction

$$I_2(g) + Br_2(g) \rightleftharpoons 2IBr(g)$$

is $1.2 \times 10^2$ at 150°C. What is the direction (forward or reverse) needed to attain equilibrium at 150°C?

**15.81** A gaseous mixture containing 1.00 mol each of CO, $H_2O$, $CO_2$, and $H_2$ is exposed to a zinc oxide–copper oxide catalyst at 1000°C. The reaction is

$$CO(g) + H_2O(g) \rightleftharpoons CO_2(g) + H_2(g)$$

and the equilibrium constant $K_c$ is 0.58 at 1000°C. What is the direction of reaction (forward or reverse) as the mixture attains equilibrium?

**15.82** A 2.0-L reaction flask initially contains 0.10 mol CO, 0.20 mol $H_2$, and 0.50 mol $CH_3OH$ (methanol). If this mixture

is brought in contact with a zinc oxide–chromium(III) oxide catalyst, the equilibrium

$$CO(g) + 2H_2(g) \rightleftharpoons CH_3OH(g)$$

is attained. The equilibrium constant $K_c$ for this reaction at 300°C is $1.1 \times 10^{-2}$. What is the direction of reaction (forward or reverse) as the mixture attains equilibrium?

**15.83** Hydrogen bromide dissociates when heated according to the equation

$$2HBr(g) \rightleftharpoons H_2(g) + Br_2(g)$$

The equilibrium constant $K_c$ equals $1.6 \times 10^{-2}$ at 200°C. What are the moles of substances in the equilibrium mixture at 200°C if we start with 0.010 mol HBr in a 1.0-L vessel?

**15.84** Iodine monobromide, IBr, occurs as brownish-black crystals that vaporize with decomposition:

$$2IBr(g) \rightleftharpoons I_2(g) + Br_2(g)$$

The equilibrium constant $K_c$ at 100°C is 0.026. If 0.010 mol IBr is placed in a 1.0-L vessel at 100°C, what are the moles of substances at equilibrium in the vapor?

**15.85** Phosgene, $COCl_2$, is a toxic gas used in the manufacture of urethane plastics. The gas dissociates at high temperature.

$$COCl_2(g) \rightleftharpoons CO(g) + Cl_2(g)$$

At 400°C, the equilibrium constant $K_c$ is $8.05 \times 10^{-4}$. Find the percentage of phosgene that dissociates at this temperature when 1.00 mol of phosgene is placed in a 25.0-L vessel.

**15.86** Dinitrogen tetroxide, $N_2O_4$, is a colorless gas (boiling point, 21°C), which dissociates to give nitrogen dioxide, $NO_2$, a reddish-brown gas.

$$N_2O_4(g) \rightleftharpoons 2NO_2(g)$$

The equilibrium constant $K_c$ at 25°C is 0.125. What percentage of dinitrogen tetroxide is dissociated when 0.0300 mol $N_2O_4$ is placed in a 1.00-L flask at 25°C?

**15.87** Suppose you start with a mixture of 1.00 mol CO and 4.00 mol $H_2$ in a 10.00-L vessel. Find the moles of substances present at equilibrium at 1200 K for the reaction

$$CO(g) + 3H_2(g) \rightleftharpoons CH_4(g) + H_2O(g); K_c = 3.92$$

You will get an equation of the form

$$f(x) = 3.92$$

where $f(x)$ is an expression in the unknown $x$ (the amount of $CH_4$). Solve this equation by guessing values of $x$, then computing values of $f(x)$. Find values of $x$ such that the values of

$f(x)$ bracket 3.92. Then choose values of $x$ to get a smaller bracket around 3.92. Obtain $x$ to two significant figures.

**15.88** What are the moles of substances present at equilibrium at 450°C if 1.00 mol $N_2$ and 4.00 mol $H_2$ in a 10.0-L vessel react according to the following equation?

$$N_2(g) + 3H_2(g) \rightleftharpoons 2NH_3(g)$$

The equilibrium constant $K_c$ is 0.153 at 450°C. Use the numerical procedure described in Problem 15.87.

**15.89** The amount of nitrogen dioxide formed by dissociation of dinitrogen tetroxide,

$$N_2O_4(g) \rightleftharpoons 2NO_2(g)$$

increases as the temperature rises. Is the dissociation of $N_2O_4$ endothermic or exothermic?

**15.90** The equilibrium constant $K_c$ for the synthesis of methanol, $CH_3OH$,

$$CO(g) + 2H_2(g) \rightleftharpoons CH_3OH(g)$$

is 4.3 at 250°C and 1.8 at 275°C. Is this reaction endothermic or exothermic?

**15.91** For the reaction

$$N_2(g) + 3H_2(g) \rightleftharpoons 2NH_3(g)$$

show that $\qquad K_c = K_p(RT)^2$

Do not use the formula $K_p = K_c(RT)^{\Delta n}$ given in the text. Start from the fact that $P_i = [i]RT$, where $P_i$ is the partial pressure of substance $i$ and $[i]$ is its molar concentration. Substitute into $K_c$.

**15.92** For the reaction

$$COCl_2(g) \rightleftharpoons CO(g) + Cl_2(g)$$

show that $\qquad K_c = K_p/(RT)$

Do not use the formula $K_p = K_c(RT)^{\Delta n}$ given in the text. See Problem 15.91.

**15.93** At high temperatures, a dynamic equilibrium exists between carbon monoxide, carbon dioxide, and solid carbon.

$$C(s) + CO_2(g) \rightleftharpoons 2CO(g); \Delta H° = 172.5 \text{ kJ}$$

At 850°C, $K_c$ is 0.153.
a. What is the value of $K_p$?
b. If the original reaction system consisted of just carbon and 1.50 atm of $CO_2$, what are the pressures of $CO_2$ and CO when equilibrium has been established?
c. How will the equilibrium pressure of CO change if the temperature is decreased?

**15.94** At high temperatures, a dynamic equilibrium exists between carbon monoxide, carbon dioxide, and solid carbon.

$$C(s) + CO_2(g) \rightleftharpoons 2CO(g); \Delta H° = 172.5 \text{ kJ}$$

At 900°C, $K_c$ is 0.238.
a. What is the value of $K_p$?
b. Some carbon dioxide is added to the hot carbon and the system is brought to equilibrium at 900°C. If the total pressure in the system is 6.40 atm, what are the partial pressures of CO and $CO_2$?
c. How will the equilibrium system respond if additional carbon at 900°C is added to the reaction system? Why?

**15.95** The equilibrium constant $K_c$ for the reaction

$$PCl_3(g) + Cl_2(g) \rightleftharpoons PCl_5(g)$$

equals 4.1 at 300°C.
a. A sample of 35.8 g of $PCl_5$ is placed in a 5.0-L reaction vessel and heated to 300°C. What are the equilibrium concentrations of all of the species?
b. What fraction of $PCl_5$ has decomposed?
c. If 35.8 g of $PCl_5$ were placed in a 1.0-L vessel, what qualitative effect would this have on the fraction of $PCl_5$ that has decomposed (give a qualitative answer only; do not do the calculation)? Why?

**15.96** At 25°C in a closed system, ammonium hydrogen sulfide exists as the following equilibrium:

$$NH_4HS(s) \rightleftharpoons NH_3(g) + H_2S(g)$$

a. When a sample of pure $NH_4HS(s)$ is placed in an evacuated reaction vessel and allowed to come to equilibrium at 25°C, total pressure is 0.660 atm. What is the value of $K_p$?
b. To this system, sufficient $H_2S(g)$ is injected until the pressure of $H_2S$ is three times that of the ammonia at equilibrium. What are the partial pressures of $NH_3$ and $H_2S$?
c. In a different experiment, 0.750 atm of $NH_3$ and 0.500 atm of $H_2S$ are introduced into a 1.00-L vessel at 25°C. How many moles of $NH_4HS$ are present when equilibrium is established?

**15.97** At moderately high temperatures, $SbCl_5$ decomposes into $SbCl_3$ and $Cl_2$ as follows:

$$SbCl_5(g) \rightleftharpoons SbCl_3(g) + Cl_2(g)$$

A 65.4-g sample of $SbCl_5$ is placed in an evacuated 5.00-L vessel and it is raised to 195°C and the system comes to equilibrium. If at this temperature 35.8% of the $SbCl_5$ is decomposed, what is the value of $K_p$?

**15.98** The following reaction is important in the manufacture of sulfuric acid.

$$SO_2(g) + \tfrac{1}{2}O_2(g) \rightleftharpoons SO_3(g)$$

At 900 K, 0.0216 mol of $SO_2$ and 0.0148 mol of $O_2$ are sealed in a 1.00-L reaction vessel. When equilibrium is reached, the concentration of $SO_3$ is determined to be 0.0175 $M$. Calculate $K_c$ for this reaction.

**15.99**    Sulfuryl chloride is used in organic chemistry as a chlorinating agent. At moderately high temperatures it decomposes as follows:

$$SO_2Cl_2(g) \rightleftharpoons SO_2(g) + Cl_2(g)$$

with $K_c = 0.045$ at 650 K.
a. A sample of 8.25 g of $SO_2Cl_2$ is placed in a 1.00-L reaction vessel and heated to 650 K. What are the equilibrium concentrations of all of the species?
b. What fraction of $SO_2Cl_2$ has decomposed?
c. If 5 g of chlorine is inserted into the reaction vessel, what qualitative effect would this have on the fraction of $SO_2Cl_2$ that has decomposed?

**15.100**    Phosgene was used as a poisonous gas in World War I. At high temperatures it decomposes as follows:

$$COCl_2(g) \rightleftharpoons CO(g) + Cl_2(g)$$

with $K_c = 4.6 \times 10^{-3}$ at 800 K.
a. A sample of 6.55 g of $COCl_2$ is placed in a 1.00-L reaction vessel and heated to 800 K. What are the equilibrium concentrations of all of the species?
b. What fraction of $COCl_2$ has decomposed?
c. If 3 g of carbon monoxide is inserted into the reaction vessel, what qualitative effect would this have on the fraction of $COCl_2$ that has decomposed?

**15.101**    Gaseous acetic acid molecules have a certain tendency to form dimers. (A *dimer* is a molecule formed by the association of two identical, simpler molecules.) The equilibrium constant $K_c$ at 25°C for this reaction is $3.2 \times 10^4$.
a. If the initial concentration of $CH_3COOH$ monomer (the simpler molecule) is $4.0 \times 10^{-4}$ M, what are the concentrations of monomer and dimer when the system comes to equilibrium? (The simpler quadratic equation is obtained by as-

suming that all of the acid molecules have dimerized and then some of it dissociates to monomer.)
b. Why do acetic acid molecules dimerize? What type of structure would you draw for the dimer?
c. As the temperature increases, would you expect the percentage of dimer to increase or decrease? Why?

**15.102**    Gaseous acetic acid molecules have a certain tendency to form dimers. (A *dimer* is a molecule formed by the association of two identical, simpler molecules.) The equilibrium constant $K_p$ at 25°C for this reaction is $1.3 \times 10^3$.
a. If the initial pressure of $CH_3COOH$ monomer (the simpler molecule) is $7.5 \times 10^{-3}$ atm, what are the pressures of monomer and dimer when the system comes to equilibrium? (The simpler quadratic equation is obtained by assuming that all of the acid molecules have dimerized and then some of it dissociates to monomer.)
b. Why do acetic acid molecules dimerize? What type of structure would you draw for the dimer?
c. As the temperature decreases, would you expect the percentage of dimer to increase or decrease? Why?

**15.103**    When 0.112 mol of NO and 18.22 g of bromine are placed in a 1.00-L reaction vessel and sealed, the mixture is heated to 350 K and the following equilibrium is established:

$$2NO(g) + Br_2(g) \rightleftharpoons 2NOBr(g)$$

If the equilibrium concentration of nitrosyl bromide is 0.0824 M, what is $K_c$?

**15.104**    When 0.0322 mol of NO and 1.52 g of bromine are placed in a 1.00-L reaction vessel and sealed, the mixture reacts and the following equilibrium is established:

$$2NO(g) + Br_2(g) \rightleftharpoons 2NOBr(g)$$

At 25°C the equilibrium pressure of nitrosyl bromide is 0.438 atm. What is $K_p$?

# Cumulative-Skills Problems

**15.105**    The following equilibrium was studied by analyzing the equilibrium mixture for the amount of $H_2S$ produced.

$$Sb_2S_3(s) + 3H_2(g) \rightleftharpoons 2Sb(s) + 3H_2S(g)$$

A vessel whose volume was 2.50 L was filled with 0.0100 mol of antimony(III) sulfide, $Sb_2S_3$, and 0.0100 mol $H_2$. After the mixture came to equilibrium in the closed vessel at 440°C, the gaseous mixture was removed, and the hydrogen sulfide was dissolved in water. Sufficient lead(II) ion was added to react completely with the $H_2S$ to precipitate lead(II) sulfide, PbS. If 1.029 g PbS was obtained, what is the value of $K_c$ at 440°C?

**15.106**    The following equilibrium was studied by analyzing the equilibrium mixture for the amount of HCl produced.

$$LaCl_3(s) + H_2O(g) \rightleftharpoons LaOCl(s) + 2HCl(g)$$

A vessel whose volume was 1.25 L was filled with 0.0125 mol of lanthanum(III) chloride, $LaCl_3$, and 0.0250 mol $H_2O$. After the mixture came to equilibrium in the closed vessel at 619°C, the gaseous mixture was removed and dissolved in more water. Sufficient silver(I) ion was added to precipitate the chloride ion completely as silver chloride. If 3.59 g AgCl was obtained, what is the value of $K_c$ at 619°C?

**15.107**    Phosphorus(V) chloride, $PCl_5$, dissociates on heating to give phosphorus(III) chloride, $PCl_3$, and chlorine.

$$PCl_5(g) \rightleftharpoons PCl_3(g) + Cl_2(g)$$

A closed 2.00-L vessel initially contains 0.0100 mol $PCl_5$. What is the total pressure at 250°C when equilibrium is achieved? The value of $K_c$ at 250°C is $4.15 \times 10^{-2}$.

**15.108**  Antimony(V) chloride, $SbCl_5$, dissociates on heating to give antimony(III) chloride, $SbCl_3$, and chlorine.

$$SbCl_5(g) \rightleftharpoons SbCl_3(g) + Cl_2(g)$$

A closed 3.50-L vessel initially contains 0.0125 mol $SbCl_5$. What is the total pressure at 248°C when equilibrium is achieved? The value of $K_c$ at 248°C is $2.50 \times 10^{-2}$.

# Media Activities

**15.109**  Open the Equilibrium Calculator *Understanding Concepts* simulation on the student web site. Input the appropriate quantities into the Equilibrium Calculator to run the hypothetical reaction $3A + 2B \longrightarrow 3X$ with the initial $[A] = [B] = [X]$, and an equilibrium constant of $1 \times 10^4$ (1E4 needs to be entered into the program for this number). Click on the Calculate Equilibrium Concentrations button.
a. Write the equilibrium-constant expression for the reaction.
b. Does this reaction favor reactants or products?
c. Change one of the concentrations to shift the equilibrium to favor reactants. Click on the Calculate Equilibrium Concentrations button to see if you caused the correct shift.
d. When you change the concentrations of reactants and products, does the equilibrium constant change?

**15.110**  Work through the Equilibrium *Understanding Concepts* activity on the student CD, including the exercises.
a. Is the iron thiocyanate reaction exothermic or endothermic?
b. Why does the addition of $AgNO_3(aq)$ cause the iron thiocyanate reaction to shift?
c. How would the iron thiocyanate reaction be affected by the addition of $NaCl(aq)$?

**15.111**  Run the Equilibrium Decomposition of $N_2O_4$ *Visualization* on the student CD.

a. Based on the molecules present in the container prior to heating, is the equilibrium constant at this temperature greater or less than 1?
b. What was the effect of increased temperature on the equilibrium constant for the reaction?
c. If the volume of the container depicted in the *Visualization* was reduced by 50%, what would you observe?
d. If the volume of the container depicted in the *Visualization* was doubled, how would this affect the equilibrium constant?

**15.112**  Watch the Effect of Temperature Change on Equilibrium *Visualization* on the student CD.
a. Is the reaction depicted in the *Visualization* exothermic or endothermic?
b. For the reaction depicted in the *Visualization*, what solution color would you observe if you added $HCl(aq)$ to the room temperature solution?
c. For the reaction depicted in the *Visualization*, what solution color would you observe if you added $HCl(aq)$ to the cold solution?

**15.113**  For review, take the ACE practice test for Chapter 15 on the student web site.

**16**

Acid–base indicator dyes in solutions of different hydronium-ion concentration.

# Acids and Bases

**A**cids and bases were first recognized by simple properties, such as taste. Acids have a sour taste; bases are bitter. Also, acids and bases change the color of certain dyes called indicators, such as litmus and phenolphthalein. Acids change litmus from blue to red and basic phenolphthalein from red to colorless. Bases change litmus from red to blue and phenolphthalein from colorless to pink. As you can see from these color changes, acids and bases neutralize, or reverse, the action of one another. During neutralization, acids and bases react with each other to produce ionic substances called salts. Acids react with active metals, such as magnesium and zinc, to release hydrogen.

The Swedish chemist Svante Arrhenius framed the first successful concept of acids and bases. He defined acids and bases in terms of the effect these substances have on water. According to Arrhenius, acids are substances that increase the concentration of $H^+$ ion in aqueous solution, and bases increase the concentration of $OH^-$ ion in aqueous solution. But many reactions that have characteristics of acid–base

**Figure 16.1**
**Reaction of HCl(*g*) and NH₃(*g*) to form NH₄Cl(*s*).**
Gases from the concentrated solutions diffuse from their watch glasses (shallow dishes) and react to give a smoke of ammonium chloride.

reactions in aqueous solution occur in other solvents or without a solvent. For example, hydrochloric acid reacts with aqueous ammonia, which in the Arrhenius view is a base because it increases the concentration of $OH^-$ ion in aqueous solution. The reaction can be written

$$HCl(aq) + NH_3(aq) \longrightarrow NH_4Cl(aq)$$

The product is a solution of $NH_4Cl$—that is, a solution of $NH_4^+$ and $Cl^-$ ions. A very similar reaction occurs between hydrogen chloride and ammonia dissolved in benzene, $C_6H_6$. The product is again $NH_4Cl$, which in this case precipitates from the solution.

$$HCl(benzene) + NH_3(benzene) \longrightarrow NH_4Cl(s)$$

Hydrogen chloride and ammonia react even in the gas phase. If watch glasses (shallow glass dishes) of concentrated hydrochloric acid and concentrated ammonia are placed next to each other, dense white fumes of $NH_4Cl$ form where HCl gas and $NH_3$ gas come into contact (Figure 16.1).

$$HCl(g) + NH_3(g) \longrightarrow NH_4Cl(s)$$

These reactions of HCl and $NH_3$, in benzene and in the gas phase, are similar to the reaction in aqueous solution but cannot be explained by the Arrhenius concept. Broader acid–base concepts are needed.

In the first part of this chapter, we will discuss the Arrhenius, the Brønsted–Lowry, and the Lewis concepts of acids and bases. The Brønsted–Lowry and Lewis concepts apply to nonaqueous as well as aqueous solutions and also enlarge on the Arrhenius concept in other ways. This chapter expands on what you learned in Chapter 4 about acids and bases.

# Acid–Base Concepts

Antoine Lavoisier was one of the first chemists to try to explain what makes a substance acidic. In 1777 he proposed that oxygen was an essential element in acids. (*Oxygen,* which he named, means "acid-former" in Greek.) But in 1808 Humphry Davy showed that hydrogen chloride, which dissolves in water to give hydrochloric acid, contains only hydrogen and chlorine. Although some chemists argued that chlorine was a compound of oxygen, chlorine was eventually proved to be an element. Chemists then noted that hydrogen, not oxygen, must be the essential constituent of acids. The cause of acidity and basicity was first explained in 1884 by Svante Arrhenius.

## 16.1 Arrhenius Concept of Acids and Bases

We can state the Arrhenius concept of an acid as follows: *An acid is a substance that, when dissolved in water, increases the concentration of hydronium ion, $H_3O^+(aq)$.* For simplicity, chemists frequently use the notation $H^+(aq)$ for the $H_3O^+(aq)$ ion, and call it the hydrogen ion. Remember, however, that the aqueous hydrogen ion is not a bare proton in water, but a proton chemically bonded to water—that is, $H_3O^+(aq)$. The $H_3O^+(aq)$ ion is itself associated through hydrogen bonding with a variable number of water molecules. One such species is shown in Figure 16.2. *A base, in the Arrhenius concept, is a substance that, when dissolved in water, increases the concentration of hydroxide ion, $OH^-(aq)$.* ◀

▶ **This section summarizes the discussion in Section 4.4.**

The special role of the hydronium ion (or hydrogen ion) and the hydroxide ion in aqueous solutions arises from the following reaction:

$$H_2O(l) + H_2O(l) \rightleftharpoons H_3O^+(aq) + OH^-(aq)$$

The addition of acids and bases alters the concentrations of these ions in water.

In Arrhenius's theory, a *strong acid* is a substance that completely ionizes in aqueous solution to give $H_3O^+(aq)$ and an anion. An example is perchloric acid, $HClO_4$.

$$HClO_4(aq) + H_2O(l) \longrightarrow H_3O^+(aq) + ClO_4^-(aq)$$

Other examples of strong acids are $H_2SO_4$, HI, HBr, HCl, and $HNO_3$. A *strong base* completely ionizes in aqueous solution to give $OH^-$ and a cation. Sodium hydroxide is an example of a strong base.

$$NaOH(s) \xrightarrow{\text{H}_2\text{O}} Na^+(aq) + OH^-(aq)$$

**Figure 16.2**
**The hydronium ion, $H_3O^+$.**
The species is shown here hydrogen-bonded to three water molecules. The positive charge shown is actually distributed over the ion.

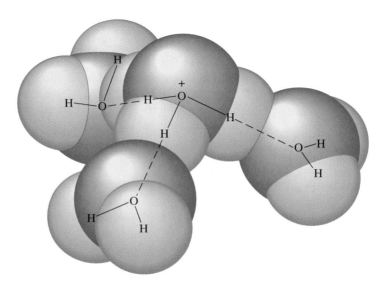

► We introduced strong and weak acids and bases in Section 4.4.

**Table 16.1**
**Common Strong Acids and Bases**

| Strong Acids | Strong Bases* |
|---|---|
| $HClO_4$ | LiOH |
| $H_2SO_4$ | NaOH |
| HI | KOH |
| HBr | $Ca(OH)_2$ |
| HCl | $Sr(OH)_2$ |
| $HNO_3$ | $Ba(OH)_2$ |

*In general, the Group IA and Group IIA hydroxides (except beryllium hydroxide) are strong bases.

The principal strong bases are the hydroxides of Group IA elements and Group IIA elements (except Be). See Table 16.1. ◄

Most of the other acids and bases you encounter are *weak*. They are not completely ionized in solution and exist in reversible reaction with the corresponding ions. Consider acetic acid, $HC_2H_3O_2$. The reaction is

$$HC_2H_3O_2(aq) + H_2O(l) \rightleftharpoons H_3O^+(aq) + C_2H_3O_2^-(aq)$$

Evidence for the Arrhenius theory comes from the heat of reaction, $\Delta H°$, for the neutralization of a strong acid by a strong base. This neutralization is essentially the reaction of $H_3O^+(aq)$ with $OH^-(aq)$ and should therefore always give the same $\Delta H°$ per mole of water formed. For example, if you write the neutralization of $HClO_4$ with NaOH in ionic form, you have

$$H_3O^+(aq) + ClO_4^-(aq) + Na^+(aq) + OH^-(aq) \longrightarrow Na^+(aq) + ClO_4^-(aq) + 2H_2O(l)$$

After canceling, you get the net ionic equation

$$H_3O^+(aq) + OH^-(aq) \longrightarrow 2H_2O(l)$$

Experimentally, it is found that all neutralizations involving strong acids and bases have the same $\Delta H°$: $-55.90$ kJ per mole of $H^+$. This indicates that the same reaction occurs in each neutralization, as Arrhenius's theory predicts.

Despite its successes, the Arrhenius concept is limited. In addition to looking at acid–base reactions only in aqueous solutions, it singles out the $OH^-$ ion as the source of base character, when other species can play a similar role. Broader definitions of acids and bases are described in the next sections.

# 16.2 Brønsted–Lowry Concept of Acids and Bases

► Animation: Proton Transfer.

► To be precise, we should say *hydrogen nucleus* instead of proton (because natural hydrogen contains some $^2H$ as well as $^1H$). The term *proton* is conventional in this context, however.

In 1923 the Danish chemist Johannes N. Brønsted (1879–1947) and, independently, the British chemist Thomas M. Lowry (1874–1936) pointed out that acid–base reactions can be seen as proton-transfer reactions and that acids and bases can be defined in terms of this proton ($H^+$) transfer. According to the Brønsted–Lowry concept, an **acid** is *the species donating a proton in a proton-transfer reaction.* A **base** is *the species accepting the proton in a proton-transfer reaction.* ◄

Consider, for example, the reaction of hydrochloric acid with ammonia, which was mentioned in the chapter opening. Writing it as an ionic equation, you have

$$H_3O^+(aq) + Cl^-(aq) + NH_3(aq) \longrightarrow H_2O(l) + NH_4^+(aq) + Cl^-(aq)$$

After canceling $Cl^-$, you obtain the net ionic equation.

$$H_3O^+(aq) + NH_3(aq) \longrightarrow H_2O(l) + NH_4^+(aq)$$

In this reaction in aqueous solution, a proton, $H^+$, is transferred from the $H_3O^+$ ion to the $NH_3$ molecule, giving $H_2O$ and $NH_4^+$ (Figure 16.3). Here $H_3O^+$ is the proton

**Figure 16.3**
**Representation of the reaction $H_3O^+ + NH_3 \rightleftharpoons H_2O + NH_4^+$.** Note the transfer of a proton, $H^+$, from $H_3O^+$ to $NH_3$. The charges indicated for ions are *overall* charges. They are not to be associated with specific locations on the ions.

$H_3O^+$ + $NH_3$ ⇌ $H_2O$ + $NH_4^+$

donor, or acid, and $NH_3$ is the proton acceptor, or base. Note that in the Brønsted–Lowry concept, acids (and bases) can be ions as well as molecular substances.

You can also apply the Brønsted–Lowry concept to the reaction of HCl and $NH_3$ dissolved in benzene, $C_6H_6$, which was mentioned in the chapter opening. In benzene, HCl and $NH_3$ are not ionized. The equation is

$$\text{(H)Cl}(benzene) + \overset{\frown}{N}H_3(benzene) \longrightarrow NH_4Cl(s)$$
$$\qquad\quad \text{acid} \qquad\qquad\quad \text{base}$$

Here the HCl molecule is the proton donor, or acid, and the $NH_3$ molecule is the proton acceptor, or base.

In any reversible acid–base reaction, both forward and reverse reactions involve proton transfers. Consider the reaction of $NH_3$ with $H_2O$.

$$\overset{\frown}{N}H_3(aq) + \text{(H)}_2O(l) \rightleftharpoons N\text{(H)}_4{}^+(aq) + OH^-(aq)$$
$$\quad \text{base} \qquad\quad \text{acid} \qquad\qquad \text{acid} \qquad\quad \text{base}$$

In the forward reaction, $NH_3$ accepts a proton from $H_2O$. Thus, $NH_3$ is a base and $H_2O$ is an acid. In the reverse reaction, $NH_4{}^+$ donates a proton to $OH^-$. $NH_4{}^+$ ion is the acid and $OH^-$ is the base.

Note that $NH_3$ and $NH_4{}^+$ differ by a proton. That is, $NH_3$ becomes the $NH_4{}^+$ ion by gaining a proton, whereas the $NH_4{}^+$ ion becomes the $NH_3$ molecule by losing a proton. The species $NH_4{}^+$ and $NH_3$ are a conjugate acid–base pair. A **conjugate acid–base pair** consists of *two species in an acid–base reaction, one acid and one base, that differ by the loss or gain of a proton*. The acid in such a pair is called the *conjugate acid* of the base, whereas the base is the *conjugate base* of the acid. Here $NH_4{}^+$ is the conjugate acid of $NH_3$, and $NH_3$ is the conjugate base of $NH_4{}^+$.

**Understanding Concepts: Weak and Strong Acids**

▶ An alternate example is provided in the Instructor's Resource Manual.

▶ Video no. 36 (Series D), "Bicarbonate/Hydrochloric Acid Volcano."

---

### EXAMPLE 16.1

## Identifying Acid and Base Species

In the following equations, label each species as an acid or a base. Show the conjugate acid–base pairs.

a. $HCO_3{}^-(aq) + HF(aq) \rightleftharpoons H_2CO_3(aq) + F^-(aq)$

b. $HCO_3{}^-(aq) + OH^-(aq) \rightleftharpoons CO_3{}^{2-}(aq) + H_2O(l)$

**◆ PROBLEM STRATEGY**

Recall that a Brønsted–Lowry acid is a proton donor, and the base is a proton acceptor. Examine each equation to find the proton donor on each side. Then label the acids and bases.

**SOLUTION**

a. On the left, HF is the proton donor; on the right, $H_2CO_3$ is the proton donor. The proton acceptors are $HCO_3{}^-$ and $F^-$. Once the proton donors and acceptors are identified, the acids and bases can be labeled.

$$HCO_3{}^-(aq) + \text{(H)}F(aq) \rightleftharpoons \text{(H)}_2CO_3(aq) + F^-(aq)$$
$$\quad \text{base} \qquad\qquad \text{acid} \qquad\qquad \text{acid} \qquad\quad \text{base}$$

In this reaction, $H_2CO_3$ and $HCO_3^-$ are a conjugate acid–base pair, as are HF and $F^-$.

b. You have

$$\text{H}CO_3^-(aq) + OH^-(aq) \rightleftharpoons CO_3^{2-}(aq) + \text{H}_2O(l)$$
$$\quad\text{acid} \qquad\qquad \text{base} \qquad\qquad\quad \text{base} \qquad\qquad \text{acid}$$

Here $HCO_3^-$ and $CO_3^{2-}$ are a conjugate acid–base pair, as are $H_2O$ and $OH^-$. Note that although $HCO_3^-$ functions as an acid in this reaction, it functions as a base in a.

---

### EXERCISE 16.1

For the reaction

$$H_2CO_3(aq) + CN^-(aq) \rightleftharpoons HCN(aq) + HCO_3^-(aq)$$

label each species as an acid or a base. For the base on the left, what is the conjugate acid?

---

See Problems 16.27 and 16.28.

The Brønsted–Lowry concept defines a species as an acid or a base according to its function in the acid–base, or proton-transfer, reaction. As you saw in Example 16.1, some species can act as either an acid or a base. An **amphiprotic species** is *a species that can act as either an acid or a base (it can lose or gain a proton)*, depending on the other reactant. ◀ For example, $HCO_3^-$ acts as an acid in the presence of $OH^-$ but as a base in the presence of HF. Anions with ionizable hydrogens, such as $HCO_3^-$, and certain solvents, such as water, are amphiprotic.

The amphiprotic characteristic of water is important in the acid–base properties of aqueous solutions. Consider, for example, the reactions of water with the base $NH_3$ and with the acid $HC_2H_3O_2$ (acetic acid).

▶ *Amphoteric* **is a general term referring to a species that can act as an acid or a base. The species need not be amphiprotic, however. For example, aluminum oxide is an amphoteric oxide, because it reacts with acids and bases. It is not amphiprotic, because it has no protons.**

$$NH_3(aq) + \text{H}_2O(l) \rightleftharpoons NH_4^+(aq) + OH^-(aq)$$
$$\quad\text{base} \qquad\qquad \text{acid} \qquad\qquad\quad \text{acid} \qquad\qquad \text{base}$$

$$\text{H}C_2H_3O_2(aq) + H_2O(aq) \rightleftharpoons C_2H_3O_2^-(aq) + H_3O^+(aq)$$
$$\qquad\text{acid} \qquad\qquad\quad \text{base} \qquad\qquad\quad\quad \text{base} \qquad\qquad\quad \text{acid}$$

In the first case, water reacts as an acid with the base $NH_3$. In the second case, water reacts as a base with the acid $HC_2H_3O_2$.

You have now seen several ways in which the Brønsted–Lowry concept of acids and bases has greater scope than the Arrhenius concept. In the Brønsted–Lowry concept:

1. A base is a species that accepts protons; $OH^-$ is only one example of a base.

2. Acids and bases can be ions as well as molecular substances.

3. Acid–base reactions are not restricted to aqueous solution.

4. Some species can act as either acids or bases, depending on what the other reactant is.

**CONCEPT CHECK 16.1**

Chemists in the seventeenth century discovered that the substance that gives red ants their irritating bite is an acid with the formula $HCHO_2$. They called this substance formic acid after the ant, whose Latin name is *formica rufus*. Formic acid has the following structural formula and molecular model:

$$H-\overset{\displaystyle \overset{O}{\|}}{C}-O-H$$

Write the acid–base equilibria connecting all components in the aqueous solution. Now list all of the species present.

## 16.3 Lewis Concept of Acids and Bases

▶ **Acidic and basic oxides were discussed in Section 8.7.**

Certain reactions have the characteristics of acid–base reactions but do not fit the Brønsted–Lowry concept. An example is the reaction of the basic oxide $Na_2O$ with the acidic oxide $SO_3$ to give the salt $Na_2SO_4$. ◀

$$Na_2O(s) + SO_3(g) \longrightarrow Na_2SO_4(s)$$

G. N. Lewis, who proposed the electron-pair theory of covalent bonding, realized that the concept of acids and bases could be generalized to include reactions of acidic and basic oxides and many other reactions, as well as proton-transfer reactions. According to this concept, a **Lewis acid** is *a species that can form a covalent bond by accepting an electron pair from another species;* a **Lewis base** is *a species that can form a covalent bond by donating an electron pair to another species.* The Lewis and the Brønsted–Lowry concepts are simply different ways of looking at certain chemical reactions. Such different views are often helpful in devising new reactions.

Consider again the neutralization of $NH_3$ by HCl in aqueous solution, mentioned earlier. It consists of the reaction of a proton from $H_3O^+$ with $NH_3$:

$$H^+ \quad + \quad :\!\overset{\displaystyle H}{\underset{\displaystyle H}{N}}\!:\!H \quad \longrightarrow \quad \left[ H\!:\!\overset{\displaystyle H}{\underset{\displaystyle H}{N}}\!:\!H \right]^+$$

electron-pair          electron-pair
acceptor                  donor

Here the red arrow shows the proton accepting an electron pair from $NH_3$ and an H—N bond being formed. The proton is an electron-pair acceptor, so it is a Lewis acid. Ammonia, $NH_3$, which has a lone pair of electrons, is an electron-pair donor and therefore a Lewis base.

Now let us look at the reaction of $Na_2O$ with $SO_3$. It involves the reaction of the oxide ion, $O^{2-}$, from the ionic solid, $Na_2O$, with $SO_3$.

$$:\!\overset{..}{\underset{..}{O}}\!:^{2-} \quad + \quad \overset{\displaystyle :\overset{..}{O}:}{\underset{\displaystyle :\overset{..}{O}:}{S::O}} \quad \longrightarrow \quad \left[ \overset{\displaystyle :\overset{..}{O}:}{\underset{\displaystyle :\overset{..}{O}:}{:\overset{..}{O}:S:\overset{..}{O}:}} \right]^{2-}$$

Lewis          Lewis
base            acid

Here $SO_3$ accepts the electron pair from the $O^{2-}$ ion. At the same time, an electron pair from the $S=O$ bond moves to the O atom. Thus, $O^{2-}$ is the Lewis base and $SO_3$ is the Lewis acid.

The Lewis concept embraces many reactions that we might not think of as acid–base reactions. The reaction of boron trifluoride with ammonia is an example.

In this reaction, the $NH_3$ molecule donates the lone pair of electrons on the nitrogen atom to the boron atom of $BF_3$. Boron trifluoride accepts the electron pair, so is a Lewis acid. Ammonia donates the electron pair and so is a Lewis base. The electron pair originally on the nitrogen atom is now shared between the nitrogen and boron atoms, forming a B–N bond.

The formation of *complex ions* can also be looked at as Lewis acid–base reactions. Complex ions are formed when a metal ion bonds to electron pairs from molecules such as $H_2O$ or $NH_3$ or from anions such as $:C{\equiv}N:^-$. An example of a complex ion is $Al(H_2O)_6^{3+}$. Hydrated ions like $Al(H_2O)_6^{3+}$ are present in compounds (hydrates) and in aqueous solution. The formation of a hydrated metal ion, such as $Al(H_2O)_6^{3+}$, involves a Lewis acid–base reaction.

$Al(OH_2)_6^{3+}$

► An alternate example is provided in the Instructor's Resource Manual.

## EXAMPLE 16.2

## Identifying Lewis Acid and Base Species

In the following reactions, identify the Lewis acid and the Lewis base.

a. $Ag^+ + 2NH_3 \rightleftharpoons Ag(NH_3)_2^+$

b. $B(OH)_3 + H_2O \rightleftharpoons B(OH)_4^- + H^+$

### PROBLEM STRATEGY

Write the equations using Lewis electron-dot formulas. Then identify the electron-pair acceptor, or Lewis acid, and the electron-pair donor, or Lewis base.

### SOLUTION

a. The silver ion, $Ag^+$, forms a complex ion with two $NH_3$ molecules.

b. The reaction is

EXERCISE 16.2

Identify the Lewis acid and the Lewis base in each of the following reactions. Write the chemical equations using electron-dot formulas.

a. $BF_3 + CH_3OH \longrightarrow F_3B:\overset{\overset{\displaystyle H}{|}}{\underset{..}{O}}CH_3$

b. $O^{2-} + CO_2 \longrightarrow CO_3{}^{2-}$

**See Problems 16.31, 16.32, 16.33, and 16.34.**

# Acid and Base Strengths

The Brønsted–Lowry concept considers an acid–base reaction as a proton-transfer reaction. It is useful to consider such acid–base reactions as a competition between species for protons. From this point of view, you can order acids by their relative strengths as proton donors. The stronger acids are those that lose their protons more easily than other acids. Similarly, the stronger bases are those that hold on to protons more strongly than other bases.

## 16.4 Relative Strengths of Acids and Bases

▶ Experiments in General Chemistry, 16, "The Relative Strengths of Some Acids."

By comparing various acid–base reactions, you can construct a table of relative strengths of acids and bases (see Table 16.2). To see how you might do this, look again at what we have called a strong acid. Recall that an acid is strong if it completely ionizes in water. In the reaction of hydrogen chloride with water, for example, water acts as a base, accepting the proton from HCl.

$$\underset{\text{acid}}{HCl(aq)} + \underset{\text{base}}{H_2O(l)} \longrightarrow \underset{\text{base}}{Cl^-(aq)} + \underset{\text{acid}}{H_3O^+(aq)}$$

The reverse reaction occurs only to an extremely small extent. Because the reaction goes almost completely to the right, you say that HCl is a strong acid. However, even though the reaction goes almost completely to products, you can consider the reverse reaction. In it, the $Cl^-$ ion acts as the base, accepting a proton from the acid $H_3O^+$.

# Taking Your Medicine

**LIFE SCIENCE**

The expressions "take your medicine" and "it's a bitter pill" are metaphors that refer to doing what's necessary to solve a difficult personal problem. To take your medicine might be unpleasant if it is bitter—and medicines do frequently taste bitter. Why?

A bitter taste appears to be a common feature of a base. It is a fact that many medicinal substances are nitrogen bases, substances that organic chemists call *amines*. Such substances are considered derivatives of ammonia, in which one or more hydrogen atoms have been substituted by carbon-containing groups. Here, for example, are the structures of some amines:

$$H-\underset{..}{N}-H \qquad H_3C-\underset{..}{\overset{\overset{\displaystyle CH_3}{|}}{N}}-H \qquad H_3C-\underset{..}{\overset{\overset{\displaystyle CH_3}{|}}{N}}-CH_3$$

The nitrogen atom in an amine has a lone pair of electrons that can be donated to form a covalent bond. So, an amine is a Lewis base. But, like ammonia, an amine accepts a hydrogen ion from an acid to form an amine ion, so it is also a Brønsted–Lowry base:

$$H-\underset{..}{\overset{\overset{\displaystyle CH_3}{|}}{N}}-H + H_3O^+ \rightleftharpoons \left[H-\overset{\overset{\displaystyle CH_3}{|}}{\underset{\underset{\displaystyle H}{|}}{N}}-H\right]^+ + H_2O$$

Some of our most important medicinal drugs have originated from plants. Lewis and Clark took Peruvian bark, or cinchona bark, with them as a medicine on their 1804 expedition from the eastern United States to the Pacific Coast and back. The bitter essence of cinchona bark is quinine, an amine drug that has been used to combat malaria. Quinine is responsible for the bitter taste of tonic water, a carbonated beverage (Figure 16.4). Some other amines that come from plants are caffeine (from coffee, a stimulant), atropine (from the deadly nightshade, used to dilate the pupil of the eye for eye exams), and codeine (from the opium poppy, used as a painkiller).

**Quinine (antimalarial)**

**Figure 16.4**
**Tonic water.** *Left:* This carbonated beverage contains quinine, which accounts for its bitter taste. *Right:* Structural formula of quinine.

Look at this reaction in terms of the relative strengths of the two acids, HCl and $H_3O^+$. Because HCl is a strong acid, it must lose its proton readily, more readily than $H_3O^+$ does. You would say that HCl is a stronger acid than $H_3O^+$, or that, of the two, $H_3O^+$ is the weaker acid.

$$\underset{\substack{\text{stronger}\\\text{acid}}}{HCl(aq)} + H_2O(l) \rightleftharpoons Cl^-(aq) + \underset{\substack{\text{weaker}\\\text{acid}}}{H_3O^+(aq)}$$

**Table 16.2**
**Relative Strengths of Acids and Bases**

| | Acid | Base | |
|---|---|---|---|
| Strongest acids | $HClO_4$ | $ClO_4^-$ | Weakest bases |
| | $H_2SO_4$ | $HSO_4^-$ | |
| | HI | $I^-$ | |
| | HBr | $Br^-$ | |
| | HCl | $Cl^-$ | |
| | $HNO_3$ | $NO_3^-$ | |
| | $H_3O^+$ | $H_2O$ | |
| | $HSO_4^-$ | $SO_4^{2-}$ | |
| | $H_2SO_3$ | $HSO_3^-$ | |
| | $H_3PO_4$ | $H_2PO_4^-$ | |
| | $HNO_2$ | $NO_2^-$ | |
| | HF | $F^-$ | |
| | $HC_2H_3O_2$ | $C_2H_3O_2^-$ | |
| | $Al(H_2O)_6^{3+}$ | $Al(H_2O)_5OH^{2+}$ | |
| | $H_2CO_3$ | $HCO_3^-$ | |
| | $H_2S$ | $HS^-$ | |
| | HClO | $ClO^-$ | |
| | HBrO | $BrO^-$ | |
| | $NH_4^+$ | $NH_3$ | |
| | HCN | $CN^-$ | |
| | $HCO_3^-$ | $CO_3^{2-}$ | |
| | $H_2O_2$ | $HO_2^-$ | |
| | $HS^-$ | $S^{2-}$ | |
| Weakest acids | $H_2O$ | $OH^-$ | Strongest bases |

It is important to understand that the terms *stronger* and *weaker* are used here only in a comparative sense. The $H_3O^+$ ion is a relatively strong acid.

An acid–base reaction normally goes in the direction of the weaker acid. You can use this fact to compare the relative strengths of any two acids, as we did in comparing the relative strengths of HCl and $H_3O^+$. As another example, look at the ionization of acetic acid, $HC_2H_3O_2$, in water.

$$HC_2H_3O_2(aq) + H_2O(l) \rightleftharpoons C_2H_3O_2^-(aq) + H_3O^+(aq)$$

Experiment reveals that in a $0.1\ M$ acetic acid solution, only about 1% of the acetic acid molecules have ionized by this reaction. This implies that $HC_2H_3O_2$ is a weaker acid than $H_3O^+$. If you look at the similar situation for $0.1\ M$ HF, you find that about

▶ Demonstration of relative acid and base strengths: Bassam Z. Shakhashiri, *Chemical Demonstrations,* Vol. 3 (Madison: University of Wisconsin Press, 1989), no. 8.25, pp. 158–161.

3% of the HF molecules have dissociated. Thus, HF is a weaker acid than $H_3O^+$ but a stronger acid than $HC_2H_3O_2$. You have already seen that HCl is stronger than $H_3O^+$. Thus, you have determined that the acid strengths for these four acids are in the order $HCl > H_3O^+ > HF > HC_2H_3O_2$.

This procedure of determining the relative order of acid strengths by comparing their relative ionizations in water cannot be used to obtain the relative strengths of two strong acids such as HCl and HI. When these acids are dissolved in water, they are essentially 100% ionized. However, if you look at solutions of equal concentrations of these acids in another solvent that is less basic than water—say, pure acetic acid—you do see a difference. Neither acid is completely ionized, but a greater fraction of HI molecules is found to be ionized. Thus, HI is a stronger acid than HCl. In water, the acid strengths of the strong acids appear to be the same; that is, they are "leveled out." We say that water exhibits a *leveling effect* on the strengths of the strong acids.

The first column of Table 16.2 lists acids by their strength; the strongest is at the top of the table. Note that the arrow down the left side of the table points toward the weaker acid and is in the direction the reaction goes. For example, in the reaction involving HCl and $H_3O^+$, the arrow points from HCl toward $H_3O^+$ (the weaker acid) and is in the direction in which the reaction occurs.

You can also view this same reaction in terms of the bases, $H_2O$ and $Cl^-$. A stronger base picks up a proton more readily than does a weaker one. Water has greater base strength than the $Cl^-$ ion; that is, $H_2O$ picks up protons more readily than $Cl^-$ does. In fact, $Cl^-$ has little attraction for protons. (If it had more, HCl would not lose its proton so readily.) Because $H_2O$ molecules compete more successfully for protons than $Cl^-$ ions do, the reaction goes almost completely to the right. (Note that the arrow on the right in Table 16.2 points from $H_2O$ to $Cl^-$.) That is, at the completion of the reaction, concentrations of product species ($Cl^-$ and $H_3O^+$) are much greater than concentrations of reactant species.

$$HCl(aq) + H_2O(l) \longrightarrow Cl^-(aq) + H_3O^+(aq)$$

stronger         weaker
base           base

The reaction goes in the direction of the weaker base.

By comparing reactions between different pairs of bases, you can arrive at a relative order for base strengths, just as for acids. A definite relationship exists between acid and base strengths. When you say an acid loses its proton readily, you can also say that its conjugate base does not hold the proton very tightly. *The strongest acids have the weakest conjugate bases, and the strongest bases have the weakest conjugate acids.* This means that a list of conjugate bases of the acids in Table 16.2 will be in order of increasing base strength. The weakest bases will be at the top of the table and the strongest at the bottom.

You can use Table 16.2 to predict the direction of an acid–base reaction. The direction for an acid–base reaction always favors the weaker acid and weaker base; that is, the normal direction of reaction is from the stronger acid and base to the weaker acid and base. For the reaction we have been discussing, you have

$$HCl(aq) + H_2O(l) \longrightarrow Cl^-(aq) + H_3O^+(aq)$$

stronger   stronger     weaker    weaker
acid       base        base      acid

The reaction follows the direction of the arrows at the left and right of Table 16.2.

▶ An alternate example is provided in the Instructor's Resource Manual.

## EXAMPLE 16.3

### Deciding Whether Reactants or Products Are Favored in an Acid–Base Reaction

For the following reaction, decide which species (reactants or products) are favored at the completion of the reaction.

$$SO_4^{2-}(aq) + HCN(aq) \rightleftharpoons HSO_4^-(aq) + CN^-(aq)$$

 **PROBLEM STRATEGY**

Use Table 16.2 to compare the relative strengths of acids and bases.

**SOLUTION**

If you compare the relative strengths of the two acids HCN and $HSO_4^-$, you see that HCN is weaker. Or, comparing the bases $SO_4^{2-}$ and $CN^-$, you see that $SO_4^{2-}$ is weaker. Hence, the reaction would normally go from right to left.

$$SO_4^{2-}(aq) + HCN(aq) \longleftarrow HSO_4^-(aq) + CN^-(aq)$$

| weaker base | weaker acid | stronger acid | stronger base |

**The reactants are favored.**

### EXERCISE 16.3

Determine the direction of the following reaction from the relative strengths of acids and bases.

$$H_2S(aq) + C_2H_3O_2^-(aq) \rightleftharpoons HC_2H_3O_2(aq) + HS^-(aq)$$

See Problems 16.37, 16.38, 16.39, and 16.40.

 **CONCEPT CHECK 16.2**

Formic acid, $HCHO_2$, is a stronger acid than acetic acid, $HC_2H_3O_2$. Which is the stronger base, formate ion, $CHO_2^-$, or acetate ion, $C_2H_3O_2^-$?

## 16.5 Molecular Structure and Acid Strength

The strength of an acid depends on how easily the proton, $H^+$, is lost or removed from an H—X bond in the acid species. By understanding the factors that determine the ease of proton loss, you will be able to predict the relative strengths of similar acids.

Two factors are important in determining relative acid strengths. One is the polarity of the bond to which the H atom is attached. The H atom should have a positive partial charge:

$$\overset{\delta+ \quad \delta-}{H—X}$$

The more polarized the bond is in this direction, the more easily the proton is removed and the greater the acid strength. Figure 16.5 dramatically illustrates this idea. The

**Figure 16.5**
**Electron density plots of acetic acid and sulfuric acid molecules.** High electron densities are shown in red and low densities are shown in blue, with intermediate colors of the spectrum for intermediate densities. Note the hydrogen atoms at the right ends of each molecule (blue colors, indicating low electron densities). These atoms have partial positive charges and thus are acidic. (Electron density figure taken from G.D. Shusterman, A.J. Shusterman, *J. Chem. Ed.*, 74, 771, 1997.)

CH₃COOH

H₂SO₄

figure shows the electron density distributions of acetic acid, $HC_2H_3O_2$ or $CH_3COOH$, and of sulfuric acid, $H_2SO_4$:

Acetic acid          Sulfuric acid

High electron densities are shown in red (giving a partial negative charge, $\delta-$) and low densities are shown in blue (giving a partial positive charge, $\delta+$). Intermediate densities are shown in colors between red and blue. A hydrogen atom bonded to an oxygen atom generally has a partial positive charge, but in comparing such an atom in acetic acid (light blue atom at the right end of the molecule) and sulfuric acid (dark blue atom at the right end of the molecule), you can see that the hydrogen atom in sulfuric acid is much more positive. You would expect the hydrogen atom in sulfuric acid to be more acidic—and it certainly is.

The second factor determining acid strength is the strength of the bond—that is, how tightly the proton is held. This, in turn, depends on the size of atom X. The larger atom X, the weaker is the bond and the greater the acid strength.

Consider a series of binary acids, HX, formed from a given column of elements of the periodic table. The acids would be compounds of these elements with hydrogen, such as the binary acids of the Group VIIA elements: HF, HCl, HBr, and HI. As you go down the column of elements, each time adding a shell of electrons to the atom, the radius increases markedly. For this reason, the size of atom X is the dominant factor in determining the acid strength. *In going down a column of elements of the periodic table, the size of atom X increases, the H—X bond strength decreases, and the strength of the binary acid increases.* You predict the following order of acid strength:

$$HF < HCl < HBr < HI$$

This is the same order shown in Table 16.2.

As you go across a row of elements of the periodic table, the atomic radius decreases slowly. For this reason, the relative strengths of the binary acids of these elements are less dependent on the size of atom X. Now the polarity of the H—X bond becomes the dominant factor in determining acid strength. *Going across a row of elements of the periodic table, the electronegativity increases, the H—X bond polarity increases, and the acid strength increases.* For example, the binary acids of the last two elements of the second period are $H_2O$ and HF. The acid strengths are

$$H_2O < HF$$

HClO

$HClO_2$

$HClO_3$

$HClO_4$

▶ **The formulas of these acids may be written HXO or HOX, depending on the convention used. Formulas of oxoacids are generally written with the acidic H atoms first, followed by the characteristic element (X), then O atoms. However, the formulas of molecules composed of three atoms are often written in the order in which the atoms are bonded, which in this case is HOX.**

This again is the order shown in Table 16.2. Hydrogen fluoride, HF, is a weak acid, and $H_2O$ is a very weak acid.

Now consider the oxoacids. An oxoacid has the structure

$$H-O-Y-$$

The acidic H atom is always attached to an O atom, which, in turn, is attached to an atom Y. Other groups, such as O atoms or O—H groups, may also be attached to Y. Bond polarity appears to be the dominant factor determining relative strengths of the oxoacids. This, in turn, depends on the electronegativity of atom Y. If the electronegativity of atom Y is large, the H—O bond is relatively polar and the acid strength large. *For a series of oxoacids of the same structure, differing only in the atom Y, the acid strength increases with the electronegativity of Y.* Consider, for example, the acids HClO, HBrO, and HIO. ◀ The structures are

$$H-\overset{..}{\underset{..}{O}}-\overset{..}{\underset{..}{Cl}}: \quad H-\overset{..}{\underset{..}{O}}-\overset{..}{\underset{..}{Br}}: \quad H-\overset{..}{\underset{..}{O}}-\overset{..}{\underset{..}{I}}:$$

The electronegativity of Group VIIA elements decreases going down the column of elements, so the electronegativity decreases from Cl to Br to I. The order of acid strengths is

$$HIO < HBrO < HClO$$

*For a series of oxoacids, $(HO)_mYO_n$, the acid strength increases with n, the number of O atoms bonded to Y (excluding O atoms in OH groups).* The oxoacids of chlorine provide an example:

$$H-\overset{..}{\underset{..}{O}}-\overset{..}{\underset{..}{Cl}}: \quad H-\overset{..}{\underset{..}{O}}-\overset{..}{\underset{..}{Cl}}-\overset{..}{\underset{..}{O}}: \quad H-\overset{..}{\underset{..}{O}}-\overset{\overset{\displaystyle :\overset{..}{O}:}{|}}{\underset{..}{Cl}}-\overset{..}{\underset{..}{O}}: \quad H-\overset{..}{\underset{..}{O}}-\overset{\overset{\displaystyle :\overset{..}{O}:}{|}}{\underset{\underset{\displaystyle :\overset{..}{O}:}{|}}{Cl}}-\overset{..}{\underset{..}{O}}:$$

With each additional O atom, the Cl atom becomes effectively more electronegative. As a result, the H atom becomes more acidic. The acid strengths increase in the following order:

$$HClO < HClO_2 < HClO_3 < HClO_4$$

Before we leave the subject of molecular structure and acid strength, let us look at the relative acid strengths of a polyprotic acid and its corresponding acid anions. For example, $H_2SO_4$ ionizes by losing a proton to give $HSO_4^-$, which, in turn, ionizes to give $SO_4^{2-}$. $HSO_4^-$ can lose a proton, so it is acidic. However, because of the negative charge of the ion, which tends to attract protons, its acid strength is reduced from that of the uncharged species. That is, the acid strengths are in the order

$$HSO_4^- < H_2SO_4$$

We conclude that *the acid strength of a polyprotic acid and its anions decreases with increasing negative charge* (see Table 16.2).

---

**EXERCISE 16.4**

Which member of each of the following pairs is the stronger acid? a. $NH_3$, $PH_3$; b. HI, $H_2Te$; c. $HSO_3^-$, $H_2SO_3$; d. $H_3AsO_4$, $H_3AsO_3$; e. $HSO_4^-$, $HSeO_4^-$

---

**See Problems 16.43 and 16.44.**

# Self-Ionization of Water and pH

In aqueous solutions, two ions have dominant roles. These ions, the hydronium ion, $H_3O^+$ (or hydrogen ion, $H^+$), and the hydroxide ion, ($OH^-$), are available in any aqueous solution as a result of the *self-ionization* of water, a reaction of water with itself, which we will describe in the next section. This will also give us some background to acid–base equilibrium calculations, which we will discuss in Chapter 17.

## 16.6 Self-Ionization of Water

**Visualization: Self-Ionization of Water to Form $H^+$ and $OH^-$ in Equilibrium**

▶ Animation: Self-ionization of water to form $H^+$ and $OH^-$ in equilibrium.

Although pure water is often considered a nonelectrolyte (nonconductor of electricity), precise measurements do show a very small conduction. This conduction results from **self-ionization** (or **autoionization**), *a reaction in which two like molecules react to give ions.* In the case of water, a proton from one $H_2O$ molecule is transferred to another $H_2O$ molecule, leaving behind an $OH^-$ ion and forming a hydronium ion, $H_3O^+(aq)$.

$$H_2O(l) + H_2O(l) \rightleftharpoons H_3O^+(aq) + OH^-(aq)$$

You can see the slight extent to which the self-ionization of water occurs by noting the small value of its equilibrium constant $K_c$.

$$K_c = \frac{[H_3O^+][OH^-]}{[H_2O]^2}$$

Because the concentration of ions formed is very small, the concentration of $H_2O$ remains essentially constant, about 56 $M$ at 25°C. If you rearrange this equation, placing $[H_2O]^2$ with $K_c$, the ion product $[H_3O^+][OH^-]$ equals a constant.

$$\underbrace{[H_2O]^2 K_c}_{\text{constant}} = [H_3O^+][OH^-]$$

▶ The thermodynamic equilibrium constant is defined in terms of activities. The activity of water is essentially 1, so we do not include water explicitly in the equilibrium expression.

We call *the equilibrium value of the ion product* $[H_3O^+][OH^-]$ the **ion-product constant for water,** which is written $K_w$. ◀ At 25°C, the value of $K_w$ is $1.0 \times 10^{-14}$. Like any equilibrium constant, $K_w$ varies with temperature. At body temperature (37°C), $K_w$ equals $2.5 \times 10^{-14}$.

$$K_w = [H_3O^+][OH^-] = 1.0 \times 10^{-14} \text{ at } 25°C$$

Because we often write $H^+(aq)$ for $H_3O^+(aq)$, the ion-product constant for water can be written $K_w = [H^+][OH^-]$.

Using $K_w$, you can calculate the concentrations of $H_3O^+$ and $OH^-$ ions in pure water. These ions are produced in equal numbers in pure water, so their concentrations are equal. Let $x = [H_3O^+] = [OH^-]$. Then, substituting into the equation for the ion-product constant,

$$K_w = [H_3O^+][OH^-]$$

you get, at 25°C,

$$1.0 \times 10^{-14} = x^2$$

Hence, $x$ equals $1.0 \times 10^{-7}$. Thus, the concentrations of $H_3O^+$ and $OH^-$ are both $1.0 \times 10^{-7}$ $M$ in pure water.

If you add an acid or a base to water, the concentrations of $H_3O^+$ and $OH^-$ will no longer be equal. The equilibrium-constant equation $K_w = [H_3O^+][OH^-]$ will still hold, however, as we will discuss in the next section.

## 16.7 Solutions of a Strong Acid or Base

Consider an aqueous solution of a strong acid or base. Suppose you dissolve 0.10 mol HCl in 1.0 L of aqueous solution, giving 0.10 $M$ HCl. You would like to know the concentration of $H_3O^+$ ion in this solution. In addition to the self-ionization of water, you have the reaction of HCl with water, which also produces $H_3O^+$ ion. A strong acid such as hydrochloric acid, HCl($aq$), essentially reacts completely with water.

$$HCl(aq) + H_2O(l) \longrightarrow H_3O^+(aq) + Cl^-(aq)$$

Because you started with 0.10 mol HCl in 1.0 L of solution, the reaction will produce 0.10 mol $H_3O^+$, so the concentration of $H_3O^+$ ion from HCl is 0.10 $M$.

Now consider the concentration of $H_3O^+$ ion produced by the self-ionization of water. In pure water, the concentration of $H_3O^+$ produced is $1.0 \times 10^{-7}$ $M$; in an acid solution, the contribution of $H_3O^+$ from water will be even smaller. You can see this by applying Le Chatelier's principle to the self-ionization reaction. When you increase the concentration of $H_3O^+$ in water by adding an acid, the self-ionization of water reverses until a new equilibrium is obtained.

$$H_2O(l) + H_2O(l) \longleftarrow H_3O^+(aq) + OH^-(aq)$$

Consequently, the concentration of $H_3O^+$ produced by the self-ionization of water ($<1 \times 10^{-7}$ $M$) is negligible in comparison with that produced from HCl (0.10 $M$). So, 0.10 $M$ HCl has a concentration of $H_3O^+$ ion equal to 0.10 $M$.

In a solution of a strong acid, you can normally ignore the self-ionization of water as a source of $H_3O^+$. The $H_3O^+$ concentration is usually determined by the strong acid concentration. (This is not true when the acid solution is extremely dilute, however. In a solution that is $1.0 \times 10^{-7}$ $M$ HCl, the self-ionization of water produces an amount of $H_3O^+$ comparable with that produced by HCl.)

Although you normally ignore the self-ionization of water in calculating the $H_3O^+$ concentration in a solution of a strong acid, the self-ionization equilibrium still exists and is responsible for a small concentration of $OH^-$ ion. You can use the ion-product constant for water to calculate this concentration. As an example, calculate the concentration of $OH^-$ ion in 0.10 $M$ HCl. You substitute $[H_3O^+] = 0.10$ $M$ into the equilibrium equation for $K_w$ (for 25°C).

$$K_w = [H_3O^+][OH^-]$$
$$1.0 \times 10^{-14} = 0.10 \times [OH^-]$$

Solving for $[OH^-]$,

$$[OH^-] = \frac{1.0 \times 10^{-14}}{0.10} = 1.0 \times 10^{-13}$$

The $OH^-$ concentration is $1.0 \times 10^{-13}$ $M$.

Now consider a solution of a strong base, such as 0.010 $M$ NaOH. What are the $OH^-$ and $H_3O^+$ concentrations in this solution? Because NaOH is a strong base, all of the NaOH is present in the solution as ions. One mole of NaOH dissolves in water as one mole of $Na^+$ and one mole of $OH^-$. Therefore, the concentration of $OH^-$ obtained from NaOH in 0.010 $M$ NaOH solution is 0.010 $M$. The concentration of $OH^-$ produced from the self-ionization of water in this solution ($< 1 \times 10^{-7}$ $M$) is negligible and can be ignored. Therefore, the concentration of $OH^-$ ion in the solution is 0.010 $M$. Hydronium ion, $H_3O^+$, is produced by the self-ionization of water. To obtain its concentration, you substitute into the equilibrium equation for $K_w$ (for 25°C).

$$K_w = [H_3O^+][OH^-]$$

$$1.0 \times 10^{-14} = [H_3O^+] \times 0.010$$

Solving for $H_3O^+$ concentration,

$$[H_3O^+] = \frac{1.0 \times 10^{-14}}{0.010} = 1.0 \times 10^{-12}$$

The $H_3O^+$ concentration is $1.0 \times 10^{-12}$ $M$.

The following example further illustrates the calculation of $H_3O^+$ and $OH^-$ concentrations in solutions of a strong acid or base.

▶ An alternate example is provided in the Instructor's Resource Manual.

## EXAMPLE 16.4

### Calculating Concentrations of $H_3O^+$ and $OH^-$ in Solutions of a Strong Acid or Base

Calculate the concentrations of hydronium ion and hydroxide ion at 25°C in:
a. 0.15 $M$ $HNO_3$, b. 0.010 $M$ $Ca(OH)_2$.

 **PROBLEM STRATEGY**

A strong acid (or base) is fully ionized in solution, so you can use its formula and molar concentration to determine the $H_3O^+$ (or $OH^-$) ion concentration. (Any contribution to this ion concentration from the self-ionization of water is negligible.) $K_w$ relates the $H_3O^+$ and $OH^-$ ion concentrations; therefore, you use this relation to obtain the $OH^-$ concentration, given the $H_3O^+$ concentration, or vice versa.

### SOLUTION

a. Every mole of $HNO_3$ contributes one mole of $H_3O^+$ ion, so the $H_3O^+$ concentration is **0.15 $M$.** The $OH^-$ concentration is obtained from the equation for $K_w$.

$$K_w = [H_3O^+][OH^-]$$

$$1.0 \times 10^{-14} = 0.15 \times [OH^-]$$

so

$$[OH^-] = \frac{1.0 \times 10^{-14}}{0.15} = 6.7 \times 10^{-14}$$

The concentration of $OH^-$ in 0.15 $M$ $HNO_3$ is **$6.7 \times 10^{-14}$ $M$.**

b.

$$Ca(OH)_2(s) \xrightarrow{H_2O} Ca^{2+}(aq) + 2OH^-(aq)$$

Every mole of $Ca(OH)_2$ that dissolves yields two moles of $OH^-$. Therefore, $0.010\ M\ Ca(OH)_2$ contains $2 \times 0.010\ M\ OH^- = \textbf{0.020}\ \textbf{\textit{M}}\ \textbf{OH}^-$. The $H_3O^+$ concentration is obtained from

$$K_w = [H_3O^+][OH^-]$$

$$1.0 \times 10^{-14} = [H_3O^+] \times 0.020$$

so

$$[H^+] = \frac{1.0 \times 10^{-14}}{0.020} = 5.0 \times 10^{-13}$$

The hydronium-ion concentration is $\textbf{5.0} \times \textbf{10}^{-13}\ \textbf{\textit{M}}$.

---

**EXERCISE 16.5**

A solution of barium hydroxide at 25°C is $0.125\ M\ Ba(OH)_2$. What are the concentrations of hydronium ion and hydroxide ion in the solution?

---

See Problems 16.45 and 16.46.

By dissolving substances in water, you can alter the concentrations of $H_3O^+$ and $OH^-$ ions. In a *neutral* solution, the concentrations of $H_3O^+$ and $OH^-$ remain equal, as they are in pure water. In an *acidic* solution, the concentration of $H_3O^+$ is greater than that of $OH^-$. In a *basic* solution, the concentration of $OH^-$ is greater than that of $H_3O^+$. At 25°C, you observe the following conditions:

> In an acidic solution, $[H_3O^+] > 1.0 \times 10^{-7}\ M$.
>
> In a neutral solution, $[H_3O^+] = 1.0 \times 10^{-7}\ M$.
>
> In a basic solution, $[H_3O^+] < 1.0 \times 10^{-7}\ M$.

---

**EXERCISE 16.6**

A solution has a hydroxide-ion concentration of $1.0 \times 10^{-5}\ M$ at 25°C. Is the solution acidic, neutral, or basic?

---

See Problems 16.51 and 16.52.

# 16.8 The pH of a Solution

▶ The Danish biochemist S. P. L. Sørensen devised the pH scale while working on the brewing of beer.

You see that whether an aqueous solution is acidic, neutral, or basic depends on the hydronium-ion concentration. You can quantitatively describe the acidity by giving the hydronium-ion concentration. But because these concentration values may be very small, it is often more convenient to give the acidity in terms of **pH,** which is defined as *the negative of the logarithm of the molar hydronium-ion concentration:* ◀

**Figure 16.6**
**The pH scale.** Solutions having pH less than 7 are acidic; those having pH more than 7 are basic.

$$pH = -\log [H_3O^+]$$

(You can also write $pH = -\log [H^+]$.) For a solution in which the hydronium-ion concentration is $1.0 \times 10^{-3} M$, the pH is

$$pH = -\log(1.0 \times 10^{-3}) = 3.00$$

Note that the number of places after the decimal point in the pH equals the number of significant figures reported in the hydronium-ion concentration. ◄

► **See Appendix A for a discussion of logarithms.**

A neutral solution, whose hydronium-ion concentration at 25°C is $1.0 \times 10^{-7} M$, has a pH of 7.00. For an acidic solution, the hydronium-ion concentration is greater than $1.0 \times 10^{-7} M$, so the pH is less than 7.00. Similarly, a basic solution has a pH greater than 7.00. Figure 16.6 shows a diagram of the pH scale and lists pH values of some common solutions.

---

### EXAMPLE 16.5

## Calculating the pH from the Hydronium-Ion Concentration

A sample of orange juice has a hydronium-ion concentration of $2.9 \times 10^{-4} M$. What is the pH? Is the solution acidic?

### PROBLEM STRATEGY

pH is the negative logarithm of the molar hydronium-ion concentration. Determine the log by using either an electronic calculator or log tables (see Appendix A for a discussion of logarithms). On a calculator, enter the $H_3O^+$ concentration, press *log*, and change the sign of the result. You should write the pH with the same number of decimal places as there are significant figures in $[H_3O^+]$.

**Problem Solving:**
**Calculating pH and pOH**

**SOLUTION**

$$pH = -\log[H_3O^+] = -\log(2.9 \times 10^{-4}) = \mathbf{3.54}$$

The pH is less than 7.00, so the solution is **acidic** (as you expect for orange juice).

---

**EXERCISE 16.7**

What is the pH of a sample of gastric juice (digestive juice in the stomach) whose hydronium-ion concentration is 0.045 $M$?

---

See Problems 16.59 and 16.60.

You can find the pH of a solution of known hydroxide-ion concentration by first solving for the hydronium-ion concentration, as shown in Example 16.4 b. However, you can also find the pH simply from the *pOH*, a measure of hydroxide-ion concentration similar to the pH:

$$pOH = -\log[OH^-]$$

▶ **Taking the logarithm of both sides of the equation**

$$[H_3O^+][OH^-] = 1.0 \times 10^{-14}$$

**you get**

$$\log[H_3O^+] + \log[OH^-]$$
$$= -14.00$$

**or**

$$(-\log[H_3O^+]) + (-\log[OH^-])$$
$$= 14.00$$

**Hence, pH + pOH = 14.00.**

Then, because $K_w = [H_3O^+][OH^-] = 1.0 \times 10^{-14}$ at 25°C, you can show that ◀

$$pH + pOH = 14.00$$

You can rearrange this equation to obtain the pH from the value of pOH. For example, suppose you wish to find the pH of an ammonia solution whose hydroxide-ion concentration is $1.9 \times 10^{-3}$ $M$. You first calculate the pOH.

$$pOH = -\log(1.9 \times 10^{-3}) = 2.72$$

Then the pH is

$$pH = 14.00 - pOH$$
$$= 14.00 - 2.72 = 11.28$$

---

**EXERCISE 16.8**

A saturated solution of calcium hydroxide has a hydroxide-ion concentration of 0.025 $M$. What is the pH of the solution?

---

See Problems 16.65 and 16.66.

The usefulness of the pH scale depends on the ease with which you can interconvert pH and hydronium-ion concentrations. In the next example, you will find the hydronium-ion concentration, given the pH value.

---

**EXAMPLE 16.6**

## Calculating the Hydronium-Ion Concentration from the pH

The pH of human arterial blood is 7.40. What is the hydronium-ion concentration?

**PROBLEM STRATEGY**

The hydronium-ion concentration is

$$[H_3O^+] = antilog(-pH) = 10^{-pH}$$

On some calculators, you obtain the antilogarithm by taking the *inverse* of the logarithm. On others, you use a $10^x$ key.

**SOLUTION**

The result is

$$[H_3O^+] = antilog(-7.40) = 10^{-7.40} = \mathbf{4.0 \times 10^{-8}\, M}$$

Because the pH was given to two decimal places, you write the hydronium-ion concentration with two significant figures.

---

**EXERCISE 16.9**

A brand of carbonated beverage has a pH of 3.16. What is the hydronium-ion concentration of the beverage?

---

See Problems 16.67 and 16.68.

---

**EXERCISE 16.10**

A 0.010 $M$ solution of ammonia, $NH_3$, has a pH of 10.6 at 25°C. What is the concentration of hydroxide ion? (One way to solve this problem is to find the pOH first and then calculate the hydroxide-ion concentration.)

---

See Problems 16.69 and 16.70.

The pH of a solution can be accurately measured by a pH meter (Figure 16.7). This instrument consists of specially designed electrodes that are dipped into the solution. A voltage, which depends on the pH, is generated between the electrodes and is read on a meter calibrated directly in pH. (The principle behind the operation of a pH meter is discussed in Chapter 20.)

Although less precise, acid–base indicators are often used to measure pH, because they usually change color within a small pH range. The color change of an indicator involves establishment of an equilibrium between an acid form and a base form that have different colors. If we denote the acid form as HIn, then the base form is In⁻ and the equilibrium is

$$HIn(aq) + H_2O(l) \rightleftharpoons H_3O^+(aq) + In^-(aq)$$

Consider the case of phenolphthalein. The acid form is colorless and the base form is pink. When a base is added to an acidic solution of phenolphthalein, OH⁻ ion from the base reacts with $H_3O^+$ in the solution. According to Le Chatelier's principle, the equilibrium is shifted to the right (to replenish $H_3O^+$). Thus, the colorless acid form of the indicator, HIn, is converted to the pink base form, In⁻. A solution of phenolphthalein begins to turn pink at about pH 8.0. By pH 9.7, the color change

**Figure 16.7**
**A digital pH meter.** The experimenter places a probe containing electrodes into the solution and reads the pH on the meter.

▶ Litmus contains several substances and changes color over a broader range, from about pH 5 to pH 8.

is essentially complete. ◀ Figure 16.8 shows the color changes for various acid–base indicators. (See also Figure 16.9.) Paper strips impregnated with several indicators are often used to measure pH values. Such "pH paper" gives a definite color for different pH ranges and can give the pH to the nearest integer value or better.

 **CONCEPT CHECK 16.3**

You have solutions of $NH_3$, HCl, NaOH, and $HC_2H_3O_2$ (acetic acid), all with the same solute concentrations. Rank these solutions in order of pH, from the highest to the lowest.

**Figure 16.8**
**Color changes of some acid–base indicators.**
Acid–base indicators are dyes whose acid form has one color and whose base form has another color. Thymol blue is a diprotic acid and has two different color changes (listed here as the acidic range and the basic range).

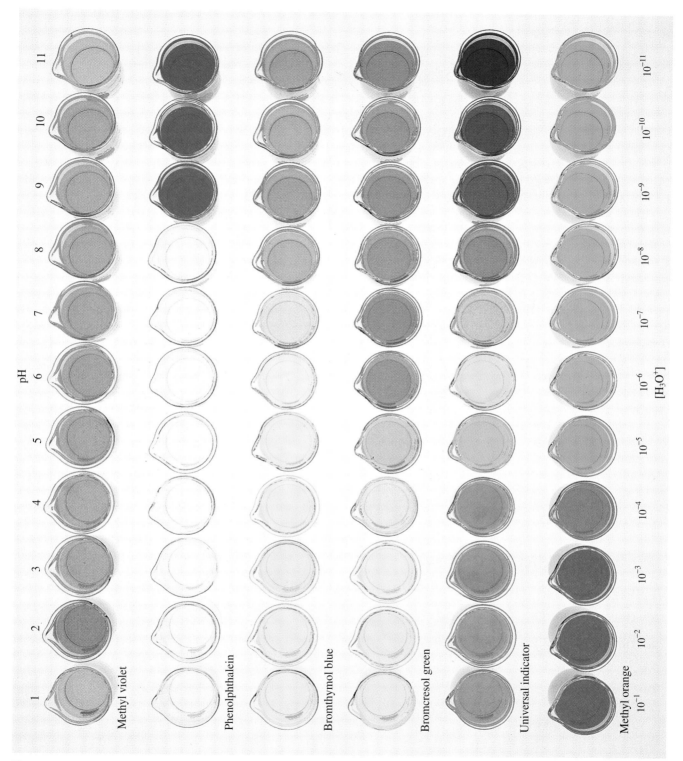

**Figure 16.9**
**Some acid–base indicators in solutions of different H$_3$O$^+$ concentrations.**

# Unclogging the Sink and Other Chores

**EVERYDAY LIFE**

When the sink drain is clogged, you might need a strong arm, but a strong base might work just as well. Drain cleaner solutions, available in grocery stores, are simply solutions of sodium hydroxide, a strong base. Such solutions work by chemically reacting with fat and with hair, the usual ingredients of a stopped drain. When fat reacts with a strong base such as sodium hydroxide, it forms the salt of a fatty acid, a product otherwise known as soap. Hair is a protein material, and in the presence of a strong base, the protein breaks up into the salts of its constituent amino acids. So a wad of greasy hair packed in a drain, in the presence of sodium hydroxide, becomes a soapy solution of amino acid salts. That solution washes easily down the drain. Oven cleaner is a paste made of sodium hydroxide. It cleans by reacting with the fatty deposits and other residues in the oven.

Sodium hydroxide is produced commercially by sending a direct current through an aqueous solution of sodium chloride, a process referred to as electrolysis (Figure 16.10). Hydrogen gas is released at the negative pole and chlorine gas is released at the positive pole. At the end of the electrolysis, the solution contains sodium hydroxide. The overall reaction is

$$2NaCl(aq) + 2H_2O(l) \xrightarrow{\text{electrolysis}} H_2(g) + Cl_2(g) + 2NaOH(aq)$$

Chlorine is prepared commercially in this same process.

**Figure 16.10**
**Preparation of sodium hydroxide by hydrolysis.**
A solution of sodium chloride is electrolyzed. Note the pink color at the negative electrode. Phenolphthalein indicator turns pink from the hydroxide ion formed at the electrode. Later, if we evaporate the solution, sodium hydroxide crystallizes out.

## A Checklist for Review

### Important Terms

acid (Brønsted–Lowry) (16.2)
base (Brønsted–Lowry) (16.2)

conjugate acid–base pair (16.2)
amphiprotic species (16.2)
Lewis acid (16.3)

Lewis base (16.3)
self-ionization (autoionization) (16.6)

ion-product constant for water ($K_w$) (16.6)
pH (16.8)

### Key Equations

$$K_w = [H_3O^+][OH^-]$$
$$= 1.0 \times 10^{-14} \text{ at } 25°C$$

$$pH = -\log[H_3O^+]$$
$$pH + pOH = 14.00$$

### Summary of Facts and Concepts

The *Arrhenius concept* was the first successful theory of acids and bases. Then in 1923, Brønsted and Lowry characterized

acid–base reactions as proton-transfer reactions. According to the *Brønsted–Lowry concept,* an acid is a proton donor and a base is a proton acceptor. The *Lewis concept* is even more general than the Brønsted–Lowry concept. A Lewis acid is an electron-pair acceptor and a Lewis base is an electron-pair donor. Reactions of acidic and basic oxides and the formation of complex ions, as well as proton-transfer reactions, can be described in terms of the Lewis concept.

Acid–base reactions can be viewed as a competition for protons. Using this idea, you can construct a table of *relative strengths of acids and bases.* Reaction favors the weaker acid and the weaker base. Acid strength depends on the polarity and the strength of the bond involving the acidic hydrogen atom. Therefore, you can *relate molecular structure and acid strength.* The text gives a number of rules that allow you to predict the relative strengths of similar acids based on molecular structure.

Water ionizes to give hydronium ion and hydroxide ion. The concentrations of these ions in aqueous solution are related by the *ion-product constant for water ($K_w$).* Thus, you can describe the acidity or basicity of a solution by the hydronium-ion concentration. You often use the pH, which is the negative logarithm of the hydronium-ion concentration, as a measure of acidity.

## Operational Skills

1. **Identifying acid and base species** Given a proton-transfer reaction, label the acids and bases, and name the conjugate acid–base pairs. **(EXAMPLE 16.1)**
2. **Identifying Lewis acid and base species** Given a reaction involving the donation of an electron pair, identify the Lewis acid and the Lewis base. **(EXAMPLE 16.2)**
3. **Deciding whether reactants or products are favored in an acid–base reaction** Given an acid–base reaction and the relative strengths of acids (or bases), decide whether reactants or products are favored. **(EXAMPLE 16.3)**
4. **Calculating concentrations of $H_3O^+$ and $OH^-$ in solutions of a strong acid or base** Given the concentration of a strong acid or base, calculate the hydronium-ion and hydroxide-ion concentrations. **(EXAMPLE 16.4)**
5. **Calculating the pH from the hydronium-ion concentration, or vice versa** Given the hydronium-ion concentration, calculate the pH **(EXAMPLE 16.5)**; or given the pH, calculate the hydronium-ion concentration **(EXAMPLE 16.6)**.

# Review Questions

**16.1** Which of the following are strong acids? Which are weak acids? (a) $HC_2H_3O_2$; (b) $HClO$; (c) $HCl$; (d) $HNO_3$; (e) $HNO_2$; (f) $HCN$

**16.2** Describe any thermochemical (heat of reaction) evidence for the Arrhenius concept.

**16.3** Define an acid and a base according to the Brønsted–Lowry concept. Give an acid–base equation and identify each species as an acid or a base.

**16.4** What is meant by the conjugate acid of a base?

**16.5** Write an equation in which $H_2PO_3^-$ acts as an acid and another in which it acts as a base.

**16.6** Describe four ways in which the Brønsted–Lowry concept enlarges on the Arrhenius concept.

**16.7** Define an acid and a base according to the Lewis concept. Give a chemical equation to illustrate.

**16.8** Explain why an acid–base reaction favors the weaker acid.

**16.9** Give two important factors that determine the strength of an acid. How does an increase in each factor affect the acid strength?

**16.10** What is meant by the self-ionization of water? Write the expression for $K_w$. What is its value at 25°C?

**16.11** What is meant by the pH of a solution? Describe two ways of measuring pH.

**16.12** Which is more acidic, a solution having a pH of 4 or one having a pH of 5?

**16.13** What is the pH of a neutral solution at 37°C, where $K_w$ equals $2.5 \times 10^{-14}$?

**16.14** What is the sum of the pH and the pOH for a solution at 37°C, where $K_w$ equals $2.5 \times 10^{-14}$?

# Conceptual Problems

**16.15** Aqueous solutions of ammonia, $NH_3$, were once thought to be solutions of an ionic compound ammonium hydroxide, $NH_4OH$, in order to explain how the solutions could contain hydroxide ion. Using the Brønsted–Lowry concept, show how $NH_3$ yields hydroxide ion in aqueous solution without involving the species $NH_4OH$.

**16.16** Blood contains several substances that minimize changes in its acidity by reacting with either an acid or a base. One of these is the hydrogen phosphate ion, $HPO_4^{2-}$. Write one equation showing this species acting as a Brønsted–Lowry acid and another in which the species acts as a Brønsted–Lowry base.

**16.17**  Self-contained environments, such as that of a space station, require that the carbon dioxide exhaled by people be continuously removed. This can be done by passing the air over solid alkali hydroxide, in which carbon dioxide reacts with hydroxide ion. What ion is produced by the addition of $OH^-$ ion to $CO_2$? Use the Lewis concept to explain this.

**16.18**  Compare the structures of $HNO_2$ and $H_2CO_3$. Which would you expect to be the stronger acid? Explain your choice.

**16.19**  The value of the ion-product constant for water, $K_w$, increases with temperature. What will be the effect of lowering the temperature on the pH of pure water?

**16.20**  You make solutions of ammonia and sodium hydroxide by adding the same moles of each solute to equal volumes of water. Which solution would you expect to have the higher pH?

## Practice Problems

### Brønsted–Lowry Concept

**16.21**  Write the balanced reaction of hypobromous acid, HOBr, with water to form hydronium ion and the conjugate base of HOBr. Identify each species as either an acid or a base.

**16.22**  Write the balanced reaction of hydroxide ion with hydrofluoric acid, HF, to form fluoride ion and water. Identify each species as either an acid or a base.

**16.23**  Give the conjugate base to each of the following species regarded as acids.
a. $H_2PO_4^-$    b. $H_2Se$    c. $HNO_2$    d. $HAsO_4^{2-}$

**16.24**  Give the conjugate base to each of the following species regarded as acids.
a. $HSO_4^-$    b. $PH_4^+$    c. $HS^-$    d. $HOCl$

**16.25**  Give the conjugate acid to each of the following species regarded as bases.
a. $ClO^-$    b. $AsH_3$    c. $H_2PO_4^-$    c. $TeO_3^{2-}$

**16.26**  Give the conjugate acid to each of the following species regarded as bases.
a. $HS^-$    b. $NH_2^-$    c. $BrO_2^-$    d. $N_2H_4$

**16.27**  For the following reactions, label each species as an acid or a base. Indicate the species that are conjugates of one another.
a. $HSO_4^- + NH_3 \rightleftharpoons SO_4^{2-} + NH_4^+$
b. $HPO_4^{2-} + NH_4^+ \rightleftharpoons H_2PO_4^- + NH_3$
c. $Al(H_2O)_6^{3+} + H_2O \rightleftharpoons Al(H_2O)_5OH^{2+} + H_3O^+$
d. $SO_3^{2-} + NH_4^+ \rightleftharpoons HSO_3^- + NH_3$

**16.28**  For the following reactions, label each species as an acid or a base. Indicate the species that are conjugates of one another.
a. $H_2PO_4^- + HCO_3^- \rightleftharpoons HPO_4^{2-} + H_2CO_3$
b. $F^- + HSO_4^- \rightleftharpoons HF + SO_4^{2-}$
c. $HSO_4^- + H_2O \rightleftharpoons SO_4^{2-} + H_3O^+$
d. $H_2S + CN^- \rightleftharpoons HS^- + HCN$

### Lewis Acid–Base Concept

**16.29**  The following shows ball-and-stick models of the reactants in a Lewis acid–base reaction.

Write the complete equation for the reaction, including the product. Identify each reactant as a Lewis acid or a Lewis base.

**16.30**  The following shows ball-and-stick models of the reactants in a Lewis acid–base reaction.

Write the complete equation for the reaction, including the product. Identify each reactant as a Lewis acid or a Lewis base.

**16.31**  Complete each of the following equations. Then write the Lewis formulas of the reactants and products and identify each reactant as a Lewis acid or a Lewis base.
a. $I^- + I_2 \longrightarrow$
b. $GaBr_3 + Br^- \longrightarrow$

**16.32**  Complete each of the following equations. Then write the Lewis formulas of the reactants and products and identify each reactant as a Lewis acid or a Lewis base.
a. $BF_3 + F^- \longrightarrow$
b. $BBr_3 + AsH_3 \longrightarrow$

**16.33**  In the following reactions, identify each reactant as a Lewis acid or a Lewis base.
a. $Cr^{3+} + 6H_2O \longrightarrow Cr(H_2O)_6^{3+}$
b. $BF_3 + (C_2H_5)_2\overset{..}{O}: \longrightarrow F_3B:\overset{..}{O}(C_2H_5)_2$

**16.34** In the following reactions, label each reactant as a Lewis acid or a Lewis base.
a. $BeF_2 + 2F^- \longrightarrow BeF_4{}^{2-}$
b. $SnCl_4 + 2Cl^- \longrightarrow SnCl_6{}^{2-}$

**16.35** Natural gas frequently contains hydrogen sulfide, $H_2S$. $H_2S$ is removed from natural gas by passing it through aqueous ethanolamine, $HOCH_2CH_2NH_2$ (an ammonia derivative), which reacts with the hydrogen sulfide. Write the equation for the reaction. Identify each reactant as either a Lewis acid or a Lewis base. Explain how you arrived at your answer.

**16.36** Coal and other fossil fuels usually contain sulfur compounds that produce sulfur dioxide, $SO_2$, when burned. One possible way to remove the sulfur dioxide is to pass the combustion gases into a tower packed with calcium oxide, $CaO$. Write the equation for the reaction. Identify each reactant as either a Lewis acid or a Lewis base. Explain how you arrived at your answer.

## Acid and Base Strengths

**16.37** Complete the following equation. Using Table 16.2, predict whether you would expect the reaction to occur to any significant extent or whether the reaction is more likely to occur in the opposite direction.

$$HSO_4{}^-(aq) + ClO^-(aq) \longrightarrow$$

**16.38** Complete the following equation. Using Table 16.2, predict whether you would expect the reaction to occur to any significant extent or whether the reaction is more likely to occur in the opposite direction.

$$HCN(aq) + SO_4{}^{2-}(aq) \longrightarrow$$

**16.39** Use Table 16.2 to decide whether the species on the left or those on the right are favored by the reaction.
a. $NH_4{}^+ + H_2PO_4{}^- \rightleftharpoons NH_3 + H_3PO_4$
b. $HCN + HS^- \rightleftharpoons CN^- + H_2S$
c. $HCO_3{}^- + OH^- \rightleftharpoons CO_3{}^{2-} + H_2O$
d. $Al(H_2O)_6{}^{3+} + OH^- \rightleftharpoons Al(H_2O)_5OH^{2+} + H_2O$

**16.40** Use Table 16.2 to decide whether the species on the left or those on the right are favored by the reaction.
a. $NH_4{}^+ + CO_3{}^{2-} \rightleftharpoons NH_3 + HCO_3{}^-$
b. $HCO_3{}^- + H_2S \rightleftharpoons H_2CO_3 + HS^-$
c. $CN^- + H_2O \rightleftharpoons HCN + OH^-$
d. $CN^- + H_2CO_3 \rightleftharpoons HCN + HCO_3{}^-$

**16.41** In the following reaction of trichloroacetic acid, $HC_2Cl_3O_2$, with formate ion, $CHO_2{}^-$, the formation of trichloroacetate ion, $C_2Cl_3O_2{}^-$, and formic acid, $HCHO_2$ is favored.

$$HC_2Cl_3O_2 + CHO_2{}^- \longrightarrow C_2Cl_3O_2{}^- + HCHO_2$$

Which is the stronger acid, trichloroacetic acid or formic acid? Explain.

**16.42** In the following reaction of tetrafluoroboric acid, $HBF_4$, with the acetate ion, $C_2H_3O_2{}^-$, the formation of tetrafluoroborate ion, $BF_4{}^-$, and acetic acid, $HC_2H_3O_2$ is favored.

$$HBF_4 + C_2H_3O_2{}^- \longrightarrow BF_4{}^- + HC_2H_3O_2$$

Which is the weaker base, $BF_4{}^-$ or acetate ion?

**16.43** For each of the following pairs, give the stronger acid. Explain your answer.
a. $H_2S$, $HS^-$  b. $H_2SO_3$, $H_2SeO_3$  c. $HBr$, $H_2Se$
d. $HIO_4$, $HIO_3$  e. $H_2S$, $H_2O$

**16.44** Order each of the following pairs by acid strength, giving the weaker acid first. Explain your answer.
a. $HNO_3$, $HNO_2$  b. $HCO_3{}^-$, $H_2CO_3$  c. $H_2S$, $H_2Te$
d. $HCl$, $H_2S$  e. $H_3PO_4$, $H_3AsO_4$

## Solutions of a Strong Acid or Base

**16.45** What are the concentrations of $H_3O^+$ and $OH^-$ in each of the following?
a. 2.5 $M$ HCl  b. 0.35 $M$ NaOH
c. 0.065 $M$ Ca(OH)$_2$  d. 0.30 $M$ HNO$_3$

**16.46** What are the concentrations of $H_3O^+$ and $OH^-$ in each of the following?
a. 1.50 $M$ KOH  b. 0.25 $M$ Ba(OH)$_2$
c. 0.040 $M$ HClO$_4$  d. 0.300 $M$ HCl

**16.47** A solution of hydrochloric acid is 0.050 $M$ HCl. What is the hydronium-ion concentration at 25°C? What is the hydroxide-ion concentration at 25°C?

**16.48** A solution is 0.020 $M$ HNO$_3$ (nitric acid). What is the hydronium-ion concentration at 25°C? What is the hydroxide-ion concentration at 25°C?

**16.49** What are the hydronium-ion and the hydroxide-ion concentrations of a solution at 25°C that is 0.0050 $M$ strontium hydroxide, Sr(OH)$_2$?

**16.50** A saturated solution of magnesium hydroxide is $3.2 \times 10^{-4}$ $M$ Mg(OH)$_2$. What are the hydronium-ion and hydroxide-ion concentrations in the solution at 25°C?

**16.51** The following are solution concentrations. Indicate whether each solution is acidic, basic, or neutral.
a. $5 \times 10^{-6}$ $M$ $H_3O^+$  b. $5 \times 10^{-9}$ $M$ $OH^-$
c. $1 \times 10^{-7}$ $M$ $OH^-$  d. $2 \times 10^{-9}$ $M$ $H_3O^+$

**16.52** The following are solution concentrations. Indicate whether each solution is acidic, basic, or neutral.
a. $2 \times 10^{-11}$ $M$ $OH^-$  b. $2 \times 10^{-6}$ $M$ $H_3O^+$
c. $6 \times 10^{-10}$ $M$ $OH^-$  d. $6 \times 10^{-3}$ $M$ $H_3O^+$

**16.53** A shampoo solution at 25°C has a hydroxide-ion concentration of $1.5 \times 10^{-9}$ $M$. Is the solution acidic, neutral, or basic?

**16.54** An antiseptic solution at 25°C has a hydroxide-ion concentration of $8.4 \times 10^{-5}\ M$. Is the solution acidic, neutral, or basic?

## Calculations Involving pH

**16.55** Which of the following pH values indicate an acidic solution at 25°C? Which are basic and which are neutral?
a. 10.7  b. 1.9  c. 7.0  d. 4.3

**16.56** Which of the following pH values indicate an acidic solution at 25°C? Which are basic and which are neutral?
a. 1.9  b. 4.3  c. 8.7  d. 13.8

**16.57** For each of the following, state whether the solution at 25°C is acidic, neutral, or basic: (a) A beverage solution has a pH of 3.5. (b) A 0.50 $M$ solution of potassium bromide, KBr, has a pH of 7.0. (c) A 0.050 $M$ solution of pyridine, $C_5H_5N$, has a pH of 9.0. (d) A solution of iron(III) chloride has a pH of 5.5.

**16.58** For each of the following, state whether the solution at 25°C is acidic, neutral, or basic: (a) A 0.1 $M$ solution of trisodium phosphate, $Na_3PO_4$, has a pH of 12.0. (b) A 0.1 $M$ solution of calcium chloride, $CaCl_2$, has a pH of 7.0. (c) A 0.2 $M$ solution of copper(II) sulfate, $CuSO_4$, has a pH of 4.0. (d) A sample of rainwater has a pH of 5.7.

**16.59** Obtain the pH corresponding to the following hydronium-ion concentrations.
a. $1.0 \times 10^{-8}\ M$  b. $5.0 \times 10^{-12}\ M$
c. $7.5 \times 10^{-3}\ M$  d. $6.35 \times 10^{-9}\ M$

**16.60** Obtain the pH corresponding to the following hydronium-ion concentrations.
a. $1.0 \times 10^{-4}\ M$  b. $3.2 \times 10^{-10}\ M$
c. $2.3 \times 10^{-5}\ M$  d. $2.91 \times 10^{-11}\ M$

**16.61** A sample of vinegar has a hydronium-ion concentration of $7.5 \times 10^{-3}\ M$. What is the pH of the vinegar?

**16.62** Some lemon juice has a hydronium-ion concentration of $5.0 \times 10^{-3}\ M$. What is the pH of the lemon juice?

**16.63** Obtain the pH corresponding to the following hydroxide-ion concentrations.
a. $5.25 \times 10^{-9}\ M$  b. $8.3 \times 10^{-3}\ M$
c. $3.6 \times 10^{-12}\ M$  d. $2.1 \times 10^{-8}\ M$

**16.64** Obtain the pH corresponding to the following hydroxide-ion concentrations.
a. $4.83 \times 10^{-11}\ M$  b. $3.2 \times 10^{-5}\ M$
c. $2.7 \times 10^{-10}\ M$  d. $5.0 \times 10^{-4}\ M$

**16.65** A solution of washing soda (sodium carbonate, $Na_2CO_3$) has a hydroxide-ion concentration of 0.0040 $M$. What is the pH at 25°C?

**16.66** A solution of lye (sodium hydroxide, NaOH) has a hydroxide-ion concentration of 0.050 $M$. What is the pH at 25°C?

**16.67** The pH of a cup of coffee (at 25°C) was found to be 5.12. What is the hydronium-ion concentration?

**16.68** A wine was tested for acidity, and its pH was found to be 3.85 at 25°C. What is the hydronium-ion concentration?

**16.69** A detergent solution has a pH of 11.63 at 25°C. What is the hydroxide-ion concentration?

**16.70** Morphine is a narcotic that is used to relieve pain. A solution of morphine has a pH of 9.61 at 25°C. What is the hydroxide-ion concentration?

**16.71** A 1.00-L aqueous solution contained 5.80 g of sodium hydroxide, NaOH. What was the pH of the solution at 25°C?

**16.72** A 1.00-L aqueous solution contained 6.78 g of barium hydroxide, $Ba(OH)_2$. What was the pH of the solution at 25°C?

**16.73** A certain sample of rainwater gives a yellow color with methyl red and a yellow color with bromthymol blue. What is the approximate pH of the water? Is the rainwater acidic, neutral, or basic? (See Figure 16.8.)

**16.74** A drop of thymol blue gave a yellow color with a solution of aspirin. A sample of the same aspirin solution gave a yellow color with bromphenol blue. What was the pH of the solution? Was the solution acidic, neutral, or basic? (See Figure 16.8.)

## General Problems

**16.75** Identify each of the following as an acid or a base in terms of the Arrhenius concept. Give the chemical equation for the reaction of the substance with water, showing the origin of the acidity or basicity.
a. BaO  b. $H_2S$  c. $CH_3NH_2$  d. $SO_2$

**16.76** Which of the following substances are acids in terms of the Arrhenius concept? Which are bases? Show the acid or base character by using chemical equations.
a. $P_4O_{10}$  b. $K_2O$  c. $N_2H_4$  d. $H_2Se$

**16.77** Write a reaction for each of the following in which the species acts as a Brønsted acid. The equilibrium should favor the product side.
a. $H_2O_2$     b. $HCO_3^-$     c. $NH_4^+$     d. $H_2PO_4^-$

**16.78** Write a reaction for each of the following in which the species acts as a Brønsted base. The equilibrium should favor the product side.
a. $H_2O$     b. $HCO_3^-$     c. $NH_3$     d. $H_2PO_4^-$

**16.79** For each of the following, write the complete chemical equation for the acid–base reaction that occurs. Describe each using Brønsted language (if appropriate) and then using Lewis language (show electron-dot formulas).
a. The $ClO^-$ ion reacts with water.
b. The reaction of $NH_4^+$ and $NH_2^-$ in liquid ammonia to produce $NH_3$.

**16.80** For each of the following, write the complete chemical equation for the acid–base reaction that occurs. Describe each using Brønsted language (if appropriate) and then using Lewis language (show electron-dot formulas).
a. The $HS^-$ ion reacts in water to produce $H_2S$.
b. Cyanide ion, $CN^-$, reacts with $Fe^{3+}$.

**16.81** Complete the following proton-transfer reaction. Decide on the basis of relative acid strengths whether the reaction is more likely to go in the direction written or in the opposite direction.

$$HNO_2 + F^- \rightleftharpoons$$

**16.82** Complete the following reaction. Decide on the basis of relative acid strengths whether the reaction is more likely to go in the direction written or in the opposite direction.

$$HS^- + HCN \rightleftharpoons$$

**16.83** List the following compounds in order of increasing acid strength: $HBr$, $H_2Se$, $H_2S$.

**16.84** List the following compounds in order of increasing acid strength: $HBrO_2$, $HClO_2$, $HBrO$.

**16.85** A solution is 0.25 $M$ KOH. What are the concentrations of $H_3O^+$ and $OH^-$ in this solution?

**16.86** A solution is 0.35 $M$ $Sr(OH)_2$. What are the concentrations of $H_3O^+$ and $OH^-$ in this solution?

**16.87** A wine has a hydronium-ion concentration equal to $1.5 \times 10^{-3}$ $M$. What is the pH of this wine?

**16.88** A sample of lemon juice has a hydronium-ion concentration equal to $2.5 \times 10^{-2}$ $M$. What is the pH of this sample?

**16.89** A sample of apple cider has a pH of 3.15. What is the hydroxide-ion concentration of this solution?

**16.90** A sample of grape juice has a pH of 4.05. What is the hydroxide-ion concentration of this solution?

**16.91** A 2.500-g sample of a mixture of sodium hydrogen carbonate and potassium chloride is dissolved in 25.00 mL of 0.437 $M$ $H_2SO_4$. Some acid remains after treatment of the sample.
a. Write both the net ionic and the molecular equations for the complete reaction of sodium hydrogen carbonate with sulfuric acid.
b. If 35.4-mL of 0.108 $M$ NaOH were required to titrate the excess sulfuric acid, how many moles of sodium hydrogen carbonate were present in the original sample?
c. What is the percent composition of the original sample?

**16.92** A 2.500-g sample of a mixture of sodium carbonate and sodium chloride is dissolved in 25.00 mL of 0.798 $M$ HCl. Some acid remains after the treatment of the sample.
a. Write the net ionic equation for the complete reaction of sodium carbonate with hydrochloric acid.
b. If 28.7 mL of 0.108 $M$ NaOH were required to titrate the excess hydrochloric acid, how many moles of sodium carbonate were present in the original sample?
c. What is the percent composition of the original sample?

**16.93** The bicarbonate ion has the ability to act as an acid in the presence of a base and as a base in the presence of an acid, so it is said to be amphiprotic. Illustrate this behavior with water by writing Brønsted–Lowry acid–base reactions. Also illustrate this property by selecting a common strong acid and base to react with the bicarbonate ion.

**16.94** The dihydrogen phosphate ion has the ability to act as an acid in the presence of a base and as a base in the presence of an acid. What is this property called? Illustrate this behavior with water by writing Brønsted–Lowry acid–base reactions. Also illustrate this property by selecting a common acid and base to react with the dihydrogen phosphate ion.

**16.95** The hydride ion does not exist in water because it has a greater attraction for the hydronium ion than the hydroxide ion does. Write the equation for the reaction that occurs when calcium hydride is added to water. Which is the stronger base, the hydride ion or the hydroxide ion? What is the meaning of the statement that the strongest base that can exist in water is the hydroxide ion?

**16.96** The nitride ion and the amide ion, $NH_2^-$, have greater attractions for the hydronium ion than the hydroxide ion does. Write the equations for the reactions that occur when calcium nitride and sodium amide are added to water. Which is the stronger base, the nitride ion or the amide ion? Why? What is the meaning of the statement that the hydroxide ion is the strongest base that can exist in water?

**16.97** Pure liquid hydrogen fluoride ionizes in a way similar to that of water.

a. Write the equilibrium reaction for the self-ionization of liquid hydrogen fluoride.
b. Will sodium fluoride be an acid or a base in liquid hydrogen fluoride? Why?
c. Perchloric acid is a strong acid in liquid hydrogen fluoride. Write the chemical equation for the ionization reaction. What is the conjugate acid in this medium?

**16.98**   Pure liquid ammonia ionizes in a way similar to that of water.
a. Write the equilibrium reaction of liquid ammonia.
b. Will sodium amide, $NaNH_2$, be an acid or a base in liquid ammonia? Why?
c. Ammonium chloride can be used to titrate sodium amide in liquid ammonia. Write the chemical equation for the titration of sodium amide with ammonium chloride in liquid ammonia.

**16.99**   A solution contains 4.25 g of ammonia per 250.0 mL of solution. Electrical conductivity measurements at 25°C show that 0.42% of the ammonia has reacted with water. Write the equation for this reaction and calculate the pH of the solution.

**16.100**   A solution contains 0.675 g of ethylamine, $C_2H_5NH_2$, per 100.0 mL of solution. Electrical conductivity measurements at 20°C show that 0.98% of the ethylamine has reacted with water. Write the equation for this reaction. Calculate the pH of the solution.

## Cumulative-Skills Problems

**16.101**   Phosphorous acid, $H_3PO_3$, and phosphoric acid, $H_3PO_4$, have approximately the same acid strengths. From this information, and noting the possibility that one or more hydrogen atoms may be directly bonded to the phosphorus atom, draw the structural formula of phosphorous acid. How many grams of sodium hydroxide would be required to completely neutralize 1.00 g of this acid?

**16.102**   Hypophosphorous acid, $H_3PO_2$, and phosphoric acid, $H_3PO_4$, have approximately the same acid strengths. From this information, and noting the possibility that one or more hydrogen atoms may be directly bonded to the phosphorus atom, draw the structural formula of hypophosphorous acid. How many grams of sodium hydroxide would be required to completely neutralize 1.00 g of this acid?

**16.103**   Boron trifluoride, $BF_3$, and ammonia, $NH_3$, react to produce $BF_3 \cdot NH_3$. A coordinate covalent bond is formed between the boron atom on $BF_3$ and the nitrogen atom on $NH_3$. Write the equation for this reaction, using Lewis electron-dot formulas. Label the Lewis acid and the Lewis base. Determine how many grams of $BF_3 \cdot NH_3$ are formed when 10.0 g $BF_3$ and 10.0 g $NH_3$ are placed in a reaction vessel, assuming that the reaction goes to completion.

**16.104**   Boron trifluoride, $BF_3$, and diethyl ether, $(C_2H_5)_2O$, react to produce a compound with the formula $BF_3 \cdot (C_2H_5)_2O$. A coordinate covalent bond is formed between the boron atom on $BF_3$ and the oxygen atom on $(C_2H_5)_2O$. Write the equation for this reaction, using Lewis electron-dot formulas. Label the Lewis acid and the Lewis base. Determine how many grams of $BF_3 \cdot (C_2H_5)_2O$ are formed when 10.0 g $BF_3$ and 20.0 g $(C_2H_5)_2O$ are placed in a reaction vessel, assuming that the reaction goes to completion.

## Media Activities

**16.105**   Work through the Acids and Bases *Problem Solving* activity on the student web site. Complete the following reactions and identify the acid, base, conjugate acid, and conjugate base.
a. $HIO_2 + OH^- \longrightarrow$
b. $NH_3 + H_2O \longrightarrow$
c. $HCN + CO_3{}^{2-} \longrightarrow$

**16.106**   Run the Self-Ionization of Water to Form $H^+$ and $OH^-$ at Equilibrium *Visualization* on the student CD.
a. Write the reaction for the self-ionization of water.
b. How does the addition of acid to water affect the concentration of $OH^-$?
c. Calculate the number of $OH^-$ anions present in 1.0 L of water at 25°C.

**16.107**   Work through the pH *Problem Solving* activity on the student web site. What volumes of 12.0 *M* HCl and water would be used to prepare 2.00 L of solution with a pH of 1.30?

**16.108**   Watch the Addition of HCl to Water *Visualization* on the student CD.
a. After the addition of the HCl to the water, how do the concentrations of $Cl^-$ and $H_3O^+$ compare?
b. Is the $Cl^-$ in the solution depicted at the end of the *Visualization* a strong base?
c. Would you expect the solution formed at the end of the *Visualization* to have a pH greater or less than 7? To calculate the pH of this solution, what information would you need?
d. What would be different about the solution if the acid was $HNO_3$ instead of HCl? What would be the same?

**16.109**   For review, take the ACE practice test for Chapter 16 on the student web site.

Limestone building ornaments (gargoyles) damaged by acid rain.

# Acid–Base Equilibria

**M**any well-known substances are weak acids or bases. The following are weak acids: aspirin (acetylsalicylic acid, a headache remedy), phenobarbital (a sedative), saccharin (a sweetener), and niacin (nicotinic acid, a B vitamin). That these are weak acids means that their reactions with water do not go to completion. To discuss such acid–base reactions, you need to look at the equilibria involved and be able to calculate the concentrations of species in a reaction mixture.

Consider, for example, how you could answer the following questions: What is the hydronium-ion concentration of 0.10 *M* niacin (nicotinic acid)? What is the hydronium-ion concentration of the solution obtained by dissolving one 5.00-grain tablet of aspirin (acetylsalicylic acid) in 0.500 L of water? If these were solutions of strong acids, the calculations would be simple; 0.10 *M* monoprotic acid would yield 0.10 *M* $H_3O^+$ ion. However, because niacin is a weak monoprotic acid, the $H_3O^+$ concentration is less than 0.10 *M*. To find the concentration, you need the equilibrium constant for the reaction involved, and you need to solve an equilibrium problem.

## CONTENTS

**Solutions of a Weak Acid or Base**

**Solutions of a Weak Acid or Base with Another Solute**

**723**

A similar process is involved in finding the hydronium-ion concentration of $0.10\,M$ sodium nicotinate (sodium salt of niacin). In this case, you need to look at the acid–base equilibrium of the nicotinate ion. Let us see how you answer questions of this sort.

# Solutions of a Weak Acid or Base

▶ **Experiments in General Chemistry, 17A, "Equilibria with Weak Acids and Weak Bases."**

The simplest acid–base equilibria are those in which a single acid or base solute reacts with water. We will look first at solutions of weak acids, then at solutions of weak bases. Then we will consider solutions of salts, which can have acidic or basic properties as a result of reactions of their ions with water.

## 17.1 Acid-Ionization Equilibria

An acid reacts with water to produce hydronium ion (hydrogen ion) and the conjugate base ion. The process is called *acid ionization* or *acid dissociation*. Consider an aqueous solution of the weak acid acetic acid, $HC_2H_3O_2$, the sour constituent of vinegar (Figure 17.1). When the acid is added to water, it reacts according to the equation

$$HC_2H_3O_2(aq) + H_2O(l) \rightleftharpoons H_3O^+(aq) + C_2H_3O_2^-(aq)$$

Because acetic acid is a weak electrolyte, the acid *ionizes* to a small extent in water (about 1% or less, depending on the concentration of acid).

For a strong acid, which ionizes completely in solution, the concentrations of ions are determined by the stoichiometry of the reaction from the initial concentration of acid. However, for a weak acid such as acetic acid, the concentrations of ions in solution are determined from the **acid-ionization constant** (also called the acid-dissociation constant), which is *the equilibrium constant for the ionization of a weak acid.*

To find the acid-ionization constant, write HA for the general formula of a weak, monoprotic acid. The acid-ionization equilibrium in aqueous solution is

$$HA(aq) + H_2O(l) \rightleftharpoons H_3O^+(aq) + A^-(aq)$$

The corresponding equilibrium constant is

$$K_c = \frac{[H_3O^+][A^-]}{[HA][H_2O]}$$

Assuming that this is a dilute solution and that the reaction occurs to only a small extent, the concentration of water will be nearly constant. Rearranging this equation gives

$$K_a = [H_2O]K_c = \frac{[H_3O^+][A^-]}{[HA]}$$

Thus, $K_a$, the acid-ionization constant, equals the constant $[H_2O]K_c$. ◀

$$K_a = \frac{[H_3O^+][A^-]}{[HA]}$$

**Figure 17.1**
**The structure of acetic acid.** *Top:* Structural formula. Only the hydrogen atom (in color) attached to the oxygen atom is acidic. The hydrogen atoms attached to carbon atoms are not acidic. The acidic group

$$-\overset{\overset{\displaystyle O}{\|}}{C}-O-H$$

(often written —COOH and called the carboxylic acid group) occurs in many organic acids. *Bottom:* Model of acetic acid.

▶ **In terms of the thermodynamic equilibrium constant, the activity of $H_2O$ is nearly constant and essentially 1, so it does not appear explicitly in the equilibrium constant.**

**Figure 17.2**
**The structure of nicotinic acid.** *Top:* The structural formula of nicotinic acid (niacin); note the acidic hydrogen atom (in color). Nicotinic acid was first prepared by oxidizing nicotine, a basic substance present in tobacco. *Bottom:* Molecular model of nicotinic acid.

## Experimental Determination of $K_a$

The ionization constant for a weak acid is usually determined experimentally by one of two methods. In one method, the electrical conductivity or some colligative property of a solution of the acid is measured to obtain its degree of ionization. The **degree of ionization** of a weak electrolyte is *the fraction of molecules that react with water to give ions.* This may also be expressed as a percentage, giving the *percent ionization.* In the other method, the pH of a solution of the weak acid is determined. From the pH one finds the concentration of $H_3O^+$ ion and then the concentrations of other ions. The following example shows how to calculate $K_a$ from the pH of a solution of nicotinic acid (Figure 17.2). It also shows how to calculate the degree of ionization and the percent ionization.

---

### EXAMPLE 17.1

## Determining $K_a$ from the Solution pH

Nicotinic acid (niacin) is a monoprotic acid with the formula $HC_6H_4NO_2$. A solution that is 0.012 $M$ in nicotinic acid has a pH of 3.39 at 25°C. What is the acid-ionization constant, $K_a$, for this acid at 25°C? What is the degree of ionization of nicotinic acid in this solution?

 **PROBLEM STRATEGY**

It is important to realize that when we say the solution is 0.012 $M$, this refers to *how the solution is prepared.* The solution is made up by adding 0.012 mol of substance to enough water to give a liter of solution. Once the solution is prepared, some molecules ionize, so the actual concentration is somewhat less than 0.012 $M$. To solve for $K_a$, you follow the three steps outlined in Chapter 15 for solving equilibrium problems.

**SOLUTION**

**STEP 1** Abbreviate the formula for nicotinic acid as HNic. Then 1 L of solution contains 0.012 mol HNic and 0 mol Nic$^-$, the acid anion, before ionization. The $H_3O^+$ concentration at the start is that from the self-ionization of water. It is usually much smaller than that obtained from the acid (unless the solution is extremely dilute or $K_a$ is quite small), so you can write $[H_3O^+]$ = ~0 (meaning approximately zero). If $x$ mol HNic ionizes, $x$ mol each of $H_3O^+$ and Nic$^-$ is formed, leaving $(0.012 - x)$ mol HNic in solution. You can summarize the situation as follows:

| Concentration ($M$) | $HNic(aq) + H_2O(l) \rightleftharpoons$ | $H_3O^+(aq) +$ | $Nic^-(aq)$ |
|---|---|---|---|
| Starting | 0.012 | ~0 | 0 |
| Change | $-x$ | $+x$ | $+x$ |
| Equilibrium | $0.012 - x$ | $x$ | $x$ |

Thus, the molar concentrations of HNic, $H_3O^+$, and Nic$^-$ at equilibrium are $(0.012 - x)$, $x$, and $x$, respectively.

**STEP 2** The equilibrium-constant equation is

$$K_a = \frac{[H_3O^+][Nic^-]}{[HNic]}$$

When you substitute the expressions for the equilibrium concentrations, you get

$$K_a = \frac{x^2}{(0.012 - x)}$$

**STEP 3** The value of $x$ equals the numerical value of the molar hydronium-ion concentration, and can be obtained from the pH of the solution.

$$x = [H_3O^+] = \text{antilog}(-pH) = \text{antilog}(-3.39) = 4.1 \times 10^{-4} = 0.00041$$

You can substitute this value of $x$ into the equation obtained in Step 2. Note first, however, that

$$0.012 - x = 0.012 - 0.00041 = 0.01159 \simeq 0.012 \qquad \text{(to two significant figures)}$$

This means that the concentration of undissolved acid is equal to the original concentration of the acid within the precision of the data. (We will make use of this type of observation in later problem solving.) Therefore,

$$K_a = \frac{x^2}{(0.012 - x)} \simeq \frac{x^2}{0.012} \simeq \frac{(0.00041)^2}{0.012} = \mathbf{1.4 \times 10^{-5}}$$

To obtain the degree of ionization, note that $x$ mol out of 0.012 mol of nicotinic acid ionizes. Hence,

$$\text{Degree of ionization} = \frac{x}{0.012} = \frac{0.00041}{0.012} = \mathbf{0.034}$$

The percent ionization is obtained by multiplying this by 100, which gives 3.4%.

---

### EXERCISE 17.1

Lactic acid, $HC_3H_5O_3$, is found in sour milk, where it is produced by the action of lactobacilli on lactose, or milk sugar. A 0.025 $M$ solution of lactic acid has a pH of 2.75. What is the ionization constant $K_a$ for this acid? What is the degree of ionization? ◄

---

See Problems 17.27 and 17.28.

▶ **Lactic acid has the structural formula**

**and the molecular model**

Table 17.1 lists acid-ionization constants for various weak acids. The weakest acids have the smallest values of $K_a$.

## Calculations with $K_a$

Once you know the value of $K_a$ for an acid HA, you can calculate the equilibrium concentrations of species HA, $A^-$, and $H_3O^+$ for solutions of different molarities. The general method for doing this was discussed in Chapter 15. Here we illustrate the use of a simplifying approximation that can often be used for weak acids. In the next example, we will look at an amplification of the first question posed in the chapter opening.

---

### EXAMPLE 17.2

## Calculating Concentrations of Species in a Weak Acid Solution Using $K_a$ (Approximation Method)

What are the concentrations of nicotinic acid, hydrogen ion, and nicotinate ion in a solution of 0.10 $M$ nicotinic acid, $HC_6H_4NO_2$, at 25°C? What is the pH of the solution? What is the degree of ionization of nicotinic acid? The acid-ionization constant, $K_a$, was determined in the previous example to be $1.4 \times 10^{-5}$.

**Table 17.1**
**Acid-Ionization Constants at 25°C***

| Substance | Formula | $K_a$ |
|---|---|---|
| Acetic acid | $HC_2H_3O_2$ | $1.7 \times 10^{-5}$ |
| Benzoic acid | $HC_7H_5O_2$ | $6.3 \times 10^{-5}$ |
| Boric acid | $H_3BO_3$ | $5.9 \times 10^{-10}$ |
| Carbonic acid | $H_2CO_3$ | $4.3 \times 10^{-7}$ |
| | $HCO_3^-$ | $4.8 \times 10^{-11}$ |
| Cyanic acid | HOCN | $3.5 \times 10^{-4}$ |
| Formic acid | $HCHO_2$ | $1.7 \times 10^{-4}$ |
| Hydrocyanic acid | HCN | $4.9 \times 10^{-10}$ |
| Hydrofluoric acid | HF | $6.8 \times 10^{-4}$ |
| Hydrogen sulfate ion | $HSO_4^-$ | $1.1 \times 10^{-2}$ |
| Hydrogen sulfide | $H_2S$ | $8.9 \times 10^{-8}$ |
| | $HS^-$ | $1.2 \times 10^{-13\dagger}$ |
| Hypochlorous acid | HClO | $3.5 \times 10^{-8}$ |
| Nitrous acid | $HNO_2$ | $4.5 \times 10^{-4}$ |
| Oxalic acid | $H_2C_2O_4$ | $5.6 \times 10^{-2}$ |
| | $HC_2O_4^-$ | $5.1 \times 10^{-5}$ |
| Phosphoric acid | $H_3PO_4$ | $6.9 \times 10^{-3}$ |
| | $H_2PO_4^-$ | $6.2 \times 10^{-8}$ |
| | $HPO_4^{2-}$ | $4.8 \times 10^{-13}$ |
| Phosphorous acid | $H_2PHO_3$ | $1.6 \times 10^{-2}$ |
| | $HPHO_3^-$ | $7 \times 10^{-7}$ |
| Propionic acid | $HC_3H_5O_2$ | $1.3 \times 10^{-5}$ |
| Pyruvic acid | $HC_3H_3O_3$ | $1.4 \times 10^{-4}$ |
| Sulfurous acid | $H_2SO_3$ | $1.3 \times 10^{-2}$ |
| | $HSO_3^-$ | $6.3 \times 10^{-8}$ |

*The ionization constants for polyprotic acids are for successive ionizations. For example, for $H_3PO_4$, the equilibrium is $H_3PO_4 + H_2O \rightleftharpoons H_3O^+ + H_2PO_4^-$. For $H_2PO_4^-$, the equilibrium is $H_2PO_4^- + H_2O \rightleftharpoons H_3O^+ + HPO_4^{2-}$.
†This value is in doubt. Some evidence suggests that it is about $10^{-19}$. See R.J. Myers, *J. Chem. Educ.*, **63**, 687 (1986).

 **PROBLEM STRATEGY**

You follow the three steps for solving equilibrium problems that were introduced in Example 15.7. In the last step, you solve the equilibrium-constant equation for the equilibrium concentrations. The resulting equation is quadratic, but because the equilibrium concentration of a weak acid is usually negligibly different from its starting value, the equation simplifies so that it involves only the square of the unknown, which is easily solved by taking the square root. (You will need to check that this assumption is valid.)

## SOLUTION

**STEP 1** At the start (before ionization), the concentration of nicotinic acid, HNic, is 0.10 $M$ and that of its conjugate base, $Nic^-$, is 0. The concentration of $H_3O^+$ is essentially zero ($\sim 0$), assuming that the contribution from the self-ionization of water can be neglected. In 1 L of solution, the nicotinic acid ionizes to give $x$ mol $H_3O^+$ and $x$ mol $Nic^-$, leaving $(0.10 - x)$ mol of nicotinic acid. These data are summarized in the following table:

| Concentration ($M$) | $HNic(aq) + H_2O(l) \rightleftharpoons$ | $H_3O^+(aq) +$ | $Nic^-(aq)$ |
|---|---|---|---|
| Starting | 0.10 | $\sim 0$ | 0 |
| Change | $-x$ | $+x$ | $+x$ |
| Equilibrium | $0.10 - x$ | $x$ | $x$ |

The equilibrium concentrations of HNic, $H_3O^+$, and $Nic^-$ are $(0.10 - x)$, $x$, and $x$, respectively.

**STEP 2** Now substitute these concentrations and the value of $K_a$ into the equilibrium-constant equation for acid ionization:

$$\frac{[H_3O^+][Nic^-]}{[HNic]} = K_a$$

You get

$$\frac{x^2}{(0.10 - x)} = 1.4 \times 10^{-5}$$

**STEP 3** Now solve this equation for $x$. This is actually a quadratic equation, but it can be simplified so that the value of $x$ is easily found. Because the acid-ionization constant is small, the value of $x$ is small. Assume that $x$ is much smaller than 0.10, so that

$$0.10 - x \simeq 0.10$$

You will need to check that this assumption is valid after you obtain a value for $x$. The equilibrium-constant equation becomes

$$\frac{x^2}{0.10} \simeq 1.4 \times 10^{-5}$$

or

$$x^2 \simeq 1.4 \times 10^{-5} \times 0.10 = 1.4 \times 10^{-6}$$

Hence,

$$x \simeq 1.2 \times 10^{-3} = 0.0012$$

At this point, you should check to make sure that the assumption that $0.10 - x \simeq 0.10$ is valid. You substitute the value obtained for $x$ into $0.10 - x$.

$$0.10 - x = 0.10 - 0.0012 = 0.10 \quad \text{(to two significant figures)}$$

The assumption is indeed valid.

Now you can substitute the value of $x$ into the last line of the table written in Step 1 to find the concentrations of species. The concentrations of nicotinic acid, hydronium ion, and nicotinate ion are **0.10 *M*, 0.0012 *M*, and 0.0012 *M*,** respectively.

The pH of the solution is

$$pH = -\log [H_3O^+] = -\log (0.0012) = \mathbf{2.92}$$

The degree of ionization equals the amount per liter of nicotinic acid that ionizes ($x = 0.0012$) divided by the total amount per liter of nicotinic acid initially present (0.10). Thus, the degree of ionization is $0.0012/0.10 = \mathbf{0.012}$.

---

**Problem Solving:**
**The pH of Weak Acids**

### EXERCISE 17.2

What are the concentrations of hydrogen ion and acetate ion in a solution of 0.10 *M* acetic acid, $HC_2H_3O_2$? What is the pH of the solution? What is the degree of ionization? See Table 17.1 for the value of $K_a$.

---

See Problems 17.29 and 17.30.

In Example 17.2, the degree of ionization of nicotinic acid (0.012) is relatively small in a 0.10 *M* solution. That is, only 1.2% of the molecules ionize. It is the small value of the degree of ionization that allows you to neglect $x$ in the term $0.10 - x$ and thereby simplify the calculation.

The degree of ionization of a weak acid depends on both $K_a$ and the concentration of the acid solution. For a given concentration, the larger the $K_a$, the greater is the degree of ionization. For a given value of $K_a$, however, the more dilute the solution, the greater is the degree of ionization. Figure 17.3 shows how the percent ionization (degree of ionization $\times$ 100) varies with the concentration of solution.

How do you know when you can use the simplifying assumption of Example 17.2, where you neglected $x$ in the denominator of the equilibrium equation? *It can be shown that this simplifying assumption gives an error of less than 5% if the concentration of acid, $C_a$, divided by $K_a$ equals 100 or more.* For example, for an acid concentration of $10^{-2}$ *M* and $K_a$ of $10^{-5}$, $C_a/K_a = 10^{-2}/10^{-5} = 10^3$, so the assumption is valid. It would not be valid, however, if the concentration of the same acid were $10^{-4}$ *M*.

**Figure 17.3**
**Variation of percent ionization of a weak acid with concentration.**
On this curve for nicotinic acid, note that the percent ionization is greatest for the most dilute solutions.

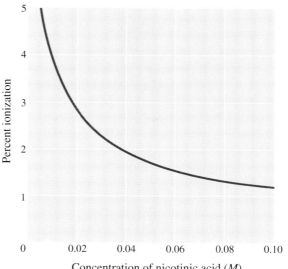

Concentration of nicotinic acid (*M*)

If the simplifying assumption of Example 17.2 is not valid, you can solve the equilibrium equation exactly by using the *quadratic formula*. We illustrate this in the next example with a solution of aspirin, $HC_9H_7O_4$, a common headache remedy. (Figure 17.4 shows the structural formula.) ◀

▶ You might find it helpful to compare the results of some equilibrium calculations using the simplifying assumption with exact results from the "Equilibrium Calculator (Equilib)," by Robert Allendorfer, JCE: Software, Madison, WI.

**Figure 17.4**
**The structure of aspirin.**
*Top:* Structural formula of aspirin (acetylsalicylic acid), a well-known headache remedy. Note the acidic hydrogen atom (in color). *Bottom:* Molecular model of aspirin.

## EXAMPLE 17.3

## Calculating Concentrations of Species in a Weak Acid Solution Using $K_a$ (Quadratic Formula)

What is the pH at 25°C of the solution obtained by dissolving a 5.00-grain tablet of aspirin (acetylsalicylic acid) in 0.500 L of water? The tablet contains 5.00 grains, or 0.325 g, of acetylsalicylic acid, $HC_9H_7O_4$. The acid is monoprotic, and $K_a$ equals $3.3 \times 10^{-4}$ at 25°C.

### PROBLEM STRATEGY

You follow the three steps for equilibrium problems. However, the approximation used in the previous example to solve the equilibrium-constant equation is not valid here. Verify this by calculating $C_a/K_a$; it is not greater than 100. You solve the equation exactly using the quadratic formula.

### SOLUTION

The molar mass of $HC_9H_7O_4$ is 180.2 g. From this you find that an aspirin tablet contains 0.00180 mol of the acid. Hence, the concentration of the aspirin solution is 0.00180 mol/0.500 L, or 0.0036 M. (Retain two significant figures, the number of significant figures in $K_a$.) $C_a/K_a = 0.0036/3.3 \times 10^{-4} = 11$, which is less than 100, so we must solve the equilibrium equation exactly.

**STEP 1** Abbreviate the formula for acetylsalicylic acid as HAcs and let $x$ be the amount of $H_3O^+$ ion formed in 1 L of solution. The amount of acetylsalicylate ion formed is also $x$ mol; the amount of nonionized acetylsalicylic acid is $(0.0036 - x)$ mol. These data are summarized in the following table:

| Concentration (M) | HAcs(aq) + H₂O(l) ⇌ H₃O⁺(aq) + Acs⁻(aq) | | |
|---|---|---|---|
| Starting | 0.0036 | ~0 | 0 |
| Change | −x | +x | +x |
| Equilibrium | 0.0036 − x | x | x |

**STEP 2** If you substitute the equilibrium concentrations and the value of $K_a$ into the equilibrium-constant equation, you get

$$\frac{[H_3O^+][Acs^-]}{[HAcs]} = K_a$$

$$\frac{x^2}{0.0036 - x} = 3.3 \times 10^{-4}$$

**STEP 3** You can solve the equation exactly using the quadratic formula. Rearrange the preceding equation to put it in the form $ax^2 + bx + c = 0$. You get

$$x^2 = (0.0036 - x) \times 3.3 \times 10^{-4} = 1.2 \times 10^{-6} - 3.3 \times 10^{-4}x$$

or

$$x^2 + 3.3 \times 10^{-4}x - 1.2 \times 10^{-6} = 0$$

Now substitute into the quadratic formula.

$$x = \frac{-b \pm \sqrt{b^2 - 4ac}}{2a}$$

$$= \frac{-3.3 \times 10^{-4} \pm \sqrt{(3.3 \times 10^{-4})^2 + 4(1.2 \times 10^{-6})}}{2}$$

$$= \frac{-3.3 \times 10^{-4} \pm 2.2 \times 10^{-3}}{2}$$

▶ Pyruvic acid is an important biochemical intermediate formed during the breakdown of glucose by a cell. In muscle tissue, glucose is broken down to pyruvic acid, which is eventually oxidized to carbon dioxide and water. The structural formula of pyruvic acid is

and the molecular model is

If you take the lower sign in $\pm$, you will get a negative value for $x$. But $x$ equals the concentration of $H_3O^+$ ion, so it must be positive. Therefore you ignore the negative solution. Taking the upper sign, you get

$$x = [H_3O^+] = 9.4 \times 10^{-4}$$

Now you can calculate the pH.

$$pH = -\log [H_3O^+] = -\log(9.4 \times 10^{-4}) = \textbf{3.03}$$

### EXERCISE 17.3

What is the pH of an aqueous solution that is 0.0030 $M$ pyruvic acid, $HC_3H_3O_3$? ◀

See Problems 17.35 and 17.36.

### CONCEPT CHECK 17.1

You have prepared dilute solutions of equal molar concentrations of $HC_2H_3O_2$ (acetic acid), $HNO_2$, HF, and HCN. Rank the solutions from the highest pH to the lowest pH. (Refer to Table 17.1.)

## 17.2  Polyprotic Acids

In the preceding section, we dealt only with acids releasing one $H_3O^+$ ion or proton. Some acids, however, have two or more such protons; these acids are called *polyprotic acids*. Sulfuric acid, for example, can lose two protons in aqueous solution. One proton is lost completely to form $H_3O^+$ (sulfuric acid is a strong acid).

$$H_2SO_4(aq) + H_2O(l) \longrightarrow H_3O^+(aq) + HSO_4^-(aq)$$

The hydrogen sulfate ion, $HSO_4^-$, may then lose the other proton. In this case, an equilibrium exists.

$$HSO_4^-(aq) + H_2O(l) \rightleftharpoons H_3O^+(aq) + SO_4^{2-}(aq)$$

For a weak diprotic acid like carbonic acid, $H_2CO_3$, there are two simultaneous equilibria to consider.

$$H_2CO_3(aq) + H_2O(l) \rightleftharpoons H_3O^+(aq) + HCO_3^-(aq)$$

$$HCO_3^-(aq) + H_2O(l) \rightleftharpoons H_3O^+(aq) + CO_3^{2-}(aq)$$

Carbonic acid is in equilibrium with the hydrogen carbonate ion, $HCO_3^-$, which is in turn in equilibrium with the carbonate ion, $CO_3^{2-}$. Each equilibrium has an associated acid-ionization constant. For the loss of the first proton,

$$K_{a1} = \frac{[H_3O^+][HCO_3^-]}{[H_2CO_3]} = 4.3 \times 10^{-7}$$

and for the loss of the second proton,

$$K_{a2} = \frac{[H_3O^+][CO_3^{2-}]}{[HCO_3^-]} = 4.8 \times 10^{-11}$$

Note that $K_{a1}$ for carbonic acid is much larger than $K_{a2}$ (by a factor of about $1 \times 10^4$). This indicates that carbonic acid loses the first proton more easily than the second one, because the first proton separates from an ion of single negative charge, whereas the second proton separates from an ion of double negative charge. This double negative charge strongly attracts the proton back to it.

> In general, the second ionization constant, $K_{a2}$, of a polyprotic acid is much smaller than the first ionization constant, $K_{a1}$. In the case of a triprotic acid, the third ionization constant, $K_{a3}$, is much smaller than the second one, $K_{a2}$.

See the values for phosphoric acid, $H_3PO_4$, in Table 17.1.

Calculating the concentrations of various species in a solution of a polyprotic acid might appear complicated, because several equilibria occur at once. However, reasonable assumptions can be made that simplify the calculation, as we show in the next example.

## EXAMPLE 17.4

### Calculating Concentrations of Species in a Solution of a Diprotic Acid

Ascorbic acid (vitamin C) is a diprotic acid, $H_2C_6H_6O_6$ (Figure 17.5). What is the pH of a 0.10 $M$ solution? What is the concentration of ascorbate ion, $C_6H_6O_6^{2-}$? The acid ionization constants are $K_{a1} = 7.9 \times 10^{-5}$ and $K_{a2} = 1.6 \times 10^{-12}$.

 **PROBLEM STRATEGY**

Diprotic acids have two ionization constants, one for the loss of each proton. To be exact, you should account for both reactions. However, $K_{a2}$ is so much smaller than $K_{a1}$ that the smaller amount of hydronium ion produced in the second reaction can be neglected. Follow the three steps for solving equilibrium problems.

Considering only the first ionization, set up a table of concentrations (starting, change, and equilibrium). Substitute these values into the equilibrium equation for $K_{a1}$, solve for $x = [H_3O^+]$, and then solve for pH (as in Example 17.2). Ascorbate ion is produced in the second ionization; its concentration, as we show, equals $K_{a2}$.

### SOLUTION

**CALCULATION OF pH**    Abbreviate the formula for ascorbic acid as $H_2Asc$. Hydronium ions are produced by two successive acid ionizations.

$$H_2Asc(aq) + H_2O(l) \rightleftharpoons H_3O^+(aq) + HAsc^-(aq); K_{a1} = 7.9 \times 10^{-5}$$

$$HAsc^-(aq) + H_2O(l) \rightleftharpoons H_3O^+(aq) + Asc^{2-}(aq); K_{a2} = 1.6 \times 10^{-12}$$

**Figure 17.5**
**The structure of ascorbic acid.** *Top:* Structural formula of ascorbic acid (vitamin C); note the acidic hydrogen atoms (in color). *Bottom:* Molecular model of ascorbic acid.

If you let $x$ be the amount of $H_3O^+$ formed in the first ionization, you get the following results:

| Concentration ($M$) | $H_2Asc(aq) + H_2O(l) \rightleftharpoons H_3O^+(aq) + HAsc^-(aq)$ | | |
|---|---|---|---|
| Starting | 0.10 | ~0 | 0 |
| Change | $-x$ | $+x$ | $+x$ |
| Equilibrium | $0.10 - x$ | $x$ | $x$ |

Now substitute into the equilibrium-constant equation for the first ionization.

$$\frac{[H_3O^+][HAsc^-]}{[H_2Asc]} = K_{a1}$$

$$\frac{x^2}{0.10 - x} = 7.9 \times 10^{-5}$$

Assuming $x$ to be much smaller than 0.10, you get

$$\frac{x^2}{0.10} \simeq 7.9 \times 10^{-5}$$

or

$$x^2 \simeq 7.9 \times 10^{-5} \times 0.10$$

$$x \simeq 2.8 \times 10^{-3} = 0.0028$$

(Note that $0.10 - x = 0.10 - 0.0028 = 0.10$, correct to two significant figures. Therefore, the assumption that $0.10 - x \simeq 0.10$ is correct.)

The hydronium-ion concentration is 0.0028 $M$, so

$$pH = -\log [H_3O^+] = -\log(0.0028) = \textbf{2.55}$$

**ASCORBATE-ION CONCENTRATION** Ascorbate ion, $Asc^{2-}$, which we will call $y$, is produced only in the second reaction. Assume the starting concentrations of $H_3O^+$ and $HAsc^-$ for this reaction to be those from the first equilibrium. The amounts of each species in 1 L of solution are as follows:

| Concentration ($M$) | $HAsc^-(aq) + H_2O(l) \rightleftharpoons H_3O^+(aq) + Asc^{2-}(aq)$ | | |
|---|---|---|---|
| Starting | 0.0028 | 0.0028 | 0 |
| Change | $-y$ | $+y$ | $+y$ |
| Equilibrium | $0.0028 - y$ | $0.0028 + y$ | $y$ |

Now substitute into the equilibrium-constant equation for the second ionization.

$$\frac{[H_3O^+][Asc^{2-}]}{[HAsc^-]} = K_{a2}$$

$$\frac{(0.0028 + y)y}{0.0028 - y} = 1.6 \times 10^{-12}$$

This equation can be simplified if you assume that $y$ is much smaller than 0.0028. (Again, this assumes that the reaction occurs to only a small extent, as you expect from the magnitude of the equilibrium constant.) That is,

$$0.0028 + y \simeq 0.0028$$

$$0.0028 - y \simeq 0.0028$$

# A CHEMIST LOOKS AT

## Acid Rain

**ENVIRONMENT**

*Acid rain* means rain having a pH lower than that of natural rain (pH 5.6). Natural rain dissolves carbon dioxide from the atmosphere to give a slightly acidic solution. The pH of rain in eastern North America and western Europe, however, is approximately 4 and sometimes lower. This acidity is primarily the result of the dissolving in rainwater of sulfur oxides and nitrogen oxides from human activities. In the northeastern United States, the strong acid components in acid rain are about 62% sulfuric acid, 32% nitric acid, and 6% hydrochloric acid.

The sulfuric acid in acid rain has been traced to the burning of fossil fuels and to the burning of sulfide ores in the production of metals, such as zinc and copper. Coal, for example, contains some sulfur mainly as pyrite, or iron(II) disulfide, $FeS_2$. When this burns in air, it produces sulfur dioxide.

$$4FeS_2(s) + 11O_2(g) \longrightarrow 2Fe_2O_3(s) + 8SO_2(g)$$

In the presence of dust particles and other substances in polluted air, the sulfur dioxide oxidizes further to give sulfur trioxide, which reacts with water to form sulfuric acid.

$$2SO_2(g) + O_2(g) \longrightarrow 2SO_3(g)$$

$$SO_3(g) + H_2O(l) \longrightarrow H_2SO_4(aq)$$

Almost all of the sulfuric acid dissociates into the ions $H_3O^+$, $HSO_4^-$, and $SO_4^{2-}$. The first ionization of $H_2SO_4$ is complete; the second ionization (that of $HSO_4^-$) is partial ($K_{a2} = 1.1 \times 10^{-2}$).

Acid rain can be harmful to some plants, to fish (by changing the pH of lake water), and to structural materials and monuments (Figure 17.6). Marble, for example, is composed of calcium carbonate, $CaCO_3$, which dissolves in water of low pH.

$$H_3O^+(aq) + CaCO_3(s) \longrightarrow Ca^{2+}(aq) + HCO_3^-(aq) + H_2O(l)$$

It is instructive to calculate the pH of natural rain—that is, water saturated with carbon dioxide from the atmosphere. For this, you need to solve two equilibrium problems. First, you need the equilibrium for the solution of $CO_2$ in water.

$$CO_2(g) + H_2O(l) \rightleftharpoons H_2CO_3(aq)$$

$$K = \frac{[H_2CO_3]}{P_{CO_2}} = 3.5 \times 10^{-2}$$

Here $[H_2CO_3]$ represents the concentration of the molecular species formed when $CO_2$ dissolves in water. The partial pressure of $CO_2$ in the atmosphere is 0.00033 atm. If you substitute this for $P_{CO_2}$ and solve for $[H_2CO_3]$, you obtain $1.2 \times 10^{-5}$ M. (Note that the equilibrium-constant equation involving $P_{CO_2}$ is essentially Henry's law, discussed in Section 12.3.) Now you need to account for the acid ionization of $H_2CO_3$. You

Then the equilibrium equation reads

$$\frac{(0.0028)y}{0.0028} \simeq 1.6 \times 10^{-12}$$

Hence,

$$y \simeq 1.6 \times 10^{-12}$$

(Note that $1.6 \times 10^{-12}$ is indeed much smaller than 0.0028, as you assumed.) The concentration of ascorbate ion equals $K_{a2}$, or **$1.6 \times 10^{-12}$ M.**

---

### EXERCISE 17.4

Sulfurous acid, $H_2SO_3$, is a diprotic acid with $K_{a1} = 1.3 \times 10^{-2}$ and $K_{a2} = 6.3 \times 10^{-8}$. The acid forms when sulfur dioxide (a gas with a suffocating odor) dissolves in water. What is the pH of a 0.25 M solution of sulfurous acid? What is the concentration of sulfite ion, $SO_3^{2-}$, in the solution? Note that $K_{a1}$ is relatively large.

---

See Problems 17.39 and 17.40.

look only at the first ionization, which produces almost all of the $H_3O^+$.

$$H_2CO_3(aq) + H_2O(l) \rightleftharpoons H_3O^+(aq) + HCO_3^-(aq)$$

$$K_{a1} = \frac{[H_3O^+][HCO_3^-]}{[H_2CO_3]} = 4.3 \times 10^{-7}$$

Denoting $[H_3O^+] = [HCO_3^-]$ by $x$, you can write the equilibrium equation

$$\frac{x^2}{1.2 \times 10^{-5} - x} = 4.3 \times 10^{-7}$$

Making the usual assumption about neglecting $x$ in the denominator, you obtain $x = [H_3O^+] = 2.3 \times 10^{-6}$. This assumption and the neglect of the self-ionization of water affect the answer in the second significant figure. Retaining one significant figure, pH $= -\log(2 \times 10^{-6})$ $= 5.7$, which is essentially what we gave earlier for the pH of natural rain.

**Figure 17.6**
**Effect of acid rain on marble.** These photographs of a marble statue of George Washington, located in New York, were taken 60 years apart. The more recent photograph shows that much of the detail of the statue has been dissolved by acid rain. Marble ($CaCO_3$) reacts with $H_3O^+$ ion to form soluble $Ca(HCO_3)_2$.

Example 17.4 shows that $[H_3O^+]$ and $[HA^-]$ in a solution of a diprotic acid $H_2A$ can be calculated from the first ionization constant, $K_{a1}$.

The concentration of the ion $A^{2-}$ equals the second ionization constant, $K_{a2}$.

# 17.3 Base-Ionization Equilibria

Equilibria involving weak bases are treated similarly to those for weak acids. Ammonia, for example, ionizes in water as follows:

$$NH_3(aq) + H_2O(l) \rightleftharpoons NH_4^+(aq) + OH^-(aq)$$

The corresponding equilibrium constant is

$$K_c = \frac{[NH_4^+][OH^-]}{[NH_3][H_2O]}$$

**735**

**Table 17.2**
**Base-Ionization Constants at 25°C**

| Substance | Formula | $K_b$ |
|-----------|---------|-------|
| Ammonia | $NH_3$ | $1.8 \times 10^{-5}$ |
| Aniline | $C_6H_5NH_2$ | $4.2 \times 10^{-10}$ |
| Dimethylamine | $(CH_3)_2NH$ | $5.1 \times 10^{-4}$ |
| Ethylamine | $C_2H_5NH_2$ | $4.7 \times 10^{-4}$ |
| Hydrazine | $N_2H_4$ | $1.7 \times 10^{-6}$ |
| Hydroxylamine | $NH_2OH$ | $1.1 \times 10^{-8}$ |
| Methylamine | $CH_3NH_2$ | $4.4 \times 10^{-4}$ |
| Pyridine | $C_5H_5N$ | $1.4 \times 10^{-9}$ |
| Urea | $NH_2CONH_2$ | $1.5 \times 10^{-14}$ |

▶ In terms of the thermodynamic equilibrium constant, the activity of $H_2O$ is nearly constant and essentially 1, so it does not appear explicitly in the equilibrium constant.

Because the concentration of $H_2O$ is nearly constant, you can rearrange this equation as you did for acid ionization. ◀

$$K_b = [H_2O]K_c = \frac{[NH_4^+][OH^-]}{[NH_3]}$$

In general, a weak base B with the *base ionization*

$$B(aq) + H_2O(l) \rightleftharpoons HB^+(aq) + OH^-(aq)$$

has a **base-ionization constant,** $K_b$ (*the equilibrium constant for the ionization of a weak base*), equal to

$$K_b = \frac{[HB^+][OH^-]}{[B]}$$

Table 17.2 lists ionization constants for some weak bases. All of these bases are similar to ammonia in that they have an N atom with a lone pair of electrons that accepts an $H^+$ ion from water. For example, the base ionization of methylamine, $CH_3NH_2$, is

$$H\!-\!\overset{\displaystyle H}{\underset{\displaystyle H}{C}}\!-\!\overset{\displaystyle H}{\underset{\displaystyle H}{N}}\!: + H_2O \rightleftharpoons H\!-\!\overset{\displaystyle H}{\underset{\displaystyle H}{C}}\!-\!\overset{\displaystyle H}{\underset{\displaystyle H}{\overset{+}{N}}}\!-\!H + OH^-$$

Many drugs and plant substances such as caffeine are bases. Often a prescription or over-the-counter drug is a salt of a base, because the salt is more soluble in water (Figure 17.7). These salts are analogous to the salts of ammonia, such as $NH_4Cl$. Other common bases are anions of weak acids. Examples are $F^-$ and $C_2H_3O_2^-$ (acetate ion). We will discuss these in the next section.

In the following exercise, you will determine the $K_b$ of a molecular base from the pH of a solution. Note that this is essentially the problem you looked at in Example 17.1.

**Figure 17.7**
**Many over-the-counter drugs are salts of bases.** These salts are often chlorides (frequently called hydrochlorides) and are analogous to ammonium chloride. You might find it instructive to read the ingredients from the labels for these products.

▶ Some other well-known alkaloids and their plant sources are nicotine (tobacco), cocaine (coca bush), mescaline (species of cactus), and caffeine (coffee and tea).

### EXERCISE 17.5

Quinine is an alkaloid, or naturally occurring base, used to treat malaria. A 0.0015 *M* solution of quinine has a pH of 9.84. The basicity of alkaloids is due to a nitrogen atom that picks up protons from water in the same manner as ammonia does. ◀ What is $K_b$?

See Problems 17.43 and 17.44.

▶ An alternate example is provided in the Instructor's Resource Manual.

### EXAMPLE 17.5

## Calculating Concentrations of Species in a Weak Base Solution Using $K_b$

Morphine, $C_{17}H_{19}NO_3$, is administered medically to relieve pain. It is a naturally occurring base, or alkaloid. What is the pH of a 0.0075 *M* solution of morphine at 25°C? The base-ionization constant, $K_b$, is $1.6 \times 10^{-6}$ at 25°C.

 **PROBLEM STRATEGY**

The calculation for the ionization of a weak base parallels that used with weak acids in Examples 17.2 and 17.3: you write the equation, make a table of concentrations (Step 1), set up the equilibrium-constant equation for $K_b$ (Step 2), and solve for $x = [OH^-]$ (Step 3). Assume the self-ionization of water can be neglected. You obtain $[H_3O^+]$, and then the pH, by solving $K_w = [H_3O^+][OH^-]$.

**SOLUTION**

Abbreviate the formula for morphine as Mor. Morphine ionizes by picking up a proton from water (as does ammonia).

$$Mor(aq) + H_2O \rightleftharpoons HMor^+(aq) + OH^-(aq)$$

**STEP 1** Summarize the concentrations in a table. For the *change* values, note that the morphine in 1 L of solution ionizes to give $x$ mol $HMor^+$ and $x$ mol $OH^-$.

| Concentration ($M$) | $Mor(aq)$ + $H_2O(l)$ | $\rightleftharpoons$ | $HMor^+(aq)$ + | $OH^-(aq)$ |
|---|---|---|---|---|
| Starting | 0.0075 | | 0 | ~0 |
| Change | $-x$ | | $+x$ | $+x$ |
| Equilibrium | $0.0075 - x$ | | $x$ | $x$ |

**STEP 2** Substituting into the equilibrium equation

$$\frac{[HMor^+][OH^-]}{[Mor]} = K_b$$

gives

$$\frac{x^2}{0.0075 - x} = 1.6 \times 10^{-6}$$

**STEP 3** If you assume that $x$ is small enough to neglect compared with 0.0075, you have

$$0.0075 - x \simeq 0.0075$$

and you can write the equilibrium-constant equation as

$$\frac{x^2}{0.0075} \simeq 1.6 \times 10^{-6}$$

Hence,

$$x^2 \simeq 1.6 \times 10^{-6} \times 0.0075 = 1.2 \times 10^{-8}$$

$$x = [OH^-] \simeq 1.1 \times 10^{-4}$$

You can now calculate the hydronium-ion concentration. The product of the hydronium-ion concentration and the hydroxide-ion concentration equals $1.0 \times 10^{-14}$ at 25°C:

$$[H_3O^+][OH^-] = 1.0 \times 10^{-14}$$

Substituting into this equation gives

$$[H_3O^+] \times 1.1 \times 10^{-4} = 1.0 \times 10^{-14}$$

Hence,

$$[H_3O^+] = \frac{1.0 \times 10^{-14}}{1.1 \times 10^{-4}} = 9.1 \times 10^{-11}$$

The pH of the solution is

$$pH = -\log [H_3O^+] = -\log(9.1 \times 10^{-11}) = \mathbf{10.04}$$

Note that you could also calculate the pH using the formula $pH + pOH = 14.00$. You would first calculate the pOH; then $pH = 14.00 - pOH$.

**Problem Solving:
The pH of Weak Bases**

### EXERCISE 17.6

What is the hydronium-ion concentration of a 0.20 $M$ solution of ammonia in water? See Table 17.2 for $K_b$.

See Problems 17.45 and 17.46.

# 17.4 Acid–Base Properties of Salt Solutions

A salt may be regarded as an ionic compound obtained by a neutralization reaction in aqueous solution. The resulting salt solution may be neutral, but often it is acidic or basic (Figure 17.8). One of the successes of the Brønsted–Lowry concept of acids and bases was in pointing out that some ions can act as acids or bases. The acidity or basicity of a salt solution is explained in terms of the acidity or basicity of individual ions in the solution.

Consider a solution of sodium cyanide, NaCN. A 0.1 $M$ solution has a pH of 11.1 and is therefore fairly basic. Sodium cyanide dissolves in water to give $Na^+$ and $CN^-$ ions.

$$NaCN(s) \xrightarrow{H_2O} Na^+(aq) + CN^-(aq)$$

Sodium ion, $Na^+$, is unreactive with water, but the cyanide ion, $CN^-$, reacts to produce HCN and $OH^-$.

$$CN^-(aq) + H_2O(l) \rightleftharpoons HCN(aq) + OH^-(aq)$$

From the Brønsted–Lowry point of view, the $CN^-$ ion acts as a base, because it accepts a proton from $H_2O$. You can also see, however, that $OH^-$ ion is a product, so you expect the solution to have a basic pH. This explains why solutions of NaCN are basic.

The reaction of the $CN^-$ ion with water is referred to as the *hydrolysis* of $CN^-$. The **hydrolysis** of an ion is *the reaction of an ion with water to produce the conjugate acid and hydroxide ion or the conjugate base and hydronium ion.* The $CN^-$ ion hydrolyzes to give the conjugate acid and $OH^-$ ion. As another example, consider the ammonium ion, $NH_4^+$, which hydrolyzes as follows:

$$NH_4^+(aq) + H_2O(l) \rightleftharpoons NH_3(aq) + H_3O^+(aq)$$

The ammonium ion acts as an acid, donating a proton to $H_2O$ and forming $NH_3$. Note that this equation has the form of an acid ionization, so you could write the $K_a$ expression for it. Similarly, the hydrolysis reaction for the $CN^-$ ion has the form of a base ionization, so you could write the $K_b$ expression for it.

▶ Animation: Hydrolysis.

**Figure 17.8**
**Salt solutions are acidic, neutral, or basic.** A pH meter measures the pH of solutions of the salts $NH_4Cl$, $NaCl$, and $Na_2CO_3$.

$NH_4Cl$

NaCl

$Na_2CO_3$

Two questions become apparent. First, how can you predict whether a particular salt solution will be acidic, basic, or neutral? Then, how can you calculate the concentrations of $H_3O^+$ ion and $OH^-$ ion in the salt solution (or equivalently, how do you predict the pH of the salt solution)? We will now look at the first of these questions.

## Prediction of Whether a Salt Solution Is Acidic, Basic, or Neutral

When the $CN^-$ ion hydrolyzes, it produces the conjugate acid, HCN. Hydrogen cyanide, HCN, is a weak acid. That means that it tends to hold on to the proton strongly (does not ionize readily). In other words, the cyanide ion, $CN^-$, tends to pick up a proton easily, so it acts as a base. This argument can be generalized: *The anions of weak acids are basic. On the other hand, the anions of strong acids have hardly any basic character; that is, these ions do not hydrolyze.* For example, the $Cl^-$ ion, which is conjugate to the strong acid HCl, shows no appreciable reaction with water.

$$Cl^-(aq) + H_2O(l) \longrightarrow \text{no reaction}$$

Now consider a cation conjugate to a weak base. The simplest example is the $NH_4^+$ ion, which we just discussed. It behaves like an acid. *The cations of weak bases are acidic. On the other hand, the cations of strong bases (metal ions of Groups IA and IIA elements—except Be) have hardly any acidic character; that is, these ions do not hydrolyze.* For example,

$$Na^+(aq) + H_2O(l) \longrightarrow \text{no reaction}$$

Aqueous metal ions, other than the cations of the strong bases, usually hydrolyze by acting as acids. These ions usually form hydrated metal ions. For example, the aluminum ion $Al^{3+}$ forms the hydrated ion $Al(H_2O)_6^{3+}$. The bare $Al^{3+}$ ion acts as a Lewis acid, forming bonds to the electron pairs on O atoms of $H_2O$ molecules. Because the electrons are drawn away from the O atoms by the Al atom, the O atoms in turn tend to draw electrons from the O—H bonds, making them highly polar. As a result, the $H_2O$ molecules in $Al(H_2O)_6^{3+}$ are acidic. The $Al(H_2O)_6^{3+}$ ion hydrolyzes in aqueous solution by donating a proton (from one of the acidic $H_2O$ molecules on the ion) to water molecules in the solvent. After the proton leaves the hydrated metal ion, the ion charge is reduced by 1 and an $H_2O$ molecule on the ion becomes a hydroxide ion bonded to the metal. Thus, the formula of the ion becomes $Al(H_2O)_5(OH)^{2+}$.

$$Al(H_2O)_6^{3+}(aq) + H_2O(l) \rightleftharpoons Al(H_2O)_5(OH)^{2+}(aq) + H_3O^+(aq)$$

To predict the acidity or basicity of a salt solution, you need to examine the acidity or basicity of the ions composing the salt. Consider potassium acetate, $KC_2H_3O_2$. The ions are $K^+$ and $C_2H_3O_2^-$ (acetate ion). Potassium is a Group IA element, so $K^+$ does not hydrolyze. However, the acetate ion is the conjugate of acetic acid, a weak acid. Therefore, the acetate ion is basic.

$$K^+(aq) + H_2O(l) \longrightarrow \text{no reaction}$$
$$C_2H_3O_2^-(aq) + H_2O(l) \rightleftharpoons HC_2H_3O_2(aq) + OH^-(aq)$$

A solution of potassium acetate is predicted to be basic.

From this discussion, we can derive a set of rules for deciding whether a salt solution will be neutral, acidic, or basic. These rules apply to normal salts (those in which the anion has no acidic hydrogen atoms).

1. **A salt of a strong base and a strong acid.** The salt has no hydrolyzable ions and so gives a neutral aqueous solution. An example is NaCl.

2. **A salt of a strong base and a weak acid.** The anion of the salt is the conjugate of the weak acid. It hydrolyzes to give a basic solution. An example is NaCN.

3. **A salt of a weak base and a strong acid.** The cation of the salt is the conjugate of the weak base. It hydrolyzes to give an acidic solution. An example is $NH_4Cl$.

4. **A salt of a weak base and a weak acid.** Both ions hydrolyze. Whether the solution is acidic or basic depends on the relative acid–base strengths of the two ions. To determine this, you need to compare the $K_a$ of the cation with the $K_b$ of the anion. If the $K_a$ is larger, the solution is acidic. If the $K_b$ is larger, the solution is basic. Consider solutions of ammonium formate, $NH_4CHO_2$. These solutions are slightly acidic, because the $K_a$ for $NH_4^+$ ($= 5.6 \times 10^{-10}$) is somewhat larger than the $K_b$ for formate ion, $CHO_2^-$ ($= 5.9 \times 10^{-11}$).

The next example illustrates the application of these rules.

### EXAMPLE 17.6

► An alternate example is provided in the Instructor's Resource Manual.

## Predicting Whether a Salt Solution Is Acidic, Basic, or Neutral

Decide whether aqueous solutions of the following salts are acidic, basic, or neutral: a. KCl; b. NaF; c. $Zn(NO_3)_2$; d. $NH_4CN$.

**SOLUTION**

a. KCl is a salt of a strong base (KOH) and a strong acid (HCl), so none of the ions hydrolyze, and a solution of KCl is **neutral.**

b. NaF is a salt of a strong base (NaOH) and a weak acid (HF), so a solution of NaF is **basic** because of the hydrolysis of $F^-$. (You assume HF is a weak acid, because it is not one of the common strong acids listed in Table 16.1.)

c. $Zn(NO_3)_2$ is the salt of a weak base ($Zn(OH)_2$) and a strong acid ($HNO_3$), so a solution of $Zn(NO_3)_2$ is **acidic.** Note also that $Zn^{2+}$ is not an ion of a Group IA or Group IIA element, so you can expect it to form a metal hydrate ion that hydrolyzes to give an acidic solution.

d. $NH_4CN$ is a salt of a weak base ($NH_3$) and a weak acid (HCN). Therefore, to answer the question, you would need the $K_a$ for $NH_4^+$ and the $K_b$ for $CN^-$. Although these are listed in some tables, more frequently they are not. However, as you will see later in this section, values are easily calculated from the molecular substances that are conjugates ($NH_3$ and HCN). The $K_a$ for $NH_4^+$ is $5.6 \times 10^{-10}$, and the $K_b$ for $CN^-$ is $2.0 \times 10^{-5}$. From these values, you conclude that a solution of $NH_4CN$ is **basic.**

---

### EXERCISE 17.7

Consider solutions of the following salts: a. $NH_4NO_3$; b. $KNO_3$; c. $Al(NO_3)_3$. For each, state whether the solution is acidic, basic, or neutral.

---

See Problems 17.51 and 17.52.

## The pH of a Salt Solution

Suppose you wish to calculate the pH of 0.10 $M$ NaCN. As you have seen, the solution is basic because of the hydrolysis of the $CN^-$ ion. The calculation is essentially the same as that for any weak base (Example 17.5). You would require the base-ionization constant of the cyanide ion. However, frequently the ionization constants for ions are not listed directly in tables, because the values are easily related to those for the conjugate molecular species. Thus, the $K_b$ for $CN^-$ is related to the $K_a$ for HCN.

To see this relationship between $K_a$ and $K_b$ for conjugate acid–base pairs, consider the acid ionization of HCN and the base ionization of $CN^-$. When these two reactions are added, you get the ionization of water.

$$
\begin{array}{ll}
HCN(aq) + H_2O(l) \rightleftharpoons H_3O^+(aq) + CN^-(aq) & K_a \\
\underline{CN^-(aq) + H_2O(l) \rightleftharpoons HCN(aq) + OH^-(aq)} & K_b \\
2H_2O(l) \rightleftharpoons H_3O^+(aq) + OH^-(aq) & K_w
\end{array}
$$

The equilibrium constants for the reactions are shown symbolically at the right. In Section 15.2, it was shown that when two reactions are added, their equilibrium constants are multiplied. Therefore, ◄

$$\boxed{K_a K_b = K_w}$$

This relationship is general and shows that the product of acid- and base-ionization constants in aqueous solution for conjugate acid–base pairs equals the ion-product constant for water, $K_w$.

▶ Another way to obtain this result is by multiplying the expression for $K_b$ by $[H_3O^+]/[H_3O^+]$ and rearranging.

$$K_b = \frac{[HCN][OH^-]}{[CN^-]} \times \frac{[H_3O^+]}{[H_3O^+]}$$

$$= [H_3O^+][OH^-] \Big/ \frac{[H_3O^+][CN^-]}{[HCN]}$$

This is equivalent to

$$K_b = K_w/K_a \quad \text{or} \quad K_a K_b = K_w$$

---

### EXAMPLE 17.7

## Obtaining $K_a$ from $K_b$ or $K_b$ from $K_a$

Use Tables 17.1 and 17.2 to obtain the following at 25°C: a. $K_b$ for $CN^-$; b. $K_a$ for $NH_4^+$.

### SOLUTION

a. The conjugate acid of $CN^-$ is HCN, whose $K_a$ is $4.9 \times 10^{-10}$ (Table 17.1). Hence,

$$K_b = \frac{K_w}{K_a} = \frac{1.0 \times 10^{-14}}{4.9 \times 10^{-10}} = 2.0 \times 10^{-5}$$

Note that $K_b$ is approximately equal to $K_b$ for ammonia ($1.8 \times 10^{-5}$). This means that the base strength of $CN^-$ is comparable with that of $NH_3$.

b. The conjugate base of $NH_4^+$ is $NH_3$, whose $K_b$ is $1.8 \times 10^{-5}$ (Table 17.2). Hence,

$$K_a = \frac{K_w}{K_b} = \frac{1.0 \times 10^{-14}}{1.8 \times 10^{-5}} = \mathbf{5.6 \times 10^{-10}}$$

$NH_4^+$ is a relatively weak acid. Acetic acid, by way of comparison, has $K_a$ equal to $1.7 \times 10^{-5}$.

**Problem Solving:
Salts as Weak Bases**

### EXERCISE 17.8

Calculate the following, using Tables 17.1 and 17.2: a. $K_b$ for $F^-$; b. $K_a$ for $C_6H_5NH_3^+$ (conjugate acid of aniline, $C_6H_5NH_2$).

**See Problems 17.55 and 17.56.**

For a solution of a salt in which only one of the ions hydrolyzes, the calculation of the concentrations of species present follows that for solutions of weak acids or bases. The only difference is that you must first obtain $K_a$ or $K_b$ for the ion that hydrolyzes. The next example illustrates the reasoning and calculations involved.

### EXAMPLE 17.8

## Calculating Concentrations of Species in a Salt Solution

What is the pH of 0.10 $M$ sodium nicotinate at 25°C (a problem posed in the chapter opening)? The $K_a$ for nicotinic acid was determined in Example 17.1 to be $1.4 \times 10^{-5}$ at 25°C.

### SOLUTION

Sodium nicotinate gives $Na^+$ and nicotinate ions in solution. Only the nicotinate ion hydrolyzes. Write HNic for nicotinic acid and $Nic^-$ for the nicotinate ion. The hydrolysis of nicotinate ion is

$$Nic^-(aq) + H_2O(l) \rightleftharpoons HNic(aq) + OH^-(aq)$$

Nicotinate ion acts as a base, and you can calculate the concentration of species in solution as in Example 17.5. First, however, you need $K_b$ for the nicotinate ion. This is related to $K_a$ for nicotinic acid by the equation $K_aK_b = K_w$. Substituting, you get

$$K_b = \frac{K_w}{K_a} = \frac{1.0 \times 10^{-14}}{1.4 \times 10^{-5}} = 7.1 \times 10^{-10}$$

Now you can proceed with the equilibrium calculation. We will only sketch this calculation, because it is similar to that in Example 17.5. You let $x = [HNic] = [OH^-]$, then substitute into the equilibrium-constant equation.

$$\frac{[HNic][OH^-]}{[Nic^-]} = K_b$$

This gives

$$\frac{x^2}{0.10 - x} = 7.1 \times 10^{-10}$$

Benzoic acid

Solving this equation, you find that $x = [OH^-] = 8.4 \times 10^{-6}$. Hence,

$$pH = 14.00 - pOH = 14.00 + \log(8.4 \times 10^{-6}) = \mathbf{8.92}$$

As expected, the solution has a pH greater than 7.00.

### EXERCISE 17.9

Benzoic acid, $HC_7H_5O_2$, and its salts are used as food preservatives. What is the concentration of benzoic acid in an aqueous solution of 0.015 $M$ sodium benzoate? What is the pH of the solution? $K_a$ for benzoic acid is $6.3 \times 10^{-5}$.

See Problems 17.57, 17.58, 17.59, and 17.60.

### CONCEPT CHECK 17.2

Which of the following aqueous solutions has the highest pH and which has the lowest: a. 0.1 $M$ NH$_3$; b. 0.1 $M$ NH$_4$Br; c. 0.1 $M$ NaF; d. 0.1 $M$ NaCl?

# Solutions of a Weak Acid or Base with Another Solute

In the preceding sections, we looked at solutions that contained either a weak acid, a weak base, or a salt of a weak acid or base. In the remaining sections of this chapter, we will look at the effect of adding another solute to a solution of a weak acid or base. The solutes we will look at are those that significantly affect acid or base ionization—that is, strong acids and bases, and salts that contain an ion that is produced in the acid or base ionization. These solutes affect the equilibrium through the *common-ion effect.*

## 17.5 Common-Ion Effect

The **common-ion effect** is *the shift in an ionic equilibrium caused by the addition of a solute that provides an ion that takes part in the equilibrium.* Consider a solution of acetic acid, $HC_2H_3O_2$, in which you have the following acid-ionization equilibrium:

$$HC_2H_3O_2(aq) + H_2O(l) \rightleftharpoons C_2H_3O_2^-(aq) + H_3O^+(aq)$$

Suppose you add HCl($aq$) to this solution. What is the effect on the acid-ionization equilibrium? Because HCl($aq$) is a strong acid, it provides $H_3O^+$ ion, which is present on the right side of the equation for acetic acid ionization. According to Le Chatelier's principle, the equilibrium composition should shift to the left. ◄

▶ Le Chatelier's principle was applied in Section 15.7 to the problem of adding substances to an equilibrium mixture.

$$HC_2H_3O_2(aq) + H_2O(l) \longleftarrow C_2H_3O_2^-(aq) + H_3O^+(aq)$$
$$\text{added}$$

The degree of ionization of acetic acid is decreased by the addition of a strong acid. This repression of the ionization of acetic acid by HCl($aq$) is an example of the common-ion effect.

## EXAMPLE 17.9

## Calculating the Common-Ion Effect on Acid Ionization (Effect of a Strong Acid)

The degree of ionization of acetic acid, $HC_2H_3O_2$, in a $0.10\,M$ aqueous solution at 25°C is 0.013. $K_a$ at this temperature is $1.7 \times 10^{-5}$. Calculate the degree of ionization of $HC_2H_3O_2$ in a $0.10\,M$ solution at 25°C to which sufficient HCl is added to make it $0.010\,M$ HCl. How is the degree of ionization affected?

 **PROBLEM STRATEGY**

This is an acid-ionization problem, but it differs from the simple ionization illustrated by Example 17.2. Here you have a starting concentration of $H_3O^+$ ($= 0.010\,M$) from the addition of a strong acid (HCl). This gives a different type of equation in Step 2, but you solve it in Step 3 by using a similar approximation method. You assume that $x$ is small compared with starting concentrations of acid and $H_3O^+$, so that the resulting equation is linear, rather than quadratic. The calculation of degree of ionization follows that in Example 17.2.

**SOLUTION**

**STEP 1**   Starting concentrations are $[HC_2H_3O_2] = 0.10\,M$, $[H_3O^+] = 0.010\,M$ (from HCl), and $[C_2H_3O_2^-] = 0$. The acetic acid ionizes to give an additional $x$ mol/L of $H_3O^+$ and $x$ mol/L of $C_2H_3O_2^-$. The table is

| Concentration ($M$) | $HC_2H_3O_2(aq)$ | $+\ H_2O(l)$ | $\rightleftharpoons$ | $H_3O^+(aq)$ | $+\ C_2H_3O_2^-(aq)$ |
|---|---|---|---|---|---|
| Starting | 0.10 | | | ~0 | 0.20 |
| Change | $-x$ | | | $+x$ | $+x$ |
| Equilibrium | $0.10 - x$ | | | $x$ | $0.20 + x$ |

**STEP 2**   You substitute into the equilibrium-constant equation

$$\frac{[H_3O^+][C_2H_3O_2^-]}{[HC_2H_3O_2]} = K_a$$

obtaining

$$\frac{(0.010 + x)x}{0.10 - x} = 1.7 \times 10^{-5}$$

**STEP 3**   To solve this equation, assume that $x$ is small compared with 0.010. Then

$$0.010 + x \simeq 0.010$$

$$0.10 - x \simeq 0.10$$

The equation becomes

$$\frac{0.010x}{0.10} \simeq 1.7 \times 10^{-5}$$

Solving for $x$, you get

$$x = 1.7 \times 10^{-5} \times \frac{0.10}{0.010} = 1.7 \times 10^{-4}$$

(Check that $x$ can indeed be neglected in $0.010 + x$ and $0.10 - x$.) The degree of ionization of $HC_2H_3O_2$ is $x/0.10 = \mathbf{0.0017}$. This is much smaller than the value for $0.10\ M\ HC_2H_3O_2$ (0.013), because **the addition of HCl represses the ionization of $HC_2H_3O_2$.**

---

### EXERCISE 17.10

What is the concentration of formate ion, $CHO_2^-$, in a solution at 25°C that is $0.10\ M\ HCHO_2$ and $0.20\ M$ HCl? What is the degree of ionization of formic acid, $HCHO_2$?

---

*See Problems 17.61 and 17.62.*

As you will see in the following section, solutions that contain a weak acid or a weak base and a corresponding salt are especially important. Therefore, you will need to be able to calculate the concentrations of species present in such solutions. A solution of acetic acid and sodium acetate is an example. From Le Chatelier's principle, you can see that the addition of acetate ion (from sodium acetate) represses the ionization of the acetic acid by the common-ion effect:

$$HC_2H_3O_2(aq) + H_2O(l) \longleftarrow \underset{\text{added}}{C_2H_3O_2^-}(aq) + H_3O^+(aq)$$

The next example illustrates the calculation of pH, which is similar to that of the previous example.

---

### EXAMPLE 17.10

▶ An alternate example is provided in the Instructor's Resource Manual.

## Calculating the Common-Ion Effect on Acid Ionization (Effect of a Conjugate Base)

A solution is prepared to be $0.10\ M$ acetic acid, $HC_2H_3O_2$, and $0.20\ M$ sodium acetate, $NaC_2H_3O_2$. What is the pH of this solution at 25°C? $K_a$ for acetic acid at 25°C is $1.7 \times 10^{-5}$.

**SOLUTION**

**STEP 1**   Consider the equilibrium

$$HC_2H_3O_2(aq) + H_2O(l) \rightleftharpoons H_3O^+(aq) + C_2H_3O_2^-(aq)$$

Initially, 1 L of solution contains 0.10 mol of acetic acid. Sodium acetate is a strong electrolyte, so 1 L of solution contains 0.20 mol of acetate ion. When the acetic acid ionizes, it gives $x$ mol of hydronium ion and $x$ mol of acetate ion. This is summarized in the following table:

| Concentration ($M$) | $HC_2H_3O_2(aq) + H_2O(l) \rightleftharpoons$ | $H_3O^+(aq) +$ | $C_2H_3O_2^-(aq)$ |
|---|---|---|---|
| Starting | 0.10 | ~0 | 0.20 |
| Change | $-x$ | $+x$ | $+x$ |
| Equilibrium | $0.10 - x$ | $x$ | $0.20 + x$ |

**STEP 2**   The equilibrium-constant equation is

$$\frac{[H_3O^+][C_2H_3O_2^-]}{[HC_2H_3O_2]} = K_a$$

Substituting into this equation gives

$$\frac{x(0.20 + x)}{0.10 - x} = 1.7 \times 10^{-5}$$

**STEP 3**  To solve the equation, assume that $x$ is small compared with 0.10 and 0.20. Then

$$0.20 + x \simeq 0.20$$

$$0.10 - x \simeq 0.10$$

The equilibrium equation becomes

$$\frac{x(0.20)}{0.10} \simeq 1.7 \times 10^{-5}$$

Hence,

$$x \simeq 1.7 \times 10^{-5} \times \frac{0.10}{0.20} = 8.5 \times 10^{-6}$$

(Note that $x$ is indeed much smaller than 0.10 or 0.20.) The hydronium-ion concentration is $8.5 \times 10^{-6} M$, and

$$\text{pH} = -\log [\text{H}_3\text{O}^+] = -\log(8.5 \times 10^{-6}) = \textbf{5.07}$$

For comparison, the pH of 0.10 $M$ acetic acid is 2.88.

### EXERCISE 17.11

One liter of solution was prepared by dissolving 0.025 mol of formic acid, $\text{HCHO}_2$, and 0.018 mol of sodium formate, $\text{NaCHO}_2$, in water. What was the pH of the solution? $K_a$ for formic acid is $1.7 \times 10^{-4}$.

See Problems 17.63 and 17.64.

The previous examples considered the common-ion effect in solutions of weak acids. You also encounter the common-ion effect in solutions of weak bases. For example, the base ionization of $\text{NH}_3$ is

$$\text{NH}_3(aq) + \text{H}_2\text{O}(l) \Longrightarrow \text{NH}_4^+(aq) + \text{OH}^-(aq)$$

This suggests that you could repress, or decrease the effect of, the base ionization by adding an ammonium salt (providing $\text{NH}_4^+$ ion) or a strong base (providing $\text{OH}^-$ ion). The calculations are similar to those for acid ionization, except that you use the appropriate base ionization.

# 17.6 Buffers

▶ Demonstration of buffer action: Summerlin et al., *Chemical Demonstrations,* Vol. 2 (Washington, D.C.: Am. Chem. Soc., 1987), no. 91, pp. 172–173.

A **buffer** is *a solution characterized by the ability to resist changes in pH when limited amounts of acid or base are added to it.* If 0.01 mol of hydrochloric acid is added to 1 L of pure water, the pH changes from 7.0 to 2.0—a pH change of 5.0 units. By contrast, the addition of this amount of hydrochloric acid to 1 L of buffered solution might change the pH by only 0.1 unit. Biological fluids, such as blood, are usually buffer solutions; the control of pH is vital to proper functioning of these fluids. The

**Figure 17.9**
**Some laboratory buffers.**
These commercially prepared buffer solutions have different proportions of the same conjugate acid and base.

▶ Video no. 25 (Series A), "Buffers."

▶ Video Series C, "Buffering."

▶ Animations: HCl added to water; HCl added to buffer.

oxygen-carrying function of blood depends on its being maintained very near a pH of 7.4. If the pH were to change by a tenth of a unit, the capacity of the blood to carry oxygen would be lost.

Buffers contain either a weak acid and its conjugate base or a weak base and its conjugate acid. Thus, a buffer solution contains both an acid species and a base species in equilibrium. To understand the action of a buffer, consider one that contains approximately equal molar amounts of a weak acid HA and its conjugate base $A^-$. When a strong acid is added to the buffer, it supplies hydronium ions that react with the base $A^-$.

$$H_3O^+(aq) + A^-(aq) \longrightarrow HA(aq) + H_2O(l)$$

On the other hand, when a strong base is added to the buffer, it supplies hydroxide ions. These react with the acid HA.

$$OH^-(aq) + HA(aq) \longrightarrow H_2O(l) + A^-(aq)$$

A buffer solution resists changes in pH through its ability to combine with both $H_3O^+$ and $OH^-$ ions.

Blood, as a buffer solution, contains $H_2CO_3$ and $HCO_3^-$, as well as other conjugate acid–base pairs. A buffer frequently used in the laboratory contains varying proportions of the conjugate acid–base pair $H_2PO_4^-$ and $HPO_4^{2-}$ (Figure 17.9).

Buffers also have commercial applications. For example, the label on a package of artificial fruit-juice mix says that it contains "citric acid to provide tartness (acidity) and sodium citrate to regulate tartness." A solution of citric acid and its conjugate base, citrate ion (provided by sodium citrate), functions as an acid–base buffer, which is what "to regulate tartness" means. The pH of the buffer is in the acid range.

Two important characteristics of a buffer are its pH and its *buffer capacity,* which is the amount of acid or base the buffer can react with before giving a significant pH change. Buffer capacity depends on the amount of acid and conjugate base in the solution. Figure 17.10 illustrates the change in pH of a buffer solution containing 1.0 mol of acetic acid and 1.0 mol of acetate ion to which varying amounts of $H_3O^+$ and $OH^-$ are added. This buffer changes less than 0.5 pH units as long as no more than 0.5 mol $H_3O^+$ or $OH^-$ ion is added. Note that this is one-half (or less than one-half) the

**Figure 17.10**
**Effect of added acid or base on a buffer solution.**
The buffer contains 1.0 mol of acetic acid and 1.0 mol of acetate ion in 1.00 L of solution. Note that the addition of 0.5 mol or less of strong acid or base gives only a small change of pH.

amounts of acid and conjugate base in the solution. The ratio of amounts of acid and conjugate base is also important. Unless this ratio is close to 1 (between 1:10 and 10:1), the buffer capacity will be too low to be useful.

The other important characteristic of a buffer is its pH. Let us now look at how to calculate the pH of a buffer.

### The pH of a Buffer

The solution described in Example 17.10 is a buffer, because it is a solution of a weak acid (0.10 $M$ acetic acid) and its conjugate base (0.20 $M$ acetate ion from sodium acetate). The example described how to calculate the pH of such a buffer (we obtained a pH of 5.07). The next example is similar, except now you are given a recipe for making up a buffer from volumes of solutions and are asked to calculate the pH. In this example, the buffer consists of a molecular base ($NH_3$) and its conjugate acid ($NH_4^+$).

---

**EXAMPLE 17.11**

## Calculating the pH of a Buffer from Given Volumes of Solution

Instructions for making up a buffer say to mix 60. mL (0.060 L) of 0.100 $M$ $NH_3$ with 40. mL (0.040 L) of 0.100 $M$ $NH_4Cl$. What is the pH of this buffer?

 **PROBLEM STRATEGY**

The buffer contains a base and its conjugate acid in equilibrium. The equation is

$$NH_3(aq) + H_2O(l) \rightleftharpoons NH_4^+(aq) + OH^-(aq)$$

The problem is to obtain the concentration of $H_3O^+$ or $OH^-$ in this equilibrium mixture. To do the equilibrium calculation, you need the starting concentrations in the solution obtained by mixing the $NH_3$ and $NH_4Cl$ solutions. For this, you calculate the moles of $NH_3$ and moles of $NH_4^+$ added to the buffer solution, then divide by the total volume of buffer. Then you are ready to do the equilibrium calculation.

**SOLUTION**

How many moles of $NH_3$ are added? Recall that

$$\text{Molarity } NH_3 = \frac{\text{moles } NH_3}{\text{liters } NH_3 \text{ solution}}$$

Note that the instructions say to add 60. mL (or 0.060 L) of 0.100 $M$ $NH_3$. So

$$\text{Moles } NH_3 = \text{molarity } NH_3 \times \text{liters } NH_3 \text{ solution}$$
$$= 0.100 \frac{\text{mol}}{\text{L}} \times 0.060 \text{ L} = 0.0060 \text{ mol } NH_3$$

In the same way, you find that you have added 0.0040 mol $NH_4^+$ (from $NH_4Cl$). Assume that the total volume of buffer equals the sum of the volumes of the two solutions.

$$\text{Total volume buffer} = 60. \text{ mL} + 40. \text{ mL} = 100. \text{ mL (0.100 L)}$$

Therefore, the concentrations of base and conjugate acid are

$$[NH_3] = \frac{0.0060 \text{ mol}}{0.100 \text{ L}} = 0.060 \ M$$

$$[NH_4^+] = \frac{0.0040 \text{ mol}}{0.100 \text{ L}} = 0.040 \ M$$

**STEP 1**   Fill in the concentration table for the acid–base equilibrium (base ionization of $NH_3$).

| Concentration (M) | $NH_3(aq)$ | $+ H_2O(l)$ | $\rightleftharpoons$ | $NH_4^+ (aq)$ | $+ OH^-(aq)$ |
|---|---|---|---|---|---|
| Starting | 0.060 | | | 0.040 | 0 |
| Change | $-x$ | | | $+x$ | $+x$ |
| Equilibrium | $0.060 - x$ | | | $0.040 + x$ | $x$ |

**STEP 2**   You substitute the equilibrium concentrations into the equilibrium-constant equation.

$$\frac{[NH_4^+][OH^-]}{[NH_3]} = K_b$$

$$\frac{(0.040 + x)x}{0.060 - x} = 1.8 \times 10^{-5}$$

**STEP 3**   To solve this equation, you assume that $x$ is small compared with 0.040 and 0.060. So this equation becomes

$$\frac{0.040x}{0.060} = 1.8 \times 10^{-5}$$

Therefore, $x = (0.060/0.040) \times (1.8 \times 10^{-5}) = 2.7 \times 10^{-5}$. (Check that $x$ can be neglected in $0.040 + x$ and $0.060 - x$.) Thus, the hydroxide-ion concentration is $2.7 \times 10^{-5} \ M$. The pH of the buffer is

$$pH = 14.00 - pOH = 14.00 + \log(2.7 \times 10^{-5}) = \textbf{9.43}$$

(If you have $2.7 \times 10^{-5}$ in your calculator from the previous calculation, you obtain the pH by pressing the *log* button, then adding 14.)

---

**EXERCISE 17.12**

What is the pH of a buffer prepared by adding 30.0 mL of 0.15 $M$ $HC_2H_3O_2$ (acetic acid) to 70.0 mL of 0.20 $M$ $NaC_2H_3O_2$ (sodium acetate)?

---

See Problems 17.67 and 17.68.

**Problem Solving:**
**The pH of a Buffered Solution**

**Understanding**
**Concepts: Buffers**

## Adding an Acid or a Base to a Buffer

We noted earlier that a buffer resists changes in pH, and we also described qualitatively how this occurs. What we want to do here is to show quantitatively that a buffer does indeed tend to resist pH change when a small quantity of an acid or a base is added to it.

All of the previous acid–base examples dealt with situations in which an acid and its conjugate base were close to equilibrium. In the following example, we consider the addition of a strong acid to an acetic acid/acetate ion buffer. Hydronium ion reacts with the conjugate base (acetate ion) in the buffer, so that the concentrations of species in solution change markedly from their initial value. (If you add a strong base to this buffer, hydroxide ion reacts with acetic acid in the buffer.)

### EXAMPLE 17.12

## Calculating the pH of a Buffer When a Strong Acid or Strong Base Is Added

Calculate the pH of 75 mL of the buffer solution described in Example 17.10 (0.10 $M$ $HC_2H_3O_2$ and 0.20 $M$ $NaC_2H_3O_2$) to which 9.5 mL of 0.10 $M$ hydrochloric acid is added. Compare the pH change with what would occur if this amount of acid were added to pure water.

 **PROBLEM STRATEGY**

Do the problem in two parts. First, you assume that the $H_3O^+$ ion from the strong acid and the conjugate base from the buffer react completely. This is a stoichiometric calculation. Actually, the $H_3O^+$ ion and the base from the buffer reach equilibrium just before complete reaction. So you now solve the equilibrium problem using concentrations from the stoichiometric calculation. Because these concentrations are not far from equilibrium, you can use the usual simplifying assumption about $x$.

**SOLUTION**

When hydronium ion (from hydrochloric acid) is added to the buffer, it reacts with acetate ion.

$$H_3O^+(aq) + C_2H_3O_2^-(aq) \longrightarrow HC_2H_3O_2(aq) + H_2O(l)$$

Because acetic acid is a weak acid, you can assume as a first approximation that the reaction goes to completion. This part of the problem is simply a stoichiometric calculation. Then you assume that the acetic acid ionizes slightly. This part of the problem involves an acid-ionization equilibrium.

**STOICHIOMETRIC CALCULATION**   You must first calculate the amounts of hydrogen ion, acetate ion, and acetic acid present in the solution before reaction. The molar amount of hydrogen ion will equal the molar amount of hydrochloric acid added, which you obtain by converting the volume of hydrochloric acid, HCl, to moles of HCl. To do this, you note for 0.10 $M$ HCl that 1 L HCl is equivalent to 0.10 mol HCl. Hence, to convert 9.5 mL (= $9.5 \times 10^{-3}$ L) of hydrochloric acid, you have

$$9.5 \times 10^{-3} \; \text{L HCl} \times \frac{0.10 \; \text{mol HCl}}{1 \; \text{L HCl}} = 0.00095 \; \text{mol HCl}$$

Hydrochloric acid is a strong acid, so it exists in solution as the ions $H_3O^+$ and $Cl^-$. Therefore, the amount of $H_3O^+$ added is 0.00095 mol.

The amounts of acetate ion and acetic acid are found in a similar way. The buffer in Example 17.10 contains 0.20 mol of acetate ion and 0.10 mol of acetic acid in 1 L of solution. The amounts in 75 mL (= 0.075 L) of solution are obtained by converting to moles.

$$0.075 \; \text{L soln} \times \frac{0.20 \; \text{mol} \; C_2H_3O_2^-}{1 \; \text{L soln}} = 0.015 \; \text{mol} \; C_2H_3O_2^-$$

$$0.075 \; \text{L soln} \times \frac{0.10 \; \text{mol} \; HC_2H_3O_2}{1 \; \text{L soln}} = 0.0075 \; \text{mol} \; HC_2H_3O_2$$

You now assume that all of the hydronium ion added (0.00095 mol) reacts with acetate ion. Therefore, 0.00095 mol of acetic acid is produced and 0.00095 mol of acetate ion is used up. Hence, after reaction you have

Moles of acetate ion = (0.015 − 0.00095) mol $C_2H_3O_2^-$ = 0.014 mol $C_2H_3O_2^-$

Moles of acetic acid = (0.0075 + 0.00095) mol $HC_2H_3O_2$ = 0.0085 mol $HC_2H_3O_2$

**EQUILIBRIUM CALCULATION** You first calculate the concentrations of $HC_2H_3O_2$ and $C_2H_3O_2^-$ present in the solution before you consider the acid-ionization equilibrium. Note that the total volume of solution (buffer plus hydrochloric acid) is 75 mL + 9.5 mL, or 85 mL (0.085 L). Hence, the starting concentrations are

$$[HC_2H_3O_2] = \frac{0.0085 \; \text{mol}}{0.085 \; \text{L}} = 0.10 \; M$$

$$[C_2H_3O_2^-] = \frac{0.014 \; \text{mol}}{0.085 \; \text{L}} = 0.16 \; M$$

From this, you construct the following table:

| Concentration ($M$) | $HC_2H_3O_2(aq) +$ | $H_2O(l) \rightleftharpoons$ | $H_3O^+(aq) +$ | $C_2H_3O_2^-(aq)$ |
|---|---|---|---|---|
| Starting | 0.10 | | ~0 | 0.16 |
| Change | $-x$ | | $+x$ | $+x$ |
| Equilibrium | $0.10 - x$ | | $x$ | $0.17 + x$ |

The equilibrium-constant equation is

$$\frac{[H_3O^+][C_2H_3O_2^-]}{[HC_2H_3O_2]} = K_a$$

Substituting, you get

$$\frac{x(0.16 + x)}{0.10 - x} = 1.7 \times 10^{-5}$$

If you assume that $x$ is small enough that $0.16 + x \approx 0.16$ and $0.10 - x \approx 0.10$, this equation becomes

$$\frac{x(0.16)}{0.10} = 1.7 \times 10^{-5}$$

or

$$x = 1.7 \times 10^{-5} \times \frac{0.10}{0.16} = 1.1 \times 10^{-5}$$

Note that $x$ is indeed small, so the assumptions you made earlier are correct. The $H_3O^+$ concentration is $1.0 \times 10^{-5}\,M$. The pH is

$$pH = -\log[H_3O^+] = -\log(1.1 \times 10^{-5}) = \mathbf{4.96}$$

Because the pH of the buffer was 5.07 (see Example 17.10), the pH has changed by $5.07 - 4.96 = 0.11$ units.

**ADDING HCl TO PURE WATER** If 9.5 mL of $0.10\,M$ hydrochloric acid were added to 75 mL of pure water, the hydronium-ion concentration would change to

$$[H_3O^+] = \frac{\text{amount of } H_3O^+ \text{ added}}{\text{total volume of solution}}$$

$$[H_3O^+] = \frac{0.00095\ \text{mol } H_3O^+}{0.085\ \text{L solution}} = 0.011\,M$$

(The total volume is 75 mL of water plus 9.5 mL HCl, assuming no change of volume on mixing.) The pH is

$$pH = -\log[H_3O^+] = -\log(0.011) = 1.96$$

**The pH of pure water is 7.00, so the change in pH is $7.00 - 1.96 = 5.04$ units, compared with 0.11 units for the buffered solution.**

**Problem Solving:
pH Changes in a
Buffered Solution**

### EXERCISE 17.13

What is the pH of the solution described in Exercise 17.11 if 50.0 mL of $0.10\,M$ sodium hydroxide is added to 1 L of solution?

**See Problems 17.69 and 17.70.**

## Henderson–Hasselbalch Equation

How do you prepare a buffer of given pH? We can show that the buffer must be prepared from a conjugate acid–base pair in which the acid-ionization constant is approximately equal to the desired $H_3O^+$ concentration. To illustrate, consider a buffer made up of a weak acid HA and its conjugate base $A^-$. The acid-ionization equilibrium is

$$HA(aq) + H_2O(l) \rightleftharpoons H_3O^+(aq) + A^-(aq)$$

and the acid-ionization constant is

$$K_a = \frac{[H_3O^+][A^-]}{[HA]}$$

By rearranging, you get an equation for the $H_3O^+$ concentration.

$$[H_3O^+] = K_a \times \frac{[HA]}{[A^-]}$$

This equation expresses the $H_3O^+$ concentration in terms of the $K_a$ for the acid and the ratio of concentrations of HA and $A^-$. This equation was derived from the equilibrium constant, so the concentrations of HA and $A^-$ should be equilibrium values. But because the presence of $A^-$ represses the ionization of HA, these concentrations do not differ significantly from the values used to prepare the buffer. If [HA] and

[A$^-$] are approximately equal, the hydronium-ion concentration of the buffer is approximately equal to $K_a$.

You can use the preceding equation to derive an equation for the pH of a buffer. Take the negative logarithm of both sides of the equation. That is,

$$-\log [H_3O^+] = -\log\left(K_a \times \frac{[HA]}{[A^-]}\right) = -\log K_a - \log\frac{[HA]}{[A^-]}$$

The left side equals the pH. You can also simplify the right side. The p$K_a$ of a weak acid is defined in a manner similar to pH and pOH. ◀

▶ Acid- and base-ionization constants are often listed as p$K_a$s and p$K_b$s (p$K_b$ = −log $K_b$).

$$pK_a = -\log K_a$$

The previous equation can be written

$$pH = pK_a - \log\frac{[HA]}{[A^-]} = pK_a + \log\frac{[A^-]}{[HA]}$$

More generally, you can write

$$pH = pK_a + \log\frac{[base]}{[acid]}$$

This is *an equation relating the pH of a buffer for different concentrations of conjugate acid and base;* it is known as the **Henderson–Hasselbalch equation.** By substituting the value of p$K_a$ for the conjugate acid and the ratio [base]/[acid], you obtain the pH of the buffer.

The question we asked earlier was how to prepare a buffer of a given pH—for example, pH 4.90. You can see that you need to find a conjugate acid–base pair in which the p$K_a$ of the acid is close to the desired pH. $K_a$ for acetic acid is $1.7 \times 10^{-5}$, and its p$K_a$ is $-\log(1.7 \times 10^{-5}) = 4.77$. You can get a pH somewhat higher by increasing the ratio [base]/[acid].

Consider the calculation of the pH of a buffer containing 0.10 $M$ NH$_3$ and 0.20 $M$ NH$_4$Cl. The conjugate acid is NH$_4^+$, whose $K_a$ you can calculate from $K_b$ for NH$_3$ ($= 1.8 \times 10^{-5}$). $K_a$ for NH$_4^+$ is $5.6 \times 10^{-10}$, and the p$K_a$ is $-\log(5.6 \times 10^{-10}) = 9.25$. Hence,

$$pH = 9.25 + \log\frac{0.10}{0.20} = 8.95$$

The solution is basic, as you could have guessed, because the buffer contains the weak base NH$_3$ and the very weak acid NH$_4^+$.

## CONCEPT CHECK 17.3

You add 1.5 mL of 1 $M$ HCl to each of the following solutions. Which one will show the least change of pH?

a. 15 mL of 0.1 $M$ NaOH

b. 15 mL of 0.1 $M$ HC$_2$H$_3$O$_2$

c. 30 mL of 0.1 $M$ NaOH and 30 mL of 0.1 $M$ HC$_2$H$_3$O$_2$

d. 30 mL of 0.1 $M$ NaOH and 60 mL of 0.1 $M$ HC$_2$H$_3$O$_2$

# 17.7   Acid–Base Titration Curves

An acid–base titration is a procedure for determining the amount of acid (or base) in a solution by determining the volume of base (or acid) of known concentration that will completely react with it. ◀ An **acid–base titration curve** is *a plot of the pH of a solution of acid (or base) against the volume of added base (or acid).* Such curves are used to gain insight into the titration process. You can use the titration curve to choose an indicator that will show when the titration is complete.

▶ The technique of titration was discussed in Section 4.10.

Visualization: Neutralization of a Strong Acid by a Strong Base

▶ Experiments in General Chemistry, 17B, "An Acid–Base Titration Curve."

**Figure 17.11**
**Curve for the titration of a strong acid by a strong base.**  Here 25.0 mL of 0.100 *M* HCl is titrated by 0.100 *M* NaOH. The portions of the curve where indicators bromcresol green and phenolphthalein change color are shown. Note that both indicators change color where the pH changes rapidly (the nearly vertical part of the curve).

## Titration of a Strong Acid by a Strong Base

Figure 17.11 shows a curve for the titration of 25.0 mL of 0.100 *M* HCl by 0.100 *M* NaOH. Note that the pH changes slowly at first until the molar amount of base added nearly equals that of the acid—that is, until the titration is near the equivalence point. The **equivalence point** is *the point in a titration when a stoichiometric amount of reactant has been added.* At the equivalence point, the pH of this solution of NaOH and HCl is 7.0, because it contains a salt, NaCl, that does not hydrolyze. However, the pH changes rapidly near the equivalence point, from a pH of about 3 to a pH of about 11. To detect the equivalence point, you add an indicator that changes color within the pH range 3–11. Phenolphthalein can be used, because it changes from colorless to pink in the pH range 8.2–10.0. (Figure 16.8 showed the pH ranges for the color changes of indicators.) Even though this color change occurs on the basic side, only a fraction of a drop of base is required to change the pH several units when the titration is near the equivalence point. The indicator bromcresol green, whose color changes in the pH range 3.8–5.4, would also work. Because the pH change is so large, many other indicators could be used.

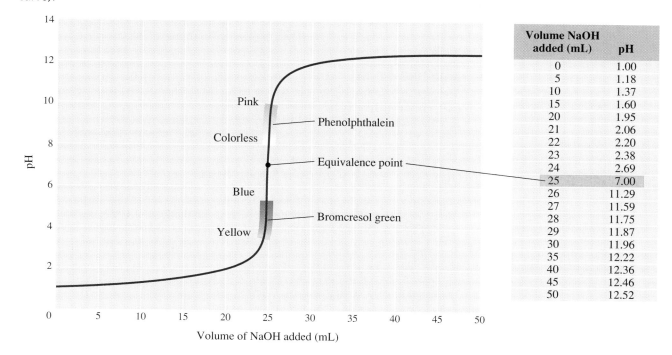

| Volume NaOH added (mL) | pH |
|---|---|
| 0 | 1.00 |
| 5 | 1.18 |
| 10 | 1.37 |
| 15 | 1.60 |
| 20 | 1.95 |
| 21 | 2.06 |
| 22 | 2.20 |
| 23 | 2.38 |
| 24 | 2.69 |
| 25 | 7.00 |
| 26 | 11.29 |
| 27 | 11.59 |
| 28 | 11.75 |
| 29 | 11.87 |
| 30 | 11.96 |
| 35 | 12.22 |
| 40 | 12.36 |
| 45 | 12.46 |
| 50 | 12.52 |

Volume of NaOH added (mL)

The following example shows how to calculate a point on the titration curve of a strong acid and a strong base.

▶ **Animation: Neutralization of a strong acid by a strong base.**

### EXAMPLE 17.13

## Calculating the pH of a Solution of a Strong Acid and a Strong Base

Calculate the pH of a solution in which 10.0 mL of 0.100 $M$ NaOH is added to 25.0 mL of 0.100 $M$ HCl.

**SOLUTION**

Because the reactants are a strong acid and a strong base, the reaction is essentially complete. The equation is

$$H_3O^+(aq) + OH^-(aq) \longrightarrow H_2O(l) + H_2O(l)$$

You get the amounts of reactants by multiplying the volume (in liters) of each solution by its molar concentration.

$$Mol\ H_3O^+ = 0.0250\ L \times 0.100\ mol/L = 0.00250\ mol$$

$$Mol\ OH^- = 0.0100\ L \times 0.100\ mol/L = 0.00100\ mol$$

All of the $OH^-$ reacts, leaving an excess of $H_3O^+$.

$$Excess\ H_3O^+ = (0.00250 - 0.00100)\ mol = 0.00150\ mol\ H_3O^+$$

You obtain the concentration of $H_3O^+$ by dividing this amount of $H_3O^+$ by the total volume of solution (= 0.0250 L + 0.0100 L = 0.0350 L).

$$[H_3O^+] = \frac{0.00150\ mol}{0.0350\ L} = 0.0429$$

Hence,

$$pH = -\log\,[H_3O^+] = -\log(0.0429) = \mathbf{1.368}$$

### EXERCISE 17.14

What is the pH of a solution in which 15 mL of 0.10 $M$ NaOH has been added to 25 mL of 0.10 $M$ HCl?

See Problems 17.77 and 17.78.

## Titration of a Weak Acid by a Strong Base

The titration of a weak acid by a strong base gives a somewhat different curve. Figure 17.12 shows the curve for the titration of 25.0 mL of 0.100 $M$ nicotinic acid, $HC_6H_4NO_2$, by 0.100 $M$ NaOH. The titration starts at a higher pH than the titration of HCl, because nicotinic acid is a weak acid. As before, the pH changes slowly at first, then rapidly near the equivalence point. The pH range in which the rapid change is seen occurs from about pH 7 to pH 11. Note that the pH range is shorter than that for the titration of a strong acid by a strong base. This means that the choice of an indicator is more critical. Phenolphthalein would work; it changes color in the range 8.2–10.0. Bromcresol green would not work because it changes color in the range 3.8–5.4, which occurs before the titration curve rises steeply.

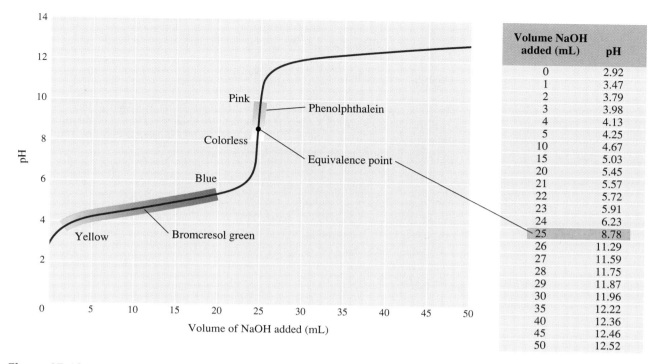

| Volume NaOH added (mL) | pH |
|---|---|
| 0 | 2.92 |
| 1 | 3.47 |
| 2 | 3.79 |
| 3 | 3.98 |
| 4 | 4.13 |
| 5 | 4.25 |
| 10 | 4.67 |
| 15 | 5.03 |
| 20 | 5.45 |
| 21 | 5.57 |
| 22 | 5.72 |
| 23 | 5.91 |
| 24 | 6.23 |
| 25 | 8.78 |
| 26 | 11.29 |
| 27 | 11.59 |
| 28 | 11.75 |
| 29 | 11.87 |
| 30 | 11.96 |
| 35 | 12.22 |
| 40 | 12.36 |
| 45 | 12.46 |
| 50 | 12.52 |

**Figure 17.12**
**Curve for the titration of a weak acid by a strong base.** Here 25.0 mL of 0.100 $M$ nicotinic acid, a weak acid, is titrated by 0.100 $M$ NaOH. Note that bromcresol green changes color during the early part of the titration, well before the equivalence point. Phenolphthalein changes color where the pH changes rapidly (near the equivalence point). Thus, phenolphthalein could be used as an indicator for the titration, whereas bromcresol green could not.

▶ An alternate example is provided in the Instructor's Resource Manual.

Note also that the equivalence point for the titration curve of nicotinic acid occurs on the basic side. This happens because at the equivalence point the solution is that of the salt, sodium nicotinate, which is basic from the hydrolysis of the nicotinate ion. The optimum choice of indicator would be one that changes color over a range that includes the pH of the equivalence point.

The following example shows how to calculate the pH at the equivalence point in the titration of a weak acid and a strong base.

## EXAMPLE 17.14

## Calculating the pH at the Equivalence Point in the Titration of a Weak Acid by a Strong Base

Calculate the pH of the solution at the equivalence point when 25 mL of 0.10 $M$ nicotinic acid is titrated by 0.10 $M$ sodium hydroxide. $K_a$ for nicotinic acid equals $1.4 \times 10^{-5}$.

 **PROBLEM STRATEGY**

At the equivalence point, equal molar amounts of nicotinic acid and sodium hydroxide react to give a solution of sodium nicotinate. You first calculate the concentration of nicotinate ion (stoichiometry problem). Then you find the pH of this solution (hydrolysis problem).

**SOLUTION**

**CONCENTRATION OF NICOTINATE ION**  Assume that the reaction of the base with the acid is complete. In this case, 25 mL of 0.10 $M$ sodium hydroxide is needed to react with 25 mL of 0.10 $M$ nicotinic acid. The molar amount of nicotinate ion formed equals the initial molar amount of nicotinic acid.

$$25 \times 10^{-3} \, \text{L soln} \times \frac{0.10 \, \text{mol nicotinate ion}}{1 \, \text{L soln}} = 2.5 \times 10^{-3} \, \text{mol nicotinate ion}$$

The total volume of solution is 50 mL (25 mL NaOH solution plus 25 mL of nicotinic acid solution, assuming there is no change in volume on mixing). Dividing the molar amount of nicotinate ion by the volume of solution in liters gives the molar concentration of nicotinate ion.

$$\text{Molar concentration} = \frac{2.5 \times 10^{-3} \, \text{mol}}{50 \times 10^{-3} \, \text{L}} = 0.050 \, M$$

**HYDROLYSIS OF NICOTINATE ION**    This portion of the calculation follows the method given in Example 17.8. You find that $K_b$ for nicotinate ion is $7.1 \times 10^{-10}$ and that the concentration of hydroxide ion is $6.0 \times 10^{-6} \, M$. The pH is **8.78.**

---

**EXERCISE 17.15**

What is the pH at the equivalence point when 25 mL of 0.10 $M$ HF is titrated by 0.15 $M$ NaOH?

---

See Problems 17.79 and 17.80.

**Figure 17.13**
**Curve for the titration of a weak base by a strong acid.**    Here 25.0 mL of 0.100 $M$ NH$_3$ is titrated by 0.100 $M$ HCl. Methyl red can be used as an indicator for the titration.

## Titration of a Weak Base by a Strong Acid

When you titrate a weak base by a strong acid, you get a titration curve similar to that obtained when a weak acid is titrated by a strong base. Figure 17.13 shows the

| Volume HCl added (mL) | pH |
|---|---|
| 0 | 11.13 |
| 1 | 10.64 |
| 2 | 10.32 |
| 3 | 10.13 |
| 4 | 9.98 |
| 5 | 9.86 |
| 10 | 9.44 |
| 15 | 9.08 |
| 20 | 8.66 |
| 21 | 8.54 |
| 22 | 8.39 |
| 23 | 8.20 |
| 24 | 7.88 |
| 25 | 5.28 |
| 26 | 2.70 |
| 27 | 2.40 |
| 28 | 2.22 |
| 29 | 2.10 |
| 30 | 2.00 |
| 35 | 1.70 |
| 40 | 1.52 |
| 45 | 1.40 |
| 50 | 1.30 |

pH changes during the titration of 25.0 mL of 0.100 $M$ $NH_3$ by 0.100 $M$ HCl. In this case, the pH declines slowly at first, then falls abruptly from about pH 7 to pH 3. Methyl red, which changes color from yellow at pH 6.0 to red at pH 4.8, is a possible indicator for this titration. Note that phenolphthalein could not be used to find the equivalence point.

---

### EXERCISE 17.16

What is the pH at the equivalence point when 35 mL of 0.20 $M$ ammonia is titrated by 0.12 $M$ hydrochloric acid? $K_b$ for ammonia is $1.8 \times 10^{-5}$.

---

See Problems 17.81 and 17.82.

# A Checklist for Review

## Important Terms

acid-ionization
constant (17.1)
degree of ionization (17.1)

base-ionization
constant (17.3)
hydrolysis (17.4)

common-ion effect (17.5)
buffer (17.6)
Henderson–Hasselbalch
equation (17.6)

acid–base titration
curve (17.7)
equivalence point (17.7)

## Key Equations

$$K_a = \frac{[H_3O^+][A^-]}{[HA]}$$

$$K_b = \frac{[HB^+][OH^-]}{[B]}$$

$$K_a K_b = K_w$$

$$pH = pK_a + \log\frac{[base]}{[acid]}$$

## Summary of Facts and Concepts

When a weak acid dissolves in water, it ionizes to give hydronium ions and the conjugate base ions. The equilibrium constant for this *acid ionization* is $K_a = [H_3O^+][A^-]/[HA]$, where HA is the general formula for the acid. The constant $K_a$ can be determined from the pH of an acid solution of known concentration. Once obtained, the acid-ionization constant can be used to find the concentrations of species in any solution of the acid. In the case of a *diprotic* acid, $H_2A$, the concentration of $H_3O^+$ and $HA^-$ are calculated from $K_{a1}$, and the concentration of $A^{2-}$ equals $K_{a2}$. Similar considerations apply to *base ionizations.*

Solutions of salts may be acidic or basic because of *hydrolysis* of the ions. The equilibrium constant for hydrolysis equals $K_a$ for a cation or $K_b$ for an anion. Calculation of the pH of a solution of a salt in which one ion hydrolyzes is fundamentally the same as calculation of the pH of a solution of an acid or base. However, $K_a$ or $K_b$ for an ion is usually obtained from the conjugate base or acid by applying the equation $K_a K_b = K_w$.

A *buffer* is a solution that can resist changes in pH when small amounts of acid or base are added to it. A buffer contains either a weak acid and its conjugate base or a weak base and its conjugate acid. The concentrations of acid and base conjugates are approximately equal. Different pH ranges are possible for acid–base conjugates, depending on their ionization constants.

An *acid–base titration curve* is a plot of the pH of the solution against the volume of reactant added. During the titration of a strong acid by a strong base, the pH changes slowly at first. Then, as the amount of base nears the stoichiometric value—that is, nears the *equivalence point*—the pH rises abruptly, changing by several units. The pH at the equivalence point is 7.0. A similar curve is obtained when a weak acid is titrated by a strong base. However, the pH changes less at the equivalence point. Moreover, the pH at the equivalence point is greater than 7.0 because of hydrolysis of the salt produced. An indicator must be chosen that changes color within a pH range near the equivalence point, where the pH changes rapidly.

## Operational Skills

1. **Determining $K_a$ (or $K_b$) from the solution pH** Given the molarity and pH of a solution of a weak acid, calculate $K_a$ for the acid **(EXAMPLE 17.1)**. The $K_b$ for a base can be determined in a similar way (see Exercise 17.5).

2. **Calculating concentrations of species in a weak acid solution using $K_a$** Given $K_a$, calculate the hydronium-ion concentration and pH of a solution of a weak acid of known molarity **(EXAMPLES 17.2, 17.3)**. Given $K_{a1}$, $K_{a2}$, and the molarity of a diprotic acid solution, calculate the pH and the concentrations of $H_3O^+$, $HA^-$, and $A^{2-}$ **(EXAMPLE 17.4)**.

3. **Calculating concentrations of species in a weak base solution using $K_b$** Given $K_b$, calculate the hydronium-ion concentration and pH of a solution of a weak base of known molarity. **(EXAMPLE 17.5)**

4. **Predicting whether a salt solution is acidic, basic, or neutral** Decide whether an aqueous solution of a given salt is acidic, basic, or neutral. **(EXAMPLE 17.6)**

5. **Obtaining $K_a$ from $K_b$ or $K_b$ from $K_a$** Calculate $K_a$ for a cation or $K_b$ for an anion from the ionization constant of the conjugate base or acid. **(EXAMPLE 17.7)**

6. **Calculating concentrations of species in a salt solution** Given the concentration of a solution of a salt in which one ion hydrolyzes, and given the ionization constant of the conjugate acid or base of this ion, calculate the $H_3O^+$ concentration. **(EXAMPLE 17.8)**

7. **Calculating the common-ion effect on acid ionization** Given $K_a$ and the concentrations of weak acid and strong acid in a solution, calculate the degree of ionization of the weak acid **(EXAMPLE 17.9)**. Given $K_a$ and the concentrations of weak acid and its salt in a solution, calculate the pH **(EXAMPLE 17.10)**.

8. **Calculating the pH of a buffer from given volumes of solution** Given concentrations and volumes of acid and conjugate base from which a buffer is prepared, calculate the buffer pH. **(EXAMPLE 17.11)**

9. **Calculating the pH of a buffer when a strong acid or strong base is added** Calculate the pH of a given volume of buffer solution (given the concentrations of conjugate acid and base in the buffer) to which a specified amount of strong acid or strong base is added. **(EXAMPLE 17.12)**

10. **Calculating the pH of a solution of a strong acid and a strong base** Calculate the pH during the titration of a strong acid by a strong base, given the volumes and concentrations of the acid and base. **(EXAMPLE 17.13)**

11. **Calculating the pH at the equivalence point in the titration of a weak acid by a strong base** Calculate the pH at the equivalence point for the titration of a weak acid by a strong base **(EXAMPLE 17.14)**. Be able to do the same type of calculation for the titration of a weak base by a strong acid.

## Review Questions

**17.1** Write an equation for the ionization of hydrogen cyanide, HCN, in aqueous solution. What is the equilibrium expression $K_a$ for this acid ionization?

**17.2** Which of the following is the weakest acid: $HClO_4$, HCN, or $HC_2H_3O_2$? See Table 16.1 and Table 17.1.

**17.3** Briefly describe two methods for determining $K_a$ for a weak acid.

**17.4** Describe how the degree of ionization of a weak acid changes as the concentration increases.

**17.5** Consider a solution of 0.0010 $M$ HF ($K_a = 6.8 \times 10^{-4}$). In solving for the concentrations of species in this solution, could you use the simplifying assumption in which you neglect $x$ in the denominator of the equilibrium equation? Explain.

**17.6** Phosphorous acid, $H_2PHO_3$, is a diprotic acid. Write equations for the acid ionizations. Write the expressions for $K_{a1}$ and $K_{a2}$.

**17.7** What is the concentration of oxalate ion, $C_2O_4^{2-}$, in 0.10 $M$ oxalic acid, $H_2C_2O_4$? $K_{a1}$ is $5.6 \times 10^{-2}$, and $K_{a2}$ is $5.1 \times 10^{-5}$.

**17.8** Write the equation for the ionization of aniline, $C_6H_5NH_2$, in aqueous solution. Write the expression for $K_b$.

**17.9** Which of the following is the strongest base: $NH_3$, $C_6H_5NH_2$, or $CH_3NH_2$? See Table 17.2.

**17.10** Do you expect a solution of anilinium chloride (aniline hydrochloride), $C_6H_5NH_3Cl$, to be acidic or basic? (Anilinium chloride is the salt of aniline and hydrochloric acid.) Write the equation for the reaction involved. What is the equilibrium expression? How would you obtain the value for the equilibrium constant from Table 17.2?

**17.11** What is meant by the common-ion effect? Give an example.

**17.12** The pH of 0.10 $M$ $CH_3NH_2$ (methylamine) is 11.8. When the chloride salt of methylamine, $CH_3NH_3Cl$, is added to this solution, does the pH increase or decrease? Explain, using Le Chatelier's principle and the common-ion effect.

**17.13** Define the term *buffer*. Give an example.

**17.14** What is meant by the capacity of a buffer? Describe a buffer with low capacity and the same buffer with greater capacity.

**17.15** Describe the pH changes that occur during the titration of a weak base by a strong acid. What is meant by the term *equivalence point*?

**17.16** If the pH is 8.0 at the equivalence point for the titration of a certain weak acid with sodium hydroxide, what indicator might you use? See Figure 16.8. Explain your choice.

# Conceptual Problems

**17.17** You have 0.10-mol samples of three acids identified simply as HX, HY, and HZ. For each acid, you make up 0.10 $M$ solutions by adding sufficient water to each of the acid samples. When you measure the pH of these samples, you find that the pH of HX is greater than the pH of HY, which in turn is greater than the pH of HZ.
a. Which of the acids is the least ionized in its solution?
b. Which acid has the largest $K_a$?

**17.18** What reaction occurs when each of the following is dissolved in water?
a. HF    b. NaF    c. $C_6H_5NH_2$    d. $C_6H_5NH_3Cl$

**17.19** You have the following solutions, all of the same molar concentrations: KBr, HBr, $CH_3NH_2$, and $NH_4Cl$. Rank them from the lowest to the highest hydroxide-ion concentrations.

**17.20** Rantidine is a nitrogen base that is used to control stomach acidity by suppressing the stomach's production of hydrochloric acid. The compound is present in Zantac® as the chloride salt (rantidinium chloride; also called rantidine hydrochloride). Do you expect a solution of rantidine hydrochloride to be acidic, basic, or neutral? Explain by means of a general chemical equation.

**17.21** A chemist prepares dilute solutions of equal molar concentrations of $NH_3$, $NH_4Br$, NaF, and NaCl. Rank these solutions from highest pH to lowest pH.

**17.22** You want to prepare a buffer solution that has a pH equal to the $pK_a$ of the acid component of the buffer. If you have 100 mL of a 0.10 $M$ solution of the acid HA, what volume

and concentration of NaA solution could you use in order to prepare the buffer?

**17.23** A friend of yours has performed three titrations: strong acid with a strong base, weak acid with a strong base, and a weak base with a strong acid. He hands you the three titration curves, saying he has forgotten which is which. What attributes of the curves would you look at to correctly identify each curve?

**17.24** You are given the following acid–base titration data, where each point on the graph represents the pH after adding a given volume of titrant (substance being added during the titration).

Volume of Titrant (mL)

a. What substance is being titrated, a strong acid, strong base, weak acid, or weak base?
b. What is the pH at the equivalence point of the tiration?
c. What indicator might you use to perform this titration? Explain.

# Practice Problems

*Note:* For values of $K_a$ and $K_b$ that are not given in the following problems, see Tables 17.1 and 17.2.

## Acid Ionization

**17.25** Write chemical equations for the acid ionizations of each of the following weak acids (express these in terms of $H_3O^+$).
a. HOCN (cyanic acid)    b. HF (hydrofluoric acid)
c. HIO (hypoiodous acid)    d. $HCO_2H$ (formic acid)

**17.26** Write chemical equations for the acid ionizations of each of the following weak acids (express these in terms of $H_3O^+$).

a. $HN_3$ (hydrazoic acid)    b. $HClO_2$ (chlorous acid)
c. $HNO_2$ (nitrous acid)    d. HCN (hydrocyanic acid)

**17.27** Acrylic acid, whose formula is $HC_3H_3O_2$ or $HO_2CCH{=}CH_2$, is used in the manufacture of plastics. A 0.10 $M$ aqueous solution of acrylic acid has a pH of 2.63. What is $K_a$ for acrylic acid?

**17.28** Heavy metal azides, which are salts of hydrazoic acid, $HN_3$, are used as explosive detonators. A solution of 0.20 $M$ hydrazoic acid has a pH of 3.21. What is the $K_a$ for hydrazoic acid?

**17.29** Boric acid, $B(OH)_3$, is used as a mild antiseptic. What is the pH of a 0.025 $M$ aqueous solution of boric acid?

What is the degree of ionization of boric acid in this solution? The hydronium ion arises principally from the reaction

$$B(OH)_3(aq) + 2H_2O(l) \rightleftharpoons B(OH)_4^-(aq) + H_3O^+(aq)$$

The equilibrium constant for this reaction is $5.9 \times 10^{-10}$.

**17.30**    Formic acid, $HCHO_2$, is used to make methyl formate (a fumigant for dried fruit) and ethyl formate (an artificial rum flavor). What is the pH of a 0.12 $M$ solution of formic acid? What is the degree of ionization of $HCHO_2$ in this solution? See Table 17.1 for $K_a$.

**17.31**    $C_6H_4NH_2COOH$, *para*-aminobenzoic acid (PABA), is used in some sunscreen agents. Calculate the concentrations of hydronium ion and *para*-aminobenzoate ion, $C_6H_4NH_2 COO^-$, in a 0.050 $M$ solution of the acid. The value of $K_a$ is $2.2 \times 10^{-5}$.

**17.32**    Barbituric acid, $HC_4H_3N_2O_3$, is used to prepare various barbiturate drugs (used as sedatives). Calculate the concentrations of hydronium ion and barbiturate ion in a 0.20 $M$ solution of the acid. The value of $K_a$ is $9.8 \times 10^{-5}$.

**17.33**    A solution of acetic acid, $HC_2H_3O_2$, on a laboratory shelf was of undetermined concentration. If the pH of the solution was found to be 2.68, what was the concentration of the acetic acid?

**17.34**    A chemist wanted to determine the concentration of a solution of lactic acid, $HC_3H_5O_3$. She found that the pH of the solution was 2.51. What was the concentration of the solution? The $K_a$ of lactic acid is $1.4 \times 10^{-4}$.

**17.35**    Hydrofluoric acid, HF, unlike hydrochloric acid, is a weak electrolyte. What is the hydronium-ion concentration and the pH of a 0.040 $M$ aqueous solution of HF?

**17.36**    Chloroacetic acid, $HC_2H_2ClO_2$, has a greater acid strength than acetic acid, because the electronegative chlorine atom pulls electrons away from the O—H bond and thus weakens it. Calculate the hydronium-ion concentration and the pH of a 0.015 $M$ solution of chloroacetic acid. $K_a$ is $1.3 \times 10^{-3}$.

**17.37**    What is the hydronium-ion concentration of a 2.00 $M$ solution of 2,6-dinitrobenzoic acid, $(NO_2)_2C_6H_3COOH$, for which $K_a = 7.94 \times 10^{-2}$?

**17.38**    What is the hydronium-ion concentration of a $2.00 \times 10^{-4}$ $M$ solution of *p*-bromobenzoic acid, $BrC_6H_4COOH$, for which $K_a = 1.00 \times 10^{-4}$?

**17.39**    Phthalic acid, $H_2C_8H_4O_4$, is a diprotic acid used in the synthesis of phenolphthalein indicator. $K_{a1} = 1.2 \times 10^{-3}$, and $K_{a2} = 3.9 \times 10^{-6}$. (a) Calculate the hydronium-ion concentration of a 0.015 $M$ solution. (b) What is the concentration of the $C_8H_4O_4^{2-}$ ion in the solution?

**17.40**    Carbonic acid, $H_2CO_3$, can be found in a wide variety of body fluids (from dissolved $CO_2$). (a) Calculate the hydronium-ion concentration of a $5.45 \times 10^{-4}$ $M$ $H_2CO_3$ solution. (b) What is the concentration of $CO_3^{2-}$?

## Base Ionization

**17.41**    Write the chemical equation for the base ionization of methylamine, $CH_3NH_2$. Write the $K_b$ expression for methylamine.

**17.42**    Write the chemical equation for the base ionization of aniline, $C_6H_5NH_2$. Write the $K_b$ expression for aniline.

**17.43**    Ethanolamine, $HOC_2H_4NH_2$, is a viscous liquid with an ammonialike odor; it is used to remove hydrogen sulfide from natural gas. A 0.15 $M$ aqueous solution of ethanolamine has a pH of 11.34. What is $K_b$ for ethanolamine?

**17.44**    Trimethylamine, $(CH_3)_3N$, is a gas with a fishy, ammonialike odor. An aqueous solution that is 0.25 $M$ trimethylamine has a pH of 11.63. What is $K_b$ for trimethylamine?

**17.45**    What is the concentration of hydroxide ion in a 0.065 $M$ aqueous solution of methylamine, $CH_3NH_2$? What is the pH?

**17.46**    What is the concentration of hydroxide ion in a 0.20 $M$ aqueous solution of hydroxylamine, $NH_2OH$? What is the pH?

## Acid–Base Properties of Salt Solutions; Hydrolysis

**17.47**    Note whether hydrolysis occurs for each of the following ions. If hydrolysis does occur, write the chemical equation for it. Then write the equilibrium expression for the acid or base ionization (whichever occurs).
a. $I^-$    b. $CHO_2^-$    c. $CH_3NH_3^+$    d. $IO^-$

**17.48**    Note whether hydrolysis occurs for each of the following ions. If hydrolysis does occur, write the chemical equation for it. Then write the equilibrium expression for the acid or base ionization (whichever occurs).
a. $NO_2^-$    b. $Br^-$    c. $NO_3^-$    d. $NH_2NH_3^+$

**17.49** Write the equation for the acid ionization of the $Zn(H_2O)_6^{2+}$ ion.

**17.50** Write the equation for the acid ionization of the $Cu(H_2O)_6^{2+}$ ion.

**17.51** For each of the following salts, indicate whether the aqueous solution will be acidic, basic, or neutral.
a. $Fe(NO_3)_3$    b. $Na_2CO_3$
c. $Ca(CN)_2$    d. $NH_4ClO_4$

**17.52** Note whether aqueous solutions of each of the following salts will be acidic, basic, or neutral.
a. $Na_2S$    b. $Cu(NO_3)_2$
c. $KClO_4$    d. $CH_3NH_3Cl$

**17.53** Decide whether solutions of the following salts are acidic, neutral, or basic.
a. ammonium acetate    b. anilinium acetate

**17.54** Decide whether solutions of the following salts are acidic, neutral, or basic.
a. ammonium cyanate    b. pyridinium cyanate

**17.55** Obtain (a) the $K_b$ value for $NO_2^-$; (b) the $K_a$ value for $C_5H_5NH^+$ (pyridinium ion).

**17.56** Obtain (a) the $K_b$ value for $ClO^-$; (b) the $K_a$ value for $NH_3OH^+$ (hydroxylammonium ion).

**17.57** What is the pH of a 0.025 $M$ aqueous solution of sodium propionate, $NaC_3H_5O_2$? What is the concentration of propionic acid in the solution?

**17.58** Calculate the $OH^-$ concentration and pH of a 0.010 $M$ aqueous solution of sodium cyanide, NaCN. Finally, obtain the hydronium-ion, $CN^-$, and HCN concentrations.

**17.59** Calculate the concentration of pyridine, $C_5H_5N$, in a solution that is 0.15 $M$ pyridinium bromide, $C_5H_5NHBr$. What is the pH of the solution?

**17.60** What is the pH of a 0.35 $M$ solution of methylammonium chloride, $CH_3NH_3Cl$? What is the concentration of methylamine in the solution?

## Common-Ion Effect

**17.61** Calculate the degree of ionization of (a) 0.80 $M$ HF (hydrofluoric acid); (b) the same solution that is also 0.10 $M$ HCl.

**17.62** Calculate the degree of ionization of (a) 0.20 $M$ $HCHO_2$ (formic acid); (b) the same solution that is also 0.10 $M$ HCl.

**17.63** What is the pH of a solution that is 0.10 $M$ $KNO_2$ and 0.15 $M$ $HNO_2$ (nitrous acid)?

**17.64** What is the pH of a solution that is 0.20 $M$ KOCN and 0.10 $M$ HOCN (cyanic acid)?

**17.65** What is the pH of a solution that is 0.10 $M$ $CH_3NH_2$ (methylamine) and 0.15 $M$ $CH_3NH_3Cl$ (methylammonium chloride)?

**17.66** What is the pH of a solution that is 0.15 $M$ $C_2H_5NH_2$ (ethylamine) and 0.10 $M$ $C_2H_5NH_3Br$ (ethylammonium bromide)?

## Buffers

**17.67** A buffer is prepared by adding 45.0 mL of 0.15 $M$ NaF to 35.0 mL of 0.10 $M$ HF. What is the pH of the final solution?

**17.68** A buffer is prepared by adding 115 mL of 0.30 $M$ $NH_3$ to 145 mL of 0.15 $M$ $NH_4NO_3$. What is the pH of the final solution?

**17.69** What is the pH of a buffer solution that is 0.10 $M$ $NH_3$ and 0.10 $M$ $NH_4^+$? What is the pH if 12 mL of 0.20 $M$ hydrochloric acid is added to 125 mL of buffer?

**17.70** A buffer is prepared by mixing 525 mL of 0.50 $M$ formic acid, $HCHO_2$, and 475 mL of 0.50 $M$ sodium formate, $NaCHO_2$. Calculate the pH. What would be the pH of 85 mL of the buffer to which 8.5 mL of 0.15 $M$ hydrochloric acid has been added?

**17.71** What is the pH of a buffer solution that is 0.15 $M$ chloroacetic acid and 0.10 $M$ sodium chloroacetate? $K_a = 1.3 \times 10^{-3}$.

**17.72** What is the pH of a buffer solution that is 0.10 $M$ propionic acid and 0.20 $M$ sodium propionate?

**17.73** What is the pH of a buffer solution that is 0.15 $M$ pyridine and 0.10 $M$ pyridinium bromide?

**17.74** What is the pH of a buffer solution that is 0.20 $M$ methylamine and 0.15 $M$ methylammonium chloride?

**17.75** How many moles of sodium acetate must be added to 2.0 L of 0.10 $M$ acetic acid to give a solution that has a pH equal to 5.00? Ignore the volume change due to the addition of sodium acetate.

**17.76** How many moles of hydrofluoric acid, HF, must be added to 500.0 mL of 0.30 $M$ sodium fluoride to give a buffer of pH 3.50? Ignore the volume change due to the addition of hydrofluoric acid.

## Titration Curves

**17.77** What is the pH of a solution in which 15 mL of 0.10 $M$ NaOH is added to 25 mL of 0.10 $M$ HCl?

**17.78** What is the pH of a solution in which 40 mL of 0.10 $M$ NaOH is added to 25 mL of 0.10 $M$ HCl?

**17.79**   A 1.24-g sample of benzoic acid was dissolved in water to give 50.0 mL of solution. This solution was titrated with 0.180 $M$ NaOH. What was the pH of the solution when the equivalence point was reached?

**17.80**   A 0.400-g sample of propionic acid was dissolved in water to give 50.0 mL of solution. This solution was titrated with 0.150 $M$ NaOH. What was the pH of the solution when the equivalence point was reached?

**17.81**   Find the pH of the solution obtained when 32 mL of 0.087 $M$ ethylamine is titrated to the equivalence point with 0.15 $M$ HCl.

**17.82**   What is the pH at the equivalence point when 22 mL of 0.20 $M$ hydroxylamine is titrated with 0.15 $M$ HCl?

**17.83**   Calculate the pH of a solution obtained by mixing 500.0 mL of 0.10 $M$ NH$_3$ with 200.0 mL of 0.15 $M$ HCl.

**17.84**   Calculate the pH of a solution obtained by mixing 35.0 mL of 0.15 $M$ acetic acid with 25.0 mL of 0.10 $M$ sodium acetate.

## General Problems

**17.85**   Salicylic acid, $C_6H_4OHCOOH$, is used in the manufacture of acetylsalicylic acid (aspirin) and methyl salicylate (wintergreen flavor). A saturated solution of salicylic acid contains 2.2 g of the acid per liter of solution and has a pH of 2.43. What is the value of $K_a$?

**17.86**   Cyanoacetic acid, $CH_2CNCOOH$, is used in the manufacture of barbiturate drugs. An aqueous solution containing 5.0 g in a liter of solution has a pH of 1.89. What is the value of $K_a$?

**17.87**   A 0.050 $M$ aqueous solution of sodium hydrogen sulfate, NaHSO$_4$, has a pH of 1.73. Calculate $K_{a2}$ for sulfuric acid. Sulfuric acid is a strong electrolyte, so you can ignore hydrolysis of the HSO$_4^-$ ion.

**17.88**   A 0.10 $M$ aqueous solution of sodium dihydrogen phosphate, NaH$_2$PO$_4$, has a pH of 4.10. Calculate $K_{a2}$ for phosphoric acid. You can ignore hydrolysis of the H$_2$PO$_4^-$ ion.

**17.89**   Calculate the base-ionization constants for $CN^-$ and $CO_3^{2-}$. Which ion is the stronger base?

**17.90**   Calculate the base-ionization constants for $PO_4^{3-}$ and $SO_4^{2-}$. Which ion is the stronger base?

**17.91**   Calculate the pH of a 0.15 $M$ aqueous solution of aluminum chloride, AlCl$_3$. The acid ionization of hydrated aluminum ion is

$$Al(H_2O)_6^{3+}(aq) + H_2O(l) \rightleftharpoons$$
$$Al(H_2O)_5OH^{2+}(aq) + H_3O^+(aq)$$

and $K_a$ is $1.4 \times 10^{-5}$.

**17.92**   Calculate the pH of a 0.15 $M$ aqueous solution of zinc chloride, ZnCl$_2$. The acid ionization of hydrated zinc ion is

$$Zn(H_2O)_6^{2+}(aq) + H_2O(l) \rightleftharpoons$$
$$Zn(H_2O)_5OH^-(aq) + H_3O^+(aq)$$

and $K_a$ is $2.5 \times 10^{-10}$.

**17.93**   An artificial fruit beverage contains 11.0 g of tartaric acid, $H_2C_4H_4O_6$, and 20.0 g of its salt, potassium hydrogen tartrate, per liter. What is the pH of the beverage? $K_{a1} = 1.0 \times 10^{-3}$.

**17.94**   A buffer is made by dissolving 13.0 g of sodium dihydrogen phosphate, NaH$_2$PO$_4$, and 15.0 g of disodium hydrogen phosphate, Na$_2$HPO$_4$, in a liter of solution. What is the pH of the buffer?

**17.95**   Blood contains several acid–base systems that tend to keep its pH constant at about 7.4. One of the most important buffer systems involves carbonic acid and hydrogen carbonate ion. What must be the ratio of [HCO$_3^-$] to [H$_2$CO$_3$] in the blood if the pH is 7.40?

**17.96**   Codeine, $C_{18}H_{21}NO_3$, is an alkaloid ($K_b = 6.2 \times 10^{-9}$) used as a painkiller and cough suppressant. A solution of codeine is acidified with hydrochloric acid to pH 4.50. What is the ratio of the concentration of the conjugate acid of codeine to that of the base codeine?

**17.97**   Calculate the pH of a solution obtained by mixing 456 mL of 0.10 $M$ hydrochloric acid with 285 mL of 0.15 $M$ sodium hydroxide. Assume the combined volume is the sum of the two original volumes.

**17.98**   Calculate the pH of a solution made up from 2.0 g of potassium hydroxide dissolved in 115 mL of 0.19 $M$ perchloric acid. Assume the change in volume due to adding potassium hydroxide is negligible.

**17.99**   Find the pH of the solution obtained when 25 mL of 0.065 $M$ benzylamine, $C_7H_7NH_2$, is titrated to the equivalence point with 0.050 $M$ hydrochloric acid. $K_b$ for benzylamine is $4.7 \times 10^{-10}$.

**17.100**   What is the pH of the solution obtained by titrating 1.24 g of sodium hydrogen sulfate, NaHSO$_4$, dissolved in 50.0 mL of water with 0.180 $M$ sodium hydroxide until the equivalence point is reached? Assume that any volume change due to adding the sodium hydrogen sulfate or to mixing the solutions is negligible.

**17.101** Ionization of the first proton from $H_2SO_4$ is complete ($H_2SO_4$ is a strong acid); the acid-ionization constant for the second proton is $1.1 \times 10^{-2}$. (a) What would be the approximate hydronium-ion concentration in 0.100 $M$ $H_2SO_4$ if ionization of the second proton were ignored? (b) The ionization of the second proton must be considered for a more exact answer, however. Calculate the hydronium-ion concentration in 0.100 $M$ $H_2SO_4$, accounting for the ionization of both protons.

**17.102** Ionization of the first proton from $H_2SeO_4$ is complete ($H_2SeO_4$ is a strong acid); the acid-ionization constant for the second proton is $1.2 \times 10^{-2}$. (a) What would be the approximate hydronium-ion concentration in 0.150 $M$ $H_2SeO_4$ if ionization of the second proton were ignored? (b) The ionization of the second proton must be considered for a more exact answer, however. Calculate the hydronium-ion concentration in 0.150 $M$ $H_2SeO_4$, accounting for the ionization of both protons.

**17.103** Methylammonium chloride is a salt of methylamine, $CH_3NH_2$. A 0.10 $M$ solution of this salt has a pH of 5.82.
a. Calculate the value for the equilibrium constant for the reaction

$$CH_3NH_3^+ + H_2O \rightleftharpoons CH_3NH_2 + H_3O^+$$

b. What is the $K_b$ value for methylamine?
c. What is the pH of a solution in which 0.450 mole of solid methylammonium chloride is added to 1.00 L of a 0.250 $M$ solution of methylamine? Assume no volume change.

**17.104** Sodium benzoate is a salt of benzoic acid, $C_6H_5COOH$. A 0.15 $M$ solution of this salt has a pOH of 5.31 at room temperature.
a. Calculate the value for the equilibrium constant for the reaction

$$C_6H_5COO^- + H_2O \rightleftharpoons C_6H_5COOH + OH^-$$

b. What is the $K_a$ value for benzoic acid?
c. Benzoic acid has a low solubility in water. What is its molar solubility if a saturated solution has a pH of 2.83 at room temperature?

**17.105** Each of the following statements concerns a 0.010 $M$ solution of a weak acid, HA. Briefly describe why each statement is either true or false.
a. [HA] is approximately equal to 0.010 $M$.
b. [HA] is much greater than $[A^-]$.
c. $[OH^-]$ is approximately equal to $[H_3O^+]$.
d. The pH is 2.
e. The $H_3O^+$ concentration is 0.010 $M$.
f. $[H_3O^+]$ is approximately equal to $[A^-]$.

**17.106** Each of the following statements concerns a 0.10 $M$ solution of a weak organic base, B. Briefly describe why each statement is either true or false.
a. [B] is approximately equal to 0.10 $M$.
b. [B] is much greater than $[HB^+]$.
c. $[H_3O^+]$ is greater than $[HB^+]$.
d. The pH is 13.
e. $[HB^+]$ is approximately equal to $[OH^-]$.
f. $[OH^-]$ equals 0.10 $M$.

**17.107** A 0.288-g sample of an unknown monoprotic organic acid is dissolved in water and titrated with a 0.115 $M$ sodium hydroxide solution. After the addition of 17.54 mL of base, a pH of 4.92 is recorded. The equivalence point is reached when a total of 33.83 mL of NaOH is added.
a. What is the molar mass of the organic acid?
b. What is the $K_a$ value for the acid? The $K_a$ value could have been determined very easily if a pH measurement had been made after the addition of 16.92 mL of NaOH. Why?

**17.108** A 0.239-g sample of unknown organic base is dissolved in water and titrated with a 0.135 $M$ hydrochloric acid solution. After the addition of 18.35 mL of acid, a pH of 10.73 is recorded. The equivalence point is reached when a total of 39.24 mL of HCl is added. The base and acid combine in a 1:1 ratio.
a. What is the molar mass of the organic base?
b. What is the $K_b$ value for the base? The $K_b$ value could have been determined very easily if a pH measurement had been made after the addition of 19.62 mL of HCl. Why?

**17.109** (a) Draw a pH titration curve that represents the titration of 50.0 mL of 0.10 $M$ $NH_3$ by the addition of 0.10 $M$ HCl from a buret. Label the axes and put a scale on each axis. Show where the equivalence point and the buffer region are on the titration curve. You should do calculations for the 0%, 30%, 50%, and 100% titration points.
(b) Is the solution neutral, acidic, or basic at the equivalence point? Why?

**17.110** (a) Draw a pH titration curve that represents the titration of 25.0 mL of 0.15 $M$ propionic acid, $CH_3CH_2COOH$, by the addition of 0.15 $M$ KOH from a buret. Label the axes and put a scale on each axis. Show where the equivalence point and the buffer region are on the titration curve. You should do calculations for the 0%, 50%, 60%, and 100% titration points.
(b) Is the solution neutral, acidic, or basic at the equivalence point? Why?

**17.111** The equilibrium equations and $K_a$ values for three reaction systems are given below.

$$H_2C_2O_4(aq) + H_2O \rightleftharpoons$$
$$H_3O^+(aq) + HC_2O_4^-(aq); K_a = 5.6 \times 10^{-2}$$

$$H_3PO_4(aq) + H_2O \rightleftharpoons$$
$$H_3O^+(aq) + H_2PO_4^-(aq); K_a = 6.9 \times 10^{-3}$$

$$HCOOH(aq) + H_2O \rightleftharpoons$$
$$H_3O^+(aq) + HCOO^-(aq); K_a = 1.7 \times 10^{-4}$$

a. Which conjugate pair would be best for preparing a buffer with a pH of 2.88?
b. How would you prepare 50 mL of a buffer with a pH of 2.88 assuming that you had available 0.10 $M$ solutions of each pair?

**17.112**    The equilibrium equations and $K_a$ values for three reaction systems are given below.

$$NH_4^+(aq) + H_2O \rightleftharpoons$$
$$H_3O^+ + NH_3(aq); K_a = 5.6 \times 10^{-10}$$

$$H_2CO_3(aq) + H_2O \rightleftharpoons$$
$$H_3O^+(aq) + HCO_3^-(aq); K_a = 4.3 \times 10^{-7}$$

$$H_2PO_4^-(aq) + H_2O \rightleftharpoons$$
$$H_3O^+(aq) + HPO_4^{2-}(aq); K_a = 6.2 \times 10^{-8}$$

a. Which conjugate pair would be best for preparing a buffer with a pH of 6.96? Why?
b. How would you prepare 100 mL of a buffer with a pH of 6.96 assuming that you had available 0.10 $M$ solutions of each pair?

**17.113**    A 25.0-mL sample of hydroxylamine is titrated to the equivalence point with 35.8 mL of 0.150 $M$ HCl.
a. What was the concentration of the original hydroxylamine solution?
b. What is the pH at the equivalence point?
c. Which indicators, bromphenol blue, methyl red, or phenolphthalein, should be used to detect the end point of the titration? Why?

**17.114**    A 25.00-mL sample contains 0.562 g of NaHCO_3. This sample is used to standardize an NaOH solution. At the equivalence point, 42.36 mL of NaOH has been added.
a. What was the concentration of the NaOH?
b. What is the pH at the equivalence point?
c. Which indicator, bromthymol blue, methyl violet, or alizarin yellow R, should be used in the titration? Why?

**17.115**    A solution made up of 1.0 $M$ NH_3 and 0.50 $M$ (NH_4)_2SO_4 has a pH of 9.26.
a. Write the net ionic equation that represents the reaction of this solution with a strong acid.
b. Write the net ionic equation that represents the reaction of this solution with a strong base.
c. To 100. mL of this solution, 10.0 mL of 1.00 $M$ HCl are added. How many moles of NH_3 and NH_4^+ are present in the reaction system before and after the addition of the HCl? What is the pH of the resulting solution?
d. Why did the pH change only slightly upon the addition of HCl?

**17.116**    A solution is prepared from 0.150 mol of formic acid and enough water to make 0.425 L of solution.

a. Determine the concentrations of H_3O^+ and HCOO^- in this solution.
b. Determine the H_3O^+ concentration that would be necessary to decrease the HCOO^- concentration above by a factor of ten. How many milliliters of 2.00 $M$ HCl would be required to produce this solution? Consider that the solution was made by combining the HCl, the HCOOH, and enough water to make 0.425 L of solution.
c. Qualitatively, how can you account for the differences in the percentage dissociation of formic acid in parts a and b of this problem?

**17.117**    An important component of blood is the buffer combination of dihydrogen phosphate ion and the hydrogen phosphate ion. Consider blood with a pH of 7.44.
a. What is the ratio of $\dfrac{[H_2PO_4^-]}{[HPO_4^{2-}]}$ ?
b. What does the pH become if 25% of the hydrogen phosphate ions are converted to dihydrogen phosphate ion?
c. What does the pH become if 15% of the dihydrogen phosphate ions are converted to hydrogen phosphate ions?

**17.118**    An important component of blood is the buffer combination of bicarbonate ion and carbonic acid. Consider blood with a pH of 7.42.
a. What is the ratio of $\dfrac{[H_2CO_3]}{[HCO_3^-]}$ ?
b. What does the pH become if 15% of the bicarbonate ions are converted to carbonic acid?
c. What does the pH become if 25% of the carbonic acid molecules are converted to bicarbonate ions?

**17.119**    Tartaric acid is a weak diprotic fruit acid with $K_1 = 1.0 \times 10^{-3}$ and $K_2 = 4.6 \times 10^{-5}$.
a. Letting the symbol H_2A represent tartaric acid, write the chemical equations that represent $K_1$ and $K_2$. Write the chemical equation that represents $K_1 \times K_2$.
b. Qualitatively describe the relative concentrations of H_2A, HA^-, A^{2-}, and H_3O^+ in a solution that is about 0.5 $M$ in tartaric acid.
c. Calculate the pH of a 0.0250 $M$ tartaric acid solution and the equilibrium concentration of [H_2A].
d. What is the A^{2-} concentration?

**17.120**    Malic acid is a weak diprotic organic acid with $K_1 = 4.0 \times 10^{-4}$ and $K_2 = 9.0 \times 10^{-6}$.
a. Letting the symbol H_2A represent malic acid, write the chemical equations that represent $K_1$ and $K_2$. Write the chemical equation that represents $K_1 \times K_2$.
b. Qualitatively describe the relative concentrations of H_2A, HA^-, A^{2-}, and H_3O^+ in a solution that is about one molar in malic acid.
c. Calculate the pH of a 0.0100 $M$ malic acid solution and the equilibrium concentration of [H_2A].
d. What is the A^{2-} concentration?

**17.121** A quantity of 0.25 *M* sodium hydroxide is added to a solution containing 0.15 mole of acetic acid. The final volume of the solution is 375 mL and the pH of this solution is 4.45.
a. What is the molar concentration of the sodium acetate?
b. How many milliliters of sodium hydroxide were added to the original solution?
c. What was the original concentration of the acetic acid?

**17.122** A quantity of 0.15 *M* hydrochloric acid is added to a solution containing 0.10 mole of sodium acetate. Some of the sodium acetate is converted to acetic acid, resulting in a final volume of 650 mL of solution. The pH of the final solution is 4.56.
a. What is the molar concentration of the acetic acid?
b. How many milliliters of hydrochloric acid were added to the original solution?
c. What was the original concentration of the sodium acetate?

## Cumulative-Skills Problems

**17.123** The pH of a white vinegar solution is 2.45. This vinegar is an aqueous solution of acetic acid with a density of 1.09 g/mL. What is the mass percentage of acetic acid in the solution?

**17.124** The pH of a household cleaning solution is 11.87. This cleanser is an aqueous solution of ammonia with a density of 1.00 g/mL. What is the mass percentage of ammonia in the solution?

**17.125** What is the freezing point of 0.92 *M* aqueous acetic acid? The density of this solution is 1.008 g/cm$^3$.

**17.126** What is the freezing point of 0.87 *M* aqueous ammonia? The density of this solution is 0.992 g/cm$^3$.

**17.127** A chemist needs a buffer with pH 4.35. How many milliliters of pure acetic acid (density = 1.049 g/mL) must be added to 465 mL of 0.0941 *M* NaOH solution to obtain such a buffer?

**17.128** A chemist needs a buffer with pH 3.50. How many milliliters of pure formic acid (density = 1.220 g/mL) must be added to 325 mL of 0.0857 *M* NaOH solution to obtain such a buffer?

## Media Activities

**17.129** Work through the Weak and Strong Acids *Understanding Concepts* activity on the student CD. At the end of this activity, choose HF and $H_2C_2O_4$ (oxalic acid) and answer the on-screen questions. To verify your answers, look up the $K_a$ values (Appendix E) and calculate the pH of a 0.050 *M* solution of HF and of a 0.050 *M* $H_2C_2O_4$.

**17.130** Run the Neutralization of a Strong Acid with a Strong Base *Visualization* on the student CD. For the solution depicted at the end of the *Visualization*, answer the following questions:
a. What is the pH?
b. What is the concentration of $H_3O^+$?
c. How can you be certain that the number of moles of acid added to the solution was equivalent to the number of moles of base?
d. What would be different if the acid used for the titration were $HNO_3$ instead of HCl? What would be the same?

**17.131** Work through the Buffers *Understanding Concepts* activity on the student CD. Without performing calculations, predict which of the following solutions you would expect to be a buffer.
a. 20 mL of 0.1 *M* HCl in 50.0 mL of $H_2O$
b. 0.01 mol of NaCN and 0.02 mol of HCN in 2.00 L of $H_2O$
c. 0.20 mol of $NaNO_3$ and 0.20 mol of $HNO_3$ in 4.00 L of $H_2O$

**17.132** Step through the Buffers *Problem Solving* activity on the student web site.
a. If a solution were prepared with 90. mL of 0.10 *M* formic acid and 10. mL of 0.75 *M* sodium formate, estimate how the pH of this solution compares to that of the solution described in the activity without performing a calculation.
b. Calculate the pH of the solution described in part a.
c. If 1.0 L of 0.10 *M* HCl were added to the solution in part a, estimate the pH of the resulting solution.

**17.133** For review, take the ACE practice test for Chapter 17 on the student web site.

# 18

Formation of the violet copper–ammonia complex ion.

# Solubility and Complex-Ion Equilibria

**M**any natural processes depend on the precipitation or the dissolving of a slightly soluble salt. For example, caves form in limestone (calcium carbonate) over thousands of years as groundwater seeps through cracks, dissolving out cavities in the rock. Kidney stones form when salts such as calcium phosphate or calcium oxalate slowly precipitate in the kidney (Figure 18.1). Poisoning by oxalic acid is also explained by the precipitation of the calcium salt. If oxalic acid is accidentally ingested, the oxalate-ion concentration in the blood may increase sufficiently to precipitate calcium oxalate. Calcium ion, which is needed for proper muscle control, is then removed from the blood, and muscle tissues go into spasm.

To understand such phenomena quantitatively, you must be able to solve problems in solubility equilibria. Calcium oxalate kidney stones form when the concentrations of calcium ion and oxalate ion are sufficiently great. What is the relationship between the concentrations of ions and the solubility of a salt? What is the minimum concentration of oxalate ion that gives a precipitate of the calcium salt from a 0.0025 $M$ solution of $Ca^{2+}$ (the approximate concentration of calcium ion in blood plasma)? What is the effect of pH on the solubility of this salt? We will look at questions such as these in this chapter.

# Solubility Equilibria

To deal quantitatively with an equilibrium, you must know the equilibrium constant. In the next section, we will look at the equilibria of slightly soluble, or nearly insoluble, ionic compounds and show how you can determine their equilibrium constants. Once you find these values for various ionic compounds, you can use them to answer solubility or precipitation questions.

## 18.1  The Solubility Product Constant

**Figure 18.1**
**A kidney stone.**  Kidney stones are usually calcium phosphate or calcium oxalate that precipitates as a crystalline mass.

▶ An ionic equilibrium is affected to a small extent by the presence of ions not directly involved in the equilibrium. We will ignore this effect here.

When an ionic compound is dissolved in water, it usually goes into solution as the ions. When an excess of a slightly soluble ionic compound is mixed with water, an equilibrium occurs between the solid compound and the ions in the saturated solution. For the salt calcium oxalate, $CaC_2O_4$, you have the following equilibrium:

$$CaC_2O_4(s) \xrightarrow{H_2O} Ca^{2+}(aq) + C_2O_4^{2-}(aq)$$

The equilibrium constant for this solubility process is called the *solubility product constant* of $CaC_2O_4$. It is written

$$K_{sp} = [Ca^{2+}][C_2O_4^{2-}]$$

In general, the **solubility product constant ($K_{sp}$)** is *the equilibrium constant for the solubility equilibrium of a slightly soluble (or nearly insoluble) ionic compound*. It equals the product of the equilibrium concentrations of the ions in the compound, each concentration raised to a power equal to the number of such ions in the formula of the compound. Like any equilibrium constant, $K_{sp}$ depends on the temperature, but at a given temperature it has a constant value for various concentrations of the ions. ◀

Lead(II) iodide, $PbI_2$, is another slightly soluble salt. The equilibrium in water is

$$PbI_2(s) \rightleftharpoons Pb^{2+}(aq) + 2I^-(aq)$$

and the expression for the equilibrium constant, or solubility product constant, is

$$K_{sp} = [Pb^{2+}][I^-]^2$$

**EXAMPLE 18.1**

## Writing Solubility Product Expressions

Write the solubility product expressions for the following salts: a. AgCl; b. $Hg_2Cl_2$; c. $Pb_3(AsO_4)_2$.

**SOLUTION**

The equilibria and solubility product expressions are

a. $AgCl(s) \rightleftharpoons Ag^+(aq) + Cl^-(aq)$; $K_{sp} = [Ag^+][Cl^-]$

b. $Hg_2Cl_2(s) \rightleftharpoons Hg_2^{2+}(aq) + 2Cl^-(aq)$; $K_{sp} = [Hg_2^{2+}][Cl^-]^2$

Note that the mercury(I) ion is $Hg_2^{2+}$. (Mercury also has salts with the mercury(II) ion, $Hg^{2+}$.)

c. $Pb_3(AsO_4)_2(s) \rightleftharpoons 3Pb^{2+}(aq) + 2AsO_4^{3-}(aq)$; $K_{sp} = [Pb^{2+}]^3[AsO_4^{3-}]^2$

---

### EXERCISE 18.1

Give solubility product expressions for the following: a. barium sulfate; b. iron(III) hydroxide; c. calcium phosphate.

---

See Problems 18.21 and 18.22.

The solubility product constant, $K_{sp}$, of a slightly soluble ionic compound is expressed in terms of the molar concentrations of ions in the saturated solution. These ion concentrations are in turn related to the *molar solubility* of the ionic compound, which is the moles of compound that dissolve to give a liter of saturated solution. The next two examples show how to determine the solubility product constant from the solubility of a slightly soluble ionic compound.

---

### EXAMPLE 18.2

## Calculating $K_{sp}$ from the Solubility (Simple Example)

A liter of a solution saturated at 25°C with calcium oxalate, $CaC_2O_4$, is evaporated to dryness, giving a 0.0061-g residue of $CaC_2O_4$. Calculate the solubility product constant for this salt at 25°C.

 **PROBLEM STRATEGY**

The grams of residue from a liter of saturated solution equals the solubility in grams per liter. Convert this to molar solubility (solubility in moles per liter). Then follow the three steps similar to those for an equilibrium calculation: set up a table of ion concentrations (starting, change, and equilibrium) and calculate $K_{sp}$ from the concentrations. The starting concentrations are those before any solid has dissolved (= 0), and the change values are obtained from the molar solubility.

**SOLUTION**

The solubility of calcium oxalate is 0.0061 g/L of *solution*, and the formula weight of $CaC_2O_4$ is 128 amu. Convert grams per liter to moles per liter:

$$\text{Molar solubility of } CaC_2O_4 = 0.0061 \text{ g } CaC_2O_4\text{/L} \times \frac{1 \text{ mol } CaC_2O_4}{128 \text{ g } CaC_2O_4}$$

$$= 4.8 \times 10^{-5} \text{ mol } CaC_2O_4\text{/L}$$

Now look at the equilibrium problem.

**STEP 1** Suppose you mix solid $CaC_2O_4$ in a liter of solution. Of this solid, $4.8 \times 10^{-5}$ mol will dissolve to form $4.8 \times 10^{-5}$ mol of each ion. The results are summarized in the following table. (Because the concentration of the solid does not appear in $K_{sp}$, we do not include it in the table.)

| Concentration ($M$) | $CaC_2O_4(s) \rightleftharpoons$ | $Ca^{2+}(aq)$ | $+$ | $C_2O_4^{2-}(aq)$ |
|---|---|---|---|---|
| Starting | | 0 | | 0 |
| Change | | $+4.8 \times 10^{-5}$ | | $+4.8 \times 10^{-5}$ |
| Equilibrium | | $4.8 \times 10^{-5}$ | | $4.8 \times 10^{-5}$ |

**STEP 2** You now substitute into the equilibrium-constant equation:

$$K_{sp} = [Ca^{2+}][C_2O_4^{2-}] = (4.8 \times 10^{-5})(4.8 \times 10^{-5}) = \mathbf{2.3 \times 10^{-9}}$$

### EXERCISE 18.2

Silver ion may be recovered from used photographic fixing solution by precipitating it as silver chloride. The solubility of silver chloride is $1.9 \times 10^{-3}$ g/L. Calculate $K_{sp}$.

*See Problems 18.23 and 18.24.*

▶ An alternate example is provided in the Instructor's Resource Manual.

### EXAMPLE 18.3

## Calculating $K_{sp}$ from the Solubility (More Complicated Example)

By experiment, it is found that $1.2 \times 10^{-3}$ mol of lead(II) iodide, $PbI_2$, dissolves in 1 L of aqueous solution at 25°C. What is the solubility product constant at this temperature?

 **PROBLEM STRATEGY**

This is similar to the previous example, except that every mole of the compound ($PbI_2$) dissolves to form *two* moles of the anion ($I^-$). This difference affects the calculation in two places: it affects the molar concentration of the ion (the molar concentration of the anion is *twice* that of the molar solubility), and it affects the equilibrium-constant expression (which contains the anion concentration *squared*).

### SOLUTION

**STEP 1** Suppose the solid lead(II) iodide is mixed into 1 L of solution. You find that $1.2 \times 10^{-3}$ mol dissolves to form $1.2 \times 10^{-3}$ mol $Pb^{2+}$ and $2 \times (1.2 \times 10^{-3})$ mol $I^-$, as summarized in the following table:

| Concentration ($M$) | $PbI_2(s) \rightleftharpoons$ | $Pb^{2+}(aq)$ | $+$ | $2I^-(aq)$ |
|---|---|---|---|---|
| Starting | | 0 | | 0 |
| Change | | $+1.2 \times 10^{-3}$ | | $+2 \times (1.2 \times 10^{-3})$ |
| Equilibrium | | $1.2 \times 10^{-3}$ | | $2 \times (1.2 \times 10^{-3})$ |

**STEP 2** You substitute into the equilibrium-constant equation:

$$K_{sp} = [Pb^{2+}][I^-]^2 = (1.2 \times 10^{-3}) \times (2 \times 1.2 \times 10^{-3})^2 = \mathbf{6.9 \times 10^{-9}}$$

### EXERCISE 18.3

Lead(II) arsenate, $Pb_3(AsO_4)_2$, has been used as an insecticide. It is only slightly soluble in water. If the solubility is $3.0 \times 10^{-5}$ g/L, what is the solubility product constant? Assume that the solubility equilibrium is the only important one.

**See Problems 18.25 and 18.26.**

Table 18.1 lists the solubility product constants for various ionic compounds. If the solubility product constant is known, the solubility of the compound can be calculated.

**Table 18.1**
**Solubility Product Constants, $K_{sp}$, at 25°C**

| Substance | Formula | $K_{sp}$ |
|---|---|---|
| Aluminum hydroxide | $Al(OH)_3$ | $4.6 \times 10^{-33}$ |
| Barium chromate | $BaCrO_4$ | $1.2 \times 10^{-10}$ |
| Barium fluoride | $BaF_2$ | $1.0 \times 10^{-6}$ |
| Barium sulfate | $BaSO_4$ | $1.1 \times 10^{-10}$ |
| Cadmium oxalate | $CdC_2O_4$ | $1.5 \times 10^{-8}$ |
| Cadmium sulfide | $CdS$ | $8 \times 10^{-27}$ |
| Calcium carbonate | $CaCO_3$ | $3.8 \times 10^{-9}$ |
| Calcium fluoride | $CaF_2$ | $3.4 \times 10^{-11}$ |
| Calcium oxalate | $CaC_2O_4$ | $2.3 \times 10^{-9}$ |
| Calcium phosphate | $Ca_3(PO_4)_2$ | $1 \times 10^{-26}$ |
| Calcium sulfate | $CaSO_4$ | $2.4 \times 10^{-5}$ |
| Cobalt(II) sulfide | $CoS$ | $4 \times 10^{-21}$ |
| Copper(II) hydroxide | $Cu(OH)_2$ | $2.6 \times 10^{-19}$ |
| Copper(II) sulfide | $CuS$ | $6 \times 10^{-36}$ |
| Iron(II) hydroxide | $Fe(OH)_2$ | $8 \times 10^{-16}$ |
| Iron(II) sulfide | $FeS$ | $6 \times 10^{-18}$ |
| Iron(III) hydroxide | $Fe(OH)_3$ | $2.5 \times 10^{-39}$ |
| Lead(II) arsenate | $Pb_3(AsO_4)_2$ | $4 \times 10^{-36}$ |
| Lead(II) chloride | $PbCl_2$ | $1.6 \times 10^{-5}$ |
| Lead(II) chromate | $PbCrO_4$ | $1.8 \times 10^{-14}$ |
| Lead(II) iodide | $PbI_2$ | $6.5 \times 10^{-9}$ |
| Lead(II) sulfate | $PbSO_4$ | $1.7 \times 10^{-8}$ |
| Lead(II) sulfide | $PbS$ | $2.5 \times 10^{-27}$ |
| Magnesium arsenate | $Mg_3(AsO_4)_2$ | $2 \times 10^{-20}$ |
| Magnesium carbonate | $MgCO_3$ | $1.0 \times 10^{-5}$ |
| Magnesium hydroxide | $Mg(OH)_2$ | $1.8 \times 10^{-11}$ |
| Magnesium oxalate | $MgC_2O_4$ | $8.5 \times 10^{-5}$ |
| Manganese(II) sulfide | $MnS$ | $2.5 \times 10^{-10}$ |

**Table 18.1 (continued)**

| Substance | Formula | $K_{sp}$ |
|---|---|---|
| Mercury(I) chloride | $Hg_2Cl_2$ | $1.3 \times 10^{-18}$ |
| Mercury(II) sulfide | HgS | $1.6 \times 10^{-52}$ |
| Nickel(II) hydroxide | $Ni(OH)_2$ | $2.0 \times 10^{-15}$ |
| Nickel(II) sulfide | NiS | $3 \times 10^{-19}$ |
| Silver acetate | $AgC_2H_3O_2$ | $2.0 \times 10^{-3}$ |
| Silver bromide | AgBr | $5.0 \times 10^{-13}$ |
| Silver chloride | AgCl | $1.8 \times 10^{-10}$ |
| Silver chromate | $Ag_2CrO_4$ | $1.1 \times 10^{-12}$ |
| Silver iodide | AgI | $8.3 \times 10^{-17}$ |
| Silver sulfide | $Ag_2S$ | $6 \times 10^{-50}$ |
| Strontium carbonate | $SrCO_3$ | $9.3 \times 10^{-10}$ |
| Strontium chromate | $SrCrO_4$ | $3.5 \times 10^{-5}$ |
| Strontium sulfate | $SrSO_4$ | $2.5 \times 10^{-7}$ |
| Zinc hydroxide | $Zn(OH)_2$ | $2.1 \times 10^{-16}$ |
| Zinc sulfide | ZnS | $1.1 \times 10^{-21}$ |

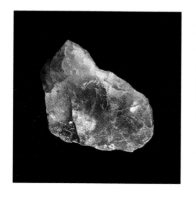

**EXAMPLE 18.4**

## Calculating the Solubility from $K_{sp}$

The mineral fluorite is calcium fluoride, $CaF_2$. Calculate the solubility (in grams per liter) of calcium fluoride in water from the solubility product constant ($3.4 \times 10^{-11}$).

 **PROBLEM STRATEGY**

This problem is the reverse of the preceding ones; instead of finding $K_{sp}$ from the solubility, here you calculate solubility from the $K_{sp}$. You follow the three steps for equilibrium problems, but since the molar solubility is not immediately known, you assign it the value $x$. For Step 1, you obtain the concentration of each ion by multiplying $x$ by the coefficient of the ion in the chemical equation. In Step 2, you obtain $K_{sp}$ as a cubic in $x$. In Step 3, you solve the equilibrium-constant equation for $x$, which will involve taking a cube root.

**SOLUTION**

**STEP 1** Let $x$ be the molar solubility of $CaF_2$. When solid $CaF_2$ is mixed into a liter of solution, $x$ mol dissolves, forming $x$ mol $Ca^{2+}$ and $2x$ mol $F^-$.

| Concentration ($M$) | $CaF_2(s) \rightleftharpoons Ca^{2+}(aq) + 2F^-(aq)$ | |
|---|---|---|
| Starting | 0 | 0 |
| Change | $+x$ | $+2x$ |
| Equilibrium | $x$ | $2x$ |

STEP 2 You substitute into the equilibrium-constant equation.

$$[Ca^{2+}][F^-]^2 = K_{sp}$$

$$(x) \times (2x)^2 = 3.4 \times 10^{-11}$$

$$4x^3 = 3.4 \times 10^{-11}$$

STEP 3 You now solve for $x$.

$$x = \sqrt[3]{\frac{3.4 \times 10^{-11}}{4}} = 2.0 \times 10^{-4}$$

The molar solubility is $2.0 \times 10^{-4}$ mol $CaF_2$ per liter. To get the solubility in grams per liter, you convert, using the molar mass of $CaF_2$ (78.1 g/mol).

$$\text{Solubility} = 2.0 \times 10^{-4} \text{ mol } CaF_2/L \times \frac{78.1 \text{ g } CaF_2}{1 \text{ mol } CaF_2} = 1.6 \times 10^{-2} \text{ g } CaF_2/L$$

---

**EXERCISE 18.4**

Anhydrite is a calcium sulfate mineral deposited when seawater evaporates. What is the solubility of calcium sulfate, in grams per liter? Table 18.1 gives the solubility product constant for calcium sulfate.

---

See Problems 18.29, 18.30, 18.31, and 18.32.

These examples illustrate the relationship between the solubility of a slightly soluble ionic compound in pure water and its solubility product constant. In the next section, you will see how the solubility product constant can be used to calculate the solubility in the presence of other ions. $K_{sp}$ is also useful in deciding whether to expect precipitation under given conditions.

 **CONCEPT CHECK 18.1**

Lead compounds have been used as paint pigments, but because the lead(II) ion is toxic, the use of lead paints in homes is now prohibited. Which of the following lead(II) compounds would yield the greatest number of lead(II) ions when added to the same quantity of water (assuming that some undissolved solid always remains): $PbCrO_4$, $PbSO_4$, or $PbS$?

## 18.2 Solubility and the Common-Ion Effect

The importance of the solubility product constant becomes apparent when you consider the solubility of one salt in the solution of another salt having the same cation or anion. For example, suppose you wish to know the solubility of calcium oxalate in a solution of calcium chloride. Each salt contributes the same cation ($Ca^{2+}$). The effect of the calcium ion provided by the calcium chloride is to make calcium oxalate less soluble than it would be in pure water.

You can explain this decrease in solubility in terms of Le Chatelier's principle. Suppose you first mix crystals of calcium oxalate in a quantity of pure water to establish the equilibrium

$$CaC_2O_4(s) \rightleftharpoons Ca^{2+}(aq) + C_2O_4^{2-}(aq)$$

**Figure 18.2**
**Demonstration of the common-ion effect.**
When the experimenter adds lead(II) nitrate solution (colorless) from the dropper to the saturated solution of lead(II) chromate (pale yellow), a yellow precipitate of lead chromate forms.

▶ An alternate example is provided in the Instructor's Resource Manual.

Now imagine that you add some calcium chloride. Calcium chloride is a soluble salt, so it dissolves to give an increase in calcium-ion concentration. You can regard this increase as a stress on the original equilibrium. According to Le Chatelier's principle, the ions will react to remove some of the added calcium ion.

$$CaC_2O_4(s) \longleftarrow Ca^{2+}(aq) + C_2O_4^{2-}(aq)$$

In other words, some calcium oxalate precipitates from the solution. The solution now contains less calcium oxalate. You conclude that calcium oxalate is less soluble in a solution of calcium chloride than in pure water.

A solution of soluble calcium chloride has an ion *in common* with the slightly soluble calcium oxalate. The decrease in solubility of calcium oxalate in a solution of calcium chloride is an example of the *common-ion effect* (first discussed in Section 17.5). In general, any ionic equilibrium is affected by a substance producing an ion involved in the equilibrium, as you would predict from Le Chatelier's principle. Figure 18.2 illustrates the principle of the common-ion effect for lead(II) chromate, $PbCrO_4$, which is only slightly soluble in water at 25°C. When the very soluble $Pb(NO_3)_2$ is added to a saturated solution of $PbCrO_4$, the concentration of the common ion, $Pb^{2+}$, increases and $PbCrO_4$ precipitates. The equilibrium is

$$PbCrO_4(s) \Longrightarrow Pb^{2+}(aq) + CrO_4^{2-}(aq)$$

The next example shows how you can calculate the solubility of a slightly soluble salt in a solution containing a substance with a common ion.

### EXAMPLE 18.5

## Calculating the Solubility of a Slightly Soluble Salt in a Solution of a Common Ion

What is the molar solubility of calcium oxalate in 0.15 *M* calcium chloride? Compare this molar solubility with that found earlier (Example 18.2) for $CaC_2O_4$ in pure water ($4.8 \times 10^{-5}$ *M*). The solubility product constant for calcium oxalate is $2.3 \times 10^{-9}$.

 **PROBLEM STRATEGY**

The two salts have a common ion ($Ca^{2+}$): one salt ($CaCl_2$) is soluble and provides $Ca^{2+}$ ion that suppresses the solubility of the slightly soluble salt ($CaC_2O_4$). You solve the solubility equilibrium for the slightly soluble salt, noting that when you set up the concentration table, you have a starting concentration of $Ca^{2+}$ ion from $CaCl_2$. The equilibrium-constant equation you obtain is a quadratic, which you can reduce to a linear equation using the same approximation used in Chapter 17; you assume *x* is small compared with the number you add it to. (That is, the starting $Ca^{2+}$ concentration is not much affected by the dissolution of the slightly soluble salt.) The value of *x* equals the molar solubility of the slightly soluble salt in the presence of the common ion.

### SOLUTION

**STEP 1**  Suppose solid $CaC_2O_4$ is mixed into 1 L of 0.15 *M* $CaCl_2$ and that the molar solubility of the $CaC_2O_4$ is *x M*. At the start (before $CaC_2O_4$ dissolves), there is 0.15 mol $Ca^{2+}$ in the solution. Of the solid $CaC_2O_4$, *x* mol dissolves to give $x$ mol of additional $Ca^{2+}$ and $x$ mol $C_2O_4^{2-}$. The following table summarizes these results:

| Concentration ($M$) | $CaC_2O_4(s) \rightleftharpoons$ | $Ca^{2+}(aq)$ | $+$ | $C_2O_4^{2-}(aq)$ |
|---|---|---|---|---|
| Starting | | 0.15 | | 0 |
| Change | | $+x$ | | $+x$ |
| Equilibrium | | $0.15 + x$ | | $x$ |

**STEP 2** You substitute into the equilibrium-constant equation:

$$[Ca^{2+}][C_2O_4^{2-}] = K_{sp}$$

$$(0.15 + x)x = 2.3 \times 10^{-9}$$

**STEP 3** Now rearrange this equation to give

$$x = \frac{2.3 \times 10^{-9}}{0.15 + x}$$

Because calcium oxalate is only slightly soluble, you might expect $x$ to be negligible compared with 0.15. In that case,

$$0.15 + x \simeq 0.15$$

and the previous equation becomes

$$x \simeq \frac{2.3 \times 10^{-9}}{0.15} = 1.5 \times 10^{-8}$$

Note that $x$ is indeed much smaller than 0.15, so your assumption was correct. Therefore, the molar solubility of calcium oxalate in 0.15 $M$ $CaCl_2$ is **1.5 × $10^{-8}$ $M$.** In pure water, the molar solubility is $4.8 \times 10^{-5}$ $M$, which is over **3000 times greater.**

---

### EXERCISE 18.5

a. Calculate the molar solubility of barium fluoride, $BaF_2$, in water at 25°C. The solubility product constant for $BaF_2$ at this temperature is $1.0 \times 10^{-6}$.

b. What is the molar solubility of barium fluoride in 0.15 $M$ NaF at 25°C? Compare the solubility in this case with that of $BaF_2$ in pure water.

---

See Problems 18.33, 18.34, 18.35, and 18.36.

 **CONCEPT CHECK 18.2**

Suppose you have equal volumes of saturated solutions of $NaNO_3$, $Na_2SO_4$, and PbS. Which solution would dissolve the most lead(II) sulfate, $PbSO_4$?

## 18.3 Precipitation Calculations

In the chapter opening, we mentioned that calcium oxalate precipitates to form kidney stones. The same salt would precipitate in the body if oxalic acid (a poison) were accidentally ingested, because $Ca^{2+}$ ion is present in the blood. To understand processes such as these, you must understand the conditions under which precipitation occurs.

Precipitation is merely another way of looking at a solubility equilibrium. Rather than ask how much of a substance will dissolve in a solution, you ask: Will precipitation occur for given starting ion concentrations?

## Criterion for Precipitation

The question just asked can be stated more generally: Given the concentrations of substances in a reaction mixture, will the reaction go in the forward or the reverse direction? To answer this, you evaluate the *reaction quotient* $Q_c$ and compare it with the equilibrium constant $K_c$. ◀ The reaction quotient has the same form as the equilibrium-constant expression, but the concentrations of substances are not necessarily equilibrium values. Rather, they are concentrations at the start of a reaction. To predict the direction of reaction, you compare $Q_c$ with $K_c$.

▶ **The reaction quotient was discussed in Section 15.5.**

If $Q_c < K_c$, the reaction should go in the forward direction.

If $Q_c = K_c$, the reaction mixture is at equilibrium.

If $Q_c > K_c$, the reaction should go in the reverse direction.

▶ **Animation: Precipitation**

Suppose you add lead(II) nitrate, $Pb(NO_3)_2$, and sodium chloride, NaCl, to water to give a solution that is 0.050 $M$ $Pb^{2+}$ and 0.10 $M$ $Cl^-$. Will lead(II) chloride, $PbCl_2$, precipitate? To answer this, you first write the solubility equilibrium.

$$PbCl_2(s) \xrightleftharpoons{H_2O} Pb^{2+}(aq) + 2Cl^-(aq)$$

The reaction quotient has the form of the equilibrium-constant expression, which in this case is the $K_{sp}$ expression, but the concentrations of the products are starting values, denoted by the subscript $i$.

$$Q_c = [Pb^{2+}]_i[Cl^-]_i^2$$

Here $Q_c$ for a solubility reaction is often called the **ion product** (rather than reaction quotient), because it is *the product of ion concentrations in a solution, each concentration raised to a power equal to the number of ions in the formula of the ionic compound.*

To evaluate the ion product $Q_c$, you substitute the concentrations of $Pb^{2+}$ and $Cl^-$ ions in the solution at the start of the reaction. You have 0.050 $M$ $Pb^{2+}$ and 0.10 $M$ $Cl^-$. Substituting, you find that

$$Q_c = (0.050)(0.10)^2 = 5.0 \times 10^{-4}$$

$K_{sp}$ for $PbCl_2$ is $1.6 \times 10^{-5}$ (Table 18.1), so $Q_c$ is greater than $K_{sp}$. Therefore, the reaction goes in the reverse direction. That is,

$$PbCl_2(s) \longleftarrow Pb^{2+}(aq) + 2Cl^-(aq)$$

In other words, $Pb^{2+}$ and $Cl^-$ react to precipitate $PbCl_2$. As precipitation occurs, the ion concentrations, and hence the ion product, decrease. Precipitation ceases when the ion product equals $K_{sp}$. The reaction mixture is then at equilibrium.

We can summarize our conclusions in terms of the following criterion for precipitation.

▶ **Precipitation may not occur even though the ion product has been exceeded. In such a case, the solution is supersaturated. Usually a small crystal forms after a time, and then precipitation occurs rapidly.**

Precipitation is expected to occur if the ion product for a solubility reaction is greater than $K_{sp}$. If the ion product is less than $K_{sp}$, precipitation will not occur (the solution is unsaturated with respect to the ionic compound). If the ion product equals $K_{sp}$, the reaction is at equilibrium (the solution is saturated with the ionic compound). ◀

▶ An alternate example is provided in the Instructor's Resource Manual.

## EXAMPLE 18.6

### Predicting Whether Precipitation Will Occur (Given Ion Concentrations)

The concentration of calcium ion in blood plasma is 0.0025 $M$. If the concentration of oxalate ion is $1.0 \times 10^{-7}$ $M$, do you expect calcium oxalate to precipitate? $K_{sp}$ for calcium oxalate is $2.3 \times 10^{-9}$.

### SOLUTION

The ion product for calcium oxalate is

$$\text{Ion product} = [\text{Ca}^{2+}]_i[\text{C}_2\text{O}_4^{2-}]_i = (0.0025) \times (1.0 \times 10^{-7}) = 2.5 \times 10^{-10}$$

This value is smaller than the solubility product constant, so **you do not expect precipitation to occur.**

### EXERCISE 18.6

Anhydrite is a mineral composed of $\text{CaSO}_4$ (calcium sulfate). An inland lake has $\text{Ca}^{2+}$ and $\text{SO}_4^{2-}$ concentrations of 0.0052 $M$ and 0.0041 $M$, respectively. If these concentrations were doubled by evaporation, would you expect calcium sulfate to precipitate?

*See Problems 18.41 and 18.42.*

Precipitation is an important industrial and laboratory process. Figure 18.3 shows the vats used to precipitate magnesium ion from seawater, which is a source of magnesium metal. The following example is a typical problem in precipitation. You are given the volumes and concentrations of two solutions and asked whether a precipitate will form when the solutions are mixed.

## EXAMPLE 18.7

### Predicting Whether Precipitation Will Occur (Given Solution Volumes and Concentrations)

Sulfate ion, $\text{SO}_4^{2-}$, in solution is often determined quantitatively by precipitating it as barium sulfate, $\text{BaSO}_4$. The sulfate ion may have been formed from a sulfur compound. Analysis for the amount of sulfate ion then indicates the percentage of sulfur in the compound. Is a precipitate expected to form at equilibrium when 50.0 mL of 0.0010 $M$ $\text{BaCl}_2$ is added to 50.0 mL of 0.00010 $M$ $\text{Na}_2\text{SO}_4$? The solubility product constant for barium sulfate is $1.1 \times 10^{-10}$. Assume that the total volume of solution, after mixing, equals the sum of the volumes of the separate solutions.

###  PROBLEM STRATEGY

You need to calculate $Q_c$, the ion product, to compare it with $K_{sp}$, the solubility product. For $Q_c$, you need to obtain the concentrations of ions in the solution just after mixing, but before precipitation, if any occurs. First, calculate the moles of each ion added. Then, using the total volume of solution (assumed to

**Figure 18.3**
**Vats for the precipitation of magnesium hydroxide from seawater.** Seawater contains magnesium ion (in addition to other ions). When base is added, magnesium hydroxide precipitates. This is the source of magnesium metal.

equal the sum of the two added solutions), calculate the molar concentration of each ion present. Finally, calculate $Q_c$ and compare it with $K_{sp}$.

**SOLUTION**

The molar amount of $Ba^{2+}$ present in 50.0 mL (= 0.0500 L) of 0.0010 $M$ $BaCl_2$ is

$$\text{Amount of } Ba^{2+} = \frac{0.0010 \text{ mol } Ba^{2+}}{1 \text{ L soln}} \times 0.050 \text{ L soln} = 5.0 \times 10^{-5} \text{ mol } Ba^{2+}$$

The molar concentration of $Ba^{2+}$ in the total solution equals the molar amount of $Ba^{2+}$ divided by the total volume (0.0500 L $BaCl_2$ + 0.0500 L $Na_2SO_4$ = 0.1000 L).

$$[Ba^{2+}] = \frac{5.0 \times 10^{-5} \text{ mol}}{0.1000 \text{ L soln}} = 5.0 \times 10^{-4} M$$

Similarly, you find $\quad [SO_4^{2-}] = 5.0 \times 10^{-5} M$

(Try the calculations for $SO_4^{2-}$.) The ion product is

$$Q_c = [Ba^{2+}]_i[SO_4^{2-}]_i = (5.0 \times 10^{-4}) \times (5.0 \times 10^{-5}) = 2.5 \times 10^{-8}$$

Because the ion product is greater than the solubility product constant (1.1 × $10^{-10}$), **you expect barium sulfate to precipitate.**

---

**EXERCISE 18.7**

A solution of 0.00016 $M$ lead(II) nitrate, $Pb(NO_3)_2$, was poured into 456 mL of 0.00023 $M$ sodium sulfate, $Na_2SO_4$. Would a precipitate of lead(II) sulfate, $PbSO_4$, be expected to form if 255 mL of the lead nitrate solution were added?

---

See Problems 18.43 and 18.44.

### Fractional Precipitation

**Fractional precipitation** is *the technique of separating two or more ions from a solution by adding a reactant that precipitates first one ion, then another, and so forth.* For example, suppose a solution is 0.10 $M$ $Ba^{2+}$ and 0.10 $M$ $Sr^{2+}$. As we will show in the following paragraphs, when you slowly add a concentrated solution of potassium chromate, $K_2CrO_4$, to the solution of $Ba^{2+}$ and $Sr^{2+}$ ions, barium chromate precipitates first. After most of the $Ba^{2+}$ ion has precipitated, strontium chromate begins to come out of solution. It is therefore possible to separate $Ba^{2+}$ and $Sr^{2+}$ ions from a solution by fractional precipitations using $K_2CrO_4$.

To understand why $Ba^{2+}$ and $Sr^{2+}$ ions can be separated in this way, calculate (1) the concentration of $CrO_4^{2-}$ necessary to just begin the precipitation of $BaCrO_4$, and (2) the concentration of $CrO_4^{2-}$ necessary to just begin the precipitation of $SrCrO_4$. Ignore any volume change in the solution of $Ba^{2+}$ and $Sr^{2+}$ ions resulting from the addition of the concentrated $K_2CrO_4$ solution. To calculate the $CrO_4^{2-}$ concentration when $BaCrO_4$ begins to precipitate, you substitute the initial $Ba^{2+}$ concentration into the solubility product equation. $K_{sp}$ for $BaCrO_4$ is $1.2 \times 10^{-10}$.

$$[Ba^{2+}][CrO_4^{2-}] = K_{sp} \text{ (for } BaCrO_4\text{)}$$

$$(0.10)[CrO_4^{2-}] = 1.2 \times 10^{-10}$$

$$[CrO_4^{2-}] = \frac{1.2 \times 10^{-10}}{0.10} = 1.2 \times 10^{-9} \, M$$

In the same way, you can calculate the $CrO_4^{2-}$ concentration when $SrCrO_4$ begins to precipitate. $K_{sp}$ for $SrCrO_4$ is $3.5 \times 10^{-5}$.

$$[Sr^{2+}][CrO_4^{2-}] = K_{sp} \text{ (for } SrCrO_4\text{)}$$

$$(0.10)[CrO_4^{2-}] = 3.5 \times 10^{-5}$$

$$[CrO_4^{2-}] = \frac{3.5 \times 10^{-5}}{0.10} = 3.5 \times 10^{-4} \, M$$

Note that $BaCrO_4$ precipitates first, because the $CrO_4^{2-}$ concentration necessary to form the $BaCrO_4$ precipitate is smaller.

These results reveal that as the solution of $K_2CrO_4$ is slowly added to the solution of $Ba^{2+}$ and $Sr^{2+}$, barium chromate begins to precipitate when the $CrO_4^{2-}$ concentration reaches $1.2 \times 10^{-9} \, M$. Barium chromate continues to precipitate as $K_2CrO_4$ is added. When the concentration of $CrO_4^{2-}$ reaches $3.5 \times 10^{-4} \, M$, strontium chromate begins to precipitate.

What is the percentage of $Ba^{2+}$ ion remaining just as $SrCrO_4$ begins to precipitate? First calculate the concentration of $Ba^{2+}$ at this point. You write the solubility product equation and substitute $[CrO_4^{2-}] = 3.5 \times 10^{-4}$, the concentration of chromate ion when $SrCrO_4$ begins to precipitate.

$$[Ba^{2+}][CrO_4^{2-}] = K_{sp} \text{ (for } BaCrO_4\text{)}$$

$$[Ba^{2+}](3.5 \times 10^{-4}) = 1.2 \times 10^{-10}$$

$$[Ba^{2+}] = \frac{1.2 \times 10^{-10}}{3.5 \times 10^{-4}} = 3.4 \times 10^{-7} \, M$$

To calculate the percentage of $Ba^{2+}$ ion remaining, you divide this concentration of $Ba^{2+}$ by the initial concentration (0.10 $M$), and multiply by 100%.

$$\frac{3.4 \times 10^{-7}}{0.10} \times 100\% = 0.00034\%$$

**Figure 18.4**
**Titration of chloride ion by silver nitrate using potassium chromate as an indicator.** *Left:* A small amount of $K_2CrO_4$ (yellow) has been added to a solution containing an unknown amount of $Cl^-$ ion. *Center:* The solution is titrated by $AgNO_3$ solution, giving a white precipitate of AgCl. *Right:* When nearly all of the $Cl^-$ ion has precipitated as AgCl, silver chromate begins to precipitate. Silver chromate, $Ag_2CrO_4$, has a red-brown color, and the appearance of this color signals the end of the titration. An excess of $Ag^+$ was added to show the color of $Ag_2CrO_4$ more clearly.

The percentage of $Ba^{2+}$ ion remaining in solution is quite low, so most of the $Ba^{2+}$ ion has precipitated by the time $SrCrO_4$ begins to precipitate. You conclude that $Ba^{2+}$ and $Sr^{2+}$ can indeed be separated by fractional precipitation. (Another application of fractional precipitation is shown in Figure 18.4.)

# 18.4 Effect of pH on Solubility

▶ Video Series C, "Sodium Arsenate and Silver Nitrate."

In discussing solubility, we have assumed that the only equilibrium of interest was the one between the solid ionic compound and its ions in solution. Sometimes, however, it is necessary to account for other reactions the ions might undergo. For example, if the anion is the conjugate base of a weak acid, the anion reacts with $H_3O^+$ ion. You should expect the solubility to be affected by pH. We will look into this possibility now.

## Qualitative Effect of pH

Consider the equilibrium between solid calcium oxalate, $CaC_2O_4$, and its ions in aqueous solution:

$$CaC_2O_4(s) \rightleftharpoons Ca^{2+}(aq) + C_2O_4^{2-}(aq)$$

Because the oxalate ion is conjugate to a weak acid (hydrogen oxalate ion, $HC_2O_4^-$), you would expect it to react with any $H_3O^+$ ion that is added—say, from a strong acid:

$$C_2O_4^{2-}(aq) + H_3O^+(aq) \rightleftharpoons HC_2O_4^-(aq) + H_2O(l)$$

According to Le Chatelier's principle, as $C_2O_4^{2-}$ ion is removed by reaction with $H_3O^+$ ion, more calcium oxalate dissolves to replenish some of the $C_2O_4^{2-}$ ion.

$$CaC_2O_4(s) \longrightarrow Ca^{2+}(aq) + C_2O_4^{2-}(aq)$$

Therefore, you expect calcium oxalate to be more soluble in acidic solution (low pH) than in pure water.

In general, salts of weak acids should be expected to be more soluble in acidic solutions. As an example, consider the process of tooth decay. Bacteria on the teeth produce an acidic medium as a result of the metabolism of sugar. Teeth are normally composed of a calcium phosphate mineral hydroxyapatite, which you can denote as either $Ca_5(PO_4)_3OH$ or $3Ca_3(PO_4)_2 \cdot Ca(OH)_2$. This mineral salt of the weak acid $H_2O$ (conjugate to $OH^-$) dissolves in the presence of the acid medium, producing cavities in the teeth. Fluoride toothpastes provide $F^-$ ion, which gradually replaces the $OH^-$ ion in the teeth to produce fluorapatite, $Ca_5(PO_4)_3F$ or $3Ca_3(PO_4)_2 \cdot CaF_2$, which is much less soluble than hydroxyapatite.

▶ An alternate example is provided in the Instructor's Resource Manual.

---

### EXAMPLE 18.8

## Determining the Qualitative Effect of pH on Solubility

Consider two slightly soluble salts, calcium carbonate and calcium sulfate. Which of these would have its solubility more affected by the addition of HCl, a strong acid? Would the solubility of the salt increase or decrease?

 **PROBLEM STRATEGY**

A strong acid affects the solubility of a salt of a weak acid by providing hydronium ion that reacts with the anion of the salt in the solution. As the anion is removed by this reaction, more salt dissolves to replenish the anion concentration. You need to determine which salt has the anion corresponding to the weaker acid.

**SOLUTION**

Calcium carbonate gives the solubility equilibrium

$$CaCO_3(s) \rightleftharpoons Ca^{2+}(aq) + CO_3^{2-}(aq)$$

When a strong acid is added, the hydrogen ion reacts with carbonate ion, because it is conjugate to a weak acid ($HCO_3^-$).

$$H_3O^+(aq) + CO_3^{2-}(aq) \rightleftharpoons H_2O(l) + HCO_3^-(aq)$$

As carbonate ion is removed, calcium carbonate dissolves. Moreover, the hydrogen carbonate ion itself is removed in further reaction.

$$H_3O^+(aq) + HCO_3^-(aq) \longrightarrow H_2O(l) + H_2CO_3(aq) \longrightarrow 2H_2O(l) + CO_2(g)$$

Bubbles of carbon dioxide gas appear as more calcium carbonate dissolves. For calcium sulfate, the corresponding equilibria are

$$CaSO_4(s) \rightleftharpoons Ca^{2+}(aq) + SO_4^{2-}(aq)$$

$$H_3O^+(aq) + SO_4^{2-}(aq) \rightleftharpoons H_2O(l) + HSO_4^-(aq)$$

Again, the anion of the insoluble salt is removed by reaction with hydronium ion. You would expect calcium sulfate to become more soluble in strong acid.

However, $HSO_4^-$ is a much stronger acid than $HCO_3^-$, as you can see by comparing acid-ionization constants. (The values $K_{a2}$ for $H_2CO_3$ and $K_a$ for $HSO_4^-$ in Table 17.1 are $4.8 \times 10^{-11}$ and $1.1 \times 10^{-2}$, respectively.) **Thus, calcium carbonate is much more soluble in acidic solution, whereas the solubility of calcium sulfate is only slightly affected.**

### EXERCISE 18.8

Which salt would have its solubility more affected by changes in pH, silver chloride or silver cyanide?

See Problems 18.53 and 18.54.

 **CONCEPT CHECK 18.3**

If you add a dilute acidic solution to a mixture containing magnesium oxalate and calcium oxalate, which of the two compounds is more likely to dissolve?

## Separation of Metal Ions by Sulfide Precipitation

Many metal sulfides are insoluble in water but dissolve in acidic solution. The sulfide scheme discussed later in Section 18.7 uses this change in solubility of the metal sulfides with pH, or hydronium-ion concentration, to separate a mixture of metal ions. Here is how the separation scheme works.

Consider a solution that is 0.10 $M$ in zinc ion and in lead(II) ion. You add hydrogen sulfide, $H_2S$, to this solution. Hydrogen sulfide ionizes in water as a diprotic acid:

$$H_2S(aq) + H_2O(l) \rightleftharpoons H_3O^+(aq) + HS^-(aq)$$

$$HS^-(aq) + H_2O(l) \rightleftharpoons H_3O^+(aq) + S^{2-}(aq)$$

The ionization forms sulfide ion, $S^{2-}$, which can combine with the metal ions to precipitate the sulfides. Note that by increasing the hydronium-ion concentration, you can reverse both of the previous equilibria, and in that way reduce the sulfide-ion concentration. This allows you to control the sulfide-ion concentration of a hydrogen sulfide solution by varying its pH; the lower the pH, the lower the sulfide-ion concentration.

Whether lead(II) sulfide and zinc sulfide precipitate from the hydrogen sulfide solution depends on the solubility product constants of the sulfides, and on the sulfide-ion concentration, which in turn depends on the pH of the solution. Here are the solubility equilibria:

$$Zn^{2+}(aq) + S^{2-}(aq) \rightleftharpoons ZnS(s); \ K_{sp} = 1.1 \times 10^{-21}$$

$$Pb^{2+}(aq) + S^{2-}(aq) \rightleftharpoons PbS(s); \ K_{sp} = 2.5 \times 10^{-27}$$

By adjusting the pH of the solution, you adjust the sulfide-ion concentration in order to precipitate the least soluble metal sulfide while maintaining the other metal ion in solution. The solubility product constant of lead(II) sulfide is much smaller than that of zinc sulfide, so it is the least soluble sulfide. When a solution that is 0.10 $M$ in each metal ion and 0.30 $M$ in hydronium ion is saturated with hydrogen sulfide, lead(II) sulfide precipitates, but zinc ion remains in solution, You can now filter off the precipitate of lead(II) sulfide, leaving a solution containing the zinc ion.

# Limestone Caves

**ENVIRONMENT**

On the side of a hill, covered by tall grass and shrubs, the cave entrance might easily be missed. It is barely large enough for a person to enter, even if hunched over, but the space soon expands into a chamber that one can easily stand in. Rock columns rise to an arched ceiling, and there are caverns leading mysteriously away on both sides of the chamber. Limestone caves, like the one described here, which is in Tennessee, dot the landscapes of the eastern and southwestern United States and of Europe and other places as well. Though many limestone caves are inaccessible, a few, like Mammouth Cave in Kentucky and Carlsbad Caverns in New Mexico, are spectacular tourist attractions with multicolored rock draperies and icicles hanging from the ceilings of large underground rooms (Figure 18.5). How do such caves form?

Limestone itself formed millions of years ago. Seashells accumulated on ocean floors, and then layers of these shells were subsequently compacted by great pressure from overlying sediment. Later movements of the earth's crust pushed the limestone layers out of the sea, often leaving them well inland.

Seashells and limestone are primarily calcium carbonate. Calcium carbonate is fairly insoluble, as you would expect from its solubility product constant ($3.8 \times 10^{-9}$). But it dissolves readily in acidic solution. Water that has filtered through decomposing vegetation contains carbonic acid, as well as other acids. When such an acidic solution comes into contact with limestone, it carves out caverns. The water is now a solution of calcium hydrogen carbonate.

$$CaCO_3(s) + H_2O(l) + CO_2(aq) \longrightarrow Ca^{2+}(aq) + 2HCO_3^-(aq)$$

This process is reversible, which you would predict from Le Chatelier's principle. As the calcium hydrogen car-

**Figure 18.5**
**A limestone cave (Luray Caverns, Virginia).** Such caves are formed by the action of acidic groundwater on limestone deposits. The icicle-like formations (stalagmites and stalactites) in the caves are caused by the reprecipitation of calcium carbonate as carbon dioxide in the solution equilibrates with the surrounding air.

bonate solution percolates through holes in the limestone and drips into caverns, water evaporates and carbon dioxide gas escapes; calcium carbonate precipitates onto icicle-shaped *stalactites*.

$$Ca^{2+}(aq) + 2HCO_3^-(aq) \longrightarrow CaCO_3(s) + H_2O(l) + CO_2(g)$$

Excess solution drips onto the cavern floor, where calcium carbonate precipitates on upward-growing *stalagmites*. Iron compounds in the solution add yellow and brown colors to the columns of rock. Stalactites and stalagmites grow at the rate of about a centimeter every 300 years.

# Complex-Ion Equilibria

Many metal ions, especially those of the transition elements, form coordinate covalent bonds with molecules or anions having lone pairs of electrons. For example, the silver ion, $Ag^+$, can react with $NH_3$ to form the $Ag(NH_3)_2^+$ ion. The lone pair of electrons on the N atom of $NH_3$ forms a coordinate covalent bond to the silver ion to form what is called a complex ion. (See Figure 18.6.)

$$Ag^+ + 2(:NH_3) \longrightarrow (H_3N:Ag:NH_3)^+$$

▶ The Lewis concept of acids and bases was discussed in Section 16.3.

Note that the metal ion acts as a Lewis acid, and the molecule or anion having the lone pair of electrons acts as a Lewis base. ◀

A **complex ion** is an *ion formed from a metal ion with a Lewis base attached to it by a coordinate covalent bond.* A *complex* is a compound containing complex ions. A **ligand** is *a Lewis base that bonds to a metal ion to form a complex ion.* For example, $Ag(NH_3)_2^+$ is a complex ion formed from the $Ag^+$ ion and two $NH_3$ molecules. The $NH_3$ molecules are the ligands.

# 18.5 Complex-Ion Formation

The aqueous silver ion forms a complex ion with ammonia by reacting with $NH_3$ in steps.

$$Ag^+(aq) + NH_3(aq) \rightleftharpoons Ag(NH_3)^+(aq)$$

$$Ag(NH_3)^+(aq) + NH_3(aq) \rightleftharpoons Ag(NH_3)_2^+(aq)$$

When you add these equations, you get the overall equation for the formation of the complex ion $Ag(NH_3)_2^+$.

$$Ag^+(aq) + 2NH_3(aq) \rightleftharpoons Ag(NH_3)_2^+(aq)$$

The **formation constant**, or **stability constant, $K_f$,** of a complex ion is *the equilibrium constant for the formation of the complex ion from the aqueous metal ion and the ligands.* The formation constant of $Ag(NH_3)_2^+$ is

$$K_f = \frac{[Ag(NH_3)_2^+]}{[Ag^+][NH_3]^2}$$

The value of $K_f$ for $Ag(NH_3)_2^+$ is $1.7 \times 10^7$. Its large value means that the complex ion is quite stable. When a large amount of $NH_3$ is added to a solution of $Ag^+$, you expect most of the $Ag^+$ ion to react to form the complex ion. Table 18.2 lists values for the formation constants of some complex ions.

The **dissociation constant ($K_d$)** for a complex ion is *the reciprocal, or inverse, value of $K_f$.* The equation for the dissociation of $Ag(NH_3)_2^+$ is

$$Ag(NH_3)_2^+(aq) \rightleftharpoons Ag^+(aq) + 2NH_3(aq)$$

and its equilibrium constant is

$$K_d = \frac{1}{K_f} = \frac{[Ag^+][NH_3]^2}{[Ag(NH_3)_2^+]}$$

### Equilibrium Calculations with $K_f$

The next example shows how to calculate the concentration of an aqueous metal ion in equilibrium with a complex ion.

**Figure 18.6**
**Complex ions of nickel(II).**
The test tube on the left contains a solution of $Ni(H_2O)_6^{2+}$ ion (green). Adding ammonia to a similar test tube containing this ion, on the right, yields the ammonia complex ion $Ni(NH_3)_6^{2+}$ (purple).

---

**EXAMPLE 18.9**

## Calculating the Concentration of a Metal Ion in Equilibrium with a Complex Ion

What is the concentration of $Ag^+(aq)$ ion in 0.010 $M$ AgNO$_3$ that is also 1.00 $M$ NH$_3$? $K_f$ for $Ag(NH_3)_2^+$ ion is $1.7 \times 10^7$.

## Table 18.2
## Formation Constants of Complex Ions at 25°C

| Complex Ion | $K_f$ |
|---|---|
| $Ag(CN)_2^-$ | $5.6 \times 10^{18}$ |
| $Ag(NH_3)_2^+$ | $1.7 \times 10^7$ |
| $Ag(S_2O_3)_2^{3-}$ | $2.9 \times 10^{13}$ |
| $Cd(NH_3)_4^{2+}$ | $1.0 \times 10^7$ |
| $Cu(CN)_2^-$ | $1.0 \times 10^{16}$ |
| $Cu(NH_3)_4^{2+}$ | $4.8 \times 10^{12}$ |
| $Fe(CN)_6^{4-}$ | $1.0 \times 10^{35}$ |
| $Fe(CN)_6^{3-}$ | $9.1 \times 10^{41}$ |
| $Ni(CN)_4^{2-}$ | $1.0 \times 10^{31}$ |
| $Ni(NH_3)_6^{2+}$ | $5.6 \times 10^8$ |
| $Zn(NH_3)_4^{2+}$ | $2.9 \times 10^9$ |
| $Zn(OH)_4^{2-}$ | $2.8 \times 10^{15}$ |

▶ Video Series C, "Reactions of Silver(I); Reactions of Nickel(II)."

### PROBLEM STRATEGY

The formation constant for $Ag(NH_3)_2^+$ is large, so silver exists primarily as this ion. This suggests that you do this problem in two parts. First you do the *stoichiometry calculation,* in which you assume that $Ag^+(aq)$ reacts completely to form $Ag(NH_3)_2^+(aq)$. Then you do the *equilibrium calculation,* in which this complex ion dissociates to give a small amount of $Ag^+(aq)$.

### SOLUTION

### Stoichiometry Calculation

In 1 L of solution, you initially have 0.010 mol $Ag^+(aq)$ from $AgNO_3$. This reacts to give 0.010 mol $Ag(NH_3)_2^+$, leaving $(1.00 - 2 \times 0.010)$ mol $NH_3$, which equals 0.98 mol $NH_3$ to two significant figures. You now look at the equilibrium for the dissociation of $Ag(NH_3)_2^+$.

### Equilibrium Calculation

**STEP 1**    One liter of the solution contains 0.010 mol $Ag(NH_3)_2^+$ and 0.98 mol $NH_3$. The complex ion dissociates slightly, so that 1 L of solution contains $x$ mol $Ag^+$. These data are summarized in the table:

| Concentration (M) | $Ag(NH_3)_2^+ (aq)$ | $\rightleftharpoons$ | $Ag^+(aq)$ | + | $2NH_3(aq)$ |
|---|---|---|---|---|---|
| Starting | 0.010 | | 0 | | 0.98 |
| Change | $-x$ | | $+x$ | | $+2x$ |
| Equilibrium | $0.010 - x$ | | $x$ | | $0.98 + 2x$ |

**STEP 2**    The formation constant is

$$K_f = \frac{[Ag(NH_3)_2^+]}{[Ag^+][NH_3]^2} = 1.7 \times 10^7$$

From this you can write the dissociation constant, which corresponds to the reaction as shown in the table.

$$\frac{[Ag^+][NH_3]^2}{[Ag(NH_3)_2^+]} = K_d = \frac{1}{K_f}$$

Substituting into this equation gives

$$\frac{x(0.98 + 2x)^2}{(0.010 - x)} = \frac{1}{1.7 \times 10^7}$$

**STEP 3**    The right-hand side of the equation equals $5.9 \times 10^{-8}$. If you assume $x$ to be small compared with 0.010,

$$\frac{x(0.98)^2}{0.010} \simeq 5.9 \times 10^{-8}$$

and

$$x \simeq 5.9 \times 10^{-8} \times 0.010/(0.98)^2 = 6.1 \times 10^{-10}$$

The silver-ion concentration is **$6.1 \times 10^{-10}\,M$** (over 10 million times smaller than its value in 0.010 $M$ $AgNO_3$ that does not contain ammonia).

## EXERCISE 18.9

What is the concentration of $Cu^{2+}(aq)$ in a solution that was originally 0.015 $M$ $Cu(NO_3)_2$ and 0.100 $M$ $NH_3$? The $Cu^{2+}$ ion forms the complex ion $Cu(NH_3)_4^{2+}$. Its formation constant is given in Table 18.2.

See Problems 18.57 and 18.58.

### Amphoteric Hydroxides

An **amphoteric hydroxide** is *a metal hydroxide that reacts with both bases and acids.* Zinc is an example of a metal that forms such a hydroxide. Zinc hydroxide, $Zn(OH)_2$, is an insoluble hydroxide, as are most of the metal hydroxides except those of Groups IA and IIA elements. Zinc hydroxide reacts with a strong acid, as you would expect of a base, and the metal hydroxide dissolves.

$$Zn(OH)_2(s) + 2H_3O^+(aq) \longrightarrow Zn^{2+}(aq) + 4H_2O(l)$$

With a base, however, $Zn(OH)_2$ reacts to form the complex ion $Zn(OH)_4^{2-}$.

$$Zn(OH)_2(s) + 2OH^-(aq) \longrightarrow Zn(OH)_4^{2-}(aq)$$

In this way the hydroxide also dissolves in a strong base.

When a strong base is slowly added to a solution of $ZnCl_2$, a white precipitate of $Zn(OH)_2$ first forms.

$$Zn^{2+}(aq) + 2OH^-(aq) \longrightarrow Zn(OH)_2(s)$$

But as more base is added, the white precipitate dissolves, forming the complex ion $Zn(OH)_4^{2-}$. (See Figure 18.7.)

Other common amphoteric hydroxides are those of aluminum, chromium(III), lead(II), tin(II), and tin(IV). The amphoterism of $Al(OH)_3$ is commercially used to separate aluminum oxide from the aluminum ore bauxite. Bauxite contains hydrated $Al_2O_3$ plus impurities, such as silica sand ($SiO_2$) and iron oxide ($Fe_2O_3$). The aluminum oxide dissolves in NaOH solution as the $Al(OH)_4^-$ ion. After the insoluble impurities are filtered off, the solution of $Al(OH)_4^-$ can be slightly acidified to precipitate pure $Al(OH)_3$.

$$Al(OH)_4^-(aq) + H_3O^+(aq) \longrightarrow Al(OH)_3(s) + 2H_2O(l)$$

**Figure 18.7**
**Demonstration of the amphoteric behavior of zinc hydroxide.** *Left:* The beaker on the left contains $ZnCl_2$; the one on the right contains NaOH. *Center:* As some of the solution of NaOH is added to the solution of $ZnCl_2$, a white precipitate of $Zn(OH)_2$ forms. *Right:* After more NaOH is added, the precipitate dissolves by forming the hydroxo complex $Zn(OH)_4^{2-}$.

## 18.6 Complex Ions and Solubility

Example 18.9 demonstrates that the formation of the complex ion $Ag(NH_3)_2^+$ reduces the concentration of silver ion, $Ag^+(aq)$, in solution. The solution was initially $0.010\,M$ $AgNO_3$, or $0.010\,M$ $Ag^+(aq)$. When this solution is made $1.00\,M$ in $NH_3$, the $Ag^+(aq)$ ion concentration decreases to $6.1 \times 10^{-10}\,M$. You can see that the ion product for a slightly soluble silver salt could be decreased to below $K_{sp}$. Then the slightly soluble salt might not precipitate in a solution containing ammonia, whereas it otherwise would.

---

### EXAMPLE 18.10

### Predicting Whether a Precipitate Will Form in the Presence of the Complex Ion

a. Will silver chloride precipitate from a solution that is $0.010\,M$ $AgNO_3$ and $0.010\,M$ NaCl? b. Will silver chloride precipitate from this solution if it is also $1.00\,M$ $NH_3$?

 **PROBLEM STRATEGY**

Part a. is a simple precipitation problem in which you compare $Q_c$ with $K_{sp}$. See Example 18.6. Part b. differs only in that you need to know the $Ag^+$ concentration in equilibrium with a complex ion.

**SOLUTION**

a. To determine whether a precipitate should form, you calculate the ion product and compare it with $K_{sp}$ for AgCl ($1.8 \times 10^{-10}$).

$$\text{Ion product} = [Ag^+]_i[Cl^-]_i = (0.010)(0.010) = 1.0 \times 10^{-4}$$

This is greater than $K_{sp} = 1.8 \times 10^{-10}$, so **a precipitate should form.**

b. You first need to calculate the concentration of $Ag^+(aq)$ in a solution containing $1.00\,M$ $NH_3$. We did this in Example 18.9 and found that $[Ag^+]$ equals $6.1 \times 10^{-10}$. Hence,

$$\text{Ion product} = [Ag^+]_i[Cl^-]_i = (6.1 \times 10^{-10})(0.010) = 6.1 \times 10^{-12}$$

Because the ion product is smaller than $K_{sp} = 1.8 \times 10^{-10}$, **no precipitate should form.**

---

### EXERCISE 18.10

Will silver iodide precipitate from a solution that is $0.0045\,M$ $AgNO_3$, $0.15\,M$ NaI, and $0.20\,M$ KCN?

---

See Problems 18.59 and 18.60.

In Example 18.10, we determined whether a precipitate of AgCl is expected to form from a solution made up with given concentrations of $Ag^+$, $Cl^-$, and $NH_3$. The problem involved the $Ag(NH_3)_2^+$ complex in equilibrium. From it, we determined the concentration of free $Ag^+(aq)$ and then calculated the ion product of AgCl.

Suppose you want to find the solubility of AgCl in a solution of aqueous ammonia of given concentration. This problem involves the solubility equilibrium for AgCl, in addition to the complex-ion equilibrium. As silver chloride dissolves to give ions, the $Ag^+$ ion reacts with $NH_3$ to give the complex ion $Ag(NH_3)_2^+$. The equilibria are

$$AgCl(s) \rightleftharpoons Ag^+(aq) + Cl^-(aq)$$

$$Ag^+(aq) + 2NH_3(aq) \rightleftharpoons Ag(NH_3)_2^+(aq)$$

When $Ag^+(aq)$ reacts to give the complex ion, more AgCl dissolves to partially replenish the $Ag^+(aq)$ ion, according to Le Chatelier's principle. Therefore, silver chloride is more soluble in aqueous ammonia than in pure water. The next example shows how you can calculate this solubility.

► An alternate example is provided in the Instructor's Resource Manual.

---

**EXAMPLE 18.11**

## Calculating the Solubility of a Slightly Soluble Ionic Compound in a Solution of the Complex Ion

Calculate the molar solubility of AgCl in 1.0 $M$ $NH_3$.

 **PROBLEM STRATEGY**

If you add the solubility equilibrium and the complex-ion equilibrium (the two equations just before this example), you obtain the overall equilibrium for the dissolving of AgCl in $NH_3$. The equilibrium constant for this reaction equals the product of the equilibrium constants for solubility and complex-ion formation. (We will show that in the solution to this example; however, we discussed the general principle at the end of Section 15.2.) Once you have the equilibrium constant, you can solve the equilibrium problem.

**SOLUTION**

You obtain the overall equation for the process by adding the solubility and complex-ion equilibria.

$$AgCl(s) \rightleftharpoons \cancel{Ag^+(aq)} + Cl^-(aq)$$
$$\cancel{Ag^+(aq)} + 2NH_3(aq) \rightleftharpoons Ag(NH_3)_2^+(aq)$$
$$\overline{AgCl(s) + 2NH_3(aq) \rightleftharpoons Ag(NH_3)_2^+(aq) + Cl^-(aq)}$$

The equilibrium constant for the overall equation is

$$K_c = \frac{[Ag(NH_3)_2^+][Cl^-]}{[NH_3]^2}$$

Because the overall equation is the sum of the solubility and complex-ion equilibria, the equilibrium constant $K_c$ equals the product of the equilibrium constants for the solubility and complex-ion equilibria. You can easily verify this.

$$K_c = K_{sp}K_f = \cancel{[Ag^+]}[Cl^-]\frac{[Ag(NH_3)_2^+]}{\cancel{[Ag^+]}[NH_3]^2}$$

The value of $K_c = K_{sp}K_f$ can be obtained from $K_{sp}$ for AgCl in Table 18.1 and $K_f$ for $Ag(NH_3)_2^+$ in Table 18.2.

$$K_{sp}K_f = (1.8 \times 10^{-10})(1.7 \times 10^7) = 3.1 \times 10^{-3}$$

You now solve the equilibrium problem. The concentration table follows:

| Concentration ($M$) | $AgCl(s)$ + | $2NH_3(aq)$ | $\rightleftharpoons$ | $Ag(NH_3)_2^+(aq)$ | + | $Cl^-(aq)$ |
|---|---|---|---|---|---|---|
| Starting | | 1.0 | | 0 | | 0 |
| Change | | $-2x$ | | $+x$ | | $+x$ |
| Equilibrium | | $1.0 - 2x$ | | $x$ | | $x$ |

Substituting into the equilibrium-constant equation

$$\frac{[Ag(NH_3)_2^+][Cl^-]}{[NH_3]^2} = K_{sp}K_f$$

gives

$$\frac{x^2}{(1.0 - 2x)^2} = 3.1 \times 10^{-3}$$

You solve this equation by taking the square root of both sides.

$$\frac{x}{1.0 - 2x} = 0.056$$

Rearranging,

$$x = 0.056(1.0 - 2x) = 0.056 - 0.11x$$

Hence,

$$x = \frac{0.056}{1.11} = 0.050$$

Note that the molar solubility of AgCl equals the molar concentration of silver in the solution. Because of the stability of $Ag(NH_3)_2^+$, most of the silver in solution will be in the form of this complex ion. Therefore, because $x$ equals the concentration of $Ag(NH_3)_2^+$, the molar solubility of AgCl equals **0.050 $M$.**

---

**EXERCISE 18.11**

What is the molar solubility of AgBr in 1.0 $M$ $Na_2S_2O_3$ (sodium thiosulfate)? Silver ion forms the complex ion $Ag(S_2O_3)_2^{3-}$. See Tables 18.1 and 18.2 for data.

---

See Problems 18.61 and 18.62.

# An Application of Solubility Equilibria

**Qualitative analysis** involves *the determination of the identity of the substances present in a mixture.* One scheme for the qualitative analysis of a solution of metal ions is based on the relative solubilities of the metal sulfides. This scheme is no longer widely used in practical analysis, having been replaced largely by modern instrumental methods. But it is still taught in general chemistry laboratory as a way of imparting some knowledge of inorganic chemistry while illustrating the concepts of solubility, buffers, and so forth. We will look at the main outline of the sulfide scheme.

# 18.7 Qualitative Analysis of Metal Ions

In the qualitative analysis scheme for metal ions, a cation is usually detected by the presence of a characteristic precipitate. For example, silver ion gives a white precipitate with chloride ion. But other ions also give a white precipitate with chloride ion. Therefore, you must first subject a mixture to a procedure that separates individual ions before applying a precipitation test to any particular ion.

Figure 18.8 shows a flowchart illustrating how the metal ions in an aqueous solution are first separated into five analytical groups. The Analytical Group I ions, $Ag^+$, $Hg_2^{2+}$, and $Pb^{2+}$, are separated from a solution of other ions by adding dilute hydrochloric acid. The ions are precipitated as the chlorides $AgCl$, $Hg_2Cl_2$, and $PbCl_2$, which are removed by filtering.

The other metal ions remain in the filtrate, the solution that passes through the filter. Many of these ions can be separated by precipitating them as metal sulfides with $H_2S$. This separation takes advantage of the fact that only the least soluble metal sulfides precipitate in acidic solution. After these are removed and the solution is made basic, other metal sulfides precipitate. ◄

► The solubility of the metal sulfides at different pH was discussed in Section 18.4.

**Figure 18.8**
**Flowchart of the qualitative analysis scheme for the separation of metal ions.** The first group of ions consists of those precipitated by HCl. The separation of the next two groups is based on the differences in solubilities of the metal sulfides. The fourth group consists of ions of the alkaline earth elements, which are precipitated as carbonates or phosphates. The final group of ions is in the filtrate after these precipitations.

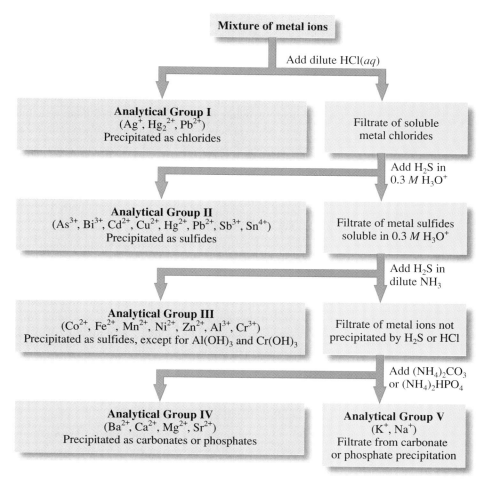

► Hydrogen sulfide is often prepared in the solution by adding thioacetamide, $CH_3CSNH_2$, and warming.

$$CH_3CSNH_2 + H_2O \longrightarrow$$
$$CH_3CONH_2 + H_2S$$

Analytical Group II consists of metal ions precipitated as metal sulfides from a solution that is 0.3 $M$ $H_3O^+$ and is saturated with $H_2S$. ◄ The ions are $As^{3+}$, $Bi^{3+}$, $Cd^{2+}$, $Cu^{2+}$, $Hg^{2+}$, $Pb^{2+}$, $Sb^{3+}$, and $Sn^{4+}$. Lead(II) ion can appear in Analytical Group II as well as in Group I, because $PbCl_2$ is somewhat soluble and therefore $Pb^{2+}$ may not be completely precipitated by HCl.

Analytical Group III consists of the metal ions in the filtrate from Group II that are precipitated in weakly basic solution with $H_2S$. The ions $Co^{2+}$, $Fe^{2+}$, $Mn^{2+}$, $Ni^{2+}$, and $Zn^{2+}$ are precipitated as the sulfides. The ions $Al^{3+}$ and $Cr^{3+}$ are precipitated as hydroxides.

The filtrate obtained after the Group III metal ions have precipitated contains the alkali metal ions and the alkaline earth ions. The Analytical Group IV ions, $Ba^{2+}$, $Ca^{2+}$, $Mg^{2+}$, and $Sr^{2+}$, are precipitated as carbonates or phosphates by adding $(NH_4)_2CO_3$ or $(NH_4)_2HPO_4$. The filtrate from this separation contains the Analytical Group V ions, $K^+$ and $Na^+$.

To illustrate the separation scheme, suppose you have a solution containing the metal cations $Ag^+$, $Cu^{2+}$, $Zn^{2+}$, $Ca^{2+}$, and $Na^+$. When you add dilute HCl($aq$) to this solution, $Ag^+$ precipitates as AgCl.

$$Ag^+(aq) + Cl^-(aq) \longrightarrow \underset{\text{white}}{AgCl(s)}$$

You filter off the precipitate of AgCl. The filtrate contains the remaining ions.

When you make the filtrate 0.3 $M$ $H_3O^+$ and then saturate it with $H_2S$, the $Cu^{2+}$ ion precipitates as CuS.

$$Cu^{2+}(aq) + S^{2-}(aq) \longrightarrow \underset{\text{black}}{CuS(s)}$$

Zinc(II) ion also forms a sulfide, but it is soluble in acidic solution. After removing the precipitate of CuS, the solution is made weakly basic with $NH_3(aq)$. This increases the $S^{2-}$ concentration sufficiently to precipitate ZnS.

$$Zn^{2+}(aq) + S^{2-}(aq) \longrightarrow \underset{\text{white}}{ZnS(s)}$$

The filtrate from this separation contains $Ca^{2+}$ and $Na^+$. Calcium ion can be precipitated as the phosphate or the carbonate. For example, by adding $(NH_4)_2CO_3$, you get the reaction

$$Ca^{2+}(aq) + CO_3^{2-}(aq) \longrightarrow \underset{\text{white}}{CaCO_3(s)}$$

The filtrate from this solution contains $Na^+$ ion. Now the five ions have been separated.

In the complete analysis, once the ions are separated into analytical groups, these are further separated into the individual ions. We can illustrate this with Analytical Group I, which is precipitated as a mixture of the chlorides AgCl, $Hg_2Cl_2$, and $PbCl_2$. Of these, only $PbCl_2$ is significantly soluble in hot water. Thus, if the Analytical Group I precipitate is mixed with hot water, lead chloride dissolves and the silver and mercury(I) chlorides can be filtered off. Any lead(II) ion is in the filtrate and can be revealed by adding potassium chromate. Lead(II) ion gives a bright yellow precipitate of lead(II) chromate, $PbCrO_4$.

$$Pb^{2+}(aq) + CrO_4^{2-}(aq) \longrightarrow \underset{\text{yellow}}{PbCrO_4(s)}$$

If there is precipitate remaining after the extraction with hot water, it consists of $AgCl$, $Hg_2Cl_2$, or both. However, only silver ion forms a stable ammonia complex ion. Therefore, silver chloride dissolves in ammonia.

$$AgCl(s) + 2NH_3(aq) \longrightarrow Ag(NH_3)_2{}^+(aq) + Cl^-(aq)$$

Mercury(I) chloride is simultaneously oxidized and reduced in ammonia solution, giving a precipitate of mercury(II) amido chloride, $HgNH_2Cl$, and mercury metal, which appears black because of its finely divided state.

$$Hg_2Cl_2(s) + 2NH_3(aq) \longrightarrow \underbrace{HgNH_2Cl(s) + Hg(l)}_{\text{black or gray}} + NH_4{}^+(aq) + Cl^-(aq)$$

The presence of silver ion in the filtrate is revealed by adding hydrochloric acid, which combines with the $NH_3$, releasing $Ag^+(aq)$ ion. The $Ag^+(aq)$ ion reacts with $Cl^-(aq)$ to give a white precipitate of $AgCl$.

$$Ag(NH_3)_2{}^+(aq) + 2H_3O^+(aq) + Cl^-(aq) \longrightarrow AgCl(s) + 2NH_4{}^+(aq) + 2H_2O(l)$$

The other analytical groups are separated in similar fashion. Once a given ion has been separated, its presence is usually confirmed by a particular reactant—perhaps one that gives a distinctive precipitate.

# A Checklist for Review

## Important Terms

solubility product constant ($K_{sp}$) (18.1)
ion product (18.3)
fractional precipitation (18.3)

complex ion (page 785)
ligand (page 785)
formation (stability) constant ($K_f$) (18.5)

dissociation constant (of a complex ion) ($K_d$) (18.5)
amphoteric hydroxide (18.5)

qualitative analysis (page 790)

## Summary of Facts and Concepts

The equilibrium constant for the equilibrium between a slightly soluble ionic solid and its ions in solution is called the *solubility product constant, $K_{sp}$*. Its value can be determined from the solubility of the solid. Conversely, when the solubility product constant is known, the solubility of the solid can be calculated. The solubility is decreased by the addition of a soluble salt that supplies a common ion. Qualitatively, this can be seen to follow from Le Chatelier's principle. Quantitatively, the *common-ion effect* on solubility can be obtained from the solubility product constant.

Rather than look at the solubility process as the dissolving of a solid in a solution, you can look at it as the *precipitation* of the solid from the solution. You can decide whether precipitation will occur by computing the *ion product*. Precipitation occurs when the ion product is greater than $K_{sp}$.

*Solubility is affected by the pH* if the compound supplies an anion conjugate to a weak acid. As the pH decreases ($H_3O^+$

ion concentration increases), the anion concentration decreases because the anion forms the weak acid. Therefore, the ion product decreases and the solubility increases.

The concentration of a metal ion in solution is decreased by *complex-ion formation*. The equilibrium constant for the formation of the complex ion from the aqueous metal ion and the ligands is called the *formation constant* (or *stability constant*), $K_f$. Because complex-ion formation reduces the concentration of aqueous metal ion, an ionic compound of the metal is more soluble in a solution of the ligand.

The sulfide scheme of *qualitative analysis* separates the metal ions in a mixture by using precipitation reactions. Variation of solubility with pH and with complex-ion formation is used to aid in the separation.

## Operational Skills

1. **Writing solubility product expressions**   Write the solubility product expression for a given ionic compound. **(EXAMPLE 18.1)**
2. **Calculating $K_{sp}$ from the solubility, or vice versa**   Given the solubility of a slightly soluble ionic compound, calculate $K_{sp}$ **(EXAMPLES 18.2, 18.3)**. Given $K_{sp}$, calculate the solubility of an ionic compound **(EXAMPLE 18.4)**.
3. **Calculating the solubility of a slightly soluble salt in a solution of a common ion**   Given the solubility product constant, calculate the molar solubility of a slightly soluble ionic compound in a solution that contains a common ion.   **(EXAMPLE 18.5)**
4. **Predicting whether precipitation will occur**   Given the concentrations of ions originally in solution, determine whether a precipitate is expected to form **(EXAMPLE 18.6)**. Determine whether a precipitate is expected to form when two solutions of known volume and molarity are mixed **(EXAMPLE 18.7)**. For both problems, you will need the solubility product constant.

5. **Determining the qualitative effect of pH on solubility**   Decide whether the solubility of a salt will be greatly increased by decreasing the pH.   **(EXAMPLE 18.8)**
6. **Calculating the concentration of a metal ion in equilibrium with a complex ion**   Calculate the concentration of aqueous metal ion in equilibrium with the complex ion, given the original metal-ion and ligand concentrations **(EXAMPLE 18.9)**. The formation constant $K_f$ of the complex ion is required.
7. **Predicting whether a precipitate will form in the presence of the complex ion**   Predict whether an ionic compound will precipitate from a solution of known concentrations of cation, anion, and ligand that complexes with the cation **(EXAMPLE 18.10)**. $K_f$ and $K_{sp}$ are required.
8. **Calculating the solubility of a slightly soluble ionic compound in a solution of the complex ion**   Calculate the molar solubility of a slightly soluble ionic compound in a solution of known concentration of a ligand that complexes with the cation **(EXAMPLE 18.11)**. $K_{sp}$ and $K_f$ are required.

## Review Questions

**18.1**   Suppose the molar solubility of nickel hydroxide, $Ni(OH)_2$, is $x$ $M$. Show that $K_{sp}$ for nickel hydroxide equals $4x^3$.

**18.2**   Explain why calcium sulfate is less soluble in sodium sulfate solution than in pure water.

**18.3**   What must be the concentration of silver ion in a solution that is in equilibrium with solid silver chloride and that is 0.10 $M$ in $Cl^-$?

**18.4**   Discuss briefly how you could predict whether a precipitate will form when solutions of lead nitrate and potassium iodide are mixed. What information do you need to have?

**18.5**   Explain why barium fluoride dissolves in dilute hydrochloric acid but is insoluble in water.

**18.6**   Explain how metal ions such as $Pb^{2+}$ and $Zn^{2+}$ are separated by precipitation with hydrogen sulfide.

**18.7**   Lead chloride at first precipitates when sodium chloride is added to a solution of lead nitrate. Later, when the solution is made more concentrated in chloride ion, the precipitate dissolves. Explain what is happening. What equilibria are involved? Note that lead ion forms the complex ion $PbCl_4^{2-}$.

**18.8**   A precipitate forms when a small amount of sodium hydroxide is added to a solution of aluminum sulfate. This precipitate dissolves when more sodium hydroxide is added. Explain what is happening.

**18.9**   Describe how you would separate the metal ions in a solution containing silver ion, copper(II) ion, and nickel(II) ion, using the sulfide scheme of qualitative analysis.

**18.10**   A solution containing calcium ion and magnesium ion is buffered with ammonia–ammonium chloride. When carbonate ion is added to the solution, calcium carbonate precipitates but magnesium carbonate does not. Explain.

## Conceptual Problems

**18.11**   Which compound in each of the following pairs of compounds is the more soluble one?
a. silver chloride or silver iodide
b. magnesium hydroxide or copper(II) hydroxide

**18.12**   You are given two mineral samples: halite, which is NaCl, and fluorite, which is $CaF_2$. Describe a simple test you could use to discover which mineral is fluorite.

**18.13** You are given a saturated solution of lead(II) chloride. Which one of the following solutions would be most effective in yielding a precipitate when added to the lead(II) chloride solution?
a. 0.1 *M* NaCl(*aq*)
c. saturated PbS(*aq*)
b. 0.1 *M* $Na_2SO_4$(*aq*)

**18.14** Which of the following pictures best represents a solution made by adding 10 g of silver chloride, AgCl, to a liter of water? In these pictures, the dark spheres represent $Ag^+$ ions and the light spheres represent chloride ions. For clarity, water molecules are not shown.

**18.15** Which of the following pictures best represents an unsaturated solution of sodium chloride, NaCl? In these pictures, the dark spheres represent $Na^+$ ions and the light spheres

represent chloride ions. For clarity, water molecules are not shown.

**18.16** When ammonia is first added to a solution of copper(II) nitrate, a pale blue precipitate of copper(II) hydroxide forms. As more ammonia is added, however, this precipitate dissolves. Describe what is happening.

**18.17** You are given a solution of the ions $Mg^{2+}$, $Ca^{2+}$, and $Ba^{2+}$. Devise a scheme to separate these ions using sodium sulfate. Note that magnesium sulfate is soluble.

**18.18** You add dilute hydrochloric acid to a solution containing a metal ion. No precipitate forms. After the acidity is adjusted to 0.3 *M* hydronium ion, you bubble hydrogen sulfide into the solution. Again no precipitate forms. Is it possible that the original solution contained silver ion? Could it have contained copper(II) ion?

## Practice Problems

### Solubility and $K_{sp}$

**18.19** Use the solubility rules (Table 4.1) to decide which of the following compounds are expected to be soluble and which insoluble.
a. $NaNO_3$     b. $SrCO_3$     c. $PbI_2$     d. $NH_4Cl$

**18.20** Use the solubility rules (Table 4.1) to decide which of the following compounds are expected to be soluble and which insoluble.
a. AgI     b. $CaBr_2$     c. $SrSO_4$     d. NaI

**18.21** Write solubility product expressions for the following compounds.
a. $SrSO_4$     b. $Mg(OH)_2$     c. $Ca_3(PO_4)_2$     d. $Ag_2S$

**18.22** Write solubility product expressions for the following compounds.
a. $MgC_2O_4$     c. $Zn(OH)_2$
b. $Al(OH)_3$     d. $Ca_3(AsO_4)_2$

**18.23** The solubility of silver bromate, $AgBrO_3$, in water is 0.0072 g/L. Calculate $K_{sp}$.

**18.24** The solubility of magnesium oxalate, $MgC_2O_4$, in water is 0.0093 mol/L. Calculate $K_{sp}$.

**18.25** Calculate the solubility product constant for copper(II) iodate, $Cu(IO_3)_2$. The solubility of copper(II) iodate in water is 0.13 g/100 mL.

**18.26** The solubility of silver chromate, $Ag_2CrO_4$, in water is 0.022 g/L. Calculate $K_{sp}$.

**18.27** The pH of a saturated solution of magnesium hydroxide (milk of magnesia) was found to be 10.52. From this, find $K_{sp}$ for magnesium hydroxide.

**18.28** A solution saturated in calcium hydroxide (limewater) has a pH of 12.35. What is $K_{sp}$ for calcium hydroxide?

**18.29** Strontianite (strontium carbonate) is an important mineral of strontium. Calculate the solubility of strontium carbonate, $SrCO_3$, from the solubility product constant (see Table 18.1).

**18.30** Magnesite (magnesium carbonate, $MgCO_3$) is a common magnesium mineral. From the solubility product constant (Table 18.1), find the solubility of magnesium carbonate in grams per liter of water.

**18.31** What is the solubility of $PbF_2$ in water? The $K_{sp}$ for $PbF_2$ is $2.7 \times 10^{-8}$.

**18.32** What is the solubility of $MgF_2$ in water? The $K_{sp}$ for $MgF_2$ is $7.4 \times 10^{-11}$.

## Common-Ion Effect

**18.33** What is the solubility (in grams per liter) of strontium sulfate, $SrSO_4$, in 0.20 $M$ sodium sulfate, $Na_2SO_4$? See Table 18.1.

**18.34** What is the solubility (in grams per liter) of lead(II) chromate, $PbCrO_4$, in 0.15 $M$ potassium chromate, $K_2CrO_4$? See Table 18.1.

**18.35** The solubility of magnesium fluoride, $MgF_2$, in water is 0.016 g/L. What is the solubility (in grams per liter) of magnesium fluoride in 0.015 $M$ sodium fluoride, NaF?

**18.36** The solubility of silver sulfate, $Ag_2SO_4$, in water has been determined to be 8.0 g/L. What is the solubility in 0.65 $M$ sodium sulfate, $Na_2SO_4$?

**18.37** What is the solubility (in grams per liter) of magnesium oxalate, $MgC_2O_4$, in 0.020 $M$ sodium oxalate, $Na_2C_2O_4$? Solve the equation exactly. See Table 18.1 for $K_{sp}$.

**18.38** Calculate the molar solubility of strontium sulfate, $SrSO_4$, in 0.0015 $M$ sodium sulfate, $Na_2SO_4$. Solve the equation exactly. See Table 18.1 for $K_{sp}$.

## Precipitation

**18.39** From each of the following ion concentrations in a solution, predict whether a precipitate will form in the solution.
a. $[Ba^{2+}] = 0.020\ M$, $[F^-] = 0.015\ M$
b. $[Mg^{2+}] = 0.0012\ M$, $[CO_3^{2-}] = 0.041\ M$

**18.40** From each of the following ion concentrations in a solution, predict whether a precipitate will form in the solution.
a. $[Sr^{2+}] = 0.012\ M$, $[CrO_4^{2-}] = 0.0015\ M$
b. $[Pb^{2+}] = 0.0035\ M$, $[Cl^-] = 0.15\ M$

**18.41** Lead(II) chromate, $PbCrO_4$, is used as a yellow paint pigment ("chrome yellow"). When a solution is prepared that is $5.0 \times 10^{-4}\ M$ in lead ion, $Pb^{2+}$, and $5.0 \times 10^{-5}\ M$ in chromate ion, $CrO_4^{2-}$, would you expect some of the lead(II) chromate to precipitate? See Table 18.1.

**18.42** Lead sulfate, $PbSO_4$, is used as a white paint pigment. When a solution is prepared that is $5.0 \times 10^{-4}\ M$ in lead ion, $Pb^{2+}$, and $1.0 \times 10^{-5}\ M$ in sulfate ion, $SO_4^{2-}$, would you expect some of the lead sulfate to precipitate? See Table 18.1.

**18.43** The following solutions are mixed: 1.0 L of 0.00010 $M$ NaOH and 1.0 L of 0.0020 $M$ $MgSO_4$. Is a precipitate expected? Explain.

**18.44** A 45-mL sample of 0.015 $M$ calcium chloride, $CaCl_2$, is added to 55 mL of 0.010 $M$ sodium sulfate, $Na_2SO_4$. Is a precipitate expected? Explain.

**18.45** A 45.0-mL sample of 0.0015 $M$ $BaCl_2$ was added to a beaker containing 75.0 mL of 0.0025 $M$ KF. Will a precipitate form?

**18.46** A 65.0-mL sample of 0.010 $M$ $Pb(NO_3)_2$ was added to a beaker containing 40.0 mL of 0.035 $M$ KCl. Will a precipitate form?

**18.47** How many moles of calcium chloride, $CaCl_2$, can be added to 1.5 L of 0.020 $M$ potassium sulfate, $K_2SO_4$, before a precipitate is expected? Assume that the volume of the solution is not changed significantly by the addition of calcium chloride.

**18.48** Magnesium sulfate, $MgSO_4$, is added to 456 mL of 0.040 $M$ sodium hydroxide, NaOH, until a precipitate just forms. How many grams of magnesium sulfate were added? Assume that the volume of the solution is not changed significantly by the addition of magnesium sulfate.

**18.49** What is the $I^-$ concentration just as AgCl begins to precipitate when 1.0 $M$ $AgNO_3$ is slowly added to a solution containing 0.015 $M$ $Cl^-$ and 0.015 $M$ $I^-$?

**18.50** What is the $Cl^-$ concentration just as $Ag_2CrO_4$ begins to precipitate when 1.0 $M$ $AgNO_3$ is slowly added to a solution containing 0.015 $M$ $Cl^-$ and 0.015 $M$ $CrO_4^{2-}$?

## Effect of pH on Solubility

**18.51** Write the net ionic equation in which the slightly soluble salt barium fluoride, $BaF_2$, dissolves in dilute hydrochloric acid.

**18.52** Write the net ionic equation in which the slightly soluble salt lead(II) carbonate, $PbCO_3$, dissolves in dilute hydrochloric acid.

**18.53** Which salt would you expect to dissolve readily in acidic solution, barium sulfate or barium fluoride? Explain.

**18.54** Which salt would you expect to dissolve readily in acidic solution, calcium phosphate, $Ca_3(PO_4)_2$, or calcium sulfate, $CaSO_4$? Explain.

## Complex Ions

**18.55** Write the chemical equation for the formation of the $Cu(CN)_2^-$ ion. Write the $K_f$ expression.

**18.56** Write the chemical equation for the formation of the $Ni(NH_3)_6^{2+}$ ion. Write the $K_f$ expression.

**18.57** Sufficient sodium cyanide, NaCN, was added to 0.015 $M$ silver nitrate, $AgNO_3$, to give a solution that was initially 0.100 $M$ in cyanide ion, $CN^-$. What is the concentration of silver ion, $Ag^+$, in this solution after $Ag(CN)_2^-$ forms? The formation constant $K_f$ for the complex ion $Ag(CN)_2^-$ is $5.6 \times 10^{18}$.

**18.58** The formation constant $K_f$ for the complex ion $Zn(OH)_4^{2-}$ is $2.8 \times 10^{15}$. What is the concentration of zinc ion, $Zn^{2+}$, in a solution that is initially 0.20 $M$ in $Zn(OH)_4^{2-}$?

**18.59** Predict whether cadmium oxalate, $CdC_2O_4$, will precipitate from a solution that is 0.0020 $M$ $Cd(NO_3)_2$, 0.010 $M$ $Na_2C_2O_4$, and 0.10 $M$ $NH_3$. Note that cadmium ion forms the $Cd(NH_3)_4^{2+}$ complex ion.

**18.60** Predict whether nickel(II) hydroxide, $Ni(OH)_2$, will precipitate from a solution that is 0.0020 $M$ $NiSO_4$, 0.010 $M$ $NaOH$, and 0.10 $M$ $NH_3$. Note that nickel(II) ion forms the $Ni(NH_3)_6^{2+}$ complex ion.

**18.61** What is the molar solubility of $CdC_2O_4$ in 0.10 $M$ $NH_3$?

**18.62** What is the molar solubility of NiS in 0.10 $M$ $NH_3$?

### Qualitative Analysis

**18.63** Describe how you could separate the following mixture of metal ions: $Cd^{2+}$, $Pb^{2+}$, and $Sr^{2+}$.

**18.64** Describe how you could separate the following mixture of metal ions: $Na^+$, $Hg^{2+}$, and $Ca^{2+}$.

**18.65** A student dissolved a compound in water, then added hydrochloric acid. No precipitate formed. Then she bubbled $H_2S$ into this solution, but again no precipitate formed. However, when she made the solution basic with ammonia and bubbled in $H_2S$, a precipitate formed. Which of the following are possible as the cation in the compound?
a. $Ag^+$     b. $Ca^{2+}$     c. $Mn^{2+}$     d. $Cd^{2+}$

**18.66** A student was asked to identify a compound. In an effort to do so, he first dissolved the compound in water. He found that no precipitate formed when hydrochloric acid was added, but when $H_2S$ was bubbled into this acidic solution, a precipitate formed. Which one of the following could be the precipitate?
a. $PbCrO_4$     b. $CdS$     c. $MnS$     d. $Ag_2S$

## General Problems

**18.67** Lead(II) sulfate is often used as a test for lead(II) ion in qualitative analysis. Using the solubility product constant (Table 18.1), calculate the molar solubility of lead(II) sulfate in water.

**18.68** Mercury(II) ion is often precipitated as mercury(II) sulfide in qualitative analysis. Using the solubility product constant (Table 18.1), calculate the molar solubility of mercury(II) sulfide in water, assuming that no other reactions occur.

**18.69** Mercury(I) chloride, $Hg_2Cl_2$, is an unusual salt in that it dissolves to form $Hg_2^{2+}$ and $2Cl^-$. Use the solubility product constant (Table 18.1) to calculate the following: (a) the molar solubility of $Hg_2Cl_2$; (b) the solubility of $Hg_2Cl_2$ in grams per liter.

**18.70** Magnesium ammonium phosphate is an unusual salt in that it dissolves to form $Mg^{2+}$, $NH_4^+$, and $PO_4^{3-}$ ions. $K_{sp}$ for magnesium ammonium phosphate equals $2.5 \times 10^{-13}$. Calculate: (a) its molar solubility; (b) its solubility in grams per liter.

**18.71** For cerium(III) hydroxide, $Ce(OH)_3$, $K_{sp}$ equals $2.0 \times 10^{-20}$. Calculate: (a) its molar solubility (recall that taking the square root twice gives the fourth root); (b) the pOH of the saturated solution.

**18.72** Copper(II) ferrocyanide, $Cu_2Fe(CN)_6$, dissolves to give $Cu^{2+}$ and $[Fe(CN)_6]^{4-}$ ions; $K_{sp}$ for $Cu_2Fe(CN)_6$ equals $1.3 \times 10^{-16}$. Calculate: (a) the molar solubility, and (b) the solubility in grams per liter of copper(II) ferrocyanide.

**18.73** What is the solubility of magnesium hydroxide in a solution buffered at pH 8.80?

**18.74** What is the solubility of silver oxide, $Ag_2O$, in a solution buffered at pH 10.50? The equilibrium is $Ag_2O(s) + H_2O(l) \rightleftharpoons 2Ag^+(aq) + 2OH^-(aq)$; $K_c = 2.0 \times 10^{-8}$.

**18.75** What is the molar solubility of $Mg(OH)_2$ in a solution containing $1.0 \times 10^{-1}$ $M$ NaOH? See Table 18.1 for $K_{sp}$.

**18.76** What is the molar solubility of $Al(OH)_3$ in a solution containing $1.0 \times 10^{-3}$ $M$ NaOH? See Table 18.1 for $K_{sp}$.

**18.77** What must be the concentration of sulfate ion in order to precipitate calcium sulfate, $CaSO_4$, from a solution that is 0.0030 $M$ $Ca^{2+}$?

**18.78** What must be the concentration of chromate ion in order to precipitate strontium chromate, $SrCrO_4$, from a solution that is 0.0025 $M$ $Sr^{2+}$? $K_{sp}$ for strontium chromate is $3.5 \times 10^{-5}$.

**18.79** A 3.20-L solution of $1.25 \times 10^{-3}$ $M$ $Pb(NO_3)_2$ is mixed with a 0.80-L solution of $5.0 \times 10^{-1}$ $M$ NaCl. Calculate $Q_c$ for the dissolution of $PbCl_2$. No precipitate has formed. Is the solution supersaturated, saturated, or unsaturated?

**18.80** A 0.150-L solution of $2.4 \times 10^{-5}$ $M$ $MgCl_2$ is mixed with 0.050 L of $4.0 \times 10^{-3}$ $M$ NaOH. Calculate $Q_c$ for the dissolution of $Mg(OH)_2$. No precipitate has formed. Is the solution supersaturated, saturated, or unsaturated?

**18.81** How many grams of sodium chloride can be added to 785 mL of 0.0015 $M$ silver nitrate before a precipitate forms?

**18.82** How many grams of sodium sulfate can be added to 435 mL of 0.0028 $M$ barium chloride before a precipitate forms?

**18.83** Solid KSCN was added to a 2.00 $M$ $Fe^{3+}$ solution so that it was also initially 2.00 $M$ $SCN^-$. These ions then reacted to give the complex ion $Fe(SCN)^{2+}$, whose formation constant was $9.0 \times 10^2$. What is the concentration of $Fe^{3+}(aq)$ at equilibrium? Be sure to check any simplifying assumption you make.

**18.84**    Solid KSCN was added to a 2.00 $M$ $Co^{2+}$ solution so that it was also initially 2.00 $M$ $SCN^-$. These ions then reacted to give the complex ion $Co(SCN)^+$, whose formation constant was $1.0 \times 10^2$. What is the concentration of $Co^{2+}(aq)$ at equilibrium? Be sure to check any simplifying assumption you make.

**18.85**    Calculate the molar solubility of silver bromide, AgBr, in 5.0 $M$ $NH_3$.

**18.86**    Calculate the molar solubility of silver iodide, AgI, in 2.0 $M$ $NH_3$.

**18.87**    The solubility of zinc oxalate, $ZnC_2O_4$, in 0.0150 $M$ ammonia is $3.6 \times 10^{-4}$ mol/L. What is the oxalate-ion concentration in the saturated solution? If the solubility product constant for zinc oxalate is $1.5 \times 10^{-9}$, what must be the zinc-ion concentration in the solution? Now calculate the formation constant for the complex ion $Zn(NH_3)_4{}^{2+}$.

**18.88**    The solubility of cadmium oxalate, $CdC_2O_4$, in 0.150 $M$ ammonia is $6.1 \times 10^{-3}$ mol/L. What is the oxalate-ion concentration in the saturated solution? If the solubility product constant for cadmium oxalate is $1.5 \times 10^{-8}$, what must be the cadmium-ion concentration in the solution? Now calculate the formation constant for the complex ion $Cd(NH_3)_4{}^{2+}$.

**18.89**    Ammonia, $NH_3$, is a base that ionizes to give $NH_4{}^+$ and $OH^-$ ($K_b = 1.8 \times 10^{-5}$). You add magnesium sulfate to an ammonia solution. Calculate the concentration of $Mg^{2+}$ ion when magnesium hydroxide, $Mg(OH)_2$, just begins to precipitate from 0.10 $M$ $NH_3$. $K_{sp}$ for $Mg(OH)_2$ is $1.8 \times 10^{-11}$.

**18.90**    Hydrazine, $N_2H_4$, is a base that ionizes to give $N_2H_5{}^+$ and $OH^-$ ($K_b = 1.7 \times 10^{-6}$). You add magnesium sulfate to a hydrazine solution. Calculate the concentration of $Mg^{2+}$ ion when magnesium hydroxide, $Mg(OH)_2$, just begins to precipitate from 0.20 $M$ $N_2H_4$. $K_{sp}$ for $Mg(OH)_2$ is $1.8 \times 10^{-11}$.

**18.91**    A saturated solution of copper(II) iodate in pure water has a copper-ion concentration of $2.7 \times 10^{-3}$ $M$.
a. What is the molar solubility of copper iodate in a 0.35 $M$ potassium iodate solution?
b. What is the molar solubility of copper iodate in a 0.35 $M$ copper nitrate solution?
c. Should there be a difference in the answers to parts a and b? Why?

**18.92**    A saturated solution of lead iodate in pure water has an iodate-ion concentration of $8.0 \times 10^{-5}$ $M$.
a. What is the molar solubility of lead iodate in a 0.15 $M$ lead nitrate solution at the same temperature?

b. Should the molar solubility of lead iodate in part a be the same as, greater than, or less than that of lead iodate in pure water? Why?

**18.93**    A solution contains 0.0150 $M$ lead(II) ion. A concentrated sodium iodide solution is added to precipitate lead iodide (assume no volume change).
a. At what concentration of $I^-$ does precipitate start to form?
b. When $[I^-] = 2.0 \times 10^{-3}$ $M$, what is the lead-ion concentration? What percentage of the lead(II) originally present remains in solution?

**18.94**    A solution contains 0.00750 $M$ calcium ion. A concentrated sodium fluoride solution is added to precipitate calcium fluoride (assume no volume change).
a. At what concentration of $F^-$ does precipitate start to form?
b. When $[F^-] = 9.5 \times 10^{-4}$ $M$, what is the calcium-ion concentration? What percentage of the calcium ion has precipitated?

**18.95**    a. If the molar solubility of cobalt(II) hydroxide is $5.4 \times 10^{-6}$ mol/L in pure water, what is its $K_{sp}$ value?
b. What is the molar solubility of $Co(OH)_2$ in a buffered solution that has a pH of 10.43?
c. Account for the differences in molar solubility in parts a and b.

**18.96**    a. If the molar solubility of beryllium(II) hydroxide is $8.6 \times 10^{-7}$ $M$ in pure water, what is its $K_{sp}$ value?
b. What is the molar solubility of beryllium(II) hydroxide in a solution that is 1.50 $M$ in $NH_3$ and 0.25 $M$ in $NH_4Cl$?
c. Account for the differences in molar solubility in parts a and b.

**18.97**    Although silver chloride is insoluble in water, it readily dissolves upon the addition of ammonia.

$$AgCl(s) + 2NH_3(aq) \rightleftharpoons Ag(NH_3)_2{}^+ (aq) + Cl^-(aq)$$

a. What is the equilibrium constant for this dissolving process?
b. Ammonia is added to a solution containing excess $AgCl(s)$. The final volume is 1.00 L and the resulting equilibrium concentration of $NH_3$ is 0.80 $M$. Calculate the number of moles of AgCl dissolved, the molar concentration of $Ag(NH_3)_2{}^+$, and the number of moles of $NH_3$ added to the original solution.

**18.98**    Crystals of AgBr can be removed from black-and-white photographic film by reacting the AgBr with sodium thiosulfate.

$$AgBr(s) + 2S_2O_3{}^{2-}(aq) \rightleftharpoons [Ag(S_2O_3)_2]^{3-}(aq) + Br^-(aq)$$

a. What is the equilibrium constant for this dissolving process?
b. In order to dissolve 2.5 g of AgBr in 1.0 L of solution how many moles of $Na_2S_2O_3$ must be added?

**18.99**    A 1.0-L solution that is 4.2 $M$ in ammonia is mixed with 26.7 g of ammonium chloride.
a. What is the hydroxide-ion concentration of this solution?

b. 0.075 mol of $MgCl_2$ is added to the above solution. Assume that there is no volume change. After $Mg(OH)_2$ has precipitated, what is the molar concentration of magnesium ion? What percent of the $Mg^{2+}$ is removed from solution?

**18.100**   A 1.0-L solution that is 1.6 $M$ in ammonia is mixed with 75.8 g of ammonium sulfate.

a. What is the hydroxide-ion concentration of this solution?
b. 0.058 mol of $MnCl_2$ is added to the above solution. Assume that there is no volume change. After $Mn(OH)_2$ has precipitated, what is the molar concentration of manganese ion? What percent of the $Mn^{2+}$ is removed from solution ($K_{sp}$ of $Mn(OH)_2$ is $4.6 \times 10^{-14}$)?

## Cumulative-Skills Problems

**18.101**   A solution is $1.5 \times 10^{-4} M$ $Zn^{2+}$ and 0.20 $M$ $HSO_4^-$. The solution also contains $Na_2SO_4$. What should be the minimum molarity of $Na_2SO_4$ to prevent the precipitation of zinc sulfide when the solution is saturated with hydrogen sulfide (0.10 $M$ $H_2S$)?

**18.102**   A solution is $1.8 \times 10^{-4} M$ $Co^{2+}$ and 0.20 $M$ $HSO_4^-$. The solution also contains $Na_2SO_4$. What should be the minimum molarity of $Na_2SO_4$ to prevent the precipitation of cobalt(II) sulfide when the solution is saturated with hydrogen sulfide (0.10 $M$ $H_2S$)?

**18.103**   What is the solubility of calcium fluoride in a buffer solution containing 0.45 $M$ $HCHO_2$ (formic acid) and 0.20 $M$ $NaCHO_2$? (*Hint:* Consider the equation $CaF_2(s) + 2H^+(aq) \rightleftharpoons Ca^{2+}(aq) + 2HF(aq)$, and solve the equilibrium problem.)

**18.104**   What is the solubility of magnesium fluoride in a buffer solution containing 0.45 $M$ $HC_2H_3O_2$ (acetic acid) and 0.20 $M$ $NaC_2H_3O_2$? The $K_{sp}$ for magnesium fluoride is $6.5 \times 10^{-9}$. (See the hint for Problem 18.103.)

**18.105**   A 67.0-mL sample of 0.350 $M$ $MgSO_4$ is added to 45.0 mL of 0.250 $M$ $Ba(OH)_2$. What is the net ionic equation for the reaction that occurs? What are the concentrations of ions in the mixture at equilibrium?

**18.106**   A 50.0-mL sample of 0.0150 M $Ag_2SO_4$ is added to 25.0 mL of 0.0100 M $PbCl_2$. What is the net ionic equation for the reaction that occurs? What are the concentrations of ions in the mixture at equilibrium?

 ## Media Activities

**18.107**   Step through the Predicting Precipitation *Problem Solving* activity on the student web site. Will a precipitate form from a solution that has $1.62 \times 10^{-5} M$ $Ca(NO_3)_2$ and $4.25 \times 10^{-1} M$ $H_2SO_4$? The $K_{sp}$ of $CaSO_4$ is $2.4 \times 10^{-5}$. If no precipitate forms, what concentration of $Ca(NO_3)_2$ would allow precipitation to occur?

**18.108**   Watch the Dynamic Equilibrium of Solubility Process *Visualization*.
a. Describe, on a molecular level, what you would observe if you added additional NaCl to the solution depicted in the *Visualization*.

b. If you added additional NaCl($s$) to the solution depicted in the *Visualization*, what would happen to the concentration of $Cl^-$?
c. Describe, on a molecular level, what you would observe if you added $Pb(NO_3)_2(aq)$ to the solution depicted in the *Visualization*.

**18.109**   For review, take the ACE practice test for Chapter 18 on the student web site.

Rust forms in a spontaneous reaction of iron with oxygen in air.

# Thermodynamics and Equilibrium

U rea, $NH_2CONH_2$, is an important industrial chemical. It is used to make synthetic resins for adhesives and melamine plastics (Figure 19.1). Its major use, however, is as a nitrogen fertilizer for plants. Urea is produced by reacting ammonia with carbon dioxide.

$$2NH_3(g) + CO_2(g) \longrightarrow NH_2CONH_2(aq) + H_2O(l)$$

Suppose you want to determine whether this or some other reaction could be useful for the industrial preparation of urea. Some of your immediate questions might be the following: Does the reaction naturally go in the direction it is written? Will the reaction mixture contain a sufficient amount of the product at equilibrium? We addressed these questions in the preceding chapters by looking at the equilibrium constant. Now we want to discuss these questions from the point of view of thermodynamics.

*Thermodynamics* is the study of the relationship between heat and other forms of energy involved in a chemical or physical process. With only heat measurements of substances, you can answer the questions just posed. You can predict the natural direction of a chemical reaction, and you can also determine the composition of a reaction mixture at equilibrium. Consider, then, the possibility of reacting $NH_3$ and $CO_2$ to give urea. Just how do you apply thermodynamics to such a reaction?

# 19.1    First Law of Thermodynamics; Enthalpy

We introduced the thermodynamic property of enthalpy, *H*, in Chapter 6. There we noted that the change in enthalpy equals the heat of reaction at constant pressure. Now we want to look at this property again, but define it more precisely, in terms of the energy of the system. We begin by discussing the first law of thermodynamics.

## First Law of Thermodynamics

Thermodynamics is normally described in terms of three laws. The first law of thermodynamics is essentially the law of conservation of energy applied to thermodynamic systems. ◄ To state the law, you need to understand what the internal energy of a system is and how you can change the internal energy.

► According to the law of conservation of energy, the total energy remains constant; energy is neither created nor destroyed.

The **internal energy, *U*,** is *the sum of the kinetic and potential energies of the particles making up the system.* The kinetic energy includes the energy of motion of electrons, nuclei, and molecules. The potential energy results from the chemical bonding of atoms and from the attractions between molecules.

Internal energy is a **state function.** That is, it is *a property of a system that depends only on its present state, which is completely determined by variables such as temperature and pressure.* Thus, 1 mol of water at 0°C and 1 atm pressure has a definite quantity of energy. When a system changes from one state to another, its internal energy changes from one definite value to another. You can calculate the change in internal energy, $\Delta U$, from the initial value of the internal energy, $U_i$, and the final value of the internal energy, $U_f$.

$$\Delta U = U_f - U_i$$

Normally you are interested in changes in internal energy rather than in absolute values. You measure these changes for a thermodynamic system by noting the exchanges of energy between the system and its surroundings. These exchanges of energy are of two kinds: heat and work. Heat, you will recall, is energy that moves into or out of the system because of a temperature difference between the system and its surroundings. **Work,** on the other hand, is *the energy exchange that results when a force F moves an object through a distance d; work (w) equals $F \times d$.*

The distinction between heat and work is illustrated in the system shown in Figure 19.2. This system consists of a gas in a vessel equipped with a movable piston. On top of the piston is a weight, which you can consider as part of the surroundings. You momentarily fix the position of the piston so it does not move. Suppose the temperature of the surroundings is raised, and as a result, heat passes from the surroundings to the vessel. If you find that the energy of the surroundings decreases by 165 J in this way, you know from the law of conservation of energy that the internal energy of the system must have increased by just this quantity. You write $q = +165$ J, using the sign convention that heat (denoted by the symbol *q*) absorbed by the system is positive (energy is added to the system) and that heat evolved by the system is negative (energy is subtracted from the system).

**Figure 19.1**
**Urea.**    *Top:* Urea is used as a plant fertilizer because it slowly decomposes in the soil to provide ammonia. *Bottom:* A molecular model of urea.

As the temperature of the gas in the vessel increases, the gas pressure increases at fixed volume. You now allow the piston to move, so the gas expands and lifts the piston and the weight on top of it. In lifting the weight, the system does work. The energy gained by the weight equals the force of gravity on the weight times the height to which the weight was raised (the distance it was moved). Suppose this energy is 92 J. Because the surroundings, which include the weight, have gained 92 J of energy, the system must have lost 92 J of energy (energy cannot be created or destroyed). You write $w = -92$ J, adhering to the sign convention that work done *on* the system is positive (energy is added to the system) and work done *by* the system is negative (energy is subtracted from the system). ◄

▶ The convention given here is that adopted by IUPAC (International Union of Pure and Applied Chemistry).

The system in Figure 19.2 gains internal energy from the heat absorbed and loses internal energy via the work done. In general, the net change of internal energy equals heat plus work.

$$\Delta U = q + w$$

The **first law of thermodynamics** states that *the change in internal energy of a system, $\Delta U$, equals $q + w$.* For the system shown in Figure 19.2,

$$\Delta U = (+165 \text{ J}) + (-92 \text{ J}) = +73 \text{ J}$$

### EXERCISE 19.1

A gas is enclosed in a system similar to that shown in Figure 19.2. More weights are added to the piston, giving a total mass of 2.20 kg. As a result, the gas is compressed and the weights are lowered 0.250 m. At the same time, 1.50 J of heat evolves from (leaves) the system. What is the change in internal energy of the system, $\Delta U$? The force of gravity on a mass $m$ is $mg$, where $g$ is the constant acceleration of gravity ($g = 9.80 \text{ m/s}^2$).

**See Problems 19.23 and 19.24.**

Work

Weight

Piston

Heat

Gas

**Figure 19.2**
**Exchanges of heat and work with the surroundings.** A gas is enclosed in a vessel with a piston. Heat flows into the vessel from the surroundings, which are at a higher temperature. As the temperature of the gas increases, the gas expands, lifting the weight (doing work).

### Heat of Reaction and Internal Energy

Now consider a chemical system—for example, the reaction of zinc metal with hydrochloric acid. The reaction is

$$Zn(s) + 2H_3O^+(aq) \longrightarrow Zn^{2+}(aq) + 2H_2O(l) + H_2(g)$$

When this reaction is carried out in a beaker open to the atmosphere, the reaction is exothermic, evolving 152.4 kJ of heat per mole of zinc consumed. You write $q_p = -152.4$ kJ, where the subscript $p$ indicates that the process occurs at constant pressure (the pressure of the atmosphere).

The hydrogen gas that is produced increases the volume of the system. As hydrogen is evolved, work must be done by the system to push back the atmosphere. How can you calculate this work? Imagine for the moment that the atmosphere is replaced by a piston and weights, whose downward force from gravity $F$ creates a pressure on the gas equivalent to that of the atmosphere. The pressure $P$ equals $F$ divided by the cross-sectional area of the piston, $A$. (See Figure 19.3.)

Suppose the increase in volume of the system due to the production of hydrogen is $\Delta V$. Because the volume of a cylinder equals its height $h$ times its cross-sectional area $A$, the change in volume is $\Delta V = A \times h$, and $h = \Delta V/A$. The work

**Figure 19.3**
**Reaction of zinc metal with hydrochloric acid at constant pressure.**  *(A)* The beginning of the reaction. The constant pressure of the atmosphere has been replaced by a piston and weight to give an equivalent pressure. *(B)* The reaction produces hydrogen gas. This increases the volume of the system, so that the piston and weight are lifted upward. Work is done by the system on the piston and weight.

done by the system in expanding equals the force of gravity times the distance the piston moves.

$$w = -F \times h = -F \times \frac{\Delta V}{A} = -\frac{F}{A} \times \Delta V$$

The negative sign is given because $w$ is work done by the system and represents energy lost by it. Note that $F/A$ is the pressure, $P$, which equals that of the atmosphere. Therefore,

$$w = -P\Delta V$$

This formula tells you that you can calculate the work done by a chemical reaction carried out in an open vessel by multiplying the atmospheric pressure $P$ by the change in volume of the chemical system, $\Delta V$. For example, when 1.00 mol Zn reacts with excess hydrochloric acid, 1.00 mol $H_2$ is produced. At 25°C and 1.00 atm (= $1.01 \times 10^5$ Pa), this amount of $H_2$ occupies 24.5 L (= $24.5 \times 10^{-3}$ m$^3$). The work done by the chemical system in pushing back the atmosphere is

$$w = -P\Delta V = -(1.01 \times 10^5 \text{ Pa}) \times (24.5 \times 10^{-3} \text{ m}^3)$$
$$= -2.47 \times 10^3 \text{ J, or } -2.47 \text{ kJ}$$

If you apply the first law to this chemical system, you can relate the change in internal energy of the system to the heat of reaction. You have

$$\Delta U = q_p + w = q_p - P\Delta V$$

For the reaction of Zn with HCl, $q_p = -152.4$ kJ and $w = -P\Delta V = -2.47$ kJ, so

$$\Delta U = -152.4 \text{ kJ} - 2.47 \text{ kJ} = -154.9 \text{ kJ}$$

We can now summarize what happens when 1.00 mol Zn reacts with excess hydrochloric acid. When the reaction occurs, the internal energy changes as the kinetic and potential energies change in going from reactants to products. This energy change, $\Delta U$, equals $-154.9$ kJ. Energy leaves the system mostly as heat ($q_p = -152.4$ kJ) but partly as expansion work ($w = -2.47$ kJ).

### EXERCISE 19.2

Consider the combustion (burning) of methane, $CH_4$, in oxygen.

$$CH_4(g) + 2O_2(g) \longrightarrow CO_2(g) + 2H_2O(l)$$

The heat of reaction at 25°C and 1.00 atm is $-890.2$ kJ. What is the change in volume when 1.00 mol $CH_4$ reacts with 2.00 mol $O_2$? (You can ignore the volume of liquid water, which is insignificant compared with volumes of gases.) What is $w$ for this change? Calculate $\Delta U$ for the change indicated by the chemical equation.

See Problems 19.25 and 19.26.

### Enthalpy and Enthalpy Change

In Chapter 6, we defined enthalpy tentatively in terms of the relationship of $\Delta H$ to the heat of reaction at constant pressure, $q_p$. We now define **enthalpy, $H$,** precisely as *the quantity $U + PV$.* Because $U$, $P$, and $V$ are state functions, $H$ is also a state function. This means that for a given temperature and pressure, a given amount of a substance has a definite enthalpy. Therefore, if you know the enthalpies of substances, you can calculate the change of enthalpy, $\Delta H$, for a reaction.

Let us now show the relationship of $\Delta H$ to the heat of reaction $q_p$. Note that $\Delta H$ is the final enthalpy $H_f$ minus the initial enthalpy $H_i$.

$$\Delta H = H_f - H_i$$

Now, substitute defining expressions for $H_f$ and $H_i$, using the subscripts $f$ and $i$ to indicate *final* and *initial*, respectively. Note, however, that the pressure is constant.

$$\Delta H = (U_f + PV_f) - (U_i + PV_i) = (U_f - U_i) + P(V_f - V_i) = \Delta U + P\Delta V$$

Earlier we found that $\Delta U = q_p - P\Delta V$. Hence,

$$\Delta H = (q_p - P\Delta V) + P\Delta V = q_p$$

This shows that $\Delta H = q_p$, as noted in Chapter 6.

In practice, what you do is measure certain heats of reaction and use them to tabulate enthalpies of formation, $\Delta H_f^\circ$. The standard enthalpy change for a reaction is

$$\Delta H^\circ = \Sigma\, n\Delta H_f^\circ(\text{products}) - \Sigma\, m\Delta H_f^\circ(\text{reactants})$$

Standard enthalpies of formation for selected substances and ions are listed in Table 6.2 and in Appendix C.

Suppose you would like to know how much heat is absorbed or evolved in the reaction between $NH_3$ and $CO_2$ to produce urea and water, which is the reaction given in the chapter opening. The following are standard enthalpies of formation at 25°C for the substances indicated (in kJ/mol): $NH_3(g)$, $-45.9$; $CO_2(g)$, $-393.5$; $NH_2CONH_2(aq)$, $-319.2$; $H_2O(l)$, $-285.8$. Substituting into the previous equation yields

$$\Delta H^\circ = [(-319.2 - 285.8) - (-2 \times 45.9 - 393.5)]\ \text{kJ} = -119.7\ \text{kJ}$$

From the minus sign in the value of $\Delta H^\circ$, you conclude that heat is evolved. The reaction is exothermic.

Spontaneous process

Nonspontaneous process

**Figure 19.4**
**Examples of a spontaneous and a nonspontaneous process.**
*Top:* The rolling of a rock downhill is a spontaneous process. The rock eventually comes to equilibrium at the bottom of the hill. *Bottom:* The rolling of a rock uphill is a nonspontaneous process.

# Spontaneous Processes and Entropy

Why does a chemical reaction go naturally in a particular direction? To answer this question, you need to look at spontaneous processes. A **spontaneous process** is *a physical or chemical change that occurs by itself.* It requires no continuing outside agency to make it happen. A rock at the top of a hill rolls down (Figure 19.4, top).

Heat flows from a hot object to a cold one. An iron object rusts in moist air (Figure 19.5). These processes occur spontaneously, or naturally, without requiring an outside force or agency. They continue until equilibrium is reached. If these processes were to go in the opposite direction, they would be *nonspontaneous*. The rolling of a rock uphill by itself is not a natural process; it is nonspontaneous (see Figure 19.4, bottom). The rock could be moved to the top of the hill, but work would have to be expended. Heat can be made to flow from a cold to a hot object, but a heat pump or refrigerator is needed. Rust can be converted to iron, but the process requires chemical reactions used in the manufacture of iron from its ore (iron oxide).

## 19.2   Entropy and the Second Law of Thermodynamics

When you ask whether a chemical reaction goes in the direction in which it is written, as we did in the chapter opening, you are asking whether the reaction is spontaneous in this direction. The first law of thermodynamics cannot help you answer such a question. It does help you keep track of the various forms of energy in a chemical change, using the constancy of total energy (conservation of energy). But although at one time it was thought that spontaneous reactions must be exothermic ($\Delta H < 0$), many spontaneous reactions are now known to be endothermic ($\Delta H > 0$); recall the reaction shown in Figure 6.1.

The second law of thermodynamics, which we will discuss in this section, provides a way to answer questions about the spontaneity of a reaction. The second law is expressed in terms of a quantity called *entropy*.

### Entropy

**Entropy, S,** is *a thermodynamic quantity that is a measure of the randomness or disorder in a system.* As we will discuss later, the SI unit of entropy is joules per kelvin (J/K). Entropy, like enthalpy, is a state function. That is, the quantity of entropy in a given amount of substance depends only on variables, such as temperature and pressure, that determine the state of the substance. If these variables, are fixed, the quantity of entropy is fixed. For example, 1 mol of ice at 0°C and 1 atm pressure has an entropy that has been determined experimentally to be 41 J/K. One mole of liquid water at 0°C and 1 atm has an entropy of 63 J/K. (In the next section, we will describe how entropy values are measured.)

You would expect the entropy (disorder) to increase when ice melts to the liquid. Ice has an ordered crystalline structure (see Figure 11.51). When ice melts to liquid water, the crystalline structure breaks down, and a less-ordered liquid structure results. In ice, the $H_2O$ molecules occupy regular, fixed positions in a crystal lattice; that is, ice has a relatively low entropy. But in the liquid, the molecules move about freely, giving a disordered structure—one having greater entropy. You can see this in the entropy values for these states.

You calculate the entropy change, $\Delta S$, for a process similarly to the way you calculate $\Delta H$. If $S_i$ is the initial entropy and $S_f$ is the final entropy, the change in entropy is

$$\Delta S = S_f - S_i$$

For the melting of ice to liquid water,

$$H_2O(s) \longrightarrow H_2O(l)$$
$$\Delta S = (63 - 41) \text{ J/K} = 22 \text{ J/K}$$

**Figure 19.5**
**The rusting of iron in moist air is a spontaneous reaction.** This sculpture by Picasso in Chicago was allowed to rust to give a pleasing effect.

When 1 mol of ice melts at 0°C, the water increases in entropy by 22 J/K. The entropy increases, as you expect, because a substance becomes more disordered when it melts.

 **CONCEPT CHECK 19.1**

You have a sample of 1.0 mg of solid iodine at room temperature. Later, you notice that the iodine has sublimed (passed into the vapor state). What can you say about the change of entropy of the iodine?

## Second Law of Thermodynamics

A process occurs naturally as a result of an *overall* increase in disorder of a system plus its surroundings. There is a natural tendency for things to mix and to break down, events that represent increasing disorder. Where you see order or structure being built, it results from the use of greater order elsewhere. Order in one place is used to build order in another. Thus, a house is built (and order produced) at the expense of order in the food molecules of workers and of order in substances needed to construct building materials. Moreover, during the overall process of building the house, some order is degraded to disorder. The *net* result, as in any natural process, is to increase disorder.

We can put this more precisely in terms of the **second law of thermodynamics,** which states that *the total entropy of a system and its surroundings always increases for a spontaneous process.* Note that entropy is quite different from energy. Energy can be neither created nor destroyed during chemical change (law of conservation of energy). But entropy is created during a spontaneous, or natural, process.

For a spontaneous process carried out at a given temperature, the second law can be restated in a form that refers only to the system. To arrive at that form, consider the changes in entropy that occur *in the system.* As the process takes place, entropy is created. At the same time, heat flows into or out of the system, and entropy accompanies that heat flow. Thus, when heat flows into the system, entropy flows into the system. In general, the entropy change associated with heat $q$ at absolute temperature $T$ can be shown to equal $q/T$. The net change in entropy of the system, $\Delta S$, equals the sum of the entropy created during the spontaneous process and the change in entropy associated with the heat flow (entropy flow). ◄

▶ Consider a simple analogy. A baker makes (creates) cookies. She also buys cookies from a wholesaler and sells cookies at her store. The change in number of cookies in her store in any time interval equals the number created plus the flow of cookies in and out of her store (number bought minus number sold). The analogy breaks down if someone eats cookies in the store. Whereas entropy can only be created, cookies can be created and destroyed (eaten).

$$\Delta S = \text{entropy created} + \frac{q}{T} \qquad \text{(spontaneous process)}$$

Normally the quantity of entropy created during a spontaneous process cannot be directly measured. Because it is a positive quantity, however, if you delete it from the right side of the equation, you can conclude that $\Delta S$ must be greater than $q/T$ for a spontaneous process.

$$\Delta S > \frac{q}{T} \qquad \text{(spontaneous process)}$$

The restatement of the second law is as follows: *for a spontaneous process at a given temperature, the change in entropy of the system is greater than the heat divided by the absolute temperature.*

### Entropy Change for a Phase Transition

Certain processes occur at equilibrium or, more precisely, very close to equilibrium. For example, ice at 0°C is in equilibrium with liquid water at 0°C. If heat is slowly absorbed by the system, it remains very near equilibrium, but the ice melts. Under these essentially equilibrium conditions, no significant amount of entropy is created. The entropy change results entirely from the absorption of heat. Therefore,

$$\Delta S = \frac{q}{T} \quad \text{(equilibrium process)}$$

▶ When a system is at equilibrium, a small change in a condition can make the process go in one direction or the other. The process is said to be *reversible*. Think of a two-pan balance in which the weight on each pan has been adjusted to make the balance beam level—that is, so the beam is at equilibrium. A small weight on either pan will tip the scale one way or the other.

Other phase changes, such as the vaporization of a liquid, also occur under equilibrium conditions. ◀

You can use the previous equation to obtain the entropy change for a phase change. Consider the melting of ice. The heat absorbed is the heat of fusion, $\Delta H_{fus}$, which is known from experiment to be 6.0 kJ for 1 mol of ice. You get the entropy change for melting by dividing $\Delta H_{fus}$ by the absolute temperature of the phase transition, 273 K (0°C). Because entropy changes are usually expressed in joules per kelvin, you convert $\Delta H_{fus}$ to $6.0 \times 10^3$ J.

$$\Delta S = \frac{\Delta H_{fus}}{T} = \frac{6.0 \times 10^3 \text{ J}}{273 \text{ K}} = 22 \text{ J/K}$$

Note that this is the $\Delta S$ value we obtained earlier for the conversion of ice to water.

---

### EXAMPLE 19.1

## Calculating the Entropy Change for a Phase Transition

The heat of vaporization, $\Delta H_{vap}$, of carbon tetrachloride, $CCl_4$, at 25°C is 43.0 kJ/mol.

$$CCl_4(l) \longrightarrow CCl_4(g); \Delta H_{vap} = 43.0 \text{ kJ/mol}$$

If 1 mol of liquid carbon tetrachloride at 25°C has an entropy of 214 J/K, what is the entropy of 1 mol of the vapor in equilibrium with the liquid at this temperature?

 **PROBLEM STRATEGY**

The entropy change for this equilibrium vaporization is $\Delta H_{vap}/T$. The entropy of the vapor equals the entropy of the liquid plus the entropy change.

**SOLUTION**

When liquid $CCl_4$ evaporates, it absorbs heat: $\Delta H_{vap} = 43.0$ kJ/mol ($43.0 \times 10^3$ J/mol) at 25°C, or 298 K. The entropy change, $\Delta S$, is

$$\Delta S = \frac{\Delta H_{vap}}{T} = \frac{43.0 \times 10^3 \text{ J/mol}}{298 \text{ K}} = 144 \text{ J/(mol} \cdot \text{K)}$$

In other words, 1 mol of carbon tetrachloride increases in entropy by 144 J/K when it vaporizes. The entropy of 1 mol of the vapor equals the entropy of 1 mol of liquid (214 J/K) plus 144 J/K.

$$\text{Entropy of vapor} = (214 + 144) \text{ J/(mol} \cdot \text{K)} = \textbf{358 J/(mol} \cdot \textbf{K)}$$

**EXERCISE 19.3**

Liquid ethanol, $C_2H_5OH(l)$, at 25°C has an entropy of 161 J/(mol · K). If the heat of vaporization, $\Delta H_{vap}$, at 25°C is 42.3 kJ/mol, what is the entropy of the vapor in equilibrium with the liquid at 25°C?

*See Problems 19.27 and 19.28.*

Now you can see how thermodynamics is applied to the question of whether a reaction is spontaneous. Suppose that you propose a chemical reaction for the preparation of a substance. You also assume that the reaction occurs at constant temperature and pressure. For urea, you propose the reaction

$$2NH_3(g) + CO_2(g) \longrightarrow NH_2CONH_2(aq) + H_2O(l)$$

Is this a spontaneous reaction? That is, does it go left to right as written? You can use the second law in the form $\Delta S > q/T$ to answer this question if you know both $\Delta H$ and $\Delta S$ for the reaction.

Recall that the heat of reaction at constant pressure, $q_p$, equals the enthalpy change $\Delta H$. The second law for a spontaneous reaction at constant temperature and pressure becomes

$$\Delta S > \frac{q_p}{T} = \frac{\Delta H}{T} \qquad \text{(spontaneous reaction, constant } T \text{ and } P)$$

$\Delta S$ is greater than $\Delta H/T$ for a spontaneous reaction at constant temperature and pressure. Therefore, if you subtract $\Delta S$ from $\Delta H/T$, you get a negative quantity. That is,

$$\frac{\Delta H}{T} - \Delta S < 0 \qquad \text{(spontaneous reaction, constant } T \text{ and } P)$$

Multiplying each term of this inequality by the positive quantity $T$, you get

$$\Delta H - T\Delta S < 0 \qquad \text{(spontaneous reaction, constant } T \text{ and } P)$$

▶ Demonstration of a spontaneous reaction (crystallization from supersaturated solution): Bassam Z. Shakhashiri, *Chemical Demonstrations*, Vol. 1 (Madison: University of Wisconsin Press, 1983), no. 1.11, pp. 27–30. Note that the reaction involves a decrease in entropy of the system ($\Delta S < 0$, because of the formation of crystals) but is sufficiently exothermic that $\Delta H - T\Delta S < 0$.

This inequality is important. If you have a table of entropies of substances, you can calculate $\Delta S$ for the proposed reaction. From Table 6.2, you can also calculate $\Delta H$. If $\Delta H - T\Delta S$ is negative for the reaction, you would predict that it is spontaneous left to right, as written. However, if $\Delta H - T\Delta S$ is positive, you would predict that the reaction is nonspontaneous in the direction written but spontaneous in the opposite direction. If $\Delta H - T\Delta S$ is zero, the reaction is at equilibrium.

## 19.3 Standard Entropies and the Third Law of Thermodynamics

▶ Heat capacity was discussed in Section 6.6.

To determine experimentally the entropy of a substance, you first measure the heat absorbed by the substance by warming it at various temperatures. That is, you find the heat capacity at different temperatures. ◀ You then calculate the entropy as we will describe. This determination of the entropy is based on the third law of thermodynamics.

# Entropy and Disorder

**EVERYDAY LIFE**

In the text, we noted that entropy is related to the disorder of a system. Let us look more closely at this notion. The order or disorder of a system is itself related to the relative number of arrangements of the system. As an analogy, consider a pack of playing cards. A new deck is arranged in a particular way, with cards arranged by suit, and the cards within a suit arranged from the ace to the king. We will call this the "ordered" state of the cards. Of course, there are many other possible arrangements of the cards. When you shuffle this ordered deck, the chances are you will end up with one of these other arrangements. When you shuffle cards, you obtain a "disordered" state. The chance of shuffling the cards and obtaining the ordered state is very slim, because there are so many possible arrangements of the disordered state and only one of the ordered state.

The idea that entropy is a measure of randomness or disorder was first advanced by the Austrian physicist Ludwig Boltzmann (1844–1906). Boltzmann related entropy, $S$, to a quantity, $W$, which is the number of different possible arrangements of molecules of the system, each arrangement having the same energy:

$$S = k \ln W$$

Here $k$ is Boltzmann's constant and equals the gas constant divided by Avogadro's number ($k = 1.38 \times 10^{-23}$ J/K). A system with large $W$ is one that can occur in many different molecular arrangements and is one that is "disordered." A disordered state has large $W$ and large entropy.

To understand this equation, consider a simplified model of a gas at first confined to a single glass bulb and later allowed to expand into a similar bulb containing a vacuum, as in Figure 19.6A. For the model, imagine two containers, each having three numbered boxes (Figure 19.6B). All possible arrangements of two balls, representing gas molecules, in the boxes of the two containers are shown in Figure 19.7. The first three arrangements represent the gas filling only the left container ($W = 3$). The next nine represent the gas distributed evenly between the two containers ($W = 9$), and the last three represent the gas filling only the right container ($W = 3$). The state in which the system is confined to one container, say, the left one, is the more ordered state and has lower entropy ($S = k \ln 3$) than the state in which gas is evenly distributed between containers ($S = k \ln 9$). In the equal distribution between the containers, a gas becomes more disordered; that is, the gas changes so that the entropy increases (at a fixed energy).

*(continued)*

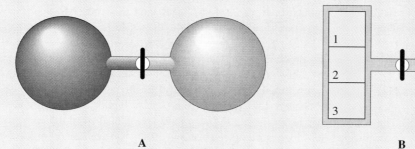

**Figure 19.6**
**A gas experiment.** *(A)* A gas is confined to the glass bulb on the left; the other bulb contains a vacuum. Later, the stopcock will be opened and the gas will expand into the vacuum. *(B)* A model of the gas experiment. It consists of two containers, each holding three numbered boxes.

**Figure 19.7**

**Arrangements of balls in two containers.** The first three arrangements represent a gas confined to the left container; the last three represent a gas confined to the right container. The middle nine arrangements represent a gas evenly distributed between the containers.

$W = 3$

$W = 9$

$W = 3$

### Third Law of Thermodynamics

The **third law of thermodynamics** states that *a substance that is perfectly crystalline at 0 K has an entropy of zero.* This seems reasonable. A perfectly crystalline substance at 0 K should have perfect order. When the temperature is raised, however, the substance becomes more disordered as it absorbs heat.

> ► The process just described is essentially the numerical evaluation of an integral, which you can obtain as follows: The heat absorbed for temperature change $dT$ is $C_p(T)dT$, and the entropy change is $C_p(T)dT/T$. The standard entropy at temperature $T$ is
>
> $$\int_0^T \frac{C_p(T)dT}{T}$$

You can determine the entropy of a substance at a temperature other than 0 K—say, at 298 K (25°C)—by slowly heating the substance from near 0 K to 298 K. Recall that the entropy change $\Delta S$ that occurs when heat is absorbed at a temperature $T$ is $q/T$. Suppose you heat the substance from near 0.0 K to 2.0 K, and the heat absorbed is 0.19 J. You find the entropy change by dividing the heat absorbed by the average absolute temperature $[\frac{1}{2}(0.0 + 2.0)K = 1.0$ K$]$. Therefore, $\Delta S$ equals 0.19 J/1.0 K = 0.19 J/K. This gives you the entropy of the substance at 2.0 K. Now you heat the substance from 2.0 K to 4.0 K, and this time 0.88 J of heat is absorbed. The average temperature is $\frac{1}{2}(2.0 + 4.0)K = 3.0$ K, and the entropy change is 0.88 J/3.0 K = 0.29 J/K. The entropy of the substance at 4.0 K is $(0.19 + 0.29)$ J/K = 0.48 J/K. Proceeding this way, you can eventually get the entropy at 298 K. ◄

Figure 19.8 shows how the entropy of a substance changes with temperature. Note that the entropy increases gradually as the temperature increases. But when there is a phase change (for example, from solid to liquid), the entropy increases sharply. The entropy change for the phase transition is calculated from the enthalpy of the phase transition, as described earlier (see Example 19.1). The **standard entropy** of a substance

> ► Entropies of substances must be positive. Those for ions, however, can be negative because they are derived in a different way by arbitrarily setting $S°$ for $H_3O^+(aq)$ equal to zero.

or ion (Table 19.1), also called its **absolute entropy,** $S°$, is *the entropy value for the standard state of the species* (indicated by the superscript degree sign). For a substance, the standard state is the pure substance at 1 atm pressure. For a species in solution, the standard state is the 1 $M$ solution. Table 19.1 gives standard entropies of various substances at 25°C and 1 atm. Note that the elements have nonzero values, unlike standard enthalpies of formation, $\Delta H_f°$, which by convention are zero. The symbol $S°$, rather than $\Delta S°$, is chosen for standard entropies to emphasize that they originate from the third law. ◄

**Figure 19.8**
**Standard entropy of methyl chloride, CH₃Cl, at various temperatures (approximate schematic graph).**   The entropy rises gradually as the temperature increases, but jumps sharply at each phase transition.

## Entropy Change for a Reaction

Once you determine the standard entropies of all substances in a reaction, you can calculate the change of entropy, $\Delta S°$, for the reaction. A sample calculation is described in Example 19.3. Even without knowing values for the entropies of substances, you can sometimes predict the sign of $\Delta S°$ for a reaction. The entropy usually increases in the following situations:

1.  A reaction in which a molecule is broken into two or more smaller molecules.

2.  A reaction in which there is an increase in moles of gas. (This may result from a molecule breaking up, in which case Rules 1 and 2 are related.)

3.  A process in which a solid changes to liquid or gas or a liquid changes to a gas.

▶ An alternate example is provided in the Instructor's Resource Manual.

**EXAMPLE 19.2**
## Predicting the Sign of the Entropy Change of a Reaction

a.  The following equation represents the essential change that takes place during the fermentation of glucose (grape sugar) to ethanol (ethyl alcohol).

$$\underset{\text{glucose}}{C_6H_{12}O_6(s)} \longrightarrow \underset{\text{ethanol}}{2C_2H_5OH(l)} + 2CO_2(g)$$

Is $S°$ positive or negative? Explain.

b.  Do you expect the entropy to increase or decrease in the preparation of urea from $NH_3$ and $CO_2$?

$$2NH_3(g) + CO_2(g) \longrightarrow NH_2CONH_2(aq) + H_2O(l)$$

Explain.

c.  What is the sign of $\Delta S°$ for the following reaction?

$$CO(g) + H_2O(g) \longrightarrow CO_2(g) + H_2(g)$$

**Table 19.1**
**Standard Entropies (at 25°C)**

| Formula | $S°$, J/(mol · K)* | Formula | $S°$, J/(mol · K) | Formula | $S°$, J/(mol · K) |
|---|---|---|---|---|---|
| **Hydrogen** | | **Carbon (continued)** | | **Sulfur** | |
| $H_3O^+(aq)$ | 0 | $CS_2(l)$ | 151.0 | $S_2(g)$ | 228.1 |
| $H_2(g)$ | 130.6 | $HCN(g)$ | 201.7 | S(rhombic) | 31.9 |
| **Sodium** | | $HCN(l)$ | 112.8 | S(monoclinic) | 32.6 |
| $Na^+(aq)$ | 60.2 | $CCl_4(g)$ | 309.7 | $SO_2(g)$ | 248.1 |
| $Na(s)$ | 51.4 | $CCl_4(l)$ | 214.4 | $H_2S(g)$ | 205.6 |
| $NaCl(s)$ | 72.1 | $CH_3CHO(g)$ | 266 | **Fluorine** | |
| $NaHCO_3(s)$ | 102 | $C_2H_5OH(l)$ | 161 | $F^-(aq)$ | −9.6 |
| $Na_2CO_3(s)$ | 139 | **Silicon** | | $F_2(g)$ | 202.7 |
| **Calcium** | | $Si(s)$ | 18.0 | $HF(g)$ | 173.7 |
| $Ca^{2+}(aq)$ | −55.2 | $SiO_2(s)$ | 41.5 | **Chlorine** | |
| $Ca(s)$ | 41.6 | $SiF_4(g)$ | 285 | $Cl^-(aq)$ | 55.1 |
| $CaO(s)$ | 38.2 | **Lead** | | $Cl_2(g)$ | 223.0 |
| $CaCO_3(s)$ | 92.9 | $Pb(s)$ | 64.8 | $HCl(g)$ | 186.8 |
| **Carbon** | | $PbO(s)$ | 66.5 | **Bromine** | |
| C(graphite) | 5.7 | $PbS(s)$ | 91.3 | $Br^-(aq)$ | 80.7 |
| C(diamond) | 2.4 | **Nitrogen** | | $Br_2(l)$ | 152.2 |
| $CO(g)$ | 197.5 | $N_2(g)$ | 191.5 | **Iodine** | |
| $CO_2(g)$ | 213.7 | $NH_3(g)$ | 193 | $I^-(aq)$ | 109.4 |
| $HCO_3^-(aq)$ | 95.0 | $NO(g)$ | 210.6 | $I_2(s)$ | 116.1 |
| $CH_4(g)$ | 186.1 | $NO_2(g)$ | 239.9 | **Silver** | |
| $C_2H_4(g)$ | 219.2 | $HNO_3(aq)$ | 146 | $Ag^+(aq)$ | 73.9 |
| $C_2H_6(g)$ | 229.5 | **Oxygen** | | $Ag(s)$ | 42.7 |
| $C_6H_6(l)$ | 172.8 | $O_2(g)$ | 205.0 | $AgF(s)$ | 84 |
| $HCHO(g)$ | 219 | $O_3(g)$ | 238.8 | $AgCl(s)$ | 96.1 |
| $CH_3OH(l)$ | 127 | $OH^-(aq)$ | −10.5 | $AgBr(s)$ | 107.1 |
| $CS_2(g)$ | 237.8 | $H_2O(g)$ | 188.7 | $AgI(s)$ | 114 |
| | | $H_2O(l)$ | 69.9 | | |

*See Appendix C for additional values.

 **PROBLEM STRATEGY**

Compare the reactants and products in each reaction. Look for the following: molecules breaking into smaller parts or aggregating into larger parts, moles of gas increasing or decreasing, and changes of state. Then apply the three rules preceding this example.

**SOLUTION**

a. A molecule (glucose) breaks into smaller molecules ($C_2H_5OH$ and $CO_2$). More-over, this results in a gas being released. You predict that $\Delta S°$ for this reaction is **positive.** That is, the entropy increases.

b. In this reaction, the moles of gas decrease (by 3 mol), which would decrease the entropy. You predict that the entropy should **decrease.** That is, $\Delta S°$ is negative.

c. Because there is no change in the number of moles of gas, **you cannot predict the sign of $\Delta S°$ from the rules given.**

---

**EXERCISE 19.4**

Predict the sign of $\Delta S°$ for each of the following reactions.

a. $CaCO_3(s) \longrightarrow CaO(s) + CO_2(g)$

b. $CS_2(l) \longrightarrow CS_2(g)$

c. $2Hg(l) + O_2(g) \longrightarrow 2HgO(s)$

d. $2Na_2O_2(s) + 2H_2O(l) \longrightarrow 4NaOH(aq) + O_2(g)$

---

See Problems 19.31 and 19.32.

It is useful to be able to predict the sign of $\Delta S°$. You gain some understanding of the reaction, and you can use the prediction for qualitative work. For quantitative work, however, you need to find the value of $\Delta S°$. You can find the standard change of entropy, $\Delta S°$, for a reaction by subtracting the standard entropies of reactants from the standard entropies of products, similar to the way you obtained $\Delta H°$.

$$\Delta S° = \Sigma\, nS°(\text{products}) - \Sigma\, mS°(\text{reactants})$$

The next example illustrates the calculations.

---

**EXAMPLE 19.3**

## Calculating $\Delta S°$ for a Reaction

Calculate the change of entropy, $\Delta S°$, at 25°C for the reaction in which urea is formed from $NH_3$ and $CO_2$.

$$2NH_3(g) + CO_2(g) \longrightarrow NH_2CONH_2(aq) + H_2O(l)$$

The standard entropy of $NH_2CONH_2(aq)$ is 174 J/(mol · K). See Table 19.1 for other values.

 **PROBLEM STRATEGY**

The calculation is similar to that used to obtain $\Delta H°$ from $\Delta H_f°$ values.

**SOLUTION**

It is convenient to put the standard entropy values multiplied by stoichiometric coefficients below the formulas in the balanced equation.

$$2NH_3(g) + CO_2(g) \longrightarrow NH_2CONH_2(aq) + H_2O(l)$$
$$S°: 2 \times 193 \quad 214 \qquad\qquad 174 \qquad\quad 70$$

You calculate the entropy change by subtracting the entropy of the reactants from the entropy of the products.

$$\Delta S° = \Sigma \, nS°(\text{products}) - \Sigma \, mS°(\text{reactants})$$
$$= [(174 + 70) - (2 \times 193 + 214)] \text{ J/K} = -356 \text{ J/K}$$

Note that the sign of $\Delta S°$ agrees with the solution of Example 19.2b.

**EXERCISE 19.5**

Calculate the change of entropy, $\Delta S°$, for the reaction given in Example 19.2a. The standard entropy of glucose, $C_6H_{12}O_6(s)$, is 212 J/(mol · K). See Table 19.1 for other values.

See Problems 19.33 and 19.34.

# Free-Energy Concept

At the end of Section 19.2, you saw that the quantity $\Delta H - T\Delta S$ can serve as a criterion for spontaneity of a reaction at constant temperature and pressure. If the value of this quantity is negative, the reaction is spontaneous. If it is positive, the reaction is nonspontaneous. If it equals zero, the reaction is at equilibrium.

As an application of this criterion, consider the reaction described in the chapter opening, in which urea is prepared from $NH_3$ and $CO_2$. The heat of reaction $\Delta H°$ was calculated from enthalpies of formation in Section 19.1, where you obtained $-119.7$ kJ. Then, in Example 19.3, you calculated the entropy of reaction $\Delta S°$ and found a value of $-356$ J/K, or $-0.356$ kJ/K. Substitute these values and $T = 298$ K (25°C) into the expression $\Delta H° - T\Delta S°$.

$$\Delta H° - T\Delta S° = (-119.7 \text{ kJ}) - (298 \text{ K})(-0.356 \text{ kJ/K}) = -13.6 \text{ kJ}$$

You see that $\Delta H° - T\Delta S°$ is a negative quantity, from which you conclude that the reaction is spontaneous under standard conditions.

## 19.4 Free Energy and Spontaneity

It is very convenient to define a new thermodynamic quantity in terms of $H$ and $S$ that will be directly useful as a criterion of spontaneity. For this purpose, the American physicist J. Willard Gibbs (1839–1903) introduced the concept of **free energy, G,** which is *a thermodynamic quantity defined by the equation $G = H - TS$.* This quantity gives a direct criterion for spontaneity of reaction.

As a reaction proceeds at a given temperature and pressure, reactants form products and the enthalpy $H$ and entropy $S$ change. These changes in $H$ and $S$, denoted $\Delta H$ and $\Delta S$, result in a change in free energy, $\Delta G$, given by the equation

$$\Delta G = \Delta H - T\Delta S$$

Note that the change in free energy, $\Delta G$, equals the quantity $\Delta H - T\Delta S$ that you just saw serves as a criterion for spontaneity of a reaction. Thus, if you can show that $\Delta G$ for a reaction at a given temperature and pressure is negative, you can predict that the reaction will be spontaneous.

## Standard Free-Energy Change

Recall that for purposes of tabulating thermodynamic data, certain *standard states* are chosen, which are indicated by a superscript degree sign on the symbol of the quantity. The standard states are as follows: for pure liquids and solids, 1 atm pressure; for gases, 1 atm partial pressure; for solutions, 1 $M$ concentration. The temperature is the temperature of interest, usually 25°C (298 K).

The standard free-energy change, $\Delta G°$, is the free-energy change that occurs when reactants in their standard states are converted to products in their standard states. Example 19.4 illustrates the calculation of the standard free-energy change, $\Delta G°$, from $\Delta H°$ and $\Delta S°$.

$$\Delta G° = \Delta H° - T\Delta S°$$

---

### EXAMPLE 19.4

## Calculating $\Delta G°$ from $\Delta H°$ and $\Delta S°$

What is the standard free-energy change, $\Delta G°$, for the following reaction at 25°C?

$$N_2(g) + 3H_2(g) \longrightarrow 2NH_3(g)$$

Use values of $\Delta H_f°$ and $S°$, from Tables 6.2 and 19.1.

 **PROBLEM STRATEGY**

Calculate $\Delta H°$ and $\Delta S°$, then substitute these values into $\Delta H° - T\Delta S°$ to obtain $\Delta G°$.

**SOLUTION**

Write the balanced equation and place below each formula the values of $\Delta H_f°$ and $S°$ multiplied by stoichiometric coefficients.

| | $N_2(g) +$ | $3H_2(g)$ | $\longrightarrow$ | $2NH_3(g)$ |
|---|---|---|---|---|
| $\Delta H_f°$: | 0 | 0 | | $2 \times (-45.9)$ kJ |
| $S°$: | 191.5 | $3 \times 130.6$ | | $2 \times 193$ J/K |

You calculate $\Delta H°$ and $\Delta S°$ by taking values for products and subtracting values for reactants.

$$\Delta H° = \Sigma\, n\Delta H_f°(\text{products}) - \Sigma\, m\Delta H_f°(\text{reactants})$$
$$= [2 \times (-45.9) - 0]\ \text{kJ} = -91.8\ \text{kJ}$$

$$\Delta S° = \Sigma\, nS°(\text{products}) - \Sigma\, mS°(\text{reactants})$$
$$= [2 \times 193 - (191.5 + 3 \times 130.6)]\ \text{J/K} = -197\ \text{J/K}$$

You now substitute into the equation for $\Delta G°$ in terms of $\Delta H°$ and $\Delta S°$. Note that you substitute $\Delta S°$ in units of kJ/K.

$$\Delta G° = \Delta H° - T\Delta S° = -91.8\ \text{kJ} - (298\ \text{K})(-0.197\ \text{kJ/K}) = \textbf{-33.1 kJ}$$

**Problem Solving:**
**Calculating $\Delta H°$, $\Delta S°$, and $\Delta G°$**

**EXERCISE 19.6**

Calculate $\Delta G°$ for the following reaction at 25°C. Use data given in Tables 6.2 and 19.1.

$$CH_4(g) + 2O_2(g) \longrightarrow CO_2(g) + 2H_2O(g)$$

See Problems 19.37 and 19.38.

## Standard Free Energies of Formation

The **standard free energy of formation, $\Delta G_f°$,** of a substance is defined similarly to the standard enthalpy of formation. That is, $\Delta G_f°$ is *the free-energy change that occurs when 1 mol of substance is formed from its elements in their stablest states at 1 atm and at a specified temperature* (usually 25°C). For example, the standard free energy of formation of $NH_3(g)$ is the free-energy change for the reaction

$$\tfrac{1}{2}N_2(g) + \tfrac{3}{2}H_2(g) \longrightarrow NH_3(g)$$

The reactants, $N_2$ and $H_2$, each at 1 atm, are converted to the product, $NH_3$, at 1 atm pressure. In Example 19.4, you found $\Delta G°$ for the formation of 2 mol $NH_3$ from its elements to be −33.1 kJ. Hence, $\Delta G_f°$ (NH$_3$) = −33.1 kJ/2 mol = −16.6 kJ/mol.

As in the case of standard enthalpies of formation, the standard free energies of formation of elements in their stablest states are assigned the value zero. By tabulating $\Delta G_f°$ for substances, as in Table 19.2, you can easily calculate $\Delta G°$ for any reaction involving those substances. You simply subtract the standard free energies of reactants from the standard free energies of products:

$$\Delta G° = \Sigma\, n\Delta G_f°(\text{products}) - \Sigma\, m\Delta G_f°(\text{reactants})$$

The next example illustrates the calculations.

▶ An alternate example is provided in the Instructor's Resource Manual.

**EXAMPLE 19.5**

## Calculating $\Delta G°$ from Standard Free Energies of Formation

Calculate $\Delta G°$ for the combustion of 1 mol of ethanol, $C_2H_5OH$, at 25°C.

$$C_2H_5OH(l) + 3O_2(g) \longrightarrow 2CO_2(g) + 3H_2O(g)$$

Use the standard free energies of formation given in Table 19.2.

 **PROBLEM STRATEGY**

Calculate $\Delta G°$ from $\Delta G_f°$ values, similar to the way you calculate $\Delta H°$ from $\Delta H_f°$.

**SOLUTION**

Write the balanced equation with values of $\Delta G_f°$ multiplied by stoichiometric coefficients below each formula.

$$C_2H_5OH(l) + 3O_2(g) \longrightarrow 2CO_2(g) + 3H_2O(g)$$

$\Delta G_f°$:    −174.8        0        2(−394.4)    3(−228.6) kJ

**Table 19.2**
**Standard Free Energies of Formation (at 25°C)**

| Formula | $\Delta G_f^\circ$, kJ/mol* | Formula | $\Delta G_f^\circ$, kJ/mol | Formula | $\Delta G_f^\circ$, kJ/mol |
|---|---|---|---|---|---|
| *Hydrogen* | | *Carbon (continued)* | | *Sulfur* | |
| $H_3O^+(aq)$ | 0 | $CS_2(l)$ | 63.6 | $S_2(g)$ | 80.1 |
| $H_2(g)$ | 0 | $HCN(g)$ | 125 | $S(rhombic)$ | 0 |
| *Sodium* | | $HCN(l)$ | 121 | $S(monoclinic)$ | 0.10 |
| $Na^+(aq)$ | −261.9 | $CCl_4(g)$ | −53.7 | $SO_2(g)$ | −300.2 |
| $Na(s)$ | 0 | $CCl_4(l)$ | −68.6 | $H_2S(g)$ | −33 |
| $NaCl(s)$ | −384.0 | $CH_3CHO(g)$ | −133.7 | *Fluorine* | |
| $NaHCO_3(s)$ | −851.9 | $C_2H_5OH(l)$ | −174.8 | $F^-(aq)$ | −276.5 |
| $Na_2CO_3(s)$ | −1048.1 | *Silicon* | | $F_2(g)$ | 0 |
| *Calcium* | | $Si(s)$ | 0 | $HF(g)$ | −275 |
| $Ca^{2+}(aq)$ | −553.0 | $SiO_2(s)$ | −856.5 | *Chlorine* | |
| $Ca(s)$ | 0 | $SiF_4(g)$ | −1506 | $Cl^-(aq)$ | −131.2 |
| $CaO(s)$ | −603.5 | *Lead* | | $Cl_2(g)$ | 0 |
| $CaCO_3(s)$ | −1128.8 | $Pb(s)$ | 0 | $HCl(g)$ | −95.3 |
| *Carbon* | | $PbO(s)$ | −189.2 | *Bromine* | |
| $C(graphite)$ | 0 | $PbS(s)$ | −96.7 | $Br^-(aq)$ | −102.8 |
| $C(diamond)$ | 2.9 | *Nitrogen* | | $Br_2(l)$ | 0 |
| $CO(g)$ | −137.2 | $N_2(g)$ | 0 | *Iodine* | |
| $CO_2(g)$ | −394.4 | $NH_3(g)$ | −16 | $I^-(aq)$ | −51.7 |
| $HCO_3^-(aq)$ | −587.1 | $NO(g)$ | 86.60 | $I_2(s)$ | 0 |
| $CH_4(g)$ | −50.8 | $NO_2(g)$ | 51 | *Silver* | |
| $C_2H_4(g)$ | 68.4 | $HNO_3(aq)$ | −110.5 | $Ag^+(aq)$ | 77.1 |
| $C_2H_6(g)$ | −32.9 | *Oxygen* | | $Ag(s)$ | 0 |
| $C_6H_6(l)$ | 124.5 | $O_2(g)$ | 0 | $AgF(s)$ | −185 |
| $HCHO(g)$ | −110 | $O_3(g)$ | 163 | $AgCl(s)$ | −109.7 |
| $CH_3OH(l)$ | −166.2 | $OH^-(aq)$ | −157.3 | $AgBr(s)$ | −95.9 |
| $CS_2(g)$ | 66.9 | $H_2O(g)$ | −228.6 | $AgI(s)$ | −66.3 |
| | | $H_2O(l)$ | −237.2 | | |

*See Appendix C for additional values.

The calculation is

$$\Delta G^\circ = \Sigma\, n\Delta G_f^\circ(\text{products}) - \Sigma\, m\Delta G_f^\circ(\text{reactants})$$
$$= [2(-394.4) + 3(-228.6) - (-174.8)] \text{ kJ}$$
$$= \mathbf{-1299.8 \text{ kJ}}$$

---

**EXERCISE 19.7**

Calculate $\Delta G°$ at 25°C for the following reaction, using values of $\Delta G_f°$.

$$CaCO_3(s) \longrightarrow CaO(s) + CO_2(g)$$

---

**See Problems 19.41 and 19.42.**

### $\Delta G°$ as a Criterion for Spontaneity

You have already seen that the quantity $\Delta G = \Delta H - T\Delta S$ can be used as a criterion for the spontaneity of a reaction. The change of free energy, $\Delta G$, should be calculated for the conditions at which the reaction occurs. If the reactants are at standard conditions and give products at standard conditions, the free-energy change you need to look at is $\Delta G°$. The calculation is simple, as shown in Example 19.5. For other conditions, you should look at the appropriate $\Delta G$ value. This would be a more complicated calculation. Nevertheless, the standard free-energy change $\Delta G°$ is still a useful guide to the spontaneity of reaction in these cases. The following rules are useful in judging the spontaneity of a reaction:

1. When $\Delta G°$ is a large negative number (more negative than about $-10$ kJ), the reaction is spontaneous as written, and reactants transform almost entirely to products when equilibrium is reached.

2. When $\Delta G°$ is a large positive number (larger than about 10 kJ), the reaction is nonspontaneous as written, and reactants do not give significant amounts of products at equilibrium.

3. When $\Delta G°$ has a small negative or positive value (less than about 10 kJ), the reaction gives an equilibrium mixture with significant amounts of both reactants and products.

---

> ▶ An alternate example is provided in the Instructor's Resource Manual.

**EXAMPLE 19.6**

### Interpreting the Sign of $\Delta G°$

Calculate $\Delta H°$ and $\Delta G°$ for the reaction

$$2KClO_3(s) \longrightarrow 2KCl(s) + 3O_2(g)$$

Interpret the signs obtained for $\Delta H°$ and $\Delta G°$. Values of $\Delta H_f°$ (in kJ/mol) are as follows: $KClO_3(s)$, $-391.2$; $KCl(s)$, $-436.7$. Similarly, values of $\Delta G_f°$ (in kJ/mol) are as follows: $KClO_3(s)$, $-289.9$; $KCl(s)$, $-408.8$. Note that $O_2(g)$ is the reference form of the element, so $\Delta H_f° = \Delta G_f° = 0$ for it.

 **PROBLEM STRATEGY**

Calculate $\Delta H°$ and $\Delta G°$; then interpret the signs of these quantities. The interpretation of the sign of $\Delta H$ was discussed in Chapter 6. As long as $\Delta G°$ is not close to zero, you can interpret its sign as follows: a negative $\Delta G°$ means the reactants tend to go mostly to products; a positive $\Delta G°$ means that the reaction is nonspontaneous as written. When the value of $\Delta G°$ is close to zero, the reaction gives an equilibrium mixture.

**SOLUTION**

The problem is set up as follows:

$$2KClO_3(s) \longrightarrow 2KCl(s) + 3O_2(g)$$

$\Delta H_f^\circ$: $2 \times (-391.2)$    $2 \times (-436.7)$    0 kJ

$\Delta G_f^\circ$: $2 \times (-289.9)$    $2 \times (-408.8)$    0 kJ

Then,

$$\Delta H^\circ = [2 \times (-436.7) - 2 \times (-391.2)] \text{ kJ} = \mathbf{-91.0\ kJ}$$
$$\Delta G^\circ = [2 \times (-408.8) - 2 \times (-289.9)] \text{ kJ} = \mathbf{-237.8\ kJ}$$

**The reaction is exothermic, liberating 91.0 kJ of heat. The large negative value for $\Delta G^\circ$ indicates that the equilibrium composition is mostly potassium chloride and oxygen.**

**EXERCISE 19.8**

Which of the following reactions are spontaneous in the direction written? See Table 19.2 for data.

a. $C(graphite) + 2H_2(g) \longrightarrow CH_4(g)$

b. $2H_2(g) + O_2(g) \longrightarrow 2H_2O(l)$

c. $4HCN(g) + 5O_2(g) \longrightarrow 2H_2O(l) + 4CO_2(g) + 2N_2(g)$

d. $Ag^+(aq) + I^-(aq) \longrightarrow AgI(s)$

**See Problems 19.45 and 19.46.**

**CONCEPT CHECK 19.2**

Consider the reaction of nitrogen, $N_2$, and oxygen, $O_2$, to form nitric oxide, NO:

$$N_2(g) + O_2(g) \longrightarrow 2NO(g)$$

From the standard free energy of formation of NO, what can you say about this reaction?

# 19.5 Interpretation of Free Energy

You have seen that the free-energy change serves as a criterion for spontaneity of a chemical reaction. This gives you some idea of what free energy is and how you interpret it. In this section, we will look more closely at the meaning of free energy.

## Maximum Work

Theoretically, spontaneous reactions can be used to obtain useful work. By useful work, we mean energy that can be used directly to move objects of normal size. We use the combustion of gasoline to move an automobile, and we use a reaction in a battery to generate electricity to drive a motor. Similarly, biochemical reactions in muscle tissue occur in such a way as to contract muscle fibers and lift a weight.

# Coupling of Reactions

**LIFE SCIENCE**

Many important processes are thermodynamically unfavorable, or *nonspontaneous*. Assembling many small amino acid molecules into a large protein molecule is an example of a nonspontaneous process. Yet biological organisms do assemble proteins from amino acids. Biological organisms accomplish many similar nonspontaneous processes. How do they do this?

Let's look at a simpler example. We would like to obtain iron metal from iron ore. Here is the equation for the direct decomposition of iron(III) oxide, the iron-containing substance in the iron ore hematite:

$$2Fe_2O_3(s) \longrightarrow 4Fe(s) + 3O_2(g); \quad \Delta G° = +1487 \text{ kJ}$$

This reaction is clearly nonspontaneous, because $\Delta G°$ is a large positive quantity. The nonspontaneity of this reaction is in agreement with common knowledge. Iron tends to rust (form iron oxide) in air, by combining with oxygen. You do not expect a rusty wrench to turn spontaneously into shiny iron and oxygen. But this does not mean you cannot change iron(III) oxide to iron metal. It merely means that you have to do work on the iron(III) oxide to reduce it to the metal. You must, in effect, find a way to couple this nonspontaneous reaction to one that is sufficiently spontaneous; that is, you must couple this reaction to one having a more negative $\Delta G°$.

Consider the spontaneous reaction of carbon monoxide with oxygen (the burning of CO):

$$2CO(g) + O_2(g) \longrightarrow 2CO_2(g); \quad \Delta G° = -514.4 \text{ kJ}$$

For 3 mol $O_2$, $\Delta G°$ is $-1543$ kJ, which is more negative than for the direct decomposition of 2 mol $Fe_2O_3$ to its elements. Let us add the two reactions:

$$2Fe_2O_3(s) \longrightarrow 4Fe(s) + 3O_2(g);$$
$$\Delta G° = 1487 \text{ kJ}$$
$$6CO(g) + 3O_2(g) \longrightarrow 6CO_2(g);$$
$$\Delta G° = -1543 \text{ kJ}$$

$$\overline{2Fe_2O_3(s) + 6CO(g) \longrightarrow 4Fe(s) + 6CO_2(g);}$$
$$\Delta G° = -56 \text{ kJ}$$

The net result is to reduce iron(III) oxide to the metal by reacting the oxide with carbon monoxide. This is the reaction that occurs in a blast furnace, where iron ore is commercially reduced to iron.

The concept of the coupling of two chemical reactions, a nonspontaneous reaction with a spontaneous one, to give an overall spontaneous change is a very useful one, especially in biochemistry. Adenosine triphosphate, or ATP, is a molecule containing a triphosphate group. It reacts with water, in the presence of an enzyme (biochemical catalyst), to give adenosine diphosphate, ADP, and a phosphate ion. (Figure 19.11 shows the structure of ATP and ADP.) The reaction has a negative $\Delta G°$ (a spontaneous reaction.)

$$ATP + H_2O \longrightarrow ADP + \text{phosphate ion}; \quad \Delta G° = -31 \text{ kJ}$$

ATP is first synthesized in a living organism; the energy for this synthesis is obtained from food. The spontaneous reaction of ATP to give ADP is then coupled to various nonspontaneous reactions in the organism to accomplish necessary life processes (see Figure 19.12).

Often reactions are not carried out in a way that does useful work. The reactants are simply poured together in a reaction vessel, and products are separated from the mixture. As the reaction occurs, the free energy of the system decreases and entropy is produced, but no useful work is obtained.

In principle, if a reaction is carried out to obtain the maximum useful work, no entropy is produced. It can be shown that the maximum useful work, $w_{max}$, for a spontaneous reaction is $\Delta G$.

$$w_{max} = \Delta G$$

The term *free energy* comes from this result. *The free-energy change is the maximum energy available, or free, to do useful work.* As a reaction occurs in such a way as to give the maximum useful work, the free energy decreases and a corresponding quantity of useful work is obtained. ◄

▶ The sign of *w* (work) is defined so that a negative value means work obtained *from* the system (that is, energy is subtracted from the system). You can obtain work from a reaction if its $\Delta G$ is negative.

**Figure 19.11**
**Structural formulas of ATP and ADP.** Shown are ionic forms of adenosine triphosphate (ATP) and adenosine diphosphate (ADP). The rings have carbon atoms at the junctures.

**Figure 19.12**
**Coupling of reactions in a biological cell.** The mitochondria are parts of a cell that obtain energy in the form of ATP from glucose (from food), $C_6H_{12}O_6$. ATP molecules are used in various cell processes, such as protein synthesis from amino acids, which occurs in the ribosomes.

The concept of maximum work from a chemical reaction is an idealization. In any real situation, less than this maximum work is obtained and some entropy is created. When this work is eventually expended, it appears in the environment as additional entropy.

## Free-Energy Change During Reaction

You have seen that the free-energy change is related to the work done during a chemical reaction. Consider the combustion of gasoline in $O_2$. The reaction is spontaneous, so the free-energy change is negative. That is, the free energy of the system changes to a lower value as the reactants are converted to products. Figure 19.9 shows the free-energy change that occurs during this reaction. When the gasoline is burned in a gasoline stove, the decrease in free energy shows up as an increase in entropy of the

**Figure 19.9**
**Free-energy change during a spontaneous reaction (combustion of gasoline).**    The free energy decreases as the reaction proceeds. At equilibrium, the free energy becomes a minimum, and the equilibrium mixture is mostly products.

▶ **What happens in the gasoline engine is complicated, but you need look at only the initial and final states. Reactants at the surrounding temperature and pressure go into the engine, which produces work and heat, and the products leave the engine, eventually reaching the initial temperature and pressure of the reactants.**

system and the surroundings. However, when the gasoline is burned in an automobile engine, some of the decrease in free energy shows up as work done. Theoretically, all of the free-energy decrease can be used to do work. This gives the maximum work. In practice, less work is obtained, and the difference appears as an increase of entropy. Ultimately, the work is itself used up—that is, changed to entropy. ◀

Look at Figure 19.9 again. At the start of the reaction, the system contains only reactants: gasoline and $O_2$. The free energy has a relatively high value. This decreases as the reaction proceeds. The decrease in free energy appears either as an increase in entropy or as work done. Eventually, the free energy of the system reaches its minimum value. Then the net reaction stops; it comes to equilibrium.

Before we leave this section, consider a reaction in which $\Delta G°$ is positive. You would predict that the reaction is nonspontaneous. Figure 19.10 shows the free-energy change as the reaction proceeds. Note that there is a small decrease in free energy as the system goes to equilibrium. Some reaction occurs in order to give the equilibrium mixture. But this mixture consists primarily of reactants, because the reaction does not go very far before coming to equilibrium. To change reactants completely to products is a nonspontaneous reaction (shown by the arrow along the curve in Figure 19.10).

**Figure 19.10**
**Free-energy change during a nonspontaneous reaction.**    The free energy decreases until equilibrium, at which the minimum value is reached. There is very little reaction, however, because the equilibrium mixture is mostly reactants. To change reactants to mostly products is to undergo a nonspontaneous reaction.

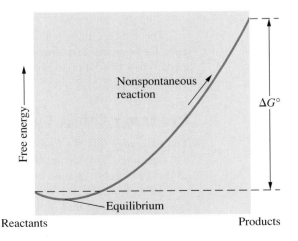

# Free Energy and Equilibrium Constants

In the previous section, you saw that for any spontaneous reaction, the total free energy of the substances in the reaction mixture decreases as the reaction proceeds (that is, the free-energy change is negative). Earlier, you saw that the standard free-energy change, $\Delta G°$, can be used as a criterion for the spontaneity of a reaction. When $\Delta G°$ is negative, the reaction is spontaneous. We will now describe how the standard free-energy change is related to the equilibrium constant.

## 19.6  Relating $\Delta G°$ to the Equilibrium Constant

One of the most important results of chemical thermodynamics is an equation relating the standard free-energy change for a reaction to the equilibrium constant. Before we look at this equation, we must discuss the thermodynamic form of the equilibrium constant that occurs in that equation.

The **thermodynamic equilibrium constant, K,** is *the equilibrium constant in which the concentrations of gases are expressed in partial pressures in atmospheres, whereas the concentrations of solutes in liquid solutions are expressed in molarities.* For a reaction involving only solutes in liquid solution, $K$ is identical to $K_c$; for reactions involving only gases, $K$ equals $K_p$. ◄

▶ To be precise, $K$ is defined in terms of activities, which are dimensionless quantities numerically equal to "effective" concentrations and pressures.

### EXAMPLE 19.7

### Writing the Expression for a Thermodynamic Equilibrium Constant

Write expressions for the thermodynamic equilibrium constants for each of the following reactions:

a. The reaction given in the chapter opening,

$$2NH_3(g) + CO_2(g) \rightleftharpoons NH_2CONH_2(aq) + H_2O(l)$$

b. The solubility process,

$$AgCl(s) \rightleftharpoons Ag^+(aq) + Cl^-(aq)$$

**SOLUTION**

a. Note that $H_2O$ is a solvent, so it does not appear explicitly in $K$. The gases appear in $K$ as partial pressures, and the solute appears as a molar concentration.

$$K = \frac{[NH_2CONH_2]}{P_{NH_3}^2 P_{CO_2}}$$

b. Note that AgCl is a solid, so it does not appear explicitly in $K$. The solutes appear as molar concentrations.

$$K = [Ag^+][Cl^-]$$

This is identical to $K_{sp}$.

---

**EXERCISE 19.9**

Give the expression for $K$ for each of the following reactions.

a. $CaCO_3(s) \rightleftharpoons CaO(s) + CO_2(g)$

b. $PbI_2(s) \rightleftharpoons Pb^{2+}(aq) + 2I^-(aq)$

c. $H^+(aq) + HCO_3^-(aq) \rightleftharpoons H_2O(l) + CO_2(g)$

---

See Problems 19.51 and 19.52.

The standard free-energy change, $\Delta G°$, for a reaction can be calculated using data from thermodynamic tables. If you want the free-energy change when reactants in nonstandard states are changed to products in nonstandard states ($\Delta G$), you can obtain it from the standard free-energy change $\Delta G°$ using the following equation:

$$\Delta G = \Delta G° + RT \ln Q$$

Here $Q$ is the thermodynamic form of the reaction quotient. It has the same general appearance as the thermodynamic equilibrium constant $K$, but the concentrations and partial pressures are those for a mixture at some instant, perhaps at the beginning of a reaction. You obtain $\Delta G$ from $\Delta G°$ by adding $RT \ln Q$, where $\ln Q$ is the natural logarithm of $Q$ ($\ln Q = 2.303 \log Q$).

You can derive the equation relating $\Delta G°$ to the equilibrium constant $K$ from the preceding equation. In the previous section, you saw that as a chemical reaction approaches equilibrium, the free energy decreases and continues to decrease until equilibrium is reached. At equilibrium, the free energy ceases to change; then $\Delta G = 0$. Also, the reaction quotient $Q$ becomes equal to the equilibrium constant $K$. If you substitute $\Delta G = 0$ and $Q = K$ into the preceding equation, you obtain

$$0 = \Delta G° + RT \ln K$$

This result easily rearranges to give the basic equation relating the standard free-energy change to the equilibrium constant.

$$\Delta G° = -RT \ln K$$

The next example illustrates the calculation of a thermodynamic equilibrium constant from the standard free-energy change, $\Delta G°$.

---

▶ An alternate example is provided in the Instructor's Resource Manual.

**EXAMPLE 19.8**

## Calculating $K$ from the Standard Free-Energy Change (Molecular Equation)

Find the value of the equilibrium constant $K$ at 25°C (298 K) for the reaction

$$2NH_3(g) + CO_2(g) \rightleftharpoons NH_2CONH_2(aq) + H_2O(l)$$

The standard free-energy change, $\Delta G°$, at 25°C equals $-13.6$ kJ. (We calculated this value of $\Delta G° = \Delta H° - T\Delta S°$ just before Section 19.4.)

**PROBLEM STRATEGY**

Rearrange the equation $\Delta G° = -RT \ln K$ to give

$$\ln K = \frac{\Delta G°}{-RT}$$

$\Delta G°$ and $R$ must be in compatible units. You normally express $\Delta G°$ in joules and set $R$ equal to 8.31 J/(K · mol).

**SOLUTION**

Substituting numerical values into this equation,

$$\ln K = \frac{-13.6 \times 10^3}{-8.31 \times 298} = 5.49$$

Hence,

$$K = e^{5.49} = 2.42 \times 10^2$$

Note that although the value of $K$ indicates that products predominate at equilibrium, $K$ is only moderately large. You would expect that the composition could be easily shifted toward reactants if you could remove either $NH_3$ or $CO_2$ (according to Le Chatelier's principle). This is what happens when urea is used as a fertilizer. As $NH_3$ is used up, more $NH_3$ is produced by the decomposition of urea.

**EXERCISE 19.10**

Use the data from Table 19.2 to obtain the equilibrium constant $K_p$ at 25°C for the reaction

$$CaCO_3(s) \rightleftharpoons CaO(s) + CO_2(g)$$

Note that values of $\Delta G_f°$ are needed for $CaCO_3$ and $CaO$, even though the substances do not appear in $K_p = P_{CO_2}$.

See Problems 19.53, 19.54, 19.55, and 19.56.

The method of calculating $K$ from $\Delta G°$ also works for net ionic equations. In Example 19.9, you obtain a solubility product constant from thermodynamic data.

**EXAMPLE 19.9**

## Calculating $K$ from the Standard Free-Energy Change (Net Ionic Equation)

Calculate the equilibrium constant $K_{sp}$ at 25°C for the reaction

$$AgCl(s) \rightleftharpoons Ag^+(aq) + Cl^-(aq)$$

using standard free energies of formation.

 **PROBLEM STRATEGY**

This is similar to the preceding example, except that you must first calculate $\Delta G_f°$ for the ionic equation. Note that $K_{sp}$ equals $K$, the thermodynamic equilibrium constant.

**SOLUTION**

You first calculate $\Delta G°$. Writing $\Delta G_f°$ values below the formulas in the equation gives

$$AgCl(s) \rightleftharpoons Ag^+(aq) + Cl^-(aq)$$
$$\Delta G_f°: \quad -109.7 \qquad\quad 77.1 \qquad -131.2 \text{ kJ}$$

Hence,

$$\Delta G° = [(77.1 - 131.2) - (-109.7)] \text{ kJ}$$
$$= 55.6 \text{ kJ (or } 55.6 \times 10^3 \text{ J)}$$

You now substitute numerical values into the equation relating $\ln K$ and $\Delta G°$.

$$\ln K = \frac{\Delta G°}{-RT} = \frac{55.6 \times 10^3}{-8.31 \times 298} = -22.45$$

Therefore,

$$K = e^{-22.45} = \mathbf{1.8 \times 10^{-10}}$$

**EXERCISE 19.11**

Calculate the solubility product constant for $Mg(OH)_2$ at 25°C. The $\Delta G_f°$ values (in kJ/mol) are as follows: $Mg^{2+}(aq)$, $-456.0$; $OH^-(aq)$, $-157.3$; $Mg(OH)_2(s)$, $-933.9$.

*See Problems 19.57 and 19.58.*

You can understand the use of $\Delta G°$ as a criterion for spontaneity by looking at its relationship to the equilibrium constant, $\Delta G° = -RT \ln K$. When the equilibrium constant is greater than 1, $\ln K$ is positive and $\Delta G°$ is negative. Similarly, when the equilibrium constant is less than 1, $\ln K$ is negative and $\Delta G°$ is positive. This agrees with the first two rules listed at the end of Section 19.4. You can get the third rule by substituting $\Delta G° = \pm 10 \times 10^3$ J into $\Delta G° = -RT \ln K$ and solving for $K$. You find that $K$ is between 0.018 and 57. In this range, the equilibrium mixture contains significant amounts of reactants as well as products.

 **CONCEPT CHECK 19.3**

The following reaction is spontaneous in the drection given.

$$A(g) + B(g) \longrightarrow C(g) + D(g)$$

Suppose you are given a vessel containing an equilibrium mixture of A, B, C, and D, and you increase the concentration of C by increasing its partial pressure.

a. How is the value of $\Delta G°$ affected by the addition of C to the vessel?

b. How is the value of $\Delta G$ affected by the addition of C to the vessel?

# 19.7 Change of Free Energy with Temperature

In the previous sections, you obtained the free-energy change and equilibrium constant for a reaction at 25°C, the temperature at which the thermodynamic data were given. How do you find $\Delta G°$ or $K$ at another temperature? Precise calculations are possible but are rather involved. Instead, we will look at a simple method that gives approximate results.

In this method, you assume that $\Delta H°$ and $\Delta S°$ are essentially constant with respect to temperature. (This is only approximately true.) You get the value of $\Delta G_T°$ at any temperature $T$ by substituting values of $\Delta H°$ and $\Delta S°$ at 25°C (obtained from appropriate tables) into the following equation. ◀

▶ This approximation is most accurate for temperatures not too different from the temperature for which the $\Delta H°$ and $\Delta S°$ values are obtained. Much different temperatures give greater error.

$$\Delta G_T° = \Delta H° - T\Delta S° \qquad \text{(approximation for } \Delta G_T°)$$

Remember that the superscript degree sign (°) refers to substances in standard states, that is, at 1 atm and at the *specified temperature*. Although until now this was 25°C (298 K), we now consider other temperatures. Note that in general, $\Delta G°$ depends strongly on temperature.

## Spontaneity and Temperature Change

All of the four possible choices of signs for $\Delta H°$ and $\Delta S°$, listed in Table 19.3, give different temperature behaviors for $\Delta G°$. Consider the case in which $\Delta H°$ is negative and $\Delta S°$ is positive. An example is the reaction

$$\underset{\text{glucose}}{C_6H_{12}O_6(s)} \longrightarrow \underset{\text{ethanol}}{2C_2H_5OH(l)} + 2CO_2(g)$$

This represents the overall change of glucose (grape sugar) to ethanol (ethyl alcohol). The signs of $\Delta H°$ and $\Delta S°$ are easily explained. The formation of more stable bonds, such as occur in $CO_2$, releases energy as heat. The reaction is exothermic and $\Delta H°$ is negative. As explained in Example 19.2, the breaking up of a molecule ($C_6H_{12}O_6$) into smaller ones and the formation of a gas are expected to increase the entropy, so $\Delta S°$ is positive.

When $\Delta H°$ is negative and $\Delta S°$ is positive, both terms in $\Delta G°$ (that is, $\Delta H°$ and $-T\Delta S°$) are negative. Therefore, $\Delta G°$ is always negative and the reaction is spontaneous whatever the temperature.

If the signs of $\Delta H°$ and $\Delta S°$ are reversed (that is, if $\Delta H°$ is positive, or endothermic, and $\Delta S°$ is negative), $\Delta G°$ is always positive. Thus, the reaction is nonspontaneous at all temperatures. An example is the reaction in which oxygen gas is converted to ozone.

$$3O_2(g) \longrightarrow 2O_3(g)$$

### Table 19.3
### Effect of Temperature on the Spontaneity of Reactions

| $\Delta H°$ | $\Delta S°$ | $\Delta G°$ | Description* | Example |
|---|---|---|---|---|
| − | + | − | Spontaneous at all $T$ | $C_6H_{12}O_6(s) \longrightarrow 2C_2H_5OH(l) + 2CO_2(g)$ |
| + | − | + | Nonspontaneous at all $T$ | $3O_2(g) \longrightarrow 2O_3(g)$ |
| − | − | + or − | Spontaneous at low $T$; nonspontaneous at high $T$ | $2NH_3(g) + CO_2(g) \longrightarrow NH_2CONH_2(aq) + H_2O(l)$ |
| + | + | + or − | Nonspontaneous at low $T$; spontaneous at high $T$ | $Ba(OH)_2 \cdot 8H_2O(s) + 2NH_4NO_3(s) \longrightarrow$ $Ba(NO_3)_2(aq) + 2NH_3(g) + 10H_2O(l)$ |

*The terms *low temperature* and *high temperature* are relative. For a particular reaction, high temperature could mean room temperature.

To accomplish this conversion, oxygen is passed through a tube in which an electrical discharge occurs. The electrical discharge supplies the necessary free energy for this otherwise nonspontaneous change.

In the reaction described in the chapter opening,

$$2NH_3(g) + CO_2(g) \longrightarrow NH_2CONH_2(aq) + H_2O(l)$$

both $\Delta H°$ and $\Delta S°$ are negative. In this case, the sign of $\Delta G°$ depends on the relative magnitudes of the terms $\Delta H°$ and $-T\Delta S°$, which have opposite signs. At some temperature, these terms just cancel and $\Delta G°$ equals zero. Below this temperature, $\Delta G°$ is negative. Above it, $\Delta G°$ is positive. Therefore, this reaction is spontaneous at low temperatures but nonspontaneous at sufficiently high temperatures. This particular reaction is spontaneous at 25°C but nonspontaneous at about 60°C.

A reaction in which both $\Delta H°$ and $\Delta S°$ are positive is the one described in the opening to Chapter 6. ◄

▶ **The reaction mixture spontaneously cools enough to freeze water. See Figure 6.1.**

$$Ba(OH)_2 \cdot 8H_2O(s) + 2NH_4NO_3(s) \longrightarrow Ba(NO_3)_2(aq) + 2NH_3(g) + 10H_2O(l)$$

The reaction is endothermic, and because crystalline solids change to a solution and a gas, the entropy increases. Again, the sign of $\Delta G°$ depends on the relative magnitudes of the terms $\Delta H°$ and $-T\Delta S°$. The reaction is spontaneous at room temperature but would be nonspontaneous at a sufficiently low temperature. Table 19.3 summarizes this discussion.

## CONCEPT CHECK 19.4

Consider the decomposition of dinitrogen tetroxide, $N_2O_4$, to nitrogen dioxide, $NO_2$:

$$N_2O_4(g) \longrightarrow 2NO_2(g)$$

How would you expect the spontaneity of the reaction to behave with temperature change?

## Calculation of $\Delta G°$ at Various Temperatures

As an application of the method of calculating $\Delta G°$ at various temperatures, assuming $\Delta H°$ and $\Delta S°$ are constant, look at the following reaction:

$$CaCO_3(s) \longrightarrow CaO(s) + CO_2(g)$$

At 25°C, $\Delta G°$ equals $+130.9$ kJ, and the equilibrium partial pressure of $CO_2$ calculated from this is $1.1 \times 10^{-23}$ atm. The very small value of this partial pressure shows that $CaCO_3$ is quite stable at room temperature. In the next example, you will see how $\Delta G°$ and $K_p$ for this reaction change at higher temperature.

---

**EXAMPLE 19.10**

### Calculating $\Delta G°$ and $K$ at Various Temperatures

a. What is $\Delta G°$ at 1000°C for the following reaction?

$$CaCO_3(s) \rightleftharpoons CaO(s) + CO_2(g)$$

Is this reaction spontaneous at 1000°C and 1 atm?

b. What is the value of $K_p$ at 1000°C for this reaction? What is the partial pressure of $CO_2$?

 **PROBLEM STRATEGY**

a. Calculate $\Delta H°$ and $\Delta S°$ at 25°C using standard enthalpies of formation and standard entropies. Then substitute into the equation for $\Delta G_T°$.

b. Use the value of $\Delta G_T°$ to find $K$ ($= K_p$), as in Example 19.8.

**SOLUTION**

a. From Tables 6.2 and 19.1, you have

$$CaCO_3(s) \rightleftharpoons CaO(s) + CO_2(g)$$

|  | $CaCO_3(s)$ | $CaO(s)$ | $CO_2(g)$ |
|---|---|---|---|
| $\Delta H_f°$: | $-1206.9$ | $-635.1$ | $-393.5$ kJ |
| $S°$: | $92.9$ | $38.2$ | $213.7$ J/K |

You calculate $\Delta H°$ and $\Delta S°$ from these values.

$$\Delta H° = [(-635.1 - 393.5) - (-1206.9)] \text{ kJ} = 178.3 \text{ kJ}$$

$$\Delta S° = [(38.2 + 213.7) - (92.9)] \text{ J/K} = 159.0 \text{ J/K}$$

Now you substitute $\Delta H°$, $\Delta S°$ ($= 0.1590$ kJ/K), and $T$ ($= 1273$ K) into the equation for $\Delta G_T°$.

$$\Delta G_T° = \Delta H° - T\Delta S° = 178.3 \text{ kJ} - (1273 \text{ K})(0.1590 \text{ kJ/K}) = -24.1 \text{ kJ}$$

Because $\Delta G_T°$ is negative, the reaction should be **spontaneous** at 1000°C and 1 atm

b. Substitute the value of $\Delta G°$ at 1273 K, which equals $-24.1 \times 10^3$ J, into the equation relating $\ln K$ and $\Delta G°$.

$$\ln K = \frac{\Delta G°}{-RT} = \frac{-24.1 \times 10^3}{-8.31 \times 1273} = 2.278$$

$$K = K_p = e^{2.278} = \textbf{9.76}$$

$K_p = P_{CO_2}$, so the partial pressure of $CO_2$ is **9.76 atm.**

---

**EXERCISE 19.12**

The thermodynamic equilibrium constant for the vaporization of water,

$$H_2O(l) \rightleftharpoons H_2O(g)$$

is $K_p = P_{H_2O}$. Use thermodynamic data to calculate the vapor pressure of water at 45°C. Compare your answer with the value given in Table 5.6.

---

See Problems 19.59 and 19.60.

You can use the ideas described in this section to find the temperature at which a reaction such as the decomposition of $CaCO_3$ changes from nonspontaneous to spontaneous under standard conditions (1 atm for reactants and products). At this temperature, $\Delta G°$ equals zero.

$$\Delta G° = 0 = \Delta H° - T\Delta S°$$

Solving for $T$ gives

$$T = \frac{\Delta H^\circ}{\Delta S^\circ}$$

For the decomposition of $CaCO_3$, using values obtained in Example 19.10, you get

$$T = \frac{178.3 \text{ kJ}}{0.1590 \text{ kJ/K}} = 1121 \text{ K } (848°C)$$

Thus, $CaCO_3$ should be stable to thermal decomposition to CaO and $CO_2$ at 1 atm until heated to 848°C. This is only approximate, of course, because you assumed that $\Delta H^\circ$ and $\Delta S^\circ$ are constant with temperature.

---

### EXERCISE 19.13

To what temperature must magnesium carbonate be heated to decompose it to MgO and $CO_2$ at 1 atm? Is this higher or lower than the temperature required to decompose $CaCO_3$? Values of $\Delta H_f^\circ$ (in kJ/mol) are as follows: MgO(s), −601.2; $MgCO_3$(s), −1112. Values of $S°$ (in J/K) are as follows: MgO(s), 26.9; $MgCO_3$(s), 65.9. Data for $CO_2$ are given in Tables 6.2 and 19.1.

See Problems 19.61 and 19.62.

---

# A Checklist for Review

## Important Terms

internal energy, $U$ (19.1)
state function (19.1)
work (19.1)
first law of
    thermodynamics (19.1)
enthalpy, $H$ (19.1)

spontaneous
    process (p. 804)
entropy, $S$ (19.2)
second law of
    thermodynamics (19.2)

third law of thermody-
    namics (19.3)
standard (absolute)
    entropy (19.3)
free energy, $G$ (19.4)

standard free energy of
    formation, $\Delta G_f^\circ$ (19.4)
thermodynamic
    equilibrium constant,
    $K$ (19.6)

## Key Equations

$\Delta U = q + w$

$w = -P\Delta V$

$\Delta S > \dfrac{q}{T}$  (spontaneous process)

$\Delta S = \dfrac{q}{T}$  (equilibrium process)

$\Delta S^\circ = \Sigma \, nS°(\text{products}) - \Sigma \, mS°(\text{reactants})$

$\Delta G^\circ = \Delta H^\circ - T\Delta S^\circ$

$\Delta G^\circ = \Sigma \, n\Delta G_f^\circ(\text{products}) - \Sigma \, m\Delta G_f^\circ(\text{reactants})$

$\Delta G = \Delta G^\circ + RT \ln Q$

$\Delta G^\circ = -RT \ln K$

$\Delta G_T^\circ = \Delta H^\circ - T\Delta S^\circ$  (approximation for $\Delta G_T^\circ$)

## Summary of Facts and Concepts

The *first law of thermodynamics* states that $\Delta U = q + w$. For a chemical reaction at constant pressure, this becomes $\Delta U = q_p - P\Delta V$. If you define *enthalpy, H*, as $U + PV$, you can show that for a reaction at constant pressure, $\Delta H = \Delta U + P\Delta V = q_p$. *Entropy, S*, is a thermodynamic measure of randomness or disorder in a system. According to the *second law of thermody-*

*namics*, the total entropy of a system and its surroundings increases for a *spontaneous process*, one that occurs of its own accord. For a spontaneous reaction at constant $T$ and $P$, you can show that the entropy change $\Delta S$ is greater than $\Delta H/T$. For an equilibrium process, such as a phase transition, $\Delta S = \Delta H/T$. The standard entropy $S°$ of a substance is determined by measuring the heat absorbed in warming it at increasing temperatures at 1 atm. The method depends on the *third law of thermodynamics,* which states that perfectly crystalline substances at 0 K have zero entropy.

*Free energy, G,* is defined as $H - TS$. The change in free energy, $\Delta G$, for a reaction at constant $T$ and $P$ equals $\Delta H - T\Delta S$, which is negative for a spontaneous change. *Standard free-energy changes,* $\Delta G°$, can be calculated from $\Delta H_f°$ and $S°$ values or from *standard free energies of formation,* $\Delta G_f°$. $\Delta G_f°$ for a substance is the standard free-energy change for the formation of the substance from the elements in their stablest states. The standard free-energy change, $\Delta G°$, serves as a *criterion for spontaneity* of a reaction. A negative value means the reaction is spontaneous. A value of zero means the reaction is at equilibrium. A positive value means the reaction is nonspontaneous.

The free-energy change, $\Delta G$, is related to the *maximum work* that can be done by a reaction. This maximum work equals $\Delta G$. A nonspontaneous reaction can be made to go by coupling it with a reaction that has a negative $\Delta G$, so that $\Delta G$ for the overall result is negative. The reaction with the negative $\Delta G$ does work on the nonspontaneous reaction.

The *thermodynamic equilibrium constant, K,* can be calculated from the standard free-energy change by the relationship $\Delta G° = -RT \ln K$. $\Delta G°$ at various temperatures can be obtained by using the approximation that $\Delta H°$ and $\Delta S°$ are constant. Therefore, $\Delta G°$, which equals $\Delta H° - T\Delta S°$, can be easily calculated for any $T$.

## Operational Skills

1. **Calculating the entropy change for a phase transition** Given the heat of phase transition and the temperature of the transition, calculate the entropy change, $\Delta S$, of the system. **(EXAMPLE 19.1)**

2. **Predicting the sign of the entropy change of a reaction** Predict the sign of $\Delta S°$ for a reaction to which the rules listed in the text can be clearly applied. **(EXAMPLE 19.2)**

3. **Calculating $\Delta S°$ for a reaction** Given the standard entropies of reactants and products, calculate the entropy of reaction, $\Delta S°$. **(EXAMPLE 19.3)**

4. **Calculating $\Delta G°$ from $\Delta H°$ and $\Delta S°$** Given enthalpies of formation and standard entropies of reactants and products, calculate the standard free-energy change $\Delta G°$ for a reaction. **(EXAMPLE 19.4)**

5. **Calculating $\Delta G°$ from standard free energies of formation** Given the free energies of formation of reac-

tants and products, calculate the standard free-energy change $\Delta G°$ for a reaction. **(EXAMPLE 19.5)**

6. **Interpreting the sign of $\Delta G°$** Use the standard free-energy change to determine the spontaneity of a reaction. **(EXAMPLE 19.6)**

7. **Writing the expression for a thermodynamic equilibrium constant** For any balanced chemical equation, write the expression for the thermodynamic equilibrium constant. **(EXAMPLE 19.7)**

8. **Calculating $K$ from the standard free-energy change** Given the standard free-energy change for a reaction, calculate the thermodynamic equilibrium constant. **(EXAMPLES 19.8, 19.9)**

9. **Calculating $\Delta G°$ and $K$ at various temperatures** Given $\Delta H°$ and $\Delta S°$ at 25°C, calculate $\Delta G°$ and $K$ for a reaction at a temperature other than 25°C. **(EXAMPLE 19.10)**

# Review Questions

**19.1** What is a spontaneous process? Give three examples of spontaneous processes. Give three examples of nonspontaneous processes.

**19.2** Which contains greater entropy, a quantity of frozen benzene or the same quantity of liquid benzene at the same temperature? Explain in terms of the degree of order of the substance.

**19.3** State the second law of thermodynamics.

**19.4** The entropy change $\Delta S$ for a phase transition equals $\Delta H/T$, where $\Delta H$ is the enthalpy change. Why is it that the entropy change for a system in which a chemical reaction occurs spontaneously does not equal $\Delta H/T$?

**19.5** Describe how the standard entropy of hydrogen gas at 25°C can be obtained from heat measurements.

**19.6** Describe what you would look for in a reaction involving gases in order to predict the sign of $\Delta S°$. Explain.

**19.7** Define the free energy $G$. How is $\Delta G$ related to $\Delta H$ and $\Delta S$?

**19.8** What is meant by the standard free-energy change $\Delta G°$ for a reaction? What is meant by the standard free energy of formation $\Delta G_f°$ of a substance?

**19.9** Explain how $\Delta G°$ can be used to decide whether a chemical equation is spontaneous in the direction written.

**19.10** What is the useful work obtained in the ideal situation in which a chemical reaction with free-energy change $\Delta G$ is run so that it produces no entropy?

**19.11** Give an example of a chemical reaction used to obtain useful work.

**19.12** How is the concept of coupling of reactions useful in explaining how a nonspontaneous change could be made to occur?

**19.13** Explain how the free energy changes as a spontaneous reaction occurs. Show by means of a diagram how $G$ changes with the extent of reaction.

**19.14** Explain how an equilibrium constant can be obtained from thermal data alone (that is, from measurements of heat only).

**19.15** Discuss the different sign combinations of $\Delta H°$ and $\Delta S°$ that are possible for a process carried out at constant temperature and pressure. For each combination, state whether the process must be spontaneous or not, or whether both situations are possible. Explain.

**19.16** Consider a reaction in which $\Delta H°$ and $\Delta S°$ are positive. Suppose the reaction is nonspontaneous at room temperature. How would you estimate the temperature at which the reaction becomes spontaneous?

## Conceptual Problems

**19.17** For each of the following statements, indicate whether it is true or false.
a. A spontaneous reaction always releases heat.
b. A spontaneous reaction is always a fast reaction.
c. The entropy of a system always increases for a spontaneous change.
d. The entropy of a system and its surroundings always increases for a spontaneous change.
e. The energy of a system always increases for a spontaneous change.

**19.18** Which of the following are spontaneous processes?
a. A cube of sugar dissolves in a cup of hot tea.
b. A rusty crowbar turns shiny.
c. Butane from a lighter burns in air.
d. A clock pendulum, initially stopped, begins swinging.
e. Hydrogen and oxygen gases bubble out from a glass of pure water.

**19.19** For each of the following series of pairs, indicate which one of each pair has the greater quantity of entropy.
a. 1.0 mol of carbon dioxide gas at 20°C, 1 atm, or 2.0 mol of carbon dioxide gas at 20°C, 1 atm
b. 1.0 mol of butane liquid at 20°C, 10 atm, or 1.0 mol of butane gas at 20°C, 10 atm
c. 1.0 mol of solid carbon dioxide at −80°C, 1 atm, or 1.0 mol of solid carbon dioxide at −90°C, 1 atm
d. 25 g of solid bromine at −7°C, 1 atm, or 25 g of bromine vapor at −7°C, 1 atm

**19.20** Predict the sign of the entropy change for each of the following processes.
a. A drop of food coloring diffuses throughout a glass of water.
b. A tree leafs out in the spring.
c. Flowers wilt and stems decompose in the fall.
d. A lake freezes over in the winter.
e. Rainwater on the pavement evaporates.

**19.21** Here is a simple experiment. Take a rubber band and stretch it. (Is this a spontaneous process? How does the Gibbs free-energy change?) Place the rubber band against your lips; note how warm the rubber band has become. (How does the enthalpy change?) According to polymer chemists, the rubber band consists of long, coiled molecules. On stretching the rubber band, these long molecules uncoil and align themselves in a more ordered state. Show how the experiment given here is in accord with this molecular view of the rubber band.

**19.22** Describe how you would expect the spontaneity ($\Delta G°$) for each of the following reactions to behave with a change of temperature.
a. Phosgene, $COCl_2$, the starting material for the preparation of polyurethane plastics, decomposes as follows:

$$COCl_2(g) \longrightarrow CO(g) + Cl_2(g)$$

b. Chlorine adds to ethylene to produce dichloroethane, a solvent:

$$Cl_2(g) + C_2H_4(g) \longrightarrow C_2H_4Cl_2(l)$$

## Practice Problems

### First Law of Thermodynamics

**19.23** A gas is cooled and loses 82 J of heat. The gas contracts as it cools, and work done on the system equal to 29 J is exchanged with the surroundings. What are $q$, $w$, and $\Delta U$?

**19.24** An ideal gas is cooled isothermally (at constant temperature). The internal energy of an ideal gas remains constant during an isothermal change. If $q$ is −76 J, what are $\Delta U$ and $w$?

**19.25** What is $\Delta U$ when 1.00 mol of liquid water vaporizes at 100°C? The heat of vaporization, $\Delta H°_{vap}$, of water at 100°C is 40.66 kJ/mol.

**19.26** What is $\Delta U$ for the following reaction at 25°C?

$$N_2(g) + 3H_2(g) \longrightarrow 2NH_3(g); \Delta H° = -91.8 \text{ kJ}$$

## Entropy Changes

**19.27**  Chloroform, $CHCl_3$, is a solvent and has been used as an anesthetic. The heat of vaporization of chloroform at its boiling point (61.2°C) is 29.6 kJ/mol. What is the entropy change when 1.20 mol $CHCl_3$ vaporizes at its boiling point?

**19.28**  Diethyl ether (known simply as ether), $(C_2H_5)_2O$, is a solvent and anesthetic. The heat of vaporization of diethyl ether at its boiling point (35.6°C) is 26.7 kJ/mol. What is the entropy change when 1.34 mol $(C_2H_5)_2O$ vaporizes at its boiling point?

**19.29**  The enthalpy change when liquid methanol, $CH_3OH$, vaporizes at 25°C is 37.4 kJ/mol. What is the entropy change when 1.50 mol of vapor in equilibrium with liquid condenses to liquid at 25°C? The entropy of the vapor at 25°C is 252 J/(mol · K). What is the entropy of the liquid at this temperature?

**19.30**  The heat of vaporization of carbon disulfide, $CS_2$, at 25°C is 29 kJ/mol. What is the entropy change when 1.85 mol of vapor in equilibrium with liquid condenses to liquid at 25°C? The entropy of the vapor at 25°C is 248 J/(mol · K). What is the entropy of the liquid at this temperature?

**19.31**  Predict the sign of $\Delta S°$, if possible, for each of the following reactions. If you cannot predict the sign for any reaction, state why.
a. $C_2H_2(g) + 2H_2(g) \longrightarrow C_2H_6(g)$
b. $N_2(g) + O_2(g) \longrightarrow 2NO(g)$
c. $2C_2H_2(g) + 3O_2(g) \longrightarrow 4CO(g) + 2H_2O(g)$
d. $2C(s) + O_2(g) \longrightarrow 2CO(g)$

**19.32**  Predict the sign of $\Delta S°$, if possible, for each of the following reactions. If you cannot predict the sign for any reaction, state why.
a. $2SO_3(g) \longrightarrow 2SO_2(g) + O_2(g)$
b. $C_2H_5OH(l) + 3O_2(g) \longrightarrow 2CO_2(g) + 3H_2O(l)$
c. $P_4(g) \longrightarrow P_4(s)$
d. $2NaHCO_3(s) \longrightarrow Na_2CO_3(s) + H_2O(g) + CO_2(g)$

**19.33**  Calculate $\Delta S°$ for the following reactions, using standard entropy values.
a. $2Na(s) + Cl_2(g) \longrightarrow 2NaCl(s)$
b. $NaCl(s) \longrightarrow NaCl(aq)$
c. $CS_2(l) + 3O_2(g) \longrightarrow CO_2(g) + 2SO_2(g)$
d. $2CH_3OH(l) + 3O_2(g) \longrightarrow 2CO_2(g) + 4H_2O(g)$

**19.34**  Calculate $\Delta S°$ for the following reactions, using standard entropy values.
a. $2Ca(s) + O_2(g) \longrightarrow 2CaO(s)$
b. $CaCl_2(s) \longrightarrow CaCl_2(aq)$
c. $CS_2(g) + 4H_2(g) \longrightarrow CH_4(g) + 2H_2S(g)$
d. $C_2H_4(g) + 3O_2(g) \longrightarrow 2CO_2(g) + 2H_2O(g)$

**19.35**  Calculate $\Delta S°$ for the reaction

$$CH_4(g) + 2O_2(g) \longrightarrow CO_2(g) + 2H_2O(l)$$

See Table 19.1 for values of standard entropies. Does the entropy of the chemical system increase or decrease as you expect? Explain.

**19.36**  What is the change in entropy, $\Delta S°$, for the reaction

$$CaCO_3(s) + 2H^+(aq) \longrightarrow Ca^{2+}(aq) + H_2O(l) + CO_2(g)$$

See Table 19.1 for values of standard entropies. Does the entropy of the chemical system increase or decrease as you expect? Explain.

## Free-Energy Change and Spontaneity

**19.37**  Using enthalpies of formation (Appendix C), calculate $\Delta H°$ for the following reaction at 25°C. Also calculate $\Delta S°$ for this reaction from standard entropies at 25°C. Use these values to calculate $\Delta G°$ for the reaction at this temperature.

$$2CH_3OH(l) + 3O_2(g) \longrightarrow 2CO_2(g) + 4H_2O(l)$$

**19.38**  Using enthalpies of formation (Appendix C), calculate $\Delta H°$ for the following reaction at 25°C. Also calculate $\Delta S°$ for this reaction from standard entropies at 25°C. Use these values to calculate $\Delta G°$ for the reaction at this temperature.

$$4HCN(l) + 5O_2(g) \longrightarrow 2H_2O(g) + 4CO_2(g) + 2N_2(g)$$

**19.39**  The free energy of formation of one mole of compound refers to a particular chemical equation. For each of the following, write that equation.
a. $KBr(s)$     b. $CH_3Cl(l)$     c. $H_2S(g)$     d. $AsH_3(g)$

**19.40**  The free energy of formation of one mole of compound refers to a particular chemical equation. For each of the following, write that equation.
a. $MgO(s)$     b. $COCl_2(g)$     c. $CF_4(g)$     d. $PCl_5(g)$

**19.41**  Calculate the standard free energy of the following reactions at 25°C, using standard free energies of formation.
a. $CH_4(g) + 2O_2(g) \longrightarrow CO_2(g) + 2H_2O(g)$
b. $CaCO_3(s) + 2H^+(aq) \longrightarrow Ca^{2+}(aq) + H_2O(l) + CO_2(g)$

**19.42**  Calculate the standard free energy of the following reactions at 25°C, using standard free energies of formation.
a. $C_2H_4(g) + 3O_2(g) \longrightarrow 2CO_2(g) + 2H_2O(g)$
b. $Na_2CO_3(s) + H^+(aq) \longrightarrow 2Na^+(aq) + HCO_3^-(aq)$

**19.43**  On the basis of $\Delta G°$ for each of the following reactions, decide whether the reaction is spontaneous or nonspontaneous as written. Or, if you expect an equilibrium mixture with significant amounts of both reactants and products, say so.
a. $SO_2(g) + 2H_2S(g) \longrightarrow 3S(s) + 2H_2O(g)$; $\Delta G° = -91$ kJ
b. $2H_2O_2(aq) \longrightarrow O_2(g) + 2H_2O(l)$; $\Delta G° = -211$ kJ
c. $HCOOH(l) \longrightarrow CO_2(g) + H_2(g)$; $\Delta G° = 119$ kJ
d. $I_2(s) + Br_2(l) \longrightarrow 2IBr(g)$; $\Delta G° = 7.5$ kJ
e. $NH_4Cl(s) \longrightarrow NH_3(g) + HCl(g)$; $\Delta G° = 92$ kJ

**19.44**  For each of the following reactions, state whether the reaction is spontaneous or nonspontaneous as written or is

easily reversible (that is, is a mixture with significant amounts of reactants and products).
a. $HCN(g) + 2H_2(g) \longrightarrow CH_3NH_2(g); \Delta G° = -92$ kJ
b. $N_2(g) + O_2(g) \longrightarrow 2NO(g); \Delta G° = 173$ kJ
c. $2NO(g) + 3H_2O(g) \longrightarrow 2NH_3(g) + \frac{5}{2}O_2(g);$
   $\Delta G° = 479$ kJ
d. $H_2(g) + Cl_2(g) \longrightarrow 2HCl(g); \Delta G° = -191$ kJ
e. $H_2(g) + I_2(s) \longrightarrow 2HI(g); \Delta G° = 2.6$ kJ

**19.45** Calculate $\Delta H°$ and $\Delta G°$ for the following reactions at 25°C, using thermodynamic data from Appendix C; interpret the signs of $\Delta H°$ and $\Delta G°$.
a. $Al_2O_3(s) + 2Fe(s) \longrightarrow Fe_2O_3(s) + 2Al(s)$
b. $COCl_2(g) + H_2O(l) \longrightarrow CO_2(g) + 2HCl(g)$

**19.46** Calculate $\Delta H°$ and $\Delta G°$ for the following reactions at 25°C, using thermodynamic data from Appendix C; interpret the signs of $\Delta H°$ and $\Delta G°$.
a. $2PbO(s) + N_2(g) \longrightarrow 2Pb(s) + 2NO(g)$
b. $CS_2(l) + 2H_2O(l) \longrightarrow CO_2(g) + 2H_2S(g)$

## Maximum Work

**19.47** Consider the reaction of 2 mol $H_2(g)$ at 25°C and 1 atm with 1 mol $O_2(g)$ at the same temperature and pressure to produce liquid water at these conditions. If this reaction is run in a controlled way to generate work, what is the maximum useful work that can be obtained? How much entropy is produced in this case?

**19.48** Consider the reaction of 1 mol $H_2(g)$ at 25°C and 1 atm with 1 mol $Cl_2(g)$ at the same temperature and pressure to produce gaseous HCl at these conditions. If this reaction is run in a controlled way to generate work, what is the maximum useful work that can be obtained? How much entropy is produced in this case?

**19.49** What is the maximum work that could be obtained from 4.85 g of zinc metal in the following reaction at 25°C?
$$Zn(s) + Cu^{2+}(aq) \longrightarrow Zn^{2+}(aq) + Cu(s)$$

**19.50** What is the maximum work that could be obtained from 3.65 g of zinc metal in the following reaction at 25°C?
$$Zn(s) + 2H^+(aq) \longrightarrow Zn^{2+}(aq) + H_2(g)$$

## Calculation of Equilibrium Constants

**19.51** Give the expression for the thermodynamic equilibrium constant for each of the following reactions.
a. $CO(g) + H_2O(g) \rightleftharpoons CO_2(g) + H_2(g)$
b. $Mg(OH)_2(s) \rightleftharpoons Mg^{2+}(aq) + 2OH^-(aq)$
c. $2Li(s) + 2H_2O(l) \rightleftharpoons 2Li^+(aq) + 2OH^-(aq) + H_2(g)$

**19.52** Write the expression for the thermodynamic equilibrium constant for each of the following reactions.
a. $CO(g) + 2H_2(g) \rightleftharpoons CH_3OH(g)$
b. $2Ag^+(aq) + CrO_4^{2-}(aq) \rightleftharpoons Ag_2CrO_4(s)$
c. $CaCO_3(s) + 2H^+(aq) \rightleftharpoons Ca^{2+}(aq) + H_2O(l) + CO_2(g)$

**19.53** What is the standard free-energy change $\Delta G°$ at 25°C for the following reaction? Obtain necessary information from Appendix C.
$$H_2(g) + Br_2(l) \longrightarrow 2HBr(g)$$
What is the value of the thermodynamic equilibrium constant $K$?

**19.54** What is the standard free-energy change $\Delta G°$ at 25°C for the following reaction? See Table 19.2 for data.
$$C(graphite) + 2Cl_2(g) \longrightarrow CCl_4(l)$$
Calculate the value of the thermodynamic equilibrium constant $K$.

**19.55** Calculate the standard free-energy change and the equilibrium constant $K_p$ for the following reaction at 25°C. See Table 19.2 for data.
$$CO(g) + 3H_2(g) \rightleftharpoons CH_4(g) + H_2O(g)$$

**19.56** Calculate the standard free-energy change and the equilibrium constant $K_p$ for the following reaction at 25°C. See Appendix C for data.
$$CO(g) + 2H_2(g) \rightleftharpoons CH_3OH(g)$$

**19.57** Obtain the equilibrium constant $K_c$ at 25°C from the free-energy change for the reaction
$$Fe(s) + Cu^{2+}(aq) \rightleftharpoons Fe^{2+}(aq) + Cu(s)$$
See Appendix C for data.

**19.58** Calculate the equilibrium constant $K_c$ at 25°C from the free-energy change for the following reaction:
$$Zn(s) + 2Ag^+(aq) \rightleftharpoons Zn^{2+}(aq) + Ag(s)$$
See Appendix C for data.

## Free Energy and Temperature Change

**19.59** Use data given in Tables 6.2 and 19.1 to obtain the value of $K_p$ at 1000°C for the reaction
$$C(graphite) + CO_2(g) \rightleftharpoons 2CO(g)$$
Carbon monoxide is known to form during combustion of carbon at high temperatures. Do the data agree with this? Explain.

**19.60** Use data given in Tables 6.2 and 19.1 to obtain the value of $K_p$ at 2000°C for the reaction
$$N_2(g) + O_2(g) \rightleftharpoons 2NO(g)$$
Nitric oxide is known to form in hot flames in air, which is a mixture of $N_2$ and $O_2$. It is present in auto exhaust from this reaction. Are the data in agreement with this result? Explain.

**19.61** Sodium carbonate, $Na_2CO_3$, can be prepared by heating sodium hydrogen carbonate, $NaHCO_3$.

$$2NaHCO_3(s) \longrightarrow Na_2CO_3(s) + H_2O(g) + CO_2(g)$$

Estimate the temperature at which $NaHCO_3$ decomposes to products at 1 atm. See Appendix C for data.

**19.62** Oxygen was first prepared by heating mercury(II) oxide, HgO.

$$2HgO(s) \longrightarrow 2Hg(g) + O_2(g)$$

Estimate the temperature at which HgO decomposes to $O_2$ at 1 atm. See Appendix C for data.

## General Problems

**19.63** Find the sign of $\Delta S°$ for the reaction

$$2N_2O_5(s) \longrightarrow 4NO_2(g) + O_2(g)$$

The reaction is endothermic and spontaneous at 25°C. Explain the spontaneity of the reaction in terms of enthalpy and entropy changes.

**19.64** The combustion of acetylene, $C_2H_2$, is a spontaneous reaction given by the equation

$$2C_2H_2(g) + 5O_2(g) \longrightarrow 4CO_2(g) + 2H_2O(l)$$

As expected for a combustion, the reaction is exothermic. What is the sign of $\Delta H°$? What do you expect for the sign of $\Delta S°$? Explain the spontaneity of the reaction in terms of the enthalpy and entropy changes.

**19.65** Estimate the value of $\Delta H°$ for the following reaction from bond energies (Table 9.5).

$$H_2(g) + Cl_2(g) \longrightarrow 2HCl(g)$$

Is the reaction exothermic or endothermic? Note that the reaction involves the breaking of symmetrical molecules ($H_2$ and $Cl_2$) and the formation of a less symmetrical product (HCl). From this, would you expect $\Delta S°$ to be positive or negative? Comment on the spontaneity of the reaction in terms of the changes in enthalpy and entropy.

**19.66** Compare the energies of the bonds broken and those formed (see Table 9.5) for the reaction

$$HCN(g) + 2H_2(g) \longrightarrow CH_3NH_2(g)$$

From this, conclude whether the reaction is exothermic or endothermic. What is the sign of $\Delta S°$? Explain. The reaction is spontaneous at 25°C. Explain this in terms of the enthalpy and entropy changes.

**19.67** Acetic acid, $CH_3COOH$, freezes at 16.6°C. The heat of fusion, $\Delta H_{fus}$, is 69.0 J/g. What is the change of entropy, $\Delta S$, when 1 mol of liquid acetic acid freezes to the solid?

**19.68** Acetone, $CH_3COCH_3$, boils at 56°C. The heat of vaporization of acetone at this temperature is 29.1 kJ/mol. What is the entropy change when 1 mol of liquid acetone vaporizes at 56°C?

**19.69** Without doing any calculations, decide what the sign of $\Delta S°$ will be for each of the following reactions.
a. $2LiOH(aq) + CO_2(g) \longrightarrow Li_2CO_3(aq) + H_2O(l)$
b. $(NH_4)_2Cr_2O_7(s) \longrightarrow N_2(g) + 4H_2O(g) + Cr_2O_3(s)$
c. $2N_2O_5(g) \longrightarrow 4NO_2(g) + O_2(g)$
d. $O_2(g) + 2F_2(g) \longrightarrow 2OF_2(g)$

**19.70** For each of the following reactions, decide whether there is an increase or a decrease in entropy. Why do you think so? (No calculations are needed.)
a. $N_2(g) + 3H_2(g) \longrightarrow 2NH_3(g)$
b. $NH_4Cl(s) \longrightarrow NH_3(g) + HCl(g)$
c. $CO(g) + 2H_2(g) \longrightarrow CH_3OH(l)$
d. $Li_3N(s) + 3H_2O(l) \longrightarrow 3LiOH(aq) + NH_3(g)$

**19.71** The following equation shows how nitrogen dioxide reacts with water to produce nitric acid:

$$3NO_2(g) + H_2O(l) \longrightarrow 2HNO_3(l) + NO(g)$$

Predict the sign of $\Delta S°$ for this reaction.

**19.72** Ethanol burns in air or oxygen according to the equation

$$C_2H_5OH(l) + 3O_2(g) \longrightarrow 2CO_2(g) + 3H_2O(g)$$

Predict the sign of $\Delta S°$ for this reaction.

**19.73** Acetic acid in vinegar results from the bacterial oxidation of ethanol.

$$C_2H_5OH(l) + O_2(g) \longrightarrow CH_3COOH(l) + H_2O(l)$$

What is $\Delta S°$ for this reaction? Use standard entropy values. (See Appendix C for data.)

**19.74** Methanol is produced commercially from carbon monoxide and hydrogen.

$$CO(g) + 2H_2(g) \longrightarrow CH_3OH(l)$$

What is $\Delta S°$ for this reaction? Use standard entropy values.

**19.75** Is the following reaction spontaneous as written? Explain. Do whatever calculation is needed to answer the question.

$$SO_2(g) + H_2(g) \longrightarrow H_2S(g) + O_2(g)$$

**19.76** Is the following reaction spontaneous as written? Explain. Do whatever calculation is needed to answer the question.

$$CH_4(g) + N_2(g) \longrightarrow HCN(g) + NH_3(g)$$

**19.77** The reaction

$$CO_2(g) + H_2(g) \longrightarrow CO(g) + H_2O(g)$$

is nonspontaneous at room temperature but becomes spontaneous at a much higher temperature. What can you conclude from this about the signs of $\Delta H^\circ$ and $\Delta S^\circ$, assuming that the enthalpy and entropy changes are not greatly affected by the temperature change? Explain your reasoning.

**19.78** The reaction

$$N_2(g) + 3H_2(g) \longrightarrow 2NH_3(g)$$

is spontaneous at room temperature but becomes nonspontaneous at a much higher temperature. From this fact alone, obtain the signs of $\Delta H^\circ$, and $\Delta S^\circ$, assuming that $\Delta H^\circ$ and $\Delta S^\circ$ do not change much with temperature. Explain your reasoning.

**19.79** Calculate $\Delta G^\circ$ at 25°C for the reaction

$$CaF_2(s) \rightleftharpoons Ca^{2+}(aq) + 2F^-(aq)$$

The value of $\Delta G_f^\circ$ at 25°C for $CaF_2(s)$ is $-1162$ kJ/mol. See Table 19.2 for other values. What is the value of the solubility product constant, $K_{sp}$, for this reaction at 25°C?

**19.80** Calculate $\Delta G^\circ$ at 25°C for the reaction

$$BaSO_4(s) \rightleftharpoons Ba^{2+}(aq) + SO_4^{2-}(aq)$$

The values of $\Delta G_f^\circ$ at 25°C are (in kJ/mol): $BaSO_4(s)$, $-1353$; $Ba^{2+}(aq)$, $-561$; $SO_4^{2-}(aq)$, $-742$. What is the value of the solubility product constant, $K_{sp}$, for this reaction at 25°C?

**19.81** Consider the decomposition of phosgene, $COCl_2$.

$$COCl_2(g) \longrightarrow CO(g) + Cl_2(g)$$

Calculate $\Delta H^\circ$, and $\Delta S^\circ$, at 25°C for this reaction. See Appendix C for data. What is $\Delta G^\circ$ at 25°C? Assume that $\Delta H^\circ$ and $\Delta S^\circ$ are constant with respect to a change of temperature. Now calculate $\Delta G^\circ$ at 800°C. Compare the two values of $\Delta G^\circ$. Briefly discuss the spontaneity of the reaction at 25°C and at 800°C.

**19.82** Consider the reaction

$$CS_2(g) + 4H_2(g) \rightleftharpoons CH_4(g) + 2H_2S(g)$$

Calculate $\Delta H^\circ$, $\Delta S^\circ$, and $\Delta G^\circ$ at 25°C for this reaction. Assume $\Delta H^\circ$ and $\Delta S^\circ$ are constant with respect to a change of temperature. Now calculate $\Delta G^\circ$ at 650°C. Compare the two values of $\Delta G^\circ$. Briefly discuss the spontaneity of the reaction at 25°C and at 650°C.

**19.83** a. From a consideration of the following reactions, calculate $\Delta H_f^\circ$ for methane, $CH_4(g)$.

$$CO_2(g) + 2H_2(g) \longrightarrow CH_2O(g) + H_2O(g); \Delta H^\circ(kJ) = 35.5$$
$$CO_2(g) \longrightarrow C(s) + O_2(g); \Delta H^\circ(kJ) = 393$$
$$CH_2O(g) + 2H_2(g) \longrightarrow CH_4(g) + H_2O(g); \Delta H^\circ(kJ) = -201$$
$$2H_2(g) + O_2(g) \longrightarrow 2H_2O(g); \Delta H^\circ(kJ) = -484$$

b. From appropriate entropy data, calculate $\Delta S^\circ$ for the formation of methane from the elements.

c. Using the information from above, calculate $\Delta G_f^\circ$ for methane at 25°C.

**19.84** a. From a consideration of the following reactions, calculate $\Delta H_f^\circ$ for ethane, $C_2H_6(g)$.

$$CH_3CHO(g) + 2H_2(g) \longrightarrow C_2H_6(g) + H_2O(l); \Delta H^\circ(kJ) = -205$$
$$2H_2(g) + O_2(g) \longrightarrow 2H_2O(g); \Delta H^\circ(kJ) = -484$$
$$2C_2H_5OH(l) + O_2(g) \longrightarrow 2CH_3CHO(g) + 2H_2O(l); \Delta H^\circ(kJ) = -348$$
$$H_2O(l) \longrightarrow H_2O(g); \Delta H^\circ(kJ) = 44$$
$$2C_2H_5OH(l) \longrightarrow 4C(s) + 6H_2(g) + O_2(g); \Delta H^\circ(kJ) = 556$$

b. From appropriate entropy data, calculate $\Delta S^\circ$ for the formation of ethane from the elements.

c. Using the information from above, calculate $\Delta G_f^\circ$ for ethane at 25°C.

**19.85** For the reaction

$$CH_3OH(l) + \tfrac{3}{2}O_2(g) \longrightarrow 2H_2O(l) + CO_2(g)$$

the value of $\Delta G^\circ$ is $-702.6$ kJ at 25°C. Other data are as follows:

| | $\Delta H_f^\circ$ (kJ/mol) at 25°C | $S^\circ$ (J/mol · K) at 25°C |
|---|---|---|
| $CH_3OH(l)$ | $-238.6$ | 127 |
| $H_2O(l)$ | $-285.8$ | 70 |
| $CO_2(g)$ | $-393.5$ | 214 |

Calculate the absolute entropy, $S^\circ$, per mole of $O_2(g)$.

**19.86** For the reaction

$$CH_2O(g) + \tfrac{2}{3}O_3(g) \longrightarrow CO_2(g) + H_2O(g)$$

the value of $\Delta G^\circ$ is $-621.7$ kJ/mol at 25°C. Other data are as follows:

| | $\Delta H_f^\circ$ (kJ/mol) at 25°C | $S^\circ$ (J/mol · K) at 25°C |
|---|---|---|
| $CH_2O(g)$ | $-116.0$ | 219 |
| $H_2O(g)$ | $-241.8$ | 189 |
| $CO_2(g)$ | $-393.5$ | 214 |
| $O_3(g)$ | 143 | ? |

Calculate the absolute entropy, $S^\circ$, per mole of $O_3(g)$.

**19.87** Tin(IV) oxide can be reacted with either hydrogen or carbon to form tin and water vapor or carbon dioxide, respectively. Data not available in the text are $\Delta H_f^\circ$ ($SnO_2(s)$) = −580.7 kJ/mol and $S^\circ$ ($SnO_2(s)$) = 52.3 J/mol · K.

a. Calculate $\Delta H^\circ$ and $\Delta S^\circ$ at 25°C for the reaction of $SnO_2$ with $H_2$ and for the reaction of $SnO_2$ with C.

b. At what temperature will each of these processes become spontaneous?

c. Industrially, which process is preferred? Why?

**19.88** Tungsten is usually produced by the reduction of $WO_3$ with hydrogen

$$WO_3(s) + 3H_2(g) \longrightarrow W(s) + 3H_2O(g)$$

Consider the following data:

|  | **$WO_3(s)$** | **$H_2O(g)$** |
|---|---|---|
| $\Delta H_f^\circ$ (kJ/mol) | −839.9 | −241.8 |
| $\Delta G_f^\circ$ (kJ/mol) | −763.1 | −228.6 |

a. Is $K > 1$ or $< 1$ at 25°C? Explain your answer.

b. What is the value of $\Delta S^\circ$ at 25°C?

c. What is the temperature at which $\Delta G^\circ$ equals zero for this reaction at 1 atm pressure?

d. What is the driving force of this reaction?

**19.89** For the decomposition of formic acid,

$$HCOOH(l) \longrightarrow H_2O(l) + CO(g)$$

$\Delta H^\circ$ = +13 kJ/mol at 25°C.

a. Does the tendency of this reaction to proceed to a state of minimum energy favor the formation of water and carbon monoxide or formic acid? Explain.

b. Does the tendency of this reaction to proceed to a state of maximum entropy favor the formation of products or reactants? Explain.

**19.90** a. Why are some reactions exothermic and others endothermic?

b. Discuss the driving force in a spontaneous reaction that is highly exothermic and in one that is endothermic.

**19.91** For the reaction

$$2Cu(s) + S(s) \longrightarrow Cu_2S(s)$$

$\Delta H^\circ$ and $\Delta G^\circ$ are negative and $\Delta S^\circ$ is positive.

a. At equilibrium, will reactants or products predominate? Why?

b. Why must the reaction system be heated in order to produce copper(I) sulfide?

**19.92** For the reaction

$$C_6H_6(l) + Br_2(l) \longrightarrow C_6H_5Br(l) + HBr(g)$$

the products have a lower energy than the reactants.

a. Predict the sign of the entropy change for this reaction. Explain.

b. Predict whether reactants or products would predominate in an equilibrium mixture. Why?

c. How can you account for the fact that a mixture of the reactants must be heated in the presence of a catalyst in order to produce products?

**19.93** When 1.000 g of gaseous butane, $C_4H_{10}$, is burned at 25°C and 1.00 atmosphere, $H_2O(l)$ and $CO_2(g)$ are formed with the evolution of 49.57 kJ of heat.

a. Calculate the molar enthalpy of formation of butane. (It will be necessary to use data from Appendix C.)

b. $\Delta G_f^\circ$ of butane is −15.7 kJ/mol. What is $\Delta G^\circ$ for the combustion of 1 mol butane?

c. What is $\Delta S^\circ$ for the combustion of 1 mol butane?

**19.94** When 1.000 g of ethylene glycol, $C_2H_6O_2$, is burned at 25°C and 1.00 atmosphere pressure, $H_2O(l)$ and $CO_2(g)$ are formed with the evolution of 19.18 kJ of heat.

a. Calculate the molar enthalpy of formation of ethylene glycol. (It will be necessary to use data from Appendix C.)

b. $\Delta G_f^\circ$ of ethylene glycol is −322.5 kJ/mol. What is $\Delta G^\circ$ for the combustion of 1 mol ethylene glycol?

c. What is $\Delta S^\circ$ for the combustion of 1 mol ethylene glycol?

**19.95** a. Calculate $K_1$ at 25°C for phosphoric acid:

$$H_3PO_4(aq) \rightleftharpoons H^+(aq) + H_2PO_4{}^-(aq)$$

|  | **$H_3PO_4(aq)$** | **$H^+(aq)$** | **$H_2PO_4{}^-(aq)$** |
|---|---|---|---|
| $\Delta H_f^\circ$ (kJ/mol) | −1289 | 0 | −1302 |
| $S^\circ$ (J/mol · K) | 176 | 0 | 89 |

b. Which thermodynamic factor is the most significant in accounting for the fact that phosphoric acid is a weak acid? Why?

**19.96** a. Calculate $K_1$ at 25°C for sulfurous acid:

$$H_2SO_3(aq) \rightleftharpoons H^+(aq) + HSO_3{}^-(aq)$$

|  | **$H_2SO_3(aq)$** | **$H^+(aq)$** | **$HSO_3{}^-(aq)$** |
|---|---|---|---|
| $\Delta H_f^\circ$ (kJ/mol) | −608 | 0 | −635 |
| $S^\circ$ (J/mol · K) | 234 | 0 | 109 |

b. Which thermodynamic factor is the most significant in accounting for the fact that sulfurous acid is a weak acid? Why?

## Cumulative-Skills Problems

**19.97** Hydrogen bromide dissociates into its gaseous elements, $H_2$ and $Br_2$, at elevated temperatures. Calculate the percent dissociation at 375°C and 1.00 atm. What would be the percent dissociation at 375°C and 10.0 atm? Use data from Appendix C and any reasonable approximation to obtain $K$.

**19.98** Hydrogen gas and iodine gas react to form hydrogen iodide. If 0.500 mol $H_2$ and 1.00 mol $I_2$ are placed in a closed 10.0-L vessel, what is the mole fraction of HI in the mixture when equilibrium is reached at 205°C? Use data from Appendix C and any reasonable approximations to obtain $K$.

**19.99** A 20.0-L vessel is filled with 1.00 mol of ammonia, $NH_3$. What percent of ammonia dissociates to the elements if equilibrium is reached at 345°C? Use data from Appendix C and make any reasonable approximation to obtain $K$.

**19.100** A 25.0-L vessel is filled with 0.0100 mol CO and 0.0300 mol $H_2$. How many moles of $CH_4$ and how many moles of $H_2O$ are produced when equilibrium is reached at 785°C? Use data from Appendix C and make any reasonable approximation to obtain $K$.

**19.101** $K_a$ for acetic acid at 25.0°C is $1.754 \times 10^{-5}$. At 50.0°C, $K_a$ is $1.633 \times 10^{-5}$. What are $\Delta H°$ and $\Delta S°$ for the ionization of acetic acid?

**19.102** $K_{sp}$ for silver chloride at 25.0°C is $1.782 \times 10^{-10}$. At 35.0°C, $K_{sp}$ is $4.159 \times 10^{-10}$. What are $\Delta H°$ and $\Delta S°$ for the reaction?

## Media Activities

**19.103** Watch the Work vs. Energy Flow *Visualization* on the student web site. Make sure that you can answer the questions posed in the *Visualization*.
a. If you were to run the reaction depicted in the *Visualization* in an insulated container with a fixed volume, would you expect an increase or decrease in the internal energy?
b. If you ran the reaction depicted in the *Visualization* where no heat was allowed to flow from the container, how would this affect the amount of work?

**19.104** Watch the Spontaneous Reactions *Visualization* on the student web site. The reaction that you observed at the end of the *Visualization* is $2Na(s) + 2H_2O(l) \longrightarrow 2NaOH(aq) + H_2(g)$.
a. How did the internal energy of the reaction system change in going from products to reactants?
b. Predict the signs of $\Delta S$, $\Delta H$, and $\Delta G$ for the reaction.

**19.105** Work through the $\Delta G_f$ *Problem Solving* activity on the student web site.
a. If $H_2O(l)$ were produced in the reaction instead of $H_2O(g)$, how would this affect $\Delta G°$? Would the reaction still be spontaneous?
b. Calculate $\Delta G°$ for the reaction with $H_2O(l)$ and $CO_2(g)$ as products. Do the results of this calculation support your answer for part a?

**19.106** For review, take the ACE practice test for Chapter 19 on the student web site.

Electrolysis of a potassium iodide solution producing iodine and hydrogen gas.

# Electrochemistry

T he first battery was invented by Alessandro Volta about 1800. He assembled a pile consisting of pairs of zinc and silver disks separated by paper disks soaked in salt water. With a tall pile, he could detect a weak electric shock when he touched the two ends of the pile. Later Volta showed that any two different metals could be used to make such a voltaic pile (Figure 20.1).

A battery cell that became popular during the nineteenth century was constructed in 1836 by the English chemist John Frederick Daniell. This cell used zinc and copper. The basic principle was that of Volta's battery pile, but each metal was surrounded by a solution of the metal ion, and the solutions were kept separate by a porous ceramic barrier. Each metal with its solution was a half-cell; a zinc half-cell and a copper half-cell made up one voltaic cell. This construction became the standard form of such cells, which exploit the spontaneous chemical reaction

$$Zn(s) + Cu^{2+}(aq) \longrightarrow Zn^{2+}(aq) + Cu(s)$$

to generate electric energy.

**839**

> The student should be familiar with half-reactions, introduced in Section 4.6.

Today many different kinds of batteries are in use. They include miniature button batteries for pocket calculators and watches, as well as larger batteries to power electric cars or to store the energy from the solar collectors of communications satellites. But all such batteries operate on the same general principles of Volta's pile or Daniell's cell. In this chapter, we will look at the general principles involved in setting up a chemical reaction as a battery. We will answer such questions as the following: What voltage can you expect from a particular battery? How can you relate the battery voltage to the equilibrium constant for the reaction?

# Half-Reactions

> You may wish to review Sections 4.5 and 4.6.

A voltaic cell employs a spontaneous oxidation–reduction reaction as a source of energy. It is constructed to separate the reaction physically into two half-reactions, one involving oxidation and the other reduction, with electrons moving from the oxidation half-reaction through the external circuit to the reduction half-reaction. ◄

## 20.1 Balancing Oxidation–Reduction Reactions in Acidic and Basic Solutions

> Experiments in General Chemistry, 20A, "Oxidation–Reduction Reactions."

In Chapter 4 (Section 4.6) we introduced the *half-reaction method* of balancing simple oxidation–reduction reactions. We now extend this method to reactions that occur in acidic or basic solution. The steps used to balance these equations successfully are the same as those presented in Chapter 4, so much of what is presented here will probably be familiar. Keep in mind that oxidation–reduction reactions involve a transfer of electrons from one species to another. For example, in the reaction described in the chapter opener, zinc metal becomes zinc(II) ion: each zinc atom loses two electrons, and copper(II) ion becomes copper metal (each copper ion gains two electrons).

In this chapter, we will represent the hydronium ion, $H_3O^+(aq)$, by its simpler notation $H^+(aq)$, which chemists call the hydrogen ion. In the chapters that focused on acid–base reactions, we stressed proton transfer in solution, so we adopted the hydronium-ion notation. Here, we shift our focus to the electron-transfer aspect of certain reactions, so we adopt the simpler hydrogen-ion notation. Remember, however, that this is simply a notation change. It does not change the chemistry.

## Skeleton Oxidation–Reduction Equations

▶ Video no. 25 (Series D), "Permanganate/Paper Towel Reaction," and no. 26 (Series D), "The Chemical Traffic Light."

To tackle more complex oxidation–reduction reactions in acidic and basic solutions, we need to review and discuss the essential information required to describe an oxidation–reduction reaction, which is called a *skeleton equation*. To set up the skeleton equation and then balance it, you need answers to the following questions:

1. What species is being oxidized (or, what is the reducing agent)? What species is being reduced (or, what is the oxidizing agent)?

2. What species result from the oxidation and reduction?

3. Does the reaction occur in acidic or basic solution?

For example, iron(II) ion may be oxidized easily by a number of oxidizing agents to yield the iron(III) ion. Permanganate ion in acidic, aqueous solution is a strong oxidizing agent and can oxidize $Fe^{2+}$ to $Fe^{3+}$. In the process of oxidizing the iron(II) ion, the permanganate ion ($MnO_4^-$) is reduced to manganese(II) ion ($Mn^{2+}$). We can express these facts in the following skeleton equation:

$$\overset{+2}{Fe^{2+}}(aq) + \overset{+7}{MnO_4^-}(aq) \longrightarrow \overset{+3}{Fe^{3+}}(aq) + \overset{+2}{Mn^{2+}}(aq) \qquad \text{(acidic solution)}$$

We have written oxidation numbers over the appropriate atoms. Note that the skeleton equation is not balanced. Nor is it complete; in an acidic, aqueous solution $H^+(aq)$ and $H_2O(l)$ may be possible reactants or products in the equation. (If this were a basic solution, $OH^-$ ion would replace the $H^+$ ion.)

## Half-Reaction Method in Acidic and Basic Solutions

Although the preceding skeleton equation is not complete, it does give the essential information about the oxidation–reduction reaction. Moreover, given the skeleton equation, you can complete and balance the equation using the half-reaction method. Let us see how to do that. We first look at balancing oxidation–reduction equations in acidic solution. To balance such equations in basic solution requires additional steps.

## Steps in Balancing Oxidation–Reduction Equations in Acidic Solution

▶ Demonstration of oxidation by dichromate ion in acid solution: Summerlin et al., Chemical Demonstrations, Vol. 2 (Washington, D.C.: Am. Chem. Soc., 1987), no. 30, pp. 56–57.

1. Assign oxidation numbers to each atom so you know what is oxidized and what is reduced.

2. Split the skeleton equation into two half-reactions, proceeding as follows. Note the species containing the element that increases in oxidation number and write those species to give the oxidation half-reaction. Similarly, note the species containing the element that decreases in oxidation number and write the reduction half-reaction.

3. Complete and balance each half-reaction.

   a. Balance all atoms except O and H.

   b. Balance O atoms by adding $H_2O$'s to one side of the equation.

   c. Balance H atoms by adding $H^+$ ions to one side of the equation.

   d. Balance electric charge by adding electrons ($e^-$) to the more positive side.

4. Combine the two half-reactions to obtain the final balanced oxidation–reduction equation.

    a. Multiply each half-reaction by a factor such that when the half-reactions are added, the electrons cancel. (Electrons cannot appear in the final equation.)

    b. Simplify the balanced equation by canceling species that occur on both sides and reduce the coefficients to smallest whole numbers. Check that the equation is indeed balanced.

Let us see how we would apply this method to the skeleton equation we wrote for the oxidation of iron(II) ion by permanganate ion in acidic solution. We have already determined the elements that change oxidation number (Step 1). We now write the incomplete equation representing the oxidation half-reaction. Iron(II) is oxidized to iron(III) ion. So, following Step 2, we write

$$\overset{+2}{Fe^{2+}}(aq) \longrightarrow \overset{+3}{Fe^{3+}}(aq)$$

Note that the equation is balanced in Fe as it stands (Step 3a). Steps 3b and 3c require that the half-reaction be balanced in O and H. However, since these atoms do not appear in the half-reaction, we can ignore those steps. Then (Step 3d) you need to balance the half-reaction in electric charge. You do that by adding one electron to the more positive side.

$$Fe^{2+}(aq) \longrightarrow Fe^{3+}(aq) + e^- \qquad \text{(oxidation half-reaction)}$$

Check that the half-reaction is balanced in Fe and in electric charge. Note that electrons are lost, as you expect for oxidation.

Now let us look at the reduction half-reaction. From the skeleton equation, you write the species containing the atom for which the oxidation number decreases (Step 2):

$$\overset{+7}{MnO_4^-}(aq) \longrightarrow \overset{+2}{Mn^{2+}}(aq)$$

The equation is balanced in Mn as it stands (Step 3a). For Step 3b, you add four $H_2O$'s to the right side of the equation to balance the O atoms. (Water is present, since this is an aqueous solution.)

$$MnO_4^-(aq) \longrightarrow Mn^{2+}(aq) + 4H_2O(l)$$

Step 3c says to balance the H atoms by adding $H^+$ ions. (Hydrogen ions are present, since this is an acidic solution.) You add eight $H^+$ ions to the left side of the equation.

$$MnO_4^-(aq) + 8H^+(aq) \longrightarrow Mn^{2+}(aq) + 4H_2O(l)$$

Finally, you balance the half-reaction in electric charge by adding electrons to the more positive side of the equation (Step 3d). Note that the left side has a charge of $(-1 + 8) = +7$, whereas the right side has a charge of $+2$. So you add five electrons to the left side:

$$MnO_4^-(aq) + 8H^+(aq) + 5e^- \longrightarrow Mn^{2+}(aq) + 4H_2O(l) \qquad \text{(reduction half-reaction)}$$

Check that this half-reaction is balanced in the different kinds of atoms and in electric charge. Note that electrons are gained, as you expect for reduction.

In Step 4a, you need to multiply each half-reaction by a factor such that when the two half-reactions are added, the electrons cancel. You multiply the oxidation half-reaction by 5 (the number of electrons in the reduction half-reaction) and multiply the reduction half-reaction by 1 (the number of electrons in the oxidation half-reaction). Then you add the two half-reactions:

$$5 \times (Fe^{2+} \longrightarrow Fe^{3+} + e^-)$$
$$1 \times (MnO_4^- + 8H^+ + 5e^- \longrightarrow Mn^{2+} + 4H_2O)$$
$$\overline{5Fe^{2+} + MnO_4^- + 8H^+ + 5e^- \longrightarrow 5Fe^{3+} + Mn^{2+} + 4H_2O + 5e^-}$$

After canceling electrons, the equation is in simplest form (Step 4b). The final balanced equation is

$$5Fe^{2+}(aq) + MnO_4^-(aq) + 8H^+(aq) \longrightarrow 5Fe^{3+}(aq) + Mn^{2+}(aq) + 4H_2O(l)$$

You can write this equation in terms of the hydronium-ion notation, $H_3O^+(aq)$, if you wish, by changing $8H^+(aq)$ to $8H_3O^+(aq)$ and adding $8H_2O(l)$ to the other side of the equation to give a total of $12H_2O(l)$.

The following example further illustrates the half-reaction method.

▶ An alternate example is provided in the Instructor's Resource Manual.

## EXAMPLE 20.1

# Balancing Equations by the Half-Reaction Method (Acidic Solution)

Zinc metal reacts with nitric acid, $HNO_3$, to produce a number of products depending on how dilute the acid solution is. In a concentrated solution, zinc reduces nitrate ion to ammonium ion; zinc is oxidized to zinc ion, $Zn^{2+}$. Write the net ionic equation for this reaction.

 **PROBLEM STRATEGY**

From the word statement of the problem, you write the skeleton equation representing the essential features of the oxidation–reduction reaction. Then, assign oxidation numbers to the atoms (Step 1) and separate the skeleton equation into two incomplete half-reactions (Step 2). Now complete and balance each half-reaction (Step 3). (Balance the half-reaction in all elements except O and H, then balance O atoms, then H atoms, and finally electric charge.) In Step 4, you combine the half-reactions to eliminate electrons and simplify the result.

### SOLUTION

Note that nitric acid is a strong acid, so it exists in solution as $H^+$ and $NO_3^-$ ions. For the skeleton equation, write just the $NO_3^-$ ion, because you will be adding $H^+$ later as needed to balance the equation. The skeleton equation, with oxidation numbers for those atoms that change values (following Step 1), is

$$\overset{0}{Zn}(s) + \overset{+5}{NO_3^-}(aq) \longrightarrow \overset{+2}{Zn^{2+}}(aq) + \overset{-3}{NH_4^+}(aq)$$

Now you separate this equation into two incomplete half-reactions (Step 2). Note that zinc is oxidized (increases in oxidation number), and nitrogen is reduced (decreases in oxidation number). The two incomplete half-reactions are

$$Zn(s) \longrightarrow Zn^{2+}(aq)$$
$$NO_3^-(aq) \longrightarrow NH_4^+(aq)$$

The oxidation half-reaction is balanced in Zn, as it stands (Step 3a). There are no O or H atoms, so you can ignore Steps 3b and 3c. Following Step 3d, you add electrons to the more positive side to balance the charge.

$$Zn(s) \longrightarrow Zn^{2+}(aq) + 2e^- \qquad \text{(oxidation half-reaction)}$$

The reduction half-reaction is balanced in N (Step 3a), but is not complete in either O or H. Therefore, you add three $H_2O$'s to the right side to balance O atoms (Step 3b), and add ten $H^+$ ions to the left side to balance the four H atoms in the $NH_4^+$ ion and the six H atoms in the $H_2O$'s you just added (Step 3c).

$$10H^+(aq) + NO_3^-(aq) \longrightarrow NH_4^+(aq) + 3H_2O(l)$$

Now, you add electrons to the more positive side to balance the charge.

$$10H^+(aq) + NO_3^-(aq) + 8e^- \longrightarrow NH_4^+(aq) + 3H_2O(l)$$
$$\text{(reduction half-reaction)}$$

Check that the half-reaction is balanced in charge, as well as in the different kinds of atoms.

Finally (Step 4a), you multiply the two half-reactions by factors so that when added, the electrons cancel. You multiply the oxidation half-reaction by 4 and the reduction half-reaction by 1:

$$4 \times (Zn \longrightarrow Zn^{2+} + 2e^-)$$
$$\underline{10H^+ + NO_3^- + 8e^- \longrightarrow NH_4^+ + 3H_2O}$$
$$4Zn + 10H^+ + NO_3^- + \cancel{8e^-} \longrightarrow 4Zn^{2+} + NH_4^+ + 3H_2O + \cancel{8e^-}$$

The equation requires no further simplification beyond canceling electrons (Step 4b). The net ionic equation is

$$\mathbf{4Zn(s) + 10H^+(aq) + NO_3^-(aq) \longrightarrow 4Zn^{2+}(aq) + NH_4^+(aq) + 3H_2O(l)}$$

### EXERCISE 20.1

Iodic acid, $HIO_3$, can be prepared by reacting iodine, $I_2$, with concentrated nitric acid. The skeleton equation is

$$I_2(s) + NO_3^-(aq) \longrightarrow IO_3^-(aq) + NO_2(g)$$

Balance this equation.

**See Problems 20.27 and 20.28.**

To balance an oxidation–reduction equation in basic solution, you begin by balancing the equation *as if it were a reaction in acidic solution*. Then, you add the following steps. ◄

▶ You can apply Steps 5 and 6 to the individual half-reactions, rather than to the overall equation. Although this is more work, it does yield the complete half-reactions, which can be useful in discussing batteries (later in this chapter).

## Additional Steps for Balancing Oxidation–Reduction Equations in Basic Solution

5. Note the number of $H^+$ ions in the equation. Add this number of $OH^-$ ions to both sides of the equation.

6. Simplify the equation by noting that $H^+$ reacts with $OH^-$ to give $H_2O$. Cancel any $H_2O$'s that occur on both sides of the equation and reduce the equation to simplest terms.

The following example illustrates how to apply these rules.

▶ An alternate example is provided in the Instructor's Resource Manual.

### EXAMPLE 20.2

### Balancing Equations by the Half-Reaction Method (Basic Solution)

Permanganate ion oxidizes sulfite ion in basic solution according to the following skeleton equation:

$$MnO_4^-(aq) + SO_3^{2-}(aq) \longrightarrow MnO_2(s) + SO_4^{2-}(aq)$$

Use the half-reaction method to complete and balance this equation.

 **PROBLEM STRATEGY**

First complete and balance this equation as if it were in acid solution, as in the previous example. Then, following Step 5, add as many $OH^-$ ions to both sides of the equation as there are $H^+$ ions. After noting that each $H^+$ ion reacts with an $OH^-$ ion to give $H_2O$, simplify the equation (Step 6).

**SOLUTION**

After balancing the equation as if it were in acid solution, you obtain the following:

$$2MnO_4^- + 3SO_3^{2-} + 2H^+ \longrightarrow 2MnO_2 + 3SO_4^{2-} + H_2O$$

Following Step 5, you add $2OH^-$ to both sides of the equation.

$$2MnO_4^- + 3SO_3^{2-} + \underbrace{2H^+ + 2OH^-}_{2H_2O} \longrightarrow 2MnO_2 + 3SO_4^{2-} + H_2O + 2OH^-$$

You replace $2H^+ + 2OH^-$ by $2H_2O$ on the left side, then cancel one of these with the $H_2O$ on the right side. The balanced equation for the reaction in basic solution is

$$\mathbf{2MnO_4^-(aq) + 3SO_3^{2-}(aq) + H_2O(l) \longrightarrow 2MnO_2(s) + 3SO_4^{2-}(aq) + 2OH^-(aq)}$$

### EXERCISE 20.2

Balance the following equation using the half-reaction method.

$$H_2O_2 + ClO_2 \longrightarrow ClO_2^- + O_2 \qquad \text{(basic solution)}$$

■ **See Problems 20.29 and 20.30.**

# Voltaic Cells

The next several sections describe battery cells, or voltaic cells (also called galvanic cells). These are a kind of electrochemical cell. An **electrochemical cell** is *a system consisting of electrodes that dip into an electrolyte and in which a chemical reaction* *either uses or generates an electric current.* A **voltaic,** or **galvanic, cell** is *an electrochemical cell in which a spontaneous reaction generates an electric current.* An **electrolytic cell** is *an electrochemical cell in which an electric current drives an otherwise nonspontaneous reaction.* In the next sections, we will discuss the basic principles behind voltaic cells and then explore some of their commercial uses.

## 20.2 Construction of Voltaic Cells

A voltaic cell consists of two *half-cells* that are electrically connected. Each **half-cell** is *the portion of an electrochemical cell in which a half-reaction takes place.* A simple half-cell can be made from a metal strip that dips into a solution of its metal ion. An example is the zinc–zinc ion half-cell (often called simply a zinc electrode), which consists of a zinc metal strip dipping into a solution of a zinc salt. Another simple half-cell consists of a copper metal strip dipping into a solution of a copper salt (copper electrode).

In a voltaic cell, two half-cells are connected in such a way that electrons flow from one metal electrode to another through an external circuit, and ions flow from one half-cell to another through an internal cell connection. Figure 20.2 illustrates an atomic view of a voltaic cell consisting of a zinc electrode and a copper electrode. As long as there is an external circuit, electrons can flow through it from one electrode to another. Because zinc tends to lose electrons more readily than copper, zinc atoms in the zinc electrode lose electrons to produce zinc ions. These electrons flow through the external circuit to the copper electrode, where they react with the copper ions in that half-cell to deposit copper metal atoms. The net result is that zinc metal reacts with copper ions to produce zinc ions and copper metal, and an electric current flows through the external circuit.

The two half-cells must be connected internally to allow ions to flow between them. As zinc ions continue to be produced, the zinc ion solution begins to build up a positive charge. Similarly, as copper ions plate out as copper, the solution builds up a negative charge. The half-cell reactions will stop unless positive ions can move from the zinc half-cell to the copper half-cell, and negative ions from the copper half-cell can move to the zinc half-cell. It is necessary that these ion flows occur without mixing of the zinc ion and copper ion solutions. If copper ion were to come in contact

**Figure 20.2**
**Atomic view of a voltaic cell.** In the half-cell on the left, a zinc metal atom loses two electrons. These flow through the zinc electrode to the external circuit, then to the copper electrode in the half-cell on the right. Negative charge on the copper electrode (from the electrons) attracts a copper ion, which reacts with the electrons on the electrode to form a copper metal atom. The internal connection, discussed later in the text, allows ions to flow between the two half-cells; it is required to maintain charge balance.

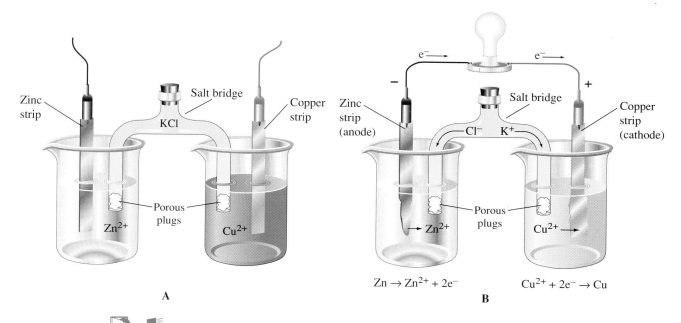

Zinc strip — Salt bridge KCl — Copper strip

Zn$^{2+}$   Cu$^{2+}$   Porous plugs

**A**

e$^-$ →   e$^-$ →

−   Salt bridge   +

Zinc strip (anode) — Cl$^-$   K$^+$ — Copper strip (cathode)

Zn$^{2+}$   Porous plugs   Cu$^{2+}$ →

$$Zn \rightarrow Zn^{2+} + 2e^-$$   $$Cu^{2+} + 2e^- \rightarrow Cu$$

**B**

**Figure 20.3**
**A zinc–copper voltaic cell.**
(A) A zinc electrode and a copper electrode, without an external circuit; there is no cell reaction. (B) When the two electrodes are connected by an external circuit (with a lightbulb), chemical reaction occurs. (C) A similar cell, with the salt bridge replaced by a porous glass plate and the bulb replaced by a voltmeter. Note the blue color of $Cu^{2+}(aq)$.

▶ **Animation: Electrochemical half-reactions in a galvanic cell.**

▶ **The porous barrier in the original Daniell cell serves the same purpose as the salt bridge.**

with the zinc metal, for example, direct reaction would occur without an electric current being generated. The voltage would drop, and the battery would run down quickly.

Figure 20.3A shows the two half-cells of a voltaic cell connected by a salt bridge. A **salt bridge** is *a tube of an electrolyte in a gel that is connected to the two half-cells of a voltaic cell; the salt bridge allows the flow of ions but prevents the mixing of the different solutions that would allow direct reaction of the cell reactants.* In Figure 20.3B, the half-cells are connected externally so that an electric current flows. Figure 20.3C shows an actual setup of the zinc–copper cell, but in this case the salt bridge is replaced by a porous glass plate, which accomplishes a similar purpose. ◀

The two half-cell reactions, as noted earlier, are

$$Zn(s) \longrightarrow Zn^{2+}(aq) + 2e^- \qquad \text{(oxidation half-reaction)}$$
$$Cu^{2+}(aq) + 2e^- \longrightarrow Cu(s) \qquad \text{(reduction half-reaction)}$$

▶ Video no. 44 (Series D), "Clock with Orange Juice."

The first half-reaction, in which a species loses electrons, is the oxidation half-reaction. *The electrode at which oxidation occurs* is called the **anode.** The second half-reaction, in which a species gains electrons, is the reduction half-reaction. *The electrode at which reduction occurs* is called the **cathode.** These definitions of anode and cathode hold for all electrochemical cells, including electrolytic cells.

Note that the sum of the two half-reactions

$$Zn(s) + Cu^{2+}(aq) \longrightarrow Zn^{2+}(aq) + Cu(s)$$

is *the net reaction that occurs in the voltaic cell;* it is called the **cell reaction.**

▶ Animations: Anode reaction; Cathode reaction.

Once you know which electrode is the anode and which is the cathode, you can determine the direction of electron flow in the external portion of the circuit. Electrons are given up by the anode (from the oxidation half-reaction) and thus flow from it, whereas electrons are used up by the cathode (by the reduction half-reaction) and so flow into this electrode. The anode in a voltaic cell has a negative sign, because electrons flow from it. The cathode in a voltaic cell has a positive sign. Look again at Figure 20.3*B* and note the labeling of the electrodes as anode and cathode; also note the direction of electron flow in the external circuit and the signs of the electrodes. Note too the migration of ions in the solutions. The following example further illustrates these points about a voltaic cell.

▶ An alternate example is provided in the Instructor's Resource Manual.

**EXAMPLE 20.3**

## Sketching and Labeling a Voltaic Cell

A voltaic cell is constructed from a half-cell in which a cadmium rod dips into a solution of cadmium nitrate, $Cd(NO_3)_2$, and another half-cell in which a silver rod dips into a solution of silver nitrate, $AgNO_3$. The two half-cells are connected by a salt bridge. Silver ion is reduced during operation of the voltaic cell. Draw a sketch of the cell. Label the anode and cathode, showing the corresponding half-reactions at these electrodes. Indicate the electron flow in the external circuit, the signs of the electrodes, and the direction of cation migration in the half-cells.

**SOLUTION**

Because silver ion is reduced at the silver electrode, the silver electrode is the cathode. The half-reaction is

$$Ag^+(aq) + e^- \longrightarrow Ag(s)$$

The cadmium electrode must be the anode (electrode at which oxidation occurs); the half-reaction is

$$Cd(s) \longrightarrow Cd^{2+}(aq) + 2e^-$$

Now that you have labeled the electrodes, you can see that the electron flow in the external circuit is from the cadmium electrode (anode) to the silver electrode (cathode). Positive ions will flow in the solution portion of the circuit opposite to the direction of the electrons. **The sketch for this cell, including the labeling, is given in Figure 20.4.**

**Figure 20.4**
**Cell described in the example.** A voltaic cell consisting of cadmium and silver electrodes.

Cd (anode) − Salt bridge + Ag (cathode)

Cd²⁺ Ag⁺

$$Cd \rightarrow Cd^{2+} + 2e^-  \quad  Ag^+ + e^- \rightarrow Ag$$

**EXERCISE 20.3**

A voltaic cell consists of a silver–silver ion half-cell and a nickel–nickel(II) ion half-cell. Silver ion is reduced during operation of the cell. Sketch the cell, labeling the anode and cathode and indicating the corresponding electrode reactions. Show the direction of electron flow in the external circuit and the direction of cation movement in the half-cells.

See Problems 20.35 and 20.36.

 **CONCEPT CHECK 20.1**

If you were to construct a wet cell and decided to replace the salt bridge with a piece of copper wire, would the cell produce a sustainable current? Explain your answer.

## 20.3  Notation for Voltaic Cells

It is convenient to have a shorthand way of designating particular voltaic cells. The cell described earlier, consisting of a zinc metal–zinc ion half-cell and a copper metal–copper ion half-cell, is written

$$Zn(s)|Zn^{2+}(aq)\,\|\,Cu^{2+}(aq)|Cu(s)$$

In this notation, the anode, or oxidation half-cell, is always written on the left; the cathode, or reduction half-cell, is written on the right. The two electrodes are electrically connected by means of a salt bridge, denoted by two vertical bars.

$$\underset{\text{anode}}{Zn(s)|Zn^{2+}(aq)} \quad \underset{\text{salt bridge}}{\|} \quad \underset{\text{cathode}}{Cu^{2+}(aq)|Cu(s)}$$

— Electrode
connection

$H_2$                    $H_2$

Salt
bridge

— Platinum
plate

$H^+$

**Figure 20.5**
**A hydrogen electrode.**
Hydrogen gas bubbles over a
platinum surface, where the
half-reaction $2H^+(aq) + 2e^-$
$\rightleftharpoons H_2(g)$ occurs.

The cell terminals are at the extreme ends in this cell notation, and a single vertical bar indicates a phase boundary—say, between a solid terminal and the electrode solution. For the anode of the same cell, you have

$$Zn(s) \qquad | \qquad Zn^{2+}(aq)$$

anode terminal  phase boundary  anode electrolyte

When the half-reaction involves a gas, an inert material such as platinum serves as a terminal and as an electrode surface on which the half-reaction occurs. The platinum catalyzes the half-reaction but otherwise is not involved in it. Figure 20.5 shows a hydrogen electrode; hydrogen bubbles over a platinum plate that is immersed in an acidic solution. The cathode half-reaction is

$$2H^+(aq) + 2e^- \rightleftharpoons H_2(g)$$

The notation for the hydrogen electrode, written as a cathode, is

$$H^+(aq)|H_2(g)|Pt$$

To write such an electrode as an anode, you simply reverse the notation:

$$Pt|H_2(g)|H^+(aq)$$

Here are several additional examples of this notation for electrodes (written as cathodes). A comma separates ions present in the same solution. We will write the cathode with oxidized species before reduced species, in the same order as in the half-reaction.

| **Cathode** | **Cathode Reaction** |
|---|---|
| $Cl_2(g)|Cl^-(aq)|Pt$ | $Cl_2(aq) + 2e^- \rightleftharpoons 2Cl^-(aq)$ |
| $Fe^{3+}(aq), Fe^{2+}(aq)|Pt$ | $Fe^{3+}(aq) + e^- \rightleftharpoons Fe^{2+}(aq)$ |
| $Cd^{2+}(aq)|Cd(s)$ | $Cd^{2+}(aq) + 2e^- \rightleftharpoons Cd(s)$ |

**EXERCISE 20.4**

Write the notation for a cell in which the electrode reactions are

$$2H^+(aq) + 2e^- \longrightarrow H_2(g)$$
$$Zn(s) \longrightarrow Zn^{2+}(aq) + 2e^-$$

**See Problems 20.41 and 20.42.**

You can write the overall cell reaction from the cell notation by first writing the appropriate half-cell reactions, multiplying as necessary, and then summing so that the electrons cancel. Example 20.4 shows how this is done.

▶ An alternate example is
provided in the Instructor's
Resource Manual.

**EXAMPLE 20.4**

**Writing the Cell Reaction from the Cell Notation**

a. Write the cell reaction for the voltaic cell

$$Tl(s)|Tl^+(aq)\,\|\,Sn^{2+}(aq)|Sn(s)$$

b. Write the cell reaction for the voltaic cell

$$Zn(s)|Zn^{2+}(aq)\,\|\,Fe^{3+}(aq), Fe^{2+}(aq)|Pt$$

**PROBLEM STRATEGY**

The cell notation gives the species involved in each half-reaction. Complete and balance each half-reaction, then multiply by factors so that when you add the half-reactions, the electrons cancel. The result is the cell reaction.

**SOLUTION**

a. The half-cell reactions are

$$Tl(s) \longrightarrow Tl^+(aq) + e^-$$
$$Sn^{2+}(aq) + 2e^- \longrightarrow Sn(s)$$

Multiplying the anode reaction by 2 and then summing the half-cell reactions gives

$$\textbf{2Tl}(s) + \textbf{Sn}^{2+}(aq) \longrightarrow \textbf{2Tl}^+(aq) + \textbf{Sn}(s)$$

b. The half-cell reactions are

$$Zn(s) \longrightarrow Zn^{2+}(aq) + 2e^-$$
$$Fe^{3+}(aq) + e^- \longrightarrow Fe^{2+}(aq)$$

and the cell reaction is

$$\textbf{Zn}(s) + \textbf{2Fe}^{3+}(aq) \longrightarrow \textbf{Zn}^{2+}(aq) + \textbf{2Fe}^{2+}(aq)$$

---

**EXERCISE 20.5**

Give the overall cell reaction for the voltaic cell

$$Cd(s)\,|\,Cd^{2+}(aq)\,\|\,H^+(aq)\,|\,H_2(g)\,|\,Pt$$

---

**See Problems 20.45 and 20.46.**

To fully specify a voltaic cell, it is necessary to give the concentrations of solutions or ions and the pressure of gases. In the cell notation, these are written within parentheses for each species. For example,

$$Zn(s)\,|\,Zn^{2+}(1.0\ M)\,\|\,H^+(1.0\ M)\,|\,H_2(1.0\ \text{atm})\,|\,Pt$$

# 20.4 Electromotive Force

Work is needed to move electrons in a wire or to move ions through a solution to an electrode. The situation is analogous to pumping water from one point to another. Work must be expended to pump the water. The analogy may be extended. Water moves from a point at high pressure to a point at low pressure. Thus, a pressure difference is required. The work expended in moving water through a pipe depends on the volume of water and the pressure difference. The situation with electricity is similar. An electric charge moves from a point at high electric potential (high electrical pressure) to a point at low electric potential (low electrical pressure). The work needed to move an electric charge through a conductor depends on the total charge moved and the potential difference. **Potential difference** is *the difference in electric potential (electrical pressure) between two points.* You measure this quantity in volts.

**Figure 20.6**
**A digital voltmeter.**
A digital voltmeter draws negligible current, so it can be used to measure cell emf.

The **volt, V,** is *the SI unit of potential difference.* The electrical work expended in moving a charge through a conductor is

$$\text{Electrical work} = \text{charge} \times \text{potential difference}$$

Corresponding SI units for the terms in this equation are

$$\text{Joules} = \text{coulombs} \times \text{volts}$$

The **Faraday constant, $F$,** is *the magnitude of charge on one mole of electrons;* *it equals $9.65 \times 10^4$ C* (96,500 coulombs). The *faraday* is a unit of charge equal to $9.65 \times 10^4$ C. In moving this quantity of charge (one faraday of charge) from one electrode to another, the numerical value of the work done by a voltaic cell is the product of the faraday constant $F$ times the potential difference between the electrodes. The work $w$ is the negative of this, because the voltaic cell loses energy as it does work on the surroundings.

$$w = -F \times \text{potential difference}$$

In the normal operation of a voltaic cell, the potential difference (voltage) across the electrodes is less than the maximum possible voltage of the cell. One reason for this is that it takes energy or work to drive a current through the cell itself. The decrease in cell voltage as current is drawn reflects this energy expenditure within the cell; and the greater the current, the lower the voltage. Thus, the cell voltage has its maximum value only when no current flows. The situation is analogous to measuring the difference between the pressure of water in a faucet and that of the outside atmosphere. You expect this difference to be higher if you measure it by a device that does not require you to run water from the faucet. When you do run water, the pressure difference drops.

*The maximum potential difference between the electrodes of a voltaic cell* is referred to as the **electromotive force (emf)** of the cell, or $E_{\text{cell}}$. It can be measured by an electronic digital voltmeter (Figure 20.6), which draws negligible current.

You can also determine which electrode of a cell is the anode and which is the cathode from this measurement. The digital voltmeter gives the reading with a sign attached, from which you can deduce the sign, or polarity, of each electrode. The anode of a voltaic cell has negative polarity, and the cathode has positive polarity.

We can now write an expression for the maximum work obtainable from a voltaic cell. Let $n$ be the number of electrons transferred in the overall cell equation. The maximum electrical work of a voltaic cell for molar amounts of reactants (according to the cell equation *as written*) is

$$\boxed{w_{\text{max}} = -nFE_{\text{cell}}}$$

Here $E_{\text{cell}}$ is the cell emf, and $F$ is the Faraday constant, $9.65 \times 10^4$ C.

▶ An alternate example is provided in the Instructor's Resource Manual.

**EXAMPLE 20.5**

## Calculating the Quantity of Work from a Given Amount of Cell Reactant

The emf of a particular voltaic cell with the cell reaction

$$\text{Hg}_2{}^{2+}(aq) + \text{H}_2(g) \rightleftharpoons 2\text{Hg}(l) + 2\text{H}^+(aq)$$

is 0.650 V. Calculate the maximum electrical work of this cell when 0.500 g $\text{H}_2$ is consumed.

**PROBLEM STRATEGY**

The maximum work of a cell equals $-nFE_{cell}$. To get $n$, you split the cell reaction into half-reactions, multiplying each by a factor so that when added together they give the cell reaction. Then $n$ equals the number of electrons in either half-reaction.

**SOLUTION**

The half-reactions are

$$Hg_2^{2+}(aq) + 2e^- \rightleftharpoons 2Hg(l)$$

$$H_2(g) \rightleftharpoons 2H^+(aq) + 2e^-$$

$n$ equals 2, and the maximum work for the reaction as written is

$$w_{max} = -nFE_{cell} = -2 \times 9.65 \times 10^4 \text{ C} \times 0.650 \text{ V}$$
$$= -1.25 \times 10^5 \text{ V} \cdot \text{C}$$
$$= -1.25 \times 10^5 \text{ J}$$

(Remember that a joule is equal to a volt-coulomb. Also, note that the negative sign means work is done by the cell—that is, energy is lost by the cell.) For 0.500 g $H_2$, the maximum work is

$$0.500 \text{ g } H_2 \times \frac{1 \text{ mol } H_2}{2.02 \text{ g } H_2} \times \frac{-1.25 \times 10^5 \text{ J}}{1 \text{ mol } H_2} = -3.09 \times 10^4 \text{ J}$$

Note that the conversion factor $-1.25 \times 10^5$ J/1 mol $H_2$ is determined by the chemical equation as written. In the equation, 1 mol $H_2$ reacts, and the maximum work produced is $-1.25 \times 10^5$ J.

---

**EXERCISE 20.6**

What is the maximum electrical work that can be obtained from 6.54 g of zinc metal that reacts in a Daniell cell, described in the chapter opening, whose emf is 1.10 V? The overall cell reaction is

$$Zn(s) + Cu^{2+}(aq) \longrightarrow Zn^{2+}(aq) + Cu(s)$$

---

See Problems 20.49, 20.50, 20.51, and 20.52.

# 20.5 Standard Cell emf's and Standard Electrode Potentials

A cell emf is a measure of the driving force of the cell reaction. This reaction occurs in the cell as separate half-reactions: an oxidation half-reaction and a reduction half-reaction. The general forms of these half-reactions are

Reduced species $\longrightarrow$ oxidized species + $ne^-$      (oxidation/anode)

Oxidized species + $ne^- \longrightarrow$ reduced species      (reduction/cathode)

You can imagine that the cell emf is composed of a contribution from the anode (whose value depends on the ability of the oxidation half-reaction to lose electrons)

and a contribution from the cathode (whose value depends on the ability of the reduction half-reaction to gain electrons). We call these contributions the *oxidation potential* and the *reduction potential*, respectively. Then

$$E_{cell} = \text{oxidation potential} + \text{reduction potential}$$

A reduction potential is a measure of the tendency for an oxidized species to gain electrons in the reduction half-reaction (in other words, a measure of the ability of the species to act as an oxidizing agent). If you can construct a table of reduction potentials, you will have a list of strengths of oxidizing agents, in addition to having a way of calculating cell emf's.

You can look at an oxidation half-reaction as the reverse of a corresponding reduction half-reaction. The oxidation potential for an oxidation half-reaction equals the negative of the reduction potential for the reverse half-reaction (which is a reduction).

Oxidation potential for a half-reaction = −reduction potential for the reverse half-reaction

This means that in practice you need to tabulate only oxidation potentials or reduction potentials. The choice, by convention, is to tabulate reduction potentials. We call these *electrode potentials,* and we denote them by the symbol $E$ (without the subscript cell, as in $E_{cell}$).

Consider the zinc–copper cell described earlier.

$$Zn(s)\,|\,Zn^{2+}(aq)\,\|\,Cu^{2+}(aq)\,|\,Cu(s)$$

The half-reactions are

$$Zn(s) \longrightarrow Zn^{2+}(aq) + 2e^-$$
$$Cu^{2+}(aq) + 2e^- \longrightarrow Cu(s)$$

The first half-reaction is an oxidation. If you write $E_{Zn}$ for the electrode potential corresponding to the reduction half-reaction $Zn^{2+}(aq) + 2e^- \longrightarrow Zn(s)$, then $-E_{Zn}$ is the potential for the oxidation half-reaction $Zn(s) \longrightarrow Zn^{2+}(aq) + 2e^-$. The copper half-reaction is a reduction. Write $E_{Cu}$ for the electrode potential.

The cell emf is the sum of the potentials for the reduction and oxidation half-reactions. For the cell we have been describing, the emf is the sum of the reduction potential (electrode potential) for the copper half-cell and the oxidation potential (negative of the electrode potential) for the zinc half-cell.

$$E_{cell} = E_{Cu} + (-E_{Zn}) = E_{Cu} - E_{Zn}$$

Note that the cell emf equals the difference between the two electrode potentials. You can think of the electrode potential as the electric potential on the electrode, and you obtain the cell emf as a potential difference in which you subtract the anode potential from the cathode potential.

$$E_{cell} = E_{cathode} - E_{anode}$$

The electrode potential is an *intensive property*. This means that its value is independent of the amount of species in the reaction. Thus, the electrode potential for the half-reaction

$$2Cu^{2+}(aq) + 4e^- \longrightarrow 2Cu(s)$$

is the same as for

$$Cu^{2+}(aq) + 2e^- \longrightarrow Cu(s)$$

## Tabulating Standard Electrode Potentials

The emf of a voltaic cell depends on the concentrations of substances and the temperature of the cell. For purposes of tabulating electrochemical data, it is usual to choose thermodynamic standard-state conditions for voltaic cells. The **standard emf,** $E°_{cell}$, is *the emf of a voltaic cell operating under standard-state conditions (solute concentrations are each 1 M, gas pressures are each 1 atm, and the temperature has a specified value—usually 25°C).* Note the superscript degree sign (°), which signifies standard-state conditions. ◄

If you can derive a table of electrode potentials, you can calculate cell emf's from them. This provides a great advantage over tabulating cell emf's. From a small table of electrode potentials, you can obtain the emf's of all the cells that you could construct from pairs of electrodes. For instance, a table of 40 electrode potentials would give the emf's of nearly 800 voltaic cells.

However, it is not possible to measure the potential of a single electrode; only emf's of cells can be measured. What you can do is measure the emf's of cells constructed from various electrodes connected in turn to one particular electrode, which you choose as a reference. You arbitrarily assign this reference electrode a potential equal to zero and obtain the potentials for the other electrodes by measuring the emf's. By convention, the reference chosen for comparing electrode potentials is the *standard hydrogen electrode.* This is an electrode like that pictured in Figure 20.5 operating under standard-state conditions.

The **standard electrode potential,** $E°$, is *the electrode potential when the concentrations of solutes are 1 M, the gas pressures are 1 atm, and the temperature has a specified value (usually 25°C).* The superscript degree sign (°) signifies standard-state conditions.

To understand how standard electrode potentials are obtained, look at how you would find the standard electrode potential, $E°$, for the zinc electrode. You connect a standard zinc electrode to a standard hydrogen electrode. When you measure the emf of the cell with a voltmeter, you obtain 0.76 V, with the zinc electrode acting as the anode.

Now write the cell emf in terms of the electrode potentials. The cell is

$$Zn(s)\,|\,Zn^{2+}(aq)\,\|\,H^+(aq)\,|\,H_2(g)\,|\,Pt$$

and the half-reactions with corresponding half-cell potentials (oxidation or reduction potentials) are

$$Zn(s) \longrightarrow Zn^{2+}(aq) + 2e^-; \ -E°_{Zn}$$
$$2H^+(aq) + 2e^- \longrightarrow H_2(g); \ E°_{H_2}$$

The cell emf is the sum of the half-cell potentials.

$$E_{cell} = E°_{H_2} + (-E°_{Zn})$$

Substitute 0.76 V for the cell emf and 0.00 V for the standard hydrogen electrode potential. This gives $E°_{Zn} = -0.76$ V.

Proceeding in this way, you can obtain the electrode potential for a series of half-cell reactions. Table 20.1 lists standard electrode potentials for selected half-cells at 25°C.

## Strengths of Oxidizing and Reducing Agents

Standard electrode potentials are useful in determining the strengths of oxidizing and reducing agents under standard-state conditions. Because electrode potentials are reduction potentials, those reduction half-reactions in Table 20.1 with the larger (that

► To be precise, the standard-state conditions are those in which the activities (ideal concentrations) of solutes and gases equal 1. This is equivalent to 1 *M* only for ideal solutions. Because of the strong attractions between ions, substantial deviations from ideal conditions exist in electrolyte solutions, except when very dilute. We will ignore these deviations here.

**TABLE 20.1**
**Standard Electrode (Reduction) Potentials in Aqueous Solution at 25°C\***

| Cathode (Reduction) Half-Reaction | Standard Potential, $E°$(V) |
|---|---|
| $Li^+(aq) + e^- \rightleftharpoons Li(s)$ | $-3.04$ |
| $Na^+(aq) + e^- \rightleftharpoons Na(s)$ | $-2.71$ |
| $Mg^{2+}(aq) + 2e^- \rightleftharpoons Mg(s)$ | $-2.38$ |
| $Al^{3+}(aq) + 3e^- \rightleftharpoons Al(s)$ | $-1.66$ |
| $2H_2O(l) + 2e^- \rightleftharpoons H_2(g) + 2OH^-(aq)$ | $-0.83$ |
| $Zn^{2+}(aq) + 2e^- \rightleftharpoons Zn(s)$ | $-0.76$ |
| $Cr^{3+}(aq) + 3e^- \rightleftharpoons Cr(s)$ | $-0.74$ |
| $Fe^{2+}(aq) + 2e^- \rightleftharpoons Fe(s)$ | $-0.41$ |
| $Cd^{2+}(aq) + 2e^- \rightleftharpoons Cd(s)$ | $-0.40$ |
| $Ni^{2+}(aq) + 2e^- \rightleftharpoons Ni(s)$ | $-0.23$ |
| $Sn^{2+}(aq) + 2e^- \rightleftharpoons Sn(s)$ | $-0.14$ |
| $Pb^{2+}(aq) + 2e^- \rightleftharpoons Pb(s)$ | $-0.13$ |
| $Fe^{3+}(aq) + 3e^- \rightleftharpoons Fe(s)$ | $-0.04$ |
| $2H^+(aq) + 2e^- \rightleftharpoons H_2(g)$ | $0.00$ |
| $Sn^{4+}(aq) + 2e^- \rightleftharpoons Sn^{2+}(aq)$ | $0.15$ |
| $Cu^{2+}(aq) + e^- \rightleftharpoons Cu^+(aq)$ | $0.16$ |
| $Cu^{2+}(aq) + 2e^- \rightleftharpoons Cu(s)$ | $0.34$ |
| $IO^-(aq) + H_2O(l) + 2e^- \rightleftharpoons I^-(aq) + 2OH^-(aq)$ | $0.49$ |
| $Cu^+(aq) + e^- \rightleftharpoons Cu(s)$ | $0.52$ |
| $I_2(s) + 2e^- \rightleftharpoons 2I^-(aq)$ | $0.54$ |
| $Fe^{3+}(aq) + e^- \rightleftharpoons Fe^{2+}(aq)$ | $0.77$ |
| $Hg_2^{2+}(aq) + 2e^- \rightleftharpoons 2Hg(l)$ | $0.80$ |
| $Ag^+(aq) + e^- \rightleftharpoons Ag(s)$ | $0.80$ |
| $Hg^{2+}(aq) + 2e^- \rightleftharpoons Hg(l)$ | $0.85$ |
| $ClO^-(aq) + H_2O(l) + 2e^- \rightleftharpoons Cl^-(aq) + 2OH^-(aq)$ | $0.90$ |
| $2Hg^{2+}(aq) + 2e^- \rightleftharpoons Hg_2^{2+}(aq)$ | $0.90$ |
| $NO_3^-(aq) + 4H^+(aq) + 3e^- \rightleftharpoons NO(g) + 2H_2O(l)$ | $0.96$ |
| $Br_2(l) + 2e^- \rightleftharpoons 2Br^-(aq)$ | $1.07$ |
| $O_2(g) + 4H^+(aq) + 4e^- \rightleftharpoons 2H_2O(l)$ | $1.23$ |
| $Cr_2O_7^{2-}(aq) + 14H^+(aq) + 6e^- \rightleftharpoons 2Cr^{3+}(aq) + 7H_2O(l)$ | $1.33$ |
| $Cl_2(g) + 2e^- \rightleftharpoons 2Cl^-(aq)$ | $1.36$ |
| $MnO_4^-(aq) + 8H^+(aq) + 5e^- \rightleftharpoons Mn^{2+}(aq) + 4H_2O(l)$ | $1.49$ |
| $H_2O_2(aq) + 2H^+(aq) + 2e^- \rightleftharpoons 2H_2O(l)$ | $1.78$ |
| $S_2O_8^{2-}(aq) + 2e^- \rightleftharpoons 2SO_4^{2-}(aq)$ | $2.01$ |
| $F_2(g) + 2e^- \rightleftharpoons 2F^-(aq)$ | $2.87$ |

\*See Appendix I for a more extensive table.

is, more positive) electrode potentials have the greater tendency to go left to right as written. A reduction half-reaction has the general form

$$\text{Oxidized species} + ne^- \longrightarrow \text{reduced species}$$

The oxidized species acts as an oxidizing agent. Consequently, *the strongest oxidizing agents in a table of standard electrode potentials are the oxidized species corresponding to half-reactions with the largest (most positive) $E°$ values.*

Those reduction half-reactions with lower (that is, more negative) electrode potentials have a greater tendency to go right to left. That is,

$$\text{Reduced species} \longrightarrow \text{oxidized species} + ne^-$$

The reduced species acts as a reducing agent. Consequently, *the strongest reducing agents in a table of standard electrode potentials are the reduced species corresponding to half-reactions with the smallest (most negative) $E°$ values.*

The first two and last two entries in Table 20.1 are as follows:

$$\text{Li}^+(aq) + e^- \longrightarrow \text{Li}(s)$$
$$\text{Na}^+(aq) + e^- \longrightarrow \text{Na}(s)$$
$$\cdots$$
$$\cdots$$
$$\text{S}_2\text{O}_8^{2-}(aq) + 2e^- \longrightarrow 2\text{SO}_4^{2-}(aq)$$
$$\text{F}_2(g) + 2e^- \longrightarrow 2\text{F}^-(aq)$$

The strongest oxidizing agents are the species on the left side in the last two entries in the table (given here in red). The strongest reducing agents are the species on the right side in the first two entries in the table (given here in blue).

▶ An alternate example is provided in the Instructor's Resource Manual.

## EXAMPLE 20.6

### Determining the Relative Strengths of Oxidizing and Reducing Agents

a. Order the following oxidizing agents by increasing strength under standard-state conditions: $Cl_2(g)$, $H_2O_2(aq)$, $Fe^{3+}(aq)$.

b. Order the following reducing agents by increasing strength under standard-state conditions: $H_2(g)$, $Al(s)$, $Cu(s)$.

**PROBLEM STRATEGY**

a. Read down the table of electrode potentials, picking out the reduction half-reactions in which the species of interest occur as reactants. The order from *top to bottom* in which these species occur in the table is the order of increasing oxidizing power.

b. Read down the table of electrode potentials, picking out the reduction half-reactions in which the species of interest occur as products. The order from *bottom to top* in which these species occur in the table is the order of increasing reducing strength.

**SOLUTION**

a. The half-reactions and corresponding electrode potentials are as follows:

$$Fe^{3+}(aq) + e^- \longrightarrow Fe^{2+}(aq) \qquad 0.77 \text{ V}$$
$$Cl_2(g) + 2e^- \longrightarrow 2Cl^-(aq) \qquad 1.36 \text{ V}$$
$$H_2O_2(aq) + 2H^+(aq) + 2e^- \longrightarrow 2H_2O(l) \qquad 1.78 \text{ V}$$

The order by increasing oxidizing strength is **$Fe^{3+}(aq)$, $Cl_2(g)$, $H_2O_2(aq)$.**

b. The half-reactions and corresponding electrode potentials are

$$Al^{3+}(aq) + 3e^- \longrightarrow Al(s) \qquad -1.66 \text{ V}$$
$$2H^+(aq) + 2e^- \longrightarrow H_2(g) \qquad 0.00 \text{ V}$$
$$Cu^{2+}(aq) + 2e^- \longrightarrow Cu(s) \qquad 0.34 \text{ V}$$

The order by increasing reducing strength is **$Cu(s)$, $H_2(g)$, $Al(s)$.**

---

**EXERCISE 20.7**

Which is the stronger oxidizing agent, $NO_3^-(aq)$ in acidic solution (to NO) or $Ag^+(aq)$?

---

See Problems 20.53, 20.54, 20.55, and 20.56.

You can use a table of electrode potentials to predict the direction of spontaneity of an oxidation–reduction reaction. You need note only the relative strengths of the oxidizing agents on the left and right sides of the equation. The stronger oxidizing agent will be on the reactant side when the equation is written as a spontaneous reaction. (Alternatively, you can look at the reducing agents; the stronger reducing agent will be on the reactant side of the spontaneous reaction.)

---

▶ An alternate example is provided in the Instructor's Resource Manual.

**EXAMPLE 20.7**

## Determining the Direction of Spontaneity from Electrode Potentials

Consider the reaction

$$Zn^{2+}(aq) + 2Fe^{2+}(aq) \longrightarrow Zn(s) + 2Fe^{3+}(aq)$$

Does the reaction go spontaneously in the direction indicated, under standard conditions?

 **PROBLEM STRATEGY**

Find the oxidizing agents in the equation; one is on the left side and the other on the right side. Locate these oxidizing agents in a table of electrode potentials. (The oxidizing agent is on the left side of the reduction half-reaction.) The stronger oxidizing agent is the one involved in the half-reaction with the more positive standard electrode potential.

**SOLUTION**

In this reaction, $Zn^{2+}$ is the oxidizing agent on the left side; $Fe^{3+}$ is the oxidizing agent on the right side. The corresponding standard electrode potentials are

$$Zn^{2+}(aq) + 2e^- \longrightarrow Zn(s); \ E° = -0.76 \text{ V}$$

$$Fe^{3+}(aq) + e^- \longrightarrow Fe^{2+}(aq); E° = 0.77 \text{ V}$$

The stronger oxidizing agent is the one involved in the half-reaction with the more positive standard electrode potential, so $Fe^{3+}$ is the stronger oxidizing agent. The reaction is **nonspontaneous** as written.

### EXERCISE 20.8

Does the following reaction occur spontaneously in the direction indicated, under standard conditions?

$$Cu^{2+}(aq) + 2I^-(aq) \longrightarrow Cu(s) + I_2(s)$$

*See Problems 20.57 and 20.58.*

## Calculating Cell emf's from Standard Potentials

The emf of a voltaic cell constructed from standard electrodes is easily calculated using a table of electrode potentials. As an example, consider the cell constructed from a cadmium electrode and a silver electrode having the following reduction half-reactions and corresponding standard electrode potentials (reduction potentials):

$$Cd^{2+}(aq) + 2e^- \longrightarrow Cd(s); E°_{Cd} = -0.40 \text{ V}$$
$$Ag^+(aq) + e^- \longrightarrow Ag(s); E°_{Ag} = 0.80 \text{ V}$$

You will need to reverse one of these half-reactions to obtain the oxidation part of the cell reaction. The cell reaction is spontaneous with the stronger reducing agent on the left. This will be Cd, because it is the reactant in the half-reaction with the more negative electrode potential. Therefore, you reverse the first half-reaction, as well as the sign of the half-cell potential (to obtain the oxidation potential, which equals $-E°_{Cd}$).

$$Cd(s) \longrightarrow Cd^{2+}(aq) + 2e^-; -E°_{Cd} = 0.40 \text{ V}$$
$$Ag^+(aq) + e^- \longrightarrow Ag(s); E°_{Ag} = 0.80 \text{ V}$$

To obtain the cell reaction, you must multiply the half-reactions by factors so that when the half-reactions are added together, the electrons cancel. This does not affect the half-cell potentials, which do not depend on amount of substance. You multiply the second half-reaction by 2, then add.

| | |
|---|---|
| $Cd(s) \longrightarrow Cd^{2+}(aq) + 2e^-$ | $-E°_{Cd} = 0.40 \text{ V}$ |
| $2Ag^+(aq) + 2e^- \longrightarrow 2Ag(s)$ | $E°_{Ag} = 0.80 \text{ V}$ |
| $Cd(s) + 2Ag^+(aq) \longrightarrow Cd^{2+}(aq) + 2Ag(s)$ | $E°_{Ag} - E°_{Cd} = 1.20 \text{ V}$ |

The cell emf, $E°_{cell}$, is 1.20 V. The cell diagram is

$$Cd(s)|Cd^{2+}(aq)\|Ag^+(aq)|Ag(s)$$

Suppose that at the beginning of this problem, you had mistakenly reversed the silver half-reaction, instead of correctly reversing the cadmium half-reaction. Then your work would have looked like the following:

| | |
|---|---|
| $Cd^{2+}(aq) + 2e^- \longrightarrow Cd(s)$ | $E°_{Cd} = -0.40 \text{ V}$ |
| $2Ag(s) \longrightarrow 2Ag^+(aq) + 2e^-$ | $-E°_{Ag} = -0.80 \text{ V}$ |
| $Cd^{2+}(aq) + 2Ag(s) \longrightarrow Cd(s) + 2Ag^+(aq)$ | $E°_{Cd} - E°_{Ag} = -1.20 \text{ V}$ |

The corresponding cell notation would have been

$$Ag(s)|Ag^+(aq)\|Cd^{2+}(aq)|Cd(s)$$

Everything is simply the reverse of the correct result. At this point, you realize your mistake, because the cell emf is found to be negative. A negative emf merely indicates that the cell reaction is nonspontaneous as written. To obtain the spontaneous reaction and a positive emf, you simply reverse both half-reactions and corresponding half-cell potentials. (You would also reverse the cell notation.) This changes the sign of the emf.

Note that the emf of the cell equals the standard electrode potential of the cathode minus the standard electrode potential of the anode. So, alternatively, you can calculate cell emf's from the equation

$$E^\circ_{cell} = E^\circ_{cathode} - E^\circ_{anode}$$

You obtain

$$E^\circ_{cell} = E^\circ_{Ag} - E^\circ_{Cd} = 0.80\ V - (-0.40\ V) = 1.20\ V$$

▶ An alternate example is provided in the Instructor's Resource Manual.

### EXAMPLE 20.8

## Calculating the emf from Standard Potentials

Calculate the standard emf of the following voltaic cell at 25°C using standard electrode potentials.

$$Al(s)\,|\,Al^{3+}(aq)\,\|\,Fe^{2+}(aq)\,|\,Fe(s)$$

What is the cell reaction?

 **PROBLEM STRATEGY**

From a table of electrode potentials, write the two reduction half-reactions and standard electrode potentials for the cell. The cell notation assumes that the anode (the oxidation half-cell) is on the left. Change the direction of this half-reaction and the sign of its electrode potential. (Assuming that the cell notation was written correctly, you change the direction of the half-reaction corresponding to the smaller, or more negative, electrode potential.) Multiply the half-reactions (but not the electrode potentials) by factors so that when the half-reactions are added, the electrons cancel. The sum of the half-reactions is the cell reaction. Add the electrode potentials to get the cell emf.

### SOLUTION

The reduction half-reactions and standard electrode potentials are

$$Al^{3+}(aq) + 3e^- \longrightarrow Al(s);\ E^\circ_{Al} = -1.66\ V$$
$$Fe^{2+}(aq) + 2e^- \longrightarrow Fe(s);\ E^\circ_{Fe} = -0.41\ V$$

You reverse the first half-reaction and its half-cell potential to obtain

$$Al(s) \longrightarrow Al^{3+}(aq) + 3e^-;\ -E^\circ_{Al} = 1.66\ V$$
$$Fe^{2+}(aq) + 2e^- \longrightarrow Fe(s);\ E^\circ_{Fe} = -0.41\ V$$

You obtain the cell emf by adding the half-cell potentials. Because you also want the cell reaction, you multiply the first half-reaction by 2 and the second half-reaction by 3, so that when the half-reactions are added, the electrons cancel. The addition of half-reactions is displayed below. The cell emf is **1.25 V**.

$$2Al(s) \longrightarrow 2Al^{3+}(aq) + 6e^- \qquad -E^\circ_{Al} = \quad 1.66 \text{ V}$$
$$\underline{3Fe^{2+}(aq) + 6e^- \longrightarrow 3Fe(s) \qquad\qquad E^\circ_{Fe} = -0.41 \text{ V}}$$
$$\textbf{2Al(s) + 3Fe}^{2+}\textbf{(aq)} \longrightarrow \textbf{2Al}^{3+}\textbf{(aq) + 3Fe(s)} \qquad E^\circ_{Fe} - E^\circ_{Al} = \quad 1.25 \text{ V}$$

Note that the cell emf is the standard electrode potential for the cathode minus the standard electrode potential for the anode. You could have calculated the cell emf from the formula

$$E^\circ_{cell} = E^\circ_{cathode} - E^\circ_{anode}$$

You obtain

$$E^\circ_{cell} = E^\circ_{Fe} - E^\circ_{Al} = -0.41 \text{ V} - (-1.66 \text{ V}) = 1.25 \text{ V}$$

### EXERCISE 20.9

Using standard electrode potentials, calculate $E^\circ_{cell}$ at 25°C for the following cell.

$$Zn(s)\,|\,Zn^{2+}(aq)\,\|\,Cu^{2+}(aq)\,|\,Cu(s)$$

See Problems 20.61 and 20.62.

**Problem Solving:
Cell Potential**

 **CONCEPT CHECK 20.2**

Let us define the reduction of $I_2$ to $I^-$ ions, $I_2(s) + 2e^- \longrightarrow 2I^-(aq)$, as the standard reduction reaction with $E^\circ = 0.00$ V. We then construct a new standard reduction table based on this definition.

a. What would be the new standard reduction potential of $H^+$?

b. Would using a new standard reduction table change the *measured* value of a freshly prepared voltaic cell made from Cu and Zn? (Assume you have the appropriate solutions and equipment to construct the cell.)

c. Would the calculated voltage for the cell in part b be different if you were using the values presented in Table 20.1? Do the calculations to justify your answer.

## 20.6   Equilibrium Constants from emf's

Some of the most important results of electrochemistry are the relationships among cell emf, free-energy change, and equilibrium constant. Recall that the free-energy change $\Delta G$ for a reaction equals the maximum useful work of the reaction (Section 19.5).

$$\Delta G = w_{max}$$

For a voltaic cell, this work is the electrical work, $-nFE_{cell}$ (where $n$ is the number of electrons transferred in a reaction), so when the reactants and products are in their standard states, you have

$$\Delta G^\circ = -nFE^\circ_{cell}$$

With this equation, emf measurements become an important source of thermo-dynamic information. Alternatively, thermodynamic data can be used to calculate cell emf's. These calculations are shown in the following examples.

▶ An alternate example is provided in the Instructor's Resource Manual.

### EXAMPLE 20.9

## Calculating the Free-Energy Change from Electrode Potentials

Using standard electrode potentials, calculate the standard free-energy change at 25°C for the reaction

$$Zn(s) + 2Ag^+(aq) \longrightarrow Zn^{2+}(aq) + 2Ag(s)$$

 **PROBLEM STRATEGY**

Substitute into $\Delta G° = -nFE°_{cell}$. Use a table of standard electrode potentials to obtain $E°_{cell}$. The cell reaction equals the sum of the half-reactions after they have been multiplied by factors so that the electrons cancel in the summation. Note that $n$ is the number of electrons involved in each half-reaction.

**SOLUTION**

The half-reactions, corresponding half-cell potentials, and their sums are displayed below.

$$
\begin{array}{ll}
Zn(s) \longrightarrow Zn^{2+}(aq) + 2e^- & -E° = 0.76 \text{ V} \\
2Ag^+(aq) + 2e^- \longrightarrow 2Ag(s) & E° = 0.80 \text{ V} \\
\hline
Zn(s) + 2Ag^+(aq) \longrightarrow Zn^{2+}(aq) + 2Ag(s) & E°_{cell} = 1.56 \text{ V}
\end{array}
$$

Note that each half-reaction involves two electrons; hence, $n = 2$. Also, $E°_{cell} = 1.56$ V, and the faraday constant, $F$, is $9.65 \times 10^4$ C. Therefore,

$$\Delta G° = -nFE°_{cell} = -2 \times 9.65 \times 10^4 \text{ C} \times 1.56 \text{ V} = -3.01 \times 10^5 \text{ J}$$

Recall that (coulombs) × (volts) = joules. Thus, the standard free-energy change is **−301 kJ.**

**Problem Solving: Calculating ΔG°**

### EXERCISE 20.10

What is $\Delta G°$ at 25°C for the reaction

$$Sn^{2+}(aq) + 2Hg^{2+}(aq) \longrightarrow Sn^{4+}(aq) + Hg_2^{2+}(aq)$$

For data, see Table 20.1.

See Problems 20.65 and 20.66.

▶ An alternate example is provided in the Instructor's Resource Manual.

### EXAMPLE 20.10

## Calculating the Cell emf from Free-Energy Change

Suppose the reaction of zinc metal and chlorine gas is utilized in a voltaic cell in which zinc ions and chloride ions are formed in aqueous solution.

$$Zn(s) + Cl_2(g) \xrightarrow{H_2O} Zn^{2+}(aq) + 2Cl^-(aq)$$

Calculate the standard emf for this cell at 25°C from standard free energies of formation (see Appendix C).

 **PROBLEM STRATEGY**

Calculate $\Delta G°$ and substitute it and the value of $n$ into the equation $\Delta G° = -nFE°_{cell}$. Solve for $E°_{cell}$.

**SOLUTION**

Write the equation with $\Delta G°_f$'s beneath.

$$Zn(s) + Cl_2(g) \longrightarrow Zn^{2+}(aq) + 2Cl^-(aq)$$
$$\Delta G°_f: \quad 0 \qquad 0 \qquad\qquad -147 \qquad 2 \times (-131)\ kJ$$

Hence,

$$\Delta G° = \Sigma\, n\Delta G°_f(\text{products}) - \Sigma\, m\Delta G°_f(\text{reactants})$$
$$= [-147 + 2 \times (-131)]\ kJ$$
$$= -409\ kJ = -4.09 \times 10^5\ J$$

You obtain $n$ by splitting the reaction into half-reactions.

$$Zn(s) \longrightarrow Zn^{2+}(aq) + 2e^-$$
$$Cl_2(g) + 2e^- \longrightarrow 2Cl^-(aq)$$

Each half-reaction involves two electrons, so $n = 2$. Now you substitute into

$$\Delta G° = -nFE°_{cell}$$
$$-4.09 \times 10^5\ J = -2 \times 9.65 \times 10^4\ C \times E°_{cell}$$

Solving for $E°_{cell}$, you get $E°_{cell} = 2.12\ V.$

---

**EXERCISE 20.11**

Use standard free energies of formation (Appendix C) to obtain the standard emf of a cell at 25°C with the reaction

$$Mg(s) + Cu^{2+}(aq) \longrightarrow Mg^{2+}(aq) + Cu(s)$$

---

See Problems 20.69 and 20.70.

The measurement of cell emf's gives you yet another way to obtain equilibrium constants. Combining the previous equation, $\Delta G° = -nFE°_{cell}$, with the equation $\Delta G° = -RT \ln K$ from Section 19.6, you get

$$nFE°_{cell} = RT \ln K$$

or

$$E°_{cell} = \frac{RT}{nF} \ln K = \frac{2.303\ RT}{nF} \log K$$

Substituting values for the constants $R$ and $F$ at 25°C gives the equation

$$E°_{cell} = \frac{0.0592}{n} \log K \qquad \text{(values in volts at 25°C)}$$

**Figure 20.7**
**The relationships among**
$K$, $\Delta G°$, and $E°_{cell}$.   The
diagram shows how
composition data, calorimetric
data, and electrochemical data
are related.

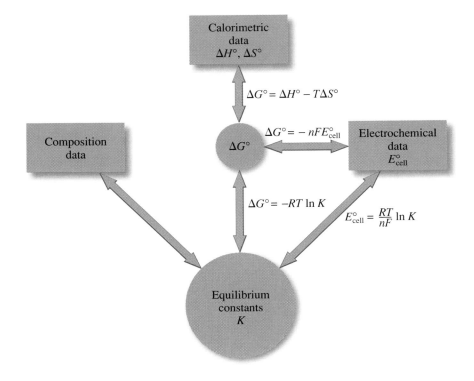

▶ **This formula is typically written
using the common logarithm
(base 10) function as it is
shown here. The factor 2.303
allows for conversion between
the common logarithm and
base e logarithm (2.303 log
$x$ = ln $x$).**

▶ **An alternate example is
provided in the Instructor's
Resource Manual.**

Figure 20.7 summarizes the various relationships among $K$, $\Delta G°$, and $E°_{cell}$. ◀
   The next example illustrates calculation of the thermodynamic equilibrium constant from the cell emf.

## EXAMPLE 20.11

## Calculating the Equilibrium Constant from Cell emf

The standard emf for the following voltaic cell is 1.10 V:

$$Zn(s)|Zn^{2+}(aq)\|Cu^{2+}(aq)|Cu(s)$$

Calculate the equilibrium constant $K_c$ for the reaction

$$Zn(s) + Cu^{2+}(aq) \rightleftharpoons Zn^{2+}(aq) + Cu(s)$$

 **PROBLEM STRATEGY**

Substitute the standard emf into the equation relating this quantity to the thermodynamic equilibrium constant, $K$. Solve for $K$. Note that $K = K_c$.

**SOLUTION**

The reaction corresponds to the one for the voltaic cell. Note that $n = 2$. Substituting into the equation relating $E°_{cell}$ and $K$ gives

$$1.10 = \frac{0.0592}{2} \log K_c$$

Solving for $\log K_c$, you find

$$\log K_c = 37.2$$

Now take the antilog of both sides:

$$K_c = \text{antilog } (37.2) = 1.6 \times 10^{37}$$

The number of significant figures in the answer equals the number of decimal places in 37.2 (one). Thus

$$K_c = 2 \times 10^{37}$$

---

### EXERCISE 20.12

Calculate the equilibrium constant $K_c$ for the following reaction from standard electrode potentials.

$$\text{Fe}(s) + \text{Sn}^{4+}(aq) \rightleftharpoons \text{Fe}^{2+}(aq) + \text{Sn}^{2+}(aq)$$

---

See Problems 20.73 and 20.74.

## 20.7   Dependence of emf on Concentration

The emf of a cell depends on the concentrations of ions and on gas pressures. For that reason, cell emf's provide a way to measure ion concentrations. The pH meter, for example, depends on the variation of cell emf with hydrogen-ion concentration. You can relate cell emf's for various concentrations of ions and various gas pressures to standard electrode potentials by means of an equation first derived by the German chemist Walther Nernst (1864–1941). ◄

▶ Nernst also formulated the third law of thermodynamics, for which he received the Nobel Prize in 1920.

### Nernst Equation

Recall that the free-energy change, $\Delta G$, is related to the standard free-energy change, $\Delta G°$, by the following equation (Section 19.6):

$$\Delta G = \Delta G° + RT \ln Q$$

Here $Q$ is the thermodynamic reaction quotient. The reaction quotient has the form of the equilibrium constant, except that the concentrations and gas pressures are those that exist in a reaction mixture at a given instant. You can apply this equation to a voltaic cell. In that case, the concentrations and gas pressures are those that exist in the cell at a particular instant. If you substitute $\Delta G = -nFE_{\text{cell}}$ and $\Delta G° = -nFE°_{\text{cell}}$ into this equation, you obtain

$$-nFE_{\text{cell}} = -nFE°_{\text{cell}} + RT \ln Q$$

This result rearranges to give the **Nernst equation,** *an equation relating the cell emf to its standard emf and the reaction quotient.*

$$E_{\text{cell}} = E°_{\text{cell}} - \frac{RT}{nF} \ln Q \qquad \text{or} \qquad E_{\text{cell}} = E°_{\text{cell}} - \frac{2.303 \, RT}{nF} \log Q$$

If you substitute 298 K (25°C) for the temperature in the Nernst equation and put in values for $R$ and $F$, you get (using common logarithms)

$$E_{\text{cell}} = E°_{\text{cell}} - \frac{0.0592}{n} \log Q \qquad \text{(values in volts at 25°C)}$$

You can show from the Nernst equation that the cell emf, $E_{cell}$, decreases as the cell reaction proceeds. As the reaction occurs in the voltaic cell, the concentrations of products increase and the concentrations of reactants decrease. Therefore, $Q$ and $\log Q$ increase. The second term in the Nernst equation, $(0.0592/n) \log Q$, increases, so that the difference $E°_{cell} - (0.0592/n) \log Q$ decreases. Thus, the cell emf, $E_{cell}$, becomes smaller. Eventually the cell emf goes to zero, and the cell reaction comes to equilibrium.

As an example of the computation of the reaction quotient, consider the following voltaic cell:

$$Cd(s)|Cd^{2+}(0.0100\ M)\,\|\,H^+(1.00\ M)|H_2(1.00\ atm)|Pt$$

The cell reaction is

$$Cd(s) + 2H^+(aq) \rightleftharpoons Cd^{2+}(aq) + H_2(g)$$

and the expression for the equilibrium constant is

$$K = \frac{[Cd^{2+}]P_{H_2}}{[H^+]^2}$$

Note that the hydrogen-gas concentration is given here in terms of the pressure (in atmospheres). The expression for the reaction quotient has the same form as $K$, except that the values for the ion concentrations and hydrogen-gas pressures are those that exist in the cell. ◀ Hence,

▶ We have omitted subscript $i$'s on the concentrations and pressures in $Q$ to simplify the notation.

$$Q = \frac{[Cd^{2+}]P_{H_2}}{[H^+]^2} = \frac{0.0100 \times 1.00}{(1.00)^2} = 0.0100$$

The next example illustrates a complete calculation of emf from the ion concentrations in a voltaic cell.

---

▶ An alternate example is provided in the Instructor's Resource Manual.

### EXAMPLE 20.12

## Calculating the Cell emf for Nonstandard Conditions

What is the emf of the following voltaic cell at 25°C?

$$Zn(s)|Zn^{2+}(1.00 \times 10^{-5}M)\,\|\,Cu^{2+}(0.100\ M)|Cu(s)$$

The standard emf of this cell is 1.10 V.

 **PROBLEM STRATEGY**

From the cell reaction, deduce the value of $n$; then calculate $Q$. Now substitute $n$, $Q$, and $E°_{cell}$ into the Nernst equation to obtain $E_{cell}$.

**SOLUTION**

The cell reaction is

$$Zn(s) + Cu^{2+}(aq) \rightleftharpoons Zn^{2+}(aq) + Cu(s)$$

The number of electrons transferred is two; hence, $n = 2$. The reaction quotient is

$$Q = \frac{[Zn^{2+}]}{[Cu^{2+}]} = \frac{1.00 \times 10^{-5}}{0.100} = 1.00 \times 10^{-4}$$

The standard emf is 1.10 V, so the Nernst equation becomes

$$E_{cell} = E^{\circ}_{cell} - \frac{0.0592}{n} \log Q$$

$$= 1.10 - \frac{0.0592}{2} \log (1.0 \times 10^{-4})$$

$$= 1.10 - (-0.12) = 1.22 \text{ V}$$

The cell emf is **1.22 V.** This result is qualitatively what you would expect. Because the concentration of product ($Zn^{2+}$) is much less than the standard value (1 $M$), whereas the concentration of reactant ($Cu^{2+}$) is 0.100 $M$, the spontaneity of the reaction as measured by $E_{cell}$ is greater than the standard value.

**Problem Solving:
The Nernst Equation**

---

**EXERCISE 20.13**

What is the emf of the following voltaic cell at 25°C?

$$Zn(s)|Zn^{2+}(0.200\ M)\,\|\,Ag^+(0.00200\ M)|Ag(s)$$

---

See Problems 20.77 and 20.78.

### Determination of pH

The pH of a solution can be obtained very accurately from emf measurements, using the Nernst equation to relate emf to pH. To see how this is done, suppose you have a test solution whose pH you would like to determine. You set up a voltaic cell as follows: You use the test solution as the electrolyte for a hydrogen electrode and bubble in hydrogen gas at 1 atm. (The hydrogen electrode is shown in Figure 20.5.) Now connect this hydrogen electrode to a standard zinc electrode to give the following cell:

$$Zn|Zn^{2+}(1\ M)\,\|\,H^+(\text{test solution})|H_2(1\ atm)|Pt$$

The cell reaction is

$$Zn(s) + 2H^+(\text{test solution}) \longrightarrow Zn^{2+}(1\ M) + H_2(1\ atm)$$

The emf of this cell depends on the hydrogen-ion concentration of the test solution, according to the Nernst equation. The standard emf of the cell equals 0.76 V, and

$$Q = \frac{[Zn^{2+}]P_{H_2}}{[H^+]^2} = \frac{1}{[H^+]^2}$$

Substituting into the Nernst equation, you obtain

$$E_{cell} = 0.76 - \frac{0.0592}{2} \log \frac{1}{[H^+]^2} = 0.76 + 0.0592 \log [H^+]$$

where $[H^+]$ is the hydrogen-ion concentration of the test solution.

To obtain the relationship between the cell emf ($E_{cell}$) and pH, ◀ you substitute the following into the preceding equation:

▶ Keeping in mind that $H^+$ is a shorthand representation of $H_3O^+$, the equation for pH used here is the same as presented in Chapter 16 (pH = $-\log$ [$H_3O^+$]).

$$pH = -\log [H^+]$$

The result is

$$E_{cell} = 0.76 - 0.0592\ pH$$

which you can rearrange to give the pH directly in terms of the cell emf:

$$pH = \frac{0.76 - E_{cell}}{0.0592}$$

In this way, measurement of the cell emf gives you the pH of the solution.

The hydrogen electrode is seldom employed in routine laboratory work, because it is awkward to use. It is often replaced by a *glass electrode*. This compact electrode (see Figure 20.8) consists of a silver wire coated with silver chloride immersed in a solution of dilute hydrochloric acid. The electrode solution is separated from the test solution by a thin glass membrane, which develops a potential across it depending on the hydrogen-ion concentrations on its inner and outer surfaces. A mercury–mercury(I) chloride (calomel) electrode is often used as the other electrode. The emf of the cell depends linearly on the pH. In a common arrangement, the emf is measured with a voltmeter that reads pH directly (see Figure 16.7).

The glass electrode is an example of an *ion-selective electrode*. Many electrodes have been developed that are sensitive to a particular ion, such as $K^+$, $NH_4^+$, $Ca^{2+}$, or $Mg^{2+}$. They can be used to monitor solutions of that ion. It is even possible to measure the concentration of a nonelectrolyte. To measure urea, $NH_2CONH_2$, in solution, you can use an electrode selective to $NH_4^+$ that is coated with a gel containing the enzyme urease. (An enzyme is a biochemical catalyst.) The gel is held in place around the electrode by means of a nylon net. Urease catalyzes the decomposition of urea to ammonium ion, whose concentration is measured.

$$NH_2CONH_2(aq) + 2H_2O(l) + H^+(aq) \xrightarrow{\text{urease}} 2NH_4^+(aq) + HCO_3^-(aq)$$

**Figure 20.8**
**A glass electrode.** *Left:* A small, commercial glass electrode. *Right:* A sketch showing the construction of a glass electrode for measuring hydrogen-ion concentrations.

Ag–AgCl electrode

HCl

Thin glass membrane

**EXERCISE 20.14**

What is the nickel(II)-ion concentration in the voltaic cell

$$Zn(s)\,|\,Zn^{2+}(1.00\,M)\,\|\,Ni^{2+}(aq)\,|\,Ni(s)$$

if the emf is 0.34 V at 25°C?

See Problems 20.81 and 20.82.

 **CONCEPT CHECK 20.3**

Consider a voltaic cell, $Fe(s)\,|\,Fe^{2+}(aq)\,\|\,Cu^{2+}(aq)\,|\,Cu(s)$, being run under standard conditions.

a. Is $\Delta G°$ positive or negative for this process?

b. Change the concentrations from their standard values in such a way that $E_{cell}$ is reduced. Write your answer using the shorthand notation of Section 20.3.

## 20.8  Some Commercial Voltaic Cells

► Demonstration of a dry cell: Bassam Z. Shakhashiri, *Chemical Demonstrations,* Vol. 4 (Madison: University of Wisconsin Press, 1992), no. 11.4, pp. 111–114.

We commonly use voltaic cells as convenient, portable sources of energy. Flashlights and radios are examples of devices that are often powered by the **zinc–carbon,** or **Leclanché, dry cell** (Figure 20.9). *This voltaic cell has a zinc can as the anode; a graphite rod in the center, surrounded by a paste of manganese dioxide, ammonium and zinc chlorides, and carbon black, is the cathode.* The electrode reactions are complicated but are approximately these:

$$Zn(s) \longrightarrow Zn^{2+}(aq) + 2e^- \qquad \text{(anode)}$$
$$2NH_4^+(aq) + 2MnO_2(s) + 2e^- \longrightarrow Mn_2O_3(s) + H_2O(l) + 2NH_3(aq) \qquad \text{(cathode)}$$

**Figure 20.9** **Leclanché dry cell.** The cell has a zinc anode and a graphite rod with a paste of $MnO_2$, $NH_4Cl$, $ZnCl_2$, and C as the cathode.

- Positive terminal
- Air space
- Asphalt seal
- Porous paper
- Paste of $MnO_2$, $ZnCl_2$, $NH_4Cl$, and C
- Graphite rod (cathode)
- Zinc can (anode)
- Metal jacket
- Insulator
- Negative terminal

**Figure 20.10**
**A small alkaline dry cell.**
The anode is zinc powder and the cathode is $MnO_2$, as in the Leclanché cell; however, the electrolyte is KOH.

Anode (negative): Zn powder + KOH electrolyte

Anode cap

Gasket

Cathode (positive): $MnO_2$ + KOH electrolyte

Cell can

Separator

Negative pole

Li metal anode

Nickel mesh

$I_2$ complex cathode

LiI crystals

**Figure 20.11**
**A solid-state lithium–iodine battery.**
The anode is lithium metal and the cathode is a complex of iodine, $I_2$; the electrodes are separated by a thin crystalline layer of lithium iodide. The battery consists of two cells enclosed in a titanium shell.

The voltage of this dry cell is initially about 1.5 V, but it decreases as current is drawn off. The voltage also deteriorates rapidly in cold weather.

An **alkaline dry cell** (Figure 20.10) is *similar to the Leclanché cell, but it has potassium hydroxide in place of ammonium chloride.* This cell performs better under current drain and in cold weather. The half-reactions are

$$Zn(s) + 2OH^-(aq) \longrightarrow Zn(OH)_2(s) + 2e^- \quad \text{(anode)}$$
$$2MnO_2(s) + H_2O(l) + 2e^- \longrightarrow Mn_2O_3(s) + 2OH^-(aq) \quad \text{(cathode)}$$

A dry cell is not truly "dry," because the electrolyte is an aqueous paste. Solid-state batteries have been developed, however. One of these is a **lithium–iodine battery,** *a voltaic cell in which the anode is lithium metal and the cathode is an $I_2$ complex.* These solid-state electrodes are separated by a thin crystalline layer of lithium iodide (Figure 20.11). Current is carried through the crystal by diffusion of $Li^+$ ions. Although the cell has high resistance and therefore low current, the battery is very reliable and is used to power heart pacemakers. The battery is implanted within the patient's chest and lasts about ten years before it has to be replaced.

Once a dry cell is completely discharged (has come to equilibrium), the cell is not easily reversed, or recharged, and is normally discarded. Some types of cells are rechargeable after use, however. An important example is the **lead storage cell.** *This voltaic cell consists of electrodes of lead alloy grids; one electrode is packed with a spongy lead to form the anode, and the other electrode is packed with lead dioxide to form the cathode* (see Figure 20.12). Both are bathed in an aqueous solution of sulfuric acid, $H_2SO_4$. The half-cell reactions during discharge are

$$Pb(s) + HSO_4^-(aq) \longrightarrow PbSO_4(s) + H^+(aq) + 2e^- \quad \text{(anode)}$$

**Figure 20.12**
**A lead storage battery.**
Each cell delivers about 2 V, and a battery consisting of six cells in series gives about 12 V.

Anode (negative): lead grids filled with spongy lead

Cathode (positive): lead grids filled with $PbO_2$

$H_2SO_4$ electrolyte fills the battery case

**Figure 20.13** (left)
**A maintenance-free battery.** This lead storage battery contains calcium–lead alloy grids for electrodes. This alloy resists the decomposition of water, so the battery can be sealed.

**Figure 20.14** (right)
**Nicad storage batteries.** These rechargeable cells have a cadmium anode and a hydrated nickel oxide cathode.

▶ Demonstration of a lead storage battery: Bassam Z. Shakhashiri, *Chemical Demonstrations,* Vol. 4 (Madison: University of Wisconsin Press, 1992), no. 11.5, pp. 115–118.

▶ Demonstration of a fuel cell: Bassam Z. Shakhashiri, *Chemical Demonstrations,* Vol. 4 (Madison: University of Wisconsin Press, 1992), no. 11.7, pp. 123–129.

$$PbO_2(s) + 3H^+(aq) + HSO_4^-(aq) + 2e^- \longrightarrow PbSO_4(s) + 2H_2O(l) \quad \text{(cathode)}$$

White lead(II) sulfate coats each electrode during discharge, and sulfuric acid is consumed. Each cell delivers about 2 V, and a battery consisting of six cells in series gives about 12 V.

After the lead storage battery is discharged, it is recharged from an external electric current. The previous half-reactions are reversed. Some water is decomposed into hydrogen and oxygen gas during this recharging, so more water may have to be added at intervals. However, newer batteries use lead electrodes containing some calcium metal; the calcium–lead alloy resists the decomposition of water. These *maintenance-free* batteries are sealed (Figure 20.13).

The **nickel–cadmium cell** (nicad cell) is a common storage battery. It is *a voltaic cell consisting of an anode of cadmium and a cathode of hydrated nickel oxide (approximately NiOOH) on nickel; the electrolyte is potassium hydroxide.* Nicad batteries are used in calculators, portable power tools, shavers, and toothbrushes. The half-cell reactions during discharge are

$$Cd(s) + 2OH^-(aq) \longrightarrow Cd(OH)_2(s) + 2e^- \quad \text{(anode)}$$
$$NiOOH(s) + H_2O(l) + e^- \longrightarrow Ni(OH)_2(s) + OH^-(aq) \quad \text{(cathode)}$$

These half-reactions are reversed when the cell is recharged. Nicad batteries can be recharged and discharged many times (see Figure 20.14).

A **fuel cell** is *essentially a battery, but it differs by operating with a continuous supply of energetic reactants, or fuel.* Figure 20.15 shows a fuel cell that uses hydrogen and oxygen. ◀ At one electrode, oxygen passes through a porous material that catalyzes the following reaction:

$$O_2(g) + 2H_2O(l) + 4e^- \longrightarrow 4OH^-(aq) \quad \text{(cathode)}$$

At the other electrode, hydrogen reacts.

$$2H_2(g) + 4OH^-(aq) \longrightarrow 4H_2O(l) + 4e^- \quad \text{(anode)}$$

The sum of these half-cell reactions is

$$2H_2(g) + O_2(g) \longrightarrow 2H_2O(l)$$

▶ The first fuel cell was invented in 1839 by William Groves. He obtained the cell by electrolyzing water and collecting the hydrogen and oxygen in small, inverted test tubes. After removing the battery current from the electrolysis cell, he found that a current would flow from the cell. In effect, the electrolysis charged the cell, after which it operated as a fuel cell.

which is the net reaction in the fuel cell. Such cells are used in the space shuttle orbiters to supply electric energy. Other types of cells employing hydrocarbon fuels have been constructed.

**Figure 20.15**
**A hydrogen–oxygen fuel cell.** Hydrogen gas passes into a chamber where the gas is in contact with a porous material in contact with a hot aqueous solution of potassium hydroxide. This forms the anode at which hydrogen is oxidized to water. Oxygen gas enters a similar electrode and is reduced to hydroxide ion. The net chemical change is the reaction of hydrogen with oxygen to form water.

Another use of voltaic cells is to control the corrosion of underground pipelines and tanks. Such pipelines and tanks are usually made of steel, an alloy of iron, and their corrosion or rusting is an electrochemical process.

Consider the rusting that occurs when a drop of water is in contact with iron. The edge of the water drop exposed to the air becomes one pole of a voltaic cell (see Figure 20.16). At this edge, molecular oxygen from air is reduced to hydroxide ion in solution.

$$O_2(g) + 2H_2O(l) + 4e^- \longrightarrow 4OH^-(aq)$$

The electrons for this reduction are supplied by the oxidation of metallic iron at the center of the drop, which acts as the other pole of the voltaic cell.

$$Fe(s) \longrightarrow Fe^{2+}(aq) + 2e^-$$

These electrons flow from the center of the drop through the metallic iron to the edge of the drop. The metallic iron functions as the external circuit between the cell poles.

Ions move within the water drop, completing the electric circuit. Iron(II) ions move outward from the center of the drop, and hydroxide ions move inward from the edge. The two ions meet in a doughnut-shaped region, where they react to precipitate iron(II) hydroxide.

$$Fe^{2+}(aq) + 2OH^-(aq) \longrightarrow Fe(OH)_2(s)$$

**Figure 20.16**
**The electrochemical process involved in the rusting of iron.** Here a single drop of water containing ions forms a voltaic cell in which iron is oxidized to iron(II) ion at the center of the drop (this is the anode). Oxygen gas from air is reduced to hydroxide ion at the periphery of the drop (the cathode). Hydroxide ions and iron(II) ions migrate together and react to form iron(II) hydroxide. This is oxidized to iron(III) hydroxide by more $O_2$ that dissolves at the surface of the drop. Iron(III) hydroxide precipitates, and this settles to form rust on the surface of the iron.

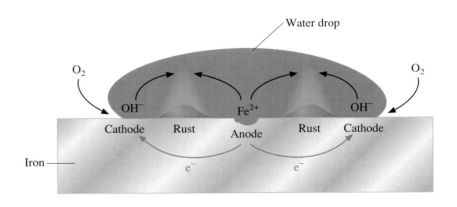

**Figure 20.17**
**Cathodic protection of a buried steel pipe.** Iron in the steel becomes the cathode in an iron–magnesium voltaic cell. Magnesium is then oxidized in preference to iron.

Magnesium rod

$O_2 + 2H_2O + 4e^- \longrightarrow 4OH^-$

Steel pipe

Moist soil

$Mg \longrightarrow Mg^{2+} + 2e^-$

This precipitate is quickly oxidized by oxygen to rust (approximated by the formula $Fe_2O_3 \cdot H_2O$).

$$4Fe(OH)_2(s) + O_2(g) \longrightarrow 2Fe_2O_3 \cdot H_2O(s) + 2H_2O(l)$$

If a buried steel pipeline (Figure 20.17) is connected to an active metal (that is, a highly electropositive substance) such as magnesium, a voltaic cell is formed; the active metal is the anode and iron becomes the cathode. Wet soil forms the electrolyte, and the electrode reactions are

$$Mg(s) \longrightarrow Mg^{2+}(aq) + 2e^- \qquad \text{(anode)}$$
$$O_2(g) + 2H_2O(l) + 4e^- \longrightarrow 4OH^-(aq) \qquad \text{(cathode)}$$

As the cathode, the iron-containing steel pipe is protected from oxidation. Of course, the magnesium rod is eventually consumed and must be replaced, but this is cheaper than digging up the pipeline. This use of an active metal to protect iron from corrosion is called *cathodic protection*. See Figure 20.18 for a laboratory demonstration of cathodic protection.

**Figure 20.18**
**A demonstration of cathodic protection.** The nails are in a gel containing phenolphthalein indicator and potassium ferricyanide. Iron corrosion yields $Fe^{2+}$, which reacts with ferricyanide ion to give a dark blue precipitate. Where $OH^-$ forms, phenolphthalein appears pink. *Left:* The unprotected nail. *Right:* The nail has magnesium wrapped around the center. Note that no corrosion (dark blue or black region) appears on it.

> **CONCEPT CHECK 20.4**
>
> Keeping in mind that seawater contains a number of ions, explain why seawater corrodes iron much faster than freshwater.

# Electrolytic Cells

<image ▶ **Video no. 45 (Series D), "Hoffman Electrolysis Apparatus."**

An *electrolytic cell,* you may recall, is an electrochemical cell in which an electric current drives an otherwise nonspontaneous reaction. *The process of producing a chemical change in an electrolytic cell* is called **electrolysis.** Many important substances, including aluminum and chlorine, are produced commercially by electrolysis. We will begin by looking at the electrolysis of molten salts.

## 20.9 Electrolysis of Molten Salts

Figure 20.19 shows a simple electrolytic cell. Wires from a battery are connected to electrodes that dip into molten sodium chloride. (NaCl melts at 801°C.) At the electrode connected to the negative pole of the battery, globules of sodium metal form; chlorine gas evolves from the other electrode. The half-reactions are

$$Na^+(l) + e^- \longrightarrow Na(l)$$
$$Cl^-(l) \longrightarrow \tfrac{1}{2}Cl_2(g) + e^-$$

**Figure 20.19**
**Electrolysis of molten sodium chloride.** Sodium metal forms at the cathode from the reduction of $Na^+$ ion; chlorine gas forms at the anode from the oxidation of $Cl^-$ ion. Sodium metal is produced commercially this way, although the commercial cell must be designed to collect the products and to keep them away from one another.

As noted earlier (Section 20.2), the *anode* is the electrode at which oxidation occurs, and the *cathode* is the electrode at which reduction occurs (these definitions hold for electrolytic cells as well as for voltaic cells). Thus, during the electrolysis of molten NaCl, the reduction of $Na^+$ to Na occurs at the cathode, and the oxidation of $Cl^-$ to $Cl_2$ occurs at the anode (note the labeling of electrodes in Figure 20.19).

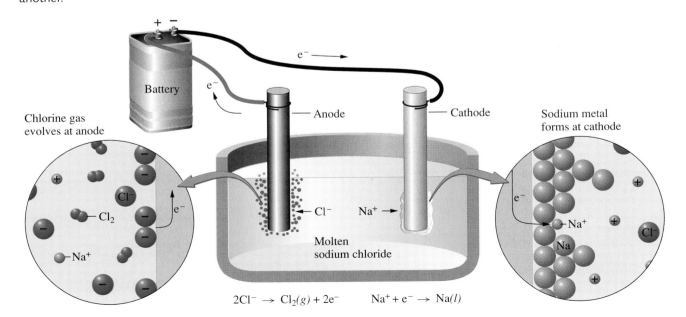

Chlorine gas evolves at anode

Anode

Cathode

Sodium metal forms at cathode

Molten sodium chloride

$$2Cl^- \rightarrow Cl_2(g) + 2e^- \qquad Na^+ + e^- \rightarrow Na(l)$$

**Figure 20.20**
**A Downs cell for the preparation of sodium metal.** In this commercial cell, sodium is produced by the electrolysis of molten sodium chloride. The salt contains calcium chloride, which is added to lower the melting point of the mixture. Liquid sodium forms at the cathode, where it rises to the top of the molten salt and collects in a tank. Chlorine gas is a by-product.

Chlorine gas

Add sodium chloride

Molten sodium chloride

Iron cathode

Liquid sodium

Graphite anode        +

This electrolysis of molten NaCl is used commercially to obtain sodium metal from sodium chloride. A **Downs cell** is *a commercial electrochemical cell used to obtain sodium metal by the electrolysis of molten sodium chloride* (Figure 20.20). The cell is constructed to keep the products of the electrolysis separate, because they would otherwise react. Calcium chloride is added to the sodium chloride to lower the melting point from 801°C for NaCl to about 580°C for the mixture. (Remember that the melting point, or freezing point, of a substance is lowered by the addition of a solute.) You obtain the cell reaction by adding the half-reactions.

$$Na^+(l) + e^- \longrightarrow Na(l)$$
$$\underline{\qquad\qquad Cl^-(l) \longrightarrow \tfrac{1}{2}Cl_2(g) + e^-}$$
$$Na^+(l) + Cl^-(l) \longrightarrow Na(l) + \tfrac{1}{2}Cl_2(g)$$

A number of other reactive metals are obtained by the electrolysis of a molten salt or ionic compound. Lithium, magnesium, and calcium metals are all obtained by the electrolysis of the chlorides. The first commercial preparation of sodium metal adapted the method used by Humphry Davy when he discovered the element in 1807. Davy electrolyzed molten sodium hydroxide, NaOH, whose melting point (318°C) is relatively low for an ionic compound. The half-reactions are

$$Na^+(l) + e^- \longrightarrow Na(l) \qquad\qquad\text{(cathode)}$$
$$4OH^-(l) \longrightarrow O_2(g) + 2H_2O(g) + 4e^- \qquad\qquad\text{(anode)}$$

Many of the commercial uses of electrolysis involve aqueous solutions. We will look at electrolysis in aqueous solution in the next section.

**EXERCISE 20.15**

Write the half-reactions for the electrolysis of the following molten compounds:
a. KCl; b. KOH.

See Problems 20.83 and 20.84.

## 20.10 Aqueous Electrolysis

▶ Video no. 28 (Series A), "Electrolysis of Water."

▶ Video Series C, "Electrolysis of Water."

In the electrolysis of a molten salt, the possible half-reactions are usually limited to those involving ions from the salt. When you electrolyze an aqueous solution of an ionic compound, however, you must consider the possibility that water is involved at one or both electrodes. Let us look at the possible half-reactions involving water.

Water can be reduced or oxidized in half-reactions, and you can easily obtain these half-reactions. To do this, first note the species likely to be involved. In addition to $H_2O$, these are $H_2$, $O_2$, $H^+$, and $OH^-$. Only $H_2$ and $O_2$ involve a change of oxidation state. Hydrogen in $H_2$ has a lower oxidation state (0) than in $H_2O$ (+1), whereas the oxygen in $O_2$ has a higher oxidation state (0) than in $H_2O$ (−2). Therefore, you can reduce water to $H_2$ or oxidize it to $O_2$.

Consider the reduction half-reaction. It must involve the reduction of $H_2O$ to $H_2$. You need to balance the half-reaction by putting an oxygen-containing species on the right. The only species in which there are no changes in oxidation states from those in $H_2O$ is $OH^-$. The balanced half-reaction is

$$2H_2O(l) + 2e^- \longrightarrow H_2(g) + 2OH^-(aq)$$

You can obtain the oxidation half-reaction for water in a similar way. It involves the oxidation of $H_2O$ to $O_2$. You need to put a hydrogen-containing species on the right side of the equation to balance it. The only species in which there is no change in oxidation state is $H^+$. The balanced half-reaction is

$$2H_2O(l) \longrightarrow O_2(g) + 4H^+(aq) + 4e^-$$

Now let us consider the electrolysis of different aqueous solutions and try to decide what the half-reactions might be. Once you have the half-reactions, you can obtain the overall chemical change due to the electrolysis.

### Electrolysis of Sulfuric Acid Solutions

To decide what is likely to happen during the electrolysis of a solution of sulfuric acid, $H_2SO_4$, you must consider the half-reactions involving ionic species from $H_2SO_4$, in addition to those involving water. Sulfuric acid is a strong acid and ionizes completely into $H^+$ and $HSO_4^-$. The $HSO_4^-$ ion is relatively strong and for the most part ionizes into $H^+$ and $SO_4^{2-}$. Therefore, the species you have to look at are $H^+$, $SO_4^{2-}$, and $H_2O$.

At the cathode, the possible reduction half-reactions are

$$2H^+(aq) + 2e^- \longrightarrow H_2(g)$$
$$2H_2O(l) + 2e^- \longrightarrow H_2(g) + 2OH^-(aq)$$

At the anode, the possible oxidation half-reactions are

$$2SO_4^{2-}(aq) \longrightarrow S_2O_8^{2-}(aq) + 2e^-$$
$$2H_2O(l) \longrightarrow O_2(g) + 4H^+(aq) + 4e^-$$

In the first half-reaction, the sulfate ion, $SO_4^{2-}$, is oxidized to the peroxydisulfate ion, $S_2O_8^{2-}$. The $S_2O_8^{2-}$ ion contains the peroxy group —O—O— and therefore has O in the −1 oxidation state. Oxygen in this half-reaction is oxidized from oxidation state −2 to −1.

Consider the cathode reaction. As you can easily show, the two reduction half-reactions (reduction of $H^+$ and reduction of $H_2O$) in acidic solution are essentially equivalent, although this is not immediately obvious. Consider the reduction of water to give $H_2$ and $OH^-$. In acidic solution, $OH^-$ reacts with $H^+$ to give $H_2O$. The net result is obtained by adding the two reactions.

$$\begin{array}{r} 2H_2O(l) + 2e^- \longrightarrow H_2(g) + 2OH^-(aq) \\ 2H^+(aq) + 2OH^-(aq) \longrightarrow 2H_2O(l) \\ \hline 2H^+(aq) + 2e^- \longrightarrow H_2(g) \end{array}$$

You can consider the cathode reaction to be the reduction of $H^+$.

For the anode, consider the oxidation half-reactions and their oxidation potentials (which equal the electrode potentials with signs reversed).

$$2SO_4^{2-}(aq) \longrightarrow S_2O_8^{2-}(aq) + 2e^-; \; -E° = -2.01 \text{ V}$$

$$2H_2O(l) \longrightarrow O_2(g) + 4H^+(aq) + 4e^-; \; -E° = -1.23 \text{ V}$$

The species whose oxidation half-reaction has the larger (less negative) oxidation potential is the more easily oxidized. Therefore, under standard conditions, you expect $H_2O$ to be oxidized in preference to $SO_4^{2-}$.

A note of caution must be added here. Electrode potentials are measured under conditions in which the half-reactions are at or very near equilibrium. In electrolysis, in which the half-reactions can be far from equilibrium, you may have to supply a larger voltage than predicted from electrode potentials. This additional voltage, called *overvoltage*, can be fairly large (several tenths of a volt), particularly for forming a gas. This means that in trying to predict which of two half-reactions is the one that actually occurs at an electrode, you must be especially careful if the electrode potentials are within several tenths of a volt of one another. For example, the standard oxidation potential for $O_2$ is $-1.23$ V, but because of overvoltage, the effective oxidation potential may be several tenths of a volt lower (more negative). The sulfate ion is so difficult to oxidize ($-E° = -2.01$ V), however, that the previous conclusion about the anode half-reaction in electrolyzing aqueous sulfuric acid is unaltered.

You obtain the cell reaction by adding the half-reactions that occur at the electrodes.

$$\begin{array}{lr} 2 \times [2H^+(aq) + 2e^- \longrightarrow H_2(g)] & \text{(cathode)} \\ 2H_2O(l) \longrightarrow O_2(g) + 4H^+(aq) + 4e^- & \text{(anode)} \\ \hline 2H_2O(l) \longrightarrow 2H_2(g) + O_2(g) & \end{array}$$

The net cell reaction is simply the *electrolysis of water*. Where electricity is cheap, hydrogen can be prepared commercially by the electrolysis of water. Here $H_2SO_4$ was used as an electrolyte; however, various electrolytes can be used, including NaCl.

## Electrolysis of Sodium Chloride Solutions

When you electrolyze an aqueous solution of sodium chloride, NaCl, the possible species involved in half-reactions are $Na^+$, $Cl^-$, and $H_2O$. The possible cathode half-reactions are

$$Na^+(aq) + e^- \longrightarrow Na(s); \; E° = -2.71 \text{ V}$$

$$2H_2O(l) + 2e^- \longrightarrow H_2(g) + 2OH^-(aq); \; E° = -0.83 \text{ V}$$

▶ The $OH^-$ concentration is actually $1 \times 10^{-7}$ *M*. From Nernst's equation, *E* for the reduction of $H_2O$ is $-0.41$ V; so you still conclude that $H_2O$ is reduced in preference to $Na^+$.

Under standard conditions, you expect $H_2O$ to be reduced in preference to $Na^+$, which agrees with what you observe. Hydrogen gas evolves at the cathode. ◀

The possible anode half-reactions are

$$2Cl^-(aq) \longrightarrow Cl_2(g) + 2e^-; \; -E° = -1.36 \text{ V}$$

$$2H_2O(l) \longrightarrow O_2(g) + 4H^+(aq) + 4e^-; \; -E° = -1.23 \text{ V}$$

**Figure 20.21**
**Electrolysis of aqueous potassium iodide.** Iodine forms at the anode by the oxidation of $I^-$ ion. Iodine reacts with iodide ion to form $I_3^-$ ion, which is red-brown. Hydrogen bubbles and hydroxide ion form at the cathode by the reduction of water.

Under standard-state conditions, you might expect $H_2O$ to be oxidized in preference to $Cl^-$. However, the potentials are close and overvoltages at the electrodes could alter this conclusion.

It is possible, nevertheless, to give a general statement about the product expected at the anode. Electrode potentials, as you have seen, depend on concentrations. It turns out that when the solution is concentrated enough in $Cl^-$, $Cl_2$ is the product; but in dilute solution, $O_2$ is the product. To see this, you would simply apply the Nernst equation to the $Cl^-|Cl_2$ half-reaction.

$$2Cl^-(aq) \longrightarrow Cl_2(g) + 2e^-$$

Starting with very dilute NaCl solutions, you would find that the oxidation potential of $Cl^-$ is very negative, so $H_2O$ is reduced in preference to $Cl^-$. But as you increase the NaCl concentration, you would find that the oxidation potential of $Cl^-$ increases until eventually $Cl^-$ is oxidized in preference to $H_2O$. The product changes from $O_2$ to $Cl_2$.

The half-reactions and cell reaction for the electrolysis of aqueous sodium chloride to chlorine and hydroxide ion are as follows:

$$\begin{array}{ll} 2H_2O(l) + 2e^- \longrightarrow H_2(g) + 2OH^-(aq) & \text{(cathode)} \\ 2Cl^-(aq) \longrightarrow Cl_2(g) + 2e^- & \text{(anode)} \\ \hline 2H_2O(l) + 2Cl^-(aq) \longrightarrow H_2(g) + Cl_2(g) + 2OH^-(aq) & \end{array}$$

Because the electrolysis started with sodium chloride, the cation in the electrolyte solution is $Na^+$. When you evaporate the electrolyte solution at the cathode, you obtain sodium hydroxide, NaOH. Figure 20.21 shows the similar electrolysis of aqueous potassium iodide, KI.

The electrolysis of aqueous sodium chloride is the basis of the *chlor-alkali indus-try,* the major commercial source of chlorine and sodium hydroxide. Commercial cells are of several types, but in each the main problem is to keep the products separate, because chlorine reacts with aqueous sodium hydroxide.

Figure 20.22 shows a modern **chlor-alkali membrane cell,** *a cell for the electrolysis of aqueous sodium chloride in which the anode and cathode compartments*

**Figure 20.22**
**A chlor-alkali membrane cell.** Sodium chloride solution enters the anode compartment, where $Cl^-$ ion is oxidized to $Cl_2$. Sodium ion migrates from the anode compartment through the membrane to the cathode compartment. Here water is reduced to hydrogen and $OH^-$ ion. Sodium hydroxide solution is removed at the bottom of the cathode compartment.

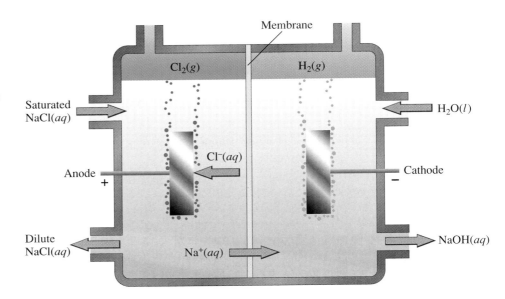

**Figure 20.23
A chlor-alkali mercury
cell.** Chloride ion oxidizes
to chlorine gas at the graphite
anodes (+ electrodes),
and sodium deposits at
the mercury cathode.
Sodium amalgam (a liquid
sodium–mercury alloy)
circulates to the amalgam
decomposer, where sodium
reacts with water to form
NaOH(aq) and $H_2$.

are separated by a special plastic membrane that allows only cations to pass through it. Sodium chloride solution is added to the anode compartment, where chloride ion is oxidized to chlorine. The sodium ions carry the current from the anode to the cathode by passing through the membrane. Water is added at the top of the cathode compartment, where it is reduced to hydroxide ion, and sodium hydroxide solution is removed at the bottom of the cathode compartment.

The older **chlor-alkali mercury cell** is *a cell for the electrolysis of aqueous sodium chloride in which mercury metal is used as the cathode* (Figure 20.23). At the mercury cathode, sodium ion is reduced in preference to water. Sodium ion is reduced to sodium to form a liquid sodium–mercury alloy called sodium amalgam. (An *amalgam* is an alloy of mercury with any of various other metals.)

$$Na^+(aq) + e^- \longrightarrow Na(amalgam)$$

Sodium amalgam circulates from the electrolytic cell to the amalgam decomposer. The amalgam and graphite particles in the decomposer form the electrodes of many voltaic cells, in which sodium reacts with water to give sodium hydroxide solution and hydrogen gas. Loss of mercury from these cells has been a source of mercury pollution of waterways. Under environmental regulations, the discharge water has been filtered to reduce the mercury losses.

## Electroplating of Metals

Many metals are protected from corrosion by plating them with other metals. Zinc coatings are often used to protect steel, because the coat protects the steel by cathodic protection even when the zinc coat is scratched (see end of Section 19.8). A thin zinc coating can be applied to steel by *electrogalvanizing,* or zinc electroplating. (Galvanized steel has a thick zinc coating obtained by dipping the object in molten zinc.) The steel object is placed in a bath of zinc salts and made the cathode in an electrolytic cell. The cathode half-reaction is

$$Zn^{2+}(aq) + 2e^- \longrightarrow Zn(s)$$

**Figure 20.24**
**Purification of copper by electrolysis.** *Left:* Copper(II) ion leaves the anode and plates out on the cathode. Reactive ions, such as $Zn^{2+}$, remain in solution, and unreactive substances (including gold) collect as a mud under the anode. *Right:* Very pure copper sheets alternate with impure copper slabs in this electrolytic tank. The pure copper sheets grow in size and are removed in about a month.

Electrolysis is also used to purify some metals. For example, copper for electrical use, which must be very pure, is purified by electrolysis. Slabs of impure copper serve as anodes, and pure copper sheets serve as cathodes; the electrolyte bath is copper(II) sulfate, $CuSO_4$ (Figure 20.24). During the electrolysis, copper(II) ions leave the anode slabs and plate out on the cathode sheets. Less reactive metals, such as gold, silver, and platinum, that were present in the impure copper form a valuable mud that collects on the bottom of the electrolytic cell. Metals more reactive than copper remain as ions in the electrolytic bath. After about a month in the electrolytic cell, the pure copper cathodes are much enlarged and are removed from the cell bath.

▶ An alternate example is provided in the Instructor's Resource Manual.

**EXAMPLE 20.13**

**Predicting the Half-Reactions in an Aqueous Electrolysis**

What do you expect to be the half-reactions in the electrolysis of aqueous copper(II) sulfate?

**SOLUTION**

The species you should consider for half-reactions are $Cu^{2+}(aq)$, $SO_4^{2-}(aq)$, and $H_2O$.

Possible cathode half-reactions are

$$Cu^{2+}(aq) + 2e^- \longrightarrow Cu(s); \quad E° = 0.34 \text{ V}$$

$$2H_2O(l) + 2e^- \longrightarrow H_2(g) + 2OH^-(aq); \quad E° = -0.83 \text{ V}$$

Because the copper electrode potential is much larger than the reduction potential of water, you expect $Cu^{2+}$ to be reduced.

Possible anode half-reactions are

$$2SO_4{}^{2-}(aq) \longrightarrow S_2O_8{}^{2-}(aq) + 2e^-; \quad -E° = -2.01 \text{ V}$$

$$2H_2O(l) \longrightarrow O_2(g) + 4H^+(aq) + 4e^-; \quad -E° = -1.23 \text{ V}$$

You expect $H_2O$ to be oxidized.

The expected half-reactions are

$$\mathbf{Cu^{2+}(aq) + 2e^- \longrightarrow Cu(s)}$$

$$\mathbf{2H_2O(l) \longrightarrow O_2(g) + 4H^+(aq) + 4e^-}$$

---

**EXERCISE 20.16**

Give the half-reactions that occur when aqueous silver nitrate is electrolyzed. Nitrate ion is not oxidized during the electrolysis.

---

See Problems 20.85 and 20.86.

## 20.11 Stoichiometry of Electrolysis

In 1831 and 1832, the British chemist and physicist Michael Faraday showed that the amounts of substances released at the electrodes during electrolysis are related to the total charge that has flowed in the electric circuit. ◀ If you look at the electrode reactions, you see that the relationship is stoichiometric.

When molten sodium chloride is electrolyzed, sodium ions migrate to the cathode, where they react with electrons.

$$Na^+ + e^- \longrightarrow Na(l)$$

Similarly, chloride ions migrate to the anode and release electrons.

$$Cl^- \longrightarrow \tfrac{1}{2}Cl_2(g) + e^-$$

Therefore, when one mole of electrons reacts with sodium ions, one faraday of charge (the magnitude of charge on one mole of electrons) passes through the circuit. One mole of sodium metal is deposited at one electrode, and one-half mole of chlorine gas evolves at the other electrode.

What is new in this type of stoichiometric problem is the measurement of numbers of electrons. You do not weigh them as you do substances. Rather, you measure the quantity of electric charge that has passed through the circuit. You use the following fact:

One faraday ($9.65 \times 10^4$ C) is equivalent to the charge on one mole of electrons.

If you know the current in a circuit and the length of time it has been flowing, you can calculate the electric charge.

$$\text{Electric charge} = \text{electric current} \times \text{time lapse}$$

Corresponding units are

$$\text{Coulombs} = \text{amperes} \times \text{seconds}$$

▶ Faraday's results were summarized in two laws: (1) The quantity of a substance liberated at an electrode is directly proportional to the quantity of electric charge that has flowed in the circuit. (2) For a given quantity of electric charge, the amount of any metal deposited is proportional to its equivalent weight (atomic weight divided by the charge on the metal ion). These laws follow directly from the stoichiometry of electrolysis.

▶ Experiments in General Chemistry, 20B, "Electrochemistry," and 20C, "Estimating an Electrode Potential."

The **ampere (A)** is *the base unit of current in the International System (SI).* The coulomb (C), the SI unit of electric charge, is equivalent to an ampere-second. So a current of 0.50 amperes flowing for 84 seconds gives a charge of 0.50 A × 84 s = 42 A · s, or 42 C.

If you are given the amount of substance produced at an electrode and the time of electrolysis, you can determine the current. If you are given the current and the time of electrolysis, you can calculate the amount of substance produced at an electrode. The next two examples illustrate these calculations.

▶ An alternate example is provided in the Instructor's Resource Manual.

## EXAMPLE 20.14

### Calculating the Amount of Charge from the Amount of Product in an Electrolysis

When an aqueous solution of copper(II) sulfate, $CuSO_4$, is electrolyzed, copper metal is deposited.

$$Cu^{2+}(aq) + 2e^- \longrightarrow Cu(s)$$

(The other electrode reaction gives oxygen: $2H_2O \longrightarrow O_2 + 4H^+ + 4e^-$.) If a constant current was passed for 5.00 h and 404 mg of copper metal was deposited, what was the current?

**🧩 PROBLEM STRATEGY**

The electrode equation for copper says that 1 mol Cu is equivalent to 2 mol $e^-$. You can use this in the conversion of grams of Cu to moles of electrons needed to deposit this amount of copper. The Faraday constant (which says that one mole of electrons is equivalent to $9.65 \times 10^4$ C) converts moles of electrons to coulombs. The conversions are

$$g\ Cu \longrightarrow mol\ Cu \longrightarrow mol\ e^- \longrightarrow coulombs\ (C)$$

The current in amperes (A) equals the charge in coulombs divided by the time in seconds.

**SOLUTION**

The conversion of grams of Cu to coulombs required to deposit this amount of copper is

$$0.404\ g\ Cu \times \frac{1\ mol\ Cu}{63.6\ g\ Cu} \times \frac{2\ mol\ e^-}{1\ mol\ Cu} \times \frac{9.65 \times 10^4\ C}{1\ mol\ e^-} = 1.2\underline{2}6 \times 10^3\ C$$

The time lapse, 5.00 h, equals $1.80 \times 10^4$ s. Thus,

$$Current = \frac{charge}{time} = \frac{1.226 \times 10^3\ C}{1.80 \times 10^4\ s} = \mathbf{6.81 \times 10^{-2}\ A}$$

## EXERCISE 20.17

A constant electric current deposits 365 mg of silver in 216 min from an aqueous silver nitrate solution. What is the current?

See Problems 20.87 and 20.88.

◄ An alternate example is provided in the Instructor's Resource Manual.

## EXAMPLE 20.15

### Calculating the Amount of Product from the Amount of Charge in an Electrolysis

When an aqueous solution of potassium iodide is electrolyzed using platinum electrodes, the half-reactions are

$$2I^-(aq) \longrightarrow I_2(aq) + 2e^-$$
$$2H_2O(l) + 2e^- \longrightarrow H_2(g) + 2OH^-(aq)$$

How many grams of iodine are produced when a current of 8.52 mA flows through the cell for 10.0 min?

 **PROBLEM STRATEGY**

The current in amperes multiplied by the time in seconds equals the charge in coulombs. Convert the charge in coulombs to moles of electrons. Then, noting that each mole of iodine produced requires 2 mol $e^-$, convert mol $e^-$ to mol $I_2$. Finally, convert mol $I_2$ to g $I_2$.

$$\text{Coulombs} \longrightarrow \text{mol } e^- \longrightarrow \text{mol } I_2 \longrightarrow \text{g } I_2$$

**SOLUTION**

When the current flows for $6.00 \times 10^2$ s (10.0 min), the amount of charge is

$$8.52 \times 10^{-3} \text{ A} \times 6.00 \times 10^2 \text{ s} = 5.11 \text{ C}$$

Note that two moles of electrons are equivalent to one mole of $I_2$. Hence,

$$5.11 \text{ C} \times \frac{1 \text{ mol } e^-}{9.65 \times 10^4 \text{ C}} \times \frac{1 \text{ mol } I_2}{2 \text{ mol } e^-} \times \frac{254 \text{ g } I_2}{1 \text{ mol } I_2} = \mathbf{6.73 \times 10^{-3} \text{ g } I_2}$$

## EXERCISE 20.18

How many grams of oxygen are liberated by the electrolysis of water with a current of 0.0565 A after $1.85 \times 10^4$ s?

See Problems 20.89 and 20.90.

# A Checklist for Review

## Important Terms

electrochemical cell (p. 845)
voltaic (galvanic)
  cell (p. 846)
electrolytic cell (p. 846)
half-cell (20.2)
salt bridge (20.2)
anode (20.2)
cathode (20.2)
cell reaction (20.2)

potential difference (20.4)
volt (V) (20.4)
Faraday constant (F) (20.4)
electromotive force
  (emf) (20.4)
standard emf (20.5)
standard electrode
  potential (20.5)

Nernst equation (20.7)
zinc–carbon (Leclanché)
  dry cell (20.8)
alkaline dry cell (20.8)
lithium–iodine
  battery (20.8)
lead storage cell (20.8)
nickel–cadmium cell (20.8)

fuel cell (20.8)
electrolysis (p. 874)
Downs cell (20.9)
chlor-alkali membrane
  cell (20.10)
chlor-alkali mercury
  cell (20.10)
ampere (A) (20.11)

## Key Equations

$$w_{\max} = -nFE_{\text{cell}}$$

$$E^{\circ}_{\text{cell}} = \frac{0.0592}{n} \log K \quad \text{(values in volts at } 25°C\text{)}$$

$$E^{\circ}_{\text{cell}} = E^{\circ}_{\text{cathode}} - E^{\circ}_{\text{anode}}$$

$$\Delta G^{\circ} = -nFE^{\circ}_{\text{cell}}$$

$$E_{\text{cell}} = E^{\circ}_{\text{cell}} - \frac{0.0592}{n} \log Q \quad \text{(values in volts at } 25°C\text{)}$$

## Summary of Facts and Concepts

*Oxidation–reduction reactions* involve a transfer of electrons from one species to another. The *half-reaction method* can be applied to balancing oxidation–reduction reactions in acidic and basic solutions. Many of the principles required for balancing these reactions were presented in Chapter 4.

Electrochemical cells are of two types: voltaic and electrolytic. A *voltaic cell* uses a spontaneous chemical reaction to generate an electric current. It does this by physically separating the reaction into its oxidation and reduction half-reactions. These half-reactions take place in *half-cells*. The half-cell in which reduction occurs is called the *cathode;* the half-cell in which oxidation occurs is called the *anode*. Electrons flow in the external circuit from the anode to the cathode.

The *electromotive force (emf)* is the maximum voltage of a voltaic cell. It can be directly related to the maximum work that can be done by the cell. A *standard electrode potential,* or reduction potential, refers to the potential of an electrode in which molar concentrations and gas pressures (in atmospheres) have unit values. A table of standard electrode potentials is useful for establishing the direction of spontaneity of an oxidation–reduction reaction and for calculating the *standard emf* of a cell.

The standard free-energy change, standard emf, and equilibrium constant are all related. Knowing one, you can calculate the others. Electrochemical measurements can therefore provide equilibrium or thermodynamic information.

An electrode potential depends on concentrations of the electrode substances, according to the *Nernst equation*. Because of this relationship, cell emf's can be used to measure ion concentrations. This is the basic principle of a pH meter, a device that measures the hydrogen-ion concentration.

Voltaic cells are used commercially as portable energy sources (batteries). In addition, the basic principle of the voltaic cell is employed in the *cathodic protection* of buried pipelines and tanks.

*Electrolytic cells* represent another type of electrochemical cell. They use an external voltage source to push a reaction in a nonspontaneous direction. The electrolysis of an aqueous solution often involves the oxidation or reduction of water at the electrodes. Electrolysis of concentrated sodium chloride solution, for example, gives hydrogen at the cathode. The amounts of substances released at an electrode are related to the amount of charge passed through the cell. This relationship is stoichiometric and follows from the electrode reactions.

## Operational Skills

1. **Balancing equations in acidic and basic solutions by the half-reaction method** Given the skeleton equation for an oxidation–reduction equation, complete and balance it. **(EXAMPLES 20.1, 20.2)**
2. **Sketching and labeling a voltaic cell** Given a description of a voltaic cell, sketch the cell, labeling the anode and cathode, and give the directions of electron flow and ion migration. **(EXAMPLE 20.3)**
3. **Writing the cell reaction from the cell notation** Given the notation for a voltaic cell, write the overall cell reaction **(EXAMPLE 20.4)**. Alternatively, given the cell reaction, write the cell notation.
4. **Calculating the quantity of work from a given amount of cell reactant** Given the emf and overall reaction for a voltaic cell, calculate the maximum work that can be obtained from a given amount of reactant. **(EXAMPLE 20.5)**
5. **Determining the relative strengths of oxidizing and reducing agents** Given a table of standard electrode potentials, list oxidizing or reducing agents by increasing strength. **(EXAMPLE 20.6)**

6. **Determining the direction of spontaneity from electrode potentials** Given standard electrode potentials, decide the direction of spontaneity for an oxidation–reduction reaction under standard conditions. **(EXAMPLE 20.7)**
7. **Calculating the emf from standard potentials** Given standard electrode potentials, calculate the standard emf of a voltaic cell. **(EXAMPLE 20.8)**
8. **Calculating the free-energy change from electrode potentials** Given standard electrode potentials, calculate the standard free-energy change for an oxidation–reduction reaction. **(EXAMPLE 20.9)**
9. **Calculating the cell emf from free-energy change** Given a table of standard free energies of formation, calculate the standard emf of a voltaic cell. **(EXAMPLE 20.10)**
10. **Calculating the equilibrium constant from cell emf** Given standard potentials (or standard emf), calculate the equilibrium constant for an oxidation–reduction reaction. **(EXAMPLE 20.11)**

**11. Calculating the cell emf for nonstandard conditions** Given standard electrode potentials and the concentrations of substances in a voltaic cell, calculate the cell emf. **(EXAMPLE 20.12)**

**12. Predicting the half-reactions in an aqueous electrolysis** Using values of electrode potentials, decide which electrode reactions actually occur in the electrolysis of an aqueous solution. **(EXAMPLE 20.13)**

**13. Relating the amounts of charge and product in an electrolysis** Given the amount of product obtained by electrolysis, calculate the amount of charge that flowed **(EXAMPLE 20.14)**. Given the amount of charge that flowed, calculate the amount of product obtained by electrolysis **(EXAMPLE 20.15)**.

## Review Questions

**20.1** Describe the difference between a voltaic cell and an electrolytic cell.

**20.2** Define *cathode* and *anode*.

**20.3** What is the SI unit of electrical potential?

**20.4** Define the *faraday*.

**20.5** Why is it necessary to measure the voltage of a voltaic cell when no current is flowing to obtain the cell emf?

**20.6** How are standard electrode potentials defined?

**20.7** Express the SI unit of energy as the product of two electrical units.

**20.8** Give the mathematical relationships between the members of each possible pair of the three quantities $\Delta G°$, $E°_{cell}$, and $K$.

**20.9** Using the Nernst equation, explain how the corrosion of iron would be affected by the pH of a water drop.

**20.10** Describe the zinc–carbon, or Leclanché, dry cell and the lead storage battery.

**20.11** What is a fuel cell? Describe an example.

**20.12** Explain the electrochemistry of rusting.

**20.13** Iron may be protected by coating with tin (tin cans) or with zinc (galvanized iron). Galvanized iron does not corrode as long as zinc is present. By contrast, when a tin can is scratched, the exposed iron underneath corrodes rapidly. Explain the difference between zinc and tin as protective coatings against iron corrosion.

**20.14** The electrolysis of water is often done by passing a current through a dilute solution of sulfuric acid. What is the function of the sulfuric acid?

**20.15** Describe a method for the preparation of sodium metal from sodium chloride.

**20.16** Potassium was discovered by the British chemist Humphry Davy when he electrolyzed molten potassium hydroxide. What would be the anode reaction?

**20.17** Briefly explain why different products are obtained from the electrolysis of molten NaCl and the electrolysis of a dilute aqueous solution of NaCl.

**20.18** Write the Nernst equation for the electrode reaction $2Cl^-(aq) \longrightarrow Cl_2(g) + 2e^-$. With this equation, explain why the electrolysis of concentrated sodium chloride solution might be expected to release chlorine gas rather than oxygen gas at the anode.

## Conceptual Problems

**20.19** Keeping in mind that aqueous $Cu^{2+}$ is blue and aqueous $Zn^{2+}$ is colorless, predict what you would observe over a several-day period if you performed the following experiments.
a. A strip of Zn is placed into a beaker containing aqueous $Zn^{2+}$.
b. A strip of Cu is placed into a beaker containing aqueous $Cu^{2+}$.
c. A strip of Zn is placed into a beaker containing aqueous $Cu^{2+}$.
d. A strip of Cu is placed into a beaker containing aqueous $Zn^{2+}$.

**20.20** You are working at a plant that manufactures batteries. A client comes to you and asks for a 6.0-V battery that is made

from silver and cadmium. Assuming that you are running the battery under standard conditions, how should it be constructed?

**20.21** The composition of the hull of a submarine is mostly iron. Pieces of zinc, called "zincs," are placed in contact with the hull throughout the inside of the submarine. Why is this done?

**20.22** You place a battery in a flashlight in which all of the electrochemical reactions have reached equilibrium. What do you expect to observe when you turn on the flashlight? Explain your answer.

**20.23** The difference between a "heavy-duty" and a regular zinc–carbon battery is that the zinc can in the heavy-duty battery is thicker walled. What makes this battery heavy-duty in terms of output?

**20.24** From an electrochemical standpoint, what metal, other than zinc, would be a reasonable candidate to coat a piece of iron to prevent corrosion (oxidation)?

**20.25** Household bleach is a solution containing $NaClO$, which is a very strong oxidizing agent. Provide a brief explanation as to how bleach is able to "whiten" stains in clothing.

**20.26** The development of lightweight batteries is an ongoing research effort combining many of the physical sciences.

You are a member of an engineering team trying to develop a lightweight battery that will effectively react with $O_2(g)$ from the atmosphere as an oxidizing agent. A reducing agent must be chosen for this battery that will be lightweight, have nontoxic products, and react spontaneously with oxygen. Using data from Appendix I, suggest a likely reducing agent, being sure the above conditions are met. Are there any drawbacks to your selection?

## Practice Problems

### Balancing Oxidation–Reduction Equations

**20.27** Balance the following oxidation–reduction equations. The reactions occur in acidic solution.
a. $Cr_2O_7^{2-} + C_2O_4^{2-} \longrightarrow Cr^{3+} + CO_2$
b. $Cu + NO_3^- \longrightarrow Cu^{2+} + NO$
c. $MnO_2 + HNO_2 \longrightarrow Mn^{2+} + NO_3^-$
d. $PbO_2 + Mn^{2+} + SO_4^{2-} \longrightarrow PbSO_4 + MnO_4^-$
e. $HNO_2 + Cr_2O_7^{2-} \longrightarrow Cr^{3+} + NO_3^-$

**20.28** Balance the following oxidation–reduction equations. The reactions occur in acidic solution.
a. $Mn^{2+} + BiO_3^- \longrightarrow MnO_4^- + Bi^{3+}$
b. $Cr_2O_7^{2-} + I^- \longrightarrow Cr^{3+} + IO_3^-$
c. $MnO_4^- + H_2SO_3 \longrightarrow Mn^{2+} + SO_4^{2-}$
d. $Cr_2O_7^{2-} + Fe^{2+} \longrightarrow Cr^{3+} + Fe^{3+}$
e. $As + ClO_3^- \longrightarrow H_3AsO_3 + HClO$

**20.29** Balance the following oxidation–reduction equations. The reactions occur in basic solution.
a. $Mn^{2+} + H_2O_2 \longrightarrow MnO_2 + H_2O$
b. $MnO_4^- + NO_2^- \longrightarrow MnO_2 + NO_3^-$
c. $Mn^{2+} + ClO_3^- \longrightarrow MnO_2 + ClO_2$
d. $MnO_4^- + NO_2 \longrightarrow MnO_2 + NO_3^-$
e. $Cl_2 \longrightarrow Cl^- + ClO_3^-$

**20.30** Balance the following oxidation–reduction equations. The reactions occur in basic solution.
a. $Cr(OH)_4^- + H_2O_2 \longrightarrow CrO_4^{2-} + H_2O$
b. $MnO_4^- + Br^- \longrightarrow MnO_2 + BrO_3^-$
c. $Co^{2+} + H_2O_2 \longrightarrow Co(OH)_3 + H_2O$
d. $Pb(OH)_4^{2-} + ClO^- \longrightarrow PbO_2 + Cl^-$
e. $Zn + NO_3^- \longrightarrow NH_3 + Zn(OH)_4^{2-}$

**20.31** Balance the following oxidation–reduction equations. The reactions occur in acidic or basic aqueous solution, as indicated.
a. $H_2S + NO_3^- \longrightarrow NO_2 + S_8$ (acidic)
b. $NO_3^- + Cu \longrightarrow NO + Cu^{2+}$ (acidic)
c. $MnO_4^- + SO_2 \longrightarrow SO_4^{2-} + Mn^{2+}$ (acidic)
d. $Bi(OH)_3 + Sn(OH)_3^- \longrightarrow Sn(OH)_6^{2-} + Bi$ (basic)

**20.32** Balance the following oxidation–reduction equations. The reactions occur in acidic or basic aqueous solution, as indicated.
a. $Hg_2^{2+} + H_2S \longrightarrow Hg + S_8$ (acidic)
b. $S^{2-} + I_2 \longrightarrow SO_4^{2-} + I^-$ (basic)
c. $Al + NO_3^- \longrightarrow Al(OH)_4^- + NH_3$ (basic)
d. $MnO_4^- + C_2O_4^{2-} \longrightarrow MnO_2 + CO_2$ (basic)

**20.33** Balance the following oxidation–reduction equations. The reactions occur in acidic or basic aqueous solution, as indicated.
a. $MnO_4^- + I^- \longrightarrow MnO_2 + IO_3^-$ (basic)
b. $Cr_2O_7^{2-} + Cl^- \longrightarrow Cr^{3+} + Cl_2$ (acidic)
c. $S_8 + NO_3^- \longrightarrow SO_2 + NO$ (acidic)
d. $H_2O_2 + MnO_4^- \longrightarrow O_2 + MnO_2$ (basic)
e. $Zn + NO_3^- \longrightarrow Zn^{2+} + N_2$ (acidic)

**20.34** Balance the following oxidation–reduction equations. The reactions occur in acidic or basic aqueous solution, as indicated.
a. $Cr_2O_7^{2-} + H_2O_2 \longrightarrow Cr^{3+} + O_2$ (acidic)
b. $CN^- + MnO_4^- \longrightarrow CNO^- + MnO_2$ (basic)
c. $Cr(OH)_4^- + OCl^- \longrightarrow CrO_4^{2-} + Cl^-$ (basic)
d. $Br_2 + SO_2 \longrightarrow Br^- + SO_4^{2-}$ (acidic)
e. $CuS + NO_3^- \longrightarrow Cu^{2+} + NO + S_8$ (acidic)

### Electrochemical Cells

**20.35** A voltaic cell is constructed from the following half-cells: a magnesium electrode in magnesium sulfate solution and a nickel electrode in nickel sulfate solution. The half-reactions are

$$Mg(s) \longrightarrow Mg^{2+}(aq) + 2e^-$$
$$Ni^{2+}(aq) + 2e^- \longrightarrow Ni(s)$$

Sketch the cell, labeling the anode and cathode (and the electrode reactions), and show the direction of electron flow and the movement of cations.

**20.36** Half-cells were made from a nickel rod dipping in a nickel sulfate solution and a copper rod dipping in a copper sulfate solution. The half-reactions in a voltaic cell using these half-cells were

$$Cu^{2+}(aq) + 2e^- \longrightarrow Cu(s)$$
$$Ni(s) \longrightarrow Ni^{2+}(aq) + 2e^-$$

Sketch the cell and label the anode and cathode, showing the corresponding electrode reactions. Give the direction of electron flow and the movement of cations.

**20.37** Zinc reacts spontaneously with silver ion.

$$Zn(s) + 2Ag^+(aq) \longrightarrow Zn^{2+}(aq) + 2Ag(s)$$

Describe a voltaic cell using this reaction. What are the half-reactions?

**20.38** Iron reacts spontaneously with copper(II) ion.

$$Fe(s) + Cu^{2+}(aq) \longrightarrow Fe^{2+}(aq) + Cu(s)$$

Obtain half-reactions for this, then describe a voltaic cell using these half-reactions.

**20.39** A silver oxide–zinc cell maintains a fairly constant voltage during discharge (1.60 V). The button form of this cell is used in watches, hearing aids, and other electronic devices. The half-reactions are

$$Zn(s) + 2OH^-(aq) \longrightarrow Zn(OH)_2(s) + 2e^-$$
$$Ag_2O(s) + H_2O(l) + 2e^- \longrightarrow 2Ag(s) + 2OH^-(aq)$$

Identify the anode and the cathode reactions. What is the overall reaction in the voltaic cell?

**20.40** A mercury battery, used for hearing aids and electric watches, delivers a constant voltage (1.35 V) for long periods. The half-reactions are

$$HgO(s) + H_2O(l) + 2e^- \longrightarrow Hg(l) + 2OH^-(aq)$$
$$Zn(s) + 2OH^-(aq) \longrightarrow Zn(OH)_2(s) + 2e^-$$

Which half-reaction occurs at the anode and which occurs at the cathode? What is the overall cell reaction?

## Voltaic Cell Notation

**20.41** Write the cell notation for a voltaic cell with the following half-reactions.

$$Ni(s) \longrightarrow Ni^{2+}(aq) + 2e^-$$
$$Pb^{2+}(aq) + 2e^- \longrightarrow Pb(s)$$

**20.42** Write the cell notation for a voltaic cell with the following half-reactions.

$$Al(s) \longrightarrow Al^{3+}(aq) + 3e^-$$
$$2H^+(aq) + 2e^- \longrightarrow H_2(g)$$

**20.43** Give the notation for a voltaic cell constructed from a hydrogen electrode (cathode) in 1.0 $M$ HCl and a nickel electrode (anode) in 1.0 $M$ NiSO$_4$ solution. The electrodes are connected by a salt bridge.

**20.44** A voltaic cell has an iron rod in 0.30 $M$ iron(III) chloride solution for the cathode and a zinc rod in 0.40 $M$ zinc sulfate solution for the anode. The half-cells are connected by a salt bridge. Write the notation for this cell.

**20.45** Write the overall cell reaction for the following voltaic cell.

$$Fe(s) | Fe^{2+}(aq) \| Ag^+(aq) | Ag(s)$$

**20.46** Write the overall cell reaction for the following voltaic cell.

$$Pt | H_2(g) | H^+(aq) \| Br_2(l) | Br^-(aq) | Pt$$

**20.47** Consider the voltaic cell

$$Cd(s) | Cd^{2+}(aq) \| Ni^{2+}(aq) | Ni(s)$$

Write the half-cell reactions and the overall cell reaction. Make a sketch of this cell and label it. Include labels showing the anode, cathode, and direction of electron flow.

**20.48** Consider the voltaic cell

$$Zn(s) | Zn^{2+}(aq) \| Cr^{3+}(aq) | Cr(s)$$

Write the half-cell reactions and the overall cell reaction. Make a sketch of this cell and label it. Include labels showing the anode, cathode, and direction of electron flow.

## Electrode Potentials and Cell emf's

**20.49** A voltaic cell whose cell reaction is

$$2Fe^{3+}(aq) + Zn(s) \longrightarrow 2Fe^{2+}(aq) + Zn^{2+}(aq)$$

has an emf of 0.72 V. What is the maximum electrical work that can be obtained from this cell per mole of iron(III) ion?

**20.50** A particular voltaic cell operates on the reaction

$$Zn(s) + Cl_2(g) \longrightarrow Zn^{2+}(aq) + 2Cl^-(aq)$$

giving an emf of 0.853 V. Calculate the maximum electrical work generated when 20.0 g of zinc metal is consumed.

**20.51** What is the maximum work you can obtain from 30.0 g of nickel in the following cell when the emf is 0.97 V?

$$Ni(s) | Ni^{2+}(aq) \| Ag^+(aq) | Ag(s)$$

**20.52** Calculate the maximum work available from 25.0 g of aluminum in the following cell when the emf is 1.15 V.

$$Al(s) | Al^{3+}(aq) \| H^+(aq) | O_2(g) | Pt$$

Note that O$_2$ is reduced to H$_2$O.

**20.53**   Order the following oxidizing agents by increasing strength under standard-state conditions: $O_2(g)$; $MnO_4^-(aq)$; $NO_3^-(aq)$ (in acidic solution).

**20.54**   Order the following oxidizing agents by increasing strength under standard-state conditions: $Ag^+(aq)$; $I_2(aq)$; $MnO_4^-(aq)$ (in acidic solution).

**20.55**   Consider the reducing agents $Cu^+(aq)$, $Zn(s)$, and $Fe(s)$. Which is strongest? Which is weakest?

**20.56**   Consider the reducing agents $Sn^{2+}(aq)$, $Cu(s)$, and $I^-(aq)$. Which is strongest? Which is weakest?

**20.57**   Consider the following reactions. Are they spontaneous in the direction written, under standard conditions at 25°C?
a.   $Sn^{4+}(aq) + 2Fe^{2+}(aq) \longrightarrow Sn^{2+}(aq) + 2Fe^{3+}(aq)$
b.   $4MnO_4^-(aq) + 12H^+(aq) \longrightarrow$
$\phantom{xxxxxxxx} 4Mn^{2+}(aq) + 5O_2(g) + 6H_2O(l)$

**20.58**   Answer the following questions by referring to standard electrode potentials at 25°C.
a.   Will dichromate ion, $Cr_2O_7^{2-}$, oxidize iron(II) ion in acidic solution under standard conditions?
b.   Will copper metal reduce $1.0\,M\ Ni^{2+}(aq)$ to metallic nickel?

**20.59**   What would you expect to happen when chlorine gas, $Cl_2$, at 1 atm pressure is bubbled into a solution containing $1.0\,M\ F^-$ and $1.0\,M\ Br^-$ at 25°C? Write a balanced equation for the reaction that occurs.

**20.60**   Dichromate ion, $Cr_2O_7^{2-}$, is added to an acidic solution containing $Br^-$ and $Mn^{2+}$. Write a balanced equation for any reaction that occurs. Assume standard conditions at 25°C.

**20.61**   Calculate the standard emf of the following cell at 25°C.
$$Cr(s)\,|\,Cr^{3+}(aq)\,\|\,Hg_2^{2+}(aq)\,|\,Hg(l)$$

**20.62**   Calculate the standard emf of the following cell at 25°C.
$$Sn(s)\,|\,Sn^{2+}(aq)\,\|\,Cu^{2+}(aq)\,|\,Cu(s)$$

**20.63**   What is the standard emf you would obtain from a cell at 25°C using an electrode in which $I^-(aq)$ is in contact with $I_2(s)$ and an electrode in which a chromium strip dips into a solution of $Cr^{3+}(aq)$?

**20.64**   What is the standard emf you would obtain from a cell at 25°C using an electrode in which $Hg_2^{2+}(aq)$ is in contact with mercury metal and an electrode in which an aluminum strip dips into a solution of $Al^{3+}(aq)$?

## Relationships Among $E°_{cell}$, $\Delta G°$, and $K$

**20.65**   Calculate the standard free-energy change at 25°C for the following reaction.
$$3Cu(s) + 2NO_3^-(aq) + 8H^+(aq) \longrightarrow$$
$$\phantom{xxxxxx} 3Cu^{2+}(aq) + 2NO(g) + 4H_2O(l)$$
Use standard electrode potentials.

**20.66**   Calculate the standard free-energy change at 25°C for the following reaction.
$$4Al(s) + 3O_2(g) + 12H^+(aq) \longrightarrow 4Al^{3+}(aq) + 6H_2O(l)$$
Use standard electrode potentials.

**20.67**   What is $\Delta G°$ for the following reaction?
$$2I^-(aq) + Cl_2(g) \longrightarrow I_2(s) + 2Cl^-(aq)$$
Use data given in Table 20.1.

**20.68**   Using electrode potentials, calculate the standard free-energy change for the reaction
$$Na(s) + \tfrac{1}{2}Br_2(g) \xrightarrow{\ H_2O\ } Na^+(aq) + Br^-(aq)$$

**20.69**   Calculate the standard emf at 25°C for the following cell reaction from standard free energies of formation (Appendix C).
$$Mg(s) + 2Ag^+(aq) \longrightarrow Mg^{2+}(aq) + 2Ag(s)$$

**20.70**   Calculate the standard emf at 25°C for the following cell reaction from standard free energies of formation (Appendix C).
$$2Al(s) + 3Cu^{2+}(aq) \longrightarrow 2Al^{3+}(aq) + 3Cu(s)$$

**20.71**   Calculate the standard emf of the lead storage cell whose overall reaction is
$$PbO_2(s) + 2HSO_4^-(aq) + 2H^+(aq) + Pb(s) \longrightarrow$$
$$\phantom{xxxxxxxx} 2PbSO_4(s) + 2H_2O(l)$$
See Appendix C for free energies of formation.

**20.72**   Calculate the standard emf of the cell corresponding to the oxidation of oxalic acid, $H_2C_2O_4$, by permanganate ion, $MnO_4^-$.
$$5H_2C_2O_4(aq) + 2MnO_4^-(aq) + 6H^+(aq) \longrightarrow$$
$$\phantom{xxxxxx} 10CO_2(g) + 2Mn^{2+}(aq) + 8H_2O(l)$$
See Appendix C for free energies of formation; $\Delta G_f°$ for $H_2C_2O_4(aq)$ is $-698$ kJ.

**20.73**   Calculate the equilibrium constant $K$ for the following reaction at 25°C from standard electrode potentials.
$$Fe^{3+}(aq) + Cu(s) \longrightarrow Fe^{2+}(aq) + Cu^{2+}(aq)$$
The equation is not balanced.

**20.74** Calculate the equilibrium constant $K$ for the following reaction at 25°C from standard electrode potentials.

$$Sn^{4+}(aq) + Hg(l) \longrightarrow Sn^{2+}(aq) + Hg_2^{2+}(aq)$$

The equation is not balanced.

**20.75** Copper(I) ion can act as both an oxidizing agent and a reducing agent. Hence, it can react with itself.

$$2Cu^+(aq) \longrightarrow Cu(s) + Cu^{2+}(aq)$$

Calculate the equilibrium constant at 25°C for this reaction, using appropriate values of electrode potentials.

**20.76** Use electrode potentials to calculate the equilibrium constant at 25°C for the reaction

$$ClO_4^-(aq) + ClO_2^-(aq) \rightleftharpoons 2ClO_3^-(aq)$$

See Appendix I for data.

### Nernst Equation

**20.77** Calculate the emf of the following cell at 25°C.

$$Cr(s)\,|\,Cr^{3+}(1.0 \times 10^{-3}\,M)\,\|\,Ni^{2+}(1.5\,M)\,|\,Ni(s)$$

**20.78** What is the emf of the following cell at 25°C?

$$Ni(s)\,|\,Ni^{2+}(1.0\,M)\,\|\,Sn^{2+}(1.0 \times 10^{-4}\,M)\,|\,Sn(s)$$

**20.79** Calculate the emf of a cell operating with the following reaction at 25°C, in which $[MnO_4^-] = 0.010\,M$, $[Br^-] = 0.010\,M$, $[Mn^{2+}] = 0.15\,M$, and $[H^+] = 1.0\,M$.

$$2MnO_4^-(aq) + 10Br^-(aq) + 16H^+(aq) \longrightarrow$$
$$2Mn^{2+}(aq) + 5Br_2(l) + 8H_2O(l)$$

**20.80** Calculate the emf of a cell operating with the following reaction at 25°C, in which $[Cr_2O_7^{2-}] = 0.020\,M$, $[I^-] = 0.015\,M$, $[Cr^{3+}] = 0.40\,M$, and $[H^+] = 0.50\,M$.

$$Cr_2O_7^{2-}(aq) + 6I^-(aq) + 14H^+(aq) \longrightarrow$$
$$2Cr^{3+}(aq) + 3I_2(s) + 7H_2O(l)$$

**20.81** The voltaic cell

$$Cd(s)\,|\,Cd^{2+}(aq)\,\|\,Ni^{2+}(1.0\,M)\,|\,Ni(s)$$

has an electromotive force of 0.240 V at 25°C. What is the concentration of cadmium ion? ($E^\circ_{cell} = 0.170$ V.)

**20.82** The emf of the following cell at 25°C is 0.475 V.

$$Zn\,|\,Zn^{2+}(1\,M)\,\|\,H^+(\text{test solution})\,|\,H_2(1\text{ atm})\,|\,Pt$$

What is the pH of the test solution?

### Electrolysis

**20.83** What are the half-reactions in the electrolysis of (a) $CaS(l)$; (b) $CsOH(l)$?

**20.84** What are the half-reactions in the electrolysis of (a) $MgBr_2(l)$; (b) $Ca(OH)_2(l)$?

**20.85** Describe what you expect to happen when the following solutions are electrolyzed: (a) aqueous $Na_2SO_4$; (b) aqueous KBr. That is, what are the electrode reactions? What is the overall reaction?

**20.86** Describe what you expect to happen when the following solutions are electrolyzed: (a) aqueous $CuCl_2$; (b) aqueous $Cu(NO_3)_2$. That is, what are the electrode reactions? What is the overall reaction? Nitrate ion is not oxidized.

**20.87** In the commercial preparation of aluminum, aluminum oxide, $Al_2O_3$, is electrolyzed at 1000°C. (The mineral cryolite is added as a solvent.) Assume that the cathode reaction is

$$Al^{3+} + 3e^- \longrightarrow Al$$

How many coulombs of electricity are required to give 3.61 kg of aluminum?

**20.88** Chlorine, $Cl_2$, is produced commercially by the electrolysis of aqueous sodium chloride. The anode reaction is

$$2Cl^-(aq) \longrightarrow Cl_2(g) + 2e^-$$

How long will it take to produce 1.18 kg of chlorine if the current is $5.00 \times 10^2$ A?

**20.89** When molten lithium chloride, LiCl, is electrolyzed, lithium metal is liberated at the cathode. How many grams of lithium are liberated when $5.00 \times 10^3$ C of charge passes through the cell?

**20.90** How many grams of cadmium are deposited from an aqueous solution of cadmium sulfate, $CdSO_4$, when an electric current of 1.51 A flows through the solution for 221 min?

## General Problems

**20.91** Some metals, such as iron, can be oxidized to more than one oxidation state. Obtain the balanced net ionic equations for the following oxidation–reduction reactions, in which nitric acid is reduced to nitric oxide, NO.

a. Oxidation of iron metal to iron(II) ion by nitric acid.
b. Oxidation of iron(II) ion to iron(III) ion by nitric acid.
c. Oxidation of iron metal to iron(III) by nitric acid. [Consider adding a and b equations.]

**20.92**    Some metals, such as thallium, can be oxidized to more than one oxidation state. Obtain the balanced net ionic equations for the following oxidation–reduction reactions, in which nitric acid is reduced to nitric oxide, NO.
a. Oxidation of thallium metal to thallium(I) ion by nitric acid.
b. Oxidation of thallium(I) ion to thallium(III) ion by nitric acid.
c. Oxidation of thallium metal to thallium(III) by nitric acid. [Consider adding a and b equations.]

**20.93**    Balance the following skeleton equations. The reactions occur in acidic or basic aqueous solution, as indicated.
a.  $MnO_4^- + S^{2-} \longrightarrow MnO_2 + S_8$     (basic)
b.  $IO_3^- + HSO_3^- \longrightarrow I^- + SO_4^{2-}$     (acidic)
c.  $Fe(OH)_2 + CrO_4^{2-} \longrightarrow Fe(OH)_3 + Cr(OH)_4^-$     (basic)
d.  $Cl_2 \longrightarrow Cl^- + ClO^-$     (basic)

**20.94**    Balance the following skeleton equations. The reactions occur in acidic or basic aqueous solution, as indicated.
a.  $MnO_4^- + H_2S \longrightarrow Mn^{2+} + S_8$     (acidic)
b.  $Zn + NO_3^- \longrightarrow Zn^{2+} + N_2O$     (acidic)
c.  $MnO_4^{2-} \longrightarrow MnO_4^- + MnO_2$     (basic)
d.  $Br_2 \longrightarrow Br^- + BrO_3^-$     (basic)

**20.95**    Iron(II) hydroxide is a greenish precipitate that is formed from iron(II) ion by the addition of a base. This precipitate gradually turns to the yellowish-brown iron(III) hydroxide from oxidation by $O_2$ in the air. Write a balanced equation for this oxidation by $O_2$.

**20.96**    A sensitive test for bismuth(III) ion consists of shaking a solution suspected of containing the ion with a basic solution of sodium stannite, $Na_2SnO_2$. A positive test consists of the formation of a black precipitate of bismuth metal. Stannite ion is oxidized by bismuth(III) ion to stannate ion, $SnO_3^{2-}$. Write a balanced equation for the reaction.

**20.97**    Give the notation for a voltaic cell that uses the reaction

$$Ca(s) + Cl_2(g) \longrightarrow Ca^{2+}(aq) + 2Cl^-(aq)$$

What is the half-cell reaction for the anode? for the cathode? What is the standard emf of the cell?

**20.98**    Give the notation for a voltaic cell whose overall cell reaction is

$$Mg(s) + 2Ag^+(aq) \longrightarrow Mg^{2+}(aq) + 2Ag(s)$$

What are the half-cell reactions? Label them as anode or cathode reactions. What is the standard emf of this cell?

**20.99**    Use electrode potentials to answer the following questions. (a) Is the oxidation of nickel by iron(III) ion a spontaneous reaction under standard conditions?

$$Ni(s) + 2Fe^{3+}(aq) \longrightarrow Ni^{2+}(aq) + 2Fe^{2+}(aq)$$

(b) Will iron(III) ion oxidize tin(II) ion to tin(IV) ion under standard conditions?

$$2Fe^{3+}(aq) + Sn^{2+}(aq) \longrightarrow 2Fe^{2+}(aq) + Sn^{4+}(aq)$$

**20.100**    Use electrode potentials to answer the following questions, assuming standard conditions. (a) Do you expect permanganate ion ($MnO_4^-$) to oxidize chloride ion to chlorine gas in acidic solution? (b) Will dichromate ion ($Cr_2O_7^{2-}$) oxidize chloride ion to chlorine gas in acidic solution?

**20.101**    Determine the emf of the following cell.

$$Pb\,|\,PbSO_4(s),\ SO_4^{2-}(1.0\ M)\,\|\,H^+(1.0\ M)\,|\,H_2(1.0\ atm)\,|\,Pt$$

The anode is essentially a lead electrode, $Pb\,|\,Pb^{2+}(aq)$. However, the anode solution is saturated with lead sulfate, so that the lead(II)-ion concentration is determined by the solubility product of $PbSO_4$ ($K_{sp} = 1.7 \times 10^{-8}$).

**20.102**    Determine the emf of the following cell.

$$Pt\,|\,H_2(1.0\ atm)\,|\,H^+(1.0\ M)\,\|\,Cl^-(1.0\ M),\ AgCl(s)\,|\,Ag$$

The cathode is essentially a silver electrode, $Ag^+(aq)\,|\,Ag$. However, the cathode solution is saturated with silver chloride, so that the silver-ion concentration is determined by the solubility product of $AgCl$ ($K_{sp} = 1.8 \times 10^{-10}$).

**20.103**    a. Calculate the equilibrium constant for the following reaction at 25°C.

$$Sn(s) + Pb^{2+}(aq) \rightleftharpoons Sn^{2+}(aq) + Pb(s)$$

The standard emf of the corresponding voltaic cell is 0.010 V.
b. If an excess of tin metal is added to 1.0 $M$ $Pb^{2+}$, what is the concentration of $Pb^{2+}$ at equilibrium?

**20.104**    a. Calculate the equilibrium constant for the following reaction at 25°C.

$$Ag^+(aq) + Fe^{2+}(aq) \rightleftharpoons Ag(s) + Fe^{3+}(aq)$$

The standard emf of the corresponding voltaic cell is 0.030 V.
b. When equal volumes of 1.0 $M$ solutions of $Ag^+$ and $Fe^{2+}$ are mixed, what is the equilibrium concentration of $Fe^{2+}$?

**20.105**    How many faradays are required for each of the following processes? How many coulombs are required?
a.  Reduction of 1.0 mol $Na^+$ to Na
b.  Reduction of 1.0 mol $Cu^{2+}$ to Cu
c.  Oxidation of 1.0 g $H_2O$ to $O_2$
d.  Oxidation of 1.0 g $Cl^-$ to $Cl_2$

**20.106**    How many faradays are required for each of the following processes? How many coulombs are required?
a.  Reduction of 1.0 mol $Fe^{3+}$ to $Fe^{2+}$
b.  Reduction of 1.0 mol $Fe^{3+}$ to Fe
c.  Oxidation of 1.0 g $Sn^{2+}$ to $Sn^{4+}$
d.  Reduction of 1.0 g $Au^{3+}$ to Au

**20.107** In an analytical determination of arsenic, a solution containing arsenious acid, $H_3AsO_3$, potassium iodide, and a small amount of starch is electrolyzed. The electrolysis produces free iodine from iodide ion, and the iodine immediately oxidizes the arsenious acid to hydrogen arsenate ion, $HAsO_4^{2-}$.

$$I_2(aq) + H_3AsO_3(aq) + H_2O(l) \longrightarrow$$
$$2I^-(aq) + HAsO_4^{2-}(aq) + 4H^+(aq)$$

When the oxidation of arsenic is complete, the free iodine combines with the starch to give a deep blue color. If, during a particular run, it takes 65.4 s for a current of 10.5 mA to give an endpoint (indicated by the blue color), how many grams of arsenic are present in the solution?

**20.108** The amount of lactic acid, $HC_3H_5O_3$, produced in a sample of muscle tissue was analyzed by reaction with hydroxide ion. Hydroxide ion was produced in the sample mixture by electrolysis. The cathode reaction was

$$2H_2O(l) + 2e^- \longrightarrow H_2(g) + 2OH^-(aq)$$

Hydroxide ion reacts with lactic acid as soon as it is produced. The endpoint of the reaction is detected with an acid–base indicator. It required 115 s for a current of 15.6 mA to reach the endpoint. How many grams of lactic acid (a monoprotic acid) were present in the sample?

**20.109** A standard electrochemical cell is made by dipping a silver electrode into a 1.0 $M$ $Ag^+$ solution and a cadmium electrode into a 1.0 $M$ $Cd^{2+}$ solution.
a. What is the spontaneous chemical reaction, and what is the maximum potential produced by this cell?
b. What would be the effect on the potential of this cell if sodium sulfide were added to the $Cd^{2+}$ half-cell and CdS were precipitated? Why?
c. What would be the effect on the potential of the cell if the size of the silver electrode were doubled?

**20.110** A standard electrochemical cell is made by dipping an iron electrode into a 1.0 $M$ $Fe^{2+}$ solution and a copper electrode into a 1.0 $M$ $Cu^{2+}$ solution.
a. What is the spontaneous chemical reaction, and what is the maximum potential produced by this cell?
b. What would be the effect on the potential of this cell if sodium carbonate were added to the $Cu^{2+}$ half-cell and $CuCO_3$ were precipitated? Why?
c. What would be the effect on the potential of the cell if the size of the iron electrode were halved?

**20.111** A solution of copper(II) sulfate is electrolyzed by passing a current through the solution using inert electrodes. Consequently, there is a decrease in the $Cu^{2+}$ concentration and an increase in the hydronium ion concentration. Also, one electrode increases in mass and a gas evolves at the other electrode. Write half-reactions that occur at the anode and at the cathode.

**20.112** A potassium chloride solution is electrolyzed by passing a current through the solution using inert electrodes. A gas evolves at each electrode, and there is a large increase in pH of the solution. Write the half-reactions that occur at the anode and at the cathode.

**20.113** A constant current of 1.50 amp is passed through an electrolytic cell containing a 0.10 $M$ solution of $AgNO_3$ and a silver anode and a platinum cathode until 2.48 g of silver is deposited.
a. How long does the current flow to obtain this deposit?
b. What mass of chromium would be deposited in a similar cell containing 0.10 $M$ $Cr^{3+}$ if the same amount of current were used?

**20.114** A constant current of 1.25 amp is passed through an electrolytic cell containing a 0.050 $M$ solution of $CuSO_4$ and a copper anode and a platinum cathode until 2.10 g of copper is deposited.
a. How long does the current flow to obtain this deposit?
b. What mass of silver would be deposited in a similar cell containing 0.10 $M$ $Ag^+$ if the same amount of current were used?

**20.115** An aqueous solution of an unknown salt of gold is electrolyzed by a current of 2.75 amps for 3.50 hours. The electroplating is carried out with an efficiency of 90.0%, resulting in a deposit of 21.221 g of gold.
a. How many faradays are required to deposit the gold?
b. What is the charge on the gold ions (based on your calculations)?

**20.116** An aqueous solution of an unknown salt of vanadium is electrolyzed by a current of 2.50 amps for 1.90 hours. The electroplating is carried out with an efficiency of 95.0%, resulting in a deposit of 2.850 g of vanadium.
a. How many faradays are required to deposit the vanadium?
b. What is the charge on the vanadium ions (based on your calculations)?

**20.117** An electrochemical cell is made by placing a zinc electrode in 1.00 L of 0.200 $M$ $ZnSO_4$ solution and a copper electrode in 1.00 L of 0.0100 $M$ $CuCl_2$ solution.
a. What is the initial voltage of this cell when it is properly constructed?
b. Calculate the final concentration of $Cu^{2+}$ in this cell if it is allowed to produce an average current of 1.00 amp for 225 s.

**20.118** An electrochemical cell is made by placing an iron electrode in 1.00 L of 0.15 $M$ $FeSO_4$ solution and a copper electrode in 1.00 L of 0.036 $M$ $CuSO_4$ solution.
a. What is the initial voltage of this cell when it is properly constructed?
b. Calculate the final concentration of $Cu^{2+}$ in this cell if it is allowed to produce an average current of 1.25 amp for 335 s.

**20.119**    a. Calculate $\Delta G°$ for the following cell reaction:

$$Cd(s)\,|\,Cd^{2+}(aq)\,\|\,Co^{2+}(aq)\,|\,Co(s)$$

b. From $\Delta G°$, calculate the standard emf for the cell reaction and from this, determine the standard potential for $Co^{2+}(aq) + 2e^- \longrightarrow Co(s)$.

**20.120**    a. Calculate $\Delta G°$ for the following cell reaction:

$$Tl(s)\,|\,Tl^+(aq)\,\|\,Pb^{2+}(aq)\,|\,Pb(s)$$

The $\Delta G_f°$ for $Tl^+(aq)$ is $-32.4$ kJ/mol.

b. From $\Delta G°$, calculate the standard emf for the cell reaction and from this, determine the standard potential for $Tl^+(aq) + e^- \longrightarrow Tl(s)$.

## Cumulative-Skills Problems

**20.121**    Consider the following cell reaction at 25°C.

$$2Cr^{3+}(aq) + 3Zn(s) \longrightarrow 3Zn^{2+}(aq) + 2Cr(s)$$

Calculate the standard emf of this cell from the standard electrode potentials, and from this obtain $\Delta G°$ for the cell reaction. Use data in Appendix C to calculate $\Delta H°$; note that $Cr(H_2O)_6^{3+}(aq)$ equals $Cr^{3+}(aq)$. Use these values of $\Delta H°$ and $\Delta G°$ to obtain $\Delta S°$ for the cell reaction.

**20.122**    Consider the following cell reaction at 25°C.

$$2Cr(s) + 3Fe^{2+}(aq) \longrightarrow 2Cr^{3+}(aq) + 3Fe(s)$$

Calculate the standard emf of this cell from the standard electrode potentials, and from this obtain $\Delta G°$ for the cell reaction. Use data in Appendix C to calculate $\Delta H°$; note that $Cr(H_2O)_6^{3+}(aq)$ equals $Cr^{3+}(aq)$. Use these values of $\Delta H°$ and $\Delta G°$ to obtain $\Delta S°$ for the cell reaction.

**20.123**    Under standard conditions for all concentrations, the following reaction is spontaneous at 25°C.

$$O_2(g) + 4H^+(aq) + 4Br^-(aq) \longrightarrow 2H_2O(l) + 2Br_2(l)$$

If $[H^+]$ is decreased so that the pH = 3.60, what value will $E_{cell}$ have, and will the reaction be spontaneous at this $[H^+]$?

**20.124**    Under standard conditions for all concentrations, the following reaction is spontaneous at 25°C.

$$O_3(g) + 2H^+(aq) + 2Co^{2+}(aq) \longrightarrow$$
$$O_2(g) + H_2O(l) + 2Co^{3+}(l)$$

If $[H^+]$ is decreased so that the pH = 9.10, what value will $E_{cell}$ have, and will the reaction be spontaneous at this $[H^+]$?

**20.125**    Under standard conditions for all concentrations, the following reaction is spontaneous at 25°C.

$$O_2(g) + 4H^+(aq) + 4Br^-(aq) \longrightarrow 2H_2O(l) + 2Br_2(l)$$

If $[H^+]$ is adjusted by adding a buffer of 0.10 $M$ NaOCN and 0.10 $M$ HOCN ($K_a = 3.5 \times 10^{-4}$), what value will $E_{cell}$ have, and will the reaction be spontaneous at this $[H^+]$?

**20.126**    Under standard conditions for all concentrations, the following reaction is spontaneous at 25°C.

$$O_3(g) + 2H^+(aq) + 2Co^{2+}(aq) \longrightarrow$$
$$O_2(g) + H_2O(l) + 2Co^{3+}(aq)$$

If $[H^+]$ is adjusted by adding a buffer of 0.10 $M$ NaClO and 0.10 $M$ HClO ($K_a = 3.5 \times 10^{-8}$), what value will $E_{cell}$ have, and will the reaction be spontaneous at this $[H^+]$?

**20.127**    An electrode is prepared by dipping a silver strip into a solution saturated with silver thiocyanate, AgSCN, and containing 0.10 $M$ SCN$^-$. The emf of the voltaic cell constructed by connecting this electrode as the cathode to the standard hydrogen half-cell as the anode is 0.45 V. What is the solubility product of silver thiocyanate?

**20.128**    An electrode is prepared from liquid mercury in contact with a saturated solution of mercury(I) chloride, $Hg_2Cl_2$, containing 1.00 $M$ Cl$^-$. The emf of the voltaic cell constructed by connecting this electrode as the cathode to the standard hydrogen half-cell as the anode is 0.268 V. What is the solubility product of mercury(I) chloride?

## Media Activities

**20.129**    Work through the Balancing Redox Equations *Problem Solving* activity on the student web site. Use the half-reaction method to balance the skeleton oxidation-reduction reaction that occurs in acidic solution:

$$MnO_2(s) + Br^-(aq) \longrightarrow Br_2(aq) + MnBr_2(aq)$$

**20.130**    Open the Balancing Redox Equations *Understanding Concepts* simulation on the student web site. Balance one of the listed reactions for an acidic solution and a basic solution.

a. What are the differences between balancing a reaction in an acidic solution versus a basic solution?

b. Does the fact that the solution is acidic or basic have any impact on the number of electrons lost or gained during an oxidation-reduction reaction?

**20.131** Work through the Spontaneity from Potential *Problem Solving* activity on the student web site.

a. What values of $\Delta G$ indicate a spontaneous reaction?

b. Would you expect a piece of pure silver to react in an aqueous 1.0 $M$ solution of $Cu(NO_3)_2$ under standard conditions?

c. Calculate $\Delta G°$ for the reaction of $Cu^{2+}(aq)$ and silver described in part b. Is this calculated value of $\Delta G°$ consistent with your answer for part b?

**20.132** Work through the Nernst Equation *Problem Solving* activity on the student web site. Answer the following questions that are based on the reaction depicted in this activity.

a. How would $E_{cell}$ and $E°_{cell}$ compare if the concentrations of $Zn^{2+}$ and $Ag^+$ were equal?

b. If you were to increase the concentration of $Ag^+$, how would this affect the values of $E_{cell}$ and $E°_{cell}$?

c. If you were to decrease the concentration of $Zn^{2+}$, how would this affect the values of $E_{cell}$ and $E°_{cell}$?

**20.133** For review, take the ACE practice test for Chapter 20 on the student web site.

Image of healthy human heart produced by recording gamma emissions of the radionuclide, thallium-201. In order to produce this image, the radioactive thallium-210 was injected into the patient.

# Nuclear Chemistry

Technetium is an unusual element. Although a *d*-transition element (under manganese in Group VIIB) with a small atomic number ($Z = 43$), it has no stable isotopes. The nucleus of every technetium isotope is radioactive and decays, or disintegrates, to give an isotope of another element. Many of the technetium isotopes decay by emitting an electron from the nucleus.

Because of its nuclear instability, technetium is not found naturally on earth. Nevertheless, it is produced commercially in kilogram quantities from other elements by nuclear reactions, processes in which nuclei are transformed into different nuclei. Technetium (from the Greek *tekhnetos,* meaning "artificial") was the first new element produced in the laboratory from another element. It was discovered in 1937 by Carlo Perrier and Emilio Segrè when the element molybdenum was bombarded with deuterons (nuclei of hydrogen, each having one proton and one neutron). Later, technetium was found to be a product of the fission, or splitting, of uranium nuclei. Technetium is produced in nuclear fission reactors used to generate electricity.

**Figure 21.1**
**Image of a person's skeleton obtained using an excited form of technetium-99.**   A technetium compound was injected into the body, where it concentrated in bone tissue. Gamma rays (similar to x rays) emitted by technetium were detected by special equipment to produce this image.

Technetium is one of the principal isotopes used in medical diagnostics based on radioactivity. A compound of technetium is injected into a vein, where it concentrates in certain body organs. The energy emitted by technetium nuclei is detected by special equipment and gives an image of these body organs. Figure 21.1 shows the image of a person's skeleton obtained from technetium administered in this manner. The technetium is eliminated by the body after several hours.

In this chapter, we will look at nuclear processes such as those we have described for technetium. We will answer such questions as the following: How do you describe the radioactive decay of technetium? How do you describe the transformation of a molybdenum nucleus into technetium? How is technetium produced from uranium by nuclear fission, or splitting? What are some practical applications of nuclear processes?

# Radioactivity and Nuclear Bombardment Reactions

▶ You can cover this chapter early in the course if you wish. The kinetics equations in Section 21.4 can be used without referring to chemical kinetics as background.

In chemical reactions, only the outer electrons of the atoms are disturbed. The nuclei of the atoms are not affected. In nuclear reactions, however, the nuclear changes that occur are independent of the chemical environment of the atom. For example, the nuclear changes in a radioactive $^3_1H$ atom are the same if the atom is part of an $H_2$ molecule or incorporated into $H_2O$.

We will look at two types of nuclear reactions. One type is **radioactive decay,** *the process in which a nucleus spontaneously disintegrates, giving off radiation.* The radiation consists of one or more of the following, depending on the nucleus: electrons, nuclear particles (such as neutrons), smaller nuclei (usually helium-4 nuclei), and electromagnetic radiation.

The second type of nuclear reaction is a **nuclear bombardment reaction,** *a nuclear reaction in which a nucleus is bombarded, or struck, by another nucleus or by a nuclear particle.* If there is sufficient energy in this collision, the nuclear particles of the reactants rearrange to give a product nucleus or nuclei. First, we will look at radioactive decay.

# 21.1 Radioactivity

The phenomenon of radioactivity was discovered by Antoine Henri Becquerel in 1896. He discovered that photographic plates develop bright spots when exposed to uranium minerals, and he concluded that the minerals give off some sort of radiation.

The radiation from uranium minerals was later shown to be separable by electric (and magnetic) fields into three types, alpha ($\alpha$), beta ($\beta$), and gamma ($\gamma$) rays (Figure 21.2). *Alpha rays* bend away from a positive plate and toward a negative plate, indicating that they have a positive charge; they are now known to consist of helium-4 nuclei (nuclei with two protons and two neutrons). *Beta rays* bend in the opposite direction, indicating that they have a negative charge; they are now known to consist of high-speed electrons. *Gamma rays* are unaffected by electric and magnetic fields: they have been shown to be a form of electromagnetic radiation that is similar to x rays, except they are higher in energy with shorter wavelengths (about 1 pm, or $1 \times 10^{-12}$ m). Uranium minerals contain a number of radioactive elements, each emitting one or more of these radiations. Uranium-238, the main uranium isotope in uranium minerals, emits alpha rays and thereby decays, or disintegrates, to thorium-234 nuclei.

A sample of uranium-238 decays, or disintegrates, spontaneously over a period of billions of years. After about 30 billion years, the sample would be nearly gone. Strontium-90, formed by nuclear reactions that occur in nuclear weapons testing and nuclear power reactors, decays more rapidly. A sample of strontium-90 would be nearly gone after several hundred years. In either case, it is impossible to know when a particular nucleus will decay, although, as you will see in Section 21.4, precise information can be given about the rate of decay of any radioactive sample.

## Nuclear Equations

You can write an equation for the nuclear reaction corresponding to the decay of uranium-238 much as you would write an equation for a chemical reaction. You represent the uranium-238 nucleus by the *nuclide symbol* $^{238}_{92}$U. ◀ The radioactive decay of $^{238}_{92}$U by alpha-particle emission (loss of a $^{4}_{2}$He nucleus) is written

▶ Nuclide symbols were introduced in Section 2.3. For uranium-238, you write

mass number ⟶ $^{238}_{92}$U
atomic number ⟶

$$^{238}_{92}\text{U} \longrightarrow {}^{234}_{90}\text{Th} + {}^{4}_{2}\text{He}$$

The product, in addition to helium-4, is thorium-234. This is an example of a **nuclear equation,** which is *a symbolic representation of a nuclear reaction*. Normally, only the nuclei are represented. It is not necessary to indicate the chemical compound or

**Figure 21.2**
**Separation of the radiation from a radioactive material (uranium mineral).** The radiation separates into alpha ($\alpha$), beta ($\beta$), and gamma ($\gamma$) rays when it passes through an electric field.

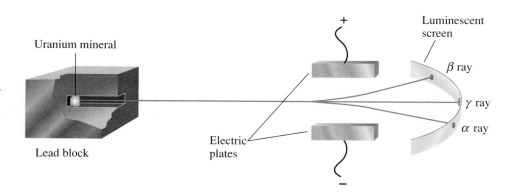

the electron charges for any ions involved, because the chemical environment has no effect on nuclear processes.

Reactant and product nuclei are represented in nuclear equations by their nuclide symbols. Other particles are given the following symbols, in which the subscript equals the charge and the superscript equals the total number of protons and neutrons in the particle (mass number):

| | | | |
|---|---|---|---|
| Proton | $^1_1\text{H}$ | or | $^1_1\text{p}$ |
| Neutron | $^1_0\text{n}$ | | |
| Electron | $^{\phantom{-}0}_{-1}\text{e}$ | or | $^{\phantom{-}0}_{-1}\beta$ |
| Positron | $^0_1\text{e}$ | or | $^0_1\beta$ |
| Gamma photon | $^0_0\gamma$ | | |

The decay of a nucleus with the emission of an electron, $^{\phantom{-}0}_{-1}\text{e}$, is usually called beta emission, and the emitted electron is sometimes labeled $^{\phantom{-}0}_{-1}\beta$. A **positron** is *a particle similar to an electron, having the same mass but a positive charge*. A **gamma photon** is *a particle of electromagnetic radiation of short wavelength (about 1 pm, or $10^{-12}$ m) and high energy*.

▶ An alternate example is provided in the Instructor's Resource Manual.

## EXAMPLE 21.1

### Writing a Nuclear Equation

Write the nuclear equation for the radioactive decay of radium-226 by alpha decay to give radon-222. A radium-226 nucleus emits one alpha particle, leaving behind a radon-222 nucleus.

 **PROBLEM STRATEGY**

Use the list of elements on the inside back cover to obtain the atomic numbers of radium and radon, so you can write their nuclide symbols for the nuclear equation. The nuclide symbol of the alpha particle is $^4_2\text{He}$.

**SOLUTION**

The nuclear equation is

$$^{226}_{88}\text{Ra} \longrightarrow {}^{222}_{86}\text{Rn} + {}^4_2\text{He}$$

## EXERCISE 21.1

Potassium-40 is a naturally occurring radioactive isotope. It decays to calcium-40 by beta emission. When a potassium-40 nucleus decays by beta emission, it emits one beta particle and gives a calcium-40 nucleus. Write the nuclear equation for this decay.

See Problems 21.27, 21.28, 21.29, and 21.30.

The total charge is conserved, or remains constant, during a nuclear reaction. This means that the sum of the subscripts (number of protons, or positive charges, in the nuclei) for the products must equal the sum of the subscripts for the reactants. For the equation in Example 21.1, the subscript for the reactant $^{226}_{88}\text{Ra}$ is 88. For the products, the sum of the subscripts is $86 + 2 = 88$.

Similarly, the total number of *nucleons* (protons and neutrons) is conserved, or remains constant, during a nuclear reaction. This means that the sum of the superscripts (the mass numbers) for the reactants equals the sum of the superscripts for the products. For the equation in Example 21.1, the superscript for the reactant nucleus is 226. For the products, the sum of the superscripts is $222 + 4 = 226$.

Note that if all reactants and products but one are known in a nuclear equation, the identity of that one nucleus or particle can be easily obtained. This is illustrated in the next example.

▶ An alternate example is provided in the Instructor's Resource Manual.

### EXAMPLE 21.2

### Deducing a Product or Reactant in a Nuclear Equation

Technetium-99 is a long-lived radioactive isotope of technetium. Each nucleus decays by emitting one beta particle. What is the product nucleus?

 **PROBLEM STRATEGY**

First, write the nuclear equation using the symbol $_Z^A X$ for the unknown nuclide or particle. Then, solve for $A$ and $Z$ using the following relations: (1) the sum of the subscripts for the reactants equals the sum of the subscripts for the products, and (2) the sum of the superscripts for the reactants equals the sum of the superscripts for the products.

**SOLUTION**

Technetium-99 has the nuclide symbol $_{43}^{99} Tc$. A beta particle is an electron; the symbol is $_{-1}^{0} e$. The nuclear equation is

$$_{43}^{99} Tc \longrightarrow {}_Z^A X + {}_{-1}^{0} e$$

From the superscripts, you can write

$$99 = A + 0, \text{ or } A = 99$$

Similarly, from the subscripts, you get

$$43 = Z - 1, \text{ or } Z = 43 + 1 = 44$$

Hence $A = 99$ and $Z = 44$, so the product is $_{44}^{99} X$. Because element 44 is ruthenium, symbol Ru, you write the product nucleus as $_{44}^{99} \textbf{Ru.}$

### EXERCISE 21.2

Plutonium-239 decays by alpha emission, with each nucleus emitting one alpha particle. What is the other product of this decay?

See Problems 21.31, 21.32, 21.33, and 21.34.

## Nuclear Stability

At first glance, the existence of several protons in the small space of a nucleus is puzzling. Why wouldn't the protons be strongly repelled by their like electric charges? The existence of stable nuclei with more than one proton is due to the nuclear force. The **nuclear force** is *a strong force of attraction between nucleons that acts only at very short distances (about $10^{-15}$ m).* Beyond nuclear distances, these nuclear forces become negligible. Therefore, two protons that are much farther apart than $10^{-15}$ m repel one another by their like electric charges. Inside the nucleus, however, two protons are close enough together for the nuclear force between them to be effective.

This force in a nucleus can more than compensate for the repulsion of electric charges and thereby give a stable nucleus.

The protons and neutrons in a nucleus appear to have energy levels much as the electrons in an atom have energy levels. The **shell model of the nucleus** is *a nuclear model in which protons and neutrons exist in levels, or shells, analogous to the shell structure that exists for electrons in an atom.* Recall that in an atom, filled shells of electrons are associated with the special stability of the noble gases. The total numbers of electrons for these stable atoms are 2 (for He), 10 (for Ne), 18 (for Ar), and so forth. Experimentally, note that nuclei with certain numbers of protons or neutrons appear to be very stable. These numbers, called *magic numbers* and associated with specially stable nuclei, were later explained by the shell model. According to this theory, a **magic number** is *the number of nuclear particles in a completed shell of protons or neutrons.* Because nuclear forces differ from electrical forces, these numbers are not the same as those for electrons in atoms. For protons, the magic numbers are 2, 8, 20, 28, 50, and 82. Neutrons have these same magic numbers, as well as the magic number 126. For protons, calculations show that 114 should also be a magic number.

Some of the evidence for these magic numbers, and therefore for the shell model of the nucleus, is as follows. Many radioactive nuclei decay by emitting alpha particles, or $^4_2$He nuclei. There appears to be special stability in the $^4_2$He nucleus. It contains two protons and two neutrons; that is, it contains a magic number of protons (2) and a magic number of neutrons (also 2).

Another piece of evidence is seen in the final products obtained in natural radioactive decay. For example, uranium-238 decays to thorium-234, which in turn decays to protactinium-234, and so forth. Each product is radioactive and decays to another nucleus until the final product, $^{206}_{82}$Pb, is reached. This nucleus is stable. Note that it contains 82 protons, a magic number. Other radioactive decay series end at $^{207}_{82}$Pb and $^{208}_{82}$Pb, each of which has a magic number of protons. Note that $^{208}_{82}$Pb also has a magic number of neutrons (208 − 82 = 126).

Evidence also points to the special stability of pairs of protons and pairs of neutrons, analogous to the stability of pairs of electrons in molecules. Table 21.1 lists the number of stable isotopes that have an even number of protons and an even number of neutrons (157). By comparison, only 5 stable isotopes have an odd number of protons and an odd number of neutrons.

Finally, when you plot each stable nuclide on a graph with the number of protons ($Z$) on the horizontal axis and the number of neutrons ($N$) on the vertical axis, these stable nuclides fall in a certain region, or band, of the graph. The **band of stability** is *the region in which stable nuclides lie in a plot of number of protons against number of neutrons.* Figure 21.3 shows the band of stability; the rest of the figure is

**Table 21.1**
**Number of Stable Isotopes with Even and Odd Numbers of Protons and Neutrons**

|  | Number of Stable Isotopes | | | |
| --- | --- | --- | --- | --- |
|  | 157 | 52 | 50 | 5 |
| Number of protons | Even | Even | Odd | Odd |
| Number of neutrons | Even | Odd | Even | Odd |

**Figure 21.3**
**Band of stability.** The stable nuclides, indicated by black dots, cluster in a band. Nuclides to the left of the band of stability usually decay by beta emission, whereas those to the right usually decay by positron emission or electron capture. Nuclides of Z > 83 often decay by alpha emission.

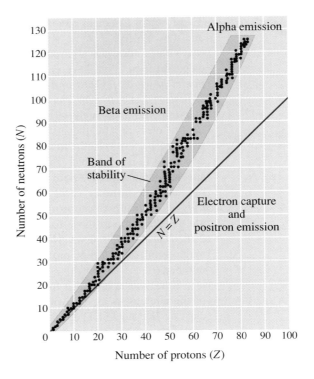

explained later in this section. For nuclides up to $Z = 20$, the ratio of neutrons to protons is about 1.0 to 1.1. As $Z$ increases, however, the neutron-to-proton ratio increases to about 1.5. This increase in neutron-to-proton ratio with increasing $Z$ is believed to result from the increasing repulsions of protons from their electric charges. More neutrons are required to give attractive nuclear forces to offset these repulsions.

It appears that when the number of protons becomes very large, the proton–proton repulsions become so great that stable nuclides are impossible. No stable nuclides are known with atomic numbers greater than 83. On the other hand, all elements with $Z$ equal to 83 or less have one or more stable nuclides, with the exception of technetium ($Z = 43$), as noted in the chapter opening, and promethium ($Z = 61$).

▶ An alternate example is provided in the Instructor's Resource Manual.

### EXAMPLE 21.3

### Predicting the Relative Stabilities of Nuclides

One of the nuclides in each of the following pairs is radioactive; the other is stable. Which is radioactive and which is stable? Explain.

a. $^{208}_{84}Po$, $^{209}_{83}Bi$    b. $^{39}_{19}K$, $^{40}_{19}K$    c. $^{71}_{31}Ga$, $^{76}_{31}Ga$

 **PROBLEM STRATEGY**

You must decide which nuclide of each pair is more likely to be stable, based on the general principles stated in the preceding text.

### SOLUTION

a. Polonium has an atomic number greater than 83, so $^{208}_{84}Po$ **is radioactive.** Bismuth-209 has 126 neutrons (a magic number), so $^{209}_{83}Bi$ **is expected to be stable.**

b. Of these two isotopes, $^{39}_{19}$K has a magic number of neutrons (20), so $^{39}_{19}$**K is expected to be stable.** The isotope $^{40}_{19}$K has an odd number of protons (19) and an odd number of neutrons (21). Because stable odd–odd nuclei are rare, **you might expect $^{40}_{19}$K to be radioactive.**

c. Of the two isotopes, $^{76}_{31}$Ga lies farther from the center of the band of stability, so it is more likely to be radioactive. For this reason, **you expect $^{76}_{31}$Ga to be radioactive and $^{71}_{31}$Ga to be stable.**

### EXERCISE 21.3

Of the following nuclides, two are radioactive. Which are radioactive and which is stable? Explain. a. $^{118}_{50}$Sn; b. $^{76}_{33}$As; c. $^{227}_{89}$Ac.

See Problems 21.35 and 21.36.

## Types of Radioactive Decay

There are six common types of radioactive decay; the first five are listed in Table 21.2.

1. **Alpha emission** (abbreviated $\alpha$): *emission of a $^4_2$He nucleus, or alpha particle, from an unstable nucleus.* An example is the radioactive decay of radium-226.

$$^{226}_{88}\text{Ra} \longrightarrow {}^{222}_{86}\text{Rn} + {}^4_2\text{He}$$

The product nucleus has an atomic number that is two less, and a mass number that is four less, than that of the original nucleus.

▶ Demonstration of beta rays: See the Instructor's Resource Manual.

2. **Beta emission** (abbreviated $\beta$ or $\beta^-$): *emission of a high-speed electron from an unstable nucleus.* Beta emission is equivalent to the conversion of a neutron to a proton.

$$^1_0\text{n} \longrightarrow {}^1_1\text{p} + {}^{\ 0}_{-1}\text{e}$$

An example of beta emission is the radioactive decay of carbon-14.

$$^{14}_6\text{C} \longrightarrow {}^{14}_7\text{N} + {}^{\ 0}_{-1}\text{e}$$

The product nucleus has an atomic number that is one more than that of the original nucleus. The mass number remains the same.

**Table 21.2**
**Types of Radioactive Decay**

| Type of Decay | Radiation | Equivalent Process | Resulting Nuclear Change | | Usual Nuclear Condition |
| --- | --- | --- | --- | --- | --- |
| | | | Atomic Number | Mass Number | |
| Alpha emission ($\alpha$) | $^4_2$He | — | $-2$ | $-4$ | $Z > 83$ |
| Beta emission ($\beta$) | $^{\ 0}_{-1}$e | $^1_0\text{n} \longrightarrow {}^1_1\text{p} + {}^{\ 0}_{-1}\text{e}$ | $+1$ | $0$ | $N/Z$ too large |
| Positron emission ($\beta^+$) | $^0_1$e | $^1_1\text{p} \longrightarrow {}^1_0\text{n} + {}^0_1\text{e}$ | $-1$ | $0$ | $N/Z$ too small |
| Electron capture (EC) | x rays | $^1_1\text{p} + {}^{\ 0}_{-1}\text{e} \longrightarrow {}^1_0\text{n}$ | $-1$ | $0$ | $N/Z$ too small |
| Gamma emission ($\gamma$) | $^0_0\gamma$ | — | $0$ | $0$ | Excited |

3. **Positron emission** (abbreviated $\beta^+$): *emission of a positron from an unstable nucleus.* A positron, denoted in nuclear equations as $^0_1e$, is a particle identical to an electron in mass but having a positive instead of a negative charge. Positron emission is equivalent to the conversion of a proton to a neutron. ◀

$$^1_1p \longrightarrow {}^1_0n + {}^0_1e$$

The radioactive decay of technetium-95 is an example of positron emission.

$$^{95}_{43}Tc \longrightarrow {}^{95}_{42}Mo + {}^0_1e$$

The product nucleus has an atomic number that is one less than that of the original nucleus. The mass number remains the same.

▶ Positrons are annihilated as soon as they encounter electrons. When a positron and an electron collide, both particles vanish with the emission of two gamma photons that carry away the energy.

$$^0_1e + {}^0_{-1}e \longrightarrow 2^0_0\gamma$$

4. **Electron capture** (abbreviated EC): *the decay of an unstable nucleus by capturing, or picking up, an electron from an inner orbital of an atom.* In effect, a proton is changed to a neutron, as in positron emission.

$$^1_1p + {}^0_{-1}e \longrightarrow {}^1_0n$$

An example is given by potassium-40, which has a natural abundance of 0.012%. ◀ Potassium-40 can decay by electron capture, as well as by beta and positron emissions. The equation for electron capture is

$$^{40}_{19}K + {}^0_{-1}e \longrightarrow {}^{40}_{18}Ar$$

▶ Most of the argon in the atmosphere is believed to have resulted from the radioactive decay of $^{40}_{19}K$.

The product nucleus has an atomic number that is one less than that of the original nucleus. The mass number remains the same. When another orbital electron fills the vacancy in the inner-shell orbital created by electron capture, an x-ray photon is emitted.

5. **Gamma emission** (abbreviated $\gamma$): *emission from an excited nucleus of a gamma photon, corresponding to radiation with a wavelength of about $10^{-12}$ m.* In many cases, radioactive decay results in a product nucleus that is in an excited state. As in the case of atoms, the excited state is unstable and goes to a lower-energy state with the emission of electromagnetic radiation. For nuclei, this radiation is in the gamma-ray region of the spectrum.

Often gamma emission occurs very quickly after radioactive decay. In some cases, however, an excited state has significant lifetime before it emits a gamma photon. A **metastable nucleus** is *a nucleus in an excited state with a lifetime of at least one nanosecond ($10^{-9}$ s).* In time, the metastable nucleus decays by gamma emission. An example is metastable technetium-99, denoted $^{99m}_{43}Tc$, which is used in medical diagnosis, as discussed in Section 21.5.

$$^{99m}_{43}Tc \longrightarrow {}^{99}_{43}Tc + {}^0_0\gamma$$

The product nucleus is simply a lower-energy state of the original nucleus, so there is no change of atomic number or mass number.

6. **Spontaneous fission:** *the spontaneous decay of an unstable nucleus in which a heavy nucleus of mass number greater than 89 splits into lighter nuclei and energy is released.* For example, a uranium-236 atom can spontaneously undergo the following nuclear reaction:

$$^{236}_{92}U \longrightarrow {}^{96}_{39}Y + {}^{136}_{53}I + 4^1_0n$$

This process will be discussed further in Section 21.6.

Nuclides outside the band of stability (Figure 21.3) are generally radioactive. Nuclides to the left of the band of stability have a neutron-to-proton ratio ($N/Z$) larger than that needed for stability. These nuclides tend to decay by beta emission. Beta emission reduces the neutron-to-proton ratio, because in this process a neutron is changed to a proton. The product is a stabler nuclide. In contrast, nuclides to the right of the band of stability have a neutron-to-proton ratio smaller than that needed for stability. These nuclides tend to decay by either positron emission or electron capture. Both processes convert a proton to a neutron, increasing the neutron-to-proton ratio and giving a stabler product nuclide. The types of radioactive decay expected of unstable nuclides are noted in Figure 21.3.

Consider a series of isotopes of a given element, such as carbon. Carbon-12 and carbon-13 are stable isotopes, whereas the other isotopes of carbon are radioactive. The isotopes of mass number smaller than 12 decay by positron emission. For example, carbon-11 decays by positron emission to boron-11.

$$\ce{^{11}_{6}C} \longrightarrow \ce{^{11}_{5}B} + \ce{^{0}_{1}e}$$

Carbon-11 has a neutron-to-proton ratio of 5/6 ($= 0.8$), which increases in the product boron-11 to 6/5 (1.2). The isotopes of carbon with mass number greater than 13 decay by beta emission. Carbon-14 decays by beta emission to produce nitrogen-14.

$$\ce{^{14}_{6}C} \longrightarrow \ce{^{14}_{7}N} + \ce{^{0}_{-1}e}$$

Carbon-14 has a neutron-to-proton ratio of 8/6 (1.3), which decreases in the product nitrogen-14 to 7/7 (1.0).

Now consider the radioactive isotope phosphorus-30. You can predict the expected type of radioactive decay of this isotope by noting whether the mass number is less than or greater than the mass number of stable isotopes. Generally, the mass numbers of stable isotopes will be close to the numerical value of the atomic weight of the element. The atomic weight of phosphorus is 31.0 amu, so you might expect phosphorus-31 to be a stable isotope (which it is). Phosphorus-30 has a mass number less than that of the stable isotope phosphorus-31. Therefore, you expect that phosphorus-30 will decay by either positron emission or electron capture. Positron emission is actually observed.

Positron emission and electron capture are competing radioactive decay processes, and what is observed depends on the relative rates of the two processes. The rate of electron capture increases with atomic number of the decaying nuclide and therefore becomes important in heavier elements. Positron emission is generally seen in lighter elements (recall that phosphorus-30 decays by positron emission). However, in the very heavy elements, especially those with $Z$ greater than 83, radioactive decay is often by alpha emission (noted in Figure 21.3 at the top of the figure). $\ce{^{238}_{92}U}$, $\ce{^{226}_{88}Ra}$, and $\ce{^{232}_{90}Th}$ are examples of alpha emitters.

◀ An alternate example is provided in the Instructor's Resource Manual.

---

**EXAMPLE 21.4**

## Predicting the Type of Radioactive Decay

Predict the expected type of radioactive decay for each of the following radioactive nuclides: a. $\ce{^{47}_{20}Ca}$; b. $\ce{^{25}_{13}Al}$.

### ▶ PROBLEM STRATEGY

Compare each nuclide with the stable nuclides of the same element. A nuclide with an $N/Z$ ratio greater than that of the stable nuclides is expected to exhibit beta emission. A nuclide with an $N/Z$ ratio less than that of the stable nuclides is expected to exhibit positron emission or electron capture; electron capture is important with heavier elements. Since you are comparing nuclides of the same $Z$, you can compare mass numbers ($= N + Z$), rather than $N/Z$.

**SOLUTION**

a. The atomic weight of calcium is 40.1 amu, so you expect calcium-40 to be a stable isotope. Calcium-47 has a mass number greater than that of the stable isotope, so you expect it to decay by **beta emission.** (This is the observed behavior of calcium-47.)

b. The atomic weight of aluminum is 27.0 amu, so you expect aluminum-27 to be a stable isotope. The mass number of aluminum-25 is less than 27, so you expect aluminum-25 to decay by either **positron emission or electron capture.** Positron emission is actually observed.

---

**EXERCISE 21.4**

Predict the type of decay expected for each of the following radioactive nuclides: a. $^{13}_{7}N$; b. $^{26}_{11}Na$.

---

**See Problems 21.37 and 21.38.**

## Radioactive Decay Series

All nuclides with atomic number greater than $Z = 83$ are radioactive, as we have noted. Many of these nuclides decay by alpha emission. Alpha particles, or $^{4}_{2}He$ nuclei, are especially stable and are formed in the radioactive nucleus at the moment of decay. By emitting an alpha particle, the nucleus reduces its atomic number, becoming more stable. However, if the nucleus has a very large atomic number, the product nucleus is also radioactive. Natural radioactive elements, such as uranium-238, give a **radioactive decay series,** *a sequence in which one radioactive nucleus decays to a second, which then decays to a third, and so forth.* Eventually, a stable nucleus, which is an isotope of lead, is reached.

Three radioactive decay series are found naturally. One of these series begins with uranium-238. Figure 21.4 shows the sequence of nuclear decay processes. In the first step, uranium-238 decays by alpha emission to thorium-234.

$$^{238}_{92}U \longrightarrow {}^{234}_{90}Th + {}^{4}_{2}He$$

This step is represented in Figure 21.4 by a red arrow labeled $\alpha$. Each alpha decay reduces the atomic number by 2 and the mass number by 4. The atomic number and mass number of each nuclide are given by its horizontal and vertical position on the graph. Thorium-234 in turn decays by beta emission to protactinium-234, which decays by beta emission to uranium-234.

$$^{234}_{90}Th \longrightarrow {}^{234}_{91}Pa + {}^{0}_{-1}e$$
$$^{234}_{91}Pa \longrightarrow {}^{234}_{92}U + {}^{0}_{-1}e$$

Beta emission is represented in Figure 21.4 by a blue arrow labeled $\beta$. Each beta decay increases the atomic number by one but has no effect on the mass number. After the decay of protactinium-234 and the formation of uranium-234, there are a number of alpha-decay steps. The final product of the series is lead-206.

Natural uranium is 99.28% $^{238}_{92}U$, which decays as we have described. However, the natural element also contains 0.72% $^{235}_{92}U$. This isotope starts a second radioac-

**Figure 21.4**
**Uranium-238 radioactive decay series.**  Each nuclide occupies a position on the graph determined by its atomic number and mass number. Alpha decay is shown by a red diagonal line. Beta decay is shown by a short blue horizontal line.

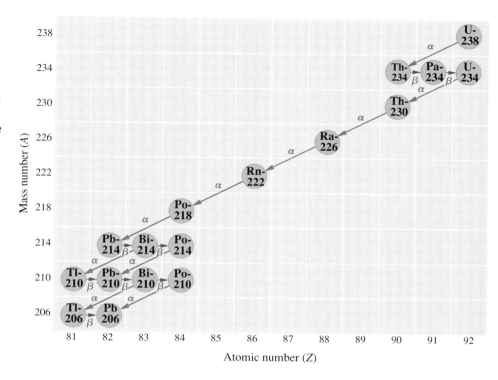

tive decay series, which consists of a sequence of alpha and beta decays, ending with lead-207. The third naturally occurring radioactive decay series begins with thorium-232 and ends with lead-208. All three radioactive decay series found naturally end with an isotope of lead.

## CONCEPT CHECK 21.1

You have two samples of water, each made up of different isotopes of hydrogen: one contains $_1^1H_2O$ and the other, $_1^3H_2O$.

a.  Would you expect these two water samples to be chemically similar?

b.  Would you expect these two water samples to be physically the same?

c.  Which one of these water samples would you expect to be radioactive?

# 21.2  Nuclear Bombardment Reactions

The nuclear reactions discussed in the previous section are radioactive decay reactions, in which a nucleus spontaneously decays to another nucleus and emits a particle, such as an alpha or beta particle. In 1919, Ernest Rutherford discovered that it is possible to change the nucleus of one element into the nucleus of another element by processes that can be controlled in the laboratory. **Transmutation** is *the change of one element to another by bombarding the nucleus of the element with nuclear particles or nuclei.*

## Transmutation

Rutherford used a radioactive element as a source of alpha particles and allowed these particles to collide with nitrogen nuclei. He discovered that protons are ejected in the process. The equation for the nuclear reaction is

$$^{14}_{7}N + ^{4}_{2}He \longrightarrow ^{17}_{8}O + ^{1}_{1}H$$

The experiments were repeated on other light nuclei, most of which were transmuted to other elements with the ejection of a proton. These experiments yielded two significant results. First, they strengthened the view that all nuclei contain protons. Second, they showed for the first time that it is possible to change one element into another under laboratory control.

When beryllium is bombarded with alpha particles, a penetrating radiation is given off that is not deflected by electric or magnetic fields. Therefore, the radiation does not consist of charged particles. The British physicist James Chadwick (1891–1974) suggested in 1932 that the radiation from beryllium consists of neutral particles, each with a mass approximately that of a proton. The particles are called neutrons. The reaction that resulted in the discovery of the neutron is

$$^{9}_{4}Be + ^{4}_{2}He \longrightarrow ^{12}_{6}C + ^{1}_{0}n$$

In 1933, a nuclear bombardment reaction was used to produce the first artificial radioactive isotope. Irène and Frédéric Joliot-Curie found that aluminum bombarded with alpha particles produces phosphorus-30, which decays by emitting positrons. The reactions are

$$^{27}_{13}Al + ^{4}_{2}He \longrightarrow ^{30}_{15}P + ^{1}_{0}n$$
$$^{30}_{15}P \longrightarrow ^{30}_{14}Si + ^{0}_{1}e$$

Phosphorus-30 was the first radioactive nucleus produced in the laboratory. Since then over a thousand radioactive isotopes have been made. ◄

► Some of the uses of radioactive isotopes are discussed in Section 21.5.

Nuclear bombardment reactions are often referred to by an abbreviated notation. For example, the reaction

$$^{14}_{7}N + ^{4}_{2}He \longrightarrow ^{17}_{8}O + ^{1}_{1}H$$

is abbreviated $^{14}_{7}N(\alpha, p)^{17}_{8}O$. In this notation, you first write the nuclide symbol for the original nucleus (target), then in parentheses you write the symbol for the projectile particle (incoming particle), followed by a comma and the symbol for the ejected particle. After the last parenthesis, you write the nuclide symbol for the product nucleus. The following symbols are used for particles:

| | |
|---|---|
| Neutron | n |
| Proton | p |
| Deuteron, $^{2}_{1}H$ | d |
| Alpha, $^{4}_{2}He$ | $\alpha$ |

► An alternate example is provided in the Instructor's Resource Manual.

---

**EXAMPLE 21.5**

## Using the Notation for a Bombardment Reaction

a. Write the abbreviated notation for the following bombardment reaction, in which neutrons were first discovered.

$$^{9}_{4}Be + ^{4}_{2}He \longrightarrow ^{12}_{6}C + ^{1}_{0}n$$

b. Write the nuclear equation for the bombardment reaction denoted $^{27}_{13}\text{Al}(\text{p, d})^{26}_{13}\text{Al}$.

**SOLUTION**

a. The notation is $^{9}_{4}\text{Be}(\alpha, \text{n})^{12}_{6}\text{C}$.

b. The nuclear equation is

$$^{27}_{13}\text{Al} + {}^{1}_{1}\text{H} \longrightarrow {}^{26}_{13}\text{Al} + {}^{2}_{1}\text{H}$$

---

**EXERCISE 21.5**

a. Write the abbreviated notation for the reaction

$$^{40}_{20}\text{Ca} + {}^{2}_{1}\text{H} \longrightarrow {}^{41}_{20}\text{Ca} + {}^{1}_{1}\text{H}$$

b. Write the nuclear equation for the bombardment reaction $^{12}_{6}\text{C}(\text{d, p})^{13}_{6}\text{C}$.

---

**See Problems 21.41, 21.42, 21.43, and 21.44.**

Elements of large atomic number merely scatter, or deflect, alpha particles from natural sources, rather than giving a transmutation reaction. These elements have nuclei of large positive charge, and the alpha particle must be traveling very fast in order to penetrate the nucleus and react. Alpha particles from natural sources do not have sufficient kinetic energy. To shoot charged particles into heavy nuclei, it is necessary to accelerate the charged particles.

A **particle accelerator** is *a device used to accelerate electrons, protons, and alpha particles and other ions to very high speeds.* The basis of a particle accelerator is the fact that a charged particle will accelerate toward a plate having a charge opposite in sign to that of the particle. It is customary to measure the kinetic energies of these particles in units of electron volts. An **electron volt (eV)** is *the quantity of energy that would have to be imparted to an electron (whose charge is $1.602 \times 10^{-19}$ C) to accelerate it by one volt potential difference.*

$$1 \text{ eV} = (1.602 \times 10^{-19} \text{ C}) \times (1 \text{ V}) = 1.602 \times 10^{-19} \text{ J}$$

Typically, particle accelerators give charged particles energies of millions of electron volts (MeV). To keep the accelerated particles from colliding with molecules of gas, the apparatus is enclosed and evacuated to low pressures, about $10^{-6}$ mmHg or less.

A charged particle can be accelerated in stages to very high kinetic energy. Figure 21.5 shows a diagram of a **cyclotron,** *a type of particle accelerator consisting of two hollow, semicircular metal electrodes, called dees (because the shape resembles the letter D), in which charged particles are accelerated by stages to higher and higher kinetic energies.* Ions introduced at the center of the cyclotron are accelerated in the space between the two dees. Magnet poles (not shown in the figure) above and below the dees keep the ions moving in an enlarging spiral path. The dees are connected to a high-frequency electric current that changes their polarity so that each time an ion moves into the space between the dees, it is accelerated. Thus, the ion is continually accelerated until it finally leaves the cyclotron at high speed. Outside the cyclotron, the ions are directed toward a target element so that investigators may study nuclear reactions or prepare isotopes.

Technetium was first prepared by directing **deuterons,** or *nuclei of hydrogen-2 atoms,* from a cyclotron to a molybdenum target. ◀ The nuclear reaction is $^{96}_{42}\text{Mo}(\text{d, n})^{97}_{43}\text{Tc}$, or

$$^{96}_{42}\text{Mo} + {}^{2}_{1}\text{H} \longrightarrow {}^{97}_{43}\text{Tc} + {}^{1}_{0}\text{n}$$

▶ The hydrogen-2 atom is often called deuterium and given the symbol D. It is a stable isotope with a natural abundance of 0.015%. Deuterium was discovered in 1931 by Harold Urey and coworkers and was prepared in pure form by G. N. Lewis.

**Figure 21.5**
**A cyclotron.** Positive ions are introduced at the center of the cyclotron. A fraction of the ions will be crossing the gap between the dees when the electric polarities are just right to accelerate them. (The positive ions are accelerated toward the negative dee and away from the positive dee.) The dees alternate in polarities, so that these ions continue to be accelerated each time they pass between the gap in the dees. Magnet poles above and below the dees produce a magnetic field that keeps the ions moving in a spiral path within the dees. At the end of their path, the ions encounter a negative electrode that deflects them to a target material.

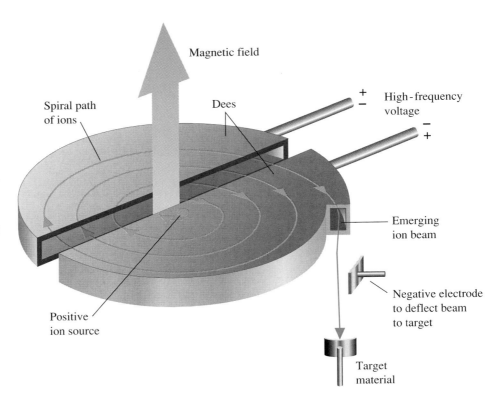

The energies of particles obtained from a cyclotron are limited (to about 20 MeV for deuterons, for example). For higher energies, more sophisticated accelerators are required. Figure 21.6 shows the accelerator at Fermilab in Batavia, Illinois. Larger and larger particle accelerators are being built to study the basic constituents of matter.

### Transuranium Elements

The **transuranium elements** are *elements with atomic numbers greater than that of uranium (Z = 92), the naturally occurring element of greatest Z.* In 1940, E. M. McMillan and P. H. Abelson, at the University of California at Berkeley, discovered the first transuranium element. They produced an isotope of element 93, which they

**Figure 21.6**
**The Fermilab accelerator in Batavia, Illinois.** The tunnel of the main accelerator is shown here. Protons are accelerated in the upper ring of conventional magnets (red and blue) to 400 billion electron volts. These protons are then injected into the lower ring of superconducting magnets (yellow), where they are accelerated to almost a trillion electron volts.

named neptunium, by bombarding uranium-238 with neutrons. This gave uranium-239, by the capture of a neutron, and this nucleus decayed in a few days by beta emission to neptunium-239.

$$^{238}_{92}U + ^{1}_{0}n \longrightarrow ^{239}_{92}U$$
$$^{239}_{92}U \longrightarrow ^{239}_{93}Np + ^{0}_{-1}e$$

The next transuranium element to be discovered was plutonium ($Z = 94$). Deuterons, the positively charged nuclei of hydrogen-2, were accelerated by a cyclotron and directed at a uranium target to give neptunium-238, which decayed to plutonium-238.

$$^{238}_{92}U + ^{2}_{1}H \longrightarrow ^{238}_{93}Np + 2^{1}_{0}n$$
$$^{238}_{93}Np \longrightarrow ^{238}_{94}Pu + ^{0}_{-1}e$$

Another isotope of plutonium, plutonium-239, is now produced in large quantity in nuclear reactors, as described in Section 21.7. Plutonium-239 is used for nuclear weapons.

The discovery of the next two transuranium elements, americium ($Z = 95$) and curium ($Z = 96$), depended on an understanding of the correct positions in the periodic table of the elements beyond actinium ($Z = 89$). It had been thought that these elements should be placed after actinium under the $d$-transition elements. So uranium was placed in Group VIB under tungsten. However, Glenn T. Seaborg, then at the University of California, Berkeley, postulated a second series of elements to be placed at the bottom of the periodic table, under the lanthanides, as shown in modern tables (see inside front cover). These elements, the actinides, would be expected to have chemical properties similar to those of the lanthanides. Once they understood this, Seaborg and others were able to use the predicted chemical behaviors of the actinides to separate americium and curium.

Figure 21.7 shows the apparatus used by Albert Ghiorso at the Lawrence Berkeley Laboratory in Berkeley, California, for the discovery of elements 104, 105, and 106. A team of scientists led by Peter Armbruster in Darmstadt, Germany, created elements 107

**Figure 21.7**
**The vertical wheel, an apparatus used in the discovery of elements 104, 105, and 106.** The apparatus was designed by Albert Ghiorso and colleagues. They synthesized element 106 by bombarding californium-249 with oxygen-18. The products were whisked by gas jets to the vertical wheel, where they were deposited (at the top). Detectors around the periphery of the wheel monitored the alpha activity to identify the products. Binary-coded holes label the positions around the wheel for computer collection of data.

**Figure 21.8**
**The High Flux Isotope Reactor at Oak Ridge National Laboratory.** The reactor in which transuranium isotopes are produced lies beneath a protecting pool of water. Visible light is emitted by high-energy particles moving through the water.

▶ An alternate example is provided in the Instructor's Resource Manual.

to 109 in the early 1980s by bombarding other elements with heavy-metal ions. During the 1994–1996 period, they reported the discovery of elements 110, 111, and 112. Recent work, described in the essay at the end of Section 2.5, has yielded other elements including the heaviest to date, 118.

The transuranium elements have a number of commercial uses. Plutonium-238 emits only alpha radiation, which is easily stopped by shielding. The isotope has been used as a power source for space satellites, navigation buoys, and heart pacemakers. Americium-241 is both an alpha-ray and a gamma-ray emitter. The gamma rays are used in devices that measure the thickness of materials such as metal sheets. Americium-241 is also used in home smoke detectors, in which the alpha radiation ionizes the air in a chamber within the detector and renders it electrically conducting. Smoke reduces the conductivity of the air, and this reduced conductivity is detected by an alarm circuit.

Figure 21.8 shows the High Flux Isotope Reactor at Oak Ridge National Laboratory, where technicians produce transuranium elements by bombarding plutonium-239 with neutrons. Neutrons are produced by nuclear fission, a process discussed in Section 21.7.

---

**EXAMPLE 21.6**

### Determining the Product Nucleus in a Nuclear Bombardment Reaction

Plutonium-239 was bombarded by alpha particles. Each $^{239}_{94}$Pu nucleus was struck by one alpha particle and emitted one neutron. What was the product nucleus?

**SOLUTION**

This is similar to Example 21.2. You can write the nuclear equation as follows:

$$^{239}_{94}\text{Pu} + ^4_2\text{He} \longrightarrow ^A_Z\text{X} + ^1_0\text{n}$$

To balance this equation, write

$$239 + 4 = A + 1 \quad \text{(from superscripts)}$$
$$94 + 2 = Z + 0 \quad \text{(from subscripts)}$$

Hence,
$$A = 239 + 4 - 1 = 242$$
$$Z = 94 + 2 = 96$$

The product is $^{242}_{96}$**Cm.**

---

**EXERCISE 21.6**

Carbon-14 is produced in the upper atmosphere when a particular nucleus is bombarded with neutrons. A proton is ejected for each nucleus that reacts. What is the identity of the nucleus that produces carbon-14 by this reaction?

See Problems 21.47, 21.48, 21.49, and 21.50.

---

## 21.3 Radiations and Matter: Detection and Biological Effects

Radiations from nuclear processes affect matter in part by dissipating energy in it. An alpha, beta, or gamma particle traveling through matter dissipates energy by ionizing atoms or molecules, producing positive ions and electrons. In some cases, these radi-

**Figure 21.9**
**A Geiger counter.**  A particle of radiation enters the thin window and passes into the gas. Energy from the particle ionizes gas molecules, giving positive ions and electrons, which are accelerated to the electrodes. The electrons, which move faster, strike the wire anode and create a pulse of current. Pulses from radiation particles are counted.

ations may also excite electrons in matter. When these electrons undergo transitions back to their ground states, light is emitted. The ions, free electrons, and light produced in matter can be used to detect nuclear radiations. Because nuclear radiations can ionize molecules and break chemical bonds, they adversely affect biological organisms. We will first look at the detection of nuclear radiations and then briefly discuss biological effects and radiation dosage in humans.

### Radiation Counters

Two types of devices—*ionization counters* and *scintillation counters*—are used to count particles emitted from radioactive nuclei and other nuclear processes. Ionization counters detect the production of ions in matter. Scintillation counters detect the production of scintillations, or flashes of light.

A **Geiger counter** (Figure 21.9), *a kind of ionization counter used to count particles emitted by radioactive nuclei, consists of a metal tube filled with gas, such as argon.* The tube is fitted with a thin glass or plastic window through which radiation enters. A wire runs down the tube's center, from which the wire is insulated. The tube and wire are connected to a high-voltage source so that the tube becomes the negative electrode and the wire the positive electrode. Normally the gas in the tube is an insulator and no current flows through it. However, when radiation, such as an alpha particle, $_2^4\text{He}^{2+}$, passes through the window of the tube and into the gas, atoms are ionized. Free electrons are quickly accelerated to the wire. As they are accelerated to the wire, additional atoms may be ionized from collisions with these electrons and more electrons set free. An avalanche of electrons is created, and this gives a pulse of current that is detected by electronic equipment. The amplified pulse activates a digital counter or gives an audible "click."

Alpha and beta particles can be detected directly by a Geiger counter. ◀ To detect neutrons, boron trifluoride is added to the gas in the tube. Neutrons react with boron-10 nuclei to produce alpha particles, which can then be detected.

$$_0^1\text{n} + {}_5^{10}\text{B} \longrightarrow {}_3^7\text{Li} + {}_2^4\text{He}$$

A **scintillation counter** (Figure 21.10) is *a device that detects nuclear radiation from flashes of light generated in a material by the radiation.* A *phosphor* is a substance that emits flashes of light when struck by radiation. Rutherford used zinc sulfide as a

▶ Demonstration of a cloud chamber, a simple device used to detect radiation: see the Instructor's Resource Manual.

▶ Gamma and low-energy alpha particles are better detected by scintillation counters. These counters are filled with solids or liquids, which are more likely to stop gamma rays. Also, low-energy alpha particles may be absorbed by the window of the Geiger counter and go undetected.

**Figure 21.10**
**A scintillation counter probe.** *Left:* Radiation passes through the window into a phosphor (here, NaI). Flashes of light from the phosphor fall on the photocathode, which ejects electrons by the photoelectric effect. The electrons are accelerated to electrodes of increasing positive voltage, each electrode ejecting more electrons, so that the original signal is magnified. *Right:* Photograph of a scintillation probe (the window is on the left).

▶ The photoelectric effect was discussed in Section 7.2.

▶ The curie was originally defined as the number of disintegrations per second from 1.0 g of radium-226.

▶ Sources of alpha radiation outside the body are relatively harmless, because the radiation is absorbed by the skin. Internal sources, however, are very destructive.

phosphor to detect alpha particles. A sodium iodide crystal containing thallium(I) iodide is used as a phosphor for gamma radiation. (Excited technetium-99 emits gamma rays and is used for medical diagnostics. The gamma rays are detected by a scintillation counter.)

The flashes of light from the phosphor are detected by a *photomultiplier.* A photon of light from the phosphor hits a photoelectric-sensitive surface (the photocathode). ◀ This emits an electron, which is accelerated by a positive voltage to another electrode, from which several electrons are emitted. These electrons are accelerated by a higher voltage to the next electrode, from which more electrons are emitted, and so forth. The result is that a single electron may produce a million electrons and therefore a detectable pulse of electric current.

A radiation counter can be used to measure the rate of nuclear disintegrations in a radioactive material. The **activity of a radioactive source** is *the number of nuclear disintegrations per unit time occurring in a radioactive material.* A **curie (Ci)** is *a unit of activity equal to 3.700 $\times$ $10^{10}$ disintegrations per second.* ◀ For example, a sample of technetium having an activity of $1.0 \times 10^{-2}$ Ci is decaying at the rate of $(1.0 \times 10^{-2}) \times (3.7 \times 10^{10}) = 3.7 \times 10^{8}$ nuclei per second.

### Biological Effects and Radiation Dosage

Although the quantity of energy dissipated in a biological organism from a radiation dosage might be small, the effects can be quite adverse because important chemical bonds may be broken. DNA in the chromosomes of the cell is especially affected, which interferes with cell division. Cells that divide the fastest, such as those in the blood-forming tissue in bone marrow, are most affected by nuclear radiations.

To monitor the effect of nuclear radiations on biological tissue, it is necessary to have a measure of radiation dosage. The **rad** (from *r*adiation *a*bsorbed *d*ose) is *the dosage of radiation that deposits 1 $\times$ $10^{-2}$ J of energy per kilogram of tissue.* However, the biological effect of radiation depends not only on the energy deposited in the tissue but also on the type of radiation. For example, neutrons are more destructive than gamma rays of the same radiation dosage measured in rads. A **rem** is *a unit of radiation dosage used to relate various kinds of radiation in terms of biological destruction.* It equals the rad times a factor for the type of radiation, called the relative biological effectiveness (RBE).

$$\text{rems} = \text{rads} \times \text{RBE}$$

Beta and gamma radiations have an RBE of about 1, whereas neutron radiation has an RBE of about 5 and alpha radiation an RBE of about 10. ◀

The effects of radiation on a person depend not only on the dosage but also on the length of time in which the dose was received. A series of small doses has less overall effect than these dosages given all at once. A single dose of about 500 rems is fatal to most people, and survival from a much smaller dose can be uncertain or leave the person chronically ill. Detectable effects are seen with doses as low as 30 rems. Continuous exposure to such low levels of radiation may result in cancer or leukemia. At even lower levels, the answer to whether the radiation dose is safe depends on the possible genetic effects of the radiation. Because radiation can cause chromosome damage, heritable defects are possible.

Safe limits for radiations are much debated. Although there may not be a strictly safe limit, it is important to have in mind the magnitude of the radiations humans may be subjected to. The background radiation we all receive results from cosmic rays (radiations from space) and natural radioactivity. This averages about 0.1 rem per year but varies considerably with location. Uranium and its decay products in the soil are an important source of background radiation. ◀ Another source is potassium-40, a radioactive isotope with a natural abundance of 0.012%. In addition to natural background radiation, we may receive radiation from other fairly common sources. The most important of these are x rays used in medical diagnosis. The average person receives a radiation dose from this source that is about equal to that of the natural background. Very small radiation sources include consumer items such as television sets and luminous watches.

The background radiation to which we are all subjected has increased slightly since the advent of nuclear technology. Fallout from atmospheric testing of nuclear weapons increased this background by several percent, but it has decreased since atmospheric testing was banned. The radiation contributed by nuclear power plants is only a fraction of a percent of the natural background.

▶ Soils contain varying amounts of uranium-238, which decays in several steps to radium-226, then to radon-222, a gas (see Figure 21.4). Some homes situated in areas of high uranium content have been found to accumulate radon gas. Radon has a half-life of 3.8 days, decaying by alpha emission to radioactive lead, bismuth, and polonium. These decay products can remain in the lungs and may lead to lung cancer.

## CONCEPT CHECK 21.2

If you are internally exposed to 10 rads of $\alpha$, $\beta$, and $\gamma$ radiation, which form of radiation will cause the greatest biological damage?

# 21.4 Rate of Radioactive Decay

Although technetium-99 is radioactive and decays by emitting beta particles (electrons), it is impossible to say when a particular nucleus will disintegrate. A sample of technetium-99 continues to give off beta rays for millions of years. Thus, a particular nucleus might disintegrate the next instant or several million years later. The rate of this radioactive decay cannot be changed by varying the temperature, the pressure, or the chemical environment of the technetium nucleus. Radioactivity, whether from technetium or some other nucleus, is not affected by the variables that affect the rate of a chemical reaction. In this section, we will look at how to express quantitatively the rate of radioactive decay.

### Rate of Radioactive Decay and Half-Life

The rate of radioactive decay—that is, the number of nuclei disintegrating per unit time—is found to be proportional to the number of radioactive nuclei in the sample. You can express this rate mathematically as

$$\text{Rate} = kN_t$$

▶ **The rate equation for radioactive decay has the same form as the rate law for a first-order chemical reaction. Indeed, radioactive decay is a first-order rate process, and the mathematical relationships used in Chapter 14 for first-order reactions apply here also.**

Here $N_t$ is the number of radioactive nuclei at time $t$, and $k$ is the **radioactive decay constant,** or *rate constant for radioactive decay.* This rate constant is a characteristic of the radioactive nuclide, each nuclide having a different value. ◀

You can obtain the decay constant for a radioactive nucleus by counting the nuclear disintegrations over a period of time. The original definition of the curie (now $3.7 \times 10^{10}$ disintegrations per second) was the activity or decay rate of 1.0 g of radium-226. You can use this with the rate equation to obtain the decay constant of radium-226. Radium-226 has a molar mass of 226 g. A 1.0-g sample of radium-226 contains the following number of nuclei:

$$1.0 \text{ g Ra-226} \times \frac{1 \text{ mol Ra-226}}{226 \text{ g Ra-226}} \times \frac{6.02 \times 10^{23} \text{ Ra-226 nuclei}}{1 \text{ mol Ra-226}} = 2.7 \times 10^{21} \text{ Ra-226 nuclei}$$

This equals the value of $N_t$. When you solve Rate $= kN_t$ for $k$, you get

$$k = \frac{\text{rate}}{N_t}$$

Substituting into this gives

▶ **Recall from Chapter 7 that /s is equivalent to the unit s⁻¹.**

$$k = \frac{3.7 \times 10^{10} \text{ nuclei/s}}{2.7 \times 10^{21} \text{ nuclei}} = 1.4 \times 10^{-11}\text{/s} \quad ◀$$

▶ **An alternate example is provided in the Instructor's Resource Manual.**

## EXAMPLE 21.7

### Calculating the Decay Constant from the Activity

A 1.0-mg sample of technetium-99 has an activity of $1.7 \times 10^{-5}$ Ci (Ci = curies), decaying by beta emission. What is the decay constant for $^{99}_{43}\text{Tc}$?

 **PROBLEM STRATEGY**

Solve the equation Rate $= kN_t$ for $k$; obtain Rate and $N_t$ from values in the problem statement. The rate of decay (Rate) of Tc-99 equals the activity in curies expressed as nuclei decaying per second, noting that 1 Ci $= 3.7 \times 10^{10}$ nuclei/s. Obtain the number of Tc-99 nuclei, $N_t$, by converting $1.0 \times 10^{-3}$ g of Tc-99 as follows:

$$\text{g Tc-99} \longrightarrow \text{mol Tc-99} \longrightarrow \text{number Tc-99 nuclei}$$

For the first conversion, note that the numerical value of the molar mass of Tc-99 is approximately equal to the mass number of Tc(99).

### SOLUTION

The rate of decay in this sample is

$$\text{Rate} = 1.7 \times 10^{-5} \text{ Ci} \times \frac{3.7 \times 10^{10} \text{ nuclei/s}}{1.0 \text{ Ci}} = 6.3 \times 10^5 \text{ nuclei/s}$$

The number of nuclei in this sample of $1.0 \times 10^{-3}$ g $^{99}_{43}\text{Tc}$ is

$$1.0 \times 10^{-3} \text{ g Tc-99} \times \frac{1 \text{ mol Tc-99}}{99 \text{ g Tc-99}} \times \frac{6.02 \times 10^{23} \text{ Tc-99 nuclei}}{1 \text{ mol Tc-99}}$$

$$= 6.1 \times 10^{18} \text{ Tc-99 nuclei}$$

The decay constant is

$$k = \frac{\text{rate}}{N_t} = \frac{6.3 \times 10^5 \text{ nuclei/s}}{6.1 \times 10^{18} \text{ nuclei}} = \mathbf{1.0 \times 10^{-13}/s}$$

### EXERCISE 21.7

The nucleus $^{99m}_{43}\text{Tc}$ is a metastable nucleus of technetium-99; it is used in medical diagnostic work. Technetium-99$m$ decays by emitting gamma rays. A 2.5-$\mu$g (microgram) sample has an activity of 13 Ci. What is the decay constant (in units of /s)?

See Problems 21.51, 21.52, 21.53, and 21.54.

▶ **The half-life of a reaction was discussed in Section 14.4.**

▶ **Iodine is attracted to the thyroid gland, which incorporates the element into the growth hormone thyroxine. Radiation from iodine-131 kills cancer cells in the thyroid gland.**

We define the **half-life** of a radioactive nucleus as *the time it takes for one-half of the nuclei in a sample to decay.* ◀ The half-life is independent of the amount of sample. In some cases, you can find the half-life of a radioactive sample by directly observing how long it takes for one-half of the sample to decay. For example, you find that 1.000 g of iodine-131, an isotope used in treating thyroid cancer, decays to 0.500 g in 8.07 days. ◀ Thus, its half-life is 8.07 days. In another 8.07 days, this sample would decay to one-half of 0.500 g, or 0.250 g, and so forth. Figure 21.11 shows this decay pattern.

Although you might be able to obtain the half-life of a radioactive nucleus by direct observation in some cases, this is impossible for many nuclei because they decay too quickly or too slowly. Uranium-238, for example, has a half-life of 4.51 billion years, much too long to be directly observed! The usual method of determining the half-life is by measuring decay rates and relating them to half-lives.

You relate the half-life for radioactive decay, $t_{1/2}$, to the decay constant $k$ by the equation

$$t_{1/2} = \frac{0.693}{k}$$

**Figure 21.11**
**Radioactive decay of a 1.000-g sample of iodine-131.** The sample decays by one-half in each half-life of 8.07 days.

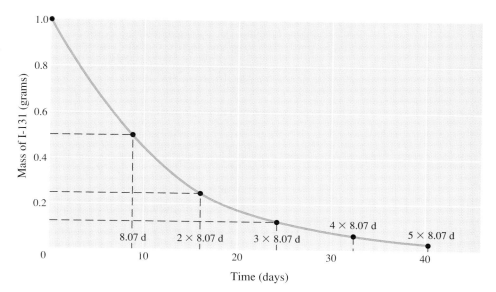

The next example illustrates the use of this relation.

▶ An alternate example is provided in the Instructor's Resource Manual.

---

### EXAMPLE 21.8

## Calculating the Half-Life from the Decay Constant

The decay constant for the beta decay of $^{99}_{43}$Tc was obtained in Example 21.7. We found that $k$ equals $1.0 \times 10^{-13}$/s. What is the half-life of this isotope in years?

**SOLUTION**

Substitute the value of $k$ into the preceding equation.

$$t_{1/2} = \frac{0.693}{k} = \frac{0.693}{1.0 \times 10^{-13}/s} = 6.9 \times 10^{12}\ s$$

Then you convert this half-life in seconds to years.

$$6.9 \times 10^{12}\ s \times \frac{1\ min}{60\ s} \times \frac{1\ h}{60\ min} \times \frac{1\ d}{24\ h} \times \frac{1\ y}{365\ d} = \mathbf{2.2 \times 10^{5}\ y}$$

---

### EXERCISE 21.8

Cobalt-60, used in cancer therapy, decays by beta and gamma emission. The decay constant is $4.18 \times 10^{-9}$/s. What is the half-life in years?

---

**See Problems 21.55 and 21.56.**

Tables of radioactive nuclei often list the half-life. When you want the decay constant or the activity of a sample, you can calculate them from the half-life. This calculation is illustrated in the next example.

▶ An alternate example is provided in the Instructor's Resource Manual.

---

### EXAMPLE 21.9

## Calculating the Decay Constant and Activity from the Half-Life

Tritium, $^{3}_{1}$H, is a radioactive nucleus of hydrogen. It is used in luminous watch dials. Tritium decays by beta emission with a half-life of 12.3 years. What is the decay constant (in /s)? What is the activity (in Ci) of a sample containing 2.5 $\mu$g of tritium? The atomic mass of $^{3}_{1}$H is 3.02 amu.

 **PROBLEM STRATEGY**

▶ Experiments in General Chemistry, 21, "Natural Radioactivity."

First convert the half-life to its value in seconds and then calculate $k$. Then you can use the rate equation to find the rate of decay of the sample (in nuclei/s) and finally the activity.

**SOLUTION**

The conversion of the half-life to seconds gives

$$12.3\ y \times \frac{365\ d}{1\ y} \times \frac{24\ h}{1\ d} \times \frac{60\ min}{1\ h} \times \frac{60\ s}{1\ min} = 3.88 \times 10^{8}\ s$$

Because $t_{1/2} = 0.693/k$, you solve this for $k$ and substitute the half-life in seconds.

$$k = \frac{0.693}{t_{1/2}} = \frac{0.693}{3.88 \times 10^8 \text{ s}} = \textbf{1.79} \times \textbf{10}^{-9}\textbf{/s}$$

Before substituting into the rate equation, you need to know the number of tritium nuclei in a sample containing $2.5 \times 10^{-6}$ g of tritium. You get

$$2.5 \times 10^{-6} \text{ g H-3} \times \frac{1 \text{ mol H-3}}{3.02 \text{ g H-3}} \times \frac{6.02 \times 10^{23} \text{ H-3 nuclei}}{1 \text{ mol H-3}} = 5.0 \times 10^{17} \text{ H-3 nuclei}$$

Now you substitute into the rate equation.

$$\text{Rate} = kN_t = 1.79 \times 10^{-9}\text{/s} \times 5.0 \times 10^{17} \text{ nuclei} = 9.0 \times 10^8 \text{ nuclei/s}$$

You obtain the activity of the sample by converting the rate in disintegrations of nuclei per second to curies.

$$\text{Activity} = 9.8 \times 10 \text{ nuclei/s} \times \frac{1.0 \text{ Ci}}{3.7 \times 10^{10} \text{ nuclei/s}} = \textbf{0.024 Ci}$$

### EXERCISE 21.9

Strontium-90, $^{90}_{38}\text{Sr}$, is a radioactive decay product of nuclear fallout from nuclear weapons testing. Because of its chemical similarity to calcium, it is incorporated into the bones if present in food. The half-life of strontium-90 is 28.1 y. What is the decay constant of this isotope? What is the activity of a sample containing 5.2 ng ($5.2 \times 10^{-9}$ g) of strontium-90?

**See Problems 21.57, 21.58, 21.59, and 21.60.**

Once you know the decay constant for a radioactive isotope, you can calculate the fraction of the radioactive nuclei that remains after a given period of time by the following equation. ◄

► A similar equation is given in Chapter 14 for the reactant concentration at time $t$, $[\text{A}]_t$:

$$\ln \frac{[\text{A}]_t}{[\text{A}]_0} = -kt$$

where $[\text{A}]_0$ is the concentration of A at $t = 0$.

$$\ln \frac{N_t}{N_0} = -kt$$

Here $N_0$ is the number of nuclei in the original sample ($t = 0$). After a period of time $t$, the number of nuclei decreases by decay to the number $N_t$. The fraction of nuclei remaining after time $t$ is $N_t/N_0$. The next example illustrates the use of this equation.

► An alternate example is provided in the Instructor's Resource Manual.

### EXAMPLE 21.10

## Determining the Fraction of Nuclei Remaining after a Specified Time

Phosphorus-32 is a radioactive isotope with a half-life of 14.3 d (d = days). A biochemist has a vial containing a compound of phosphorus-32. If the compound is used in an experiment 5.5 d after the compound was prepared, what fraction of the radioactive isotope originally present remains? Suppose the sample in the vial originally contained 0.28 g of phosphorus-32. How many grams remain after 5.5 d?

**SOLUTION**

If $N_0$ is the original number of P-32 nuclei in the vial and $N_t$ is the number after 5.5 d, the fraction remaining is $N_t/N_0$. You can obtain this fraction from the equation

$$\ln \frac{N_t}{N_0} = -kt$$

You substitute $k = 0.693/t_{1/2}$.

$$\ln \frac{N_t}{N_0} = \frac{-0.693t}{t_{1/2}}$$

Because $t = 5.5$ d and $t_{1/2} = 14.3$ d, you obtain

$$\ln \frac{N_t}{N_0} = \frac{-0.693 \times 5.5 \text{ d}}{14.3 \text{ d}} = -0.267$$

(An additional digit has been retained for further calculation.) Hence,

$$\text{Fraction nuclei remaining} = \frac{N_t}{N_0} = e^{-0.267} = \mathbf{0.77}$$

(Note that 0.766 is rounded to two significant figures.) Thus, 77% of the nuclei remain. The mass of $^{32}_{15}\text{P}$ in the vial after 5.5 d is

$$0.28 \text{ g} \times 0.766 = \mathbf{0.21 \text{ g}}$$

**Problem Solving:
Kinetics of Nuclear Decay**

**EXERCISE 21.10**

A nuclear power plant emits into the atmosphere a very small amount of krypton-85, a radioactive isotope with a half-life of 10.76 y. What fraction of this krypton-85 remains after 25.0 y?

See Problems 21.63 and 21.64.

## Radioactive Dating

Fixing the dates of relics and stone implements or pieces of charcoal from ancient campsites is an application based on radioactive decay rates. Because the rate of radioactive decay of a nuclide is constant, this rate can serve as a clock for dating very old rocks and human implements. Dating wood and similar carbon-containing objects that are several thousand to fifty thousand years old can be done with radioactive carbon, carbon-14, which has a half-life of 5730 y.

Carbon-14 is present in the atmosphere as a result of cosmic-ray bombardment of earth. Cosmic rays are radiations from space that consist of protons and alpha particles, as well as heavier ions. These radiations produce other kinds of particles, including neutrons, as they bombard the upper atmosphere. The collision of a neutron with a nitrogen-14 nucleus (the most abundant nitrogen nuclide) can produce a carbon-14 nucleus.

$$^{14}_{7}\text{N} + ^{1}_{0}\text{n} \longrightarrow ^{14}_{6}\text{C} + ^{1}_{1}\text{H}$$

Carbon dioxide containing carbon-14 mixes with the lower atmosphere. Because of the constant production of $^{14}_{6}\text{C}$ and its radioactive decay, a small, constant fractional abundance of carbon-14 is maintained in the atmosphere.

Living plants, which continuously use atmospheric carbon dioxide, also maintain a constant abundance of carbon-14. Similarly, living animals, by feeding on plants, have a constant fractional abundance of carbon-14. But once an organism dies, it is no longer in chemical equilibrium with atmospheric $CO_2$. The ratio of carbon-14 to carbon-12 begins to decrease by radioactive decay of carbon-14. In this way, this ratio of carbon isotopes becomes a clock measuring the time since the death of the organism.

▶ Analyses of tree rings have shown that this assumption is not quite valid. Before 1000 B.C., the levels of carbon-14 were somewhat higher than they are today. Moreover, recent human activities (burning of fossil fuels and atmospheric nuclear testing) have changed the fraction of carbon-14 in atmospheric $CO_2$.

If you assume that the ratio of carbon isotopes in the lower atmosphere has remained at the present level for the last 50,000 years (presently 1 out of $10^{12}$ carbon atoms is carbon-14), you can deduce the age of any dead organic object by measuring the level of beta emissions that arise from the radioactive decay of carbon-14. ◀ This is the decay reaction:

$$^{14}_{6}\text{C} \longrightarrow {}^{14}_{7}\text{N} + {}^{0}_{-1}\text{e}$$

In this way, bits of campfire charcoal, parchment, jawbones, and the like have been dated.

## EXAMPLE 21.11

## Applying the Carbon-14 Dating Method

A piece of charcoal from a tree killed by the volcanic eruption that formed the crater in Crater Lake (in Oregon) gave 7.0 disintegrations of carbon-14 nuclei per minute per gram of total carbon. Present-day carbon (in living matter) gives 15.3 disintegrations per minute per gram of total carbon. Determine the date of the volcanic eruption. Recall that the half-life of carbon-14 is 5730 y.

 **PROBLEM STRATEGY**

You substitute $k = 0.693/t_{1/2}$ into the equation for the number of nuclei in a sample after time $t$.

$$\ln \frac{N_t}{N_0} = -kt = \frac{-0.693t}{t_{1/2}}$$

Hence,

$$t = \frac{t_{1/2}}{0.693} \ln \frac{N_0}{N_t}$$

To get $N_0/N_t$, you assume that the ratio of $^{14}_{6}\text{C}$ to $^{12}_{6}\text{C}$ in the atmosphere has remained constant. Then you can say that 1.00 g of total carbon from the living tree gave 15.3 disintegrations per minute. The ratio of the number of $^{14}_{6}\text{C}$ nuclei originally present to the number that existed at the time of dating equals the ratio of rates of disintegration.

### SOLUTION

The ratio of rates of disintegration is

$$\frac{N_0}{N_t} = \frac{15.3}{7.0} = 2.2$$

Substituting this value of $N_0/N_t$ and $t_{1/2} = 5730$ y into the equation developed in the Problem Strategy gives

$$t = \frac{t_{1/2}}{0.693} \ln \frac{N_0}{N_t} = \frac{5730 \text{ y}}{0.693} \ln 2.2 = 6.5 \times 10^3 \text{ y}$$

▶ An alternate example is provided in the Instructor's Resource Manual.

**Problem Solving:**
**¹⁴C Dating**

Thus, the date of the eruption was about **4500 B.C.,** 6500 years before the present.

### EXERCISE 21.11

A jawbone from the archaeological site at Folsom, New Mexico, was dated by analysis of its radioactive carbon. The activity of the carbon from the jawbone was 4.5 disintegrations per minute per gram of total carbon. What was the age of the jawbone? Carbon from living material gives 15.3 disintegrations per minute per gram of carbon.

**See Problems 21.69 and 21.70.**

For the age of rocks and meteorites, other similar methods of dating have been used. One method depends on the radioactivity of naturally occurring potassium-40, which decays by positron emission and electron capture (as well as by beta emission).

$$\ce{^{40}_{19}K -> ^{40}_{18}Ar + ^{0}_{1}e}$$

$$\ce{^{40}_{19}K + ^{0}_{-1}e -> ^{40}_{18}Ar}$$

Potassium occurs in many rocks. Once such a rock forms by solidification of molten material, the argon from the decay of potassium-40 is trapped. To obtain the age of a rock, you first determine the number of $\ce{^{40}_{19}K}$ atoms and the number of $\ce{^{40}_{18}Ar}$ atoms in a sample. The number of $\ce{^{40}_{19}K}$ atoms equals $N_t$. The number originally present, $N_0$, equals $N_t$ plus the number of argon atoms, because each argon atom resulted from the decay of a $\ce{^{40}_{19}K}$ nucleus. You then calculate the age of the rock from the ratio $N_t/N_0$.

The oldest rocks on earth have been dated at $3.8 \times 10^9$ y. The rocks at the earth's surface have been subjected to extensive weathering, so even older rocks may have existed. This age, $3.8 \times 10^9$ y, therefore represents the minimum possible age of the earth—the time since the solid crust first formed. Ages of meteorites, assumed to have solidified at the same time as other solid objects in the solar system, including earth, have been determined to be $4.4 \times 10^9$ y to $4.6 \times 10^9$ y. It is now believed from this and other evidence that the age of the earth is $4.6 \times 10^9$ y.

 **CONCEPT CHECK 21.3**

Why do you think that carbon-14 dating is limited to less than 50,000 years?

## 21.5   Applications of Radioactive Isotopes

We have already described two applications of nuclear chemistry. One was the preparation of elements not available naturally. We noted that the discovery of the transuranium elements clarified the position of the heavy elements in the periodic table. In the section just completed, we discussed the use of radioactivity in dating objects. We will discuss practical uses of nuclear energy later in the chapter. Here we will look at the applications of radioactive isotopes to chemical analysis and to medicine.

### Chemical Analysis

A **radioactive tracer** is *a very small amount of radioactive isotope added to a chemical, biological, or physical system to study the system.* The advantage of a radioactive tracer is that it behaves chemically just as a nonradioactive isotope does, but it can be detected in exceedingly small amounts by measuring the radiations emitted.

As an illustration of the use of radioactive tracers, consider the problem of establishing that chemical equilibrium is a dynamic process. Let us look at the equilibrium of solid lead(II) iodide and its saturated solution, containing $Pb^{2+}(aq)$ and $I^-(aq)$. The equilibrium is

$$PbI_2(s) \xrightleftharpoons{H_2O} Pb^{2+}(aq) + 2I^-(aq)$$

In two separate beakers, you prepare saturated solutions of $PbI_2$ in contact with the solid. One beaker contains only natural iodine atoms with nonradioactive isotopes. The other beaker contains radioactive iodide ion, $^{131}I^-$. Some of the solution, but no solid, containing the radioactive iodide ion is now added to the other beaker containing nonradioactive iodide ion. Both solutions are saturated, so the amount of solid in this beaker remains constant. Yet after a time the solid lead iodide, which was originally nonradioactive, becomes radioactive. This is evidence for a dynamic equilibrium, in which radioactive iodide ions in the solution substitute for nonradioactive iodide ions in the solid.

With only naturally occurring iodine available, it would have been impossible to detect the dynamic equilibrium. By using $^{131}I^-$ as a radioactive tracer, you can easily follow the substitution of radioactive iodine into the solid by measuring its increase in radioactivity.

A series of experiments using tracers was carried out in the late 1950s by Melvin Calvin at the University of California, Berkeley, to discover the mechanism of photosynthesis in plants. ◄ The overall process of photosynthesis involves the reaction of $CO_2$ and $H_2O$ to give glucose, $C_6H_{12}O_6$, and $O_2$. Energy for photosynthesis comes from the sun.

$$6CO_2(g) + 6H_2O(l) \xrightarrow{sunlight} C_6H_{12}O_6(aq) + 6O_2(g)$$

This equation represents only the net result of photosynthesis. As Calvin was able to show, the actual process consists of many separate steps. In several experiments, algae (single-celled plants) were exposed to carbon dioxide containing much more radioactive carbon-14 than occurs naturally. Then the algae were extracted with a solution of alcohol and water. The various compounds in this solution were separated by chromatography and identified. ◄ Compounds that contained radioactive carbon were produced in the different steps of photosynthesis. Eventually, Calvin was able to use tracers to show the main steps in photosynthesis.

Another example of the use of radioactive tracers in chemistry is **isotope dilution,** *a technique to determine the quantity of a substance in a mixture or of the total volume of solution by adding a known amount of an isotope to it.* After removal of a portion of the mixture, the fraction by which the isotope has been diluted provides a way of determining the quantity of substance or volume of solution.

As an example, suppose you wish to obtain the volume of water in a tank but are unable to drain the tank. You add 100 mL of water containing a radioactive isotope. After allowing this to mix completely with the water in the tank, you withdraw 100 mL of solution from the tank. You find that the activity of this solution in curies is 1/1000 that of the original radioactive solution. The isotope has been diluted by a factor of 1000, so the volume of the tank is $1000 \times 100$ mL = 100,000 mL (100 L).

A typical chemical example of isotope dilution is the determination of the amount of vitamin $B_{12}$, a cobalt-containing substance, in a sample of food. Although part of the vitamin in food can be obtained in pure form, not all of the vitamin can

► Melvin Calvin received the Nobel Prize in chemistry in 1961 for his work on photosynthesis.

► Chromatography was discussed in the essay at the end of Section 1.4.

be separated. Therefore, you cannot precisely determine the quantity of vitamin $B_{12}$ in a sample of food by separating the pure vitamin and weighing it. But you can determine the amount of vitamin $B_{12}$ by isotope dilution. Suppose you add $2.0 \times 10^{-7}$ g of vitamin $B_{12}$ containing radioactive cobalt-60 to 125 g of food and mix well. You then separate $5.4 \times 10^{-7}$ g of pure vitamin $B_{12}$ from the food and find that the activity in curies of this quantity of the vitamin contains 5.6% of the activity added from the radioactive cobalt. This indicates that you have recovered 5.6% of the total amount of vitamin $B_{12}$ in the sample. Therefore, the mass of vitamin $B_{12}$ in the food, including the amount added ($2.0 \times 10^{-7}$ g), is

$$5.4 \times 10^{-7} \text{ g} \times \frac{100}{5.6} = 9.6 \times 10^{-6} \text{ g}$$

Subtracting the amount added in the analysis gives

$$9.6 \times 10^{-6} \text{ g} - 2.0 \times 10^{-7} \text{ g} = 9.4 \times 10^{-6} \text{ g}$$

**Neutron activation analysis** is *an analysis of elements in a sample based on the conversion of stable isotopes to radioactive isotopes by bombarding a sample with neutrons.* Human hair contains trace amounts of many elements. By determining the exact amounts and the position of the elements in the hair shaft, you can identify whom the hair comes from (assuming you have a sample known to be that person's hair for comparison). Consider the analysis of human hair for arsenic, for example. When the natural isotope $^{75}_{33}\text{As}$ is bombarded with neutrons, a metastable nucleus $^{76m}_{33}\text{As}$ is obtained.

$$^{75}_{33}\text{As} + {}^{1}_{0}\text{n} \longrightarrow {}^{76m}_{33}\text{As}$$

A metastable nucleus is in an excited state. It decays, or undergoes a transition, to a lower state by emitting gamma rays. The frequencies, or energies, of the gamma rays emitted are characteristic of the element and serve to identify it. Also, the intensities of the gamma rays emitted are proportional to the amount of the element present. The method is very sensitive; it can identify as little as $10^{-9}$ g of arsenic. ◀

▶ **Neutron activation analysis has been used to authenticate oil paintings by giving an exact analysis of pigments used. Pigment compositions have changed, so it is possible to detect fraudulent paintings done with more modern pigments. The analysis can be done without affecting the painting.**

## Medical Therapy and Diagnosis

The use of radioactive isotopes has had a profound effect on the practice of medicine. Radioisotopes were first used in medicine in the treatment of cancer. This treatment is based on the fact that rapidly dividing cells, such as those in cancer, are more adversely affected by radiation from radioactive substances than are cells that divide more slowly. Radium-226 and its decay product radon-222 were used for cancer therapy a few years after the discovery of radioactivity. Today gamma radiation from cobalt-60 is more commonly used.

Cancer therapy is only one of the ways in which radioactive isotopes are used in medicine. The greatest advances in the use of radioactive isotopes have been in the diagnosis of disease. Radioactive isotopes are used for diagnosis in two ways. They are used to develop images of internal body organs so that their functioning can be examined. And they are used as tracers in the analysis of minute amounts of substances, such as a growth hormone in blood, to deduce possible disease conditions.

Technetium-99*m* is the radioactive isotope used most often to develop images (pictures) of internal body organs. It has a half-life of 6.02 h, decaying by gamma emission to technetium-99 in its nuclear ground state. The image is prepared by scanning part of the body for gamma rays with a scintillation detector. (Figure 21.1 shows the image of a person's skeleton obtained with technetium-99*m*.)

**Figure 21.12**
**A technetium-99$m$ generator.** *Left:* A conceptual view of the generator. Molybdenum-99, in the form of $MoO_4^{2-}$ ion adsorbed on alumina, decays to technetium-99$m$. The technetium is leached from the generator with a salt solution (saline charge) as $TcO_4^-$. *Right:* Photograph of a technetium-99$m$ generator, with lead case and vials of saline charge.

The technetium isotope is produced in a special container, or generator, shown in Figure 21.12. The generator contains radioactive molybdate ion, $MoO_4^{2-}$, adsorbed on alumina granules. Radioactive molybdenum-99 is produced from uranium-235 nuclear fission products (see Section 21.7). This radioactive molybdenum, adsorbed on alumina, is placed in the generator and sent to the hospital. Pertechnetate ion is obtained when the molybdenum-99 nucleus in $MoO_4^{2-}$ decays. The nuclear equation is

$$_{42}^{99}\text{Mo} \longrightarrow _{43}^{99m}\text{Tc} + _{-1}^{0}\text{e}$$

Each day pertechnetate ion, $TcO_4^-$, is leached from the generator with a salt solution whose osmotic pressure is the same as that of blood. One use of this pertechnetate ion is in assessing the performance of a patient's heart. The physician injects tin(II) ion into a vein and, a few minutes later, administers a similar injection of the pertechnetate ion. In the presence of tin(II) ion, the pertechnetate ion binds to the red blood cells. The heart then becomes visible in gamma-ray imaging equipment, because of the quantity of blood in the heart. Technetium-99$m$ pyrophosphate ($Tc_2P_2O_7$) is another technetium species useful for gamma-ray imaging. This compound binds especially strongly to recently damaged heart muscle, so these gamma-ray images can be used to assess the extent of damage from a heart attack.

Thallium-201 is a radioisotope used to determine whether a person has heart disease (caused by narrowing of the arteries to the heart). This isotope decays by electron capture and emits x rays and gamma rays, which can be used to obtain images similar to those obtained from technetium-99$m$ (Figure 21.13). Thallium-201 injected into the blood binds particularly strongly to heart muscle. Diagnosis of heart disease depends on the fact that only tissue that receives sufficient blood flow binds thallium-201. When someone exercises strenuously, some part of the person's heart tissue may not receive sufficient blood because of narrowed arteries. These areas do not bind thallium-201 and show up on an image as dark spots.

**Figure 21.13**
**Using thallium-201 to diagnose heart disease.**   *Left:* A patient is undergoing a heart scan using a portable thallium-201 scintillation counter (on the pivoted arm). Gamma-ray scintillations are counted and collected by a computer (in the foreground) and then presented on the screen as an image. *Right:* A series of cross-sectional images of a patient's heart after exercise (labeled "stress") and then some time afterward (labeled "rest"). By comparing the stress and rest images, a physician can see if there is impaired blood flow to an area of the heart (the area is dark in the stress image but light in the rest image) or if the heart muscle has been damaged through a heart attack (the area is dark in both stress and rest images).

More than a hundred different radioactive isotopes have been used in medicine. Besides technetium-99*m* and thallium-201, other examples include iodine-131, used to measure thyroid activity; phosphorus-32, used to locate tumors; and iron-59, used to measure the rate of formation of red blood cells.

*Radioimmunoassay* is a technique for analyzing blood and other body fluids for very small quantities of biologically active substances. The technique depends on the reversible binding of the substance to an antibody. Antibodies are produced in animals as protection against foreign substances. They protect by binding to the substance and countering its biological activity. Consider, for example, the analysis for insulin in a sample of blood from a patient. Before the analysis, a solution of insulin-binding antibodies has been prepared from laboratory animals. This solution is combined with insulin containing a radioactive isotope, in which the antibodies bind with radioactive insulin. Now the blood sample containing an unknown amount of insulin is added to the antibody–radioactive insulin mixture. The nonradioactive insulin replaces some of the radioactive insulin bound to the antibody. As a result, the antibody loses some of its radioactivity. The loss in radioactivity can be related to the amount of insulin in the blood sample.

# Energy of Nuclear Reactions

Nuclear reactions, like chemical reactions, involve changes of energy. However, the changes of energy in nuclear reactions are enormous by comparison with those in chemical reactions. The energy released by certain nuclear reactions is used in nuclear

# Positron Emission Tomography (PET)

**LIFE SCIENCE**

Positron emission tomography (PET) is a technique for following biochemical processes within the organs (brain, heart, and so forth) of the human body. Like magnetic resonance imaging (see essay at the end of Section 8.1), a PET scan produces an image of a two-dimensional slice through a body organ of a patient. The image shows the distribution of some positron-emitting isotope present in a compound that was administered earlier by injection. By comparing the PET scan of the patient with that of a healthy subject, a physician can diagnose the presence or absence of disease (Figure 21.14). The PET scan of the brain of an Alzheimer's patient differs markedly from that of a healthy subject. Similarly, the PET scan of a patient with a heart damaged by a coronary attack clearly shows the damaged area. Moreover, the PET scan may help the physician assess the likelihood of success of by-pass surgery.

Some isotopes used in PET scans are carbon-11, nitrogen-13, oxygen-15, and fluorine-18. All have short

**Figure 21.15**
**A patient undergoing a PET scan of the brain.**
Note the circular bank of gamma-ray detectors.

**Figure 21.14**
**PET scans of normal and diseased patients.**
The scans are of the brains of a normal patient and a schizophrenic patient. The scans show differences in brain activities measured by glucose usages—yellow-to-red colors show high values; green-to-blue are low values.

half-lives, so the radiation dosage to the patient is minimal. However, because of the short half-life, a chemist must prepare the diagnostic compound containing the radioactive nucleus shortly before the physician administers it. The preparation of this compound requires a cyclotron, whose cost (several million dollars) is a major deterrent to the general use of PET scans.

Figure 21.15 shows a patient undergoing a PET scan of the brain. The instrument actually detects gamma radiation. When a nucleus emits a positron within the body, the positron travels only a few millimeters before it reacts with an electron. This reaction is an example of the annihilation of matter (an electron) by antimatter (a positron). Both the electron and the positron disappear and produce two gamma photons. The gamma photons easily pass through human tissue, so they can be recorded by scintillation detectors placed around the body. You can see the circular bank of detectors in Figure 21.15. The detectors record the distribution of gamma radiation, and from this information a computer constructs images that can be used by the physician.

power reactors and to provide the energy for nuclear weapons. In the case of a power plant, the energy from the reactions is released in controlled, small amounts by the fission of uranium-235. With a nuclear weapon, the objective is to release the energy as rapidly as possible and produce a nuclear explosion. Although the source of energy for nuclear power plants and weapons can be the same, a typical nuclear power plant does not contain enough fissionable material in high enough concentration to produce a nuclear explosion.

# 21.6 Mass–Energy Calculations

When nuclei decay, they form products of lower energy. The change of energy is related to changes of mass, according to the mass–energy equivalence relation derived by Albert Einstein in 1905. Using this relation, you can obtain the energies of nuclear reactions from mass changes.

## Mass–Energy Equivalence

One of the conclusions from Einstein's theory of special relativity is that the mass of a particle changes with its speed: the greater the speed, the greater the mass. Or, because kinetic energy depends on speed, the greater the kinetic energy of a particle, the greater its mass. This result, according to Einstein, is even more general. Energy and mass are equivalent and are related by the equation

$$E = mc^2$$

Here $c$ is the speed of light, $3.00 \times 10^8$ m/s.

The meaning of this equation is that to any mass there is an associated energy, or to any energy there is an associated mass. If a system loses energy, it must also lose mass. For example, when carbon burns in oxygen, it releases heat energy.

$$C(graphite) + O_2(g) \longrightarrow CO_2(g); \Delta H = -393.5 \text{ kJ}$$

► The change in mass gives the change in internal energy, $\Delta U$. The enthalpy change, $\Delta H$, equals $\Delta U + P\Delta V$. For the reaction given in the text, $P\Delta V$ is essentially zero.

Because this chemical system loses energy, it should also lose mass. In principle, you could obtain $\Delta H$ for the reaction by measuring the change in mass and relating this by Einstein's equation to the change in energy. ◄ However, weight measurements are of no practical value in determining heats of reaction—the changes in mass are simply too small to detect.

Calculation of the mass change for a typical chemical reaction, the burning of carbon in oxygen, will show just how small this quantity is. When the energy changes by an amount $\Delta E$, the mass changes by an amount $\Delta m$. You can write Einstein's equation in the form

$$\Delta E = (\Delta m)c^2$$

The change in energy when 1 mol of carbon reacts is $-3.935 \times 10^5$ J, or $-3.935 \times 10^5$ kg · m²/s². Hence,

$$\Delta m = \frac{\Delta E}{c^2} = \frac{-3.935 \times 10^5 \text{ kg} \cdot m^2/s^2}{(3.00 \times 10^8 \text{ m/s})^2} = -4.37 \times 10^{-12} \text{ kg}$$

For comparison, a good analytical balance can detect a mass as small as $1 \times 10^{-7}$ kg, but this is ten thousand times greater than the mass change caused by the release of heat during the combustion of 1 mol of carbon.

By contrast, the mass changes in nuclear reactions are approximately a million times larger per mole of reactant than those in chemical reactions. Consider the alpha decay of uranium-238 to thorium-234. The nuclear equation is

$$^{238}_{92}\text{U} \longrightarrow ^{234}_{90}\text{Th} + ^{4}_{2}\text{He}$$

$$238.05078 \qquad 234.04359 \qquad 4.00260 \text{ amu}$$

**Table 21.3**
**Masses of Some Elements and Other Particles**

| Symbol | Z | A | Mass (amu) | Symbol | Z | A | Mass (amu) |
|--------|-----|-----|------------|--------|-----|-----|------------|
| $e^-$ | $-1$ | 0 | 0.000549 | Co | 27 | 59 | 58.93320 |
| n | 0 | 1 | 1.008665 | Ni | 28 | 58 | 57.93535 |
| p | 1 | 1 | 1.00728 | | 28 | 60 | 59.93079 |
| H | 1 | 1 | 1.00783 | Pb | 82 | 206 | 205.97444 |
| | 1 | 2 | 2.01400 | | 82 | 207 | 206.97587 |
| | 1 | 3 | 3.01605 | | 82 | 208 | 207.97663 |
| He | 2 | 3 | 3.01603 | Po | 84 | 208 | 207.98122 |
| | 2 | 4 | 4.00260 | | 84 | 210 | 209.98285 |
| Li | 3 | 6 | 6.01512 | Rn | 86 | 222 | 222.01757 |
| | 3 | 7 | 7.01600 | Ra | 88 | 226 | 226.02540 |
| Be | 4 | 9 | 9.01218 | Th | 90 | 230 | 230.03313 |
| B | 5 | 10 | 10.01294 | | 90 | 234 | 234.04359 |
| | 5 | 11 | 11.00931 | Pa | 91 | 234 | 234.04330 |
| C | 6 | 12 | 12.00000 | U | 92 | 233 | 233.03963 |
| | 6 | 13 | 13.00336 | | 92 | 234 | 234.04095 |
| O | 8 | 16 | 15.99492 | | 92 | 235 | 235.04392 |
| Cr | 24 | 52 | 51.94051 | | 92 | 238 | 238.05078 |
| Fe | 26 | 56 | 55.93494 | Pu | 94 | 239 | 239.05216 |

Here we have written the atomic mass (in amu) beneath each nuclide symbol. (Table 21.3 lists masses of some elements and other particles.) The change in mass for this nuclear reaction, starting with molar amounts, is

$$\Delta m = (234.04359 + 4.00260 - 238.05078)\ \text{g} = -0.00459\ \text{g}$$

As in calculating $\Delta H$ and similar quantities, you subtract the value for the reactant from the sum of the values for the products. The minus sign indicates a loss of mass. This loss of mass is clearly large enough to detect.

From a table of atomic masses, such as Table 21.3, you can use Einstein's equation to calculate the energy change for a nuclear reaction. This is illustrated in the next example. Recall from Section 21.2 that $1\ \text{eV} = 1.602 \times 10^{-19}$ J. Therefore, 1 MeV equals $1.602 \times 10^{-13}$ J. ◀

▶ From Einstein's equation, one can show that 1 amu is equivalent to 931 MeV. This number can be used to obtain energies of nuclear reactions. For the reaction given in Example 21.12, the mass change is $-0.01960$ amu. Multiplying this by 931 MeV gives the energy change, $-18.2$ MeV, which agrees with the answer given to three significant figures.

**EXAMPLE 21.12**

## Calculating the Energy Change for a Nuclear Reaction

a. Calculate the energy change in joules (four significant figures) for the following nuclear reaction per mole of $_1^2\text{H}$:

$$_1^2\text{H} + _2^3\text{He} \longrightarrow _2^4\text{He} + _1^1\text{H}$$

Atomic masses are given in Table 21.3; the speed of light is given on the inside back cover of this book.

b. What is the energy change in MeV for one $^2_1H$ nucleus?

**SOLUTION**

a. You write the atomic masses below each nuclide symbol and then calculate $\Delta m$. Once you have $\Delta m$, you can obtain $\Delta E$.

$$^2_1H + {}^3_2He \longrightarrow {}^4_2He + {}^1_1H$$
$$2.01400 \quad 3.01603 \qquad 4.00260 \quad 1.00783 \text{ amu}$$

Hence,

$$\Delta m = (4.00260 + 1.00783 - 2.01400 - 3.01603) \text{ amu}$$
$$= -0.01960 \text{ amu}$$

To obtain the energy change for molar amounts, note that the molar mass of an atom in grams is numerically equal to the mass of a single atom in amu. Therefore, the mass change for molar amounts in this nuclear reaction is $-0.01960$ g, or $-1.960 \times 10^{-5}$ kg. The energy change is

$$\Delta E = (\Delta m)c^2 = (-1.960 \times 10^{-5} \text{ kg})(2.998 \times 10^8 \text{ m/s})^2$$
$$= -1.762 \times 10^{12} \text{ kg} \cdot \text{m}^2/\text{s}^2 \quad \text{or} \quad \mathbf{-1.762 \times 10^{12} \text{ J}}$$

b. The mass change for the reaction of one $^2_1H$ atom is $-0.01960$ amu. First change this to grams. Recall that 1 amu equals 1/12 the mass of a $^{12}_6C$ atom, whose mass is $12 \text{ g}/6.022 \times 10^{23}$. Thus, 1 amu $= 1 \text{ g}/6.022 \times 10^{23}$. Hence, the mass change in grams is

$$\Delta m = -0.01960 \text{ amu} \times \frac{1 \text{ g}}{1 \text{ amu} \times 6.022 \times 10^{23}}$$
$$= -3.255 \times 10^{-26} \text{ g} \quad (\text{or } -3.255 \times 10^{-29} \text{ kg})$$

Then,

$$\Delta E = (\Delta m)c^2 = (-3.255 \times 10^{-29} \text{ kg})(2.998 \times 10^8 \text{ m/s})^2$$
$$= -2.926 \times 10^{-12} \text{ J}$$

Now convert this to MeV:

$$\Delta E = -2.926 \times 10^{-12} \text{ J} \times \frac{1 \text{ MeV}}{1.602 \times 10^{-13} \text{ J}} = \mathbf{-18.26 \text{ MeV}}$$

**EXERCISE 21.12**

a. Calculate the energy change in joules when 1.00 g $^{234}_{90}Th$ decays to $^{234}_{91}Pa$ by beta emission.

b. What is the energy change in MeV when one $^{234}_{90}Th$ nucleus decays? Use Table 21.3 for these calculations.

**See Problems 21.75 and 21.76.**

## Nuclear Binding Energy

The equivalence of mass and energy explains the otherwise puzzling fact that the mass of an atom is always less than the sum of the masses of its constituent particles. For example, the helium-4 atom consists of two protons, two neutrons, and two electrons, giving the following sum:

$$
\begin{aligned}
\text{Mass of 2 electrons} &= 2 \times 0.000549 \text{ amu} = 0.00110 \text{ amu} \\
\text{Mass of 2 protons} &= 2 \times 1.00728 \text{ amu} = 2.01456 \text{ amu} \\
\text{Mass of 2 neutrons} &= 2 \times 1.00867 \text{ amu} = \underline{2.01734 \text{ amu}} \\
\text{Total mass of particles} &= 4.03300 \text{ amu}
\end{aligned}
$$

The mass of the helium-4 atom is 4.00260 amu (see Table 21.3), so the mass difference is

$$\Delta m = (4.00260 - 4.03300) \text{ amu} = -0.03040 \text{ amu}$$

This mass difference is explained as follows. When the nucleons come together to form a nucleus, energy is released. (The nucleus has lower energy and is therefore more stable than the separate nucleons.) According to Einstein's equation, there must be an equivalent decrease in mass.

The **binding energy** of a nucleus is *the energy needed to break a nucleus into its individual protons and neutrons.* Thus, the binding energy of the helium-4 nucleus is the energy change for the reaction

$$_2^4\text{He} \longrightarrow 2\,_1^1\text{p} + 2\,_0^1\text{n}$$

The **mass defect** of a nucleus is *the total nucleon mass minus the atomic mass.* In the case of helium-4, the mass defect is 4.03300 amu − 4.00260 amu = 0.03040 amu. (This is the positive value of the mass difference we calculated earlier.) Both the binding energy and the corresponding mass defect are reflections of the stability of the nucleus.

To compare the stabilities of various nuclei, it is useful to compare binding energies per nucleon. Figure 21.16 shows values of this quantity (in MeV per nucleon) plotted against the mass number for various nuclides. Most of the points lie near the smooth curve drawn on the graph.

**Figure 21.16**
**Plot of binding energy per nucleon versus mass number.**  The binding energy of each nuclide is divided by the number of nucleons (total of protons and neutrons), then plotted against the mass number of the nuclide. Note that nuclides near mass number 50 have the largest binding energies per nucleon. Thus, heavy nuclei are expected to undergo fission to approach this mass number, whereas light nuclei are expected to undergo fusion.

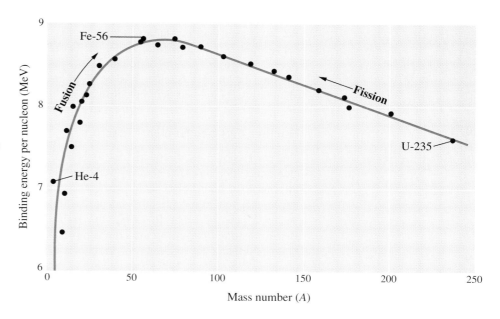

Note that nuclides near mass number 50 have the largest binding energies per nucleon. This means that a group of nucleons would tend to form those nuclides, because they would form nuclei of the lowest energy. For this reason, heavy nuclei might be expected to split to give lighter nuclei, and light nuclei might be expected to react, or combine, to form heavier nuclei.

**Nuclear fission** is *a nuclear reaction in which a heavy nucleus splits into lighter nuclei and energy is released.* For example, californium-252 decays both by alpha emission (97%) and by spontaneous fission (3%). There are many possible ways in which the nucleus can split. One way is represented by the following equation:

$$^{252}_{98}\text{Cf} \longrightarrow ^{142}_{56}\text{Ba} + ^{106}_{42}\text{Mo} + 4^1_0\text{n}$$

In some cases, a nucleus can be induced to undergo fission by bombarding it with neutrons. An example is the nuclear fission of uranium-235. When a neutron strikes the $^{235}_{92}\text{U}$ nucleus, the nucleus splits into roughly equal parts, giving off several neutrons. Three possible splittings are shown in the following equations:

$$^1_0\text{n} + ^{235}_{92}\text{U} \left\{ \begin{array}{l} \longrightarrow ^{142}_{54}\text{Xe} + ^{90}_{38}\text{Sr} + 4^1_0\text{n} \\ \longrightarrow ^{139}_{56}\text{Ba} + ^{94}_{36}\text{Kr} + 3^1_0\text{n} \\ \longrightarrow ^{144}_{55}\text{Cs} + ^{90}_{37}\text{Rb} + 2^1_0\text{n} \end{array} \right.$$

Nuclear fission is discussed further in Section 21.7.

**Nuclear fusion** is *a nuclear reaction in which light nuclei combine to give a stabler, heavier nucleus plus possibly several neutrons, and energy is released.* An example of nuclear fusion is

$$^2_1\text{H} + ^3_1\text{H} \longrightarrow ^4_2\text{He} + ^1_0\text{n}$$

Even though a nuclear reaction is energetically favorable, the reaction may be imperceptibly slow unless the proper conditions are present (see Section 21.7).

## 21.7 Nuclear Fission and Nuclear Fusion

We have seen that the stablest nuclei are those of intermediate size (with mass numbers around 50). Nuclear fission and nuclear fusion are reactions in which nuclei attain sizes closer to this intermediate range. In doing so, these reactions release tremendous amounts of energy. Nuclear fission of uranium-235 is employed in nuclear power plants to generate electricity. Nuclear fusion may supply us with energy in the future.

**Visualization: Nuclear Fission**

### Nuclear Fission; Nuclear Reactors

Nuclear fission was discovered as a result of experiments to produce transuranium elements. Soon after the neutron was discovered in 1932, experimenters realized that this particle, being electrically neutral, should easily penetrate heavy nuclei. They began using neutrons in bombardment reactions, hoping to produce isotopes that would decay to new elements. In 1938, Otto Hahn, Lise Meitner, and Fritz Strassmann in Berlin identified barium in uranium samples that had been bombarded with neutrons. Soon afterward, the presence of barium was explained as a result of fission of the uranium-235 nucleus. When this nucleus is struck by a neutron, it splits into two nuclei. Fissions of uranium nuclei produce approximately 30 different elements of intermediate mass, including barium.

**Figure 21.17**

**Representation of a chain reaction of nuclear fissions.**   Each nuclear fission produces two or more neutrons, which can in turn cause more nuclear fissions. At each stage, a greater number of neutrons are produced, so that the number of nuclear fissions multiplies quickly. Such a chain reaction is the basis of nuclear power and nuclear weapons. The original nucleus splits into pieces of varying mass number.

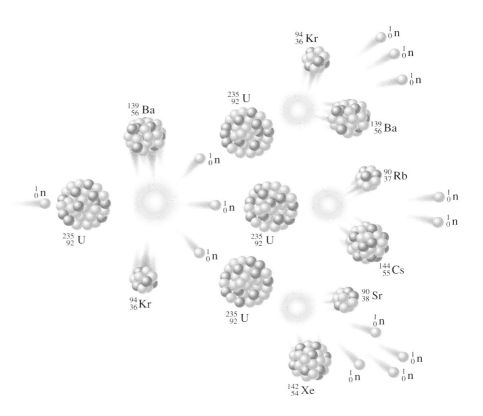

When the uranium-235 nucleus splits, approximately two or three neutrons are released. If the neutrons from each nuclear fission are absorbed by other uranium-235 nuclei, these nuclei split and release even more neutrons. In this way, a chain reaction can occur. A nuclear **chain reaction** is *a self-sustaining series of nuclear fissions caused by the absorption of neutrons released from previous nuclear fissions.* The number of nuclei that split quickly multiply as a result of the absorption of neutrons released from previous nuclear fissions. Figure 21.17 shows how such a chain reaction occurs. The chain reaction of nuclear fissions is the basis of nuclear power and nuclear weapons.

To sustain a chain reaction in a sample of fissionable material, a nucleus that splits must give an average of one neutron that results in the fission of another nucleus, and so on. If the sample is too small, many of the neutrons leave the sample before they have a chance to be absorbed. There is thus a **critical mass** for a particular fissionable material, which is *the smallest mass of fissionable material in which a chain reaction can be sustained.* If the mass is much larger than this (a *supercritical* mass), the number of nuclei that split multiply rapidly. An atomic bomb is detonated with a small amount of chemical explosive that pushes together two or more masses of fissionable material to get a supercritical mass (Figure 21.18). A rapid chain reaction results in the splitting of most of the fissionable nuclei—and in the explosive release of an enormous amount of energy.

A **nuclear fission reactor** is *a device that permits a controlled chain reaction of nuclear fissions.* In power plants, a nuclear reactor is used to produce heat, which in turn is used to produce steam to drive an electric generator. A nuclear reactor consists of fuel rods alternating with control rods contained within a vessel. The **fuel rods**

**Figure 21.18**
**An atomic bomb.**   In the gun-type bomb, one piece of uranium-235 of subcritical mass is hurled into another piece by a chemical explosive. The two pieces together give a supercritical mass, and a nuclear explosion results. Bombs using plutonium-239 require an implosion technique, in which wedges of plutonium arranged on a spherical surface are pushed into the center of the sphere by a chemical explosive, where a supercritical mass of plutonium results in a nuclear explosion.

are *the cylinders that contain fissionable material.* In the light-water (ordinary water) reactors commonly used in the United States (see Figure 21.19), these fuel rods contain uranium dioxide pellets in a zirconium alloy tube. Natural uranium contains only 0.72% uranium-235, which is the isotope that undergoes fission. The uranium used for fuel in these reactors is "enriched" so that it contains about 3% of the uranium-235 isotope. **Control rods** are *cylinders composed of substances that absorb neutrons, such as boron and cadmium, and can therefore slow the chain reaction.* By varying the depth of the control rods within the fuel-rod assembly (reactor core), one can increase or decrease the absorption of neutrons. If necessary, these rods can be dropped all the way into the fuel-rod assembly to stop the chain reaction.

A **moderator,** *a substance that slows down neutrons,* is required if uranium-235 is the fuel and this isotope is present as a small fraction of the total fuel. The neu-

**Figure 21.19**
**Light-water nuclear reactor.**   The nuclear reactor consists of fuel rods with interspersed control rods. By raising or lowering the control rods, an operator can increase or decrease the rate of energy release from the fuel rods. Heat from the fuel rods raises the temperature of the liquid water in the reactor. A pump circulates the hot water to a steam generator, and the resulting steam passes through a turbine that operates an electric generator. The steam leaves the turbine and goes into the condenser, where it liquefies on the cooling coil. A nearby river or lake provides the cold water for the condenser.

trons that are released by the splitting of uranium-235 nuclei are absorbed more readily by uranium-238 than by other uranium-235 nuclei. However, when the neutrons are slowed down by a moderator, they are more readily absorbed by uranium-235, so it is possible to sustain a chain reaction with low fractional abundance of this isotope. Commonly used moderators are heavy water ($_1^2H_2O$), light water, and graphite.

In the light-water reactor, ordinary water acts as both a moderator and a coolant. Figure 21.19 shows a pressurized-water design of this type of reactor. Water in the reactor is maintained at about 350°C under high pressure (150 atm) so it does not boil. The hot water is circulated to a heat exchanger, where the heat is used to produce steam to run a turbine and generate electricity.

After a period of time, fission products that absorb neutrons accumulate in the fuel rods. This interferes with the chain reaction, so eventually the fuel rods must be replaced. Originally, the intention was to send these fuel rods to *reprocessing plants,* where fuel material could be chemically separated from the radioactive wastes. Opposition to constructing these plants has been intense, however. Plutonium-239 would be one of the fuel materials separated from the spent fuel rods. This isotope is produced during the operation of the reactor when uranium-238 is bombarded with neutrons. It is fissionable and can be used to construct atomic bombs. For this reason, many people believe that the availability of this element in large quantities would increase the chance that many countries and terrorist groups could divert enough plutonium to produce atomic bombs.

Whether or not the spent fuel rods are reprocessed, a pressing problem facing the nuclear power industry is how to dispose of radioactive wastes safely. One of many proposals is to encase the waste in a ceramic material and store the solids deep in the earth, perhaps in salt mines.

## Nuclear Fusion

As we noted in Section 21.6, energy can be obtained by combining light nuclei into a heavier nucleus by nuclear fusion. Such fusion reactions have been observed in the laboratory by means of bombardment using particle accelerators. Deuterons ($_1^2H$ nuclei), for example, can be accelerated toward targets containing deuterium ($_1^2H$ atoms) or tritium ($_1^3H$ atoms). The reactions are

$$_1^2H + {_1^2H} \longrightarrow {_2^3He} + {_0^1n}$$
$$_1^2H + {_1^3H} \longrightarrow {_2^4He} + {_0^1n}$$

▶ **Natural hydrogen contains 0.015% deuterium.**

To get the nuclei to react, the bombarding nucleus must have enough kinetic energy to overcome the repulsion of electric charges of the nuclei. The first reaction uses only deuterium, which is present in ordinary water. ◀ It is therefore very attractive as a source of energy. But, as we will discuss, the second reaction is more likely to be used first.

Energy cannot be obtained in a practical way using particle accelerators. Another way to give nuclei sufficient kinetic energy to react is by heating the nuclear materials to a sufficiently high temperature. The reaction of deuterium, $_1^2H$, and tritium, $_1^3H$, turns out to require the lowest temperature of any fusion reaction. For this reason it is likely to be the first fusion reaction developed as an energy source. For practical purposes, the temperature will have to be about 100 million °C. At this temperature, all atoms have been stripped of their electrons so that a plasma results. A **plasma** is *an electrically neutral gas of ions and electrons.* At 100 million °C, the plasma is essentially separate nuclei and electrons. Thus, the development of nuclear fusion requires study of the properties of plasmas at high temperatures.

# The Chernobyl Nuclear Accident

**ENVIRONMENT**

The nuclear accident that occurred at the Chernobyl reactor, north of Kiev, Ukraine, on April 26, 1986, renewed fears in some about the safety of nuclear reactors. It is important to understand the nature of the accident at Chernobyl. A nuclear reactor using normal fuel elements cannot become an atomic bomb. However, without proper design and safeguards, it is possible for a malfunction of a reactor to disperse dangerous radioactivity over a populated area. This is in fact what occurred at Chernobyl.

The Chernobyl reactor had serious design problems. Unlike U.S. reactors, the Chernobyl reactor had no confinement shell surrounding it to contain radioactive spills. Another problem was the design of the cooling system. The fuel core had a graphite moderator with water cooling in which some of the liquid water changed to steam. Liquid water is a good absorber of neutrons, whereas steam is not. This meant that as the reactor ran hotter, producing a greater percentage of steam, more neutrons became available for nuclear fission. Unless these extra neutrons were absorbed, the reactor would become even hotter and then would run out of control. Normally, as the reactor became hotter, control rods were automatically pushed in to absorb the extra neutrons. In a U.S. light-water reactor, the water coolant is always under pressure to maintain the liquid phase. As long as coolant water is present, the reactor is under control.

On the day of the Chernobyl accident, operators disabled the safety system on the reactor while they pro-

▶ Even though the plasma is nearly 100 million °C above the melting point of any material, the total quantity of heat that could be transferred from the plasma is very small, because its concentration is extremely low. Therefore, if the plasma were to touch the walls of the reactor, the plasma would cool but the walls would not melt.

It is now believed that the energy of stars, including our sun, where extremely high temperatures exist, derives from nuclear fusion. The hydrogen bomb also employs nuclear fusion for its destructive power. High temperature is first attained by a fission bomb. This then ignites fusion reactions in surrounding material of deuterium and tritium.

The main problem in developing controlled nuclear fusion is how to heat a plasma to high temperature and maintain those temperatures. When a plasma touches any material whatsoever, heat is quickly conducted away, and the plasma temperature quickly falls. ◄ A *magnetic fusion reactor* uses a magnetic field to hold the plasma

**Figure 21.20**
**Torus Fusion Reactor.**
The National Spherical Torus Experiment at the Princeton Plasma Physics Laboratory uses this reactor to test an advanced magnetic field configuration to confine a compact, spherically shaped plasma.

ceeded with an experimental test of the reactor. During the test, the reactor cooled excessively and threatened to shut down. If this happened, the reactor could not be restarted for a long period. Therefore, the operators removed most of the control rods (counter to safety instructions). Then the reactor began to overheat. With the safety system disabled, the operators were unable to work fast enough to correct the overheating, and the reactor went out of control.

The fuel rods melted and spilled their hot contents into the superheated water, which flashed into steam. The sudden pressure increase blew the top off the reactor and the roof off the building, spewing radioactive material skyward (see Figure 21.21). Hot steam reacted with the zirconium shell of the fuel rods and with the graphite moderator to produce hydrogen gas, which ignited. The graphite moderator burned for a long period, spreading more radioactivity.

The cost of the Chernobyl accident was enormous. Many people died, and several hundred were hospitalized. Thousands of people had to be evacuated and

**Figure 21.21**
**View of the Chernobyl nuclear plant.** This photograph was taken in 1987, one year after the nuclear accident.

resettled. The accident was the direct result of a faulty reactor coupled with a disregard of reactor safety procedures.

away from any material (see Figure 21.20). A *laser fusion reactor* employs a bank of lasers aimed at a single point. Pellets containing deuterium and tritium would drop into the reactor, where they would be heated to 100 million °C by bursts of laser light.

## A Checklist for Review

### Important Terms

radioactive decay (p. 895)
nuclear bombardment
  reaction (p. 895)
nuclear equation (21.1)
positron (21.1)
gamma photon (21.1)
nuclear force (21.1)
shell model of the
  nucleus (21.1)
magic number (21.1)
band of stability (21.1)
alpha emission (21.1)
beta emission (21.1)
positron emission (21.1)

electron capture (21.1)
gamma emission (21.1)
metastable nucleus (21.1)
spontaneous fission (21.1)
radioactive decay
  series (21.1)
transmutation (21.2)
particle accelerator (21.2)
electron volt (eV) (21.2)
cyclotron (21.2)
deuterons (21.2)
transuranium
  elements (21.2)
Geiger counter (21.3)

scintillation counter (21.3)
activity of a radioactive
  source (21.3)
curie (Ci) (21.3)
rad (21.3)
rem (21.3)
radioactive decay
  constant (21.4)
half-life (21.4)
radioactive tracer (21.5)
isotope dilution (21.5)
neutron activation
  analysis (21.5)
binding energy (21.6)

mass defect (21.6)
nuclear fission (21.6)
nuclear fusion (21.6)
chain reaction (21.7)
critical mass (21.7)
nuclear fission
  reactor (21.7)
fuel rods (21.7)
control rods (21.7)
moderator (21.7)
plasma (21.7)

### Key Equations

$$\text{Rate} = kN_t$$

$$t_{1/2} = \frac{0.693}{k}$$

$$\ln \frac{N_t}{N_0} = -kt$$

$$\Delta E = (\Delta m)c^2$$

## Summary of Facts and Concepts

Nuclear reactions are of two types, *radioactive decay* and *nuclear bombardment*. Such reactions are represented by nuclear equations, each nucleus being denoted by a nuclide symbol. The equations must be balanced in charge (subscripts) and in nucleons (superscripts).

According to the *nuclear shell model,* the nucleons are arranged in shells. *Magic numbers* are the numbers of nucleons in a completed shell of protons or neutrons. Nuclei with magic numbers of protons or neutrons are especially stable. Pairs of protons and pairs of neutrons are also especially stable. When placed on a plot of $N$ versus $Z$, stable nuclei fall in a *band of stability.* Those radioactive nuclides that fall to the left of the band of stability in this plot usually decay by beta emission. Those radioactive nuclides that fall to the right of the band of stability usually decay by positron emission or electron capture. However, nuclides with $Z > 83$ often decay by alpha emission. Uranium-238 forms a *radioactive decay series.* In such series, one element decays to another, which decays to another, and so forth, until a stable isotope is reached (lead-206, in the case of the uranium-238 series).

*Transmutation* of elements has been carried out in the laboratory by bombarding nuclei with various atomic particles. Alpha particles from natural sources can be used as reactants with light nuclei. For heavier nuclei, positive ions such as alpha particles must first be accelerated in a *particle accelerator.* Many of the *transuranium elements* have been obtained by bombardment of elements with accelerated particles. For example, plutonium was first made by bombarding uranium-238 with deuterons ($^2_1H$ nuclei) from a *cyclotron,* a type of particle accelerator.

Particles of radiation from nuclear processes can be counted by Geiger counters or scintillation counters. In a *Geiger counter,* the particle ionizes a gas, which then conducts a pulse of electricity between two electrodes. In a *scintillation counter,* the particle hits a phosphor, and this emits a flash of light that is detected by a photomultiplier tube. The activity of a radioactive source, or the number of nuclear disintegrations per unit time, is measured in units of *curies* ($3.700 \times 10^{10}$ disintegrations per second).

Radiation affects biological organisms by breaking chemical bonds. The *rad* is the measure of radiation dosage that deposits $1 \times 10^{-2}$ J of energy per kilogram of tissue. A *rem* equals the number of rads times a factor to account for the relative biological effectiveness of the radiation.

Radioactive decay is a first-order rate process. The rate is characterized by the *decay constant, k,* or by the *half-life, $t_{1/2}$.* The quantities $k$ and $t_{1/2}$ are related. Knowing one or the other, you can calculate how long it will take for a given radioactive sample to decay by a certain fraction. Methods of *radioactive dating* depend on determining the fraction of a radioactive isotope that has decayed and, from this, the time that has elapsed.

Radioactive isotopes are used as *radioactive tracers* in chemical analysis and medicine. *Isotope dilution* is one application of radioactive tracers in which the dilution of the tracer can be related to the original quantity of nonradioactive isotope. *Neutron activation analysis* is a method of analysis that depends on the conversion of elements to radioactive isotopes by neutron bombardment.

According to Einstein's *mass–energy equivalence,* mass is related to energy by the equation $E = mc^2$. A nucleus has less mass than the sum of the masses of the separate nucleons. The positive value of this mass difference is called the *mass defect;* it is equivalent to the *binding energy* of the nucleus. Nuclides having mass numbers near 50 have the largest binding energies per nucleon. It follows that heavy nuclei should tend to split, a process called *nuclear fission,* and that light nuclei should tend to combine, a process called *nuclear fusion.* Tremendous amounts of energy are released in both processes. Nuclear fission is used in conventional nuclear power reactors. Nuclear fusion reactors are in the experimental stage.

## Operational Skills

1. **Writing a nuclear equation**   Given a word description of a radioactive decay process, write the nuclear equation. **(EXAMPLE 21.1)**
2. **Deducing a product or reactant in a nuclear equation** Given all but one of the reactants and products in a nuclear reaction, find that one nuclide. **(EXAMPLES 21.2, 21.6)**
3. **Predicting the relative stabilities of nuclides** Given a number of nuclides, determine which are most likely to be radioactive and which are most likely to be stable. **(EXAMPLE 21.3)**
4. **Predicting the type of radioactive decay** Predict the type of radioactive decay that is most likely for given nuclides. **(EXAMPLE 21.4)**
5. **Using the notation for a bombardment reaction** Given an equation for a nuclear bombardment reaction, write the abbreviated notation, or vice versa. **(EXAMPLE 21.5)**
6. **Calculating the decay constant from the activity** Given the activity (disintegrations per second) of a radioactive isotope, obtain the decay constant. **(EXAMPLE 21.7)**
7. **Relating the decay constant, half-life, and activity** Given the decay constant of a radioactive isotope, obtain the half-life **(EXAMPLE 20.8)**, or vice versa **(EXAMPLE 20.9)**. Given the decay constant and mass of a radioactive isotope, calculate the activity of the sample **(EXAMPLE 21.9)**.

**8. Determining the fraction of nuclei remaining after a specified time** Given the half-life of a radioactive isotope, calculate the fraction remaining after a specified time. **(EXAMPLE 21.10)**

**9. Applying the carbon-14 dating method** Given the disintegrations of carbon-14 nuclei per gram of carbon in a dead organic object, calculate the age of the object—that is, the time since its death. **(EXAMPLE 21.11)**

**10. Calculating the energy change for a nuclear reaction** Given nuclear masses, calculate the energy change for a nuclear reaction **(EXAMPLE 21.12)**. Obtain the answer in joules per mole or MeV per particle.

## Review Questions

**21.1** What are the two types of nuclear reaction? Give an example of a nuclear equation for each type.

**21.2** What are *magic numbers?* Give several examples of nuclei with magic numbers of protons.

**21.3** List characteristics to look for in a nucleus to predict whether it is stable.

**21.4** What are the six common types of radioactive decay? What condition usually leads to each type of decay?

**21.5** What are the isotopes that begin each of the naturally occurring radioactive decay series?

**21.6** Give equations for (a) the first transmutation of an element obtained in the laboratory by nuclear bombardment, and for (b) the reaction that produced the first artificial radioactive isotope.

**21.7** What is a particle accelerator, and how does one operate? Why are they required for certain nuclear reactions?

**21.8** In what major way has the discovery of transuranium elements affected the form of modern periodic tables?

**21.9** Describe how a Geiger counter works. How does a scintillation counter work?

**21.10** Define the units *curie, rad,* and *rem.*

**21.11** The half-life of cesium-137 is 30.2 y. How long will it take for a sample of cesium-137 to decay to 1/8 of its original mass?

**21.12** What is the age of a rock that contains equal numbers of $^{40}_{19}K$ and $^{40}_{18}Ar$ nuclei? The half-life of $^{40}_{19}K$ is $1.28 \times 10^9$ y.

**21.13** What is a radioactive tracer? Give an example of the use of such a tracer in chemistry.

**21.14** Isotope dilution has been used to obtain the volume of blood supply in a living animal. Explain how this could be done.

**21.15** Briefly describe neutron activation analysis.

**21.16** The deuteron, $^2_1H$, has a mass that is smaller than the sum of the masses of its constituents, the proton plus the neutron. Explain why this is so.

**21.17** Certain stars obtain their energy from nuclear reactions such as

$$^{12}_6C + ^{12}_6C \longrightarrow ^{23}_{11}Na + ^1_1H$$

Explain in a sentence or two why this reaction might be expected to release energy.

**21.18** Briefly describe how each of the following operates: (a) nuclear fission reactor; (b) tokamak fusion reactor.

## Conceptual Problems

**21.19** When considering the lifetime of a radioactive species, a general rule of thumb is that after 10 half-lives have passed, the amount of radioactive material left in the sample is negligible. The disposal of some radioactive materials is based on this rule.
a. What percentage of the original material is left after 10 half-lives?
b. When would it be a bad idea to apply this rule?

**21.20** Identify the following reactions as fission, fusion, a transmutation, or radioactive decay.
a. $4^1_1H \longrightarrow ^4_2He + 2^0_1e$
b. $^{14}_6C \longrightarrow ^{14}_7N + ^0_{-1}e$
c. $^1_0n + ^{235}_{92}U \longrightarrow ^{140}_{56}Ba + ^{93}_{36}Kr + 3^0_1e$
d. $^{14}_7N + ^4_2He \longrightarrow ^{17}_8O + ^1_1H$

**21.21** Sodium has only one naturally occurring isotope, sodium-23. Using the data presented in Table 21.3, explain how the molecular weight of sodium is 22.98976 amu and not the sum of the masses of the protons, neutrons, and electrons.

**21.22** You have a mixture that contains 10 g of Pu-239 with a half-life of $2.4 \times 10^4$ years and 10 g of Np-239 with a half-life of 2.4 days. Estimate how much time must elapse before the quantity of radioactive material is reduced by 50%.

**21.23** Come up with an explanation as to why $\alpha$ radiation is easily blocked by materials such as a piece of wood, whereas $\gamma$ radiation easily passes through those same materials.

**21.24** You have an acquaintance who tells you that he is going to reduce his radiation exposure to zero. What examples could you present that would illustrate this to be an impossible goal?

**21.25** In Chapter 7 (p. 286, A Chemist Looks At: Zapping Hamburger with Gamma Rays) there is a discussion of how gamma radiation is used to kill bacteria in food. As indicated in the feature, there is concern on the part of some people that the irradiated food is radioactive. Why is this not the case? If you wanted to make the food radioactive, what would you have to do?

**21.26** You have a pile of I-131 atoms with a half-life of 8 days. A portion of the solid I-131 is represented below. Can you predict how many half-lives will occur before the green I-131 atom undergoes decay?

## Practice Problems

### Radioactivity

**21.27** Rubidium-87, which forms about 28% of natural rubidium, is radioactive, decaying by the emission of a single beta particle to strontium-87. Write the nuclear equation for this decay of rubidium-87.

**21.28** Write the nuclear equation for the decay of phosphorus-32 to sulfur-32 by beta emission. A phosphorus-32 nucleus emits a beta particle and gives a sulfur-32 nucleus.

**21.29** Thorium is a naturally occurring radioactive element. Thorium-232 decays by emitting a single alpha particle to produce radium-228. Write the nuclear equation for this decay of thorium-232.

**21.30** Radon is a radioactive noble gas formed in soil containing radium. Radium-226 decays by emitting a single alpha particle to produce radon-222. Write the nuclear equation for this decay of radium-226.

**21.31** Fluorine-18 is an artificially produced radioactive isotope. It decays by emitting a single positron. Write the nuclear equation for this decay.

**21.32** Scandium-41 is an artificially produced radioactive isotope. It decays by emitting a single positron. Write the nuclear equation for this decay.

**21.33** Polonium was discovered in uranium ores by Marie and Pierre Curie. Polonium-210 decays by emitting a single alpha particle. Write the nuclear equation for this decay. ○

**21.34** Actinium was discovered in uranium ore residues by André-Louis Debierne. Actinium-227 decays by emitting a single alpha particle. Write the nuclear equation for this decay.

**21.35** From each of the following pairs, choose the nuclide that is radioactive. (One is known to be radioactive, the other stable.) Explain your choice.
a. $^{122}_{51}Sb$, $^{136}_{54}Xe$    b. $^{204}_{82}Pb$, $^{204}_{85}At$    c. $^{87}_{37}Rb$, $^{80}_{37}Rb$

**21.36** From each of the following pairs, choose the nuclide that is radioactive. (One is known to be radioactive, the other stable.) Explain your choice.
a. $^{102}_{47}Ag$, $^{109}_{47}Ag$    b. $^{25}_{12}Mg$, $^{24}_{10}Ne$    c. $^{203}_{81}Tl$, $^{223}_{90}Th$

**21.37** Predict the type of radioactive decay process that is likely for each of the following nuclides.
a. $^{228}_{92}U$    b. $^{8}_{5}B$    c. $^{68}_{29}Cu$

**21.38** Predict the type of radioactive decay process that is likely for each of the following nuclides.
a. $^{60}_{30}Zn$    b. $^{10}_{6}C$    c. $^{241}_{93}Np$

**21.39** Four radioactive decay series are known—three naturally occurring, and one beginning with the synthetic isotope $^{241}_{94}Pu$. To which of these decay series does the isotope $^{219}_{86}Rn$ belong? To which series does $^{220}_{86}Rn$ belong? Each isotope in these series decays by either alpha emission or beta emission. How do these decay processes affect the mass number?

**21.40** Four radioactive decay series are known—three naturally occurring, and one beginning with the synthetic isotope $^{241}_{94}Pu$. To which of these decay series does the isotope $^{227}_{89}Ac$ belong? To which series does $^{225}_{89}Ac$ belong? Each isotope in these series decays by either alpha emission or beta emission. How do these decay processes affect the mass number?

### Nuclear Bombardment Reactions

**21.41** Write the abbreviated notations for the following bombardment reactions.
a. $^{26}_{12}Mg + ^{2}_{1}H \longrightarrow ^{24}_{11}Na + ^{4}_{2}He$
b. $^{16}_{8}O + ^{1}_{0}n \longrightarrow ^{16}_{7}N + ^{1}_{1}p$

**21.42** Write abbreviated notations for the following bombardment reactions.
a. $^{14}_{7}N + ^{1}_{0}n \longrightarrow ^{14}_{6}C + ^{1}_{1}H$
b. $^{63}_{29}Cu + ^{4}_{2}He \longrightarrow ^{66}_{31}Ga + ^{1}_{0}n$

**21.43** Write the nuclear equations for the following bombardment reactions.

a. $^{27}_{13}Al(d, \alpha)^{25}_{12}Mg$     b. $^{10}_{5}B(\alpha, p)^{13}_{6}C$

**21.44** Write the nuclear equations for the following bombardment reactions.

a. $^{45}_{21}Sc(n, \alpha)^{42}_{19}K$     b. $^{63}_{29}Cu(p, n)^{63}_{30}Zn$

**21.45** A proton is accelerated to 14.3 MeV per particle. What is this energy in kJ/mol?

**21.46** An alpha particle is accelerated to 23.1 MeV per particle. What is this energy in kJ/mol?

**21.47** Fill in the missing parts of the following reactions.

a. $^{6}_{3}Li + ^{1}_{0}n \longrightarrow ? + ^{3}_{1}H$

b. $^{232}_{90}Th(?, n)^{235}_{92}U$

**21.48** Fill in the missing parts of the following reactions.

a. $^{27}_{13}Al + ^{3}_{1}H \longrightarrow ^{27}_{12}Mg + ?$

b. $^{12}_{6}C(^{3}_{1}H, ?)^{14}_{6}C$

**21.49** Curium was first synthesized by bombarding an element with alpha particles. The products were curium-242 and a neutron. What was the target element?

**21.50** Californium was first synthesized by bombarding an element with alpha particles. The products were californium-245 and a neutron. What was the target element?

## Rate of Radioactive Decay

**21.51** Tritium, or hydrogen-3, is prepared by bombarding lithium-6 with neutrons. A 0.250-mg sample of tritium decays at the rate of $8.94 \times 10^{10}$ disintegrations per second. What is the decay constant (in /s) of tritium, whose atomic mass is 3.02 amu?

**21.52** The first isotope of plutonium discovered was plutonium-238. It is used to power batteries for heart pacemakers. A sample of plutonium-238 weighing $2.8 \times 10^{-6}$ g decays at the rate of $1.8 \times 10^{6}$ disintegrations per second. What is the decay constant of plutonium-238 in reciprocal seconds (/s)?

**21.53** Sulfur-35 is a radioactive isotope used in chemical and medical research. A 0.48-mg sample of sulfur-35 has an activity of 20.4 Ci. What is the decay constant of sulfur-35 (in /s)?

**21.54** Sodium-24 is used in medicine to study the circulatory system. A sample weighing $5.2 \times 10^{-6}$ g has an activity of 45.3 Ci. What is the decay constant of sodium-24 (in /s)?

**21.55** Tellurium-123 is a radioactive isotope occurring in natural tellurium. The decay constant is $1.7 \times 10^{-21}$/s. What is the half-life in years?

**21.56** Neptunium-237 was the first isotope of a transuranium element to be discovered. The decay constant is $1.03 \times 10^{-14}$/s. What is the half-life in years?

**21.57** Carbon-14 has been used to study the mechanisms of reactions that involve organic compounds. The half-life of carbon-14 is $5.73 \times 10^{3}$ y. What is the decay constant (in /s)?

**21.58** Promethium-147 has been used in luminous paint for dials. The half-life of this isotope is 2.5 y. What is the decay constant (in /s)?

**21.59** Gold-198 has a half-life of 2.69 d. What is the activity (in curies) of a 0.86-mg sample?

**21.60** Cesium-134 has a half-life of 2.05 y. What is the activity (in curies) of a 0.25-mg sample?

**21.61** A sample of a phosphorus compound contains phosphorus-32. This sample of radioactive isotope is decaying at the rate of $6.0 \times 10^{12}$ disintegrations per second. How many grams of $^{32}P$ are in the sample? The half-life of $^{32}P$ is 14.3 d.

**21.62** A sample of sodium thiosulfate, $Na_2S_2O_3$, contains sulfur-35. Determine the mass of $^{35}S$ in the sample from the decay rate, which was determined to be $7.7 \times 10^{11}$ disintegrations per second. The half-life of $^{35}S$ is 88 d.

**21.63** A sample of sodium-24 was administered to a patient to test for faulty blood circulation by comparing the radioactivity reaching various parts of the body. What fraction of the sodium-24 nuclei would remain undecayed after 12.0 h? The half-life is 15.0 h. If a sample contains 5.0 $\mu$g of $^{24}Na$, how many micrograms remain after 12.0 h?

**21.64** A solution of sodium iodide containing iodine-131 was given to a patient to test for malfunctioning of the thyroid gland. What fraction of the iodine-131 nuclei would remain undecayed after 7.0 d? If a sample contains 2.5 $\mu$g of $^{131}I$, how many micrograms remain after 7.0 d? The half-life of I-131 is 8.07 d.

**21.65** If 28.0% of a sample of nitrogen-17 decays in 1.97 s, what is the half-life of this isotope (in seconds)?

**21.66** If 18.0% of a sample of zinc-65 decays in 69.9 d, what is the half-life of this isotope (in days)?

**21.67** A sample of iron-59 initially registers 125 counts per second on a radiation counter. After 10.0 d, the sample registers 107 counts per second. What is the half-life (in days) of iron-59?

**21.68** A sample of copper-64 gives a reading of 88 counts per second on a radiation counter. After 9.5 h, the sample gives a reading of 53 counts per second. What is the half-life (in hours) of copper-64?

**21.69** Carbon from a cypress beam obtained from the tomb of Sneferu, a king of ancient Egypt, gave 8.1 disintegrations of

$^{14}$C per minute per gram of carbon. How old is the cypress beam? Carbon from living material gives 15.3 disintegrations of $^{14}$C per minute per gram of carbon.

**21.70** Carbon from the Dead Sea Scrolls, very old manuscripts found in Israel, gave 12.1 disintegrations of $^{14}$C per minute per gram of carbon. How old are the manuscripts? Carbon from living material gives 15.3 disintegrations of $^{14}$C per minute per gram of carbon.

**21.71** Several hundred pairs of sandals found in a cave in Oregon were found by carbon-14 dating to be $9.0 \times 10^3$ years old. What must have been the activity of the carbon-14 in the sandals in disintegrations per minute per gram? Assume the original activity was 15.3 disintegrations per minute per gram.

**21.72** Some mammoth bones found in Arizona were found by carbon-14 dating to be $1.13 \times 10^4$ years old. What must have been the activity of the carbon-14 in the bones in disintegrations per minute per gram? Assume the original activity was 15.3 disintegrations per minute per gram.

## Mass–Energy Equivalence

**21.73** Find the change of mass (in grams) resulting from the release of heat when 1 mol C reacts with 1 mol $O_2$.

$$C(s) + O_2(g) \longrightarrow CO_2(g); \Delta H = -393.5 \text{ kJ}$$

**21.74** Find the change of mass (in grams) resulting from the release of heat when 1 mol $SO_2$ is formed from the elements.

$$S(s) + O_2(g) \longrightarrow SO_2(g); \Delta H = -297 \text{ kJ}$$

**21.75** Calculate the energy change for the following nuclear reaction (in joules per mole of $^2_1$H).

$$^2_1H + ^3_1H \longrightarrow ^4_2He + ^1_0n$$

Give the energy change in MeV per $^2_1$H nucleus. See Table 21.3.

**21.76** Calculate the change in energy, in joules per mole of $^1_1$H, for the following nuclear reaction.

$$^1_1H + ^1_1H \longrightarrow ^2_1H + ^0_1e$$

Give the energy change in MeV per $^1_1$H nucleus. See Table 21.3.

**21.77** Obtain the mass defect (in amu) and binding energy (in MeV) for the $^6_3$Li nucleus. What is the binding energy (in MeV) per nucleon? See Table 21.3.

**21.78** Obtain the mass defect (in amu) and binding energy (in MeV) for the $^{58}_{28}$Ni nucleus. What is the binding energy (in MeV) per nucleon? See Table 21.3.

## General Problems

**21.79** Sodium-23 is the only stable isotope of sodium. Predict how sodium-20 will decay and how sodium-26 will decay.

**21.80** Aluminum-27 is the only stable isotope of aluminum. Predict how aluminum-24 will decay and how aluminum-30 will decay.

**21.81** A uranium-235 nucleus decays by a series of alpha and beta emissions until it reaches lead-207. How many alpha emissions and how many beta emissions occur in this series of decays?

**21.82** A thorium-232 nucleus decays by a series of alpha and beta emissions until it reaches lead-208. How many alpha emissions and how many beta emissions occur in this series of decays?

**21.83** A bismuth-209 nucleus reacts with an alpha particle to produce an astatine nucleus and two neutrons. Write the complete nuclear equation for this reaction.

**21.84** A bismuth-209 nucleus reacts with a deuteron to produce a polonium nucleus and a neutron. Write the complete nuclear equation for this reaction.

**21.85** Complete the following equation by filling in the blank.

$$^{238}_{92}U + ^{12}_6C \longrightarrow \underline{\hspace{1cm}} + 4^1_0n$$

**21.86** Complete the following equation by filling in the blank.

$$^{246}_{96}Cm + ^{12}_6C \longrightarrow \underline{\hspace{1cm}} + 4^1_0n$$

**21.87** Tritium, or hydrogen-3, is formed in the upper atmosphere by cosmic rays, similar to the formation of carbon-14. Tritium has been used to determine the age of wines. A certain wine that has been aged in a bottle has a tritium content only 72% of that in a similar wine of the same mass that has just been bottled. How long has the aged wine been in the bottle? The half-life of tritium is 12.3 y.

**21.88** The naturally occurring isotope rubidium-87 decays by beta emission to strontium-87. This decay is the basis of a method for determining the ages of rocks. A sample of rock contains 102.1 $\mu$g $^{87}$Rb and 5.3 $\mu$g $^{87}$Sr. What is the age of the rock? The half-life of rubidium-87 is $4.8 \times 10^{10}$ y.

**21.89** When a positron and an electron collide, they are annihilated and two gamma photons of equal energy are emitted. Calculate the wavelength corresponding to this gamma emission.

**21.90** When technetium-99$m$ decays to technetium-99, a gamma photon corresponding to an energy of 0.143 MeV is emitted. What is the wavelength of this gamma emission? What is the difference in mass between Tc-99$m$ and Tc-99?

**21.91** Calculate the energy released when 3.00 kg of uranium-235 undergoes the following fission process.

$$_{0}^{1}n + _{92}^{235}U \longrightarrow _{53}^{136}I + _{39}^{96}Y + 4_{0}^{1}n$$

The masses of $_{53}^{136}I$ and $_{39}^{96}Y$ nuclei are 135.8401 amu and 95.8629 amu, respectively. Other masses are given in Table 21.3. Compare this energy with the heat released when 3.00 kg C(graphite) burns to $CO_2(g)$.

**21.92** Calculate the energy released when 1.00 kg of hydrogen-1 undergoes fusion to helium-4, according to the following reaction.

$$4_{1}^{1}H \longrightarrow _{2}^{4}He + 2_{1}^{0}e$$

This reaction is one of the principal sources of energy from the sun. Compare the energy released by 1.00 kg of $_{1}^{1}H$ in this reaction to the heat released when 1.00 kg of C(graphite) burns to $CO_2(g)$. See Table 21.3 for data.

**21.93** The half-life of calcium-47 is 4.536 days and it decays by the emission of a beta particle.
a. Write a balanced equation for the decay of Ca-47.
b. If 10.0 $\mu$g of Ca-47 is needed for an experiment, what mass of $^{47}CaSO_4$ must be ordered if it takes 48 h for it to arrive from the supplier?

**21.94** The radioactive isotope phosphorus-32 is often used in biochemical research. Its half-life is 14.28 days and it decays by beta emission.
a. Write a balanced equation for the decomposition of P-32.
b. If the original sample were 275 mg of $K_3^{32}PO_4$, what amount remains after 45.0 d? What percent of the sample has undergone decay?

**21.95** The half-life of $_{35}^{82}Br$ is 1.471 days. This isotope decays by the emission of a beta particle.
a. Gaseous HBr is made with Br-82. When the bromine isotope decays, the HBr produces $H_2$ and the bromine decay product. Write a balanced equation for the decay of Br-82. Now, write a balanced equation for the decomposition of $H^{82}Br$.
b. If a pure sample of 0.0150 mol of HBr made entirely with Br-82 is placed in an evacuated 1.00-L flask, how much HBr remains after 12.0 h? If the temperature is 22°C, what is the pressure in the flask?

**21.96** The half-life of $_{52}^{132}Te$ is 3.26 days. This isotope decays by beta emission to a highly unstable intermediate that decays rapidly to a stable product by beta emission.
a. What is the ultimate product obtained from Te-132? Write a balanced equation for this reaction.
b. Gaseous $H_2Te$ is made with Te-132. When the tellurium isotope decays, the $H_2Te$ produces $H_2$ and the ultimate decay product of Te-132. Write a balanced equation for the formation of stable products from $H_2^{132}Te$.
c. If a pure sample of 0.0125 mol of $H_2Te$ made entirely of Te-132 is placed in an evacuated 1.50-L flask, how much $H_2Te$ remains after 99.0 h? If the temperature is 25°C, what is the pressure in the flask?

## Cumulative-Skills Problems

**21.97** A sample of sodium phosphate, $Na_3PO_4$, weighing 54.5 mg contains radioactive phosphorus-32 (with mass 32.0 amu). If 15.6% of the phosphorus atoms in the compound is phosphorus-32 (the remainder is naturally occurring phosphorus), how many disintegrations of this nucleus occur per second in this sample? Phosphorus-32 has a half-life of 14.3 d.

**21.98** A sample of sodium thiosulfate, $Na_2S_2O_3$, weighing 38.1 mg contains radioactive sulfur-35 (with mass 35.0 amu). If 22.3% of the sulfur atoms in the compound is sulfur-35 (the remainder is naturally occurring sulfur), how many disintegrations of this nucleus occur per second in this sample? Sulfur-35 has a half-life of 87.9 d.

**21.99** Polonium-210 has a half-life of 138.4 days, decaying by alpha emission. Suppose the helium gas originating from the alpha particles in this decay were collected. What volume of helium at 25°C and 735 mmHg could be obtained from 1.0000 g of polonium dioxide, $PoO_2$, in a period of 48.0 h?

**21.100** Radium-226 decays by alpha emission to radon-222, a noble gas. What volume of pure radon-222 at 23°C and 785 mmHg could be obtained from 543.0 mg of radium bromide, $RaBr_2$, in a period of 37.5 y? The half-life of radium-226 is 1602 years.

**21.101** What is the energy (in joules) evolved when 1 mol of helium-4 nuclei is produced from protons and neutrons? How many liters of ethane, $C_2H_6(g)$, at 25°C and 725 mmHg are needed to evolve the same quantity of energy when the ethane is burned in oxygen to $CO_2(g)$ and $H_2O(g)$? See Table 21.3 and Appendix C for data.

**21.102**    Plutonium-239 has been used as a power source for heart pacemakers. What is the energy obtained from the following decay of 215 mg of plutonium-239?

$$^{239}_{94}\text{Pu} \longrightarrow \ ^{4}_{2}\text{He} + \ ^{235}_{92}\text{U}$$

Suppose the electric energy produced from this amount of plutonium-239 is 25.0% of this value. What is the minimum grams of zinc that would be needed for the standard voltaic cell $\text{Zn}|\text{Zn}^{2+}\|\text{Cu}^{2+}|\text{Cu}$ to obtain the same electric energy?

 **Media Activities**

**21.103**    Watch the Radiation of Everyday Objects *Visualization* on the student CD.
a. Which isotope of uranium makes the orange plate radioactive? What type of emission would you predict would occur with uranium?
b. If the thorium in the lantern mantle undergoes alpha emission, what is the product nucleus?

**21.104**    Open the Band of Stability *Understanding Concepts* activity on the student CD. Work through the Self-Assessment and write the nuclear decay reactions for C-11, P-30, B-8, Cu-68, C-10, and Zn-66.

**21.105**    Watch the Nuclear Fission *Visualization* on the student CD. If the fission of one U-235 nucleus only produced a single neutron, would it be possible to have a sustainable chain reaction?

**21.106**    Work through the Radioactive Half-Life *Problem Solving* activity on the student web site. Use the information presented in the activity to solve the following problem: If a wooden spoon were 8300 years old, what would be the carbon-14 activity of the spoon—in disintegrations per minute?

**21.107**    For review, take the ACE practice test for Chapter 21 on the student web site.

The exothermic reaction of magnesium and silicon dioxide forming magnesium oxide and silicon.

# Chemistry of the Main-Group Elements

**F**igure 22.1 is an abridged periodic table showing the main-group elements but omitting the transition elements. Many of these elements are of interest because they are naturally abundant and, when combined with each other, form many of the necessary ingredients for life and living on our planet. For example, consider oxygen, the most abundant element on the surface of the earth: in its elemental form as $O_2$, it is the airborne fuel for the animal kingdom; when combined with carbon to form $CO_2$ gas, carbon can be delivered to plants for photosynthesis; and when combined with hydrogen to form water, a substance that is essential for all forms of life results. Many of the other main-group elements play equally important roles. In this chapter you will have the opportunity to explore the properties, chemistry, and commercial applications of several of these elements.

**22**

**Figure 22.1**
**An abridged periodic table showing the main-group elements.** Elements to the left of the heavy staircase line are largely metallic in character; those to the right are largely nonmetallic.

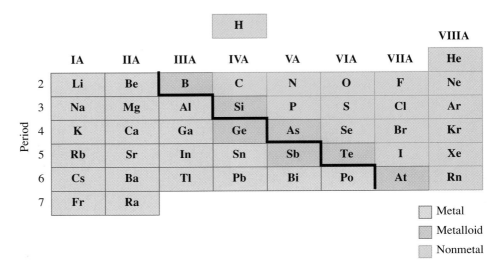

## 22.1 General Observations About the Main-Group Elements

In this section, we want to consider some periodic trends that you will find useful as you study the chemistry of specific elements. The elements on the left side of the periodic table in Figure 22.1 are largely metallic in character; those on the right side are largely nonmetallic. Table 22.1 compares the characteristics of metallic and non-metallic elements, and the following discussion describes some of these characteristics in greater detail.

The metallic elements generally have low ionization energies and low electronegativities compared with the nonmetallic elements. As a result, the metals tend to lose their valence electrons to form cations ($Na^+$, $Ca^{2+}$, $Al^{3+}$) in compounds or in

**Table 22.1**
**Comparison of Metallic and Nonmetallic Elements**

| Metals | Nonmetals |
|---|---|
| Lustrous | Nonlustrous |
| Solids at 20°C (except Hg, which is a liquid) | Solids or gases at 20°C (except $Br_2$, which is a liquid) |
| Solids are malleable and ductile | Solids are usually hard and brittle |
| Conductors of heat and electricity | Nonconductors of heat and electricity (except graphite, an allotrope of C) |
| Low ionization energies | Moderate to high ionization energies |
| Low electronegativities | Moderate to high electronegativities |
| Form cations | Form monatomic anions or oxoanions |
| Oxides are basic (unless metal is in a high oxidation state) | Oxides are acidic |

aqueous solution. Nonmetals, on the other hand, form monatomic anions ($O^{2-}$, $Cl^-$) and oxoanions ($NO_3^-$, $SO_4^{2-}$).

Also, as noted earlier (Section 8.7), the oxides of the metals are usually basic. The oxides of the most reactive metals react with water to give basic solutions. For example,

$$CaO(s) + H_2O(l) \longrightarrow Ca^{2+}(aq) + 2OH^-(aq)$$

[Oxides of metals in high oxidation states, which you encounter in some transition metals, can be acidic; chromium(VI) oxide, $CrO_3$, is an acidic oxide.] The oxides of the nonmetals are acidic. Sulfur dioxide, an oxide of a nonmetal, dissolves in water to form an acidic solution.

$$SO_2(g) + 2H_2O(l) \rightleftharpoons H_3O^+(aq) + HSO_3^-(aq)$$

Silicon dioxide (silica sand) does not dissolve in water, but it does react when melted with basic oxides such as calcium oxide to form salts.

$$SiO_2(l) + CaO(s) \longrightarrow CaSiO_3(l)$$

Table 22.2 shows the oxidation states displayed by compounds of the main-group elements. The principal conclusions are as follows: the metallic elements generally have oxidation states equal to the group number (the Roman numeral for a column), representing a loss of the valence electrons in forming compounds. Some of the metallic elements in the fifth and sixth periods also have oxidation states equal to the group number minus two (for example, $Pb^{2+}$). However, the nonmetals (except for the most electronegative elements fluorine and oxygen) have a variety of oxidation states extending from the group number (the most positive value) to the group number minus eight (the most negative value). For example, chlorine (a Group VIIA element) has the following oxidation states in compounds: $+7$, $+5$, $+3$, $+1$, $-1$.

The metallic–nonmetallic characteristics of the elements change in definite ways as you move from left to right across a period or down a column of the periodic table. *The metallic characteristics of the main-group elements in the periodic table generally decrease in going across a period from left to right.* Figure 22.1 illustrates this in a broad way. In any period, the elements at the far left are metals and those at the far right are nonmetals. The trend of decreasing metallic character can be seen clearly in the third-period elements. Sodium, magnesium, and aluminum are metals. Silicon is a metalloid, whereas phosphorus, sulfur, and chlorine are nonmetals. The oxides of these elements show the expected trend from basic to acidic. Sodium oxide and magnesium oxide are basic. Aluminum oxide is amphoteric (it has basic and acidic character), as we noted in Section 13.2. The oxides of the elements silicon to chlorine are acidic.

*The metallic characteristics of the main-group elements in the periodic table become more important going down any column (group).* This trend is more pronounced in the middle groups of the periodic table (Groups IIIA to VA). For example, in Group IVA, carbon is a nonmetal, silicon and germanium are metalloids, and tin and lead are metals. The metallic elements tend to become more reactive as you progress down a column. You can see this most clearly in the Group IIA elements; beryllium is much less reactive than strontium and barium.

*A second-period element is often rather different from the remaining elements in its group.* The second-period element generally has a small atom that tends to hold electrons strongly, giving rise to a relatively high electronegativity (electron-withdrawing power). For example, nitrogen has an electronegativity of 3.0, but the other Group VA elements have electronegativities between 1.9 and 2.1. Another reason for the difference in behavior between a second-period element and the other

**Table 22.2**
**Oxidation States in Compounds of the Main-Group Elements\***

| Period | Group | | | | | | |
|--------|-------|------|------|------|------|------|------|
| | IA | IIA | IIIA | IVA | VA | VIA | VIIA |
| 2 | Li +1 | Be +2 | B +3 | C +4 +2 −4 | N +5 +4 +3 +2 +1 −3 | O −1 −2 | F −1 |
| 3 | Na +1 | Mg +2 | Al +3 | Si +4 −4 | P +5 +3 −3 | S +6 +4 +2 −2 | Cl +7 +5 +3 +1 −1 |
| 4 | K +1 | Ca +2 | Ga +3 | Ge +4 +2 | As +5 +3 −3 | Se +6 +4 −2 | Br +7 +5 +1 −1 |
| 5 | Rb +1 | Sr +2 | In +3 +1 | Sn +4 +2 | Sb +5 +3 −3 | Te +6 +4 −2 | I +7 +5 +1 −1 |
| 6 | Cs +1 | Ba +2 | Tl +3 +1 | Pb +4 +2 | Bi +5 +3 | Po +4 +2 | At +5 −1 |

\*The most common oxidation state is shown in color. Some uncommon oxidation states are not shown.

elements of the same group has to do with the fact that bonding in the second-period elements involves only $s$ and $p$ orbitals, whereas the other elements may use $d$ orbitals. This places a limit on the types of compounds formed by the second-period elements. For example, although nitrogen forms only the trihalides (such as $NCl_3$), phosphorus has both trihalides ($PCl_3$) and pentahalides ($PCl_5$), which it forms using $3d$ orbitals.

# Chemistry of the Main-Group Metals

In the first half of this chapter, we will focus on the chemistry of the Groups IA, IIA, and IIIA metals. The sources and preparation of several of these elements were covered in Sections 13.1 and 13.2 of Chapter 13. Here we will address these questions: What are the chemical and physical properties of some of the main-group metallic elements and their compounds? What are some of the commercial uses of these substances?

# 22.2   Group IA: The Alkali Metals

The alkali metals, the elements in the first column of the periodic table except hydrogen; are all soft, silvery metals. Figure 22.2 shows sodium metal being cut with a knife. The alkali metals are the most reactive of all metals, reacting readily with air and water. Because the valence configuration of these elements is $ns^1$, the alkali metals usually react by losing this electron to form the $+1$ ions, such as $Li^+$ and $Na^+$. Most of the compounds of these ions are soluble in water.

Because of their chemical reactivity, the Group IA elements never occur as free metals in nature. They do occur extensively in silicate minerals, which weather to form soluble compounds of the elements (particularly of sodium and potassium). These soluble compounds eventually find their way to landlocked lakes and oceans, where they concentrate. Enormous underground beds of sodium and potassium compounds formed when lakes and seas became isolated by geologic events; the water eventually evaporated, leaving solid deposits of alkali metal compounds. Commercially, sodium and potassium compounds are common, and both sodium and lithium metals are available in quantity.

## Lithium

In recent years, the commercial importance of lithium metal has risen markedly. Among the alkali metals, only sodium metal is more important.

As you might expect for an alkali metal, lithium is chemically reactive. Like the other alkali metals, it is a relatively soft metal, although the hardest of the Group IA elements. Lithium does exhibit properties that are somewhat different from those of the lower members of the Group IA elements. Many of its ionic compounds are much less soluble than are similar compounds of the other alkali metals. Lithium carbonate, $Li_2CO_3$, for example, is only slightly soluble in water at room temperature, whereas sodium carbonate, $Na_2CO_3$, is soluble.

**Lithium Metallurgy**   The commercial source of lithium metal is the ore spodumene, which is a lithium aluminum silicate mineral, $LiAl(SiO_3)_2$. The ore is heated and then washed with sulfuric acid to obtain a solution of lithium ion, which is precipitated as lithium carbonate. Lithium carbonate from lithium ore is the primary starting substance for the production of lithium metal and lithium compounds. Lithium metal is obtained by electrolysis of the chloride, which was described in Section 13.2.

The use of lithium metal has greatly expanded in recent years (Figure 22.3). A major use is in the production of low-density aluminum alloy for aircraft construction. Batteries with lithium metal anodes have also become common (Figure 22.4). The anode reaction is the oxidation of lithium to the ion:

$$Li(s) \longrightarrow Li^+ + e^-$$

The cathode material varies, but the cathodes in batteries used in cameras, calculators, and similar devices consist of manganese(II) oxide, the same substance used in common dry cells. The electrolyte in this $Li/MnO_2$ cell is a solution of lithium perchlorate, $LiClO_4$, dissolved in an organic solvent.

**Reactions of Lithium Metal**   Lithium, like the other alkali metals, reacts readily with water and with moisture in the air to produce lithium hydroxide and hydrogen gas:

$$2Li(s) + 2H_2O(l) \longrightarrow 2LiOH(aq) + H_2(g)$$

**Figure 22.2**
**Cutting of sodium metal.**
The metal is easily cut with a sharp knife.

**Figure 22.3** *(left)*
**A roll of lithium metal for batteries.** The metal must be handled in humidity-free rooms to prevent corrosion.

**Figure 22.4** *(right)*
**Lithium battery.** Lithium batteries are used wherever a reliable current is required for a lengthy period. The photograph shows a desktop computer circuit board with lithium battery.

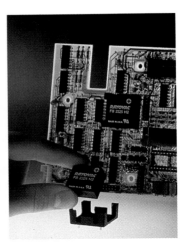

When a pellet of lithium is placed in water, the pellet spins around on the surface of the water, evolving bubbles of hydrogen gas. (Lithium floats because it is less dense than water.) The reaction is not as vigorous, however, as the similar reaction of sodium and water.

Lithium burns in air to produce lithium oxide, $Li_2O$, a white powder.

$$4Li(s) + O_2(g) \longrightarrow 2Li_2O(s)$$

When heated with nitrogen gas, lithium reacts to form lithium nitride, $Li_3N$, the only stable alkali-metal nitride.

$$6Li(s) + N_2(g) \longrightarrow 2Li_3N(s)$$

Lithium nitride is an ionic compound of the nitride ion, $N^{3-}$.

**Lithium Compounds** Lithium carbonate, which is obtained from lithium ore, is the primary source of other lithium compounds. Significant quantities of lithium carbonate are also used in the preparation of porcelain enamels, glazes, and special glass. When purified, the carbonate is used as a source of lithium ion for the treatment of manic-depressive disorders. The physiological action of lithium ion in this treatment is not completely understood, although lithium ion is known to inhibit certain biochemical reactions involved in hormone action.

Lithium hydroxide, LiOH, is a strong base. It is produced by reaction of lithium carbonate with calcium hydroxide in a precipitation reaction. Calcium carbonate is much less soluble than lithium carbonate:

$$Li_2CO_3(aq) + Ca(OH)_2(aq) \longrightarrow 2LiOH(aq) + CaCO_3(s)$$

The calcium carbonate is filtered off, and lithium hydroxide is recovered from the filtrate. Lithium hydroxide is used in the production of lithium soaps, which are used in making lubricating greases from oil. (Lithium soap thickens the oil.) Soaps are salts of fatty acids; they are prepared by heating a fat with a strong base.

Lithium hydroxide is used to remove carbon dioxide from the air in spacecraft and submarines. Humans exhale carbon dioxide that is produced by the normal bio-chemical reactions of the body. For people to live in a closed space, it is necessary to remove this carbon dioxide. Although carbon dioxide is not normally toxic, increasing concentrations do have a physiological effect by interfering with the blood's capacity to carry oxygen, and high concentrations of carbon dioxide are hazardous. Like the other alkali-metal hydroxides, lithium hydroxide absorbs carbon dioxide from air by forming the carbonate and hydrogen carbonate:

$$2LiOH(s) + CO_2(g) \longrightarrow Li_2CO_3(s) + H_2O(l)$$

$$Li_2CO_3(s) + H_2O(l) + CO_2(g) \longrightarrow 2LiHCO_3(s)$$

Sodium hydroxide could be used to remove carbon dioxide from air, since it undergoes the same reactions. However, the formula weight of LiOH (24 amu) is much lower than that of NaOH (40 amu). Where weight is at a premium, lithium hydroxide is preferred, even though it is more expensive than sodium hydroxide.

Table 22.3 lists some uses of lithium and other alkali metal compounds.

**Table 22.3**
**Uses of Alkali Metal Compounds**

| Compound | Use |
| --- | --- |
| $Li_2CO_3$ | Preparation of porcelain, glazes, special glasses |
| | Preparation of LiOH |
| | Treatment of manic-depressive disorders |
| LiOH | Manufacture of lithium soaps for lubricating greases |
| | In air-regeneration systems |
| LiH | Reducing agent in organic syntheses |
| $LiNH_2$ | Preparation of antihistamines and other pharmaceuticals |
| NaCl | Source of sodium and sodium compounds |
| | Condiment and food preservative |
| | Soap manufacture (precipitates soap from reaction mixture) |
| NaOH | Pulp and paper industry |
| | Extraction of aluminum oxide from ore |
| | Manufacture of viscose rayon |
| | Petroleum refining |
| | Manufacture of soap |
| $Na_2CO_3$ | Manufacture of glass |
| | In detergents and water softeners |
| $Na_2O_2$ | Textile bleach |
| $NaNH_2$ | Preparation of indigo dye for denim (blue jeans) |
| KCl | Fertilizer |
| | Source of other potassium compounds |
| KOH | Manufacture of soft soap |
| | Manufacture of other potassium compounds |
| $K_2CO_3$ | Manufacture of glass |
| $KNO_3$ | Fertilizers |
| | Explosives and fireworks |

## Sodium and Potassium

Sodium compounds are of enormous economic importance. Common table salt, which is sodium chloride, has been an important article of commerce since earliest recorded history. Salt was of such importance in Roman times that a specific allowance of salt was part of the soldiers' pay. The word *salary* derives from the Latin word for this salt allowance (*salarium,* which comes from the Latin word *sal,* for "salt"). Today, we manufacture sodium metal as well as other sodium compounds from sodium chloride.

The economic importance of potassium stems in large part from its role as a plant nutrient. Enormous quantities of potassium chloride from underground deposits are used as fertilizer to increase the world's food supply. In fact, plants were an early source of potassium compounds. Wood and other plant materials were burned in pots to give ashes (potash) that consist primarily of potassium carbonate, from which other potassium compounds could be manufactured. The name potassium has its origin in the word *potash* from this earlier source of the element.

Sodium metal is more reactive than lithium metal, and potassium metal is still more reactive. In general, as you move down the column of alkali metals in Group IA of the periodic table, you find that the metals become more chemically reactive (forming the +1 ions). This is partly a result of the decrease in ionization energy of the atom (from 520 kJ/mol for the first ionization of lithium to 376 kJ/mol for that of cesium). As the atomic size increases going from the top of the column to the bottom, the valence electrons are held less strongly and so are lost more easily.

The Group IA metals also increase in softness going from the top of the column to the bottom. Lithium is a moderately firm metal, whereas potassium is soft enough to be cut with a butter knife. This increasing softness of the metals is in part a result of increasing atomic size, which results in decreasing strength of bonding by the valence-shell *s* electrons.

**Sodium Metallurgy**   Sodium metal is obtained by the electrolysis of molten, or fused, sodium chloride. Sodium chloride is mined from huge underground deposits. The other source of sodium chloride is seawater, which is a solution of many dissolved substances, but sodium chloride is the principal one. Sodium chloride melts at 801°C, but commercial electrolysis employs a mixture of NaCl and $CaCl_2$, which melts at 580°C. We discussed the electrolysis in some detail in Section 20.9. ◀

▶ **Figure 20.20 depicts the commercial Downs cell for the preparation of sodium from molten sodium chloride.**

Sodium metal is a strong reducing agent, and this accounts for many of its major uses. For example, it is used to obtain metals such as titanium and zirconium by reduction of their compounds. Titanium is a strong metal used in airplane and spacecraft manufacture. It is produced by reduction of titanium tetrachloride, $TiCl_4$, obtained by chemical processing of titanium ores. The overall process can be written as follows:

$$TiCl_4(g) + 4Na(l) \longrightarrow Ti(s) + 4NaCl(s)$$

Sodium is also employed as a reducing agent in the production of a number of organic compounds, including dyes and pharmaceutical drugs.

**Reactions of Sodium Metal**   Like lithium, sodium metal reacts with water, but with even greater vigor. The reaction is sufficiently exothermic to ignite the hydrogen gas that is a product of the reaction:

$$2Na(s) + 2H_2O(l) \longrightarrow 2NaOH(aq) + H_2(g)$$

**Figure 22.5**
**Oxygen compounds of alkali metals.** Sodium peroxide, $Na_2O_2$ (yellowish-white), and potassium superoxide, $KO_2$ (orange-yellow). Potassium superoxide is the principal product formed when potassium metal burns in air.

Sodium burns in air, producing some sodium oxide, $Na_2O$, but mainly sodium peroxide, $Na_2O_2$, a compound containing the peroxide ion $O_2^{2-}$ (Figure 22.5).

$$2Na(s) + O_2(g) \longrightarrow Na_2O_2(s)$$

The peroxide ion acts as an oxidizing agent (Figure 22.6).

▶ Demonstration of the reaction of sodium peroxide and sulfur: Bassam Z. Shakhashiri, *Chemical Demonstrations,* Vol. 1 (Madison: University of Wisconsin Press, 1983), no. 1.23, pp. 57–58.

**Sodium Compounds** Sodium hydroxide, NaOH, is among the top ten industrial chemicals. It is produced by the electrolysis of aqueous sodium chloride. The overall electrolysis, which was described in Section 20.10, can be written as

$$2NaCl(aq) + 2H_2O(l) \xrightarrow{\text{electrolysis}} H_2(g) + Cl_2(g) + 2NaOH(aq)$$

▶ Figures 20.22 and 20.23 show two different commercial cells for the electrolysis of aqueous sodium chloride.

Both chlorine gas and sodium hydroxide are major products of this electrolysis. ◀

Sodium hydroxide is a strong base, and this property is useful in many applications. It is used in large quantities in aluminum production (which depends on its reaction with the amphoteric aluminum hydroxide) and in the refining of petroleum (where it chemically reacts with acid constituents during the refining process). Large quantities of sodium hydroxide are also used to produce sodium compounds, which occur in a wide variety of commercial products.

For example, soap is made by heating a fat with sodium hydroxide solution. The product is a mixture of sodium salts of fatty acids, obtained from the fat. Fatty acids are chemically similar to acetic acid, but they consist of long chains of carbon atoms.

**Figure 22.6**
**Sodium peroxide as an oxidizing agent.** *Left:* A couple of drops of water are added to a mixture of sodium peroxide and sulfur to initiate a reaction. *Right:* During the violent reaction, sulfur is oxidized to sulfur dioxide.

▶ The action of soap in dispersing oil in water is described in Section 12.9.

**Figure 22.7**
**Sodium compounds available at the grocery store.** *Left to right:* Salt (NaCl), baking soda (NaHCO$_3$), washing soda (Na$_2$CO$_3$ · 10H$_2$O), drain opener (NaOH), and oven cleaner (NaOH).

▶ Soda ash decomposes to sodium oxide when heated. Sodium and calcium oxides, which are basic oxides, react with fused silicon dioxide, SiO$_2$, an acidic oxide, to produce silicate glass.

Compare the structure of the acetate ion with that of the stearate ion. (Stearic acid is a typical fatty acid.) ◀

acetate ion

stearate ion

Sodium hydroxide is commonly known as lye or caustic soda. You can buy it in a grocery store as a drain opener and as an oven cleaner (Figure 22.7). Both usages depend on the reaction of sodium hydroxide with organic materials, such as fats and hair, to produce soluble materials. Because of these reactions, sodium hydroxide and its solutions require careful handling.

Sodium carbonate is another important compound of sodium. The anhydrous compound, Na$_2$CO$_3$, is known commercially as *soda ash*. Large quantities of soda ash are consumed with sand and lime in making glass. ◀ The decahydrate, Na$_2$CO$_3$ · 10H$_2$O, is added to many detergent preparations and is sold commercially as washing soda. In the United States, most sodium carbonate is produced from the mineral trona, whose chemical formula is approximately Na$_2$CO$_3$ · NaHCO$_3$ · 2H$_2$O. There are large deposits of trona in southwestern Wyoming.

Worldwide, a large fraction of sodium carbonate is still produced by the Solvay process. The **Solvay process** is *an industrial method for obtaining sodium carbonate from sodium chloride and limestone.* In the main step of this process, ammonia is first dissolved in a saturated solution of sodium chloride. Then carbon dioxide is bubbled in, and sodium hydrogen carbonate (baking soda) precipitates. Figure 22.8 shows a laboratory demonstration of this reaction.

$$NH_3(g) + \underbrace{H_2O(l) + CO_2(g)}_{H_2CO_3(aq)} + NaCl(aq) \longrightarrow NaHCO_3(s) + NH_4Cl(aq)$$

**Figure 22.8**
**Laboratory demonstration of the Solvay process.**
*Left:* A concentrated solution of ammonia is saturated with sodium chloride. When pieces of dry ice (solid CO$_2$) are added, a cloud of cooled water vapor forms. *Right:* Sodium hydrogen carbonate precipitates from the cold reaction mixture.

You can think of this as an exchange reaction of $NH_4HCO_3$ (from $NH_3$ + $H_2CO_3$) with NaCl, to give the products $NaHCO_3$ and $NH_4Cl$. The sodium hydrogen carbonate is filtered from the solution. When heated to 175°C, it decomposes to sodium carbonate.

$$2NaHCO_3(s) \xrightarrow{\Delta} Na_2CO_3(s) + CO_2(g) + H_2O(g)$$

The reaction to produce sodium hydrogen carbonate uses relatively expensive ammonia. The industrial Solvay process illustrates how you can affect the economics of a process by the ingenious use of raw materials and by recycling intermediate products. Carbon dioxide for the industrial process is obtained by heating limestone ($CaCO_3$):

$$CaCO_3(s) \xrightarrow{\Delta} CaO(s) + CO_2(g)$$

The calcium oxide obtained is used to recycle the ammonia from ammonium chloride for use in the preparation of sodium hydrogen carbonate.

$$CaO(s) + 2NH_4Cl(aq) \xrightarrow{\Delta} 2NH_3(g) + CaCl_2(aq) + H_2O(l)$$

The carbon dioxide obtained when sodium hydrogen carbonate is heated to form sodium carbonate is also recycled to prepare more sodium hydrogen carbonate. The overall result of these steps in the industrial Solvay process is the reaction of limestone ($CaCO_3$) and salt (NaCl) to produce sodium carbonate and calcium chloride:

$$CaCO_3 + 2NaCl \longrightarrow Na_2CO_3 + CaCl_2$$

The raw materials (limestone and salt) are cheap, and for many years the Solvay process was the principal source of sodium carbonate. Unfortunately, there is not sufficient demand for the by-product calcium chloride. Some is used for deicing of roads, but much of it has to be disposed of as waste. The environmental cost of this waste disposal has made the Solvay process increasingly less attractive.

**Potassium and Potassium Compounds** The principal source of potassium and potassium compounds is potassium chloride, KCl, which is obtained from underground deposits. Small amounts of this salt are used to prepare the metal. Potassium is prepared by the chemical reduction of potassium chloride rather than by electrolysis of the molten chloride, as in the preparation of sodium. In the commercial process, potassium chloride is melted with sodium metal by heating to 870°C.

$$Na(l) + KCl(l) \longrightarrow NaCl(l) + K(g)$$

At this temperature, potassium forms as a vapor. The reaction goes in the direction written because potassium vapor leaves the reaction chamber and is condensed.

Almost all of the potassium metal produced is used in the preparation of potassium superoxide, $KO_2$, for self-contained breathing apparatus used in situations such as fire fighting where toxic fumes may be present. When potassium burns in air, it produces potassium superoxide:

$$K(s) + O_2(g) \longrightarrow KO_2(s)$$

(The superoxide ion is $O_2^-$, and the oxidation state of oxygen in this ion is $-\frac{1}{2}$.) In a self-contained breathing apparatus, the potassium superoxide is contained in a canister through which one's breath passes. Moisture in the breath attacks the superoxide, releasing oxygen.

$$4KO_2(s) + 2H_2O(l) \longrightarrow 4KOH(s) + 3O_2(g)$$

The potassium hydroxide produced in this reaction removes the carbon dioxide from the exhaled air. The net effect of the self-contained breathing apparatus is to remove moisture and carbon dioxide from exhaled air and provide oxygen.

More than 90% of the potassium chloride that is mined is used directly as a plant fertilizer. The rest is used in the preparation of potassium compounds. Potassium hydroxide, used in the manufacture of liquid soaps, is obtained by the electrolysis of aqueous potassium chloride. Potassium nitrate is prepared by the reaction of potassium hydroxide and nitric acid. This compound is used in fertilizers and for explosives and fireworks. (Table 22.3 summarizes the major uses of potassium compounds, as well as those of lithium and sodium.)

## 22.3 Group IIA: The Alkaline Earth Metals

▶ Video Series C, "Dry Ice and Magnesium."

**Figure 22.9**
**Magnesium and barium metals.** Magnesium metal turnings are in the beaker (left). Barium metal (right) is much more reactive than magnesium and must be stored in a bottle of kerosene to exclude moisture and oxygen, with which barium reacts.

The Group IIA elements are also known as the alkaline earth metals. These metals illustrate the expected periodic trends. If you compare an alkaline earth metal with the alkali metal in the same period, you see that the alkaline earth metal is less reactive and a harder metal. For example, lithium reacts readily with water, but beryllium reacts hardly at all even with steam. Lithium is a soft metal, whereas beryllium is hard enough to scratch glass. You also see the expected trend within the column of alkaline earth metals; the elements at the bottom of the column are more reactive and are softer metals than those at the top of the column (Figure 22.9). Barium, in the sixth period, is a very reactive metal and, when placed in water, reacts much like an alkali metal. It is also a soft metal, much like the alkali metals. Magnesium, in the third period, is much less reactive and a harder metal (comparable to aluminum).

As we noted in Section 22.1, a second-period element is often considerably different from the other elements in its column. Those differences become more pronounced as you progress to the right in the periodic table. We briefly noted differences in properties of lithium from those of the other alkali metals, though those differences are not great. Beryllium, however, shows rather marked differences from the other alkaline earth elements. We have already noted that beryllium differs in its lack of reactivity compared with the other Group IIA metals. Another notable difference is in the properties of the hydroxides. Whereas those of the elements magnesium to barium are basic, beryllium hydroxide is amphoteric, reacting with both acids and bases.

Like the alkali metals, the Group IIA elements occur in nature as silicate rocks. They also occur as carbonates and sulfates, and many of these are commercial sources of alkaline earth metals and compounds.

Calcium and magnesium are the most common and commercially useful of the alkaline earth elements, and we will consider them in some detail in the next two sections. As can be seen from Table 13.1, calcium is the fifth and magnesium the eighth most abundant element in the earth's crust.

### Magnesium

The name magnesium comes from the name of the mineral magnesite, which in turn is believed to stem from the name Magnesia, a site in northern Greece where various minerals, including those of magnesium, have been mined since ancient times. The British chemist Humphry Davy discovered the pure element magnesium in 1808. Davy had already discovered the alkali metals potassium and sodium in late 1807. Several months later, he managed to isolate in quick succession the alkaline earth metals barium, strontium, calcium, and finally magnesium. He obtained magnesium by a

**Figure 22.10**
**Magnesium in motorcycles.** Magnesium alloys are used in many motorcycle parts. Shown here is a closeup of the engine area of a Harley-Davidson motorcycle. The V engine (upper right), engine block (lower right), and transmission case (lower left) are made of an aluminum–magnesium alloy.

▶ Magnesium metal is isolated from seawater by the Dow process, which is described in Section 13.2.

procedure similar to the one he used for the other alkaline earths. He electrolyzed a moist mixture of magnesium oxide and mercury(II) oxide, from which he obtained magnesium amalgam (an alloy of magnesium dissolved in mercury). To obtain pure magnesium, Davy distilled the mercury from the amalgam.

**Magnesium Metallurgy**   Magnesium has become increasingly important as a structural metal. Its great advantages are its low density ($1.74$ g/cm$^3$, which compares with $7.87$ g/cm$^3$ for iron and $2.70$ g/cm$^3$ for aluminum, the other important structural metals) and the relative strength of its alloys.

Some important commercial sources of magnesium are the minerals dolomite, $CaCO_3 \cdot MgCO_3$, and magnesite, $MgCO_3$. However, magnesium ion, $Mg^{2+}$, is the third most abundant dissolved ion in the oceans, after $Cl^-$ and $Na^+$. The oceans, therefore, are an essentially inexhaustible supply of the ion, from which the metal can be obtained. ◀

Pure magnesium metal is a relatively reactive element. Its alloys, however, contain aluminum and small quantities of other metals to impart both strength and corrosion resistance. Increasing quantities of magnesium alloy are used to make automobile and aircraft parts, as well as consumer materials such as power-tool and lawn-mower housings (Figure 22.10). Most commercial aluminum metal contains some percentage of magnesium, which improves the hardness and corrosion resistance of the aluminum.

Magnesium metal is also used as a reducing agent in the manufacture of titanium and zirconium from their tetrachlorides. (Sodium is also used as the reducing agent in the production of titanium, as we mentioned earlier.)

$$ZrCl_4(g) + 2Mg(l) \longrightarrow Zr(s) + 2MgCl_2(s)$$

Zirconium metal is used to make containment vessels for uranium-235 fuel rods for nuclear power reactors, because of its low absorption of the neutrons needed in the nuclear fission of uranium-235.

**Reactions of Magnesium Metal**   Once magnesium metal has been ignited, it burns vigorously in air to form the oxide.

$$2Mg(s) + O_2(g) \longrightarrow 2MgO(s)$$

The metal powder and fine wire burn readily. Burning magnesium metal gives off an intense white light; this is the white light you see in the burning of some fireworks and flares.

Magnesium also burns in carbon dioxide, producing magnesium oxide and soot, or carbon (Figure 22.11).

$$2Mg(s) + CO_2(g) \longrightarrow 2MgO(s) + C(s)$$

**Figure 22.11**
**Reaction of magnesium and carbon dioxide.**
When a glowing ribbon of magnesium metal is thrust into a beaker of carbon dioxide (from the sublimation of the dry ice at the bottom of the beaker), the metal bursts into a bright flame, producing a smoke of magnesium oxide and carbon.

Many fire extinguishers contain carbon dioxide, which normally smothers a fire by preventing oxygen from air getting to combustible materials. Carbon dioxide, however, cannot be used to extinguish magnesium fires; burning magnesium must be smothered with sand.

The pure metal reacts slowly with water. (Magnesium alloys are even less reactive.) Magnesium does react readily with steam, however.

$$Mg(s) + H_2O(g) \longrightarrow MgO(s) + H_2(g)$$

**Magnesium Compounds**   When magnesite, a magnesium carbonate mineral, is heated above 350°C, it decomposes to the oxide, a white solid:

$$MgCO_3(s) \xrightarrow{\Delta} MgO(s) + CO_2(g)$$

**Figure 22.12**
**Milk of magnesia.** Milk of magnesia is a suspension of $Mg(OH)_2$ in water. It is used as an antacid and a laxative.

Careful heating at low temperature results in a powdery form of the oxide that is relatively reactive. It reacts slowly with water to produce magnesium hydroxide but reacts readily with acids to yield the corresponding salts:

$$MgO(s) + H_2O(l) \longrightarrow Mg(OH)_2(s)$$
$$MgO(s) + 2HCl(aq) \longrightarrow MgCl_2(aq) + H_2O(l)$$

Large quantities of magnesium oxide are used in animal feed supplements, because magnesium ion is an important nutrient for animals. (The ion is also important for human nutrition.) Firebricks are produced by strongly heating magnesite (to about 1400°C) to give a hard, relatively inert form of magnesium oxide. Magnesium oxide is quite stable at high temperatures and is a good thermal insulator.

Magnesium hydroxide is only slightly soluble in water. (The solubility product, $K_{sp}$, is $1.8 \times 10^{-11}$.) A suspension of the white compound in water is called *milk of magnesia* (Figure 22.12). This suspension has a pH of about 10, which means the solution is mildly basic. Milk of magnesia is sold as an antacid, because the magnesium hydroxide reacts to neutralize excess hydrochloric acid in the stomach. Table 22.4 gives a summary list of the uses of compounds of magnesium and other Group IIA elements.

**Table 22.4**
**Uses of Alkaline Earth Compounds**

| Compound | Use |
| --- | --- |
| MgO | Refractory bricks (for furnaces) |
| | Animal feeds |
| $Mg(OH)_2$ | Source of magnesium for the metal and compounds |
| | Milk of magnesia (antacid and laxative) |
| $MgSO_4 \cdot 7H_2O$ | Fertilizer |
| | Medicinal uses (laxative and analgesic) |
| | Mordant (used in dyeing fabrics) |
| CaO and $Ca(OH)_2$ | Manufacture of steel |
| | Neutralizer for chemical processing |
| | Water treatment |
| | Mortar |
| | Stack-gas scrubber (to remove $H_2S$ and $SO_2$) |
| $CaCO_3$ | Paper coating and filter |
| | Antacids, dentifrices |
| $CaSO_4$ | Plaster, wallboard |
| | Portland cement |
| $Ca(H_2PO_4)_2$ | Soluble phosphate fertilizer |
| $BaSO_4$ | Oil-well drilling mud |
| | Gastrointestinal x-ray photography |
| | Paint pigment |

### CONCEPT CHECK 22.1

As discussed in the text, Zr is used as cladding for nuclear fuel rods in power plants. Why wouldn't you want to use a fuel rod cladding material that absorbs neutrons?

## Calcium

Calcium is a common element. It is present in the earth's crust as silicates, which weather to give free calcium ion, $Ca^{2+}$. The ion is about as abundant in seawater as magnesium ion. Calcium ion is an important nutrient for living organisms. As noted in the preceding section, seashells are principally calcium carbonate. Corals are marine organisms that grow in colonies; their calcium carbonate skeletons eventually form enormous coral reefs in warm waters. The Bahamas and the Florida Keys originated as coral reefs. Deposits of limestone (mostly $CaCO_3$) formed in earlier times as sediments of seashells and coral and by the direct precipitation of calcium carbonate from seawater. Gypsum, $CaSO_4 \cdot 2H_2O$, is another important mineral of calcium; deposits originated from evaporation of inland lakes and seas.

**Calcium Metallurgy**   Most calcium metal is obtained by the reduction of calcium oxide with aluminum. Calcium metal is used mainly in some alloys. For example, the addition of a small quantity of calcium to the lead used for the electrodes of lead storage batteries substantially reduces the decomposition of water into its elements during the recharging of the battery. Since gases are not produced during recharging, the battery can be sealed so it requires less maintenance.

**Reactions of Calcium Metal**   Calcium is a soft, reactive metal. It reacts with water, like the alkali metals do, to produce the metal hydroxide and hydrogen; the reaction, however, is much less vigorous.

$$Ca(s) + 2H_2O(l) \longrightarrow Ca(OH)_2(aq) + H_2(g)$$

Calcium burns in air to produce the oxide, CaO, and with chlorine to produce calcium chloride. Calcium also reacts directly with hydrogen to give the hydride. Calcium hydride, $CaH_2$, is a convenient source of hydrogen, because it reacts easily with water to yield $H_2$.

$$CaH_2(s) + 2H_2O(l) \longrightarrow Ca(OH)_2(aq) + 2H_2(g)$$

**Calcium Compounds**   Calcium compounds are extremely important commercially (see Table 22.4). Enormous quantities of them are used in metallurgy, in building materials, in the making of glass and paper, and in other products. Limestone ($CaCO_3$), dolomite ($CaCO_3 \cdot MgCO_3$), anhydrite ($CaSO_4$), and gypsum ($CaSO_4 \cdot 2H_2O$) are all important sources of calcium compounds. An interesting use of ground limestone is shown in Figure 22.13; this application depends on the basic character of the carbonate ion. The mineral gypsum, $CaSO_4 \cdot 2H_2O$, forms plaster of Paris, $CaSO_4 \cdot \frac{1}{2}H_2O$, when heated to 100°C. Plaster of Paris is used to make wallboard.

Calcium oxide, CaO, is among the top ten industrial chemicals. It is prepared by calcining calcium carbonate. Limestone and seashells are all possible starting materials.

$$CaCO_3(s) \xrightarrow[900°C]{\Delta} CaO(s) + CO_2(g)$$

**Figure 22.13**
**Raising the pH of a lake with ground limestone.** A specially designed barge applies a slurry of finely ground limestone ($CaCO_3$) to a lake to neutralize its acidity, caused by acid-rain pollution. Acid rain results from the burning of sulfur-containing materials (coal and sulfide ores) and from nitrogen oxide pollution from automobiles.

▶ Iron metallurgy is discussed in Section 13.2.

▶ Video Series C, "Limewater."

**Figure 22.14**
**Some antacids contain calcium carbonate.** These antacid preparations consist of purified calcium carbonate with flavorings and a binder. Such antacids are sometimes prescribed as a nutrient calcium supplement.

Calcium oxide is known commercially as *quicklime,* or simply *lime.* Much of the calcium oxide produced is used in the manufacture of iron from its ores. Iron is obtained by reducing iron ore (containing iron oxides with silicate impurities) in a blast furnace. A charge of iron ore, carbon, and calcium oxide is added at the top of the furnace, and a stream of compressed air flows in from the bottom. The purpose of calcium oxide is to combine with silicon dioxide (sand) and silicate impurities in the iron ore to produce a glassy material (called *slag*). The slag is molten at the temperatures of the blast furnace, and it flows to the bottom of the furnace and is withdrawn. The essential reaction is that between a basic oxide (calcium oxide) and an acidic oxide (silicon dioxide). ◀

$$\underset{\substack{\text{basic} \\ \text{oxide}}}{CaO(s)} + \underset{\substack{\text{acidic} \\ \text{oxide}}}{SiO_2(s)} \longrightarrow \underset{\substack{\text{calcium silicate} \\ \text{slag}}}{CaSiO_3(l)}$$

Calcium oxide reacts exothermically with water to produce calcium hydroxide. The considerable heat released is apparent when you add a few drops of water to a pile of calcium oxide powder; there is a hiss, and puffs of steam rise from the mixture.

$$CaO(s) + H_2O(l) \longrightarrow Ca(OH)_2(s); \ \Delta H^\circ = -65.7 \ kJ$$

Commercially, calcium hydroxide is referred to as *slaked lime.* Calcium hydroxide is a strong base, and many of its uses depend on this fact. For example, hydrochloric and sulfuric acids are used to remove rust from steel products. After the steel has been cleaned this way, calcium hydroxide is used to neutralize the excess acid.

Calcium hydroxide solutions react with gaseous carbon dioxide (the acid oxide of carbonic acid) to give a white, milky precipitate of calcium carbonate.

$$Ca(OH)_2(aq) + CO_2(g) \longrightarrow CaCO_3(s) + H_2O(l)$$

This reaction, with its formation of a milky precipitate, is the basis of a simple test for carbon dioxide. Commercially, the reaction is important as a method of preparing precipitated calcium carbonate, a pure form of the finely divided compound. An important use of precipitated calcium carbonate is as a filler for paper. (The purpose of a filler is to improve the paper's characteristics, such as brightness and inking ability.) Precipitated calcium carbonate is also used in toothpowders, antacids, and nutritional supplements (Figure 22.14).

**Figure 22.15**
**Reaction of carbon dioxide with calcium hydroxide solution.** *Left:* Calcium hydroxide solution with bromthymol-blue indicator. *Center:* Carbon dioxide from dry ice reacts with calcium ion to precipitate calcium carbonate. *Right:* In an excess of carbon dioxide, the calcium carbonate dissolves to form a solution of calcium ion and hydrogen carbonate ion, $HCO_3^-$. In this acidic solution, bromthymol blue changes to a yellow color.

▶ **Mortar was used by the ancient Romans for buildings and roads.**

Bricklaying mortar is made by mixing slaked lime, $Ca(OH)_2$, with sand and water. The mortar hardens as the mixture dries and calcium hydroxide crystallizes. Over time, however, the mortar sets to a harder solid as calcium hydroxide reacts with carbon dioxide in the air to form calcium carbonate crystals that intertwine with the sand particles. ◀

$$Ca(OH)_2(s) + CO_2(g) \longrightarrow CaCO_3(s) + H_2O(l)$$

Large amounts of quicklime, CaO, and slaked lime, $Ca(OH)_2$, are used to *soften* municipal water supplies. This may seem paradoxical at first, because the process of "softening" water refers to the removal of certain metal ions, particularly calcium ions, from the water. Water that is "hard" contains these metal ions. When soap, which is a sodium salt of a fatty acid, is added to hard water, a curdy precipitate of the calcium salt of the fatty acid forms. (When this precipitate adheres to the tub, it is called bathtub ring.) The calcium ion in hard water results when water containing carbon dioxide from air (and from rotting leaves and other organic matter) passes through limestone ($CaCO_3$). The calcium carbonate dissolves in this water to produce soluble calcium hydrogen carbonate. ◀

▶ **A brief discussion of hard water and water softening appears in the essay on water at the end of Chapter 11.**

$$\underset{\text{limestone}}{CaCO_3(s)} + H_2O(l) + CO_2(g) \longrightarrow \underset{\text{calcium hydrogen carbonate}}{Ca^{2+}(aq) + 2HCO_3^-(aq)}$$

Figure 22.15 illustrates the dissolution of calcium carbonate by carbon dioxide. The apparent paradox is solved when we observe that the addition of a stoichiometric amount of calcium hydroxide to hard water containing calcium hydrogen carbonate can precipitate all of the calcium ions as calcium carbonate.

$$Ca^{2+}(aq) + 2HCO_3^-(aq) + Ca(OH)_2(aq) \longrightarrow 2CaCO_3(s) + 2H_2O(l)$$

## 22.4 Group IIIA and Group IVA Metals

The Group IIIA elements clearly show the trend of increasing metallic character in going down any column of elements in the periodic table. Boron, at the top of Group IIIA, is a metalloid, and its chemistry is typical of a nonmetal. The compound $B(OH)_3$ is actually acidic (boric acid). The rest of the elements (aluminum, gallium, indium, and thallium) are metals, but their hydroxides change from amphoteric (acidic and basic) for aluminum and gallium to basic for indium and thallium.

Bonding to boron is covalent; the atom shares its $2s^2 2p^1$ valence electrons to give the $+3$ oxidation state. Although aluminum has many covalent compounds, it also has definitely ionic ones, such as $AlF_3$. The cation $Al^{3+}(aq)$ is present in aqueous solutions of aluminum salts. Gallium, indium, and thallium also have ionic compounds and give cations in aqueous solution. The $+3$ oxidation state is the only important one for aluminum, but some $+1$ compounds are known for gallium and indium, and in the case of thallium, both $+1$ and $+3$ states are important.

The Group IVA elements also show the trend of greater metallic character in going down the column of elements. The last two elements of the column, tin and lead, are well-known metals, and we will discuss them in this section. Tin and lead, in their compounds, exist in the $+2$ and $+4$ oxidation states. In tin, both oxidation states are common. In most lead compounds, lead is in the $+2$ oxidation state. Lead(IV) compounds, such as $PbO_2$, are strong oxidizing agents, indicating their tendency to revert to lead(II).

## Aluminum

Aluminum is the third most abundant element in the earth's crust (after oxygen and silicon). It occurs primarily in aluminosilicate minerals. The weathering of these rocks results in aluminum-containing clays, which are an essential part of most soils. Further weathering of clay yields bauxite, the principal ore of aluminum, which contains aluminum hydroxide, $Al(OH)_3$, and aluminum oxide hydroxide, $AlO(OH)$. Deposits of bauxite occur throughout the world but are particularly common in tropical and subtropical regions. Corundum is a hard mineral of aluminum oxide, $Al_2O_3$. The pure oxide is colorless, but the presence of impurities can give various colors to it. Sapphire (usually blue) and ruby (deep red) are gem-quality corundum (Figure 22.16).

**Aluminum Metallurgy** We discussed the details of aluminum metallurgy in Section 13.2. Bauxite, the ore of aluminum, is treated chemically in the Bayer process to obtain aluminum hydroxide, which when calcined yields aluminum oxide, $Al_2O_3$ (referred to commercially as *alumina*). Aluminum metal is obtained by electrolysis of aluminum oxide dissolved in fused cryolite, $Na_3AlF_6$.

Aluminum is the most important commercial metal after iron. Despite the fact that pure aluminum is soft and chemically reactive, the addition of a small quantity of other metals, such as copper and magnesium, yields hard, corrosion-resistant alloys. The corrosion resistance of these alloys, together with their relatively low densities, allows their use in structural applications and for containers and packaging (Figure 22.17). Aluminum is also a very good conductor of electricity. Because of this property and its low density, aluminum is used to make electrical transmission wire.

**Reactions of Aluminum Metal** Aluminum is a chemically reactive metal, although much less reactive than the alkali metals and alkaline earth metals. It does not react at an appreciable rate with water at room temperature. In air, aluminum metal reacts readily with oxygen, but the aluminum oxide that forms gives an adherent coat, which protects the underlying metal from further reaction. (The metal will burn vigorously once started, however.) Thus, unlike iron, which is chemically less reactive than aluminum but corrodes quickly in a moist environment, aluminum is corrosion resistant. The demonstration depicted in Figure 22.18 shows that aluminum does corrode quickly in air in the absence of an oxide coating. That aluminum metal does not nor-

▶ Video no. 29 (Series A), "An Exothermic Reaction: The Goldschmidt Process."

▶ Demonstration of the thermite reaction: Bassam Z. Shakhashiri, *Chemical Demonstrations*, Vol. 1 (Madison: University of Wisconsin Press, 1983), no. 1.36, pp. 85–89.

**Figure 22.16**
**Gem-quality corundum.**
This is the Logan sapphire, on exhibit at the Smithsonian Institution in Washington, D.C. It is the largest sapphire on public display in the United States.

**Figure 22.17**
**Aluminum products.**
Common items made from aluminum metal.

▶ The ruby laser is described in the essay at the end of Section 7.3.

**Figure 22.18**
**Protective oxide coating on aluminum.** In air, aluminum metal forms an adherent coating of aluminum oxide that protects the metal from further oxidation. Note the shiny surface of the metal plate on the top. When a similar plate of aluminum on the bottom is coated with mercury, however, needle-like crystals of aluminum oxide form at the aluminum–mercury surface. Instead of adhering to the metal, the aluminum oxide flakes away from the metal, and the aluminum continuously oxidizes in air.

mally corrode, or rust, makes the metal extremely useful in many practical applications. However, this property is also the source of an environmental problem. Tin cans (nowadays mostly steel) disintegrate quickly in the environment through rusting, but aluminum cans remain intact for decades. The solution to this problem is to recycle aluminum cans, which also saves on the energy otherwise required in the electrolytic production of the metal.

Aluminum metal is also used to produce other metals. The **Goldschmidt process** is *a method of preparing a metal by reduction of its oxide with powdered aluminum.* Chromium metal is obtained this way; the reaction is highly exothermic, because of the large negative heat of formation of $Al_2O_3$:

$$Cr_2O_3(s) + 2Al(s) \longrightarrow Al_2O_3(l) + 2Cr(l); \ \Delta H° = -536 \text{ kJ}$$

A similar reaction with a mixture of iron(III) oxide and aluminum powder (called *thermite*) produces iron for certain kinds of welding. Once the thermite powder is ignited, the reaction is self-sustaining and gives a spectacular incandescent shower. The reaction is also the basis of certain kinds of incendiary bombs.

**Aluminum Compounds** The most important compound of aluminum is aluminum oxide, $Al_2O_3$, or alumina. It is prepared by heating aluminum hydroxide, obtained from bauxite, at low temperature (550°C). Alumina is a white powder or porous solid. Although most alumina is used to make aluminum metal, large quantities are used for other purposes. For example, alumina is used as a carrier, or support, for many of the heterogeneous catalysts required in chemical processes, including those used in the production of gasoline (Figure 22.19).

When aluminum oxide is fused (melted) at high temperature (2045°C), it forms corundum, one of the hardest materials known. Corundum is used as an abrasive for grinding tools. When aluminum oxide is fused with small quantities of other metal oxides, synthetic sapphires and rubies are obtained. Synthetic ruby, for example, contains about 2.5% chromium oxide, $Cr_2O_3$. Ruby is used in fine instrument bearings ("jewel" bearings) and to make lasers. ◀

Aluminum oxide is used in the manufacture of industrial ceramics. Industrial ceramics are materials made by high-temperature firing (heating) of minerals or inorganic substances. (The term *ceramics* derives from the Greek word *kerimikos*, which means "of pottery," referring to objects made by firing clay.) Ceramics made from aluminum oxide are used to line metallurgical furnaces, and the white ceramic material in automobile spark plugs is made from aluminum oxide. Ceramic fibers composed of aluminum oxide with other metal oxides have been developed for special applications, including ceramic-fiber-reinforced aluminum.

Aluminum sulfate octadecahydrate, $Al_2(SO_4)_3 \cdot 18H_2O$, is the most common soluble salt of aluminum. It is prepared by dissolving bauxite in sulfuric acid. The salt is acidic in aqueous solution. In water, aluminum ion forms a strong hydration complex, $Al(H_2O)_6^{3+}$, and this ion in turn hydrolyzes.

$$Al(H_2O)_6^{3+}(aq) + H_2O(l) \rightleftharpoons Al(H_2O)_5OH^{2+}(aq) + H_3O^+(aq)$$

Large quantities of aluminum sulfate are used in the paper industry. Printing papers require the addition of various materials, such as clay and rosin (a tree resin), to improve the capacity of the paper to hold ink without spreading. Colloidal suspensions (suspensions of extremely fine particles) of clay and rosin in water are coagulated, or aggregated, onto the paper fibers with aluminum sulfate. The aluminum ion does make the papers acidic, however, so paper made by this process deteriorates over time (Figure 22.20). ◀

**Figure 22.19**
**Heterogeneous catalysts.**
Many heterogeneous catalysts use aluminum oxide as a carrier, or support.

▶ Coagulation of colloids by multiply charged ions is discussed in Section 12.9.

**Figure 22.20**
**Deterioration of paper in books.** Aluminum sulfate and similar acidic compounds mixed with materials such as clay and rosin are added to paper to improve its printing characteristics. However, the acids decompose the cellulose fibers in paper, causing the paper to deteriorate over time. The Library of Congress estimates that 25% of its collection is brittle from such deteriorating paper. Calcium carbonate, a basic substance, has been suggested as an alternative in preparing paper.

Aluminum sulfate is also used to treat the wastewater obtained from the process of making paper pulp, a water slurry of fibers obtained from wood. Aluminum sulfate and a base, such as calcium hydroxide, are added to the wastewater, and a gelatinous precipitate of aluminum hydroxide forms.

$$Al^{3+}(aq) + 3OH^-(aq) \longrightarrow Al(OH)_3(s)$$

Colloidal particles of clay and other substances adhere to the precipitate, which is then filtered from the water. The same procedure is one of the steps in the purification of municipal water supplies. The process removes colloidal particles of clay and some bacteria.

Aluminum hydroxide is amphoteric. With acids, aluminum hydroxide acts as a base, as most metal hydroxides do. The reaction is simply a neutralization.

$$Al(OH)_3(s) + 3H_3O^+(aq) \longrightarrow Al^{3+}(aq) + 6H_2O(l)$$

With bases, aluminum hydroxide forms a hydroxo ion (tetrahydroxoaluminate ion, often simply called the aluminate ion). The acidic behavior of the hydroxide is expected of a nonmetal hydroxide.

$$Al(OH)_3(s) + OH^-(aq) \longrightarrow Al(OH)_4^-(aq)$$

The amphoteric character of aluminum hydroxide is a reflection of the partial nonmetallic character of aluminum. Table 22.5 lists the major uses of aluminum compounds.

## Tin and Lead

Tin and lead were both known in ancient times. The Egyptians used lead coins and made lead sculptures perhaps as early as 5000 B.C. The use of tin in the form of bronze, an alloy of copper and tin, dates from about 3500 B.C.

Tin is a relatively rare element (ranking 50th or so in abundance in the earth's crust). However, the element occurs in localized deposits of the tin ore cassiterite ($SnO_2$), so the metal is much more common than you might expect from its abundance in the earth's crust. Lead is more abundant than tin. Its most important ore is galena, a lead(II) sulfide mineral (PbS). Galena is a common mineral and often occurs in association with other important metallic elements, such as silver.

**Table 22.5**
**Uses of Aluminum Compounds**

| Compound | Use |
| --- | --- |
| $Al_2O_3$ | Source of aluminum and its compounds |
| | Abrasive |
| | Refractory bricks and furnace linings |
| | Synthetic sapphires and rubies |
| $Al_2(SO_4)_3 \cdot 18H_2O$ | Making of paper |
| | Water purification |
| $AlCl_3$ | Preparation of aluminum |
| | Catalyst in organic reactions |
| $AlCl_3 \cdot 6H_2O$ | Antiperspirant |

**Figure 22.21**
**Allotropes of tin.** A bar of metallic tin (white tin) was cooled to −45°C in a solution of tin(IV) ion catalyst. After several hours, an area of gray tin formed on the bar, then grew rapidly until the bar broke apart.

▶ **Experiments in General Chemistry, 21, "Qualitative Analysis of $Mg^{2+}$, $Ca^{2+}$, $Ba^{2+}$, and $Al^{3+}$ Ions."**

As noted earlier, metallic character increases in moving down a group of elements. Tin, which is between the metalloid germanium and the metal lead in Group IVA, illustrates this periodic trend in a very interesting way: it has two different forms, or allotropes; one is a metal and the other is a nonmetal.

The nonmetallic form of tin, called *gray tin*, is a brittle, gray powder. The metallic form of tin is called *white tin*. White tin is stable above 13°C, but at lower temperatures white tin slowly undergoes a transition to gray tin.

$$\text{White tin} \underset{}{\overset{13°C}{\rightleftharpoons}} \text{gray tin}$$

The transition from white tin to gray tin becomes more rapid if the temperature is much lower than 13°C. During a cold winter in the 1850s, the tin pipes of some church organs in Russia and other parts of Europe began crumbling from what was described then as "tin disease." Tin disease, as we now know, is simply the transition from white to gray tin (Figure 22.21).

**Metallurgy of Tin and Lead**  Tin metal is obtained from cassiterite. Purified tin(IV) oxide, $SnO_2$, from tin ore is reduced to the metal by heating with carbon in a furnace.

$$SnO_2(s) + 2C(s) \overset{\Delta}{\longrightarrow} Sn(l) + 2CO(g)$$

Lead metal is obtained from ores containing galena. The ore is first concentrated in the lead(II) sulfide mineral using flotation. The concentrated ore is then roasted; that is, the sulfide ore is burned in air to yield lead(II) oxide.

$$2PbS(s) + 3O_2(g) \longrightarrow 2PbO(s) + 2SO_2(g)$$

The fused mass from the roasting is broken up, mixed with coke (carbon), and fed into a blast furnace. Here the lead(II) oxide is reduced with carbon monoxide produced in the blast furnace by partial oxidation of the carbon.

$$PbO(s) + CO(g) \longrightarrow Pb(l) + CO_2(g)$$

Tin is used to make tin plate, which is steel (iron alloy) sheeting with a thin coating of tin. Tin plate is used for food containers ("tin cans"). The tin coating protects the iron from reaction with air and food acids. Tin is also used to make a number of alloys. Solder is a low-melting alloy of tin and lead; bronze is an alloy of copper and tin (Figure 22.22).

▶ **Lead storage batteries are discussed in Section 20.8.**

More than half of the lead produced is used to make electrodes for lead storage batteries. ◀ The manufacture of military and sporting ammunition also consumes a significant fraction of the lead produced. Because lead resists attack from many corrosive substances, the metal is also used to make chemical plant equipment.

**Reactions of Tin and Lead Metals**  Tin and lead are much less reactive than the metals of Groups IA, IIA, and IIIA. Whereas aluminum reacts vigorously with dilute hydrochloric and sulfuric acids, tin reacts only slowly with these acids. Tin reacts more rapidly with the concentrated acids. The products are tin(II) ion and hydrogen:

$$Sn(s) + 2HCl(aq) \longrightarrow SnCl_2(aq) + H_2(g)$$
$$Sn(s) + H_2SO_4(aq) \longrightarrow SnSO_4(aq) + H_2(g)$$

Lead metal reacts with these acids, but the products $PbCl_2$ and $PbSO_4$ are insoluble and adhere to the metal. As a result, the reaction soon stops.

$$Pb(s) + 2HCl(aq) \longrightarrow PbCl_2(s) + H_2(g)$$
$$Pb(s) + H_2SO_4(aq) \longrightarrow PbSO_4(s) + H_2(g)$$

**Figure 22.22**
**Tin alloys.** Shown here are solder (and a soldering iron) and bronze.

**Table 22.6**
**Uses of Tin and Lead Compounds**

| Compound | Use |
|---|---|
| $SnO_2$ | Manufacture of tin compounds |
| | Glazes and enamels |
| $SnCl_2$ | Reducing agent in preparing organic compounds |
| $PbO$ | Lead storage batteries |
| | Lead glass |
| $PbO_2$ | Cathode in lead storage batteries |
| $Pb_3O_4$ | Pigment for painting structural steel |

**Tin and Lead Compounds** Tin(II) chloride, $SnCl_2$, is used as a reducing agent in the preparation of dyes and other organic compounds. In reactions where tin(II) ion acts as a reducing agent, tin(II) ion oxidizes to tin(IV) species. Tin(II) compounds are commonly referred to as *stannous compounds,* using an older naming system. Thus, tin(II) chloride is commonly called stannous chloride. Tin also has a number of tin(IV), or stannic, compounds. Tin(IV) oxide, $SnO_2$, is, as we have seen, the chemical substance in the mineral cassiterite. Tin(IV) chloride, $SnCl_4$, is a liquid; it freezes at $-33°C$. (These properties indicate that the substance is molecular, rather than ionic.)

Lead also exists in compounds in the $+2$ and $+4$ oxidation states, although lead(II) compounds are the more common. The starting compound for preparing most lead compounds is lead(II) oxide, $PbO$. This is a reddish-yellow solid, commercially called *litharge.* Lead(II) oxide is prepared by exposing molten lead to air. Lead(IV) oxide, $PbO_2$, is a dark brown or black powder; it forms the cathode of lead storage batteries. The cathode is made by packing a paste of lead(II) oxide (litharge) into a lead metal grid. When the battery is charged, the lead(II) oxide is oxidized to lead(IV) oxide. Table 22.6 gives a summary list of the uses of tin and lead compounds.

# Chemistry of the Nonmetals

The nonmetals are elements that do not exhibit the characteristics of a metal. Nearly half of them are colorless gases; and until the eighteenth century, gases were referred to as "airs" and were not well differentiated. In contrast, the special characteristics of metals held special fascination for early humans—only fire was more fascinating. And of the two nonmetals known to the ancients, carbon and sulfur, both are associated with fire.

Carbon was known in the form of charcoal and lampblack, or soot, which are products of fire. Charcoal may have been so common that it was hardly noticed at first, until its role in the reduction of metal ores was discovered. Lampblack was used by the ancient Egyptians to produce ink for writing on papyrus.

Free sulfur was less widely available than carbon, although it was probably well known because of its ready occurrence in volcanic areas. No doubt its yellow color made it stand out among other rocks. That sulfur also burned with a beautiful blue flame made it especially distinctive. The old English name of sulfur was *brimstone,* which means "a stone that burns." This term survives today in the expression "fire and brimstone."

**Figure 22.23
Some nonmetals.** *Left to right:* Sulfur, bromine, white phosphorus, and carbon.

These two nonmetals have rather distinctive physical characteristics, as do some of the nonmetals discovered later. For example, white phosphorus is a waxy, white solid; bromine is a reddish-brown liquid; chlorine is a greenish-yellow gas. Some nonmetallic elements are shown in Figure 22.23.

In the remainder of this chapter, we look at the chemistry of the nonmetals. Some questions we will address are these: What are the chemical and physical properties of the more important nonmetallic elements? What are some of the commercial uses of these substances and their compounds?

## 22.5 Hydrogen

Hydrogen is the most abundant element in the universe, comprising nearly 90% of all atoms and is the third most abundant element on the surface of the earth. (Oxygen and silicon are the most abundant.) Most stars, including our sun, consist primarily of hydrogen. The hydrogen in our sun is the fuel for the fusion reactions that produce the life-sustaining energy that reaches our planet. On earth, the majority of hydrogen is found in oceans combined with oxygen as water. Hydrogen occurs in a variety of compounds that you have had the opportunity to study as a part of your general chemistry course. These include the organic compounds (Chapter 24), biologically important compounds (Chapter 25) and acids and bases (Chapters 4 and 15).

### Properties and Preparation of Hydrogen

Hydrogen was first isolated and identified by Henry Cavendish in 1766. His experiments consisted of reacting iron, zinc, and tin with several different binary acids. For example, $H_2(g)$ can easily be produced on a small scale according to the reaction

$$2HCl(g) + Zn(s) \longrightarrow ZnCl_2(aq) + H_2(g)$$

Hydrogen is a colorless, odorless gas that is less dense than air. Even though hydrogen is often placed in Group IA of the periodic table, aside from its valence electron configuration, it has little in common with the alkali metals that make up the rest of Group IA. It is much less likely to form a cation than any of the alkali metals, with a first ionization energy of 1312 kJ/mol versus 520 kJ/mol for lithium. Because of this, hydrogen generally combines with nonmetallic elements to form covalent compounds such as $CH_4$, $H_2S$, and $PH_3$.

There are three isotopes of hydrogen: *protium*, $_1^1H$ or H, which is the most abundant; *deuterium*, $_1^2H$ or D; and *tritium*, $_1^3H$ or T. All three isotopes are naturally occurring; however, only 0.0156% is D and a trace is T. Because an atom of D has about twice the mass of protium, compounds that contain deuterium often have different properties than those that contain only protium. For example, the normal boiling point of $D_2O$ is 101.42°C versus 100.00°C for $H_2O$. Tritium is hydrogen produced naturally in the upper atmosphere by nuclear reactions that are induced by cosmic rays or in a nuclear reactor by bombarding lithium-6 with neutrons:

$$_3^6Li + _0^1n \longrightarrow _1^3H + _2^4He$$

Tritium is radioactive with a half-life of 12.3 years, hence very little of what is naturally produced in the upper atmosphere reaches the surface of the earth.

The isotopes of hydrogen find many applications. They are used as markers or labels that can be followed during chemical reactions. For example, if a chemist were interested in determining if hydrogen atoms move between water molecules, he could make a solution that contains DOD and HOH. If hydrogen atom transfer takes place (which is the case in this example), then the solution, after a period of time, would be expected to contain DOD, HOH, and the new compound DOH. Information about the exchange rate of H with D can also be obtained from the experiment by measuring the rate at which the DOH forms. Because so many compounds contain hydrogen, the hydrogen isotopes are widely used in this manner.

The elemental form of hydrogen is a diatomic molecule having a bond dissociation energy of 432 kJ/mol. This is a large value when compared to chlorine at 240 kJ/mol. This relatively high bond dissociation energy indicates why hydrogen is less reactive than its halogen counterparts. However, with the addition of heat or light, or in the presence of a suitable catalyst, hydrogen can be induced to react.

Hydrogen is currently produced on a massive industrial scale, with an annual U.S. production on the order of $10^{10}$ m$^3$. Approximately 40% of this production is used to manufacture ammonia via the following chemical reaction with nitrogen at high temperature and pressure:

$$N_2(g) + 3H_2(g) \longrightarrow 2NH_3(g)$$

The hydrogen for this reaction is generally prepared using the **steam-reforming process** *where steam and hydrocarbons from natural gas or petroleum react at high temperature and pressure in the presence of a catalyst to form carbon monoxide and hydrogen.* For example:

$$C_3H_8(g) + 3H_2O(g) \xrightarrow[\Delta]{Ni} 3CO(g) + 7H_2(g)$$

Another viable route for hydrogen production is the *water–gas reaction,* which is no longer used commercially but may become important in the future as natural gas and petroleum become more expensive and scarce. In this reaction, steam is passed over red-hot coke or coal.

$$C(s) + H_2O(g) \longrightarrow CO(g) + H_2(g)$$

Both of these reactions produce a mixture of hydrogen and carbon monoxide. Such mixtures are used to produce various organic compounds, but to obtain pure hydrogen the carbon monoxide must be removed. First the carbon monoxide is reacted with steam in the presence of a catalyst to give carbon dioxide and more hydrogen.

$$CO(g) + H_2O(g) \xrightarrow[\Delta]{catalyst} CO_2(g) + H_2(g)$$

The carbon dioxide is then removed by passing the mixture of gases through a basic aqueous solution.

Hydrogen can also be produced via the electrolysis of water. Water that contains a small amount of an acid can undergo the following reaction with an $E_{cell}^{\circ} = -1.23$ V.

$$2H_2O(l) \longrightarrow 2H_2(g) + O_2(g)$$

Due to the cost of electricity, this reaction is not economically feasible when more than a very small amount of hydrogen is required. However, electrolysis does become economically viable on an industrial scale when the hydrogen is produced as a by-product of the electrolysis of concentrated aqueous NaCl solutions during the production of NaOH and chlorine. ◄

► See Section 20.10 for a discussion of the electrolysis of sodium chloride solutions.

## Hydrogen Reactions and Compounds

In addition to the preparation of ammonia, the other major use of hydrogen is in the petrochemical industry. In many cases the reaction is one where hydrogen is added to hydrocarbon compounds containing carbon–carbon double bonds to produce compounds that contain carbon–carbon single bonds. For example, 1-butene can be reacted with hydrogen using a platinum or palladium catalyst to produce butane.

$$CH_3CH_2CH{=}CH_2 + H_2 \xrightarrow{\text{Pt}} CH_3CH_2CH_2CH_3$$

This process, called *hydrogenation,* is used in the food processing industry where oils (liquids) that contain many carbon–carbon double bonds are converted to fats (solids) that contain few or no carbon–carbon double bonds. Another important process that requires hydrogen is the cobalt-catalyzed *synthesis gas* reaction with carbon monoxide to produce methyl alcohol.

$$CO(g) + 2H_2(g) \xrightarrow{\text{cobalt catalyst}} CH_3OH(g)$$

Hydrogen is also used to reduce metal oxides to extract pure metals. For example, molybdenum and tungsten oxides can be reduced at high temperatures via the reaction

$$MO_3(s) + 3H_2(g) \longrightarrow M(s) + 3H_2O(g)$$

Industrial processes are emerging that allow this reaction to be used for iron oxides as well.

When hydrogen combines with another element it forms a **binary hydride**—that is, *a compound that contains hydrogen and one other element.* There are three categories of binary hydrides: ionic hydrides, covalent hydrides, and metallic hydrides.

*Ionic hydrides,* which contain the hydride ion, $H^-$, can be directly formed via the reaction of an alkali metal or the larger Group IIA metals (Ca, Sr, and Ba) with hydrogen gas near 400°C.

$$2Li(s) + H_2(g) \longrightarrow 2LiH(s)$$
$$Ba(s) + H_2(g) \longrightarrow BaH_2(s)$$

These hydrides are white crystalline compounds in which the H atoms have an oxidation state of $-1$. Ionic hydrides can undergo an oxidation–reduction reaction with water to produce hydrogen and a basic solution. For example:

$$LiH(s) + H_2O(l) \longrightarrow H_2(g) + LiOH(aq)$$

Because of this, hydrides can be used as a source of hydrogen gas where transportation of $H_2(g)$ is impractical, such as for inflating weather balloons. Ionic hydrides are also used as reducing agents (a source of electrons) during chemical reactions.

*Covalent hydrides* are molecular compounds in which hydrogen is covalently bonded to another element. Examples of these compounds are $NH_3$, $H_2O$, $H_2O_2$, and HF. Some of these compounds often can be formed from the direct reaction of the elements. If the nonmetal reacting with hydrogen is reactive, the reaction will readily occur without the need for elevated temperatures or a catalyst:

$$F_2(g) + H_2(g) \longrightarrow 2HF(g)$$

The reaction of hydrogen with oxygen to form water is an example of a reaction that requires the input of energy to get started; however, once it does, the reaction is rapid and exothermic.

$$2H_2(g) + O_2(g) \longrightarrow 2H_2O(g) \qquad \Delta H = -484 \text{ kJ}$$

Because it is such an exothermic reaction and the product is a gas, it is an ideal rocket fuel (Figure 22.24).

The combustion of hydrogen produces more heat per gram than any other fuel (120 kJ/g). Unlike hydrocarbons, it is a "clean fuel" because the product (water) is environmentally benign. These features, in the face of a dwindling supply of hydrocarbons, indicate that hydrogen gas may become the favorite fuel of the twenty-first century.

*Metallic hydrides* are compounds containing a transition metal and hydrogen. Generally, the formula of these compounds is $MH_x$, where $x$ is often not an integer. These compounds contain hydrogen atoms that are spread throughout a metal crystal occupying the holes in the crystal lattice. Often, hydrogen atoms enter the holes in nonstoichiometric amounts. The result is that the composition of the metallic hydride is variable. For example, under one pressure of $H_2$ the composition of the metallic hydride $MH_x$ might be $MH_{0.4}$, whereas at a higher pressure of $H_2$ it might be $MH_{0.5}$. One way to think about these compounds is to consider the metal crystal to be similar to an apartment building with the lattice holes representing the individual apartments. The occupancy of an apartment building is not fixed; it is instead a function of factors like the rent, the condition of the building, and so on. Examples of these compounds are $TiH_{1.7}$ and $ZrH_{1.9}$.

**Figure 22.24**
**A liquid-hydrogen storage tank.** Liquid hydrogen is used as a rocket fuel.

## 22.6 Group IVA: The Carbon Family

The elements of Group IVA show in more striking fashion than the previous groups the normal periodic trend of greater metallic character going down a column. Carbon, the first element in the group, is distinctly nonmetallic. Silicon and germanium are metalloids, or semimetals, although their chemical properties are primarily those of nonmetals. Tin and lead, the last two elements in the group, are metals.

In the following subsections, we will look at carbon and silicon. Both elements have many important compounds that exhibit tetrahedral ($sp^3$) bonding. Carbon, however, also has many compounds that contain multiple bonds in which $sp$ and $sp^2$ bonding occurs; silicon has very few such compounds. Another difference in the bonding characteristics of these elements is silicon's ability to use $sp^3d^2$ hybrid orbitals in octahedral bonding, which is not possible with carbon. The hexafluorosilicate ion, $SiF_6^{2-}$, is an example of such bonding.

### Carbon

▶ Due to the extensive coverage of these elements elsewhere in the text (carbon in Chapters 13 and 24, and silicon in Chapter 13), this discussion focuses on the properties and chemistry not presented in other chapters.

One of the most important features of the carbon atom is its ability to bond to other carbon atoms to form chains and rings of enormous variety. *The covalent bonding of two or more atoms of the same element to one another* is referred to as **catenation.** Although other elements display catenation, none show it to the same degree as carbon. Millions of carbon compounds are known, most classified as *organic.* These can be thought of as derivatives of hydrocarbons, each consisting of a chain (or more complicated arrangement) of carbon atoms to which hydrogen atoms are bonded. In this section, we will confine ourselves to the element and its oxides and carbonates.

**Allotropes of Carbon**   Until recently, carbon was thought to occur in only two principal allotropic forms: diamond and graphite. Both allotropes are covalent network solids, whose structures we discussed in detail in Section 11.8. In diamond, each carbon atom is tetrahedrally ($sp^3$) bonded to four other carbon atoms. To move one plane of atoms in the diamond crystal relative to another requires the breaking of many strong carbon–carbon bonds. Because of this, diamond is one of the hardest substances known. As a pure substance, diamond is colorless, although natural diamond may be colored by impurities.

Graphite is a black substance having a layer structure. Each layer consists of carbon atoms bonded to three other carbon atoms to give a hexagonal pattern of carbon atoms arranged in a plane. The bonding involves $sp^2$ hybridization of the carbon atoms with delocalized $\pi$ electrons. You can also describe the bonding in terms of resonance formulas with alternating single and double bonds. One layer of carbon atoms in graphite is held to another layer only by van der Waals forces. Because of the relative weakness of these forces, the layers in graphite easily slide over one another, resulting in a substance that is soft and slippery. Graphite, unlike diamond, is a good electrical conductor, because of the delocalized bonding within layers.

In 1985 a third allotropic form of carbon, known as buckminsterfullerene ($C_{60}$), was discovered. This molecular form of carbon has been studied intensively. The molecule has a stable "soccer-ball" structure, described in Chapter 13.

**Carbon Black**   The form of carbon known as *carbon black* is composed of extremely small crystals of carbon having an amorphous, or imperfect, graphite structure. It is produced in large quantities by burning natural gas ($CH_4$) or petroleum hydrocarbons in a limited supply of air so that heat "cracks" (or breaks bonds in) the hydrocarbon. Lampblack, mentioned earlier, is a form of carbon black.

$$CH_4(g) \xrightarrow{\Delta} \underset{\text{carbon black}}{C(s)} + 2H_2(g)$$

Carbon black is used in the manufacture of rubber tires (to increase wear) and as a pigment in black printing inks. Coke is an amorphous carbon obtained by heating coal in the absence of air; it is used in large quantities in the making of iron.

**Oxides of Carbon**    Carbon has two principal oxides: carbon monoxide, CO, and carbon dioxide, $CO_2$. Carbon and organic compounds burn in an excess of oxygen to give carbon dioxide, $CO_2$. However, an equilibrium exists among carbon, carbon monoxide, and carbon dioxide that favors carbon monoxide above 700°C.

$$CO_2(g) + C(s) \rightleftharpoons 2CO(g)$$

For this reason, carbon monoxide is almost always one of the products of combustion of carbon and organic compounds, unless an excess of oxygen is present, in which case the carbon monoxide burns to carbon dioxide.

These two oxides of carbon are quite different in their chemical and physiological properties. Carbon monoxide is a colorless, odorless gas that burns in air with a blue flame. It is a toxic gas, which poisons by attaching strongly to iron atoms in the hemoglobin of red blood cells, preventing them from carrying out their normal oxygen-carrying function. As a result, the cells of the body become starved for oxygen.

Carbon monoxide is manufactured industrially from natural gas ($CH_4$) and petroleum hydrocarbons, either by reaction with steam or by partial oxidation. For example,

$$CH_4(g) + H_2O(g) \xrightarrow{Ni} CO(g) + 3H_2(g)$$
$$2CH_4(g) + O_2(g) \longrightarrow 2CO(g) + 4H_2(g)$$

The product in these reactions is a mixture of carbon monoxide and hydrogen, which is called *synthesis gas*. Synthesis gas can yield any of a number of organic products depending on the reaction conditions and catalyst. Methanol ($CH_3OH$), for example, is produced in large quantities from synthesis gas.

$$CO(g) + 2H_2(g) \xrightarrow{\text{catalyst}} CH_3OH(l)$$

Carbon dioxide is a colorless, odorless gas with a faint acid taste. Under normal circumstances, the gas is nontoxic, although at high concentrations it interferes with respiration. Carbon dioxide does not support most combustions, which is why $CO_2$ fire extinguishers are used in many fire-fighting situations.

Carbon dioxide is produced whenever carbon or organic materials are burned. For example,

$$CH_4(g) + 2O_2(g) \longrightarrow CO_2(g) + 2H_2O(g)$$
$$CS_2(l) + 3O_2(g) \longrightarrow CO_2(g) + 2SO_2(g)$$
$$C_2H_5OH(l) + 3O_2(g) \longrightarrow 2CO_2(g) + 3H_2O(g)$$

Carbon dioxide is obtained commercially as a by-product in the production of ammonia (see Section 22.7) and in the calcining (strong heating) of limestone to give calcium oxide. Liquid carbon dioxide and solid carbon dioxide (dry ice) are used in large quantities as refrigerants. Carbonated beverages are made by dissolving carbon dioxide gas under pressure in an aqueous solution of sugar and flavorings. Carbonated water is acidic as the result of the formation of carbonic acid,

**Figure 22.25**
**Test for carbon dioxide.**
When carbon dioxide is bubbled into a solution of calcium hydroxide (limewater), a milky white precipitate of calcium carbonate forms. This is the basis of a test for carbon dioxide. The reaction is also used to manufacture pure calcium carbonate. (The needle at the right is to provide an escape for excess gas.)

although carbonated beverages frequently also contain fruit acids and phosphoric acid.

**Carbonates** Carbon dioxide dissolves in water to form an aqueous solution of carbonic acid.

$$CO_2(g) + H_2O(l) \rightleftharpoons H_2CO_3(aq)$$

The acid is diprotic (has two acidic H atoms per molecule) and dissociates to form hydrogen carbonate ion and carbonate ion.

$$H_2CO_3(aq) + H_2O(l) \rightleftharpoons H_3O^+(aq) + HCO_3^-(aq)$$
$$HCO_3^-(aq) + H_2O(l) \rightleftharpoons H_3O^+(aq) + CO_3^{2-}(aq)$$

Carbonic acid has never been isolated from solution, but its salts, hydrogen carbonates and carbonates, are well known. When you bubble carbon dioxide gas into an aqueous solution of calcium hydroxide, a milky white precipitate of calcium carbonate forms (Figure 22.25).

$$CO_2(g) + Ca(OH)_2(aq) \longrightarrow CaCO_3(s) + H_2O(l)$$

This is a standard test for carbon dioxide. The reaction is also used to manufacture a pure calcium carbonate for antacids and other products.

Carbonate minerals are very common and many are of commercial importance. Limestone contains the mineral calcite, which is calcium carbonate, $CaCO_3$. Much of this was formed by marine organisms, although some limestone was also formed by direct precipitation from water solution.

Table 22.7 lists uses of some compounds of carbon (and of silicon, discussed in the next section).

### Silicon

Silicon is the second most abundant element in the earth's crust, following only oxygen. Almost 95% of the crustal rocks and materials that make up the earth contain silicon. It has been important to humans ever since flint, a silicon dioxide mineral, was first used to fashion tools and weapons by early civilizations. The Romans took advantage of silicon-containing minerals when they discovered how to use them

**Table 22.7**
**Uses of Some Compounds of Carbon and of Silicon**

| Compound | Use |
| --- | --- |
| CO | Fuel; reducing agent |
| | Synthesis of methanol, $CH_3OH$ |
| $CO_2$ | Refrigerant |
| | Carbonation of beverages |
| $SiO_2$ | Source of silicon and its compounds |
| | Abrasives |
| | Glass |
| $(CH_3)_2SiCl_2$ | Manufacture of silicones (used as lubricants, hydraulic fluids, caulking compounds, and medical implants) |

**Figure 22.26**
**Molecular model of a silicone.** This model shows a fragment of the polydimethylsiloxane molecule. Note how the chain structure is created by linking $(CH_3)_2SiO_2$ tetrahedra through Si—O bonds.

to make concrete. Today, our computer industry depends on obtaining pure silicon for the manufacture of integrated circuits and semiconductors.

**Compounds That Contain Silicon and Oxygen**  The variety and complexity of silicon-containing compounds are immense. However, unlike carbon, which involves a profusion of compounds with atoms of the element bonding to one another, compounds containing chains and networks of silicon–oxygen bonds dominate silicon chemistry. Chapter 13 contains a discussion of the element and many of the important Si—O compounds, which include silica and the silicates. The chemical and physical stability of these compounds can be attributed to the strong Si—O bonds, which have an average bond dissociation energy of 460 kJ/mol. In contrast, a single C—O bond is about 20% weaker.

An important class of compounds that contain silicon and oxygen are the **silicones,** which are *synthetic compounds that contain chains or rings of Si—O bonds with one or more of the bonding positions on each Si atom occupied by an organic group*. Typically, silicones are large molecules that consist of a chain of alternating Si—O bonds with two organic groups bonded to each silicon atom in the chain. This structure is similar to the structure of the silicates where two Si—O silicate bonds of each $SiO_4$ unit are replaced by two Si—C bonds. Figure 22.26 depicts a portion of a silicone chain. The preparation of many silicone compounds begins with the reaction of Si with methyl chloride at elevated temperature in the presence of a copper catalyst.

$$Si(s) + 2CH_3Cl(g) \xrightarrow[300°C]{\text{catalyst}} (CH_3)_2SiCl_2(g)$$

If a long chain of Si—O bonds is desired, the product of this reaction is reacted with water.

$$(CH_3)_2SiCl_2(l) + 2H_2O(l) \longrightarrow (CH_3)_2Si(OH)_2(l) + 2HCl(g)$$

The silicon-containing product of this reaction then undergoes a condensation reaction to form a silicone oil (polydimethylsiloxane) and water.

$$n(CH_3)Si(OH)_2 \longrightarrow \begin{bmatrix} CH_3 \\ | \\ Si—O— \\ | \\ CH_3 \end{bmatrix}_n + nH_2O$$

Depending on the application for the silicone, the chain length can be varied to make different weight oils, or the chains can be linked together to form *elastomers* (rubbers). The methyl group, $CH_3$, can also be replaced by a different organic group, which can lead to the formation of hard materials called resins.

Silicones have a wide variety of applications (Figure 22.27). Due to the strong Si—O and Si—C bonds, silicones are generally unreactive and can be synthesized to

**Figure 22.27**
**Products that contain silicone.**

**Figure 22.28**
**Silane bonding.**  The Lewis structures of three different compounds with the formula $Si_5H_{12}$, each with a different bonding arrangement of the silicon atoms.

perform in temperature ranges of $-75°F$ to $400°F$. Silicone oils can be found in cosmetics, car wax, and hydraulic fluids. The silicone elastomers are used for medical tubing, heart valve implants, electrical tape, caulk, and gaskets. Silicone resins are used to insulate electrical equipment and in the molding process of electronic circuit boards. These examples represent only a small portion of the uses for silicone compounds. Because of this extensive use, several hundred thousand tons of silicones are manufactured each year.

**Silicon Hydrides (Silanes) and Silicon Halides**  Silicon hydrides and silicon halides are synthetic silicon compounds that have the distinction of not containing Si—O bonds. Like carbon, which forms hydrocarbons, silicon can be made to form compounds with hydrogen called *silanes*. Silanes are straight chain and branched chain compounds with the general formula $Si_nH_{2n+2}$ with $n$ being a number in the range of 1 to 8. Figure 22.28 illustrates the bonding arrangements in these compounds. Silanes can also form rings, making $Si_5H_{10}$ and $Si_6H_{12}$. All of the silanes are colorless gases or liquids that explode or ignite when they come in contact with air. This is in contrast to the hydrocarbons, which only combust in air if ignited. In all cases, the silanes are much more reactive than their carbon analogs.

Silicon reacts with all of the halogens to form silicon halides. For example, $SiCl_4$ can be formed by the reaction

$$Si(s) + 2Cl_2(g) \longrightarrow SiCl_4(g)$$

This reaction is of particular use to the semiconductor industry because the $SiCl_4$ product can be used as a source of ultrapure silicon. All of the silicon halides are very reactive and volatile.

## 22.7  Group VA: Nitrogen and the Phosphorus Family

Like the carbon family of elements, the Group VA elements show the distinct trend of increasing metallic character as you go from top to bottom of the column. The first members, nitrogen and phosphorus, are nonmetallic; arsenic and antimony are metalloids; bismuth is a metal.

As expected for a second-period element, nitrogen is in many respects different from the other elements in its group. You can see this in the formulas of the elements and compounds. Elementary nitrogen is $N_2$, while white phosphorus is $P_4$. Similarly, the common +5 oxoacid of nitrogen is $HNO_3$, that of phosphorus is $H_3PO_4$, and that of arsenic is $H_3AsO_4$. [Nitrogen exists in many molecular compounds in a variety of oxidation states. Phosphorus exists in many compounds in the +5 and +3 oxidation states. Many phosphorus compounds have P—O—P bonding.]

### Nitrogen

The element nitrogen is crucial to life: it is a component of all proteins, which are involved in almost every biochemical process that occurs in living organisms. Most of the available nitrogen on earth, however, is present as nitrogen gas (dinitrogen, $N_2$) in the atmosphere, which consists of 78.1% $N_2$, by mass. Dinitrogen, also simply called "nitrogen," has collected in the atmosphere because of its relative chemical unreactivity. Most organisms cannot use dinitrogen from the atmosphere as their source of the element. However, certain soil bacteria, as well as bacteria that live in nodules on the roots of beans, clover, and similar plants, can "fix" nitrogen; that is,

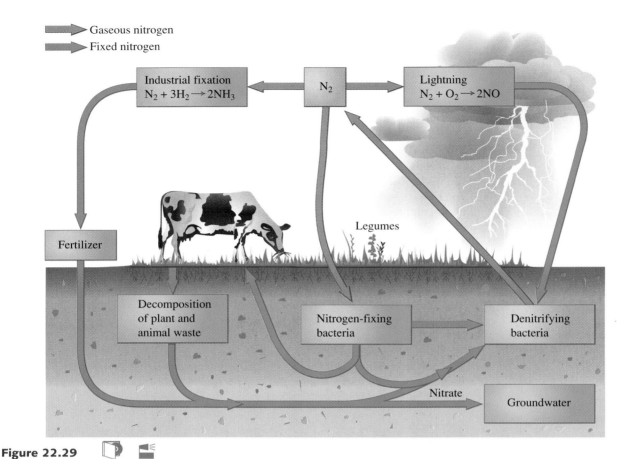

**Figure 22.29**
**The nitrogen cycle.**
Nitrogen, $N_2$, is fixed (converted to compounds) by bacteria, by lightning, and by the industrial synthesis of ammonia. Fixed nitrogen is used by plants and enters the food chain of animals. Later, plant and animal wastes decompose. Denitrifying bacteria complete the cycle by producing free nitrogen again.

they convert dinitrogen to ammonium and nitrate compounds. Plants use these simple nitrogen compounds to make proteins and other complex nitrogen compounds. Animals eat these plants, and other animals eat those animals. Finally, bacteria in decaying organic matter convert the nitrogen compounds back to dinitrogen. In this way, nitrogen in our environment continually cycles from dinitrogen to living organisms and back. Figure 22.29 depicts this *nitrogen cycle.*

**Properties and Uses of Nitrogen**   Daniel Rutherford, a chemist and physician, discovered nitrogen, $N_2$, in air in 1772. In his experiments, he removed the oxygen from air by burning a substance in it. When burning carbon-containing substances, he removed the carbon dioxide that formed by reacting it with aqueous potassium hydroxide. He then showed that the residual gas would no longer support either combustion or living organisms. Although this residual gas is mostly $N_2$, it does contain small amounts of noble gases.

Nitrogen, $N_2$, is a relatively unreactive element because of the stability of the nitrogen–nitrogen triple bond. (The $N{\equiv}N$ bond energy is 942 kJ/mol, compared with 167 kJ/mol for the $N{-}N$ bond energy.) When substances burn in air, they generally react with oxygen, leaving the nitrogen unreacted. Some very reactive metals do react directly with nitrogen, however. For example, when magnesium metal burns in air, it forms the nitride, as well as the oxide.

$$3Mg(s) \ + \ N_2(g) \longrightarrow Mg_3N_2(s)$$

Because the nitride ion, $N^{3-}$, is a very strong base, ionic nitrides react with water, producing ammonia.

$$N^{3-}(aq) + 3H_2O(l) \longrightarrow NH_3(g) + 3OH^-(aq)$$

Magnesium nitride reacts with water to give magnesium hydroxide and ammonia.

$$Mg_3N_2(s) + 6H_2O(l) \longrightarrow 3Mg(OH)_2(s) + 2NH_3(g)$$

Air is the major commercial source of nitrogen. The components of air are separated by liquefaction, followed by distillation. Nitrogen is the most volatile component in liquid air, so it is the first to distill off, leaving behind a liquid that is primarily oxygen with a small amount of noble gases (mostly argon).

Liquid nitrogen is used as a refrigerant to freeze foods, to freeze soft or rubbery materials prior to grinding them, and to freeze biological materials (Figure 22.30). Large quantities of nitrogen are also used as a *blanketing gas,* whose purpose is to protect a material from oxygen during processing or storage. Thus, electronic components are often made under a nitrogen atmosphere. The other principal use of nitrogen is to prepare nitrogen compounds.

**Nitrogen Compounds**   Ammonia, $NH_3$, is the most important commercial compound of nitrogen. A colorless gas with a characteristic irritating or pungent odor, it is prepared commercially from $N_2$ and $H_2$ by the *Haber process.* Figure 22.31 shows a flowchart of the industrial preparation of ammonia from natural gas, steam, and air. Small amounts can be prepared in the laboratory by heating a solution of an ammonium salt with a strong base, such as NaOH or $Ca(OH)_2$ (Figure 22.32).

$$NH_4Cl(aq) + NaOH(aq) \xrightarrow{\Delta} NH_3(g) + H_2O(l) + NaCl(aq)$$

Ammonia is easily liquefied, and the liquid is used as a nitrogen fertilizer. Ammonium salts, such as the sulfate and nitrate, are also sold as fertilizers. Large quantities

**Figure 22.30**
**Liquid nitrogen.** When liquid nitrogen is poured on a table (which, although at room temperature, is about 220°C above the boiling point of nitrogen), the liquid sizzles and boils away violently.

**Figure 22.31**
**Industrial preparation of ammonia (flowchart).** The raw materials are natural gas, water, and air. Hydrogen for the Haber process is obtained by reacting natural gas with steam to give carbon monoxide and hydrogen. In the next step, carbon monoxide is reacted with steam to give carbon dioxide and additional hydrogen. The carbon dioxide is removed by dissolving it in water solution.

**Figure 22.32**
**Preparation of ammonia from an ammonium salt.**
Sodium hydroxide solution was added to ammonium chloride. Ammonia gas, formed in the reaction, turns colorless phenolphthalein in the moist paper to a bright pink.

▶ The biological importance of NO was discussed in an essay in Chapter 5.

of ammonia are converted to urea, $NH_2CONH_2$, which is used as a fertilizer, as a livestock feed supplement, and in the manufacture of urea–formaldehyde plastics. Ammonia is also the starting compound for the preparation of most other nitrogen compounds.

Nitrous oxide, $N_2O$, is a colorless gas with a sweet odor. It can be prepared by careful heating of molten ammonium nitrate. (If heated strongly, it explodes.)

$$NH_4NO_3(s) \xrightarrow{\Delta} N_2O(g) + 2H_2O(g)$$

Nitrous oxide, or laughing gas, is used as a dental anesthetic. It is also useful as a propellant in whipped-cream dispensers. The gas dissolves in cream under pressure. When the cream is dispensed, the gas bubbles out, forming a foam.

Nitric oxide, NO, is a colorless gas that is of great industrial and biological importance. ◀ Although it can be prepared by the direct combination of the elements at elevated temperatures, large amounts are prepared from ammonia as the first step in the commercial preparation of nitric acid. The ammonia is oxidized in the presence of a platinum catalyst.

$$4NH_3(g) + 5O_2(g) \xrightarrow{Pt} 4NO(g) + 6H_2O(g)$$

Nitric acid, $HNO_3$, is the third most important industrial acid (after sulfuric and phosphoric acids). It is used to prepare explosives, nylon, and polyurethane plastics. Nitric acid is produced commercially by the **Ostwald process,** which is *an industrial preparation of nitric acid starting from the catalytic oxidation of ammonia to nitric oxide.* In this process, ammonia is burned in the presence of a platinum catalyst to give NO, which is then reacted with oxygen to form $NO_2$. The $NO_2$ is dissolved in water, where it reacts to form nitric acid and nitric oxide.

$$4NH_3(g) + 5O_2(g) \xrightarrow{Pt} 4NO(g) + 6H_2O(g)$$
$$2NO(g) + O_2(g) \longrightarrow 2NO_2(g)$$
$$3NO_2(g) + H_2O(l) \longrightarrow 2HNO_3(aq) + NO(g)$$

The nitric oxide produced in the last step is recycled for use in the second step.

Nitric acid is a strong oxidizing agent. Although copper metal is unreactive to most acids, it is oxidized by the nitrate ion in acid solution. In dilute acid, nitric oxide is the principal reduction product (Figure 22.33).

$$3Cu(s) + 8H_3O^+(aq) + 2NO_3^-(aq) \longrightarrow 3Cu^{2+}(aq) + 2NO(g) + 12H_2O(l)$$

**Figure 22.33**
**Reaction of copper metal with dilute nitric acid.**
*Left:* Copper metal is oxidized by nitric acid to $Cu^{2+}(aq)$ ion, which has a blue color. The principal reduction product from dilute nitric acid is nitric oxide, NO, a colorless gas.
*Right:* When the stopper is lifted from the flask, air enters, and the NO reacts with $O_2$ to produce $NO_2$, a reddish-brown gas.

**Figure 22.34**
**Reaction of copper metal with concentrated nitric acid.**   The principal reduction product from concentrated nitric acid is nitrogen dioxide, $NO_2$ (reddish-brown gas).

▶ Demonstration of reaction of nitric acid and copper: Summerlin et al., *Chemical Demonstrations*, Vol. 2 (Washington, D.C.: Am. Chem. Soc., 1987), no. 2, pp. 4–5.

With concentrated nitric acid, nitrogen dioxide is obtained (Figure 22.34).

$$Cu(s) + 4H_3O^+(aq) + 2NO_3^-(aq) \longrightarrow Cu^{2+}(aq) + 2NO_2(g) + 6H_2O(l)$$

When certain nitrates, such as sodium nitrate, are heated, they decompose to the nitrites.

$$2NaNO_3(s) \xrightarrow{\Delta} 2NaNO_2(s) + O_2(g)$$

The corresponding acid, nitrous acid, is unstable and is usually prepared when needed as an aqueous solution. By adding the stoichiometric amount of sulfuric acid to an aqueous solution of barium nitrite, $Ba(NO_2)_2$, barium sulfate precipitates, leaving a solution of nitrous acid.

$$Ba(NO_2)_2(aq) + H_2SO_4(aq) \longrightarrow BaSO_4(s) + 2HNO_2(aq)$$

Table 22.8 lists uses of some compounds of nitrogen (and of phosphorus, discussed in the next section).

## CONCEPT CHECK 22.2

Considering the fact that $N_2$ makes up about 80% of the atmosphere, why don't animals use the abundant $N_2$ instead of $O_2$ for biological reactions?

**Table 22.8**
**Uses of Some Compounds of Nitrogen and of Phosphorus**

| Compound | Use |
| --- | --- |
| $NH_3$ | Nitrogen fertilizer |
| | Manufacture of nitrogen compounds |
| $N_2H_4$ | Blowing agent for foamed plastics |
| | Water treatment |
| $HNO_3$ | Explosives |
| | Polyurethane plastics |
| $Ca(H_2PO_4)_2 \cdot H_2O$ | Phosphate fertilizer |
| | Baking powder |
| $CaHPO_4 \cdot 2H_2O$ | Animal feed additive |
| | Toothpowder |
| $H_3PO_4$ | Manufacture of phosphate fertilizers |
| | Soft drinks |
| $PCl_3$ | Manufacture of $POCl_3$ |
| | Manufacture of pesticides |
| $POCl_3$ | Manufacture of plasticizers (substances that keep plastics pliable) |
| | Manufacture of flame retardants |
| $Na_5P_3O_{10}$ | Detergent additive |

**Figure 22.35**
**Allotropes of phosphorus.**
*Left:* White phosphorus. *Right:*
Red phosphorus.

## Phosphorus

▶ Demonstration of reaction of white phosphorus and potassium chlorate: Summerlin and Ealy, *Chemical Demonstrations,* Vol. 1 (Washington, D.C.: Am. Chem. Soc., 1985), no. 103, p. 166.

▶ Video Series C, "Barking Dogs (White Phosphorus)."

**Figure 22.36**
**Structure of the $P_4$ molecule.** *Top:* The reactivity of white phosphorus results from the small P—P—P angle (60°). *Bottom:* Space-filling molecular model.

Phosphorus, the most abundant of the Group VA elements, occurs in phosphate minerals, such as fluorapatite, whose formula is written either $Ca_5(PO_4)_3F$ or $3Ca_3(PO_4)_2 \cdot CaF_2$. Unlike nitrogen, which exists in important compounds with oxidation states between $-3$ and $+5$, the most important oxidation states of phosphorus are $+3$ and $+5$. Like nitrogen, however, phosphorus is an important element in living organisms. DNA (deoxyribonucleic acid), a chainlike biological molecule in which information about inheritable traits resides, contains phosphate groups along the length of its chain. Similarly, ATP (adenosine triphosphate), the energy-containing molecule of living organisms, contains phosphate groups.

**Allotropes of Phosphorus** Phosphorus has two common allotropes: white phosphorus and red phosphorus (Figure 22.35). White phosphorus, a waxy, white solid, is very poisonous and very reactive. If white phosphorus is left exposed to air, it bursts spontaneously into flame. Because of its reactivity with oxygen, white phosphorus is stored under water, in which it is insoluble. As you might expect from its low melting point (44°C), white phosphorus is a molecular solid, with the formula $P_4$.

The phosphorus atoms in the $P_4$ molecule are arranged at the corners of a regular tetrahedron such that each atom is single-bonded to the other three (Figure 22.36). The experimentally determined P—P—P bond angle is 60°. Because this is considerably smaller than the normal bond angle, the bonding in $P_4$ is strained and therefore weaker. This accounts for the reactivity seen in this phosphorus allotrope; chemical reactions of $P_4$ replace its weak bonds by stronger ones.

White phosphorus is a major industrial chemical and is prepared by heating phosphate rock with coke (C) and quartz sand ($SiO_2$) in an electric furnace. The overall reaction is

$$2Ca_3(PO_4)_2(s) + 6SiO_2(s) + 10C(s) \xrightarrow{1500°C} 6CaSiO_3(l) + 10CO(g) + P_4(g)$$
$$\text{calcium silicate}$$

The gases from the furnace are cooled by water to condense the phosphorus vapor to the liquid; the carbon monoxide gas is used as a fuel. The other product, calcium silicate glass (called *slag*), is drained periodically from the bottom of the furnace.

Most of the white phosphorus produced is used to manufacture phosphoric acid, $H_3PO_4$. Some white phosphorus is converted to red phosphorus, which has a chain structure (Figure 22.37). Red phosphorus is much less reactive than white phospho-

**Figure 22.37**
**Chain structure of red phosphorus.** The structure is obtained by linking $P_4$ tetrahedra together after breaking a bond in each tetrahedron.

rus and can be stored in the presence of air. This phosphorus allotrope is relatively nontoxic and is used in the striking surface for safety matches. Red phosphorus is made by heating white phosphorus at about 400°C for several hours.

**Phosphorus Oxides** Phosphorus has two common oxides, $P_4O_6$ and $P_4O_{10}$. Their common names are phosphorus trioxide and phosphorus pentoxide, respectively. The names are at odds with present rules of nomenclature but stem from their empirical formulas, $P_2O_3$ and $P_2O_5$. We will refer to these compounds as phosphorus(III) oxide ($P_4O_6$) and phosphorus(V) oxide ($P_4O_{10}$), using the Stock system of nomenclature, in which the oxidation states are given by Roman numerals.

These oxides have interesting structures (Figure 22.38). Phosphorus(III) oxide has a tetrahedron of phosphorus atoms, as in $P_4$, but with oxygen atoms between each pair of phosphorus atoms. Phosphorus(V) oxide is similar, except that each phosphorus atom has an additional oxygen atom bonded to it. These phosphorus–oxygen bonds are much shorter than the ones in the P—O—P bridges (139 pm versus 162 pm); hence, they are best represented as double bonds.

Phosphorus(III) oxide is a low-melting solid (m.p. 23°C) and the anhydride of phosphorous acid, $H_3PO_3$. Phosphorus(V) oxide is the most important oxide; it is a white solid that sublimes at 360°C. This oxide is the anhydride of phosphoric acid, $H_3PO_4$. The reaction with water is quite vigorous, making phosphorus(V) oxide useful in the laboratory as a drying agent. It is prepared by burning white phosphorus in air.

$$P_4(s) + 5O_2(g) \longrightarrow P_4O_{10}(s)$$

Phosphorus(V) oxide is used to manufacture phosphoric acid, an oxoacid.

**Oxoacids of Phosphorus** Phosphorus has many oxoacids, but the most important of these can be thought of as derivatives of orthophosphoric acid (often called simply phosphoric acid), $H_3PO_4$. Orthophosphoric acid is a colorless solid, melting at

**Figure 22.38**
**Structures of the phosphorus oxides.**
*Left:* The phosphorus atoms in $P_4O_6$ have tetrahedral positions, as in $P_4$; however, the phosphorus atoms are bonded to oxygen atoms, forming P—O—P bridges between each pair of phosphorus atoms. *Right:* The $P_4O_{10}$ molecule is similar, except that an additional oxygen atom is bonded to each phosphorus atom. Both ball-and-stick and space-filling models are shown.

$P_4O_6$

$P_4O_{10}$

42°C when pure. It is usually sold as an aqueous solution, however. Orthophosphoric acid is triprotic (three acidic H atoms per molecule); possible sodium salts are sodium dihydrogen phosphate ($NaH_2PO_4$), disodium hydrogen phosphate ($Na_2HPO_4$), and trisodium phosphate ($Na_3PO_4$). The acid has the following electron-dot formula:

$$\begin{array}{c} :O: \\ \| \\ H-O-P-O-H \\ | \\ :O: \\ | \\ H \end{array}$$

Orthophosphoric acid is produced in enormous quantity either directly from phosphate rock or from phosphorus(V) oxide, which in turn is obtained by burning white phosphorus. In the direct process, phosphate rock is treated with sulfuric acid, from which a solution of phosphoric acid is obtained by filtering off the calcium sulfate and other solid materials. The product is an impure phosphoric acid, which is used in the manufacture of phosphate fertilizers. ◄

▶ Because phosphate rock contains $CaF_2$, hydrofluoric acid is a by-product in the preparation of phosphoric acid. HF is used in aluminum production.

$$Ca_3(PO_4)_2(s) + 3H_2SO_4(aq) \longrightarrow 3CaSO_4(s) + 2H_3PO_4(aq)$$

A purer grade of orthophosphoric acid is obtained by reacting phosphorus(V) oxide with water. It is used in soft drinks for tartness and also in making phosphates for detergent formulations.

Orthophosphoric acid, $H_3PO_4$, undergoes condensation reactions to form other phosphoric acids. For example, two orthophosphoric acid molecules condense to form diphosphoric acid (also called pyrophosphoric acid). ◄

▶ Condensation reactions are discussed in Section 13.6.

$$\begin{array}{c} O \\ \| \\ H-O-P-O-H \\ | \\ O \\ | \\ H \end{array} + \begin{array}{c} O \\ \| \\ H-O-P-O-H \\ | \\ O \\ | \\ H \end{array} \longrightarrow \begin{array}{c} O \quad\quad O \\ \| \quad\quad \| \\ H-O-P-O-P-O-H \\ | \quad\quad | \\ O \quad\quad O \\ | \quad\quad | \\ H \quad\quad H \end{array} + H_2O$$

Two series of these phosphoric acids exist (all having phosphorus in the +5 oxidation state). One series consists of the linear **polyphosphoric acids,** which are *acids with the general formula* $H_{n+2}P_nO_{3n+1}$ *formed from linear chains of P—O bonds.*

$$\begin{array}{c} O \quad\quad O \\ \| \quad\quad \| \\ H-O-P-O-P-O-H \\ | \quad\quad | \\ O \quad\quad O \\ | \quad\quad | \\ H \quad\quad H \end{array} \quad\quad \begin{array}{c} O \quad\quad O \quad\quad O \\ \| \quad\quad \| \quad\quad \| \\ H-O-P-O-P-O-P-O-H \\ | \quad\quad | \quad\quad | \\ O \quad\quad O \quad\quad O \\ | \quad\quad | \quad\quad | \\ H \quad\quad H \quad\quad H \end{array}$$
diphosphoric acid      triphosphoric acid

The other series consists of the **metaphosphoric acids,** which are *acids with the general formula* $(HPO_3)_n$. Figure 22.39 shows the structure of a cyclic, or ring, metaphosphate anion. When a linear polyphosphoric acid chain is very long, the formula becomes $(HPO_3)_n$, with $n$ very large, and the acid is called *polymetaphosphoric acid.*

The polyphosphates and metaphosphates are used in detergents, where they act as water softeners by complexing with metal ions in the water. Sodium triphosphate, $Na_5P_3O_{10}$, one of the most commonly used polyphosphates, is manufactured by adding sufficient sodium carbonate to orthophosphoric acid to give a solution of the salts $NaH_2PO_4$ and $Na_2HPO_4$. When this solution is sprayed into a hot kiln, the orthophosphate ions condense to give sodium triphosphate.

**Figure 22.39**
**The structure of trimetaphosphate ion, $P_3O_9^{3-}$.** A cyclic metaphosphate ion; the general formula of the metaphosphate ions is $(PO_3^-)_n$.

The use of phosphates in detergents has been criticized for contributing to the overfertilization of plants and algae in lakes (a process referred to as *eutrophication*). Such lakes become oxygen-deficient from decomposing plants and algae, and the fish die. In the United States, the use of phosphates in laundry detergents has been banned in some states.

Table 22.8 lists some uses of compounds of phosphorus.

# 22.8  Group VIA: Oxygen and the Sulfur Family

Group VIA, like the preceding groups, shows the trend from nonmetallic to metallic as you proceed from top to bottom of the column of elements. Oxygen and sulfur are strictly nonmetallic. Although the chemical properties of selenium and tellurium are predominantly those of nonmetals, they do have semiconducting allotropes as expected of metalloids. Polonium is a radioactive metal.

Oxygen, a second-period element, has rather different properties from those of the other members of Group VIA. It is very electronegative and bonding involves only $s$ and $p$ orbitals. For the other members of Group VIA, $d$ orbitals become a factor in bonding. Oxygen exists in compounds mainly in the $-2$ oxidation state. The other Group VIA elements exist in compounds in this state also, but the $+4$ and $+6$ states are common. (Oxygen is the most abundant element on earth, making up 48% by mass of the earth's crust. Sulfur, although not as abundant as oxygen, is a common element. Both oxygen and sulfur are important to living organisms.)

## Oxygen

Oxygen in the form of dioxygen, $O_2$, makes up 20.9 mole percent of the atmosphere. Although this might seem to constitute a considerable quantity of the element, most of the oxygen on earth is present as oxide and oxoanion minerals (in silicates, carbonates, sulfates, and so forth). Indeed, oxygen combines with almost every element—only the noble gases helium, neon, and argon have no known compounds. The chemistry of oxygen is therefore very important, though we frequently discuss it in the context of other elements. In fact, much of the chemistry of the elements discussed in this chapter concerns either their reactions with oxygen or the properties of their oxides and oxoacids.

### Properties and Preparation of Oxygen   The common form of the element oxygen is dioxygen, $O_2$. (The element also exists as the allotrope ozone, $O_3$.) Dioxygen, usually called simply oxygen, is a colorless, odorless gas under standard conditions. The critical temperature is $-118°C$. Therefore, you can liquefy oxygen if you first cool the gas below this temperature and then compress it. Both liquid and solid $O_2$ have a pale blue color. The melting point of the solid is $-218°C$, and the boiling point at 1 atm is $-183°C$. ◄

▶ **Ozone and its presence in the atmosphere are discussed in an essay at the end of Chapter 10.**

Oxygen is produced in enormous quantity from air. As described in the discussion of nitrogen, air is first liquefied, then distilled. Nitrogen and argon are more volatile components of air and distill off, leaving liquid oxygen behind.

Oxygen can be prepared in small quantities by decomposing certain oxygen-containing compounds. Both the Swedish chemist Karl Wilhelm Scheele and the British chemist Joseph Priestley are credited with the discovery of oxygen. Priestley obtained the gas in 1774 by heating mercury(II) oxide.

$$2HgO(s) \xrightarrow{\Delta} 2Hg(l) + O_2(g)$$

In one laboratory preparation, potassium chlorate, $KClO_3$, is heated with pure manganese(IV) oxide, $MnO_2$, as a catalyst.

$$2KClO_3(s) \xrightarrow[MnO_2]{\Delta} 2KCl(s) + 3O_2(g)$$

**Reactions of Oxygen**   Molecular oxygen is a very reactive gas and combines directly with many substances. The products are usually oxides. An **oxide** is *a binary compound with oxygen in the −2 oxidation state.*

Most metals react readily with oxygen to form oxides, especially if the metal is in a form that exposes sufficient surface area. For example, magnesium wire and iron wool burn brightly in air to yield the oxides.

$$2Mg(s) + O_2(g) \longrightarrow 2MgO(s)$$
$$3Fe(s) + 2O_2(g) \longrightarrow Fe_3O_4(s)$$

**Figure 22.40**
**Chromium oxides.**  *Left:* Chromium(III) oxide, a basic oxide. *Right:* Chromium(VI) oxide, an acidic oxide.

The resulting oxides MgO and $Fe_3O_4$ are basic oxides, as is true of most metal oxides. If the metal is in a high oxidation state, however, the oxide may be acidic. For example, chromium(III) oxide, $Cr_2O_3$, is a basic oxide, but chromium(VI) oxide, $CrO_3$, is an acidic oxide (Figure 22.40).

The alkali metals form an interesting series of binary compounds with oxygen. When an alkali metal burns in air, the principal product with oxygen depends on the metal. With lithium, the product is the basic oxide, $Li_2O$. With the other alkali metals, the product is predominantly the peroxide and superoxide. A **peroxide** is *a compound with oxygen in the −1 oxidation state.* (Peroxides contain either the $O_2^{2-}$ ion or the covalently bonded group —O—O—.) A **superoxide** is *a binary compound with oxygen in the $-\frac{1}{2}$ oxidation state;* superoxides contain the superoxide ion, $O_2^-$. Sodium metal burns in air to give mainly the peroxide.

$$2Na(s) + O_2(g) \longrightarrow Na_2O_2(s)$$

Potassium and the other alkali metals form mainly the superoxides.

$$K(s) + O_2(g) \longrightarrow KO_2(s)$$

Nonmetals react with oxygen to form covalent oxides, most of which are acidic. For example, carbon burns in an excess of oxygen to give carbon dioxide, which is the acid anhydride of carbonic acid (that is, carbon dioxide produces carbonic acid when it reacts with water). Sulfur, $S_8$, burns in oxygen to give sulfur dioxide, $SO_2$, the acid anhydride of sulfurous acid.

$$S_8(s) + 8O_2(g) \longrightarrow 8SO_2(g)$$

Sulfur forms another oxide, sulfur trioxide, $SO_3$, but only small amounts are obtained during the burning of sulfur in air. Sulfur trioxide is the acid anhydride of sulfuric acid.

Compounds in which at least one element is in a reduced state are oxidized by oxygen, giving compounds that would be expected to form when the individual elements are burned in oxygen. For example, a hydrocarbon such as octane, $C_8H_{18}$, burns to give carbon dioxide and water.

$$2C_8H_{18}(l) + 25O_2(g) \longrightarrow 16CO_2(g) + 18H_2O(g)$$

Some other examples are given in the following equations:

$$2H_2S(g) + 3O_2(g) \longrightarrow 2H_2O(g) + 2SO_2(g)$$
$$CS_2(l) + 3O_2(g) \longrightarrow CO_2(g) + 2SO_2(g)$$
$$2ZnS(s) + 3O_2(g) \longrightarrow 2ZnO(s) + 2SO_2(g)$$

Note the products that are formed; sulfur compounds usually form $SO_2$.

**Figure 22.41**
**Sulfur obtained from underground deposits.**
Sulfur is obtained from these deposits by pumping in superheated water; molten sulfur is then pumped to the earth's surface. (Details of this *Frasch process* are described later.) Here, molten sulfur from a well is being directed to an area where it can cool and solidify.

### CONCEPT CHECK 22.3

Why do we need such low temperatures to liquefy gases such as nitrogen, oxygen and helium?

## Sulfur

Sulfur occurs in sulfate minerals, such as gypsum ($CaSO_4 \cdot 2H_2O$), and in sulfide and disulfide minerals, many of which are important metal ores. Pyrite is iron(II) disulfide, $FeS_2$; it consists of $Fe^{2+}$ and $S_2^{2-}$ ions. This mineral is sometimes called "fool's gold," because its golden color often fooled novice miners into thinking they had found gold. Sulfur is also present in coal and petroleum as organic sulfur compounds and in natural gas as hydrogen sulfide, $H_2S$. Free sulfur occurs in some volcanic areas, perhaps formed by the reaction of hydrogen sulfide and sulfur dioxide, which are present in volcanic gases.

$$16H_2S(g) + 8SO_2(g) \longrightarrow 16H_2O(l) + 3S_8(s)$$

Commercial deposits of free sulfur also occur in the rock at the top of salt domes, which are massive columns of salt embedded in rock a hundred meters or more below the earth's surface. These deposits are believed to have formed by bacterial action involving calcium sulfate minerals. Figure 22.41 shows free sulfur obtained from such deposits that occur in the United States along the Gulf of Mexico.

Sulfur also occurs in several amino acids, which are the building blocks of the proteins in living organisms. Plants are able to use sulfate ion as a source of sulfur for the synthesis of amino acids. Animals and decay bacteria derive most of their nutritional sulfur from organic sources.

**Figure 22.42**
**Structure of the $S_8$ molecule.** *Top:* Each molecule consists of eight S atoms arranged in a ring (in the shape of a crown). *Bottom:* Space-filling molecular model.

**Allotropes of Sulfur**   Sulfur has a fascinating array of allotropes, including two common crystal forms, rhombic sulfur and monoclinic sulfur (see Figure 6.14). *Rhombic sulfur* is the stablest form of the element under normal conditions; natural sulfur is rhombic sulfur. It is a yellow, crystalline solid with a lattice consisting of crown-shaped $S_8$ molecules (Figure 22.42). The relative stability of this molecule results in part from the ability of sulfur atoms to undergo catenation—that is, to form stable bonds to other sulfur atoms. Rhombic sulfur melts at 113°C to give an orange-colored liquid.

When this liquid is cooled, it crystallizes to give *monoclinic sulfur*. This allotrope also consists of $S_8$ molecules; it differs from rhombic sulfur only in the way the molecules are packed to form crystals. Monoclinic sulfur melts at 119°C. It is unstable below 96°C, and in a few weeks at room temperature it reverts to rhombic sulfur.

**Figure 22.43**
**Sulfur at various temperatures.** *Left:* Solid rhombic sulfur at 20°C. *Center:* Orange-colored liquid sulfur at 120°C. *Right:* Viscous liquid sulfur at 200°C.

If, instead of cooling the liquid sulfur, you raise its temperature above 160°C, the sulfur begins to darken and at somewhat higher temperatures changes to a dark reddish-brown, viscous liquid. The original melt consists of $S_8$ molecules, but these rings of sulfur atoms open up, and the fragments join to give long spiral chains of sulfur atoms. The spiral chains have unpaired electrons at their ends, and these unpaired electrons are responsible for the color. The viscosity of the liquid increases as compact $S_8$ molecules are replaced by long spiral chains that can intertwine. At temperatures greater than 200°C, the chains begin to break apart and the viscosity decreases. Figure 22.43 shows the appearance of sulfur at various temperatures.

When molten sulfur above 160°C (but below 200°C) is poured into water, the liquid changes to a rubbery mass, called *plastic sulfur.* Plastic sulfur is an amorphous mixture of sulfur chains. The rubberiness of this sulfur results from the ability of the spiral chains of sulfur atoms to stretch and then relax to their original length. Plastic sulfur reverts to rhombic sulfur after a period of time.

Sulfur boils at 445°C, giving a vapor of $S_8$, $S_6$, $S_4$, and $S_2$ molecules. Measurements of vapor density depend on the temperature, as a result of these different species of molecular sulfur in the vapor. (The relationship between gas density and molecular weight is described in Section 5.3.)

▶ Demonstration of the preparation of plastic sulfur: Bassam Z. Shakhashiri, *Chemical Demonstrations,* Vol. 1 (Madison: University of Wisconsin Press, 1983), no. 3.12, pp. 243–244.

**Production of Sulfur** Free sulfur that occurs in deep underground deposits is mined by the **Frasch process,** *a mining procedure in which underground deposits of solid sulfur are melted in place with superheated water, and the molten sulfur is forced upward as a froth using air under pressure* (see Figure 22.44). A sulfur well is similar to an oil well but consists of three concentric pipes. Superheated water flows down the outer pipe, and compressed air flows down the inner pipe. The superheated water melts the sulfur, which is then pushed up the middle pipe by the compressed air. Molten sulfur spews from the well and solidifies in large storage areas. Sulfur obtained this way is 99.6% pure.

Hydrogen sulfide, $H_2S$, recovered from natural gas and petroleum is also a source of free sulfur. The sulfur is obtained from the hydrogen sulfide gas by the **Claus process,** *a method of obtaining free sulfur by the partial burning of hydrogen sulfide.* The partial burning of hydrogen sulfide produces some sulfur, as well as sulfur dioxide.

$$8H_2S(g) + 4O_2(g) \longrightarrow S_8(s) + 8H_2O(g)$$
$$2H_2S(g) + 3O_2(g) \longrightarrow 2SO_2(g) + 2H_2O(g)$$

The sulfur dioxide that forms reacts with the hydrogen sulfide to produce more sulfur.

$$16H_2S(g) + 8SO_2(g) \longrightarrow 3S_8(s) + 16H_2O(g)$$

Most of the sulfur produced (almost 90%) is used to make sulfuric acid. The remainder has a wide variety of uses, including the vulcanization of rubber (sulfur converts the initially tacky material into useful rubber), the production of carbon disulfide (to make cellophane), and the preparation of sulfur dioxide for bleaching.

**Sulfur Oxides and Oxoacids** Sulfur dioxide, $SO_2$, is a colorless, toxic gas with a characteristic suffocating odor. Its presence in polluted air (from the burning of fossil fuels) is known to cause respiratory ailments. The $SO_2$ molecule has a bent geometry with a bond angle of 119.5°, very close to that predicted by the VSEPR model (Figure 22.45).

Sulfur dioxide gas is very soluble in water, producing acidic solutions. Although these solutions are often referred to as solutions of *sulfurous acid*, they appear to be composed primarily of hydrated species of $SO_2$; the acid $H_2SO_3$ has never been isolated. An aqueous solution of sulfur dioxide, $SO_2(aq)$, does apparently contain small amounts of the ions $HSO_3{}^-$ and $SO_3{}^{2-}$, which would be expected to be produced by the ionization of $H_2SO_3$.

$$H_2O(l) + SO_2(g) \rightleftharpoons H_2SO_3(aq)$$

$$H_2SO_3(aq) + H_2O(l) \rightleftharpoons H_3O^+(aq) + HSO_3{}^-(aq)$$

$$HSO_3{}^-(aq) + H_2O(l) \rightleftharpoons H_3O^+(aq) + SO_3{}^{2-}(aq)$$

When an appropriate amount of base is added to an aqueous solution of sulfur dioxide, the corresponding hydrogen sulfite salt or sulfite is obtained. Sodium hydrogen sulfite (also called sodium bisulfite) and sodium sulfite are produced this way commercially using sodium carbonate as the base:

$$Na_2CO_3(aq) + 2SO_2(aq) + H_2O(l) \longrightarrow 2NaHSO_3(aq) + CO_2(g)$$

$$Na_2CO_3(aq) + SO_2(aq) \longrightarrow Na_2SO_3(aq) + CO_2(g)$$

Sulfites and hydrogen sulfites decompose when treated with acid to give sulfur dioxide. For example,

$$NaHSO_3(aq) + HCl(aq) \longrightarrow NaCl(aq) + H_2O(l) + SO_2(g)$$

This reaction can be used to prepare small amounts of sulfur dioxide in the laboratory.

Sulfur dioxide is produced on a large scale by burning sulfur, $S_8$. It is also obtained as a by-product of the roasting of sulfide ores (such as $FeS_2$, $CuS$, $ZnS$, and $PbS$). Most of this sulfur dioxide is used to prepare sulfuric acid. Some is used as a bleach for wood pulp and textiles (Figure 22.46) and as a disinfectant and food preservative (for example, in wine and dried fruit). Its use as a food preservative depends on the fact that sulfur dioxide is especially toxic to yeasts, molds, and certain bacteria. Because some people are allergic to sulfur dioxide, foods containing it must be properly labeled.

Sulfur trioxide is a liquid at room temperature. The liquid actually consists of $S_3O_9$ molecules in equilibrium with $SO_3$ molecules (Figure 22.47). The vapor-phase molecule is $SO_3$, which has a planar triangular geometry.

Sulfur trioxide is formed in small amounts when sulfur is burned in air, although the principal product is sulfur dioxide. Thermodynamically, sulfur trioxide is the preferred product of sulfur and oxygen. Sulfur dioxide does react slowly with oxygen in air to produce sulfur trioxide, but the reaction is much faster in the presence of a catalyst, such as platinum. Sulfur trioxide is produced commercially by the oxidation of sulfur dioxide in the presence of vanadium(V) oxide catalyst.

$$2SO_2(g) + O_2(g) \xrightarrow{V_2O_5} 2SO_3(g)$$

Sulfur trioxide reacts vigorously and exothermically with water to produce sulfuric acid.

$$SO_3(g) + H_2O(l) \longrightarrow H_2SO_4(aq)$$

The **contact process** is an industrial method for the manufacture of sulfuric acid. It consists of the reaction of sulfur dioxide with oxygen to form sulfur trioxide using a

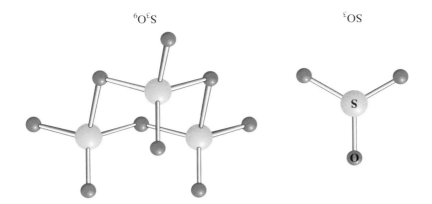

**Figure 22.47**
**Structures of $SO_3$ and $S_3O_9$ molecules.** These molecules are in equilibrium in liquid sulfur trioxide. The vapor consists mostly of $SO_3$.

SO₃

S₃O₉

**Figure 22.46**
**Bleaching of a rose by sulfur dioxide.** The dye in the rose is reduced by sulfur dioxide (contained in the beaker) to a colorless substance.

**Table 22.9**
**Uses of Some Sulfur Compounds**

| Compound | Use |
|---|---|
| $CS_2$ | Manufacture of rayon and cellophane |
| | Manufacture of $CCl_4$ |
| $SO_2$ | Manufacture of $H_2SO_4$ |
| | Food preservative |
| | Textile bleach |
| $H_2SO_4$ | Manufacture of phosphate fertilizers |
| | Petroleum refining |
| | Manufacture of various chemicals |
| $Na_2S_2O_3$ | Photographic fixer |

*catalyst of vanadium(V) oxide, followed by the reaction of sulfur trioxide with water.* Because the direct reaction of sulfur trioxide with water produces mists that are unmanageable, the sulfur trioxide is actually dissolved in concentrated sulfuric acid, which is then diluted with water.

◀ **Sulfuric acid is discussed in an essay at the end of Chapter 6, and acid rain is discussed in an essay at the end of Section 17.2.**

Sulfuric acid is a component of acid rain and forms from sulfur dioxide. Atmospheric sulfur dioxide has both natural and human origins. Natural sources include plant and animal decomposition and volcanic emissions. However, the burning of coal, oil, and natural gas has been identified as a major source of acid pollution. After persisting in the atmosphere for some time, sulfur dioxide is oxidized to sulfur trioxide, which dissolves in rain to give $H_2SO_4(aq)$.

Concentrated sulfuric acid is a viscous liquid and a powerful dehydrating agent. The concentrated acid is also an oxidizing agent. Copper is not dissolved by most acids. $E°$ for $Cu^{2+}|Cu$ is positive, so Cu is not oxidized by $H^+(H_3O^+)$. It is, however, dissolved by hot, concentrated sulfuric acid. In this reaction, sulfate ion in acid solution is reduced to sulfur dioxide:

$$\overset{0}{Cu}(s) + 2H_2\overset{+6}{S}O_4(l) \longrightarrow \overset{+2}{Cu}SO_4(aq) + 2H_2O(l) + \overset{+4}{S}O_2(g)$$

◀ **Video Series C, "Sulfuric Acid and Sugar."**

More sulfuric acid is made than any other industrial chemical. Most of this acid is used to make soluble phosphate and ammonium sulfate fertilizers. Sulfuric acid is also used in petroleum refining and in the manufacture of many chemicals. Table 22.9 lists uses of some sulfur compounds.

## 22.9  Group VIIA: The Halogens

The Group VIIA elements, called the halogens, have very similar properties, or at least they have properties that change smoothly in progressing down the column. All are reactive nonmetals, except perhaps for astatine, whose chemistry is not well known.

As a second-period element, fluorine does exhibit some differences from the other elements of Group VIIA, although these are not so pronounced as those of the second-period elements in Groups IIIA to VIA. The solubilities of the fluorides in water, for

◀ **Video Series C, "Zinc and Iodine."**

example, are often quite different from those of the chlorides, bromides, and iodides. Calcium chloride, bromide, and iodide are very soluble in water. Calcium fluoride, however, is insoluble. Silver chloride, bromide, and iodide are insoluble, but silver fluoride is soluble.

All of the halogens form stable compounds in which the element is in the $-1$ oxidation state. In fluorine compounds, this is the only oxidation state. Chlorine, bromine, and iodine also have compounds in which the halogen is in one of the positive oxidation states $+1$, $+3$, $+5$, or $+7$. The higher positive oxidation states ($> +1$) are due to the involvement of $d$ orbitals in bonding.

## Chlorine

Chlorine gas, $Cl_2$, was discovered in 1774 by the Swedish chemist Karl Wilhelm Scheele by heating hydrochloric acid with manganese dioxide.

$$4HCl(aq) + MnO_2(s) \longrightarrow MnCl_2(aq) + Cl_2(g) + 2H_2O(l)$$

► The preparation of chlorine using this reaction is shown in Figure 3.16.

He immediately noted the suffocating odor of the gas. Scheele also discovered that chlorine solutions could bleach cotton cloth. Within a few decades, chlorine-based bleaches became major items of commerce. Currently, chlorine is the most commercially important halogen.

**Properties of Chlorine** Chlorine gas has a pale greenish-yellow color. It is a very reactive oxidizing agent and supports the combustion of many substances in a manner similar to oxygen. We discussed the reactions of chlorine with sodium and antimony in Chapter 2.

► See the introduction to Chapter 2.

All of the halogens are oxidizing agents, though the oxidizing power decreases from fluorine to iodine. Thus, chlorine is a stronger oxidizing agent than either bromine or iodine. When chlorine gas is bubbled into a bromide solution, free bromine is obtained. Similarly, chlorine oxidizes iodide ion to iodine.

$$Cl_2(g) + 2KBr(aq) \longrightarrow 2KCl(aq) + Br_2(aq)$$
$$Cl_2(g) + 2KI(aq) \longrightarrow 2KCl(aq) + I_2(aq)$$

These reactions can be used as a test for bromide and iodide ions. Suppose an aqueous solution of chlorine is added to a test tube containing either bromide or iodide ion. The corresponding free halogen is formed in the water solution. It is readily identified by adding the organic solvent methylene chloride, $CH_2Cl_2$, which dissolves the halogen, forming a colored layer at the bottom of the test tube. Bromide ion gives an orange layer; iodide ion gives a violet layer. Of course, neither bromine nor iodine is strong enough to oxidize chloride ion.

Chlorine reacts with water by disproportionating (that is, it is both oxidized and reduced):

$$\overset{0}{Cl_2}(g) + H_2O(l) \rightleftharpoons \overset{+1}{H}ClO(aq) + H\overset{-1}{Cl}(aq)$$

In an aqueous solution of chlorine at 25°C, about two-thirds of the chlorine is present as $Cl_2(aq)$; the rest is HClO and HCl.

**Preparation and Uses of Chlorine** Chlorine is a major industrial chemical. It is prepared commercially by the electrolysis of aqueous sodium chloride.

► The electrolysis of aqueous sodium chloride is discussed in Section 20.10.

Chlorine can be prepared in small amounts for laboratory use by the reaction of chloride ion with a strong oxidizing agent, such as potassium dichromate or man-

**Figure 22.48**
**The hydrogen chloride fountain.** The flask contains hydrogen chloride gas. When water is added to the flask from the dropper, the hydrogen chloride dissolves in it, reducing the pressure in the flask. Atmospheric pressure pushes the water in the beaker into the flask, forming a stream or fountain. Water in the beaker has bromthymol-blue indicator in it. The indicator solution in the flask is yellow because it is acidic.

▶ Demonstration of the HCl fountain: Bassam Z. Shakhashiri, *Chemical Demonstrations,* Vol. 2 (Madison: University of Wisconsin Press, 1985), no. 6.24, pp. 205–206.

ganese dioxide. However, chlorine is readily available in steel cylinders for laboratory work.

The principal use of chlorine is in the preparation of chlorinated hydrocarbons, such as vinyl chloride, $CH_2=CHCl$ (for polyvinyl chloride plastics), and methyl chloride, $CH_3Cl$ (for the manufacture of silicones, polymers with Si—O bonds and organic groups). Various insecticides are also chlorinated hydrocarbons; many of these (such as DDT) are now restricted in their use because of possible environmental damage.

Other major uses of chlorine are as a bleaching agent for textiles and paper pulp and as a disinfectant. Not long after the discovery of chlorine, chlorine bleaches became available commercially. Chlorine solutions were used as disinfectants early in the nineteenth century. Today, chlorine gas is commonly used for disinfecting municipal water supplies.

**Hydrogen Chloride**   Hydrogen chloride, HCl, is a colorless gas with a sharp, penetrating odor. The gas is very soluble in water (Figure 22.48), and the water solution is commonly referred to as hydrochloric acid. The molecular species HCl ionizes nearly completely in aqueous solution:

$$HCl(g) + H_2O(l) \longrightarrow H_3O^+(aq) + Cl^-(aq)$$

Hydrogen chloride can be produced by heating sodium chloride with concentrated sulfuric acid.

$$NaCl(s) + H_2SO_4(l) \xrightarrow{\Delta} NaHSO_4(s) + HCl(g)$$

On stronger heating, the sodium hydrogen sulfate reacts with sodium chloride to produce additional hydrogen chloride.

$$NaCl(s) + NaHSO_4(s) \xrightarrow{\Delta} Na_2SO_4(s) + HCl(g)$$

Hydrogen bromide and hydrogen iodide can also be produced from their salts by a similar replacement reaction, but in these cases phosphoric acid is used instead of sulfuric acid, which tends to oxidize the bromide and iodide ions to the respective elements (Figure 22.49).

Most of the hydrogen chloride available commercially is obtained as a by-product in the manufacture of chlorinated hydrocarbons. In these reactions, hydrogen bonded

**Figure 22.49**
**The action of concentrated sulfuric acid on halide salts.**
Concentrated sulfuric acid was added to watch glasses containing, from left to right, NaCl, NaBr, and NaI. Note the formation of some $Br_2$ in the center watch glass (brown) and the formation of $I_2$ vapor (purple) over the watch glass on the right. These halogens form when concentrated sulfuric acid oxidizes the corresponding halide ions. Chloride ion is not oxidized by $H_2SO_4$.

**Table 22.10**
**Halogen Oxoacids**

| Oxidation State | Fluorine Oxoacids | Chlorine Oxoacids | Bromine Oxoacids | Iodine Oxoacids | General Name |
|---|---|---|---|---|---|
| +1 | HFO* | HClO† | HBrO† | HIO† | Hypohalous acid |
| +3 | — | HClO₂† | HBrO₂† | — | Halous acid |
| +5 | — | HClO₃† | HBrO₃† | HIO₃ | Halic acid |
| +7 | — | HClO₄ | HBrO₄† | HIO₄ | Perhalic acid |
| | | | | H₅IO₆ | |

*The oxidation state of F in HFO is −1.
†These acids are known only in aqueous solution.

to carbon is replaced by chlorine, forming the chlorinated compound and HCl. An example is the preparation of methyl chloride, $CH_3Cl$, from methane.

$$CH_4(g) + Cl_2(g) \longrightarrow CH_3Cl(g) + HCl(g)$$

Hydrochloric acid is the fourth most important industrial acid (after sulfuric, phosphoric, and nitric acids). It is used to clean metal surfaces of oxides (a process called *pickling*) and to extract certain metal ores, such as those of tungsten.

**Oxoacids of Chlorine**    The halogens form a variety of oxoacids (Table 22.10). Figure 22.50 shows the structures of the chlorine oxoacids. The acidic character of these acids increases with the number of oxygen atoms bonded to the halogen atom—that is, in the order $HClO$, $HClO_2$, $HClO_3$, $HClO_4$. (See Section 16.5 for a discussion of molecular structure and acid strength.) Perchloric acid, $HClO_4$, is the strongest of the common acids. Of the chlorine oxoacids, only perchloric acid is stable; the other oxoacids have never been isolated and are known only in aqueous solution.

Hypochlorous acid, $HClO$, is produced when chlorine disproportionates (is oxidized and reduced) in water. In basic solution, the equilibrium is very far toward the acid anions:

$$Cl_2(g) + 2OH^-(aq) \longrightarrow Cl^-(aq) + ClO^-(aq) + H_2O(l)$$

Solutions of sodium hypochlorite are manufactured by allowing the chlorine gas released by the electrolysis of aqueous sodium chloride to mix with the cold solution

▶ Video Series C, "Sugar, Potassium Chlorate, and Sulfuric Acid."

▶ Experiments in General Chemistry, 22, "The Strength of a Laundry Bleach."

**Figure 22.50**
**Structures of the chlorine oxoacids.** The models of the oxoacids also include lone pairs on the Cl atom.

HClO

HClO₂

HClO₃                    HClO₄

**Figure 22.51**
**Solution of sodium hypochlorite bleach.** The solution is manufactured by allowing chlorine and sodium hydroxide solution (from the electrolysis of aqueous sodium chloride) to react.

of sodium hydroxide that is also obtained in the electrolysis. These solutions are sold as a bleach (Figure 22.51).

Hypochlorite ion itself is unstable, disproportionating into chlorate ion, $ClO_3^-$, and chloride ion.

$$3ClO^-(aq) \longrightarrow ClO_3^-(aq) + 2Cl^-(aq)$$

At room temperature the reaction is slow; but in hot solution in the presence of base, the reaction is fast. Therefore, when chlorine reacts with hot sodium hydroxide solution, sodium chlorate is the product instead of sodium hypochlorite.

$$3Cl_2(g) + 6NaOH(aq) \longrightarrow NaClO_3(aq) + 5NaCl(aq) + 3H_2O(l)$$

Sodium chlorate can be crystallized from the solution. Solutions of chloric acid, $HClO_3$, can be prepared, although the pure acid cannot be isolated.

Sodium perchlorate and potassium perchlorate are produced commercially by the electrolysis of a saturated solution of the corresponding chlorate. The anode reaction is

$$ClO_3^-(aq) + 3H_2O(l) \longrightarrow ClO_4^-(aq) + 2H_3O^+(aq) + 2e^-$$

Hydrogen evolves at the cathode. Perchloric acid can be prepared by treating a perchlorate salt with sulfuric acid.

$$KClO_4(s) + H_2SO_4(l) \longrightarrow KHSO_4(s) + HClO_4(l)$$

The perchloric acid is distilled from the mixture at reduced pressure (to keep the temperature below 92°C, where perchloric acid decomposes explosively).

Table 22.11 lists uses of some halogen compounds.

**Table 22.11**
**Uses of Some Halogen Compounds**

| Compound | Use |
| --- | --- |
| AgBr, AgI | Photographic film |
| $CCl_4$ | Manufacture of fluorocarbons |
| $CH_3Br$ | Pesticide |
| $C_2H_4Cl_2$ | Manufacture of vinyl chloride (plastics) |
| HCl | Metal treating |
| | Food processing |
| NaClO | Household bleach |
| | Manufacture of hydrazine for rocket fuel |
| $NaClO_3$ | Paper pulp bleaching (with $ClO_2$) |
| KI | Human nutritional and animal feed supplement |

# 22.10   Group VIIIA: The Noble Gases

In our discussions of bonding, we pointed out the relative stability of the electron configurations of the Group VIIIA noble gases. For many years, it was thought that because the atoms of these elements had completed octets, the noble gases would be

| IA | | | | | | | | | VIIIA |
|----|----|----|----|----|----|----|----|----|----|
| H | IIA | | | IIIA IVA VA VIA VIIA | | | | | He |
| Li | Be | | | B | C | N | O | F | Ne |
| Na | Mg | IIIB | IIB | Al | Si | P | S | Cl | Ar |
| K | Ca | Sc | Zn | Ga | Ge | As | Se | Br | Kr |
| Rb | Sr | Y | Cd | In | Sn | Sb | Te | I | Xe |
| Cs | Ba | La | Hg | Tl | Pb | Bi | Po | At | Rn |
| Fr | Ra | Ac | | | | | | | |

completely unreactive. Consequently, these elements were known as *inert gases.* Compounds of krypton, xenon, and radon have since been prepared, however, so the term is not quite appropriate.

## Helium and the Other Noble Gases

The noble gases were not known until 1894. A couple of years earlier, the English physicist Lord Rayleigh discovered that the density of nitrogen gas obtained from air (1.2561 g/L at STP) was noticeably greater than the density of nitrogen obtained by decomposition of nitrogen compounds (1.2498 g/L at STP). He concluded that one of these two nitrogen sources was contaminated with another substance.

Soon after this, Rayleigh began collaborating with the Scottish chemist William Ramsay. Ramsay passed atmospheric nitrogen over hot magnesium to remove the nitrogen as magnesium nitride, $Mg_3N_2$, and obtained a nonreactive residual gas. He placed this gas in a sealed glass tube and passed a high-voltage electrical discharge through it to observe the emission spectrum of the corresponding atom. The spectrum showed a series of red and green lines and was unlike that of any known element. Ramsay and Rayleigh concluded that they had discovered a new element, which they called argon (from the Greek word *argos*, meaning "lazy"—referring to argon's lack of chemical reactivity). They also surmised that argon was a member of a new column of elements in the periodic table.

Before the discovery of argon, some lines in the spectrum of the sun were ascribed to an element not yet known on earth. This element was called helium (from the Greek *helios*, meaning "sun"). In 1895, Ramsay and the Swedish chemist Per Theodor Cleve (working independently) announced the discovery of helium gas in the mineral cleveite (a uranium ore) by identifying the spectrum of helium. Several years later, Ramsay discovered neon, krypton, and xenon by fractional distillation of liquid air. Radon was discovered in 1900 as a gaseous decay product of radium.

**Preparation and Uses of the Noble Gases** Commercially, all the noble gases except helium and radon are obtained by the distillation of liquid air. The principal sources of helium are certain natural-gas wells. Helium has the lowest boiling point ($-268.9°C$) of any substance and is very important in low-temperature research. The major use of argon is as a blanketing gas (inert gas) in metallurgical processes. It is also used as a mixture with nitrogen to fill incandescent lightbulbs. In these bulbs the gas mixture conducts heat away from the hot tungsten filament, without reacting with it. All the noble gases are used in gas discharge tubes (containing gas through which a high-voltage electric current can be discharged, giving light from atomic emission). Neon gives a highly visible red-orange emission and has long been used in advertising signs. The noble gases are also used in a number of lasers. The helium–neon laser was the first continuously operating gas laser. It emits red light at a wavelength of 632.8 nm.

**Compounds of the Noble Gases** Neil Bartlett, working at the University of British Columbia, prepared the first noble-gas compound after he discovered that molecular oxygen reacts with platinum hexafluoride, $PtF_6$, to form the ionic solid $[O_2^+][PtF_6^-]$. Because the ionization energy of xenon ($1.17 \times 10^3$ kJ/mol) is close to that of molecular oxygen ($1.21 \times 10^3$ kJ/mol), Bartlett reasoned that xenon should also react with

▶ The product actually has variable composition and can be represented by the formula $Xe(PtF_6)_n$, where $n$ is between 1 and 2.

## Table 22.12
### Some Compounds of Xenon

| Compound | Formula | Description |
|---|---|---|
| Xenon difluoride | $XeF_2$ | Colorless crystals |
| Xenon tetrafluoride | $XeF_4$ | Colorless crystals |
| Xenon hexafluoride | $XeF_6$ | Colorless crystals |
| Xenon trioxide | $XeO_3$ | Colorless crystals, explosive |
| Xenon tetroxide | $XeO_4$ | Colorless gas, explosive |

**Figure 22.52**
**Crystals of xenon tetrafluoride.** This photomicrograph shows crystals obtained in the experiment that first produced a binary compound of xenon.

platinum hexafluoride. In 1962 he reported the synthesis of an orange-yellow compound with the approximate formula $XePtF_6$. ◄ Later in the same year, chemists from Argonne National Laboratory near Chicago reported that xenon reacts directly with fluorine at 400°C to give the tetrafluoride. The elements react even at room temperature when exposed to sunlight.

$$Xe(g) + 2F_2(g) \longrightarrow XeF_4(s)$$

The product is a volatile, colorless solid (Figure 22.52). Since then, a number of noble-gas compounds have been prepared, all involving bonds to the highly electronegative elements fluorine and oxygen. Most are compounds of xenon (see Table 22.12), but a few are compounds of krypton and radon. Recently, an argon compound, HArF, was synthesized at −446°C.

## A Checklist for Review

### Important Terms

Solvay process (22.2)
Goldschmidt process (22.4)
steam-reforming process (22.5)

binary hydride (22.5)
catenation (22.6)
silicones (22.6)
Ostwald process (22.7)

polyphosphoric acids (22.7)
metaphosphoric acids (22.7)
oxide (22.8)
peroxide (22.8)

superoxide (22.8)
Frasch process (22.8)
Claus process (22.8)
contact process (22.8)

### Summary of Facts and Concepts

Several general observations can be made about the main-group elements. First, the metallic characteristics of these elements generally decrease across a period from left to right in the periodic table. Second, metallic characteristics of the main-group elements become more pronounced going down any column (group). Finally, a second-period element is usually rather different from the other elements in its group.

The Group IA metals (alkali metals) are soft, chemically reactive elements. Lithium, sodium, and potassium are important alkali metals. In recent years, the commercial uses of lithium have grown markedly. The metal is obtained by the electrolysis of molten lithium chloride and is used in the production of low-density alloys and as a battery anode. Lithium hydroxide is used to make lithium soap for lubricating greases; it is produced by the reaction of lithium carbonate and calcium hydroxide.

Sodium metal is prepared in large quantities. It is used as a reducing agent in the preparation of other metals, such as titanium and zirconium, and in the preparation of dyes and pharmaceuticals. Sodium compounds are of enormous economic importance. Sodium chloride is the source of sodium and most of its compounds. Sodium hydroxide is prepared by the electrolysis of aqueous sodium chloride; as a strong base, it has many useful commercial applications, including aluminum production and petroleum refining. Sodium carbonate is obtained from the mineral trona, which contains sodium carbonate and sodium hydrogen carbonate, and by the *Solvay process* from salt (NaCl) and limestone ($CaCO_3$). Sodium carbonate is used to make glass. Potassium metal is produced in relatively small quantities, but potassium compounds are important. Large quantities of potassium chloride are used as a plant fertilizer.

Magnesium and calcium are the most important of the Group IIA (alkaline earth) metals. Magnesium and its alloys are important structural metals. Calcium is important primarily as its compounds, which are prepared from natural carbonates, such as limestone, and the sulfates, such as gypsum. When limestone is heated strongly, it decomposes to calcium oxide (lime). Enormous quantities of lime are used in the production of iron from its ores.

Of the Group IIIA and Group IVA metals, aluminum, tin, and lead are especially important. Aluminum is the third most abundant element in the earth's crust. It is obtained commercially from bauxite; through chemical processing bauxite yields pure aluminum oxide. Most of this aluminum oxide is used in the production of aluminum by electrolysis. Some aluminum oxide is used as a carrier for heterogeneous catalysts and in manufacturing industrial ceramic materials.

Tin is normally a metal (called white tin) but does undergo a low-temperature conversion to a nonmetallic form (called gray tin). Tin is obtained by reduction of cassiterite, a mineral form of $SnO_2$. Tin is used to make tin plate, bronze, and solder. Lead is obtained from galena, which is a sulfide ore, $PbS$. More than half of the lead produced is used to make electrodes for lead storage batteries. Litharge, or lead(II) oxide, is an important lead compound from which other lead compounds are prepared.

Carbon and tin are the least metallic of the Group IVA elements. *Catenation* is an important feature of carbon chemistry and is responsible for the enormous number of organic compounds. Carbon has several allotropes, the principal ones being diamond and graphite, which are covalent-network solids, and buckminsterfullerene, which is molecular ($C_{60}$). The element has important industrial uses, including carbon black for rubber tires. The principal oxides of carbon are $CO$ and $CO_2$. Mixtures of carbon monoxide and hydrogen (synthesis gas) are used to prepare various organic compounds. Liquid and solid carbon dioxide are used as refrigerants, and the gas is used to make carbonated beverages.

Hydrogen is the most abundant element in the universe and is the third most abundant element on the surface of the earth. Most of the hydrogen on earth is found in water. Hydrogen has three isotopes: *protium, deuterium,* and *tritium.* Protium is the most abundant, with less than 0.02% being deuterium and only a trace being radioactive tritium. Deuterium and tritium isotopes can be substituted for protium in chemical compounds in order to provide markers that can be followed during a chemical reaction, or to change the chemical and physical properties of the compound. Elemental hydrogen is produced on an industrial scale by the *steam-reforming process* in which a hydrocarbon is reacted with water in the presence of a catalyst at high temperature. The bulk of the hydrogen produced in this manner is used to make organic compounds including methanol. Hydrogen forms three classes of binary compounds called binary hydrides: ionic hydrides, covalent hydrides, and metallic hydrides. The *ionic hydrides* are reactive solids formed either by the reaction of hydrogen with an alkali metal to form compounds with the general formula $MH$,

or with larger alkaline earth metals to form $MH_2$. The *covalent hydrides* are compounds in which hydrogen is covalently bonded to another element. The *metallic hydrides* contain a transition metal element and hydrogen. In these compounds, the lattice of metal atoms forms a porous structure that allows hydrogen atoms to enter and bond. Metallic hydrides are often nonstoichiometric, meaning that the ratio of hydrogen atoms to metal atoms is not a whole number.

Silicon is the second most abundant element in the crust of the earth. The majority of silicon-containing compounds consist of chains and networks of silicon–oxygen bonds. Sharing some of the structural attributes of the network solids discussed in Chapter 13 are the *silicones:* materials that contain chains or rings of Si—O bonds with organic groups bonded to the silicon atoms. Depending on the amount and type of bonding between the rings or chains, the silicones can be formulated to be oils, *elastomers,* or resins. Due to their low reactivity and thermal stability, the silicones find many commercial and industrial applications ranging from cosmetics to hydraulic fluids. The *silicon hydrides* (also known as *silanes*) and *silicon halides* are silicon-containing compounds that do not contain Si—O bonds. The silanes consist of straight chains or branched chains of silicon atoms combined with hydrogen atoms with the general formula $Si_nH_{n+2}$ or as rings with the formula $Si_5H_{10}$ or $Si_6H_{12}$. The silicon halides can be formed by the direct reaction of Si with a halogen. Both the silanes and silicon halides are very reactive.

Of the Group VA elements, nitrogen and phosphorus are particularly important. Nitrogen, $N_2$, is obtained from liquid air by fractional distillation; liquid nitrogen is used as a refrigerant. Ammonia, $NH_3$, is the most important compound of nitrogen. It is prepared from the elements and is used as a fertilizer. Ammonia is also the starting compound for the manufacture of other nitrogen compounds. For example, in the *Ostwald process* for the preparation of nitric acid, ammonia is burned in the presence of a catalyst to nitric oxide, $NO$. The nitric oxide reacts with oxygen to give nitrogen dioxide, which reacts with water to give nitric acid.

Phosphorus has two common allotropes, white phosphorus ($P_4$) and red phosphorus (chain structure). White phosphorus is obtained by heating a phosphate mineral with sand and coke in an electric furnace. When phosphorus burns in air, it forms phosphorus(V) oxide, $P_4O_{10}$. This oxide reacts with water to give orthophosphoric acid, $H_3PO_4$. Phosphorus has many oxoacids; most are obtained by condensation reactions with orthophosphoric acid. One series is called the *polyphosphoric acids;* they have the general formula $H_{n+2}P_nO_{3n+1}$. Triphosphoric acid is an example; sodium triphosphate, $Na_5P_3O_{10}$, is used in detergents. The *metaphosphoric acids* have the general formula $(HPO_3)_n$.

Oxygen, a Group VIA element, occurs in the atmosphere (as $O_2$), but mostly it is present on earth as oxide and oxoanion minerals. Oxygen has two allotropes: dioxygen, $O_2$, and ozone, $O_3$. Dioxygen, usually called simply oxygen, is obtained commercially from liquid air. Oxygen reacts with almost all elements to give *oxides* or, in some cases, *peroxides* or *super-oxides.*

Sulfur, another Group VIA element, occurs in sulfate and sulfide minerals. Free sulfur, $S_8$, occurring in deep underground deposits is mined by the *Frasch process*. Sulfur is also produced by the *Claus process,* in which hydrogen sulfide (obtained from natural gas and petroleum) is partially burned. Most of the sulfur is used to prepare sulfuric acid by the *contact process*. In this process, sulfur is burned to sulfur dioxide, $SO_2$, which in the presence of a catalyst and oxygen forms sulfur trioxide, $SO_3$. This oxide dissolves in concentrated sulfuric acid, which when diluted with water gives additional sulfuric acid. Sulfuric acid is the most important compound of sulfur.

The Group VIIA elements, or halogens, are reactive. Chlorine ($Cl_2$), a pale greenish-yellow gas, is prepared commercially by the electrolysis of aqueous sodium chloride. Its principal uses are in the preparation of chlorinated hydrocarbons and as a bleaching agent and disinfectant. Hydrogen chloride, HCl, is one of the most important compounds of chlorine; aqueous solutions of HCl are known as hydrochloric acid.

The Group VIIIA elements, the noble gases, were discovered at the end of the nineteenth century. Although the noble gases were at first thought to be unreactive, compounds of xenon, krypton, radon, and argon are now known.

## Operational Skills

*Note:* The problem-solving skills used in this chapter are discussed in previous chapters.

# Review Questions

**22.1** Give the commercial source of each of the following metals: Li, Na, Mg, Ca, Al, Sn, Pb.

**22.2** Give equations for the reactions of lithium and sodium metals with water and with oxygen in air.

**22.3** Write the equation for the reaction of lithium carbonate with barium hydroxide.

**22.4** Which do you expect to be more reactive, lithium or potassium? Explain.

**22.5** Ethanol, $C_2H_5OH$, reacts with sodium metal because the hydrogen atom attached to the oxygen atom is slightly acidic. Write a balanced equation for the reaction of sodium with ethanol to give the salt sodium ethoxide, $NaOC_2H_5$.

**22.6** (a) Write electrode half-reactions for the electrolysis of fused sodium chloride. (b) Do the same for fused sodium hydroxide. The hydroxide ion is oxidized to oxygen and water.

**22.7** How is sodium hydroxide manufactured? What is another product in this process?

**22.8** Describe common uses of the following sodium compounds: sodium chloride, sodium hydroxide, and sodium carbonate.

**22.9** Describe the main step in the Solvay process. Give the balanced equation for this reaction.

**22.10** Because magnesium reacts with oxygen, steam, and carbon dioxide, magnesium fires are extinguished only by smothering the fire with sand. Write balanced equations for the reactions of magnesium with $O_2$, $H_2O(g)$, and $CO_2$.

**22.11** (a) Calcium oxide is prepared industrially from what calcium compound? Write the chemical equation for the reaction. (b) Write the equation for the preparation of calcium hydroxide.

**22.12** Calcium carbonate is used in some antacid preparations to neutralize the hydrochloric acid in the stomach. Write the equation for this neutralization.

**22.13** Purified calcium carbonate is prepared by precipitation. Write the equation for the commercial process using carbon dioxide. You can also prepare calcium carbonate by precipitation using sodium carbonate. Write the equation.

**22.14** Write the chemical equation of the thermite reaction, in which iron(III) oxide is reduced by aluminum.

**22.15** What are some major uses of aluminum oxide?

**22.16** How is aluminum sulfate used to purify municipal water supplies?

**22.17** Lead(IV) oxide forms the cathode of lead storage batteries. How is this substance produced for these batteries?

**22.18** Why are lead pigments no longer used in house paints?

**22.19** The yellow paint pigment chrome yellow is produced by a precipitation reaction. Write an equation in which chrome yellow is produced from lead(II) nitrate.

**22.20** Describe the reactions that are used in the steam-reforming process for the production of hydrogen.

**22.21** Give the names and symbols for the three isotopes of hydrogen. Which isotope is radioactive?

**22.22** Explain why hydrogen has the potential to be widely used as a fuel.

**22.23** Give an example of a compound of each of the binary hydrides.

**22.24** What is meant by the term *catenation*? Give an example of a carbon compound that displays catenation.

**22.25** Carbon monoxide is a poisonous gas. What is the mechanism of this poisoning?

**22.26** Write equations for the equilibria involving carbon dioxide in water.

**22.27** Describe some of the desirable properties of silicones that have led to their widespread use.

**22.28** Give the molecular formula for two *silanes* and two hydrocarbon analogs for these compounds.

**22.29** Write the balanced equation for the reaction of a halogen with silicon to form a *silicon halide*.

**22.30** Describe the natural cycle of nitrogen from the atmosphere to biological organisms and back to the atmosphere.

**22.31** Describe Rutherford's preparation of nitrogen from air. Was the gas he obtained pure nitrogen? Explain.

**22.32** List the different nitrogen oxides. What is the oxidation number of nitrogen in each?

**22.33** In your own words, describe the manufacture of ammonia from natural gas, steam, and air.

**22.34** Describe the steps in the Ostwald process for the manufacture of nitric acid from ammonia.

**22.35** Describe the structure of white phosphorus. How does the structure account for the reactivity of this substance?

**22.36** Give chemical equations for the reaction of white phosphorus in an excess of air and for the reaction of the product with water.

**22.37** Describe two different methods used to manufacture phosphoric acid, $H_3PO_4$, starting from $Ca_3(PO_4)_2$.

**22.38** By means of an equation, show how triphosphoric acid could be formed from orthophosphoric acid and diphosphoric acid.

**22.39** What is the purpose of adding polyphosphates to a detergent?

**22.40** What reaction was used by Priestley in preparing pure oxygen?

**22.41** What is the most important commercial means of producing oxygen?

**22.42** Define *oxide, peroxide,* and *superoxide.* Give an example of each.

**22.43** Give an example of an acidic oxide and a basic oxide.

**22.44** List three natural sources of sulfur or sulfur compounds.

**22.45** Describe the structure of the rhombic sulfur molecule.

**22.46** Discuss the various allotropes of sulfur and describe how they can be prepared from rhombic sulfur.

**22.47** Describe the Frasch process for mining sulfur.

**22.48** With the aid of chemical equations, describe the Claus process for the production of sulfur from hydrogen sulfide.

**22.49** Give equations for preparing each of the following:
a. $H_2S$  b. $SO_2$

**22.50** Give equations for the steps in the contact process for the manufacture of sulfuric acid from sulfur.

**22.51** Write equations for each of the following.
a. reaction of $H_2S$ with $SO_2$
b. oxidation of $SO_2$ with $Cr_2O_7{}^{2-}(aq)$ to $SO_4{}^{2-}$
c. reaction of hot, concentrated $H_2SO_4$ with Cu
d. reaction of sulfur with $Na_2SO_3$

**22.52** Give the equation for the reaction used by the Swedish chemist Scheele to prepare chlorine.

**22.53** Complete and balance the following equations. Write *NR* if no reaction occurs.
a. $I_2(aq) + Cl^-(aq) \longrightarrow$
b. $Cl_2(aq) + Br^-(aq) \longrightarrow$
c. $Br_2(aq) + I^-(aq) \longrightarrow$
d. $Br_2(aq) + Cl^-(aq) \longrightarrow$

**22.54** A test tube contains a solution of one of the following salts: NaCl, NaBr, NaI. Describe a single test that can distinguish among these possibilities.

**22.55** What are some major commercial uses of chlorine?

**22.56** How is sodium hypochlorite prepared? Give the balanced chemical equation.

**22.57** Describe the preparation of perchloric acid from sodium chloride.

**22.58** What was the argument used by Bartlett that led him to the first synthesis of a noble-gas compound?

# Conceptual Problems

**22.59** When producing coke, why is the coal heated in the absence of air? Write the chemical reaction for what would happen if it were to be heated in air.

**22.60** Even though hydrogen isn't a metal, why is it in Group IA of most periodic tables?

**22.61** What happens to the metallic character of the main-group elements as you move left to right across any row of the periodic table? What happens to the metallic character of the main-group elements as you move down a column (group)?

**22.62** Lithium hydroxide, like sodium hydroxide, becomes contaminated when exposed to air. What is the source of this contamination? What reactions take place?

**22.63** Aluminum hydroxide is an amphoteric substance. What does this mean? Write equations to illustrate.

**22.64** Tin metal would not make a very good structural metal in cold climates. Why?

**22.65** Oxygen, like other second-period elements, is somewhat different from the other elements in its group. List some of these differences.

**22.66** Given that the reaction $Cl_2(g) + 2KBr(aq) \longrightarrow 2KCl(aq) + Br_2(aq)$ readily occurs, would you expect the reaction $I_2(s) + 2KCl(aq) \longrightarrow 2KI(aq) + Cl_2(aq)$ to occur?

**22.67** Hydrogen chloride can be prepared by heating NaCl with concentrated sulfuric acid. Why is substituting NaBr for NaCl in this reaction not a satisfactory way to prepare HBr?

**22.68** Do you expect an aqueous solution of sodium hypochlorite to be acidic, neutral, or basic? What about an aqueous solution of sodium perchlorate?

## Practice Problems

### Group IA: The Alkali Metals

**22.69** Caustic soda, NaOH, can be manufactured from sodium carbonate in a manner similar to the preparation of lithium hydroxide. Write balanced equations (in three steps) for the preparation of NaOH from slaked lime, $Ca(OH)_2$, salt (NaCl), carbon dioxide, ammonia, and water.

**22.70** Sodium phosphate, $Na_3PO_4$, is produced by the neutralization reaction. Phosphoric acid, $H_3PO_4$, is obtained by burning phosphorus to $P_4O_{10}$, then reacting the oxide with water to give the acid. Write balanced equations (in four steps) for the preparation of $Na_3PO_4$ from $P_4$, $H_2O$, air, and NaCl.

**22.71** Complete and balance the following equations.
a. $K(s) + Br_2(l) \longrightarrow$
b. $K(s) + H_2O(l) \longrightarrow$
c. $NaOH(s) + CO_2(g) \longrightarrow$
d. $Li_2CO_3(aq) + HNO_3(aq) \longrightarrow$
e. $K_2SO_4(aq) + Pb(NO_3)_2(aq) \longrightarrow$

**22.72** Complete and balance the following equations.
a. $LiHCO_3(s) \xrightarrow{\Delta}$
b. $Na_2SO_4(aq) + BaCl_2(aq) \longrightarrow$
c. $K_2CO_3(aq) + Ca(OH)_2(aq) \longrightarrow$
d. $Li(s) + HCl(aq) \longrightarrow$
e. $Na(s) + ZrCl_4(g) \longrightarrow$

**22.73** Francium was discovered as a minor decay product of actinium-227. Write the nuclear equation for the decay of actinium-227 by alpha emission.

**22.74** Francium-223 is a radioactive alkali metal that decays by beta emission. Write the nuclear equation for this decay process.

### Group IIA: The Alkaline Earth Metals

**22.75** Sodium hydroxide and calcium hydroxide are strong bases. What simple chemical test could you use to distinguish between solutions of these two bases?

**22.76** Potassium hydroxide and barium hydroxide are strong bases. What simple chemical test could you use to distinguish between solutions of these two bases?

**22.77** Devise a chemical method for separating a solution containing $MgCl_2$ and $BaCl_2$ to give two solutions or compounds each containing only one of the metal ions.

**22.78** Devise a chemical method for separating a solution containing NaCl and $MgCl_2$ to give two solutions or compounds each containing only one of the metal ions.

**22.79** Thorium-230, which occurs in uranium minerals, decays by alpha emission to radium. Write the nuclear equation for this decay process.

**22.80** A short-lived isotope of radium decays by alpha emission to radon-219. Write the nuclear equation for this decay process.

**22.81** Complete and balance the following equations.
a. $BaCO_3(s) \xrightarrow{\Delta}$
b. $Ba(s) + H_2O(l) \longrightarrow$
c. $Mg(OH)_2(s) + HNO_3(aq) \longrightarrow$
d. $Mg(s) + NiCl_2(aq) \longrightarrow$
e. $NaOH(aq) + MgSO_4(aq) \longrightarrow$

**22.82** Complete and balance the following equations.
a. $KOH(aq) + MgCl_2(aq) \longrightarrow$
b. $Mg(s) + CuSO_4(aq) \longrightarrow$
c. $Sr(s) + H_2O(l) \longrightarrow$
d. $SrCO_3(s) + HCl(aq) \longrightarrow$
e. $Ba(OH)_2(aq) + CO_2(g) \longrightarrow$

**22.83** You have a 0.12 $M$ solution of calcium hydrogen carbonate. How many grams of calcium hydroxide must be added to 25.0 mL of the solution to precipitate all of the calcium?

**22.84** You have a 0.21 $M$ solution of magnesium ion. How many grams of calcium hydroxide must be added to 50.0 mL of the solution to precipitate all of the magnesium ion?

### Group IIIA and Group IVA Metals

**22.85** Baking powders contain sodium (or potassium) hydrogen carbonate and an acidic substance. When water is added to a baking powder, carbon dioxide is released. One kind of baking powder contains $NaHCO_3$ and sodium aluminum sulfate, $NaAl(SO_4)_2$. Write the net ionic equation for the reaction that occurs in water solution.

**22.86** When aluminum sulfate is dissolved in water, it produces an acidic solution. Suppose the pH of this solution is raised by the dropwise addition of aqueous sodium hydroxide. (a) Describe what you would observe as the pH continues to rise. (b) Write balanced equations for any reactions that occur.

**22.87** The following solid substances are in separate but unlabeled test tubes: $Al_2(SO_4)_3 \cdot 18H_2O$, $BaCl_2 \cdot 2H_2O$, KOH. Describe how you could identify the compounds by chemical tests using only these substances and water.

**22.88** Unlabeled test tubes contain solid $AlCl_3 \cdot 6H_2O$ in one, $Ba(OH)_2 \cdot 8H_2O$ in another, and $MgSO_4 \cdot 7H_2O$ in the other. How could you find out what is in each test tube, using chemical tests that involve only these compounds plus water?

**22.89** The $Sn^{2+}(aq)$ ion can be written in more detail as $Sn(H_2O)_6^{2+}$. This ion is acidic by hydrolysis. Write a possible equation for this hydrolysis.

**22.90** Lead(II) nitrate, one of the few soluble lead salts, gives a solution with a pH of about 3 to 4. Write a chemical equation to explain why the solution is acidic.

**22.91** Lead(IV) oxide is a strong oxidizing agent. For example, lead(IV) oxide will oxidize hydrochloric acid to chlorine, $Cl_2$. Write the balanced equation for this reaction.

**22.92** Lead(IV) oxide can be prepared by oxidizing plumbite ion, $Pb(OH)_3^-$, which exists in a basic solution of $Pb^{2+}$. Write the balanced equation for this oxidation by $OCl^-$ in basic solution.

**22.93** Complete and balance the following equations.
a. $Al_2O_3(s) + H_2SO_4(aq) \longrightarrow$
b. $Al(s) + AgNO_3(aq) \longrightarrow$
c. $Pb(NO_3)_2(aq) + NaI(aq) \longrightarrow$
d. $Al(s) + Mn_3O_4(s) \longrightarrow$
e. $Ga(OH)_3(s) \xrightarrow{\Delta}$

**22.94** Complete and balance the following equations.
a. $Pb(NO_3)_2(aq) + Al(s) \longrightarrow$
b. $Pb(NO_3)_2(aq) + Na_2CrO_4(aq) \longrightarrow$
c. $Al_2(SO_4)_3(aq) + \text{dilute } LiOH(aq) \longrightarrow$
d. $Al(s) + HCl(aq) \longrightarrow$
e. $Sn(s) + HBr(aq) \longrightarrow$

## Hydrogen

**22.95** Calculate the amount of heat evolved when $2.5 \times 10^4$ kg of hydrogen is combusted.

$$2H_2(g) + O_2(g) \longrightarrow 2H_2O(g) \qquad \Delta H = -484 \text{ kJ}$$

**22.96** How much heat will be evolved when 20.0 grams of the binary covalent hydride HF is produced via the following reaction?

$$F_2(g) + H_2(g) \longrightarrow 2HF(g) \qquad \Delta H = -546 \text{ kJ}$$

**22.97** Indicate the oxidation state for the element noted in each of the following:
a. H in $CaH_2$    b. H in $H_2O$
c. C in $CH_4$    d. S in $H_2SO_4$

**22.98** Indicate the oxidation state for the element noted in each of the following:
a. H in $H_2$    c. Si in $SiH_4$
b. H in $C_2H_4$    d. N in $HNO_3$

## Group IVA: The Carbon Family

**22.99** Describe the bonding (using valence bond theory) of the Group IVA atoms in each of the following:
a. $C_2H_6$    b. $SiF_6^{2-}$
c. $CH_3CH{=}CH_2$    d. $SiH_4$

**22.100** Describe the bonding (using valence bond theory) of the Group IVA atoms in each of the following:
a. $CCl_4$    b. HCN
c. $SiF_4$    d. $CH_3COOH$ (acetic acid)

**22.101** Calculate the standard enthalpy change for the following "cracking" reactions.
a. $CH_4(g) \longrightarrow C(\text{graphite}) + 2H_2(g)$
b. $C_2H_6(g) \longrightarrow C_2H_4(g) + H_2(g)$

**22.102** Calculate the standard enthalpy change for the following reactions involving synthesis gas.
a. $CO(g) + 2H_2(g) \longrightarrow CH_3OH(g)$
b. $CO(g) + 3H_2(g) \longrightarrow CH_4(g) + H_2O(g)$

**22.103** Complete and balance the following equations.
a. $CO_2(g) + Ba(OH)_2(aq) \longrightarrow$
b. $MgCO_3(s) + HBr(aq) \longrightarrow$

**22.104** Complete and balance the following equations.
a. $NaHCO_3(aq) + HC_2H_3O_2(aq) \longrightarrow$
b. $Ca(HCO_3)_2(aq) + Ca(OH)_2(aq) \longrightarrow$

**22.105** Use balanced equations to show how you could prepare $Na_2CO_3$ from carbon, NaOH, air, and $H_2O$.

**22.106** Use balanced equations to show how you could prepare methanol, $CH_3OH$, from ethane, $C_2H_6$, and water.

## Group VA: Nitrogen and the Phosphorus Family

**22.107** Magnesium nitride, $Mg_3N_2$, reacts with water to produce magnesium hydroxide and ammonia. How many grams of ammonia can you obtain from 7.50 g of magnesium nitride?

**22.108** Ammonia reacts with oxygen in the presence of a platinum catalyst to give nitric oxide, NO. How many grams of oxygen are required in this reaction to give 5.00 g NO?

**22.109** You have the following substances: $NH_3$, $O_2$, Pt, and $H_2O$. Write equations for the preparation of $N_2O$ from these substances.

**22.110** Give equations for the preparation of $N_2O$. You can use NaOH, $NaNO_3$, $H_2SO_4$, and $(NH_4)_2SO_4$ (plus $H_2O$). Several steps are required.

**22.111** Zinc metal reacts with concentrated nitric acid to give zinc ion and ammonium ion. Write the balanced equation for this reaction.

**22.112** Silver metal reacts with nitric acid to give silver ion and nitric oxide. Write the balanced equation for this reaction.

**22.113** Although phosphorus pentabromide exists in the vapor phase as $PBr_5$ molecules, in the solid phase the substance is ionic and has the structure $[PBr_4^+]Br^-$. What is the expected geometry of $PBr_4^+$? Describe the bonding to phosphorus in $PBr_4^+$.

**22.114** Although phosphorus pentachloride exists in the vapor phase as $PCl_5$ molecules, in the solid phase the substance is ionic and has the structure $[PCl_4^+][PCl_6^-]$. What is the expected geometry of $PCl_6^-$? Describe the bonding to phosphorus in $PCl_6^-$.

**22.115** Phosphorous acid, $H_3PO_3$, is oxidized to phosphoric acid, $H_3PO_4$, by hot, concentrated sulfuric acid, which is reduced to $SO_2$. Write the balanced equation for this reaction.

**22.116** Phosphorous acid, $H_3PO_3$, is oxidized to phosphoric acid, $H_3PO_4$, by nitric acid, which is reduced to nitric oxide, NO. Write the balanced equation for this reaction.

**22.117** According to an analysis, a sample of phosphate rock contains 74.6% $Ca_3(PO_4)_2$, by mass. How many grams of phosphoric acid, $H_3PO_4$, can be obtained from 15.0 g of the phosphate rock according to the following reaction?

$$Ca_3(PO_4)_2(s) + 3H_2SO_4(aq) \longrightarrow 3CaSO_4(s) + 2H_3PO_4(aq)$$

**22.118** According to an analysis, a sample of phosphate rock contains 71.2% $Ca_3(PO_4)_2$, by mass. How many grams of calcium dihydrogen phosphate, $Ca(H_2PO_4)_2$, can be obtained from 10.0 g of phosphate rock from the following reaction?

$$Ca_3(PO_4)_2(s) + 4H_3PO_4(aq) \longrightarrow 3Ca(H_2PO_4)_2(aq)$$

## Group VIA: Oxygen and the Sulfur Family

**22.119** Write an equation for each of the following.
a. burning of lithium metal in oxygen
b. burning of methylamine, $CH_3NH_2$, in excess oxygen (N ends up as $N_2$)
c. burning of diethyl sulfide, $(C_2H_5)_2S$, in excess oxygen

**22.120** Write an equation for each of the following.
a. burning of calcium metal in oxygen
b. burning of phosphine, $PH_3$, in excess oxygen
c. burning of ethanolamine, $HOCH_2CH_2NH_2$, in excess oxygen (N ends up as $N_2$)

**22.121** What are the oxidation numbers of sulfur in each of the following?
a. $SF_6$    b. $SO_3$    c. $H_2S$    d. $CaSO_3$

**22.122** What are the oxidation numbers of sulfur in each of the following?
a. $S_8$    b. CaS    c. $CaSO_4$    d. $SCl_4$

**22.123** Selenous acid, $H_2SeO_3$, is reduced by $H_2S$ to sulfur, $S_8$, and selenium, $Se_8$. Write the balanced equation for this reaction.

**22.124** Concentrated sulfuric acid oxidizes iodide ion to iodine, $I_2$. Write the balanced equation for this reaction.

**22.125** Sodium hydrogen sulfite is prepared from sodium carbonate and sulfur dioxide:

$$Na_2CO_3(s) + 2SO_2(g) + H_2O(l) \longrightarrow \\ 2NaHSO_3(aq) + CO_2(g)$$

How many grams of $NaHSO_3$ can be obtained from 25.0 g of $Na_2CO_3$?

**22.126** Sodium thiosulfate, $Na_2S_2O_3$, is prepared from sodium sulfite and sulfur:

$$8Na_2SO_3(aq) + S_8(s) \longrightarrow 8Na_2S_2O_3(aq)$$

How many grams of $Na_2S_2O_3$ can be obtained from 25.0 g of sulfur?

## Group VIIA: The Halogens

**22.127** A solution of chloric acid may be prepared by reacting a solution of barium chlorate with sulfuric acid. Barium sulfate precipitates. Write the balanced equation for the reaction.

**22.128** A solution of chlorous acid may be prepared by reacting a solution of barium chlorite with sulfuric acid. Barium sulfate precipitates. Write the balanced equation for the reaction.

**22.129** Chlorine can be prepared by oxidizing chloride ion (from hydrochloric acid) with potassium dichromate, $K_2Cr_2O_7$, which is reduced to $Cr^{3+}$. Write the balanced equation for the reaction.

**22.130** Iodic acid, $HIO_3$, can be prepared by oxidizing elemental iodine with concentrated nitric acid, which is reduced to nitrogen dioxide, $NO_2(g)$. Write the balanced equation for the reaction.

**22.131** Discuss the bonding in each of the following molecules or ions. What is the expected molecular geometry?
a. $Cl_2O$    b. $BrO_3^-$    c. $BrF_3$

**22.132** Discuss the bonding in each of the following molecules or ions. What is the expected molecular geometry?
a. HClO    b. $ClO_4^-$    c. $ClF_5$

**22.133** Write balanced equations for each of the following.
a. Bromine reacts with aqueous sodium hydroxide to give hypobromite and bromide ions.

b. Hydrogen bromide gas forms when sodium bromide is heated with phosphoric acid.

**22.134** Write balanced equations for each of the following.
a. Solid calcium fluoride is heated with sulfuric acid to give hydrogen fluoride vapor.
b. Solid potassium chlorate is carefully heated to yield potassium chloride and potassium perchlorate.

**22.135** By calculating the standard emf, decide whether aqueous sodium hypochlorite solution will oxidize $Fe^{2+}(aq)$ to $Fe^{3+}(aq)$ in basic solution under standard conditions. See Appendix I for data.

**22.136** Using standard electrode potentials, decide whether aqueous sodium hypochlorite solution will oxidize $Br^-$ to $Br_2$ in basic solution under standard conditions. See Appendix I for data.

## Group VIIIA: The Noble Gases

**22.137** Xenon tetrafluoride, $XeF_4$, is a colorless solid. Give the Lewis formula for the $XeF_4$ molecule. What is the hybridization of the xenon atom in this compound? What geometry is predicted by the VSEPR model for this molecule?

**22.138** Xenon tetroxide, $XeO_4$, is a colorless, unstable gas. Give the Lewis formula for the $XeO_4$ molecule. What is the hybridization of the xenon atom in this compound? What geometry would you expect for this molecule?

**22.139** Xenon difluoride, $XeF_2$, is hydrolyzed (broken up by water) in basic solution to give xenon, fluoride ion, and $O_2$ as products. Write a balanced equation for the reaction.

**22.140** Xenon trioxide, $XeO_3$, is reduced to xenon in acidic solution by iodide ion. Iodide ion is oxidized to iodine, $I_2$. Write a balanced chemical equation for the reaction.

# General Problems

**22.141** A 50.00-mL volume of 0.4987 $M$ HCl was added to a 5.436-g sample of milk of magnesia. This solution was then titrated with 0.2456 $M$ NaOH. If it required 39.42 mL of NaOH to reach the endpoint, what was the mass percentage of $Mg(OH)_2$ in the milk of magnesia?

**22.142** An antacid tablet consists of calcium carbonate with other ingredients. The calcium carbonate in a 0.9863-g sample of the antacid was dissolved in 50.00 mL of 0.5068 $M$ HCl, then titrated with 41.23 mL of 0.2601 $M$ NaOH. What was the mass percentage of $CaCO_3$ in the antacid?

**22.143** Calculate the standard enthalpy change, $\Delta H^\circ$, for the following reaction at 25°C.

$$Fe_2O_3(s) + 2Al(s) \longrightarrow 2Fe(s) + Al_2O_3(s)$$

What is the enthalpy change per mole of iron?

**22.144** Calculate the standard enthalpy change, $\Delta H^\circ$, for the following reaction at 25°C.

$$3CaO(s) + 2Al(s) \longrightarrow 3Ca(s) + Al_2O_3(s)$$

What is the enthalpy change per mole of calcium?

**22.145** A sample of limestone was dissolved in hydrochloric acid, and the carbon dioxide gas that evolved was collected. If a 0.1662-g sample of limestone gave 34.56 mL of dry carbon dioxide gas at 745 mmHg and 21°C, what was the mass percentage of $CaCO_3$ in the limestone?

**22.146** A sample of rock containing magnesite, $MgCO_3$, was dissolved in hydrochloric acid, and the carbon dioxide gas that evolved was collected. If a 0.1504-g sample of the rock gave 37.71 mL of dry carbon dioxide gas at 758 mmHg and 22°C, what was the mass percentage of $MgCO_3$ in the rock?

**22.147** How many grams of sodium chloride are required to produce 10.00 g of $NaHCO_3$ by the Solvay process?

**22.148** How many grams of aluminum are required to react with 15.00 g of chromium(III) oxide by the Goldschmidt process for the production of chromium metal?

**22.149** Estimate the temperature at which strontium carbonate begins to decompose to strontium oxide and $CO_2$ at 1 atm.

$$SrCO_3(s) \longrightarrow SrO(s) + CO_2(g)$$

Use thermodynamic data in Appendix C.

**22.150** Estimate the temperature at which barium carbonate decomposes to barium oxide and $CO_2$ at 1 atm.

$$BaCO_3(s) \longrightarrow BaO(s) + CO_2(g)$$

Use thermodynamic data in Appendix C.

**22.151** Calculate $E^\circ$ for the disproportionation of $In^+(aq)$.

$$3In^+(aq) \rightleftharpoons 2In(s) + In^{3+}(aq)$$

(*Disproportionation* is a reaction in which a species undergoes both oxidation and reduction.) Use the following standard potentials:

$$In^+(aq) + e^- \rightleftharpoons In(s); E^\circ = -0.21 \text{ V}$$
$$In^{3+}(aq) + 2e^- \rightleftharpoons In^+(aq); E^\circ = -0.40 \text{ V}$$

From $E^\circ$, calculate $\Delta G^\circ$ for the disproportionation (in kilojoules). Does this reaction occur spontaneously?

**22.152** Calculate $E^\circ$ for the disproportionation of $Tl^+(aq)$.

$$3Tl^+(aq) \rightleftharpoons 2Tl(s) + Tl^{3+}(aq)$$

(*Disproportionation* is a reaction in which a species undergoes both oxidation and reduction.) Use the following standard potentials:

$$Tl^+(aq) + e^- \rightleftharpoons Tl(s); \quad E° = -0.34 \text{ V}$$
$$Tl^{3+}(aq) + 2e^- \rightleftharpoons Tl^+(aq); \quad E° = 1.25 \text{ V}$$

From $E°$, calculate $\Delta G°$ for the disproportionation (in kilojoules). Does this reaction occur spontaneously?

**22.153** Lithium hydroxide has been used in spaceships to absorb carbon dioxide exhaled by astronauts. Assuming that the product is lithium carbonate, determine what mass of lithium hydroxide is needed to absorb the carbon dioxide from 1.00 L of air containing 30.0 mmHg partial pressure of $CO_2$ at 25°C.

**22.154** Potassium chlorate, $KClO_3$, is used in fireworks and explosives. It can be prepared by bubbling chlorine into hot aqueous potassium hydroxide; $KCl(aq)$ and $H_2O$ are the other products in the reaction. How many grams of $KClO_3$ can be obtained from 156 L of $Cl_2$ whose pressure is 784 mmHg at 25°C?

**22.155** The main ingredient in many phosphate fertilizers is $Ca(H_2PO_4)_2 \cdot H_2O$. If a fertilizer is 17.1% P (by mass), and all of this phosphorus is present as $Ca(H_2PO_4)_2 \cdot H_2O$, what is the mass percentage of this salt in the fertilizer?

**22.156** A fertilizer contains phosphorus in two compounds, $Ca(H_2PO_4)_2 \cdot H_2O$ and $CaHPO_4$. The fertilizer contains 30.0% $Ca(H_2PO_4)_2 \cdot H_2O$ and 10.0% $CaHPO_4$ (by mass). What is the mass percentage of phosphorus in the fertilizer?

**22.157** NaClO solution is made by electrolysis of $NaCl(aq)$ by allowing the products NaOH and $Cl_2$ to mix. How long must a cell operate to produce $1.00 \times 10^3$ L of 5.25% NaClO solution (density = 1.00 g/mL) if the cell current is $3.00 \times 10^3$ A?

**22.158** Sodium perchlorate, $NaClO_4$, is produced by electrolysis of sodium chlorate, $NaClO_3$. If a current of $2.50 \times 10^3$ A passes through an electrolytic cell, how many kilograms of sodium perchlorate are produced per hour?

**22.159** The amount of sodium hypochlorite in a bleach solution can be determined by using a given volume of bleach to oxidize excess iodide ion to iodine, because the reaction goes to completion. The amount of iodine produced is then determined by titration with sodium thiosulfate, $Na_2S_2O_3$, which is oxidized to sodium tetrathionate, $Na_2S_4O_6$. Potassium iodide was added in excess to 5.00 mL of bleach (density = 1.00 g/mL). This solution, containing the iodine released in the reaction, was titrated with 0.100 $M$ $Na_2S_2O_3$. If 34.6 mL of sodium thiosulfate was required to reach the endpoint (detected by disappearance of the blue color of the starch–iodine complex), what was the mass percentage of NaClO in the bleach?

**22.160** Ascorbic acid, $C_6H_8O_6$ (vitamin C), is a reducing agent. The amount of this acid in solution can be determined quantitatively by a titration procedure involving iodine, $I_2$, in which ascorbic acid is oxidized to dehydroascorbic acid, $C_6H_6O_6$.

$$C_6H_8O_6(aq) + I_2(aq) \longrightarrow C_6H_6O_6(aq) + 2HI(aq)$$

A 30.0-g sample of an orange-flavored beverage mix was placed in a flask to which 10.00 mL of 0.0500 $M$ $KIO_3$ and excess KI were added. The $IO_3^-$ and $I^-$ ions react in acid solution to give $I_2$, which then reacts with ascorbic acid. Excess iodine is titrated with sodium thiosulfate (see Problem 22.159). If 31.2 mL of 0.0300 $M$ $Na_2S_2O_3$ is required to titrate the excess $I_2$, how many grams of ascorbic acid are there in 100.0 g of beverage mix?

 ## Media Activities

**22.161** Watch the Reaction of Potassium in Water *Visualization* on the student CD.
a. If cesium metal were reacted with water, how would it compare with the reaction shown in the *Visualization?* Write the balanced equation for the reaction of Cs(*s*) with water; be sure to include all state symbols in your equation.
b. What are the signs of: $\Delta H$, $\Delta S$, and $\Delta G$ for the reaction of sodium with water?

**22.162** Watch the Burning of a Match in $CO_2$ and the Reaction of Magnesium Metal with $CO_2$ *Visualizations* on the student web site.
a. Why did the match go out in the $CO_2$ atmosphere?
b. Why did the magnesium metal react with the $CO_2$?
c. If you assume that the products of the magnesium and carbon dioxide reaction are carbon and magnesium oxide, identify the oxidizing agent and the reducing agent.

**22.163** Watch the Barking Dogs (White Phosphorus) *Visualization* on the student web site.
a. Using the tabulated values from Appendix C, calculate $\Delta G°$ for the reaction of white phosphorus, $P_4$, with oxygen, $O_2$, to produce $P_4O_{10}(s)$. (Don't forget to balance the equation.) Do the results of this calculation agree with what you observed in the *Visualization?*
b. Using the tabulated values from Appendix C, calculate $\Delta S°$ and $\Delta H°$ for the reaction: $N_2(g) + 2O_2(g) \longrightarrow 2NO_2(g)$. Based on the results of these calculations, would you expect this reaction to be spontaneous?
c. Why do you think $N_2$ is much less reactive than $P_4$?

**22.164** For review, take the ACE practice test for Chapter 22 on the student web site.

Heating pink [CO(NH₃)5H₂O]Cl₃(s) drives off water molecules, forming blue [CO(NH₃)5Cl]Cl₂(s).

# The Transition Elements and Coordination Compounds

I n the previous chapter, we studied the main-group elements—the A groups in the periodic table. Between columns IIA and IIIA are ten columns of the transition elements—the B groups. Among these elements are metals with familiar commercial applications: iron tools, copper wire, silver jewelry and coins. Many catalysts for important industrial reactions involve transition elements. Examples of these are found in petroleum refining and in the synthesis of ammonia from $N_2$ and $H_2$.

In addition to their commercial usefulness, many transition elements have biological importance. Iron compounds, for example, are found throughout the plant and animal kingdoms. Iron is present in hemoglobin, the molecule in red blood cells that is responsible for the transport of oxygen, $O_2$, from the lungs to other body tissue. Myoglobin, found in muscle tissue, is a similar molecule containing iron. Myoglobin takes oxygen from hemoglobin, holding it until it is required by the muscle cells. Cytochromes, iron-containing compounds within each cell, are involved in the oxidation

▶ You can tailor the material in this chapter to the particular needs of your course. For example, if you want to concentrate on coordination compounds, you can begin with Section 23.3. You may also want to limit your discussion of bonding to either Section 23.6 or 23.7.

of food molecules. In these cases, the transition element is central to the structure and function of the biological molecule. Hemoglobin and myoglobin are examples of *metal complexes,* or *coordination compounds,* in which the metal atom is surrounded by other atoms bonded to it by the electron pairs these atoms donate. In hemoglobin and myoglobin, the $O_2$ molecule bonds to the iron atom.

We will look at the general properties of transition metals in the first section of the chapter; in the second section we will investigate the chemistry of two transition elements, chromium and copper. The remaining sections of the chapter will cover the structure, naming, and bonding of complex ions and coordination compounds.

# Properties of the Transition Elements

Often, all of the B elements of the periodic table are classified as transition elements, though strictly speaking, the **transition elements** are defined as *those metallic elements that have an incompletely filled d subshell or easily give rise to common ions that have incompletely filled d subshells.* Iron, whose electron configuration is $[Ar]3d^6 4s^2$ is an example of such an element; iron has an incompletely filled $3d$ subshell. Another example is copper. Although the free element has the configuration $[Ar]3d^{10}4s^1$, in which the $3d$ subshell is complete, copper readily forms the copper(II) ion, whose configuration $[Ar]3d^9$ has an incomplete $3d$ subshell. The Group IIB elements zinc, cadmium, and mercury have filled $d$ subshells in the element and in the common ions, so in the strict sense these are not transition elements, although they are often included with them.

Sometimes, too, chemists include the two rows of elements at the bottom of the periodic table with the transition elements. These two rows, often referred to as the *inner-transition elements,* have partially filled $f$ subshells in common oxidation states. The elements in the first row are called the *lanthanides,* or *rare earths,* and the elements in the second row are called the *actinides.* Figure 23.1 shows the divisions of the transition elements. The B columns of transition elements, as well as the inner-transition elements, frequently form complex ions and coordination compounds.

**Figure 23.1**
**Classification of the transition elements.** The classification into the B groups of transition elements and the inner-transition elements.

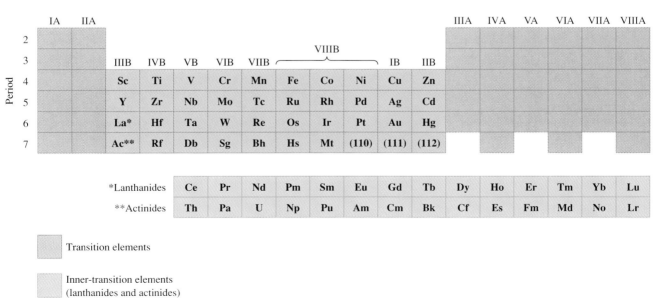

# 23.1  Periodic Trends in the Transition Elements

**Understanding Concepts:
Periodic Trends**

The transition elements have a number of characteristics that set them apart from the main-group elements:

1. All of the transition elements are metals and, except for the IIB elements, have high melting points and high boiling points and are hard solids. For instance, of the fourth-period elements from scandium to copper, the lowest-melting metal is copper (1083°C) and the highest-melting metal is vanadium (1890°C). Of the main-group metals, only beryllium melts above 1000°C; the rest melt at appreciably lower temperatures.

2. With the exception of the IIIB and IIB elements, each transition element has several oxidation states. Vanadium, for example, exists as aqueous ions in all oxidation states from +2 to +5 (Figure 23.2). Of the main-group metals, only the heavier ones display several oxidation states. Because of their multiplicity of oxidation states, the transition elements are often involved in oxidation–reduction reactions.

3. Transition-metal compounds are often colored, and many are paramagnetic. Most compounds of the main-group metals are colorless and diamagnetic.

▶ **Demonstration of oxidation
states of vanadium: See the
Instructor's Resource Manual.**

We will look at points 1 and 2 in some detail in this section. In particular, we will examine trends in melting points, boiling points, hardness, and oxidation states. We will also look at trends in covalent radii and ionization energies, which we can relate to chemical properties. We will discuss the color and paramagnetism of transition-metal complexes later in the chapter.

## Electron Configurations

Electronic structure is central to any discussion of the transition elements. Table 23.1 lists the ground-state electron configurations of the fourth-period transition elements. For the most part, these configurations are predicted by the building-up principle. Following the building-up principle, the $3d$ subshell begins to fill after calcium (configuration $[Ar]4s^2$). Thus, scandium has the configuration $[Ar]3d^14s^2$, and, as you go across the period, additional electrons go into the $3d$ subshell. You get the configuration $[Ar]3d^24s^2$ for titanium, and $[Ar]3d^34s^2$ for vanadium. Then in chromium the

**Figure 23.2
Oxidation states of
vanadium.** The test tubes
contain, left to right: $VO_3^-(aq)$,
pale yellow; $VO^{2+}(aq)$,
bright blue; $V^{3+}$, gray-blue;
$V^{2+}$, violet. The oxidation
states of vanadium are, left to
right, +5, +4, +3, and +2.

**TABLE 23.1**
**Properties of the Fourth-Period Transition Elements**

| Property | Scandium | Titanium | Vanadium | Chromium | Manganese |
|---|---|---|---|---|---|
| Electron configuration | $[Ar]3d^14s^2$ | $[Ar]3d^24s^2$ | $[Ar]3d^34s^2$ | $[Ar]3d^54s^1$ | $[Ar]3d^54s^2$ |
| Melting point, °C | 1541 | 1660 | 1890 | 1857 | 1244 |
| Boiling point, °C | 2831 | 3287 | 3380 | 2672 | 1962 |
| Density, g/cm$^3$ | 3.0 | 4.5 | 6.0 | 7.2 | 7.2 |
| Electronegativity (Pauling scale) | 1.3 | 1.5 | 1.6 | 1.6 | 1.5 |
| Covalent radius, pm | 144 | 132 | 122 | 118 | 117 |
| Ionic radius (for $M^{2+}$), pm | — | 100 | 93 | 87 | 81 |

| Property | Iron | Cobalt | Nickel | Copper | Zinc |
|---|---|---|---|---|---|
| Electron configuration | $[Ar]3d^64s^2$ | $[Ar]3d^74s^2$ | $[Ar]3d^84s^2$ | $[Ar]3d^{10}4s^1$ | $[Ar]3d^{10}4s^2$ |
| Melting point, °C | 1535 | 1495 | 1453 | 1083 | 420 |
| Boiling point, °C | 2750 | 2870 | 2732 | 2567 | 907 |
| Density, g/cm$^3$ | 7.9 | 8.9 | 8.9 | 8.9 | 7.1 |
| Electronegativity (Pauling scale) | 1.8 | 1.8 | 1.8 | 1.9 | 1.6 |
| Covalent radius, pm | 117 | 116 | 115 | 117 | 125 |
| Ionic radius (for $M^{2+}$), pm | 75 | 79 | 83 | 87 | 88 |

configuration predicted by the building-up principle is $[Ar]3d^44s^2$, but the actual configuration is $[Ar]3d^54s^1$. The configurations for the rest of the elements are those predicted by the building-up principle, until you get to copper. Here, the predicted configuration is $[Ar]3d^94s^2$, but the actual configuration is $[Ar]3d^{10}4s^1$.

## Melting Points and Boiling Points

Table 23.1 reveals that the melting points of the transition metals increase from 1541°C for scandium to 1890°C for vanadium and 1857°C for chromium, then decrease from 1535°C for iron to 1083°C for copper and 420°C for zinc. The same pattern is observed in the fifth-period and sixth-period elements. As you read across a row of transition elements, the melting points increase, reaching a maximum at the Group VB or VIB elements, after which the melting points decrease. For instance, tungsten in Group VIB has the highest melting point (3410°C) of any metallic element, and mercury in Group IIB has the lowest (−39°C). Similar trends can be seen in the boiling points of the metals. These properties depend on the strength of metal bonding, which in turn depends roughly on the number of unpaired electrons in the metal atoms. At the beginning of a period of transition elements, there is one unpaired $d$ electron. The number of unpaired $d$ electrons increases across a period until Group VIB, after which the electrons begin to pair.

## Atomic Radii

Trends in atomic radii are of concern because chemical properties are determined in part by atomic size. Looking at the fourth-period covalent radii (one measure of atomic size) in Table 23.1, you see that they decrease quickly from scandium (144 pm) to titanium

**Figure 23.3**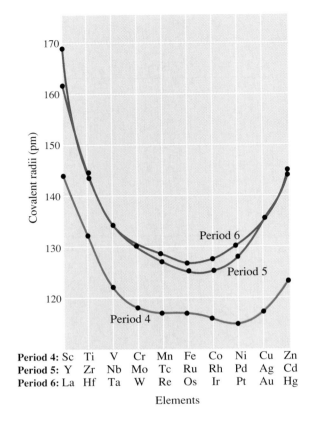
**Covalent radii for transition elements of Periods 4 to 6.** Note that elements of the fifth and sixth periods of the same column have approximately the same radii.

| Period 4: | Sc | Ti | V | Cr | Mn | Fe | Co | Ni | Cu | Zn |
| Period 5: | Y | Zr | Nb | Mo | Tc | Ru | Rh | Pd | Ag | Cd |
| Period 6: | La | Hf | Ta | W | Re | Os | Ir | Pt | Au | Hg |

Elements

(132 pm) and vanadium (122 pm). This decrease in atomic size across a row is also observed in the main-group elements. It is due to an increase in *effective nuclear charge* that acts on the outer electrons and pulls them in more strongly. The effective nuclear charge is the positive charge "felt" by an electron; it equals the nuclear charge minus the shielding or screening of the positive charge by intervening electrons. After vanadium, the covalent radii decrease slowly from 118 pm for chromium to 115 pm for nickel. Then the covalent radii increase slightly to 117 pm for copper and 125 pm for zinc. ◀ The relative constancy in covalent radii for the later elements is partly responsible for the similarity in properties of the Group VIIIB elements iron, cobalt, and nickel.

▶ **The small increase in covalent radius for copper and zinc has no simple explanation.**

Figure 23.3 compares the covalent radii of the transition elements. The atomic radii increase in going from a fourth-period to a fifth-period element within any column. For example, reading down the Group IIIB elements, you find that the covalent radius of scandium is 144 pm and of yttrium is 162 pm. You would expect an increase of radius from scandium to yttrium because of the addition of a shell of electrons. Continuing down the column of IIIB elements, you find a small increase of covalent radius to 169 pm in lanthanum. But all of the remaining elements of the sixth period have nearly the same covalent radii as the corresponding elements in the fifth period.

This similarity of radii in the fifth- and sixth-period transition elements can be explained in terms of the *lanthanide contraction*. Between lanthanum and hafnium in the sixth period are 14 lanthanide elements (cerium to lutetium), in which the 4*f* subshell fills. The covalent radius decreases slowly from cerium (165 pm) to lutetium (156 pm), but the total decrease is substantial. By the time the 4*f* subshell is complete, the covalent radii of the transition elements from hafnium on are similar to those of

**TABLE 23.2**
**First Ionization Energies of the Transition Elements (kJ/mol)**

| Period | IIIB | IVB | VB | VIB | VIIB | VIIIB | | | IB | IIB |
|--------|------|-----|-----|------|------|-----|-----|-----|------|------|
| Fourth | Sc 631 | Ti 658 | V 650 | Cr 652 | Mn 717 | Fe 759 | Co 758 | Ni 737 | Cu 745 | Zn 906 |
| Fifth | Y 616 | Zr 660 | Nb 664 | Mo 685 | Tc 702 | Ru 711 | Rh 720 | Pd 805 | Ag 731 | Cd 868 |
| Sixth | La 538 | Hf 680 | Ta 761 | W 770 | Re 760 | Os 840 | Ir 880 | Pt 870 | Au 890 | Hg 1007 |

the elements in the preceding row of the periodic table. Therefore, the covalent radius of hafnium (146 pm) is approximately the same as that of zirconium (145 pm).

The chemical properties of the transition elements parallel the pattern seen in covalent radii. That is, the fourth-period elements have substantially different properties from the elements in the same group in the fifth and sixth periods. However, elements in the same group of the fifth and sixth periods are very much alike. Hafnium, for example, is so much like zirconium that it remained undiscovered until it was identified in 1923 in zirconium ores from its x-ray spectrum. ◄ Zirconium (mixed with hafnium as an impurity) was discovered more than a hundred years earlier, in 1787.

► The atomic number of an atom can be obtained from its x-ray spectrum, which therefore gives unequivocal identification of an element. See the essay at the end of Section 8.2.

## Ionization Energies

Looking at the first ionization energies of the fourth-period transition elements (Table 23.2), you see that although they vary somewhat irregularly, they tend to increase from left to right. The other rows of transition elements behave similarly. Most noteworthy, however, is that all of the sixth-period elements after lanthanum have ionization energies higher than those of the fourth-period and fifth-period transition elements in the same group. This behavior is opposite to that of the main-group elements, where the ionization energies decrease regularly down a column. The high ionization energies of the sixth-period elements from osmium to mercury are no doubt one determinant of the relative unreactivity of these elements.

## Oxidation States

Table 23.3 gives oxidation states in compounds of the fourth-period transition elements. Scandium occurs as the 3+ ion and zinc as the 2+ ion. The other elements exhibit several oxidation states. This multiplicity of oxidation states is due to the varying involvement of the $d$ electrons in bonding.

Most of the transition elements have a doubly filled $ns$ orbital. Because the $ns$ electrons ionize before the $(n - 1)d$ electrons, you might expect the +2 oxidation state to be common. This oxidation state is in fact seen in all of the fourth-period elements except scandium, where the 3+ ion with an Ar configuration is especially stable.

► The maximum oxidation state is generally found in compounds of the transition elements with very electronegative elements, such as F and O (in oxides and oxoanions).

The maximum oxidation state possible equals the number of $s$ and $d$ electrons in the valence shell (which equals the group number for the elements up to iron). Titanium in Group IVB has a maximum oxidation state of +4. Similarly, vanadium (Group VB), chromium (Group VIB), and manganese (Group VIIB) exhibit maximum oxidation states of +5, +6, and +7, respectively. ◄

**TABLE 23.3**
**Oxidation States of the Fourth-Period Transition Elements**

| IIIB | IVB | VB | VIB | VIIB | VIIIB | | | IB | IIB |
|------|-----|-----|-----|------|------|-----|-----|-----|-----|
| Sc | Ti | V | Cr | Mn | Fe | Co | Ni | Cu | Zn |
|  |  |  |  |  |  |  | +1 | **+1** |  |
|  | **+2** | **+2** | **+2** | **+2** | **+2** | **+2** | **+2** | **+2** | **+2** |
| **+3** | **+3** | **+3** | **+3** | **+3** | **+3** | **+3** | +3 | **+3** |  |
|  | **+4** | **+4** | +4 | **+4** | +4 | +4 | +4 |  |  |
|  |  | **+5** | +5 | +5 | +5 |  |  |  |  |
|  |  |  | **+6** | +6 | +6 |  |  |  |  |
|  |  |  |  | **+7** |  |  |  |  |  |

*Key:* Common oxidation states are in boldface. Additional oxidation states, particularly zero and negative values, may be observed in complexes with CO and with organic compounds.

The total number of oxidation states actually observed increases from scandium ($+3$) to manganese (all states from $+2$ to $+7$). From iron on, however, the maximum oxidation state is not attained, so the number of observed states decreases. The highest oxidation state seen in iron is $+6$. ◀ Thereafter, the highest observed oxidation number decreases until, for zinc compounds, you find only the $+2$ oxidation state. The $Zn^{2+}$ ion has a stable $[Ar]3d^{10}$ configuration.

▶ Presumably, Fe(VIII) would oxidize any element to which it would bond.

## 23.2    The Chemistry of Two Transition Elements

▶ Experiments in General Chemistry, 23A, "Qualitative Analysis of Chromium, Iron, and Copper."

In this section, we look at the chemical properties of two transition elements, chromium and copper. Both elements are well known and commercially important; in addition, they provide many examples of colorful compounds.

### Chromium

A fascinating feature of chromium chemistry is the many colorful compounds of this element. In fact, this feature is the origin of the name chromium, which derives from the Greek word *khroma,* meaning "color." The common oxidation states of chromium in compounds are $+2$, $+3$, and $+6$.

The chromium(II), or chromous, ion, $Cr(H_2O)_6^{2+}(aq)$, has a bright blue color. The ion is obtained when an acid, such as $HCl(aq)$, reacts with chromium metal (see Figure 23.4, left).

$$Cr(s) + 2HCl(aq) + 6H_2O(l) \longrightarrow Cr(H_2O)_6^{2+}(aq) + 2Cl^-(aq) + H_2(g)$$

The chromium(II) ion is easily oxidized to chromium(III) ion by $O_2$ (Figure 23.4, right). Therefore, to prepare chromium(II) salts, we must carry out this reaction in the absence of air.

The most common oxidation state of chromium is $+3$. Chromium metal burns in air to give chromium(III) oxide, $Cr_2O_3$ (also called chromic oxide). This oxide is a dark green solid, which has been used as a paint pigment (chrome green). Chromium(III) oxide dissolves in acid solution to form the violet-colored ion

**Figure 23.4**
**Aqueous chromium ion.**
*Left:* Chromium metal reacts with hydrochloric acid to produce $Cr(H_2O)_6^{2+}$, which is blue. *Right:* $Cr(H_2O)_6^{2+}$ oxidizes with air to give a chromium(III) species, which is green.

$Cr(H_2O)_6^{3+}(aq)$. (Hydrochloric acid solution, however, yields green-colored complex ions of $Cr^{3+}$ with $Cl^-$.) Potassium chromium(III) sulfate (or chrome alum), $KCr(SO_4)_2 \cdot 12H_2O$, is a common chromium(III) salt. It is a deep purple compound used for leather tanning.

The +6 oxidation state is represented by the chromate ion, $CrO_4^{2-}$, and the dichromate ion, $Cr_2O_7^{2-}$. Sodium chromate, $Na_2CrO_4$, which is yellow, and sodium dichromate, $Na_2Cr_2O_7$, which is orange, are the primary sources of compounds of chromium. Sodium chromate, $Na_2CrO_4$, is prepared from chromite, $FeCr_2O_4(s)$, the principal chromium ore, by strongly heating it with sodium carbonate in air. Chromite is an iron(II) chromium(III) oxide, and both iron and chromium are oxidized in this reaction.

$$4FeCr_2O_4(s) + 8Na_2CO_3(s) + 7O_2(g) \xrightarrow{\;1100°C\;} 8Na_2CrO_4(s) + 2Fe_2O_3(s) + 8CO_2(g)$$

Sodium chromate is soluble and can be leached from the mixture with water.

Sodium chromate can be easily converted to sodium dichromate by treating it with an acid. When the yellow solution of a chromate salt is acidified, it turns orange from the formation of dichromate ion, $Cr_2O_7^{2-}(aq)$ (Figure 23.5).

$$\underset{\text{yellow}}{2CrO_4^{2-}(aq)} + 2H^+(aq) \rightleftharpoons \underset{\text{orange}}{Cr_2O_7^{2-}(aq)} + H_2O(l)$$

The chromate and dichromate ions are in equilibrium, which is sensitive to pH changes; lower pH favors dichromate ion. (Increasing the $H^+$ concentration pushes the equilibrium to the right, according to Le Chatelier's principle.)

The dichromate ion is a strong oxidizing agent in acid solution.

$$Cr_2O_7^{2-}(aq) + 14H^+(aq) + 6e^- \longrightarrow 2Cr^{3+}(aq) + 7H_2O(l); \; E° = 1.33 \text{ V}$$

The color of the chromium(III) species depends on anions in solution that can form complexes with $Cr^{3+}$. Frequently the ion is green, so there is a definite color change from orange to green during the reaction.

Chromium(VI) oxide (also called chromium trioxide, $CrO_3$) is a red, crystalline compound. It precipitates when concentrated sulfuric acid is added to concentrated solutions of a dichromate salt. ◄

▶ Chromium trioxide is a strong oxidizing agent. Vapors of ethanol, $C_2H_5OH$, spontaneously ignite in the presence of the solid oxide.

$$K_2Cr_2O_7(aq) + 2H_2SO_4(l) \longrightarrow 2KHSO_4(aq) + 2CrO_3(s) + H_2O(l)$$

**Figure 23.5**
**Chromate–dichromate equilibrium.** The beaker on the left contains $CrO_4^{2-}$ (yellow). When the experimenter adds sulfuric acid to a similar solution in the beaker on the right, $CrO_4^{2-}$ is converted to $Cr_2O_7^{2-}$ (orange).

$CrO_4^{2-}$

$Cr_2O_7^{2-}$

Chromium trioxide is an acidic oxide, and its aqueous solutions are referred to as chromic acid. Chromic acid is used in chromium plating.

Pure chromium metal is prepared by the exothermic reaction of chromium(III) oxide, obtained from chromite ore, with aluminum (the *Goldschmidt process*). Once the reaction mixture is ignited, the large heat of reaction produces molten chromium.

$$Cr_2O_3(s) + 2Al(s) \longrightarrow 2Cr(l) + Al_2O_3(s); \Delta H° = -488 \text{ kJ}$$

One of the chief uses of chromium is in steel making. For this purpose, chromium is usually prepared directly as an alloy of iron and chromium, called ferrochromium, by reducing chromite ore with carbon in an electric furnace.

$$FeCr_2O_4(s) + 4C(s) \xrightarrow{\Delta} Fe(l) + 2Cr(l) + 4CO(g)$$

### Copper

Copper and the other Group IB elements, silver and gold, can be found in nature as the free metals. This is a reflection of the stability of their 0 oxidation states. For example, copper is not attacked by most acids.

▶ Demonstration of precipitates and complexes of copper(II): Bassam Z. Shakhashiri, *Chemical Demonstrations,* Vol. 1 (Madison: University of Wisconsin Press, 1983), no. 4.8, pp. 318–323.

Copper metal does react with hot, concentrated sulfuric acid and with nitric acid. In these cases, the anion of the acid acts as the oxidizing agent (rather than $H^+$, the usual oxidizing agent in acids). Sulfuric acid is reduced to sulfur dioxide:

$$Cu(s) + 2H_2SO_4(l) \longrightarrow Cu^{2+}(aq) + SO_4^{2-}(aq) + SO_2(g) + 2H_2O(l)$$

Dilute nitric acid is reduced to NO; concentrated nitric acid, to $NO_2$:

$$3Cu(s) + 2NO_3^-(aq) + 8H^+(aq) \longrightarrow 3Cu^{2+}(aq) + 2NO(g) + 4H_2O(l)$$
$$Cu(s) + 2NO_3^-(aq) + 4H^+(aq) \longrightarrow Cu^{2+}(aq) + 2NO_2(g) + 2H_2O(l)$$

The copper(II) ion, or cupric ion, given in these equations as $Cu^{2+}(aq)$, is written more precisely as $Cu(H_2O)_6^{2+}$. It has a bright blue color. Hydrated copper(II)

**Figure 23.6**
**Benedict's test for glucose.** The experimenter adds glucose solution to a basic solution of copper(II) ion (blue-green solution in left test tube); the solution also contains citrate ion to complex copper(II) ion, so that it will not precipitate as copper(II) hydroxide. When this solution is heated, the glucose reduces copper(II) ion to copper(I) oxide (brick-red precipitate in right test tube).

▶ Flotation is discussed in Section 13.2.

salts, such as copper(II) sulfate pentahydrate, $CuSO_4 \cdot 5H_2O$, also have a blue color. Four of the water molecules of copper(II) sulfate pentahydrate are associated with $Cu^{2+}$, and the fifth is hydrogen-bonded to the sulfate ion as well as to the water molecules on the copper ion. You can write the formula as $[Cu(H_2O)_4]SO_4 \cdot H_2O$ to better represent its structure. When this salt is heated, it loses its water of hydration in stages. The hydrate $CuSO_4 \cdot H_2O$ and the anhydrous salt $CuSO_4$ are nearly colorless substances; the complexing of water molecules to the copper ion is responsible for the blue color of the pentahydrate.

Although most of the aqueous chemistry of copper involves the +2 oxidation state, there are a number of important compounds of copper(I). When copper is heated in oxygen below 1000°C, it forms the black copper(II) oxide, CuO. But above this temperature, it forms the brick-red copper(I) oxide, $Cu_2O$. This oxide is found naturally as the mineral cuprite.

The formation of copper(I) oxide by the reduction of copper(II) ion in basic solution forms the basis of a common test for glucose in urine.

$$RCHO(aq) + 2Cu^{2+}(aq) + 5OH^-(aq) \longrightarrow RCOO^-(aq) + Cu_2O(s) + 3H_2O(l)$$
blue                                  brick-red

When the copper(II) ion is heated in basic solution with a reducing sugar, such as glucose (represented here by the general formula RCHO), the ion is reduced to copper(I) oxide. The blue-green solution of $Cu^{2+}$ gives a brick-red precipitate of $Cu_2O$ (Figure 23.6).

The principal commercial use of copper metal is as an electrical conductor. Common ores of copper are native copper (the free metal), copper oxides, and copper sulfides. Most copper is presently obtained by open-pit mining of low-grade rock containing only a small percentage of copper as copper sulfides. The ore is concentrated in copper by flotation. In this process, a slurry of the crushed ore is agitated with air, and the copper sulfides are carried away in the froth. This concentrated ore is then treated in several steps that result in the production of molten copper(I) sulfide, $Cu_2S$. This molten material, called *matte*, is reduced to copper by blowing air through it.

$$Cu_2S(l) + O_2(g) \longrightarrow 2Cu(l) + SO_2(g)$$

The metal produced is called *blister copper* and is about 99% pure. For electrical use, the copper must be further purified or refined by electrolysis (see Figure 20.24).

# Complex Ions and Coordination Compounds

▶ Experiments in General Chemistry, 23B, "Thermochemistry and Complex Ions." 🔺

As shown in the previous section, ions of the transition elements exist in aqueous solution as *complex ions.* Iron(II) ion, for example, exists in water as $Fe(H_2O)_6^{2+}$. The water molecules in this ion are arranged about the iron atom with their oxygen atoms bonded to the metal by donating electron pairs to it. Replacing the $H_2O$ molecules by six $CN^-$ ions gives the $Fe(CN)_6^{4-}$ ion. Some of the transition elements have biological activity and their role in human nutrition (Table 23.4) depends in most cases on the formation of *complexes,* or *coordination compounds,* which exhibit the type of bonding that occurs in $Fe(H_2O)_6^{2+}$ and $Fe(CN)_6^{4-}$. In the chapter opening, we saw that hemoglobin, a complex of iron, is vital to the transport of oxygen by the red blood cells.

**Table 23.4**
**Transition Elements Essential to Human Nutrition**

| Element | Some Biochemical Substances | Function |
|---|---|---|
| Chromium | Glucose tolerance factor | Utilization of glucose |
| Manganese | Isocitrate dehydrogenase | Cell energetics |
| Iron | Hemoglobin and myoglobin | Transport and storage of oxygen |
|  | Cytochrome $c$ | Cell energetics |
|  | Catalase | Decomposition of hydrogen peroxide |
| Cobalt | Cobalamin (vitamin $B_{12}$) | Development of red blood cells |
| Copper | Ceruloplasmin | Synthesis of hemoglobin |
|  | Cytochrome oxidase | Cell energetics |
| Zinc | Carbonic anhydrase | Elimination of carbon dioxide |
|  | Carboxypeptidase A (pancreatic juice) | Protein digestion |
|  | Alcohol dehydrogenase | Oxidation of ethanol |

## 23.3   Formation and Structure of Complexes

▶ Lewis acid–base reactions are discussed in Section 16.3.

A metal atom, particularly a transition-metal atom, often functions in chemical reactions as a Lewis acid, accepting electron pairs from molecules or ions. ◀ For example, $Fe^{2+}$ and $H_2O$ can bond to one another in a Lewis acid–base reaction.

$$Fe^{2+} \ + \ :\!\overset{..}{\underset{..}{O}}\!-\!H \longrightarrow \left[ Fe:\overset{..}{\underset{..}{O}}\!-\!H \right]^{2+}$$

coordinate covalent bond

Lewis acid   Lewis base

A pair of electrons on the oxygen atom of $H_2O$ forms a coordinate covalent bond to $Fe^{2+}$. In water, the $Fe^{2+}$ ion ultimately bonds to six $H_2O$ molecules to give the $Fe(H_2O)_6{}^{2+}$ ion.

The $Fe^{2+}$ ion also undergoes a similar Lewis acid–base reaction with cyanide ions, $CN^-$. In this case, the $Fe^{2+}$ ion bonds to the electron pair on the carbon atom of $CN^-$.

$$Fe^{2+} + (:C{\equiv}N:)^- \longrightarrow (Fe:C{\equiv}N:)^+$$

▶ Although the ion $Fe(CN)_6{}^{4-}$ is a complex of $Fe^{2+}$ and $CN^-$ ions, a solution of $Fe(CN)_6{}^{4-}$ contains negligible concentration of $CN^-$. Therefore, a substance such as $K_4[Fe(CN)_6]$ is relatively nontoxic, even though the free cyanide ion is a potent poison.

Finally, a very stable ion $Fe(CN)_6{}^{4-}$ is obtained with six cyanide ions. ◀ Note that the charge on the $Fe(CN)_6{}^{4-}$ ion equals the sum of the charges on the ions from which it is formed: $+2 + 6(-1) = -4$.

In some cases, a neutral species is produced from a metal ion and anions. Cis-platin, the anticancer drug discussed in the opening of Chapter 1, has the structure

$$H_3N:\overset{\displaystyle :\!\overset{..}{\underset{..}{Cl}}\!:}{\underset{\displaystyle NH_3}{Pt}}:\overset{..}{\underset{..}{Cl}}\!:$$

**Table 23.5**
**Examples of Complexes of Various Coordination Numbers**

| Complex | Coordination Number |
|---|---|
| $Ag(NH_3)_2{}^+$ | 2 |
| $HgI_3{}^-$ | 3 |
| $PtCl_4{}^{2-}$, $Ni(CO)_4$ | 4 |
| $Fe(CO)_5$, $Co(CN)_5{}^{3-}$ | 5 |
| $Co(NH_3)_6{}^{3+}$, $W(CO)_6$ | 6 |
| $Mo(CN)_7{}^{3-}$ | 7 |
| $W(CN)_8{}^{4-}$ | 8 |

It consists of $Pt^{2+}$ with two $NH_3$ molecules (neutral) and two $Cl^-$ ions, giving a neutral species. Iron pentacarbonyl, $Fe(CO)_5$, is an example of a neutral species formed from a neutral iron atom and CO molecules.

## Basic Definitions

A **complex ion** is *a metal ion with Lewis bases attached to it through coordinate covalent bonds.* A **complex** (or **coordination compound**) is *a compound consisting either of complex ions and other ions of opposite charge* (for example, the compound $K_4[Fe(CN)_6]$ of the complex ion $Fe(CN)_6{}^{4-}$ and four $K^+$ ions) *or of a neutral complex species* (such as cisplatin).

▶ **A cation in which the positive charge is far removed from an electron pair that could be donated can function as a ligand. An example is the pyrazinium ion:**

▶ **Very high coordination numbers (9 to 12) are known for some complex ions of the lanthanide elements.**

**Ligands** are *the Lewis bases attached to the metal atom in a complex.* They are electron-pair donors, so ligands may be neutral molecules (such as $H_2O$ or $NH_3$) or anions (such as $CN^-$ or $Cl^-$) that have at least one atom with a lone pair of electrons. Cations only rarely function as ligands. You might expect this, because an electron pair on a cation is held securely by the positive charge, so it would not be involved in coordinate bonding. ◀

The **coordination number** of a metal atom in a complex is *the total number of bonds the metal atom forms with ligands.* In $Fe(H_2O)_6{}^{2+}$, the iron atom bonds to each oxygen atom in the six water molecules. Therefore, the coordination number of iron in this ion is 6, by far the most common coordination number. Coordination number 4 is also well known, and many examples of number 5 have been discovered. Table 23.5 gives some examples of complexes for the coordination numbers 2 to 8. The coordination number for an atom depends on several factors, but size of the metal atom is important. For example, coordination numbers 7 and 8 are seen primarily in fifth- and sixth-period elements, whose atoms are relatively large. ◀

## Polydentate Ligands

The ligands we have discussed so far bond to the metal atom through one atom of the ligand. For instance, ammonia bonds through the nitrogen atom. This type of bonding indicates a **monodentate ligand** (meaning "one-toothed" ligand)—that is, *a ligand that bonds to a metal atom through one atom of the ligand.* A **bidentate ligand**

▶ Video no. 26 (Series A), "Aqueous Nickel(II) Chemistry." The video illustrates the formation of complexes of polydentate ligands.

▶ Demonstration of nickel complexes: Summerlin et al., *Chemical Demonstrations,* Vol. 2 (Washington, D.C.: Am. Chem. Soc., 1987), no. 40, pp. 73–74.

("two-toothed" ligand) is *a ligand that bonds to a metal atom through two atoms of the ligand.* Ethylenediamine is an example.

ethylenediamine

Nitrogen atoms at the ends of the molecule have lone pairs of electrons that can form coordinate covalent bonds. In forming a complex, the ethylenediamine molecule bends around so that both nitrogen atoms coordinate to the metal atom, M.

Because ethylenediamine is a common bidentate ligand, it is frequently abbreviated in formulas as "en." Figure 23.7 shows the structure of the stable ion $Co(en)_3^{3+}$. The oxalate ion, $C_2O_4^{2-}$, is another common bidentate ligand.

oxalate ion

$C_2O_4^{2-}$

The hemoglobin molecule in red blood cells is an example of a complex with a *quadridentate ligand*—one that bonds to the metal atom through four ligand atoms. Hemoglobin consists of the protein *globin* chemically bonded to *heme,* whose struc-

**Figure 23.7 Structure of tris(ethylenediamine) cobalt(III) ion, $Co(en)_3^{3+}$.** *(A)* Ball-and-stick model. *(B)* Shorthand notation. Note that N⌒N represents ethylenediamine.

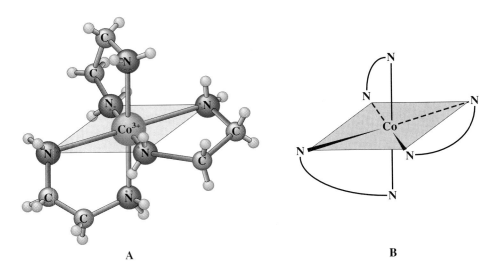

A                                    B

**Figure 23.8**
**Complexes with ligands that bond with more than one atom to the metal atom (polydentate ligands).** *Left:* Structure of heme. *Right:* Complex of $Fe^{2+}$ and ethylenediaminetetraacetate ion (EDTA). Note how the EDTA ion envelops the metal ion.

**Heme**

**Complex of $Fe^{2+}$ and EDTA**

EDTA

▶ The term *chelate* is derived from the Greek *chele* for "claw," because a polydentate ligand appears to attach itself to the metal atom like crab claws to some object.

ture is shown in Figure 23.8 (left). Heme is a planar molecule consisting of iron(II) to which a quadridentate ligand is bonded through its four nitrogen atoms.

Ethylenediaminetetraacetate ion (EDTA) is a ligand that bonds through six of its atoms.

It can completely envelop a metal atom, simultaneously occupying all six positions in an octahedral geometry (Figure 23.8, right).

A **polydentate ligand** ("having many teeth") is *a ligand that can bond with two or more atoms to a metal atom. A complex formed by polydentate ligands* is frequently quite stable and is called a **chelate** (pronounced "key-late"). ◀ Because of the stability of chelates, polydentate ligands (also called *chelating agents*) are often used to remove metal ions from a chemical system. EDTA, for example, is added to certain canned foods to remove transition-metal ions that can catalyze the deterioration of the food. The same chelating agent has been used to treat lead poisoning because it binds $Pb^{2+}$ ions as the chelate, forming a substance that can then be excreted by the kidneys.

## Discovery of Complexes; Formula of a Complex

In 1798 B. M. Tassaert found that a solution of cobalt(II) chloride in aqueous ammonia exposed to air (an oxidizing agent) deposits orange-yellow crystals. He assigned the formula $CoCl_3 \cdot 6NH_3$ to these crystals. Later, a similar compound of platinum was assigned the formula $PtCl_4 \cdot 6NH_3$. Because these formulas suggested that the substances were somehow composed of two stable compounds—platinum(IV)

**Table 23.6**
**Some Platinum(IV) Complexes Studied by Werner**

| Old Formula | Modern Formula | Number of Ions | Number of Free $Cl^-$ Ions |
|---|---|---|---|
| $PtCl_4 \cdot 6NH_3$ | $[Pt(NH_3)_6]Cl_4$ | 5 | 4 |
| $PtCl_4 \cdot 4NH_3$ | $[Pt(NH_3)_4Cl_2]Cl_2$ | 3 | 2 |
| $PtCl_4 \cdot 3NH_3$ | $[Pt(NH_3)_3Cl_3]Cl$ | 2 | 1 |
| $PtCl_4 \cdot 2NH_3$ | $[Pt(NH_3)_2Cl_4]$ | 0 | 0 |

chloride and ammonia in the latter case—they were called complex compounds, or simply complexes.

The basic explanation for the structure of these complexes was given by the Swiss chemist Alfred Werner in 1893. According to Werner, a metal atom exhibits two kinds of valences, a primary valence and a secondary valence. The primary valence is what we now call the oxidation number of the metal. The secondary valence corresponds to what we now call the coordination number, which is often 6. In Werner's view, the substance previously represented by the formula $PtCl_4 \cdot 6NH_3$ is composed of the ion $Pt(NH_3)_6^{4+}$, with six ammonia molecules directly attached to the platinum atom. The charge of this ion is balanced by four $Cl^-$ ions, giving a neutral compound with the structural formula $[Pt(NH_3)_6]Cl_4$.

Table 23.6 lists a series of platinum(IV) complexes studied by Werner and both the old and Werner's modern formulas. (Note that square brackets are used in modern formulas to separate the metal and its associated groups from the rest of the formula.) According to Werner's theory, these complexes should dissolve to give different numbers of ions per formula unit. For example, $[Pt(NH_3)_4Cl_2]Cl_2$ dissolves to give three ions: $Pt(NH_3)_4Cl_2^{2+}$ and two $Cl^-$ ions. Werner was able to show that the electrical conductances of solutions of these complexes were equal to what was expected for the number of ions predicted by his formulas. He also demonstrated that the chloride ions in the platinum complexes were of two kinds: those that could be precipitated from solution as AgCl using silver nitrate and those that could not. He explained that the chloride ions within the platinum complex ion are securely attached to the metal atom and that only those outside the complex ion can be precipitated with silver nitrate. The number of free $Cl^-$ ions (those not attached to platinum) determined this way agreed exactly with his formulas. Over a period of 20 years, Werner carried out many experiments on complexes, the results of all of them in basic agreement with his theory. He received the Nobel Prize for his work in 1913.

## CONCEPT CHECK 23.1

Another complex studied by Werner had a composition corresponding to the formula $PtCl_4 \cdot 2KCl$. From electrical-conductance measurements, he determined that each formula unit contained three ions. He also found that silver nitrate did not give a precipitate of AgCl with this complex. Write a formula for this complex that agrees with this information.

# Salad Dressing and Chelate Stability

EVERYDAY LIFE

The list of ingredients for a particular mayonnaise reads: vegetable oil, eggs, vinegar, and calcium disodium EDTA (ethylenediaminetetraacetate). EDTA? Yes, commercial mayonnaise and salad dressings use EDTA to remove traces of metal ions. Metal ions can catalyze undesirable reactions or else provide nutrient for bacteria, resulting in off-flavors and spoilage of the product. EDTA is a polydentate ligand that forms particularly stable chelates with many metal ions, effectively removing those ions from the product. Many commercial products contain chelating agents such as EDTA (Figure 23.9).

What accounts for the special stability of chelates? Their stability stems from the additional entropy obtained when they are formed. This leads to a large negative $\Delta G°$, which is equivalent to a large equilibrium constant for the formation of the chelate. To understand how this happens, consider the formation of the chelate $Co(en)_3^{3+}$ from the complex ion $Co(NH_3)_6^{3+}$, with monodentate ligands $NH_3$.

$$Co(NH_3)_6^{3+} + 3en \rightleftharpoons Co(en)_3^{3+} + 6NH_3$$

Each en molecule replaces two $NH_3$ molecules. Therefore, the number of particles in the reaction mixture is increased when the reaction goes to the right. In most cases, an increase in number of particles increases the possibilities for randomness or disorder. Therefore, when the reaction goes to the right, there is an increase in entropy; that is, $\Delta S°$ is positive.

At the same time, the reaction involves very little change of internal energy ($\Delta U°$) or enthalpy ($\Delta H° = \Delta U° + P\Delta V$), because the bonds are similar; all consist of a nitrogen atom coordinated to a cobalt atom. Six nitrogen-metal bonds in $Co(NH_3)_6^{3+}$ are broken, and six new nitrogen-metal bonds of about the same energy are formed in $Co(en)_3^{3+}$. Therefore, $\Delta H° \simeq 0$.

**Figure 23.9**
**Some products containing EDTA.** Mayonnaise and salad dressings contain EDTA to reduce the concentrations of certain metal ions. Some plant fertilizers contain EDTA chelates of copper and other metals, providing the metal in a soluble form that plants can utilize.

The spontaneity of a reaction depends on $\Delta G°$, which equals $\Delta H° - T\Delta S°$. But because $\Delta H° \simeq 0$,

$$\Delta G° = \Delta H° - T\Delta S°$$
$$= -T\Delta S°$$

The entropy change is positive, so $\Delta G°$ is negative, and the reaction is spontaneous from left to right.

The fact that the equilibrium for the reaction favors the chelate $Co(en)_3^{3+}$ means that the chelate has thermodynamic stability. A similar argument could be made for any equilibrium involving the replacement of monodentate ligands by polydentate ligands. Reaction tends to favor the chelate.

# 23.4 Naming Coordination Compounds

Thousands of coordination compounds are now known. A systematic method of naming such compounds, or *nomenclature*, needs to provide basic information about the structure of a coordination compound. What is the metal in the complex? Does the metal atom occur in the cation or the anion? What is the oxidation state of the metal? What are the ligands? You can answer these questions by following the rules of nomenclature agreed upon by the International Union of Pure and Applied Chemistry (IUPAC). These rules are essentially an extension of those originally given by Werner.

1. In naming a salt, the name of the cation precedes the name of the anion. (This is a familiar rule.) For example,

$$K_4[Fe(CN)_6] \text{ is named } \underbrace{\text{potassium}}_{\text{cation}} \underbrace{\text{hexacyanoferrate(II)}}_{\text{anion}}$$

$$[Co(NH_3)_6]Cl_3 \text{ is named } \underbrace{\text{hexaamminecobalt(III)}}_{\text{cation}} \underbrace{\text{chloride}}_{\text{anion}}$$

2. The name of the complex—whether anion, cation, or neutral species—consists of two parts written together as one word. Ligands are named first, and the metal atom is named second. For example,

$$Fe(CN)_6{}^{4-} \text{ is named } \underbrace{\text{hexacyanoferrate(II)}}_{\substack{\text{ligand} \\ \text{name}}} \underbrace{\text{ion}}_{\substack{\text{metal} \\ \text{name}}}$$

$$Co(NH_3)_6{}^{3+} \text{ is named } \underbrace{\text{hexaamminecobalt(III)}}_{\substack{\text{ligand} \\ \text{name}}} \underbrace{\text{ion}}_{\substack{\text{metal} \\ \text{name}}}$$

3. The complete ligand name consists of a Greek prefix denoting the number of ligands, followed by the specific name of the ligand. When there are two or more ligands, the ligands are written in alphabetical order (disregarding Greek prefixes).

a. Anionic ligands end in -o. Some examples:

| Anion Name | Ligand Name |
|---|---|
| Bromide, $Br^-$ | Bromo |
| Carbonate, $CO_3{}^{2-}$ | Carbonato |
| Chloride, $Cl^-$ | Chloro |
| Cyanide, $CN^-$ | Cyano |
| Fluoride, $F^-$ | Fluoro |
| Hydroxide, $OH^-$ | Hydroxo |
| Oxalate, $C_2O_4{}^{2-}$ | Oxalato |
| Oxide, $O^{2-}$ | Oxo |
| Sulfate, $SO_4{}^{2-}$ | Sulfato |

b. Neutral ligands are usually given the name of the molecule. There are, however, several important exceptions:

| Molecule | Ligand Name |
|---|---|
| Ammonia, $NH_3$ | Ammine |
| Carbon monoxide, CO | Carbonyl |
| Water, $H_2O$ | Aqua |

c. The prefixes used to denote the number of ligands are *mono-* (1, usually omitted); *di-* (2); *tri-* (3); *tetra-* (4); *penta-* (5); *hexa-* (6); and so forth. To see how the ligand name is formed, consider the complex ions

$$Fe(CN)_6{}^{4-} \text{ or } \underbrace{\text{hexacyanoferrate(II)}}_{6 \ CN^- \text{ ligands}} \text{ ion}$$

$$Co(NH_3)_6{}^{3+} \text{ or } \underbrace{\text{hexaamminecobalt(III)}}_{6 \ NH_3 \text{ ligands}} \text{ ion}$$

d. When the name of the ligand also has a number prefix, the number of ligands is denoted with *bis* (2), *tris* (3), *tetrakis* (4), and so forth. The name of the ligand follows in parentheses. For example, the complex [Co(en)$_3$]Cl$_3$ is named as follows:

$$\underbrace{\text{tris}}_{3}(\underbrace{\text{ethylenediamine}}_{\text{ligand name}})\text{cobalt(III) chloride}$$

4. The complete metal name consists of the name of the metal, followed by *-ate* if the complex is an anion, followed by the oxidation number of the metal as a Roman numeral in parentheses. (An oxidation state of zero is indicated by 0 in parentheses.) When there is a Latin name for the metal, it is used to name the anion (except for mercury). These names are:

| English Name | Latin Name | Anion Name |
|---|---|---|
| Copper | Cuprum | Cuprate |
| Gold | Aurum | Aurate |
| Iron | Ferrum | Ferrate |
| Lead | Plumbum | Plumbate |
| Silver | Argentum | Argentate |
| Tin | Stannum | Stannate |

Examples are

hexacyano**ferrate**(II)    hexaamminecobalt(III)

ferrum | oxidation            metal   oxidation
= iron | number 2             name    number 3
        indicates
        an anion

▶ An alternate example is provided in the Instructor's Resource Manual.

### EXAMPLE 23.1

## Writing the IUPAC Name Given the Structural Formula of a Coordination Compound

Give the IUPAC name of each of the following coordination compounds:
a. [Pt(NH$_3$)$_4$Cl$_2$]Cl$_2$; b. [Pt(NH$_3$)$_2$Cl$_2$]; c. K$_2$[PtCl$_6$].

**SOLUTION**

a. The cation is listed first in the formula.

$$\underbrace{[\text{Pt(NH}_3)_4\text{Cl}_2]}_{\text{cation}}\underbrace{\text{Cl}_2}_{\text{anions}}$$

There are two Cl$^-$ anions, so the charge on the cation is +2: Pt(NH$_3$)$_4$Cl$_2{}^{2+}$. The oxidation number of platinum plus the sum of the charges on the ligands (−2) equals the cation charge +2. Therefore, the oxidation number of Pt is +4. Hence, the name of the compound is **tetraamminedichloroplatinum(IV) chloride.** Note that the ligands are listed in alphabetical order (that is, ammine before chloro).

b. This is a neutral complex species. The oxidation number of platinum must balance that of the two chloride ions. The name of the compound is **diamminedichloroplatinum(II).**

c. The complex anion is PtCl$_6{}^{2-}$. The oxidation number of platinum is 4, and the name of the compound is **potassium hexachloroplatinate(IV).**

---

**EXERCISE 23.1**

Give the IUPAC names of a. $[Co(NH_3)_5Cl]Cl_2$; b. $K_2[Co(H_2O)(CN)_5]$; c. $Fe(H_2O)_5(OH)^{2+}$.

---

See Problems 23.43, 23.44, 23.45, and 23.46.

▶ An alternate example is provided in the Instructor's Resource Manual.

---

**EXAMPLE 23.2**

**Writing the Structural Formula Given the IUPAC Name of a Coordination Compound**

Write the structural formula corresponding to each of the following IUPAC names: a. hexaaquairon(II) chloride; b. potassium tetrafluoroargentate(III); c. pentachlorotitanate(II) ion.

**SOLUTION**

a. The complex cation hexaaquairon(II) is $Fe^{2+}$ with six $H_2O$ ligands: $Fe(H_2O)_6^{2+}$. The formula of the compound is $[Fe(H_2O)_6]Cl_2$. (Remember to enclose the formula of the complex ion in square brackets.)

b. The compound contains the complex anion tetrafluoroargentate(III)—that is, $Ag^{3+}$ with four $F^-$ ligands. The formula of the ion is $AgF_4^-$. Hence, the formula of the compound is $K[AgF_4]$.

c. The ion contains $Ti^{2+}$ and five $Cl^-$ ligands. The charge on the ion is $+2 + 5(-1) = -3$. The formula of the complex ion is $TiCl_5^{3-}$.

---

**EXERCISE 23.2**

Write structural formulas for each of the following: a. potassium hexacyanoferrate(II), b. tetraamminedichlorocobalt(III) chloride, c. tetrachloroplatinate(II) ion.

---

See Problems 23.47 and 23.48.

# 23.5 Structure and Isomerism in Coordination Compounds

Although we described the formation of a complex as a Lewis acid–base reaction (Section 23.3), we did not go into any details of structure. We did not look at the geometry of complex ions or inquire about the precise nature of the bonding. Three properties of complexes have proved pivotal in determining these details.

1. *Isomerism*   Isomers are compounds with the same molecular formula (or the same simplest formula, in the case of ionic compounds) but with different arrangements of atoms. Because their atoms are differently arranged, isomers have different properties. There are many possibilities for isomerism in coordination compounds. The study of isomerism can lead to information about atomic arrangement in coordination compounds. Werner's research on the isomerism of coordination compounds finally convinced others that his views were essentially correct.

**Figure 23.10**
**Gouy balance for measuring the paramagnetism of a substance.** If the sample is attracted into the field of the magnet, the left pan of the balance will be subjected to a downward force. This force is balanced by weights added to the right pan.

Glass tube

Sample

Magnet

2. *Paramagnetism*   Paramagnetic substances are attracted to a strong magnetic field. Paramagnetism is due to unpaired electrons in a substance. (*Ferromagnetism* in solid iron is also due to unpaired electrons, but in this case the magnetism of many iron atoms is aligned, giving a magnetic effect perhaps a million times stronger than that seen in paramagnetic substances.) Many complex compounds are paramagnetic. The magnitude of this paramagnetism can be measured with a *Gouy balance,* in which the force of magnetic attraction is balanced with weights (Figure 23.10). Because paramagnetism is related to the electron configuration of the complex, these measurements can give information about the bonding.

3. *Color*   A substance is colored because it absorbs light in the visible region of the spectrum. The absorption of visible light is due to a transition between closely spaced electronic energy levels. As you have seen, many coordination compounds are highly colored. The color is related to the electronic structure of the compounds. ◄

▶ Recall that the wavelength of light absorbed by a substance is related to the difference between two energy levels in the substance. See Section 7.3.

In this section, we will investigate the relationship between structure and isomerism. In the following sections, we will look into explanations of the paramagnetism and color of coordination compounds.

There are two major kinds of isomers. **Constitutional isomers** are *isomers that differ in how the atoms are joined together*—that is, in the order in which the atoms are bonded to each other. (For example, H—N≡C═O and N≡C—O—H are constitutional isomers because the H atom is bonded to N in one case and to O in the other.) **Stereoisomers,** on the other hand, are *isomers in which the atoms are bonded to each other in the same order but differ in the precise arrangement of the atoms in space.*

## Constitutional Isomerism

Coordination compounds provide many special types of constitutional isomers. Here, for example, are two cobalt isomers:

$$[Co(NH_3)_5(SO_4)]Br \qquad \text{a red compound}$$
$$[Co(NH_3)_5Br]SO_4 \qquad \text{a violet compound}$$

In the first isomer, the sulfate ion is attached to the cobalt atom, and the bromide ion is in the crystal lattice as the anion countering the metal complex cation. In the second isomer, the sulfate ion is in the crystal lattice, with the bromide ion attached to the cobalt atom. ◄

► **This type of constitutional isomerism is called *coordination isomerism.***

Both the cation and the anion can be metal complex ions, yielding another type of isomer. The following copper–platinum compounds provide an example:

$$[Cu(NH_3)_4][PtCl_4] \qquad \text{and} \qquad [Pt(NH_3)_4][CuCl_4]$$

In the first compound, the $NH_3$ ligands are associated with a copper atom and the $Cl^-$ ligands with a platinum atom. In the second compound, the ligands are transposed.

A more subtle type of constitutional isomer is displayed by the following compounds:

$$[Co(NH_3)_5(ONO)]Cl_2$$
$$[Co(NH_3)_5(NO_2)]Cl_2$$

► **This type of constitutional isomerism is called *linkage isomerism.***

In the first compound, the nitrite ligand, $NO_2^-$, bonds to the cobalt atom through an electron pair on an oxygen atom. This is a red substance that slowly changes to the second compound. In this yellow-brown compound, the nitrite ligand bonds to the cobalt atom through an electron pair on the nitrogen atom. ◄

 **CONCEPT CHECK 23.2**

A complex has the composition $Co(NH_3)_4(H_2O)BrCl_2$. Conductance measurements show that there are three ions per formula unit, and precipitation of AgCl with silver nitrate shows that there are two $Cl^-$ ions not coordinated to cobalt. What is the structural formula of the compound?

## Stereoisomerism

The existence of various types of structural isomers is strong evidence for the view that complexes consist of groups directly bonded to a central metal atom. The existence of stereoisomers not only strengthens this view but also helps explain how the groups are arranged about the central atom.

**Geometric isomers** are *isomers in which the atoms are joined to one another in the same way but differ because some atoms occupy different relative positions in space.* Consider the complex diamminedichloroplatinum(II), $[Pt(NH_3)_2Cl_2]$. Two compounds of this composition are known. One is an orange-yellow compound with solubility at 25°C of 0.252 g per 100 g of water. The other compound is pale yellow and much less soluble (0.037 g per 100 g of water at 25°C). ◄ How do you explain the occurrence of these two isomers?

► **The isomers of [Pt(NH₃)₂Cl₂] also differ in their biological properties, as noted in Chapter 1. The orange-yellow compound acts as an anticancer drug; the other isomer does not.**

There are two symmetrical geometries of complexes of four ligands: tetrahedral and square planar. Let us write $MA_2B_2$ for a complex with ligands A and B about the

**Figure 23.11**
**The tetrahedral complex MA$_2$B$_2$ has no geometric isomers.** The molecule on the left can be rotated to look like the molecule on the right. A and B are any two ligands.

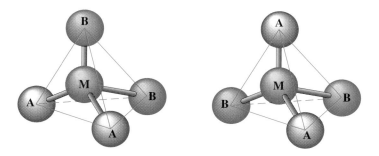

metal atom M. As Figure 23.11 shows, a tetrahedral geometry for MA$_2$B$_2$ allows only one arrangement of ligands. The square planar geometry for MA$_2$B$_2$, however, gives two possible arrangements. (The two arrangements for [Pt(NH$_3$)$_2$Cl$_2$] are shown in Figure 23.12.) One, labeled *cis,* has the two A ligands on one side of the square and the two B ligands on the other side. The other arrangement, labeled *trans,* has A and B ligands across the square from one another. The *cis* and *trans* arrangements of MA$_2$B$_2$ are examples of geometric isomers.

**Figure 23.12**
**Geometric isomers of the square planar complex diamminedichloroplatinum(II).** The existence of two isomers of [Pt(NH$_3$)$_2$Cl$_2$] is evidence of a square planar geometry. (The tetrahedral geometry would not give isomers.)

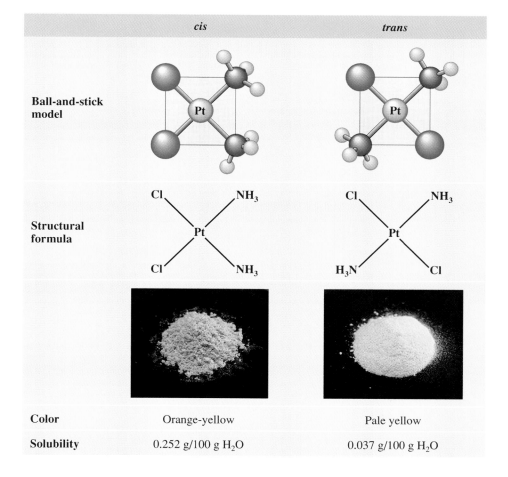

| | *cis* | *trans* |
|---|---|---|
| **Ball-and-stick model** | Pt | Pt |
| **Structural formula** | Cl, Pt, NH$_3$ / Cl, NH$_3$ | Cl, Pt, NH$_3$ / H$_3$N, Cl |
| **Color** | Orange-yellow | Pale yellow |
| **Solubility** | 0.252 g/100 g H$_2$O | 0.037 g/100 g H$_2$O |

**Figure 23.13**
**Octahedral geometry.**
(A) Each position in this geometry is equivalent. (B) Octahedral geometry is often represented like this.

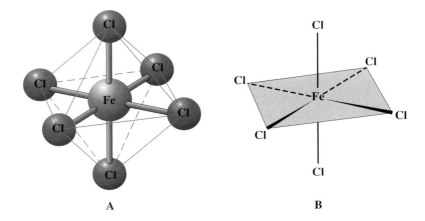

A                                    B

That there are two isomers of $[Pt(NH_3)_2Cl_2]$ is evidence for the square planar geometry in this complex. But how do you identify the substances and their properties with the *cis* and *trans* arrangements? You can distinguish between them by predicting their polarity. The *trans* arrangement, being completely symmetrical, is nonpolar. The *cis* arrangement, with electronegative Cl atoms on one side of the platinum atom, is polar. This difference between *cis* and *trans* isomers should be reflected in their solubilities in water, because polar substances are more soluble in water (itself a polar substance) than are nonpolar substances. Thus, you would expect the more soluble isomer to have the *cis* arrangement.

Although the difference in solubility is revealing, the most direct evidence of polarity in a molecule comes from measurements of dipole moment. (The dipole moment is a measure of charge separation in a molecule.) These values show that the less soluble platinum(II) complex has no dipole moment and must be *trans* and that the other isomer does have a dipole moment and must be *cis*. Figure 23.12 shows the two isomers.

Six-coordinate complexes have only one symmetrical geometry: octahedral (Figure 23.13). Geometric isomers are possible for this geometry also. Consider the complex $MA_4B_2$ in which two of the A ligands occupy positions just opposite one another. The other four ligands have a square planar arrangement in which *cis–trans* isomers are possible. Tetraamminedichlorocobalt(III) chloride, $[Co(NH_3)_4Cl_2]Cl$, is an example of a complex with such geometric isomers. The *cis* compound is purple; the *trans* compound is green. See Figure 23.14.

▶ An alternate example is provided in the Instructor's Resource Manual.

### EXAMPLE 23.3

### Deciding Whether Geometric Isomers Are Possible

Are there any geometric isomers of the complex cation $Co(NH_3)_4(NO_2)_2^+$? If so, draw them.

#### SOLUTION

**Yes,** there are geometric isomers. This is easily seen if you first draw two $NH_3$ ligands opposite one another in an octahedral geometry. Then the $NO_2^-$ ligands

**Figure 23.14**
**Geometric isomers of tetraamminedichlorocobalt (III) ion, $Co(NH_3)_4Cl_2^+$.** The two isomers have different properties. For example, the *cis* compound is purple; the *trans* compound is green.

can have *cis* or *trans* arrangements on the plane perpendicular to the axis of the first $NH_3$ ligands. See Figure 23.15.

**Figure 23.15**
**Cis–trans isomers of $Co(NH_3)_4(NO_2)_2^+$.** Note the arrangements of groups on the gray planes.

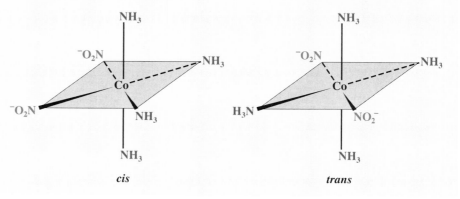

---

**EXERCISE 23.3**

Do any of the following stable octahedral complexes have geometric isomers? If so, draw them.

a. $Co(NH_3)_5Cl^{2+}$       b. $Co(NH_3)_4(H_2O)_2^{3+}$

---

See Problems 23.49 and 23.50.

Bidentate ligands increase the potential for isomerism in octahedral complexes. Figure 23.16 shows the isomers of the dichlorobis(ethylenediamine)cobalt (III) ion, $CoCl_2(en)_2^+$. Isomer A is the *trans* isomer (it has a green color). Both B and C are *cis* isomers (both have a violet color). Yet isomers B and C are not identical molecules. They are **enantiomers,** or **optical isomers;** that is, they are *isomers that are nonsuperimposable mirror images of one another.*

To better understand the nature of enantiomers, note that the two *cis* isomers have the same relationship to each other as the relationship of your left hand and right hand. The mirror image of a left hand looks like the right hand, and vice versa (Figure 23.17). But neither the right hand nor the mirror image of the left hand can be turned in any way to look exactly like the left hand; that is, the right and left hands cannot be *superimposed* on one another. (Remember that a left-handed glove does not fit a right hand.) Any physical object *possessing the quality of handedness*—whose mirror image is not identical with itself—is said to be **chiral** (from the Greek *cheir,* meaning "hand"). A glove is chiral. However, a pencil, whose mirror image looks identical to the real pencil, is *achiral.* Isomer A in Figure 23.16 is achiral; its mirror image is identical to itself. Isomer B is chiral; its mirror image, isomer C, is not superimposable on B.

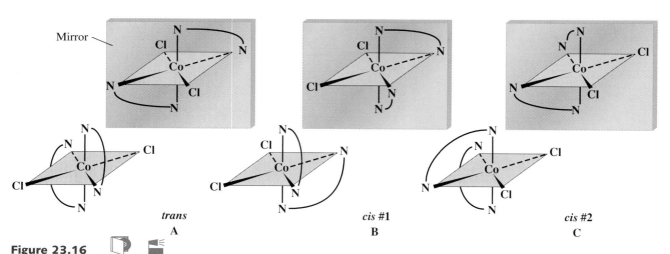

**Figure 23.16**
**Isomers of dichlorobis(ethylenediamine)cobalt(III) ion, CoCl₂(en)₂⁺.** Each isomer is shown with its mirror image. N⌒N represents ethylenediamine. *(A)* The *trans* isomer and its mirror image are superimposable, so they represent the same molecule. You can see this by rotating the original structure 90° counterclockwise (about the axis perpendicular to the gray plane). *(B, C)* Cis #1 and *cis* #2 are enantiomers (mirror-image, or optical, isomers) of the *cis* form. You cannot rotate one molecule to superimpose it on the other. However, the mirror image of *cis* #1 is identical to *cis* #2. To see this, rotate the mirror image of *cis* #1 by 180° and compare it with *cis* #2.

**Figure 23.17**
**Nonsuperimposable mirror images.** The mirror image of the left hand looks like the right hand, but the left hand itself cannot be superimposed on the right hand.

Enantiomers (optical isomers) have identical properties in a symmetrical environment. They have identical melting points and the same solubilities and colors. Enantiomers are usually differentiated by the manner in which they affect *plane-polarized* light. Normally, a light beam consists of electromagnetic waves vibrating in all possible planes about the direction of the beam (Figure 23.18). One of these planes may be selected out by passing a beam of light through a *polarizer* (say through a Polaroid lens). When this plane-polarized light is passed through a solution containing an enantiomer, such as one isomer of *cis*-[CoCl$_2$(en)$_2$]Cl, the plane of the light wave is twisted. One of the enantiomers twists the plane to the right (as the light comes out toward you); the other isomer twists the plane to the left by the same angle. Because of *the ability to rotate the plane of light waves, either as pure substances or in solution,* enantiomers are said to be **optically active.** (This is also the origin of the term *optical isomer.*) Figure 23.19 shows a sketch of a *polarimeter,* an instrument that determines the angular change in the plane of a light wave made by an optically active compound.

*A compound whose solution rotates the plane of polarized light to the right (when looking toward the source of light)* is called **dextrorotatory** and is labeled *d*. *A compound whose solution rotates the plane of polarized light to the left (when looking toward the source of light)* is called **levorotatory** and is labeled *l*. Thus, the dextrorotatory isomer of *cis*-[CoCl$_2$(en)$_2$]$^+$ is *d*-*cis*-dichlorobis(ethylenediamine)cobalt(III).

A chemical reaction normally produces *a mixture of equal amounts of optical isomers,* called a **racemic mixture.** Because the two isomers rotate the plane of polarized light in equal but opposite directions, a racemic mixture has no net effect on polarized light.

**Figure 23.18**
**Polarization of light.** Light from the source consists of waves vibrating in various planes along any axis. The polarizer filters out all waves except those vibrating in a particular plane.

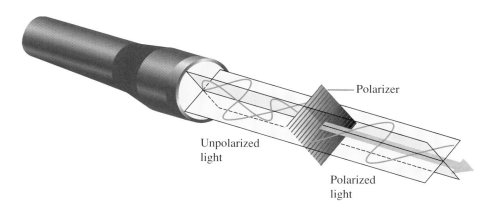

Polarizer

Unpolarized light

Polarized light

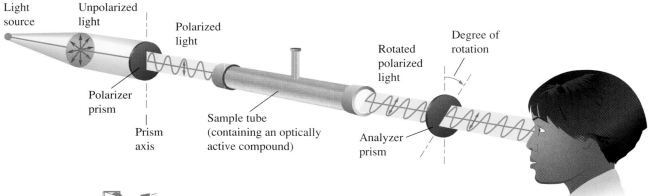

Light source
Unpolarized light
Polarized light
Polarizer prism
Prism axis
Sample tube (containing an optically active compound)
Rotated polarized light
Analyzer prism
Degree of rotation

**Figure 23.19**
**Sketch of a polarimeter.**
Light from a source (at the left) enters a polarizer prism, which splits the light into polarized components. The component that is polarized vertically enters the sample tube containing a solution of an optically active compound, which rotates the plane of the polarized light by an angle that depends on the compound and its concentration. The experimenter can see the light coming through the analyzer prism only if that prism is rotated by an equivalent angle (degree of rotation). [From Hart, *Organic Chemistry: A Short Course,* Ninth Edition. Copyright © 1995 by Houghton Mifflin Company. Used with permission.]

▶ An alternate example is provided in the Instructor's Resource Manual.

To show that optical isomers exist, a racemic mixture must be separated into its *d* and *l* isomers; that is, the racemic mixture must be *resolved.* One way to resolve a mixture containing *d* and *l* complex ions is to prepare a salt with an optically active ion of opposite charge. For example, the tartaric acid, $H_2C_4H_4O_6$, prepared from the white substance in wine vats is the optically active isomer *d*-tartaric acid. When the racemic mixture of *cis*-[CoCl$_2$(en)$_2$]Cl is treated with *d*-tartaric acid, the *d*-tartrate salts of *d*- and *l*-*cis*-[CoCl$_2$(en)$_2$]$^+$ may be crystallized. These salts will no longer be optical isomers of one another and will have different solubilities.

### EXAMPLE 23.4

### Deciding Whether Optical Isomers Are Possible

Are there optical isomers of the complex [Co(en)$_3$]Cl$_3$? If so, draw them.

**SOLUTION**

Yes, there are optical isomers, because the complex Co(en)$_3{}^{3+}$ has nonsuperimposable mirror images. See Figure 23.20.

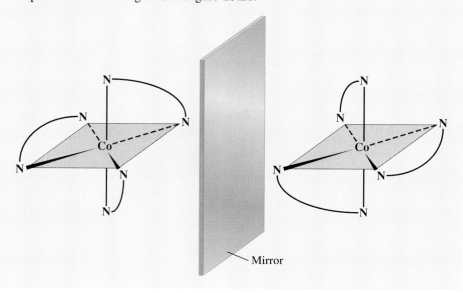

Mirror

**Figure 23.20**
**Optical isomers of the complex ion Co(en)$_3{}^{3+}$.**
Note that the isomers are nonsuperimposable mirror images of one another.

**EXERCISE 23.4**

Do any of the following have optical isomers? If so, draw them.

a. *trans*-Co(en)$_2$(NO$_2$)$_2$$^+$    b. *cis*-Co(en)$_2$(NO$_2$)$_2$$^+$

c. Ir(en)$_3$$^{3+}$                        d. [Ir(H$_2$O)$_3$Cl$_3$]

See Problems 23.51 and 23.52.

 **CONCEPT CHECK 23.3**

a. Which of the following molecular models of octahedral complexes are mirror images of the complex X? Keep in mind that you can rotate the models when performing comparisons.

b. Which complexes are optical isomers of molecule X?

c. Identify the distinct geometric isomers of complex X. (Note that some of the models may represent the same molecule.)

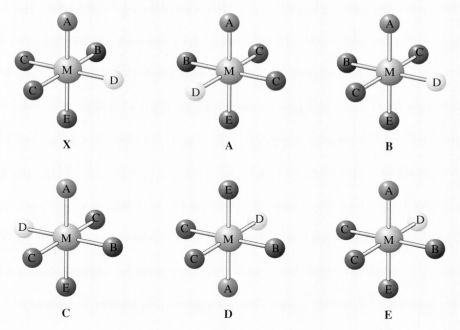

## 23.6  Valence Bond Theory of Complexes

A complex is formed when electron pairs from ligands are donated to a metal ion. This does not explain how the metal ion can accept electron pairs. Nor does it explain the paramagnetism often observed in complexes. *Valence bond theory* provided the first detailed explanation of the electronic structure of complexes. This explanation is essentially an extension of the view of covalent bonding we described earlier. ◄ According to this view, a covalent bond is formed by the overlap of two orbitals, one

▶ See Section 10.3.

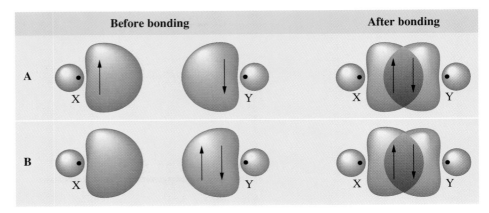

**Figure 23.21**
**Covalent bond formation between atoms X and Y.** *(A)* In the usual case, each overlapping orbital contains one electron. *(B)* When a coordinate covalent bond forms, one orbital containing a lone pair of electrons overlaps an empty orbital.

from each bonding atom. In the usual covalent bond formation, each orbital originally holds one electron, and after the orbitals overlap, a bond is formed that holds two electrons. In the formation of a coordinate covalent bond in a complex, however, a ligand orbital containing two electrons overlaps an unoccupied orbital on the metal atom. Figure 23.21 diagrams these two bond formations.

As an example of complex formation in a transition metal ion, let's look at the formation of the yellow $Cr(NH_3)_6^{3+}$ ion. This ion is known from experiment to be paramagnetic. We can think of this ion as formed by complexing $:NH_3$ ligands with the free chromium(III) ion. First, we need the configuration of the free ion. The chromium atom configuration is $[Ar]3d^5 4s^1$. In the formation of a transition-metal ion, the outer $s$ electrons (here, $4s^1$) are lost first, then the outer $d$ electrons (here, two of the $3d^5$ electrons). Therefore, $Cr^{3+}$ has the configuration $[Ar]3d^3$. The orbital diagram is

$$Cr^{3+}: \quad [Ar] \; \underset{3d}{\uparrow \; \uparrow \; \uparrow \; \bigcirc \; \bigcirc} \quad \underset{4s}{\bigcirc} \quad \underset{4p}{\bigcirc \; \bigcirc \; \bigcirc} \quad \underset{4d}{\bigcirc \; \bigcirc \; \bigcirc \; \bigcirc \; \bigcirc}$$

Note that the $3d^3$ electrons are placed in separate orbitals with their spins in the same direction, following Hund's rule. Two empty $3d$ orbitals, in addition to orbitals of the $n = 4$ shell, can be used for bonding to ligands.

To bond electron pairs from six $:NH_3$ ligands to $Cr^{3+}$, forming six equivalent bonds, octahedral hybrid orbitals are required. These hybrid orbitals will use two $d$ orbitals, the $4s$ orbital, and the three $4p$ orbitals. The $d$ orbitals could be either $3d$ or $4d$; the two available $3d$ orbitals are used because they have lower energy. We can now write the orbital diagram for the metal atom in the complex:

$$Cr \text{ in } Cr(NH_3)_6^{3+}: \quad [Ar]$$

$d^2sp^3$ bonds to ligands

Electron pairs donated from ligands are shown in color. Note that there are three unpaired electrons in $3d$ orbitals on the chromium atom, which explains the paramagnetism of the complex ion. The bonding in other complexes can be explained in a similar fashion.

# 23.7  Crystal Field Theory

Although valence bond theory explains the bonding and magnetic properties of complexes, it is limited in two important ways. First, the theory cannot easily explain the color of complexes. Second, the theory is difficult to extend quantitatively. Consequently, another theory—crystal field theory—has emerged as the prevailing view of transition-metal complexes.

**Crystal field theory** is *a model of the electronic structure of transition-metal complexes that considers how the energies of the d orbitals of a metal ion are affected by the electric field of the ligands.* According to this theory, the ligands in a transition-metal complex are treated as point charges. So a ligand anion becomes simply a point of negative charge. A neutral molecule, with its electron pair that it donates to the metal atom, is replaced by a partial negative charge, representing the negative end of the molecular dipole. In the electric field of these negative charges, the five *d* orbitals of the metal atom no longer have exactly the same energy. The result, as you will see, explains both the paramagnetism and the color observed in certain complexes.

The simplifications used in crystal field theory are drastic. Treating the ligands as point charges is essentially the same as treating the bonding as ionic. However, it turns out that the theory can be extended to include covalent character in the bonding. This simple extension is usually referred to as *ligand field theory,* but after including several levels of refinements, the theory becomes equivalent to molecular orbital theory.

## Effect of an Octahedral Field on the *d* Orbitals

All five *d* orbitals of an isolated metal atom have the same energy. But if the atom is brought into the electric field of several point charges, these *d* orbitals may be affected in different ways and therefore may have different energies. To understand how this can happen, you must first see what these *d* orbitals look like. You will then be able to picture what happens to them in the crystal field theory of an octahedral complex.

Figure 23.22 shows the shapes of the five *d* orbitals. The orbital labeled $d_{z^2}$ has a dumbbell shape along the *z*-axis, with a collar in the *x*−*y* plane surrounding this dumbbell. Remember that this shape represents the volume most likely to be occupied by an electron in this orbital. The other four *d* orbitals have "cloverleaf" shapes, each differing from one another only in the orientation of the orbitals in space. The "cloverleaf" orbital $d_{x^2-y^2}$ has its lobes along the *x*-axis and the *y*-axis. Orbitals $d_{xy}$, $d_{yz}$, and $d_{xz}$ have their lobes directed between the two sets of axes designated in the orbital label. Orbital $d_{xy}$, for example, has its lobes lying between the *x*- and *y*-axes.

A complex ion with six ligands will have the ligands arranged octahedrally about the metal atom to reduce mutual repulsion. Imagine that the ligands are replaced by negative charges. If the ligands are anions, they are replaced by the anion charge. If the ligands are neutral molecules, they are replaced by the partial negative charge from the molecular dipole. The six charges are placed at equal distances from the metal atom, one charge on each of the positive and negative sides of the *x*-, *y*-, and *z*-axes. (See Figure 23.22.)

Fundamentally, the bonding in this model of a complex is due to the attraction of the positive metal ion for the negative charges of the ligands. However, an electron in a *d* orbital of the metal atom is repelled by the negative charge of the ligands. This repulsion alters the energy of the *d* orbital depending on whether it is directed *toward* or *between* ligands. For example, consider the difference in the repulsive effect of ligands on metal-ion electrons in the $d_{z^2}$ and the $d_{xy}$ orbitals. Because the $d_{z^2}$ orbital

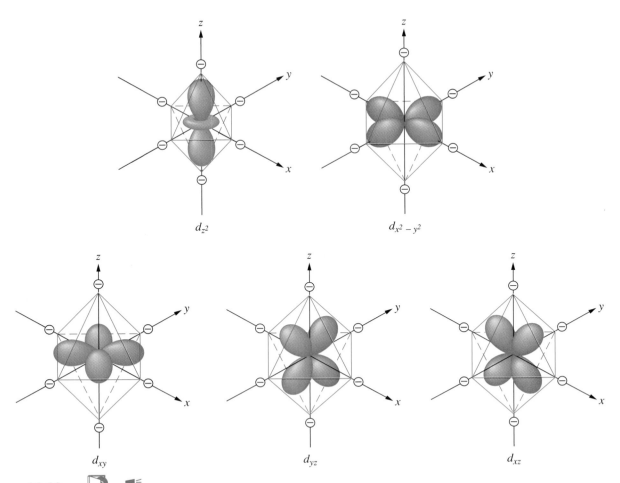

**Figure 23.22**
**The five $d$ orbitals.** The $d_{z^2}$ orbital has a dumbbell shape with a collar; the other orbitals have cloverleaf shapes. In an isolated atom, these orbitals have the same energy. However, in an octahedral complex, the orbitals split into two sets, with the $d_{z^2}$ and $d_{x^2-y^2}$ orbitals having higher energy than the other three. Note that the lobes of the $d_{z^2}$ and $d_{x^2-y^2}$ orbitals point toward the ligands (represented here by negative charges), whereas the lobes of the other orbitals point between ligands. The repulsion is greater in the case of the $d_{z^2}$ and $d_{x^2-y^2}$ orbitals.

is directed at the two ligands on the $z$-axis (one on the $-z$ side and the other on the $+z$ side), an electron in the orbital is rather strongly repelled by them. The energy of the $d_{z^2}$ orbital becomes greater. Similarly, an electron in the $d_{xy}$ orbital is repelled by the negative charge of the ligands, but because the orbital is not pointed directly at the ligands, the repulsive effect is smaller. The energy is raised, but less than the energy of the $d_{z^2}$ orbital is raised.

   If you look at the five $d$ orbitals in an octahedral field (electric field of octahedrally arranged charges), you see that you can divide them into two sets. Orbitals $d_{z^2}$ and $d_{x^2-y^2}$ are directed toward ligands, and orbitals $d_{xy}$, $d_{yz}$, and $d_{xz}$ are directed between ligands. The orbitals in the first set ($d_{z^2}$ and $d_{x^2-y^2}$) have higher energy than those in the second set ($d_{xy}$, $d_{yz}$, and $d_{xz}$). Figure 23.23 shows the energy levels of the $d$ orbitals in an octahedral field. *The difference in energy between the two sets of d*

**Figure 23.23**
**Energy levels of *d* orbitals in an octahedral field.**
The positive metal ion is attracted to the negative charges (ligands), but electrons in the *d* orbitals are repelled by them. Thus, although there is an overall attraction, the *d* orbitals no longer have the same energy.

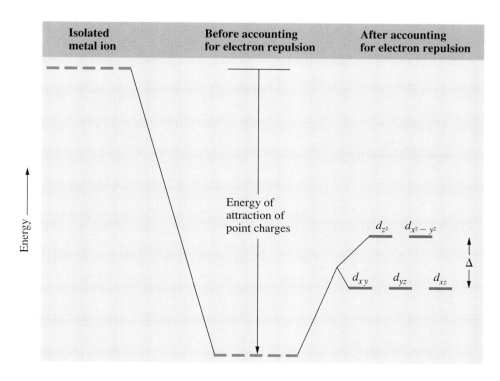

*orbitals on a central metal ion that arises from the interaction of the orbitals with the electric field of the ligands is called the* **crystal field splitting, $\Delta$.**

## High-Spin and Low-Spin Complexes

Once you have the energy levels for the *d* orbitals in an octahedral complex, you can decide how the *d* electrons of the metal ion are distributed in them. Knowing this distribution, you can predict the magnetic characteristics of the complex.

Consider the complex ion $Cr(NH_3)_6^{3+}$. According to crystal field theory, it consists of the $Cr^{3+}$ ion surrounded by $NH_3$ molecules treated as partial negative charges. The effect of these charges is to split the *d* orbitals of $Cr^{3+}$ into two sets as shown in Figure 23.23. The question now is how the *d* electrons are distributed among the *d* orbitals of the $Cr^{3+}$ ion. Because the electron configuration of $Cr^{3+}$ is $[Ar]3d^3$, there are three *d* electrons to distribute. They are placed in the *d* orbitals of lower energy, following Hund's rule (Figure 23.24). You see that the complex ion $Cr(NH_3)_6^{3+}$ has three unpaired electrons and is therefore paramagnetic.

If you examine any transition-metal ion that has configurations $d^4$, $d^5$, $d^6$, or $d^7$, you will find two bonding possibilities, which yield high-spin complexes in one case and low-spin complexes in the other case. A **high-spin complex** is *a complex in which there is minimum pairing of electrons in the d orbitals of the metal atom.* A **low-spin complex** is *a complex in which there is more pairing of electrons in the d orbitals of the metal atom than in a corresponding high-spin complex.*

Consider the complex ion $Fe(H_2O)_6^{2+}$. What are its magnetic characteristics? Remember, you need look at only the *d* electrons of the metal ion, $Fe^{2+}$. The electron configuration of the ion is $[Ar]3d^6$. You distribute six electrons among the *d* orbitals of the complex in such a way as to get the lowest total energy. If you place all six electrons into the lower three *d* orbitals, you get the distribution shown by the

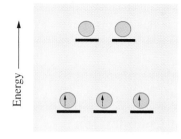

**Figure 23.24**
**Occupation of the 3*d* orbitals in an octahedral complex of $Cr^{3+}$.** The electrons occupy different lower-energy orbitals but have the same spin (Hund's rule).

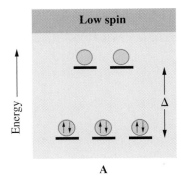

**Low spin**

Energy

Δ

A

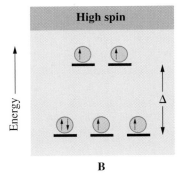

**High spin**

Energy

Δ

B

**Figure 23.25**
**Occupation of the 3*d*** **orbitals in complexes of** **Fe²⁺.** *(A)* Low spin. *(B)* High spin.

energy-level diagram in Figure 23.25*A*. All of the electrons are paired, so you would predict that this distribution gives a diamagnetic complex. The $Fe(H_2O)_6^{2+}$ ion, however, is paramagnetic. Therefore, the distribution given in Figure 23.25*A* is not correct. What was the mistake?

The mistake was to ignore the **pairing energy, *P***, *the energy required to put two electrons into the same orbital*. When an orbital is already occupied by an electron, it requires energy to put another electron into that orbital because of their mutual repulsion. Suppose that this pairing energy is greater than the crystal field splitting; that is, suppose $P > \Delta$. In that case, once the first three electrons have singly occupied the three lower-energy *d* orbitals, the fourth electron will go into one of the higher-energy *d* orbitals. It will take less energy to do that than to pair up with an electron in one of the lower-energy orbitals. Similarly, the fifth electron will go into the last empty *d* orbital. The sixth electron must pair up, so it goes into one of the lower-energy orbitals. Figure 23.25*B* shows this electron distribution. In this case, there are four unpaired electrons and the complex is paramagnetic.

We see that crystal field theory predicts two possibilities: a *low-spin complex* when $P < \Delta$ and a *high-spin complex* when $P > \Delta$. The value of $\Delta$, as we will explain later, can be obtained from the spectrum of a complex, and the value of $P$ can be calculated theoretically. Even in the absence of these numbers, however, the theory predicts that a paramagnetic octahedral complex of $Fe^{2+}$ should have a magnetism equal to that of four unpaired electrons. This is what you find for the $Fe(H_2O)_6^{2+}$ ion.

You would expect low-spin diamagnetic $Fe^{2+}$ complexes to occur for ligands that bond strongly to the metal ion—that is, for those giving large $\Delta$. Ligands that might give a low-spin complex are suggested by a look at the spectrochemical series. The **spectrochemical series** is *an arrangement of ligands according to the relative magnitudes of the crystal field splittings they induce in the d orbitals of a metal ion*. The following is a short version of the spectrochemical series:

weak-bonding ligands                                                  strong-bonding ligands

$$I^- < Br^- < Cl^- < F^- < OH^- < H_2O < NH_3 < en < NO_2^- < CN^- < CO$$

increasing $\Delta$ ⟶

From this series, you see that the $CN^-$ ion bonds more strongly than $H_2O$, which explains why $Fe(CN)_6^{4-}$ is a low-spin complex ion and $Fe(H_2O)_6^{2+}$ a high-spin complex ion. You can also see why carbon monoxide might be expected to be poisonous. You know that $O_2$ bonds reversibly to the Fe(II) atom of hemoglobin, so the bonding is only moderately strong. According to the spectrochemical series, however, carbon monoxide, CO, forms a strong bond. The bonding in this case is irreversible (or practically so). It results in a very stable complex of CO and hemoglobin, which cannot function then as a transporter of $O_2$.

▶ An alternate example is provided in the Instructor's Resource Manual.

**EXAMPLE 23.5**

**Describing the Bonding in an Octahedral Complex Ion (Crystal Field Theory)**

Describe the distribution of *d* electrons in the complex ion $Co(H_2O)_6^{2+}$, using crystal field theory. The hexaaquacobalt(II) ion is a high-spin complex ion. What would be the distribution of *d* electrons in an octahedral cobalt(II) complex ion that is low spin? How many unpaired electrons are there in each ion?

**SOLUTION**

The electron configuration of $Co^{2+}$ is $[Ar]3d^7$. The high-spin and low-spin distributions in the $d$ orbitals are

**High-spin**      **Low-spin**

$Co(H_2O)_6{}^{2+}$, a high-spin complex, has three unpaired electrons. A low-spin complex would have one unpaired electron.

**EXERCISE 23.5**

Describe the distribution of $d$ electrons in $Ni(H_2O)_6{}^{2+}$, using crystal field theory. How many unpaired electrons are there in this ion?

See Problems 23.53 and 23.54.

## Tetrahedral and Square Planar Complexes

When a metal ion bonds with tetrahedrally arranged ligands, the $d$ orbitals of the ion split to give two $d$ orbitals at lower energy and three $d$ orbitals at higher energy (just the opposite of what is found for an octahedral field). See Figure 23.26A. In the field of ligands in a square planar arrangement, the $d$ orbitals split as shown in Figure 23.26B.

The observed splittings, $\Delta$, in a tetrahedral field are approximately one-half the size of those in comparable octahedral complexes. Only high-spin complexes are observed, because the pairing energy is always greater than $\Delta$. In the square planar case, only low-spin complexes have been found.

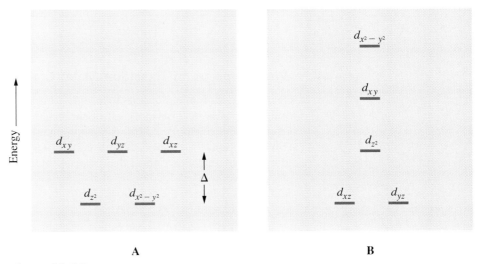

A                                    B

**Figure 23.26**
**Energy splittings of the $d$ orbitals in a complex with four ligands.** *(A)* A tetrahedral field. The crystal field splitting, $\Delta$, is smaller than in a comparable octahedral complex. *(B)* A square planar field.

▶ An alternate example is provided in the Instructor's Resource Manual.

**EXAMPLE 23.6**

## Describing the Bonding in a Four-Coordinate Complex Ion (Crystal Field Theory)

Describe the $d$-electron distributions of the complexes $Ni(NH_3)_4^{2+}$ and $Ni(CN)_4^{2-}$, according to crystal field theory. The tetraamminenickel(II) ion is paramagnetic, and the tetracyanonickelate(II) ion is diamagnetic.

**SOLUTION**

You expect the tetrahedral field to give high-spin complexes and the square planar field to give low-spin complexes. Therefore, the geometry of the $Ni(NH_3)_4^{2+}$ ion, which is paramagnetic, is probably tetrahedral. The distribution of $d$ electrons in the $Ni^{2+}$ ion (configuration $d^8$) is

$$Ni(NH_3)_4^{2+} \text{(tetrahedral)}$$

The geometry of $Ni(CN)_4^{2-}$, which is diamagnetic, is probably square planar; the distribution of $d$ electrons is

$$Ni(CN)_4^{2-} \text{(square planar)}$$

**EXERCISE 23.6**

Describe the distribution of $d$ electrons in the $CoCl_4^{2-}$ ion. The ion has a tetrahedral geometry. Assume a high-spin complex.

*See Problems 23.55 and 23.56.*

## Visible Spectra of Transition-Metal Complexes

Frequently, substances absorb light only in regions outside the visible spectrum and reflect, or pass on (transmit), all of the visible wavelengths. As a result, these substances appear white or colorless. (White light is a mixture of all visible wavelengths.) However, some substances absorb certain wavelengths in the visible spectrum and transmit the remaining ones; they appear colored. Many transition-metal complexes, as we have noted, are colored substances. The color results from electron jumps, or transitions, between the two closely spaced $d$ orbital energy levels that come from the crystal field splitting. ◀

▶ The colors of many gemstones are due to transition-metal-ion impurities in the mineral. Ruby has $Cr^{3+}$ in alumina, $Al_2O_3$, and emerald has $Cr^{3+}$ in beryl, $Be_3Al_2(SiO_3)_6$.

The spectrum of a $d^1$ configuration complex is particularly simple. Hexaaquatitanium(III) ion, $Ti(H_2O)_6^{3+}$, is an example. Titanium has the configuration $[Ar]3d^24s^2$, and $Ti^{3+}$ has the configuration $[Ar]3d^1$. According to crystal field theory, the $d$ electron occupies one of the lower-energy $d$ orbitals of the octahedral complex. Figure 23.27 shows the visible spectrum of $Ti(H_2O)_6^{3+}$. It results from a transition, or jump, of the $d$ electron from a lower-energy $d$ orbital to a higher-energy $d$ orbital,

**Figure 23.27**
*(Right)*
**Color and visible spectrum of $Ti(H_2O)_6^{3+}$.**
*Left:* A test tube containing a solution of $Ti(H_2O)_6^{3+}$. *Right:* Visible spectrum of $Ti(H_2O)_6^{3+}$. Unlike atomic spectra, which show absorption lines, spectra of ions and molecules in solution yield broad bands resulting from changes in nuclear motion that accompany the electronic transitions.

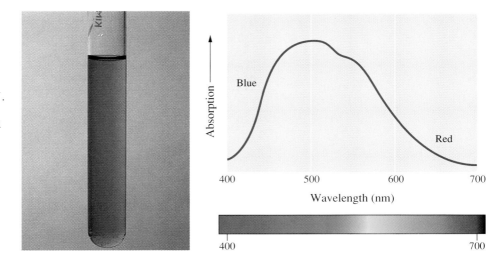

▶ Recall from Section 7.3 that the energy change during a transition equals $h\nu$.

as shown in Figure 23.28. Note that the energy change equals the crystal field splitting, $\Delta$. Consequently, the wavelength, $\lambda$, of light absorbed is related to $\Delta$. ◀

$$\Delta = h\nu = \frac{hc}{\lambda}$$

or

$$\lambda = \frac{hc}{\Delta}$$

When white light, which contains all visible wavelengths (400 nm to 750 nm), falls on a solution containing $Ti(H_2O)_6^{3+}$, blue-green light is absorbed. (The maximum light absorption is observed at 500 nm, which is blue-green light. See Table 23.7.) The other wavelengths of visible light, including red and some blue light, pass through the solution, giving it a red-purple color.

When the ligands in the $Ti^{3+}$ complex are changed, $\Delta$ changes and therefore the color of the complex changes. For example, replacing $H_2O$ ligands by weaker $F^-$ ligands should give a smaller crystal field splitting and therefore an absorption at longer wavelengths. The absorption of $TiF_6^{3-}$ is at 590 nm, in the yellow rather than the blue-green, and the color observed is violet-blue.

From this discussion, you see that the visible spectrum can be related to the crystal field splitting, and values of $\Delta$ can be obtained by spectroscopic analysis. However,

**Figure 23.28**
**The electronic transition responsible for the visible absorption in $Ti(H_2O)_6^{3+}$.**
An electron undergoes a transition from a lower-energy $d$ orbital to a higher-energy $d$ orbital. The energy change equals the crystal field splitting, $\Delta$.

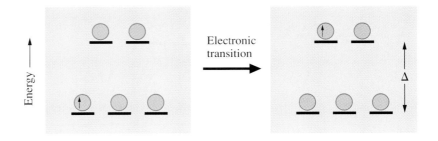

**Table 23.7**
**Color Observed for Given Absorption of Light by an Object**

| Wavelength Absorbed (nm) | Color Absorbed | Approximate Color Observed* |
|---|---|---|
| 410 | Violet | Green-yellow |
| 430 | Violet-blue | Yellow |
| 480 | Blue | Orange |
| 500 | Blue-green | Red |
| 530 | Green | Purple |
| 560 | Green-yellow | Violet |
| 580 | Yellow | Violet-blue |
| 610 | Orange | Blue |
| 680 | Red | Blue-green |
| 720 | Purple-red | Green |

*The exact color depends on the relative intensities of various wavelengths coming from the object and on the response of the eye to those wavelengths.

when there is more than one $d$ electron, several excited states can be formed. Consequently, the spectrum generally consists of several lines, and the analysis is more complicated than for $Ti^{3+}$.

► An alternate example is provided in the Instructor's Resource Manual.

**EXAMPLE 23.7**

## Predicting the Relative Wavelengths of Absorption of Complex Ions

When water ligands in $Ti(H_2O)_6^{3+}$ are replaced by $CN^-$ ligands to give $Ti(CN)_6^{3-}$, the maximum absorption shifts from 500 nm to 450 nm. Is this shift in the expected direction? Explain. What color do you expect to observe for this ion?

**SOLUTION**

According to the spectrochemical series, $CN^-$ is a more strongly bonding ligand than $H_2O$. Consequently, $\Delta$ should increase, and the wavelength of the absorption ($\lambda = hc/\Delta$) should decrease. **So the shift of the absorption is in the expected direction.** Because the absorbed light is between blue and violet-blue (see Table 23.7), the observed color is **orange-yellow** (this is the *complementary color* of the color between blue and violet-blue).

**EXERCISE 23.7**

The $Fe(H_2O)_6^{3+}$ ion has a pale purple color, and the $Fe(CN)_6^{3-}$ ion has a ruby-red color. What are the approximate wavelengths of the maximum absorption for each ion? Is the shift of wavelength in the expected direction? Explain.

See Problems 23.57, 23.58, 23.59, and 23.60.

# The Cooperative Release of Oxygen from Oxyhemoglobin

**LIFE SCIENCE**

Hemoglobin is an iron-containing substance in red blood cells responsible for the transport of $O_2$ from the lungs to various parts of the body. Myoglobin is a similar substance in muscle tissue, acting as a reservoir for the storage of $O_2$ and as a transporter of $O_2$ within muscle cells. The explanation for the different actions of these two substances involves some fascinating transition-metal chemistry.

Myoglobin consists of heme—a complex of Fe(II) bonded to a quadridentate ligand (Figure 23.8)—and globin. Globin, a protein, is attached through a nitrogen atom to one of the octahedral positions of Fe(II). The sixth position is vacant in free myoglobin but is occupied by $O_2$ in oxymyoglobin. Hemoglobin is essentially a four-unit structure of myoglobinlike units—that is, a *tetramer* of myoglobin.

Myoglobin and hemoglobin exist in equilibrium with the oxygenated forms oxymyoglobin and oxyhemoglobin, respectively. For example, hemoglobin (Hb) and $O_2$ are in equilibrium with oxyhemoglobin ($HbO_2$).

$$Hb + O_2 \rightleftharpoons HbO_2$$

Although hemoglobin is a tetramer of myoglobin, it does not function simply as four independent units of myoglobin. For it to function efficiently as a transporter of $O_2$ from the lungs and then be able to release that $O_2$ easily to myoglobin, hemoglobin must be less strongly attached to $O_2$ in the vicinity of a muscle cell than is myoglobin. In hemoglobin, the release of $O_2$ from one heme group triggers the release of $O_2$ from another heme group of the same molecule. In other words, there is a *cooperative release* of $O_2$ from hemoglobin that makes it possible for it to give up its $O_2$ to myoglobin.

The mechanism postulated for this cooperative release of $O_2$ depends on a change of iron(II) from a low-spin to a high-spin form, with a corresponding change in radius of the iron atom. In oxyhemoglobin, iron(II) exists in the low-spin form. When $O_2$ leaves, the iron atom goes to a high-spin form with two electrons in the higher-energy $d$ orbital. These higher-energy orbitals are somewhat larger than the lower-energy $d$ orbitals.

Low-spin Fe(II)
(smaller radius)

High-spin Fe(II)
(larger radius)

When an $O_2$ molecule leaves a heme group, the radius of the iron atom increases, and the atom pops out of the heme plane by about 70 pm. In hemoglobin, this change triggers the cooperative release of another $O_2$ molecule. As the iron atom moves, the attached globin group moves with it. This motion of one globin group causes an adjacent globin group in the tetramer to alter its shape, which in turn makes possible the easy release of an $O_2$ molecule from its heme unit.

## A Checklist for Review

### Important Terms

transition elements (p. 1003)
complex ion (23.3)
complex (coordination compound) (23.3)
ligands (23.3)
coordination number (23.3)
monodentate ligand (23.3)

bidentate ligand (23.3)
polydentate ligand (23.3)
chelate (23.3)
constitutional isomers (23.5)
stereoisomers (23.5)
geometric isomers (23.5)

enantiomers (optical isomers) (23.5)
chiral (23.5)
optically active (23.5)
dextrorotatory (23.5)
levorotatory (23.5)
racemic mixture (23.5)

crystal field theory (23.7)
crystal field splitting, $\Delta$ (23.7)
high-spin complex (23.7)
low-spin complex (23.7)
pairing energy, $P$ (23.7)
spectrochemical series (23.7)

### Summary of Facts and Concepts

The *d-block transition elements* are defined as those elements having a partially filled *d* subshell in any common oxidation state. They have a number of characteristics, including high melting points and a multiplicity of oxidation states. Compounds of transition elements are frequently colored and many are paramagnetic. These properties are due to the participation of $d$ orbitals in bonding. We described the chemical properties of two transition elements: Cr and Cu. Chromium metal reacts with acids to give $Cr^{2+}$ ion, which is readily oxidized to $Cr^{3+}$. The $+3$ state is the most common oxidation state; chromium(III) oxide is a green pigment. The $+6$ oxidation state is represented by chromates and dichromates. The dichromate ion, $Cr_2O_7^{2-}$, in acid solution is a strong oxidizing agent. Chromium metal is obtained by reduction of the ore chromite, $FeCr_2O_4$. Copper metal reacts only with acids having strongly oxidizing anions, such as $HNO_3$; it gives $Cu^{2+}$ ion. Copper(I) oxide, $Cu_2O$, occurs naturally as a copper mineral. This oxide also forms as a brick-red precipitate when copper(II) ion is reduced in basic solution. The reaction is used as a test for glucose.

Transition-metal atoms often function as Lewis acids, reacting with groups called *ligands* by forming coordinate covalent bonds to them. The metal atom with its ligands is a *complex ion* or neutral *complex*. Ligands that bond through more than one atom are called *polydentate,* and the complex formed is called a *chelate*. The IUPAC has agreed on a nomenclature of complexes that gives basic structural information about the species. The presence of isomers in coordination compounds is evidence for particular geometries. For example, $[Pt(NH_3)_2Cl_2]$ has two isomers, which is evidence for a square planar geometry having *cis–trans isomers*. Octahedral complexes, often those that have bidentate ligands, may have *optical isomers*— that is, isomers that are mirror images of one another.

*Valence bond theory* gave the earliest theoretical description of the electronic structure of a complex. According to this theory, a complex forms when doubly occupied ligand orbitals overlap unoccupied orbitals of the metal atom.

*Crystal field theory* treats the ligands as electric charge points that affect the energy of the $d$ orbitals of the metal ion. In an octahedral complex, two of the $d$ orbitals have higher energy than the other three. A high-spin complex forms when the *pairing energy* is greater than the *crystal field splitting,* so that electrons would "prefer" to occupy a higher-energy $d$ orbital rather than pair up with an electron in a lower-energy orbital. When the pairing energy is smaller than the crystal field splitting, the $d$ orbitals are occupied in the normal fashion, giving a low-spin complex. Color in transition-metal complexes is explained as due to a jump of an electron from the lower-energy to the higher-energy $d$ orbitals. The crystal field splitting can be obtained experimentally from the visible spectrum of a complex.

## Operational Skills

1. **Writing the IUPAC name given the structural formula of a coordination compound, and vice versa** Given the structural formulas of coordination compounds, write the IUPAC names **(EXAMPLE 23.1)**. Given the IUPAC names of complexes, write the structural formulas **(EXAMPLE 23.2)**.
2. **Deciding whether isomers are possible** Given the formula of a complex, decide whether geometric isomers are possible and, if so, draw them **(EXAMPLE 23.3)**. Given the structural formula of a complex, decide whether enantiomers (optical isomers) are possible and, if so, draw them **(EXAMPLE 23.4)**.
3. **Describing the bonding in a complex ion** Given a transition-metal complex ion, describe the bonding types (high-spin and low-spin, if both exist), using crystal field theory. **(EXAMPLES 23.5, 23.6)**
4. **Predicting the relative wavelengths of absorption of complex ions** Given two complexes that differ only in the ligands, predict, on the basis of the spectrochemical series, which complex absorbs at higher wavelength. Given the absorption maxima, predict the colors of the complexes. **(EXAMPLE 23.7)**

## Review Questions

**23.1** What characteristics of the transition elements set them apart from the main-group elements?

**23.2** According to the building-up principle, what is the electron configuration of the ground state of the technetium atom (atomic number 43)?

**23.3** The highest melting point for metals in the fifth period occurs for molybdenum. Explain why this is expected.

**23.4** Iron, cobalt, and nickel are similar in properties and are sometimes studied together as the "iron triad." For example, each is a fairly active metal that reacts with acids to give hydrogen and the $+2$ ions. In addition to the $+2$ ions, the $+3$ ions of the metals also figure prominently in the chemistries of the elements. Explain why these elements are similar.

**23.5** Palladium and platinum are very similar to one another. Both are unreactive toward most acids. However, nickel, which is in the same column of the periodic table, is an active metal. Explain why this difference exists.

**23.6** Write balanced equations for the reactions of Cr and Cu metals with HCl(*aq*). If no reaction occurs, write *NR*.

**23.7** Write balanced equations for the reactions of chromium(III) oxide with a strong acid.

**23.8** Describe the structure of copper(II) sulfate pentahydrate. What color change occurs when the salt is heated? What causes the color change?

**23.9** Copper(II) ion in basic solution is reduced by formaldehyde, HCHO, to copper(I) oxide. Formaldehyde is oxidized to formate ion, $HCOO^-$. Write the balanced equation for this reaction.

**23.10** What evidence did Werner obtain to show that the platinum complex $PtCl_4 \cdot 4NH_3$ has the structural formula $[Pt(NH_3)_4Cl_2]Cl_2$?

**23.11** Define the terms *complex ion, ligand,* and *coordination number*. Use an example to illustrate the use of these terms.

**23.12** Define the term *bidentate ligand*. Give two examples.

**23.13** Rust spots on clothes can be removed by dissolving the rust in oxalic acid. The oxalate ion forms a stable complex with $Fe^{3+}$. Using an electron-dot formula, indicate how an oxalate ion bonds to the metal ion.

**23.14** What three properties of coordination compounds have been important in determining the details of their structure and bonding?

**23.15** Define each of the following and give an example of each: (a) ionization isomerism, (b) hydrate isomerism, (c) coordination isomerism, (d) linkage isomerism.

**23.16** Define the terms *geometric isomerism* and *optical isomerism* and give an example of each.

**23.17** Explain the difference in behavior of *d* and *l* isomers with respect to polarized light.

**23.18** What is a racemic mixture? Describe one method of resolving a racemic mixture.

**23.19** Describe the formation of a coordinate covalent bond between a metal-ion orbital and a ligand orbital.

**23.20** (a) Describe the steps in the formation of a high-spin octahedral complex of $Fe^{2+}$ in valence bond terms. (b) Do the same for a low-spin complex.

**23.21** Explain why *d* orbitals of a transition-metal atom may have different energies in the octahedral field of six negative charges. Describe how each of the *d* orbitals is affected by the octahedral field.

**23.22** (a) Use crystal field theory to describe a high-spin octahedral complex of $Fe^{2+}$. (b) Do the same for a low-spin complex.

**23.23** What is meant by the term *crystal field splitting*? How is it determined experimentally?

**23.24** What is the spectrochemical series? Use the ligands $CN^-$, $H_2O$, $Cl^-$, and $NH_3$ to illustrate the term. Then arrange them in order, describing the meaning of this order.

**23.25** What is meant by the term *pairing energy*? How do the relative values of pairing energy and crystal field splitting determine whether a complex is low-spin or high-spin?

**23.26** A complex absorbs red light from a single electron transition. What color is this complex?

# Conceptual Problems

**23.27** A cobalt complex whose composition corresponded to the formula $Co(NO_2)_2Cl \cdot 4NH_3$ gave an electrical conductance equivalent to two ions per formula unit. Excess silver nitrate solution immediately precipitates 1 mol AgCl per formula unit. Write a structural formula consistent with these results.

**23.28** For the following coordination compounds, identify the geometric isomer(s) of compound X.

**23.29** Describe step by step how the name potassium hexacyanoferrate(II) leads one to the structural formula $K_4[Fe(CN)_6]$.

**23.30** Compounds A and B are known to be stereoisomers of one another. Compound A has a violet color; compound B has a green color. Are they geometric or optical isomers?

**23.31** A complex has a composition corresponding to the formula $CoBr_2Cl \cdot 4NH_3$. What is the structural formula if conductance measurements show two ions per formula unit? Silver nitrate solution gives an immediate precipitate of AgCl but no AgBr. Write the structural formula of an isomer.

**23.32** For the complexes shown here, which would have the *d* electron distribution shown in the diagram below: $MF_6{}^{3-}$, $M(CN)_6{}^{3-}$, $MF_6{}^{4-}$, $M(CN)_6{}^{4-}$? Note that the neutral metal atom, M, in each complex is the same and has the ground state electron configuration $[Ar]4s^2 3d^6$.

## Practice Problems

### Properties of the Transition Elements

**23.33** Find the oxidation numbers of the transition metal in each of the following compounds:
a. $FeCO_3$    b. $MnO_2$    c. $CuCl$    d. $CrO_2Cl_2$

**23.34** Find the oxidation numbers of the transition metal in each of the following compounds:
a. $CoSO_4$    b. $Ta_2O_5$    c. $Cu_2(OH)_3Cl$

**23.35** Write the balanced equation for the reaction of iron(II) ion with nitrate ion in acidic solution. Nitrate ion is reduced to NO.

**23.36** Write the balanced equation for the reaction of sulfurous acid with dichromate ion.

### Structural Formulas and Naming of Complexes

**23.37** Give the coordination number of the transition-metal atom in each of the following complexes.
a. $Au(CN)_4^-$    b. $[Co(NH_3)_4(H_2O)_2]Cl_3$
c. $[Au(en)_2]Cl_3$    d. $Cr(en)_2(C_2O_4)^+$

**23.38** Give the coordination number of the transition element in each of the following complexes.
a. $[Ni(NH_3)_6](ClO_3)_2$    b. $[Cu(NH_3)_4]SO_4$
c. $[Cr(en)_3]Cl_3$    d. $K_2[Ni(CN)_4]$

**23.39** Determine the oxidation number of the transition element in each of the following complexes.
a. $K_2[Ni(CN)_4]$    b. $Mo(en)_3^{3+}$
c. $Cr(C_2O_4)_3^{3-}$    d. $[Co(NH_3)_5(NO_2)]Cl_2$

**23.40** For each of the following complexes, determine the oxidation state of the transition-metal atom.
a. $[CoCl(en)_2(NO_2)]NO_2$    b. $PtCl_4^{2-}$
c. $K_3[Cr(CN)_6]$    d. $Fe(H_2O)_5(OH)^{2+}$

**23.41** Consider the complex ion $Cr(NH_3)_2Cl_2(C_2O_4)^-$.
a. What is the oxidation state of the metal atom?
b. Give the formula and name of each ligand in the ion.
c. What is the coordination number of the metal atom?
d. What would be the charge on the complex if all ligands were chloride ions?

**23.42** Consider the complex ion $Mn(NH_3)_2(H_2O)_3(OH)^{2+}$.
a. What is the oxidation state of the metal atom?
b. Give the formula and name of each ligand in the ion.
c. What is the coordination number of the metal atom?
d. What would be the charge on the complex if all ligands were chloride ions?

**23.43** Write the IUPAC name for each of the following coordination compounds.
a. $K_3[FeF_6]$    b. $Cu(NH_3)_2(H_2O)_2^{2+}$
c. $(NH_4)_2[Fe(H_2O)F_5]$    d. $Ag(CN)_2^-$

**23.44** Name the following complexes, using IUPAC rules.
a. $K_4[Mo(CN)_8]$    b. $CrF_6^{3-}$
c. $V(C_2O_4)_3^{2-}$    d. $K_2[FeCl_4]$

**23.45** Give the IUPAC name for each of the following.
a. $Fe(CO)_5$    b. $Rh(CN)_2(en)_2^+$
c. $[Cr(NH_3)_4SO_4]Cl$    d. $MnO_4^-$

**23.46** Give the IUPAC name for each of the following.
a. $W(CO)_8$    b. $[Co(H_2O)_2(en)_2](SO_4)_3$
c. $K[Mo(CN)_8]$    d. $CrO_4^{2-}$

**23.47** Write the structural formula for each of the following compounds.
a. potassium hexacyanomanganate(III)
b. sodium tetracyanozincate(II)
c. tetraamminedichlorocobalt(III) nitrate
d. hexaamminechromium(III) tetrachlorocuprate(II)

**23.48** Give the structural formula for each of the following complexes.
a. diaquadicyanocopper(II)
b. potassium hexachloroplatinate(IV)
c. tetraamminenickel(II) perchlorate
d. tetraammineplatinum(II) tetrachlorocuprate(II)

### Isomerism

**23.49** Draw *cis–trans* structures of any of the following square planar or octahedral complexes that exhibit geometric isomerism. Label the drawings *cis* or *trans*.
a. $[Pd(NH_3)_2Cl_2]$    b. $Pd(NH_3)_3Cl^+$
c. $Pd(NH_3)_4^{2+}$    d. $Ru(NH_3)_4Br_2^+$

**23.50** If any of the following octahedral complexes display geometric isomerism, draw the structures and label them *cis* or *trans*.
a. $Co(NO_2)_4(NH_3)_2^-$    b. $Co(NH_3)_5(NO_2)^{2+}$
c. $Co(NH_3)_6^{3+}$    d. $Cr(NH_3)_5Cl^{2+}$

**23.51** Determine whether there are optical isomers of any of the following. If so, sketch the isomers.
a. *cis*-$Co(NH_3)_2(en)_2^{3+}$    b. *trans*-$IrCl_2(C_2O_4)_2^{3-}$

**23.52** Sketch mirror images of each of the following. From these sketches, determine whether optical isomers exist and note this fact on the drawings.
a. $Rh(en)_3^{3+}$    b. *cis*-$Cr(NH_3)_2(SCN)_4^-$

### Crystal Field Theory

**23.53** Using crystal field theory, sketch the energy-level diagram for the *d* orbitals in an octahedral field; then fill in the electrons for the metal ion in each of the following complexes. How many unpaired electrons are there in each case?
a. $V(CN)_6^{3-}$    b. $Co(C_2O_4)_3^{4-}$ (high-spin)
c. $Mn(CN)_6^{3-}$ (low-spin)

**23.54** Using crystal field theory, sketch the energy-level diagram for the $d$ orbitals in an octahedral field; then fill in the electrons for the metal ion in each of the following complexes. How many unpaired electrons are there in each case?
a. $ZrCl_6^{4-}$
b. $OsCl_6^{2-}$ (low-spin)
c. $MnCl_6^{4-}$ (high-spin)

**23.55** Obtain the distribution of $d$ electrons in the complex ions listed below, using crystal field theory. Each ion is either tetrahedral or square planar. On the basis of the number of unpaired electrons (given in parentheses), decide the correct geometry.
a. $Pt(NH_3)_4^{2+}$ (0)    (b) $Co(en)_2^{2+}$ (1)
c. $FeCl_4^-$ (5)    (d) $Co(NCS)_4^{2-}$ (3)

**23.56** Obtain the distribution of $d$ electrons in the complex ions listed below, using crystal field theory. Each ion is either tetrahedral or square planar. On the basis of the number of unpaired electrons (given in parentheses), decide the correct geometry.
a. $Pt(NH_3)_2(NO_2)_2^{2+}$ (2)    b. $MnCl_4^{2-}$ (5)
c. $NiCl_4^{2-}$ (2)    d. $AuF_4^-$ (0)

## Color

**23.57** The $Co(SCN)_4^{2-}$ ion has a maximum absorption at 530 nm. What color do you expect for this ion?

**23.58** The $Co(en)_3^{3+}$ ion has a maximum absorption at 470 nm. What color do you expect for this ion?

**23.59** The $Co(NH_3)_6^{3+}$ ion has a yellow color, but when one $NH_3$ ligand is replaced by $H_2O$ to give $Co(NH_3)_5(H_2O)^{3+}$, the color shifts to red. Is this shift in the expected direction? Explain.

**23.60** The $Co(en)_3^{3+}$ ion has a yellow color, but the $CoF_6^{3-}$ ion has a blue color. Is the shift from yellow to blue expected when ethylenediamine ligands are replaced by $F^-$ ligands? Explain.

**23.61** What is the value of $\Delta$ (in kJ/mol) when $\lambda = 500$ nm, corresponding to an electron jump between $d$-orbital levels in a complex with $d^1$ configuration?

**23.62** What is the value of $\Delta$ (in kJ/mol) when $\lambda = 680$ nm, corresponding to an electron jump between $d$-orbital levels in a complex with $d^1$ configuration?

## General Problems

**23.63** The hexaaquascandium(III) ion, $Sc(H_2O)_6^{3+}$, is colorless. Explain why this might be expected.

**23.64** The tetraaquazinc(II) ion, $Zn(H_2O)_4^{2+}$, is colorless. Explain why this might be expected.

**23.65** There are only two geometric isomers of the tetraamminedichlorocobalt(III) ion, $Co(NH_3)_4Cl_2^+$. How many geometric isomers would be expected for this ion if it had a regular planar hexagonal geometry? Give drawings for them. Does this rule out a planar hexagonal geometry for $Co(NH_3)_4Cl_2^+$? Explain.

**23.66** There are only two geometric isomers of the triamminetrichloroplatinum(IV) ion, $Pt(NH_3)_2Cl_4^+$. How many geometric isomers would be expected for this ion if it had a regular planar hexagonal geometry? Give drawings for them. Does this rule out a planar hexagonal geometry for $Pt(NH_3)_2Cl_4^+$? Explain.

**23.67** Find the concentrations of $Cu^{2+}(aq)$, $NH_3(aq)$, and $Cu(NH_3)_4^{2+}(aq)$ at equilibrium when 0.10 mol $Cu^{2+}(aq)$ and 0.40 mol $NH_3(aq)$ are made up to 1.00 L of solution. The dissociation constant, $K_d$, for the complex $Cu(NH_3)_4^{2+}$ is $2.1 \times 10^{-13}$.

**23.68** Find the concentrations of $Ag^+(aq)$, $NH_3(aq)$, and $Ag(NH_3)_2^+(aq)$ at equilibrium when 0.10 mol $Ag^+(aq)$ and 0.20 mol $NH_3(aq)$ are made up to 1.00 L of solution. The dissociation constant, $K_d$, for the complex $Ag(NH_3)_2^+$ is $5.9 \times 10^{-8}$.

##  Media Activities

**23.69** Watch the Thermite Reaction *Visualization* on the student web site.
a. What is the oxidation state of the iron before and after the thermite reaction?
b. Write the electron configuration of iron before and after the thermite reaction.

**23.70** Complete the $d$ orbital Splitting in an Octahedral Field *Understanding Concepts* activity on the student web site. For a square planar complex:

a. Explain why the splitting of the $d$ orbitals is not the same as the splitting of the octahedral complex.
b. Explain why the $d_{z^2}$ orbital is higher in energy than the $d_{xz}$ and $d_{yz}$ orbitals.

**23.71** For review, take the ACE practice test for Chapter 23 on the student web site.

Sugar crystals viewed with a microscope under polarized light.

# Organic Chemistry

Organic chemistry is the chemistry of compounds containing carbon. As discussed in Chapter 2, carbon-containing compounds make up the majority of known compounds. Organic chemistry plays a central role in most of the substances that you encounter and use every day: the food you ate this morning, the shampoo that cleans your hair, the fuel that you use to heat your house and generate electricity, and the list goes on. Also, your life and that of every living organism on earth depends on the chemical reactions of organic molecules.

In this chapter, we introduce this fascinating area of chemistry with a discussion of the structural features of organic molecules, nomenclature, and a few important chemical reactions.

**Figure 24.1**
**Petroleum-based products contain long chains of carbon atoms.** *Left:* The black, tarry substance in asphalt consists of molecules with 30 or more carbon atoms bonded together. *Right:* Polyethylene contains chains of many thousands of carbon atoms bonded together. Polyethylene is used to create a variety of everyday materials.

## 24.1   The Bonding of Carbon

Because carbon is in Group IVA of the periodic table, it has four valence electrons. To fill its octet, it requires four additional electrons, which can be obtained through the formation of four covalent bonds. Carbon forms single, double, and triple bonds to achieve a filled octet. Therefore, the possible bonding combinations for carbon are as follows.

$$
-\overset{|}{\underset{|}{C}}- \qquad -\overset{|}{C}= \qquad =C= \qquad -C\equiv
$$

Recall from the VSEPR model (Chapter 10) that the molecular geometry around an atom is dictated by the number of regions of electron density. Keeping in mind that double and triple bonds count as one area of electron density, carbon can have a tetrahedral, trigonal planar, or linear geometry.

A unique feature of carbon is its ability to bond with other carbon atoms to form chains and rings of various lengths. Several other elements have limited ability to form such chains or rings of like atoms, but only carbon does this with more than a few atoms. Some petroleum-containing products, such as asphalt, consist of molecules with 30 or more carbon atoms bonded together; the molecules that make up polyethylene can have chains with thousands of carbon atoms (Figure 24.1).

## Hydrocarbons

The simplest organic compounds are **hydrocarbons,** *compounds containing only carbon and hydrogen.* All other organic compounds, for example, those containing O, N, and the halogen atoms, are classified as being derived from hydrocarbons. At

**Figure 24.2**
**Molecular models for the different hydrocarbons.**
*(A)* Molecular model of propane, $C_3H_8$, a saturated, acyclic hydrocarbon.
*(B)* Molecular model of cyclohexane, $C_6H_{12}$, a saturated, cyclic hydrocarbon.
*(C)* Molecular model of acetylene, $C_2H_2$, an unsaturated hydrocarbon.
*(D)* Molecular model of benzene, $C_6H_6$, an aromatic hydrocarbon.

A          B          C          D

first glance, you might think that the hydrocarbons represent a very limited set of molecules; however, several hundred thousand molecules exist that contain only hydrogen and carbon atoms.

Hydrocarbons can be separated into three main groups:

1. **Saturated hydrocarbons** are *hydrocarbons that contain only single bonds between the carbon atoms.* Saturated hydrocarbon molecules can be *cyclic* or *acyclic.* A cyclic hydrocarbon is one in which a chain of carbon atoms has formed a ring. An acyclic hydrocarbon is one that does not contain a ring of carbon atoms.

2. **Unsaturated hydrocarbons** are *hydrocarbons that contain double or triple bonds between carbon atoms.*

3. **Aromatic hydrocarbons** are *hydrocarbons that contain benzene rings or similar features.*

The saturated and unsaturated hydrocarbons are often referred to as the *aliphatic* hydrocarbons (Figure 24.2).

## 24.2   Alkanes and Cycloalkanes

The **alkanes** are *acyclic saturated hydrocarbons,* and the **cycloalkanes** are *cyclic saturated* hydrocarbons. The simplest hydrocarbon, an alkane called methane, consists of one carbon atom to which four hydrogen atoms are bonded in a tetrahedral arrangement. You can represent methane by its *molecular formula,* $CH_4$, which gives the number and kind of atoms in the molecule, or by its *structural formula,* which shows how the atoms are bonded to one another

$$CH_4$$

molecular formula
of methane

$$H-\underset{\underset{H}{|}}{\overset{\overset{H}{|}}{C}}-H$$

structural formula
of methane

109.5°

**Figure 24.3**
**Three-dimensional models of methane, $CH_4$.**
*Left:* Space-filling model of methane. *Right:* Ball-and-stick model of methane with bond angle.

Note that the structural formula does not convey information about the three-dimensional arrangement of the atoms. To do this, you would need to draw the three dimensional formula depicting the *molecular geometry* (see Figure 24.3). Methane is a very important molecule since it is the principal component of natural gas. In 1998, more than $2 \times 10^{11}$ ft$^3$ of natural gas were consumed in the United States to supply heating, transportation, and industrial needs.

## The Alkane Series

▶ The term *paraffin* comes from the Latin *parum affinus*, meaning "little affinity." The alkanes do not react with many reagents.

The alkanes, also called *paraffins*, ◀ have the general formula $C_nH_{2n+2}$. For $n = 1$, methane, the formula is $CH_4$; for $n = 2$, $C_2H_6$; for $n = 3$, $C_3H_8$; and so on. Note that the general formula conveys no information about how the atoms are connected. For now, we will assume that the carbon atoms are bonded together in a straight chain with hydrogen atoms completing the four required bonds to each carbon atom; these are called *straight-chain* or *normal* alkanes. The structural formulas for the first four straight chain alkanes are shown.

Because carbon atoms typically have four bonds, chemists often write the structures of the parts of organic compounds using **condensed structural formulas,** or con- **densed formulas,** *where the bonds around each carbon atom in the compound are not explicitly written.* For example,

Condensed formulas of the first four alkanes ($n = 1$ to 4) are

$$CH_4 \quad CH_3CH_3 \quad CH_3CH_2CH_3 \quad CH_3CH_2CH_2CH_3$$
methane      ethane         propane            butane

Note that the condensed formula of an alkane differs from that of the preceding alkane ($n - 1$) by a $—CH_2—$ group. The alkanes constitute a **homologous series,** which is *a series of compounds in which one compound differs from a preceding one by a fixed group of atoms.* Members of a homologous series have similar chemical properties, and their physical properties change throughout the series in a regular way. Table 24.1 lists the melting points and boiling points of the first ten straight-chain alkanes ($n = 1$ to $n = 10$). ◀ Note that the melting points and boiling points generally increase in the series with an increase in the number of carbon atoms in the chain. This is a result of increasing intermolecular forces, which increase with molecular weight.

## Constitutional Isomerism and Branched-Chain Alkanes

▶ Formerly, the names of the straight-chain alkanes were distinguished from branched-chain isomers by the prefix *n-* for *normal.* This designation is still common (butane is called *n*-butane) but is not used in the IUPAC name, which we will discuss in Section 24.5.

In addition to the straight-chain alkanes, *branched-chain* alkanes are possible. For example, isobutane (or 2-methylpropane) has the structure

**Table 24.1**
**Physical Properties of Straight-Chain Alkanes**

| Name | Number of Carbons | Formula | Melting Point (°C) | Boiling Point (°C) |
|---|---|---|---|---|
| Methane | 1 | $CH_4$ | −183 | −162 |
| Ethane | 2 | $CH_3CH_3$ | −172 | −89 |
| Propane | 3 | $CH_3CH_2CH_3$ | −187 | −42 |
| Butane | 4 | $CH_3(CH_2)_2CH_3$ | −138 | 0 |
| Pentane | 5 | $CH_3(CH_2)_3CH_3$ | −130 | 36 |
| Hexane | 6 | $CH_3(CH_2)_4CH_3$ | −95 | 69 |
| Heptane | 7 | $CH_3(CH_2)_5CH_3$ | −91 | 98 |
| Octane | 8 | $CH_3(CH_2)_6CH_3$ | −57 | 126 |
| Nonane | 9 | $CH_3(CH_2)_7CH_3$ | −54 | 151 |
| Decane | 10 | $CH_3(CH_2)_8CH_3$ | −30 | 174 |

Adapted from Robert D. Whitaker et al., *Concepts of General, Organic, and Biological Chemistry*, p. 231. Copyright © 1981 by Houghton Mifflin Company. Used with permission.

Isobutane, $C_4H_{10}$, has the same molecular formula as butane, the straight-chain hydrocarbon. However, isobutane and butane have different structural formulas and, therefore, different molecular structures. Butane and isobutane are **constitutional** (or **structural**) **isomers,** *compounds with the same molecular formula but different structural formulas*. Figure 24.4 depicts molecular models of isobutane and butane. Because these isomers have different structures, they have different properties. For example, isobutane boils at −12°C whereas butane boils at 0°C. Here the difference in boiling point can be attributed to the fact that isobutane has a more compact molecular structure than butane, which results in weaker intermolecular interactions between isobutane molecules.

The number of constitutional isomers rapidly increases with the number of carbons in the series. For example, there are three constitutional isomers with the molecular formula $C_5H_{12}$ (pentanes; see Figure 24.5), five of $C_6H_{14}$ (hexanes), and 75 of $C_{10}H_{22}$ (decanes).

Branched alkanes are usually written using condensed structural formulas. The following example and exercise will help you make the transition to writing hydrocarbons in this way.

**Figure 24.4**
**Constitutional isomers of butane.** Ball-and-stick models of isobutane (2-methylpropane) and butane.

Isobutane                Butane

pentane, b.p. 36°C

2-methylbutane, b.p. 28°C

2,2-dimethylpropane, b.p. 9.5°C

**Figure 24.5**
**Isomers of pentane.** Note that each isomer is a different compound with a different boiling point.

**EXAMPLE 24.1**
# Writing Condensed Structural Formulas

Write the condensed structural formula of each of the following alkanes.

**SOLUTION**

a. To write the condensed structural formula, it helps to identify the carbon backbone, or the skeleton of carbon atoms in the molecule. It is good practice to write the longest chain of carbon atoms in the molecule in a straight line. For this case, the skeleton looks like:

Next, write the appropriate number of H atoms next to each carbon atom.

$$CH_3-\overset{\underset{\displaystyle CH_3}{|}}{C}H-\overset{\underset{\displaystyle CH_3}{|}}{C}H-CH_3$$

Finally, find the straight chain(s) of carbon atoms that contain more than two carbon atoms and remove only the bonds (dashes) that connect those carbon atoms.

$$CH_3CHCHCH_3$$ with $CH_3$ above and $CH_3$ below

After a little practice you will be able to combine these steps.

b. Following the steps outlined above, the condensed structural formula for the part b molecule is

$$CH_3CH_2CHCH_2CH_3$$ (with $CH_3$/$CH_2$ branch)  or  $$CH_3CH_2CHCH_2CH_3$$ (with $CH_2CH_3$ branch)

Either of the above structures is correct; the choice of which structure you use depends on how much structural information you want to convey. Following the steps in the solution for part a, and recognizing that the —$CH_2$—$CH_3$ fragment in the structure on the left is a chain of carbons, you should be able to arrive at the structure on the right.

### EXERCISE 24.1

Write the condensed structural formula for the following alkane.

H<br>
|<br>
H—C—H  H  H<br>
|  |  |<br>
H—C ——— C—C—H<br>
|  |  |<br>
H—C—H  H  H<br>
|<br>
H

See Problems 24.23 and 24.24.

## Cycloalkanes

The general formula for cyclic cycloalkanes is $C_nH_{2n}$. Figure 24.6 gives the names and structural formulas for the first four members of the cycloalkane series. In the condensed structural formulas, a carbon atom and its attached hydrogen atoms are assumed to be at each corner.

| Molecular formula | $C_3H_6$ | $C_4H_8$ | $C_5H_{10}$ | $C_6H_{12}$ |
|---|---|---|---|---|
| Ball-and-stick model | | | | |
| Full structural formula | | | | |
| Condensed structural formula | △ | □ | ⬠ | ⬡ |
| Name | Cyclopropane | Cyclobutane | Cyclopentane | Cyclohexane |

**Figure 24.6**
**First four members of the cycloalkane series.**  These are saturated aliphatic hydrocarbons characterized by carbon-atom rings.

**CONCEPT CHECK 24.1**

In the model shown here, C atoms are black and H atoms are light blue:

a. Write the molecular formula.

b. Write the condensed structural formula.

## Sources and Uses of Alkanes and Cycloalkanes

Fossil fuels (natural gas, petroleum, and coal) are the principal sources of all types of organic chemicals. Natural gas is a mixture of low molecular weight hydrocarbons made up primarily of methane, $CH_4$, with lesser amounts of ethane, $C_2H_6$, propane, $C_3H_8$, and butane, $C_4H_{10}$. Petroleum, or crude oil, is the raw material extracted from a well. It is a mixture of alkanes and cycloalkanes with small amounts of aromatic hydrocarbons. The composition of petroleum is not consistent; it is dependent on geologic location and the organic matter present during oil formation. Because of this, a barrel of oil from Saudi Arabia contains a much different mixture of organic materials than a barrel of oil from the North Slope of Canada. The Saudi crude oil might be made up primarily of molecules with 5 to 20 carbon atoms, whereas the Canadian crude might consist mainly of molecules containing 20 to 40 carbon atoms. Crude oils that contain a majority of the lower molecular weight hydrocarbons are often more desirable because they are easier to transport and require less refining to convert to high-demand products like gasoline. Tables 24.2 and 24.3 show, respectively, the world's largest oil-producing countries and those with the largest reserves. Current world oil consumption is more than 70 million barrels per day.

Because fossil fuels are extracted from their source as mixtures of hydrocarbons, it is usually necessary to separate these mixtures into various components. Such separations are most easily performed by distillating the mixture into fractions that contain mixtures of different molecular weight compounds. As you can see from

**Table 24.2**
**Top Oil-Producing Countries**

| Country | 1996 Production (millions of barrels per day) |
|---|---|
| Saudi Arabia | 8.1 |
| Former Soviet Union | 6.9 |
| United States | 6.5 |
| I.R. Iran | 3.6 |
| China | 3.2 |

From *OPEC Annual Statistical Bulletin 1996.*

**Table 24.3**
**Countries Having the Largest Oil Reserves**

| Country | 1996 Reserves (millions of barrels) |
|---|---|
| Saudi Arabia | 261,444 |
| Iraq | 112,000 |
| United Arab Emirates | 97,800 |
| Kuwait | 96,500 |
| I.R. Iran | 92,600 |

From *OPEC Annual Statistical Bulletin 1996.*

Table 24.4, common names have been given to some of these fractions such as gasoline ($C_5$ to $C_{12}$) and kerosene ($C_{12}$ to $C_{15}$).

Often, there is a need to further separate the fractions listed in Table 24.4 into molecules with the same molecular weight, molecular formula, or structure. Through chemical processes that usually involve catalysts, small molecules can be combined into larger molecules, and large molecules can be broken apart. The processing of petroleum via distillation or chemical reactions is called *petroleum refining.* The type and extent of the petroleum refining that is performed is highly dependent on the type of crude oil that is available and the demand for a particular type of product. One chemical process, called *catalytic cracking,* involves passing hydrocarbon vapor over a heated catalyst of alumina ($Al_2O_3$) and silica ($SiO_2$) to break apart, or "crack," high molecular weight

**Table 24.4**
**Fractions from the Distillation of Petroleum**

| Boiling Range (°C) | Name | Range of Carbon Atoms per Molecule | Use |
|---|---|---|---|
| Below 20 | Gases | $C_1$ to $C_4$ | Heating, cooking, and petrochemical raw material |
| 20–200 | Naphtha; straight-run gasoline | $C_5$ to $C_{12}$ | Fuel; lighter fractions (such as petroleum ether, b.p. 30°C–60°C) are also used as laboratory solvents |
| 200–300 | Kerosene | $C_{12}$ to $C_{15}$ | Fuel |
| 300–400 | Fuel oil | $C_{15}$ to $C_{18}$ | Heating homes, diesel fuel |
| Over 400 | | Over $C_{18}$ | Lubricating oil, greases, paraffin waxes, asphalt |

From Harold Hart, *Organic Chemistry: A Short Course,* Eighth Edition, p. 102. Copyright © 1991 by Houghton Mifflin Company. Used with permission.

**Figure 24.7**
**Consumer products derived from petroleum.**
All of these common items contain organic compounds that used petroleum as a starting material.

hydrocarbons to produce hydrocarbons of low molecular weight. For example, this process can be used to convert the fuel oil fraction of petroleum to gasoline.

Petroleum refining also involves the conversion of the relatively abundant alkanes to unsaturated hydrocarbons, aromatic hydrocarbons, and hydrocarbon derivatives. The alkanes serve as the starting point for the majority of organic compounds, including plastics and pharmaceutical drugs. It is amazing to think about the number of everyday materials we use that started out as an alkane or cycloalkane (Figure 24.7).

### Reactions of Alkanes with Oxygen

From your experiences with alkanes, it is probably apparent that they are not particularly reactive molecules at normal temperatures. But imagine if this were not the case; you would have to be extremely careful about how you filled your automobile with fuel and you couldn't safely use natural gas in your home for heating or cooking. Also, the refining process could not involve the separation of the crude oil components at high temperatures by distillation. However, we do react alkanes every day through combustion with $O_2$; all hydrocarbons burn (combust) in an excess of $O_2$ at elevated temperatures to produce carbon dioxide, water, and heat. For example, a propane gas grill uses the reaction

$$C_3H_8(g) + 5O_2(g) \longrightarrow 3CO_2(g) + 4H_2O(l); \Delta H^\circ = -2220 \text{ kJ/mol}$$

The large negative $\Delta H^\circ$ value for this reaction and all hydrocarbon reactions with oxygen demonstrates why we rely on these molecules to meet our energy needs.

### Substitution Reactions of Alkanes

Under the right conditions, alkanes can react with other molecules. An important example is the reaction of alkanes with the halogens $F_2$, $Cl_2$, and $Br_2$. Reaction with $Cl_2$, for example, requires light (indicated by $hv$) or heat.

This is an example of a substitution reaction. A **substitution reaction** is *a reaction in which a part of the reacting molecule is substituted for an H atom on a hydrocarbon or hydrocarbon group.* All of the H atoms of an alkane may undergo substitution, leading to a mixture of products.

$$CH_3Cl + Cl_2 \xrightarrow{hv} CH_2Cl_2 + HCl$$

$$CH_2Cl_2 + Cl_2 \xrightarrow{hv} CHCl_3 + HCl$$

$$CHCl_3 + Cl_2 \xrightarrow{hv} CCl_4 + HCl$$

The $CCl_4$ product can be reacted with HF in the presence of a $SbCl_5$ catalyst to produce trichlorofluoromethane, $CCl_3F$, also known as CFC-11.

$$CCl_4 + HF \xrightarrow{SbCl_5} CCl_3F + HCl$$

$CCl_3F$

$CF_3CH_2F$

This compound is one of a number of chlorofluorocarbons (CFCs) that have been used for much of the twentieth century as a refrigerant. Data obtained in the 1970s revealed that these compounds survived long enough to travel to the stratospheric region of our atmosphere, where they facilitate the destruction of the ozone layer. (See the essay in Chapter 10 on this topic.) Recent refrigerants that do not as readily contribute to ozone destruction include hydrofluorocarbons (HFCs), such as $CF_3CH_2F$, which do not contain chlorine atoms.

### CONCEPT CHECK 24.2

For gasoline to function properly in an engine, it should not begin to burn before it is ignited by the spark plug. If it does, it makes the noise we think of as engine "knock." The octane-number scale rates the anti-knock characteristics of a gasoline. This linear scale is based on heptane, given an octane number of 0, and on 2,2,4-trimethylpentane (an octane constitutional isomer), given an octane number of 100. The higher the octane number, the better the anti-knock characteristics. If you had a barrel of heptane and a barrel of 2,2,4-trimethylpentane, how would you blend these compounds to come up with a 90 octane mixture?

## 24.3 Alkenes and Alkynes

Alkenes and alkynes are unsaturated hydrocarbons. Because they contain carbon–carbon multiple bonds, they are typically much more reactive than alkanes.

Under the proper conditions, molecular hydrogen can be added to an alkane or alkyne to produce a saturated compound in a process called *catalytic hydrogenation.* For example, ethylene adds hydrogen to give ethane.

$$
\underset{\text{ethylene}}{\begin{array}{c} H \\ \diagdown \\ H \end{array} C = C \begin{array}{c} H \\ \diagup \\ H \end{array}} + H_2 \xrightarrow{\text{Ni catalyst}} \underset{\text{ethane}}{H - \overset{\overset{\displaystyle H}{|}}{\underset{\underset{\displaystyle H}{|}}{C}} - \overset{\overset{\displaystyle H}{|}}{\underset{\underset{\displaystyle H}{|}}{C}} - H}
$$

Catalytic hydrogenation is also used in the food industry to convert carbon–carbon double bonds to carbon–carbon single bonds. For example, margarine can be manufactured by hydrogenating some of the double bonds present in corn oil to change it from oil to a solid (fat).

### Alkenes and Geometric Isomerism

**Alkenes** are *hydrocarbons that have the general formula $C_nH_{2n}$ and contain a carbon–carbon double bond.* (These compounds are also called *olefins.*) The simplest alkene, ethylene, has the condensed formula $CH_2=CH_2$ and the structural formula

$$
\begin{array}{c} H \\ \diagdown \\ H \end{array} C = C \begin{array}{c} H \\ \diagup \\ H \end{array}
$$

**Figure 24.8**
**Representations of the ethylene molecule.** *Left:* Pi bonding forces the two CH$_2$ groups in ethylene to bond so as to give a planar structure. *Right:* A space-filling model shows the planar structure of the molecule.

Ethylene is a gas with a sweet odor. It is obtained from the refining of petroleum and is an important raw material in the chemical industry. For example, when ethylene molecules are linked they make polyethylene (see Section 25.1), which is commonly used to make soda bottles and milk jugs. Plants also produce small amounts of ethylene. Fruit suppliers have found that exposure of fruit to ethylene speeds ripening.

In ethylene and other alkenes, all of the atoms connected to the two carbon atoms of the double bond lie in a single plane, as Figure 24.8 shows. This is due to the need for maximum overlap of 2*p* orbitals on the carbon atoms to form a pi ($\pi$) bond. As a result, rotation about a carbon–carbon double bond cannot occur without breaking the $\pi$ bond. For rotation to occur, a significant amount of energy must be supplied to break the $\pi$ bond, so rotation does not normally occur. (This is in contrast to carbon–carbon single bonds, which have very low energy requirements for rotation so they freely rotate under most conditions.) This inability to rotate gives rise to geometric isomers in certain alkenes. **Geometric isomers** are *isomers in which the atoms are joined to one another in the same way but differ because some atoms occupy different relative positions in space.* For example, 2-butene, C$_4$H$_8$, has two geometric isomers, called *cis*-2-butene and *trans*-2-butene.

*cis*-2-butene
b.p. 3.7°C

*trans*-2-butene
b.p. 0.9°C

The different boiling points confirm that they are different compounds. Note that in the *cis* isomer the two CH$_3$ groups are attached to the same side of the double bond, whereas in the *trans* isomer the two CH$_3$ groups are attached on opposite sides of the double bond. An alkene with the general formula

$$\begin{array}{c} A \qquad C \\ \diagdown \diagup \\ C{=}C \\ \diagup \diagdown \\ B \qquad D \end{array}$$

exists as geometric isomers only if groups A and B are different and groups C and D are different. For instance, geometric isomers do not exist for propene, CH$_2$=CHCH$_3$.

▶ An alternate example is
provided in the Instructor's
Resource Manual.

---

**EXAMPLE 24.2**

## Predicting *cis–trans* Isomers

For each of the following alkenes, decide whether *cis–trans* isomers are possible. If so draw structural formulas of the isomers.

a. $CH_3CH_2CH=C(CH_3)_2$     b. $CH_3CH=CHCH_2CH_3$

**SOLUTION**

a. Writing the structure, you have

Because two methyl groups are attached to the second carbon atom of the double bond, **geometric isomers are not possible.** b. **Geometric isomers are possible.** They are

*cis*-2-pentene          *trans*-2-pentene

---

**EXERCISE 24.2**

Decide whether *cis–trans* isomers are possible for each of the following compounds. If isomers are possible, draw the structural formulas.

a. $CH_3CH=CHCH_2CH_2CH_3$     b. $CH_3CH_2CH=CH_2$

See Problems 24.25 and 24.26.

---

*Cis* and *trans* isomers of dietary fats and oils are suspected to play a role in human health. All unprocessed fats and oils contain only *cis* isomers. During processing, some of the *cis* isomers are converted to *trans* isomers, called *trans* fatty acids. *Trans* fatty acids are suspected of raising the amount of serum cholesterol in the blood stream, which can cause health problems.

### Oxidation Reactions of Alkenes

Because alkenes are hydrocarbons, they undergo complete combustion reactions with oxygen at high temperatures to produce carbon dioxide and water. Unsaturated hydrocarbons can also be partially oxidized under relatively mild conditions. For example, when aqueous potassium permanganate, $KMnO_4(aq)$, is added to an alkene (or alkyne), the purple color of $KMnO_4$ fades and a brown precipitate of manganese dioxide forms (Figure 24.9).

$$3C_4H_9CH=CH_2 + 2MnO_4^-(aq) + 4H_2O \longrightarrow$$

1-hexene

**Figure 24.9**
**Test for unsaturation
using KMnO₄(aq).** *Left:* A
purple KMnO₄ solution prior
to the addition of an alkene.
*Right:* KMnO₄ solution that
has turned brown due to the
formation of MnO₂(s) due to
reaction with an alkene.

## Addition Reactions of Alkenes

Alkenes are more reactive than alkanes because of the presence of the double bond. Many reactants *add* to the double bond. A simple example is the addition of a halogen, such as $Br_2$, to propene.

$$CH_3CH=CH_2 + Br_2 \longrightarrow CH_3CH-CH_2$$
$$\underset{\text{propene}}{} \qquad\qquad \underset{\overset{|}{Br} \quad \overset{|}{Br}}{}$$

An **addition reaction** is *a reaction in which parts of a reactant are added to each carbon atom of a carbon–carbon double bond, which converts to a carbon–carbon single bond.* (Addition to triple bonds is also possible, giving a product with a double bond.) The addition of $Br_2$ to an alkene is fast. In fact, it occurs so readily that bromine dissolved in carbon tetrachloride, $CCl_4$, is a useful reagent to test for unsaturation. When a few drops of the solution are added to an alkene, the red-brown color of the $Br_2$ disappears immediately.

Unsymmetrical reagents, such as HCl and HBr, add to unsymmetrical alkenes to give two products that are constitutional isomers. For example,

$$CH_3CH=CH_2 + HBr \rightarrow \overset{3}{CH_3}\overset{2}{CH}-\overset{1}{CH_2}$$
$$\underset{\text{2-bromopropane}}{\overset{|}{Br} \quad \overset{|}{H}}$$

and

$$CH_3CH=CH_2 + HBr \rightarrow \overset{3}{CH_3}\overset{2}{CH}-\overset{1}{CH_2}$$
$$\underset{\text{1-bromopropane}}{\overset{|}{H} \quad \overset{|}{Br}}$$

In one case, the hydrogen atom of HBr adds to carbon atom 1, giving 2-bromopropane; in the other case, the hydrogen atom of HBr adds to carbon atom 2, giving 1-bromopropane. (The name 1-bromopropane means that a bromine atom is substituted for a hydrogen atom at carbon atom 1.) However, the two products are not formed in equal amounts; one is more likely to form. **Markownikoff's rule** is *a generalization that states that the major product formed by the addition of an unsymmetrical reagent such as H—Cl, H—Br, or H—OH is the one obtained when the H atom of the reagent* *adds to the carbon atom of the multiple bond that already has the greater number of hydrogen atoms attached to it.* In the preceding example, the H atom of HBr should add preferentially to carbon atom 1, which has two hydrogen atoms attached to it. The major product then is 2-bromopropane.

▶ An alternate example is provided in the Instructor's Resource Manual.

---

**EXAMPLE 24.3**

### Predicting the Major Product of an Addition Reaction

What is the major product of the following reaction?

$$CH_3-\underset{1}{C}\overset{CH_3}{\underset{2}{=}}\underset{3}{\overset{H}{C}}-\underset{4}{CH_3} + HCl \longrightarrow$$

**SOLUTION**

One H atom is attached to carbon 3 and none to carbon 2. Therefore the major product is

$$CH_3-\overset{CH_3}{\underset{Cl}{C}}-\overset{H}{\underset{H}{C}}-CH_3$$

**EXERCISE 24.3**

Predict the main product when HBr adds to 1-butene.

See Problems 24.31 and 24.32.

---

▶ Demonstration of the preparation of acetylene: See the Instructor's Resource Manual.

### Alkynes

**Alkynes** are *unsaturated hydrocarbons containing a carbon–carbon triple bond.* The general formula is $C_nH_{2n-2}$. The simplest alkyne is acetylene (ethyne), a linear molecule.

$$H-C\equiv C-H$$

Acetylene is a very reactive gas that is used to produce a variety of other chemical compounds. It burns with oxygen in the oxyacetylene torch to give a very hot flame (about 3000°C). Acetylene is produced commercially from methane.

$$2CH_4 \xrightarrow{1600°C} CH\equiv CH + 3H_2$$

Acetylene is also prepared from calcium carbide, $CaC_2$. Calcium carbide is obtained by heating calcium oxide and coke (carbon) in an electric furnace:

$$CaO(s) + 3C(s) \xrightarrow{2000°C} CaC_2(l) + CO(g)$$

The calcium carbide is cooled until it solidifies. The carbide ion, $C_2^{2-}$, is strongly basic and reacts with water to produce acetylene (Figure 24.10).

$$CaC_2(s) + 2H_2O(l) \longrightarrow Ca(OH)_2(aq) + C_2H_2(g)$$

**Figure 24.10**
**Preparation of acetylene gas.** Here acetylene is prepared by the reaction of water with calcium carbide. The acetylene burns with a sooty flame.

The alkynes, like the alkenes, undergo addition reactions, usually adding two molecules of the reagent for each C≡C bond. The major product is the isomer predicted by Markownikoff's rule, as shown in the following reaction.

$$CH_3{-}C{\equiv}C{-}H + 2HCl \longrightarrow CH_3{-}\underset{\underset{\displaystyle Cl}{|}}{\overset{\overset{\displaystyle Cl}{|}}{C}}{-}\underset{\underset{\displaystyle H}{|}}{\overset{\overset{\displaystyle H}{|}}{C}}{-}H$$

## 24.4  Aromatic Hydrocarbons

Aromatic hydrocarbons usually contain benzene rings—six membered rings of carbon atoms with alternating carbon–carbon single and carbon–carbon double bonds. The electronic structure of benzene can be represented by resonance formulas. For benzene,

This electronic structure can also be described using molecular orbitals (Figure 24.11). In this description, π molecular orbitals encompass the entire carbon-atom ring, and the π electrons are said to be delocalized. Delocalization of π electrons means that the double bonds in benzene do not behave as isolated double bonds. Two condensed formulas for benzene are

and

**Figure 24.11**
**The molecular orbitals of benzene.** The sigma bond framework (in green) forms by the overlap of $sp^2$ hybrid orbitals on each carbon atom; H atom 1s orbitals overlap these hybrid orbitals to form C—H bonds. Delocalized pi orbitals form by the overlap of carbon atom 2p orbitals that are perpendicular to the plane of the molecule; they give orbitals with lobes above and below this plane. Only the lowest-energy pi orbital is shown (yellow).

**Figure 24.12**
**Bond delocalization in benzene.** The space-filling model of benzene shows that all carbon–carbon bond distances are identical.

where the circle in the formula at right represents the delocalization of the $\pi$ electrons (and therefore the double bonds). Although you will often encounter benzene represented with alternating double and single bonds as shown on the left, the better representation is the one that indicates the bond delocalization. The space-filling model of benzene is shown in Figure 24.12.

Aromatic compounds are found everywhere. The term *aromatic* implies that compounds that contain a benzene ring have aromas and, indeed, this is the case. Flavoring agents that can be synthesized in the laboratory or found in nature include the flavor and aroma of cinnamon, cinnamaldehyde, and the wintergreen flavor of candies and gum, methyl salicylate (Figure 24.13).

Benzene rings also exist in the pain relievers acetylsalicylic acid (aspirin), acetaminophen (Tylenol), and the illicit drug mescaline.

**Figure 24.13**
**Examples of aromatic compounds.** *Left:* A molecular model of cinnamaldehyde, the molecule responsible for the taste and smell of cinnamon. *Right:* A molecular model of methylsalicylate, the molecule that produces the flavor of wintergreen.

Acetylsalicylic acid (aspirin)

Acetaminophen (Tylenol)

Mescaline

Cinnamaldehyde

Methyl salicylate

Other examples include the active ingredients in some sunscreens, *p*-aminobenzoic acid (often abbreviated as PABA on the label) and oxybenzone.

*p*-Aminobenzoic acid
(PABA)

Oxybenzone

Benzene rings can also fuse together to form polycyclic aromatic hydrocarbons in which two or more rings share carbon atoms. Figure 24.14 gives formulas for some polycyclic aromatic hydrocarbons. Naphthalene is one of the compounds that gives mothballs their characteristic odor.

## Substitution Reactions of Aromatic Hydrocarbons

Although benzene, $C_6H_6$, is an unsaturated hydrocarbon, it does not usually undergo addition reactions because the delocalized $\pi$ electrons of benzene are more stable than the localized $\pi$ electrons. Because of this, benzene does not undergo an addition reaction

**Figure 24.14**
**Some polycyclic aromatic hydrocarbons.**
Naphthalene is the simplest member of this series. It is a white, crystalline substance used in manufacturing plastics and plasticizers (to keep plastics pliable). Small amounts are used for mothballs.

with $Br_2$ in carbon tetrachloride like alkenes do. The usual reactions of benzene are substitution reactions. For example, in the presence of an iron(III) bromide catalyst, a hydrogen atom on the benzene ring is substituted for a bromine atom.

Similarly, benzene undergoes substitution with nitric acid in the presence of sulfuric acid to give nitrobenzene.

## 24.5  Naming Hydrocarbons

A nomenclature for organic compounds has developed over the years as a way of understanding and classifying their structures. This nomenclature is now formulated in rules agreed upon by the International Union of Pure and Applied Chemistry (IUPAC).

### Nomenclature of Alkanes

The first four straight-chain alkanes (methane, ethane, propane, and butane) have long-established names. Higher members of the series are named from the Greek words indicating the number of carbon atoms in the molecule, plus the suffix *-ane*. For example, the straight-chain alkane $C_5H_{12}$ is named pentane. Table 24.5 gives the formulas, names, and structures of the first ten straight-chain alkanes.

The following four IUPAC rules are applied in naming the branched-chain alkanes:

1. Determine the longest continuous (not necessarily straight) chain of carbon atoms in the molecule. The base name of the branched-chain alkane is that of the straight-chain alkane (Table 24.5) corresponding to the number of carbon atoms in this longest chain. For example, in

the longest continuous carbon chain, shown in color, has seven carbon atoms, giving the base name heptane. The full name for the alkane includes the names of any branched chains. These names are placed in front of the base name, as described in the remaining rules.

**Table 24.5**
**Formulas, Names, and Structures of the First Ten Straight-Chain Alkanes**

| Name | Elemental Formula | Condensed Structural Formula | Molecular model |
|---|---|---|---|
| Methane | $CH_4$ | $CH_4$ | |
| Ethane | $C_2H_6$ | $CH_3CH_3$ | |
| Propane | $C_3H_8$ | $CH_3CH_2CH_3$ | |
| Butane | $C_4H_{10}$ | $CH_3CH_2CH_2CH_3$ | |
| Pentane | $C_5H_{12}$ | $CH_3CH_2CH_2CH_2CH_3$ | |
| Hexane | $C_6H_{14}$ | $CH_3CH_2CH_2CH_2CH_2CH_3$ | |
| Heptane | $C_7H_{16}$ | $CH_3CH_2CH_2CH_2CH_2CH_2CH_3$ | |
| Octane | $C_8H_{18}$ | $CH_3CH_2CH_2CH_2CH_2CH_2CH_2CH_3$ | |
| Nonane | $C_9H_{20}$ | $CH_3CH_2CH_2CH_2CH_2CH_2CH_2CH_2CH_3$ | |
| Decane | $C_{10}H_{22}$ | $CH_3CH_2CH_2CH_2CH_2CH_2CH_2CH_2CH_2CH_3$ | |

2. Any chain branching off the longest chain is named as an alkyl group. An alkyl group is an alkane less one hydrogen atom. (Table 24.6 lists some alkyl groups.) When a hydrogen atom is removed from an end carbon atom of a straight-chain alkane, the resulting alkyl group is named by changing the suffix -*ane* of the alkane to -*yl*. For example, removing a hydrogen atom from methane gives the

**Table 24.6
Important Alkyl Groups**

| Original Alkane | Structure of Alkyl Group | Name of Alkyl Group |
|---|---|---|
| Methane, $CH_4$ | $CH_3-$ | Methyl |
| Ethane, $CH_3CH_3$ | $CH_3CH_2-$ | Ethyl |
| Propane, $CH_3CH_2CH_3$ | $CH_3CH_2CH_2-$ | Propyl |
| Propane, $CH_3CH_2CH_3$ | $CH_3\underset{\mid}{C}HCH_3$ | Isopropyl |
| Butane, $CH_3CH_2CH_2CH_3$ | $CH_3CH_2CH_2CH_2-$ | Butyl |
| Isobutane, $CH_3\underset{\underset{CH_3}{\mid}}{C}HCH_3$ | $CH_3\underset{\underset{CH_3}{\mid}}{\overset{\mid}{C}}CH_3$ | *Tertiary*-butyl (*t*-butyl) |

methyl group, $-CH_3$. The structure shown in Rule 1 has a methyl group as a branch on the heptane chain.

3. The complete name of a branch requires a number that locates the branch on the longest chain. For this purpose, you number each carbon atom on the longest chain in the direction that gives the smaller numbers for the locations of all branches. The structural formula in Rule 1 is numbered as shown:

$$\underset{7\ \ \ 6\ \ \ \ 5\ \ \ \ 4}{CH_3CH_2CH_2CH_2}-\overset{\overset{H}{\underset{3}{|}}}{\underset{\underset{\underset{1}{CH_3}}{\underset{2}{CH_2}}}{C}}-CH_3 \qquad \left( not\ \ \underset{1\ \ \ \ \ 2\ \ \ \ \ 3\ \ \ \ \ 4}{CH_3CH_2CH_2CH_2}-\overset{\overset{H}{\underset{5}{|}}}{\underset{\underset{\underset{7}{CH_3}}{\underset{6}{CH_2}}}{C}}-CH_3 \right)$$

The methyl branch is located at carbon 3 of the heptane chain (not carbon 5). The complete name of the branch is 3-methyl, and the compound is named 3-methyl-heptane. Note that the branch name and the base name are written as a single word, and a hyphen follows the number.

4. When there are more than one alkyl branch of the same kind (say, two methyl groups), this number is indicated by a prefix, such as *di-, tri-,* or *tetra-,* used with the name of the alkyl group. The position of each group on the longest chain is given by numbers. For example,

$$\underset{7\ \ \ 6\ \ \ 5}{CH_3CH_2CH_2}-\overset{\overset{H}{\underset{4}{|}}}{\underset{\underset{CH_3}{|}}{C}}-\overset{\overset{H}{\underset{3}{|}}}{\underset{\underset{\underset{1}{CH_3}}{\underset{2}{CH_2}}}{C}}-CH_3 \qquad\qquad \underset{7\ \ \ 6\ \ \ 5\ \ \ 4}{CH_3CH_2CH_2CH_2}-\overset{\overset{CH_3}{\underset{3}{|}}}{\underset{\underset{\underset{1}{CH_3}}{\underset{2}{CH_2}}}{C}}-CH_3$$

$$\text{3,4-dimethylheptane} \qquad\qquad\qquad \text{3,3-dimethylheptane}$$

Note that the position numbers are separated by a comma and are followed by a hyphen.

When there are two or more different alkyl branches, the name of each branch, with its position number, precedes the base name. The branch names are placed in alphabetical order. For example,

$$\underset{\text{3-ethyl-2-methylpentane}}{\overset{\overset{\displaystyle CH_3}{\underset{1\ \ \ 2\ |\ \ \ 3\ \ \ 4\ \ \ 5}{CH_3CHCHCH_2CH_3}}}{\underset{\displaystyle CH_2CH_3}{|}}}$$

Note the use of hyphens.

► An alternate example is provided in the Instructor's Resource Manual.

---

**EXAMPLE 24.4**

## Writing the IUPAC Name of an Alkane Given the Structural Formula

Give the IUPAC name for each of the following compounds.

a.  $CH_3CH_2CH_2$
      $\phantom{CH_3CH_2CH_2}\diagdown$
      $\phantom{CH_3CH_2CH_2CH_2}CHCH_2CH_2CH_3$
      $\phantom{CH_3CH}\diagup$
    $CH_3CH_2CH_2CH_2$

b.  $\phantom{xxx}CH_3$
      $\phantom{xxxx}|$
    $CH_3-C-CH_3$
      $\phantom{xxxx}|$
      $\phantom{xxx}CH_2$
      $\phantom{xxxx}|$
      $\phantom{xxx}CH_2$
      $\phantom{xxxx}|$
      $\phantom{xxx}CH_2$
      $\phantom{xxxx}|$
      $\phantom{xxx}CH_3$

**SOLUTION**

a. The longest continuous chain is numbered as follows:

$CH_3CH_2CH_2$ $\underset{8\ \ \ 7\ \ \ 6\ \ \ \ 5}{\phantom{CH_3CH_2CH_2}}$ $\overset{4\ \ \ 3\ \ \ \ 2\ \ \ 1}{\phantom{xxx}}$
      $\phantom{CH_3CH_2CH_2}\diagdown$
      $\phantom{CH_3CH_2CH_2CH_2}CHCH_2CH_2CH_3$
      $\phantom{CH_3CH}\diagup$
    $CH_3CH_2CH_2CH_2$

The name of the compound is **4-propyloctane.** If the longest chain had been numbered in the opposite direction, you would have the name 5-propyloctane. But because 5 is larger than 4, this name is unacceptable. b. The numbering of the longest chain is

$$\underset{}{\overset{\overset{1}{\displaystyle CH_3}}{\underset{\overset{3}{\displaystyle CH_2}}{\underset{\overset{4}{\displaystyle CH_2}}{\underset{\overset{5}{\displaystyle CH_2}}{\underset{\overset{6}{\displaystyle CH_3}}{}}}}}}$$

$$\overset{2}{CH_3-C-CH_3}$$

Any of the three clustered methyl carbon atoms could be given the number 1, and the other methyl groups branch off carbon atom 2. Hence, the name is **2,2-dimethylhexane.**

---

**EXERCISE 24.4**

What is the IUPAC name for each of the following hydrocarbons?

a.
$$CH_3$$
$$|$$
$$CH_3CHCHCH_3$$
$$|$$
$$CH_3$$

b.
$$CH_2CH_2CH_3$$
$$|$$
$$CH_3CHCHCH_2CH_3$$
$$|$$
$$CH_3$$

---

See Problems 24.33 and 24.34.

---

► An alternate example is provided in the Instructor's Resource Manual.

**EXAMPLE 24.5**

## Writing the Structural Formula of an Alkane Given the IUPAC Name

Write the condensed structural formula of 4-ethyl-3-methylheptane.

**SOLUTION**

First write the carbon skeleton for heptane.

$$\begin{array}{ccccccc} 1| & 2| & 3| & 4| & 5| & 6| & 7| \\ -C- & C- & C- & C- & C- & C- & C- \\ | & | & | & | & | & | & | \end{array}$$

Then attach the alkyl groups.

$$CH_2CH_3$$
$$\begin{array}{ccccccc} & & & | & & & \\ 1| & 2| & 3| & 4| & 5| & 6| & 7| \\ -C- & C- & C- & C- & C- & C- & C- \\ | & | & | & | & | & | & | \\ & & CH_3 & & & & \end{array}$$

After filling out the structure with H atoms, you have

$$CH_2CH_3$$
$$|$$
$$CH_3CH_2CHCHCH_2CH_2CH_3$$
$$|$$
$$CH_3$$

---

**EXERCISE 24.5**

Write the condensed structural formula of 3,3-dimethyloctane.

---

See Problems 24.35 and 24.36.

---

### Nomenclature of Alkenes and Alkynes

The IUPAC name for an alkene is determined by first identifying the longest chain containing the double bond. As with the alkanes, the longest chain provides the stem name, but the suffix is -*ene* rather than -*ane*. The carbon atoms of the longest chain are then numbered from the end nearer the carbon–carbon double bond, and the position of the double bond is given the number of the first carbon atom of that bond (the smaller number). This number is written in front of the stem name of the alkene. Branched chains are named the same way as the alkanes. The simplest alkene, $CH_2{=}CH_2$, is called ethene, although the common name is ethylene.

<table>
<tr><td>► An alternate example is provided in the Instructor's Resource Manual.</td></tr>
</table>

**EXAMPLE 24.6**

## Writing the IUPAC Name of an Alkene Given the Structural Formula

What is the IUPAC name of each of the following alkenes?

a.  $CH_2{=}CHCHCH_2CH_3$
    $\qquad\qquad |$
    $\qquad\quad CH_3$

b.  $CH_3CH_2CH_2CH_2CHCH{=}CHCH_3$
    $\qquad\qquad\qquad\quad |$
    $\qquad\qquad\qquad CH_2$
    $\qquad\qquad\qquad\quad |$
    $\qquad\qquad\qquad CH_2$
    $\qquad\qquad\qquad\quad |$
    $\qquad\qquad\qquad CH_2$
    $\qquad\qquad\qquad\quad |$
    $\qquad\qquad\qquad CH_2$
    $\qquad\qquad\qquad\quad |$
    $\qquad\qquad\qquad CH_3$

**SOLUTION**

a. The numbering of the carbon chain is

$$\overset{1}{C}H_2{=}\overset{2}{C}H\overset{3}{C}H\overset{4}{C}H_2\overset{5}{C}H_3$$
$$\qquad\qquad |$$
$$\qquad\quad CH_3$$

Because the longest chain containing a double bond has five carbon atoms, this is a pentene. It is a 1-pentene, because the double bond is between carbon atoms 1 and 2. The name of the compound is **3-methyl-1-pentene.** Note the placement of hyphens. If the numbering had been in the opposite direction, you would have named the compound as a 4-pentene. But this is unacceptable, because 4 is greater than 1. b. The numbering of the longest chain containing the double bond is

$$CH_3CH_2CH_2\overset{4}{C}H_2\overset{3}{C}HCH{=}\overset{2}{C}H\overset{1}{C}H_3$$
$$\qquad\qquad\qquad \overset{5}{|}$$
$$\qquad\qquad\qquad CH_2$$
$$\qquad\qquad\qquad \overset{6}{|}$$
$$\qquad\qquad\qquad CH_2$$
$$\qquad\qquad\qquad \overset{7}{|}$$
$$\qquad\qquad\qquad CH_2$$
$$\qquad\qquad\qquad \overset{8}{|}$$
$$\qquad\qquad\qquad CH_2$$
$$\qquad\qquad\qquad \overset{9}{|}$$
$$\qquad\qquad\qquad CH_3$$

This gives the name **4-butyl-2-nonene.** There is a longer chain (of ten carbon atoms), but it does not contain the double bond.

---

**EXERCISE 24.6**

Give the IUPAC name for each of the following compounds.

a.  CH$_3$C=CHCHCH$_3$
     |     |
    CH$_3$  CH$_2$
          |
         CH$_3$

b.  CH$_3$CH$_2$CH$_2$CHCH$_2$CH$_2$CH$_3$
               |
              CH
               ‖
             CH$_2$

*See Problems 24.37 and 24.38.*

---

**EXERCISE 24.7**

Write the condensed structural formula of 2,5-dimethyl-2-heptene.

*See Problems 24.39 and 24.40.*

---

Recall that some alkenes also exhibit *cis* and *trans* isomerism. In these cases, either *cis* or *trans* may be included as a prefix, separated by a hyphen, to the name.

---

**EXERCISE 24.8**

Give the IUPAC names, including the *cis* or *trans* label, for each of the geometric isomers of CH$_3$CH$_2$CH=CHCH$_2$CH$_3$.

*See Problems 24.41 and 24.42.*

---

Alkynes are named using the IUPAC rules in the same way as the alkenes, except that the stem name is determined from the longest chain that contains the carbon–carbon triple bond. The suffix for the stem name is *-yne*.

---

**EXERCISE 24.9**

Give the IUPAC name for each of the following alkynes.

a.  CH$_3$C≡CH

b.  CH≡CCHCH$_3$
          |
        CH$_2$CH$_3$

*See Problems 24.43 and 24.44.*

---

## Nomenclature of Aromatic Hydrocarbons

Simple benzene containing hydrocarbons that have one group substituted on a benzene ring use the name *benzene* as the suffix and the name of the group as the prefix. For example, the following compound is named ethylbenzene.

CH$_2$CH$_3$

ethylbenzene

When two groups are on the benzene ring, three isomers are possible. The isomers may be distinguished by using the prefixes *ortho-* (*o-*), *meta-* (*m-*), and *para-* (*p-*). For example,

o-dimethylbenzene (o-xylene)     m-dimethylbenzene (m-xylene)     p-dimethylbenzene (p-xylene)

A numbering system is also used to show the positions of two or more groups.

3-ethyl-1-methylbenzene          1,3,5-trimethylbenzene

It is sometimes preferable to name a compound containing a benzene ring by regarding the ring as a group in the same manner as alkyl groups. Pulling a hydrogen atom from benzene leaves the phenyl group, $C_6H_5$—. For example, you name the following compound diphenylmethane by using methane as the stem name.

diphenylmethane

---

**EXERCISE 24.10**

Write the structural formula of a. ethylbenzene; b. 1,2-diphenylethane.

---

See Problems 24.45 and 24.46.

 **CONCEPT CHECK 24.3**

In the model shown here, C atoms are black and H atoms are light blue:

a. Write the molecular formula.

b. Write the condensed structural formula.

c. Write the IUPAC name.

# Derivatives of Hydrocarbons

Certain groups of atoms in organic molecules are particularly reactive and have characteristic chemical properties. A **functional group** is *a reactive portion of a molecule that undergoes predictable reactions.* As discussed in Section 24.3, the C=C bond in a compound reacts readily and predictably with the reagents $Br_2$ and HBr in addition reactions. Thus, the C=C bond is a functional group. Many functional groups contain an atom other than carbon that has lone pair(s) of electrons. These lone pairs of electrons contribute to the reactivity of the functional group. Other functional groups, such as C=O, have multiple bonds that are reactive. Table 24.7 lists some common functional groups.

In the previous sections of the chapter, we discussed the hydrocarbons and their reactions. All other organic compounds can be considered to be derivatives of hydrocarbons. In these compounds, one or more hydrogen atoms of a hydrocarbon have been replaced by atoms other than carbon to give a functional group.

**Table 24.7
Some Organic Functional Groups**

| Structure of General Compound* (Functional Group in Color) | Name of Functional Group |
|---|---|
| R—Cl:      R—Br: | Organic halide |
| R—O—H | Alcohol |
| R—O—R′ | Ether |
| :O: ‖ R—C—H | Aldehyde |
| :O: ‖ R—C—R′ | Ketone |
| :O: ‖ R—C—O—H | Carboxylic acid |
| :O: ‖ R—C—O—R′ | Ester |
| R—N—H      R—N—H      R—N—R″ <br>   H          R′         R′ | Amine |
| :O: ‖ R—C—N—R′ <br>         H | Amide |

*R, R′, and R″ are general hydrocarbon groups.

# 24.6  Organic Compounds Containing Oxygen

Many of the important functional groups in organic compounds contain oxygen. Examples are alcohols, ethers, aldehydes, ketones, carboxylic acids, and esters. In this section, we will look at characteristics of these compounds.

## Alcohols and Ethers

Structurally, you may think of an **alcohol** as *a compound obtained by substituting a hydroxyl group (—OH) for an —H atom on a tetrahedral (sp³ hybridized) carbon atom of a hydrocarbon group.* Some examples are

$$CH_3OH \qquad CH_3CH_2OH \qquad \overset{\displaystyle OH}{\underset{\displaystyle |}{CH_3CHCH_3}}$$

methanol       ethanol       2-propanol
(methyl alcohol)    (ethyl alcohol)    (isopropyl alcohol)

Alcohols are named by IUPAC rules similar to those for the hydrocarbons, except that the stem name is determined from the longest chain containing the carbon atom to which the —OH group is attached. The suffix for the stem name is *-ol*. The position of the —OH group is indicated by a number preceding the stem name. (The number is omitted if unnecessary, as in ethanol.)

Alcohols are usually classified by the number of carbon atoms attached to the carbon atom to which the —OH group is bonded. A *primary alcohol* has one such carbon atom, a *secondary alcohol* has two, and a *tertiary alcohol* has three. The following are examples:

1-butanol        2-butanol        2-methyl-2-propanol
(a primary alcohol)    (a secondary alcohol)    (a tertiary alcohol)

Methanol, ethanol, ethylene glycol, and glycerol are some common alcohols. Methanol, $CH_3OH$, was at one time separated from the liquid distilled from sawdust—hence the common name wood alcohol. It is a toxic liquid prepared in large quantities by reacting carbon monoxide with hydrogen at high pressure in the presence of a catalyst. It is used as a solvent and as the starting material for the preparation of formaldehyde. Ethanol is manufactured by the fermentation of glucose (a sugar) or by the addition of water to the double bond of ethylene. The latter

**Figure 24.15**
**Gasoline containing ethanol.** The gasoline in this pump was formulated with ethanol. Ethanol is added to boost octane and oxygenate the fuel.

▶ Polyesters are discussed in Chapter 25.

▶ Video no. 18 (Series B), "Alcohol Explosions: The Chemistry of Engines."

reaction is carried out by heating ethylene with water in the presence of sulfuric acid.

$$CH_2{=}CH_2 + HOH \xrightarrow{\text{H}_2\text{SO}_4} \begin{array}{c} CH_2{-}CH_2 \\ | \quad\quad | \\ H \quad\quad OH \end{array}$$

ethylene                                          ethanol

Alcoholic beverages contain ethanol. Ethanol is also a solvent and a starting material for many organic compounds. It is mixed with gasoline and sold as gasohol, an automotive fuel (Figure 24.15). Ethylene glycol (IUPAC name: 1,2-ethanediol) and glycerol (1,2,3-propanetriol) are alcohols containing more than one hydroxyl group.

$$\begin{array}{cc} CH_2{-}CH_2 & CH_2{-}CH{-}CH_2 \\ | \quad\quad | & | \quad\quad | \quad\quad | \\ OH \quad OH & OH \quad OH \quad OH \end{array}$$

ethylene glycol                     glycerol

Ethylene glycol is a liquid prepared from ethylene and is used as an antifreeze agent. It is also used in the manufacture of polyester plastics and fibers. ◀ Glycerol is a nontoxic, sweet-tasting liquid obtained from fats during the making of soap. It is used in foods and candies to keep them soft and moist.

Just as an alcohol may be thought of as a derivative of water in which one H atom of $H_2O$ has been replaced by a hydrocarbon group, R, an **ether** is *a compound formally obtained by replacing both H atoms of $H_2O$ by the hydrocarbon groups R and R′.*

$$\underset{\text{water}}{H{-}O{-}H} \quad\quad \underset{\text{an alcohol}}{R{-}O{-}H} \quad\quad \underset{\text{an ether}}{R{-}O{-}R'}$$

Common names for ethers are formed from the hydrocarbon groups plus the word *ether*. For example, $CH_3OCH_2CH_2CH_3$ is called methyl propyl ether. By IUPAC rules, the ethers are named as derivatives of the longest hydrocarbon chain. For example, $CH_3OCH_2CH_2CH_3$ is 1-methoxypropane; the methoxy group is $CH_3O{-}$. The best-known ether is diethyl ether, $CH_3CH_2OCH_2CH_3$, often called simply ether, a volatile liquid used as a solvent and as an anesthetic.

---

**EXERCISE 24.11**
Give the IUPAC name of the following compound.

$$\begin{array}{c} OH \\ | \\ CH_3CH_2CH_2CCH_2CH_3 \\ | \\ CH_2CH_3 \end{array}$$

---

See Problems 24.49 and 24.50.

---

**EXERCISE 24.12**
Give the common name of each of the following compounds: a. $CH_3OCH_3$; b. $CH_3OCH_2CH_3$.

---

See Problems 24.53 and 24.54.

## Aldehydes and Ketones

Aldehydes and ketones are compounds containing a *carbonyl group.*

Ethanal

An **aldehyde** is *a compound containing a carbonyl group with at least one H atom attached to it.*

Here (H)R indicates a hydrocarbon group or H atom. The aldehyde function is usually abbreviated —CHO, and the structural formula of acetaldehyde is written $CH_3CHO$.

A **ketone** is *a compound containing a carbonyl group with two hydrocarbon groups attached to it.*

Propanone

$$
\underset{\text{a ketone}}{R-\overset{\overset{\textstyle O}{\|}}{C}-R'} \qquad \underset{\substack{\text{propanone}\\(\text{acetone})}}{CH_3-\overset{\overset{\textstyle O}{\|}}{C}-CH_3} \qquad \underset{\substack{\text{2-butanone}\\(\text{methyl ethyl ketone})}}{CH_3-\overset{\overset{\textstyle O}{\|}}{C}-CH_2CH_3}
$$

The ketone functional group is abbreviated —CO—, and the structural formula of acetone is written $CH_3COCH_3$.

Aldehydes and ketones are named according to IUPAC rules similar to those for naming alcohols. You first locate the longest carbon chain containing the carbonyl group to get the stem hydrocarbon name. Then you change the *-e* ending of the hydrocarbon to *-al* for aldehydes and *-one* for ketones. In the case of aldehydes, the carbon atom of the —CHO group is always the number-1 carbon. In ketones, however, the carbonyl group may occur in various nonequivalent positions on the carbon chain. For ketones, the position of the carbonyl group is indicated by a number before the stem name, just as the position of the hydroxyl group is indicated in alcohols. The carbon chain is numbered to give the smaller number for the position of the carbonyl group.

The aldehydes of lower molecular weight have sharp, penetrating odors. Formaldehyde (methanal), HCHO, and acetaldehyde (ethanal), $CH_3CHO$, are examples. With increasing molecular weight, the aldehydes become more fragrant. Some aldehydes of aromatic hydrocarbons have especially pleasant odors (see Figure 24.16). Formaldehyde is a gas produced by the oxidation of methanol. The gas is very soluble in water, and a 37% aqueous solution called Formalin is marketed as a disinfectant and as a preservative of biological specimens. The main use of formaldehyde is in the manufacture of plastics and resins. Acetone, $CH_3COCH_3$, is the simplest ketone. It is a liquid with a fragrant odor. The liquid is an important solvent for lacquers, paint removers, and nail polish removers.

**Figure 24.16**
**Some aldehydes of aromatic hydrocarbons.**
The name *aromatic* in aromatic hydrocarbons derives from the aromatic odors of many derivatives of these hydrocarbons, including the aldehydes shown here.

| Benzaldehyde (oil of almonds) | Vanillin (extract of vanilla) | Cinnamaldehyde (oil of cinnamon) |
|---|---|---|

---

**EXERCISE 24.13**

Name the following compounds by IUPAC rules:

$$\text{a. } CH_3CH_2CH_2 \overset{\overset{\textstyle O}{\|}}{-C} -CH_3 \qquad \text{b. } H \overset{\overset{\textstyle O}{\|}}{-C} -CH_2CH_2CH_3$$

---

See Problems 24.55 and 24.56.

### Carboxylic Acids and Esters

A **carboxylic acid** is *a compound containing the carboxyl group,* —COOH.

Ethanoic acid

$$-\overset{\overset{\textstyle O}{\|}}{C}-O-H \qquad R-\overset{\overset{\textstyle O}{\|}}{C}-O-H \qquad CH_3-\overset{\overset{\textstyle O}{\|}}{C}-O-H$$

carboxyl group       a carboxylic acid       ethanoic acid (acetic acid)

These compounds are named by IUPAC rules like those for the aldehydes, except that the ending on the stem name is *-oic* followed by the word *acid*. Many carboxylic acids have been known for a long time and are usually referred to by common names (see Table 24.8). The carboxylic acids are weak acids because of the acidity of the H atom on the carboxyl group. Acid-ionization constants are about $10^{-5}$.

An **ester** is *a compound formed from a carboxylic acid, RCOOH, and an alcohol, R'OH.* The general structure is

$$\overset{\overset{\textstyle :O:}{\|}}{RC}-\overset{..}{\underset{..}{O}}-R'$$

## 24.7  Organic Compounds Containing Nitrogen

Alcohols and ethers, you may recall, can be considered derivatives of $H_2O$, where one or both H atoms are replaced by hydrocarbon groups, R. Thus, the general formula of an alcohol is ROH, and that of an ether is ROR'. Another important class of organic compounds is obtained by similarly substituting R groups for the H atoms of ammonia, $NH_3$.

**Table 24.8**
**Common Carboxylic Acids**

| Carbon Atoms | Formula | Source | Common Name | IUPAC Name |
|---|---|---|---|---|
| 1 | HCOOH | Ants (Latin, *formica*) | Formic acid | Methanoic acid |
| 2 | CH₃COOH | Vinegar (Latin, *acetum*) | Acetic acid | Ethanoic acid |
| 3 | CH₃CH₂COOH | Milk (Greek, *protos pion*, "first fat") | Propionic acid | Propanoic acid |
| 4 | CH₃(CH₂)₂COOH | Butter (Latin, *butyrum*) | Butyric acid | Butanoic acid |
| 5 | CH₃(CH₂)₃COOH | Valerian root (Latin, *valere*, "to be strong") | Valeric acid | Pentanoic acid |
| 6 | CH₃(CH₂)₄COOH | Goats (Latin, *caper*) | Caproic acid | Hexanoic acid |
| 7 | CH₃(CH₂)₅COOH | Vine blossom (Greek, *oenanthe*) | Enanthic acid | Heptanoic acid |
| 8 | CH₃(CH₂)₆COOH | Goats (Latin, *caper*) | Caprylic acid | Octanoic acid |
| 9 | CH₃(CH₂)₇COOH | Pelargonium (an herb with stork-shaped capsules; Greek, *pelargos*, "stork") | Pelargonic acid | Nonanoic acid |
| 10 | CH₃(CH₃)₈COOH | Goats (Latin, *caper*) | Capric acid | Decanoic acid |

From Harold Hart, *Organic Chemistry: A Short Course*, Eighth Edition, p. 272. Copyright © 1991 by Houghton Mifflin Company. Used with permission.

Most organic bases are **amines,** which are *compounds that are structurally derived by replacing one or more hydrogen atoms of ammonia with hydrocarbon groups.*

$$R—N—H \qquad R—N—H \qquad R—N—R''$$
$$\underset{\text{a primary amine}}{|\,H} \qquad \underset{\text{a secondary amine}}{|\,R'} \qquad \underset{\text{a tertiary amine}}{|\,R'}$$

Table 24.9 lists some common amines.

Amines are bases, because the nitrogen atom has an unshared electron pair that can accept a proton to form a substituted ammonium ion. For example,

$$\underset{\text{methylamine}}{H—\overset{\displaystyle H}{\underset{\displaystyle CH_3}{N}}\!:} \;+\; H_2O \;\rightleftharpoons\; \underset{\text{methylammonium ion}}{\left[\,H—\overset{\displaystyle H}{\underset{\displaystyle CH_3}{N}}—H\,\right]^{+}} \;+\; OH^-$$

Like ammonia, amines are weak bases. Table 16.2 gives base-ionization constants of some amines.

**Amides** are *compounds derived from the reaction of ammonia, or of a primary or secondary amine, with a carboxylic acid.* For example, when ammonia is strongly heated with acetic acid, they react to give the amide acetamide. ◄

$$\underset{\text{ammonia}}{H—\overset{\displaystyle H}{N}—H} \;+\; \underset{\text{acetic acid}}{HO—\overset{\displaystyle}{\underset{\displaystyle O}{C}}—CH_3} \;\longrightarrow\; \underset{\text{acetamide}}{H—\overset{\displaystyle H}{N}—\overset{\displaystyle}{\underset{\displaystyle O}{C}}—CH_3} \;+\; H_2O$$

► The condensation to give an amide occurs under milder conditions if the ammonia or amine is reacted with the acid chloride, a derivative of the carboxylic acid obtained by replacing

$$\overset{\displaystyle O}{\underset{\displaystyle \|}{}}—C—OH \quad \text{by} \quad \overset{\displaystyle O}{\underset{\displaystyle \|}{}}—C—Cl$$

In this reaction, HCl is a product.

Methylamine and acetic acid react to give *N*-methylacetamide.

$$CH_3-\overset{\overset{\displaystyle H}{|}}{N}-H \ + \ HO-\overset{\overset{\displaystyle}{\underset{\underset{\displaystyle O}{\|}}{C}}}{}-CH_3 \ \longrightarrow \ CH_3-\overset{\overset{\displaystyle H}{|}}{N}-\overset{\overset{\displaystyle}{\underset{\underset{\displaystyle O}{\|}}{C}}}{}-CH_3 \ + \ H_2O$$

methylamine          acetic acid               *N*-methylacetamide

► **Experiments in General Chemistry, 24, "Molecular Models of Organic Molecules."**

**Table 24.9
Some Common Amines**

| Name | Formula | Boiling Point (°C) |
|------|---------|--------------------|
| Methylamine | $CH_3-NH_2$ | −6.5 |
| Dimethylamine | $CH_3-\overset{\displaystyle}{\underset{\underset{\displaystyle H}{|}}{N}}-CH_3$ | 7.4 |
| Trimethylamine | $CH_3-\overset{\displaystyle}{\underset{\underset{\displaystyle CH_3}{|}}{N}}-CH_3$ | 3.5 |
| Ethylamine | $CH_3CH_2-NH_2$ | 16.6 |
| Piperidine | ⬡N—H | 106 |
| Aniline | ⬡—$NH_2$ | 184 |

From Robert D. Whitaker et al., *Concepts of General, Organic, and Biological Chemistry*, p. 343. Copyright © 1981 by Houghton Mifflin Company. Used with permission.

---

## A Checklist for Review

### Important Terms

hydrocarbons (24.1)
saturated hydrocarbons (24.1)
unsaturated hydrocarbons (24.1)
aromatic hydrocarbons (24.1)
alkanes (24.2)

cycloalkanes (24.2)
condensed structural formula (24.2)
homologous series (24.2)
constitutional (structural) isomer (24.2)
substitution reaction (24.2)
alkenes (24.3)

geometric isomers (24.3)
addition reaction (24.3)
Markownikoff's rule (24.3)
alkynes (24.3)
functional group (p. 1070)
alcohol (24.6)
ether (24.6)
aldehyde (24.6)

ketone (24.6)
carboxylic acid (24.6)
ester (24.6)
amines (24.7)
amides (24.7)

### Summary of Facts and Concepts

*Organic compounds* are hydrocarbons or derivatives of hydrocarbons. Hydrocarbons contain only H and C atoms. The three main groups of hydrocarbons are *saturated hydrocarbons*, hydrocarbons with only single bonds between the car-

bon atoms; *unsaturated hydrocarbons*, hydrocarbons that contain double or triple bonds between carbon atoms; and *aromatic hydrocarbons*, hydrocarbons that contain a benzene ring (a six-membered ring of carbon atoms with alternating single and double carbon–carbon bonds described by resonance formulas).

The *alkanes* are *acyclic,* saturated hydrocarbons that form a *homologous series* of compounds, with the general formula $C_nH_{2n+2}$. The *cycloalkanes* are *cyclic,* saturated hydrocarbons that form another homologous series with the general formula $C_nH_{2n}$ in which the carbon atoms are joined in a ring. The *alkenes* and *alkynes* are unsaturated hydrocarbons (cyclic or acyclic) that contain carbon–carbon double or triple bonds.

The alkanes and aromatic hydrocarbons usually undergo *substitution reactions.* Alkenes and alkynes undergo *addition reactions. Markownikoff's rule* predicts the major product in the addition of an unsymmetrical reagent to an unsymmetrical alkene.

A *functional group* is a portion of an organic molecule that reacts readily in predictable ways. Important functional groups containing oxygen are *alcohols* (ROH), *aldehydes* (RCHO), *ketones* (RCOR′), and *carboxylic acids* (RCOOH). When a functional group is present in an organic compound, the compound is considered to be a *hydrocarbon derivative. Amines* are organic derivatives of ammonia. They react with carboxylic acids to give *amides.*

## Operational Skills

1. **Writing a condensed structural formula**  Given the structural formula of a hydrocarbon, write the condensed structural formula.  **(EXAMPLE 24.1)**
2. **Predicting *cis–trans* isomers**  Given a condensed structural formula of an alkene, decide whether *cis* and *trans* isomers are possible, and, if so, draw the structural formulas.  **(EXAMPLE 24.2)**
3. **Predicting the major product of an addition reaction**  Predict the major product in the addition of an unsymmetrical reagent to an unsymmetrical alkene.  **(EXAMPLE 24.3)**
4. **Writing the IUPAC name of a hydrocarbon given the structural formula, and vice versa**  Given the structure of a hydrocarbon, state the IUPAC name **(EXAMPLES 24.4, 24.6)**. Given the IUPAC name of a hydrocarbon, write the structural formula **(EXAMPLE 24.5)**.

## Review Questions

**24.1**  Give the molecular formula of an alkane with 30 carbon atoms.

**24.2**  Draw structural formulas of the five isomers of $C_6H_{14}$.

**24.3**  Draw structural formulas of an alkane, a cycloalkane, an alkene, and an aromatic hydrocarbon, each with seven carbon atoms.

**24.4**  Explain why there are two isomers of 2-butene. Draw their structural formulas and name the isomers.

**24.5**  Draw structural formulas for the isomers of ethylmethylbenzene.

**24.6**  Fill in the following table.

| Hydrocarbon | Source | Use |
|---|---|---|
| Methane | | |
| Octane | | |
| Ethylene | | |
| Acetylene | | |

**24.7**  Give condensed structural formulas of all possible substitution products of ethane and $Cl_2$.

**24.8**  Define the terms *substitution reaction* and *addition reaction.* Give examples of each.

**24.9**  What would you expect to be the major product when two molecules of HCl add successively to acetylene? Explain.

**24.10**  Write the structure of propylbenzene. Write the structure of *p*-dichlorobenzene.

**24.11**  What is a functional group? Give an example and explain how it fits this definition.

**24.12**  An aldehyde contains the carbonyl group. Ketones, carboxylic acids, and esters also contain the carbonyl group. What distinguishes these latter compounds from an aldehyde?

**24.13**  Fill in the following table.

| Compound | Source | Use |
|---|---|---|
| Methanol | | |
| Ethanol | | |
| Ethylene glycol | | |
| Glycerol | | |
| Formaldehyde | | |

**24.14**  Identify and name the functional group in each of the following.
a. $CH_3COCH_3$
b. $CH_3OCH_2CH_3$
c. $CH_3CH{=}CH_2$
d. $CH_3CH_2COOH$
e. $CH_3CH_2CHO$
f. $CH_3CH_2CH_2OH$

# Conceptual Problems

**24.15** You are distilling a barrel of oil that contains the hydrocarbons listed in Table 24.4. You heat the contents of the barrel to 200°C.
a. What molecules will no longer be present in your sample of oil?
b. What molecules will be left in the barrel?
c. Provide an explanation for your answers in parts a and b.
d. Which molecule would boil off at a lower temperature, hexane or 2,3-dimethylbutane?

**24.16** A classmate tells you that the following compound has the name 3-propylhexane.

$$CH_2CH_2CH_3$$
$$|$$
$$CH_3CH_2CHCH_2CH_2CH_3$$

a. Is he right? If not, what error did he make and what is the correct name?
b. How could you redraw the condensed formula to better illustrate the correct name?

**24.17** Explain why you wouldn't expect to find a compound with the formula $CH_5$.

**24.18** Catalytic cracking is an industrial process used to convert high-molecular-weight hydrocarbons to low-molecular-weight hydrocarbons. A petroleum company has a huge supply of heating oil stored as straight-chain $C_{17}H_{36}$, and demand has picked up for shorter chain hydrocarbons to be used in formulating gasoline. The company uses catalytic cracking to create the shorter chains necessary for gasoline. If they produce two molecules in the cracking, and 1-octene is one of them, what is the formula of the other molecule produced? As part of your answer, draw the condensed structural formula of the 1-octene.

**24.19** In the models shown here, C atoms are black, H atoms are light blue, O atoms are red, and N atoms are dark blue:

a. Write the molecular formula of each molecule.
b. Write the condensed structural formula for each molecule.
c. Identify the functional group for each molecule.

**24.20** In the models shown here, C atoms are black and H atoms are light blue:

a. Write the molecular formula of each molecule.
b. Write the condensed structural formula for each molecule.
c. Give the IUPAC name of each molecule.

**24.21** Why would you expect the melting points of the alkanes to increase in the series methane, ethane, propane, and so on?

**24.22** Consider the following formulas of two esters:

$$CH_3CH_2-\overset{\overset{\displaystyle O}{\|}}{C}-O-CH_3 \qquad CH_3CH_2-O-\overset{\overset{\displaystyle O}{\|}}{C}-CH_3$$

One of these is ethyl ethanoate (ethyl acetate) and one is methyl propanoate (methyl propionate). Which is which?

## Practice Problems

### Condensed Structural Formulas

**24.23** Write the condensed structural formula of the following alkane.

$$
\begin{array}{c}
\phantom{H-}\overset{\displaystyle H}{\phantom{|}} \\
H \quad H \quad H-\overset{\displaystyle |}{C}-H \\
H-\overset{\displaystyle |}{\underset{\displaystyle |}{C}}-\overset{\displaystyle |}{\underset{\displaystyle |}{C}}-\overset{\displaystyle |}{\underset{\displaystyle |}{C}}-H \\
H \quad H \quad H
\end{array}
$$

**24.24** Write the condensed formula of the following alkane.

$$
\begin{array}{c}
H \qquad\qquad H \\
H \quad H-\overset{}{C}-H \quad H \quad H-\overset{}{C}-H \quad H \\
H-C-\quad C-\quad C-\quad C-\quad C-H \\
H \quad H \quad H-C-H \quad H \quad H \\
\phantom{H-C-}H
\end{array}
$$

### Geometric Isomers

**24.25** If there are geometric isomers for the following, draw structural formulas showing the isomers.
a. $CH_3CH_2CH{=}CHCH_2CH_3$
b. $CH_3C{=}CHCH_2CH_3$
    $\phantom{CH_3}|$
    $\phantom{CH_3}CH_2CH_3$

**24.26** One or both of the following have geometric isomers. Draw the structures of any geometric isomers.
a. $CH_3CHCH{=}CHCH_3$
    $\phantom{CH_3}|$
    $\phantom{CH_3}CH_3$

b. $CH_3C{=}CHCH_2CH_3$
    $\phantom{CH_3}|$
    $\phantom{CH_3}CH_3$

### Reactions of Hydrocarbons

**24.27** Complete and balance the following equations. Note any catalyst used.
a. $C_2H_4 + O_2 \longrightarrow$
b. $CH_2{=}CH_2 + MnO_4^- + H_2O \longrightarrow$
c. $CH_2{=}CH_2 + Br_2 \longrightarrow$
d. 

e. 

**24.28** Complete and balance the following equations. Note any catalyst used.
a. $C_4H_{10} + O_2 \longrightarrow$
b.

cyclohexene

c. $CH_2\!\!=\!\!CH_2 + HBr \longrightarrow$

d.

$CH_2CH_3$

$+$ $\longrightarrow$

$CH_2CH_3$

$NO_2$

e.

$+ Cl_2 \xrightarrow{FeCl_3}$

---

**24.29** Write an equation for a possible substitution reaction of ethane, $C_2H_6$, with $Cl_2$.

**24.30** Write an equation for a possible substitution reaction of propane, $C_3H_8$, with $Br_2$.

---

**24.31** What is the major product when HBr is added to methylpropene?

**24.32** Complete the following equation, giving only the main product.

$$CH_2\!\!=\!\!CHCH_3 + H\!\!-\!\!OH \xrightarrow[\text{catalyst}]{H_2SO_4}$$

## Naming Hydrocarbons

---

**24.33** Give the IUPAC name for each of the following hydrocarbons.

a.
$$\begin{array}{c} CH_3 \\ | \\ CH_3CHCH_2CHCH_3 \\ | \quad\quad | \\ CH_3 \quad CH_3 \end{array}$$

b.
$$\begin{array}{c} CH_3 \quad\quad CH_3 \\ | \quad\quad\quad | \\ CH_3CCH_2CH_2CH_2CCH_3 \\ | \quad\quad\quad | \\ CH_3 \quad\quad CH_3 \end{array}$$

c.
$$\begin{array}{c} CH_3CH_2CHCH_2CH_2CH_2CH_3 \\ | \\ CH_2CH_2CH_3 \end{array}$$

d.
$$\begin{array}{c} CH_3 \\ | \\ CH_3CHCHCH_2CH_2CH_2 \\ | \quad\quad\quad\quad | \\ CH_2CH_3 \quad\quad CH_3 \end{array}$$

**24.34** What is the IUPAC name of each of the following compounds?

a.
$$\begin{array}{c} CH_3 \\ | \\ CH_3CH_2CHCHCH_3 \\ | \\ CH_3 \end{array}$$

b.
$$\begin{array}{c} CH_2CH_3 \\ | \\ CH_3CH_2CH_2CHCH_2CH_3 \end{array}$$

c.
$$\begin{array}{c} CH_3CH_2CH_2CHCH_2CH_2CH_3 \\ | \\ CH_3CHCH_3 \\ | \\ CH_3 \end{array}$$

d.
$$\begin{array}{c} CH_3CHCH_2CH_2CHCH_2CH_3 \\ | \\ CH_2CH_3 \end{array}$$

**24.35** Write the condensed structural formula for each of the following compounds.
a. 2,3-dimethylhexane
b. 3-ethylhexane
c. 2-methyl-4-isopropylheptane
d. 2,2,3,3-tetramethylpentane

**24.36** Write the condensed structural formula for each of the following compounds.
a. 2,2-dimethylpentane
b. 3-isopropylhexane
c. 3-ethyl-4-methyloctane
d. 3,4,4,5-tetramethylheptane

---

**24.37** Give the IUPAC name of each of the following.
a. $CH_3CH\!\!=\!\!CHCH_2CH_3$
b.
$$\begin{array}{c} CH_3C\!\!=\!\!CHCH_2CHCH_3 \\ | \quad\quad\quad\quad | \\ CH_3 \quad\quad\quad CH_3 \end{array}$$

**24.38** For each of the following, write the IUPAC name.
a.
$$\begin{array}{c} CH_3CH_2 \\ \phantom{CH_3CH_2}\searrow \\ \phantom{CH_3CH_2xx}C\!\!=\!\!CHCH_2CH_3 \\ \phantom{CH_3CH_2}\nearrow \\ CH_3CH_2 \end{array}$$
b.
$$\begin{array}{c} CH_3CH_2CCH_2CH_2CH_3 \\ \| \\ CH_2 \end{array}$$

**24.39** Give the condensed structural formula for each of the following compounds.
a. 3-ethyl-2-pentene
b. 4-ethyl-2-methyl-2-hexene

**24.40** Write the condensed structural formula for each of the following compounds.
a. 2,3-dimethyl-2-pentene
b. 2-methyl-4-propyl-3-heptene

---

**24.41** Give the IUPAC names and include the *cis* or *trans* label for each of the isomers of $CH_3CH\!\!=\!\!CHCH_2CH_3$.

**24.42** Give the IUPAC names and include the *cis* or *trans* label for each of the isomers of $CH_3CH\!\!=\!\!CHCH_3$.

---

**24.43** Give the IUPAC name of each of the following compounds.
a. $CH_3C\!\!\equiv\!\!CCH_3$
b.
$$\begin{array}{c} CH\!\!\equiv\!\!CCHCHCH_3 \\ | \\ CH_3 \end{array}$$

**24.44** Write the IUPAC name of each of the following hydrocarbons.

a. CH₃CHC≡CH
    |
    CH₃

b. CH₃C≡CCH₂CH₃

**24.45** Write structural formulas for (a) 1,1,1-triphenylethane; (b) *o*-ethylmethylbenzene.

**24.46** Write structural formulas for (a) 1,2,3-trimethylbenzene; (b) *p*-diethylbenzene.

## Naming Oxygen-Containing Organic Compounds

**24.47** Circle and name the functional group in each compound.

          O
          ‖
a. CH₃—C—CH₂CH₂CH₃

          OH
          |
b. CH₃—C—CH₂CH₃
          |
          H

          O
          ‖
c. HO—C—CH₂CH₃

        O
        ‖
d. H—C—CH₂CH₃

**24.48** Circle and name the functional group in each compound.

          O
          ‖
a. CH₃—C—OH

b. O=C—OH
       |
    CH₃CHCH₃

            CH₃
            |
c. HO—CH₂CH
            |
            CH₃

d. O=C—CH₃
       |
    CH₃CHCH₃

**24.49** Give the IUPAC name for each of the following.

a. HOCH₂CH₂CH₂CH₂CH₃

b. CH₃CHCH₂CH₂CH₃
       |
       OH

c. CH₃CH₂CH₂CHCH₂CH₂CH₃
             |
        H—C—OH
            |
            H

        OH     CH₃
        |        |
d. CH₃CH₂CH₂CHCH₂CHCH₂CH₃

**24.50** Write the IUPAC name for each of the following.

a. HOCH₂CHCH₂CH₃
         |
        CH₂CH₂CH₃

b. HOCH₂CH₂CH₂CHCH₃
                |
                CH₃

c. CH₃CHCH₂CH₃
       |
  H—C—OH
       |
       CH₃

d. HOCH₂CHCH₂CH₃
         |
         CH₃

**24.51** State whether each of the following alcohols is primary, secondary, or tertiary.

a. CH₃CHCH₂CH₃    b.  CH₂CH₃
      |                 |
      OH          HCCH₂CH₃
                      |
                    OH

c.       CH₃     d.     CH₃
        |               |
  HOCH₂CH        HOCH₂CCH₃
        |               |
        CH₃            CH₃

**24.52** Classify each of the following as a primary, secondary, or tertiary alcohol.

a.    CH₃  CH₃    b.        CH₃
        |    |                 |
  CH₃CCH₂CHOH    CH₃CH₂CH₂COH
       |                      |
     CH₃                   CH₃

c.       CH₂OH    d.    CH₂CH₃
         |                 |
  CH₃CH₂CCH₃    CH₃CH₂COH
         |                 |
       CH₃            CH₂CH₃

**24.53** What is the common name of each of the following compounds?

a. CH₃CH₂OCH₂CH₂CH₃

b.       CH₃
          |
   H—COCH₃
          |
         CH₃

**24.54** What is the common name of each of the following compounds?

a.       CH₃
         |
  CH₃OCCH₃
         |
        CH₃

b.

**24.55**   According to IUPAC rules, what is the name of each of the following compounds?

a. $CH_3COCH_2CH_3$

b. $CH_3CH_2CH_2CHO$

c.
$$\underset{\overset{|}{CH_3}}{H-\underset{}{\overset{\overset{O}{\|}}{C}}CH_2CH_2\overset{\overset{O}{\|}}{C}CH_3}$$

d.
$$CH_3\underset{\overset{|}{CH_2CH_3}}{CH}\overset{\overset{O}{\|}}{C}CH_3$$

**24.56**   Write the IUPAC name of each of the following compounds.

a. $CH_3\underset{\overset{|}{CHO}}{CH}CH_3$

b. $CH_3\underset{\overset{|}{COCH_3}}{CH}CH_3$

c. $CH_3\underset{\overset{|}{CH_2\overset{\overset{O}{\|}}{C}-H}}{CH_2}$

d. $CH_3\underset{\overset{|}{CH_2-\overset{\overset{O}{\|}}{C}-CH_2CH_3}}{CH}CH_3$

## General Problems

**24.59**   Give the IUPAC name of each of the following compounds.

a. $CH_3\underset{\overset{|}{CH_3}}{CH}CH_2COOH$

b. $CH_3\underset{\overset{|}{CH_3}}{CH}CH_2\underset{\overset{}{H}}{\overset{H}{C}}=\underset{\overset{}{CH_3}}{\overset{H}{C}}$

c. $CH_3\underset{\overset{|}{\underset{\overset{|}{CH_3}}{CH_2}}}{CH}\overset{\overset{O}{\|}}{C}CH_2\underset{\overset{|}{CH_3}}{CH}$

d. $CH_3\underset{\overset{|}{CH_3}}{CH}C\equiv CCH_3$

**24.60**   Give the IUPAC name of each of the following compounds.

a. $CH_3\underset{\overset{|}{CH_2CH_3}}{\overset{\overset{|}{CHO}}{C}}CH_3$

b. $CH_3\underset{\overset{|}{CH_3}}{\overset{\overset{|}{COOH}}{C}}CH_2CH_3$

c. $CH_3\underset{\overset{|}{CH_2CH_3}}{\overset{\overset{|}{CH_3}}{C}}COCH_3$

d. $CH_3\underset{\overset{|}{CH_3}}{\overset{\overset{|}{CH_3}}{C}}CH_2CH_3$

**24.61**   Write the structural formula for each of the following compounds.

a. isopropyl propionate

b. *t*-butylamine

c. 2,2-dimethylhexanoic acid

d. *cis*-3-hexene

**24.62**   Write the structural formula for each of the following compounds.

a. 3-ethyl-1-pentene

b. 1,1,2,2-tetraphenylethane

## Organic Compounds Containing Nitrogen

**24.57**   Identify each of the following compounds as a primary, secondary, or tertiary amine, or as an amide.

a. $CH_3NH$ (with benzene ring attached)

b. $CH_3CH_2NHCH_2CH_3$

**24.58**   Identify each of the following compounds as a primary, secondary, or tertiary amine, or as an amide.

a. $O=C-NH_2$ (with benzene ring attached)

b. $CH_3CH_2CH_2NH_2$

c. 1-phenyl-2-butanone

d. cyclopentanone

**24.63**   Describe chemical tests that could distinguish between:

a. propionaldehyde (propanal) and acetone (propanone)

b. $CH_2=CH-C\equiv C-CH=CH_2$ and benzene

**24.64**   Describe chemical tests that could distinguish between:

a. acetic acid and acetaldehyde (ethanal)

b. toluene (methylbenzene) and 2-methylcyclohexene

**24.65**   Identify each of the following compounds.

a. A gas with a sweetish odor that promotes the ripening of green fruit.

b. An unsaturated compound of the formula $C_7H_8$ that gives a negative test with bromine in carbon tetrachloride.

c. A compound with an ammonia-like odor that acts as a base; its molecular formula is $CH_5N$.

d. An alcohol used as a starting material for the manufacture of formaldehyde.

**24.66**   Identify each of the following compounds.

a. An acidic compound that also has properties of an aldehyde; its molecular formula is $CH_2O_2$.

b. A compound used as a preservative for biological specimens and as a raw material for plastics.

c. A saturated hydrocarbon boiling at 0°C that is liquefied and sold in cylinders as a fuel.

d. A saturated hydrocarbon that is the main constituent of natural gas.

**24.67**   A compound that is 85.6% C and 14.4% H and has a molecular weight of 56.1 amu reacts with water and sulfuric acid to produce a compound that reacts with acidic potassium dichromate solution to produce a ketone. What is the name of the original hydrocarbon?

**24.68** A compound with a fragrant odor reacts with dilute acid to give two organic compounds, A and B. Compound A is identified as an alcohol with a molecular weight of 32.0 amu. Compound B is identified as an acid. It can be reduced to give a compound whose composition is 60.0% C, 13.4% H, and 26.6% O and whose molecular weight is 60.1 amu. What is the name of the original compound?

## Media Activities

**24.69** Watch the C$=$C Bond Formation *Visualization* on the student web site.

a. What is the molecular geometry of each carbon atom in $H_2O=CH_2$?

b. What is the hybridization of each carbon atom in $H_2O=CH_2$?

c. How many pi and sigma bonds are between the carbon atoms in acetylene, HCCH?

**24.70** Watch the Addition Reaction *Visualization* on the student web site.

a. What is the product of the following reaction?

$$Cl_2 + CH_3CHCHCH_2CH_3$$

b. What is the IUPAC name of the product from the reaction in part a?

**24.71** Work through the Organic Nomenclature *Problem Solving* activity on the student CD. Name the following compounds:

a.
$$\begin{array}{c} CH_3 \\ | \\ CH_3CHCHCH_3 \\ | \\ CH_3 \end{array}$$

b.
$$\begin{array}{c} OH \quad\quad CH_2CH_3 \\ | \quad\quad\quad | \\ CH_3CHCH_2CH \\ | \\ CH_3 \end{array}$$

**24.72** For review, take the ACE practice test for Chapter 24 on the student web site.

Pulling nylon from a reaction mixture (blue dye added).

# Polymer Materials: Synthetic and Biological

Celluloid was the first commercial plastic, discovered around 1850. Initially it was used as a cheap substitute for ivory in making billiard balls, but it soon became the material of choice for many consumer items such as combs and toothbrushes. Perhaps its best known application, though, was in producing photographic film for motion pictures. Celluloid is a tough plastic, but it has one nasty property—it catches fire and burns explosively, a property that stems from celluloid's main ingredient, cellulose nitrate. Cellulose nitrate (also known as guncotton) is made by soaking a cellulose material such as cotton in a mixture of nitric and sulfuric acids. Soon after celluloid billiard balls came on the market, there were reports of billiard balls that vanished with a flash and a pop when touched accidentally by a person's cigar. There were also a number of serious theater fires. By the early twentieth century, celluloid was displaced in most of its applications by Bakelite, a hard plastic produced from formaldehyde, HCHO, and phenol, $C_6H_5OH$.

Plastics, such as celluloid and Bakelite, are *polymers,* which are giant molecules constructed by covalently bonding together many small molecules. The cellulose

**Figure 25.1**
**Products made of synthetic polymers.** *Front:* Nylon rope and polyethylene bags. *Rear (left to right):* Thread (polyester core), eye drop lubricant (hydroxypropyl methylcellulose), paper glue (polyvinyl acetate), latex paint (acrylic polymers), and polyethylene bottle.

nitrate of celluloid is a chemically modified form of the natural polymer cellulose, which plants make from the sugar glucose $C_6H_{12}O_6$. Bakelite was the first completely synthetic polymer, produced commercially from nonpolymeric chemicals. Since then, many different plastics with various properties have been developed (Figure 25.1).

In this chapter, we will look at some basic features of synthetic and biological, or natural, polymers. In the part concerning biological polymers, we will concentrate on proteins and how the genetic code, constructed from the polymer DNA, leads to the synthesis of proteins.

# Synthetic Polymers

A **polymer** is *a chemical species of very high molecular weight made up from many units of low molecular weight covalently linked together. A compound used to make a polymer (and from which the polymer's unit arises) is called a* **monomer.** A simple example is polyethylene, which consists of many ethylene units, $-CH_2CH_2-$, covalently bonded to one another to form long-chain molecules consisting of thousands of ethylene units. The name *polyethylene* stems from the name of the monomer (ethylene) plus the prefix *poly-*, meaning "many." Shown here is a short piece of a polyethylene molecule.

$$
\begin{array}{cccccccc}
H & H & H & H & H & H & H & H \\
| & | & | & | & | & | & | & | \\
-C- & C- & C- & C- & C- & C- & C- & C- \\
| & | & | & | & | & | & | & | \\
H & H & H & H & H & H & H & H \\
\end{array}
$$

One ethylene unit is shown in color.

Polyethylene is an example of a synthetic polymer, one made in the laboratory or commercially from nonpolymeric substances. We will look at synthetic polymers in the next two sections. The first polymers chemists studied, however, were natural or biological polymers, including rubber, cellulose, starch, and proteins. Proteins and deoxyribonucleic acids (DNA) are polymers that form the basis of life on earth. We will examine these in the last part of the chapter.

## 25.1 Synthesis of Organic Polymers

The idea that certain natural products such as rubber were composed of giant molecules, or polymers, consisting of many repeating units linked by covalent bonds was due largely to the work of the German chemist Hermann Staudinger (1881–1965) in the early 1920s. He convinced skeptical chemists of this idea, in part, by linking organic monomers one at a time through addition reactions to produce larger and larger molecules. His work ushered in an era of systematic study of polymers.

It would be difficult to overemphasize the importance of synthetic polymers to present-day technology. Perhaps you are seated in front of a computer monitor, whose casing is made of hard plastic. The desk on which it sits most likely has a coating of a polymer resin, such as polyurethane. Your clothing might well be made of fibers composed of synthetic polymers, such as polyester or nylon. And, any elastic bands, say, on your socks, are made from *elastomers,* an elastic or rubbery type of polymer.

In this section, we describe several synthetic polymers. Polymers are classified as *addition polymers* or *condensation polymers,* depending on the type of reaction used in forming them.

### Addition Polymers

An **addition polymer** is *a polymer formed by linking together many molecules by addition reactions.* The monomers must have multiple bonds that will undergo addition reactions. For example, when propylene (IUPAC name, propene) is heated under pressure with a catalyst, it forms polypropylene.

$$\ldots + \underset{\mathrm{CH_3}}{\mathrm{CH}}{=}\mathrm{CH_2} + \underset{\mathrm{CH_3}}{\mathrm{CH}}{=}\mathrm{CH_2} + \underset{\mathrm{CH_3}}{\mathrm{CH}}{=}\mathrm{CH_2} + \ldots \longrightarrow$$

$$-\underset{\mathrm{CH_3}}{\mathrm{CH}}-\mathrm{CH_2}-\underset{\mathrm{CH_3}}{\mathrm{CH}}-\mathrm{CH_2}-\underset{\mathrm{CH_3}}{\mathrm{CH}}-\mathrm{CH_2}-$$
polypropylene

Note that the $\pi$ electrons in the double bonds of the propylene molecules form new $\sigma$ bonds that unite the monomers.

► Demonstration of the preparation of polystyrene: Bassam Z. Shakhashiri, *Chemical Demonstrations,* Vol. 1 (Madison: University of Wisconsin Press, 1983), no. 3.11, pp. 241–242.

**Free-Radical Addition** The preparation of an addition polymer is often induced by an *initiator,* a compound that produces free radicals (species having an unpaired electron). Organic peroxides (organic compounds with the —O—O— group) are frequently used as initiators. If you write R for an organic group, the general formula of such a peroxide is R—O—O—R. When heated, the —O—O— bond in the organic peroxide breaks, and free radicals form. (The unpaired electron is symbolized by a dot.) Figure 25.2 illustrates the preparation of polystyrene using benzoyl peroxide as an initiator.

$$\mathrm{R{-}O{-}O{-}R} \xrightarrow{\Delta} \mathrm{R{-}O \cdot} + \mathrm{\cdot O{-}R}$$

The free radical reacts with the alkene monomer to produce a new free-radical species, which in turn reacts with another monomer molecule. For example, the polymerization of propylene occurs as follows:

$$\mathrm{R{-}O \cdot} + \underset{\underset{\mathrm{H}}{|}}{\overset{\overset{\mathrm{CH_3}}{|}}{\mathrm{C}}}{=}\underset{\underset{\mathrm{H}}{|}}{\overset{\overset{\mathrm{H}}{|}}{\mathrm{C}}} \longrightarrow \mathrm{R{-}O{-}}\underset{\underset{\mathrm{H}}{|}}{\overset{\overset{\mathrm{CH_3}}{|}}{\mathrm{C}}}{-}\underset{\underset{\mathrm{H}}{|}}{\overset{\overset{\mathrm{H}}{|}}{\mathrm{C}}} \cdot$$

**Figure 25.2**
**Preparation of polystyrene.** When styrene, $C_6H_5CH=CH_2$, is heated with a small amount of benzoyl peroxide, ROOR ($R = C_6H_5CO—$), it yields a viscous liquid. After some time, this liquid sets to a hard plastic (sample shown at left).

$$R—O—\overset{\overset{\displaystyle CH_3}{|}}{\underset{\underset{\displaystyle H}{|}}{C}}—\overset{\overset{\displaystyle H}{|}}{\underset{\underset{\displaystyle H}{|}}{C}}\cdot \ + \ \overset{\overset{\displaystyle CH_3}{|}}{\underset{\underset{\displaystyle H}{|}}{C}}=\overset{\overset{\displaystyle H}{|}}{\underset{\underset{\displaystyle H}{|}}{C}} \ \longrightarrow \ R—O—\overset{\overset{\displaystyle CH_3}{|}}{\underset{\underset{\displaystyle H}{|}}{C}}—\overset{\overset{\displaystyle H}{|}}{\underset{\underset{\displaystyle H}{|}}{C}}—\overset{\overset{\displaystyle CH_3}{|}}{\underset{\underset{\displaystyle H}{|}}{C}}—\overset{\overset{\displaystyle H}{|}}{\underset{\underset{\displaystyle H}{|}}{C}}\cdot$$

At each addition step, the unpaired electron from the free radical pairs with one of the $\pi$ electrons of the alkene monomer to form a $\sigma$ bond. One end of the chain retains an unpaired electron and can add another monomer molecule. The reaction terminates when the free electron at the end of the chain reacts with another free radical. Some addition polymers are listed in Table 25.1.

**Natural and Synthetic Rubber**  Natural rubber is an addition polymer of isoprene (formal name, 2-methyl-1,3-butadiene). Rubber has been reproduced in the laboratory by heating isoprene with a catalyst:

$$n\,CH_2{=}\underset{\underset{\displaystyle\text{isoprene}}{}}{\overset{\overset{\displaystyle CH_3}{|}}{C}}{-}CH{=}CH_2 \ \longrightarrow \ \underset{\text{poly-}cis\text{-isoprene (rubber)}}{\overset{\displaystyle CH_3 \qquad H \qquad CH_3 \qquad H}{\cdots}}$$

isoprene                    poly-*cis*-isoprene (rubber)

**Table 25.1**
**Some Addition Polymers**

| Polymer | Monomer | Uses |
|---|---|---|
| Polyethylene | $CH_2{=}CH_2$ | Bottles, plastic tubing |
| Polypropylene | $CH_2{=}\overset{\overset{\displaystyle CH_3}{|}}{CH}$ | Bottles, carpeting, textiles |
| Polytetrafluoroethylene (Teflon®) | $CF_2{=}CF_2$ | Nonstick surface for frying pans |
| Poly(vinyl chloride) (PVC) | $CH_2{=}\overset{\overset{\displaystyle Cl}{|}}{CH}$ | Plastic pipes, floor tile |
| Polyacrylonitrile (Orlon®, Acrilan®) | $CH_2{=}\overset{\overset{\displaystyle CN}{|}}{CH}$ | Carpets, textiles |
| Polystyrene | $CH_2{=}\overset{\overset{\displaystyle C_6H_5}{|}}{CH}$ | Plastic foam insulation, cups |

**Figure 25.3**
**Collecting rubber latex.**
The outer bark of the rubber tree (*Hevea brasiliensis*) and the layer just beneath it are cut diagonally so that the latex runs into a collecting cup.

Rubber is obtained from the milky sap, or latex, of the rubber tree (Figure 25.3). After the latex is collected, it is coagulated by the addition of formic or acetic acid.

The elasticity of rubber results from its long, coiled polymer chains. When you stretch a piece of rubber, the long coils unwind. When you let go, the molecules move back to roughly their original positions. Actually, the molecular strands do move relative to one another, which accounts for rubber's pliability and tackiness. Early rubber goods were sticky in the summer and brittle in the winter. In 1839, Charles Goodyear discovered that crude rubber heated with sulfur remained elastic but lacked these undesirable characteristics. Goodyear's process, called *vulcanization,* results in cross-linking of the polymer chains, which anchors them relative to one another.

Natural rubber is a **homopolymer,** *a polymer whose monomer units are all alike.* Most of the plastics we discussed earlier are also homopolymers. On the other hand, some synthetic rubbers (as well as many plastics) are copolymers. A **copolymer** is *a polymer consisting of two or more different monomer units.* Styrene–butadiene rubber (SBR) is an important synthetic rubber; it is a copolymer made from about 25% styrene and 75% 1,3-butadiene. Its approximate structure is

$$-CH_2-CH=CH-CH_2-CH_2-CH=CH-CH_2-CH_2-CH=CH-CH_2-CH_2-\overset{\overset{\displaystyle C_6H_5}{|}}{C}H-$$

| 1,3-butadiene unit | 1,3-butadiene unit | 1,3-butadiene unit | styrene unit |

**EXERCISE 25.1**

An addition polymer is prepared from vinylidene chloride, $CH_2=CCl_2$. Write the structure of the addition polymer.

See Problems 25.23 and 25.24.

## Condensation Polymers

A **condensation polymer** is *a polymer formed by linking together many molecules by condensation reactions.* Wallace Carothers (1896–1937), a chemist at E. I. du Pont de Nemours and Company, realized that polymers could be prepared by condensation reactions. He and his coworkers produced a number of polymers this way, including

**Figure 25.4**
**Nylon being pulled from a reaction mixture.**
Nylon is formed by reacting hexamethylene diamine (bottom water layer) with adipic acid or, as in this experiment, with adipoyl chloride, $ClCO(CH_2)_4COCl$ (top hexane layer). Nylon, formed at the interface of the two layers, is continuously removed.

polyamides (nylon) and polyesters. The first synthetic fibers, produced from nylon, were announced by du Pont in 1939.

**Polyesters**  A substance with two alcohol groups reacts with a substance with two carboxylic acid groups to form a *polyester,* a polymer whose repeating units are joined by ester groups. The reactant molecules join as links of a chain to form a very long molecule. The polyester Dacron, used as a textile fiber, is prepared from ethylene glycol and terephthalic acid.

$$\cdots + HOCH_2CH_2OH + HO-\overset{O}{\underset{}{C}}\!\!-\!\!\bigcirc\!\!-\!\!\overset{O}{\underset{}{C}}\!\!-OH + HOCH_2CH_2OH + HO-\overset{O}{\underset{}{C}}\!\!-\!\!\bigcirc\!\!-\!\!\overset{O}{\underset{}{C}}\!\!-OH + \cdots \longrightarrow$$

ethylene glycol        terephthalic acid        ethylene glycol        terephthalic acid

$$\sim OCH_2CH_2O-\overset{O}{\underset{}{C}}\!\!-\!\!\bigcirc\!\!-\!\!\overset{O}{\underset{}{C}}\!\!-OCH_2CH_2O-\overset{O}{\underset{}{C}}\!\!-\!\!\bigcirc\!\!-\!\!\overset{O}{\underset{}{C}}\!\!\sim + nH_2O$$

Dacron (a polyester)

A water molecule is split out during the formation of each ester linkage.

**Polyamides**  When a compound containing two amine groups reacts with a compound containing two carboxylic acid groups, a condensation polymer called a *polyamide* is formed (Figure 25.4). Nylon-6,6 is an example. It is prepared by heating hexamethylene diamine (1,6-diaminohexane) and adipic acid (hexanedioic acid).

$$\cdots + H-\overset{H}{\underset{}{N}}(CH_2)_6\overset{H}{\underset{}{N}}-H + HO-\underset{\underset{O}{\|}}{C}(CH_2)_4\underset{\underset{O}{\|}}{C}-OH + H-\overset{H}{\underset{}{N}}(CH_2)_6\overset{H}{\underset{}{N}}-H + HO-\underset{\underset{O}{\|}}{C}(CH_2)_4\underset{\underset{O}{\|}}{C}-OH + \cdots \xrightarrow{\Delta}$$

hexamethylene diamine        adipic acid        hexamethylene diamine        adipic acid

$$\sim N(CH_2)_6\overset{H}{\underset{}{N}}-\underset{\underset{O}{\|}}{C}(CH_2)_4\underset{\underset{O}{\|}}{C}-\overset{H}{\underset{}{N}}(CH_2)_6\overset{H}{\underset{}{N}}-\underset{\underset{O}{\|}}{C}(CH_2)_4\underset{\underset{O}{\|}}{C}\sim + nH_2O$$

nylon-6,6 (a polyamide)

## 25.2 Electrically Conducting Polymers

Sometime during the early 1970s, a graduate student in the laboratory of Professor Hideki Shirakawa in Japan mistakenly added 1000 times too much catalyst in a reaction to polymerize acetylene to polyacetylene. (The basic reaction is similar to the one in which ethylene is polymerized to polyethylene; see the previous section.) To his surprise, the plastic film coating the inside of the vessel, which in previous experiments was black, now had a silvery sheen to it. The silvery film was *trans*-polyacetylene, which has the following Lewis resonance formulas:

The immediate question was this: If it looks like a metal, does it conduct electricity? In fact, the material had the electrical conductivity of a pure semiconductor, which is to say it was slightly conducting.

You can explain the slight conductivity of polyacetylene from its electronic structure. The Lewis resonance formulas of the long-chain molecule have alternating double and single bonds, indicative of delocalized bonding. In a molecular orbital description, the orbitals would encompass the entire length of the molecular chain. Thus, an electron would be able to move easily from one end of the molecule to the other, which means a current could travel the length of the molecule. Pure polyacetylene plastic is only slightly conductive, however, because electric current does not easily pass from one molecule to another.

Sometime later, Shirakawa collaborated with two scientists from the United States, Alan MacDiarmid, a chemist, and Alan Heeger, a physicist. MacDiarmid and Heeger had been doing experiments in which they "doped" (added small amounts of particular substances to) an inorganic polymer to increase its electrical conductivity. In experiments in 1977 with polyacetylene, they found that the electrical conductivity of the material increased dramatically when it was doped with iodine, $I_2$.

How does doping increase the conductivity of the polymer? Polyacetylene consists of entwined strands of individual polymer molecules. Experiments show that iodine inserts itself between polyacetylene molecules, where it forms triiodide ion, $I—I—I^-$. Each ion carries a negative charge that was abstracted from a $\pi$ orbital of the polymer molecule. The polymer molecule is left with a positive charge, or "hole." The triiodide ion can facilitate the movement of an electron from an uncharged polymer molecule to a positively charged polymer molecule (a hole). In effect, the hole moves (Figure 25.5). The movement of holes through the material constitutes an electric current. Doped polyacetylene was only the first conductive polymer material to be discovered. Since 1977, other conducting polymers have also been developed (Figure 25.6).

During experiments to find alternate ways of doping polymers, a graduate student of Professor MacDiarmid discovered one of the first practical applications of polyacetylene. He placed two strips of polyacetylene into a solution of doping ions and connected the strips to an electric current. The hope was that the ions would penetrate the polymer under the action of the external electric current. But after the student removed the external electric current, the polymer strips retained a charge and

**Figure 25.5**
**Movement of charge in polyacetylene.** The diagram shows the movement of electrons by arrows. When an electron moves away from the polymer chain (perhaps going to an electrode) at the far right, it leaves a hole, or positive charge, in the double bond. An electron from the adjacent double bond can then move into this hole, leaving a hole further to the left. In this way, the hole moves right to left. The triiodide ion facilitates electron transfer between the polymer chains. (Hydrogen atoms in polyacetylene were omitted for clarity.)

acted as a battery. Since then an all-plastic battery has been developed. Its electrodes are made of conductive polymers, and the electrolyte is a polymer gel.

Scientists have succeeded in making various electronic devices, including transistors and light-emitting diodes (LEDs), from conducting polymers. Commercial applications are just now coming to market. A conductive polymer acting as an LED is being used in a display for a mobile phone. Alan Heeger, Alan MacDiarmid, and Hideki Shirakawa won the Nobel Prize in chemistry for the year 2000 for their discovery and development of conductive polymers.

**Figure 25.6**
**Other conducting polymers.** Polypyrrole has been used in the construction of an all-plastic battery. Polyaniline has become a popular conducting polymer.

Polypyrrole

Polyaniline

Polyparaphenylene vinylene (PPV)

Plastics have traditionally been thought of as insulators. In fact, electric wires are generally covered with a coating of polyethylene insulation. Now, some plastics have been found to be as conductive as metals. In chemistry, certainly, it pays to keep an open mind!

# Biological Polymers

The molecules of biological organisms fall into four groups: proteins, carbohydrates, nucleic acids, and lipids (nonpolymeric materials soluble in organic solvents, such as fats and oils). Proteins and nucleic acids, and many carbohydrates, are polymers. In this chapter, we will limit our discussion to proteins and nucleic acids. Proteins are the basic constituents of living organisms; we discuss these substances in the next section. Nucleic acids are important in the synthesis of proteins; we discuss these in the final section of the chapter.

## 25.3   Proteins

**Proteins** are *biological polymers of small molecules called amino acids.* Their molecular weights range from about 6000 amu to 250,000 amu, so proteins can be very large molecules. Many proteins also contain non–amino-acid components such as metal ions (for example, $Fe^{2+}$, $Zn^{2+}$, $Cu^+$, $Mg^{2+}$) or certain complex organic molecules usually derived from vitamins. (Vitamins are organic molecules necessary in very small quantities for normal cell structure and metabolism.)

Proteins are very important molecules in cells and organisms, playing both structural and functional roles. The proteins of bone and connective tissue, for example, are of major structural importance. Some proteins are enzymes, catalyzing specific biochemical reactions; some transport materials in the blood stream or across biological membranes; and some (hormones) carry chemical messages to coordinate the body's activities. *Insulin* and *glucagon,* for example, are protein hormones made in the pancreas and secreted to regulate the body's blood-sugar level.

### Amino Acids

An **amino acid** is *a compound containing an amino group (—NH$_2$) and a carboxyl group (—COOH).* The building blocks of proteins are alpha-amino acids ($\alpha$-amino acids). An $\alpha$-amino acid has the general structure

$$\text{R}-\underset{\underset{\text{NH}_2}{|}}{\overset{\overset{\text{H}}{|}}{\text{C}}}-\overset{\overset{\text{O}}{\|}}{\text{C}}-\text{O}-\text{H}$$

The carbon atom next to the carboxyl carbon is called the $\alpha$ carbon atom. In an $\alpha$-amino acid the amino group is attached to the $\alpha$ carbon atom. Glycine, $NH_2CH_2COOH$, is the simplest $\alpha$-amino acid; R is an H atom.

$$\text{H}-\underset{\underset{\text{NH}_2}{|}}{\overset{\overset{\text{H}}{|}}{\text{C}}}-\overset{\overset{\text{O}}{\|}}{\text{C}}-\text{O}-\text{H}$$

glycine

Table 25.2 lists the 20 amino acids from which most proteins are composed. Under the name of each amino acid, its common three-letter abbreviation is shown. For instance, the abbreviation for glycine is gly.

**Table 25.2**
**The Amino Acids Found in Most Proteins**

Nonpolar Side Chain

glycine
gly

L-alanine
ala

L-valine
val

L-leucine
leu

L-isoleucine
ile

L-proline
pro

L-phenylalanine
phe

L-tryptophan
trp

L-methionine
met

Polar Side Chain

L-serine
ser

L-cysteine
cys

L-threonine
thr

L-aspartic acid
asp

L-glutamic acid
glu

L-tyrosine
tyr

L-asparagine
asn

L-glutamine
gln

L-histidine
his

L-lysine
lys

L-arginine
arg

Each amino acid has a different R group, or *side chain*. The side chains of the amino acids in a protein determine the protein's properties. In Table 25.2 the side chains of the amino acids are shown within a color screen. Nine of the amino acids have nonpolar, or hydrocarbon, side chains (left side of table). The remaining 11 amino acids have polar side chains (right side of table), capable of ionizing or forming hydrogen bonds with other amino acids or with water.

At the near-neutral pH of a biological system, the amino acid groups in a protein are generally in the form of *zwitterions,* which are species having both a positive and a negative charge. The carboxyl group is acidic, so its proton ionizes:

$$-COOH \rightleftharpoons -COO^- + H^+$$

The amino group is basic, so it tends to pick up a proton:

$$H^+ + -NH_2 \rightleftharpoons -NH_3^+$$

At neutral pH, the net result is that the carboxyl group loses $H^+$ to the amino group, yielding the zwitterion:

$$R-\underset{\underset{NH_3^+}{|}}{\overset{\overset{H}{|}}{C}}-\overset{\overset{O}{\|}}{C}-O^-$$

Except for the simplest $\alpha$-amino acid, glycine, $NH_2CH_2COOH$, the $\alpha$-amino acids exist as *enantiomers,* or optical isomers. Such isomers are mirror images. Any molecule having one tetrahedral carbon atom bonded to four different groups of atoms exhibits optical isomerism. The mirror-image isomers are referred to as the D-isomer and the L-isomer; their three-dimensional structures are represented by the following formulas: ◄

► Optical isomerism is discussed in Section 23.5.

$$\underset{\text{D-amino acid}}{H-\underset{\underset{R}{|}}{\overset{\overset{COOH}{|}}{C}}-NH_2} \qquad \underset{\text{L-amino acid}}{H_2N-\underset{\underset{R}{|}}{\overset{\overset{COOH}{|}}{C}}-H}$$

These two-dimensional structural formulas use a standard convention to represent the three-dimensional molecules. The groups attached to the central carbon atom along horizontal bonds are understood to bend toward you. The groups attached along vertical bonds bend away from you. For instance, in the L-amino acid, the $H_2N-$ group and the H atom bend toward you, and the $-COOH$ and the R groups bend away from you. All of the amino acids that occur naturally in proteins are L-amino acids.

The amino acids in a protein are linked together by peptide bonds. A **peptide** (or **amide**) **bond** is *the C—N bond resulting from a condensation reaction between the carboxyl group of one amino acid and the amino group of a second amino acid:*

$$H-\underset{\underset{H}{|}}{\overset{\overset{R_1}{|}}{N}}-\underset{\underset{H}{|}}{\overset{\overset{O}{\|}}{C}}-\overset{\overset{}{}}{C}-O-H + H-\underset{\underset{H}{|}}{\overset{\overset{R_2}{|}}{N}}-\underset{\underset{H}{|}}{\overset{}{C}}-\overset{\overset{O}{\|}}{C}-O-H \longrightarrow$$

$$H-\underset{\underset{H}{|}}{\overset{\overset{R_1}{|}}{N}}-\underset{\underset{H}{|}}{\overset{}{C}}-\overset{\overset{O}{\|}}{C}-\underset{\underset{H}{|}}{\overset{\overset{}{}}{N}}-\underset{\underset{H}{|}}{\overset{\overset{R_2}{|}}{C}}-\overset{\overset{O}{\|}}{C}-O-H + H_2O$$

peptide bond

The product in this example is a *dipeptide*, a molecule formed by linking together two amino acids. Similarly, a *tripeptide* is formed by linking together three amino acids. A **polypeptide** is *a polymer formed by the linking of many amino acids by peptide bonds*. It may or may not have a biological function. A *protein* is a polypeptide that has a biological function.

 **CONCEPT CHECK 25.1**

Two common amino acids are

$$
\begin{array}{cc}
\underset{\substack{| \\ H}}{\overset{\substack{CH_3 \\ |}}{H_2N-C-COOH}}
&
\underset{\substack{| \\ H}}{\overset{\substack{H \\ | \\ HO-C-CH_3 \\ |}}{H_2N-C-COOH}}
\\[1em]
\text{alanine} & \text{threonine}
\end{array}
$$

Write the structural formulas of all the dipeptides that they could form with each other.

## Protein Primary Structure

The **primary structure** of a protein refers to *the order, or sequence, of the amino-acid units in the protein*. This order of amino acids is conveniently shown by denoting the amino acids by their three-letter abbreviations, separated by a dash. In this notation, it is understood that the amino group is on the left and the carboxyl group is on the right. For example, the dipeptide formed from alanine and glycine is written ala–gly.

$$
\underset{\text{ala}}{\overset{\substack{CH_3 \quad O}}{H-N-C-\!\!-\!C-O-H}} + \underset{\text{gly}}{\overset{\substack{H \quad O}}{H-N-C-C-O-H}} \longrightarrow
$$

$$
\underset{\text{ala–gly}}{\overset{\substack{CH_3 \quad O \qquad H \quad O}}{H-N-C-\!\!-\!C-N-C-C-O-H}} + H_2O
$$

Consider a polypeptide containing the following five amino-acid units: glycine, alanine, valine, histidine, and serine. These five units can be ordered in many different ways to make a polypeptide. Three possible primary structures are

gly–ser–ala–val–his

his–ala–ser–val–gly

ser–ala–gly–his–val

Each of the possible sequences would produce a different polypeptide with different properties.

In any protein, which may contain as few as 50 or more than 1000 amino-acid units, the amino acids are arranged in one unique sequence, the primary structure of that protein molecule. Figure 25.7 diagrams the primary structure of one form of the hormone insulin.

**Figure 25.7**
**Primary structure of human proinsulin.** The diagram shows the order of the amino acids in the structure of the protein molecule proinsulin. Also shown are the positions of three disulfide linkages (discussed below). The actual three-dimensional shape of the molecule is not represented here.

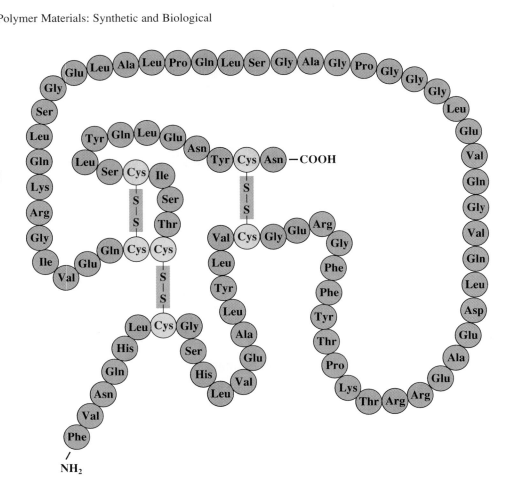

## Shapes of Proteins

A protein molecule of a unique amino-acid sequence spontaneously folds and coils into a characteristic three-dimensional conformation in aqueous solution (Figure 25.8). The side chains of the amino-acid units, with their different chemical properties (nonpolar or polar), determine the characteristic shape of the protein. For an energetically stable shape, the parts of the chain with nonpolar amino-acid side chains are buried within the structure away from water, because nonpolar groups are hydrophobic (not attracted to water). Conversely, most polar groups are stablest on the surface, where they can hydrogen bond with water or with other polar side chains. Occasionally, side chains form ionic bonds.

One type of covalent linkage, *the disulfide linkage* (—S—S—), is especially important to protein shape. The amino acid cysteine, with its thiol (—SH) side chain, is able to react with a second cysteine in the presence of an oxidizing agent, as follows:

$$\text{cys—SH} + \text{HS—cys} + \text{(O)} \longrightarrow \text{cys—S—S—cys} + \text{H}_2\text{O}$$

When two cysteine side chains in a polypeptide are brought close together by folding of the molecule, then oxidized, they form a disulfide linkage, which helps anchor the folded chain into position (Figure 25.7). ◄

▶ Hair can be permanently waved by the use of disulfide linkage. Hair proteins have many disulfides that can be chemically reduced and reformed again while the hair is rolled on curlers. The new disulfide links hold the hair shafts in their new, wavy conformation.

**Figure 25.8**
**Three-dimensional structure of myoglobin.** Myoglobin is present in muscle tissue, and is a reservoir for $O_2$. Note that many sections of the protein have a helical, or coiled, structure, which results from hydrogen bonding. The helical sections fold so that the nonpolar side chains are inside the globular molecule, and the polar side chains point outward. The planar structure in the upper right is called a heme group; it contains an iron ion that can bind to an $O_2$ molecule. (After R.E. Dickerson, *The Proteins*, H. Neurath, ed., 2nd ed., Academic Press, 1964.)

Proteins may be classed as fibrous or globular on the basis of their arrangements in space. **Fibrous proteins** are *proteins that form long coils or align themselves in parallel to form long, water-insoluble fibers. The relatively simple coiled or parallel arrangement of a protein molecule* is called its **secondary structure. Globular proteins** are *proteins in which long coils fold into compact, roughly spherical shapes.* In addition to their secondary structure (long coil), globular proteins have a **tertiary structure,** or *structure associated with the way the protein coil is folded.* Most globular proteins are water soluble, because they are relatively small and have hydrophilic ("water-loving") surfaces that bind water molecules.

The characteristic coiling or folding pattern of protein molecules produces unique surface features, such as grooves or indentations, and hydrophobic or charged areas. These surface features are fundamental to protein function. They allow protein molecules to interact with other molecules in specific ways. For example, certain protein molecules characteristically aggregate because of complementary surface features, forming tubules or filaments or other organized structures. Particular surface features of enzyme molecules are responsible for their ability to bind specific molecules and catalyze specific reactions. The surface feature of an enzyme where catalysis occurs is referred to as its *active site* (Figure 25.9). Enzyme catalysis is discussed in Section 14.9.

## 25.4  Nucleic Acids

▶ *Genetic engineering* refers to the manipulation of nucleic acids to change the characteristics of organisms (for example, to correct genetic flaws).

*Nucleic acids* are vital to the life cycles of cells, because they are the carriers of species inheritance. There are two types of nucleic acids: *deoxyribonucleic acid (DNA)* and *ribonucleic acid (RNA)*. Both are polymers of *nucleotide* building blocks. ◀

**Figure 25.9**

**Active site on an enzyme.**
This is a computer-generated model of a portion of thymidylate synthase, an enzyme used in the synthesis of thymine (one of the bases in DNA, which is discussed in Section 25.4). The main body of the enzyme is represented by colored coils and ribbons (in green and dark blue), but the amino-acid groups on the active site are shown by ball-and-stick models (in light purple with numbers showing the places of the amino acids in the protein sequence). The model shows the active site containing a molecule (in black) on which the enzyme is acting. The molecule is attached to the active site by hydrogen bonds (yellow dashes).

## Sugar Constituents of Nucleic Acid

The building blocks of nucleic acids are themselves composed of certain organic bases linked to either ribose or deoxyribose, both of which are sugar molecules. In this short subsection, we will briefly discuss the structure of *monosaccharides,* or sugars, before we launch into the structure of nucleic acids.

Monosaccharides are simple sugars, each containing three to nine carbon atoms, generally all but one of which bear a hydroxyl group. The remaining carbon atom is part of a carbonyl group, either an aldehyde or a ketone. The monosaccharides exist as D- and L-isomers, because each molecule has at least one carbon atom with four different groups attached. Only a few of the many known monosaccharides are of major biological importance. All are D-sugars, and most are 5- and 6-carbon sugars (pentoses and hexoses). D-Glucose, D-fructose, D-ribose, and 2-deoxy-D-ribose are by far the most important sugars. D-Glucose is common blood sugar, an important energy source for cell function. D-Fructose is a common sugar in fruits and is involved in cell metabolism (the chemical processes occurring in the cell). D-Ribose and 2-deoxy-D-ribose are parts of nucleic acids, as we noted earlier, where we referred to them simply as ribose and deoxyribose. You could regard 2-deoxy-D-ribose as derived from D-ribose by replacing the hydroxyl group on carbon 2 by a hydrogen atom.

D-glucose          D-fructose          D-ribose          2-deoxy-D-ribose

Although one often draws the simple sugars as straight-chain molecules, they do not exist predominantly in that form. The chain curls up and the carbonyl group reacts with an alcohol group. With aldehydes, alcohols give *hemiacetals*, or compounds in which an —OH group, an —OR group, and an H atom are attached to the same carbon atom.

$$R'-\overset{\overset{\textstyle O}{\|}}{C}-H \;+\; \overset{\textstyle H}{\underset{\textstyle \text{alcohol}}{:O-R}} \;\rightleftharpoons\; \underset{\textstyle \text{hemiacetal}}{R'-\overset{\overset{\textstyle OH}{|}}{\underset{\underset{\textstyle OR}{|}}{C}}-H}$$

carbonyl

A ketone and an alcohol give a *hemiketal*, which is similar to a hemiacetal except that the H atom is replaced by an R″ group.

The straight-chain form of D-glucose, whose formula we gave earlier, is a flexible molecule that can bend around to give a cyclic hemiacetal.

| α-D-glucose (hemiacetal) | D-glucose (straight-chain form) | β-D-glucose (hemiacetal) |

The straight-chain form, in the center, is shown bent around so that an —OH group and the —CHO group can react to give a cyclic hemiacetal. A new —OH group is obtained (on the right in each hemiacetal) and can point either down or up. Thus, there are two isomers of the hemiacetal—one labeled α (—OH down) and the other labeled β (—OH up).

Although for simplicity we have drawn the six-membered rings of the hemiacetals as flat, they are in fact puckered, as shown in these more accurate three-dimensional formulas:

α-D-glucose          β-D-glucose

In these three-dimensional formulas, it is understood that there are carbon and hydrogen atoms at each of the corners in the ring, except where there is an oxygen atom.

The structures of the sugars we will need in this section are:

β-D-ribose          2-deoxy-β-D-ribose

These sugars have five-membered rings. Four of the ring atoms are planar, but either carbon atom 2 or carbon atom 3 (with attached H and OH) points upward or downward from the planar ring atoms.

## Nucleotides

**Nucleotides** are *the building blocks of nucleic acids*. There are two types: *ribonucleotides* and *deoxyribonucleotides*. The general structure of nucleotides consists of an organic base (described below) linked to carbon 1 of a pentose (5-carbon sugar), which is linked in turn at carbon 5 to a phosphate group. The only difference between the two types of nucleotides is in the sugar. Ribonucleotides contain $\beta$-D-ribose; deoxyribonucleotides contain 2-deoxy-$\beta$-D-ribose. Both sugar groups have furanose rings.

a ribonucleotide

a deoxyribonucleotide

The carbon atoms of ribose are labeled with numbers 1', 2', ... 5'. (Numbers without primes are used for carbon atoms in the organic base.) Note that the phosphate group is shown in ionized form—$PO_4{}^{2-}$, as this form predominates near neutral pH in cells.

Five organic bases (amines) are most often found in nucleotides. The bases are *adenine, guanine, cytosine, uracil,* and *thymine.* They link to the sugars through the indicated nitrogen (in color) in each of the following structural formulas:

adenine       guanine       cytosine       uracil       thymine

The base plus the sugar group (without the phosphate group) is called a *nucleoside*. Nucleosides are named from the bases. For example, the nucleoside composed of adenine with $\beta$-D-ribose is called adenosine. The nucleoside composed of adenine with 2-deoxy-$\beta$-D-ribose is called deoxyadenosine. A nucleotide is named by adding monophosphate (or diphosphate or triphosphate) after the nucleoside name. A number with a prime indicates the position of the phosphate group on the ribose ring. Thus, adenosine-5'-monophosphate is a nucleotide composed of adenine, $\beta$-D-ribose, and a phosphate group at the 5' position of $\beta$-D-ribose.

## Polynucleotides and Their Conformations

A **polynucleotide** is *a linear polymer of nucleotide units* linked from the hydroxyl group at the 3' carbon of the pentose of one nucleotide to the phosphate group of the other nucleotide. For example,

Just as the unique sequence of amino acids in a protein determines the protein's nature, so the sequence of nucleotides determines the particular properties and functions of a polynucleotide. **Nucleic acids** are *polynucleotides folded or coiled into specific three-dimensional shapes.*

**Complementary bases** are *nucleotide bases that form strong hydrogen bonds with one another.* Adenine and thymine are complementary bases, as are adenine and uracil, and guanine and cytosine. Hydrogen bonding of complementary bases, called *base pairing,* is the key to nucleic acid structure and function (see Figure 25.10).

**Deoxyribonucleic acid (DNA)** is *the hereditary constituent of cells and consists of two polymer strands of deoxyribonucleotide units.* One of the strands consists of nucleotide bases that are complementary to those of the other strand. The two strands coil about each other in a double helix, with base pairing along the entire lengths of the strands (Figure 25.11). **Ribonucleic acid (RNA)** is *a constituent of cells that is*

**Figure 25.10**
**Hydrogen-bonded complementary bases.**
Each base hydrogen bonds strongly to only its complementary base. Adenine hydrogen bonds to thymine or to uracil; guanine hydrogen bonds to cytosine.

**Figure 25.11**
**Double helix of the DNA molecule.** The structure was first deduced by James Watson, an American scientist, and Francis Crick, a British scientist, in the early 1950s.

*used to manufacture proteins from genetic information. It is a polymer of ribonucleotide units.* Both DNA and RNA have nucleotides with the bases adenine, guanine, and cytosine. In addition, DNA contains thymine but not uracil, and RNA contains uracil but not thymine. It is often convenient in writing the bases in DNA and RNA sequences to use the abbreviations A = adenine, C = cytosine, G = guanine, T = thymine, and U = uracil.

## DNA and the Nature of the Genetic Code

Photomicrographs of dividing cells show structures, called chromosomes, as dense, thick rods (Figure 25.12). *Chromosomes are cell structures that contain DNA and proteins;* the DNA contains the genetic inheritance of the cell and organism. Before cell division, the cell synthesizes a new and identical set of chromosomes or, more particularly, a new and identical set of DNA molecules—the genetic information—to be transmitted to the new cell. Thus, the new cell will have all the necessary instructions for normal structure and function.

What is the nature of this genetic information? The genetic information is coded into the linear sequence of nucleotides in the DNA molecule. Each DNA molecule is composed of hundreds of genes. A **gene** is *a sequence of nucleotides in a DNA molecule that codes for a given protein.* The nucleotides in a gene are grouped in sets of three, or triplets. Each triplet codes for one amino acid in a protein.

New DNA must be synthesized during cell division, as we mentioned. This synthesis, or DNA replication, proceeds as follows. The two strands of DNA unravel at one end of the helix to form two single strands. As the two strands unravel, and with the aid of an enzyme called *polymerase*, complementary nucleotides are attached one after the other to these single strands to form two new double strands of DNA. At the completion of the process, two double-stranded DNA chains exist where previously there was one (Figure 25.13).

A similar laboratory technique, based on the *polymerase chain reaction* (PCR), is used to amplify the quantity of DNA in a sample. In the PCR technique, a sample of DNA is heated with polymerase to separate the DNA into single strands. The polymerase then catalyzes the formation of new double-stranded DNA by adding complementary nucleotides to the original single strands. The process is continued, separating double-stranded DNA into single strands and forming new double-stranded DNA from these. Within hours, the original sample of DNA can be amplified a millionfold. Using PCR, researchers can diagnose diseases by identifying the DNA of infectious agents originally present in minute quantities in blood, food, and other samples. The PCR technique also makes it possible to replicate the DNA from Egyptian mummies and ancient bone samples.

**Figure 25.12**
**Chromosomes in a dividing cell.** The chromosomes appear as dense, thick rods.

**Figure 25.13**
**DNA replication.** The first stage of replication involves the unraveling of the two chains of the DNA double helix. As the chains begin to unravel, the exposed bases on each chain attract the complementary bases. The process is finished when there are two DNA double helixes where there had been only one.

Occasionally, an error is made during the synthesis of new DNA. Such a change in the genetic information, or genetic error, is called a *mutation*. One possible consequence could be the synthesis of a faulty or inactive protein. We know of various genetic diseases that result from a single error leading to the change of an amino acid in a protein sequence that has a profound effect on the protein's activity and so on the life of the organism.

## RNA and the Transmission of the Genetic Code

RNA is used to translate the genetic information stored in DNA into protein structure. There are three classes of RNA: ribosomal RNA, messenger RNA, and transfer RNA.

**Ribosomes** are *tiny cellular particles on which protein synthesis takes place.* They are constructed of numerous proteins plus three or four RNA molecules. *The RNA in a ribosome* is called **ribosomal RNA.** Ribosomes provide a surface on which to organize the process of protein synthesis, and they also contain enzymes that catalyze the process.

**Figure 25.14**
**Formation of messenger RNA (transcription).** A portion of a DNA double helix corresponding to a single protein unravels, and complementary ribonucleotides line up, attracted by hydrogen bonding to one of the DNA chains. The ribonucleotides are then linked to form messenger RNA. After being released from the DNA, the messenger RNA migrates from the cell nucleus into the cytoplasm to a ribosome, where protein biosynthesis occurs.

A **messenger RNA** (mRNA) molecule is *a relatively small RNA molecule that can diffuse about the cell and attach itself to a ribosome, where it serves as a pattern for protein biosynthesis.* The first step in protein biosynthesis, called *transcription*, is synthesis of a messenger RNA molecule that has a sequence of bases complementary to that of a portion of a DNA strand corresponding to a single protein (Figure 25.14).

*A sequence of three bases in a messenger RNA molecule that serves as the code for a particular amino acid* is called a **codon.** When the messenger RNA molecule attaches itself to a ribosome, its codons provide the sequence of amino acids in a protein that will be synthesized. *Translation* is the biosynthesis of protein using messenger RNA codons. Table 25.3, the complete messenger RNA code-

**Table 25.3**
**Genetic Code Dictionary***

|  | U | | C | | A | | G | |
|---|---|---|---|---|---|---|---|---|
| **U** | UUU | Phe | UCU | Ser | UAU | Tyr | UGU | Cys |
|  | UUC | Phe | UCC | Ser | UAC | Tyr | UGC | Cys |
|  | UUA | Leu | UCA | Ser | UAA | End | UGA | End |
|  | UUG | Leu | UCG | Ser | UAG | End | UGG | Trp |
| **C** | CCU | Leu | CCU | Pro | CAU | His | CGU | Arg |
|  | CUC | Leu | CCC | Pro | CAC | His | CGC | Arg |
|  | CUA | Leu | CCA | Pro | CAA | Gln | CGA | Arg |
|  | CUG | Leu | CCG | Pro | CAG | Gln | CGG | Arg |
| **A** | AUU | Ile | ACU | Thr | AAU | Asn | AGU | Ser |
|  | AUC | Ile | ACC | Thr | AAC | Asn | AGC | Ser |
|  | AUA | Ile | ACA | Thr | AAA | Lys | AGA | Arg |
|  | AUG | Met | ACG | Thr | AAG | Lys | AGG | Arg |
| **G** | GUU | Val | GCU | Ala | GAU | Asp | GGU | Gly |
|  | GUC | Val | GCC | Ala | GAC | Asp | GGC | Gly |
|  | GUA | Val | GCA | Ala | GAA | Glu | GGA | Gly |
|  | GUG | Val | GGG | Ala | GAG | Glu | GGG | Gly |

*The four nucleotide bases of RNA—U, C, A, and G—are arranged along the left side of the table and along the top. Combining any two of them and then adding a third (again, U, C, A, or G) gives one of the 64 three-nucleotide codons shown in capital letters in the table. Next to each codon appears the abbreviation for the amino acid it codes for during protein biosynthesis. In three cases, the word *End* appears because each of these codons serves as a signal to terminate protein biosynthesis. [From Robert D. Whitaker et al., *Concepts of General, Organic, and Biological Chemistry*, p. 697. Copyright © 1981 by Houghton Mifflin Company. Used with permission.]

**Figure 25.15**
**Synthesis of protein molecules (translation).** Messenger RNA provides the code needed to determine the sequence of amino acids to be added to the protein. Transfer RNA brings the necessary amino acids to the mRNA in the ribosome. The tRNA bonds to the mRNA, and the amino acid is transferred to the growing protein chain. After transfer of the amino acid, the tRNA is released from the mRNA.

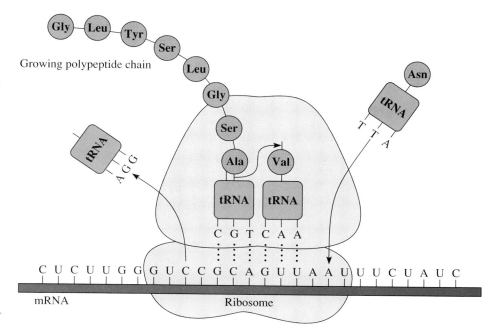

word dictionary, shows the specific amino acids coded for by messenger RNA codons. There are 64 possible arrangements of the four RNA bases, so there are 64 codons. Three of the codons do not signify amino acids but signify the end of a message and thus are called *termination codons.* The remaining 61 codons signify particular amino acids. Because only 20 different amino acids exist in proteins, there are a number of instances in which 2, 3, 4, or even 6 codons translate to the same amino acid.

A **transfer RNA** (tRNA) molecule is *a small RNA molecule that binds to a particular amino acid and carries it to a ribosome.* At the ribosome, the tRNA molecule attaches itself through base pairing to an mRNA codon. The amino acid is then transferred to a growing protein chain. The process of protein biosynthesis stops when a termination codon is reached on the mRNA.

Figure 25.15 illustrates this process of protein biosynthesis. It shows a tRNA molecule about to add to the mRNA codon AAU. Note that the tRNA molecule has a triplet base sequence TTA at one end of the molecule and the amino acid asparagine (Asn) attached to the other end. The triplet TTA, called an *anticodon,* is the complementary base sequence to the codon AAU. The tRNA anticodon pairs up with the mRNA codon. The amino acid asparagine is then transferred to the end of the protein chain, where it links to valine (Val). Once the amino acid has been transferred, the tRNA molecule is released from mRNA.

 **CONCEPT CHECK 25.2**

Write the RNA sequence complementary to the following DNA sequence:

ATGCTACGGATTCAA

# Tobacco Mosaic Virus and Atomic Force Microscopy

**LIFE SCIENCE**

The amount of detail provided by the latest scanning probe microscopes is breathtaking. Figure 25.16 shows what looks like a tumble of logs on a forest floor, each log covered with mossy growth. The image, provided by an atomic force microscope, is of rods of tobacco mosaic virus, a disease agent that infects tobacco and many other crops. Each virus rod is covered by molecules of bovine serum albumin (from the blood serum of cows). Chemists obtained this image as part of a study of the interaction of the albumin protein molecule with the virus. The virus itself consists of a strand of RNA, containing the genetic material of the virus, surrounded by a protective protein coat.

The atomic force microscope (Figure 25.17) is a cousin of the scanning tunneling microscope, which we discussed in an essay at the end of Section 7.4. Both microscopes use a probe to scan a surface; but, whereas the scanning tunneling microscope measures an electric current between the probe tip and the sample, the atomic

**Figure 25.16**
**Tobacco mosaic virus rods.** The virus rods are covered with adsorbed protein molecules (bovine serum albumin). The protein molecules show up as bright bumps on the dark yellow color of the virus. A computer adds color to the image to create contrast. (The image was obtained by an atomic force microscope produced in the laboratory of Gang-Yu Liu at Wayne State University.)

microscope measures the attractive van der Waals force between the probe tip and the sample. The advantage of the atomic force microscope is that it can be used with almost any surface, whereas a scanning tunneling microscope requires a conductive surface.

**Figure 25.17**
**Atomic force microscope.** The sample moves by a scanner under a fixed probe. The probe consists of a silicon nitride tip attached to a cantilever, which can move up and down. A laser beam is reflected from the back of the cantilever to a photodiode divided into four segments (quadrant photodiode). As the cantilever bends up and down (or left and right) from surface forces acting on the probe tip, it deflects the laser beam to different quadrants of the photodiode. A computer coordinates the output from the photodiode with the sample position to create an image that appears on the computer screen. (From an illustration provided by Gang-Yu Liu of Wayne State University.)

## A Checklist for Review

### Important Terms

polymer (p. 1085)
monomer (p. 1085)
addition polymer (25.1)
homopolymer (25.1)
copolymer (25.1)
condensation polymer (25.1)
proteins (25.3)
amino acid (25.3)

peptide (amide) bonds (25.3)
polypeptide (25.3)
primary structure (of a protein) (25.3)
fibrous proteins (25.3)
secondary structure (of a protein) (25.3)
globular proteins (25.3)

tertiary structure (of a protein) (25.3)
nucleotides (25.4)
polynucleotide (25.4)
nucleic acids (25.4)
complementary bases (25.4)
deoxyribonucleic acid (DNA) (25.4)

ribonucleic acid (RNA) (25.4)
gene (25.4)
ribosomes (25.4)
ribosomal RNA (25.4)
messenger RNA (25.4)
codon (25.4)
transfer RNA (25.4)

### Summary of Facts and Concepts

*Polymers* are species of very high molecular weight consisting of many units *(monomers)*. They are classified as *addition polymers* or *condensation polymers,* depending on the type of reaction used in forming them. Rubber is an addition polymer. Natural rubber is a *homopolymer* of isoprene units; the synthetic rubber SBR (styrene-butadiene rubber) is a *copolymer.* Nylon-6,6 is an example of a condensation polymer; it is produced by condensing hexamethylene diamine with adipic acid.

Certain polymers, such as *trans*-polyacetylene, are semiconductors. By doping them with substances such as iodine, their electrical conductivities can be greatly increased. One of the first applications of a conducting polymer was to produce a battery. These polymers have also been used to produce electronic devices, including transistors and LEDs.

Proteins are polymers of *α-amino acids* linked by *peptide bonds.* The 20 common amino acids of protein have different *side chains,* which may be *polar* or *nonpolar.* The unique *primary structure* of a polypeptide (the amino-acid sequence) is responsible for spontaneous folding into a unique shape maintained by interaction between side chains and disulfide bonds.

This folding produces surface features responsible for the protein's function. An enzyme has an *active site* where catalysis takes place.

Nucleic acids are polymers of *nucleotides.* Nucleotides are base–sugar–phosphate compounds. Conformations of nucleic acids are the result of hydrogen bonding, called base pairing, between *complementary bases. DNA* is the genetic material of chromosomes. DNA consists of two complementary polynucleotides coiled into a *double helix;* each has all the information necessary to direct synthesis of a new complementary strand. The genetic code is the relationship between nucleotide sequence and protein amino-acid sequence. Sixty-four *codons* (three-nucleotide sequences) are the basis of the code. *Transcription* is the synthesis of a *messenger RNA* molecule, which represents a copy of the information in a DNA *gene.* The messenger RNA binds to a *ribosome,* where *translation* of the message into an amino-acid sequence occurs. *Transfer RNA* molecules attach to amino acids and bind to the messenger RNA by *codon–anticodon* pairing, bringing each amino acid in turn into the position specified by the codon sequence.

### Operational Skills

*Note:* The problem-solving skills used in this chapter are discussed in previous chapters.

## Review Questions

**25.1** What is the difference between an addition polymer and a condensation polymer? Give an example of each, writing the equation for its formation.

**25.2** Write the equation for the addition of one isoprene molecule to another.

**25.3** Give the structural formula for a portion of the chain polymer obtained by the reaction of ethylene glycol, $CH_2OH—CH_2OH$ with malonic acid, $HOOC—CH_2—COOH$.

**25.4** Show how the pi electrons in one resonance formula of polyacetylene would have to shift to obtain the other resonance formula.

**25.5** Explain how the doping of polyacetylene increases its electrical conductivity.

**25.6** Describe the primary structure of protein. What makes one protein different from another protein of the same size? What is the basis of the unique conformation of a protein?

**25.7** Distinguish between secondary and tertiary structures of protein.

**25.8** What is an enzyme?

**25.9** What is the difference in structure between cellulose and amylose?

**25.10** What are the structural forms of D-glucose present in the blood?

**25.11** Name the complementary base pairs. Describe the DNA double helix.

**25.12** How do ribonucleotides and deoxyribonucleotides differ in structure? Do they form polymers in the same way?

**25.13** Explain the nature of the genetic code.

**25.14** Define *codon* and *anticodon*. How do they interact?

**25.15** Show mathematically why there are 64 possible triplet codons.

**25.16** Outline how the genetic message in a gene is translated to produce a polypeptide. Include the roles of the gene, messenger RNA, ribosomes, and transfer RNA.

 ## Conceptual Problems

**25.17** It is your job to manufacture polymers from a series of monomer units. These monomer units are called A, B, and C. In this problem you need to "build" polymers by linking the monomer units. Represent the polymer linkages using dashes. For example, −A−B−C− represents a polymer unit made from linking monomer units A, B, and C.
a. Build two different homopolymers from your monomer units.
b. Build a copolymer from A and B.
c. Build a copolymer that is about 33% C and 66% A.

**25.18** Use resonance formulas to explain why polyacetylene has delocalized molecular orbitals extending over the length of the molecule, whereas the following molecule does not.

**25.19** A common amino acid in the body is ornithine. It is involved in the excretion of excess nitrogen into the urine. The structural formula of ornithine is

$$H_2NCH_2CH_2CH_2-\overset{\overset{\displaystyle H}{|}}{\underset{\underset{\displaystyle NH_2}{|}}{C}}-COOH$$

Write the fully ionized form of the molecule. (Note that there are two ionizable amino groups.)

**25.20** Write the structural formulas of all possible tripeptides with the composition of two glycines and one serine. (See the structural formulas in Table 25.2.)

**25.21** What amino-acid sequence would result if the following messenger RNA sequence were translated from left to right?

AGAGUCCGAGACUUGACGUGA

**25.22** Give one of the nucleotide sequences that would translate to the peptide lys−pro−ala−phe−trp−glu−his−gly.

## Practice Problems

### Synthetic Polymers

**25.23** Teflon is an addition polymer of 1,1,2,2-tetrafluoroethene. Write the equation for the formation of the polymer.

**25.24** Poly(vinyl chloride) (PVC) is an addition polymer of vinyl chloride, $CH_2{=}CHCl$. Write the equation for the formation of the polymer.

**25.25** Consider the following polymer:

$$-OCH_2CH_2O\overset{O}{\overset{||}{C}}CH_2CH_2\overset{O}{\overset{||}{C}}OCH_2CH_2O\overset{O}{\overset{||}{C}}CH_2CH_2\overset{O}{\overset{||}{C}}-$$

From which two monomers is the polymer made?

**25.26** Consider the following polymer:

$$-\overset{O}{\overset{||}{C}}(CH_2)_{10}\overset{O}{\overset{||}{C}}NH(CH_2)_6NH\overset{O}{\overset{||}{C}}(CH_2)_{10}\overset{O}{\overset{||}{C}}NH(CH_2)_6NH-$$

From which two monomers is the polymer made?

### Conducting Polymers

**25.27** Which of the following monomers might you expect would lead to a conducting polymer?
a. $H-C{\equiv}C-CH_3$
b. $H_2C{=}CH-CH_3$

**25.28** Which of the following monomers might you expect would lead to a conducting polymer?
a. $CH_3—CH=CH—CH_3$     b. $CH_3—C\equiv C—CH_3$

## Amino Acids and Primary Structure

**25.29** Alanine has the structure

$$CH_3—\overset{\overset{\displaystyle H}{|}}{\underset{\underset{\displaystyle NH_2}{|}}{C}}—COOH$$

Draw the zwitterion that would exist at neutral pH.

**25.30** Valine has the structure

$$CH_3CH—\overset{\overset{\displaystyle H}{|}}{\underset{\underset{\displaystyle NH_2}{|}}{\underset{\underset{\displaystyle CH_3}{}}{C}}}—COOH$$

Draw the zwitterion that would exist at neutral pH.

**25.31** Write the structural formula of a dipeptide formed from the reaction of L-alanine and L-histidine. How many dipeptides are possible?

**25.32** Write the structural formulas of two tripeptides formed from the reaction of L-tryptophan, L-glutamate, and L-tyrosine. How many tripeptides are possible?

## Nucleotides and Polynucleotides

**25.33** If adenine, thymine, guanine, and cytosine were each analyzed separately in a sample of DNA, what molar ratios of A : T and G : C would you expect to find?

**25.34** If a sample of DNA isolated from a microorganism culture were analyzed and found to contain 1.5 mol of cytosine nucleotides and 0.5 mol of adenosine nucleotides, what would be the amounts of guanine and thymine nucleotides in the sample?

**25.35** Write a structural formula for the nucleotide adenosine-5′-monophosphate.

**25.36** Write a structural formula for the nucleotide deoxyadenosine-5′-monophosphate.

**25.37** How many hydrogen bonds link a guanine–cytosine base pair? an adenine–uracil base pair? Would you expect any difference in strength between guanine–cytosine bonding and adenine–uracil bonding? Explain.

**25.38** How many hydrogen bonds link an adenine–thymine base pair? Would there be any difference in strength between adenine–thymine bonding and adenine–uracil bonding? between adenine–thymine and cytosine–guanine bonding? Explain.

## DNA, RNA, and Protein Synthesis

**25.39** If a codon consists of two nucleotides, how many codons would be possible? Would this be a workable code for the purpose of protein synthesis?

**25.40** If a codon consists of four nucleotides, how many codons would be possible? Would this be workable as an amino-acid code?

**25.41** Nucleic acids can be denatured by heat, as proteins can. What bonds are broken when a DNA molecule is denatured? Would DNA of greater percent composition of guanine and cytosine denature more or less readily than DNA of greater percent composition of adenine and thymine?

**25.42** Would RNA of a certain percent composition of guanine and cytosine denature more or less readily than RNA with a lower percent of guanine and cytosine? Why?

**25.43** Consulting Table 25.3, write the amino-acid sequence resulting from left-to-right translation of the messenger RNA sequence

GGAUCCCGCUUUGGGCUGAAAUAG

**25.44** Write the amino-acid sequence obtained from left-to-right translation of the messenger RNA sequence

AUUGGCGCGAGAUCGAAUGAGCCCAGU

See Table 25.3.

# General Problems

**25.45** List the codons to which the following anticodons would form base pairs:

Anticodon: GAC UGA GGG ACC
Codon:

**25.46** List the anticodons to which the following codons would form base pairs:

Codon: UUG CAC ACU GAA
Anticodon:

**25.47** Give one of the nucleotide sequences that would translate to

leu–ala–val–glu–asp–cys–met–trp–lys

**25.48** Write one nucleotide sequence that would translate to

tyr–ile–pro–his–leu–his–thr–ser–phe–met

**25.49** Write the structural formula of the addition polymer made from *trans*-1,2-dichloroethene:

**25.50** Neoprene is an addition polymer of chloroprene:

Write the structural formula of neoprene.

**25.51** Kevlar® is a polymer fiber used to make bulletproof vests.

What are the structures of the monomer units in this polymer?

**25.52** Nylon 6 has the following structure:

What is the structure of the monomer unit in this polymer? Show the reaction corresponding to the formation of the polymer.

**25.53** Which of the following amino acids has a nonpolar side chain?

**25.54** Which of the following amino acids has a polar side chain?

**25.55** Draw the zwitterion structure for the amino acid serine.

**25.56** Draw the zwitterion structure for the amino acid leucine.

**25.57** Write the structural formula of one of the two dipeptides that could be formed from the amino acids in Problem 25.53.

**25.58** Write the structural formula of one of the two peptides that could be formed by the amino acids in Problem 25.54.

**25.59** A peptide contains six amino acids: L-arginine, L-proline, L-glutamate, L-glycine, L-asparagine, and L-glutamine. How many different peptides of this amino-acid composition are possible?

**25.60** Using abbreviations for the amino acids, give three possible sequences of the peptide described in Problem 25.59.

**25.61** Name each of the three major constituents of the nucleotide adenosine-5′-monophosphate.

**25.62** Write the structural formula of deoxyguanosine-5′-monophosphate.

## Media Activities

**25.63** Work through the Amino Acids *Problem Solving* activity on the student CD. Consult Table 25.2 for amino acid structures.
a. Write the structure of the zwitterion of the amino acid L-leucine.
b. Write the structures of all the possible dipeptides that could form from L-leucine and L-valine.

**25.64** Watch the Hemiacetal Formation in the D-Glucose *Visualization* on the student web site. Write the structure of one of the cyclic hemiacetals that can form from the sugar D-ribose. (See section 25.4 for the structure of D-ribose.)

**25.65** Step through the Synthesis of Protein Molecules (Translation) *Visualization* on the student web site.
a. For the ribosome depicted in the *Visualization*, what are the six complementary base pairs of the first six nucleic acids (read from left to right)?
b. How many bases are required to provide the code for the synthesis of a particular amino acid?
c. What is the anticodon of the mRNA codon AAU?

**25.66** For review, take the ACE practice test for Chapter 25 on the student web site.

# Appendixes

## Appendix A — Mathematical Skills

Only a few basic mathematical skills are required for the study of general chemistry. But to concentrate your attention on the concepts of chemistry, you will find it necessary to have a firm grasp of these basic mathematical skills. In this appendix, we will review scientific (or exponential) notation, logarithms, simple algebraic operations, the solution of quadratic equations, and the plotting of straight-line graphs.

### A.1 Scientific (Exponential) Notation

In chemistry, you frequently encounter very large and very small numbers. For example, the number of molecules in a liter of air at 20°C and normal barometric pressure is 25,000,000,000,000,000,000,000, and the distance between two hydrogen atoms in a hydrogen molecule is 0.000,000,000,074 meters. In these forms, such numbers are inconvenient to write and difficult to read. For this reason, you normally express them in scientific, or exponential, notation. Scientific calculators also use this notation.

In scientific notation, a number is written in the form $A \times 10^n$. $A$ is a number greater than or equal to 1 and less than 10, and the exponent $n$ (the $n$th power of ten) is a positive or negative integer. For example, 4853 is written in scientific notation as $4.853 \times 10^3$, which is 4.853 multiplied by three factors of 10:

$$4.853 \times 10^3 = 4.853 \times 10 \times 10 \times 10 = 4853$$

The number 0.0568 is written in scientific notation as $5.68 \times 10^{-2}$, which is 5.68 divided by two factors of 10:

$$5.68 \times 10^{-2} = \frac{5.68}{10 \times 10} = 0.0568$$

Any number can be conveniently transformed to scientific notation by moving the decimal point to obtain a number, $A$, greater than or equal to 1 and less than 10. If the decimal point is moved to the left, you multiply $A$ by $10^n$, where $n$ equals the number of places moved. If the decimal point is moved to the right, you multiply $A$ by $10^{-n}$. Consider the number 0.00731. You must move the decimal point to the right three places. Therefore, 0.00731 equals $7.31 \times 10^{-3}$. To transform a number written in scientific notation to the usual form, the process is reversed. If the exponent is positive, the decimal point is shifted right. If the exponent is negative, the decimal point is shifted left.

---

**EXAMPLE 1**

### Expressing Numbers in Scientific Notation

Express the following numbers in scientific notation:

a. 843.4     b. 0.00421     c. 1.54

**SOLUTION**

Shift the decimal point to get a number between 1 and 10; count the number of positions shifted.

a. $8\underset{\frown}{4}\underset{\frown}{3}.4 = \mathbf{8.434 \times 10^2}$

b. $0.\underset{\frown}{0}\underset{\frown}{0}\underset{\frown}{4}21 = \mathbf{4.21 \times 10^{-3}}$

c. **leave as is or write $1.54 \times 10^0$**

---

**EXERCISE 1**

Express the following numbers in scientific notation:

a.  4.38     b.  4380     c.  0.000483

---

**EXAMPLE 2**

## Converting Numbers in Scientific Notation to Usual Form

Convert the following numbers in scientific notation to usual form:

a.  $6.39 \times 10^{-4}$     b.  $3.275 \times 10^2$

**SOLUTION**

a. $0.\underset{\frown}{0}\underset{\frown}{0}\underset{\frown}{0}\underset{\frown}{6}.39 \times 10^{-4} = \mathbf{0.000639}$

b. $3.\underset{\frown}{2}\underset{\frown}{7}5 \times 10^2 = \mathbf{327.5}$

---

**EXERCISE 2**

Convert the following numbers in scientific notation to usual form:

a.  $7.025 \times 10^3$     b.  $8.97 \times 10^{-4}$

## Addition and Subtraction

Before adding or subtracting two numbers written in scientific notation, it is necessary to express both to the same power of ten. After adding or subtracting, it may be necessary to shift the decimal point to express the result in scientific notation.

---

**EXAMPLE 3**

## Adding and Subtracting in Scientific Notation

Carry out the following arithmetic; give the result in scientific notation.

$$(9.42 \times 10^{-2}) + (7.6 \times 10^{-3})$$

**SOLUTION**

You can shift the decimal point in either number to obtain both to the same power of ten. For example, to get both numbers to $10^{-2}$, you shift the decimal point one place to the left and add 1 to the exponent in the expression $7.6 \times 10^{-3}$.

$$7.6 \times 10^{-3} = 0.76 \times 10^{-2}$$

Now you can add the two numbers.

$$(9.42 \times 10^{-2}) + (0.76 \times 10^{-2}) = (9.42 + 0.76) \times 10^{-2} = 10.18 \times 10^{-2}$$

Since 10.18 is not between 1 and 10, you shift the decimal point to express the final result in scientific notation.

$$10.18 \times 10^{-2} = \mathbf{1.018 \times 10^{-1}}$$

### EXERCISE 3

Add the following, and express the sum in scientific notation:

$$(3.142 \times 10^{-4}) + (2.8 \times 10^{-6})$$

## Multiplication and Division

To multiply two numbers in scientific notation, you first multiply the two powers of ten by adding their exponents. Then you multiply the remaining factors. Division is handled similarly. You first move any power of ten in the denominator to the numerator, changing the sign of the exponent. After multiplying the two powers of ten by adding their exponents, you carry out the indicated division.

### EXAMPLE 4

### Multiplying and Dividing in Scientific Notation

Do the following arithmetic, and express the answers in scientific notation.

a. $(6.3 \times 10^2) \times (2.4 \times 10^5)$

b. $\dfrac{6.4 \times 10^2}{2.0 \times 10^5}$

### SOLUTION

a. $(6.3 \times 10^2) \times (2.4 \times 10^5) = (6.3 \times 2.4) \times 10^7 = 15.12 \times 10^7 = \mathbf{1.512 \times 10^8}$

b. $\dfrac{6.4 \times 10^2}{2.0 \times 10^5} = \dfrac{6.4}{2.0} \times 10^2 \times 10^{-5} = \dfrac{6.4}{2.0} \times 10^{-3} = \mathbf{3.2 \times 10^{-3}}$

### EXERCISE 4

Perform the following operations, expressing the answers in scientific notation:

a. $(5.4 \times 10^{-7}) \times (1.8 \times 10^8)$   b. $\dfrac{5.4 \times 10^{-7}}{6.0 \times 10^{-5}}$

## Powers and Roots

A number $A \times 10^n$ raised to a power $p$ is evaluated by raising $A$ to the power $p$ and multiplying the exponent in the power of ten by $p$:

$$(A \times 10^n)^p = A^p \times 10^{n \times p}$$

You extract the $r$th root of a number $A \times 10^n$ by first moving the decimal point in $A$ so that the exponent in the power of ten is exactly divisible by $r$. Suppose this has been done, so that $n$ in the number $A \times 10^n$ is exactly divisible by $r$. Then

$$\sqrt[r]{A \times 10^n} = \sqrt[r]{A} \times 10^{n/r}$$

---

### EXAMPLE 5

### Finding Powers and Roots in Scientific Notation

Evaluate the following expressions:

a. $(5.29 \times 10^2)^3$

b. $\sqrt{2.31 \times 10^7}$

**SOLUTION**

a. $(5.29 \times 10^2)^3 = (5.29)^3 \times 10^6 = 148 \times 10^6 = \mathbf{1.48 \times 10^8}$

b. $\sqrt{2.31 \times 10^7} = \sqrt{23.1 \times 10^6} = \sqrt{23.1} \times 10^{6/2} = \mathbf{4.81 \times 10^3}$

---

### EXERCISE 5

Obtain the values of the following, and express them in scientific notation:

a. $(3.56 \times 10^3)^4$     b. $\sqrt[3]{4.81 \times 10^2}$

## Electronic Calculators

Scientific calculators will perform all of the arithmetic operations we have just described (as well as those discussed in the next section). On most calculators, the basic operations of addition, subtraction, multiplication, and division are similar and straightforward. However, more variation exists in raising a number to a power and extracting a root. If you have the calculator instructions, by all means read them. Otherwise, the following information may help.

Squares and square roots are usually obtained with special keys, perhaps labeled $x^2$ and $\sqrt{x}$. Thus, to obtain $(5.15)^2$, you enter 5.15 and press $x^2$. To obtain $\sqrt{5.15}$, you enter 5.15 and press $\sqrt{x}$ (or perhaps INV, for inverse, and $x^2$). Other powers and roots require a $y^x$ (or $a^x$) key. The answer to Example 5(a), which is $(5.29 \times 10^2)^3$, would be obtained by a sequence of steps such as the following: enter $5.29 \times 10^2$, press the $y^x$ key, enter 3, and press the = key.

The same sequence can be used to extract a root. Suppose you want $\sqrt[5]{2.18 \times 10^6}$. This is equivalent to $(2.18 \times 10^6)^{1/5}$ or $(2.18 \times 10^6)^{0.2}$. If the calculator has a $1/x$ key, the sequence would be as follows: enter $2.18 \times 10^6$, press the $y^x$ key, enter 5, press the $1/x$ key, then press the = key. Some calculators have a $\sqrt[x]{y}$ key, so this can be used to extract the $x$th root of $y$, using a sequence of steps similar to that for $y^x$.

## A.2   Logarithms

The *logarithm* to the *base a* of a number $x$, denoted $\log_a x$, is the exponent of the constant $a$ needed to equal the number $x$. For example, suppose $a$ is 10, and you would like the logarithm of 1000; that is, you would like the value of $\log_{10} 1000$. This is the exponent, $y$, of 10 such that $10^y$ equals 1000. The value of $y$ here is 3. Thus, $\log_{10} 1000 = 3$.

*Common logarithms* are logarithms in which the base is 10. The common logarithm of a number $x$ is often denoted simply as $\log x$. It is easy to see how to obtain the common logarithms of 10, 100, 1000, and so forth. But logarithms are defined for all positive numbers, not just the powers of ten. In general, the exponents, or values, of the logarithm will be decimal numbers. To understand the meaning of a decimal exponent, consider $10^{0.400}$. This is equivalent to $10^{400/1000} = 10^{2/5} = \sqrt[5]{10^2} = 2.51$. Therefore, log 2.51 = 0.400. Any decimal exponent is essentially a fraction, $p/r$, so by evaluating the expressions $10^{p/r} = \sqrt[r]{10^p}$ one could construct a table of logarithms. In practice, power series or other methods are used.

The following are fundamental properties of all logarithms:

$$\log_a 1 = 0 \tag{1}$$

$$\log_a(A \times B) = \log_a A + \log_a B \tag{2}$$

$$\log_a \frac{A}{B} = \log_a A - \log_a B \tag{3}$$

$$\log_a A^p = p \log_a A \tag{4}$$

$$\log_a \sqrt[r]{A} = \frac{1}{r} \log_a A \tag{5}$$

These properties are very useful in working with logarithms.

Electronic calculators that evaluate logarithms are now available for about $10 or so. Their simplicity of operation makes them well worth the price. To obtain the logarithm of a number, you enter the number and press the LOG key.

---

### EXERCISE 6

Find the values of (a) log 0.00582; (b) log 689.

## Antilogarithm

The antilogarithm (abbreviated antilog) is the inverse of the common logarithm. Antilog $x$ is simply $10^x$. If your electronic calculator has a $10^x$ key, you obtain the antilogarithm of a number by entering the number and pressing the $10^x$ key. (It may be necessary to press an inverse key before pressing a $10^x$/LOG key.) If your calculator has a $y^x$ (or $a^x$) key, you enter 10, press $y^x$, enter $x$, then press the = key.

---

### EXERCISE 7

Evaluate (a) antilog 5.728; (b) antilog $(-5.728)$.

## Natural Logarithms and Natural Antilogarithms

The mathematical constant $e = 2.71828$, like $\pi$, occurs in many scientific and engineering problems. It is frequently seen in the natural exponential function $y = e^x$. The inverse function is called the *natural logarithm*, $x = \ln y$, where $\ln y$ is simplified notation for $\log_e y$.

Finding the natural logarithm with a calculator is analogous to finding the common logarithm, only in this case you enter the number and press the LN key. For example, you may obtain the natural logarithm of 6.2 by entering 6.2 and pressing the LN key (ln 6.2 = 1.82). The natural antilogarithm of a number is the constant $e$ raised to a power equal to the number. Therefore, the expression for natural antilogarithm of 1.34 is written as

$$\text{natural antilog } 1.34 = e^{1.34}$$

In order to evaluate the expression $e^{1.34}$ on your calculator, you can use the $e^x$ key, or you may have to use the INV key in combination with the LN key ($e^{1.34} = 3.8$). 

There are instances where you may want to relate the natural logarithm to the common logarithm; the expression is

$$\ln x = 2.303 \log x$$

---

**EXERCISE 8**

Evaluate (a) $\ln 9.93$; (b) $e^{1.10}$.

---

## A.3  Algebraic Operations and Graphing

Often you are given an algebraic formula that you need to rearrange to solve for a particular quantity. As an example, suppose you would like to solve the following equation for $V$:

$$PV = nRT$$

You can eliminate $P$ from the left-hand side by dividing by $P$. But to maintain the equality, you must perform the same operation on both sides of the equation:

$$\frac{PV}{P} = \frac{nRT}{P}$$

or

$$V = \frac{nRT}{P}$$

### Quadratic Formula

A quadratic equation is one involving only powers of $x$ in which the highest power is two. The general form of the equation can be written

$$ax^2 + bx + c = 0$$

where $a$, $b$, and $c$ are constants. For given values of these constants, only certain values of $x$ are possible. (In general, there will be two values.) These values of $x$ are said to be the solutions of the equation.

These solutions are given by the *quadratic formula:*

$$x = \frac{-b \pm \sqrt{b^2 - 4ac}}{2a}$$

In this formula, the symbol $\pm$ means that there are two possible values of $x$—one obtained by taking the positive sign, the other by taking the negative sign.

---

**EXAMPLE 6**

### Obtaining the Solutions of a Quadratic Equation

Obtain the solutions of the following quadratic equation:

$$2.00x^2 - 1.72x - 2.86 = 0$$

**SOLUTION**

Using the quadratic formula, you substitute $a = 2.00$, $b = -1.72$, and $c = -2.86$. You get

$$x = \frac{1.72 \pm \sqrt{(-1.72)^2 - 4 \times 2.00 \times (-2.86)}}{2 \times 2.00} = \frac{1.72 \pm 5.08}{4.00} = \textbf{-0.840 and +1.70}$$

Mathematically there are two solutions, but in any real problem one solution may not be allowed. For example, if the solution is some physical quantity that can have only positive values, a negative solution must be rejected.

**EXERCISE 9**

Find the positive solution (or solutions) to the following equation:

$$1.80x^2 + 0.850x - 9.50 = 0$$

## The Straight-Line Graph

A graph is a visual means of representing a mathematical relationship or physical data. Consider the following data in which values of $y$ from some experiment are given for four values of $x$.

| $x$ | $y$ |
|-----|-----|
| 1 | $-1$ |
| 2 | 1 |
| 3 | 3 |
| 4 | 5 |

By plotting these $x$, $y$ points on a graph (Figure A.1), you see that they fall on a straight line. This suggests (but does not prove) that other points from this type of

**Figure A.1**
A straight-line plot of some data.

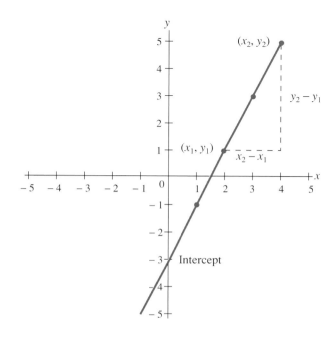

experiment might fall on the same line. It would be useful to have the mathematical equation for this line.

The general equation form of a straight line is

$$y = mx + b$$

The constant m is called the *slope* of the straight line. It is obtained by dividing the vertical distance between any two points on the line by the horizontal distance. If the two points are $(x_1, y_1)$ and $(x_2, y_2)$, the slope is given by the following formula:

$$\text{slope} = \frac{y_2 - y_1}{x_2 - x_1}$$

Suppose you choose the points (2, 1) and (4, 5) from the data. Then

$$\text{slope} = \frac{5 - 1}{4 - 2} = 2$$

For the straight line in Figure A.1, $m = 2$.

The constant b is called the *intercept*. It is the value of y at $x = 0$. From Figure A.1, you see that the intercept is $-3$. Therefore, $b = -3$. Hence, the equation of the straight line is

$$y = 2x - 3$$

## Appendix B    Vapor Pressure of Water at Various Temperatures

| Temperature (°C) | Pressure (mmHg) | Temperature (°C) | Pressure (mmHg) |
|---|---|---|---|
| 0 | 4.6 | 27 | 26.7 |
| 5 | 6.5 | 28 | 28.3 |
| 10 | 9.2 | 29 | 30.0 |
| 11 | 9.8 | 30 | 31.8 |
| 12 | 10.5 | 35 | 42.2 |
| 13 | 11.2 | 40 | 55.3 |
| 14 | 12.0 | 45 | 71.9 |
| 15 | 12.8 | 50 | 92.5 |
| 16 | 13.6 | 55 | 118.0 |
| 17 | 14.5 | 60 | 149.4 |
| 18 | 15.5 | 65 | 187.5 |
| 19 | 16.5 | 70 | 233.7 |
| 20 | 17.5 | 75 | 289.1 |
| 21 | 18.7 | 80 | 355.1 |
| 22 | 19.8 | 85 | 433.6 |
| 23 | 21.1 | 90 | 525.8 |
| 24 | 22.4 | 95 | 633.9 |
| 25 | 23.8 | 100 | 760.0 |
| 26 | 25.2 | 105 | 906.1 |

# Appendix C

# Thermodynamic Quantities for Substances and Ions at 25°C

| Substance or Ion | $\Delta H^\circ_f$ (kJ/mol) | $\Delta G^\circ_f$ (kJ/mol) | $S^\circ$ (J/mol · K) |
|---|---|---|---|
| $e^-(g)$ | 0 | 0 | 20.87 |
| $H^+(g)$ | 1536.3 | 1517.1 | 108.83 |
| $H^+(aq)$ | 0 | 0 | 0 |
| $H(g)$ | 218.0 | 203.30 | 114.60 |
| $H_2(g)$ | 0 | 0 | 130.6 |
| **Group IA** | | | |
| $Li^+(g)$ | 687.163 | 649.989 | 132.91 |
| $Li^+(aq)$ | −278.46 | −293.8 | 14 |
| $Li(g)$ | 161 | 128 | 138.67 |
| $Li(s)$ | 0 | 0 | 29.10 |
| $LiF(s)$ | −616.9 | −588.7 | 35.66 |
| $LiCl(s)$ | −408 | −384 | 59.30 |
| $LiBr(s)$ | −351 | −342 | 74.1 |
| $LiI(s)$ | −270 | −270 | 85.8 |
| $Na^+(g)$ | 609.839 | 574.877 | 147.85 |
| $Na^+(aq)$ | −239.66 | −261.87 | 60.2 |
| $Na(g)$ | 107.76 | 77.299 | 153.61 |
| $Na(s)$ | 0 | 0 | 51.446 |
| $NaF(s)$ | −575.4 | −545.1 | 51.21 |
| $NaCl(s)$ | −411.1 | −384.0 | 72.12 |
| $NaBr(s)$ | −361 | −349 | 86.82 |
| $NaI(s)$ | −288 | −285 | 98.5 |
| $NaHCO_3(s)$ | −947.7 | −851.9 | 102 |
| $Na_2CO_3(s)$ | −1130.8 | −1048.1 | 139 |
| $K^+(g)$ | 514.197 | 481.202 | 154.47 |
| $K^+(aq)$ | −251.2 | −282.28 | 103 |
| $K(g)$ | 89.2 | 60.7 | 160.23 |
| $K(s)$ | 0 | 0 | 64.672 |
| $KF(s)$ | −568.6 | −538.9 | 66.55 |
| $KCl(s)$ | −436.68 | −408.8 | 82.55 |
| $KBr(s)$ | −394 | −380 | 95.94 |
| $KI(s)$ | −328 | −323 | 106.39 |
| $Rb^+(g)$ | 495.04 | | |
| $Rb^+(aq)$ | −246 | −282.2 | 124 |
| $Rb(g)$ | 85.81 | 55.86 | 169.99 |
| $Rb(s)$ | 0 | 0 | 69.5 |
| $RbF(s)$ | −549.28 | | |
| $RbCl(s)$ | −430.58 | | |
| $RbBr(s)$ | −389.2 | −378.1 | 108.3 |

| Substance or Ion | $\Delta H_f^\circ$ (kJ/mol) | $\Delta G_f^\circ$ (kJ/mol) | $S^\circ$ (J/mol · K) |
|---|---|---|---|
| RbI(s) | −328 | −326 | 118.0 |
| $Cs^+(g)$ | 458.5 | 427.1 | 169.72 |
| $Cs^+(aq)$ | −248 | −282.0 | 133 |
| Cs(g) | 76.7 | 49.7 | 175.5 |
| Cs(s) | 0 | 0 | 85.15 |
| CsF(s) | −554.7 | −525.4 | 88 |
| CsCl(s) | −442.8 | −414 | 101.18 |
| CsBr(s) | −395 | −383 | 121 |
| CsI(s) | −337 | −333 | 130 |
| **Group IIA** | | | |
| $Mg^{2+}(g)$ | 2351 | | |
| $Mg^{2+}(aq)$ | −461.96 | −456.01 | −118 |
| $Mg^+(g)$ | 894.1 | | |
| Mg(g) | 150 | 115 | 148.55 |
| Mg(s) | 0 | 0 | 32.69 |
| $MgCl_2(s)$ | −641.6 | −592.1 | 89.630 |
| MgO(s) | −601.2 | −569.0 | 26.9 |
| $Mg_3N_2(s)$ | −461 | −401 | 88 |
| $MgCO_3(s)$ | −1112 | −1028 | 65.86 |
| $Ca^{2+}(g)$ | 1934.1 | | |
| $Ca^{2+}(aq)$ | −542.96 | −553.04 | −55.2 |
| $Ca^+(g)$ | 788.6 | | |
| Ca(g) | 192.6 | 158.9 | 154.78 |
| Ca(s) | 0 | 0 | 41.6 |
| $CaF_2(s)$ | −1215 | −1162 | 68.87 |
| $CaCl_2(s)$ | −795.0 | −750.2 | 114 |
| CaO(s) | −635.1 | −603.5 | 38.2 |
| $CaCO_3(s)$ | −1206.9 | −1128.8 | 92.9 |
| $CaSO_4(s)$ | −1432.7 | −1320.3 | 107 |
| $Ca_3(PO_4)_2(s)$ | −4138 | −3899 | 263 |
| $Sr^{2+}(g)$ | 1784 | | |
| $Sr^{2+}(aq)$ | −545.51 | −557.3 | −39 |
| $Sr^+(g)$ | 719.6 | | |
| Sr(g) | 164 | 110 | 164.54 |
| Sr(s) | 0 | 0 | 54.4 |
| $SrCl_2(s)$ | −828.4 | −781.2 | 117 |
| SrO(s) | −592.0 | −562.4 | 55.5 |
| $SrCO_3(s)$ | −1218 | −1138 | 97.1 |
| $SrSO_4(s)$ | −1445 | −1334 | 122 |
| $Ba^{2+}(g)$ | 1649.9 | | |
| $Ba^{2+}(aq)$ | −538.36 | −560.7 | 13 |
| $Ba^+(g)$ | 684.6 | | |

| Substance or Ion | $\Delta H_f^\circ$ (kJ/mol) | $\Delta G_f^\circ$ (kJ/mol) | $S^\circ$ (J/mol · K) |
|---|---|---|---|
| $Ba(g)$ | 175.6 | 144.8 | 170.28 |
| $Ba(s)$ | 0 | 0 | 62.5 |
| $BaCl_2(s)$ | −806.06 | −810.9 | 126 |
| $BaO(s)$ | −548.1 | −520.4 | 72.07 |
| $BaCO_3(s)$ | −1219 | −1139 | 112 |
| $BaSO_4(s)$ | −1465 | −1353 | 132 |
| **Group IIIA** | | | |
| $B(\beta\text{-rhombohedral})$ | 0 | 0 | 5.87 |
| $B_2O_3(s)$ | −1272 | −1193 | 53.8 |
| $Al(s)$ | 0 | 0 | 28.3 |
| $Al^{3+}(aq)$ | −524.7 | −481.2 | −313 |
| $Al_2O_3(s)$ | −1676 | −1582 | 50.94 |
| **Group IVA** | | | |
| $C(g)$ | 715.0 | 669.6 | 158.0 |
| $C(\text{graphite})$ | 0 | 0 | 5.686 |
| $C(\text{diamond})$ | 1.896 | 2.866 | 2.439 |
| $CO(g)$ | −110.5 | −137.2 | 197.5 |
| $CO_2(g)$ | −393.5 | −394.4 | 213.7 |
| $CO_2(aq)$ | −412.9 | −386.2 | 121 |
| $CO_3^{2-}(aq)$ | −676.26 | −528.10 | −53.1 |
| $HCO_3^{-}(aq)$ | −691.11 | −587.06 | 95.0 |
| $H_2CO_3(aq)$ | −698.7 | −623.42 | 191 |
| $CH_4(g)$ | −74.87 | −50.81 | 186.1 |
| $C_2H_2(g)$ | 227 | 209 | 200.85 |
| $C_2H_4(g)$ | 52.47 | 68.36 | 219.22 |
| $C_2H_6(g)$ | −84.667 | −32.89 | 229.5 |
| $C_6H_6(l)$ | 49.0 | 124.5 | 172.8 |
| $CH_3OH(g)$ | −201.2 | −161.9 | 238 |
| $CH_3OH(l)$ | −238.6 | −166.2 | 127 |
| $HCHO(g)$ | −116 | −110 | 219 |
| $HCOO^{-}(aq)$ | −410 | −335 | 91.6 |
| $HCOOH(l)$ | −409 | −346 | 129.0 |
| $HCOOH(aq)$ | −410 | −356 | 164 |
| $C_2H_5OH(l)$ | −277.63 | −174.8 | 161 |
| $CH_3CHO(g)$ | −166 | −133.7 | 266 |
| $CH_3COOH(l)$ | −487.0 | −392 | 160 |
| $CN^{-}(aq)$ | 151 | 166 | 118 |
| $HCN(g)$ | 135 | 125 | 201.7 |
| $HCN(l)$ | 105 | 121 | 112.8 |
| $HCN(aq)$ | 105 | 112 | 129 |
| $CS_2(g)$ | 117 | 66.9 | 237.79 |
| $CS_2(l)$ | 87.9 | 63.6 | 151.0 |

| Substance or Ion | $\Delta H_f^\circ$ (kJ/mol) | $\Delta G_f^\circ$ (kJ/mol) | $S^\circ$ (J/mol · K) |
|---|---|---|---|
| $CH_3Cl(g)$ | $-83.7$ | $-60.2$ | 234 |
| $CH_2Cl_2(l)$ | $-117$ | $-63.2$ | 179 |
| $CHCl_3(l)$ | $-132$ | $-71.5$ | 203 |
| $CCl_4(g)$ | $-96.0$ | $-53.7$ | 309.7 |
| $CCl_4(l)$ | $-139$ | $-68.6$ | 214.4 |
| $COCl_2(g)$ | $-220$ | $-206$ | 283.74 |
| $Si(s)$ | 0 | 0 | 18.0 |
| $SiO_2(s)$ | $-910.9$ | $-856.5$ | 41.5 |
| $Sn(gray)$ | 3 | 4.6 | 44.8 |
| $Sn(white)$ | 0 | 0 | 51.5 |
| $SnCl_4(l)$ | $-545.2$ | $-474.0$ | 259 |
| $Pb^{2+}(aq)$ | 1.6 | $-24.3$ | 21 |
| $Pb(s)$ | 0 | 0 | 64.785 |
| $PbO(s, red)$ | $-219.0$ | $-189.2$ | 66.5 |
| $PbO_2(s)$ | $-276.6$ | $-219.0$ | 76.6 |
| $PbS(s)$ | $-98.3$ | $-96.7$ | 91.3 |
| $PbCl_2(s)$ | $-359$ | $-314$ | 136 |
| $PbSO_4(s)$ | $-918.39$ | $-811.24$ | 147 |
| **Group VA** | | | |
| $N(g)$ | 473 | 456 | 153.2 |
| $N_2(g)$ | 0 | 0 | 191.5 |
| $NO(g)$ | 90.29 | 86.60 | 210.65 |
| $NO_2(g)$ | 33.2 | 51 | 239.9 |
| $N_2O_4(g)$ | 9.16 | 97.7 | 304.3 |
| $N_2O_5(g)$ | 11 | 118 | 346 |
| $NH_3(g)$ | $-45.9$ | $-16$ | 193 |
| $NH_3(aq)$ | $-80.83$ | 26.7 | 110 |
| $NO_3^-(aq)$ | $-206.57$ | $-110.5$ | 146 |
| $HNO_3(l)$ | $-173.23$ | $-79.914$ | 155.6 |
| $HNO_3(aq)$ | $-206.57$ | $-110.5$ | 146 |
| $NH_4NO_3(s)$ | $-365.6$ | $-184.0$ | 151.1 |
| $P(g)$ | 333.9 | 292.0 | 163.1 |
| $P(red)$ | 0 | 0 | 22.8 |
| $P_4(white)$ | 69.8 | 48.0 | 164 |
| $P_2(g)$ | 146.2 | 103.8 | 218.0 |
| $P_4(g)$ | 128.9 | 72.4 | 128.9 |
| $PCl_3(g)$ | $-287$ | $-267.8$ | 311.7 |
| $PCl_5(g)$ | $-343$ | $-278$ | 364.5 |
| $P_4O_{10}(s)$ | $-2940$ | $-2675$ | 228.9 |
| $PO_4^{3-}(aq)$ | $-1277$ | $-1018$ | $-222$ |
| $HPO_4^{2-}(aq)$ | $-1292.1$ | $-1089.3$ | $-33$ |
| $H_2PO_4^-(aq)$ | $-1296.3$ | $-1130.4$ | 90.4 |
| $H_3PO_4(aq)$ | $-1288.3$ | $-1142.7$ | 158 |

| Substance or Ion | $\Delta H_f^\circ$ (kJ/mol) | $\Delta G_f^\circ$ (kJ/mol) | $S^\circ$ (J/mol · K) |
|---|---|---|---|
| **Group VIA** | | | |
| O(g) | 249.2 | 231.7 | 160.95 |
| $O_2(g)$ | 0 | 0 | 205.0 |
| $O_3(g)$ | 143 | 163 | 238.82 |
| $OH^-(aq)$ | −229.94 | −157.30 | −10.54 |
| $H_2O(g)$ | −241.826 | −228.60 | 188.72 |
| $H_2O(l)$ | −285.840 | −237.192 | 69.940 |
| $H_2O_2(l)$ | −187.8 | −120.4 | 110 |
| $H_2O_2(aq)$ | −191.2 | −134.1 | 144 |
| S(g) | 279 | 239 | 168 |
| $S_2(g)$ | 129 | 80.1 | 228.1 |
| $S_8(g)$ | 101 | 49.1 | 430.211 |
| S(rhombic) | 0 | 0 | 31.9 |
| S(monoclinic) | 0.30 | 0.096 | 32.6 |
| $S^{2-}(aq)$ | 41.8 | 83.7 | 22 |
| $HS^-(aq)$ | −17.7 | 12.6 | 61.1 |
| $H_2S(g)$ | −20.2 | −33 | 205.6 |
| $H_2S(aq)$ | −39 | −27.4 | 122 |
| $SO_2(g)$ | −296.8 | −300.2 | 248.1 |
| $SO_3(g)$ | −396 | −371 | 256.66 |
| $SO_4^{2-}(aq)$ | −907.51 | −741.99 | 17 |
| $HSO_4^-(aq)$ | −885.75 | −752.87 | 126.9 |
| $H_2SO_4(l)$ | −813.989 | −690.059 | 156.90 |
| $H_2SO_4(aq)$ | −907.51 | −741.99 | 17 |
| **Group VIIA** | | | |
| F(g) | 78.9 | 61.8 | 158.64 |
| $F^-(g)$ | −255.6 | −262.5 | 145.47 |
| $F^-(aq)$ | −329.1 | −276.5 | −9.6 |
| $F_2(g)$ | 0 | 0 | 202.7 |
| HF(g) | −273 | −275 | 173.67 |
| Cl(g) | 121.0 | 105.0 | 165.1 |
| $Cl^-(g)$ | −234 | −240 | 153.25 |
| $Cl^-(aq)$ | −167.46 | −131.17 | 55.10 |
| $Cl_2(g)$ | 0 | 0 | 223.0 |
| HCl(g) | −92.31 | −95.30 | 186.79 |
| HCl(aq) | −167.46 | −131.17 | 55.06 |
| Br(g) | 111.9 | 82.40 | 174.90 |
| $Br^-(g)$ | −218.9 | | |
| $Br^-(aq)$ | −120.9 | −102.82 | 80.71 |
| $Br_2(g)$ | 30.91 | 3.13 | 245.38 |
| $Br_2(l)$ | 0 | 0 | 152.23 |
| HBr(g) | −36 | −53.5 | 198.59 |
| I(g) | 106.8 | 70.21 | 180.67 |

| Substance or Ion | $\Delta H_f^\circ$ (kJ/mol) | $\Delta G_f^\circ$ (kJ/mol) | $S^\circ$ (J/mol · K) |
|---|---|---|---|
| $I^-(g)$ | $-194.7$ | | |
| $I^-(aq)$ | $-55.94$ | $-51.67$ | 109.4 |
| $I_2(g)$ | 62.442 | 19.38 | 260.58 |
| $I_2(s)$ | 0 | 0 | 116.14 |
| $HI(g)$ | 25.9 | 1.3 | 206.33 |
| **Group IB** | | | |
| $Cu^+(aq)$ | 51.9 | 50.2 | $-26$ |
| $Cu^{2+}(aq)$ | 64.39 | 64.98 | $-98.7$ |
| $Cu(g)$ | 341.1 | 301.4 | 166.29 |
| $Cu(s)$ | 0 | 0 | 33.1 |
| $Ag^+(aq)$ | 105.9 | 77.111 | 73.93 |
| $Ag(g)$ | 289.2 | 250.4 | 172.892 |
| $Ag(s)$ | 0 | 0 | 42.702 |
| $AgF(s)$ | $-203$ | $-185$ | 84 |
| $AgCl(s)$ | $-127.03$ | $-109.72$ | 96.11 |
| $AgBr(s)$ | $-99.50$ | $-95.939$ | 107.1 |
| $AgI(s)$ | $-62.38$ | $-66.32$ | 114 |
| $Ag_2S(s)$ | $-31.8$ | $-40.3$ | 146 |
| **Group IIB** | | | |
| $Zn^{2+}(aq)$ | $-152.4$ | $-147.21$ | $-106.5$ |
| $Zn(g)$ | 130.5 | 94.93 | 160.9 |
| $Zn(s)$ | 0 | 0 | 41.6 |
| $ZnO(s)$ | $-348.0$ | $-318.2$ | 43.9 |
| $ZnS(s$, zinc blende$)$ | $-203$ | $-198$ | 57.7 |
| $Cd^{2+}(aq)$ | $-72.38$ | $-77.74$ | $-61.1$ |
| $Cd(g)$ | 112.8 | 78.20 | 167.64 |
| $Cd(s)$ | 0 | 0 | 51.5 |
| $CdS(s)$ | $-144$ | $-141$ | 71 |
| $Hg^{2+}(aq)$ | | 164.8 | |
| $Hg_2^{2+}(aq)$ | | 153.9 | |
| $Hg(g)$ | 61.30 | 31.8 | 174.87 |
| $Hg(l)$ | 0 | 0 | 76.027 |
| $HgCl_2(s)$ | $-230$ | $-184$ | 144 |
| $Hg_2Cl_2(s)$ | $-264.9$ | $-210.66$ | 196 |
| $HgO(s)$ | $-90.79$ | $-58.50$ | 70.27 |
| **Group VIB** | | | |
| $[Cr(H_2O)_6]^{3+}(aq)$ | $-1971$ | | |
| $Cr(s)$ | 0 | 0 | 23.8 |
| $CrO_4^{2-}(aq)$ | $-863.2$ | $-706.3$ | 38 |
| $Cr_2O_7^{2-}(aq)$ | $-1461$ | $-1257$ | 214 |

| Substance or Ion | $\Delta H_f^\circ$ (kJ/mol) | $\Delta G_f^\circ$ (kJ/mol) | $S^\circ$ (J/mol · K) |
|---|---|---|---|
| **Group VIIB** | | | |
| $Mn^{2+}(aq)$ | −219 | −223 | −84 |
| $Mn(s)$ | 0 | 0 | 31.8 |
| $MnO_2(s)$ | −520.9 | −466.1 | 53.1 |
| $MnO_4^-(aq)$ | −518.4 | −425.1 | 190 |
| **Group VIIIB** | | | |
| $Fe^{3+}(aq)$ | −47.7 | −10.5 | −293 |
| $Fe^{2+}(aq)$ | −87.9 | −84.94 | 113 |
| $Fe(s)$ | 0 | 0 | 27.3 |
| $Fe(l)$ | 13.13 | 11.05 | 34.29 |
| $FeO(s)$ | −272.0 | −251.4 | 60.75 |
| $Fe_2O_3(s)$ | −825.5 | −743.6 | 87.400 |
| $Fe_3O_4(s)$ | −1121 | −1018 | 145.3 |
| $Co^{2+}(aq)$ | −67.4 | −51.5 | −155 |
| $Co(s)$ | 0 | 0 | 30 |
| $Ni^{2+}(aq)$ | −64.0 | −46.4 | −159 |
| $Ni(s)$ | 0 | 0 | 30.1 |

# Appendix D

# Electron Configurations of Atoms in the Ground State

| Z | Element | Configuration | Z | Element | Configuration |
|---|---|---|---|---|---|
| 1 | H | $1s^1$ | 20 | Ca | $[Ar]4s^2$ |
| 2 | He | $1s^2$ | 21 | Sc | $[Ar]3d^14s^2$ |
| 3 | Li | $[He]2s^1$ | 22 | Ti | $[Ar]3d^24s^2$ |
| 4 | Be | $[He]2s^2$ | 23 | V | $[Ar]3d^34s^2$ |
| 5 | B | $[He]2s^22p^1$ | 24 | Cr | $[Ar]3d^54s^1$ |
| 6 | C | $[He]2s^22p^2$ | 25 | Mn | $[Ar]3d^54s^2$ |
| 7 | N | $[He]2s^22p^3$ | 26 | Fe | $[Ar]3d^64s^2$ |
| 8 | O | $[He]2s^22p^4$ | 27 | Co | $[Ar]3d^74s^2$ |
| 9 | F | $[He]2s^22p^5$ | 28 | Ni | $[Ar]3d^84s^2$ |
| 10 | Ne | $[He]2s^22p^6$ | 29 | Cu | $[Ar]3d^{10}4s^1$ |
| 11 | Na | $[Ne]3s^1$ | 30 | Zn | $[Ar]3d^{10}4s^2$ |
| 12 | Mg | $[Ne]3s^2$ | 31 | Ga | $[Ar]3d^{10}4s^24p^1$ |
| 13 | Al | $[Ne]3s^23p^1$ | 32 | Ge | $[Ar]3d^{10}4s^24p^2$ |
| 14 | Si | $[Ne]3s^23p^2$ | 33 | As | $[Ar]3d^{10}4s^24p^3$ |
| 15 | P | $[Ne]3s^23p^3$ | 34 | Se | $[Ar]3d^{10}4s^24p^4$ |
| 16 | S | $[Ne]3s^23p^4$ | 35 | Br | $[Ar]3d^{10}4s^24p^5$ |
| 17 | Cl | $[Ne]3s^23p^5$ | 36 | Kr | $[Ar]3d^{10}4s^24p^6$ |
| 18 | Ar | $[Ne]3s^23p^6$ | 37 | Rb | $[Kr]5s^1$ |
| 19 | K | $[Ar]4s^1$ | 38 | Sr | $[Kr]5s^2$ |

| Z | Element | Configuration | Z | Element | Configuration |
|---|---------|---------------|---|---------|---------------|
| 39 | Y | $[Kr]4d^15s^2$ | 78 | Pt | $[Xe]4f^{14}5d^96s^1$ |
| 40 | Zr | $[Kr]4d^25s^2$ | 79 | Au | $[Xe]4f^{14}5d^{10}6s^1$ |
| 41 | Nb | $[Kr]4d^45s^1$ | 80 | Hg | $[Xe]4f^{14}5d^{10}6s^2$ |
| 42 | Mo | $[Kr]4d^55s^1$ | 81 | Tl | $[Xe]4f^{14}5d^{10}6s^26p^1$ |
| 43 | Tc | $[Kr]4d^55s^2$ | 82 | Pb | $[Xe]4f^{14}5d^{10}6s^26p^2$ |
| 44 | Ru | $[Kr]4d^75s^1$ | 83 | Bi | $[Xe]4f^{14}5d^{10}6s^26p^3$ |
| 45 | Rh | $[Kr]4d^85s^1$ | 84 | Po | $[Xe]4f^{14}5d^{10}6s^26p^4$ |
| 46 | Pd | $[Kr]4d^{10}$ | 85 | At | $[Xe](4f^{14}5d^{10}6s^26p^5)$ |
| 47 | Ag | $[Kr]4d^{10}5s^1$ | 86 | Rn | $[Xe]4f^{14}5d^{10}6s^26p^6$ |
| 48 | Cd | $[Kr]4d^{10}5s^2$ | 87 | Fr | $[Rn](7s^1)$ |
| 49 | In | $[Kr]4d^{10}5s^25p^1$ | 88 | Ra | $[Rn]7s^2$ |
| 50 | Sn | $[Kr]4d^{10}5s^25p^2$ | 89 | Ac | $[Rn]6d^16s^2$ |
| 51 | Sb | $[Kr]4d^{10}5s^25p^3$ | 90 | Th | $[Rn]6d^27s^2$ |
| 52 | Te | $[Kr]4d^{10}5s^25p^4$ | 91 | Pa | $[Rn](5f^26d^17s^2)$ |
| 53 | I | $[Kr]4d^{10}5s^25p^5$ | 92 | U | $[Rn](5f^36d^17s^2)$ |
| 54 | Xe | $[Kr]4d^{10}5s^25p^6$ | 93 | Np | $[Rn](5f^46d^17s^2)$ |
| 55 | Cs | $[Xe]6s^1$ | 94 | Pu | $[Rn]5f^67s^2$ |
| 56 | Ba | $[Xe]6s^2$ | 95 | Am | $[Rn]5f^77s^2$ |
| 57 | La | $[Xe]5d^16s^2$ | 96 | Cm | $[Rn](5f^76d^17s^2)$ |
| 58 | Ce | $[Xe]4f^15d^16s^2$ | 97 | Bk | $[Rn](5f^97s^2)$ |
| 59 | Pr | $[Xe]4f^36s^2$ | 98 | Cf | $[Rn](5f^{10}7s^2)$ |
| 60 | Nd | $[Xe]4f^46s^2$ | 99 | Es | $[Rn](5f^{11}7s^2)$ |
| 61 | Pm | $[Xe]4f^56s^2$ | 100 | Fm | $[Rn](5f^{12}7s^2)$ |
| 62 | Sm | $[Xe]4f^66s^2$ | 101 | Md | $[Rn](5f^{13}7s^2)$ |
| 63 | Eu | $[Xe]4f^76s^2$ | 102 | No | $[Rn](5f^{14}7s^2)$ |
| 64 | Gd | $[Xe]4f^75d^16s^2$ | 103 | Lr | $[Rn](5f^{14}6d^17s^2)$ |
| 65 | Tb | $[Xe]4f^96s^2$ | 104 | Rf | $[Rn](5f^{14}5d^27s^2)$ |
| 66 | Dy | $[Xe]4f^{10}6s^2$ | 105 | Db | $[Rn](5f^{14}6d^37s^2)$ |
| 67 | Ho | $[Xe]4f^{11}6s^2$ | 106 | Sg | $[Rn](5f^{14}6d^47s^2)$ |
| 68 | Er | $[Xe]4f^{12}6s^2$ | 107 | Bh | $[Rn](5f^{14}6d^57s^2)$ |
| 69 | Tm | $[Xe]4f^{13}6s^2$ | 108 | Hs | $[Rn](5f^{14}6d^67s^2)$ |
| 70 | Yb | $[Xe]4f^{14}6s^2$ | 109 | Mt | $[Rn](5f^{14}6d^77s^2)$ |
| 71 | Lu | $[Xe]4f^{14}5d^16s^2$ | (110) | Uun | $[Rn](5f^{14}6d^87s^2)$ |
| 72 | Hf | $[Xe]4f^{14}5d^26s^2$ | (111) | Uuu | $[Rn](5f^{14}6d^97s^2)$ |
| 73 | Ta | $[Xe]4f^{14}5d^36s^2$ | (112) | Uub | $[Rn](5f^{14}6d^{10}7s^2)$ |
| 74 | W | $[Xe]4f^{14}5d^46s^2$ | (114) | Uuq | $[Rn](5f^{14}6d^{10}7s^27p^2)$ |
| 75 | Re | $[Xe]4f^{14}5d^56s^2$ | (116) | Uuh | $[Rn](5f^{14}6d^{10}7s^27p^4)$ |
| 76 | Os | $[Xe]4f^{14}5d^66s^2$ | (118) | Uuo | $[Rn](5f^{14}6d^{10}7s^27p^6)$ |
| 77 | Ir | $[Xe]4f^{14}5d^76s^2$ | | | |

Configurations in this table, except those in parentheses, are taken from Charlotte E. Moore, *National Standard Reference Data Series*, National Bureau of Standards, 34 (September 1970). Those in parentheses were obtained on the basis of their assumed position in the periodic table.

# Appendix E

# Acid-Ionization Constants at 25°C

| Substance | Formula | $K_a$ |
|---|---|---|
| Acetic acid | $HC_2H_3O_2$ | $1.7 \times 10^{-5}$ |
| Arsenic acid* | $H_3AsO_4$ | $6.5 \times 10^{-3}$ |
| | $H_2AsO_4^-$ | $1.2 \times 10^{-7}$ |
| | $HAsO_4^{2-}$ | $3.2 \times 10^{-12}$ |
| Ascorbic acid* | $H_2C_6H_6O_6$ | $6.8 \times 10^{-5}$ |
| | $HC_6H_6O_6^-$ | $2.8 \times 10^{-12}$ |
| Benzoic acid | $HC_7H_5O_2$ | $6.3 \times 10^{-5}$ |
| Boric acid | $H_3BO_3$ | $5.9 \times 10^{-10}$ |
| Carbonic acid* | $H_2CO_3$ | $4.3 \times 10^{-7}$ |
| | $HCO_3^-$ | $4.8 \times 10^{-11}$ |
| Chromic acid* | $H_2CrO_4$ | $1.5 \times 10^{-1}$ |
| | $HCrO_4^-$ | $3.2 \times 10^{-7}$ |
| Cyanic acid | $HOCN$ | $3.5 \times 10^{-4}$ |
| Formic acid | $HCHO_2$ | $1.7 \times 10^{-4}$ |
| Hydrocyanic acid | $HCN$ | $4.9 \times 10^{-10}$ |
| Hydrofluoric acid | $HF$ | $6.8 \times 10^{-4}$ |
| Hydrogen peroxide | $H_2O_2$ | $1.8 \times 10^{-12}$ |
| Hydrogen sulfate ion | $HSO_4^-$ | $1.1 \times 10^{-2}$ |
| Hydrogen sulfide* | $H_2S$ | $8.9 \times 10^{-8}$ |
| | $HS^-$ | $1.2 \times 10^{-13}$ |
| Hypochlorous acid | $HClO$ | $3.5 \times 10^{-8}$ |
| Lactic acid | $HC_3H_5O_3$ | $1.3 \times 10^{-4}$ |
| Nitrous acid | $HNO_2$ | $4.5 \times 10^{-4}$ |
| Oxalic acid* | $H_2C_2O_4$ | $5.6 \times 10^{-2}$ |
| | $HC_2O_4^-$ | $5.1 \times 10^{-5}$ |
| Phenol | $C_6H_5OH$ | $1.1 \times 10^{-10}$ |
| Phosphoric acid* | $H_3PO_4$ | $6.9 \times 10^{-3}$ |
| | $H_2PO_4^-$ | $6.2 \times 10^{-8}$ |
| | $HPO_4^{2-}$ | $4.8 \times 10^{-13}$ |
| Phosphorous acid* | $H_2PHO_3$ | $1.6 \times 10^{-2}$ |
| | $HPHO_3^-$ | $7 \times 10^{-7}$ |
| Propionic acid | $HC_3H_5O_2$ | $1.3 \times 10^{-5}$ |
| Pyruvic acid | $HC_3H_3O_3$ | $1.4 \times 10^{-4}$ |
| Sulfurous acid* | $H_2SO_3$ | $1.3 \times 10^{-2}$ |
| | $HSO_3^-$ | $6.3 \times 10^{-8}$ |

*The ionization constants for polyprotic acids are for successive ionizations. For example, for $H_3PO_4$, the equilibrium is $H_3PO_4 \rightleftharpoons H^+ + H_2PO_4^-$. For $H_2PO_4^-$, the equilibrium is $H_2PO_4^- \rightleftharpoons H^+ + HPO_4^{2-}$.

## Appendix F    Base-Ionization Constants at 25°C

| Substance | Formula | $K_b$ |
|---|---|---|
| Ammonia | $NH_3$ | $1.8 \times 10^{-5}$ |
| Aniline | $C_6H_5NH_2$ | $4.2 \times 10^{-10}$ |
| Dimethylamine | $(CH_3)_2NH$ | $5.1 \times 10^{-4}$ |
| Ethylamine | $C_2H_5NH_2$ | $4.7 \times 10^{-4}$ |
| Ethylenediamine | $NH_2CH_2CH_2NH_2$ | $5.2 \times 10^{-4}$ |
| Hydrazine | $N_2H_4$ | $1.7 \times 10^{-6}$ |
| Hydroxylamine | $NH_2OH$ | $1.1 \times 10^{-8}$ |
| Methylamine | $CH_3NH_2$ | $4.4 \times 10^{-4}$ |
| Pyridine | $C_5H_5N$ | $1.4 \times 10^{-9}$ |
| Trimethylamine | $(CH_3)_3N$ | $6.5 \times 10^{-5}$ |
| Urea | $NH_2CONH_2$ | $1.5 \times 10^{-14}$ |

## Appendix G    Solubility Product Constants at 25°C

| Substance | Formula | $K_{sp}$ |
|---|---|---|
| Aluminum hydroxide | $Al(OH)_3$ | $4.6 \times 10^{-33}$ |
| Barium chromate | $BaCrO_4$ | $1.2 \times 10^{-10}$ |
| Barium fluoride | $BaF_2$ | $1.0 \times 10^{-6}$ |
| Barium sulfate | $BaSO_4$ | $1.1 \times 10^{-10}$ |
| Cadmium oxalate | $CdC_2O_4$ | $1.5 \times 10^{-8}$ |
| Cadmium sulfide | $CdS$ | $8 \times 10^{-27}$ |
| Calcium carbonate | $CaCO_3$ | $3.8 \times 10^{-9}$ |
| Calcium fluoride | $CaF_2$ | $3.4 \times 10^{-11}$ |
| Calcium oxalate | $CaC_2O_4$ | $2.3 \times 10^{-9}$ |
| Calcium phosphate | $Ca_3(PO_4)_2$ | $1 \times 10^{-26}$ |
| Calcium sulfate | $CaSO_4$ | $2.4 \times 10^{-5}$ |
| Cobalt(II) sulfide | $CoS$ | $4 \times 10^{-21}$ |
| Copper(II) hydroxide | $Cu(OH)_2$ | $2.6 \times 10^{-19}$ |
| Copper(II) sulfide | $CuS$ | $6 \times 10^{-36}$ |
| Iron(II) hydroxide | $Fe(OH)_2$ | $8 \times 10^{-16}$ |
| Iron(II) sulfide | $FeS$ | $6 \times 10^{-18}$ |
| Iron(III) hydroxide | $Fe(OH)_3$ | $2.5 \times 10^{-39}$ |
| Lead(II) arsenate | $Pb_3(AsO_4)_2$ | $4 \times 10^{-36}$ |
| Lead(II) chloride | $PbCl_2$ | $1.6 \times 10^{-5}$ |
| Lead(II) chromate | $PbCrO_4$ | $1.8 \times 10^{-14}$ |
| Lead(II) iodide | $PbI_2$ | $6.5 \times 10^{-9}$ |
| Lead(II) sulfate | $PbSO_4$ | $1.7 \times 10^{-8}$ |
| Lead(II) sulfide | $PbS$ | $2.5 \times 10^{-27}$ |

| | | |
|---|---|---|
| Magnesium arsenate | $Mg_3(AsO_4)_2$ | $2 \times 10^{-20}$ |
| Magnesium carbonate | $MgCO_3$ | $1.0 \times 10^{-5}$ |
| Magnesium hydroxide | $Mg(OH)_2$ | $1.8 \times 10^{-11}$ |
| Magnesium oxalate | $MgC_2O_4$ | $8.5 \times 10^{-5}$ |
| Manganese(II) sulfide | $MnS$ | $2.5 \times 10^{-10}$ |
| Mercury(I) chloride | $Hg_2Cl_2$ | $1.3 \times 10^{-18}$ |
| Mercury(II) sulfide | $HgS$ | $1.6 \times 10^{-52}$ |
| Nickel(II) hydroxide | $Ni(OH)_2$ | $2.0 \times 10^{-15}$ |
| Nickel(II) sulfide | $NiS$ | $3 \times 10^{-19}$ |
| Silver acetate | $AgC_2H_3O_2$ | $2.0 \times 10^{-3}$ |
| Silver bromide | $AgBr$ | $5.0 \times 10^{-13}$ |
| Silver chloride | $AgCl$ | $1.8 \times 10^{-10}$ |
| Silver chromate | $Ag_2CrO_4$ | $1.1 \times 10^{-12}$ |
| Silver iodide | $AgI$ | $8.3 \times 10^{-17}$ |
| Silver sulfide | $Ag_2S$ | $6 \times 10^{-50}$ |
| Strontium carbonate | $SrCO_3$ | $9.3 \times 10^{-10}$ |
| Strontium chromate | $SrCrO_4$ | $3.5 \times 10^{-5}$ |
| Strontium sulfate | $SrSO_4$ | $2.5 \times 10^{-7}$ |
| Zinc hydroxide | $Zn(OH)_2$ | $2.1 \times 10^{-16}$ |
| Zinc sulfide | $ZnS$ | $1.1 \times 10^{-21}$ |

# Appendix H

# Formation Constants of Complex Ions at 25°C

| Complex Ion | $K_f$ |
|---|---|
| $Ag(CN)_2^-$ | $5.6 \times 10^{18}$ |
| $Ag(NH_3)_2^+$ | $1.7 \times 10^7$ |
| $Ag(S_2O_3)_2^{3-}$ | $2.9 \times 10^{13}$ |
| $Cd(NH_3)_4^{2+}$ | $1.0 \times 10^7$ |
| $Cu(CN)_2^-$ | $1.0 \times 10^{16}$ |
| $Cu(NH_3)_4^{2+}$ | $4.8 \times 10^{12}$ |
| $Fe(CN)_6^{4-}$ | $1.0 \times 10^{35}$ |
| $Fe(CN)_6^{3-}$ | $9.1 \times 10^{41}$ |
| $Ni(CN)_4^{2-}$ | $1.0 \times 10^{31}$ |
| $Ni(NH_3)_6^{2+}$ | $5.6 \times 10^8$ |
| $Zn(NH_3)_4^{2+}$ | $2.9 \times 10^9$ |
| $Zn(OH)_4^{2-}$ | $2.8 \times 10^{15}$ |

# Appendix I

# Standard Electrode (Reduction) Potentials in Aqueous Solution at 25°C

| Cathode (Reduction) Half-Reaction | Standard Potential, $E°$ (V) |
|---|---|
| $Li^+(aq) + e^- \rightleftharpoons Li(s)$ | $-3.04$ |
| $K^+(aq) + e^- \rightleftharpoons K(s)$ | $-2.92$ |
| $Ca^{2+}(aq) + 2e^- \rightleftharpoons Ca(s)$ | $-2.76$ |
| $Na^+(aq) + e^- \rightleftharpoons Na(s)$ | $-2.71$ |
| $Mg^{2+}(aq) + 2e^- \rightleftharpoons Mg(s)$ | $-2.38$ |
| $Al^{3+}(aq) + 3e^- \rightleftharpoons Al(s)$ | $-1.66$ |
| $2H_2O(l) + 2e^- \rightleftharpoons H_2(g) + 2OH^-(aq)$ | $-0.83$ |
| $Zn^{2+}(aq) + 2e^- \rightleftharpoons Zn(s)$ | $-0.76$ |
| $Cr^{3+}(aq) + 3e^- \rightleftharpoons Cr(s)$ | $-0.74$ |
| $Fe^{2+}(aq) + 2e^- \rightleftharpoons Fe(s)$ | $-0.41$ |
| $Cd^{2+}(aq) + 2e^- \rightleftharpoons Cd(s)$ | $-0.40$ |
| $Ni^{2+}(aq) + 2e^- \rightleftharpoons Ni(s)$ | $-0.23$ |
| $Sn^{2+}(aq) + 2e^- \rightleftharpoons Sn(s)$ | $-0.14$ |
| $Pb^{2+}(aq) + 2e^- \rightleftharpoons Pb(s)$ | $-0.13$ |
| $Fe^{3+}(aq) + 3e^- \rightleftharpoons Fe(s)$ | $-0.04$ |
| $2H^+(aq) + 2e^- \rightleftharpoons H_2(g)$ | $0.00$ |
| $Sn^{4+}(aq) + 2e^- \rightleftharpoons Sn^{2+}(aq)$ | $0.15$ |
| $Cu^{2+}(aq) + e^- \rightleftharpoons Cu^+(aq)$ | $0.16$ |
| $ClO_4^-(aq) + H_2O(l) + 2e^- \rightleftharpoons ClO_3^-(aq) + 2OH^-(aq)$ | $0.17$ |
| $AgCl(s) + e^- \rightleftharpoons Ag(s) + Cl^-(aq)$ | $0.22$ |
| $Cu^{2+}(aq) + 2e^- \rightleftharpoons Cu(s)$ | $0.34$ |
| $ClO_3^-(aq) + H_2O(l) + 2e^- \rightleftharpoons ClO_2^-(aq) + 2OH^-(aq)$ | $0.35$ |
| $IO^-(aq) + H_2O(l) + 2e^- \rightleftharpoons I^-(aq) + 2OH^-(aq)$ | $0.49$ |
| $Cu^+(aq) + e^- \rightleftharpoons Cu(s)$ | $0.52$ |
| $I_2(s) + 2e^- \rightleftharpoons 2I^-(aq)$ | $0.54$ |
| $ClO_2^-(aq) + H_2O(l) + 2e^- \rightleftharpoons ClO^-(aq) + 2OH^-(aq)$ | $0.59$ |
| $Fe^{3+}(aq) + e^- \rightleftharpoons Fe^{2+}(aq)$ | $0.77$ |
| $Hg_2^{2+}(aq) + 2e^- \rightleftharpoons 2Hg(l)$ | $0.80$ |
| $Ag^+(aq) + e^- \rightleftharpoons Ag(s)$ | $0.80$ |
| $Hg^{2+}(aq) + 2e^- \rightleftharpoons Hg(l)$ | $0.85$ |
| $ClO^-(aq) + H_2O(l) + 2e^- \rightleftharpoons Cl^-(aq) + 2OH^-(aq)$ | $0.90$ |
| $2Hg^{2+}(aq) + 2e^- \rightleftharpoons Hg_2^{2+}(aq)$ | $0.90$ |
| $NO_3^-(aq) + 4H^+(aq) + 3e^- \rightleftharpoons NO(g) + 2H_2O(l)$ | $0.96$ |
| $Br_2(l) + 2e^- \rightleftharpoons 2Br^-(aq)$ | $1.07$ |
| $O_2(g) + 4H^+(aq) + 4e^- \rightleftharpoons 2H_2O(l)$ | $1.23$ |
| $Cr_2O_7^{2-}(aq) + 14H^+(aq) + 6e^- \rightleftharpoons 2Cr^{3+}(aq) + 7H_2O(l)$ | $1.33$ |
| $Cl_2(g) + 2e^- \rightleftharpoons 2Cl^-(aq)$ | $1.36$ |
| $Ce^{4+}(aq) + e^- \rightleftharpoons Ce^{3+}(aq)$ | $1.44$ |

$$MnO_4^-(aq) + 8H^+(aq) + 5e^- \rightleftharpoons Mn^{2+}(aq) + 4H_2O(l) \qquad 1.49$$

$$H_2O_2(aq) + 2H^+(aq) + 2e^- \rightleftharpoons 2H_2O(l) \qquad 1.78$$

$$Co^{3+}(aq) + e^- \rightleftharpoons Co^{2+}(aq) \qquad 1.82$$

$$S_2O_8^{2-}(aq) + 2e^- \rightleftharpoons 2SO_4^{2-}(aq) \qquad 2.01$$

$$O_3(g) + 2H^+(aq) + 2e^- \rightleftharpoons O_2(g) + H_2O(l) \qquad 2.07$$

$$F_2(g) + 2e^- \rightleftharpoons 2F^-(aq) \qquad 2.87$$

# Answers to Exercises

**NOTE:** Your answers may differ from those given here in the last significant figure.

## Chapter 1

**1.1** 11.02 g    **1.2** Physical properties: soft; silver color; melts at 64°C. Chemical properties: metal, reacts with water; reacts with oxygen; reacts with chlorine.    **1.3** (a) 4.9    (b) 2.48    (c) 0.08    (d) 3    **1.4** (a) 1.84 nm    (b) 5.67 ps    (c) 7.85 mg    (d) 9.7 km    (e) 0.732 ms or 732 $\mu$s    (f) 0.154 nm or 154 pm    **1.5** (a) 39.2°C    (b) 195 K    **1.6** 7.87 g/cm$^3$; the object is made of iron    **1.7** 38.4 cm$^3$    **1.8** $1.21 \times 10^{-7}$ mm    **1.9** $6.76 \times 10^{-26}$ dm$^3$    **1.10** 3.24 m

## Chapter 2

**2.1** $^{35}_{17}$Cl    **2.2** 35.453 amu    **2.3** (a) Se: Group VIA, Period 4; nonmetal    (b) Cs: Group IA, Period 6; metal    (c) Fe: Group VIIIB, Period 4; metal    (d) Cu: Group IB, Period 4; metal    (e) Br: Group VIIA, Period 4; nonmetal    **2.4** $K_2CrO_4$    **2.5** (a) calcium oxide    (b) lead(II) chromate    **2.6** $Tl(NO_3)_3$    **2.7** (a) dichlorine hex(a)oxide    (b) phosphorus trichloride    (c) phosphorus pentachloride    **2.8** (a) $CS_2$    (b) $SO_3$    **2.9** (a) boron triflouride    (b) hydrogen selenide    **2.10** perbromate ion, $BrO_4^-$    **2.11** sodium carbonate decahydrate    **2.12** $Na_2S_2O_3 \cdot 5H_2O$    **2.13** (a) $O_2 + 2PCl_3 \longrightarrow 2POCl_3$    (b) $P_4 + 6N_2O \longrightarrow P_4O_6 + 6N_2$    (c) $2As_2S_3 + 9O_2 \longrightarrow 2As_2O_3 + 6SO_2$    (d) $Ca_3(PO_4)_2 + 4H_3PO_4 \longrightarrow 3Ca(H_2PO_4)_2$

## Chapter 3

**3.1** (a) 46.0 amu    (b) 180 amu    (c) 40.0 amu    (d) 58.3 amu    **3.2** (a) $SO_3$, 80.1 amu    (b) $H_2SO_4$, 98.1 amu    **3.3** (a) $6.656 \times 10^{-23}$ g    (b) $7.65 \times 10^{-23}$ g    **3.4** 30.9 g $H_2O_2$    **3.5** 0.452 mol $HNO_3$    **3.6** $1.2 \times 10^{21}$ HCN molecules    **3.7** 35.0% N, 60.0% O, 5.04% H    **3.8** 17.0 g N    **3.9** 40.9% C, 4.57% H, 54.5% O    **3.10** $SO_3$    **3.11** $C_7H_6O_2$    **3.12** $C_2H_4O$
**3.13**

| $H_2$ | + | $Cl_2$ | → | 2HCl |
|---|---|---|---|---|
| 1 molecule $H_2$ | + | 1 molecule $Cl_2$ | → | 2 molecules HCl |
| 1 mol $H_2$ | + | 1 mol $Cl_2$ | → | 2 mol HCl |
| 2.016 g $H_2$ | + | 70.9 g $Cl_2$ | → | $2 \times 36.5$ g HCl |

**3.14** 178 g Na    **3.15** 2.46 kg $O_2$    **3.16** 81.1 g Hg    **3.17** 0.12 mol $AlCl_3$    **3.18** 11.0 g ZnS    **3.19** 21.4 g; 89.1%

## Chapter 4

**4.1** (a) soluble    (b) soluble    (c) insoluble
**4.2** (a) $2H^+(aq) + 2NO_3^-(aq) + Mg(OH)_2(s) \longrightarrow$
$$2H_2O(l) + Mg^{2+}(aq) + 2NO_3^-(aq)$$
$2H^+(aq) + Mg(OH)_2(s) \longrightarrow 2H_2O(l) + Mg^{2+}(aq)$

(b) $Pb^{2+}(aq) + 2NO_3^-(aq) + 2Na^+(aq) + SO_4^{2-}(aq) \longrightarrow$
$$PbSO_4(s) + 2Na^+(aq) + 2NO_3^-(aq)$$
$Pb^{2+}(aq) + SO_4^{2-}(aq) \longrightarrow PbSO_4(s)$
**4.3** $2NaI(aq) + Pb(C_2H_3O_2)_2(aq) \longrightarrow$
$$PbI_2(s) + 2NaC_2H_3O_2(aq)$$
$2I^-(aq) + Pb^{2+}(aq) \longrightarrow PbI_2(s)$
**4.4** (a) weak acid    (b) weak acid    (c) strong acid    (d) strong base
**4.5** $HCN(aq) + LiOH(aq) \longrightarrow H_2O(l) + LiCN(aq)$
$HCN(aq) + OH^-(aq) \longrightarrow H_2O(l) + CN^-(aq)$
**4.6** $H_2SO_4(aq) + KOH(aq) \longrightarrow H_2O(l) + KHSO_4(aq)$
$KHSO_4(aq) + KOH(aq) \longrightarrow H_2O(l) + K_2SO_4(aq)$
$H^+(aq) + OH^-(aq) \longrightarrow H_2O(l)$
$HSO_4^-(aq) + OH^-(aq) \longrightarrow H_2O(l) + SO_4^{2-}(aq)$
**4.7** $CaCO_3(s) + 2HNO_3(aq) \longrightarrow$
$$Ca(NO_3)_2(aq) + CO_2(g) + H_2O(l)$$
$CaCO_3(s) + 2H^+(aq) \longrightarrow Ca^2(aq) + CO_2(g) + H_2O(l)$
**4.8** (a) K: +1, Cr: +6, O: −2    (b) Mn: +7, O: −2
**4.9** $Ca + Cl_2 \longrightarrow Ca^{2+} + 2Cl^-$    **4.10** 0.0464 $M$ NaCl
**4.11** 10.1 mL    **4.12** 0.0075 mol NaCl; 0.44 g NaCl
**4.13** 12 mL    **4.14** 34.19%    **4.15** 48.4 mL $NiSO_4$    **4.16** 5.07%

## Chapter 5

**5.1** 0.56 atm, $4.3 \times 10^2$ mmHg    **5.2** 24.1 L    **5.3** 4.47 dm$^3$    **5.4** 5.54 dm$^3$    **5.5** Use $PV = nRT$. Solve for $n$: $n = PV/RT = P(V/RT)$. Everything within parentheses is constant; therefore, $n = $ constant $\times P$ (or $\propto P$).    **5.6** 46.0 atm    **5.7** density of He = 0.164 g/L; difference in mass between 1 L of air and 1 L of He = 1.024 g    **5.8** 64.1 amu    **5.9** 2.00 L    **5.10** $P_{O_2} = 0.0769$ atm; $P_{CO_2} = 0.0310$ atm; $P = 0.1080$ atm; $X_{O_2} = 0.712$    **5.11** 0.01591 mol $O_2$; $V = 0.406$ L    **5.12** 219 m/s    **5.13** $T = 52.4$ K, 728 K    **5.14** 9.96 s    **5.15** 44.0 amu    **5.16** $P_{vdW} = 0.992$ atm; $P_{ideal} = 1.000$ atm (larger)

## Chapter 6

**6.1** $1.1 \times 10^{-17}$ J, $2.7 \times 10^{-18}$ cal
**6.2** exothermic; $q = -1170$ kJ
**6.3** $2N_2H_4(l) + N_2O_4(l) \longrightarrow 3N_2(g) + 4H_2O(g)$;
$$\Delta H = -1049 \text{ kJ}$$
**6.4** (a) $N_2H_4(l) + \frac{1}{2}N_2O_4(l) \longrightarrow \frac{3}{2}N_2(g) + 2H_2O(g)$;
$$\Delta H = -5.245 \times 10^2 \text{ kJ}$$
(b) $4H_2O(g) + 3N_2(g) \longrightarrow N_2O_4(l) + 2N_2H_4(l)$; $\Delta H = 1049$ kJ
**6.5** −164 kJ    **6.6** $1.80 \times 10^2$ J    **6.7** −54 kJ;
$HCl(aq) + NaOH(aq) \longrightarrow NaCl(aq) + H_2O(l)$; $\Delta H = -54$ kJ
**6.8** $\Delta H = -1789$ kJ    **6.9** $\Delta H°_{vap} = 44.0$ kJ
**6.10** $\Delta H° = -136.7$ kJ    **6.11** $\Delta H° = 62.0$ kJ

## Chapter 7

**7.1** 767 nm  **7.2** $6.58 \times 10^{14}$/s  **7.3** $2.0 \times 10^{-19}$ J, $2.0 \times 10^{-17}$ J, and $2.0 \times 10^{-15}$ J; greatest energy: x-ray region; least energy: infrared region  **7.4** 103 nm  **7.5** $3.38 \times 10^{-19}$ J  **7.6** $\lambda = 332$ pm  **7.7** (a) The value of $n$ must be a positive whole number.  (b) The values of $l$ range from 0 to $n - 1$. Here, $l$ has a value greater than $n$.  (c) The values for $m_l$ range from $-l$ to $+l$. Here, $m_l$ has a value greater than that of $l$.  (d) The values for $m_s$ are $+\frac{1}{2}$ or $-\frac{1}{2}$, not 0.

## Chapter 8

**8.1** (a) possible  (b) possible  (c) impossible. There are two electrons in a $2p$ orbital with the same spin.  (d) possible  (e) impossible. Only two electrons are allowed in an $s$ subshell.  (f) impossible. Only six electrons are allowed in a $p$ subshell.  **8.2** $1s^2 2s^2 2p^6 3s^2 3p^6 3d^5 4s^2$  **8.3** $4s^2 4p^3$  **8.4** Group IVA, Period 6; main-group element

**8.5** 

1s   2s   2p   3s   3p

**8.6** In order of increasing radius: Be, Mg, Na  **8.7** It is more likely that 1000 kJ/mol is the ionization energy for iodine because ionization energies tend to decrease as atomic number increases in a column.  **8.8** F

## Chapter 9

**9.1** 

**9.2** The electron configuration for $Ca^{2+}$ is [Ar] and its Lewis symbol is $Ca^{2+}$. The electron configuration for $S^{2-}$ is $[Ne]3s^2 3p^6$ and its Lewis symbol is 

.  **9.3** The electron configurations of Pb and $Pb^{2+}$ are $[Xe]4f^{14}5d^{10}6s^2 6p^2$ and $[Xe]4f^{14}5d^{10}6s^2$, respectively.  **9.4** $[Ar]3d^5$  **9.5** One would expect $S^{2-}$, because it has two additional electrons.  **9.6** In order of increasing ionic radius: $Mg^{2+}$, $Ca^{2+}$, $Sr^{2+}$  **9.7** In order of increasing ionic radius: $Ca^{2+}$, $Cl^-$, $P^{3-}$  **9.8** C—O is the most polar bond.  **9.9** 

**9.10** 

**9.11** (a) 

(b) 

**9.12** 

**9.13** 

**9.14** 

**9.15** 

**9.16** 103 pm  **9.17** 123 pm  **9.18** $-1304$ kJ

## Chapter 10

**10.1** (a) trigonal pyramidal (b) bent (c) tetrahedral  **10.2** T-shaped  **10.3** (b), (c)  **10.4** (b); the F atoms are distributed in a tetrahedral arrangement, to give a molecule with a dipole moment of zero.  **10.5** For the N atom:

N (ground state) = [He] 

and

2s   2p

N (hybridized) = [He] 

$sp^3$

Each N—H bond is formed by the overlap of a $1s$ orbital of a hydrogen atom with one of the $sp^3$ hybrid orbitals of the nitrogen atom.  **10.6** Each P—Cl bond is formed by the overlap of a phosphorus $sp^3d$ hybrid orbital with a singly occupied $3p$ chlorine orbital.  **10.7** The C atom is $sp$ hybridized. A carbon–oxygen bond is double and is composed of a $\sigma$ bond and a $\pi$ bond. The $\sigma$ bond is formed by the overlap of a hybrid C orbital with a $2p$ orbital of O that lies along the axis. The $\pi$ bond is formed by the sidewise overlap of a $2p$ on C with a $2p$ on O.  **10.8** The structural formulas for the isomers are as follows:

cis          trans

These compounds exist as separate isomers. If they were interchangeable, one end of the molecule would have to remain fixed as the other rotated. This would require breaking the $\pi$ bond and a considerable expenditure of energy.

**10.9** KK

$\sigma_{2s}$  $\sigma_{2s}^*$  $\pi_{2p}$  $\sigma_{2p}$  $\pi_{2p}^*$  $\sigma_{2p}^*$

$KK(\sigma_{2s})^2(\sigma_{2s}^*)^2(\pi_{2p})^4$; diamagnetic; bond order = 2

**10.10** KK

$\sigma_{2s}$  $\sigma_{2s}^*$  $\pi_{2p}$  $\sigma_{2p}$  $\pi_{2p}^*$  $\sigma_{2p}^*$

$KK(\sigma_{2s})^2(\sigma_{2s}^*)^2(\pi_{2p})^4(\sigma_{2p})^2$; bond order = 3; diamagnetic

## Chapter 11

**11.1** $1.37 \times 10^3$ kJ, $4.12 \times 10^3$ g    **11.2** $5.30 \times 10^2$ mmHg    **11.3** 43.0 kJ/mol    **11.4** (a) Methyl chloride can be liquefied by sufficiently increasing the pressure as long as the temperature is below 144°C.   (b) Oxygen can be liquefied by compression as long as the temperature is below $-119$°C.   **11.5** (a) hydrogen bonding, dipole–dipole forces, and London forces   (b) London forces   (c) dipole–dipole forces and London forces   **11.6** In order of increasing vapor pressure we have: butane ($C_4H_{10}$), propane ($C_3H_8$), and ethane ($C_2H_6$). London forces increase with increasing molecular weight. Therefore, one would expect the lowest vapor pressure to correspond to the highest molecular weight.   **11.7** Hydrogen bonding is negligible in methyl chloride but strong in ethanol, which explains the lower vapor pressure of ethanol.   **11.8** (a) metallic solid   (b) ionic solid   (c) covalent network solid   (d) molecular solid   **11.9** $C_2H_5OH$, molecular; $CH_4$, molecular; $CH_3Cl$, molecular; $MgSO_4$, ionic. In order of increasing melting point: $CH_4$, $CH_3Cl$, $C_2H_5OH$, and $MgSO_4$.   **11.10** 1 atom   **11.11** $6.02 \times 10^{23}$/mol   **11.12** 533 pm

## Chapter 12

**12.1** A dental filling, made up of liquid mercury and solid silver, is a solid solution.   **12.2** $C_4H_9OH$, because of hydrogen bonding.   **12.3** $Na^+$   **12.4** $8.45 \times 10^{-3}$ g $O_2$/L   **12.5** 7.07 g HCl; 27.9 g $H_2O$   **12.6** 3.09 $m$ $C_6H_5CH_3$   **12.7** mole fraction toluene = 0.194; mole fraction benzene = 0.806   **12.8** mole fraction methanol = 0.00550; mole fraction ethanol = 0.995   **12.9** 7.24 $m$ $CH_3OH$   **12.10** 2.96 $M$ $(NH_2)_2CO$   **12.11** 2.20 $m$ $(NH_2)_2CO$   **12.12** $\Delta P = 1.22$ mmHg; $P = 155$ mmHg   **12.13** $1.89 \times 10^{-1}$ g $CH_2OHCH_2OH$   **12.14** 176 amu   **12.15** 124 amu; $P_4$   **12.16** $\pi = 3.5$ atm   **12.17** 100.077°C   **12.18** $AlCl_3$

## Chapter 14

**14.1** $\dfrac{\Delta[NO_2F]}{\Delta t} = -\dfrac{\Delta[NO_2]}{\Delta t}$

**14.2** $1.1 \times 10^{-4}$ $M$/s   **14.3** zero order in [CO]; second order in [$NO_2$]; second order overall   **14.4** rate = $k[NO_2]^2$; $k =$ 0.71 L/(mol · s)   **14.5** (a) $[N_2O_5]_t = 1.24 \times 10^{-2}$ mol/L (b) $t = 4.80 \times 10^3$ s   **14.6** half-life = 0.075 s; decreases to 50% of initial concentration in 0.075 s, to 25% in 0.151 s   **14.7** $Ea = 2.07 \times 10^5$ J/mol, $k = 5.80 \times 10^{-2}/(M^{1/2} \cdot$ s)   **14.8** $2H_2O_2 \longrightarrow 2H_2O + O_2$   **14.9** bimolecular   **14.10** rate = $k[NO_2]^2$   **14.11** rate = $k_1[H_2O_2][I^-]$   **14.12** rate = $k_2(k_1/k_{-1})[NO]^2[O_2]$

## Chapter 15

**15.1** 0.57 mol CO, 0.57 mol $H_2O$, 0.43 mol $CO_2$, and 0.43 mol $H_2$   **15.2** $K_c = \dfrac{[NH_3]^2[H_2O]^4}{[NO_2]^2[H_2]^7}$; $K_c = \dfrac{[NH_3][H_2O]^2}{[NO_2][H_2]^{7/2}}$

**15.3** $CO(g) + H_2O(g) \rightleftharpoons H_2(g) + CO_2(g)$; 0.57   **15.4** $K_c = 2.3 \times 10^{-4}$   **15.5** $K_p = 1.24$   **15.6** $K_c = \dfrac{[Ni(CO)_4]}{[CO]^4}$   **15.7** products; 0.018 $M$   **15.8** $Q_c = 0.67$; more CO will form   **15.9** 0.0096 moles of $PCl_5$   **15.10** 0.11 mol $H_2$, 0.11 mol $I_2$, 0.78 mol HI   **15.11** 0.86 $M$ $PCl_5$, 0.135 $M$ $PCl_3$, 0.135 $M$ $Cl_2$   **15.12** (a) in the reverse direction   (b) in the reverse direction   **15.13** (a) no; pressure has no effect.   (b) no   (c) yes   **15.14** high temperature   **15.15** High temperatures and low pressure would give the best yield of product.

## Chapter 16

**16.1** $H_2CO_3(aq) + CN^-(aq) \rightleftharpoons HCN(aq) + HCO_3^-(aq)$
acid        base                acid            base
HCN is the conjugate acid of $CN^-$.

**16.2** (a)

(b)

**16.3** The reactants are favored.   **16.4** (a) $PH_3$   (b) HI   (c) $H_2SO_3$   (d) $H_3AsO_4$   (e) $HSO_4^-$   **16.5** 0.250 $M$ $OH^-$, $4.0 \times 10^{-14}$ $M$ $H^+$   **16.6** basic   **16.7** 1.35   **16.8** 12.40   **16.9** $6.9 \times 10^{-4}$ $M$   **16.10** $4 \times 10^{-4}$ $M$

## Chapter 17

**17.1** $1.4 \times 10^{-4}$, 0.071   **17.2** $[H^+] = [C_2H_3O_2^-] = 1.3 \times 10^{-3}$ $M$; pH = 2.89; 0.013   **17.3** 3.24   **17.4** pH = 1.29; $[SO_3^{2-}] = 6.3 \times 10^{-8}$ $M$   **17.5** $3.3 \times 10^{-6}$   **17.6** $5.3 \times 10^{-12}$ $M$   **17.7** (a) acidic   (b) neutral   (c) acidic   **17.8** (a) $1.5 \times 10^{-11}$ (b) $2.4 \times 10^{-5}$   **17.9** $1.5 \times 10^{-6}$; 8.19   **17.10** $8.5 \times 10^{-5}$ $M$, $8.5 \times 10^{-4}$   **17.11** 3.63   **17.12** 5.26   **17.13** 3.83   **17.14** 1.60   **17.15** 7.97   **17.16** 5.19

## Chapter 18

**18.1** (a) $K_{sp} = [Ba^{2+}][SO_4^{2-}]$   (b) $K_{sp} = [Fe^{3+}][OH^-]^3$   (c) $K_{sp} = [Ca^{2+}]^3[PO_4^{3-}]^2$   **18.2** $1.8 \times 10^{-10}$   **18.3** $4.5 \times 10^{-36}$   **18.4** 0.67 g/L   **18.5** (a) $6.3 \times 10^{-3}$ mol/L (b) $4.4 \times 10^{-5}$ mol/L   **18.6** Precipitation is expected.   **18.7** No precipitate will form.   **18.8** silver cyanide   **18.9** $1.2 \times 10^{-9}$ $M$   **18.10** No precipitate will form.   **18.11** 0.44 mol/L

## Chapter 19

**19.1** 3.89 J   **19.2** $-48.9$ L; 4.96 kJ; $-885.2$ kJ
**19.3** 303 J/(mol · K)   **19.4** (a) positive   (b) positive
(c) negative   (d) positive   **19.5** 537 J/K   **19.6** $-800.7$ kJ
**19.7** 130.9 kJ   **19.8** All four reactions are spontaneous in the
direction written.   **19.9** (a) $K = P_{CO_2}$   (b) $K = [Pb^{2+}][I^-]^2$
(c) $K = \dfrac{P_{CO_2}}{[H^+][HCO_3^-]}$   **19.10** $1.1 \times 10^{-23}$   **19.11** $2.4 \times 10^{-29}$
**19.12** $1 \times 10^{-1}$ atm   **19.13** $T = 671$ K, which is lower than
that for $CaCO_3$.

## Chapter 20

**20.1** $I_2(g) + 10NO_3^-(aq) + 8H^+(aq) \longrightarrow$
$\qquad\qquad\qquad 4H_2O(l) + 2IO_3^-(aq) + 10NO_2(g)$
**20.2** $H_2O_2(aq) + 2ClO_2(aq) + 2OH^-(aq) \longrightarrow$
$\qquad\qquad\qquad 2ClO_2^-(aq) + O_2(g) + 2H_2O(l)$
**20.3** anode: $Ni(s) \longrightarrow Ni^{2+}(aq) + 2e^-$; cathode:
$Ag^+(aq) + e^- \longrightarrow Ag(s)$; electrons flow from anode to
cathode; $Ag^+$ flows to Ag, $Ni^{2+}$ flows away from Ni.
**20.4** $Zn(s)|Zn^{2+}(aq)\|H^+(aq)|H_2(g)|Pt$
**20.5** $Cd(s) + 2H^+(aq) \longrightarrow Cd^{2+}(aq) + H_2(g)$
**20.6** $-2.12 \times 10^4$ J   **20.7** $NO_3^-$   **20.8** No   **20.9** 1.10 V
**20.10** $-1.4 \times 10^2$ kJ   **20.11** 2.699 V   **20.12** $1 \times 10^{19}$
**20.13** 1.42 V   **20.14** $4 \times 10^{-7}$ M $Ni^{2+}$
**20.15** (a) $K^+(l) + e^- \longrightarrow K(l)$
$2Cl^-(l) \longrightarrow Cl_2(g) + 2e^-$
(b) $K^+(l) + e^- \longrightarrow K(l)$
$4OH^-(l) \longrightarrow O_2(g) + 2H_2O(g) + 4e^-$
**20.16** $Ag^+(aq) + e^- \longrightarrow Ag(s)$ (cathode)
$2H_2O(l) \longrightarrow O_2(g) + 4H^+(aq) + 4e^-$ (anode)
**20.17** $2.52 \times 10^{-2}$ A   **20.18** $8.66 \times 10^{-2}$ g

## Chapter 21

**21.1** $^{40}_{19}K \longrightarrow {}^{40}_{20}Ca + {}^{0}_{-1}e$   **21.2** $^{235}_{92}U$   **21.3** (a) is stable;
(b) and (c) are radioactive.   **21.4** (a) positron emission
(b) beta emission   **21.5** (a) $^{40}_{20}Ca(d,p)^{41}_{20}Ca$
(b) $^{12}_6C + {}^2_1H \longrightarrow {}^{13}_6C + {}^1_1H$   **21.6** $^{14}_7N$   **21.7** $3.2 \times 10^{-5}$/s
**21.8** 5.26 y   **21.9** $7.82 \times 10^{-10}$/s; $7.4 \times 10^{-7}$ Ci
**21.10** 0.200   **21.11** $1.0 \times 10^4$ y   **21.12** $-1.1 \times 10^8$ J;
$-0.27$ MeV

## Chapter 23

**23.1** (a) pentaamminechlorocobalt(III) chloride   (b) potassium
aquapentacyanocobaltate(III)   (c) pentaaquahydroxoiron(III) ion
**23.2** (a) $K_4[Fe(CN)_6]$   (b) $[Co(NH_3)_4Cl_2]Cl$   (c) $PtCl_4^{2-}$

**23.3** (a) no geometric isomers possible   (b) geometric isomers
possible (see Fig. 23.15 for similar example)   **23.4** Only (b) and
(c); (b) is similar to Figure 23.16; (c) is similar to Figure 23.20.
**23.5**

$$\underset{\text{2 unpaired electrons}}{\uparrow\downarrow \quad \uparrow\downarrow \quad \uparrow\downarrow} \quad \uparrow \quad \uparrow$$

**23.6**

$$\uparrow \quad \uparrow \quad \uparrow$$
$$\uparrow\downarrow \quad \uparrow\downarrow$$

**23.7** 530 nm, 500 nm, yes; $CN^-$ is a more strongly bonding
ligand than $H_2O$.

## Chapter 24

**24.1**
$$CH_3CHCH_2CH_3$$
with $CH_3$ branch

**24.2** (a)

$$\overset{H}{\underset{CH_3}{}}C=C\overset{H}{\underset{CH_2CH_2CH_3}{}} \qquad \overset{H}{\underset{CH_3}{}}C=C\overset{CH_2CH_2CH_3}{\underset{H}{}}$$

(b) none   **24.3** 2-bromobutane   **24.4** (a) 2,3-dimethylbutane
(b) 3-ethyl-2-methylhexane

**24.5** $CH_3CH_2\overset{\displaystyle CH_3}{\underset{\displaystyle CH_3}{-\,C\,-}}CH_2CH_2CH_2CH_3$

**24.6** (a) 2,4-dimethyl-2-hexene   (b) 3-propyl-1-hexene
**24.7** $CH_3\,\overset{}{-}\,C=CH\,-\,CH_2\,-\,CH\,-\,CH_2CH_3$
with $CH_3$ substituents
**24.8** *cis*-3-hexene, *trans*-3-hexene   **24.9** (a) propyne
(b) 3-methyl-1-pentyne
**24.10** (a)

**24.11** 3-ethyl-3-hexanol   **24.12** (a) dimethyl ether
(b) methyl ethyl ether   **24.13** (a) 2-pentanone   (b) butanal

# Answers to Concept Checks

## Chapter 1

**1.1** (A) element (B) compound (C) mixture. **1.2** (a) two if you weigh under 100 lb, three if you weigh 100 lb or more (b) A 165-lb person would be reported as weighing $1.7 \times 10^2$ lb. (c) A 74.8-kg or 165-lb male would weigh 50 kg or 200 lb on the truck scale. **1.3** (a) Meters, decimeters, or centimeters would be appropriate units. (b) approximately 9 m (c) 39°C would mean that you had a moderate fever. (d) You would probably be comfortable in a 23°C room. **1.4** You need to perform an experiment that will give you the density of the marcasite and then compare that to the density of gold.

## Chapter 2

**2.1** The molecular model on the far right that contains 3 atoms with a ratio of 1:2 **2.2** One would conclude that the atom is made up primarily of a large impenetrable mass with a positive charge. **2.3** They must have similar chemical and/or physical properties. **2.4** (a) ether (b) alcohol (c) carboxylic acid (d) hydrocarbon **2.5** Q is likely to be an element in Group VIA.

## Chapter 3

**3.1** (a) 1.5 mol (b) 4.5 mol (c) 3.0 mol of $OH^-$ since there are 2 mol of $OH^-$ per mol of $Mg(OH)_2$ **3.2** (a) The total mass of C and H that you collect will be less than the total mass of material you started with. (b) The calculated percent C and H would be less than the real value. **3.3** (a) No, the empirical formula needs to be the smallest whole-number ratio of subscripts. (b) No, the subscripts are not whole numbers. (c) Yes, the empirical and molecular formulas can be the same. **3.4** (a) correct (b) incorrect (c) incorrect (d) correct (e) correct (f) incorrect (g) incorrect **3.5** (a) $X_2(g) + 2Y(g) \longrightarrow 2XY(g)$ (b) container 1 (c) $X_2(g)$

## Chapter 4

**4.1** The left beaker contains the strong electrolyte $LiI(aq)$. The right beaker contains the nonelectrolyte $CH_3OH(aq)$.
**4.2** (a) $Na_2SO_4(aq) + Sr(C_2H_3O_2)_2(aq) \longrightarrow$
$$SrSO_4(s) + 2NaC_2H_3O_2(aq)$$
(b) $2Na^+(aq) + SO_4{}^{2-}(aq) + Sr^{2+}(aq) + 2C_2H_3O_2{}^-(aq) \longrightarrow$
$$SrSO_4(s) + 2Na^+(aq) + 2C_2H_3O_2{}^-(aq)$$
(c) $Sr^{2+}(aq) + SO_4{}^{2-}(aq) \longrightarrow SrSO_4(s)$
**4.3** (a) $MOH(s)$ is a base. M could be any Group I metal. (b) HA is an acid. A is an anion with a $-1$ charge. (c) $H_2A$ is an acid. A is an anion with a $-2$ charge. (d) $M \longrightarrow Na^+$, $A \longrightarrow C_2H_3O_2{}^-$, $A \longrightarrow SO_3{}^{2-}$.

**4.4** Concentration order: $B > C = D > A$. Leave A alone, add water to fill beaker B, add water to double the volumes of D and C. **4.5** (a) The acid in flask C has three times as many acidic protons as the acid in flask A. The acid in flask B has two times as many acidic protons as the acid in flask A. (b) Yes, if we assume that flask A contains a monoprotic acid.

## Chapter 5

**5.1** Since the height of the column is related to the density of the liquid, the column containing the $H_2O$ would be higher. **5.2** (a) Step 1, pressure decrease; Step 2, pressure decrease; total change, final pressure less than starting pressure. (b) Step 1, pressure increase; Step 2, pressure increase; total change, final pressure greater than starting pressure. **5.3** (a) All contain same amount of gas. (b) Xe flask (c) He flask (d) They would all still have the same number of moles. **5.4** (a) No change in $H_2$ pressure. (b) They are equal. (c) The total pressure is the sum of the pressures of the $H_2$ and Ar in the container **5.5** (a) He (b) Raise the temperature of the Ar. **5.6** (a) Pressure would be greater for ideal gas. (b) Pressure would be less for ideal gas. (c) Cannot make a determination because the effect of molecular volume and attractions on the pressure of a real gas is opposite.

## Chapter 6

**6.1** Sun's energy is changed to electric energy, to chemical energy (of the battery), and back to electric energy, then to kinetic energy of motor and water, and to potential energy of water. **6.2** c and d are exothermic; d is most exothermic **6.3** $\Delta H_{sub} = \Delta H_{fus} + \Delta H_{vap}$

## Chapter 7

**7.1** The wavelength is halved; ultraviolet. **7.2** UV **7.3** The proton speed would be about 2000 times smaller.

## Chapter 8

**8.1** 4 **8.2** Mg and Al **8.3** IIA **8.4** As

## Chapter 9

**9.1** c, e **9.2** b **9.3** a

## Chapter 10

**10.1** The four pairs have a tetrahedral arrangement; the molecular geometry is trigonal pyramidal.   **10.2** Molecule Y is likely to be trigonal planar, but trigonal pyramidal or T-shaped geometries are possible. Molecule Z cannot have a trigonal planar geometry, but must be either trigonal pyramidal or T-shaped.   **10.3** $sp$

## Chapter 11

**11.1** (a) (i) slightly lower (ii) higher (iii) more molecules (iv) evaporation is greater
(v)

$t = 1$

(b) (i) slightly lower (ii) higher (iii) more molecules (iv) they will be equal
(v)

$t = 2$

**11.2** Cook it longer because the $H_2O$ temperature would be lower.   **11.3** (a) The hydrogen and oxygen would form an explosive mixture.   (b) Many times greater (more positive) for the wrong reaction.   (c) Apply Hess's law.   **11.4** (a) AB (b) face-centered cubic

## Chapter 12

**12.1** (a) Mo solute, Cr solvent   (b) $MgCl_2$ solute, $H_2O$ solvent (c) $O_2$ and $N_2$ solutes, Ar solvent   **12.2** The $A^{2+}$ ion must be smaller so it has a greater energy of hydration.   **12.3** At high elevations, the partial pressure of $O_2$ is so low that not enough dissolves in the water to sustain fish.   **12.4** A soluble, chemically similar liquid with a higher vapor pressure than water **12.5** To prevent water from flowing into the pickle via osmosis. The pickles would be huge and would probably burst. **12.6** Hightest to lowest: $MgCl_2 > KBr > Na_2SO_4 > NaCl$. **12.7** The iron(III) hydroxide precipitates because the electrons from the negative electrode neutralize the positive charge on the iron(III) hydroxide.

## Chapter 13

**13.1** Metals in ores and minerals are in an oxidized form in which the free metal form is the neutral, elemental state. **13.2** More energy because you would need to give the electrons enough energy to jump the gap into the unoccupied orbitals that make up the unfilled band.   **13.3** The same as in glass: curved downward from the walls of the container. It attains the shape because of hydrogen bonding interactions between the O atoms in the $SiO_2$ tetrahedra and the H atoms of the water molecules.

## Chapter 14

**14.1** (a) Point A has a faster instantaneous rate.   (b) No, it decreases with time.   **14.2** (a) $[Q] = 0.0\,M$ slowest (no reaction); the other two are equal in rate because they are zero order with respect to $[Q]$.   (b) Rate $= k[R]^2$   **14.3** (a) Rate of aging $= (\text{diet})^w(\text{exercise})^x(\text{sex})^y(\text{occupation})^z$   (b) Gather a sample of people that have all of the factors the same except one. For example, using the equation given in part (a), you could determine the effect of diet if you had a sample of people who were the same sex, exercised the same amount, and had the same occupation. You would need to isolate each factor in this fashion to determine the exponent on each factor.   (c) 2   **14.4** Second order   **14.5** (a) the A + B reaction   (b) the E + F reaction (c) exothermic   **14.6** (a) It will increase the rate since the activation energy, $E_a$, is inversely related to the rate constant, $k$. A decrease in $E_a$ results in an increase in the value of $k$.   (b) yes (c) No, it should be rate $= k[Y]^2$.

## Chapter 15

**15.1** $2A + B \longrightarrow 2C$   **15.2** Left to right.   **15.3** It is quadrupled.

## Chapter 16

**16.1** $HCHO_2(aq) + H_2O(l)$   $CHO_2^-(aq) + H_3O^+(aq)$; $HCHO_2(aq)$, $H_2O(l)$, $CHO_2^-(aq)$, and $H_3O^+(aq)$
**16.2** Acetate ion   **16.3** NaOH, $NH_3$, $HC_2H_3O_2$, HCl

## Chapter 17

**17.1** HCN, $HC_2H_3O_2$, $HNO_2$, HF   **17.2** a, b   **17.3** d

## Chapter 18

**18.1** $PbSO_4$   **18.2** $NaNO_3$   **18.3** Magnesium oxalate

## Chapter 19

**19.1** It has increased.   **19.2** The NO concentration is low at equilibrium.   **19.3** a. $\Delta G^\circ$ is not changed, b. $\Delta G$ increases. **19.4** The concentration of $NO_2$ increases with temperature.

## Chapter 20

**20.1** No sustainable current would flow. The wire does not contain mobile positive and negatively charged species, which are

necessary to balance the accumulation of charges in each of the half-cells.   **20.2** (a) $-0.54$ V   (b) No   (c) They both would be 1.10 V.   **20.3** (a) Negative   (b) Change the concentrations in a manner to increase $Q$, where $Q = \dfrac{[Fe^{2+}]}{[Cu^{2+}]}$.
For example: $Fe(s)|Fe^{2+}(1.10\,M)\|Cu^{2+}(0.50\,M)|Cu(s)$.
**20.4** Many of the ions contained in seawater have very high reduction potentials—higher than $Fe(s)$. This means that spontaneous electrochemical reactions will occur with the $Fe(s)$, causing the iron to form ions and go into solution while at the same time, the ions in the sea are reduced and plate out on the surface of the iron.

## Chapter 21

**21.1** (a) Yes   (b) No, since the $_1^3H_2O$ molecule is more massive than $_1^1H_2O$.   (c) $_1^3H_2O$   **21.2** The $\alpha$ particle, since it has the highest RBE.   **21.3** After 50,000 years, enough half-lives have passed (about 10) that there would be almost no carbon-14 present to detect and measure.

## Chapter 22

**21.1** Neutrons cause fission and a sustainable nuclear reaction. If the neutrons were absorbed by the cladding, no nuclear reaction would occur.   **22.2** Given the high energy demands of animals to move and maintain body temperature, breaking the very strong triple bond of $N_2$ requires too much energy when compared to the lower-energy double bond of $O_2$.   **22.3** Because the only intermolecular forces in these materials are very weak van der Waals forces.

## Chapter 23

**23.1** $K_2[PtCl_6]$   **23.2** $[Co(NH_3)_4(H_2O)Br]Cl_2$, $[Co(NH_3)_4H_2OCl]BrCl$   **23.3** (a) Complexes A, D, and E. (*Note:* B and C are the same complex, and A and E are the same complex.)   (b) None   (c) A and E, B and C.

## Chapter 24

**24.1** (a) $C_7H_{14}$   (b)

$$CH_3CHCHCH_2CH_3$$

with $CH_3$ groups attached

**24.2** Blend of 90% 2,2,4-trimethylpentane and 10% heptane

**24.3** (a) $C_5H_{10}$   (b)

(c) *cis*-2-pentene

## Chapter 25

**25.1**

**25.2** UACGAUGCCUAAGUU

# Answers to Odd-Numbered Problems

## Chapter 1

**1.21** (a) Immerse the thermometer in a solution of ice and liquid water and when the mercury stops moving, mark that point. Then immerse the thermometer in boiling water, repeat the procedure, and mark that point. (b) Take the thermometer that you marked up and make 19 evenly spaced marks between those two original points.

**1.23** (a) $°F = °YS \left( \dfrac{120°F}{100°YS} \right) - 100°F$ (b) $-21°F$

**1.25** The box that contains the cubes has a greater density.
**1.27** (a) A paper clip has the closest mass to 1 g. (b) Your answers will vary depending on sample: grain of sand $1 \times 10^{-5}$ kg, paper clip $1 \times 10^{-3}$ kg, nickel $5 \times 10^{-3}$ kg, 5.0-gallon bucket of water $2.0 \times 10^1$ kg, a brick 3 kg, a car $1 \times 10^3$ kg.
**1.29** 6.6 g **1.31** 29.8 g **1.33** (a) solid (b) liquid (c) gas (d) solid **1.35** (a) physical (b) physical (c) chemical (d) physical **1.37** heating of mercury(II) oxide = chemical change; solidification of mercury = physical change; burning of wood = chemical change **1.39** (a) physical (b) chemical (c) physical (d) physical (e) chemical **1.41** Physical properties: lustrous; blue-black crystals that vaporize readily to a violet gas. Chemical properties: combines with many metals, including aluminum. **1.43** (a) physical (b) chemical (c) physical (d) chemical (e) physical **1.45** (a) solution (b) substance (c) substance (d) heterogeneous mixture **1.47** (a) pure substance; bromine liquid, bromine gas (b) mixture; liquid solution and solid (c) pure substance; nonmolten portion (solid), liquid (d) mixture; solid sodium hydrogen carbonate and solid potassium hydrogen tartrate **1.49** (a) 6 (b) 3 (c) 4 (d) 5 (e) 3 (f) 4 **1.51** $4.0 \times 10^4$ km **1.53** (a) 4.7 (b) 82.5 (c) 111 (d) $2.3 \times 10^3$ **1.55** 32 cm$^3$ **1.57** (a) 5.89 ps (b) 20.1 cm (c) 2.560 ng (d) 6.05 km **1.59** (a) $6.15 \times 10^{-12}$ s (b) $3.781 \times 10^{-6}$ m (c) $1.546 \times 10^{-10}$ m (d) $9.7 \times 10^{-3}$ g **1.61** (a) 20°C (b) $-24°C$ (c) 79°F (d) $-94°F$ **1.63** $-6.0°F$ **1.65** 7.56 g/cm$^3$ **1.67** ethanol **1.69** $1.3 \times 10^2$ g **1.71** 25.1 cm$^3$ **1.73** $4.80 \times 10^5$ mg **1.75** $5.55 \times 10^{-4}$ mm **1.77** $3.73 \times 10^{17}$ m$^3$; $3.73 \times 10^{20}$ L **1.79** $3.25 \times 10^6$ g **1.81** $4.435 \times 10^3$ m **1.83** 0.86 g **1.85** 2.53 g **1.87** 114.5 g **1.89** (a) bromine (b) phosphorus (c) gold (d) carbon **1.91** no **1.93** $6.07 \times 10^4$ cm$^3$ **1.95** 50.4 gal **1.97** $9.9 \times 10^1$ cm$^3$ **1.99** (a) $7.6 \times 10^{-1}$ (b) $1.63 \times 10^1$ (c) $4.76 \times 10^2$ (d) $1.12 \times 10^{-1}$ **1.101** (a) 9.12 cg/mL (b) 66 pm (c) 7.1 $\mu$m (d) 56 nm **1.103** (a) $1.07 \times 10^{-12}$ s (b) $5.8 \times 10^{-6}$ m (c) $3.19 \times 10^{-7}$ m (d) $1.53 \times 10^{-2}$ s **1.105** 6170°F **1.107** $1.52 \times 10^3$°F **1.109** 303.0 K, 85.6°F **1.111** 907.8°C, 1180.9 K **1.113** $1.74 \times 10^3$ kg/m$^3$ **1.115** 2.65 g/mL **1.117** chloroform **1.119** $1.82 \times 10^3$ g **1.121** 33.24 mL **1.123** (a) $8.45 \times 10^9$ $\mu$g (b) $3.18 \times 10^{-1}$ ms (c) $9.3 \times 10^{13}$ nm (d) 3.71 cm **1.125** (a) $5.91 \times 10^6$ mg (b) $7.53 \times 10^5$ $\mu$g (c) $9.01 \times 10^4$ kHz (d) $4.98 \times 10^{-4}$ kJ **1.127** $1.2230 \times 10^{16}$ L **1.129** $2.8 \times 10^4$ L **1.131** 55.0 g **1.133** 2.4 L **1.135** $2.13 \times 10^3$ g **1.137** $2.6 \times 10^{21}$ g **1.139** 34%, 79.3 proof **1.141** 8.10 g/cm$^3$ **1.143** 2.4 g/cm$^3$

## Chapter 2

**2.21** Since there is no massive, positive nucleus, one would expect no or very little deflection of the alpha particles when they are shot at the gold foil. **2.23** Group them by physical properties (density, mass, color, conductivity, etc.) or by chemical properties (reaction with air, reaction with water, etc.). **2.25** (a) $X_2SO_4$, $XSO_4$, and $X_2(SO_4)_5$ (b) exy(I) sulfate, exy(II) sulfate, exy(V) sulfate **2.27** 19 protons, 20 neutrons, and 38 electrons **2.29** (a) neon (b) zinc (c) silver (d) magnesium **2.31** (a) K (b) S (c) Fe (d) Mn **2.33** $9.05 \times 10^{-31}$ kg **2.35** D,C **2.37** Cl-35: 17 protons, 18 neutrons, 17 electrons; Cl-37: 17 protons, 20 neutrons, 17 electrons **2.39** $^{28}_{14}Si$ **2.41** 13.90 **2.43** 10.812 amu **2.45** 39.10 amu, potassium **2.47** 24.615 amu **2.49** (a) Group IVA, Period 2, nonmetal (b) Group VIA, Period 6, metal (c) Group VIB, Period 4, metal (d) Group IIA, Period 3, metal (e) Group IIIA, Period 2, metalloid **2.51** (a) Te (b) Al **2.53** (a) O, oxygen (b) Na, sodium (c) Fe, iron (d) Ce, cerium **2.55** Differences: as a solid, each molecule contains 8 sulfur atoms, whereas gas molecules contain various numbers of sulfur atoms. Alike: molecules of solid and gas contain only S—S bonds. **2.57** $4.10 \times 10^{22}$ N atoms, $2.73 \times 10^{22}$ N atoms **2.59** $1.1 \times 10^{21}$ NH$_3$ molecules **2.61** (a) N$_2$H$_4$ (b) H$_2$O$_2$ (c) C$_3$H$_8$O (d) PCl$_3$ **2.63** (a) PCl$_5$ (b) NO$_2$ (c) CH$_3$CH$_2$COOH **2.65** 1 Fe to 9 O **2.67** (a) Fe(CN)$_3$ (b) K$_2$SO$_4$ (c) Li$_3$N (d) Ca$_3$P$_2$ **2.69** (a) sodium sulfate (b) calcium oxide (c) copper(I) chloride (d) chromium(III) oxide **2.71** (a) PbCr$_2$O$_7$ (b) Ba(HCO$_3$)$_2$ (c) Cs$_2$S (d) Fe(C$_2$H$_3$O$_2$)$_2$ **2.73** (a) molecular (b) ionic (c) molecular (d) ionic **2.75** (a) dinitrogen monoxide (b) tetraphosphorus decoxide (c) arsenic trichloride (d) dichlorine heptoxide **2.77** (a) NBr$_3$ (b) XeO$_4$ (c) OF$_2$ (d) Cl$_2$O$_5$ **2.79** (a) selenium trioxide (b) disulfur dichloride (c) carbon monoxide **2.81** (a) bromic acid HBrO$_3$ (b) hyponitrous acid H$_2$N$_2$O$_2$ (c) disulfurous acid H$_2$S$_2$O$_5$ (d) arsenic acid H$_3$AsO$_4$ **2.83** sodium sulfate decahydrate **2.85** FeSO$_4 \cdot$7H$_2$O **2.87** 9 **2.89** (a) Sn + 2NaOH $\longrightarrow$ Na$_2$SnO$_2$ + H$_2$ (b) 8Al + 3Fe$_3$O$_4$ $\longrightarrow$ 4Al$_2$O$_3$ + 9Fe

(c) $2CH_3OH + 3O_2 \longrightarrow 2CO_2 + 4H_2O$   (d) $P_4O_{10} + 6H_2O$ $\longrightarrow 4H_3PO_4$   (e) $PCl_5 + 4H_2O \longrightarrow H_3PO_4 + 5HCl$

**2.91** $Ca_3(PO_4)_2(s) + 3H_2SO_4(aq) \longrightarrow$
$$2H_3PO_4(aq) + 3CaSO_4(s)$$

**2.93** $2NH_4Cl(aq) + Ba(OH)_2(aq) \xrightarrow{\Delta}$
$$2NH_3(g) + BaCl_2(aq) + 2H_2O(l)$$

**2.95** The ratio of O masses is 5/4 (B to A).   **2.97** $-1.6 \times 10^{-19}$ C; 2, 4, 6, 7   **2.99** 63, 60   **2.101** $^{80}_{35}Br^-$   **2.103** Ag-107: 0.518; Ag-109: 0.482   **2.105** (a) bromine, Br   (b) hydrogen, H (c) niobium, Nb   (d) fluorine, F   **2.107** (a) lead(II) ion (b) lead(IV) ion   (c) copper(I) ion   (d) copper(II) ion **2.109** $Na_2SO_4$, NaCl, $NiSO_4$, $NiCl_2$   **2.111** (a) tin(II) phosphate (b) ammonium nitrite   (c) magnesium hydroxide (d) chromium(II) sulfate   **2.113** (a) $Hg_2S$   (b) $Co_2(SO_3)_3$ (c) $(NH_4)_2Cr_2O_7$   (d) $Al_2S_3$   **2.115** (a) arsenic tribromide (b) dihydrogen selenide   (c) diphosphorus pentoxide (d) silicon dioxide   **2.117** (a) $2C_2H_6 + 7O_2 \longrightarrow 4CO_2 + 6H_2O$ (b) $P_4O_6 + 6H_2O \longrightarrow 4H_3PO_3$   (c) $4KClO_3 \longrightarrow$
$$3KClO_4 + KCl$$
(d) $(NH_4)_2SO_4 + 2NaOH \longrightarrow 2NH_3 + 2H_2O + Na_2SO_4$ (e) $2NBr_3 + 3NaOH \longrightarrow N_2 + 3NaBr + 3HOBr$ **2.119** 26, nickel   **2.121** 12.502 amu   **2.123** $6.06 \times 10^9$ miles **2.125** $NiSO_4 \cdot 7H_2O$, $NiSO_4 \cdot 6H_2O$, 4.826 g **2.127** 0.8000 g, 63.54 amu, Cu

## Chapter 3

**3.13** (a) $3H_2(g) + N_2(g) \longrightarrow 2NH_3(g)$   (b) $H_2(g)$ (c) six moles of $H_2$   **3.15** (a) unreasonable   (b) unreasonable (c) reasonable   (d) unreasonable   **3.17** (a) charcoal   (b) Mg (c) $H_2$   **3.19** (a) 32.0 amu   (b) 137 amu   (c) 138 amu (d) 366 amu   **3.21** (a) 64.1 amu   (b) 137 amu **3.23** 80.05 g/mol   **3.25** (a) $3.818 \times 10^{-23}$ g/atom (b) $5.325 \times 10^{-23}$ g/atom   (c) $8.383 \times 10^{-23}$ g/molecule (d) $5.390 \times 10^{-22}$ g/unit   **3.27** $1.231 \times 10^{-22}$ g/molecule **3.29** (a) 3.4 g Na   (b) 19.0 g S   (c) 236 g $CH_2Cl_2$   (d) $4.8 \times 10^3$ g $Na_2SO_3$   **3.31** 33.6 g   **3.33** (a) 0.238 mol   (b) $9.94 \times 10^{-2}$ mol   (c) 1.3 mol   (d) 0.355 mol   **3.35** $5.81 \times 10^{-3}$ mol $CaSO_4$; $1.16 \times 10^{-2}$ mol $H_2O$; for every mole of $CaSO_4$ there are two moles of $H_2O$.   **3.37** (a) $7.12 \times 10^{23}$ atoms (b) $2.41 \times 10^{23}$ atoms   (c) $2.2 \times 10^{24}$ molecules   (d) $3.75 \times 10^{23}$ units   (e) $6.59 \times 10^{22}$ $SO_4^{2-}$ ions   **3.39** $2.97 \times 10^{19}$ molecules   **3.41** 86.27%   **3.43** 20.2%   **3.45** $6.56 \times 10^{-1}$ kg **3.47** 72.3% Al, 27.7% Mg   **3.49** (a) 42.9% C, 57.1% O (b) 27.3% C, 72.7% O   (c) 19.2% Na, 1.68% H, 25.8% P, 53.3% O   (d) 32.2% Co, 15.3% N, 52.5% O   **3.51** 91.2% C, 8.8% H   **3.53** Ethanol   **3.55** 38.7% C, 9.79% H, 51.5% O **3.57** $OsO_4$   **3.59** $K_2MnO_4$   **3.61** $C_3H_4O_2$   **3.63** (a) CH (b) $C_4H_4$, $C_6H_6$   **3.65** $C_4H_{12}N_2$

**3.67** $C_2H_2O_4$

**3.69**

| $C_2H_4$ | $+$ | $3O_2$ | $\longrightarrow$ |
|---|---|---|---|
| 1 molecule $C_2H_4$ | $+$ | 3 molecules $O_2$ | $\longrightarrow$ |
| 1 mole $C_2H_4$ | $+$ | 3 moles $O_2$ | $\longrightarrow$ |
| 28.054 g $C_2H_4$ | $+$ | $3 \times 32.00$ g $O_2$ | $\longrightarrow$ |

| $2CO_2$ | $+$ | $2H_2O$ |
|---|---|---|
| 2 molecules $CO_2$ | $+$ | 2 molecules $H_2O$ |
| 2 moles $CO_2$ | $+$ | 2 moles $H_2O$ |
| $2 \times 44.01$ g $CO_2$ | $+$ | $2 \times 18.016$ g $H_2O$ |

**3.71** 1.2 mol $CO_2$   **3.73** 7.82 mol $O_2$   **3.75** 8.21 g $NO_2$ **3.77** $1.46 \times 10^5$ g W   **3.79** 22.4 g $CS_2$   **3.81** 7.563 g $NO_2$ **3.83** $KO_2$ is the limiting reactant; 0.19 moles of oxygen **3.85** 40.5 g $CH_3OH$; $H_2$ remains; 5.1 g $H_2$   **3.87** 9.07 g **3.89** 2.61 g is the theoretical yield; 75.9%   **3.91** 49.5% C, 5.19% H, 28.9% N, 16.5% O   **3.93** $C_6H_4Cl_2$   **3.95** $C_4H_4S$, $C_4H_4S$   **3.97** 84.2%   **3.99** 59.5%   **3.101** 60.3 g **3.103** $1.31 \times 10^3$ g $CaC_2$   **3.105** 16.2%   **3.107** 0.21 g CuO **3.109** 616 amu   **3.111** Ag and Cl   **3.113** $1.45 \times 10^{26}$ Fe atoms

## Chapter 4

**4.15** (a) Any soluble salt that will form a precipitate when reacted with $Pb^{2+}$ ions in solution will work; for example, $CaCl_2$, $Na_2S$, and $(NH_4)_2CO_3$.   (b) No precipitate would form. (c) You would underestimate the amount of lead present in the solution.   **4.17** (a) $3Ca(C_2H_3O_2)_2(aq) + 2(NH_4)_3PO_4(aq) \longrightarrow$
$$Ca_3(PO_4)_2(s) + 6NH_4C_2H_3O_2(aq)$$
(b) $3Ca^{2+}(aq) + 6C_2H_3O_2^-(aq) + 6NH_4^+(aq) + 2PO_4^{3-}(aq)$
$$\longrightarrow Ca_3(PO_4)_2(s) + 6NH_4^+(aq) + 6C_2H_3O_2^-(aq)$$
(c) $3Ca^{2+}(aq) + 2PO_4^{3-}(aq) \longrightarrow Ca_3(PO_4)_2(s)$
**4.19** Probably not, because the ionic compound that is a nonelectrolyte is not soluble.   **4.21** (a) insoluble   (b) soluble (c) soluble   (d) soluble   **4.23** (a) insoluble   (b) soluble; $Li^+$, $SO_4^{2-}$   (c) insoluble   (d) soluble; $Na^+$, $CO_3^{2-}$ **4.25** (a) $H^+(aq) + OH^-(aq) \longrightarrow H_2O(l)$
(b) $Ag^+(aq) + Br^-(aq) \longrightarrow AgBr(s)$
(c) $S^{2-}(aq) + 2H^+(aq) \longrightarrow H_2S(g)$
(d) $OH^-(aq) + NH_4^+(aq) \longrightarrow NH_3(g) + H_2O(l)$
**4.27** $Pb(NO_3)_2(aq) + Na_2SO_4(aq) \longrightarrow$
$$PbSO_4(s) + 2NaNO_3(aq);$$
$Pb^{2+}(aq) + SO_4^{2-}(aq) \longrightarrow PbSO_4(s)$
**4.29** (a) $FeSO_4(aq) + NaCl(aq) \longrightarrow NR$
(b) $Na_2CO_3(aq) + MgBr_2(aq) \longrightarrow MgCO_3(s) + 2NaBr(aq)$;
$CO_3^{2-}(aq) + Mg^{2+}(aq) \longrightarrow MgCO_3(s)$
(c) $MgSO_4(aq) + 2NaOH(aq) \longrightarrow$
$$Mg(OH)_2(s) + Na_2SO_4(aq);$$
$Mg^{2+}(aq) + 2OH^-(aq) \longrightarrow Mg(OH)_2(s)$
(d) $NiCl_2(aq) + NaBr(aq) \longrightarrow NR$
**4.31** (a) $Ba(NO_3)_2(aq) + Li_2SO_4(aq) \longrightarrow$
$$BaSO_4(s) + 2LiNO_3(aq);$$
$Ba^{2+}(aq) + SO_4^{2-}(aq) \longrightarrow BaSO_4(s)$
(b) $NaBr(aq) + Ca(NO_3)_2(aq) \longrightarrow NR$

(c) $Al_2(SO_4)_3(aq) + 6NaOH(aq) \longrightarrow$
$$2Al(OH)_3(s) + 3Na_2SO_4(aq);$$
$Al^{3+}(aq) + 3OH^-(aq) \longrightarrow Al(OH)_3(s)$
(d) $3CaBr_2(aq) + 2Na_3PO_4(aq) \longrightarrow$
$$Ca_3(PO_4)_2(s) + 6NaBr(aq);$$
$3Ca^{2+}(aq) + 2PO_4^{3-}(aq) \longrightarrow Ca_3(PO_4)_2(s)$
**4.33** (a) weak acid   (b) strong base   (c) strong acid
(d) weak acid
**4.35** (a) $NaOH(aq) + HNO_3(aq) \longrightarrow NaNO_3(aq) + H_2O(l);$
$H^+(aq) + OH^-(aq) \longrightarrow H_2O(l)$
(b) $2HCl(aq) + Ba(OH)_2(aq) \longrightarrow 2H_2O(l) + BaCl_2(aq);$
$H^+(aq) + OH^-(aq) \longrightarrow H_2O(l)$
(c) $2HC_2H_3O_2(aq) + Ca(OH)_2(aq) \longrightarrow$
$$Ca(C_2H_3O_2)_2(aq) + 2H_2O(l);$$
$HC_2H_3O_2(aq) + OH^-(aq) \longrightarrow H_2O(l) + C_2H_3O_2^-(aq)$
(d) $NH_3(aq) + HNO_3(aq) \longrightarrow NH_4NO_3(aq);$
$H^+(aq) + NH_3(aq) \longrightarrow NH_4^+(aq)$
**4.37** (a) $2HBr(aq) + Ca(OH)_2(aq) \longrightarrow CaBr_2(aq) + 2H_2O(l);$
$H^+(aq) + OH^-(aq) \longrightarrow H_2O(l)$
(b) $Al(OH)_3(s) + 3HNO_3(aq) \longrightarrow Al(NO_3)_3(aq) + 3H_2O(l);$
$3H^+(aq) + Al(OH)_3(s) \longrightarrow 3H_2O(l) + Al^{3+}(aq)$
(c) $2HCN(aq) + Ca(OH)_2(aq) \longrightarrow Ca(CN)_2(aq) + 2H_2O(l);$
$HCN(aq) + OH^-(aq) \longrightarrow H_2O(l) + CN^-(aq)$
(d) $LiOH(aq) + HCN(aq) \longrightarrow H_2O(l) + LiCN(aq);$
$HCN(aq) + OH^-(aq) \longrightarrow H_2O(l) + CN^-(aq)$
**4.39** (a) $2KOH(aq) + H_3PO_4(aq) \longrightarrow$
$$2H_2O(l) + K_2HPO_4(aq);$$
$H_3PO_4(aq) + 2OH^-(aq) \longrightarrow 2H_2O(l) + HPO_4^{2-}(aq)$
(b) $3H_2SO_4(aq) + 2Al(OH)_3(s) \longrightarrow Al_2(SO_4)_3(aq) + 6H_2O(l);$
$3H^+(aq) + Al(OH)_3(s) \longrightarrow Al^{3+}(aq) + 3H_2O(l)$
(c) $2HC_2H_3O_2(aq) + Ca(OH)_2(aq) \longrightarrow$
$$2H_2O(l) + Ca(C_2H_3O_2)_2(aq);$$
$HC_2H_3O_2(aq) + OH^-(aq) \longrightarrow H_2O(l) + C_2H_3O_2^-(aq)$
(d) $H_2SO_3(aq) + NaOH(aq) \longrightarrow NaHSO_3(aq) + H_2O(l);$
$H_2SO_3(aq) + OH^-(aq) \longrightarrow HSO_3^-(aq) + H_2O(l)$
**4.41** $2H_2SO_3(aq) + Ca(OH)_2(aq) \longrightarrow$
$$Ca(HSO_3)_2(aq) + 2H_2O(l);$$
$Ca(HSO_3)_2(aq) + Ca(OH)_2(aq) \longrightarrow 2CaSO_3(s) + 2H_2O(l);$
$H_2SO_3(aq) + OH^-(aq) \longrightarrow H_2O(l) + HSO_3^-(aq);$
$Ca^{2}(aq) + HSO_3^-(aq) + OH^-(aq) \longrightarrow CaSO_3(s) + H_2O(l)$
**4.43** (a) $CaS(aq) + 2HBr(aq) \longrightarrow H_2S(g) + CaBr_2(aq);$
$2H^+(aq) + S^{2-}(aq) \longrightarrow H_2S(g)$
(b) $MgCO_3(s) + 2HNO_3(aq) \longrightarrow$
$$CO_2(aq) + H_2O(l) + Mg(NO_3)_2(aq);$$
$2H^+(aq) + MgCO_3(s) \longrightarrow CO_2(aq) + H_2O(l) + Mg^{2+}(aq)$
(c) $K_2SO_3(aq) + H_2SO_4(aq) \longrightarrow$
$$SO_2(g) + H_2O(l) + K_2SO_4(aq);$$
$2H^+(aq) + SO_3^{2-}(aq) \longrightarrow SO_2(g) + H_2O(l)$
**4.45** $FeS(s) + 2HCl(aq) \longrightarrow H_2S(g) + FeCl_2(aq);$
$FeS(s) + 2H^+(aq) \longrightarrow Fe^{2+}(aq) + H_2S(g)$   **4.47** (a) +3
(b) +4   (c) +7   (d) +6   **4.49** (a) −3   (b) +5   (c) +3
(d) +5   **4.51** (a) Mn: +2, Cl: +5, O: −2   (b) Fe: +3,
Cr: +6, O: −2   (c) Hg: +2, Cr: +6, O: −2   (d) Co: +2,
P: +5, O: −2

**4.53** (a) $P_4(s)$ + $5O_2(g)$ $\longrightarrow$ $P_4O_{10}(s)$
       reducing    oxidizing
        agent       agent
(b) $Co(s)$ + $Cl_2(g)$ $\longrightarrow$ $CoCl_2(s)$
   reducing    oxidizing
    agent       agent
**4.55** (a) $2Al(s)$ + $3F_2(g)$ $\longrightarrow$ $2AlF_3(s)$
       reducing    oxidizing
        agent       agent
(b) $Hg^{2+}(aq)$ + $NO_2^-(aq)$ + $H_2O(l)$ $\longrightarrow$
  oxidizing      reducing
   agent          agent
$$Hg(l) + 2H^+(aq) + NO_3^-(aq)$$
**4.57** (a) $3Cu^{2+}(aq) + 2Al(s) \longrightarrow 2Al^{3+}(aq) + 3Cu(s)$
(b) $2Cr^{3+}(aq) + 3Zn(s) \longrightarrow 2Cr(s) + 3Zn^{2+}(aq)$
**4.59** 1.02 $M$   **4.61** 0.101 $M$   **4.63** 1.25 L   **4.65** 35.8 mL
**4.67** $1.44 \times 10^{-4}$ mol   **4.69** 0.33 g   **4.71** 7.6 mL
**4.73** 0.302 g; 65.9%   **4.75** (a) 35.81%   (b) CuCl
**4.77** $FeCl_2$   **4.79** 50.9 mL nitric acid   **4.81** 163 mL
**4.83** 3.19%   **4.85** $Mg(s) + 2HBr(aq) \longrightarrow H_2(g) +$
$MgBr_2(aq); Mg(s) + 2H^+(aq) \longrightarrow H_2(g) + Mg^{2+}(aq)$
**4.87** $NiSO_4(aq) + 2LiOH(aq) \longrightarrow Ni(OH)_2(s) + Li_2SO_4(aq);$
$Ni^{2+}(aq) + 2OH^-(aq) \longrightarrow Ni(OH)_2(s)$
**4.89** (a) $LiOH(aq) + HCN(aq) \longrightarrow H_2O(l) + LiCN(aq);$
$OH^-(aq) + HCN(aq) \longrightarrow H_2O(l) + CN^-(aq)$
(b) $Li_2CO_3(aq) + 2HNO_3(aq) \longrightarrow$
$$2LiNO_3(aq) + H_2O(l) + CO_2(g)$$
$CO_3^{2-}(aq) + 2H^+(aq) \longrightarrow H_2O(l) + CO_2(g)$
(c) $LiCl(aq) + AgNO_3(aq) \longrightarrow LiNO_3(aq) + AgCl(s)$
$Cl^-(aq) + Ag^+(aq) \longrightarrow AgCl(s)$
(d) $LiCl(aq) + MgSO_4(aq) \longrightarrow NR$
**4.91** (a) $Sr(OH)_2(aq) + 2HC_2H_3O_2(aq) \longrightarrow$
$$Sr(C_2H_3O_2)_2(aq) + 2H_2O(l);$$
$OH^-(aq) + HC_2H_3O_2(aq) \longrightarrow H_2O(l) + C_2H_3O_2^-(aq)$
(b) $NH_4I(aq) + CsCl(aq) \longrightarrow NR$
(c) $NaNO_3(aq) + CsCl(aq) \longrightarrow NR$
(d) $NH_4I(aq) + AgNO_3(aq) \longrightarrow AgI(s) + NH_4NO_3(aq)$
$I^-(aq) + Ag^+(aq) \longrightarrow AgI(s)$
**4.93** (a) $BaCl_2(aq) + CuSO_4(aq) \longrightarrow BaSO_4(s) + CuCl_2(aq);$
$BaSO_4$ is filtered off, and the solution evaporated to give solid
$CuCl_2$
(b) $CaCO_3(s) + 2HC_2H_3O_2(aq) \longrightarrow$
$$CO_2(g) + H_2O(l) + Ca(C_2H_3O_2)_2(aq);$$
evaporate $H_2O$ to leave $Ca(C_2H_3O_2)_2(s)$
(c) $Na_2SO_3(s) + 2HNO_3(aq) \longrightarrow$
$$SO_2(g) + H_2O(l) + 2NaNO_3(aq);$$
evaporate $H_2O$ to leave $NaNO_3(s)$
(d) $Mg(OH)_2(s) + 2HCl(aq) \longrightarrow 2H_2O(l) + MgCl_2(aq);$
evaporate $H_2O$ to leave $MgCl_2(s)$
**4.95** (a) decomposition   (b) decomposition   (c) combination
(d) displacement   **4.97** (a) $Pb(NO_3)_2 + Ba(OH)_2$, $Pb(NO_3)_2 +$
$H_2SO_4$, $Pb(NO_3)_2 + MgSO_4$   (b) $Ba(OH)_2 + MgSO_4$
(c) $Ba(OH)_2 + H_2SO_4$   **4.99** 0.0405 $M$ $CaCl_2$, 0.0405 $M$ $Ca^{2+}$,
0.0811 $M$ $Cl^-$   **4.101** 329 mL   **4.103** 0.610 $M$   **4.105** 0.0967 $M$
**4.107** 21.30%   **4.109** $CuSO_4 \cdot 2H_2O$   **4.111** 35.0% Cl, $AuCl_3$
**4.113** 9.66%   **4.115** $Pb(NO_3)_2$; $Cs_2SO_4$;

$Pb(NO_3)_2(aq) + Cs_2SO_4(aq) \longrightarrow PbSO_4(s) + 2CsNO_3(aq)$;
$Pb^{2+}(aq) + SO_4^{2-}(aq) \longrightarrow PbSO_4(s)$;
lead(II) sulfate; cesium nitrate;
$Pb(NO_3)_2(aq) + Na_2SO_4(aq) \longrightarrow PbSO_4(s) + 2NaNO_3(aq)$
**4.117** $CaBr_2(aq) + Cl_2(g) \longrightarrow Br_2(l) + CaCl_2(aq)$;
$2Br^-(aq) + Cl_2(g) \longrightarrow Br_2(l) + 2Cl^-(aq)$; $5.67 \times 10^3$ g
**4.119** $Hg(NO_3)_2(aq) + H_2S(aq) \longrightarrow 2HNO_3(aq) + HgS(s)$,
$Hg^{2+}(aq) + H_2S(aq) \longrightarrow 2H^+(aq) + HgS(s)$, nitric acid,
mercury(II) sulfide, 581.5 g   **4.121** $Fe^{2+}:Fe^{3+}$ 5:2
**4.123** 0.0121 L   **4.125** 0.930 mol/L   **4.127** 69.72 g/mol,
gallium   **4.129** $P_4O_{10} + 6H_2O \longrightarrow 4H_3PO_4$; 55.7 g $P_4O_{10}$
**4.131** 85.23%   **4.133** 0.194 L   **4.135** 61.7%

## Chapter 5

**5.23** (a) At constant volume, the pressure of a gas is directly
proportional to temperature ($P \propto T$).   (b) At constant volume,
the pressure of a gas is directly proportional to temperature
($P \propto T$).   (c) The vapor pressure of the $H_2O$ increases with
increasing temperature (Table 5.6).   (d) At constant temperature, the pressure of a gas is inversely proportional to the volume
($P \propto 1/V$).   **5.25** (a) increase by 2   (b) decrease by 2   (c)
increase by 1.5   (d) $P$ no change; $V$ increase by 2   **5.27** (a) $O_2$
(b) $H_2$   (c) Both the same   (d) There will be no pressure change.
(e) 1/4   **5.29** Double the kelvin temperature or reduce the
pressure by 1/2.   **5.31** 36 mmHg   **5.33** 2.32 L   **5.35** $1.02 \times$
$10^3$ L   **5.37** $3.64 \times 10^{-4}$ kPa   **5.39** 2.50 mL   **5.41** 0.65 L
**5.43** 156 K   **5.45** 31.6 mL   **5.47** 1 volume   **5.49** $PV = nRT$;
$V = nRT/P$; if temperature and moles remain constant, $V =$
constant $\times 1/P$   **5.51** 8.02 atm   **5.53** 29.4 L   **5.55** 143°C
**5.57** 0.675 g/L   **5.59** 2.15 g/L   **5.61** 47.7 g/mol
**5.63** $1.60 \times 10^2$ g/mol   **5.65** Gas density depends on molecular
weight or average molecular weight of a mixture. Thus, the
density of a gas of $NH_4Cl$ would be greater than that of a mixture
of $NH_3$ and HCl, since $NH_3$ and HCl have lower molecular
weights than $NH_4Cl$.   **5.67** 1.4 L   **5.69** 160. L   **5.71** 493 L
**5.73** $2NH_3(g) + H_2SO_4(aq) \longrightarrow (NH_4)_2SO_4(aq)$; 46.7 L
**5.75** 0.279 atm   **5.77** $P(O_2) = 3.80 \times 10^{-3}$ atm; $P(He) =$
0.017 atm; P(total) = 0.020 atm   **5.79** $P(CO_2) = 494$ mmHg;
$P(H_2) = 190.$ mmHg; P(HCl) = 41 mmHg; P(HF) = 21 mmHg;
$P(SO_2) = 13$ mmHg; $P(H_2S) = 0.8$ mmHg   **5.81** 6.34 g
**5.83** At $T = 298$ K, $u = 5.15 \times 10^2$ m/s; at $T = 398$ K, $u = 5.95$
$\times 10^2$ m/s; graph as in Figure 5.20.   **5.85** $1.53 \times 10^2$ m/s
**5.87** $6.50 \times 10^3$ K ($6.23 \times 10^3$°C)   **5.89** rate $N_2$/rate $O_2 =$
1.069   **5.91** 4.6 s   **5.93** 146 amu   **5.95** 0.9605 atm (0.9716
atm from ideal gas law)   **5.97** At 1.00 atm: 22.4 L from ideal
gas law, 22.2 L from van der Waals equation; at 10.0 atm: 2.24 L
from ideal gas law, 2.08 L from van der Waals equation.
**5.99** 80.7 cm³   **5.101** 142 mL   **5.103** 6.5 dm³
**5.105** $3.01 \times 10^{20}$ atoms   **5.107** 28.979 amu   **5.109** $1.2 \times$
$10^3$ g LiOH   **5.111** 20.4 atm   **5.113** 0.449 $M$   **5.115** $H_2SO_4$
**5.117** −95°C   **5.119** 1.0043   **5.121** 28.10 g/mol
**5.123** $P(O_2) = 0.167$ atm; $P(CO_2) = 0.333$ atm
**5.125** 0.732 g/L   **5.127** 0.194 g $Na_2O_2$   **5.129** 76% $CaCO_3$,
24% $MgCO_3$   **5.131** 8.81 g/L   **5.133** $3.00 \times 10^2$ m/s

## Chapter 6

**6.25** small car   **6.27** $\Delta H = \Delta H_f - \Delta H_{vap}$   **6.29** (a) b   (b) a
**6.31** Equate $\Delta H$ for the burning to $\Delta H_f$ (products) minus $\Delta H_f$
(reactants), then solve for $\Delta H_f$ of $P_4S_3$. You will need $\Delta H_f$ of
$P_4O_{10}$ and $SO_2$.   **6.33** −106.4 kcal   **6.35** $7.2 \times 10^5$ J, $1.7 \times$
$10^5$ cal   **6.37** $5.24 \times 10^{-21}$ J/molecule   **6.39** positive; cold
**6.41** endothermic; +66.4 kJ   **6.43** $2Fe(s) + 6HCl(aq) \longrightarrow$
$2FeCl_3(aq) + 3H_2(g)$; $\Delta H = -175.8$ kJ   **6.45** $+2.940 \times 10^3$ kJ
**6.47** −452.8 kJ
**6.49** −1.90 kJ/g   **6.51** $-6.66 \times 10^2$ kJ   **6.53** 7.96 g
**6.55** $5.8 \times 10^4$ J   **6.57** 5.81°C   **6.59** 20.5 kJ   **6.61** $-1.36 \times$
$10^3$ kJ   **6.63** 50.6 kJ   **6.65** −906.3 kJ   **6.67** −137 kJ
**6.69** +42.2 kJ   **6.71** −1125 kJ   **6.73** −23.5 kJ   **6.75** −75.2 kJ
**6.77** 13.9 kJ   **6.79** kg · m²/s²   **6.81** $1.59 \times 10^3$ J/g
**6.83** 226 J, 31.6 m/s   **6.85** 38.0 kJ   **6.87** −255 kJ/mol
**6.89** 0.128 J/g°C   **6.91** 0.383 J/g°C   **6.93** −23.6 kJ/mol
**6.95** −871.7 kJ/mol   **6.97** −20 kJ   **6.99** 206.2 kJ
**6.101** 178 kJ   **6.103** −39.7 kJ   **6.105** −2225 kJ/mol
**6.107** 35.5 g   **6.109** 37.4°C   **6.111** −22 kJ   **6.113** 30.3 g
CO, 40.4 g $CO_2$   **6.115** 113 kJ   **6.117** (a) 1.88 kJ
(b) 11.9%   **6.119** $1.07 \times 10^3$ g LiOH

## Chapter 7

**7.19** frequency-doubled beam   **7.21** one   **7.23** $x$   **7.25** (a)
**7.27** 219.6 m   **7.29** $6.28 \times 10^{14}$/s   **7.31** $1.9 \times 10^2$ s
**7.33** $6.05802106 \times 10^{-7}$ m, $4.959 \times 10^{14}$/s   **7.35** $9.045 \times$
$10^{-28}$ J   **7.37** $3.72 \times 10^{-19}$ J   **7.39** just barely; red
**7.41** $2.34 \times 10^{14}$/s   **7.43** $1.22 \times 10^{-7}$ m, near UV
**7.45** 93.9 nm   **7.47** $4.699 \times 10^{-19}$ J   **7.49** 95.4 pm
**7.51** $7.28 \times 10^7$ m/s   **7.53** $1.52 \times 10^{-22}$ pm, much smaller
**7.55** $l = 0, 1, 2, 3$; $m_l = -3, -2, -1, 0, 1, 2, 3$   **7.57** 3, 7
**7.59** (a) 6$d$   (b) 5$g$   (c) 4$f$   (d) 6$p$   **7.61** (a) $m_s$ can only be
$+\frac{1}{2}$ or $-\frac{1}{2}$   (b) $l$ can only be as large as $n - 1$   (c) $m_l$ cannot
exceed +2 in magnitude   (d) $n$ cannot be 0
(e) $m_s$ can only be $+\frac{1}{2}$ or $-\frac{1}{2}$   **7.63** $6.51 \times 10^{14}$/s, $4.31 \times$
$10^{-19}$ J   **7.65** 485 nm, blue-green   **7.67** $6.55 \times 10^{14}$/s
**7.69** $5.59 \times 10^5$ m/s   **7.71** $4.35 \times 10^{-7}$ m   **7.73** 7
**7.75** $1.64 \times 10^{-7}$ m; near UV   **7.77** $1.50 \times 10^4$ eV
**7.79** (a) five   (b) seven   (c) three   (d) one   **7.81** 6$s$, 6$p$,
6$d$, 6$f$, 6$g$, 6$h$   **7.83** $5.01 \times 10^{-7}$ m   **7.85** $5.26 \times 10^{28}$
**7.87** $3.60 \times 10^{-19}$ J, $2.16 \times 10^2$ kJ/mol   **7.89** $1.94 \times 10^{-11}$ m

## Chapter 8

**8.25** $1s^1, 1s^2, 1s^3, 1s^4, 1s^42s^1, 1s^42s^2$   **8.27** Sr, Y   **8.29** Z
would form a cation of charge 2+.   **8.31** Ge, Sn   **8.33** (a) not
allowed; 2$s$ electrons should have opposite spins.   (b) allowed;
$1s^22s^22p^3$   (c) not allowed; second 2$p$ electrons should have
opposite spins.   (d) allowed; $1s^22s^22p^4$   **8.35** (a) possible
(b) possible   (c) impossible; 2$p$ subshell can hold no more than
six electrons.   (d) impossible; 3$d$ subshell can hold no more
than ten electrons.

**8.37**

1s    2p

**8.39** $1s^2 2s^2 2p^6 3s^2 3p^5$    **8.41** $1s^2 2s^2 2p^6 3s^2 3p^6 3d^3 4s^2$
**8.43** $4s^2 4p^5$    **8.45** $4d^{10} 5s^2$    **8.47** Group IIIA, period = 6, main-group element
**8.49** [Ar] ⇅⇅⇅↑↑  ⇅
3d        4s

**8.51** ⇅  ⇅  ⇅⇅⇅  ⇅  ⇅⇅⇅  ↑
1s    2s    2p      3s    3p      4s

Paramagnetic
**8.53** S, Se, As    **8.55** Na, Al, Cl, Ar    **8.57** (a) Br    (b) F
**8.59** $LiBrO_4$    **8.61** $1s^2 2s^2 2p^6 3s^2 3p^6 3d^{10} 4s^2 4p^6 5s^2$
**8.63** $6s^2 6p^4$
**8.65** ⇅  ⇅  ⇅⇅⇅  ⇅  ⇅⇅⇅  ⇅⇅⇅⇅⇅
1s    2s    2p      3s    3p      3d

⇅  ↑↑↑
4s    4p

**8.67** $[Rn]5f^{14} 6d^{10} 7s^2 7p^2$, metal, eka-PbO or eka-$PbO_2$
**8.69** 370 kJ/mol
**8.71** [Kr] ↑↑↑↑↑  ↑
4d        5s

**8.73** (a) $Cl_2$    (b) Na    (c) Sb    (d) Ar
**8.75** $1s^2 2s^2 2p^6 3s^2 3p^6 3d^3 4s^2$, Group VB, Period 4, *d*-block transition element
**8.77** $Ba(s) + 2H_2O(l) \longrightarrow H_2(g) + Ba(OH)_2(aq)$; 447 mL
**8.79** RaO, 93.4% Ra    **8.81** 108 J    **8.83** $1.313 \times 10^3$ kJ/mol H
**8.85** −639 kJ/mol

## Chapter 9

**9.19** $XY_3$    **9.21** b    **9.23** (a) Bond Cl and H to C.    (b) Bond O atoms to N, then bond H to one O.    (c) Bond O and F to N. (d) Place N atoms in center with two O atoms bonded to each N.
**9.25** (a) $Ca_3X_2$    (b) ionic

**9.27** (a) ·In·    (b) $In^{3+}$    (c) ·P·    (d) $\left[:\ddot{P}:\right]^{3-}$

**9.29** (a) Na· + ·Ï: ⟶ $Na^+$ + $\left[:\ddot{I}:\right]^-$

(b) :Ċl· + ·Ca· + ·Ċl: ⟶

$\left[:\ddot{Cl}:\right]^-$ + $Ca^{2+}$ + $\left[:\ddot{Cl}:\right]^-$

**9.31** (a) $1s^2 2s^2 2p^6 3s^2 3p^6$, $Ca^{2+}$

(b) $1s^2 2s^2 2p^6 3s^2 3p^6 3d^{10} 4s^2 4p^4$, :S̈e·

(c) $1s^2 2s^2 2p^6 3s^2 3p^6 3d^{10} 4s^2 4p^6$, $\left[:\ddot{Se}:\right]^{2-}$

(d) $1s^2 2s^2 2p^6 3s^2 3p^6 3d^{10} 4s^2 4p^6$, $\left[:\ddot{Br}:\right]^-$

**9.33** (a) $[Xe]4f^{14} 5d^{10} 6s^2 6p^3$    (b) $[Xe]4f^{14} 5d^{10} 6s^2$
**9.35** (a) $[Ar]3d^8$    (b) $[Ar]3d^7$    **9.37** (a) I, $I^-$; electron repulsion makes $I^-$ larger.    (b) $Ca^{2+}$, Ca; loss of valence electrons in Ca makes $Ca^{2+}$ smaller.    **9.39** $S^{2-}$, $Se^{2-}$, $Te^{2-}$. The radius increases with increasing number of filled shells.
**9.41** $Na^+$, $F^-$, $N^{3-}$. The ions are isoelectronic, so the atomic radius increases with decreasing nuclear charge.

**9.43** 3 H· + :Ṗ· ⟶ [P bonded to three H atoms showing bonding electron pairs and lone pair]
Bonding electron pairs
Lone pair

**9.45** $AsH_3$    **9.47** (a) P, N, O    (b) Na, Mg, Al
(c) Al, Si, C    **9.49** C—Br, As—Br, Si—O

**9.51**
$\overset{\delta+\ \ \delta-}{Si-O}$,  $\overset{\delta+\ \ \delta-}{C-Br}$,  $\overset{\delta+\ \ \delta-}{As-Br}$

**9.53** (a) :Br—Br:    (b) H—S̈e—H

(c) :F̈:N̈:
    :F̈:

**9.55** (a) :P≡P:    (b) :Ö=C—Br:
(with Br above carbon)

(c) H—Ö—N=Ö:

**9.57** (a) $\left[:\ddot{Cl}—\ddot{O}:\right]^-$    (b) $\left[:\ddot{Cl}—\underset{}{\overset{:\ddot{Cl}:}{Sn}}—\ddot{Cl}:\right]^-$

(c) $\left[:\ddot{S}—\ddot{S}:\right]^{2-}$

**9.59** (a) $\left[:\ddot{O}—\ddot{N}=\ddot{O}\right]^- \longleftrightarrow \left[\ddot{O}=\ddot{N}—\ddot{O}:\right]^-$

(b) H—Ö—N with :O: and ⟷ H—Ö—N with O and O:

**9.61** H—C—N (with H atoms on C, :O: and O) ⟷ H—C—N (with H atoms on C, O: and O)

**9.63** (a) :F—Xe—F:    (b) :F—Se—F: with :F: above and :F: below

(c) Te with F atoms (F, F top; F, F bottom) :F—Te—F:    (d) [Xe with F atoms]$^+$

**9.65** (a) :Cl—B—Cl: with :Cl: above    (b) [:Cl—Tl—Cl:]$^+$

(c) :Br—Be—Br:

**9.67** (a) O=O—O: with charges (−) (+)    (b) :C≡O: with (−)(+)

(c) H—O—N with O (+) and O: (−)

**9.69** (a) S with =O above, :F: :F: below    (b) S with =O above, O—H, O—H below

(c) O=Cl—O—H

**9.71** 174 pm    **9.73** (a) 114 pm    (b) 203 pm    (c) 213 pm
(d) 183 pm    **9.75** 147 pm, methylamine; 116 pm, acetonitrile
**9.77** −78 kJ    **9.79** (a) ionic, SrO, strontium oxide
(b) covalent, $CBr_4$, carbon tetrabromide    (c) ionic, $GaF_3$, gallium(III) fluoride    (d) covalent, $NBr_3$, nitrogen tribromide

**9.81** [:O—As—O: with :O: above and :O: below]$^{3-}$  or  [:O—As—O: with :O: above (double bond) and :O: below]$^{3-}$ , $Pb_3(AsO_4)_2$

**9.83** H—O—I—O: with :O: below  or  H—O—I=O with :O: below    **9.85** [H—N—H]$^-$

**9.87** [:O=N=O:]$^+$    **9.89** (a) :Cl—Se—Cl: or

:Cl—Se—Cl: with =O above (b) :Se=C=Se:

(c) [:Cl—Ga—Cl: with :Cl: above and :Cl: below]$^-$    (d) [:C≡C:]$^{2-}$

**9.91** (a) :Cl—Sb—Cl: with :Cl: below    (b) :I—C≡N:

(c) :Cl—I—Cl: with :Cl: below    (d) I with F, F top and F, F bottom and —F

**9.93** (a) :O=Se—O: ⟷ :O—Se=O:

(b) resonance structures of N—N with O groups (four structures shown)

**9.95** resonance structures with S and N in ring (four structures shown)

None of the resonance forms has formal charges of zero for all atoms.

**9.97** resonance structures of N—O—N with O groups (four structures shown)

The outer N—O bonds are 118 pm, and the inner N—O bonds are 136 pm.    **9.99** −134 kJ    **9.101** $-1.60 \times 10^2$ kJ
**9.103** $H_3PO_3(aq) + 2NaOH(aq) \longrightarrow Na_2HPO_3(aq) + 2H_2O(l)$, 0.007031 M
**9.105** $Mg(ClO_4)_2$, magnesium perchlorate, $Mg^{2+}$, and

[:O—Cl—O: with :O: above and :O: below]$^-$  or  [O=Cl=O with :O: above and :O: below]$^-$

**9.107**

**9.109** 261 amu; the compound is volatile, so is molecular;

**9.111** 1271 kJ/mol; 860 kJ/mol compared to table value of 887 kJ/mol   **9.113** 3.4   **9.115** 3.47

## Chapter 10

**10.17** (a) ii   (b)  i   (c) iv   (d) iii   **10.19** 120°; the O—C—O angle would be larger, the other angles are smaller than 120°
**10.21**

**10.23** (a) angular   (b) tetrahedral   (c) trigonal planar
(d) trigonal pyramidal   **10.25** (a) trigonal planar   (b) tetrahedral   (c) linear   (d) tetrahedral   **10.27** (a) 109°, less
(b) 109°, same   (c) 109°, less   (d) 120°; the two F—C=O bond angles should be larger than 120°, and the F—C—F angle should be smaller than 120°   **10.29** (a) trigonal bipyramidal
(b) T-shaped   (c) square pyramidal   (d) distorted tetrahedral
**10.31** (a) trigonal bipyramidal   (b) octahedral   (c) linear
(d) square planar   **10.33** (a) trigonal pyramidal and T-shaped
(b) bent   **10.35** $XeF_4$ has zero dipole by symmetry (square planar)   **10.37** (a) uses $sp^3$ orbitals, (b) uses $sp^3$ orbitals
**10.39** (a) $sp^2$   (b) $sp^2$   **10.41** (a) Two single bonds and no lone pair on Hg suggests $sp$ hybridization. An Hg—Cl bond is formed by overlapping an Hg hybrid orbital with a $3p$ orbital of Cl.
(b) The presence of three single bonds and one lone pair on P suggests $sp^3$ hybridization. Each hybrid orbital overlaps a $3p$ orbital of a Cl atom to form a P—Cl bond. The fourth hybrid orbital contains a lone pair.   **10.43** (a) $sp^3d^2$   (b) $sp^3d$   (c) $sp^3d$
(d) $sp^3d$   **10.45** The P atom in $PCl_6^-$ has 6 single bonds around it and no lone pairs. This suggests $sp^3d^2$ hybridization. Each bond is a $\sigma$ bond formed by overlap of an $sp^3d^2$ hybrid orbital on P with a $3p$ orbital on Cl.   **10.47** (a) The C atom is bonded to three other atoms; therefore, we assume $sp^2$ hybrid orbitals. One $2p$ orbital remains unhybridized. Each carbon–hydrogen bond is a $\sigma$ bond formed by the overlap of an $sp^2$ hybrid orbital on C with a $1s$ orbital on H. The remaining $sp^2$ hybrid orbital on C overlaps with a $2p$ orbital on O to form a $\sigma$ bond. The unhybridized $2p$ orbital on C overlaps with a parallel $2p$ orbital on O to form a $\pi$ bond. Together, the $\sigma$ and $\pi$ bonds describe the C=O bond.   (b) The nitrogen atoms are $sp$ hybridized, and a $\sigma$ bond is formed by the overlap of an $sp$ hybrid orbital from each N. The remaining $sp$ hybrid orbitals contain lone pairs of elec-

trons. The two unhybridized $2p$ orbitals on one N overlap with the unhybridized $2p$ orbitals on the other N to form two $\pi$ bonds.
**10.49**

Each of the N atoms has a lone pair of electrons and is bonded to two atoms. The N atoms are $sp^2$ hybridized. The two possible arrangements of the oxygen atoms relative to one another are shown above. Since the $\pi$ bond between the nitrogen atoms must be broken to interconvert these forms, it is expected that the hyponitrite ion will exhibit *cis–trans* isomerization.   **10.51**
(a) $KK(\sigma_{2s})^2(\sigma_{2s}^*)^2(\pi_{2p})^2$, bond order = 1, stable, paramagnetic
(b) $KK(\sigma_{2s})^2(\sigma_{2s}^*)^2(\pi_{2p})^1$, bond order = 1/2, stable, paramagnetic   (c) $KK(\sigma_{2s})^2(\sigma_{2s}^*)^2(\pi_{2p})^4(\sigma_{2p})^2(\pi_{2p}^*)^3$, bond order = 3/2, stable, paramagnetic
**10.53** $KK(\sigma_{2s})^2(\sigma_{2s}^*)^2(\pi_{2p})^4(\sigma_{2p})^2$, bond order = 3, diamagnetic
**10.55** (a) bent   (b) trigonal planar   (c) linear   (d) octahedral
**10.57** none   **10.59** Left:  Both carbon atoms with the double bonds are $sp^2$ hybridized. The carbon atom with the OH group is $sp^3$ hybridized. Right:  Both carbon atoms are $sp$ hybridized.
**10.61** The *trans* isomer is expected to have a zero dipole moment, whereas the *cis* isomer is not.   **10.63** The center carbon forms $\pi$ bonds with its neighbors, which are perpendicular to each other. Each $-CH_2$ is perpendicular to the adjacent $\pi$ bond and so must be perpendicular to each other.   **10.65** $(\sigma_{1s})^2$, stable
**10.67** $KK(\sigma_{2s})^2(\sigma_{2s}^*)^2(\pi_{2p})^4(\sigma_{2p})^2$; 3   **10.69** $O_2^+$ has one less antibonding electron than $O_2$; $O_2^-$ has a longer bond because it has one more antibonding electron.   **10.71** The $N_2$ ground state is $KK(\sigma_{2s})^2(\sigma_{2s}^*)^2(\pi_{2p})^4(\sigma_{2p})^2$, the first excited state is $KK(\sigma_{2s})^2(\sigma_{2s}^*)^2(\pi_{2p})^4(\sigma_{2p})^1(\pi_{2p}^*)^1$. The transition decreases the bond order from 3 to 2, and the ground state is diamagnetic while the excited state is paramagnetic.

**10.73** $XeO_3F_2$;

trigonal bipyramidal, $sp^3d$ hybrid
**10.75** $ClF_3$; the Cl atom has $sp^3d$ hybrid orbitals, giving three Cl—F bonds and two lone pairs.   **10.77** (a) $N_2$: 110 pm, linear geometry, $sp$ hybrids   (b) $N_2F_2$: 122 pm, trigonal planar geometry, $sp^2$ hybrids   (c) $N_2H_4$: 145 pm, tetrahedral geometry, $sp^3$ hybrids

**10.79**

trigonal planar geometry, $sp^2$ hybrids, $\Delta H_f^\circ = -40$ kJ/mol, resonance energy = 95 kJ

## Chapter 11

**11.21** (a) C   (b) A   (c) A   **11.23** The heat released when the liquid to solid phase change occurs prevents the fruit from freezing.   **11.25** (a) $AB_2$   (b) yes, body-centered cubic   **11.27** The molecules with the higher kinetic energy evaporate, leaving behind the lower energy molecules. Because the cup is insulated, the energy lost with the evaporated molecules is not rapidly replaced.   **11.29** (a) vaporization   (b) freezing of eggs and sublimation of ice   (c) condensation   (d) condensation (deposition)   (e) freezing   **11.31** 10°C   **11.33** 15.5 kJ/mol   **11.35** 3.50 kJ   **11.37** 1.25 g   **11.39** 27 g   **11.41** 264 mmHg   **11.43** 53.3 kJ/mol   **11.45** (a) gas   (b) gas   (c) liquid   (d) no **11.49** $SO_2$ and $C_2H_2$ can be lique-fied at 25°C; to liquefy $CH_4$, reduce temperature below $-82$°C, then apply pressure greater than 46 atm; to liquefy CO, reduce temperature below $-140$°C, then apply pressure greater than 35 atm.   **11.51** (a) solid   (b) liquid   **11.53** Values increase with molecular weight as expected.   **11.55** (a) London forces   (b) dipole–dipole forces, London forces, hydrogen bonding   (c) dipole–dipole forces, London forces   (d) London forces   **11.57** $CCl_4$, $SiCl_4$, $GeCl_4$   **11.59** $CCl_4$; highest molecular weight and greatest London forces.   **11.61** $HOCH_2CH_2OH$, $FCH_2CH_2OH$, $FCH_2CH_2F$; hydrogen bonding decreases in magnitude from left to right.   **11.63** $CH_4$, $C_2H_6$, $CH_3OH$, $CH_2OHCH_2OH$   **11.65** (a) metallic   (b) metallic   (c) covalent network   (d) molecular   (e) ionic   **11.67** (a) metallic   (b) covalent network   (c) molecular   (d) molecular   **11.69** $C_2H_5OC_2H_5$, $C_4H_9OH$, KCl, CaO   **11.71** (a) low-melting, brittle   (b) high-melting, hard, and brittle   (c) malleable, electrically conducting   (d) high-melting, hard   **11.73** (a) LiCl   (b) SiC   (c) $CHI_3$   (d) Co   **11.75** one   **11.77** $9.26 \times 10^{-23}$ g; $9.274 \times 10^{-23}$ g from AW   **11.79** 361 pm   **11.81** 4, fcc   **11.83** 19.25 g/cm³   **11.85** $1.72 \times 10^{-23}$ cm³, 160 pm   **11.87** Water vapor deposits as frost (deposition), frost melts (melting), liquid water evaporates (vaporization)   **11.89** 68.4%   **11.91** 80°C   **11.93** (a) condenses to liquid (and to solid at high pressure)   (b) condenses to solid   (c) remains a gas   **11.95** Hydrogen bonds form between H atom of OH group on one molecule and O atom on another molecule.   **11.97** Hydrogen bonding between ethylene glycol molecules leads to higher boiling point and greater viscosity.   **11.99** Al, metallic; Si, covalent network; P, molecular; S, molecular.   **11.101** (a) KCl; smaller charges on ions   (b) $CCl_4$; lower molecular weight and smaller London forces   (c) Zn; Group IIB metal is lower melting   (d) $C_2H_5Cl$; no hydrogen bonding.   **11.103** $3.171 \times 10^{-22}$ g, 191.0 amu   **11.105** 128 pm   **11.107** 68.0%   **11.109** (a) The boiling point increases with molecular size.   (b) Hydrogen bonding occurs between HF molecules.   (c) The hydrogen halides also have dipole–dipole interactions.   **11.111** (a) Diamond and silicon carbide are covalent network solids with strong covalent bonds; graphite is a layered structure with weak forces between layers.   (b) Silicon dioxide is a covalent network solid; carbon dioxide is a small, discrete molecule.   **11.113** (a) $CO_2$ consists of nonpolar mole-cules; $SiO_2$ is a covalent network solid.   (b) HF has hydrogen bonding.   (c) $SiF_4$ is a larger molecule.   **11.115** 0.220 mol $N_2$, 0.01023 mol $C_3H_8O$, 0.0444, 33.1 mmHg, 33.1 mmHg

**11.117** 7.6 kJ   **11.119** 449 J   **11.121** 0.42 g monomer, 1.42 g dimer, 1.83 g/L

# Chapter 12

**12.21** There is not enough dissolved oxygen in the warm water to support trout, which need more $O_2$.   **12.23** The lattice energy must be less for the XZ compound.   **12.25** The charged smoke particles are neutralized by the electric current, which allows them to aggregate into large particles. These large particles are too big to be carried out the stack.   **12.27** Because the salt concentration outside the lettuce leaf is higher, water will pass out of the lettuce leaf into the dressing via osmosis.   **12.29** Aqueous ammonia   **12.31** Ethanol   **12.33** $H_2O$, $CH_2OHCH_2OH$, $C_{10}H_{22}$   **12.35** $Al^{3+}$   **12.37** $Ba(IO_3)_2$, $Sr(IO_3)_2$, $Ca(IO_3)_2$, $Mg(IO_3)_2$. Because $IO_3^-$ is large, the change in lattice energy in these iodates is small. Therefore, the hydra-tion energy of the cation dominates the solubility trend.   **12.39** 0.886 g/100 mL   **12.41** Dissolve 1.81 g KI in 70.7 g $H_2O$   **12.43** 5.16 g   **12.45** 1.52 $m$   **12.47** 155 g   **12.49** 0.358, 0.642   **12.51** 0.0116   **12.53** 0.232, 16.8 $m$   **12.55** 0.563 $M$   **12.57** 0.796 $m$   **12.59** 41.6 mmHg, 0.631 mmHg   **12.61** $T_b = 100.0481$°C, $T_f = -0.175$°C   **12.63** $4.6 \times 10^{-2}$ $m$   **12.65** 122 amu   **12.67** 163 amu   **12.69** $6.85 \times 10^4$ amu   **12.71** $-0.050$°C   **12.73** $2.78 \approx 3$   **12.75** (a) aerosol   (b) sol   (c) foam   (d) sol   **12.77** $Al_2(SO_4)_3$   **12.79** 0.671 mol fraction $N_2$, 0.329 mol fraction $O_2$   **12.81** 1.74 $m$, 0.0303, 1.63 $M$   **12.83** 33 g of propane, 42 g of butane   **12.85** 139 mmHg   **12.87** (a) 2.500 $\times 10^{-4}$ $M$   (b) $5.000 \times 10^{-4}$ $M$   (c) $2.50 \times 10^{-4}$ $M$   **12.89** $-24$°C   **12.91** 0.30 $M$   **12.93** 0.10 $m$ $CaCl_2$   **12.95** 1.8 g/mL; $5.0 \times 10^2$ $m$   **12.97** $Mn_2C_{10}O_{10}$   **12.99** (a) $C_6H_{12}O_4$   (b) 148.2 g/mol   **12.101** 30. atm   **12.103** $Ca^{2+}(aq) + CO_3{}^{2-}(aq) \longrightarrow CaCO_3(s)$, $2Ag^+(aq) + CO_3{}^{2-}(aq) \longrightarrow Ag_2CO_3(s)$, 0.0750 $M$ $CO_3{}^{2-}$, 0.375 $M$ $Na^+$, 0.225 $M$ $NO_3{}^-$   **12.105** $-783$ kJ/mol, $-445$ kJ/mol   **12.107** 0.565 $m$   **12.109** 0.701 $M$   **12.111** 3.1%   **12.113** $C_3H_8O_3$

# Chapter 13

**13.31** No. Most aluminum is available in clays and aluminosili-cate minerals, which cannot be used economically as ores to obtain aluminum metal.   **13.33** Boron acts as a dopant (doping agent), which yields positive holes that become current carriers.   **13.35** Examples: $MgCaSi_3O_8$ and $KNaMgSi_3O_8$   **13.37** The impure lead metal serves as the anode and pure lead serves as the cathode. Impurities either remain at the anode or go into solution. Anode: $Pb(s) \longrightarrow Pb^{2+}(aq) + 2e^-$; Cathode: $PbSiF_6(aq) + 2e^- \longrightarrow Pb(s) + SiF_6{}^{2-}(aq)$   **13.39** $Fe_2O_3 + 3H_2 \longrightarrow 2Fe + 3H_2O$   **13.41** $3.69 \times 10^4$ g   **13.43** $-417.5$ kJ; exothermic   **13.45** The diagram is similar to the one for Na.   **13.47** $2.62 \times 10^{19}$   **13.49** Diamond   **13.51** 0.00187 L   **13.53** The hybrid orbitals as $sp^2$; the other $2p$

orbital on each C atom forms pi orbitals. **13.55** (b), (d) **13.57** Each Si atom has four tetrahedral bonds to other Si atoms, with an $sp^3$ orbital on one Si atom overlapping an $sp^3$ orbital on another Si atom. **13.59** 10.7 kg **13.61** Spodumene consists of a long chain of $SiO_4$ tetrahedra linked together, with the negative charge of this chain balanced by lithium and aluminum cations. Figure 13.31 shows the linking of two $SiO_4$ tetrahedra. Note that in this linkage, one O atom is shared by the two tetrahedra. The repeating unit of the $SiO_4$ chain of spodumene consists of an Si atom with two unshared O atoms and two shared O atoms (each shared atom counting as $\frac{1}{2}$ O atom to each repeating unit). The repeating unit also has a charge of 2− on it. Thus, the formula of the repeating unit is $SiO_3{}^{2-}$. Two of these repeating units would have a 4− charge, which in spodumene is balanced by an $Li^+$ ion and an $Al^{3+}$ ion, giving the formula $LiAl(SiO_3)_2$.
**13.63** $2ZnS(s) + 3O_2(g) \longrightarrow 2ZnO(s)\ 2SO_2(g)$
$ZnO(s) + C(s) \longrightarrow Zn(g) + CO(g)$
**13.65** 0.584 metric tons **13.67** Buckminsterfullerene is similar to graphite in being constructed from carbon-atom hexagons. However, buckminsterfullerene also contains carbon-atom pentagons, which allow the molecule to curl around to give a cage structure. **13.69** 689.9 kJ

# Chapter 14

**14.23** (a) Rate of depletion of $A = -\dfrac{\Delta [A]}{\Delta t}$ or Rate of formation of $B = +\dfrac{\Delta [B]}{\Delta t}$. (b) No, the rate of depletion of A would be faster than the rate of formation of B.
(c) $-\dfrac{\Delta [A]}{3\Delta t} = \dfrac{\Delta [B]}{2\Delta t}$
**14.25** (a) $x = 0$ (b) $x = 1$ (c) $x = 3$ **14.27** (a) region C (b) region A **14.29** A number of answers will work as long as you match one of the existing concentrations of A or B. For example, $[A] = 2.0\ M$ with $[B] = 2.0\ M$, or $[A] = 1.0\ M$ with $[B] = 2.0\ M$. **14.31** $-\frac{1}{2}\Delta[NO_2]/\Delta t = \Delta[O_2]/\Delta t$
**14.33** $\frac{1}{5}\Delta[Br^-]/\Delta t = \Delta[BrO_3{}^-]/\Delta t$ **14.35** $2.3 \times 10^{-2}\ M/hr$
**14.37** $1.2 \times 10^{-6}\ M/s$ **14.39** 1, 1, 2 **14.41** Orders with respect to $MnO_4{}^-$, $H_2C_2O_4$, and $H^+$ are 1, 1, 0, respectively. Overall order is 2. **14.43** rate = $k[CH_3NNCH_3]$, $k = 2.5 \times 10^{-4}/s$ **14.45** rate = $k[NO]^2[H_2]$, $k = 2.9 \times 10^2/(M^2 \cdot s)$
**14.47** rate = $k[ClO_2]^2[OH^-]$, $k = 2.3 \times 10^2/(M^2 \cdot s)$
**14.49** $2.1 \times 10^{-2}\ M$ **14.51** $5.2 \times 10^{-4}\ M$ **14.53** $1.1 \times 10^3$ s, $1.1 \times 10^3$ s, $2.2 \times 10^3$ s **14.55** 96 hr, $1.9 \times 10^2$ hr, $2.9 \times 10^2$ hr, $3.9 \times 10^2$ hr, $4.8 \times 10^2$ hr **14.57** $2.6 \times 10^2$ hr
**14.59** First order, $k = 0.101/s$ **14.61** $E_a = 210$ kJ
**14.63** $1.0 \times 10^5$ J/mol, $1.7 \times 10^{-3}/s$ **14.65** $8.38 \times 10^4$ J/mol
**14.67** $1.1 \times 10^2$ kJ/mol **14.69** $NOCl_2$; $2NO + Cl_2 \longrightarrow$ $2NOCl$ **14.71** (a) bimolecular (b) bimolecular (c) unimolecular (d) termolecular **14.73** (a) rate = $k[O_3]$ (b) rate = $k[NOCl_2][NO]$ **14.75** rate = $k[C_3H_6]^2$
**14.77** rate = $(k_2k_1/k_{-1}) \times [H_2][I_2] = k[H_2][I_2]$
**14.79** $2H_2O_2 \longrightarrow 2H_2O + O_2$; $Br^-$; yes ($BrO^-$) **14.81** $3.5 \times$

$10^{-6}\ M/s$, $3.2 \times 10^{-6}\ M/s$, $2.5 \times 10^{-6}\ M/s$ **14.83** Average $k = 2.5 \times 10^{-4}/s$ **14.85** $8.33 \times 10^3$ s **14.87** $5.50 \times 10^3$ s **14.89** 0.22 $M$ **14.91** $2.5 \times 10^{-4}/s$ **14.93** $1.14 \times 10^5$ J/mol, $5 \times 10^9$, 14 $M/s$ **14.95** rate = $k[NO_2][CO]$ **14.97** rate = $k_1[NO_2Br]$ **14.99** rate = $(k_2k_1/k_{-1}) \times [NH_4{}^+][OCN^-] = k[NH_4{}^+][OCN^-]$ **14.101** (a) rate = $k[NO]^2[O_2]$ (b) 0.12 mol/L $\cdot$ s **14.103** (a) i. decreases ii. decreases (b) i. increases ii. decreases **14.105** A catalyst provides another pathway with lower activation energy.
**14.107** (a) Rate is the change in concentration with time. (b) It changes because the concentration of reactant changes. (c) It equals $k$ when all reactants are 1.00 $M$.
**14.109** 6.8 L **14.111** 0.210 kJ/s
**14.113** $6.60 \times 10^{-8}$ mmHg/s

# Chapter 15

**15.13** $\frac{2}{3}x$ **15.15** a **15.17** The equilibrium composition is increased by increasing the pressure and decreasing the temperature. The rate of reaction is increased by increasing the temperature and using the proper catalyst. The choice of temperature is a compromise between increased equilibrium product concentration and rate of reaction **15.19** $K_c = 0.25$ **15.21** 1.05 mol $N_2O_3$, 0.45 mol $NO_2$, 0.45 mol NO **15.23** 0.0266 mol NO, 0.0134 mol $Br_2$, 0.0389 mol NOBr **15.25** 0.1187 mol CO, 0.2374 mol $H_2$, 0.0313 mol $CH_3OH$

**15.27** (a) $K_c = \dfrac{[NO_2][NO]}{[N_2O_3]}$

(b) $K_c = \dfrac{[H_2]^2\,[S_2]}{[H_2S]^2}$ (c) $K_c = \dfrac{[NO_2]^2}{[NO]^2\,[O_2]}$

(d) $K_c = \dfrac{[P(NH_2)_3]\,[HCl]^3}{[PCl_3]\,[NH_3]^3}$

**15.29** $2H_2S(g) + 3O_2(g) \rightleftharpoons 2H_2O(g) + 2SO_2(g)$

**15.31** $K_c = \dfrac{[N_2O_5]}{[NO_2]^2\,[O_2]^{1/2}}$

**15.33** $K_c = 0.543$ **15.35** 49.1

**15.37** 10.5 **15.39** $1.0 \times 10^{-3}$ **15.41** (a) $K_p = \dfrac{P_{HBr}{}^2}{P_{H_2}P_{Br_2}}$

(b) $K_p = \dfrac{P_{CH_4}P_{H_2S}{}^2}{P_{CS_2}P_{H_2}{}^4}$ (c) $K_p = \dfrac{P_{H_2O}{}^2P_{Cl_2}{}^2}{P_{HCl}{}^4P_{O_2}}$

(d) $K_p = \dfrac{P_{CH_3OH}}{P_{CO}P_{H_2}{}^2}$ **15.43** $3.0 \times 10^{-5}$ **15.45** 56.3

**15.47** (a) $K_c = \dfrac{[CO]^2}{[CO_2]}$ (b) $K_c = \dfrac{[CO_2]}{[CO]}$ (c) $K_c =$

$\dfrac{[CO_2]}{[SO_2][O_2]^{1/2}}$ (d) $K_c = [Pb^{2+}][I^-]^2$ **15.49** (a) not

complete (b) nearly complete **15.51** no, $3.2 \times 10^{-48}$ mol,

yes   **15.53** (a) goes to left   (b) goes to right   (c) equilibrium
(d) goes to left   **15.55** goes to left   **15.57** 0.37 $M$
**15.59** $[I_2] = [Br_2] = 4.7 \times 10^{-5} M$, $[IBr] = 5.1 \times 10^{-4} M$
**15.61** 0.18 mol $PCl_3$, 0.18 mol $Cl_2$, 0.32 mol $PCl_5$
**15.63** $[CO] = 0.0613 M$, $[H_2] = 0.1839 M$, $[CH_4] =$
0.0387 $M$, $[H_2O] = 0.0387 M$   **15.65** forward   **15.67** (a) no
effect   (b) no effect   (c) goes to left   **15.69** No, the fraction
of methanol would decrease.   **15.71** decrease   **15.73** low
temperature and low pressure   **15.75** 4.3 **15.77** 0.153
**15.79** reverse   **15.81** reverse   **15.83** 0.008 mol HBr,
0.0010 mol $H_2$, 0.0010 mol $Br_2$   **15.85** 13.2%
**15.87** 0.48 mol CO, 2.44 mol $H_2$, 0.52 mol $CH_4$, 0.52 mol $H_2O$
**15.89** endothermic

**15.91** $K_p = \dfrac{P_{NH_3}^2}{P_{N_2}P_{H_2}^3} = \dfrac{[NH_3]^2(RT)^2}{[N_2](RT)[H_2]^3(RT)^3} = K_c/(RT)^2$, so

$K_c = K_p(RT)^2$   **15.93** (a) 14.1   (b) 0.36 atm, 2.28 atm
(c) decreases   **15.95** (a) $[Cl_2] = [PCl_3] = 0.0306 M$, $[PCl_5] =$
0.0038 $M$   (b) 0.89   (c) Less $PCl_5$ would decompose.
**15.97** 0.335   **15.99** (a) $[SO_2] = [Cl_2] = 0.0346$ mol/L,
$[SO_2Cl_2] = 0.0266$ mol/L   (b) 0.565   (c) decreases
**15.101** (a) $7.2 \times 10^{-5}$, $1.6 \times 10^{-4}$   (b) hydrogen bonding
occurs   (c) decrease   **15.103** 106   **15.105** 0.430
**15.107** 0.408 atm

# Chapter 16

**16.15** Hydroxide ion forms in the reaction $NH_3(aq) + H_2O(l)$
$\rightleftharpoons NH_4^+(aq) + OH^-(aq)$   **16.17** The hydroxide ion acts as a
base and donates a pair of electrons on the O atom, forming a
bond with $CO_2$ to give the $HCO_3^-$ ion.   **16.19** The pH
increases.

**16.21** $\underset{\text{acid}}{HOBr} + \underset{\text{base}}{H_2O} \rightleftharpoons \underset{\text{acid}}{H_3O^+} + \underset{\text{base}}{OBr^-}$

**16.23** (a) $HPO_4^{2-}$   (b) $HSe^-$   (c) $NO_2^-$   (d) $AsO_4^{3-}$
**16.25** (a) HClO   (b) $AsH_4^+$   (c) $H_3PO_4$   (d) $HTeO_3^-$

**16.27** (a) $\underset{\text{acid}}{HSO_4^-} + \underset{\text{base}}{NH_3} \rightleftharpoons \underset{\text{base}}{SO_4^{2-}} + \underset{\text{acid}}{NH_4^+}$

(b) $\underset{\text{base}}{HPO_4^{2-}} + \underset{\text{acid}}{NH_4^+} \rightleftharpoons \underset{\text{acid}}{H_2PO_4^-} + \underset{\text{base}}{NH_3}$

(c) $\underset{\text{acid}}{Al(H_2O)_6^{3+}} + \underset{\text{base}}{H_2O} \rightleftharpoons \underset{\text{base}}{Al(H_2O)_5OH^{2+}} + \underset{\text{acid}}{H_3O^+}$

(d) $\underset{\text{base}}{SO_3^{2-}} + \underset{\text{acid}}{NH_4^+} \rightleftharpoons \underset{\text{acid}}{HSO_3^-} + \underset{\text{base}}{NH_3}$

**16.29** $BF_3 + :AsF_3 \longrightarrow BF_3:AsF_3$
   $BF_3$, Lewis acid; $AsF_3$, Lewis base.
**16.31** (a) $I^- + I_2 \longrightarrow I_3^-$

$$\underset{\text{base}}{\left[\ddot{\underset{..}{I}}:\right]^-} + \underset{\text{acid}}{:\ddot{\underset{..}{I}}\!:\!\ddot{\underset{..}{I}}\!:} \longrightarrow \left[\ddot{\underset{..}{I}} - \ddot{\underset{..}{I}} - \ddot{\underset{..}{I}}\!:\right]^-$$

(b) $GaBr_3 + Br^- \longrightarrow GaBr_4^-$

**16.33** (a) $\underset{\substack{\text{Lewis}\\\text{acid}}}{Cr^{3+}} + \underset{\substack{\text{Lewis}\\\text{base}}}{6H_2O} \longrightarrow Cr(H_2O)_6^{3+}$

(b) $\underset{\substack{\text{Lewis}\\\text{acid}}}{BF_3} + \underset{\substack{\text{Lewis}\\\text{base}}}{(C_2H_5)_2\ddot{O}:} \longrightarrow F_3B:\ddot{O}(C_2H_5)_2$

**16.35** $\underset{\substack{\text{Lewis}\\\text{acid}}}{H_2S} + \underset{\substack{\text{Lewis}\\\text{base}}}{HOCH_2CH_2NH_2} \longrightarrow$

$HS^- + HOCH_2CH_2NH_3^+$

The $H^+$ ion from $H_2S$ accepts a pair of electrons from the N atom
in $HOCH_2CH_2NH_2$.   **16.37** $HSO_4^- + ClO^- \longrightarrow SO_4^{2-} +$
HClO; reaction goes to the right because HClO is a weaker
acid than $HSO_4^-$.   **16.39** (a) left   (b) left   (c) right
(d) right   **16.41** trichloroacetic acid; an acid–base reaction
normally goes in the direction of the weaker acid.
**16.43** (a) $H_2S$; acid strength decreases with negative anion
charge   (b) $H_2SO_3$; acid strength increases with
electronegativity   (c) HBr; acid strength increases with
electronegativity   (d) $HIO_4$; acid strength increases with
number of O atoms   (e) $H_2S$; acid strength increases with
central-atom size   **16.45** (a) $[H_3O^+] = 2.5 M$,
$[OH^-] = 4.0 \times 10^{-15} M$   (b) $[H_3O^+] = 2.9 \times 10^{-14} M$,
$[OH^-] = 0.35 M$   (c) $[H_3O^+] = 7.7 \times 10^{-14} M$,
$[OH^-] = 0.13 M$   (d) $[H_3O^+] = 0.30 M$, $[OH^-] = 3.3 \times$
$10^{-14} M$   **16.47** $[H_3O^+] = 0.050 M$, $[OH^-] = 2.0 \times 10^{-13} M$
**16.49** $[H_3O^+] = 1.0 \times 10^{-12} M$, $[OH^-] = 0.010 M$
**16.51** (a) acidic   (b) acidic   (c) neutral   (d) basic
**16.53** acidic   **16.55** (a) basic   (b) acidic   (c) neutral
(d) acidic   **16.57** (a) acidic   (b) neutral   (c) basic
(d) acidic   **16.59** (a) 8.00   (b) 11.30   (c) 2.12   (d) 8.197
**16.61** 2.12   **16.63** (a) 5.72   (b) 11.92   (c) 2.56   (d) 6.32
**16.65** 11.60   **16.67** $7.6 \times 10^{-6} M$   **16.69** $4.3 \times 10^{-3} M$
**16.71** 13.16   **16.73** 5.5 to 6.5, acidic   **16.75** (a) BaO is a
base; $BaO + H_2O \longrightarrow Ba^{2+} + 2OH^-$   (b) $H_2S$ is an acid; $H_2S$
$+ H_2O \longrightarrow H_3O^+ + HS^-$   (c) $CH_3NH_2$ is a base; $CH_3NH_2 +$
$H_2O \longrightarrow CH_3NH_3^+ + OH^-$   (d) $SO_2$ is an acid; $SO_2 + 2H_2O$
$\longrightarrow H_3O^+ + HSO_3^-$   **16.77** (a) $H_2O_2(aq) + S^{2-}(aq) \longrightarrow$
$HO_2^-(aq) + HS^-(aq)$   (b) $HCO_3^-(aq) + OH^-(aq) \longrightarrow$
$CO_3^{2-}(aq) + H_2O(l)$   (c) $NH_4^+(aq) + CN^-(aq) \longrightarrow NH_3(aq)$
$+ HCN(aq)$
(d) $H_2PO_4^-(aq) + OH^-(aq) \longrightarrow HPO_4^{2-}(aq) + H_2O(l)$

**16.79** (a) $\underset{\text{base}}{ClO^-} + \underset{\text{acid}}{H_2O} \rightleftharpoons \underset{\text{acid}}{HClO} + \underset{\text{base}}{OH^-}$

$$\left[:\ddot{\underset{..}{Cl}} - \ddot{\underset{..}{O}}:\right]^- + H - \ddot{\underset{..}{O}} - H \longrightarrow :\ddot{\underset{..}{Cl}} - \ddot{\underset{..}{O}} - H + \left[:\ddot{\underset{..}{O}} - H\right]^-$$

(b) $NH_4^+ + NH_2^- \rightleftharpoons 2NH_3$

   acid      base

$$\left[\begin{matrix} & H & \\ & | & \\ H\!-\!\!&N&\!\!-\!H \\ & | & \\ & H & \end{matrix}\right]^+ + \left[\begin{matrix} & H & \\ & | & \\ & :\!N\!: & \\ & | & \\ & H & \end{matrix}\right]^- \longrightarrow 2\begin{matrix} & H & \\ & | & \\ :&N&\!\!-\!H \\ & | & \\ & H & \end{matrix}$$

**16.81** $HNO_2 + F^- \rightleftharpoons NO_2^- + HF$, should go in the direction written because HF is a slightly weaker acid than $HNO_2$
**16.83** $H_2S$, $H_2Se$, HBr  **16.85** $[H_3O^+] = 4.0 \times 10^{-14}$ M, $[OH^-] = 0.25$ M  **16.87** 2.82  **16.89** $1.4 \times 10^{-11}$ M
**16.91** (a) $H_3O^+(aq) + HCO_3^-(aq) \longrightarrow CO_2(g) + 2H_2O(l)$
$H_2SO_4(aq) + 2NaHCO_3(aq) \longrightarrow$
$$Na_2SO_4(aq) + CO_2(g) + H_2O(l)$$
(b) 0.0180 mol  (c) 60.5% $NaHCO_3$
**16.93** $HCO_3^-(aq) + H_2O(l) \rightleftharpoons H_3O^+(aq) + CO_3^{2-}(aq)$
$HCO_3^-(aq) + H_2O(l) \rightleftharpoons H_2CO_3(aq) + OH^-(aq)$
$HCO_3^-(aq) + \cancel{Na^+(aq)} + OH^-(aq) \rightleftharpoons$
$$\cancel{Na^+(aq)} + CO_3^{2-}(aq) + H_2O$$
$HCO_3^-(aq) + H^+(aq) + \cancel{Cl^-(aq)} \rightleftharpoons H_2O + CO_2 + \cancel{Cl^-(aq)}$
**16.95** $CaH_2(s) + 2H_2O(l) \longrightarrow Ca(OH)_2(s) + H_2(g)$; hydride ion is the stronger base  **16.97** (a) $2HF(l) \rightleftharpoons$
$H_2F^+ + F^-$  (b) base  (c) $HClO_4 + HF \longrightarrow H_2F^+ + ClO_4^-$;
$H_2F^+$  **16.99** 11.62

**16.101**
$$\begin{matrix} & OH & \\ & | & \\ HO\!-\!\!&P&\!\!=\!O \\ & | & \\ & H & \end{matrix}$$
; 0.976 g NaOH

**16.103** $BF_3 + :NH_3 \longrightarrow F_3B:NH_3$; $BF_3$ is the Lewis acid, and $NH_3$ is the Lewis base; 12.5 g

## Chapter 17

**17.17** HX, HZ  **17.19** HBr, $NH_4Cl$, KBr, $CH_3NH_2$
**17.21** $NH_3$, NaF, NaCl, $NH_4Br$  **17.23** Look at the pH corresponding to the middle of the vertical portion of the titration curve. Titration of a strong acid with a strong base (pH 7); titration of a weak acid with a strong base (pH > 7); titration of a weak base with a strong acid (pH < 7).  **17.25** (a) HOCN(aq) $+ H_2O(l) \rightleftharpoons OCN^-(aq) + H_3O^+(aq)$
(b) $HF(aq) + H_2O(l) \rightleftharpoons H_3O^+(aq) + F^-(aq)$
(c) $HIO(aq) + H_2O(l) \rightleftharpoons H_3O^+(aq) + IO^-(aq)$
(d) $HCO_2H(aq) + H_2O(l) \rightleftharpoons CO_2H^-(aq) + H_3O^+(aq)$
**17.27** $5.6 \times 10^{-5}$  **17.29** 5.42, $1.5 \times 10^{-4}$
**17.31** $[H_3O^+] = [C_6H_4NH_2COO^-] = 1.0 \times 10^{-3}$ M
**17.33** 0.26 M  **17.35** $4.9 \times 10^{-3}$ M, 2.31  **17.37** 0.361 M
**17.39** (a) $3.7 \times 10^{-3}$ M  (b) $3.9 \times 10^{-6}$ M
**17.41** $CH_3NH_2(aq) + H_2O(l) \rightleftharpoons$
$$CH_3NH_3^+(aq) + OH^-(aq)$$
$K_b = \dfrac{[CH_3NH_3^+][OH^-]}{[CH_3NH_2]}$
**17.43** $3.2 \times 10^{-5}$  **17.45** $5.1 \times 10^{-3}$ M, 11.71
**17.47** (a) No hydrolysis

(b) $CHO_2^-(aq) + H_2O(l) \rightleftharpoons HCHO_2(aq) + OH^-(aq)$
$K_b = \dfrac{[HCHO_2][OH^-]}{[CHO_2^-]}$
(c) $CH_3NH_3^+(aq) + H_2O(l) \rightleftharpoons H_3O^+(aq) + CH_3NH_2(aq)$
$K_a = \dfrac{[H_3O^+][CH_3NH_2]}{[CH_3NH_3^+]}$
(d) $IO^-(aq) + H_2O(l) \rightleftharpoons HIO(aq) + OH^-(aq)$
$K_b = \dfrac{[HIO][OH^-]}{[IO^-]}$
**17.49** $Zn(H_2O)_6^{2+}(aq) + H_2O(l) \rightleftharpoons$
$$Zn(H_2O)_5(OH)^+(aq) + H_3O^+(aq)$$
**17.51** (a) acidic  (b) basic  (c) basic  (d) acidic
**17.53** (a) nearly neutral  (b) acidic  **17.55** (a) $2.2 \times 10^{-11}$
(b) $7.1 \times 10^{-6}$  **17.57** 8.64, $4.4 \times 10^{-6}$ M  **17.59** $1.0 \times 10^{-3}$
M, 2.99  **17.61** (a) 0.029  (b) 0.0065  **17.63** 3.17
**17.65** 10.47  **17.67** 3.45  **17.69** 9.26, 9.09  **17.71** 2.68
**17.73** 5.32  **17.75** 0.34 mol  **17.77** 1.60  **17.79** 8.59
**17.81** 5.97  **17.83** 9.08  **17.85** $1.1 \times 10^{-3}$
**17.87** $1.1 \times 10^{-2}$  **17.89** $K_b(CN^-) = 2.0 \times 10^{-5}$;
$K_b(CO_3^{2-}) = 2.1 \times 10^{-4}$; $CO_3^{2-}$  **17.91** 2.84  **17.93** 3.16
**17.95** 11/1  **17.97** 2.4  **17.99** 3.11  **17.101** (a) 0.100 M
(b) 0.11 M  **17.103** (a) $2.3 \times 10^{-11}$  (b) $4.3 \times 10^{-4}$
(c) 10.38  **17.105** (a) true  (b) true  (c) false  (d) false
(e) false  (f) true  **17.107** (a) 74.0 g/mol  (b) $1.3 \times 10^{-5}$
**17.109** (a) Initially: pH = 11.13; 30% titration: pH = 9.62; 50% titration: pH = 9.26; 100% titration: pH = 5.28  (b) acidic
**17.111** (a) $H_3PO_4$ and $H_2PO_4^-$  (b) 8 mL $H_3PO_4(aq)$ and 42 mL $H_2PO_4^-(aq)$  **17.113** (a) 0.215 M  (b) 3.55
(c) bromphenol blue
**17.115** (a) $NH_3(aq) + H_3O^+(aq) \longrightarrow NH_4^+(aq) + H_2O(l)$
(b) $NH_4^+(aq) + OH^-(aq) \longrightarrow NH_3(aq) + H_2O(l)$
(c) moles $NH_3$ before/after: 0.10/0.090; moles $NH_4^+$ before/after: 0.10/0.11; pH 9.2  (d) this is a buffer system
**17.117** (a) 0.59  (b) 7.16  (c) 7.55  **17.119** (a) $H_2A + 2H_2O \rightleftharpoons 2H_3O^+ + A^{2-}$  (b) $[H_2A] \gg [H_3O^+] = [HA^-] \gg [A^{2-}]$  (c) pH = 2.34; $[H_2A] = 0.0205$ M  (d) $4.6 \times 10^{-5}$ M
**17.121** (a) 0.13 M  (b) $1.9 \times 10^2$ mL  (c) 0.83 M
**17.123** 4.1%  **17.125** $-1.8°C$  **17.127** 9.1 mL

## Chapter 18

**18.11** From the $K_{sp}$ values, these are the more soluble: (a) silver chloride (b) magnesium hydroxide  **18.13** a  **18.15** middle beaker  **18.17** Add just enough $Na_2SO_4$ to precipitate all $Ba^{2+}$; filter off $BaSO_4$; add more $Na_2SO_4$ to precipitate all $Ca^{2+}$; filter off $CaSO_4$; $Mg^{2+}$ remains in solution.  **18.19** (a) soluble (b) insoluble  (c) insoluble  (d) soluble  **18.21** (a) $K_{sp} = [Sr^{2+}][SO_4^{2-}]$  (b) $K_{sp} = [Mg^{2+}][OH^-]^2$  (c) $K_{sp} = [Ca^{2+}]^3[PO_4^{3-}]^2$  (d) $K_{sp} = [Ag^+]^2[S^{2-}]$  **18.23** $9.3 \times 10^{-10}$
**18.25** $1.2 \times 10^{-7}$  **18.27** $1.8 \times 10^{-11}$  **18.29** 0.0045 g/L
**18.31** $1.9 \times 10^{-3}$ M  **18.33** $2.3 \times 10^{-4}$ g/L  **18.35** $1.9 \times$

$10^{-5}$ g/L   **18.37** 0.4 g/L   **18.39** (a) yes   (b) yes
**18.41** Lead chromate will precipitate.   **18.43** no precipitate
**18.45** no precipitate   **18.47** 0.0018 mol
**18.49** $6.9 \times 10^{-9} M$   **18.51** $BaF_2(s) + 2H_3O^+(aq) \longrightarrow$
$Ba^{2+}(aq) + 2HF(aq) + 2H_2O(l)$   **18.53** $BaF_2$; $F^-$ is the conjugate base of the weak acid HF.   **18.55** $Cu^+(aq) + 2CN^-(aq)$
$\longrightarrow Cu(CN)_2^-(aq)$

$$K_f = \frac{[Cu(CN)_2^-]}{[Cu^+][CN^-]^2}$$

**18.57** $5.5 \times 10^{-19} M$   **18.59** will precipitate   **18.61** $3.0 \times 10^{-3} M$   **18.63** Add HCl; $Pb^{2+}$ will be precipitated as the chloride. Filter off $PbCl_2$; add $H_2S$ in 0.3 $M$ $H^+$. $Cd^{2+}$ will precipitate as the sulfide; filter off CdS. The filtrate contains $Sr^{2+}$.
**18.65** $Mn^{2+}$   **18.67** $1.3 \times 10^{-4} M$   **18.69** (a) $6.9 \times 10^{-7} M$;
(b) $3.2 \times 10^{-4}$ g/L   **18.71** (a) $5.2 \times 10^{-6} M$;
(b) pOH = 4.81   **18.73** 26 g/L   **18.75** $1.8 \times 10^{-9} M$
**18.77** $8.0 \times 10^{-3} M$   **18.79** $1.0 \times 10^{-5}$; unsaturated
**18.81** $5.5 \times 10^{-6}$ g   **18.83** $4.7 \times 10^{-2} M$   **18.85** $1.4 \times 10^{-2} M$   **18.87** $3.6 \times 10^{-4} M$, $4.2 \times 10^{-6} M$, $2.5 \times 10^9$
**18.89** $1.0 \times 10^{-5} M$   **18.91** (a) $6.4 \times 10^{-7} M$ (b) $2.4 \times 10^{-4}$ $M$ (c) yes   **18.93** (a) $6.6 \times 10^{-4} M$   (b) 11%
**18.95** (a) $6.3 \times 10^{-16}$   (b) $8.7 \times 10^{-9} M$   (c) common ion effect   **18.97** (a) $3.1 \times 10^{-3}$   (b) 0.044 mol; 0.044 $M$; 0.89 mol   **18.99** (a) $1.5 \times 10^{-4} M$   (b) 99.0%   **18.101** 0.18 $M$
**18.103** $2.7 \times 10^{-4} M$   **18.105** $Ba^{2+}(aq) + 2OH^-(aq) +$
$Mg^{2+}(aq) + SO_4^{2-}(aq) \longrightarrow BaSO_4(s) + Mg(OH)_2(s)$; $[SO_4^{2-}]$
$= 0.109\ M$; $[Ba^{2+}] = 1.0 \times 10^{-9} M$; $[Mg^{2+}] = 0.109\ M$; $[OH^-]$
$= 1.3 \times 10^{-5} M$

## Chapter 19

**19.17** (a) false   (b) false   (c) false   (d) true   (e) false
**19.19** (a) 2.0 mol $CO_2$   (b) butane gas   (c) $CO_2$ at $-80°C$
(d) bromine vapor   **19.21** The process is not spontaneous; $\Delta G$ is positive; $\Delta H$ is negative; this implies that $\Delta S$ is negative, which is consistent with an increase in order.   **19.23** $-82$ J,
29 J, $-53$ J   **19.25** 37.58 kJ   **19.27** 106 J/K   **19.29** $-188$ J/K,
$1.90 \times 10^2$ J/K for 1.50 mol   **19.31** (a) negative   (b) not predictable   (c) positive   (d) positive   **19.33** (a) $-181.6$ J/K
(b) 43.2 J/K   (c) $-56.1$ J/K   (d) 313 J/K
**19.35** $-242.6$ J/K; entropy decreases, as expected, since moles of gas decrease.   **19.37** $-1453.2$ kJ, $-162$ J/K, $-1404.9$ kJ
**19.39** (a) $K(s) + \frac{1}{2}Br_2(l) \longrightarrow KBr(s)$
(b) $C(graphite) + \frac{3}{2}H_2(g) + \frac{1}{2}Cl_2(g) \longrightarrow CH_3Cl(l)$
(c) $H_2(g) + \frac{1}{8}S_8$ (rhombic) $\longrightarrow H_2S(g)$
(d) $As(s) + \frac{3}{2}H_2(g) \longrightarrow AsH_3(g)$   **19.41** (a) $-800.8$ kJ
(b) $-55.8$ kJ   **19.43** (a) spontaneous   (b) spontaneous
(c) nonspontaneous   (d) equilibrium mixture with significant amounts of reactants and products   (e) nonspontaneous
**19.45** (a) 851 kJ; 838 kJ; endothermic reaction with mainly reactants at equilibrium.   (b) $-72$ kJ; $-142$ kJ; exothermic reaction with mainly products at equilibrium.
**19.47** $-474.4$ kJ; 0   **19.49** $-15.7$

**19.51** (a) $K = \dfrac{P_{CO_2}P_{H_2}}{P_{CO}P_{H_2O}}$   (b) $K = [Mg^{2+}][OH^-]^2$
(c) $K = [Li^+]^2[OH^-]^2 P_{H_2}$   **19.53** $-107$ kJ; $6 \times 10^{18}$
**19.55** $-142.2$ kJ; $8.3 \times 10^{24}$   **19.57** $-149.92$ kJ; $1.84 \times 10^{26}$
**19.59** $1.2 \times 10^2$; yes   **19.61** 383 K   **19.63** Reaction is spontaneous because $\Delta S°$ is large and positive.   **19.65** $-184$ kJ;
exothermic; positive; both $\Delta H$ and $-T\Delta S$ are negative, so reaction is spontaneous at all $T$.   **19.67** $-14.3$ J/K
**19.69** (a) negative   (b) positive   (c) positive   (d) negative
**19.71** negative   **19.73** $-136$ J/K   **19.75** no, $\Delta G°$ is $+267$ kJ
**19.77** $\Delta H°$ is positive; $\Delta S°$ is positive   **19.79** 56 kJ; $2 \times 10^{-10}$
**19.81** $1.10 \times 10^2$ kJ; 137 J/K; 69 kJ; $-37$ kJ;
nonspontaneous at 25°C, but spontaneous at 800°C
**19.83** (a) $-74.5$ kJ   (b) $-80.8$ J/mol $\cdot$ K   (c) $-50.4$ kJ/mol
**19.85** 205 J/mol $\cdot$ K   **19.87** (a) 97.1 kJ; 116 J/K; 187.2 kJ;
208 J/K   (b) 837 K; 900 K   (c) carbon   **19.89** (a) formic acid   (b) products   **19.91** (a) products   (b) Need to heat to increase rate of reaction.   **19.93** (a) $-122$ kJ/mol
(b) $-2748$ kJ/mol   (c) $-446$ J/mol $\cdot$ K
**19.95** (a) $5.4 \times 10^{-3}$   (b) change in entropy
**19.97** 0.0040%, same; $K = 4.0 \times 10^{-10}$   **19.99** 87%
**19.101** $-2.29$ kJ/mol; $-98.7$ J/(mol $\cdot$ K)

## Chapter 20

**20.19** (a) No change   (b) No change   (c) The Zn strip would dissolve, the blue color of the solution would fade, and solid copper would precipitate out of the solution.   (d) No change   **20.21** The Zn is a sacrificial electrode that keeps the hull from undergoing oxidation by the dissolved ions in seawater. Zinc works because it is more easily oxidized than Fe.
**20.23** Because more zinc is present, the oxidation–reduction reactions in the battery will run for a longer period of time. This assumes that the zinc is the limiting reactant.   **20.25** A stain is visible in clothing because it is absorbing only a portion of the visible light striking the material (which means some light is reflected off of the cloth, giving the stain color). When you bleach the stain, the compounds that make up the stain are oxidized. This oxidation often changes the electronic properties of the stain to make it absorb no colors in the visible region, thus reflecting back all wavelengths of visible light, making the stain "disappear."   **20.27** (a) $Cr_2O_7^{2-} + 14H^+ + 3C_2O_4^{2-} \longrightarrow$
$2Cr^{3+} + 6CO_2 + 7H_2O$
(b) $3Cu + 2NO_3^- + 8H^+ \longrightarrow 3Cu^{2+} + 2NO + 4H_2O$
(c) $MnO_2 + H^+ + HNO_2 \longrightarrow Mn^{2+} + H_2O + NO_3^-$
(d) $5PbO_2 + 2Mn^{2+} + 5SO_4^{2-} + 4H^+ \longrightarrow$
$5PbSO_4 + 2MnO_4^- + 2H_2O$
(e) $3HNO_2 + Cr_2O_7^{2-} + 5H^+ \longrightarrow 2Cr^{3+} + 3NO_3^- + 4H_2O$
**20.29** (a) $Mn^{2+} + H_2O_2 + 2OH^- \longrightarrow MnO_2 + 2H_2O$
(b) $2MnO_4^- + 3NO_2^- + H_2O \longrightarrow 2MnO_2 + 3NO_3^- + 2OH^-$
(c) $Mn^{2+} + 2ClO_3^- \longrightarrow MnO_2 + 2ClO_2$
(d) $MnO_4^- + 3NO_2 + 2OH^- \longrightarrow MnO_2 + 3NO_3^- + H_2O$
(e) $3Cl_2 + 6OH^- \longrightarrow 5Cl^- + ClO_3^- + 3H_2O$
**20.31** (a) $8H_2S + 16NO_3^- + 16H^+ \longrightarrow 16NO_2 + S_8 + 16H_2O$

(b) $2NO_3^- + 3Cu + 8H^+ \longrightarrow 2NO + 3Cu^{2+} + 4H_2O$

(c) $2MnO_4^- + 5SO_2 + 2H_2O \longrightarrow 5SO_4^{2-} + 2Mn^{2+} + 4H^+$

(d) $2Bi(OH)_3 + 3Sn(OH)_3^- + 3OH^- \longrightarrow 3Sn(OH)_6^{2-} + 2Bi$

**20.33** (a) $2MnO_4^- + I^- + H_2O \longrightarrow 2MnO_2 + IO_3^- + 2OH^-$

(b) $Cr_2O_7^{2-} + 6Cl^- + 14H^+ \longrightarrow 2Cr^{3+} + 3Cl_2 + 7H_2O$

(c) $3S_8 + 32NO_3^- + 32H^+ \longrightarrow 24SO_2 + 32NO + 16H_2O$

(d) $3H_2O_2 + 2MnO_4^- \longrightarrow 3O_2 + 2MnO_2 + 2H_2O + 2OH^-$

(e) $5Zn + 2NO_3^- + 12H^+ \longrightarrow 5Zn^{2+} + N_2 + 6H_2O$

**20.35** Anode: $Mg \longrightarrow Mg^{2+} + 2e^-$

Cathode: $Ni^{2+} + 2e^- \longrightarrow Ni$

Electron flow in circuit is from Mg to Ni.

**20.37** $Zn(s) \longrightarrow Zn^{2+}(aq) + 2e^-$

$Ag^+(aq) + e^- \longrightarrow Ag(s)$

**20.39** Anode: $Zn(s) + 2OH^-(aq) \longrightarrow Zn(OH)_2(s) + 2e^-$

Cathode: $Ag_2O(s) + H_2O(l) + 2e^- \longrightarrow 2Ag(s) + 2OH^-(aq)$

Overall: $Zn(s) + Ag_2O(s) + H_2O(l) \longrightarrow 2Ag(s) + Zn(OH)_2(s)$

**20.41** $Ni(s)|Ni^{2+}(aq)\|Pb^{2+}(aq)|Pb(s)$

**20.43** $Ni(s)|Ni^{2+}(1\ M)\|H^+(1\ M)|H_2(g)|Pt$

**20.45** $2Ag^+(aq) + Fe(s) \longrightarrow Fe^{2+}(aq) + 2Ag(s)$

**20.47** Anode: $Cd(s) \longrightarrow Cd^{2+}(aq) + 2e^-$

Cathode: $Ni^{2+}(aq) + 2e^- \longrightarrow Ni(s)$

Overall: $Cd(s) + Ni^{2+}(aq) \longrightarrow Cd^{2+}(aq) + Ni(s)$

**20.49** $-69$ kJ **20.51** $-95$ kJ **20.53** $NO_3^-(aq)$, $O_2(g)$, $MnO_4^-(aq)$ **20.55** strongest is $Zn(s)$; weakest is $Cu^+(aq)$

**20.57** (a) not spontaneous (b) spontaneous **20.59** $Cl_2$ will oxidize $Br^-$; $Cl_2(g) + 2Br^-(aq) \longrightarrow Br_2(l) + 2Cl^-(aq)$

**20.61** 1.54 V **20.63** 1.28 V **20.65** $-3.6 \times 10^5$ J

**20.67** $-1.6 \times 10^5$ J **20.69** 3.56 V **20.71** 1.93 V

**20.73** $3.4 \times 10^{14}$ **20.75** $1.2 \times 10^6$ **20.77** 0.57 V

**20.79** 0.29 V **20.81** $4 \times 10^{-3}\ M$ **20.83** (a) Cathode:

$Ca^{2+}(l) + 2e^- \longrightarrow Ca(l)$; anode: $S^{2-}(l) \longrightarrow S(l) + 2e^-$

(b) $Cs^+(l) + e^- \longrightarrow Cs(l)$

$4OH^-(l) \longrightarrow O_2(g) + 2H_2O(g) + 4e^-$

**20.85** (a) Anode: $2H_2O(l) \longrightarrow O_2(g) + 4H^+(aq) + 4e^-$;

cathode: $2H_2O(l) + 2e^- \longrightarrow H_2(g) + 2OH^-(aq)$;

overall: $2H_2O(l) \longrightarrow 2H_2(g) + O_2(g)$

(b) Anode: $2Br^-(aq) \longrightarrow Br_2(l) + 2e^-$; cathode: $2H_2O(l) + 2e^- \longrightarrow H_2(g) + 2OH^-(aq)$; overall:

$Br^-(aq) + 2H_2O(l) \longrightarrow Br_2(l) + H_2(g) + 2OH^-(aq)$

**20.87** $3.87 \times 10^7$ C **20.89** 0.360 g

**20.91** (a) $3Fe + 2HNO_3 + 6H^+ \longrightarrow 3Fe^{2+} + 2NO + 4H_2O$

(b) $3Fe^{2+} + HNO_3 + 3H^+ \longrightarrow 3Fe^{3+} + NO + 2H_2O$

(c) $Fe + HNO_3 + 3H^+ \longrightarrow Fe^{3+} + NO + 2H_2O$

**20.93** (a) $16MnO_4^- + 24S^{2-} + 32H_2O \longrightarrow$
$$16MnO_2 + 3S_8 + 64OH^-$$

(b) $IO_3^- + 3HSO_3^- \longrightarrow I^- + 3SO_4^{2-} + 3H^+$

(c) $3Fe(OH)_2 + CrO_4^{2-} + 4H_2O \longrightarrow$
$$3Fe(OH)_3 + Cr(OH)_4^- + OH^-$$

(d) $Cl_2 + 2OH^- \longrightarrow Cl^- + ClO^- + H_2O$

**20.95** $4Fe(OH)_2 + O_2 + 2H_2O \longrightarrow 4Fe(OH)_3$

**20.97** $Ca(s)|Ca^{2+}(aq)\|Cl^-(aq)|Cl_2(g)|Pt(s)$; anode:

$Ca(s) \longrightarrow Ca^{2+}(aq) + 2e^-$; cathode: $Cl_2(g) + 2e^- \longrightarrow$

$2Cl^-(aq)$; 4.12 V **20.99** (a) spontaneous (b) $Fe^{3+}$ will oxidize $Sn^{2+}$ to $Sn^{4+}$. **20.101** 0.36 V **20.103** (a) 2.2 (b) 0.3 $M$ **20.105** (a) 1.0 F, $9.6 \times 10^4$ C (b) 2.0 F, $1.9 \times 10^5$ C

(c) 0.11 F, $1.1 \times 10^{-4}$ C (d) $2.8 \times 10^{-2}$ F, $2.7 \times 10^3$ C

**20.107** $2.83 \times 10^{-4}$ g As **20.109** (a) $Cd(s) + 2Ag^+(aq) \rightleftharpoons$

$2Ag(s) + Cd^{2+}(aq)$; 1.20 V (b) It would increase. (c) no

effect **20.111** Anode: $2H_2O(l) \longrightarrow O_2(g) + 4H^+(aq) + 4e$;

cathode: $Cu^{2+}(aq) + 2e \longrightarrow Cu(s)$ **20.113** (a) $1.48 \times 10^3$ s

(b) 0.399 g **20.115** (a) 0.323 faraday (b) 3+

**20.117** (a) 1.06 V (b) $8.8 \times 10^{-3}\ M$ **20.119** (a) $-26.2$ kJ

(b) $-0.26$ V for $Co^{2+} + 2e \longrightarrow Co(s)$ **20.121** 11.7 kJ/K

**20.123** $-0.05$ V; not spontaneous **20.125** $-0.04$ V; not spontaneous **20.127** $K_{sp} = 1 \times 10^{-7}$

## Chapter 21

**21.19** (a) 0.1% (b) If you had a large quantity of material, 0.1% still would be a significant quantity. Also, if the material were particularly toxic in addition to being radioactive.

**21.21** Some of the expected mass is in the form of energy: the mass defect. **21.23** The large, positively charged He nucleus that makes up alpha ($\alpha$) radiation is unable to pass through the atoms that make up solid materials such as wood without coming into contact or being deflected by the nuclei. Gamma ($\gamma$) radiation, however, with its small wavelength and high energy, can pass through large amounts of material without interaction, just like x rays can pass through skin and other soft tissue.

**21.25** Irradiation of an atom with gamma radiation does not result in the creation of radioactive elements; therefore, the food cannot become radioactive. The addition of a radioactive element directly to the meat would make the meat radioactive. Under the right conditions, nuclear bombardment could also lead to the production of radioactive elements. **21.27** $^{87}_{37}Rb \longrightarrow {}^{87}_{38}Sr +$

$^{0}_{-1}e$ **21.29** $^{232}_{90}Th \longrightarrow {}^{4}_{2}He + {}^{228}_{88}Ra$ **21.31** $^{18}_{9}F \longrightarrow {}^{18}_{8}O +$

$^{0}_{1}e$ **21.33** $^{210}_{84}Po \longrightarrow {}^{4}_{2}He + {}^{206}_{82}Pb$

**21.35** (a) $^{122}_{51}Sb$ (b) $^{204}_{85}At$ (c) $^{80}_{37}Rb$ **21.37** (a) $\alpha$ emission

(b) positron emission or electron capture (c) $\beta$ emission

**21.39** $^{235}_{92}U$ decay series; $^{232}_{90}Th$ decay series. Alpha emission decreases the mass number by 4. Beta emission has no effect upon the mass number. **21.41** (a) $^{26}_{12}Mg(d, \alpha)^{24}_{11}Na$

(b) $^{16}_{8}O(n, p)^{16}_{7}N$ **21.43** (a) $^{27}_{13}Al + {}^{1}_{1}H \longrightarrow {}^{25}_{12}Mg + {}^{4}_{2}He$

(b) $^{10}_{5}B + {}^{4}_{2}He \longrightarrow {}^{1}_{1}H + {}^{13}_{6}C$ **21.45** $1.38 \times 10^9$ kJ/mol

**21.47** (a) $^{4}_{2}He$ (b) $\alpha$ **21.49** plutonium-239 **21.51** $1.79 \times 10^{-9}\ s^{-1}$ **21.53** $9.1 \times 10^{-8}\ s^{-1}$ **21.55** $1.3 \times 10^{13}$ years

**21.57** $3.84 \times 10^{-12}$/s **21.59** $2.1 \times 10^2$ Ci **21.61** $5.7 \times 10^{-4}$ g **21.63** 0.574, 2.9 $\mu$g **21.65** 4.16 s **21.67** 44.6 d

**21.69** $5.3 \times 10^3$ y **21.71** 5.2 disintegrations/(min · g)

**21.73** $4.37 \times 10^{-9}$ g **21.75** $-1.688 \times 10^{12}$ J; $-17.50$ MeV

**21.77** 0.032715 amu; 30.0489 MeV; 5.082 MeV

**21.79** Na-20 is expected to decay by electron capture or positron emission. Na-26 is expected to decay by beta emission.

**21.81** 7 alpha emissions and 4 beta emissions

**21.83** $^{209}_{83}Bi + {}^{4}_{2}He \longrightarrow {}^{211}_{85}At + 2{}^{1}_{0}n$ **21.85** $^{246}_{98}Cf$

**21.87** 5.8 y **21.89** 2.42 pm **21.91** $-3.61 \times 10^{14}$ J;

$9.84 \times 10^4$ kJ **21.93** (a) $^{47}_{20}Ca \longrightarrow {}^{47}_{21}Sc + {}^{0}_{-1}\beta$

(b) 46.1 $\mu$g **21.95** (a) $^{82}_{35}Br \longrightarrow {}^{82}_{36}Kr + {}^{0}_{-1}\beta$; $2H^{82}Br(g)$

$\longrightarrow H_2 + 2Kr$ (b) 0.0118 mol; 0.32 atm **21.97** $1.75 \times 10^{13}$ nuclei/s **21.99** 0.001 L **21.101** $2.63 \times 10^9$ kJ; $4.73 \times 10^7$ L

## Chapter 22

**22.59** It would react with the oxygen by the following equation: $C(s) + O_2(g) \longrightarrow CO_2(g)$.    **22.61** The metallic character decreases from left to right and increases going down a column.    **22.63** This means that aluminum hydroxide reacts with both acids and bases. For example,

$Al(OH)_3(s) + 3HCl(aq) \longrightarrow 3H_2O(l) + AlCl_3(aq)$

$Al(OH)_3(s) + NaOH(aq) \longrightarrow Na^+(aq) + Al(OH)_4^-(aq)$

**22.65** Oxygen is a very electronegative element, and its bonding involves only the $s$ and $p$ orbitals, in contrast to bonding using the $d$ orbitals in sulfur, etc. Molecular oxygen is a reactive gas but forms mainly compounds in which its oxidation state is $-2$, compared to compounds of sulfur, etc., which exhibit positive oxidation states as well as the $-2$ state.    **22.67** HBr cannot be prepared by adding sulfuric acid to NaBr because the hot concentrated acid will oxidize the bromide ion to bromine.

**22.69** $CO_2(g) + NH_3(g) + NaCl(aq) + H_2O(l) \longrightarrow$
$$NaHCO_3(s) + NH_4Cl(aq)$$

$2NaHCO_3(s) \overset{\Delta}{\longrightarrow} Na_2CO_3(s) + CO_2(g) + H_2O(g)$

$Ca(OH)_2(aq) + Na_2CO_3(aq) \longrightarrow 2NaOH(aq) + CaCO_3(s)$

**22.71** (a) $2K(s) + Br_2(l) \longrightarrow 2KBr(s)$

(b) $2K(s) + 2H_2O(l) \longrightarrow 2KOH(aq) + H_2(g)$

(c) $2NaOH(s) + CO_2(g) \longrightarrow Na_2CO_3(s) + H_2O(l)$

(d) $Li_2CO_3(aq) + 2HNO_3(aq) \longrightarrow$
$$2LiNO_3(aq) + H_2O(l) + CO_2(g)$$

(e) $K_2SO_4(aq) + Pb(NO_3)_2(aq) \longrightarrow PbSO_4(s) + 2KNO_3(aq)$

**22.73** $^{227}_{89}Ac \longrightarrow ^4_2He + ^{223}_{87}Fr$    **22.75** Add $CO_3^{2-}$ to each solution. Calcium carbonate will precipitate from the calcium hydroxide solution.    **22.77** Add $Na_2SO_4$ to the solution. $BaSO_4(s)$ can then be filtered from the solution. The filtrate will contain $Mg^{2+}$ ions.    **22.79** $^{230}_{90}Th \longrightarrow ^{226}_{88}Ra + ^4_2He$

**22.81** (a) $BaCO_3(s) \overset{\Delta}{\longrightarrow} BaO(s) + CO_2(g)$

(b) $Ba(s) + 2H_2O(l) \longrightarrow Ba(OH)_2(aq) + H_2(g)$

(c) $Mg(OH)_2(s) + 2HNO_3(aq) \longrightarrow Mg(NO_3)_2(aq) + 2H_2O(l)$

(d) $Mg(s) + NiCl_2(aq) \longrightarrow MgCl_2(aq) + Ni(s)$

(e) $2NaOH(aq) + MgSO_4(aq) \longrightarrow Mg(OH)_2(s) + Na_2SO_4(aq)$

**22.83** 0.22 g $Ca(OH)_2$

**22.85** $Al(H_2O)_6^{3+}(aq) + HCO_3^-(aq) \longrightarrow$
$$Al(H_2O)_5OH^{2+}(aq) + H_2O(l) + CO_2(g)$$

**22.87** Test portions of solutions of each compound with the others. If one solution is poured into one of the other solutions and no precipitate forms, that means that $BaCl_2$ was mixed with KOH and the third solution is $Al_2(SO_4)_3$. Adding KOH to $Al_2(SO_4)_3$ will form a precipitate of $Al(OH)_3$, which will dissolve when an excess of KOH is added. $BaCl_2$ and $Al_2(SO_4)_3$ will form a precipitate of $BaSO_4$.    **22.89** $Sn(H_2O)_6^{2+}(aq) + H_2O(l) \longrightarrow$
$$Sn(H_2O)_5(OH)^+(aq) + H_3O^+(aq)$$

**22.91** $PbO_2 + 4HCl \longrightarrow PbCl_2 + Cl_2 + 2H_2O$

**22.93** (a) $Al_2O_3(s) + 3H_2SO_4(aq) \longrightarrow$
$$Al_2(SO_4)_3(aq) + 3H_2O(l)$$

(b) $Al(s) + 3AgNO_3(aq) \longrightarrow Al(NO_3)_3(aq) + 3Ag(s)$

(c) $Pb(NO_3)_2(aq) + 2NaI(aq) \longrightarrow PbI_2(s) + 2NaNO_3(aq)$

(d) $8Al(s) + 3Mn_3O_4(s) \longrightarrow 4Al_2O_3(s) + 9Mn(s)$

(e) $2Ga(OH)_3(s) \overset{\Delta}{\longrightarrow} Ga_2O_3(s) + 3H_2O(l)$

**22.95** $-5.8 \times 10^9$ kJ    **22.97** (a) $-1$    (b) $+1$    (c) $-4$    (d) $+6$    **22.99** (a) Each C atom has $sp^3$ hybridization and forms four $\sigma$ bonds, one bond to each of three H atoms, and one bond to the other C atom.    (b) Since the Si atom is bonded to six F atoms, Si has six $sp^3d^2$ hybrid orbitals.    (c) The double bond between two of the C atoms consists of a $\sigma$ bond and a $\pi$ bond. The $\sigma$ bond is formed by the overlap of an $sp^2$ hybrid orbital from each of the two C atoms. The $\pi$ bond is formed by the overlap of the unhybridized $2p$ orbital from each of the two C atoms. The single bond between the center C atom and the other C atom is a $\sigma$ bond formed by the overlap of an $sp^2$ hybrid orbital from the center C atom with an $sp^3$ hybrid orbital from the outer C atom.    (d) Since the Si atom is bonded to four H atoms, Si has four $sp^3$ hybrid orbitals.    **22.101** (a) 74.87 kJ    (b) 137.14 kJ

**22.103** (a) $CO_2(g) + Ba(OH)_2(aq) \longrightarrow BaCO_3(s) + H_2O(l)$

(b) $MgCO_3(s) + 2HBr(aq) \longrightarrow$
$$MgBr_2(aq) + H_2O(l) + CO_2(g)$$

**22.105** $C + O_2$ (from air) $\longrightarrow CO_2$

$NaOH + CO_2 \longrightarrow NaHCO_3$

$2NaHCO_3 \overset{\Delta}{\longrightarrow} Na_2CO_3 + H_2O + CO_2$

**22.107** 2.53 g $NH_3$

**22.109** $4NH_3(g) + 5O_2(g) \overset{Pt}{\longrightarrow} 4NO(g) + 6H_2O(g)$

$2NO(g) + O_2(g) \longrightarrow 2NO_2(g)$

$3NO_2(g) + H_2O(l) \longrightarrow 2HNO_3(aq) + NO(g)$

$HNO_3(g) + NH_3(aq) \longrightarrow NH_4NO_3(aq)$

$NH_4NO_3(aq) \longrightarrow NH_4NO_3(s)$

$NH_4NO_3(s) \overset{\Delta}{\longrightarrow} N_2O(g) + 2H_2O(g)$

**22.111** $4Zn + 10H^+ + NO_3^- \longrightarrow 4Zn^{2+} + NH_4^+ + 3H_2O$

**22.113** Tetrahedral. Each bond is formed by the overlap of a $4p$ orbital from Br with an $sp^3$ orbital from P.

**22.115** $H_3PO_3 + H_2SO_4 \longrightarrow H_3PO_4 + SO_2 + H_2O$

**22.117** 7.07 g $H_3PO_4$    **22.119** (a) $4Li + O_2 \longrightarrow 2Li_2O$

(b) $4CH_3NH_2 + 9O_2 \longrightarrow 4CO_2 + 10H_2O + 2N_2$

(c) $2(C_2H_5)_2S + 15O_2 \longrightarrow 8CO_2 + 10H_2O + 2SO_2$

**22.121** (a) $+6$    (b) $+6$    (c) $-2$    (d) $+4$

**22.123** $8H_2SeO_3 + 16H_2S \longrightarrow Se_8 + 2S_8 + 24H_2O$

**22.125** 49.1 g

**22.127** $Ba(ClO_3)_2(aq) + H_2SO_4(aq) \longrightarrow$
$$2HClO_3(aq) + BaSO_4(s)$$

**22.129** $8H^+ + 6HCl + K_2Cr_2O_7 \longrightarrow$
$$3Cl_2 + 2Cr^{3+} + 7H_2O + 2K^+$$

**22.131** (a) $sp^3$; bent    (b) $sp^3$; trigonal pyramidal
(c) $sp^3d$; T-shaped

**22.133** (a) $Br_2(l) + 2NaOH(aq) \longrightarrow$
$$BrO^-(aq) + Br^-(aq) + H_2O(l) + 2Na^+(aq)$$

(b) $NaBr(s) + H_3PO_4(aq) \overset{\Delta}{\longrightarrow}$
$$HBr(g) + Na^+(aq) + H_2PO_4^-(aq)$$

**22.135** 0.13 V; reaction occurs

**22.137** $\ddot{:}\ddot{F}$ $\ddot{F}\ddot{:}$ $sp^3d^2$, square planar
Xe
$:\ddot{F}$ $\ddot{F}:$

**22.139** $2XeF_2 + 4OH^- \longrightarrow 2Xe + 4F^- + O_2 + 2H_2O$
**22.141** 8.184%  **22.143** $-851$ kJ; $-425$ kJ/mol Fe  **22.145**
84.5%  **22.147** 6.957 g  **22.149** $1.35 \times 10^3$ K  **22.151** $E° =$
0.19 V; $-37$ kJ; yes  **22.153** $7.73 \times 10^{-2}$ g  **22.155** 70.0%
**22.157** 12.6 hr  **22.159** 2.58%

## Chapter 23

**23.33** (a) $+2$  (b) $+4$  (c) $+1$  (d) $+6$
**23.35** $3Fe^{2+} + NO_3^- + 4H^+ \longrightarrow 3Fe^{3+} + NO + 2H_2O$
**23.37** (a) 4  (b) 6  (c) 4  (d) 6  **23.39** (a) $+2$  (b) $+3$
(c) $+3$  (d) $+3$
**23.41** (a) $+3$  (b) *Name*   *Formula*  (c) 6  (d) $-3$
ammine   $NH_3$
chloro   $Cl^-$
oxalato   $C_2O_4^{2-}$
**23.43** (a) potassium hexafluoroferrate(III)
(b) diamminediaquacopper(II) ion
(c) ammonium aquapentafluoroferrate(III)
(d) dicyanoargentate(I) ion
**23.45** (a) pentacarbonyliron(0)
(b) dicyanobis(ethylenediamine)rhodium(III) ion
(c) tetraaminesulfatochromium(III) chloride
(d) tetraoxomanganate(VII) ion
**23.47** (a) $K_3[Mn(CN)_6]$  (b) $Na_2[Zn(CN)_4]$
(c) $[Co(NH_3)_4Cl_2]NO_3$  (d) $[Cr(NH_3)_6]_2[CuCl_4]_3$

**23.49** (a)
*cis*   *trans*

(b) no geometric isomerism  (c) no geometric isomerism
(d)

*cis*

*trans*

**23.51** (a) Has optical isomers, similar to those in Figure 23.16 B
and C.  (b) No
**23.53** (a) 2 unpaired electrons  (b) 3 unpaired electrons
(c) 2 unpaired electrons  **23.55** (a) 8 electrons in square planar
field (low spin)  (b) 7 electrons in square planar field (low spin)
(c) 5 electrons in tetrahedral field (high spin)  (d) 7 electrons in
tetrahedral field (high spin)  **23.57** purple
**23.59** Yes, $H_2O$ is a weaker bonding ligand.  **23.61** 239 kJ/mol
**23.63** no $d$ electrons  **23.65** 2  **23.67** $[Cu^{2+}] = 6.1 \times 10^{-4}$
$M$; $[NH_3] = 2.4 \times 10^{-3}$ $M$; $[Cu(NH_3)_4^{2+}] = 0.10$ $M$

## Chapter 24

**24.15** (a) The molecules with carbon chains in the $C_5$–$C_{12}$
range.  (b) The molecules with carbon chains greater than $C_{11}$.
(c) Low molecular weight hydrocarbons have fewer polarizable
electrons; therefore, they have weaker London forces than the
larger chains and boil at a lower temperature.  (d) The more
compact 2,3-dimethylbutane would boil at a lower temperature.
**24.17** This compound would violate the octet rule.
**24.19** (a) trimethylamine, $C_3H_9N$; acetaldehyde, $C_2H_4O$;
2-propanol, $C_3H_8O$; acetic acid, $C_2H_4O_2$

(b) trimethylamine:

$CH_3$
|
$CH_3NCH_3$

acetaldehyde:

$O$
‖
$CH_3CH$

2-propanol:

$OH$
|
$CH_3CHCH_3$

acetic acid:

$O$
‖
$CH_3COH$

(c) trimethylamine: amine (tertiary); acetaldehyde: aldehyde;
2-propanol: alcohol; acetic acid: carboxylic acid
**24.21** The molecules increase regularly in molecular weight.
Therefore, you expect their intermolecular forces, and thus their
melting points, to increase.  **24.23** $CH_3CH_2CH_2CH_3$

**24.25** (a)

$CH_3CH_2$   $CH_2CH_3$
C=C
$H$   $H$
*cis*-3-hexene

$CH_3CH_2$   $H$
C=C
$H$   $CH_2CH_3$
*trans*-3-hexene

(b)

$CH_3$   $H$
C=C
$CH_3CH_2$   $CH_2CH_3$
*cis*-3-methyl-3-hexene

$CH_3$   $CH_2CH_3$
C=C
$CH_3CH_2$   $H$
*trans*-3-methyl-3-hexene

**24.27** (a) $C_2H_4 + 3O_2 \longrightarrow 2CO_2 + 2H_2O$
(b) $3CH_2{=}CH_2 + 2MnO_4^- + 4H_2O \longrightarrow$

$OH$  $OH$
|  |
$3CH_2{-}CH_2 + 2MnO_2 + 2OH^-$

(c) $CH_2{=}CH_2 + Br_2 \longrightarrow$

$Br$  $Br$
|  |
$CH_2{-}CH_2$

(d)

$$CH_3 \text{(toluene)} + Br_2 \xrightarrow{FeBr_3} CH_3 \text{(with Br)} + HBr$$

(e)

$$CH_3 \text{(toluene)} + HNO_3 \xrightarrow{H_2SO_4} CH_3 \text{(with } NO_2\text{)} + H_2O$$

**24.29** One possibility: $C_2H_6 + Cl_2 \longrightarrow C_2H_5Cl + HCl$

**24.31**

$$CH_3-\underset{\underset{CH_3}{|}}{\overset{\overset{Br}{|}}{C}}-CH_3$$

**24.33** (a) 2,3,4-trimethylpentane
(b) 2,2,6,6-tetramethylheptane   (c) 4-ethyloctane
(d) 3,4-dimethyloctane

**24.35** (a) $CH_3\underset{\underset{CH_3}{|}}{\overset{\overset{CH_3}{|}}{CH}}CHCH_2CH_2CH_3$

(b) $CH_3CH_2\underset{\underset{CH_3}{|}}{CH}CH_2CH_2CH_3$

(c) $CH_3\underset{\underset{CH_3CHCH_3}{|}}{\overset{\overset{CH_2CH_3}{|}}{CH}}CH_2CHCH_2CH_2CH_3$

(d) $CH_3\underset{\underset{CH_3}{|}}{\overset{\overset{CH_3}{|}}{C}}\;\;\;\;\underset{\underset{CH_3}{|}}{\overset{\overset{CH_3}{|}}{C}}CH_2CH_3$

**24.37** (a) 2-pentene   (b) 2,5-dimethyl-2-hexene
**24.39** (a) $CH_3CH{=}\underset{\underset{CH_2CH_3}{|}}{C}CH_2CH_3$

(b) $CH_3\underset{\underset{CH_3}{|}}{C}{=}CH\underset{\underset{CH_2CH_3}{|}}{CH}CH_2CH_3$

**24.41** *cis*-2-pentene, *trans*-2-pentene   **24.43** (a) 2-butyne
(b) 3-methyl-1-pentyne
**24.45** (a)

(b)

**24.47** (a) $CH_3\overset{\overset{\textstyle O}{\|}}{C}CH_2CH_2CH_3$ ketone   (b) $CH_3\overset{\overset{\textstyle OH}{|}}{CH}CH_2CH_3$ alcohol

(c) $HOC\overset{\overset{\textstyle O}{\|}}{}CH_2CH_3$ carboxylic acid   (d) $H-\overset{\overset{\textstyle O}{\|}}{C}CH_2CH_3$ aldehyde

**24.49** (a) 1-pentanol   (b) 2-pentanol   (c) 2-propyl-1-pentanol   (d) 6-methyl-4-octanol   **24.51** (a) secondary alcohol
(b) secondary alcohol   (c) primary alcohol   (d) primary alcohol   **24.53** (a) ethyl propyl ether   (b) methyl isopropyl ether   **24.55** (a) butanone   (b) butanal
(c) 4,4-dimethylpentanal   (d) 3-methyl-2-pentanone
**24.57** (a) secondary amine   (b) secondary amine
**24.59** (a) 3-methylbutanoic acid   (b) *trans*-5-methyl-2-hexene
(c) 2,5-dimethyl-4-heptanone   (d) 4-methyl-2-pentyne

**24.61** (a) $CH_3CH_2\overset{\overset{\textstyle O}{\|}}{C}-O-\underset{\underset{CH_3}{|}}{CH}$   (b) $CH_3-\underset{\underset{CH_3}{|}}{\overset{\overset{CH_3}{|}}{C}}-NH_2$

(c) $CH_3CH_2CH_2CH_2\underset{\underset{CH_3}{|}}{\overset{\overset{CH_3}{|}}{C}}COOH$

(d) $\underset{H}{\overset{CH_3CH_2}{}}C{=}C\underset{H}{\overset{CH_2CH_3}{}}$

**24.63** (a) Addition of dichromate ion in acidic solution to propionaldehyde will cause the reagent to change from orange to green as the aldehyde is oxidized. Under similar conditions, acetone would not react.   (b) Add $Br_2$ in $CCl_4$ to each compound. First compound reacts; the color of $Br_2$ fades. Benzene does not react.   **24.65** (a) ethylene   (b) toluene
(c) methylamine   (d) methanol   **24.67** 1-butene or 2-butene

## Chapter 25

**25.17** (a) —A—A—A—A—A—A— and
—B—B—B—B—B—B—
(b) —A—B—A—B—A—B—
(c) —A—A—C—A—A—C—

**25.19** $^+H_3NCH_2CH_2CH_2-\underset{\underset{NH_3{}^+}{|}}{\overset{\overset{H}{|}}{C}}-COO^-$

**25.21** The corresponding amino-acid sequence is Arg-Val-Arg-Asp-Leu-Thr. The triplet UGA is the code to end the sequence.
**25.23** $nCF_2{=}CF_2 \longrightarrow$
$-CF_2-CF_2-CF_2-CF_2-CF_2-CF_2-$
**25.25** $HOCH_2CH_2OH$ and $HOOCCH_2CH_2COOH$   **25.27** (a) would give a polymer having alternating single and double bonds, in which resonance might be expected to give a conduct-

ing polymer.

**25.29** $CH_3 - \overset{\overset{\displaystyle H}{|}}{\underset{\underset{\displaystyle NH_2}{|}}{C}} - COOH$

**25.31** 2 dipeptides are possible:

$H_2N - \overset{\overset{\displaystyle CH_3}{|}}{\underset{\underset{\displaystyle H}{|}}{C}} - \overset{\overset{\displaystyle O}{\|}}{C} - \overset{\overset{\displaystyle }{|}}{\underset{\underset{\displaystyle H}{|}}{N}} - \overset{\overset{\displaystyle CH_2}{|}}{\underset{\underset{\displaystyle H}{|}}{C}} - COOH$  and

$H_2N - \overset{\overset{\displaystyle CH_2}{|}}{\underset{\underset{\displaystyle H}{|}}{C}} - \overset{\overset{\displaystyle O}{\|}}{C} - \overset{\overset{\displaystyle }{|}}{\underset{\underset{\displaystyle H}{|}}{N}} - \overset{\overset{\displaystyle CH_3}{|}}{\underset{\underset{\displaystyle H}{|}}{C}} - COOH$

**25.33** A : T, 1 : 1, G : C, 1 : 1

**25.35**

**25.37** 3; 2; guanine-cytosine bonding is stronger.
**25.39** 16; no    **25.41** hydrogen bonds; less readily
**25.43** gly-ser-arg-phe-gly-leu-lys-end
**25.45** CUG, ACU, CCC, UGG
**25.47** CUU GCU GUU GAA GAU UGU AUG UGG AAA

**25.49** $-\overset{\overset{\displaystyle H}{|}}{\underset{\underset{\displaystyle Cl}{|}}{C}} - \overset{\overset{\displaystyle H}{|}}{\underset{\underset{\displaystyle Cl}{|}}{C}} - \overset{\overset{\displaystyle H}{|}}{\underset{\underset{\displaystyle Cl}{|}}{C}} - \overset{\overset{\displaystyle H}{|}}{\underset{\underset{\displaystyle Cl}{|}}{C}} -$

**25.51** $HOOC - C_6H_5 - COOH$, $NH_2 - C_6H_5 - NH_2$

**25.53** $CH_3SCH_2CH_2\overset{\overset{\displaystyle NH_2}{|}}{CH} - COOH$ is nonpolar

**25.55** $\overset{\overset{\displaystyle COOH}{|}}{H_2N - \underset{\underset{\underset{\underset{\displaystyle OH}{|}}{\displaystyle CH_2}}{|}}{C} - H}$

**25.57** $CH_3SCH_2CH_2\overset{\overset{\displaystyle NH_2}{|}}{CH} - \overset{\overset{\displaystyle O}{\|}}{C} - NH - \overset{\overset{\displaystyle }{}}{\underset{\underset{\displaystyle CH_2SH}{|}}{CH}} - COOH$

$HSCH_2\overset{\overset{\displaystyle NH_2}{|}}{CH} - \overset{\overset{\displaystyle O}{\|}}{C} - NH - \overset{\overset{\displaystyle }{}}{\underset{\underset{\underset{\underset{\displaystyle CH_2SCH_3}{|}}{\displaystyle CH_2}}{|}}{CH}} - COOH$

**25.59** 720    **25.61** adenine, ribose, phosphate ion

# Glossary

The number given in blue at the end of a definition indicates the section (or page) where the term was introduced. In a few cases, several sections are indicated.

**Absolute entropy** see *Standard entropy.*

**Absolute temperature scale** a temperature scale in which the lowest temperature that can be attained theoretically is zero. (1.6)

**Accuracy** the closeness of a single measurement to its true value. (1.5)

**Acid** (Arrhenius definition) a substance that produces hydrogen ions, $H^+$, (hydronium ion, $H_3O^+$) when it dissolves in water. (4.4 and 16.1)     (Brønsted–Lowry definition) the species (molecule or ion) that donates a proton to another species in a proton-transfer reaction. (4.4 and 16.2)

**Acid–base indicator** a dye used to distinguish between acidic and basic solutions by means of the color changes it undergoes in these solutions. (4.4)

**Acid–base titration curve** a plot of the pH of a solution of acid (or base) against the volume of added base (or acid). (17.7)

**Acidic oxide** an oxide that reacts with bases. (8.7)

**Acid-ionization (or acid-dissociation) constant ($K_a$)** the equilibrium constant for the ionization of a weak acid. (17.1)

**Acid rain** rain having a pH lower than that of natural rain, which has a pH of 5.6. (p. 734)

**Acid salt** a salt that has an acidic hydrogen atom and can undergo neutralization with bases. (4.4)

**Actinides** elements in the last of the two rows at the bottom of the periodic table; the 14 elements following actinium in the periodic table, in which the $5f$ subshell is filling. (p. 1003)

**Activated complex (transition state)** an unstable grouping of atoms that can break up to form products. (13.5)

**Activation energy ($E_a$)** the minimum energy of collision required for two molecules to react. (14.5)

**Activity of a radioactive source** the number of nuclear disintegrations per unit time occurring in a radioactive material. (21.3)

**Activity series** a listing of the elements in order of their ease of losing electrons during reactions in aqueous solution. (4.5)

**Addition polymer** a polymer formed by linking together many molecules by addition reactions. (25.1)

**Addition reaction** a reaction in which parts of a reactant are added to each carbon atom of a carbon–carbon double bond, which then becomes a C—C single bond. (24.3)

**Adsorption** the binding or attraction of molecules to a surface. (14.9) See *Chemisorption* and *Physical adsorption.*

**Aerosol** a colloid consisting of liquid droplets or solid particles dispersed throughout a gas. (12.9)

**Alcohol** a compound obtained by substituting a hydroxyl group (—OH) for an —H atom on a tetrahedral ($sp^3$ hybridized) carbon atom of a hydrocarbon group. (24.6)

**Aldehyde** a compound containing a carbonyl group with at least one H atom attached to it. (24.6)

**Aliphatic hydrocarbon** a saturated or unsaturated hydrocarbon; a hydrocarbon that does not contain benzene rings. (24.1)

**Alkali metals** the Group IA elements; they are reactive metals. (2.5 and 8.7)

**Alkaline dry cell** a voltaic cell that is similar to the Leclanché dry cell but uses potassium hydroxide in place of ammonium chloride. (20.8)

**Alkaline earth metals** the Group IIA elements; they are reactive metals, though less reactive than the alkali metals. (8.7)

**Alkane** an acyclic saturated hydrocarbon; a saturated hydrocarbon with the general formula $C_nH_{2n+2}$. (24.2)

**Alkene** a hydrocarbon that has the general formula $C_nH_{2n}$ and contains a carbon–carbon double bond. (24.3)

**Alkyl group** an alkane less one hydrogen atom. (24.5)

**Alkyne** an unsaturated hydrocarbon containing a carbon–carbon triple bond. The general formula is $C_nH_{2n-2}$. (24.3)

**Allotrope** one of two or more distinct forms of an element in the same physical state. (6.8)

**Alloy** a material with metallic properties that is either a compound or a mixture. (p. 554)

**Alpha emission** emission of a $^4_2He$ nucleus, or alpha particle, from an unstable nucleus. (21.1)

**Aluminosilicate mineral** a mineral consisting of silicate sheets or three-dimensional networks in which some of the $SiO_4$ tetrahedra of a silicate structure have been replaced by $AlO_4$ tetrahedra. (13.6)

**Ambidentate ligand** a ligand that can bond to a metal atom through one atom or another. (23.5)

**Amide bonds** see *Peptide bonds.*

**Amide** a compound derived from the reaction of ammonia, or a primary or secondary amine, with a carboxylic acid. (24.7)

**Amine** a compound that is structurally derived by replacing one or more hydrogen atoms of ammonia with hydrocarbon groups. (24.7)

**Amino acid** a compound containing an amino group (—$NH_2$) and a carboxyl group (—COOH). (25.3)

**Amorphous solid** a solid that has a disordered structure; it lacks the well-defined arrangement of basic units (atoms, molecules, or ions) found in a crystal. (**11.7**)

**Ampere (A)** the base unit of current in the International System (SI). (**20.11**)

**Amphiprotic species** a species that can act as either an acid or a base (that is, it can lose or gain a proton). (**16.2**)

**Amphoteric hydroxide** a metal hydroxide that reacts with both bases and acids. (**18.5**)

**Amphoteric oxide** an oxide that has both acidic and basic properties. (**8.7**)

**Angstrom (Å)** a non-SI unit of length; $1\ \text{Å} = 10^{-10}\text{m}$. (**1.6**)

**Angular geometry** see *Bent geometry.*

**Angular momentum quantum number** ($l$) also known as the azimuthal quantum number. The quantum number that distinguishes orbitals of given $n$ having different shapes; it can have any integer value from 0 to $n - 1$. (**7.5**)

**Anion** a negatively charged ion. (**2.6**)

**Anode** the electrode at which oxidation occurs. (**20.2**)

**Antibonding orbitals** molecular orbitals that are concentrated in regions other than between two nuclei. (**10.5**)

**Aromatic hydrocarbon** a hydrocarbon that contains a benzene ring or similar structural feature. (**24.1**)

**Arrhenius equation** the mathematical equation $k = Ae^{-E_a/RT}$, which expresses the dependence of the rate constant on temperature. (**14.6**)

**Association colloid** a colloid in which the dispersed phase consists of micelles. (**12.9**)

**Atmosphere (atm)** a unit of pressure equal to exactly 760 mmHg; 1 atm = 101.325 kPa (exact). (**5.1**)

**Atom** an extremely small particle of matter that retains its identity during chemical reactions. (**2.1**)

**Atomic mass unit (amu)** a mass unit equal to exactly one-twelfth the mass of a carbon-12 atom. (**2.4**)

**Atomic number (Z)** the number of protons in the nucleus of an atom. (**2.3**)

**Atomic orbital** a wave function for an electron in an atom; pictured qualitatively by describing the region of space where there is a high probability of finding the electron. (**7.5**)

**Atomic symbol** a one- or two-letter notation used to represent an atom corresponding to a particular element. (**2.1**)

**Atomic theory** an explanation of the structure of matter in terms of different combinations of very small particles (atoms). (**2.1**)

**Atomic weight** the average atomic mass for the naturally occurring element, expressed in atomic mass units. (**2.4**)

**Aufbau principle** see *Building-up principle.*

**Autoionization** see *Self-ionization.*

**Avogadro's law** a law stating that equal volumes of any two gases at the same temperature and pressure contain the same number of molecules. (**5.2**)

**Avogadro's number** ($N_A$) the number of atoms in a 12-g sample of carbon-12, equal to $6.02 \times 10^{23}$ to three significant figures. (**3.2**)

**Axial direction** one of two directions pointing from the center of a trigonal bipyramid along its axis. Compare *Equatorial direction.* (**10.1**)

**Azimuthal quantum number** see *Angular momentum quantum number.*

**Band of stability** the region in which stable nuclides lie in a plot of number of protons against number of neutrons. (**21.1**)

**Band theory** molecular orbital theory of metals. (**13.3**)

**Barometer** a device for measuring the pressure of the atmosphere. (**5.1**)

**Base** (Arrhenius definition) a substance that produces hydroxide ions, $OH^-$, when it dissolves in water. (**4.4 and 16.1**) (Brønsted–Lowry definition) the species (molecule or ion) that accepts a proton in a proton-transfer reaction. (**4.4 and 16.2**)

**Base-ionization (or base-dissociation) constant** ($K_b$) the equilibrium constant for the ionization of a weak base. Thus $K_b$ for $NH_3$ is $1.8 \times 10^{-5}$. (**17.3**)

**Base pairing** the hydrogen bonding of complementary bases. (**25.4**)

**Basic oxide** an oxide that reacts with acids. (**8.7**)

**Bayer process** a chemical procedure in which purified aluminum oxide, $Al_2O_3$, is separated from the aluminum ore bauxite. (**13.2**)

**Bent geometry** nonlinear molecular geometry, in the case of a molecule of three atoms. (**10.1**)

**Beta emission** emission of a high-speed electron from an unstable nucleus. (**21.1**)

**Bidentate ligand** a ligand that bonds to a metal atom through two atoms of the ligand. (**23.3**)

**Bimolecular reaction** an elementary reaction that involves two reactant molecules. (**14.7**)

**Binary compound** a compound composed of only two elements. (**2.8**)

**Binary hydride** a compound that contains hydrogen and one other element. (**25.5**)

**Binding energy (of a nucleus)** the energy needed to break a nucleus into its individual protons and neutrons. (**21.6**)

**Body-centered cubic unit cell** a cubic unit cell in which there is a lattice point at the center of the unit cell as well as at the corners. (**11.7**)

**Boiling point** the temperature at which the vapor pressure of a liquid equals the pressure exerted on the liquid (atmospheric pressure, unless the vessel containing the liquid is closed). (**11.2**)

**Boiling-point elevation** a colligative property of a solution equal to the boiling point of the solution minus the boiling point of the pure solvent. (12.6)

**Bond distance** see *Bond length.*

**Bond energy** the average enthalpy change for the breaking of a bond in a molecule in the gas phase. (9.11)

**Bonding orbitals** molecular orbitals that are concentrated in regions between nuclei. (10.5)

**Bonding pair** an electron pair shared between two atoms. (9.4)

**Bond length (bond distance)** the distance between the nuclei in a bond. (9.10)

**Bond order** in a Lewis formula, the number of pairs of electrons in a bond. (9.10)     In molecular orbital theory, one-half the difference between the number of bonding electrons and the number of antibonding electrons. (10.5)

**Boyle's law** the volume of a sample of gas at a given temperature varies inversely with the applied pressure. (5.2)

**Brønsted–Lowry concept** a concept of acids and bases in which an acid is the species donating a proton in a proton-transfer reaction, whereas a base is the species accepting a proton in such a reaction. (4.4 and 16.2)

**Buckminsterfullerene** a molecular form of carbon ($C_{60}$) informally called "buckyball," referring to its soccer-ball shape; a fullerene. (13.4, p. 574)

**Buffer** a solution characterized by the ability to resist changes in pH when limited amounts of acid or base are added to it. (17.6)

**Building-up principle (Aufbau principle)** a scheme used to reproduce the electron configurations of the ground states of atoms by successively filling subshells with electrons in a specific order (the building-up order). (8.2)

**Calorie (cal)** a non-SI unit of energy commonly used by chemists, originally defined as the amount of energy needed to raise the temperature of 1 g of water by 1°C; now defined as 1 cal = 4.184 J (exact). (6.1)

**Calorimeter** a device used to measure the heat absorbed or evolved during a physical or chemical change. (6.6)

**Carboxylic acid** a compound containing the carboxyl group, —COOH. (24.6)

**Catalysis** the increase in rate of a reaction as the result of the addition of a catalyst. (14.9)

**Catalyst** a substance that increases the rate of reaction without being consumed in the overall reaction. (2.9 and p. 591)

**Catenation** the covalent bonding of two or more atoms of the same element to one another. (22.6)

**Cathode** the electrode at which reduction occurs. (20.2)

**Cathode rays** the rays emitted by the cathode (negative electrode) in a gas discharge tube (tube of low-pressure gas through which an electric current is discharged). (2.2)

**Cation** a positively charged ion. (2.6)

**Cation exchange resin** a resin (organic polymer) that removes cations from solution and replaces them with hydrogen ions. (p. 493)

**Cell reaction** the net reaction that occurs in a voltaic cell. (20.2)

**Celsius scale** the temperature scale in general scientific use. There are exactly 100 units between the freezing point and the normal boiling point of water. (1.6)

**Ceramics** nonmetallic, inorganic solids that are hard and brittle and usually produced at elevated temperature. (13.7)

**Chain reaction, nuclear** a self-sustaining series of nuclear fissions caused by the absorption of neutrons released from previous nuclear fissions. (21.7)

**Change of state (phase transition)** a change of a substance from one state to another. (p. 442)

**Charles's law** the volume occupied by any sample of gas at a constant pressure is directly proportional to the absolute temperature. (5.2)

**Chelate** a complex formed by polydentate ligands. (23.3)

**Chemical bond** a strong attractive force that exists between certain atoms in a substance. (p. 348)

**Chemical change** see *Chemical reaction.*

**Chemical equation** the symbolic representation of a chemical reaction in terms of chemical formulas. (2.9)

**Chemical equilibrium** the state reached by a reaction mixture when the rates of forward and reverse reactions have become equal. (15.1)

**Chemical formula** a notation that uses atomic symbols with numerical subscripts to convey the relative proportions of atoms of the different elements in a substance. (2.6)

**Chemical kinetics** the study of how reaction rates change under varying conditions and of what molecular events occur during the overall reaction. (p. 590)

**Chemical nomenclature** the systematic naming of chemical compounds. (2.8)

**Chemical property** a characteristic of a material involving its chemical change. (1.4)

**Chemical reaction (chemical change)** a change in which one or more kinds of matter are transformed into a new kind of matter or several new kinds of matter. (1.4)     The rearrangement of the atoms present in the reacting substances to give new chemical combinations present in the substances formed by the reaction. (2.1)

**Chemical vapor deposition** a process in which a chemical reaction in a gas or vapor produces a product that can be deposited onto a solid material, or substrate. (13.4)

**Chemisorption** the binding of a species to a surface by chemical bonding forces. (14.9)

**Chiral** possessing the quality of handedness. A chiral object has a mirror image that is not identical to the object. (23.5)

**Chlor−alkali membrane cell** a cell for the electrolysis of aqueous sodium chloride in which the anode and cathode compartments are separated by a special plastic membrane that allows only cations to pass through it. (20.10)

**Chlor−alkali mercury cell** a cell for the electrolysis of aqueous sodium chloride in which mercury metal is used as the cathode. (20.10)

**Chromatography** a name given to a group of similar separation techniques that depend on how fast a substance moves, in a stream of gas or liquid, past a stationary phase to which the substance may be slightly attracted. (p. 15)

**Clausius−Clapeyron equation** an equation that expresses the relation between the vapor pressure $P$ of a liquid and the absolute temperature $T$: $\ln P = -\Delta H_{vap}/RT + B$, where $B$ is a constant. (11.2)

**Claus process** a method of obtaining free sulfur by the partial burning of hydrogen sulfide. (22.8)

**Coagulation** the process by which the dispersed phase of a colloid is made to aggregate and thereby separate from the continuous phase. (12.9)

**Codon** a sequence of three bases in a messenger RNA molecule that serves as the code for a particular amino acid. (25.4)

**Colligative properties** properties that depend on the concentration of solute molecules or ions in a solution but not on the chemical identity of the solute. (p. 517)

**Collision theory** a theory that assumes that, in order for reaction to occur, reactant molecules must collide with an energy greater than some minimum value and with proper orientation. (13.5)

**Colloid** a dispersion of particles of one substance (the dispersed phase) throughout another substance or solution (the continuous phase). (12.9)

**Combination reaction** a reaction in which two substances combine to form a third substance. (4.5)

**Combustion reaction** a reaction of a substance with oxygen, usually with the rapid release of heat to produce a flame. (4.5)

**Common-ion effect** the shift in an ionic equilibrium caused by the addition of a solute that provides an ion that takes part in the equilibrium. (17.5 and 18.2)

**Complementary bases** nucleotide bases that form strong hydrogen bonds with one another. (25.4)

**Complete ionic equation** a chemical equation in which strong electrolytes (such as soluble ionic compounds) are written as separate ions in the solution. (4.2)

**Complex (coordination compound)** a compound consisting either of complex ions and other ions of opposite charge or of a neutral complex species. (23.3)

**Complex ion** an ion formed from a metal ion with a Lewis base attached to it by a coordinate covalent bond. (p. 785 and 23.3)

**Composite** a material constructed of two or more different kinds of materials in separate phases. (13.8)

**Compound** a substance composed of two or more elements chemically combined. (1.4) A type of matter composed of atoms of two or more elements chemically combined in fixed proportions. (2.1)

**Concentration** a general term referring to the quantity of solute in a standard quantity of solvent or solution. (4.7 and 12.4)

**Condensation** the change of a gas to either the liquid or the solid state. (11.2)

**Condensation polymer** a polymer formed by linking many molecules together by condensation reactions. (25.1)

**Condensation reaction** a reaction in which two molecules or ions are chemically joined by the elimination of a small molecule such as water. (13.6 and 24.6)

**Condensed (structural) formula** a structural formula in which the bonds around each carbon atom is not explicitly written. (24.2)

**Conjugate acid** in a conjugate acid−base pair, the species that can donate a proton. (16.2)

**Conjugate acid−base pair** two species in an acid−base reaction, one acid and one base, that differ by the loss or gain of a proton. (16.2)

**Conjugate base** in a conjugate acid−base pair, the species that can accept a proton. (16.2)

**Constitutional (structural) isomers** isomers that differ in how the atoms are joined together (23.5); compounds with the same molecular formula but different structural formulas. (24.2)

**Contact process** an industrial method for the manufacture of sulfuric acid. It consists of the reaction of sulfur dioxide with oxygen to form sulfur trioxide using a catalyst of vanadium(V) oxide, followed by the reaction of sulfur trioxide with water. (22.8)

**Continuous spectrum** a spectrum containing light of all wavelengths. (7.3)

**Control rods** cylinders composed of substances that absorb neutrons, such as boron and cadmium, and can therefore slow a nuclear chain reaction. (21.7)

**Conversion factor** a factor equal to 1 that converts a quantity expressed in one unit to a quantity expressed in another unit. (1.8)

**Coordinate covalent bond** a bond formed when both electrons of the bond are donated by one atom. (9.4)

**Coordination compound** a compound consisting either of complex ions and other ions of opposite charge or of a neutral complex species. (23.3)

**Coordination isomers** isomers consisting of complex cations and complex anions that differ in the way the ligands are distributed between the metal atoms. (23.5)

**Coordination number** in a crystal, the number of nearest-neighbor atoms of an atom. (11.8) In a complex, the total number of bonds the metal atom forms with ligands. (23.3)

**Copolymer** a polymer consisting of two or more different monomer units. (25.1)

**Covalent bond** a chemical bond formed by the sharing of a pair of electrons between atoms. (p. 362)

**Covalent network solid** a solid that consists of atoms held together in large networks or chains by covalent bonds. (11.6)

**Covalent radii** values assigned to atoms in such a way that the sum of covalent radii of atoms A and B predicts an approximate A—B bond length. (9.10)

**Critical mass** the smallest mass of fissionable material in which a chain reaction can be sustained. (21.7)

**Critical pressure** the vapor pressure at the critical temperature; it is the minimum pressure that must be applied to a gas at the critical temperature to liquefy it. (11.3)

**Critical temperature** the temperature above which the liquid state of a substance no longer exists regardless of the pressure. (11.3)

**Crystal** a kind of solid having a regular three-dimensional arrangement of atoms, molecules, or ions. (2.6)

**Crystal field splitting (Δ)** the difference in energy between the two sets of $d$ orbitals on a central metal ion that arises from the interaction of the orbitals with the electric field of the ligands. (23.7)

**Crystal field theory** a model of the electronic structure of transition-metal complexes that considers how the energies of the $d$ orbitals of a metal ion are affected by the electric field of the ligands. (23.7)

**Crystal lattice** the geometric arrangement of lattice points of a crystal, in which we choose one lattice point at the same location within each of the basic units of the crystal. (11.7)

**Crystalline solid** a solid composed of one or more crystals; each crystal has a well-defined ordered structure in three dimensions. (11.7)

**Crystal systems** the seven basic shapes possible for unit cells; a classification of crystals. (11.7)

**Cubic close-packed structure (ccp)** a crystal structure composed of close-packed atoms (or other units) with the stacking ABCABCABCA · · ·. (11.8)

**Curie (Ci)** a unit of activity, equal to $3.700 \times 10^{10}$ disintegrations per second. (21.3)

**Cycloalkane** a cyclic saturated hydrocarbon; that is, a saturated hydrocarbon in which the carbon atoms form a ring; the general formula is $C_nH_{2n}$. (24.2)

**Cyclotron** a type of particle accelerator consisting of two hollow, semicircular metal electrodes, called dees (because the shape resembles the letter D), in which charged particles are accelerated by stages to higher and higher kinetic energies. Ions introduced at the center of the cyclotron are accelerated in the space between the two dees. (21.2)

**Dalton's law of partial pressures** the sum of the partial pressures of all the different gases in a mixture is equal to the total pressure of the mixture. (5.5)

**_d_-Block transition elements** those transition elements with an unfilled $d$ subshell in common oxidation states. (8.2)

**de Broglie relation** the equation $\lambda = h/mv$. (7.4)

**Decomposition reaction** a reaction in which a single compound reacts to give two or more substances. (4.5)

**Degree of ionization** the fraction of molecules that react with water to give ions. (17.1)

**Delocalized bonding** a type of bonding in which a bonding pair of electrons is spread over a number of atoms rather than localized between two. (9.7)

**Denitrifying bacteria** bacteria that use nitrate ion, $NO_3^-$, as a source of energy; they convert the ion to gaseous nitrogen. (22.7)

**Density** the mass per unit volume of a substance or solution. (1.7)

**Deoxyribonucleic acid (DNA)** the hereditary constituent of cells; it consists of two polymer strands of deoxyribonucleotide units. (25.4)

**Deposition** the change of a vapor to a solid. (11.2)

**Derived unit** see _SI derived unit._

**Desalinate** remove ions from brackish (slightly salty) water or seawater, to make drinkable or industrially usable water. (12.7)

**Deuterons** nuclei of hydrogen-2 atoms. (21.2)

**Dextrorotatory** refers to a compound whose solution rotates the plane of polarized light to the right (when looking toward the source of light). (23.5)

**Diamagnetic substance** a substance that is not attracted by a magnetic field or is very slightly repelled by such a field. This property generally means that the substance has only paired electrons. (8.4)

**Diffusion** the process whereby a gas spreads out through another gas to occupy the space uniformly. (5.7)

**Dimensional analysis (factor-label method)** the method of calculation in which one carries along the units for quantities. (1.8)

**Dipole–dipole force** an attractive intermolecular force resulting from the tendency of polar molecules to align themselves such that the positive end of one molecule is near the negative end of another. (11.5)

**Dipole moment** a quantitative measure of the degree of charge separation in a molecule. (10.2)

**Dispersion forces** see _London forces._

**Displacement reaction (single-replacement reaction)** a reaction in which an element reacts with a compound, displacing an element from it. (4.5)

**Dissociation constant of a complex ion ($K_d$)** the reciprocal, or inverse, value of the formation constant. (18.5)

**Distillation** the process in which a liquid is vaporized and condensed; used to separate substances of different volatilities. (1.4)

**Distorted tetrahedral geometry** see _Seesaw geometry._

**Disulfide linkage** a covalent linkage formed between two cysteine groups within a protein molecule by oxidation of the two groups. (25.3)

**Double bond** a covalent bond in which two pairs of electrons are shared by two atoms. (9.4)

**Downs cell** a commercial electrochemical cell used to obtain sodium metal by the electrolysis of molten sodium chloride. (20.9)

**Dow process** a commercial method for isolating magnesium from seawater. (13.2)

**Dry cell** see *Alkaline dry cell* and *Zinc–carbon dry cell.*

**Effective nuclear charge** the positive charge that an electron experiences from the nucleus, equal to the nuclear charge but reduced by any shielding or screening from any intervening electron distribution. (8.6)

**Effusion** the process in which a gas flows through a small hole in a container. (5.7)

**Elastomer** elastic or rubbery type of polymer. (25.1)

**Electrochemical cell** a system consisting of electrodes that dip into an electrolyte and in which a chemical reaction either uses or generates an electric current. (p. 845)

**Electrolysis** the process of producing a chemical change in an electrolytic cell. (p. 874)

**Electrolyte** a substance, such as sodium chloride, that dissolves in water to give an electrically conducting solution. (4.1)

**Electrolytic cell** an electrochemical cell in which an electric current drives an otherwise nonspontaneous reaction. (p. 846)

**Electromagnetic spectrum** the range of frequencies or wavelengths of electromagnetic radiation. (7.1)

**Electromotive force (emf)** the maximum potential difference between the electrodes of a voltaic cell. (20.4)

**Electron** a very light, negatively charged particle that exists in the region around the atom's positively charged nucleus. (2.2)

**Electron affinity** the energy change for the process of adding an electron to a neutral atom in the gaseous state to form a negative ion. (8.6)

**Electron capture** the decay of an unstable nucleus by capturing, or picking up, an electron from an inner orbital of an atom. (21.1)

**Electron configuration** the particular distribution of electrons among available subshells. (8.1)

**Electronegativity** a measure of the ability of an atom in a molecule to draw bonding electrons to itself. (9.5)

**Electron spin quantum number ($m_s$)** see *Spin quantum number.*

**Electron volt (eV)** the quantity of energy that would have to be imparted to an electron (whose charge is $1.602 \times 10^{-19}$ C) to accelerate it by one volt potential difference. (21.2)

**Element** a substance that cannot be decomposed by any chemical reaction into simpler substances. (1.4)     A type of matter composed of only one kind of atom, each atom of a given kind having the same properties. (2.1)     A substance whose atoms all have the same atomic number. (2.3)

**Elementary reaction** a single molecular event, such as a collision of molecules, resulting in a reaction. (14.7)

**Empirical formula (simplest formula)** the formula of a substance written with the smallest integer (whole number) subscripts. (3.5)

**Emulsion** a colloid consisting of liquid droplets dispersed throughout another liquid. (12.9)

**Enantiomers (optical isomers)** isomers that are nonsuperimposable mirror images of one another. (23.5)

**Endothermic process** a chemical reaction or physical change in which heat is absorbed. (6.2)

**Energy** the potential or capacity to move matter. (6.1)

**Energy levels** specific energy values in an atom. (7.3)

**Enthalpy ($H$)** an extensive property of a substance that can be used to obtain the heat absorbed or evolved in a chemical reaction at constant pressure. (6.3)     It equals the quantity $U + PV$. (19.1)

**Enthalpy of formation** see *Standard enthalpy of formation.*

**Enthalpy of reaction ($\Delta H$)** the change in enthalpy for a reaction at a given temperature and pressure; it equals the heat of reaction at constant pressure. (6.3)

**Entropy ($S$)** a thermodynamic quantity that is a measure of the randomness or disorder in a system. (19.2)

**Enzyme** a protein that catalyzes a biochemical reaction. (25.3)

**Equatorial direction** one of the three directions pointing from the center of a trigonal bipyramid to a vertex other than one on the axis. Compare *Axial direction.* (10.1)

**Equilibrium** see *Chemical equilibrium.*

**Equilibrium constant $K_c$** the value obtained for the equilibrium-constant expression when equilibrium concentrations are substituted. (15.2)

**Equilibrium constant $K_p$** an equilibrium constant for a gas reaction, similar to $K_c$, but in which concentrations of gases are replaced by partial pressures (in atm). (15.2)

**Equilibrium-constant expression** an expression obtained for a reaction by multiplying the concentrations of products, dividing by the concentrations of reactants, and raising each concentration term to a power equal to the coefficient in the chemical equation. (15.2)

**Equivalence point** the point in a titration when a stoichiometric amount of reactant has been added. (17.7)

**Ester** a compound formed from a carboxylic acid, RCOOH, and an alcohol, R'OH. (24.6)

**Ether** a compound formally obtained by replacing both H atoms of $H_2O$ by hydrocarbon groups R and R'. (24.6)

**Exact number** a number that arises when you count items or sometimes when you define a unit. (1.5)

**Exchange (metathesis) reaction** a reaction between compounds that, when written as a molecular equation, appears to involve the exchange of parts between the two reactants. (4.3)

**Excited state** a quantum-mechanical state of an atom or molecule associated with any energy level except the lowest, which is the ground state. (8.2)

**Exclusion principle** see *Pauli exclusion principle.*

**Exothermic process** a chemical reaction or physical change in which heat is evolved. (6.2)

**Experiment** an observation of natural phenomena carried out in a controlled manner so that the results can be duplicated and rational conclusions obtained. (1.2)

**Face-centered cubic unit cell** a cubic unit cell in which there are lattice points at the centers of each face of the unit cell in addition to those at the corners. (11.7)

**Factor-label method** see *Dimensional analysis.*

**Faraday constant (*F*)** the magnitude of charge on one mole of electrons, equal to $9.65 \times 10^4$ C. (20.4)

***f*-Block transition elements (inner-transition elements)** the elements with a partially filled *f* subshell in common oxidation states. (8.2)

**Fibrous proteins** proteins that form long coils or align themselves in parallel to form long, water-insoluble fibers. (25.3)

**First ionization energy (first ionization potential)** the minimum energy needed to remove the highest-energy (that is, the outermost) electron from a neutral atom in the gaseous state. (8.6)

**First law of thermodynamics** the change in internal energy of a system, $\Delta U$, equals $q + w$. (19.1)

**Flotation** a physical method of separating a mineral from the gangue that depends on differences in their wettabilities by a liquid solution. (13.2)

**Formal charge** (of an atom in a Lewis formula) the hypothetical charge you obtain by assuming that bonding electrons are equally shared between bonded atoms and that electrons of each lone pair belong completely to one atom. (9.9)

**Formation constant (stability constant) of a complex ion ($K_f$)** the equilibrium constant for the formation of a complex ion from the aqueous metal ion and the ligands. (18.5)

**Formula unit** the group of atoms or ions explicitly symbolized in the formula. (2.6)

**Formula weight (FW)** the sum of the atomic weights of all atoms in a formula unit of a compound. (3.1)

**Fractional (isotopic) abundance** the fraction of the total number of atoms that is composed of a particular isotope. (2.4)

**Fractional precipitation** the technique of separating two or more ions from a solution by adding a reactant that precipitates first one ion, then another, and so forth. (18.3)

**Frasch process** a mining procedure in which underground deposits of solid sulfur are melted in place with superheated water,

and the molten sulfur is forced upward as a froth using air under pressure. (22.8)

**Free energy (*G*)** a thermodynamic quantity defined by the equation $G = H - TS$. (19.4)

**Free energy of formation** see *Standard free energy of formation.*

**Freezing** the change of a liquid to the solid state. (11.2)

**Freezing point** the temperature at which a pure liquid changes to a crystalline solid, or freezes. (11.2)

**Freezing-point depression** a colligative property of a solution equal to the freezing point of the pure solvent minus the freezing point of the solution. (12.6)

**Frequency (*ν*)** the number of wavelengths of a wave that pass a fixed point in one unit of time (usually one second). (7.1)

**Frequency factor** the symbol *A* in the Arrhenius equation, assumed to be a constant. (14.6)

**Fuel cell** essentially a battery, but it differs by operating with a continuous supply of energetic reactants, or fuel. (20.8)

**Fuel rods** the cylinders that contain fissionable material for a nuclear reactor. (21.7)

**Fullerene** a family of molecules consisting of a closed cage of carbon atoms arranged in pentagons as well as hexagons. (13.4)

**Functional group** a reactive portion of a molecule that undergoes predictable reactions. (p. 1070)

**Fusion** see *Melting.*

**Galvanic cell** see *Voltaic cell.*

**Gamma emission** emission from an excited nucleus of a gamma photon, corresponding to radiation with a wavelength of about $10^{-12}$ m. (21.1)

**Gamma photon** a particle of electromagnetic radiation of short wavelength (about 1 pm, or $10^{-12}$ m) and high energy. (21.1)

**Gangue** the worthless portion of an ore. (13.2)

**Gas** the form of matter that is an easily compressible fluid; a given quantity of gas will fit into a container of almost any size and shape. (1.4)

**Gas chromatography** a chromatographic separation method in which a gaseous mixture of vaporized substances is separated into its components by passing the mixture through a column of packing material. Substances in the gaseous mixture are attracted to the packing material to different extents and thus move through the column at different rates. (p. 16)

**Geiger counter** a kind of ionization counter used to count particles emitted by radioactive nuclei. It consists of a metal tube filled with gas, such as argon. (21.3)

**Gene** a sequence of nucleotides in a DNA molecule that codes for a given protein. (25.4)

**Geometric isomers** isomers in which the atoms are joined to one another in the same way but differ because some atoms occupy different relative positions in space. (10.4, 23.5, and 24.3)

**Glass electrode** a compact electrode used to determine pH by emf measurements. (20.7)

**Globular proteins** proteins in which long coils fold into compact, roughly spherical shapes. (25.3)

**Goldschmidt process** a method of preparing a metal by reduction of its oxide with powdered aluminum. (22.4)

**Graham's law of effusion** the rate of effusion of gas molecules from a particular hole is inversely proportional to the square root of the molecular weight of the gas at constant temperature and pressure. (5.7)

**Gravimetric analysis** a type of quantitative analysis in which the amount of a species in a material is determined by converting the species to a product that can be isolated completely and weighed. (4.9)

**Ground state** a quantum-mechanical state of an atom or molecule associated with the lowest energy level. States associated with higher energy levels are called *excited states*. (8.2)

**Group (of the periodic table)** the elements in any one column of the periodic table. (2.5)

**Haber process** an industrial process for the preparation of ammonia from nitrogen and hydrogen with a specially prepared catalyst, high temperature, and high pressure. (3.6 and 22.7)

**Half-cell** the portion of an electrochemical cell in which a half-reaction takes place. (20.2)

**Half-life ($t_{1/2}$)** the time it takes for the reactant concentration to decrease one-half of its initial value. (14.4)     The time it takes for one-half of the nuclei in a sample to decay. (21.4)

**Half-reaction** one of two parts of an oxidation–reduction reaction, one part of which involves a loss of electrons (or increase of oxidation number) and the other a gain of electrons (or decrease of oxidation number). (4.5 and 20.1)

**Hall–Héroult process** the commercial method for producing aluminum by the electrolysis of a molten mixture of aluminum oxide in cryolite, $Na_3AlF_6$. (13.2)

**Halogens** the Group VIIA elements; they are reactive nonmetals. (2.5 and 8.7)

**Heat** the energy that flows into or out of a system because of a difference in temperature between the thermodynamic system and its surroundings. (6.2)

**Heat capacity ($C$)** the quantity of heat needed to raise the temperature of a sample of substance one degree Celsius (or one kelvin). (6.6)

**Heat of formation** see *Standard enthalpy of formation*.

**Heat of fusion (enthalpy of fusion)** the heat needed for the melting of a solid. (11.2)

**Heat of reaction** the heat absorbed (or evolved) during a chemical reaction; it equals the value of $q$ required to return the chemical system to a given temperature at the completion of the reaction. (6.2)

**Heat of solution** the heat absorbed (or evolved) when an ionic substance dissolves in water. (12.3)

**Heat of vaporization (enthalpy of vaporization)** the heat needed for the vaporization of a liquid. (11.2)

**Henderson–Hasselbalch equation** an equation relating the pH of a buffer for different concentrations of conjugate acid and base: pH = $pK_a$ + log [base]/[acid]. (17.6)

**Henry's law** the solubility of a gas is directly proportional to the partial pressure of the gas above the solution. (12.3)

**Hess's law of heat summation** for a chemical equation that can be written as the sum of two or more steps, the enthalpy change for the overall equation equals the sum of the enthalpy changes for the individual steps. (6.7)

**Heterogeneous catalysis** the use of a catalyst that exists in a different phase from the reacting species, usually a solid catalyst in contact with a gaseous or liquid solution of reactants. (14.9)

**Heterogeneous equilibrium** an equilibrium involving reactants and products in more than one phase. (15.3)

**Heterogeneous mixture** a mixture that consists of physically distinct parts, each with different properties. (1.4)

**Heteronuclear diatomic molecules** molecules composed of two different nuclei. (10.6)

**Hexagonal close-packed structure (hcp)** a crystal structure composed of close-packed atoms (or other units) with the stacking ABABABA · · ·; the structure has a hexagonal unit cell. (11.8)

**High-spin complex ion** a complex ion in which there is minimum pairing of electrons in the orbitals of the metal atom. (23.6)

**Homogeneous catalysis** the use of a catalyst in the same phase as the reacting species. (14.9)

**Homogeneous equilibrium** an equilibrium that involves reactants and products in a single phase. (15.3)

**Homogeneous mixture (solution)** a mixture that is uniform in its properties throughout given samples. (1.4)

**Homologous series** a series of compounds in which one compound differs from a preceding one by a fixed group of atoms, for example, a —$CH_2$— group. (24.2)

**Homonuclear diatomic molecules** molecules composed of two like nuclei. (10.6)

**Homopolymer** a polymer whose monomer units are all alike. (25.1)

**Hund's rule** the lowest-energy arrangement of electrons in a subshell is obtained by putting electrons into separate orbitals of the subshell with the same spin before pairing electrons. (8.4)

**Hybrid orbitals** orbitals used to describe bonding that are obtained by taking combinations of atomic orbitals of the isolated atoms. (10.3)

**Hydrate** a compound that contains water molecules weakly bound in its crystals. (2.8)

**Hydrate isomers** isomers of a complex that differ in the placement of water molecules in the complex. (23.5)

**Hydration** the attraction of ions for water molecules. (12.2)

**Hydrocarbons** compounds containing only carbon and hydrogen. (p. 1045)

**Hydrogen bonding** a weak to moderate attractive force that exists between a hydrogen atom covalently bonded to a very electronegative atom and a lone pair of electrons on another small, electronegative atom. It is represented in formulas by a series of dots. (11.5)

**Hydrologic cycle** the natural cycle of water from the oceans to fresh water sources and back to the oceans. (p. 492)

**Hydrolysis** the reaction of an ion with water to produce the conjugate acid and hydroxide ion or the conjugate base and hydrogen ion. (16.4)

**Hydronium ion** the $H_3O^+$ ion; also called the hydrogen ion and written $H^+(aq)$. (4.4)

**Hydrophilic colloid** a colloid in which there is a strong attraction between the dispersed phase and the continuous phase (water). (12.9)

**Hydrophobic colloid** a colloid in which there is a lack of attraction between the dispersed phase and the continuous phase (water). (12.9)

**Hypothesis** a tentative explanation of some regularity of nature. (1.2)

**Ideal gas law** the equation $PV = nRT$, which combines all of the gas laws. (5.3)

**Ideal solution** a solution of two or more substances each of which follows Raoult's law. (12.5)

**Immiscible fluids** fluids that do not mix but form separate layers. (12.1)

**Inert gases** see *Noble gases.*

**Initiator** a compound that produces free radicals in a reaction for the preparation of an addition polymer. (25.1)

**Inner-transition elements** the two rows of elements at the bottom of the periodic table (2.5); the elements with a partially filled *f* subshell in common oxidation states. (8.2 and p. 1003)

**Inorganic compounds** compounds composed of elements other than carbon. A few simple compounds of carbon, including carbon monoxide, carbon dioxide, carbonates, and cyanides, are generally considered to be inorganic. (2.8)

**Intermolecular forces** the forces of interaction between molecules. (11.5)

**Internal energy (*U*)** the sum of the kinetic and the potential energies of the particles making up a system. (6.1 and 19.1)

**International System of Units (SI)** a particular choice of metric units that was adopted by the General Conference of Weights and Measures in 1960. (1.6)

**Ion** an electrically charged particle obtained from an atom or a chemically bonded group of atoms by adding or removing electrons. (2.6)

**Ion exchange** a process in which a water solution is passed through a column of a material that replaces one kind of ion in solution with another kind. (p. 493)

**Ionic bond** a chemical bond formed by the electrostatic attraction between positive and negative ions. (9.1)

**Ionic compound** a compound composed of cations and anions. (2.6)

**Ionic equation** see *Complete ionic equation.*

**Ionic radius** a measure of the size of the spherical region around the nucleus of an ion within which the electrons are most likely to be found. (9.3)

**Ionic solid** a solid that consists of cations and anions held together by the electrical attraction of opposite charges (ionic bonds). (11.6)

**Ionization energy** the energy needed to remove an electron from an atom (or molecule). Often used to mean *first ionization energy* (which see). (8.6)

**Ionization isomers** isomers of a complex that differ in the anion that is coordinated to the metal atom. (23.5)

**Ion product (*Q_c*)** the product of ion concentrations in a solution, each concentration raised to a power equal to the number of ions in the formula of the ionic compound. (18.3)

**Ion-product constant for water (*K_w*)** the equilibrium value of the ion product $[H_3O^+][OH^-]$. (16.6)

**Ion-selective electrode** an electrode whose emf depends on the concentration of a particular ion in solution. (20.7)

**Isoelectronic** refers to different species having the same number and configuration of electrons. (9.3)

**Isomers** compounds of the same molecular formula but with different arrangements of the atoms. (10.4)

**Isotope dilution** a technique to determine the quantity of a substance in a mixture or the total volume of solution by adding a known amount of an isotope to it. (21.5)

**Isotopes** atoms whose nuclei have the same atomic number but different mass numbers; that is, the nuclei have the same number of protons but different numbers of neutrons. (2.3)

**Joule (J)** the SI unit of energy; $1 \text{ J} = 1 \text{ kg} \cdot \text{m}^2/\text{s}^2$. (6.1)

**Kelvin (K)** the SI base unit of temperature; a unit on an absolute temperature scale. (1.6)

**Ketone** a compound containing a carbonyl group with two hydrocarbon groups attached to it. (24.6)

**Kilogram (kg)** the SI base unit for mass; equal to about 2.2 pounds. (1.6)

**Kinetic energy** the energy associated with an object by virtue of its motion. (6.1)

**Kinetic-molecular theory (kinetic theory)** the theory that a gas consists of molecules in constant random motion. (p. 210)

**Kinetics, chemical** see *Chemical kinetics.*

**Lanthanides** the first of the two rows of elements at the bottom of the periodic table; the 14 elements following lanthanum in the periodic table, in which the $4f$ subshell is filling. (p. 995)

**Laser** a source of intense, highly directed beam of monochromatic light; the word *laser* is an acronym meaning *l*ight *a*mplification by *s*timulated *e*mission of *r*adiation. (p. 292)

**Lattice energy** the change in energy that occurs when an ionic solid is separated into isolated ions in the gas phase. (9.1 and 12.2)

**Law** a concise statement or mathematical equation about a fundamental relationship or regularity of nature. (1.2)

**Law of combining volumes** a relation stating that gases at the same temperature and pressure react with one another in volume ratios of small whole numbers. (5.2)

**Law of conservation of energy** energy may be converted from one form to another, but the total quantity of energy remains constant. (6.1)

**Law of conservation of mass** the total mass remains constant during a chemical change (chemical reaction). (1.3)

**Law of definite proportions (law of constant composition)** a pure compound, whatever its source, always contains definite or constant proportions of the elements by mass. (1.4)

**Law of effusion** see *Graham's law of effusion.*

**Law of heat summation** see *Hess's law of heat summation.*

**Law of mass action** the values of the equilibrium-constant expression $K_c$ are constant for a particular reaction at a given temperature, whatever equilibrium concentrations are substituted. (15.2)

**Law of multiple proportions** when two elements form more than one compound, the masses of one element in these compounds for a fixed mass of the other element are in ratios of small whole numbers. (2.1)

**Law of partial pressures** see *Dalton's law of partial pressures.*

**Lead storage cell** a voltaic cell that consists of electrodes of lead alloy grids; one electrode is packed with a spongy lead to form the anode, and the other is packed with lead dioxide to form the cathode. (20.8)

**Le Chatelier's principle** when a system in equilibrium is disturbed by a change of temperature, pressure, or concentration variable, the system shifts in equilibrium composition in a way that tends to counteract this change of variable. (12.3 and 15.7)

**Leclanché dry cell** see *Zinc–carbon dry cell.*

**Levorotatory** refers to a compound whose solution rotates the plane of polarized light to the left (when looking toward the source of light). (23.5)

**Lewis acid** a species that can form a covalent bond by accepting an electron pair from another species. (16.3)

**Lewis base** a species that can form a covalent bond by donating an electron pair to another species. (16.3)

**Lewis electron-dot formula** a formula using dots to represent valence electrons. (9.4)

**Lewis electron-dot symbol** a symbol in which the electrons in the valence shell of an atom or ion are represented by dots placed around the letter symbol of the element. (9.1)

**Ligand** a Lewis base that bonds to a metal ion to form a complex ion. (p. 785 and 23.3)

**Limiting reactant (limiting reagent)** the reactant that is entirely consumed when a reaction goes to completion. (3.8)

**Linear geometry** a molecular geometry in which all atoms line up along a straight line. (10.1)

**Line spectrum** a spectrum showing only certain colors or specific wavelengths of light. (7.3)

**Linkage isomers** isomers of a complex that differ in the atom of a ligand that is bonded to the metal atom. (23.5)

**Lipids** biological substances like fats and oils that are soluble in organic solvents, such as chloroform and carbon tetrachloride. (p. 543)

**Liquefaction** the process in which a substance that is normally a gas changes to the liquid state. (11.2)

**Liquid** the form of matter that is a relatively incompressible fluid; a liquid has a fixed volume but no fixed shape. (1.4)

**Liter (L)** a unit of volume equal to a cubic decimeter (equal to approximately one quart). (1.7)

**Lithium–iodine battery** a voltaic cell in which the anode is lithium metal and the cathode is an $I_2$ complex. (20.8)

**London forces (dispersion forces)** the weak attractive forces between molecules resulting from the small, instantaneous dipoles that occur because of the varying positions of the electrons during their motion about nuclei. (11.5)

**Lone pair (nonbonding pair)** an electron pair that remains on one atom and is not shared. (9.4)

**Low-spin complex ion** a complex ion in which there is more pairing of electrons in the orbitals of the metal atom than in a corresponding high-spin complex ion. (23.6)

**Magic number** the number of nuclear particles in a completed shell of protons or neutrons. (21.1)

**Magnetic quantum number ($m_l$)** the quantum number that distinguishes orbitals of given $n$ and $l$—that is, of given energy and shape—but having a different orientation in space; the allowed values are the integers from $-l$ to $+l$. (7.5)

**Main-group element (representative element)** an element in an A column of the periodic table, in which an outer $s$ or $p$ subshell is filling. (2.5 and 8.2)

**Manometer** a device that measures the pressure of a gas or liquid in a sealed vessel. (5.1)

**Markownikoff's rule** a generalization stating that the major product formed by the addition of an unsymmetrical reagent such as H—Cl, H—Br, or H—OH is the one obtained when the H atom of the reagent adds to the carbon atom of the multiple bond that already has the greater number of hydrogen atoms attached to it. (24.3)

**Mass** the quantity of matter in a material. (1.3)

**Mass defect** the total nucleon mass minus the atomic mass of a nucleus. (21.6)

**Mass number (A)** the total number of protons and neutrons in a nucleus. (2.3)

**Mass percentage** parts per hundred parts of the total, by mass. (3.3)

**Mass percentage of solute** the percentage by mass of solute contained in a solution. (12.4)

**Mass spectrometer** an instrument, such as one based on Thomson's principles, that measures the mass-to-charge ratios of atoms. (2.4)

**Material** any particular kind of matter. (p. 2)

**Matter** all of the objects around you (p. 2) whatever occupies space and can be perceived by our senses. (1.3)

**Maxwell's distribution of molecular speeds** a theoretical relationship that predicts the relative number of molecules at various speeds for a sample of gas at a particular temperature. (5.7)

**Melting (fusion)** the change of a solid to the liquid state. (11.2)

**Melting point** the temperature at which a crystalline solid changes to a liquid, or melts. (11.2)

**Messenger RNA** a relatively small RNA molecule that can diffuse about the cell and attach itself to a ribosome, where it serves as a pattern for protein synthesis. (25.4)

**Metal** a substance or mixture that has a characteristic luster, or shine, is generally a good conductor of heat and electricity, and is malleable and ductile. (2.5 and p. 554)

**Metallic solid** a solid that consists of positive cores of atoms held together by a surrounding "sea" of electrons (metallic bonding). (11.6)

**Metalloid (semimetal)** an element having both metallic and nonmetallic properties. (2.5)

**Metallurgy** the scientific study of the production of metals from their ores and the making of alloys having various useful properties. (13.2)

**Metal refining** in metallurgy, the purification of a metal. (21.2)

**Metaphosphoric acids** acids with the general formula $(HPO_3)_n$. (22.7)

**Metastable nucleus** a nucleus in an excited state with a lifetime of at least one nanosecond ($10^{-9}$ s). (21.1)

**Metathesis reaction** see *Exchange reaction.*

**Meter (m)** the SI base unit of length. (1.6)

**Micelle** a colloidal-sized particle formed in water by the association of molecules or ions each of which has a hydrophobic end and a hydrophilic end. (12.9)

**Millimeters of mercury (mmHg)** a unit of pressure also known as the *torr.* A unit of pressure equal to that exerted by a column of mercury 1 mm high at 0.00°C. (5.1)

**Mineral** a naturally occurring inorganic solid substance or solid solution with definite crystalline form. (13.1)

**Miscible fluids** fluids that mix with or dissolve in each other in all proportions. (12.1)

**Mixture** a material that can be separated by physical means into two or more substances. (1.4)

**Moderator** a substance that slows down neutrons in a nuclear fission reactor. (21.7)

**Molality** the moles of solute per kilogram of solvent. (12.4)

**Molar concentration (molarity), M** the moles of solute dissolved in one liter (cubic decimeter) of solution. (4.7)

**Molar gas constant (R)** the constant of proportionality relating the molar volume of a gas to $T/P$. (5.3)

**Molar gas volume** the volume of one mole of gas. (5.2)

**Molarity** see *Molar concentration.*

**Molar mass** the mass of one mole of substance. In grams, it is numerically equal to the formula weight in atomic mass units. (3.2)

**Mole (mol)** the quantity of a given substance that contains as many molecules or formula units as the number of atoms in exactly 12 g of carbon-12. The amount of substance containing Avogadro's number of molecules or formula units. (3.2)

**Molecular equation** a chemical equation in which the reactants and products are written as if they were molecular substances, even though they may actually exist in solution as ions. (4.2)

**Molecular formula** a chemical formula that gives the exact number of different atoms of an element in a molecule. (2.6)

**Molecular geometry** the general shape of a molecule, as determined by the relative positions of the atomic nuclei. (p. 396)

**Molecularity** the number of molecules on the reactant side of an elementary reaction. (14.7)

**Molecular orbital theory** a theory of the electronic structure of molecules in terms of molecular orbitals, which may spread over several atoms or the entire molecule. (p. 423)

**Molecular solid** a solid that consists of atoms or molecules held together by intermolecular forces. (11.6)

**Molecular substance** a substance that is composed of molecules, all of which are alike. (2.6)

**Molecular weight (MW)** the sum of the atomic weights of all the atoms in a molecule. (3.1)

**Molecule** a definite group of atoms that are chemically bonded together—that is, tightly connected by attractive forces. (2.6)

**Mole fraction** the fraction of moles of a component in the total moles of a mixture. (5.5) The moles of a component substance divided by the total moles of solution. (12.4)

**Monatomic ion** an ion formed from a single atom. (2.8)

**Mond process** a chemical procedure for the purification of nickel that depends on the formation and later decomposition of a volatile compound of the metal (nickel tetracarbonyl). (13.2)

**Monodentate ligand** a ligand that bonds to a metal atom through one atom of the ligand. (23.3)

**Monomer** the small molecules that are linked together to form a polymer (2.6); a compound used to make a polymer (and from which the polymer's repeating unit arises). (p. 1085)

**Monoprotic acid** an acid that yields one acidic hydrogen per molecule. (4.4)

**Monosaccharides** simple sugars, each containing three to nine carbon atoms, generally all but one of which bear a hydroxyl group, the remaining one being part of a carbonyl group. (25.4)

**Nanotechnology** an area of technology in which one manipulates materials on a molecular scale to create useful devices. (13.4)

**Nernst equation** an equation relating the cell emf, $E_{cell}$, to its standard emf, $E^{\circ}_{cell}$, and the reaction quotient, $Q$. At 25°C, the equation is $E_{cell} = E^{\circ}_{cell} - (0.0592/n)\log Q$. (19.7)

**Net ionic equation** an ionic equation from which spectator ions have been canceled. (4.2)

**Neutralization reaction** a reaction of an acid and a base that results in an ionic compound and possibly water. (4.4)

**Neutron** a particle found in the nucleus of an atom; it has a mass almost identical to that of the proton but no electric charge. (2.3)

**Neutron activation analysis** an analysis of elements in a sample based on the conversion of stable isotopes to radioactive isotopes by bombarding a sample with neutrons. (21.5)

**Nickel–cadmium cell** a voltaic cell consisting of an anode of cadmium and a cathode of hydrated nickel oxide (approximately NiOOH) on nickel; the electrolyte is potassium hydroxide. (19.8)

**Nitrogen cycle** the circulation of the element nitrogen in the biosphere, from nitrogen fixation to the release of free nitrogen by denitrifying bacteria. (22.7)

**Nitrogen-fixing bacteria** bacteria that can produce nitrogen compounds in the soil from atmospheric nitrogen. (22.7)

**Noble-gas core** an inner-shell configuration corresponding to one of the noble gases. (8.2)

**Noble gases (inert gases; rare gases)** the Group VIIIA elements; all are gases consisting of uncombined atoms. They are relatively unreactive elements. (8.7)

**Nomenclature** see *Chemical nomenclature*.

**Nonbonding pair** an electron pair that remains on one atom and is not shared. (9.4)

**Nonelectrolyte** a substance, such as sucrose, or table sugar $(C_{12}H_{22}O_{11})$, that dissolves in water to give a nonconducting or very poorly conducting solution. (4.1)

**Nonmetal** an element that does not exhibit the characteristics of a metal. (2.5)

**Nonstoichiometric compound** a compound whose composition varies from its idealized formula. (11.7)

**Nuclear bombardment reaction** a nuclear reaction in which a nucleus is bombarded, or struck, by another nucleus or by a nuclear particle. (p. 895)

**Nuclear equation** a symbolic representation of a nuclear reaction. (21.1)

**Nuclear fission** a nuclear reaction in which a heavy nucleus splits into lighter nuclei and energy is released. (21.6)

**Nuclear fission reactor** a device that permits a controlled chain reaction of nuclear fissions. (21.7)

**Nuclear force** a strong force of attraction between nucleons that acts only at very short distances (about $10^{-15}$ m). (21.1)

**Nuclear fusion** a nuclear reaction in which light nuclei combine to give a stabler, heavier nucleus plus possibly several neutrons, and energy is released. (21.6)

**Nucleic acids** polynucleotides folded or coiled into specific three-dimensional shapes. (25.4)

**Nucleotides** the building blocks of nucleic acids. (25.4)

**Nucleus** the atom's central core; it has most of the atom's mass and one or more units of positive charge. (2.2)

**Nuclide** a particular atom characterized by a definite atomic number and mass number. (2.3)

**Nuclide symbol** a symbol for a nuclide, in which the atomic number is given as a left subscript and the mass number is given as a left superscript to the symbol of the element. (2.3)

**Number of significant figures** the number of digits reported for the value of a measured or calculated quantity, indicating the precision of the value. (1.5)

**Octahedral geometry** the geometry of a molecule in which six atoms occupy the vertices of a regular octahedron (a figure with eight faces and six vertices) with the central atom at the center of the octahedron. (10.1)

**Octet rule** the tendency of atoms in molecules to have eight electrons in their valence shells (two for hydrogen atoms). (9.4)

**Optical isomers (enantiomers)** isomers that are nonsuperimposable mirror images of one another. (23.5)

**Optically active** having the ability to rotate the plane of light waves, either as a pure substance or in solution. (23.5)

**Orbital diagram** a diagram to show how the orbitals of a subshell are occupied by electrons. (8.1)

**Ore** a rock or mineral from which a metal or nonmetal can be economically produced. (13.1)

**Organic compounds** compounds that contain carbon combined with other elements, such as hydrogen, oxygen, and nitrogen. (2.7)

**Osmosis** the phenomenon of solvent flow through a semipermeable membrane to equalize the solute concentrations on both sides of the membrane. (12.7)

**Osmotic pressure** a colligative property of a solution equal to the pressure that, when applied to the solution, just stops osmosis. (12.7)

**Ostwald process** an industrial preparation of nitric acid starting from the catalytic oxidation of ammonia to nitric oxide. (22.7)

**Overall order of a reaction** the sum of the orders of the reactant species in the rate law. (14.3)

**Oxidation** the part of an oxidation–reduction reaction in which there is a loss of electrons by a species (or an increase in the oxidation number of an atom). (4.5)

**Oxidation number (oxidation state)** either the actual charge on an atom in a substance, if the atom exists as a monatomic ion, or a hypothetical charge assigned by simple rules. (4.5)

**Oxidation potential** the negative of the standard electrode potential. (19.5)

**Oxidation–reduction reaction (redox reaction)** a reaction in which electrons are transferred between species or in which atoms change oxidation number. (4.5)

**Oxide** a binary compound with oxygen in the −2 oxidation state. (22.8)

**Oxidizing agent** a species that oxidizes another species; it is itself reduced. (4.5)

**Oxoacid** an acid containing hydrogen, oxygen, and another element. (2.7)    A substance in which O atoms (and possibly other electronegative atoms) are bonded to a central atom, with one or more H atoms usually bonded to the O atoms. (9.6)

**Pairing energy ($P$)** the energy required to put two electrons into the same orbital. (23.7)

**Paramagnetic substance** a substance that is weakly attracted by a magnetic field; this attraction generally results from unpaired electrons. (8.4)

**Partial pressure** the pressure exerted by a particular gas in a mixture. (5.5)

**Particle accelerator** a device used to accelerate electrons, protons, and alpha particles and other ions to very high speeds. (21.2)

**Pascal (Pa)** the SI unit of pressure; 1 Pa = 1 kg/(m · s$^2$). (5.1)

**Pauli exclusion principle** no two electrons in an atom can have the same four quantum numbers. It follows from this that an orbital can hold no more than two electrons and can hold two only if they have different spin quantum numbers. (8.1)

**Peptide (amide) bonds** the C—N bonds resulting from a condensation reaction between the carboxyl group of one amino acid and the amino group of a second amino acid. (25.3)

**Percentage composition** the mass percentages of each element in a compound. (p. 98)

**Percentage yield** the actual yield (experimentally determined) expressed as a percentage of the theoretical yield (calculated). (3.8)

**Period (of the periodic table)** the elements in any one horizontal row of the periodic table. (2.5)

**Periodic law** when the elements are arranged by atomic number, their physical and chemical properties vary periodically. (8.6)

**Periodic table** a tabular arrangement of elements in rows and columns, highlighting the regular repetition of properties of the elements. (2.5)

**Peroxide** a compound with oxygen in the −1 oxidation state. (22.8)

**pH** the negative of the logarithm of the molar hydrogen-ion concentration. (16.8)

**Phase** one of several different homogeneous materials present in the portion of matter under study. (1.4)

**Phase diagram** a graphical way to summarize the conditions under which the different states of a substance are stable. (11.3)

**Phase transition** see *Change of state.*

**Phospholipid** a type of lipid that contains a phosphate group. These molecules have hydrophobic and hydrophilic ends, and they aggregate to give a layer structure. The process is similar to the formation of micelles. (p. 543)

**Phospholipid bilayer** a part of a biological membrane consisting of two layers of phospholipid molecules. (p. 543)

**Photoelectric effect** the ejection of electrons from the surface of a metal or other material when light shines on it. (7.2)

**Photon** particle of electromagnetic energy with energy $E$ proportional to the observed frequency of light: $E = h\nu$. (7.2)

**Physical adsorption** adsorption in which the attraction is provided by weak intermolecular forces. (14.9)

**Physical change** a change in the form of matter but not in its chemical identity. (1.4)

**Physical property** a characteristic that can be observed for a material without changing its chemical identity. (1.4)

**Pi ($\pi$) bond** a bond that has an electron distribution above and below the bond axis. (10.4)

**Planck's constant ($h$)** a physical constant with the value $6.63 \times 10^{-34}$ J · s. It is the proportionality constant relating the frequency of light to the energy of a photon. (7.2)

**Plasma** an electrically neutral gas of ions and electrons. (21.7)

**Polar covalent bond** a covalent bond in which the bonding electrons spend more time near one atom than near the other. (9.5)

**Polyamide** a polymer formed by reacting a substance containing two amine groups with a substance containing two carboxylic acid groups. (25.1)

**Polyatomic ion** an ion consisting of two or more atoms chemically bonded together and carrying a net electric charge. (2.8)

**Polydentate ligand** a ligand that can bond with two or more atoms to a metal atom. (23.3)

**Polyester** a polymer formed by reacting a substance containing two alcohol groups with a substance containing two carboxylic acid groups. (25.1)

**Polymer** a very large molecule made up of a number of smaller molecules repeatedly linked together. (2.6); a chemical species of very high molecular weight that is made up from many repeating units of low molecular weight. (p. 1085)

**Polymerase chain reaction (PCR)** a technique used to amplify the quantity of DNA in a sample. (25.4)

**Polynucleotide** a linear polymer of nucleotide units. (25.4)

**Polypeptide** a polymer formed by the linking of many amino acids by peptide bonds. (25.3)

**Polyphosphoric acids** acids with the general formula $H_{n+2}P_nO_{3n+1}$ formed from linear chains of P—O bonds. (22.7)

**Polyprotic acid** an acid that yields two or more acidic hydrogens per molecule. (4.4 and 17.2)

**Positron** a particle that is similar to an electron and has the same mass but a positive charge. (20.1)

**Positron emission** emission of a positron from an unstable nucleus. (20.1)

**Potential difference** the difference in electrical potential (electrical pressure) between two points. (19.4)

**Potential energy** the energy an object has by virtue of its position in a field of force. (6.1)

**Precipitate** an insoluble solid compound formed during a chemical reaction in solution. (4.3)

**Precision** the closeness of the set of values obtained from identical measurements of a quantity. (1.5)

**Pressure** the force exerted per unit area of surface. (5.1)

**Primary alcohol** an alcohol in which the hydroxyl group is attached to a carbon atom that is itself bonded to only one other carbon atom. (24.6)

**Primary structure (of a protein)** the order, or sequence, of the amino-acid units in the protein. (25.3)

**Principal quantum number ($n$)** the quantum number on which the energy of an electron in an atom principally depends; it can have any positive value: 1, 2, 3, $\cdots$. (7.5)

**Product** a substance that results from a chemical reaction. (2.9)

**Protein** a biological polymer of small molecules called amino acids. (25.3)

**Proton** a particle found in the nucleus of the atom; it has a positive charge equal in magnitude, but opposite in sign, to that of the electron and a mass 1836 times that of the electron. (2.3)

**Pseudo-noble-gas core** the noble-gas core together with $(n-1)d^{10}$ electrons. (8.2)

**Qualitative analysis** the determination of the identity of substances present in a mixture. (p. 790)

**Quantitative analysis** the determination of the amount of a substance or species present in a material. (p. 167)

**Quantum (wave) mechanics** the branch of physics that mathematically describes the wave properties of submicroscopic particles. (7.4)

**Racemic mixture** a mixture of equal amounts of optical isomers. (23.5)

**Rad** the dosage of radiation that deposits $1 \times 10^{-2}$ J of energy per kilogram of tissue. (20.3)

**Radioactive decay** the process in which a nucleus spontaneously disintegrates, giving off radiation. (p. 852)

**Radioactive decay constant ($k$)** rate constant for radioactive decay. (20.4)

**Radioactive decay series** a sequence in which one radioactive nucleus decays to a second, which then decays to a third, and so on. (20.1)

**Radioactive tracer** a very small amount of radioactive isotope added to a chemical, biological, or physical system to facilitate study of the system. (20.5)

**Radioactivity** spontaneous radiation from unstable elements. (20.1)

**Raoult's law** the partial pressure of solvent, $P_A$, over a solution equals the vapor pressure of the pure solvent, $P_A^\circ$, times the mole fraction of the solvent, $X_A$, in solution: $P_A = P_A^\circ X_A$. (12.5)

**Rare gases** see *Noble gases.*

**Rate constant** a proportionality constant in the relationship between rate and concentrations. (14.3)

**Rate-determining step** the slowest step in a reaction mechanism. (14.8)

**Rate law** an equation that relates the rate of a reaction to the concentrations of reactants (and catalyst) raised to various powers. (14.3)

**Reactant** a starting substance in a chemical reaction. (2.9)

**Reaction intermediate** a species produced during a reaction that does not appear in the net equation because it reacts in a subsequent step in the mechanism. (14.7)

**Reaction mechanism** the set of elementary reactions whose overall effect is given by the net chemical equation. (14.7)

**Reaction order** the exponent of the concentration of a given reactant species in the rate law, as determined experimentally. (14.3) See also *Overall reaction order.*

**Reaction quotient ($Q_c$)** an expression that has the same form as the equilibrium-constant expression but whose concentration values are not necessarily those at equilibrium. (15.5)

**Reaction rate** the increase in molar concentration of product of a reaction per unit time or the decrease in molar concentration of reactant per unit time. (14.1)

**Redox reaction** see *Oxidation–reduction reaction.*

**Reducing agent** a species that reduces another species; it is itself oxidized. (4.5)

**Reduction** the part of an oxidation–reduction reaction in which there is a gain of electrons by a species (or a decrease of oxidation number of an atom). (4.5)

**Reduction potential** see *Standard electrode potential.*

**Reference form** the stablest form (physical state and allotrope) of an element under standard thermodynamic conditions. (6.8)

**Rem** a unit of radiation dosage used to relate various kinds of radiation in terms of biological destruction. It equals the rad times a factor for the type of radiation, called the relative biological effectiveness (RBE): rems = rads × RBE. (21.3)

**Refining** see *Metal refining.*

**Representative element** see *Main-group element.*

**Resonance description** a representation in which you describe the electron structure of a molecule having delocalized bonding by writing all possible electron-dot formulas. (9.7)

**Reverse osmosis** a process in which a solvent, such as water, is forced by a pressure greater than the osmotic pressure to flow through a semipermeable membrane from a concentrated solution to a more dilute one. (12.7)

**Ribonucleic acid (RNA)** a constituent of cells that is used to manufacture proteins from genetic information. It is a polymer of ribonucleotide units. (25.4)

**Ribosomal RNA** the RNA in a ribosome. (25.4)

**Ribosomes** tiny cellular particles on which protein synthesis takes place. (25.4)

**Roasting** the process of heating a mineral in air to obtain the oxide. (13.2)

**Root-mean-square (rms) molecular speed** a type of average molecular speed, equal to the speed of a molecule having the average molecular kinetic energy. It equals $\sqrt{3RT/M_m}$, where $M_m$ is the molar mass. (5.7)

**Rounding** the procedure of dropping nonsignificant digits in a calculation result and adjusting the last digit reported. (1.5)

**Salt** an ionic compound that is a product of a neutralization reaction. (4.4)

**Salt bridge** a tube of an electrolyte in a gel that is connected to the two half-cells of a voltaic cell; it allows the flow of ions but prevents the mixing of the different solutions that would allow direct reaction of the cell reactants. (20.2)

**Saponification** the hydrolysis of an ester in the presence of a base. (24.6)

**Saturated hydrocarbon** a hydrocarbon that has only single bonds between carbon atoms; all carbon atoms are bonded to the maximum number of hydrogen atoms (that is, the hydrocarbon is *saturated* with hydrogen). A saturated hydrocarbon molecule can be cyclic or acyclic. See *cyclic hydrocarbon* and *acyclic hydrocarbon.* (24.1)

**Saturated solution** a solution that is in equilibrium with respect to a given dissolved substance. (12.2)

**Scientific method** the general process of advancing scientific knowledge through observation, the framing of laws, hypotheses, or theories, and the conducting of more experiments. (1.2)

**Scientific notation** the representation of a number in the form $A \times 10^n$, where $A$ is a number with a single nonzero digit to the left of the decimal point and $n$ is an integer, or whole number. (1.5)

**Scintillation counter** a device that detects nuclear radiations from flashes of light generated in a material by the radiation. (21.3)

**Second (s)** the SI base unit of time. (1.6)

**Secondary alcohol** an alcohol in which the hydroxyl group is attached to a carbon atom that is itself bonded to two carbon atoms. (24.6)

**Secondary structure (of a protein)** the relatively simple coiled or parallel arrangement of a protein molecule. (25.3)

**Second law of thermodynamics** the total entropy of a system and its surroundings always increases for a spontaneous process. Also, for a spontaneous process at a given temperature, the change in entropy of the system is greater than the heat divided by the absolute temperature. (19.2)

**Seesaw geometry** the geometry of a molecule having four atoms bonded to a central atom, in which two of these outer atoms occupy axial positions of a trigonal bipyramid and the other two occupy equatorial positions. (10.1)

**Self-ionization (autoionization)** a reaction in which two like molecules react to give ions. (16.6)

**Semiconducting element** an element that exhibits very low electrical conductivity at room temperature when pure, but whose electrical conductivity rises with temperature or with the addition of certain other elements. The process of adding small quantities of these other elements to a semiconducting element to increase its conductivity is called *doping.* (13.5)

**Semimetal** see *Metalloid.*

**Shell model of the nucleus** a nuclear model in which protons and neutrons exist in levels, or shells, analogous to the shell structure that exists for electrons in an atom. (21.1)

**SI** see *International System.*

**SI base units** the SI units of measurement from which all others can be derived. (1.6)

**SI derived unit** a unit derived by combining SI base units. (1.7)

**SI prefix** a prefix used in the International System to indicate a power of ten. (1.6)

**Sigma ($\sigma$) bond** a bond that has a cylindrical shape about the bond axis. (10.4)

**Significant figures** those digits in a measured number (or result of a calculation with measured numbers) that include all certain digits plus a final one having some uncertainty. (1.5)

**Silane** a hydride of silicon that contains chains of single-bonded silicon atoms. (22.6)

**Silica** a covalent network solid of $SiO_2$ in which each silicon atom is covalently bonded in tetrahedral directions to four oxygen atoms; each oxygen atom is in turn bonded to another silicon atom. (13.6)

**Silicate** a compound of silicon and oxygen (with one or more metals) that may be formally regarded as a derivative of silicic acid, $H_4SiO_4$ or $Si(OH)_4$. (13.6)

**Silicone** a compound that contains chains or rings of Si—O with one or more of the bonding positions on each Si atom occupied by an organic group. (22.6)

**Simple cubic unit cell** a cubic unit cell in which lattice points are situated only at the corners of the unit cell. (11.7)

**Simplest formula** see *Empirical formula.*

**Single bond** a covalent bond in which a single pair of electrons is shared by two atoms. (9.4)

**Single-replacement reaction** see *Displacement reaction.*

**Sol** a colloid that consists of solid particles dispersed in a liquid. (12.9)

**Solid** the form of matter characterized by rigidity; a solid is relatively incompressible and has fixed shape and volume. (1.4)

**Solubility** the amount of a substance that dissolves in a given quantity of solvent (such as water) at a given temperature to give a saturated solution. (12.2)

**Solubility-product constant ($K_{sp}$)** the equilibrium constant for the solubility equilibrium of a slightly soluble (or nearly insoluble) ionic compound. (18.1)

**Solute** in the case of a solution of a gas or solid dissolved in a liquid, the gas or solid; in other cases, the component in smaller amount. (12.1)

**Solution** see *Homogeneous mixture.*

**Solvay process** an industrial method for obtaining sodium carbonate from sodium chloride and limestone. (22.2)

**Solvent** in a solution of a gas or solid dissolved in a liquid, the liquid; in other cases, the component in greater amount. (12.1)

**Specific heat capacity (specific heat)** the quantity of heat required to raise the temperature of one gram of a substance by one degree Celsius (or one kelvin) at constant pressure. (6.6)

**Spectator ion** an ion in an ionic equation that does not take part in the reaction. (4.2)

**Spectrochemical series** an arrangement of ligands according to the relative magnitudes of the crystal field splittings they induce in the *d* orbitals of a metal ion. (23.7)

**Spin quantum number ($m_s$)** the quantum number that refers to the two possible orientations of the spin axis of an electron; possible values are $+\frac{1}{2}$ and $-\frac{1}{2}$. (7.5)

**Spontaneous fission** the spontaneous decay of an unstable nucleus in which a heavy nucleus of mass number greater than 89 splits into lighter nuclei and energy is released. (21.1)

**Spontaneous process** a physical or chemical change that occurs by itself. (p. 804)

**Square planar geometry** the geometry of a molecule in which a central atom is surrounded by four other atoms arranged in a square and in a plane containing the central atom. (10.1)

**Square pyramidal geometry** the geometry of a molecule in which a central atom is at the apex of a pyramid and four other atoms form the square base of the pyramid. (10.1)

**Stability constant (of a complex)** see *Formation constant.*

**Standard electrode potential ($E°$)** the electrode potential when the concentrations of solutes are 1 *M*, the gas pressures are 1 atm, and the temperature has a specified value—usually 25°C. (20.5)

**Standard emf ($E°_{cell}$)** the emf of a voltaic cell operating under standard-state conditions (solute concentrations are 1 *M*, gas pressures are 1 atm, and the temperature has a specified value—usually 25°C). (20.5)

**Standard enthalpy of formation (standard heat of formation), $\Delta H_f°$** the enthalpy change for the formation of one mole of a substance in its standard state from its elements in their reference forms and in their standard states. (6.8)

**Standard entropy ($S°$)** the entropy value for the standard state of a species. (19.3)

**Standard free energy of formation ($\Delta G_f°$)** the free-energy change that occurs when one mole of substance is formed from its elements in their stablest states at 1 atm and at a specified temperature (usually 25°C). (19.4)

**Standard heat of formation** see *Standard enthalpy of formation.*

**Standard state** the standard thermodynamic conditions (1 atm and usually 25°C) chosen for substances when listing or comparing thermochemical data. (6.8)

**Standard temperature and pressure (STP)** the reference conditions for gases chosen by convention to be 0°C and 1 atm. (5.2)

**State function** a property of a system that depends only on its present state, which is determined by variables such as temperature and pressure, and is independent of any previous history of the system. (6.3 and 19.1)

**States of matter** the three forms that matter can assume—solid, liquid, and gas. (1.4)

**Steam-reforming process** an industrial preparation in which steam and hydrocarbons from natural gas or petroleum react at high temperature and pressure in the presence of a catalyst to form carbon monoxide and hydrogen. (22.5)

**Stereoisomers** isomers in which the atoms are bonded to each other in the same order but that differ in the precise arrangement of the atoms in space. (23.5)

**Stock system** a system of chemical nomenclature in which the charge on a metal atom or oxidation number of an atom is denoted by a Roman numeral in parentheses following the element name. (2.8)

**Stoichiometry** the calculation of the quantities of reactants and products involved in a chemical reaction. (p. 108)

**Stratosphere** the region of the atmosphere that lies just above the troposphere and wherein the temperature increases with increasing altitude. (p. 430)

**Strong acid** an acid that ionizes completely in water; it is a strong electrolyte. (4.4 and 16.1)

**Strong base** a base that is present in aqueous solution entirely as ions, one of which is $OH^-$; it is a strong electrolyte. (4.4 and 16.1)

**Strong electrolyte** an electrolyte that exists in solution almost entirely as ions. (4.1)

**Structural formula** a chemical formula that shows how the atoms are bonded to one another in a molecule. (2.6)

**Structural isomers** isomers that differ in how the atoms are joined together. (23.5)

**Sublimation** the change of a solid directly to the vapor. (11.2)

**Substance** a kind of matter that cannot be separated into other kinds of matter by any physical process. (1.4)

**Substitution reaction** a reaction in which a part of the reacting molecule is substituted for an H atom on a hydrocarbon or a hydrocarbon group. (24.2)

**Substrate** the substance whose reaction an enzyme catalyzes. (13.9)

**Superoxide** a binary compound with oxygen in the $-\frac{1}{2}$ oxidation state; it contains the superoxide ion, $O_2^-$. (22.8)

**Supersaturated solution** a solution that contains more dissolved substance than does a saturated solution; the solution is not in equilibrium with the solid substance. (12.2)

**Surface tension** the energy required to increase the surface area of a liquid by a unit amount. (11.4)

**Surroundings** everything in the vicinity of a thermodynamic system. (6.2)

**Synthesis gas** a mixture of CO and $H_2$ gases used in the industrial preparation of a number of organic compounds, including methanol, $CH_3OH$. (p. 639)

**System (thermodynamic)** the substance or mixture of substances under study in which a change occurs. (6.2)

**Termolecular reaction** an elementary reaction that involves three reactant molecules. (14.7)

**Tertiary alcohol** an alcohol in which the hydroxyl group is attached to a carbon atom that is itself bonded to three carbon atoms. (24.6)

**Tertiary structure (of a protein)** the structure associated with the way the protein coil is folded. (25.3)

**Tetrahedral geometry** the geometry of a molecule in which four atoms bonded to a central atom occupy the vertices of a tetrahedron with the central atom at the center of this tetrahedron. (10.1)

**Theoretical yield** the maximum amount of product that can be obtained by a reaction from given amounts of reactants. (3.8)

**Theory** a tested explanation of basic natural phenomena. (1.2)

**Thermal equilibrium** a state in which heat does not flow between a system and its surroundings because they are both at the same temperature. (6.2)

**Thermochemical equation** the chemical equation for a reaction (including phase labels) in which the equation is given a molar interpretation, and the enthalpy of reaction for these molar amounts is written directly after the equation. (6.4)

**Thermochemistry** the study of the quantity of heat absorbed or evolved by chemical reactions. (p. 236)

**Thermodynamic equilibrium constant ($K$)** the equilibrium constant in which the concentrations of gases are expressed in partial pressures in atmospheres, whereas the concentrations of solutes in liquid solutions are expressed in molarities. (19.6)

**Thermodynamics** the study of the relationship between heat and other forms of energy involved in a chemical or physical process. (p. 236 and p. 758)

**Thermodynamic system** see *System.*

**Third law of thermodynamics** a substance that is perfectly crystalline at 0 K has an entropy of zero. (19.3)

**Titration** a procedure for determining the amount of substance A by adding a carefully measured volume of a solution with known concentration of B until the reaction of A and B is just complete. (4.10)

**Torr** see *Millimeters of mercury.*

**Transfer RNA (tRNA)** a small RNA molecule; it binds to a particular amino acid, carries it to a ribosome, and then attaches itself (through base pairing) to a messenger RNA codon. (25.4)

**Transition element** the B columns of elements in the periodic table. (2.5); the *d*-block transition elements in which a *d* subshell is being filled. (8.2); Those metallic elements that have an incompletely filled *d* subshell or easily give rise to common ions that have incompletely filled *d* subshells. (p. 1003)

**Transition-state theory** a theory that explains the reaction resulting from the collision of two molecules in terms of an activated complex (transition state). (14.5)

**Transmutation** the change of one element to another by bombarding the nucleus of the element with nuclear particles or nuclei. (21.2)

**Transuranium elements** those elements with atomic numbers greater than that of uranium ($Z = 92$), the naturally occurring element of greatest $Z$. (21.2)

**Trigonal bipyramidal geometry** the geometry of a molecule in which five atoms bonded to a central atom occupy the vertices of a trigonal bipyramid (formed by placing two trigonal pyramids base to base) with the central atom at the center of this trigonal bipyramid. (10.1)

**Trigonal planar geometry** the geometry of a molecule in which a central atom is surrounded by three other atoms arranged in a triangle and in a plane containing the central atom. (10.1)

**Trigonal pyramidal geometry** the geometry of a molecule in which a central atom is at the apex of a pyramid and three other atoms form the triangular base of the pyramid. (10.1)

**Triple bond** a covalent bond in which three pairs of electrons are shared by two atoms. (9.4)

**Triple point** the point on a phase diagram representing the temperature and pressure at which three phases of a substance coexist in equilibrium. (11.3)

**T-shaped geometry** the geometry of a molecule in which three atoms are bonded to a central atom to form a T. (10.1)

**Tyndall effect** the scattering of light by colloidal-size particles. (12.9)

**Uncertainty principle** a relation stating that the product of the uncertainty in position and the uncertainty in momentum (mass times speed) of a particle can be no smaller than Planck's constant divided by $4\pi$. (7.4)

**Unimolecular reaction** an elementary reaction that involves one reactant molecule. (14.7)

**Unit** a fixed standard of measurement. (1.5)

**Unit cell** the smallest boxlike unit (each box having faces that are parallelograms) from which you can imagine constructing a crystal by stacking the units in three dimensions. (11.7)

**Unsaturated hydrocarbon** a hydrocarbon that has at least one double or triple bond between carbon atoms; all carbon atoms are not bonded to the maximum number of hydrogen atoms (that is, the hydrocarbon is *unsaturated* with hydrogen). (24.1)

**Unsaturated solution** a solution that is not in equilibrium with respect to a given dissolved substance and in which more of the substance can dissolve. (12.2)

**Valence bond theory** an approximate theory to explain the electron pair or covalent bond in terms of quantum mechanics. (10.3)

**Valence electron** an electron in an atom outside the noble-gas or pseudo-noble-gas core. (8.2)

**Valence-shell electron-pair repulsion (VSEPR) model** predicts the shapes of molecules and ions by assuming that the valence-shell electron pairs are arranged about each atom so that electron pairs are kept as far away from one another as possible, thus minimizing electron-pair repulsions. (10.1)

**van der Waals forces** a general term for those intermolecular forces that include dipole–dipole and London forces. (11.5)

**van der Waals equation** an equation similar to the ideal gas law, but includes two constants, *a* and *b*, to account for deviations from ideal behavior. (5.8)

**Vapor** the gaseous state of any kind of matter that normally exists as a liquid or solid. (1.4)

**Vaporization** the change of a solid or a liquid to the vapor. (11.2)

**Vapor pressure** the partial pressure of the vapor over the liquid, measured at equilibrium at a given temperature. (11.2)

**Vapor-pressure lowering** a colligative property equal to the vapor pressure of the pure solvent minus the vapor pressure of the solution. (12.5)

**Viscosity** the resistance to flow that is exhibited by all liquids and gases. (11.4)

**Volatile** refers to a liquid or solid having a relatively high vapor pressure at normal temperatures. (11.2)

**Volt (V)** the SI unit of potential difference. (20.4)

**Voltaic cell (galvanic cell)** an electrochemical cell in which a spontaneous reaction generates an electric current. (p. 846)

**Volumetric analysis** a method of analysis based on titration. (4.10)

**Wavelength ($\lambda$)** the distance between any two adjacent identical points of a wave. (7.1)

**Wave mechanics** see *Quantum mechanics.*

**Weak acid** an acid that is only partly ionized in water; it is a weak electrolyte. (4.4)

**Weak base** a base that is only partly ionized in water; it is a weak electrolyte. (4.4)

**Weak electrolyte** an electrolyte that dissolves in water to give a relatively small percentage of ions. (4.1)

**Work** the energy exchange that results when a force *F* moves an object through a distance *d*; it equals $F \times d$. (19.1)

**Zinc–carbon (Leclanché) dry cell** a voltaic cell that has a zinc can as the anode; a graphite rod in the center, surrounded by a paste of manganese dioxide, ammonium and zinc chlorides, and carbon black, is the cathode. (20.8)

**Zwitterion** an amino acid in the doubly ionized form, in which the carboxyl group has lost an $H^+$ to give —$COO^-$ and the amino group has gained an $H^+$ to give —$NH_3^+$. (25.3)

# Photo Credits

**Computer-generated molecular models by Frank Cox:** Pages xviii (bottom), 362 (upper left and upper right), 1102 (top)

**Computer-generated molecular models by Frank Cox in association with Julie Newdoll:** Pages 983 (bottom), 985

**Photographs by American Color:** Pages v (two photos), ix (bottom), x (two photos), xii (bottom), xiii (bottom), xiv (top), xv (bottom), xvi (bottom), xvii (bottom), xviii (top), 3, 8 (top right), 12, 13 (top), 19, 27 (bottom right), 44 (two photos), 78, 90, 94, 98 (two photos), 102 (right), 106, 112 (top), 129, 132 (two photos), 133 (two photos), 135, 141, 143, 150, 152 (three photos), 157 (two photos), 159, 164 (three photos), 167, 171 (three photos), 183, 193 (two photos), 212 (two photos), 217 (two photos), 218, 236 (two photos), 240, 242 (two photos), 248 (right), 278, 340, 342 (bottom), 357, 362 (two photos), 406, 407 (two photos), 441, 457, 458 (bottom), 485 (bottom), 506, 515, 516, 529, 538, 559 (top two photos), 562, 564, 574 (right), 580 (three photos), 589, 597 (three photos), 612 (two photos), 648, 661 (right), 673 (three photos), 677, 692, 701, 714, 715, 737, 739 (three photos), 773, 775, 781 (three photos), 785, 839, 843, 847, 852, 871 (two photos), 873 (two photos), 878, 943, 951 (top), 958 (bottom), 961 (bottom), 972, 976 (top), 982, 984 (three photos), 989 (two photos), 1002, 1004, 1009 (two photos), 1011, 1017, 1023 (two photos), 1025 (two photos), 1045 (right), 1053, 1054, 1057 (two photos), 1060 (two photos), 1072, 1084, 1085, 1089

**Photographs by James Scherer:** Pages vii (top), xiv (bottom), 8 (bottom two photos), 13 (bottom two photos), 15 (three photos), 27 (top and bottom left), 49 (three photos), 64 (two photos), 76, 91, 96 (two photos), 112 (bottom), 117, 130, 144 (two photos), 153, 154, 158, 160, 161, 168 (two photos), 184, 190 (two photos), 195, 201, 259, 279 (four photos), 341, 358, 384 (three photos), 443 (two photos), 459, 472, 508 (three photos), 541 (two photos), 542, 554 (right), 556 (right), 570, 590 (two photos), 591 (two photos), 680 (two photos), 681, 693, 716, 748, 787 (three photos), 801, 840, 947, 951 (bottom two photos), 952 (three photos), 954, 955 (bottom), 956, 959 (three photos), 963 (two photos), 965, 971, 975, 976 (bottom two photos), 977, 978, 986, 991, 1010, 1037, 1058, 1087

**Credits for all other photos are listed below, with page numbers in boldface.**

**vii (bottom)** Photo Courtesy of Digital Instruments; **viii (top)** © 2001 PhotoDisc, Inc., **(bottom)** © Corbis; **ix (top)** © Corbis; **xi (bottom)** NASA; **xii (top)** Courtesy: O.D. Lavrentovich, Liquid Crystal Institute, Kent State University; **xiii (top)** Argonne National Laboratory; **xvii (top)** Siemens Medical Systems, Inc.; **1** © Yoav Levy/Phototake NYC; **2** © Douglas Elbinger; **4 (top left)** Courtesy of Corning, Incorporated; **4 (top right)** Courtesy of International Business Machines Corporation. Unauthorized use not permitted; **16** Courtesy of PerkinElmer Instruments, Shelton, Connecticut; **23** © Geoffrey Wheeler; **32** © Spencer Grant/PhotoEdit; **43** IBM Almaden Research Center. Research done by Dr. Don Eigler and coworkers; **45** Photo Courtesy of Digital Instruments; **48** The Cavendish Laboratory; **53** © Deep Light Productions/Photo Researchers; **56** Edgar Fahs Smith Collection, Van Pelt-Dietrich Library, University of Pennsylvania; **59** Lawrence Berkeley National Laboratory; **62 (right)** Teflon® is a registered trademark of DuPont; **63 (left)** © E.R. Degginger; **102 (left)** © 2001 PhotoDisc, Inc.; **138** © Corbis; **196** Schwarz Pharma; **197** © AP/Wide World Photos; **199** © Corbis; **204** Courtesy of Chrysler Corporation; **220** Frank Hoffmann/Department of Energy Photo Department; **235** © Joseph P. Sinnot/Fundamental Photographs, New York; **237 (left)** University of Missouri–Rolla Solar Car Team; **237 (right)** Michael Siluk/The Image Works; **239** Tennessee Valley Authority; **248 (left)** Ken O'Donoghue; **268** NASA; **274 (left)** Corbis; **274 (right)** NASA; **280** (no credit); **286** Photo courtesy of MDS Nordion; **287** © AP/Wide World Photos; **292** © Robert Holmgren/Peter Arnold, Inc.; **295** Carl Zeiss, Inc., Thornwood, NY; **296** © Biophoto Associates/Photo Researchers; **299** IBM Almaden Research

Center; **300** IBM Almaden Research Center. Research done by Dr. Don Eigler and coworkers; **311** © Richard Megna/Fundamental Photographs, New York; **312** Culver Pictures; **316** © Paul Shambroom/Photo Researchers; **330** Courtesy of A. Geim, University of Nijmegen; **342 (top)** Ken O'Donoghue; **348** © Clyde H. Smith/Peter Arnold, Inc.; **349** © E.R. Degginger; **355** Reprinted with permission from *Chemical Engineering News,* March 30, 1998, 76(13), p. 33. Copyright 1998 American Chemical Society; **365** © Craig Hammell/The Stock Market; **385 (two photos)** Nicolet Instrument Corporation; **394** © Alfred Pasieka/Science Photo Library/Photo Researchers; **399** (no credit); **417** Zuo, Kim, O'Keeffe, Spence, Arizona State University. From *Nature* magazine, September 1999. Reprinted with permission; **423** © Yoav Levy/Phototake NYC; **434** NASA; **440** Courtesy: O.D. Lavrentovich, Liquid Crystal Institute, Kent State University; **458 (top)** © Biophoto Associates/Photo Researchers; **466** Reprinted with permission from the *Journal of Chemical Education,* Vol. 74, No. 7, 1997, cover. Copyright © 1997, Division of Chemical Education, Inc.; **485 (top)** UCAR Carbon Company; **488** From Preston, Proceedings of the Royal Society, A, Volume 172, plate 4, figure 5A; **491** Courtesy of Bruker Analytical X-Ray Systems, Inc., Madison, Wisconsin, USA; **504** © Richard Megna/Fundamental Photographs, New York; **513 (two photos)** © Bill Longcore/Photo Researchers; **518** © Bill Stanton/Rainbow; **535 (three photos)** © David Phillips/Photo Researchers; **536** Seawater Systems, Inc.; **552** © IBM Research/Peter Arnold, Inc.; **553 (top)** NASA Ames Research Center; **553 (bottom)** Courtesy of Derby Cycle Corporation/Michael Hammond, photographer; **554 (left)** Alcoa Photo; **555** National Park Service; **556 (left)** © M. Claye/Jacana/Photo Researchers; **559 (bottom)** © Hank Morgan/Photo Researchers; **568** Argonne National Laboratory; **571** © Corbis; **572 (top right)** Courtesy of Sony Electronics, Inc.; **572 (bottom)** Copyright St. Jude Medical, Inc. 2001. This image is provided courtesy of St. Jude Medical, Inc. All rights reserved; **574 (left)** By permission of Dr. Richard Smalley, Rice University; **576** © Charles Falco/Photo Researchers; **577** © Dick Luria/Photo Researchers; **581–582** © Joel DeGrand 1999; **583 (top)** Stuart Gray; **583 (bottom)** Visions glass-ceramic cookware is made by World Kitchen, Inc., Elmira, NY; **584 (top)** Wright Medical Technology; **584 (bottom)** Courtesy of Textron Systems; **632 (two photos)** Engelhard Corporation; **633** Delphi Automotive Systems; **636** California Institute of Technology; **649** Courtesy © Eastman Kodak Company; **661 (left)** E.R. Degginger; **663** American Gas Association; **705** Reprinted with permission from the *Journal of Chemical Education,* Vol. 74, No. 7, 1997, cover. Copyright © 1997, Division of Chemical Education, Inc.; **723** © Richard Megna/Fundamental Photographs, New York; **735 (left)** © NYC Photo Archive c/o Fundamental Photographs, New York; **735 (right)** © Kristen Brochman/Fundamental Photographs, New York; **768** © Richard Megna/Fundamental Photographs, New York; **769** © C. Allan Morgan/Peter Arnold, Inc.; **779** © Barrie Rokeach/The Image Bank; **784** © E.R. Degginger; **800** © Paul Silverman/Fundamental Photographs, New York; **805** © Nelson Morris/Photo Researchers; **868** Corning Science Products; **880** Kennecott Corporation; **894** © CNRI/Science Photo Library/Photo Researchers; **895** © Alexander Tsiaras/Photo Researchers; **908** Fermilab Visual Media Services; **909** Lawrence Berkeley Laboratory; **910** Oak Ridge National Laboratories; **912** Courtesy of Bicron, division of Saint-Gobain Industrial Ceramics, Inc.; **923** DuPont Pharmaceuticals Company; **924 (two photos)** Siemens Medical Systems, Inc.; **925 (left)** © NIH/Science Source/Photo Researchers; **925 (right)** © Hank Morgan/Science Source/Photo Researchers; **934** Princeton Plasma Physics Laboratory; **935** © Novosti/Science Photo Library/Photo Researchers; **948 (left)** DuPont Magazine; **948 (right)** Rayovac Corporation; **955 (top)** © E.R. Degginger; **958 (top)** Sweetwater Technology; **960** Smithsonian Institution Photo No. 77-10626P; **961 (top)** © E.R. Degginger; **962 (top)** Engelhard Corporation; **962 (bottom)** © E.R. Degginger; **968** NASA; **983 (top)** © Farrell Frehan/Photo Researchers; **993** Argonne National Laboratory; **1044** © Bruce Iverson; **1045 (left)** © Arthur R. Hill/Visuals Unlimited; **1088** © P&G Bowater/The Image Bank; **1098** By permission of Robert M. Stroud, University of California, San Francisco; **1102 (bottom)** © Manfred Kage/Peter Arnold, Inc.; **1106** By permission of Gang-Yu Liu, Wayne State University

# Index

# Table of Atomic Numbers and Atomic Weights

| Name | Symbol | Atomic number | Atomic weight | Name | Symbol | Atomic number | Atomic weight |
|------|--------|---------------|---------------|------|--------|---------------|---------------|
| Actinium | Ac | 89 | (227) | Molybdenum | Mo | 42 | 95.94 |
| Aluminum | Al | 13 | 26.981538 | Neodymium | Nd | 60 | 144.24 |
| Americium | Am | 95 | (243) | Neon | Ne | 10 | 20.1797 |
| Antimony | Sb | 51 | 121.760 | Neptunium | Np | 93 | (237) |
| Argon | Ar | 18 | 39.948 | Nickel | Ni | 28 | 58.6934 |
| Arsenic | As | 33 | 74.92160 | Niobium | Nb | 41 | 92.90638 |
| Astatine | At | 85 | (210) | Nitrogen | N | 7 | 14.00674 |
| Barium | Ba | 56 | 137.327 | Nobelium | No | 102 | (259) |
| Berkelium | Bk | 97 | (247) | Osmium | Os | 76 | 190.23 |
| Beryllium | Be | 4 | 9.012182 | Oxygen | O | 8 | 15.9994 |
| Bismuth | Bi | 83 | 208.98038 | Palladium | Pd | 46 | 106.42 |
| Bohrium | Bh | 107 | (264) | Phosphorus | P | 15 | 30.973762 |
| Boron | B | 5 | 10.811 | Platinum | Pt | 78 | 195.078 |
| Bromine | Br | 35 | 79.904 | Plutonium | Pu | 94 | (244) |
| Cadmium | Cd | 48 | 112.411 | Polonium | Po | 84 | (209) |
| Calcium | Ca | 20 | 40.078 | Potassium | K | 19 | 39.0983 |
| Californium | Cf | 98 | (251) | Praseodymium | Pr | 59 | 140.90765 |
| Carbon | C | 6 | 12.0107 | Promethium | Pm | 61 | (145) |
| Cerium | Ce | 58 | 140.116 | Protactinium | Pa | 91 | 231.03588 |
| Cesium | Cs | 55 | 132.90545 | Radium | Ra | 88 | (226) |
| Chlorine | Cl | 17 | 35.4527 | Radon | Rn | 86 | (222) |
| Chromium | Cr | 24 | 51.9961 | Rhenium | Re | 75 | 186.207 |
| Cobalt | Co | 27 | 58.933200 | Rhodium | Rh | 45 | 102.90550 |
| Copper | Cu | 29 | 63.546 | Rubidium | Rb | 37 | 85.4678 |
| Curium | Cm | 96 | (247) | Ruthenium | Ru | 44 | 101.07 |
| Dubnium | Db | 105 | (262) | Rutherfordium | Rf | 104 | (261) |
| Dysprosium | Dy | 66 | 162.50 | Samarium | Sm | 62 | 150.36 |
| Einsteinium | Es | 99 | (252) | Scandium | Sc | 21 | 44.955910 |
| Erbium | Er | 68 | 167.26 | Seaborgium | Sg | 106 | (263) |
| Europium | Eu | 63 | 151.964 | Selenium | Se | 34 | 78.96 |
| Fermium | Fm | 100 | (257) | Silicon | Si | 14 | 28.0855 |
| Fluorine | F | 9 | 18.9984032 | Silver | Ag | 47 | 107.8682 |
| Francium | Fr | 87 | (223) | Sodium | Na | 11 | 22.989770 |
| Gadolinium | Gd | 64 | 157.25 | Strontium | Sr | 38 | 87.62 |
| Gallium | Ga | 31 | 69.723 | Sulfur | S | 16 | 32.066 |
| Germanium | Ge | 32 | 72.61 | Tantalum | Ta | 73 | 180.9479 |
| Gold | Au | 79 | 196.96655 | Technetium | Tc | 43 | (98) |
| Hafnium | Hf | 72 | 178.49 | Tellurium | Te | 52 | 127.60 |
| Hassium | Hs | 108 | (265) | Terbium | Tb | 65 | 158.92534 |
| Helium | He | 2 | 4.002602 | Thallium | Tl | 81 | 204.3833 |
| Holmium | Ho | 67 | 164.93032 | Thorium | Th | 90 | 232.0381 |
| Hydrogen | H | 1 | 1.00794 | Thulium | Tm | 69 | 168.93421 |
| Indium | In | 49 | 114.818 | Tin | Sn | 50 | 118.710 |
| Iodine | I | 53 | 126.90447 | Titanium | Ti | 22 | 47.867 |
| Iridium | Ir | 77 | 192.217 | Tungsten | W | 74 | 183.84 |
| Iron | Fe | 26 | 55.845 | Ununbium | Uub | 112 | (277) |
| Krypton | Kr | 36 | 83.80 | Unununilium | Uun | 110 | (269) |
| Lanthanum | La | 57 | 138.9055 | Unununium | Uuu | 111 | (272) |
| Lawrencium | Lr | 103 | (262) | Uranium | U | 92 | 238.0289 |
| Lead | Pb | 82 | 207.2 | Vanadium | V | 23 | 50.9415 |
| Lithium | Li | 3 | 6.941 | Xenon | Xe | 54 | 131.29 |
| Lutetium | Lu | 71 | 174.967 | Ytterbium | Yb | 70 | 173.04 |
| Magnesium | Mg | 12 | 24.3050 | Yttrium | Y | 39 | 88.90585 |
| Manganese | Mn | 25 | 54.938049 | Zinc | Zn | 30 | 65.39 |
| Meitnerium | Mt | 109 | (268) | Zirconium | Zr | 40 | 91.224 |
| Mendelevium | Md | 101 | (258) | | | | |
| Mercury | Hg | 80 | 200.59 | | | | |

A value in parentheses is the mass number of the isotope of longest half-life.

Atomic weights in this table are from the IUPAC report "Atomic Weights of the Elements 1993," *Pure and Applied Chemistry,* Vol. 68, No. 12 (1996), pp. 2339–2359.